SCANNING ELECTRON MICROSCOPY/1979/II

An
INTERNATIONAL REVIEW
of
ADVANCES
in
TECHNIQUES AND APPLICATIONS
of the
SCANNING ELECTRON MICROSCOPE
PART II

Published by

SCANNING ELECTRON MICROSCOPY, Inc.,
P.O. Box 66507
AMF O'Hare, IL 60666, USA

For all additional information or inquiries contact:

        Dr. Om Johari
        SEM Inc.,
        P.O. Box 66507
        AMF O'Hare, IL 60666, U.S.A.

Information about obtaining Scanning Electron Microscopy/
1978 or 1979 as well as regarding Scanning Electron Microscopy/
1980 meetings scheduled from April 21-25, 1980 in Chicago,
IL is available from the above address.

LIBRARY OF CONGRESS CATALOGUE NO. 72-626068

ISBN: 0-931288-00-2   for
SCANNING ELECTRON MICROSCOPY/Year...series from 1978 on.

ISBN---SCANNING ELECTRON MICROSCOPY/1978/I: 0-931288-01-0
       SCANNING ELECTRON MICROSCOPY/1978/II: 9-931288-02-9
       SCANNING ELECTRON MICROSCOPY/1979/I: 0-931288-04-5
       SCANNING ELECTRON MICROSCOPY/1979/II: 0-931288-05-3
       SCANNING ELECTRON MICROSCOPY/1979/III: 0-931288-06-1

LIBRARIANS AND OTHERS PLEASE NOTE:

    For this publication the volume number and the year are the
    same. Individual papers should be referred to by the year and
    the part, (Roman Numeral included after the year). Thus,
    SEM/1979 has three parts: I, II, and III; and SEM/1978 had
    two parts: I and II.

                    PRINTED IN THE UNITED STATES OF AMERICA.

SCANNING ELECTRON MICROSCOPY/1979/II

EDITORS	Om Johari
	Robert P. Becker
ASSOCIATE EDITOR	Irene Corvin Pontarelli
EDITORIAL ASSISTANCE	Jan Elder and Joseph Staschke

EXPLANATION OF THE TYPE OF PAPERS IN THIS VOLUME:*

TUTORIAL   Presentation of established material in teaching lecture type format with emphasis on techniques and "how to do it".

REVIEW   A review of the chosen subject with emphasis on author's own work, placing it in context with relevant literature, and putting the topic in perspective.

* For identification of these papers see Table of Contents or Major Subject Index.

Most of the papers included in this volume were presented at the Scanning Electron Microscopy/1979 meetings, held April 16-20, 1979, at Sheraton Park Hotel, Washington, D.C.. The program for various parts of the meeting was prepared and conducted by the advisors listed above. Om Johari was the director of the meetings.

The division of papers in three parts of the Scanning Electron Microscopy/1979 volume is somewhat arbitrary. In general, the editors attempted to follow the scheme given below. We realize that such a scheme is not satisfactory in all cases; thus biological microanalysis papers are placed in part II because of the increasing interaction between physical and biological scientists in that field. Similarly, many papers under mineralized tissue originated from geology departments, and so we elected to place them in part II.

Part I   Papers primarily of interest to physical scientists.

Part III   Papers primarily of interest to biological scientists.

Part II   Remaining papers, presumably of interest to both - physical and biological scientists.

A major subject index has been included in parts II and III.

Additionally, a major subject index of SEM/1978/I and SEM/1978/II has been included in SEM/1979/II and SEM/1979/III respectively.

TABLE OF CONTENTS

Organization   iii

Discussion with Reviewers   x

Reviewers List   xi

Major Subjects Index:
  General Interest Papers, Electron Sources and Analytical Electron
  Microscopy   xv
  Particulate Matter, Sediments, Archaeology, SEM & the Law   xvi
  Materials Characterization, Failure Analysis, Semiconductor
  Applications, Surface Analysis   xvii

General Information about SEM Inc. and Its activities   xviii

KEYNOTE PAPER: MICROANALYSIS IN THE SCANNING ELECTRON MICROSCOPE:
PROGRESS AND PROSPECTS
        D.E. Newbury
                                                                                      1

THIN FILMS FOR HIGH RESOLUTION CONVENTIONAL SCANNING ELECTRON MICROSCOPY
        P. Echlin and G. Kaye
                                                                                      21

CHARGE NEUTRALIZATION USING VERY LOW ENERGY IONS
        C.K. Crawford
                                                                                      31

ON-LINE TOPOGRAPHIC ANALYSIS IN THE SEM
        D.M. Holburn and K.C.A. Smith
                                                                                      47

A PRECISION SEM IMAGE ANALYSIS SYSTEM WITH FULL FEATURE EDXA CHARACTERIZATION
        W.R. Stott and E.J. Chatfield
                                                                                      53

USE OF MICROTOPOGRAPHY CAPABILITY IN THE SEM FOR ANALYSING FRACTURE SURFACES
        J. Lebiedzik, J. Lebiedzik, R. Edwards and B. Phillips   61

THE PERCEPTION AND MEASUREMENT OF DEPTH IN THE SEM   (TUTORIAL PAPER)
        A. Boyde
                                                                                      67

THE OBSERVATION AND MEASUREMENT OF DISPLACEMENTS AND STRAIN BY STEREOIMAGING
        D.L. Davidson
                                                                                      79

A THEORY OF SURFACE-ORIGINATING CONTAMINATION AND A METHOD FOR ITS
ELIMINATION   (REVIEW PAPER)
        J.T. Fourie
                                                                                      87

REDUCTION OF CONTAMINATION IN ANALYTICAL ELECTRON MICROSCOPY
        Y. Harada, T. Tomita, T. Watabe, H. Watanabe and T. Etoh   103

ELECTRON-SPECIMEN INTERACTIONS   (TUTORIAL PAPER)
        L. Reimer
                                                                                      111

APPLICATIONS OF THE CATHODOLUMINESCENCE METHOD IN BIOLOGY
AND MEDICINE   (REVIEW PAPER)
        W. Bröcker and G. Pfefferkorn
                                                                                      125

SCANNING ELECTRON MICROSCOPY AT MACROMOLECULAR RESOLUTION IN LOW ENERGY MODE ON
BIOLOGICAL SPECIMENS COATED WITH ULTRA THIN METAL FILMS   (REVIEW PAPER)
        K. -R. Peters
                                                                                      133

FURTHER DEVELOPMENT OF THE CONVERTED BACKSCATTERED ELECTRON DETECTOR
        S.H. Moll, F. Healey, B. Sullivan and W. Johnson   149

CONVERSION OF EXISTING SEM COMPONENTS TO FORM AN EFFICIENT BACKSCATTERED ELECTRON
DETECTOR, AND ITS FORENSIC APPLICATIONS
        N. Zeldes and M. Tassa
                                                                                      155

THE APPLICATION OF BACKSCATTERED ELECTRON IMAGE TO
FORENSIC SCIENCE   (REVIEW PAPER)
        V. R. Matricardi
                                                                                      159

FIREARM IDENTIFICATION BY EXAMINATION OF BULLET FRAGMENTS:
AN SEM/EDS STUDY   (REVIEW PAPER)
        R. L. Taylor, M. S. Taylor and T. T. Noguchi   167

CHARACTERIZATION OF GUNSHOT RESIDUE PARTICLES BY LOCALIZATIONS
OF THEIR CHEMICAL CONSTITUENTS
        M. Tassa and N. Zeldes
                                                                                      175

A METHOD FOR PREPARING FIREARMS RESIDUE SAMPLES
FOR SCANNING ELECTRON MICROSCOPY
     J.S. Wallace and R.H. Keeley   179

FORENSIC APPLICATIONS OF SEM/EDX   (REVIEW PAPER)
     E.H. Sild and S. Pausak   185

QUANTITATIVE INVESTIGATION OF SULPHUR AND CHLORINE IN HUMAN HEAD HAIRS
BY ENERGY DISPERSIVE X-RAY MICROANALYSIS
     S. Seta, H. Sato and M. Yoshino   193

THE CHARACTERIZATION OF DYNAMITE WRAPPING PAPER BY SEM:
A PRELIMINARY REPORT
     M. Ueyama, I. Suzuki, R.L. Taylor and T.T. Noguchi   203

APPLICATIONS OF SEM AND X-RAY MICROANALYSIS IN CRIMINALISTICS:
AN INDEXED BIBLIOGRAPHY
     R.L. Taylor and T.T. Noguchi   209

TEACHING SCANNING ELECTRON MICROSCOPY
     G.R. Hooper, K.K. Baker and S.L. Flegler   217

USE OF A TABLE-TOP SEM IN UNDERGRADUATE TEACHING
     D.B. Williams J.I. Goldstein   221

COORDINATED, MULTIDISCIPLINE COURSES IN SEM
     K.S. Howard, M.D. Socolofsky, R.L. Chapman and C.H. Moore, Jr.   225

APPLICATION OF THE UNIVERSITY OF ILLINOIS COMPUTER BASED PLATO IV
NETWORK TO THE TEACHING OF ELECTRON MICROSCOPY
     R.L. McConville and F. Scheltens   231

CRITICAL POINT DRYING OF BIOLOGICAL SPECIMENS
     W.J. Humphreys and W.G. Henk   235

AN EVALUATION OF VIDEO TAPES IN TEACHING SEM
     E.L. Thurston   241

THE USE OF SLIDE/TAPE MODULES IN TEACHING SEM
     K.K. Baker, S.L. Flegler and G.R. Hooper   243

TEACHING METHODS FOR SEM SHORT COURSES
     O.C. Wells   247

THE DEVELOPMENT AND IMPLEMENTATION OF AUTOTUTORIAL METHODS FOR
ELECTRON MICROSCOPY UTILIZING SOUND ON SLIDE PRESENTATIONS
     Judith A. Murphy   253

SCANNING AUGER MICROSCOPY - AN INTRODUCTION FOR BIOLOGISTS   (TUTORIAL PAPER)
     A.P. Janssen and J.A. Venables   259

LASER MICROPROBE MASS ANALYSIS: ACHIEVEMENTS AND ASPECTS   (REVIEW PAPER)
     R. Kaufmann, F. Hillenkamp, R. Wechsung, H.J. Heinen and M. Schürmann   279

MASS MEASUREMENTS WITH THE ELECTRON MICROSCOPE   (REVIEW PAPER)
     J.S. Wall   291

CRITICAL POINT DRYING -- PRINCIPLES AND PROCEDURE   (TUTORIAL PAPER)
     A.L. Cohen   303

A COMPREHENSIVE FREEZING, FRACTURING AND COATING SYSTEM FOR
LOW TEMPERATURE SCANNING ELECTRON MICROSCOPY   (REVIEW PAPER)
     A.W. Robards and P. Crosby   325

SCANNING ELECTRON MICROSCOPY OF BIOLOGICAL SPECIMENS SURFACE-ETCHED BY
AN OXYGEN PLASMA
     W.J. Humphreys, W.G. Henk and D.B. Chandler   345

COATING BY ION SPUTTERING DEPOSITION FOR ULTRAHIGH RESOLUTION SEM
     J.D. Geller, T. Yoshioka and D.A. Hurd   355

ASBESTOS FIBRE COUNTING BY AUTOMATIC IMAGE ANALYSIS
     R.N. Dixon and C.J. Taylor   361

COMPUTER GRAPHICS ANALYSIS OF STEM IMAGES
     R. Llinás, R. Spitzer, D. Hillman and M. Chujo   367

PREPARATION AND OBSERVATION OF VERY THIN, VERY CLEAN SUBSTRATES FOR
SCANNING TRANSMISSION ELECTRON MICROSCOPY
     M. Ohtsuki, M.S. Isaacson and A.V. Crewe   375

RAMAN MICROPROBE STUDIES OF TWO MINERALIZING TISSUES:
ENAMEL OF THE RAT INCISOR AND THE EMBRYONIC CHICK TIBIA
      F.S. Casciani, E.S. Etz, D.E. Newbury and S.B. Doty   383

ESTIMATION OF THE SIZE OF RESORPTION LACUNAE IN MAMMALIAN CALCIFIED
TISSUES USING SEM STEREOPHOTOGRAMMETRY
      A. Boyde and S.J. Jones   393

ULTRASTRUCTURAL STUDIES ON CALCIFICATION IN VARIOUS ORGANISMS   (REVIEW PAPER)
      N. Watabe and D.G. Dunkelberger   403

COCCOLITH MORPHOLOGY AND PALEOCLIMATOLOGY...II. CELL ULTRASTRUCTURE AND
FORMATION OF COCCOLITHS IN *CYCLOCOCCOLITHINA LEPTOPORA* (MURRAY AND BLACKMAN)
WILCOXON AND *GEPHYROCAPSA OCEANICA* KAMPTNER
      P.L. Blackwelder, L.E. Brand and R.L. Guillard   417

THE ORGANIZATION OF A STRUCTURAL ORGANIC MATRIX WITHIN THE SKELETON
OF A REEF-BUILDING CORAL   (REVIEW PAPER)
      I.S. Johnston   421

STATOLITH SYNTHESIS AND EPHYRA DEVELOPMENT IN *AURELIA* METAMORPHOSING IN
STRONTIUM   AND LOW CALCIUM CONTAINING SEA WATER
      D.B. Spangenberg   433

COMPARATIVE SHELL MICROSTRUCTURE OF THE MOLLUSCA, BRACHIOPODA AND BRYOZOA
      J.G. Carter   439

THE EFFECTS OF MARINE MICROPHYTES ON CARBONATE SUBSTRATA   (REVIEW PAPER)
      K.J. Lukas   447

ON THE INTERNAL STRUCTURES OF THE NACREOUS TABLETS IN MOLLUSCAN SHELLS
      H. Mutvei   457

SHELL DISSOLUTION: DESTRUCTIVE DIAGENESIS IN A METEORIC ENVIRONMENT
      B.M. Walker   463

FORMATION OF A DISSOLUTION LAYER IN MOLLUSCAN SHELLS
      D.A. Wilkes and M.A. Crenshaw   469

ULTRASTRUCTURAL RELATIONSHIPS OF MINERAL AND ORGANIC MATTER IN AVIAN EGGSHELLS
      A.S. Pooley   475

SEM OF ENAMEL LAYER IN ORAL TEETH OF FOSSIL AND EXTANT
CROSSOPTERYGIAN AND DIPNOAN FISHES
      M. M. Smith   483

A METHOD OF INTERPRETING ENAMEL PRISM PATTERNS
      D. G. Gantt   491

THE ARRANGEMENT OF PRISMS IN THE ENAMEL OF THE ANTERIOR TEETH OF THE AYE-AYE
      R. P. Shellis and D. F. G. Poole   497

COMPARATIVE INVESTIGATIONS ON FLUOROSED ENAMEL
      M. Triller, A. Bouratbine, D. Guillaumin and R. Weill   507

SCANNING ELECTRON MICROSCOPY OF RACHITIC RAT BONE
      D.W. Dempster, H.Y. Elder and D.A. Smith   513

THE ARCHITECTURE OF METAPHYSEAL BONE HEALING
      K. Draenert and Y. Draenert   521

COLONIZATION OF VARIOUS NATURAL SUBSTRATES BY OSTEOBLASTS *IN VITRO*
      S.J. Jones and A. Boyde   529

USE OF SEM FOR THE STUDY OF THE SURFACE RECEPTORS OF OSTEOCLASTS *IN SITU*
      I.M. SHAPIRO, S.J. Jones, N.M. Hogg, M. Slusarenko and A. Boyde   539

STRUCTURAL CONVERGENCES BETWEEN ENAMELOID OF ACTINOPTERYGIAN TEETH
AND OF SHARK TEETH
      W.-E. Reif   547

APPLICATION OF ELECTRON PROBE X-RAY MICROANALYSIS TO CALCIFICATION STUDIES
OF BONE AND CARTILAGE   (REVIEW PAPER)
      W.J. Landis   555

REPLICATION TECHNIQUES WITH NEW DENTAL IMPRESSION MATERIALS IN COMBINATION
WITH DIFFERENT NEGATIVE IMPRESSION MATERIALS   (REVIEW PAPER)
      C.H. Pameijer   571

INLAYED TEETH OF ANCIENT MAYANS: A TRIBOLOGICAL STUDY USING THE SEM
      A.J. Gwinnett and L. Gorelick   575

OPTIMIZATION OF AN ANALYTICAL ELECTRON MICROSCOPE FOR X-RAY MICROANALYSIS:
INSTRUMENTAL PROBLEMS   (TUTORIAL PAPER)
    J. Bentley, N. J. Zaluzec, E. A. Kenik and R. W. Carpenter   581

PRINCIPLES OF X-RAY MICROANALYSIS IN BIOLOGY   (KEYNOTE PAPER)
    J. A. Chandler   595

GENERAL CONSIDERATIONS OF X-RAY MICROANALYSIS OF
FROZEN HYDRATED TISSUE SECTIONS   (TUTORIAL PAPER)
    A. J. Saubermann   607

QUANTIFICATION OF ELECTROLYTES IN FREEZE-DRIED CRYOSECTIONS BY
ELECTRON MICROPROBE ANALYSIS   (REVIEW PAPER)
    R. Rick, A. Dörge, R. Bauer, K. Gehring and K. Thurau   619

ORIGIN OF ARTIFACTUAL QUANTITATION OF ELECTROLYTES IN MICROPROBE
ANALYSIS OF FROZEN SECTIONS OF ERYTHROCYTES
    J.M. Tormey and R.M. Platz   627

NON-FREEZING TECHNIQUES OF PREPARING BIOLOGICAL SPECIMENS FOR ELECTRON
MICROPROBE X-RAY MICROANALYSIS   (REVIEW PAPER)
    A.J. Morgan   635

STANDARDS FOR X-RAY MICROANALYSIS OF BIOLOGICAL SPECIMENS   (REVIEW PAPER)
    G.M. Roomans   649

QUANTITATIVE METHODS IN BIOLOGICAL X-RAY MICROANALYSIS   (REVIEW PAPER)
    N.C. Barbi   659

ANALYSIS OF ELEMENTAL RATIOS IN "THIN" SAMPLES   (TUTORIAL PAPER)
    J.C. Russ   673

PREVENTION OF LOSSES DURING X-RAY MICROANALYSIS   (REVIEW PAPER)
    T. Bistricki   683

IDENTIFICATION AND PREVENTION OF ARTIFACTS IN BIOLOGICAL X-RAY MICROANALYSIS
                                                    (TUTORIAL PAPER)
    B.J. Panessa-Warren   691

MICROANALYSIS IN BIOLOGY AND MEDICINE. A REVIEW OF RESULTS OBTAINED WITH
THREE MICROANALYTICAL METHODS   (REVIEW PAPER)
    P. Galle, J.P. Berry and R. Lefevre   703

ELECTRON PROBE ANALYSIS OF MUSCLE AND X-RAY MAPPING OF BIOLOGICAL SPECIMENS
WITH A FIELD EMISSION GUN   (REVIEW PAPER)
    A.P. Somlyo, A.V. Somlyo, H. Shuman and M. Stewart   711

ANALYTICAL ELECTRON MICROSCOPIC STUDIES OF ISCHEMIC AND HYPOXIC MYOCARDIAL INJURY
    H.K. Hagler, K.P. Burton, L. Sherwin, C. Greico, A. Siler, L. Lopez
                                   and L.M. Buja   723

INTRACELLULAR ELECTROLYTE CONCENTRATIONS IN EPITHELIAL TISSUE DURING
VARIOUS FUNCTIONAL STATES   (REVIEW PAPER)
    K. Thurau, A. Dörge, R. Rick, Ch. Roloff, F. Beck, J. Mason and R. Bauer   733

X-RAY ANALYSIS OF PATHOLOGICAL CALCIFICATIONS INCLUDING URINARY STONES
    K.M. Kim   739

APPLICATION OF SCANNING AND TRANSMISSION ELECTRON MICROSCOPY, X-RAY ENERGY
SPECTROSCOPY, AND X-RAY DIFFRACTION TO CALCIUM PYROPHOSPHATE CRYSTAL FORMATION
IN VITRO
    S.A. Omar, P.-T. Cheng, S.C. Nyburg and K.P.H. Pritzker   745

DOCUMENTATION OF ENVIRONMENTAL PARTICULATE EXPOSURES IN HUMANS
USING SEM AND EDXA   (REVIEW PAPER)
    J.L. Abraham   751

TECNIQUES FOR QUANTITATIVE ORGANIC ANALYSIS IN MICRODROPLETS (TUTORIAL PAPER)
    R. Beeuwkes III   767

ELECTRON PROBE STUDIES OF $Na+-K+-ATPase$ (REVIEW PAPER)
    S. Rosen and R. Beeuwkes III   773

X-RAY MICROANALYSIS OF PYROANTIMONATE-PRECIPITABLE CATIONS   (REVIEW PAPER)
    J.A.V. Simson, H.L. Bank and S.S. Spicer   779

CYTOCHEMICAL LOCALIZATION OF CATIONS IN MYELINATED NERVE USING TEM, HVEM,
SEM AND ELECTRON PROBE MICROANALYSIS
    M.H. Ellisman, P.L. Friedman and W.J. Hamilton   793

METAL BINDING BY INTESTINAL MUCUS
        J.R. Coleman and L.B. Young   801

PROGRESS IN THE DEVELOPMENT OF THE PEAK-TO- BACKGROUND METHOD FOR THE
QUANTITATIVE ANALYSIS OF SINGLE PARTICLES WITH THE ELECTRON PROBE
        J.A. Small, K.F.J. Heinrich, D.E. Newbury and R.L. Myklebust   807

PROGRESS IN THE QUANTITATION OF ELECTRON ENERGY-LOSS SPECTRA   (REVIEW PAPER)
        D.C. Joy, R.F. Egerton and D.M. Maher   817

ELECTRON ENERGY ANALYSIS IN A VACUUM GENERATORS HB5 STEM
        M.T. Browne   827

VISUALIZATION OF SUBSURFACE STRUCTURES IN CELLS AND TISSUES BY
BACKSCATTERED ELECTRON IMAGING   (TUTORIAL PAPER)
        R.P. Becker and M. Sogard   835

EVALUATION OF TISSUE-RESPONSE TO HYDROGEL COMPOSITE MATERIALS
        R.T. Greer, R.L. Knoll and B.H. Vale   871

PRACTICAL EXPERIENCES WITH SCANNING ELECTRON MICROSCOPY IN A
FORENSIC SCIENCE LABORATORY   (REVIEW PAPER)
        J. Andrasko, S. Bendtz and A.C. Maehly   879

SEM AND X-RAY MICROANALYSIS OF ARTIFACTS RETRIEVED FROM MARINE
ARCHAEOLOGICAL EXCAVATIONS IN NEWFOUNDLAND
        V.C. Barber   885

DEHYDRATION OF SCANNING ELECTRON MICROSCOPY SPECIMENS:
A BIBLIOGRAPHY (1974-1978)
        K.S. Howard and M.T. Postek   892

Classification of papers by major subjects from SCANNING ELECTRON MICROSCOPY/1978/I 903

ERRATA/SUGGESTION REQUEST   906

SUBJECT INDEX   907

AUTHOR INDEX   909

DISCUSSION WITH REVIEWERS

Each paper in this volume contains a discussion with reviewers. This discussion follows the text and references and should be read with the paper. Each paper submitted to SEM Inc. for publication was reviewed by at least three (and often more, on the average five) reviewers. The reviewers were asked to separate their comments and questions. Comments were useful in determining the acceptability of the paper as submitted. Of the over 315 papers submitted, nearly 10% are not being published; while almost all were subjected to minor, and, in nearly half the cases, major revisions. Although the comments required no written responses, in several cases, the authors have included responses to comments, or to questions phrased from or based on comments(either due to editorial suggestions along these lines or on the author's own initiative).

The questions for the most part originated as a result of the following statements in the cover letter accompanying each paper sent to the reviewers: "...You can ask relevant questions by supposing that you are an attendee at a conference where this paper, as written, is presented. Please write the questions which you would ask the author...." From the questions so asked, some are not included because the authors attended to them by revising the text. In a few cases, editorial and/or space considerations excluded publication of all questions asked by several reviewers for a paper. This year the authors prepared their own discussion with reviewers in the ready-to-print format, i.e., they retyped the questions and gave the answers, hopefully following the instructions we sent them. In some instances the authors edited the questions and/or combined several similar questions from different reviewers to provide one answer. While all efforts were made to check that the questions in the printed version faithfully represented the views of the specific reviewer, the editors apologize to the reviewers, if in some cases, the actual meaning and/or emphasis may have been changed by the author.

The cover letter to the reviewers also stated (in red ink):
"1. Your name will be conveyed to the author with your review UNLESS YOU ASK US NOT TO.
2. The questions published in the proceedings will be identified as coming from you UNLESS YOU ADVISE OTHERWISE...."

In all cases we made genuine efforts to respect and honor reviewers' wishes when they requested that they be not identified. But, in nearly 99% of the cases, the reviewers did not instruct us otherwise and their names were conveyed to the authors and are included with the questions printed with each paper. A general list of reviewers, containing names of all those who let us include their names, is provided in the beginning pages of each volume. We apologize for any errors/omissions.

Finally the readers are urged to be cautious on the weight they attach to authors replies, since the answers to the questions repre- sent the authors unchallenged views (since, except for minor editorial changes, the authors generally had the last word). Also, please remember that the questions were, in almost all instances, raised on the originally submitted paper and may not have the same significance/meaning for the revised final paper being published.

With grateful thanks to the authors and reviewers who make prompt publication in this format possible, we invite your comments on the ways to improve our procedure and seek volunteers to help us with the reviewing of papers in the future.

The Editors.

REVIEWERS LIST

The list below includes names of reviewers related to this part.

Abraham, J.L.	Univ. California, San Diego
Adar, F.	Instruments SA, NJ
Allen, C.W.	Univ. Notre Dame, IN
Allen, R.D.	Dartmouth College, Hanover, NH
Anderson, R.F.	Bureau of Mint, Washington, DC
Anderson, R.M.	IBM, East Fishkill, NY
Arima, E.	Kagoshima University, Japan
Arnott, H.J.	Univ. Texas, Arlington, TX
Arteaga, M.	Hoffman-LaRoche Inc., Nutley, NJ
Artz, B.E.	Ford Motor Company, Dearborn, MI
Ashraf, M.	Univ. Cincinnati, OH
Bahr, G.F.	Armed Forces Inst. Pathology, Washington, DC
Baker, K.K.	Michigan State University, E. Lansing
Balk, L.	Gesamthochschule Duisburg, West Germany
Banfield, W.G.	National Institute of Health, Bethesda, MD
Barber, V.C.	Memorial University, Newfoundland, Canada
Barbi, N.C.	Princeton-Gamma Tech, NJ
Barnard, T.	Wenner-Gren Institute, Stockholm, Sweden
Barrnett, R.J.	Yale University Sch of Medicine, CT
Beaman, D.R.	Dow Chemical Company, Midland, MI
Beer, M.	Johns Hopkins Univ., Baltimore, MD
Bentley, J.	Oak Ridge National Laboratory, TN
Berezesky, I.K.	Univ. Maryland School Medicine, Baltimore
Berry, V.K.	Univ. Texas, San Antonio
Bistricki, T.	Canada Ctr. Inland Watres, Burlington, Ont.
Blanquet, R.	Georgetown Univ., Wash. DC
Bolon, R.B.	General Electric Co, Corp. R&D, NY
Boyde, A.	Univ. College, London, U.K.
Brody, A.R.	National Institute Environmental Heal. Sci.
Brown, J.A.	McCrone Associates, Chicago, IL
Brown, W.A.S.	King's College, Univ. London, U.K.
Buja, L.M.	Univ. Texas, Dallas
Burge, R.E.	Queen Elizabeth College, U.K.
Carlysle, T.C.	USDA-ARS-Attractants Lab., FL
Carpenter, R.W.	Oak Ridge National Laboratory, TN
Carriker, M.R.	Univ. Delaware, Lewes, DE
Carter, J.G.	Univ. North Carolina, Chapel Hill, NC
Carter, H.W.	St. Barnabus Medical Center, Livingston, NJ
Chandler, J.A.	Tenovus Institute, Heath, Cardiff, U.K.
Chatfield, E.J.	Ontario Research Foundation, Canada
Clark, M.A.	Naval Medical Research Institute, Bethesda
Cohen, A.L.	Washington State University, Pullman, WA
Cohen, S.H.	Army Natick R&D Lab., MA
Coleman, J.R.	Univ. Rochester Sch. Med., Rochester, NY
Colliex, C.	Univ. Paris Sud, Orsay, France
Condrate, R.A., Sr.	NYS College of Ceramics, Alfred, NY
Conru, H,W,	IBM, Burlington, VT
Costa, J.L.	National Inst. Health, Bethesda, MD
Cotton, W.R.	Naval Medical Reserach Inst., Bethesda, MD
Cowley, J.M.	Arizona State University, Tempe
Crenshaw, M.A.	Univ. North Carolina, Chapel Hill, NC
Culbreth, K.L.	NC State Bureau of Investigation, Raleigh
Daniel, J.L.	Battelle Northwest, Richland, WA
Das, S.	Argonne National Lab., IL
Davidson, D.L.	Southwest Research Institute, San Antonio, TX
de Bruyn, W.C.	Univ. Leiden, Netherlands
de Mets, M.	Univ. Ghent, Belgium
DeNee, P.B.	Inhalation Tech. Res. Inst., Albuquerque, NM
Dillaman, R.M.	Duke University, Durham, NC
Dische, F.E.	Dulwich Hospital, London, U.K.
Draenert, K. & Y.	Univ. Munich, West Germany
Dronzek, B.	Univ. Manitoba, Winnipeg, Canada
Eades, J.L.	Univ. Florida, Gainesville, FL

Eades, J.A.	Univ. Bristol, U.K.
Echlin, P.	Univ. Cambridge, U.K.
Ekelund, S.	Swedish Institute for Metals Research
Elliott, J.C.	London Hosp. Medical College, U.K.
Engel, A.	Univ. Basel, U.K.
Eurell, J.A.	Texas A&M University, College Station, TX
Fairing, J.D.	Monsanto Company, St. Louis
Falk, R.H.	Univ. California, Davis
Farber, P.A.	Temple Univ. Dental School, Philadelphia
Fenoglio, C.M.	College of Physicians and Surgeons, NYC
Fernquist, R.G.	Xerox Corporation, Webster, NY
Fiori, C.E.	Natl. Inst. Health, Bethesda, MD
Fuchs, W.	Univ. Saarlands, West Germany
Galil, K.A.	Univ. Western Ontario, London, Canada
Gantt, D.G.	Senckenberg Museum, Frankfurt, W. Germany
Gedcke, D.A.	EG&G Ortec, Oak Ridge, TN
Geiss, R.H.	IBM Research Labs., San Jose, CA
Geller, J.D.	JEOL, USA, Inc.
Germinario, L.T.	Case Western Reserve Univ., Cleveland, OH
Goldstein, J.I.	Lehigh Univ., Bethlehem, PA
Gould, K.G.	Emory University, Atlanta, GA
Greday, T.	Ctr. Rech. Metall., Leige, Belgium
Grimes, G.W.	Hofstra Univ., Hempstead, NY
Gwinnett, A.J.	SUNY at Stony Brook, NY
Haggis, G.H.	Canada Dept. Agriculture, Ottawa
Hagler, H.K.	Univ. Texas, Dallas, TX
Hall, E.L.	Mass. Inst. Technology, Cambridge, MA
Harasaki, H.	Cleveland Clinic, OH
Hart, R.K.	Pasat Research Associates, Atlanta, GA
Haudenschild, C.C.	Mallory Inst. Pathology, Boston, MA.
Hawkes, P.W.	Lab. D'Optique Electronique, Toulouse, France
Hayes, R.L.	Morehouse College, Atlanta, GA
Heinrich, K.F.J.	National Bureau Standards, Wash., DC
Holland, V.F.	Monsanto Dev. Ctr., RTP, NC
Hooper, G.R.	Michigan State Univ., East Lansing, MI
Hren, J.J.	Univ. Florida, Gainesville, FL
Hutchinson, T.E.	Univ. Washington, Seattle
Hylander, W.	Duke University, Durham, NC
Ichinokawa, T.	Waseda University, Tokyo, Japan
Ingram, P.	Research Triangle Inst., North Carolina
Jakstys, B.P.	Univ. Illinois, Urbana
Jalanti, T.	Univ. Lausanne, Switzerland
Janossy, A.	Inst. Biophysics, Szeged, Hungary
Jessen, H.	Univ. Copenhagen, Denmark
Johnson, D.E.	Univ. Washington, Seattle
Johnston, I.S.	Univ. Calif., Los Angeles
Jones, A.L.	VA Medical Center, San Francisco, CA
Jones, M.P.	Imperial College, London, U.K.
Jones, S.J.	Univ. College, London, U.K.
Jongebloed, W.L.	Univ. Groningen, Netherlands
Joy, D.C.	Bell Laboratories, Murray Hill, NJ
Judd, G.	Rensselear Polutech. Inst., Troy, NY
Kaufmann, R.	Univ. Dusseldorf, West Germany
Keeley, R.H.	Metropolitan Police Forensic Sci. Lab., U.K.
Kim, K.M.	V.A. Hospital, Baltimore
Kimzey, S.L.	NASA- Johnson Space Ctr., Houston, TX
Kuhn, C.	Washington Univ., St. Louis, MO
Kuptsis, J.D.	IBM-Watson Research Ctr., Hopewell Jct., NY
Kyser, D.F.	IBM Reserach Labs, San Jose, CA
Lamvik, M.K.	Brandeis Univ., Waltham, MA
Lane, W.C.	Fremont, California
Langer, A.M.	Mt. Sinai School of Medicine, NYC
Lankford, J.	Southwest Research Inst., San Antonio, TX
Larrabee, G.B.	Texas Instruments, Dallas
LeCampion, T.	Stat. Marine d'Endoume, Marseille, France
Lee, R.J.	U.S. Steel Research Laboratory, Monroeville
Ledbetter, M.C.	Brookhaven National Lab., Upton, NY
LeFurgey, A.	Duke Univ. Med. Ctr., Durham, NC
LeGressus, C.	CEN-Saclay, France

Lieberman, A.R.	Univ. College, London, U.K.
Lifshin, E.	General Electric Co.- Corp. R&D, Schenectady
Lucas, D.M.	Canada Ctr. Forensic Sci., Toronto
Maggiore, C.J.	Los Alamos Sci. Lab., Los Alamos, NM
Makita, T.	Albany Medical College, NY
Margolis, S.V.	Univ. Hawaii, Honolulu
Marks. S.C., Jr.,	Univ. Mass. Med. Ctr., Worcester, MA
Marshall, A.T.	La Trobe Univ., Australia
Maser, M.D.	Marine Biol. Lab., Woods Hole, MA
Massover, W.H.	Brown Univ., Providence, RI
Matthews, J.L., W. Davis, and R. Jones	Univ. Texas, Dallas
Matricardi, V.R.	FBI Laboratory, Washington DC
McCarthy, J.	Tracor-Northern Inc., Middleton, WI
McConville, R.L.	Univ. Illinois, Urbana
McKinney, W.R.	Brookhaven National Lab., Upton, NY
Meuzelaar, H.L.C.	Univ. Utah, Salt Lake City
Miller, S.C.	Univ. Utah, Salt Lake City
Mills, J.W.	Mass. Gen. Hospital, Boston
Mizuhira, V.	Tokyo Medical and Dental Univ., Japan
Moll, S.H.	AMRAY Corp., Bedford, MA
Morton, R.	Bausch and Lomb, Rochester, NY
Mundell, R.D.	Univ. Pittsburgh, PA
Murphy, J.A.	Univ. Southern Illinois, Carbondale, IL
Nagatani, T.	Hitachi Ltd., Naka Works, Japan
Newbury, D.E.	National Bureau of Standards, Wash., D.C.
Newesely, H.	Free Univ. Berlin, W. Germany
Nielsen, N.A.	Dupont Experimental Station, Wilmington,DE
Nowell, J.	Univ. Calif., Santa Cruz
Okagaki, T.	Univ. Minnesota, Minneapolis,MN
Pallaghy, C.K.	La Trobe Univ., Australia
Pameijer, C.H.	Boston Univ., MA
Pande, S.C.	Brookhaven National Lab., Upton, NY
Panessa-Warren, B.J.	SUNY at Stony Brook, NY
Pappas, G.	Univ. Illinois Medical Center, Chicago
Parikh, M.	IBM Research Lab., San Jose, Calif.
Parham, R.A.	ITT Rayonier Inc., Shelton, WA
Parsons, D.F.	NY State Dept. Health, Albany
Pawley, J.B.	Univ. Wisconsin, Madison
Pearson, E.F.	Metropolitan Police Forensic Sci. Lab., U.K.
Pearson, C.	Canberra College of Advanced Education, Austr.
Perkins, R.D.	Duke University, Durham, NC
Perlaki, F.	Hewlett-Packard, Palo Alto, CA
Peters, K.-R.	Yale Univ. Medical School, CT
Pfefferkorn, G.E.	University of Munster, West Germany
Pickering, N.E.	Rome Air Development Ctr., Bedford, MA
Pritzker, K.P.H.	Mount Sinai Hosp., Toronto, Canada
Powell, C.J.	National Bureau of Standards, Washington, DC
Ratner, B.D.	Univ. Washington, Seattle, WA
Reed, S.J.B.	Cambridge Univ., U.K.
Reid, A.F.	CSIRO Mineral Technology, Australia
Reimer, L.	Univ. Munster, West Germany
Reiter, R.J.	Univ. Washington, Seattle
Rice, R.W.	Texas A&M University, College Station, TX
Rick, R.	Univ. Munich, West Geramny
Robards, A.W.	Univ. York, U.K.
Robinson, V.N.E.	Univ. New South Wales, Australia
Roinel, N.	CEN Saclay, France
Roomans, G.M.	Wenner-Gren Inst., Stockholm, Sweden
Rose, J.C.	Univ. Arkansas, Fayetteville
Rosen, J.J.	Univ. Washington, Seattle
Rosen, S.	Beth Israel Hospital, Boston
Ruffolo, J.J.	Medical College Virginia, Richmond
Russ, J.C.	EDAX International, Prairie View, IL
Ryder, P.L.	Univ. Bremen, West Germany
Sachs, I.B.	Forest Products Labotarory, Madison, WI
Saleuddin, A.S.M.	York Univ., Downsview, Ont., Canada
Saubermann, A.J.	Beth Israel Hospital, Boston, MA

Schidlovsky, G.   Brookhaven National Lab., Upton, NY
Schneider, G.B.   Univ. Mass. Medical Center, Worcester, MA
Sevely, J.C.   Lab. D'Optique Electron., Toulouse, France
Shelburne, J.D.   Duke University Med. Ctr., Durham, NC
Shellis, R.P.   MRC Dental Unit., Bristol, U.K.
Shih, C.Y.   Univ. Iowa, Iowa City
Siegesmund, K.A.   Medical Coll. Wisconsin, Milwaukee
Simkiss, K.   Univ. Reading, U.K.
Simson, J.A.V.   Medical Univ. South Carolina, Charleston
Sinclair, R.   Stanford University, CA
Sjöström, M.   Univ. Umea, Sweden
Skobe, Z.   Forsyth Dental Center, Boston, MA
Small, J.A.   National Bureau of Standards, Wash., D.C.
Smith, M.M.   St. George's Hosp. Med. School, London, U.K.
Sogard, M.   Univ. Chicago, IL
Sorauf, J.E.   SUNY at Binghamton, NY
Spector, M.   Medical Univ. South Carolian, Charleston
Spence, J.C.H.   Arizona State Univ., Tempe
Spurr, A.R.   Univ. Calif., Davis
Statham, P.J.   Link Systems Ltd., U.K.
Sturcken, E.F.   Savanah River Labs., Aiken, SC
Swift, J.A.   Unilever Research Labs., Isleworth, U.K.
Tannenbaum, M. & S.   College of Physicians and Surgeons, NYC
Taylor, R.L.   Los Angeles County Coroner, CA
Thompson, M.N.   Philips, Netherlands
Thomas, R.S.   USDA-ARS, West. Reg. Res. Lab., Berkeley
Thrower, P.A.   Penn. State Univ., Univ. Park, PA
Thurston, E.L.   Texas A&M Univ., College Station, TX
Todd, G.L.   Univ. Nebraska Med. Ctr., Omaha
Tompa, A.   Univ. Michigan, Ann Arbor, MI
Towe, K.M.   Smithsonian Institution, Washington, DC
Trump. B.F.   Univ. Maryland Med. School, Baltimore, MD
VanderSande, J.B.   Mass. Inst. Technology, Cambridge, MA
Vassamillet, L.F.   Carnegie-Mellon Univ., Pittsburgh
Venables, J.A.   Univ. Sussex, Brighton, Sussex, U.K.
von Koenigswald, W.   Frankfurt, West Germany
Wall, J.S.   Brookhaven National Laboratory, Upton, NY
Ward, C.M.   Lawrence Livermore Laboratory, Livermore, CA
Watabe, N.   Univ. South Carolina, Columbia, SC
Weiss, R.E.   National Inst. Health, Bethesda, MD
Wells, O.C.   IBM Watson Research Laboratory, NY
Whittaker, D.K.   Dental School, Heath, Cardiff, U.K.
Wiggins, W.   Johns Hopkins Univ., Baltimore
Wilbur, K.M.   Duke Univ., Durham, NC
Williams, D.B.   Lehigh Univ., Bethlehem, PA
Wise, S.W., Jr.   Florida State University, Tallahassee
Wolbarsht, M.L.   Duke Univ. Medical Center, Durham, NC
Wroblewski, R.   Wenner-Gren Inst., Stockholm, Sweden
Zimny, M.L.   Louisiana State Univ. Med. Ctr., New Orleans

Major Subjects Index

For

NONBIOLOGICAL PAPERS

This index should be used in conjunction with either Table of Contents, or Author Index, or Subject Index of Scanning Electron Microscopy/1979/II or I as applicable. Biological papers have been classified in a similar index included at pages xi-xvi of Scanning Electron Microscopy/1979/III.

The index below contains only an abridged title, name of the first author, and the page number where a paper begins.

PAPERS OF GENERAL INTEREST, ELECTRON-SPECIMEN INTERACTIONS, IMAGE PROCESSING, TECHNIQUES ETC.

KEYNOTE PAPER: MICROANALYSIS IN THE SEM: PROGRESS AND PROSPECTS; D.E. Newbury (1/II

For other papers on microanalysis see papers under Analytical Electron Microscopy.

TUTORIAL: ELECTRON-SPECIMEN INTERACTIONS; L. Reimer (111/II

REVIEW: A TRANSPORT EQUATION THEORY OF ELECTRON BACKSCATTERING; D.J. Fathers (55/I

CHARGE NEUTRALIZATION USING VERY LOW ENERGY IONS; C.K. Crawford (31/II

TUTORIAL: VISUALIZATION OF SUBSURFACE STRUCTURES IN CELLS AND TISSUES BY BACKSCATTERED ELECTRON IMAGING; R.P. Becker (835/II

FURTHER DEVELOPMENT OF THE CONVERTED BSE DETECTOR; S.H. Moll (149/II

See also: Zeldes (155/II), Matricardi (159/II), Moza (473/I), Green (495/I)

THIN FILMS FOR HIGH RESOLUTION CONVENTIONAL SEM; P. Echlin (21/II

COATING BY ION SPUTTERING DEPOSITION FOR ULTRAHIGH RESOLUTION SEM; J.D. Geller (355/II

REVIEW: SEM AT MACROMOLECULAR RESOLUTION IN LOW ENERGY MODE ON BIOLOGICAL SPECIMENS COATED WITH ULTRA THIN METAL FILMS; K.-R. Peters (133/II

See papers under biological specimen preparation, Chatfield (563/I).

REVIEW: A THEORY OF SURFACE-ORIGINATING CONTAMINATION AND A METHOD FOR ITS ELIMINATION; J.T. Fourie (87/II

REDUCTION OF CONTAMINATION IN ANALYTICAL EM; Y. Harada (103/II

SECONDARY ELECTRON EMISSION DEPENDENCE ON ELECTRON BEAM DENSITY DOSE AND SURFACE INTERACTIONS FROM AES AND ELS IN ULTRA HIGH VACUUM SEM; C. Le Gressus (161/I

ON-LINE TOPOGRAPHIC ANALYSIS IN THE SEM; D.M. Holburn (47/II

A PRECISION SEM IMAGE ANALYSIS SYSTEM WITH FULL FEATURE EDXA CHARACTERIZATION; W.R. Stott (53/II

See also: On contamination - Bentley (581/II) On image processing - Lebiedzik (61/II), Dixon (361/II), Llinas (367/II), Moza (473/I), McAlear (729/III)

ELECTRON SOURCES, ELECTRON OPTICS

REVIEW: THERMAL CATHODE ILLUMINATION SYSTEMS FOR ROUND BEAM ELECTRON PROBE SYSTEMS; A.N. Broers (1/I

BRIGHTNESS OF SINGEL CRYSTAL LaB_6 CATHODES..; R. Shimizu (11/I

MOUNTING METHODS AND OPERATING CHARACTERISTICS FOR LaB_6 CATHODES; C.K. Crawford (19/I

BASIC STUDY OF TF EMISSION; N. Tamura (31/I

...SOME ELECTROSTATIC GUN LENSES FOR FIELD EMISSION; J. Orloff (39/I

EMISSION CHARACTERISTICS OF A LIQUID GALLIUM ION SOURCE; L.W. Swanson (45/I

NOTE ON SEM ELECTRON OPTICS; O.C. Wells (52/I

See also: Oppolzer (111/I), Carpenter (153/I)

STEREO TECHNIQUES

TUTORIAL: THE PERCEPTION AND MEASUREMENT OF DEPTH IN SEM; A. Boyde (67/II

THE OBSERVATION AND MEASUREMENT OF DISPLACEMENTS AND STRAIN BY STEREOIMAGING; D.L. Davidson (79/II

See also: Holburn (47/II), Lebiedzik (61/II)

ANALYTICAL ELECTRON MICROSCOPY

See also: Newbury, keynote paper (1/II), papers classified under biological microanalysis in the index in SEM/1979/III (xii/III)

TUTORIAL: OPTIMIZATION OF AN AEM FOR X-RAY MICRO-ANALYSIS..; J. Bentley (581/II

...QUANTITATIVE ANALYSIS OF SINGLE PARTICLES..; J.A. Small (807/II

ADVANTAGES OF A STEM WITH A FIELD EMISSION GUN FOR X-RAY ANALYSIS OF INHOMOGENEITIES IN METALS AND CERAMICS; H. Oppolzer (111/I

FACTORS AFFECTING THE MEASUREMENT OF COMPOSITION PROFILES IN STEM; A.M. Ritter (121/I

TUTORIAL: RAMAN MICROPROBE ANALYSIS: PRINCIPLES AND APPLICATIONS; E.S. Etz (67/I

APPLICATIONS OF THE MOLECULAR OPTICAL LASER EXAMINER ..; F. Adar (83/I

See also: Blaha (93/I), Casciani (383/II), Vuhas (103/I). For Auger and other surface analytical techniques, see Semiconductor applications.

REVIEW: LASER MICROPROBE MASS ANALYSIS..; R. Kaufmann (279/II

REVIEW: PROGRESS IN THE QUANTITATION OF ELECTRON ENERGY-LOSS SPECTRA;   D.C. Joy   (817/II

ELECTRON ENERGY ANALYSIS IN A VACUUM GENERATORS HB5 STEM;   M.T. Browne   (827/II

See also: Le Gressus (161/I), Galle (703/II)

COMPUTER GRAPHICS ANALYSIS OF STEM IMAGES;   R. Llinas   (367/II

STEM IMAGES AT HIGH RESOLUTION: THE INFLUENCE OF DETECTOR GEOMETRY;   R.E. Burge   (127/I

DATA RECORDING AND REPLAY IN STEM;   S. Lackovic   (137/I

REVIEW: MASS MEASUREMENTS WITH THE ELECTRON MICROSCOPE;   J.S. Wall   (291/II

PREPARATION AND OBSERVATION OF VERY THIN, VERY CLEAN SUBSTRATES FOR STEM;   M. Ohtsuki   (375/II

ELECTRON CHANNELING AND MICRODIFFRACTION FROM CRYSTAL SURFACES;   G.G. Hembree   (145/I

ON THE PERFORMANCE OF A FIELD EMISSION GUN TEM/ STEM;   R.W. Carpenter   (153/I

See also: Fourie (87/II), Harada (103/II), and other papers under General Interest Papers.

PARTICULATES EMPHASIZING ASBESTOS

See also: Papers under General Interest Papers, Analytical Electron Microscopy, Surface analysis under Semiconductor Applications and Boyde (67/II), Gibbon (501/I), Pierce (555/I), Tassa (175/II), Wallace (179/II), Collins (439/I), Nolan (449/I).

TUTORIAL: PREPARATION AND ANALYSIS OF PARTICULATE SAMPLES BY ELECTRON MICROSCOPY, WITH SPECIAL REFERENCE TO ASBESTOS; E.J. Chatfield   (563/I

CHRYSOTILE, PALYGORSKITE AND HALLOYSITE IN DRINKING WATER;   J.R. Millette   (579/I

See: Small (807/II)

MOLECULAR ANALYSIS OF MICROSCOPIC SAMPLES WITH A RAMAN MICROPROBE:...PARTICULATE CHARACTERIZATION;   J.J. Blaha   (93/I

DEVELOPMENT OF METHODS TO ISOLATE ASBESTOS FROM SPIKED BEVERAGES AND FOODS FOR SEM CHARACTERIZATION;   J.T. Stasny   (587/I

ASBESTOS FIBRE COUNTING BY AUTOMATIC IMAGE ANALYSIS;   R.N. Dixon   (361/II

SEDIMENTS

TUTORIAL: PETROGRAPHIC ANALYSIS OF BITUMINOUS COAL: OPTICAL AND SEM IDENTIFICATION OF CONSTITUENTS;   R.W. Stanton   (465/I

INORGANIC ELEMENT ANALYSIS OF COAL PARTICLES USING COMPUTER EVALUATION OF SEM IMAGES;   A.K. Moza   (473/I

ORGANIC AND INORGANIC SULFUR IN COAL;   R.T. Greer   (477/I

SEM STUDY OF SOLUBILIZATION OF COAL UNDER MILD CONDITIONS;   J.H. Shinn   (487/I

THE APPLICATION OF HEAVY METAL STAINING (OsO_4) AND BSE IMAGING FOR THE DETECTION OF ORGANIC MATERIAL IN GAS AND OIL SHALES;   D.A. Green   (495/I

REVIEW: MICROCHARACTERIZATION OF FLY-ASH AND ANALOGS:..;   D.L. Gibbon   (501/I

VOLCANIC ASH: SOME EXAMPLES OF DEVITRIFICATION AND EARLY DIAGENESIS;   S.W. Wise, Jr.   (511/I

PARTICULATE MATERIAL SUSPENDED IN ESTUARINE AND OCEANIC WATERS;   J.W. Pierce   (555/I

CLAY FABRIC AND RELATED PORE GEOMETRY OF SELECTED SUBMARINE SEDIMENTS..; R.H. Bennett   (519/I

...STUDY OF THE TEXTURAL RELATIONSHIP OF KAOLINITE CLAY FLOCCULATED IN INCREASINGLY SALINE SOLUTIONS;   W.P. Lanier   (525/I

SEM STUDIES OF TRIASSIC RESERVOIR SANDSTONES..;   V.S. Colter   (531/I

MORPHOLOGY AND CHEMISTRY OF MANGANESE MICRONODULES ..;   V.K. Berry   (539/I

QUARTZ SILT PRODUCTION AND SAND GRAIN SURFACE TEXTURES FROM FLUVIAL AND GLACIAL ENVIRONMENTS;   W.B. Whalley   (547/I

See also: Papers under General Interest Papers, Analytical Electron Microscopy, Particulates, Mineralized Tissue (xvi/III), Boyde (67/II), Charola (379/I), Gouda (387/I), Yuhas (103/I), and Surface Analysis Papers under Semiconductor Applications.

ARCHAELOGY AND ART HISTORY

REVIEW: REPLICATION TECHNIQUES WITH NEW DENTAL IMPRESSION MATERIALS IN COMBINATION WITH DIFFERENT NEGATIVE IMPRESSION MATERIALS.   C.H. Pameijer   (571/II

FUNCTIONAL ANALYSIS OF DRILLING ON ANCIENT NEAR EASTERN SEALS USING SEM;   L. Gorelick   (405/I

REVIEW: DAGUERREOTYPES: A STUDY OF THE PLATES AND THE PROCESS;   A. Swan   (411/I

SEM AND X-RAY MICROANALYSIS OF ARTIFACTS RETRIEVED FROM MARINE ARCHAEOLOGICAL EXCAVATIONS..;   V.C. Barber   (885/II

INLAYED TEETH OF ANCIENT MAYANS: A TRIBOLOGICAL STUDY..;   A.J. Gwinnett   (575/II

SEM AND THE LAW

APPLICATIONS OF SEM AND X-RAY MICROANALYSIS IN CRIMINALISTICS: AN INDEXED BIBLIOGRAPHY;   R.L. Taylor   (209/II

FIREARM IDENTIFICATION BY EXAMINATION OF BULLET FRAGMENTS:..;   R.L. Taylor   (167/II

REVIEW: PRACTICAL EXPERIENCES WITH SEM IN A FORENSIC SCIENCE LABORATORY;   J. Andrasko (879/II

REVIEW: FORENSIC APPLICATIONS OF SEM/EDX;   E.H. Sild   (185/II

REVIEW: THE APPLICATION OF BSE IMAGE TO FORENSIC SCIENCE;   V.R. Matricardi   (159/II

CONVERSION OF EXISTING SEM COMPONENTS TO FORM AN EFFICIENT BSE DETECTOR AND ITS FORENSIC APPLICA- TIONS;   N. Zeldes   (155/II

THE IDENTIFICATION OF TYPES AND BRANDS OF COATED ARC ELECTRODES..;   K. Mogami   (433/I

IDENTIFICATION OF DEBRIS FROM THE OXYGEN OR ABRASIVE CUTTING OF SAFES;   B. Collins   (439/I

...IDENTIFICATION OF FOREIGN TOOL MARKS IN THE INTERIOR OF PICKED CYLINDER LOCKS;   M. Tassa   (445/I

COMPARISION AND CLASSIFICATION OF SMALL PAINT
FRAGMENTS..; P.J. Nolan (449/I

ANALYSIS OF COUNTERFEIT GOLD COINS..;
J. Andrasko (455/I

CLASSIFICATION OF SMALL GLASS FRAGMENTS..;
R.H. Keeley (459/I

THE CHARACTERIZATION OF DYNAMITE WRAPPING PAPER..;
M. Ueyama (203/II

QUANTITATIVE INVESTIGATION OF SULPHUR AND CHLORINE
IN HUMAN HEAD HAIRS..; S. Seta (193/II

A METHOD FOR PREPARING FIREARM RESIDUE SAMPLES FOR
SEM; J.S. Wallace (179/II

CHARACTERIZATION OF GUNSHOT RESIDUE PARTICLES..;
M. Tassa (175/II

See also: Papers listed under various catagories
as applicable to individual situations relating to
application of SEM to Law.

MATERIAL CHARACTERIZATION

See also: Papers under General Interest Papers,
Analytical Electron Microscopy, Particulates,
SEM and the Law, Stereo Techniques, and Surface
Analysis papers under Semiconductor Applications.
Also See: Failure Analysis, Pameijer (571/II),
Swan (411/I)

SURFACE CHARACTERIZATION OF FILM GROWTH ON LITHIUM
METAL; K.M. Black (363/I

SEM CHARACTERIZATION OF ALUMINUM SURFACES ANODIZED
IN PHOSPHORIC ACID; T.P. Remmel (369/I

SCANNING AUGER MICROSCOPY OF ISOLATED METALLIC
PARTICLES ON A POORLY CONDUCTING SUBSTRATE;
G. Todd (207/I

DIRECT OBSERVATION OF THE GROWTH OF VOIDS IN
MULTIFILAMENTARY SUPERCONDUCTING MATERIALS VIA
HOT STAGE SEM; J.L.-F. Wang (399/I

EFFLORESCENCES ON BUILDING STONES..;
A.E. Charola (379/I

CLINKER CHARACTERIZATION BY SEM; G.R, Gouda (387/I

ACOUSTIC MICROSCOPY, SEM AND OPTICAL MICROSCOPY:
CORRELATIVE INVESTIGATIONS IN CERAMICS;
D.E. Yuhas (103/I

MORPHOLOGICAL IDENTITY OF POROSITY IN CHEMICALLY
ACTIVATED ACRYLIC CEMENTS; J.C. Keller (425/I

FAILURE ANALYSIS

TUTORIAL: FAILURE ANALYSIS- HOW TO CHOOSE THE
RIGHT TOOL; B.E. Boardman (339/I

TUTORIAL: PRESERVATION AND CLEANING OF FRACTURES
FOR FRACTOGRAPHY; R.D. Zipp (355/I

For Replicas: see Pameijer (571/II)
For Auger Electrons: see papers under surface
analysis under Semiconductor Applications.
For Legal aspects; see papers under SEM & the Law
See also General Interest Papers, Analytical
Electron Microscopy, Particulates, and Materials
Characterization.

USE OF MICROTOPOGRAPHY CAPABILITY IN THE SEM FOR
ANALYSING FRACTURE SURFACES;
J. Lebiedzik (61/II

SEMICONDUCTOR APPLICATIONS INCLUDING SURFACE
ANALYSIS

See also General Interest Papers, Electron Sources,
Analytical Electron Microscopy.

TUTORIAL:.. X-RAY PHOTOELECTRON AND AUGER ELECTRON
SPECTROSCOPIES TO EVALUATE MICROELECTRONIC
PROCESSING; G.E. McGuire (173/I

SOME RECENT APPLICATIONS OF THE SCANNING AUGER
MICROPROBE TO SEMICONDUCTORS;
T.J. Shaffner (183/I

REVIEW: INTERDIFFUSION IN Au REFRACTORY THIN FILM
SYSTEMS STUDIED BY A COMBINATION OF SCATTERING
TECHNIQUES; A. Christou (191/I

SURFACE FAILURES RELATED TO CARBON CONTAMINATED
LAYERS; G. Tissier (203/I

See also: Le Gressus (161/I), Todd (207/I), Janssen
(259/II), Kaußmann (279/II), Newbury (1/II), Etz
(67/I), Adar (83/I), Bentley (581/II).

SOME SIGNAL TO NOISE CONSIDERATIONS FOR SCANNING
AUGER MICROSCOPY; C.T. Hovland (213/I

PROGRESS WITH IMAGE PROCESSING ON A SCANNING AUGER
MICROPROBE; T.J. Shaffner (219/I

...TAILORED MODULATION TECHNIQUES TO IMAGE BSE IN A
SCANNING AUGER MICROPROBE; G.L. Jones (225/I

SPATIALLY RESOLVED CL STUDIES OF GaP LED's..;
K. Löhnert (229/I

ELECTRON BEAM EFFECTS OBSERVED IN CL AND AES IN
NATURAL MATERIALS..; G. Remond (237/I

APPLICATIONS OF STROBOSCOPIC CL MICROSCOPY
G.V. Saparin (267/I

APPLICATIONS OF CL METHOD IN BIOLOGY AND MEDICINE
(REVIEW); W. Bröcker (125/II

REVIEW: ELECTRON BEAM DEPTH PROFILING IN SEMI-
CONDUCTORS; G.E. Possin (245/I

REVIEW: CONTRAST FORMATION IN SEM-CHARGE COLLECTION
IMAGES OF SEMICONDUCTOR DEFECTS;
C. Donolato (257/I

TUTORIAL: LITHOGRAPHIC PROCESSES IN VLSI CIRCUIT
FABRICATION; M. Hatzakis (275/I

REVIEW: VLSI TESTING USING THE ELECTRON PROBE;
H.-P. Feuerbaum (285/I

ELECTRON BEAM TEST SYSTEM FOR VLSI CIRCUIT
INSPECTION; E. Menzel (297/I

REVIEW: ELECTRON BEAM CHOPPING SYSTEMS IN THE SEM;
E. Menzel (305/I

VOLTAGE CONTRAST OBSERVATIONS OF SURFACE ACOUSTIC
WAVES; N.D. Wittles (319/I

...VISUALIZATION AND MEASUREMENTS OF SURFACE
POTENTIALS..; E.I. Rau (325/I

STUDIES OF SEM SURFACE PATTERNS OF OSCILLATING QUAR
QUARTZ CRYSTALS..; H. Bahadur (333/I

TEACHING SEM

For papers on teaching SEM (of interest to both
biologists and nonbiologists) see pages 217-258
of SEM/1979/II, and an index of these papers in
the Table of Contents of SEM/1979/II.

General Information

▶ SCANNING ELECTRON MICROSCOPY, Inc.

    SEM Inc. is a non-profit organization devoted to promotion and advancement of the science of SEM
and related material characterization techniques; to promote applications of these techniques in existing
and new areas of investigations; and to promote these techniques so that users of these instruments
obtain maximum information of highest quality from their instruments.

    SEM Inc. sponsors the annual SEM meetings and publishes the annual journal "SCANNING ELECTRON
MICROSCOPY/year" devoted to advances in instrumentation, techniques, theory, interpretation and
applications of the SEM and related techniques. Sponsorship of other activities to promote the aims of
SEM Inc. are being continuously explored. Please send your suggestions to one of the following officers:

John D. Fairing, President
809 Westwood Drive
Ballwin, MO 63011
Phone: 314-694-5007

Robert P. Becker, Vice-President
1S640 Brook Court,
Glen Ellyn, IL 60137
Phone: 312-996-6791

**Om Johari, Secretary-Treasurer
P.O.Box 66507
AMF O'Hare, IL 60666
Phone 312-529-6677

Irving B. Sachs, Asst. Director
SEM Meetings
4402 Vale Circle,
Madison, WI 53705
Phone: 608-257-2211 Ext. 249

** General contact

▶ SCANNING ELECTRON MICROSCOPY/1980

                           April, 21-25, 1980 --- CHICAGO, ILLINOIS

    The planning for SEM/1980 has already started. Your suggestions for review and tutorial papers,
and workshop topics are earnestly requested; please forward them at the earliest to Om Johari. The
tentative outline of important dates is as follows:

CALL FOR PAPERS ISSUED:	Sept. 1, 1979
Abstracts Due	Nov. 1, 1979
Program Outline available	Dec. 10, 1979
Manuscripts of accepted abstracts due	Jan. 15, 1980
Meeting Dates	Apr. 21-25, 1980

    The meetings will take place at McCormick Inn, 23rd and Lake Shore Drive; Chicago, IL 60616, USA.
For room reservations, contact the Inn directly; those in USA may call 800-621-6909.Economy rooms will be
available at McCormick Inn's affiliated motels (on a regular bus route from the Inn).

    For all information about the meetings contact Om Johari.

▶ MAILING LIST INFORMATION:

    Immediately after the SEM/1979 meetings we intend to start computerizing our mailing lists. All those
who have registered for the 1979 meetings, or ordered the SEM volume during the last 12 months or wrote
to us to get on our lists will be automatically added to the list. Please send name, complete postal
address (including zip or other postal codes) and telephone number to Om Johari at the above address.
Our lists are organized by zip/postal codes, therefore we cannot add or retain your name on our lists if
the zip code information is missing.

▶ SEM Inc. Publications:*

Scanning Electron Microscopy/1978/I	Physical, technical and General interest papers	$37.00*
Scanning Electron Microscopy/1978/II	Biological techniques and applications	$40,50*
SEM/1978/set of above two parts		$67.50*
SEM of Brain Ventricular Surfaces-- Republished from SEM/1978/II		$14.95*

Scanning Electron Microscopy/1979 is being published in three parts:

          Part I :   Papers primarily of interest to physical scientists
          Part II :   Technical and other papers of general interest
          Part III: Papers primarily of interest to biologists.

One must order at least part II and one other part, current prices for SEM/1979 are as follows:

Scanning Electron Microscopy/1979/P parts I and II	(2 parts)	$65.50*
Scanning Electron Microscopy/1979/B parts II and III	(2 parts)	$67.00*
Scanning Electron Microscopy/1979/S set of parts I,II and III	(3 parts)	$84.00*

These prices are in effect from April 1, 1979 and subject to change without notice.

Illinois residents must add 5% state sales tax.

*Those from outside the United States must add $2.00 per part for postage.

SCANNING ELECTRON MICROSCOPY/1979/II
SEM Inc., AMF O'Hare, IL 60666, USA

KEYNOTE PAPER: MICROANALYSIS IN THE SCANNING ELECTRON MICROSCOPE:
PROGRESS AND PROSPECTS

D. E. Newbury

National Bureau of Standards
Analytical Chemistry Division
Washington DC 20234

Abstract

Many techniques of analysis are available
for the characterization of a sample on the micrometer scale in the scanning electron microscope.
Directly available as a result of the electron
bombardment are x-rays, Auger electrons, and long
wavelength electromagnetic radiation (cathodoluminescence). Auxiliary excitation sources
which can be added to the SEM for analysis include
energetic ion beams for secondary ion mass
spectrometry and ion scattering spectrometry,
x-rays for x-ray-induced characteristic x-ray
emission (x-ray fluorescence), and monochromatic
photons from laser sources for Raman spectroscopy
and mass spectrometry. The information available
from these techniques includes elemental and
molecular composition, structural information from
crystals, and a variety of semiconductor properties. The information can be obtained from
regions with lateral dimensions of the order of
1 µm and with depth resolutions ranging from one
atom layer to several micrometers. Qualitative
and quantitative analysis can be carried out for
all chemical elements, with relative errors in
the best cases being less than five percent, and
with limits of detection in the range 1-100 ppm
under optimum conditions. No one technique
provides all of the information required for
total characterization of a sample, and combinations of techniques are required.

KEY WORDS: Microanalysis, Auger Electron
Spectroscopy, Cathodoluminescence, X-ray
Microanalysis, Ion Scattering Spectrometry,
Laser Microprobe Mass Analyser, Laser Raman
Microprobe, Secondary Ion Mass Spectrometry

Introduction

A constant theme in the literature of scanning electron microscopy (SEM) in recent years
has been the joint application of microanalysis
techniques with SEM imaging techniques to greatly
enhance the capabilities for specimen characterization. Historically, the techniques of microanalysis and scanning electron microscopy tended
to develop separately until about 1968. The
roots of both techniques go well back into the
1930's, but the development of the modern instruments can be firmly dated from the initiation of
Prof. Oatley's research group in scanning electron
microscopy at Cambridge University in 1948 and
the publication of Prof. Castaing's thesis
describing the electron probe microanalyzer in
1951.[1,2] During the 1950's, scanning electron
microscopes were developed with the principal
goal of imaging with electron signals, while
electron probe microanalyzers were developed with
static beams to measure x-ray signals exclusively.
In the late 1950's and early 1960's, experiments
were carried out to employ scanning techniques in
the electron probe microanalyzer, primarily to
provide x-ray intensity maps showing elemental
distributions.[3,4] These scanning electron microprobes could also make use of the specimen
current signal for electron imaging, although the
probe size limited the resolution to the range
0.1-1 micrometers. Finally, the development of
an energy dispersive x-ray spectrometer based on
the lithium-drifted silicon detector provided an
eminently suitable means for measuring x-ray
spectra in the scanning electron microscope.[5]
Since the publication of the first paper
describing the union of such a spectrometer to an
analytical electron beam instrument in 1968, much
progress has been made in combining microanalysis
with microscopy.[6] The classical electron probe
microanalyzer has been designed to include electron optics and electron detectors comparable to
a high performance scanning electron microscope.
Meanwhile, x-ray spectrometers of both the wavelength dispersive and energy dispersive types
have been attached to conventional scanning electron microscopes. Indeed, the only features
which distinguish the classical electron probe
microanalyzer configuration are a high x-ray
emergence angle and the inclusion of an optical
microscope for accurate placement of the sample

at the focussing position of the wavelength spectrometers. These requirements have now been incorporated with high resolution electron imaging to produce an instrument with uncompromised performance both as a scanning electron microscope and as an electron probe microanalyzer.

To augment and extend the compositional information available in the x-ray signal, developments have taken place to employ other compositionally-related signals simultaneously generated during electron bombardment, including Auger electrons and long wavelength electromagnetic radiation (cathodoluminescence). Finally, in the past several years, additional analytical techniques which employ auxiliary excitation sources have been added to the SEM. Examples include secondary ion mass spectrometry from ion bombardment and x-ray fluorescence induced from x-rays produced in a target foil by the electron beam.

In this paper, the present state of development of each of these techniques will be briefly reviewed with special emphasis on spatial and depth resolution, information derived from a specimen, accuracy of qualitative and quantitative analysis, sensitivity, and limitations. In addition, several new techniques which might be included in SEM systems of the future, such as the laser Raman microprobe, laser microprobe mass analyzer, and ion scattering spectrometer, will also be discussed.

Definitions and General Concepts

For the purposes of the following discussion, it is important to establish the meanings of a number of important terms within the context with which they are used in this field:

(1) Microanalysis refers to the identification and chemical analysis of very small volumes of matter. For the techniques of microanalysis which we will consider in this paper, the analyzed sample mass is of the order of 10^{-12}g and the volume sampled is of the order of 1 μm^3. Note that the term microanalysis implies nothing about the limits of detection within the mass which is sampled.

(2) Interaction volume: The region of a sample excited by the primary radiation. The size of the interaction volume is typically determined by the elastic and inelastic scattering characteristics of the primary radiation in the target.

(3) Sampling volume: The region of the sample from which the secondary radiation employed for analysis is generated. The absorption and scattering characteristics of the secondary radiation determine what fraction of the interaction volume is represented by the sampling volume. In some cases, the sampling volume may exceed the size of the interaction volume of the primary radiation, e.g., when effects such as x-ray induced fluorescence occur.

(4) Limit of detection (sensitivity): The limit of detection refers to the mass fraction of a given constituent in the sample mass which can be distinguished by an analytical technique. From counting statistics, the limit of detection is given by that characteristic signal which is three times the standard deviation of the background at the peak position.[7] By such a definition, the probability of the background count exceeding three standard deviations of the average of adjacent points is 0.2 percent. It is important to note that the limit of detection is frequently a function of the instrumental operating parameters, e.g., beam energy, and specimen characteristics, such as the composition of the matrix in which the element of interest is dispersed. The limit of detection is therefore listed in the following discussion with the caution that a particular analytical situation might deviate substantially from the value given.

(5) Trace analysis - trace analysis refers to the measurement of a minute fraction of a constituent within the sample being analyzed. The level of a component of a sample which constitutes a "trace amount" (or a "minor" or "major" component) is somewhat subjective, but a practical scale of these terms for microanalysis is:

major constituent: 10 percent or greater
minor constituent: 0.1 to 10 percent
trace constituent: less than 0.1 percent (1000 parts per million (ppm))

Symbol Table

A = atomic weight (g/mole)

B = background

C = concentration

E = beam energy

E_c = critical excitation energy

EI = exponential integral

E_0 = initial beam energy

i = measured signal

J = mean ionization energy

k = intensity ratio between sample and standard

k_A = absorption correction

k_c = correction for continuum fluorescence

k_F = correction for characteristic fluorescence

k_Z = atomic number correction

P = peak

R = electron range

S_j = absolute sensitivity factor for constituent j

$S_{X/R}$ = relative sensitivity factor for X relative to R

U = overvoltage

ϵ = SIMS matrix correction factor

ρ = density

σ = cross section

Approaches to Quantitative Analysis

In carrying out quantitative analysis by any technique, three levels of sophistication exist: 1) direct use of calibration standards; 2) use of relative or absolute sensitivity factors; and 3) a "first principles" approach, in which the necessary corrections are calculated from a knowledge of the physics of the interaction of the primary and secondary radiation with matter.

Direct Use of Standards

The simplest approach to quantitative analysis involves the direct use of standards which are nearly identical to the unknown in composition. A suite of standards is employed which spans the composition range of interest. From this suite of standards, a "working curve" is determined which relates the analytical signal to the concentration for the constituent of interest. Analysis of the unknown is accomplished by comparing the measured intensity from the unknown with the working curve. In principle, this empirical approach to quantitative analysis will always give the best possible analytical accuracy relative to standards of known composition (as determined by an independent reference analytical technique) since there is no change in matrix (i.e., the chemical environment) between the specimen and the standards, thus eliminating matrix effects. The accuracy is limited by the number of standards available to define the working curve, especially in the case where the analyte is present at high concentrations, greater than one percent, where the working curve is likely to become non-linear due to inter element effects. The working curve method is the least flexible approach to quantitative analysis, since a new set of standards is required for each change in matrix of the unknown. As a general observation, the preparation of standards which are homogeneous on a micrometer scale and therefore suitable for microanalysis is quite difficult. Moreover, when surface analysis techniques are considered, the requirement for standards with a known surface composition further complicates the situation.

Sensitivity Factors

An absolute sensitivity factor, S_j, for the element j can be defined as

$$S_j = i_j/C_j \qquad (1)$$

where i is the measured signal and C_j is the known concentration. For analysis, S_j values are determined from standards and applied to unknowns. In general, absolute sensitivity factors are a function of the matrix and the instrument, and the accuracy of the quantitative analysis based upon such factors will be limited by matrix effects. Instrumental effects are minimized by working under standardized operating conditions.

Relative elemental sensitivity factors, $S_{x/R}$, are defined as:

$$S_{x/R} = (i_x/C_x)/(i_R/C_R) \qquad (2)$$

where x and R denote any two elements. Relative elemental sensitivity factors can be determined from standards and applied to the analysis of an unknown in the following way. A suite of sensitivity factors for all elements in a sample relative to a reference element R must be available. From equation (2), the concentration of an element x in terms of element R is given by:

$$C_x = (i_x/i_R)(C_R/S_{x/R}) \qquad (3)$$

If the set of values C_x represents all elements in the analyzed volume, or a known fraction thereof, then:

$$\sum_x C_x + C_R = \sum_x (i_x/i_R)(C_R/S_{x/R}) + C_R = 1 \qquad (4)$$

In equation (4), only C_R is an unknown and can therefore be immediately determined. From this value of C_R, the values of C_x for all remaining elements can be obtained from equation (3).

The advantage of the relative sensitivity factor approach to analysis is that by taking the response of one element relative to another in the same sample, the relative sensitivity factor is somewhat less dependent on matrix effects than the absolute sensitivity factor. A second advantage is that by determining the sensitivity factor on the same instrument as that used for the analysis of the unknown, local instrument factors which influence response are automatically included in the final analysis.

An example of a method which incorporates some aspects of the sensitivity factor approach in microanalysis is the alpha coefficient ("empirical method") which is widely used in electron probe microanalysis.[8,9]

"First Principles Analysis"

If the physics of the interaction of the primary and secondary radiation with matter is well known and suitable functions are available to describe these phenomena accurately, then a "first principles" approach to quantitative analysis can be developed. Such an approach should offer the maximum flexibility for analysis, since the need for standards is obviated by the capability of calculating all necessary quantities. However, such an approach is rarely practical in microanalysis. It is necessary to have accurate models not only for the interactions in the specimen, but also in the subsequent measuring system. Often the measuring system, including the processes of collection and detection, is difficult to describe with sufficient accuracy to permit a "pure" first principles approach. In such circumstances, it is common to fall back on a hybrid of the sensitivity factor method and the first principles method. By the use of a ratio of absolute sensitivity factors, comparing the unknown sample to a known standard, instrumental effects tend to cancel. The specimen matrix effects may then be calculated separately from a first principles approach. An example of this hybrid approach is the well known "ZAF" method used in x-ray microanalysis.[7,9]

Error Distributions

In discussing the accuracy of methods of quantitative analysis, it is important to realize that there is no unique error associated with a set of results. Rather, there exists a distribution of errors, such as that illustrated in

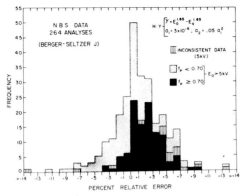

Figure 1. Histogram of errors observed in analyzing NBS certified standard alloys (SRM 479, 480, 481, 482, 483) by x-ray microanalysis with ZAF quantitative analysis. From Ref. (19).

Figure 1. In general, the errors associated with uncertainties in matrix corrections may be difficult to predict, although error propagation has been studied in detail in some cases.[10] Therefore in assessing the possible uncertainty to be associated with a final result, an experimentally determined error distribution like that in Figure 1 is invaluable. The determination of such an error distribution, however, requires the analysis of numerous samples of known composition to build up the histogram. Such error distributions are only available for a few techniques at the present time, and as development continues, analytical experience should allow the preparation of additional distributions. In applying the experience contained in such a histogram to the analysis of an unknown, it is necessary to be aware of the analytical conditions which were used to establish the error distributions and the consequences of deviations from those conditions. For example, in x-ray microanalysis, the ideal conditions for analysis require a flat, polished specimen set normal to the beam. Error distributions derived for this case cannot be presumed to apply to rough samples where the beam penetration and the angles of beam incidence and the x-ray emergence are poorly defined.

Microanalysis by Electron Bombardment

The Electron Interaction Volume

Energetic electrons which strike a target undergo elastic and inelastic scattering events, causing deviations in the direction of travel of the electrons and a loss of energy. A good qualitative description of the appearance of the interaction volume which develops from scattering can be obtained from the Monte Carlo electron trajectory simulation technique.[11] In Figure 2, despite the fact that the incident probe confines the beam electrons to an area of 10 nm diameter, scattering within the sample leads to the development of an interaction volume in excess of one micrometer in dimensions. A first estimate of the size of the interaction volume can be obtained from the Bethe range, which gives the total length of path in the solid. A useful formulation of this range is given by:[11]

$$R(cm) = \frac{J^2}{7.85 \times 10^4} \frac{A}{\rho Z} \left[EI(2 \log(1.166 \frac{E_o}{J})) - EI(2 \log(1.166 \frac{E_J}{J})) \right] \tag{5}$$

where A is the atomic weight (g/mole), ρ is the density (g/cm^3), Z is the atomic number, E_o is the beam energy (keV), E_J = 1.03 J, and J is the mean ionization potential:[9]

$$J = (9.76 Z + 58.5 Z^{-0.19}) \times 10^{-3} \text{ (keV)} \tag{6}$$

EI is the exponential integral given by:

$$EI(x) = \int_{-\infty}^{x} \frac{e^t}{t} dt = \log_e x + x + \frac{x^2}{2 \cdot 2!} + \dots + \frac{x^n}{n \cdot n!} + 0.5772$$

The maximum dimension of the interaction volume from the surface of the target will be less than the Bethe range due to the irregular nature of the trajectories, Figure 2(a), which results from elastic scattering. An example of an electron range which considers elastic and inelastic scattering, such as that determined by Kanaya and Okayama[12], is useful:

$$R = 2.76 \times 10^{-6} E_o^{1.67} A/(Z^{0.889} \rho) \text{ (cm)} \tag{7}$$

Throughout the interaction volume, inelastic scattering processes can take place which lead to the formation of several analytical signals, including x-rays, Auger electrons, and cathodoluminescence, providing the electrons have sufficient energy to excite the process of interest.

X-ray Microanalysis

Sampling volume. The cross section for inner shell ionization which leads to the formation of characteristic x-rays and Auger electrons, has the form

$$\sigma = \frac{7.92 \times 10^{-20}}{E_c^2} \frac{\log U}{U} \text{ (cm}^2) \tag{8}$$

where E_c is the critical excitation energy (keV) and U is the overvoltage, $U = E/E_c$.[13]

For a 20 keV beam incident upon a copper target, the distribution of K-shell ionization sites, as calculated by the Monte Carlo simulation is shown in Figure 2(b). Although the electron energy must be greater than 8.98 keV in order to cause ionization of the copper K-shell, the distribution of such ionization sites is nearly as large as the electron interaction volume. This is a result of the fact that as the electron energy decreases, elastic scattering becomes dominant, and the trajectory "curls up", limiting further penetration of the solid. Heinrich has derived a useful equation for assessing the size of the sampling volume in which the x-rays of interest are generated:[14]

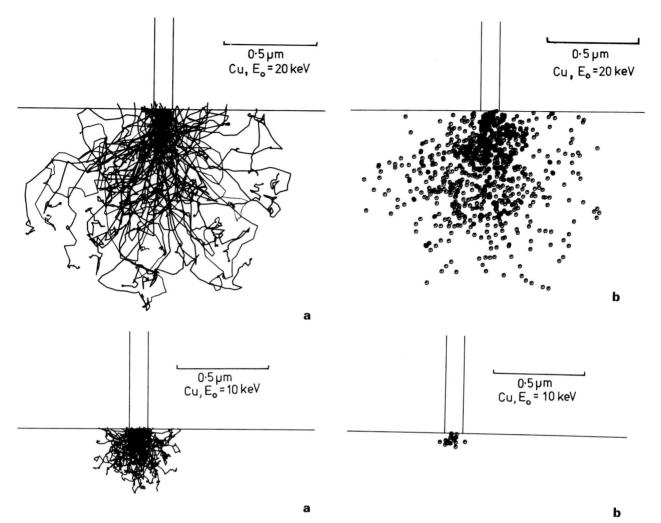

Figure 2. Monte Carlo electron trajectory simulations in copper at 10 and 20 keV: (a) electron trajectories; (b) sites of K-shell ionization events.

$$R = \frac{7 \times 10^{-6}}{\rho} (E_o^{1.65} - E_c^{1.65}) \ (cm) \qquad (9)$$

Choice of x-ray spectrometer. The analyst may be fortunate in having both an energy dispersive spectrometer (EDS) and a wavelength dispersive spectrometer (WDS) available for spectral measurement. An example of a spectrum measured by each type of spectrometer is shown in Figure 3. These two different types of spectrometers are properly regarded as complementary rather than competitive. The relative merits of these spectrometers have been discussed at length elsewhere.[7,9] Important points about the two spectrometers include:

(a) The WDS has superior resolution, 10 eV compared to 150 eV at the energy of manganese, which reduces the number of interferences of one peak upon another.

(b) The WDS has a superior peak-to-background ratio by a factor of 10 to 50 compared to EDS, which improves the limit of detection.

(c) The nearly simultaneous collection of all x-ray energies of interest by the EDS allows a faster qualitative analysis for major or minor elements, perhaps by as much as a factor of 20 over WDS.

(d) The overall efficiency of x-ray measurement is superior for the EDS because of the larger solid angle of collection (typically 0.01-0.1 ster for EDS and 0.001 ster for WDS) and the greater quantum efficiency (100% for EDS in the 3-15 keV range compared to 30% for WDS). Hence, the EDS can operate more effectively with the low currents typical of SEM imaging modes. A minimum of 10^{-10} A is required for the EDS, and 10^{-8} A for WDS.

Quantitative X-ray Microanalysis. Quantitative x-ray microanalysis proceeds by the measurement of the intensity for an element A in the unknown relative to the intensity emitted from a standard containing A under the same operating conditions.[7,10] This value, the k-ratio, is to a first approximation, proportional to the ratio of concentrations between the unknown and the standard:

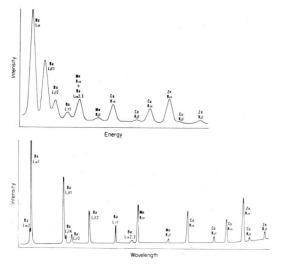

Figure 3. Comparison of energy dispersive and wavelength dispersive scans from a glass excited with a 20 keV electron beam. The constituents include barium, manganese, cobalt, and zinc. (Courtesy C. E. Fiori.)

$$k = \frac{I_A}{I_A^s} \simeq \frac{C_A}{C_A^s} \qquad (10)$$

The use of a relative intensity rather than an absolute intensity immediately compensates for instrumental effects, such as detector efficiency. The constant of proportionality in equation (10) is dependent on three distinct matrix effects, which are usually separately calculated and combined in a multiplicative fashion to yield a final concentration:

$$C_A = C_A^s\, k\, k_Z\, k_A\, k_F\, k_c \qquad (11)$$

k_Z is a correction factor for the electron retardation and backscattering which depend strongly on the atomic number of the target.

k_A is a correction for the absorption of x-rays, which are produced over a large fraction of the interaction volume, during their passage through the solid.

k_F and k_c are corrections for x-ray-induced ionization of element A. The x-ray-induced fluorescence may be caused by characteristic x-rays of energy greater than the critical excitation energy of A (k_F) or from continuous x-rays (k_c), with both effects leading to enhanced x-ray emission of A above that due to electron-induced ionization.

Since all of the matrix correction factors are themselves compositionally dependent, the corrections must be calculated in an iterative fashion. Fortunately for the analyst, a number of computer procedures are available to implement the corrections.[15,16,17] Recently, these procedures have been advanced to consider the particular requirements of quantitative EDS analysis. Additional corrections must be applied for peak overlaps and for accurate background subtraction.[18]

Except in cases of extreme overlap, the accuracy is not compromised in EDS analysis.

While other techniques such as the empirical method are available, the approach to quantitative x-ray microanalysis given by equation (11) provides the greatest flexibility for analysis. The ZAF procedure allows the use of a pure element or compound standards which are readily available for the analysis of a wide array of unknowns, since the matrix correction factors are capable of compensating for large changes in composition between the standard and the unknown. A great deal of experience has been accumulated in the application of the ZAF method to known samples from which the magnitude of errors likely to be encountered in analyzing unknowns can be estimated. For flat, polished samples, normal beam incidence, and for elements with $Z \geq 11$, the error distribution suggests:[19]

relative error	percentage of analyses
1%	30%
2.5%	50%
5%	90%

Limits of detection. Following Reed, an estimate of the limit of detection can be made as follows.[7] Consider that the peak-to-background ratio for the best case in a wavelength spectrometer is 1000:1. If the background correction, made by tuning the spectrometer off the peak, can be made with an error of one percent, then the minimum detectable signal above background will be 1:100,000 or 10 ppm. To discern this level above the background would require 100,000 counts from both the peak and the background. The practical detection limits actually observed are closer to 100 ppm. Geller gives the following results in a recent comparison of the WDS and EDS:[20]

Line	Limit of Detection (ppm)	
	WDS	EDS
Na K	210	1950
Al K	80	690
Ca K	90	850

These values pertain only to a particular instrument and operating conditions, but are a good guide for estimation. The limit of detection is poorer for the energy dispersive spectrometer because of its inherently poorer peak-to-background.

Areas of Further Development. Quantitative electron probe microanalysis has been developed to a high degree, but there remain a number of problem areas, including 1) the analysis of small particles; 2) the analysis of rough solid targets for which the x-ray emergence angle is poorly defined; and 3) the analysis of light elements, particularly carbon, nitrogen, and oxygen. The analysis of particles and rough surfaces is of special interest in the scanning electron microscope, since the typical SEM specimen surface has rough topography and does not meet the requirements for the ideal specimen for x-ray microanalysis. For particles, a number of procedures have been developed over the past several years, but each of these methods is very sensitive to the specimen size and shape, requiring subjective operator input and extensive calculations.[21,22] Recently, significant progress has been reported in this area by making use of simultaneous measurements of the characteristic and the continuum at each energy of interest.

The X-ray Continuum as an Aid to Quantitative Analysis. In the past, the x-ray continuum has rightfully been regarded as an unfortunate nuisance for quantitative analysis. The occurrence of x-rays at the energy of interest from a source other than inner shell ionization is clearly a major limitation to accurate quantitative analysis, and consequently, a method for accurately estimating the background under a peak must be incorporated in a quantitative analysis procedure. For energy dispersive analysis, the peaks are so broad that the technique of extrapolating the background under a peak from measurements of nearby energies is difficult to perform accurately. Mathematical fitting procedures are frequently used to overcome this problem.[23] The x-ray continuum does, however, carry some valuable information about the target. The continuum at any energy of interest varies as a function of the atomic number and is a measure of the total mass of atoms excited by the electron beam. Measurements of the x-ray continuum to correct for specimen mass effects have been previously used by Hall in a quantitative analysis procedure for thin films.[24] At the 1978 SEM Symposium, it was independently reported by Small, et al. and by Statham and Pawley that the peak-to-background measured on a bulk solid and on a particle of the same composition were nearly identical for a wide range of particle sizes and x-ray energies, despite the fact that the ratio of characteristic intensities varied strongly as a function of particle size.[25,26] Based on these observations, Small et al. developed an argument demonstrating how the peak-to-background measurement could be incorporated into a conventional ZAF matrix correction to yield absolute concentrations from particles and rough objects.[25,27] The x-ray signal from an irregular target as compared to a flat, bulk standard set normal to the beam varies because of 1) differences in the composition between the unknown and the standard which the analyst wishes to measure and 2) effects of the specimen geometry, including differences in the beam incidence and x-ray emergence angles and beam penetration in particles. Ideally, a method is needed to separate these two effects. Given a series of characteristic x-ray peak intensities, P_i, measured from a particle or a rough target, we need the intensities P^* which would be measured from the unknown if it could be converted into a flat, bulk target with no geometrical effects. From the experimental observation that

$$(P/B)_{particle} = (P/B)_{bulk} \qquad (12)$$

a modified intensity for the unknown P^* can be obtained:

$$P^* = P_{bulk} = \frac{P_{particle}}{B_{particle}} \times B_{bulk} \qquad (13)$$

In general, the analyst will not have a standard with the same composition as the bulk material from which to measure B_{bulk}. In the method of Small et al.[25], B_{bulk} is estimated from the background measured on a known standard, B_{std}, modified by an appropriate function of atomic number:

$$B_{bulk} = B_{std} \frac{f(Z_{bulk})}{f(Z_{std})} \qquad (14)$$

Z_{bulk} is determined from the current estimate of the concentrations of the elements in the sample. The modified peak intensities from equations (13) and (14) are used to generate a set of k values

$$k^* = P^*/P_{std} \qquad (15)$$

which are used for input to a conventional ZAF matrix correction procedure. Analyses of particles by this procedure have been shown to yield quantitative values which generally lie within 10 percent relative of known values for standard particles.[27,28] Moreover, the procedure is capable of handling particles of simple shape such as spheres as well as complex, irregularly shaped particles. An example of the analysis of particles by this method compared to a conventional ZAF analysis with no particle corrections is given in Table 1. Since rough, irregular particles can be analyzed by the peak-to-background method, the extension of this method to the analysis of rough solid specimens also shows considerable promise.

Analysis of Light Elements. The analysis of elements with $Z < 11$ is difficult because the low energy of the characteristic x-rays of these elements leads to high absorption effects both in the specimen and in the window material of the EDS or WDS x-ray spectrometer. The development of reliable windowless energy dispersive x-ray spectrometers has significantly increased the interest in performing analysis in the energy range 0.25-1.0 keV.[29] Unfortunately, x-ray mass absorption coefficients are poorly known for x-ray energies in this range, and this source of error, compounded by absorption which often exceeds 90 percent of the x-rays generated, leads to large relative errors. To minimize x-ray absorption, operation at a low beam energy, 5 keV or less, would be desirable. For significant progress in this area, a number of points must be resolved: 1) an improvement in the accuracy of absorption expressions for low incident beam energy; 2) accurate measurements of x-ray absorption coefficients for low energy x-rays; and 3) development of accurate background subtraction schemes for the low energy region which take into account detector noise as well as the x-ray continuum.

Auger Electron Microanalysis

During the electron transitions which occur following the ionization of an inner shell of an atom, the Auger process can take place.[30] The transition of an electron from a filled outer

Table 1. X-ray Microanalysis of Particles

| Element | True Composition (weight percent) | Sphere, 2 μm | | Irregular Particle | |
		P/B Method	Conventional ZAF	P/B Method	ZAF
Al	0.079	0.079	0.011	0.084	0.034
Si	0.187	0.185	0.025	0.208	0.079
Ca	0.107	0.103	0.013	0.101	0.039
Ba	0.134	0.130	0.017	0.128	0.050
Fe	0.105	0.110	0.014	0.104	0.043

Sample: NBS Glass K309

Beam: 20 keV

shell to the vacancy in an inner shell is accompanied by the ejection of an outer shell electron, leaving the atom doubly-ionized. The emitted electron has a characteristic energy related to the energy levels of the atom, and by measurement of this energy, qualitative analysis is possible for all elements with $Z \geq 3$ (Li). Auger electron emission is particularly efficient for the light elements where the fluorescence yield of x-rays is poor. The forms of an Auger electron spectrum in both the direct (number of electrons as a function of energy, N(E) vs. E) and the differentiated (N'(E) vs. E) modes are illustrated in Figure 4.

Sampling volume. The map of the distribution of inner shell ionization sites, Figure 2(b), reveals the distribution of the generation of both x-rays and Auger electrons. Although the generation volumes are the same for the two signals, the sampling volumes of the two techniques differs markedly. While x-rays can undergo absorption while escaping from the solid, inelastic scattering of x-rays has a very low probability, so that the x-rays which do escape still retain their characteristic energy. Auger electrons undergo inelastic scattering with a mean free path of the

order of a nanometer. Auger electrons generated at depths exceeding about 1 nm lose their characteristic energy and contribute to the background rather than the peak. The sampling volume for Auger microanalysis thus has the shape of a disc with a diameter determined by the distribution of backscattered electrons, about 0.1-1 μm, and a thickness equal to the sampling depth, about 1 nm.

Quantitative Analysis. Quantitative Auger analysis techniques have been extensively described by a number of authors, with two excellent papers in recent SEM proceedings.[31,32] By preparing working curves from standards, Morabito has demonstrated that quantitative analyses can be carried out with relative errors in the range 5-10 percent.[33] To extend the range of compositions which can be studied from a limited set of standards, the use of relative elemental sensitivity factors has been advocated. With relative sensitivity factors derived from standard Auger spectra recorded from pure elements, Morabito reports relative errors in the range 25-50 percent.[31]

Considerable progress has been reported in the advancement of a "first principles" method for Auger analysis.[31,32] The factors which must be taken into account in such a scheme include 1) matrix insensitive factors, e.g., ionization cross section, transition probability, and instrumental factors such as transmission, and 2) matrix sensitive factors, e.g., atom density, Auger electron escape depth, and the effect of electron backscattering. By developing analytical expressions involving these factors, Morabito and Hall have demonstrated quantitative analysis of simple binary systems with relative errors of 5-30 percent.

Morabito identifies four effects which limit the accuracy of quantitative Auger analysis: 1) the use of sputtering to remove surface contaminants or for depth profiling which can lead to compositional changes of the surface layer sampled by the Auger electrons due to preferential sputtering; 2) the necessity for backscattering and escape depth corrections, both of which depend on the composition of the target; 3) spectral interference problems, particularly for the peaks of elements present at low concentrations;

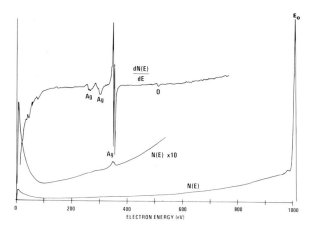

Figure 4. Auger electron spectrum from silver excited with a 10 keV electron beam, shown in both the direct (N(E)) and differentiated (dN(E)/dE) forms. (Courtesy N. C. MacDonald.)

4) chemical effects on peak position and shape, and related chemical effects on relative sensitivity factors.[31]

Limits of detection. The limits of detection by Auger microanalysis are relatively poor compared to other techniques such as x-ray microanalysis because of the low peak-to-background, as can be observed in the direct N(E) vs. E spectrum, Figure 4. The background is formed by backscattered electrons, Auger electrons which have undergone inelastic scattering, and secondary electrons arising from other processes. Signal differentiation is frequently employed as a technique to enhance the peaks, acting as a high pass frequency filter which suppresses the low frequency background component, producing the spectrum shown in Figure 4. However, differentiation also acts to enhance the noise in the signal, and there is no gain over the information available in the original N(E) vs. E spectrum. The limit of detection by Auger microanalysis is in the range 0.1-0.5 percent.

Chemical Information. In addition to the qualitative and quantitative analysis of elemental constituents, the Auger signal can also be used to deduce information on the chemical state of the sample surface. Chemical bonding can lead to measurable changes in the Auger peak shape and position which can be correlated with compound formation, e.g., the state of oxidation.[33] While not as specific as the molecular information available from other electron spectroscopy, chemical effects in Auger spectroscopy are of particular interest because of the possibility of both lateral and depth specificity.

Cathodoluminescence

Cathodoluminescence is the phenomenon of the emission of long wavelength electromagnetic radiation by matter under electron bombardment. In ionically or covalently bonded solids, there are several mechanisms which result in cathodoluminescence:[34] 1) an electron-hole pair, formed when inelastic scattering of the beam electron promotes an electron from a filled valence band to an empty conduction band state, can recombine, emitting radiation with an energy equal to the band gap. 2) the electron and hole may be bound into an energetic uncharged state called an "exciton". The "exciton" can interact with chemical impurities, leading to the emission of radiation upon the recombination of the exciton. 3) additional mechanisms of generation involve electron transitions to energy levels associated with impurity atoms. 4) in amorphous systems, such as biological targets, cathodoluminescence originates from intra-molecular electron transitions involving excited singlet and triplet states and the ground state.[36]

Examples of a cathodoluminescence spectra obtained at high spatial resolution are shown in Figure 5, taken from the work of Petroff et al.[35] The spectra, recorded from a gallium-aluminum-arsenic target, correspond to: 1) a thick region of the sample; 2) a thin (200 nm) region of the sample which was defect-free; and 3) a thin region containing dislocation loops and defect centers. The intensity of the principal emission is significantly lower in the region of the

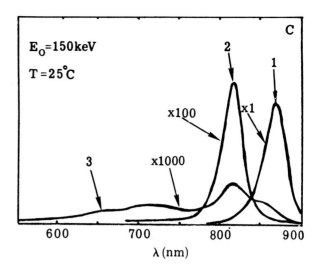

Figure 5. Cathodoluminescence spectra excited from $Ga_{1-x}Al_xAs$ with a 150 keV electron beam. Curve 1 - thick sample; curve 2 - thin (200 nm) sample which is defect free); curve 3 - thin sample is an area which includes dislocation loops. (Courtesy P. Petroff.)

defect, as a result of associated non-radiative centers.

To make use of the weak cathodoluminescence signal generated in the SEM, numerous designs of efficient collection systems have been described.[34,37] In addition, for analytical work, the response of the spectrometer must be carefully measured. For detection, the photomultiplier must be cooled to reduce the dark current, and single photon counting techniques are usually employed.

The information potentially available from the cathodoluminescence signal is quite extensive. For example, Kniseley and Laabs list the following possibilities for ionically or covalently bonded materials, particularly semiconductors:[38]

1) Spectral distribution of cathodoluminescence: composition, band structure, dopant-activation characterization.

2) Spatial distribution (mapping): electrical and compositional homogeneity, defect distributions, diffusion profiles.

3) Intensity: dopant concentration, diffusion length, surface recombination velocity, and internal absorption of light.

4) Rise and decay times of emission: trapping density, temperature effects, and efficiency of radiative recombination.

Herbst and Hoder have reviewed the information available from cathodoluminescence signals in biological targets.[36] Characteristic spectra can be obtained from many significant organic compounds which can be used for qualitative identification. These compounds can then be used as characteristic stains to identify particular cell structures and processes. In extending immunofluorescence studies to include cathodoluminescent stains, higher spatial resolution can be achieved as compared to light microscopy.

Quantitative analysis with cathodoluminescence is dependent on the generation of working curves from standards. The complexity and nature of the electron interaction and relaxation processes makes it difficult to calculate generated and/or emitted intensities from first principles. Where working curves are available, quantitative results have been achieved with relative errors of 10 percent. The limit of detection depends greatly on the system being studied, but in cases where the cathodoluminescence effect is strong, such as semiconductor materials where the electronic structure is markedly affected by trace levels of an impurity, the limit of detection can be as low as 1 ppm.[39]

The spatial resolution of cathodoluminescence is also strongly dependent on the properties of the solid. Since the energy necessary to create an electron-hole pair is small, cathodoluminescence can be created throughout most of the interaction volume. The sampling volume is then dependent on the absorption of the radiation in the solid. In a transparent solid, the sampling volume will be of the order of one micrometer, while in opaque solids, the sampling volume can be much smaller. A major limitation to high resolution cathodoluminescence studies is the quenching of the effect caused by electron beam damage which is dose dependent.

Microanalysis Based on Auxiliary Excitation

Ion Bombardment

By attaching a source of energetic ions (1-25 keV) and suitable focussing lenses to the sample chamber of the SEM, a beam of ions can be made to impinge upon a specimen. The interaction of ions with a solid target is illustrated schematically in Figure 6.[40] A small fraction of the primary beam is backscattered, providing the basis for ion scattering spectrometry (ISS), while most of the primary ions lose all of their energy in collisions with specimen atoms, becoming implanted in the specimen. The momentum transferred to specimen atoms in the collision cascade is frequently adequate to dislodge them from their lattice sites. Atoms lying in the first several layers at the surface of the target may be dislodged with a component of motion sufficient to carry them out of the specimen; this process is known as sputtering. Sputtered particles include single atoms, clusters, and complex molecules, mostly in a neutral charge state. A small fraction, ranging from 10^{-3} to 10^{-1}, of the sputtered particles are emitted in an ionized state, the so-called secondary ions, which form the basis for secondary ion mass spectrometry (SIMS).

Secondary Ion Mass Spectrometry. In secondary ion mass spectrometry, the secondary ions are collected and passed through a mass spectrometer to provide dispersion according to the mass-to-charge ratio. Several investigators have described SIMS experiments in the scanning electron microscope.[41,42] In all cases, the mass spectrometer employed was of the quadrupole type, which uses electric fields for mass/charge dispersion, and has the advantage of offering rapid scanning and peak switching capabilities.

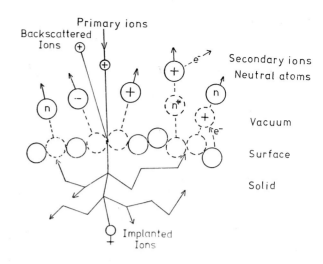

Figure 6. Schematic illustration of energetic ion interaction with a solid target, illustrating primary ion backscattering, implantation, sputtering, and secondary ion production and neutralization.

The spectrum shown in Figure 7 will serve to highlight the outstanding features of SIMS which include:[43]

1) SIMS offers the capability of detecting all of the elements, including hydrogen.

2) The isotopes of an element can be measured, offering the possibilities of isotopic ratio determination for diffusion studies or geological age dating of a selected region of a sample.

3) Molecular signals can be detected, so that studies of chemical compounds are possible. Detection of molecular species is best made under conditions of low primary ion dose; even complicated organic molecules can be studied.[44]

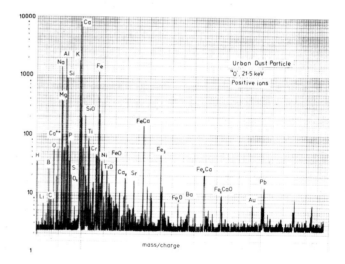

Figure 7. Positive secondary ion mass spectrum of an individual urban airborne dust particle, diameter 10 μm. Substrate — gold.

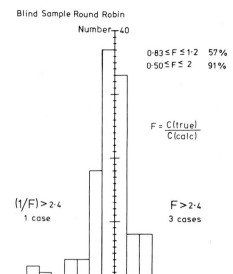

Blind Sample Round Robin

$0.83 \leq F \leq 1.2$ 57%
$0.50 \leq F \leq 2$ 91%

$F = \dfrac{C(true)}{C(calc)}$

$(1/F) > 2.4$
1 case

$F > 2.4$
3 cases

Figure 8. Distribution of errors observed in a blind sample round robin analysis of glasses analyzed by the method of relative sensitivity factors. Ref. (46).

4) Through the use of negative primary ion bombardment and/or charge neutralization by simultaneous electron bombardment, SIMS spectra can be readily obtained from insulators.

Quantitation of SIMS data has proven to be difficult because of strong matrix effects arising from the electronic character of matrix which determines the availability of free electrons for the neutralization of secondary ions. A great deal of the difficulty observed in performing quantitative analysis with physical models for secondary ion emission has recently been traced to instrumental effects on relative elemental sensitivity.[45] When quantitative analysis is performed by the method of relative sensitivity factors, where the factors are determined from standards on the same instrument used for the analysis of the unknowns, the error distribution which results is shown in Figure 8.[46] It should be noted that the analytical performance demonstrated on blind unknowns in Figure 8 is actually quite competitive with x-ray microanalysis, due to the low levels of some of the elements and superior response of SIMS in the measurement of light elements. To increase the analytical flexibility, future developments of quantitative procedures in SIMS may make use of an approach similar to ZAF x-ray microanalysis corrections. A relative sensitivity factor measured in matrix A will be modified through matrix correction factors, ε_i, calculated from physical theories in order to yield appropriate factors for a matrix B:

$$(S_{X/M})_B = (S_{X/M})_A \, \varepsilon_1 \varepsilon_2 \cdots \qquad (16)$$

In this approach, local instrumental sensitivity effects are directly accounted for in the original

Table 2. Limits of Detection Observed in Secondary Ion Mass Spectrometry

Element	c/s/nA/atom Percent	Detection Limit
Li	7.2×10^4	0.42 ppm
B	4.7×10^3	6.4
Mg	4.7×10^4	0.64
Al	6.3×10^4	0.47
P	1.3×10^3	23.
Ti	5.4×10^4	0.56
Cr	3.0×10^4	1.0
Fe	1.8×10^4	1.7
Ni	1.1×10^4	2.7
Ge	7.5×10^3	4.0
Zr	4.2×10^4	0.71
Ba	7.4×10^4	4.0
Ce	1.9×10^4	1.6
Eu	6.5×10^4	4.6
Ta	2.5×10^3	12.
Th	1.4×10^4	2.2
U	1.4×10^4	2.2

Matrix: Silicon-lead-oxygen glass
Beam: $^{16}O^-$ ions at 21.5 keV

measurement of the relative sensitivity factor, while matrix effects are calculated separately.

Despite the fact that only a small fraction of the sputtered particles are ionized, the limit of detection by SIMS is normally in the trace region. Some typical limits of detection for species in a lead-silicate glass are listed in Table 2. The sensitivity variation which is observed for the different elements is a result of the influence of chemical effects on ion emission. An important concept in SIMS is the mass of sample which must be consumed in order to detect a given level of a constituent, since the process is destructive.[43] Since most SIMS spectrometers allow only one or at most a few channels of simultaneous detection, sample consumption results in data collection from dissimilar regions, even in the same spectrum. This non-simultaneous data collection is the major limitation in applying SIMS to a number of problems. The possible development of parallel recording instruments would be of special utility in SIMS.

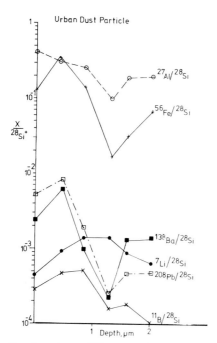

Figure 9. Depth profile of selected elements in an individual urban dust particulate, diameter 10 μm. The signals are normalized with $^{28}Si^+$ signal to minimize matrix effects. Beam $^{16}O^-$ ions at 21.5 keV.

Although the sample must be worn away to be analyzed, this liability can be turned into an asset in the technique of depth profiling. Since the analytical signal emerges from a thin layer of 1-2 nm near the surface, controlled sputtering to carefully erode the sample can lead to depth profiles with a depth resolution as fine as 5 nm.[43] As an example of this powerful technique, a multi-element depth profile of a single 10 μm diameter particle is shown in Figure 9.[47] Preferential surface concentrations of several elements, including the heavy elements (barium, lead), are observed.

The lateral resolution of dedicated ion microprobes is of the order of 1 μm in commercial instruments, with beam sizes as small as 0.1 μm in instruments currently under development. High resolution ion imaging with both secondary ions and secondary electrons is also possible in these instruments. In SIMS systems directly attached to the SEM, beam sizes have been considerably larger, in the range 100-1000 μm.

Ion Scattering Spectrometry. The backscattering of noble gas primary ions from a solid forms the basis for the technique of ion scattering spectrometry (ISS), Figure 10.[48] For a particular scattering angle, the ratio of the mass of the target atom M_s to the incident ion M_o is related to the fraction of the incident energy which the primary ion retains:

$$\frac{M_s}{M_o} = f\left(\frac{E_{out}}{E_{in}}\right) \qquad (17)$$

By measuring the energy of the scattered ions in an energy spectrometer, such as a cylindrical mirror analyzer, the species present can be identified, for all elements with $Z \geq 3$ (Li), e.g., Figure 11.

The most remarkable aspect of ISS is its shallow sampling depth, generally regarded to be one atomic layer. Although backscattering can occur from deeper layers, the incident ions are virtually completely neutralized by charge exchange with the solid, so that by using a detection system which is only sensitive to ions, a shallow sampling depth is obtained.

The salient features of ISS include:
1) Elements with $Z > 3$ can be detected, with little variation in relative elemental sensitivity.
2) Mass resolution is about five percent of the detected mass number.
3) Isotopic ratio measurements are possible for light elements.
4) The extraordinary surface sensitivity makes possible the characterization of the outermost layer of a sample and makes possible depth profiles with extremely fine depth resolution.

Quantitative analysis in ISS is based primarily upon the direct use of standards, where relative accuracies of the order of 20 percent can be achieved. The problem of standards is especially acute for ISS, since there are virtually no reference techniques which can independently analyze with the same sampling depth and sufficient accuracy. In developing "first principles" approaches to quantitation, the most serious difficulty is the existence of a matrix dependent neutralization effect which can affect the relative ion scattering yields.

ISS has not yet been directly included with the SEM, primarily for reasons of vacuum requirements. Because of the extreme surface sensitivity, contamination from a poor vacuum is especially troublesome in ISS. Dedicated ISS/SIMS combination systems typically operate with clean vacua and residual pressures below 10^{-7} Pa. Experiments have demonstrated that ISS can be carried out with focussed ion beams less than 100 μm in diameter, and therefore the possible combination of high spatial resolution ISS and the SEM in an ultrahigh vacuum system is attractive.[49]

X-ray Bombardment

The well known technique of x-ray fluorescence analysis can be adapted to the SEM through the use of a thin foil target of a metal such as molybdenum above the specimen.[50,51] The electron beam is stopped in the foil, generating characteristic and continuum x-rays. A pure element is relatively transparent to its own characteristic radiation. By proper selection of the foil thickness, the electron beam can be totally contained within the foil while a significant fraction of the x-rays generated will be transmitted through the foil to irradiate the specimen. The primary x-rays striking the specimen can ionize all electron shells for which the critical excitation energy is exceeded, leading to the emission of secondary x-rays. In the application to the SEM, x-ray detection is normally carried out with an energy dispersive x-ray spectrometer.

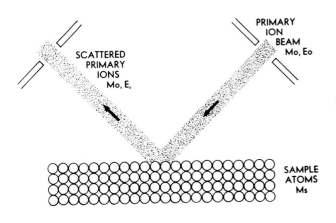

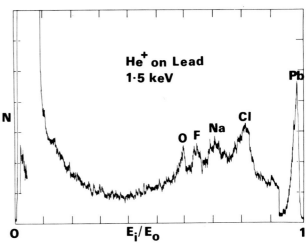

He⁺ on Lead
1·5 keV

Figure 10. Schematic diagram of a low energy ion scattering experiment. (Courtesy — G. Sparrow.)

Figure 11. Ion scattering spectrum obtained with 1.5 keV He⁺ ions on lead sheet. (Courtesy — G. Sparrow.

Since the primary x-rays diverge from a point source above the specimen, a large region of the specimen (and possibly the surrounding stage) can be excited. Moreover, the primary x-rays can cause fluorescence to a considerable depth in the sample, frequently to 100 μm or more. Thus, x-ray fluorescence in the SEM is more of a macroscopic or bulk analysis technique, and it is thus somewhat out of context in the present discussion of microanalysis. However, this procedure can be a useful auxiliary method to conventional electron-excited x-ray microanalysis. The principal advantage of x-ray fluorescence can be readily seen in a comparison of electron-excited and x-ray-excited spectra from the same target, Figure 12(a),(b). The peak-to-background is higher under x-ray excitation compared to electron excitation. A great reduction in the background is realized in the x-ray fluorescence spectrum because of the elimination of most of the bremsstrahlung from inelastic electron scattering. The background in the x-ray fluorescence spectrum originates from 1) inelastic scattering of the primary and secondary x-rays, 2) elastic scattering of the primary x-rays (which in the SEM case, include both characteristic and continuum x-rays at all energies), and 3) bremsstrahlung created by inelastic scattering of photoelectrons produced during absorption of the primary x-rays. The background is sufficiently low to realize detection limits in the ppm range. For example, in a study of mercury in coal, Myklebust et al. demonstrated detection limits of 2 ppm in a difficult situation in which inelastic x-ray scattering from the light element matrix caused a relatively high background.[52]

Techniques of quantitative x-ray fluorescence rely primarily on the use of standards similar to the unknown to minimize matrix effects.[50] Preliminary investigations of a fundamental parameters method have been described, but further development and testing is necessary before such an approach can be widely applied.[53]

For the situation of x-ray fluorescence in the SEM, in which both continuum and characteristic radiation irradiate the specimen, the use of standards would be the most accurate approach to quantitation. With close standards and good control of the electron beam current, relative errors of five percent or less should be readily attainable.

Laser Bombardment

Laser Microprobe Mass Analyzer. In the laser microprobe mass analyzer (LAMMA), a focussed laser beam is used to couple sufficient energy to the specimen to evaporate and partially ionize a small volume of the specimen.[54] These ions are then mass analyzed by a time-of-flight mass spectrometer to produce a spectrum such as that shown in Figure 13. If the target is in the form of a thin foil of the order of 0.5 μm thick, a spatial resolution of the order of 1 μm can be achieved. The technique can be applied to a bulk target with some degradation in spatial resolution due to the evaporation of a larger volume of material caused by the complete absorption of the laser pulse.

The particular advantages of this technique are: 1) all elements in the target region are measured with a single laser pulse; 2) the limits of detection for metals in an organic matrix have been shown to lie in the range 0.2-20 ppm; 3) both positive and negative ions can be measured simultaneously, allowing detection of both electropositive and electronegative species; 4) since the ionization process takes place in a vapor phase or a plasma, matrix effects are much less significant than for the corresponding SIMS case; and 5) since a neutral beam is employed, charging is not a problem in the analysis of insulating samples.

Laser Raman Microprobe. When monochromatic photons of frequency ν_o in the visible range strike a solid, scattering processes can take place involving the molecular bonding energy levels in the solid (Raman scattering). If the polarizability of a molecule changes while under-

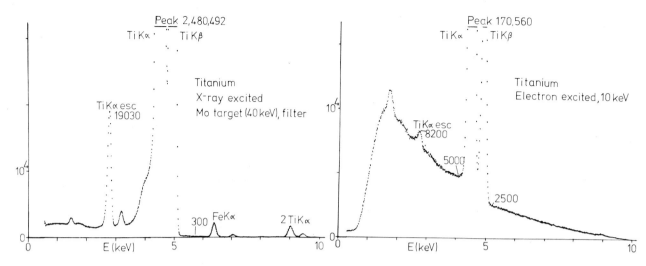

Figure 12 (a) and (b). Comparison of energy dispersive x-ray spectra of titanium excited by: (a) x-rays from a molybdenum target and filter (40 keV) and (b) electrons at 10 keV.

going rotation or vibration at a characteristic frequency ν_1, then a fraction of the scattered photons will be shifted in frequency to $\nu_o \pm \nu_1$. Measurements of these characteristic frequency shifts can be employed in studying the structure of molecules. Because of the specificity of the set of vibrational, rotational, and stretching modes of a particular molecule, observation of the Raman spectrum can be a very effective tool for the determination of the compounds present in a sample.

Recently, instrumentation has been developed by two independent groups which enables micro-Raman spectroscopy to be carried out on targets as small as 1 μm in lateral dimension.[55,56,57,58] Both microprobe systems employ a laser as the source of high brightness radiation and utilize an optical system producing a diffraction-limited

probe size of the order of 1 μm in diameter. An efficient collection optic is used to transfer the radiation scattered from a target into a scanning monochromator. Simultaneous viewing of the target with an optical microscope permits accurate location of the probe on the sample. Experiments have demonstrated that Raman spectra obtained from individual micrometer-sized particles are similar to spectra observed from macroscopic targets, thus enabling the analyst to make use of the extensive literature of Raman spectra.

As an illustration of the technique, spectra obtained from various sulfate-bearing particles are shown in Figure 14. The micro-Raman technique is easily able to distinguish the various chemical forms and can follow reactions from one form to another.

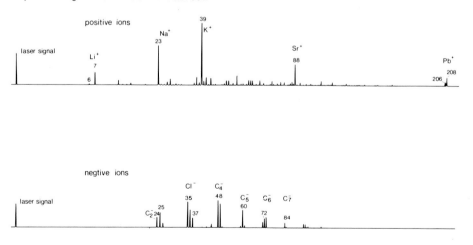

Figure 13. Laser microprobe mass analyzer spectra of epon film doped with Na, K, Rb, and Cs. (Courtesy R. Wechsung.)

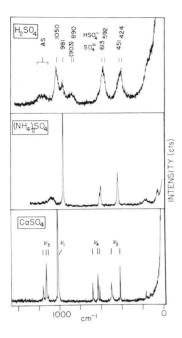

Figure 14. Laser Raman microprobe spectra of individual microparticles of H_2SO_4, $(NH_4)_2SO_4$, and $CaSO_4$, illustrating the capability of identifying and differentiating similar compounds. (Courtesy E. Etz.)

Micro-Raman spectroscopy offers a number of powerful capabilities to the analyst:

1) Characterization and identification of all types of molecular structures.

2) Information on the lattice structure of crystalline materials with ionic and/or covalent bonds.

3) The opportunity of analyzing samples in ambient conditions or in special reaction cells which can produce a wide range of temperature/pressure/chemical environment conditions. For a great many samples, analysis without exposure to vacuum conditions is clearly a great advantage.

4) Images with the characteristic Raman signals can be formed, which allows for the selective mapping of the distribution of molecular species embedded in a heterogeneous sample.

At the present time, Raman microprobe analysis is primarily used for qualitative analysis, e.g., the identification of compounds present in an unknown by comparison with spectra from prepared standards or literature spectra. In this regard, the lack of elemental information in the Raman spectrum is a liability in searching the vast number of molecular possibilities when a total unknown is examined. Techniques for quantitative analysis are currently under development. Experiments indicate that the limits of detection within the analyzed volume are limited to major and minor molecular constituents.

Synergism of Microanalysis Techniques

The general features of the technique of microanalysis which have been discussed in this paper are collected in Table 3. It is readily apparent that no single technique is capable of fully characterizing a sample. In order to determine fine scale morphology, elemental constituents, molecular species, and lateral and depth distributions, a variety of microanalytical techniques must be employed. While it is certainly possible to move a sample from one instrument to another, this is invariably tedious, and it is frequently difficult to locate the same region of interest when the features are on a micrometer scale. Moreover, the sample may undergo substantial changes over the period of time necessary to make all of the measurements. The trend to combine several microanalytical techniques in the same vacuum system, which began with the combination of electron imaging and x-ray microanalysis, can thus be expected to continue. Collections of several analytical techniques in the same instrument have already proven to be especially effective in surface analysis, and similar synergism can be expected in future microanalytical instrumentation. To successfully implement some useful combinations of techniques, it may be necessary to alter the philosophy of the design of the instrumentation. In the past, microanalytical instrumentation has been primarily treated as an "add-on" to the SEM column and vacuum system. However, to gain the maximum advantage from the microanalysis techniques, it will be necessary to design the instrumentation around the characteristics of the microanalysis techniques and "add-on." the SEM! An example of this approach can be seen in Auger spectroscopy coupled with the SEM. Early investigators were able to produce satisfactory results for certain types of problems by attaching an Auger spectrometer directly on an SEM with conventional oil pumping.[59] To make full use of the surface analysis capabilities, however, a clean vacuum is necessary, and so the ultrahigh vacuum scanning Auger microprobe has been developed in which the SEM is miniaturized and placed inside the bore of a cylindrical mirror analyzer, providing optimum illumination and collection for Auger spectroscopy in the proper vacuum environment.

The complexity of the microanalysis techniques is such that it would probably be impossible to combine all techniques in a single instrument. It would also be undesirable, since they would probably never all be working simultaneously! However, several combinations of microanalytical techniques and scanning electron microscopy suggest themselves as being particularly powerful.

Surface and Bulk Microanalysis

It is frequently of great interest to simultaneously determine both the surface and the bulk composition of a sample with a lateral resolution of 1 μm. Such a requirement could be met by a combined analytical system consisting of an Auger electron spectrometer and a windowless energy dispersive x-ray spectrometer. Independent and simultaneous measurements of all elements with $Z \geq 5$ (boron) could be achieved with sampling depths of 1 nm and 1000 nm. By adding an ion gun/lens system and a quadrupole mass spectrometer, SIMS measurements would further extend capabilities to high sensitivities and depth profiling in both the Auger and SIMS modes.

Table 3. Summary of Characteristics of Microanalysis Techniques

Technique	Primary Radiation	Secondary Radiation	Constituents Measured	Lateral Resolution	Depth Resolution	Typical Detection Limit
(1) X-ray Microanalysis	Electron	X-ray	Elements $Z \geq 11$ (EDS) $Z \geq 4$ (WDS)	1 μm	1 μm	750 ppm (EDS) 100 ppm (WDS)
(2) Auger Microanalysis	Electron	Electron	Elements $Z \geq 3$	0.1-1 μm	1 nm	0.1 percent
(3) Cathodo-luminescence	Electron	Photon	Molecules, other	1 μm	1 μm	1 - 1000 ppm varies strongly
(4) Secondary Ion Mass Spectrometry	Ion	Ion	All elements Molecules	1 μm	1 nm	1 ppm
(5) Ion Scattering Spectrometry	Ion	Ion	Elements $Z \geq 3$	100 μm	1 atom layer	100 ppm - 0.1 percent
(6) X-ray Fluorescence	X-ray	X-ray	Elements $Z \geq 11$ (EDS)	1 mm - 1 cm	100 μm	1 - 10 ppm
(7) Laser Raman Micro-probe	Photons	Photons	Molecules	1 μm	1 μm	1 percent
(8) Laser Microprobe Mass Analyzer	Photons	Ions	All elements (Molecules)	1 μm	1 - 10 μm	1 ppm

Table 3. (Continuation)

	Quantitation Technique	Relative Accuracy	Characteristic Signal Images	Non-Conducting Samples	Vacuum Requirements
(1)	ZAF	1 - 5 percent $(Z \geq 11)$	Yes	Yes, coated	$<10^{-2}$ Pa
(2)	Sensitivity Factors	10 - 25 percent	Yes	Difficult	$<10^{-6}$ Pa
(3)	Working Curve	10 percent	Yes	Yes, coated	$<10^{-2}$ Pa
(4)	Sensitivity Factors	20 percent	Yes	Yes	$<10^{-5}$ Pa
(5)	Working Curve	20 percent	Yes	Yes	$<10^{-7}$ Pa
(6)	Working Curve	5 percent	No	Yes	$<10^{-1}$ Pa
(7)	Under Development	-	Yes	Yes	Atmospheric Pressure
(8)	Working Curve	5 percent	No	Yes	$<10^{-3}$ Pa

Elemental and Molecular Microanalysis

Knowledge of the elements present in a specimen and the molecules which they form is invaluable in studying the course of chemical reactions. An obvious complementary combination is x-ray microanalysis and laser Raman microanalysis. X-ray microanalysis for qualitative identification of the elements present would greatly simplify the task of qualitative identification of the Raman spectral features by limiting the possible reference spectra which would have to be searched. Since the sampling volumes of the two techniques are similar, quantitative x-ray microanalysis would also aid quantitative Raman microanalysis by providing some indication of the fractions of the constituents associated with each compound. Furthermore, the collection optic and optical spectrometer used for the Raman signal would also be ideal for the collection and analysis of the electron-excited cathodoluminescence signal.

X-ray Microanalysis and the LAMMA

The LAMMA-type instrument has the advantage of outstanding sensitivity and the capability for measurement of all elements and isotopes. By its very nature, it must destroy the sample during analysis, preventing a repetitive analysis. Moreover, it is restricted to the point analysis mode; scanning images for mapping purposes are impractical, and therefore a survey of a complicated specimen is tedious. A combination of an SEM with EDS and/or WDS x-ray microanalysis with the LAMMA would provide high resolution imaging, nondestructive analysis, and scanning x-ray images for survey purposes to augment the mass spectrometry. X-ray microanalysis would be invaluable in selecting suitable sites for the destructive mass analysis and in providing a comparison analysis.

Summary

In this brief survey, a number of important aspects of the wealth of microanalysis techniques available to the analyst have been described. The exact choice of a technique to solve a particular problem is dependent on the nature of the problem. In many cases, several techniques, applied alone or in combination, will be necessary to completely characterize a sample. We are indeed fortunate to have such a variety of powerful instrumentation available to us, for which we owe a debt of gratitude to the instrumentation developers. We can look forward to more developments in the future, particularly involving combinations of analytical techniques in a single instrument, as a perusal of the other papers in this volume attests.

Acknowledgements

The author wishes to thank his many colleagues who have made valuable suggestions and aided in the preparation of this paper, including K. F. J. Heinrich, R. L. Myklebust, J. A. Small, E. Etz, C. E. Fiori, H. Yakowitz, P. Petroff, G. Sparrow, and N. C. MacDonald.

References

1. C. W. Oatley, The Scanning Electron Microscope, Part 1: The Instrument (University Press, Cambridge, 1972).

2. R. Castaing, "Application of Electron Probes to Local Chemical and Crystallographic Analysis", Ph.D. Diss., Univ. Paris (1951).

3. V. E. Cosslett and P. Duncumb, "Microanalysis by a Flying Spot X-ray Method", Nature, 177 (1956) 1172-1173.

4. K. F. J. Heinrich, "Concentration Mapping Device for the Scanning Electron Probe Microanalyzer", Rev. Sci. Instrum., 33 (1962) 884.

5. H. R. Bowman, E. K. Hyde, S. G. Thompson, and R. C. Jared, "Application of High Resolution Semiconductor Detectors in X-ray Emission Spectrography", Science, 151 (1966) 562-568.

6. R. Fitzgerald, K. Keil, and K. F. J. Heinrich, "Solid-State Energy-Dispersion for Electron Microprobe X-ray Analysis", Science, 159 (1968) 528-529.

7. S. J. B. Reed, Electron Microprobe Analysis (University Press, Cambridge, 1975).

8. T. O. Ziebold and R. E. Ogilvie, "An Empirical Method for Electron Microanalysis", Analyt. Chem., 36 (1964) 322-327.

9. J. I. Goldstein, H. Yakowitz, D. E. Newbury, E. Lifshin, J. W. Colby, and J. R. Coleman, Practical Scanning Electron Microscopy (Plenum, New York, 1975).

10. K. F. J. Heinrich and H. Yakowitz, "Propagation of Errors in Correction Models for Quantitative Electron Probe Microanalysis", in 5th Int'l Conf. X-ray Optics and Microanalysis, ed. G. Mollenstedt and K. H. Gaukler (Springer - Verlag, Berlin, 1969) 151-159.

11. For a recent comprehensive review of Monte Carlo techniques, see Monte Carlo Calculations in Scanning Electron Microscopy and Electron Probe Microanalysis, ed. K. F. J. Heinrich, D. E. Newbury, and H. Yakowitz, National Bureau of Standards Special Publication 460 (Washington, 1976).
J. Henoc and F. Maurice, "Characteristics of a Monte Carlo Program for Microanalysis: Study of Energy Losses", in NBS SP 460, ibid., 61-95.

12. K. Kanaya and S. Okayama, "Penetration and Energy-loss Theory of Electrons in Solid Targets", J. Phys. D: Appl. Phys., 5 (1972) 43-58.

13. C. J. Powell, "Evaluation of Formulas for Inner Shell Ionization Cross Sections", in NBS SP 460, ibid., 97-104.

14. K. F. J. Heinrich, quoted by H. Yakowitz and D. E. Newbury, "A Simple Analytical Method for Thin Film Analysis with Massive Pure Element Standards", SEM/1976/I (IITRI, Chicago, 1976) 151-162.

15. J. W. Colby, "Quantitative Microprobe Analysis of Thin Insulating Films" in Advances in X-ray Analysis, ed. J. B. Newkirk, G. R. Mallett, and H. G. Pfeiffer (Plenum, New York, 1968) 287-305.

16. H. Yakowitz, R. L. Myklebust, and K. F. J. Heinrich, "FRAME: An On-line Correction Procedure for Quantitative Electron Probe Microanalysis" National Bureau of Standards Technical Note 796 (Washington, 1973).

17. J. Henoc, K. F. J. Heinrich, and R. L. Myklebust, "A Rigorous Correction Procedure for Quantitative Electron Probe Microanalysis (COR 2)", National Bureau of Standards Technical Note 769 (Washington, 1973).

18. R. L. Myklebust, C. E. Fiori, and K. F. J. Heinrich, FRAME C: A Compact Procedure for Quantitative Energy Dispersive Electron Probe X-ray Analysis", NBS Technical Note, (in press).

19. H. Yakowitz, "X-ray Microanalysis in Scanning Electron Microscopy", SEM/1974, ed. O. Johari (IITRI, Chicago, 1974) 1029-1042.

20. J. D. Geller, "A Comparison of Minimum Detection Limits Using Energy and Wavelength Dispersive Spectrometers", SEM/1977/I, ed. O. Johari, (IITRI, Chicago, 1977) 281-288.

21. J. Armstrong, "Methods of Quantitative Analysis of Individual Microparticles with Electron Beam Instruments", SEM/1978/I, ed. O. Johari (SEM, Inc., AMF O'Hare, IL, 1978) 455-468.

22. N. C. Barbi, M. A. Giles, and D. P. Skinner, "Estimating Elemental Concentrations in Small Particles Using X-ray Analysis in the Electron Microscope", SEM/1978/I, ed. O. Johari (SEM, Inc., AMF O'Hare, IL, 1978) 193-200.

23. C. E. Fiori, R. L. Myklebust, K. F. J. Heinrich, and H. Yakowitz, "Prediction of Continuum Intensity in X-ray Dispersive Microanalysis", Analyt. Chem., 48 (1976) 172-176.

24. T. Hall, "Some Aspects of the Microprobe Analysis of Biological Specimens" in Quantitative Electron Probe Microanalysis, ed. K. F. J. Heinrich, National Bureau of Standards Special Publication 298 (Washington, 1968) 269-299.

25. J. A. Small, K. F. J. Heinrich, D. E. Newbury, R. L. Myklebust, C. E. Fiori, and M. F. Dilmore, "The Production and Characterization of Glass Fibers and Spheres for Microanalysis", SEM/1978/I, ed. O. Johari (SEM/Inc., AMF O'Hare, IL, 1978) 445-454.

26. P. J. Statham and J. B. Pawley, "A New Method for Particle X-ray Microanalysis Based on Peak-to-Background Measurements", SEM/1978/I, ed., O. Johari (SEM/Inc., AMF O'Hare, IL, 1978) 469-478.

27. J. A. Small, K. F. J. Heinrich, C. E. Fiori, D. E. Newbury, and R. L. Myklebust, "Progress in the Quantitation of Single-Particle Analysis with the Electron Probe" in Proc. 13th Ann. Conf., Microbeam Analysis Society (1978) 56A-56K.

28. J. A. Small, K. F. J. Heinrich, D. E. Newbury, and R. L. Myklebust, "Progress in the Development of the Peak-to-Background Method for the Quantitative Analysis of Single Particles with the Electron Probe", SEM/1979/II - this volume.

29. J. C. Russ, "Procedures for Quantitative Ultralight element analysis" SEM/1977/I, ed. O. Johari, (IITRI, Chicago, 1977) 289-296.

30. A. Joshi, J. E. Davis, and P. W. Palmberg, "Auger Electron Spectroscopy" in Methods of Surface Analysis, ed. A. W. Czanderna (Elsevier, Amsterdam, 1975) 159-222.

31. J. M. Morabito and P. M. Hall, "Quantitative Auger Electron Spectroscopy" SEM/1976/I, ed. O. Johari (IITRI, Chicago, 1976) 221-230.

32. P. H. Holloway, "Quantitative Auger Electron Spectroscopy - Problems and Prospects", SEM/1978/I, ed. O. Johari (SEM, Inc., AMF O'Hare, IL, 1978) 361-374.

33. J. M. Morabito, "Quantitative Analysis of Light Elements (N, C, and O) in Sputtered Tantalum Films by Auger Electron Spectroscopy and Secondary Ion Mass Spectrometry", Analyt. Chem., 46 (1974) 189-196

34. M. D. Muir and D. B. Holt, "Analytical Cathodoluminescence Mode Scanning Electron Microscopy", SEM/1974, ed. O. Johari (IITRI, Chicago, 1974) 135-142.

35. P. M. Petroff, D. V. Lang, J. L. Strudel, and R. A. Logan, "Scanning Transmission Electron Microscopy Techniques for Simultaneous Electronic Analysis and Observation of Defects in Semiconductors", SEM/1978/I, ed. O. Johari (SEM, Inc., AMF O'Hare, IL, 1978) 352-332.

36. R. Herbst and D. Hoder, "Cathodoluminescence in Biological Studies", Scanning, 1 (1978) 35-41.

37. M. de Mets, "Improved Cathodoluminescence Detection System", J. Phys. E., 7 (1974) 91-97.

38. R. N. Kniseley and F. C. Laabs, "Applications of Cathodoluminescence in Electron Microprobe Analysis" in Microprobe Analysis, ed. C. A. Anderson (Wiley, New York, 1973) 371-382.

39. H. C. Casey, Jr. and R. H. Kaiser, "Analysis of n-type GaAs with Electron Beam Excited Radiative Recombination", J. Electrochem. Soc., 114 (1967) 149-153.

40. G. K. Wehner, "The Aspects of Sputtering in Surface Analysis Methods" in Methods of Surface Analysis, ed. A. W. Czanderna (Elsevier, Amsterdam, 1975) 5-37.

41. W. C. Lane and N. C. Yew, "Ion Analysis in the SEM", in SEM/1973, ed. O. Johari, (IITRI, Chicago, 1973) 81-88.

42. J. A. Leys and J. T. McKinney, "Surface Analysis in a SEM with SIMS Imaging", in SEM/1976/I, ed. O. Johari (IITRI, Chicago, 1976) 231-238.

43. J. A. McHugh, "Secondary Ion Mass Spectrometry" in Methods of Surface Analysis, ed. A. W. Czanderna (Elsevier, Amsterdam, 1975) 223-278.

44. A. Benninghoven and W. K. Sichtermann, "Detection, Identification, and Structural Investigation of Biologically Important Compounds by Secondary Ion Mass Spectrometry", Analyt. Chem., 50 (1978) 1180-1184.

45. D. E. Newbury, "Report on the United States - Japan Cooperative Analysis of Glasses by Secondary Ion Mass Spectrometry", 2nd US-Japan Joint Seminar on Secondary Ion Mass Spectrometry (Osaka, Japan, Oct. 23-27, 1978), available from author.

46. D. E. Newbury, "On the Accuracy of Quantitative Analysis in Secondary Ion Mass Spectrometry - Round Robin Results" in Proc. 13th Ann. Conf., Microbeam Analysis Society (1978) 6A-6I.

47. D. E. Newbury, "Secondary Ion Mass Spectrometry for Particulate Analysis", in Environmental Pollutants, ed. T. Y. Toribara (Plenum, New York, 1978) 317-348.

48. T. M. Buck, "Low Energy Ion Scattering Spectrometry", in Methods of Surface Analysis (Elsevier, Amsterdam, 1975) 75-102.

49. J. A. Leys and J. T. McKinney, "SEM+ SIMS: A Unique Combination for Surface Characterization" in Proc. 10th Ann. Conf., Microbeam Analysis Society (1975) 59A-59E.

50. R. O. Muller, Spectrochemical Analysis by X-ray Fluorescence (Plenum, New York, 1972).

51. L. M. Middleman and J. D. Geller, "Trace Element Analysis Using X-ray Excitation with an Energy Dispersive Spectrometer on a Scanning Electron Microscope", SEM/1976/I, ed. O. Johari (IITRI, Chicago, 1976) 171-178.

52. R. L. Myklebust, M. M. Darr, and K. F. J. Heinrich, "Evaluation of X-ray Fluorescence Analysis for the Determination of Mercury in Coal", National Bureau of Standards Interim Report 75-675 (Washington, 1974).

53. R. L. Myklebust, C. E. Fiori, D. N. Breiter, and K. F. J. Heinrich, "Data Reduction Procedure for Monochromatic X-ray Fluorescence Analysis with a Si(Li) Detector", in Proc. 3rd Ann. Conf., Federation of Analytical Chemistry and Spectroscopy Societies (1976) paper 241.

54. R. Wechsung, F. Hillenkamp, R. Kaufmann, R. Nitsche, and H. Vogt, "LAMMA - A New Laser Microprobe Mass Analyzer", SEM/1978/I, ed. O. Johari, (SEM, Inc., AMF O'Hare, IL, 1978) 611-620.

55. G. J. Rosasco, E. S. Etz, and W. A. Cassatt, "The Analysis of Discrete Fine Particles by Raman Spectroscopy", Appl. Spectrosc. 29 (1975) 396-404.

56. G. J. Rosasco and E. S. Etz, "The Raman Microprobe: A New Analytical Tool", Res. and Dev., 28 (1977) 20-35.

57. M. Delhaye and P. Dhamelincourt, "Raman Microprobe and Microscope with Laser Excitation", J. Raman Spectrosc., 3 (1975) 33-43.

58. M. Delhaye, E. DaSilva, and G. S. Hayat, "Molecular Microprobe", Amer. Lab., (1977) 83-86.

59. E. K. Brandis, "High Spatial Resolution Auger Electron Spectroscopy in an 'Ordinary' Diffusion Pumped SEM", SEM/1975/I, ed. O. Johari (IITRI, Chicago, 1975) 141-148.

Note: Proceedings of the Microbeam Analysis Society can be obtained from Robert Myklebust, Secretary, MAS 222/A121, National Bureau of Standards, Washington, D.C. 20234.

Discussion with Reviewers

Om Johari: What future trends do you predict in microanalysis?

Author: I believe the principal developments in microanalysis will occur in three areas:

(1) Synergistic combinations of existing techniques. As mentioned in the text, most microanalysis techniques have limitations of one sort or another. Combinations of carefully chosen techniques, such as those examples noted in the text, could provide powerful analytical capabilities for the characterization of specimens. In particular, quantitative analysis would be advanced for many techniques if some compositional information could be obtained by an independent technique operating simultaneously on the same region of the sample. To make the most efficient use of multiple analytical techniques, it will be necessary to carefully design the instrumentation around the particular requirements for optimum analysis conditions.

(2) Other advances can be expected on the basis of a full examination of the information available in the various spectra which can now be recorded. Thus, in x-ray spectrometry, the bremsstrahlung which forms the background has recently been found to offer extremely useful information for the analysis of particles and rough surfaces. An example of particle analyses is given in Table 1. The same approach can be applied to rough samples. As an example, a rough surface was prepared by ductile fracture of the homogeneous 60 Au-40 Cu (nominal) NBS standard reference material. Electron-excited EDS spectra were analyzed by a conventional ZAF scheme (FRAME C) and the modified ZAF method which employs peak-to-background measurements (FRAME P). An analysis of the region shown in Figure D1 revealed:

	Actual Wt%	FRAME C Wt%	Relative Error
Au	60.3	28.7	-52%
Cu	39.6	28.2	-29%

	Normalized FRAME C	Relative Error	FRAME P Wt%	Relative Error
Au	50.5	-16%	58.0	-3.8%
Cu	49.4	+25%	44.0	+11%

Other advances can be expected in the future on the basis of a re-examination of existing spectrometry techniques.

(3) On-line computerization - The extraordinary development of powerful laboratory computers providing on-line, dedicated computation facilities offers tremendous promise for enhancing the rate of information acquisition as well as the quality of that information in all types of microanalysis. Laboratory computers are typically incorporated in "smart" multichannel analyzers for x-ray microanalysis. However, the same computer can be used

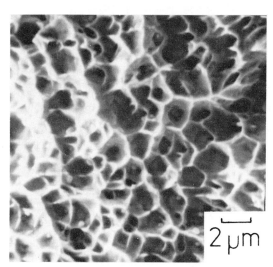

Figure D1. Fracture surface of NBS Standard Reference Material 482. 60Au-40Cu (Microprobe Standard).

expected, particularly as a result of the rapid development of solid state backscattered electron detectors.

for other tasks. The possibility exists of interfacing the computer to other operations of interest, such as image analysis. Smith et al. (K. C. A. Smith, B. M. Unitt, D. M. Holburn, and W. J. Tee, "Gradient Image Processing Using an On-Line Digital Computer" SEM/1977/1 (Chicago, IITRI, 1977) 49-56) have demonstrated on-line image analysis with such a system to characterize surface topography. Many more developments of this type are to be expected as laboratory computer systems advance.

David Kyser: What are the possibilities for the use of backscattered electrons for microanalysis? Author: Backscattered electrons can be used directly for microanalysis. Heinrich, (K. F. J. Heinrich, "Electron Probe Microanalysis of Specimen Current Measurement", in 4th Int'l Conf. on X-ray Optics and Microanalysis", ed. R. Castaing, P. Deschamps, and J. Philibert, (Hermann, Paris, 1966) 159-167) has demonstrated that accurate quantitative microanalysis of binary alloys can be carried out with the backscattered electron signal (as indirectly measured with specimen current), providing the secondary electron current is suppressed by biasing. Niedrig (H. Niedrig, "Backscattered Electrons as a Tool for Film Thickness Determination", SEM/1978/1, ed. O. Johari (SEM, Inc., AMF O'Hare, IL, 1978) 841-858) has demonstrated the utility of the backscattered electron signal for assessing the mass thickness of films. DeNee (P. B. DeNee, "Measurement of Mass and Thickness of Respirable Size Dust Particles by SEM Backscattered Electron Imaging", SEM/1978/1 ed. O. Johari (SEM, Inc., AMF O'Hare, IL, 1978) 741-746) has extended this technique to particles. There is potentially much information available in the backscattered electron signal, particularly when the energy and angular distributions of the backscattered electrons are considered. Further progress in the application of this signal to analysis can be

SCANNING ELECTRON MICROSCOPY/1979/II
SEM Inc., AMF O'Hare, IL 60666, USA

THIN FILMS FOR HIGH RESOLUTION CONVENTIONAL SCANNING ELECTRON MICROSCOPY

P. Echlin   and   G. Kaye

Botany School   Polaron Equipment Ltd.
University of Cambridge   Holywell Industrial Estate
Cambridge CB2 3EA   Watford, WD1 8XG
U.K.   U.K.

Abstract

The use of high-brightness electron sources on conventional scanning electron microscopes and the introduction of scanning attachments on transmission electron microscopes means that it is now possible to obtain a spatial resolution of 3 nm in the reflective mode of instrument operation. At this resolution it is possible to distinguish sub-structure in the thin films normally used for coating non-conductive specimens.

A study has been made using transmission and conventional scanning electron microscopy of thin metallic films prepared from both noble and refractory metals. Electron beam evaporation and resistive heating evaporation appear to result in a smaller particle size and these techniques must remain the optimal method of preparing thin films for high resolution (2-3 nm) scanning transmission microscopy. It has been possible to sputter some of the refractory metals at ambient and sub-ambient temperatures, and provided this is done in a clean environment, the resultant films have a smaller particle size than some of the noble metal thin films. However, although the back-scattered electron yield shows a significant increase with atomic number, the secondary electron yield does not show such a dramatic improvement. The most convenient technique for preparing thin films for medium resolution (5.0-8.0 nm) scanning electron microscopy is to use sputter coated films of the noble metals and their alloys other than gold on specimens maintained at below ambient temperatures. Advantage can also be made of the conductivity of discontinuous metal films which can provide an effective coating layer for scanning electron microscopy when only a few nanometres thick.

/S.I. units have been used throughout this paper. For readers unfamiliar with these terms 273 K = 0° etc and 7 Pa = 5 x 10^{-2} t, 2 mPa = 1 x 10^{-5} t and 260 μPa = 2 x 10^{-6} t./

KEY WORDS: Coatings, Specimen Preparation, Non-conductive Specimens, Sputter Coating, Evaporative Coatings, Thin Films, Noble Metals, Refractory Metals

Introduction

Most modern scanning electron microscopes are capable of resolving 5-8 nm and scanning attachments on transmission electron microscopes can easily halve this figure. This increase in spatial resolution means that greater attention must be paid to the coating techniques which form a central part of specimen preparation for scanning electron microscopy. A recent paper[1] briefly discussed some of the problems associated with high resolution coating methods, and this present paper will consider these factors in greater detail. We have chosen to emphasise the practical aspects of the subject as applied to conventional SEM using readily available equipment, and only passing reference will be given to the theoretical aspects of thin film formation. We will attempt to relate substrate temperature, target composition and film thickness to the production of thin films with a minimum sub-structure using sputter and evaporation techniques.

Methods

Transmission electron microscopy of thin films (Figs. 1-32)

The metal layers were deposited by different means onto carbon films supported on 400 mesh copper grids. One half of the grid was covered by a cleaned razor blade or glass coverslip before being placed in the coating unit. In Figs. 1-14 different metals were sputtered at 7 Pa and 2.5 kV using a 8 mA argon plasma at a substrate temperature of 283 K in a Polaron E5100 cool coater to give a final thickness of 15 nm as measured by a Polaron E5500 thin film monitor placed to one side of the specimens. The thin film monitor had been previously calibrated by measuring varying film thicknesses of different metals deposited on a glass slide using an interference microscope. The same method was used to prepare the thin films shown in Figs. 21-32 except that the substrate temperatures were varied by simply adjusting the electrical input into the Peltier cooled specimen table. Figs. 15 and 16 were prepared by evaporating 15 nm of gold as measured on a Balzers QSG-201 thin film monitor from a heated tungsten wire at 260 μPa in a Balzers BAE 120 vacuum evaporator. Figs. 17-20 were prepared by evaporating 5 nm of material using an electron gun at 1.8-2.0 kV and 80-170 μA at 2 mPa in a Balzers freeze etch unit.

All the samples were examined and photographed at 80 kV in an AEI EM6B transmission electron microscope fitted with a liquid nitrogen cooled anti-contamination plate. A whole series of photographs were taken of each sample ranging from the thickest to the thinnest layer of material. It was relatively easy to judge the thickest layer as this was in the region far away from the part of the grid which remained covered during coating. The regions of the grid which were well covered received no coating. This was checked by using an energy dispersive X-ray spectrometer which gave no signal for the metal concerned from well covered regions and an appreciable signal from uncovered regions. The intersection of the covered and uncovered regions was not clear cut and it was possible to find a gradual gradation of particle sizes. The thinnest region of any film sample was taken as the point when it was just possible to see particles at a magnification of 113,000 using the transmission electron microscope.

Scanning electron microscopy of bulk specimens (Figs. 33-64)

Three different test specimens were used for this aspect of the work. Glutaraldehyde and osmium fixed, alcohol dehydrated and CO_2 critical point dried leaves of Lemna minor were dry-fractured after the method of Flood[2]. The fracture faces were mounted on specimen stubs using colloidal silver and kept at 313 K for 16 h before coating. Aqueous suspensions of cleaned diatom frustules and 109 nm diameter polystyrene latex spheres were placed on cleaned specimen stubs and air dried at 313 K for 16 h before coating. The specimens shown in Figs. 33-48 were sputter-coated with 15 nm of material as described previously, while the specimens shown in Figs. 49-64 were coated with different thicknesses of gold-palladium using the same method.

All the samples were examined and photographed at 20 kV in a Cambridge Instruments S-4 scanning electron microscope fitted with liquid nitrogen cooled baffles.

Great care was taken throughout the preparation and examination of all the samples to minimise contaminations. The protocol and procedures used for coating are described in an earlier paper[1].

Results

Transmission microscopy of metal films prepared at ambient temperature

It is first necessary to relate the transmitted electron image of a metal film to its surface features as viewed in the scanning electron microscope. It is well known that metal films build up from little islands which gradually coalesce to form a continuous layer[3]. The thickness at which the first continuous

layer of metal is formed varies with different elements. Thus with carbon, a continuous layer is formed when only 1.0 nm of material has been deposited, whereas with gold, at least 45 nm of material must be deposited before a continuous film is formed. The island growth occurs laterally as well as vertically, and the end result is a continuous layer with a variously irregular surface. It is the surface irregularity of the film which is of interest to the scanning microscopist, because large particulate aggregations may well obscure the surface detail.

As can be seen in the images of the thick films of the seven different metals (Figs. 1-20) there is considerable range in both the average particle size and the film integrity. Most of the films show the characteristic cracking of the surface, a feature of most thick continuous metal films. This is due in part to the variable thickness of the film as viewed in a transmitted electron image and to the way the film is being formed once the first continuous layer has been established. Hodgkin and Murr[4] have compared the sub-structure of thin films by TEM and SEM and show that nuclei appear the same size when viewed by the two methods. This finding thus validates our assumption that the TEM images of thin films (Figs. 1-20) are representative of the particle size on the top of the film layer.

Consider first the sputter coated films (Figs. 1-14) where it is clear that platinum (Figs. 5,6), molybdenum (Figs. 9,10) and possibly tungsten (Figs. 13,14) have a smaller particle size than the other metals. It is interesting to note that gold/palladium has a smaller particle size than gold alone. In the one experiment we did with thermally evaporated gold (Figs. 15,16) there was only a marginal reduction in particle size of the thin film over that seen in the sputtered film (Figs. 1,2). Fig. 16 very nicely demonstrates why gold alone is not a good coating material for even medium-high resolution SEM. The thin film has formed an irregular polycrystalline layer, similar to the structures shown earlier by Pashley et al.[5]. However, Hodgkin and Murr[4] have shown that pre-coating with carbon enhances nucleation density, but not to an extent to make it useful for medium-high resolution SEM.

There is however a considerable diminution in the particle size of the films which are prepared by electron beam evaporation (Figs. 17-20) and there is little difference in the size of particles between the tantalum tungsten film (Figs. 17,18) and the platinum-carbon film (Figs. 19,20) when examined at high magnification in the transmission electron microscope.

Transmission microscopy of metal films prepared at low and high temperatures

Although the temperature differential is only 35°C there is a difference in the appearance of films prepared at 265 K and 300 K (Figs. 21-32). The average particle size of thin gold-palladium (Figs. 21,22) and gold (Figs. 25,26) films appears smaller, although this difference is not so apparent in the platinum films (Figs. 29,30). The differences in film structure are also seen in the thicker films, which more closely resemble the layers one uses in scanning electron microscopy. This is more obvious in the gold-palladium and gold films and suggests that keeping the forming film layer at as little as only a few degrees below ambient temperature limits the amount of lateral movement in atoms arriving at the surface[3]. It is important to carry this out in a clean environment.

Scanning microscopy of metal films on biological material

The earlier results of this study would suggest that platinum, molybdenum and possibly tungsten might be suitable for high resolution SEM. To a certain extent this is true and in the series of images shown in Figs. 33-48 it is clear that platinum (Figs. 37,38) and gold-palladium (Figs. 33,34) give good secondary electron images. Tungsten (Figs. 43-45) and molybdenum (Figs. 46-48) are less satisfactory for secondary electron images because of their lower emissivity but are probably better suited for low loss back-scattered images. Tantalum films (Figs. 39,40) although of small particle size have a low secondary electron emissivity and diminished conductivity and appear less suitable for scanning electron microscopy. It was hoped that nickel films (Figs. 41,42) would have a high secondary electron emission but the images were generally of rather poor quality. At low magnifications gold (Figs. 35,36) is satisfactory, but the larger particle size precludes its use at magnifications much above 10,000-15,000 in the conventional SEM.

Scanning microscopy of films of varying thickness

The results given in Figs. 49-64 show that it is possible to obtain high resolution images from non-conductors with thin layers of metal coating. The thicker layers of coating do not appear to significantly improve the image quality, and increase the diameter of the polystyrene latex spheres. In the specimens coated with 50 nm of gold-palladium (Figs. 49,50), the average diameter of the latex sphere is 133 nm, with a 20 nm layer (Figs. 51, 52) it is 126 nm, with a 10 nm layer (Figs. 53, 54) it is 119 nm, and with a 7.5 nm layer (Figs. 55,56) it is 111 nm. At film thickness below 7.5 nm the diameter of the spheres remains relatively constant at between 107 and 110 nm. The specimens with greatest contrast are those which have been coated with 7.5 nm of gold-palladium. Below this thickness the image quality begins to fall off, but is still surprisingly good even with film thickness as thin as 1 nm (Figs. 61,62).

Discussion

The results of these experiments show that it is possible to obtain 5-8 nm resolution images of non-conducting biological material in the conventional SEM after careful application of metal films to the specimen.

For medium resolution SEM where one is working up to 20,000 diameters magnification, sputter coating with either gold-palladium or platinum on a cooled specimen generally gives satisfactory results. Neugebauer[3] had already shown that lowering the substrate temperature leads to small grain size and shortens the time for the formation of a continuous film. At higher magnifications, the particle size of gold-palladium becomes visible and specimens should be coated with platinum alone. Slayter[6] had shown that platinum metal evaporated at 1-5 μPa is capable of resolving features as small as 2 nm. The quality of the coating layer is however very dependent on carrying out the process in a clean vacuum[7] and if platinum is to be sputtered in a conventional cool sputter coater, great care must be taken to avoid back-streaming of the rotary pump oil and to flush the chamber many times with clean dry argon before coating.

For studies using scanning attachments on transmission microscopes or high resolution SEM's fitted with field emission sources it is necessary to resort to high vacuum electron beam evaporation in order to apply the metal film. This is the basis of the technique which is used routinely to produce the high resolution replicas of freeze-fractured surfaces and shows no film sub-structure up to magnifications of 150,000 diameters. If the image is to be formed by secondary electrons, then carbon/platinum should be the material of choice, whereas tungsten or a tantalum/tungsten mixture would be better suited for images formed by low loss back-scattered primary electrons. Provided the vacuum system is clean, lowering the substrate temperature results in small critical nuclei. Although we have not tried sub-ambient specimens in vacuum evaporated systems, there is no reason why this same effect should not be observed. Further work is needed to show if it is better to sputter or evaporate metal films. Much seems to depend on the cleanliness of the vacuum and it may be necessary to only sputter coat in a system which has previously been pumped to at least 100 μPa before being back-filled to working pressure with clean dry argon. Alternatively, we should perhaps further investigate the cryo-absorption pumped system proposed by de Harven and colleagues[8]. Adachi et al.[9] have shown that high resolution films can be made by ion-beam sputtering refractory metals such as tungsten and tantalum, which allow resolution down to the 1.0 nm level in the TEM. However, such films have a poor secondary electron emission and should only be considered for low loss back-scattered electron imaging.

It is also apparent that in many instances we are putting far too much metal on our non-conducting samples. As we have shown in this paper, metallic films as thin as 2-5 nm give surprisingly good results even though such thicknesses are unlikely to be forming continuous layers. However as Morris and Coutts[10] have shown, discontinuous metal films are conducting even if the mechanism for conductance is not understood.

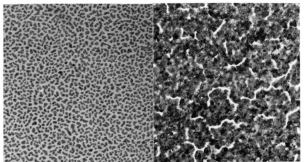

Fig. 1. Thin layer of Au sputtered at 285 K. PW = 465 nm.

Fig. 2. Thick layer of Au sputtered at 285 K. PW = 465 nm.

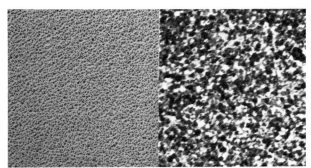

Fig. 3. Thin layer of Au/Pd sputtered at 285 K. PW = 465 nm.

Fig. 4. Thick layer of Au/Pd sputtered at 285 K. PW = 465 nm.

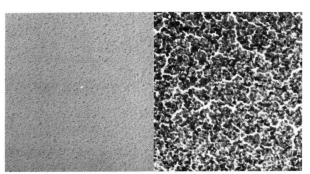

Fig. 5. Thin layer of Pt sputtered at 285 K. PW = 465 K.

Fig. 6. Thick layer of Pt sputtered at 285 K. PW = 465 nm.

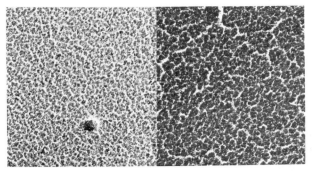

Fig. 7. Thin layer of Ni sputtered at 285 K. PW = 465 nm.

Fig. 8. Thick layer of Ni sputtered at 285 K. PW = 465 nm.

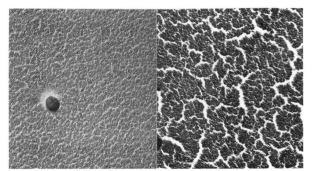

Fig. 9. Thin layer of Mo sputtered at 285 K. PW = 465 nm.

Fig. 10. Thick layer of Mo sputtered at 285 K. PW = 465 nm.

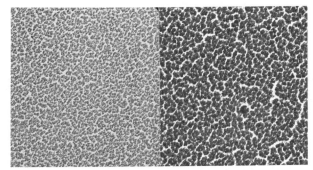

Fig. 11. Thin layer of Ta sputtered at 285 K. PW = 465 nm.

Fig. 12. Thick layer of Ta sputtered at 285 K. PW = 465 nm.

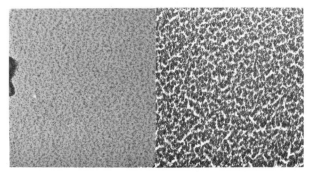

Fig. 13. Thin layer of W sputtered at 285 K. PW = 465 nm.

Fig. 14. Thick layer of W sputtered at 285 K. PW = 465 nm.

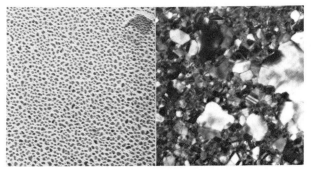

Fig. 15. Thin layer of Au thermal evaporation at 295 K. PW = 465 nm.

Fig. 16. Thick layer of Au thermal evaporation at 295 K. PW = 465 nm.

PW = Photo Width

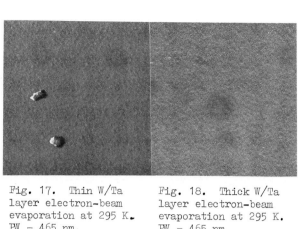 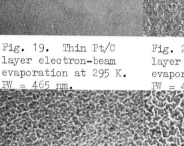

Fig. 17. Thin W/Ta
layer electron-beam
evaporation at 295 K.
PW = 465 nm.

Fig. 18.   Thick W/Ta
layer electron-beam
evaporation at 295 K.
PW = 465 nm.

Fig. 19. Thin Pt/C
layer electron-beam
evaporation at 295 K.
PW = 465 nm.

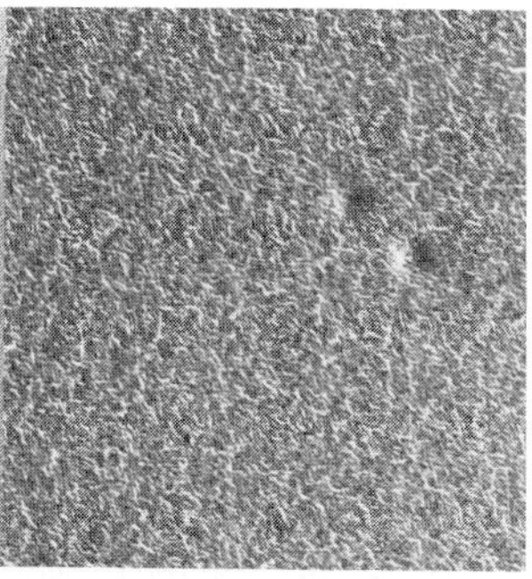

Fig. 20.   Thick Pt/C
layer electron-beam
evaporation at 295 K.
PW = 465 nm.

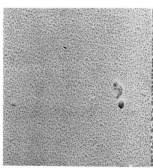 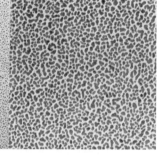 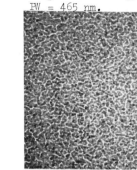 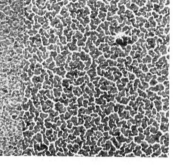

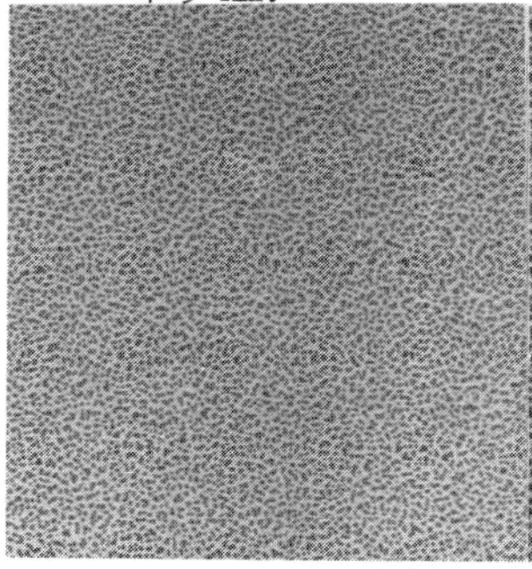

Fig. 21. Thin Au/Pd
layer sputtered at
265 K. PW = 465 nm.

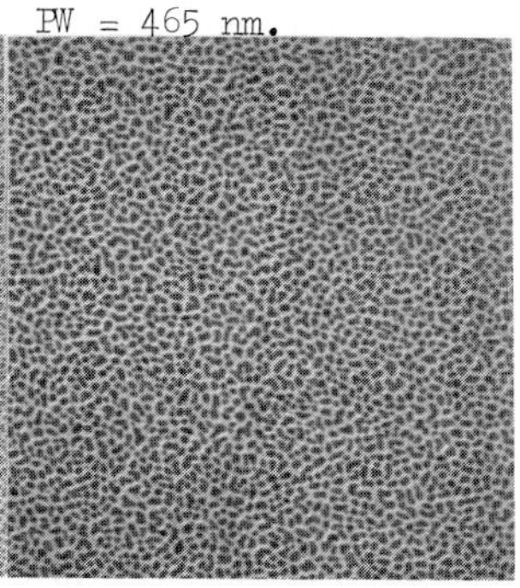

Fig. 22. Thin Au/Pd
layer sputtered at
300 K. PW = 465 nm.

Fig. 23. Thick Au/Pd
layer sputtered at
265 K. PW = 465 nm.

Fig. 24. Thick Au/Pd
layer sputtered at
300 K. PW = 465 nm.

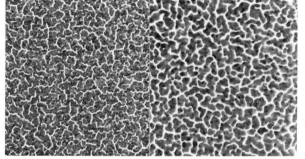

Fig. 25. Thin Au
layer sputtered at
265 K. PW = 465 nm.

Fig. 26. Thin Au
layer sputtered at
300 K. PW = 465 nm.

Fig. 27. Thick Au
layer sputtered at
265 K. PW = 465 nm.

Fig. 28. Thick Au
layer sputtered at
300 K. PW = 465 nm.

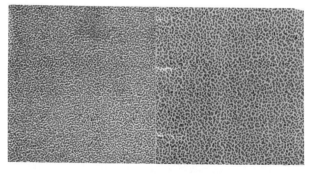 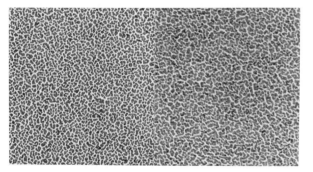

Fig. 29. Thin Pt
layer sputtered at
265 K. PW = 465 nm.

Fig. 30.   Thin Pt
layer sputtered at
300 K. PW = 465 nm.

Fig. 31. Thick Pt
layer sputtered at
265 K. PW = 465 nm.

Fig. 32.   Thick Pt
layer sputtered at
300 K. PW = 465 nm.

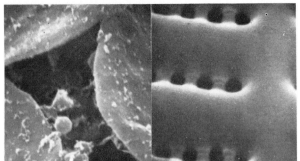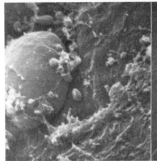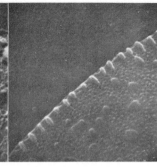

Fig. 33. <u>Lemna</u>
sputtered with 15 nm
Au/Pd at 285 K.
PW = 5.3 µm.

Fig. 34. Diatom
sputtered with 15 nm
Au/Pd at 285 K.
PW = 1.8 µm.

Fig. 35. <u>Lemna</u>
sputtered with 15 nm
Au at 285 K.
PW = 9.3 µm.

Fig. 36. Diatom
sputtered with 15 nm
Au at 285 K.
PW = 4.4 µm.

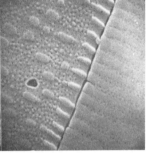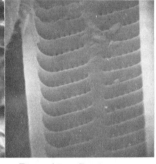

Fig. 37. <u>Lemna</u>
sputtered with 15 nm
Pt at 285 K.
PW = 13.5 µm.

Fig. 38. Diatom
sputtered with 15 nm
Pt at 285 K.
PW = 4.4 µm.

Fig. 39. <u>Lemna</u>
sputtered with 15 nm
Ta at 285 K.
PW = 27.3 µm.

Fig. 40. Diatom
sputtered with 15 nm
Ta at 285 K.
PW = 8.8 µm.

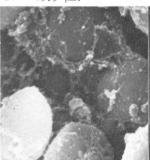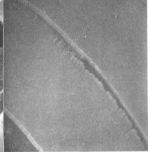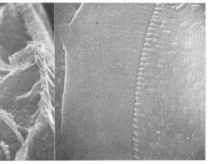

Fig. 41. <u>Lemna</u>
sputtered with 15 nm
Ni at 285 K.
PW = 11.4 µm.

Fig. 42. Diatom
sputtered with 15 nm
Ni at 285 K.
PW = 8.8 µm.

Fig. 43. <u>Lemna</u>
sputtered with 15 nm
W at 285 K.
PW = 25 µm.

Fig. 44. Diatom
sputtered with 15 nm
W at 285 K.
PW = 8.8 µm.

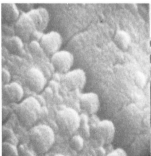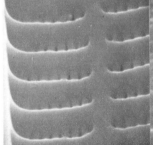

Fig. 45. Diatom
sputtered with 15 nm
W at 285 K.
PW = 1.8 µm.

Fig. 46. <u>Lemna</u>
sputtered with 15 nm
Mo at 285 K.
PW = 10 µm.

Fig. 47. Diatom
sputtered with 15 nm
Mo at 285 K.
PW = 4.4 µm.

Fig. 48. Diatom
sputtered with 15 nm
Mo at 285 K.
PW = 1.8 µm.

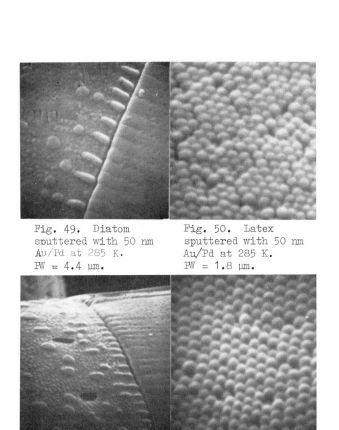

Fig. 49. Diatom
sputtered with 50 nm
Au/Pd at 285 K.
PW = 4.4 µm.

Fig. 50. Latex
sputtered with 50 nm
Au/Pd at 285 K.
PW = 1.8 µm.

Fig. 51. Diatom
sputtered with 20 nm
Au/Pd at 285 K.
PW = 4.4 µm.

Fig. 52. Latex
sputtered with 20 nm
Au/Pd at 285 K.
PW = 1.8 µm.

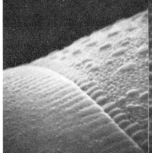

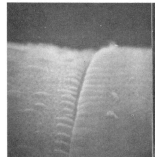

Fig. 53. Diatom
sputtered with 10 nm
Au/Pd at 285 K.
PW = 4.4 µm.

Fig. 54. Latex
sputtered with 10 nm
Au/Pd at 285 K.
PW = 1.8 µm.

Fig. 55. Diatom
sputtered with 7.5 nm
Au/Pd at 285 K.
PW = 4.4 µm.

Fig. 56. Latex
sputtered with 7.5 nm
Au/Pd at 285 K.
PW = 1.8 µm.

Fig. 57. Diatom
sputtered with 5 nm
Au/Pd at 285 K.
PW = 4.4 µm.

Fig. 58. Latex
sputtered with 5 nm
Au/Pd at 285 K.
PW = 1.8 µm.

Fig. 59. Diatom
sputtered with 2 nm
Au/Pd at 285 K.
PW = 4.4 µm.

Fig. 60. Latex
sputtered with 2 nm
Au/Pd at 285 K.
PW = 1.8 µm.

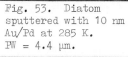

Fig. 61. Diatom
sputtered with 1 nm
Au/Pd at 285 K.
PW = 4.4 µm.

Fig. 62. Latex
sputtered with 1 nm
Au/Pd at 285 K.
PW = 1.8 µm.

Fig. 63. Diatom
uncoated. PW = 4.4 µm.

Fig. 64. Latex
uncoated. PW = 1.8 µm.

Some of our work corroborates the findings of earlier studies. Thus Braten[11] had shown using a STEM instrument that sputter coating with 25 nm of gold-palladium produced smaller particles than sputter coating a similar thickness of gold alone. Of the two metals used in the study by Braten, thermal evaporation of gold-palladium with or without carbon gave the best results, although particle sub-structure was visible at high magnifications. Katoh and Nakazuka[12] using a notched wire flashing method have produced high resolution films of tungsten, tantalum and molybdenum which have a grain size as small as conventionally prepared carbon films. Maeki and Benoki[13] have obtained good results in a STEM instrument using evaporated chromium. The particle size was small enough to give good secondary electron images at 30,000x with a useful contrast range. Watt[14] has recently shown that thin films of platinum prepared in a diode sputter coater have a smaller grain size than gold films prepared by the same technique.

Although conventional sputter and evaporative coaters are capable of providing the medium resolution (5-8 nm) coating layers which modern scanning microscopes require such methods do not give the high resolution films required for scanning transmission microscopy. Sputter coating at high vacuum, the use of electron beam evaporant sources and a careful look at the nature of the target material is urgently required to provide thin conducting films 2-3 nm thick.

Acknowledgements

We are grateful to Brian Chapman for carrying out the transmission electron microscopy, to Sylvia Dalton for her photographic expertise, and to Ruth Hockaday for carefully and patiently typing the manuscript.

References

1.P. Echlin. Coating techniques for scanning electron microscopy and X-ray microanalysis. SEM/1978/I 109-132 SEM Inc. AMF O'Hare Ill. 60666.

2.P.R. Flood. Dry fracturing techniques for the study of soft internal biological tissue in the SEM. SEM/1975 287-294 IIT Research Institute Chicago Ill. 60616.

3.C.A. Neugebauer. Condensation, nucleation and growth of thin films. Chapter 8 in Handbook of Thin Film Technology, ed. L.I. Maissel and R. Glang. McGraw Hill, New York, 1970.

4.N.M. Hodgkin and L.E. Murr. Quantitative study of vapour-deposited metal coatings for SEM. Microstructural Science 2, 129-146, 1974.

5.D.W.M. Pashley, M.J. Stowell, M.H. Jacobs and T.J. Law. Growth and structure of gold and silver deposits formed by evaporation inside an electron microscope. Phil. Mag. 10, 127, 1964.

6.H.S. Slayter. High resolution metal replication of macromolecules. Ultramicroscopy 1, 341-357, 1976.

7. L.I. Maissel. Application of sputtering to the deposition of films. Chapter 4 in Handbook of Thin Film Technology, ed. L.I. Maissel and R. Glang. McGraw Hill, New York, 1970.

8.E. de Harven, N. Lampen and D. Pla. Sputter coating in oil contamination free vacuum for SEM. SEM/1978/I 167-169 SEM Inc. AMF O'Hare Ill. 60666.

9.K. Adachi, K. Jojou, M. Katoh and K. Kanaya. High resolution shadowing for electron microscopy by sputter deposition. Ultramicroscopy 2, 17-29, 1976.

10.J.E. Morris and T.J. Coutts. Electrical conductance in discontinuous metal films: A discussion. Thin-solid Films 47, 1-65, 1977.

11.T. Braten. High resolution scanning electron microscopy in biology: artefacts caused by the nature and mode of application of the coating material. J. of Microscopy 113, 53-59, 1978.

12.M. Katoh and H. Nakazuka. A simple vacuum evaporation method for high melting point metals and its applications. J. Electron Micros. 26, 219-222, 1977.

13.G. Maeki and M. Benoki. Effect of chromium shadowing for the improvement of the SEM image of biological specimens. J. Electron Micros. 26, 223-224, 1977.

14.I.M. Watt. A comparison of gold and platinum sputtered coatings for SEM. Electron Microscopy 1978, Vol. II, Biology 94-95. Microscopical Society of Canada, 150 College St., Toronto, Ontario.

Discussion with Reviewers

Reviewer I and J. Geller: What steps were taken to minimize contamination during sputtering?
Authors: We have experimented with cold traps working at 272 K and 193 K and found they made little difference to either the pump down time or contamination rates. Liquid nitrogen traps cannot be used because the argon condenses on the baffles and eventually clogs up the system. The sputter coater is pumped using a ballasted two stage rotary pump and we are careful never to run it at its maximal obtainable vacuum. The argon gas is anhydrous, and the chamber was flushed ten times before turning on the high voltage supply. We only noticed a small amount of contamination after prolonged sputtering, i.e. 30 mins.

K.R. Peters: What was the ratio of elements in the alloys used for coating?
Authors: The Au/Pd ratio was 80:20, the W/Ta ratio was between 70/85:30/15 and the Pt/C ratio was 85:15. Such figures are of little value for evaporated materials as few alloys evaporate congruently because of differences in their vapour pressure. Sputter coating on the other hand does maintain the alloy ratios in the film which are present in the target.

K.R. Peters: Did you observe any evidence for the influence of contamination on the thickness, fine structure and secondary emission of the metal films?
Authors: As indicated earlier, a small amount of contamination was seen after prolonged sputtering. No evidence of contamination was seen on the evaporated films.

K.R. Peters: Why does the Ta film obtained by cool sputter coating (Figs. 11 and 12) show discontinuities?
Authors: We presume this is a feature of the way the film is formed, and the uneven features and large grain size of the sputtered Ta film make it a less useful coating material for scanning electron microscopy.

Reviewer I and K.R. Peters: Are the settings for signal generation, amplification, signal recording and image processing kept constant to allow a quantitative comparison of image quality of Figs. 33-64?
Authors: As far as possible these parameters were kept constant during examination and recording. But in all cases the final image quality was adjusted to give the best picture as judged by focus, resolution, contrast and information content.

Reviewer I and K.R. Peters: How long does it take to obtain the various coatings described in the paper?
Authors: Because the coating thicknesses were all monitored using quartz thin film monitors, no accurate record was made of the time taken to apply each coating layer. However, the evaporative coatings were completed within 30 secs, the variable thickness Au/Pd coatings (Figs. 49-64) ranged from 10 mins to 25 secs, and the various noble metal sputter coatings were completed within 3-5 mins. Nickel and molybdenum sputter coating took between 10 and 15 mins while the tantalum and tungsten coating usually took approximately 20 mins. Occasionally, Ta and W took much longer, but such films were found to be contaminated and were discarded.

K.R. Peters: What are the distances between the source and specimen in the various pieces of equipment used?
Authors: These details may be found in the paper given by one of us (P.E.) at the 1978 SEM meeting - see reference number one. The source to specimen distance in the electron beam evaporation equipment is 150 nm.

K.R. Peters: How did you estimate the lowest value of the applied substrate temperature for obtaining minimal grain size?
Authors: The temperature of 265 K was the lowest we could obtain on the Peltier cooled table on the model sputter coater we used for our experiments. We are at present carrying out sputter coating at much lower temperatures, but

as Slayter has recently shown /H.S. Slayter, Electron microscopy of glycoproteins by high resolution metal replication, Chapter 6 of Principles and Techniques of Electron Microscopy, Vol. 9, ed. M.A. Hayat, Van Nostrand, New York, 1978/, it is necessary to ensure that low temperature coating is only carried out in a clean environment.

K.R. Peters: How accurate is the monitoring of the thickness of thin films described as just visible in the TEM?
Authors: We do not consider that the assessment of film thickness by visibility in the TEM at a fixed high magnification as a reliable method. It would be wrong to make comparisons of different film thicknesses by this method. In Figs. 1-32 we put down a known film thickness and then examined the extinction point where we could just see particles at 113,000 magnification which should allow us to resolve 2-3 nm. The technique was used as a means of measuring the size of metal particles. Ideally these measurements should also be carried out in a high resolution STEM instrument, but we have as yet not had an opportunity of carrying out such investigations.

K.R. Peters: What reasons do you give for the diminished conductivity of the 15 nm tantalum film?
Authors: Tantalum bulk metal has a much poorer conductivity (0.575 W.cm^{-1}) than say the noble metals (i.e. gold 3.17 W.cm^{-1}) used in sputter coating and it may well be that a continuous film is not formed at a thickness of 15 nm. Tantalum and tungsten were only used in this work because they have been shown to give high resolution images of replicas examined in the TEM. It is clear that these two metals have little use as coating material for conventional SEM.

K.R. Peters: Is the poor image of the Lemna specimen coated with 15 nm of Ta due to insufficient specimen mounting or a non-representative picture?
Authors: Tantalum invariably gave poor SEM images of biological material and we consider this is related to its poor conductivity and lower secondary electron emissivity.

K.R. Peters: What is the reference giving the secondary electron emission coefficient of different metals?
Authors: General references are given in the paper by Echlin[1] and by Goldstein /J.I. Goldstein, Electron beam-specimen interaction, Chapter III in Practical Scanning Electron Microscopy, ed. J.I. Goldstein and H. Yakowitz, Plenum Press, New York, 1975/. A more detailed study is given in the paper by Wittry /D.B. Wittry in X-ray Optics and X-ray Analysis, 168-180, ed. R. Castaing et al., Herman, Paris, 1966/.

J. Geller: Why are there no SEM micrographs of material coated using electron beam evaporated W/Ta and Pt/C?

Authors: After the disappointing results with sputter coated W and Ta material we decided it was not worth further investigation at this time. One of the problems is that the electron beam technique is usually used to put on thin c. 1 nm films, and we would like to investigate a range of film thicknesses. We intend carrying out further experiments using electron beam evaporation and hope to be able to examine specimens in the JEOL high resolution STEM instrument.

K.R. Peters: The change of latex sphere diameters with increased thickness of Au/Pd cool coated films (Figs. 49-64) is not linear. Do these results represent decoration artefacts rather than altered coating thickness?

Authors: Although decoration artefacts may well play a small part in the changes seen in the images, we consider that the major change is due to differences in film thickness, particularly as we were careful to minimize specimen contamination during the coating procedure.

P.B. DeNee: Do you agree that your results corroborate our earlier findings /P.B. DeNee and E.R. Walker, Specimen coating technique for SEM, SEM/1975 225-232 IIT Research Institute Chicago Ill 60616/ that both Au and Au/Pd sputter coating consists mainly of nearly spherical particles and that a 10 nm Au film has a larger particle size than a 10 nm Au/Pd film?

Authors: Our results appear to be in close agreement with your findings and demonstrate the added advantage of using a cool sputter coater.

P.B. DeNee: Assuming that a 15 nm coating layer consists of 10-15 nm spherical islands and that these islands cover about 50% of the specimen surface, would it be correct to conclude that this is in fact equivalent to a measured film of only 5-8 nm, and that it is this film thickness which we should consider to be the maximum coating we should apply to specimens?

Authors: We were careful to calibrate the thickness monitor using interferometer measurements on thin films. However, this only measures the film thickness on a flat surface, and you may well be correct in assuming that a proportionally thinner layer lands on a rough sample.

J. Geller: Why does a thin layer of evaporated Au/Pd have a finer grain size than sputtered Au/Pd whereas a thick layer of evaporated Au/Pd is coarser than sputtered Au/Pd?

Authors: The heat input to the specimen by radiation probably affects crystallite growth. The higher heat input from the evaporative source would be likely to result in increased crystallite growth and diminished nucleation frequency.

J. Geller: Surely the image contrast is a function of coating thickness and resolution not of coating thickness alone?

Authors: You are correct, it is a function of coating thickness and resolution.

K.R. Peters: What is your opinion of the use of W coating in high resolution work?

Authors: From our limited experience of using it for preparing sputter coated films, we would not recommend using it for high resolution secondary emission SEM. However, its proven small particle size from the extensive work on high resolution TEM replicas makes it worth investigating for coating specimens to be examined by low loss back-scattered electrons.

SCANNING ELECTRON MICROSCOPY/1979/II
SEM Inc., AMF O'Hare, IL 60666, USA

CHARGE NEUTRALIZATION USING VERY LOW ENERGY IONS

C. K. Crawford

KIMBALL PHYSICS Inc., Kimball Hill Road
Wilton, NH 03086

Abstract

A new method of controlling insulating specimen charge-up in scanning electron microscopes
is proposed and demonstrated. The method consists
of neutralizing the charge buildup as needed, in
real time, with a beam of very low energy ions.
The ions stabilize the surface at the ion zero
kinetic-energy point independent of the nature of
the insulating surface; they cause no sputtering
because their energy is very low; they are not
attracted to the secondary collector because of
their positive charge.

The method is very successful for eliminating
most charge-up effects. These include primary
electron trajectory effects, secondary electron
trajectory effects, secondary yield effects, and
breakdown effects. The method is not as effective
for buried charge effects, however these are
usually minor. It is now possible to see atomic
number contrast with insulating specimens. And
the technique will also permit easier x-ray,
SEM/SIMS, and Auger analysis of insulators. In a
few cases the charging effects themselves may even
be used to obtain useful information, now that
they can be controlled.

A theory is presented for this kind of surface
potential control; its limitations are explored.
It will not be appropriate for all specimens at
all magnifications. Results are presented for
semiconductor, ceramic, plastic and biological
samples.

KEY WORDS: Specimen Charging Effects, Charge
Neutralization, Low Energy Ions, Insulators,
Specimen Preparation

Introduction

Insulating specimen charge-up is a serious
problem in scanning electron microscopes and many
similar instruments. Deleterious effects due to
charging include primary electron beam trajectory
effects, secondary electron trajectory and yield
effects, vacuum, surface, and bulk breakdown effects, specimen conduction effects, and even
specimen mechanical motion effects. Many of these
phenomena have been well studied.[1-6]

Two main groups of methods have been used to
control specimen charging of insulating specimens.
By far the most common group involves improving the
specimen surface or bulk conductivity, usually by
the addition of a highly conductive layer to the
surface.[7] The second group of methods involves no
specimen conductivity modification. Instead the
net arrival and departure rates of charged particles are controlled such that surface neutrality
(or some approximation thereto) is achieved
without need for either surface or bulk conductive
paths.[8-11]

This paper concerns a new method in the second
group. Positive ions with energies in the few eV
range continuously neutralize electron charge
buildup at each point on the surface of an insulating specimen. The ion beam is adjusted to have
a current significantly higher than the primary
scanning electron beam. Each point of the surface
then charges until the potential of that point is
just sufficiently positive to cut off further ion
bombardment. The resulting surface potential is
very close to the potential of the ion generation
region, which can in turn be adjusted to be very
close to ground. The process is inherently stable
since the ions and electrons are of opposite charge.

Charge-Up Theory

The theory of insulator charge-up under electron bombardment has been studied for many
years.[12] For scanning microscopes and similar
instruments, an appropriate model is that depicted
in Fig. 1. Here the scanning microscope column
is modeled simply as the generator of a constant
primary bombarding electron current i_{PO} (i_{PO} is
defined as the electron current leaving the final
column aperture and is always a negative number,
i.e., negative particles flowing in the direction
shown). A fraction of this current, denoted as
current i_P strikes the insulator surface (normally
$i_P = i_{PO}$, but this need not always be the case). The

primary electrons strike the insulating specimen, in an elemental area at a point where its potential is V_X (note that all potentials are measured with respect to ground, normally the specimen chamber wall), and cause the release of a secondary electron current i_S (another negative current). Since the primary electrons originate from a cathode at potential V_{EC} (a large negative number numerically equal to the instrument KeV setting), their kinetic energy when they strike the specimen is simply $e(V_X - V_{EC})$, where e is the magnitude of the electron charge. The secondary electrons are attracted into a secondary collector at positive potential V_{SC}, their kinetic energy is $e(V_{SC} - V_X)$ when they reach the collector.

If ion charge neutralization is to be used to control the surface potential V_X, then there is an ion current I_{NO} which is directed at the surface. Of this, a portion I_N actually impinges on the surface. The ions strike the surface with kinetic energy $q(V_{IS} - V_X)$, where V_{IS} is the potential (positive) of the ion generation region (i.e. the potential where the ions have zero kinetic energy), and q is the ion charge.

Finally, if the specimen has a non-infinite resistivity a conduction current i_C (positive for positive V_X) will flow to ground.

The general case is complex. The specimen surface potential V_X is a function of both spacial position on the surface and of time. The surface shape may be highly irregular. The resistivity is often non-constant, nonlinear, time-varying, and a function of position on the surface. Even if the resistivity itself is a constant, an irregular geometry will make the resistance to ground a complex function of position. The secondary emission properties of the surface are in general nonuniform. Finally a variety of sporadic emission and breakdown phenomena may be superimposed. For purposes of analysis it is easiest to take the simplest case, i.e., a planar, homogeneous sample, with constant resistivity, uniform secondary properties, and which charges uniformly. This avoids what for most samples would be impossible integrations. In spite of the grossness of these approximations, this simple analysis can accurately describe what is actually taking place in the case where ion neutralization dominates. This is because of the internal feedback stabilization inherent in the ion neutralization technique. For unneutralized insulating specimens however, this analysis is far too crude, and should be regarded only as a very poor first order approximation to what is really going on.

Proceed as follows: Let Q_X be the uniformly distributed surface charge on the idealized specimen surface, C be the capacitance (also assumed linear) of the front specimen surface to ground (the specimen holder), and I_X be the net change in surface charge per unit time.

Then:
$$Q_X = CV_X, \qquad\qquad [1]$$

and:
$$I_X = \frac{dQ_X}{dt} = C\frac{dV_X}{dt}. \qquad [2]$$

From the conservation of charge:
$$I_X = i_P - i_S - i_C + I_N \qquad [3]$$

and:
$$\frac{dV_X}{dt} = \frac{1}{C}\left[\, i_P - i_S - i_C + I_N \,\right]. \qquad [4]$$

Symbols Used:

C Capacitance of specimen surface to ground (farads)

e Magnitude of electronic charge (coulombs)

eV_P Primary electron energy (electron volts)

eV_S Secondary electron energy (electron volts)

eV_5 Primary energy where secondary suppression begins (electron volts)

i_C Specimen conduction current to ground (amperes)

i_{PO} Primary electron column current leaving last aperture (amperes)

i_P Primary current reaching specimen surface (amperes)

i_S Total secondary electron current leaving surface, includes reflected primaries (amperes)

I_{NO} Ion source emitted current (amperes)

I_N Ion current reaching specimen surface (amperes)

I_X Rate of change of specimen surface charge, equal to net current to specimen surface (amperes)

q Charge on a single ion (coulombs)

Q_X Specimen surface charge (coulombs)

R Resistance of specimen surface to ground (ohms)

V_{EC} Primary-electron cathode potential (volts)

V_{EC1} Primary-electron cathode potential (volts)

V_{EC2} Neutralizing-electron cathode potential (volts)

V_{IS} Ion source potential (volts)

V_{SC} Secondary electron collector potential (volts)

V_X Specimen surface potential (volts)

V_1 First unity secondary emission potential (volts)

V_2 Second unity secondary emission potential (volts)

V_3 Surface potential without neutralization (volts)

V_4 Surface potential with ion neutralization (volts)

δ True low energy secondary yield

η Reflected primary yield

σ Total secondary electron yield

The desired goal is to keep the surface potential constant and near zero, that is:

$$\frac{dV_X}{dt} = 0, \text{ and } V_X = 0 . \qquad [5]$$

One way to do this is to start with an uncharged specimen and somehow maintain a balance such that

$$i_P - i_S - i_C + I_N = 0 . \qquad [6]$$

The currents in Eq. 6 however, are not only functions of V_X, but different functions of V_X. It is worthwhile to examine these functions.

To a good approximation the function $i_P(V_X)$ looks electrically like a current source of value i_{PO} which cuts off sharply at $V_X = V_{EC}$ (ignoring the energy spread and trajectory effects which are not very important for this analysis). The current i_P must cut off at V_{EC} because the electron kinetic energy changes sign at this point, and of course negative kinetic energies are impossible. This function is graphed in Fig. 2a.

The secondary current i_S is easily calculated from the well known total secondary yield curve as shown in Fig. 2b (recall that the primary energy is just $e(V_X - V_{EC})$). It is necessary to (temporarily) make the further assumption that V_{SC} is sufficiently positive to collect all secondaries emitted and that none are lost due to other causes. Since by definition

$$\sigma = \frac{i_S}{i_P} \qquad [7]$$

it follows that

$$i_P - i_S = i_P(1 - \sigma). \qquad [8]$$

This function has several interesting points. First, for the case where $i_C=0$ and $I_N=0$, I_X is equal to i_P-i_S as graphed in Fig. 2c. The net current to the specimen, I_X, is just the difference between the incoming primary beam current and the departing secondary current; its value can be either positive or negative. Now from Eq. 2, I_X and dV_X/dt are linearly proportional to each other and always have the same sign. Thus when I_X is positive, dV_X/dt is positive, i.e. V_X is increasing; likewise when I_X is negative, V_X is decreasing. Hence when the system is described by an operating point above the V_X axis *(I_X positive)*, this operating point will move to the right; an operating point below the axis will move to the left. An operating point on the axis, corresponding to $I_X = dV_X/dt = 0$ will not move. For example, an initially uncharged specimen starts off with $V_X = 0$ and I_X equal to the negative value at which the curve crosses the I_X axis. Hence the specimen will charge negatively at a rate determined by the magnitude of this intercept and the magnitude of the specimen to ground capacitance C. As the operating point approaches the point $I_X = 0$, $V = V_2$, it will slow down and stop. If somehow the operating point lies somewhere along the curve between V_1 and V_2, it will move to the right, again coming to rest at point $0,V_2$. For this curve there are two points, $0,V_1$ and $0,V_2$ and a line, $V_X < V_{EC}$ (which corresponds to an infinity of points) where $I_X = 0$. All of these

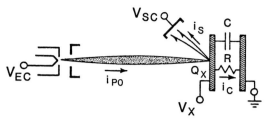

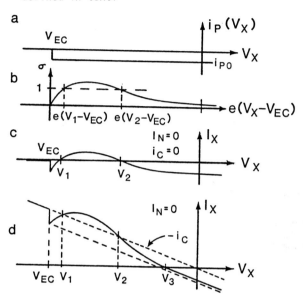

Fig. 1 Model for electron charge-up of an insulating specimen in a SEM. Symbols are defined in text.

Fig. 2 Voltage-current relationships for unneutralized insulator charge-up. I_X is the sum of all currents arriving at the insulator surface. Insulator surface potential is V_X. Curves are highly idealized.

points are equilibrium points. Point $0,V_2$ is a stable equilibrium point, the infinity of points to the left of V_{EC} are neutral equilibrium points, and point $0,V_1$ is an unstable equilibrium point. In general, if dI_X/dV_X is negative the point is stable, if dI_X/dV_X is positive the point is unstable, and if dI_X/dV_X is zero the point is neutral.

Consider now the change which results if the specimen has a conduction path to ground. Assume initially that the resistivity is linear, homogeneous, and constant. Then

$$i_C = \frac{V_X}{R} , \qquad [9]$$

where R, the resistance to ground, can in principle be calculated from the resistivity and the specimen geometry. Subtracting this current from the characteristic shown in Fig. 2c results in the characteristic shown in Fig. 2d. There is now only one equilibrium point, the point $0,V_3$, which is a stable equilibrium point located between the point at V_2 and the origin.

Suppose the resistivity changes. For a decrease in resistivity V_3 moves toward the origin, and, in fact, the entire characteristic rotates clockwise about the point where the characteristic intercepts the I_X axis. For a metallic specimen, or a specimen coated with a thin metallic layer, the characteristic is essentially vertical, and the specimen potential is stabilized at ground. Having a metallic specimen is clearly the simplest way to avoid specimen charging.

With a real specimen additional complexities arise. If a high primary energy is used, breakdowns may occur before the specimen charges negatively enough to reach the points at V_2 or V_3. If the specimen does not have a region where $\sigma > 1$ (many don't), the points at V_1 and V_2 do not exist, and the specimen will try to charge negatively to the full cathode potential, V_{EC}. Normally there is no mechanism to charge a specimen more negatively than V_{EC}; however the operator can sometimes cause this by making a step reduction in V_{EC}. If this should occur, the specimen can sit for an indefinite period at any of these points, depending on surface leakage and other effects.

A major additional complexity in SEM's is that the beam is scanned in a raster. Thus each point on the sample is charged separately at a rate determined by the magnitude of the primary bombardment current, the scanning rate, and the surface properties and topography at that point. The importance of the scanning rate is often such that satisfactory images can be obtained at fast scan rates where none at all is possible at slower rates.[8]

Ion Neutralization Theory

Explore now the possibility of neutralizing the negative surface charge using a current of the opposite sign, such as a beam of very low energy positive ions. The addition of a small ion beam generator directly in the specimen chamber, as shown in Fig. 3, will result in the idealized ion current characteristic shown in Fig. 4a. This characteristic is very similar to the electron gun characteristic of Fig. 2a. except for three important differences. First, the signs are all reversed because the ions are positive. Second, the ion source potential, V_{IS}, is chosen to be very near ground (typically a few volts positive), whereas the electron cathode potential, V_{EC}, is far from ground (typically many kilovolts negative). Third, the ion current is chosen to be much larger than the primary electron current. The reasons for these choices will appear presently.

Consider first a hypothetical case where only i_P and I_N exist, i.e., $i_S=0$ and $i_C=0$. I_X then equals i_P+I_N. Adding the curves of Figures 2a and 4a results in a characteristic of the type shown in Fig. 4b. There is one stable equilibrium point, point O, V_4, which is located approximately at the ion source potential, V_{IS}. Note that the reason that the stable point is near V_{IS} rather than V_{EC} is that the magnitude of I_{NO} is greater than that of i_{PO}. If this is not the case, then a characteristic such as that shown in Fig. 4c results. The analysis up to here is completely symmetric with respect to positive and negative. It should now be clear that the major reason for choosing V_{IS} close to ground, i.e., using very low energy ions, is to guarantee that the surface potential will be stabilized close to ground. A large ion current is

chosen to insure that the ions dominate in controlling the surface potential.

Consider next a more realistic case where the secondary current, i_S, and the conduction current, i_C, are nonzero . Equation 3 then has all nonzero terms, and a characteristic such as shown in Fig. 5 results. The important idea here is that there is still only one equilibrium point, a stable point, O, V_4, located almost exactly at O, V_{IS}, the same as before. Within very broad limits the specimen secondary emission properties and the specimen conductivity no longer matter. This is exactly the desired result: the insulating surface potential is now uniform, and fixed close to ground, independent of the specimen properties. In fact, since the ion source potential is easily varied, the insulator surface potential is now under direct operator control.

If the resistivity becomes lower the curve again rotates clockwise toward the vertical, about the I_X axis intercept point. (Note that the vertical curve segments remain vertical, constant in length, and located at the same values of V_X.) When the knee of the ion current characteristic drops down far enough to pass the point O, V_{IS}, an effect similar to that observed in Fig. 2d occurs. The stable point O, V_4 starts moving away from O, V_{IS} toward the origin along the V_X axis. For low enough resistivities, the stable point again approaches the origin as the curve becomes nearly vertical.

In practice, with a charge neutralizer working properly, a real system will almost always be found stabilized either near point O, V_{IS}, or very near the origin; only rarely will it be found at an in-between point. This is because real specimen conductivities vary numerically over about twenty-five orders of magnitude, from $10^{-6} \Omega$cm for metals like copper and gold, up to $10^{19} \Omega$cm for paraffin and some plastics. But the changeover from near O, V_{IS} to near the origin takes only about one order of magnitude; hence the probability of the specimen resistivity being just right for an intermediate point is small. The situation is almost, but not quite, like binary stable points with the system sitting either at V_{IS} or ground.

A specimen with mixed conductivities, such as a semiconductor circuit, presents an interesting case. Those conductor portions of the circuit which have low impedances to ground will remain at ground potential. But the insulator surfaces and various electrically isolated conductor portions will stabilize at V_{IS}. Mobility effects are important. Bombardment induced conductivity effects, where isolated surfaces are effectively connected to underlying grounds or connected to each other, are also important.

Note that both the position of the stable point and the magnitude of the ion neutralization current are separately under operator control. This means for example, that the point O, V_{IS} can be moved so close to the origin that everything is stabilized at ground potential; or alternatively, the ion source potential and current can be set such that interesting voltage contrast effects occur. In principle, where the specimen geometry can be calculated, ion neutralization techniques could be used to measure specimen conductivity or other parameters on a microscopic scale.

The magnitudes of currents and voltages need to be discussed. Primary beam currents in scanning microscopes range from about 10^{-12}A for very high resolution, up to about 10^{-7}A for x-ray and Auger work, with currents of 10^{-11}A being typical. In order to guarantee that the ion neutralization current can control the surface potential, the ion current should be chosen 2 to 3 orders of magnitude larger. Hence values of I_{NO} of about 10^{-8}A are appropriate for normal SEM work. Note however that no ion current of this magnitude actually bombards the specimen. From either Fig. 4b or Fig. 5, it can be seen that the ion current, I_N, which actually lands on the surface is approximately equal to the electron current, i_P, landing in the same general vicinity. This conclusion also results directly from the fact that the total current to the surface must be zero. Hence nearly all the ions approach the insulator surface, turn around, and actually land somewhere else. They either return to the ion generator (landing predominately on the outer housing), or become neutralized on the specimen chamber walls. The ions cause no problem wherever they land because their energy is low, typically chosen to be between a few eV and 100 eV as they leave the source. Since the sample charges up to the ion generation region potential, those ions which actually reach the specimen have even lower energy, in some cases a fraction of an eV. The actual landing energy depends on the available ion current density, the ion energy spread of the ion source, and primary beam parameters such as the primary electron density and the raster scan rate. One place ions can't go (fortunately) is to the electron collector; being positive they have the wrong sign.

The type of ion is another variable under operator control, and is an important choice for some specimens. Almost any type of ion can be used for ion neutralization, though some types are much more convenient to generate and control than others.

Comparison with Electron Neutralization

It is instructive to compare the ion neutralization technique with electron neutralization, which is the only similar technique thus far proposed for neutralization of negative charge buildup. Electron neutralization has a long but sparse history of use in analytical instruments. Columbe,[9] Spivak et.al.,[10] and Morin,[11] among others, all used electron neutralization for this purpose. Similar techniques have been used for many years in storage tubes and in television camera tubes. And of course electrons have been regularly used with good success to neutralize positive charge buildup in many positive ion instruments. The use of electron neutralization in scanning microscopes however has not been very successful. This is because the conditions here are too variable, and the attempt to control negative charge buildup with negative particle bombardment lacks the inherent stability required to overcome this variability. Because none of the proponents of the electron method seem to have presented a complete theory; an abbreviated theory is presented here.

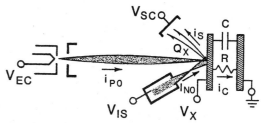

Fig. 3 Model for ion charge neutralization of insulating specimen. Miniature ion beam generation delivers weak flux of ions to specimen surface to continuously neutralize charge buildup. Ion Neutralizer generates ions with zero-kinetic-energy points at the ion source potential, V_{IS}.

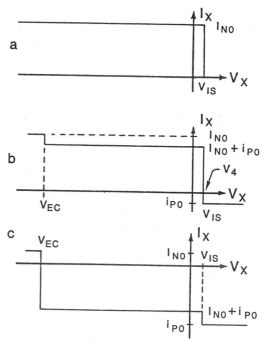

Fig. 4 Specimen voltage-current characteristics for ion neutralization considering only primary electron and neutralizing ion currents. Normal operation corresponds to (b), where the magnitude of I_{NO} is much larger than that of i_{PO}.

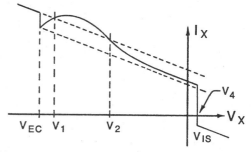

Fig. 5 Voltage-current characteristic for normal ion neutralization including secondary electron and conduction currents. Specimen surface is stabilized at potential V_4, which is very nearly the same as V_{IS}, and which can be adjusted to be very close to ground.

Since it is the most recent and the most advanced, Morin's work will be used as the starting point. A standard scanning microscope is used, except that an adjustable 100 eV to 3000 eV auxiliary electron gun and an auxiliary optical microscope are added as shown schematically in Fig. 6.

To understand the process it is necessary to look at secondary emission in more detail. The kinetic energy distribution of the secondary electrons leaving the specimen surface looks roughly like that shown in Fig. 7a, where the area under the curve is σ by definition. The area associated with the Auger peaks is negligible, hence to an excellent approximation $\sigma=\delta+\eta$, where δ is the true energy secondary yield, and η is the primary backscatter coefficient. The curve of Fig. 7a can be drawn for any given primary energy. Hence it is possible to divide the secondary yield versus energy curve into δ and η components, as shown in Fig. 7b. Note that the η curve is always less than unity, since one cannot have more reflected primaries than one has primaries. At low primary energies, where insulators typically have yields well above unity, δ is several times η. Thus for low primary energies, the surface is emitting mostly low energy secondaries. For high primary energies, the surface is emitting mostly backscattered primaries. Now consider what happens to these electrons: The backscattered primaries, having energies nearly equal to the original primary energy, leave the surface with essentially straight line trajectories; surface electric fields near the sample have little effect on them. The low energy secondaries however, are very susceptible to surface fields. Unless there is an electrode more positive than the specimen surface in the vicinity of the specimen, such that there is an electric field to draw away the low energy secondaries, the secondaries have difficulty permanently escaping from the surface. They may redistribute surface charge by making short hops; but permanent escape is energetically impossible when the specimen is positive by an amount equal to the initial kinetic energies. An interesting effect occurs if the reason for a primary energy change is a specimen surface-potential change. An increase in primary energy implies the surface becoming more positive. And at some point the surface electric field reverses sign, whereupon the low energy secondary emission starts being cut off. This point, labeled eV_5 in Fig. 7b, may for example occur when the surface becomes more positive than the secondary electron collector in a SEM. The result is an effective secondary emission yield curve like the solid line in Fig. 7b. The shape of the change-over from the "everything-escapes" curve to the "only-high-energies-escape" curve is determined by the width and shape of the δ energy distribution, and the geometry of the surface electric fields.

In scanning microscopes this low-energy-secondary cutoff effect is observed on charged insulating specimens as the "black cloud" effect. A portion of the insulating surface is charged a few volts positive with respect to the remainder of the specimen, by an amount sufficient to prevent the escape of the low energy secondaries. The dearth of secondaries causes a dark area in the image. This dark area has image information at less than half normal intensity, from those backscattered electrons which can reach the secondary collector. Note that although the area is uniformly dark, the potential is not uniform. It does not take much charging to cause this effect since the width of the δ distribution is only a few electron volts.

To explain the electron neutralization of Morin, it is necessary to combine a primary SEM beam characteristic such as that shown in Fig. 2c, with a low energy neutralization electron gun characteristic, using the detail of Fig. 7b for the low energy gun. A schematic characteristic is shown in Fig. 8; V_{EC1} and V_{EC2} refer to the primary beam and neutralization beam cathode potentials respectively. This drawing is not to scale. The magnitude of the discharging neutralization-beam current is necessarily much larger than the primary beam (several thousand times larger in Morin's apparatus), while the magnitude of V_{EC2} is typically about 1000 eV. The characteristic shows that there is a stable surface potential somewhat more positive than ground, assuming that there is enough grounded surface in the vicinity of the specimen. In this case the nearby grounded surface holds the surface potential close to ground by controlling the escape of the low energy secondaries. If, on the other hand, the electron collector is nearby and at a positive potential V_{SC}, the surface potential will move close to V_{SC}, as shown by the dashed line in Fig. 8.

This method is an interesting one, and it has provided some good pictures; however there are some problems.

First, because the neutralizing current consists of electrons, it is necessary to prevent the electrons from swamping the secondary electron detector. This is done by pulsing the neutralization gun and the SEM primary beam alternately (thus Fig. 8 must be regarded as a time average as well as being out of scale), and additional complex electronics is required. It is necessary to turn the auxiliary gun off very solidly to prevent electron leakage which would upset operation of the SEM.

Second, because the neutralizing beam is composed of negative particles, and it is a negative charge which is to be neutralized, there is no tendency for the (relatively high energy) neutralizing particles to seek out the specimen surface charge. The neutralizing particles move in approximately straight lines and must be carefully aimed at the area of primary beam illumination. This is accomplished using the auxiliary optical microscope and a small piece of phosphor type material, a somewhat bothersome procedure which requires extra apparatus. A concomitant problem is that if the specimen has complex surface relief, then the neutralizing beam will not get into shadow areas and full surface discharging will not occur.

Third, the neutralization process depends on the properties of the secondary emission curve for each specific area of specimen surface being stabilized. And the process requires a secondary emission yield greater than unity in order to work. Most clean metals do not have yields greater than unity, hence isolated metal areas in (or on) an insulating specimen cannot be neutralized. On specimens where the method works, not all areas of a mixed material specimen necessarily stabilize at the same potential.

Fourth, one of the charge-up effects which is important for some specimens, is the charge-up of nearby insulators associated with the specimen, but not in the area of observation. For example the charge-up (and subsequent breakdown) of seals on microcircuit headers by backscattered electrons. Unlike ion neutralization, the electron method cannot cope with this difficulty.

Finally, the nature of the curve in Fig. 8 allows operation at more than one value of surface potential. For example, if the surface charge-up is allowed to become too large before the neutralizing gun is turned on, (i.e., more negative than the unstable zero-crossing point which lies somewhere between V_{EC2} and the origin), the surface will attempt to stabilize way out at V_2. And once a surface potential starts to run away, the neutralizing gun is powerless to bring it back.

Ion neutralization does not have any of these problems. That is not to say that ion neutralization is problem free; the ion method has its own set of non-trivial problems. Unfortunately for the electron neutralization method however, the above problems have been serious enough to prevent any widespread adoption.

Experimental Results

Returning now to the ion method, let us review some of the experimental results to date. The results are encouraging, but problems remain.

Several ion beam generators of a couple of different types were designed and fabricated to test the ion neutralization theory. The two most convenient generation methods appear to be first: thermal ionization of alkali metals, particularly lithium; and second: electron impact ionization of background gas, usually water vapor.

One of the neutralizers constructed is shown in Fig. 9. The overall length is just over 5cm, and it is completely self-contained; thus it fits easily in almost any scanning microscope type instrument. The source contains electrodes for ion generation, extraction, focusing, deflection, and stray particle suppression. All required voltages are brought to the source via a single 3mm diameter teflon-insulated multi-conductor cable. A multi-conductor feedthrough must be added somewhere through the specimen chamber wall; an existing hole is normally used. By the change of a source cartridge, which plugs in at the rear of the neutralizer, it is possible to generate either alkali metal or background gas ions all in the same unit; the ions are extracted, focused and deflected by an ion optical column which takes up about 60% of the neutralizer length. This design should not be regarded as an optimum; it is possible to make smaller sources and obtain higher currents.

Figures 10 through 15 are pairs of photographs showing results obtained with the ion charge neutralization method. Most of these pictures were taken with 8 eV Li[7] ions. For each photo, brightness and contrast were adjusted to give the best possible picture, but no other parameters were changed when the preadjusted ion source was

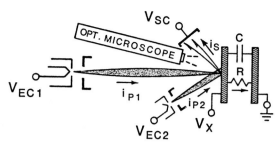

Fig. 6 Model for surface potential control using the electron neutralization method. An auxilliary electron gun generates a flood beam of roughly 1 kV electrons. Optical microscope is used for alignment.

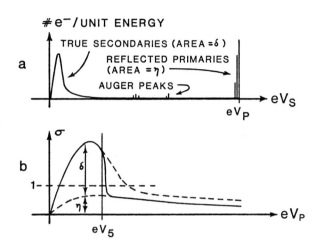

Fig. 7 (a) Energy distribution of secondary electrons, reflected primaries, and Auger electrons emitted from surface bombarded by electrons of energy eV_P. Area under this curve is σ by definition.
  (b) Total yield vs. primary energy curve showing transition from everything-escapes curve to only-high-energies-escape curve at energy slightly higher than eV_5.

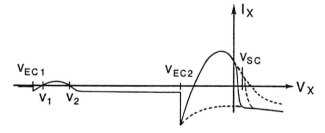

Fig. 8 Specimen voltage-current characteristic for electron neutralization. Curve is a time average and not to scale. Primary electron and neutralizing electron beams are not operated simultaneously. Specimen surface potential may vary according to surface electric field geometry.

turned on with a single switch. The pictures were taken on several different instruments made by various manufacturers. None of the specimens were given any pre-treatment. They were simply attached to a specimen stub and inserted into the respective microscopes uncoated. All the photographs shown are low energy secondary images; however they also all contain some scattered primaries, as is typical for the standard Everhart-Thornley detectors used.

Figure 10 is a Q-tip*at very low magnification. A Q-tip is an attractive example for ion neutralization because it is hard to do by coating. The black cloud effect is strongly in evidence in the ions-off photo, but completely disappears when the ions are turned on. Note also that while using ions it is possible to see deep down between fibres. The ions appear willing to execute contorted trajectories to seek out and neutralize electron charge buildup, even down deep holes. In this case the ions are mainly neutralizing electrons on the outer fibres: In the ions-off photo, it is the surface which is charging negative with respect to the interior which is closer to neutral, rather than the other way around. These photos, and most of the others shown, were taken at 25 kV.

Figure 11 is a somewhat higher magnification of the same Q-tip which shows three or four more effects which can be alleviated by ion charge neutralization. The worst of these is the swimming effect where the specimen charges so badly that the incoming primary trajectories are deflected. The distorted picture continuously swims around because the charge distribution slowly changes during the picture frame time. When the neutralizer is turned on, all picture motion

Fig. 9 Miniature ion charge neutralizer unit attached to mounting bracket. Neutralizer is just over 5 cm long. Ion source cartridge plugs into rear (upper right); drive cable also plugs into rear (not shown). Ions exit from hole in deflection/focusing unit (lower center).

*Cotton swab

immediately stops. Also visible in Fig. 11 are image shifts due to rapid discharge of some charged-up area not necessarily in the field of view (note that the scan lines are horizontal as printed in these pictures). Also visible is a contrast change on the left side compared to the rest of the picture. Finally the ions-off photo contains hundreds of little white specks apparently due to a different type of breakdown (these may not be easily visible in the printed version). All these effects disappear with the turn on of ions.

Figure 12 shows a crack in a piece of plastic. Note that with the ions on, it is possible to see down into the crack, with the ions off it is not. Again this is because without neutralization, the surface charges negative with respect to the bottom of the crack, which prevents the escape of low energy secondaries. Apparently the reflected primaries are also undetected in this case, because the crack is not orientated toward the electron collector. Also note in the Figure 12 ions-on photo, that some very bright areas remain, presumably a charge-up effect. This often happens, and the reason is not completely clear. Perhaps the mechanism is the anomalously-high secondary-yield effect described in the next section.

Figure 13 is a piece of high grade alumina at medium magnification. The ions-off picture shows light bands and thousands of tiny breakdown spots. The ions-on picture is much better, but notice that the white bands are still slightly visible. This memory-like effect has also been seen several times. Also note in this picture the sidewise image shift of about 10μm when the ions are turned on. This was caused by a small stray magnetic field generated by the particular ion source used, a problem which has since been largely corrected.

Figure 14 is a low magnification picture of fiberglass of a type commonly used for house insulation. Glass has the capability of supporting large voltages, and the effects of primary beam trajectory distortion and of image shifts due to breakdowns are again clearly visible.

Figure 15 shows a semiconductor circuit which was difficult to examine in its uncoated state. Ion neutralization eliminated the high secondary emission area. Note the atomic number contrast which would not have been as visible had the specimen been coated with gold.

Problems and Subsidiary Effects

The introduction of a miniature ion beam generator into the specimen chamber of a complex instrument such as a SEM might be expected to cause a panoply of new effects. It does. These can be divided into two catagories: unwanted instrumental interactions where the ion column disrupts the operation of the SEM or vice versa, and ion specimen interactions which include the desired charge neutralization effect along with several others.

Fortunately the instrumental interactions seem to be easily controllable. Ion sources can produce not only the desired ions, but also electric and magnetic fields, light, stray electrons, neutral particles, ions of the wrong kind, and even soft x-rays. We ran resolution tests to determine if presence or operation of a

Fig. 10 Low magnification comparison of Q-tip with ions off and
ions on. Black cloud effect visible on left of ions-off photo.

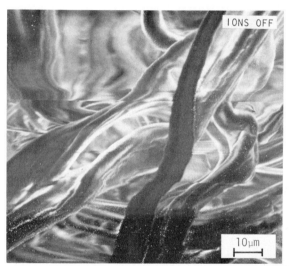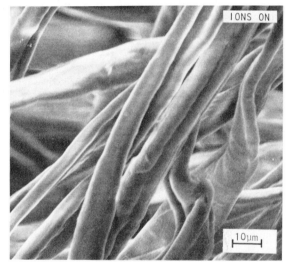

Fig. 11 Higher magnification of Q-tip. Swimming effect,
and various breakdown effects all stop with turn on of ions.

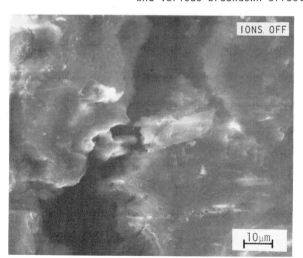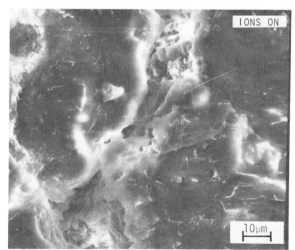

Fig. 12 Crack in a piece of plastic; note new capability to see
down into crack. Also note that not all charge-up effects disappear.

neutralization ion source in any way degraded the resolution of the instrument. Comparison photographs were taken of magnetic tape resolution standards at magnifications of 60,000 diameters. No change in resolution could be detected. A slight DC image shift was observed. It is caused by a residual milligauss-level magnetic field generated by the source when on. If the source power leads are not properly designed, this effect can be large, as seen in Fig. 13. Additionally, the power supplies driving the source, particularly the heater supply, must be well regulated to prevent the slight DC shift from turning into ripple. The sources are constructed from non-magnetic materials; and it has not been necessary to incorporate any magnetic shielding.

A more difficult problem in the design of background gas sources, where electron bombardment is used to generate ions in the source cartridge, is the prevention of stray bombardment electrons from reaching the secondary electron collector. The scale of the problem is best visualized by noting that the desired signal at the detector is eight orders of magnitude less than the bombardment current. Careful shielding design is required; a similar problem was solved before;[13] the solution lies in the suppression of small effects.

A small specimen chamber pressure rise can be noted when operating any of the ion sources. This results from the desorption of background gas (mostly water vapor), by the couple of watts of source input power. The effect is similar to that occurring in a gun chamber when running a filament, noticeable, but not normally objectionable.

An important factor to consider when using very low energy ions is the effect of all electrostatic fields near the sample. The ions are very easy to deflect, attract, or sweep away. Their ability to follow small fields is exactly the reason they neutralize electron charge build-up so well, they will follow the weak force lines of an incipient charge build-up. We have been able to obtain satisfactory operation with an ion neutralizer deliberately misaligned, such that it was aimed at the specimen more than 15 mm away from the primary SEM beam. It is possible for the ions to execute multiple 90° trajectory bends in reaching the electron charge. However they will follow any electrostatic field, not just the desired ones. Most existing instruments use a secondary collector draw-in field which is stronger than necessary. It is helpful to reduce this field to a minimum. It is also helpful to place the neutralization source on the same side of the specimen as the secondary collector; that way the collector field tends to push ions toward the sample. We have observed with the source and collector at right angles, that the ions may be driven away from the collector side such that only half the sample can be neutralized. A surface field control grid cage as proposed by Moll[14] for other purposes would presumably be a substantial help here; however such a device has not yet been tried.

Surface fields from other sources must also be controlled. The electrical impedance of the source driving the field is a critical factor.

High impedance fields (such as those due to electron charge-up) are easily handled by the ion generator; it is the low impedance fields which represent an uncontrollable infinite sink for ions. Experimentally we have sometimes observed that it is easier to get good neutralization on an "impossible" specimen than on some specimens which are almost all right. This leads to an interesting conclusion. One way to improve ion neutralization may be to increase the amount of insulating material near the specimen (contrary to all prior good SEM practice). The low energy ions can control the surface potentials of insulators, but they cannot control the surface potential of a metal connected to a low impedance source. It is only possible to pull ions a long way to a charged-up area if they have no easy way of getting lost en route.

Another way of losing ions before neutralization is complete, is to use ions of too high an energy, say 100eV, with respect to specimen chamber ground. If higher energy ions are used, the ion deflection plates must be adjusted such that the ions fall accurately in the right area. Otherwise the majority of the ions may be wasted. By contrast, when using low energies it is frequently unnecessary to use the deflection or focusing controls at all. The ions automatically go to the right place, wherever they are needed most. Note that higher energy ions are needed in the presence of electrostatic fields which cannot be reduced or shielded in some other way. Higher energy ions can also be focused into higher current densities if required. However, they stabilize the specimen potential farther from ground.

Turning now to ion interactions with the specimen, there are again several interesting effects:

The obvious effect, sputtering, normally does not occur. The ion energies are too low. There are two exceptions however. First, if sputtering is desired, the ion energy can be deliberately increased. Second, if a specimen has already charged to high negative potential before ion source turn on, sputtering can occur while the surface is being brought back to ground potential. On occasion sputtering may even occur continuously if the ion source does not have the capacity to bring the surface back to ground. We have yet to positively identify this effect, because the currents are also too small.

A potentially more detrimental effect involves the current density balance between electrons and ions arriving at the surface. For satisfactory neutralization, the ions and electrons must get together. Fortunately the current densities need not be equal because the electrons spread out significantly in the sample surface. Suppose however, that there are grossly too many electrons in one spot. An excess electron charge starts building up, and this charge generates a field which starts to pull in more ions. However this field is required in the steady state, and a potential difference is required to support the field. Hence not all of the surface is stabilized accurately at the ion source potential. This effect is worse at high magnifications (where fortunately it seems that charge-up problems are generally less severe), but it also can show up at

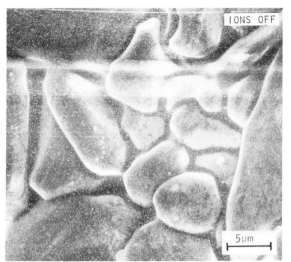 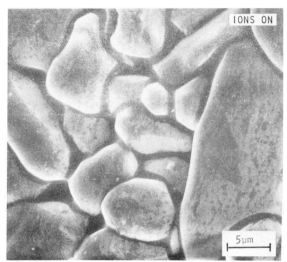

Fig. 13 High quality aluminum oxide at medium magnification. Note
memory effect, where horizontal bright band does not completely disappear.

Fig. 14 Fiberglass insulation showing image shifts and distortions without ions. Specimens such as this are very hard to do by any other technique.

Fig. 15 Semiconductor specimen, typical of another class of specimen
where coating is often undesirable. Note atomic number contrast.

low magnifications as a frame-rate dependent effect. We have several times observed that an ion density was sufficient for a certain frame rate, but insufficient for a slower rate. The general solution is probably higher ion densities, and more care not to waste ion current via some of the effects already discussed. This problem is the most significant encountered to date. Ion space charge plays a role. Modeling and calculations are now in progress.

Another effect, which we believe we have seen but cannot prove, is what might be called anomalously-high secondary yield. At typical SEM electron energies the primary electrons end up buried a micron or so below the surface. For some insulating samples a good fraction of this charge stays right there. The electric field from this buried charge draws in sufficient neutralizing ions to the surface to force the vacuum electric field to zero. But a powerful subsurface field remains. New primaries impinging on this thin-layer charged capacitor release secondaries which are strongly accelerated toward the surface, and thus have an enhanced probability of escape. The result appears to be very high secondary emission from certain localized areas. One way to test for this effect would be to examine the energy spectrum of the electrons emitted from the bright areas. The subsurface field also exists in coated specimens and presumably results in the same electron acceleration. However the mean free path in a metal coating is too short for electrons to be transmitted. Hence only the ordinary metal surface secondaries are seen.

Another class of effects, on which we have done no work at all, is the chemical interaction of low energy ions with the surface. Both of the ions used most, lithium and water vapor, can chemically react with many surfaces. Lithium deposits as a solid. The rates are low because the ion arrival rate is so low - only about 10^8 ions/sec. But these may still be significant at high magnification where the area involved is also very small. No such effects have been identified so far. If a problem should develop with some samples, it is possible to use other ions, such as argon (with a slightly more complex apparatus).

One final effect which also has not been observed as yet, is the Auger ion neutralization process, whereby a low energy ion landing on a surface can release a low energy electron. The energy for the electron emission comes from the potential energy of the ion, and the condition for this effect is that the ionization energy of the ion must be greater than twice the surface work function. A possible result in the present experiments would be the release of unwanted electrons from the specimen chamber walls by the unused neutralizing ions. If this should become a problem, the solution would be to avoid use of high ionization potential ions.

Conclusions

A new method for controlling electron charge buildup on insulating specimens in scanning electron microscopes and similar instruments has been proposed and demonstrated. The technique permits easier sample preparation for many samples, and

new types of experiments with some samples. Contrast from atomic number effects, crystallographic properties, and surface potential differences can now be observed on insulating specimens; SEM/SIMS, x-ray, and Auger analysis of insulators should all be made easier. Gold surface coating is frequently no longer required. The apparatus is easily adapted to fit virtually all scanning microscopes. Not all the changes are necessarily improvements however, and the new method will not be appropriate in every case. For example, in addition to improving surface conductivity, a gold coating usually improves the signal (because of the high secondary yield due to its high atomic number), and improves resistance to primary electron radiation damage (because of its thermal conductivity). These advantages of gold coating are lost if one switches to ion neutralization.

In addition, a great deal is still not understood about the panoply of new effects introduced by use of ion neutralization. Neutralization will not be appropriate for all specimens at all magnifications. There is an obvious need for much more experimentation, with more comparison photos at high magnification, comparisons between ion neutralization and gold coating using the same sample, and studies on many other types of samples. Effort is needed for even more miniaturized ion neutralizers, with higher currents and higher current densities. For some applications, such as higher current Auger, gas feed ion sources will probably be required. The theory developed to date is too simplistic; more effort is also needed here.

Perhaps now that the worst charging effects can be eliminated, it may be possible to perform controlled experiments to study those which remain. Some charging effects may represent a source of potentially useful sample information.

References

1.   D.B. Wittry and C.J. Wu, "The Charging of Semi-Insulating Specimens in Electron Microprobe Instruments", SEM/1975, IIT Research Institute, Chicago, IL 60616, 441-446.
2.   T.J. Shaffner, and J.W.S. Hearle, "Recent Advances in Understanding Specimen Charging", SEM/1976/I, IIT Research Institute, Chicago, IL 60616, 61-82.
3.   G.E. Pfefferkorn, H. Gruter and M. Pfautsch, "Observations on the Prevention of Specimen Charging", SEM/1972, IIT Research Institute, Chicago, IL 60616, 147-152.
4.   J.B. Pawley, "Charging Artifacts in the Scanning Electron Microscope", SEM/1972, ITT Research Institute, Chicago, IL 60616, 153-160.
5.   W. Fuchs, J.D. Brombach and W. Trösch II, "Charging Effect in Electron-Irradiated Ice", J. of Microscopy, 112 (1), 1978, 63-74.
6.   D.M. Taylor, "Electron-beam Charging of Polyethylene Terephthalate Films", J. Phys. D 9, 1976, 2269-2279.
7.   P. Echlin and P.J.W. Hyde, "The Rationale and Mode of Application of Thin Films to Non-Conducting Materials", SEM/1972, IIT Research Institute, Chicago, IL 60616, 137-146.

8.   L.M. Welter and A.N. McKee, "Observations On Uncoated, Nonconducting or Thermally Sensitive Specimens Using a Fast Scanning Field Emission Source SEM", SEM/1972, IIT Research Institute, Chicago, IL 60616, 161-168.
9.   M.J. Columbe, "System for Charge Neutralization", United States Patent No. 2,890,342, (1959).
*10. G.V. Spivak, E.I. Rau, A.E. Lukianov, "Des Images Non Alterees Des Isolants Dans Un Microscope Electronique a Balayage", Proc. 5th European Cong. on Electron Microscopy, 1972, 492-493.
11. P. Morin, M. Pitavel and E. Vicario, "Direct Observation of Insulators with a Scanning Electron Microscope", J. Phys. E. 9, 1976, 1017-1020.
12. M.D. Hare, "The Stable Potential Attained by an Essentially-Floating Electron Emitter Bombarded by Electrons", Report on 20th Annual Conf. on Physical Electronics, M.I.T., 1960, 37-43.
13. C.K. Crawford, "Electron Impact Ionization Cross Sections", M.I.T. Particle Optics Laboratory Technical Report ARML-TR-67-376, Wright Patterson Air Force Base, 1967, 59-67.
14. S.H. Moll, F. Healey, B. Sullivan and W. Johnson, "A High Efficiency, Nondirectional Backscattered Electron Detection Mode for SEM", SEM/1978/I, SEM Inc., AMF O'Hare, IL 60666, 303-307.

Discussion with Reviewers

S. Moll: It should be noted that atomic number effects are quite commonly visible in so-called secondary electron images even when samples are coated. The reemerging BSE flux from beneath the surface contributes to the signal information directly and indirectly (by producing SE at the specimen surface). In fact, at low to medium magnifications, even though the detector is in the SE mode, the image contrast on relatively flat specimens is essentially that due to BSE interactions and light surface coatings make no significant difference if the kV is high enough and the chemistry effects extend into the bulk of the specimen a micron or so.
Author: We agree.

S. Moll: Another significant advantage of gold coating is that of improving spacial resolution due to topographic contrast at high magnification. Since the escape depth of SE in gold is only about 30A (metal conduction), while it is some 300 A in nonconductors, the ability to image surface details smaller than 100 A is improved by light gold coating even when the obscuration effects of coatings are considered. Would the author comment on the reviewer's statement that the charge neutralization method, while yielding stable images, does not improve spatial resolution, as coatings do, since the secondary electron escape depth would still be that of the sample examined (conductor, semi-conductor or insulator).
Author: The statement is correct. To the extent that a specific coating improves contrast or resolution, and that this improvement is desired, coating should of course be considered. The existence of ion charge neutralization allows this decision to be made on its own merits.

*Published by Institute of Physics, Bristol, England

S. Moll: What effect would charge neutralization have in maintaining incident electron beam potential in the case of charging samples undergoing x-ray analysis? This could be quite important. It is well known that charge buildup on the specimen surface can reduce the effective incident electron energy (kV) and thus introduce error into quantitative ZAF calculations of chemistry.
Author: Charge neutralization should in principle completely solve this problem. Under good conditions, with adequate ion density and a low ion energy spread, surface potential control to better than one volt is possible. At the present time, however, the miniature ion sources developed by us do not provide sufficient total ion current to completely neutralize the typically large electron probe currents. The ion current must be many times the electron current for good surface potential control. The problem should be regarded as of a temporary engineering nature, but effort required for solution is not clear. For probe currents comparable to SEM currents, the method should work right now.

S. Moll: Would the author please estimate the rate of surface build-up for alkali ions for the purpose of x-ray analysis considerations? Any typical case would be useful.
Author: For particles which condense on the surface, the surface build-up rate is

$$\frac{ds}{dt} = \frac{m}{\rho a q} I_N,$$

where s is the layer thickness, I_N is the ion current which actually lands on the surface, m is the ion mass, ρ is the density of the surface layer produced, a is the bombarded area, and q is the ion charge. To preserve charge neutrality I_N must be equal to the net electron current. Using singly charged lithium ions (m = 1.17×10^{-23}gm, ρ = 0.534gm/cm^3), a magnification of 1000 (a = 10^{-4}cm2), and picking I_N = 10^{-8}A (a high value) the build-up rate calculates out to 1.37×10^{-8}cm/sec, or a bit more than an angstrom per second. Several comments: First, the rate scales up and down rapidly for changes in electron beam current and magnification. For SEM primary currents and low magnifications, the effect is very small; for x-ray work at high magnifications, it would be large. Second, these calculations ignore possible effects such as active metal gettering of background gas, or surface chemical reactions. Third, if build-up becomes a problem, the appropriate solution is to switch to gaseous ions which do not build up. For sensitive specimens such as semiconductors, gaseous non-reactive ions may be needed at all times.

J. Pawley: How do your observations of subsurface charging differ from those of Shaffner and Hearle (Fig. 12)? (your paper seems to indicate that they do.) (text ref.2)
Author: The observation of subsurface charging effects appears to be the same, it is the explanation which is different. Shaffner and Hearle used a simple linear theory, and describe their effect as sub-surface mirroring. Primary electrons are turned around and re-emitted from the surface by the subsurface electric field. This theory is a

step in the right direction (which is all Shaffner and Hearle intended). However, since the vacuum effects can now be eliminated by ion charge neutralization, it is worthwhile to examine subsurface effects in greater detail. These include what might be called anomalously high secondary yield, the subsurface mirroring effect, charge storage effects, conductivity effects, field dependent mobility effects, temperature effects, Malter effect, subsurface field emission, and perhaps others. The following comments are only a starting point, with regard to anomalously high secondary yield:

(1) Some insulators can support large subsurface fields, in excess of 10^6 V/cm.

(2) Most insulators allow long mean free paths for low energy electrons, in some cases more than $1\mu m$. This is the reason for the large escape depth sometimes observed with a SE detector. (The long mean paths have also been studied extensively by UV photoelectron spectroscopy using synchrotron radiation.)

(3) Bombardment by high energy primary electrons generates large numbers of excited electrons below the surface. In the absence of a subsurface electric field, most of these electrons never reach the surface.

(4) The existance of an intense subsurface field must accelerate the low energy electrons toward the surface. Presumably a large number escape.

(5) This effect can occur at much lower subsurface field strengths than can subsurface mirroring. Therefore it presumably occurs sooner during charge-up and is more frequently observed.

(6) The electrons emitted from the surface presumably have a broader energy distribution than normal low energy secondaries. The angular distribution will not be isotropic, and should be dependent on surface topography.

(7) Which electrons are detected using various detectors in an SEM will depend on the point of emission from the surface, the direction of emission, and the emission energy. Additionally, detection will depend on the physical location and applied potentials of the detector.

Hard work will be required to understand just this one effect, even without considering the others.

S. Moll: Can the electrostatic field of the collector of the SE detector be used to direct ions to the specimen if the geometry is proper? What sort of geometry would be recommended?
J. Pawley: I can see the reasons for using as low an ion beam voltage as possible. Could you outline those conditions when one might wish to use a higher voltage? (ie. 100eV)
Author: In general, the most uniform neutralization has been obtained using the lowest energy ions. The minimum energy is that necessary to overcome the effects of the SE detector field, which thus far has been nothing but a hinderance. It is difficult to see a way of using the detector field to help direct ions, since the ions will always gain unwanted energy falling through this field. An optimum geometry is probably a field free region close to the sample (into which the ions are injected), with a SE collection field outside this region.

Higher energies may have other advantages. For example it might be desired to stabilize a specimen surface at a potential far from ground. It is also easier to control and extract ions when they have higher kinetic energies. (For this reason higher voltages are used inside the source, the ions being decelerated as they leave the source.)

S. Moll: The image shift in Fig. 13 is proposed as a turn-on magnetic field effect. Is this also true for Fig. 11 where the shift is twice as large?
Author: No, the effect in Fig. 11 results from the electrostatic displacement of the primary beam, due to the large charge-up fields present. It is easy to distinguish charge neutralizer effects from charging effects by testing with a metallic specimen which does not charge.

J. Pawley: Could you please provide more details on the construction and operation of your source? Where do the ions come from (manufacturer of sources)? What type of ions? How is plasma maintained? Total power needed? Mechanical layout?
Author: All the sources used to date were designed and built at Kimball Physics. As far as we know, no one else has had a need to build ion sources which meet neutralizer type requirements: low energy, low energy spread, small physical size, low power dissipation, very low stray particle output, and small stray effects. It would appear that ions of virtually any species can accomplish the neutralization function. Thus secondary considerations such as ease of generation, required energy spread, and surface effects, determine what ion is to be used. To date we have used lithium ions, which are easy to generate in a surface ionization type ion source, and water vapor ions, which are easy to generate from background gas in an electron bombardment type source. The surface ionization sources have advantages in an inherently low energy spread (which leads to very uniform surface potential stabilization), in that they have negligible stray particle problems, and in that they can generate ions in the absence of background gas. By contrast, the electron impact sources have advantages in that background gas (if present) is inexhaustible, the ions do not condense on surfaces, the ions do not disrupt the operation of semiconductors, and there are minimal chemical reaction problems. The basic principles of source design for both these source types have been the subject of many books and papers. Neither type involves a plasma to generate the required ions. Power dissipation (for present sources) is about two watts for the electron bombardment type (where the bombardment filament must be heated to electron emission temperatures), and about four watts for the surface ionization type (where the ionization surface temperature must be raised to ionization temperatures). Both types require several potentials for extraction and focusing, however these electrodes require no power. The dimensions of the source can be seen in Fig. 9, where the ion exit aperture (bottom center of picture) is 1 cm in diameter.

It should be pointed out that the sources used thus far are not well optimized. Smaller, higher current, lower power, sources are under development. The biggest improvement available in the near term, will be the addition of a gas feed to the electron bombardment type. This will provide much higher ion currents, as well as allowing user choice of the ion species.

Editor: Since your paper was submitted for publication, a paper by Monirieff, Robinson, and Harris has appeared, J. Phys. D. Appl. Phys., 11 (1978) 2315-2325. It describes a new theory for another type of charge neutralization, using ions in environmental cells. Can you put your work in perspective with relation to this work?
Author: The environmental-cell charge-neutralization technique is an interesting method which should have wide application, particularly to biological specimens. We were not aware of this theory prior to seeing the new article.

A comparison between ion charge neutralization and environmental-cell charge neutralization is instructive. Both work on the same basic principle: positive ions combining with electrons on an insulating specimen surface to prevent excessive negative charge buildup. The ICN technique obtains the ions from a small auxiliary ion source as described in our paper. The environmental cell technique obtains ions from an electron assisted Townsend discharge between the specimen and the specimen chamber. Here the primary beam provides both the electrons and the energy to initiate the discharge, while the specimen chamber pressure is raised to about 10 Pa to provide the necessary gas.

The advantages of the environmental cell method are several. First, the method is simple; little complex equipment is required. Second, the method compliments well the major purpose of the environmental cell, that of providing a water vapor partial pressure for the observation of biological specimens. If an experiment already requires the cell to meet specimen requirements, there is no additional cost for charge neutralization. Third, the ion current scales with the primary beam current, and the ions are generated close to where they are needed. In contrast to the ICN method, there are always enough ions (assuming adequate cell pressure), and no effort is required to focus or transport ions. Fourth, because a backscattered electron detector is used, the low-voltage charge-up effects (which the technique does not treat) are not visible.

The comparative disadvantages of the cell method are also several. The requirement of high gas pressure means that only backscatter and x-ray detectors can be used to obtain sample information. Low energy secondary electron detectors, and techniques like Auger, SIMS, and channelling are all precluded. Thus significantly less sample information is available. The high cell pressure also precludes the use of bright electron sources, and may be hard on vacuum systems which must handle a couple of grams of water per hour. In order to obtain a BSE signal which compares favorably with that easily obtainable from a SE detector, it is necessary to use a BSE detector with a large solid angle. This consumes valuable space in the neighborhood of the specimen. Next, the use of a Townsend discharge generates ions with a very large energy spread; thus accurate surface potential control is impossible. Also the residual surface potentials (in the hundreds-of-volts range) depend on both the surface properties and the gas pressure. The errors are large enough to cause difficulty in x-ray work using the ZAF method. This contrasts poorly with ICN neutralization where stabilization can be as good as a fraction of a volt. The cell method also generates such high energy ions that the possibility of unwanted sputtering still exists. It is also more difficult to control the ion species.

Neither method can recover the resolution and contrast advantages which a heavy atomic number coating gives to a low atomic number specimen. Both methods quite adequately deal with high voltage charge-up effects. Neither method stops effects such as the anomalously high secondary yield. The cell method tends to see it less because of the use of only a backscatter detector; however, it is always possible to switch to backscatter mode while using the ICN method.

V.N.E. Robinson: Over the past few years, a lot of work has been published showing the elimination of charging artefacts using a backscattered electron detector. Your technique has not eliminated charging artefacts using secondary electron imaging. Why not try it with backscattered electron imaging?
Author: This question misses two major points. First, the technique has eliminated surface charge-up (and the artifacts resulting therefrom) for some very large classes of samples. Backscattered electron imaging, by contrast, simply fails to see most of the effects. Second, the intent of modern surface analysis has been to employ the largest possible number of techniques, to gain the maximum information from each sample. Over-reliance on any single technique, or single kind of imaging, is unwise.

V.N.E. Robinson: Your technique for neutralizing charge, to avoid charging artefacts using a secondary electron detector, does not appear to be any better than a good backscattered electron detector without any other effects, see for example J. Phys. E., 7 (1974) 650-652, or J. Phys. E. 8 (1975), 607-610, your reference No. 2, or Zeldes and Tassa, in this volume. Have you any evidence to show that your technique is better than an efficient backscattered electron detector, as far as studying surface topography is concerned?
Author: Yes, for example, the high voltage charge-up effects shown in Fig. 11 and Fig. 14 of this paper are beyond the capacity of a BSE detector. See also the answers for the preceding two questions.

Additional discussion with reviewers of the paper "On-Line Topographic Analysis in the SEM" by
D.M. Holburn et al, continued from page 52.

B. Artz: What is the theoretical justification
for using the maximum of the total surface gradient as a measure of the in-focus condition?
Authors: A full two-dimensional treatment leads
to an analytically intractable integral. However,
the case of a one-dimensional electron beam incident on an idealised specimen yields readily to a
mathematical approach. An expression may be
written down for the digitally recorded intensity
and used to derive an equation which represents a
one-dimensional intensity gradient of the form:

$$G \propto \frac{\sin \theta}{\theta}$$

where G is the intensity gradient and θ is linearly proportional to the ratio d_o/D, where d_o
is the probe diameter and D is the period of the
variations in the secondary emission coefficient.
G may be shown to increase monotonically as θ
decreases, provided $d_o < D/4$. Thus G will reach a
peak when the probe size is minimised, which is
an accepted criterion for focus, and is the condition attained when adjustments are made in the
orthodox way by maximising the steepness of the
oscilloscope trace of the video signal.

B. Artz: What is the purpose of the signal
conditioner, and how does it affect the focus
determination?
Authors: The signal conditioner performs two
functions. Firstly, it consists of a linear
buffer amplifier whose purpose is to scale the
video signal available from the SEM to a range
acceptable to the computer. Secondly, it re-
strictsthe bandwidth of the video signal in
order to reduce the effects of high frequency
noise that may be present, and it fulfills the
requirements imposed by Shannon's sampling
theorem.
     The effects of maladjustment on the condit-
ioner are similar to those which arise on the
conventionally observed image when the gain or
rise-time of the video amplifier are incorrectly
set.

B. Artz: What type of sample microstructure is
required? For example, will this method work on
a slowly varying smooth sample surface?
Authors: The only requirement is that there be
a substantial amount of detail in the field under
inspection, to allow the generation of contrast
which is dependent on the degree of focus. Provided these conditions are fulfilled, the method
will work on a slowly varying surface.

B. Artz: The example shown in figure 5 has topography variations two orders of magnitude above
your theoretical vertical limitations. Are there
any examples that show anywhere near a 1 μm resolution in the vertical direction?

Authors: With the system described, the resolution in the vertical sense is in practice
determined by the current increment, which in
the example given corresponded to 4.22 μm. This
may readily be reduced by a factor of three,
giving approximately 1.5 μm resolution; however,
with a sample possessing the wide range of feature heights of the specimen used in this run,
this would have required that each point be
scanned many more times, and the increased scanning time would have been unacceptable when considered in combination with the drift characteristics of our lens supply. We hope to improve
these aspects of the system by attention to the
programming algorithms and to the specifications
of the power supplies.

J.B. Pawley: A piece of phonograph record might
make a good test specimen for this method; have
you considered making a comparison of a Y - modulated picture and a set of profiles across such
a test specimen?
Authors: The idea of comparing line profiles
and Y - modulation profiles is certainly an interesting one, and we should like to take up this
suggestion in our future investigations.

SCANNING ELECTRON MICROSCOPY/1979/II
SEM Inc., AMF O'Hare, IL 60666, USA

ON-LINE TOPOGRAPHIC ANALYSIS IN THE SEM

D. M. Holburn and K. C. A. Smith

Cambridge University Engineering Dept.
Trumpington St.
Cambridge CB2 1PZ   U.K.

Abstract

A novel method for on-line topographic analysis of rough surfaces in the SEM has been investigated. It utilises a digital minicomputer configured to act as a programmable scan generator and automatic focusing unit. The computer is coupled to the microscope through digital-to-analogue converters which enable it to generate ramp waveforms allowing the beam to be scanned over a small sub-region of the field under program control. A further digital-to-analogue converter regulates the current supply to the objective lens of the microscope. The video signal is sampled by means of an analogue-to-digital converter and the resultant binary code stored in the computer's memory as an array of numbers describing relative image intensity. Computations based on the intensity gradient of the image allow the objective lens current to be found for the in-focus condition, which may be related to the working distance through a previous calibration experiment. The sensitivity of the method for detecting small height changes is theoretically of the order of 1 μm.

In practice the operator specifies features of interest by means of a mobile spot cursor injected into the SEM display screen, or he may scan the specimen at sub-regions corresponding to pre-determined points on a regular grid defined by him. The operation then proceeds under program control.

KEY WORDS: Computer, Microtopography, Topographic Analysis, Height Measurement, Automatic Focusing

Introduction

Ever since its inception the SEM has proved to be an invaluable tool for the examination of rough surfaces. However, the quantitative interpretation of the image produced by it is not to be undertaken lightly; seldom is it possible to make direct measurements of the heights and depths of protuberances and depressions from a single two-dimensional image, and for this reason, even inferences of distances apparently in the plane of the image may be of doubtful accuracy. Several techniques have gradually become established to circumvent these difficulties. Hoover[1] and Swift[2] have described methods that involve marking the surface of interest with a thin film of a heavy metal, or simply a line of organic contamination, then observing the local displacements of the marks after tilting through a known angle. Brandis[3] has suggested a technique for the measurement of convex features in which measurements are taken of the shadows cast when metal atoms are incident on the surface at an oblique angle from a remote evaporating source. Boyde[4] has shown how the depths of regular concavities may be deduced from careful measurements of width followed by a controlled tilt operation. Lebiedzik and White[5] have used an array of electron detectors to determine the orientation, and hence the heights of zones of a gold-coated specimen. Perhaps the most comprehensive approach to the problem is that of stereoscopic measurement from a pair of images. This has been described in detail by Boyde[6].

None of the techniques described may be universally applied, with the possible exception of stereometry; this is a serious shortcoming. Moreover, most require some prior knowledge of the nature of the specimen surface, which may not be available in every case. In addition, the majority require the attentions of a skilled operator and do not lend themselves to automation.

A method of a rather different nature, and which has already found some limited applications in optical microscopy involves the measurement of heights by adjustment of the optical system, using a calibrated control to achieve a precisely focused image of the feature of interest; with the optical microscope, very fine discrimination is possible owing to the small depth of focus afforded by the large numerical apertures which apply. With the SEM the same procedure is possible in principle, but owing to the larger depth of focus which accrues from the use under normal circumstances of relatively fine beam-defining apertures, the sensitivity of the image to small changes in relief is much less pronounced. Yew[7] has suggested that a depth of focus of 1 μm should normally be obtainable using a 200 μm aperture, at a numerical magnification of 1000. For many applications, a spatial resolution of this order would be sufficient to allow useful information to be obtained. In this connection, the ease with which the SEM may be controlled electronically, and the convenience with which the image may be examined by processing of the video signal are particularly advantageous. Hersener and Ricker[8] have described a system in which these principles are utilised to measure heights in the SEM; here the amplitude of the high frequency content of the signal is taken to indicate the relative sharpness of the image, the value reaching a peak as the system passes through focus. By careful calibration of the objective lens supply in terms of working distance, and using a line scan to focus the electron probe accurately by making manual adjustments, they were able to gauge to a precision of $\pm$ 2 μm the altitude of the line feature being scanned. However, their apparatus was not really capable of making measurements of small, isolated features on a large scale. This degree of elaboration demands the facilities for independent programmed control afforded by a computer. Digital computers have been used in conjunction with the SEM to perform a variety of functions, such as processing of X-ray data (Yakowitz[9]), distortion analysis (Maune[10]), on-line image processing (Unitt[11], Oron and Gilbert[12], Smith et. al.[13]), and automatic control of various functions. A digital focusing unit for the SEM has been proposed by Tee et. al.[14].

The system described in this paper provides automatic measurement in the SEM of the relative heights of features of a specimen, using a method of focusing the instrument controlled by an on-line minicomputer.

Instrumentation

The computing installation consists of a 32 kiloword 16-bit minicomputer (Computer Automation, LSI 2-20G), equipped with a keyboard, storage display, floppy disk bulk data storage system, and a high speed printer. Modifications to the scanning electron microscope (Cambridge Instruments, Stereoscan IIA) have been kept to the minimum, and involve the provision of an interface which furnishes:-
(i) An externally controlled objective lens power supply, governed by a 15-bit digital to analogue converter (DAC) coupled to the computer.
(ii) Buffers to allow a pair of computer-driven 12-bit DACs to determine the resolution and position of the scanning raster by feeding suitable waveforms to the microscope scan amplifiers, and to permit the computer to blank or unblank the image display units.

(iii) A signal conditioner, consisting of a buffer amplifier and variable low pass filter to adjust the amplitude and bandwidth of the video signal to allow it to be sampled by a computer-controlled 8-bit analogue to digital converter (ADC). (See figure 1.)

The video signal obtained from the SEM is digitised in the interface, and used to calculate a parameter related to the sharpness of a selected part of the image. Subroutines written in the computer's own assembly language to maximise speed of execution synthesise ramp waveforms from sequences of discrete integer values and a suitable bias, allowing the DAC and scan amplifier combination to generate a sub-raster of the required size, positioned anywhere within the image field of up to 4096x4096 points. The sub-raster is arranged so that it coincides with the feature whose height is to be measured. At each picture point within the sub-raster, the ADC produces an output proportional to the local image intensity, and this value is linearly combined with others in order to calculate an approximation to the spatial rate of change of image intensity (that is, the intensity gradient) at that point. The disposition of the points used for computation is shown in figure 2. If the picture element currently being scanned is P, then the quantities evaluated are

$$\Delta I_h = I_P - I_Q$$

$$\Delta I_v = I_P - I_R$$

representing respectively the horizontal and vertical changes in image intensity; the computed intensity gradient is then defined by

$$G = |\Delta I_h| + |\Delta I_v|$$

Hence it is necessary to maintain only two complete lines simultaneously in computer memory during the scanning operation, and not the entire frame as might first be supposed. A running total of the intensity gradients is accumulated as each point is scanned, and at the end of the frame, the gradient sum provides the necessary measure of the sharpness of the image. In order to focus the microscope the value of the gradient sum is determined for a sequence of different objective lens currents, determined by the 15-bit DAC. To guard against inconsistencies which might arise from hysteresis of the ferromagnetic materials used in the lens, the current is made to describe exactly the same series of values for each sequence, by setting it first to zero and ramping it smoothly up to the starting value for the sequence; during the focusing run itself the current is varied monotonically throughout. The previous magnetic history of the lens is erased by arranging for the hysteresis loop so defined to be traversed a few times before any measurements are taken. The size of the current increment depends to some extent on the resolution required; the smallest increment obtainable with the hardware described is 30 μA,

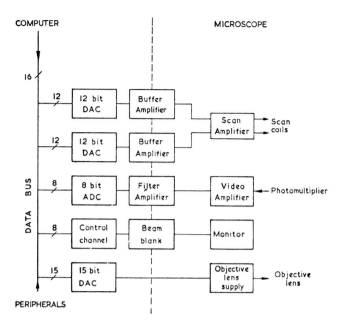

Figure 1. Essential system components.

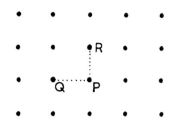

Figure 2. Image points used for computation of intensity gradient.

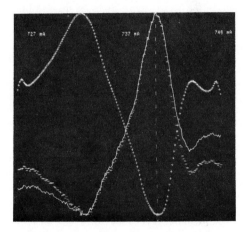

Figure 3. Graphical output for continuous monitoring of performance. Functions of image intensity gradient plotted against lens current: Continuous line represents raw gradient total; '-' symbols show action of background stripping; '+' symbols show least squares errors for fitted Gaussian. Vertical line shows computer-determined focus.

corresponding to a change of approximately 1 μm. in a working distance of 5mm. The exact size of the sub-raster employed may be specified by the operator, but is typically 4 x 4 points, this being found to give a satisfactory compromise between immunity to noise, and smallness, to permit discrimination of small features (approximately 1 mm square at a magnification of 100 times, positioned with a resolution of 0.2 mm).

The program finds the position of a peak in the array of gradient totals by a computational algorithm which is designed to take into account the presence in the data of electron beam noise, and possibly a superimposed gradual variation due to the change of probe current with lens excitation. Any sloping background present is first removed from the data array, leaving a focusing peak which is approximately Gaussian in shape. An array containing a Gaussian function of predetermined width is then subtracted from the data, the relative positions of the two arrays being progressively altered, and the total error between the Gaussian and the data computed for each case; this error reaches a clearly defined minimum at the objective setting which corresponds to optimum focus. Figure 3 shows the graphical output obtained at the end of a focusing run.

The number of frames scanned, and hence the number of current increments depends on the required resolution and the range of heights present in the field of view, since this governs the spread of objective currents required to ensure location of a focus. It takes 0.3 seconds to scan each frame and prepare for the next.

In practice it is necessary to perform a preliminary scan in order to assess the approximate range of objective currents required. This is arranged to occur at the start of the program when sequences of 64 frames are scanned using larger rasters with fairly coarse current increments to determine approximate foci. The information derived from this operation is used to establish a suitable value for the current increment in the later stages of the program.

The current value corresponding to the best

focus setting for the point under consideration
is inserted into a third-order polynomial expression which computes the difference in heights
between this point and a datum, which is normally
taken as the feature at the centre of the image.
A preliminary calibration experiment was performed to derive the polynomial coefficients, in
which the operation of the stage Z-motion was
investigated using a travelling microscope,
allowing a resolution of 10 μm. A specimen with
plenty of fine detail was then installed in the
SEM, and a conspicuous feature was accurately
focused at different heights. In this way the
relationship between lens excitation and working
distance was ascertained.

Operation

The operator may choose one of two options
for the selection of the points at which he
wishes to measure the height. Firstly one may
define a hypothetical rectangular grid by specifying the pitch, and have the program perform
scan sequences at each of the mesh points. Alternatively, one may prefer to choose features
of particular interest, and distributed at random
over the field of view. This is facilitated by
an additional refinement to the program, which
produces a fairly rapid but coarse raster (about
four frames per second) covering the entire image
field and displaying the detected signal on the
visual monitor. A 'bright-up' cursor is injected into the frame by means of manipulation of
the blanking signal; for the majority of the frame the picture information is displayed with the
beam only partially unblanked, giving an image of
low brightness, while over the small range of co-
ordinates corresponding to the cursor, the beam
is fully enabled, and produces a bright spot.
This may be guided about the screen by simple
keyboard commands which allow the operator to
position it at or near the feature of interest
(see figure 4); further keystrokes allow the
resolution to be increased by stages, at the
expense of an increased frame time, in order to
allow the cursor to be brought progressively
closer to and ultimately coincident with the
feature. Finally the X and Y coordinates of the
centre of the cursor may be retained in special
data arrays for reference during the scan
sequences; up to 200 points may be randomly selected in this way. When coordinate definition
has been satisfactorily completed, the program
requests information about various salient microscope parameters; these are magnification range
(the exact magnification is calculated from a
polynomial function of the lens current at focus),
operator-defined mean focus current and current
spread. Finally the operator may define the size
and resolution of the subraster. For all parameters except the magnification range, the processor is perfectly capable of determining suitable values without the operator's assistance;
indeed, in a system designed for completely unattended operation it would be required to do so,
and it is only for the convenience of instrumental
development that provision has been made for modification of the various parameters. Operation

100 μm.

Figure 4. Micrograph showing cursor
stationed at the point (3534,3214).
Specimen: fractured steel rod.

then continues completely automatically and without intervention. The necessary heights are
determined, and may be displayed by means of an
XY recorder in the form of a profile along a
single plane through the specimen, or alternatively, as a chart similar to those produced by
hydrographic surveys with 'soundings' plotted in
appropriate units at the correct coordinates
referred to cartesian axes. A complete log is
maintained by the high speed printer of all
points inspected (see figure 5), and this may be
used in conjunction with the chart or profile
information produced by the XY recorder.

Resolution

Two main factors set limits to the best
height resolution obtainable. They are the depth
of focus of the SEM, and the precision to which
the objective lens may be set.
The depth of focus δ is given by

$$\delta = \frac{d}{\alpha} \qquad (1)$$

where d is the probe diameter and α is the beam
semi-angular aperture. The probe diameter under
normal working conditions in the SEM is limited
by spherical aberration so that the probe diameter is approximately

$$d = \tfrac{1}{2} C_s \alpha^3 \qquad (2)$$

where C_s is the spherical aberration coefficient.
Substituting (2) in (1) gives

$$\delta = \tfrac{1}{2} C_s \alpha^2 \qquad (3)$$

This expression shows that if the probe is
spherical aberration limited, the depth of focus
decreases when α is decreased - contrary to
normal experience in light optics.
Taking typical values: $C_s = 20$ mm, $\alpha = 10^{-2}$
radian (corresponding to an aperture of 200 μm
and a focal length of 10 mm), equation (3) gives

```
ALTIMETRIC SCAN PROGRAM                    D. M. HOLBURN    01 03 1979

          8 DISCRETE SCAN LOCATION COORDINATES

          K                   X                    Y

          1                  592                  3216
          2                 1104                  3216
          3                 1680                  3216
          4                 1936                  3216
          5                 2320                  3216
          6                 2640                  3216
          7                 3088                  3216
          8                 3536                  3216


MEAN FOCUS CURRENT       =   735.95 mA
CENTRAL FOCUS CURRENT    =   739.98 mA
MIN. FOCUS CURRENT       =   731.68 mA
MAX. FOCUS CURRENT       =   742.42 mA
CURRENT SPREAD           =    10.74 mA


NUMBER OF DISCRETE SCANNED POINTS =      8
MEAN MAGNIFICATION                =   133.2   NUMBER OF FRAMES PER POINT =    200


RASTER DEFINITION              =     4  BY    4    RASTER INCREMENT  =      1 PP
MEAN ALTIMETRIC CURRENT        =   737.00 mA
MEAN WORKING DISTANCE          =     5.49 mm
ALTIMETRIC CURRENT SPREAD      = +  300   DAC units  = +  9.16 mA   = +  421.77 micron
ALTIMETRIC CURRENT INCREMENT   =     3   DAC units  =    .09 mA/frame  =    4.22 micron/frame

  LOCATION   X COORD.   Y COORD.      CURRENT    WKG DIST    INCREM WD        PROFILE

       1        590       3214        740.47      5.33        -158.
       2       1102       3214        740.20      5.34        -146.
       3       1678       3214        741.57      5.28        -207.
       4       1934       3214        740.84      5.31        -174.
       5       2318       3214        739.46      5.38        -113.
       6       2638       3214        737.08      5.48          -4.
       7       3086       3214        733.97      5.63         141.
       8       3534       3214        731.04      5.77         280.
```

Figure 5. Printer output from system. 'WKG DIST' represents distance from objective to feature in mm; 'INCREM WD' expresses the height of the feature in micrometers relative to the datum; 'PROFILE' is a log of the image intensity gradient plotted against objective excitation.

δ = 1 μm. Thus we might expect that the method would provide a resolution in terms of height of the order of 1 μm.

The smallest incremental change, ΔI, in the objective lens current, I, required to achieve this order of resolution may be found by considering the expression for the focal length:

$$f = \frac{k}{I^2} \qquad (4)$$

where k is a constant
Then differentiating

$$\frac{\Delta f}{f} = \frac{2\Delta I}{I} \qquad (5)$$

For a change Δf = 1 μm and for f = 10 mm, we obtain from (5) $\Delta I/I = 0.5 \times 10^{-4}$. In our instrument the objective lens current is about 0.7A thus the lens current must be set to a precision of 0.03 mA or better. This requires a DAC of 13 or more bits.

Accuracy

A preliminary experiment to assess the accuracy of the method has been carried out. A standard specimen grid was imaged in the microscope at various nominal tilt angles in the range 13 to 55 degrees. A previous experiment had established the true values of tilt corresponding to these settings to within 0.05 degrees. For each orientation, a series of topographic determinations was performed along a line perpendicular to the tilt axis. Feature heights were then plotted as a function of position for all points measured, and profiles were then constructed graphically to allow the tilt angles to be determined independently. With height differences between separate features of about 700 μm, the uncertainty in the spot heights was found to be no worse than 10 μm, and the computed angles of inclination agreed with those indicated by the tilt control to within 3%.

Acknowledgements

This work was supported by grants from the Science Research Council. The loan of a scanning electron microscope by the Cambridge Instrument Company is gratefully acknowledged.

References

1. R.A. Hoover, "Measuring Surface Variations with the SEM using Deposited Contamination Lines", J. Phys.E: Sci. Instrum., 4, 1971, 747-749.
2. J.A. Swift, "Measuring Surface Variations with the SEM using Lines of Evaporated Metal", J. Phys.E.: Sci. Instrum, 9, 1976, 803-804.
3. E.K. Brandis, "Comparison of Height and Depth Measurements using a Shadow Casting Technique", SEM/1972/I, IIT Research Institute, Chicago, IL, 60616, 241-248.
4. A. Boyde, "Photogrammetry and the Scanning Electron Microscope", The Photogrammetric Record, 8, 1975, 442-443.
5. J. Lebiedzik and E.W. White, "Multiple Detector Method for Quantitative Determination of Microtopography", SEM/1975/I, IIT Research Institute, Chicago, IL, 181-188.
6. A. Boyde "Quantitative Photogrammetric Analysis of SEM Images", J. Microscopy, 98, (3), 1973, 452-471.
7. N.C. Yew, "Dynamic Focusing Technique for Tilted Samples", SEM/1971/I, IIT Research Institute, Chicago, IL, 60616, 33-40.
8. J. Hersener and Th Ricker, "Eine automatische Fokussierungseinrichtung fur Rasterelektronenmikroskope", Beiträge zur elektronenmikroskopischen Direktabbildung von Oberflächen, 5, 1972, 337-387.
9. H. Yakowitz, "X-ray Analysis with Small Computers", in Practical Scanning Electron Microscopy, J. Goldstein and H. Yakowitz (ed.),Plenum, New York, U.S.A., 1975, 373-400.
10. D.F. Maune, "Photogrammetric Self-calibration of SEMs", Photogrammetric Engineering and Remote Sensing, 42, (9), 1976, 1161-1172.
11. B.M. Unitt, "On-line Digital Image Processing for the SEM", Ph.D. Dissertation, University of Cambridge, 1976 (copies available from: The Librarian, Cambridge University, Cambridge, England.)
12. M. Oron and D. Gilbert, "Combined SEM-Minicomputer System for Digital Image Processing", SEM/1976/I, IIT Research Institute, Chicago, IL, 60616, 121-128.
13. K.C.A. Smith, B.M. Unitt, D.M. Holburn and W.J. Tee, "Gradient Image Processing using an On-line Digital Computer", SEM/1977/I, IIT Research Institute, Chicago, IL, 60616, 49-56.
14. W.J. Tee, K.C.A. Smith and D.M. Holburn, "An Automatic Focusing and Stigmating System for the SEM", J. Phys.E: Sci. Instrum., 12, 1979, 35-38.

DISCUSSION WITH REVIEWERS

J.B. Pawley: Have you attempted to demonstrate the repeatability of the system described?
Authors: Yes, but we have not so far obtained reliable measurements for two reasons: firstly, repeated scanning of a sub-region of the field causes contamination of the surface: secondly, the programmable lens supply that we are using has insufficient stability in terms of drift.

D.E. Newbury: In your paper in SEM/1977 (Smith et al., p 49) the definition of the image gradient was $G = a(p+q)$, where p and q are the partial derivatives and a is a constant. This seems to be a slightly different definition than

is used in this paper. Please explain the current equation and how it relates to your earlier work.
Authors: In the paper referred to it was explained that the true intensity gradient, as defined rigorously by $G = \sqrt{(p^2 + q^2)}$, requires an inconveniently long computational time; the approximation $G = a(p+q)$, may be computed much more rapidly. The approximation to the gradient used in this paper corresponds to $G = p + q$, that is, we have set the constant $a = 1$.

D.E. Newbury: Again referring to the 1977 paper, you gave an error analysis for the effect of not taking a rigorous numerical solution for the gradient. In that analysis, you concluded that a maximum error of 24% could be observed between the approximate value and the rigorous value of the gradient. If the current definition of the gradient is subject to the same error due to the approximation, how does this error affect the accuracy with which a depth measurement can be made?
Authors: We have here necessarily taken the simplest possible algorithm to provide a measure of the sharpness of focus because of the limited number of the picture elements available and because of the requirements of computational speed. The fact that this measure of sharpness does not correspond to the true intensity gradient does not affect accuracy as far as vertical resolution is concerned.

J.B. Pawley: What is the time required for carrying out the procedures described?
D.E. Newbury: How much time is required to obtain a profile like figure 5?
Authors: The time required to measure 8 points, as in figure 5, is approximately 4 minutes. The program is not optimised for speed since the method is still under development.

W.C. Lane: As the objective lens current changes, the image also rotates about the electron optical axis of the SEM. How does this image shift phenomenon affect your system? Can the computer be used to correct for this effect?
Authors: We have measured the way in which the orientation of the scanning raster varies according to the objective excitation: at a working distance of 5 mm, a change of 1 mA in the lens current rotates the image field through 5 minutes of arc. Since the image field consists of a matrix of 4096 separate addressable points, the shift in a feature located at the periphery of the field produced by such a current increment is approximately:

$$2048 \times \frac{5}{60} \times \frac{\pi}{180} = 3 \text{ picture points}$$

The shift is of less significance for features closer to the optic axis.

We think that the computer could be used to correct for this effect since the relationship between rotation and objective excitation is known and is available as data to the computer.

For additional discussion see page 46.

SCANNING ELECTRON MICROSCOPY/1979/II
SEM Inc., AMF O'Hare, IL 60666, USA

A PRECISION SEM IMAGE ANALYSIS SYSTEM WITH FULL-FEATURE EDXA CHARACTERIZATION

W. R. Stott and E. J. Chatfield

Electron Optical Laboratory
Ontario Research Foundation
Sheridan Park, Mississauga,
Ontario, Canada L5K 1B3

Abstract

An image-processing microanalyzer system has been developed which is able to perform image processing and X-ray compositional analysis of features in an SEM image. The major components of the system are an SEM, an energy dispersive X-ray analyzer and a minicomputer which controls both the SEM raster and the X-ray spectrometer. The system is able to classify all features of an image according to their geometrical parameters and to categorize them by their chemical compositions. A new image analysis computer program has been developed which identifies data from a raster image and assigns information in each scan line to the discrete features present. A binary representation of the image is obtained which identifies the edges of features intersected by each scan line. Images containing features of great complexity are analyzed correctly as to convoluted feature shapes, holes within features and interdigitation with adjacent features. The location co-ordinates, area, perimeter and the appropriate edge information are computed for each of the features encountered. In the elemental analysis section of the system the SEM electron beam is scanned uniformly over each feature, using the edge information to obtain an X-ray spectrum characteristic of the whole area of each feature. Thus in a field of view of the sample object, all features observed are characterized by their size, geometry, abundance and chemical composition. The new system has been used to investigate mineral particles in coke concretions, particulate in urban air samples, and a number of industrial powders. The system has considerable flexibility, so that individual requirements of an analysis, such as the facility to search for only one particle geometry or composition, can be easily accommodated.

KEY WORDS: Automatic Image Analysis, Image Analysis, Backscattered Electrons, X-Ray Microanalysis, Particle Classification, Fibre Analysis, Digital Scanning, Computer

Introduction

The value of an image analyzing system for the SEM is well recognized and this paper presents a new approach to the analysis of the features present in an image. Once the position, area and perimeter information on each of the discrete features in the image has been determined, an X-ray analysis process is begun which obtains an elemental analysis of each feature. The SEM electron beam is scanned over the full surface of each feature. The scanning of the beam over the individual features is accomplished using information derived from the data collected during the original complete scan of the field of view. Using the energies and the intensities of the X-ray lines excited, a comparison is made with a stored file of energies and relative intensities of elemental X-ray lines, in order to determine the elements present in each feature.

The major components of the system consist of an SEM, a Si(Li) X-ray detector with associated multi-channel analyzer and a minicomputer. Through a hardware interface, the minicomputer controls the SEM and the X-ray energy spectrometer system, in addition to performing the computation for the image and X-ray analysis. A block diagram is shown in Figure 1.

The system would be best applied to situations in which repetitive determinations are required on a large number of similar samples. Work of this nature has traditionally been accomplished manually by a highly trained operator who counts and classifies particles. Hard-wired image analysis devices are available,[1] but lack the versatility of software-based systems. The use of computers to accomplish the controlling, acquisition and processing of images is now much more prevalent while incorporating modal, intercept length and X-ray analysis.[2,3,4,5] The present system has the advantage of great flexibility which allows for a broad range of analytical requirements and in addition incorporates X-ray elemental classification.

Hardware

The computer currently in use for controlling the digital scanning and also the image and X-ray analysis is a Data General Nova 1200. This

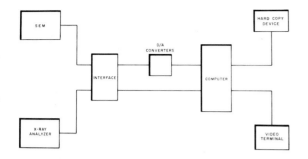

Fig. 1. Block diagram of hardware of image processing microanalyzer.

---

has 32K of 16-bit words of core storage with an associated hard disk on which the programs and the data arrays are stored. Two 10-bit digital-to-analog converters were assembled to direct the electron beam of the SEM in both x and y directtions with these being connected between the computer and the interface unit. The interface unit (Figure 2) contains circuits which control the beam deflection on the Cambridge S4 SEM, the blanking of the screen on the SEM display and operation of the Kevex 5100 X-ray analyzer. The unit also processes the SEM video signal such that when the signal rises above or falls below a certain threshold voltage a signal is sent to the computer to indicate that a feature edge has been encountered. The setting of this threshold is accomplished through a variable resistance adjustable by the operator of the SEM. The video terminal of the computer and the interface unit are both located close to the SEM, such that the operator is able to maintain direct communication with the computer.

System Operation

This image processing microanalyzer operates in a series of more refined analytical steps, each step being a reduction of data from the preceding step. The initial acquisition of image data is performed by the computer which controls the scanning of the focussed electron beam through two 10-channel digital-to-analog converters. For the present device the D/A converters divide each of 1024 lines into 1024 elements such that the scanning image consists of a 1024 x 1024 matrix of points. The beam moves from left to right in a line and then moves down one unit to scan across a new line.

To store all of the matrix points would be both impractical and unnecessary.[6,7] For each scan line, only those points which identify feature edges are stored. When a scan line crosses an image feature there will in general be a point where the beam of scanned electron enters the feature and a point where it leaves the feature. By setting an electronic threshold on the SEM video signal, these points of entry and exit

are determined and the image is represented in a binary form. The locations of all the feature edges are stored in a file on a disk of the computer, typically of size 4096 words or less for a reasonably filled field of view. All subsequent image analysis and beam control is performed using data derived from this file. Once the complete image has been acquired, the binary image is displayed on the screen of the computer video terminal.

The actual feature analysis is accomplished in a line-by-line manner in which the line segment between each pair of entry and exit points is checked to determine if it is contiguous with (overlaps) a similar line segment on the preceding line. If contiguity is established then it is known that both line segments belong to the same feature. There are six possible situations that can occur in this line segment contiguity checking procedure. These are summarized in Figure 3. Position (a) represents the first scan line to intercept the feature. This is recognized as a new feature, since there are no similar contiguous line segments on the line immediately above it. Since at this point in the analysis there is no reason to believe otherwise, this feature is treated separately, even though it may actually be one lobe of a multiply-lobed feature. Position (b) represents a condition where there is no change in the status from that in the previous line. The scan line at position (c) is at a low point in the feature where it is recognized that Features I and II are one and the same. This is recognized since the line segment is contiguous with line segments of both Features I and II. The scan line at position (d) has no line segment contiguous with any part of the feature being considered and so the feature is considered complete. At position (e) two segments are contiguous with one line segment of the previous line. This is the case where a feature is splitting into lobes. The situation represented in position (f) is similar to that of position (d). In this case, however, termination of only one of the lobes of the feature has occurred on the scan line.

The six situations shown in Figure 3 are the only ones which can occur in line-by-line analysis of a feature. Analysis of the most complicated individual feature can be reduced to a combination of these six events no matter how convoluted it is in shape, and even if interior holes are present. Features that are interdigitated are properly separated by this line-by-line analysis since no contiguity exist to connect them.

The appropriate area and perimeter is determined for each of the features present on a scan line after contiguity of the line segments has been established. The number of picture elements in a line segment gives the area to be summed while the perimeters are calculated by computing the vector distance between the adjacent endpoints of contiguous line segments. Line by line, a running calculation of area and perimeter for each feature is thus made which takes into proper account the joining, separating and terminating of lobes including the internal holes

Fig. 2. Photograph of the interface unit for the system.

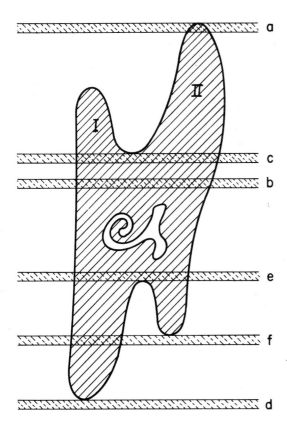

Fig. 3. Illustration of the six possible situations encountered by a scan line intercepting a feature.

encountered. Thus at the termination of a feature the area and perimeter are known and these, along with the lowest left-hand point of the feature are stored in a file on the computer.

Some features extend beyond the field of view and thus they have edges which are seen as being at the perimeter of the field of view. Consequently all of the information on these features is not available and so they are flagged to indicate that they extend beyond the edge. In the output file the complete features are numbered and those that intercept edges are assigned negative numbers (e.g. see Table 1).

#	X	Y	Area	Perim.
1	943	101	88.0	37.5
2	415	130	807.0	115.5
-3	1	160	14.0	18.0
4	584	161	179.0	52.6
-5	1020	162	41.0	28.3
6	633	165	171.0	51.8
7	771	165	53.0	30.5
8	541	176	2339.0	290.4
9	92	195	244.0	65.1
10	94	201	23.0	23.4
11	370	202	63.0	33.8
12	367	212	46.0	27.5
13	828	226	31.0	23.9
14	804	244	696.0	108.5
15	365	296	182.0	57.4
16	116	346	40.0	25.5
17	806	366	3.0	8.0
18	946	386	253.0	78.4
19	579	392	894.0	122.1
20	815	507	235.0	62.0
21	916	518	818.0	122.5
22	414	524	1009.0	130.7
23	623	555	469.0	87.2
24	634	569	21.0	22.1
25	522	601	2955.0	241.9
26	414	612	11.0	14.2
27	597	621	572.0	92.3
28	607	655	841.0	130.8
29	224	660	233.0	60.6
30	56	683	1179.0	143.2
31	814	732	635.0	99.2
32	123	745	35.0	25.7
33	32	749	331.0	79.8
34	555	753	21.0	21.0
35	15	796	170.0	72.6
36	544	820	234.0	58.8
37	667	848	34.0	24.1
38	616	885	159.0	52.3
39	492	894	602.0	97.4
40	797	912	531.0	95.2
41	618	927	869.0	146.2
42	969	927	2.0	6.0
43	802	937	58.0	29.6
44	317	947	124.0	48.0
45	221	956	36.0	28.5
46	536	1006	100.0	39.9

PARTICLE ANALYSIS

X	Y	Z		INTENSITY
415	130			
		14	SI	11573.
		13	AL	2132.
		12	MG	2273.
		26	FE	496.
		20	CA	491.
541	176			
		26	FE	770.
		14	SI	517.
		24	CR	245.
		13	AL	380.
		16	S	127.
579	392			
		26	FE	3954.
		14	SI	1210.
		13	AL	444.
		12	MG	204.
916	518			
		26	FE	1246.
		24	CR	723.
		14	SI	701.
		13	AL	419.
		16	S	185.
414	524			
		26	FE	4898.
		14	SI	1102.
		24	CR	340.
		13	AL	507.
522	601			
		26	FE	6503.
		24	CR	3272.
		14	SI	1492.
		13	AL	585.
		23	V	208.
607	655			
		26	FE	8636.
		14	SI	1479.
		13	AL	550.
56	683			
		26	FE	3859.
		14	SI	1530.
		13	AL	500.
		24	CR	164.
618	927			
		26	FE	4710.
		24	CR	1162.
		14	SI	1417.
		13	AL	625.
		25	MN	272.

Table 1. Analysis of the SEM image and the elemental analysis of some of the features.

Individual Feature Analysis

Once all of the features have been characterized as to area and perimeter, it is possible to analyze a set of features based on geometrical factors such as area, perimeter or shape. To do an analysis of an individual feature, all the line segments belonging to that feature are required. By returning to the original data of the full scan, and knowing the lowest point of these features, an analysis similar to the original process can be made which will yield the appropriate line segments for each feature. Once this information is at hand a complete characterization of the feature is made possible. There is a fundamental difference between the original analysis of the data and this individual feature analysis in that the feature's line segment information is stored during the second pass through the data whereas this information was lost in first pass.

X-Ray Elemental Analysis

To further categorize the features beyond size, geometry and abundance, a chemical identification is made possible using EDXA analysis of the features. Using the line segment information already obtained, the electron beam is driven uniformly over the feature in question with the correct shape being completely scanned. The scan of each feature lasts for a defined period, (currently 20 sec.). While the beam of electrons is scanning over the feature, the X-ray analyzer acquires the emitted X-ray spectrum. After the counting time has expired, the spectrum is automatically transferred to the computer for immediate analysis.

The first step in the analysis of the X-ray spectrum is to extract the energies and the intensities of the peaks. The spectrum is first smoothed[8] and then the background is subtracted. The intensity and centroid energy of the highest peak is determined[8] and then the peak is subtracted from the spectrum. The next highest peak is then found and subtracted until all the peaks in the spectrum have been accounted for.[9] Because each element has a recognizable fingerprint of lines with specific energies and relative intensities, the identification of elements is made by matching the measured lines and intensities to the appropriate elemental lines and intensities. Along with the elemental identification, the measured intensity of the strongest line for each of the elements is stored. Though this information is not currently used, it will be incorporated in the near future into a semi-quantitative interpretation of the features.

System Performance

As mentioned previously, one of the advantages of the line-by-line approach to image analysis is that no large storage arrays are needed to allow a comprehensive study to be made of the field of view. This reduced array storage size has not resulted in the sacrifice of speed of analysis. Analysis of the initial complete scan as to area, perimeter and the locations of all features takes approximately 30 seconds, but of course, this time depends upon the complexity and abundance of features within the scan frame. The time taken to complete the computer-driven 1024 x 1024 scan is about 70 seconds, this being the minimum time possible currently if there is to be no loss of instrumental resolution. To extract the list of individual feature line segments takes typically a few seconds per feature, again depending upon the complexity of the feature involved. During X-ray acquisition in the current program, the beam scans over a feature for 20 seconds. This time can be changed and in the future the integrated beam current on the specimen will be the operational parameter. The time to transmit and analyze the X-ray spectrum is 10 to 15 seconds. The total time to perform the operations and analysis of the example given in Table 1 was 8.5 minutes.

An important consideration in the reliability of the system is the capability of accurately directing the electron beam back to a feature after a period of time. The electronic drift of the system will be the sum of drifts in the D/A converter drivers, the interface circuitry and the SEM itself. For this system the drift at worst is ±3 units of the 1024 unit divisions either in the horizontal or vertical directions, regardless of magnification.

Figures 4 and 5 illustrate the manner in which the system processes an SEM image. Figure 4 is an SEM micrograph of an air pollution sample taken from the vicinity of a steel mill. Figure 5 is a plot of the stored binary representation of the specimen image. The intensity of the backscattered electron signal varies considerably since the composition and surface structure of the particulates vary. Some of the particles seen in the micrograph have significant topographical detail such as the large composite of particles near the top. The surface structure is not represented in the computer image. This is because at no point within the feature did the signal fall below the set threshold until the beam had passed completely outside of the particle. There are also some particulates which are not represented at all, since the signal originating from them never did rise above the threshold voltage. However, if the threshold had been reduced these too would have been seen in the computer image. Similarly if the threshold had been raised then some particles seen here would not appear at all. Thus the threshold level offers a powerful tool in discriminating particles especially since the backscattered signal is dependent on the average atomic number of the feature involved. Currently, the backscattered signal is under utilized since only one grey-level is being used per image representation in the on-off threshold approach. In the near future representations of the field of view will be stored either via grey-level storage of the image or via successive acquisitions with automatically set thresholds.

Table 1 contains the results of the initial analysis of the full field of view and also the X-ray analysis of some features. The image analysis consists of the list of particles including the x and y locations, areas and perimeters. The edge particles are designated by negative numbering. From this list those features with areas greater than 800 square units were chosen. The analysis is shown including the intensity of the largest peak belonging to that element, whether it is a K, L or M line. This most intense line and the relative elemental line intensities have been determined empirically using pure element standards. As a consequence of inadequate collimation of the X-ray detector, X-ray emission from specimen chamber components and the silicon backscattered electron detector (Fe Kα, Al Kα, Si Kα) forms a part of the X-ray spectrum (e.g. Si and Al lines of particle (916, 518) in Table 1). Elimination of the majority of these spurious X-ray lines will be achieved by further shielding, and the residuals will be taken into account by the particle species identification logic.

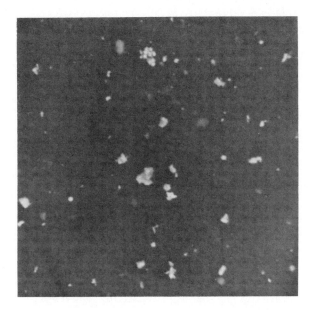

Fig. 4. SEM micrograph of a steel mill air pollution sample.

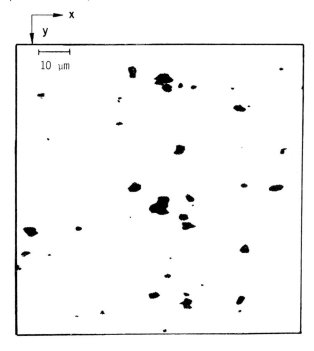

Fig. 5. The computer binary image of the SEM image.

Future Considerations

Because this image analyzer is based on a raster scan of the sample, the process lends itself to optical systems through the use of a television camera. Implementation of this system to optical microscopes and transmission electron microscopes will be done and should be relatively simple and direct. Another area to which this system is being directed is in the field of fibre counting, where overlapping fibres pose a serious problem in conventional image analysis. The separation of crossing fibres is made possible, since during line-by-line analysis it is possible to determine where contiguous lobes join and separate. The result of fibre analysis will be presented as lengths and widths of individual fibres which form an aggregate of crossing fibres.

The nature of line-by-line analysis used in this system allows a sequence of progressively more detailed analyses to be achieved using small computer array sizes and without seriously compromising the speed of operation. The initial raster scan first acquires the binary image representation. The next step is the extraction from this file of the individual features present and their geometrical characteristics. Line segment data belonging to an individual feature can then be isolated and used for more sophisticated procedures, such as scanning the electron beam over the feature during X-ray acquisition. Furthermore, the collected line segment data may be processed again to achieve other results, such as the case of overlapping fibre analysis.

Acknowledgements

The authors wish to acknowledge contributions from members of the Department of Applied Physics, Ontario Research Foundation. Funding of this instrumental development was provided by the Research Council of Alberta.

References

1. M. Cole "Instrument Errors in Quantitative Image Analysis", The Microscope 19, 1971 p.87-112.
2. Grant, G., Hall, J.S., Reid, A.F. and Zuiderwyk, M., "Multicompositional Particle Characterization Using the SEM-Microprobe", in SEM/1976/1, IITRI, Chicago, Il, 60616, p.401-408.
3. Ekelund, S. and Werlefors, T., "A System for the Quantitative Characterization of Micro-structures by Combined Image Analysis and X-ray Discrimination in the Scanning Electron Microscope", in SEM/1976/1, IITRI, Chicago, Il, 60616, p.417-424.
4. Lee, R.J., Huggins, F.E. and Huffman, G.P., "Correlated Mossbauer - SEM Studies of Coal Mineralogy", in SEM/1978/1, SEM. Inc., AMF O'Hare, Il, 60666, p.561-568.
5. Dinger, D.R. and White, E.W., "Analysis of Polished Sections as a Method for the Quantitative 3-D Characterization of Particulate Materials", in SEM/1976/1, IITRI, Chicago, Il, 60616, p.409-416.
6. Rink, M., "A Computerized Quantitative Image Analysis Procedure for Investigating Features and an Adapted Image Process", in J. of Microscopy 107 (Pt. 3), 1976, p.267-286.
*7. Grant, G., Hall, J.S., Reid, A.F. and Zuiderwyk, M., "Characterization of Particulate and Composite Mineral Grains by On-Line Computer Processing of SEM Images", 15th. APCOM Symposium, Brisbane, 1977.
8. Savitzky, A. and Golay, M.J.E., "Smooth-

ing and Differentiation of Data by Simplified Least Squares Procedures", Analytical Chemistry 36, 1964, p.1627-1639.

*9. Statham, P.J., "A Comparison of Some Quantitative Techniques for Treating Energy Dispersive X-ray Spectra", Ninth Annual Conference, Microbeam Analysis Society, No. 21, 1974.

Discussion with Reviewers

S.Ekelund: Please estimate the total measuring time to do a size distribution and elemental analysis for some 500 particles in the range of 0.5 - 5 microns.

Authors: The answer to this question can be divided into two parts. One deals with the geometrical analysis of the features while the other deals with the X-ray elemental analysis. The array sizes for the toal number of particles in a field of view in the current software are 256. However, there is no reason why this size could not be doubled to take into account 500 particles. For the system as it exists now consider only 250 particles rather than 500. No matter how many particles there are in a scan field, the acquisition time is about 70 seconds. For 250 particles the acquisition of the image and the geometrical analysis will take about 2.5 minutes. If a complete EDXA characterization were to be done on all 250 particles the total time would take about 85 minutes. About 90% of this time period is for the electron beam bombardment of each of the feature for 20 seconds. However, an analyst is not likely to do all features in this manner as only a selected subset of features will require the full detailed analysis. Also the 20 second electron excitation of each feature may be altered for the purposes required by a specific analytical approach.

J.D. Fairing: It is proposed to use this technique for fiber measurements with an optical microscope to resolve the problems of crossed fibers. How is it proposed to eliminate the depth-of-field problem so that all fibers will be sharply imaged?

Authors: In the collection of fibres from air or liquid they are passed through filters which have a significant thickness. Thus when the microscopical analysis is done the fibres are not all lying in the same plane and may be at oblique angles with respect to the surface of the filter. The way to attack this problem is not at the instrumental measurement end but to collect and prepare the fibre sample such that the fibres are all lying in a common plane (e.g. using Nuclepore filters). If this is done the depth-of-field problem would be resolved.

J.A. Small: You mention that the threshold adjustment is important. I would like to know if there is any method available which will automatically and correctly adjust the threshold on a field-by-field basis?

Authors: Manual global thresholding that is currently being used presents serious problems as to the accuracy of the measurements, c.f. reference 1. Shading within a field of view, variability of brigthness between fields of view and variations in brightness of image features are considerations for an automatic system. There are sophisticated techniques available to process grey level information of digital images (e.g. "Digital Image Processing", William K. Pratt, John Wiley and Sons, 1978) which require extensive amounts of computation. The approach we are undertaking is to differentiate a single scan line in order to accentuate the edges of features. The feature edges then can be separated from slowly varying signal levels and the changes in absolute brightness.

A.F. Reid: Can the system described be made to distinguish between more than one phase per particle?

Authors: The system as described acquires a binary representation of the SEM image. Features or phases with signal intensities greater than the threshold level are recognized but they are not distinguished. Grey-level storage of the image or multiple acquisitions of image representations with automatically set thresholds is the major hardware improvement planned.

T. Greday: What is the minimum size of the particles correctly assessable with the system in a case such as the steel mill air pollution sample presented?

Authors: The minimum size of particle correctly assessable with respect to its geometry and its area is directly related to the electronics of the SEM and the edge detection. The scan lines are stable. The y-axis resolution is thus essentially one scan line or one part in 1024. Because of the crude threshold method of detecting feature edges the resolution in the x or horizontal direction is approximately three parts in 1024. Accordingly if particle sizes are approaching this limit of resolution, an increase in magnification of the SEM is necessary in order to correctly determine the geometrical shape of the particles. With planned improvements in the image acquisition electronics, the resolution in the horizontal direction will improve to the one picture element level.

J.A. Small: Can you explain how you handle the statistical analysis for a sample which must be run at different magnifications as a result of a wide range in particle sizes?

Authors: For a sample with a wide size distribution of particles, increasing the magnification to obtain accurate geometrical characterization is necessary. In order to be confident in the analysis two conditions must be met. Firstly, the ratio of the magnifications must be well known in order that a precise measure of the fraction of the original field of view is known. Secondly, the small area being viewed at the higher magnification must be representative of the full view of the sample, that is the particulate must be uniformly distributed. If these conditions are met, then the large size particulate can be measured at the low magnification, while the higher magnification measurements of the smaller particle may be scaled to correspond to the full scan field.

T. Greday: *The way in which the features extending beyond the field of view are treated introduces an important bias in the statistical value of the image analysis. How do the authors think they could resolve that problem?*
Authors: The problem of obtaining statistically valid results in particle counting for limited fields of view is a difficult problem in microscopy. One approach which would be applicable for this system is to create an imaginary window smaller than the actual scan field. If a particle intersects either of two adjacent sides of the window then they will be counted, but if they intersect either of the other two sides they will not be counted. Thus if enough scan fields are counted then a reliable statistical analysis may be done. Another approach that can be followed because of the line-by-line nature of this image analysis process is to acquire a long continuous strip of scan lines across the sample by means of a mechanical drive on the specimen stage. Essentially only two sides of the field of view then present a problem; these are treated as rejection and acceptance sides in a similar manner to the above procedure.

J.A. Small and T. Greday: *You count X-rays for a 20 second time period. What is a typical value for the integrated counts over the entire spectrum and how does this value compare for two runs on the same area? Many particles contain elements such as S at the one percent level; can these elements be detected? How is it intended to measure the integrated beam current on the specimen and will the dead time of the spectrometer be taken into account?*
Authors: During the scanning of individual features with the electron beam the X-ray count rate is typically 1000-2000 counts per second. At the time of the writing of this paper only qualitative analysis has been used although semi-quantitative analysis is being incorporated into the system. This is being accomplished by measuring the current collected by the electrically isolated final aperture. This current is directly related to the specimen current and will be used to normalize the X-ray peak intensities. Using the live time as calculated by the spectrometer and the ratio of X-ray peak counts to the measured aperture current, a standardized X-ray measurement can be achieved. At present, without such normalization, the actual X-ray intensities vary considerably since the SEM beam current varies with time. The peak-to-background ratio of 1% sulfur in a stainless steel specimen was measued and found to be about 15%, which is detectable.

A.F. Reid: *To what extent are the EDXA spectra of standard mineral or phase compositions used to identify or verify the compositions observed during image analysis.*

Authors: As yet no mineralogical, phase or chemical identifications through comparison with standards have been incorporated into the analysis. Nonetheless, once the semi-quantitative elemental analysis is fully functional, comparison with known standards will be implemented in aiding identification.

S. Ekelund: *In the authors' opinion, what is the most significant advantage in combining image analysis and EDX-techniques? What is the most important field of applications?*
Authors: Image and EDXA analysis are both significant analytical tools by themselves and the combination of the two in one system gives the analyst a great deal of flexibility in the approach to his problems. Discrimination of features on the basis of the image analysis may result in a drastic reduction in the number of features requiring X-ray analysis with subsequent reduction in time and cost of analysis. Because the system is a general tool it has a broad range of applicability. For instance in air pollution studies which require analysis of a large assortment of particle types, this system would be ideally suited especially if classification into types is needed. In the field of metallurgy, for which inclusion and phase analysis as to type and amount is extremely important, the image and X-ray analysis gives a powerful evaluation of the material composition. The line-by-line analysis lends itself to analysis of specimens with a sparse population of features, since instead of analyzing a series of square images a large continuous strip of scan lines may be analyzed.

J.D. Fairing: *The images in Figures 4 and 5 do not correspond vertically or horizontally with respect to the picture boundaries. Is there an inherent shift in the system?*
Authors: The shift between the computer representation and the micrograph is not significant since the horizontal and vertical shift and gain of the representation is merely a function of the D/A converter voltages. The representation could however be made to overlap the micrograph if it were critical.

J.D. Fairing: *Will the authors make available, upon request, details of the interface wiring and system connections as well as listings of the computer programs and/or algorithms used?*
Authors: Assistance from the authors to others with regards to hardware connections, interfacing and algorithms used will be available. Since financial support was received from another organization with a view to possible commercial development of the system, the depth of detail that will be available is subject to discussion with them.

* *Editor:* *How may a reader obtain references 7 & 9?*
Authors: For Ref.7 write to Dr. A.F.Reid, Inst. Earth Resources, CSIRO, PO Box 124, Port Melbourne, Vic. 3207, Australia. For Ref. 9 contact NBS, Analytical Chemistry Div., Washington DC 20234.

Additional discussion with reviewers of the paper "Use of Microtopographhy Capability in the SEM..." by J. Lebiedzik et al, continued from page 66.

R.J. Lee: Are the D.C. Balance, and BSE Intensity checks required before each analysis, and is a reference sample required?
Authors: It is necessary to adjust the D.C. balance and BSE Intensities only once. However, if a variety of materials with large differences in atomic number are to be analyzed, it is good practice to check the raw intensities before analysis to make sure that operation is within the range of ADC.

J. Pawley: How would a measure of the surface roughness of the alumina ordinarily be obtained and how do the results compare with the SEM topography? (Percentage error?)
J.L. Daniel: In both sets of data (Table 1 and Figure 3) considerable overlap occurs in the measured values for different samples, although the RMS values seem to indicate different roughness. What is the precision of replicate measurements, not including the contribution from random surface variations? Couldn't some of the small differences simply be the result of sampling statistics? Do you have profilometer data available for the same samples?
C.M. Ward: What is the potential measurement accuracy of this technique?
Authors: Alumina samples are usually done by mechanical profilometry. Profilometer values were not available on the studied samples. Comparison of methods using a surface standard was done (Figure 1). The spread of RMS values in Table 1 originates from roughness differences within a sample. Each sample was measured in 24 locations in each direction (longitudinal and transverse). In the graphite samples (Table 1), the fractured surface was measured. It is quite possible that the relatively large spread of RMS values originates more from the fracture than from the oxidation, thus somewhat masking the desired results. This is one of the typical cases where surface roughness expressed in simple RMS values is insufficient and where a more complex expression would better characterize the surface, separating general textures from unusual ones.

A second measurement at the same location shows a precision of better than 1%. However because of field drift in the SEM, the reproducibility is strongly influenced by this drift since it is nearly impossible to scan again along the same line. The measurement accuracy by this technique is expected to reach a range of $\pm$ 1nm.

D.L. Davidson & J. Lankford: If one considers the surface roughnesses of the three aluminas (for either removal rate), they do not correlate with the fracture stresses or grain size, although the hardnesses of the materials should be approximately equal. Why not?

Material	Fracture Stress	Roughness	Grain Size(μm)
Al 300	32,200	4.33	25
Al 600	51,300	3.72	10
Al 995	39,500	1.81	12(25)

Authors: This is a typical example of surface roughness measurement where the RMS or AA value is not sufficient to fully characterize the roughness. S.E.I. photomicrographs of sample Al 600 and Al 995 reveal that Al 995 is smoother than Al 600 except where occasional large grains are removed by the grinding process. Although larger grains were removed, their low number contributed little to overall roughness.

J.L. Daniel: The illustrations you use are all for structures which have relatively smooth or rounded corners and edges. What errors are introduced when the surface contains sharp thin edges with their characteristic bright emissions independent of angle or orientation?
Authors: The "Edge Effect", so characteristic to Secondary Electron Imaging is negligible in BSE Imaging. However if it is a problem, equation (1) usually compensates for it. But there are special cases where stray BSE are directed by a unique topography to a preferred detector. The edge effect may also be reduced greatly by operating at lower Accelerating Potential. An error analysis for such an edge effect was not performed.

SCANNING ELECTRON MICROSCOPY/1979/II
SEM Inc., AMF O'Hare, IL 60666, USA

USE OF MICROTOPOGRAPHY CAPABILITY IN THE SEM FOR ANALYZING FRACTURE SURFACES

J. Lebiedzik, J. Lebiedzik, R. Edwards and B. Phillips

LeMont Scientific, Inc.
1359 E. College Ave.
State College, PA 16801

Abstract

A computer-controlled four-detector method was used to make multiple scans of surfaces to determine roughness values on samples for which fracture data were available. Fractured surfaces of graphites which were equilibrium oxidized to various levels of weight loss and ground surfaces of aluminas affecting strength were studied. The roughness values obtained are consistent with the other physical data in both cases. The method described to do microtopography in the SEM is accurate, consistent in result with current methods and because of computer involvement, both rapid and reliable.

KEY WORDS: Backscattered Electrons, Microtopography, Surface Roughness, Computer, Image Processing, Graphite, Alumina, Fracture

Introduction

At any instant, various electron and x-ray signals are generated from a sample exposed to the electron beam in the scanning electron microscope. From these various signals, the physical and chemical nature of most samples can be derived. The x-rays are necessary to establish the chemistry, but much of the surface characterization of the sample is generally made through secondary or backscattered electron images. The secondary electrons, because they are largely influenced by the configuration, and because they can be collected efficiently (close to 100%), produce images which show very sharp edges, slopes, ridges, etc. that are present on the surface, without obvious relation to the chemistry. On the other hand, backscattered electrons (BSE), which result in comparatively weaker video signals due to inefficient collection, are strongly influenced by variations in average atomic number of the sample. This makes BSE images particularly useful for observing structural differences in samples containing two or more phases. With a good BSE detector system, phases with average atomic number differences less than 0.5 can readily be contrasted. This capability, plus the fact that the emission of back scattered electrons is also influenced by the surface features, led to the development of the multiple detector method for measuring microtopography of SEM-Analyzed samples.[1] Further improvements in this method, along with better computer analysis of the results, make the technique valid for analyzing the microtopography of fractured surfaces, surfaces which influence fractures, and surfaces which cannot be analyzed by other techniques.

Generalities of the Multiple Detector Method

The multiple detector method, as it is now used, employs four BSE detectors in the determination of the microtopography. In scanning the sample in the north-south-direction, for example, the two opposing detectors are also N-S slightly above and on opposite sides of the sample, and positioned at a take off angle of 45°. Analysis in the east-west direction can be done then with a second set of two opposing detectors. The detected and amplified signals are used to generate a signal by difference (N minus S) which is a function of a combination of the surface slope and surface orientation. Using the mathematics outlined in the next section, the results can be used to calculate root mean square (RMS) or arithmetic average (AA) roughness values. In addition, the computer can generate directly on the pictured image, plots of the variation in the Z direction along the analyzed line, as well as a plot of deviation versus length in standard print-out style for any analyzed line.

Pertinent Mathematics and Computer Calculations

The detailed theory of this approach is described elsewhere.[1,2] In short, the original method involved one secondary electron detector for slope determination and four backscattered electron detectors for orientation of the slope determination in x-y plane. From these two values and a point to point spacing, the incremental change of surface elevations were calculated. Although the original method proved to be useful and practical, additional work yielded an improved, reliable and easy-to-use technique for routine analyses that requires no special sample preparation or system calibration.

The original work was modified as follows:

a. The secondary electron signal was eliminated because of its poor reliability as a slope indicator. Alternatively, the slope at each electron beam spot was determined from backscattered electron signals only.

b. Relative BSE intensities per total BSE yield were used instead of absolute intensities of a pre-calibrated system. This minimizes the need for calibration and makes the system insensitive to atomic number contrast and beam intensity variations.

c. Two opposing detectors were used to measure and reconstruct a scan line vs. elevation profile and surface roughness calculation for one direction.

d. The calibration consisted of a D.C. balance check between the two opposing BSE detectors to prevent any systematic error leading to artificial sloping of the surface plus a check on the BSE intensities, keeping them within the range of the Analog To Digital Converter.

The following equations were derived for the surface roughness determination from Detector D_0 and D_{180} located at 6 and 12 o'clock respectively around the objective lens aperture and having a take-off angle of 45°.

$$I_B = \frac{I_0 - I_{180}}{I_0 + I_{180}} \qquad (1)$$

where I_B - Normalized difference of BSE signals
I_0 - BSE signal from Detector D_0 range (0-255)
I_{180} - BSE signal from Detector D_{180} range (0-255)

Then for a slope of the interacting solid surface $\emptyset$

$$\sin\emptyset = I_B \qquad (2)$$

$$\text{or} \quad \emptyset = \text{Arc sin } I_B \qquad (3)$$

The incremental elevation:

$$\Delta Z = -C \cdot \text{Tan } \emptyset \quad [\mu m/\text{point}] \quad (4)$$

where C = point to point spacing

$$C = \frac{\text{scan length}}{\text{magnif. x point density}} \quad [\mu m/\text{point}] \quad (5)$$

For a nth point on a scan line the actual elevation with respect to the starting point will be given by

$$Z_n = Z_{n-1} + \Delta Z \qquad (6)$$

From a series of Z values one can readily proceed to calculate surface roughness as arithmetic average deviation or root mean square (RMS).

$$AA = \int_{X=0}^{X=L} \frac{1}{L} |z-\overline{z}| dx \qquad (7)$$

where L = length of the line scan

$$\text{and} \quad \overline{z} = \int_{X=0}^{X=L} \frac{1}{L} z dx \qquad (8)$$

$$RMS = \left[\frac{1}{L} \int_{X=0}^{X=L} (z-\overline{z})^2 dx\right]^{1/2} \qquad (9)$$

For discrete values of z, the integrals may be replaced by summations, thus:

$$z = \sum_{i=1}^{i=n} \frac{1}{n} z_i \qquad (10)$$

where n = number of points/line it follows then that:

$$AA = \frac{1}{n} \sum_{i=1}^{i=n} |z_i - \overline{z}| \qquad (11)$$

$$\text{and} \quad RMS = \left[\frac{1}{n} \sum_{i=1}^{i=n} (z_i - \overline{z})^2\right]^{1/2} . \qquad (12)$$

Experimental Procedures

The multiple-detector method employed in this study is automated; the scans are controlled and all information is gathered through a minicomputer. The microtopography for two sets of samples (graphites and aluminas) was determined and roughness values (RMS & AA) were calculated. For one set of samples (graphites), the roughness data were fitted to eliminate any slope effect caused by slanted sample mountings.

For the graphite samples (uncoated - as received) six N-S scans were made at four locations on each of the controlled-fracture surfaces from eight different samples. The scans were made at a rate of 256 points/line with a signal averaging of four readings/point.

For the alumina samples (gold coated), two each of three types, six N-S scans were made at eight locations on the preground surface of each tested sample. The scans were made at the same rate and condition as above.

To confirm that the measurements made were accurate, a standard roughness surface (GAR 63 ST) was then scanned along six lines and the AA value calculated. Scan rates and signal averaging were the same as those used on samples. The image, scan lines, and six profile lines are shown in Figure 1. The standard has an AA roughness value of 1.8μm. Values obtained by the described method are in the range

1.54 - 2.1μm and vary with location. This confirmed that the method in use is accurate and capable of measuring very small Z values.

Results and Discussion

A. Graphites

The graphites used in this study were plates cut from one block of material, which was made from 60-70% coke flakes and 30-40% coal tar pitch binder and subsequently pressed and graphitized. The coke flakes (~75μm in the flat dimension) are somewhat oriented by the processing and the samples were made in regard to that orientation. Plates referred to as longitudinal were cut in a way that controlled fracture could supposedly occur between flakes, i.e. the flatness of the flakes was the same as the fracture direction and perpendicular to plate flatness. Those referred to as transverse were cut in a way that the flatness of the flakes was the same direction as the flatness of the sample plate and fracture through the plate must cause flake fracture.

In addition to those variables, some samples were equilibrium oxidized at 500° C to 5%, 10%, and 20% weight loss. Beyond 5% weight loss, in addition to dimensional change, porosity increased. Table 1 combines all of the significant results.

Aside from porosity, which is oxidation dependent, the other variables are generally higher for the transverse samples than for the longitudinal samples. Fracture toughness, fracture surface energy and roughness of the fractured surface are all higher for those samples where flakes were fractured. There seems to be little difference in roughness as a function of oxidation-induced porosity. This could be expected in that oxidation increases the number and size of pores but should not change significantly the texture of the fracture surface between pores, until perhaps much greater losses of binder are caused by the oxidation process.

Figure 2 shows an example of the fractured surface of as-received graphite (longitudinal variety). The roughness results fit the data from the other studies and are consistent with expected results as a function of the oxidation.

B. Aluminas

The Aluminas used in this study were control specimens used in a study on the effect of surface grinding rate on the sample strength. The six samples were of three different materials, fine grain (600), coarse grain (300), and a mixture of coarse and fine grain (995) material, two each with finish removal rates differing by a factor of ten--6.35 vs. 0.635 μm/pass. The grinding direction was perpendicular to the long direction of the tensile test specimen, thus producing the most severe flaws which could affect the strength.

Results of the strength tests and the average of eight roughness measurements on the ground surface are summarized in Table 2, and all of the roughness measurements as a function of surface removal rate are shown in Figure 3.

As expected, the strengths of the samples with faster removal rates are less, and the roughness values are greater. This is clear in every case, but a most impressive fact is that out of forty-eight measurements on these six samples by this technique, none is visibly "in error", and we're looking at very small roughness values where inconsistent data would be apt to occur. This is strong evidence for the accuracy of the technique and should lead to other similar studies and additional pertinent applications where minor differences in surface roughness are critical or have large effect.

Figure 4 is an example of the ground alumina surface as seen in the SEM (BSE) signal with scan lines and surface profile lines superimposed. An added benefit of computer control, data storage and calculation, is that any profile line may be reproduced by the printer showing actual Z-values for each picture point as a percentage of maximum elevation. An example is given in Figure 5, which is the profile for the line most to the right in Figure 4.

Conclusions

This fully automatic method of microtopography measurement, which yields rapid results including permanent data storage for further analysis, is superior to a mechanical profilometer in many aspects. Speed, no moving parts, no wear or damage to the surface and extreme sensitivity are better when compared to profilometer performance. Another distinct advantage of the method is that it can be used on spherical or cylindrical surfaces. The curvature effect is easily removed by curve fitting and the results appear as if the contour was flat. The only disadvantage of this approach is that it cannot be used on all surface types, such as those with a very steep slope or a break in the surface normally invisible from above. This results from the fact that slopes greater than 60° are not detected correctly. Other disadvantages of the original approach were solved by eliminating the secondary electron signal and using relative BSE intensities instead of absolute calibrated ones. No special sample preparation or calibration is necessary. The sample need not be coated as long as it is conductive. The measurement is fast, accurate, consistent in result with other methods, and versatile in output via computer involvement. As seen in this study, the method can be used to look at previously fractured surfaces to aid in understanding the fracture, at surfaces which have an effect on the strength of materials, and on materials too soft to be scanned by a stylus type profilometer. These latter materials include soft inorganics and organics of all types, from graphites through living tissues. We expect that the method will be established among the techniques commonly employed to get more meaningful data from the SEM.

TABLE 1

Oxidation Wt. Loss, %	LONGITUDINAL				TRANSVERSE			
	K_{I_c}	G_{I_c}	Porosity %	Fitted (1st Order) RMS	K_{I_c}	G_{I_c}	Porosity %	Fitted (1st Order) RMS
0	1.00	95.1	24.2	43-49	1.20	160.2	24.2	52-55
5	.81	71.5	23.9	39-48	.88	92.0	23.9	49-53
10	.57	52.1	27.3	47-53	.63	59.0	27.3	48-52
20	.27	32.0	34.2	48-51	.30	43.0	34.2	50-57

K_{I_c} - Fracture Toughness in meganewtons/meter 3/2

G_{I_c} - Fracture Surface Energy in Joules/meter^2

RMS - Range established from four separate measurements on each fracture surface in µm.

TABLE 2

Sample	Average Grain Size,(µm)	Specific Gravity,g/cc	Removal Rate 6.35µm/Pass		Removal Rate 0.635µm/Pass	
			Fracture Stress,Kg/cm^2	Average Roughness, RMSµm	Fracture Stress,Kg/cm^2	Average Roughness RMSµm
Al 300	25	3.74	2190	4.33	2510	4.18
Al 600	10	3.71	3490	3.72	3763	1.77
Al 995	12(25)*	3.85	2687	1.81	2749	1.52

* Average fines (coarse)

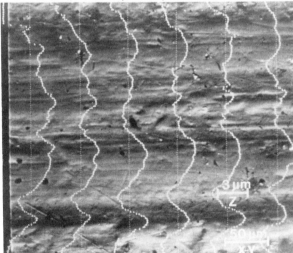

Fig.1. SEM Picture Showing Surface Roughness Standard (GAR 63 ST) Scan and Profile Lines as Obtained by the Method Herein Described. Image: BSE single detector 6 o'clock. Standard profil-ometer value AA=1.8µm. AA range obtained by the SEM method=1.54-2.1µm. In this and subsequent figures, raised surfaces are indicated by varia-tions to the right, and depressions to the left.

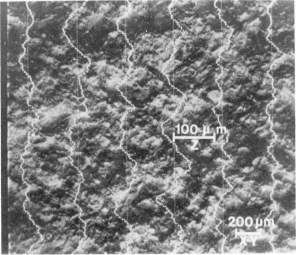

Fig.2 SEM Picture of Fracture Surface of Graphite As Received. Image: BSE single detector located 6 o'clock. The surface topography curves indicate only localized changes in elevation, since overall slope and large surface curvature were elimi-nated by a curve lifting procedure.

Roughness measurements of ground surfaces of various aluminas as a
function of surface removal rate (lower to the left and higher to
the right).

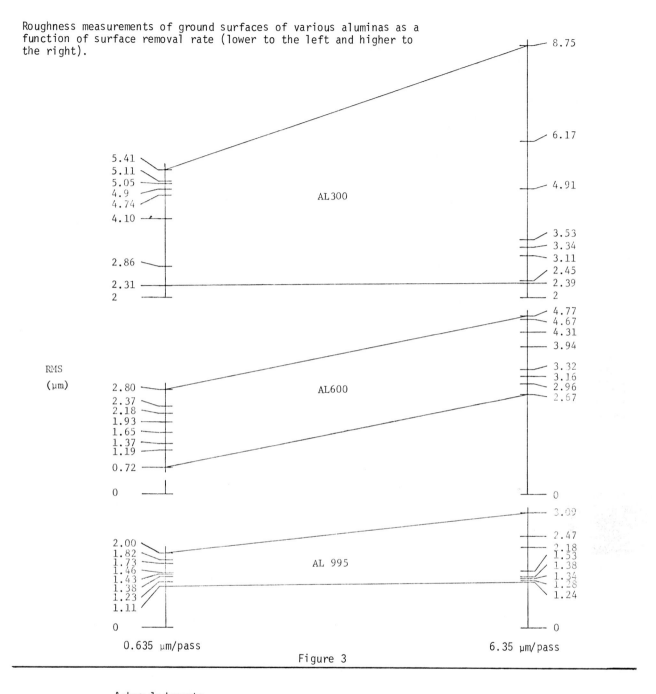

Figure 3

Acknowledgments

The graphites and aluminas used in this
study along with the characterizing and strength
information are from works by Gerald L. Wood and
T.E. Easler respectively, and are bases for
theses in preparation for Ph.D. and M.S. degrees
in Ceramic Science at the Pennsylvania State
University. We gratefully acknowledge the
cooperation and assistance of these students in
the use of the samples and data in this work.
The comments and suggestions of the reviewers
add significantly to the matter of this paper
and they are much appreciated.

References

1. Lebiedzik, J., "Multiple Electron Detector
Method for Quantitative Microtopographic
Characterization in the SEM," Ph.D. Thesis, The
Pennsylvania State University, 1975.
2. Lebiedzik, J., and White, E.W., "Multiple
Detector Method for Quantitative Determination
of Microtopography in the SEM," Scanning
Electron Microscopy/1975,   IIT Res. Inst.,
Chicago, 181-188.

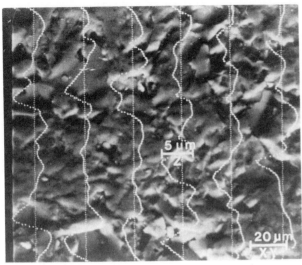

Fig.4. SEM Picture of Al 995 with Surface Removal Rate of 6.35 μm/pass with Scan Lines and Surface Profile Lines superimposed. Again the Z-direction is small in magnitude, yet very clear. Image: BSE single detector located @ 6 o'clock. Micron markers are indicated for x-y and Z direction.

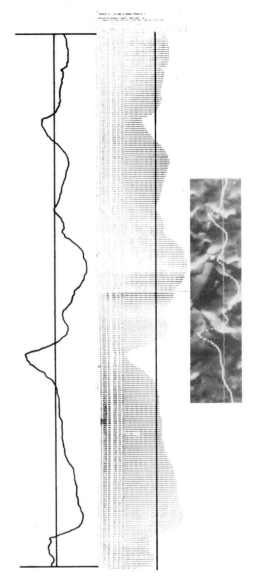

Fig.5. Computer Printout Showing Surface Deviation Versus Length for the Scan Line Most Right in Fig. 4.

Discussion with Reviewers

C.M. Ward: This technique has substantial benefits over stylus profilometry in addition to those mentioned in the paper. It is not sensitive to the tip radius which should result in much more accurate measurements on samples with narrowly spaced ridges and also allow a more accurate total profile measurement.

C.M. Ward: Reference is made to this being a "rapid" technique. How much time is required for the measurement of each point and the calculation of both the Z dimension and the average values?

R.J. Lee: How do the area total line length scanned and time per measurement compare with other methods?

Authors: The length of the line scan is typically shorter than those covered by stylus profilometry. Maximum length for a typical SEM scan would be 10mm. Line scan time varies with the chosen point density, but at 7 ms/point, simultaneous Z, AA and RMS calculations at 256 points/line takes approximately 1.8 seconds/scan. With additional data processing, such as curvature correction, from 10 to 30 ms/point may be required.

C.M. Ward: What beam current was used and how is the accuracy of the measurement influenced by operating parameters such as time per point, kV and magnification?

J. Pawley: Could you give us the beam voltage + currents?

J. Pawley: The backscatter coefficient of graphite is notably low. Did this limit your results and could they become biased by contamination buildup on the scanned line?

Authors: For graphite samples, an absorbed current of 5×10^{-7} A and an acceleration potential of 25kV was used. The relatively high absorbed current was necessary because of the very low BSE yield from graphite. A major change of operating parameters has some influence on results but minor changes are typically negligible. An error analysis related to change of operating parameters, surface texture, edge effect, etc., has not been performed.

The contamination build-up did not cause a noticeable problem since only relative BSE intensities were used for analyses (see Equation 1).

For additional discussion see page 60.

SCANNING ELECTRON MICROSCOPY/1979/II
SEM Inc., AMF O'Hare, IL 60666, USA

THE PERCEPTION AND MEASUREMENT OF DEPTH IN THE SEM

A. *Boyde*

Anatomy Department
University College, London
Gower Street
London WC1E 6BT, U.K.

Abstract

This tutorial reviews various aspects of the appreciation and measurement of three dimensional (3-D) depth in the scanning electron microscope (SEM). These two facets of this general problem area may or may not be linked. It is very useful to make 3-D SEM images in order to obtain a better understanding of a complex object. It is imperative to make such stereo-pair images under close-ly controlled conditions in order to be able to make 3-D co-ordinate measurements (contour maps, line profiles, etc.). The stereo-pairs for measurement purposes can be measured, one image at a time, to obtain the 3-D co-ordinate data, i.e. they need not be viewed stereoscopically. However, 3-D viewing is a great aid to parallax measurement, enabling much greater precision and productivity.

The tutorial considers how to obtain the stereo-pair images with specimen tilt and beam tilt methods; how to view them stereoscopically for one observer and multi-observer display; what to measure and how to measure the pictures separately to make 3-D analyses from the 2-D data; and what to measure and how to combine stereo-viewing and parallax and X,Y measurements from the SEM micrographs in order to obtain X, Y and Z specimen co-ordinates (stereophotogrammetry). For most practical applications in SEM photogrammetry it is sufficient to assume a parallel projection system. The geometrical relationships are then simple. The SEM user is strongly recommended and encouraged to use stereo-measurement methods in furthering the solution and understanding of his own problem areas.

Key words. SEM/stereophotogrammetry/stereoviewing/Real-time TV beam tilt stereo/simple photogrammetric instrumentation

Introduction

Seeing and understanding 3-D depth in SEM micrographs may not be the same thing. If you see a 3-D image, i.e. a stereoscopic pair of images with one member presented to each eye, then you will certainly understand that there is depth in the field of view. But you may know that there is depth there without a stereo-pair since some kinds of SEM contrast are due to surface topography and strongly suggest the nature of the surface relief of the specimen. You may also measure parameters from which you can calculate the 3-D depth in a stereo-pair without actually viewing the image stereoscopically[1]. You may view a stereo-pair and have a false psychological impression of depth if too little or too much tilt angle was used, i.e. if the convergence angle between the two images is not close to the normal physiological values for objects viewed at close distances, about 8 degrees.

In order to remove the incipient confusion which may therefore exist for the stereo beginner, it is useful to divide the subject matter up into various levels, so that the user can decide what will be good for him to know.

Non-stereoscopic and non-stereometric depth analysis.

That a great many SEM users are happy in their ignorance of practical stereoscopic SEM procedures is witness to the fact that there are ways of obtaining enough information about a specimen surface topography, to satisfy a particular usage, which do not require the specimen in the SEM to be carefully positioned and the image measured and/or recorded on at least two occasions. The striking similarity of optical images of combined diffusely and directionally illuminated surfaces to the conventional secondary mode image in the SEM provides part of the explanation for the apparent ability to appreciate 3-D in one 2-D image. However, the operation of searching the specimen with movement, rotation and tilting controls means that large numbers of images from different viewpoints have been processed by the SEM operator's brain, and these all add subliminally to our understanding of the surface under study in the SEM. (Note the implicit message that the image interpreter was the operator.) With TV speed scanning and rapid search movements, motion parallax can contribute strongly to understanding topography.

Foreknowledge of the likely surface topography is another factor of great importance in understanding single or multiple non-stereoscopic images of rough surfaces. None of the above came within the compass of the present tutorial.

Stereoscopy (seeing 3-D, stereovision)

The first level to which most normal SEM users with normal stereoscopic binocular vision should wish to aspire is to see depth in the SEM picture. You merely require (a) stereo-pair images, (b) means of viewing them and (c) stereopsis, i.e. the individual ability to fuse images and see them as 3-D.

Stereoscopic parallaxes viewed monocularly

One (good)-eyed users and those with squints do not despair. You can appreciate that stereo-depth is present in a stereo-pair if the two images are presented in such a way that they are over-lapped as if for stereoviewing by projection on to a common viewing plane but only viewed with one (good) eye. (Figure 1). If the two members of the stereo-pair are printed in black and white and the details that they contain are too repetitious, then it may be difficult to distinguish the separate doubled features (repeated once for each part of the pair). This problem of recognising the two images is reduced if they are displayed in contrasting colours. In many cases, therefore, it is possible to recognise exactly overlapping features. The overlapped features lie in the plane of the image, the others either above or below the paper (or in front of or behind the projection screen). Relative displacement in opposite senses means above or below the mean plane (e.g. the surface of the paper) of the image.

Simple measurement for 3-D XYZ co-ordinate reconstruction. (Stereoscopic parallaxes measured, but not viewed).

Referring to the foregoing case (Figure 1), the separation distance between doubled features is the parallax, which can be measured. Twice as much parallax means twice as much height difference. 3-D XYZ co-ordinates can also be reconstructed from measurements made singly in each member of the stereo-pair. In addition to distance measurements in X only or in X and Y, the magnification, M; the working distance from the raster perspective center to the mean datum plane of the specimen, D; and either the base-shift distance, B, or the tilt angle difference, α, needs also to be measured.

Stereophotogrammetry (Stereometry) and Stereoplotting

The final evolution of stereopsis and geometry combined gives rise to the possibility of using stereoscopic binocular vision with its attendant exactitude of judgment of relative distance as an aid in the precise measurement of both distance and distance differences or parallax. Measuring and plotting instruments of various degrees of complexity, from the simple parallax bar or its analog (e.g. the SB 180 and SB 185 series instruments [2,3] manufactured by Cartographic Engineering Ltd., Landford Manor, Nr. Salisbury, Wiltshire, England) through special SEM and TEM plotters [4-9], have been evolved as aids to the partial or complete mechanical solution of stereo-pair geometrics. The limited involvement of most SEM laboratories in the serious measurement of unknown objects has meant that the cost of a medium level special plotter for SEM stereo-pairs is considered to be too expensive. However, the volume of sales of simple mirror stereoscopes with parallax measuring facilities (e.e. SB 185 [2,3]) indicates that large numbers of SEM users have appreciated the usefulness of being able to make the occasional exact and and exacting measurement. Coupling such simple optics and mechanics to digitising beds and rotary digital encoders will surely mean that computers and software will become more popular means of improving efficiency, accuracy and productivity in 3-D mapping, contouring and profiling in the future.

Other things that this tutorial is not about: Stereology and signal level contouring and Y modulation.

Stereoscopy and stereometry are not to be confused with stereology even though the latter science may have an important place in SEM studies in material science and metallurgy. Stereology deals with plane cut or flat surfaces which would be totally uninteresting to us because they are only 2-D. (The Journal of Microscopy is the official journal of the International Society for Stereology and can be consulted for papers dealing with the theory and practice of computations of 3-D structure based on 1- and 2-D measurements of intercept and area, etc.)

Signal intensity is not simply related to specimen topography in the SEM, even under the best circumstances in which a combination of detectors is used to sort out variations in signal level due to slope from those due to material contrast [10,11,12]. Y modulation displays, therefore, do not represent height profiles across the specimen and contour maps showing equal signal levels (grey scale boundaries) do not show lines of equal height.

STAGE I STEREOSCOPY

(a) Stereo-pair images

Stereo-pair images can be generated by either (i) shifting or (ii) tilting a specimen with respect to a stationary electron beam raster[13-17]; or (iii) by tilting the beam with respect to a stationary specimen [18-21].

(i) Shifting the specimen obviously only works at relatively low magnifications - best within the range of magnifications possible with stereoscopic binocular optical microscopes. Above say 50X, one runs out of overlapped area in the two fields of view (the two members of the stereo-pair) if the specimen has been shifted far enough to generate enough stereo-parallax for viewing and/or measurement. (ii) Tilting the specimen can be used at any and all magnifications, and nearly all SEM stages are equipped with the facility of tilting the specimen so that it faces more towards or more away from the electron collector system. This facility can be used to give two views of the same area of the specimen separated by a small angle of about 8°. If one could choose, one would opt for a stage in which the stub height can be adjusted by a delta-Z control so that the specimen area of interest can be brought to lie on the level of the tilt axis, so that the specimen does

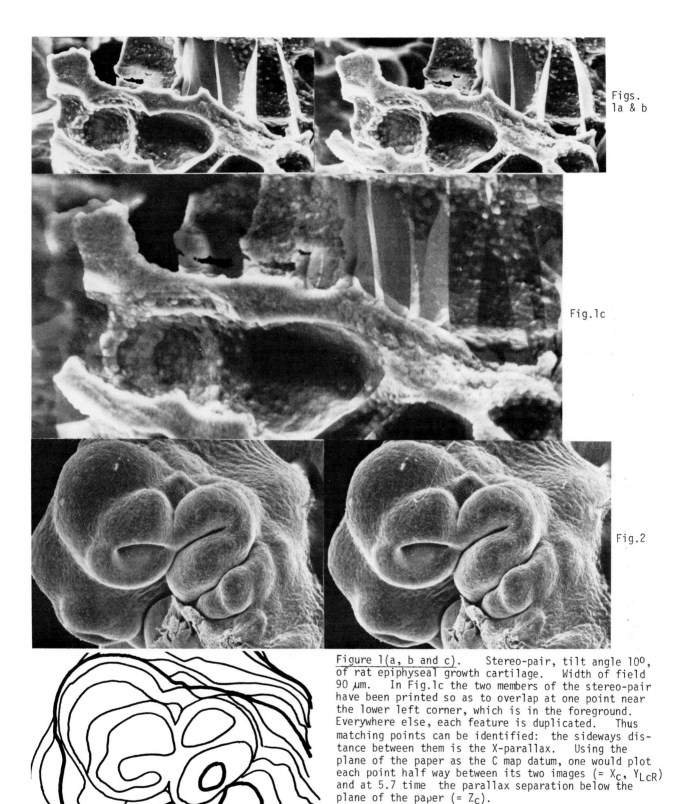

Figure 1(a, b and c).   Stereo-pair, tilt angle 10°, of rat epiphyseal growth cartilage.   Width of field 90 µm.   In Fig.1c the two members of the stereo-pair have been printed so as to overlap at one point near the lower left corner, which is in the foreground. Everywhere else, each feature is duplicated.   Thus matching points can be identified: the sideways dis-tance between them is the X-parallax.   Using the plane of the paper as the C map datum, one would plot each point half way between its two images (= X_c, Y_{LcR}) and at 5.7 time the parallax separation below the plane of the paper (= Z_c).

Figure 2.   Stereo-pair of mouse embryo (courtesy of Prof. Arnold Tamarin, Seattle).   Width of field = 1.9 mm.   10° tilt angle difference.   Contour map is drawn at 1 mm parallax intervals at the original 50X magnification scale = 5.7 mm height intervals at that working scale = 110 µm at the true scale.

Figs. 1a & b

Fig.1c

Fig.2

not move when it is tilted. A trifling additional advantage would ensue if the stereo tilt axis direction were perpendicular rather than parallel to the face of the electron collector, so that the "illumination intensity" (signal level) would not change significantly when the specimen was tilted, and the images would not have to be rotated through 90^0 for stereo viewing.

Optimal stages would have, therefore, two tilt axes - one for altering the mean specimen attitude (basic tilt) and the other designed to give the stereo tilt - in addition to rotation, Z, delta-Z, and X and Y shifts. So many of us, however, must make do with second best that we will digress to see how to proceed in order that the results need not be one jot less valuable (or exact, if measurements are to be attempted). We presume that the stage possesses basic tilt (rotation),X, Y and Z controls.   We will presume also that the specimen does not lie exactly in the tilt axis, so that, when tilted, it moves in X (the direction in which dimensions change, i.e. in which stereoscopic parallaxes are generated) and Z - the vertical, electron beam parallel direction. Even for simple stereoscopy, it is as well to keep the magnification of both members of the stereo-pair the same.   The change of focus due to the latter, Z, movement should therefore be corrected with the Z control and the final lens control, usually used for refocusing, left well alone.   The movement in X should be corrected with the X control so that one has the same field of view after tilting (plus or minus) a little bit at the top and bottom of the field according to whether you increased or decreased the mean tilt of the specimen with respect to the electron beam).   Marking a cover on the visual CRT with a wax pencil is a simple means of aiding the memory about what was on the middle of the field of view when you recorded the first member of the pair.   The movement in X on tilting, incidentally, also defines the X direction for you - the control necessary to correct the tilt generated movement is called X.   If you use this, the X shift control, the picture should move parallel with the side of the CRT. If it does not, then you have an image rotation problem.   You can (i) ignore it, and mount the photos for stereo viewing with this direction parallel to the interocular axis; (ii) rotate the scan coils manually, until X movement is parallel to the edge of the frame; (iii) adjust the working distance by altering the height of the specimen until X is the frame direction   or (iv) use a scan rotator unit to adjust the rotation of the raster.   In the last three cases, the photos can be mounted square on the stereo viewing apparatus.

How much to tilt the specimen? is a question frequently asked. To get a correct, to-scale impression of depth (as checked by measuring heights/depths in stereo-models against computed heights/depths from parallax mensuration) one should use a tilt angle of about 8^0 [20,22].   This value might vary slightly from observer to observer, according to variations in the interocular base. This variation, however, should be of no practical significance when judged against the likely uses of analysis by simple 3-D viewing!

More and less tilt and stereo parallax

For both perception and measurement purposes, however, it may be useful to exaggerate[23] or diminish[24] stereo-parallax.   To increase apparent Z differences use more tilt[23] - more parallax is also easier to measure, and one will have a better proportional accuracy.   If the 3-D depth of the field of view is significantly greater than its width in X and Y, then one may wish to reduce the stereo tilt angle (though not too much, as in ref. [24]), to ensure that the excessive stereo parallax does not prevent one from fusing the two images as a stereo-pair, i.e. to prevent the two images from looking too dissimilar.

Influence of stage design on choice of tilt angle

Due to simple mechanical design problems, it is often better to choose 10^0 for the tilt angle difference (alpha) between the two specimen stage positions used when the stereo-pair photographs are recorded.   Many stages are calibrated so that it is easier to read every 10^0 of movement. Others only give reliable readings when the set values are 10^0 apart.   Unless the stage is coupled directly to the tilt control, it is important to approach both tilt angle settings from the same direction[25].   One only needs to worry about the measured value of the tilt angle if one wants to measure depth, and there the golden rule is to calibrate the stage unless it is obvious that you need not.

Beam tilt system

Beam tilt stereo can be added as an afterthought to many standard commercial conventional SEM's.   In our own system[21], the additional beam deflections are given by a special set of scanning coils which fit in the usual position in the final lens base.   The usual final beam defining aperture at the exit of the final lens is replaced by one just before it.

Post lens deflection coils[18-20] have the advantage that the usual scanning coils and the final aperture position do not need to be modified. but they are extra clutter inside the specimen chamber.   As such they could conceivably interfere with ancilliary devices, such as micromanipulation and dissection apparatus[26] and EDX detectors in the specimen chamber.

Some of the main advantages of both systems are connected with the means of viewing (to be considered in the next section).   However, we should note that real time stereo at TV speed scanning of, perhaps, a changing object - changing as we heat, cool, stretch, compress or cut it - could not realistically be obtained by mechanical specimen tilting.   Beam tilt stereo also allows stereo-pairs to be taken of specimens which would be unmanageably large for mechanical tilting, or which could not be tilted because of some other problem, for example, because we were cutting it with another device in the SEM chamber.   Beam tilt stereo for recording purposes is achieved at the flick of a switch.   We can "hold" the left or right eye views whilst they are recorded.

Beam tilt stereo systems are simple to operate. One will find knobs, (A) to control the left-right switching speed ((1) every other line, (2) every other half-TV frame interface, (3) every full TV frame, (4) hold left (5) hold right); (B) to adjust the convergence of the two beams so that they overlap somewhere near the centre of the field of view in the line scan direction (one or two knobs, depending on the design); (C) to overlap the two images on the frame direction; (D) ideally, there should be separate stigmators for the left and right deflected beam positions; (E) there must be a control to set the magnitude of the beam tilt, i.e. to adjust the tilt angle. In commercial apparatus, this will have been calibrated by the manufacturer. But the user must note that the beam tilt stereo angle will vary with the working distance and the accelerating voltage[21,27] unless some complex compensatory circuitry is used.

However, it is a simple matter to calibrate the beam tilt angle for any one working distance and accelerating voltage[27]. One uses an object with a known height step, Z, measures the stereo parallax, P, and hence calculates the tilt angle, α, from the relationship $\alpha = 2 \ (\text{arc sin} \ (\frac{P}{2Z})) \ (1)$.

Systems for viewing the stereo-pair

(i) One observer at a time Most practised stereo observers can fuse a correctly mounted stereo-pair without any viewing aid. It is worth noting that the correct way of doing this is to uncross one's eyes, as if gazing to infinity . In this case one can view any normal stereo-pairs mounted for viewing with a simple lens viewer and see the object the right way out. If the eyes are crossed then a pseudoscopic inversion results, each eye seeing the incorrect image so that the object is seen from inside out. The functions of focusing and convergence, normally closely coupled, are uncoupled in order to achieve unaided stereo-viewing.

Simple lens viewers have two lenses separated by the interocular width and held at their focal distance from the images. Each half of the stereo-pair should be printed as the average interocular width (i.e. 68 mm.) wide. Since the lenses give a slight magnification, printers should be asked to use a finer screen size when printing stereo-pairs, using a half tone process. Journals vary widely in the quality (fineness) of the screen they use, and would-be authors are advised to look very closely at this problem to avoid disappointment.

Prism viewers. One (or two) prisms may be used to allow pictures mounted side by side to be viewed stereoscopically. In the system Nesh [28] viewer, the two photos are mounted vertically one above the other, and any size of photo may be used and accommodated merely by adjusting the viewing distance. This cheaply manufactured viewer therefore has the advantage that published stereo-pairs can be of a conventional picture size, and the use of a conventional screen does not give poor results.

Two-mirror "Wheatstone" Stereoscopes[29]. The Wheatstone Stereoscope is contructed from two mirrors mounted at right angles to each other, each facing one eye and one photo at 45°. The images are mirror inversions, but this can be of little practical significance. Most practical photograph sizes can be accommodated by adjusting the eye-mirror-photo distances appropriately.

Four mirror stereoscopes. In the commonest type of mirror stereoscope in use to-day, a second pair of mirrors performs an additional inversion bringing the photos the "right way round again"

Like the Wheatstone stereoscope, these can be used with or without additional lenses to increase magnification or to remove the need to focus at a near distance. They have the advantage that larger photographs can be used than with a simple lens viewer - anything up to half the distance between the centers of the second pair of mirrors being accommodated. The photos can be mounted permanently side by side and there are several other points of convenience over the Wheatstone[29] Stereoscope.

Flicker display. If the two members of the stereo-pair are displayed alternately on the same surface, and a shutter mechanism operates so as to allow each eye only to see its appropriate image, then a stereoscopic fusion of the images will take place if the flicker speed is not too slow. A recent example of a flicker shutter mechanism is that using PZLT liquid crystal material in spectacles[23]. This material could be made opaque or transparent to allow the wearer to see the stereo-pair images displayed on a black and white TV monitor.

All the foregoing systems for stereo viewing are suitable for use with any form of image, i.e. whether the image is a negative or positive print or transparency, on paper, glass, plastic or on a live TV display.

(ii) Systems for several observers.

Lenticular overlays. If the image is broken up into a series of lines, each of which is laid in the proper position behind a linear prismatic overlay, then the image will only be seen by one eye. The opposite image fills the lines in between and is seen through the other side of the prismatic overlay. This technique could be used for a real time TV display of SEM stereo images, but is more commonly seen in 3-D postcards.

Twin optical paths, polaroid filters and silvered screen. The polaroid projection system is the best for projecting stereo images to be viewed by audiences. Each image is projected separately through a 045° or 135° oriented polaroid filter on to a silvered (Al spray paint) reflecting screen. The images are overlapped at the center of the field of view. Viewers have 045° and 135° oriented polaroid filters in a spec-tacle frame. Special lantern slides are in common use which hold the two members of the stereo-pair ready oriented. The special twin projectors have controls for vertical and horizontal overlap of the two images.

However, any pair of lantern slide projectors (e.g. Carousels) can be used to project the images so long as they give the same magnification, aim at the same screen, and are appropriately equipped with filters. Most multi-user laboratories will have two identical projectors, so why not have a go? Scrap polaroid to make filters and spectacles can be obtained very cheaply. Some of the advantages of Twin-Carousel display are that 1) one can use the same slides in mono: special mounts are not needed; 2) one still has automatic advance and remote focus so that the speaker can remain near the screen; 3) the low cost ($150 for a Carousel compared with $500-$800 for a stereo projector); 4) the stereo slides are more easily integrated with the mono slides in a presentation; 5) if no aluminized screen is available, it is easier to convert to anaglyph by fitting color filters instead.

The disadvantages of the polaroid projection system are that a special set-up or system is required, and that the polaroid acts as a neutral density filter, thus severely reducing the intensity of the projected image. Depending on the size of the projection screen, it is therefore sometimes necessary to use high intensity projector bulbs and adequate heat filters to project the the lantern slide emulsions.

The anaglyph system. In the anaglyph system [18,21,30], the two members of the stereo-pair are again projected or printed on to the same surface, but here they are coded in colors. The usual code is red for the port eye and green for the starboard, appropriate filters being worn in spectacle frames for viewing. It works with color printing in publications, colored photographic prints, transparencies and TV sets[21]. It is merely important, as in all stereo work, to overlap the pictures very precisely in the vertical direction, and to have some point of overlap horizontally - at least, do not displace the two images so far sideways that it is difficult to accommodate this "false parallax".

Anaglyph lantern slides are handy when one wants to insert one or two stereo slides in the middle of other monoscopic images - the red and green viewers are cheap and/or easy to make. Anaglyph slides can be made directly at the SEM by putting color negative film in the camera and inverting the display contrast[31]. Alternatively, the separate black and white images can be printed as diazo-positives in red and blue, and these are clamped together in the correctly overlapped position in a single lantern slide frame[30].

The anaglyph system also lends itself to real-time stereo TV display[18,21]. The alternating left-red:right-green in formation is fed into the appropriate color channels of a color TV monitor. Since the display can be sited anywhere convenient and does not have to be viewed from within a confined range of angles, the system lends itself well to small groups of individuals participating in 3-D operations at the SEM, as, for example, in using the SEM to dissect small specimens[32].

STAGE II MEASUREMENT WITHOUT STEREOPSIS

What we have to measure to recover X, Y and Z co-ordinates for specimen surface features varies according to the means of generating the stereo parallaxes, and whether we are justified in using simplifying assumptions about the projective geometry in the SEM[33-35]. When working at low magnifications, it cannot be justifiable to deal with the perspective geometry as anything other than that! Therefore, the length of the principal projector, i.e. the distance from the raster perspective center (normally, the final aperture) to the principal point at the specimen must be known. If the specimen in shifted horizontally in X to generate the stereo-pair, the base shift distance, B, through which it was moved must also be known. It can be measured with the stage X shift control, or from the images. The

height difference, Z, is expressed in a direction parallel to that of the principal projector, which is perpendicular to the plane of both the photographs. P is the parallax, which is the left photo X-distance X_L minus the right photo X-distance X_R between any pair of points. $Z=PD/B$ (2) all measurements made in the micrograph being multiplied by the magnification M.

If more than a simple two point height difference is required, then a three-dimensional co-ordinate system must be constructed taking the principal point, whose position (where the electron beam is perpendicular to the image) must therefore be known, as the origin of the co-ordinate system

$$X_L = x_L \frac{(D - Z)}{D} \qquad (3)$$

$$\text{and} \quad Y_L = y_L \frac{(D - Z)}{D} \qquad (4)$$

where large X and Y are calculated and small x and y are measured co-ordinates, and L is a subscript referring to one of the photographs, for example, the one which would be viewed by the left eye if the images were viewed as a stereo-pair[33].

In addition to the foregoing requirements of known true working distance, D, and the position of the principal point in each image, from either of which an X, Y, Z co-ordinate frame can be erected, one needs to know the magnification M and the tilt angle difference, α, for the case where the stereo-pair is generated by tilting the specimen. The working formulae (see refs.33 and 35) are too complex to be workable without a programmable calculator and there is little point in reproducing them here, since a derivation has recently been published[35] and other derivations will be found elsewhere[13-17].

When working at high magnifications it is obvious that the angle included in the raster scanned across the specimen is small. As the magnification (or the working distance) increases one increasingly approximates a parallel projective system. The assumption of parallel projection is thoroughly justified in theory and in practice at magnifications above 1000X - the "errors" involved in making this assumption have recently been tabulated[36]. If the specimen is only tilted about the Y axis to generate X parallaxes, then Y co-ordinates are unaltered and are the same in both images. The X distances (suffixed L for the photo of the specimen in the less tilted position, to be viewed by the left eye, and R for the opposite, the right photo[33,34]) between corresponding pairs of points in the images are used together to compute the Z co-ordinate (again suffixed L or R, relative to the L or R photo, either of which may be chosen). Since, however, the photo of the specimen in the more tilted position will show parts of the specimen which cannot be "heighted" because they are not represented in both photos, I believe that it is sensible to choose the less tilted photo as the one which to use as a map. X_L co-ordinates are then used, uncorrected, like Y_L

$$\text{and} \quad Z_L = (X_L \cos \alpha - X_R)/\sin \alpha \qquad (5)$$

$$\text{or} \quad Z_L = (X_L/\tan \alpha) - (X_R/\sin \alpha) \qquad (6)$$

all distances measured in the micrographs being divided by the magnification, M, or the final answer divided by M, as suits one's practice.

It is also sometimes convenient to use an imaginary central photograph (suffixed C) taken at a tilt angle half way intermediate between the less and more tilted positions as an X, Y, datum plane against which to erect the Z direction.

$$Y_C = Y_L = Y_R$$

$$X_C = \frac{(X_L + Y_R)}{2}$$

$$Z_C = \frac{X_L + X_R}{2 \sin \alpha/2} = p.k. \qquad (7)$$

where k = a constant = $\frac{1}{2 \sin \alpha/2}$

Again, all measured values may be divided by the magnification, or the final answer divided by the magnification, as seems convenient to the user.

The simplicity of the above relationships holds great appeal; the drawback is that central photos do not normally exist. Neither do plotting machines which give this solution mechanically, although they would be easily constructed.

To measure the magnification, M: use a cross ruled diffraction grating standard[14,15] photographed and processed under the same conditions as the stereo-pair micrographs. This procedure has the advantage that it can be used to control the distortion of the raster in the SEM, the CRT display raster, and any due to the camera and enlarger optics. The stage X, Y controls can also be used to calibrate the magnification[37].

To measure the principal distance, D: using a suitable test object and the stage Z control, make a plot of the width of the field, W, scanned on the specimen as its distance from the final lens is changed either side of a mean focus level, but do not alter the focus, i.e. do not change the excitation of the final lens as you perform this procedure[38]. Naturally, your image will be out of focus either side of focus, but with a small final aperture and at a low magnification you will be able to measure over a large range of Z movement. Extrapolating the W plot line back to zero on the Z axis (the perspective center) gives the distance from the chosen focus level to the raster center. Note that D is only required for low magnification stereophotogrammetric work and the number of such applications so far published in the literature is vanishingly small.

Measurement of the stage tilt angle difference, α, is only a problem because most specimen stages are not designed to cope with producing an accurately measured difference value. They are not even equipped with verniers! If you have a good stage, no need to worry; but the proportional accuracy of your Z calculation is the same as your proportional accuracy in tilt angle measurement. Probably, therefore, most users would like to aim for better than 1%, or better than a tenth of a degree over ten degrees. Both set values of the tilt angle should be approached from the same direction to avoid hysteresis and slack

problems[25]. If the stage tilt control knob has to drive a separate tilt indicator, then avoid using numbers ending with the digit 0, since extra strain (and hence mechanical variability) has to be used to turn over two digits at once . It is therefore sensible to use tilt values ending in the digit 5 with this kind of stage.

If you have to resort to calibrating the stage tilt, remember that with certain stage designs the X and Z controls which one uses to reposition the specimen after tilting may also introduce tilts around the Y axis, so these need to be calibrated too[25]. Methods for calibration include
(1) optical lever methods[25],
(2) crystallographic orientation, e.g. Kikuchi[39] or chaneling pattern analysis, and
(3) using an object of known height (see equation (1)).
In the latter case, the method due to M. Bode (personal communication) has a lot to recommend it. If spherical objects are mounted on a flat ground(as long as the sphere has not settled into a groove or pit!) the height is equal to the width of the sphere. If the size of the sphere is also known, it can also serve as the magnification calibration standard.

Measurement of the image rotation[40,41]: it is easier to choose a standard working distance at which the raster is square with the X and Z stage axes than to have to mount the stereo micrographs in a rotated position for measurement. If this is not convenient, and no scan rotation facility is available, then one should calibrate the rotation of the image against the final lens current (focus) setting using a straight edged object set parallel with the tilt axis[41].

Measurement of X_L and X_R: can be simply performed with a ruler, and the beginner is advised to practise such measurements as an aid to understanding what stereophotogrammetry is all about. It is only necessary to remember to measure X coordinates as such, i.e. use only the X component of a distance between two points when entering X in the simple formulae in equations 5, 6 and 7.

One of the bugbears of SEM images is the distortion[42-46] of the display - which is mainly due to the distortion of the display CRT but may be contributed to by distortion of the raster scanned across the specimen in the SEM, increasingly with decreasing magnification, i.e. with increasing angle of scan across the specimen[43,46]. A normal incidence photograph of a cross ruled grating standard taken at the same magnification as the stereo-pairs can be used as an X,Y co-ordinate and measurement system which eliminates distortion problems[14,34,46]. Measurements are made in terms of the number of grid intervals at the scale of the photograph of the grid. At high magnifications, it is safe to ignore the SEM raster distortion, and to assume that all the distortion is due to the non-linearity of the record CRT[46]. If the record CRT is driven by a digital scan generator, or the line direction can be varied to give either horizontal or vertical lines, it is possible to write a self-distortion grid[34,42] on each photograph and to use this grid spacing as a basis for measurement[34]. If the SEM scan can be driven by the scan generator of an image analysing computer, linear measurements can be obtained directly from each image in terms of the number of picture points or pixels[23,47]. A recent evolution in this direction is described by K.C.A Smith and D. Holburn[48] (this conference). They have constructed an arrangement whereby the two stereo-pair images can be stored, pixel by pixel, on floppy discs, for re-display at any later time. To facilitate the process of memorising which features are to be measured, the images can be interchanged between two display CRTs, one of which is equipped with a cursor whose X,Y co-ordinates can be fed to a computer.

STAGE III. STEREOPHOTOGRAMMETRY, STEREOPLOTTING AND STEREOMODELLING

In this section I propose to deal with instru-ments which produce digital or analog analytical data from stereo-pair photographs and involve the essential intervention of the stereoscopic observation process. All such instruments and methods make use of the principle of the floating mark or dot. The way in which the floating mark works will be clearly understood following a demonstration.

To demonstrate the floating mark, use 3 projector channels, 2 of which are the twin projector channels of the polaroid projector which display the stereo-pair images, coded by the 90° crossed vibration directions of the 045° and 135° oriented polaroid filters. The third projector is used to project a light dot on to the same screen. The light dot can be made by making a pin hole in a piece of Al foil bound in a 35 mm lantern slide mount. This projector has no polaroid filter, so that the dot image is seen equally through both of the polaroid filters of the viewing spectacles, and it appears to be in the plane of the metal-lised viewing screen. It will be found that, if the light dot is shone on to the same, overlapped feature area, then it will appear to lie in the surface of the stereoscopic optical model which results from the fusion of the stereo-projector, the light dot can be made to float in front of, or behind the surface of the stereoscopic optical model. (In borrowing from the jargon of air survey photogrammetry, we say that the floating dot "can be brought into contact with the ground"). In fact, it is easy to see that in this case all over-lapped features of the left and right images lie in the plane of the projection screen, and that what we are doing is merely pinpointing one of the overlapped areas. We can move the light dot around to join up all the features which overlap because they have the same height in the stereo-optical model - and thus produce a contour line. By changing the separation of the two images by regular increments, we could trace out a family of such contours to make a contour map. By measuring the lateral shift applied to the X-separation screw whilst moving the light dot in a straight line across the screen, we have a parallax value which can be used to draw out a profile line.

The "floating dot" is merely a device by which overlapping identical features can be identified with great precision. In its more u-sual form, in most photogrammetric instruments, it consists of two components: either black marks or light dots superimposed upon each member of the stereo-pair or injected into the optical train through which the images are viewed[2,3]. Each "half of the dot" can be moved with respect to the image to which it relates, and the two halves can be moved with respect to one another so that they can overlie features which lie at different separation distances.

Specimen X,Y and Z co-ordinates can be measured directly using the specimen stage X,Y and Z micrometers and without knowing the magnification principal distance or tilt angle if one uses a beam-tilt stereo system. This, conceptually, is the simplest photogrammetric instrument for SEM purposes which could exist - the SEM itself! In principle, the "left" and "right" beams in the SEM cross each other at a fixed point in space. This point can be made to correspond to the place on the specimen surface under observation and which is to be "heighted". The point is identified by introducing a floating mark. This mark may consist of two components - a bright cursor mark of equal intensity displayed in the same position in both the red and the green images[1,27]; or, it may be a single black mark placed in the center of the color TV monitor. Features are brought into register with the floating mark. Even without stereopsis, it is possible to adjust the specimen height (Z) and position (X,Y) so that the feature, in both the red and green images, is in register with the mark[1]. Using stereoscopic observation, this adjustment is achieved more rapidly and more reliably. The co-ordinates of the point brought to register with the floating mark are the X,Y,X specimen stage readings. The X,Y,Z controls could easily be attached to digitizing encoders, and the co-ordinates fed, at the press of a switch, to computer memory and/or tape store for later processing. Such a system would save greatly on photographic and accessory instrumental costs.

Stereo-pairs can be made into stereo models if the stereoscopic optical model (the 3-D image that you can see when a stereo-pair is properly "fused") and a real modelling material can be seen in the same 3-D space[2]. To understand how this is possible, think back to the example of a stereo pair projected via the polaroid system: the optical model usually lies half in front of, and half behind the screen. This modelling method is brought into reality by constructing a four mirror stereoscope (like the Stereosketch[2], now manufactured by Cartographic Engineering Ltd., Landford Manor, nr. Salisbury. Wiltshire, England) in which the first pair are semi-transarent so that one can see through them into the space where the stereoscopic optical model is found. This modelling space is illuminated so that one's hands, modelling tools and modelling clay, for example, can be seen at the appropriate light intensity compared with the 3-D image. Modelling clay is then formed to reproduce the optical model[2]. Alternatively, simple models of complex shapes such as filaments, for example, can be made by bending and twisting an appropriate filamentous modelling material, like soft wire, or a cylinder of modelling clay[49].

Mirror stereoscopes with parallax measuring facilities, and a combined photo-mounting platten attached to a parallel guidance mechanism so that all areas of the stereo model can be inspected and measured[2,3,50] are to be highly recommended as the simplest and most economical aid to making accurate "stereo" measurements. They are used for simple stereoviewing and for parallax measurements using the floating mark technique - the best kind is that in which the two components of the mark, two light dots, are injected through the semi-silvered second pair of mirrors of a four mirror stereoscope system (as in the SB 180 and SB 190 series instruments[2,3] now manufactured by Cartographic Engineering Ltd., Landford Manor, nr. Salisbury, Wiltshire, England, and obtainable through many different outlets worldwide). Simple height differences, like the depths of pits, grooves or depressions, are obtained with the unmodified instrument. However, two simple additions to such an instrument increase its usefulness enormously. Firstly, one may add a drawing pencil to the photoplatten assembly[2], when it is possible to trace out lines of equal parallax difference (height) by moving the photo-plattens by hand so as to keep the floating mark in contact with the ground.* (Contour maps produced in this way contain an inbuilt minor error - hence the reason for the evolution of more complex instruments - see next section). Secondly, one may add an X_L, Y_{LCR} co-ordinate measuring system, which may be as simple as a piece of graph-paper fastened underneath the plattens or as complex as a vernier mechanical stage[41]. It is then possible to produce $X_L + p = X_R$, to be entered into equations 1,2,5,6,7, to derive co-ordinates with reference to the L, C, or R photo datum planes. Profiles in the X direction are simply computed by moving along one line of the graph paper and recording the parallax screw reading at, say, 1 mm intervals[41]. These values are most simply plotted by multiplying them all by k in equation 7, to give the Z_C values; the parallax is added to each X_L point to give X_C.

Special plotting instruments. It is probable that most complex air survey photogrammetric plotting instruments could be used to obtain the necessary 3-D data from SEM images. Such instruments have been used for this purpose[6,9] but they are in the same range of cost as electron microscopes. Even special EM stereoplotters may be very expensive and complicated[50,51]. Some years ago, therefore, we proposed to develop a solution to the simple deficiency of the SB 185 type of instrument, the defect being that one ends up with parallax values only. By adding an X, Y co-ordinate measuring facility one only derives X_L and Y_{LCR}. By adding a mapping pencil one is drawing Z_C contours against the left datum plane, which is tilted by $\alpha/2$ against the desired center datum plane.

If one examines the parallel projection equations for, for example, the left photo datum plane

$$Z_L = \frac{X_L \cos \alpha - X_R}{\sin \alpha} \qquad (5)$$

one finds that there is no simple relationship

between height and parallax which one can use in a mechanical plotting instrument. If, however, we are to modify all the X_L co-ordinates by multiplying them, mechanically or optically, by cos α, then we have a modified parallax which is in constant relationahip to height differences. We achieved this solution by placing a cylindrical lens close to one of the images to apply the scale correction. The resulting instruments (EMPD2[4], [5]) were designed to deal with a constant tilt angle difference ($\alpha = 10°$ for EMPD1 and $9°34'$ for EMPD2) stereo-pairs and to give (1) immediate digital read-out of height differences in millimetres at the plot scale, (2) contour maps, and (3) profile sections in any chosen direction through the 3-D image, the contour interval being chosen in mm at the plot scale, and the profiles to true height scale for stereo-pairs taken with the designed tilt angle. The instruments were also suitable for TEM and SEM stereo work. The astonishing development of cheap computing facilities has probably been the reason why such opto-mechanical instruments, which are comparatively ever more expensive to produce, have not caught on. I would predict that the future development of instrumentation for SEM photogrammetry will be towards hybrid systems, which will feed out digitized positional information to a programmable calculator and plotter.

ACKNOWLEDGEMENTS

I am very grateful for the financial support of the SRC in the development of EMPD1 and for the particular assistance of Peter Howell, James B. Pawley and Hamish Ross.

REFERENCES

1. A. Boyde. Measurement of specimen height differences and beam tilt angles in anaglyph TV stereo SEM systems. SEM/1975, IIT Research Institute, Chicago, IL. 60616, pp.
2. A. Boyde. A single-stage carbon replica method and some related techniques for the analysis of the electron microscope image. J.roy. microsc.Soc. 86, 1967, 359-370.
3. D.B. Martin: A new mirror stereoscope with optical means of measuring parallax. Proc. Int. Soc. Photogrammetry Commission VII, Paris Sept. 1966.
4. A. Boyde, H.F. Ross and W.B. Bucknall. Plotting instruments for use with images produced by scanning electron microscopes, pp 483-493 in Biostereometrics 74, Amer.Soc.Photogrammetry, Falls Church, Va., 1974.
5. A. Boyde and H.F. Ross. Photogrammetry and the SEM. Photogrammetric Record 8, 1975, 408-457.
6. J.D. Eick, L.N. Johnson and R.F. McGivern. Stereoscopic measurements on the microscale by combining scanning electron microscopy and photogrammetry, pp 89-110 in Close Range Photogrammetry, pubd. by American Society of Photogrammetry, Falls Church, Virginia 22046, U.S.A., 1971.
7. D.F. Maune. Photogrammetric self-calibration of a scanning electron microscope. Ph.D.

*See Figure 2.

Dissertation, Ohio State University Aug. 1973.

8. D.F. Maune. SEM photogrammetric calibration. SEM/IITRI/1975, 207-215.

9. R. Wood. The modification of a topographic plotter and its application in the three dimensional plotting of stereomicrographs. Photogrammetric Record 7, 1972, 454-465.

10. S. Kimoto, H. Hashimoto and T. Suganuma. Stereoscopic observation in scanning microscopy using multiple detectors. Proc. Electron Microprobe Symp. Washington D.C. 1964, 480-489, Wiley and Sons, New York, 1966.

11. E.P. George and V.N.E. Robinson. The influence of electron scattering on the detection of fine topographic detail in the SEM. SEM/IITRI/1977 Vol.1, 63-70.

12. J. Lebiedzik and E.W. White. Multiple detector method for quantitative determination of microtopography in the SEM. SEM/IITRI/1975, 181-188.

13. J.B.F. Cripps and H. Sang. Stereo height measurements in scanning electron microscopy. Rev.Sci.Instr. 41, 1970, 1825-1827.

14. J.E. Hilliard. Quantitative analysis of scanning electron micrographs. J.Microscopy, 95, 1972, 45-58.

15. G.S. Lane. Dimensional measurements. Chap.11 pp 219-238 in J.W.S. Hearle, J.T. Sparrow and P.M. Cross (Eds), The use of the scanning electron microscope, Pergamon Press, Oxford, 1972.

16. G. Piazzesi. Photogrammetry with the scanning electron microscope. J.Phys.(E). Sci.Instr. 6, 1973, 392-396.

17. N.K. Tovey. A general photogrammetric method for the analysis of scanning electron micrographs, pp 82-87 in Scanning electron microscopy: systems and applications 1973. Institute of Physics, London and Bristol, 1973

18. E.J. Chatfield, J. More and V.H. Nielsen. Stereoscopic scanning electron microscopy at T.V.scan rates. SEM/IITRI/1974, 117-124.

19. A.R. Dinnis. After lens deflection and its uses. SEM/IITRI/1971. 41-48.

20. A.R. Dinnis. Limiting factors in direct stereo viewing. pp 76-81 in Scanning electron microscopy: systems and applications. Institute of Physics, London and Bristol, 1973.

21. A. Boyde. A stereo-plotting device for SEM micrographs: and a real time 3-D system for the SEM. SEM/IITRI/1974, 93-100.

22. J.A. Westfall and J.W. Townsend. Stereo SEM applied to the study of feeding behaviour in Hydra. SEM/IITRI/1976 Vo.II, 563-568.

23. Y. Kato, S. Fukuhara and T. Komoda. Stereoscopic observation and three dimensional measurement for SEM. SEM/IITRI/1977 Vol.I, 41-48

24. J.A. Westfall and J.W. Townsend. Scanning electron stereomicroscopy of the gastrodermis of Hydra. SEM/IITRI/1977, Vol.II, 623-629.

25. A. Boyde. Calibration of the specimen stage of Stereoscan. BEDO 3 Remy Verlag, Munster, 403-410, 1970.

26. J.B. Pawley and A.Boyde. A robust micromanipulator for the SEM. J.Microscopy 103,1975, 265-270.

27. J.B. Pawley. Design and performance of presently available TV-rate stereo SEM systems. SEM/SEM Inc./1978 VolI, pp 157-166.

28. G. Neubauer and A. Schnitger: A new sterepscope for stereoscopic pictures of all sizes. BEDO 3. Remy Verlag, Munster, 411-414, 1970.

29. H.B. Haanstra. Stereophotography with the electron microscope. Philips Technical Review 27, 1966, 231-237.

30. M. Nemanic. Preparation of stereo slides from electron micrograph stereo-pairs. pp.135-147 in M.A. Hayat ed., Principles and Techniques of SEM Vol.I, Van Nostrand Reinhold, New York, 1974.

31. A. Boyde. Direct recording of stereoscopic images with the SEM by the anaglyph colour technique. Med.and Biol.Illustration 21, 1971, 130-133.

32. A. Tamarin and A. Boyde. 3-D anatomy of the 8-day mouse conceptus. J.Embryol.exp.Morph. 36, 1976, 157-196.

33. P.G.T. Howell and A. Boyde. Comparison of various methods for reducing measurements from stereo-pair scanning electron micrographs to "real 3-D data". SEM/IITRI/1972, 233-240.

34. A. Boyde. Quantitative photogrammetric analysis and qualitative stereoscopic analysis of SEM images. J.Microscopy 98, 1973, 452-471

35. P.G.T. Howell. The derivation of working formulae for SEM photogrammetry. Scanning 1, 1978, 230-232.

36. P.G.T. Howell. A theoretical approach to the errors in SEM photogrammetry. Scanning 1, 1978, 118-124.

37. C.W. Oatley. The scanning electron microscope. Cambridge University Press, Cambridge 1972.

38. A. Boyde. Determination of the principal distance and the location of the perspective centre in low magnification SEM photogrammetry. J.Microscopy 105, 97-105.

39. W. Kleinn. Quantitative electron microscopy. Part 2. Establishment of control of scaling the micrograph. Photogrammetric Engineering 31, 1965, 800-802.

40 R. Christenhuss and G. Pfefferkorn. Bild-Drehung und -Verzerrung beim Raster-Elektronenmikroskop Stereoscan. BEDO 1, Remy Verlag Munster, 129-140, 1968.

41. A. Boyde. Height measurements from stereo-pair scanning electron micrographs. BEDO 1, Remy Verlag Munster, 97-105, 1968.

42. O.C. Wells. The contruction of a SEM and its application to the study of fibers. Ph.D. Thesis University of Cambridge Sept. 1957.

43. G. Owen and W.C. Nixon. The effects of dynamic correction on post lens deflection in a scanning electron microscope. pp 22-27 in Scanning electron microscopy: systems and applications 1973. The Institute of Physics, London and Bristol.

44. S. Murray and A.H. Windle. Characterisation and correction of distortions in SEM micrographs. pp 88-93, in Scanning electron microscopy: systems and applications, 1973. Institute of Physics, London and Bristol.

45. S.A. Bradley, P.N. Thielen and J.E. Hilliard. The use of moire patterns to determine the source and degree of distortion in scanning

electron microscopes. J.Microscopy 103(1), 1975, 25-31.

46. P.G.T. Howell. A practical method for correction of distortions in SEM photogrammetry. SEM/IITRI/1975, 199-206.
47. A. Boyde.   Photogrammetry of stereo-pair SEM images using separate measurements from the two images. SEM/IITRI/1974, 101-108.
48. K.C.A. Smith and D.M.Holburn . On-line topographic analysis in the SEM. SEM/SEM Inc. Part II, this volume, p 47-52
49. A. Boyde and R.A.D. Williams. Estimation of the volume of bacterial cells by scanning electron microscopy. Archs.oral Biol. 16, 1971, 259-267.
50. R. Burkhardt. Quantitative electron microscopy. Part 3. Determination of the third dimension of objects by stereoscopy. Photogrammetric Engineering 31, 1965, 802-806.
51. R. Burkhardt and J.G. Helmcke. Der Elmigraph I, ein photogrammetrisches Auswertgerät mit selbstkartier ng für elektronenmikroskopische Stereo aufnahmen   Proc.Int.Conf.Electron Microscopy, London, 1954, pp 651-658.

Appendix
SOME USEFUL SOURCES OF EQUIPMENT
(Names and addresses of U.S. suppliers, courtesy Dr. James B. Pawley)

SB 185 and SB 190 mirror stereoscope with parallax measuring facility, (text refs.2,3,41) and the Stereosketch (text refs.3,34), formerly manufactured by Hilger and Watts Ltd. and later by Rank Precision Industries Ltd., is now manufactured by Cartographic Engineering Ltd., Landford Manor, nr. Salisbury, Wiltshire SP5 2EW, England. The latter firm also manufactures the Electron Micrograph Plotting Device Mk.II. (EMPD2 text refs.4,5,21,34).

The above equipment is available in the U.K. from the manufacturers or from Agar Aids 66a Cambridge Rd., Stansted, Essex CM24 8DA, England; or in the U.S.A. from 1) PSI Instruments 7840 Airpark Rd, Gaithersburg, Maryland 20760, 2) Commonwealth Scientific Corporation, 500 Pendleton Street, Alexandria, Virginia 22314, or 3) Ladd Research Industries Inc., P.O. Box 901, Burlington, Vermont 05401.

Other suppliers of stereo (measuring) equipment in the U.S.A. are 1) Eugene Dietzgen Co., 50 West 44th St., N.W., New York, N.Y. 10036, 2) Wild Heerbrugg Instruments, 465 Smith Street, Farmingdale, N.Y. 11735, and 3) Alan Gordon Enterprises, Inc., 5362 Cahuenga Blvd.. North Hollywood, CA. 91601.

Old aerial survey plotting machines suitable for micrograph analysis are often available at nominal cost to educational institutions from the U.S. Geological Survey in the U.S.A.

In the U.K. the following firm specialises in supplying digitizing equipment for photogrammetric instruments: Surveying and Scientific Instruments Ltd., 1 Moot Lane, Downton, Wiltshire, England.

Stereo slides, cameras, new and secondhand stereoprojectors are available from Duval Studios, 217 High Rd., Chiswick, London, W.4, England.

Various stereo supplies in the U.S.A. and

details for the preparation of stereo slides are given by M. Nemanic, Preparation of stereo slides from electron micrograph stereo-pairs, pp 135-148 in (M. Hayat ed.) Principles and techniques of SEM Van Nostrand Reinhold, New York, 1974; and see Frontispiece of same volume for printed anaglyphs.

Discussion with Reviewers

G.F. Bahr: It is my experience from showing stereo-pairs that about one fourth of the audience and more of spectators cannot see stereo; even after a time of accommodation to the relatively new visual situation.   That is most obvious when the person looking through a stereoviewer, with interpupillary distance well adjusted, moves two loose pairs back and forth and finally states that he sees 3-D.   In fact he has given up.
Author: It is generally accepted that the non-stereo proportion of the population is about 10%, but I agree with you, it may be higher.

G.F. Bahr: How does change of the signal intensity upon stereo tilt influence the ability to measure i.e. to discern structural detail?   Measurements are based on visual judgment of parallax by the person measuring.   Furthermore, tilting the specimen may result in an out-of-focus condition. Refocusing at a different level, may change image properties.   Please clarify how and if measurements are influenced.
Author: The change of signal intensity due to tilting the specimen towards or away from the electron collector in the SEM is rarely significant for practical purposes.   This is because one usually operates at near normal incidence to the mean surface plane, and only tilts through a few degrees.   If the mean surface were tilted at, say, 70° and 78° or at 0° and 30° for the two members of the stereo-pair, one would have a problem. It is obviously preferable to have a tilting stage which tilts the specimen so that the mean signal level will not change at all, i.e. the tilt axis should point to the electron collector and not lie perpendicular to it.   Again, it is obvious that one cannot measure what one cannot see: therefore it is important to make sure that features seen in one view are not eclipsed by other features in the second member of the stereo-pair.

Measurements are not based on visual judgment of parallax.   They are either based on the observer's ability to recognise matching surface details in two images if, for example, the measurements are being made with a ruler on the two separate images: or they are based on the principle that the floating mark of the stereophotogrammetric instrument appears to lie in the plane of the ground.   If the two components of the floating mark lie in the plane of the photo, there will be no discrepancy between observers.   There is no discrepancy between observers using properly designed photogrammetric equipment: we would have the ridiculous situation that contour maps based on air survey photographs would differ from one machine operator to the next, if this were the case.

If tilting the specimen causes portions of it to go out of focus, it is certainly the case that the specimen does not satisfy the condition of lying nearly perpendicular to the beam. However, it rarely matters that the top and bottom fringes are slightly out of focus. Many authors have noted that one member of a stereo-pair can be significantly blurred without that affecting stereoscopic acuity. Furthermore, one can always change to a smaller final aperture to increase the depth of focus. If the worst comes to the worst, one could use an auto focus system, although it must be noted that the apparent perspective center is then changed (see Figs.1,2,3 in text ref.38).

Refocusing at a different level is naturally avoided like the plague because one would then have used a different focal length lens for the second image. Refocusing must be done if the specimen area does not lie exactly in the stage tilt axis, and it must be done by a mechanical Z shift using the stage controls, as described in the text. This creates no problems: most people are used to focusing optical microscopes with an analogous mechanism.

G.F. Bahr: For the observation of a continuously variable portion of the specimen, I am reminded of an ingenious system in which the frequency (about 60 Hz) at which the beam was tilted between the two views is matched by the opening and closure of the respective optical path in a binocular viewer. This concept goes back to Dr. M. Renfrew of the University of Oregon, I believe. The method appears to be a cousin of the flicker display.
Author: You describe a flicker display. The concept dates back before the turn of the century.

G.F. Bahr: The distortion (of whatever source) in SEM can be shown by a cross-ruled grating as is mentioned in the text. An old method of correcting linear distortions is the very practical and inexpensive method of photocopying grating and images onto a tiltable easel or with a tiltable projection system in an enlarger. Many make of enlarger are provided by the manufacturers with a facility to tilt.
Author: The simple distortions you describe are those in the image of a plane surface due to the fact that it is tilted. These are not a problem. It is by measuring them that we derive the 3-D coordinates. The important corrections to be applied are those for pincushion, barrel and other distortions catalogued by S. Murray and A.H. Windle (text ref.44). These are best dealt with by storing a matrix of corrections for all parts of the record CRT for particular combinations of kV, working distance, and magnification (see text refs.44 and 46).

G.F. Bahr: How is resolution limited by the discrete steps of a digitizing potentiometer?
Author: The resolution of commercially available X,Y digitizing beds is ca. 0.1 mm or the same as the line width of a 100 mm wide, 1000 line image, and considerably less than the width of a drawn line. This limit could have no practical effect on resolution, assuming that the parallax values were read separately. There would, however, be a practical limit to the "height" resolution given by a digital encoder for the parallax reading.

For example, suppose we work with 10° tilted pairs (a usual compromise for measurement work) and we want a height resolution at the photo scale of 0.1 mm ($\pm$ 0.05 mm), then the step size of the digital encoder should be one sixth of this, i.e. less than 16.6 μm.

If X_L and X_R are measured in discrete units like picture points (pixels), and the parallaxes derived as $X_L - X_R$ from these data, serious problems may arise. (See p.104 in ref.47 and Fig. 17 in ref.5)

R.H. Geiss: Where can one find a diagram showing the location of the perspective center in the SEM i.e. the point from which working distance (D, in your text) is to be measured?
Author: See figures 1 and 2 in text, ref.35 and Figs.1,2 and 3 in text ref.38.

Previous stereo tutorials might also help, e.g. Fig.2 in P.G.T. Howell, Taking, presenting and treating stereo data from the SEM, pp 698-706 in SEM/IITRI/1975: A. Boyde and P.G.T. Howell, Taking, presenting and treating stereo data from the SEM II, pp 571-579 in SEM/IITRI/1977, Vol.I.

SCANNING ELECTRON MICROSCOPY/1979/II
SEM Inc., AMF O'Hare, IL 60666, USA

THE OBSERVATION AND MEASUREMENT OF DISPLACEMENTS AND STRAIN BY STEREOIMAGING

D. L. Davidson

Southwest Research Institute
6220 Culebra Road
P.O. Brawer 28510
San Antonio, TX 78284

Abstract

The human visual system has the ability to image a localized area of displacements in an otherwise undisturbed field as the third dimension. This ability is utilized in imaging the displacement fields associated with a crack tip from photographs made in the unloaded and loaded states. A special loading stage for the SEM is used to make these comparison photographs. When a photograph of each condition of the crack tip area is placed in the stereoviewer, displacements in the eye axis can be imaged; therefore, it is necessary to view the photographs in two orientations 90° apart to obtain the total displacement magnitude and direction. By parallax bar measurement of the relative motion of points, these visualized displacements may be quantified and resultant , or total, displacement diagrams produced. Values of engineering strain may be deduced from these displacement diagrams by differentiation.

KEY WORDS: Stereopsis, Fatigue Crack, Crack Tip Displacements, Special Loading Stage, Strain, Photogrammetry

Introduction

In the study of cracking phenomena, such as are encountered in fatigue, creep, and stress corrosion, there is a need for studying the region very near the crack tip in as much detail as possible. Very near the tips of many cracks, there is a small region where the material deforms nonelastically which is called the plastic zone. During the study of fatigue crack plastic zones, a new technique for visualizing and measuring the plasticity caused by cyclic loading of the crack has been developed. The method involves viewing two photographs of the same area, but made under different conditions, in a stereo viewer. When this is done, displacements in the plane of the photograph (x,y) may be visualized as changes in the third (z-axis) dimension.

The Origins of Three-Dimensional Vision

Isaac Newton, in 1704, first proposed the basic notion as to how various biological systems have the ability to see depth. The process is very complex insofar as the details of how the brain perceives depth, and is still being elucidated, but Newton did correctly predict that part of the information gathered from one eye goes via fibers of the optic nerve (the contralateral fibers) to the optic chiasm where it comes together with fibers of the optic nerve from the other eye (the ipsilateral fibers) and proceeds to a part of the brain known as the lateral geniculate nucleus.[1] Pettigrew[1,2] gives more complete information on how the brain detects this information and processes it to construct the three-dimensional images we "see."

The study of this process of three-dimensional vision, or stereopsis, led Julesz[3] to the generation of random dot stereograms, which have regions that stand out clearly as floating above (or below) the plane of the figure when viewed stereographically, but which are invisible to monocular inspection. In this type of image, the sole basis of stereopsis is the disparity, along the axis of the eyes, between portions of the figures. This disparity between figures is translated into the three-dimensional images we "see" by the neurophysiology of the brain.

The ability to image material displacements due to localized plasticity at a crack tip has its origins in this unique capability of the eye-brain system to detect disparities in the same direction as the eye axis and reveal these disparities as the third dimension.

The utilization of stereopsis in the study of localized plasticity has the significant advantage that it allows a rapid visualization of:

(1) the location of where the plasticity is occurring
(2) the relative magnitude of the displacements
(3) the direction of the displacements.

It is possible to determine the direction of displacements because the mechanism of stereopsis is operative only in the direction (horizontal) of the axis of the eyes (a line connecting the eye centers). Displacements perpendicular to this direction (vertical) cannot be imaged, and may cause difficulties in imaging, although the visual system does have some tolerance for vertical displacements. Thus, by rotating the figures relative to the axis of the eyes, the maximum displacement magnitude may be determined.

Visualization of Localized Plasticity

The use of stereopsis for determining the displacements associated with localized plasticity has thus far been employed in the analysis of photographs made while using a cyclic loading stage designed for the scanning electron microscope.[4] This facility is capable of a loading magnitude of up to 4450 N statically, or 3780 N cyclically. It is hydraulically powered and controlled by a closed loop servo-system. Observed resolution is about 50 nm (500 Å) for static observations and 100 nm (1000 Å) for dynamic observations made at television scan rates. Although most operation is at 0.5 Hz or less, the stage can cycle up to 5 Hz. Real time observations of fatigue crack propagation in metals and polymers have been made up to 10,000 times magnification.

This loading facility may be used in essentially three ways:

(1) Displacements of the loaded specimen may be determined relative to the unloaded specimen
(2) Displacements of the loaded or unloaded specimen may be determined relative to the original specimen condition. For example, the displacements caused by the existence of the crack relative to the uncracked specimen may be determined.
(3) Displacements associated with time dependent deformation may be determined.

To visualize the displacements discussed above, place the two photographs to be compared in the stereoscope as illustrated in Figure 1(a). After adjusting the photographs for individual visual characteristics, it should be possible to image displacements in the plane of the photographs and along the eye axis as third dimension, or z-axis, values. How the eye-brain visual system accomplishes this imaging is shown in Figure 1b which indicates the relative motion

of points on either side of a crack in the loaded and unloaded conditions.

Interpretation of Images

Although the technique described provides a remarkably simple method of visualizing localized plasticity, interpretation of what is imaged is more complex. Whether the displacement of a particular point imaged moves up (toward the viewer) or down depends on which of the two photographs is placed on the left. Whereas this is not important in the simple imaging of displacements, operationally it is convenient to standardize on one way of comparison so that a sense of relative displacements can be generated through experience. It should also be noted again that the displacements imaged as the third dimension are in-plane displacements. Actual z-axis displacements cannot be imaged except by the usual use of stereopairs. It also is not possible to image total displacements of all points on the photographs simultaneously unless all motion has been along one axis of the photographs, which would be an unlikely finding for the case of localized plasticity. Engineering strain, the change in length per unit length, is often the desired quantity in plasticity studies, and this can be inferred while imaging the displacements by observing the rate of change of the displacements as the eyes move across the photographs.

Quantification

The ability to image localized plasticity rapidly has been a great aid in assessing microplasticity while conducting an experiment, but it is also highly desirable to be able to quantify what can be visualized. Because the stereopsis capability of the visual system is being utilized in a different sense than it is in normal stereographic photography, it is necessary to adopt a different method of analysis. Fortunately, the analysis is much simpler than in normal stereophotography, although the basic measurements are the same: A parallax bar is used to measure the distance between the same point on the two photographs, Figure 1(b), along one axis. If the plasticity is localized, there should be sets of points on the two photographs which have the same distance measurement, and which can provide a reference distance (x_0). By subtracting this distance from measurements made in the zone of plasticity, the displacements (Δx) may be measured, and their sense determined. Since plasticity is usually not specific to one axis, it is necessary to measure the displacements of a point along two orthogonal axes (the photographs are reoriented 90°, y_0 is established and Δy is measured). From these two measurements, the resultant in-plane displacement can be determined in both magnitude and direction. For almost any case of localized plasticity, there will almost surely be an out-of-plane, or z-axis, component of displacement, but this component cannot be measured without analysis of a true stereopair.

The measurement of displacements from a loaded/unloaded pair of photographs of a fatigue crack is shown in Figure 2. If these displacements

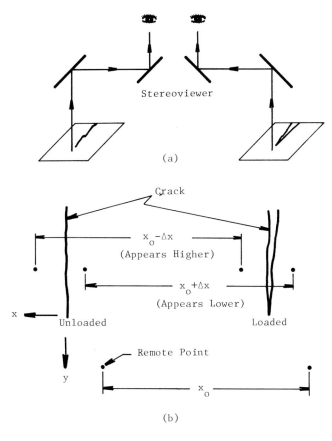

Stereoviewer

(a)

Crack

$x_o - \Delta x$

(Appears Higher)

$x_o + \Delta x$

(Appears Lower)

Unloaded   Loaded

Remote Point

x_o

(b)

Figure 1. (a) Positioning of a loaded/unloaded pair of photographs in the stereoviewer for visualizing localized plasticity. (b) The z-axis visualization of displacements is generated by comparison of the movements of points near the crack, either closer together or farther apart relative to some point remote to the crack which has not experienced any displacement.

are imaged in a stereoscope, the displacements seen in 2(c) and 2(d) will of course, be visualized; the resultants cannot. In understanding the deformation being studied, it is often useful to examine the displacements along various lines ahead and behind the crack tip. The total (or resultant) displacement as a function of x is shown for two positions in y in Figure 3(b) and 3(c). Since engineering strain is defined as the change in length, or displacement, per unit of original length, values of engineering strain may be deduced from these diagrams by differentiation. Figure 3(d) and 3(e) may be deduced from these diagrams by differentiation. Figure 3(d) and 3(e) show the result of graphical differentiation of 3(b) and 3(c), respectively.

Interpretation of the results of these displacement and strain values is complex. If D = resultant displacement,

$$D = \sqrt{D_x^2 + D_y^2},$$

then the tensor values of strains (ε) could be defined

$$\varepsilon_{xx} = \frac{dD_x}{dx} \; , \; \varepsilon_{yy} = \frac{dD_y}{dx} \; , \; \varepsilon_{xy} = \frac{\partial D_x}{\partial y}$$

The principal strain

$$\varepsilon_1 = \frac{dD}{dr}$$

where r is the direction of D, could also be defined. At this point in time, only a few cases have been examined using the methods illustrated in Figures 2 and 3, and these have been relatively simple cases to interpret; considerably more effort on interpreting the obtainable results is surely required.

To obtain results such as shown in Figures 2 and 3, a laborious point by point analysis of photographs is required. The data is then hand plotted and graphically differentialted to obtain strains. The accuracy of the results, particularly of the differentials, depends upon the quantity of data taken, i.e., the density of measured points, which is a direct function of the time and labor available for analysis. It is possible to use equipment designed for reducing aerial photographs to topographic maps to digitize the position and displacement of all points, and then to handle the plotting and differentiation digitally, and efforts to implement this degree of automation are underway at this time. Difficulties in finding proper analysis equipment have been encountered because of the general case where displacements can occur in both the x and y axes. A aircraft moving along a track making overlapping photographs to be used in aerial photogrammetry can yaw, wing dip, climb or dive between photographs, and most aerial stereographic plotters are designed to compensate for these motions. A few instruments exist which can translate one photograph in the x or y direction relative to the other, which is what is needed for the displacement analysis.

Display of the data derived from pairs of photographs, as shown in Figure 2 and 3, can become quite complex if used in conjunction with normal photogrammetric measurements of out-of plane displacements. For this case, it would be necessary to derive a display more sophisticated than anything yet considered.

Discussion

The technique described in the foregoing sections has its advantages and limitations. So far, only photographs taken while using the SEM cyclic loading stage[4] have been subjected to analysis using the technique, but other uses are also conceivable. Non-axial displacements in a tensile sample have been visualized from light-optical photographs, thus it is possible to use optical microscopy as well as electron microscopy. On the other end of the scale, there should be no reason why the deformation of macrostructures, such as dams, bridges, mine mouths, earthquake fault lines, etc should not be imaged, so long as the displacements are localized. Conversely, it may also be possible to use the technique on microstructures of specimens which will not fit into an SEM through the techniques of replication.

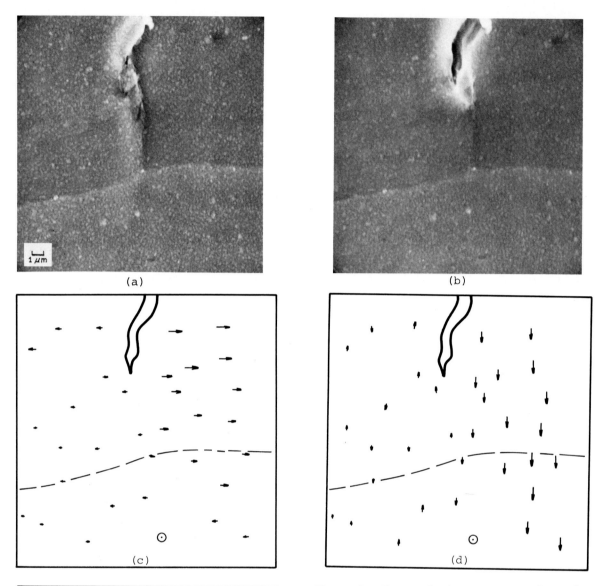

Figure 2. The crack tip at a stress intensity of (a) K = 2MN/m$^{3/2}$ and (b) K = 9.7 MN/M$^{3/2}$. The displacement diagrams for (c) Δx and (d) Δy, and (e) the resultant diagram. 2024-T4 aluminum alloy, crack propagation in vacuum of 10^{-4} torr.

It may be possible to examine replicas in the transmission electron microscope at higher resolution than in the SEM, or positive replicas[5] may be made for the SEM which have as good as, or better, resolution than the original specimen. This latter use would allow small areas of very large structures suspected of deformation to be examined.

The sensitivity of the stereoimaging and measurement technique described is dependent upon a number of factors:

(1) The accuracy to which parallax may be measured, which is currently approximately 0.05 mm;

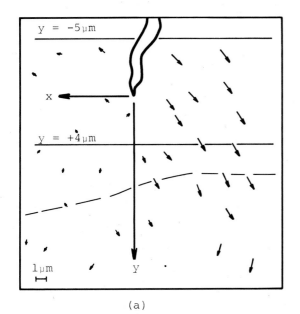

(a)

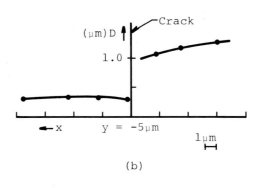

(b)

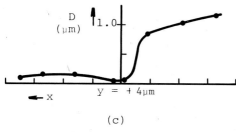

(c)

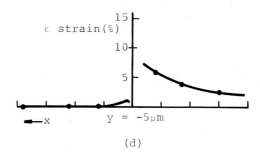

(d)

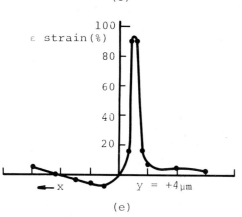

(e)

Figure 3. The resultant displacements D in (a) are shown as a function of x for two values of y: (b) for y = -5 μm (relative to the crack tip) and (c) y = +4 μm. By differentiating, a strain $(\varepsilon) = \frac{dD}{dx}$ may be derived : (d) from (b) and (e) from (c). 2024-T4 aluminum alloy, crack propagating in vacuum.

(2) The magnification of the photographs being used, which is currently limited by the spatial extent of the displacements of interest, since it is necessary to include an undeformed region, within the photographic field for reference when making quantative measurements;

(3) The fineness of resolvable points on the photographs being analyzed will limit the spatial access of data, so that if there are only course features to observe, the density of quantifiable displacements will be limited;

(4) The distortion of the scanning and display systems of the SEM may introduce errors in the analysis.

For very small displacement fields, magnifications of 8×10^3 have been achieved, and parallax differences of 0.05 mm have been measured, giving a displacement resolution of 70 angstroms, although it is unlikely that this magnitude of resolution is actually useful, and on most specimens it will not be possible to create a fine enough density of sufficiently small topographic features to allow such small parallax measurements.

All SEM scanning and display systems are likely to have distortions, and care must be exercised so that these distortions do not affect the displacement measurements; in practice, it may not be possible to achieve this goal. If the area coverage of the two photographs is correlated very carefully, then it is at least possible to minimize these distortional effects. On the SEM in use with the cyclic loading facility described above, a specimen photographed, translated 2 to 3 cm and rephotographed, evidences a small displacement field, therefore distortion, which could only be entering through the scanning and recording system, but so far, this effect is very small, compared to the displacements of interest, and no effort has as yet been expended for devising a way of compensating for it.

When using the described technique with replicas, distortions resulting from the replicating process may occur, and it will be necessary to devise methods of identifying artifacts. If displacements in macrostructures are to be derived, imperfections in photographic lenses and enlargers might also be sources of distortion, and these must be considered.

Interpretation of the displacement diagrams generated may be difficult because of the new opportunities to visualize and measure complex conditions hitherto unseen, and it may be a challenge to future users of the technique described to explain what may now be imaged and measured.

Acknowledgements

I appreciate the better understanding of stereopsis achieved through a discussion with Prof. John D. Pettigrew of California Institute of Technology, and the assistance of Mr. Bill French of the American Society of Photogrammetry in the fruitless search of that literature for references to previous use of the technique described herein. The support of the Air Force Office of Scientific Research for research on fatigue crack propagation, during which the stereoimaging technique was discovered, is also gratefully acknowledged.

References

1. J. D. Pettigrew, The Neurophysiology of Binocular Vision, Scientific American, 227(2), 1972, pp 84-95.
2. J. D. Pettigrew, Stereo Visual Processing, Nature, 273 (4 May), 1978, pp 9-11.
3. B. Julesz, Texture and Visual Perception, Scientific American, 212(2), 1965, pp 38-48.
4. D. L. Davidson and A. Nagy, A Low-Frequency Cyclic Loading Stage for the SEM, J. Phys. E. Sci. Instruments, 11, 1978, pp 207-210.
5. J. Lankford and J. G. Barbee, SEM Characterization of Fatigue Crack Tip Deformation in Stainless Steel Using a Positive Replica Technique, J. of Materials Science, 9, 1974, pp 1906-1908.

DISCUSSION WITH REVIEWERS

Reviewer III: Have you considered graphic display of your data, such as discussed in the paper "Computer Films for Research" (Physics Today, January 1979, pp. 46-52)?

Author: The mathematical description of crack tip plasticity is still poorly developed, thus it is not possible to generate computer-based films. We are making real-time observations of crack propagation using more conventional techniques of television*, and have hopes of using the stereoimaging technique to produce animation sequences to go along with the dynamic observations so that we may come to understand the sequence of events which causes fatigue crack propagation.

(* See D. L. Davidson and J. Lankford, "Dynamic, Real-Time Fatigue Crack Propagation at High Resolution as Observed in the Scanning Electron Microscope," ASTM STP-675, J. Fong, editor, Am. Soc. Testing Materials, Philadelphia, 1979, in press.)

Reviewer I: The interpretation of an apparent z-axis displacement as a lateral deformation by stereoimaging assumes that the pair of photographs are identical except for such lateral deformations. Changes in specimen tilt, specimen orientation towards the collector or significant translation with respect to the optical axis of the microscope can produce stereoimages that reflect the z-axis contour of the specimen. What precautions are taken to assure that none of these effects contribute significantly to the interpretation of lateral deformations?

Author: With the in situ fixturing we are using (see text ref. 4), only x and y translations of any magnitude can occur; tilt and rotation are fixed. Occasionally a small amount of z-motion occurs upon changing the magnitude of the load, which requires refocusing slightly. Duplicate photographs have been taken with an intermediate translation between them in both the x and y directions, and it is possible to detect distortions in the scanning and display system this way. Thus, the procedure established for making photographs for quantitative displacement measurements is to fix the position of the feature of importance (the crack tip in our case) on the display screen at the same position before photographing. This does not completely eliminate errors due to display distortion, but it certainly minimizes them. For the system we have in use, display system distortions, although detectible, are very small in comparison to the displacement magnitudes generally being measured.

Reviewer II: What is the optimum distance between photographs for making displacement measurements on the stereoviewer? Should it be the separation of the eyes in the head of the viewer?

Author: Stereoimaging conditions depend upon the physiology, training and experience of the individual, and so far as I know, it is not possible to arrive at a standard positioning of the photographs for using the methods I have described.

Reviewer II: How much time was required to make the experimental measurements for Figures 2 and 3?

Author: Although it depends on what equipment is available for making the measurements, I can obtain the data like that shown in Figures 2 and 3, excluding preparation of the graphics, in about 4 hours (now that I have experience on the equipment). Persons reducing aerial survey photographs can do the work much faster. Soon we shall have our processing of data much more automated, so that the strain values are done by computational procedures, increasing both the accuracy and speed of the results.

Regardless of the equipment used for quantification, the information which can be derived easily by qualitatively visualizing the displacements, can only be laboriously quantified; the technique is rather equipment intensive if accurate work is to be done.

Reviewers II and III: Concerning the interpretation of the strain values, is it possible that one could assume linear homogeneous deformation for a small element of area and identify its strain state simply by its principal strains?
Author: Yes, in general. The photographs can be rotated until the maximum displacements are visualized in any one area of the photographs, and this will essentially give the maximum strain direction. One note of caution: whereas the eye-brain combination does a magnificent job of detecting changes in depth, the quantification of this information is perceived very poorly (see text ref. 1); in short, you can see it well, but you can measure it only poorly, without the aid of instrumentation. Also note that maximum strain detection by rotation can only be done one point at a time.

Reviewer III: You state in the Discussion that "the magnification of the photographs being used is limited by the spatial extent of the displacements of interst, since it is necessary to include an undeformed region..." How can this be true? Masking all but a portion of the photographs presented still allows the three-dimensional effect to be seen.
Author: Yes, it is true that the extent of coverage of the photograph is unrelated to the ability to visualize displacements. But quantitative measurements cannot be carried out without a reference state, and that is provided by portions of the crack tip region which have not deformed, which can only be observed by using a low enough magnification photograph to show an area outside the displacement field.

Reviewer III: You state in the discussion that "the fineness of the resolvable points on the photographs being analyzed will limit the spatial access of data, so that if there are only coarse features to observe, the density of quantifiable displacements will be limited." A stereo pair need not be in focus to be seen in stereo. In fact, whatever method the brain uses to visualize the 3-dimensional world, it has little to do with resolution. You can prove this to yourself by taking two stereo pairs of some surface-- say a fracture surface. Set up for the left and right pictures in the normal way except that after you take the in-focus picture, defocus the condenser lens by a large amount so that the image is just an unintelligible blur, and take a second photo. View the totally out-of-focus stereo pair first, and you will see that the general 3-dimensional character of the surface is easily seen. The in-focus picture allows for a more accurate determination of the surface, but stereo vision does not depend on resolution or the position of discrete points, as does photogrammetry.
Author: My statement in no way relates resolution to the ability to visualize 3-dimensionally; it was included to give a relatively common-sense notion: you cannot visualize or measure those things you cannot see; therefore, you have to prepare specimens in such a way that the surface has a texture. Experience has shown that a "granular" texture (points of resolution) is better than a cross grid of scratches (lines of resolution) because there is more opportunity to see the surface; you cannot see anything you cannot develop contrast from, or features on smooth surfaces are difficult to see. Specimen preparation is therefore important to the success of using the technique.

Reviewer III: Won't it be difficult to use the technique you describe on macrostructures because of the necessity to take two photographs of the structure from the same position?
Author: I can think of several problems with looking at macrostructures, but taking two photographs from the same place is not uppermost on the list. For example, the Rephotographic Project (see Life Magazine March 1979) was able to take photographs in 1977 from the exact place and even under the same lighting conditions as was used when the same place was photographed in 1877. To use a hypothetical situation: taking a picture from a ship in San Francisco harbor before and after earthquake damage to the Golden Gate Bridge might prove difficult because of the reason you state.

Reviewer III: Could it be that the pseudo surface that is imaged by your method is, in fact, an image of the strain tensor associated with a progressive crack?
Author: Figure DR-1 shows two photographs of a fatigue crack made at the same load magnitude but 174 cycles apart. What may be visualized using this pair is the displacement field caused by the crack progressing, in contrast to the effect of load magnitude (text Fig. 2). Unfortunately, what can be imaged in either case is not the strain tensor, unless a particular viewer can somehow differentiate the displacements visualized (see text section on Quantification) in his head. Arriving at the elements of a strain tensor, except ε_{zz}, ε_{xz} and ε_{yz}, is the principal use for which the technique was developed. This information cannot be directly visualized and requires quantification of the displacements, as explained in the text.

REVIEWERS: I   J. D. Fairing
             II E. F. Sturcken
             III W. C. Lane

See also page 86.

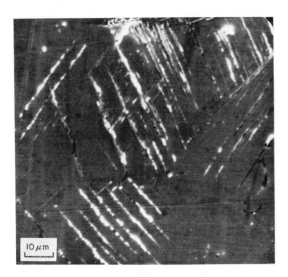

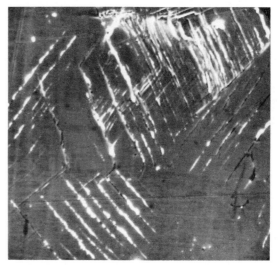

Figure DR-1. Two views of the crack tip region of a brass specimen at the same load, with the right photograph being taken 174 cycles after the left one. Stereoimaging allows visualization of the damage caused by the cyclic loading.

Author's Late Addition: Dr. W.C. Lane suggested an alternative method of mounting stereopairs for publication: "Mount the pairs in the right-hand column on adjoining odd numbered pages. Then the book can be placed under a stereo viewer with the left hand photo positioned by folding back the first page....". A discussion with the editor indicated that this was not possible at present. The stereo photos are not mounted in the usual way because of the restriction of stereo effect being realized in one direction only. To appreciate the total effect described in the text it is necessary to have the ability to rotate the photographs also and view the stereo effect at different orientations to obtain the information described. The interested reader may obtain copies of photos by contact the author.

SCANNING ELECTRON MICROSCOPY/1979/II
SEM Inc., AMF O'Hare, IL 60666, USA

A THEORY OF SURFACE-ORIGINATING CONTAMINATION AND A METHOD FOR ITS ELIMINATION

J. T. Fourie

National Physical Research Laboratory,
P. O. Box 395
Pretoria, 0001
South Africa

Abstract

Experiments are described which support the theory of the generation of an electric field by the interaction of primary electrons with thin foils. It is shown that the electric field is due to a positive charge on the foil resulting from the ejection of secondary electrons by the transmitted primary electrons. It is further shown that the electric field strength increases with the primary current, I_p. Experiments which indicate that secondary electron generation occurs up to about 1,5 µm from the primary beam perimeter, are discussed.

The theory of contamination by surface diffusion is formulated. It is indicated that the electric field around the illuminated region can exercise a drift force on molecules, thus influencing local concentrations of adsorbed hydrocarbon molecules. The theory indicates that the nature of this force is fundamentally different for small beam diameters when compared with large beam diameters, and that this could explain the high rates of contamination in narrow beams.

A low energy anti-contamination electron (LEACE) gun is described. It is shown that contamination can be eliminated totally with the use of this gun and an efficient cryo anti-contamination device.

KEY WORDS: Contamination, Scanning Transmission Electron Microscope, Secondary Electrons, Low Energy Anti-contamination Electron Gun, Surface Diffusion

Introduction

The total elimination of contamination in electron microscopes has become increasingly important with modern developments in the field of electron microscopy[1]. This is particularly true for small probe techniques such as scanning transmission electron microscopy (STEM), electron energy loss spectroscopy (EELS), so-called micro-micro diffraction and convergent beam diffraction. Recent articles[2,3] in this journal have been devoted to the subject of contamination and these are recommended for important considerations which are not dealt with here.

In all electron microscopes there are two means by which hydrocarbon molecules responsible for contamination can reach the electron-illuminated region. The first is by volume diffusion in the vacuum environment, leading to direct impingement in the illuminated region. The second is by the process of surface diffusion of hydrocarbon molecules already adsorbed on the specimen surface. It is very difficult to deal with the problem of contamination either theoretically or experimentally when both of these transport processes operate simultaneously (see e.g. Hart, Kassner and Maurin[4]). Therefore, it is essential to eliminate firstly the simplest component, namely, that of volume diffusion. This can be achieved by straightforward means of having either an ultra-high vacuum environment or an efficient cryo anti-contamination device (ACD) surrounding the sample. The surface diffusion component, which is the most difficult to eliminate as its nature varies with different samples and microscopes, represents the final problem in contamination and is dealt with in this paper.

The paper is divided into three main sections. Firstly, some unpublished experiments which have a direct bearing on the theory are discussed. Secondly, the theory, which will refer only to the surface diffusion component of contamination and which will further be limited to those aspects which relate specifically to small electron probes, is formulated. Thirdly, a method[5] for the total elimination of contamination by means of a low energy anti-contamination electron (LEACE) gun is described.

Experiments

Evidence of positive charge creation in thin foils by high energy electrons

When high energy electrons impinge on a thin foil, the majority are transmitted, but, in the process, secondary electrons are excited and escape from the foil, leaving behind a net positive charge[6]. For a conductor foil the charge would be neutralized by conduction of electrons from the grid bar. For an insulator foil or a conductor foil in which the surface of the irradiated region is covered with an insulating layer (such as contamination), the charge density associated with this phenomenon will continue to increase until a steady state condition is reached where the rate of production of positive charge is balanced by the flux of electrons from the grid bar to the illuminated region.

In practice, this effect can be demonstrated on an insulator foil, such as silicon monoxide, by means of STEM. The phenomenon shown in Fig. 1 was obtained by scanning a small central region of the

Fig. 1. STEM micrograph of region of a silicon monoxide foil which has been charged by ejection of secondary electrons from it.

----

foil-covered grid square for about 5 minutes, blanking the beam to allow setting of the magnification to a low value, and then, in a single 10 second scan, recording the micrograph. The fact that there is a charged region is indicated by the bright central spot in a totally dark background. This can be explained with reference to Fig. 2. A high charge density at the centre of the foil would have associated with it a series of concentric equipotential surfaces which would refract electrons according to Snell's Law in electron optics[7]. If $V_n > V_0$, where V_n is the potential at the centre of the foil and V_0 is the potential to which the primary electrons are accelerated (from -20 kV to ground potential V_0), the charge is positive and the scanning beam would be refracted as indicated in Fig. 2. That is, when the beam is in position a or to the left of it, the electrons in the beam would be refracted through an angle larger than that subtended by the aperture at the sample (one degree for the JSM-U3 microscope used in this experiment). Thus, the scanned region to the left of

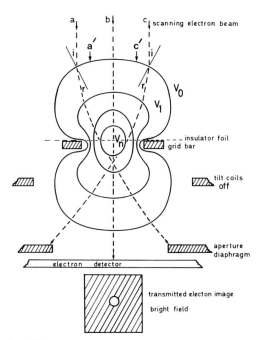

Fig. 2. Electrons are refracted at equipotential surfaces around the charged foil.

----

position a would appear dark. In the region a to c, the electrons would be striking the equipotential surfaces at angles sufficiently close to normal to be transmitted through the aperture. To the right of c, by the same argument, electrons would again, after refraction, be cut off by the aperture diaphragm, and that region of scan would appear dark.

The fact that the charge is positive can be proved by tilting the beam below the specimen as shown in Fig. 3. Using the same arguments as before, it follows that if the beam is tilted below the specimen as shown (i.e. anti-clockwise around an axis into the paper), beams toward the right hand of the scanned area would be transmitted and a pattern as in Fig. 4 would be obtained. If the charge were negative, the bright region would appear on the left-hand side for the tilt as specified. However, this is never observed.

Evidence of dynamic electric field effects

With further reference to Fig. 2 and for convenience of argument, assuming that there is only one equipotential surface which separates two spaces at potentials V_0 and V_n, it is obvious that if V_n were increased, the quantity $(V_n/V_0)^{\frac{1}{2}}$ in the expression for Snell's Law,

$$(V_n/V_0)^{\frac{1}{2}} = \sin i / \sin r \qquad (1)$$

would increase in magnitude. Therefore, for a fixed angle of incidence, i, the angle of refraction, r, would decrease to maintain the equality. Thus, the total deviation in the incident beam direction would increase. The beam at position a, for example, would then strike the aperture diaphragm at a position further out from the centre after refraction. The position of the beam where transmission through the aperture would *just* occur

scanning electron beam
i
i
r
r
V_0
V_1
V_n
insulator foil
grid bar
tilt coils
aperture diaphragm
electron detector
transmitted electon image
bright field

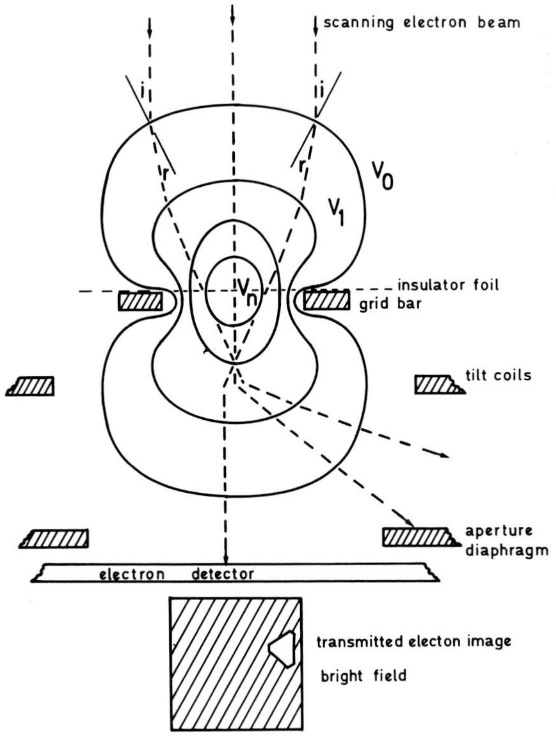

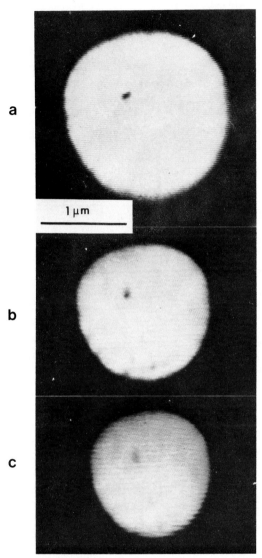

Fig. 3. Tilt of the beam after refraction at equipotential surfaces to distinguish between positive and negative charge.

10μm

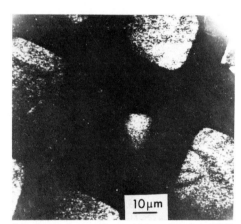

Fig. 4. Experimental verification by STEM of the existence of a positive charge on the foil surface.

would thus shift inwards from a to a', for example, i.e. to smaller i; and likewise from c to c'. This would mean that the diameter of the bright region would shrink if the potential V_n were increased.

An increase in V_n would occur if the primary beam current, I_p, and thus the rate of secondary electron production in the thin foil were increased. According to the argument, above, it can therefore be expected that the central bright region would shrink with increased I_p. This is shown to occur in Fig. 5 which represents electric field effects

a
1 μm
b
c

Fig.5. Refraction of electrons within the aperture below the sample (STEM mode) is limited to a central disc due to electric field effects. The diameter of the disc decreases with increasing I_p: a) $6{,}7 \cdot 10^{-12}$ A; b) $24 \cdot 10^{-12}$ A; c) $31 \cdot 10^{-12}$ A.

in a silicon monoxide foil, scanned at relatively high magnification and photographed during uninterrupted scanning. Here the sample was scanned at 20 kV in the STEM mode. The beam currents indicated were obtained by altering the strength of the condenser lens. The gun bias and apertures were retained unchanged throughout. The aperture between the sample and the primary electron (STEM) detector subtended an angle of 1° at the sample. The detector is situated about 400 mm below the sample.

Since V_n increases with I_p, and since the distance from the grid bar, from which the neutralising charge must flow in steady state, is constant, it follows that -dV/dr in the plane of the foil must increase with I_p. By definition, the electric intensity in the plane of the foil is given by the relation

$$E_r = -dV/dr \qquad (2)$$

Hence E_r will increase with I_p. From known parameters and equations (1) and (2) it can be estimated that V_n for the examples in Fig. 5 has a value of the order of $V_0 + 10^3$ volts, and that E_r is of the order of 10^9 volt/metre.

Evidence for secondary emission from a region extending further than 1 μm outside the primary electron beam

From theoretical considerations of experimental contamination configurations, it became apparent that secondary emission in thin foils probably occurred within a region ranging from within the primary beam to a considerable distance outside it. This fact was also noted recently by Broers et al.[8]. The situation was thus investigated experimentally as follows:

A JSM-U3 SEM operated in the secondary electron line-scan mode was used to record the secondary electron signal as a function of position on an insulator foil. The position of scan was across the edge of a double layer where the foil had been folded over onto a single layer region, as shown in a 100 kV TEM micrograph of the region, taken after the experiment, in Fig. 6. The edge region of importance for this description, across which the beam was traversed, is indicated by P in Fig. 6. This is a composite figure by necessity, because of the difference in scattering of

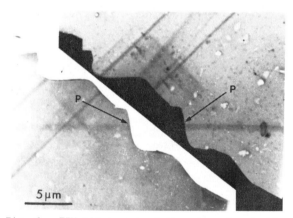

Fig. 6. TEM micrograph of the edge of a double layer of a silicon monoxide foil.
----

primary electrons within the double and single layers in TEM imaging. Thus two different negatives were exposed in photographing the edge. The exposure which was correct for the double layer (on the left) produced a totally overexposed negative for the single layer and this is the white region in the positive print. Similarly, correct exposure for the single layer (on the right) produced an underexposed negative for the double layer and this is the black region in the positive print. The two prints were mounted together, but displaced horizontally along the line P-P, so that

the edge detail remained fully visible. An edge on view, with the direction of the edge of the double layer being into the paper, at the position P is given diagramatically in Figs. 7a and 7b together with the corresponding experimental secondary signal recorded as a function of position. These figures will now be discussed.

1. For Fig. 7a it is clear that as the scanning beam approaches P from the left, the secondary signal begins to increase from a distance greater than 1 μm from the edge. It is also evident that the secondary signal reaches its lowest value after the primary beam moves from the double layer onto the single layer. However, the signal increases with distance to the right of P to a distance of about 1,3 μm. This result could be explained if high energy secondary electrons (referred to as tertiary electrons by Thornton[9]) were moving away from the scanning beam within the bulk of the foil and roughly parallel to the foil surface. These high energy secondaries would create many low energy secondaries close to the surface from which they could escape relatively easily. For the primary beam in a position to the left of P, the escape of low energy secondaries from the top layer would be enhanced as soon as the edge came within the range of high energy secondaries. Escape of secondaries from the lower layer would also be greatly enhanced when the range of high energy secondaries moved into the single layer. This is because the low energy secondaries produced by high energy secondaries in the region below the top layer would have a low probability of reaching the detector, but a high probability when they are produced in the single layer. Thus the signal would rise.

Immediately the primary beam moves to the right of P, a large proportion of high energy secondaries moving within and parallel to the foil, would move into the region to the left of P which is covered by the top layer of foil. Low energy secondaries generated there would have difficulty in escaping at the top surface of the second layer. Thus the signal would reach its lowest value. As the beam moves further to the right of P, conditions for secondaries to the left of the beam to reach the detector, would improve and the signal would rise.

2. For Fig. 7b, where the foil configuration has been inverted with respect to the primary beam and the secondary detector, a similar explanation can be given. As the primary beam approaches P from the left the rise in the signal would be less pronounced. This is because the secondaries which escape from the edge have to pass through the upper foil to reach the detector and only those with relatively high energies would be able to do so. When the beam is immediately to the right of P, some of the scattered primaries, or of the high energy secondaries, travelling at small angles to the primary beam (i.e. out of the foil) could impinge on the edge of the lower foil and generate secondaries with a resultant slight increase in the secondary signal due to those that penetrate the upper layer. This is observed, for as the beam moves further to the right there is a slight drop in signal.

3. Figures 8a and 8b depict the same type of

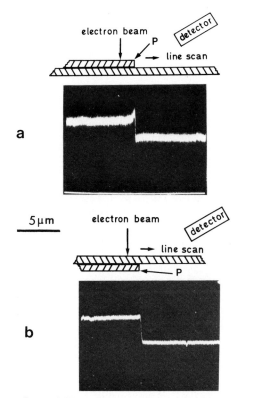

Figs. 7a and 7b. The secondary electron emission signal in the line scan mode; the scan is across the edge of the double layer in the orientation shown diagrammatically. (Silicon monoxide foil about 30 nm thick.)

----

experiment except that a grid bar, which would be totally opaque to secondary electrons, is used in place of the second layer. The observed effects are in agreement with those described in (1) and (2) above and can be explained similarly.

The ratio of secondary electrons reaching the detector from outside the primary beam to those from within

Calculations, according to the theory, developed in the following section, of contamination rate distribution profiles, and comparison with experimental profiles, indicated that, in thin foils, the required ratio of secondary electrons created *outside* the primary beam to those *within*, was about 5:1. The validity of this postulated ratio was therefore examined with respect to the measurements in Fig. 8. The solid curve in Fig. 9 is the re-plotted equivalent of the curve in Fig. 8a. The dotted curve was calculated on the basis of a simple model which assumes that the primary beam is surrounded by a circular region with a radius equal to the range of high energy secondaries (in this case 1,3 μm), in which low energy secondaries are generated uniformly. The fact that the experimental curve decreases by only about half the required amount as the grid bar is approached, could be attributed to extraneous effects. The most obvious of these is that when the beam is near the grid bar, secondaries and back scattered primaries would strike it and

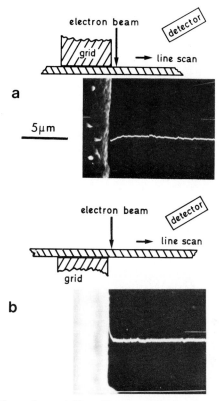

Figs. 8a and 8b. The secondary electron emission signal in the line scan mode; the scan is across the edge of the grid bar, with the foil in the orientation shown diagrammatically. (Evaporated carbon foil about 10 nm thick.)

----

generate secondaries which would increase the signal. Thus the decrease in signal in Fig. 9 is superimposed upon a general rise in signal with decreasing distance from the grid bar. This rise is observable in Fig. 7a and especially in Fig. 8a, where a slight rise beginning already at the right-hand edge of the photograph can clearly be discerned if the upper trace is compared with the lower trace for zero signal. The measured decrease

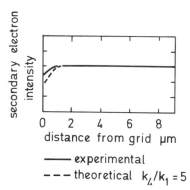

Fig. 9. Experimental measurements replotted from Fig. 8a together with a calculated curve of the secondary electron signal as a function of distance from the grid bar.

in signal would thus always be less than that actually due to the specific effect under consideration. It can therefore be concluded that the experimental measurements are not necessarily at variance with the ratio of 5:1 mentioned earlier.

Table of Symbols

Symbol	Meaning	Units
r	radial coordinate	μm
I_p	primary electron beam current	A
σ_p	charge generated per unit time by primary beam within radius r	Cs^{-1}
σ_s	charge generated per unit time by high energy secondaries within radius r	Cs^{-1}
σ	$\sigma_p + \sigma_s$	Cs^{-1}
I	surface current at r required to neutralize σ	A
i_r	surface current at r required to neutralize σ_p	A
i_s	surface current at r required to neutralize σ_s	A
E_r	radial electric field intensity in surface plane of sample	Vm^{-1}
j_r	specific two-dimensional surface current at radius r	Am^{-1}
ρ	two-dimensional radial surface resistivity	ohm
r_1	radius of primary electron beam	μm
σ_{p1}	charge generated per unit time by primary beam within radius r_1	Cs^{-1}
i_{r1}	surface current at r_1 required to neutralise σ_{p1}	A
k_1	secondary electron generation coefficient relevant to direct primary electron interaction	-
k_4	secondary electron generation coefficient relevant to the interaction of high energy secondary electrons with the sample while moving transversely to the primary beam within the bulk of the sample	-
r_3	maximum range, within the sample, of high energy secondary electrons	μm
i_{s3}	the magnitude of i_s at r_3	A
W_1 to W_5	constants	not relevant
m, b_1 and b_2	constants in equation for parabola	not relevant
F	electric drift force on polarised hydrocarbon molecules	N
α	polarizability of hydrocarbon molecules	$C^2\,mN^{-1}$
c	surface concentration of hydrocarbon molecules	m^{-2}
U	rate of contamination	$m^{-2}s^{-1}$
J	current density in primary beam	Am^{-2}
ϕ	reaction cross-section for electron-molecule interaction	$m^2\,C^{-1}$

Theory

In those electron microscopes where the vacuum is sufficiently high for the mean free path of volume diffusing molecules to be larger than the distance from the sample to the ACD, contamination which originates from the sample can *only* occur by surface diffusion. This is because molecules which escape from the attractive surface forces of the sample, travel in straight lines, thereafter, and thus cannot return to the sample. An understanding of contamination by surface diffusion is therefore applicable to all forms of sample-originating contamination.

Surface-diffusing hydrocarbon molecules can become polarised in a strong electric field. If the field is inhomogeneous, a drift force will act on the molecules driving them in the direction of the force. When the force is radial, local concentrations of molecules and, hence, rates of contamination can be affected significantly. In the following, the theory of these effects is considered with special emphasis on their relation to small beam diameters as used in scanning electron microscopy and related stationary electron probe techniques.

Model of theory

The theory which is derived with reference to Fig. 10, is based on the following model, many aspects of which are supported by experimental measurements:

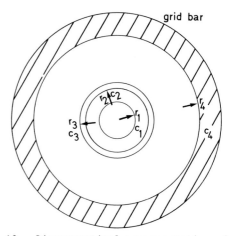

Fig. 10. Diagrammatical representation of important zones around an electron beam of radius r_1 incident on a thin foil surrounded by a conducting grid bar at radius r_4. Secondary electrons are produced in the region $r = 0$ to r_3.

A high energy electron beam, of diameter $2r_1$, is incident on, and transmitted through, an insulater thin foil. None of the primary electrons are retained in the foil but they generate low energy secondary electrons within a region of radius r_1. A proportion of these escape from the foil and this leads to a given rate of positive charge production at the surface of the foil. The intensity distribution in the primary beam and thus the rate of charge production per unit area is assumed to be uniform over the radius r_1. In addition to low energy secondary electrons, the primary electrons also create high energy secondary electrons. About 1 % of these will be travelling sufficiently parallel to the foil surface to travel a distance of about 1,5 μm within the bulk of the foil. In the process they will create many low energy secondaries which will have a high probability of escaping from the surface and leaving behind a positive charge. Thus, there will be this second mode of positive charge production which will extend as far as r_3, the maximum range of high energy secondary electrons.

After the primary beam has been stationary in one position for a given period, steady state conditions will exist. That is, the rate of charge production *within* a given radius, r, will be matched, at r, by an equivalent two-dimensional surface flux of electrons from the grid bar at r_4. The flux of electrons across the surface is assumed to occur after the electric field has become high enough to cause electrical break-down across the surface.

Derivation of theory

It was shown with reference to Fig. 5, that the electric field intensity in the surface plane, E_r, increased with I_p. Therefore, it can be postulated that E_r will be proportional to the *specific* surface current, j_r, where,

$$j_r = I/(2\pi r) \qquad (3)$$

Here I is the surface current required to neutralize the secondary electron associated charge, σ, produced per unit time within the radius r by the two mechanisms. Thus,

$$I = i_r + i_s \qquad (4)$$

where i_r is the surface current required to neutralize σ_p, the charge generated per unit time within the radius r by direct primary electron interaction, and i_s is the surface current required to neutralize σ_s, the charge generated per unit time within the radius r by high energy secondaries moving transversely to the primary beam.

The postulated proportionality between E_r and j_r, leads to

$$E_r = \rho j_r \qquad (5)$$

where the proportionality constant ρ is the two-dimensional electrical resistivity constant of the surface. If k_1 is the secondary emission coefficient relating to primary-secondary excitation for the region r = o to r_1, the current required to neutralize the total secondary charge created by primaries within that region per unit time is

$$i_{r_1} = k_1 I_p \qquad (6)$$

For $r < r_1$, it follows that

$$i_r = k_1 I_p(r^2/r_1^2) \qquad (7)$$

The formal calculation of i_s requires not only a knowledge of secondary electron production parameters which are not available at present, but also complicated mathematical procedures which are beyond the scope of the present study. However, expressions based on physically intuitive considerations can be used, and these are discussed below:

It is to be expected that high energy secondary electrons moving transversely from the centre of the primary beam at r = o, would immediately begin to create low energy secondaries. Since the path length of such high energy secondaries inside the foil is two orders of magnitude larger than that of the primaries which pass directly through the foil, there should be a corresponding proportionality in the number of low energy secondaries produced per electron by the two mechanisms. In addition, the efficiency with which the high energy secondaries produce other secondaries should increase as they decelerate in the foil. .

The foregoing would also apply to high energy secondaries produced at $r = r_1/2$ and $r = r_1$, for example. If the charge contributions of all the secondaries produced along the entire paths of high energy secondaries which originate between r = o to $r = r_1$ are summed, then clearly the highest value of di_s/dr would occur well outside r_1, namely at r_2, providing r_1 is in the μm and sub-μm range. For mathematical convenience it is assumed that i_s, the current required to neutralize σ_s, can be described by a parabolic relation

$$i_s = mr^2 + b_1r + b_2 \qquad (8)$$

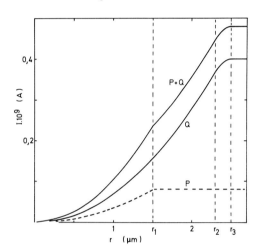

Fig. 11. The curves P and Q represent the surface current required to neutralize σ_p and σ_s respectively. The curve P + Q is the sum of the two curves and represents the surface current I required to neutralize σ. $r_1 = 1,5$ μm.

With reference to curve Q, Fig. 11, it can be seen that i_s is described by *two* parabolas, one for the zone $o \leqq r \leqq r_2$, where

$$d^2i_s/dr^2 > o \qquad (9),$$

and another for the zone $r_2 \leqq r \leqq r_3$, where

$$d^2i_s/dr^2 < o \qquad (10).$$

The constants m, b_1 and b_2 for both parabolas are obtained by specifying the values of r_2 and r_3, and i_{s_3} (the value of i_s at r_3). The magnitude of i_{s_3} is usually expressed in terms of i_{r_1}, equation (6). For example, for the present calculations, it was assumed that

$$i_{s_3} = 5i_{r_1} \qquad (11)$$

93

Further, if

$$i_{s_3} = k_4 I_p \qquad (12)$$

where k_4 is a constant, it follows from equations (6), (11) and (12) that for the present calculations,

$$k_4/k_1 = 5 \qquad (13),$$

as for the curves in Figs. 9, 11, 12 and 13.

It is assumed that the intervals (r_2-r_1) and (r_3-r_2) are constant regardless of the magnitude of r_1. An obvious improvement in this aspect of the theory would be to make r_2 and r_3 functions of r_1. However, because of the uncertainty concerning the actual function, this was not considered worthwhile at the present stage. It is not expected that the conclusions would be affected significantly by such further sophistication in the theory.

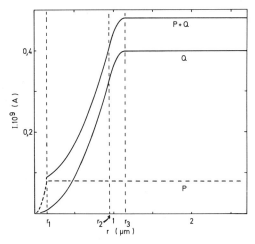

Fig. 12. The same as for Fig. 11 except that $r_1 = 0,15$ µm.

As illustration, curves of i_r, i_s and I as a function of r were calculated for primary beam radii of 1,5 µm and 0,15 µm. The results are shown in Figs. 11 and 12 where the currents required for the the two different modes of charge production (curves P and Q) and for the combined modes (curves (P + Q)) are plotted as a function of r. The curve E_r is shown in Fig. 13, where curves A and B have been derived from Figs. 11 and 12 respectively according to equations (3), (4) and (5).

The analytical function for E_r has the following form in the first two regions shown in Fig. 10.
For $o \leq r \leq r_1$,

$$E_r = w_1 r \qquad (14)$$

where w_1 is a constant. It follows that

$$dE_r/dr = w_1 \qquad (15)$$

that is, the gradient is constant and positive.
For $r_1 \leq r \leq r_2$
From (4),

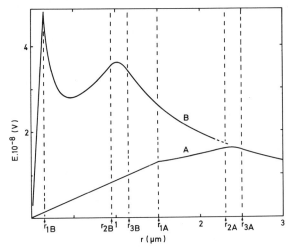

Fig. 13. The electric field intensity as a function of r for two beam diameters. Curves A and B are derived from Figs. 11 and 12 respectively according to the formula $E_r = \rho j_r$ (see text). For the same I_p, curve B becomes coincident with curve A at $r \geq r_{3A}$, i.e. beyond the radius of charge generation.

$$I = i_{r_1} + i_s \qquad (16)$$

Thus, from (6) and (8), (16) becomes

$$I = k_1 I_p + m r^2 + b_1 r + b_2 \qquad (17),$$

where $b_1 = o$ and $b_2 = o$.
Thus, from (3)

$$j_r = (1/2\pi)\{k_1 I_p/r + mr\} \qquad (18).$$

It follows from (5), that

$$E_r = w_2/r + w_3 r \qquad (19),$$

where w_2 and w_3 are constants. Thus,

$$dE_r/dr = -w_2/(r^2) + w_3 \qquad (20)$$

For $r_2 \leq r \leq r_3$
As in (17)

$$I = k_1 I_p + m r^2 + b_1 r + b_2 \qquad (21),$$

but here $b \neq o$ and $c \neq o$.
Thus

$$j_r = (1/2\pi)\{k_1 I_p/r + mr + b_1 + b_2/r\} \qquad (22)$$

i.e.

$$E = w_4/r + w_3 r + w_5 \qquad (23),$$

where w_3, w_4 and w_5 are constants, and

$$dE_r/dr = -w_4/(r^2) + w_3 \qquad (24).$$

In both (20) and (24) the gradient is determined by two terms. If r_1 were large, r could only assume large values and the negative term would be small. Thus, dE_r/dr would be positive as for curve A, Fig. 13. However, for small r_1, r could assume small values, and the negative term could thus dominate, as for curve B, Fig. 13, where dE_r/dr is negative immediately outside the primary beam.

The drift force on polarised molecules in the electric field E_r is given by

$$F = \alpha E_r(dE_r/dr) \qquad (25)$$

where α is the polarisability of the molecule. Clearly then, with reference to curve B, Fig. 13, there would be a drift force in the negative r-direction, forcing the molecules toward the primary beam and thus increasing the concentration of molecules right up to the primary beam perimeter. It is this property of the interaction of small diameter beams with thin foils which could be responsible for the high rate of contamination in such beams. For the beam which is an order of magnitude larger (curve A), the region adjoining the beam perimeter has only a repulsive drift force, thus reducing the concentration of molecules near the beam, even though there is a long range attraction for molecules.

When molecular drift due to electrical forces and molecular diffusion down concentration gradients are in a state of equilibrium, the Nernst-Einstein equation and Fick's First Law can be used to obtain a differential equation which describes the equilibrium. This equation has the following form:

$$-\partial c/\partial r + cF/(kT) = o \qquad (26)$$

where c is the concentration of molecules per unit surface area, k is Boltzmann's constant and T the temperature in Kelvin.

The equation can be solved to obtain an expression

$$c = c_n \exp(-1/(kT) \int_r^{r_n} Fdr) \qquad (27)$$

where n = 1, 2, 3 or 4.

By assigning an appropriate value to c_4, it is possible to compute in turn c_3, c_2, c_1 and finally c as a function of r within the primary beam.

For an electric field intensity distribution such as in curve A (Fig. 13), the highest surface concentration at equilibrium will be at the apex of the curve between r_{2A} and r_{3A}. This is because the drift force to the left of the apex tends to establish a positive concentration gradient (i.e. $dc/dr > o$) and that to the right a negative concentration gradient (i.e. $dc/dr < o$). If the parameters which determine the relative magnitudes of the drift forces to the left and right of the apex are chosen so that $(dc/dr)_{left} > (dc/dr)_{right}$, an increase in I_p would produce a decrease in concentration at r_{1A} notwithstanding an increase in concentration at the apex.

Similarly, with reference to curve B (Fig. 13), if the parameters are chosen so that the magnitude of E at the apex between r_{2B} and r_{3B} lies below the value of E at r_{1B}, the concentration gradients would be so matched that the highest concentration is at r_{1B}. An increase in I_p would then result in an increased concentration at r_{1B} because of the inter-relation of dc/dr immediately to the right of r_{1B} and to the left and right of the apex between r_{2B} and r_{3B}. These arguments are expressed more exactly in equation (27) from which the predicted phenomena can be calculated. It

should be emphasized that curves A and B (Fig.13) were calculated without any changes in parameters except for the parameter r_1. The rate of contami-

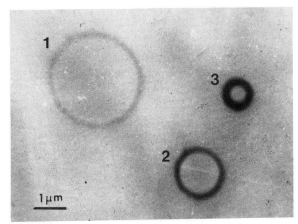

Fig. 14a. Experimental demonstration of beam size effect on contamination rates at constant J.

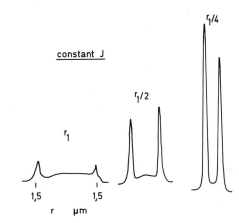

Fig. 14b. Densitometer profiles of the contamination configurations in Fig. 14a. Density of contamination measured along vertical axis.

----

nation, U, per unit area within the primary beam is given by

$$U = \phi Jc \qquad (28),$$

where, ϕ is the reaction cross-section for an electron-molecule interaction leading to precipitation of a carbon radical, and J is the current density in the primary beam.

The experimental observations in Figs. 14(a) and 14(b) are compared with a calculation based on equations (27) and (28) for various beam sizes but *constant* J in Fig. 14(c). It is clear that the rate of contamination in the small beam greatly exceeds that in the larger beams even though J is constant for both experiment and calculation. This result thus supports the theory.

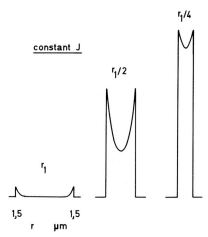

Fig. 14c. Densitometer profiles corresponding to those in Fig. 14b but calculated according to theory. Density of contamination measured along vertical axis.

The total elimination of contamination by means of a low energy anti-contamination electron (LEACE) gun

In a recent experiment[10] evidence was found for the drastic reduction in contamination in the presence of low energy electron illumination generated at a grid bar by the primary beam (Fig. 15). It was concluded that in the case of b, Fig. 15, the low energy electrons had impinged on the positively charged region of the foil in and around the beam, thus eliminating the drift force F, discussed in the foregoing. For this reason it was considered justifiable to construct a low energy electron gun which could be inserted through the X-ray take-off port of a Philips EM-200 CTEM. With this gun the specimen could be subjected to indirect bombardment with low energy electrons, as shown schematically in Fig. 16. The success of the LEACE gun was immediately apparent and a reduction of two orders of magnitude in contamination rates was achieved[5].

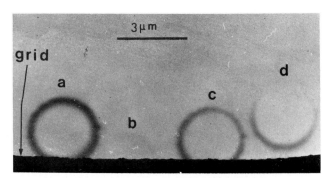

Fig. 15. The effect of secondary electron production at the grid bar in reducing contamination at b. Courtesy reference[10].

The gun was then installed in a standard JSM-U3 SEM through a suitable hole in the left-

hand side plate, on the same level as the sample, with the anode of the gun about 10 mm away from it. As the experiments were conducted in STEM, and therefore collection of the secondary signal was not required, the microscope could be equipped with a cryo-ACD similar to the one described by Rackham and Eades[11]. The ACD formed a totally

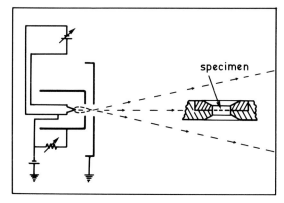

Fig. 16. A diagrammatical representation of the low energy anti-contamination electron (LEACE) gun. Courtesy reference[5].

----

enclosed volume around the sample except for 0,5 mm holes, top and bottom, to provide passage for the primary electron beam, and suitable front and side openings, for the insertion of the sample and the injection of electrons from the LEACE gun, respectively. Without this sophisticated ACD the LEACE gun was only partially effective, since, presumably, the flux of directly impinging hydrocarbon molecules from the microscope environment was of the same magnitude as the flux from surface diffusion. In the following, some examples of the effectiveness of the LEACE gun are given as well as an example of the quality of image obtainable in STEM with simultaneous operation of the LEACE gun.

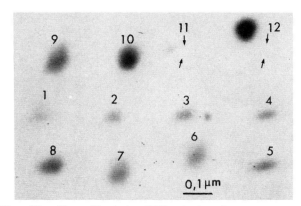

Fig. 17. The effect in STEM of the electric field around the primary beam in increasing the rate of contamination from positions 1-10. Note the total elimination of contamination at positions 11 and 12. The sample is a 4 nm carbon foil.

The history of the specimen represented in Fig. 17 is that it had been left in the evacuated microscope with the vacuum system isolated, for 15 hours. Thereafter the vacuum system was re-connected to the microscope which was then pumped at high vacuum ($6 \cdot 10^{-3}$ Pa) for 2 hours. The cryo-ACD had also been cooled for a period of 2 hours. The total system, including the specimen, should thus have been in a condition close to equilibrium. The specimen, a 4 nm thick carbon foil covered with MoO_3 particles for focusing purposes, was then illuminated for 60 seconds with a 25 kV electron beam of 30 nm diameter at position 1, Fig. 17. At 2 minute intervals the same experiment was repeated at positions 2-9. The beam was then blanked electrically for 30 minutes, during which time all vacuum conditions were maintained unchanged. Hereafter, the contamination spot in position 10 was produced under the same conditions used for spots 1-9. Immediately following this, the LEACE gun was switched on at 225 V and a beam current of about 25 μA, after which the focused high energy primary beam was stopped at position 11 for 60 seconds and then at position 12 for 120 seconds. Clearly there is no contamination whatsoever in either case. The micrograph was then produced during a 50 second single scan while the LEACE gun remained switched on.

The increase in contamination rate shown progressively from spots 1-9 in Fig. 17, is a clear indication of the electric field effect on the surface concentration of molecules. This aspect of the experiment has been repeated, with the same result. It can be deduced that the drift force predicted by theory continued to cause an increase in the concentration of molecules around the beam from the time it was stationary at position 1, until it was blanked at position 9. The fact that a tremendous increase in the surface concentration of molecules appears to have occurred during the time elapse between 9 and 10, indicates that, even with the beam blanked, the remaining charge continued to exercise a drift force on surface diffusing molecules.

Figure 18 is an example of the LEACE gun's effectiveness when applied to metal foils. The sample is a thin polycrystalline gold foil prepared by vacuum evaporation. A focused 10 nm beam at 45 kV and a current of 10^{-11} A was used to produce contamination spots 1, 2 and 3 by stopping the beam in those positions for 60 seconds. The LEACE gun was then switched on at 225 V and a beam current of about 25 μA. Immediately afterwards, the high energy beam was stopped at position 4 for a total of 12 minutes. Clearly there is no contamination.

The resolution of high magnification scans of samples during LEACE illumination is apparently not degraded. This is illustrated in Fig. 19, where the same area of a carbon replica, shadowed with gold-palladium, of slip lines on a deformed copper single crystal, is imaged, (a) without and (b) with LEACE illumination. The resolution in both images is better than 10 nm. The image in Fig. 19(b) is somewhat distorted, which indicates that the LEACE current might cause an accumulation of negative charge on the surface. The associated equipotential surfaces appear to deflect the beam slightly when I_p is low, as required for high resolution. When I_p is high, as for high resolution imaging with field-emission or LaB_6 guns, or for

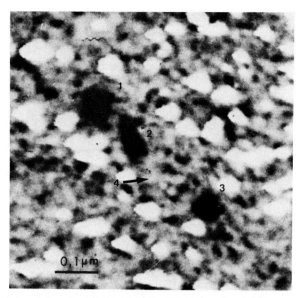

Fig. 18. Contamination in STEM due to surface diffusion of hydrocarbons on a polycrystalline gold foil (1, 2 and 3). Note total elimination of contamination at 4.

----

low resolution imaging with standard electron guns, the high rate of positive charge production around the beam, as discussed previously, would reduce considerably the equilibrium negative surface charge due to the LEACE current. No distortion of the image should then be observed, as is shown to be the case in a low resolution image of slip lines in Fig. 20, produced by means of a standard electron gun at a current of 10^{-10} A.

There are situations under which LEACE illumination is not possible during the imaging process, such as when the specimen is in the magnetic gap between the pole pieces of a high resolution lens. In such cases, it has been demonstrated[5] that LEACE illumination prior to imaging, with the magnetic lens switched off, can also reduce contamination drastically. It is clear from Fig. 16, that the LEACE current can only reach the sample via a process of reflection from surrounding surfaces, such as the surface of the ACD. The current density of the incident low energy electrons will consequently be very low. There was no evidence that LEACE illumination caused any radiation damage in formvar films which showed immediate damage when subjected to 60 kV electrons.

The simplicity of the LEACE gun method for preventing contamination is to be found in the fact that direct illumination of the sample is not essential. This greatly simplifies the equipment required and thus reduces the associated cost. A related technique of "flooding" the specimen by high energy electrons has been discussed by Rackham and Eades[11], Lempfuhl[12], Le Poole[13] and others. However, these high energy electrons would cause prior damage to the sample and secondly the technique requires readjustment of the electron optical system and could not be used simultaneously with imaging or during analysis. In the case of the LEACE gun a

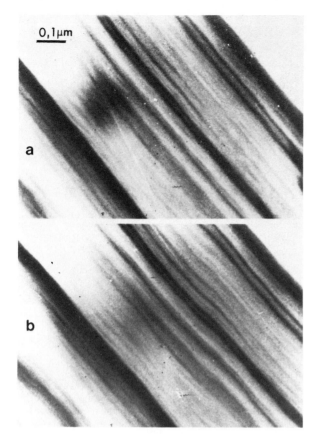

Fig. 19. Imaging in STEM with simultaneous LEACE illumination does not produce loss in resolution; (a) without LEACE illumination, (b) with LEACE illumination. Sample is a carbon-gold-palladium replica of slip lines in deformed copper. Resolution is better than 10 nm in both a) and b) but there is distortion present in b). $I_p \cong 10^{-11}$ A.

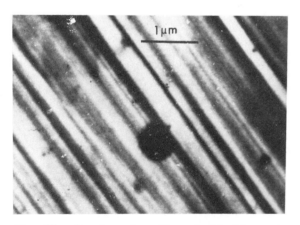

Fig. 20. Imaging of replica of slip lines at $I_p \cong 10^{-10}$ A, i.e. at low resolution, simultaneous with LEACE illumination. There is no distortion present.

----

single switch is all that is required to activate the system.

The fact that the resolution of high magnification scans of specimens during LEACE illumination is apparently not degraded, is very important in regard to many aspects of scanning electron microscopy and also in regard to stationary probe techniques such as in electron energy loss spectroscopy and localised convergent beam diffraction. It follows that the ability to image or analyse the specimen by primary high energy electrons without the interference of contamination can be prolonged usefully.

Summary

The most important conclusions from this paper are:

1. The electric field around the illuminated region in a sample can influence contamination.
2. Secondary electron emission in thin foils occurs over distances of at least 1 μm from the perimeter of the primary beam.
3. The distribution of the electric field strength outside a small diameter beam (<100 nm) is fundamentally different from that in a large diameter beam (>1000 nm). This difference is responsible for the high rate of contamination in narrow beams.
4. Contamination can be prevented totally by indirect illumination of the sample with low energy electrons, a procedure which may be carried out simultaneously with imaging.

Acknowledgements

The author is greatly indebted to Professor F R L Schöning of the University of the Witwatersrand for numerous discussions. He is also indebted to Dr N R Comins, J T Thirlwall and J L Crawford for discussion and assistance with experiments, and to M Hengstberger for preparing the gold foils.

References

1. D.C. Joy, Scanning electron microscopy - where next? SEM/1977/I, IIT Research Institute, Chicago, IL, 60616, pp. 1-8.
2. P. Echlin, Contamination in the scanning electron microscope, SEM/1975, IIT Research Institute, Chicago, IL, 60616, pp. 679-686.
3. D.E. Miller, SEM vacuum techniques and contamination management, SEM/1978/I, SEM Inc., AMF O'Hare, IL, 60666, pp. 513-528.
4. R.K. Hart, T.F. Kassner and J.K. Maurin, The contamination of surfaces during high-energy electron irradiation, Phil. Mag. 21, 1970, 453-467.
5. J.T. Fourie, The elimination of surface-originating contamination in electron microscopes, Optik, 1979, in press.
6. J.T. Fourie, Contamination phenomena in cryo-pumped TEM and ultra-high vacuum field-emission STEM systems, SEM/1976/I, IIT Research Institute, Chicago, IL. 60616, pp. 53-60.
7. V.E. Cosslett, *Introduction to electron optics*, Oxford University Press, London, England, 1950, pp. 53-60.

8. A.N. Broers, J. Cuomo and J. Harper, High resolution electron beam fabrication using STEM, Ninth Internat. Electron Microscopy Conf., J.M. Sturgess (ed.), Microscopical Society of Canada, Toronto, Canada, 1978, pp. 343-354.
9. P.R. Thornton, *Scanning electron microscopy*, Chapman and Hall, London, England, 1968, pp. 97-98.
10. J.T. Fourie, High contamination rates from strongly adsorbed hydrocarbon molecules and a suggested solution, Optik 52(1), 1978, 91-95.
11. G.M. Rackham and J.A. Eades, Specimen contamination in the electron microscope when small probes are used, Optik 47(2), 1977, 227-232..
12. G. Lempfuhl, Convergent beam electron diffraction, Ninth Internat. Electron Microscopy Conf., J.M. Sturgess (ed.), Microscopical Society of Canada, Toronto, Canada, 1978, pp. 304-315.
13. J.B. Le Poole, Trends in instrumentation, *Developments in electron microscopy and analysis*, J.A. Venables (ed.), Academic Press, London, England, 1976, pp. 79-81.

Discussion with Reviewers

Reviewer I: Your model for the sideways flux of electrons from the grid bar to the charged zone to maintain a steady state condition is unconfirmed by any other theoretical or experimental work, and yet would surely produce other detectable effects.
Author: Some detectable effects are shown in Fig. 21. It is believed that the streaks emanating radially from the primary-beam irradiated

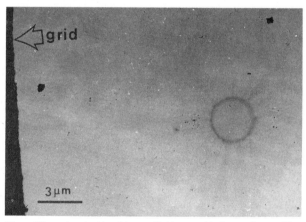

Fig. 21. An indication of the radial surface flux of electrons toward the primary beam (10 nm SiO foil).

----

zone are due to concentrated regions of electron flux across the surface. Some of the electrons moving in these regions would interact with hydrocarbon molecules on the surface with subsequent deposition of these molecules on the foil.

J. Bentley: Can the author briefly describe the application of his model to the case of conducting metal foils? Is some sort of insulating surface film on the metal foils required? How does the LEACE gun work when conducting specimens are being examined (e.g., the Au foil in Fig. 18)?
W. Wiggins. The initial section of the paper pre-

sents evidence that an insulating film, SiO, will charge due to secondary emission. The idea of charging is then used to explain contamination of films that are expected to be highly conducting. Charging effects of the sort illustrated in Fig. 1 are not usually seen in carbon and gold films. What justifies the application of a nonconducting film theory to contamination on conducting films?
R. Anderson: You require a positive charge at the beam site to attract contamination. Why does contamination build up with similiar speed and to a similiar extent on conducting metal specimens where there can be no positive charge?
Author: The original model (text reference[6]) took into account the need for an insulating layer on conductor foils and postulated that a thin contamination layer was formed initially on the conductor film to give its surface within the illuminated region insulating properties.

In applying the model to conductor foils and if we assume that some insulating contamination layer is present due to secondary electron-hydrocarbon interaction up to r_3 (Fig. 10), the grid bar in Fig. 10 effectively contracts to the position where $r_4 = r_3$, because of the radial conductivity of the foil beyond r_3, where no insulating layers are present. With reference to Fig. 13, curve A, for example, this would cause a rapid attenuation of the field for $r > r_{3A}$. Thus dE/dr in this region would be large, resulting in a strong drift force towards the beam and consequently an increase in the concentration of molecules. Thus, for the same electric field a much higher rate of contamination is predicted for a conductor foil.

The mechanism by which the LEACE gun works is not clearly understood at present either for insulator foils or conductor foils. It has been shown experimentally (text reference[5]), that the degree of reduction in contamination rate produced by the LEACE-gun is *time* dependent rather than total exposure (current·time) dependent. This suggests that an effect other than simple electron-molecule collision is involved since such a model would lead to an exposure-dependent phenomenon.

If the LEACE illumination produced an electric field at the surface, this could have the effect of accelerating the normal thermal desorption rate of adsorbed hydrocarbons when the specimen is placed in a high vacuum. That there might be a field present in the case of a conducting film during LEACE-illumination, is demonstrated in Fig. 19b, where the distortion is probably due to refraction of the imaging beam by equipotential surfaces as discussed in the text.

Reviewer II. Can the grid bar effect at d in Fig. 15 be explained by a temperature gradient due to electron beam heating?
Author: It should be emphasized that this phenomenon can be explained only on the basis of an electric field theory. Postulating a temperature gradient near the grid bar is not acceptable since a, b and c in Fig. 15 should then have had similar configurations to d. On the basis of the theory outlined here, the effect seen in d can be predicted by calculation as shown in Fig. 22. For an insulator foil, the proximity of the grid bar

would have the effect of rapidly attenuating the
electric field from a point where the field still
has a large value. As for the conductor foil, a
large drift force on hydrocarbons towards the
beam would result, thus increasing the rate of
contamination on the side of the grid bar.

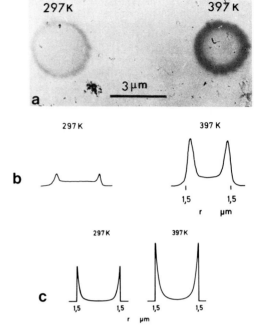

Fig. 23: The temperature effect in contamination.
Contaminating species: silicone high vacuum
grease. (a) micrograph of contamination at
(left to right) 297 and 397 K; (b) experimental
densitometer profiles. Density of contamination
is measured along vertical axis; (c) Calculations according to theory.

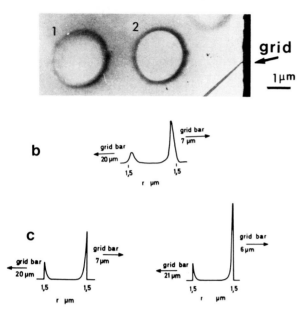

Fig. 22. (a) Grid bar effect on contamination
rates for insulator foil (measurements and calculations refer only to 1. Contamination 2
carried out at smaller J; (b) Experimental
densitometer profile with contamination
density measured along vertical axis;
(c) Calculations according to theory.

J.J. Hren: What local temperature would be expected in the thin film SiO specimens due to beam
heating? Please comment on how this would affect
the expected surface diffusivity desorption rate
of hydrocarbons, and electrical conductivity of
the SiO.
Author: For the beam currents of about 10^{-10} A
used in most of the TEM experiments on silicon
monoxide the temperature rise should not be more
than a few degrees. It is unlikely that beam
heating under these conditions has a significant
effect on the desorption rate of hydrocarbons or
the electrical conductivity of silicon monoxide.
    The effect of temperature on contamination
is taken into account in the theory (see equation (26)). The accuracy with which the theory
can predict the observed experimental phenomena is
shown in Fig. 23 (a) - (c). Clearly, from experiment and theory, an increased sample temperature
for large diameter beams results in increased
rates of contamination due to increased surface
diffusivity of the molecules. Obviously, if the
temperature was increased to the point where rapid
desorption occurred, the rate of contamination
would decrease due to a decreased concentration
of hydrocarbons.

Reviewer II: Why was it necessary to postulate

a ratio $k_4/k_1 = 5$?
Author: In addition to beam size effects (Fig. 14)
and temperature effects (Fig. 23), a further experimental observation on primary current-density
effects (Fig. 24) was made. It was found by experiment that in 3 μm diameter beams, an increasing
J would result in the rate of contamination passing through a maximum as indicated in Fig. 24.
In order to predict the effects of r_1 (Fig. 14),
T (Fig. 23) and J (Fig. 24) on the contamination
rate, by calculation, using unaltered parameters
except for those above, it was required that a
k_4/k_1 ratio of about 5:1 be postulated.

J.J. Hren: There is strong evidence to suggest
that hydrocarbons are readily polymerized by even
low energy electron beams (e.g. Christy, *J. Appl.
Phys.*, 31 (1960) 1680). Could the LEACE gun be
simply immobilizing all mobile hydrocarbons?
R. Anderson: You say your low energy electron
beam eliminates contamination by cancelling the
positive charge on the specimen. An alternative
explanation would be that the flood of low energy
electrons from your LEACE gun polymerizes the
mobile surface ions - fixing them in place. Did
you test for this possibility by flooding a
specimen with electrons prior to applying the
instruments electron beam, immobilizing the contamination before a positive charge can be built
up?
Author: Firstly, we agree that the low energy
beam would be effective in polymerising hydrocarbons. However, because of the indirect LEACE-

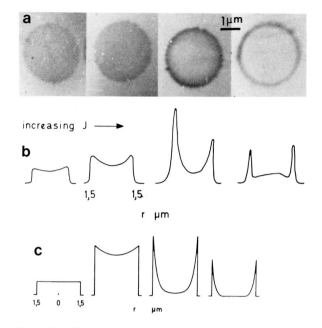

increasing J ⟶

b

1,5   1,5

r μm

c

1,5 0 1,5

r   μm

Fig. 24: The rate of contamination passes through
a maximum with increased J. (a) Experimental
observations, (b) Densitometer profiles of
micrographs in (a). Density of contamination
measured along vertical axis; (c) Calculations
according to theory.

----

illumination (see Fig. 16), the current density
of electrons reaching the sample must be extremely
small and perhaps not capable of immobilizing all
mobile hydrocarbons. Secondly, during the present
STEM experiments with the LEACE-gun, the gun was
on occasion operated without the sophisticated
ACD device being present. LEACE-illumination was
carried out over a period of many hours in this
condition. Had it been an effective polymeriser
of hydrocarbons on the sample, the electron trans-
missibility of the sample should have decreased
dramatically over such a period, since the vacuum
environment was high in hydrocarbon content.
However, such an effect was not observed. Thirdly,
as mentioned elsewhere, the LEACE-effect is not
proportional to prior electron exposure.

It is believed that the LEACE-action is twofold. Firstly, it produces an accelerated desorption of hydrocarbon molecules from the surface
and, secondly, eliminates whatever effects the
electric field around the beam had exercised.
Obviously, some hydrocarbon molecules would be
immobilized by direct low-energy electron-molecule
collisions. However, this is probably a peripheral effect.

W. Wiggins: Accepting the fact that "LEACE" irradiation reduces contamination, it might be by any
of several mechanisms. Could it be that the LEACE
gun warms the sample and drives off contaminating
molecules? Could it be that the diffuse beam
deposits the contaminating molecules over a wide
area before they come into the field of view?
Author: Because of the technique of indirect
LEACE-irradiation the current density of electrons

incident on the sample is extremely low and heating of the sample by electron irradiation can be
ignored. Two much more likely aspects have been
considered. Firstly, because of the small distance between the sample holder and the heated
LEACE gun filament (about 10 mm), some sample
heating might be expected, even though the sample
has no direct sight of the filament. Secondly,
because of the high temperature of the filament
a considerable emission of ultra-violet light
could be expected. It was found experimentally,
however, that the LEACE gun was ineffective while
the filament heating current *only* was on. As soon
as the accelerating voltage was switched on and an
electron current registered, the LEACE gun became
an effective anti-contamination device. Consequently, effects due to heating or ultra-violet
radiation could be discounted. The last point of
your question is dealt with in the previous reply.

J. Bentley: Can the author explain how LEACE
illumination *prior* to imaging can reduce contamination?
Author: It is believed that the LEACE gun action
greatly accelerates the desorption of hydro-carbon
molecules from the sample. Hence the concentration of such molecules could be reduced to a negligible level by LEACE-illumination prior to
imaging. Further contamination would then only be
possible once re-adsorption of hydrocarbons from
the vacuum environment had occurred. Continuous
LEACE-illumination prevents such re-adsorption.

J.J. Hren: Why don't the transversely scattered,
high energy secondaries (and their subsequent products) immobilize the contaminant over a diameter
comparable to that of the charged region? This
mechanism would, of course, deplete the region of
mobile hydrocarbons.
Author: We expect that some hydrocarbon molecules
are immobilized in the manner you suggest and evidence for this can be seen around some irradiated
regions. However, the density of the "products"
of high energy secondaries should be low compared
to the current density in the direct beam. Therefore most of the polymerization of hydrocarbons
would occur within the beam.

J. Bentley: How does the total volume of material
deposited in Fig. 14a change as a function of beam
size?
W. Wiggins: In Fig. 14, has it been quantitatively shown that the total contamination around the
small spot is greater than that around the large
spot or is the contamination around the large spot simply dimmer because it is spread over a larger area?
Author: The total volume of material in regions
1, 2 and 3 can be calculated from the densitometer
profiles in Fig. 14b on the assumption that the
peak heights are proportional to the density of
the deposit. It follows, then, that if V_1, V_2 and
V_3 are the volumes, $V_1:V_2:V_3 = 3,4:6,6:10$. Thus
the volume deposited by the narrow beam was about
3 times the volume deposited by the wide beam for
the same current density. Since the total electron exposure in the small beam was accordingly
only 1/16 of that in the wide beam, the true rate
of contamination in the small beam was about 48
times that in the wide beam.

J.J. Hren: If the ratio of secondary electrons emitted outside the irradiated zone is greater than that within (5:1 or 3:1), why is the instrumental resolution so good in this mode of operation (particularly in thin films)?

R. Anderson: The secondary electrons emitted are the same electrons used to produce a conventional SEM image. If they come from a 2,6 micron diameter area, how do you account for SEM resolutions less than 10 nm?

Author: Images produced by surface scanning in SEM are inherently poor in resolution due to an unfavourable peak-to-background ratio, as would be expected if the secondaries were coming from a 2,6 μm diameter area as postulated in the present paper. However, improved peak-to-background ratios would be obtained in regions of the sample where sharp discontinuities, either geometrical or chemical are present, such as for example at the edge P in Fig. 7a. For the same reason, the familiar magnetic tape sample can be used as a means of measuring the probe width of an SEM but not the actual resolution capability in normal surface scanning. When, for the magnetic tape sample, the beam scans across the *surface* of a particle, the resolution is poor, and very little surface detail of the particle can be discerned. Nevertheless, if a narrow gap between two particles is approached, the secondary signal, even if it is produced by electrons coming from a 2,6 μm diameter around the primary beam, would register a sharp decrease as the beam moved into the crevice between the particles and, further, a sharp increase as it reached the adjacent particle.

J. Bentley: What was the thickness of SiO foil used in Figs. 6-8? How might the k_4/k_1 ratio be expected to change with foil thickness?

Author: The thickness was not determined accurately but from experience can be estimated to have been about 30 nm.

A moderate change of foil thickness is not expected to alter the k_4/k_1 ratio appreciably. For example, if the foil were made thicker, a greater percentage of high energy secondaries would travel within the foil to a radius of r_3 (Fig. 10) thus producing an increased number of low energy secondaries in the process. However, the number of low energy secondaries excited directly by the primary beam would also increase with thickness. Thus the k_4/k_1 ratio should remain fairly constant.

If the thickness of the sample were increased to the point where low energy secondaries excited directly by the primary beam towards the centre of the foil could no longer reach the surface, the k_4/k_1 ratio should increase.

W. Wiggins: How thick was the SiO foil referred to in Fig. 1? Would the charging effect be seen with a very thin foil?

Author: This foil was about 10 nm thick, The rate of charge production due to the interaction of primary electrons with the foil will decrease with decreasing thickness. Therefore, for foils thinner than 10 nm an increased primary-beam current would be required to attain the same equilibrium configurations as in Fig. 5, for example. Some thinner foils were studied and these did show charging effects.

J. Bentley: In Fig. 3 and 4, how do you know the sense of the tilt relative to the STEM image? What angle of tilt was used to obtain Fig. 4?

Author: The sense of the tilt for a given sense of turn on the control knob can be determined by viewing the transmitted beam directly on a fluorescent screen below the sample.

The angle of tilt was not measured accurately but can be estimated to be about 1-2°.

J.J. Hren: The build-up of positive charge should decrease the (low energy) secondary yield. Is this observed? Please cite any experiments.

Author: When a thin foil is irradiated with high energy electrons, the majority are transmitted and will not register a specimen current. The ejected secondaries, however, will produce a potential difference between the grid bar and the irradiated zone and a specimen current will be registered. Such measurements have been made but due to complicating factors, such as simultaneous contamination, these cannot be interpreted reliably at this stage.

I agree that the build-up of positive charge should decrease the low energy secondary yield. For example, in Figs. 1-5 where the potential difference between the irradiated zone and the grid bar (ground potential) is estimated to be about 1 kV, one would expect that only secondary electrons with energies above 1 keV would still escape from the foil.

J.J. Hren: Has the author performed any experiments as a function of accelerating voltage on SiO and if so what effects were observed?

Author: The experiments on charging effects were all performed at 20 kV only. The experiments on secondary generation outside the irradiated zone (Figs. 6, 7 and 8) were carried out at 25 kV. Some similar experiments were done at 50 kV and a preliminary conclusion from these would be that the zone where secondary electrons were produced outside the primary beam, increased in radius with accelerating voltage.

J.J. Hren: Have experiments with metallic specimens containing oxide films on the surface been studied? If so, please cite briefly the results.

Author: No, such specimens have not been studied. However, they would represent a composite insulator-conductor foil which is essentially similar to an oxide free metal foil (or an evaporated carbon foil) covered with an (insulator) contamination layer. It is believed that results on contamination in such specimens would coincide to a large extent with results described here.

J. Bentley: Were the specimens used in this work pre-treated in any way to ensure reproducibility of contamination rate?

Author: Considerable reproducibility of contamination results and also the selection of a particular species of contaminating molecule are possible by dipping specimens in a dilute solution of the oil or grease in toluene. Suitable concentrations are about 0,7 gL^{-1}. This procedure was used for many of the systematic experiments in the present work. For most of the LEACE-gun experiments in the SEM, however, samples prepared in the usual manner were used to investigate the gun's general applicability, i.e. no contaminant was purposely added.

SCANNING ELECTRON MICROSCOPY/1979/II
SEM Inc., AMF O'Hare, IL 60666, USA

REDUCTION OF CONTAMINATION IN ANALYTICAL ELECTRON MICROSCOPY

Y. Harada, T. Tomita, T. Watabe, H. Watanabe and T. Etoh

Technical Operations Division, Scientific Instrument Project
JEOL Ltd.
Akishima, Tokyo 196
Japan

Abstract

This paper describes the method to measure quantitatively the contamination caused by the use of a small electron probe and instrumental improvements for reducing it. These improvements include the adoption of a new type of vacuum grease and the development of a new type anti-contamination device. Since the weight of contamination deposits on the carbon film changes linearly with the probe irradiation time, a highly accurate method of contamination measurement has been established by indicating the contamination rate in g/min. Thus, it has become possible to correctly evaluate instrumental improvements for reducing contamination. The partial pressure of hydrocarbon gases in the specimen chamber was reduced by approximately one order, through various instrumental improvements. This resulted in a drastic reduction of the contamination rate and therefore in the performance improvement of the analytical electron microscope. In order to further reduce contamination, some specimen pretreatments that eliminate contamination source existing in a specimen itself, were carried out. The effects of these pretreatments are discussed.

KEY WORDS: Contamination, Contamination Rate, Analytical Electron Microscopy, X-ray Microanalysis, Micro-diffraction, Specimen Pretreatment, Anti Contamination Device

Introduction

In the advanced electron microscopes, various improvements have been made on the vacuum system, and a clean high vacuum specimen chamber (10^{-7} Torr) has been realized[1]. As a result, it is possible to take TEM performance micrographs (1.4 Å gold lattice images) without using the anti-contamination device (ACD).

On the other hand, in the analytical electron microscope (which is capable of elemental analysis of submicron areas using both the energy dispersive X-ray spectrometer and electron energy analyzer, and micro-diffraction and convergent beam diffraction) a small electron probe with a high current density is frequently used, and in many cases the problem of contamination arises as a result of keeping the focused electron probe at one spot on the specimen. Therefore, the ACD is generally used with other attempts to reduce the contamination.

Specimen contamination usually builds up in the following process: when an incident electron beam strikes hydrocarbon gas molecules on the specimen surface, the bonds of hydrogen and carbon are broken, resulting in the carbon deposition on the specimen surface[2]. This contamination is classified roughly into two types depending on the source of hydrocarbon gas molecules: the contamination inherent in the instrument and the contamination resulting from the specimen itself[3,4]. The former is caused when hydrocarbon gas molecules coming from the inner wall of the instrument, O-rings, grease and other components are adsorbed on the specimen surface. This contamination can be greately reduced by improving the vacuum system and carefully eliminating the contamination source in the process of instrument assembly. The latter is caused when hydrocarbon gas molecules adsorbed on the surface of specimen during specimen preparation are brought into the specimen chamber. This type of contamination cannot be reduced by means of instrumental improvements alone, therefore hydrocarbon gas molecules must be eliminated by carrying out specimen pretreatment in a special manner.

This paper describes a method to measure quantitatively the contamination caused by the use of a small electron probe (several tens to hundreds of Å), and instrumental improvements for reducing it. These improvements include the adoption of a new type of vacuum grease and the development of a new type ACD. In order to further reduce contamination, some specimen pretreatments were carried out that eliminated contamination source existing in the specimen itself. The effects of these pretreatments are discussed.

Method of Contamination Measurement

In order to evaluate the instrumental improvements for reducing the specimen contamination, it is important to establish an accurate method of contamination measurement. Our experiment was conducted using a TEM equipped with a scanning attachment, the JEM-100CX with ASID-4D. The electron guns used were thermal emission and field emission types, which were operated at 100 kV and 80 kV. Electron probe less than 100 Å was formed on the specimen surface and was kept at one spot for a certain period of time (called the spot mode), then the contamination on the specimen surface was observed. The specimens used for the experiment were thin films of carbon, aluminium and silicon. They were not pretreated at all with regard to the contamination sources adsorbed on their surfaces.

It is generally known that when a small electron probe irradiates a thin film specimen, a cone-shaped contamination deposits nearly symmetrically on both the upper and lower sides of the specimen[5], as shown schematically in Fig. 1. Fig. 2 shows a TEM image of a cone-shaped contamination deposit on a thin carbon film of 700 Å thickness. The specimen was tilted at 45° for photographing after it was irradiated with a small electron probe of 30 Å diameter for 100 seconds without the ACD.

In order to evaluate such cone-shaped contamintion deposits quantitatively, the small electron probe was successively impinged on different positions on the specimen surface for different times, and then secondary electron images of the cone-shaped deposits thus produced were taken at a tilting angle of 45° (see Fig. 3). By measuring the diameter and height of the contamination deposits from these micrographs, it is possible to estimate, by means of cone approximation, the volumes of the contamination deposits produced on the upper side of the specimen. Furthermore, it is possible to calculate the weight of the contamination deposits by assuming that its density[6] is 2.3 g/cm^3, because this deposit is carbon[7].

Figs. 4, 5 and 6 show changes with irradiation times, of the bottom diameters, heights and weights, respectively, of cone-shaped contamination deposits produced on thin film specimens of carbon, aluminium and silicon irradiated under conditions of probe diameter 30 Å, and probe current 2×10^{-10} A. These data were obtained after instrumental improvements described below. The diameter and height of the deposit rapidly increase up to one minute after the start of electron probe irradiation and slowly increase after that. These results agree almost with those obtained by other researchers[5, 8]. The weight of

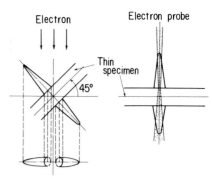

Fig. 1 Schematic cross section of cone-shaped contamination deposit formed symmetrically on both surfaces of thin film in spot mode, where the electron ray path is depicted before contamination builds-up.

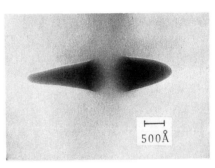

Fig.2 Cone-shaped contamination deposit on C thin film. The specimen was tilted at 45°, after it was irradiated with electron probe of 30 Å in diameter for 100 seconds without the ACD.

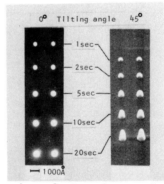

Fig.3 Secondary electron images of cone-shaped deposits on Aℓ thin film for different irradiation times without ACD, photographed at 0° and 45° tilt.

the deposit increases linearly with the time[5] for the carbon films, though the weight increases of deposits for Aℓ and Si do not necessarily show linearity. It is possible to measure the contamination rate (g/min) from its gradient for the carbon films. Defining the contamination rate in the spot mode in this way allows not only the accurate quantitative evaluation of contamination, but also the proper evaluation of instrumental improvements.

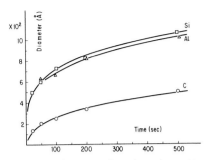

Fig.4 Diameter of contamination deposits as function of the probe irradiation time for C, Aℓ and Si thin films (probe diameter: 30 Å, probe current: 2 × 10⁻¹⁰ A):

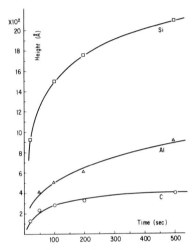

Fig.5 Height of contamination deposits as function of the probe irradiation time for C, Aℓ and Si thin films (probe diameter: 30 Å, probe current: 2 × 10⁻¹⁰ A).

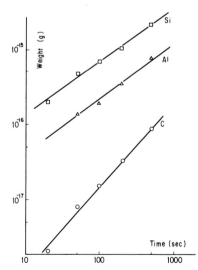

Fig.6 Weight of contamination deposits as function of the probe irradiation time for C, Aℓ and Si thin films (probe diameter: 30 Å, probe current: 2 × 10⁻¹⁰ A).

Instrumental Improvements: Reduction of partial pressure of hydrocarbon gases

In order to reduce the partial pressure of residual hydrocarbon gases in the specimen chamber, the following steps were taken:
(1) Santovac-5[9] was used as the oil for two diffusion pumps.
(2) All the O-rings used in the column were fully baked out in vacuum in advance.
(3) Fomblin RT-15[10] which contains no hydrocarbon molecules was used as the vacuum grease.
Table 1 tabulates these improvements.

	Before improve.	After improve.
DP₁ Oil	Lion – S	Santovac – 5
DP₂ Oil	Santovac – 5	Santovac – 5
O-ring pretreat -ment	None	Done
Detergent	Trichloroethane	Trichrotrifluro ethane
Vacuum Grease	Apiezon L	Fomblin RT-l5

Table 1 The differences of vacuum systems before and after instrumental improvements

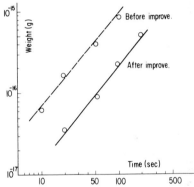

Fig.7 Weight of contamination deposits as function of the probe irradiation time for C thin film before and after instrumental improvements (probe diameter: 30 Å, probe current: 2 × 10⁻¹¹ A)

Fig.7 compares the contaminations in the spot mode before and after these improvements. The experimental conditions were probe diameter 30 Å, probe current 2 × 10⁻¹¹ A, and carbon thin film as the specimen. By reducing the partial pressure of hydrocarbon gas, the contamination rate was found to be reduced to 1/4.

Instrumental Improvements: Improvements on anti-contamination device

In order to have hydrocarbon gas molecules coming from the parts around the specimen, trapped effectively by the ACD before reaching the specimen surface, the following improvements were made on the ACD:
(1) Improvement of thermal conduction system to further lower the cooling temperature of the cold trap of the ACD.

(2) Improvement of the position and shape of the cold trap.

Even with the same shape of the cold trap, the weight of contamination varies because the adsorption efficiency for hydrocarbon gas molecules varies with the cooling temperature of the trap. Fig. 8 shows the dependency of contamination rate (g/min) upon the cooling temperature of the trap. The experimental conditions were electron probe diameter 60 Å, probe current 4×10^{-10} A, and the electron probe was kept on a carbon thin film specimen for 60 sec. It is clear that the lower the trap cooling temperature, the smaller is the contamination rate. How to lower the temperature of the cold trap is, therefore, an important problem. Conventionally, the trap is cooled by positioning the copper rod as a thermal conductor, between the cold trap of the ACD and the liquid nitrogen tank. Extreme lowering of the trap temperature by the copper rod is not possible. Thus, the heat pipe[11], using nitrogen gas as its working fluid, was employed. This had ten times the cooling capability of the copper rod. The heat pipe conducts heat in the form of latent heat of evaporation, through the processes of condensation, liquid transition, evaporation and vapor transition. The thermal characteristics of the heat pipe depend upon the pressure of the enclosed nitrogen gas at room temperature. After a basic experiment on it a pressure of approximately 20 kg/cm^2 was decided upon.

Fig 9 shows changes-with-time of the cooling temperature of the cold traps, one with a heat pipe and the other with a copper rod of the same size as the thermal conductor. In the case of the copper rod, the attainable temperature of the trap is -135°C, but in the case of the heat pipe, the trap is rapidly cooled down to -150°C. Thus the heat pipe was superior to the copper rod in cooling capability.

Described below are improvements on the position and shape of the ACD. In general, an analytical electron microscope incorporates, as a standard component, a side entry goniometer affording superb versatility. However, the hydrocarbon gas molecules released from the O-rings and vacuum grease used in the movable parts of the goniometer cause contamination. In order to have these hydrocarbon gas molecules adsorbed effectively near their sources, as shown in Fig. 10, a cylindrical cold trap (called the outer trap) was provided on the outer circumference of objective lens pole piece, in addition to the conventional cold trap surrounding the specimen (called the inner trap). Fig. 11 shows the dependency of the weight of the contamination deposit upon the irradiation time for the cases of not using the ACD, using the inner trap only, and using both the inner and outer traps. The experimental conditions were probe diameter 30 Å, probe current 2×10^{-10} A, and thin carbon film as the specimen. The contamination rates are 4×10^{-15} g/min, 3×10^{-17} g/min, and 8×10^{-18} g/min when not using the ACD, using the inner trap only, and using both the inner and outer traps, respectively. The outer trap, therefore, is seen to be very effective.

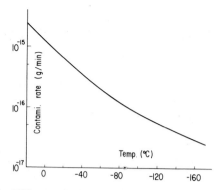

Fig.8 Effect of the cooling temperature of trap on the contamination rate (g/min) for C thin film (probe diameter: 60 Å, probe current: 4×10^{-10} A)

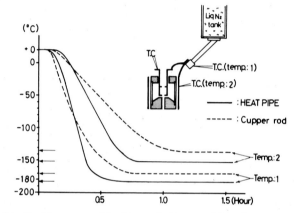

Fig.9 Changes-with-time of the cooling temperature of the cold traps for thermal conductors with heat pipe and copper rod

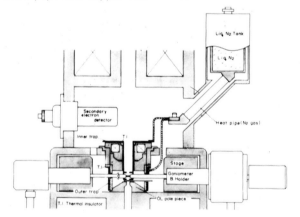

Fig.10 Special designed ACD assembly

Effect of Instrumental Improvements

By lowering the partial pressure of hydrocarbon gas and by making instrumental improvements including the improvement of the ACD, as mentioned above, the contamination rate has greately reduced. As an example of the effects, the changes-with-time of the micro-diffraction pattern and the weight of the contamination

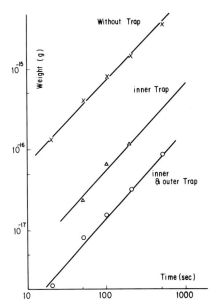

Fig.11 Weight of contamination deposits as function of the probe irradiation time with C thin film for the cases of not using ACD, using the inner trap only and using both the inner and outer traps (probe diameter: 30 Å, probe current: 2×10^{-10} A)

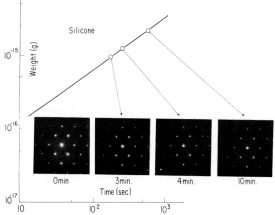

Fig.12 Weight of contamination deposits and micro diffraction pattern as function of the probe irradiation time for Si thin film when using special designed ACD (probe diameter: 300 Å, probe current: 2×10^{-9} A)

deposit are shown in Fig. 12, in which the electron probe is kept on a thin Si film for 3 min, 4 min and 10 min. The experimental conditions were probe diameter 300 Å, and probe current 2×10^{-9}A. Thus, diffraction patterns, which vanish in several tens of seconds with a conventional instrument, can clearly be observed even after ten minutes.

Effect of Specimen Pretreatments

To further reduce the contamination, another contamination source - which exists on and in a specimen itself - must be eliminated. In this case, contamination occurs because the specimen is brought into the specimen chamber with hydrocarbon gas molecules adsorbed on and stored in the specimen itself during the process of specimen preparation, etc. To prevent this, hydrocarbon gas molecules must be removed from the specimen by a special pretreatment. Several pretreatment methods are now available, some of which have produced good results[6, 12, 13].

We used glow discharge and ethanol immersion as specimen pretreatment methods. The specimen holder was not treated. In the former method, specimens were placed in the environment of stable glow discharge in a sputtering device. The sputtering condition was 500 V/2mA in the 0.2 Torr air. In the latter method, specimens were immersed in pure ethanol for 18 hours. Fig. 13 shows the effect of the specimen pretreatment on the weight of contamination deposits, as a function of the probe irradiation time. The experimental conditions were probe diameter 60 Å, and probe current 4.5×10^{-11} A. The specimens used were evaporated carbon film and electrolytically

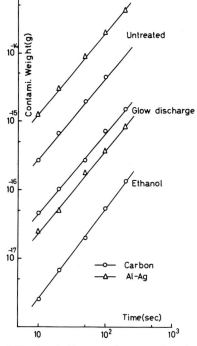

Fig.13 Effect of the specimen pretreatment methods on the weight of contamination deposits as function of the probe irradiation time (probe diameter: 60 Å, probe current: 4.5×10^{-11} A)

polished Aℓ-5.5 Wt% Ag thin foil. Both treatment methods, as was expected, were effective for reducing the contamination rate in the spot mode. In the case of carbon film pretreated with the glow discharge and the ethanol immersion method, the contamination rates were found to be reduced to 1/7 and 1/80 respectively that of the untreated specimens. Fig. 14 shows the contamination rate as a function of the time of specimen pretreatment, i.e., immersing a carbon film in ethanol. Although this pretreatment time was

18 hours for Fig. 13, a sufficient effect was attained by treating only for 10 min. The same results as in Fig. 13 and Fig. 14 were also confirmed by using organic solvents[3] such as methanol and acetone.

It is important to establish and use these specimen pretreatment methods for reducing contamination rate in analytical electron microscopy, since the contamination source of the instrument itself has already been extensively reduced.

Discussions

As shown, even in the diffusion pump evacuation system, contamination can be negligibly reduced, by making instrumental improvements such as employment of a new type of vacuum grease, modification of the ACD structure, lowering of the cooling temperature of the cold trap and by paying through consideration to contamination in the manufacturing process of an instrument. Fig. 15 represents a result of residual gas analysis around the specimen, obtained after these instrumental improvements. The total pressure is 8×10^{-8} Torr, approximately 40 % of which is the partial pressure of H_2O. The partial pressure of hydrocarbon gases is in the order of 10^{-10} Torr, an extremely small value. This is one order better than that obtained before such improvements. Consequently, the experiment shows that lowering the hydrocarbon partial pressure is reducing the contamination rate. Besides, while the partial pressure of H_2O mostly determines the total pressure, it may be possible to lower the partial pressure of H_2O by baking out the column at higher temperature, because the vacuum grease used in this experiment has a lower pressure. Fig. 16 visually shows changes-with-time of the contamination deposit both in the case of the specimen chamber of a conventional instrument being evacuated simply by an ion pump and a sublimation pump (Fig. 16a), and in the case of the specimen chamber being evacuated by a diffusion pump with such improvements as mentioned above (Fig. 16b). The experimental conditions are, in the both cases, probe diameter 20 Å, probe current 2×10^{-10} A, total pressure of specimen chamber 3×10^{-7} Torr, and thin carbon film as the specimen. As seen, there is a great difference between both contamination rates; the contamination rates in Fig. 16a and Fig. 16b are 1.5×10^{-15} g/min and 1.5×10^{-17} g/min, respectively. Thus, contamination cannot be reduced effectively by employing such a dry pumping system. More important, than the discussion on whether to use a dry or wet pumping system, is to improve the column and particularly the specimen chamber into those with less hydrocarbon outgassing. When measuring the weight of a cone-shaped contamination deposit, the following were assumed:
(i) The deposit is cone-shaped
(ii) The deposit is carbon, whose density is 2.3 g/cm³.
With respect to assumption (i), the deposit should be considered to be a truncated cone, but

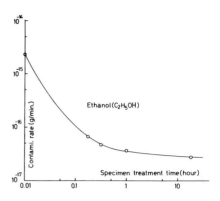

Fig.14 Contamination rate as function of the specimen pretreatment time for carbon film

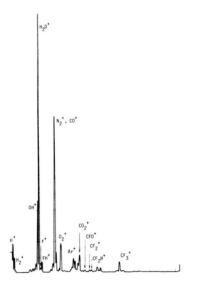

Fig.15 Mass spectrum of residual gases around the specimen (total pressure is 8×10^{-8} Torr)

the precision is considered to be within the error range of ±5 % since the probe diameter used is, in most cases, as small as tens of angstroms.

In regard to supposition (ii), the volume of contamination is expressed as an equivalent weight of carbon in grams (g). This is intuitively easier to comprehend plus the detectability limits of energy analyses are usually expressed in gram weight. Then, the density of solid carbon, i.e., 2.3 g/cm³ was used. It will be necessary to measure it with a suitable method such as a quartz oscillator.

Conclusions

(1) Since the weight of the contamination deposits on the carbon film changes linearly with time, a very intuitive and highly accurate method of contamination measurement has been established by indicating the contamination rate by g/min. Thus, it has become possible to correctly evaluate instrumental

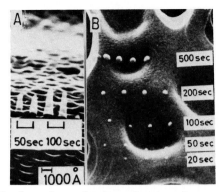

Fig.16 Secondary electron images of contamination deposits both for the case of specimen chamber of a conventional instrument being evacuated simply by dry pump system (Fig. 16a) and for the case of improved specimen chamber being evacuated by diffusion pump (Fig. 16b)

improvements and effects of specimen pretreatment methods for reducing contamination.

(2) The partial pressure of hydrocarbon gases in the specimen chamber was reduced by approximately one order through various instrumental improvements. This resulted in a drastic reduction of the contamination rate and therefore improved performance of the analytical electron microscope.

(3) Even with a diffusion pump evacuation system, the contamination source inherent in the instrument was reduced to a negligible amount through anti-contamination considerations.

(4) The contamination rate in the spot mode was decreased effectively by using specimen pretreatment methods, i.e., placing the specimen in the environment of glow discharge or immersing the specimen in organic solvents.

References

1. N. Yoshimura, K. Shirota and T. Etoh, "The clean high vacuum system of JEM-100CX", JEOL news, 14e, No. 2, 1977, 2-8
2. A. E. Ennos, "The origin of specimen contamination in the electron microscope", Brit. J. Appl. Phys., 4, 1953, 101-106
3. J. T. Fourie, "Contamination phenomena in cryopumped TEM and ultrahigh vacuum field-emission STEM systems", Scanning Electron Microscopy/1976/Vol I, O. Johari (ed.), IIT Research Institute, Chicago, IL, 1976, 53-60
4. J. J. Hren, "Specimen contamination in analytical electron microscopy: Sources and solutions", Ultramicroscopy, 3, 1979, 375-380
5. H. W. Conru and P. C. Laberge, "Oil contamination with the SEM operated in the spot scanning mode", J. Phys. E., 8, 1975, 136-138
6. L. Reimer and M. Wüchter, "Contribution to the contamination problem in transmission electron microscopy", Ultramicroscopy, 3, 1978, 169-174
7. H. König, "Verkohlung organizcher Objecte durch Elektronen", Z. Physik., 129, 1951, 483-490
8. W. A. Knox, "Contamination formed around a very narrow electron beam", Ultramicroscopy, 1, 1976, 175-180
9. J. S. Cleaver and O. N. Fiveash, "A study of some diffusion pump oils using a six-inch radius 60° sector field mass spectrometer", Vacuum, 20, 1970, 49-54
10. "Fomblin, Fluorinated Fluids in the Vacuum Industry", Informazioni Techniche (Technical Information), Montedison S.p.A., No. 1096E, 1973
11. S. W. Chi, Heat Pipe Theory and Practice, Hemisphere Publishing Corpo., 1967, Chap. 2
12. E. K. Brandis, F. W. Anderson and R. Hoover, "Reduction of carbon contamination in the SEM", Scanning Electron Microscopy/1971, O. Johari (ed.), IIT Research Institute, Chicago, IL, 505-510
13. B. Bauer and R. Speidel, "Herabsetzung der Kontaminationsrate im STEM bei einem Druck von 10^{-5} Torr", Optik, 48, 1977, 237-246

Discussion With Reviewers

R. Anderson: You state that the Al and Si were thin foils. Most Al and Si foil preparation procedures require rinses in a solvent following etching or electropolishing. If such rinses were employed, how can you state that the specimens were not pretreated? Your figure 14 shows that very short immersion in a solvent will drastically alter contamination build-up.

Authors: Fig. 14 shows that very short immersion in ethanol reduces the contamination rate for evaporated carbon films. The same effect seems to be attained also for metallic foils prepared by etching or electropolishing, if the metallic foils are immersed in ethanol at the final preparation stage.

H.W. Conru: Figures 5 and 6 illustrate an increase in contamination rate with increasing atomic number. My experimental results (data unpublished) showed to a 1st order approximation a linear dependence of contamination rate to atomic number. I attributed this to an increase in backscattered electrons with increasing atomic number. What are your comments?

Authors: We do not have enough data to explain the dependence of contamination rate to atomic number. That is, the specimens used are only light elements and their thicknesses are not equal. However, the fact is that backscattered electrons as well as primary electrons affect the contamination rate. To check the dependence, it will be important for us to make a precise experiment including specimen pretreatment, etc.

H.W. Conru: The text states that the temperature of the improved cold trap is reduced to -150°C as compared to -135°C with a conventional Cu rod. From Figure 8 this temperature shows only an incremental reduction in contamination rate. The text further states a reduction in contamination rate by almost two orders of magnitude. Explain?

Authors: We stated the two improvements on the ACD. One is the adoption of the cooling system employing the heat pipe and the other is the provision of an additional outer trap. The former gives a 10 % reduction in the contamination rate by lowering the trap temperature from -135°C to -150°C (Fig. 8), and it has an additional effect of shortening the time of cooling down to -150°C (Fig. 9). The latter gives a four times larger reduction in contamination rate than when using the conventional inner trap only (Fig. 11). You state that the reduction in contamination rate by almost two orders of magnitude. This seems to imply a comparison of contamination rate when not using the ACD and that when using both the inner and outer traps (Fig. 11).

H.W. Conru: *You state that the contamination rate of an ethanol pretreated sample is 1/80 that of an untreated sample? Was this with the vacuum improvements in place?*
Authors: We obtained the above result in three attempts under the same vacuum conditions. This pretreatment method probably dissolved the hydrocarbon molecules in carbon films, as pointed out by Reimer et al.[6]

H.W. Conru: *The micro-diffraction patterns illustrated in Figure 12 were obtained with a probe current of 2×10^{-9}A. Specimen contamination is very low at such a high current. How long would the diffraction pattern remain visible under a probe current of 5×10^{-11}A?*
Authors: We have no data of micro-dffraction obtained under a probe current of 5×10^{-11}A. However, judging from the convergent diffraction patterns shown in Fig. 17 we suppose micro diffraction pattern would remain visible for 10 min under the same probe current.

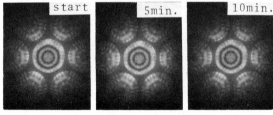

Fig.17 Convergent diffraction patterns obtained at varying probe irradiation times for Si foil by use of a specially designed ACD (probe diameter: 150 Å, probe current: 2×10^{-11}A).

H.W. Conru: *The carbon deposits shown in Figure 16a appear to be at least 2000 Å high (allowing for foreshortening) even though the chamber was ion/sublimation pumped. What type of pump was used to rough the system? Are you implying that most of the contamination shown is from the sample itself? If this is so, then Auger analyses of many samples would be impossible which is not the case.*
Authors: An oil rotary and a diffusion pump were used to rough the system to a pressure of 10^{-5} Torr. We are implying that most of the contamination shown is not from the sample itself but from the system. Therefore, we emphasized that it is important to improve the specimen chamber into one with less hydrocarbon outgassing.

H.W. Conru: *What is the estimated total reduction in contamination achieved by employing all the improvements and sample pretreatment described?*
Authors: It is difficult to estimate the total reduction in contamination rate, because the probe irradiation and specimen conditions are not equal between Fig. 11 and Fig. 13.

S. Das: *At what temperature and in what vacuum were the O-rings baked out? Was it done in a separate vacuum chamber outside the microscope?*
Authors: The O-rings were baked out at 120°C in a separate vacuum chamber held at 10^{-6} Torr.

S. Das: *Will the authors comment on whether the contamination deposit rate in the improved system is sufficiently low for micro-micro-analysis (in spot mode) using energy dispersive x-ray analysis, where long counting times are needed.*
Authors: The contamination rate in the improved system is sufficiently low for micro-micro-analysis, as seen from Fig. 12 and Fig. 17.

J.J. Hren: *Have the authors attempted to calculate the extent of beam spreading anticipated by passing through 2000 Å of carbon?*
Authors: No, we have not, but according to a paper by J. I. Goldstein et al. ("Quantitative X-ray analysis in the electron microscope" Scanning Electron Microscopy/1977/Vol. I, O. Johari (ed.), IIT Research Institute, Chicago, IL, 1977, 315-321), the extent of beam spreading is 100 Å by passing through 2000 Å of carbon at 100 kV.

J.J. Hren & D.C. Joy: *Have you made any attempt to verify by direct microanalysis that the contamination deposit is, in fact, pure carbon?*
Authors: We conducted a direct microanalysis by energy loss spectroscopy (ELS) and energy dispersivie X-ray spectroscopy (EDS), as shown in Fig. 18a and b, respectively. A contamination deposit on thin boron-nitride (BN) was analyzed. Fig. 18a shows ELS spectra of BN before and after the contamination has built up. It is seen that a carbon peak in addition to the BN peaks appears after the contamination build-up. Fig. 18b shows an EDS spectrum from the same region. There are no remarkable peaks, though background increases appreciably due to the contamination. The contamination deposit is probably pure carbon, judging from these data.

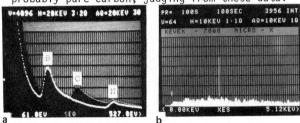

Fig.18 Spectra of contamination deposit on BN, obtained by ELS (Fig. 18a) and EDS (Fig. 18b)

SCANNING ELECTRON MICROSCOPY/1979/II
SEM Inc., AMF O'Hare, IL 60666, USA

ELECTRON - SPECIMEN INTERACTIONS

L. Reimer

Universität Münster
Physikalisches Institut
Schloszplatz 7
D-4400 Munster, West Germany

Abstract

The discussion of electron-specimen interactions starts with the scattering processes by a single atom. One can distinguish between elastic scattering resulting in large-angle scattering and bending of the electron trajectories and inelastic scattering resulting in low-angle scattering and energy losses.

The multiple scattering of electrons results in the formation of an electron range and a diffusion cloud which limits the resolution of BSE images and x-ray analysis.

The backscattered electron (BSE) coefficient η depends on atomic number, tilt angle, take-off angle, crystal orientation, direction of magnetization and is less influenced by variations of electron beam energy. This can be used for the atomic number (Z) contrast with high take-off angles and topographic contrast with low take-off angles, crystal orientation contrast, electron channelling patterns and magnetic contrast type II. Energy filtering will be a further possibility to get better information from the BSE signal (e.g. low-loss electrons).

The secondary electron (SE) yield δ shows a less pronounced variation with atomic number and depends on tilt angle and electron energy. It can be used for resolving the surface topography with a resolution in the order of the beam diameter. SE are influenced by external magnetic stray fields (magnetic contrast type I).

Ionisation processes in the specimen generate not only secondary electrons, however, also heat, radiation damage, light (cathodoluminescence), electron-hole pairs (electron beam induced current, EBIC) and characteristic x-ray quanta for elemental analysis.

KEY WORDS: Electron Scattering, Electron Diffusion, Backscattered Electrons, Secondary Electrons, Contrast, X-ray Microanalysis

Introduction

Everybody working with an SEM has to know as much as possible about electron-specimen interactions. This is important for the optimum specimen preparation, for the interpretation of specimen damage and image contrast, and for the optimum detector strategy.

It is impossible to cover all these aspects in one tutorial. We restrict ourself, therefore, to a discussion of some of those most important physical results which are of interest in practical work.

Fig.1 shows a schematic view of what happens inside the specimen and of what comes out of the specimen while irradiating a specimen with a beam of primary electrons (PE) of energy E_0. The electrons do not penetrate along straight lines into the material but instead are scattered by about 20-50 elastic, large-angle scattering processes (with scattering angles $\theta=10-180^0$) and a few hundred small-angle scattering processes. Fig.1 contains three examples of electron trajectories computed by the Monte-Carlo method [1-4] which give an impression of the bending of the electron trajectories. Some of the electrons can escape as backscattered electrons (BSE). The electrons continuously lose energy on their path through the material by inelastic scattering processes. Most of the energy losses ΔE in the order of 5-50 eV cause excitation and ionisation in the outer electron shells or in the broadened energy levels (band structure) of a solid. This deceleration process limits the electron range R and the diameter of the diffusion cloud. Most of the energy losses are converted into heat [5].

Some of the low energy electrons from the ionisation processes can leave the specimen as secondary electrons (SE) from a small exit depth t_{SE}. A charge is transfered to the specimen if more PE are stopped inside the specimen than SE and BSE can escape. This causes a negative charging of insulating material or can be used as the specimen current signal for conductive specimens. Most of the

Table of symbols

a   Auger electron yield
a_H Bohr radius
A   atomic mass
e   elementary charge
E_0 primary electron energy (keV)
$E<E_0$ electron energy after interaction
E_K ionisation energy of K shell
E_m mean electron energy
$d\sigma/d\Omega$ differential scattering cross section
$f(\theta)$ atomic scattering amplitude
J   mean ionisation energy of an atom (eV)
m_0c^2 relativistic electron rest energy
N_A Avogadro constant
p   impact parameter
R   practical electron range
R_B Bethe range
R   backscattering correction factor for x-ray microanalysis
s   coordinate along the bended electron trajectory
t_{SE} exit depth of secondary electrons
u   overvoltage ratio E_0/E_K
z   distance below the surface
Z   atomic number

β ratio of mean SE yield of one BSE to the mean yield of one primary electron
δ secondary electron yield
ε_0 vacuum permittivity
η backscattering coefficient
θ scattering angle
θ_0 screening angle of elastic scattering
λ wavelength of electron
Λ mean free path between scattering processes
ρ density
$\sigma = \eta + \delta$ total electron yield
σ_{el}, σ_{inel} total elastic (inelastic) scattering cross section
φ azimut angle of surface facets
ϕ tilt angle of specimen between surface normal and primary beam
$\phi(z)$ depth distribution of ionisation
ω x-ray fluorescence yield
Ω solid angle
ψ take-off angle between direction of observation and surface plane
ψ amplitude of electron wave

---

ionisation processes are reversible, but irreversible ionisation damage can occure in organic materials[6-8] and some semiconductor structures.[9] The generation of electron-hole pairs in semiconductors can be used for the electron beam induced current (EBIC) mode if the charge carriers are separated by a depletion layer.[10-12] One can also observe the emitted light quanta from the recombination of excited electrons in cathodoluminescent material.[6,13-15]

Inner atomic shells are ionized with low probability, but the refilling of a vacancy in an inner shell by an electron from an outer shell produces an Auger electron with a discrete energy or a characteristic x-ray quantum, both of which are important for elemental analysis. The characteristic x-ray quanta and the x-ray continuum can also penetrate deeper into the specimen and produce inner shell ionisation with a following emission of characteristic x-ray quanta in a larger volume than the electron diffusion cloud (x-ray fluorescence).

The dependence of the different types of signals on the electron energy E_0, the atomic number Z, the tilt angle ϕ and the take-off angle ψ determines the measureable information.

Cross sections for scattering and ionisation

One can describe a scattering process by a cross section, which will be introduced using the example of elastic large-angle scattering (Fig.2a). Electrons are attracted by the positive charge of the nucleus and travel along hyperbolic trajectories. The scattering angle θ decreases with increasing impact parameter p. Electrons of the parallel beam of energy E_0 within an area $d\sigma$ are scattered into a solid angle $d\Omega$. The ratio $d\sigma/d\Omega$ is called the differential cross section and depends on the scattering angle. In quantum mechanics (Fig.2b) the scattering centre (nucleus) produces a secondary spherical wave with an amplitude $A_0 f(\theta)$ with the correlation $d\sigma/d\Omega = |f(\theta)|^2$. Alpha particles with a charge +2e show similar trajectories bent in opposite direction. Rutherford investigated in 1911 the elastic scattering of alpha particles on thin foils and concluded that the positive charge and the mass of an atom is concentrated in a very small volume. Therefore, this interaction is called 'Rutherford elastic scattering' with

$$\left(\frac{d\sigma}{d\Omega}\right)_R = \frac{Z^2 e^4}{(8\pi\varepsilon_0)^2 E^2}\left[\frac{E+m_0c^2}{E+2m_0c^2}\right]^2 \frac{1}{\left[\sin^2\frac{\theta}{2} + \frac{\theta_0^2}{4}\right]^2} \tag{1}$$

The dependence of $d\sigma/d\Omega$ on Z^2 and $\csc^4(\theta/2)$ is characteristic for this type of scattering. The additional term with $\sin\theta_0 \simeq \theta_0$ in the last denominator considers the increased screening of the nuclear charge by the inner atomic shells with increasing p ($\theta_0 = \lambda/2\pi r_0$ with $r_0 \simeq a_H Z^{-1/3}$). By integrating over all solid angle elements $d\Omega = 2\pi \sin\theta\, d\theta$ one obtains the total elastic cross section

$$\sigma_R = \frac{Z^2 e^4}{4\pi\varepsilon_0^2 E^2}\left[\frac{E+m_0c^2}{E+2m_0c^2}\right]^2 \frac{1}{\theta_0^2(1+\theta_0^2/4)} \tag{2}$$

with the dimensions of an area. One can represent each atom by a circular target of area σ_R. There are $N=N_A\rho/A$ atoms per unit volume and Nz atoms per unit area in a layer of thickness z. If all atomic cross sections σ do not overlap, one needs $1/\sigma$ atoms for complete scattering corresponding to a layer of thickness $z=\Lambda$ (mean free path length). In reality some of the cross sections σ overlap and only a fraction $1/e$ is scattered, and the number of unscattered electrons decreases as $\exp(-z/\Lambda)$.

Bethe's continuous slowing down approximation

Most of the inelastic scattering processes result from interaction with the electrons in the outer shells or energy levels, and the energy transfer or energy loss of the primary electron is in the order of $\Delta E=5-5o$ eV. Because of this deceleration of electrons in small steps, one can use the Bethe relation [16]

$$\frac{dE_m}{ds}= - \frac{\rho N_A}{2\pi\varepsilon_o^2} \frac{e^4}{A} \frac{Z}{E} \ln \frac{1.66\ E}{J} \ ; \quad J=13,5\ Z \quad (3)$$

for describing the mean energy loss dE_m per element ds along the path length of an electron trajectory. The 'mass thickness' ρs in $g.cm^{-2}$ is a more useful quantity for describing scattering processes than the geometrical path length s or specimen thickness because then the relations do not depend on the density ρ of the material.

However, one has to consider in a multiple scattering theory the statistics of the single energy losses and also larger energy losses by ionisation of inner atomic shells. This causes a 'straggling'[17] of the energy distribution. The Landau and Blunck and Leisegang[18] formulas are approximations to describe the energy distribution of transmitted electrons through transparent films (for comparison with experiments see[19]). One should not use furthermore the 'historical' Thomson-Whiddington law which cannot describe the complex nature of electron diffusion.

Starting with $E_m=E_o$ at $s=o$ (electron entrance) one can calculate the slowing down of E_m as a function of s by integrating equation (3) (Fig.3). The maximum value of ρs for total slowing down ($E_m=o$) is called the 'Bethe range R_B' which increases with increasing atomic number. This Bethe range can be much larger than the practical range R due to the increasing bending of electron trajectories with increasing atomic number.

Electron range and diffusion

The penetration depth of electrons in a solid - the practical electron range R -

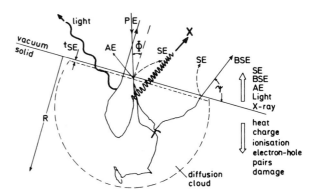

Fig.1 Electron-specimen interactions with three examples of electron trajectories obtained by Monte Carlo calculation (t_{SE}= exit depth of SE, R = electron range)

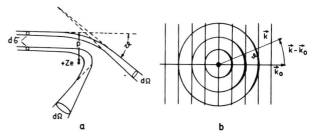

Fig.2 a) Classical model of Rutherford scattering of an electron at a nucleus of charge +Ze to explain the differential cross section $d\sigma/d\Omega$. b) Wave optical model of scattering with a scattered spherical wave with amplitude $A_o f(\theta)$.

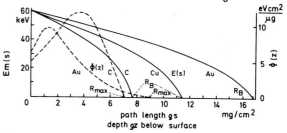

Fig.3 Decrease of $E_m(s)$ of 6o keV primary electrons in C, Cu and Au with increasing path length s calculated with the Bethe continuous slowing down approximation, and depth distribution $\Phi(z)$ of ionisation as a function of depth z (in $\mu g.cm^{-2}$) below the surface for Au and C (Monte Carlo calculations)

depends on the definition used. Experimental values of R can be obtained by measuring the electron transmission into the exit half space as a function of film thickness. Figs.4 show transmission curves for Al and Au with typical differences in shape[20]. If one defines a maximum range, one gets difficulties because of the long tail in the transmission of Au. One can use a definition of R showing 1o% transmission or one can extrapolate the last straight part of the transmission curve to zero (Fig.4). There are only small differences between such definitions of a practical electron range. In contrast to the Bethe range R_B, the practical electron range R is approximately independent of the material, and depends only on electron energy E_0. Such range-energy relations can be approximated by powers of E_0 [20-23]. We use the rule of thumb

$$R = 10 \ E_0^{1.43} \quad (R \text{ in } \mu g.cm^{-2}, E_0 \text{ in keV} \quad (4)$$

to estimate electron ranges for E_0=1-100 keV.

For low Z material the practical range R is of the same order as the Bethe range R_B (see depth distribution of ionisation $\Phi(z)$ for carbon in Fig.3). The probability for large-angle elastic scattering is small due to Z^2 in the Rutherford formula (1). A large fraction of electrons can then pass through the material with only a large number of inelastic small angle scattering events. For carbon the total inelastic cross section $\sigma_{inel} \simeq 3 \ \sigma_{el}$. Increasing Z results in a larger probability for large angle scattering, e.g. σ_{inel}=0.4 σ_{el} for platinum. The electron trajectories are bent more strongly and no electron can penetrate to a depth which is in the order of R_B (compare depth distribution of ionisation $\Phi(z)$ for C and Au with $E_m(s)$ in Fig.3). This diffusion model also explains the increase of the backscattering with increasing Z.

Sometimes one characterizes the shape of the electron diffusion cloud with the shape of a fruit. One can say that for low Z material it has the shape of a pear and for high Z that of an apple or plum. Fig.5 demonstrates this with photographs of the light emission of a diffusion cloud generated by 100 keV electrons penetrating into air and argon at atmospheric pressure (real diameter $\simeq$ 8 cm).

This light emission as well as the generation of heat [5] and the density of electron-hole pairs as a function of depth z is called the 'depth distribution of ionisation' $\Phi(z)$. It shows, at normal beam incidence, a maximum below the surface. Fig.3 contains two $\Phi(z)$ curves for C and Au. Monte Carlo calculations for increasing foil thicknesses show how the ionisation increases with increasing depth due to the decrease of the mean free path for multiple scattering and due to the increase of the ionisation probability of decelerated electrons (Fig.6). The surface value $\Phi(0)$ for small thicknesses increases due to the contribution of backscattered electrons.

Depth distributions and transmission curves show similar shapes for E_0=10-100 keV when reduced lengths are used in units of the electron range R (see the examples for transmission in Fig.4 which are normalized at a transmission of 50%). This also results in approximate energy independent values of the backscattering coefficient. Electron energies E_0>100 keV result in an increased forward scattering, below 10 keV the differences in the backscattering coefficient of low and high Z material becomes smaller (see next section).

Backscattered electrons (BSE)

Backscattering coefficient

The backscattering coefficient η is defined as the ratio of the number (current) of backscattered electrons to the number of the incident electrons, without considering their energy and angular distribution. A comprehensive collection of results and literature about electron backscattering has been published by Niedrig [25]. Fig.7a shows an experimental arrangement [26] to realize exact values of this quantity and the secondary electron yield δ. The outer collector electrode C is always positively biased (+50 V) relative to the grid G of high transparency. This prevents the escape of SE generated by the BSE at the collector. One measures $\eta+\delta$ with a positively biased grid (+50 V) relative to the specimen and only η with a negative bias (-50 V). It is important that the specimen dimension be small compared to the collector dimension so that only a small fraction of multiply backscattered electrons at the collector can hit the specimen again. One never obtains accurate values of η by the specimen current mode in SEM without simulating the above conditions with additional electrodes and grids.

Fig.8 shows the energy dependence of η for different atomic numbers at normal incidence for electron energies E_0<10 keV. The monotonic increase of η with increasing Z (see also Fig.13) is decreased in magnitude for E_0 of few keV [27]. One has to look for contrast effects in low energy SEM which can be explained by this result. In addition, η increases with increasing tilt angle ϕ (Fig.9) [26,28]. One can see the reason from the angular characteristics of BSE (Fig.10) discussed below.

Exit depth of BSE

The exit depth of BSE can be determined by two methods. One can use the increase and saturation of η with increasing foil thickness [26] or one can observe the disappearance of SEM contrast when a material is covered with an evaporated layer of different Z [29]. Both methods result in the order of half the electron range for normal incidence. Similar experiments for tilted specimens and energy filtering will be of interest.

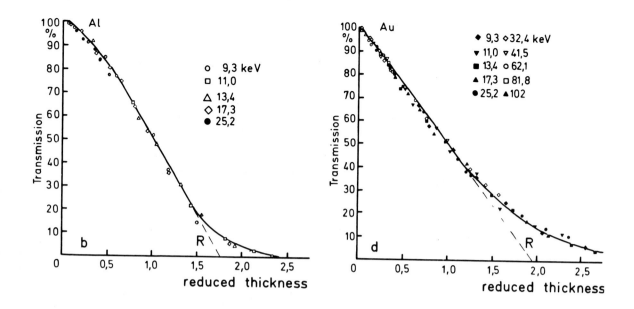

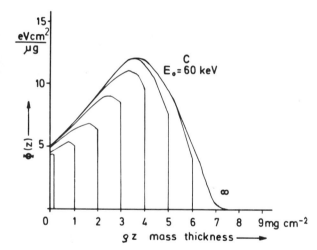

Fig.4 Transmission $T=I/I_0$ of a) Al and b) Au films as a function of a reduced thickness t/t_H with $T(t_H)=50$ %.

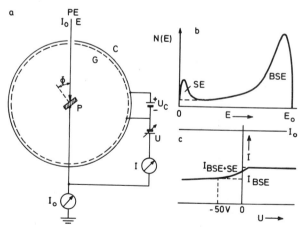

Fig.5 Photographs of the light emission from diffusion clouds of 100 keV electrons in a) air and b) argon gas at atmospheric pressure.

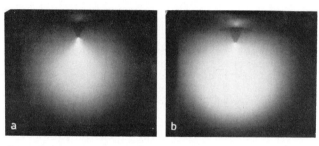

Fig.7 a) Experimental arrangement for measuring exact values of η and δ. b) Schematical energy distribution of SE and BSE, c) retarding field curve.

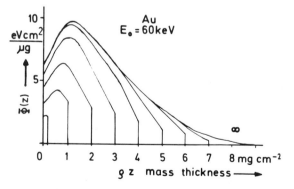

Fig.6 Depth distribution $\Phi(z)$ of ionisation generated by 60 keV electrons in a) C and b) Au as a function of depth z for increasing film thicknesses.

Angular and energy distribution

The angular distribution of BSE at normal incidence can approximately be described by a sin ψ-law (ψ=take-off angle). With increasing tilt angle ϕ one observes a reflection-like angular characteristic (Fig.1o) [28,30]. For large ϕ and low ψ the differential backscattering coefficient $d\eta/d\Omega$ is larger for low Z material and a directional sensitive BSE detector shows a contrast reversal [30].

There are only few experiments reported about the energy distribution of BSE [31,32]. The results in Fig.11 are obtained by a retarding field method. For SEM work more accurate results about the energy distribution for different tilt and take-off angles will be of interest. Especially low-loss electrons with $\Delta E < 10-500$ eV at high ϕ are concentrated at the maximum of the angular characteristic. However, the maximum of low-loss electron characteristic does not coincide with that one without energy filtering [33].

Consequences for SEM imaging with BSE

Because of the angular and energy characteristics, the BSE signal depends on the position and solid angle of the BSE detector. One also has to consider the dependence of the efficiency of the detector on the BSE energy. The signal of scintillators and semiconductor detectors is proportional to E (BSE energy) and shows a low energy cut-off of some keV (dead layer). If one records the BSE by BSE/SE conversion [34] the signal becomes proportional to $E^{-0.8}$ (see energy dependence of SE emission). One of the advantages of BSE imaging is the increase of η with increasing Z, resulting in the well-known Z contrast. There are few experiments which show how the BSE contrast depends on Z for various positions and types of BSE detectors. One has to look for conditions which show a stronger increase of BSE signal, especially for higher Z, or if one wants to get only topographic contrast for those without a significant Z dependence. The use of two BSE detectors [35] or a movable small detector [30,36] seems to be the first attempts to make the best use of BSE. Energy filtering is another experimental possibility for considering the physical laws of BSE emission. The low-loss electrons are scattered in a very small exit depth and can show resolution comparable to that one of the SE [37]. However, selection of solid angle and energy is limited by the decreasing signal-to-noise ratio.

Special contrast effects with BSE

In all experiments and calculations on electron diffusion one does not consider the crystalline state of a solid, and indeed there is no observable difference in the electron range for amorphous and crystalline material. However, one has to consider electron diffraction effects (electron channelling) in a surface layer of some tens of nm. A plane incident electron wave is modified inside the crystal and the exponentially decreasing amplitude ψ of the Bloch wave field shows a periodicity perpendicular to the lattice planes with increasing or decreasing $\psi\psi^*$ at the nuclei positions on the lattice planes depending on the sign of the tilt out of the Bragg position. The large-angle scattering needs a small impact parameter p (Fig.2a) or a large probability $\psi\psi^*$ of the electron wave field at the position of a nucleus. Therefore, backscattering is modulated proportional $\psi\psi^*$ and is larger for a negative tilt and lower for a positive tilt (reciprocal lattice point inside the Ewald sphere).

This dependence of η on crystal orientation can be used for the determination of crystal structures by illuminating with a rocking probe (electron channelling pattern, ECP) [38] or one can detect different crystal orientations in polycrystalline or cold worked material (crystal orientation contrast). The variation $\Delta\eta$ of the backscattering coefficient due to channelling effects is only in the order of some percent and decreases with increasing electron energy [39,40]. Fig.12 shows the variation $\Delta\eta_{max}$ across a 111 pole. There can also be a contrast reversal when changing the tilt and take-off angles. Variations of the BSE signal due to crystal orientation are not only observed when tilting the electron beam but also by changing the take-off angle [30]. The angular distribution of BSE in Fig.10 are modulated by Kikuchi lines and bands. This can be observed as electron back-scattering patterns (EBSP) [41].

The crystal orientation contrast or contrast in ECP increases with increasing high-pass filtering of the BSE because a larger part of the high energy BSE is directly scattered out of the Bloch wave field by Rutherford single and multiple scattering. It will be possible to detect single dislocations by the distortion of the Bloch wave field and the corresponding changes in BSE intensity if energy filtering, black level and field emission guns are used. [42]

The BSE are also influenced by the internal magnetic field in ferromagnetic material due to the Lorentz force (magnetic contrast type II). Maximal variations of η are only of the order of one percent. The largest contrast is observed for tilt angles $\phi \simeq 40-50°$ and increases proportional to $E_0^{1.5}$ [43]. There is also an optimum take-off angle at $\psi \simeq 60°$, whereas the maximum of the topographic contrast is observed at $30°$ [36].

These last examples of special BSE contrast effects demonstrate the importance of a good detector strategy based

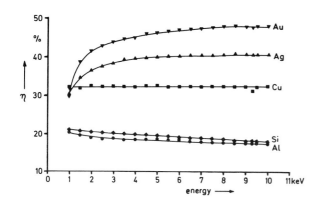

Fig.8 BSE coefficient η as a function of electron energy in the range E_o=1-10 keV

Fig.10 Angular characteristic of BSE for E_o=100 keV (full curves) and E_o=9 keV (dotted lines) at different angles of incidence: a) φ=0 (normal incidence, b) φ=60° and c) φ=80°

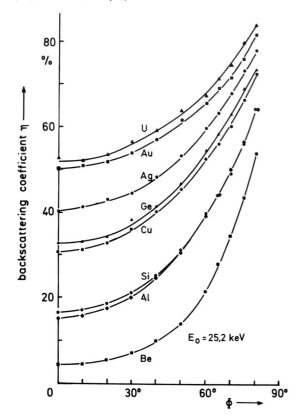

Fig.9 Dependence of η on the specimen tilt angle φ for 25 keV electrons

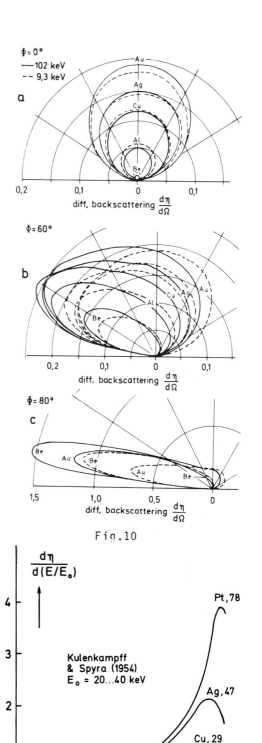

Fig.10

Fig.11. Energy spectrum of BSE (Kulenkampff and Spyra [32])

on the knowledge of electron-specimen interactions.

Secondary electrons

SE yield

The SE energy distribution shows a most probable energy at 3-5 eV [44]. One cannot find a clear limit between SE and BSE in the high energy tail of the SE energy distribution. To use 50 eV is only an arbitrary convention. Indeed one cannot see further changes when filtering an SE image by a negative retarding voltage larger than 10-20 V.

The SE yield δ for normal incidence does not show a monotonic increase with increasing Z like the BSE (Fig.13) [45-47]. There is only an increasing tendency which is caused by the contribution of the BSE to the SE yield. One has to consider that δ can be influenced by a variation of the work function due to different surface contaminations. Results of δ are, therefore, not very reproducible.

One can assume that the SE have a sin ψ-like angular distribution. There are no findings that the energy and angular distributions change with increasing tilt angle.

Exit depth of SE

The exit depth t_{SE} is in the order of 1-10 nm [48]. One knows the exit depth of Auger electrons in the range 10-200 eV. Several experiments confirm a minimum of t in the order of 5 nm at E=100 eV [49]. However, one has to look at the definition of this exit depth. This value for the Auger electrons means the mean free path without a plasmon loss. It is difficult to measure t_{SE} directly. Experiments with evaporated films of increasing thickness [27] show that one gets more information about the growth of the film than on the exit depth.

Dependence of SE yield on ϕ and E_0

δ shows a maximum >1 at primary electron energies in the order of $E_0=E_{max}=$ 500-800 eV [44]. The decrease at higher electron energies can be approximated by δ proportional to $E^{-0.8}$ [10]. The critical energy $E_0=E_c$ which shows $\sigma=\eta+\delta \lessgtr 1$ for $E_0 \lessgtr E_c$ respectively is important for the avoiding of specimen charging. If $\sigma>1$ for $E_0<E_c$ one expects a positive charging of the specimen. SE of lower energies are then attracted, and one observes a kind of self-biased charge neutralization. E_c is only in the order of a few keV. One can increase E_c by tilting the specimen (increase of η and δ) [50]. Another possibility of avoiding charging is the intermediate bombardement with low energy electrons ($E_0 \simeq E_{max}$) during the sweep of the electron beam from one line to the next or during every second or third line [51,52].

One can expect δ proportional to sec ϕ because of the increasing path length of the primary beam in the surface layer of thickness t_{SE}. However, the contribution of the BSE causes a systematic deviation because δ is influenced by the dependence of the number of the BSE and their angular and energy distributions on the tilt angle ϕ. Experimental results (Fig.14) which are normalized to 1 at $\phi=0$ show that typical differences in the $\delta(\phi)$ curves for low and high atomic numbers can be explained by Monte Carlo calculations considering the BSE contribution with a model discussed below [53]. The curve δ_0 represents the contribution of the primary electrons and δ_{BSE} that one of the BSE ($\delta=\delta_0+\delta_{BSE}$).

Contribution of the BSE to the SE yield

The SE yield consists of two contributions, one due to the SE emission of the primary beam and the other one by the BSE on their trajectories through the surface layer of thickness t_{SE}. Experimental results [20] of $\delta(E_0)$ for E=10-100 keV can be approximated by δ proportional to $E_0^{-0.8}$, whereas the dependence $E_0^{-1}\ln(E_0/J)$ in the Bethe law (3) results in $|dE_m/ds|$ proportional $E_0^{-0.84}$ for Be and $E_0^{-0.73}$ for U. These exponents are obtained from the slope of an approximate straight line when plotting this quantity in the same energy interval (10-100 keV) in a double-logarithmic diagram. One can postulate, therefore, that the rate of SE generation is proportional to the rate of energy loss $|dE_m/ds|$ in the exit depth t_{SE}. This means that a BSE with a relative energy E/E_0 and a take-off angle ψ contributes proportional $(E/E_0)^{-0.8}\sec \psi$ to the SE yield. One has to average over the energy and angular distributions of BSE to obtain the mean value of BSE contribution to δ [20]. This results in a $\beta=2-3$ times larger mean value per one BSE at normal incidence and one can write

$$\delta = \delta_0 + \delta_{BSE}= \delta_0(1 + \beta \eta) \qquad (5)$$

with $\delta_0=$ mean value of SE yield per one primary electron. All experimental results, also those for δ on the lower side of a transparent film [54] are in complete agreement to this model. This means, if there is any prefered direction of the primary SE excitation due to a conservation of momentum in the initial ionisation process [55], this is completely changed to an isotropic distribution of the SE trajectories due to multiple scattering before the SE leave the specimen.

For SEM work one has to consider that there is a third source of SE emission by the BSE at the other parts of the specimen chamber. This group can be decreased by a suitable arrangement of carbon foils (low η) and negatively biased grids at the lower pole piece plate or one records only this latter group by using a negatively biased grid over the specimen [20,34].

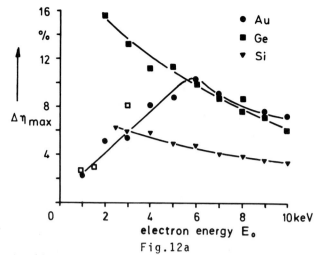

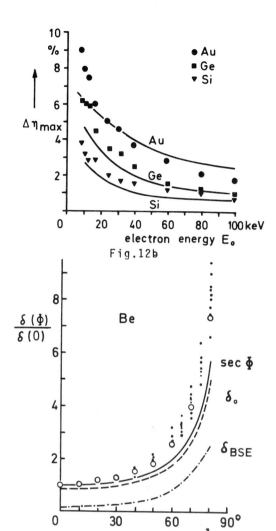

Fig.12a

Fig.12b

Fig.12. Dependence of maximal variation $\Delta\eta_{max}$ of the backscattering coefficient near the 111 pole of 111 oriented Au, Ge and Si single crystals for a) E_o=1-10 keV and b) 10-100 keV

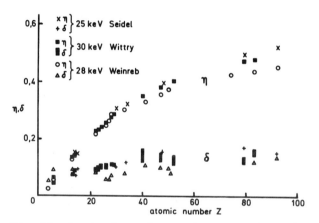

Fig.13. Variation of η and δ as a function of atomic number Z

Fig.14. Measurement of the variation of δ with ϕ (small points) and Monte Carlo calculations (open circles) with the contributions δ_0 of the primary electrons and δ_{BSE} of the backscattered electrons for a) Be and b) U

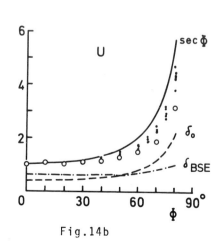

Fig.14a

• experiments

⊙ } Monte Carlo calculations

-·-·-

Fig.14b

Consequences for the imaging with SE

The generation of a larger part of the SE signal inside the electron probe area and the low exit depth demonstrates that the SE mode is the most important one for topography and resolution. Surface facets with different tilt angles ϕ are imaged with different intensities due to $\delta(\phi)$ (Fig.14). The contribution of BSE to the SE signal cannot be avoided. However, one should do more to suppress multiple scattering of BSE in the specimen chamber and their SE contribution by negatively biased grids especially at the lower lens plate. One cannot distinguish between facets of equal ϕ but different azimut angles ψ of the surface normal if all SE are collected. One obtains a better impression of topography if only about half of all emitted SE at $\phi=0$ are detected. One observes more shadow effects if using directional sensitive detector arrangements [56] which will be important not only for increasing topographic contrast but also for the observation of magnetic contrast type I due to the influence of external magnetic stray fields on the trajectories of the SE.

The recording of the SE energy distribution by an electron spectrometer [57,58] allows the determination of surface potentials which is important for the investigation of semiconductor devices.

X-ray emission

Ionisation of inner shells

One needs electron energies $E_0 \geq E_I$ with E_I= ionisation energy of the I=K,L or M shell for ionizing. The primary electron is predominately scattered into small angles θ of some few degrees and the energy loss spectrum shows a steep increase for an energy loss $\Delta E = E_I$ (for further details about energy loss spectroscopy see [59]). The total cross section of K shell ionisation can be described by [60]

$$\sigma_K = \frac{e^4}{8\pi\varepsilon_0^2 u E_K^2} \, b_K \, \ln \frac{4uE_K}{B_K} \qquad (6)$$

with $b_K=0.35$ and $B_K=\{1.65+2.35 \exp(1-u)\}E_K$ and the 'overvoltage ratio' $u=E_0/E_K$ which results in $\sigma_K E_K^2$ const for u=const.

The vacancy in the inner shell can be filled by an electron from an outer shell. The difference of the initial and final energy states either is emitted with a probability ω (fluorescence yield) as an x-ray quantum or is directly transfered with a probability 'a' to another electron which leave the atom as an Auger electron of discrete energy ($\omega+a=1$). The fluorescence yield is very low for low $Z(\omega \simeq 2\times10^{-2}$ for Na K) and increases with increasing Z ($\simeq 0.97$ for Au K). The Auger electrons with energies of some hundred or thousand eV can leave the specimen without energy loss (e.g. plasmon loss) only from a small exit depth.

X-ray emission of a solid

If one wants to obtain the number of emitted x-ray quanta in a solid, one has to calculate the number of ionisation processes on the electron trajectory by considering the Bethe continuous slowing down approximation. The number of emitted x-ray quanta per path element $d(\rho s)$ and per unit of solid angle for an element of concentration c_a will be [61]

$$dN_K = \frac{1}{4\pi} c_a \omega_K \sigma_K p_K \frac{c_a N_A}{A} \, d(\rho s)$$

$$= \frac{1}{4\pi} \omega_K p_K \frac{c_a N_A}{A} \, \frac{\sigma_K}{dE/d(\rho s)} \, dE \qquad (7)$$

(p_K= ratio of intensity of the observed characteristic line to that one of all lines of the K series).

The last denominator in (7) can be obtained from the Bethe law (3) and one has to integrate over all electron energies from E_0 down to $E_m = E_I$

$$N_K = \frac{1}{4\pi} \omega_K p_K \frac{c_a N_A}{A} R \int_{E_K}^{E_0} \frac{\sigma_K}{dE/d(\rho s)} \, dE \qquad (8)$$

The factor R<1 considers the loss in the total number of generated x-ray quanta due to electron backscattering, and depends on the energy distribution N(E) of the BSE (Fig.11). Analogous to the depth distribution of ionisation, there is a depth distribution of x-ray emission $\phi(z)$ with a similar shape but reduced dimensions. Only a fraction

$$f(\chi) = \int_0^\infty \phi(z) \, \exp\{-\chi \rho z\} \, dz \; / \int_0^\infty \phi(z) \, dz \qquad (9)$$

with $\chi = (\mu/\rho) \csc\psi$ leaves the specimen without absorption.

The continuous background of the bremsstrahlung can be calculated by similar methods. Of interest is the peak-to-background ratio using an energy dispersive x-ray detector. Fig.15 shows the increase of this ratio with increasing electron energy and a comparison with a theoretical approximation [61]. However, one works normally with overvoltages u=2-3 to keep small the volume of x-ray emission and the $f(\chi)$-correction (9)

Quantitative x-ray analysis for tilted specimens

For x-ray analysis in an SEM it is important to know the dependence of the quantities in (8) and (9) on the tilt angle. Normally one uses an angle of 90° between electron incidence and take-off direction ($\phi=\psi$). An example of the dependence of x-ray intensity for equal probe current on the tilt angle ϕ is shown in Fig.16. For low ϕ or ψ the decrease is caused by increasing absorption due to (9) and for large ϕ by the decrease of R in (8) due to increasing backscattering and a shift of the mean BSE energy to larger values. It will be necessary, for quantitative analysis, to evaluate analytic approximations for the angular dependence of the correction factors. With the existing formula [63] it is possible to obtain quantitative results of element concentrations by comparison with a pure element standard for tilt angles $\phi = 30-50°$. [64]

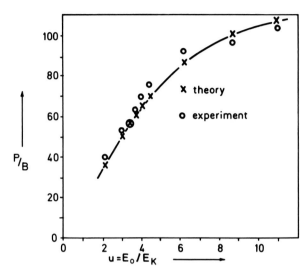

Fig.15 Increase of peak-to-background ratio of the Cu Kα line (E_K=8.047 keV) with increasing overvoltage ratio u (full curve: theory of Albert [62])

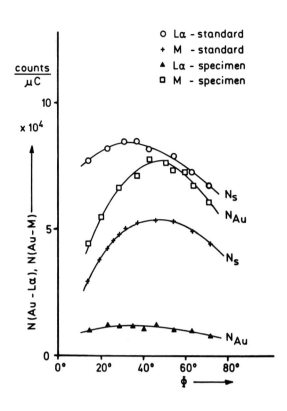

Fig.16 Dependence of energy dispersive x-ray signal on the tilt angle ϕ of the specimen (Au-Cu alloy) and the pure element standard (Au).

References

1. H.E.Bishop,"Monte Carlo calculation on the scattering of electrons in copper", Proc.Phys.Soc. 85, 1965, 855-866
2. L.Reimer,"Monte Carlo-Rechnungen zur Elektronendiffusion", Optik 27, 1968, 86-98
3. K.Murata, T.Matsukawa and R.Shimizu, "Monte Carlo calculations on electron scattering in a solid target", Jap.J.Appl. Phys. 10, 1971, 678-686
4. K.F.J.Heinrich, D.E.Newbury and H.Yakowitz (eds.),"Use of Monte Carlo calculations in electron probe microanalysis and SEM", NBS special Publ. 460, 1976
5. R.Christenhusz and L.Reimer,"Schichtdickenabhängigkeit der Wärmeerzeugung durch Elektronenbestrahlung im Energiebereich 9 bis 100 keV",Z.angew.Phys. 23, 1967, 397-404
6. M.DeMets and A.Lagasse,"An investigation of some organic chemicals as cathodoluminescent dyes in the SEM",J.Micr. 94, 1971, 151-156
7. R.M.Glaeser,"Radiation damage and biological electron microscopy",in Physical Aspects of Electron Microscopy and Microbeam Analysis, p.205-230, John Wiley &Sons New York 1975

8. L.Reimer,"Review of the radiation damage problem of organic specimens", in Physical Aspects of Electron Microscopy and Microbeam Analysis, p.231-246, John Wiley & Sons, New York 1975
9.W.J.Keery, K.O.Leedy and K.F.Galloway, "Electron beam effects on microelectronic devices", SEM/1976/I, IIT Research Institute, Chicago, IL, 60616, p.507-514
10.W.Czaja and J.R.Patel,"Observation of individual dislocations and oxygen precipitates in Si with a scanning electron beam method",J.Appl.Phys. 36, 1965, 1476-1482
11.K.V.Ravi, V.J.Varker and C.E.Volk, "Electrically active stacking faults in Si", J.Electrochem.Soc.120, 1973,533-541
12.H.J.Leamy, L.C.Kimerling and S.D.Ferris "Silicon single crystal characterization by SEM", SEM/1976/I, IIT Research Institute, Chicago, IL, 60616, p.529-538
13. W.Bröcker and G.Pfefferkorn,"Bibliography on cathodoluminescence,part I", SEM/1976/I, p.725-737, "part II", SEM/ 1977/I, IIT Research Institute, Chicago, IL, 60616, p.455-461
14.W.Bröcker and G.Pfefferkorn,"Application of the cathodoluminescent method in biology and medicine", SEM/1979/II, SEM Inc., this volume

15.L.J.Balk and E.Kubalek,"Micron scaled cathodoluminescence of semiconductors", SEM/1977/I, IIT Research Institute, Chicago,IL, 60616, p.739-746

16.H,Bethe,"Quantenmechanik der Ein- und Zwei-Elektronenprobleme", Handbuch d.Physik 24,273-560, Springer, Berlin 1933

17.L.Landau,"On the energy loss of fast electrons by ionization",J.Phys.USSR 8, 1944, 201

18.O.Blunck and S.Leisegang,"Zum Energieverlust schneller Elektronen in dünnen Schichten",Z.Physik 128, 1950, 500-505

19.L.Reimer, K.Brockmann and U.Rhein," Energy losses of 20-40 keV electrons in 150-650 µg cm^{-2} metal films", J.Phys.D 11, 1978, 2151-2155

20.H.Drescher, L. Reimer and H.Seidel," Rückstreukoeffizient und Sekundärelektronenausbeute von 10-100 keV-Elektronen und Beziehungen zur Rasterelektronenmikroskopie", Z.angew,Phys. 29, 1970, 331-336

21.R.O.Lane and D.J.Zaffarano,"Transmission of 0-40 keV electrons by thin films with applications to beta-ray spectroscopy", Phys.Rev. 94, 1954, 960-964

22.J.R.Young,"Penetration of electrons and ions in aluminum," J.Appl.Phys. 27,1956, 1-4

23.V.E.Cosslett and R.N.Thomas,"Multiple scattering of 5-30 keV electrons in evaporated metal films II. Range-energy relations",Brit.J.Appl.Phys. 15, 1964, 1283-1300

24."D.J.Fathers and P.Rez," A transport equation theory of electron backscattering in SEM/1979/I, O. Johari, ed., SEM Inc., AMF O'Hare, IL 60666, USA, this publication

25.H.Niedrig,"Physical background of electron backscattering", Scanning 1,1978, 17-34

26.L.Reimer, H.Seidel and H.Gilde,"Einfluß derElektronendiffusion auf die Bildentstehung im Raster-Elektronenmikroskop", Beitr.elektr.mikr.Direktabb.Oberfl. 1, 53-65, Remy-Verlag, Münster 1968

27.L.Reimer,"Contrast in the different modes of SEM", Scanning Electron Microscopy: systems and applications, 120-125, Institute of Physics, London-Bristol 1973

28.H.Kanter,"Zur Rückstreuung von Elektronen im Energiebereich von 10-100 keV," Ann.d.Phys. 20, 1957, 144-166

29.H.Seiler,"Determination of the "information depth" in the SEM", SEM/1976/I, IIT Research Institute, Chicago, IL, 60616, p.9-16

30.L.Reimer, W.Pöpper and W.Bröcker," Experiments with a small solid angle detector for BSE", SEM/1978/I, SEM.Inc,, AMF O'Hare, IL, 60666, p.705-710

31.E.J.Sternglass,"Backscattering of kilovolt electrons from solids", Phys.Rev. 95, 1954, 345-358

32.H.Kulenkampff and W.Spyra,"Energieverteilung rückdiffundierter Elektronen", Z.Physik 137, 1954, 416-425

33.O.C.Wells,"Measurements of low-loss electron emission from amorphous targets", SEM/1975, IIT Research Institute, Chicago, IL, 60616, p. 43-50

34.S.H.Moll et al,"A high efficiency nondirectional backscattered electron detection mode for SEM",SEM/1978/I, SEM.Inc., AMF O'Hare, IL, 60666, p. 303-310

35.S.Kimoto, H.Hashimoto and T.Sugmana," Stereoscopic observation in SEM using multiple detectors", Proc.Electron Microprobe Symp. 1964, p.480-489, John Wiley & Sons, New York 1966

36. O.C.Wells,"Effect of collection position on type-2 magnetic contrast in SEM", SEM/1978/I, SEM.Inc., AMF O'Hare, Chicago, IL, 60666, p.293-298

37.O.C.Wells, A.N.Broers and C.G.Bremer, "Method for examining solid specimens with improved resolution in the SEM".Appl.Phys. Letters 23, 1973, 353-355

38.G.R.Booker,"Electron channelling effects", in Modern Diffraction and Imaging Techniques in Material Science, p.597-653 , North Holland Publ.Comp., Amsterdam-London 1970

39.H.Seiler and G.Kuhnle,"Zur Anisotropie der Elektronenausbeute in Abhängigkeit von der Energie der auslösenden Primärelektronen von 5-50 keV", Z.angew.Phys. 29, 1970, 254-260

40.H.Drescher et al.,"The orientation dependence of the electron backscattering coefficient of gold single crystal films", Z.Naturforschg. 29a, 1974, 833-837

41.S.A.Venables and C.J.Harland,"Electron backscattering patterns- a new technique for obtaining crystallographic information in the SEM", Phil.Mag. 27, 1973, 1193-1200

42.M.Pitaval et al.,"Advances in crystalline contrast from defects", SEM/1977/I, IIT Research Institute, Chicago, IL, 60616, p.439-444

43.D.J.Fathers et al.,"A new method of observing magnetic domains by SEM", Phys. stat.sol.(a) 20, 1973, 535-544

44.R.Kollath,"Sekundärelektronen-Emission fester Körper bei Bestrahlung mit Elektronen", Handbuch d.Physik 21, 232-303, Springer, Berlin 1956

45.E.Weinryb, Thesis, University Paris 1965

46.D.B.Wittry, in Optique des rayon X et microanalyse (eds.Castaing, Descamps, Philibert), p.168, Hermann, Paris 1966

47.H.Seidel, in Rasterelektronenmikroskopie, 2.edition (eds. Reimer and Pfefferkorn), p.43, Springer, Berlin 1977

48. H.Seiler,"Einige aktuelle Probleme der Sekundärelektronenemission", Z.angew. Phys. 22, 1967, 249-263

49.C.R.Brundle,"The application of electron spectroscopy to surface studies", J.Vac.Sci.Technol. 11, 1974, 212-224

50.T.Ichinokawa et al.,"Charging effects of specimen in SEM"Jap.J.Appl.Phys. 13, 1974, 1272-1277

51.G.V.Spivak et al.,"Des images non alte-
rees des isolants dans une microscope
electronique a balayage", Electron
Microscopy 1972, p.492-493, Institute of
Physics, London-Bristol 1972
52.P.Morin, M.Pitaval and E.Vicario,
"Direct observation of insulators by
SEM",Developments in Electron Microscopy
and Analysis, p.115-118. Academic Press,
London 1976
53.L.Reimer and H.Seidel,"Messung der
Elektronenemission zur Deutung des Kon-
trastes im Raster-Elektronenmikroskop",
Electron Microscopy 1968, Vol.I, p.79-80
Tipografia poliglotta Vaticana, Rome 1968
54. L.Reimer and H.Drescher,"Secondary
electron emission of 10-100 keV electrons
from transparent films of Al and Au",
J.Phys.D 10, 1977, 805-815
55.V.N.E.Robinson,"The dependence of
emitted secondary electrons upon the
direction of travel of the exciting
electron", J.Phys.D 8, 1975, L74-L76
56.J.R.Banbury and W.C.Nixon," A high
contrast directional detector for the
SEM", J.Phys.E 2, 1968, 1055-1059
57.W.J.Tee and A.Gopinath,"Improved
voltage measurement system using the SEM",
Rev.Sci.Instr. 48, 1977, 350-355
58. L.J.Balk et al.,"Quantitative voltage
contrast at high frequencies in the SEM",
SEM/1976/I, IIT Research Institute, Chicago, IL, 60616, p. 615-624
59.M.Isaacson,"All you might want to know
about ELS (but are afraid to ask):A
Tutorial",SEM/1978/I, SEM.Inc., AMF O'Hare
IL, 60666, p.763-776
60.C.R.Worthington and S.G.Tomlin,"The
intensity of emission of characteristic
x-radiation",Proc.Phys.Soc.A 69, 1956,
401-412
61.S.J.B.Reed,"Electron Microprobe Analysis", Cambridge University Press,
Cambridge 1975
62.L.Albert,"Photonen-Ausbeute, Signal/Un-
tergrund-Verhältnis und Nachweisgrenzen
elektronenangeregter Kα-Linien", Beitr.
elektr.mikr.Direktabb.Oberfl. 5, 109-
128, Remy-Verlag, Münster 1972
63.G.Love, M.G.Cox and V.D.Scott,"A versatile atomic number correction for
electron-probe microanalysis", J.Phys.D
11, 1978, 7-21
64.B.Lödding and L.Reimer,"Energy dispersive x-ray microanalysis on tilted specimens using a modified ZAF correction",
Scanning 1, 1978, 225-229

Discussion with Reviewers

D.F.Kyser: What are the physical limitations in the Bethe equation (2) for the
application you discuss ? Are there any
good alternative theories for energy loss
to be used ?
Author: The physical limitation of the
Bethe equation is that it does not consider the 'straggling' of the energy distribution. It is an approximation for the
mean energy remaining after travelling a
distance s. Especially in high Z material
one has to consider the increase of path
length in a layer of thickness z. This
effect decreases the mean energy (see text
reference 19).This reference also contains
comparisons of measured transmission
energy spectra with the Landau and Blunck
and Leisegang formulas. However, there
exist no good formula to describe the
energy distribution of transmitted electrons.

D.C.Joy: For the benefit of those people
who do not work in this area could you
please explain what is implied by Monte
Carlo, and Transport Theory methods of
modelling electron-solid interactions ?
Author: Monte Carlo methods use the
atomic scattering cross sections and
energy losses for single atoms. The large-
angle scattering is considered by single
Rutherford scattering. The low-angle
scattering is considered by a multiple
scattering formula and the energy losses
by the Bethe continuous slowing down
approximation. Scattering angles and
mean free pathes between scattering processes are calculated by correlating a
random number with the probability function of angle and path length. One calculates some thousands of electron trajectories. One single trajectory as shown in
Fig.1 has no practical meaning, because
never an electron will travel the same
way, but the statistics over some thousands of trajectories show comparable
results if asking for backscattering,
energy distribution of BSE etc.

The Transport Theory starts with the
Boltzmann transport equation. The solution
can be reduced to a system of first-order
coupled differential equations and solving
an eigenwert problem (diagonalisation of
a matrix)(for further details see the
contribution of Fathers and Rez in this
proceeding). If one want to make accurate
calculations it seems to be also a time-
consuming method like the Monte Carlo-
method. However, this method is an alternative one has to look for in the future
to solve simple problems in a short
computation time.

Both methods have the disadvantage
that they do not result in analytic
solutions. One obtains a "collection of
stamps" and has to recalculate each new
problem.

V.N.E. Robinson: Would you please explain the Gunshot Residue (GSR) Particle Technique and give a greater description of the advantages to be gained from BEI, particularly in terms of time factor savings?

V.R. Matricardi: What magnification is used for the search of GSR particles?

Authors: When a firearm is discharged, the hands and clothes of the shooter are sprayed with microscopic particles originating in the primer of the cartridge. These particles, usually 1-50μm in diameter, are identified as GSR by the combination of their morphology and chemical composition. The hands of a suspect are sampled by an adhesive-coated aluminum disk, which is then examined in the SEM. The disk is searched at a magnification of X800-1000, and particles of suitable morphology are analyzed by X-ray EDS. Since GSR particles are at best a tiny fraction of the overall population, the search might cover hundreds of particles before one GSR particle is found; a single sample could demand a full SEM workday before a conclusion was reached. However, practically all GSR compositions contain one or more of the heavy elements lead, barium and antimony; while the "innocent" particles are mostly made of silicon, calcium, chlorine and similarly light elements. Using BEI, one can adjust the SEM so that only elements above atomic number 25 appear on the screen, so that GSR appears bright on a completely black background. Provided this can be done at TV rate, the search becomes faster by a factor of perhaps 20, as well as more reliable (i.e., the probability of GSR passing undetected is practically eliminated). The introduction of the detector described has indeed permitted us to distribute sampling kits to many more police units than before, and to cope with the increased number of required examinations within the same SEM time allocation.

V.R. Matricardi: Would it be possible to design two light guides which simultaneously impinge on the PMT? This would allow viewing the specimen by BEI (by turning off the bias) and by SEI and BEI when the bias is turned on; all without altering the physical structure.

Authors: This undeniably elegant solution has been considered, and it was decided not to attempt it, for the following reasons:
a. The proposed method would not permit using SEI alone; since SEI still has its advantages (e.g., at very high magnifications), it should not be permanently cancelled.
b. In our SEM, physical alterations in the SEI system would be necessary, and might result in reduced image quality in the SEI mode. It was our intention to leave the SEI system unchanged in order to ensure against any unpredictable side-effects.
c. The SEI light guide would have to retain its cage, forcing us to route the BEI guide around it; this would cause losses of the BEI signal in the longer, curved path.

R. Keeley: Backscattered imaging requires higher probe currents than are used for SEI. Do you experience difficulties in EDX analysis due to excessively high count rates?

Authors: With the detector described, BEI does not require higher probe currents. We use precisely the same beam parameters (KV, probe size and current) for BEI as we do for SEI, and the X-ray count rate is quite adequate.

V.N.E. Robinson: You indicate that conversion from SEI to BEI and vice versa is quite quick. How often do you use SEI compared to BEI in your forensic studies?

Authors: We use BEI in cases involving the applications described, which form perhaps 50% of our SEM casework. The change between BEI and SEI is made once or twice a day.

SCANNING ELECTRON MICROSCOPY/1979/II
SEM Inc., AMF O'Hare, IL 60666, USA

APPLICATIONS OF THE CATHODOLUMINESCENCE METHOD IN BIOLOGY
AND MEDICINE

W. Bröcker and G. Pfefferkorn

Institut für Medizinische Physik
Universität Münster
Hüfferstr. 68
D-4400 Münster
West-Germany

Abstract

Applications of cathodoluminescence (CL) in scanning electron microscopy (SEM) for biomedical problems can be categorised as being produced by autoluminescence or by staining methods. In the first, one uses direct investigation of luminescent specimens or parts of a specimen. Examples include localization of herbicides on leaves and the distribution of calcified regions in human aortae.

CL dyes have been added to filling materials to study their spread in tooth root canals. Immunocathodoluminescence permits investigations similar to those of immunofluorescence microscopy. Antigen-antibody reactions can be demonstrated with the CL mode in the SEM. The passage of marked macromolecules from capillaries into the extravascular space can also be shown.

Problems of the CL mode derive from backscattered electrons that can produce light in parts of the CL equipment and thus disturb the CL picture. Beam damage causes many of the luminescent materials which have been used in biomedical investigations to lose some of their CL intensity under electron bombardment. The application of a coating layer reduces the detected light. Some new preparation methods may therefore need to be developed.

However the CL method in SEM has the advantage that sensitive detector systems can be used. It is also possible to combine the CL mode with other imaging or analytical equipment of the SEM.

KEY WORDS: Cathodoluminescence, Autoluminescence, Calcifications, Staining Methods, Immunocathodoluminescence, Microcirculation, Fluorescence Microscopy, Biology, Medicine

Introduction

In the last few years cathodoluminescence (CL) has become more and more successful in the scanning electron microscope (SEM) even in the field of biomedical applications. The first CL micrograph of a biological sample was published by Pease and Hayes (1966).[1] They treated spinach leaves with the dye thioflavine T and could observe a selective CL in the cell walls. Since then more new different kinds of applications have been published (see e.g., the different review papers Bröcker (1976)[2], Herbst and Hoder (1978)[3] and our CL bibliography in the last proceedings[4]). In this paper we present a review of this method by showing some selected applications and discussing some problems.

CL may be defined as light emission in the visible, near ultraviolet (UV) and infrared (IR) regions due to excitation by electrons. This effect is similar to excitation by UV light and can also be distinguished as fluorescence and phosphorescence (afterglow). There are some advantages of the CL mode in comparison to UV fluorescence microscopy:
a) Some substances exhibit luminescence only under electron irradiation.
b) Even weak CL intensities can be considerably amplified.
c) The CL signal can be better measured quantitatively.
d) The CL image can be correlated with secondary electron (SE), backscattered electron (BE) or transmission electron (TE) images or with X-ray analysis. SEM images have a higher resolution than light microscopical images.
e) The depth of focus and the resolution in the CL mode are higher than in the fluorescence microscope.

The wavelength of the emitted light can also be analysed. This would be useful for some autoluminescence problemes, for double staining techniques and for separating staining from autoluminescence.

The advantages of the CL mode should make it useful in the study of certain biomedical problems.

Detector Systems

In early applications of the CL mode the emitted light was collected by an optical microscope in an electron micro-probe. The first detector systems developed especially for the SEM consisted of a light pipe near the specimen which was connected to a photomultiplier. Commercially availabe scanning electron microscopes often have a lens for focussing the emitted light onto a light guide and thus onto the entrance of a photomultiplier.

These systems are not very efficient for biomedical specimens, which often demonstrate very weak luminescence. Several different mirror systems have been proposed in the last few years, of which the most recent is very effective for thin transparent specimens.[5] A good summary of the different photon collecting systems will be found in the paper by Hörl.[6] In most of our works we have used the elliptical mirror developed by Hörl and Mügschl,[7] because this detector system is very efficient for compact specimens and for thin specimens mounted on polished aluminium stubs which reflect the CL light.

For analysing the wavelength of the emitted light filters or monochromators can be used. The high spectral resolution of a monochromator is often not necessary for biological specimens. Therefore an investigation with a continuous interference filter can be sufficient. An example of this is shown in Fig. 1.[8] The CL spectra of fluoresceinisothyiocyanate (FITC) labelled serum (see below) and an un-labelled serum are investigated with a continuous interference filter (Fig. 1a) and with a monochromator (Fig. 1b). There are no significant differences between the two diagrams.

For future work two other parts of the whole detector system should be improved. The photomultiplier should have a high sensitivity all over the regions of interesting wavelengths. Furthermore cooling the photomultiplier from room temperature to -25°C leads to a 100 to 1000 times reduction of the dark current.[9]

The Lock-In technique as used successfully for semiconductor application[10] seems to be a very useful amplifier system for some biomedical problems. First experiences in our laboratory confirm this. Some of the advantages of this system will be discussed in a later section.

Preparation Techniques

The accommodation of the specimen preparation technique to biomedical CL investigations is very important, because the preparation steps must not change any luminescence (autoluminescence or staining) in the specimen.

The staining process is often the first preparation step. Most specimens, however, have been fixed, dehydrated, dried and coated as is usual in SEM work.

Fixation

Since certain fixatives can, among other things, increase the autolumines-cence of a tissue or decrease the CL of a dye[11], the best fixation method must be found empirically in each new experiment. In some investigations, different fixation techniques have been tested. Glutaraldehyde seems to be the best fixative for CL investigations of vaginal cells.[11,12] This finding corresponds with results obtained with mesenteric tissues.[13] These conclusions, however, can not automatically be applied to all specimens.

Dehydration and drying

Air drying does not change dye distribution, but the conservation of the tissue structure is not good. Critical point drying is also successful in CL work, and allows the detection of fluorescein labelled anti-IgG-induced caps on mouse B lymphocytes and brilliant sulfoflavine staining of mesenteric tissue from rabbits.[13,14]

Using cryostat sections air drying or drying in the vacuum of the SEM is not so serious. Therefore in our early CL studies we used cryostat sections of mouse kidney mounted directly on aluminium stubs.[15,16] These stubs were polished so that no surface details of the stub could impress through the section and so that the light emitted downwards could be reflected into the CL detector. The penetration of most of the electron beam (at a primary energy of 20 keV or more) led to a good conductivity without the need to apply any coating. For other problems cryostat sections were mounted on nylon foils (stretched over an 1cm diameter aluminium ring), which were then dried in the cryostat and subsequently coated with aluminium or carbon. This specimen preparation technique was designed for electron microprobe investigations, but has also been successfully applied to the CL method.[17,18]

Another direct method is to use a cold stage in combination with a transfer line and a cold finger, when no preparation steps except the rapid freezing of the specimen are necessary.

No chemical fixation or dehydration is necessary if freeze drying technique is used. Specimens are rapidly frozen in liquid nitrogen at its melting point and freeze dried at -80°C.

Embedding

Usually sections were cut from embedded specimens. But unfortunately all of the well known embedding compounds luminesce more or less. Epon has the lowest CL intensity in comparison with most of the other important compounds and

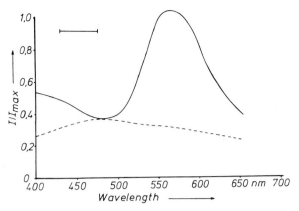
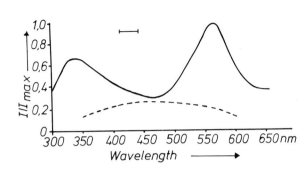

Fig. 1: CL spectra of a FITC-serum and a pure serum without FITC (dashed line), a: measured with a continuous interference filter, b: with monochromator, ├─────┤ = spectral response of the filter and monochromator respectively.[8]

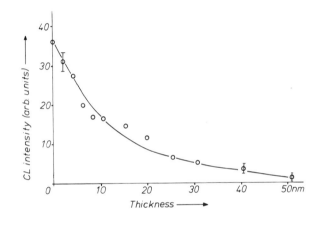

Fig. 2: CL intensity as a function of the thickness of a gold evaporation layer.[8]

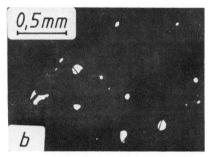

Fig. 3: SEM micrographs of a turkey mullein leaf surface treated with the herbicide MCPA (from Hess et al., 1974).[20] a: SE image, b: CL image and c: composite micrograph of SE and CL showing the distribution of MCPA on leaf surface.

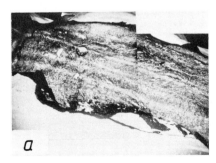

Fig. 4: About 5µm thick unstained cryostat section of human aorta with stenosis of the isthmus (air dried).[17,18] a: TE image, b: CL image. The bright regions in (b) are calcified.

0,5 mm

Table 1: Possible types of applications of the cathodoluminesce method in biology and medicine

1. Autoluminescence: Direct investigations of luminescent specimens without staining (e.g., herbicides, lichens, calcifications or nucleic acids)[5,17,18,20,21,22]

2. Stainings Methods:

    a) Demonstration of clefts, splits, cracks or fissures   } e.g. with

    b) Examination of the depth of penetration of certain substances[23]   } optical brightener

    c) Production of luminescent precipitates (e.g., detection of SO_4^{2-}; not yet realized)

    d) Immunocathodoluminescence (e.g., FITC)[14,15,16]

    e) Staining of cells, tissues and organs by direct addition of specific luminescent dyes (e.g., for staining of nuclei)[1,11,12,13,24]

---

is therefore most suitable.[8] The beam damage effect will be discussed in a later section.

Coating

Most of the dried biomedical specimens must be coated to avoid charging in the SEM. This is necessary for high resolution SE images. Coating layers diminish, however, more or less the emitted CL. Fig. 2 shows the CL intensity as a function of the thickness of a gold evaporation layer.[8] The application of a 10nm thick layer reduces the CL intensity to about 50%. Using a carbon film, however, the thickness can be twice as much, but a double thickness of carbon does not reduce charging as effectively as a single thickness of gold, and therefore offers no advantage.[19] Charging is not often an important problem for CL image alone. If it is possible to use TE or BE, rather than SE images for comparison purposes, a thinner coating layer can be used.[13,17] On the other hand one can make first the CL pictures of the uncoated specimen, then coat the specimen and make the high resolution SE pictures.

Selected biomedical applications

Table 1 is a summary of different types of biomedical application of the CL method in SEM. In the following we shall discuss a few applications and point out the advantages of the method in each case.

Autoluminescence

Autoluminescence means that specimens or parts of specimens luminesce under electron bombardment without the application of any dye. This technique could become increasingly important for substances which luminesce under electron but not under UV light irradiation. An earlier example of this phenomenon is shown by the herbicides like MCPA ([(4-chloro otolyl)oxy]acetic acid).[20] Fig. 3a shows the SEM image of a turkey mullein leaf surface treated with MCPA and Fig. 3b the corresponding CL image. Fig. 3c is a composite micrograph of both SE and CL images showing the distribution of the herbicide on the leaf surface.[20]

In a different application, the autoluminescence of substances in lichens was studied.[21] The CL mode allows one to localize these substances (e.g., xanthones), which have also been analysed by chemical means and by X-ray microanalysis. It has been possible to discriminate the CL of different crystals inside the lichens using different colour filters.[21]

Fig. 4 shows two SEM micrographs of an unstained cryostat section of a human aorta from a patient suffering from a stenosis of the isthmus.[17,18] The specimens were air dried and than coated with Al (see "Preparation techniques"). They were prepared for a later quantitative analysis of Ca, P and other important elements in an electron microprobe. Fig. 4a shows a TE image and Fig. 4b the corresponding CL image of the same section. The autoluminescence of the calcified regions is clearly shown. Our investigations show that the CL of calcium-phosphate compounds is a more sensitive indicator for these substances than the normally used stain alizarin red S. Moreover, the CL studies can be made using the same specimen as is used in an electron microprobe. It is an advantage that no parallel sections and no staining are necessary. Furthermore, we can get corresponding pictures of the tissue surface with the BE mode or pictures of the mass distribution with the TE mode (Fig. 4a).[17]

Another important example of autoluminescence is given by the nucleic acids.[5,22]

Staining Methods

We distinguish five different staining methods in Table 1. It is possible to localize clefts, splits, cracks

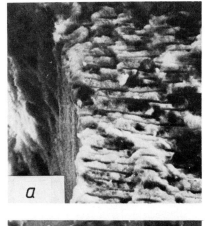

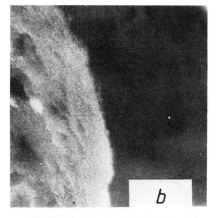

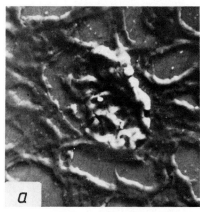

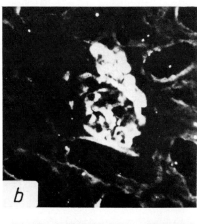

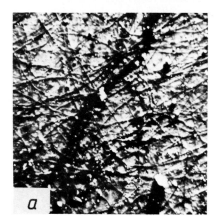

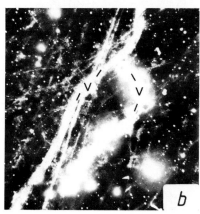

or fissures in certain kinds of specimens. The specimen is saturated with a solution which contains a strongly luminescent substance (e.g., optical brightener, Bayer AG, Leverkusen, W. Germany) and then dried. CL then only appears in the clefts or cracks which existed before the staining. If cracks or clefts are generated in later preparation procedures, they demonstrate no CL.

In other cases, penetration of certain substances in naturally preexisting spaces can be demonstrated. For example, a viscous dental root canal filling material was labelled with an optical brightener and the failure of this material to penetrate dentine tubules could be shown (Fig. 5a and b).[23] The specimen is a fractured tooth, carbon coated.

Soni et al. have been able to demonstrate CL of fluorescein labelled anti-IgG-induced caps on mouse B lymphocytes. The resolution obtained in the CL mode was the same as that using UV fluorescence microscopy. However, only the CL mode allowed the cap areas to be distinguished with certainty. This is not possible with the normally used SE or BE modes in the SEM.[14]

We have examined the deposition of fibrin in rat kidneys after the infusion of an endotoxin.[15,16] The fibrin was made visible in the SEM by treatment with fluoresceinisothiocyanate (FITC) labelled anti-rat-fibrinogen serum, which luminesces very strongly. The strong CL of the fibrin deposits in the glomeruli can be seen in the CL picture of an uncoated cryostat section of rat kidney prepared in this way (Fig. 6). The last two examples were called immunocathodoluminescence similar to the immunofluorescence microscopy.

As already mentioned, thioflavin T was used to stain spinach leaves in the very first CL investigation in the biomedical field.[1] Since then, many different dyes have been investigated for CL. Quinacrine and acridine orange have been successfully used for staining cells in vaginal smears.[11,12] Herbst et al. could prove that the cell nuclei were clearly apparent in liver sections stained with fluorescein sodium in CL, whereas there was no emission under UV light.[24]

The following new application of CL in our laboratory is founded on studies using blue light vital fluorescence microscopy. These studies showed that there is a movement of dye labelled proteins from capillaries into the extravascular space in rabbit mesentery, followed by its movement along the connective tissue fibres. For corresponding CL studies, proteins labelled with the dye brilliant sulfoflavine were injected intravenously. After in situ fixation and removal, by dissection, of the mesenteric tissue, these were dehydrated, critical point dried and gold coated. Fig. 7a shows the BE image and Fig. 7b the corresponding CL image of a rabbit mesenteric tissue sample prepared in this way.[13] The passage of the labelled protein into the extravascular space is shown by the secondary luminescence around the microvessels. The autoluminescence of the tissue constitues a problem, but this luminescent halo distinguishes CL pictures of stained tissues from unstained controls. Another new result is the finding that autoluminescence seems to be seen only from the elastic fibres and not from the collagen fibres.[13]

Problems of the CL Mode

The above mentioned applications extend the analytical methods in the biomedical field available in an SEM. However some problems of this method must be discussed.

Problem of intensity

A technical problem of the CL mode is that the intensities available from biomedical specimens are often very low, and the signal to noise ratio is therefore very small. New detector systems mentioned above have diminished these difficulties. Future improved amplifier systems, e.g., using Lock-In techniques, could improve the CL signal again.

Stray light

Light from the heated tungsten filament can pass down the column and be reflected from the specimen into the CL detector.[25] Reflection varies with the orientation and the reflective properties of the specimen surface. Using polished aluminium stubs, the intensity of the reflected light from the filament can be twice as high as the normal background noise from the amplifier system.[8] If this effect proves to be important in CL investigations, this background can be removed by tilting the complete gun or for low magnifications putting a thin aluminium foil in one of the apertures so that light cannot pass through.[25,8] Furthermore this problem is completely eliminated using the Lock-In technique or by using a cool field emission gun.

Backscattered electrons

Primary electrons can be backscattered from the specimen into the lens or the light guide of the CL equipment. They generate photons which contribute to the CL signal. This effect does not normally lead to a constant background, but to a so called "CL" image which appears to be similar to the BE image from the same area, and, as such, it may be detected. To avoid this effect lenses and light guides should be placed as far as possible from the specimen, and the beam current and the acceleration voltage should be as low as possible. Using the detector developed by Hörl, this effect is small because electrons have to be reflected twice before they can enter the light guide.[8,15] Using only mirrors without any light guides much less CL can be produced by BE.

Beam damage

Negative oxygen ions can pass from the electron gun through the column to the specimen. Reduction of CL intensity due to interaction with negative oxygen ions, however, could only be demonstrated in semiconductor material.[26] In comparison with this most of the organic specimens are damaged by the scanning beam in such a way that the CL intensity is diminished after a short electron irradiation time. This is a real disadvantage of the CL method. For example with plastic scintillators, the CL intensity is half of its original value after 5.10^{-5} Amp seconds per cm^2.[27] On the other hand CL intensity increases with irradiation time with certain embedding compounds.[8] Beam damage effects can somethimes be delayed or diminished by cooling the specimen to a very low temperature. However, these effects must be tested anew for every new kind of specimen because beam damage and cooling effects differ for different specimens and cannot presently be reliably predicted.

Summary

Often weak CL intensities, coating layers, stray light, reflected fast electrons and beam damage diminish the resolution which is obtainable using the CL mode. The known investigations of the different laboratories show, however, that the possible resolution of the CL mode can be better than in a fluorescence microscope. Despite of the above numbered disadvantages the CL method in an SEM is an useful adjunct for solving problems in the biomedical field.

Acknowledgements

We thank Dr. E. Wolfgang, Siemens Forschungslaboratorien München, for the measurements using a monochromator and the Deutsche Forschungsgemeinschaft for support.

References

1. R.F.W. Pease, T.L. Hayes "Scanning electron microscopy of biological material" Nature 210, 1966, 1049.
2. W. Bröcker "Biologisch-medizinische Anwendungen der Kathodolumineszenz am Raster-Elektronenmikroskop" Microsc. Acta 78, 1976, 105-117.
3. R. Herbst, D. Hoder "Cathodoluminescence in Biological Studies" Scanning 1, 1978, 35-41.
4. W. Bröcker, G. Pfefferkorn "Bibliography on cathodoluminescence" in SEM/1978/I, SEM Inc., AMF O'Hare, IL, 60666, 33-351.
5. W.R. McKinney, P.V.C. Hough "A new detector system for cathodoluminescence microscopy" SEM/1977/I, IIT Research Institute, Chicago, IL, 60616, 251-256.
6. E.M. Hörl "Cathodoluminescence - actual state of instrumentation" Microsc. Acta, Suppl. 2, 1978, 236-248.
7. E.M. Hörl, E. Mügschl "Scanning electron microscopy of metals using light emission" in Proc. 5th Europ. Congr. Electron Microsc., Inst.of Physics, London, 1972, 502-503.
8. W. Bröcker "Kathodolumineszenz im Raster-Elektronenmikroskop" Thesis, Universität Münster, 1977.
9. M. Cole, D. Ryer "Cooling PM tubes for best spectral response" Electro Optical Systems Design, Milton S. Kiver Publications, Inc., 1972, 16-19.
10. L.J. Balk, E. Kubalek "Micron scaled cathodoluminescence of semiconductors" in SEM/1977/I, IIT Research Institute, Chicago, IL, 60616, 739-746.
11. E.H. Schmidt, W. Bröcker, M. Pfautsch "Cathodoluminescence properties of frequently used fluorescent stains in vaginal smears" in SEM/1976/II, IIT Research Institute, Chicago, IL, 60616, 327-334.
12. W.A. Barnett, E.L. Jones, M.L.H. Wise "Observations on natural and induced cathodoluminescence from vaginal epithelial cells" Micron 6, 1975, 93-100.
13. W. Bröcker, G. Hauck.,R. Blaschke et al. "Cathodoluminescence technique combined with blue light fluorescence microscopy for the investigation of substance transport through the extravascular space of mammalians" Microsc. Acta, Suppl. 2, 1978, 260-270.
14. S.L. Soni, V.I. Kalnis, G.H. Haggis "Localisation of caps on mouse B lymphocytes by scanning electron microscopy" Nature, 255, 1975, 717-719.
15. W. Bröcker, E.H. Schmidt, G. Pfefferkorn, F.K. Beller "Demonstration of cathodoluminescence in fluorescein marked biological tissues" in SEM/1975, IIT Research Institute, Chicago, IL, 60616, 243-250.
16. E.H. Schmidt, W. Bröcker, W. Wagner et al. "Glomerular microcapillary thrombosis demonstrated by the new technique of immunocathodoluminescence" Amer.J. Pathology 81, 1975, 43-48.
17. W. Bröcker, H.J. Höhling, W.A.P. Nicholson et al. "Comparison of the methods of cathodoluminescence, electron probe microanalysis and calcium staining applied to human aorta with isthmus stenosis" Path.Res.Pract.163, 1978, 310-322.
18. W. Schlake, W. Bröcker., H.J.Höhling, B. Drüen "Zur Methodik kombinierter Kathodolumineszenz-Mikrosonden-Untersuchungen und Calciumfärbungen an Aorten mit Stenose" Verh.Dtsch.Ges.Path.60, 1976,446.
19. M.D. Muir, P.R. Grand, G. Hubbard, J. Mundell "Cathodoluminescence spectra" in SEM/1971, IIT Research Institute, Chicago IL, 60616, 401-408.
20. F.D. Hess, D.E. Bayer, R.H. Falk "Herbicide dispersal pattern.I. As a function of leaf surface" Weed. Sci., 22, 1974, 394-401.
21. D. Hoder, A. Mathey "Cathodoluminescence of the lichens Lecanora, Buellia and Laurera in the SEM" Microsc.Acta, Suppl 2, 1978, 271-280.
22. P.V.C. Hough, W.R. McKinney, M.C. Ledbetter et al. "Identification of biological molecules in situ at high resolution via the fluorescence excited by a scanning electron beam" Proc.Nat. Acad.Sci. USA, 73, 1976, 317-321.
23. W. Kovermann "Untersuchung des Wurzelfüllmaterials Biocalex mit dem Raster-Elektronenmikroskop" Thesis, Universität Münster, 1973.
24. R. Herbst, E.M. Hörl, A.M. Multier-Lajous "Scanning electron microscopy of biological material using the luminescent mode" Beitr.elektronenmikr.Direktabb. Oberfl. 6, 1973, 169-176.
25. G.H. Haggis, E.F. Bond, R.G. Fulcher "Improved resolution in cathodoluminescence microscopy of biological material" J.Microscopy 108, 1976, 177-184.
26. W. Bröcker, L. Reimer "Specimen Damage by Negative Oxygen Ions from the

SEM Cathode Detected by CdS Cathodoluminescence" Scanning, 1, 1978, 60-62.
27. W. Bröcker, E.R. Krefting, L. Reimer
"Beobachtung der Strahlenschädigung während des Abrastvorganges im Raster-
Elektronenmikroskop mit Hilfe der Kathodolumineszenz" Beitr.elektronenmikroskop.
Direktabb.Oberfl. 7, 1974, 75-88.

Discussion with Reviewers

G.H. Haggis: For progress in biomedical CL
work it seems vitally important that we
should gain more theoretical understanding
of the reasons some fluorescent materials
give better light emission than others,
under electron beam excitation, and more
understanding of the molecular mechanisms,
for example, by which aldehyde fixatives
might be used to enhance and stabilize
the light emission of CL stains. Do you
have further comments on this aspect?
Also could you give references to the
latest reviews and research papers which
contribute to greater theoretical understanding of CL emission?
Authors: There are indeed rather few
papers in which the theoretical background
of cathodoluminescence of organic molecules is discussed, because this is a very
difficult problem. Furthermore only practical investigations exist which fixatives
stabilize the CL stains and which not (see
text references 11, 12 und 13). M. De Mets
gives some explanations in his papers why
some certain organic compounds emit CL
(e.g.: M. De Mets "Relationship between
cathodoluminescence and molecular structure
of organic compounds" Microsc. Acta, 76,
1975, 405-414).

M. De Mets: Present CL work is to be done
on adapted instruments. CL results would
be much better if SEMs especially built
for this type of work were availabe. Can
you agree with that?
Authors: Yes, we fully agree with your
statement! This special instrument should
be a high resolution SEM with some built-
in additional equipments, e.g. a high
sensitive CL detector and a cooled photomultiplier, a lock-in amplifier system,
two visual screens and a signal mixing
unit.

W.R. McKinney: Do you have any other
criteria for choosing accelerating voltage
besides the specimen charging and backscattered electron CL effect?
Authors: Yes, we do: The acceleration
voltage alters the thickness of the surface layer from which CL is emitted. Lower
voltage leads more to a surface analysis
than higher voltage does. However, the
higher the acceleration voltage, the higher
the CL intensity from a bulk specimen (see
text reference 8).

W.R. McKinney: In your studies of frozen
cryostat sections, have you ever detected
CL from ice crystals?
Authors: The cryostat sections which are
shown in this paper (Fig. 4 and Fig. 6)
are dried, and therefore no ice crystals
could be observed. In other investigations with a cold stage we could not
detect CL from ice crystals.

W.R. McKinney: Does the amount of dehydration effect the intensity of your
CL signals for any cathodoluminescent
substance that you have used?
Authors: No, we did not observe such an
effect.

H.W. Carter: Have decalcified specimens
been observed with CL? Does the CL
persist?
Authors: No, but we investigated unmineralized specimen under the same
conditions as the specimens with
stenosis. There exists no significant CL.

W.R. McKinney: Does the autoluminescence
from the elastic fibers decrease with
exposure to the elctron beam? Do you have
any idea which compounds are luminescing?
Authors: Yes, the autoluminescence
decreases.So far we have no idea which
compounds luminesce. Our investigations
in this field are in progress. The
spectrum of the autoluminescence of the
elastic fibres in the mesentery of
rabbits shows a very broad band with a
maximum between 450 - 500 nm.

W.R. McKinney: Have you tried other
protein-dye combinations than FITC and
sulfoflavine? What are the relative
resistances to beam damage of those
that you have tried?
Authors: We have only used FITC and
brilliant sulfoflavine for protein-dye
combination. Brilliant sulfoflavine
seems to be somewhat more stable under
electron bombardment.

M. De Mets: What is, in your opinion,the
most important problem concerning present
CL methodology?
Authors: We think that the most important
problems of the CL method are the often
weak CL intensities and the electron beam
damage.

H.W. Carter: What is the best resolution
you have achieved with CL?
Authors: The best resolution we achieved
in the CL mode ranges from 100   to
50 nm. 40 - 50 nm seems to be the best
resolution in bulk specimens, which can
be obtained today.

SCANNING ELECTRON MICROSCOPY/1979/II
SEM Inc., AMF O'Hare, IL 60666, USA

SCANNING ELECTRON MICROSCOPY AT MACROMOLECULAR RESOLUTION IN LOW ENERGY
MODE ON BIOLOGICAL SPECIMENS COATED WITH ULTRA THIN METAL FILMS

K. - R. Peters

Section of Cell Biology
Yale University School of Medicine
333 Cedar Street
New Haven, CT 06510

Abstract

In this report, conditions for attaining high resolution in scanning electron microscopy with soft biological specimens are described using the currently available high resolution scanning electron microscopes in emission mode of low energy electrons (secondary and charging electrons). Retinal rod outer segments, red blood cells, intestinal mucosa, and ferritin molecules were all used as biological test specimens. From uncoated specimens a new source of signal, referred to as a discharge signal, can provide a high yield of low energy electrons from an excitation area approximately the size of the beam's cross section. Additionally, under these conditions sufficient topographic contrast can be achieved by applying ultra thin metal coatings. A 0.5 nm thick gold film is found sufficient for generating the total signal, whereas increased coating thickness causes additional topographic background signal. However, a 2.0 nm film is needed for imaging surface details with the present instrument. Ultra thin, even, and grainless tantalum films have been found effective in eliminating the charging artifacts caused by external fields, and the decoration artifacts caused by crystal growth as seen in gold films. To improve, in high magnification work on ultra thin coated specimen, signal-to-noise ratio, methods for obtaining saturation of the signal with discharge electrons are shown. The necessity of confirming the information obtained in SEM by independent techniques (TEM of stereo-replicas or ultra thin sections) is discussed.

KEY WORDS: High Resolution; Field Emission; Biological Specimen Preparation, Specimen Coating, Metal Film, Discharge Contrast; Decoration Artifact; Topographic Contrast; Stereo-Replica; Signal-to-Noise Ratio.

Introduction

In electron microscopy three dimensional fine structures of surfaces of biological specimens are visualized with transmission or scanning electron microscopes (TEM and SEM). Depending on the microscopic method, either the three dimensional representation or the fine structural resolution will be optimal. Thus the proper imaging method must be chosen for the actual problem.

Preshadow-casted "stereo-replicas"[1] allow the visualization of fine structures on very complex three dimensional cell surfaces or subcellular components with the highest resolution obtained by TEM. But the grain size of metal used for generating the contrast limits the attainable resolution to approximately 2 nm. Surface scanning electron microscopy, however, allows more detailed three dimensional analyses due to increased depth of focus and extensive possible tilting of the specimen. Highest resolution in the range of 2.0 nm is now obtainable with SEM using instruments in the "low energy loss mode"[2]. But the geometrical conditions for generating low loss electrons limit tilting of the specimen. The "emission mode", on the other hand, which conventionally uses mainly the low energy secondary electrons (SE), allows the most extensive topographic analysis. But until recently this mode was limited, on soft biological specimen, to the very low resolution of about 15 to 20 nm, due to both large probe sizes and insufficient specimen preparation. Since "high resolution" instruments with probe diameters smaller than 5.0 nm are available, resolution was expected to be much improved. First results, however, are discouraging, and improvement seems difficult, because there is not very much known about image generation in high resolution scanning electron microscopy.

However, most of the work on image formation and resolution was done on inorganic, low resistance specimens ranging from carbon to gold. For these specimens the dependence of topographic resolution on the interaction of the probe with the specimen is still under discussion[3,4]. A new type of SE electron emission generated on the specimen's surface is observed on metal specimens and their edges in areas approximately the diameter of field emission-generated electron probes[5]. Also, increased SE emission has been described for non-coated biological specimens[6] as well as for specimens coated with discontinous metal films, emphasizing details in the range of 3 to 30 nm[7]. Nevertheless, for metal coated soft biological specimens the parameters required for obtaining highest resolution have yet to be defined.

Even the specimen preparation required for high resolution microscopy, is not yet optimal[2]. Two problems are involved. First, the dimensions of the specimen's fine structure must be preserved during dehydration and drying (reviewed in refs. 8 and 9). Second, metal coating (reviewed in ref. 10) must be applied in the right way. The kind and thickness of metal deposition needed to increase the contrast, but not to obscure fine surface details, is of primary interest. Most investigators use, for high resolution analysis, "semi-thin" coatings in the range of 5 to 10 nm thickness on bulk specimens[6,7,11,12]. In low magnification work, however, a much thicker coating is used to prevent charging artifacts[13]. Usually, along with the signal-to-noise ratio, the conductivity of the specimen's surface has to be increased to avoid distortion of the final high magnification image[6,10]. To insure conductivity and prevent image distortion, noncoating techniques (review 1978: 14) have also been applied for high resolution work[15,16].

Regarding the sometimes contradictory reports for improving resolution, this paper will try to emphasize those parameters involved in high resolution scanning electron microscopy of "soft biological" specimens. "High resolution scanning electron microscopy" will be taken here as meaning microscopy performed with high resolution instruments[4] at magnifications where the image element represents a specimen area of almost the beam diameter, i.e. 50,000 to 100,000 magnification. Furthermore, only the "emission mode" using "low energy electrons" will be used for analyzing "topographic" features of the specimens. For this mode (most commonly used at lower magnification on biological specimens) image formation and interpretation, particularly as related to metallic coats, will be discussed here.

Material and Methods

Instruments. A JEOL JFSM 30 with cold field emission, provided with a Everhart-Thornley detector, and a JEOL JEM 100 CX with an ASID scanning attachment (JEOL Comp., Boston, Ma) were used.

Calibration of magnification. Calibration of the magnification was done at a magnification of 100,000 times in TEM as well as in SEM by using grating fragments (Grating suspension, # 1038, E. Fullam Inc., Schenectady, NY), prepared by air drying a drop of suspension on a holey grid or a carbon support.

Biological specimens. Rod outer segments of frog retina were isolated by gentle shaking, and fragments of appropriate size were separated by differential centrifugation[17]. Red blood cells (rat) were washed five times in 0.9% NaCl, and then fixed for 1 hr in 2% glutaraldehyde in saline at 4° C. Small intestine (rat) was washed with buffered saline and fixed with Karnovsky's fixative for 1 to 10 hr, then postfixed in 1% Osmium tetroxide at 20° C. Cationized ferritin was used

(#5753; Sigma Chem. Co., Saint Louis, Mo).

Specimen preparation. Except for the intestine, all specimens were premounted on their support and then (including the intestine) processed in a simple semi-automatic glass apparatus (Exchange apparatus: H. Hert, München, Fed.Rep. Germany; or Polaron Instr. Inc., Warrington, Pa), for washing, fixing, and dehydration in a linear gradient between buffer and ethanol[18,19]. For critical point drying, carbon dioxide was used in a E 3000 Critical point apparatus (Polaron Instr. Inc.).

Metal coating. DSM 1 diode sputtering module (Tousimis Res. Corp., Rockville, Md) was used in a vacuum evaporator Edwards 306 coater (Edwards High Vacuum Inc., Grand island, NY). The penning sputter coating system PSC 1a (Zentrum für Elektronenmikroskopie, Graz, Austria) was attached to a Balzers high vacuum coating plant system BAE 120 (Balzers High Vac. Corp., Nashua, NH). Metal thickness measurements were done with a Infinction XTM quartz monitor (Infincton, East Syracuse, NY) with an accuracy of 0.01 nm/20 g/ ccm.

Replicas. Pt-C shadow casting was performed in a Balzers freeze-etch unit BAE 300 (Balzers High Vacc. Corp.) equipped with a Pt-C electron gun. Carbon deposition for the replica matrix was obtained with a Polaron diode sputter system and replicas were cleaned in HF and hypochlorite as described previously[1].

Results and their Interpretation

In order to attain the greatest resolution of topographical details with the scanning electron microscope (SEM), the operational conditions of the instrument must be optimized. The probe diameter should be minimize[4,20] and the interaction of the electron probe with the specimen must be controlled so that the highest possible signal is generated from the "smallest area at the specimen's surface". To rule out the possible existence of artifacts, the validity of the image must be checked by other independent methods. In the following sections, the important phenomena involved in high resolution analysis of surface structures, will be discussed in the context of simple comparative experiments, performed on biological specimens.

I. Instrumental parameters for high resolution scanning electron microscopy.

High resolution SEM's (see ref. 4) have construction features that reduce the spherical aberration of the optical system which generates the electron probe. Either the electron beam leaves the electron gun at such a small aperture that only the central parts of the optical system are used for demagnification, as in the JFSM 30 with cold field emission gun or the specimen is placed near or within a strongly excited final lens, as in the JEM 100 CX ASID. The first arrangement allows a long working distance and makes possible a long depth of focus.

Minimizing the probe diameter. Operational parameters in high resolution work are designed to achieve the smallest electron probe diameter. From the physical point of view[4], this means, working at the highest possible incident beam current and acceleration voltage. According to the manufacturer a beam diameter of 2.5 nm can be obtained at 100 KV in the JEM 100 CX ASID and 1.0 nm for the JFSM 30 at 35 KV.

Minimizing the excitation volume of the specimen. In a low resistance specimen of uniform atomic-number composition, interaction with the incident beam produces mainly SE and back scattered electrons (BSE). In high resolution work, where the picture element represents an area almost the size of the probe diameter, only those electrons collected from the surface area excitated by the primary electrons (PE) of the probe, contain information of fine structures. Biological specimens (of low atomic-number) impregnated by atoms of high atomic-number for purpose of fixation, staining or enhanced conductivity, potentially lose topographic resolution due to increase in BSE generated in the depth of the specimen. When metal coating is applied, to increase SE emission out of the excited surface area, the film should be as thin as possible to reduce BSE yield. As will be shown later, these BSE cause an unspecific topographic signal.

Increasing signal-to-noise ratio. The resolution is also dependent on the signal-to-noise (S/N) ratio, in that a detail must be represented by a minimum amount of electrons against the background[20,21]. Because in high resolution work the collecting time (frame time) is limited due to instabilities of beam generation or to fluctuations in the electrical fields of the specimen, variation of beam current must achieve sufficient signals. For field emission SEM's there is a wide range for variation of the incident beam current without change of the beam diameter.

II. Signal generation in structurally homogeneous biological specimens.

To study signal generation in dried soft biological specimens, fragments of retinal rod outer segments (ROS) were used as test on account of their structural homogenity as a regular stacking of discs demonstrated in ultra thin sections (fig.

1). These reveal a very flat surface as shown by stereo-replica method (fig. 2). The fragments have a uniform diameter of 6 to 8 μm but varying heights depending on the number of discs in the stack (fig. 3).

Signal generation. Signal strength depends not only on the incident beam, but also on the atomic composition, structural features and electrical resistance of the specimen. In low resistance specimens, the emission signal is characterized by the BSE coefficient and the total SE yield[4]. Variations in electron emission and detection related to the structural or atomic features of the specimen, generates the well known types of contrast important for biological analysis: material, thickness and topographic contrast. Fig. 4 shows, for example, a good material contrast between the Si support which generates more electrons than the flat disc membranes (mostly C), which appear dark on account of differences in atomic-number. For high S/N ratio, high resolution work is best done with low energy electrons.

Charging artifacts. Signal generation on high resistance specimens is also well understood[13]. External electrical fields are created by charges buidling up at the surface of the specimen, whereas internal fields are located within the specimen at maximal penetration depth of the PE, because the implanted electrons (charging electrons) have restricted mobility in an environment of high resistance. Both fields can deflect the incident beam or the emitted electrons, leading to trajectory artifacts in the image. On the other hand, positive external fields will absorb emitted SE, whereas strong negative internal fields can accelerate SE or lead to field emission of electrons, known as bright bursts. These charging artifacts can also be observed on low resistance specimens when the latter are not connected to the electrical ground. In that case, if the electrons are mobile, as in metals, the absorbed electrons will create an electrical field resulting in a voltage contrast or in charging artifact if the latter becomes too high. If the absorbed electrons have restricted mobility, charging

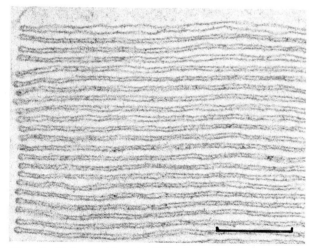

Fig. 1 Thin section in TEM of rod outer segment (ROS). Homogeneous internal structure of stacked membrane discs. Bar 0.2 μm.

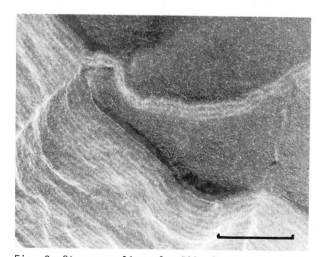

Fig. 2 Stereo-replica of a ROS fragment in TEM. Disc surface is flat and even with well defined edges. Bar 0.2 μm.

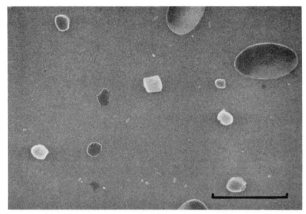

Fig. 3 Fragmented frog rod outer segment (ROS) and blood cells. Uncoated, Si support. Contrast is related to heights of specimens. Bar 25 μm.

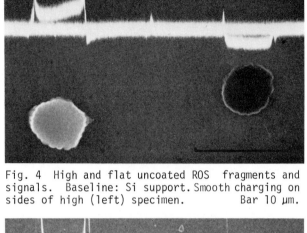

Fig. 4 High and flat uncoated ROS fragments and signals. Baseline: Si support. Smooth charging on sides of high (left) specimen. Bar 10 μm.

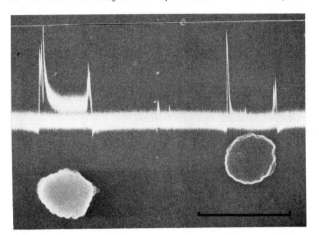

Fig. 5 Same specimen as in fig. 4, but diode sputter coated with 0.5 nm gold. Strong charging on sides of left specimen. Bar 10 μm.

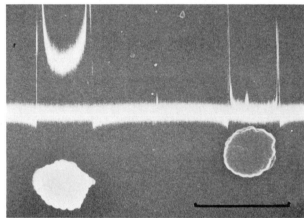

Fig. 6 Same specimen as in fig. 4, recoated to a total of 2.0 nm gold. Left specimen's signal superposed by additional topogr. signal. Bar 10 μm.

artifacts will occur regulary as seen on non-grounded biological specimens.

Discharge contrast. In SEM, biological specimens mounted on grounded, low resistance supports show a very special behavior (fig. 3 and 4) which applies to both high and low resistance specimens. Although all ROS fragments consist of the same material with the same electron emission coefficient, they show, depending on their height, variations in contrast from less to more than the Si support. The remarkable contrast shown by the high fragments is not thickness contrast, since the atomic-number and the emission coefficients of the specimens are lower than that of the silicon. Fig. 4 shows in detail a high (about 6 μm) and a short ROS fragment (about 100nm). The signal difference between the two is demonstrated by the line scan in the same figure. Only the bright particle shows a slight charging effect, that appears as dark areas on both their sides (figs. 3 and 4).

To investigate the origin of this contrast phenomenon, metal coatings of increasing thickness were applied with diode sputtering. Fig. 5 shows the same particles with 0.5 nm gold coating. The flat particle now shows the same signal as the support, indicating that the total signal on the particle as well as on the support is generated by the ultra thin gold film. Very strong and narrow edge effects indicate increased SE emission related to the beam diameter [5]. The charging effect at the high particle is highly increased, but disappears after 2.0 nm total coating (fig. 6). This indicates that the isolated gold crystals of the 0.5 nm gold film capture electrons which generate a strong external electrical field but at 2.0 nm Au the distance between gold islands become close enough for the electrons to flow to ground. At the 2.0 nm coating the bright particle shows an enormous increase of signal which no longer forms a plateau. These findings are indicative of a typical topographic contrast generated by PE scattered in the thick gold layer and generate SE over the whole surface of the large particle. As shown in the curve in fig. 7, which relates signal obtained to the increased thickness of the metal coating, this topographic yield is generated at Au films of more than 0.5 nm thickness and increases rapidly. On the thin specimen and the support, the signal increases much more slowly reflecting the thickness contrast of the gold layer up to the signal obtainable with solid gold. The thin specimen shows additionally a small topographic fraction in its total signal.

The origin of the additional signal emitted by the high ROS fragment can not be an increased SE or BSE emission, because the emission coefficients of the ROS components are expected to be constant, and because the signal from the high uncoated specimen exceeds the signal from solid Si (fig. 8) It produces only a few electrons at the edges indicating a small BSE and SE yield. It is assumed, that due to the high acceleration voltage of the beam, PE are absorbed by the specimen which thereby becomes negatively charged[4]. If the specimens are mounted on a low resistance support, the absorbed electrons must flow to ground during the scan but in a restricted manner, because only a very light and stable charging effect is visible. Thus the absorbed electrons have a restricted mobility in the specimen and are accessible to the probe if their pathway through the specimen is above approximately 100 nm (height of the flat specimen). The horizontal diameter of the excitation volume of the probe is very narrow as seen by the narrow edge effect on the uncoated high specimen and the plateau of its signal (fig. 4). Although the 0.5 nm Au coating increases the edge effects, the plateau is still maintained. This proves that the additional contrast in the high specimen after this coating is still generated by discharging of the absorbed electons. The decrease of contrast between both particles after 0.5 nm coating obviously also proves a decrease of the discharge signal, due to scattering of the PE in the metallic coat, thus reducing the depth of excitation volume. If the observed contrast is suspected to be a "topographic contrast", it should increase with the coating. Such topographic contrast is seen only after a "thicker" coating than 0.5 nm Au, due to its dependency on sufficient scattering of PE. Probably, the charging electrons leave the specimen through a channel of altered resistance (created by the incident beam) due to the internal field, which is built up by the charging electrons themselves. The internal field is homogeneous as indicated by the plateaus in figs. 4 and 5, and is also very weak, producing only a smooth charging artifact on the uncoated specimen (figs. 3, 4 and 8). The contrast generated by the charging electrons will be referred hereafter as "discharge contrast".

The discharge contrast is observed only in objects of intermediate resistance such as biological specimens. To some extent the contrast seems proportional to the height of the specimen (fig. 3) and the maximum will be attainable for specimens approximately as high as the ionization distance of the PE. On specimens coated with more than 0.5 nm Au, no measurements of the discharge yield could be made because of superimposed topographic signal (fig. 7). But it can be assumed, that the discharge yield decreases rapidly due to increased scattering of the electrons in the metal coat.

Properties of the discharge signal. Beside the specific resistance of the specimen material, its structural composition may also influence the adsorption and distribution of charging electrons as well as their retarded flow to ground. To establish the properties of the discharge signal, high ROS fragments were analyzed at high magnification after applying Au coatings of different thickness by diode sputtering. At the periphery of the rod, single discs may overlap each other, building an ideal biological step specimen, if somewhat moderately displaced. Fig. 9 shows an uncoated specimen. As seen in fig. 4 only a small edge effect is visible, revealing an extremely good step resolution. This means that the signal is generated in a surface equal to the cross section of the incident beam. Because of the small area from which it is generated, the discharge signal does not lower the resolution and appears to be ideally suited for increasing the S/N ratio, thereby contributing substantially to high resolution image.

However, because of the low electron emission

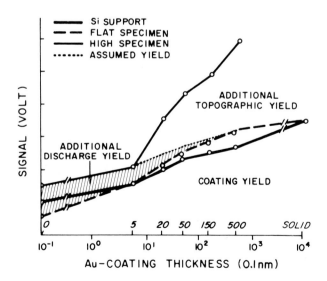

Fig. 7 Diagramm of signal obtained for specimens in fig. 4 at different coating thicknes.

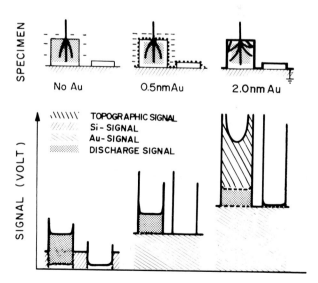

Fig. 8 Diagram of signal values shown in fig. 7 and their assumed relation to the specimens.

coefficient of the low atomic-number biological specimens, only poor topographic contrast is visible on the flat surface of the uncoated discs at high magnification (fig. 9). To increase this contrast, a very light coating with metal is sufficient (as shown in fig. 8). However, 1.0 nm Au coating results in a loss of resolution caused by external fields created by isolated gold crystals (fig. 10). A 2.0 nm Au film (fig. 11), eliminates this artifact and enables good visualization of details on the flat surface of the same specimen as fig. 10. These details may represent slightly protruding areas of the membrane as can be seen in ultra thin sections (fig. 1). However, this same metal coating, causes strong edge contrast, obscuring the details at the periphery of the stacks. Thicker coating (fig. 12) allows the visualization of single Au crystals due to their increased material contrast. The strong topographic signal produced at the periphery of the specimen (fig. 6), lowers the visibility of the gold crystals in the outer region of the disc (fig. 12) by increasing locally the S/N ratio of the crystals.

III. Properties of ultra thin metal coating films.

As shown on the ROS discs, ultra thin metal coatings with very low thickness of below 2.0 nm improve the topographic signal, but can cause with increased thickness a high nonspecific background signal (figs. 8, 11, 12). Yet, they have several other negative aspects for high resolution work. Since they are applied on top of the specimen's surface, the metal film can obscure surface details; in addition metal crystals can create strong external fields on the surface; the material contrast component of the metal crystal can cover up the topographic contrast of small structures, thereby altering the final image, and crystal formation can result in extreme decoration artifacts. Other important points have to do with artifacts produced by the coating technique itself due to energy transfer or contamination depositions. Therefore the parameters involved in metal

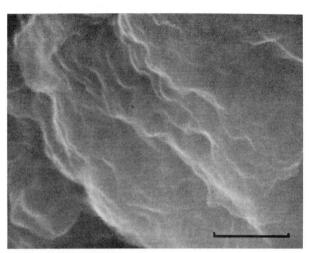

Fig. 9 ROS fragment from the same sample as in fig. 3, uncoated; 35 KV, 3 µA emission current. High step resolut.at displaced discs. Bar 0.2 µm.

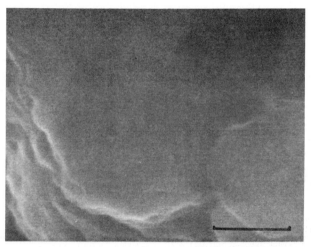

Fig. 10 Similar specimen and imaging conditions as in fig. 9, but 1.0 nm Au diode sputter coated. Resolut. loss due to external fields. Bar 0.2 µm.

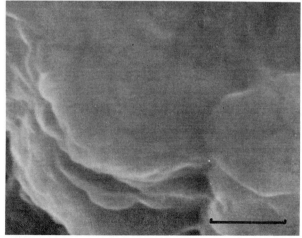

Fig. 11 Same specimen as in fig. 10, but coating increased to 2.0 nm. Topographic features of flat surface are well represented.   Bar 0.2 µm.

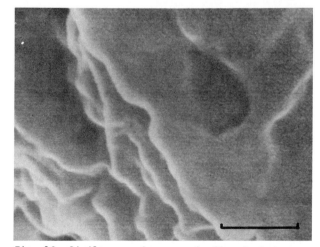

Fig. 12 Similar specimen as in fig. 9, but coated with 5.0 nm Au. Gold crystals and strong edge effects become visible.   Bar 0.2 µm.

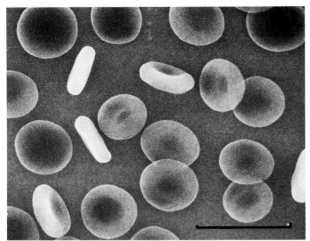

Fig. 13 Rat red blood cells on Si support; glutaraldehyde fixed; no coating. Smooth internal field charging artifacts are visible. Bar 10 μm.

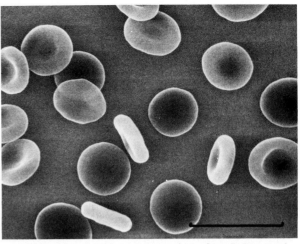

Fig. 14 Same preparation as fig. 13; 7.5 nm Au diode sputter coated. Strong external field charging visible as dark flags. Bar 10 μm.

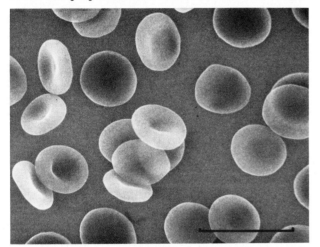
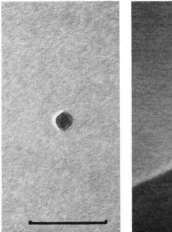
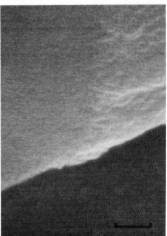

Fig. 15 Same preparation as fig. 13; 2.0 nm Ta coating applied by penning sputtering, no charging occurs. Bar 10 μm.

Fig. 16 Left: 2.0 nm thick Ta film on carbon. Right: Same prep. as fig. 13, shadow-casted with 4.0 nm Ta. No external fields occur. Bars 100 nm.

coating have to be well understood and controlled in order to eliminate from the beginning each of these possibilities for fine structural artifacts.

Deposition of ultra thin metal films for SEM. Techniques for metal deposition have been reviewed[10]. Most common among these procedures are metal evaporation and sputter techniques including diode -, triode -, and ion beam sputtering. A new technique, recently introduced, is penning sputtering[22]. The ions are produced in a low pressure penning gas discharge chamber between two cathodes. Large deflecting electrodes prevent the escape of charged particles from the gun. Moreover, the very low partial pressure of argon inside the discharge chamber and the separation of the chamber from the specimen by the second cathode reduces the number of escaping neutral argon atoms. With this sputter technique in a 2.0 nm film of gold on carbon smallest grain sizes are obtained, when compared to the other deposition techniques. These smallest grains result from sufficient high energy of the target atoms (primary energy), generated by 4 KV argon ions.

Other aspects, beside reducing crystal formation, are important in metal deposition. High secondary energy, transferred by photons, electrons, ions, or neutral particles, may be very destructive to the structures of the specimen surface as may high primary energy, demonstrated by the 20 KeV ion beam used to etch biological specimens[23]. Also undesirable is the formation of contamination products by the high energy electrons used in some of the low vacuum sputtering techniques. In diode sputtering, under conditions where an aluminium target will not be sputtered, its use instead of a gold target, under the same conditions as for gold deposition, will produce a yellow contamination layer a few nm thick[1], as in diode and cool diode sputtering. Low voltage, as applied in cool triode sputtering, does not generate a visible contamination layer. In all high vacuum techniques, contamination can be almost eliminated.

Thus the best method of metal deposition for high resolution work seems to involve a high primary energy of the metal atoms for restricting their mobility on the specimen surface in a contamination free environment, while avoiding any secondary energy transfer to the specimen surface.

Prevention of external field charging artifacts on biological specimens. When the discharge signal is used for high resolution microscopy, any external fields on thin metal coated specimens will reduce resolution (see figs. 5, 10). One way to avoid this effect is to reduce the distance between all metal islands to allow electron flow to ground. As shown in fig. 6, 2.0 nm Au diode sputtered will fulfill these conditions for a very flat membrane specimen. If, however, the specimen has a structurally uneven surface, more gold deposition will be needed. This is demonstrated on red blood cells as test specimen. Fig. 13 shows a specimen without coating on a low resistance support; the cells, especially those standing on edge, show mainly discharge contrast. On the same specimen (but in a different area) diodic coated with 7.5 nm gold (fig. 14) charging artifacts generated by external fields are still seen. On all specimens with rough surface and deep cavities, shadowed sites will be less well coated. With all crystal forming metals, external fields will develop at such sites due to the mobility of the charging electrons in the specimen.

The ability of metal atoms to form crystals in thin films is closely related to their boiling point. Metals with boiling points over 3,000° C, such as tantalum and tungsten, are preferable for coating techniques[10]. Unfortunately high energy is needed to sputter or evaporate them, which is mostly combined with a high secondary energy release. Therefore, penning sputtering is used to generate Ta atoms. Such a 2.0 nm Ta film, applied on red blood cells (fig. 15) does not show any external fields, and thus seems to be grainless. As proven by TEM, a 2.0 nm Ta film on carbon shows no crystals, neither in its substructure (fig. 16 - left) nor by electron diffraction. However, problems are encountered when a point source is used for thin metal coating of rough surfaces. Even with elaborated tumbling movements of the specimen during metal deposition, shadowed and uncoated areas on the specimen will always remain. If metal islands are present at a fuzzy border between shadowed and coated areas, external field will develop along these borders. Only very narrow point sources are able to form sharp shadows and avoid an extended diffuse border where the metal film is discontinous. The satisfactory performance of the penning sputtering system is proven in fig. 16 (right part) which shows that no external field builds up along the border of the coating, even when a 4.0 nm Ta film has been applied on the specimen. Using the technique described above, the kind of movement of the specimen during metal deposition and even the absence of coating in shadowed surface areas becomes irrelevant for avoiding charging artifacts.

Prevention of decoration artifact. In high resolution work where the shape and dimensions of the smallest details are considered relevant, the changes of these features caused by metal coating have to be well understood. The main problems are: the formation of decoration artifacts, due to the high mobility of the metal atoms on the specimen's surface; the production of uniform films; and the

determination of their thickness. To investigate these points, ferritin molecules have been used. They consist of a 5.5 nm iron core and an outer protein shell; the overall diameter of the molecule is 11 to 12 nm. Because of the high concentration of iron in the core, they give a good contrast in TEM as well as in SEM. Although it seems difficult to show single molecules on cell surfaces, they are easily visible with high contrast after light gold coating of 2.0 to 5.0 nm[11,24,25].

The ferritin solution used here also contains some proteins of smaller sizes, fragments of ferritin molecules. In SEM, on low resistance carbon support without coating, only the ferritin cores are visible (fig. 17a) as well as in TEM on a 5.0 nm thin carbon film (fig. 17b). The iron core, gives in high resolution SEM, as well as in TEM, a good contrast against a low atomic-number support. With both methods and without metal coating, the protein shell of the ferritin and the protein molecules are not visible due to their low electron emission or low electron scattering. However, Ta rotory shadow casting (fig. 17b, insert) reveals the shape of the particles. The ferritin show an outer diameter of approximately 11 nm. In SEM, as well as in TEM, the the outer contours are made visible by thin metal coating. Ferritin on silicon support needs a 2.0 to 5.0 nm Au coating (diode sputtered) to visualize the molecules in SEM[11] in a reasonable contrast, as shown in fig. 17c with 2.0 nm Au coating. Although the material contrast of the iron core against the silicon support becomes negligible, contrast is now established by decoration of ferritin molecules by gold crystals, revealing a very irregular shape and accumulations of bright crystals. This is shown in TEM with a replica (preventing any loss or rearrangement of crystals) of the preparation used for fig. 17c. Fig. 17d shows that the crystals of the 2.0 nm Au coat have a very irregular distribution. Stereo analysis reveals, that most crystals are located at different heights above the support plane. At certain spots, very high piles of crystals are built up, the heights of the crystals corresponding to the height of the "coated" particles. This decoration increases the material contrast of the gold by a micro roughness effect, emphazising small particles, but thus does not directly correspond to the shape of the molecules. In thick Au coating this piling effect (shown here for 2.0 nm average thickness), may produce much more highly pronounced artifacts[26].

A more accurate representation of the shape and size of the molecule is obtained when ferritin is coated with less mobile atoms. This is demonstrated in fig. 17e, which shows ferritin molecules coated with a 2.0 nm Ta film applied by penning sputtering under 0 to 90° tilt rotation. The higher noise in this micrograph is caused by the very thin metal layer. Figs. 17a, c and e are taken under very similar conditions (no tilt, 30 KV acceleration voltage, and 4 μA emission current) and are printed on the same grade paper. The 2.0 nm average thick gold film in fig. 17b is composed of isolated crystals of greater thickness, which results in a higher signal, similar to that of a continuous film of increased thickness.

It is easy to see, that the 5.5 nm iron-rich core (Fe has a total yield of SE similar to Au and Ta) gives as good a contrast as the gold crystals, but that the Ta layer produces a poorer signal. Through the Ta film the iron cores are still visible. The main contrast on the particle surface in this case is no longer material contrast, it is a "true" topographic contrast, as is shown by its light-and-shadow effect in direction to the detector.

It should be noted, that the size of the particle in fig. 17e seems uniformly increased. To demonstrate more clearly the surface of the specimen after Ta coating, a replica was made of the preparation and the metal dissolved in hydrofluoric acid. Then the negative side of the matrix of the specimen surface thus obtained was shadowcasted with Ta penning sputtering at 30° rotation with a 1.0 nm Ta deposition (measured on particles surfaces facing perpendicular to the source). In such preparations (fig. 17f) the outer diameter of the contrasted negative matrix represents the actual outer surface of the Ta coating on the particles. The smooth surface proves an even coat. Measurement of the diameter of the imprint of the Ta coated ferritin gives an average value of 15.0 nm, and thereby independently confirms a coating thickness of 2.0 nm. Since the original Ta film was very homogeneous, as seen in fig.16, it can be

concluded that the light contrast of the molecules in fig. 17e is primarily topographic.

The determination of the actual metal deposit on a specimen during tilt-rotation is very easily done by setting the tooling factor (ratio of the deposits at the sites of monitor and specimen) semi-empirically by using gold films on a test specimen as a standard. Gold films reveal distinct colors at a thickness of 0.5 to 2.5 nm. Because the actual coating thickness on the specimen surface is very dependent on regional orientation to the metal gun, the gold colors on the side of a small 3 mm cube gives a good approximation of the coating. With 0 to 90° tilt-rotation and low sputter rates, a very uniform coating on all sides is obtainable on small particles positioned on a flat plane. Of course, sites which appear partially or totally shadowed, when seen from the direction of the metal gun, receive directionally-oriented coating or none at all, thus fine structures' details should be analyzed only on specimen sites totally exposed during coating when a point source is used. All low vacuum sputtering techniques with stationary specimen position give a very uneven coating, which is less than one third as thick near the base of perpendicular planes of the test cube.

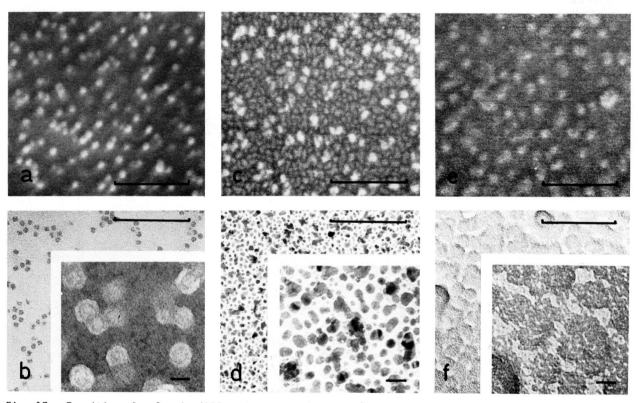

Fig. 17. Ferritin molecules in different preparations.   a) Ferritin on bulk carbon in SEM, no metal, only core visible in material contrast; b) Ferritin on thin carbon film in TEM. Insert: shadow-casted with l.0 nm tantalum; c) Ferritin on silicon support. 2.0 nm Au diode sputtered. Au crystals obscure particles;   d) Replica of the same specimen as in c, preserving sputtered Au film. Insert: showing piled Au crystals indicating decorated particles; e) Ferritin on bulk carbon 2.0 nm Ta film penning sputtered, seen with low S/N ratio; f) Surface of specimen in fig. e., TEM of negative matrix, rotat. shadowed (30°) with Ta. Even coating represented by regular round prints.   Bars 100 nm, 10 nm Inserts.

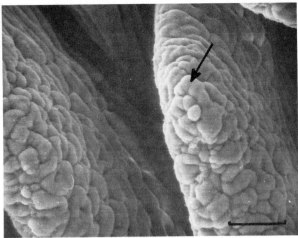

Fig. 18 Rat intestine, mucosal surface, 5.0 nm Au applied by diode sputtering. Strong charging visible. Bar 0.1 mm.

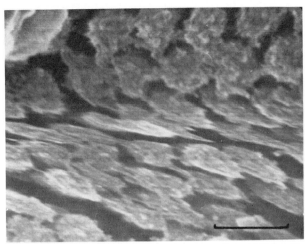

Fig. 19 Detail of fig. 18 showing microvilli. At high magnification internal fields cause trajectory artifacts. Bar 200 nm.

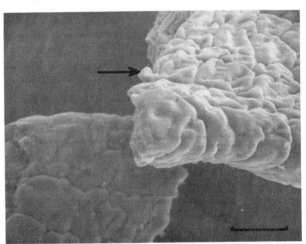

Fig. 20 Isolated villi coated with 7.5 nm Au with diode sputtering. One villus retains discharge signal. Bar 50 µm.

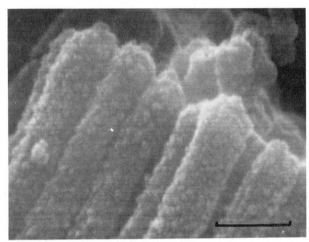

Fig. 21 Detail of fig. 20, gold decoration on microvillar surface is seen as single crystals surrounding dark spots. Bar 200 nm.

IV. Signal generation on structural nonhomogeneous biological specimens.

As demonstrated above, there are well defined conditions which permit the attainment of high resolution on biological specimens. The use of the smallest spot size and the generation of a pure topographic contrast on specimens coated with thin, even, and grainless metal film are the technical preconditions for microscopy. Obtaining a high S/N ratio in generating a discharge signal is the operator's contribution to microscopy. As seen in fig. 17e, the S/N ratio of the applied thin Ta film is very low in conventional operation mode. Although the bulk carbon support, due to its low resistance, will give a lower background than the bulk biological specimen, the lateral resolution in topographic contrast will not go far beyond the dimensions of the ferritin particles. Thus the increase of the S/N ratio is an absolute prerequisite for improving topographical resolution on thin coated specimen. Using a high incident beam current, the capacity of the specimen for charging electrons has to be saturated for

generating the additional discharge signal. Good contact of the specimen to its support in a large area is enough to control the internal fields in structurally homogeneous specimens like the ROS fragments. Most of the bulk biological specimens are, however, structurally nonhomogeneous, thus they require a special kind of signal generation.

Bulk massive specimens. In bulk massive specimens, like tissues, structural nonhomogeneity will always create strong internal fields. Thus only thick metal coatings should be applied to suppress charging artifacts in low magnifications. High resolution, however, requires thinner coatings. Therefore coating thickness must be adapted to the information desired. Intestinal microvilli have been used as specimens to explore the possibility of fine structural analysis. First, a semithin Au film of 5.0 nm is applied. At low magnifications, high internal fields are observed as bright areas, but at intermediate magnifications (fig. 18) a smaller scanning field can be placed over an area at the tips of the villi from which the flow of charging electrons is retarded. After a few scans in a slow scan mode of 5 to 10 sec per frame, a saturation of the discharge signal is usually obtained and can be recognized as a stable bright area in the image. Several such areas must be examined to check their suitability for work at high magnification. Often strong internal fields lead to charging artifacts. These are seen as beam trajectory deflections (fig. 19). If no stable conditions can be obtained on such a specimen, internal fields can be reduced by two methods: heavy metal coating, or the reduction of the distance to ground. Both allow a better flow of electrons out of the sample. Intestinal villi can be easily shaved from the mucosa and mounted on low resistance supports. Depending on the extent of their connection to ground and their internal resistance, some villi will show a good discharge contrast and thus will appear much brighter than the others (fig. 20). In such small specimens, internal fields are more homogeneous, so work at low magnification can be done. The increased S/N ratio obtained under these conditions also permits

work at high resolution. If gold is applied as a 7.5 nm thick coating on the sample, small particles can be seen in high contrast (fig. 21). Most of them show a dark center surrounded by a bright, irregular halo. These structures appear to be decoration artifacts produced by gold crystals, thus they do not represent the surface fine structure.

Signal saturation. When ultra thin metal coatings are applied on bulk massive specimens, internal fields interfere more severely with the formation of the image. Reducing the specimen height is the best way eliminate this problem. On a specimen coated with a 2.0 nm Ta film (fig. 22), heavy charging is seen at low magnification on the villus at its sectioned base. But other surfaces reveal stable contrast after a few scans at low magnification. A small scan frame is placed on those areas where the brightest signal is obtained without any distortion. If the beam current applied is not variable, several spots must be tried in order to find conditions where signal

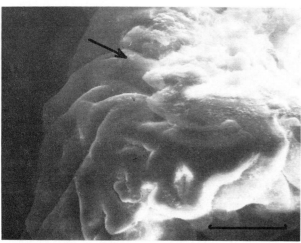

Fig. 22 Villi of same preparation as fig. 20, 2.0 nm Ta penning sputtering coated. Strong charging, but also stable areas (arrow). Bar 20 μm.

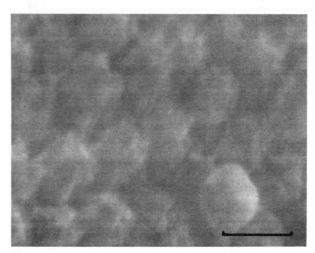

Fig. 23 Detail of fig. 22 (arrow), emission current applied 1 μA. Insufficient signal with low contrast. Bar 200 nm.

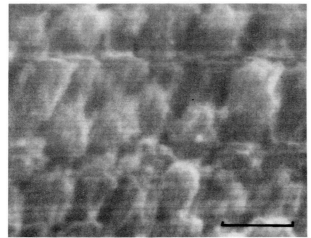

Fig. 24 Same specimen as in fig. 23, emission current 4 μA. Strong internal fields distort the image. Bar 200 nm.

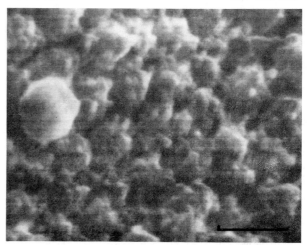

Fig. 25 Same specimen as in fig. 23, emission current 3 μA, signal saturation. Particles visible in "true" topographic contrast. Bar 200 nm.

saturation can be obtained. With variable emission current, incident beam current can be adapted to an area selected on account of other parameters. On an area fully exposed during metal coating (fig. 22 - arrow), a beam current as low as 1 μA will not saturate the signal, resulting in poor contrast and a fuzzy image after focussing (fig. 23). Too high a beam current, such as 4 μA, leads to oversaturation and creates nonhomogeneous internal fields and charging artifacts (fig. 24). These charging artifacts are seen as trajectory deflections of the electron probe, deflections of the emitted low energy electrons, distortion of the image, reduction of the contrast, or bright areas caused by electron emission from internal fields. Signal saturation on the other hand is easily recognized as the condition which gives an image high contrast and fine detail. This condition was achieved on the area in fig. 25 with an emission current of 3 μA. Except for the beam current, figs. 23. 24 and 25 are identical in instrumental and photographic conditions.

On the surfaces of the microvilli (fig. 25), clusters of small particles are seen in typical topographic contrast with light-and-shadow effects. Every bright spot can be interpreted as representing particles emerging from the surface due to the evenness of the metal film applied. Comparison of images obtained by conventional scanning mode (fig. 17e) emphazises the improvement obtained in using the discharge signal for increasing the S/N ratio. Thus the topographic features of the surface are represented in a more uniform and accurate manner than after coating with discontinuously thin gold films (figs. 17c, 19, and 21), where bright spots represent single crystals. However, the existence and characteristic features of the fine structures observed must be proven with independent methods. This critically important comparison has to take into account possible artifacts produced by the preparation and imaging methods applied in each individual procedure. Because of possible beam damage caused by the probe, the existence of the fine

structures observed should be confirmed by other methods which do not produce artifacts of the type encountered in SEM (like stereo-replicas in TEM).

Tissue specimens, processed as for SEM up to the common dehydration step, were embedded and then examined in TEM. Due to the orientation of the sectioning plane ultra thin sections show only isolated particles attached to the surface of the microvilli viewed at a few places (fig. 26). Stereo-replicas, however, prepared from the critical point-dried villi show the same type of fine structures on the tips of the microvilli (fig. 27), as visualized in high resolution scanning electron microscopy, thus confirming the information obtained.

Discussion

Of all the SEM modes, low energy electron emissions, induced by an electron probe of high acceleration voltage, offer the best conditions for analyzing at high magnification the three dimensional features of soft biological specimens. However, even with the high brightness of modern electron sources, the S/N ratio remains unsatisfactory, because of the insufficient SE emission from specimens of low atomic-number. This low S/N ratio prevents the achievement of theoretically possible resolutions in the range of the probe's diameter. Up to now, the only way to increase the topographic signal of the specimen was to apply metal coatings to the surface. But metal coatings must be both thick enough to generate sufficient signal from the topographic fine structure, and thin enough so as not to obscure the details of this fine structure or generate too high a nonspecific signal from randomly scattered electrons. As shown, a total specific signal available from the specimen surface can be generated from samples coated with as little as 0.5 nm metal. Thicker coatings contribute only an unspecific signal, which leads to lowered resolution. This is why the resolution of topographic features with a conventional 5 to 15 nm metal coating is always so

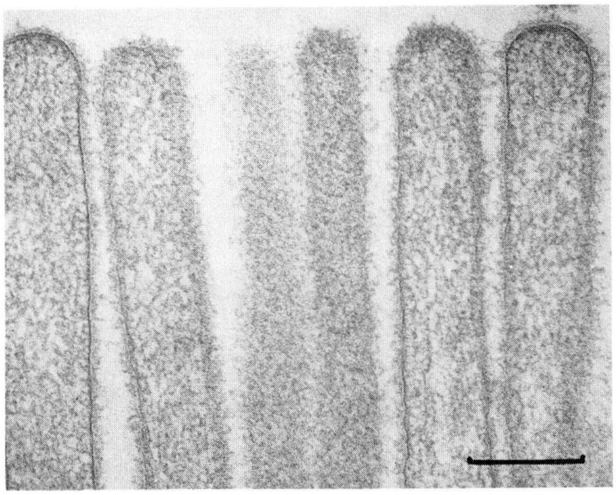

Fig. 26 Ultra thin section of rat intestine, mucosal surface. Particles infrequently seen, due to orientation of section plane. Bar 200 nm.

Fig. 27 Stereo-replica of intestine mucosal surface. Particles of similar size and distribution seen as in SEM. Bar 200 nm.

disappointingly low. However, when using discontinuous metal films, the single metal particles on the soft biological specimen surface are visualized mainly by their material contrast, though of course, in much higher resolution. Thus gold decoration may be used to show the distribution of surface features but not fine structures, as in the case of the smallest spherical markers used for labelling, where the fine structure and the kind of decoration obtained are known from other high resolution imaging methods.

Metal coatings of excessive thickness are often required to prevent charging when specimens are mounted on high resistance supports (such as mica, glass, formvar, plastic, etc.), and when metals which give an uneven and grainy film are used for coating. The use of grainless metal coatings and of low resistance supports (such as aluminium, silicon or carbon) eliminates the need for thick metal coats to lead to ground. Another reason for these thick metal coats is the reduction of charging within the specimen itself. Charging on soft biological specimen is caused by excessive accumulation of charging electrons. In low magnification work, variations in the yield of charging electrons from the specimen result in an uneven, and therefore unreliable, topographic representation of the specimen's surface. Thick metal coatings can be applied to prevent the emission of these electrons, or the internal conductivity of the specimen can be increased to allow these electrons to flow through the specimen to ground. However, both methods interfere with high resolution work. Fine details are either blanketed by metal or lost in the increased background noise caused by the impregnation of the specimen with metals. An example of the latter is the OTO method in which the fine structures on the intestinal microvilli cannot be visualized[15]. Because in most soft biological specimens variations in the yield of charging electrons are negligible for small surface areas, it is possible to use the charging to increase the low S/N ratio of ultra thin metal coatings. Thus the generation and control of the charging electrons by specimen preparation becomes an essential part of high resolution scanning electron microscopy and makes use of the full resolution potential of modern instruments.

For this new kind of high resolution microscopy, the preservation of the resolvable fine structures now becomes a problem of high priority. Specimen distortions are known or expected to occur during preparation for microscopy (dehydration and drying artifacts) as well as during imaging (beam damage). The increased resolution of scanning electron microscopy now makes this method useful, in addition to the TEM stereo-replica method, as a monitor of those alterations.

Acknowlegments

This work has been supported by NIH Grant Number GM 21714. The author wishes to express his grateful acknowledgments to Ms. Barbara Schneider, Ms. Pamela Ossorio, Ms. Linda Mazzacane, and Mr. H. Stukenbrok for assistance in ultra thin sectioning, photographic work and preparation of the manuscript. The results involving biological specimens come from investigations in progress carried out in collaboration with Dr. G. E. Palade (intestine mucosa) and Dr. D. S. Papermaster(rod).

References

1. K.-R. Peters. Stereo surface replicas of culture cells for high resolution electron microscopy. J. Ultrastruct. Res. 61,1977:115-123.
2. A.N. Broers, B.P. Panessa, and J.F. Gennaro. High resolution SEM of biological specimen. SEM/1975 , IIT Res. Inst., Chicago, IL:233-237.
3. E.P. George and V.N.E. Robinson. The influ. of electron scattering on the detection of fine topographic detail... SEM/1977/I, IIT Res. Inst., Chicago, IL:63-69.
4. L. Reimer. Scanning electron microscopy: Present state and trends. Scanning 1,1978:3-16.
5. T. Matsukawa and R. Shimizu. A new type edge effect in high resolution scanning electron microscopy. Jap. J. Appl. Physics 13,1974:583-586.
6. S-D. Lin and M.K. Lamvik. High resolution scanning electron microscopy at the subcellular level. J. Microscopy 103,1974:249-257.
7. R.H, Kirschner and M. Rusli. Identification and characterization of isolated cell organelles.. SEM/1976/II, IIT Res. Inst., Chicago, Il:153-162.
8. A. Boyde, E. Bailey, S.J. Jones,et al. Dimensional changes during specimen preparation for scanning electron microscopy. SEM/1977/I, IIT Res. Inst., Chicago, IL:507-515.
9. A. Boyde. Pros and cons of critical point drying and freeze drying for SEM. SEM/1978/II, SEM Inc., AMF O'Hare, IL:303-312.
10. P. Echlin. Coating techniques for scanning electron microscopy and x-ray microanalysis. SEM/1978/I, SEM Inc., AMF O'Hare, IL:109-132.
11. P.M. Male and D. Biemersdorfer. Silicon wafers as support for biological macromolecules... SEM/1978/II, SEM Inc., AMF O'Hare, IL:643-648.
12. R.H. Kirschner, M. Ruslin, and T.E. Martin. Characterization of the nuclear envelope, pore compl. and dense lamina of mouse liver nuclei... J. Cell Biol. 72,1977:118-132.
13. T.J. Shaffner and J.W.S. Herle. Recent advances in understanding specimen charging. SEM/1976/I, IIT Res. Inst., Chicago, IL:61-68.
14. J.A. Murphy. Non-coating techniques to render biological specimens conductive. SEM/1978/II, SEM Inc., AMF O'Hare, IL:175-193.
15. G. Takahashi. OsO_4-Tannin-OsO_4 methods for transmission and scanning electron microscopy... Electron Microscopy,1978, II:56-57.
16. T. Murakami. Tannin-Osmium conductive staining of biological specimens for non-coated scanning el. micr. Scanning 1,1978: 127-129.
17. D.S. Papermaster and W.J. Dreyer. Rhodopsin content in the outer segment membranes of bovine and frog ret. rods. Biochem. 13,1974:2438-2444.
18. K.-R. Peters and G. Rutter. Veränderungen an HeLa-Zellen (Deckglaskulturen) während der Präparation für REM-Untersuchungen. Beitr. elektronenmikroskop. Direktabb. Oberfl. 7,1974:465-482.
19. K.-R. Peters, H.H. Gschwender, W. Haller, and G. Rutter. Utilization of high resolution spherical marker for labeling of virus antigens... SEM/1976/II, IIT Res. Inst., Chicago, IL:75-84.
20. O.C. Wells. Resolution of the topographic image in the SEM. SEM/1976/I, IIT Res. Inst.,

Chicago, IL:61-68.
21. D. Joy. The scanning electron microscope - principles and application. SEM/1973 , IIT Res. Inst., Chacago, IL:743-749.
22. E. Jakopic, A. Brunegger, R. Essl, and G. Windisch. A sputter source for electron microscopic pre.. Electron Microscopy,1978, I:150-151.
23. K. Hoju, H. Fujita, M. Ito et al. Etching of red blood cells by means of ion beam sputtering. Electron Microscopy, 1978, II:38-39.
24. M.M.B. Kay. Mechanism of removal of renescent cells by human macrophages. Proc. Nat. Acad. Sci. USA 72, 1975:3521-3525.
25. T. Tokunaga, T. Fujita, T. Habbori, et al. Scanning electron microscopic observation of immunoreactions on the cell surface: Analysis of Candida albicans cell wall antigens by immunofer.. SEM/1976/I, IIT Res. Inst., Chicago, IL:301-310.
26. V.F. Holland. Some artifacts associated with sputter-coated samples observed at high mag... SEM/1976/ I, IIT Res. Inst., Chicago, IL:71-74.

Discussion with Reviewers

J.B. Pawley: In your study you take as given the primary requirement that the spot size must be minimized at all cost. As the ability to resolve small details is limited by both probe size and signal contrast and as the resolution you show appears to be at least several times larger than the 1 nm spot size you claim, might it not have been worthwhile to operate the instrument at a lower voltage where the contrast would be expected to be several times greater? At 30 kV, secondaries can emerge from Si microns from the entry point, giving rise to the low image contrast you mention.
Author: You are right, in practice the acceleration voltage should be adapted to the specimen. Unforunately due to a lack of a simple test, we have to trust the manufacturers and physicists, to tell us the conditions for a small spot size and how small it should be. If you, however, consider the image element divided by the magnification as the operational spot size in high resolution work (comparable to the metal grain in replicas) then it should be as small as possible and the specimen preparation should be adapted to the imaging requirements. As shown the main source for low contrast is "thick" coating of 2.0 nm or more, not the support if bulk specimens are used. For small particles, like ferritin, the oposite is the case. To "show resolution" is also dependent on the specimen. As long as we don't have a test specimen with well-defined topographic features and with different well-defined atomic composition, there is more speculation then "truth" about resolution on soft biol. specimen, as you can see even by the interpretation of ferritin images, a particle of 11 nm in size.

J.B. Pawley: How do you discount the following explanation of figs. 4 - 6? The high particle in fig. 5 retains some small negative charges and accompanying field while the beam strikes it (a dark shadow on either side of the particle but not above or below it could be negative (reduced) collection trajectory contrast caused by such a field). The coating is not continuous in this sample and the field of this internal charge is only partially shielded by it. The bright edge on the sides, but not on the top and bottom are simply positive (increased) collection contrast (see Pawley, SEM/1972:153-159). Large number of secondaries are produced when the beam strikes the sample at glanzing incidence. Usually only a few of these are collected but when they are produced from a negatively charged surface a greater proportion are collected ie: the secondary electron coefficient remains the same but the collection efficiency is increased. What you refer to as discharge contrast is merely a change in collection efficiency caused by an incompletely shielded subsurface charge.
Author: This is a possible explanation for the edge effect in fig. 4 (bright particle) but the chraging/discharging phenomena may be different. The experiments refered to were made on metal and latex, etc. As you mentioned specimen collection-related and collection efficiency-related phenomena may be frequently observed also on biological specimen, as shown under conditions like in fig. 22. However, on well grounded homogeneous soft biological specimens or in a small frame at high magnification high voltage microscopy with high brightness sources may produce other phenomena. On the specimens, which this paper deals with, the probe may easily penetrate the smaller one, so that the channel of altered resistivity easily reaches the ground, even when only ultra thin coatings is applied (fig. 4,5), preventing accumulation of high charges. In a bulk specimen at high magnification the locally injected electrons diffuse in the depth of the specimen, if a flow to ground is possible. The discontinuity of the metal film produces an additional external field, but its influence on the contrast described here, is negligible. First this external field (fig. 5, 14) is much stronger than the internal and is to be suspected to be the same on the entire surface, creating dark flags, as you describe. But the metal film, even if continous does not prevent the bright signal (fig. 6, 15). On the other hand, in the uncoated or ultra thin coated specimen at high magnification, the signal obtained is very closely related to the excitation surface area which represents the outside of the channel of altered resistance, due to the electron probe. This very close connection (indicated by structural resolution at 100,000 magnification in fig. 9, 11, 16, 21 and 25) withstands the interpretation of specimen collection alterations, because the latter is induced by accumulated surface charges spatially distinct from the incident beam.

J.B. Pawley: Could you give us the beam current and scan times used for your high resolution micrographs?
Author: With the JFSM 30 only emission current (used between 2 and 4 uA) and specimen current (in the range of 5 10^{-11} A) is indicated. No easy way for proving incident beam current is available from the manufactorer.Frame time was 50 seconds.

J.B. Pawley: You allude briefly to the problem of radiation damage. I estimate the doses used to make your micrographs to be 10^4-10^5 times those required for complete protein denaturation. Do you observe any visible changes while viewing these samples?

Author: Using ultra thin metal coatings, which only scatter the PE slightly and are not good thermal conductors even when continous, beam-damage is a serious problem. Scanning of a ferritin sample on bulk support five times (at 200,000 mag.) will alter the core from 5.5 nm to about 8.0 nm, if only aldehyde fixed. On bulk specimens, mass loss is visible as seen on the RBC in fig. 16 (only aledyde fixed). Thus, stereo-replicas are the only proof for recognizing beam damages, if the preshadow casting is done without any secondary energy load. To control the artifacts produced by the replication, ultra thin sections should be made. Using ultra thin metal coats to improve the resolution on the specimen surface, beam damages are now as serious for SEM as for TEM samples. In both methods, a stabilization of the specimen should be found by a compound of low atomic-number and low resistance. SEM, using thicker specimens than TEM, will require the most attention in stabilizing bulk specimens. As long, as there are no other ways for reduction of beam damages except low voltage, cool stages, osmification or semi thin metal coating, compromises between resolution and specimen alterations must be made.

J.B. Pawley: Published TEM micrographs of brush border microvilli (Tilney and Moosiker, J. Cell Biol., 1977 and other) show many microfilaments running axially inside them. As these are not visible in Fig. 26, is it possible that your surface micrograph of this sample are also lacking in certain structures?

Author: Alterations of (internal and external) fine structures are always taken into account in any structural analysis. Due to the lack of a more defined test specimen with properties of soft biological material, I chose the intestinal mucosal surface here like other authors (ref. 15). A fine structure analysis would not be covered by this paper.

W.H. Massover: Couldn't the multi-layer osmication procedure developed by Kelley et al. (1973, J. Ultrastruct. Res. 45:254-258) permit you to totally eliminate all resolution-destroying coating of the specimen?

Author: If you refer to "coating" as coat on the surface, the advantage of ultra thin metal coating (if measured accurate, deposited as decribed and decorations prevented) is that it produces more signal in a thinner layer without emphazising the background noise. Also, with controlled deposition as with 0 to 90° tilt rotation, analysis of the image information is much easyer, because the size increase of structures is known.

W.H. Massover: What "new details" are seen in your specimens by virtue of using the coating and observational procedures described?

Author: "True" three dimensional arrangement of particles, not crystal piles, on the surface of biological soft specimens at high magnifications.

W.H. Massover: How does the resolution measured with your specimens in the most favorable case compare with (A) the resolution measured with in-

organic specimens using the same microscope, and (B) the details that can be resolved with surface replicas by TEM?

Author: Gold crystals on bulk carbon will be resolved "center-to-center" above 5.0 nm, visualized in material contrast. On soft biological specimens, single crystals can be seen, if discharge signal is used, with a center-to-center spacing above 3.0 nm. Surface replicas will resolve ca. 2 to 5 nm, depending on the metal deposition method. These data are not comparable, because they are obtained with very different contrast mechanisms and different microscopic methods, and thus represent very different information.

V.N.E. Robinson: I would like to suggest, that your "discharge contrast" is just a "topographic signal". It is really just another case where primary electrons are incident at the top of the specimen and are scattered out of the side of the specimen. As they pass out of the side of the specimen they release many more secondary electrons than were released as the beam entered the surface, because of the forward dependence of SE emission upon the direction of travel of the exciting PE (Robinson, J. Phys. D: Appl. Phys. 8,1975:74-76). The theory of this forward dependence of SE emission has been well established. It has the same features as displayed by your "discharge signal", but does not depend upon the conductivity of the specimen.

Author: In fact, the generation of "topographic contrast", as you mention, is the reason why variations of the "surface area" of the excitation volume create a signal related to the topography. Two properties of this signal are important for biologists, who have to apply metal films on soft specimens for sufficient "specific" signal generation: the edge effect and the increase of the signal with the thickness of the metal film. Both effects are easily seen with thick coatings (figs. 7, 8). However, on non-coated (fig. 4) or ultra thin coated specimens (fig. 5) both effects are missing. The signal is a plateau along the surface of the ROS, which is very smooth, and even after a metal film of 0.5 nm Au is applied, the signal doesn't increase, still preserving its plateau. Instead of generating a topographic signal on the flat specimen, the metal film generates a thickness contrast, as supposed; but on the high specimen, there is no increase whatsoever. Thus the high specimen shows a decrease of its signal arrived at by a different contrast mechanism. Thicker metal films generate enough scattered PEs, which are then recognizable as an increase of a "topographic signal", as you describe it.

Reviewer V: Why is it that the individual microvilli cannot be as clearly seen in fig. 27 as they can in fig. 25?

Author: Because the replica is made from an other part of the specimen, they may be longer or shorter, thus may be arranged in a different way after the preparation. Also in the area represented by the replica, some material seems to be between the microvilli. More improtant, both methods establish contrast in different ways. Whereas in SEM the surface is seen from the incident beam in the "light" of the detector, in the replica the

representation of structures is only dependent
upon the glancing angle of the metal deposition.

W.H. Massover: What are the improvements regarding ferritin, since the images published from
JEOL (Watabe et al. 1978, Ultramicroscopy 3:19-27)
appear to be better?
Author: Referring to the mentioned work in Jeol
News 14, 1976,11-14, sent to me by JEOL as a
reference for ferritin images obtained, you raise
an interesting question about comparison of images
obtained under different conditions. Thin films
give very favorable imaging conditions for small
particles because the exitation volume for the
signal is extremly small (related to film thickness and probe diameter). Thus uncoated ferritin
can be easily visualized at high voltage without
coating, but of course, only the iron core is seen
in material contrast. The images, referred to as
ferritin of 7.0 to 12.0 nm in size, represent the
cores with increased size, due to superposition or
beam damages. 5.0 nm Au/Pd coating of ferritin on
film or bulk Al (foil) generates heavy decoration,
thus revealing mainly visualization of crystals.
Similar effects are obtained by heavy Pt-C shadow-
cating (which obscure the image of the iron core).
In biology, we are mostly concerned with bulk
specimens. Single macromolecules or macromolecular
complexes are easily seen in TEM. Thus in this
work, ferritin absorbed on bulk carbon is used as
a topographic test specimen. After ultra thin Ta
coating (2.0 nm thick surface coat) this specimen
is not visualized as on thin film. Figs. 5 and 6
prove that even an ultra thin metal film on a bulk
specimen scatter a serious number of PEs, generating a strong nonspecific noise.

SCANNING ELECTRON MICROSCOPY/1979/II
SEM Inc., AMF O'Hare, IL 60666, USA

FURTHER DEVELOPMENT OF THE CONVERTED BACKSCATTERED ELECTRON DETECTOR

S. H. Moll, F. Healey, B. Sullivan and W. Johnson

AMRAY, Inc.
160 Middlesex Turnpike
Bedford, MA 01730

Abstract

The Converted Backscattered Electron Detector introduced in 1978 has been further characterized. Comparisons of images obtained with both the CBSE detector and a solid state diode array have been carried out under identical conditions of sample excitation and sample geometry. Previous measurements of relative signal contributions for various detection modes, carried out in terms of signal modulation depth, have been remeasured in terms of total average signal level for specimens of varying atomic number and a simplified theoretical treatment has been developed. The original grid mesh structure of the detector has been alternatively replaced with a simple ring structure whose performance is essentially identical. The ring detector allows an unimpeded view of the sample surface down to the lowest possible magnifications and introduces virtually no artifacts in the case of x-ray analysis.

KEY WORDS: Backscattered Electrons, Electron Detector, Atomic Number Contrast, Topographic Contrast, Stained Biological Tissue, Metallurgical Specimens

Introduction

A backscattered electron detection system involving the suppression of secondary electrons emitted by the specimen, followed by conversion of BSE to SE at the pole piece of the final lens and collection of these by the conventional scintillator-photomultiplier detector developed by Everhart and Thornley[1] has been described previously[2]. It has been shown that the collection efficiency of the Converted Backscattered Electron Detector (CBSE) is excellent. Since the video signal is generated by all BSE leaving the specimen over essentially 2π steradians, BSE images may easily be formed at low and high magnification utilizing incident currents as low as 10^{-11} amps. Spatial resolution of uncoated samples has been shown to be better than $100\overset{\circ}{A}$. Frequency response of the system is determined by the conventional detector employed to gather the SE produced by the BSE conversion and is well known to exceed that necessary for TV rate imaging.

The functional components of the CBSE detector are shown in Figure 1. The trajectories of various electrons are indicated in accordance with the mechanisms we believe to be operating in producing the observed signals. In the CBSE mode, the grid (or retarding ring) is biased with a negative potential on the order of -50 to -100 volts, the specimen is grounded and the collector is at $+300$ volts. Conventional SE images are obtained with a small positive potential applied to the grid ($+10$ to $+100$ volts) while the collector remains at $+300$ volts and the sample is grounded. If the collector is held at -100 volts, all SE of any origin are rejected and the typical, directionally shadowed, low solid angle collection efficiency BSE images result. Low voltage biasing of the CBSE grid does not affect this image.

Reviews[2] of the original presentation of results obtained with the CBSE system indicated that further investigation of the system was of interest. Measurements of relative signal strengths in the various CBSE detection modes were carried out previously in terms of the signal modulation depth or peak to peak excursion of the video signal. In the current work, relative signal strengths were measured in terms of

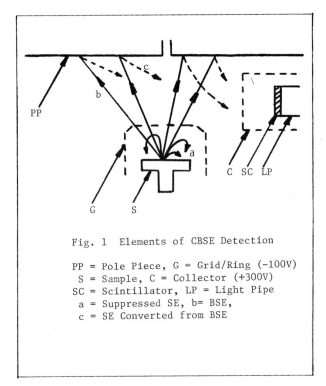

Fig. 1   Elements of CBSE Detection

PP = Pole Piece, G = Grid/Ring (-100V)
S = Sample, C = Collector (+300V)
SC = Scintillator, LP = Light Pipe
a = Suppressed SE, b= BSE,
c = SE Converted from BSE

the absolute DC level of the total video signal (brightness) for various geometries and sample atomic number. A theoretical treatment of the electron detection mechanism was also indicated to be of interest.

It has also been suggested by reviewers that the CBSE detector should be more sensitive to low energy BSE (<10 keV) since the SE yield increases at the pole piece as the exciting electron energy decreases, whereas with solid state BSE detectors the higher energy BSE will contribute more to signal output. Therefore, comparisons of images obtained with the CBSE detector to images obtained with solid state detectors at high incident beam energy, where the solid state detector is most efficient, was recommended.

Relative Signal Strengths and Electron Contributions

Relative signal strengths for various CBSE detector operating modes were measured in terms of the absolute DC level of the total video signal. This is easily done using the line scan mode (or any DC level measuring meter). Careful attention must be paid to the linearity of the measuring system, to fixing any variable gain control in the video chain, et cetera. Data were obtained for specimen tilts of 0° and 45° for samples of Au, Cu and Al. Measurements were made at 20kV, 200 micron final lens aperture, incident currents in the 10^{-11} amp range, at magnifications on the order of 500X, and at 12mm W. D. (working distance).

Table 1 summarizes the results of this investigation. The data are reported relative to the highest video signal measured in the data

set, i.e., that for the secondary electron mode signal measured with the Au sample tilted at 45°.

The new data relating to absolute DC signal level is in good agreement with the data previously obtained for signal modulation depth[2]. The BSE signal obtained with the CBSE detector is about fifteen to thirty percent of the SE signal under the conditions used. In all cases the signal contribution of the BSE collected directly by the scintillator is on the order of two to six percent, depending upon the geometry and atomic number of the sample. Specimen tilts increase the BSE flux coming from the specimen as does an increased atomic number of the specimen itself in accordance with known electron-specimen interaction data.

As previously discussed by Drescher[3], Everhart[4], Moll[2], and more recently by Moncrieff[5], when the SEM is operated in the SE detection mode, the large majority of SE collected by the scintillator-photomultiplier detector is not generated within the sample inside a localized volume around the point of excitation. Rather, significant numbers of SE emanate from areas far removed from the point of impact of the electron beam due to the re-emerging BSE flux; many are generated by excitation of the pole piece, specimen chamber, et cetera, and some may also be collected from excitations at the final lens aperture, depending upon design and geometry.

It is clear, based on the CBSE detector measurements, that some twenty percent of the total SE signal is generated by the BSE striking the pole piece, et cetera. It may also be speculated that the SE produced by the BSE emerging from the specimen are at least as abundant as those produced by these same BSE striking the pole piece, et cetera. It would appear that more than fifty percent of the SE signal collected by the conventional detector operating in the SE mode are the result of these excitations.

Calculation of Relative Signal Strengths

Based on published values for SE and BSE yields, and the geometrical collection efficiency of the various structures used, it should be possible to calculate the ratios of CBSE to SE signals as reported in Table 1. Certain assumptions are made in the following simplified, preliminary derivation.

(1) The collection efficiency of the SE detector is the same for SE produced at the sample and at the pole piece surface.

(2) The signal contribution of BSE striking the scintillator directly (on the order of a few percent of the total SE signal) may be neglected. The contribution of BSE scattered from the pole piece is also neglected.

(3) The secondary electrons produced at the pole piece may be assumed to be excited by the peak energy of the BSE energy distribution leaving the specimen.

(4) The efficiency of collection of SE from the pole piece by the SE collector does not vary with the location of SE produced.

Table 1
Relative Signal Strengths
Total DC Signal Level

	45° Tilt			0° Tilt		
	Au	Cu	Al	Au	Cu	Al
Standard SE Signal Grid removed Collector = +300 volts	1.00	0.80	0.72	0.46	0.40	0.24
CBSE Signal Grid = -100 volts Collector = +300 volts	0.26	0.16	0.14	0.087	0.060	0.036
Standard BSE Signal Grid = 0 volts Collector = -100 volts	0.060	0.030	0.030	0.020	0.010	0.005

The signal which results from the collection of SE produced at the sample (S_s) is proportional to the secondary yield of the specimen and the collection efficiency of the SE detector where δ_s is the appropriate SE emission coefficient.

$$S_s = K \delta_s \qquad (1)$$

The CBSE signal which results from the collection of SE produced at the pole piece (S_p) is proportional to the backscatter yield of the specimen (η_s) times the secondary yield of the pole piece (δ_p) corrected for the collection efficiency of the pole piece (C_1) (intercepted solid angle compared to full 2π surface) and the attenuation of the grid mesh situated between the sample and pole piece in the CBSE mode (C_2). The secondary yield at the pole piece (δ_p) is taken at the most probable (peak) energy of the BSE energy distribution leaving the specimen. It should be noted that $C_2 = 1$ if the mesh is removed.

$$S_p = K \delta_p \eta_s C_1 C_2 \qquad (2)$$

The total SE signal generated at the detector is thus the sum of (1) and (2) and the ratio of the CBSE signal to the total SE signal, with the grid mesh removed, is given by...

$$S_p / S_p + S_s = \frac{\delta_p \eta_s C_1 C_2}{\delta_s + \delta_p \eta_s C_1} \qquad (3)$$

For the case of a Au specimen excited at 0° tilt with 20kV electrons and with a Ag plated pole piece, the values of δ_s, δ_p, and η_s are reported to be 0.27, 0.23 and 0.48, respectively, from the data of Moncrieff[5] and review by Niedrig[6]. The values of C_1 and C_2 are taken as 0.9 and 0.7, respectively, based on calculations. The calculated ratio from (3) is then 0.19 in excellent agreement with the experimental value of 0.19. Similarly, calculated values for Cu and Al are 0.16 and 0.14, respectively, while the experimental values are 0.15 and 0.15, respectively.

The CBSE signals for Au, Cu and Al for the flat specimen are in the ratio of 1.0, 0.69 and 0.41, respectively, while calculated values are in excellent agreement at 1.0, 0.63 and 0.39, respectively.

While the simplified analysis presented seems to corroborate the signal data for perpendicular electron beam incidence, an analytical treatment in the case of tilted specimens has not yet been developed. It is clear that tilting the specimen increases the ratio of CBSE signal to the total SE signal. This would indicate that the BSE coefficient tends to increase with respect to the SE coefficient as the sample is tilted. Further experimental work as well as analytical treatment appears to be necessary to explain these data.

Comparison of the CBSE Detector with a Solid State Diode Detector

Electron micrographs utilizing BSE were generated under identical conditions for both a solid state diode detector (SS) and the CBSE detector. A solid state diode detector array was positioned just under the pole piece and was arranged symmetrically around the bore. Followed by a high gain, low noise amplifier, the diode detector subtended a solid angle of collection of about 1-2 steradians for working distances of about 12mm. A CBSE detector was installed simultaneously. It is estimated that the solid angle of excitation on the pole piece for BSE was on the order of five to six steradians. Comparison micrographs could be obtained under identical conditions, by simply selecting the video signal of each detector system alternately. In all cases, electron optical conditions were optimized for the SS detector, the micrograph obtained, and the CBSE detector image obtained immediately thereafter.

Four different types of specimens were examined in this way. These included a metallographically prepared multiphase alloy of Si, Fe, Cu and Ag, a Ag stained human kidney specimen, a silicon single crystal examined in the SACP mode, and a metallic iron base alloy particle. In most cases, the micrographs obtained were so similar that it would not be justified to present them here, since the differences could not

Fig. 2. Metallic powder particle, SS image,
i = 8X10⁻¹¹ amps, 0° tilt, 60 sec. scan.

Fig. 3. Metallic powder particle, CBSE image,
i = 8X10⁻¹¹ amps, 0° tilt, 60 sec. scan.

be observed within the constraints of the reproduction process. In the case of the metallo-
graphic specimen and the SACP image, the micrographs were essentially identical but some small
differences could be noted in the case of the
other two.

   Figures 2 and 3 are BSE images of a metallic
iron base alloy powder particle. The image formation mechanism would be expected to be primarily
that of topographic contrast. The CBSE image
reveals a more developed three dimensional quality than the SS image which appears "flattened"
in contrast, as if a nonlinear "gamma" amplifier
had been used. Slight variations in gray level
in the video signal appear to be more faithfully
reproduced in the CBSE image. All parts of the
roughly spherical object, including the "pit"
at middle right, are better illuminated in the
CBSE image. Note the presence of the black
shadow in the upper right hand corner of the SS
image which is absent in the CBSE image as well

as the presence of picture information in the
lower right hand corner of the CBSE image which
is absent in the SS image. The lack of "omni-
directional illumination" in the SS image
probably results from its lower collection
efficiency (1-2 steradians compared to five
to six for the CBSE detector). On the other
hand, the better developed topographic detail
in the CBSE image may result from a small
directionality due to enhanced collection of
converted BSE at the portion of the pole piece
which is closer to the collector.

   A human kidney section, mounted on glass,
selectively Ag stained and carbon coated, was
contributed by Dr. P. DeNee. Complete preparation details are reported elsewhere[7]. Figures
4 and 5 are the BSE images obtained with the SS
and CBSE detectors, respectively. Surface and
subsurface cell nuclei are rich with the Ag
stain allowing their visualization due to an
atomic number contrast mechanism. The images

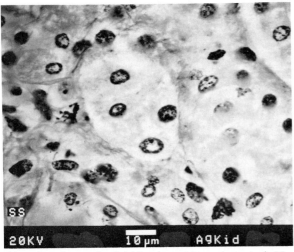

Fig. 4. Ag stained kidney section, inverted SS
image, i = 7X10⁻¹¹ amps, 0° tilt, 60 sec. scan.

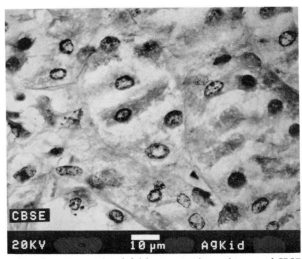

Fig. 5. Ag stained kidney section, inverted CBSE
image, i = 7X10⁻¹¹ amps, 0° tilt, 60 sec. scan.

are video inverted and thus the Ag rich areas are dark. The CBSE image reveals somewhat finer detail and exhibits a gray scale in the image which reveals the lightly stained areas also. In general, however, the images are very similar.

Comparison of the two detector systems thus reveals no large differences in BSE imaging capability. There appears to be a small tendency for a more faithful linearity in signal amplification with the CBSE detector compared to the SS detector, but this supposition would require further investigation. Observations of the line scan of the video signal revealed a better signal to noise ratio with the CBSE detector, and the frequency response of the SS system was inferior to that of the CBSE system.[8]

Alternative Geometry for the CBSE Detector

The wire grid cage utilized in the original design[2] to develop the appropriate electrostatic field gradient in the vicinity of the specimen may be replaced by other structures which function just as well in producing CBSE images. These may be left to the imagination of the designer, but one geometry with certain advantages over the original grid mesh was investigated.

A wire ring structure consisting of two or three loops which surrounded the sample stub was used. The uppermost wire ring was only about 5mm above the sample surface and had an open diameter on the order of 15mm. The significant advantage of this geometry is the fact that with untilted or only slightly tilted specimens, extremely low magnifications down to about 10X are possible without interference of the biasing structure. In the case of the original grid mesh, the grid wires were visible in images below about 80X magnification. Relative signal strengths measured with the ring geometry were virtually identical to those reported in Table 1 and this "detector" was actually used to obtain silicon channeling patterns. Figure 6 is a micrograph of the ring structure operating in the SE mode. The ring and stub have been purposely

tilted to about 50° to reveal the construction. The SE image is uniformly illuminated.

Figure 7 is a comparison image with the ring structure biased to -100 volts (CBSE mode). The specimen stub within the ring is much darker since SE are suppressed. Note that the ring itself is still bright as are some areas remote to the ring, due to SE emission. It should also be pointed out that although in the construction used by the authors the specimen is always grounded while the grid or ring structure is negatively biased for CBSE images, grounding of the grid and positive biasing of the specimen would probably work just as well.

X-Ray Artifacts

It was anticipated that the grid or ring structure would contribute spurious peaks to EDX spectra unless care was taken to paint the structures with carbon paint or to use Be as a construction material. Artifactual peaks produced by the wire mesh or ring structure were virtually undetectable, however. Using a carbon planchet and 30kV incident beam energy, spectra were collected for long period of time with and without the grid or ring present. In addition to a very weak Al peak due to excitation of the specimen stage, a very weak Cu peak was evident when the Cu wire mesh grid was installed, but not when the ring structure was present. It was noted that a Si peak from the SS detector was more intense than any peaks from the CBSE detector structure. The general conclusion was that the introduction of spurious peaks was of no great significance for normal EDX studies.

Conclusion

Measurements of absolute signal strength (DC level) with the CBSE detector are in good agreement with similar measurements made previously in terms of signal modulation depth (contrast). A simple theoretical treatment derived to predict the relative SE and CBSE signal strengths appears to be in accord with the

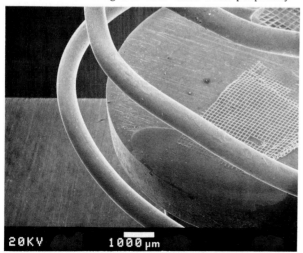

Fig. 6. Biased ring geometry, SE image, i = 1X10^{-10} amps, 50° tilt, 60 sec. scan.

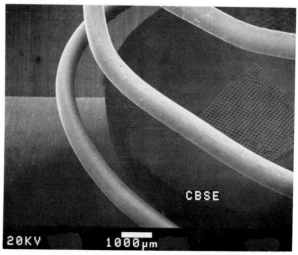

Fig. 7. Biased ring geometry, CBSE image, i = 1X10^{-10} amps, 50° tilt, 60 sec. scan.

experimental data in the case of perpendicular electron beam incidence on the sample. It is again noted that a high level of SE contribution arises from BSE excitations when operating SEM's in the conventional SE mode.

Comparisons of BSE images obtained for a variety of specimens involving atomic number, topographic and electron channeling pattern contrasts and, using both a solid state diode detector (SS) and the CBSE detector, revealed no significant differences in their ability to form acceptable images. Conditions were chosen to optimize performance of the SS system compared to the CBSE system (20kV, higher incident currents). Further work comparing SE, SS and CBSE images at low beam potentials in the range of 1-10kV would be of interest.

To form CBSE images, it is possible to employ many forms of biasing structures which will generate an appropriate field gradient in the vicinity of the specimen surface. The ring structure described in this work has the advantage that the low magnification capability is not limited as compared to the wire mesh grid originally utilized. X-ray artifacts produced by the biasing structures are insignificant.

References

1. T.E. Everhart and R.F.M. Thornley, "Wide Band Detector for Micro-microampere Low Energy Currents," J. Sci. Inst., 37, 1960, 246-248.
2. S.H. Moll, F. Healey, B. Sullivan and W. Johnson, "A High, Energy, Nondirectional Backscattered Electron Detection Mode for SEM," SEM/1978/I, SEM, Inc., AMF O'Hare, Il., 60666, 303-310.
3. H. Drescher, L. Reimer, H. Seidel, "Ruckstreukoeffizient und Sekundarelectronenausbeute von 10-100 keV-Elektronen und Beziehungen zur Raster-Elektronenmikroskopie." Z. angew. Phys. 29, 331-336 (1970).
4. T.E. Everhart, "Contrast Formation in the Scanning Electron Microscope," Ph.D. Dissertation, Cambridge University, (1958).
5. D.A. Moncrieff and P.R. Barker, "Secondary Electron Emission in the SEM," Scanning, Vol. 1, 3, (1978) 195-197.
6. H. Niedrig, "Physical Background of Electron Backscattering," Scanning, 1, No. 1, (1978), 17-34.
7. P.B. DeNee, R.G. Frederickson and R.S. Pope, Heavy Metal Staining of Paraffin, Epoxy and Glycol Methacrylate Embedded Biological Tissue for SEM Histology," SEM/1977/II, ITT Research Institute, Chicago, Il., 60616, 83-92.
8. D.A. Gedcke, J.B. Ayers, P.B. DeNee, "A Solid State Backscatter Detector Capable of Operating at T.V. Scan Rates," SEM/1978/I, SEM, Inc., AMF O'Hare, Il., 60666, 581-594.

DISCUSSION WITH REVIEWERS

M. Sogard: Can you provide any examples of the CBSE picture quality as a function of beam voltage? Since the SE coefficient decreases with increasing energy, what is the highest energy for which you can obtain a decent image?
Authors: CBSE micrographs published to date have been concerned primarily with comparisons to other types of BSE detectors which operate most efficiently above 15kV (scintillator and solid state). However, we have operated the CBSE detector using beam voltages down to the 1-5kV range and intend to report on these studies in the future. With respect to high beam voltages, we have obtained[2] micrographs of uncoated magnetic tape at 30kV exhibiting spatial resolution better than 100Å.

D. Davidson, M. Sogard: Have you tried enhancing the CBSE signal still more by covering the pole piece with something like Au to enhance the SE coefficient?
Authors: We have addressed this question previously (Reference 2, questions by Reviewers III, IV, VI). The basic signal to noise ratio in the observed CBSE signal is felt to be determined by the shot-noise of the emerging BSE flux and not the converting of BSE electrons to SE electrons. On the other hand, the pole piece in the SEM employed is Ag plated in the standard design. We have not attempted to change its surface chemistry.

L. Reimer: Can you show retarding field curves (CBSE signal versus grid bias) for the two detector" types (grid and wire ring)?
Authors: We have observed that when adjusting the voltage applied to the grid or ring, that the values for either CBSE detection (negative voltages) or for SE detection (positive voltages) will vary depending upon the sample tilt, working distance, grid or ring size or position and we simply tune until the signal no longer changes. We agree that such data could be informative for a few selected configurations.

L. Reimer: Has a +100V biasing an effect on astigmatism at high magnifications?
Authors: Yes. In our earliest efforts with the grid "detector" we have observed small but significant changes depending upon the sense of the applied voltage. We actually observe essentially no astigmatism to be corrected for under certain conditions of geometry and bias.

M. Sogard: Your model ignores contributions from secondary electrons produced in the aperture. Moncrieff and Barker (text ref. 5) estimated this contribution to be 20-40% of the secondary electron yield. Also, you neglect the standard BSE signal, which according to your table is more than 20% of the CBSE signal in the case for gold. Please comment.
Author: The derivation of relative signal strengths was basically carried out to suggest, in a simple way, those physical parameters which control CBSE signal contribution such as SE and BSE yields and the collection efficiency of the pole piece. The close agreement with the experimental data is perhaps fortuitous. The contribution of BSE striking the scintillator would tend to reduce the calculated ratio as would the effect of SE produced at the final aperture. The contribution of SE produced at the final aperture may be reduced by appropriate geometry. In the SEM employed measurements of SE signal at the detector with the primary beam entering a Faraday Cage indicate this effect to be less than 2% of the total normal SE signal.

SCANNING ELECTRON MICROSCOPY/1979/II
SEM Inc., AMF O'Hare, IL 60666, USA

CONVERSION OF EXISTING SEM COMPONENTS TO FORM AN EFFICIENT BACKSCATTERED
ELECTRON DETECTOR, AND ITS FORENSIC APPLICATIONS

N. Zeldes and M. Tassa

Criminal Identification Division
Israel National Police Headquarters
Jerusalem, Israel

Abstract

Most SEMs are not equipped with a BE detector
suitable for forensic work. A detector is described, which is eminently suited to forensic
needs, very inexpensive and simple in construction. This is achieved by converting existing
components of the Secondary Electron Imaging (SEI)
system of the SEM. A full technical description
is followed by performance data and some examples
of the uses to which the detector is being put in
routine case work.

KEY WORDS: Backscattered Electron Imaging,
Detector, Gunshot Residue, Atomic Number Contrast,
Paint Smears, Specimen Charging

Introduction

Over the past few years interest in Backscattered Electron Imaging (BEI)[1,2] among
forensic scientists has been rapidly growing.[3]
This interest was in many instances triggered by
the recently introduced Gunshot Residue (GSR)
Particle Technique. For all its power, this
method cannot be used on any large scale without
a fast search procedure for detecting these Pb -
Sb - Ba rich particles on hand-samples containing
very large populations of lighter dust particles.
BEI, which provides for Atomic Number Contrast,
neatly solves this acute problem; it also has the
added advantages of freedom from charging artefacts[1,4] and edge highlighting.

Achieving these advantages can, however, be
a painstaking problem. Some SEMs do not provide
for BE detection. Those that do (usually as an
option) mostly use silicon solid-state detectors,
which preclude imaging at TV scan rates—a limitation which all but rules out effective GSR
search. Furthermore, commercially available BE
systems are rather expensive.

Confronted by these problems, the authors
were led to attempt the development of a detector designed to fulfill the specific needs of the
forensic laboratory.

Design Considerations

The following design goals were outlined
for this project:
a. Acceptable image quality at TV scan rate.
b. Unobstructed specimen motion of ±15mm on x
and y axes (to permit the scan of an entire
specimen stub 25mm in diameter), at a working
distance suitable for efficient X-ray analysis.
c. Simple and inexpensive construction, using
easily available facilities.
d. A minimum of irreversible alterations in SEM
mechanical structure.

Since the Backscattered Electrons are very
energetic (10-20 KeV for a 25 KeV primary beam),[5]
it is impractical to curve their divergent trajectories towards a small area detector, as is
done with the <10eV Secondary Electrons. Instead, it is necessary to detect them by intercepting their straight trajectories. A BE
detector should therefore subtend a solid angle
at the specimen sufficiently large to give an

acceptable S/N ratio, subject to the constraints of specimen free motion.

At present one finds in use two main categories of BEI detectors: those using silicon solid-state detectors,[6,7] and photomultiplier (PMT)/scintillator systems, successfully adapted to BE Imaging by Robinson in 1973.[8] The former are very compact, but their slow response makes TV rate operation impossible. The latter type, while more cumbersome, is free of this limitation and was thus indicated for the present project.

In the design to be described, use has been made of most of the existing parts of the SEI system. The scintillator is positioned near the beam axis, facing the specimen, and its light output is routed to the SEI PMT. The coupling is achieved by a curved light guide, which carries the scintillator on its far end. The other end is mounted onto the bare end of the SEI detector's light guide, which has been exposed by removing the cage and SEI scintillator tip.

This approach has three attractive advantages:
a. By avoiding the use of a separate PMT (or P-N junction detectors, for that matter) the interface between the BE detector and the SEM is purely optical rather than electronic. No critical amplifiers are involved, and no mismatches can exist. In fact, the designer's job is reduced to getting as much light as possible to the PMT photocathode; the rest is done by the tried and proven electronics provided by the SEM manufacturer for his SEI signals.
b. There is no need to make any permanent alterations in the SEM proper, such as drilling holes, modifying circuits, etc.
c. By dispensing with a separate PMT and its associated circuitry system cost is greatly reduced.

The efficiency of such a detector depends chiefly on the geometry of the scintillator, its position and the design of the extension light guide. A curved scintillator enclosing the specimen, as proposed by Robinson,[8] is best from signal-to-noise considerations. However, it is impractical for common forensic work since it severely limits specimen size and motion and precludes the introduction of an X-ray detector. After some trial and error, the geometry presented below was chosen, which permits using the SEM's original scintillator disc and is very easy to make. Although it intercepts only part of the BE flux, it provides an adequate S/N ratio and permits X-ray analysis and free specimen motion.

Technical Details

The detector is designed for a Cambridge Scanning Co. Model Cam Scan III S SEM. The SEI detector of this SEM is shown in Fig. 1. The Installed BE detector is shown in Fig. 2, where the cage has been removed and the extension light guide is mounted on the SEI guide instead of the SEI scintillator tip, using the latter's mounting ring for support. Details of the detector are shown in Fig. 3.

The light guide extension was made from a 19.5mm diameter perspex rod, curved under heat and given a fine polish. The outside was coated with

Fig. 1. Original SEI detector system. 1 - Final Lens. 2 - Cage. 3 - Light Guide. 4 - PMT. 5 - X-ray Detector. (Column in this SEM is horizontal.)

Fig. 2. BEI detector installed. 1 - Extension Light Guide. 2 - SEI Scintillator Mounting Ring. 3 - Mounting Ring. 4 - Grounding Wire.

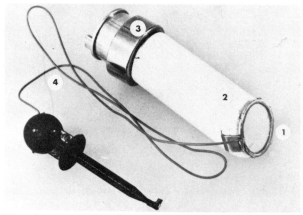

Fig. 3. Extension light guide Assembly. 1 - Scintillator disk in clip ring. 2 - Perspex Light Guide. 3 - Mounting Ring. 4 - Grounding Wire.

a white TiO_2 water-based paint. (Although recommended, this is not essential for effective results.) One end of the guide is equipped with a threaded brass ring and flange, which fit the original scintillator mounting hardware on the SEI guide. The curved end accepts the SEM's standard scintillator tip, a glass disc coated with an inorganic phosphor layer. This is clipped to the perspex by a ring which also makes contact to the conductive outer layer of the scintillator. A wire leads accumulated charge from this ring to an earthed clip.

Applying this design, care must be taken of the high voltage connection on the SEI light guide tip, which should be kept insulated.

Performance

Performance was found to depend on the scintillator's position and angle relative to the specimen. The geometry shown in Fig. 3 performs very satisfactorily at TV scan rate, giving an image quality and resolution nearly equal to those obtained at SEI at the same working conditions. In practice, this means that structural details on a typical GSR particle can be comfortably observed, at a magnification of about x10,000, at TV rate.

The advantages of BEI, namely Atomic Number Contrast, reduced Edge Highlighting and freedom from Charging Artefacts are fully achieved.

To convert the SEM from SEI to BEI or back, one merely opens the chamber and replaces the cage structure with the extension guide, or vice versa; the whole procedure takes about 1 minute.

Overall expenditure for this project, including early prototypes, was about $100, consisting mainly of cost of machine shop labour.

Limitations and Planned Improvements

The main limitation of this configuration is that since the detector is offset to one side of the beam, there is some shadowing of details on the "far" side of curved specimen features. This is not critical in normal use, but work is now in progress on a new model with a larger scintillator surrounding the beam axis, in which the scintillating layer is deposited directly onto the perspex surface.

Topography enhancement, as achieved with split-detector operation, is inherently impossible in a PMT/scintillator configuration; however, it is the authors' opinion that this is a small price in return for the much more useful capability of TV scan rate operation.

Forensic Applications

The BEI detector is now in routine use on forensic case work. Some typical examples of its applications follow:

a. GSR Search. This previously laborious task has been revolutionized. Where hours were needed to find a GSR particle, now minutes suffice. Contrast is simply adjusted so that heavy particles appear white on a completely black background (Fig. 4). Although some "false alarms" still occur (Barium Sulphate particles, for

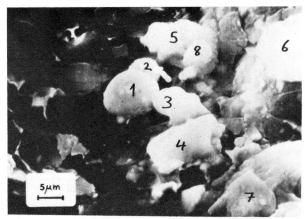

Fig. 4a   Particle population from suspect hand sample; SE Image. 1,2 are GSR particles (Pb, Sb, Ba). 3-7 are Si, Ca, Cl particles.

Fig. 4b. The same population in a BE Image. GSR particles stand out clearly.

Fig. 5. BE Image of some paint smears on an uncoated cloth specimen.

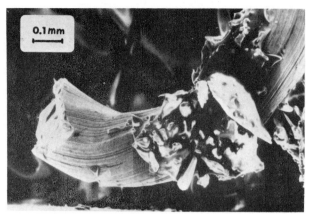

Fig. 6a. SE Image of tungsten lamp filament with molten glass fragments.

Fig. 6b. BE Image. Charging artefacts are eliminated, making glass surface details clearly visible.

instance), the search is faster by at least an order of magnitude.

    b. Cloth and Paper examination. These specimens are hard to coat properly, and charging problems can make it impossible to locate and analyze small paint particles located between the fibers. Using BEI, not only is the charging rendered harmless, but the paint stands out clearly due to atomic number contrast (Fig. 5).

    c. Examination of uncoated specimens. A good example is the case of molten glass fragments on tungsten lamp filaments, which would be blurred under an aura of charging artefacts in SEI, but can be observed very clearly in BEI (Fig. 6).

    d. Further applications arise continuously, usually in cases where specimens must be searched for smears or particles distinguishable by their average atomic number.

Summary

    The suitability of the detector described to forensic work is proved by its performance in routine case work. Although this model was designed specifically for a Cam Scan SEM, the authors are confident that any competent SEM user can implement the principle to convert his own type of instrument to BEI with equal success. It should be noted that although performance does depend on a wise choice of geometry, experience shows that this dependence is not too critical.

References

1. V.N.E. Robinson, A Re-appraisal of the Complete Electron Emission Spectrum in Scanning Electron Microscopy, J. Phys. D:Appl. Phys. 6, 1973, L105-7.
2. V.N.E. Robinson, Backscattered Electron Imaging, SEM/1975, Ed. O. Johari & I. Corvin, IIT Research Institute, Chicago, U.S.A., pp. 51-60.
3. V.R. Matricardi, The Application of the Back-Scattered Electron Image to Forensic Science. SEM/1979/II, SEM Inc, AMF O'Hare, IL 60666, This Volume.
4. T.J. Shaffner & J.W.S. Hearle, Recent Advances in Understanding Specimen Charging. SEM/1976/I, Ed. O. Johari, IIT Res. Inst. Chicago 60666, pp. 61-70.
5. P.R. Thornton, Scanning Electron Microscopy, Chapman & Hall, London, 1968, pp. 85-95.
6. A.V. Crewe & P.S.D. Lin, The Use of Back-scattered Electrons for Imaging Purposes in a SEM, Ultramicroscopy 1, 1976, pp. 231-8.
7. J. Stephen, B.J. Smith et al., Applications of a Semi-conductor Backscattered Electron Detector in a SEM, J. Phys. E: Sci. Instrum. 8, 1975, pp. 607-10.
8. V.N.E. Robinson, The Construction and Uses of an Efficient Backscattered Electron Detector for SEM, J. Phys. E:Sci. Instrum. 7, 1974, pp. 650-2.

Discussion with Reviewers

R. Keeley: How do you propose to increase the solid angle of collection of your detector without obstructing sample movement and X-ray collection?
Authors: Obviously, a tradeoff is necessary between detector area on one hand, and sample movement and X-ray collection on the other. However, once the limitation of using the flat scintillator disk is removed, we can use irregular surfaces which permit a better compromise than that presently achieved. The shape of the optimal solution would vary from one SEM to another, and we intend to describe our choice at the conclusion of the research now under way.

V.R. Matricardi: What is the scintillator material which you are using for your larger detector? Is it available commercially?
Authors: We are expecting delivery of P47 phosphor, which is available commercially. Details, including deposition technique, can be found in an excellent article by Comins et al., J. Phys. E:Sci. Instrum., 11, 1978, pp. 1041-7.

For additional discussion see p.124.

SCANNING ELECTRON MICROSCOPY/1979/II
SEM Inc., AMF O'Hare, IL 60666, USA

THE APPLICATION OF THE BACKSCATTERED ELECTRON IMAGE TO FORENSIC SCIENCE

V. R. Matricardi

Elemental Analysis Unit,
FBI Laboratory
Washington, DC 20535

Abstract

The backscattered electron image (BSI) has been used extensively in our Laboratory to analyze specimens which originate during criminal acts. The ability of the BSI to discriminate between elements of appreciably different atomic numbers and to view insulating specimens without severe charging artifacts makes it a useful addition to the SEM in a crime laboratory.

Often the examiner has no previous information about the specimen which is being examined; thus an examination procedure is required which will allow the maximum information to be gathered without destroying or altering the specimen. The light microscope is, of course, the first instrument of choice. Additional analytical and topographical information is obtained from the BSI with the SEM. Prior to X-ray analysis the BSI is used to localize areas of inhomogeneity or inclusions on the surface of the specimen which are often not detectable with the light microscope or the secondary electron image. Typical applications would include the localization of smears (on bone, car bumpers, tools, weapons, etc.), identification of foreign particles (precious metal chips in a salted ore, rare earth powders used as tags in illegal transfers or thefts), and identification of gunshot residue particles from the hands of a suspected shooter.

KEY WORDS: Backscattered Electrons, Atomic Number Contrast, Forensic Applications, Gun Shot Residue

Introduction

For the last four years, the backscattered electron detector has been used routinely to examine criminal evidence in the author's laboratory. Next to the energy dispersive X-ray analyzer (EDX) it has been one of the more useful attachments to the scanning electron microscope (SEM). This paper is meant to provide the forensic scientist with an overview of the backscattered electron image (BSI), the general methods used in specimen preparation prior to obtaining the BSI and some relevant applications to criminal case work.

In a crime laboratory the criminalist is often asked to obtain as much information as possible from a microscopic specimen without altering or consuming the specimen. After a thorough examination and classification by light microscopy the SEM-EDX can be used to further classify the material. Initially an EDX spectrum is obtained from the whole specimen. The secondary electron image (SEI) provides further information about the outline and surface of the specimen. The SEI is the image normally observed in SEMs without attachments. This has been used to good advantage in some casework. However the SEI by itself is of limited use in a crime laboratory. Some of the early papers dealing with SEM in criminalistics (including a paper by this author) emphasized the SEI. In the author's opinion, it is the analytical capability available with the EDX which really makes the SEM-EDX useful to criminalists.

The backscattered electron image (BSI) provides some additional useful information and should be considered during the examination by SEM-EDX. Before going into specific examples, the origin of the BSI will be presented.

Characteristics of the BSI

The BSI is the result of the collection and amplification of some of the backscattered electrons (BSE) which are arbitrarily defined as those electrons which, due to electron excitation by the beam, leave the specimen with an energy greater than 50 electron volts. For recent reviews of the BSI the reader is referred to Wells [1] and Niedrig [2].

The forensic applications of the BSI are subdivided in this paper into three subjective categories:

A) Localization of areas of the specimen surface having an atomic number different from that of the matrix (Z-contract)
B) Determination of the three-dimensional shape of the surface of the specimen (topographic-contrast)
C) Viewing of insulating specimens.

Note that electron-channeling contrast and magnetic contrast have been omitted because of a lack of known applications in forensic examinations. The author feels confident that the users would be able to expand on these categories if, indeed, they have not done so already.

Given a flat clean specimen, the intensity of one picture point of the BSI will be a function of the average atomic number of the specimen. In particular, the BSI intensity increases as the atomic number [3] of the specimen is increased. Thus if the cross section of a gold coated brass wire (85%Cu, 15%Zn) is imaged by BSI, the gold layer (Z=79) will appear more intense than the brass (Z (effective) = 29.15).

Unfortunately the criminalist cannot always afford the luxury of having clean flat specimens because physical evidence has to be preserved in its original state or not enough material is present to embed and polish. For a rough dirty specimen, Z-contrast from the surface is modulated by surface contamination and the three dimensional geometry of the surface (topographic contrast). The contribution of the contamination layer (dirt) can be compensated by scraping the surface lightly (without making deep grooves) and comparing the intensity of the cleaned area to that of the original surface. The topographic contrast effects are due to changes in the emission of BSE with the angle between beam and surface and geometrical beam-specimen-detector considerations based on the directional nature of BSE emission. As a result, some specimen surfaces which would be expected to provide a high intensity by Z-contrast appear dark in the BSI because of orientation [3]. However it should be emphasized that topographic contrast has not, in the author's opinion, appreciably reduced the effectiveness of the search for foreign particles. For a very rough specimen surface the examination is more difficult.

There are also disadvantages to the BSI; especially from images obtained with older detectors. Generally a BSI has lower resolution than the corresponding SEI; it requires a higher beam intensity incident on the specimen; it requires direct line of sight between specimen and detector and the amplifiers which are used cause smearing of the image at rates of 100 frames per minute or faster. Some manufacturers now provide efficient BSE detectors and amplifiers that can operate at (or near) T.V. rates and there are recent modifications which are continuing the evolution toward an improved and less costly system (4-8).

Topographic contrast can, however, provide useful information about the three-dimensional shape of the specimen's surface (2, 9). This topic will be described briefly here since the author has not done any work in this area; however, the technique is a promising one for a criminalist to increase the morphological information available from a surface. Using topographical contrast, depth profiling of striations and fracture topography can be obtained. Fracture topography would be a useful extension of present fracture matching (10). A practical comparison of the application of topographical contrast and quantitative stereo viewing to surface topography would be useful.

Viewing of an insulating specimen is the last of the categories which was listed at the beginning of this section. The charging artifacts normally encountered in the SEI are not as pronounced in the BSI. Thus the requirement for metallic coating on an insulating specimen is relaxed to a great extent. The author does not coat specimens during examinations unless high magnification micrographs are required.

Experimental Procedures

The BSE detector used in this study consists of 3 diode detectors (36 mm ^{2}, total) located on one side of the objective lens pole piece (Etec Corp.). The detectors and amplifiers used in this study are over 6 years old and require higher beam currents and slower scan rates than more recent equivalents.

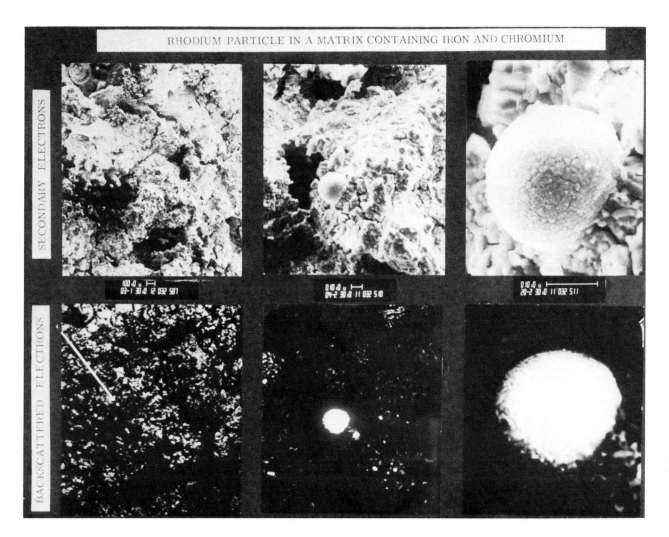

SECONDARY ELECTRONS

BACKSCATTERED ELECTRONS

Figure 1. Composite micrograph showing the SEI (top) and the BSI (bottom) of an iron-chromium sponge being searched for particles containing platinum and/or rhodium. The left BSI (marker corresponds to 100 um) shows an arrow pointing to a bright spot. Middle and right micrographs show increasing magnifications (marker corresponds to 10 um) of area indicated by arrow.

As an example, an undistorted image of the surface is obtained at 10 frames per minute; however, at 120 frames per minute, smearing of the image in the direction of scan occurs elongating a sphere into an ellipse. As was mentioned previously, recent developments have improved on these shortcomings remarkably (4-8). Specimen preparation is held to a minimum, mainly consisting of scraping a small area of the surface to determine bulk properties. The specimen is placed on a freshly sanded carbon surface and held there by gravity or a small adhesive layer (Scotch brand XBA 721510 or equivalent). The surface is tilted 10 to 30 degrees to allow line-of-sight access to the EDX collimator (which in this case has a horizontal axis). The distance between the specimen and the BSE detector is reduced to improve the signal collection and the voltage is normally set at 30kV.

Applications

A case which shows the ability of the BSI to isolate particles of interest is one of an iron (Z=26) - chromium (Z=24) matrix containing particles of platinum (Z=78) and rhodium (Z=45). The number of the precious particles was low thus bulk X-ray analysis provided marginal results. Figure 1 is a composite of SEI and BSI micrographs taken at 3 different magnifications. The BSI at the bottom left corner shows topographic contrast and Z-contrast. The former exhibits broad dark areas while the latter isolates the brighter particles. At higher magnifications the intensity of the particle shows it to be easily distinguished from the background. In the right micrograph the particle is seen to be a spheroid possessing a mottled surface. Subsequent EDX analysis showed the particle to be composed of rhodium. Other particles having similar appearance consisting of Pt and Rh were quickly found both in the source material and that found in the suspect's possession. The examiner was thus able to state that the Pt and Rh was localized in particles which were spherical and had a mottled surface appearance.

The number of cases involving a search for particles or smears on a surface is very large. A case in point was a blackened area on a broken bone found in a floating, decayed corpse. The BSI of the bone is shown in figure 2 where bright areas are clearly visible. When EDX was used with the electron beam focused on the high intensity areas, lead was detected. The possibility that this lead originated from a bullet must be considered.

Another method often used is the tape dabbing of materials which are supposed to have come in contact with stolen objects having precious metal components or constructed of unique materials. A suitcase belonging to a suspect was supposed to have been used to carry a valuable gold statuette. The interior surface of the suitcase was dabbed with tape and the tape examined using the BSI. Figure 3 shows the SEI and BSI of an area of the tape. The bright object in the BSI was one of many found, and most were gold.

The examination for particulate gunshot residue (GSR) by SEM-EDX was introduced recently [11] and a number of forensic laboratories have adopted this procedure. The technique consists of the examination of an adhesive surface which had been dabbed on the back of the hands of a suspected shooter. After coating with carbon, the surface is scanned by SEI for spheroidal or otherwise suspect particles [11] and the EDX spectra of these is obtained to search for the presence of antimony, barium and/or lead. Some or all of these three elements are present in primer mixtures of U.S. manufacture. The search for GSR particles is simplified appreciably by the introduction of the BSI [12]. Figure 4 shows the comparison of the SEI and the BSI. In this case, the bright particles found in the BSI were GSR as determined by EDX and by morphology obtained by viewing the particles at higher magnification. Figure 4 shows that suspected GSR particles can be easily recognized using the BSI.

The number of examples where the BSI could be used is indeed very large.

For example, the search for precious metal granules in ore samples, the search for lead on the victim's clothing to determine grazing by a bullet, the localization of aluminum smears on a knife blade used in an entry made through an aluminum screen door, the search for a rare-earth tag on suspected stolen bills, etc.

The BSI was used effectively in the previous examples to localize the particles of interest. Instead, EDX could have been used for the search by applying the elemental mapping technique. In elemental mapping, the collected X-rays having an energy falling within a predetermined range (window) are used to produce dot patterns on the image. For example, if the gold M-line is straddled by the selecting window, the image of any gold particles will be covered with dots. This technique is thus able to specifically isolate the element of interest. A reason why elemental mapping was not chosen as the method of search for the previous examples was that it takes longer to obtain an elemental map than it does to obtain a BSI and the resolution is poorer.

The following is an example in point. Figure 5 shows a shiny metallic fragment found at a murder scene. The BSI in the middle micrograph shows a dark strip folded on itself enclosing a bright area. In addition there are bright striations on the surface. The electron beam was focused on the dark (low Z?) area, and the EDX spectrum showed the presence of Al. When the bright (high Z?) areas of the specimen were selected the EDX spectrum showed lead to be present. These results supported the theory that a so called "silver tip" bullet (actually a lead bullet clad by aluminum) had been used in the crime. The right hand micrograph of figure 5 shows the elemental distribution of Pb on this specimen. What is noticed is that the central Pb area is plainly visible, however the Pb deposit on the striations are not visible. Because of the low resolution of elemental mapping, fine details (roughly less than one tenth of the picture width) cannot be detected above background. In addition, elemental mapping (like the BSI) requires that the X-ray detector have a line-of-sight view of the area in question. Thus the deep valleys of a specimen with a rough surface may not be analyzed. Elemental mapping is an excellent tool to determine the extent of a particular elemental distribution (if the area of interest has been localized) but it is a poor and slow tool for the search of areas of interest. It is in the search that BSI excels.

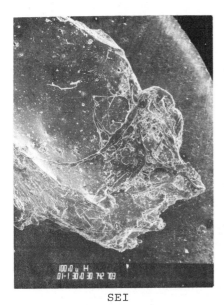  

SEI   BSI   BSI

LOW CONTRAST   HIGH CONTRAST

Figure 2. Bone fragment taken from cadaver. Bright areas seen in the two BSIs proved to contain lead (marker corresponds to 100 μm).

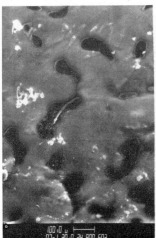 

SEI   BSI

Figure 3. SEI (left) and BSI (right) of tape lift from suitcase (marker corresponds to 100 μm). Bright particles in the BSI were shown to be gold by EDX.

Figure 4. BSI (left) and SEI (right) of a tape lift taken from the hand of a shooter (marker corresponds to 10 μm). The two particles of suspect gun shot residue are clearly distinguished from the other particles in the field.

Another area where the BSI is useful is in the determination of the uniformity (homogeneity) of the specimen. Because the criminalist often works with unknown specimens, he must be wary of bulk analysis. That is, if a small metallic fragment is composed of iron, chromium and zinc the BSI of the fragment may show two distinct areas of intensity. On further, more selective, EDX analysis the zinc may be associated with one of the areas and the iron-chromium with the other indicating that a zinc plating may have been present on the original material from which this fragment originated.

Conclusions

In this paper the reasons were presented why the backscattered electron image (BSI) is very useful in the examination of physical evidence. With the new backscatter detectors which are now available, the BSI will be even more effective in the search for particles, smears and inhomogeneities having different atomic number than the background.

However, because of topographic contrast and because the criminalist must often deal with rough and unprepared specimens, the backscatter detector cannot be thought of as a replacement for the energy dispersive X-ray analyzer (EDX). In the author's laboratory the BSI has been used routinely and to good advantage in the analysis of physical evidence for over four years.

Bibliography

1. O.C. Wells, "BSI in the SEM", SEM/1977/I, IIT Research Institute, Chicago, IL 747-777.
2. H. Niedrig, "...Electron Backscattering", Scanning, 1 (1) 1978, 17-34.
3. D.E. Newbury in Practical SEM, J. I. Goldstein and H. Yakowitz (Eds.), Plenum Press, N.Y., 1975, 107-110.
4. D.A. Gedcke et al., "Solid State BSE Detector... at TV Rates", SEM/1978/I, SEM Inc., AMF O'Hare, IL., 581-594.
5. S.H. Moll et al., "...BSE Detection Mode for SEM", Ibid, 303-310.
6. V.N.E. Robinson and B.W. Robinson, "Material Characterization in a SEM Environmental Cell", Ibid, 595-602.
7. D.E. Newbury, "The Utility of Specimen Current Imaging in the SEM", SEM/1976/I, IIT Research Institute, Chicago, IL., 111-120.
8. N. Zeldes and M. Tassa, "Conversion... to Efficient BSE Detector", SEM/1979/II - This volume, 154-158.
9. J. Lebiedzik and E.W. White, "...Microtopography in the SEM", SEM/1975/I, IIT Research Institute, Chicago, IL., 181-188.
10. V.R. Matricardi, "Matching of Surfaces", Ibid, 503-510.
11. R.S. Nesbitt, J.E. Wessel and P.E. Jones, "...Detection of GSR By the Use of Particle Analysis", J. Forensic Sciences, 21, 1976, 595-610.
12. V.R. Matricardi and J.W. Kilty, "Detection of GSR Particles From the Hands of a Shooter", J. Forensic Sciences, 22, 1977, 725-738.

Discussion with Reviewers

R. Keeley and M. Sogard: What advantage does backscattered electron topographic contrast have over secondary electron contrast, other than eliminating the need for specimen coating?
Author: Topographic contrast in the BSI provides quantitative information about the topography of the specimen surface because of the directional nature of BSE emission and geometrical beam-specimen-detector considerations. Given the signals from multiple BSE detectors and proper data processing a full topographic map of a homogenous specimen can be obtained. These results can be used for 3-dimensional striation and fracture matching which can only be approximated by SEI.

P.B. DeNee: Do you do any work at voltages below 30kV? What voltage range can be used for the type of forensic work which you do?
Author: Most of the initial analyses are performed at 30kV. However when thin platings, small particles, low Z materials or insulating specimens are examined lower voltages (from 1 to 20kV) are used.

P.B. DeNee: Why do you use a freshly sanded carbon surface?
Author: The SEM facility at the FBI Laboratory examines about 300 cases per year, where each case usually requires more than one specimen stub. Given this large throughput of specimens, carbon cylinders (15 mm diameter, 10 mm high) are used which have been drilled to accept a 3 mm (1/8 in.) rod. By sanding the surface these can be reused many times as long as we're aware that particles containing only silicon could have originated from the sandpaper. The advantages are that the specimen cylinder can sit flat on a work surface, indexing is convenient since we have numbered the cylinders, and low cost.

M. Sogard: The coefficient for the production of BSE is larger than that for secondary electrons, so when you say a BSI requires a higher beam intensity you are presumably referring to the small solid angles subtended by past BSE detectors, and therefore low collection efficiency, rather than any intrinsic inefficiency in the BSE process?
Author: I agree with you that the solid angle subtended by the detector is a factor as is detector collection efficiency, amplifier chain efficiency, SE production by BSEs incident on the chamber walls, etc. My statement was based on experimental fact for the system used and the conditions employed.

R. Keeley: Do you have any procedure for setting the brightness and contrast of the display when searching for high atomic number particles?
Author: Yes, both the brightness and contrast can be adjusted to fit the requirements of the specimen and instrumental operating conditions such as beam current and specimen tilt.

J. L. Abraham: Please provide more details on the actual instrument conditions used for each figure, such as beam current and recording frame time.

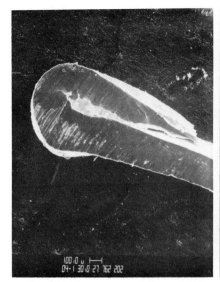

Figure 5. Composite micrograph of bright metallic object found at crime scene. The SEI is shown at left, the BSI in the middle and the elemental map (dot pattern) for lead is shown on the right. Marker in these micrographs corresponds to 100 μm. Localized EDX analysis showed that all bright areas seen in the BSI are composed of lead.

Were the same conditions used for secondary electron imaging as for backscattered electron imaging? It is sometimes useful in my experience to get a much better secondary picture by reducing the beam current, especially with poorly conducting samples.

P.B. DeNee: What beam current do you normally use for your BSI work? How does it compare with the beam current used to obtain adequate secondary electron images?

Author: The beam current used for BSI work will vary with the specimen composition, working distance (which in turn depends on the bulk of the specimen) and the X-ray production capability of the items which are being sought (in order not to have high or low dead-times in the EDX when suspect particles are found). Generally 0.5 to 5×10^{-9} amps are used for BSI work (as measured by a Faraday cup with an uncalibrated meter) and about 0.8×10^{-10} amps are used for SEI work. To obtain high resolution SEIs it is better to reduce the beam current, I agree. The recording frame time for all the micrographs in this paper was 20 seconds since resolution was of minor concern.

J.L. Abraham: What kind of tape was used in preparation of the sample you illustrated in Figure 3?

Author: In this case the tape was desk-top transparent tape. In most similar cases "Scotch" brand no. 465

adhesive layer is used or a liquid adhesive is spread on the stub.

J.L. Abraham: In Figure 5 was the lead L line used for the map? The lead M line overlaps with the sulfur K line, thus, caution must be used in interpreting these two elements.

Author: The L line was used. Your comment is well taken.

J.L. Abraham: The use of these techniques is very valuable in forensic examination of tissues as well as samples such as you have illustrated. Comment on your experience in this area.

Author: The examination of tissue is a relatively infrequent experience for the SEM in the FBI Laboratory. Most of the tissue specimens received were examined for the presence of contaminant particles or smears on the victim's skin surface (for example the search for foreign particles on the scalp of a victim having a crushed skull to determine the composition of the weapon used).

P.B. DeNee: Is SEM evidence such as you show here acceptable in a court of law?

Author: FBI Laboratory evidence examiners have given testimony regarding information obtained through the use of the SEM in criminal courts. In the case of People v. Palmer, 80 Cal. App. 3d 234, 145 Cal. Rptr. 466 (1978), the use of the SEM was specifically approved by the California Court of Appeal.

Additional discussion of the paper "A Method of Preparing Firearms Residue..." by J.S. Wallace et al, continued from page 184.

K.L. Culbreth: Could you elaborate on the results from single and multiple firings?

Authors: There were no differences between recovery rates from single and multiple firings. Larger amounts of lead and barium were measured from multiple firings, but the measurements were less than would be expected from multiplying the value for a single firing by the number of rounds fired. Saturation of the sampling material may be responsible for this but we have no definite explanation.

K.L.Culbreth: Have you utilized your filtration technique in casework? If so have you encountered any difficulty in searching for particles with the SEM/EDX?

J.A. Brown: You imply that a good secondary electron image is necessary for good searching. Don't you consider that the backscattered image is equally important as well as being necessary for rapid searching?

Authors: We have only recently obtained an efficient backscattering detector and are very pleased with the great reduction in searching time which has resulted from its use. The filtration method has been used in casework on samples from clothing; the reduction in extraneous material together with the improved backscattered image have made searching much easier. Positive results have been obtained in two out of the eight cases where this method has been used; all involved clothing collected more than five hours after firing.

V.R. Matricardi: Was there a reason why vacuum rather than pressure from the syringe at the top of your apparatus was used for filtration.

Authors: Application of pressure from the syringe produced airlocks and incomplete filtration. Vacuum filtration was much more satisfactory.

SCANNING ELECTRON MICROSCOPY/1979/II
SEM Inc., AMF O'Hare, IL 60666, USA

FIREARM IDENTIFICATION BY EXAMINATION OF BULLET FRAGMENTS:
AN SEM/EDS STUDY

R. L. Taylor, M. S. Taylor[*] and T. T. Noguchi

Forensic Science Center
Dept. of Chief Medical Examiner-Coroner
1104 N. Mission Street
Los Angeles, CA 90033

[*]American Inst. of Forensic Science
PO Box 33213
Los Angeles, CA 90033

Abstract

Deaths by gunshot frequently occur where the lethal bullet is not found and the identity of the responsible firearm is unknown. Either no weapon was found at the crime scene or two or more known weapons were involved, and it isn't known which one fired the lethal bullet.

Important clues--clues generally overlooked--as to the identity of the responsible weapon may be found in the gunshot residues and bullet fragments left in the victim's body and clothing. Two famous cases are presented (the SLA 'Symbionese Liberation Army' Shootout and the Oscar Bonavena homicide) in which examination of gunshot residues and/or bullet fragments by scanning electron microscopy/energy dispersive spectroscopy facilitated the identification of the responsible weapon.

Introduction

Last year in Los Angeles County, there were 749 homicides due to firearms. A significant number of these deaths involved "through and through" gunshot wounds in which the responsible bullet was not recovered. The most a medical examiner can generally say in such cases is that the wound was caused by a bullet of either "large caliber" or "small caliber," and perhaps indicate the possible range and direction of firing. This is all valuable information, but it doesn't go very far in helping the police identify the ammunition or firearm which was used.

Not infrequently, the police take into evidence two or more firearms, one of which was responsible for the killing- but which one is the question? The answer can, of course, be most critical. Our laboratory has been involved in answering this question in several important cases.

The SLA Shootout

Donald DeFreeze

On May 17, 1974 six members of the Symbionese Liberation Army (SLA) died in the now famous encounter with the Los Angeles Police Department. The cause and mode of death was not particularly difficult to determine in five of the six deaths. The sixth, however, Donald DeFreeze ("Field Marshal Cinque") presented special problems, largely because the lethal bullet was not recovered. Although, there was no question as to the cause of DeFreeze's death--a bullet had entered his right temple and exited his left--the question we faced was whether or not the death was suicide or homicide.

DeFreeze was found with a .38 cal. revolver beneath him. It was a five-chamber, Smith and Wesson, two-inch barrel, "Chief's Special." It isn't known if he was holding this gun at the time of his death, but if his death was

KEY WORDS: Firearms Identification, Gunshot Residues, X-ray Microanalysis, Bullet Fragments due to a self-inflicted gunshot wound, then this is presumably the weapon he would have used. There was no other weapon within his reach that could have produced the gunshot wound through his head. The bullet trajectory was consistent with a self-inflicted gunshot wound, and the appearance of the wound was consistent with a .38 cal. gunshot wound to the head. However, "death at the hands of another" was a distinct possibility given the fact of the gun battle itself.

The gun described above, hereafter referred to as "DeFreeze's gun," held five empty cartridges, four with firing pin marks and a fifth without which had apparently discharged in the heat of the fire. All were standard factory loads, four marked ".38 SPL, S & W-F," and the fifth simply ".38 SPL." Because of their position in the cylinder, the cartridge that killed DeFreeze, if the gunshot wound was self-inflicted, would have had to have been one of the S & W-F's. The problem then, restated, was to analyze and compare trace materials found along the wound tract with the S & W-F's (both the remaining cartridge cases in the gun and additional ammunition from the SLA's arsenal collected near DeFreeze's body).

To address this problem, we took two main approaches: neutron activation analysis and SEM/EDS (scanning electron microscopic/energy dispersive spectroscopy). Through the use of multiple radiographs and triangulation, small bullet fragments along the wound track were located and removed. These fragments, together with ammunition of the types used in DeFreeze's gun and that used by the law enforcement agencies, were sent to Dr. Vincent P. Guinn at the University of California, Irvine, for the neutron activation analysis. The SEM/EDS studies were performed on samples of ammunition used by DeFreeze and on microscopic gunshot residue particles adhering to the skin, bone, and dura, of both DeFreeze's gunshot wound and control wounds produced in biological material (hog's heads).

Materials and Methods. DeFreeze's scalp, bone, and dura mater at both the entrance and exit wounds were removed at the time of autopsy and placed in 12.5% formalin. Specimens for SEM/EDS analysis were subsequently dehydrated via a graded series of acetone. The bone and dura were allowed to air dry whereas the skin lesions were critical point dried utilizing carbon dioxide as the transitional fluid.

Test firings were performed on freshly killed hog's heads. The ammunition used at contact range and at a distance of 10 feet was .22 cal., .38 cal., and 9 mm. Specimens of bone surrounding the entrance gunshot wound were removed and processed as described above for the decedent's bone.

DeFreeze's ammunition, i.e., the .38 SPL, S & W-F's, were "pulled" and appropriate samples of the bullet jacket, lead projectile, cartridge case, anvil, and primer cap were mounted for SEM/EDS analysis.

All specimens were mounted on carbon substrates attached to aluminum studs, and the tissue specimens were coated with carbon. When necessary colloidal graphite was used to increase the conductivity between the specimen and the stud. All specimens were examined at 25kV in a Cambridge Stereoscan S-4 equipped with an EDAX x-ray analyzer (707A) interfaced with the EDAX-Edit computerized data system. Studs were positioned facing the x-ray collector at a 45^{0} angle from the electron collector, and specimen tilt and count rate (ca. 3200 counts per second) were optimized for x-ray collection. Spectra were collected for 200 seconds at 20eV/channel. Generally, a minimum of three spectra were generated, each from a different region on the specimen, and the data averaged.

Results. From the test firings on hog's heads, we learned that each of the following cartridge components can be deposited in a contact gunshot wound and detected by SEM/EDS analysis: (1) primer residue, (2) brass--copper (Cu) and zinc (Zn)--particles derived from the casing components, i.e., cartridge case, anvil, or primer cap (Fig. 1), (3) particles derived from the bullet jacket (Fig. 2), and (4) in most cases lead (Pb) from the projectile. The Cu and Zn composition of cartridge casing and various jacket brasses was determined by microprobe analyses and compared to the Cu:Zn ratios, based on integrated counts after background subtraction, of the same brasses as determined by SEM/EDS analysis. From these Cu:Zn ratios we were able to determine whether we were looking at "brass" from the cartridge case (70% Cu, 30% Zn) or jacket (generally 90% Cu, 10% Zn; 95% Cu, 5% Zn; or 100% Cu) (Fig. 3).

These studies told us that if Donald DeFreeze killed himself, we should find in his wound track particles derived from the jacket of the ammunition he was using. The casing components in the decedent's ammunition consisted of 70% Cu, 30% Zn (Fig. 3a), and the jackets consisted of 95% Cu, 5% Zn (Fig. 3c). Since none of the ammunition used by the police had jackets of this composition, and since our test firings into hog's heads told us that we should find particles of jacket material in a suicidal gunshot wound, we now had all the information we needed to determine if

Fig. 1. Particles of cartridge casing found bordering contact gunshot wounds in hog skulls. a) Particle derived from a .38 caliber cartridge casing. b) Cu Kα x-ray distribution image of particle in Fig. 1a. c) Particle derived from a .22 caliber cartridge casing. d) Cu Kα x-ray distribution image of particle in Fig. 1c. The Zn Kα x-ray distribution images of these particles are not shown here, but they overlap perfectly with the Cu Kα images indicating that copper and zinc are alloyed and therefore that the particles are brass. The Cu:Zn ratios are equivalent to a brass of 70% Cu, 30% Zn, indicating that these particles are derived from cartridge casing (See Fig. 3a).

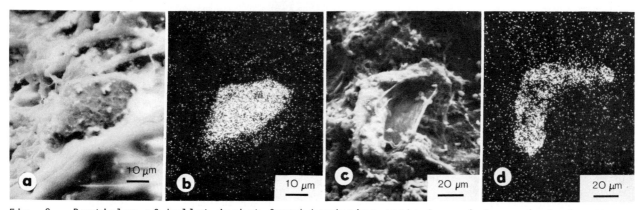

Fig. 2. Particles of bullet jacket found bordering contact gunshot wounds in hog skulls. a) Particle derived from a .22 caliber bullet "jacket." b) Cu Kα x-ray distribution image of particle in Fig. 2a. On EDS analysis this particle was shown to consist of copper only, and therefore represents the copper "plating" or "flash" commonly present on .22 caliber ammunition and which was present on the ammunition used (See Fig. 3d). c) Particle derived from a .38 caliber bullet jacket. d) Cu Kα x-ray distribution image of particle in Fig. 2c. EDS analysis showed this particle to be a brass of 90% Cu, 10% Zn--identical with the bullet jacket of the ammunition used (See Fig. 3b).

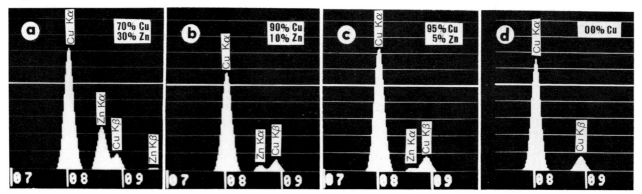

Fig. 3. EDS x-ray spectra of cartridge brasses. a) Casing brass of 70% Cu, 30% Zn. b) Jacket brass of 90% Cu, 10% Zn. c) Jacket brass of 95% Cu, 5% Zn. d) Jacket of 100% Cu.

DeFreeze killed himself or died "at the hands of another." We then examined the tissues along the wound track for jacket particles with a composition of 95% Cu, 5% Zn.

SEM/EDS examination of the tissues along the wound track revealed the following elements: Cu, Zn, iron (Fe), manganese (Mn), nickel (Ni), and Pb. Since these elements were present in unique form in comparison to control tissue, they were presumably ejected from the weapon which discharged the lethal bullet.

Elemental mapping showed the Cu and Zn to be alloyed in particles of approximately 70% Cu, 30% Zn (Fig. 4). These particles ranged in size from 5 to 100 μm. Since brass of these compositions does not occur in bullet jackets, the particles are most likely derived from casing, primer cap, or anvil. Some of these brass particles appear to be essentially discrete whereas others appear in association (not alloyed) with Fe/Mn, Ni, or Pb in varying combinations.

Significantly, nowhere within DeFreeze's wound track were particles found with the jacket composition of DeFreeze's ammunition.

Elemental mapping showed the Fe and Mn to be alloyed and therefore to probably represent particles of steel. Nickel was frequently juxtaposed with the Fe/Mn alloy in particulate form. Occasionally Cu was found associated with these particles. These findings suggest a nickel-plated, steel-jacketed bullet.

Lead was present in the wound also, and was apparently derived from both the projectile (Fig. 5) and the primer powder.

Discussion. The absence in DeFreeze's wound of any brass particles consistent with the jacket brass of his ammunition suggests that the lethal bullet was not fired from his weapon. Further, the SEM/EDS studies of the trace materials found in DeFreeze's gunshot would suggest that he was killed by a steel-jacketed bullet, plated with nickel. There is 9 mm European military ammunition dating from World War II which fits this description and which is available in this country. Investigation on our part revealed that such ammunition was being used by one member of the LAPD SWAT team - the one officer who because of his firing position was most likely to deliver a lethal bullet to Donald DeFreeze. Unfortunately, all ammunition of this type was "used up" in the Shootout and neither intact cartridges, projectiles, nor spent casings were made available to us for analysis.

The results of the neutron activation analysis when taken in conjunction with the SEM/EDS results support the conclusion that Donald DeFreeze died "at the hands of another." (V.P. Guinn, private communication).

Patricia Soltyzik

SEM/EDS played a role in answering a second question which arose in the course of our investigation of the SLA Shootout. Routine radiographs of the body of Patricia Soltyzik revealed the presence of a partly deformed bullet and a partly deformed jacket in her forearm. Upon removal from the tissue, it was noted that one side of the bullet and jacket had been scraped or shaved off and was missing. Most unusual, however, was the absence of any rifling, an observation which suggested that the projectile was probably from a live cartridge which discharged due to the heat of the fire and did not pass through the barrel of a gun. The projectile was subsequently identified as an unfired .38 special, jacketed, hollow-point bullet. The question we were asked was, "What was the source of this bullet"?

SEM/EDS analysis of the .38 cal. copper jackets used by all parties in the Shootout showed only one with the same jacket composition as that in Soltyzik's elbow, and that was the .38 SPL, S & W-F's used by DeFreeze which, as already noted, were 95% Cu, 5% Zn. Also noted earlier was the fact that DeFreeze's gun held five empty cartridge cases, one of which was an S & W-F lacking a firing pin impression. It was positioned in the cylinder such that if discharged due to the heat, the bullet could have escaped out the side of the gun, one portion of it being scraped away by the gun frame which partially covered the chamber. This is obviously what happened. These observations as well as the relative position of Soltyzik's body and the revolver in question support the conclusion that the cartridge which "heat-discharged" in DeFreeze's gun is the source of the projectile found in Soltyzik's forearm.

Oscar Bonavena

On May 22, 1976 Argentine heavyweight boxer Oscar Bonavena was shot and killed at the Mustang Ranch, the biggest and most celebrated legal brothel in the United States.

The bullet entered the left side of the chest, shattered the heart, lacerated the liver, grazed the lower lobe of the right lung, exited the back slightly to the right of the midline, and continued into the desert and was never recovered.

According to the Storey County Sheriff's Office, Bonavena was killed either by Ross Brymer, bodyguard to the whorehouse's owner, using a thirty-ought-six rifle (.30-06 cal. ammunition) or by an unknown individual firing an AR-15 rifle (.223 cal. ammunition) from a gun tower overlooking the "Ranch's" grounds. We were asked if we could determine which ammunition killed Oscar Bonavena. We accepted the case and the body was flown to Los Angeles for our examination.

Fig. 4. Particles of cartridge casing found in DeFreeze's gunshot wound. a) Particle on skull at entrance wound. b) Cu Kα x-ray distribution image of particle in Fig. 4a. c) Particle on dura at exit wound. d) Cu Kα x-ray distribution image of particle in Fig. 4c. The Zn Kα x-ray distribution images of these particles are not shown here but overlap perfectly with the Cu Kα images indicating that the Cu and Zn are alloyed and therefore that the particles are brass. Their Cu and Zn composition (70% Cu, 30% Zn) as determined by EDS analysis indicate that they are derived from cartridge casing and not bullet jacket.

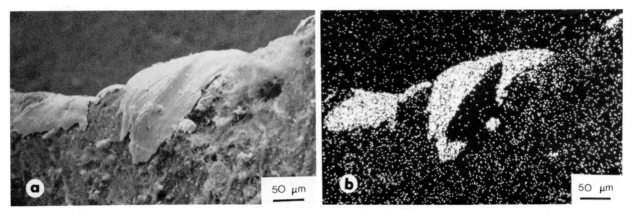

Fig. 5. a) Lead on the skull at the edge of the entrance gunshot wound. b) X-ray distribution image for lead in Fig. 5a.

Materials and Methods

Four .30-06 cal. cartridges and two .223 cal. cartridges removed from the two suspect weapons were provided for our examination. The projectiles were "pulled" from the cartridges and samples of the lead projectile and jacket brass were removed and mounted for SEM/EDS analysis.

Radiographs of the clothing at the exit wound and of the wound track revealed the presence of numerous small radiodense particles (Fig. 6). Over 28 of these particles--ranging in weight from under 4 mg to about 58 mg--were isolated, cleaned by brief sonication in detergent solution followed by distilled water, and examined under the dissecting microscope. All appeared to be lead. Fragments from the clothing and one from the body were analyzed by SEM/EDS for comparison with the .30-06 cal. and .223 cal. ammunition provided. One fragment removed from the liver contained a microscopic particle of what appeared to be brass. This fragment was also analyzed by SEM/EDS (Fig. 7). The conditions for SEM/EDS analysis were the same as given above for the SLA

investigation.

Results

SEM/EDS analysis of the lead fragments removed from the wound track and those from the clothing revealed detectable levels of antimony (Sb) (Fig. 8). All four .30-06 cal. bullets contained similar Sb levels, whereas neither of the .223 cal. bullets contained Sb, at least not at the level of sensitivity of the EDS (Table 1).

SEM/EDS analysis of the brass-appearing particle on the lead fragment removed from the decedent's liver gave a copper to zinc ratio equivalent to a brass of 95% Cu, 5% Zn (Fig. 3c). Analysis of the jackets of the four .30-06 cal. bullets gave the same copper to zinc ratio, whereas the jackets of the .223 cal. bullets gave ratios equivalent to a brass of 90% Cu, 10% Zn (Fig. 3b) (Table 1).

Discussion

The elemental composition of the bullet fragments removed from Bonavena's wound track and clothing, as determined by SEM/EDS analysis, is consistent with the .30-06 cal. bullets and not the .223

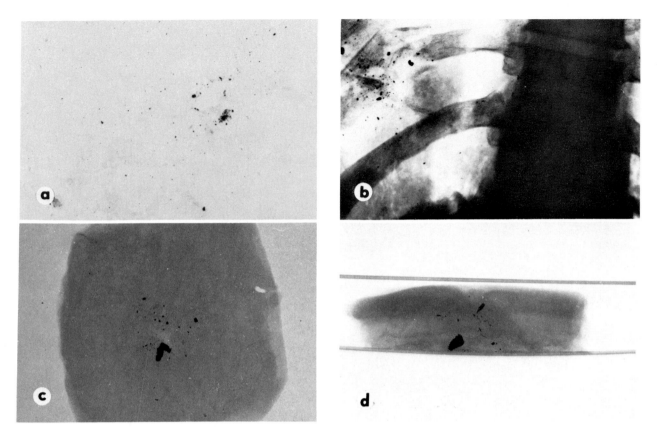

Fig. 6. Radiographs of the clothing at the exit wound and of portions of the wound track showing the presence of numerous small bullet fragments. a) Radiograph of Bonavena's shirt at exit wound (1X). b) Radiograph of wound track in thoracic region (.75X). c) Radiograph of skin at exit wound (2X). d) Same as Fig. 6c after rotating tissue 90° (2X). Bullet fragments are often difficult to locate and remove from the tissue when radiographs are taken in only one plane.

TABLE 1. SEM/EDS ANALYSIS OF SUSPECT CARTRIDGES AND BULLET FRAGMENTS FROM DECEDENT

| | PROJECTILE | | JACKET |
SPECIMEN	Pb	Sb	% Cu,Zn
Suspect Cartridges:			
.30-06	+	+	95,5
.30-06	+	+	95,5
.30-06	+	+	95,5
.30-06	+	+	95,5
.223	+	-	90,10
.223	+	-	90,10
Bullet Fragments Removed from Decedent:			
From clothing	+	+	
From liver	+	+	95,5

+ = element detected
- = element not detected (if present, it is below the level of sensitivity of the instrumentation.

cal. bullets. Given that Bonavena could only have been killed by one of these two types, the lethal bullet could only have been one of the .30-06 cal. bullets.

Lead fragments from the wound track as well as lead from the projectiles were submitted to Dr. Vincent P. Guinn, University of California, Irvine, for neutron activation analysis. The results of his examination fully support the SEM/EDS findings.

Ross Brymer, the bodyguard, was charged with the murder of Oscar Bonavena. He subsequently pleaded guilty to a reduced charge of voluntary manslaughter, and was sentenced to two years in the Nevada State Prison.

Conclusion

As a result of cases such as reported above, we've learned that in "through and through" gunshot wounds valuable physical evidence may be deposited along the wound track. And in cases where the bullet isn't recovered and there are questions as to the identity of the ammunition or firearm used in the shooting, this physical

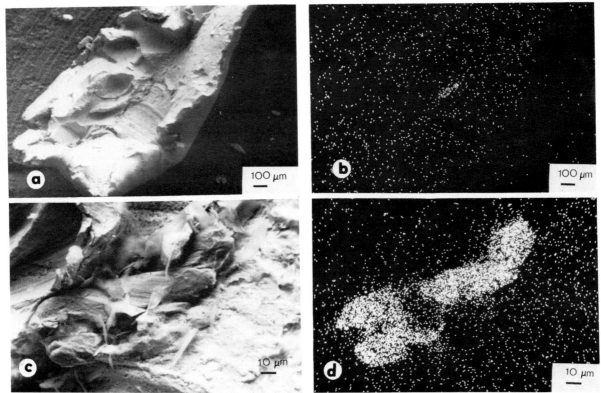

Fig. 7. SEM/EDS micrographs of bullet fragment which contains a microscopic particle of jacket brass. a) Low magnification micrograph of fragment. b) Cu Kα x-ray distribution image of fragment in Fig. 7a pinpointing the location of the jacket brass. c) Higher magnification of region containing the jacket brass. d) Cu Kα x-ray distribution image of Fig. 7c, clearly delineating the jacket fragment. The Zn Kα x-ray distribution image is not shown but overlaps with the Cu Kα image indicating that the Cu and Zn are alloyed. EDS analysis resulted in a Cu:Zn ratio equivalent to a brass of 95% Cu, 5% Zn.

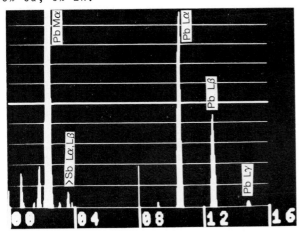

Fig. 8. Typical EDS x-ray spectrum of lead fragment removed from the wound track and clothing of Bonavena. Sb Lα can be clearly seen.

evidence may be crucial to solving the case. Therefore, it should be collected and not buried or burned with the body as is the common practice. We recommend that whenever similar cases arise those tissues containing the fragments be saved--no matter how small the fragments are. If and when suspect ammunition turns up, then the fragments can be isolated and a proper forensic comparison can be made. That is, a comparison by SEM/EDS. When appropriate subsequent analyses, such as neutron activation, can be performed also. We would advise that if a metal probe is used by the medical examiner to determine the trajectory of the bullet through the body, that it be taken into evidence. Ideally, a plastic or wooden probe would be used to avoid any possibility of metallic contamination.

Had the techniques reported here been applied at the time of the Robert Kennedy assassination, some of the controversy regarding the source of the bullets might have been averted. In any unsolved cases where tissues containing bullet fragments are still available, and there are questions as to the identity of the responsible ammunition, these analytical procedures should be considered.

Acknowledgements

We are grateful to Mr. Chris Sloan and Mr. Bill Lystrup for their excellent technical assistance with various phases of this work, and we thank Vera Wrobel for her patience in typing the manuscript.

Discussion with Reviewers

I.C. Pontarelli: What is a typical composition of a primer powder, which you suggest might be an additional source of lead found in DeFreeze's tissue?
Authors: The precise composition of primer powders varies to some extent with the manufacturers and is generally regarded as proprietary information. A typical composition, however, would be as follows: 60% Basic Lead styphnate, 5% Tetracene, 25% Barium nitrate, and 10% Antimony sulfide.

V.R. Matricardi: In the Soltyzik case, were there obvious metal smears in the DeFreeze gun where bullet grazing had occurred?
Authors: The gun was never in our possession, but in the hands of the Los Angeles Police Department. We do not know if it was examined for metal smears where bullet grazing would have had to have occurred.

V. Reeve: Was antimony detected by examining the L lines in the x-ray spectra?
Authors: Yes.

V.R. Matricardi: From your experience, can any conclusions be drawn from the absence along the wound tract of particles characteristic of the suspect bullet?
Authors: No. Unless, of course, particles are found which are totally inconsistent with the suspect bullet and therefore point to a different bullet altogether.

V.R. Matricardi: Does analysis of particles performed in the "as-is" condition provide a surface for analysis which may have had a different thermal history than the bulk specimen?
Authors: Yes. But, in our experience, the SEM/EDS analysis of cartridge casing and jacket materials is unaffected by the temperatures experienced during the firing of the ammunition.

V.R. Matricardi: Should the fragments be cut to expose clean, unaffected areas?
Authors: It is, of course, always preferable to analyze a clean surface. However, none of the jacket or casing particles removed from the decedent's in the cases discussed here were large enough to be cut. We do not believe that our data would have improved significantly had we been able to cut the particles.

R.H. Keeley: How many different types of bullet jacketing are in general use?
Authors: Personal communication with a number of ammunition experts throughout the United States has revealed the following different types of bullet jacketing in general use: (1) brass jackets; most are 90% Cu, 10% Zn or 95% Cu, 5% Zn, (2) steel jackets, and (3) copper, nickel, cupronickel, or lubaloy plating. These materials may be present directly on the lead projectile or on the surface of brass or steel jackets.

R.H. Keeley: What is the maximum range of firing that would result in primer residues and cartridge case fragments entering the wound?
Authors: This is an important question, and one which hopefully will be answered by future research. From our experience the answer will be complex depending on the caliber, velocity, and composition of the ammunition; the characteristics of the weapon, e.g. barrel length and mechanism ; and the environmental conditions.

R.H. Keeley: In cases where it is not possible to distinguish between bullets by SEM/EDS are there any other methods which might be used?
Authors: Yes, assuming that the specimen is large enough to be processed by other analytical methods, such as neutron activation analysis, atomic absorption analysis, x-ray induced EDS or WDS, electron probe microanalysis, etc. However, in our experience specimens are often too small to be located and analyzed without the use of the SEM.

SCANNING ELECTRON MICROSCOPY/1979/II
SEM Inc., AMF O'Hare, IL 60666, USA

CHARACTERIZATION OF GUNSHOT RESIDUE PARTICLES BY LOCALIZATIONS OF THEIR CHEMICAL CONSTITUENTS

M. Tassa and N. Zeldes

Criminal Identification Division
Israel National Police Headquarters
Jerusalem, Israel

Abstract

Gunshot Residue (GSR) Particles are usually identified by their morphology and overall chemical composition. In many cases these criteria do not suffice for a definite identification. A further criterion is presented, namely, the distribution of the chemical elements over the morphological features of the particle's surface. A study of this distribution has been made for particles originating in various types of ammunition, using SEM and X-ray microanalysis together with a digital multi-colour mapping technique. The results described indicate that a correlation does exist between morphological and chemical details in a variety of primer compositions. This correlation is suggested as an additional parameter in the characterization of GSR.

KEY WORDS: Gunshot Residue, Particle Analysis, X-Ray Digital Mapping

Introduction

The technique for gunshot residue (GSR) detection by particle analysis is now being used by a growing number of forensic laboratories. Of major importance for this method is the ability to positively identify a microscopic particle sampled from a suspect's hand as being of gunshot origin. So far, this has been done by using the criteria of morphology and overall chemical composition, as formulated in the authoritative work of Wolten et al.[1] While in some cases these parameters suffice to give a positive ("unique") identification, in others identification by these criteria is limited to "characteristic, but not unique". For example, some common primer formulas produce particles containing lead and barium; however, the same composition exists in "innocent" particles derived from certain paint materials. In such a case, positive differentiation of GSR from other particles of the same composition would necessitate some further criterion.

Any GSR examiner is probably familiar with the phenomenon of GSR particles having a variety of lumps and nodules on their surface. Matricardi[2] reported the observation of particles where the nodules were richer in lead relative to the rest of the particle's surface. Data accumulated from the authors' routine case work shows that this and similar phenomena are typical of GSR in general. It was then theorized that since this segregation of the particle's constituents must take place during its formation in the firing chamber, in a complex dynamical process, it should be strongly characteristic of GSR. In order to evaluate this proposition, a study was made of the aforementioned phenomenon, its occurrence and properties.

Experimental Procedure

The particles studied were derived from a variety of sources:
a. Particle samples from the hands of suspects, accumulated from routine case work. The sampling is done with adhesive-coated aluminium disks.
b. Similar hand samples from indoor range test firings.
c. Simulations performed by directly shooting primers onto sampling disks.
d. Samples collected from the interior of fired cartridge cases.

Each sample was examined thoroughly to determine the presence of GSR particles with round nodules or lumps attached to them. Where found, their composition and the distribution of the various elements on the particle's surface were determined.

The SEM used is a Cambridge Scanning Co. model CamScan III S, equipped with an Elscint Proxan III X-ray analyzer. Backscattered Electron (BE) Imaging was used to locate GSR particles on hand samples. The BE system developed by the authors mainly for that purpose is described elsewhere in SEM/1979.

Study of the elemental distribution in a sample was greatly simplified by use of the Elscint Element Distribution Processor (EDP). This microprocessor-based system accumulates counts from the X-ray pulse processor into a 128x128 element digital memory array. The array is scanned in synchronization with the electron beam's scan of the sample. By selecting the appropriate energy window on the pulse processor, one can thus obtain in the EDP's memory array a digital X-ray map of a given element's distribution in the sample. The EDP can store these maps on a flexible disk, process them (e.g., by background rejection or smoothing), and display them on a colour TV monitor. A scale of 16 different colours is used to indicate the number of counts in each picture point. Acquisition of a map might take 5-10 minutes; one element is mapped at a time, but the processor can combine two such maps to display a composite image where one element's concentration is represented in shades of red, and the other's in blue.

Results

Fig. 1 shows clearly the morphology with which this study is concerned. The bulges on a particle can vary in size from small dots to hemispherical domes half the particle's diameter in size; they shall be termed, collectively, "nodules" in what follows.

Ten representative types of cartridges of different calibers and primer compositions have been studied. Table 1 gives a condensed summary of the experimental data obtained by examining

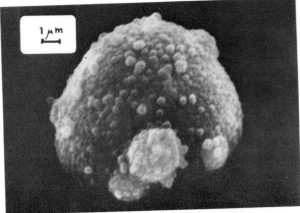

Fig. 1. A GSR particle exhibiting typical nodule morphology. (Dominion .38 S&W cartrdige).

Ammunition	Overall	Body	Nodules	Remarks
Winchester Western .30 Carbine	Pb,Sb, Ba,Al	Ba,Al	Pb,Sb	Unusually high fraction of particles have nodules.
FN 9mm Parabellum	Pb,Sn, Ca,Ba	Ba,Ca	Pb,Sn	
IMI 9mm Parabellum	Pb,Cl,K Sb,S	--	--	No nodules observed.
Hirtenberg 9mm Parabellum	Pb,Ca, Ba,S,Si	Pb,Ca, Ba,S	Pb,Ca, Ba,S, Si	Nodules & bulk of same composition
BPD 9mm Parabellum	Hg,S, Cl,K,Sb	K,Cl	Hg,Sb, S	Relatively few nodules
Winchester Western 9mm Parabellum	Pb,Ba,Sb	Ba	Pb,Sb	
Eley .22 LR	Pb,Ba	Ba	Pb	
Remington .22 LR	Pb,Ba	Ba	Pb	
Peters .22 LR	Pb,Si,Ca	Pb,Si, Ca	Pb	
Dominion .38 S&W	Pb,Ba, Ca,Si,Cu	Ba,Ca, Si	Pb,Cu	

Table 1. Representative nodule-bearing particles of various ammunitions.

two to six samples of each ammunition. The Table lists, for each ammunition, the overall composition of the GSR particles found, the composition of the nodules and that of the particle's surface in nodule-free regions, which hereafter shall be termed "body". (In all three cases, minor or insignificant constituents have been disregarded). The compositions listed in Table 1 are typical rather than exclusive: The quantitative ratios of the elements vary among individual particles of the same ammunition and even among two nodules on the same particle. One can also expect particles exhibiting some intermixture of the body and nodule compositions given in the Table. This should come as no surprise when one keeps in mind that GSR particles are formed from a vapor phase in a violent, non-equilibrium process. Despite this inherent variation, for each ammunition, the typical composition in Table 1 is characteristic of the particle population as a whole.

Discussion

The detailed results of the observations made have led the authors to the following conclusions:
a. In a majority of the ammunitions studied (9 out of 10) nodules of various sizes were present on a significant fraction (between 10 and 90%) of the particles.

b. The nodule-bearing particles are normally spheroidal in shape, and larger than 10μm in diameter.

c. In most cases the composition of the nodules differs from that of the "body" of the particle. When this happens, the elements lead and antimony (either or both) are found concentrated in the nodules, whereas barium, calcium and silicon tend to concentrate in the body of the particle. A typical example is vividly displayed by the use of elemental X-ray mapping in Fig. 2.

The above is seen to hold true for a variety of cartridge types and primer compositions. It can therefore be considered typical of the formation process of GSR, rather than of a particular type of ammunition. This suggests that particles formed in different processes, e.g., paint particles, shall not display such nodules even if their overall composition is consistent with GSR (e.g., lead and barium). The Authors' experience with such particles does indeed confirm this suggestion.

The above results, as well as the experience accumulated from numerous cases solved, have led the authors to conclude that the additional criterion derived above should complement and strengthen identification in cases where the overall composition criterion would give uncertain results. Explicitly, when a particle is found to have a composition "consistent with GSR, but not unique",[1] the presence of nodules of the previously listed characteristics would strongly indicate GSR origin; otherwise the possibility of other origins (e.g., paint) cannot be excluded.

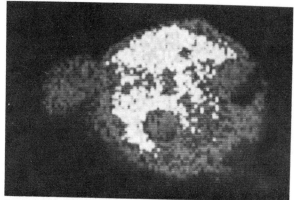

Fig. 2b, above.   Fig. 2c, below.

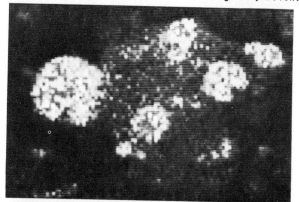

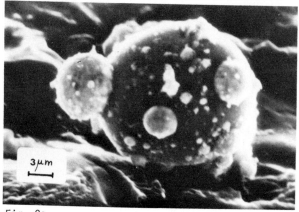

Fig. 2a

Fig. 2. Element distribution across a GSR particle. (Winchester-Western .30 Carbine cartridge).
a.   SEM Image of particle
b,c,d. EDP X-ray maps of Ba, Sb and Pb concentrations, respectively. In these B&W reproductions of colour images, brighter areas correspond to higher concentration (Originally a 16-colour scale was used).
e.   EDP 2-colour map of barium (shades of blue) and lead (shades of red) concentrations. In B&W, lead appears white and barium grey.

Fig. 2d, above.   Fig. 2e, below.

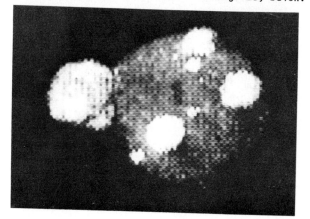

References

1. G.M. Wolten, R.S. Nesbitt, A.R. Calloway et al., Final Report on Particle Analysis for Gunshot Residue Detection, Report No. ATR-77 (7915)-3, Aerospace Corp., El Segundo, Calif., 1977, pp. 13-22.
2. V.R. Matricardi and J.W. Kilty, Detection of Gunshot Residue Particles from the Hands of a Shooter, J. Forensic Sciences, 22 (4), 1977, pp. 725-738.

Discussion with Reviewers

R. Keeley: Do you find as high a proportion of nodule-bearing particles in breech deposits from hands and clothing as occurs in samples collected from muzzle blast and cartridge residues? Since the majority of nodule-bearing particles are larger than 10μm, are they not rapidly lost from the surfaces of skin and clothing under casework conditions?

D.M. Lucas: In case work, do you find particles greater than 10μm in diameter where more than two hours had elapsed between the time of the firing and the collection of the samples?

Authors: Hand deposits are indeed poorer in nodule-bearing particles than are muzzle deposits, and that may well be caused by the larger size of such particles. In our casework experience, nodules were detected with time lapses up to and including 2 hours between firing and sampling; longer lapses produced small particles only.

We must stress, however, that the purpose of this preliminary study was to characterize nodules when they are found. The questions of when, why and under what conditions of firing and sampling this happens require a larger data base, which we are at present accumulating.

D.M. Lucas: Have you seen any "innocent" particles derived from certain paint materials that were spherical and contained lead and barium?

Authors: We have not yet seen such particles, but we do not preclude their existence because we have encountered spheroidal particles derived from paints, albeit of other compositions.

R. Keeley: Could your colour mapping system be used for searching?

R.L. Taylor: Do you see any applications of the colour mapping mode in routine GSR analysis?

Authors: The acquisition of a colour map requires 5-10 minutes, which is prohibitively long for a rapid search routine. In the present study EDP was used to evaluate and demonstrate the phenomenon of segregation in GSR particles. For routine casework, however, "point" analyses from different parts of the particle are faster and quite satisfactory.

SCANNING ELECTRON MICROSCOPY/1979/II
SEM Inc., AMF O'Hare, IL 60666, USA

A METHOD FOR PREPARING FIREARMS RESIDUE SAMPLES FOR
SCANNING ELECTRON MICROSCOPY

J. S. Wallace   and   R. H. Keeley

Northern Ireland Forensic Science
  Laboratory
180 Newtownbreda Road
Belfast, BT8 4QR
Northern Ireland, U.K.

Metropolitan Police Forensic Science
  Laboratory
109 Lambeth Road
London SE1 7LP
U.K.

Abstract

A serious disadvantage to the use of the Scanning Electron Microscope for the detection of firearms residue particles is the excessively long time required to search for and locate the particles.

A simple method is described for the removal of extraneous material from the sample and concentration of particles onto a small area, thereby substantially reducing the searching time. Samples are suspended in a non-polar solvent and filtered through a two stage filtration apparatus. The first filter removes most of the extraneous material and the second filter retains small particles including firearms residues. The surface of the second filter can then be searched in the SEM.

Introduction

The identification of firearms residue has always been a difficult problem in forensic science. Until recently all methods have been based on the qualitative or quantitative analysis of samples for lead, antimony and barium, some or all of which are present in most firearms ammunition primers. It is not possible to identify primer residue conclusively by elemental analysis alone because of the normal occurence of these elements in the environment.

A method has been developed which combines the visual and analytical capabilities of the SEM equipped with an X-ray spectrometer to enable firearms residues to be identified conclusively by their morphology and composition[1-3]. This greatly enhances the quality of scientific evidence which can be presented in court. Particle analysis is the best method available for the identification of residues but it suffers from the fact that searching for particles can be a long and tedious process.

The lengthy searching is due to (a) the large sample area, (b) the time spent examining, analysing and rejecting a large number of non-firearms particles and (c) extraneous material such as fibres and skin debris which may make the particles difficult to find or even conceal them from view. The areas on a suspect that are normally sampled are the hands, face, outer clothing and the insides of the pockets; the problem is most acute in samples from clothing where much unwanted material is collected. The filtration method described here will remove most of this material and concentrate the remaining sample into a small area. The procedure is fast, simple, reliable, gives high percentage recoveries and does not alter the appearance or composition of the residue particles.

KEY WORDS: Sample Preparation, Firearms Residue, Gun Shot Residue, Filtration, Forensic Applications

Firearms Residue Particles

Firearms residue particles are composed of lead and inorganic compounds of heavy elements which are insoluble in non-polar organic solvents. They vary in size from less than 0.5 μm to more than 200 μm, though in casework the size range is typically 0.5 μm to 10 μm. Larger particles are seldom detected on a suspect because they are very rapidly lost from the surfaces of skin and clothing. The particles may be spherical or irregular in shape and sometimes the spherical particles may have smaller spherical nodules on their surfaces.

We have tried to separate residue particles from unwanted material by sedimentation but this method has yielded poor percentage recoveries. Although the residue particles are denser than most of the other materials in the sample, they do not separate out from a suspension. This may be due to viscosity effects and the retention of particles by adsorption onto the surfaces of the apparatus. Filtration of the suspension and washings from the sedimentation tube gives good yields of particles. A two stage filtration apparatus has been used to recover particles in the size range 2-50 μm. Measurement of percentage recovery by examining unfiltered and filtered samples with the SEM/EDX is tedious so we have used atomic absorption spectrophotometry (AA) for speed and convenience. Filtered samples have been examined with the SEM/EDX to check for any changes in the appearance and composition of particles.

Apparatus

The double filtration system (Figure 1) consists of two filter holders (Gelman-Swinny hypodermic adaptors) joined by a Luer-Lok connector. The sample is introduced through a 50 cm^3 syringe barrel and the whole apparatus fits onto a 100 cm^3 conical filtering flask which is connected to a rotary pump via a cold trap. All of the screw threads of the joints are sealed with PTFE tape. The first filter is either a 50 μm nickel mesh or porous polythene (pore size 50 μm) cut from sheet with a cork borer. The final filter is a membrane with a pore size of 0.2 μm (Gelman Metricel Alpha 8,13 mm diameter). Recovery was measured with a Perkin-Elmer 603 atomic absorption spectrophotometer equipped with a HGA 76 graphite furnace. The SEM was a Camscan 330E with a Link Systems EDX system.

Sample Collection

Residues were obtained from indoor firings of several types of handgun and ammunition. The number of rounds of ammunition discharged was varied so that some samples had a large deposit of residues and others a small amount. Samples were taken from the back and palm of the firing hand and the cuff of the outer garment immediately after firing; adhesive coated SEM stubs (13mm diameter), flexible acrylic strips and cotton swabs were used. Stubs and acrylic strip were pressed onto the relevant area until it was covered or the stickiness of the stub diminished. Cotton handswabs were moistened with a non-polar solvent (120-160 bp petroleum ether or trichlorethylene), and rubbed over areas where particles might be found. The choice of adhesive is important since solutions of the adhesive from some adhesive tapes (and Scotch Spraymount) will not pass through fine membrane filters. A number of adhesives have been tested and the most satisfactory of these was Vistanex LH-HM, a polyisobutylene. Concentrated Vistanex is difficult to work with since it is very viscous; for this work a 30% solution was made in trichlorethylene and a thin coating applied to the surfaces of the stubs and acrylic strip. They were then warmed in an oven to evaporate the solvent.

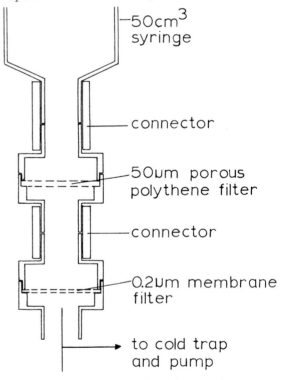

Figure 1. Filtration Apparatus.

20 cm^3 petroleum ether

containing firearms residue particles and extraneous material

Mix well and divide immediately

10 cm^3	10 cm^3
Evaporated to dryness with a hair dryer and residue taken up in 5 cm^3 20% HCl. Aliquots taken for lead and barium determination.	Filtered through the apparatus. Final filter placed in 5 cm^3 20% HCl. Aliquots taken for lead and barium determination.

$$\% \text{ Recovery} = \frac{\text{Final Concentration}}{\text{Initial Concentration}} \times 100$$

Figure 2. Procedure for measuring percentage recovery.

Element	Initial Concentration ng/cm^3	First Filter ng/cm^3	Second Filter ng/cm^3	% Recovery
Lead	2.0 X 10^3	1.4 X 10^2	1.8 X 10^3	90
Barium	6.0 X 10^2	2.0 X 10^1	5.5 X 10^2	91
Lead	1.7 X 10^3	1.1 X 10^2	1.4 X 10^3	82
Barium	1.0 X 10^3	1.0 X 10^2	8.5 X 10^2	85
Lead	3.3 X 10^3	2.0 X 10^2	2.8 X 10^3	85
Barium	7.6 X 10^2	8.0 X 10^1	6.4 X 10^2	82
Lead	4.2 X 10^3	3.7 X 10^2	3.7 X 10^3	88
Barium	7.6 X 10^2	6.0 X 10^1	6.8 X 10^2	90

Table 1. Percentage recovery of residues.

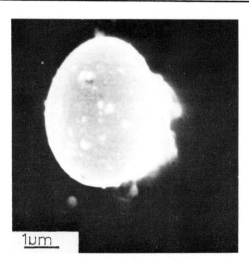

Figure 3. Spherical particle with surface features, after filtration.

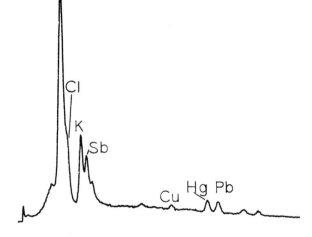

Figure 4. Spectrum of particle in Figure 3; a chlorate primer residue.

Procedure

Flexible strips and sample stubs with residue collected on the sticky surface were placed in a plastic beaker and covered with petroleum ether (boiling range 120-160°C) which dissolved the adhesive leaving the solid material in suspension. Cotton swabs were vigorously agitated in the solvent in an ultra-sonic bath to dislodge particles from the fibres. Suspensions were filtered in the filtration apparatus; the apparatus was then flushed with more solvent. The extraction, filtration and washing was repeated a further six times. After the liquid from the final washing had passed through suction was continued for a few minutes to evaporate any residual solvent. The final filter was carefully removed, attached to a sample stub and carbon coated for examination in the SEM. The scheme shown in Figure 2 was used for the measurement of percentage recovery. The solvent extract was divided and half used to estimate initial concentration of lead and barium. The remainder was filtered and the initial and final filters extracted for the measurement of final concentration.

All reagents and both filters were checked for their lead and barium content before starting. Neither of these elements were found in the membrane filter or on the nickel mesh, but both were detected in the porous polythene. This probably came from contamination during manufacture and storage and was easily removed by washing the filters in 20% HCl. The filtration apparatus was carefully cleaned in a similar manner between samples.

Results

The results from four different filtrations are shown in Table 1; the average percentage recovery for lead is 87% and 86% for barium. Most of the losses occur in the first filter, but there is a small discrepancy between the initial concentration measured in the unfiltered extract and the sum of the amounts detected in the two filters, indicating that losses also occur elsewhere. Neither lead nor barium were detected in the filtrate. Some material particularly very small particles will be lost by adsorption onto the surfaces of the containers used and the differ - ence between the two totals is probably due to further adsorption from the filtered portion onto the surfaces of the filtration apparatus. The nickel and porous polythene filters were examined in the SEM for evidence of residue particles trapped in the fibres

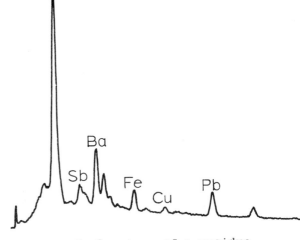

Figure 5. Spectrum of a residue particle from barium nitrate primer, after filtration.

retained on the surfaces of the filters but none were observed.

The barium levels are fairly constant but there is a twofold difference between the lead recovered in samples 1 and 4. The firings and samplings were conducted under conditions as close as possible to those which would be experienced in case-work; some of the lead and barium probably comes from the normal environmental deposits on the surfaces of skin and clothing. Lead particles from bullets and variations in primer composition also contribute to the differences.

There are often large differences in composition between residue particles from the same firing and it is not possible to compare untreated to extracted and filtered samples directly. However particles observed on the final filters were of normal appearance and composition. Figure 3 shows a spherical particle with some surface features; this particle is from a chlorate primer. The residues from this type of primer contain potassium chloride which is very soluble in water. The presence of potassium and chlorine in the spectrum (Figure 4) shows that the particles are not attacked by non-polar solvents. The composition of the more common barium nitrate primer (Figure 5) is similarily unchanged.

Various sizes of filters and holder have been investigated and all gave high percentage recoveries; the small area of the 13 mm filters makes them the obvious choice. We used sintered glass primary filters for our first attempts and achieved very low recoveries because of retention of particles in the pores. Particles are most clearly visible on membrane filters with smooth surfaces; Figure 6a shows a filter where the

Figure 6a. Secondary electron image; Figure 6b. Back-scattered electron image of residue particles on a filter with a rough surface and closely spaced pores.

Figure 7a-b. Secondary electron image of a residue particle on a smooth filter with wide pore spacing.

contrast between the particles and the surface is low, except in the back-scattered electron image (Figure 6b), making searching difficult. The surface of the filter in Figure 7a-b is smooth and the pores widely spaced so that particles are clearly visible in the secondary electron image. The effect of filtration in removing much of the material which obscures particles and makes searching difficult is also noticeable in Figure 6a.

The extraction and filtration method has been developed in response to a need to speed up the location and identification of residue particles. Although it represents an additional stage in the examination, it has reduced searching times by a factor of five, particularly when samples have come from clothing. The method can be adapted to any form of sampling as long as the particles can be got into suspension and, coupled with improvements in searching techniques, should enable a forensic science laboratory with a heavy firearms case-load to use the SEM/EDX as a routine method.

Acknowledgements

The authors wish to thank Dr. W.H.D. Morgan and Dr. R.L. Williams, Directors of the Northern Ireland Forensic Science Laboratory and the Metropolitan Police Laboratory; Mr.V. Beavis and Mr. W. McDowell.

References

1. G.M. Wolten et al. Final Report on Particle Analysis for Gunshot Residues. ATR-77 (7915)-3. The Aerospace Corporation, El Segundo, CA.,1977.

2. J. Andrasko and A.C. Maehly.Detect-
ion of Gunshot Residues on Hands by
SEM. J. For. Sciences, 22(2) 1977.
pp. 279-287.
3. R.H. Keeley. Applications of Electron Probe Instruments in Forensic
Science. Proc. Anal. Div. of the Chemical Society, 13(6) 1976. pp. 178-181.

Discussion with Reviewers

J.A.Brown: Are you satisfied that the
80-90% recovery of particles represents
the composition of all particles as
recovered?

K.L.Culbreth; Did your examinations
include antimony and other elements?

Authors: The values given for percentage recovery for lead and barium are
relevant to only part of the particle
composition, since they may also contain antimony, calcium, silicon,copper
and possibly other elements. Although
we have only measured lead and barium
directly, we have observed no changes
in appearance and composition of individual particles and are satisfied that
the particles collected on the filter
are representative of the untreated
sample.

V.R. Matricardi: What precautions are
taken to prevent contamination of the
stem and backside of the stub when
being handled by the arresting officer?
Is the bottom of the stub washed prior
to immersion in the solvent?

R.L. Taylor: Contamination of the filtration apparatus with GSR would appear
to be a potential problem. How would
you convince the court that the single
unique GSR particle found in a particular case was not the result of a
contaminated apparatus?

Authors: Contamination is a serious
problem, at the time of arrest and
during subsequent examination in the
laboratory. The standard sampling kit
contains several pairs of surgical
gloves and officers are instructed to
wear these while taking samples. Accidental contamination of the hands and
clothing of a suspect before sampling
is a much more serious problem and has
been used as a defence in court in the
U.K. in a number of recent cases. In one
of these cases it was possible to show
that primer residue found on the de-
fendant could not have come from ammunition used by the arresting officers,
but where this is not possible, it can
not be categorically stated that contamination did not occur. All clothing
and hand samples submitted to the laboratory are unpacked and examined in a
dust free cabinet in the SEM section
which is some distance away from the
firearms section. The filtration
apparatus is cleaned with HCl between
samples and we are satisfied that this
cleaning is effective. Cases in which
a single particle was found would only
be reported as positive if only one
weapon was involved. The police in the
U.K. (except Northern Ireland) are normally unarmed.

V.R. Matricardi: You mention cotton
swabs; did the cotton fibres retain
some of the particles through physical
entrapment? Did loose fibres from the
swab hamper filtering?

K.L. Culbreth: The swab technique is
utilized in many laboratories; could
you elaborate on their purpose in your
examination and the results of their
use?

R.L. Taylor: Have you compared the
stickiness of the Vistanex LH-HM
adhesive with 3M adhesive transfer tape
465? Wouldn't collection efficiency
improve with a stub of larger surface
area?

Authors: The SEM/EDX method has been
used routinely for residue identification on clothing submitted to the laboratory but we have not yet decided on
the most suitable method for sampling
hands. At present hands are sampled
with cotton swabs moistened with dilute
HCl and these are examined by AA, since
they are not suitable for SEM/EDX. In
this work cotton swabs moistened with
a non-polar solvent gave results comparable with other methods of sampling
and we experienced no difficulty in
removing particles from the fibres and
filtering the suspension. We find that
sample stubs coated with adhesive are
difficult to use and do not recommend
them; if an adhesive method is introduced operationally it will be on
flexible acrylic strips. These have
two separate sampling areas, the second
is used when all the stickiness on the
first has disappeared. Large areas of
adhesive improve collection efficiency
but unless a method such as that described here is used to reduce sample
area, any improvement is offset by the
increase in time needed to search large
area samples. We have not investigated
3M 465 tape; any adhesive tape used for
this method must have an adhesive which
dissolves and passes through the filter.

K.L. Culbreth: Could you clarify your
concern about the twofold difference
between the lead recovered in samples
1 and 4? Should one really be more
concerned with the percent recovery,
which is relatively constant?

Authors: We agree that percent recovery is most important but were drawing
attention to the variation between
samples from apparently similar firings.

For additional discussion see page. 166

SCANNING ELECTRON MICROSCOPY/1979/II
SEM Inc., AMF O'Hare, IL 60666, USA

FORENSIC APPLICATIONS OF SEM/EDX

E. H. Sild and S. Pausak

Centre of Forensic Sciences
25 Grosvenor Street
Toronto, Canada M7A 2G8

Abstract

Scanning Electron Microscopy-Energy Dispersive X-ray analysis has been applied in case work for over two years. Three cases are described where analysis without SEM-EDX would have been very difficult. Two of the cases are examples of physical matching of broken surfaces and the third is SEM-EDX analysis of a small amount of an unusual alloy.

Discussion is also presented on plasma ashing of gunshot residue samples removed from hands by double sided tape. A suction device suitable for collecting gunshot residue samples from hands or clothing is also described.

KEY WORDS: EDX, Forensic Applications, Plasma Ashing, Suction Collection Device

Introduction

The Centre of Forensic Sciences has been using scanning electron microscopy in combination with energy dispersive x-ray analysis since the fall of 1975. In that time, approximately 100 cases have involved some aspect of SEM or x-ray examination. The types of examination have ranged from the identification of arsenic on the copper wire used in the Reisch test to the examination of fracture surfaces in failure analysis. The equipment used consists of a Cambridge S-4-10 SEM with a LaB_6 gun system, interfaced with a Kevex 5100 x-ray energy spectrometer. When not involved in case work, a considerable amount of time has been devoted to the development of an acceptable method for the identification of gunshot residues (GSR) by SEM-EDX.

For the purposes of this paper, three cases have been selected which illustrate forensic applications of SEM-EDX analysis. In each of these cases, all of the physical evidence resulted from SEM-EDX examination. Also presented will be the approaches that have been taken to optimize the SEM GSR detection technique.

Case Examples

The first two cases involve the physical matching of broken surfaces as suggested by Matricardi[1].

Case 1

Two suspects were detained on suspicion of theft from parking meters. In their possession were found a number of parking meter keys, seven lock picks and what appeared to be a dental pick with both ends broken. The local parking authority repair shop had recently been broken into and parking meter keys reported missing. The lock on the side door of the repair shop was found to be jammed. Examination of the keyway revealed a small piece of metal. This piece, the dental pick, the lock cylinder and the lock picks were submitted for examination.

Using plasticine as a support medium, one of the ends of the dental pick and the piece from the lock were carefully mounted, under a stereo microscope, into complementary positions. The plasticine was then painted with colloidal graphite and the mounting was examined in the SEM. Fig. 1

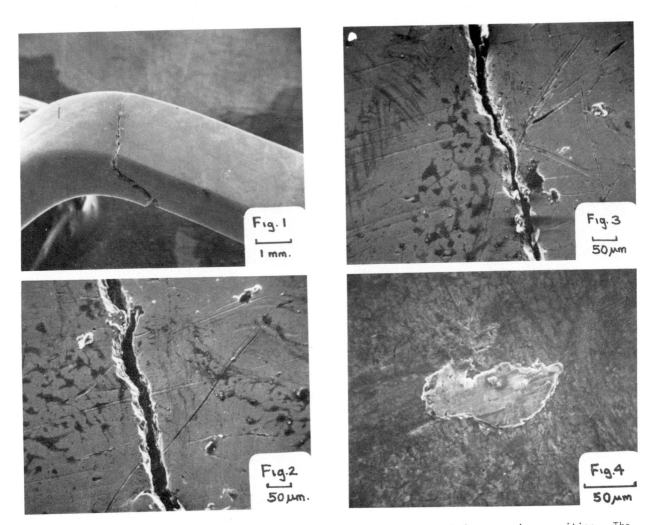

Fig. 1   Low magnification view of dental pick and broken piece from lock in comparison position.   The
          dental pick is on the left hand side.
Fig. 2 and 3.   Higher magnification views of two of the areas used in matching the configuration shown
          in Fig. 1.
Fig. 4.   One of the two brass specks that was found to be similar in composition to the lock cylinder
          interior.

shows a low magnification view of the mount. Since, due to damage and distortion, the break surfaces were not suitable for matching, comparison was made on the basis of common accidental surface scratch marks. Fig. 2 and 3 show a higher magnification view of two of the several areas used to make a positive comparison.

One of the lock picks had smears and specks of brass at the fine tip end. Using x-ray analysis two of the larger specks were found to have a chemical composition similar to the lock cylinder interior. Fig. 4 shows one of these specks.

Case 2

A battery was reported stolen from a truck. The battery leads had been cut and the clamps were missing. However, one of the leads had four strands of wire which appeared to have been broken in tension. A suspect was found with a battery of a similar type in his possession and also a battery clamp with four strands of wire. The short-

est strand from the battery clamp and the longest from the lead in the truck were selected for comparison. Under a stereo microscope these two wires were positioned side by side and clamped between two strips of lead. Fig. 5 shows a low magnification view of this configuration. The comparison was made by taking photographs of the central areas of the two broken surfaces at increasing magnifications. By noting distinctive features on the two broken surfaces and matching the corresponding negatives, a successful match was made. One such matching area is illustrated in Fig. 6 and 7. The details labelled A, B, C and D were used for the comparison in this area.

Case 3

Part of the sprinkler system in a warehouse was set off. No fire could be found that could have caused this. A suspect was found drenched in rusty water but he denied any knowledge of why the system was activated. A cigarette lighter was found in his possession and on the sparkwheel of the lighter were tiny blobs of a foreign material. The trigger on the sprinkler head consisted of two brass plates held together by a thin layer of a low melting point alloy (71°C). The alloy consisted of lead, tin, bismuth and cadmium. Fig. 8 shows a low magnification view of the sparkwheel with the blobs of material. Fig. 9 is a higher magnification view of one of the blobs showing a molten appearance. X-ray analysis of the material revealed the presence of lead, tin, bismuth and cadmium.

Gunshot Residues

No discussion of forensic applications of SEM-EDX would be complete without some mention of GSR. In this field, we have followed the work of Jones[2] and his group at the Aerospace Corporation. The GSR samples are collected with a 2.5 cm diameter aluminum stub covered with double sided tape (3M Type 465) and then coated with carbon. Examination in the SEM is done at a magnification of approximately 1000X and at TV scanning rates. All bright spherical particles are analyzed by EDX for the presence of lead, antimony and barium. To date, we have found that if the elapsed time is more than four hours between the firing of a weapon and the sampling, the search time required to locate distinctive particles becomes prohibitively long. In an attempt to solve this problem, we have taken two approaches. The first is to "clean" the sample by removing obstructing debris; the second is to concentrate the sample, thereby decreasing the search area.

Low Temperature Plasma Ashing

After examining a number of hand samples, it became evident a large number of GSR particles were likely hidden from view, covered by debris that was also removed in the sampling process. The most common obstructing material is skin debris. Since skin is essentially organic in nature, it was felt that low temperature plasma ashing could remove a substantial amount of it. Fig. 10 shows a low magnification view of a GSR hand sample. The two spheres in the central region are GSR particles each containing lead, antimony and barium. Fig. 11 shows the same area after three hours ashing. Although no new GSR particles are evident in Fig. 11, the presence of the two spheres indicates that the process does not damage or dislodge GSR particles. Fig. 12 and 13 show views of two different areas of another ashed sample. In Fig. 13 the bright spherical particles are lead GSR particles 1-2 μm in diameter. As in Fig. 11, Fig. 13 also illustrates the supposition that ashing "shrinks" skin debris, thereby exposing previously covered areas of tape.

To date, our experience with ashing is somewhat limited and we have not determined the optimum operating conditions. The samples that have been ashed were done by an outside agency and thus we did not have full control over the operating conditions. Presently, we are awaiting arrival of our own equipment and will then more actively investigate this technique.

GSR Collection by Suction

Since the first publication[3] on detection of gunshot residues (GSR) by SEM much has been

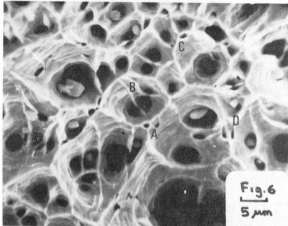

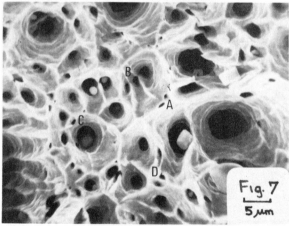

Fig. 5. Low magnification view of the two wires held in position between two lead strips. This is at 0° tilt.

Fig. 6. View of a small area of the central region of the break surface of the wire from the battery clamp. Details labelled A, B, C, D were used as a starting point for comparison.

Fig. 7. View of a small area of the central region of the break surface of the wire from the battery lead. Details labelled A, B, C, D were compared to details labelled A, B, C, D in Fig. 6.

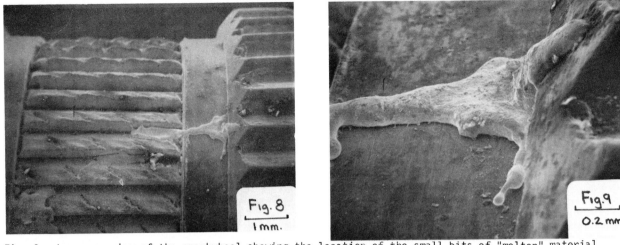

Fig. 8. Low mag. view of the sparkwheel showing the location of the small bits of "molten" material.
Fig. 9. A higher magnification view of one of the bits of "molten" material.

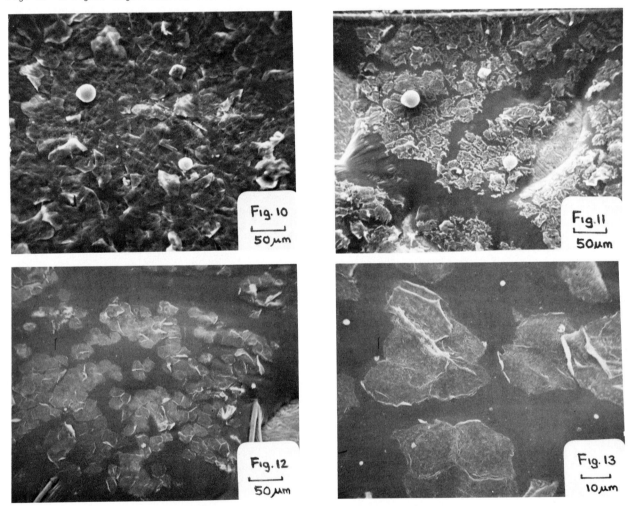

Fig. 10. View of double sided tape used to remove gunshot residues from a hand. The two bright spheres
          in the central region are GSR particles each containing Pb, Sb and Ba.
Fig. 11. Same as Fig. 10 after three hours of plasma ashing.
Fig. 12. View of another plasma ashed taped GSR hand sample.
Fig. 13. Higher magnification view of another area of the sample shown in Fig. 12. The bright spherical
          particles are Pb GSR particles.

188

learned about the properties of GSR particles: size, distribution, elemental content, morphology and other physical and chemical properties. Various GSR particles are considered to be unique[2-5], i.e. they have not been found as environmental particles. Thus, the SEM detection of GSR is much more specific than the "standard" analysis by neutron activation or atomic absorption of complete or partial hand washes or swabs[6].

A major problem of the SEM technique lies in the collection method. Double sided adhesive tape is used to collect GSR by multiple dabbings on the hand. In this process the adhesive loses its "stickiness" by collecting too much skin debris and dirt in addition to GSR. This results firstly in incomplete pickup of GSR and, secondly, many GSR particles are covered by debris and cannot be detected by SEM. Further, the area of the tape, about 2.5 cm^2, requires too much time for a complete search for small particles.

In order to diminish the problem of long search time and excessive debris a miniature suction device has been designed. Fig. 14 is a schematic of this device. The stub holder is attached to a vacuum line and the flat end of the glass nozzle is moved in parallel strokes over the sampling area (hand, clothing, etc.). The air flow passes through the glass nozzle slits, picks up the particles from the surface, and brings them to the sieve. Sieves with various pore sizes, 5 μm, 20 μm, 40 μm and 90 μm, have been tested. The most effective sieve appears to be one with a pore size of 20 μm; it retains the majority of dirt and skin debris particles and allows through almost all GSR particles. The air jet formed in the aluminum nozzle shoots the small particles onto the SEM stub with the adhesive tape on the top of it. After the sampling is completed, the stub is coated with a thin carbon layer and analyzed by SEM. The material retained on the sieve is examined under a stereo microscope and any large GSR particles are separated for chemical analysis and transferred to the SEM stub. An acetone wash of the material retained by the sieve is analyzed by gas chromatography for nitroglycerine traces.

Fig. 15 shows the small SEM stub with adhesive tape. The stub is 5 mm in diameter and the majority of the particles are concentrated in an area 2 mm in diameter. This means a 100X reduction in the sample area and consequently a large reduction in the search time. Fig. 16 shows a small area at a magnification of 1800X. A half dozen spherical lead particles appear as bright spots.

The suction device has been tested for collection of GSR from hands and clothing. Although we have experienced a number of technical problems the apparatus works well and the results so far are satisfactory. Further improvements of the apparatus are in progress. Table I is a list of eleven tests in which the suction apparatus was used.

Acknowledgements

The authors wish to thank D.M. Lucas, L. Allair and L. Blunt for their assistance in preparing this paper.

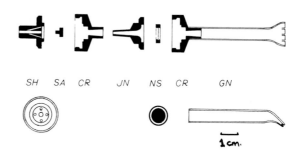

FIG. 14 THE EXPLODED VIEW OF THE SUCTION APPARATUS
SH - STUB HOLDER   JN - JET NOZZLE
SA - STUB AND ADHESIVE   NS - NICKEL SIEVE
CR - COUPLING RING   GN - GLASS NOZZLE

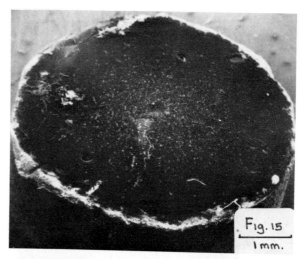

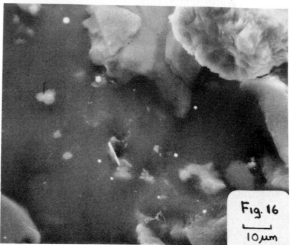

Fig. 15. Low magnification view of the sample stub used in the suction apparatus.

Fig. 16. View of a small area of the sample stub shown in Fig. 15. The bright spherical particles are Pb GSR particles.

TABLE I

Test No.	Firearm	Firing Hand	No. of Shots	Delay - Firing Sampling	Area Sampled	SEM Findings Pb	SEM Findings Pb+Sb+Ba	G.C. Test for Nitroglycerine
1	.38 handgun	Right	2	15 min.	R.H.	>>20	>20	not done
					R.S.	>20	2	+
2	.38 handgun	Right	2	1 hour	R.H.	>>20	7	+
					R.S.	>>20	0	+
		Left	2	1 hour	L.H.	>>20	>20	+
					L.S.	>>20	>20	+ (?)
					F.	>>20	5	+
3	.38 handgun	Right	2	30 min.	R.H.	>20	3	+
					R.S.	10	1	+
					F.	>20	0	−
4	.38 handgun	Right	2	10 min.	R.H.	>20	8	+
					R.S.	>20	5	+
5	.38 handgun	Right	2	1 hour	R.H.	0	1	−
					R.S.	>20	2	+
6	.38 handgun	Right	2	1½ hours	R.H.	>20	8	+
					R.S.	>20	4	+
					F.	>20	2	+
					P.	>20	3	+
7	.38 handgun	Right	2	2 days	F.	>20	1	−
8	.38 handgun	Right	4	4 hours	R.S.	>20	0	−
9	shotgun 12 gauge	Right	1	½ hour	F.	>>>20	>20	+
					R.G.	>20	3	+
					L.G.	>20	3	−
10	.303 rifle	Right	2	2 hours	R.S.	>>>20	0	−
					L.S.	>>>20	1	−
11	.22 rifle	Right	2	1½ hours	R.S.	>20	0	−

LEGEND:

R.H. - Right Hand R.G. - Right Glove
L.H. - Left Hand   L.G. - Left Glove
R.S. - Right Sleeve F. - Front of Lab Coat
L.S. - Left Sleeve   P. - Pocket of Lab Coat

> - the whole stub contains more than 20 particles
>> - a single sweep (magnification 1000X) reveals more than 20 particles
>>> - a single viewing field (magnification 1000X) has more than 20 particles

References

1. Matricardi, V.R., Clark, M.S., Defonja, F.S., The Comparison of Broken Surfaces: A Scanning Electron Microscope Study, J. Forensic Sciences, 20 (1975), p. 507-523.
2. Nesbitt, R.S., Wessel, J.E., and Jones, P.F., Detection of Gunshot Residue by Use of the Scanning Electron Microscope, J. Forensic Sciences, 21 (1976), p. 595-610.
3. Boehm, E., Application of the SEM in Forensic Medicine, Proceedings of the 4th Annual Scanning Electron Microscopy Symposium, 1971, p. 553-562.
4. Andrasko, J., Maehly, A.C., Detection of GSR on Hands by Scanning Electron Microscopy, J. Forensic Sciences, 22 (1977), p. 279-287.
5. Matricardi, V.R., Kilty, J.W., Detection of Gunshot Residue Particles from the Hands of the Shooter, J. Forensic Sciences, 22 (1977), p. 725-738.
6. Krishnan, S.S., Detection of Gunshot Residues on the Hands by Activation and Atomic Absorption Analysis, J. Forensic Sciences, 19 (1974), p. 780-797.

Discussion with Reviewers

Reviewer II: The suction method would appear to have two serious drawbacks: (1) The need for a vacuum source in the field and (2) Possible contamination of the vacuum device from prior usage. How could you convince the court that the single, unique GSR particle you found didn't arise from an improperly cleaned suction device?

Reviewer I: Is there any evidence of small particles being trapped on the internal surfaces of your suction apparatus?

Authors: The suction apparatus described is a prototype model. A suction device for field use is under design. It will have its own vacuum source and a detachable part consisting of a nozzle, sieve and sample stub with adhesive. The detachable part would be used once and then sent to the laboratory for analysis and decontamination. The suction device can be cleaned efficiently.

Reviewer I: Do you have any explanation for the generally low recovery of particles from clothing when the results from Table I (7, 10) and elsewhere indicate that residues persist for longer periods on clothing than on hands?
Authors: Since we do not know the number of GSR particles initially deposited in the area sampled, we cannot agree with the statement "generally low recovery of particles". In a test hand sampling, the first suctioning was positive for GSR and nitroglycerine. However, immediate re-suctioning resulted in no GSR particles being found, and a negative test for nitroglycerine.

Reviewer II: Methodology for "speeding up" GSR analysis by SEM is much needed. Of the two procedures you discuss--plasma ashing and suction-concentration--which appears to be the most promising at this time?
Authors: At present we do not have enough data to knowledgeably answer this question. However, it appears suctioning would be the method of choice for removing particles from clothing.

Reviewer III: In the matching of fracture surfaces do you always work with the as-is images or at times with the mirror image of one of the surfaces?
Authors: In the comparison of fracture surfaces as shown in Fig. 6 and 7 we would overlay and reverse the negatives to find comparison points (Fig. 17, 18). Another simple way of doing this directly on the viewing screen is to place transparent plastic on the screen and outline distinctive features in one colour. Then the plastic is flipped over, similar distinctive features are located on the surface to be compared and new features outlined in another colour. Repeating this process results in a two colour montage of complementary detail.

Reviewer III: The adhesive layer is a source of problems in the search for gunshot residue with the SEM/EDX. Would it be possible for the adhesive surface in your device to be replaced by a 0.2 μm filter membrane?
Authors: We agree the adhesive layer is a source of problems. Without an adhesive layer we anticipate a loss of particles through subsequent handling and coating.

Reviewer III: Would you have an approximate value for the airflow required for your pickup device? Has this been optimized?
Reviewer IV: Please provide some quantitation of your suction technique for collecting samples. Have you prepared any dispersions of known concentrations and sampled them?
Authors: The airflow that we use at present is 150-200 cc per sec., but this has not been optimized. We have no data at this time on the efficiency of the apparatus. We would appreciate any

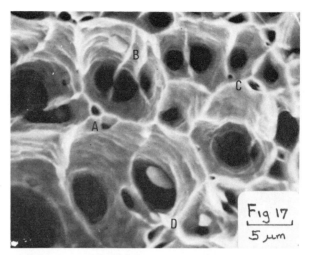

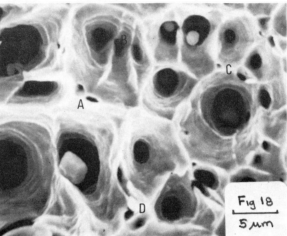

Fig. 17. Higher magnification view of central area of Fig. 6.
Fig. 18. Mirror image higher magnification view of central area of Fig. 7.

suggestions on the preparation of surface dispersions of known concentration.

Reviewer I: When does it become necessary to use the SEM rather than well established and more convenient optical methods for examining and comparing surface detail?
Authors: The choice of SEM over optical methods is dependent on the type of sample features to be examined and the magnification required to clearly interpret these. Generally, if magnifications much over 100X are required, lack of depth of field and lighting effects such as glare or shadowing impose severe limitations on optical methods. Obviously at the limits of optical magnification, SEM becomes necessary. For example, the comparison in Case 1 (Fig. 1, 2, 3) possibly could have been made by optical means, however, the comparison in Case 2 (Fig. 5, 6, 7) would have been impossible optically.

Reviewer III: Would you evaluate, if possible, the relative importance to your laboratory of

imaging versus analytical capabilities of the
SEM/EDX.
<u>Authors</u>: Our experience is that the combination
of SEM and EDX is so complementary that it becomes
futile to assess them individually. The examination of a large majority of the items would have
been meaningless without the analytical capabilities of EDX. However, when imaging alone has been
used, e.g. Case 2 in text of paper, or in fracture
analysis, the information has been crucial.

<u>REVIEWERS</u>:   I   R. Keeley
                II   R. L. Taylor
               III   V. R. Matricardi
                IV   J. L. Abraham

SCANNING ELECTRON MICROSCOPY/1979/II
SEM Inc., AMF O'Hare, IL 60666, USA

QUANTITATIVE INVESTIGATION OF SULFUR AND CHLORINE IN HUMAN HEAD HAIRS
BY ENERGY DISPERSIVE X-RAY MICROANALYSIS

S. Seta, H. Sato and M. Yoshino

National Research Institute of Police Science
6 Sanban-Cho
Chiyoda-Ku
Tokyo 102
Japan

Abstract

Criminal identification of human head hairs
has been mainly led by morphological examinations.
Analytical investigation of elements of head hairs
coupled with morphological findings would encourage the accuracy of the identification. Although
many analytical techniques principally led by neutron activation analysis have been used to study
the trace element contents in human head hairs,
they are not necessarily acceptable as the routine
use in forensic laboratories, especially because
of the difficulty in the access to the instrument.
So, energy dispersive X-ray microanalyzer in a SEM
was applied to the quantitative analysis of human
head hairs in order to routinely introduce the
analytical data to the criminal identification of
head hairs in forensic laboratories.

Investigated elements were limited to sulfur
and chlorine. Sulfur is the most aboundant of the
elements measured in hairs. Chlorine has been
taken to show some variation in its content, which
presumably originated in the status of individual
constitution. Hair samples were collected from 74
Japanese (48 males and 26 females) ranging from
newborns to 50 years of age. After rinsing in
ether-alcohol (1:1), the proximal end of about one
cm long from each sample was mounted on carbon
specimen holder and coated with carbon. Pure sulfur and sodium chloride of analytical grade were
used as standard samples for the quantitative analysis. Each X-ray spectrum from samples on a
scale of 0-10 kev under the same operating conditions was stored in a computer interfaced energy
dispersive X-ray detection system.

Sulfur contents showed relatively constant
values with 5-6 % as the mean value, whereas chlorine contents showed some variation, presumably
because of age and sex. The values from young age
groups below 13 years of age were 0.93±0.39 % in
males and 1.09±0.62 % in females. The values from
adult groups ranging from 20 to 50 years of age
were 0.67±0.20 % in males and 0.45±0.12 % in females. Sulfur and chlorine contents obtained in
the present investigation approximately coincide
with those by the chemical analysis and neutron
activation analysis.

KEY WORDS: Sulfur, Chlorine, Hair, X-ray Microanalysis, Quantitative Analysis, Forensic
Applications

Introduction

Criminal identification of human head hairs
in Japanese police laboratories has been practised mainly by morphological comparison and
blood grouping[1].

If hair specimens collected from an actual
crime scene are small or little, hair discrimination may not be effectively carried out only by
the method mentioned above. In such a case,
using chemical analysis in addition to those
methods will be expected to increase the probability of identification. Up to now analytical
examinations of human head hairs have been made
mainly by neutron activation analysis (NAA)[2,3,4] but
recently proton induced or radioisotope excited
X-ray emission method has been introduced to the
head hair analysis[5,6]. These analytical methods
seem not convenient to use together with
morphological examinations in forensic science
laboratories because of their technical complexities. From this point of view we have been
trying such method for examination of head hairs
as hair specimens can be conveniently analyzed in
laboratories and submitted to other methods of
identification after analytical examinations. In
a previous report we discussed the probability of
applying a wave length dispersive X-ray microanalyzer (WDX) to analysis of human head hairs[7].
Despite the higher resolving power of wave length
systems, we used an energy dispersive X-ray microanalyzer (EDX) in a SEM for analysis of human
head hairs because the analysis by WDX damages
the hair surface severely and demands long time
as compared with EDX.

In order to discriminate head hair specimens effectively by comparing the patterns of
characteristic X-ray spectra obtained with EDX,
we attempted in this paper to determine conveniently chlorine content in human head hairs
which has been shown different among individuals
according to the investigations using NAA and WDX
in previous papers. Stimultaneously with the
measurement of chlorine content, sulfur which is
the most abundant of elements contained in hair
specimens was also investigated in order to evaluate the precision of EDX in analyzing hair
specimens. Also, the practical application of
EDX to actual case works and some technical problems in the analysis of hair specimens have been
described.

Materials and Methods

Before carrying out quantitative analysis of head hair specimens, a few factors which might influence the characteristic X-ray intensity from hair specimens with EDX were investigated; that is, the difference in the X-ray intensity between a spot and a square analyses, the effects of unevenness of specimens, the direction of specimen alignment to the detector axis in SEM sample chamber and the washing solution such as ether-alcohol (1:1) solution and distilled water, and the positional difference such as cuticle, cortex and medulla.

Spot and square analyses

A head hair used was collected from the parietal part of a 35-year-old Japanese male. One cm in length near to the hair root was washed for 15 minutes in each of three changes of ether-alcohol (1:1) solution bottles. The specimen was placed on a carbon specimen mount pasted with double-face tape and coated with carbon in *vacuo*. The specimen was aligned to be parallel to the detector axis and the whole part of specimen holder in SEM sample chamber was deviced to be shielded with carbon in order to eliminate the generation of unexpected X-ray peaks (Fig. 1). Four positions on the cuticle surface, illustrated in Fig. 1, were analyzed for 200 seconds in spot and square modes at ×3,000 magnification, thus the characteristic X-ray intensities of sulfur and chlorine at the center of peak were measured. When analyzing in the spot mode the electron beam spot was always fixed at the center of CRT.

Unevenness of specimen

Apart from a polished metal specimen, the surface of hair specimen is uneven in external form, therefore, it is naturally presumed that a characteristic X-ray intensity fluctuates depending on different analyzing positions. Thus, using a head hair from the same person as above the effect of unevenness was investigated. The specimen was prepared in the same method as above, and as shown in Fig. 2 it was aligned to be in the direction perpendicular to the detector axis in SEM sample chamber. Six positions on the scale surface, as shown in Fig. 2, were adopted for the analysis by fixing a electron beam spot at the center on CRT at ×3,000 magnification.

Washing

Ether-alcohol (1:1) solution and distilled water were used for washing solutions of hair specimens. From the parietal part of a 40-year-old Japanese male ten head hairs were collected, from which specimens of one cm each in length near to the hair root were used. Each specimen was divided into two pieces and one was washed in ether-alcohol (1:1) solution and another in distilled water for 15 minutes in each of three changes of washing solutions in both cases. Moreover, a dorsal hair from horse which is known to perspire profusely was simultaneously added to this investigation. All the specimens were aligned to be parallel to the detector axis in the SEM specimen chamber, and analyzed by fixing a electron beam spot on the middle position of cuticle surface at ×3,000 magnification.

Positional differences in a cross section

A cross section of 25μm in thickness of a head hair from the parietal part of 45-year-old Japanese male was made by embedding it in araldite. In this case a cross section of a dorsal hair from horse was also investigated. Cuticle, cortex and medulla were analyzed in the same manner as mentioned above.

After these investigations the quantitative analysis of sulfur and chlorine in head hairs was carried out. Head hair specimens were collected from 25 males and 11 females from 0 to 13 year-old Japanese, and 23 males and 15 females from 20 to 50 year-old Japanese. From the parietal part of each person, five head hairs were collected and one cm in length near to the hair root from each specimen was analyzed. Each specimen was washed in ether-alcohol (1:1) solution for 15 minutes in each of three changes of washing bottles and coated with carbon in *vacuo*. They were aligned on a carbon specimen mount pasted with a double-face tape, to be in the direction parallel to the detector axis in the SEM specimen chamber. The analysis was carried out on the middle position of cuticle surface by fixing a electron beam spot at the center of CRT at ×3,000 magnification. For quantitative analysis a pure sulfur powder for sulfur and a sodium chloride for chlorine were used as standard samples, respectively. A 0-10 kev spectrum from each specimen under the same operating conditions was recorded in a computer interfaced X-ray detection system. K ratios in the intensities of sulfur and chlorine of hair specimens to those of standard samples were at the same time calculated in the computer with respect to K lines of elements. The arithmetic mean of K ratios from five specimens from each person was used as the value of the individual. In the case of calculating of K ratios of chlorine responses, it was corrected based on the theoretical weight ratio of sodium chloride, because it is not representative of pure sample with respect to chlorine content (Table 1). The sulfur and chlorine contents in head hairs were determined by multiplying K ratio by the correction coefficients for quantitative analysis such as atomic number, absorption, and fluorescence effects.

In all the present investigations specimens were analyzed for 200 seconds under the same operating conditions of 25 kv as a accelerating voltage and 1.5×10^{-10}A as a specimen current, and the intensities were measured as the X-ray counts at the center of the peak of element. In all cases the measurements were carried out at 45° of take-off angle and with a 30 mm specimen-detector distance. A computer interfaced, energy dispersive X-ray detection system (NS 880, Northern Scientific, U.S.A.) equipped to SEM (JEOL-U3) and a Si (Li) semiconductor detector (having 10 mm^2 surface, 3 mm thickness, 158 ev resolving power, Nuclear Semiconductor, U.S.A.) were employed in the present investigations.

Results

Characteristic X-ray intensities of sulfur and chlorine measured in each position in spot and square modes on the cuticle surface are shown in Fig. 1 and Table 2. The intensities of chlorine were hardly variable in every analyzing position and in both analyzing modes, showing a

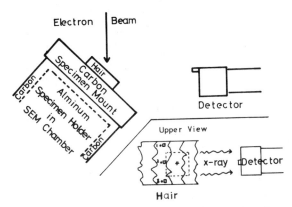

Fig. 1. Illustration of the alignment of hair specimens in SEM specimen chamber and the analyzing positions in spot and square modes. Each number indicates the analyzing position. Small squares were about $3.3\mu m^2$ and large one was about $33\mu m^2$.

Table 1. Procedure in the calculation of the correction coefficient for Cl K ratio obtained between NaCl and hair specimen.

	K	Z	A	F	ZAF
NA	-0.393	0.982	1.500	0.996	1.468
CL	-0.606	1.013	1.100	1.000	1.115

K value was derived from the theoretical weight ratio of NaCl. Z: atomic number effect, A: absorption effect, F: fluorescence effect. Correction coefficient was calculated from K/ZAF.

slight increase in intensity at position 3 in the spot mode. Intensities of sulfur showed a slight variation according to the analyzing position, showing the intensity at position 3 was slightly higher than at other positions in both analyzing modes. Although it is impossible to ellucidate the cause of the slight positional variations in intensities, it could be presumed that they are due to the constitutional difference in the distribution of elements on the hair scale surface, or in the mechanical relationship between analyzing specimen and detector.

When hair specimens were aligned in the direction perpendicular to the detector axis in SEM specimen chamber (Fig. 2), characteristic X-ray intensities both of sulfur and chlorine showed marked variation according to the analyzing positions. As shown in Table 3, intensities measured at the opposite side of the detector were much far reduced as considered to be nearly unable to detect. And as the analyzing position moved towards the side of the detector intensities were increased. These marked variations in intensities according to analyzing positions clearly indicate that they are not due to the constitutional difference but the direct effect of specimen unevenness on intensities. Therefore, in the analytical investigation of hair specimens with EDX in a SEM, specimens should be aligned in the direction parallel to the detector axis in SEM specimen chamber in order to eliminate the effect of specimen unevenness. Even after washing hair

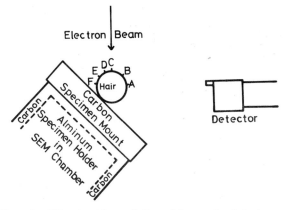

Fig. 2. Illustration of the alignment of hair specimens and the analyzing positions indicated as A, B, C, D, E and F.

Table 2. Intensity at the center of peaks of S and Cl on head hair cuticle surface measured in spot and square modes with EDX in a SEM.

Position number in Fig. 1	Spot mode		Square mode	
	S	Cl	S	Cl
1	1193	205	1179	195
2	1274	201	1290	210
3	1301	223	1303	207
4	—	—	1163	208

Intensity was showed as X-ray counts measured in 200 seconds under the operating conditions of 25 kv in accelerating voltage and 1.5×10^{-10}A in specimen current. Specimen was aligned in the direction parallel to the axis of detector in the specimen chamber in a SEM.

specimens in ether-alcohol solution, minute contaminations persisted in adhering on the scale surface (Fig. 3). In order to eliminate the X-ray generation originating from such contaminations a spot mode would be preferable, as intensities measured in spot mode were almost of the same value as those in square mode on the scale surface.

Characteristic X-ray intensity after washing in ether-alcohol solution and in distilled water are shown in Table 4. Intensities of sulfur hardly varied between both washing treatments, whereas those of chlorine varied, showing that the value after washing in distilled water was nearly 50 % (160.6 ± 5.1) of that in ether-alcohol solution (322.9 ± 11.2). X-ray spectra of dorsal hairs from horse after both washing treatments (in Fig.4), show that the peak of chlorine after washing in ether-alcohol solution is very conspicuous, and after washing in distilled water it almost disappeared. Also in the case of horse, intensities of sulfur hardly varied between both washing treatments.

Intensities of sulfur and chlorine measured on the cross section of a head hair are shown in Table 5. A tendency was seen that the intensity of sulfur was slightly higher in cuticle and lower in medulla and that of chlorine was highest in medulla. It is interesting that the

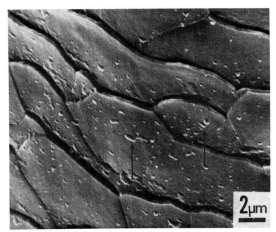

Fig. 3. Cuticle surface of a head hair after washing for 15 minutes in each of three changes of ether-alcohol (1;1) solution bottles. Arrows indicated some contaminations persisted in adhering on the scale surface.

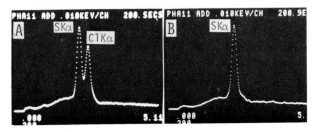

Fig. 4. EDX spectra of dorsal hairs of a horse after washing in ether-alcohol (1:1) solution (A) and distilled water (B). In photograph B, Cl peak nearly disappeared.

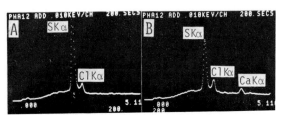

Fig. 5. EDX spectra of a cross section from a human head hair. A was from cuticle and B was from medulla. In the same vertical full scale S intensity in cuticle overflowed and Ca peak was very clear in medulla.

characteristic X-ray peak of calcium which was not clear in cuticle and cortex appeared clearly in medulla (Fig. 5). As shown in Fig. 7, peak heights of sulfur on the cross section of dorsal hair from horse (Fig. 6) were nearly of the same value in cuticle and cortex as were those of chlorine. Peak height of sulfur in medulla decreased remarkably, while the intensity of chlorine varied extremely depending on different analyzing positions. As shown in Fig. 6, dorsal hair from horse has wide medulla in which many vacuoles are seen. Some vacuoles may be packed with coarse materials. Analysis of this material

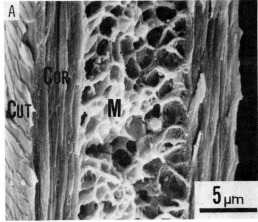

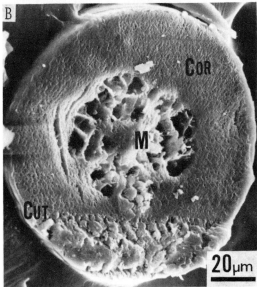

Fig. 6. Longitudinal surface (A) and cross sec- of dorsal hairs of horse medulla was very wide and filled with many vacuoles. Some vacuoles were packed with coarse materials. Cut: Cuticle, Cor: Cortex, M: Medulla.

Table 3. Effect of the unevenness of the head hair on the intensity at the center of peaks of S and Cl of the cuticle surface with EDX in a SEM.

Analyzing position in Fig. 2	S	Cl
A	1231	231
B	1056	175
C	716	129
D	577	115
E	117	39
F	60	20

Intensity was shown as X-ray counts measured in spot mode of 200 seconds under the operating conditions of 25 kv in accelerating voltage and 1.5×10^{-10}A in specimen current. Specimen was aligned in the direction perpendicular to the axis of the detector in the sample chamber in a SEM.

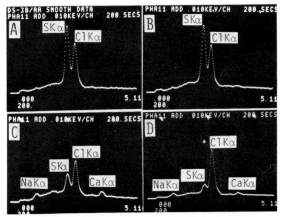

Fig. 7. EDX spectra of a cross section from a horse. A and B were from cuticle and cortex, respectively. C and D were from medulla, in which Ca peak were very clear. The peak heights of Cl were variable depending on analyzing positions and D obtained in analyzing coarse materials in the vacuole gave highest peak of Cl. The counts of vertical full scale of D were two times as many as others.

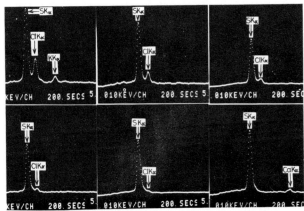

Fig. 8. Several patterns of EDX spectra from human head hairs which were assorted by the peak height of Cl. Some specimen generated the peak of K or Ca.

showed an extremely high peak of chlorine which was approximately twice as high as the peak from the cuticle, and also showed a clear peak of sodium (Fig. 7). The peak height of sulfur in medulla was much lower than that in cuticle or cortex. Also in horse, a peak of calcium was very clearly detected only in the medulla in the same manner as in human. From these results it would be presumed that chlorine and calcium were distributed in hair medulla in higher levels than in cuticle and cortex.

After these investigations, quantitative analysis of human head hair specimens was carried out in such a way that specimens were aligned in the direction parallel to the detector axis in SEM specimen chamber. As shown in Fig. 8, X-ray spectra in an energy range from 0 to 10 kev obtained in the analysis of head hair specimens

Table 4. Washing effect of ether-alcohol solution and distilled water on the intensity at the center of peaks S and Cl of the head hair cuticle surface with EDX in a SEM.

Specimen number	Ether-alcohol S	Cl	Distilled water S	Cl
1	1072	335	1072	158
2	1030	327	1021	170
3	997	333	1074	159
4	1011	307	1103	163
5	1031	342	1028	158
6	1074	319	1049	153
7	1026	328	1078	164
8	1001	309	1034	154
9	1052	313	1061	160
10	1056	316	1053	167
M.±S.D.	1035.0 ± 26.3	322.9 ± 11.2	1057.3 ± 24.2	160.6 ± 5.1

Intensity was showed as X-ray counts in spot mode of 200 seconds under the operating conditions of 25 kv in accelerating voltage and 1.5 ± 10^{-10}A in specimen current. Ten head hairs collected from the same individual were examined. (Mean±Standard deviation)

Table 5. Positional difference of the intensity at the center of the peaks of S, Cl and Ca examined in the cross section of the head hair with EDX in a SEM.

Position	S	Cl	Ca
Cuticle	1359	214	—
Cortex	1192	238	—
Medulla	885	276	144

Intensity was showed as X-ray counts in spot mode of 200 seconds under the operating conditions of 25 kv in accelerating voltage and 1.5 ± 10^{-10}A in specimen current.

Table 6. Content of S and Cl in human head hairs measured with EDX.

Age	Sex	Number of specimen	S %	Cl %
0 — 13	Male	25	5.04±1.02	0.93±0.39
	Female	11	5.99±1.00	1.09±0.62
20 — 50	Male	23	6.09±0.93	0.67±0.20
	Female	15	5.99±0.91	0.45±0.12
			Mean±Standard deviation	

with EDX were classified into several patterns according to peak height of chlorine. Although of rare occurrence, potassium and/or calcium peaks were detected in some specimens. The heights of chlorine varied from the maximum peak (about 400 counts) to the minimum peak which could be hardly observed as a peak height under the analyzing conditions described in this paper.

Table 6 shows results of quantitative analysis of sulfur and chlorine contents in human head hair specimens. The sulfur contents of hairs from 0-13 year old individuals were 5.04±1.02 %

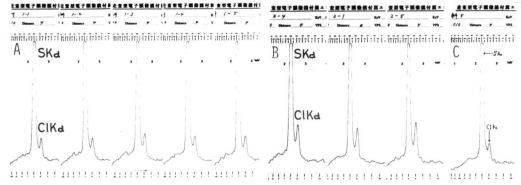

Fig. 9. Excerpted figure from a case work report of hair identification. A: EDX spectra of 5 single head hairs from a suspect, B: EDX spectra of 3 single axillary hairs from the same person as A, C: EDX spectra of a hair specimen from a crime scene.

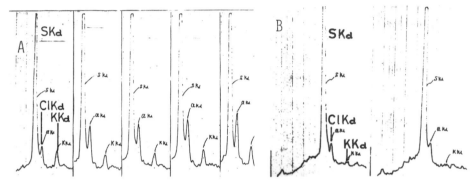

Fig. 10. Excerpted figure from a case work report of hair identification. A: EDX spectra of 5 single head hairs from a suspect, B: EDX spectra of 2 single hair specimens from a crime scene.

and 5.99±1.00 %, and those of 20-50 year old were 6.09±0.93 % and 5.99±0.91 %, for male and female, respectively. Chlorine contents of 0-13 year old individuals were 0.93±0.39 % and 1.09±0.62 %, and those of 20-50 year old were 0.67±0.20 % and 0.45±0.12 %, for male and female, respectively. A tendency for higher chlorine contents was seen in the younger age groups including both sexes. At the same time, chlorine contents of the adult age groups had a tendency to be higher in the male group than in the female group.

Figure 9 was excerpted from a case work report, in which A and B show spectra of 5 single human head hairs and 3 single axillary hairs from a suspect, respectively. Figure 9C shows the spectrum of a hair specimen from the crime scene. Simultaneously with the visual comparison of X -ray spectra among specimens, the confidence interval of chlorine counts of the suspect was estimated from the counts of 5 single head hairs. In this case the confidence limits of the chlorine counts from the suspect were 147.9-168.1 and 141.3-174.6 in the confidence coefficient of 95 % and 99 %, respectively. The chlorine counts of 3 single axillary hairs from the suspect were 210, 225 and 235 which were in higher values than head hairs. The chlorine count of a hair specimen from the crime scene was 134. The counts were deviated from the upper and lower limits of the confidence interval of the suspect in 99 % as well as in 95 % confidence coefficient.

Therfore, the hair specimen from the crime scene was identified to be differ from hairs of the suspect. Figure 10 was excerpted from figures of another case work report. Figure 10A shows EDX spectra of 5 single head hairs from a suspect, and Fig. 10B shows EDX spectra of 2 single human head hairs from a crime scene. In this case work visual comparison of spectra was fairly effective in differentiating among specimens, because spectra from the suspect showed commonly higher peak of potassium as well as chlorine than those from the crime scene.

Discussion

The determination of chlorine content in human head hairs has been mainly investigated by NAA[2,3,4]. Kozuka et al.[3] reported that chlorine contents in head hairs in Japanese population fluctuate between 12,000 ppm (1.2 %) maximum and 80 ppm (0.008 %) minimum with the mean value of 2010 ppm (0.201 %), while it has recently reported by Takeuchi et al.[4] that the chlorine contents fluctuate between 2500 ppm (0.25 %) maximum and 3.6 ppm (0.00036 %) minimum with the mean value of 540±550 ppm (0.054±0.055 %). This discrepancy in chlorine content values would originate in the difference of hair washings prior to the measurement of the element. Kozuka et al.[3] used hair specimens washed for 10 minutes in each of three changes of alcohol

-acetone (1:1) solution bottles, while Takeuchi et al.[4] used hair specimens washed in sequence of acetone, water, water, water and acetone. Kanda et al.[8] reported that sodium and chlorine were remarkably eluted from the hair, when the hair was washed in deionized water. In fact, the clear chlorine peak, which was observed in a horse hair specimen washed in ether-alcohol solution, nearly disappeared in the specimen washed in water as described in this paper. Therefore, in discussing the analytical values of chlorine contents of hair specimens among individuals, special attention should be paid to what solutions were used for the hair washing prior to the measurment of the element. Even if the mean values of chlorine contents obtained by Takeuchi et al.[4] is left out of consideration, the mean values in this paper are rather higher than those by Kozuka et al.[3]. It would be partially due to it that the part near to the hair root was examined in the present investigation, because it was suggested in a previous paper that the characteristic X-ray counts of chlorine measured with WDX decreased with ascending from the hair root to tip[7].

In the determination of sulfur contents of head hairs from 55 Japanese persons by a wet chemical analysis, Sakata[9] reported that the mean values of sulfur contents were 46.8±2.0 % in males and 43.6±2.0 % in females with a very slight difference in the values between both sexes. Clay et al.[10] also suggested that sulfur contents of human light and dark head hairs were within the range of 4.54-5.67 % in males and 4.76-5.78 % in females by the same method as above, with no variations according to age and sex. Basco et al.[6] measured the sulfur contents of human head hairs with radioisotope excited X-ray emission method and showed that the sulfur contents were found to fluctuate between 6±1 % with relatively constant values among 204 samples. The mean values of sulfur contents in the present investigation were 5.04 %, 5.99 %, 6.09 % and 5.99 % in each age and sex group described here. On the whole it may safely be said that there are no individual variations according to age and sex in the sulfur contents of human head hairs. The slight discrepancy in the sulfur contents among investigators would be due to the differences in the analytical methods and specimen sources. Moreover, in the case of the analysis of hair specimens with EDX, it is necessary to consider which part of the hair analyzed is. As caluculating based on amino acid compositions of the head hairs from Caucasian by Wolfram et al.[11] the sulfur content of the hair cuticle is estimated 6.33 % and the value of the cuticle is a little higher than that of the cortex. Considering the penetration depth of the electron beam into the hair specimen which was suggested by Brown et al.[12], it is presumed that the mean values of sulfur contents in the present investigation is obtained from the hair cuticle.

It has been suggested by Coleman[2] and Cornells et al.[13] that the sodium and chlorine are present as NaCl mainly originating from perspiration with, possibly, a certain amount of extra chlorine in some other chemical form. Basing on the suggestion, the perspiration rate of the individual could be mentioned with relation to the chlorine contents of the human head hairs. Koyama[14] reported that the numbers of sweat gland pores per one cm^2 in Japanese parietal skin decrease with age from infancy with about 716 in 0-1 age and 223 in 30-60 age. With respect to the perspiration rate, it has been suggested by Kuno[15] that the sweat volume per one hour is likely to be twice as high for infants as it is in adults and the perspiration rate of women is much lower than that of men because of the lower activity of basal metabolism in womwen. These anatomical and physiological findings on the perspiration could explain the general trend in the currently observed chlorine contents of human head hairs. Recently, Gaudette[16] has discussed the value of human head hairs. According to Gaudette, the greatest value of trace element analysis is through multielement analysis to determine individuals with abnormally high or low contents of certain elements. Although the chlorine contents of human head hairs do not imply the abnormal characteristics of the individual, it might be used as one of the factors showing the status of the individual constitution because of its wide variation among individuals presumably based on perspiration rate.

References

1. Sudo, T. and S. Seta: Individual identification of hair samples in criminalistics. Proc. 1st Int. Symp. on Biol. and Disease of Hair, 1, 543-553, 1975.
2. Coleman, R. F.: The application of neutron activation analysis to forensic sciences. J. For. Sci. Soc., 6, 19-27, 1966.
3. Kozuka, H., H. Isono and N. Tsunoda et al.: Activation analysis of human head hair. J. Hyg. Chem., 18, 1-6, 1972.
4. Takeuchi, T., H. Kozuka and S. Ohmori et al.: Survey of the trace elements in hairs of the normal Japanese. Annu. Rep. Res. Reactor Inst. Kyoto Univ., 11, 177-185, 1978.
5. Cookson, J. A. and A. D. Pilling: Trace element distribution across the diameter of human hair. Phys. Med. Biol., 20, 1015-1020, 1975.
6. Basco, J. P., P. Kovacs and S. Horrath: Investigation of some inorganic compounds in human hair. Radiochem. Radioanal. Letters, 33, 273-280, 1978.
7. Seta, S., H. Kozuka and T. Sudo: Criminal investigation of human head hairs by means of SEM and EPMA with special reference to age and sex differences. SEM/1975/II. IIT Research Institute, Chicago, IL, 60616, 579-587.
8. Kanda, Y., H. Isono and H. Kozuka: Elution behaviors of sodium, chlorine and bromine from the hair. Radioisotop., 23, 398-401, 1974.
9. Sakata, S.: On the amounts of cystine and total sulfur in human hairs. J. Hiroshima Med. Ass., 4, 1358-1365, 1956.
10. Clay, R. C., K. Cook and J. I. Routh: Studies in the composition of human hair. J. Am. Chem. Soc., 62, 2709-2710, 1940.
11. Wolfram, L. J. and M. O. Lindemann: Some observations on the hair cuticle. J. Soc. Cosmet. Chem., 22, 839-850, 1971.
12. Brown, A. C., R. J. Gerdes and J. Johnson:

Scanning electron microscopy and electron probe analysis of congenital hair defectss. SEM/1971/I. IIT Research Institute, Chicago, IL, 60616, 371-376.

13. Cornells, R. and A. Seecke. Neutron activation analysis of human hair collected at regular intervals for 25 years. J. For. Sci. Soc., 11, 29-46, 1971.
14. Koyama, K.: Über die Verteilung der Schweissdrüsen bei den Japanern. Fol. Anat. Jap., 15, 571-594, 1937.
15. Kuno, Y.: In "Physiology of human perspiration " in japanese. pp. 72-75, Koseikan, Tokyo, 1970.
16. Gaudette, B. D.: Some further thoughts on probabilities and human hair comparisons. J. For. Sci., 23, 758-763, 1978.

Discussion with Reviewers

J.A. Swift: What were the angles of incidence and the take-off angles for this work?
Authors: The angle of incidence is one between the incident electron beam and the specimen surface which is inclined at an 45° in SEM specimen chamber. The take-off angle is one between the normal line of the specimen surface and the detector axis.

Reviewer I: What specific ether-alcohol solutions were used and in what ratio, i. e., concentrations, for the washing of hairs?
Authors: We used the solution which consists of 1:1 mixture of ethyl ether and ethyl alcohol of reagent grade. In the present investigation all hair specimens were washed for 15 minutes in each of three changes of the solution bottles.

R.K. Hart: Why were pure sulfur and sodium chloride used for standards instead of chemically analyzed hair? Also, how many determinations were made on each hair sample, and at what positions along the shaft from the proximal end were these recordings made?
Authors: It has been generally accepted that it is very difficult to obtain the suitable standard samples for the quantitative analysis of the trace bioelements such as chlorine, potassium and calcium. So, we used pure sulfur and sodium chloride of analytical grade. Of course, the K ratio of chlorine should be corrected basing on the theoretical weight ratio of sodium chloride. I think the calculation of K ratio between the specimen and the standard sample should be carried out very seriously and theoretically. Moreover, the positional difference in the element content along the hair shaft should be taken into consideration. I think that the chemically analyzed hair is able to be used as the control specimen of the individual, but it can not be used as the standard sample for the quantitative analysis with EDX. The part of one cm in length near to the hair root was used as a specimen. The analysis was carried out once at the middle position of the specimen.

K.A. Siegesmund: Double back scotch tape has chlorine in it. Was this placed far enough from the area analyzed? Also the araldite section would

have chlorine in it. To get any meaningful value for chlorine a background analysis should be done.
Authors: This basic problem has been solved.

J.A. Swift: Please comment on the reasons for the apparent difference in sulphur content of human and equine medulla in the light of the fact that wet chemical analyses of medulla isolated from both these animal species (Bradbury, Adv. Protein Chem., 1973, 27, 111-211) contain sulphur at the same low level of concentration? In relation to this do you not think that whereas analyses in equine medulla are confined to this histological component, because of the smaller dimensions of human medulla there is electron beam spreading into and therefore analytical contributions from the adjacent cortex?
Authors: We have not refered to the difference in the sulfur contents between human and equine medulla, and we have merely showed that in medulla of both human and equine the surfur intensities measured with EDX were of lower levels than those in both the cuticle and the cortex. In analyzing at ×3,000 magnification the additional X-ray from the cortex is not seemed to be generated. Of course, we think, the electron beam spreading to adjacent area should be seriously taken into consideration in analyzing a very minute object.

J.A. Swift: What for the various elements you have analysed are the means and standard deviations between hairs from the same subject as compared with those between subjects. In other words what is the statistical significance of the data you show for example in figure 9.
Authors: Figure 9A shows the spectra of five single hairs from a suspect. We calculated the confidence interval (upper and lower limits) from the chlorine counts of five single hairs (see text). When the chlorine count of the hair specimen from the crime scene deviates from the confidence intervals calculated in the confidence coefficients of 95 % and 99 %, the hair specimen from the crime scene is statistically identified not to originate from the suspect.

J.A. Swift: Will you comment upon the exogenous and endogenous origins of Cl, K and Ca that may contribute to the variations in concentration of these elements you found in human hair?
Authors: The chlorine detected on the cuticle surface is present as NaCl mainly originated from the perspiration. As the chlorine intensity decreases remarkably after washing hair specimen in water, the chlorine is considered exogeneous. In analyzing sweat with EDX, the potassium is always detected along with the main peak of chlorine. However, while some hair specimens which have high chlorine peaks give no potassium peak, some specimens which have no chlorine peak give high potassium peaks. From these observations, the potassium is presumed endogeneous, and of cource, contaminated hair specimens often give the potassium peaks. The calcium of the hair specimen would be of both exogeneous and endogeneous origins. The exogeneous calcium is generally considered to originate from surface contaminations, and after the permannent wave treatment the cal-

cium content increase remarkably. The endogeneous
calcium is well explained in "text reference 6"
in which it has been suggested that the calcium
content of human head hair was dependent on age
and sex and presumably fluctuates during certain
illness, e. g. ischemic heart disease. In our
investigations the calcium is mainly detected in
higher level in medulla than in cuticle and cortex, which would indicate the endogeneous origin
of calcium.

Reviewer I: In figure 3 and the text, your "evidence" for contamination seems inconclusive.
Do you have addtional evidence or information that
these areas are indeed contamination, and, if so,
what is the contamination?
Authors: In our routine examinations such elements as silicon, chlorine, potassium and calcium
are often detected from these areas. However, the
surface contaminations sometimes come from fat
and detergent, and some damaged cuticle surfaces
are likely to be misread as contaminations.

K.A. Siegesmund: If the purpose of this study, as
stated, was to add analytical data on hair composition to methods of criminal identification,
why was it restricted to s and cl identifications?
R.K. Hart: This paper implies that the analysis
of indigenoue trace elements in human head hair
cannot be used for positive identification of an
individual. However, specific toxin responses
together with the spectrum of normal human head
hair inorganics can be used to identify a suspect.
Authors: It is very often that in actual criminal case works the hair specimens from the crime
scene are very small in amount. Although there
are many reports with respect to the contents of
trace elements of human head hairs with NAA and
chemical analyses, it would be rather difficult
to put them into practical use in the criminal
identification of head hairs. Therfore, we have
tried to apply the EDX system to the elemental
analysis of the hair specimens in small amount.
Even if 10 single hairs are collected from the
crime scene, we can not submit them in a lump to
the elemental analysis with NAA or other chemical
analyses, because they are hardly of the same
origin. In the analysis of hair specimens with
EDX, the element peaks detected in a rang of 0
-20 kev are generally silicon, sulfur, chlorine,
potassium and calcium, and rarely zinc. Depending on specimens, potassium or calcium peak is
not detected. Other trace elements investigated
with NAA can not be detected with EDX. It would
be due to the low contents of the elements in
head hairs. Since 12 years we have carried out
the trace element analysis with NAA and accumulated analytical data of the exogeneous and endogeneous elements. Chlorine content variations
among individuals are originally based on the
analytical data with NAA in our institute. At
present we have started to investigate the endogeneous calcium contents of human head hairs with
EDX, AA and NAA with respect to individual differences in its contents. We would like to investigated detecting abnormally high element
caused by specific toxin with EDX.

REVIEWER I: N. E. Pickering

Additional discussion with reviewers of the paper "The Characterization of Dynamite Wrapping Paper by SEM...." by M. Ueyama et al continued from page 208.

I.B. Sachs: The authors state that they presume that the particles on the surface of the remnant papers following the explosion of dynamite are melted paraffin droplets. Did the authors attempt to drop some of the paper residues into hot water or heat the residues in a solution of alkanin in 50% alcohol?
Authors: No, although we did place some, the fragments bearing "spherical" particles, in an oven at 100°C for a few minutes. Subsequent SEM examination revealed that the "spherical" particles had disappeared.

V. Matricardi: Is it correct to assume that the "spherical" features observed did not provide any characteristic peaks when examined by WDS?
Authors: Yes. Characteristic peaks were not obtained.

R.A. Parham: How closely do various commercial papers with wax treatments resemble dynamite paper? Would these papers complicate your separation of dynamite paper from all other possibilities?
Authors: We have only examined so far one commercial paper treated with wax--the Towa paper discussed in this study--and it was readily distinguished from the dynamite paper. However, it is possible that there are other commercial papers treated with wax that might be distinguished only with difficulty from dynamite paper.

I.B. Sachs: How can the authors distinguish the elements listed for ink from those that may be attributed to paper fillers or dynamite powder?
Authors: WDS analysis was performed on the ink and on the paper separate from the ink. Only when the analysis was performed on the ink did the elements listed as characteristic of the ink emerge.

I.B. Sachs: What is the size distribution of fibers examined before and after explosion of the dynamite?
Authors: The fibers ranged in diameter from 8 μm to 150 μm both before and after the explosion of the dynamite.

V. Matricardi: After the "spherical" features have been identified and the inked lines analyzed, would low temperature ashing be able to remove the paraffin layer thereby exposing the paper for analysis? Would there be any advantage in doing this?
Authors: Possibly. At least it would be interesting to try.

SCANNING ELECTRON MICROSCOPY/1979/II
SEM Inc., AMF O'Hare, IL 60666, USA

THE CHARACTERIZATION OF DYNAMITE WRAPPING PAPER BY SEM:
A PRELIMINARY REPORT

M. Ueyama*, I. Suzuki, R. L. Taylor** and T. T. Noguchi**

National Res. Inst.   **Forensic Science Center
of Police Science   Dept of Chief Medical Examiner-Coroner
6. Sanban-cho, Chiyoda-ku   1104 N. Mission Street
Tokyo, Japan   Los Angeles, CA 90033

Abstract

Since fragments of exploded dynamite paper as well as other papers may be collected from a bomb scene, a SEM study was conducted whereby unexploded dynamite paper, exploded dynamite paper, and two common commercial wrapping papers were compared. In addition, the printing ink on the dynamite paper was analyzed by wavelength-dispersive x-ray analysis. Exploded dynamite paper fragments, when dynamite itself was used as the explosive material, were shown to have a unique morphology, i.e., "spherical" particles, presumably paraffin and ranging in size from 0.4 μm to 7.5 μm, were present on the surface of the paper. These particles were not observed on the exploded dynamite paper when ammonium nitrate was used as the explosive material.

Two actual bombing cases are reported. The potential usefulness of SEM and x-ray microanalysis of paper fragments recovered from bombing scenes was demonstrated, particularly in cases where sufficient explosive residues cannot be collected for chemical analysis.

*Presently at same address as Taylor and Noguchi.

KEY WORDS: X-ray Microanalysis, Dynamite, Paper, Explosives, Bomb Scene Investigation

Introduction

The number of terrorist bombings in Japan is small but appears to be increasing. The Japanese public is naturally alarmed and outraged by the indiscriminate killings resulting from the bombings. As a result, all bombings are rigorously and thoroughly investigated. As part of this process, every bomb scene is thoroughly searched for physical evidence, and each and every potentially important fragment is collected, including any explosive residues; these are then submitted to the crime laboratory for examination and identification.

Of paramount importance in the bombing investigations is the determination of the type of explosive used. Many chemical tests are currently in use to screen and identify explosive residues.[1-4] Unfortunately, however, most explosives are consumed on detonation, and only minute traces of the explosive material can subsequently be found. It is therefore important for the laboratory investigator to be familiar with and to be able to identify the various other materials (e.g. blasting cap, dynamite paper, timer, etc.) often associated with bombs. To achieve the necessary degree of familiarity, it is important to have available an adequate reference collection of such materials.[5-7] The National Research Institute of Police Science in Japan is currently engaged in a long range program to determine the origin of all fragments produced in various types of controlled bombings and to assemble just such a reference collection.

Since fragments of exploded dynamite paper as well as other paper may be collected from a bomb scene, a reference collection of the type discussed above would by necessity contain dynamite papers as well as newspapers, bag papers, wrapping papers, etc. A scanning electron microscopic (SEM) study was therefore undertaken whereby dynamite papers, both before and after explosion of the dynamite, and two wrapping papers used to make dynamite paper and which are also common in Japanese commerce were examined in a manner similar to that in other investigations concerned with SEM of paper.[8,9] In addition, the printing inks on the unexploded and exploded dynamite papers were analyzed by wavelength-dispersive x-ray spectroscopy (WDS).

Materials and Methods

Because paraffin waxes possess good barrier properties toward water, dynamite papers are coated with them. The paraffin layer is always thicker on one side of the paper than the other, and the dynamite is wrapped with the thicker layer facing the outside.

Two kinds of dynamite which are commonly used in Japan--No. 3 Kiri and No. 2 Enoki--were used in this study. The former uses Oji paper and the latter uses Towa paper, and each is made of a different explosive powder. In order to determine the effect of different explosive materials on the morphological appearance of dynamite paper, two different explosives were used - dynamite and ammonium nitrate. Test explosions were conducted in a closed steel drum measuring 60 cm high and 50 cm in diameter. After the explosion, the fragments of exploded paper were collected.

One of the commercial wrapping papers examined (Towa Paper Company #12) had been coated with wax in manufacture, whereas the other (Oji Paper Company) had not. As a result, each paper appeared different to the naked eye. All specimens were evaporated with gold in a vacuum evaporator, and their surfaces examined at 10 Kv in the Jeol JSM-U3. WDS analyses were performed at 20 Kv in the Shimadzu EMX-SM electron probe microanalyzer.

Results and Discussion

Unexploded Dynamite Paper Morphology

The paraffin on the surface of the unexploded dynamite paper was distinctive when viewed by SEM and appeared somewhat "scaly" in its surface pattern. Figures 1 and 2 are SEM micrographs of No. 3 Kiri and No. 2 Enoki dynamite papers, respectively. The paraffin layer on the outer surfaces of the papers obscured most of the paper fibers. In comparison, more fibers could be seen on the inner surfaces where the paraffin layer was thinner.

Exploded Dynamite Paper Morphology

Dynamite Explosion. The fragments of exploded dynamite paper were small in size, measuring less than 1 mm x 2 mm, and in some of them remnants of the printing on the paper (red "lines") could be seen. Both sides of several fragments were examined in the SEM. Figures 3 and 4 show the surface morphology of the two sides. The paraffin layer as such was not seen on either side of the paper following the explosion. "Spherical" particles, however, were seen, and these appeared in greater numbers on the outer surface than on the inner. These particles are presumed to be paraffin which melted and resolidified during the explosion. The surface morphology of the dynamite paper from all of the experimental explosions appeared virtually identical, and intermediate stages between the appearance of the unexploded paraffin layer and the "spherical" particles were not seen.

Ammonium Nitrate Explosion. Figure 5 shows SEM micrographs of dynamite paper exploded by ammonium nitrate. It exhibited one distinct morphological difference from the paper exploded by dynamite (See Figs. 3 and 4); "spherical" particles were not seen on either side of the paper. This difference cannot be readily evaluated due to the small number of test explosions conducted, but probably is dependent on different detonation conditions, including the explosion temperatures.

Commercial Wrapping Paper Morphology

Figure 6 is a micrograph of one of the commercial wrapping papers (Towa Paper Company). The coating layer is very thin and easily distinguished from that of the dynamite paper. Figure 7 is a micrograph of the other paper (Oji Paper Company). The morphology illustrated is typical of the fibrous nature of paper in general.

WDS Analysis of Printing Ink on Dynamite Paper

Figure 8 is an x-ray spectrum resulting from the analysis of the red ink on unexploded dynamite paper (No. 3 Kiri dynamite). Barium (Ba), Sulfur (S), and Calcium (Ca) are consistently found in the ink.

Case Histories

Case 1.

In 1975, a woman was walking in front of a railroad station in the Miyazaki Prefecture in Japan when she saw a paper bag on the sidewalk. She stopped to pick it up, and it exploded, seriously injuring her. The only significant fragments of physical evidence collected from the bombing scene were two splinters of metal and some pieces of paper. Due to insufficient material, the type of explosive used could not be determined by chemical tests performed at the crime laboratories of Miyazaki Police Headquarters. On request from the Department of Police, the crime laboratories of Kyushu Regional Police Bureau (KRPH) examined the metals and paper fragments by a variety of procedures. The paper fragments were examined by SEM and determined to be of two different kinds (Figs. 9 and 10). The one kind was similar to the large fragments of the paper bag recovered from the bomb scene and which had enclosed the explosives. The other kind had similarities to both the

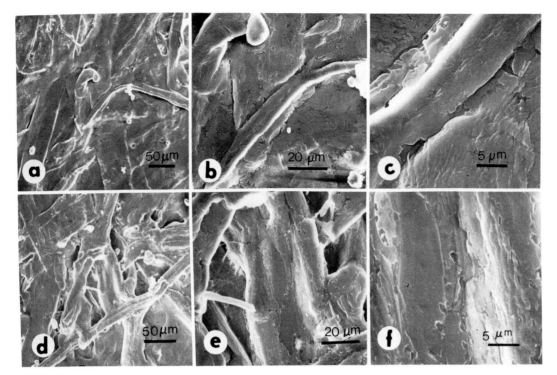

Fig. 1.   SEM micrographs of unexploded dynamite paper (No. 3
Kiri).   The paraffin layer on both sides of the paper appears
somewhat "scaly" in its surface pattern.   a-c) Outer surface
at increasing magnification; d-f) Inner surface at increasing
magnification.

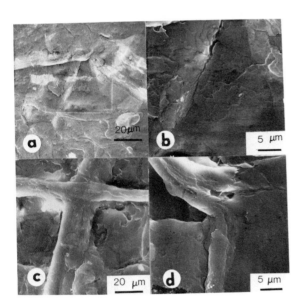

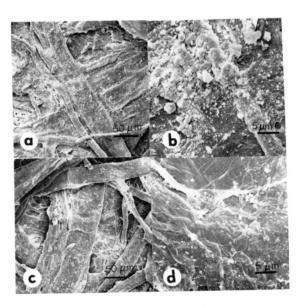

Fig. 2.   SEM micrographs of unexploded dynamite paper (No. 2 Enoki).   Like No. 3
Kiri dynamite paper, the paraffin layer
on both sides of the paper appears somewhat
"scaly" in its surface pattern.   a,b) Outer
surface.   Note the beam damage in 2a; c,d)
Inner surface.

Fig. 3.   SEM micrographs of exploded dynamite paper (No. 3 Kiri).   "Spherical" particles are present on both sides of the
paper, and in greater numbers on the outer
surface (a,b) than on the inner surface
(c,d).

unexploded and exploded dynamite papers; i.e., both a "scaly" appearing surface and "spherical" particles were present. The latter, however, appeared in fewer numbers than on the exploded dynamite paper. This appearance perhaps represents an intermediate state in the transformation of the dynamite paper from its unexploded state to its fully exploded state. It is, of course, possible that the differences observed simply reflect differences in the physical properties of the dynamite paper used in the test explosions from that used by the terrorists in the actual case. The metal fragments, incidentally, were determined to be splinters of a blasting cap.

Case 2.

In 1977, a bomb exploded at the main gate of a prominent citizen's home in Kagoshima Prefecture. The significant physical evidence collected from the bomb scene included ground samples, a splinter of metal, and a piece of paper.

By gas chromatography, nitroglycerine and nitrocompounds were detected in large volumes in the ground samples. These large volumes are apparently due to an incomplete explosion having taken place. The crime laboratories of the KRPB examined the metal and paper fragments. The metal fragment was determined to be a splinter of a blasting cap. The paper fragment, measuring 2 mm x 2 mm, was brown in color similar to dynamite paper, and, on one side, had some faint red "lines" on it. Both sides of the paper appeared "scaly" (Fig. 11), and their surface morphology was more similar to that of unexploded dynamite paper (See Figs. 1 and 2) than it was to exploded dynamite paper (Figs. 3-5). "Spherical" particles were not observed, possibly for the same reasons given for the differences observed in Case #1.

In addition, the red "lines" were analyzed by WDS. Figure 12 shows an x-ray spectrum resulting from such an analysis. The major elements are the same as previously noted for the printing ink on the unexploded dynamite paper.

Conclusion

The identification of the type of explosive used is often extremely difficult. Therefore, every fragment of physical evidence has to be thoroughly examined by various means. By SEM the morphology of exploded dynamite papers was observed to be unique and distinctly different from unexploded dynamite paper and commercial wrapping papers. In addition, x-ray microanalysis promises to be a useful technique for the characterization of the printing inks on the fragments of exploded dynamite paper. Like any other paper, the "identification" of dynamite paper as to source and type could involve fiber species identification and pulp type

identification, pigmentation (crystalline and amorphous), and analysis of other paper additives. And, in any case, the "matching" of two specimens can only be taken to the point where one could state that "there is a high probability" that they came from the same source, and not that they unequivocally came from the same source. It should be noted that the study reported here is only in its earliest stages. Additional test explosions and examinations are planned.

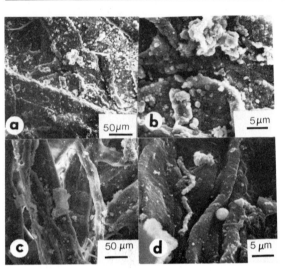

Fig. 4. SEM micrographs of exploded dynamite paper (No. 2 Enoki). Like Figure 3, the "spherical" particles are present on both sides of the paper, and appear in greater numbers on the outer surface (a,b) than on the inner surface (c,d).

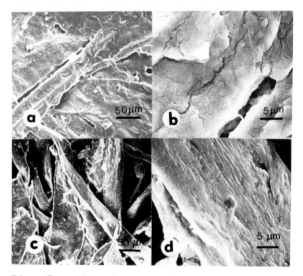

Fig. 5. SEM micrographs of ammonium nitrate-exploded dynamite paper. Note that the morphology of the fragments is very different from that of the dynamite-exploded dynamite paper (See Figs. 3 and 4). a,b) Outer surface; c,d) Inner surface.

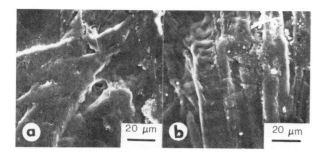

Fig. 6. SEM micrographs of one of the commercial wrapping papers (Towa Paper Company #12). This paper has a wax surface. The surface morphology is easily distinguished from that of the dynamite papers (See Figs. 1 and 2). a) Outer surface; b) Inner surface.

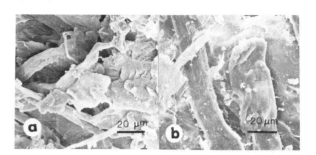

Fig. 9. SEM micrograph of one of the paper fragments from Case #1. This fragment appears similar to the paper bag recovered from the bomb scene. a) One surface; b) Opposite surface.

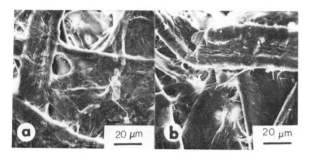

Fig. 7. SEM micrographs of the second commercial wrapping paper (Oji Paper Company). The fibrous surface of uncoated papers is demonstrated well in these micrographs. a) Outer surface; b) Inner surface.

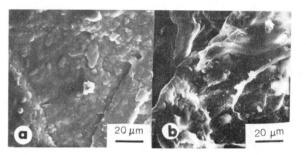

Fig. 10. SEM micrographs of a second paper fragment from Case #1. This fragment appears to be of different origin from that illustrated in Figure 9. It has similarities to both the unexploded and exploded dynamite papers. a) One surface; b) Opposite surface.

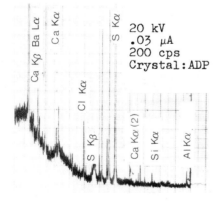

Fig. 8. Typical WDS x-ray spectrum of the red ink on unexploded dynamite paper (No. 3 Kiri). The major peaks are BaLα, SKα, and CaKα. The source of the Al, Cl, and Si is unknown but is possibly derived from the paper filler.

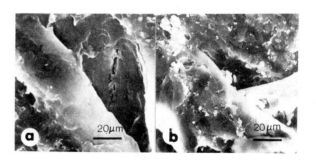

Fig. 11. SEM micrographs of one of the paper fragments from Case #2. Both sides of the paper appear somewhat "scaly." a) One surface; b) Opposite surface.

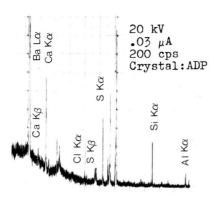

20 kV
.03 μA
200 cps
Crystal:ADP

Fig. 12. Typical WDS x-ray spectrum of the red "lines" on the paper fragments collected from Case #2. The major peaks are BaLα, SKα, CaKα, and SiKα.

References

1. R.G. Parker, M.O. Stephenson, J.M. McOwen, and J.A. Cherolis, "Analysis of explosives and explosive residues. Part 2: Thin-layer chromatography," J. Forensic Sci., 20(2), 1975, 254-256.
2. A.D. Beveridge, S.F. Payton, S.F. Audette, A.J. Lambrtus, and R.D. Shaddick, "Systematic analysis of explosive residues, J. Forensic Sci., 20(3), 1975, 431-545.
3. J.E. Chrostowski, R.N. Holmes, and B.W. Rehn, "The collection and determination of ethylene glycol, dinitrate, nitroglycerine, and trinitrotoluene explosive vapors," J. Forensic Sci., 21(3), 1976, 611-615.
4. H. Fukuda, J. Nakamura, and T. Yamada, "The analysis of nitroglycerine and nitrocompounds by gas chromatography electron capture detector," National Research Institute of Police Science Report, 30, 1977, 29-32.
5. D.G. Townshend, "Identification of electric blasting caps manufacture," J. Forensic Sci., 18(2), 1973, 405-409.
6. C.M. Hoffman and E.B. Byall, "Identification of explosive residues in bomb scene investigations," J. Forensic Sci., 18(1), 1973, 54-63.
7. H.J. Yallop, "Breaking offences with explosive - the techniques of the criminal and the scientist," J. Forensic Sci. Soc., 14, 1974, 99-102.
8. R.A. Parham, "On the use of SEM/x-ray technology for identification of paper components," SEM/1975, IIT Research Institute, Chicago, IL, 60616, 511-518.
9. J.A.W. Barnard, D.E. Polk, and B.C. Giessen, "Forensic identification of papers by elemental analysis using scanning electron microscopy," SEM/1975, IIT Research Institute, Chicago, IL, 60616, 519-527.

Discussion with Reviewers

I.B. Sachs: Can cross-sections of the paraffin paper be prepared and measured?
Authors: Yes. Such measurements might provide additional information that would assist in the identification of dynamite papers.

R.A. Parham: Was there any potential for artifact creation in the vacuum evaporator due to the melting temperatures of the waxes on the dynamite papers in this study?
Authors: Yes, the potential would appear to exist. However, we made no observations that suggested that the waxes were in any way altered by our preparatory procedures. The "spherical" particles clearly did not form as a result of the temperatures reached in the vacuum evaporator.

I.B. Sachs: Were the studies the authors quoted carried out predominately at one beam voltage--10 kV--or were there problems encountered at higher kV?
Authors: All SEM studies reported in this paper were carried out at 10 kV. At higher kV beam damage occasionally occurred.

I.B. Sachs: Did the authors experience any problems due to beam damage or contamination?
Authors: No significant beam damage or contamination occurred under the conditions these studies were conducted.

R.A. Parham: Did you dewax dynamite paper and examine it unexploded to see if any of your "particles" could have been due to paper pigmentation?
Authors: No. Although we did examine unwaxed, unexploded dynamite paper (Oji paper), and no "spherical" particles--paper pigments or otherwise--were observed.

V. Matricardi: Could a light microscope be used to distinguish most of the "spherical" features which are evident in the exploded dynamite wrapping paper?
Authors: The "spherical" particles could be observed by reflected light under a compound light microscope, but not distinctly due to the shallow depth of focus and low magnification. The SEM is required to study satisfactorily the microscopic features of the paper surface.

For additional discussion see page 202

SCANNING ELECTRON MICROSCOPY/1979/II
SEM Inc., AMF O'Hare, IL 60666, USA

APPLICATIONS OF SEM AND X-RAY MICROANALYSIS IN CRIMINALISTICS:
AN INDEXED BIBLIOGRAPHY

R. L. Taylor and T. T. Noguchi

Forensic Science Center
Department of Chief Medical Examiner-Coroner
Los Angeles County
1104 N. Mission Street
Los Angeles, CA 90033

Introduction

In the last few years the scanning electron microscope (SEM) equipped with x-ray microanalysis capability has become one of the most sought after new investigatory tools for the forensic science laboratory, and the number of laboratories acquiring such equipment has increased dramatically. It was only eight years ago that the first papers appeared reporting preliminary results and possible applications of SEM in criminalistics. Since that time many papers have been published on the application of SEM to specific forensic problems and, more recently, to the routine use of the instrument for the examination of relatively common items of physical evidence. The literature relating to the applications of SEM and x-ray microanalysis in criminalistics is now large and diverse, and it was with the intention of making this literature available and therefore more useful that this indexed bibliography was prepared.

All of the references included here have been selected because of their potential interest to individuals concerned with the applications of SEM and/or x-ray microanalysis to the field of criminalistics. Papers in the field of forensic medicine are not included. Whereas the majority of the papers are written by forensic scientists and directly address criminalistics' problems, a few are authored by individuals in other disciplines. Papers in the latter category do not directly concern forensic matters, but they do discuss evidentiary items of interest to the criminalist, and for that reason, have been included here.

KEY WORDS: X-ray Microanalysis, Criminalistics, Forensic Science

Topic Index

The evidentiary items listed below are organized in a manner which we have found useful in our work. It is admittedly arbitrary, but certainly functional. Each topic is followed by one or more numbers which refer directly to particular references in the bibliography. For example, the reader who wants to determine what is known about the SEM and x-ray microanalysis of bullet jackets would look up "Bullet Jackets" in the index and then look up the references which follow--30, 93, and 129--in the bibliography.

Key to the Use of Symbols

Most of the references in the bibliography are followed by one or more of the following symbols in parentheses: SEM, E/EDS, E/WDS, X/EDS, and X/WDS. "SEM" means that the paper contains scanning electron micrographs and/or that the use of the SEM by itself--separate from any x-ray microanalysis that may also be reported in the paper--is important to the subject matter of the paper. All of the remaining symbols mean that the paper involves some form of x-ray microanalysis. Specifically, "E" means that the fluoresced x-rays were excited by electrons either from a microprobe or an SEM, and "X" means that they were excited by x-rays from an x-ray tube. "EDS" means that the fluoresced x-rays were analyzed by energy dispersive spectroscopy, and "WDS" means that they were analyzed by wavelength dispersive spectroscopy.

A single asterisk (*) before a reference means that we have not personally seen the reference and therefore cannot validate its accuracy, have not indexed it to topic, and have not indicated at the end of the reference the instrumentation/analysis used nor any foreign languages that may be involved. A double asterisk (**) means that although we have the article, we do not have an English translation of it, and therefore have neither indexed it to topic nor indicated the instrumentation used.

Languages

If a paper is written in a language other than English, that information is placed in parentheses at the very end of the reference. For example, "German" means that the entire paper--abstracts and summaries included--is in German. "German; English summary" means that the paper is in German and contains an English summary. (The word "summary" is used broadly here to refer to either a summary or an abstract.) "German summary" means that the paper is in English and contains a German summary.

Physical Science Papers

Drugs, Poisons
    Arsenic--134
    LSD--4
    Marijuana, Hashish (See "Botanicals")
    Narcotics--4,109
Explosives, Fire/Arson Residues, Burnt Matches
    Burnt Matches--1
    Explosives--4,22,59
    Fire/Arson Residues--143
Fibers, Synthetic (Man-made)
    17,58,59,69,85,91,102,110,111,114,126,127,138
Firearms
    Bullet Jackets--30,93,129
    Clip, Extractor, Ejector, Breech-block Markings--66,70
    Firing Pin Impressions--29,43,44,59,61,69,99
    Gunshot Residues (GSR): Primer Composition, GSR on Hands, GSR around Bullet Holes (excluding gunshot wounds)--2,4,11,30,47,59,69,77,79,82,83,85,98,102,104,123,129,150
    Rifling Land and Groove Impressions--29,59,60,66,116,118,119
Gems, Coins, Paintings, Objets d'Art
    Coins--42,95,96,138,143
    Gems--135,146
    Paintings, Objets d'Art--4,138
Glass, Vehicle Lamp Filaments
    Glass--3,4,55,59,67,94,113,141,146
    Vehicle Lamp Filaments--41,46,130,138
Paint
    4,6,12,14,17,18,29,45,47,50,54,59,62,69,71,78,92,99,109,113,127,128,134,138,141,143,146,148,149
Particle/Metal Analysis
    59,60,62,65,89,93,128,129,132,146
Questioned Documents
    Handwritten and Printed Impressions--14,53,84,146
    Inks--4,5,53,86,97,128,138
    Paper: Money, Stamps, Documents, Other Papers; Fibers; Pigments--4,5,8,39,59,84,86,87,97,138
Soil(Soil diatoms are listed under "Botanicals--Phytoplankton")--69,89,109,113,127
Toolmarks, Serial Numbers, Fracture Surfaces, Bite Marks (See "Firearms" for toolmarks associated with firearms examinations)
    Bite Marks--7,116,117,118,119,120,121,137
    Cutting Tools (drills, hacksaws, machine tools)--126,127
    Metal Fracture Surfaces, Matching of Surfaces--75,76,127,138,146
    Pliers/Wirecutters--66
    Serial Numbers--32,39,59,88,152
    Staples--53,59
    Toolmarks in fabrics--58,85
    Toolmarks, Misc.--29,47,51
Miscellaneous Materials
    Asbestos--33
    Cement, Grease, Dirt, Rubber, Tape--59
    Glazing Putty--126
    Roofing Materials--17
    Safe Insulation--4,59,69

Biological Science Papers

Botanicals
    Marijuana, Hashish (and other plants bearing hair-like structures similar to marijuana)--4,14,17,28,29,59,80,102,109,133
    Paper (See "Questioned Documents")
    Phytoplankton: Diatoms (soil), Coccoliths (Chalk, glazing putty)--127,138
    Plant Fibers (See "Fibers,Natural")
    Pollen--89,138
    Wood--59,99,126,127
Feathers--25
Fibers, Natural (Animal hairs and plant fibers)
    36,58,59,85,91,110,111,126,127,138,145
Fingernails--66
Hair: Human; Normal, Abnormal, Diseased, Injured; Chemistry; Race, Sex, and Age Differentiation; Individualization
    4,9,10,13,15,16,19,27,36,38,56,57,69,73,81,91,100,102,106,112,122,124,125,126,127,131,138,142,144,145,151
Latent Prints--37
Serology
    Blood--31
    Semen (For an up-to-date bibliography consisting of 24 references on the SEM of spermatozoa, see volume 1 of Scanning, page 201, 1978)--35,36,101,102,107,108,126,129
    Sweat, Saliva, Urine--90,107
Teeth (Bite marks are listed under "Toolmarks")--16,52,115,116,132

Review, Applications, Technique, and Instrumentation Papers (Most of these papers either review the work of others or briefly discuss two or more applications of SEM to criminalistics problems)

4,14,16,17,24,26,29,34,40,47,48,59,61,62,65,66,68,69,99,102,103,104,109,111,116,118,119,126,127,128,129,138,140,141,143,146,147

SEM in the Courtroom

117,139

Bibliography

1. Andrasko, J. Identification of burnt matches by scanning electron microscopy. J. Forensic Sci., 23(4), 1978, 637-642. (SEM,E/EDS)

2. Andrasko, J. and A.C. Maehly. Detection of gunshot residues on hands by scanning electron microscopy. J. Forensic Sci., 22(2), 1977, 279-287. (SEM,E/EDS)

3. Andrasko, J. and A.C. Maehly. The discrimination between samples of window glass by combining physical and chemical techniques. J. Forensic Sci., 23(2), 1978, 250-262. (SEM,E/EDS)

4. Anonymous. Application of energy dispersive x-ray fluorescence analysis to forensic studies: An ORTEC Workshop. ORTEC Inc., 100 Midland Road, Oak Ridge, Tenn., 37830. (X/EDS)

5. Anonymous. Ink study suggests Vinland map fraud. Chemical & Eng. News, 52(6), 1974, 21. (SEM,E/EDS)

6. Anonymous. Application of SEM-EDS to the forensic investigation of paints (XAS 26). ORTEC Inc., 100 Midland Rd., Oak Ridge, Tenn., 37830, 1975. (SEM,E/EDS)

7. Bang, G. Analysis of tooth marks in a homicide case. ACTA Odontol. Scand., 34(1), 1976, 1-11. (SEM)

8. Barnard, J.A.W., D.E. Polk, and B.C. Giessen. Forensic identification of papers by elemental analysis using scanning electron microscopy. SEM/1975, IIT Research Institute, Chicago, IL, 60616, 519-527. (SEM,E/EDS)

9. Boehm, E. Untersuchungen an Kopfhaaren im Nahschussbereich mit dem Rasterelektronenmikroskop (Studies on the head hair after close range shooting with the scanning electron microscope). Arch. Kriminol., 149(3), 1972, 65-76. (SEM) (German)

10. Boehm, E. and H. Klingele. Einige rasterelektronenmikroskopische Befunde an menschlichen Haaren nach toedlicher Hochspannungsverletzung (SEM findings of human hair from a fatal electrocution case). Elektromedizin, 15, 1970, 141-155. (SEM) (German, English summary)

11. Boehm, E. and H. Klingele. Morphologische Untersuchungen an Pulverschmauchpartikeln verschiedener Munitionsarten (Morphological examination of powder smoke particles of several kinds of ammunition 'by SEM') Arch. Kriminol., 150(1&2), 1972, 31-43. (SEM) (German; English summary)

12. Bosch, K. and E. Boehm. Zum nachweis von Lackspuren auf Textilien (Identification of traces of varnish on textiles). Arch. Kriminol., 154(5&6), 1974, 157-160. (SEM)(German)

13. Bottoms, E., E. Wyatt, and S. Comaish. Progressive changes in cuticular pattern along the shafts of human hair as seen by scanning electron microscopy. Br. J. Dermatol., 86, 1972, 379-384. (SEM)

14. Bradford, L.W. and J. Devaney. Scanning electron microscopy applications in criminalistics. J. Forensic Sci., 15(1), 1970, 110-119. (SEM)

15. Brown, A.C., R.J. Gerdes, and J. Johnson. Scanning electron microscopy and electron probe analysis of congenital hair defects. SEM/1971, IIT Research Institute, Chicago, IL, 60616, 369-376. (SEM,E/EDS, E/WDS)

16. Brown, J.L. Applications of the Stereoscan to industrial materials. Proc. 2nd Annual Stereoscan Colloquium, 1969, pp. 1-3, 11-13. (SEM)

17. Brown, J.L. and J.W. Johnson. Electron microscopy and x-ray microanalysis in forensic science. J. Assoc. Official Anal. Chem., 56, 1973, 930-943. (SEM,E/EDS)

18. Butler, E.M. The examination of paint with the electron microprobe. J. Crim. Law, Criminology & Police Sci., 58, 1967, 596-602. (E/WDS)

19. Caputo, R. and B. Ceccarelli. Study of normal hair and of some malformations with a scanning electron microscope. Arch. Klin. Exp. Dermatol., 234, 1969, 242-249. (SEM) (German summary)

**20. Carlsson, L. and A.C. Maehly. Autoeinbruch oder Versicherungsbetrug? Identifizierund von Mikrospuren mit dem Rasterelektronenmikroskop. Arch. Kriminol, 157. 1976, 107-110. (German)

**21. Carlsson, L., A.C. Maehly, and E. Rudh. Toetungsversuch mit elektrischem Strom. Eine kriminaltechnische Untersuchung. Arch. Kriminol. 157, 1976, 30-36. (German)

22. Clancey, V.J. Comet G-ARCO: solving the riddle. New Scientist, 39, 1968, 533-537. (SEM,E/WDS)

*23. Collins, B. The evidential value of paint flakes, with special reference to electron probe microanalysis, in Third Australian National Symposium on the Forensic Sciences. The Australian Forensic Science Society, Sydney, 1973, pp. 25.1-25.15.

24. Corvin, I. and O. Johari. Forensic applications of the scanning electron microscope. Police, 16, 1971, 6-14. (SEM, E/EDS,E/WDS)

25. Davies, A. Micromorphology of feathers using the scanning electron microscope. J. Forensic Sci., 10, 1970, 165-174. (SEM)

26. Davies, G. Criminalistics--educational and scientific progress. Anal. Chem., 47, 1975, 318A-330A. (SEM,E/EDS)

27. Dawber, R. and S. Comaish. Scanning electron microscopy of normal and abnormal hair shafts. Arch. Dermatol., 101, 1970, 316-322. (SEM)

28. DeForest, P.R., C.V. Morton, and R.A. Henderson. Microscopic morphology of marijuana ash. J. Forensic Sci., 19(2), 1974, 372-378. (SEM)

29. Devaney, J.R. and L.W. Bradford. Applications of scanning electron microscopy to forensic science at Jet Propulsion Laboratory, 1969-1970. SEM/1971, IIT Research Institute, Chicago, IL, 60616, 561-568. (SEM)

30. Diederichs, R., M.J. Camp, A.E. Wilimovsky, M.A. Haas, and R.F. Dragen. Investigations into the adaptability of scanning electron microscopy and x-ray fluorescence spectroscopy to firearms related examinations. A.F.T.E. Journal, 63, 1974, 1-12. (SEM,E/EDS)

31. Dixon, T.R., A.V. Samudra, W.D. Stewart, and O. Johari. A scanning electron microscope study of dried blood. J. Forensic Sci., 21(4), 1976, 797. (SEM, E/EDS)

32. Dragen, R.F., A.E. Wilimovsky, R. Diederichs, M.J. Camp, and M.A. Haas. The application of the scanning electron microscope to the examination of toolmarks. A.F.T.E. Journal, 64, 1974, 1-16. (SEM)

33. Ferrell, Jr., R.E., G.G. Paulson, and C.W. Walker. Evaluation of an SEM-EDS method for identification of chrysotile. SEM/1975, IIT Research Institute, Chicago, IL, 60616, 537-546. (SEM,E/EDS)

34. Fox, R.H., F.R. McDaniel, and G.R. Howell. The criminalistics mission: a comment. Leg. Med. Annu., 1975, p. 103-113 (SEM,E/EDS)

35. Fujita, T., M. Miyoshi, and J. Tokunaga. Scanning and transmission electron microscopy of human ejaculate spermatozoa with special reference to their abnormal forms. Z.Zellforsch, 105, 1970, 483-497. (SEM)

36. Fujita, T., J. Tokunga, and H. Inoue. Atlas of Scanning Electron Microscopy in Medicine. Elsevier, N.Y., 1971, 54-59, 90-93. (SEM)

37. Garner, G.E., C.R. Fontan, and D.W. Hobson. Visualization of fingerprints in the scanning electron microscope. J. Forensic Sci. Soc., 15, 1975, 281-288. (SEM)

38. Gejvall, N.G. Superimposition plus SEM-comparison of hair cuticle for identification purpose. International J. Skeletal Research, 1, 1974, 99-103. (SEM)

39. Giessen, B.C., D.E. Polk, and J.A.W. Barnard. The applications of material science methods to forensic problems: Principles, serial number recovery and paper identification, in Educational and Scientific Progress in Forensic Science, G. Davies (Ed.), ACS Symposium series, 1975, p. 58-74. (SEM,E/EDS)

40. Goebel, R. Das Rasterelektronenmikroskop, ein neues Untersuchungsgeraet fuer die Kriminaltechnik (The SEM, a new device to be used for crime lab work). Kriminalistik, 9, 1973, 389-391. (SEM) (German)

41. Goebel, R. Examination of incandescent bulbs of motor vehicles after road accidents. SEM/1975, IIT Research Institute, Chicago, IL, 60616, 547-554. (SEM, E/EDS)

42. Green, L. and J.R. Moon. A microscopical examination of an alleged 1933 penny. J. Microscopy, 96(3), 1972, 381-384. (SEM/EDS)

43. Grove, C.A., G. Judd, and R. Horn. Examination of firing pin impressions by scanning electron microscopy. J. Forensic Sci., 17, 1972, 645-658. (SEM)

44. Grove, C.A., G. Judd, and R. Horn. Evaluation of SEM potential in the examination of shotgun and rifle firing pin impressions. J. Forensic Sci., 19, 1974, 441-447. (SEM)

45. Haag, L.C. Element profiles of automotive paint chips by x-ray fluorescence spectrometry. J. Forensic Sci. Soc., 16(3), 1977, 255-263. (X/EDS)

46. Haas, M.A., M.J. Camp, and R.F. Dragen. A comparative study of the applicability of the scanning electron microscope and the light microscope in the examination of vehicle light filaments. J. Forensic Sci., 20(1), 1975, 91-102. (SEM,E/EDS)

47. Hantsche, H. Zur Anwendung des Raster-Elektronenmikroskops in der Kriminaltechnik (Some applications of the scanning electron microscope in forensic work). Beitr. Elektronenmikroskop, Direktabb. Oberfl., 4(2), 1971, 641-656. (SEM,E/WDS)

48. Hantsche, H. Das Raster-Elektronenmikroskop: ein modernes Hilfsmittel der Kriminaltechnik, Part I. (SEM--a modern aid of criminalistics, Part I), VDI-Z 114(4), 1972, 221-224. (SEM) (German)

49. Hantsche, H. Das Raster-Elektronenmikroskop: ein modernes Hilfsmittel der Kriminaltechnik, Part II. (SEM--a modern aid of criminalistics, Part II), VDI-Z 114(5), 1972, 314-321. (SEM)(German)

50. Hantsche, H. and A. Schontag. Die Untersuchung von Lacksplittern mit dem Raster-Elektronenmikroskop als wichtiger Beitrag zu deren Identifizierung (Scanning electron microscopic examination and identification of paint chips). Arch. Kriminol., 147, 1971, 92-119.(SEM)(German)

51. Hantsche, H. and W. Schwarz. Das Raster-Elektronenmikroskop als Hilfsmittel zur Identifizierung von Werkzeugspuren (SEM as an aid in identifying toolmarks). Arch. Kriminol., 148, 1971, 24-32. (SEM) (German)

52. Harsanyi, L. Scanning electron microscopic investigation of thermal damage of the teeth. Acta Morphol. Acad. Sci. Hung., 23(4), 1975, 271-281. (SEM)

53. Horan, J.J. Scanning electron microscope in document examination, A.S.Q.D.E. Conference, 1975, 1-8(plus 3 plates). (SEM,E/EDS)

54. Howden, C.R., R.J. Dudley, and K.W. Smalldon. The non-destructive analysis of single layered household paints using energy dispersive x-ray fluorescence spectrometry. J. Forensic Sci. Soc., 17, 1977, 161-167. (X/EDS)

55. Howden, C.R., R.J. Dudley and K.W. Smalldon. The analysis of small glass fragments using energy dispersive x-ray fluorescence spectrometry, J. Forensic Sci.Soc., 18(1&2), 1978, 99-112. (X/EDS)

56. Ishizu, H., S. Hayakawa, H. Kaneko, H. Takata, Y. Funatsu, K. Ando, M. Seno, and T. Yoshino. Scanning electron microscopic studies on surface structure of hairs. Part 1. Scanning electron microscopy of the cuticles of animal hairs. Jap. J. Legal Med., 27(2), 1973, 113-122. (SEM) (Japanese; English summary)

57. Ishizu, H., S. Hayakawa, H. Kaneko, H. Takata, Y. Funatsu, K. Ando, M. Seno, and T. Yoshino. Scanning electron microscopic studies on surface structure of hairs. Part 2. Scanning electron microscopy of human hairs. Jap. J. Legal Med., 27(5), 1973, 337-345. (SEM) (Japanese; English summary)

58. Ishizu, H., Y. Doi, S. Hayakawa, H. Kaneko, H. Takata, Y. Funatsu, M. Seno, T. Yoshino, K. Nakanishi, and Y. Mikami. Relationship between impingement on textile fibers and causative instruments: a scanning electron microscopic observation. Jap. J. Legal Med., 28(2), 1974, 104-108. (SEM) (Japanese; English summary)

59. Judd, G., J. Sabo, and S. Ferriss. SEM applications under the FORSEM project. SEM/1975, IIT Research Institute, Chicago, IL, 60616, 487-492.(SEM,E/EDS)

60. Judd, G., J. Sabo, W. Hamilton, S. Ferriss, and R. Horn. SEM microstriation characterization of bullets and contaminant particle identification. J. Forensic Sci., 19, 1974, 798-811. (SEM,E/EDS)

61. Judd, G., R. Wilson, and H. Weiss. A topographical comparison imaging system for SEM applications. SEM/1973, IIT Research Institute, Chicago, IL, 60616, 167-172. (SEM)

62. Keeley, R.H. and M.C. Robeson. The routine use of SEM and electron probe microanalysis in forensic science. SEM/1975, IIT Research Institute, Chicago, IL, 60616, 479-486, 608. (SEM,E/EDS,E/WDS)

* 63. Keeley, R.H. Some applications of electron probe instruments in forensic science. Proc. Analytical Division of the Chemical Society, 13(6), 1976, 178-181.

* 64. Keeley, R.H. The economics of SEM and electron probe microanalysis in forensic science, Int. J. Legal Med., 131, 1976.

65. Kirk, P.L. Instrumentation in criminalistics. J. Forensic Sci., 10, 1970, 97-107. (SEM,E/WDS)

66. Korda, E.J., H.L. MacDonell, and J.P. Williams. Forensic applications of the scanning electron microscope. J. Crim. Law, Criminology, and Police Sci., 61, 1970, 453-458. (SEM)

67. Korda, E.J., L.H. Pruden, and J.P. Williams. Elemental analysis of glass and glass-ceramics with the scanning electron microscope. Ceramic Bulletin, 52(3), 1973, 279-284. (SEM,E/EDS)

68. Krishnan, S.S. An introduction to modern criminal investigation. Chas. C. Thomas, Springfield, IL., 1978, Chapt. 16, pp. 332-343. (SEM)

69. Leitner, J. Application of EMX-SM to forensic analysis. Sunland Laboratory Applications Report, Applied Research Laboratories, PO Box 129, Sunland, CA. 91040. (SEM,E/WDS)

70. MacDonell, H.L. and L.H. Pruden. Application of the scanning electron microscope to the examination of firearms markings. SEM/1971, IIT Research Institute, Chicago, IL, 60616, 569-576. (SEM)

71. MacQueen, H.R., G. Judd, and S. Ferriss. The application of scanning electron microscopy to the forensic evaluation of vehicular paint samples. J. Forensic Sci., 17, 1972, 659-667. (SEM)

** 72. Maehly, A.C. and L. Rammer. Zinkblende unter der Vorhaut: Tatortspur bei einem Sexualmord. Arch. Kriminol., 159, 1977, 139-143. (German)

73. Maes, D. and B.D. Pate. The spatial distribution of copper in individual human hairs. J. Forensic Sci., 21(1), 1976, 127-149. (SEM)

74. Matano, Y. Scanning electron microscope studies on spermatozoa of some mammals. J. Electron-Microscopy (Jap.), 20(3), 1971, 222. (SEM)

75. Matricardi, V.R. Matching of surfaces. SEM/1975, IIT Research Institute, Chicago, IL, 60616, 503-510.(SEM,E/EDS)

76. Matricardi, V.R., M.S. Clark, and F.S. DeRonja. The comparison of broken surfaces: a scanning electron microscopic study. J. Forensic Sci., 20(3), 1975, 507-523. (SEM,E/EDS,E/WDS)

77. Matricardi, V.R. and J.W. Kilty. Detection of gunshot residue particles from the hands of a shooter, J. Forensic Sci., 22(4), 1977, 725-738. (SEM,E/EDS)

78. Meinhold, R.H. and R.M. Sharp. The application of a multistylus recorder to the energy dispersive analysis of paint flakes in the scanning electron microscope. J. Forensic Sci., 23(2), 1978, 274-282. (E/EDS)

79. Messler, H.R. and W.R. Armstrong. Bullet residue as distinguished from powder pattern. J. Forensic Sci., 23(4), 1978, 687-692. (SEM,E/EDS)

80. Mitosinka, G.T., J.I. Thornton, and T.L. Hayes. The examination of cystolithic hairs of Cannabis and other plants by means of the scanning electron microscope. J. Forensic Sci. Soc., 12, 1972, 521-529. (SEM)

81. Muto, H. and I. Yoshioka. Scanning electron microscopic studies on hairs: Part 5. Scalp hair. J. AICHI Med. Univ. Ass., 2(4), 1974, 227-230 (plus 8 figs.). (SEM) (Japanese; English summary)

82. Nesbitt, R.S., J.E. Wessel, and P.F. Jones. Detection of gunshot residue by use of the scanning electron microscope. J. Forensic Sci., 21(3), 1976, 595-610. (SEM,E/EDS,E/WDS)

83. Oron, M., T. Arad, and A. Alkbir. SEM study of the effect of firing distances on bullet holes in clothing. SEM/1975, IIT Research Institute, Chicago, IL, 60616, 529-536. (SEM,E/WDS)

84. Oron, M. and V. Tamir. Development of SEM methods for solving forensic problems encountered in handwritten and printed documents. SEM/1974, IIT Research Institute, Chicago, IL, 60616, 207-214. (SEM)

85. Paplauskas, L. The Scanning electron microscope: a new way to examine holes in fabrics. J. Police Sci. and Adm., 1, 1973, 362-365. (SEM)

86. Parham, R.A. On the use of SEM/x-ray technology for identification of paper components. SEM/1975, IIT Research Institute, Chicago, IL, 60616, 511-518, 528. (SEM,E/EDS)

87. Polk, D.E., A.E. Attard, and B.C. Giessen. Forensic characterization of papers. II: Determination of batch differences by scanning electron microscopic elemental analysis of the inorganic components. J. Forensic Sci., 22(3), 1977, 524-533. (E/EDS)

88. Polk, D.E. and B.C. Giessen. A new serial number marking system applicable to firearms identification. J. Forensic Sci., 20(3), 1975, 501-506. (SEM)

89. Price, J.L. and T.L. Shirley. Identification of particulate air pollutants and contaminants by scanning electron microscopy and x-ray spectrometry. Proc. 29th EMSA, 1971, pp. 62-63. (SEM,E/WDS)

90. Quinton, E.P. Ultramicroanalysis of biological fluids with energy dispersive x-ray spectrometry. Micron, 9, 1978, 57-69. (SEM,E/EDS)

91. Rash, A.E. Identification of hairs and fibers--a review of the state of the art. Crime Laboratory-Division, State Laboratories Dept., Bismarck, ND, 58505. 1977, (SEM,E/EDS)

92. Reeve, V. and T. Keener. Programmed energy dispersive x-ray analysis of top coats of automotive paint. J. Forensic Sci. 21(4), 1976, 883-889. (X/EDS)

93. Reeve, V. and A. Lacis. An analytical approach to ultra-small physical evidence. J. Forensic Sci. Soc., 15, 1975, 235-244. (E/WDS)

94. Reeve, V., J. Mathiesen, and W. Fong. Elemental analysis by energy dispersive x-ray: a significant factor in the forensic analysis of glass. J. Forensic Sci., 21(2), 1976, 291-306. (X/EDS)

95. Rodgers, P.G., J.R.G. Jacob, P. Blais, and D.C. Harris. Scanning electron microscopy of forged dots on 1936 Canadian coins. J. Forensic Sci., 16(1), 1971, 92-102. (SEM,E/WDS)

96. Rodgers, P.G., J.R.G. Jacob, P. Blais, and D.C. Harris. The scanning electron microscopy of counterfeit coins. SEM/1971, IIT Research Institute, Chicago, IL, 60616, 585-586. (SEM)

97. Rodin, H.F. Application of modern science to the field of questioned document examination. A.S.Q.D.E. Conference, 1977, 1-8. (SEM,E/WDS)

98. Ryvarden, G. Problematik der Schussentfernungsbestimmung mit dem REM bei Distanzen uber 1m (Problems in determination of gunshot distance by means of the SEM at distances over 1 meter). Beitr. Gerichtl. Med., 34, 1976, 179-184. (SEM, E/EDS) (German, English summary)

99. Sabo, J., G. Judd, and S. Ferriss. Examples of SEM analyses in forensic evidence applications, in Educational and Scientific Progress in Forensic Science, G. Davies (Ed.), ACS Symposium series, 1975, p. 75-82. (SEM,E/EDS)

100. Schneider, V. Ueber die Untersuchung von Haaren mit dem Rasterelektronenmikroskop (On examination of hairs with SEM). Z. Rechtsmedizin, 71, 1972, 94-103. (SEM) (German, English summary)

101. Schneider, V. Rasterelektronenmikroskopische Untersuchung zur Alterung von Samenfaeden (SEM experimentation on the ageing of sperm). Beitr. Gerichtl. Med., 30, 1973, 394-399. (SEM) (German; English summary)

102. Schneider, V. Rasterelektronenmikroskopie in Gerichtsmedizin und Kriminalistik. Pressedienst Wissenschaft, 2, 1974, 101-110. (SEM,E/EDS) (German)

103. Schneider, V. Die Roentgenmikroanalyse am Rasterelektronenmikroskop: ein Routineverfahren im Rahmen Kriminaltechnischer Untersuchungen (X-ray microanalysis with a scanning electron microscope: a routine method in criminal investigations). Beitr. Gerichtl. Med., 35, 1977, 255-266. (SEM,E/EDS) (German; English summary)

104. Schneider, V. and H. Hantsche. Die energiedispersive Roentgenmikroanalyse: ein schnelles Verfahren zur Elementanalyze. Anwendungsbeispiele aus der Gerichtsmedizin und Kriminalistik (Energy dispersive x-ray microanalysis: a fast process for element analysis. Applications to legal medicine and criminalistics). Zacchia, Archivio di medicina legale, sociale e criminologica, 9, 1973, 1-20. (SEM,E/EDS) (German; English summary)

*105. Seta, S. Scanning electron microscope of some criminological specimens, Nat. Res. Inst. Police Sci. Report, 23(3), 1970, 1-11.

106. Seta, S., H. Kozuka, and T. Sudo. Criminal investigation of human head hair by means of SEM and electron probe microanalysis with special reference to age and sex differences. SEM/1975, IIT Research Institute, Chicago, IL, 60616, 579-588. (SEM,E/EDS)

107. Seta, S., H. Sato, K. Mamba, and S. Takeo. Application of the energy dispersive x-ray microanalyzer equipped with the scanning electron microscope to the criminal identification of body fluid stains. Jap. J. Legal Med., 30(5), 1976, 371-379. (E/EDS) (Japanese summary)

108. Seta, S., E. Suzuki, and M. Kitahama. Scanning electron microscopy and electron probe microanalysis of seminal stains. Jap. J. Legal Med., 26(6), 1972, 397-402. (SEM,E/WDS) (Japanese summary)

109. Siegesmund, K.A. and G.M. Hunter. Scanning electron microscopy of selected crime laboratory specimens. SEM/1971, IIT Research Institute, Chicago, IL, 60616, 577-584. (SEM)

110. Sikorski, J. Naturliche und Kunstliche technologische Fasein (Studies of fibrous structures). First International Conference on Electron Microscopy, Proc., Berlin, 1, 1958, 686-707. (SEM)

111. Sikorski, J., J.S. Moss, A. Hepworth, and T. Buckley. Specimen preparation for and dynamic experiments in the scanning electron microscope. Proc. 2nd annual Stereoscan colloquium, 1969, pp. 25-36. (SEM)

112. Singh, D.N., W.B. Greene, R.A. Schreiber, and G.R. Hennigar. Scanning electron microscopy of hair shafts from folic acid deficient mice. SEM/1973, IIT Research Institute, Chicago, IL, 60616, 646-649. (SEM)

113. Smale, D. The examination of paint flakes, glass and soils for forensic purposes, with special reference to electron probe microanalysis. J. Forensic Sci. 13, 1973, 5-15. (E/EDS)

114. Smalldon, K.W. The identification of acrylic fibers by polymer composition as determined by infrared spectroscopy and physical characteristics. J. Forensic Sci., 18, 1973, 69-81. (SEM)

115. Sognnaes, R.F. Post-mortem identification of Martin Bormann. Criminol, 9 (34), 1974, 3-28. (SEM)

116. Sognnaes, R.F. Forensic identifications aided by scanning electron microscopy of silicone-epoxy microreplicas of calcified and cornified structures. Proc. 33rd EMSA, 1975, pp. 678-679. (SEM)

117. Sognnaes, R.F. Dental science as evidence in court. Int. J. Forensic Dent., 3(9), 1976, 14-16. (SEM)

118. Sognnaes, R.F. Identification of bites and bullets. Harvard Dental Alumni Bull., 36(1), 1976, 12-19. (SEM)

119. Sognnaes, R.F. Talking Teeth. American Scientist, 64(4), 1976, 369-373.(SEM)

120. Sognnaes, R.F. Forensic stomatology (First of three parts). New England J. Med., 296(2), 1977, 79-85. (SEM)

121. Solheim, T. and T.I. Leidal. Scanning electron microscopy in the investigation of bite marks in foodstuffs. Forensic Sci., 6, 1975, 205-215. (SEM)

122. Sotonyi, P. Application of scanning electron microscopy in the medico-legal practice. Morph. es Ig. Orv. Szemle, 16 (2), 1976, 93-99. (SEM) (Hungarian; English summary)

123. Stone, I.C., V.J.M. DiMaio, and C.S. Petty. Gunshot wounds: visual and analytical procedures. J. Forensic Sci., 23(2), 1978, 361-367. (X/EDS)

124. Sudo, T. and S. Seta. Atlas of human hair by electron microscopy. Japan Human Hair Association, Tokyo, Japan, 1978, 125 p. (SEM)

125. Swift, J.A. Histological examination of keratin fibers by SEM. Applied Polymer Symp., 18, 1971, 185-192. (SEM)

126. Taylor, M.E. Forensic applications of scanning electron microscopy. SEM/1971, IIT Research Institute, Chicago, IL, 60616, 545-552. (SEM)

127. Taylor, M.E. Scanning electron microscopy in forensic science. J. Forensic Sci. Soc., 13, 1973, 269-280. (SEM,E/WDS, E/EDS)

128. Taylor, M.E. and C.G. van Essen. The uses of microanalysis in forensic science. Inst. Phys. Conf. Ser., 18, 1973, 234-237. (SEM,E/WDS,E/EDS)

129. Taylor, R.L., M.S. Taylor, and T.T. Noguchi. Applications of SEM/x-ray analysis at the Los Angeles County Forensic Science Center. SEM/1975, IIT Research Institute, Chicago, IL, 60616, 493-502. (SEM,E/EDS)

130. Thornton, J.I. and G.T. Mitosinka. Comparison of tungsten filaments by means of the scanning electron microscope. J. Forensic Sci. Soc., 11(3), 1971, 197-200. (SEM)

131. Tosti, A., S. Villardita, M.L. Fazzini, and R. Scalici. Contribution to the knowledge of dermatophytic invasion of hair. J. Investigative Dermatol., 55 (2), 1970, 123-134. (SEM)

132. Ueyama, M. The application of quantitative EPMA to the identification of the metal dental crowns of dead bodies. Shikwa Gakuho, 78(9), 1978, 1405-1416.

133. Ueyama, M., M. Io, and I. Nishioka. Observations on the leaves of Cannabis and its analogous plants by scanning electron microscopy and x-ray microanalyzer. Nat. Res. Inst. Police Sci. Report, 27(4), 1974, 208-222. (SEM,E/WDS) (Japanese; English summary)

134. Ueyama, M. and M. Io. Applications of scanning electron microscopy and wavelength dispersive x-ray analysis in forensic science. No. 2: analysis of arsenic in wheat flour. Nat. Res. Inst. Police Sci. Report. 28(2), 1975, 113-116. (SEM, E/WDS) (Japanese)

135. Ueyama, M. and M. Io. Applications of scanning electron microscopy and wavelength dispersive x-ray analysis in forensic science. No. 3: identification of imitation sapphire. Nat. Res. Inst. Police Sci. Report. 28(3), 1975, 188-189. (E/WDS) (Japanese)

136. Ueyama, M. and M. Io. Applications of scanning electron microscopy and wavelength dispersive x-ray analysis in forensic science. No. 4: identification of paint chip removed from a deceased individual. Nat. Res. Inst. Police Sci. Report, 28(3), 1975, 190-192. (E/WDS) (Japanese)

137. Vale, G.L., R.F. Sognnaes, G.N. Felando, and T.T. Noguchi. Unusual three-dimensional bite mark evidence in a homicide case. J. Forensic Sci., 21(3), 1976, 642-652. (SEM)

138. van Essen, C.G. Scanning electron microscopy in forensic science. Physics and Technology, 5, 1975, 234-243. (SEM, E/EDS)

139. van Essen, C.G. The validity of scanning electron microscopy evidence. SEM/1975, IIT Research Institute, Chicago, IL, 60616, 473-478. (SEM,E/EDS,E/WDS)

140. van Essen, C.G. and J.E. Morgan. A split-field comparison SEM. SEM/1973, IIT Research Institute, Chicago, IL, 60616, 159-166. (SEM)

141. Vasan, V.S., W.D. Stewart, and J.B. Wagner. X-ray analysis of forensic samples, using a scanning electron microscope. J. Assoc. Official Anal. Chem., 56, 1973, 1206-1222. (SEM,E/EDS,E/WDS)

142. Verhoeven, L.E. The advantages of the scanning electron microscope in the investigative studies of hair. J. Crim. Law, Criminology, and Police Sci., 63, 1972, 125-128. (SEM)

143. Visapaa, A. The use of x-ray fluorescence in crime investigation. Bull., Anal. Equip., N.V. Philips' Gloeilampenfabrieken, Eindhoven, The Netherlands. 1975, 4 p. (X/WDS)

144. Wasiutynski, A., J. Kwiatkowski, J. Wyhowski, and L. Suchcicki. The use of SEM for hair examination in the field of forensic science. Arch. Med. Sad. I Krym, 25(4), 1975, 317-322. (SEM) (Polish)

145. Wasiutynski, A., J. Kwiatkowski, J. Wyhowski, and L. Suchcicki. Scanning electron microscopy in medico-legal investigations of hair. Polish Med. Sci. Hist. Bull., 15(2), 1976, 151-159. (SEM)

146. Williams, R.L. An evaluation of the SEM with x-ray microanalyzer accessory for forensic work. SEM/1971, IIT Research Institute, Chicago, IL, 60616, 537-544. (SEM,E/EDS,E/WDS)

147. Williams, R.L. Forensic science-- the present and the future. Anal. Chem., 45, 1973, 1076A-1089A. (SEM,E/EDS,E/WDS)

148. Wilson, R. and G. Judd. The application of scanning electron microscopy and energy dispersive x-ray analysis to the examination of forensic paint samples. Adv. X-Ray Anal., 16, 1972, 19-26. (SEM, E/EDS)

149. Wilson, R., G. Judd, and S. Ferriss. Characterization of paint fragments by combined topographical and chemical electron optics techniques. J. Forensic Sci., 19, 1974, 363-371. (SEM,E/EDS)

150. Wolten, G.M., R.S. Nesbitt, A.R. Calloway, G.L. Loper, and P.F. Jones. Final report on particle analysis for gunshot residue detection. ATR-77 (7915)-3, The Aerospace Corporation, El Segundo, CA., 1977. (SEM,E/EDS)

151. Wyatt, E.H. and J.M. Riggott. Scanning electron microscopy of hair: observations on surface morphology with respect to site, sex, and age in man. Br. J. Dermatol., 96, 1977, 627-633. (SEM)

152. Young, S.G. The restoration of obliterated stamped serial numbers by ultrasonically induced cavitation in water. J. Forensic Sci., 19(4), 1974, 820-835. (SEM)

*153. Unknown. Criminal technical studies with the aid of a scanning electron microscope combined with a radiofluorescence analysis system. Polizei Technik Verkehr, 20(7), 1975, 266-270.

SCANNING ELECTRON MICROSCOPY/1979/II
SEM Inc., AMF O'Hare, IL 60666, USA

TEACHING SCANNING ELECTRON MICROSCOPY

G. R. Hooper, K. K. Baker and S. L. Flegler

Center for Electron Optics
Pesticide Research Center
Michigan State University
East Lansing, MI 48824

Abstract

Teaching SEM is an occupation that involves a very large number of individuals. It is not restricted to the academic world but is essential at all levels and across all disciplines that utilize the SEM. While some large classes exist in universities or special schools, much instruction is of the apprentice/master type. Currently formal classes use traditional lecture-laboratory formats to cover basic electron optical theory, instrument construction, preparatory skills and photographic-interpretive concepts. Educational methodology is both purposefully and inadvertently utilized in SEM teaching. Impetus for improving teaching is largely a personal matter since few disciplines actively encourage development of teaching methodology in SEM. Opportunities to exchange teaching technology are limited, particularly in regards to journal space. Despite these short-comings, increasing demand for instruction in SEM and the amenability of the subject matter to presentation via existing educational methods are creating a climate for change.

KEY WORDS: Teaching SEM

Introduction

R. T. Greer recently put teaching SEM into perspective, noting that, "An upper level student can be trained to operate an SEM and obtain photographic information in about an hour. So what is left to do in a lecture-lab sequence lasting approximately thirty sessions . . . ?"[1]

He, of course, offered some comments on what to do with that time and how to do it effectively.

The importance of teaching and learning in our discipline is incompletely understood. The importance of any subject within the scientific community may be partially evaluated by the volume of literature devoted to that subject. Teaching SEM is, by that measure, of little significance. Yet many of us teach and discussions about teaching at our meetings continue to draw interested participants.

A number of questions can be asked about our teaching activities. For example: Why do we teach SEM? Whom do we teach? How do we teach it? Is teaching SEM really different than teaching any other subjects?

In the following discussion and in the additional papers to be presented in this "Teaching Workshop" we hope to review current teaching philosophy and methodology as well as suggest some additional approaches to the subject.

Teaching Opportunities and Situations

There are just under 3,000 SEM's now in use in the U.S. and Canada*. The location and use of these instruments is extremely varied. While placement in industrial settings for use on non-biological materials may involve the largest single group of instruments, the academic area (colleges and universities) and medical situations also involve significant numbers of SEM's.

Users of each instrument vary from single, skilled operators to large numbers of inexperienced individuals with minimum supervision. Whatever the placement and use, there are clearly thousands of situations that require or allow teaching of SEM.

In many cases motivation for learning (hence teaching) is rather pragmatic--the skills of

*Phone survey, December 1978, of 7 major firms.

operating the instrument and preparing the materials are required by the job. In other cases the SEM may be a critical tool that enables a researcher, at pre- or post-graduate levels, to investigate a particular problem. For this latter group personal mastery of SEM skills both facilitates the research and provides a more accurate interpretive base than if the work was "hired out." Finally, many appear to be motivated to learn SEM simply for the pleasure of learning.

Teachers and their Approaches

In this discussion it is necessary to identify and expound upon the obvious. Perhaps by so doing we can become involved in a personal review of our own efforts and identify and correct areas of deficiency. A format of rhetorical questions will be used to present data on teachers and their methodology.

1. Who does SEM teaching?

Individuals who are formally appointed to teach SEM are a distinct minority. At many academic institutions teaching is done by professors on a special topic or individual student basis. At other institutions SEM teaching may be done as part of another, broader, course such as cell biology, pathology or general electron microscopy.

Much of the teaching of SEM operation is done by technicians in charge of instruments. This teaching occurs in industrial settings, at academic institutions and in hospitals.

A significant number of users have received most of their education in SEM use from manufacturer representatives and have learned the arts of specimen preparation via tutorials, from the literature, and by other on-the-job type experiences.

2. What is taught about SEM's?

There are several distinct SEM subject areas that are common to all the teaching situations. First, some instruction in electron optical theory is regarded as essential. This may take the form of very structured, electron optical information as a part of a complete course[1,2,3,4,5]. It may also be provided through attendance at specific papers, short courses, tutorials, manufacturer's seminars or similar special purpose events. Finally, optical theory may simply be learned piecemeal as it is needed for design or evaluation of experiments. This theoretical aspect of SEM may be considered the "science" of the discipline as opposed to the "art" aspects which are commented on below.

A second area of emphasis common to instruction concerns what has been termed SEM "anatomy."[4] The basic structure of SEM's, the location and function of controls and any specific physical peculiarities of the instrument to be used all require some explanation.

A third area, the preparative arts of SEM, is one of great diversification. Each discipline has its own dogmas and basic techniques. The biologists have specific problems centered around water removal and morphological integrity. Physical scientists may have little problem with maintaining gross morphology of their specimens but may require special preparative schemes to preserve crystal structure or facilitate x-ray analysis. Preparative methodology is broad in scope and dynamic in nature.

Teaching preparative arts may also require interjection of subject matter related to ancillary equipment, e.g., the critical point drying apparatus.

A final instructional area includes methods of data recording and interpretation of this data. Perhaps less time is spent in this subject area than in the others. In most instances some instruction is provided in the use of photographic materials. In addition, some may teach or advocate other recording methodology. Interpretation of recorded data is not generally accorded the time devoted to previously mentioned subjects. Some formal courses require students to submit term papers which may be used to evaluate and guide their research approach. In cases where students involve their own materials in the SEM training, they may make interpretations based upon principles taught elsewhere in their program.

The need for SEM students to be aware of the place of the instrument in science and the need for its users to be aware how they conduct and interpret research is implicit in most published accounts of teaching programs. It is perhaps most provocatively expressed in Hayes' 1975 discussion[2].

3. How do we teach and evaluate students in SEM?

Presenting subject matter and evaluating how well the material has been learned are almost inseparable. The presentation methods used may, in fact, depend upon how we plan to evaluate our students. For example, the traditional lecture approach is commonly evaluated by relating the performance of each individual to that of some larger group. This norm-referenced evaluation contrasts with other methods, which may be considered criterion referenced evaluations. This latter method requires establishment of distinct, more precisely defined, levels of achievement required of the students. One example of this approach is that termed "mastery" evaluation[6,7,8]. Where mastery learning concepts are applied, the use of programmed instructional methods is common. Eble[9] notes that in programmed instruction "a logical progression of sub-units leads to mastery of a total skill or body of knowledge." Programmed courses may take the form of self teaching approaches but are not necessarily the same thing.

Most SEM teachers do not consciously devise their courses based upon current educational theories or even with a knowledge of the psychology of learning. In fact, university or college level scientists seem almost disdainful of pure educators and educational jargon. Published methods for teaching electron microscopy or associated techniques are very rare (e.g., see Ref. 10).

The subject matter divisions previously cited can be divided roughly into the "science of SEM" where theoretical and background information are supplied, and "arts of SEM" where preparative skills and operational processes are involved. Both areas are being taught in most formal and informal instructional programs. There is a definite trend towards using the traditional lecture, assigned readings approach with norm-referenced evaluation for the "science" aspects of SEM. Criterion-referenced evaluation is often used for the "arts" aspects where laboratory demonstrations and programmed exercises are the general rule. Even in apprentice/master instruction of SEM arts, definite levels of expertise are required at each step before additional subject matter is introduced.

Teaching materials and their uses

Some common ground exists for teaching materials but the breadth of disciplines involved as well as some personal philosophies have resulted in considerable variability in the specimens used for laboratory exercises.

An actual SEM is considered essential to all teaching experiences. In apprentice/master situations, particularly in industrial settings, the student is expected to have extensive hands-on experience. A common problem encountered in teaching larger groups is that of how much, if any, access students should have to the instrument. Cost of SEM operation is an obvious factor as is a need, in many cases, for simultaneous access of trained researchers and untrained students to the same electron microscope. Some teachers advocate a demonstration approach to SEM experiences.[2] Others insist on at least some regular use of the microscope as essential to the learning experience[1,3,4,5]. Our own preference is for a minimum two hour weekly use of the SEM.

Students are exposed to a variety of photographs, films, video-tapes, slide/cassette lessons and other visual materials in both formal and informal teaching situations. No single textbook appears to satisfy a large number of instructors, but many books exist covering both the science and the art of SEM. At the graduate level (which includes most university or college courses) a strong preference for the use of current literature is evident. Local laboratory manuals and handouts exist but do not find a widespread market--presumably reflecting our highly individual approach to teaching. Perhaps the most often mentioned set of written materials used in teaching is the Proceedings of the IITRI/SEM - SEM INC. annual Symposia (e.g., see Ref. 1, 2, 5).

The materials used in laboratory exercises and/or demonstrations are varied not only by broad categories (biological, physical) but may depend upon local availability and suitability. Some materials, such as standards for x-ray analysis, are more universally acceptable. In much individual instruction and most of the courses surveyed, students are encouraged to learn techniques by applying them to their own research materials. In other courses a common set of laboratory materials is used to facilitate presentation of methodology and to allow for uniform evaluation of student performance.

Exchange of teaching information

The exchange of teaching philosophy and methodology is currently by "electronic grapevine." Personal conversations at meetings, letters, trial use of others' laboratory exercises, and similar exchanges constitute this most often used route to disseminate teaching methods or ideas. While some opportunity to publish papers devoted to teaching does exist (e.g., the 1975 and 1979 SEM symposia) none of our current journals solicits or readily accepts such papers. Journals associated with SEM are basically research oriented.

Journals interested in education and teaching might accept such papers but are not widely read by scanning electron microscopists. Furthermore, most of us are unsure of our ability to write in the acceptable form and jargon of these journals--to say nothing of our real lack of training in education.

The lack of interest exhibited by major publishing firms in laboratory manuals is known by all who have aspired to publish such material. This is presumably a fact motivated by economics rather than by the value of the laboratory program.

The existence of the Educational Committee of EMSA provides an additional avenue of exchange for all areas of electron microscopy. Their sponsorship of an "Educational Resource Center" has made possible exchange of at least some visual aids and may yet provide a method for greater interchange of educational methods.

Discussion

The foregoing remarks have been based, where possible, on published material or personal correspondence. The element of personal opinion, however, is evident and perhaps essential. In teaching SEM, and indeed in teaching any subject, the contribution of each unique personality is a major factor. Few would want each approach to be the same since the goals of teachers and students are so variable. Even in such cases where preparation for skill certification at local or national levels exists, it appears possible to use a variety of approaches to reach the same final level of expertise.

Consideration of increased instructional demand, high costs of individualized instruction, a need for more and more preparative skills, and related constraints do provide a powerful impetus for examination of our teaching programs. Several matters in particular seem pertinent to review or to comment upon.

1. The view that teaching is a second rate occupation.

George Bernard Shaw purportedly quipped, "He who can, does; he who cannot, teaches." It has also been noted that "college professors outside of the field of education, who venerate research and who prepare other college professors, shy away from research on human interaction and learning."[9]

While all involved in teaching are not college professors, few have escaped this basic, negative, attitude towards teaching. The lack of journal space in our profession, where educational matters may be expressed, is one form of documentation of this prevalent attitude.

2. The view that teaching methods can be used only by professional educators.

While it is perhaps heresy to say so, it appears that a strong element of common sense runs through educational philosophy and methodology. The fact that teaching and learning are basically enjoyable is evident if we review our own motivations and experiences. Much of the abundant literature on learning and teaching can be applied to passing on skills and concepts to others without regard for subject matter. We can and do have access to training in educational methods and communicative skills at almost any locality.

3. The view that programmed learning is inferior to close personal instruction.

One recent comparison[8] of traditional vs. mastery-taught subject matter indicated that 81% of the mastery taught students scored higher than those non-mastery taught. Our experience in a program that utilized programmed instruction and mastery evaluation of preparative SEM technique and instrument operation indicates a marked improvement in final product since we entered the program. It appears that at least instruction in the "arts" of SEM may be aided by incorporation of such methods.

Acceptance of teaching as a productive use of time and incorporation of existing educational methodology may well improve both quantity and quality of our students and leave the instructors with more time and resources to pursue other matters.

References

1. Greer, R. T. 1975. A multidisciplinary university course in scanning electron microscopy. In, O. Johari and I. Corvin (Eds), SEM/1975/IIT Research Institute, Chicago, IL, pp. 609-616.
2. Hayes, T. L. 1975. The scanning electron microscope. Subject and servant in teaching. In, O. Johari and I. Corvin (Eds), SEM/1975/IIT Research Institute, Chicago, IL, pp. 594-600.
3. Goldstein, J. I. 1975. The teaching of scanning electron microscopy and electron probe microanalysis. In, O. Johari and I. Corvin (Eds), SEM/1975/IIT Research Institute, Chicago, IL, pp. 601-608.
4. Rue, J. W. 1975. An approach to teaching scanning electron microscopy to multidisciplinary students. In, O. Johari and I. Corvin (Eds), SEM/1975/IIT Research Institute, Chicago, IL, pp. 625-630.
5. Thurston, E. L. 1975. Teaching scanning electron microscopy: Optimizing the compromises. In, O. Johari and I. Corvin (Eds), SEM/1975/IIT Research Institute, Chicago, IL, pp. 617-620.
6. Block, J. H. 1971. Mastery learning: Theory and Practice. Holt, Rinehart and Winston, New York. Chapters 1, 4, 5, 6. Part one.
7. Davis, H. R., L. T. Alexander and S. L. Yelon. 1974. Learning system design. McGraw Hill, New York. Chapters 1, 4, 6, 12.
8. Schom, C. B., L. Haley, O. P. Kamra, A. Hicks and R. W. Lee. 1978. Teaching techniques: Do results differ between approaches. AIBS Education Review. 7(1):1-5.
9. Eble, K. E. 1977. The craft of teaching.
Jossey-Bass, San Francisco. Chapters 1, 2, 8, 11, 14.
10. Hopkirk, J. W. and R. M. Glaeser. 1972. Electron microtomy training: A self-contained learning package. In, C. J. Arceneaux (Ed), Proc. EMSA Thirtieth Ann. Mtg. pp. 698-699.

DISCUSSION WITH REVIEWERS

Reviewer I: Do you feel that workshops such as the one in which this paper is being presented are essential to the establishment of communications between all of the pure and interdisciplinary sciences which use SEM?
Authors: It is difficult to say they are "essential." Our personal observations do support the idea that they are extremely useful to at least a large number of people in the SEM field.

Reviewer I: Is it the opinion of the authors that SEM teachers should become more involved in current educational theories and the psychology of learning? And if so, how will this help instruction concerning what has been termed SEM "anatomy"?
Authors: Some SEM teachers are so skilled that they might profit very little from such exposure. We believe that most of us would find such involvement very worthwhile. Instruction in any form requires communication and transmission of skills and concepts. An instructor with some idea of learning processes should be able to improve his/her communication skills.

Reviewer I: It is our experience that approximately 20% of students taking our courses are doing so purely for credit. How would these people fit into your overall view of SEM teaching?
Authors: The type of student you describe does exist. Course prerequisites (graduate standing, instructor permission, etc.) can eliminate some. The others could perhaps be viewed as our contribution to public relations.

Reviewer IV: Who takes SEM instruction and how should courses be designed to meet the needs of these students?
Authors: Students come from all the subject areas that utilize the SEM. They may be regularly enrolled college students at undergraduate or graduate levels or may be involved in specific instruction outside of the academic world. Professional programs (e.g., medical technology) are increasingly involving their trainees in SEM instruction.

Because of significant differences in preparation of biological and non-biological materials separate laboratory sections are being offered by many instructors to fit needs of students from these separate areas.

REVIEWERS: I.   R. L. McConville
*   IV.   G. Judd*

SCANNING ELECTRON MICROSCOPY/1979/II
SEM Inc., AMF O'Hare, IL 60666, USA

USE OF A TABLE-TOP SEM IN UNDERGRADUATE TEACHING

D. B. Williams and J. I. Goldstein

Department of Metallurgy and Materials Engineering
Lehigh University
Bethlehem, PA 18015

Abstract

Recent years have seen the advent of table-top SEMs, which not only offer many of the capabilities of advanced research-oriented SEMs, but also are simple to operate, portable and most importantly, relatively inexpensive. These instruments are ideal teaching aids and enable the introduction of SEM concepts and practice to the student even at the freshman level. In this way the SEM is seen as no more complex to operate and understand than a conventional high grade optical microscope, and finds a natural place in a wide range of undergraduate courses. Early hands-on experience facilitates the introduction of the instrument into many courses and at all times the undergraduate interfaces directly with the SEM without the need for back-up support from a technician. The portable nature, combined with video output facilities means that demonstrations within the laboratory and the lecture theater are routine.

KEY WORDS: Table-top SEM, Teaching SEM

Introduction

The scanning electron microscope (SEM) is a versatile tool for high resolution study of surfaces, and as such is indispensable in both biological and materials research. However, over the last five years we have noticed an ever increasing interest in using the SEM as a teaching aid in undergraduate laboratories. In this situation its main function is to complement and extend the role customarily played by the conventional bench optical microscope, and therefore the performance requirements and versatility of the instrument are not as stringent as for most research applications. However, standard SEMs are relatively complex and research oriented in design, and thus we have found that, although undergraduates are occasionally able to obtain "hands-on" experience on such an instrument, this generally requires the presence and overall supervision of a skilled technician, graduate student or faculty member. Therefore it is becoming increasingly difficult to expand teaching time on the SEM (which is now as much an integral part of many of our undergraduate courses as a conventional optical microscope) because of the personal supervision required, and also because of the time limitations and conflict with research projects for which use of the instrument is imperative. This situation has been alleviated by the recent development of relatively inexpensive, simple to operate table-top SEMs which are available from several manufacturers. This paper deals with the way such an instrument has been used to augment undergraduate teaching at Lehigh University.

Table-top SEMs are compact, capable of installation in a relatively small area, and powered directly from the mains supply. Controls are minimized, and although this obviously reduces the instrumental flexibility and sophisticated specimen handling capability compared with a standard research SEM (for example, in-situ X-ray microanalysis is not routinely possible), essential performance is not sacrificed. In particular, useful magnification up to 20,000X and a guaranteed resolution of ~ 10 nm is usually attainable. Operation of the instrument is simple and the only controls to be mastered are:

1. specimen change mechanism (manual vacuum control with safety interlock for possible misoperation)

2. high tension supply (on/off) (choice of up to two kV's)
3. filament emission control
4. variable spot size control (first condenser lens)
5. contrast and brightness controls for the image on the CRT monitor screen
6. image focus control (objective lens)
7. specimen manipulation controls-- traverse, tilt, height, and rotation
8. recording of the image using a polaroid camera.

Thus in essence the microscope requires only slightly more operational skill than a conventional optical microscope, yet offers all the specific advantages of a high grade SEM, namely high resolution, large depth of field and the ability to take stereo-pair images. Furthermore, the instrument is optically pre-aligned and also accepts large specimens which means that instrument and specimen preparation time is minimal. A final, but not insubstantial, point in favor of such an instrument is that the cost is less than that of a sophisticated optical microscope despite offering $\sim$ 500X improvement in resolution and $\sim$ 100X increase in depth of field.

Use in Undergraduate Laboratories

The problems of incorporating the use of such an instrument into existing undergraduate courses are the same in any particular area of study. First, the initial teaching of hands-on use of the instrument has to be carried out as soon as possible in the undergraduate curriculum. This expands the number of courses in which applications are possible. Secondly, instructors and lecturers have to be made aware of the possible applications and ease of use of an instrument which often they have not encountered before, or of which they are inherently wary.

We will discuss the way in which these problems have been overcome at Lehigh, with particular reference to the Department of Metallurgy and Materials Engineering. Obviously similar methods can be applied in any field.

Initial Introduction

If possible, freshman engineers are given the opportunities to see the instrument in action in the introductory engineering course. This is considered an essential step since the whole concept of electron microscopy is often not encountered at high school, or even worse, is dealt with superficially, hence introducing a notion of complexity to the instrument which is detrimental to the learning process. In the sophomore year, at the same time that such basic techniques as optical microscopy are taught in some depth, the hands-on use of the SEM is similarly taught (this is only possible because of the simplicity of operation and robust nature of the table-top SEM). Thus the student is encouraged to think of the SEM as an instrument complementary to the optical microscope, possessing certain specific advantages, already mentioned, as well as several drawbacks (e.g., contrast limitations, need for a conducting surface, vacuum system, water cooling requirements, etc.).

At this stage the undergraduate is as qualified to use the SEM as he is an optical microscope. The specific applications to which this knowledge is put in the Department of Metallurgy and Materials Engineering are listed below, to give some idea of the wide range of usage possible.

Materials Applications of the Table-top SEM

Met. 204 - Nonmetallic materials of construction - fracture and structural properties of materials such as concrete, glass-fiber reinforced plastics.

Met. 207 - Electronic and crystal structure- the SEM is used for the study of surfaces and their defects, such as grain boundary configurations. Crystal morphology and its relationship to atomic symmetry can be clearly demonstrated.

Met. 208 - Phase diagrams and transformations - the fine scale microstructural details of particular phase transformations (e.g., fine pearlite) and surface relief effects accompanying specific transformations (e.g., martensite) are observed in the SEM.

Met. 218 - Mechanical behavior of materials- the deformation and fracture of materials is particularly suited to SEM study. Fractography, the direct study of fracture surface characteristics, is an integral part of the course.

Met. 307 - Senior projects in structure and behavior of materials - e.g., extensive use of metallography for tool steel studies and fractography in e.g., temper embrittlement studies.

Met. 311 - Metallic materials for structures- the fracture analysis of structural steels, aluminum alloys, and welding materials (for non-metallurgical majors) requires in-situ SEM studies of engineering failure modes.

Met. 334 - Electron metallography - hands-on use of the research SEM is optimized by obtaining a basic understanding of SEM techniques from initial study of the theoretical and operational concepts using the table-top SEM. Laboratories on the table-top SEM are completed in the students' free time, scheduled simply by a sign-up list. In almost all cases no instructor is needed, after an introductory talk. Special contrast detection and X-ray analysis are taught using the research SEM.

The problems of specimen preparation are minimal, since in most cases the specimen is conducting. As yet, use of coating equipment to enable non-conducting specimens to be examined has not been taught, and such specimens have been provided in the coated state. However, over three years the student will interact extensively with the instrument and become well acquainted with its versatility, capabilities and limitations.

Overall Teaching Approach

There is no attempt at the freshman level to introduce electron optical concepts to the students. Their level of interaction with the instrument is limited to some supervised hands-on time, looking at typical materials samples such as a fracture surface or an integrated circuit. Contrast interpretation is not required at this stage since the main emphasis is on the specific advantages of SEM in terms of magnification and depth of field characteristics.

Since more time is available in the sophomore level laboratory-oriented course, it is possible to start drawing analogies between the secondary electron image and reflected light microscopy. However, in this course the prime objective is to teach the operational concepts without complicating the situation with detailed electron optical theory. It is considered that their state of knowledge is sufficient to enable the instrument to be of substantial use to them in their sophomore and junior (200-level) courses.

It is only in the senior year (Met. 334) that a concerted attempt is made to teach basic beam/specimen interactions and contrast interpretation as well as some finer points of operational technique such as astigmatism correction and high resolution imaging. With this level of knowledge it is now possible for the student to pursue an individual project (e.g., Met. 307). The electron metallography course, Met. 334, incorporates some laboratory sessions where the student is asked to bring his own specimen, and obtain satisfactory images (in terms of corrected astigmatism, optimum spot size, depth of field, etc.).

Examples of the levels of teaching are illustrated by Figs. 1-3 which are examples from the courses described above and taken by the students themselves. Figure 1 shows a low magnification integrated circuit, used in freshman level demonstrations of the wide range of materials that can be studied by the instrument. In Fig. 2 the depth of field and high magnification capability are illustrated by the ductile fracture surface of a mild steel. This is typical of the problems encountered in several metallurgical courses at sophomore, junior and senior levels involving mechanical behavior and fracture. Finally, a high resolution image of ZnO copy paper shows the limits of operational capability, reached by students who have taken the electron metallography course.

Operation of the Table-Top SEM Laboratory

Although there are specific advantages in terms of portability, as discussed below, the table-top SEM is usually kept in one room to facilitate student access, photographic processing and maintenance. Because of the high turn-around rate of specimens in class time, the instrument is pumped continuously to ensure that specimen pump-down time is minimized. The operation of the table-top SEM is made substantially easier by our experience gained in using a research oriented SEM for several years, and obviously all the back-up facilities such as specimen preparation and maintenance equipment already exist. However this is clearly not a prerequisite and the instrument maintenance is straightforward, requiring little sophisticated training.

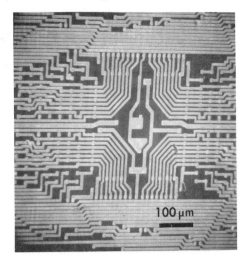

Fig. 1: Low magnification image of integrated circuit.

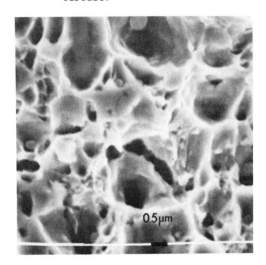

Fig. 2: Fracture surface of a mild steel specimen.

Fig. 3: High resolution specimen of ZnO particles.

The portability of the instrument is made possible by the fact that it is self-contained, apart from the mechanical vacuum pump and water recirculator, which can be moved on an adjacent trolley. It is a two-man operation to move the whole instrument from one laboratory to another, or into the lecture theater, where set-up time is less than half an hour.

For demonstration purposes, a TV scan output is essential, although this is considered secondary to the main hands-on teaching. The ability to interface with existing video display systems can also be most useful.

Finally, during times when the instrument is not in demand for teaching, it offers a very reasonable back-up research capability enabling prior specimen inspection before investigation on the research SEM.

Conclusions

The development of a low cost, robust, simple-to-operate table-top SEM can greatly increase the use of SEM as an aid to teaching. At Lehigh University, the instrument is used as a logical extension to the application of conventional optical microscopy. Student hands-on interaction is introduced at an early stage in the undergraduate curriculum enabling routine use of the instrument in many courses. The portable nature of the instrument permits it to be used directly in the lecture theater as well as in several laboratories, and the built-in TV scan simplifies the demonstration of particular points to large numbers of students. The instrument is considered essential to most of the courses in which it is used, and as such, familiarizes undergraduates with developing microstructural techniques at the earliest opportunity.

Acknowledgement

The authors wish to thank Mr. S. F. Baumann, a Lehigh graduate student, and principal table-top SEM instructor, for valuable discussions.

Discussion with Reviewers

R. G. Fernquist: We have experienced the use of a table-top SEM in an industrial complex where about a hundred users have been trained to use this SEM for materials science investigations. Some of our users (in many cases they have been exposed to SEM use in undergraduate and post-graduate courses) fail to grasp fundamental beam/sample interactions and only view the SEM as an instrument with which "to take pretty pictures." With this in mind, is there a danger of misleading students on the nature of electron microscopy in that oversimplified instrumentation might tend to delude users in thinking that 1) they understand scanning electron microscopy and have reached "expert" status after a few hours operation and, 2) they understand how to intelligently interpret the results? With nonconductive samples and powders rather than metallurgical samples, more problems are experienced, i.e., contamination of final apertures, obtaining required resolution, charging artifacts, contrast

effects from local voltage fields, and so on. Are all students exposed to a course in theory where such pitfalls are taught?
Authors: We agree that undoubtedly there is danger in oversimplifying the interpretive aspects of SEM, and with this in mind, we take care at the sophomore and junior levels to limit the specimens to those where there is little problem in interpretation, and concentrate mainly on the operational simplicity. Those students with a serious interest in the subject would take Met. 334 where all the pitfalls described above are dealt with in detail. We cannot of course guarantee that all students will not retain some delusions of their degree of expertness. However this argument could equally well apply to the casual user of the optical microscope, who thinks that he is an accomplished metallographer although lacking an understanding of the artefacts and imaging techniques commonly encountered with this technique.

R. G. Fernquist and B. Jakstys: How much mainten-ance time is required for your instrument and what is the nature of the maintenance? How does down-time of the microscope affect the productivity of your class?
Authors: Routine maintenance is minimal, primarily involving replacement of filaments, and cleaning of the column liner and its final aperture about once a week. This can be accommodated easily in the normal teaching schedule, but a substantial down-time period due to breakdown would be quite catastrophic in terms of disruption of the teaching. However the table-top SEMs appear to be reliable enough that this problem should not be faced more than a couple of times each year.

B. Jakstys: On the average, how large are your undergraduate classes in which a table-top SEM is utilized?
Authors: Courses requiring individual "hands on" experience average 12 students; other courses not requiring "hands on" vary from small groups of 3 - 5 to larger demonstrations where as many as 20 - 30 may be shown the basics of instrument operation.

B. Jakstys: Do you find that one table-top SEM is sufficient to accommodate efficiently and conveniently all of your students when "hands on" the microscope is necessary?
Authors: At present with about 12 students required to perform weekly exercises on the SEM and perhaps 5 or 6 other users one SEM is certainly sufficient. With 45 normal scheduling hours weekly, twenty hands-on students should easily be accommodated full time. Most of the courses that use the instrument do so for a relatively short time each semester. Only a couple of courses require continual hands-on use throughout the semester.

For additional discussion see page 230.

SCANNING ELECTRON MICROSCOPY/1979/II
SEM Inc., AMF O'Hare, IL 60666, USA

COORDINATED, MULTIDISCIPLINE COURSES IN SEM

K. S. Howard, M. D. Socolofsky, R. L. Chapman*, and C. H. Moore, Jr.**

Department of Microbiology
*Department of Botany
**Department of Geology
Louisiana State University
Baton Rouge, LA 70803

Abstract

Prior to 1977, electron microscopy courses on the LSU campus had developed in three separate departments (viz. Engineering, Geology and Microbiology) and included six graduate level lectures and laboratories. Early in 1977, an interest was expressed in designing a series of courses which would unify campus offerings in electron microscopy. These courses involved faculty members from the departments of Botany, Geology, Mechanical Engineering, Microbiology, and Zoology and Physiology and serve graduate students from the entire campus.

The lecture course, Electron Microscopy, emphasizes the theoretical aspects of both transmission and scanning electron microscopy as well as energy dispersive and wavelength dispersive x-ray analysis as applied to EM. Laboratory courses are not mandatory but a student may take one or more depending upon his interests. The laboratories emphasize understanding of equipment and specimen preparation techniques, as well as research application and presentation of results. Laboratories are offered at three separate sites, utilizing one TEM and two SEM's. The comprehensive nature of this new program has already resulted in particular benefits including (1) increased quality of the lecture and laboratory presentation based upon a fuller utilization of the talents, equipment, and audio-visual resources found on the campus, (2) a resulting increase in the breadth, depth, and quality of graduate study in EM, (3) fewer operational problems in the microscope facilities due to increased user competence, (4) the provision of a setting to stimulate and strengthen collaborative research and educational efforts in EM, and (5) a reduction in the duplication of faculty effort.

KEY WORDS: Teaching SEM

Introduction

In recent years, new approaches have been instituted in developing electron microscopy courses at Louisiana State University. Prior to 1977, both transmission and scanning electron microscopy facilities and courses grew independently under the direction of several faculty members in different departments. The facilities and their electron microscopy courses mirrored the unique research requirements and financial resources associated with each laboratory. Steps were undertaken to coordinate campus offerings in electron microscopy by combining the elements common to transmission and scanning electron microscopy and x-ray analysis. Such coordination provided for a more efficient use of existing staff and facilities, and broadened the electron microscopic information available to graduate-level students. This resulting group of electron microscopy courses continues to undergo evaluation and change, although a basic pattern has been established. We are finding numerous foreseen and unforeseen benefits in coordination of teaching efforts on the campus. Our experiences and course sequence may serve as a pattern of development for other campuses which lack a centralized EM facility.

Historical background

EM facilities developed in different departments and schools on the LSU campus evolving to a present series of six research laboratories, three of which house both research and teaching instruments (Table 1). During this period of development, six EM courses were offered in various disciplines.

Impetus for change

The first step toward the present coordination of general electron microscopy efforts occurred in 1976 with the establishment of a Life Sciences Electron Microscopy Discussion Group. This group of faculty, staff and graduate students realized a need to increase communication among electron microscopists in order to enhance research and teaching and to improve EM facilities on the campus. Their efforts in 1976 resulted in the establishment of the Life Sciences interdepartmental SEM facility with the purchase of a scanning electron microscope. This facility augmented the SEM-EDS system which had been available to biologists in the Department of Geology.

The group, although based primarily in the biological sciences, encouraged multidiscipline electron microscopy communication across the campus and discerned problems common to the several electron microscopy facilities. A major problem at LSU was the lack of a plan for coordinated development and efficient utilization of the various EM laboratories. As communication among LSU electron microscopists increased, it became possible to consider unification of teaching efforts using electron optics as a common theme.

TABLE 1.   Individual EM Laboratories and Equipment

Department	Instrumentation
Geology	Scanning electron microscope (SEM) with energy dispersive spectrometer (EDS)
Life Sciences (5 separate biology departments)	SEM
Geology	3-spectrometer, wavelength dispersive microprobe (WDS)
Microbiology	Transmission electron microscope (TEM)
Life Sciences (3 separate biology departments)	TEM
Mechanical Engineering	TEM

New Course Formats

Early in 1977, a coordinated series of multidiscipline electron microscopy courses was designed. The objectives of this effort were to: 1) utilize the existing facilities, funding and staff on the campus; 2) realize more fully the talents and resources at LSU; 3) broaden the electron microscopy information available to the students; 4) provide a format to encourage interdisciplinary exchange among faculty, staff and students; 5) and to reduce duplication in teaching efforts. The course formats were designed basically as follows:
1. Electron Microscopy: A two semester hour lecture course taught one time per year. This course covers fundamental theoretical information of TEM, SEM and x-ray analytical instrumentation. The major impetus of the course is to serve as background for a series of laboratory courses.
2. Three laboratory courses are currently offered:
    a. Scanning Electron Microscopy Laboratory: Biological Materials
    b. Scanning Electron Microscopy Laboratory: Geological Materials
    c. Transmission Electron Microscopy Laboratory: Biological Materials
3. Two proposed laboratory courses:
    a. Transmission Electron Microscopy Laboratory: Engineering Materials
    b. Wavelength Dispersive X-ray Analysis: Geological Materials

The basic philosophy behind this series of courses is that the similarity of theory serves as a building block for greater comprehension of each branch of electron microscopy. Students are exposed to TEM, SEM and x-ray analysis in the lecture course regardless of their ultimate choice of emphasis in the laboratories. The multidiscipline lecture and the scanning electron microscopy laboratories are of particular interest to this symposium and will be further commented upon below.

General objectives of the SEM sequence

The objectives of the courses involved in teaching scanning electron microscopy include attainment of comprehension of instrument theory, operational competency, and ability to apply instrumental capabilities to research problems. The lecture and individual laboratories are designed to work together to achieve these goals.

Specifically, the 2 semester hour lecture course, Electron Microscopy, provides background for the laboratories and also serves to expose students who do not enroll in the SEM labs to SEM theory. Four-fifths of the course is devoted to teaching theory of electron optics, vacuum systems, beam-specimen interactions, and the design of TEM, SEM, and x-ray analytical instruments (both wavelength and energy dispersive), emphasizing the similarities and uniqueness of each. One-fifth of the course is used to expose students to specimen preparation techniques, the theory of photography, and applications of specific instruments.

Scanning Electron Microscopy Laboratory: Biological Materials (2 SEM hrs)

Course objectives are: (1) to develop mastery of scanning electron microscope operation and the preparation of biological specimens for SEM observation. This includes the understanding and use of ancillary equipment and techniques such as critical point drying, sputter coating, and photography. (2) to create an ability to apply the instrument to research problems by requiring students to develop a research project. The results of research projects are prepared for formal presentation to increase the ability of the student to communicate research results.

Scanning Electron Microscopy Laboratory: Geological Materials (2 SEM hrs)

Course objectives are the same as those of the biological SEM laboratory with two major exceptions. (1) The emphasis is divided between the scanning electron microscope and energy dispersive x-ray analysis on the SEM. (2) Specimen preparation requirements stress the use of non-biological materials.

Approaches to obtaining desired objectives

The electron microscopy courses at LSU serve students with biological, physical, and earth science backgrounds. Departmental cross-listing of the EM courses (Table 2) allows students to receive credit in their home departments. Students may take none or several of the laboratories since they are taught at various times throughout the year.

Each course is sponsored by an individual department, which in part has been determined by historical precedence. Each has developed the SEM facility involved in the course and each provides staff and the primary financial support for the course. Interdepartmental assistance, although informal and limited, has been most effective in providing personnel, as indicated in Table 2. The major responsibility for each course has been placed on a specific course coordinator in the sponsoring department who is assisted by additional staff (Table 2). The function of the course coordinator is to provide not only teaching services in his area of expertise, but also to maintain continuity in the lecture and laboratories and to implement student performance evaluation procedures.

Materials used

Audio-visual aids for the lecture course include videotapes, slides, overhead transparencies, and written handouts (outlines, diagrams, references, etc.). Videotaped tours of TEM, SEM, SEM-EDS and WDS-probe laboratories were made on the campus and are presented at three points during the semester. Each tape is thirty to thirty-five minutes in length and was designed to introduce students to the instrumentation for the first time. The students responded favorably to the tapes as a means of exposing them to laboratory instruments in the absence of time-consuming visits by large numbers of people into small laboratory facilities. Handouts detailing the information in the videotapes are provided to the students so that their exposure will be a completely visual experience. One suggestion by students has been to use videotapes not only as a tool for introducing new subjects in the lecture, but as a summary of information at the end of each series of lectures on a given topic. Slides are frequently used throughout the lecture to present information. In addition to the personal collections of each of the instructors, one source of visual information which has been most helpful has been the filmstrip series Electron Microscopy: Principles & Practices by R. E. Crang and J. A. Ward available through the Electron Microscopy Society of America. These filmstrips, delivered as strips or as individual slides, utilize instrumentation common to the LSU facilities and have been especially useful in providing specimen preparation information in a short, concise manner. Additional information is provided to the students through written handouts in the form of lecture outlines and accessory explanatory

TABLE 2. Organizational Information for SEM Courses

Course:	Electron Microscopy Lecture	SEM Lab: Biological	SEM Lab: Geological
Facilities:		Life Sciences Hitachi S-500 SEM	Geology JEOL-U2 SEM and ORTEC EDS
Departmental Cross-listing:	MBIO*, BOTY*, ZOOL*, ME*, GEOL*	MBIO, BOTY, ZOOL	GEOL
Course coordinator: Additional staff:	MBIO 4 faculty-BOTY, GEOL, 2 ZOOL 2 associates-MBIO, BOTY	BOTY 1 associate-ZOOL	GEOL 2 teaching assistants-GEOL
Semesters taught:	Fall	Summer and Spring	Fall
Students/class:	1977-30 1978-25	1978-5 & 7 1979-6 & 6	1977-7 1978-6

Student performance evaluation:	3 equivalent written examinations	SEM check out P/F
		Specimen set
		  photographs 10%
		Display photo-
		  graph 5%
		Oral Presentation 5%
		Written, illustrated
		  report 80%
			Research report 50%
			Poster session 10%
			Homework 10%
			Observed
			  instrument
			  performance 30%

*MBIO: Microbiology   *ME: Mechanical Engineering
*BOTY: Botany   *GEOL: Geology
*ZOOL: Zoology and Physiology

material.

One problem as yet unsolved for the lecture course in electron microscopy deals with the lack of an appropriate text. Although several excellent volumes are available on each of the topics covered by the course, no single, affordable text over the broad range of information we cover is presently on the market. As a result, students utilize a variety of resource materials held on reserve in the LSU library. Texts that have been particularly useful include the following:

Agar, A. W., R. H. Alderson, and D. Chescoe. 1974. Principles and Practice of Electron Microscope Operation, in Practical Methods in Electron Microscopy, Vol. 2. (A. M. Glauert, ed.) North-Holland, Amsterdam.

Goldstein, J. I., H. Yakowitz, D. E. Newbury, E. Lifshin, J. W. Colby, and J. R. Coleman. 1975. Practical Scanning Electron Microscopy, Electron and Ion Microprobe Analysis (J. I. Goldstein and H. Yakowitz, ed.'s) Plenum Press, New York.

Hayat, M. A. 1970. Principles and Techniques of Electron Microscopy: Biological Applications, Vol. 1. Van Nostrand Reinhold Company, New York.

Meek, G. 1976. Practical Electron Microscopy for Biologists. Wiley, New York.

SEM/1968-1977, IIT Research Institute, Chicago, IL, 60616, and 1978, SEM. Inc., AMF O'Hare, IL, 60666.

Although an appropriate text for the lecture portion is still being sought, the manuscript Scanning Electron Microscopy: A student's handbook which has been developed at LSU (M. T. Postek, K. S. Howard, A. Johnson and K. L. McMichael) will soon be available for use in the laboratories.

Conclusions

The coordinated, multidiscipline courses in scanning electron microscopy at Louisiana State University are now into their second academic year. During their brief history they have evolved from their original design as the methods of presentation become more efficient. Specifically, we have observed that care must be taken in a multidisciplinary lecture course to select lecture material which is representative of the various disciplines but yet is stimulating to the students regardless of their professional interest. As the lecture and laboratories continue to develop, further emphasis will be placed on the utilization of visual teaching aids for the classroom and on the preparation of autotutorial aids to reinforce laboratory information.

In the past two years, the coordination of electron microscopy teaching efforts has resulted in numerous benefits to both students and staff with a few disadvantages. The following general critique of the courses was arrived at through evaluations by both students and participating staff.

Disadvantages:
1. Continuity in lecture is difficult to maintain in a team-taught series of courses. The flow of information from one topic to the next can be seriously disrupted by a change in teaching style from one instructor to another. The major responsibility of the course coordinator has been to reduce this problem by attending lectures and serving as a line of communication between the members of the teaching staff.
2. It has been necessary to teach several of the labs subsequent to the lecture course rather than concurrently. This negates the "hands on" experience with equipment which is a vital dimension in furthering student understanding of equipment design and operation. However, the use of videotaped laboratory tours and equipment demonstration has helped considerably in bridging this gap between lecture and laboratory experience.
3. The use of individual laboratories and numerous personnel requires a strong commitment on the part of all involved to the premises upon which this series of courses has been built and the goals which they seek to attain.
Advantages:
1. Team teaching has allowed the utilization of the specific areas of expertise of the instructors for the greatest benefit of the students. Each of the instructors has brought to the classroom an intimate knowledge of his subject attained through research experience with the equipment and techniques under discussion.
2. The multidiscipline nature of the introductory lecture course has resulted in broader appreciation of electron microscopy on the part of the students. This will broaden their understanding of currently developing EM concepts and their ability to apply various branches of EM to their own research.
3. Exposure to theory in the lecture course prior to entrance to the laboratory has resulted in superior performance of the students while in labs. Their comprehension of equipment and techniques as well as their ability to apply the capabilities of scanning electron microscopy to a research project is definitely enhanced by pre-exposure to the theory of electron optics and instrument design.
4. The utilization of more than one complete SEM facility for teaching laboratories permits handling approximately twenty students per year without overburdening either facility with teaching. Each of the facilities involved maintains research responsibilities within the department in which it is housed.
5. The autonomy of the individual SEM facilities has been preserved allowing for the retention of specific direction and research application of each.
6. Increased communication among the LSU electron microscopists has increased the awareness of the capabilities and inadequacies of each facility. This has resulted in increased cooperative utilization and cooperative effort toward improvement.
7. Increased communication among faculty, staff and students has also resulted in the stimulation of collaborative, multidiscipline research.

8. The higher level of competence with which operators and trainees now utilize the SEM and EDS laboratories has led to improved equipment maintenance.

9. The redundancy of teaching efforts on the LSU campus has been significantly reduced. The talents, existing resources and funding applied to electron microscopy courses have remained at their previous levels, but are now used to present a broader base of information to each student.

10. The success experienced by the campus electron microscopists in securing and maintaining new instruments has served as an impetus for undertaking other interdisciplinary ventures on campus.

In summary, the impact of improved communication and coordination between electron microscopists, and scanning electron microscopists in particular, has been multifold. It has provided broader exposure and a more unified theoretical understanding of electron microscopy in the absence of a single, centralized EM facility for the campus. However, the advantages of autonomous electron microscopy laboratories are not lost. Freedom for unique research application is maintained and accessibility for clientele at various locales is greater than would be anticipated with a centralized EM facility. The commitment to coordination has allowed LSU to enjoy the best attributes of both the individual and the centralized EM facility.

DISCUSSIONS WITH REVIEWERS

B. P. Jakstys: A two-semester hour lecture course encompassing TEM, SEM, x-ray analysis and specimen preparation seems too short a time period to cover all the topics adequately. Does this mean that your students are given heavy library assignments or do you feel that the information which the students receive during the lecture alone is sufficient?

Authors: The students do receive outside library assignments during the lecture course. In addition, specific information is given in laboratories for the equipment and techniques in use. The lecture course alone is not sufficient to cover all the necessary materials.

C. Y. Shih: On the average, how many hours per student were spent on the microscopes per semester?

Authors: In the biological SEM laboratory, each student spends 8 hours with an instructor, a minimum of 20 hours prior to checking out on the instrument in order to master preparatory techniques, and between 10 to 30 hours to complete a project, averaging 48 hours of beam time/student/semester. Students in the SEM-EDS laboratory spend approximately 40 hours/student/semester.

C. Y. Shih: What is the estimated cost per student for the SEM laboratory? How is it financed?

Authors: The estimated cost is $200.00/student. This includes equipment maintenance, expendable supplies and technician and instructor salaries. It does not include professorial salaries. The laboratories are financed through the budgets of the departments which support the two SEM facilities.

B. P. Jakstys: It appears to me that one-fifth of a two-semester hour EM lecture course devoted to specimen preparation is exceptionally short. Does it mean you also have "mini-lectures" in the laboratory?

Authors: Yes, the laboratories provide the majority of the specimen preparation information in the form of "mini-lectures," written handouts, and suggested references. The specimen preparation portion of the lecture is kept as an overview to expose all students to the basic principles and requirements of specimen preparation, while detailed, specific information is given in each laboratory course.

B. P. Jakstys: Do you find that students who have not taken the EM lecture course concurrently with the appropriate laboratories retain sufficient information to efficiently move into an EM laboratory without an extensive review of the lecture material? Certainly in this case the course coordinator does not bridge the discontinuity, or does he?

Authors: The large majority of students have no problem retaining sufficient information from the lecture. Any exceptions to this are handled by the individual course coordinator or his assisting staff.

C. Y. Shih: How many students failed in your SEM laboratory during past two years?

Authors: No students have failed the SEM laboratories. However, we have had one student withdraw from the lab, two students withdraw from the lecture and one student fail the lecture which excluded him from the laboratory.

Additional discussion with reviewers of the paper "Use of a Table-top SEM in Undergraduate Teaching", by D.B. Williams et al continued from page 224.

R. G. Fernquist and B. Jakstys: Although the number of controls are minimized on a table-top SEM, have you found problems in teaching operators to correct for astigmatism, to align the beam, and in understanding spot size and other instrument parameters? How do you decide when a student is capable of operating the instrument alone?
Authors: Every operator is given a demonstration and a short but complete "cookbook" set of instructions which, if followed, will allow smooth operation of the instrument, even for the total novice. A copy of the instructions is also attached to the instrument for ready reference. The instrument needs no prior alignment and control panel knobs are clearly labeled and numbered in agreement with the instruction sheets. Thus any literate person should be a worthy operator. Although an understanding of why a particular operation should be performed is obviously not gained until the completion of the major SEM course, this does not prevent the student from ob-obtaining reasonable images which can extend his comprehension of the particular problem under investigation.

G. Judd: How does the prior exposure to the table-top SEM affect the student's ability to work with the research level equipment?
Authors: A student does not get hands-on use on the research instrument until well into the electron metallography course (Met. 334) in his senior year. In contrast to our previous experience where only a research instrument was available, our current students who have used the table-top model have come to accept the basic simplicity of the instrument and, hence, are not as intimidated by the vast number of push buttons and dials on the research SEM. The student can then concentrate on working with the specimen rather than working the machine. He understands that relatively few of the controls are adjusted in routine operation.

Additional discussion with reviewers of the paper "Application of the University of Illinois Computer based PLATO IV...", by R.L. McConville et al, continued from page 234.

J. Murphy: If a campus has a Plato terminal, is there access to the programs you describe? If so, please identify access routes (group and name).
Authors: The Hitachi HU 11 vacuum siumlator and a short introduction to the Plato keyset can be accessed through our system router by use of the signol: user/uicem. The JEOL U3 SEM lesson is, at present, solely for "in house" teaching.

P. A. Thrower: Are programs for other SEMs planned to accommodate other users already connected to the network? If so, which ones?
Authors: No new lessons are planned for other SEMs at this date but as student needs and equipment change, so will our lessons.

APPLICATION OF THE UNIVERSITY OF ILLINOIS COMPUTER BASED PLATO IV
NETWORK TO THE TEACHING OF ELECTRON MICROSCOPY

R. L. McConville and F. Scheltens

Center for Electron Microscopy
University of Illinois
Urbana, IL 61801

Abstract

The University of Illinois' Plato IV system
is a unique visual educational aid. It allows
students the opportunity both to work individually, at their own speed, and to respond to
questions and answers via a keyboard.

The Plato IV network has been used to augment present methods of teaching in the E.M.
courses offered by the Center for Electron
Microscopy at the University of Illinois. At the
present time, the major use of the system has
been in the field of instruction on E.M. instrument control functions. Prior to "hands on" any
specific microscope, students are expected to
complete an introductory course from one of the
Plato lessons available. Encouraging results
have already been obtained using the Plato system
as a simulator. Here students are familiarized
with vacuum systems and asked to correctly valve
the microscope to attain high vacuum. Students
misvalving the E.M. simulator are informed of the
error, the trouble caused by such an action and
estimate of the cost of repairs. Currently
available and proposed SEM Plato lessons will be
described in detail.

KEY WORDS: Teaching SEM

Introduction

Technical Details of the Plato IV System

Plato,[1] which is a registered service mark
of the University of Illinois, was begun in 1960
with one terminal. The first course exclusively
taught on the Plato system occurred in the Fall
of 1965. Today there are approximately 950 terminals located at about 140 sites: twenty-six
sites on the Campus of the University of Illinois
in Urbana-Champaign, nine elementary schools, two
high schools, six community colleges, twenty
government-supported installations, twenty-seven
medical sites (fifteen at colleges or universities), thirty colleges and universities and about
twenty miscellaneous business and industrial installations. The terminals are scattered from
San Diego, California, to Boston, Massachusetts,
and from Madison, Wisconsin, to Wichita Falls,
Texas.[2]

At the present time, users have access to
about 5000 hours of instructional material in
over 140 subject areas. A partial list of programs available includes engineering, pure
sciences, foreign languages, social sciences, all
levels of mathematics, natural sciences, business, medical and health sciences, English and
psychology. Simulation studies of social, biological or technological systems abound. Over
the entire Plato network, three million contact
hours of use were logged between July 1, 1974,
and April 1, 1977.

The heart of the Plato computer system is
not one but two computers: a CDC Cyber 73-24, and
a CDC 6500. The permanent storage medium for
computer programs and lessons is a disk system
from which programs and data are transferred to
and from an Extended Core Storage (ECS) of
2,000,000 words. The programs and data in the
ECS are then swapped in and out of the central
memory for processing by the central processors
with a transfer rate several hundred times faster
than that of disks or drums and an access time of
a thousand times shorter than the retrieval time
from disks to the central computer memory.

The access time of the ECS is actually less
than five microseconds and the transfer rate is
ten million, 60 bit words per second. ECS makes
possible fractional second response times (125
milliseconds) to hundreds of graphic terminals at

relatively low cost. The electronic swapping memory (ECS) is much less expensive than a large central memory.

The communications between the computer system and the terminals is via an interface unit (CIU) onto a standard television channel or microwave link, and thence to site controllers, each handling thirty-two terminals. The information from there is distributed to each terminal via a 1200, 4800 or 9600 bit per second telephone line. Terminal-to-computer communications is via telephone line. A line concentrator at the site controller multiplexes the information from thirty-two terminals onto the single telephone line and transmits it to the computer. Each terminal is guaranteed an average of 2.3 inputs per terminal per second, which is five times more than the measured average.

Each Plato terminal, like the one shown in Fig. 1, is equipped with a plasma display panel[3] which is a flat panel also doubling as a projector screen for images stored on a microfiche. Such a display device is well suited to low cost mass production and enables economies in design and operation of the telecommunications and computer software of the system.

A plasma panel, which was invented specifically for use with the Plato system, has a grid of 512 x 512 transparent conductors at whose intersections a neon discharge can be ignited or extinguished under computer control. The writing speed for text on such a display is 180 characters per second, and graphs and line drawings are displayed at a rate of sixty connected lines per second. The panel has inherent memory and does not require repetitive replotting. Selective writing and erasing of parts of the display is possible without disturbing the rest of the display.

The principal Plato input device is a keyset which has a standard set of typewriter keys plus special keys performing different functions in different lessons.

The terminal contains character and line generators and interface electronics to a 1200 bit per second telephone line. The character generator includes 126 standard alphanumerics and 126 characters alterable under author control. The latter can be loaded from the central computer with character patterns for special alphabets or pieces of pictures which may then be presented on the plasma display panel.

Plato Lessons for Students of Electron Microscopy

The Center for Electron Microscopy is developing a repertoire of Plato lessons to augment existing teaching methods. Our teaching commitment covers instruction to approximately 100 students per year of both the physical and biological science disciplines. At the present time seven courses are offered covering SEM, TEM and special techniques. The latter involves individual instruction to qualified microscopists who may be unfamiliar with our equipment, and thus require minimal training which may take only a few hours of "hands on" microscope time.

The SEM Plato course due to be introduced in the Spring semester of 1979 will instruct students in the operation of a JEOL U3 Scanning Electron Microscope fitted with an energy

dispersive x-ray detector.

The intent of the lesson is to take students from no knowledge of the JEOL U3 SEM up to a point where their understanding allows them to both operate the microscope successfully and obtain high quality electron micrographs. The lesson begins with a display of the table of contents (Fig. 2). The student may choose to study any section or even go directly to a final quiz that will test his understanding of the material.

The table of contents contains nine headings: 1) Overall view of the microscope, 2) The control panel, 3) The specimen stage, 4) The electron optical column, 5) The energy dispersive x-ray detection system, 6) The vacuum system, 7) Operating procedures, 8) Fault finding, 9) Final quiz.

Additional flexibility in the use of the computer as an aid to teaching SEM comes from the ability to have the computer control the projection of color photographs on the plasma panel. In all sections of the lesson, full use of this facility has been made. Students are given general views and close-up views to study; since the plasma panel is unaffected by the projection of color photographs, simultaneous dynamic instruction can be presented. Various microscope function control knobs are displayed and pointed out and if he wishes he can branch from the main lesson to learn more about any particular control and then return later to the same part of the main lesson to continue his study. The random access slide selector back projects images onto the plasma panel from a 4" x 4" microfiche with a 256 image capacity. This versatility allows all important parts of the microscope to be displayed and cross referenced.

The successful application of program segments, such as those described above, to a real teaching situation involves the integration of practice problems and associated help sequence with the development of the necessary theoretical framework to assist the students in understanding the material. For example, in the section on operating procedures, a student who is either unfamiliar with or who has wrongly answered a Plato question on filament saturation can branch to a separate program segment which explains the theoretical considerations in obtaining the correct electron emission from the gun. On returning to the main program, the student is reassessed and allowed to continue only after correctly answering the given questions.

Plato itself will play a significant role in gathering both subjective and objective information on lesson functioning to aid in evaluation and revision of our SEM lessons. The computer provides an on-line message facility for communications between students and instructors, and also administers opinion questionnaires. Objective data on the performance of lessons are automatically collected as students work through lessons. Data including both tabulations of student response judged "incorrect" and statistical analyses of students' behavior with the lesson will suggest revisions which would improve the teaching effectiveness of the lesson.

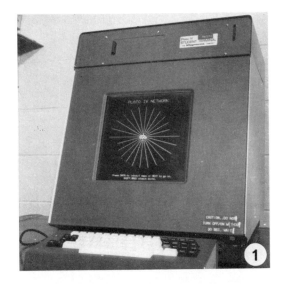

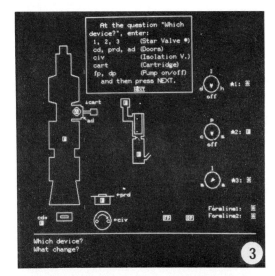

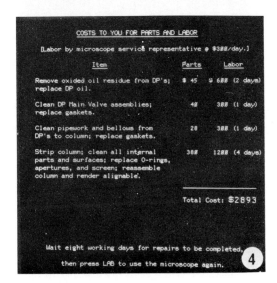

Fig. 1 One of the 950 Plato terminals. This one is located at the Center for Electron Microscopy.
Fig. 2 Introduction to Scanning Electron Microscopy. The student must choose which section he wishes to study by depressing the appropriate key.
Note: Unfortunately, the static photographs shown here do not fully convey the dynamic nature of the changing displays seen by the student.

Fig. 3 The Hitachi HU 11 series microscope shown on the Plato simulator. The student must give the computer the next instruction to proceed through the lesson.
Fig. 4 After a serious misvalving which could cause damage to the vacuum system, the student is presented with this bill itemizing the cost of repairs.

The Use of Plato as a Simulator*

In this Plato lesson students can, at their own pace, learn the necessary techniques to operate a vacuum system. When he feels competent enough, he can then proceed to the simulator. Using the simulator, he can instruct the computer to operate the vacuum valves in order to pump the system down. The simulator, which is shown in Fig. 3, currently describes a specific TEM, the Hitachi HU 11 series. However, only those students with a sound fundamental understanding of vacuum systems can be successful in its operation. The student who can master this lesson

*Program written by Daniel Davis

should have no problem operating the actual microscope valves and, with a little extra thought, successfully operate, for example, vacuum evaporators or high vacuum desiccators. Student responses are examined by the computer and judged accordingly. For example, misvalving the microscope by admitting air into the fore pump lines while the diffusion pumps are pumping the microscope and being "backed" by the rotary pumps will lead to a catastrophic failure of the vacuum system. The student is immediately informed of the error and an itemized account of charges for repairs appears on the plasma panel, as illustrated in Fig. 4. Such an irreversible mistake cannot be rectified and the student has no alternative but to either end his lesson or return to the very beginning. Less serious errors, such as leaving the specimen chamber prepump valve open while introducing the specimen into the column, are also noted by the computer. Although in many microscopes this may not be considered a very serious operating error, if this misvalving sequence is performed regularly, vacuum performance would certainly suffer. Therefore, as well as informing the student about the consequences of this action, the computer keeps count of the number of times this was performed. Eventually the computer will inform the student of a vacuum breakdown owing to his continued negligence.

Summary

The ability of the Plato system to support a one-to-one dialogue with the student offers the possibility of presenting the material in unique ways that make the student an active participant in an effective learning situation. Well-designed lesson material tends to be highly interactive and requires frequent inputs from each student in the form of answers to questions, predictions of the outcome of some experiment, parameters to be used in simulated experiments, and interpretation of a set of data or facts. In addition, since the understanding of the subject matter that a student has before starting a given program varies enormously, it is desirable to structure the program to accommodate students who just need a brief review, as well as those who are learning the material for the first time. One simple way that is employed in our Plato lessons to provide the students with flexibility in the way they study and use a given program, is to provide an index to the lesson, which allows easy access to any section.

Students' opinions about the use of Plato in our currently available lessons have been enthusiastic. Most students said that Plato helped them learn the fundamentals and expressed the opinion that Plato lessons should be a permanent part of the course.

There is no doubt that the best method of learning how to use an electron microscope and obtain high quality electron micrographs is to actually use the microscope. However, under present problems of staffing, time availability on microscopes, and financial considerations, it is not always possible to give our students free, unlimited use of our facilities. We have found that the initial ideas introduced through our varying teaching methods, significantly accelerated the learning process and made the material more interesting. In the future, the Center hopes to significantly expand its list of Plato lessons and fully integrate them in its teaching effort.

Acknowledgements

The authors wish to express their appreciation to Celia Davis of the Computer-Based Educational Research Laboratory and other members of the staff of the Plato system for their help in providing much useful information used in the preparation of this manuscript.

References

1. Lyman, E. R.; Plato Highlights; March 1978; CERL Plato Publication, University of Illinois.
2. Bitzer, D. L.; The Wide World of Computer Based Education; Advances in Computers, 15, 1976, 239-283, Morris Rubinoff and Marshall Yovits, eds., Academic Press, New York-San Francisco-London.
3. Johnson, R. L., D. L. Bitzer, and H. G. Slottow, "The Device Characteristics of the Plasma Display Element," IEEE Transactions on Electron Devices, 18-9, (Sept.), 1971, 642-749.

Discussion with Reviewers

J. Murphy: Assuming a knowledge of EM, what other background is needed to write programs for Plato (e.g., program languages, etc.)?
G. Hooper: How much instruction or background relative to computers is needed by students using Plato systems?
Authors: A working knowledge of "TUTOR", the computer language used by the Plato computer-based education system, is necessary to author Plato lessons. TUTOR, an education oriented language, differs from other business and scientific languages in that it enables the programmer to tailor the lesson to the student using it. This is accomplished through a unique series of commands in the TUTOR language which judge the student's response to a question and route him accordingly through the lesson. The Plato question-answer-judging capability is sufficiently advanced in itself so that the student may fully concentrate on his response and need not worry about computer oriented problems of any type.

Since Plato is an education based system, any preliminary instruction which is necessary, be it a short introduction to the keyset for the student or extensive training in the use of the TUTOR language for someone authoring a lesson, can be accomplished internally in the system.

G. Hooper: Do you have any cost figures for the development and operation of lessons such as you described?
P. A. Thrower: What costs are involved in being connected to the Plato network?
Authors: Due to the fact that our catalog of Plato EM lessons is limited to those described above, no cost analysis was performed. However, the College of Veterinary Medicine, which makes extensive use of the Plato system calculated their lessons development cost to be $828.00 per hour of instructional material. When this cost is distributed over a five year period, it represents a cost of $1.93 per student contact hour. The initial costs for connecting to the Plato system are:

Plato IV Terminal	$ 5035.00
Communications System	288.00
Total	$ 5323.00

For additional discussion see page 230.

SCANNING ELECTRON MICROSCOPY/1979/II
SEM Inc., AMF O'Hare, IL 60666, USA

CRITICAL POINT DRYING OF BIOLOGICAL SPECIMENS

W. J. Humphreys and W. G. Henk*

Department of Zoology and
*The Central Electron Microscopy Laboratory
The University of Georgia
Athens, GA 30602

Abstract

For teaching purposes a 20 minute (16 mm) film with an explanatory sound track was made concerning the topic of critical point drying of biological specimens. A critical point drying apparatus of our own design was used; the high pressure specimen chamber was cylindrical and horizontally mounted and it had a glass window at each end. Light was shined into the back window and events occurring in the specimen chamber were photographed through the front window. The film shows and discusses the equipment required, the proper way to fill the bomb with the transitional fluid (CO_2), disappearance of the meniscus of liquid CO_2 when the temperature of the bomb exceeds the critical temperature and recondensation of the gas to a liquid when the temperature falls below the critical temperature. A typical critical point drying run is shown. The specimen is kept in view during the exchange of intermediate fluid for transitional fluid, during the heating of the bomb and during the gradual release of the gaseous CO_2 at a temperature above its critical temperature. Special methods for handling single cell suspensions to be critical point dried are demonstrated. These include a method for drying cells directly onto specimen studs, a method for attaching cells to coverslip fragments by the use of polylysine and a method for filtering cells onto Nuclepore filters. These methods facilitate the handling of cells during critical point drying.

KEY WORDS: Teaching SEM, Biological Specimen Preparation

Introduction

The critical point method for drying biological specimens has become such an important and widely used technique for preparing specimens for scanning electron microscopy that an effective way for teaching the theory and the proper application of this technique is now an important requirement for any training program in scanning electron microscopy. The theory can be taught adequately by lecturing, but proper application of the technique is taught best by a step-by-step demonstration of the method.

A novice acquires considerably better understanding of the method if the high pressure chamber used for demonstrating critical point drying has a window that permits observation of the events taking place inside the chamber during the period when the specimen is being processed. Personal demonstrations of the method are effective only with very small groups of observers. Moreover the possibility of breaking the bomb's window under high pressure is a danger that cannot be dismissed. Even when a protective plastic shield is bolted over the window there is danger since the plastic itself could also shatter at the time the glass window breaks.

The use of a series of color transparency slides showing the sequence of steps and resulting events involved in critical point drying eliminates the danger to the audience that could result from a broken window. Color slides also can show some details more clearly than they could be seen by individuals in a small group seeing a demonstration. But events such as the disappearance of the meniscus of the transitional fluid when the bomb is heated to a temperature above the critical temperature and the dramatic recondensation of the gas into a liquid when the bomb's temperature goes below the critical temperature are dynamic events that should be witnessed and understood. Such events cannot be adequately depicted by a series of still color slides.

We decided, therefore, for purposes of teaching, to make a color film with an explanatory spoken text used for a sound track. The film was intended to demonstrate the basic physical phenomena involved in critical point drying and to show how specimens are dried by using the technique. A critical point bomb with a glass window at both ends of a horizontally mounted cylindrical chamber was used. Light was shined into the back window and events occurring in the chamber were photographed through the front window. The following written text is a transcription of the text used for the sound track.

Transcription of the Spoken Text

Equipment Required. (Film title)

Equipment required for critical point drying includes a tank of liquid CO_2 securely fastened in place. The tank must have a siphon tube which extends beneath the liquid CO_2 reaching nearly to the bottom of the tank so that it delivers the CO_2 as a liquid. The liquid CO_2 is transferred to the chamber of the high pressure vessel through high pressure tubing. The bomb is equipped with a pressure gauge designed to withstand pressures up to 3000 pounds. It also has a temperature gauge, an inlet valve, a drain valve and a vent valve.

If a window is used, it must be held in a position between resilient gaskets to make the seal. For safety a thick shield of clear tough plastic at least one-half inch thick must be securely fixed in a position between the observer and the window in case the window should shatter under the high pressures used.

The temperature of the chamber is regulated by a thermostatically controlled water bath heater-recirculator unit. Water is heated to a predetermined maximum temperature while being pumped through tubing to and through a water jacket that surrounds the chamber. The water is returned through the tubing to the water bath and recirculated.

Before the bomb is filled with liquid CO_2, the temperature of the bomb should be dropped to a temperature lower than that of the liquid CO_2 in the cylinder. To accomplish this cold water is circulated through the jacket until the temperature of the chamber is lowered to 19°C.

Filling the Bomb with Liquid CO_2. (Film title)

The temperature of the bomb is lowered to 19°C and all valves are closed. We will now see what happens when the chamber is only partially filled with liquid CO_2, and heated to a temperature above that of the critical temperature of CO_2.

The valve on the liquid carbon dioxide cylinder is opened. The inlet valve on the pressure chamber is opened. Liquid carbon dioxide enters the chamber. The pressure gauge shows that the pressure increases to about 900 pounds per square inch. When the chamber is about half full, the valve on the carbon dioxide cylinder is closed and the inlet valve is closed.

The thermostat is set to heat the water bath to a temperature of about 40°C--well above the critical temperature of carbon dioxide which is 31.1°C. As the temperature increases, the pressure inside the bomb rises, going up to as high as 2000 pounds per square inch, depending on the maximum temperature used.

Disappearance of the Meniscus. (Film title)

As the temperature increases and approaches the critical temperature of carbon dioxide

turbulence of the liquid becomes visible. The surface tension of the liquid decreases progressively until at the critical temperature it becomes zero. At about 30° centigrade the surface of the liquid becomes indistinct and at 31.1° centigrade the meniscus disappears. Above this critical temperature all the carbon dioxide will be in a dense gaseous phase.

Recondensation and Reappearance of the Meniscus. (Film title)

If the temperature of the chamber is allowed to fall below the critical temperature, the gas will recondense into minute liquid droplets. After they coalesce, the liquid surface or meniscus of the liquid phase reappears.

In order to demonstrate this--cold water is circulated through the water jacket. When the temperature falls below 31.1° centigrade, the gaseous carbon dioxide condenses into millions of minute liquid droplets. These droplets completely reflect and block out the light being shined into the back window of the chamber. As a result a very dark cloud momentarily fills the chamber. The droplets coalesce and the clear liquid phase will reappear at the bottom of the chamber. The meniscus again becomes visible.

This window shattered under a pressure of 1100 pounds per square inch. The thick plastic shield held and prevented the glass fragments from being blown from the bomb. The possibility that the plastic shield itself could have shattered is a danger, however remote, that should not be dismissed. The window is not necessary for routine critical point drying and a metal disc of the same size is routinely used in its place.

A Typical Critical Point Drying Run. (Film title)

In a typical critical point drying run the high pressure chamber is cooled to 19°C.

Ethanol, the intermediate fluid in this case, is placed in a well located just inside the opening of the chamber.

Dehydrated specimens soaking in ethanol in containers such as screen baskets or other porous containers are transferred into the fluid well. This keeps the tissue wet until the chamber is filled with the transitional fluid, liquid CO_2.

The bomb is sealed.

The tank valve is opened off screen.

The admit valve is opened.

Liquid carbon dioxide enters the chamber.

The drain valve is opened for a period of 2 minutes. Most of the ethanol in the fluid well is flushed out the drain valve during this interval.

The alcohol in the specimen is then replaced by liquid carbon dioxide:

The drain valve is closed.

The CO_2 outlet valve located at the highest part of the chamber is opened.

Liquid carbon dioxide is flushed through the chamber for 3 minutes.

Rapid expansion of the escaping carbon dioxide causes adiabatic cooling and frosting of the exhaust tubing.

The carbon dioxide outlet valve is closed. All other valves are closed and the tissue is allowed to soak in the liquid carbon dioxide for 2 minutes.

Flush 3 Minutes
Soak 2 Minutes
Repeat 3 Times. (Film title)
This sequence of flushing liquid CO_2 through the chamber for 3 minutes followed by soaking the specimens in liquid CO_2 for 2 minutes is repeated for a total of 3 times. This purges the specimens of the ethanol which is exchanged for liquid CO_2.

If amyl acetate is used as the intermediate fluid, the sequence is repeated until the odor of the amyl acetate cannot be detected in the solidified carbon dioxide collected from the exhaust tube onto a paper towel. The absence of the odor should be verified by someone who has not been recently near the fumes of amyl acetate.

The carbon dioxide outlet valve is shut first and all other valves are shut.

In order to avoid the possibility of unreliable pressure readings, which could occur if the liquid level were high enough to enter the inlet of the pressure gauge, sufficient fluid is released to lower the liquid level until a meniscus is just barely visible at the top of the chamber as seen through the viewing port.

With all valves shut the thermostat of the recirculator-heater is set for a maximum temperature of 40°C and it is turned on.

As the temperature of the recirculating water increases, the pressure inside the bomb rises. Near the critical temperature turbulence becomes optically apparent.

When the temperature reaches 40°C, the tissue is dry and in a dense gas and under high pressure.

In order to avoid turbulence and excessive adiabatic cooling, the gas is released very slowly. The temperature of the chamber is carefully kept above the critical temperature while the gas is being released over a period of from 5-10 minutes until the pressure gauge reaches zero, indicating atmospheric pressure inside the vessel. To make sure the bomb is at atmospheric pressure before opening it after all the gas is out and the pressure gauge reads zero, the drain valve should also be opened to make sure the pressure inside the vessel is at atmospheric pressure.

The chamber is opened and after the dried tissue is removed, it will be ready to be attached to the specimen stud, coated and viewed in the electron microscope.

Cell Suspensions. (Film title)
Critical point drying of small biological specimens, particularly those in liquid suspensions, requires special handling techniques.

A specimen stud was designed specifically for handling such specimens. A portion of the stud becomes the specimen carrier during critical point drying.

A retaining ring and fine mesh screen are removed and placed in 100% ethanol. The base pin is likewise removed and a hollow-ground stud disc, now acting as the specimen carrier, is placed in 100% ethanol.

Cells, fixed, dehydrated and collected by centrifugation are transferred into the carrier without allowing them to become dry. The fine mesh screen will then be placed over the shallow carrier and it is held in place by the retaining ring.

When the carrier is placed in the critical point bomb, the fine mesh screen permits fluid exchange in the specimen carrier without allowing the specimens to be lost from the carrier.

After drying, the screen is removed and the specimens, then dried onto the surface of the disc, are viewed after reassembly of the stud.

Polylysine as an Adehesive. (Film title)
Coverslip fragments coated with polylysine may also serve to retain small specimens. The fragments are cleaned and stored prior to use. A solution of 0.1% polylysine (MW 115,000) is applied and allowed to remain on the coverslip for a period of time (a few minutes is usually enough).

The coverslip is washed with distilled H_2O and then it is rinsed with an appropriate buffer.

A monolayer of polylysine remains on the glass surface and acts as an adhesive for cells in suspension. Suspended material is then added and allowed to settle (settling overnight in a moist chamber may be necessary when very small particles are used).

The coverslip with the particles or cells attached is then passed through the dehydrating sequence to 100% ethanol and then placed in a specimen carrier after which it is placed in the critical point bomb for drying.

Filtration. (Film title)
Nuclepore filters also provide a convenient means for collecting suspended material.

A Nuclepore filter selected for the appropriate porosity is wetted and placed in a "pop top" Nuclepore filter holder secured with a retaining ring.

The holder is attached to a small vacuum chamber and a syringe body is fitted to the holder.

The suspended matter is added.

The chamber is evacuated to a pressure not lower than 5 psi (gravity alone may be adequate for the filtration).

The valve to the vacuum chamber is opened. Fluid is drawn through the filter by the negative pressure.

Care is taken never to allow the filter to dry.

The next fluid in the sequence is added (it may be a fixative or a dehydrating agent) and the process is repeated. The process is repeated until the particles or cells on the filter are completely dehydrated in 100% ethanol.

The filter holder is then removed and placed in the critical point bomb. After drying, the filter is removed and mounted on a specimen stud.

Discussion of the Film

The film was not intended to be used as a self-sufficient teaching unit on the subject of critical point drying. It should be shown as an adjunct to a comprehensive lecture on this subject given prior to the film's presentation so that terms used in the spoken narrative such as critical temperature, critical pressure, critical point, intermediate fluid and transitional fluid, etc. would already be understood by the viewer at the time he first saw the film. Since the spoken text was required to coincide with the action of the film, there was no adequate opportunity for defining terms, etc. as the film progressed, the succession of scenes and their time intervals being as they were. A detailed script carefully planned to allow time for integrating more discussion of the theory of critical point drying with the demonstration of the method might be an improvement. Timing the action of the film to a spoken text would be easier than the way we chose, which was to time the spoken text to the action of the film.

The lecture preceding the film should point out variables in the technique such as the choice of different intermediate and transitional fluids and the theory behind these variables. Different types of devices used for critical point drying should also be discussed so that the students do not receive the impression that the procedure shown is the only way or the preferred way to do critical point drying.

Acknowledgements

We thank Mr. Ralph E. Morton of the University of Georgia Instrument Shop for his help in the design and fabrication of special equipment. This film was produced through the Instructional Resources Center, University of Georgia, Athens, GA 30602, USA, phone (404) 542-1528. It was photographed under the supervision of Mr. Steve A. Gamble and Mr. James W. Morgenthaler. Copies are available for purchase. Inquiries should be directed to the Instructional Resources Center at the above address.

Discussion with Reviewers

A.L. Cohen: If the window shattered at 1100 lbs/in^2, you used the wrong kind of glass! It is wrong in not being strong enough to withstand the pressure, and wrong in shattering. The proper types of windows, if they do fail, merely crack. If your bomb is "designed to withstand pressures up to 3000 pounds," this should include the window. You also remark that "the pressure may go as high as 2000 pounds per square inch." If the pressure goes this high, your 1.5X safety factor is too small. There is also no evidence of a rupture disc or other failsafe device. Whatever disclaimers are made later (e.g. abolition of the window), presentation in a teaching film is equivalent to recommendation. On safety, see the following references:

Cohen, Arthur L., 1974. "Critical point drying," in Principles and Techniques of Scanning Electron Microscopy, Vol. 1, M.A. Hayat (ed.), Van Nostrand Reinhold, NY. (p. 55)

Cohen, Arthur L., 1977. A critical look at critical point drying--theory, practice and artefacts. IITRI/SEM/Vol. I, 525-536.

Humphreys, W.J., 1977. Health and safety hazards in the SEM laboratory. IITRI/SEM/Vol. I, 537-544.

Authors: There is no safe way to predict how much pressure a piece of any glass will withstand before cracking. There are reports of glass cracking spontaneously while sitting on a bench at room temperature. The only safe assumption is that a glass window can break even at low pressures. Breakage could result from unequal pressure on the glass when it is clamped into position, from thermal stresses, and from undetected imperfections in the glass. With the exception of the windows, all components of our critical point dryer, including inlet and outlet valves are rated by our instrument shop engineers to be able to withstand 6000 p.s.i. and all components, including the windows, have been tested to withstand 3000 p.s.i. over a period of several days.

As can be seen in the film, at 40°C the pressure of our chamber reaches only about 1500 p.s.i. But it is important to know that some commercial pressure chambers used for critical point drying do sometimes have pressures in the chambers that exceed 2000 p.s.i. if liquid CO_2 is used. Bartlett and Burstyne (A.A. Bartlett and H.P. Burstyne, A review of the physics of critical point drying, SEM/1975/I, IIT Research Institute, Chicago, IL 60616, p. 305-316) in a discussion with reviewers cite data taken with the Sorvall apparatus ("Critical Point Drying Systems," Bulletin 20-72 Ivan Sorvall, Inc., Newton, Ct.) where the pressure in one chamber was 2020 p.s.i. at 46°C. In another chamber it was 2440 p.s.i. at 52°C during one run and 2300 p.s.i. at 54°C in another run. The maximum pressure for a given temperature will vary with different chambers. The larger the bubble of vapor left at the top of the chamber before heating the vessel to a temperature higher than the critical temperature, the lower the maximum pressure in the chamber will be at a given temperature.

You are correct in insisting that a safety blow-off valve be part of the chamber design. The devices we use have valves preset to blow the pressure fuse at a pressure of 1800 p.s.i. In a revised spoken text for the sound track this should be mentioned.

G.H. Haggis: If tests have shown that there is any danger of "tough plastic" ½-inch thick not standing up to a window blow-out, then a plastic cover 1 inch thick (or whatever may be necessary) should be recommended. The "tough plastic" should also be specified as to material and manufacturers.

Authors: The plastic shield used is ½-inch thick and it is made of Lexan, a polycarbonate resin. General Electric Co., Chemical Division, Plastics Avenue, Pittsfield, Mass. 01201.

A.L. Cohen: Why should there be "unreliable pressure readings" if the liquid level is so high that liquid enters the pressure gauge? A much better reason for keeping the liquid CO_2 level below the top of the chamber is to avoid the rapid increase in pressure which occurs if liquid completely fills the chamber while it is being heated.
Authors: Trapping of the liquid phase of a gas and heating it in the restricted confines of the high pressure gauge designed to read gas pressures apparently interferes with the free equilibrium between the liquid phase and the gas phase, resulting in unreliable pressure readings. Manufacturers of the Polaron critical point drying apparatus warn against heating the intermediate fluid with the liquid level above the inlet to the pressure gauge (Instruction manual, Polaron Critical Point Drying Apparatus E3000, Polaron Equipment Limited, 60/62 Greenhill Crescent, Hollywell Industrial Estate, Watford, Hertfordshire).

G.H. Haggis: I think the ideal critical point dryer is one with windows but of a design which has been very thoroughly tested for safety. How do you carry out the operation described to avoid the possibility of unreliable pressure readings when you have replaced the window with a metal disc?
Authors: Using the window, a determination is made as to how long the exhaust valve should be left fully open to bring the liquid down to the proper level.

G.H. Haggis: When the apparatus is being half filled only to illustrate certain effects, should it not be made clear that this is not the procedure used in day-to-day use of a critical point dryer?
Authors: This should be made clear.

A.L. Cohen: If cycling through or around the critical point is slow enough, you should see the critical opalescence of the vapor phase in the neighborhood of the critical point. This· is different from the dense cloud described. Is this evident?
Authors: This was not evident as observed through the front window although it might have been evident if viewed through the rear window. What one sees depends on whether the observation is made through the back window, in which case the chamber interior is illuminated by reflected light or through the front window in which case the interior is illuminated mainly by transmitted light. Thus when the gaseous CO_2 condenses into a fog when the temperature drops below the critical temperature the fog appears light when viewed by reflected light through the back window but dark as seen by transmitted light through the front window.

A.L. Cohen: Would you please supply a selected short bibliography? This is particularly important for the beginner.
Authors: For relevant references see "Dehydration of SEM Specimens, A Bibliography" by K.S. Howard and M.T. Postek in these proceedings. Also see the tutorial: "Critical Point Drying--Principles and Procedures," by A.L. Cohen in these proceedings.

A.L. Cohen: Is it necessary to carry so much ethanol into the chamber? Your procedure does not appear to involve a gradual exchange of CO_2 for ethanol, nor is ethanol so volatile that there is appreciable danger of evaporation under ordinary circumstances.
Authors: In most cases it is not necessary to carry so much ethanol into the chamber as indicated in the film. But it does no harm and some specimens such as cell suspensions or delicate mycelia of molds might dry out before being covered with the liquid CO_2 unless this kind of precaution is taken. Precautions for insuring that specimens do not dry out are more necessary if the more volatile intermediate fluids such as acetone or Freon 13 are used. Our results convince us that our exchange of ethanol with liquid CO_2 is sufficiently gradual.

A.L. Cohen: AT 19°C ethanol (0.79) is slightly denser than liquid CO_2 (0.77) and either mixes or falls to the bottom, depending on the amount of convection. There is no clear reason for using two outlet valves.
Authors: It is easier to flush out the excess ethanol from the chamber through an exhaust valve connecting the lowest part of the chamber (the fluid well) before much mixing of the ethanol and the liquid CO_2 takes place. Also this valve can be used to drain out excess ethanol before the liquid CO_2 is admitted to the chamber.

A.L. Cohen: Since ethanol seems to be satisfactory, why use (iso) amyl acetate?
Authors: Carbon dioxide snow forms when liquid CO_2 escapes from the orafice of the tubing leading away from the vent (exhaust) valve. The CO_2 is collected on a paper towel. When it no longer has the odor of amyl acetate, the exchange of CO_2 for amyl acetate is considered to be complete. Anderson (T.F. Anderson, "Techniques for the preservation of three-dimensional structure in preparing specimens for the electron microscope." Trans. N.Y. Acad. Sci., Ser. II, 13, 1951, 130-134) recommended that the flow of liquid CO_2 through the bomb continue for twice the time necessary for the odor of amyl acetate to disappear from the escaping gas. Some users prefer to use amyl acetate as the intermediate fluid because it serves as an olfactory clue as to when it is completely replaced by CO_2.

G.A. Hooper: At several points the safety of users related to the glass viewing port was discussed. Why then does the film advocate use of 2000 p.s.i. during the run when CO_2's critical pressure is well below that point (1073 p.s.i.)?
Authors: The point is made that pressures inside critical point drying vessels may reach 2000 p.s.i. It was not our intention to advocate this high a pressure. At 40°C our device registered a pressure of about 1500 p.s.i.

Discussion with Reviewers

G.W. Grimes: You make no mention in your paper regarding the background of students who are involved in your course. This might be an important factor influencing the general usefulness of the videocassette for instructional purposes. For example, if students are trained previously in TEM, then certain aspects of SEM operation essentially could be ignored. On the other hand, if students were not trained in TEM, then extensive discussion of imaging, electronoptics, column design, etc. must be considered exhaustively. On the basis of your experience, could you visualize, and would you recommend a series of cassettes which would correspond to the training level of the viewer?
Author: The individuals which use the video tapes are mostly graduate students which have completed a five semester hour course (3 hours lecture, 6 hours laboratory per week) in scanning electron microscopy. The video tapes were designed to acquaint these students to a newer scanning electron microscope. The video tape concentrates on its operation and reviews basic theory i.e., stigmation, depth of field, spot size versus resolution, etc. The approach would be similar to introducing a pilot to a new aircraft control panel via videocassettes. I would not recommend a series of videocassettes for various levels of competency.

G.W. Grimes: Do you have any concern that students learn a "stepwise mechanical" approach to E.M. using this method of instruction rather than developing a thorough understanding of the equipment? An individual should be able to use different equipment with minimal instruction if he understands the equipment, but for individuals who lack this understanding, a complete reorientation is required.
Author: The videocassettes are only used to review forgotten concepts and instrumentation dexterity. Some students haven't changed a filament in a year and often feel awkward about changing one without assistance. The video tape is used to reinforce this procedure.

R.L. McConville: What is the response of the students to the use of the videocassette program in your curriculum?
Author: Favorable.

R.L. McConville: You mention that your videocassette program has been successful in supplementing the formalized course program. Do you therefore assess your students by examination on the material they should have learned from the videocassette program? If so, what are the results?
Author: Students who employ the program as an introduction to a new instrument must prove their instrumentation competency by passing a check out examination.

R.L. McConville: What is the maximum number of students that can, at one time, usefully use the VTR?
Author: 15-20.

R.L. McConville: Have you interfaced the SEM with the VTR to illustrate such effects as depth of field, signal/noise ratio, etc.? And if this has been done, how well does the VTR reproduce such effects?
Author: No. These concepts are presented in the video tape using micrographs and diagrams. The synchronization of the SEM video generator and commercial video tape recorders can be a problem when editing is required. Such effects e.g., s/n ratio are easily demonstrated using the on-line approach, but without editing, these tapes are often lengthy and poorly narrated.

G.R. Hooper: In addition to your videotape program, are you aware of other uses of videotapes in EM instruction?
Author: Yes, there are several topics germane to electron microscopy available in videotape format (see EMSA Educational Bulletin).

G.R. Hooper: We have proposed attachment of a videorecorder to the student SEM so that "problems" occurring in the absence of instructors could be recorded for later evaluation. Do you feel such a use would be an effective teaching aid?
Author: Such information could be useful but it is my experience that a large number of problems encountered are not associated with the T.V. mode of operation but involve the slow scan mode. The latter realistically mimics the normal mode of operation. We require the student to record such problems on film, record the problem in our maintenance book along with the negative illustrating the problem. The T.V. mode of operation was designed to image a subject in real time. Resolution is usually poor at higher magnifications and noise is often a problem.

SCANNING ELECTRON MICROSCOPY/1979/II
SEM Inc., AMF O'Hare, IL 60666, USA

AN EVALUATION OF VIDEO TAPES IN TEACHING SEM

E. L. Thurston

Electron Microscopy Center
Dept. of Biology
Texas A&M University
College Station, Texas 77843

Abstract

Videocassettes represent a relatively new
format for presenting information to an academic
audience. The advantages of a compact, fool-
proof, inexpensive video-audio delivery system
are numerous, however, problems exist in the
planning and production of videotapes. In-house
production of videocassettes for use in the
teaching of electron microscopy are extremely
useful for presenting information concerning
instrumentation operation. The tapes coupled
with a portable VTR and monitor allow the operator to view a replay of the instruction while
actually operating the instrument.

Introduction

The "art" of electron microscopy has been
traditionally taught in an apprentice-like relationship. This type of individualized training
environment is perhaps the most effective and
oldest means of formal education. While it
serves its purpose well, it is seldom used in today's educational system. The "hows and whys"
of electron microscopy currently are being taught
in formalized undergraduate and graduate curricu-
la. These highly technical courses require the
melding of theory with practice on delicate instruments which cost many thousands of dollars.
Therefore, the quality and quantity of instruction must complement the nature of the "beast."

The tried and true approach of lecture-laboratory has been most satisfactory in implementing
a course syllabus. Academic programs usually
supplement theory presentation with textbooks,
selected readings, slide talks and demonstrations.
The use of 8 and 16 mm movies as supplementary
material has met with some success in specific
areas, however, the scarcity of such material on
the subject matter of electron microscopy is indicative that some problems exist in production,
duplication and cost. The advantage of the motion picture as an educational tool over 2"x2"
slides is its capability to capture the activity
of "the demonstration" over a static presentation. Video recording tape is not new to television stations, however, it has experienced
limited use in the academic environment. Indeed,
the video tape recorder (VTR) and its supportive
equipment i.e., edit machines, cameras, etc.,
usually requires a special grant to purchase, an
army of highly trained technicians to operate
and maintain the equipment and has been imprac-
ticable for most academic programs. Television
and teaching clashed in the 1960's and in most
instances, educators sen lat. abandoned this combination as being a lethal educational experience.[1,2]

The advent of the videocassette has altered
most of the financial and technical problems
associated with earlier, bulkier equipment. Videocassettes are the latest twig on the audiovisual tree and do not represent a new, separate,
different, or truly revolutionary industry.

How can videocassettes be used successfully
in the teaching of SEM? Since the teaching of

KEY WORDS: Teaching SEM, Videocassette

electron microscopy is an educational exercise and most instructors employ some form of audio visual aids, then it seems appropriate that the videocassette should be involved to some degree. The videocassette has many general advantages:

1. Easy to operate
2. Excellent picture quality
3. Reliability
4. Television delivery
5. Peripheral services

A number of topics covered in an electron microscopy program require demonstration and repetition of presentation. The quality of learning that is retained after the introduction of a subject (operation of the vacuum evaporator, changing a filament, etc.) is usually poor and often requires a great amount of reteaching if the outcome is to be achieved with minimum trauma to the pupil and equipment. The reteaching of subject matter to a large number of individuals on a one-to-one ratio is boring, time consuming and inaccurate in that it is inconsistent from session to session. Videocassettes are a means to accomplish the above teaching and overcome the mentioned pitfalls at a reasonable cost.

A videocassette program was initiated in our teaching program to test its feasibility for instruction in SEM. It has been successful in supplementing the formalized course program and a "life saver" in the continuing education program involving infrequent laboratory users. This is not an universal endorsement of videocassettes, as production of a videocassette can be time consuming and costly if production is poor due to inexperience.

There are a limited number of topics involving transmission and scanning electron microscopy that could be presented usefully via videocassette. I currently employ videocassette instruction as a means of upgrading operators to newer equipment and as a review for individuals whose instrumentation use is sporadic. Individuals who wish to acquire the skills involved in upgrading to a new instrument are required to view four 20 minute videocassettes on that instrument. Four tapes are used to present a visual demonstration of "how to" along with theory. The topics presented in the SEM tapes are as follows:

1. Introduction to the microscope-modules.
2. Start up procedure - presetting the instrument.
3. Axis alignment of the Electron Optical Column - Electron gun, condenser lens and objective aperture. Condenser spot size vs. resolution and signal/noise.
4. Astigmatism correction, depth of focus and selection of proper accelerating voltage.

A videocassette recorder player (VTR) and a television monitor can be purchased for a moderate fee and the entire unit can be housed on a portable cart. The portable nature of the equipment, its ability to operate in ambient light and its foolproof packaging make it a useful educational tool.

The advantages of videocassettes in educational programs appear attractive but let me caution zealous individuals as to some of the potential pitfalls.

Subject matter

A limited number of topics can be justifiably dealt with using videocassettes. Theory is presented poorly via this media, however, instruction involving demonstrations and equipment are excellent topics.

Script Writing

A comprehensive but terse script must accompany the tape presentation. Tapes lasting more than 20-30 minutes are monotonous to the viewer and the impact of video teaching is usually lost. In videocassette production a great amount of tape footage must be taken which will be edited at a later date. The audio portion of the tape is then superimposed and synchronized with the video in a sound studio. A great deal of expertise is required to produce a videotape which looks and sounds professional. Reducing the script to the bare essentials is similar to condensing a manuscript into an abstract without loss of information.

Visual Props

Most T.V. taping employs title cards, diagrams, micrographs, etc. which require graphic art expertise. Remember certain small details and some colors do not display well on T.V. If 2"x2" slides are to be incorporated in the presentation, they must be appropriately framed for T.V. which is not the standard 35 mm frame size.

Shooting

Camera(s) angle, lighting, available lenses, split screen, fade-ins and-outs all require skill to coordinate and a sizeable expenditure in equipment. Such equipment is usually available at the university T.V. station or through an audio visual/ multi-media center. In most cases charges are levied for equipment use.

Editing

Once the necessary footage has been taken it then becomes necessary to marry the script with the video, making sure there is no loss of information. This is one of the most time consuming and important steps in good videocassette production. Editing skills require a great deal of practice and a flair for video production.

The production of a videocassette is an educational experience to a neophyte, however, it must serve a purpose as it represents an expense ($100/min) and a great deal of planning and effort. I remain optimistic in the use of videocassette for instruction purposes involving electron microscopy.

References

1. J.H. Barwick and S. Kranz. The Complete Videocassette Users Guide. Knowledge Industry Publications, Inc. New York, U.S.A., 1973 pp. 1-169.
2. G.N. Gordon and I.A. Falk. Videocassette Technology in American Education. Educational Technology Publications. New Jersey, U.S.A., 1972 pp. 1-161.

See page 240 for discussion with reviewers.

SCANNING ELECTRON MICROSCOPY/1979/II
SEM Inc., AMF O'Hare, IL 60666, USA

THE USE OF SLIDE/TAPE MODULES IN TEACHING SEM

K. K. Baker, S. L. Flegler and G. R. Hooper

Center for Electron Optics
Pesticide Research Center
Michigan State University
East Lansing, MI 48824

Abstract

Increased student demand for instruction in SEM techniques and limited resources led to a search for economical and productive teaching techniques. A modification of the auto-tutorial approach was devised where short lessons (modules) on a 35 mm slide/cassette tape format were developed for a number of subject areas. These are used to supplement live demonstration of techniques in a preview-review manner. Student acceptance of the modules was very positive. Increases in quantity and quality of student performance per instructor hour and dollar were obtained. The concept of mastery learning with a programmed approach seems especially well suited for laboratory materials where instrument use and/or specific preparative technique is being taught.

KEY WORDS: Teaching SEM

Introduction

Between the years 1972 and 1978, the number of students taking electron microscopy courses through the Center for Electron Optics (CEO) at MSU increased from approximately 12 (No SEM) to over 100 per academic year. About 40% of these students now take "hands-on" SEM training. In addition to regularly enrolled students, a number of faculty and other staff receive basic operating instruction on one or both of the CEO's scanning electron microscopes each year.

Since only one SEM is devoted to instruction, we experienced difficulty in providing the instrument hours per student that we felt was essential for adequate instruction. We also found that the time involved in personal instruction of approximately 12-15 students each 10 week term prohibitive.

An analysis of our original teaching methods (TEM and SEM) indicated two specific areas that utilized large amounts of instrument and instructor time.

First, each student was being taught location and function of instrument controls at the SEM. This procedure tied up both instrument and instructor each time new operating principles were demonstrated.

Second, the specific preparative arts (e.g., darkroom, critical point drying) required an initial demonstration and repetition of the procedure several times for each student.

With the advice and support of Michigan State University's Learning and Evaluation Services, we sought alternative teaching methods to our current program. Many of the elements of auto-tutorial systems seemed suitable for our laboratory program. Such systems as the "Keller plan," "Postlethwait's approach"[1,2,3] and "SLATES" (Structured Learning and Teaching Environments)[4,5] were investigated. Each of these approaches is essentially programmed instruction which provides a step by step progression through related sub-units until the desired level of competency is reached.

The evaluation of the student's competency level may be determined by traditional methods of examination where each student is referenced to the performance of a selected group. Such "norm-referenced" evaluation contrasts with an approach where teachers establish a fixed reference level that students must reach. This latter approach is referred to as "mastery" or "competency based" evaluation[1,6]. The attractiveness of the mastery approach is that each student knows the minimum level of performance expected and instruction can involve manipulations of presentation, practice, feedback and evaluation at a somewhat individualized pace.

The SEM subject matter ordinarily taught in laboratory or in other practical situations does have identifiable levels of competency and lends itself well to mastery teaching.

Some of the programmed approaches cited above emphasize independent work by the students[5]. Others utilize various levels of instructor involvement. Student reaction at MSU to complete self-teaching approaches in subjects other than SEM has at times been negative. We were accordingly not inclined to follow such a system. In addition, we were concerned that safety of students and SEM's might be compromised in a largely self-teaching program. Our approach was to devise a system with elements of self-paced independent study combined with sufficient guidance to achieve our goals. The following discussion covers the approach used at MSU along with some analysis of its effectiveness and student acceptance.

Materials and Methods

Combinations of visual and audio materials were evaluated as to their suitability to provide a maximum input of data to the students. While video tapes would provide both audio and video input and can demonstrate dynamic processes in SEM, the costs of equipment and preparation seemed to us undesirable. The production of 35 mm color slide sets and cassette recordings is relatively inexpensive and simple[2,4]. This approach was chosen. Lessons of 10-20 minute duration were designed. We term each short unit a "slide/tape module." Kodachrome 40 transparency film was used exclusively. "Caramate" display units, manufactured by the Singer Corp., (Education Division, Education Systems, 3750 Monroe Avenue, Rochester, N.Y. 14603) were used in both production and actual instruction. These units are approximately 1 cubic ft. in size and very lightweight. They may be used with synchronized, automatic slide advance or manual advance modes. Earphones are available for use in crowded or distracting situations.

Basic Instructional Model

The slide/tape modules were developed primarily for use in laboratory situations. The skills in both instrument operation and specimen preparation are very amenable to evaluation by a mastery concept. In such an approach the instructors determine the level of expertise required by the student at each step and at the end of the teaching sequence. Individual modules are kept to a maximum 10 minute period and large areas of instruction are covered with a series of modules. Teaching and evaluative processes are devised to attain the planned goals. We approached

each laboratory exercise in the following sequence:

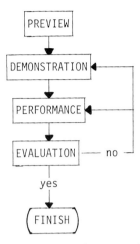

Preview. Reading assignments in texts and journals as well as the specific laboratory exercise[7] were completed by students prior to class time. Specific slide/tape modules were provided for preview by students.

Demonstration. The instructor(s) demonstrated specific preparative schemes or SEM operating principles to large groups. Emphasis was placed upon dynamic processes that were difficult to portray in the slide/tape modules.

Performance. Each student performed the exercises at their own time and pace. Instructors were on call but not involved in extensive personal instruction. The modules were available and could be used at the site of performance (e.g., the entire display unit could be taken to the laboratory bench or SEM room).

Evaluation. Since an actual product (specimen, micrograph, predicted response of the SEM) was required at the end of each unit, students could partially evaluate themselves. Instructors were available for "check out" situations. At the end of the term, a formal portfolio of work, including photographs and explanations, was submitted for grading.

Finish. Each exercise was considered finished when a successful evaluation (by student and/or instructor) was completed. Where failure occurred, the students returned to the slide/tape modules or could ask for another demonstration before repeating the performance.

Preparation of the Modules

Detailed procedures for making slide/tape modules are covered elsewhere[2,4,7]. In our preparation three areas were considered.

First, the actual use planned for the module, the basic subject matter and format and the student's preparation were determined. With these factors in mind, both general goals and specific objectives for each lesson were formulated.

Second, development of the modules proceeded in a number of steps, e.g., outline the content, sequence the content; prepare a "storyboard" and visuals; take slides; write the text; record the tape; synchronize the slides and tape. We initially approached each lesson in this systematic, formal manner. With practice, a number of shortcuts were developed.

Finally, each unit was evaluated. Student response on questionnaires and informal comments of students were invaluable. Each instructor also periodically reviewed the modules and revisions were made until the units were satisfactory. Currently a laboratory manual with accompanying slide/tape modules is being prepared for general distribution.

Discussion

The value of teaching methods can be measured by how well they aid instructors in reaching course objectives. In our case, we hoped to alleviate the problem of instructor and instrument time constraints which existed. We were also conscious of the high cost of instructing EM students. Student performance and acceptance of the teaching method was a final consideration.

We have experienced an exceptional increase in available instructor time. Where each instructor could assist 1 student at a time, he or she now works with several simultaneously. The instructors are on duty for 2-4 hour shifts in the teaching laboratory where an SEM, TEM, microtomes, critical point dryers, vacuum evaporators, etc. are being used. It is not uncommon to have as many as 8-10 students working at various stations. During periods of exceptional activity two instructors are stationed in the laboratory and are available on a demand basis.

Less time is spent with students at the SEM, when the slide-tape method is employed. During their initial session on the instrument, the instructor monitors the first part of the period. Where lack of familiarity of controls is evident, the student is asked to review the module(s) and make another appointment for his first SEM experience. The productivity of the SEM is much higher in terms of completion of experiments. Failures due to student ignorance are much reduced from previous instructional periods.

The most satisfying result of our teaching program has been a marked improvement in quality and quantity of work. Previous groups completed an average of 70% of the assigned materials. Under the current system, approximately 90% do so. The finished portfolios are uniformly better in terms of photographic documentation and textual comments.

The student attitude has been both positive and constructive. The freedom to receive instruction according to their own schedules is commented on positively in questionnaires. The fact that students can set, in part, their own pace and repeat instructions as often as desired is also considered useful by those involved.

Programmed, auto-tutorial materials have elsewhere been demonstrated as effective[3] and economical[8]. The idea of programmed instruction SEM has been mentioned in previous reports[9] and self-teaching programs have been developed for at least some EM instruction[10]. We believe they deserve further consideration.

References

1. Eble, K. E. 1977. The craft of teaching. Jossey-Bass, San Francisco. Chapters 1, 8, 11.
2. Espich, J. E. and B. Williams. 1967. Developing programmed instructional materials. Fearon Publishers, Palo Alto, California. Part One, Chapters 2, 3, 4. Part Two, Chapters 5-11.
3. Schom, C. B., L. Haley, O. P. Kamra, A. Hicks and R. W. Lee. 1978. Teaching techniques: Do results differ between approaches. AIBS Education Review. 7(1):1-5.
4. Abedor, A. J. 1977. Handbook for the Learning and Evaluation Service Workshop of self instructional module design. Learning and Evaluation Services. Michigan State University, East Lansing, MI 48824.
5. Abedor, A. J. 1977. Analysis of the SLATE instructional model. Mimeograph, 10 pp. Learning and Evaluation Services. Michigan State University, East Lansing, MI 48824.
6. Block, J. H. 1971. Mastery learning: Theory and Practice. Holt, Rinehart and Winston. New York. Chapter 1.
7. Hooper, G. R., K. K. Baker and S. L. Flegler. 1978. Exercises in Electron Microscopy - A laboratory manual. Mimeograph, Center for Electron Optics. Michigan State University, East Lansing, MI 48824. 116 pages.
8. White, J. M. and R. D. Barnes. 1978. A cost analysis of audio-tutorial and conventional instruction. AIBS Educational Review. 7(1):10-12.
9. Rue, J. W. 1975. An approach to teaching scanning electron microscopy to multidisciplinary students. In, O. Johari and I. Corvin (Eds), SEM /IIT Research Institute, Chicago, IL, pp. 625-630.
10. Hopkirk, J. W. and R. M. Glaeser. 1972. Electron microtomy training: A self contained learning package. In, C. G. Arceneaux, (Ed) Proc. Thirtieth Ann. Mtg. EMSA. p. 698-699.

DISCUSSION WITH REVIEWERS

Reviewer I: In your paper you make no mention of formal lectures on the theory of electron optics. Does this mean that you have a separate lecture course as a prerequisite or to be taken concurrently with the laboratory or, in fact, cover all the necessary information through reading assignments and the use of the slide/tape modules in the laboratory?

Authors: We have a separate lecture course (3 Quarter hours). It is required concurrently with the SEM or TEM classes.

Reviewer I: You state that the students perform various assigned exercises at their own time and pace. Do you mean time and pace per week, or throughout the ten-week period? How does student self-timing and pacing work with respect to the laboratory demonstrations given by the instructor? For instance, do you find that for some students the demonstrations are too premature, while for other students who work at a much faster pace, the demonstrations are not presented soon enough?

Authors: Students generally perform assigned exercises at their own pace throughout the 10 week period. We urge them to keep up week by week and to complete assignments as soon after the demonstration as possible. Demonstrations are given once a week with some provisions for personal demonstrations for those students working at a much faster pace.

Reviewer I: You mention that it is not uncommon to have as many as 12 to 15 students working at various stations. Am I to interpret from this statement that you have a single class time for all students, or are the students assigned different class times but are also welcome to come and work during class times other than their own but which have an instructor on hand?
Authors: There is one demonstration period per week which all students attend. In addition, students are welcome to come and work at any other time during the week on certain exercises. Where their skills are limited they may be restricted to specific time periods (see the next question).

Reviewer I: Are students permitted to work at the CEO without an instructor on the premises? How many instructors are involved in the course?
Authors: There is an equivalent of two instructors available (1full time, 2 ½time) so that an instructor is on duty most working hours.

Students are permitted to work without supervision in the teaching laboratory. However, a strict checklist procedure is used to determine which students may actually operate instruments during unsupervised hours.

Reviewer II: You indicate that a lab manual with accompanying slide/tape is being prepared for general distribution. Does this include other universities and/or industries? If so, when, at what cost and where?
Authors: The laboratory manual (syllabus, demonstrations, slide/tape modules) will be available through M.S.U. Instructional Media Center. It should be available in late 1979. At publication a complete list of available modules will be available.

Reviewer II: What texts and/or references do you find most valuable with your teaching method?
Authors: We use M. A. Hayat's "Principles and techniques of Scanning Electron microscopy" series extensively. We also assign readings in SEM/IIT Research Institute or SEM/Inc proceedings.

Reviewer II: You indicate that each lesson was 10-20 min in duration. Did you find this time from experimental results based on the student's best attention time?
Authors: The time limit was suggested by M.S.U.'s Learning and Evaluation Services. Student questionnaires confirmed a preference for several short lessons over one long lesson.

REVIEWERS: I   B. P. Jakstys
            II   J. A. Murphy

SCANNING ELECTRON MICROSCOPY/1979/II
SEM Inc., AMF O'Hare, IL 60666, USA

TEACHING METHODS FOR SEM SHORT COURSES

O. C. Wells

IBM Thomas J. Watson Research Center
PO Box 218
Yorktown Heights, NY 10598

Abstract

In an SEM short course, the emphasis is on teaching the basics of the subject, rather than on presenting the latest research results, as is the case at a conventional SEM conference. The need to provide instrument manufacturers with an opportunity to demonstrate the equipment on which the success of microscopy depends is the same in each case. Some of the experience which has been acquired during the organisation of the eight annual SUNY short courses on the SEM is summarised in this paper.

KEY WORDS: Teaching SEM

Introduction

Every year approximately 1000 scanning electron microscopes (SEMs) are sold world-wide. Modern SEMs are very simple to use - and yet, to obtain the extra measure of performance that is sometimes needed, a thorough understanding of the basic principles and a special sort of operator's diligence are both required. One of the suppliers recently told me: "We employ a training manager full-time to teach our customers how to use their instruments. There is a very significant need for extra instruction in how to use these instruments."

There are at present five main sources of information on the SEM. These are:

(1.) Instructions from the manufacturer.

(2.) Books devoted to the SEM.

(3.) Publications in the technical literature.

(4.) Conferences such as this one, which are generally organised by a scientific society or its equivalent and which have the three main aims: I. Of providing papers devoted to the most recent developments in the art, to novel applications and to instrumental improvements of an experimental nature. II. Of providing tutorial papers, which, if properly presented, will be of interest to both the novice and experienced worker in the field. III. Finally, the instrument manufacturers must be given the opportunity to show their equipment.

(5.) By attending a "short course", generally organised by a university, in which the aim is to come as close to a university course as can be done in the limited time available. As compared with a conference such as this one, the relative emphasis on the three items in the preceding paragraph must be in the order II, III, I.

It is not the purpose of this paper to discuss lecturing techniques. Rather, I shall summarize the main ways in which a short course is different from a conventional conference such as this one. It is based on the eight annual short courses organised by Dr. Angelos Patsis of the Chemistry Department at State

University of New York at New Paltz (referred to here as "the SUNY short course"). The course coordinators for the physical sciences are Dr. A.N. Broers and myself; while for the biological and medical aspects the course coordinators are H.W. Carter M.D. and Dr. B.J. Panessa. Other lecturers have included Dr. E. Lifshin on x-ray microanalysis, Prof. M.S. Isaacson on electron interactions with thin films and STEM, and Dr. N.C. MacDonald on Auger electron analysis techniques, secondary ion mass spectroscopy, and related techniques. In the most recent SUNY short course, there were 47 attendees of whom 15 had PhD degrees. 15 were medically oriented. 32 were oriented towards the physical sciences. The wide range of interests and backgrounds seemed to enhance, rather than to detract, from the success of the short course. The schedule for the practical work is described below.

Summary of the course

In choosing the material for a short course on the SEM a proper balance must be achieved between: I. Background information; II. Details of the major components of the SEM (electron gun, magnetic lenses, aperturing systems and so on); III. Practical operating procedures; and IV. Information relating to possible areas of future progress in the art. A typical day's plan is to have lectures in the morning that will be of interest to both materials scientists and health scientists, with practical laboratory sessions and lectures of specific interest to either one group or the other in the afternoon and evening.

At the SUNY short course the major points of emphasis on the five days of the course are as follows:

I. General introductory lectures by the course coordinators in the morning, followed by introductory laboratory sessions in the afternoon and some more detailed lectures and a discussion period in the evening.

II. Lecture on the major components of the SEM column in the morning, with a similar afternoon and evening program as on the first day.

III. X-ray microanalysis theory, practice and applications (all day).

IV. Some more advanced topics, including electron interactions with thin films and STEM, again with practical SEM work in the afternoon (and in the evening by private arrangement with the demonstrators).

V. Related techniques such as Auger electron analysis techniques and secondary ion mass spectroscopy, summaries, and discussion period.

Organisation of an SEM short course

The differences between a short course and a scientific conference can be summarised as follows:

Number of instructors. Since the emphasis in a short course is on teaching a subject rather than on presenting recent research results from many people, the number of instructors can (and should) be restricted to one instructor per day (or several days). This improves the continuity of the instruction and makes it easier for the participants to discuss individual problems with the instructors out of class.

Class size. Since the classes generally contain between 25 and 50 people rather than ten times those numbers, it is easier to distribute notes in advance and to entertain questions during the presentations.

Introducing concepts. It is an interesting challenge to present the material in a way which will be interesting both to the novice and to the experienced worker in the field. Many of the concepts that must be introduced are quite complicated and yet there is not enough time to introduce them with the degree of preparation that would be appropriate in a university course. Very often it is helpful to emphasize the practical usefulness of a concept rather than the theoretical background, although, of course, the underlying theory can never be entirely ignored.

New concepts can sometimes be made to be more interesting by describing how they were discovered. Consider the electron, for example. There are (in vacuum): primary electrons, secondary electrons, Auger electrons, backscattered electrons, low-loss electrons and so on. In a solid material there are: conduction electrons, inner-shell electrons, hot electrons, etc. A good way to begin might be to spend five minutes describing the "Edison effect"[1] as shown in Fig. 1, which was described by Edison[2] as follows: "I have discovered that if a conducting substance is interposed anywhere in the vacuous space within the globe of an incandescent electric lamp, and said conducting substance is connected outside of the lamp with one terminal, preferably the positive one, of the incandescent conductor, a portion of the current will, when the lamp is in operation, pass through the shunt-circuit thus formed, which shunt includes a portion of the vacuous space within the lamp. This current I have found to be proportional to the degree of incandescence of the conductor or candle-power of the lamp." Following investigations by later workers, this was explained by assuming that the heated filament emits negative "electrons" which are attracted to the positive plate.

The diagram shown in Fig. 1 can also be used to emphasize the inherent simplicity of the SEM. The basic action, in which electrons are emitted from a heated filament and are then accelerated onto a specimen or target, is the same in each case. The additional operations such as focussing the beam can be explained by comparing the diagram of the SEM column with this simplified diagram. In a case like this, it is useful to be able to refer to the diagram shown in Fig. 1 several times in the course of the introductory lecture, because there are several concepts that can be explained with reference to this single experiment.

On the subject of backscattering, a brief description of how the observation that alpha particles were backscattered from a thin metal foil led Rutherford to realise that the atoms in the foil must each contain a nucleus can also provide a useful introduction. Here, it will be useful to compare the ways in which the backscattering phenomenon is used (with electrons) in the SEM to give the backscattered electron image and the way in which it is used (with energetic helium ions) to analyse materials by the nuclear backscattering technique. Obviously, the historical approach should not be overdone, but perhaps 5 minutes in every hour can usefully be occupied in this way.

"Hands-on" experience. In any short course, a proper balance must be achieved between the practical and theoretical aspects. In the early years when such courses were offered (and this applies, I believe, to all such courses), the theoretical aspects were covered very thoroughly but it was very difficult for the student to obtain hands-on experience. Since then, there has been a steady and most welcome trend towards a better degree of balance in which every student does have the opportunity for hands-on experience in a small group of only a few people to one SEM.

At the SUNY short course, there were four SEM stations, two energy-dispersive x-ray detector systems, one specimen preparation station, and an additional lab where the column from a non-working SEM was disassembled and then reassembled by the attendees. This was organised by dividing the participants into six groups of seven or eight each which were sent on a one hour per station trip as shown in Fig. 2. Additional time was available by individual arrangement with the demonstrators.

Introductory lecture. Obviously, every lecturer will have his own ideas on how to introduce a subject, and I not wish to appear to be inflexible by saying how this should be done. The viewgraph which I use for this purpose is shown in its original state in Fig. 3(a), and in its final condition in Fig. 3(b). Copies of this viewgraph are also enclosed with the class notes so that this diagram can be similarly annotated by the participants. Additional viewgraphs can be used from time to time in order to elaborate on the details listed in this main diagram. I have found that a single diagram of this kind can provide all of the information needed for the 45 minute introductory presentation, and it seems to work better (at least for the introductory lecture) than showing a whole series of slides.

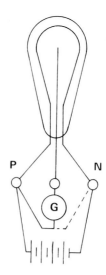

Fig. 1. Illustrating the Edison effect.[3]

	1	2	3	4	5	6
Mon.	A	B	C	D	E	F
	B	A	D	C	F	E
Tues.	E	F	A	B	C	D
	F	E	B	A	D	C
Wed.	C	D	E	F	A	B
	D	C	F	E	B	A

Fig. 2. Practical lab schedule (schematic).

It has been my impression that many lecturers whose experience has been gained at conferences will tend to show too many slides too fast. At conferences, it is usual to specify that one slide per minute is acceptable, but in a teaching environment such a rate is sometimes too high. I once attended a very effective presentation in which the lecturer, in a very large hall, used one slide for a presentation which lasted for twenty minutes. Of course, this cannot always be achieved, but there are times when this can be an effective technique to use.

Course textbook. There is a choice of books which are suitable as the textbook for an SEM short course. We have chosen to use one[4] that contains a comprehensive collection of micrographs which were taken, as far as possible, from the original publications in which the various developments of the subject were first reported, and also an extensive bibliography. Micrographs and diagrams published in the manufacturers' leaflets and newsletters can sometimes also form a useful addition to the course textbook, but here I would like to make the plea that manufacturers should give us more information relating to the specimen description, specimen preparation, instrumental details such as the beam current, incident energy, recording time, detector configuration and so forth when they publish micrographs. It has been my impression that if the lecturer is using an overhead projector then it is sometimes preferable that the participants should have copies of the micrographs in front of them, rather than changing back and forth between the overhead projector and the slides.

Conclusion

In this paper I have tried to summarize some of the main points which distinguish a short course from a conventional conference. During the past few years, there has been an increasing tendency for conventional conferences to include "teaching" sessions. However, I think that the need for a substantial amount of time at a conventional conference to be devoted to research papers and to descriptions of novel applications will tend to preserve the identity of the short course as a means by which the background of the subject may be learned.

References

(1) The expression "Edison effect" was first used by W.H. Preece, "On a peculiar behaviour of glow-lamps when raised to high incandescence," Proc. Roy. Soc. London, **38**, 219-230 (1885).

(2) T.A. Edison, "Electrical indicator," U.S. Patent 307,031 filed Nov. 15 1883, granted Oct. 21, 1884.

(3) Fig. 1 is based on E.J. Houston, "Notes on phenomena in incandescent lamps," Trans. A.I.E.E., **1**, 1-8 (1885).

(4) O.C. Wells, A. Boyde, E. Lifshin, A. Rezanowich, "Scanning Electron Microscopy," McGraw Hill, New York, 1974 (reprinted 1978).

Discussion with reviewers

T. Nagatani: It is important for beginners in scanning electron microscopy to receive lessons on the general concepts of SEMs and on the relevant basic physics. It is also necessary for them to learn about the various accessories for SEMs such as EDX spectrometers together with the appropriate scientific background.

It seems to me, however, that general tutorials, workshops or seminars for scanning electron microscopy are only to give physics oriented lessons. The reason for this is that most of those classes deal with ideal samples and give beautiful interpretations for interactions between (for example) bulk samples and electrons. There is a tendency for these classes to avoid the more complicated phenomena or problems often encountered in practical work.

Fig. 3 (on the right): Viewgraph used for introductory lecture: (a) Before use. (b) At the conclusion of the lecture.

(a) INTRODUCTION TO THE SEM:

ELECTRON-SPECIMEN INTERACTIONS:

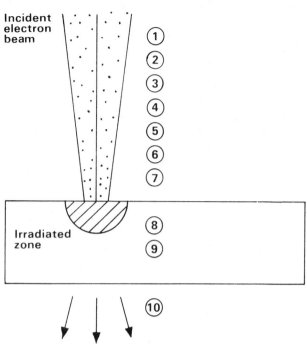

(b) INTRODUCTION TO THE SEM:

ELECTRON-SPECIMEN INTERACTIONS:

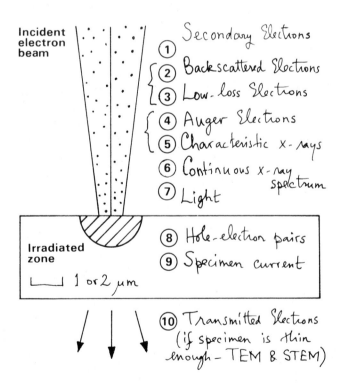

When a man wishes to use an SEM in his research or in his evaluation of some processes, it is not good enough for him simply to obtain good SEM pictures. He must also be able to understand the limitations of interpretation.

In scanning electron microscopy, image contrasts and analytical data depend greatly on the operating conditions. People must therefore understand what relations exist between the operating conditions and the needs of a particular sample. For most researchers or people working on quality control, SEM pictures or analytical data are just a few of several sorts of information available to them. In most cases, they will have already learnt something about the samples by means other than scanning electron microscopy. If an SEM picture contains artifacts or is incorrectly interpreted for some other reason, then this may lead them to a wrong conclusion.

Most SEM samples used in practical applications are not like the ones that are talked about by physicists. They are quite different from the ideal samples discussed in most classes. I would feel that true SEM schools must teach students the effects of changing the operating parameters of their SEM with these practical considerations in mind. For example, SEM images of biological samples may show the original morphological structure, additional structure caused by surface coating, carbon contamination layers and so on. Correct interpretation of this information is critical and important. I think that this is something that has to be taught in true SEM schools.

Author: Thank you for this comment. You are, of course, absolutely correct when you say that the interpretation of images which have been obtained in the course of practical investigations with the SEM should certainly be studied by the participants of an SEM course. It is also true, I believe, that sometimes SEM micrographs which are published in the literature are not accompanied by as much data as we could like (such as beam voltage, beam current, recording time, detector position and type, and so on). Might I make the plea that users of the SEM who publish their micrographs should give us more information of this type?

B.P. Jakstys: It is obvious that in a lecture course one can easily accommodate greater numbers of students than is possible in a laboratory. You, however appear to have successfully conducted SEM laboratories composed not only of large numbers of participants, but with participants whose scientific interests, accomplishments and experiences in SEM were highly varied. It would be of interest and most helpful to the readers of your paper to have a more detailed description of how your laboratory classes are conducted in order to achieve a satisfactory "hands on" experience within the span of one week.

Also, if there are any reference sources which you have found to be exceptionally useful in your short course, would you please share them with your readers.

I congratulate you in what seems to me the successful accomplishment of a herculean task.

Dr. Wells, I hope you will find it convenient to bring your teaching aids to the meeting in Washington and have them on display. I am quite sure that those interested in teaching SEM would find the materials used in your couse of great interest.

Author: Thank you for this comment. If you are agreeable, I would like to refer to these points when I answer your questions below.

B.P. Jakstys: The division of the participants of the SUNY short course into six groups of seven or eight students with a one-hour per station experience seems to be an extremely short exposure for the inexperienced user of the SEM. Does this mean that the students, in reality, get little hands-on experience but instead have different aspects of the microscope demonstrated to them?

Author: Certainly, the opening practical session begins with a description and a demonstration of the instrument. It has been our experience that many of the participants are unwilling to push themselves forward to actually operate the instruments, and a certain amount of diplomacy is needed to persuade everyone to spend some time on the machines. It was my impression that about 1 in 4 of the participants obtained more than an hour of actual operating time, while the remainder were content with less than this. A smaller number of participants find the time to spend several hours on the machines. You are perfectly correct when you draw attention to this as a sensitive issue, and we are constantly trying to find ways to improve out track record in this regard.

B.P. Jakstys: What is the average number of experienced-to-inexperienced participants attending the SUNY short course?

Author: About half of the participants have operated SEMs or some other type of electron-optical instrument before. I feel uneasy, however, in describing the remainder as "inexperienced" inasmuch as everyone has had experience in some relevant subject and can contribute usefully to the discussions.

B.P. Jakstys: Do you group the experienced SEM users with the inexperienced participants or do you group them according to experience?

Author: At first, we tried to group them closely by interest and experience. Nowadays, I think that it is better not to be too rigid in the way in which the groups are divided, because it encourages people who have different interests and backgrounds to learn from one another.

B.P. Jakstys: Would you please elucidate how a group of seven or eight participants can accomplish the mastery of a SEM within one hour? This allows the students less than ten minutes each with hands-on experience.

Author: I do not believe that it is possible to achieve "mastery" within one hour. What one **can** acquire, however, is the courage necessary to come back later and spend more time on the machine. The time spent on the machines in the organised groups is only a part of the hands-on experience. What usually happens is that the participants who become really interested in the operation of the actual machines find the time to spend more time on them in the evenings and in the afternoon break periods.

B.P. Jakstys: If it is not out of order, may I ask what SEMs do you use in the SUNY short course?

Author: Last year we had a complete range of instruments. Three of them were top of the line SEMs, one with a tungsten hairpin electron gun, one with field emission, and the other with a lanthanum hexaboride gun. There were two of the simpler instruments, one of which had a tungsten hairpin electron gun while the other had a field-emission type of gun. It is, I believe, extremely important for the participants to be shown a range of instruments, rather being shown only a single model.

G.R. Hooper: I have made a few suggestions in the text and feel that the paper is acceptable as written.

Could the author expand his own observationa and preferences to comment on the other short courses offered in North America. I feel such comment would allow many of us to get a better feel for short course structure.

Author: I am not too familiar with the other short courses which are offered. From what I have heard, the other courses would seem to offer a valuable introduction to the subject, which is our aim at the SUNY short course also.

G.R. Hooper: Is the presentation of and by instrument manufacturers really essential in your short course? It appears that it might use time needed for more basic instruction.

Author: I have come to regard the presence of instruments and the associated hands-on experience as being absolutely vital for the success of a short course. I would say that between 30 and 50 percent of the participants' time should be spent in the presence of the instruments. In my opinion, the instrument manufacturers contribute significantly to the success of short courses in exactly the same way as they contribute to the success of a conventional conference such as this one.

G.R. Hooper: Is it really true that "effectivenesss" of instruction can be judged by the number of questions elicited? A very capable instructor might not need to clarify points via questions due to a thorough and clear lecture.

Author: My comment about the "effectiveness" of a lecturer being related to the number of questions he receives was obviously not quite correct, so I deleted it from the final version of the paper. My own observation of lectures in process (either from the speaking or from the listening side) has led me to become very uneasy if the lecturer does not receive any questions atall. Personally, I find it to be very much easier to lecture if I am asked a few (but not too many) questions - but I am not sure how widespread this belief may be. Your comment about a "very capable instructor" not needing such help from his audience may very well be true - but I am not quite certain how far this ideal situation can be realised in practice, especially when there is such a wide range in the backgrounds of the participants.

SCANNING ELECTRON MICROSCOPY/1979/II
SEM Inc., AMF O'Hare, IL 60666, USA

THE DEVELOPMENT AND IMPLEMENTATION OF AUTOTUTORIAL METHODS FOR
ELECTRON MICROSCOPY UTILIZING SOUND ON SLIDE PRESENTATIONS

Judith A. Murphy

Center for Electron Microscopy
Southern Illinois University
Carbondale, IL 62901

Abstract

An effective autotutorial method involving taped slide presentations was developed for steps involved in the complex and time consuming specimen preparation and instrumentation procedures used in electron microscopy.

The system allows learning researchers to start and advance at their own level and speed until competence is achieved. Evaluation procedures are carried out at each competence level by qualified electron microscopy (EM) staff. The equipment and personnel needed to implement this autotutorial have been evaluated and are herein discussed. Efficacy studies have been successfully carried out to determine the usefulness of this type of instruction for complex procedures and instrumentation in electron microscopy.

The main objectives at the onset of this project were as follows: (1) to set up effective procedural taped slides for autotutorial instructions for instrumentation and preparation procedures in EM which would complement and expedite instruction in first experience training and refresher sessions, thus allowing more time between instructor and learning researcher for advanced discussion of theory, review of learned knowledge and interpretation of results; (2) to allow learning researchers to advance at their own pace; and (3) to increase the number of people that can be trained using the minimum number of qualified staff and at the same time increase the quality of training for each individual, thus facilitating better use of available equipment.

KEY WORDS: Teaching SEM

I. Introduction

Because of the ever growing need for people to acquire skills in Electron Microscopy for research use in multidisciplines, there was an increasing concern to develop methods which would enable more people to be trained more effectively and more efficiently.

Previous to the inception of an EM Center on our campus, there were only a few people on campus who used an electron microscope. Training was usually passed down in an apprenticeship fashion. This was ineffective because of the limited background of the trainer: merely using an electron microscope does not necessarily enable a person to train someone else in the procedures and instrumentation involved. Even with a highly qualified trainer, there is still an instructor:student ratio of 1:1, which results in overwhelming training times with much necessary repetition. Our task, therefore, involved setting up training methods for faculty senior scientists and graduate student researchers from all disciplines to familiarize them enough to effectively use the instruments and learn specimen preparation procedures and instrumentation pertinent to their particular research areas such that these researchers could utilize these valuable tools to help them solve their own research problems. This type of training is slow and time consuming.

In the learning process in this exacting field, a particular procedure or instrument operation must first be demonstrated to the learning researcher. He must then practice carrying out the procedure and/or instrument operation in a dry run fashion to ascertain that all precautions and proper sequences are understood. He then must attempt the procedure/operation in earnest. All of these steps must be under careful expert supervision to insure the correct operation of the delicate instruments involved. Beginner's mistakes, which can cause many hours of repair time and expensive part replacement cannot be eliminated but must be minimized particularly in a facility where a multitude of researchers depend on the use of the equipment, all of which makes the effective learning process even more important.

Some of the main problems facing EM instructors are (1) extensive training time is involved because of the multitude of steps involved in the operation of the various instruments and detailed procedures involved in specimen preparations which often take several days or weeks; (2) people utilizing the EM center come from all disciplines (Biological and Physical Sciences) and bring with them a multitude of backgrounds so the training methods must be extremely versatile and at the same time individualized; (3) careful and effective evaluation and follow-up training is extremely necessary, all of which is in addition to the time consuming basic training; (4) refresher lessons are often necessary for people using EM procedures and equipment at intervals with extended elapsed time between use.

As a result of these problems, the usefulness of visual aids appropriate for our EM Center was investigated.

II. Objectives

The basic objectives for utilizing audiovisual aids were as follows: (1) to increase the number of our faculty and predoctoral students who could use EM techniques in their research; (2) to increase the amount of advanced training by reducing time devoted to basic training by the EM Center staff; (3) to provide faculty and student refresher courses without retarding basic or advanced training activities; and (4) to standardize the present training program which varies in quality because of instructor fatigue.

III. Choice of Autotutorial Method

Several choices were possible. A taped instruction training can be used where the student sits at the instrument while listening to the tapes.This, however, allows no visual communication with the tapes. One can also use slides alone. These slides are then used in conjunction with a script which is read by the student. This, however, does now allow the student to participate completely in the process of learning the instrumentation as he must be concerned with reading the script at the same time. Another method is to use audio films; however, this has obvious limitations for individualized training at each instrument due to cost of production and presentation. If a portable computer terminal is available, programs can be written for EM training. This is also rather costly.

After considering the problems of the listed methods, an autotutorial instruction method was chosen for our EM Center utilizing sound on slide presentations of operational procedures for instruments and specimen preparations in electron microscopy. A picture truly can be worth a thousand words: with the use of tapes and slides together, the slide can zoom in on the point of interest and immediately focus the student's attention on the subject matter on the tape.

This method in conjunction with the fact that the student can sit directly at the instrument or prep area during the presentation was found to provide effectively and more efficiently a high quality means of transmitting the needed

procedures in a fresh and concise step-by-step manner simulating the 1:1 student:instructor relationship. This then allows the instructor with enough time for the important task of discussing theory, the review of learned knowledge and interpretation of results with the student, which is so often slighted because of the extensive time involved with basic repetition training.

For our training program, the requirements for the audio visual equipment were as follows: (1) inexpensive (approximately $1000); (2) small, lightweight and portable such that it could be placed directly by the instrument or prep table; (3) rear screen so that projection is at the instrument; (4) external speaker such that information can also be conveyed to large groups if desired; (5) instant replay to repeat audio at any time; (6) pause control to hold at any time; (7) external earphone jack for individual use; and (8) ability to add or eliminate slides without worrying about synchronization.

IV. Materials and Methods

For the initial filming of all procedures at the instruments, diagrams, and study goals and for continual updating for immediate implementataion of new methods and instrument modification, a 35 mm camera with a close up lens (e.g. 55 Macro Lens, Nikon) and a wide angle (28 or 35 mm) lens was found necessary. Two photographic flood lights with reflectors and stands were used for lighting. Kodachrome EHB or Ektachrome were used in all sequences.

The projector/recorder chosen was a unit manufactured by 3M Company (Model 625, Visual Products Division, 3M Center, St. Paul, Minn. 55101). See Fig. 1. The heart of the system is an audio diskette where each slide has its own 30 sec tape disc. The instructional message is recorded directly onto the magnetic sound disc in a frame. It works like a magnetic recording tape. See Fig. 2. One can record, erase and re-record as many times as is desired. Any 35 mm or 2x2 transparency snaps into the back of the frame. While a student is listening to the message he can study visual information provided on the slide. Picture and message are locked together with no problem of synchronization. This enables easy updating of any procedure with no loss of synchronization. This units, as most, can be operated in automatic or manual and can be reversed or repeated at any time.

This particular unit also has a responder available whereby multiple choice questions can be answered and recorded before the slide will change and an additional responder whereby duplicate additional sound slides can run automatically and unattended.

V. Experimental Design

1. Procedures/instruments are chosen which are to be used for the autotutorial method.
2. Study goals and basic concepts must be established which are necessary for understanding the operation of the instrument or procedure.

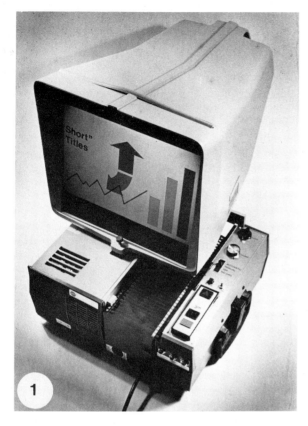

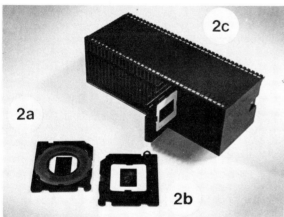

Fig.1. Projector/recorder unit by 3M, Model 625, with rear projection screen displaying 2x2 slide. Figs.2a,b. Audio diskette cartridge with slide. 2a. Tape side where each slide has its own 30 sec tape disc. 2b. Side in which slide slips into. Fig.2c. Cartridge tray for taped slide diskettes.

---

3. A script for each instrument/preparation is outlined in a fashion clearly indicating "how to do it" or "how to use it" and necessary precautions one must take when using the particular instrument or prep method. It is advisable to have 3 or 4 EM people read the script and give constructive suggestions.

4. Set ups and operation of instruments to be used during filming must be prepared.

5. Lighting set ups, filming, processing and mounting of slides can be carried out by any student familiar with 35 mm photography.

6. Slides must be edited and ordered according to procedural steps for each instrument and preparation.

7. The script used to film the sequences is recorded on the appropriate disk.

8. Preliminary evaluation is conducted to ascertain the following: (a) fluency and continuity of sound on slide presentation; (b) completeness of content for each procedure/instrumentation; (c) fulfillment of study goals by each slide sequence; (d) inclusion of adequate information in each level of a sequence so that competency can be achieved at each plateau. People were chosen in a non-science field and asked to use the sound on slide presentations for preparation equipment (i.e. critical point dryer, freeze dryer, etc.) to evaluate the completeness of the presentations. If necessary, procedures were modified for completeness.

9. Since new methods and instrument modifications are often introduced in the area of EM, camera and lighting should be on hand or easily available to update the taped slide presentations when necessary.

10. The basic staff needed to set up the sound on slide presentations is one person familiar with 35 mm phototography, and a professional EM person. It is also advisable to have 2 or 3 people familiar with EM to read the scripts as well as actually use the sound on slide presentations, thus allowing constructive feedback.

VI. Use of Sound on Slides in EM Center

Each instrument/preparation method has a cartridge(s) of taped slides which can be checked out by the learning researcher with a sound on slide projector with a rear screen. The entire set-up is easily portable on a small cart to fit next to any of the instruments or preparation areas in the EM Center. Head phones can be used if desired.

First a list of necessary accessories for the prep method is given and assembled by the student. The student can then start the taped slides and is oriented to the particular instrument/preparation. He is then instructed how to operate the instrument or carry out the process. The student sees how to operate the various controls on the slide and he is then instructed to do so in a step by step manner as the taped slides progress. Included also on the tapes are brief theoretical introductions and study goals at each level which the student must understand before attempting to further operate the instruments and/or carry out technical processing. Questions are raised during the slide program and time allowed for answering. If the proper answer is given, the student is instructed to proceed. If he does not know the correct answer, he is instructed to return to a previous slide or slide sequence and continue from there.

VII. Evaluation Procedures

At the end of each learning unit the student must review his work with the instructor, who helps him to establish whether he should continue to the next sequence or repeat a certain series. The discussion session between the instructor and student allows explanation of anything unclear as well as a review of learned knowledge. The student thus develops the correct measure of self criticism and insures the smooth functioning of the delicate instrumentation and the overall facility. At the end of the sequence for each instrument/procedure, the student is evaluated on his effective use of the instrument/preparation and his knowledge of necessary precautions as indicated on the slides. If necessary, repetition of certain slide sequences is recommended for competence to be achieved. If competency is established, then the student is allowed to sign up for the instrument use as scheduling permits. In the case of multiple sequences, especially for the electron microscopes, the student is instructed to continue on to the next slide sequence.

The time involved for each taped sequence varies with the particular instrument or procedure. The instruction for the more complicated specimen preps or equipment as well as the instruction for the electron microscopes is made in several series where each represents a certain plateau of competence which must be achieved before going on to the next plateau. Again, evaluation procedures are set up to test competence at each level. This serves not only to check the competence of the student but also to check the clarity and understanding of the sound on slide presentations. Where appropriate, lectures, hand outs, and reading assignments are used concurrently with taped slide presentations.

It should be pointed out that initially the instructor does give a general review of the instrument noting special precautions before the student is allowed to initiate the sound on slide presentation.

VIII. Advantages of System Chosen

This system has several advantages over other types of media.

1. The taped slide sequence closely simulates a 1:1 instructor: student ratio where each student receives the same high quality instruction in a fresh and concise manner.

2. With taped slides, the student can hear the technical terms pronounced. He can also hear the tone of voice which places emphasis on important points and expresses authority not sensed by reading the written word.

3. With the slides, the student's attention is immediately visually focused on the particular controls/procedural steps that are being referred to on the taped instruction.

4. The student can advance at his own pace according to his background and can break up the sessions in time allotments fitting his schedule. Information obtained can be reviewed as often as desired.

5. The instructor's time can be spent in individual review and discussion of what the student has learned as well as more time in discussion of results, which is so often slighted because of the extensive time spent in basic repetition training.

6. Electron microscopy procedures and instrumentation require careful attention to a multitude of details which frequently go unappreciated in oral instruction but which can be built into taped slides.

7. Taped slides serve handsomely for refreshers. Since most instrument procedures entail several detailed steps, which are easily forgotten, taped slide sequences serve to refresh the previously trained person as well as to teach the beginner. This is found necessary many times when a person has been away from the instruments for awhile and needs a "refresher" lesson. Many hours a week spent on this type of training can be eliminated with the use of taped slides.

8. This method allows more people to be trained more effectively in less time.

9. The particular projector/recorder system used in our EM Center has several advantages. (a).New methods are continuously reported and can be immediately implemented by inserting slides with their individual tape disc without desynchronizing the entire series making retaping necessary. (b).Going back to a previously viewed slide in a series is smoother without losing any continuity or desynchronization of tapes and slides. (c).Initial taping and synchronization is easier because of the individual disc for each slide. (d).Instantaneous re-recording of a particular slide is possible without loss of synchronization or blank spaces in the tape.(e).The unit with its rear projection screen and head phones fits neatly on a small cart which can be wheeled to any instrument or prep area in the lab.

IX. Instruments/Preps Used in Sound on Slides

The following instruments/preparations have been or are being set up for use with the autotutorial method described, each of which is divided into theory, instrumentation and technique sequences: scanning electron microscope, x-ray analysis (semi-quantitative and qualitative), transmission electron microscope, critical point drying, freeze drying, fixation procedures, preparation of substrates for mounting specimens for TEM and SEM, vacuum evaporator and the evaporation of metals and carbon on specimens for viewing, carbon rod sharpener, preparation of thin sections (obtaining samples, fixation of specimens, embedding of specimen, sectioning, and staining), knifemaking, specimen trimming by hand, Reichert trimmer, and LKB pyramitome, staining for SEM and TEM, metal shadowing, replication of specimens, autoradiography, freeze etching, photographic techniques necessary for obtaining a good quality micrograph from an electron microscope negative, slide making, etc.

When appropriate, there are several modifications suggested in the slide sequences depending on the type of sample used.

X. Conclusion

In our lab, the method described allows more people to be trained more effectively in less time. This is becoming more important as the need for training in electron microscopy skills is ever-growing and the funds for more staff are ever-decreasing. The standards of training MUST remain high which leaves only the alternative of making teaching methods more effective in less time: thus the necessity for using autotutorial instruction methods.

Acknowledgment

I wish to thank Ms. Laura Nelson for critical reading of the manuscript and Mr. Andy Piper for photographic assistance.

Discussion with Reviewers

K. Baker: What kind of feedback have you gotten from the users of your system?
Author: The feedback has been very positive. The students, i.e. anyone using the learning modules, have been able to effectively use the techniques and instruments that have been slided. There is of course still necessary instructor-student interaction, the major instruments requiring more time after the use of the sound-on-slide system than the small prep equipment.

B. P. Jakstys: Am I correct in assuming that you use your autotutorial of sound-on-slide method to teach EM to individuals in a non-structured lab type of class setting?
Author: No. The method is used in a structured EM course and in unstructured situations where individuals want to learn only microtomy or how to agar embed bacteria for EM etc.

B. P. Jakstys: In your opinion, how well would the autotutorial sound-on-slide method lend itself to be applied to a structured EM class (lab only) of 20-30 novices? What would you suggest as a recommended minimum and maximum number of instructors necessary to supplement the sound-on-slide teaching method in the above structured class system?
Author: In the size laboratory where there is enough equipment to teach 20 or 30 students per semester the system should also work well. For 20-30 students, 1 projector/recorder would be necessary to make the original tapes and 3-4 playback units for many students to use units at one time. The number of instructors necessary to supplement the sound-on-slide method in this situation would depend on how the lab sessions were organized and how much EM equipment was available at one time.

B. P. Jakstys: In a structured, large class system, would you say that lecture type of information (theory of electron optics, etc.) could also be effectively transmitted to the students using the sound-on-slide method or is this teaching approach best left to the training of EM lab procedures, methodology and instrumentation only?
Author: Certainly, lecture-type information could be effectively transmitted using this method. If one desired to use autotutorials for lectures, however, a taped lecture on a TV monitor might be easier for a large group as you mention. I myself would prefer to personally lecture to an EM class using visual aids for clarity. In this fashion the entire group can profit from questions and discussions.

E. L. Thurston: I would be very interested in the nature of the checkout procedure used to determine competency for a given instrument.
Author: Regardless how an individual learns how to use an instrument, competency check-out is the same. The instructor tests the individual on how well he can use the instrument. This can involve alignment tests, through-focal series etc.

Additional discussion with reviewers of the paper "Mass Measurement with the Electron Microscope" by J.S.Wall continued from page 302.

M. K. Lamvik: Please explain the microscope optics that make the collection efficiency of the detectors independent of specimen position.

Author: The beam below the objective lens is strongly diverging, the exit angle being a local maximum when measured as a function of lens current. Therefore small changes in specimen height which require refocusing do not appreciably affect the detector acceptance angles. This arrangement also eliminates the need to unscan the beam in normal operation.

D. F. Parsons: Since the difficulty of avoiding contamination of molecules is stressed, why is it necessary to freeze such thick (10-100 microns) layers of water? Using a humidity box it would be easy to make preparations with much thinner layers of water (and hence less contaminating material).

Author: Dr. Hainfeld in our laboratory is adapting the methodology for preparing thin frozen samples to suit our needs. We are most anxious to use this technique since, as Dr. Parsons points out, it should greatly relax the requirements for low ionic strength in the sample solution. However, for the many samples which are stable in deionized water, the present technique is adequate. Some unstable molecules have been successfully measured following crosslinking with glutaraldehyde.

M. S. Isaacson: What is the mass lost by an fd virus after irradiation of several hundred $el/\text{Å}^2$?

Author: Under our conditions, approximately 40% of the initial mass is lost before the virus reaches a stable configuration.

M. S. Isaacson & R. E. Burge: Could you estimate the error involved in Eq. 3 in the determination of the molecular weight by assuming all atoms are carbon (i.e. how much error is involved in assuming the H,N,O,P, and S atoms all scatter electrons exactly like carbon)? How are specimens stained with heavy atoms treated?

Author: Eq. 3 does not assume that all atoms scatter exactly like carbon, but rather that the scattering from an atom is proportional to its atomic weight. Theoretically[14] one expects the cross section to vary roughly as $Z^{3/2}$, so Eq. 3 may not be such a bad approximation. Experimentally one uses a mass standard which is similar in composition to the molecule under study. Therefore I would estimate that this introduces an error < 1% in a practical case where no heavy atom contaminants are present.

Specimens stained with heavy atoms can be analyzed using the same techniques if care is taken to remove unbound heavy atoms. A reference unstained particle is observed under similar conditions to determine the incremental scattering due to heavy atom binding. Alternatively one can use the different Z dependence of inelastic scattering to "subtract out" the low Z atoms and measure the remainder.

R. E. Burge: The molecular weight of the given virus example is not determined, but normalized (Figure 9) to a "known" value. What are the problems of accurate calibration of σ_c? What are the real problems involved in actually "integrating up to the radius where the particle density has fallen to background level" when minimum dose conditions (low signal to noise) are being used?

Author: The measurements presented in this paper are all absolute scattering cross sections. We have chosen not to present them as such until actual collection efficiencies are evaluated more carefully. The scattering integration is normally carried slightly past the apparent edge of the particle to avoid systematic errors. This introduces a larger error than necessary in the background subtraction, but this error is random (at low dose) and can be averaged out by considering many particles. The problem of locating the particle edge can be alleviated by recording a second scan of the same area at higher dose and using this to define the boundary.

SCANNING ELECTRON MICROSCOPY/1979/II
SEM Inc., AMF O'Hare, IL 60666, USA

SCANNING AUGER MICROSCOPY - AN INTRODUCTION FOR BIOLOGISTS

A. P. Janssen and J. A. Venables

School of Mathematical and Physical Sciences
University of Sussex
Brighton, Sussex, BN1 9QH
U.K.

Abstract

The three micro-analytical techniques of Auger electron spectroscopy, X-ray spectroscopy and electron energy loss spectroscopy are compared, with particular emphasis on Auger spectroscopy and its potential for scanning Auger microscopy of biological material. Scanning Auger microscopy combines high lateral and depth resolution, sensitivity to light elements, and the ability to observe bulk samples. It thus combines many advantages of both electron energy loss spectroscopy and X-ray spectroscopy, with a volume sensitivity much higher than for X-rays, by a factor of up to 10^5 in modern instruments.

Auger electron spectroscopy is introduced and its sensitivity is compared with the other techniques. The production efficiencies of the techniques are not radically different and the relative speeds with which analysis can be performed are largely determined by the instrumental efficiencies for detecting the relevant particles.

The types of analyser and electron gun combinations presently in use for scanning Auger microscopy are described. Projected improvements in detection techniques for Auger electrons which aim to increase the quality and speed of quantitative analysis are discussed.

Quantification methods for Auger analysis are already well developed for wide beam surface studies, with an accuracy which can approach that of X-ray analysis. The special problems of quantification on a microscopic scale are described. Recent experiments have indicated the effects of specimen topography and electron backscattering from the substrate on the Auger signal strength and spatial resolution, and have indicated the nature of some electron beam induced effects.

KEY WORDS: Auger Electron Spectroscopy, Scanning Auger Microscopy, Microanalysis, Surface Analysis, Electron Sources, Field Emission, Biological Microanalysis, Electron Energy Analyser, Ultra High Vacuum, Specimen Preparation, Quantitative Analysis

1. Introduction

The three main methods used for in situ chemical analysis in the SEM are X-ray analysis, Auger electron spectroscopy and electron energy loss spectroscopy, all of which rely on initial characteristic inner shell ionisation by the microscope's high energy electron beam.

Auger electron spectroscopy has been used extensively for surface chemical analysis from atomically clean and well prepared samples using large area, relatively low energy, electron beams. In recent years, there has been substantial effort in developing SEMs combined with AES for high spatial resolution surface chemical analysis. The technique of scanning Auger microscopy (SAM) combines a number of advantages of X-ray spectroscopy and energy loss spectroscopy such as the ability to use bulk specimens, accurate quantitative interpretation, sensitivity to light elements, and high spatial resolution. In combination with ion sputtering, depth analysis can also be performed with high resolution.

The sensitivity of AES to the top few atom layers of the surface implies that surface cleaning and the ultra-high vacuum techniques should be used. This is a major departure from traditional electron microscopy practice; however, the detection of Auger electrons from the sub-surface region may eventually de-emphasise somewhat the need for clean surfaces, especially for biological applications.

In this article we compare methods presently available for chemical analysis in the SEM with particular emphasis on AES, and explore the possibility of using SAM for analysis of biological material.

Scanning Auger microscopy has hardly been applied at all so far to biological samples. The study of such samples which damage easily and which are difficult to clean may be limited by the very surface sensitivity of Auger analysis and by the high beam current densities needed. Thus biological applications may act as a spur to develop methods of extracting Auger information from the subsurface region with greater sensitivity than is possible at present.

Symbols

E_p - primary electron energy (eV)

I_p - primary electron current (A)

Φ_k - total ionisation cross-section of K shell by electrons of energy E_p (m^2)

N - number of surface atoms (m^{-2})

γ - Auger emission probability

r - backscattering factor for a particular material

ϕ - angle of incidence of primary beam

T - analyser angular transmission factor

ϵ - analyser energy transmission function

Z - atomic number

ω - X-ray fluorescence yield

U - reduced energy = E_p/E_k for shell ionisation

λ - Auger electron inelastic mean free path (mono layer units)

Ω - analyser collection solid angle

$\Phi(\alpha)$ - ionisation cross-section integrated over scattering angle $\leq \alpha$

P_r - relative sensitivity factor for Auger intensities

η - backscattering coefficient = number of emitted backscattered electrons per incident electron

C_A - fractional concentration of element A

$f(C_A)$ - Auger correction factor for concentration C of element A

P_A - Auger derivative (dN/dE) peak intensity of element A

β - number of Auger electrons produced within the Auger escape depth per emitted backscattered electron.

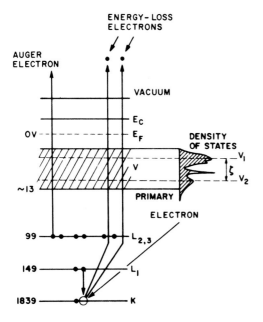

Fig.1. Diagram of the $KL_1L_{2,3}$ Auger process. Energy levels are listed on the left. The K electron loss process is also shown. (Chang ref.1).

2. What is Auger Spectroscopy? A Comparison with X-ray Analysis and Energy Loss Spectroscopy

2.1 Description of physical processes

Auger emission, X-ray emission and loss-electron emission are all initiated by ionisation of an inner shell by a sufficiently energetic primary electron. This is shown for the case of K-shell ionisation of silicon in Figure 1. Auger electron or X-ray emission are the result of de-excitation processes involving electrons from other shells and re-emission of an electron or X-ray quantum to carry away excess energy. Emission of an Auger electron leaves the atom doubly ionised and these holes are quickly filled in a free electron environment. As seen from Figure 1, Auger emission is a three stage process involving at least two energy levels. The transition shown is referred to as the $KL_1L_{2,3}$ transition (using X-ray notation) in which the Auger electron has an energy which is reasonably accurately described by

$$E_{KL_1L_{23}} = E_K(Z) - E_{L_1}(Z) - E_{L_{23}}(Z+1) - \phi_A \tag{1}$$

where ϕ_A is the analyser work-function. The shift in outer levels caused by ionisation of an inner shell is approximately compensated by replacing the outer shell energy of an atom with atomic number Z by the ion with atomic number Z + 1 (2).

It can be seen from figure 1 that there are a number of other possible paths for de-excitation involving other energy levels. The relative probability of these events and therefore the relative peak intensities are determined by the strength of electron-electron interactions between shells, being largest for electrons with more closely overlapping orbitals. As three electrons are involved, Be is the lightest detectable element. Auger peaks used for surface analysis typically have energies in the range of a few tens of eV to 2 keV; higher energy peaks are also produced in the 10 keV range, for elements of medium to high atomic number. Chemical environment may affect the shape and energy of Auger peaks somewhat (particularly those involving the valence band); valence band peaks are in general broadened by the solid state environment, and interatomic transitions are possible which involve energy levels of neighbouring atoms[3,4].

Auger, X-ray and EELS spectral energies are similar since ionisation is common to all three, and the initially ionized level is usually much more strongly bound than the other levels involved. Auger electrons and X-ray photons have a characteristic energy which is independent of the fate of the incident electron. However, in EELS it is the onset of energy loss which has a characteristic energy; higher energy losses are also possible when the

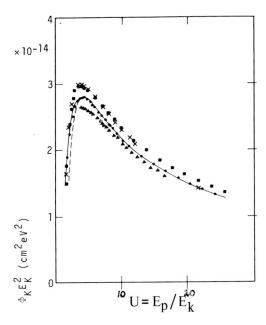

× 10⁻¹⁴ is shown at top left of plot.

y-axis: $\Phi_k E_k^2 \ (cm^2 eV^2)$

x-axis: $U = E_p / E_k$

Fig.2. Comparison of experimental total cross-section $\Phi_k E_k^2$ plotted against $U = E_p/E_k$ (Powell ref. 6)

emitted electron (which is not detected) has greater than zero energy. Typically, Auger peaks are 5-20 eV wide, whereas EELS "peaks" from inner shells are $\gtrsim$ 100 eV wide; characteristic X-ray energies are of course much better defined than either.

2.2 Auger Production Efficiency

The Auger current produced by a primary electron beam current I_p is well known to depend on the total ionisation cross-section for a particular shell Φ (which depends on the primary energy E_p), the probability of Auger emission compared to X-ray emission, the probability of a particular Auger decay occurring, and also on the backscattering factor r which accounts for additional ionisation by backscattered electrons. The basic equation for Auger production can be written.

$$I_A = I_p \cdot N \cdot [\Phi \cdot \gamma \cdot r] \sec \phi \cdot (T \cdot \epsilon) \qquad (2)$$

for a surface layer comprising N atoms/m², following a number of authors (5-8). The factors T and ϵ are instrumental angular and energy collection efficiency factors respectively for a particular analyser, and will vary with analysing energy and the particular sample and analyser configuration.

It may be useful at this stage to compare the influence of the various factors in equation (2) on the detected Auger signal with analogous factors for the EELS and X-ray techniques. Although all three techniques involve an initial ionisation event, the decay processes in Auger and X-ray emission are independent of how ionisation is produced and are dependent only on the total number of ionisations.

Φ is therefore the total cross-section integrated over all scattering angles. In EELS, however, loss electrons are detected over an angle $\alpha \sim$ 10m rad., about the forward direction and over a limited energy range Δ above the excitation edge. A differential cross-section must be used, integrated over the angle α and energy Δ. The total cross-section has been estimated using a number of formulae derived from the Bethe cross-section. For primary electron energies from 10 kV to 60 kV encountered in the SEM the ionisation cross-section may vary by up to a factor of 5 from 1.0×10^{-23} m² for K shell ionisation with E_k = 270 eV, E_p =10 keV, to 2.7×10^{-24} m², for E_k =270 eV and E_p = 60 keV, (see figure 2). The primary energy is therefore an important parameter in deciding optimum conditions for Auger sensitivity, and relatively low primary energies will be favourable especially for the low atomic number elements encountered in biological material. Beam current and spot size will however be improved by higher primary energies (see § 3.1), and an optimum primary energy will thus depend on the type of information being sought.

The Auger emission probability γ is complementary to the X-ray fluorescence yield ω: i.e. (1-ω). For low atomic number elements for which the X-ray yield is very low, the probability of Auger de-excitation is correspondingly high.

When several Auger transitions are involved the individual γ values, γ_i, sum to give $\sum \gamma_i$ =(1-ω), with the individual γ_i values essentially constant for a given element, independent of its environment.

The Auger backscattering factor arises from electrons which have undergone large angle scattering in the bulk solid and have re-emerged from the surface with sufficient energy to cause ionisation within 1 nm or so of the surface. For X-rays a similar backscattering factor can be defined, but in this case ionisation produced within the scattering volume (0.5 μm)³ below the surface contributes to the total yield[9]. The Auger backscattering factor r can vary from almost 2, i.e. equal Auger yield from primary ionisation and backscattered electrons for high Z, to about 1.2 for low Z. The backscattering measurements shown in figure 3a (10) demonstrate that r can

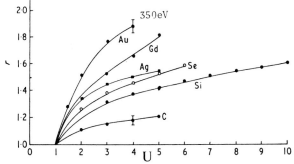

Fig.3a. Dependence of Auger back-scattering factor with $U = E_p/E_k$ Auger for a number of elements (Smith and Gallon ref.10)

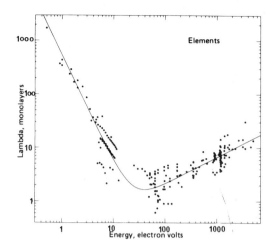

Fig.3b. Experimental values of the inelastic mean free path (IMFP) for electrons. (Seah ref. 14)

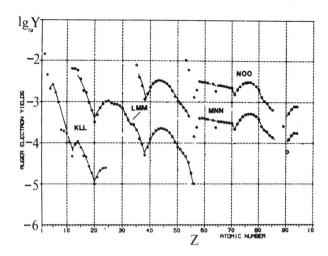

Fig.4 Auger yields for the elements for primary energy $E_p = 10$ keV. (Staib and Staudenmaier ref.19).

vary substantially over the range of reduced energy $U (= E_p/E_k)$ possible in the SEM, and will generally increase slowly with higher E_p.

The surface sensitivity of Auger spectroscopy arises from the short inelastic mean free path (IMFP) of low energy electrons. Experimental values have been compiled by a number of authors (11-14) for a range of elements. The values shown in figure 3b are in terms of monolayers. The total Auger intensity for bulk specimens is found by summing over each successive monolayer in depth with the relevant IMFP (15),

$$I_A = \sum_{n=1}^{\infty} A_n \ \exp \ \{-(n - \tfrac{1}{2})/\lambda \ \} \qquad (3)$$

where A_n is the number of electrons produced in the nth layer, and λ is the measured IMFP in monolayers (15). This description has been verified experimentally for multilayer crystal growth (16-18). Generally an effective escape depth λ_e can be used for bulk samples so that N in equation 1 is replaced by $\lambda_e \rho C$ where ρ is the atomic density (atoms m^{-3}) and C is the fractional concentration of a particular element at the surface.

The length of the primary electron path within the escape depth for Auger electrons depends on the angle of incidence ϕ; however the backscattering contribution does not precisely obey the sec ϕ dependence as assumed in equation 2 (10), so that r is itself a weak function of ϕ.

The Auger collection efficiency terms T and ϵ will be dealt with in section 3.3. In general the analogous terms are high for EELS detection, and low for AES and X-ray detection.

2.3 Comparison of sensitivities for Auger, X-ray, and EELS

The number of Auger electrons produced per incident electron can be estimated from equation 2. For example, for bulk carbon (graphite), using the KLL Auger peak at 270 eV, with a 25 keV primary beam, $\Phi = 4.6 \times 10^{-24} \ m^2$. With $\gamma = 1.2$, $\gamma' = 1$ $\rho = 1.75 \times 10^{28}$ atoms m^{-3} and $\lambda_e = 1.5$ nm., this gives an Auger yield $Y \simeq 1.5 \times 10^{-4}$ into 4π steradians. Figure 4 shows values of Y for all elements calculated by Staib and Staudenmaier (19) for $E_p = 10$ keV. It is seen that the variation in yield across the periodic table is within a factor of 5 for the most favourable transitions, and the extremes of yield are $10^{-2} > Y \gtrsim 10^{-5}$.

In comparison, it is well known that for X-rays the low fluorescence yield for low Z is a severe restriction to light element detection; the fluorescence yield is 0.004 for carbon and is greater than 0.5 only for $Z > 31$. For C, Si and Ca K shell ionisation, typically 0.015, 0.02 and 0.01 photons per incident electron are released into 4π steradians respectively for $E_p = 25$ keV; the yields decrease more strongly at higher Z. The soft X-rays for lower Z are preferentially absorbed in the material and especially in any windows placed in front of the detector.

A comparison with EELS is difficult since the number of loss electrons produced varies with the sample thickness. However for say a carbon specimen thickness of 100 nm, with $N = 1.8 \times 10^{21}$ atoms m^{-2} and for $\alpha = 20$ m rad., $\Phi_d(\alpha) = 2 \times 10^{-24} m^2$ (20) gives approximately 3.6×10^{-3} loss electrons per incident electron with $E_p = 60$ keV.

On the above figures, the three techniques have production efficiencies, Y, which are broadly similar, but with AES tending to be the smallest on account of the small escape depth λ_e. These values of Y would be the only important factors limiting data collection if ideal detectors could be designed, and noise from all other processes eliminated. In practice however, the collector will collect only a fraction of the available signal and generate noise, and all signals are superimposed on a background, which in the case of AES and EELS, is often larger than the signal. The angular distribution of particles to be detected is also quite different for EELS (which is sharply forward peaked, typically 30% being at $\alpha \lesssim 20$ m rad.) and for X-ray and Auger emission, which are more nearly isotropic. The overall sensitivity for AES is therefore rather low in comparison to EELS and X-ray techniques and particular care has to be paid to maximizing the geometric (T) and energy (ϵ) factors in equation 1, in order to obtain acceptable signal to noise ratios and data collection times. This is discussed in the next section.

3.   Instrumentation for micro-Auger analysis
3.1 Analyser-gun combinations
        Auger analysis in the SEM is now done exclusively with electrostatic analysers which must therefore possess high collection solid angle Ω, and high energy resolution, i.e. ΔE smaller than the Auger peak width. Noise introduced by either the secondary electron generation in the spectrometer, or by the detector must be much smaller than that originating in the Auger current. If the primary beam current is say 5×10^{-9} A we can expect from figure 4 between 10^{-13}A and 10^{-11}A total emitted Auger current. For practical analysers this is reduced by $T.\epsilon$. which can be as small as 10^{-3}.

        Only band-pass analysers can be considered for low current detection since high pass analysers such as the retarding field analyser (RFA) which are used extensively for wide beam surface studies at higher beam currents also detect noise from the spectrum above the Auger energy. The cylindrical mirror analyser (CMA) and concentric hemispherical

analyser (CHA) are in general use for micro-Auger analysis and a band-pass version of the RFA(19) has also been used. Although the behaviour of these analysers is fairly well understood (23-29), (for reviews see 21, 22), a choice between them depends not only on analyser operating conditions but also on specimen geometry and accessibility, particularly where the probe forming lens of the SEM occupies a large solid angle around the sample. A number of solutions to this problem are shown schematically in figure 5. (ref.30).

        The CMA has a second order focus for electrons accepted at 42.3° to its axis and usually operates with constant resolution $\Delta E/E < 1\%$, and with a maximum transmission, T, of about 12%. At this resolution ϵ is 0.05 - 0.10 depending on the details of detection scheme and Auger peak width (30). Thus the analyser factors in equation (2) are rarely greater than 5.10^{-3} and can easily be as low as 10^{-3}.

        The normal incidence geometry of fig.5a has been used in commercial SAM's (e.g. Phi and Varian) with LaB_6 or W filament guns and electrostatic lenses. These configurations however give a rather unimpressive current-spot size performance(fig 6, numbers 4, 6, 7). More recently a prototype field emission gun (FEG) system with this geometry has been developed, and the indications are that its performance is very good (31). Other machines of this type will undoubtedly be developed.

        Our own machine at Sussex uses a CMA in conjunction with a FEG and magnetic lenses in the geometry of fig 5b. Although the current-spot size behaviour is superior to other designs (fig 6, number 15) the geometry requires $\phi \gtrsim 45°$. Although the signal is then greater (due to the sec ϕ term in equation 1) the CMA collection efficiency drops away from normal incidence. We estimate $(T.\epsilon.)$ at present to be in the range $(0.75 - 1.5).10^{-3}$. (30).

        The instrument incorporates a Varian CMA which has a working distance of $\sim$ 6 mm and a sample-lens distance of 15 mm. The analyser has to be retracted on bellows to allow access to the sample for vapour deposition and other detectors (33). This is done conveniently and reproducibly using hydraulic jacks. The SEM chamber and analyser arrangements are shown schematically in figure 7. The chamber contains many other facilities besides AES (30) and this is a general feature of such "surface science" instruments. Optimising the Auger signal is only one of several, usually conflicting, design criteria. However we envisage some improvements which can increase the small analyser factors, T and ϵ. These include pre-retardation (36) so that the CMA operates in a constant energy window (ΔE) mode with ΔE freely chosen, and widening the angular acceptance to increase T (e.g. 37,38). In this way $(T.\epsilon)$ may rise to above 10^{-2} and the analyser field of view will also be improved. Different detection modes and data processing ($\S$ 3.2) may also improve matters.

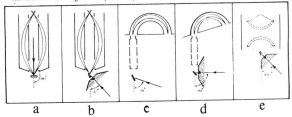

a	b	c	d	e

Fig.5.   A number of beam analyser geometries used in SAM instrument (a) CMA with electrostatic lenses, refs.(31,32). (b) CMA with magnetic lens, ref.(33). (c) CHA with electrostatic lens, ref.(34). (d) CHA with magnetic lens ref.(35). (e) Staib analyser with magnetic lens, ref.(19).

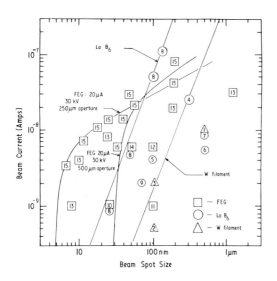

Fig. 6   A comparison of SAM electron beam current spot size performances for W, LaB_6 and F.E. emitters (ref. 30). Solid lines calculated for optimised optics at $E_p = 30$ keV. Reference to experimental points given in brackets.

W. Filament guns:	1.-Shimizu et al. (39)
	2.-Ishida et al. (40)
	3.-Staib et al. (19)
	4.-McDonald et al. (41)
	5.-Mogami et al. (42)
LaB_6 guns:	6.-Gerlach et al. (38)
	7.-McDonald et al. (32)
	8.-Christou (43)
	9.-Ishida et al. (40)
F.E. guns:	10.-Griffiths et al. (44)
	11.-Todd et al. (45)
	12.-Powell et al. (46)
	13.-Christou (43)
	14.-Browning et al. (34)
	15.-Venables et al. (33)

The CHA has the advantage of a larger working distance (~ 2 cms) and larger field of view, allowing the analyser to remain at a fixed position (34, 35). Pre-retardation with input lenses also enables the instrument to be used in a constant ΔE mode but the transmission T is only 0.8% so that $(T. \epsilon)$ is typically $\lesssim 10^{-3}$. CHA's have been used in the configurations shown in figure 5 c, d (34, 35). Notwithstanding the substantially lower T of the CHA, in practice the difference in performance from the CMA appears not to be as high as expected, possibly because of the somewhat lower take-off angle for the CHA and because field-preserving grids at the entrance and exit apertures of the CMA limit the transmission.

Figure 5 e shows a development of the retarding field analyser designed by Staib (19) combined with an SEM, with again a similar performance.

The best current-spot size characteristics are

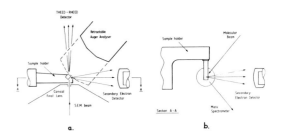

Fig. 7   Schematic arrangement of sample and detectors in the Sussex U.H.V. SEM. a) vertical and b) horizontal cross-section. (33)

provided by the field-emission gun with magnetic lenses at the high resolution end, although the LaB_6 filament would be superior above about 5×10^{-8}A. A comparison of the performance of various types of gun systems used for micro-Auger analysis is shown in Figure 6. As yet, a design involving a magnetic probe forming lens for high resolution together with an analyser which allows a low take-off angle and high transmission has not been developed, though we expect this development will not be long delayed.

3.2   Auger detection and data collection techniques

The evaluation of data collection times for Auger detection with a certain signal to noise ratio is determined not only by the Auger yield as defined in section 2.2 and the analyser transmission function $T. \epsilon$, but also by the magnitude of the secondary background from which the Auger current must be separated. The background is often up to a factor of 10 larger than the Auger intensity. Following other authors (34) we have estimated(30) that the time taken to collect spatial Auger information from M data points either in a line scan or mapping mode with a signal to (peak-to-peak) noise ratio S_n depends on the total Auger efficiency factor $\alpha = N_o \Phi . \rho . r. \sec \phi. T. \epsilon$, the monolayer concentration C and the beam current I_p according to:

$$t = \frac{36 S_n^2 M (1 + b) e}{C_1^2 C \alpha I_p} \quad (4)$$

where b is the ratio of accumulated background to Auger electrons and C_1 is the chemical contrast in the image. This is in reasonable agreement with our experience; we observe for example that a current $I_p = 10^{-8}$A is required to image 10^4 Auger image points in 200 seconds with $S_n \simeq 2$ which is approximately the minimum detection limit. Better images would obviously require either higher primary currents, or longer data collection times, or higher detection efficiency.

It is instructive to examine the form of Auger spectra to see what can be done to improve the

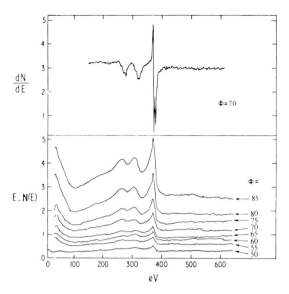

Fig. 8 dN/dE and EN(E) spectra from cadmium for various incidence angles ϕ, using a CMA. $E_p=$ 30 keV $I_p = 3.10^{-7}$A. (47)

signal to noise ratio. Figure 8 shows Auger spectra obtained in the derivative mode dN/dE and the EN(E) mode. It has been common practice to enhance Auger peaks from the slowly varying background by electronic differentiation (figure 8a). This is normally achieved by modulating the analyser pass energy and using lock-in detection. In this mode relatively good quantification is possible since only the sharp Auger peak edge is of interest, but the technique is inherently inefficient(48) and the majority of Auger electrons in the low energy tail are not detected. A number of methods for overcoming this by background subtraction offer increased sensitivity at the expense of some quantitative uncertainty and increased depth range of the information. This last point may be particularly useful for biological applications for which it may be impossible to clean the specimen surface, and where depth resolution of $\sim$ 1nm (i.e. $\sim \lambda_e$) is not of primary interest. Multiple differentiation techniques to remove low frequency Fourier components of the spectrum have had limited success(49). Spline polynomial fitting allows some control in deciding the extent of the background subtracted(50, 51).

In attempts to obtain minimum noise detection of very small Auger currents, electron multipliers, channeltrons and phosphor-photomultiplier combinations have been used in pulse counting and analogue detection modes. We have found the last to be superior as regards long term stability and background noise, compared to the electron multiplier exposed to a changing vacuum environment. The fastest useable rate for pulse counting implies the use of primary currents below about 10^{-8}A, at

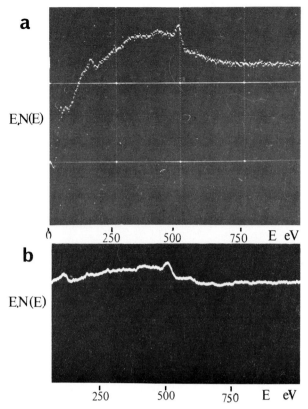

Fig. 9 A comparison of signal averaged Auger spectrum from oxidised tungsten using a multichannel analyser (MCA) with a single scan spectrum.
   a) Signal averaged EN(E) spectrum using MCA 1000 eV sweep. 1 eV per channel. 100 seconds. E_p = 60 keV. $I_p = 10^{-7}$A.
   b) Single scan spectrum. Scan time 100 secs. Details as for (a).

least for channeltrons. Primary beam modulation techniques have been used to record the N(E) spectra and to make the signal less sensitive to other current and noise sources(52, 53, 92); however, optimal use is not made of the available beam current, since the beam is off half the time. It will therefore not really be competitive with a good detector for N(E) (92).

Low frequency $1/f$ noise, which is especially high for field emission guns, may be averaged out with a multichannel analyser. This is demonstrated in figure 9 for an oxidised tungsten sample. The upper trace is a digital recording over 100 separate scans into 10^3 channels of 1eV width. The lower trace is a single analogue 1000 eV scan taken for the same data collection time. Beam current fluctuations are reproduced proportionally in the latter display which reduces the reliability of peak identification.

The above combination of reasons leads us to conclude that the full power of SAM will only be achieved if the most efficient detection scheme is used in combination with digital processing techniques to abstract the desired quantitative chemical information. This implies the use of background subtraction and other processing techniques on line, to form images and line scans, and the inevitable mini-computer or dedicated micro-processor. We are currently exploring these possibilities and feel they may be especially important for biological applications. This is because the chemical information in such material may be buried a few λ below the surface; such information is completely lost with conventional $dN_{/dE}$ detection schemes.

4. Quantitative micro-Auger analysis

4.1 Quantitative methods developed for wide beam analysis

There are a number of excellent reviews of aspects of Auger quantification which do better justice to the subject than is possible in the context of this paper, (refs.1, 5, 54-57, 63). However we would like to outline the various approaches and point out those aspects which are particularly pertinent to micro-analysis in the SEM.

There are three approaches to quantification, i) to estimate the parameters given in equation 1 from first principles and then measure I_p and I_A experimentally to give a value of N or of the concentration C; ii) to compare the Auger intensities from the sample under analysis with the intensities from a standard sample of known composition; iii) to estimate sensitivity factors using a number of known samples and infer an unknown composition from an intensity ratio using one standard. Methods ii) and iii) are widely practised in X-ray analysis.

i) The first principles approach:- The absolute Auger yield can probably be estimated to $\simeq 10\%$ accuracy although the difficulties in doing this are substantial. The first principles approach can also be used to calculate relative intensities of different Auger peaks. However, such a calculation requires an estimate of the IMFP λ, for a particular energy and material, which is subject to a very large uncertainty (see figure 3b). This will be reflected in the final analysis accuracy unless a separate determination of λ can be made, e.g. from monolayer deposition. This method is therefore likely to be difficult to use for routine quantification with high accuracy.

ii) The use of Standard samples:- The use of standards of similar composition to the unknown is probably the most accurate method of quantification. The ratio of Auger current I_A from the sample to that from the standard I_{A_o} gives a measure of concentration C_A i.e. $C_A = I_A C_{A_o}/I_{A_o}$ where C_{A_o} is the concentration of element A in the standard. A number of correction procedures are needed to allow for changes in density ρ, backscattering factor η, and possibly λ. It has been shown (57) for a number of binary alloys that the

variation in these combined factors (denoted by $f(C_B)$, $f(C_B)$) over a wide range of composition is such that the different Auger $dN_{/dE}$ peak intensities P change linearly with composition within a few percent, i.e. $P_A/P_{A_o} + P_B/P_{B_o} = 1$ and this equation implies that $f(C_A)$, $f(C_B)$ also vary linearly with composition: i.e.

$f(C_A) = 1 + a\,C_B$ and $f(C_B) = 1 + b\,C_A$ with $b \simeq -a$. This is a rather surprising fact but one which greatly simplifies quantification.

iii) Relative sensitivity factors:- A relative sensitivity factor P_r can be defined which allows the surface concentration of elements A, B to be expressed in terms of the peak heights P_A, P_B i.e.

$$P_r = P_{A_o} f(C_A) \ / \ P_{B_o} f(C_B) \text{ and for}$$

$$C_A + C_B = 1 \text{ and small } C_A, \text{ this reduces to}$$

$$P_r = P_{A_o}/P_{B_o} \ (1 + a)$$

Measured values of P_r can then be used fairly independently of material and have been estimated to yield an accuracy better than 30%. Values of sensitivity factors for a number of elements with respect to Si have been measured by Chang(58) for one apparatus. Whether these are likely to be very different in comparison with another apparatus is not certain at present.

There are a number of factors which complicate the above arguments and may provide reasons for not using "universal" factors P_r independent of sample and apparatus. Firstly, the surface concentration may not be characteristic of the bulk. This may happen by surface segregation(59) or by preferential sputtering of one species during ion sputtering which is commonly used to clean the surface before analysis(60). Secondly, anisotropic emission of Auger electrons due to diffraction on exiting the surface will decrease the accuracy of quantitative analysis in general, particularly when analysers are used with small collection solid angle. Departures in excess of 50% away from a cosine distribution in angular intensity(61, 62) have been measured in angle-resolved AES; however, such variations will be averaged out to some extent using wide angle analysers such as the CMA.

4.2 Factors affecting quantitative accuracy of micro-Auger analysis

The small probe size and large variety of specimens encountered in SAM highlight a number of factors which influence quantitative analysis to a greater extent then they would for conventional wide beam surface analysis. A number of these which we have found to be of particular importance are discussed here.

i) Topography:- The variation of secondary yield with local angle of incidence ϕ is responsible for the majority of SEM image contrast and is the primary source of SEM information. Auger analysis from a topographically varying surface, however,

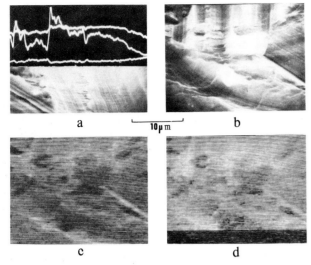

a   b

10 μm

c   d

Fig.10. Comparison of Auger line scans and maps
for conventional N'(E) signals and ratio
signals,(47).

    a) Line scan taken through bright point
shown in lower S.E. image using the
Zn 890 eV peak. i) N'(E) display showing strong influence of Topography.
ii) N'(E)/N(E) signal. (E_p= 60 kV 200 s
line scans, I_b = 5 x 10^{-8}A, τ =0.3 sec,
Vm = 10V)

    b) S.E. image of Zn surface for compa-
rison with

    c) N'(E) map and d) N'(E)/N(E) map of
the same region.

---

is complicated by the sec ϕ term in equation 1, and
Auger intensities reflect topographic detail as well as
chemical concentration. This is more marked when
the sample is examined far from normal beam incidence (§ 3.1). The Auger intensity can be corrected
for angle of incidence but this is not always easy to
measure accurately.

We have developed a dynamic correction procedure
which reduces these effects considerably. It relies on
the similar angular behaviour of the backscattered electrons; by ratioing the Auger peak intensity to the
backscattered electron intensity of the same energy,
Auger line scans and images are less sensitive to topo-
graphy(47). This is illustrated in figure 10 for a rough
zinc surface. Line scans across the sample are shown
for i) the normal derivative (dN/dE) peak intensity
and ii) which is scan i) divided by the backscattered
electron intensity. The edge highlighting which is a
feature of fig.10(c) is drastically reduced in 10(d),
and this effect has been seen on many rough samples.
This method also has the advantage of correcting for
beam current variations; however the signal to noise
is somewhat smaller than for the uncorrected signal,
given approximately by

S_n (corr) $\approx S_n$ B/(A + B), for an Auger peak
intensity A on background B. For large ratios B/A

the loss in signal to noise will be marginal.

Beam current fluctuations may also be removed
using the ratio of the Auger intensity to the secondary signal(63), but topographic and shadowing
effects are not successfully removed because of
the different detector geometries and electron energies involved. We and others(35) have also used
the current collected from the final aperture as a
means of compensating for beam current fluctuations in the unratioed dN/dE signal.

ii) Backscattering:- It has been shown that variations in the sample composition below the surface
can influence the Auger intensity due to changes in
the backscattering factor(64, 65). For instance for
an aluminium thin film on gold and silicon substrates, the signal from the gold substrate is ~ 50%
larger than from the silicon. The backscattering
coefficient η increases with atomic number and
also with increased angle of incidence, ϕ, but is
nearly independent of primary energy E_p (9, 10).
The backscattering factor $r = 1 + \eta\beta$, where
$1 \lesssim \beta \lesssim 2$ (66) also behaves similarly with E_p and
ϕ.

It may be difficult to account accurately for back
scattering effects for an unknown substrate, especially where there is a wide range of sub-surface
composition on the scale of the backscattering volume (~ 0.5 - 1μm). However, biological samples
may present a favourable case, in that the average
$\bar{Z}$ is often rather small. For example, a carbon
substrate has $\eta \simeq$ 0.08. Thus the backscattered
electrons would account for 8-16% of the Auger
production for the above values of β. If the density of the sample is known then this effect could
be estimated, but large local density excursions
will always present some difficulty.

Backscattering influences micro-Auger Analysis
in two separate ways. First, it affects quantitative
accuracy, as discussed above. Second, it affects
spatial resolution, since the Auger electrons originating from the backscattered electrons emerge
some distance from the probe.

We have determined experimentally the spatial
distribution of Auger electrons from silver layers
on tungsten(67). Silver was evaporated past 3-5 μm
nickel spheres dropped in situ on the surface, leaving a sharp shadow edge, and an Auger line scan
was made across the edge. The distinction between
the sharp drop due to the probe passing the edge,
and the longer tails due to backscattering can be
seen in figure 11a, which shows an example of an
Auger dN/dE line scan intensity for the Ag $M_{4,5}$
VV peak at 350 eV. More details, including SEM
pictures of the silver edge at 5nm resolution, are
given in ref.(67).

Monte-Carlo calculations have been made to
determine Auger profiles for particular cases(68,
69). Calculations(70) for our experimental condi-

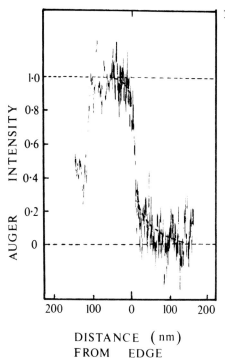

Fig.11a Auger profile for ~ 5 nm diameter beam on edge of silver layer deposited on tungsten, showing break in profile due to primary electrons and backscattered electrons. E_p =30 kV, ϕ = 50° $I_p \simeq 2 \times 10^{-9}$A τ =1 sec.(67).

INTENSITY

AUGER

DISTANCE (nm)
FROM EDGE

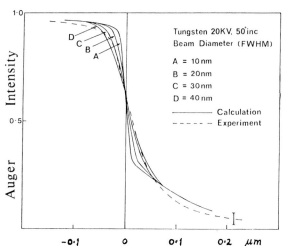

Tungsten 20KV, 50°inc
Beam Diameter (FWHM)

A = 10 nm
B = 20 nm
C = 30 nm
D = 40 nm

——— Calculation
– – – Experiment

Fig. 11b. Comparison of measured Auger profiles with Monte-Carlo calculations for the silver $M_{4,5}$ VV Auger peak from silver layer on tungsten. Angle of incidence =50°. Broken line - experimental data; full line-calculated profiles for a Gaussian beam of FWHM from 10 nm to 40 nm. (70).

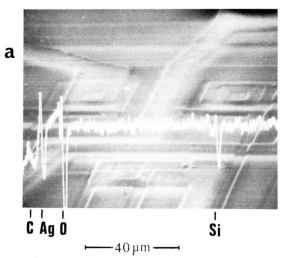

C Ag O   Si

———40 μm———

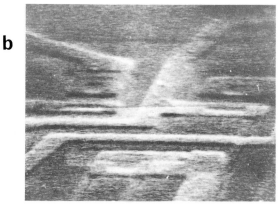

Fig. 12 Auger spectrum and map showing the effects of electron beam induced oxygen adsorption.

a) S.E. image of part of Si integrated circuit. dN/dE Auger spectrum taken from bright spot showing presence of C(270eV), Ag(350eV) 0 (510eV) and Si(1620eV). E_p =30 keV I_p =5 x 10^{-8}A.

b) Auger map using the dN/dE O KLL(510 ev) peak. Central brighter rectangle indicates increased O concentration induced by electron beam at higher magnification. E_p = 30 keV, I_p = 5 x 10^{-8}A Imaging time = 500 seconds.

tions (67) show good agreement (figure 11b). We were able to deduce a backscattering factor of 1.6 in agreement with other measurements(10) which suggests $\beta \simeq 1.2$ since $\eta \simeq 0.5$ for tungsten. Thus even in this (very severe) case, more than 60% of the Auger electrons are produced within the probe diameter. We deduce that microscopic features can be seen by SAM with the same kind of resolution as for SEM, i.e.

limited by the probe diameter; however, back-scattering affects both quantification of the signal and the effective volume of the specimen which is sampled for microanalysis. Moreover the information which can be obtained is strongly limited by the available signal to noise ratio, as seen in fig.11a.

iii) Signal to Noise Ratio:- From equation 4 and from experiments such as fig.11a the use of primary currents $I_p \leq 10^{-8}$A for a spot resolution less than ~ 20 nm implies a signal to noise ratio S_n of at most the order of ten for reasonable recording times (< 10^3 secs) and therefore a measurement error of several percent for Auger mapping.

For comparison two peak height ratios may be measured with a $S_n \gtrsim 100$ in the derivative mode for spectra taken from a spot under the same conditions At $I_p \simeq 10^{-7}A$, when the spot size is ~ 200 nm, the signal to noise ratio and the statistical accuracy is about three times better.

Any further improvements in accuracy have to be sought via analyser improvements (§3.2), or increased beam currents. But the deleterious effects of increases in beam currents are to be avoided, as discussed below.

iv) Electron beam effects:- Beam induced adsorption (71) and desorption(72, 73) have both been observed in wide-beam and micro-Auger systems. The extent of gas adsorption depends not only on the chamber pressure which should normally be in the 10^{-9}-10^{-10} torr range, but also on the overall sample cleanliness. It is our experience that these effects are not very different in SAM using high energy, high density beams from conventional low energy Auger guns.

It is widely understood that molecular species are brought to the surface either from the gas phase or by surface diffusion, and chemisorption occurs before or after cracking in the electron beam. Thus surface contamination can be seen in the SEM secondary and Auger images even under good vacuum conditions. Figure 12a shows the secondary image of part of an uncleaned integrated circuit with an overlaid Auger spectrum from a particular point of interest. Figure 12b shows the same area imaged in the Auger mode using the oxygen KLL 510 eV peak. The central area of enhanced oxygen concentration arises from scanning the beam at higher magnification for a period of time, and was not observed after thorough argon ion cleaning procedures.

It has yet to be seen to what extent these effects limit the study of biological material; our experience is that they can be fairly readily overcome on metal and semi-conductor samples.

v) Sample charging:- In comparison to EELS & X-ray, AES is very sensitive to sample charging, because it involves the detection of relatively low energy electrons, and electron energies are measured with respect to the surface potential. Severe charging is usually apparent in the SEM as a distortion of the image when the local field is sufficient to deflect the primary beam. We have experienced loss of Auger signal due to charging which produces undetectable effects in the secondary image. We are therefore careful to avoid the use of insulators in close proximity to the sample where they can be seen by primary or high energy secondary electrons. Auger analysis of some insulators e.g. glass(74), MgO (75) and ferro-electrics(76, 77) is possible for a restricted range of primary energies such that a high secondary yield stabilises the surface at a small fixed potential with respect to ground.

Samples with resistivities $> 10^8$ Ω cm will usually exhibit some degree of charging under the electron beam, which can also lead to local heating and material loss for sensitive biological tissue. A number of coating techniques used largely for biological tissue are available to increase surface conductivity, including the use of evaporation and sputter deposition of metals and carbon.

Conducting colloids and anti-static agents have been used (For a review of these techniques and materials see refs.78, 79). Althought it may be possible to detect Auger electrons from a depth of the order of the electron escape depth beneath the sample surface, many of these coating procedures developed specifically for secondary electron imaging may not be useful for AES as the layer thicknesses are likely to be too large. Any coating technique will of course also introduce quantitative uncertainty, but in situ deposition of low atomic number materials having a larger range for Auger electrons may be possible. Carbon contamination films are usually present on uncleaned material which may help to reduce charging. Whether surface layers are present unintentionally or intentionally to reduce charging, sub-surface information may have to be retrieved using data-processing techniques; it has been shown that "sub-surface" elements can be identified by a suitable background subtraction technique(93).

Other techniques used for observing insulators in the SEM are based on discharging the sample using charged particles, either positive ions or low energy electrons from another source(80). The last may be useful for SAM since no surface deposition is involved.

The extent to which sample charging will prove a serious handicap to Auger analysis of biological material is not yet clear. Charging is generally lower for thin films than for bulk samples; however, in a preliminary examination of both thin film and bulk samples we have not observed any charging effects (see § 5.2).

5.   Future developments of SAM, and outlook
                for applications in biology

We see three major areas for the future improvement of the performance of SAM. These are (i) analyser improvement (ii) digital data processing and(iii) specimen preparation, especially for biological samples.

5.1 Instrumental development

As discussed in §3 the efficiency with which Auger electrons are detected with present analysers imposes substantial upper limits on the speed of Auger analysis. Thus a significant advance would be made by improved analyser design, i.e. increasing T and ϵ . There are several possibilities including the retarding field CMA described in §3.1, and a new analyser developed by

Eastman(81) for angular ultra-violet photoelectron spectroscopy (UPS). This consists of an elliptical reflector and retarding grid, forming a band pass filter with high angular collection and with an energy resolution of ~ 1%. It might be an improvement on present analysers for SAM but the geometry and the necessary field-free regions seem difficult to combine with a probe-forming lens. Alternatively, accurately formed high transmission grids with very small spacing might also become the basis of more suitable analysers for SAM with smaller dimensions.

Although it has a low transmission T, the CHA has the advantage of an output focus plane which is nearly flat. This raises the possibility of using a position sensitive detector to collect a range of energies in parallel to improve ϵ. This has already been achieved for X-ray photoelectron spectrometers (82) and several methods of doing this are being developed(83). Such parallel detector systems increase dramatically the speed for taking spectra, though they do not markedly affect the speed of line or mapping modes. Nonetheless, it may prove convenient to have a parallel detector, and to define the energy window for these modes subsequently using electronics, as is currently done in energy dispersive detectors for X-ray analysis.

We envisage that fast 1- and 2- dimensional digital storage systems plus on-line computing will make a substantial impact on all forms of microanalysis including micro-Auger analysis. Such systems would allow simple background subtraction and other quantification routines for 1- and 2- dimensional scans, in times less than the data collection time; line scans and pictures could then be produced using quantitative measures of surface and sub-surface chemical composition and be continuously displayed in updated form.

We have recently investigated such a system(84) which is capable of collecting 2-dimensional information with 512 x 512 bit spatial resolution and 8 bit intensity resolution at up to T.V. rates, and uses a minicomputer to interact with this data store. The system could operate as a signal averager, allowing constant upgrading of the signal to noise ratio i.e. as a 2-dimensional multichannel analyser; it could also be used for many other aspects of microscopy such as particle characterisation or electron diffraction measurements. Thus fast digital data handling techniques will undoubtedly help to make quantitative micro-Auger analysis a routine technique, as well as assisting in many other areas of quantitative microscopy.

5.2 Biological Applications

Auger spectroscopy offers the ability to perform elemental analysis with high quantitative accuracy, high spatial resolution and high sensitivity for low atomic number elements. For biological material, two major problems are likely to be most significant; firstly the high surface sensi-

tivity of the technique together with the difficulty of maintaining a clean surface, and secondly the ability to control specimen charging.

Some preliminary experiments to assess the feasibility of the technique have been made by Hart(93) and recently by ourselves. Hart has shown Auger spectra from a section of rat anterior pituitary stained with vanadyl sulphate, which indicates a high surface carbon concentration, the ratio of vanadium to carbon being approximately 2×10^{-2} after ion etching.

We have made a preliminary investigation of plant tissue and human blood cells which also show a high surface carbon concentration (see discussion).

6. Comparison of SAM with X-ray and EELS techniques

Two of the most important criteria for using a particular method of micro-analysis are spatial resolution and quantitative accuracy. These aspects are discussed briefly and a summary of these and other distinctions between the techniques are outlined in table I. The data in table I should be taken as a rough guide only to the present situation. Developments in the relatively new techniques of SAM and EELS are rapid, and we can expect large improvements, particularly in quantification, in the near future.

6.1 Spatial Resolution

(i) Thin specimens:- For thin transmission specimens the resolution of any of these three techniques is expected to be less than the sample thickness, though of course somewhat larger than the beam spot size. For EELS the spatial resolution will be very close to the beam spot size since the collection angle α is normally very small, and a resolution of 10 nm has been demonstrated(85). There has been a substantial amount of work on X-ray analysis of thin films. The resolution depends on the amount of beam spreading in the sample thickness; measurements on 450 nm thick Si suggest that the spreading and thus the resolution is $\lesssim$ 20 nm for E_p =100 keV and > 400 nm for E_p = 40 keV (86). The scattering volume for thin specimens has been modelled by Monte-Carlo methods(87) which shows a scattering diameter of ~ 200 nm for 250 nm thick aluminium with E_p = 20 keV and normal incidence, which is a fairly extreme case. At higher energies the distribution is clearly narrower(88) in qualitative agreement with experiment(86). Auger electrons could be therefore produced from this wider area on the exit surface, but of course would come from essentially the beam diameter on the entrance surface.

TABLE I

	(a) SAM (AES)	(b) X-ray	(c) EELS
ELEMENTS ANALYSABLE	$Z \geq 3$	$4 \leq Z \lesssim 10$ Difficult $Z > 10$	$Z > 3$

DETECTABILITY LIMITS
  a) Minimum sample Vol. (nm^3)
  b) Mass(gm)
  c) Concn.(ppm)
	4×10^2
	10^{-19}
	10^3
		10^8
		10^{-16}
		$10^2 - 10^3$
			10^2
			10^{-19}
			(10^3)

PRESENT QUANTITATIVE ACCURACY	$\simeq 10\%$	$\gtrsim 2\%$	$\simeq 20\%$
SPATIAL RESOLUTION (nm)	$\lesssim 30$	500	$\lesssim 10$
ANALYSIS DEPTH (nm)	0.5 - 2	500	Specimen thickness $\lesssim 100$
VACUUM REQUIREMENTS (torr)	$< 10^{-8}$ (Typically $10^{-9} - 10^{-10}$)	10^{-5}	$< 10^{-5}$
SPECIAL SAMPLE REQUIREMENTS	Usually surface cleaning required	-	Thin film

(ii)   Bulk specimens:- The big advantage of both X-ray analysis and AES is that bulk specimens can be used. However, part of the signal arises from ionisation by the primary beam and part by scattered electrons in the solid. For X-rays, the signal is largely due to the latter, and since most photons produced within the scattering volume escape, the spatial resolution of X-ray analysis is therefore limited to approximately the cascade size ($\simeq 0.5\mu m$). For AES on the other hand, most of the signal arises directly from the primary beam since only those additional Auger electrons produced within the escape depth λ_e by emergent backscattered electrons will be detected. These amount to approximately 40% of the total yield for gold and $\sim$ 6-10% for carbon as outlined in sections 2.2 and 4.2(ii). The resolution will therefore be determined primarily by the beam diameter and secondarily by the backscattering factor.

The results of experiments described in §4-2 (ii) bear this out, and confirm that the Auger resolution will be approximately the beam diameter for low Z materials, and the 20-80% resolution will be better than 30 nm for high Z. For inclusions and adlayers with lateral dimensions much smaller than the cascade size ($<< 0.5\mu m$) only a small proportion of backscattered electrons will intersect them; thus, independent of the backscattering factor, we would expect the resolution of the Auger profiles to be determined almost entirely by the beam diameter.

In both X-ray and AES techniques the low detected yields mean that signal to noise ratios are important in determining resolution, especially for area mapping (§3.2 and 4.2 (iii)).

6.2 Quantitative accuracy

For X-ray analysis, the detected yield is influenced by absorption and fluorescence effects (among others) which do not appear in the case of Auger analysis, and X-ray corrections are thus more complicated. However, much effort has gone into correction procedures for X-ray micro-analysis, with the result that standard procedures are widely available which produce $\sim$ 2% accuracy. These procedures are nonetheless quite restrictive in their assumptions and the equations are difficult to calculate, and large errors may be involved when the sample is inhomogeneous on a scale comparable with the penetration distance of the electron beam. In the thin film case the quantification procedures are much simpler and the present problems are largely instrumental in nature(86).

Quantification of EELS is at present in the early stages of development(89, 90). The main problems are to determine the partial cross-section $\Phi(\alpha, \Delta)$, the behaviour of the background on which peaks are superimposed, and to account for multiple scattering in thicker samples. Comparison with standard samples may be more problematic than for X-ray or Auger because of the sensitivity to thickness variation, and the difficulty of providing homogeneous specimens with accurately known concentrations. However, EELS is presently being actively investigated in several laboratories, and encouraging results have been obtained both on quantification, with accuracies of $\sim 20\%$ claimed, and the production of standard samples by ion implantation(91).

Auger quantification is in many ways the simplest of the three techniques. It is most accurate using standard samples, for which $\simeq 10\%$ accuracy can be obtained(55). The main problems stem from the assumption that the surface layers are characteristic of the bulk composition, and some deviations are to be expected, especially after ion bombardment, as discussed in §4.1.

Acknowledgement

The authors would like to thank J.G. Duckett (Queen Mary College, University of London) and Diana M. Harvey (School of Biological Sciences, University of Sussex) for the biological specimens used in this work. The support of the Science Research Council is gratefully acknowledged.

References

1. C. C. Chang, "Auger Electron Spectroscopy " Surface Science 25 (1974) 53-74.
2. J.A.D. Matthew, "A modified $Z(Z+1)$ approximation for free atom Auger spectra " J.Phys. B10 (1977) 783-793.
3. A.P. Janssen, R. Schoonmaker, J.A.D. Matthew and A. Chambers, "Interfacial Auger transitions in oxidised Na and Mg " Solid State Commun 14 (1974) 1263-1267.
4. M. Salméron, A.M. Baró, and J. Rojo, "Interatomic Auger processes and the density of states " Surface Science 53 (1975) 688-697.
5. H.E. Bishop and J.C. Rivière, "Estimates of Efficiencies of Production and Detection of electron-excited Auger emission" J.App.Phys. 40 (1969) 1740-1744.
6. C.J. Powell, "Evaluation of formulas for inner-shell ionisation cross-sections " N.B.S.

Special Publication No.460. (1976) 97-104 and Rev. Mod. Phys. 48 (1976) 33.
7. C.J.Powell, "Relative yields of K.L.L. and L.V.V. Auger electrons from aluminum" Proc. 7th IVC, Pubs. R. Dobrozensky, postfach 300, A-1082 Vienna, Austria,(1977), 2319-2322
8. J.Kirschner, "Electron-excited core level spectroscopies" in Electron Spectroscopy for Surface Analysis, Ed.H.Ibach. Springer-Verlag, (1977), p.95
9. See S.J.B.Reed, "Electron Microprobe Analysis" Camb. Univ. Press, (1975), Chap. 12 & 13.
10. D.M.Smith and T.E.Gallon, "Auger emission from solids: the estimation of backscattering effects and ionisation cross-sections " J. Phys. D.7. (1974) 151-161.
11. C.J. Powell, R.J. Stein, P.B. Needham and T.J. Driscoll, "Attenuation lengths of low-energy electrons in solids " Phys. Rev. B16 (1977) 1370-1378.
12. C.J. Powell, "Attenuation lengths of low energy electrons " Surface Science 44 (1974) 29-46.
13. I. Lindau, and W.E. Spicer, "The probing depth in photo-emission and Auger electron spectroscopy " J. Elect.Spect.3 (1974) 409-413.
14. M.P. Seah and W.A. Dench, "Quantitative electron spectroscopy of surfaces " N.P.L. report Chem (82), April 1978. Available from N.P.L. Teddington, Middlesex, U.K.
15. M.P. Seah, "Distinction between adsorbed monolayers and thicker layers by Auger spectroscopy " J. Phys. F3 (1973) 1538-1547.
16. M.P. Seah, "Quantitative Auger electron spectroscopy and electron ranges " Surface Science 32 (1972) 703-728.
17. D.C. Jackson, T.E. Gallon and A. Chambers, "A model for the Auger electron spectroscopy of layer growth systems " Surface Science 36 (1973) 381-394.
18. C. Argile and G.E. Rhead, "Calibration in AES by means of coadsorption " Surf. Science 53 (1975) 659-674.
19. P. Staib and G. Staudenmaier, "Quantitative Auger Micro-Analysis " Proc. 7th IVC Vienna (1977) 2355-2358.
20. R.F. Egerton, "Quantitative Energy Loss Spectroscopy " SEM - (1978) I 133-142.
21. W. Steckelmacher, "Energy analysers for charged particle beams " J. Phys. E6 (1973) 1061-1071.
22. D. Roy and J.D. Carette in: "Electron Spectroscopy for Surface Analysis " Ed. H. Ibach. Springer-Verlag N.Y. (1977) p.13.
23. S. Aksela, "Analysis of the energy distribution in CMA." Rev. Sci. Instr. 42 (1971) 810-812.
24. V.V. Zashvara, M.I. Korsunskii and O.S. Kosmachev, "Focussing properties of CMA " Sov.Phys. Tech. 11 (1966) 96-99.
25. H.Z. Sar-El, "Cylindrical Capacitor as an Analyser " Rev. Sci. Instr. 38 (1967) 1210-1216.
26. C.E. Kuyatt and J.A. Simpson, "Electron Monochromator Design " Rev. Sci. Instr. 38 (1967) 103-111.
27. H. Hafner, J.A. Simpson and C.E.Kuyatt, "Comparison of the Spherical deflector and CMA "

Rev. Sci. Instr. 39 (1968) 33-35.

28.   D.W.O. Heddle, "A comparison of the étendue of electron spectrometers " J.Phys. E4 (1971) 589-592.

29.   E.N. Sickafus and D.M. Holloway, "Specimen position effects on energy shifts in CMA " Surface Science, 51 (1975) 131-139.

30.   J.A. Venables and A.P. Janssen, "Developments in Scanning Auger Microscopy " Proc. Electron Microscopy Conf. Toronto. (1978) III, (published by the Microscopical Society of Canada 150 College St. University of Toronto Ontario M58 1A1) 280-291.

31.   L.H. Veneklasen, G. Todd and H. Poppa, "An integral field emission "microsem" for UHV surface analysis " Proc. Electron Microscopy Conf. Toronto. (1978) I, 12-13.

32.   N.C. McDonald, C.J. Harland and R.L. Gerlach, "Scanning Auger microscopy for microelectronic device characterization and quality control " SEM 77. I. (1977) 201-210.

33.   J.A. Venables, A.P. Janssen, C.J. Harland and B.A. Joyce, "Scanning Auger electron microscopy at 30 nm resolution " Phil. Mag. 34 (1976) 495-500.

34.   R. Browning, P.J. Bassett, M.M. El-Gomati and M. Prutton, "A digital scanning Auger electron microscope incorporating a CHA " Proc. Roy Soc. A357 (1977) 213-230; Surf. Sci. 68 (1977) 328-337.

35.   I.R.M. Wardell and P.E. Bovey, V.G. Microscopes Ltd. East Grinstead Sussex, U.K. HB50A configuration (1978).

36.   R.L. Gerlach, "A Retarding field CMA " J. Vac. Sci. Tech. 10 (1973) 122-125.

37.   N.J. Taylor, Y.E. Strausser and T.A. Pandolfi, "A submicron automated Auger microprobe with novel data presentation capabilities " Proc. 7th I.V.C. Vienna (1977) 2621.

38.   R.L. Gerlach and N.C. MacDonald, "Recent advances in Scanning Auger Instrumentation " SEM/1976/I, IITRI, 199-203

39.   H. Shimizu, M. Ono and K. Nakayama, "Microprobe Auger analysis of heterogeneous oxidation of alloys " Proc. 7th IVC. Vienna (1977) 2359-2362.

40.   T. Ishida, T. Hayashi and Z. Oda, "High resolution Auger microanalysis by pulse counting method " Proc. 7th IVC. Vienna (1977) 2307-2313.

41.   N.C. MacDonald and J.R. Waldrop, "Auger Electron Spectroscopy in the SEM " Appl. Phys. Lett. 19 (1971) 315-318.

42.   A. Mogami and T. Sekine, "Auger electron scanning microscope with beam brightness modulation " Proc. 6th Europ. Congr. E.M. Israel. I (1976) 422-424.

43.   A. Christou, "Comparison of electron sources for high resolution   Auger spectroscopy in SEM." J. Appl.Phys.47 (1976) 5464-5466.

44.   B.W. Griffiths, A.V. Jones and I.R.M. Wardell, "An Ultra-high vacuum SEM with Auger analysis facilities " Proc. EMAG 73. (Institute of Physics) (1973) 42-45.

45.   G. Todd, H. Poppa, D. Moorhead and M. Bales "Auger electron spectroscopy at high spatial resolution " J. Vac. Sci. Tech. 12 (1975) 953-958.

46.   B.D. Powell, D.P. Woodruff and B.W. Griffiths "A scanning Auger microscope for surface studies " J. Phys. E8 (1975) 548-552.

47.   A.P. Janssen, C.J. Harland and J.A. Venables "A ratio technique for micro Auger analysis " Surface Science 62 (1977) 277-292.

48.   M. Prutton, "A theoretical comparison of electron energy analysers degraded to obtain sensitivity " J. Elect. Spect. 11 (1977) 197-204.

49.   J.E. Houston, "Dynamic background subtraction and the retrieval of threshold signals " Rev. Sci. Instr. 45 (1974) 897-903.

50.   R. Hesse, U. Littmark and P. Staib, "A method of Background Determination in Quantitative Auger Spectroscopy " Appl. Phys. 11 (1976) 233-239.

51.   P. Staib and J. Kirschner, "Absolute Atomic Densities Determined by AES " Appl. Phys. 3 (1974) 421-427.

52.   C. Le Gressus, D. Massignon and R. Sopizet, "Nouvelle methode d'analyse du spectre des electrons secondaires " C.R. Hebd. Séanc. Acad. Sci Paris. B 280 (1975) 439-442.

53.   T.J. Shaffner, "Beam brightness modulation in Auger Scanning Microscopy " SEM 78 I. (1978) 149-156.

54.   C. C. Chang in "Characterisation of Solid Surfaces " (ed. P.F. Kane and G.B. Larabee, Plenum) p.509.

55.   J.M. Morabito and P.M. Hall, " Quantitative AES," SEM/1976/I, IITRI, 221-230

56.   P.H. Holloway, "Quantitative Auger electron analysis," Surf.Sci. 66 (1977) 479-494.

57.   P.M. Hall, J.M. Morabito and D.K. Conley, "Relative sensitivity factors for quantitative Auger analysis of binary alloys " Surface Science. 62 (1977) 1-20.

58.   C. C. Chang, "General formalism for quantitative Auger analysis " Surface Science 48 (1975) 9-21.

59.   J. Erlewein and S. Hofman, "Segregation of tin on (111) and (100) surfaces of copper " Surface Science 68 (1977) 71-78.

60.   A. Jablonski, S.H. Overbury and G.A. Somorjai, "The surface composition of the gold-palladium binary alloy system " Surface Science 65 (1977) 578-592.

61.   L. McDonnell, D.P. Woodruff and B.W. Holland, "Angular dependence of Auger electron emission from Cu(111) and (100) surfaces " Surface Science 51 (1975) 249-269.

62.   D.M. Zehner, J.R. Noonan and L.H. Jenkins,

"Angular Effects in Auger electron emission from Cu(110) " Solid State Commun. 18 (1976) 483-486.

63.   P.H. Holloway, "Quantitative Auger spectroscopy - problems and prospects," SEM 78, I, (1978) 361-374.

64.   J. Kirschner, "The role of backscattered electrons in SAM " SEM/1976/I, IITRI, 215-220.

65.   L.A. Harris, "Miscellaneous topics in AES " J. Vac. Sci. Tech. 11 (1974) 23-28.

66.   H. Seiler, "Determination of the "information Depth" in the SEM " SEM 76 (1976) 10-15.

67.   A.P. Janssen and J.A. Venables, "The effect of backscattered electrons on the resolution of SAM " Surface Science 77 (1978) 351-364.

68.   J. Kirschner, "On the Influence of Backscattered electrons on the lateral resolution of SAM " App. Phys. 14 (1977) 351-354.

69.   R. Shimizu, M. Aratama, S. Ichimura, Y. Yamogaki and T. Ikuta, "Applications of Monte-Carlo calculation to fundamentals of SAM " App. Phys. Lett. 31 (1977) 692-694.

70. M.M.El-Gomati, A.P.Janssen, M.Prutton et al, "The interpretation of the spatial resolution of SAM...." Submitted to Surface Science (1979), preprint available from A.P.Janssen.

71. J.H.Neave and B.A.Joyce, "Some comments on electron beam induced adsorption" J. Phys. D10, (1977) 243-248.

72.   T.E. Madey and J.T. Yates, "Electron stimulated desorption as a tool for studies of chemisorption - a review " J. Vac. Sci. Tech. 8 (1971) 525-555.

73.   Y. Margoninski, "An AES study of electron-beam stimulated desorption of oxygen " Phys. Lett. 54A (1975) 391-392.

74.   R.A. Chappell, C.T.H. Stoddart, "An improved technique for determining surface composition by AES and ion etching applied to a multiphase glass " Proc. 7th IVC Vienna (1977) 2297-2300.

75.   A.P. Janssen, R.C. Schoonmaker, A. Chambers and M. Prutton, "Low energy Auger and loss electron spectra from magnesium and its oxide " Surface Science 45 (1974) 45-60.

76.   R. Le Bihan and M. Maussion, "Observation directe des domaines ferroelectriques du sulfate de glycocolle (T.G.S.) au microscope electronique a balayage " Journal de Physique 33 C2 (1972) 215-219.

77.   R. Le Bihan and G.Jouet, "Study of ferroelectric domains on G.A.S.H. single crystals by the SEM " Proc. Electron Microscopy Conf. Toronto I (1978) 158-159.

78.   P. Echlin, "Coating techniques for SEM and X-ray microanalysis " SEM 78 I (1978) 109-132.

79.   B.L. Munger, "The problem of specimen conductivity in electron microscopy " SEM 77 I (1977) 481-490.

80.   P. Morin, M. Pitaval and E. Vicario, "Direct observation of insulators with a scanning electron microscope " J. Phys. E 9 (1976) 1017-1020.

81.   D.L. Eastman, F.J. Himpsel and J.J.Donelon "A 2-dimensional display-type electron spectrometer for photo-emission LEED and Auger spectroscopy " Bull. Amer. Phys. Soc. 23 (1978) 363.

*82.   U. Gelius and K. Siegbahn, "New developments in electron spectroscopy " Proc. ICSFS conference Tokyo. (1978) B29.

83.   C.D. Moak, S. Datz, F. Garcia-Santibanez, and T.A. Carlson, "A position sensitive detector for electrons " J. Elec. Spec. 6 (1975) 151-156.

84.   "Intellect," manufactured by Microconsultants Ltd., Caterham, Surrey, U.K.

85.   C. Colliex and P. Trebbia, "Electron Energy Loss Spectroscopy in the electron microscope: present state of affairs " Proc. Electron Microscopy Conf. Toronto III (1978) 268-279.

86.   R. Hutchins, I.P. Jones, M.H. Loretto and R.E. Smallman, "Spatial resolution of X-ray microanalysis in thin foils " Proc. Electron Microscopy Conf. Toronto I (1978) 544-545.

87.   P. Duncumb and L. Curgenven, "Simulation of Electron trajectories in a solid target by Monte-Carlo techniques " T.I. report. 1971, Saffron Walden, Essex U.K. (1971) 303 see also R.G. Faulkner, "Prediction of X-ray spatial resolution for SEM and STEM specimens using Monte-Carlo methods " Proc. Electron Microscopy Conf. Toronto (1978) I 546-547.

88.   G. Soum, F. Arnal, J.L. Balladore, B. Jouffrey and P. Verdier, "Application of the Monte-Carlo method to high energy electron scattering " Proc. Electron Microscopy Conf. Toronto I (1978) 202-203.

89.   R.D. Leapman, P. Rez and D.F. Mayers, "A quantitative approach to inner shell losses " Proc. Electron Microscopy Conf. Toronto I (1978) 526-527.

90.   D. Joy, D. Maher and P. Mochel, "Quantitative elemental analysis by T.E.M. " Proc. Electron Microscopy Conf. Toronto I (1978) 528-529.

91.   D. Maher, D. Joy and P. Mochel, "A standard for T.E.M. " Proc. Electron Microscopy Conf. Toronto I (1978) 530-531.

92.   H.E. Bauer, P. Wiedmann and H. Seiler, "A new modulation technique for Auger electron spectroscopy " Proc. 7th I.V.C. Vienna (1977) 2371-2373.

93.   R.K. Hart, "Elemental distribution in biological materials by means of the scanning electron spectrometric microscope " Proc. 8th International Electron Microscope Conference Canberra (1974) II, 20.

---

*To be published in Surface Science

Discussion With Reviewers

<u>A.W. Robards</u>: You note that for SAM applications in biology, two major problems are likely to be encountered; surface contamination, and specimen charging. Can you provide any other experimental information on biological material and comment further on these aspects.

<u>Authors</u>: We have analysed sections of plant spermatozoid of Equisetum and Phaeoceros. These organisms were selected since they have been shown by X-ray analysis to contain high phosphorus and calcium concentrations. ("A combined ultra-structural and X-ray microanalytical study of spermatogenesis" Cytobiologie 13 (1976) 322-340. J.G. Duckett and D. Chescoe) which were prepared by fixing, embedding in epoxy resin, sectioning and then supported on carbon films over transmission grids. The samples were analysed by X-ray analysis in a CTEM prior to investigation in the Sussex UHVSEM. Initial imaging of the plant tissue specimens with a 30 keV beam resulted in noticeable brightening of the transmission image over the scanned area which we must assume is due to removal of material. After ion etching (2 minutes with 5 keV Ar^+ ions with an ion flux of $\sim 30\mu A/cm^2$) the increase in brightness from beam exposure was not as marked.

Figure 13a shows a STEM image of a sperm cell from Equisetum prepared as described together with a secondary image (Figure 13b) of the same cell taken from our UHVSEM after baking at 100°C to give a chamber pressure of 10^{-9} torr, and then ion-etching. The N(E) Auger spectra (Figure 13c, d) were also taken from a central part of the high phosphorus region and from the cell wall. These spectra result from 300 scans averaged in the MCA, with 1 second per scan. Subsequent analyses from these and other regions of the specimen after further ion etching showed only the presence of carbon. Etching was continued until the carbon support film disintegrated. The geometrical arrangement was such that although the straight through beam was stopped from entering the analyser, most of the Auger electrons released from the rear of the specimen could not be similarly stopped. Since the rear side was composed of support film, up to half the carbon Auger peak intensity would be expected to have arisen from transmitted electrons.

Human red blood cells were also studied after treating with ethanol-water solution and drying on a clean silicon surface. Figure 14 shows a secondary image together with Auger spectra taken from a particular cell and from the surrounding region after 5 minutes of ion etching. These also show an over-whelming carbon concentration on both cell and substrate.

The inability to see other elements besides carbon was disappointing. That this will not always be the case has been demonstrated by Hart (93). We have

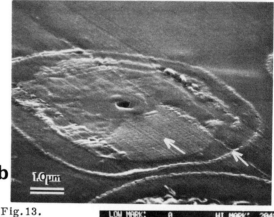

Fig.13.

Auger analysis of Equisetum sperm cell.

a) STEM image 40kV.

b) Secondary image 40kV, =70°

c) N(E) spectrum from high phosphorus area (i)

d) N(E) spectrum from cell wall (ii).

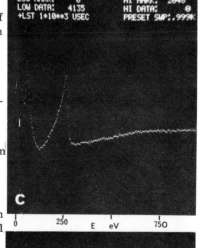

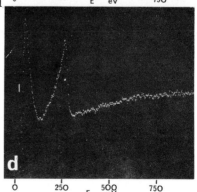

Fig. 14

a) Secondary image of blood cells on silicon substrate. 30kV

b) N(E) Auger spectra from cell (x) (upper trace), and substrate (lower trace).

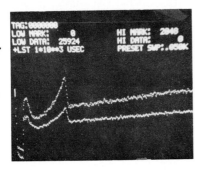

---

also tried Pt-stained material (leaf cells of Suaeda maritima) embedded in epoxy with similar result, and are of the opinion that analysis from un-embedded material may be more successful.

We are presently investigating a number of other simpler biological samples. It does seem that the success of the technique for biology will be primarily, but crucially, dependent on sample preparation; on neither the thin film samples nor the blood cells was charging observed.

C. le Gressus: Could you evaluate your experimental conditions in terms of primary beam current and energy density in the sample? Could you give some limits on the corresponding temperature increase in the irradiated spot?

Authors: This question has been asked so often that we would like to answer it in some detail. In particular, below a certain spot size ($\lesssim$ the cascade diameter) it is the total current which determines the temperature rise rather than the current density. The highest rises are produced at currents where the spot and the cascade size are of the same order, and not under conditions of highest resolution.

In our case, figure 6, No.15, shows the primary beam current for a given spot size at $E_p = 30kV$ which is the energy we choose to do much of our micro-Auger analysis. The particular operation point on the characteristic depends on whether the beam spot size is paramount, as for the data of figure 11a (5 nm dia probe, $I_p = 2 \times 10^{-9}A$, 150μm aperture), or whether Auger sensitivity dictates a higher beam current, sacrificing resolution, as in figure 12b (100 nm probe diameter, $I_p = 5 \times 10^{-8}A$

500μm aperture).

Although the current density in these two cases is high (8000 A cm^{-2} and 500 A cm^{-2}) the beam energy is dissipated over the cascade volume of $\sim 0.5\mu m^3$. The temperature rise due to an electron beam is well established by several calculations, for example those of Vine and Einstein-Heating effect of an electron beam impinging on a solid surface. Proc. IEE III $\underline{5}$ (1964) 921. They express the temperature rise ($\triangle T$) in the form $\triangle T = p.u.C.$ $E_p . I_p /K.d$, where E_p and I_p are the beam energy (V) and current (A), d is the probe diameter (cm), K is the thermal conductivity (cal. cm.$^{-1}s^{-1}. K^{-1}$) and C is a constant = 0.112. The term p accounts for power retention and is $\simeq (1 - \eta)$, with η the backscattering coefficient. The factor u depends on the ratio of the electron range to the probe diameter d, and accounts for the dissipation of energy over the cascade, which we assume to be 0.5μm. For small d, u becomes also very small.

We can estimate the values of $\triangle T$ as follows, for $E_p = 30kV$, for a range of materials illustrated in our figures, and for a range of operating conditions appropriate to our FEG machine, and also to an optimised LaB$_6$ machine, as shown in fig.6., (Tables IIa, IIb).

Table IIa

Operating conditions

Gun	d(nm)	I_p	u
FEG	5	2.10^{-9}	0.02
	20	$1.5\ 10^{-8}$	0.05
	100	4.10^{-8}	0.18
LaB$_6$	200	3.10^{-7}	0.30

From this table it is clear that severe problems can be encountered for the amorphous carbon and silicon which have low conductivities, and that the problems are most severe in the $10^{-7}A$ range, rather than at the highest resolution. Difficulties with metals and semiconductors are neither experienced, nor, from the table, are they expected. But care clearly has to be taken with mounting biological materials, composed of low conductivity materials, on high conductivity supports in bulk or thin film form. The thin film case is in general somewhat less severe (we estimate a factor of 2-3 less for C, SiO_2 and Si) since the energy losses are smaller at an energy E_p than in the bulk of the cascade.

C.J. Powell: Please discuss how the increase of the AES backscatter factor with incident electron energy (Fig. 3(b)) can be reconciled with a near energy-independent backscattering coefficient and a decreasing

Temperature rise(maximum) ΔT, for beam conditions specified in Table IIa, for four different spot sizes, 5, 20, 100 and 200 nm.

Material	K	p	ΔT for d (nm)			
			5	20	100	200
amorphous-C	.0038	1	71	330	640	(4000)
amorphous-SiO$_2$	.0032	.9	75	360	680	(4300)
Si	.358	.9	0.67	3.2	6.1	38
Ag	1.00	.65	0.18	0.82	1.6	9.8
W	.489	.6	0.33	1.6	3.0	19

cross section for inner-shell ionization as the average energy of a back-scattered electron increases.

Authors: Following Seiler's(66) arguments for secondary electrons, we can define the backscattering factor r in terms of the backscattering coefficient η and a factor $\beta = \delta_b/\delta_p$ where δ_b is the number of Auger electrons released per backscattered electron and δ_p is the number of Auger electrons released per primary electron; i.e., $r' = (1 + \eta\beta)$. The dependence of β on the primary energy is responsible for the behaviour of r(U) (Fig.3a), but the argument is somewhat subtle. From our definition of β,

$$\beta = (C/N_b) \, \Phi(E_p)) \int_{E_k}^{E_p} N_b(E) \, \Phi_b(E) \, dE$$

where $N_b(E)$ is the energy distribution of backscattered electrons and the constant C accounts for differences in the angular distribution of backscattered and primary electrons. Even if we assume that $N_b(E)$ scales with E/E_p, then r will increase slowly because $\Phi(E) \sim E^{-1} \ln(E/E_k)$ (cf C.J. Powell, ref(6), from which our fig.2 was taken!). Hence if $\overline{E}_b$ is the average energy of the backscattered distribution $\beta \simeq (C \, E_p/\overline{E}_b) \cdot (ln(E_b/E_k)/ln(E_p/E_k))$, which increases (slowly) as the logarthmic terms get bigger, and is still increasing slowly within the range of energies used in SEM (U = $E_p/E_k \simeq 60$ say, at $E_p = 30kV$).

C.J. Powell: Could you define more precisely the "instrumental angular and energy collection efficiency factors" in Eq. (2)? Please give typical values of these factors (or a range) for the CHA as well as the CMA.

Authors: The angular collection efficiency factor is used here to mean the number of electrons of energy between E and E + ΔE detected by the analyser, where ΔE is the analyser energy window, divided by the total number of electrons of the same energy produced in the surface layer. This will depend not only on the solid angle of collection but also on the angular distribution of emitted Auger electrons and the detector geometry. We have calculated the variation in angular collection efficiency factor for a range of primary beam incident angles for our particular chamber geometry (see ref.67) assuming a cosine emission distribution. For an analyser collecting a cone of emitted electrons with semiangle θ with respect to the sample normal, the angular collection efficiency is $(1 - \cos^4\theta)/2$. For a CMA collecting within $42.3 \pm 3°$, this gives a collection efficiency of 5.7% of 4π (which is normally expressed by the manufacturers as 11.4% of 2π). Increasing the collection angle to $\pm 5°$ (with consequent loss of energy resolution) would give a collection efficiency of 19% of 2π. The CHA in comparison is normally used with a collection semi-angle of $\theta < 4°$ which gives a collection efficiency of < 1% of 2π.

The energy collection efficiency factor for a particular Auger peak may be defined as the fraction of Auger electrons detected within the acceptance energy window ΔE, to the total number of Auger electrons integrated over energy and emitted within the same collection solid angle; this fraction can be small for high resolution analysers. If we consider an Auger peak having a nearly triangular loss tail extending $\sim$ 50 eV below the edge (e.g. for the cadmium N(E) peak shown in figure 8) an analyser window of 2eV at 370eV would only pass 8% off Auger electrons entering the analyser. Parallel energy detection in addition to background subtraction would in principle allow a high proportion of Auger electrons extending into the loss tail to be detected with high efficiency.

We have given a fuller discussion of these points in refs.(67) and (30).

C.J. Powell: Isn't the increase of signal predicted in Sec. 3.1 for non-normal incidence (the sec ϕ term) exactly compensated by a reduction of incident current density? Would the loss of spatial resolution in one dimension be acceptable?

Authors: Yes, but of course the total surface area irradiated by the primary beam increases as sec ϕ for a constant beam diameter. We normally aim to operate within $45° < \phi < 70°$ which would give an increase in beam width of between 1.4 and 3 times in one dimension. This is certainly acceptable for most high resolution work in both SAM and SEM. This effect has obviously to be included when measuring Auger profiles with high spatial resolution such as figure 11.

C.J. Powell: Please give an estimate of damage rates compared to signal acquisition rates (Sec. 4.2 (iv)).

Authors: The enhanced oxygen concentration seen in figure 11b represents an increase in oxygen concentration by a factor of about 2 after leaving the beam continually scanning over the central region for about 500 sec. Scanning the imaged region for the same period of time has evidently not resulted in the same enhancement of oxygen concentration (thus the contrast). The induced oxygen adsorption in this case appears to be dependent on the electron dose in the region of $0.25C.cm^{-2}$. We stress however that beam induced effects of this type are likely to depend strongly on the surface composition, cleanliness and vacuum environment: we see no effect on properly cleaned metal (Mo, W, Ag) and silicon surfaces with pressures $\sim 5 \times 10^{-10}$ torr after prolonged exposure to the beam, much in excess of the value given here.

C.J. Powell: Please comment on the relative detectability of a "peak" in EELS compared to an "edge" (for elements where sharp edges are found).

Authors: We are of the opinion that there is not much difference between the detectability of a "peak" and an "edge" for a certain data collection time with a constant detection window which is smaller than the width of the loss edge. However for parallel energy detection it would be possible to show an advantage in detecting the whole peak in the same time interval by using the information present in regions of the spectrum away from the "edge", i.e. by background subtraction.

C.J. Powell: Are the yields of X-ray photons per incident electron given in Sec. 2.3 also for a bulk sample? Are these yields calculated using the same cross sections that were used for the Auger yields?

Has account been taken of X-ray absorption for these yields?

Authors: The X-ray photon yields have been estimated by Reed (see ref.(9). p198-201) for bulk samples using the Gryzinski cross-section. However, no account has been taken of absorption.

C. le Gressus: Could you discuss the formula of Sec. 4.2.i and the limit of validity of your ratioing? Could you discuss the resulting noise of correlated noise sources, and of the validity of your ratioing technique for homogeneous and heterogeneous samples?

Authors: The details of your questions(i) and (iii) are discussed in reference(47), but we have given the general conclusions in the present paper. It is clear that the same source is used for both N(E) and N'(E) signals and therefore that the noise in each is correlated (your question(ii)). But the noise in N'(E) will be dominant, in all cases except where the background B is less than the Auger intensity A.

C. le Gressus: Could you produce some proof of your affirmation about the noise level of your phosphor detector? Could you show comparison of noise for various electron multiplier types? Could you show with your multiplier a good AES Spectrum both in En(E) and Edn/dE modes at 5.10^{-10} Amp. ?

Authors: The phosphor-photomultiplier detector was shot-noise limited for $I_p > 10^{-10}A$, which was considerably better than our previous electron multiplier, whose performance degraded with time and exposure to air. The detector is certainly capable of producing spectra of both types at $I_p = 5.10^{-10}A$, but the recording time and detection time constants have to be lengthened accordingly.

R. K. Hart: One of the most important aspects of biological sample analysis is its preparation, and this brings up a whole series of problems, most of which are of a chemical nature, for example, the type of fixative, buffering agent, and stain to be used so that their responses will have a minimum influence on the analysis. Also, one has to consider water content, organic liquids, and solution ions and what influence they have on the results.

R. K. Hart: With regard to the charge problem which is experienced when examining insulators, have the authors considered the following techniques: a) High frequency oscillation of the primary electron beam with the extraction of SEM information via a phase-locked amplifier; 2) Continually rastering the beam during data collection (See Hart, Proc. EMSA Meeting, Boston 1977, p 94); c) Biasing the sample positively above ground by a few hundred volts (see Hart, et al, Proc. 28th Symp. on Frequency Control, USAECOM, Ft. Monmouth, 1974, p 89).

Editor: No response was received from the authors in time to include with this paper.

SCANNING ELECTRON MICROSCOPY/1979/II
SEM Inc., AMF O'Hare, IL 60666, USA

LASER MICROPROBE MASS ANALYSIS: ACHIEVEMENTS AND ASPECTS

R.Kaufmann, F.Hillenkamp*, R. Wechsung**, H.J. Heinen**, and
M. Schürmann

Universität Düsseldorf
Lehrstuhl für Klinische Physiologie
Universitätsstrasse 1
D-4000, Dusseldorf
West Germany

*Universität Frankfurt
Institut für Biophysik
Theodor-Stern-Kai 7,
D-6000,Frankfurt/M.
West Germany

Abstract

After a short review of the general state
of the art of laser microprobe analysis,
recent achievements of the laser microprobe mass analyzer (LAMMA) are presented.
The Lamma-instrument, which was primarily
developed for biomedical purposes has
been recently used on a large variety of
specimens: biological, technical, organic
and inorganic in nature. Detection limits,
mass resolution and reproducibility have
been improved so that for many elements
measurements can be extended down to the
subppm range. For bulk analysis a separate version of the LAMMA-instrument has
been developed.
The examples of LAMMA-applications given
in this paper include muscle tissue (physiological cations), uterine tissue (Fe),
retina tissues (Na, Mg, K, Ca and Ba),
airborne particles (Pb, Fe, Al, Ti, Mn,
Ca, Zn), cerebral tissue (Li), a fluorinated drug, pure crystalline organic materials (including nicotinic acid) and a
selection of non biological materials
(for instance thin films).

**Leybold-Heraeus
Gaedestrasse
D-5000 Köln, West Germany

KEY WORDS: Laser Microprobe, Mass Spectrometer,
Trace Analysis, Subcellular Element Distribution,
Organic Compounds, Aerosols, Labelled Compounds,
Toxic Heavy Elements, Particle Analysis

Introduction

Radiation of a laser light source
can be easily focussed down to a diffraction limited spot size of about half
a wave-length in diameter. Therefore,
most lasers in combination with an optical focussing device can be the first
half of a laser microprobe. The analytical part of such an instrument can be
either an optical spectrometer or a mass
spectrometer depending on the mode of
interaction used to obtain the desired
analytical information from the irradiated microsample (see Tab. I). Four different secondary phenomena may be employed: (1) Raman scattering, (2) fluorescence emission from the excited solid
sample, (3) spectral emission from evaporated and excited material, and (4)
generation of ionized particles in the
laser induced microplasma.

This report will mainly deal with
recent results obtained by a laser microprobe based on mode 4 of the above list.
However, since the alternative techniques
are also available, and complementary to
some extent, a short review of the overall state of the art is in order.

History and actual state of laser microprobe techniques in general.

Microprobe analysis by Raman scattering has recently been demonstrated to be
a promising tool especially for the microchemical analysis of organic and inorganic constituents in, for instance,
airborne particles[1] or mineral samples[2].
Although instrumentation is still at its
beginning the "build-in potentialities"
of this microprobe technique are obvious
and will surely be realized in future
developments.

Laser stimulated fluorescence emission strangely enough has not been systematically exploited for microprobe analysis although some advantages may be
anticipated with respect to conventional
microfluorometry. For instance, highly
time resolved fluorescence or phosphor-

escence measurements at low temperatures may deconvolute superimposed luminescence spectra from mixed materials.

Laser microprobe analysis by emission spectroscopy, on the other hand, has a rather long history (for review see [3]). The first instruments were introduced as early as 1963 and a long list of successful applications, mainly in technical material analysis, shows the strength (and also the limitations) of this method which is comparatively cheap and straightforward. Although the detection limits of later versions of this method could be improved by several orders of magnitude[4], the ultimate sensitivity cannot compete with electron probe X-ray microanalysis (EPXM) and, even less, with ion microprobe techniques.

Mass spectrometric analysis of a laser induced microplasma should basically benefit from the high sensitivity inherent in mass spectrometry and, thus, may be expected to be a promising alternative to the other microprobe techniques. However, early attempts to construct such instruments[5,6,7,8,9] demonstrated only their feasibility, but were not convincing enough to encourage further development. It was generally felt that ion formation by laser impact necessarily implies high initial ion energies which result in multiply charged ions with a broad energy distribution which would require so much energy filtering that overall transmission (and sensitivity) would be drastically reduced.

For this reason most groups [10,11,12,13,] (for review of exceptions see [10,11]) did not continue work on laser microprobe mass analysis. Recent studies have been published by Eloy[10] who employs a magnetic sector spectrometer for laser microprobe analysis of bulk specimen including biological samples (such as dental materials). The sensitivity reported for Ca, P, O, C, Na, K, Mg, Al, F, Cl, Fe and Zn ranges from 0.1 - 0.05% for accumulated series of ten laser shots, which is clearly not sensitive enough to find trace elements in living matter.

Over the last 10 years our own work was also concentrated on the development of a laser microprobe mass analyser (called LAMMA) which was especially designed for microprobing thin sections of biomedical samples. Fortunately it turned out that the earlier predictions concerning initial energy distribution of laser induced microplasmas do not hold true for thin dielectric materials. Thus, the laser, focussing optics and a specially designed TOF (time of flight) mass spectrometer could all be optimized to obtain a spatial 0.5 μm resolution, detection limits in the sub-ppm range for many elements, and a mass resolution of (initially) $m/\Delta m = 400$. The technical

concept and some pilot results have been published [14,15,16,17,18,19]. This includes the work presented at the last SEM meeting. We report here on the recent technical improvements of the LAMMA-instrument and present a selection of applications for biomedical, enviromental, and technical microprobe analysis.

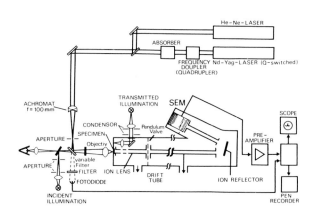

Fig. 1. Schematic diagram of the LAMMA-instrument. The specimen is a semithin section viewed in either incident or transillumination (including phase contrast). Objective lenses of the high power optical microscope are UV transparent immersion (glycerol) objectives. The area of interest is indicated in the field of view by a red spot of a Helium-Neon laser whereas the light of the impulse laser (which is triggered by a push button) is in the UV and, therefore, invisible. Laser focus (at a 100x/1.25 objective) that means the diameter of the analyzed area can be made as small as 0.5 μm. The ionized part of the evaporated material is analyzed in a time of flight (TOF) mass spectrometer which consists of a accelerating ion lens, a field free drift tube in which the ions are separated according to their (mass dependent) drift time, an ion reflector compensating for the spread of initial ion energy and an open secondary electron multiplier (SEM) for ion detection. With every laser shot a complete mass spectrum is obtained which is stored in a fast transient digital recorder for further data processing, displaying or pen recording.

The LAMMA-instrument (present state)

For extensive description of the instrument see refs.[14] and [15]. Fig. 1 gives a short description and illustrates the slightly modified basic setup. During the past year the following improvements

Tab. 1: Physical concepts of laser microprobe systems.

Probe	Interaction Process	Irradiance (Wcm^{-2})	Analytical Information	Spectrometer
LASER	Raman scattering	$10^3 - 10^6$	Molecular constituents (vibrational levels)	Optical spectrometer
	Luminescence	$10^{-2} - 10^3$	Molecular constituents (valency electrons)	"
	Spectral emission	$10^9 - 10^{13}$	Atomic constituents, (outer shell electrons)	"
	Ionization	$10^9 - 10^{10}$ (single charged particles)	Atomic and/or molecular constituents	Mass spectrometer

in the specifications have been achieved.
1. The mass resolution was increased to about m/$\triangle$m = 850 by the use of a "time focussing" ion reflector which compensates for the spread of initial ion energies (O-4OeV).
2. The detection limits were improved, especially for heavy trace elements such as Pb and U (see Tab. 2). Small atomic peaks, hidden in a background of polyatomic signals may sometimes be extracted by peak-background subtraction, due to improvements in reproducibility.
3. The reproducibility from shot to shot was improved to $\pm$ 5% SD in standard specimens.
4. Bulk specimen can be handled by a modified version of the LAMMA-instrument which is actually tested.

Quantitative analysis, comparison of LAMMA and SIMS
The linear signal intensity-concentration plots and the comparatively fair reproducibility (see Fig. 2) are already the basic prerequisites for a quantitative analysis. For absolute quantitation, however, data such as the evaporated volume, ion yield, and the (mass-dependent) ion/electron conversion efficiency of the SEM must be known. One may correct for variations of the evaporated volume by referring to an internal standard which, in the simplest case, might be the integrated organic background, whereas the other factors require either the introduction of element-specific sensitivity factors (as in ion microprobe) or the use of appropriate standard reference materials.

When comparing LAMMA and SIMS (secondary ion mass spectrometry) one must bear in mind that the formation of ionized particles is basically different in the two instruments. In the laser probe the beam energy is first transferred to the electrons, which in turn heat up the lattice atoms to the point of vaporization and plasma formation, whereas in ion probe the momentum of

primary ions is directly transferred to the superficial lattice atoms. This is the reason why, in the LAMMA-instrument, ion yield is mainly determined by the ionization potential of an element and is rather independent of the chemical nature

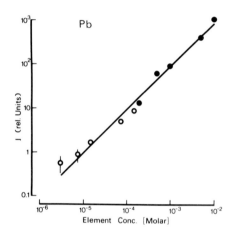

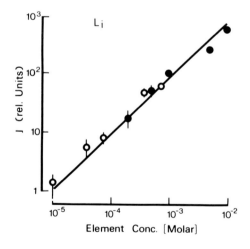

Fig. 2. Calibration plots for Li and Pb in standard specimen.

Tab. 2: Detection limits of the LAMMA-instrument (Dec. 1978). Data refer to an analyzed volume of ∼ 3 x 10^−13g of (organic matrix) material and to the condition that polyatomic fragments of the matrix do not interfere with the atomic mass signals to be measured.

	absolute (g)	relative (ppmw)
Li	2×10^{-20}	0.2
Na	2×10^{-20}	0.2
Mg	4×10^{-20}	0.4
Al	2×10^{-20}	0.2
K	1×10^{-20}	0.1
Ca	1×10^{-19}	1.0
Cu	2×10^{-18}	20.0
Rb	5×10^{-20}	0.5
Cs	3×10^{-20}	0.3
Sr	5×10^{-20}	0.5
Ag	1×10^{-19}	1.0
Ba	5×10^{-19}	0.5
Pb	1×10^{-19}	0.3
U	2×10^{-19}	2.0

of the matrix. Sensitivity for most of the ions with ionization levels (for simply charged ions) of $4eV < E_i < 8eV$ vary only over one order of magnitude in comparison to about three orders of magnitude in SIMS (see Tab. 3). This pertains particularly to trace analysis of heavy metals such as Pb or U for which the LAMMA-instrument clearly provides even better detection limits than the ion microprobe.

Biomedical LAMMA-applications

Specimen preparation
The preparation of soft biological tissues for any kind of microprobe analysis still imposes some unsolved problems. Since any

Tab. 3. Semiquantitative guide to relative SIMS and LAMMA sensitivities for various elements (normalized to Na = 1). The numbers in brackets are approximated minimum detection levels in ppm.
+) Taken from [23]
++) Preliminary estimates

	SIMS[+)	LAMMA
Li	1.7 (0.15)	0.27 (0.2)
Na	1 (0.25)	1 (0.2)
K	0.9 (0.30)	1.55 (0.1)
Sr	8×10^{-2} (3.0)	0.55 (0.5)[++]
Ba	6×10^{-2} (4.0)	0.30[++] (0.6)[++]
Pb	2.5×10^{-3} (100)	0.70 (0.3)[++]

redistribution of water soluble compounds such as the physiological cations must be avoided, conventional fixation and dehydration procedures are obsolete. For the time being cryotechniques (shock freezing, cryosectioning and/or freeze drying) are the preparation methods usually employed for microprobe analysis of hydrated biological soft tissues. For subsequent LAMMA studies the following procedure for specimen preparation was used if not otherwise indicated: The specimen was shock frozen in either melting N_2 or supercooled propane. Freeze drying was performed at -85°C in a specially designed vacuum system. Freeze dried specimens were plastic embedded under vacuum in Spurr's low viscosity medium and sectioned dry on a conventional ultramicrotome to a thickness of 0.3 - 1 μm.

Distribution of physiological cations and trace elements in soft tissues

Muscle cells
The subcellular distribution of the leading physiological cations was investigated in either single isolated skeletal muscle fibres or in heart muscle preparations. In many of the skeletal muscle fibres rather large uptake of Ca and loss of K was found as a sign of - otherwise undetectable - membrane damage during preparation. The Na/K ratio was found to be slightly different in the I- and the M-band of the sarcomer, whereas Ca was found to be rather inhomogeneously distributed (sarcoplasmic reticulum!). In healthy and ischemic myocardium the intracellular Na, K, Ca and Mg content was determined.

Retina
In a recent ion probe study M. Burns-Bellhorn detected unexpectedly large amounts of barium in the retina of cat and cow [20]. In collaboration with this author, a re-investigation was done by means of LAMMA analysis in the retina of men, cats and frogs. The presence of Ba (in association with Ca) was consistently demonstrated in all three species. As shown in Figures 3a/b the barium is strictly localized in the pigment granules of the pigment epithelium and the choroid. To a first approximation (with reference to a Ba-standard) we estimated 20-50 mM Ba in the pigment granules, whereas only trace amounts of Ba were found in other cell layers of the retina. Nothing is known so far about the physiological significance of Ba.

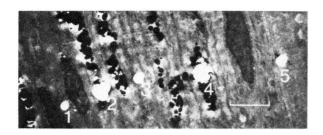

Fig. 3a. TEM picture (unstained) showing 5 analyzed areas in the choroid of a human eye specimen. Since laser light absorption in the pigment granules is higher than in the surrounding material, more laser energy is deposited and, hence more material is removed.

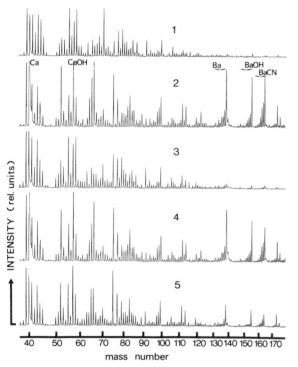

Fig. 3b. LAMMA-spectra obtained from the 5 areas of interest in Fig. 3a. Note the appearance of the Ba-signals in record 2, 4 and 5.

Detection of toxic trace elements in biological materials (Pb, Sr, Cd).

The sensitivity of EPXM is not usually sufficient for the detection of trace amounts of toxic heavy metals such as Pb, Cd or Hg. Also, ion microprobe sensitivity tends to decline by orders of magnitude with increasing Z-numbers [21,22]. Since, in LAMMA analysis, ion yield and, hence, sensitivity, is merely a function of the ionization level, heavy elements with low ionization potentials such as Pb (7.42 eV) or U (6.08 eV) can be analyzed down to sub-ppm detection limits

(see Tab. 2 and 3). This is demonstrated in the example shown in Fig. 4. For instance, the nominal (average) content of Pb in the NBS bovine liver standard material was 0.35 ppmw (or 1.5×10^{-6}M). As expected, the LAMMA analysis demonstrated a highly nonhomogeneous Pb distribution in the particulate material. Averaging the 48 measurements of the Pb-signal intensities led to the conclusion that the detection limits for Pb in the LAMMA-instrument might still be below this level.

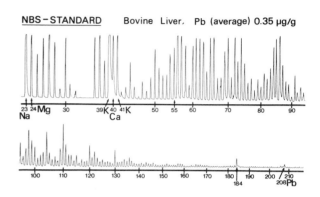

Fig. 4. LAMMA-spectrum of the NBS standard reference material 1577 (bovine liver). The Pb-signal intensity of this record corresponds to the mean value measured in 48 analyses. Analyzed volume per shot $\simeq 5 \times 10^{-13}$g. The particulate material was plastic embedded and sectioned dry (1 μm) in an ultramicrotome.

Thus the Pb-content of various fetal tissues was sufficient for detection in an incidental case of human Pb intoxication during an early stage of pregnancy (Dr. Schmidt, Institute of Physical Medicine, University of Münster). However, neither systematic distribution studies nor absolute quantification (by cross checking with chemical gross analysis) have been completed at this time. LAMMA-analysis has also revealed trace amounts of Pb in plant materials and airborne particles (see subsequent section).

Trace amounts of Sr were found in a fresh water clam (unio) specimen (see Fig. 5) which had been previously analyzed by SIMS (Dr. v. Rosenstiel, Metaalinstitut, Apeldorn). Cd could be detected in rat kidney tissues after experimental Cd-poisoning (Prof. Ohnesorge, Dept. of toxicology, University of Düsseldorf). In the latter case, however, it became clear that the LAMMA sensitivity for Cd is comparatively poor because: (1) Cd has a rather high ionization potential of 9 eV, and (2) the peaks of the various Cd isotopes are

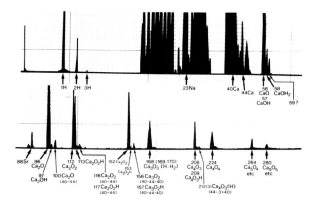

Fig. 5. Detection of trace amounts of Sr in a part of a fresh water clam (unio).

usually superimposed on strong signals of polyatomic organic fragments.

LAMMA-analysis of airborne particles

Due to the increasing pressure for environmental control, a new branch of microprobe analysis has been formed which focusses attention on the microchemical identification of aerosols. Again, EPXM has turned out to be of rather limited value for obvious reasons.

Preliminary studies with the LAMMA-instrument showed its promising capabilities to analyze and classify airborne particles of various origin and composition.

Through courtesy of Prof. Bruch (Institut für Silikoseforschung and Lufthygiene, University of Düsseldorf) we obtained a specimen of human lymphocytes, preincubated in vitro with airborne microparticles (1-2 μm), which were collected in a coal mine. Most of the particles seen in the specimen had been taken up by the tissue cultured lymphocytes via phagocytosis. Since the time required to perform a LAMMA analysis is very short, limited only by the time needed for repositioning the area of interest and for pen-recording the mass-spectrum (usually 30s per cycle), we were able to analyze 100 particles in about 1 hour. Evaluation of these spectra (see Fig. 6) revealed three major classes of particles with class I (panel c of Fig. 6) showing the typical fragmentation pattern of pit-coal, class II (panel a of Fig. 6) containing Na, Fe, Al and K, and class III (panel b of Fig.6) containing only Na, Al and K as the most abundant elements. Since these specimens had been prepared for conventional EM by the usual fixation and dehydration procedures it is, of course, not known if these results represent the true chemical nature of the particles.

Fig. 7 gives an example of a LAMMA spectrum recorded from an airborne dust particle (2-3 μm) collected in Antwerp.

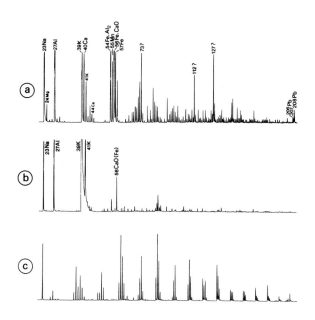

Fig. 6. LAMMA-classification of coal mine dust particles taken up by cultured human lymphocytes via phagocytosis (see text for further information).

Since this particle was neither embedded nor had passed through any chemical treatment, the spectrum reflects its natural composition. Among the major elements identified with certainty we found Li, Na, Mg, K, Ca, Ti, Mn, Co, Zn and Pb. The origin of the group of peaks at mass numbers 96,97,99 and 111, 112,114 is uncertain. If Cd were involved the natural isotopic abundances do not correspond to the observed signal intensities.

Airborne Particle

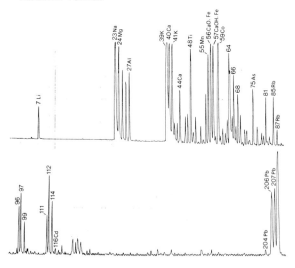

Fig. 7. LAMMA-spectrum of a native airborne particle collected in Antwerp.

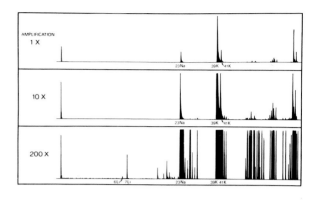

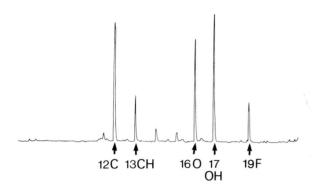

Fig. 8. Detection of a 19F-signal in the LAMMA-spectrum of negative ions recorded in a standard (epoxy-resin) specimen containing $3 \times 10^{-4}M$ of a fluorinated cardioactive drug (tri-fluor verapamil).

Fig. 9. Detection of Li in the cerebellar tissue of a rat 12 h after administration of a single therapeutic dose of 200 mg LiCl/kg. Specimen was freeze dried, embedded in Spurr's low viscosity medium and sectioned dry (1 μm thick sections).

Microprobing drugs and organic compounds

Basically the sensitivity of the LAMMA-instrument should be sufficient to search for trace amounts of labelled organic compounds or drugs. For labelling, any covalently bound atom of high ion yield may be a suitable candidate. No efforts have been made so far to produce compounds especially designed for this purpose. However, in a preliminary attempt, a fluorinated derivative of a cardioactive drug (Verapamil) was tested as a standard specimen (the drug was dissolved in an Epoxy-resin at known concentrations). In the LAMMA-spectrum of the negative ions (see Fig. 8) an F-signal was recorded down to detection limits in the range of 10^{-5} molar (the compound was a trifluorverapamil). This roughly corresponds to the sensitivity for F (2-5 ppm) found in rat tooth. Since F is certainly not the best choice for this purpose a systematic search for a more appropriate label (for instance Si instead of C!) will probably end up with much lower detection limits and, hence, will eventually produce a new tool to study subcellular pharmacokinetics.

In this context it is worth mentioning Li, which is still used as a potent drug in some mental disorders. Fig. 9 demonstrates that Li, 12 h after administration of a single therapeutic dose of 200 mg/kg, could easily be detected in the cerebellar tissues of the rat. Since the concentrations measured in the cryosectioned specimen (courtesy of Prof. Hopf, Hirnforschungsinstitut, University of Düsseldorf) were rather inhomogeneous on the 2-10 μm scale, it was concluded that Li must be highly compartmentalized on the subcellular level.

Another interesting way to use the LAMMA-instrument for mass spectrometric microanalysis of organic compounds may be based on the possibility to "desorb" organic molecules "in toto" or with only a few fragmentations. This is indeed possible, at least with microcrystallites of organic material, if the laser power in the focus is reduced to just above the threshold for visible removal of material. Fig. 10 gives some examples of how simple and easy it is to interpret those LAMMA-spectra, which closely resemble the mass spectra obtained by field desorption. The complementary nature of the negative ion spectra is also clearly demonstrated (see also 18).

LAMMA-applications in non biological materials

Although primarily developed for biomedical purposes the LAMMA-instrument has already been used for microprobe analysis of numerous technical and geological materials(ranging from dust particles of cosmic origin in the depth of the Rhone glacier to fragments of a Roman amphora).

Just to give one example which, eventually, may have some general implications for future microprocessing of thin films of dielectric material or layered sandwiches of conducting and non-conducting materials, a LAMMA-application is shown in which the problem was to produce "sensitive spots" of 2-3 μm in diameter on the surface of a pO_2 sensitive microelectrode (see Figures 11a and b). The electrode consisted of a needle-like quartz substrate on which thin films of Au, tantalum oxide, Ag and an organic insulator had been deposited. Mechanical processing to produce a small pO_2-sensi-

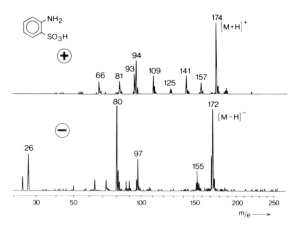

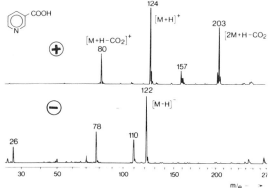

Fig. 10. LAMMA-spectra of positive and negative ions obtained from small crystallites of sulfanilic acid (upper part) and nicotinic acid (lower part).

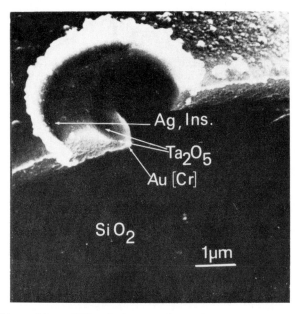

Fig. 11a. SEM picture of a microcrater obtained by repetitive "laser drilling" (see text).

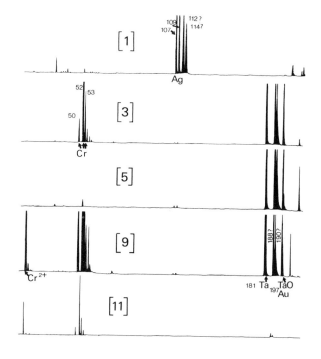

Fig. 11b. LAMMA spectra recorded during "laser drilling" of the above microcrater. Note the successive appearance of mass peaks emerging from the various layers. The numbers index the laser shot of a sequence of 12 shots.

tive spot at the tip of this electrode failed. So we attempted to locally (in an area of about 10 μm²) remove the Ag/tantalum oxide layers by repetitive, focussed laser irradiation with the LAMMA-instrument. The depth of this "laser drilling" could be nicely controlled and monitored by the mass spectrum recorded with each laser shot. The laser parameters could be adjusted such that 10 + 1 shots were needed to reach the (inner) gold layer which was freed of superimposed material to become the pO_2 sensitive spot.

The SEM picture of Fig. 11a shows the profile of such a microcrater obtained by repetitive "laser drilling" through the multi-layered sandwich consisting (in this particular case) of an insulater (1000 Å), a metallic silver coat (500 Å), a Ta_2O_5 (8000 Å) layer, a metallic gold coat (500 Å), and a (50 Å) chromium coat on a (needle shaped) quartz substrate.

Acknowledgement

This work has been supported by grants from the Stiftung Volkswagenwerk and the Federal Ministry of Research and Technology (BMFT). Some of the standards employed were prepared with the aid of organometallic complexes kindly provided by Prof. F. Vögtle and Dr. E. Buhleier, Institute for Organic Chemistry and Biochemistry, University of Bonn. The fluorinated derivative of Verapamil was kindly provided by the Knoll AG, Ludwigshafen, FRG.

References

1. E.S. Etz, G.J. Rosasco, J.J. Blaha, K.F.J. Heinrich: Particle Analysis with the Laser-Raman Microprobe. 13th annual conference of MAS, Ann Arbor, Mich.,1978, June 19-23, p 66A - 66L
2. M. Delhaye, E. DaSilva, G.S. Hayat: The Molecular Microprobe. International Laboratory, 1977, May/June, p 69-73
3. H. Moenke, L. Moenke-Blankenburg: Laser Microchemical Analysis. Crane, Russak, and Co., Inc., New York N.Y., 1973
4. K.W. Marich, P.W. Carr, W.S. Treytl, D. Glick: Effects of Matrix Material on Laser-induced Elemental Spectral Emission. Anal. Chem. Vol. 42, 1970, p. 1775-1779
5. E. Bernal, L.P. Levine, J.F. Ready: Time-of-Flight Spectrometer for Laser Surface Interaction Studies. Rev. of Sc. Instr. Vol. 37, 1966, p. 938-941
6. N.C. Fenner, N.R. Daly: Laser used for Mass Analysis. Rev. of Sc. Instr., Vol. 37, 1966, p. 1068-1070
7. J.F. Ready: Effects Due to Absorption of Laser Radiation. J. Appl. Phys., Vol. 36, 1965, p. 462-468
8. K.A. Lincoln: Mass Spectrometric Studies Applied to Laser-induced Vaporization of Polyatomic Materials. In: Pure and Applied Chemistry, Suppl.: High Temperature Technology, Butterworth, London, 1969, p 323-332
9. J.F. Eloy: Analysis Clinique Semi-quantitatives par Spectrographie de Masse a Ionisation par Bombardement Photonique (Laser a Ruby). Int. J. of Mass Spectr. and Ion Physics, 1971, p 101-115
10. J.F. Eloy: Chemical Analysis of Biological Materials with the Laser Probe Mass Spectrograph. Microscopica Acta, Suppl. 2, 1978, p 307-317
11. J. Conzemius, H.J. Svec: Scanning Laser Mass Spectrometer Milliprobe. Analytical Chemistry, Vol. 50, 1978, p 1854-1860
12. I.D. Kovalev, G.A. Maksinov, A.I. Suchkov, N.V. Larin: Analytical Capabilities of Laser-Probe Mass Spectrometry. Int. Journal Mass Spectrometry and Ion Physics. Vol. 27, 1978, p 101-137
13. B.E. Knox: In: Trace Analysis by Mass Spectrometry. A.J. Ahearn (ed.), Academic Press, London, 1972, Chapter 14, p 423-444
14. F. Hillenkamp, R. Kaufmann, R. Nitsche, E. Unsöld: A High Sensitivity Laser Microprobe Mass Analyzer. Appl. Physics, Vol. 8, 1975, p 341-348
15. R. Wechsung, F. Hillenkamp, R. Kaufmann, R. Nitsche, E. Unsöld, H. Vogt: LAMMA - A New Laser-Microprobe-Mass-Analyzer. Microscopica Acta, Suppl. 2, 1978, p 281-296
16. R. Kaufmann, F. Hillenkamp, R. Nitsche, M. Schürmann, R. Wechsung: The Laser Microprobe Mass Analyzer (LAMMA): Biomedical Applications. Microscopica Acta, Supp. 2, 1978, p 297-306
17. R. Kaufmann, F. Hillenkamp, R. Nitsche, M. Schürmann, H. Vogt, R. Wechsung: The LAMMA-instrument: A New Laser Microprobe Mass Analyzer for Biomedical Purposes. Proc. 13th annual Conf. of MAS, 1978, June 19-23, Ann Arbor, Mich., p 16A - 16E
18. H.J. Heinen, R. Wechsung, H. Vogt, F. Hillenkamp, R. Kaufmann: Laser-Mikrosonden-Massen-Analysator LAMMA. Biotechnische Umschau, Vol 2, 1978, p 346-354
19. R. Kaufmann, F. Hillenkamp, R. Wechsung: Laser Microprobe Mass Analysis. ESN-European Spectroscopy News, Vol. 20, 1978, p 41-44
20. M.S. Burns-Bellhorn, R.K. Lewis: Localization of Ions in Retina by Secondary Mass Spectrometry. Exp. Eye Res., Vol 22, 1976, p 505-518
21. M.S. Burns-Bellhorn: Review of Secondary Ion Mass Spectrometry in Biological Research. Adv. Techn. in Biomed. Electron Microscopy (ed. J. Gennaro), Masson et Cie, 1978
22. H. Liebl: Mass Spectrometry of Solids by Probe Sampling. Microchimica Acta 1978 (in press)
23. G.R. Sparrow: Quantitative SIMS Approximations for General Applications in Surface Analysis. 25th An. Conf. Mass-Spec. All. Topics, May 29 - June 3, 1977, Washington D.C.

Discussion with Reviewers

M.L. Wolbarsht: The caption to fig. 7 mentioned that the peaks at mass numbers 96,97,99,111,112, and 114 have an uncertain origin. However, those peaks are identified as Cadmium (Cd) in which the expected isotopic abundances do not correspond to the observed signal intensity. Is this comment based on your analysis of the relative isotopic abundance of Cadmium isotope or of the natural samples analyzed in the literature?

Authors: The statement is based on the expected abundances of Cd isotopes in natural samples. We do not know whether or not technical Cd-samples (such as aerosols), eventually, have different isotopic ratios.

K.F.J. Heinrich, D.E. Newbury: What is the significance of the problem of interference by molecular ions? For example, in bombarding a target of SiO_2, what would be the relative signal strengths of Si^+, Si^{++}, Si^+_2, SiO^+, and SiO_2^+?

Authors: We have not systematically investigated SiO_2 specimens except in a few (rather incidental) cases of hitting the quartz window of our specimen chamber. The main mass signals recorded were Si^+, Si_2^+, SiO^+, and SiO_2^+ with intensity ratios of about 60:10:20:20 (which in turn depends on the laser power density). A Si^{++}-signal is not observed. In this context it may be worth noting that roughly about 50% of the Ca in soft organic specimen is detected as $CaO(H)^+$ the other 50% as atomic Ca^+ whereas only minor traces (fraction of a percent) of $Ca_2O(H)^+$, $Ca_2O_2(H)^+$ and so on are seen.

K.F.J. Heinrich, D.E. Newbury: What is the expected accuracy of the isotope ratio measurements? What is the observed precision? Could you present an example of a high accuracy isotope ratio measurement?

Authors: Provided that there is no organic background interference, accuracy of isotope ratio measurements depends primarily on the concentration level and on the absolute signal intensity received. In standard specimens containing 1000 ppmw Pb we found that the precision at which the Pb isotopes could be determined was limited only by the vertical resolution of the transient recorder which is 1/256 of its dynamic range of 8 bit. Thus, accuracy of isotope ratio measurements is actually $\sim 4 \times 10^{-3}$ at the best. With future development of fast digital transient recorders featuring a better y-resolution this specification should improve accordingly.

K.F.J. Heinrich, D.E. Newbury: What are the principal limitations in applying the laser microprobe mass analyzer to solid, bulk samples of metals or ceramics?

Authors: As already mentioned in our paper a new version of the LAMMA-instrument has been specially developed for analysis of bulk samples. The main problem to be solved is to find a compromise between the required spatial resolution and the design of an appropriate ion optical device. With high magnification objectives the free working distance of only a fraction of a mm strongly limits an efficient ion extraction. Unless a solution can be found by specially designed optics (for instance a Cassegrain-optic as developed by your NBS group) serving as optical and ion optical device at the same time the only

way is to use separate long working distance objectives for both, specimen imaging and laser focussing. This limits spatial resolution to about 2-3 µm. Initial ion energy may pose another problem.

J.L. Abraham : An important element not demonstrated in the spectra you have displayed is silicon. Is there some special problem in detecting this very common element? For example, in Fig. 6 it is quite likely that most of the particles containing the elements you have listed also contain silicon as the major element, being silicates. Have you studied standards containing this element in various forms?

Authors: We have not studied silicon standards or silicates systematically. Position of Si in the periodic table indicates that it should neither easily lose nor capture an electron (ionization level 8.1 eV). Therefore, sensitivity will be probably moderate (for either negative or positive ions) in comparison with the alkali metals. The tendency for strong covalent chemical bonding let us further predict that part of the silicon will appear as for instance SiO^+, SiO_2^+. In a preliminary attempt it was easily possible to detect silicon in dust particles incorporated in lung tissue specimens of laboratory animals. See further our answer to Dr. Heinrich and Dr. Newbury.

J.L. Abraham : Can you detect beryllium as well as lithium? The ion microprobe has been especially useful in locating beryllium-containing particulates and in quantitating the beryllium concentration in tissue sections (see JL Abraham, Recent advances in pneumoconiosis, in The Lung, Thurlbeck, ad., Williams & Wilkins, 1978, pp. 96-137).

Authors: We did not have the opportunity so far to study beryllium-containing biomaterials. The only example was the demonstration of beryllium in the thin Be-Cu layer of a multiplier cathode. Detection limits for beryllium can be anticipated to be not as good as for lithium (ionization potential for Li ~ 5.4 eV, for Be ~ 9.3 eV) but should be at least fair enough to detect some ten ppmw.

J.L. Abraham : Please provide the details for each figure including how the sample was observed and the region for analysis selected, energy and time of the analysis and the spot size used. Is there some quantitation on the vertical scale of your spectra?

Authors: It is felt that inclusion of all that detail would take a bit too much space. In general the specimen is viewed in transillumination and the area to be analyzed is selected according to the structures visible in the microscopic image. In cases of doubtful structural identification the analysed section is subsequently stained or contrasted and imaged in either a light or a transmission electron microscope (see for instance Figs. 3a and b). Vertical scale calibration is always in relative units.

J.L. Abraham : How is the sample observed within the LAMMA? Is it easy to locate the areas analyzed either before and/or after LAMMA analysis?
Authors: Part of this question has been already answered in the previous statements. Since the LAMMA-instrument includes a phase contrast condenser, structural identification in the (usually unstained) specimen is facilitated to a certain extent. In cases of doubt the same or a duplicate section may be easily visualized and photographed before and/or after analysis as you suggested.

J.L. Abraham : Is there any comparison between the controlled surface analysis in an ion microprobe and the destruction burst of analyses seen with the LAMMA? The ion microprobe can be used gently to determine elemental compositions and distributions, followed by another form of observation such as SEM. Is it possible to use thicker samples of biologic tissues in the LAMMA and remove only a certain depth by controlling the analysis conditions? Have you attempted to examine for example, 5 micron paraffin sections on some solid substrate as is used in SEM and EDXA work?
Authors: The problem with thick samples in LAMMA analysis is extensively discussed in our answer to the last question of Drs. Heinrich and Newbury. The question to what extent a controlled depth profile can be obtained with the LAMMA instrument has not been systematically investigated. Laser parameters as well as physical properties of the specimen are important determinants. From our actual experience we would predict that in strongly absorbing materials (metals, ceramics) depth profiling with something of a $0.1 - 0.2$ μm resolution should be possible. This can also be concluded from our LAMMA-application in a multilayered thin film specimen shown in Figs. 11a and b.

J.L. Abraham : Please discuss the range of instrument conditions used with different kinds of samples.
Authors: In each specimen, laser parameter (irradiance, focus diameter, wave length) must be individually adjusted according to the properties of the specimen (section thickness, absorption) and the analytical problem (spatial resolution, elements and element concentration to be measured).

Only some guidelines can be given here. To reach the highest possible spatial resolution of 0.5 μm specimen thickness must be < 0.3 μm and a 100 x high resolution objective must be used. By variation of the laser power density in the focus, (by optical filter) diameter of the analyzed area can now be adjusted between say 0.5 and 1.2 μm. If larger areas (1-3 μm) or thicker specimens are laser-probed, a 32 x objective is preferred. Since in each case the ion yield for elements with higher ionization levels (C^+, Si^+, Zn^+, Hg^+) strongly depends on the irradiance, the final adjustment must take this into account.

J.L. Abraham : Is it possible to scan the laser beam over specimens at a lower energy and obtain a distribution map of different ions? What is the rate limiting part of the instrument? Would a faster spectrometer type, such as a quadrupole, be of some use?
Authors: Distribution mapping by scanning the specimen with the laser is possible. Provided that a specimen of homogeneous thickness is streched flately enough to stay in focus over the whole field of interest, one can easily move the specimen in a scanning mode. Since the rate limiting factors are data processing and repositioning of the specimen only (the analysis takes a few μs and repetition rate of the laser is in the range of 10/s), the only problem to be solved is to either store or process on line the large amount of data obtained with every laser shot. The whole thing is a computer problem rather than a problem of the instrument. Moreover, a quadrupole mass filter would be of no use, since it can be tuned for one mass only at the time and works with a transmission which is rather poor in comparison with our TOF spectrometer.

J. Ruffolo: Would you comment on the advantages of the Nd-Yag laser as compared to the ruby laser for use in a LAMMA instrument?
Authors: We have rather long experience with a LAMMA system based on a (frequency doubled) ruby laser. Provided that optical properties (divergence, TEM_{OO}-mode) and stability of the two lasers are equal, our choice in favour of the Nd-Yag laser was simply based on a slightly better cost-efficiency ratio in comparable commercial systems and on the possibility to have two laser wave lengths available, one in the near UV ($\lambda = 353$ nm) and one in the far UV ($\lambda = 265$ nm).

J. Ruffolo: With reference to your localization of barium in the human eye specimen, were you able to detect the presence of selenium?
Authors: We did not look intentionally for selenium but rechecked our recordings.

We always find signals at the relevant mass numbers of 76,77,78,80 and 82 corresponding to the main Se-isotopes. However, we cannot decide so far whether or not these (mainly molecular) peaks contain some atomic selenium. Since ion yield for Se (ionization level 9.75 eV) is rather poor we do not expect to see unequivocal Se-signals as positive atomic ions unless local concentration exceeds something around 50-100 ppmw.

H.L.C. Meuzelaar: The authors mention "generation of ionized particles in the laser-induced microplasma" as the basic phenomenon used for the LAMMA approach. However, many, if not most, specific ions pointed out in the figures and the text are obviously desorbed ion species already present in the sample. In addition the $(M+H)^+$ and $(M-H)^-$ ions observed for sulfanilic and nicotinic acid are probably formed by surface ionization phenomena, if the analogy with Field Desorption holds true. Do the authors consider it possible that in many, if not most, LAMMA applications ion generation in a microplasma plays only a minor role compared to desorption and/or surface ionization phenomena?

Authors: This is a very important question to the nature of the interaction process occurring between laser light and probed (organic) material. Current research is already extensively dealing with this problem and will be intensified in the future. We do not think that in most actual cases of LAMMA application surface ionization is the predominant process. This is already more or less excluded by the fact that we generate a physical hole in our specimen. It rather appears that electron loss or electron capture in the (non thermal equilibrium) microplasma is inducing most of the ions formed. Although occurrence of even electron systems such as $(M+H)^+$, $(M-H)^-$ does not necessarily indicate desorption, the situation may be different in laser induced organic mass spectra such as the examples show. Here, laser power density in the focus was reduced as far as possible such that, eventually, desorption may play a major role. It is felt that the choice one has got with the laser microprobe mass analyzer to select rather different ionization conditions is a unique feature and will, eventually, be of significance in mass spectrometry of organic compounds difficult to handle in conventional ion sources.

H.L.C. Meuzelaar: The authors state "EPXM turns out to be of rather limited value in this context, for obvious reasons". On what information is this statement based and what are the "obvious reasons?"

Authors: In the last MAS conference (Ann Arbor, Michigan, June 19-24, 1978) a whole section of the conference was dedicated to microprobe analysis of aerosols. It was generally felt that the analytical information needed for characterization of individual particles should not be restricted to atomic constituents of a certain range in Z-number but should include light elements and molecular constituents (organic and inorganic) as well. Special instruments such as the Raman microprobe at the NBS are mainly developed for this purpose.

SCANNING ELECTRON MICROSCOPY/1979/II
SEM Inc., AMF O'Hare, IL 60666, USA

MASS MEASUREMENTS WITH THE ELECTRON MICROSCOPE

J. S. Wall

Biology Department
Brookhaven National Laboratory
Upton, NY 11973

Abstract

The theoritical basis for performing mass measurements with the electron microscope is reviewed as well as previous experimental results. Practical errors in mass measurements are grouped into four categories: statistical errors, radiation damage induced mass loss, instrumental errors, and specimen preparation artefacts. Statistical errors are analysed in detail to show that less than 3% error should be obtained at low dose (1 electron per square Angstrom) for particles ranging in size from $2x10^6$ to $2x10^{12}$ molecular weight. Possible instrumental errors are identified and shown to be readily reducible to below 1%. Specimen preparation by freeze drying to reduce artefacts is described. Mass measurement of fd virus is presented to illustrate specimen preparation, correction of data to 0 dose condition, and practical steps in mass measurement. The standard deviation of the mass of individual virus particle obtained in this case is 4%.

Mass measurement with electron microscope promises to become a valuable technique since: minute quantities of sample are required, mass per unit length or area can be measured directly, equilibrium in aggregating systems can be studied, particles bound to substrate (e.g., DNA) can be studied in their active conformation, heavy atom binding can be quantitatively assessed at low dose, and particles can be identified in image.

KEY WORDS: Mass Measurement, Electron Scattering, Molecular Weight Measurement, Specimen Preparation, Freeze Drying, Scanning Transmission Electron Microscopy

Introduction

One of the most important quantities to be determined in characterizing a biological molecule is its molecular weight (MW). A number of physical, chemical and biophysical techniques have been developed to measure this quantity. Most techniques require large amounts of purified material (e.g. ultracentrifugation, light scattering, osmotic pressure) and are sensitive to impurities or aggregation. Data analysis requires use of theoretical models and extrapolation to zero concentration. Gel electrophoresis in the presence of SDS (sodium dodecyl sulfate) overcomes some of these difficulties, permitting analysis of complex mixtures of proteins. However non-covalent bonds are usually broken, yielding only subunits of multi-component systems. Furthermore, gel electrophoresis can only be used to compare molecules to known standards run in the same experiment and frequently produces misleading results due to presence of stable secondary structures. It is not useful for studying systems containing lipid or polysaccharide. Error of this and other techniques is usually in the range of 1 to 10%. Absolute precision can be obtained by chemical sequencing techniques or by solution of crystal structure by x-ray diffraction, however these techniques are very laborious, require large quantities of material, and become prohibitive for large structures.

A new technique capable of measuring the number of atoms in a molecule in an accurate and straightforward way regardless of chemical bonding or 3-dimensional arrangement of the atoms would find wide application. One such technique was developed 17 years ago by Bahr and Zeitler[1,2]. They employed an electron microscope operated under well controlled conditions to record images of isolated objects, both unknowns and standards. Marton and Schiff[3] had previously shown that image contrast at focus largely resulted from exclusion of elastically scattered electrons by the objective aperture, and image contrast was quantitatively related to specimen composition. Bahr and Zeitler developed a semi-quantitative technique using an integrating photometer for measuring image intensity recorded on film[1]. Routine error for their technique was $\sim$ 10% due to various nonlinearities in their system. Nevertheless their technique was widely employed and became the method of choice for particles ranging in size from 10^{-11} to 10^{-16}gm and usable to 10^{-18}gm (600,000 Dalton MW). Accuracy could be improved using a scanning microdensitometer and computer analysis, but the relatively small improvement in accuracy has not justified the increased complication, expense and time per measurement.

A major improvement in accuracy resulted from use of the scanning transmission electron microscope (STEM)[4] (see Fig. 1) for mass measurement[5,6,7]. The improvement resulted mainly from replacement of the film by a quantum efficiency electron counter with direct recording digitally on magnetic tape. Use of an annular detector with large collection angle and an electron energy loss spectrometer to produce two independent dark field signals allowed significant reduction in the dose required to produce images. Mass per unit length of DNA and fd virus was studied. These points will be discussed at greater length below.

The first biologically important result with this technique was elucidation of the disc nature of the nucleosome (as opposed to a sphere model) by Langmore and Wooley[8]. Though controversial at the time, this model is now supported by x-ray studies. Lamvik has used STEM measurements to determine the mass per unit length of muscle thick filaments, concluding on the basis of this result that there are three myosin molecules (not 4) in the 143 Å repeat[9]. Engel has recently studied the mass per unit length of F-pili and fd phage using the Basel STEM[10]. At Brookhaven National Laboratory a new STEM with a cold stage[11] is being used to study the mass per unit length of closely related filamentous viruses and monitor the mass/length change on addition of heavy atoms. Users of the Brookhaven STEM Biotechnology Resource have determined the masses of pyruvate carboxylase, of a protein covalently bound to adenovirus DNA, and of φ 6 virus (a virus with a lipid coat), previously unknown values which could not be determined reliably by other techniques. Results to date indicate that this technique will be widely applied and may become the most accurate and reliable method for molecular weight determination, short of sequencing.

It is the purpose of this paper to discuss the theory and practice of mass measurement with particular emphasis on practical limitations and means for overcoming them. The reader may find it useful to keep in mind the advantages of molecular weight measurement by electron microscopy: 1) quantity of specimen required is small ($\sim 10^{-9}$g or less), 2) high purity is not required, as long as the particle of interest is isolated and recognizable in the image, 3) direct determination of mass per unit length[10] and mass per unit area[12] is possible, 4) aggregating systems can be studied to determine relative concentrations of monomer, dimer, etc., 5) Complex systems such as protein interacting with DNA or membrane may be studied, 6) under proper conditions all carbon atoms contribute equally regardless of their surroundings, permitting determination of lipoproteins, lipids, polysaccharides, etc., 7) a considerable range of molecular weights can be determined ($\sim 10^4$ to 10^{12}Daltons) for globular particles, higher for flattened particles, 8) potential accuracy is very good, error of less than 0.1% to 1% should be obtainable.

Theory

The probability of interaction of an incident electron with an atom in the specimen can be characterized by a scattering cross section (denoted σ) which can be loosely thought of as an area blocked out by the atom such that every electron passing through that area will be scattered. Two types of interaction are observed to occur with high probability: elastic scattering causing a deviation through a relatively large angle with negligible change in energy (resulting from deflection of the electron by the screened electric field of the nucleus), and inelastic scattering causing energy loss and little angular deviation (resulting from energy transfer to electrons of the specimen). Both signals are useful for mass measurement, but only elastic scattering will be considered below. Unless stated otherwise all expressions for elastic scattering can be converted to the inelastic case by substituting the appropriate scattering cross section. The use of the energy loss signal has been described previously[5,13]. Numerical values of scattering cross sections and angular distributions in a form suitable for application to electron microscopy are available in the literature[14,15]. For the moment we need only note that the elastic scattering cross section for carbon, σ_c, is $\sim .01$Å^2 at 100 KeV. At other voltages σ is proportional to $1/\beta^2$ or roughly V_o^{-1}, where β = the ratio of the electron velocity to the velocity of light, V_o is the accelerating voltage.

If a specimen containing N_c carbon atoms uniformly distributed within a cross sectional area A (including substrate atoms) is irradiated with a dose D (electrons/unit area) the number of electrons not elastically scattered (the bright field signal), n_{un}, is given by Beer's Law:

$$n_{un} = DA \ e^{-\dfrac{N_c \sigma_c}{A}} \qquad (1)$$

Solving this equation for Nc we obtain:

$$N_c = -\dfrac{A}{\sigma_c} \ \ln\left(\dfrac{n_{un}}{DA}\right) \qquad (2)$$

n_{un} is experimentally measured as the bright field signal passing through the objective aperture in a conventional microscope or through the elastic dark field detector in a STEM. DA is the number of electrons incident on the specimen area A. This equation can be solved point by point while scanning or approximately by recording on photographic film[1]. The resulting N_c can be converted to a molecular weight using the relationship:

$$MW = 12 \cdot \left[-\dfrac{A}{\sigma_c} \ \ln\left(\dfrac{n_{un}}{DA}\right) - At/\bar{v}\right] \qquad (3)$$

where t and v are the thickness and specific volume per atom of the substrate (for carbon films $\bar{v} \simeq 9\mathring{A}^3$). When combining Eq. 2 and Eq. 3, the assumption that all atoms are carbon results in only a small error, since biological molecules are mainly composed of carbon and the Z dependence tends to cancel out. In practice, Eq. 3 (σ_c) is calibrated using standard specimens of similar composition and the background scattering is evaluated on a nearby area of the substrate devoid of specimen. The assumption of a specimen of uniform thickness was made only to clarify the presentation. It can easily be shown that Eq. 3 can be applied to specimens of any topology provided the object is divided into subareas small enough to be regarded as having constant thickness. Eq. 3 is then evaluated for each subarea and the results summed.

It is interesting at this point to calculate the statistical error inherent in using Eq. 3 (assuming all systematic errors are negligible), since this determines the accuracy as a function of dose. Statistical errors enter through n_{un}, the number of unscattered electrons recorded for a given exposure, and through the term $At/\bar{v}$ which, for non-crystalline substrates, is only an estimate of the number of substrate atoms supporting the specimen. The statistical fluctuation in these numbers will be taken to be $\sqrt{n_{un}}$ and $\sqrt{At/\bar{v}}$ respectively ($\sim$valid for n_{un}, $At/\bar{v} > 100$). The expression for background fluctuation must be regarded as a gross approximation useful only for estimating errors. Actual values of substrate noise can be determined from high dose images. The fluctuation in MW, as derived from measured n_{un} values from many different identical

Symbols

A = particle area
D = electron dose incident on a specimen, electrons per unit area
MW = molecular weight in Daltons ($= 1.67 \times 10^{-24}$ gram)
N_c = number of carbon atoms in a particle plus substrate
n_s = number of scattered electrons
n_{un} = number of unscattered electrons
STEM = scanning transmission electron microscope
t = thickness of substrate
$\bar{v}$ = specific volume per atom
V_o = accelerating voltage
Z = atomic number
β = ratio of electron velocity of velocity of light
σ = scattering cross section
σ_c = elastic scattering cross section for carbon

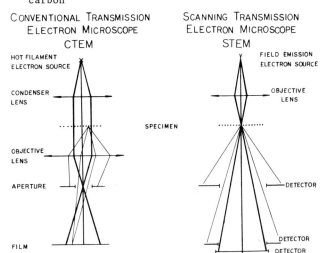

CONVENTIONAL TRANSMISSION ELECTRON MICROSCOPE CTEM

SCANNING TRANSMISSION ELECTRON MICROSCOPE STEM

Fig. 1. Comparison of electron collection in the conventional and scanning transmission electron microscopes. Scattered electrons are excluded from the image in the conventional microscope and shifted from the axial bright field detector to one of the annular detectors in the STEM. In some instruments the axial detector is replaced with an electron energy loss spectrometer for detection of inelastically scattered electrons.

particles, can be computed as:

$$\Delta MW = \dfrac{\partial MW}{\partial n_{un}} \ \Delta n_{un} + \dfrac{\partial MW}{\partial (At/\bar{v})} \ \Delta(At/\bar{v})$$

$$= \dfrac{12A}{\sigma_c \sqrt{n_{un}}} + 12 \ \sqrt{At/\bar{v}} \qquad (4)$$

and the fractional error for bright field measurement:

$$\frac{\Delta MW}{MW} = \frac{A}{\sigma_c(N_c - At/\bar{v})\sqrt{DAe^{-\frac{N_c\sigma_c}{A}}}} + \frac{\sqrt{At/\bar{v}}}{N_c - At/\bar{v}} \quad (5)$$

The dependence of $\Delta MW/MW$ on dose and molecular weight is shown in Fig. 2.

A similar expression can be derived for dark field microscopy. If we define n_s as the number of scattered electrons then:

$$n_s = DA(1-e^{\frac{-N_c\sigma_c}{A}}) \quad (6)$$

and

$$N_c = -\frac{A}{\sigma_c} \ln\left(1-\frac{n_s}{DA}\right) \quad (7)$$

For thin specimens where $N_c\sigma_c/A \ll 1$, Eq. 7 can be approximated conveniently as:

$$N_c = n_s/D\sigma_c \quad (7A)$$

Once again, the fundamental limit to the accuracy of a single measurement of a single particle is determined by the statistical fluctuations in n_s and in the substrate subtraction. Evaluating this fluctuation as before using Eqs. 3 and 7 we find:

$$\frac{\Delta MW}{MW} = \frac{\sqrt{DA(1-e^{\frac{-N_c\sigma_c}{A}})}}{D\sigma_c(N_c - At/\bar{v})e^{\frac{-N_c\sigma_c}{A}}} + \frac{\sqrt{At/\bar{v}}}{N_c - At/\bar{v}} \quad (8)$$

or

$$\frac{\Delta MW}{MW} \simeq \frac{1}{(1-At/\bar{v}N_c)\sqrt{D\sigma_c N_c}} + \frac{\sqrt{At/\bar{v}}}{N_c - At/\bar{v}} \quad (8A)$$

for thin specimens where $N_c\sigma_c \ll 1$. $\Delta MW/MW$ determined from the dark field signal (Eq. 8) is plotted in Fig. 2 for comparison to the bright field case.

Several points should be noted in comparing Eq. 5 and Eq. 8 (see Fig. 2). In each case the first term in the expression dominates at low dose and the second dominates at high dose. In both cases errors less than 3% should be achievable for compact particles ranging in size from 2×10^4 to 10^{11} molecular weight. The bright field signal is inferior for low dose, low MW measurements, as would be expected from the large error inherent in computing $\ln(n_{un}/DA)$ when $n_{un}/DA \simeq 1$. Finally, the dark field error at low dose is not sensitive to the shape of the particle, A, especially for thin specimens (see Eq. 8A).

It should be emphasized that Eq. 5 and Eq. 8 apply only to single measurements of single particles. Considerable improvement in accuracy could be realized through repeated measurements of a single particle (equivalent to higher dose) or through measurements of additional particles known to be identical and recognizable in the image (random sampling of $At/\bar{v}$). However 1% error may be adequate and should be obtainable on single large particles at very low dose. The next section of this paper will be devoted to identifying practical factors limiting accuracy and discussing means of overcoming them. Errors will be grouped into three categories: 1) radiation damage, 2) instrumental factors, and 3) specimen preparation.

Practical Limitations to Accuracy in Mass Measurement

Instrumental and specimen preparation errors can be reduced arbitrarily through careful design and suitable technique, however, radiation damage will always be with us[16]. As far as mass measurements are concerned, greatest difficulty comes from mass loss due to electron irradiation. Many biological molecules lose 50% or more of their mass with a dose of ~ 10-100 el/$\mathring{A}^2$ (at 100KV)[2,10,17,18,19]. The fractional mass loss appears to be somewhat dependent on specimen preparation conditions, however relevant variables have not been identified. Ramamurti has shown that mass loss can be dramatically reduced by cooling the specimen to $\sim 20^\circ K$[18] and several groups are attempting to reach this temperature. A second effect of radiation damage, flattening and widening of the molecule, however is easily dealt with during analysis.

Experimentally, mass loss errors are corrected by determining a dose response curve for individual molecules subjected to repeated low dose scans. On a given specimen all particles of the same type are found to give nearly identical dose response curves. These curves show an exponential fall at low dose to a relatively constant value at high dose. The dose response curve can then be used in one of two ways. A correction factor close to unity may be used to correct measurements from low dose images to zero dose conditions. Alternatively, a large correction factor (~ 1.5 to 2) may be applied to measurements from high dose images where the mass has become relatively stable. Fig. 2 indicates that for large particles it should be possible to make accurate measurements at a dose of 1 el/$\mathring{A}^2$ where minimal correction is required ($< 2\%$ correction, see Fig. 4).

Instrumental errors can severely limit accuracy in some instruments. Such errors fall into two categories: imaging errors and contrast recording errors. The parameter A in Eqs. 2 and 7 is determined from the image using the magnification (squared). An error in A can be eliminated through use of an

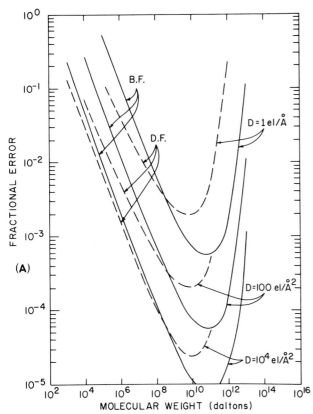

(A)

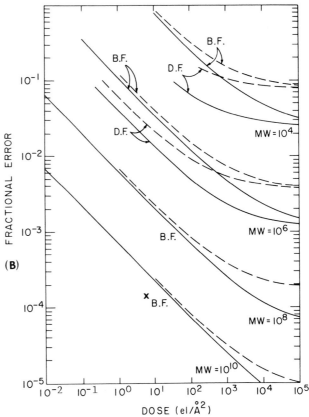

(B)

Fig. 2. Fractional error in mass
measurement $\Delta MW/MW$ as a function of molecular
weight (A) and dose (B) as predicted by Eqs. 5
and 8. A cubic particle of uniform density,
$(\bar{v} = 9\mathring{A}^3/atom)$ and composed of atoms of
uniform atomic number $(Z = 6)$ is assumed. The
scattering cross section is assumed to be
$0.01\mathring{A}^2$, roughly equal to that of carbon at
100KeV. In (A) the solid curves represent the
result of Eq. 5 for the bright field signal,
and the dashed curves the result of Eq. 8 for
the dark field signal. The substrate is
assumed to be $10\mathring{A}$ thick. In (B) bright field
and dark field curves are indicated (for
$MW \sim 10^8$ the curves are identical, for $MW =
10^{10}$ the D. F. curve is not shown). The solid
curves are calculated for a $10\mathring{A}$ thick
substrate and the dashed curves for a $100\mathring{A}$
thick substrate.

internal standard on the same specimen if
proper attention is paid to possible
distortions over the image field.

More severe errors generally result from
changes in electron collection and
nonlinearities in recording. Elastically
scattered electrons are distributed over a
wide range of scattering angles. The fraction
of scattered electrons collected by given
detector geometries with known specimen to
detector optics can be calculated from simple
relations[15]. In some systems the detector
collection efficiency may be a strong function
of operating conditions (e.g. variations in
axial specimen position). This effect can be
corrected through calibration curves[5],[10] or
internal standards[2]. Alternately it is
possible to design a system in which
collection geometry is independent of specimen
position[11]. Highest possible collection
efficiency is desirable.

The final link in the imaging chain is
the detector itself. If the detector does not
have quantum efficiency, increased dose will
be required obtain the accuracy predicted by
Eqs. 5 and 8. Actual electron counting
capability may not be required, as long as
electronic noise is substantially less than
counting noise[20]. Detector linearity with
count rate and dose must be established,
otherwise another correction factor must be
applied. Needless to say, the more correction
factors required, the more tedious and
error-prone the analysis, especially if
correction factors are large and sensitive
to experimental variables.

At this point it seems appropriate to
digress briefly to compare the conventional
microscope with the STEM for the performance
of mass measurements. Although excellent mass
measurements have been carried out with the
conventional microscope, the STEM offers
several clear advantages: 1) electrons

elastically scattered from carbon atoms can be detected at least 2 x more efficiently with the STEM (.74 collection efficiency for STEM annular detector vs .22 for tilted beam, .34 for axial beam stop conventional microscope)[14], 2) STEM can employ quantum detectors which have more linear response over a wider dynamic range than photographic film, 3) unscattered, elastically scattered, and inelastically scattered currents can be measured simultaneously on one scan, providing increased information for a given dose, 4) digital image recording can be used to preserve the inherent linearity, due to serial imaging. Two additional points, while not as important as those above, also weigh heavily in favor of the STEM. Once proper operating conditions are found, magnification is controlled very precisely ($< 0.1\%$ error) by attenuating the scan current rather than by changing lens excitation, and scan position can be offset electrically with no stage motion. Therefore large numbers of images can be recorded under nearly identical conditions. Finally, phase contrast can be an important factor in the conventional microscope. Since phase contrast changes rapidly with focus, careful attention must be paid to choice of illumination and focus conditions. With the STEM, the elastic signal obtained from an annular detector is relatively insensitive to phase contrast, especially if the detector is subdivided into two concentric annuli and the outer used[21]. Using this signal the contrast of a thin object on the substrate is always positive and has the same integrated value, only the sharpness of details depends on focal setting. Based on the above considerations, we will restrict ourselves to consideration of mass measurements performed with a STEM in further discussions.

The third class of experimental variable, specimen preparation, is often a more important limitation on accuracy than either mass loss or instrumental variables. Two types of difficulties are common: biochemical inhomogeneity, and specimen preparation artefacts. To a person with a physical background, biological molecules, (even those such as viruses which are designed to be stable) are surprisingly fragile. Small changes in pH or salt conditions, long storage, lyophilization, and bacterial or chemical attack can reduce a once homogeneous preparation to chaos. In many cases biological activity is not an adequate indicator since a substantial fraction of the population may be inactive. The safest approach is to use only freshly prepared material which has been characterized by biochemical techniques and perform a final purification immediately before preparing specimens.

Since specimens must, by definition, be unstained, a variety of new difficulties may be encountered in specimen preparation. Air drying has been found unsatisfactory for most specimens for several reasons: 1) Any residual salt is concentrated at the edges of the particle forming a low Z "negative stain", 2) unless buffers are used, pH and salt conditions are uncontrolled in the last stage of drying, frequently denaturing particles and 3) surface tension forces may flatten particles to the extent that they are not easily recognizable and do not have well defined boundaries. The use of freeze drying and vacuum transfer to overcome these problems will be described below.

Specimen Preparation for Mass Measurement

The techniques described below are being employed on a routine basis at the Brookhaven STEM Resource for preparation of specimens suitable for mass measurement. Some resuts with fd phage will be presented as an example. Freshly prepared biological molecules stored in optimum buffer conditions at $4^{\circ}C$ for less than 3 months are placed in deionized water (Millipore, Milli-Q) by passage through a Sephadex G100 (Pharmahia Fine Chemicals) column which had been previously washed with 100-1000 bed volumes of deionized water. Bed volumes typically used are ~ 0.3 ml so small quantities of samples can be used (0.01 ml at a concentration of 0.01 mg/ml). This column serves both to desalt the specimen and to remove any heavy atoms or low molecular weight impurities. The eluted sample is adjusted to the desired concentration with deionized water and a drop (0.005 ml) placed on a carbon coated holey film. After 1 min. the excess solution is wicked off, any desired washing carried out, and the grid with a layer of solution 10-100 microns thick is plunged into liquid nitrogen (or freon). The liquid nitrogen trough is actually part of the specimen exchange cartridge of the STEM, so that after freezing each specimen can be moved, under liquid nitrogen, to its slot in the exchange cartridge, ready for insertion into the STEM. After six specimens are in position, the liquid nitrogen is allowed to evaporate and the chamber pumped to $\sim 10^{-4}$ Torr in ~ 30 sec using a sorption pump. Freeze drying then occurs over a period of ~ 1 hr. as the cartridge warms to $\sim -50^{\circ}C$. The cartridge is then sealed, detached from the loading chamber, and the evacuated assembly transferred to the STEM and attached to the specimen changer which is used to insert specimens into the STEM. The specimens are maintained at a clean vacuum $< 10^{-4}$ Torr from the start of freeze drying throughout transfer to the STEM cold stage. An indicator of successful freeze drying is the presence of a network of particles which are not attached to the film (in unwashed specimens).

The rationale for this procedure is as follows: the biological specimen is maintained in deionized water for the shortest possible time (if necessary it may be fixed with glutaraldehyde without significant increase in mass). Columns must be washed extensively to remove heavy atoms and small fragments of gel which might be mistaken for molecules. The 1 minute incubation on the grid allows time for some particles to become attached to the substrate. Rapid freezing then immobilizes other particles and residual salt ions. During freeze drying, eutectics may form which concentrate residual salt into easily recognizable clumps, rather than preferentially around molecules, and pH changes, if they occur, do not appear to disrupt the specimen. Improvements in freeze drying technique are now being tested. Transfer under vacuum is judged important to prevent rehydration of hygroscopic salt clumps or freeze dried molecules.

The availability of thin clean carbon films with uniform wetting properties is essential to this preparation. Carbon is evaporated by resistance heating of spectroscopically pure carbon rods (Union Carbide) onto freshly cleaved rock salt (Harshaw Chemical Co.) in an ultra-high vacuum bell jar system. These films are intitially hydrophilic and remain so for at least 1 month as long as they are not removed from the STEM clean room (air passed through absolute and charcoal filters). Carbon films are picked up on holey carbon films supported on titanium grids as described previously[22,23]. The use of titanium grids is important for two reasons: titanium forms a chemically stable oxide which prevents heavy atom contamination observed with copper grids, and Ti grids are very sturdy and flat, providing a well defined specimen location (see above discussion of specimen axial position vs collection efficiency). The grid is then placed in a Ti ring with cap 0.25 mm thick for improved rigidity and ease of handling.

Choice of Recording Parameters

Once the specimen is transferred to the microscope and a typical area located, a dose-response curve for mass loss is determined. This can be done most conveniently with an on-line computer system[10], or by observing image intensity on a line replay, following a known number of scans. For molecules greater than 10^6 molecular weight, low dose conditions will probably give adequate accuracy. For smaller molecules a combination of high and low dose images can be used. A scan raster sufficiently fine to sample the particle at least twice across its diameter is generally required to reduce sampling error. If this results in a scan too small to include the whole particle (e.g. filamentous virus), the beam may be defocused to increase the apparent diameter of the particle (if $N_c \sigma_c / A \ll 1$). This causes little loss of accuracy for thin specimens at low dose when using a dark field signal (see Eq. 8A). Beam current is then adjusted to give the desired dose. Some care must be exercised in calculating the imaging dose if the beam size is smaller than the distance between picture elements, as damage may be non-uniformly distributed.

Choice of detector is based on specimen thickness and type of information desired. For thick specimens the unscattered electron detector (bright field) would usually be most appropriate, however, for very thick specimens ($\sigma_c N_c / A > 5$) the small and large angle detectors again become useful because of their finite outer cut off angles[24,25]. For thin specimens the large angle elastic detector will always be useful and the small angle detector can be added to improve counting statistics at the expense of increased phase contrast. The energy loss signal is mainly valuable in increasing collected scattering and in its less pronounced Z dependence, which can be used in conjunction with the elastic signal to demonstrate the presence or absence of heavy atoms in the specimen. Most STEM data aquisition systems provide for recording several signals simultaneously, so the choice of signals can be made during analysis. Since only the area scanned is damaged, it is easy to adjust all microscope parameters on an area adjacent to the one of potential interest, so that only first scan images are recorded. The Brookhaven STEM is routinely operated with the specimen at $-130^\circ C$ in order to eliminate contamination[26] which might interfere with mass measurement.

Data Analysis

Most STEM data aquisition systems provide for recording of raw image data digitally on magnetic tape. At this point the data can be analyzed using the small computer of the data aquisition system, or remotely on a large computer. Both systems have their advantages. The small computer system is ideal for providing interactive analysis with each step under the control of the operator[10]. It is also valuable for performing analysis needed for further microscope operation such as dose response or particle identification in a field of particles of different sizes. The disadvantages are: analysis and STEM operation cannot be done simultaneously, program space, speed, and accuracy are limited, as are printout capabilities, and operator subjectivity may lead to analysis errors. Using a large central computer, greater program sophistication and speed are possible. It is possible to study statistically the effect of changing analysis parameters. The price for these added capabilities is slight inconvenience, delay in getting output, and expense. A combination of both capabilities would be optimum.

The following example is presented to illustrate the steps in mass measurement as routinely performed with the Brookhaven STEM. Fig. 3 shows a 512 x 512 element scan of fd virus with a picture element size 80 x 80 Å or 4.16 micron full width recorded with a dose of 0.5 electron/Å^2 using the elastic scattering signal. Since the viruses are only 80Å wide and the beam was focused, some sampling error is evident. In the upper region of the image, charging of unattached viruses is evident. The viruses are 8830 Å long, providing a convenient internal magnification standard. A dose response curve recorded as above is presented in Fig. 4. Based on this curve the correction of first scan images to zero dose is 1.005. A line printer display is shown in Fig. 5 where intensity values above a threshold are coded first as numbers 1-9, then as letters A-Z (values greater than threshold +36 are represented as *). A histogram of intensity values in the selected area is shown in Fig. 6. Total range of possible values is 0-256 (analysis of other channels not presented). The program traces the backbone of each virus, fitting it to arcs of circles. Two printouts show the intensity as a function of distance perpendicular to the axis, and intensity along the axis. The radial distribution, in Fig. 7, is summed to give the total integrated intensity out to a given radius, after subtracting a background determined at a specified distance on each side of the virus. The mass is determined by integrating up to the radius where the particle intensity has fallen to background level. The axial distribution is displayed in several ways: the peak values of points near the axis, the difference between measured points and the value expected for a point that distance for the axis, and the actual value printed in its true relationship to the computed axis (see Fig. 8). A similar set of printouts is available for globular particles, except that intensity values around the particle center are printed in a square array. Sufficient information is available in these printouts to provide a high level of confidence that the analysis was carried out correctly. In addition the variance between actual data points and values expected from the fitted model is printed, allowing quantitative assessment of the quality of fit. For a uniform, well preserved particle this variance is close to that predicted from counting statistics and background histograms. Finally at the end of a run a summary table is printed showing important analysis parameters and integrated mass. A histogram of fd

particles in Fig. 3 is shown in Fig. 9. The observed standard deviation (4% of the mean) is in good agreement with that predicted by Eq. 8 for a particle of 16.6 x 10^{6}MW. In a larger study over 100 viruses were measured on several specimens and several areas of each specimen, with a standard deviation for the whole assembly of 8%. The cause of this systematic error is now being investigated.

For many purposes the absolute accuracy of measurement is adequate. However, for more precise measurements, fd is included as an internal standard providing both a length and mass reference, as well as a test of the quality of freeze drying. This technique is now being applied to comparison of mass per unit length of presently known filamentous viruses. Mass measurements of φ6 virus were recently carried out in collaboration with L. Mindich, NYPHS Lab giving a standard deviation of 3%, and measurements of pyruvate carboxylase carried out in collaboration with M. Utter and N. Cohen of Case Western Reserve University gave a standard deviation of 5% of the mean, both values close to those predicted by Eq. 8. It seems clear that further refinement of sample preparation technique and analysis programs will yield results predicted by Eq. 8, the quantum limit.

Presently mass measurements are performed at the Brookhaven STEM Resource as collaborative projects with STEM staff. Actual analysis requires approximately one day after suitable specimen preparation conditions are determined (usually requiring an additional day). In the future we hope to be able to offer this as one of our user services.

The state of the art in STEM mass measurements is such that it is the best available technique for determining the molecular weight of particles between $\sim 10^8$ and $\sim 10^{12}$ Daltons, competitive with other techniques between 10^5 and 10^8 Daltons, and useful down to $\sim 10^3$ Daltons or above 10^{12} in special cases. The technique has the advantage of being able to use small quantities of relatively unpurified specimens and study the equilibrium in interacting systems. It may be the only technique capable of measuring molecular weights of proteins bound to DNA or membranes, or in other complex environments. From the STEM operator's point of view it offers a solution to the request "Will the real molecule of interest please wave its flag". Finally, for heavy atom staining studies it will provide the missing link between the solution biochemistry and the high dose atom image -- a low dose determination of how many heavy atoms survived through specimen preparation.

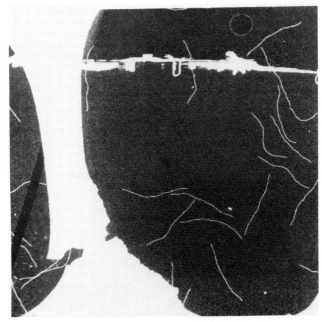

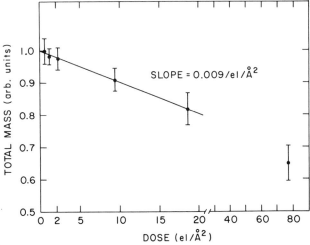

Fig. 4. Dose response curve for filamentous viruses. 19 viruses were measured on successive scans, each imparting a dose of 0.6 el/$Å^2$. Plotted points are the means of these measurements and vertical bars the standard deviations. A linear dose scale is used to demonstrate that the low dose portion of the exponential mass loss curve can be approximated by the first (linear) term. One point is shown at high dose to show that these viruses lose ~ 40% of their mass after prolonged irradiation.

Fig. 3. Dark field image of fd phage. This 512 x 512 element image was recorded at 40KeV with a dose of 0.6 el/$Å^2$ using the STEM large angle detector (40-200MRadian, 40% collection efficiency, 100% detection efficiency). fd phage were observed at -120°C supported on a 10Å thick carbon substrate stretched across a holey film (5 micron holes, edge visible in image). Note the charging artefact in the upper portion of the image due to unattached virus. 4.16 micron full scale.

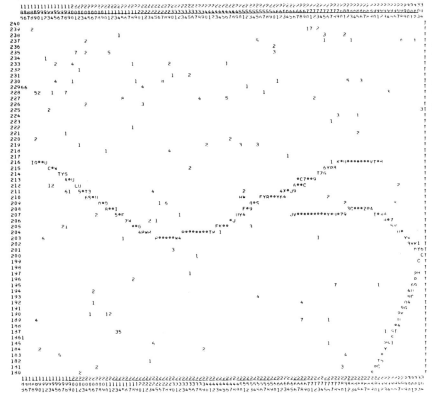

Fig. 5. Computer printout of an area of Fig. 3 slightly below and to the left of center. Individual picture elements are printed so that values below the background mean + 2 x background standard deviation are suppressed and values above this threshold (54.3 in this case) are coded first as numbers, then as letters. Values more than 36 units above the threshold are printed as *.

NUMBER PIXELS = 7138.
MEAN = 44.76
SD = 4.76

29	1	.01	I
30	1	.01	I
31	4	.06	I
32	13	.18	I
33	15	.21	I
34	30	.42	IX
35	60	.84	IXX
36	104	1.46	IXXX
37	158	2.21	IXXXX
38	239	3.35	IXXXXXXX
39	314	4.40	IXXXXXXXXX
40	400	5.60	IXXXXXXXXXXX
41	455	6.37	IXXXXXXXXXXXXXX
42	560	7.85	IXXXXXXXXXXXXXXXXX
43	564	7.90	IXXXXXXXXXXXXXXXXX
44	590	8.27	IXXXXXXXXXXXXXXXXXX
45	617	8.64	IXXXXXXXXXXXXXXXXXX
46	559	7.83	IXXXXXXXXXXXXXXXXX
47	508	7.12	IXXXXXXXXXXXXXXX
48	436	6.11	IXXXXXXXXXXXXX
49	393	5.51	IXXXXXXXXXXXX
50	316	4.43	IXXXXXXXXXX
51	227	3.18	IXXXXXXX
52	176	2.47	IXXXXXX
53	117	1.64	IXXX
54	88	1.23	IXX
55	53	.74	IX
56	46	.64	IX
57	36	.50	IX
58	16	.22	I
59	18	.25	IX
60	12	.17	I
61	12	.17	I
62	7	.10	I
63	7	.10	I
64	6	.08	I
65	2	.03	I
66	2	.03	I
67	8	.11	I
68	2	.03	I
69	1	.01	I
70	3	.04	I
71	5	.07	I
72	2	.03	I
73	4	.06	I
74	3	.04	I
75	4	.06	I
76	1	.01	I

Fig. 6. Histogram of values of picture elements in Fig. 5. Total range of values is 0-256. Background mean and standard deviation are computed by fitting the peak to a Gaussian distribution.

6 START (216.0, 292.7) END (216.0, 183.3)

	DIST	NUM	CUM NUM	DENS	BACK	INTEGRAL
	0.0	31	31	44.94	45.18	1542.4
	.1	18	49	53.06	45.18	2404.2
	.2	22	71	84.64	45.18	3272.2
	.3	12	83	78.25	45.18	3664.1
	.4	16	99	72.94	45.18	4113.2
	.5	17	116	63.76	45.18	4429.1
	.6	16	132	52.75	45.18	4550.2
	.7	16	148	49.06	45.18	4612.4
	.8	20	168	51.95	45.18	4747.8
	.9	17	185	45.76	45.18	4757.7
	1.0	60	245	45.15	45.18	4755.9
	1.1	29	274	45.79	45.18	4773.7
	1.2	11	285	44.82	45.18	4769.7
	1.3	15	300	43.53	45.18	4745.0
	1.4	18	318	43.06	45.18	4706.8
	1.5	17	335	44.41	45.18	4693.7
	1.6	12	347	42.33	45.18	4659.6
	1.7	19	366	45.42	45.18	4664.1
	1.8	15	381	45.40	45.18	4667.4
	1.9	19	400	44.21	45.18	4649.0
	2.0	57	457	44.84	45.18	4626.0
	2.1	29	486	43.84	45.18	4600.5
	2.2	17	503	48.24	45.18	4652.5
	2.3	14	517	45.00	45.18	4650.0
	2.4	23	540	43.52	45.18	4611.8
B	2.5	12	552	44.75	45.18	4606.7
B	2.6	14	566	47.14	45.18	4634.1
B	2.7	17	583	44.47	45.18	4622.1
B	2.8	16	599	43.75	45.18	4599.2
B	2.9	16	615	47.00	45.18	4628.3
B	3.0	69	684	43.96	45.18	4543.9
B	3.1	17	701	44.71	45.18	4535.8
B	3.2	19	720	46.21	45.18	4555.4
B	3.3	19	739	46.00	45.18	4571.0
B	3.4	15	754	45.00	45.18	4568.3
B	3.5	16	770	44.31	45.18	4554.4
B	3.6	15	785	43.93	45.18	4535.7
B	3.7	17	802	46.06	45.18	4550.7
B	3.8	14	816	44.93	45.18	4547.2
B	3.9	20	836	46.30	45.18	4569.6
B	4.0	64	900	44.75	45.18	4542.0
B	4.1	18	918	45.00	45.18	4538.4
B	4.2	15	933	44.93	45.18	4535.1
B	4.3	23	956	44.70	45.18	4524.0
B	4.4	14	970	47.43	45.18	4555.4
B	4.5	16	986	44.75	45.18	4548.6
B	4.6	16	1002	44.50	45.18	4537.7
B	4.7	16	1018	49.06	45.18	4599.8
B	4.8	15	1033	44.53	45.18	4590.1
B	4.9	20	1053	47.95	45.18	4645.5
B	5.0	26	1079	43.88	45.18	4611.3

Fig. 7. Printout of integrated scattering - background as a function of distance from the fitted particle axis. The virus contour is fitted to arcs of circles and the distance of picture elements from this axis is computed in units of tenths of a sampling element. Note that the integral stops changing at a distance of 0.8. Background for this integration is computed from points between 2.5 and 5.0 units from the axis. This integral is compared with a similar one computed using the background computed in Fig. 6 (not shown), any discrepancy indicating a suspicious measurement.

Acknowledgments

I would like to thank Jim Hainfeld, K. Ramamurti, D. Voreades and Elliott Shaw for advice, assistance and support; George Latham for technical help preparing specimens; and Keith Thompson for programming support. Jack Bittner, F. Kito and J. Cassidy were instrumental in the design and construction of the STEM, which would have been impossible without the professional help of many staffmembers of Brookhaven National Laboratory. The instrumentation group under the direction of G. Dimmler designed and constructed the STEM data aquisition system. This work supported by the U. S. Department of Energy and by the N.I.H. Biotechnology Resources Branch, Grant # RR-00715.

References

1. E. Zeitler, and G. F. Baher, "A photometric procedure for weight determination of submicroscopic particles. Quantitative electron microscopy," J. Appl. Phys., 33, 1962, 847-853.
2. G. F. Bahr, and E. Zeitler, "The determination of the dry mass in populations of isolated particles," in: Quantiative Electron Microscopy, (G. F. Bahr, and E. Zeitler, eds.), Williams & Wilkins Co., Baltimore, MD, 1965, 217-239.
3. L. Marton, and L. I. Schiff, "Determination of object thickness in electron microscopy," J. Appl. Phys., 12, 1941, 759-765.
4. A. V. Crewe, and J. Wall, "A scanning microscope with 5 Å resolution," J. Mol. Biol., 48, 1970, 375-393.
5. J. S. Wall, "A high resolution scanning electron microscope for the study of single biological molecules," Ph.D. dissertation, University of Chicago, Chicago, IL, 1971.

UNIT	NUM	AVE	DIF	X	Y	ANGLE	DIF	-.6-.5-.4-.3-.2-.1 0 .1 .2 .3 .4 .5
1	0	0.0	0.00	216.00	292.72	3.14	0.000	
2	1	66.0	-28.94	216.00	291.72	3.14	0.000	66
3	1	93.0	-1.94	216.00	290.72	3.14	0.000	93
4	1	84.0	-10.94	215.00	289.72	3.14	0.000	84
5	1	86.0	-8.94	216.00	288.72	3.14	0.000	86
6	1	104.0	9.06	216.00	287.72	3.14	0.000	104
7	1	95.0	.06	216.00	286.72	3.14	0.000	95
8	1	91.0	-3.94	216.00	285.72	3.14	0.000	91
9	1	98.0	3.06	216.00	284.72	3.14	0.000	98
10	1	99.0	4.06	216.00	283.72	3.14	0.000	99
11	1	96.0	1.06	216.00	282.72	3.14	0.000	96
12	1	99.0	4.06	216.00	281.72	3.14	0.000	99
13	1	85.0	-9.94	216.00	280.72	3.14	0.000	85
14	1	98.0	3.06	216.00	279.72	3.14	0.000	98
15	1	75.0	-18.06	215.92	278.73	3.23	.087	75
16	1	54.0	-30.64	215.74	277.74	3.32	.087	54
17	2	68.0	-.70	215.48	276.78	3.40	.087	80 56
18	1	89.0	4.36	215.14	275.84	3.49	.087	89
19	1	61.0	-23.64	214.77	274.91	3.52	.026	61
20	1	42.0	-10.75	214.39	273.98	3.53	.017	
21	1	90.0	17.06	213.97	273.08	3.58	.044	90
22	2	66.5	-32.99	213.59	272.15	3.53	-.052	84 49
23	1	64.0	.24	213.27	271.20	3.47	-.052	64
24	1	96.0	11.36	212.91	270.27	3.51	-.035	96
25	1	93.0	8.36	212.59	269.32	3.47	-.035	93
26	2	64.5	1.47	212.28	268.37	3.45	-.026	67 62
27	1	114.0	29.36	212.00	267.41	3.43	-.017	114
28	1	118.0	24.94	211.69	266.46	3.46	.026	118
29	2	62.5	-11.70	211.42	265.50	3.42	-.035	64 61
30	1	74.0	-4.25	211.18	264.53	3.38	-.044	74
31	1	97.0	3.94	211.00	263.54	3.32	-.052	97
32	1	88.0	-5.06	210.74	262.58	3.40	.079	88
33	2	60.0	-5.69	210.49	261.61	3.39	-.009	61 59
34	2	69.0	7.00	210.25	260.64	3.39	-.009	89
35	1	115.0	21.94	210.08	259.65	3.31	-.079	115
36	1	114.0	19.06	210.00	258.66	3.23	-.079	114
37	1	82.0	-12.94	209.97	257.66	3.17	-.061	82
38	1	89.0	-4.06	209.86	256.66	3.26	.087	89
39	1	70.0	-8.25	209.66	255.68	3.34	.087	70
40	2	67.5	-1.70	209.37	254.72	3.43	.087	83 52
41	1	107.0	13.94	209.01	253.79	3.52	.087	107
42	1	63.0	-15.25	208.56	252.90	3.60	.087	63
43	1	64.0	.24	208.08	252.02	3.65	.044	64
44	2	82.0	-1.99	207.65	251.12	3.58	-.070	94 70
45	1	59.0	6.25	207.30	250.18	3.51	-.070	59
46	1	89.0	4.36	206.90	249.26	3.54	-.035	89
47	1	85.0	.36	206.47	248.36	3.59	.044	85
48	2	59.0	-13.00	206.01	247.47	3.62	.035	74
49	1	92.0	7.36	205.61	246.56	3.56	-.061	92
50	2	69.5	13.31	205.26	245.62	3.49	-.070	92 97
51	1	97.0	3.94	205.00	244.66	3.41	-.079	97
52	1	75.0	-9.64	204.70	243.70	3.45	.035	75

Fig. 8. Printout of intensity distribution along the fitted axis. The columns printed are from left to right: distance along axis, average value of points in the interval, difference between measured and expected value (expected value from measurements of Fig. 7), X and Y coordinates of axis at that distance from end, angle of the axis at that point (in radians), difference between present and previous angle, and a display of sample points in their true relationship to the fitted axis. Not shown are means and standard deviations for columns 3, 4 and 8, together with an estimate of the sampling error.

Fig. 9. Histogram of MW measurements of viruses shown in Fig. 3. The standard deviation of these measurements is 4%. Units were chosen arbitrarily to fit the known MW of fd of 16.7 x 10^6 [27].

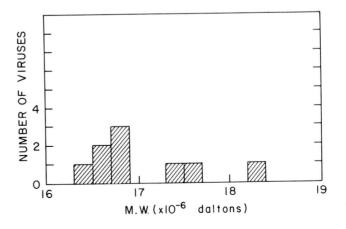

301

6. A. V. Crewe, and J. Wall, "Quantitative analysis in high resolution transmission scanning microscopy," Proc. EMSA, Claitor's Pub. Div., Baton Rouge, LA, 1971, 24-25.

7. J. Wall, "Mass and mass loss measurements on DNA and fd phage," Proc. EMSA, Claitor's Pub. Div., Baton Rouge, LA, 1972, 186-187.

8. J. P. Langmore, and J. C. Wooley, "Chromatin architecture: Investigation of a subunit of chromatin by darkfield electron microscopy," Proc. Nat. Acad. Sci. USA, 72, 1975, 2691-2695.

9. M. K., Lamvik, "Muscle thick filament mass measured by electron scattering," J. Mol. Biol., 122, 1978, 55-68.

10. A. Engel, "Molecular weight determination by scanning transmission electron microscopy," Ultramicroscopy, 3, 1978, 273-281.

11. J. Wall, and J. Hainfeld, "A new STEM capable of observing single heavy atoms in frozen biological specimens," Proc. 9th Intl. Cong. Electron Microscopy, Vol. 1, Toronto, Canada, 1978, 16-17.

12. B. P. Halloran, R. G. Kirk, and A. R. Spurr, "Quantitative electron probe microanalysis of biological thin sections: The use of the STEM for measurement of local mass thickness," Ultramicroscopy, 3, 1978, 175-184.

13. M. K. Lamvik, and J. P. Langmore, "Determination of particle mass using scanning transmission electron microscopy," IITRI/SEM, 1977/I, 401-409

14. J. P. Langmore, J. Wall, and M. S. Isaacson, "The collection of scattered electrons in dark field electron microscopy. I. Elastic scattering," Optik, 38, 1973, 335-350.

15. J. Wall, M. Isaacson, and J. P. Langmore, "The collection of scattered electrons in dark field electron microscopy. II. Inelastic scattering," Optik, 39, 1974, 359-374.

16. M. S. Isaacson, "Specimen damage in the electron microscope," in: Principles and Techniques of Electron Microscopy, Vol. 7, (M. A. Hayat, ed.), Van-Nostrand Reinhold Co., New York, 1977.

17. S. D. Lin, "Electron radiation damage of thin films of glycine, diglycine, and aromatic amino acids," Radiation Res., 59, 1974, 521-536.

18. K. Ramamurti, A. V. Crewe, and M. S. Isaacson, "Low temperature mass loss of thin films of L-phenylalanine and L-tryptophan upon electron irradiation -- a preliminary report," Ultramicroscopy 1, 1975, 156-158.

19. L. Reimer, "Irradiation changes in organic and inorganic objects," in Quantitative Electron Microscopy, (G. F. Bahr, and E. Zeitler, eds.), Williams & Wilkins Co., Baltimore, MD, 1965, 1082-1096.

20. J. W. Wiggins, "The use of scintillation detectors in the STEM, Proc. 9th Intl. Cong. Electron Microscopy/I, Toronto, Canada, 1978, 78-79.

21. J. Fertig, and H. Rose, "A reflection on partial coherence in electron microscopy," Ultramicroscopy, 2, 1977, 269-279.

22. M. Isaacson, J. Langmore, and J. Wall, "The preparation and observation of biological specimens for the high resolution scanning transmission electron microscope," IITRI/SEM/1974, 19-26.

23. J. S. Wall, J. F. Hainfeld, and J. W. Bittner, "Preliminary measurements of uranium atom motion on carbon films at low temperatures," Ultramicroscopy, 3, 1978, 81-86.

24. A. V. Crewe, and T. Groves, "Thick specimens in the CTEM and STEM. I. Contrast," J. Appl. Phys., 45, 1974, 3662-3672.

25. H. Rose, and J. Fertig, "Influence of detector geometry on image properties of the STEM for thick objects," Ultramicroscopy, 2, 1976, 77-87.

26. J. Wall, J. Bittner, and J. Hainfeld, "Contamination at low temperature," Proc. EMSA, Claitor's Pub. Div., Baton Rouge, LA, 1977, 558-559.

27. L. A. Day, in Single Stranded DNA Phages, D. T. Denhardt, (D. H. Dressler, and D. S. Ray, eds.), Cold Spring Harbor, New York, 1978.

Discussion with Reviewers

D. F. Parsons: Tracing the specimen at low temperature (-130°C) may actually enhance the contamination rate. Much depends on the surrounding geometry of cooled surfaces. What is the measured contamination rate during a molecular weight measurement?

Author: As described previously[26], the Brookhaven STEM operates at a vacuum $<$ 10^{-8}Torr in the specimen area with the specimen cooled to -130°C. Under these conditions there has been no detectable contamination on any specimen studied to date.

M. S. Isaacson: You mention that there is an 8% standard deviation on a particular set of measurements and imply that most of it is due to systematic error. Could you elaborate on the kinds of systematic errors that can arise using this technique?

Author: In this particular case the major errors arose from: 1) inadequate attention to the centering of the beam in the annular detectors, and 2) minor specimen impurities (denatured protein). Both these problems have been corrected and reproducibility of TMV particles (used as mass standards) is within $\sim 2\%$ from specimen to specimen.

For additional discussion see page 258.

SCANNING ELECTRON MICROSCOPY/1979/II
SEM Inc., AMF O'Hare, IL 60666, USA

CRITICAL POINT DRYING -- PRINCIPLES AND PROCEDURES

A. L. Cohen

Electron Microscope Center
Washington State University
Pullman, WA 99164

Abstract

This tutorial paper is an introduction to critical point drying (CPD), including preliminary processing and ancillary techniques. It gives brief descriptions of selected methods from recent literature and the author's laboratory. The topics include: 1. Surface tension and brief survey of alleviation of its damaging effects by fixation, drying from solvents, freeze drying, SEM examination of fresh and frozen material. 2. Theory and terminology of CPD. 3. Carriers and holders for small specimens. 4. Specimen processing, including selection, trimming, cleaning. 5. Fixation of small and large specimens. 6. Dehydration and intermediate fluids; selection and schedules. 7. Freon 13 and CO_2 as transitional fluids. 8. Process of CPD. 9. Variants--processing below critical pressure and temperature at reduced surface tension. 10. Safety and toxicology. 11. Conclusions--evaluations of artefacts.

KEY WORDS: Critical Point Drying, Biological Specimen Preparation, Fixation, Dehydration

Introduction

The purpose of critical point drying (CPD) is to preserve the shapes and sizes of normally hydrated specimens after they are fully dried. Although CPD has been known for 30 years[1], it had been little used, largely because the transmission electron microscope (TEM) was limited to very thin objects, such as cell or tissue sections, or to small areas of this surface replicas, or to very small specimens such as flagella, DNA threads, viruses and cell fragments (e.g. bacterial cell wall). These last maintained their shapes tolerably well when air dried or embedded in a negative stain. With the advent of the scanning electron microscope (SEM) and its revelations of surface morphology, the distortions and disruptions caused by air (evaporative) drying were too obvious, and the need to preserve surface detail became imperative. In the span of a tutorial it is not possible to discuss all the variations of specimen preparation for CPD, nor the procedure itself. Attention will be focused on common problems of specimen preparation, the most frequently used procedures, and some promising techniques whose potentialities may still be under exploration.

Nor is it possible to survey the literature exhaustively. Where feasible, references have been chosen with citations that lead back into the literature. In this paper superscript letters refer to the notes at the end of the narrative.

Surface Tension and Alleviation of its Damage

The major cause of damage on drying, as pointed out by Anderson[a][1][2] is *surface tension*, the characteristic of the surface of a liquid to behave like a stretched elastic membrane.[b] Water, the commonest intra- and extracellular environmental medium, also has the highest surface tension of any common liquid, about 73 dynes per centimeter. Water evaporating through and around a soft, hydrated specimen subjects it simultaneously to crushing and tearing forces. A rough model is given by imagining a piece of soft sticky modeling clay being pushed against a rubber membrane; the clay is both squashed and smeared as the rubber membrane, to which the clay adheres, simultaneously stretches and

*a, b, etc. in superscript refer to detailed notes included before references on p. 318 Col.2.

compresses it. Figures 1-6 illustrate the increased retention of surface characteristics of fibroblasts during drying under decreasing surface tension.

Table 1
Surface Tension Against Air at 20^0 of Common Liquids Used on Biological Electron Microscope Specimens

Name	Formula	Surface Tension (dynes/cm)[c]
Acetone	CH_3COCH	23.7
Ethanol	CH_3CH_2OH	22.8
Ethyl Ether	$C_2H_5OC_2H_5$	17.0
Methanol	CH_3OH	22.7
Freon TF (Freon 113)	$CClF_2-CCl_2F$	19.0
Water	H_2O	73.0

For decreasing surface tension, the following procedures are used:

1. Harden or toughen the material by fixation with aldehydes, osmium tetroxide, or other fixatives. Examples are listed elsewhere.[3,4] This is often combined with other methods.

2. Reduce the surface tension by substituting a liquid of lower surface tension (Table 1) (generally after fixation) as shown in Figs. 1-3 which compare the effects of evaporative drying from water, ethanol, and Freon TF. Examples using solvents of low surface tension, e.g., acetone, ethanol, and propylene oxide are cited by Cohen.[3] Thus Thornthwaite and Leif[5] claim preservation of lymphocyte microvilli in drying from xylene. However, Albrecht, *et al.*[6] in an extensive investigation of air drying from various solvents suggested caution in interpretation, as surface artefacts may be produced which are functions of solvent molecular polarity. Liepen and de Harven[7] have refined the technique by drying fixed and dehydrated specimens in Freon 113 (Freon TF) under slight vacuum. They claim results comparable to those produced by CPD.

3. Freeze the specimen quickly and dry it by evaporating (subliming) the ice under vacuum. This is a popular method exhaustively investigated by Boyde[8,9]. The techniques are discussed by Nei.[10] Freeze-drying has also been done from ethanol[11] and other solvents.[9] The method is particularly useful for specimens which have characteristic or diagnostic waxy or other cuticular coatings soluble in the treatment fluids used for CPD. The procedure is also valuable for organisms with flexible but inelastic external walls (e.g. arthropods) in which differential shrinkage of a soft interior may cause the casing partly to collapse. Within limitations the method combined with cryofracture is useful for X-ray analysis of diffusible elements (reviewed by Trump, *et al.*[12]).

4. Examine the fresh or fixed specimen before the vacuum and the beam cause obvious damage. This technique is often successful for plants or animals with cell walls or cuticles sufficiently impervious to water vapor (e.g. leaves, insects) to allow a specimen to keep its shape during observation. The high conductivity of moist tissues usually makes conductive coating unnecessary. Living insects and sprouting seeds were observed over a decade ago by Pease, Camp, and Hayes.[13] Other examples are cited by Hayat[4] (p. 83). Ledbetter's[14] striking micrographs of plant surfaces illustrate its usefulness. Fresh specimens (grass shoots) used for X-ray analysis by Dayanandan, *et al.* could be "examined for nearly 20 minutes without much damage to them."[15] The possibilities of prolonged observation are increased by the use of environmental cells.[16,17] The use of water substitutes of low vapor pressure such as glycerol[18] although very useful in some cases, has not gained extensive use for fear of contamination, although the original proponents, Panessa and Gennario, claim no such problems arose.[19]

5. Freeze the specimen and examine it frozen. Various simple and complex devices have been developed to maintain a specimen in a frozen state. The problem of preventing deposition of an obscuring coating of frost is severe. However, the method is suited for cryofracture, and X-ray analysis of a variety of objects (within limits, since soluble compounds may shift). Also, specimens may not have to be coated. Commercial devices are available for such work and for carrying out dissection or cryofracture under SEM observation.[20,21] Examination time is limited and specimens are not easily stored.

6. Critical point drying (CPD). This, the most widely used method, is explained further in theory and practice. It has the advantages over the others that for most specimens and for most workers it is the most reliable method. CPD can be particularly advocated for bulky specimens--those more than a few cubic millimeters--for preservation of internal structure at the microscopic level, for it has so far been impossible to freeze the interior of bulky specimens rapidly enough to avoid ice crystal damage. It is the one method which consistently allows the same specimen to be examined in the SEM and later impregnated with resin and sectioned for the TEM with good cytological preservation. It also is the most versatile in the variety of ways in which fractography may be practiced--dry after CPD, from frozen alcohol, from frozen resin, or cleaved wet from solvent or fixative.

All of the listed methods have been used successfully depending on the nature of the material and the goals to be achieved. No one of them is the perfect means to the microscopist's goal with its inherent contradiction of trying to stop life dead in its tracks, and keep it, like Browning's "Last Duchess," looking as if it were alive.

Theory and Terminology of Critical Point and Critical Point Drying

Suppose we approximately half fill a sturdy container (a "bomb") with a very volatile liquid, such as liquid CO_2 under pressure. The container has valved orifices for filling and venting. If the container is warmed, the liquid both expands and evaporates. If it is approximately half full, evaporation just balances

Table 2
Critical Properties of Important Fluids

Transitional Fluids	Critical Temp., T_c (degrees celsius)	Critical Pressure, P_c			Critical Density, D_c
		Atm.	lbs/in^2	kg/cm^2	
Carbon Dioxide (CO_2)	31.3	72.9	1073	75.5	0.460
Nitrous Oxide (N_2O)	26.5	71.7	1054	74.1	0.460
Freon 13 ($CClF_3$)	28.9	38.2	561	39.5	0.578
Freon 23 (CHF_3)	25.9	47.7	701	49.3	0.525
Freon 116 (CF_3-CF_3)	19.7	29.4	432	30.4	0.601
Freon TF ($CCl_2F-CClF_2$)	214.1	33.7	495	34.8	0.576
Water (H_2O)	374.0	217.7	3184	223.9	0.322

expansion and the surface of the liquid remains about in the center (Fig. 7B). However, the expanding *but nearly incompressible* liquid becomes less dense. But the *compressible* gas or vapor[d] phase becomes increasingly dense. The total density (neglecting the slight volume increase of the container) remains constant. The interface (meniscus) between the two phases becomes less distinct, and the interfacial tension (surface tension) declines as the densities and other properties of the two fluid[e] phases approach each other. At a specific temperature, the critical temperature (T_c), the densities and all other properties are identical, and the *meniscus and the surface tension vanish.* This is the critical point (CP). The pressure at the critical point is the critical pressure (P_c) and the density is the critical density (D_c). *The critical surface tension is always zero.* At the critical point, the phases are truly continuous. As the temperature is raised above T_c, the fluid behaves more and more like a gas. (The theory is discussed in greater detail elsewhere.[3,22]).

If a specimen is in the liquid, and the container is heated above T_c, the specimen is brought from a liquid to a dry vapor environment without being in contact with a surface, and it has not suffered the distorting effects of surface tension. If the valve is opened slightly while the temperature is elevated, the gas streams out until atmospheric pressure is reached, and the dry specimen may be removed. This is the essence of CPD.

In Fig. 7, the liquid is shown as dense clusters of molecules, while the vapor is shown as individual molecules. At T_c for all three columns (A, B, C), the fluid is shown as being in a somewhat intermediate condition between liquid and gas. Under a pressure of P_c or greater (Fig. B, C), this heterogeneity is coarse enough to scatter light; we get the *critical opalescence* which, in a system of decreasing temperature for a pure fluid, heralds the imminent separation of discrete liquid and vapor phases.

If the bomb is much less than half full (Fig. 7A), as the temperature is raised, evaporation outstrips expansion of the liquid, and the liquid totally evaporates before T_c is reached. Therefore, the CP is not attained. Note that the comparative pressure rise is gradual, approaching the PV curve of an ordinary gas. In Fig. 7B, the fluid is shown going through its CP; the vertical arm rises more steeply. In Fig. 7C, the bomb is initially considerably more than half filled with

the liquid phase. Now expansion outstrips evaporation, and the bomb is filled with the liquid before the critical temperature is reached. The pressure then rises abruptly. Nevertheless, the transition takes place, and above the critical temperature, only gas is present. Drying by passing the fluid through T_c at a pressure higher than P_c is called "going around the critical point." The purpose of this practice is discussed later.

The properties of transitional fluids suitable for CPD are shown in Table 2 with water listed for comparison. Table 2 shows that water has the unfortunate critical constants: T_c = 374[o] and P_c = 217 atm (3184 lbs/in^2). Therefore other fluids must be substituted. CO_2 is the most commonly used *transitional fluid*, i.e., the fluid making the transition from liquid to gas, with Freon 13[f] next. Freons 23 and 116 are rarely used.

Unfortunately, none of the extant transitional fluids are appreciably miscible with water. Therefore, we must replace water with any one of several *dehydration fluids*, usually in gradual steps; usually ethanol or acetone. Because not all of these dehydration fluids mix completely with the transitional fluids, and for other reasons, an *intermediate fluid*[f] is often used. This is usually either (iso)amyl acetate[1,2] or Freon TF.[3,22,23] The latter was selected because a small amount of Freon TF left in Freon 13 or in CO_2 (5% or less) merely raises T_c and P_c slightly. Therefore, flushing does not have to be as thorough as with the other less volatile intermediates, which if left in the bomb, evaporate at ambient pressure, and by evaporative drying defeat the purpose of the critical point technique.

Ethanol, acetone, amyl acetate, and Freon TF are completely miscible with liquid CO_2.[24] Ethanol has limited miscibility with Freon 13; the rest mix readily.[24]

Specimen Processing Before Critical Point Drying

General

Although CPD is considerably less critical than its name implies, initially it is well to proceed conservatively. This means that samples should be small, fixation times generous, dehydration steps and intermediate substitution steps fairly small, and times in each fluid substitution step more than deemed sufficient except in

Figs. 1-6: The effect of surface
tension on cultured fibroblasts
fixed in glutaraldehyde/OsO₄.
S = surface tension (dynes/cm) of
liquid. Picture width = 30 μm.

Figs. 3a-6a: Enlargement of
framed areas. Picture width = 7 μm.

Figs. 1-3: Cells dried at room
temperature and atmospheric
pressure.

Fig. 1: Dried from H₂O.*
Fig. 2: Dried from ethanol.
Fig. 3: Dried from Freon TF.

Figs. 4-6: Dried by slightly
opening chamber valve at temperatures and pressures indicated.
Pressure remains constant until
evaporation is complete. (Surface
tensions for Figs. 4 and 6 estimated.) The improvement of surface
topography with decreasing surface
tension is shown in Figs. 1-4.

*The surface tension of the water
is probably appreciably lower than
73 dynes/cm due to contamination.

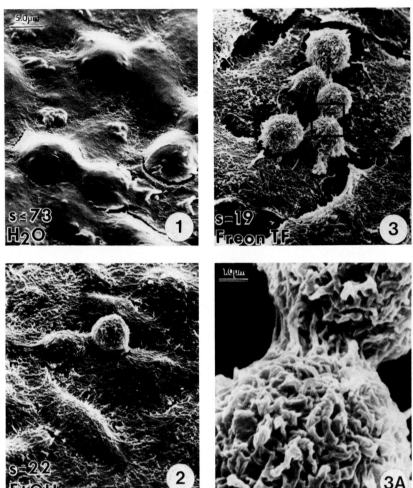

the lower (below 70%) alcohol (and possibly other dehydration fluids). Specimens should pass rapidly through these to avoid maceration and extraction. The times between flushing the bomb should be ample, and more than enough flushes used. Guidelines for each type of specimen serve to identify and quantify these factors more precisely, but nothing will do it as well as experience. Avoid short cuts at first, then any lapse or problem is usually easily identified and corrected. If possible, make pilot runs with the material to be studied. Thereby you learn the peculiarities of the material and gain expertise in its processing. If the specimens are rare, make trial preparations with as similar material as you can obtain. Time spent on practice runs is well invested. This is particularly important with seasonal or opportune material. Often the most crucial part of preparation is the initial cleaning, trimming and fixation. Suboptimal later treatment may influence results which may be only fair or good instead of excellent, but at least usable. The wrong preliminary treatment may mean failure. If possible, when you are in poorly explored technical territory, set up protocols for two or three simultaneous initial treatments through fixation. This is particularly

important if you must take fixatives into the field and depend on laboratory processing later.

As a guideline: If either microscopic or macroscopic specimens after fixation look clean, undistorted, and (except for color) "natural", your chance of success is good. *Critical point drying begins with the initial treatment of the specimen, and all steps are interrelated; arte- facts caused by earlier processing may not show up until later.*[3,25]

The following suggestions on processing are listed as closely as practicable in the order in which they would normally occur. The handling of "large" or "gross" specimens, (those being big enough to be handled individually) differs in some ways from the processing of "small" and "microscopic" specimens including those specimens which are handled *en masse*, such as cell or organelle suspensions, and cells in colonies.

Preliminaries; Specimen Selection, Trimming

Dried spores, pollen grains, lichens, mosses, etc. must be wetted before fixation if they are to have the normal vegetative appearance. Otherwise, fixation more or less perpetuates the dried collapsed state.

Beginners sometimes crowd many small organ- isms or bits of tissue into a container, and then

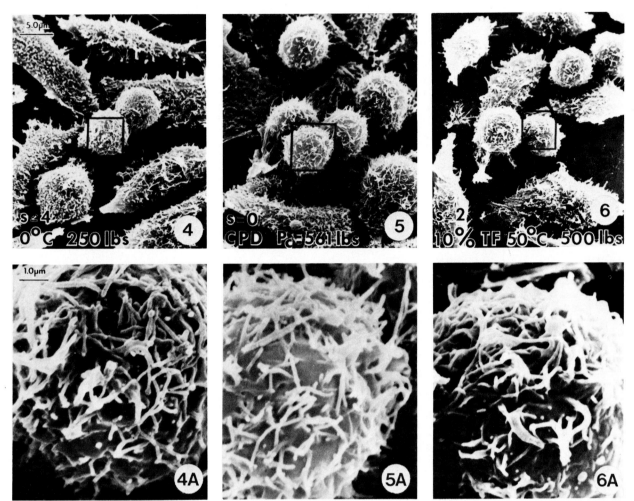

after CPD, try to separate the fragile components from a well-entwined and cemented mass. Selection and trimming may be carried out at several stages. Before fixation, initially selected specimens should be approximately trimmed to size, leaving margin for further more precise trimming later. However, if possible, particularly for animal tissues, trimming should be done in the fixative. Also when feasible, initial fixation should be by perfusing as described later (Preparation of Large Specimens). These initially cut surfaces are generally crushed and distorted. After fixation most tissues are firm and usually become progressively firmer with dehydration. Final trimming and dissection should be carried out in any appropriate liquid from buffer or water rinse after fixation through 70% ethanol. Do not attempt delicate dissection in a highly volatile or corrosive fixative (e.g. HCHO, OsO_4) or a rinse in which the fixative has not yet been diluted below irritating levels, since your eyes and nose are necessarily close to the specimens. Above 70% ethanol or its equivalent in other volatile dehydration fluids, evaporation or absorption of water vapor may compromise results. In general, higher plant tissues are trimmed easily between fixation and dehydration; agar blocks and some soft animal tissues are trimmed with cleaner surfaces and less crushing in the dehydration

fluid, usually about 70% ethanol. Razor blades and scalpels are preferable to scissors, but whatever instrument is used should be sharp, to favor cutting over crushing.

Other processes such as fracture for interior detail are discussed later.

Cleaning. It is often necessary to clean specimens of adherent secretions, for however optically transparent they are, such layers form an obscuring barrier to the SEM beam. Sometimes simple mechanical rinsing is sufficient; e.g., flushing out intestinal contents to expose the villi, or flushing out blood vessels. A few strands of mucus remaining are not undesirable, since they indicate the nature of the surface. Usually, such mechanical flushing should be done before any fixation. Enzyme treatment should be done before fixation. Certain mucoid solvents may be used after aldehyde fixation, or even as part of the fixative (e.g. sucrose). However, all cleaning should precede osmium fixation.

Mechanical cleansing can vary from gentle water rinsing as in the preparation of attached rumen bacteria[26] (also personal communications, Mr. Dan Cooper and Dr. J. Costerton) through brushing to sonication. Mechanical methods pose some danger of removing more than an obscuring layer, and specimens should be examined for damage. Bakst and Haworth[27] used sonication after aldehyde fixation for the hen's oviduct as did Behbehani, *et al.* for nematodes.[28] We have found the method very successful for removing a tough mucoid layer from aldehyde-fixed marine trematodes (unpublished). Sonication should be carried out with frequent observations. Fallah, *et al.*[29] found that any treatment, such as saline rinses, mechanical scrubbing, etc. before fixation of canine stomach mucosa could produce alterations. Therefore, they teased the mucus from the surface after glutaraldehyde fixation. Nemanic and Pitelka[30] used prolonged treatment with glycerol followed by soaking in 20% ethanol for mammary gland tissue to remove milk globules and other obscuring material; a method achieving some general use. Other methods including enzymatic treatment are reviewed by Cohen[3] (pp. 69-70). Additional examples are described by Eisenstat, *et al.*[31] for mucosa; Evans, *et al.*[32] for kidney and skin; and Mariscal, *et al.*[33] for coelenterate structures.

Cell suspensions (e.g. sperm) are not necessarily cleaned by simple centrifugation and they may be damaged by too much mechanical treatment. Atwood, Crawford and Braybrook[34] cleaned the heavily mucus-coated sperm of marine invertebrates with hyaluronidase in sea water before fixation. In our experience, best results are obtained with mammalian sperm by gentle repeated centrifugation (live) in balanced saline solution (Tyrode's); attempts to avoid ionic influence by using isotonic sucrose resulted in intractable clumping. Other methods for sperm are given by Bacetti,[35] and by Dadoune and Fain-Maurel.[36]

For cells with sufficiently tough cell walls in which wall morphology is important, more stringent methods may be used. Kinden and Brown[37] used periodic acid on osmium-fixed fungus-infected plant cells, followed by KOH treatment before dehydration and CPD. Their beautiful results showing the branching hyphae in host cells bear witness to the success of the technique.

Preservation of Coatings. There are occasions in which it is important to preserve mucoid coatings. Lutchel[38] preserved the mucus coating of trachea and bronchii by perfusing the blood vessels with glutaraldehyde, and postfixing and staining the airways with ruthenium red-OsO_4 before continuing to CPD. Thurston[39] fixed lungs intratracheally with a solution of OsO_4 in the water-immiscible perfluorobutyltetrahydrofurane, which presumably did not wash away or otherwise disturb the mucus layer.

Tissue Dissociation. By this is meant the dissolution of tissues into individual cells or cell groups. The methods for dissociating cells into their subcellular components constitute a bulky and specialized literature and will not be considered here, although the processing of these fractions for CPD is discussed further on under Processing Small Specimens. Details of the use of proteolytic and other enzymes in tissue culture technique to separate cells are given by several authors in the compendium by Kruse and Patterson.[40] A detailed procedure is given by Amsterdam and Jamieson.[41] Boyde[42] has rendered the boric acid technique more versatile by choice of fixatives, so that fractures through the cells

as well as cell separation can be achieved.

Baskets, Carriers, and Holders. Generally before or after fixation specimens should be put in or on appropriate carriers, but before they go into later highly volatile fluids or the CPD chamber. While large specimens may be carried unprotected through a series of fluids, and then after final trimming, placed in baskets, small ones must generally be placed in or on appropriate holders as soon as possible. With containers and holders, transfers are rapid, specimens are protected against loss and mechanical damage, and the chances of evaporative drying of the later solvents is much reduced. Metal or weighted holders keep specimens under the liquid surface. This is particularly important for aqueous media and for the dense Freon TF used as an intermediate fluid. Various containers for large and minute specimens and holders for grids and coverslips are reviewed, described, and illustrated by Cohen[3] and Hayat.[4] Further details on carriers for small specimens are given in the next section.

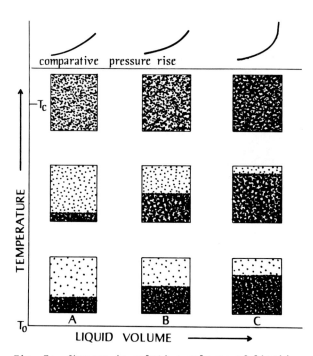

Fig. 7: Changes in relative volumes of liquid and vapor phases as the temperature is increased. T_O = initial temperature, T_C = critical temperature. A. Liquid less than half of total volume. Level decreases; evaporation is total before T_C. B. Liquid approximately half of volume, and remains at about same volume, but its density decreases and density of vapor increases with temperature rise. C. Liquid more than half of volume. Liquid expands to fill volume before T_C. Comparative pressure rise with increasing temperature, gradual for A and B, abrupt and rapid for C. This is called "going around the critical point".

Processing Small Specimens

The fixation and other procedures in processing large specimens for CPD are fairly well known. This is not the case with specimens too small to be handled individually. Many of the methods are described by Cohen,[3] Hayat,[4] Hayes and Pawley,[43] Bacetti (sperm),[35] and Kurtzman, Baker and Smiley.[44] The last authors give clear directions in a brief paper for handling a variety of organisms, and a means of transferring dried particles from membrane filters to coverslips which have a more uniform background. Because so many questions are asked in this area, this presentation will discuss the processing of small specimens in some detail.

Carriers for small specimens are basically of two types: (1) enclosed, finely porous containers such as folded filter paper packets,[3] and other fine meshed or porous capsules described below; and (2) flat surfaces such as membrane filters, glass or plastic slides and coverslips, agar plates, filmed TEM grids, and SEM stubs. These include any reasonably flat surface on which the specimens may either grow (tissue cultures, microbial colonies) or be deposited by settling, centrifugation, or section.

In general, fix particles *before* they are placed in enclosed carriers or filtered down on porous holders. Fix them *during* or *after* settling and attachment to nonporous ones and to agar.

With macroscopic mobile specimens, it is easy to see if the organism is relaxed or contracted. With microscopic plastic forms (e.g. amebae, some flagellates, lymph and blood cells) this is more difficult. It is important that these specimens be in their natural state. If they have been centrifuged or otherwise deposited on a grid or coverslip, let them assume normal shape and attachment before fixation.

Be particularly careful that organisms are not poisoned by traces of fixative, such as might be carried on forceps, needles, etc. These instruments should never be used back and forth between fixed and fresh material without rinsing. Specimens should not be downwind from fixative fumes. Rough and delayed handling can produce changes in morphology[29] (also our experience).

Porous and Fine Meshed Containers. Gershman and Rosen[45] give detailed directions for simple mesh carriers for cell aggregates and for all stages of processing from initial washing to final mounting and coating.

Teflon "Flo-thru"[g] capsules are used in the Washington State University Electron Microscope Center routinely for processing small but macroscopic specimens. We find it preferable to put the specimens in them after OsO_4 fixation and initial rinsing. Otherwise OsO_4 is continually released into the ethanol-water dehydration mixtures and darkens them. The capsules and their contents float in Freon TF unless restrained. We hold them under by pushing a sheet of filter paper into the jar over the capsules. The snugly fitting filter paper also allows the jars to be emptied quickly be decantation. The capsules slow

the diffusion of fluids in and out of the specimens and thereby moderate the osmotic changes. They protect specimens against drying during transfer through volatile fluids such as acetone, Freon TF, etc.

Marine invertebrate sperm have been processed in Flo-thru capsules which had been previously lined with membrane filter to prevent the sperm from entering and blocking the coarser pores of the capsule.[34] Many of the cells adhere to the membrane which can be mounted and examined. After CPD, any loose sperm may be shaken onto adhesive-covered stubs, but usually they are often clumped and not as satisfactory for examination as those adhering to the membrane (Mr. G. B. Braybrook, personal communication).

There are numerous descriptions of small holders made by cutting out the centers of "BEEM"[g] capsule lids and covering the holes with 3 mm grids,[4] stainless steel mesh, nylon, etc.[3] which are welded in place, or cut to fit snugly. Thomas, Wolery and Taylor[46] use packets made by folding finely perforated aluminum foil around liverwort sporophores after they had been processed through ethanol and presumably all fixative which would corrode the foil had been removed. (The foil was perforated with a needle by hand; Dr. Thomas, personal communications.) This is an elegantly simple method, adaptable to a wide variety of minute specimens. The foil is easier to manipulate than filter paper and also has sufficient density to keep specimens submerged. Gershman and Rosen[45] give a full description of a technique using screens separated by washers for handling virus-transformed cells through all stages. Their paper is valuable for its critical and comparative consideration of harvesting, fixation, and other procedures.

Membrane Holders. Similar to the holders mentioned above are W. Cohen's stacks of membrane filters for microorganisms held individually between ring magnets.[47]

Membrane filters are sometimes physically and chemically sensitive. Therefore, consult manufacturers' literature for information on the resistance to the solvents which may be encountered. During coating, these filters may be affected by the heat from a sputtering device or evaporator unless they are thermally well grounded. These filters should not be attached by the usual conducting carbon or silver paste cement, since there is danger of wetting the specimens; nor should they be attached directly to ordinary double-face tape which is a very poor thermal and electrical conductor.[43] We attach a strip of adhesive-coated copper or aluminum tape, adhesive side up, to a strip of double-faced adhesive tape on a stub. The metal tape is grounded to the stub by a dab of conductive cement. The metal foil acts as a heat sink.

OsO_4 Fume Fixation

The following general principles will cover most cases of minute specimens or cell sheets such as agar cultures, tissue cultures, etc. The method is also useful for small plant parts.

Where possible (as in animal tissue cultures, attached bacteria) grow the specimens on substrates which can serve as carriers throughout the procedure. Except for surface cultures on agar, most such preparations should be very gently washed, in water if fixed, or if unfixed, in an isotonic saline solution containing no protein or other component which can precipitate and obscure the cell surfaces.

Generally, we fix small specimens preliminarily by about 10-60 minutes exposure to the fumes of 2% OsO_4 as shown in Fig. 8. The larger, outside covered dish is a 150 mm, diameter plastic petri dish; the enclosed dish inside is a 100 mm (preferably square) dish which contains in turn a small open dish (plastic snap cap, tissue culture dish) holding 103 ml of 2% OsO_4. The double enclosure confines the fumes fairly efficiently. After the fume chamber is prepared, the tissue cultures, agar blocks, liquid suspensions on grids or coverslips are placed on a dental wax slab well away from the fume chamber. Tissue cultures should be covered with a thin but perceptible layer of fluid. Agar blocks are most easily cut out with a wet thin razor blade or scalpel and carried on a wet blade.[3] Carry the preparations to the improvised fume chamber on the slide or slab, gently but quickly lift the covers and place the preparations within the inner chamber. Close the covers as quickly as possible.

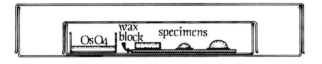

Fig. 8: OsO_4 Fume Fixation. Dish containing 2% OsO_4 and specimen carriers in double container to contain fumes. The smaller covered dish is a covered 100 mm diameter plastic petri dish. The larger is a 150 mm plastic covered petri dish.

Exposure to the concentrated fumes should be rapid so that cells are killed almost instantly. Preliminary exposure to traces of the fixative can cause surface changes as previously discussed. After approximately 10 minutes, remove the specimens and rinse them with water, or if further fixation is intended, immerse them in buffered OsO_4 fixative. The manner in which this is done is important. If the specimens are covered with a fluid film, they can be gently rinsed from a pipette or slide under the surface dish of the following liquid. If they are not so covered (e.g. colonies on agar blocks), the preparation must be placed face down in the liquid and turned right side up underneath the surface if possible without touching the bottom. This wets the surface and pushes the colony firmly against the substrate. Otherwise, if the specimen is immersed right side up, the cell layer or colony is liable to float off; an occurrence more probable with agar culture than with tissue culture lawns. These precautions are mentioned in detail because their neglect often means the loss or damage of material.

For examination of whole cells in a high voltage transmission electron microscope, Buckley and Porter[48] grew tissue cultures on Formvar film which was detached from coverslips and placed on grids. The illustrations are striking and the method may be well adapted for scanning electron microscopy. Rao[49] grew cultures directly on Formvar coated stubs, thereby minimizing possible damage caused by handling.

Suspensions Allowed to Settle on Substrates. Cell, organelle, virus, macromolecule suspensions allowed to settle on substrates and fixed before or after being placed on a smooth, structureless substrate are some of the easiest and most satisfactory of preparations. The essential conditions are: particles are well dispersed, the suspension rests undisturbed on the carrier (coverslip, filmed grid, stub, etc.) during settling, and the substrate surface is clean and hydrophilic. Particles will attach to a clean surface without further preparation, particularly if they are fixed while resting on the surface. It may be desirable to insure adhesion by preliminary coating with molecules adsorbed from solution. Polylysine, introduced by Mazia, Sale and Schatten[50] to produce positively charged surfaces is now often used. Lyon and Kramer[51] give good directions for the use of polylysine to obtain even dispersion of spores. While there are other suggested coatings ("subbings") e.g., alcian blue,[52] gelatin,[42] Masuda and Osumi[53] used poly d-glutamate to coat Nuclepore filters for the examination of polypeptide aggregates. Sarkar, Manthey and Sheffield[54] allowed suspensions of oncornaviruses to settle on grids which were then processed by CPD. They found none of the everted "tails" and other distortions prevalent on air drying and negative staining. This paper should be consulted for its detailed methodology and extensive review.

Silicon wafers "chips" from the microelectronics industry are finding use as mounts for very minute specimens. Male and Biemesderfer[55] describe the use of polylysine coated chips for supporting ferritin molecules, red cell ghosts, etc. Si wafers provide a neutral background and show less charging artefact than glass or carbon. A light metal coating may be used thereby increasing resolution of fine detail.

De Harven, et al.[56] deposited viruses on carbon coated tabbed grids (previously cleaned and rendered hydrophilic by glow discharge) and after glutaraldehyde fixation, dehydrated the specimens through an ethanol series, with uranyl acetate present in all steps. After passage through isoamyl acetate, the specimens were CPD (CO_2). Results were considerably better than with air dried specimens.

How thick should a suspension be and how long should it be allowed to settle? Differential density of medium and particles, adherence of particles to substrate, viscosity of medium, size and shape of particles, length of time suspension is in contact with substrate; all play roles. While judgment can be gained only by experience, the tolerance is remarkably broad for time and concentration.

Concern is often expressed that there will be cross-contamination between different specimens processed simultaneously. If the specimens are well dispersed so no clumps settled on the substrate to continue disintegrating, and were previously individually and gently washed to remove loose particles, the chances of cross-contamination are very slight. As a demonstration, we marked a gold coated coverslip into segments with different and easily recognizable particles in suspension in each; i.e., myxomycete spores, yeasts, bacteria, etc. The SEM showed no cross-contamination between the different segments.

More germane is the fear that some material of a heterogeneous sample may be lost selectively during treatment, either by preferential settling and adhesion or by being rinsed off, thereby distorting a differential count. Avoidance of this error is proposed by Thornthwaite, *et al.*[57] in their "Centrifugal Cytology Bucket" in which particles (blood cells) are centrifuged in a multiple well container whose common floor is a glass slide. Several preparations can thus be accommodated at once on the same slide.

Delicate Specimens Subject to Damage by Surface Tension; Plant Specimens. These include aerial hyphae of fungal colonies, sporophores, roots with root hairs, flowers with opened anthers, small leaves, etc. All of them are toughened against dispersion (spores, conidia, pollen grains) and distortion by surface tension on initial immersion. In fume fixation, at least one dimension should be small; e.g., thin petals or leaves, slender roots and stems. Such specimens may be difficult to wet. After fume fixation they may be put directly into 30% ethanol which has reduced surface tension, or before subsequent immersed fixation (e.g. glutaraldehyde-osmium) into a rinse containing a drop or two of any convenient liquid organic surfactant (Triton X-100, Aerosol OT, Teepol, commercial household detergents) before further processing.

General Preparation for Small Specimens on Surfaces. Use coverslips no greater in diameter than the SEM specimen stub surface. Clean them thoroughly, preferably by acid treatment. They should, of course, be washed completely free of acid or other cleaning material. We prefer to evaporate on a light layer of gold--golden by reflection, green by transmission. This gives a clean uniform conducting surface. Furthermore, the surface may be lightly scratched with a needle to mark off areas, put on identifying number, etc.

For transmission or scanning transmission microscopy, use grids with the smallest openings allowing unobstructed view of the particles (usually 200-300 mesh) coated with Formvar-carbon. The surface should by hydrophilic (test by putting on a drop of water and sucking it off at the edge; a film of water should be left on the grid). Non-wetting grids may be rendered hydrophilic by dipping them in 95% alcohol. They will remain hydrophilic at least an hour after drying. Put the carriers (grids, coverslips) on a wax sheet, pipette on enough suspension to make a drop not quite large enough to reach the edges. Place the laden carriers in the fume chamber as previously described, leave it at least 10 minutes. At the end of this time, gently rinse by dripping water over them.

Thereafter, coverslips or grids may be placed in one of the commercial[9] or laboratory or shop-made holders illustrated by Hayat[4] (p. 150, *et seq.*). We have not found holding them in the coils of a spring to be satisfactory. The holders described by Boublik, Jenkins and Kaback[58] for critical point drying of cell membrane fractions held tabbed grids[9] by the tabs, leaving the mesh portion free.

In summary the vapor fixation of small or minute specimens has been stressed. It provides the minimum disturbance of delicate material, though this does not mean that more common and even routine fixation procedures may not be satisfactory. Among these are the standard OsO_4 (buffered or unbuffered), glutaraldehyde and formaldehyde ("paraformaldehyde") fixations.

A few outstanding critical papers covering all stages of treatment are: Berdach (dinoflagellates; comparison of fixations, excellent stereo micrographs),[59] Biemesderfer, *et al.* (macrophage surface morphology, details of collection and treatment, use of Si wafers for mounting),[60] and Penttila, Kalimo and Trump (comparison of fixatives on cell volume, ion content; effect of osmolality).[61]

Optical and SEM Examination; Locating Particular Cells. The aspects presented by the optical microscope and the SEM may be so different that the same cell is not recognizable under both except when located with certainty. Therefore, methods have been developed to locate unambiguously the same structure under the two modes of examination. Wetzel, Erickson and Levis[62] optically examined conventionally Giemsa stained (but not dried) blood cells on coverslips and reidentified them in the SEM. Thornthwaite, *et al.*[63] photographed locator grid markings on properly prepared slides used as the floor in their centrifuge bucket[57] in order to examine the same cells by both optical and scanning microscopes.

Preparation of Large Specimens, Tissues and Organs

Standard Fixation. The general consensus favors the standard glutaraldehyde-osmium fixation: 2-4% glutaraldehyde in 0.1M phosphate or cacodylate buffer, pH 7.2-7.4/rinse in buffer/ 1% OsO_4 in same buffer. It may be modified for osmotic conditions, or specimens may require particular preparation before or between fixations to remove adherent mucus (see below). Phosphate, if not washed out, is liable to precipitate in needle clusters. Therefore, instead of buffer, water rinses are preferred between final fixation and dehydration.

The glutaraldehyde-osmium fixation has the additional advantages of being nearly universal so that it is easy to compare results with those of others. The same material, and even the same specimen, may be used for sectioning as described later.

In the conventional treatment of tissues for transmission electron microscopy, bulk specimens are cut into small pieces, usually not larger than 1 mm cubes. Scanning microscopy often requires larger pieces; therefore, complete fixation before postmortem alteration may be more difficult to attain. If perfusion (see next section) is not practicable, tissues should be cut to provide as much surface/volume ratio as possible. If 1 mm^3 pieces are too small, at least thin slices--no more than about 2 mm thick and preferably no more than 10 mm on a side--should be used. Allowing the minimum interior distance for penetration improves fixation and cuts down the time for exchange fluids in subsequent treatments. Generally such dissection should be done under the fixative.

One or two hours in glutaraldehyde at room temperature with continuous or frequent agitation (10-15 minute intervals during the first hour) is usually sufficient, longer (overnight) does no harm, but much more than this sometimes apparently allows precipitation of obscuring material on the surface.

Three rinses at 5 minute intervals with the same vehicle minus glutaraldehyde are sufficient for specimens within the 2x2x10 mm range. The subsequent osmium fixation may be from two hours to overnight. The shorter time is preferred. As is generally known, very prolonged osmium exposure can overfix with loss of internal detail. If the same specimens are desired for TEM, it may be worthwhile to remove a block from the OsO_4 fixative, rinse it, and cut it open to see if the fixative has completely penetrated as shown by blackening throughout the mass of the tissue.

A valuable comparative study of fixation and subsequent steps including freeze drying and CPD was carried out by McKee[64] on human and other tissues. The article carries a table prepared by Agnes Kormondy and provides a very useful comparative summary of techniques.

Perfusion Fixation. Large specimens are best fixed by perfusion through blood vessels or other passages. When the precise preservation of the dimensions of intercellular spaces are important then the perfusion fluid must be administered within the range of pressures normal for the organism. Suzuki, et al.[65] describe a relatively simple method of keeping a constant perfusion pressure as the volume of perfusion fluid in the reservoir diminishes. Their fine micrographs (including stereomicrographs) of the perfused and freeze-fractured splenic pulp demonstrate the success of this method. Other perfusion methods are fully described by Hayat.[4] Where circumstances are not so critical, a *small* animal (usually frog, mouse, young rat) may be anesthetized by nembutal injection, the chest cavity quickly opened, a cut made in the right ventricle and a 20 gauge needle attached to a syringe filled with fixative inserted into the left ventricle. Steady pressure is applied against the plunger to allow approximately 5 minutes to perfuse a mouse with approximately 20 ml of 2% glutaraldehyde in buffered saline. During perfusion the heart should be closely watched, and the ventricle should not be stretched by excessive fluid pressure beyond its approximately normal diastolic size. Blood followed by fixative, should flow out of the cut. The partially fixed organ (e.g. kidney, liver, gut) is removed, trimmed, rinsed, and placed in the same fixative for 1-2 hours.

Perfusion may be done through channels other than the vascular system. Thus in studies of the inner ear it is common practice to make an opening into the labyrinth through which the fixative is perfused. Hillman[66] used a glutaraldehyde-osmium fixation followed by slow dehydration through methanol/propylene oxide/Freon 113/Freon 13 CPD. The results are very successful as judged by published micrographs. Tanaka and Smith[67] in an extensive anatomical study, investigated the relations of aldehyde (HCHO and glutaraldehyde) and osmolarity in preserving the chick ear. SEM 1977 and 1978 are rich sources of papers on this subject.

Phillips, Deshmukh and Larsen[68] compared perfusion of brain lateral ventricles via the heart and by direct injection. The latter gave better preservation.

The ultimate in perfusion was done by Oishima, et al.[69] who perfused human liver biopsies by direct puncture of the samples. Preservation was good, but this is a method probably involving skill and judgment to avoid artefacts caused by puncture or excessive pressure of the perfusion fluid.

OTO (Osmium-Thiocarbohydrazide-Osmium) and Similar Methods. Thiocarbohydrazide is a ligand; it can tie ("ligate") two other atoms or molecules. By its use more osmium may be added to a cell component (C) which has osmium already attached: C-Os plus ligand (L) equals C-Os-L, plus Os equals C-Os-L-Os. Thiocarbohydrazide is such a ligand, used to attach specific cytochemical reagents to cell constituents, or to intensify a stain. Osmium increases tissue electrical conductivity. Kelley, Dekker, and Bluemink[70] used thiocarbohydrazide to add more osmium, thereby minimizing charging and and eliminating the need of a conductive coating. After a first OsO_4 fixation in the OTO technique, specimens are well washed to rid them of free osmium, then treated with aqueous thiocarbohydrazide. Following a second thorough washing, they are again exposed to OsO_4. Specimens so treated are an intense black. Also, they are generally harder and tougher than specimens which have had only the conventional glutaraldehyde-osmium fixation and in dry fracture, break with glassy surfaces.

It it extremely important that the thiocarbyhydrazide be of high quality, having a very light yellow or nearly white cast, but not dark gray. (Dr. Robert Kelley, personal communication; our own experience.) Otherwise preparations are liable to be disappointing.

The OTO treatment may not relieve all charging, in which case it may be repeated,[71] thus: C-O-T-O-T-O. These specimens, according to the several authors, show not only no charging, but better preservation of minute fragile cell surface components.

The striking micrographs of the bacteria crowding the rumen surface in the recent semi-popular article "How Bacteria Stick" by Costerton, Geesey and Cheng[26] were prepared by the repeated OTO method (personal communication, Dr. Costerton).

Tannic acid has external effects similar to the OTO method with the additional benefits of preserving interior detail and stabilizing some phospholipids (Kalina and Pease[71]). A comprehensive review of these charge-reducing methods which are also valuable fixation methods has been prepared by Murphy.[72]

Osmotic Effect of Fixatives

In aldehyde fixation, the cell and organelle membranes retain some osmotic activity (reviewed by Cohen,[3] Hayat[4] (p. 91, *et seq*.) but only the osmolarity of the vehicle needs to be taken into consideration. Within reasonable concentration (no more than 5-6%) glutaraldehyde or formaldehyde penetrates readily without usually showing osmotic effects. Good discussions and literature reviews are provided by Arborgh, *et al*. (liver cells);[73] Brunk, *et al*. (glia);[74] and, Iqbal and Weakley (hamster ovary).[75] These papers are in essential agreement that for mammalian and avian tissue at least, the osmolarity of the vehicle should be approximately 300 mOsm; about the tonicity of the normal plasma environment (for oocytes, ca. 200 mOsm is preferred).[75]

Marine animals, at least fish, seem to have somewhat different responses to osmotic effects, and the total osmolarity of the aldehyde fixative cannot be neglected. Dobbs[76] recommends different adjustments of the osmolarity for external and internal structures of osmoregulators. Thus, for external use, 2% glutaraldehyde in 77% sea water matches the osmolarity of sea water (1016 mOsm) while a balanced salt solution ("flounder saline") is used appropriately diluted for the interior tissues. We have observed pronounced, presumably osmotically generated distortion of myxomycete swarm cells on the addition of glutaraldehyde to the distilled water suspension in which they were germinated. No such distortion was observed with OsO_4.

However, using OsO_4 as the primary fixative for gills of a marine worm (*Terabella*) and a marine protozoan, Chuang[77] found that it was immaterial if sea water or distilled water were the vehicle!

Dehydration, Intermediate Fluids

No currently used transitional fluid (Table 2) is appreciably miscible with water, and any water left in the specimen means that it is subject to the distortions of evaporative drying in air. Therefore, dehydration is necessary to replace all water in the specimens with a fluid, miscible both with water and with the transitional fluid and another intermediate, as in: H_2O/ethanol/Freon TF/Freon 13→CPD.

The most commonly used dehydration fluids are acetone and ethanol. Boyde[9] lists and discusses a variety of others. Acetone has the advantages over ethanol of: it is available without special license, it is easily kept anhydrous, and it shows rapid and complete miscibility with CO_2 and the Freons. Its disadvantages are: acetone appears to cause some surface change to spores, pollen grains, and cells, possibly by dissolving structural lipids; some membrane filters and supports are either dissolved (cellulose esters) or weakened (Formvar) by it. Ethanol's advantages are: relatively low toxicity; inertness to most containers; low solvent action on cellular structural constituents, carrier filters, and membranes; and better preservation of some surface structures. Personal preference seems to play a large role.

It is usually poor practice to jump specimens from aqueous media to high concentrations of any dehydration fluid, for water often diffuses out faster than the foreign molecules can enter, with consequent collapse and distortion of the specimen. Also, as water reenters, a new equilibrium may be approached with the cells swollen or even burst. Therefore, exchanges should be gradual to maintain osmotic equilibrium. How rapid should the exchange be, and in how many steps? These are questions whose answers depend on size, shape, and permeability of the specimen, rate or diffusion or convection of fresh fluid to the specimen, and fixation history. Generally, a series of: ethanol 30/50/70/85/90/95/100% (2-3x) with 5 minutes in 30% and 50% and 10 minutes in each of the succeeding concentrations will be suitable for the majority of specimens. Specimens which are large or dense may have to spend a longer time in each, particularly in the higher ranges (70-100%). We generally use rapid (5-10 minutes for larger specimens) passage through 30% and 50%, since prolonged immersion can cause maceration, or dissolution of some components.

Small specimens such as those carried in fine meshed capsules, on filters, stubs, etc. may be processed considerably faster. The limiting exchange rate may be set by the capillary reservoirs in the spaces and crevices of holders rather than by the specimens themselves. Often 1 or 2 minutes in 30% and 50% dehydration fluid and 5 minutes in each of the later ones is sufficient.

We sometimes use a simple dehydration apparatus[3] in which a gradient flows past the specimens. Baker and Princen[78] use a series of stacked retainers past which the dehydration and other fluids flow under pressure. They claim that this "positive pressure" method gives more rapid dehydration than the conventional methods which depend on diffusion. Either method avoids the necessity of repeatedly moving the specimens in and out of liquids with consequent extensive damage to delicate structures. In another approach, carriers with ball check valves in the bottom keep specimens immersed in fluid during transfer; the valves allow drainage when the containers are immersed.[3] These devices are commercially available[g]. (It sometimes happens that at the beginning of dehydration, air bubbles are trapped by the specimen. As dehydration continues and surface tension drops, these break off and the specimen becomes uniformly wetted. Unless such bubbles occlude a large area, they seem to do no harm.)

Cohen and Shaykh[25] showed that for delicate objects (e.g. *Spirogyra*) fixation either had to be thorough (glutaraldehyde/OsO_4 instead of OsO_4 alone) or subsequent steps had to be more closely spaced than the series given above.

Other Dehydration Fluids. Recently 2,2-dimethoxypropane (DMP) has become prominent as a "chemical" dehydrant, which combines with water under slightly acid conditions to form acetone and methanol. Therefore, after fixation, it is necessary only to place the specimen in DMP for rapid and complete dehydration. The method has been used very successfully for CPD of plant tissues by Lin, Falk and Stocking.[79] Maser and Trimble[80] could detect no differences either in sections or after CPD between mammalian tissues processed through DMP and those conventionally dehydrated through ethanol. Kahn, Fromme and Cancilla,[81] in a study of fibroblasts in culture, found that either DMP or ethanol dehydration gave comparable results if used singly before CPD and for SEM examination. DMP apparently dehydrated thin layers completely in a few seconds. However, combinations of ethanol and DMP were damaging to some surface features. But in dehydrating material for thin sectioning, DMP was always inferior to ethanol, showing an apparent destruction of the plasma membrane and other structures. Boyde, et al.[8] claimed severe shrinkage in DMP with their material. DMP needs further experience.

Intermediate Fluids

Although transfer from acetone, ethanol, or DMP to either CO_2 or a Freon is possible (ethanol is incompletely soluble in Freon 13; specimen bulk must be very small or flushing numerous) the intermediates (iso)amyl acetate retain some use. Amyl acetate is useful for: low volatility which prevents specimens from drying during transfer; penetrating odor which signals in flushing that the transitional fluid is still contaminated with intermediate fluid; and less shrinkage results in the chain: isoamyl acetate/CO_2, than in ethanol/CO_2 (Boyde, et al.,[8] Wetzel, personal communication). There is some general (undocumented) evidence that isoamyl acetate as an intermediate preserves fine cellular features otherwise destroyed or distorted by direct passage from ethanol to CO_2. Freon TF (Freon 113) has the advantages: it does not have to be completely flushed out (particularly important for bulky specimens); it is less likely to attack some plastic substrates; and it very much less toxic. For 8 hours exposure, 125 parts per million of isoamyl acetate is the maximum, but 1,000 parts per million of Freon TF is allowed.[82] On the other hand, the very high volatility of Freon TF means that precautions must be taken to avoid evaporative drying of the specimen during transfer to the CPD pressure chamber.

It should be routine practice to put a few drops of Freon TF in the CPD chamber to saturate the atmosphere before transferring specimens. The surface tension of Freon TF is so low that if air drying does occur, it is probably not completely disastrous (Figs. 4, 4a, also ref. 7).

Critical Point Drying

In a sense, the term is a misnomer, for much critical point drying is done "around the critical point," i.e., at (T_c) the pressure is higher than P_c (see Fig. 7). Nevertheless, we shall continue to use CPD to mean drying through or near the critical point.

Transitional Fluids. Since drying from N_2O has not to date been practical, the discussion is confined to CO_2, Freon 13, and Freon 116. (Freon, Arcton, Genetron, etc. are the trade names of different suppliers. The number following the trade name identifies the particular fluorocarbon and is uniformly adopted.) The important properties of these compounds are given in Table 2 and are discussed by Cohen[3] and Hayat[4]. Briefly, the Freons have the advantage of critical pressures which are about half that of CO_2. Also, on venting, they do not freeze at the valves as CO_2 too often does, nor do they cause problems with some types of rubber gaskets and seals, another fault of CO_2. Preservation of delicate structure seem to be marginally or obviously better with the Freons than with CO_2[7] (various communications, our observations). However, the initial cost of Freon 13 and its decreasing availability are disadvantages. Freon 116 is less than half the cost of Freon 13, but since its T_c is only 19.7°C, it is difficult to get chambers to the 10-15° below that T_c for efficient filling and flushing. However, Freon 116 has the advantage of the lowest P_c (432 lbs/in²) of those listed.[3]

Location. The CPD instrument and its supply tank should be in a cool, dry, clean location with good ventilation preferably in a fume hood. It should be out of the way of traffic and secure from ignorant manipulation. The supply tank should be well fastened so that it cannot be inadvertently knocked over or rolled.

Supply Tank. For the expensive Freons, rather small supply tanks (7-9 lbs.) should be used; a leak in the supply line emptying a large one can be costly. For CO_2, the tank may be as large as can be conveniently handled. A pressure gauge is helpful to indicate if exhaustion is approaching, but it is not necessary. Anhydrous gases must be used, but they need not be absolutely pure; a small amount of atmospheric gases does not harm. One is commonly advised to have the supply tank upside down, or to have a siphon tube (available in CO_2 tanks) so that liquid will flow into the chamber. Whether or not there is liquid in the bomb chamber depends on its pressure and temperature. If the pressure is high enough for liquefaction, and if the chamber is below T_c, whatever fluid flows into the chamber will flow and remain as a liquid or condense to one. These conditions are easy to achieve and are in fact necessary for CPD. However, if the bomb is not at least 10°C cooler than the tank, condensation is very slow. It may take 15 minutes to fill the chamber by condensation from the overlying vapor in the supply tank, but only 2 minutes to fill it with liquid. Unfortunately, many gas tanks, particularly CO_2, have particulate and liquid impurities at the bottom, and the Freons seem to collect a slowly evaporating oily material, possible a polymer. We prefer to keep the tank (without siphon tube) upright and to cool the chamber sufficiently for rapid flow, both for cleanliness of contents, and safety in handling the tank.

Critical Point Devices (Chambers). The earliest critical point device of Anderson[1,2] was a simple sturdy chamber large enough to hold grids for the TEM. It was supplied with pressure tubing leading to the supply tank, a vent, and a pressure gauge. Immersing the pressure chamber (bomb) in buckets of cold and hot water provided the temperature cycle for CPD. Several commercial models[3,4] use essentially this arrangement but with bombs large enough to accommodate SEM specimens. Cohen, Marlow and Garner[23] improved upon the device by: (1) providing a water jacket integral with the chamber, thereby avoiding the presence of the specimens being exposed to water, (2) by providing a window or port to monitor the process, and (3) by inserting a thermometer permanently into the water jacket. The Polaron CPD device[9] is essentially an improved version of this chamber. Several CPD instruments have viewing ports and lights for observation of the specimens. The temperature cycle is performed in the Bomar[9] apparatus by a thermoelectric device which heats on one face and cools on the other when a current passes through it (Peltier effect). Reversing the current reverses the cooling and heating faces. Therefore one thermoelectric plate may serve both to heat and cool. Also no other than electric connection is necessary.

Other commercial devices depend on a resistive heater for warming, and an adiabatic expansion of the transitional fluid for cooling. The Balzers and some Tousimis instruments[9] have a separate cooling coil around the chamber through which the gas may expand. Therefore, if in the process of critical point drying (see below), it appears that all intermediate fluid has not been flushed out, the chamber may be cooled under pressure and the cycle repeated.

Several instruments with improved features have been described. Pawley and Dole[83] describe "a totally automatic critical point dryer" which can be programmed for repeated flushing, heating, cooling, and venting. Hall, Skerrett and Thomas[84] have evidence that excessively rapid heating and venting of chambers may damage some specimens. They have developed an electronically controlled instrument which automatically controls heating rate and pressure release (venting). Brown[85] describes a stainless steel chamber which is also a dehydration chamber for all fluid exchanges after fixation. This procedure eliminates the frequent transfer of specimens which can be damaging. The fluid exchange is continuous through a very simple dilution and pumping device.

The Critical Point Drying Procedure

The chamber should be dry and cooled at least 10-15° below T_c. The lower the temperature the better, up to the point that moisture may condense. During cooling, the lid should be loosely attached and the exhaust vent open. If Freon TF or acetone is used as an intermediate, and if the specimens are very thin (e.g. coverslip or grid preparations), put a few drops of the same intermediate in the chamber immediately before adding the specimens to saturate the atmosphere and retard evaporation from the specimen. This precaution is not necessary with ethanol or amyl acetate. Quickly transfer the specimens in or on their carriers, taking a moment to drain baskets containing large specimens if they were in alcohol or amyl acetate; barely drain specimens from Freon TF or acetone and do not drain small specimens which could quickly dry evaporatively in air. Seal the bomb quickly and gradually open the inlet valve. There may be problems with specimens containing air spaces (e.g. some plant tissues) which may partially collapse under the sudden external pressure. After the liquid covers the specimens, the temperature should be allowed to rise to about 20°C, i.e., about 10° lower than the T_c of Freon 13 or CO_2. The transition fluid will thus be less dense than any intermediate

fluid, and the latter will drain to the bottom. Slightly open the outlet valve and let the liquid level drop slowly until the chamber is at least 75% empty of liquid. If there is no viewing port, watch the pressure gauge of the chamber. Pressure drops slightly as soon as the valve is opened, but remains at steady pressure until the liquid is exhausted. If the pressure begins to fall again, immediately close the outlet valve and admit more fluid. Only general guidelines can be given for the number of flushes and the length of time the specimens remain in each exchange. For very small specimens, (e.g. cells on filters, monolayers on stubs, thin leaves, etc.) 2-3 changes of 5 minutes each are generally enough or more than enough with Freon TF. (We have successfully dried tissue culture monolayers on glass with no flushes, but I hesitate to recommend this practice generally.) For larger, denser ones, i.e., 2x10x10 mm, possibly 3-4 changes of 10 minutes each after the first 5 minute change are necessary.

It is a common practice to use a continuous flush with CO_2 as the transitional fluid, and if amyl acetate is the intermediate, to consider flushing complete when its odor is no longer detectable. Such a technique is extremely expensive for Freon 13, and in our opinion, not ideal for CO_2. We prefer the successive, intermittent flushes as more controllable and less wasteful. For flushing to be efficient, the outlet vent of the chamber should be near the bottom and the specimens supported slightly above it. If the temperature is no lower than about T_c - 10°, any intermediate fluid is nearly certain to be denser than any transitional fluid replacing it, and will accumulate at the bottom.[22] If the outlet is near the top, one is venting mostly entering the transitional fluid; the intermediate fluid remains.

For the drying itself, the chamber should have its free space more than half full of liquid; at T_c - 10° (about 20°C for CO_2 or Freon 13), 80% full is satisfactory. The specimens should remain covered. On heating, the fluid will pass through or around the critical point as discussed earlier. Despite the advocacy of filling the chamber completely with fluid to avoid the turbulence of boiling, it is best to avoid boiling by heating the chamber gradually, and in all cases to avoid damage to the specimens by having them in a container. The rise in pressure with temperature of a completely filled chamber is too rapid and too great for safety[22] (Fig. 7C). It is possible that rapid temperature rise, not the turbulence of boiling, does the most damage.[84]

Monitor the pressure, and if it begins to exceed a safe level, slightly vent the bomb. The temperature should be brought to at least 10° above the critical temperature of the pure transitional fluid. Small amounts of intermediate fluid left in the chamber will raise the T_c of the mixture, and the higher temperature insures that the fluid is at or near its T_c. Keeping the temperature high, the container may now be slowly vented. We find that the commonly suggested rate of 100 lbs/min. (7 atm/min.) is sufficiently slow for practically all specimens to equilibrate. We have vented as rapidly as 300 lbs/min., for many tissue specimens with no detectable adverse effects. However, it is better to be conservative, especially as recent studies[84] show damage presumably due to even moderately rapid venting.

Variants on "True" CPD. As pointed out first by Koller and Bernhard,[86] and later by Cohen,[22] and Boyde, *et al.*,[8] it is not necessary that the surface tension be reduced to zero in order to obtain results indistinguishable from true CPD. Figures 4-6 show little or no detectable difference between true CPD dried specimens and those dried at half the critical pressure with estimated surface tensions 2-4 dynes. Only in Figs. 6 and 6A there seems to be some adhesion of microvilli to each other; they may be due to coating by soluble substances not removed by flushing; but there is not the contraction of these appendages indicative of surface tension distortion.

Retrograde Condensation. It may be sometimes noted that although the chamber is being vented at a temperature remaining well above T_c, the contents become cloudy and there may actually be visible liquid condensing. This is one kind of *retrograde condensation*, a condition in which *decreasing* pressure leads to condensation, or (more rarely observed) *increasing* temperature also produces condensation.

Retrograde condensation probably means that some intermediate fluid is mixed with the transitional fluid, as it never occurs with a pure substance. A small amount does not harm, *if the intermediate is highly volatile* as is Freon TF (Fig. 6). If it is not, then on return to atmospheric pressure, the specimen dries evaporatively and the benefits of the (near) CPD are undone. In practice, as long as the condensation is not coalescing to visible drops, proceed with the venting. If liquid is obviously condensing, close the vent, cool the chamber to $T_c - 10°$ and flush again with the transitional fluid. This is not easily possible with a chamber cooled only adiabatically by expansion of gas from the chamber itself.

Mounting and Coating Specimens

The finished specimens must be mounted on a stub for observation. For very small specimens on flat supports the same carrier used for processing becomes the permanent substrate, and aside from possible conductive coating, no other attention is necessary.

Liquid adhesives applied to critical point dried specimens may be "wicked" into them, and by evaporative drying, distort them. This wicking may be avoided by using either tacky but nonliquid substrates, or thick pastes. The double-faced pressure sensitive tapes (e.g. double-face Scotch Tape) are good if they are protected from the beam, but they may be somewhat less than ideal for small specimens which sink partly into them, and they are liable to show charging. As indicated previously, we alleviate this problem by using adhesive backed copper or aluminum foil tape. Other mounting procedures are described by Cohen[3] and Hayat[4] (pp. 207-211).

Some workers affixed spores to stubs "smeared with the natural oils from one's finger,"[45] and I am assured (Dr. Thomas, personal communication) that this provides firm adhesion for selected specimens.

Coating Specimens. Unless treated by OTO or a similar method, the specimen must be coated with a conductive layer, although there are claims that some microscopic specimens do not require coating if they are attached to a conductive substrate, under low accelerating voltage, and if the beam sweeps at television rates. Most of us have to work under other conditions. Coating can be done by evaporation or by the more recent sputtering method. For dry, light specimens which are poor conductors of heat as well as electricity, there is some possibility of damage, more with the ordinary "diode" sputtering device than with evaporation. The "triode" modification alleviates this condition markedly as shown by the outstanding micrographs of euglenoid flagellates by Rosowski and Glider.[87] (This paper should also be consulted for useful preparative methods of small specimens.)

Special Applications of CPD--Immunocytochemistry, Fractography, Combined SEM and TEM Investigation

As discussed earlier, the CPD method of preparation has considerable versatility in permitting more than just surface examination. Some of these are considered briefly.

SEM Immunocytochemistry

By fixing the appropriate antibody of antigen to latex spheres, or coupling them to molecules large enough to be visualized in the SEM, specific sites or particular reactive components may be located on the surface of cells and of organelles. Molday in brief[88] and comprehensive[89] reviews discusses applications and the use of several different markers. A complete schedule is given for combined immunofluorescence, electrophoresis, and immunolatex methods by Stenman, Wartiovaara, and Vaheri.[90]

Fractography

Specimens frozen in ice and cracked have their fine structure obscured by ice crystals. Other media, such as ethanol, eliminate such damage. On cooling ethanol becomes more and more viscous, finally becoming brittle hard without any sharp solidifying temperature. Thus it is a *glass*, an amorphous material that has no crystalline ice structure to impose on any object permeated by it. Specimens that have been dehydrated to 100% ethanol may be frozen by immersion in liquid nitrogen and cracked by pressure with a precooled knife edge. The fractured specimens are then continued in processing for CPD. This method, developed by Humphreys, Spurlock and Johnson[91] provides cleanly fractured surfaces. Their paper also gives directions for sectioning such specimens for TEM.

In our procedure, small specimens are placed on top of a polished brass or aluminum block (approximately 2x2x2 cm or larger; dimensions are unimportant) projecting out of a bath of liquid N_2 in a plastic or glass dish with walls high enough to contain a layer of dense cold nitrogen vapor which blankets the block and shields it from frost deposition. A Pyrex crystallizing dish, 1 0 mm diameter, 100 mm high is very good for this purpose. Meanwhile a knife or scalpel blade with insulated handle is cooled to liquid N_2 temperature. The specimen is removed from 100% alcohol and placed on the block, and alcohol is dropped on it at once from a pipette. The alcohol freezes to form an enclosing layer which is then fractured with the knife blade, and the pieces placed in a vial of 100% ethanol at room temperature. Sometimes the specimen cracks spontaneously. For centrifuged cell pellets, Henk and Spurlock[92] have developed an elegant variation by embedding the fixed pellets in agar, dehydrating them to 100% ethanol and enclosing the specimens in parafilm strips rolled into tubes which are frozen and fractured. Such enclosures are also suitable for small or slender specimens to prevent their fragments from scattering.

Material embedded in frozen but uncured resin may be fractured similarly. After the resin is removed by solvents as described by Tanaka,[93] the specimen is subjected to CPD processing. Tanaka, Iino and Neguro[94] used a resin which could be cracked at room temperature.

Clark and Glagov in a study of aortic wall architecture perfused the vessel with 10% gelatin to maintain shape and to give it support, and after dehydration replaced 100% ethanol with glycerol.[95] The glycerol embedded specimen was frozen, fractured, and the specimen returned to 100% ethanol for subsequent CPD. (The ethanol-glycerol substitution was in 20% steps in both directions; personal communication from Dr. Clark.) The authors provide useful detailed descriptions of the appearance of the several tissues after fracturing.

Critical point dried specimens, particularly plant material, may often be fractured by hand. Sometimes the specimens are so brittle that this is harder to prevent than to accomplish. Therefore, do not throw away specimens which may have been accidentally fractured. The serendipitous view of the interior may be worth the accident!

Exposing the interior of suitable cells or tissues may be easily done by applying adhesive tape to the surface and peeling it away; a technique used to good advantage by Watson, Swedo and Page.[96,97] These and other dry fracturing techniques are reviewed by Flood.[98]

OTO treatment not only renders specimens conductive, but increases their hardness and brittleness. Tissues thus treated often fracture with a smooth, almost glassy fracture after drying, providing surfaces which are free of the ragged shreds so often seen on dry fracture. Irino, *et al.*[99] took advantage of both these properties to dissect specimens under SEM, the newly exposed surfaces being free of charging artefact. These authors also describe a very simple micromanipulator operated externally.

Sectioning Specimens After CPD for Optical and Transmission Electron Microscopy

Wickham and Worthen[100] found that material normally fixed, dehydrated, and critical point dried, could be infiltrated with epoxy resin and sectioned. The results were practically indistinguishable from those of the conventional infiltration procedure. Bucana, *et al.*[101] used the technique for combined cinematography of live cells which were then fixed and examined with the SEM after CPD and were subsequently infiltrated and sectioned for TEM observation.

Conversely, specimens already embedded may be cut into more or less thick sections which are fastened to slides or other substrates, the embedding material removed, and the section processed through CPD. It is possible to do this with conventional paraffin sections as shown by the interesting studies of Myklebust, Dalen and Saeterdal[102] who used the same heart muscle for optical microscopy, SEM, and (separate preparations) TEM study. Epon or other resin-embedded sections may have the resin removed by sodium or potassium ethylate[103] or a proprietary epoxy solvent.[104]

Artefacts

Shrinkage. The main artefact shown by CPD seems to be shrinkage,[7,8] although in many (perhaps the majority) of cases this shrinkage is quite uniform and merely results in reduction in volume without change in proportions (e.g. Gusnard and Kirschner,[105] liver nuclei; Kirschner, Rusli and Martin,[106] membrane; Waterman,[107] various embryonic tissues). Close examination of micrographs of single cells and tissues show no apparent distortion (e.g. Rosowski and Glider,[87] euglenoids; Breipohl and Fernandez,[108] embryonic olfactory epithelium). On the other hand, Schneider, Pockwinse and Billings-Gagliardi[109] found lymphocyte microvilli disproportionately narrower after CPD than after freeze drying.

Other researchers claim general better dimensional preservation with CPD than with freeze drying (Madge,[110] intestinal mucosa; Hammond and Hahlberg,[111] glandular leaf hairs; Falk, Gifford and Cutter,[112] various plant tissues). Other examples are cited by Cohen.[3,22] The micrographs of tissue and organ culture cells (transformed neuroectoderm and glioblastoma) by Copeland, *et al.*[113] show no distortion or shrinkage (confirmed by personal communication). CPD has been in general use for several years before the shrinkage was much noticed, which indicates its freedom from distortion. While 50% shrinkage in volume seems severe, it represents only 20% uniform reduction of linear dimensions. We measure magnifications linearly and we conceptualize size linearly.

Critical point drying became the most widely used method, possibly because earlier freeze drying practice had so many faults, alleviated by Boyde[8,9] who raised freeze drying to a valuable and in some cases, indispensable technique. Nermut,[103] in a balanced discussion, describes the values of both procedures.

Other Artefacts. Shrinkage results in such artefacts as cracking of cells which are stretched on a rigid substrate (e.g. bone, glass); and tautly stretched flagella, filopodia, and other cell processes, which may be normally lax. The use of flexible substrates may alleviate this condition, but the commonest materials, such as agar and membrane filters, often provide an obtrusive background. Wrinkling of the cell membrane may be due to shrinkage of internal structural components, to osmotic effects stabilized by fixation, or to loss of components by leaching, particularly in lower concentrations of dehydrating fluids. However, it is not certain that all such appearances are artefactual. Filopodia are seen stretched in normal living cells and cell membranes may have wrinkles while they are alive.

Very small holes in cell membranes which sometimes look like circular ruptures can possibly be caused by too rapid pressure changes during the CPD process. Collapse of plant cell walls may be caused by rapid pressure changes.[84] As pointed out early in this paper and elsewhere[25] CPD can develop the latent errors of previous processing. In most instances artefacts are recognizable as such, and do not occlude cell ultrastructure or lead the observer astray in conclusions.

Safety and Toxicology

Safety aspects have been touched upon in the body of this paper and are reviewed by Cohen[3,22] for the CPD technique, and by Humphreys[114] for electron microscopy in general. These papers should be considered necessary reading. Additional material is provided by Thurston.[115] No procedures involving fixatives, solvents, liquid nitrogen and other extremely cold fluids, and high pressures, are without some hazard. In low and intermediate concentrations as vapors, Freons seem to be less toxic than acetone, ethanol, or amyl acetate and not much more so than CO_2.[3,82] In high concentrations (probably 20%, or more) they sensitize the heart to the body's own adrenalin and dangerous cardiac arrhythmia may occur.

Puzzling (undocumented) reports have been made of arrythmias in laboratory workers using Freons. In old cans of Freon propellents and in tanks of Freons, we have detected slowly evaporating oily residues emitted when the container is inverted and opened. It is possible that these are derivatives (polymers?) of the Freon contents, and the aerosol droplets may be the real culprit. Therefore, until this hypothesis is proven wrong, I advise strongly against venting any inverted pressure container containing these halogenated hydrocarbons to avoid inhaling the residue from the bottom.

Acknowledgements

I am very grateful to the authors who, without exception, very willingly supplied the additional information requested. Those whose information is incorporated here are cited in context. Mr. Ted Hackstadt and Dr. James Talmadge of Washington State University supplied the tissue cultures used in preparation and illustration of this paper. Mr. David Bentley and Mr. Allen Crooker of the Electron Microscope Center helped with information and data from their experience. Mr. Robert McKenzie prepared the drawings. Mr. R.G.E. Steever assisted with all phases of photography. Miss Linda Olson showed extreme patience and skill in adapting a near infinitely amended manuscript to the Procrustean demands of this format. Miss Jan Nowell and Dr. Judith Murphy made detailed critical readings of the manuscript and provided numerous valuable suggestions.

a. Anderson wrote a number of papers on critical point drying in the period 1950-56. The 1951[1] paper is the most frequently cited. However, the two essentially equivalent editions[2] of 1956 and 1966 contain not only a clear summary of the earlier work, but very practical pointers on specimen preparation and stereography.

b. Surface tension is a particular case of interfacial tension. When two phases remain separated, there is a tension at the boundary. These boundaries may be solid/gas, solid/liquid, liquid/liquid (as in oil and water) and liquid/gas. The latter is called *surface tension*. particularly if the gas phase is air. Besides the symbol, s, used here, surface tension is often symbolized by σ or ρ. In metric notation it is given in dynes/cm; in SI notation it is in newtons/meter (N/m), or more conveniently in millinewtons/meter (mN/m) which is identical with dynes/cm.

c. Metric and the commonly used lbs/in^2 units are retained in this paper. Very few publications as yet carry surface tension in N/m or other SI units (see note b, above).

d. A *vapor* is a gas which is in actual or possible equilibrium with its liquid or solid phase (i.e. not above critical temperature). Thus, water vapor is in equilibrium with liquid water or ice. In common usage, vapor refers to the gas phase of a substance whose other phases are commonly known. Thus, we always speak of water *vapor*, even when *gas* would be better. In this article, vapor is used in the first meaning.

e. The term, *fluid*, applying to a substance which flows, gas or liquid, is deliberately used here except when distinction between liquid and gas phases is necessary. In the region of the critical point the two phases are indistinguishable.

f. The terms, *dehydration, intermediate,* and *transitional fluids,* were first applied in their current sense in 1968.[25] Current tendency is to drop the term, "dehydration fluid," and to call "intermediate" all fluids used in the chain between aqueous and transitional fluids.

g. Sources of supply:
Critical point apparatus and accessories (holders, baskets, etc.)
Balzers Corp., 99 Northeastern Blvd., Nashua, NH 03060.
The Bomar Co., P.O. Box 225, Tacoma, WA 98401.
Ladd Research Industries Inc., P.O. Box 901, Burlington, VT 05402.
Polaron Instruments Inc., 1202 Bethlehem Pike, Line Lexington, PA 18932.
Technics Inc., 5510 Vine St., Alexandria, VA 22310.
Tousimis Research Corp., P.O. Box 2189, Rockville, MD 20852.
"Flo-thru" specimen capsules--American Optical Corporation, Scientific Instrument Division, Buffalo, NY 14215.
"BEEM" capsules--Better Equipment for Electron Microscopy, Inc., P.O. Box 132, Jerome Avenue Station, Bronx, NY 10468 and various suppliers.

"*Swinney*" *adapters (filter holders)*--various suppliers of membrane filters.

"*Nuclepore*" *membrane filters*--Nuclepore Corporation, 7035 Commerce Circle, Pleasanton, CA 94566.

*Tabbed grids ("Cohen-Pelco")*Ted Pella, Co., P.O. Box 510, Tustin, CA 92680.

"*Freons*" *halogenated hydrocarbons*--E.I. Dupont de Nemours and Co., Freon Products Division, Wilmington, DE 19898.

h. Many fixative formulations call for "paraformaldehyde". This is an erroneous but now accepted terminology, for paraformaldehyde is (HCHO)n, a polymer of formaldehyde. Commercial formalin (37-40% HCHO) is often acid and contaminated with other substances which may be deleterious to fixation. Suitable ("para") formaldehyde is prepared from "true" paraformaldehyde by weighing the required amount, putting it into water (e.g. for 5%, 5 gm paraformaldehyde in 100 ml water) and gently heating to about 80°C; stir while adding dropwise 0.1N sodium or potassium hydroxide. Depolymerization is completed when all paraformaldehyde is dissolved. Sometimes a persistent faint cloudiness remains. No attempt should be made to dissolve it by adding more hydroxide, but the solution should be filtered.

References

1. Anderson, T. F. "Techniques for the preservation of three-dimensional structure in preparing specimens for the electron microscope." Trans. N. Y. Acad. Sci. 13, 1951. 130-133.
2. Anderson, T. F. "Electron microscopy of microorganisms," in Physical Techniques in Biological Research, Vol. III. G. Oster and A. Pollister (eds.) Academic Press, New York, 1956. 177-240.
_____, 2nd edition, Vol. IIIA. A. Pollister (ed.) 1966. 319-387.
3. Cohen, A. L. "Critical point drying," in Principles and Techniques of Scanning Electron Microscopy, Vol. 1. M. A. Hayat (ed.) Van Nostrand Reinhold, New York, 1974. 44-112.
4. Hayat, M. A. Introduction to Biological Scanning Electron Microscopy, University Park Press, Baltimore, 1978. 323 pp.
5. Thornthwaite, J. T. and R. C. Leif. "The plaque cytogram assay. I. Light and scanning electron microscopy of immunocompetent cells." J. Immunol. 113, 1974. 1897-1908.
6. Albrecht, R. M., et al. "Preparation of cultured cells for SEM: air drying from organic solvents." J. Microscopy 108, 1976. 21-29.
7. Liepens, A. and E. de Harven. "A rapid method for cell drying for scanning electron microscopy." SEM/1978/I, SEM. Inc., AMF O'Hare, IL. 37-44.
8. Boyde, A., et al. "Dimensional changes during specimen preparation for scanning electron microscopy." SEM/1977/I, IIT Research Institute, Chicago, IL. 507-518.
9. Boyde, A. "Pros and cons of critical point drying and freeze drying for SEM." SEM/1978/II, SEM. Inc., AMF O'Hare, IL. 303-314.
10. Nei, T. "Cryotechniques," in Principles and Techniques of Scanning Electron Microscopy,
Vol. 1. M. A. Hayat (ed.) Van Nostrand Reinhold, New York, 1974. 113-124.
11. de Harven, E., N. Lampen and D. Pla. "Alternatives to critical point drying." SEM/1977/I, IIT Research Institute, Chicago, IL. 519-524.
12. Trump, B., et al. "X-ray microanalysis of diffusible elements in scanning electron microscopy of biological thin sections. Studies of pathologically altered cells." SEM/1978/II, SEM. Inc., AMF O'Hare, IL. 1027-1040.
13. Pease, R. F. W. and T. L. Hayes. "Electron microscopy of sprouting seeds." 26th Ann. EMSA Proc., Baton Rouge, 1968. 88-89.
14. Ledbetter, M. C. "Practical problems in observation of unfixed, uncoated plant surfaces by SEM." SEM/1976/II, IIT Research Institute, Chicago, IL. 453-460.
15. Dayanandan, P. F., et al. "Structure of gravity-sensitive sheath and internodal pulvini in grass shoots." Amer. J. Bot. 64, 1972. 1189-1199.
16. Garner, G. E., A. L. Cohen and R. G. E. Steever, Jr. "An integrated device for controlling charging artifacts in the SEM." SEM/1973/I, IIT Research Institute, Chicago, IL. 189-196.
17. Parsons, D. F. "Environmental wet cells for biological medium voltage and high voltage microscopy." 31st Ann. EMSA Proc., Baton Rouge, 1973. 14-15.
18. Specian, R. D., et al. "Preparation of amyl acetate and acetone labile eggs from parasitic nematodes for scanning electron microscopy." 31st Ann. EMSA Proc., Baton Rouge, 1973. 440-441.
19. Panessa, B. J. and J. F. Gennaro, Jr. "Intracellular structures," in Principles and Techniques of Scanning Electron Microscopy, Vol. 1. M. A. Hayat (ed.) Van Nostrand Reinhold, New York, 1974. 226-241.
20. Koch, G. R. "Preparation and examination of specimens at low temperatures," in Principles and Techniques of Scanning Electron Microscopy, Vol. 4. M. A. Hayat (ed.) Van Nostrand Reinhold, New York, 1975. 1-33.
21. Pawley, J. B. and T. L. Hayes. "A freeze fracture preparation chamber attached to the SEM." 35th Ann. EMSA Proc., Baton Rouge, 1977. 588-589.
22. Cohen, A. L. "A critical look at critical point drying--theory, practice and artefacts." SEM/1977/I, IIT Research Institute, Chicago, IL. 525-536.
23. Cohen, A. L., D. P. Marlow and G. E. Garner. "A rapid critical point method using fluorocarbons ("Freons") as intermediate and transitional fluids." J. Microscopie 7(3), 1968. 331-342.
24. Lewis, E. R., L. Jackson and T. Scott. "Comparison of miscibilities and critical-point drying properties of various intermediate and transitional fluids." SEM/1975, IIT Research Institute, Chicago, IL. 317-324.
25. Cohen, A. L. and M. M. Shaykh. "Relations between fixation and dehydration in preserving cell morphology." 32nd Ann. EMSA Proc., Baton Rouge, 1974. 124-125.
26. Costerton, J. W., G. G. Geesey, and K. J. Cheng. "How bacteria stick." Scientific American 238(1), 1978. 86-95.

27. Bakst, M. and B. Howarth, Jr. "SEM preparation and observations of the hen's oviduct." Anat. Rec. 181, 1975. 211-226.

28. Behbehani, B. I., *et al.* "Scanning electron microscopy of the male nematode Physaloptera." 35th Ann. EMSA Proc., Baton Rouge, 1977. 674-675.

29. Fallah, E., *et al.* "Scanning electron microscopy of gastroscopic biopsies." Gastrointestinal Endoscopy 22(3), 1976. 137-144.

30. Nemanic, M. K. and D. R. Pitelka. "A scanning electron microscopy study of the lactating mammary gland." J. Cell Biol. 48, 1971. 410-415.

31. Eisenstat, L. F., *et al.* "A technique for removing mucus and debris from mucosal surfaces." SEM/1976/II, IIT Research Institute, Chicago, IL. 263-268.

32. Evan, A. P., *et al.* "Scanning electron microscopy of cell surfaces following removal of extracellular material." Anat. Rec. 185, 1976. 433-446.

33. Mariscal, R. N., E. J. Conklin and C. H. Bigger. "The putative sensory receptors associated with the cnidae of cnidarians." SEM/1978/II, SEM. Inc., AMF O'Hare, IL. 959-966.

34. Atwood, D. G., B. J. Crawford and G. B. Braybrook. "A technique for processing mucous coated marine invertebrate spermatozoa for scanning electron microscopy." J. Microscopy 103, 1975. 259-264.

35. Baccetti, B. "Spermatozoa," in Principles and Techniques of Scanning Electron Microscopy, Vol. 4. M. A. Hayat (ed.) Van Nostrand Reinhold, New York, 1975. 94-102.

36. Dadoune, J. and M. Fain-Maurel. "A routine technique for processing human ejaculate spermatozoa for scanning electron microscopy with special reference to their abnormal forms." Biologie Cellulaire 29, 1977. 215-218.

37. Kinden, D. A. and M. F. Brown. "Techniques for scanning electron microscopy of fungal structures within plant cells." Phytopathology 65, 1975. 74-76.

38. Lutchel, D. L. "The mucous layer of the trachea and major bronchi in the rat." SEM/1978/II, SEM. Inc., AMF O'Hare, IL. 1089-1098.

39. Thurston, R. J. "Ultrastructure of lungs fixed in inflation using a new osmium-fluorocarbon technique." J. Ultrastruct. Res. 56, 1976. 39.

40. Kruse, P. F., Jr. and M. K. Patterson, Jr. (eds.). Tissue Culture, Methods and Applications. Academic Press, New York, 1973. 3-28.

41. Amsterdam, A. and J. D. Jamieson. "Studies on dispersed pancreatic exocrine cells. I. Dissociation technique and morphologic characteristics of separated cells." J. Cell Biol. 63, 1974. 1037-1056.

42. Boyde, A. "A method for the preparation of cell surface hidden within bulk tissue for examination in the SEM." SEM/1975, IIT Research Institute, Chicago, IL. 295-304.

43. Hayes, T. L. and J. B. Pawley. "Very small biological specimens," in Principles and Techniques of Scanning Electron Microscopy, Vol. 3. M. A. Hayat (ed.) Van Nostrand Reinhold, New York, 1975. 45-81.

44. Kurtzman, C. P., F. L. Baker and M. J. Smiley. "Specimen holder to critical-point dry microorganisms for scanning electron microscopy." Applied Microbiol. 28, 1974. 708-712.

45. Gershman, H. and J. Rosen. "Cell adhesion and cell surface topography in aggregates of 3T3 and SV40-virus-transformed 3T3 cells. Visualization of interior cells by scanning electron microscopy." J. Cell Biol. 76, 1978. 639-651.

46. Thomas, R. J., M. G. Wolery and J. Taylor. "Critical point drying of liverwort spores for scanning electron microscopy." Stain Technol. 49, 1974. 261-264.

47. Cohen, W. D. "Simple magnetic holders for critical point drying of micro-specimen suspension." J. Microscopy 108, 1976. 221.

48. Buckley, I. K. and K. R. Porter. "Electron microscopy of critical point dried whole cultured cells." J. Microscopy 104, 1974. 107-120.

49. Rao, N. S. "Formvar on modified stubs for the study of cell suspensions by scanning electron microscopy." Micron. 7, 1976. 87-93.

50. Mazia, D., W. S. Sale and G. Schatten. "Polylysine as an adhesive for electron microscopy." J. Cell Biol. 63, 1974. 212a.

51. Lyon, F. L. and C. L. Kramer. "Preparing myxomycete spores for SEM." Mycologia 69, 1977. 1045-1047.

52. Sommer, J. R. "To cationize glass." J. Cell Biol. 75, 1977. 245a.

53. Masuda, Y. and M. Osumi. "Superhelical structures of polypeptide aggregates." J. Electron Micros. 23, 1974. 303-305.

54. Sarkar, N. H., W. J. Manthey and J. B. Sheffield. "The morphology of murine oncornaviruses following different methods of preparation for electron microscopy." Cancer Research 35, 1975. 740-749.

55. Male, P. M. and D. Biemesderfer. "Silicon wafers as support for biological macromolecules in high resolution cold field emission scanning electron microscopy." SEM/1978/II, SEM. Inc., AMF O'Hare, IL. 643-648.

56. de Harven, E., *et al.* "Structure of critical point dried oncornaviruses." Virology 55, 1973. 535-540.

57. Thornthwaite, J. T., *et al.* "A new method for preparing cells for critical point drying." SEM/1975, IIT Research Institute, Chicago, IL. 387-392.

58. Boublik, M., F. Jenkins and H. R. Kaback. "Use of critical point drying in the preparation of Escherichia coli membrane vesicles for electron microscopy." Cytobiologie 11, 1975. 304-308.

59. Berdach, J. T. "In situ preservation of the transverse flagellum of *Peridinium cinctum* (Dinophyceae) for scanning electron microscopy." J. Phycol. 13, 1977. 243-251.

60. Biemesderfer, D., *et al.* "Changes in macrophage surface morphology and erythrophagocytosis induced by ubiquinone-8." SEM/1978/II, SEM. Inc., AMF O'Hare, IL. 333-340.

61. Penttila, A., H. Kalimo and B. Trump. "Influence of glutaraldehyde and/or osmium tetroxide on cell volume, ion content, mechanical stability, and membrane permeability of Ehrlich ascites tumor cells." J. Cell Biol. 63, 1974. 197-214.

62. Wetzel, B., *et al.* "The need for positive

identification of leukocytes examined by SEM."
SEM/1973, IIT Research Institute, Chicago, IL.
535-542.
63. Thornthwaite, J. T., *et al.* "A technique
for combined light and scanning electron microscopy of cells." SEM/1976/II, IIT Research
Institute, Chicago, IL. 127-130.
64. McKee, A. E. "SEM in medical microbiology--
an overview." SEM/1977/II, IIT Research Institute, Chicago, IL. 239-249.
65. Suzuki, T., *et al.* "Stereoscopic scanning
electron microscopy of the red pulp of dog spleen
with special reference to the terminal structure
of the cordal capillaries." Cell Tiss. Res. 182,
1977. 441-453.
66. Hillman, D. E. "Relationship of the sensory
cell cilia to the cupula." SEM/1977/II, IIT
Research Institute, Chicago, IL. 415-420.
67. Tanaka, K. and C. Smith. "Structure of the
chicken's inner ear: SEM and TEM study." Am.
J. Anat. 153, 1978. 251-272.
68. Phillips, M. I., P. P. Deshmukh and W. Larsen.
"Morphological comparisons of the ventricular wall
of subfornical organ and organum vasculosum of
the lamina terminals." SEM/1978/II, SEM. Inc.,
AMF O'Hare, IL. 349-356.
69. Ioshima, T., *et al.* "Scanning electron
microscopy of puncture-perfused human liver biopsy
samples. Observation of bile ductules and bile
canaliculi." SEM/1978/II, SEM., Inc., AMF O'Hare,
IL. 169-174.
70. Kelley, R. O., R. A. F. Dekker and J. G.
Bluemink. "Thiocarbohydrazide-mediated osmium
binding: a technique for protecting soft biological specimens in the scanning electron microscope," in Principles and Techniques of Scanning
Electron Microscopy, Vol. 4. M. A. Hayat (ed.)
Van Nostrand Reinhold, New York, 1975. 34-44.
71. Kalina, M. and D. C. Pease. "The probable
role of phosphatidyl cholines in the tannic acid
enhancement of cytomembrane electron contrast."
J. Cell Biol. 74, 1977. 742-746.
72. Murphy, J. A. "Non-coating techniques to
render biological specimens conductive." SEM/
1978/II, SEM. Inc., AMF O'Hare, IL. 175-194.
73. Arborgh, B., *et al.* "The osmotic effect of
glutaraldehyde during fixation. A transmission
electron microscopy, scanning electron microscopy
and cytochemical study." J. Ultrastr. Res. 56,
1976. 339-350.
74. Brunk, U., *et al.* "SEM of in vitro culti-
vated cells, osmotic effects during fixation."
SEM/1975, IIT Research Institute, Chicago, IL.
379-386.
75. Iqbal, S. J. and B. S. Weakley. "The effects
of different preparative procedures on the ultrastructure of the hamster ovary. I. Effects of
various fixative solutions on ovarion oocytes
and their granulosa cells." Histochem. 38, 1974.
95-122.
76. Dobbs, G. H. "Soft tissues of marine teleosts," in Principles and Techniques of Scanning
Electron Microscopy, Vol. 2. M. A. Hayat (ed.)
Van Nostrand Reinhold, New York, 1974. 72-82.
77. Chuang, S. H. "Sea water and osmium tetroxide fixation of marine animals," in Cell Structure
and its Interpretation. S. M. McGee-Russel and
K. F. A. Ross (eds.) Edward Arnold, Ltd., London,
1968. 51-57.

78. Baker, F. L. and L. H. Princen. "Positive
displacement holder for critical point drying of
small particle materials." J. Microscopy 103(3),
1975. 383-401.
79. Lin, C. H., R. H. Falk and C. R. Stocking.
"Rapid chemical dehydration of plant material
for light and electron microscopy with 2,2-
dimethoxypropane and 2,2-diethoxypropane."
Amer. J. Bot. 64, 1977. 602-605.
80. Maser, M. D. and J. J. Trimble, III.
"Rapid chemical dehydration of mammalian tissues
for scanning electron microscopy using 2,2-di-
methoxypropane." 34th Ann. EMSA Proc., Baton
Rouge, 1976. 340-341.
81. Kahn, L. E., S. Frommes and P. Cancilla.
"Comparison of ethanol and chemical dehydration
methods for the study of cells in culture by
scanning and transmission electron microscopy."
SEM/1977/I, IIT Research Institute, Chicago, IL.
501-506.
82. OSHA (U.S. Occupational Safety and Health
Administration). 1977 Concentration Limits for
Gases (wall chart). Foxboro/Wilks Inc., South
Norwalk, Connecticut.
83. Pawley, J. and S. Dole. "A totally automatic critical point dryer." SEM/1976, IIT
Research Institute, Chicago, IL. 287-294.
84. Hall, D. J., E. J. Skerrett and W. D. E.
Thomas. "Critical point drying for scanning
electron microscopy: a semi-automatic method
of preparing biological specimens." J. Microscopy 113(3), 1978. 277-290.
85. Brown, J. N. "A simple low-cost critical
point dryer with continuous flow dehydration
attachment." J. Microscopy 111(3), 1977.
351-358.
86. Koller, T. and W. Bernhard. "Séchage de
tissus au protoxyde d'azote (N$_2$O) et coupe ultra-
fine sans matière d'inclusion." J. Microscopie
3, 1964. 589-606.
87. Rosowski, J. R. and W. V. Glider. "Comparative effects of metal coating by sputtering and
by vacuum evaporation on delicate features of
euglenoid flagellates." SEM/1977/I, IIT Research
Institute, Chicago, IL. 471-480.
88. Molday, R. S. "Cell surface markers and
labeling techniques for scanning electron
microscopy." 35th Ann. EMSA Proc., Baton Rouge,
1977. 478-481.
89. Molday, R. S. "Cell surface labeling techniques for SEM." SEM/1977/II, IIT Research Institute, Chicago, IL. 59-82.
90. Stenman, S., J. Wartiovaara and A. Vaheri.
"Changes in the distribution of a major fibroblast protein, fibronectin, during mitosis and
interphase." J. Cell Biol. 74, 1977. 453-467.
91. Humphreys, W. J., B. O. Spurlock and J. S.
Johnson. "Transmission electron microscopy of
tissue prepared for scanning electron microscopy
by ethanol-cryofracturing." Stain Tech. 50(2),
1975. 119-125.
92. Henk, W. G. and B. O. Spurlock. "Scanning
electron microscopy of ethanol infiltrated
freeze-fractured cell pellets." 32nd Ann.
EMSA Proc., Baton Rouge, 1974. 98-99.
93. Tanaka, K. "Frozen resin cracking method
and its role in cytology," in Principles and
Techniques of Scanning Electron Microscopy, Vol.
1. M. A. Hayat (ed.) Van Nostrand Reinhold,

New York, 1974. 125-134.
94. Tanaka, K., A. Iino and T. Neguro. "Styrene resin cracking method for observing biological materials by scanning electron microscopy." J. Electron Microscopy 23, 1974. 313-315.
95. Clark, J. M. and S. Glagov. "Visualization of aortic medial microarchitecture by scanning electron microscopy of frozen fractured surfaces." SEM/1978/II, SEM. Inc., AMF O'Hare, IL. 283-290.
96. Watson, J. H. L., J. L. Swedo and R. H. Page. "Observations on the nucleus of 'stripped' HeLa cells." 33rd Ann. EMSA Proc., Baton Rouge, 1975. 302-303.
97. Watson, J. H. L., R. H. Page and J. L. Swedo. "A technique for determining the interior topography of single cells (Colcemid-blocked HeLa cells)." SEM/1975, IIT Research Institute, Chicago, IL. 417-424.
98. Flood, P. R. "Dry-fracturing techniques for the study of soft internal biological tissues in the scanning electron microscope." SEM/1975, IIT Research Institute, Chicago, IL. 287-294.
99. Irino, S., et al. "Microdissection of tannin-osmium impregnated specimens in the SEM: demonstration of arterial terminals in human spleen." SEM/1978/II, SEM. Inc., AMF O'Hare, IL. 111-116.
100. Wickham, M. G. and D. M. Worthen. "Scanning and transmission electron microscopy of single tissue specimens," in Principles and Techniques of Scanning Electron Microscopy, Vol. 2. M. A. Hayat (ed.) Van Nostrand Reinhold, New York, 1974. 60-71.
101. Bucana, C., et al. "A technique for sequential examination of in vitro macrophage-tumor interaction using LM, SEM, and TEM." 34th Ann. EMSA Proc., Baton Rouge, 1976. 350-351.
102. Myklebust, R., H. Dalen and R. S. Saetersdal. "A comparative study in the transmission electron microscope and scanning electron microscope of intracellular structures in sheep heart muscle cells." J. Microscopy 105, 1975. 57-65.
103. Erlandsen, S. L., A. Thomas, and G. Wendelschafer. "A simple technique for correlating SEM with TEM on biological tissue originally embedded in epoxy resin for TEM." SEM/1973, IIT Research Institute, Chicago, IL. 349-356.
104. Winborn, W. B. and D. L. Guerrero. "The use of single tissue specimen for both transmission and scanning electron microscopy." Cytobios 10, 1974. 83-91.
105. Gusnard, D. and R. H. Kirschner. "Cell and organelle shrinkage during preparation for scanning electron microscopy: effects of fixation, dehydration and critical point drying." J. Microscopy 110, 1977. 51.
106. Kirschner, R. H., M. Rusli and T. E. Martin. "Characterization of the nuclear envelope, pore complexes, and dense lamina of mouse liver nuclei by high resolution scanning electron microscopy." J. Cell Biol. 72, 1977. 118-132.
107. Waterman, R. E. "Embryonic and fetal tissues of vertebrates," in Principles and Techniques of Scanning Electron Microscopy, Vol. 2. M. A. Hayat (ed.) Van Nostrand Reinhold, New York, 1974. 93-110.
108. Breiphol, W. and M. Fernandez. "Scanning electron microscopic investigations of olfactory epithelium in the chick embryo." Cell Tiss. Res.

182, 1977. 105-114.
109. Schneider, G. B., S. M. Pockwinse and S. Billings-Gagliardi. "Morphological changes in isolated lymphocytes during preparation for SEM: a comparative TEM/SEM study of freeze drying and critical point drying." SEM/1978/II, SEM. Inc., AMF O'Hare, IL. 77-84.
110. Madge, D. S. "Scanning electron microscopy of normal and diseased mouse small intestinal mucosa." J. Microscopie 20, 1974. 45-50.
111. Hammond, C. T. and P. G. Hahlberg. "Morphology of glandular hairs of Cannabis sativa from scanning electron microscopy." Amer. J. Bot. 60(6), 1973. 524-528.
112. Falk, R. H., E. M. Gifford, Jr. and E. G. Cutter. "The effect of various fixation schedules on the scanning electron microscopic image of Tropaeolum majus." Amer. J. Bot. 58, 1971. 676-680.
113. Copeland, D. D., et al. "Surface morphology of avian sarcoma virus and ethyl-nitrosourea transformed rat neuroectodermal cells and human glioblastoma cells in organ and monolayer culture." SEM/1976/II, IIT Research Institute, Chicago, IL. 93-100.
114. Humphreys, W. J. "Health and safety hazards in the SEM laboratory." SEM/1977/I, IIT Research Institute, Chicago, IL. 537-544.
115. Thurston, E. L. "Health and safety hazards in the SEM laboratory: update 1978." SEM/1978/II, SEM. Inc., AMF O'Hare, IL. 849-854.

Discussion with Reviewers

Reviewer I: How can you assume that cells that have been separated and mechanically placed on a substrate will assume their normal shape? They are in a foreign environment and may react quite differently.

Author: The subsection, Processing Small Specimens has details on preventing or minimizing just such changes. This is again discussed in the subsection, OsO4 Fume Fixation. As stated in the beginning, we try to make structures look as if they were alive (and normal). On the whole I think we succeed. Organisms which normally live in suspension should be fixed in suspension, and those which are normally on a solid substrate should be fixed on the substrate.

Reviewer I: I believe preferential settling and adhesion are major concerns and should be dealt with in detail. The single reference cited (57) deals with a specific example, but not the overall problem.

Author: The purpose of this paper is to alert the reader to the possibility of preferential settling and to offer a solution. The problem is touched on in Ref. 3 (p. 66) and discussed thoroughly by J. Dubochet and E. Kellenberger, Microscopica Acta. 72, 1972. 119-130.

G. H. Haggis: de Harven, et al.[11] state that their method is not freeze-drying, but "low temperature vacuum drying from ethanol". Freeze drying from alcohol is difficult, if it is possible at all, since a small amount of frost contamination is sufficient to cause the ethanol to liquify at temperatures at which the ethanol would be removed by sublimation.

Author: Your statement is supported by a comment from Dr. de Harven (personal communication). In this paper[11] the context (pp. 520-521) leads one to believe the ethanol is frozen and sublimed. As ethanol cools to a glass, and therefore has no sharp freezing temperature, this procedure may be in between solvent evaporation drying and freeze drying, with some support given to the specimen structure by the highly viscous alcohol.

J. Nowell: Can you substantiate with micrographs the statement "osmium can overfix with loss of detail"? What does the statement mean?
Author: TEM or optical microscope sections would show this, but I have not felt the need to substantiate it. It has long been accepted. Thus, Romeis states concerning OsO_4: "Overfixation can easily occur; the cells become glassy and structure is no longer apparent" (my translation) B. Romeis, Mikroskopische Technik, Oldenbourg, Munich, 1968. p. 69. See also, Pease, Histological Technique for Electron Microscopy, 2nd ed., Academic Press, New York, 1964. p. 46.

A. Boyde: Re the statement . . . "it is usually poor practice to jump specimens from aqueous media to high concentrations of any dehydration fluid, for water rushes out faster than the foreign molecules can enter." (Our) data indicate that the first stage of dehydration is a transient expansion . . .
Author: First, all dehydration results in at least brief exposure to low concentrations of dehydration fluid during the exchange process. Also the "chemical dehydration" with 2,2 dimethoxypropane will allow some time in lower concentrations. Please note that I have advocated the briefest time in these agents. The degree of swelling and whether or not there is preliminary shrinking will depend on: the nature of the cell membrane, presence and properties of cell wall or cytoskeleton, type and degree of fixation, kind and concentration of dehydrating agent. To take extremes, unfixed *Spirogyra* shrink, then swell and cells will burst in 30% methanol (class experiment!), lightly fixed ones will collapse in less than 30% ethanol (Cohen and Shaykh, IITRI/SEM, 1973, 371-378) while in 30% ethylene glycol unfixed myxomycete swarm cells first shrink and then swell again. In general our experience has indicated that without specific trial, rapid passage of fixed material through 30% and 50% ethanol is preferable to starting with higher concentrations. (See also Ref. 25.)

J. Nowell: What is the advantage of partially emptying the bomb? Would there not be less possibility of evaporative drying if the bomb is constantly filled with fresh transitional fluid as the mixture of intermediate and transitional fluid is bled off?
Author: If the intermediate fluid is denser than the transitional fluid (true in all cases above about 20° C; always true when Freon TF is the intermediate fluid), and the vent is at the bottom of the bomb, the intermediate is swept out with minimum contamination of the incoming transitional fluid. Also partially emptying the bomb to 25% will increase the dilution factor four-

fold at each "rinse". The danger of artefacts from evaporative drying is minimal. The liquid/vapor relation is near or at equilibrium, and the surface tension is negligible.

J. Nowell: By "a continuous flush with CO_2" do you mean that some investigators leave both the liquid vent valve and the gas inlet valve open until they suppose that all the intermediate fluid has been vented? Comment: If so, this probably accounts for the many poor results illustrated in the literature . . . and it certainly isn't possible to use the 'nose test' to determine when all the amyl acetate has been removed. . . . those devices . . . . in which the vent is at the top . . . can take up to 2 days or 15 lb. CO_2 of flushing for 1/2 minute every 5 minutes to completely remove amyl acetate . . ."
Author: I have quoted your comments nearly verbatim because they present a strong case for (1) venting at the bottom, and (2) intermittent flushes. I had often felt that continuous flushing might do little more than set up streams of transitional fluid from inlet to vent without having much exchange with the intermediate fluid in the specimen.

J. Nowell: How is the determination made that sufficient Freon TF has been flushed out when drying "bulky" specimens?
Reviewer II: How do you know if the intermediate fluid has not been sufficiently flushed out? What do you do if you realize you haven't flushed it out and the bomb is above T_c?
Author: If you are using an instrument with viewing port and external means of cooling such as a water jacket, thermoelectric device, or adiabatic shroud, it is easy. When you think the specimen has been purged of Freon TF, start releasing pressure above T_c. Close the vent at once if dense fog forms (disregard slight haze), or if fog forms in any part of the process above T_c. Cool the chamber to at least 10° below T_c, and flush it two or more time with the transitional fluid. If you have a window, but no easy means of cooling the chamber while it is under pressure, look for a distinct liquid layer at the bottom near or above T_c. Shaking the CPD device slightly may make it more apparent. Let the bomb cool to pressure below supply tank pressure and very carefully flush out the liquid, restoring the pressure by replenishment from the supply tank. If you have neither window nor means of cooling the bomb while it is under pressure, there is little you can do besides preventative measures such as, 4-5 flushes with the liquid transitional fluid at least 10 minutes apart.

Reviewer II: Since some accidents have occurred with windows, how may one have both windows and safety?
Author: This problem has been considered elsewhere.[3,22] Use tested instruments with pressure and temperature limiting switches, failsafe pressure release, and windows of appropriate properties. Such windows have a large safety factor, and if they do fail, they merely crack and do not shatter. Also mirrors may be used.[84] We require that Lexan face shields be worn during CPD operations for at least some protection.

Additional discussion with reviewers of the paper "A Comprehensive Freezing, Fracturing and Coating System..." by A.W. Robards continued from page 344.

J. Pawley: Radiant heat has the advantage that, at least as long as there is ice on the sample surface, the thermal energy is being directly applied where it is needed whereas when the sample is heated by conduction from the stub, the stub must be warmer than the sample and therefore the sample may be contaminated by substances desorbed from the stub. This will be particularly true when there is a long thermal path from the sample to the stub. Have you noticed this phenomenon?

Authors: Dr Pawley is referring to the question of whether ice sublimation is best carried out by radiant heating or by bringing the whole specimen (and stub) to an appropriate temperature. The attraction of the first method is that it is quick and, as stated, the heat is directed where required. Nevertheless, the degree of control using radiant heating is, we believe, less than that when the whole specimen is carefully maintained at a predetermined temperature, even though this method is lengthier. The comment that the specimen may become contaminated because the stub must be warmer could conceivably apply under the most adverse conditions but these are unlikely to prevail in our system. The heating element is effectively placed across the thermal pathway between cold sink (nitrogen dewar) and specimen surface. There will be little difference between the specimen surface and surrounding stub temperature and the influence of the closely adjacent surface of the shroud 60 K or so colder will provide an effective trap for desorbed molecules - from specimen or stub. The situation in our system is much the same as for most freeze-etching units where little trouble has been experienced from the source mentioned by the reviewer.

B. Dronzek: Have you tried your equipment with microscopes other than the Cambridge S600? What is the maximum resolution you attained using your system at low temperatures?

J. Pawley: The CSI S600 microscope should be capable of resolution of 10-20 nm. Could you please show us some pictures at >10kx demonstrating this sort of resolution, if possible as a function of observation time?

Authors: These questions largely relate to resolution which, in common with most other users of low temperature SEM equipment, we are unable to specify accurately. Dr Pawley is wrong in believing that the resolution of our CSI S600 microscope is 10-20 nm: this may apply to current models but the specification of our own early instrument (bought in 1972) was 25 nm and we have no reason to suppose that it is any better than that. It should be pointed out, however, that CSI have steadily improved the resolution of subsequent versions of the S600 since we bought ours. The question of instrumental resolution in fact begs the question of what can be obtained when using a cold stage - something that will completely change the manufacturer's specified performance. As Dr Pawley has himself pointed out[26], instrumental resolution is by no means necessarily reflected in resolution obtained from low temperature specimens. We, also, have used the fractured surface of the plasmalemma of yeast as a test specimen: this is familiar to most workers in the field of low temperature microscopy and provides a range of structures visible at different resolution. It is apparent from all the micrographs so far published of yeast plasmalemma that low temperature SEM still has some considerable way to go before it is likely to challenge freeze-etching as the method of choice for viewing cell membranes (Figs. 19,20).

Our comments about resolution must be seen against the limited performance of our S600 SEM. A commercial form of our own system has now been interfaced with an ISI System 100A SEM with resolution <10 nm (Fig. 24) and we hope to have a clearer idea of ultimate resolution in the near future.

SCANNING ELECTRON MICROSCOPY/1979/II
SEM Inc., AMF O'Hare, IL 60666, USA

A COMPREHENSIVE FREEZING, FRACTURING AND COATING SYSTEM FOR
LOW TEMPERATURE SCANNING ELECTRON MICROSCOPY

A. W. Robards and P. Crosby

Department of Biology
University of York
Heslington, York, YO1 5DD,
U.K.

Abstract

Requirements for low temperature
biological scanning electron microscopy
are discussed; some published systems are
briefly described; the latest developments of our own system are reported.

The construction of a vacuum transfer device that can be interfaced with a
Cambridge S600 SEM has already been
published. This system has been extended
so that specimens are frozen on a copper
stub, immediately withdrawn into the
vacuum transfer device, transported to
the cold table of a specially designed
fracture/coating chamber, fractured and
sputter coated, again withdrawn into the
transfer device and placed on the cold,
temperature-controlled stage of the SEM.

Specimen temperature never needs to
rise above 140 K at any time after
initial freezing. During transfer the
specimen is protected by the vacuum in
the transfer device and also by metal
shrouds that automatically close over it.
When in the fracture/coating chamber or
in the SEM the specimen is protected by
a cold anticontaminator plate. Both
cold stages cool to less than 100 K.
Using cold argon gas, sputter coating
has been accomplished without any
measurable or observable heat damage to
the specimen. In operation it is never
necessary to bring the SEM or the frac-
ture/coating chamber to atmospheric
pressure. Consequently contamination
is minimised. Specimens have been
continuously viewed at low temperatures
over many hours with no apparent change
in structure. The equipment is extremely versatile and, in principle,
compatible with microscopes other than
the Cambridge S600.

KEY WORDS: Low Temperature SEM, Cryotechnique,
Cryofracture, Cold Stage, Sputter Coating,
Contamination, Biological Specimen Preparation,
SEM Accessories

Introduction

The ability to both observe and
analyse specimens in the scanning electron microscope (SEM) potentially offers
biologists the opportunity to prepare
material for examination extremely rapidly and, at the same time, to avoid the
undesirable effects of chemical treatments. Low temperature methods are
therefore useful even for routine procedures. Moreover, frozen specimens present very considerable advantages so far
as X-ray analytical methods are concerned
and much of the impetus towards low temperature instrumentation has derived from
this fact. Further, with rapidly improving SEM resolution, it is not unthinkable
that there will come a time when the
ability to view directly a frozen-
fractured surface in the SEM will compete
with the well-established, but less
direct, technique of freeze-etching: at
the moment such resolution on low temperature biological specimens is significantly far from being demonstrated.

Technically there are a number of
problems that have to be overcome if
specimens at low temperatures are to be
observed, analysed and recorded in the
SEM. A major difficulty is the avoidance
of contamination of the specimen with
condensing water vapour, or other condensible molecules such as hydrocarbon oils,
while at the same time preventing water
from subliming from the specimen surface.
Such considerations, as well as the need
to avoid recrystallisation of ice, dic-
tate the requirement that the temperature
of the specimen at all stages during processing and observation must not go
outside prescribed limits. The precise
upper temperature limit has been arbitrarily set by many workers as -130°C (143K).
This is because recrystallisation of
pure water is considered to occur

at temperatures higher than this. Even at temperatures of 143 - 173 K ice recrystallisation in pure water would be relatively slow, and the evidence from freeze-etching biological systems appears to indicate that quite prolonged spells at 173 K do not cause problems arising from recrystallisation. Nevertheless, it has to be admitted that further work in this area would be helpful. In the absence of other criteria it is reasonable to attempt to keep the specimen surface temperature always below 143 K; temperatures between 143 and 173 K will be unlikely to have any observable adverse effects except for an increase in the sublimation rate of water which may, in any case, be a necessary part of the experimental protocol. Many workers have sought to work at as low a temperature as possible. In practice, this has usually meant close to the temperature of liquid nitrogen (-196°C - 77 K). However, such low temperatures exacerbate problems of contamination through condensation and do not seem really necessary.

In a comprehensive system for low temperature scanning electron microscopy we might expect that the specimens will be frozen; fractured or manipulated in some way; possibly etched; coated with a conducting film so that undue charging in the microscope is avoided; and observed in the microscope. At each of these stages close temperature control needs to be maintained even in the face of externally applied thermal loads such as created during the coating process, observation in the microscope, and (possibly) during fracturing. It is the purpose of this paper to review briefly the current situation concerning instrumentation available for low temperature scanning electron microscopy and to present the latest developments of our own system that was first reported at the SEM Meeting in 1978[1]. It is emphasised that this is not intended to be a comprehensive review but rather attempts to point to some of the general problems and looks at the way that some systems have sought to overcome them.

Over the past eight years or so a number of laboratories have constructed systems for low temperature SEM. With only one or two exceptions there has, until recently, been little commercially available equipment that would satisfy the stringent demands required by biological specimens. However, the situation is now rapidly changing so that, as well as systems that have been produced in the laboratories of individual scientists, a number of commercial systems are now marketed. In many cases, of course, the commercial apparatus has derived from instrumentation constructed in the first instance by scientists in their own laboratories (see Table 1). While a number

of authors have recently published on low temperature biological scanning electron microscopy, most of these papers have confined themselves to a description of the apparatus available in the author's own laboratory. Nevertheless, the following references will act as a useful guide to those who are new to this field. (Only the most recent references are given for any one research group.) The current state of work carried out in Cambridge by Echlin and his colleagues is well summarised in two recent papers[2,3] which draw from the important contribution of this group over the last 8 or 9 years. A more detailed description of some of their techniques and instrumentation is provided in an earlier reference[4]. Fuchs and Lindemann and their colleagues have published two detailed papers on the instrumentation and specimen preparation for electron beam X-ray microanalysis of frozen hydrated bulk specimens[5,6], while Koch has provided a good general review of the subject[7] and Hayes and Koch have also reviewed problems associated with low temperature micromanipulation in the SEM[8]. The work in Japan by Nei and his colleagues is presented in detail[9,10], while that of Marshall, in Melbourne, Australia, has been well summarised recently[11] and the ingenious system devised by Pawley and Norton is also fully described[12]. There are, of course, many other systems, some of which are extremely simple[13] but these will not be considered in detail here.

Of the commercial companies, Jeol have long had a system available for their JSM 35 and the latest version of their unit (known as the 'Cryoscan 35-CRU 2') is compatible with their latest JSM 35-C. Balzers have designed and built a cryo-system to be compatible with the Philips PSEM 500 scanning electron microscope. Cameca provide a low temperature system for use in conjunction with their Camebax microscope, while Etec Corporation also have comprehensive low temperature facilities (the Biosem system) for their Autoscan SEM. Cambridge Scientific Instruments Ltd. have had a heating and cooling stage available for their SEM for many years although this has not been usable for biological applications without some modification. However, this Company has recently developed a low temperature stage and transfer device suitable for allowing the observation and analysis of biological materials in the S180 or S150 microscopes (or, with some restriction on performance, in the S604, S4 and S2 stereoscans). The Bio-Chamber system designed by Pawley and Norton[12], which is attached directly to an AMR-1000A SEM, is sold by the Advanced Metals Research Company while our own system is now

Table 1

Some described equipment for biological low temperature SEM*

Described by**	Microscope	Facilities	Transfer System	Fracturing	Coating	Cold stage (Min. temp.)	Notes
Cameca	Camebax	Cold stage with sep. coating chamber on SEM	Direct into SEM via airlock	Cold knife on ball joint lead through on SEM	Carbon or metal evaporation in antechamber	Conductive cooling 121 K	Full details not avail. at time of writing
Echlin[35]	Cambridge S4	Capped stub Evaporator Cold stage	Capped stub Atmospheric pressure	-----	Carbon or metal in modified Balzers BAE 120	Copper braid 98 K	Modified Cambridge 200 Series cold stage
Fuchs[6]	Cambridge S4	Cold vac. transf. device. Balzers freeze-etch. Cold stage	Valved, cold, vacuum device	With knife in modif. Balzers freeze-etch unit	Carbon in modified Balzers freeze-etch unit	Copper braid 98 K	" " "
Cambridge Scientific Instruments	CSI S150, S180, S604, S4 & S2	Transfer device and cold stage only	Valved, cold, vacuum device	Poss. on stage or with compat. off-line unit	Possible using compatible off-line unit	Copper braid 120 K	Previously unpublished data
Jeol[23]	Jeol JSM-35C	Antechamber for SEM & cold stage	Cap. stub into antechamber at atmosph. press.	Simple cooled knife in SEM antechamber	Sputtered gold in antechamber	Copper braid 128 K	Development of earlier Jeol system
Koch[7]	Etec-Autoscan	Simple transfer device Air lock coater Cold stage	Und.liq.nitrog.at atmosph.press.into air lock coater	Micro-dissection in SEM	Carbon or metal in air lock coater	123 K Conductive cooling	Commercially available from Etec
Marshall[11]	Jeol JSM-35R	Capped stub Evaporator Cold stage	Capped stub at atmospheric pressure	Simple knife in Jeol JEE 4 evaporator	Carbon, aluminium or chromium in Jeol JEE 4	Copper braid 123 K	------
Pawley[12]	AMR-1000A	'Biochamber' Facility attached to SEM	Protect.transf.to Biochamb.at atmos. press. All other procs.und.high vac	Simple adjust-able knife in Biochamber	Carbon or metal in Biochamber	Joule-Thomson refrigerator 80 K	Commercially available from AMR
Philips[25]	PSEM-500	Balzers-designed integral unit on SEM	Single transf. into attachm. at atmos. pressure	Motorized knife	Gold sputter coating	Cold stage in cryo-system chamber 100 K	------
Saubermann[4]	Cambridge S4	Transfer device Evaporator Cold stage	'Cold tube device at atmospheric pressure	Sections (in other papers use simple cold knife in BAE 120 vacuum unit)	Aluminium in BAE 120	Copper braid 88 K	Modified Cambridge Series 200 Cold stage
Robards	Cambridge S-600	Freez.chamb. Vac. transf.device Fract./coat.chamb. Cold stage	Valved vacuum device with cold shrouded stub	Cold knife or needle on ball-joints in fract./coating chamber	Evapor. carbon or metal; sputtered gold in fract./coating chamber	Copper braid (actually, stranded copper) 98 K	Commercially avail. from Emscope. Compatible with other SEMs
Taylor[27]	Jeol JXA - 50A	Cold stage only	A detailed description of a cold stage compatible with other Cambridge equipment			Copper braid 96 K	------

* This list is by no means intended to be comprehensive. It serves more to compare some of the main contemporary approaches to the problem.

** First author only named.

327

available through Emscope Limited and has been interfaced with Cambridge S600 and S150 SEMs as well as an ISI System 100. Since the first use of cold stages for biological scanning electron microscopy (probably on a Cambridge SEM in 1970)[14,15] a number of other stages have also been described[16,17,18,19] and other full transfer systems have been reported[20,21,22]. Many other microscopists have reported the use of cold stages but the purpose of the present article is to concentrate on equipment rather than applications. We shall now consider the requirements for each part of a low temperature SEM system and examine how different approaches have sought to overcome the problems.

Design Criteria for Low Temperature SEM

a) General

Before any attempt is made to process and view biological specimens at low temperatures in an SEM a number of fundamental decisions have to be made. Among these are: the upper and lower limits of the temperature range that will be acceptable throughout the whole procedure of preparation and viewing; how the specimens should be frozen initially; whether the system should be directly attached to the SEM or separated from it (in the latter case a suitable system for transferring cold specimens without contamination will need to be developed); how the frozen specimen should be manipulated, fractured and etched; what method of coating the specimen with a conducting film (if any) should be employed; whether the stage in the SEM will be for reflection and/or transmission use; and what degrees of movement on the cold stage in the microscope are needed. The choices that are made have often depended upon the existing instrumentation available in a particular laboratory together with the specific research requirements of the individual. For example, the design of a system for allowing X-ray microanalysis of thin frozen sections in the transmission mode will have different priorities from one allowing the observation and analysis of bulk frozen specimens.

b) Some described systems

Some of the earliest systems simply allowed the transfer of a frozen specimen into the microscope followed by a period of ice sublimation to reveal underlying structures[15]; alternatively, a protective cold cap has been placed over the specimen and is only removed when ready for viewing. At all events, it must be emphasised that most bulk frozen specimens will need to be fractured and/or etched before any wet, or internal, surface will be clearly seen. Thus it is not possible to freeze wet tissues (such as many animal tissues - e.g. intestine) and view

them directly in the SEM: some sublimation of ice is necessary before the true surface is revealed.

The problem of protecting the specimen during the stages between freezing and viewing has led to the development of a number of systems that are attached directly to the SEM: in principle such systems have the advantages of relative ease of use, coupled with simple specimen movement. On the other hand, they lose some flexibility in operation; may monopolise the use of the SEM; and are usually compatible with only a single type of instrument.

One of the earlier 'dedicated' systems was that built by Jeol[20], and used by a number of workers[9,11]. The device allows the frozen specimen to be fractured with a knife, etched with a radiant heater, and coated with gold from a metal filament. The original system was found by some microscopists not to have an adequately low stage temperature and modifications to overcome this have been described (e.g.[11]); the more recent system has a lower stage temperature (see Table 1) but essentially similar facilities with the addition of gold sputter coating[23] - a feature already employed with a Jeol system for some years by K. Oates at the University of Lancaster. Nei and Fujikawa[10] have used a Jeol unit to study the controlled freeze-drying of specimens in the SEM but most workers believe that, except for specific purposes such as described by Nei, it is far better to avoid sublimation in the SEM as severe contamination of the vacuum system will inevitably occur. Such sublimation is best carried out in a separate chamber. Other dedicated systems are represented by the Etec equipment[24] with its 'airlock coater' - a system where etching in the SEM is again advocated and which has been described by Koch[7]. The Balzers system for the Philips PSEM 500[25] is also totally dedicated, allows etching in the SEM and sputter-coating with gold. Of all the dedicated systems, perhaps the best described is that designed by Pawley and Norton[12]. This system, attached to an AMR-1000A SEM is illustrated in Figs. 1 and 2. It is totally a high vacuum system, using gate-valve interlocks with the evaporators and the SEM, and allows a full range of manipulative and radiant etching facilities. It has a separate high vacuum pumping system and uses no plastics at room temperature except for Teflon wire insulation. The system has been used to study freeze-fractured yeast membranes in an attempt to compare the resolution with that obtained from the freeze-etching technique[26]: so far the results do not approach those of freeze-etching, but this interesting work to explore the

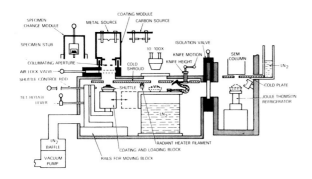

Fig. 1 Diagram of the Bio-Chamber freeze fracture system attached to the AMR-1000A SEM[12]. (Courtesy of Dr J. Pawley).

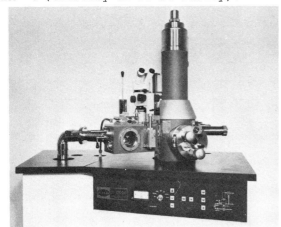

Fig. 2 AMR-1000A SEM with Bio-Chamber attached. (Courtesy of Dr J. Pawley).

full potential of the SEM in low temperature studies of membranes and molecules is well worth continuing as freeze-etching is an indirect method, whereas direct SEM observation potentially allows large continuous surfaces to be explored, X-ray analysis to be carried out, consecutive fractures to be made, and, if required, the specimen to be retrieved from the SEM after viewing. Nevertheless, for resolution of membrane particles alone, the freeze-etching technique still remains unchallenged.

Among the 'non-dedicated' systems, the work of Echlin and his colleagues is among the best-documented[2,3]. These workers have used a Cambridge Instruments S-4 SEM, and a Jeol JXA-50A electron probe microanalyser for which a redesigned cold stage has been described in detail[27]. Specimens are transferred into the SEM using a transfer device within which the frozen material is kept cold and protected; if required, the specimen is coated with a conducting metal in a Balzers BAE 120 evaporator equipped with a cold stage. Both transfers take place into the instruments at atmospheric pressure[4]. This system has been used for

the examination of a range of specimens analysed and viewed in both transmission and reflection mode[2,3,4]. Fuchs and Lindemann[5,6] have also used a Cambridge S-4 SEM as the basis of their low temperature system and they, as Echlin and his colleagues, have modified a Cambridge 200-Series heating and cooling stage for their work. Fracturing and sublimation occur in a modified Balzers BAF 300 freeze-etch apparatus; and transfer between the units is accomplished using a transportable gate-valved airlock. Cambridge Scientific Instruments have more recently designed and built a new cold stage and cooled vacuum transfer device for use with their more recent microscopes - the S180 and S150. These components are described more fully below.

It will be seen from the above discussion that a variety of approaches to the solution of the problems of viewing and analysing frozen biological specimens in the SEM has been made. The following sections will look in greater detail at the various stages in the procedure.

c) Freezing

The subject of freezing biological cells and tissues for ultrastructural examination and analysis is one that is currently receiving considerable interest and has been well reviewed[28]. In brief, only the smallest specimens can be frozen sufficiently rapidly so that ice crystal damage is minimal and ion relocation insignificant. For bulk specimens, or block material to be sectioned, it will commonly be necessary to resort to the use of cryoprotectants. Glycerol has frequently been used, but makes subsequent etching unreliable or impossible. Polymeric cryoprotectants such as polyvinylpyrrolidone (PVP) have been advocated[3], but further work is required before their true potential is known. In many cases it may well be that quite crude freezing methods will suffice for the SEM study in question, whereas other work will need the smallest possible ice crystal size.

A frequent point of discussion in low temperature SEM is whether the smallest possible specimen should be frozen alone, or whether a specimen may be frozen on some form of support. It is considered that heat transfer at the boundary of biological specimens, cooled in liquid cryogens such as the halogenated methane molecules (Freons, Arctons), propane, subcooled nitrogen etc., is a constant for the particular specimen/coolant system, and that heat flow through the specimen is a function of thermal conductivity, then it will be appreciated that the cooling rate at a given point within the specimen will be related to these two features which are, in fact, incorporated together into the

Table 2   Some representative cooling rates

		Peak Cooling Rate $(K\ s^{-1})$	Cooling Rate 272-223 K $(K\ s^{-1})$
Liquid Nitrogen	Bare T.C.	515	80
	P.B.	178	34
	P.T.F.E.	70	43
	T.C. in a droplet of yeast on a stainless steel stub		51
Subcooled Nitrogen	Bare T.C.	990	317
	P.B.	186	79
	P.T.F.E.	157	79
	T.C. in a droplet of yeast on a stainless steel stub		110
Hyperbaric Nitrogen (10 Bars)	Bare T.C.	1210	319
	P.B.	220	116
	P.T.F.E.	99	46
Arcton 12 (CCl_2F_2)	Bare T.C.	829	714
	P.B.	254	63
	P.T.F.E.	159	37

This table presents some 273-223 K cooling rates for: a bare 0.5 mm Ø copper-constantan thermocouple (T.C.); (P.B.) - similar thermocouple embedded in a 2.4 mm Ø high conductivity specimen (Phosphor-bronze - Biot number ≃ 0.002); another in a 2.4 mm Ø low conductivity specimen (P.T.F.E. - Biot number ≃ 2.5); and another in a drop of yeast on a stainless steel stub of approximately 785 mm^3. All determinations were carried out using the device described by Bald and Robards[29]; thermocouple leads were well insulated right up to the junction. The significant point is that even a small thermal resistance between the thermocouple and the cryogenic fluid serves to reduce cooling rates to the same order of magnitude. Freezing of the yeast on the surface of the large stainless steel block gave cooling rates comparable to those for the smaller specimens. Rapid cooling rates (measurable only with very small thermocouples[28]) can only be achieved at the surface of hydrated specimens. Only extremely small specimens (films - microdroplets) can be frozen sufficiently rapidly in the absence of cryoprotectants so that they have ice crystals of insignificant size. It can be seen that, although the mean 273-223 K cooling rate is given, the peak rate within this interval may be significantly higher.

dimensionless 'Biot modulus'. The Biot number of copper is approximately 0.0013 whereas that of ice is 2.16 which is close to that of P.T.F.E. (2.46) - a well known thermal insulator[30]. This indicates that the cooling rate of a watery specimen will become slower and slower as the freezing front penetrates into the specimen. The result is that the cooling rate - and hence the size of ice crystals at the surface of the specimen - is unlikely to be greatly influenced by the supporting block. This prediction is supported by the data presented in Table 2 which also illustrates some representative cooling rates achieved in different coolants under different thermodynamic conditions. It will be noted that a thermocouple mounted in a small droplet of yeast on a stainless steel stub cools from 273 to 223 K at a rate similar to that of either a small phosphor-bronze or P.T.F.E. bead when plunged into either boiling liquid nitrogen or subcooled nitrogen. The situation is not greatly different if the stub is copper. The conclusion to be drawn is that there is little point in attempting to miniaturise the specimen mount unless very high rates of freezing are required - in which case the specimen itself must be extremely small in at least one dimension and different freezing methods are likely to be used.

In fact, most workers have frozen their specimens on relatively small supports, but some have used large mounts, such as in the Jeol system[23] and in our own previous work[1]. If large stubs or holders are employed, then it is usually not difficult to freeze specimens initially on smaller supports and subsequently to mount these on the pre-cooled larger block if this is required for any special purpose.

As the cooling rate for bulk SEM specimens is in any case relatively slow, the choice of coolant is less critical

than in some other situations. In an un-
cryoprotected specimen the zone of very
small ice crystals only extends 10-15 μm
from the surface in very rapidly frozen
specimens[31] - an insignificant distance
for most SEM applications. Coolants such
as the halogenated methane molecules tend
to leave frozen residual contamination
when the specimen is transferred into
liquid nitrogen - they also give contaminating halogen peaks during microanalysis;
liquid propane is a good cryogen but is
hazardous; subcooled nitrogen gives rates
of cooling not greatly different from the
other liquid cryogens and it has no obvious disadvantages - its use is therefore recommended for routine freezing.
Other freezing methods - such as copper
block freezing[31] - may also be used but
they fall outside the scope of this
paper. It is very simple to obtain subcooled nitrogen by placing boiling nitrogen in a well insulated container and
then subjecting this, within a suitable
chamber, to the vacuum of a rotary pump.
At the triple point pressure (13.5 kNm^{-2})
the nitrogen will start to freeze and
form a 'slush'. Rapid back-leaking of
air or, preferably, an inert gas brings
the nitrogen back to ambient pressure but
in a subcooled state, at about 63 K, that
is suitable for freezing without seriously encountering the effects of the insulating jacket of nitrogen gas that forms
when freezing at 77 K. A suitable device
for freezing specimens in subcooled
nitrogen serves as the starting point for
our own low temperature instrumentation[1].

As soon as the specimen is frozen it
is usually stored in liquid nitrogen or
cooled to that temperature (77 K) prior
to transfer. At this temperature the
specimen is at extreme risk from contamination and some form of protection and/or
transfer device is necessary.

d) Transfer devices

In a sense, transfer devices are one
of the key features in successful low
temperature scanning electron microscopy.
The simplest situation is that the frozen
specimen is placed inside the SEM and observed. If this is done by withdrawing
the cold specimen from liquid nitrogen
and carrying it into the SEM (by whatever
means) two main features must be considered: the specimen may warm up to an
unacceptable level; and it will act as an
effective condensing surface for any
appropriate molecules of sufficiently
high partial pressure. It has already
been argued that the whole specimen is
best kept below 143 K and should certainly not be allowed to rise above 173 K for
more than a brief period except during
intentional etching. In principle there
is no real difficulty in keeping the specimen adequately cold during transfer;
this may be achieved either through
sufficient thermal capacity of the cold

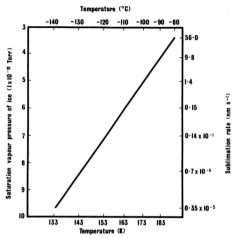

Fig. 3a Curve relating temperature, saturation vapour pressure and sublimation
rate of ice (assuming partial pressure of
water vapour in vacuum chamber is not
limiting). It will be seen that whereas
at about 170 K sublimation is relatively
slow ($\simeq$ 1.4 nm s^{-1}), even a 10 K shift
to either side results in a reduction or
increase in this rate by approximately
a factor of 10.

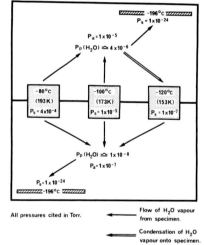

Fig. 3b Using data from Fig. 3a it is
possible to predict whether sublimation
or condensation will occur under different conditions of vacuum (partial pressure of water) and specimen temperature in
the SEM or any other vacuum chamber. The
diagram simulates the effect of placing
specimens at three different temperatures
193 K, 173 K and 153 K) into two different conditions of total pressure (P_a).
Total pressure = P_a, partial pressure =
P_p. Pressures are cited here in Torr as
this unit is probably the most familiar.
1 Torr = 1.333 millibar = 1.3 x 10^2 N
m^{-2} (Pascal). (Fig. 3 redrawn from
Robards, 1974[32]).

specimen stub or through some auxiliary
cooling system. The more difficult problem is to avoid contamination - particu-

larly with water vapour. Whether the specimen will become contaminated or not depends upon its temperature in relation to the partial pressure of the contaminating molecule. In the case of water, for example, the saturation vapour pressure at 173 K is 1×10^{-5} Torr while at 77 K it is about 1×10^{-24} Torr (Fig.3a). Within the range of these temperatures severe contamination will rapidly occur if the specimen is exposed to the normal laboratory environment. This interrelationship between specimen temperature and contamination or etching is an important one and is exemplified diagrammatically in Fig. 3b which simulates the effect of placing specimens at 3 different temperatures into two different conditions of total pressure.

In a vacuum of 1×10^{-7} Torr the partial pressure of water vapour ($P_p(H_2O)$) is approximately 1×10^{-8} Torr (own determination by mass spectrometry: this figure is extremely variable, but determinations cited here are from a well cold-trapped system and therefore represent optimistically low pressures). The saturation vapour pressure (P_S) of ice at all three temperatures is greater than the partial pressure of water in the vacuum, therefore sublimation will take place. The presence of an adjacent cold surface at 77 K, over which there will be a saturation vapour pressure for water of approximately 1×10^{-24} Torr, will lead to even more efficient trapping of water molecules. In a worse vacuum, of 1×10^{-5} Torr, the partial pressure of water is in the region of 4×10^{-6} Torr. As the saturation vapour pressure of water at 153 K is 1×10^{-7} Torr, we may expect water vapour to condense upon the specimen surface although the presence of a liquid nitrogen cooled surface will alleviate the problem, provided that it is spatially arranged so that potentially contaminating molecules are trapped before they reach the specimen. This is a grossly oversimplified picture of the interactions, but it will be clear that in conditions of poor vacuum, and/or low specimen temperature, the possibility of contamination must be foreseen and steps taken to eliminate it. The solution to the problem is to reduce the partial pressure of water (or other contaminating molecule) in the immediate environs of the specimen to a level below that of the saturation vapour pressure at that temperature. This may be accomplished in a number of ways: the specimen may be physically protected by (e.g.) immersion in liquid nitrogen during transfer; it may be protected by precluding any water vapour from approaching it by trapping such condensible molecules on a surface colder than the specimen itself; or the partial pressure of water may be reduced by transferring the specimen in an evacuated system. It is common for

two transfers to be made: firstly, from coolant to a fracturing and/or coating unit; secondly from coating unit into SEM. The first transfer is relatively uncritical with regard to contamination if the specimen is to be fractured so that 'clean' surfaces are revealed; the second transfer is always critical and the specimen must be fully protected.

Characteristic of the simple transfer devices is that described by Koch[7] where the specimen is transferred to the microscope airlock in a small 'pool' of liquid nitrogen and is surrounded by an 'atmospheric shield'. A number of workers have used a system where the specimen is protected by a 'cap' that is placed over the stub while it remains under liquid nitrogen: the cold cap then prevents water vapour from condensing on the specimen. The cap is later removed either in the microscope and/or in the fracture/coating device. Such a system was used in some of the early Cambridge work[33], as well as by other workers[11], and is incorporated into the latest Jeol system[23]. A different type of water vapour trapping transfer system has been described by Saubermann and Echlin[4] and has also been used by Hutchinson et al. [34,35]. In this type of device the cold specimen is withdrawn into a relatively long tube, the cold wall of which ensures that water molecules are trapped before they reach the specimen. Transfer is at ambient pressure with such a device and cannot be regarded as totally devoid of the possibility for contamination, particularly as the SEM and other ancillary apparatus also needs to be brought to atmospheric pressure before the transfer can be made.

More recent systems than those described above tend to give more protection to the specimen during transfer. If the system is a 'dedicated' one, then the second, critical, transfer can be made between adjacent compartments that are under vacuum and with the additional protection of cold shields: this is exemplified in the Jeol[23], Etec[24], Cameca, Philips[25], and, particularly well, AMR[12] systems. Where the fracturing and coating operations are carried out in a separate unit from the SEM, the arrangements for the second transfer become critical. For this reason many such transfer devices have valves so that the chamber can be evacuated while the specimen is carried from one position to another. Even with such a system, the vacuum provided by a rotary pump alone ($> 1 \times 10^{-3}$ Torr) will be inadequate to reduce the partial pressure of water to a sufficiently low level and additional protection of the specimen is thus usually necessary. Fuchs et al.[6] have constructed a 'transportable airlock' using a commercially available gate valve that

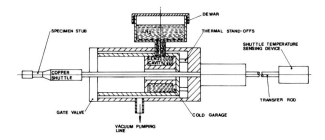

Fig. 4 Diagram of the cold vacuum transfer device designed by Cambridge Scientific Instruments. The specimen stub is
shown in the 'extended' position. When
it is retracted, the copper shuttle makes
contact with the cold garage wall - the
specimen is kept cold and protected.
(Courtesy of Cambridge Scientific Instruments).

Fig. 5 This shows the portable transfer
device shown in the diagram above with
the specimen stub fixed to the copper
shuttle which is of relatively high thermal mass. The polystyrene Dewar above
the transfer device body carries LN$_2$ for
cooling the baffle in the transfer device.
The connection for evacuating the transfer
device can also be seen. A gate valve is
housed in the transfer device body.
(Courtesy of Cambridge Scientific Instruments).

interfaces with similar valves on the SEM
and on the modified freeze-etching apparatus. When retracted into the transfer
device the specimen is completely surrounded by a liquid nitrogen cooled copper
housing, thus offering excellent protection. A device of somewhat similar principle has recently been developed by
Cambridge Scientific Instruments for use
with Cambridge S180 and S150 microscopes
(as well as the S604, S2 and S4) and is
illustrated in Figs. 4 and 5. Our own
transfer system[1] also uses an evacuated
chamber with an attached gate valve but
the stub carries its own cold shrouds
that protect the specimen during transfer.

Although the requirements of a
transfer device are well understood, and
the technical solutions are at hand, it
is important that the specimen is given
the best possible protection if contamination is to be totally eliminated.
Systems employing evacuated chambers with
additional cold shrouding of the specimen
probably offer the best likelihood of
success.

e) Specimen manipulation and fracturing

As previously stated, few specimens
can be frozen, transferred directly into
the SEM and immediately observed. It is
usually necessary to manipulate, fracture
or otherwise treat a specimen to provide
a surface of interest; and it is often
desirable to sublime away a thin layer of
ice so that underlying structures are
more clearly revealed. In some circumstances there is much to be gained from
carrying out such operations in the SEM
itself and so observing dynamic processes
as they occur but the penalty is the
likelihood of chronic contamination of
the microscope vacuum system. It is thus
preferable to perform such operations in
a separate chamber or in an ante-chamber
attached to the microscope.

Among the simplest of operations is
the crude breaking-open of a specimen or
fracturing with an elementary knife. In
either case it is necessary that the
manipulator or knife is itself pre-
cooled[8] and that the specimen is at a
temperature where it neither sublimes nor
contaminates rapidly (i.e. < 173 K). Most
described systems have embodied some form
of knife ranging from a simple scalpel on
a rod[20] to quite complex microtomes such
as used in the Balzers freeze-etching
unit[6] or, apparently, on the cryo-system
for the Philips PSEM 500[25]. Sectioning,
as such, will only be required if the
sections themselves are to be observed
and cryo-ultramicrotomy is not within the
scope of this article; in the use of bulk
specimens fracturing takes place at
temperatures below about 173 K and the
precise plane need not be determined
(and, indeed, cannot be). Very complex
knife mechanisms are thus unnecessary although it is obviously sensible to use a
cooled knife system (such as on a freeze-
etching unit) if this is available in the
laboratory. Most recent developments
have used relatively simple, but effective cold knife systems and/or simple
fracturing devices[12,23,36].

Once a suitable specimen surface has
been prepared or revealed (and it may be
an internal or external surface) it may
be required to remove some ice: either
contaminated ice on an outer surface or
ice from within the body of the frozen
tissue. Such sublimation or etching is
achieved by raising the surface temperature so that the saturation vapour pressure exceeds that of the partial pressure

of water in the surrounding vacuum (Fig. 3b). In practice (and provided that water vapour removal within the vacuum system is rapid - either through using high vacuum and/or liquid nitrogen cold shrouding) this means raising the surface temperature to about 173 K or a little warmer. This can be done in one of two ways: either by raising the temperature of the whole specimen to an appropriate level and allowing etching to occur for a known period - as in the freeze-etching technique; or by heating the surface of the specimen using a radiant heater. The advantage of the former method is that it is very controllable; the advantage of the latter is that it is quick. Many workers have opted for the radiant heat system e.g. [12,23,36], but little objective work appears to have been carried out to determine how controllable such a system is and whether the specimen <u>surface</u> temperature remains within prescribed limits (under such conditions it is very difficult to obtain true surface temperatures using thermocouples). There is much to recommend the slower, but more predictable, method of etching by controlling the whole specimen temperature. If this is to be carried out, then a temperature-controlled cold-stage and good adjacent cold-trapping surfaces are necessities.

The stage has now been reached where the frozen specimen is now almost ready for viewing in the SEM. Indeed, it may be directly viewable, but the problems of charging usually require that a conducting film is deposited.

f) Coating

Low temperature hydrated specimens develop a significant surface charge in the SEM: this can severely interfere with both observation and analysis. At low temperatures the conductivity of pure ice is only of the order of 10^{-8}ohm mm^{-1} and hence steps must be taken to conduct away the absorbed electrons so that charging cannot develop[37,38]. The nature of the coating depends in the first instance on whether the specimen is primarily for observation alone or also for analysis. If the latter, then the range of coating materials is restricted. Carbon provides a good conducting layer but its deposition can lead to substantial specimen heating[11]; aluminium and chromium have been evaporated using filament heating although the former gave inconsistent results, and consequently Marshall concluded that chromium is most satisfactory for analytical specimens[11]. While Marshall has been concerned with excessive heating during evaporation of carbon, Fuchs et al.[6] have used this element satisfactorily for their frozen specimens and Saubermann and Echlin[4] have evaporated aluminium onto their frozen sections. If specimens are not for analysis, then gold, or a gold alloy, are usually the chosen evaporants.

Most authors have evaporated conducting films onto frozen specimens by resistive heating: this appears empirically satisfactory, but the known advantages of sputter coating do not appear to have been fully explored so far as low temperature SEM specimens are concerned. The reason why sputter coating has been avoided seems in part to be because it operates in a rough vacuum (0.1 Torr) - with consequential problems of increased probability of contamination, and because the possibility of heating damage during coating may appear even more acute where low temperature specimens are concerned. However, in principle, sputter coating has much to offer in terms of improved conducting layers and lack of necessity to change evaporant between runs. Further there is no need to move the specimen during coating to ensure deposition of an even layer as there is with evaporative techniques; low vacuum need not automatically mean high partial pressure of water; and a specimen at very low temperature may well be able to stand a slight heat gain better than a sensitive specimen at ambient temperatures where changes of as little as 10 K have been argued about. One of the contributing problems is, again, the difficulty of measuring precise temperatures at the specimen surface during the coating process. Sputter coating has certainly been demonstrated by K. Oates at the university of Lancaster to be highly effective in conjunction with low temperature SEM, and is already installed as an integral component of more than one commercially available system[23,25]. So far there is no incontrovertible evidence suggesting that sputter coating is unsuitable for frozen specimens.

Once the specimen has been coated it is ready for viewing and analysis: it can now be transferred to the cold stage of the SEM.

g) Cold stages

SEM cold stages have ranged from the extremely simple[13] to relatively complicated; in some cases a true cold stage has been dispensed with altogether, the specimen merely being inserted into the SEM on a stub of high thermal capacity[39]. Some of the earliest stages were derived from equipment designed for heating and cooling specimens and were not necessarily well suited to biological SEM work[15]. When considering the requirements within the SEM itself it is not sufficient to confine attention to stage temperature alone - the whole specimen environment and positioning is important.

Many workers have attempted to obtain the lowest possible stage temperature - close to the temperature of liquid nitrogen. This can be achieved if the nitrogen is fed directly into the stage or if the Joule-Thomson system employed in the AMR system[12] is used. In this latter system dry nitrogen gas under high pressure (55kPa - 1500 p.s.i.) is fed to the refrigerator (specimen stage) where it is allowed to expand in a counter-current heat exchanger (Fig. 1). In both of these systems some vibration arising from the direct circulation of liquid or gas through the cold stage might be thought to act as a limit to resolution although both have been shown to yield good results. As emphasised by Pawley et al.[26], further work needs to be undertaken on the ultimate limitations to resolution on low temperature SEM specimens. Most other systems have extracted heat from the cold stage by some form of high conductivity metal 'braid' system. It is important to keep the number of mechanical junctions in the conduction pathway to a minimum as each junction can easily lead to a temperature difference of 10-20 K[27]; all mechanical junctions should also be lapped and/or polished to ensure the best possible fit of components and hence most efficient heat transfer. Using such an arrangement it is not difficult to cool the stage to about 100 K (Table 1) - a temperature that is, in our opinion, quite low enough for most biological work. At 100 K a well-fitting specimen stub should not be more than 10-20 K warmer and the loss across the specimen/stub interface is usually put at about 2 - 10 K. Thus, under worst conditions, the difference between stage and specimen temperature would be 30 K with the actual specimen at about 130 K - a perfectly satisfactory temperature for observation and analysis. Indeed lower temperatures encourage contamination and reduce any temperature differential between the specimen and cold plate anti-contaminators. The cold stage must be thermally insulated from the rest of the SEM (while allowing electrical coupling to avoid charging); it must have X, Y and Z movements unrestricted; it should be possible to tilt it and, ideally, some rotation should be available. These requirements mean that the braided connection must be reasonably long and flexible while retaining a relatively large cross-sectional area for heat conduction. The original Cambridge Series 200 stage was based on this principle, and has been suitably modified for biological use[4,6]; Taylor and Burgess[27] have provided a very detailed description of the construction of a transmission cold stage for the Jeol JXA 50A microanalyser; and Cambridge Scientific Instruments have recently constructed a new stage for Cambridge S180 and S150 microscopes (Table 1; Figs. 6, 7). It is not essential, but useful, to have a stage heater in the SEM: this allows specimen temperature to be raised.

Figs. 6 and 7 Two views of the new Cambridge cold stage.
Fig. 6 shows the rear view of the specimen stage/tranfer device assembly as for transferring a specimen. The rod carrying the copper shuttle and specimen stub is passed through the stage airlock to either locate or retrieve the specimen stub from the cold platform. For performing microanalysis, a standard block containing up to 7, 3mm diameter microanalysis standards replaces 2 of the specimen stubs. The copper cooling braid is clearly seen.
Fig. 7 shows the stage movements, the gravity feed liquid nitrogen (LN$_2$) Dewar, LN$_2$ feed/return pipes and copper braid assembly.

The specimen platform is shown tilted and carrying 3 specimen stubs. The optics for viewing the specimen stub during transfer can clearly be seen.

Simultaneous secondary and transmission images of thin sections can be obtained with this system when used in conjunction with the Stereoscan S180 and S150 B.(Figs.6 & 7 courtesy of C.S.I.).

If necessary, a proportional controller can be used to maintain a preset specimen temperature automatically by switching a stage heater or by actuating valves in the coolant supply line. Direct supply of coolant to the stage gives the most rapid temperature drop while the Joule-Thomson system also has excellent characteristics in this respect (Fig. 8). Cooling by long distance conduction is inevitably slower (Fig. 9) although this is not usually a significant limitation to the operation of the system. Mechanical (apart from thermal) aspects of stage design will be determined according to whether or not the stage is to be used for transmission work; whether multiple specimen loading is required; whether X-ray standards and a Faraday cage are needed; and other, similar, considerations. In general it is useful to have a stage that is large enough for modification to accommodate different requirements.

The vacuum immediately around the cold specimen in the microscope is of the utmost importance. In most microscopes it can be assumed that specimens colder than about 140 K will contaminate rather rapidly (Fig. 3b). It is usual, therefore, to fit an anticontaminator plate close to the specimen. This plate is cooled to as low a temperature as possible (usually colder than about 100 K) and must maintain a differential of at least 10 K (and normally much more) with the specimen to act effectively. Some cold stages (including commercial stages) are used without anticontaminator plates: this entails some risk. For example, a system that can cool the specimen to <110 K and for which it is said that "the clean high vacuum in the main specimen chamber ensures against condensation of contaminants on the specimen surface. A special anti-contamination device is not necessary"[25], implies that the partial pressure of water in the vacuum system must not exceed about 1×10^{-17} Torr. In a vacuum contaminated by the insertion of a frozen specimen the residual gas may well be as much as 50% H_2O, in which case the total pressure would need to be far lower than is likely to prevail - even in a clean microscope system. In general, cold shields are a simple and effective means of avoiding this problem.

Now that the general criteria for low temperature biological SEM have been considered in some detail and some of the available solutions have been briefly reviewed, we can introduce the latest stage in the development of our own system.

A new, comprehensive system for low temperature SEM

a) Introduction
In a previous paper[1] we presented a system for the freezing and protected transfer of a biological specimen into a Cambridge S600 SEM. This apparatus has now been further developed so that specimens can be frozen under controlled conditions, transferred onto a cold table in a fracturing/coating chamber, and then transferred onto a controlled temperature stage in the SEM. From the information contained in the previous section some of the design criteria will be apparent: some are mandatory; some are matters of philosophy; and others are matters of personal, subjective, choice. For our system the following criteria have been adopted.

i) After freezing the temperature of the specimen should never during any subsequent process unintentionally rise above 140 K.

ii) The preparation chamber in which fracturing and deposition of a conducting film take place should be separate from the microscope itself.

iii) A separate chamber should be available for freezing specimens in subcooled nitrogen.

iv) Transfer between freezing chamber, fracturing/coating chamber and SEM should be by evacuated transfer device with additional cold specimen shrouding.

v) The fracturing/coating chamber should allow specimens to be observed with a dissecting microscope while they are fractured or otherwise manipulated.

vi) The cold table in this chamber should be controllable in the range 173 - 108 K. The chamber should be evacuated using a rotary pump. The cold table should be enshrouded with a cold anticontaminator which remains colder than the specimen at all operational temperatures.

vii) Both evaporative and sputter coating options should be available.

viii) A simple cold stage should be constructed for the SEM; cooling should be via a copper braid. This stage should be controllable in the range 173 - 108 K.

ix) There should be an anticontaminator plate in the SEM that at all operational times remains colder than the specimen.

x) While the present system should interface with the existing Cambridge S600, provision should be made for use with other microscopes.

xi) The system should embody sufficient flexibility so that a wide range of specimens and techniques can be accommodated. Hence, large bore gate valves, large cold stages and a roomy fracturing/coating chamber are desirable. Although the present system allows scanning reflection only, subsequent development should allow scanning transmission.

The following sections deal with the different components of this system and discuss how the design criteria have been fulfilled in practice.

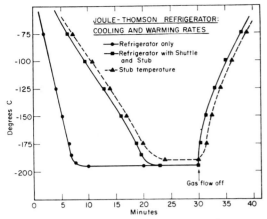

Fig. 8 Thermal performance of the Joule-Thomson refrigerator used on the AMR-1000A SEM Bio-Chamber system[12]. (Courtesy Dr J. Pawley).

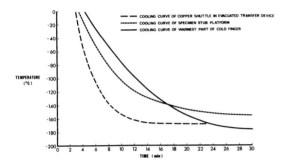

Fig. 9 Thermal performance of new Cambridge cold vacuum transfer device, SEM cold stage and cold finger. (Courtesy of Cambridge Scientific Instruments).

Fig. 10 Copper stub for York system described in this paper (side of stub measures 10 mm). The stub is shown attached to the nylon tip of the transfer rod and is equipped with the self-shrouding stainless steel 'ears'.

b) Specimen stub

The present system uses a 'self-shrouding' stub as described previously[1] and illustrated in Fig. 10. It is constructed from copper with overlapping stainless steel shrouds. When the stub is upright the shrouds automatically open; when inverted they close and so protect the specimen. During transfer the stub is always inverted so that the shrouds are closed - they are then opened by inverting the stub immediately prior to placing it on the cold stage in the SEM or fracture/coating chamber. The stub has purposely been made large - there appears to be no adverse reason - and this allows a variety of specimens and mounting/fracturing techniques to be adopted. The stub can be screwed onto either the freezing rod or the transfer rod.

c) Freezing chamber

The freezing chamber is exactly as described previously[1]. The base of the stub is screwed onto the vertical freezing rod so that it can be plunged, specimen first, into the subcooled nitrogen. When the mass of the stub has reached the temperature of liquid nitrogen (60s) the shrouds are closed by manipulation against a horizontal peg below the surface of the liquid nitrogen. The shrouded stub is then withdrawn from the cryogen, picked up on the horizontal transfer rod, released from the vertical freezing rod, and withdrawn into the transfer device. The transfer device gate valve is closed and the chamber can be evacuated using the fore-line trapped rotary pump.

d) Transfer device

This, also, is as described in the previous paper[1] (Fig. 11). A three-way vacuum valve allows either the transfer chamber, the 'dead-volume' between two adjacent gate valves, or, in the case of the freezing chamber, the chamber itself to be evacuated.

e) Fracturing/coating chamber

This component has not previously been described. It comprises a stainless steel box approximately 200 mm x 160 mm x 150 mm high with a gate valve that interfaces with the gate valve of the transfer device at one end (Figs. 12, 13). At the other end is a vacuum-walled liquid nitrogen container through the base of which is brazed a 20 mm diameter copper rod. Copper braid is clamped to this rod at one end and to the cold table at the other. The nature of the braid and its clamping is critical to the success of the cooling efficiency. After considerable experiment it was found that 'stranded' (rather than 'braided') copper wire with 100 wires, each 0.05 mm thick, in each strand gave sufficient flexibility and heat conduction when 250 such strands were used to couple the copper rod to the cold stage. (Lacquered wires must be avoided). The stranded wires were tightly clamped onto either the cold table or the clamp for the rod and were then secured in position by a small quantity of solder. This system has resulted in temperatures lower than 100 K on both cold table and SEM cold stage.

A copper shroud is positioned over the whole of the cold table with apertures in it for manipulation and for coating. This shroud is clamped directly to the end of the rod from the liquid nitrogen dewar. At all times it remains as cold as, or colder than, the table (Fig. 14). The table has a T-slot along its length (Fig. 12) in which the stub can be moved using the transfer rod. It is mounted on the base of the chamber by four nylon spacers. A resistance heater in the table allows the temperature to be raised if, for example, etching is required; the differential between table (specimen) and shroud temperature increases as the table temperature rises (Fig. 14).

The fracturing/coating chamber is sealed with a perspex (plexiglas) lid carrying an O-ring. Two manipulators on universal lead-throughs pass through the lid. These may be adapted to serve a number of different manipulating or fracturing processes. Their functional ends are cooled by placing them in suitable recesses located in the cold shroud.

A further aperture in the lid allows either a sputter head or an evaporative head to be positioned over the specimen. The special sputtering head is operated using an Emscope control system to sputter gold in an atmosphere of argon that is fed in through a narrow bore pipe in the table, so that the argon is at table temperature as it passes over the specimen.

It will be appreciated that, if sputter coating is routinely used, there is no necessity for this chamber to be brought to atmospheric pressure other than for maintenance or cleaning. This is critical to the proper operation of the system. The fracture/coating chamber is scrupulously cleaned before assembly and is then only pumped through a foreline trap of activated alumina. Purging is always with argon gas which, during sputtering, is pre-cooled. These precautions ensure that; i) contamination of the specimen in this chamber is minimal; ii) sputter coating of frozen specimens can be effected without apparent heat damage or other adverse results. As in other aspects of low temperature SEM, successful results are as much a matter of good housekeeping as of good equpment. The specimen is now once again picked up onto the transfer rod and is moved into the SEM.

f) Cold stage

The dewar and attachments for the SEM are made in the same way as for the fracturing chamber. The same amount of stranded copper connects the dewar rod to the cold stage which is positioned on a nylon pallet compatible with the normal Cambridge stage (Fig. 15); an anticontaminator plate is situated over the specimen and attached directly to the rod. The stage contains a resistance heater which allows temperature to be controlled in the range 100 - 170 K (and higher, although we have carried out no experiments above 170 K yet). The relationship between stage heating and cold plate temperature is illustrated in Fig. 16.

The S600 poses a number of particularly difficult problems for low temperature SEM: there is only one spare, available port for cooling (assuming an X-ray detector is in use); and tilt and rotate motions are removed with the standard door and are hence lost for the cold stage. We have therefore had to cool both anticontaminator and stage from a single source, but this appears to have been accomplished quite successfully. Our inability to orientate the specimen is a serious one although acceptable results have nonetheless been obtained. However, this is a limitation of the microscope rather than the system. Our low temperature equipment can easily be made compatible with other microscopes where the major specimen movements can be retained.

g) General

The freezing and fracturing chamber are mounted on the top of a trolley containing all the control equipment; rotary pump; and transfer device. The trolley can be located close to the SEM so that the transfer device (with its permanent pumping line) can easily be changed from station to station.

h) Performance and results

We have still had inadequate time to explore the full potential of our system. However, some basic data and preliminary biological results are presented.

Once the specimen has been frozen (Table 2) it is at continuous risk from sublimation and/or contamination. It is extremely difficult to follow the actual specimen temperature through all stages of the process, but we have attempted to do this and present some results in Fig. 17. This curve was obtained by placing a thermocouple in the specimen position on top of the copper stub (the worst - warmest - warmest position). Transfer times are considerably longer than normal as the trailing thermocouple cables to the specimen had to be retracted into the transfer device and then led out again with the specimen.

Fig. 11 Freezing chamber together with attached transfer device of York low temperature system. The specimen and stub are frozen in subcooled liquid nitrogen on the vertical rod. The stub is then shrouded and picked up on the transfer rod, withdrawn into the transfer device, which is evacuated and uncoupled, and carried under vacuum to the fracture/ coating unit.

Fig. 12 Top view of fracture/coating chamber (lid and cold shroud removed). The specimen stub is seen in the coating position on the cold table and is attached to the transfer rod. The cold argon inlet can be seen adjacent to the stub. The cold shroud, when in position envelops the whole of the cold table and has apertures in the manipulation and coating positions.

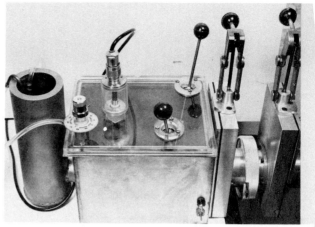

Fig. 13 External view of fracture/ coating chamber. The gate valves of the transfer device and fracture/coating chamber are coupled together. Two manipulators on ball lead-throughs pass through the perspex lid as do the sputtering head and the argon gas inlet. Under normal operational conditions this chamber is never brought to atmospheric pressure and the only gas entering it is cold argon.

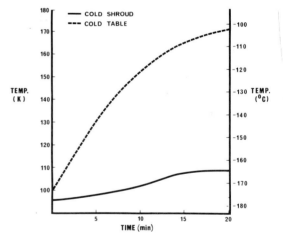

Fig. 14 Warming curves for cold table & cold plate in fracture/coating chamber. A good temperature differential is maintained between table and cold plate at all temperatures despite cooling from a common source.

Therefore Fig. 17 represents the poorest conditions for transfer that are likely to be encountered. There is a rapid warm-up of the specimen during the first transfer (2 minutes here, but usually only about half this time), but this is abruptly checked and reversed when the stub is placed on the cold table in the fracture/coating unit. Fracturing, etc., can take place at a temperature of 120 K or lower, and two minutes sputter coating has no visible effect (although using a

similar measuring system at ambient temperature small changes were readily detected[40].) The second transfer again inevitably allows some warming but equilibration on the stage (in this experiment) with a _specimen_ temperature of about 140 K is most satisfactory. From the instrumental evidence we have no reason to suppose that the specimen has either sublimed or contaminated during the run. (Note the decontaminator temperatures in relation to specimen

Fig. 15 Cold stage with anticontaminator and liquid nitrogen dewar for Cambridge S600 SEM. The clamping arrangement for the cooling connections is illustrated and allows cool-down of both components to < 100 K. (This illustration shows the prototype dewar with 12.5 mm copper rod; all later dewars have used 19.1 mm ∅ rods).

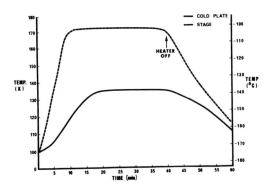

Fig. 16 Warming/cooling curves for cold stage and anticontaminatior in the SEM. It will be seen that (for example) a constant temperature of 173 K can be selected and maintained while keeping a good differential with the anticontaminator. Switching off the heater leads to steady recooling. In fact, the stage would not normally be brought so warm as the specimen temperature will be some degrees warmer than the stage itself.

temperature; when not protected by the decontaminator the specimen is protected by the stub shrouds.) This supposition is supported by the fact that frozen specimens have been viewed over periods as long as 8 hours in the SEM without any apparent change in appearance, yet warming above 170 K led to immediate, and obvious, sublimation of ice.

Some typical specimens are illustrated in Figs. 18 - 24. These show that

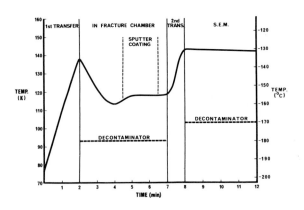

Fig. 17 Specimen temperature profile during a typical run. Final temperature stable in SEM at 143 K - specimen never warmer than this during transfer and processing. Owing to the need for thermocouples to be attached to the stub during the run, transfers were slower than normal. The first transfer usually takes approximately 1.0 minute. During transfer the specimen is protected by the cold shrouds; it is protected by the cold decontaminators in the fracture/coating chamber and in the SEM.

good results can be obtained with the system described. Still better work will come about as we gain more experience with the apparatus and, particularly, when it is combined with an SEM of better resolution and allowing a full range of specimen movements. For the present, however, we have demonstrated the effective operation of our own new low temperature preparation system.

Conclusions

This paper has sought to review briefly some of the systems described for low temperature scanning microscopy; to present our own most recent work; and to discuss some of the prerequisites for work in this area. At the present time many unknown factors remain. For example: what is the optimum stable temperature for viewing and analysis in a particular system so that neither sublimation nor contamination occur? How wide is the acceptable temperature range for viewing? How cold do stages really need to get? What is the best coating for low temperature specimens and how should it be applied? What limits attainable resolution in low temperature specimens? These, and many other questions remain to be answered. It is only by the development of suitable equipment and, importantly, by the use of different approaches to the solution of the many problems, that the best course of action can be determined. Our own system has been designed with this point very much in mind and has considerable scope for allowing further experiments to be carried out to determine whether the technique can still be improved and augmented. Finally, it must be stressed that, no matter what apparatus is available, no work in low temperature microscopy can be successful unless the most scrupulous attention is paid to cleanliness and operation within the prescibed parameters of vacuum and temperature.

Acknowledgements

The first acknowledgement goes to the many authors whose work has not been mentioned in this paper; in the main we have attempted to present a cross-section of the relatively recent developments and have not had the space to be comprehensive.

We are most grateful to those workers who have responded to our request for information at short notice and, in particular, to Mr. P.A. Jennings of Cambridge Scientific Instruments, for most helpful discussions and for allowing us to present some of his work carried out on behalf of CSI in this paper. Thanks are also due to Mr K. Oates of the University of Lancaster who has kept us informed of his innovative and successful work in the field of low temperature SEM.

References

[1] Robards, A.W. and Crosby, P. A transfer system for low temperature scanning electron microscopy. SEM/1978/II, SEM. Inc., AMF O'Hare, IL, 60666, 927-936.

[2] Echlin, P.E. Low temperature scanning electron microscopy: a review. J. Microscopy 112, 1978, 47-61.

[3] Echlin, P. Low-temperature biological scanning electron microscopy. In: Advanced Techniques in Biological Electron Microscopy II, Specific Ultrastructural Probes. Springer-Verlag, Berlin, ed. J.K. Koehler, 1978, 89-122.

[4] Saubermann, A.J. and Echlin, P. The preparation, examination and analysis of frozen hydrated tissue sections by SEM and X-ray microanalysis. J.Microscopy 105, 1975, 155-192.

[5] Fuchs, W. and Lindemann, B. Electron beam X-ray micro analysis of frozen biological bulk specimens below 130 K. I. Instrumentation and specimen preparation. J.Microsc.Biol.Cell. 22, 1975, 227-232.

[6] Fuchs, W., Lindemann, B. and Brombach, J.D. Instrumentation and specimen preparation for electron beam X-ray microanalysis of frozen hydrated bulk specimens. J.Microscopy 112, 1978, 75-87.

[7] Koch, G.R. Preparation and examination of specimens at low temperatures.

In: Principles and Techniques of Scanning Electron Microscopy. 4, Van Nostrand Reinhold, New York, ed. M.A. Hayat, 1975, 1-33.

[8] Hayes, T.L. and Koch, G. Some problems associated with low temperature micromanipulation in the scanning electron microscope. SEM/1975, IIT Research Institute, Chicago, IL, 60616, 35-42.

[9] Nei, T. Cryotechniques. In: Principles and Techniques of Scanning Electron Microscopy. 1, Van Nostrand Reinhold, New York ed. M.A. Hayat, 1974, 113-124.

[10] Nei, T. and Fujikawa, S. Freeze-drying process of biological specimens observed with a scanning electron microscope. J.Microscopy 111, 1977, 137-142.

[11] Marshall, A.T. Electron probe X-ray microanalysis of frozen-hydrated biological specimens. Microscopica Acta 79, 1977, 254-266.

[12] Pawley, J.B. and Norton, J.T. A chamber attached to the SEM for fracturing and coating frozen biological samples. J.Microscopy 112, 1978, 169-182.

[13] McAlear, J.H., Fucci, R. and Germinario, L. A simple cold stage for the scanning electron microscope. Exptl. Cell Res. 71, 1972, 235-238.

[14] Kynaston, D. and Paden, R.S. Advances in techniques and instrumentation for the stereoscan. Cambridge Stereoscan Colloquium, 1970, 109-112.

[15] Echlin, P., Paden, R. and Donzek, B. Scanning electron microscopy of labile biological material under controlled conditions. SEM/1970, IIT Research Institute, Chicago, IL, 60616, 51-64.

[16] Nei, T., Yotsumotu, H. and Hasegawa, H. Direct observation of frozen specimens with a scanning electron microscope. J.Elect.Microsc. 20, 1971, 202.

[17] Otaka, T. and Honjo, S. A new freeze-dry technique for preparation of marine biological specimens for SEM. SEM/1972, IIT Research Institute, Chicago, IL, 60616, 359-363.

*[18] Gullasch, J. Biosem-em optimates Konzept für die Analyse von Mikrostrukturen in der Biologie. In: 8th Int. Congress on Elect. Microsc. II, Canberra, 1974, 658.

*[19] Lechner, G. Experiences with a cooled transfer stage between the cryo-ultramicrotome and the SEM. In: 8th Int. Congress on Elect. Microsc. II, Canberra, 1974, 58.

[20] Hasegawa, Y., Hasegawa, M. and Auzuki, T. Soft tissue observation by cryoscan fitted with vacuum evaporating device. JEOL News, 12e, 1974, 26.

*[21] Aldrian, A. and Zedlacher, H. Transport empfindlicher elektronenmikroskopischer Proben unter Vermeidung von Kontamination. In: 8th Int. Congress on Elect. Microsc. II, Canberra, 1974.

[22] Geymayer, W. Electron microscopic observations on gels with the aid of a modified freeze-etching technique. J.Polymer Sci., Symposium 44, 1974, 25-34.

[23]Jeol instruction pamphlet for Cryoscan 35-CRU2, obtainable from JEOL Ltd., Tokyo, Japan.

[24]Etec 'Biosem' pamphlet, obtainable from Etec Corporation, Hayward, California.

[25]Philips 'Cryosystem for PSEM 500' pamphlet, obtainable from Philips, Eindhoven, The Netherlands.

[26]Pawley, J., Hayes, T.L. and Hook, G. Preliminary studies of coated complementary freeze-fractured yeast membranes viewed directly in the SEM. SEM/1978/II, SEM. Inc., AMF O'Hare, IL, 60666,683-690.

[27]Taylor, P.G. and Burgess, A. Cold stage for electron probe microanalyser. J.Microscopy 111, 1977, 51-64.

[28]Costello, M.J. and Corless, J.M. The direct measurement of temperature changes within freeze-fracture specimens during rapid quenching in liquid coolants. J.Microscopy 112, 1978, 17-37.

[29]Bald, W.B. and Robards, A.W. A device for the rapid freezing of biological specimens under precisely controlled and reproducible conditions. J.Microscopy 112, 1978, 3-15.

[30]Bald, W.B. A proposed method for specifying the temperature history of cells during the rapid cool-down of plant specimens. J.exp.Bot. 26, 1975, 103-119.

[31]Dempsey, G.P. and Bullivant, S. A copper block method for freezing non-cryoprotected tissue to produce ice-crystal free regions for electron microscopy. J.Microscopy 106, 1976, 251-271.

[32]Robards, A.W. Ultrastructural methods for looking at frozen cells. Sci.Prog. (Oxf.), 61, 1974, 1-40.

[33]Echlin, P. and Moreton, R. The preparation, coating and examination of frozen biological materials in the SEM. SEM/1973 , IIT Research Institute, Chicago, IL, 60616, 325-332.

[34]Hutchinson, T.E., Bacaner, M. and Broadhurst, J. Instrumentation for direct microscopic elemental analysis of frozen biological tissue. Rev.Sci. Instrum. 45, 1974, 252-255.

[35]Hutchinson, T.E. Energy dispersive X-ray microanalysis. In: Analytical and quantitative Methods in Microscopy. (Eds. G.A. Meek and H.Y. Elder) Cambridge University Press, Cambridge, 1977,214-226.

[36]Echlin, P. and Burgess, A. Cryofracturing and low temperature scanning electron microscopy of plant material. In: Proceedings of the Workshop on Biological Specimen Preparation Techniques. IIT Research Institute, Chicago, IL, 60616, SEM/1977/I, 491-500.

[37]Marshall, A.T. X-ray microanalysis of frozen hydrated biological specimens: the effect of charging. Micron 5, 1975, 275-280.

[38]Brombach, J.D. Electron-beam X-ray microanalysis of frozen biological bulk specimens below 130 K. II. The electrical charging of the sample in quantitative analysis. J.Microsc.Biol. Cell 22, 1975, 233-238.

[39]Turner, R.H. and Smith, C.B. A simple technique for examining fresh, frozen, biological specimens in the scanning electron microscope. J.Microscopy 102, 1974, 209-214.

[40]Robards, A.W. An introduction to techniques for scanning electron microscopy of plant cells. In: Electron Microscopy and Cytochemistry of plant cells. (ed. J.L. Hall), Elsevier/North Holland, Amsterdam, 1978, 343-415.

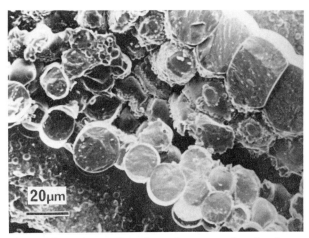

Fig. 18 Cross fractured grass leaf at temperature of approximately 140 K, coated with sputtered gold. Note ice fracture ridges in cells at top right and bottom left and empty intercellular spaces.

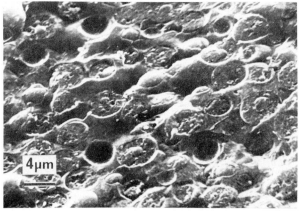

Fig. 19 Cross fractured yeast suspension prepared as Fig. 18. Although instrumental resolution is not good, the clear fracture plane and lack of etching are evident.

*Published by Australian Academy of Sci. Canberra, Auatralia

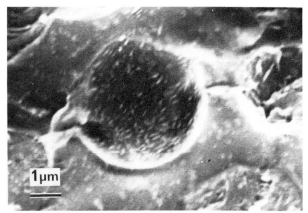

Fig. 20 High magnification image of yeast plasmalemma fracture face. The characteristic ridges are easily observable but resolution is inadequate to demonstrate further detail. This S600 micrograph compares favourably with those from better resolution instruments and augers well for use of the sputter-cryo system on a higher resolution SEM.

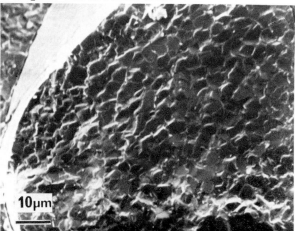

Fig. 21 Cross fractured rat skeletal muscle prepared as previous micrographs.

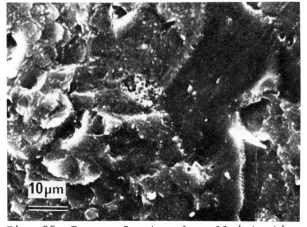

Fig. 22 Freeze fractured small intestine of rat. The brush border, lumen and a goblet cell are all obvious.

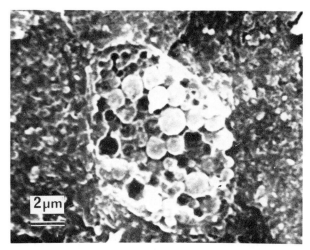

Fig. 23 High magnification view of the goblet cell in Fig. 22. Some of the vesicles have been fractured out of the cell leaving marked depressions.

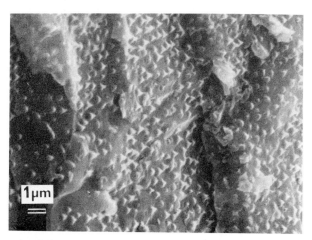

Fig. 24 Freeze-fractured 0.5 µm polystyrene latex particles. This preparation was viewed and recorded using an ISI System 100 SEM to which a commercial version (Emscope Laboratories Ltd.) of our preparation equipment had been attached.

See page 344 and 324 for discussion with reviewers.

Discussion with Reviewers

J. Pawley: While your calculations showing the cooling rate to be unaffected by size of mounting stub may be accurate for the few microns nearest the surface of a thin sample, you point out that few samples are viewed without fracturing. It is our experience that the rapid evolution of bubbles which occurs as a large metal object is put into a liquid cryogen seriously reduces the cooling rate of any object in the bubble stream and that often even the outside surface is not rapidly frozen let alone the inside revealed by fracturing. Could you comment?

Authors: The data presented in Table 2 are not calculations but actual cooling rates measured using the device described by Bald and Robards[29]. Dr Pawley's comments concerning the evolution of gas bubbles are certainly true for liquid nitrogen at its boiling point (77 K) at atmospheric pressure; they are not, in our experience, relevant to the situation that applies in our system where the specimen is plunged into subcooled nitrogen at 63 K. Under these circumstances there is a significant time of rapid cooling, without gas evolution before the enthalpy of the specimen has brought the nitrogen back to its equilibrium temperature at atmospheric pressure. The point to be drawn from Table 2 is that, whatever the _maximum_ cooling rate measured by a bare thermocouple, the cooling rate (273-223 K) in the centre of specimens of significant size is little different. We have not noticed, either in low temperature SEM or freeze-fracture preparations, poor surface cooling due to gas bubbles so long as we have frozen in subcooled nitrogen.

J. Pawley: Why did you choose PMMA (Lucite) for the lid of the York chamber given that this material is known to outgas hydrocarbons at a very high rate?

Authors: We are aware that many plastics including perspex, outgas. This is a comparative phenomenon and the degree of importance relates to the construction of the system and what is expected of it, for example, in UHV work, adsorbed water is itself often the limiting factor. Our fracturing/coating chamber operates at low vacuum and the specimen is well protected by cold shrouding: we have no evidence that any outgassing from the methacrylate lid causes problems of contamination. We are currently carrying out stringent tests to investigate whether there is contamination from _any_ source during our preparation process, but such tests may more readily be made using the better resolution of freeze-replication techniques. If we had used

a high vacuum in the fracture/coating chamber we would probably have constructed a metal lid with a glass viewing window: we do not believe this to be necessary.

T.C. Carlysle: Explain the 'Differential Increase' between the table (specimen) and shroud temperature as the table temperature rises.

Authors: This relates entirely to the heat conduction pathways between the cold shroud and the cold table. The former is always in good, direct, contact with the cold copper rod from the liquid nitrogen dewar; the latter contains a heating element which, when energised, produces a temperature gradient between the cold table and the copper rod. In this way the temperature of the shroud is maintained within narrow limits irrespective of the temperature of the heated part of the cold stage.

T.C. Carlysle: Do the functional ends of the micromanipulators remain sufficiently cold for the duration of a prolonged fracturing process or must they be re-recessed periodically for rechilling?

Authors: We have generally used rather simple, quick, fracturing procedures for which the pre-cooled manipulators are perfectly adequate. By making the manipulator ends relatively large, and of a metal of adequate thermal capacity such as stainless steel, the functional tips - whether needle-like or a scalpel blade - should stay sufficiently cold over long periods. This subject has been well discussed by Hayes and Koch[8]. We are contemplating the possibility of constructing a permanently cooled knife mechanism for fracturing but, in common with the situation in freeze-etching, do not feel that this is necessary for most preparations.

T.C. Carlysle: When the 2 element stainless steel shroud is opened to reveal the specimen is there any vibrational upset to the specimen?

Authors: The two ear-like shrouds on the specimen stub open and close quite gently and have never, in our experience, caused any damage to the specimen from vibration. Specimens for low temperature work must, in any case, be firmly attached to the stub and the opening and closing of the shrouds produces minimal disturbance.

Discussion continued on p. 324.

SCANNING ELECTRON MICROSCOPY/1979/II
SEM Inc., AMF O'Hare, IL 60666, USA

SCANNING ELECTRON MICROSCOPY OF BIOLOGICAL SPECIMENS SURFACE-ETCHED BY AN OXYGEN PLASMA

W. J. Humphreys, W. G. Henk, and D. B. Chandler

Central Electron Microscopy Laboratory and
    Department of Zoology
University of Georgia
Athens, Georgia 30602

Abstract

Tissues double fixed in glutaraldehyde and osmium tetroxide and embedded in epoxy resin were cut into sections 1 μm thick. The sections were surface-etched in a reaction chamber containing an oxygen plasma produced by exciting oxygen with a radio frequency generator. The gaseous mixture produced was highly reactive chemically causing organic bonds to be destroyed and organic material to be removed from the sample as volatile products that were pulled out with a vacuum pump. Etch-resistant residues (and possibly organo-metallic complexes) of membranes and other cell structures emerged as 3-dimensional skeletons projecting above the surrounding plastic embedment--the level of which was lowered due to its combustion and removed as volatile products. This occurs at a relatively low temperature, probably near 100°C. The depth of etching was controlled by varying the etching time. The etched specimens were metal coated and viewed in the SEM. Structural features of cell constituents that were clearly recognizable from etch patterns formed from the etch-resistant residues of structures include: profiles of plasma membranes, cytoplasmic membranes, membranes of the nuclear envelope, heterochromatin and profiles of mitochondria showing cristae and the outer limiting membranes. In the cortex of Paramecia, residues of cilia, kinetosomes, kinetodesmal fibers and trichocysts were identified. In etched thick sections of bronchial epithelium residues of microvilli were identifiable as were the residues originating from microtubule doublets of cilia and from the central pair of microtubules of the cilia.

KEY WORDS: Surface-Etching, Low Temperature Ashing, Plasma Membranes, Cytoplasmic Membranes, Oxygen Plasma, Specimen Preparation

Introduction

Numerous methods have been tried for visualizing cell organelles and their internal structural details with the scanning electron microscope (SEM) using the standard secondary electron mode. A few of the more successful methods include those used by Tanaka and Iino[1] and by Woods and Ledbetter[2]. These investigators were able to identify organelles inside cells that had been infiltrated with epoxy resin, frozen, fractured and after removal of the resin by solvents, critical point dried. The technique of freezing tissues in certain nonpolar intermediate fluids that are used for critical point drying (e.g. ethanol, acetone, Freons) followed by cryofracturing, thawing the fractured tissue in the intermediate fluid, and critical point drying it[3,4,5,6] has become a valuable technique that has provided a new and very useful way of looking inside tissues with the scanning electron microscope. Myklebust et al.[7] studied the ultrastructure of heart muscle by scanning electron microscopy of deparaffinized, critical point dried thick sections of paraffin-embedded tissue. Thick sections of tissue embedded in epoxy[8] or the smooth face of a block from which sections of embedded tissue had been cut[9] have been etched chemically with epoxy solvents to cause cell organelles such as mitochondria to appear as rounded elevations when viewed with the SEM. In none of these studies was the internal structure of organelles such as the cristae of mitochondria clearly visible. Haggis et al.[10] were able to visualize mitochondrial cristae and nuclear chromatin using the scanning electron microscope. Their specimens were freeze-fractured from an aqueous solution containing a cryoprotectant, thawed and fixed in a solution of glutaraldehyde containing a cryoprotectant, and then critical point dried.

In order to make identifications of cytoplasmic organelles unmistakable at fractured or cut surfaces of tissues, a method was sought which would etch away enough cytoplasmic matrix, mitochondrial matrix, or other matrices both inside and surrounding organellar structures so that stabilized etch-resistant membranes (and other etch-resistant constituents) would project above the surface far enough to be imaged by the SEM. It was hoped that, as a consequence of the etching, positive identification of many cytoplasmic organelles could be made by virtue of the membrane patterns or other structural patterns that would emerge.

One chemical agent that can be used for surface-etching thick sections of plastic-embedded tissue is an oxygen plasma. If radio-frequency electrical discharges are passed through a gentle stream of pure oxygen gas while it is being drawn past the specimens by a mechanical vacuum pump connected to the reaction chamber and if the oxygen is admitted at a controlled flow rate to attain an operating pressure of from 0.05 to 1.0 torr, the gas becomes highly reactive chemically. It forms a plasma that destroys organic bonds and organic material is removed from the specimen as volatile products that are swept out by the oxygen stream. Inorganic residues remain on the surface very near their position in the unreacted sample. The resulting patterns remaining on the etched surface can then be examined by SEM. This was suggested to us as a method for preparing specimens for SEM from previous studies in which excited oxygen was used for low temperature ultra-microincineration of biological specimens that were to be studied by TEM.

Fabergé[11] mechanically polished block faces of permanganate-fixed, polymer-embedded plant material (root tips) and etched the block faces in plasmas of various gases. After metal shadowing the etched surfaces were replicated and the replicas were examined in the TEM. The resolution obtained was comparable to that in ultrathin sections.

Thomas[12] pioneered the use of excited oxygen for low temperature microincineration of thin-sectioned biological materials for TEM. Thomas and Greenawalt[13] demonstrated by TEM of metal-shadowed thin sections that an oxygen plasma could be used at a low temperature (specimen temperature less than 100°C) to differentially surface-etch thin sections of Epon-embedded isolated mitochondria. This incomplete ashing caused the membrane remnants of the sectioned cristae to emerge as recognizable structures projecting upward as ridges from the surrounding embedded material some of which was volatilized and removed by the excited oxygen. An excellent review on the status of spodography (the technique of producing ash patterns) for light and electron microscopy has been published by Thomas[14].

Hohman[15] and Hohman and Schraer[16] and Hohman[17] did the first extensive studies which utilized the technique of low temperature ultramicroincineration of thin sectioned tissue. Sections of hen shell gland 100 nm-500 nm thick were completely ashed onto silicon monoxide support films and the resulting spodograms were examined by TEM. Resolution was remarkably good and it was easily possible to identify from the ash patterns such cell components as mitochondria, plasma membranes, cytoplasmic membranes and chromatin. Frazier[18] using essentially the same method did TEM of shadowed sections of odontoblasts which were etched in an oxygen plasma for varying lengths of time at a low power level. After only 30 seconds of etching, enough of the epoxy was preferentially removed to reveal ash patterns of plasma membranes mitochondria and endoplasmic reticulum with a fine

definition better than in most of the preparations shown by Hohman and Schraer[16]. Frazier's improved resolution was probably due to the use of thinner sections (about 60-100 nm).

The above results suggested to us the use of an oxygen plasma for surface-etching fractured or cut surfaces of bulk samples of tissue embedded for transmission electron microscopy. It was anticipated that a gentle, and differential surface-etching would cause the emergence of structural features within cells and within organelles that could be imaged with considerable clarity with the scanning electron microscope operated in the standard secondary electron mode.

Materials and Methods

All specimens used were fixed in 2% glutaraldehyde in 0.1M cacodylate buffer at pH 7.2 for 1 hr. and washed in 0.1M cacodylate buffer with 5% sucrose added. They were post-fixed in 1% OsO_4 in cacodylate buffer for 1 hr., washed in cacodylate buffer, dehydrated in a graded series of ethanols and embedded in Araldite-Epon[19]. In order to make sure that fixation was good, ultra-thin sections were cut, stained with aqueous uranyl acetate[20], rinsed, stained with lead citrate[21] and examined with a TEM. For SEM 1 μm thick sections were cut with a diamond knife and floated onto the surface of distilled water. Using a small wire loop the sections were transferred on a film of water to pieces of glass microscope coverslips measuring 10 X 10 mm or smaller. The sections were allowed to dry flat against the glass. The coverslip fragments bearing the 1 μm thick sections were attached to a microscope slide using double coated Scotch cellophane tape to hold them in place. Each time etching was done the slide was placed in the reaction chamber so that specimens occupied approximately the same position. We arbitrarily positioned the specimens near the end of the reaction chamber next to the operator. We used a Tegal "Plasmod" plasma generator (available from Tegal Corporation, 860 Wharf Street, Richmond, California 94804, USA). The sections were etched by an oxygen plasma following the method described by Thomas and Hollahan[22]. Briefly, the reaction chamber was evacuated to a mild vacuum by a mechanical pump. Oxygen gas obtained from standard-grade commercial oxygen bottles was admitted at a controlled flow rate (2 psi) to obtain an operating pressure in the range of 0.5 to 1.0 torr. Radio-frequency power was applied around the chamber at 13.56 MHz. This excited the oxygen molecules and changed some of them into other species such as atoms, radicals, ions and free electrons. The highly reactive gaseous plasma caused a gentle, low-temperature combustion of the organic materials in the sample. The combustion products were carried away in the gas stream leaving behind the thick sections with the exposed surfaces etched. An RF power setting of 50 watts was arbitrarily chosen and the sections were etched for a period of 60 seconds. The etched sections were shadowed by vacuum evaporation of platinum-palladium wire from an angle of 30° at a distance of 10 cm. After shadowing, the specimens were lightly sputter-coated with gold using a Hummer II sputter coater

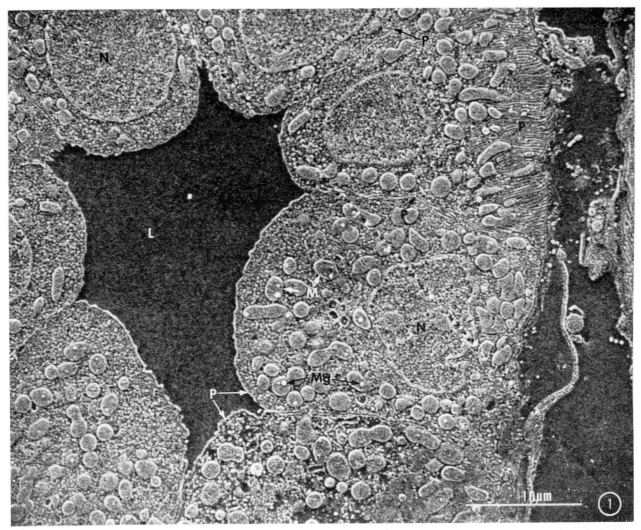

Fig. 1. Scanning electron micrograph of a 1 μm thick epoxy section of a convoluted tubule of mouse kidney, surface etched with an oxygen plasma, metal shadowed with platinum-palladium and sputter-coated with gold. Structures are identified by etch patterns that remain on the surface of the sections after etching. Plasma membranes (P) are recognizable at cell surfaces lining the lumen (L), between adjacent cells, and at the upper right of the micrograph as infoldings from the basal lamina. N, nuclei; mitochondria with mitochondrial matrix granules, Mg.

(Technics, Inc.). The etched sections were photographed with a Philips 501 scanning electron microscope at an operating voltage of 15 KV with the specimen tilted to an angle of about 25 degrees.

We sought to determine whether or not the oxygen plasma removes osmium from sectioned tissue that had been post-fixed in osmium tetroxide. Energy dispersive microanalysis was carried out on 1 μm thick sections of mouse spleen that were never exposed to the oxygen plasma and on 1 μm thick sections (cut from the same block) that were totally ashed in the oxygen plasma. In both cases sections floated onto glass distilled water were transferred onto 1/16" thick carbon planchets (E.F. Fullam, Inc.) and allowed to dry. Some of the sections were totally ashed onto the carbon planchets by exposure to the oxygen plasma for twenty minutes. Planchets were attached to the specimen

studs with a silver paint. Energy dispersive X-ray analysis was carried out with a Cambridge Mark 2A SEM equipped with an EDAX 707B X-ray analyzer. The SEM was operated at 20 KV with a beam current of 150 μA, a working distance of 12 mm, a specimen angle of 45° and lens current settings of 0.5A on the first two condenser lenses. Counting time was 400 seconds.

Observations

A low magnification SEM showing the typical appearance of a 1 μm thick epoxy section of mouse kidney after surface-etching by an oxygen plasma is shown in Fig. 1. Residues of cytoplasmic, mitochondrial, and plasma membranes show up as light lines against a dark background. The overall appearance of the etched surface resembles a TEM of a thin section printed in reverse contrast. In general the structures that would appear electron dense in typical transmission electron micrographs appear light in scanning electron micrographs of this type of preparation. Areas of the section occupied almost entirely by epoxy, such as the lumen (L) of the convoluted tubule, contain little residue after surface volatilization of the plastic and these areas appear smooth and dark in the SEM. On the other hand, after etching by the oxygen plasma, the plasma membranes (P) that border the lumen and that form the cell boundaries between cells leave prominent residue patterns in the form of ridges that appear as distinct topological structures in the micrographs. Residues of membranes of the nuclei (N) and mitochondria also form patterns that make them recognizable. Mitochondrial matrix granules (Mg) are refractory to etching by an oxygen plasma and they remain within the etch pattern of the mitochondria. A higher magnification of some of these etch-resistant structures is shown in Fig. 2. Residues originating from such structures as the nuclear membrane and the mitochondrial membranes appear to have undergone very little lateral migration as a consequence of the ashing. Shallow three-dimensional skeletons of these residues make it possible to identify such structures as perinuclear cisternae (Pc) and nuclear pores (Np). Residues of the cut and ashed mitochondrial cristae appear as discrete ridges separated from one another by spaces formed by the loss of material from between the cristae as a consequence of the etching. Residues of the outer and inner mitochondrial membranes cannot be distinguished from one another even at this higher magnification. Residues of the two membranes and the membrane space between them appear as a single structure in these preparations, probably because the metal shadowing and the subsequent metal coating caused the space between the two membranes to be bridged by the metal used. For the same reason, the double membrane structure of the cristae is not discernible from the residues of the surface-ashed cristae.

An etched section of pancreatic cells is shown in Fig. 3. This section was not shadowed with metal prior to sputter coating with gold, and resolution appears to be somewhat improved. Etch-resistant components of the endoplasmic reticulum (ER) and nuclei (N) are recognizable. The zymogen granules (G) appear to be etch-resistant to the oxygen plasma.

Fig. 4 shows a portion of the cortex of an etched section of a Paramecium. Structures recognizable from etch patterns include cilia (C), kinetosomes (K), kinetodesmosomal fibers (Kd), and trichocysts (T). Fig. 5 compares a TEM of an ultrathin section through the cilia and microvilli of mouse bronchial epithelium with a thick section cut from the same block which was etched and viewed with the SEM. The etched section has on its surface etch-resistant residues of structures that can clearly be identified as having originated from the plasma membrane of the cilia (Pc) and the microvilli (Mv). Where the cilia are cut in cross section, residues originating primarily from the microtubule doublets (that form the 9 + 2 pattern seen in TEM) form a pattern that consists of 9 separate piles of etch-resistant material

arranged in a circle with a single pile of similar material in the center (arrows Fig. 5b).

In Fig. 6 a portion of an etched section shows the macronucleus of a Paramecium. Bacterial symbionts (B) in the macronucleus are prominent. Chromatin bodies (Cb) appear to be resistant to etching. They project upward from the section as well defined bodies after the etching.

Fig. 7 compares the energy dispersive X-ray spectrum of thick unetched sections of embedded mouse spleen post-fixed with osmium tetroxide with the X-ray spectrum of similar sections totally ashed by the oxygen plasma. The unashed section exhibits peaks at 1920 eV and 8910 eV (Fig. 7a, arrows) corresponding to the M and L emissions of osmium. This clearly indicates the presence of osmium in the sections. These peaks are absent from the spectra of completely ashed sections (Fig. 7b). While it is not possible to conclude from these data that osmium is completely absent from the totally ashed sections, the concentration of osmium is reduced below detectable limits. This is in agreement with the recent more thorough study by Barnard and Thomas[23] who examined totally ashed sections of osmium tetroxide-fixed, Epon-embedded tissue before and after low-temperature oxygen plasma microincineration using a high-resolution scanning transmission electron microscope and an energy-dispersive X-ray spectrometer. Microincineration produced ash patterns that appeared to be free of osmium. As Thomas[14] had suggested earlier a likely fate of reduced osmium under these conditions is oxidation to volatile osmium tetroxide and subsequent loss from the specimen.

Discussion

Most methods used for producing ash patterns of embedded tissue that have previously been described[13,15,16,17,18] have involved the use of excited oxygen for total ashing of thin sections onto silicon monoxide films for transmission electron microscopy.

The method reported here describes an alternative procedure for producing etch patterns in which thick sections are only partially ashed or surface-etched by the use of excited oxygen and then examined in the SEM as bulk samples using the secondary electron mode. The depth of etching can be controlled by varying the length of time used for etching and by the selection of different wattages used for producing the oxygen plasma, or by an appropriate combination of these two variables. Thus etch patterns can be produced by etching thick sections to depths equal to the thicknesses of thin sections that have been totally ashed. The etched thick sections can be examined by the SEM and the totally ashed sections can be examined by TEM and the resulting etch patterns can be compared.

If such a comparison is made, the main advantage of partial etching compared to complete ashing is an apparent reduction in ash displacement and an apparent increase in resolution. Having an intact "base" under the ashed material may help to maintain the stability of the ash. The totally ashed sections studied by Hohman and Schraer[16] showed evidence of considerable ash

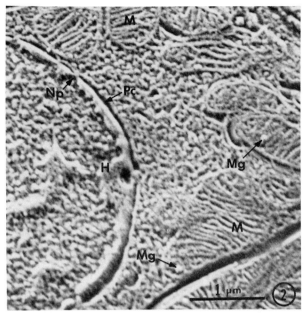

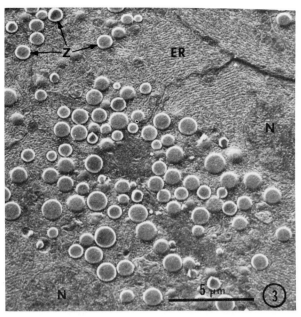

Fig. 2. A 1 μm section of mouse kidney showing a portion of a cell of a distal convoluted tubule. Etched, shadowed and coated the same as in Fig. 1. Three dimensional residues of the following partially etched cell structures are identifiable; H, heterochromatin; Np, nuclear pores; Pc, perinuclear cisterna; M, mitochondria; and Mg, mitochondrial matrix granules.

Fig. 3. SEM of a 1 μm thick section of ascinar cells of mouse pancreas. Specimen preparation the same as in Fig. 1 but without metal shadowing before coating with gold. Z, zymogen granules; N, nuclei; ER, endoplasmic reticulum.

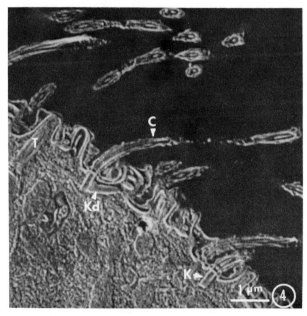

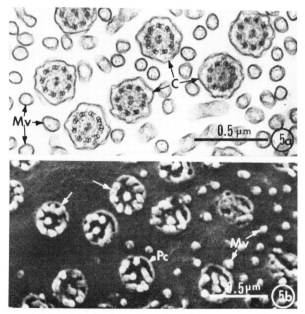

Fig. 4. SEM of a 1 μm thick section of a Paramecium showing a portion of the cortex. Etched, shadowed and coated the same as in Fig. 1. C, cilium; K, kinetosome; Kd, kinetodesmal fibers; T, trichocyst.

Fig. 5a. TEM of mouse bronchial epithelium showing cross sectional views of cilia (C) and microvilli (Mv). A 1 μm thick section of the same tissue etched, shadowed and coated the same as in Fig. 1 is shown in Fig. 5b. Pc, plasma membrane of the cilium; Mv, microvilli. Etching axonemes cut in cross section results in 9 separate piles of etch-resistant residue arranged in a circle around a central pile of residue (arrows).

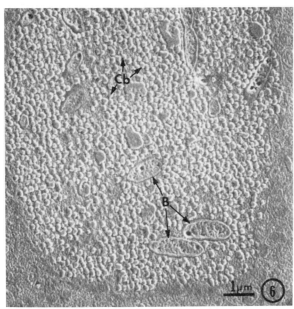

Fig. 6. A 1 μm thick section of a paramecium showing a portion of the macronucleus. Etched, shadowed and coated as in Fig. 1. Etch-resistant residues originating from bacterial endosymbionts (B) and chromatin bodies (Cb) are evident.

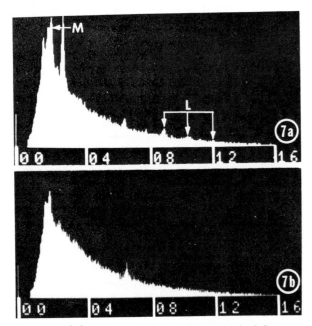

Fig. 7. (a) X-ray spectrum of an unashed 1 μm section of mouse spleen. Arrows indicate L and M peaks of osmium. (b) X-ray spectrum of totally ashed 1 μm section of the same tissue. Note absence of osmium peaks.

migration that, for example, resulted in clear zones around nuclei. This artifact was found in all his sections thicker than about 0.1 μm, the clear zones becoming wider as thicker sections were used. Such artifacts were subjectively judged to be less severe in our preparations of thick sections surface-etched for SEM. However, this could be a misleading impression since careful comparisons with regard to the depth of etching and the completeness of the ashing to the depth etched have not been made.

Recently Tanaka et al.[24] used ion-etching on critical point dried tissue that had been freeze-cracked by various methods. Such specimens when examined by SEM appeared similar to our etched tissue sections. His specimens were etched by high-energy sputter etching. Differential etching was achieved because membranous structures in the cell are generally more resistant to sputter-etching and the cytoplasmic matrix is easily etched. Structures clearly disclosed by this method of surface etching were nuclear pores, endoplasmic reticulum with ribosomes, Golgi apparatus, and mitochondrial cristae. We attempted to use excited oxygen to surface-etch freeze-fractured critical point dried tissue prepared by the method of Humphreys.[4] We were unsuccessful because the critical point dried tissue is freely permeable to the gaseous oxygen plasma which reacted chemically with the interior of the tissue at the same time as with the surface. As a consequence organic material was volatilized from the interior of the tissue at the same rate as from the tissue's surface and no differential surface-etching could be achieved.

In these preliminary studies we found that more satisfactory results were obtained when the plasma-etched sections were metal-shadowed before they were coated with metal prior to their examination in the scanning electron microscope. Theoretically this step is not necessary and higher resolution would be possible if it were dispensed with, since less metal would be present on the specimen to obscure surface detail. Our attempts to omit this step, however, usually, but not invariably, resulted in specimens that lacked adequate contrast when observed in the SEM. Better resolution could also be obtained if the etching were restricted to a depth just sufficient to generate enough variability in surface topography to permit imaging the surface after a conducting metal coat of minimal thickness was deposited onto it. Optimization of these parameters will require further study. With careful attention to such details, plasma-etched sections could conceivably be capable of showing cellular fine structure that is not now resolvable by most state-of-the-art scanning electron microscopes using the secondary electron mode.

A consistent observation has been that those cell constituents that were most electron dense in thin sections used for TEM were also the most etch-resistant structures in the plasma-etched preparations. As is well known, much of the electron density of structures in thin sections is due to the presence of reduced osmium. But this would not seem to be a significant factor in the etch-resistance since osmium appears to behave as a volatile stain in the oxygen plasma since it is re-oxidized to osmium tetroxide. However, osmium might indirectly cause retention of minerals in certain structures and these minerals would remain after ashing as structural ash residues. Hohman[17] found in ash patterns of totally low-temperature ashed thin sections of avian shell gland tissue that ash residues of certain membranes, such as membranes between cells, appeared only when glutaraldehyde fixation was followed by osmium post-fixation, although ash residues of most other membranes elsewhere in the cells could be identified when fixation was by glutaraldehyde alone.

The etching technique is new and it lacks standards that will consistently yield reproducible results. For example a slightly different position of the specimen in the reaction chamber will result in a different degree of etching. Unanswered questions such as how best to obtain the optimal amount of etching and how best to render the etched specimen electrically conductive without obscuring surface detail must be determined by further experience. Can the method be exploited for energy dispersive X-ray microanalysis? Because of these unresolved questions the use of an oxygen plasma for etching embedded tissue for observation with the SEM must be considered still in the state of technical development and exploration. Even so, certain advantages are obvious. Etching thick sections with excited oxygen for SEM is considerably less arduous than total ashing of thin sections for TEM. The need for preparing silicon monoxide films is eliminated, as well as the need to cut ultrathin sections and the concern for mounting the sections onto the film with smooth intimate contact. Resolution of the etch patterns produced by surface-etching is at least as good as in ash patterns produced by total ashing of thin sections.

Finally, as a general technique for SEM, etching thick sections of a wide variety of different types of embedded tissue by the use of an oxygen plasma yields specimens that show a resolution that is considerably better than that obtainable by most other methods of SEM currently being used for viewing the internal structure of cells and organelles in bulk samples.

Acknowledgements

We are grateful to Dr. Richard S. Thomas of the U.S. Department of Agriculture, Western Regional Research Laboratory, Berkeley, California for very helpful discussions and correspondence that provided us with a substantial amount of technical advice and guidance that made this investigation much simpler to undertake.

References

1. K. Tanaka and A. Iino. "Frozen resin cracking method for scanning electron microscopy and its application to cytology," In: Proc. 30th Annual Meeting Electron Microscopy Society of America. C.J. Arceneaux (ed.), Claitor's Publishing Div., Baton Rouge, USA. (1972) p. 408-409.
2. P.S. Woods and M.C. Ledbetter. "A method of direct visualization of plant cell organelles for scanning electron microscopy," In: Proc. 32nd Annual Meeting Electron Microscopy Society of America. C.J. Arceneaux (ed.), Claitors Publishing Div., Baton Rouge, USA (1974) p. 122-123.
3. H.D. Sybers and M. Ashraf. "Preparation of cardiac muscle for SEM," In: SEM/1973 O. Johari and I. Corvin (eds.), IIT Research Institute, Chicago, IL 60601. p. 341-348.
4. W.J. Humphreys, B.O. Spurlock and J.S. Johnson. "Critical point drying of ethanol-infiltrated, cryofractured biological specimens for scanning electron microscopy," In: SEM/1974. O. Johari and I. Corvin (eds.), IIT Research Institute, Chicago, IL 60601, p. 275-282.
5. K. Miyai, J.L. Abraham, and D.S. Lithicum. "Scanning electron microscopy of hepatic ultrastructure-secondary, backscattered, and transmitted electron imaging," Lab. Invest. 35, (1976) p. 369-376.
6. B.L. Munger and V.R. Mumaw. "Specimen preparation for SEM study of cells and cell organelles in uncoated preparations," In: SEM/1976/I O. Johari (ed.), IIT Research Institute, Chicago, IL 60601. p. 275-280.
7. R. Mykelbust, H. Dalen, and T.S. Saetersdal. "A comparative study in the transmission electron microscope of intracellular structures in sheep heart muscle cells," J. Microsc. 105 (1975) p. 57-65.
8. W.B. Winborn and D.L. Guerrero. "The use of a single tissue specimen for both transmission and scanning electron microscopy," Cytobios 10 (1974) p. 83-91.
9. S.L. Erlandsen, A. Thomas and G. Wendelschafer. "A simple technique for correlating SEM with TEM on biological tissue originally embedded in epoxy resin for TEM," In: SEM/1973 O. Johari and I. Corvin (eds.), IIT Research Institute, Chicago, IL 60601 p. 349-356.
10. G.H. Haggis, E.F. Bond and B. Phipps. "Visualization of mitochondrial cristae and nuclear chromatin by SEM," In: SEM/1976/I O. Johari (ed.), IIT Research Institute, Chicago, IL 60601 p. 281-286.
11. A.C. Fabergé. "III. Development of a Replica Process for the Electron Microscopy of Biological Material," Studies in Genetics, IV, Research Reports, Univ. of Texas at Austin, (1968) p. 21-47.
12. R.S. Thomas. "Ultrastructural localization of mineral matter in bacterial spores by microincineration," J. Cell Biol. 23 (1964) p. 113-133.
13. R.S. Thomas and J.W. Greenawalt. "Microincineration, electron microscopy, and electron diffraction of calcium phosphate-loaded mitochondria." J. Cell Biol. 39 (1968) p. 55-76.
14. R.S. Thomas. "Use of a chemically reactive gaseous plasma in preparation of specimens for microscopy," In: Techniques and Applications of Plasma Chemistry J.R. Hollahan and A.T. Bell (eds.), John Wiley and Sons Inc. N.Y., USA (1974) p. 255-346.
15. W.R. Hohman. "A study of low temperature ultramicroincineration of avian shell gland mucosa by electron microscopy," Ph.D. Thesis, Pennsylvania State Univ., University Microfilms, Ann Arbor, Mich., USA (1967) 150 pp.
16. W.R. Hohman and H. Schraer. "Low temperature ultramicroincineration of thin-sectioned tissue," J. Cell Biol. 55 (1972) p. 328-354.

17. W.R. Hohman. "Ultramicroincineration of thin-sectioned tissue," In: Principles and Techniques of Electron Microscopy--Biological Applications, Vol. 4. M.A. Hayat (ed.), Van Nostrand Reinhold, N.Y., USA. (1974) p. 129-158.

18. P.D. Frazier. "An electron microscopic investigation of mineralized tissues," Ph.D. Thesis, Washington Univ., Seattle, University Microfilms, Ann Arbor, Mich., USA (1971) 294 pp.

19. H.H. Mollenhauer. "Plastic embedding mixtures for use in electron microscopy," J. Stain Tech. 39 (1963) p. 111-114.

20. M.L. Watson. "Staining of tissue sections for electron microscopy with heavy metals," J. Biophys. Biochem. Cytol. 4 (1958) p. 475-478.

21. E.S. Reynolds. "The use of lead citrate at high pH as an electron opaque stain in electron microscopy," J. Cell. Biol. 17 (1963) p. 208-213.

22. R.S. Thomas and J.R. Hollahan. "Use of chemically-reactive gas plasmas in preparing specimens for scanning electron microscopy and electron probe microanalysis," In: SEM/1974 O. Johari and I. Corvin (eds.), IIT Research Institute, Chicago, IL 60601 p. 83-92.

23. T. Barnard and R.S. Thomas. "X-ray microanalysis of epon sections after oxygen plasma microincineration," J. Microsc. 113 (1978) p. 269-276.

24. K. Tanaka, A. Iino and T. Naguro. "Scanning electron microscopic observation on intracellular structures of ion-etched materials," Arch. Hist. Jap. 39 (1976) p. 165-175.

Discussion with Reviewers

J.A. Swift: Having had some considerable experience in the use of oxygen plasmas for etching sectioned biological materials for the SEM, I would comment that none of your micrographs are of completely ashed material. I submit that the micrographs show etching patterns relating to the susceptibility of the various microscopic components to oxidative degradation by the reactive singlet oxygen species present in the plasma. In this respect the micrographs cannot be regarded as necessarily showing ash distribution in the tissues. Do you have any SE micrographs of completely ashed tissue sections that can be shown here:

R.S. Thomas: It is probably misleading to regard the etch patterns as equivalent to ash patterns or spodograms lying on the surface. The emergent structures are simply more resistant to etching than the surrounding field and this resistance may or may not be the result of a high mineral content. The etch-resistant zymogen granules in Fig. 3 may be a case in point; do they have a high mineral content?

Authors: We do not have any micrographs of completely ashed tissue sections that can be shown here. We concede the point that our etch patterns do not represent completely ashed material and that etch-resistance may not be related to high mineral content. The zymogen granules which do not have a high mineral content are a good case in point. In recognition of this we have in the final version of this report avoided such expressions as "ash patterns" or "ash residues" when referring to our etched specimens and used

expressions such as "etch patterns" or "etch resistant residues" instead. Our method for producing etch patterns, while not necessarily showing mineral distribution in the tissues, does accomplish the main objective for which the method was employed; it allows residues of internal subcellular constituents to be resolved and identified by SEM using secondary electrons.

R.S. Thomas: You discuss your etch pattern fidelity in comparison to that for the ash patterns of Hohman and Schraer. If you etched to a depth greater than 0.5 μm (comparable to Hohman and Schraer's complete ashing of 0.5 μm thick sections) didn't you also see coarsened detail with widened cavities? This might be expected owing to lateral etching by the isotropic plasma.

Authors: This may be the case and our comparisons may not be valid. But if the ashing of our specimens is incomplete (see question above), we would expect the incompletely ashed structures to more closely resemble the structures from which they originated than if ashing were complete so that only the inorganic mineral skeleton of the structures remained. If this is so, incomplete ashing could be an advantage from the standpoint of producing etch patterns that show better structural fidelity and better resolution.

W.A. Massover: About how much material is typically removed from the surface of these sectioned specimens?

Authors: This was not measured and it is difficult to estimate since the specimens were usually metal-shadowed and all were metal-coated after etching. A guess would be 70-100 nm. A comment from a fellow reviewer, R.S. Thomas, is in order at this point:

"A reproducible internal standard by which to judge the degree of etching is provided by polystyrene latex spheres sprayed on the surface. The isotropic plasma etching causes a uniform decrease in the uniform sphere diameters which is easily measured. (See Thomas, R.S., Millard, M.M. and Scherrer, R., "Electron microscopy and photoelectron spectroscopy of oxygen plasma-etched bacterial spores and cells. in Proc. 34th Ann. Meeting Electron Microscopy Soc. Amer., G.W. Bailey, ed., Claitor's, Baton Rouge, 1976. pp. 134-135)."

W.A. Massover: How is the "low-temperature" actually known?

Authors: We have not measured the specimen temperature but we think it is not much above 100°C if that high. The microscope slide on which the specimens were etched were barely warm to the touch when removed from the chamber immediately after etching.

G.Schidlovsky: What parameters influence specimen temperature, and what is the minimum specimen temperature consistent with controlled etching of biological sections?

Authors: Thomas and Hollahan (text reference 22, in discussion with reviewers) measured surface temperatures of specimens in an oxygen plasma using an infra-red pyrometer. They report that surface temperature measurements depend on the nature of the specimen surface and the power input to the plasma. Surface heating is minimized or eliminated by etching more slowly at a reduced power level. Then the surface is no hotter than the gas which typically would be below 100°C, or perhaps even 30°C. Metal specimen substrates such as bare grids may show a substantial temperature rise in the plasma causing the specimen to heat more than if a glass support is used.

A. Boyde: Wouldn't the method used to detect residual osmium be highly suspect, since one would expect the carbon substrate to be removed from under the section so that the original situation would be severely changed?

R.S. Thomas: You indicate that sections were totally ashed while mounted on carbon planchettes. The oxygen plasma also attacks carbon. Did this cause any difficulties.

J.A. Swift: We have found that exposure of high purity amorphous carbon planchettes to oxygen plasmas in the "Plasmod" results in viscious degradation of the planchette that scatters carbon fragments over the specimen. Have you encountered this?

Authors: Degradation of the carbon planchette during exposure to the oxygen plasma presented difficulties in locating sections and in identifying fine structure within them due to the roughened surface and the presence of some carbon debris in the area of the sections. We marked the location of each section with deep scratches in the planchette surface. The scratches survived exposure to the plasma and allowed us to quickly locate the area of interest. In some cases larger structures such as nuclei were still identifiable. Regions which were indicated by the scratches and which contained some identifiable structure were subjected to X-ray analysis. In our experience the carbon planchette was not a suitable substrate for obtaining adequate micrographs of fine structure.

A. Boyde: How do you reconcile your apparent conclusion that the ash patterns are not due to osmium bound to cell membranes, etc., when the membranes are not visible unless osmium fixation is used?

Authors: Most membranes in the cell leave recognizable ash patterns after total ashing when glutaraldehyde fixation alone is used. But certain other membranes do not leave recognizable mineral residues unless they are post-fixed with osmium (text reference 17). Osmium may stabilize certain structures, phospholipids of some membranes for example, that would otherwise be leached out during alcoholic dehydration. In this case phosphate salts would remain as part of the ash residue of such membranes, even though the osmium were removed by the oxygen plasma.

G.Schidlovsky: Of the six parameters inherent in this type of etching (vacuum, RF frequency, wattage, time of exposure, gaseous environment, and location of specimen in reaction chamber), which is most critical for controlling the reaction?

Authors: We do not yet have sufficient experience with the technique to answer this question. Our approach for determining the optimal parameters

has been, and continues to be, entirely empirical.

R.S. Thomas: The image contrast between the etched tissue and the epoxy background is striking--e.g. in Figs. 1 and 4. How much of this is due to the etching and shadowing? What do unetched, unshadowed preparations look like under the same viewing conditions?

Authors: We don't know how much of the contrast is due to etching and how much is due to shadowing. Shadowing does help the contrast considerably. Unshadowed preparations such as in Fig. 3 have less contrast but they can show better resolution of the fine structure. We have not studied unetched, unshadowed preparations but we agree that this will be an interesting comparison to make.

R.S. Thomas: Oblique surface shadowing with platinum-palladium, visualized by enhanced secondary electron emission in the SEM, is clearly a valuable refinement of the present technique which should be generally useful. In the present application it evidently provides for sufficient contrast to allow the improved resolution associated with minimal etching. Couldn't this approach be improved still further by using carbon rather than gold as the over-all conductivity coating?

Authors: This is a good suggestion that we will act upon, since it will probably improve resolution.

A. Boyde: Have you tried this on unfixed thin sections?

Authors: No.

G.Schidlovsky: Do you have any experience with paraffin, methacrylate and/or other epoxy embedding media?

Authors: We tried the technique on paraffin-embedded tissue without success. The paraffin appeared to have melted. Etching more slowly at a reduced power level would cause the specimen to heat less and melting might be avoided. The methacrylate, Polyscience JB-4, etches similarly to epoxy and the etch patterns are similar to those presented here. But an increase in grain size results in some loss of detail. The etching characteristics of the epoxies, Epon, Araldite, Epon-Araldite and Maraglas seem to be identical.

W.A. Massover: Please comment on the ashed microvilli in Fig. 5b: why do they not show their limiting membrane?

Authors: It appears that the material surrounded by the limiting membrane is etch-resistant. Also the metal-shadowing and metal-coating may have been excessive in this preparation. This might have resulted in bridging of the condensed evaporated metal from one etch-resistant structure to another within this area.

W.A. Massover: HVEM can be used with specimens of the same thickness, and can give stereoscopic information on structures within such slices. Are there any anticipated advantages of using the author's SEM approach instead of HVEM?

Authors: Our approach is complementary. Thick
sections viewed in the HVEM could be surface-
etched afterwards and used to provide information
regarding the etch-resistance of certain constituents of the specimen. Or the specimen could be
used for X-ray microanalysis after etching. In
the absence of an available HVEM, etching for SEM
might be used for certain morphologic studies.
Also thick epoxy sections stained and used for
light microscopy could be surface-etched later
and used for SEM.

A. Boyde: Do you agree that the following questions are still open: 1) how much mineral is
added by binding divalent cations from the glutaraldehyde fixative; 2) how much of the original
cell mineral is removed by fixation; and 3) how
much of the visibility of the membranes is due to
the osmium added to them?
Authors: We agree. These questions are still
open.

G.Schidlovsky: In what way is RF plasma etching
more suitable to your objectives than other dry
etching techniques?
Authors: We have better control over the rate of
etching using an oxygen plasma than we would have
using ion etching. Also the specimen can be
etched more uniformly and at a lower temperature.

J.A. Swift: You raise the question 'can the
method be exploited for energy dispersive X-ray
microanalysis?" In my experience the answer to
this is that it certainly can. If thick epoxy
embedded sections are collected on titanium electron microscope grids equipped with a carbon and
silicon monoxide composite film support and then
ashed completely, there is considerable improvement in the trace element detection sensitivity
for the tissue. This is done by examining the
ash residue (now supported on the silicon monoxide film) using the transmitted electron imaging
mode of the SEM & by simultaneous nondispersive
X-ray micro-analysis. Element detection is improved roughly in the proportion of the thickness
of the original section to the thickness of the
silicon monoxide film. Using sections of 1 μm
thickness and the silicon monoxide film of 10 nm
thickness a 100 fold improvement in trace element
detection limits is thus obtained. We have had
no difficulty using this method not only in measuring zinc in human hair down to an accuracy of
5 ppm, but we are also able to observe the distribution of various levels of zinc with respect
to the morphological substructure of the hair.
Authors: Your method sounds much more practical
than what we had in mind, which may not be practical at all, and that is the use of partially
ashed thick sections used as bulk samples on a
solid substrate in the SEM for energy dispersive
X-ray microanalysis.

SCANNING ELECTRON MICROSCOPY/1979/II
SEM Inc., AMF O'Hare, IL 60666, USA

COATING BY ION SPUTTERING DEPOSITION FOR ULTRAHIGH RESOLUTION SEM

J. D. Geller, T. Yoshioka and D.A. Hurd*

JEOL U.S.A., Inc.
477 Riverside Avenue
Medford, MA 02155

*Technics Inc.
7950 Cluny Court
Springfield, VA 22153

Abstract

Specimens prepared for examination in the secondary electron imaging mode (SEI) of operation in an electron microscope that are of an insulating nature often have to be rendered conductive through the application of an electrically conductive thin film. Although there are alternate techniques to coating, such as the Os ligand process, LN_2 temperatures and low voltage microscopy, few give results that are as good as can be attained with high quality metal coatings. Thin film coatings may be deposited by vacuum evaporation, sputter coating, ion sputtering deposition (ISD), as well as many other techniques. It is important to choose the technique which approximates a perfect "mask" of coating on the specimen surface without seeing artifacts such as decoration, etching or heating effects. The grain size of the particles making up the coating, if present, must be beneath the resolution limit of the microscope.

As technology for producing high resolution images advances, so do the requirements for ancillary support equipment used to prepare the specimens. SEM's capable of producing point-to-point resolution of less than three nm (nanometers) are becoming more commonplace. Furthermore, these instruments perform at this level on a day-to-day basis.

ISD answers the need for higher resolution coatings by producing thin films which are nearly structureless in SEI images at magnifications of 100,000X and higher. It also coats at room temperatures without etching effects, under an inert gas environment. ISD is compared, using several specimens, to conventional vacuum evaporating and sputter coating techniques at appropriate magnifications, using conventional transmission (TEM) and SEI modes.

KEY WORDS: Coatings, Sputter Coating, Thin Films, Vacuum Evaporation, Ion Sputtering Deposition, Specimen Preparation

Introduction

Resolution in the secondary electron imaging mode of operation has recently been improved to the point where metallic coatings applied, which enhance the secondary electron signal and provide conductivity, are clearly resolvable with conventional coating techniques. Better than three nm point-to-point secondary resolution is near the current state of the art. That resolution capability is obtainable only with specimens which are not of an insulating nature. That is, they do not require the addition of conductive films. For secondary electron images, a thin coating is desirable on insulating specimens for high resolution to retain maximum information content in the specimen. Consequently, coatings with the finest grain structure and the absence of islands are desirable. Ultrafine coatings of carbon, with only phase-contrast structure, can be obtained by thermal deposition for such techniques as shadow casting. However, carbon is not desirable as a coating by itself, due to its low electron scattering property and secondary electron emission. Shadow casting requires fine coating to reduce measurement artifact. The need exists for high atomic number metallic coatings with a thickness of less than 15 nm to preserve detail which is resolvable.

Thermal deposition and DC plasma sputter coating are suitable for SEI images up to approximately 30,000X; however, there is a definite need for coatings with finer structure. Carbon evaporation answers this requirement, but has a very low secondary electron coefficient. In general, with pure materials, the secondary yield increases with atomic number.[1]

Metallic films were deposited by different coating techniques to provide electrical conductivity to several types of specimens. Thermal deposition (TD), DC plasma sputter coating (SC) and ion sputtering deposition (ISD) were accomplished using commercially available equipment, slightly modified in the case of ISD. Carbon coated collodion films, recording quality, resin coated, magnetic tape, silver halide crystals and biological tissue preparations were used to evaluate the techniques.

It will be shown that ISD produces a much finer grain structure with several different materials than either TD or SC. On a biological test specimen, and on silver halide grains, sputter coating leaves surface defects which are absent with ISD.[2]

It is the intended purpose of this work to illustrate the usefulness and need of ion sputtering deposition as a coating technique for ultrahigh resolution TEM and SEM.

Instrumentation

All micrographs were produced on a JEOL 100 CX/ASID TEMSCAN operated in both transmission and secondary electron imaging modes. A Tungsten filament was used at accelerating voltages of 80kV for the TEM micrographs and 40kV for the secondary electron images.

ISD coatings were made using a modified Technics, Inc. MIM-IV micro ion mill. One hollow anode TMA-1 gun of the ion accelerator type was used to generate a 1 cm ion beam at the sputtering target. The specimen was rotated to assure even coating. The sample was 2.5 cm from the sputtering target and the ion gun was operated at 5kV and 40 μa.

Sputter coating was performed with a Technics, Inc. Hummer V planar magnetron sputter coater, using targets of Au, Au/Pd and Pt in an argon atmosphere at 10–20 ma, depending on the target and at a distance of 4 cm. For comparison, Au/Pd was also sputtered in air.

A JEOL model JEE-4C vacuum evaporator with a rotating tilting specimen stage was used to evaporate different metals at 1×10^{-5} torr.

Thickness of the various films was measured first by TD to completion of a known mass of material.[3] To measure the thickness of the ISD and SC preparations X-ray energy dispersive spectroscopy was utilized.

ISD Theory

Ion sputtering deposition is widely used in the semiconductor industry for device fabrication for the deposition of high resolution, highly adhesive, uniform coatings. ISD utilizes ion beam milling, the removal of material from a substrate by an energetic ion beam. In the case of ISD, the target is placed opposing the ion beam and the removed material is subsequently deposited onto a sample to be coated.

The ion beam is generated by the ionization of an inert gas, argon, in a diode configuration with 3–5kV potential. These ions are then accelerated, collimated and directed toward the material to be deposited by the physical configuration of an anode and cathode with 13 apertures. Each ion bombards the target. The collisions result in the displacement

of the target atoms through momentum transfer. In order to dislodge the atoms from the target, the energy of the ions must exceed that of the binding energy of the target atoms. A low energy ion, typically 400 eV, at a power level of 0.5 ma/cm^2, is sufficient to dislodge surface atoms at a reasonable rate. The dislodged atoms are deposited on the sample to be coated at energies on the order of 100 eV. A schematic of the process is illustrated in figure 1.

Ion sputtering deposition has several advantages over conventional deposition processes now in use, such as vacuum evaporation and DC plasma sputtering. The plasma is generated and confined within the gun, thus the sample outside the area is not exposed to plasma. The electric field is also confined within the gun. The sample and the target are both at ground potential. This is not the case in sputter coating. Only a small number of electrons are produced from the target and little electron bombardment of the sample occurs. The heating of the substrate is substantially reduced. The heat that would be generated by direct contact with the ion beam is dissipated at the target and not in the sample. In comparison to SC, the low pressure in which the process is carried out, typically 10^{-4} to 10^{-5} torr, is seen as an advantage in that it reduces the potential of contamination of the sample and the incorporation of residual gas into the film. Kanaya, et al.,[4] researched the use of this technique as a suitable method for depositing thin films for shadowing in high resolution, transmission electron microscopy, using a modified duoplasmatron ion source. Jakopic, et al.[5] noted that since no radiation damage occurs with ISD, temperature sensitive specimens, such as biologicals, may be coated in this manner for both TEM and SEM.

Results

Collodion films, less than 30 nm thick, were prepared on TEM grids and coated with approximately 7 nm of carbon. At magnifications of 100,000X only phase contrast effects could be seen. To these preparations Au, Au/Pd and Pt thin films were added.

The purpose of the TEM images included, is to permit measurement of grain sizes and crystallite formation at magnifications that are well within the 0.3 nm point-to-point resolution capability of the TEM. As can be seen from the micrographs presented, a direct correlation between the TEM and SEI images is evident.

Figure 2 shows TEM micrographs of Au, Au/Pd (60:40) and Pt films approximately 15 nm thick, prepared by ISD, TD and SC. In all cases the TD and SC, metallic crystallites are formed in groups or islands. Thermally evaporated Pt/Pd had the smallest island size; however, it was found by X-ray analysis, that the Pt alloyed with the W basket and only Pd was deposited at about 7 nm thickness instead of the calculated 15 nm Pt/Pd.

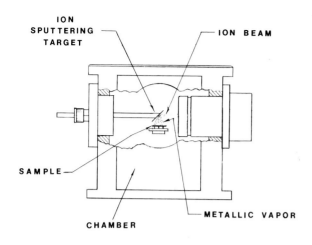

Figure 1. Ion sputtering deposition system for preparation of metallic coatings.

ISD proved superior in that no islands were observed, and in every case the grain size was significantly smaller. In figure 2 the TEM image revealed an extremely fine Pt grain. Kanaya, et al.[6] reported a 0.4 nm diameter tungsten grain size in a 15 nm thick film using ISD.

Figure 3 contains SEI's of SC and TD prepared films of Au and Au/Pd. It was not possible to resolve the surface of any ISD films in the microscope; therefore, they are not shown here. The SC films of Au/Pd, interestingly, had a larger grain size than its TD counterpart. Although there was some difference in grain size between samples sputter coated with and without argon, the grain size was still clearly resolvable in both (figs. 3b, with argon and 3c without argon). Evaporated and sputter coated Au has a very coarse structure (figs. 3d and 3e).

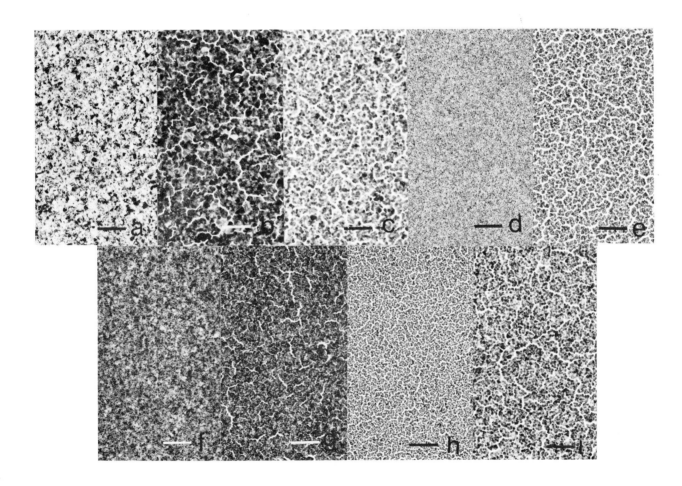

Figure 2. Transmission electron images of metallic coated collodion films; a. 12nm Au by ISD, b. 15nm Au by TD, c. 11nm Au by SC, d. 11nm Pt by ISD, e. 15nm Pt by SC, f. 15nm Au/Pd by ISD, g. 15nm Au/Pd by TD, h. 15nm Au/Pd by SC, with argon, i. 15nm Au/Pd by SC, without argon. Bar = 100nm

Figure 4 shows SEI micrographs of TKD, resin coated, cassette audio magnetic tape coated by different techniques. The Au coating by TD was clearly resolvable and the SC Au coating had extremely fine decoration present with islands of about 5 nm in diameter. The ISD Au deposition, coarsest of the ISD coatings, shows almost no inherent structure of its own, only the faithfully reproduced surface of the tape at a resolution of 3 nm in the SEM.

Figure 5 illustrates a previously reported artifact associated with sputter coating. Figure 5a is of an uncoated silver halide preparation. The surface detail is not clear. The electron beam current had to be increased approximately four times over the metal coated preparation to produce same S/N. At this higher beam current the resolution is degraded by approximately 70%. The smaller particles seen are remnants of the digestion process used to remove the grains from the photographic emulsion. The silver halide grains in Fig. 5b, which was immersed in the DC plasma field during sputter coating with Pt, the finest SC material, is bombarded by electrons, generating heat, forming artifacts which are seen as surface crystallization. The Pt ISD coated specimen (fig. 5c), Pt being the finest ISD coating, with no resolvable coating structure, clearly shows the silver grain surface detail with good contrast and free of surface crystallization.

It was observed, during the picture taking process, that the build-up of contamination during the examination of the ISD coated specimens was significantly lower than either TD or SC. A possible explanation is that the initial neutral metal atoms from the target, which are implanted in the specimen, act as a getter for surface contamination. The additional coating seals in the contamination, thereby immobilizing it. Certainly, further work must be done in this area.

Conclusions

Ion sputtering deposition produces metallic coatings with significantly smaller grain sizes and the absence of island formation when compared to conventional vacuum evaporation and DC plasma sputter coating. This should permit thinner coatings to be made while preserving electrical conductivity.

For specimens which are of a delicate nature and sensitive to the heat of thermal evaporation or the electron bombardment of sputter coating, ion sputtering deposition proves superior. It is a room temperature process with no electrical fields surrounding the specimen since the target and the specimen are both at ground potential.

The ISD Pt films had significantly smaller grain sizes than ISD prepared Au and Au/Pd. No island formation was observed with any of these coatings, but were present in every case with SC and TD using the same metals.

Acknowledgement

The authors gratefully acknowledge the assistance of David Harling in preparing the micrographs.

References

1. J. Goldstein and H. Yakowitz, Practical Scanning Electron Microscopy, Plenum Press, New York, USA, 1975, p. 65.
2. J. Franks, Properties and Applications of Saddle-Field Ion Sources, in proceedings of the 1978 American Vacuum Society, to be published in the Journal of Vacuum Science and Technology.
3. P. Echlin, Coating Techniques for Scanning Electron Microscopy and X-ray Microanalysis, in SEM/1978/I, SEM, Inc., AMF O'Hare, Il., 60666, p.125
4. K. Kanaya, K. Hojon, K. Adachi, Ion Bombardment of Suitable Targets for Atomic Shadowing for High Resolution Electron Microscopy, Micron, 3, 1974, p. 89-99.
5. E. Jakopic, A. Brunegger, R. Essl, A Sputter Source for Electron Microscopic Preparation, in 9th International Congress on Electron Microscopy, J. M. Sturgess, Microscopial Society of Canada, Toronto, Canada, 1978, p. 150-151.
6. K. Kanaya, et al. Op. Cit., p. 100
7. J. Franks, Op. Cit.

Discussion With Reviewers

Reviewer I: The authors imply that etching effects always occur in an inert gas environment (by TD or SC). Can you comment on this?

Authors: Ion sputtering deposition is a process which takes place in an inert environment, as does sputter coating. No etching effects were noted in the ion sputtering process for the range of samples covered in this manuscript. We do not mean to imply that etching effects always occur; however, Echlin indicates in reference 3, p. 126, that thermal damage does occur unless the sample is prepared by cool sputtering. Even with the cool sputter coater used in this report, we still see artifacts on the silver halide and biological specimens. The cause of these artifacts is difficult to determine.

Reviewer I: Can you comment as to the effect of using different coating thicknesses, and are your conclusions valid in view of these differences?

Authors: Film thickness does have a great deal to do with the grain size of thin films; however, the film thickness differences in the samples shown amount to 4 nm in fgs. 2-5. Unless a material has a preferential growth pattern in a lateral direction, it is hard to imagine a growth mechanism which would double the lateral size of a grain, with a change in thickness of only 4 nm. The grain must, initially, be of different sizes in each process to account for the large differences shown in the micrographs. Grain growth is not a linear function of thickness.

R. G. Fernquist: The thickness of films, sample and

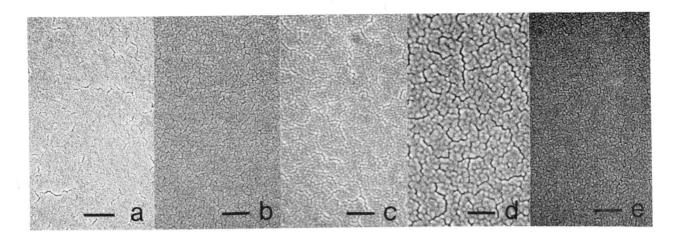

Figure 3. Secondary electron images of metallic coatings on collodion films; a. 15nm Au/Pd by TD, b. 15nm Au/Pd by SC, with argon, c. 15nm Au/Pd by SC, without argon, d. 15nm Au by TD, e. 15nm Au by SC. Bar = 100nm

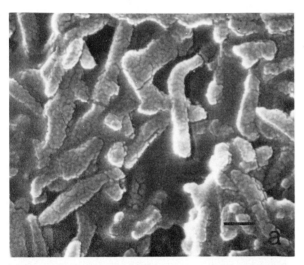

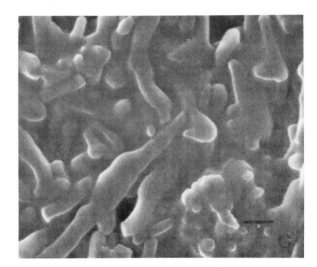

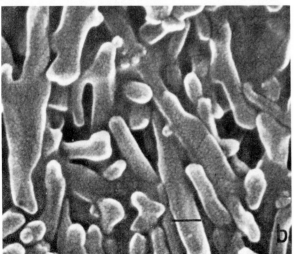

Figure 4. Secondary electron images of metallic coated magnetic tape; a. 15nm Au by TD, b. 15nm Au by SC, c. 15nm Au by ISD. Bar = 100nm

source temperatures and other variables may have considerable affect on the grain size of metallic coatings. Since the thickness of your films was measured by different techniques, do you feel that these were sufficiently accurate to enable direct comparisons, particularly between TD and SC coatings. Cross sectioned and TEM examined samples from each coating run would determine thickness accurately and eliminate any question of relative thickness effects. Authors: TD to completion of a known amount of material calculated to be 15 nm thick was used as a standard. In comparing these standards to other runs of TD to completion by X-ray analysis, they were found to be of the same thickness. The relationship was carried over to SC and ISD to determine the thickness of these films. Since it was not possible to deposit PT thermally, we used the gold calibration with X-ray analysis. We felt justified in doing this, since Au and Pt are only two atomic numbers apart and have almost the same ionization cross section. The suggestion of cross sectioned and TEM examined samples is an excellent one, and should be followed up in future studies. The temperature of the source and the sample are inherent in all of the processes used, unless some artificial cooling device is used.

R. G. Fernquist: Could you describe or reference the X-ray energy dispersive thickness measurement technique. I presume it is based on a range equation, but what is the relationship to X-ray counts, and what absorption effects from the substrate (in this instance the sample) were determined?
Authors: The method used to calculate the film thickness was by making a calibration curve of X-ray intensity as a function of film thickness in the empirical method. J. Goldstein and H. Yakowitz, Op. Cit. p. 310.

Reviewer I: Why was a cold trap not used to reduce contamination?
Authors: A cold trap is not common procedure for sputter coating or thermal deposition and, therefore, was not used in ISD.

R. G. Fernquist: Since the Pt and W from the thermally deposited film coating runs were highly reactive, do you feel it is worthwhile to attempt these coatings at a lower temperature on the source, or, alternatively, to coat Pt in the more conventional PtC shadow casting method?
Authors: If the temperature of the Pt source were lowered the Pt would not evaporate. The suggestion to evaporate Pt by the PtC method is an excellent one.

J. D. Fairing: What is meant by sputter coating with no Argon?
Authors: Instead of using a partial pressure of 100 microns of Argon, the gas bled into the vacuum chamber is air to the same pressure.

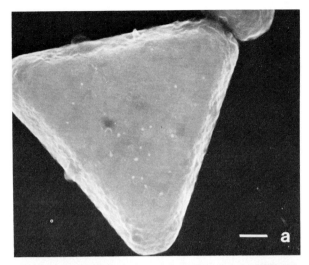

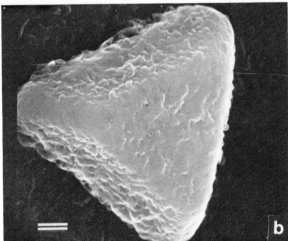

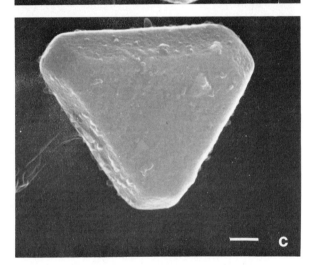

Figure 5. Secondary electron images of metallic coated silver halide grains; a. no coating, b. 11nm Pt by SC, c. 12nm Pt by ISD. Bar = 200nm

SCANNING ELECTRON MICROSCOPY/1979/II
SEM Inc., AMF O'Hare, IL 60666, USA

ASBESTOS FIBRE COUNTING BY AUTOMATIC IMAGE ANALYSIS

R. N. Dixon and C. J. Taylor

Wolfson Image Analysis Unit, Department of Medical Biophysics
Stopford Building,
University of Manchester
Oxford Road,
Manchester, M13 9PT, U.K.

Abstract

In this paper we review attempts to automate asbestos fibre counting in light microscopy and scanning electron microscopy. It is apparent that inadequate image analysis techniques have hindered the routine application of truly automatic systems.

In the second part of the paper, we develop simple but powerful image analysis algorithms that are directly suited to the problems posed by images of asbestos fibres.

Finally, we describe a practical implementation of these methods using the MAGISCAN image analyser and present some preliminary results.

Introduction

It has been standard practice for many years to characterise the fibrous content of airborne dust by using optical phase contrast microscopy on cleared membrane filters. More recently interest has been aroused in the use of scanning electron microscopy both to improve the visibility of submicron fibres and to allow the possibility of fibre identification using X-ray dispersive analysis. Manual inspection of both types of microscope image is tedious, time consuming and unreliable. Counts depend on the attentiveness of the counter, his subjective interpretation of aggregates and his subjective assessment of fibre length and width. This leads to discrepancies of over 100% between different laboratories counting the same sample[1,2] and consequently the value of an automatic image analysis system capable of reliably doing the same job has long been recognised.

The problems which arise when attempting to use automatic image analysis are similar for both optical and S.E.M. images. Fig. 1 illustrates some of the circumstances which give rise to variability in manual counting and which also make automated analysis difficult. To count the field correctly aggregates of fibres must be resolved into individual fibres. Very thin fibres, although definitely visible, are imaged with low contrast and often chance "noise" variations in background intensity may be as large as the "signal" from such fibres. Finally, single fibres which are not near the detection limit may, for various reasons, have short sections which are not imaged at high contrast and so can appear as several fibres. Harness[3] has previously reported all these problems with phase contrast optical images and Werlefors et al[4] report almost identical problems with scanning electron microscope images although, of course, the fibre diameter at which detection becomes difficult is smaller.

KEY WORDS: Asbestos Fibres, Image Analysis, Particle Sizing

Review of Previous Work

The introduction in recent years of commercial image analysers has prompted several studies of the applicability of such systems to automated asbestos fibre counting. In addition, computer-controlled scanning electron microscopes, designed for general particle analysis, have been applied to fibre counting. The development of appropriate image analysis techniques for fibre counting has, however, received scant attention from the academic community.

In this section, we review the more significant projects on automated fibre counting. Emphasis will be placed on the image analysis techniques employed and upon the aspects of hardware which influence the choice of technique. Since similar analysis methods are required for light microscope and S.E.M. images, no particular distinction will be drawn between them. All the systems discussed use television camera input for optical images.

Harness[3] in 1973 was amongst the first to investigate the suitability of commercially available instrumentation for fibre counting. A Quantimet 720 Image Analyser was used to count fibres in optical images of airborne dust samples. Local contrast thresholding was used to detect fibrous and particulate material. Fibres were distinguished from particles and aggregates on the basis of their shape. Thus, although isolated fibres of simple geometry could be counted and sized correctly, Harness reported that poor contrast, insufficient depth of focus (leading to incorrectable fragmentation of detected fibres) and the presence of aggregates of fibres caused difficulty.

The Franklin Institute, in 1975, studied automated analysis of scanning electron microscope images also using a Quantimet 720 system. The study was sponsored by the Environmental Protection Agency (EPA). Although the system was successful in counting fibres with simple geometry the EPA report[5] concluded that aggregates of fibres were likely to limit the usefulness of fully automatic measurement. Since Quantimet systems are constructed from modular, hardwired processing elements, further progress would involve the development of new hardware.

The National Institute of Occupational Safety and Health (NIOSH) commissioned a survey of potential users and suppliers of systems for automated asbestos fibre quantification. In the final report[6], November 1975, members of both the National Bureau of Standards and of the EPA were said to feel that automatic analysis of fibre aggregates was an important unresolved problem.

Two commercially available systems were proposed then by their manufacturers as being able to overcome this difficulty. Bausch and Lomb described their Pattern Analysis System as being able to distinguish between fibrillated or touching fibres and crossed fibres. More recently, they have introduced an image analyser called the Feature Analysis System - Model 2. This system is flexible since measurement of detected image features is performed by software

in a minicomputer. Fully programmable analysis of complex geometries should allow this system to meet the objections stated above. To our knowledge, the citeable literature contains no reference to either system being applied to asbestos fibre counting more recent than 1975.

In the same report, Le Mont Scientific recommended the Model B-10 Particle Analysis System as being suitable for characterising asbestos fibre samples. The system is adaptable to either S.E.M. input or optical input, using an image dissector television camera in the second case. Image scanning is controlled by a minicomputer. Since the brightness at any point in the image can be randomly accessed, it should be possible to implement sophisticated fibre detection algorithms. Furthermore, analysis of detected structure can be carried out in the minicomputer so that resolution of a fibre aggregate into its constituents is possible. Again we have found no reference more recent than 1975.

The Swedish Institute for Metals Research has developed a minicomputer-controlled scanning electron microscope and image analysis system[7]. A complete field is stored digitally as grey-level data in a Comtal 5000 special purpose image memory. The minicomputer may access this data under programme control. The system has been applied by Werlefors et al[4] to the counting of asbestos fibres. Objects in the field of view are detected by grey-level thresholding. Potentially countable fibres are discriminated from particles on the basis of shape. Fibre length is determined as the longest chord between two points on the boundary of the object. Width is measured perpendicular to this axis. Fibres with a length/width ratio greater than 3 are counted. The system requires approximately 3 minutes to process a field, of which fifty seconds is spent acquiring the image. Difficulties experienced with the image analysis techniques used were a failure to detect the finest fibres due to poor contrast, and the fragmentation of longer fibres due to incomplete detection. Since all analysis is programmed in the minicomputer, it should be possible to implement more sophisticated methods to overcome present limitations.

Pavlidis and Steiglitz[8] of Princeton University examined the problem of counting fibres in an aggregate using computational methods. Fibrous material is detected by grey-level thresholding. The resulting binary image is searched for the ends of fibres and, assuming that fibre ends are always freely visible, the number of fibres is equal to half the number of ends. Processing is simple and therefore rapid, but the approach in itself will not reveal information on fibre length and width.

All the systems reviewed can be used in a manually assisted mode, in which an operator indicates in some way what the machine is to measure. Such solutions are preferably to be avoided since one of the major reasons for introducing automation is the removal of the subjective element from fibre counting.

Summary
    Three main problems have hindered automatic
analysis of asbestos fibre images:- the poor
contrast of fine fibres, the fragmentation of
longer fibres and the presence of fibre
aggregates. These properties are common to both
optical and scanning electron microscope images.
Whilst partial success has been achieved with
automation, any or all of the above complications
may be present in a sample and systems unable to
cope reasonably well with them are unlikely to
meet the needs of the end user. Hardware of
sufficient flexibility is now available to tackle
such problems but no fully automatic system has
so far been demonstrated that has gained wide
acceptance.
    In our opinion, the failure of image
analysis to make significant impact in this field
is due to the inadequacy of the analytical
techniques brought to bear upon the subject. The
difficulties outlined above are sufficient to
require specific solutions to be developed.

An Automated Fibre Counting System

    In the remainder of this paper we describe
the image analysis methods which we have
developed to overcome the main known problems in
automating fibre counting. For each field the
task is broken down into several independent
steps. First, all fibrous material in the image
is detected. Secondly, individual fibres are
extracted by analysing the structure of detected
material. Finally, length and width are measured
for each fibre and are used as a basis for
classification. We describe also a practical
implementation of these ideas using the MAGISCAN
image analyser and present some preliminary
results.

Detection of fibrous material
    Some automatic method must be devised for
detecting all the fibrous material in any field.
This is made difficult because local background
intensity variations are often similar in size to
the signals from finer fibres (Fig. 2a).
Elementary methods of image segmentation, such as
grey-level thresholding[9] are unable to distinguish
satisfactorily between fibrous material and background (Fig. 2b). This is because the signal to
noise ratio for points taken in isolation is too
low.
    The classical method of reducing the effect
of noise is to average the data. In the case of
fine fibres it is clearly necessary to take an
average along a line locally parallel to the
fibre. To find the appropriate direction, the
mean grey-level may be calculated along lines in
several directions (Fig. 3). For fibres with
grey-level lower than background the smallest
mean will be found in the direction best aligned
with the fibre. In practice, eight line
orientations are used within a 5 x 5 picture
element region and the minimum line is sought.
Since we are concerned only with local variations
in grey-level, the difference between the mean
grey-level of the surrounding region and that of
the minimum line is taken. We can define a local
operator to perform this function and we may

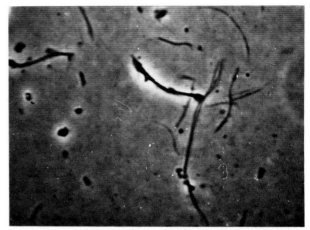

Figure 1. Optical phase contrast image of
chrysotile asbestos fibres on a membrane filter.
Picture width = 95 μm.

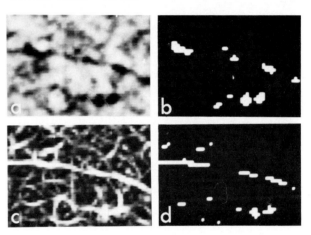

Figure 2. Detection of fibrous material using
the local operator.
(a)   a marginally visible fibre.
(b)   detection by grey-level thresholding.
(c)   local operator transform of (a)
(d)   threshold detection applied to (c).
Quadrant width = 16 μm.

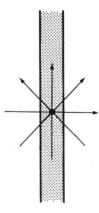

Figure 3. Principle of the local operator
used for fibre detection.

apply it at any point in the image. We would expect the output of the operator to be large at points along a fibre, where there is a linear correlation in grey-levels, but smaller at points in the background, where such a correlation is lacking. Fig. 2c is a transform of the original image, in which the brightness at each point is proportional to the output of the local operator at that point.

In practice, detection of fibrous material is performed by thresholding such a transformed image. This is illustrated in Fig. 2d where a threshold has been chosen such that the same proportion of background is detected as in the simply thresholded image (Fig. 2b). The improvement in signal to noise ratio is clearly demonstrated by the increased integrity with which the fibre has been detected.

Fibre extraction

Once the fibrous material in a field has been detected it is necessary to extract a description of each individual fibre. Fig. 4a shows a fibre complex and Fig. 4b is the detected image, which contains much irrelevant detail. It is useful as a first step to simplify this image by generating its skeleton (Fig. 4c). The skeleton consists of a set of lines of unit width which preserves the connectivity of the original image[10].

The detected image is a binary map of the original scene. Skeletonisation proceeds by removing a one picture element wide "skin" from each object in the map. However, no point is removed if by doing so the connectivity of the image would change. This test can be performed by examining the spatial relationship of points in the 3 x 3 neighbourhood of each candidate point for removal. Multiple passes through the map are made to remove successive "skins". The procedure halts after a pass in which no changes are made.

As a further simplification the skeleton is reduced to a set of simple, disjoint line segments by removing all points which have more than two neighbours. With the data in this form it is possible to deal with situations in which fibres overlap or are only partially detected. Fibre extraction proceeds by considering the likelihood that pairs of neighbouring segments derive from the same fibre.

Fig. 5 illustrates the three local geometrical parameters which are used to match segments. The first parameter is simply the distance between two end points. If two segments are distant it is unlikely that they belong (in that sequence) to the same fibre. The second parameter is a measure of path discontinuity or kink. The total angular distortion (a+b) reflects how closely each segment extrapolates to the end-point of the other. Since the size of the kink increases with distance, this function is weighted by distance. A large kink implies two separate fibres. The third parameter is the orientation of one segment with respect to the other. Large angular deviations in a continuous fibre are rarely observed.

If each of these parameters is less than a corresponding threshold value then two segments are matched. A measure of goodness of match is evaluated, based on a simple linear combination of the three parameters. This measure is used to resolve contention between several segments which may match one another. Each segment is constrained to belong to one fibre only and a logical connection is made between segments which display the best fit. Fig. 4d shows a single fibre automatically extracted from the fibre complex using this method.

Measurement and classification

In order to classify each particle as a respirable fibre or non-significant debris it is necessary to measure its length and width. For fibres which are not straight this is a non-trivial problem. The definition of length which we have chosen is that the length of a fibre is the path length of its skeleton. Width is more difficult to measure since the fibres of interest have widths comparable to the resolution limit of the microscope. The original image is scanned in a direction locally perpendicular to that of the skeleton to obtain a grey-level cross section of the fibre. The simple measure of width which we use is the distance between the first zero crossing points on either side of the fibre. This method is based on the model that a phase contrast objective produces a second derivative image of the original object. Following internationally accepted practice, fibres longer than 5 µm, with width of less than 3 µm and aspect ratio greater than 3, are included in the count of biologically hazardous, respirable fibres.

Implementation

The methods discussed so far form a logical basis for the analysis of fibre images. Practical automated image analysis demands, in addition, a machine which can implement these methods in acceptable times and at reasonable cost. This section describes briefly the important features of the hardware and the organisation of the software.

The MAGISCAN[11] is a self-contained, general purpose image analysis system. Fig. 6 shows the organisation of its main elements. The image processor performs intensive computations on bulk image data at very high speed. Normally the source image is input via a CCIR standard, closed-circuit television camera to a small cache memory at 512^2 x 6 bit resolution. On line connection to an S.E.M. is also possible. A fully programmable arithmetic unit can access this data in arbitrary sequences. A full resolution binary image plane stores intermediate processing results. The contents of this memory can be superimposed on a TV monitor display of the scene under analysis. A light-pen used on the display provides simple direct interaction. The 64k byte Data General Nova minicomputer controls the operation of the Image Processor and handles interaction with the user. The use of a standard operating system and high level language allows rapid development of software for new applications.

The image analysis software is partitioned between the two processors so as to maximise performance. Thus, the fibre detection and skeletonisation algorithms run in the image processor. Analytical functions, such as fibre extraction and width measurement, use relatively little data compared to the initial processing but require greater computational flexibility. These functions are programmed in FORTRAN to run in the minicomputer.

Results

Most of the work so far has concentrated on optical microscopy though the methods apply equally well to S.E.M. images. Table 1 summarises the results of a preliminary comparison between machine and manual counting. A test sample of airborne chrysotile asbestos dust was prepared by the health physics laboratory of a textile manufacturing plant. The filter was cleared and mounted in glycerol triacetate and was counted eight times both by experienced technicians at the laboratory and by the MAGISCAN. Details of the standard sample preparation and counting procedures that were followed are given in reference 12. For both manual and machine counts, approximately 200 fibres meeting the criteria of length > 5 μm, width < 3 μm and aspect ratio > 3, were totalled per evaluation. The machine used a square counting field of width 160 μm. Approximately eight fields were required to accumulate the count limit. Stage movement and focussing were performed by hand.

The mean of the automated counts is about 13% higher than the mean manual count and the standard deviations of the two groups of results are similar. The mean time per automated count is about a third of that for the manual counts. These preliminary results indicate reasonably good agreement although the automated counts are probably significantly higher. This may be due to some residual fragmentation of detected fibres. Although the spread of results from a single manual counting laboratory can be reasonably well controlled, inter-laboratory comparisons have revealed discrepancies of over one hundred percent. The relatively low spread of automated counts represents a real improvement because the results are free of subjective bias and could be repeated in other laboratories.

Table 1

A comparison of automated and manual counts

	Manual	Machine
Mean (fibres/ml)	4.77	5.39
Std. dev. (%)	11.7	13.5
Time/count (mins)	20	6

Summary

Light microscope and scanning electron microscope images of asbestos fibres present similar problems for automatic analysis. These problems can be summarised as: poor contrast of the finer fibres, fragmentation of longer fibres and the occurrence of fibre aggregates. The image analysis techniques previously applied have not been well matched to the complexity of such

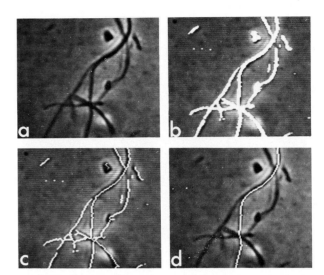

Figure 4. Analysis of an aggregate.
(a) original fibre complex.
(b) detected image.
(c) skeletonised image.
(d) single, extracted fibre.
Quadrant width = 40 μm

SEGMENT MATCHING

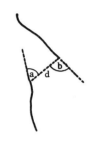

Distance = d

Kink = d (a + b)

Deviation = a - b

Figure 5. Parameters used in segment-matching.

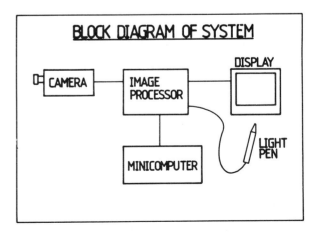

BLOCK DIAGRAM OF SYSTEM

Figure 6. Block diagram of the MAGISCAN.

images. Our approach is to analyse as fully as possible the fibre structure of each microscope field. This has necessitated the use of a local operator to stabilise detection of fibres in the presence of image noise. We have adapted simple but powerful methods to the analysis of the resulting structure to deal with fibre aggregates and fragments. Once individual fibres have been extracted, physical parameters can be measured. The implementation of these methods on the MAGISCAN Image Analyser makes full use of the complementary processors to achieve fast sample analysis whilst minimising programme development time. Initial results suggest good agreement between automatic counts and carefully controlled manual counts.

Further development will be necessary to improve fibre matching criteria and to cope with very dirty samples where there is much more irrelevant debris than fibre. With these improvements it seems possible that automated counting could be used routinely in health physics laboratories handling samples from diverse occupational environments.

References

1. Gibbs, G.W., Baron, P., Beckett, S.T. et al. A summary of asbestos fibre counting experience in seven countries. Ann. Occup. Hyg. 20 (4) 1977 321-332.
2. Duggan, M.J., Culley, E.W. The counting of small numbers of asbestos fibres on membrane filters: a comparison of results from some commercial laboratories. Ann. Occup. Hyg. 21 (1) 1978, 85-88.
3. Harness, I. Airborne asbestos dust evaluation. Ann. Occup. Hyg. 16 (4) 1973, 397-404.
4. Werlefors, T., Eskilsson, C., Ekelund, S. et al. Automated fibre measurement with a computer controlled scanning electron microscope. Paper presented at The International Symposium on the Control of Air Pollution in the Working Environment. Stockholm, 6-8 Sept., 1977. Authors address: Swedish Institute for Metals Research, S11428, Stockholm, Sweden.
5. Environmental Protection Agency, U.S.A. Development of Scanning Electron Microscopy for Measurement of Airborne Asbestos Concentrations. EPA-650/2-75-029, January, 1975.
6. Whisnant, R.A. Evaluation of Image Analysis Equipment Applied to Asbestos Fibre Counting. Final Report to the National Institute for Occupational Safety and Health, U.S.A., November, 1975. RTI No. 43U-1155-5. Research Triangle Institute, P.O. Box 12194, NC 27709, U.S.A.
7. Ekelund, S., Werlefors, T. A system for the quantitative characterisation of microstructures by combined image analysis and X-ray discrimination in the S.E.M. SEM/1976/1. IITRI, Chicago, IL, 60616, p. 417-424.
8. Pavlidis, T., Steiglitz, K. The automatic counting of asbestos fibers in air samples. IEEE Transactions on Computers, C27 (3) March 1978, 258-261.
9. Rosenfeld, A., Kak, A.C. Digital Picture Processing, Academic Press, New York, U.S.A.
1976, 258-275.
10. Hilditch, C.J. Linear skeletons from square cupboards. Machine Intelligence, 4. B. Meltzer and D. Mitchie (eds.), Edinburgh University Press, Edinburgh, 1969, 403-420.
11. Taylor, C.J. and Dixon, R.N. Quantitative image analysis using structural methods. in Proc. 7th L.H. Gray Conference. G.A. Hay (ed.), IOP/ John Wiley & Sons, Bristol, U.K., 1976, 295-308.
12. ARC - Technical Note 1. The Measurement of Airborne Asbestos Dust by the Membrane Filter Method. June, 1969. Asbestosis Research Council, P.O. Box 40, Rochdale, Lancashire, U.K.

Discussion with Reviewers

E.J. Chatfield: Beckett and Attfield (Ann. Occup. Hyg. 17, 85-96, 1974) concluded that in inter-laboratory comparisons of the counting of asbestos fibres using phase contrast optical microscopy, inexperienced counters found only one quarter of the number reported by experienced counters. They found that the principal reason for the discrepancy was the failure of the inexperienced operators to search through the full depth of focus within which fibres were present in the Millipore filter.

The system you describe obtained comparable, indeed somewhat higher results than a group of experienced fibre counters using the same sample. In view of the fact that the automatic system operates using the image obtained at a single position of focus, thus overlooking many fibres distributed at other depths, can you reconcile this with the observations of Beckett and Attfield?

Authors: It is not necessary to search many planes of focus in order to count a field correctly. Rather it is important to select the correct plane in the first place. The great majority of fine fibres in a field lie in a narrowly-defined plane a short distance ($\simeq 1$ μm) above the filter surface. Provided this plane is in focus, all fine fibres will be detectable. Selection of an appropriate plane for larger fibres however is not so critical, since their greater contrast renders them detectable over a much wider range of focus. We use this fact to control focus when running samples on the MAGISCAN. Inexperienced technicians may easily miss the plane containing the fine fibres and significant undercounting will result. This corresponds with the findings of Beckett and Attfield.

Acknowledgements

The authors gratefully acknowledge the support of the Asbestosis Research Council, U.K. We would also like to express our gratitude to Ms. Irene Hartley for her excellent preparation of the typescript.

For additional discussion see page 490.

SCANNING ELECTRON MICROSCOPY/1979/II
SEM Inc., AMF O'Hare, IL 60666, USA

COMPUTER GRAPHICS ANALYSIS OF STEM IMAGES

R. Llinás, R. Spitzer, D. Hillman and M. Chujo

Department of Physiology and Biophysics
New York University Medical Center
550 First Avenue
New York, NY 10016

Abstract

We have designed and implemented an inter-face between a PDP-15 (DEC) computer and a JEOL 100C scanning electron microscope accessory, to allow computer control of the scanning beam and direct memory storage of frames. Software con-trols the beam position during either raster scanning or searching of the field. The hardware design allows 1) transmission of analog voltages from digital-to-analog converters for positioning of the scanning beam, 2) return of intensity at each point from the SEM or STEM detector, and 3) interrupt control of both computer and STEM. Software addressing of an existing storage oscilloscope controller overrides internal scan-generating circuitry and positions the beam at any point in the frame.

Images can be scanned in a raster mode and stored as a matrix having up to 500 x 500 points, each digitized to one of 64 gray levels. Optimum density ranges are set under software control. Picture point gray levels can be windowed and a range of densities selected for display. This procedure is useful for analyzing biological specimens which have been selectively stained for structural details. In the central nervous system, myelin sheaths and synaptic structures stain significantly more densely than surrounding membrane components; thus, they can be windowed and their perimeters extracted. Software for graphics display of these stacked surface con-tours allows three-dimensional reconstruction and interactive analysis.

Alternatively, by positioning the beam and sampling the density at that point, the program can search the field and extract key features having a specific density level. For example, the program can draw perimeters and store repre-sentations of complex pictures without recording and storing the entire frame. By reducing the complexity of image acquisition, this approach yields rapid, accurate, recording and analysis of ultrastructural images.

KEY WORDS: Image Analysis, Computer, STEM, Nervous System

Introduction

Computer applications to aid study of bio-logical structures have taken diverse courses such as tomography[1], light microscopic image analysis[2-7], and ultrahigh resolution micros-copy[8-9]. In general, all input methods have utilized the common principle of density digiti-zation from images which are produced by varied forms of imaging devices[1-16]. Many computer programs are available for processing and en-hancing images once these images have been stored in the computer. These programs can ex-tract data from the image and allow processing operations such as correction of geometric dis-tortion, contrast improvement, noise reduction or improvement in spatial resolution[17]. The scanning microscope (SEM) and the scanning trans-mission electron microscope (STEM) are compatible with direct digitization of images without inter-mediate film processing[16-18]. Surprisingly, few applications have utilized the direct approach to digitization but, rather, have chosen film scanning modes (FIDAC)[8-10,13,15]. This is largely because transmission electron micro-scopic negatives have higher resolution than electron images from electron beam scanning de-vices. With the introduction of high resolution scanning attachments to transmission electron microscopes, image quality has been achieved so that direct digitization of biological STEM images more closely approaches that obtained by transmission electron microscopy. This approach eliminates the film step and allows either di-rect transfer of frames to memory or use of searching algorithms to record and analyze selected information concerning the preparation.

Approach

Direct Data Acquisition from Electron Microscope

Our approach to computerized electron mi-croscopy is based on two considerations. First, the information must be adequate to provide data for analysis at resolutions within conventional ranges of biological electron microscopical images. Second, acquisition of data must be rapid enough to allow analysis of sections at rates suitable for quantitating large areas of tissue or recording for three-dimensional reconstructions. Since these goals should be achievable in the same system, we have designed hardware which is capable of either direct digitization of images or image analysis through beam searching for predetermined density levels on this preparation.

A conventional raster scan in the direct digitization mode allows acquisition of data from the electron microscope, independent of preprocessing of information. The time for acquisition and storage of pictures of good resolution and adequate gray levels can be considerable, typically 1-5 minutes unless a direct memory picture storage is available.

In order to increase the speed of data acquisition and reduce the necessity of massive storage, a second software approach was designed. Here, recording only specific types of information served to eliminate sequential image processing and unnecessary data storage. This approach is called "software controlled vectored access (SCVA)". A similar approach has been designed for light, macro- and microscopic images as well as film[11]. This method utilizes gray level as well as shape and size to select data from images. Such as approach is desirable for quantitation and categorization of specifically stained structures which are clearly definable by density.

Criteria Considered in Designing this System

The limits of *resolution* were primarily defined by the size of the scanning beam and limits introduced by the digitizing hardware. Further restrictions are the number of points used to represent each frame and the gray levels chosen to define the image. These latter two factors determine the time for picture storage and need for storage capacity. The resolution required depends on the application, i.e. the size of objects to be studied (magnification) and the details necessary to characterize the image.

The number of *density levels* used to represent the shading of the image becomes an extremely important aspect for picture analysis. Most important in this regard is the necessity to have adequate gray levels distributed within a range where the information is contained. Limitations for gray level assigned each pixel are related to the amount of memory available and the packing of this information into each word.

Controlling of beam position could be carried out by two basic approaches: 1) the scan circuitry of the ASID could be utilized for controlling the scan or 2) the computer could provide direct control of the beam deflection circuitry. We chose the latter approach to allow the greatest range of possibilities for computer control and for simplicity of design and implementation. Software control serves to address a wide number of manipulations of beam position within the limitations of beam settling time. The STEM raster scan approach requires that every frame be scanned by line and that only the scan rate be controlled. This approach is much less flexible and should be considered only if required by the hardware, since computer control of the beam provides both raster scanning or arbitrary positioning of the beam. Software

accessing of information from images is extremely desirable and can circumvent recording the density at each picture point on the entire image. Such an approach requires a highly specialized software design which directs searching to the most likely site for data accumulation.

Other considerations were *speed of picture transmission, storage and analysis*. The limitations are in the scanning detector, the A → D and D → A converters, and the amplifiers driving the scanning coils. Amplifier capabilities and scan coil hysteresis can limit beam settling time.

Since our computer and electron microscope facility are in separate parts of the laboratory, signal transmission must be provided over a *noise-free transmitting system*. This entailed the design of a clean amplification and receiving system in order to prevent signal degradation over the length of the cable. (For reduced noise and improved speed, we are currently modifying the system to allow digital transmission and high speed conversion of the x, y axes and intensity at the microscope).

Hardware

In its simplest form, the hardware for the system may be represented by the broken line in Figure 1. A DEC PDP-15 computer is programmed to control the ASID scanning attachment of the JEOL 100C electron microscope and supplies the coordinates for beam location. At each locus, the computer examines the scanning detector output and stores a binary number corresponding to the gray level (density) for that locus. This is repeated until the entire picture is stored or, alternatively, until selected structures in the image are outlined according to software requests by interactive searching programs. These images can then be reproduced on a storage scope for examination and hard copy printout or, alternatively, they can be displayed on an interactive graphic terminal (Figure 1).

The fundamental design requires beam positioning circuitry which can be externally controlled. The D → A converters should have at least 10 bit registers each for x or y beam positioning (those addressed through our Tektronix 611 storage scope drive were utilized). Also needed is an A → D converter with 1,000 levels of digital conversions. Sixty-four gray levels are chosen for storage and optimally selected from the 1,000 levels by program software. Additionally, a display processor should be available for interactive graphics and display of extracted information.

Restrictions on the design of our system were mainly centered around the necessity to utilize equipment at hand and provide the necessary interface which would adequately carry out the desired software operations. This equipment included a scanning microscope (ASID attachment to a JEOL SEM 100C) and a DEC PDP-15 computer with 1) a fixed head disc; 2) cartridge discs; 3) 96 K of core memory; 4) an API Multi-user vector interrupt system; 5) A → D converter; 6) VT15/04 interactive graphics display processor/terminal and 7) Tektronix Model 611 storage oscilloscope with I/O bus interface.

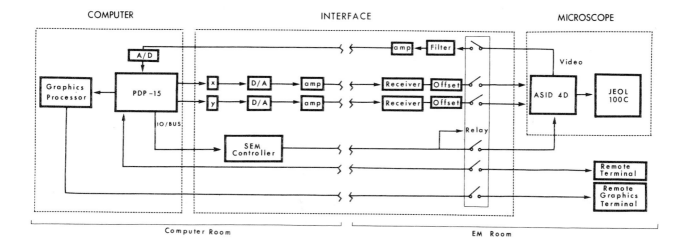

Figure 1: Diagram of Interface between Computer and Electron Microscope

The essential components of the interface are illustrated in Figure 1. Five types of interconnections constitute the fundamental interface between the computer and the electron microscope laboratory: 1) beam positioning, 2) video information, 3) digital control lines, 4) remote terminal connections and 5) remote graphics terminal.

Beam positioning is transmitted as x and y analog signals from digital to analog converters on the computer side. The signals for x and y are amplified for transmitting on 75 ohm coaxial cable to receivers in the electron microscope room (about 75 feet). In our setup, the D → A converters provide 0 to 1 volt outputs which are amplified to 0-12 volts for compatability with the scan coils of the electron microscope. An offset adjustment allows the direct superimposition of the visualized ASID frame with the computer generated scan.

The video signal is also transmitted as analog on coaxial cables to the computer. Unity gain amplifiers are employed after filtering to remove high frequency noise inherent in the scanning detector output. Variable amplification of the video signal is provided to take advantage of the full range of ± 10 volts acceptable to the A → D converter.

Digital control from the computer to the electron microscope is essential in order to supply software instructions for control of relays to engage the interface to the computer. These relays also allow isolation of the computer interface when the interface is not powered. Currently one set of digital controls activates the interface circuitry and allows the computer to take over full control of the beam positioning. This control is via a software address to the EM controller, which acts as a peripheral device on the computer I/O bus.

The fourth part of the interface consists of drivers for an oscilloscope to display the recorded information. This signal is transmitted on x, y and z (intensity) connections from either the graphics terminal or from the storage scope drivers. Unity gain amplifiers employed for driving these three lines are similar to those used for beam positioning. The fifth interconnection is to a terminal on the EM site and needs no special circuitry.

Software

Recording for direct acquisition of electron microscopical images are of two types. With the first, a rapid scan mode allows direct recording of electron images as a raster picture of specified size and resolution (Figure 2). The second method utilizes arbitrary location of the scanning beam for sensing image density at selected points. This information is used to determine the next location of the beam and thus directs the beam along lines of "isodensity". The first approach produces a complete digitization of the image while the second defines the perimeters of individual dense or light profiles. These picture or perimeter files are then processed and/or analyzed by specific programs to obtain the desired type of information.

The raster scan mode allows recording of picture points in a sequential manner, each with 64 levels of density digitization by an A → D converter. Through operator instruction, the program specifies the exact area within a maximum of 1,024 x 1,024 picture points for x and y coordinates. Also, the sequence of recorded points along each scan line can be limited in order to reduce resolution and scan time. For example, the operator can specify the starting point as x = 100 and y = 200 and then scan a box 500 points wide by 300 points high while only storing every second point. This allows selecting areas which contain information or matching the recording frame to that of the ASID.

In the raster scan mode, resolution defined by 64 gray levels is optimized by software which determines the minimum and maximum gray levels within the 1,024 range governed by the A → D converter. Thus, information which is present in the lowest part of the density scale can be recorded with sufficient gray levels to differentiate the immediate surrounding gray levels. Similarly, by selecting gray levels via a narrow density window, it is possible to favor the recording of specific information which can be isolated by, for instance, electron dense staining in the preparation.

Picture files contain 3 points per 18 bit word. This is made possible by packing coordinates and 64 gray levels together in a sequential manner. Analysis programs decode and store this information into core memory for further processing of images.

The arbitrary tracing program (SCVA) provides the first level of image analysis at a microscopic level. Essentially, a perimeter tracing program searches the specimen commencing at a specified starting point. It then outlines the perimeter of objects according to density levels specified by the operator (Figure 2). In the completely automated mode, the procedure begins as a raster scan with limited numbers of points being sensed for density levels. When a dense point is acknowledged, a perimeter trace routine begins which follows the contour of equal density. This program defines points along the perimeter by searching in localized areas for coordinates which define the boundary by density level. The relationship between the last point and the previous point provides the direction for the subsequent points to be detected. This search is programmed by the shape of a limited searching area which can be altered to accommodate characteristics of the contour. Typically, an entire frame of 500 x 800 points can be analyzed in 10 to 15 seconds. The completion of the perimeter trace by returning to the initial starting point on the contour, automatically closes the trace and the next dense structure is sought by the searching beam (usually as a raster scan of limited points which ignores regions within recorded perimeters). Coordinates for vectors along this boundary are stored as perimeter files. Each frame consists of sequential points for vectors which define perimeters of dense objects. The final length of vectors is determined by a software routine which adjusts the length according to the curvature of the boundary in order to avoid storage of every point along the perimeter.

Analysis

Picture and perimeter files are analyzed by varied programs to extract specific images and/or quantitate structures in each recorded frame (Figure 2). Picture files are analyzed for selected images by programs for picture enhancement (e.g. near neighbor analysis of picture points for removal of noise, cf. Ref. 17), or density windowing and ultimately for pattern recognition. A useful means of extracting information for picture files from raster scans is through perimeter extraction by a program which is similar to the perimeter recording approach. Here structures within the image, having density levels within specific ranges, can be selected and their perimeter defined directly in core memory. These perimeter files contain coordinates for selected images, provide necessary data for extracting perimeter length as well as area of structures within sections, or SEM images. Quantitative programs can also be used to determine shape parameters, such as short and long axes or perimeter-to-area ratio. Other parameters can be applied to quantitative stereological approaches for determining the numbers of structures as well as their volume and surface relationships within sectioned material.

Another important approach is the utilization of perimeters (extracted either directly from the microscope or from picture files) for reconstruction of surface configurations in three dimensions. Intermediate subroutines are generally necessary to allow alignment of the perimeters as well as for amending artifacts introduced in the preparation. (Alignment of sections is most effective at the electron microscope by superimposing a rapidly scanned image over a previously stored image and then correcting for rotation and x and y via the stage controls). In some cases, picture enhancement may be necessary before the perimeter routine is applied to the picture file. Programs for three-dimensional reconstruction and rotation of stacked perimeters from serial section allows viewing the structure on a graphics display with the hidden lines removed[11-14]. The rotation sequence provides motion picture recording for dynamic analysis of surface contours. In addition, the stackable picture files provide quantitative data for parameters such as volume and surface area.

Applications

Recording of selectively stained material has been found to be the most applicable for obtaining picture files or perimeter file techniques. This approach on biological specimens is presently essential, since pattern recognition of subtle differences in gray levels is not practical with limited core memory and lack of an array processor. Specific staining of structures such as synapses in the central nervous system, or cytological structures which react specifically with enzyme techniques, provides specific densities beyond that of normally stained structures and serves to select structures for recording and analysis.

Myelinated sheaths of neuronal axons are readily definable biological structures which are easily digitized from images which are routinely fixed and stained. The stacked membranes and an interposed substance define a dense mass which has the shape of a donut in a cross-section (Figure 3). These profiles are definable by their inside boundary with the axon they circumscribe. Secondary processing of the

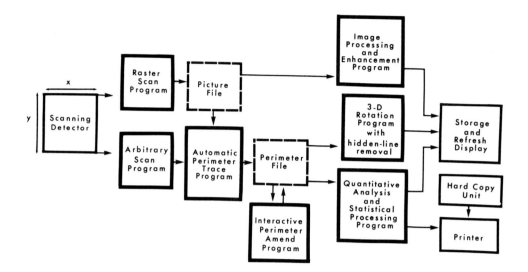

Figure 2. Software for direct electron microscopical image processing or raster scan recording and analysis.

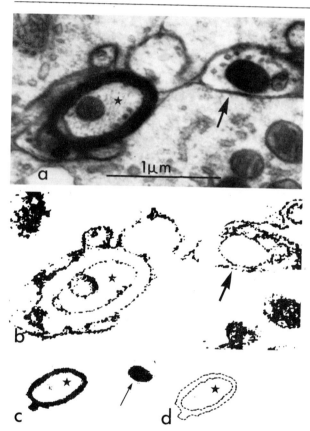

Figure 3. Raster scan digitization of STEM image from central nervous system and perimeter trace of a myelin sheath. a: Image from photographic negative; b,c: Digitized image at two density windows (b = density slice at levels 38-42, c = levels 50-64). d: Perimeter trace of myelin sheath from computer stored image.

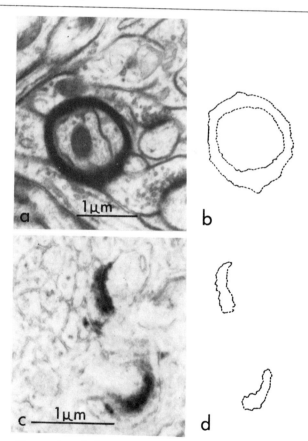

Figure 4. Software-controlled STEM perimeter trace directly on the specimen. a,b: Myelin sheath and perimeter trace. c,d: Two synaptic junctions and perimeter traces.

perimeters is required when attempting to differentiate the outer boundary except in preparations where the myelin sheaths are separated from each other.

Synaptic sites defined by pre- and post-synaptic membrane thickenings are most readily definable structures when adequate staining can be obtained. The method of post-staining of glutaraldehyde-fixed material with phosphotungstic acid has provided a means of selectively differentiating these disc-shaped structures from the surrounding neuropil[19]. As seen in Figure 3, the densities are easily selected by windowing of the gray level from raster scan picture images. These data contain coordinates for location, size and shape of the profiles in section. By extraction of perimeters from these data files, individual structures are defined and can then be easily analyzed to determine the size, shape and relative locations between structures (Figure 4).

In the direct perimeter recording approach which utilizes software controlled vector access to examine picture points, perimeter files are formed directly from the image. Here, each of the synaptic densities is outlined as in the perimeter files which are formed from picture files themselves; however, without the intermediate time consuming picture storage step. Analysis of these files subsequently provides information for quantitative analysis as well as three-dimensional reconstruction.

Conclusion

We have described hardware and software to allow a DEC PDP-15 computer to direct the beam of a JEOL 100C scanning transmission electron microscope, and store raster scanned images or preprocessed digitized representations of images directly in memory. This approach has the advantage of simplicity of design and implementation, as well as a great degree of flexibility. By allowing the operator to see results of picture processing immediately, and to interact through software with the image while at the microscope, a broad range of interactive processing becomes available. This system minimizes the need for storing large amounts of data in core computer memory and storage devices. Additional hardware and software are being developed to improve image quality and simplify system use by the non-computer specialist.

References

1. W. Wagner. Reconstruction of object layers from their x-ray projections: A simulation study. Computer Graphics and Image Processing 5, 1976, 470-483.

2. P. Coleman, C. Garvey, J. Young and W. Simon. Semi-automatic tracking of neuronal processes. In Computer Analysis of Neuronal Structures, R.D. Lindsay (ed.), Plenum, New York, 1977, 91-110.

3. E.M. Glaser and H. Van Der Loos. A semi-automatic computer microscope for the analysis of neuronal morphology. IEEE Trans.

Bio. Med. Eng. 12, 1965, 22-31.

4. C. Krieg, T. Cole, U. Deppe, E. Schierenberg, D. Schmitt, B. Yoder and G. von Ehrenstein. The cellular anatomy of embryos of the nematode *Caenorhabditis elegans*. Devel. Biol. 65, 1978, 193-215.

5. J.E. Leestma and S.S. Freeman. Computer-assisted analysis of particulate axoplasmic flow in organized CNS tissue cultures. J. Neurobiol. 8, 1977, 453-467.

6. R. Llinas and D.E. Hillman. A multipurpose tridimensional reconstruction computer system for neuroanatomy. In Golgi Centennial Symposium: Perspectives in Neurobiology, M. Santini (ed.), Raven Press, New York, 1975, 519-528.

7. J. Holmquist, E. Bengtsson, O. Eriksson and B. Stenkvist. A program system for interactive measurements on digitized cell images. J. Histochem. & Cytochem. 25, 1977, 641-654.

8. R.E. Burge, J.C. Dainty and R.F. Scott. Optical and digital image processing in high-resolution electron microscopy. Ultramicroscopy 2, 1977, 169-178.

9. P.R. Smith and J. Kistler. Surface reliefs computed from micrographs of heavy metal-shadowed specimens. J. Ultrastructure Res. 61, 1977, 124-133.

10. T. Golab, R.S. Ledley and L.S. Rotolo. FIDAC-film input to digital automatic computer. Pattern Recognition 3, 1971, 123-156.

11. D.E. Hillman, R. Llinas and M. Chujo. Automatic and semi-automatic analysis of nervous system structure. In Computer Analysis of Neuronal Structures, R.D. Lindsay (ed.), Plenum, New York and London, 1977, 73-89.

12. M. Pfoch and W. Kade. Automated classification of cells in electron microscopic images of lymphoreticular tissue. J. Histochem. & Cytochem. 25, 1977, 655-661.

13. R.S. Ledley. Automatic pattern recognition for clinical medicine. Proc. of the IEEE 57, 1969, 2017-2035.

14. M.J. Shantz and G.D. McCann. Computational morphology: Three-dimensional computer graphics for electron microscopy. IEEE Trans. Bio. Med. Eng. 25, 1978, 99-103.

15. J.A. Ungerleider, R.S. Ledley and F.E. Bloom. Automatic recognition and analysis of synapses. Comp. Biol. Med. 6, 1976, 61-66.

16. E.W. White, K. Mayberry and G.G. Johnson, Jr. Computer analysis of multichannel SEM and x-ray images from fine particles. Pattern Recognition 4, 1972, 173-174.

17. A.V. Jones and K.C.A. Smith. Image processing for scanning microscopists. In Scanning Electron Microscopy, Vol. 1/1978, O. Johari (ed.), SEM, Inc., AMF O'Hare, Chicago, 13-26.

18. B.M. Unitt and A.V. Jones. A microprocessor-based digital scan generator. In Scanning Electron Microscopy, Vol. 1/1978, O. Johari (ed.), SEM, Inc., AMF O'Hare, Chicago, 27-32.

19. F.E. Bloom and G.K. Aghajanian. Fine structural and cytochemical analysis of the staining of synaptic junctions with phosphotungstic acid. J. Ultrastruc. Res. 22, 1968, 361-375.

Acknowledgement

Electronics consultation provided by Mr. Alfred F. Benedek and Mr. Ronald K. Crank is gratefully acknowledged. Research was supported by USPHS grants NS-13742 from NINCDS and HD-10934 from NICHD.

Discussion With Reviewers

Reviewers I-II: The scan times of 1-5 minutes are extremely long testing the stage and other stabilities. Is this lengthy time the result of brightness or the detection efficiency or is it the limit imposed by the hardware?

Authors: No doubt the raster scanning speed which now requires about 3 minutes for an area of 300 x 400 pixels is considerably longer than necessary. This is because we have utilized existing converters which were developed some ten years ago. Since our goal to trace perimeters on the specimen is a viable approach, we are now modifying this system with high speed A → D and D → A converters and utilizing digital transmission between the computer and the microscope to speed analysis. There are of course limits on the rate at which the beam can be driven and an adequate density signal obtained from the detector. Ultimately the limitations are in the driving amplifiers for the scan coil as well as the hysteresis of the coils which restrict the rate of beam motion. This settling time factor is further extended by the sampling time necessary to average the high frequency noise if it is not filtered.

Reviewer III: Can you provide some justification for the use of 6 bits for picture point storage?

Authors: First of all, our computer has 18 bits per word and, therefore, this allows packing of 3 picture points per word. Secondly, the 6 bit picture point word provides 64 levels of gray. This is a favorable compromise for storage and provides a satisfactory resolution for the differentiation needed in our preparations. Since we maximally utilize the 64 points within the gray level range of the specimen, electron images have sufficient density levels for processing, windowing of specific dense structures and producing perimeter traces.

Reviewers I-III: What is the normal signal-to-noise ratio and the intensity measurement? How does the signal-to-noise ratio relate to the 64 levels of gray?

Authors: In our preparations, the minimal, traceable signal to background signal is a ratio of about 2:1 at the photomultiplier tube. This background signal contains 25% photomultiplier tube noise in our STEM preparations. High frequency filtering reduces the high frequency background and photomultiplier noise to a ratio better than 5:1 with no critical image distortion. About 8 levels of the 64 are represented by background noise without filtering and about two levels are seen after filtering. In modifications we plan to match the converter voltage range to the optimum for the photomultiplier signal range thus eliminating possible noise from amplification and overdriving of the voltage to the photomultiplier tube.

Reviewer I: How serious is the electron beam damage due to long-time scans?

Authors: During raster scan operations, the storage time is approximately triple that normally required to produce a Polaroid exposure. At completion of the scan, the scan drive is returned to the STEM circuitry and, hence, beam exposure is minimized. Automation of focusing, astigmating and stage driving will greatly reduce specimen contamination and damage. In the software controlled beam positioning mode, the sampling time is also very small as the beam rests for only a few milliseconds on each position.

Reviewer I: What are the advantages of interactive picture processing (e.g. tracing vectors) as distinct from post-processing on the raster scan image output; for example, on magnetic tape? In the latter case, criteria for accessing the image features can be readily changed in respect of a specimen that has suffered the minimum necessary beam damage for image recording. Thus, changes in criteria for interactively accessing an image will require increased electron exposure.

Authors: The principal advantage of interactive picture processing directly on a specimen is that potentially 16 million points are available for analysis without having to store the frame. (In our recent design we are utilizing 12 bit x and y converters which will allow 4,000 x 4,000 picture point frames). Since in the scanning system, the resolution is increased by greater magnification, one always desires more area for proper analysis of the structures in question. Analyzable objects which overlap the frame margins can also be traced, and thus, retrain spatial relationships. Similarly, the entire specimen is available during the analysis procedure for immediate reference at variable powers in order to identify structures and artifacts extending into the adjacent field. Other factors previously discussed include the elimination of intermediate steps, massive storage and recording-analysis turn-around time. We plan to utilize a software-directed beam to locate sites for dispersive x-ray analysis by combining image intensity with detector counts.

Reviewer III: Have you considered using a light pen to determine scan areas, perimeter traces, etc.? This would seem to be the ideal interactive approach.

Authors: This is an excellent tool and serves as a joy stick to manipulate the beam position, especially for selecting structures to be traced to the exclusion of unwanted images with the same density level. Our current approach is to utilize the light pen interactively with the display screen to delete those structures which are traced that may not be desirable.

Reviewer I: How difficult is it to close an interactive perimeter trace given variations in specimen thickness and staining?

Authors: Closing perimeters is no doubt more difficult in the interactive mode than on stored images. Essentially, the noise in beam position-

ing and intensity are a source of this problem.
This is easily remedied by altering the tolerance
in the software which checks for perimeter completion. Occasionally, perimeters which are
difficult to trace may require repeated attempts,
causing contamination and specimen heating which
alters the overall density of the area.

Reviewer III: Could you say something about the
programming language you used and comment on its
suitability for interactive uses of the kind
described here? Does it in any sense (e.g.
compilation time, cycle time, etc.) set a limit
on what you can do?

Authors: The programming language on our system
is Fortran, though we use Fortan callable machine
language subroutines to speed up processing.
This hybrid approach provides both flexibility
and efficiency. In order to more effectively
determine how various parameters of the algorithm
may affect the trace, we have implemented hardware switch register controls so that we can
directly observe which parameters are best for
certain conditions, and in which order to most
satisfactorily trace finer details on the image.
The ultimate design will no doubt be a compromise
between hardware and software for efficient and
accurate following of density levels and processing of images.

REVIEWERS: I   R.E. Burge
             II   M.S. Isaacson
             III D.C. Joy

SCANNING ELECTRON MICROSCOPY/1979/II
SEM Inc., AMF O'Hare, IL 60666, USA

PREPARATION AND OBSERVATION OF VERY THIN, VERY CLEAN SUBSTRATES FOR
SCANNING TRANSMISSION ELECTRON MICROSCOPY

M. Ohtsuki,   M.S. Isaacson,* and A.V. Crewe**

Enrico Fermi Inst.   *Also: Dept. Physics   **Also: Dept. Biophysics

University of Chicago,
Chicago, IL 60637

Abstract

The quality of STEM micrographs is dependent
upon the quality of the substrate used to support
the specimen. This dependence exists because the
intensity in the STEM dark field image (obtained
using an annular detector after the specimen) is
proportional to the mass-density of the object.
Using this mode of contrast in a STEM at a resolution of about 2.6Å, we have examined the
quality and properties of various evaporated carbon substrates.

Very thin, very clean carbon films have been
made using a conventional evaporator equipped
with an oil diffusion pump and a rotary pump.
For the fabrication of microgrids to support
these films, we have employed a modified Fukami-
Adachi method which gives very clean micro-
supporting holes without traces of heavy atom
contamination. These microgrids were placed on
acid (20% HNO$_3$) cleaned titanium grids which
have low atomic number and low solubility in
water.

Various thickness carbon films were evaporated on both mica and NaCl crystals. We have
found no apparent differences in films made on
either substrate. Carbon films with an average
thickness of 6.8Å show apparent density steps
which appear to correspond to single atomic layers. 14Å and 16Å average thickness carbon films
had grid coverages of more than 95% and were
found to be well suited both for biological
specimens (biological macromolecules in particular), and for studies of single and clustered
heavy atoms. All carbon films were shown to be
free of heavy atom and hydrocarbon contamination.

We have demonstrated the quality of these
films with dark field micrographs of specimens
such as tobacco mosaic virus, low density lipoprotein and single gold atoms.

KEY WORDS: Carbon Film, Very Thin Substrates,
Scanning Transmission Electron Microscopy,
Micro Grid, Dark Field Contrast, Single Atom
Resolution, Decontamination

Introduction

Dark field methods in electron microscopy
have been rapidly attracting interest in the
last several years since the high contrast a-
vailable with such techniques allows the visualization of individual atoms and unstained
bio-molecules[1-3]. The scanning transmission
electron microscope (STEM) is especially suited
to dark field techniques since several dark
field signals can be simultaneously available
for imaging with high collection efficiency.
The various dark field imaging modes used in the
STEM have been discussed before in the literature[1,4], but little attention has been given to
the fact that because of the high contrast a-
vailable in the dark field STEM modes of suitable
thin film substrates to support the objects
(atoms or molecules) of interest[1]. Because of
the high contrast, small imperfections of the
substrate can mask the contrast variations of
the object. Furthermore, when dealing with high
resolution scanning transmission electron microscopy, the effects of contamination on the image
become more pronounced since one must use very
high current density probes. Thus, it is essential in high resolution microscopy to reduce the
kind and number of such deleterious imperfections in the substrates, and to have them exceedingly devoid of diffusible organic contaminants which could lead to contamination build-
up effects.

It is the purpose of this paper to describe
the properties of support films which are necessary for high resolution STEM observations,
the methods we have found useful in attaining
these properties and some selected applications
of the use of these films.

Desirable Properties of Support Films

We consider the necessary requirement for
the production of suitable support films for
high resolution STEM to be:
1. They should scatter as few electrons as
possible. Since the laws of nature won't allow
supports completely transparent to electrons, we
must settle for supports which are extremely
thin and of low atomic number (since the probability of electron scattering increases with

atomic number).

2. They should be free of diffusible organic contaminants which can lead to contamination build-up under the irradiation of the electron beam. Because of the high contrast available in the dark field STEM, monolayer contamination build-up is easily detectable. This requirement is not only a necessity for high resolution STEM, but in pushing detectable limits using electron energy loss specroscopy it is helpful not to have to deal with an increase in specimen thickness due to contamination.

3. For studies involving the microscopy of heavy atoms, it is essential that the substrate be devoid of heavy atom background. Since heavy atoms are highly visible in the STEM using the annular detector signal, a background of heavy atoms which may not be obvious using bright field phase contrast may become a glaring problem in dark field. We have discussed this problem before[1,5] and have set tolerable limits as less than 2-3 background heavy atoms per $10^4 \AA^2$. This of course, depends upon the particular application.

4. Substrates should have very low structural noise. Ideally, one would like a single crystal substrate in which the substrate noise is periodic and can be filtered out using various image processing techniques. A discussion of the problems involved in reliably producing such substrates can be found in the literature[6,7]. In lieu of that, one would like substrates in which there are patches of sufficient areal extent such that the intensity modulation across the patches due to varying number of atoms under the beam "cylinder" is small compared to the contrast modulation of the object of interest.

5. The next requirement is of a more practical nature than the above four. The substrate has to be reasonably sturdy under the irradiation of the electron beam (i.e., it should not break up or structurally change upon irradiation) and the fabrication procedures have to be efficient enough so that one can routinely obtain better than 50% coverage of the grid with such films.

6. Finally, for the preparation of biological molecules, the support film must be hydrophilic so that we can get a uniform adsorption of particles. Since we want to push the limit of the thinness of our support films, they have to be supported by a "clean" microgrid network which is in turn supported by the electron microscope grid.

Preparation Methods

Fabrication of microgrids

High quality fenestrated films are required to support the very thin carbon films that we need. In an earlier paper[1], Isaacson, et. al. described a method of fabricating clean microgrids from polymethylmethacrylate. However, because the method of fabrication was not as reliable as one would like for the production of microgrids on a routine basis, we have since changed our microgrid preparation technique to one which is much more reliable and apparently free from producing heavy atom and organic contamination. In the experiments reported here, we have employed a modified version of the Fukami-Adachi method[8]. This modified method is as follows:

After cleaning a glass slide with running water which has been distilled and deionized, the glass slide is dried with a lens tissue paper (A. Rosmarin Co.). To obtain approximately $0.1 \mu m$ to $5 \mu m$ diameter holes, 0.1% (w/v) benzalkonium chloride (Tridom Chem., Inc.) in water is utilized for a water repellent treatment. The glass slide is inserted into this solution for 5 to 10 seconds, washed with the running distilled and deionized water, and then dried with the lens tissue paper. This glass slide is then placed on a precooled stainless steel block in a freezer for 5 to 10 seconds (5 seconds for a high humidity environment and 10 seconds for dry conditions). The stainless steel block should be acid cleaned with 20% HNO_3, then rinsed in acetone and then ethyl alcohol prior to placing in the freezer. After removing from the freezer, the glass slide is held until dew starts to dry from the edge. It is then inserted into a solution of 0.2% (w/v) cellulose acetate butyrate (Aldrich Chem. Co.) in ethyl acetate (reagent grade) for 5 seconds. After removing from the solution, excess solution should be drained by tilting the glass slide. To make holes in the film, one must keep the surrounding area in a high humidity condition. We have used the vapor from boiling water in a beaker placed on a hot plate by exposing the glass slide until the ethyl acetate evaporates completely. After drying, the plastic fenestrated film is checked under the optical microscope. In the areas where good holes are located, the film is scored with a clean tweezer. This film can then easily be separated from the glass slide on the surface of the water. Standard titanium grids (75 x 300 mesh, Ernest Fullam, Inc.) are then placed face side down on the floating plastic film. Newspaper quality plain white paper is placed on top of the grids. After the paper is completely soaked with water, the paper together with the grids and the plastic fenestrated film is picked up by holding the edge of the paper with clean tweezers, and placing it plastic side up on the clean glass slide. Each grid with the plastic fenestrated film is separated from the paper after it is completely dried. These grids are then placed in the vacuum evaporator for the carbon coating. Very thick (~1500Å) carbon backing was applied to the plastic by evaporation of spectroscopic grade graphite onto the plastic side. Additional carbon, about 500Å thick, was evaporated onto the other side of the grid to prevent possible charging and contamination onto the sample from the plastic support. Titanium grids are used because: (1) Titanium can be cleaned with acid (20% HNO_3) prior to acetone and then ethyl alcohol cleaning; (2) it has a moderately low atomic number; (3) it has a very low solubility in H_2O compared to that of copper or gold. Therefore, effects from possible contamination due to single heavy atoms from the metals can be minimized. Figure 1 shows the microgrid at low magnification.

Preparation of the support films

We have chosen evaporated carbon films as a supporting film for high resolution microscopy because: (1) they can be easily made into relatively uniform and sturdy thin films; (2) carbon has low electron scattering power; and (3) very clean films can be made using a conventional evaporator which employs an oil diffusion pump and a rotary pump with a liquid nitrogen cold trap located between the vacuum chamber and the diffusion pump (Varian VE-10). This last one is extremely practical in terms of cost and speed.

In our experiments, spectroscopic grade graphite rods (National L4306 AGKSP, Union Carbide) were evaporated onto either freshly cleaved NaCl crystals or mica in which the substrates were kept at room temperature. All parts inside the belljar were replaced with ones made of stainless steel (which could be acid cleaned) to prevent possible organic contamination. We found that the thickness of the evaporated carbon films can easily be controlled by sharpening a graphite rod conically to a very acute angle ($<5^{\circ}$) on one side of the rod. The other graphite electrode was flat. In addition, a 3mm diameter graphite rod gave us better control of the film thickness than did 6mm diameter rods.

The evaporated films were separated from the substrate crystals by floating them off on the surface of distilled and deionized water (Millipore Super Q filtration system). In order to observe such thin films on the surface of the water, the outside of the beaker containing the water was wrapped with black tape. Films less than 20Å average thickness could be seen by illuminating the surface of the water with a lamp at about 30° to the horizontal and viewing the surface at an angle. Such thin films appear bluish in color when viewed in this manner.

Decontamination

In the past, we have found that if proper handling techniques were observed[9], most organic contaminants could be removed by gently heating the sample to a temperature less than 50° (for biological specimens this heating temperature may be as low as 35°C) with an infra-red lamp for a period of 10 seconds to several minutes. The heating was performed while the sample was at a vacuum of 10^{-7} to 10^{-8} torr. inside of a double airlock specimen changing chamber attached to the microscope[1]. After this treatment, the sample could then be inserted into the microscope without further exposure to the air. This technique results in negligible contamination of the sample when viewed in the microscope and appears to drive off the weakly bound organic contaminants, especially those which get deposited on the specimen surface from the ambient atmosphere environment.

We observe that the gentle heating works just as well with white light as it does with infrared. The main point is to get the specimen holder warmer than the surroundings. In addition, one further step we have used which greatly increases our yield of non-contaminating support films, is that prior to depositing the

sample on the films, the grids with the microgrids and thin carbon support films are heated in air with a 100 watt lamp from a distance of 5cm for 10 minutes. This simple additional procedure results in no detectable contamination observed in dark field after an electron dose in excess of 10^{7} electrons/Å^{2}[9].

Characteristics of the Substrates

Results

Incident electrons of 37keV kinetic energy are only weakly scattered by the carbon films we generally use as support films. Typically, 2-3% are elastically scattered and 3-4% inelastically scattered. These films provide about 90-95% coverage over the holes in the plastic microgrid. We have, however, been able to fabricate films which scatter less than half that amount (on the average) although holes occupy about 50% of the area.

Figure 2 (a-d) shows some selected dark field micrographs of typical thin, very clean carbon films used in our experiments. (a) shows a low magnification micrograph of a carbon film deposited onto a NaCl crystal at room temperature. Fine decoration lines of the surface steps of the NaCl can be seen; (b) shows a higher magnification of the same film. The average thickness is 14Å; (c) shows a film of average thickness 16Å deposited on mica at room temperature. Less than 5% of the area is occupied by holes; (d) shows a carbon film of 6.8Å average thickness which had been evaporated onto a NaCl substrate at room temperature. The important point to note in figures 2 (b-d) is the negligible amount of bright spots which can be attributable to heavy atom contamination.

Discussion

In order to evaluate our films and to make quantitative comparisons with other methods, we have found it useful to evaluate the average thickness (or mass thickness) of the films from the scattered intensity. Assuming that we can treat the scattering as though the sample were incoherently illuminated, the scattered intensity we measure I_{sc} is related to the average film thickness T as:

$$I_{sc}/I_0 \stackrel{\sim}{=} nT\sigma \qquad (1)$$

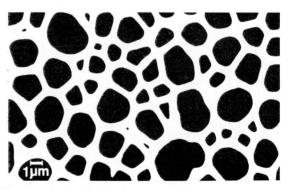

Fig 1. Dark field micrographs of typical microgrid.

where n is the number of atoms per unit volume of the film. I_0 is the incident beam current, and σ_{sc} is the collected scattering cross-section given by:

$$\sigma_{sc} = (1-f_{el}) \sigma_{el} \quad \text{elastic scattering (2A)}$$

$$= f_{in}\sigma_{in}, \quad \text{inelastic scattering} \qquad (2B)$$

where f_{el} is the fraction of electrons elastically scattered into the hole in the annular detector, f_{in} is the fraction of electrons inelastically scattered which pass through the hole in the annular detector and the transmitted by the electron spectrometer. Typically, we collect 60-80% of the elastically scattered electrons and 80-95% of those inelastically scattered. Using 37keV incident electrons, a film of 15Å average thickness and density of 0.11 atoms/$Å^3$ will scatter 4% of those electrons inelastically and 2.8% elastically (if we use the cross-sections given in references 10 and 11). Experimentally, we find the ratio of elastic to inelastic scattering to be about 0.69 after taking into account the collection efficiencies. This is in very good agreement with the ratio predicted from the cross-sections given in references 10 and 11.

If the films produced using our method were completely amorphous, we would expect the intensity fluctuations in the annular detector signal (elastic scattering signal) to be related to the number of atoms within the beam cylinder (assuming that we are not limited by beam statistics). That is, we would expect the substrate noise to vary as 1/ N where $\sqrt{N}$ is the number of atoms in the beam cylinder and thus proportional to the film thickness T. This appears to be the case when we average measurements of many areas and is consistent with earlier measurements (1). That would mean for 15Å average thickness films of density 0.11 atoms/$Å^3$ we would expect $8\pm\sqrt{8}$ atoms within a beam cylinder of 2.5Å diameter (i.e., fluctuations of $\sim$30%). On the average this is true as can be seen in the micrographs in figure 2 and line scan in figure 3. Upon close inspection, we see there are regions of about 20Å in extent that are relatively smooth (i.e., intensity fluctuations $\sim$30%). Such smoother regions are essential for visualization of smaller biological particles.

For films less than 10Å average thickness, these smooth regions are sometimes even greater in extent (see fig. 4) and as little as 0.7% of incident 37keV electrons are inelastically scattered from some of these regions. These patches appear to occur in intensity steps of multiples of some minimum value as can be seen in figure 4 and in figure 5 where we have plotted a series of measurements of the scattered signals from different patches of very thin films. In the line scan shown in figure 4 (a) and (b), such steps are clearly evident. If we deduce the mass-thickness from the thinnest regions using equations (1) and (2), we get a mass-thickness, nT, of 0.3 atoms/$Å^2$. Taking bulk graphite as having a density of 1.8 gm/cm^3 we would get a mass-thickness of 0.305 atoms/$Å^2$ per

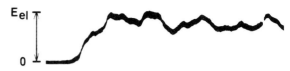

Fig 3. CRT trace of the annular detector signal obtained from a 14Å average thickness carbon film. The fluctuations in the signal (and thus, the mass-thickness) are about 30%.

layer of graphite. Thus, our results are consistent with the observations of individual graphite-like layers of carbon atoms and those smooth regions quite possibly are one, two, three, etc. layers thick. There are further subtle hints of this in that many intensity steps are bound by 120° planes. A more detailed discussion of the film structure is beyond the scope of this paper and will be dealt with in a future publication. The main point to be emphasized is that clean, thin films as thin as 1-4 atom layers thick can be reliably produced and are stable under the electron beam irradiation. We can irradiate such films over an area 200Å x 200Å with 2 x 10^{-11}A in a 2.5Å diameter probe for more than 500, 17 seconds scans and see no evidence of either contamination build-up or structural change (ref. 9). This corresponds to a dose in excess of 2 x 10^7 electrons/$Å^2$ delivered with a beam of current density of about 1/2 x 10^5 amp/cm^2. We see no evidence of thinning of the layers due to reaction with residual O_2 in the microscope column since the vacuum is four orders of magnitude better than that found in the column where such observations have been reported[12].

Applications

Biological specimens such as negatively stained (uranyl acetate) tobacco mosaic virus (TMV; see fig. 6) and unstained low density lipoprotein (see fig. 7) particles were placed on 16Å average thickness carbon films made on mica substrates; and single atoms of Au from $AuCl_3$ were deposited onto both 14Å and 6.8Å average thickness carbon films (see fig. 8,9). The micrographs shown in fig. 6-9 were obtained using the annular detector dark field mode of contrast in which both phase contrast and diffraction contrast were absent. In fig. 6, (a) shows a 1st scan micrograph of TMV, (electron does of 11e/$Å^2$) and (b) shows a micrograph of the same area after completion of 6 scans (total electron dose of 66e/$Å^2$). Hydrocarbon contamination is hardly noticed in these micrographs.[13]

For demonstration of the use of these films for single atom microscopy, $AuCl_3$ was deposited on both a 14Å and 6.8Å average thickness carbon films (see fig. 8,9). These carbon supporting films were made on NaCl substrates at room temperature. In fig. 8, comparing the micrograph of the 1st scan of the area (a) and 100th scan of the same area (b), we observe some movement of the single atoms as expected[5,14].

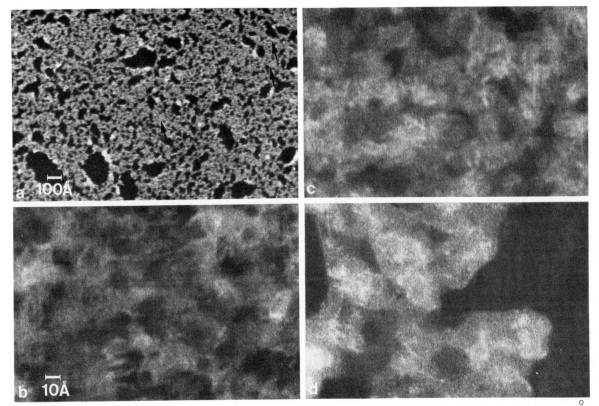

Fig 2. Dark field micrographs of typical carbon films: (a) Low magnification field of view of 6.8Å average thickness film deposited on a NaCl substrate. Arrows indicate decoration lines of the surface steps of the NaCl crystal; (b, c, d) high magnification field of view. 14Å average thickness film deposited on a NaCl substrate (b), 16Å average thickness film deposited on a mica substrate (c), and 6.8Å average thickness film deposited on a NaCl substrate **(d)** (**b–d** are shown at the same magnification).

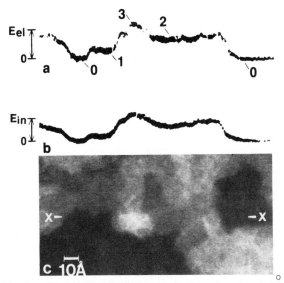

Fig 4. Possible single atom layers in the 6.8Å average thickness carbon film. Both the elastic signal (a) and the inelastic signal (b) were line-scanned at the position X shown in (c). In (a), 1, 2, 3 represent the number of atomic layers. "0" indicates the signal level from holes in the film.

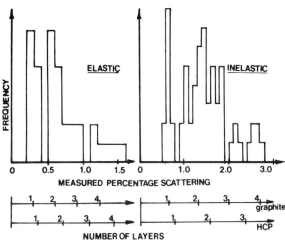

Fig 5. Histogram of the measured amount of scattering of 37 keV electrons through our thin carbon films. The measurements were made from many line scans of areas similar to that shown in Figure 4. The lower horizontal scales correspond to the predicted scattering from 1, 2, 3, 4 atomic layers arranged either with graphite packing or hexagonal close packing (HCP). The predictions take into account the detector collection efficiencies.

379

The carbon supporting film, however, is very stable under the irradiation.

Conclusions

We have described a method for reliably producing extremely thin, clean carbon film supports for high resolution scanning transmission electron microscopy. These films are reasonably uniform, have 50 to 95% open area covered by the film and are relatively free from heavy atom contamination. By utilizing appropriate preparative procedures and pre-treatment prior to insertion into the microscope, these support films are found to be essentially free from the diffusible organic contaminants which are a nuisance in high resolution and analytical electron microscopy.

From measurements of elastic and inelastic scattering, we have observed planar regions in these films that appear to be atomic layers (i.e., one atom thick). Average film thicknesses corresponding to two to four atomic layers can be produced which can easily be supported over 0.5 - 5μm holes in a plastic microgrid and are sturdy under electron beam irradiation doses in excess of $10^7 e/Å^2$ (1.6×10^4 coul/cm^2).

We have demonstrated the applicability of these films for high resolution microscopy by applying specimens of tobacco mosaic virus, low density lipoprotein and gold chloride onto these films. It is felt that if the preparation procedures discussed in this paper are followed, suitable support films for the high resolution STEM can be produced in a fairly reliable fashion using a conventional evaporator system. Thus, they should be within the reach of being prepared in most laboratories.

Acknowledgements

We thank Dr. M. P. Gordon of the University of Washington for providing us TMV and Dr. A. M. Scanu of the University of Chicago for providing low density lipoprotein. This work was supported by the Biotechnology Resources Branch of the National Institute of Health, and the United States Department of Energy. One of us (MSI) was an Alfred P. Sloan faculty fellow.

References

1. M. S. Isaacson, J. Langmore and J. Wall. "Preparation and Observation of Biological Molecules for the High Resolution Scanning Transmission Electron Microscope" SEM/1974, IITRI, Chicago, IL. 60616. (1974) pp. 19-26.
2. J. Wall, J. P. Langmore, M. S. Isaacson and A. V. Crewe. "High Resolution Scanning Transmission Electron Microscopy". Proc. Nat. Acad. Sci. (USA) 71 (1974) pp. 1-5.
3. F. P. Ottensmeyer, D. P. Baze-t-Jones and A. P. Korn. "High Resolution Structure Determination of Biological Macromolecules" in Electron Microscopy 1978. Vol. III; State of Art Symposia. (ed. J. M. Sturgiss Microscopical Society of Canada, Toronto (1978) pp. 147-159.
4. A. V. Crewe, J. P. Langmore and M. S. Isaacson. "Resolution and Contrast in the Scanning Transmission Electron Microscope" in Physical Aspects of Electron Microscopy and Microbeam Analysis (eds. B. Siegel and D. Beaman) John Wiley, New York (1975) pp. 47-62.
5. M. S. Isaacson, D. A. Kopf, M. Utlaut, N. W. Parker and A. V. Crewe. "The Direct Observation of Atomic Diffusion by Scanning Transmission Electron Microscopy". Proc. Nat. Acad. Sci. (USA). 74. (1977) pp. 1802-1806.
6. J. R. White, M. Beer and J. W. Wiggens. "Preparation of Smooth Graphite Support Films for High Resolution Electron Microscopy" Micron 2. (1971) pp. 421-427.
7. G. H. N. Riddle and B. M. Siegel. "Production of Thin Single Crystal Graphite Support Films" in Proc. 29th Annual EMSA Meeting, Boston, (ed. G. W. Bailey) Claitor's Publishing Div., Baton Rouge (1971) pp. 226-227.
8. A. Fukami and K. Adachi. "A New Method of Preparation of a Self-Perforated Micro-Plastic Grid and Its Application. I". J. Electron Microscopy (Japan) 14. (1965) pp. 112-118.
9. M. S. Isaacson, D. A. Kopf, M. Utlaut and M. Ohtsuki. "Contamination as a Psychological Problem" Ultramicroscopy 4. (1979) in press.
10. J. P. Langmore, J. Wall and M. S. Isaacson, "The Collection of Scattered Electrons in Dark Field Microscopy I. Elastic Scattering". Optik. 38. (1973) pp. 335-350.
11. J. Wall, M. S. Isaacson and J. P. Langmore. "The Collection of Scattered Electrons in Dark Field Microscopy II. Inelastic Scattering". Optik. 47. (1974) pp. 353-374.
12. S. Iijima. "High Resolution Electron Microscopy of Phase Objects: Observation of Small Holes and Steps on Graphite Crystals". Optik. 47. (1977) pp. 437-452.
13. M. Ohtsuki, M. S. Isaacson and A. V. Crewe. "Dark Field Imaging of Biological Macromolecules with the Scanning Transmission Electron Microscope". Proc. Nat. Acad. Sci. (USA). 76. (1979) appear in March issue.
14. M. S. Isaacson, M. Utlaut, M. Ohtsuki, D. A. Kopf and A. V. Crewe. "Scanning Transmission Electron Microscopy at Atomic Resolution" to be published in Science.

DISCUSSION WITH REVIEWERS

J. S. Wall: Could you describe the vacuum condition of the belljar used: number and type of o-rings, use of vacuum grease, type of diffusion pump oil, roughing trap, etc.?

Authors: The vacuum was better than 1.5 x 10^{-5}mmHg. We used a standard vacuum evaporator (varian VE-10) as noted in the text, which employs more than 20 regular o-rings. The rubber seal between the belljar and the base plate was greased with "Apiezon M" about once per 50 evaporations in such a manner that the grease was applied and then wiped off with lens tissue paper. The diffusion pump oil is "convoil 20".

W. H. Massover: What are the exact conditions for carbon evaporation (time, current, distance from ica or NaCl, indirect or direct

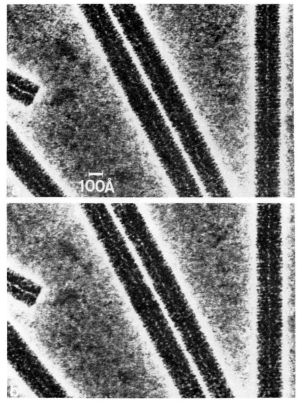

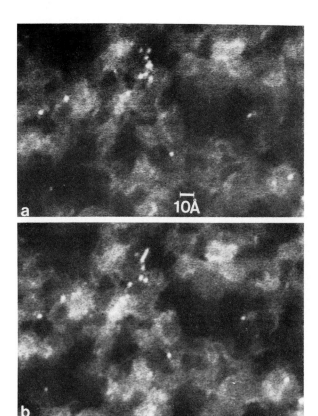

Fig 6. Micrographs of negatively stained (1% uranyl acetate) tobacco mosaic virus (TMV) placed on a 16Å average thickness carbon film (deposited on a mica substrate); (a) shows the 1st scan (electron dose of 11e/Å^2) of the area; (b) shows the same area after completion of 6 scans (total electron dose of 66e/Å^2). No hydrocarbon contamination is detectable. The specimen was dried in air.

Fig 8. Gold atoms on a 14Å average thickness carbon film (which was made on a NaCl substrate). Both the first scan (a) and 100th scan (b) of the same area are shown. The supporting film is very stable after an irradiation of more than 1.9×10^6 e/Å^2, and no hydrocarbon contamination is evident.

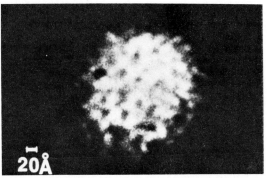

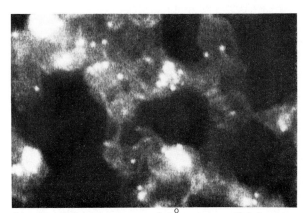

Fig 7. Unstained, freeze-dired low density lipoprotein placed on a 16Å average thickness carbon film (deposited on a mica substrate).

Fig 9. Gold atoms on a 6.8Å average thickness carbon film (made on a NaCl substrate).

evaporation, etc.)?

L.T. Germinario: Electron diffraction results on carbon films derived from vacuum evaporated graphite rods have indicated the existence of graphitic, turbostratic structures of 8 to 15Å diameter, along both a and c axis, set in a matrix of amorphous carbon (i. S. McLintock and

J. C. Orr, In. Chem. and Phys. of Carbon, Vol. II, 243 (1973) for review). These large graphitic particles are considered to have been ejected from the carbon source during evaporation. Since the ejection process commences abruptly at 2680°C, please comment on the electrode temperatures, evaporation times and

evaporation geometry (i.e., direct vs. indirect) used to fabricate the carbon substrates.
Authors: Thin carbon was directly evaporated from a distance of 10 cm with about 30 ampere and an evaporation rate of about 10Å/sec. A shutter was not used.

W. H. Massover: There appear to be many "holes" in your support films of ultra-thin carbon (e.g. Figs. 2a, 2d, 4, 9). Is this feature related to any known variable of the preparative method described?
Authors: The area of the holes in the very thin carbon film is directly related to the thickness of the film. These holes disappear when the films are thicker than 20Å.

L. T. Germinario: A most critical parameter which determines the structure and properties (i.e., smoothness, charge carrier mobility, conductivity and adsorption coefficient) of vacuum evaporated films is the temperature at which the substrate is maintained during evaporation. Strong evidence has been provided for a relationship between the boiling points (T_b) and optimum substrate temperature (T_s) for a wide range of materials with $T_s/T_b \sim 0.34$ (P. S. Vincett et al., Nature, 255, 542 (1975). What is the T_s/T_b ratio of the evaporated carbon substrates fabricated in this study? Does such a correlation exist for evaporated carbon films?
Authors: The ratio of T_s/T_b for our experiments was about 6×10^{-3}. In order to achieve a ratio of $T_s/T_b \sim .34$ the substrate would have to be at about $1600^{\circ}C$! The discrepancy exists presumably because the optimum ratio of .34 was shown to hold mainly for organic materials and photoconductive materials evaporated onto glass substrates. We do know that the substrate temperature does have an effect on the film structure and are in the process of studying this effect.

J. Spence: What determines the size of the holes in the microgrids and how can this be controlled? How reproducible is the microgrid fabrication technique?
Authors: The size of the holes can be controlled by adjusting the length of time after removing the glass slide from the freezer before insertion into the cellulose acetate butyrate. The longer the time, the smaller the holes one obtains.

J. S. Wall: Why use lens tissue instead of air drying slides? Is plastic removed with solvent after carbon coating? Is carbon coating applied using a rotary coater?
Authors: We used lens tissue paper so that the drying is more uniform which is directly related to the uniformity of the size of the hole in microgrid. The carbon was coated at a fixed position onto both sides of the plastic without removing the plastic.

J. Spence: Would you expect similar contamination rates if your method were used to prepare samples for convergent-beam work in the Philips

EM 400, fitted with field-emission gun?
Authors: We have shown that with this preparation we can eliminate contamination due to the specimen. Provided the specimen holder has been properly cleaned and no contamination comes from the optical column, there is no reason to expect that similar contamination rates wouldn't be observed in a conventional microscope.

W. H. Massover: No statement is made about hydrophilicity; are your supports hydrophilic and do they stay so, or do you use glow discharge or some other means to render them hydrophilic?
W. Wiggins: Sometimes one wishes to have hydrophilic grids for proper adsorption of biological macromolecules. Can these grids be made hydrophilic without contamination or heavy element addition?
Authors: We find that these contamination free carbon films are hydrophilic even if the films are one month old. No treatments were made on the films.

J. S. Wall: What is the substrate noise contribution in Fig. 6 relative to stain nonuniformity (i.e., what gain factors were used in Fig. 6 as compared to Fig. 2)? How was this film treated to achieve uniform spreading of uranyl acetate?
Authors: Gain factor was 10 to 20 between Fig. 6 and Fig. 2. The uniformity of staining is obtained when the protein solution wets the carbon film. This can be achieved when the concentration of the protein, in this case TMV, is somewhere between $100 \sim 500$ mg/ml. The films are not treated at all to obtain this uniform spreading.

W. Wiggins: The photos of TMV and lipoprotein look just as one would expect if they were supported on conventional carbon films. In what sense do these photos illustrate any unusual property of the very thin films?
Authors: Obviously, the less the substrate scatters the incident electrons relative to the scattering from the object, the higher the signal to noise in the image. The TMV photo was shown merely to indicate that these thin substrates can be used with conventional specimens. As for the unstained lipoprotein, the subunits are about $10-15\text{Å}$ in size, thus if the susbstrate were made thicker than this, the contrast of the subunits would be greatly reduced.

SCANNING-ELECTRON MICROSCOPY/1979/II
SEM Inc., AMF O'Hare, IL 60666, USA

RAMAN MICROPROBE STUDIES OF TWO MINERALIZING TISSUES:
ENAMEL OF THE RAT INCISOR AND THE EMBRYONIC CHICK TIBIA

F. S. Casciani, E. S. Etz*, D. E. Newbury* and S. B. Doty**

National Institute
of Dental Research
Bethesda, MD 20205

*Center for Analytical Chemistry
Nat'l Bureau of Standards
Washington, DC 20234

**Orthopedic Res. Lab.
Dept. of Anatomy
College of Physicians
and Surgeons
Columbia University
New York, NY 10032

Abstract

The laser-Raman microprobe developed at the National Bureau of Standards has been applied to the study of the mineralization process in rat incisor enamel and embryonic chick tibia. Cryostat sections were prepared from fresh frozen tissues and allowed to air dry. In these mineralizing tissues two forms of phosphorus compounds have been observed: (1) an inorganic phase identified as apatitic phosphate and (2) an organic phosphate. The distribution of these components from the mineralizing front to regions of higher mineralization has been determined with a spatial resolution of $\sim$ 15 μm. The studies suggest the existence of a carbonate, with a Raman band corresponding to that of the mineral huntite, $Mg_3Ca(CO_3)_4$, and found in regions of low phosphate mineral content.

KEY WORDS: Microanalysis, Raman Spectroscopy, Mineralizing Tissue, Raman Microprobe, Biological Mineralization, Biological Microanalysis, Dental Studies, Enamel, Bone, Dentin

Introduction

The advantages of instrumental microanalytical techniques, especially in the chemical analysis of biological specimens are well documented [1-3]. Whereas, conventional microanalysis techniques provide data on the elements present and their distribution, information related to molecular composition at the microscopic level has only recently become available with the development of Raman microprobes [4-6]. These new instruments permit direct, nondestructive examination of intact biological tissues and allow the determination of the constituent molecular species by interpretation of the Raman spectra obtained from microscopic regions of thin sections.

In this communication, we report the results of feasibility studies conducted with the NBS-developed laser-Raman microprobe. These investigations have had two primary goals: (1) to demonstrate the successful application of the Raman microprobe to microanalysis of mineralizing tissue; and (2) to determine whether Raman microprobe spectra of regions with differing degrees of mineralization would reflect changes important to the understanding of the mineralization process. The studies described herein are primarily concerned with the major organic and inorganic components of two types of mineralizing tissue, the enamel of the continually erupting rat incisor and the embryonic chick tibia, representing noncollagenous and collagenous mineralization, respectively. These two types of tissues were chosen, because both the initial site of mineralization and regions with increasing degrees of mineralization may be easily located and examined independently.

The process of mineralization in enamel is characterized by the loss of protein and water and the accumulation of mineral. Deakins[8] first reported, for pig enamel, that 80% of the organic material present in immature enamel is lost during maturation. The analyses of Glimcher et al.[9] indicate the most immature enamel contains 15-20% protein by weight, while mature enamel consists of 0.1% or less protein[8,9]. The inorganic component of rat enamel has been described both morphologically[10-14], and chemically[15,19] during mineralization and in the mature tissue as hydroxyapatite with 2-4% carbonate[16-19].

Spectroscopic studies[20,21] of mineralizing tissues have been limited by the ability to isolate samples representing single stages of development in a continuously changing tissue. However, the principles of the Raman microprobe and its microanalytical applications, which are discussed in a tutorial paper in this publication[6], eliminate these difficulties and allow chemical analysis to be correlated directly with morphology.

Experimental

The Raman Microprobe. Figure 1 illustrates the configuration of the NBS-developed Raman microprobe employed in these investigations. A complete description of the instrument[5-6] and its analytical capabilities has been described[5-7]. For the experiments described in this paper, the 514.5 nm line of the argon-krypton ion laser, operated at powers in the range 5 mW to 50 mW, was used for excitation of the spectra. Nonlasing plasma lines are removed through the use of a pre-dispersing prism. The sample is accurately and reproducibly positioned for measurement by a stage driven by remotely controlled piezo-electric translators. The microprobe is equipped with a built-in optical microscope to permit observation of microscopic detail of the specimen at various magnifications and to aid in the location and precise positioning of the sample region of interest. The laser beam is focused onto the sample with a focusing objective that provides a variable focal spot from 6-20 μm in diameter which determines the lateral spatial resolution of the probe measurement. The size of the focal spot and the incident laser power determine the power density or irradiance (watts/cm^2) incident on the sample. The effective sampling volume is determined by several factors, including laser spot size, depth resolution, scattering geometry, and optical and surface properties of the sample. We estimate that the Raman signal is typically collected from microscopic regions of mass $\sim$ 10 nanograms. The Raman microprobe employs a 180° scattering geometry. In this configuration, the total light (Mie/Rayleigh and Raman) scattered by the irradiated region is collected in the backward scattering direction (over a large solid angle) by an ellipsoidal mirror, and transfer of the scattered radiation is made to a double monochromator employing concave holographic gratings. The scattered light, at the position of the exit slit of the double monochromator, is detected by a cooled photomultiplier tube with a measured dark count of < 5 counts per second. Conventional photon counting equipment is used for amplification and monitoring of the Raman signal. Raman spectra are acquired by a strip chart recorder or by a minicomputer interfaced to the probe for data logging and total system control (e.g. sample positioning, wavelength scan, slit setting, etc.).

In these experiments, the spectroscopic measurements are performed on sections supported by a suitable sample substrate (or support). Optical quality sapphire (αAl_2O_3) and lithium fluoride (LiF) of single crystal quality are routinely used as substrate materials in the form of small rods (4 mm diameter, 6 mm long) held in a support collar. Sapphire is a weak Raman scatterer and lithium flouride lacks first order Raman activity and thus does not interfere. Both substrate materials are free from fluorescence in the spectral region of interest in this work. All the spectra presented here have been obtained with 514.5 nm excitation and 3 cm^{-1} spectral slit on samples supported on sapphire substrates.

Measurement Procedures. Examination with the optical microscope (supplemented by optical micrographs of the sample taken before and after Raman analysis) is followed by measurement in the microprobe. Spectra are obtained using irradiance levels ranging from kilowatts/cm^2. Under normal operating conditions spectra may be obtained with the following parameters: time constant, 0.5-60 seconds and scan rates 100-2 cm^{-1}/minute. With 3 cm^{-1} spectral resolution, spectra may be obtained over the spectral range 0-4000 cm^{-1} under one hour. Spectral interferences from fluorescence, or sample modification due to heating, require the use of low power densities. Under these conditions, the spectra are recorded over a limited spectral range of interest to reduce the exposure of the specimen to the laser beam. After the Raman analysis, the specimen may be carbon coated and examined in the electron or ion probe for elemental analysis of the same regions.

Various methods of sample fixation result in either spectral interference (from buffers and fixatives) or abnormally high fluorescent backgrounds. Overcoating of the tissue section with organic materials (e.g. hydrocarbon films, oils) to prevent dehydration has similar results. Freeze drying either before or after the section is on the sample support leads to poor physical contact of the sample with the sample substrate. A method which has proven to give the least amount of deleterious heating or laser induced sample modification has been to transfer frozen sections to the surface of pre-cooled (-20°C) sample substrates and allow both to come to room temperature. Thawing and air-drying of the specimen results in good physical contact between the tissue section and the sample substrate. In these studies Raman spectra have been obtained in as little as two hours after the sample is removed from the animal.

Preparation of Tissue Section. Rat Incisor. Sprague-Dawley rats, weighing from 40-100 g were sacrificed with ether. The mandibles were removed and the incisors dissected out, care being taken to avoid damaging the enamel organ. A piece 4 mm in length was cut from the apical end of the incisor and positioned in freezing water such that 10-20 μm serial saggital sections could be cut on the cryostat at -20°C. The midserial saggital sections, with the expected morphological detail were transferred to the sample support as described above.

Mature Enamel: The erupted portion of rat incisors were fractured and mounted on sample substrates with dental wax. After microscopic examination, either mature surface enamel or mature inner enamel could be identified and analysed.

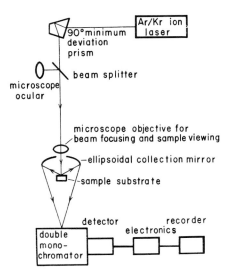

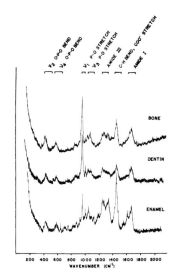

Figure 1. Schematic diagram of the Raman microprobe spectrometer employing back scattering geometry.

Chick Tibia. Tibia were removed from 13 day embryonic chicks. A 5 mm piece was removed from the distal end and prepared as was the rat incisor. The appropriate sections (10-20 μm in thickness) were transferred to the sample support as described above.

Figure 3. Raman microprobe spectra of 13 day embryonic chick tibia, mineralizing rat incisor dentin and enamel. Power, 40 mW at sample; beam, 15 μm diameter; time constant, 5 seconds; scan rate, 20 cm⁻¹/minute.

The Raman microprobe spectra in Figures 3, 4 and 5 were obtained from regions indicated in the micrographs. These spectra are plotted as scattered intensity (in arbitrary units) vs Raman shift (in units of wave number, cm⁻¹) from the exciting line and are representative of numerous measurements on each specimen obtained with normal irradiance conditions ($\sim$ 22 kW/cm^2).

Results

Optical micrographs of sections of rat incisor and chick tibia are shown in Figure 2a-c.

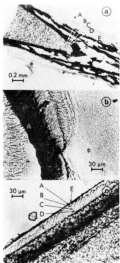

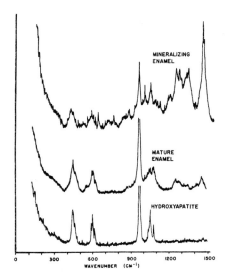

Figure 2. Light micrographs of tissue sections indicating the areas of Raman microprobe analysis (a) 13 day embryonic chick tibia (7 μm thick); spectrum in Figure 3 (bone) obtained from area E. (b) mineralizing rat incisor enamel and dentin (15 μm thick), spectra in Figure 3 (dentin and enamel) obtained from areas A and B, respectively. (c) mineralizing rat incisor enamel (20 μm thick); spectra in Figure 5 obtained from indicated areas.

Figure 4. Raman microprobe spectra of mineralizing rat incisor enamel, mature enamel and synthetic hydroxyapatite. Spectrum of mineralizing enamel obtained from area B, Figure 2. Power, 40 mW at sample; beam, 15 μm diameter; time constant, 5 seconds; 20 cm⁻¹/minute.

The Raman spectra of mineralized tissues were characterized by their two main components, an inorganic phosphate phase and a proteinaceous matrix (Table 1). The spectral bands associated with the inorganic components generally are sharper than those associated with the organic matrix.

Inorganic Component. The major inorganic component of vertebrate hard tissues is an impure hydroxyapatite, $Ca_{10}(PO_4)_6(OH)_2$. The Raman spectrum of a sample of synthetic hydroxyapatite is shown in Figure 4. The four Spectral regions specific to the internal vibrations of the orthophosphate in the apatite lattice[22,23] are the ν_1, (P-O symmetric stretch) at 960 cm^{-1}, the most intense band associated with the phosphate ion in the Raman spectrum of hydroxyapatite; the ν_2, (O-P-O symmetric bend) between 400-500 cm^{-1}; and the ν_3, (P-O asymmetric stretch) between 1000-1100 cm^{-1}; and the ν_4 (O-P-O asymmetric bend) between 560-620 cm^{-1} (see Figure 4). These four bands can be identified in the spectra of each of the tissues shown in Figures 3 and 4. The most prominent of these bands, the 960 cm^{-1} band, varies in relative intensity in the different tissues, indicating differing degrees of mineralization.

Figure 5 shows the sequential Raman microprobe spectra of rat incisor enamel in a region where the enamel is 30 μm thick (active secreting ameloblasts). These spectra illustrate the rapid increase in the relative intensity of the 960 cm^{-1} band (approximately 8x) and the development of the ν_2, ν_3, ν_4 spectral regions of the apatitic phosphate as the analysis approaches the dentino-enamel junction.

Figure 6 shows a low irradiance spectrum (∼ 2.5 kw/cm^2) obtained at the mineralizing front in rat incisor enamel. The spectrum reveals, in addition to the 960 cm^{-1} band, a band at 1003 cm^{-1} and, a sharp, relatively intense band at 1123 cm^{-1} which has only been observed in low irradiance spectra.

A similar band at 1123 cm^{-1} was also observed in low irradiance spectra of the chick tibia (Figure 7) obtained in 15 μm increments from the mineralizing front toward regions of higher mineral content as shown in Figure 2a. The intensity of this band remains constant or decreases only slightly in areas of early mineralization and low mineral content. In contrast, the intensity of the 960 cm^{-1} band increases as the mineral content increases. The 1123 cm^{-1} band was not observed in spectra of either enamel or chick tibia obtained at higher irradiance levels (22 kw/cm^2) suggesting that the component responsible for this band may be thermally unstable.

The effect of the irradiance level and exposure time on the 1123 cm^{-1} band in the chick tibia is shown in Figure 8. Spectrum A was obtained under low irradiance conditions during a two hour exposure time. Spectra, B, C, D show the progressive loss in intensity of the 1123 cm^{-1} band during a 24 minute exposure time with high irradiance conditions.

The Raman shift of the 1123 cm^{-1} band falls into the spectral range for the symmetrical stretching mode of the carbonate ion in magnesium calcium carbonate $Mg_3Ca(CO_3)_4$ (huntite), as described by Scheetz and White[24]. Figure 9 shows the coincidence (inset) of this band in the Raman spectra of both the chick tibia and huntite as

well as a reference spectrum of the mineral. The Raman spectrum of huntite was obtained from a sample particle approximately 18 μm x 12 μm in size from a reference sample of the mineral.

Organic Component. Many of the spectral features characteristic of proteins[25,26] were also observed in the spectra of the mineralizing tissues. Those associated with the peptide bond of the protein chain are (Figure 3) the Amide I, (C-O stretch) between approximately 1630-1670 cm^{-1}; Amide II, (C-N stretch and N-H bend) weak in the Raman spectra at approximately 1570 cm^{-1}. Those associated with motions of the specific amino acids (R groups) are the torsional and bending modes of CH_2 and CH_3 groups between approximately 1350-1450 cm^{-1} and C-O stretching associated with carboxyl groups at approximately 1420 cm^{-1}. The most distinctive difference in the Raman spectra of the collagenous matrix of bone and dentin and the non-collagenous matrix of mineralizing enamel were seen in the spectral regions of the Amide III and the CH_2 and CH_3 modes (Figure 3).

Table 1. Comparison of the Raman Spectra of Calcified Tissue and Collagen (in cm^{-1})

This Work		Walton et al.[26]		
Enamel*	Chick tibia*	Ox tibia	Collagen	Assignment
1675	1678	1663		Amide I
1658			1652	
	1642			
1626				
	1594	1611		
1561		1583	1571	Amide II
1458	1456	1451	1451	CH$_2$ def.
	1442		1426	
			1382	
1346	1348			
1326	1326			CH$_2$ wag.
		1309		
1278	1284		1270	
1256	1252	1256	1252	Amide III
1197			1198	
1123	1123			ν_1 CO$_3$
	1072	1071		ν_3 PO$_4$
1054				
	1036	1033		
1003	1004			ν_1 org. PO$_4$
960	962	955		ν_1 PO$_4$
888			861	
	862			
836				
		818		
		781	787	
764	760			
		713		
648			631	
		615		
594	594			
		581	579	ν_4 PO$_4$
	536		548	
450	464	448	448	
				ν_2 PO$_4$
434	434	430		
406			411	
	386			

* ± 3 cm^{-1}

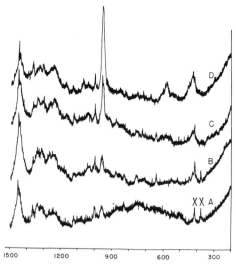

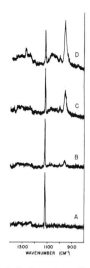

Figure 5. Sequential Raman microprobe spectra of rat incisor enamel. (A-D) obtained from areas indicated in Figure 2c, A, B, C, D areas approximately 7 μm diameter; time constant, 15 seconds; scan rate, 5 cm⁻¹/minute, X-bands due to sapphire substrate.

Figure 7. Sequential low irradiance Raman microprobe spectra (A-D) of embryonic chick tibia obtained from areas indicated in Figure 2a, A, B, C, D areas approximately 15 μm apart. Power, 5 mW at sample; beam, 15 μm diameter; time constant, 60 seconds; scan rate, 1 cm⁻¹/minute.

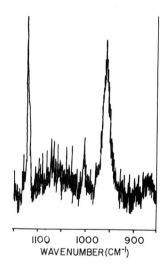

WAVENUMBER(CM⁻¹)

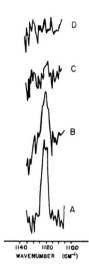

WAVENUMBER (CM⁻¹)

Figure 6. Low irradiance Raman microprobe spectrum of mineralization front in rat incisor enamel obtained from area indicated in Figure 2c, E. Power, 5 mW at sample; beam, 7 μm diameter; time constant, 50 seconds; scan rate, 1 cm⁻¹/minute.

Figure 8. Raman microprobe spectra obtained from area D in Figure 2a. Spectrum A obtained after region exposed to 5 mW for 2 hours. Spectra B, C, D from identical area D after region exposed to 60 mW for 8, 16, 24 minutes, respectively. Beam, 7 μm diameter; time constant, 20 seconds; scan rate, 5 cm⁻¹/minute.

The Raman spectra of each of the mineralized tissues have a well resolved band of weak to medium intensity at 1003 cm⁻¹. Figure 5 illustrates this band in spectra obtained at increments of 7 μm from the mineralizing front toward regions of higher mineral content. The intensity of this band at the mineralizing front is comparable to that of the 960 cm⁻¹ band. The Raman shift of this band suggests that it may be due to the ν_1 (P-O symmetric stretch) mode of the organic phosphate associated with a phosphoprotein. Raman spectra obtained from phosphorylated proteins and amino acids [e.g. phosvitin and phos-

phoserine (authors unpublished results)] and adenosine monophosphate[27], in solution or solid form, all have the most intense band of their Raman spectra in the spectral range 990-1010 cm⁻¹.

Discussion

Walton et al.[26] have discussed the advantages of Raman spectroscopy for the study of calcified tissues. This technique, however, requires irradiance levels whose effect may be unknown and our data indicate that components in our specimens are thermally unstable, requiring the use of very low irradiance levels. With suitable adjustment of experimental parameters this problem has been alleviated and good signal to noise ratios have been observed in the resulting spectra.

With this application of the Raman microprobe, we have obtained vibrational spectra of microscopic regions of intact biological sections during early stages of mineralization and mature enamel. The correlation of our data for the embryonic chick tibia with that of Walton et al.[26] on mature defattened ox tibia (cortical bone) (Table 1) indicates that the macro and micro Raman measurements are in good agreement. The slight discrepancies may be due to differences in the degree of mineralization, age and species. The great advantage of the Raman microprobe method is the ability to precisely focus pre-determined beam spot sizes on selected regions of interest.

The spectra of the mineralizing tissues are characteristic of an apatitic phosphate phase in a proteinaceous matrix. Sequential analyses from the mineralizing front to more mineralized regions all indicate a steep gradient in the accumulation of apatite phosphate mineral, while the intensities associated with the protein component appear to remain fairly constant. This suggests that in the early stages of mineralization, mineral accumulation may be accompanied by the removal of water from a very hydrated organic matrix. Numerous authors have discussed the changes which occur in mineral and water[8, 28, 29,] content and the proteins and amino acids[9, 29-34] during enamel mineralization. The spectral differences between mineralizing enamel and mature enamel (Figure 4) reflect a major protein loss and mineral accumulation described by these authors, however our spectral analyses (Figure 5) relate to the changes which occur in enamel mineralization from the mineralization front to the dentino-enamel junction where the enamel is ~ 30 μm thick. Our observations apply to relatively homogenous microregions of tissue samples which have not been previously isolated. However, the micro-dissection techniques used by Robinson[34] enabled chemical analyses of larger but comparable tissue areas. We are presently correlating their data[34] on the variation of amino acids in developing rat incisor enamel with the spectral features of each of the major amino acids in order to interpret our spectra.

The rapid increase in relative intensity of the 960 cm^{-1} band from the surface of mineralizing enamel to the dentino-enamel junction supports the data of Glick[13] in a similar region of enamel development in the rat. While our data relates specifically to mineral as determined by phosphate intensity, the correlation with the calcium analyses agrees well.

The spectrum of each mineralizing tissue has the band at approximately 1003 cm^{-1} which we assign to an organic phosphate. The characterization and identification of phosphoproteins associated with biological mineralization is presently being studied in enamel[35], predentin[36], dentin[37] and bone[38]. These phosphorylated species contain

approximately 2-6% phosphorus (approximately 6-18 % phosphate) associated primarily with the serine residues of the protein. Our observations indicate that the organic phosphate is a major component in these regions of early mineralization.

Our observations of the 1123 cm^{-1} band which we associated with an unstable carbonate, agree with the analysis of Hiller et al.[19] for magnesium and carbonate in the same regions of developing rat incisor enamel. They suggest that both magnesium and carbonate may be "diluted out" by the accumulation of calcium and phosphate. Our results indicate that the spectral feature associated with this carbonate compound is relatively intense at the mineralizing front of enamel and becomes either too weak to be observed as the mineral accumulates or the carbonate decomposes during exposure. Positive identification of this phase must include either the other spectral features associated with it or an explanation of their absence. It is interesting that Scheetz and White[24], who have performed the vibrational analysis of the mineral huntite, have observed that a particular sample of the mineral produced a very broadened lattice mode spectrum in the infrared and only the intense ν_1 mode in the Raman spectrum.

Our spectral assignment for carbonate associated it with magnesium. In both the enamel and the chick tibia samples, in which the carbonate was observed, magnesium was detectable by preliminary electron microprobe analyses (as a minor element).

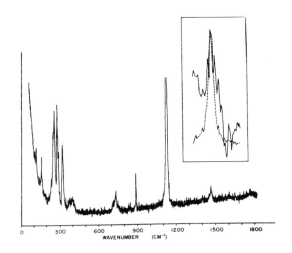

Figure 9. Raman microprobe spectrum of huntite, $Mg_3Ca(CO_3)_4$. Power, 20 mW at sample; beam, 7 μm diameter; time constant, 5 seconds; scan rate, 20 cm^{-1}/minute. Inset illustrates coincidence of ν_1 at 1123 cm^{-1} in huntite (-----) and 1123 cm^{-1} band in embryonic chick tibia (———). Inset spectra obtained at conditions described in Figure 7.

Conclusion

Raman microprobe analyses permit the molecular identification of an organic phosphate component and suggest a carbonate component, associated with calcium and magnesium, in the presence of apatitic phosphate at the mineralizing front of these tissues.

The observation of an organic phosphate component at the mineralizing front and in regions of low mineralization suggests that this component may be concentrated in these regions. Positive identification of magnesium calcium carbonate, huntite, at the mineralizing front must include either all bands in the Raman spectrum or an explanation of their absence. An important factor in the absence of the lattice modes, those bands between 200-315 cm^{-1} with an intensity approximately one-third that of the ν_1 (1123 cm^{-1}), may be the size and perfection of the crystallites at the mineralizing front.

These results suggest that the calcium and phosphate determined by elemental analysis must be partitioned among these components. The Raman, electron and ion microprobes should prove to be a powerful combination by which this partitioning may be estimated.

Acknowledgment

This research has been supported by Interagency Agreement # Y01-DE-80027 between the National Bureau of Standards and the National Institute of Dental Research.

References

(1) T. Hall, P. Echlin and R. Kaufman, Microprobe Analysis as Applied to Cells and Tissues, (ed), Academic Press Inc., London, England, 1974. See also, Biol. Microanalysis papers this Vol*
(2) J.R. Coleman, X-ray Analysis of Biological Samples, SEM/1978/II, SEM Inc., AMF O'Hara, IL. 60666, 911-926.
(3) M.B. Bellhorn and R.K. Lewis, Localization of Ions in Retina by Secondary Ion Mass Spectrometry, Expt. Eye Res., 22, 1976, 505-518.
(4) M. Delhaye and P. Dhameloncourt, Raman Microprobe and Microscope with Laser Excitation, J. Raman Spetrosc. 3, 1975, 33-43.
(5) G.J. Rasasco and E.S. Etz, The Raman Microprobe: A New Analytical Tool, Res. and Devel., 28, 1977, 20-35.
(6) E.S. Etz, Raman Microprobe Analyses-Principles and Application, SEM/1979/I, SEM, Inc., AMF O'Hare, IL. 60666.
(7) E.S. Etz and K.F.J. Heinrich, Laser Raman Probe Microanalysis of Biological Tissue, Dimensions/NBS, 62 (3), 1978, 20-21.
(8) M. Deakins, Changes in the Ash, Water and Organic Content of Pig Enamel During Calcification, J. Dent. Res. 21, 1942, 429-435.
(9) M.J. Glimcher, D. Brickley-Parsons and P.T. Levine, Studies of Enamel Proteins During Maturation. Calcif. Tiss. Res., 24, 1977, 259-270.
(10) M.U. Nylen, E.D. Eanes and K.A. Omnell, Crystal Growth in Rat Enamel, J. Cell Biol., 18, 1963, 109-123.
(11) K.A. Selvig and A. Halse, Crystal Growth in Rat Incisor Enamel, Anat. Rec., 173, 1972, 453-468.
(12) P.L. Glick and D.R. Eisenmann, Electron Microscopic and Microradiographic Investigation

*SEM/1979/II, SEM Inc., AMF O'Hare, IL 60666, USA

of a Morphological Basis for the Mineralization Pattern in Rat Incisor Enamel, Anat. Rec., 176, 1973, 289-305.
(13) P.L. Glick, Patterns of Enamel Maturation, in Proceedings of the Third International Symposium on Tooth Enamel, M.U. Nylen and J.D.Termine (ed.), J. Dent. Res., 58 (B), 1979, 883-892.
(14) H. Warshawsky, A Light and Electron Microscopic Study of the Nearly Mature Enamel of Rat Incisors, Anat. Rec., 169 (3), 1971, 559-583.
(15) J.E. Eastoe, Organic Matrix of Tooth Enamel, Nature, 187, 1960, 411-412.
(16) J.C. Elliott, The Crystallographic Structure of Dental Enamel and Related Apatites, Ph.D. Thesis, University of London, 1964.
(17) W.H. Emerson and E.E. Fischer, The Infrared Absorption Spectra of Carbonate in Calcified Tissues, Arch. Oral Biol., 7, 1962, 671-676.
(18) J.C. Elliott, The Interpretation of the Infrared Absorption Spectra of Some Carbonate Containing Apatites, In Tooth Enamel, Its Composition, Properties and Fundamental Structure, Vol.I, M.V. Stack and R.W. Fearnhead, (ed.), John Wright (Bristol), 1965, 20-28.
(19) C.R. Hiller, C. Robinson and J.A. Weatherall, Variations in the Composition of Developing Rat Incisor Enamel, Calcif. Tiss. Res., 18, 1975, 1-12.
(20) K. Tamura, M. Mori and K. Kawakatsu, A Comparative Study of X-ray Diffraction and Infrared Absorption Properties and Amino Acid Analysis of Bovine Enamel, Archs. Oral Biol., 16, 1971, 801-811.
(21) M. Suzuki, Studies on the Physicochemical Nature of Hard Tissue, n No. 230-Physico-Chimie et Crystallographie des Apatites d'Intérêt Biologique, Colloques Internationaux C.N.R.S., Paris, 1973, 77-83.
(22) K.C. Blakeslee and R.A. Condrate, Vibrational Spectra of Hydrothermally Prepared Hydroxyapatites, Jour. Amer. Cer. Soc., 54 (11), 1971, 559-563.
(23) B.O. Fowler, Infrared Studies of Apatites. I. Vibrational Assignments for Calcium, Strontium and Barium Hydroxyapatites Utilizing Isotopic Substitution, Inorg. Chem., 13, 1973, 194-207.
(24) B.E. Scheetz and W.B. White, Vibrational Spectra of the Alkaline Earth Double Carbonates, Amer. Miner., 62, 1977, 36-50.
(25) A.G. Walton and J. Blackwell, Biopolymers, Academic Press, New York and London, 1973, 168-192.
(26) A.G. Walton, M.J. Deveney and J.L. Koenig, Raman Spectroscopy of Calcified Tissue, Calcif. Tiss. Res., 6, 1970, 162-167.
(27) L. Rimai, T. Cole and J.L. Parsons et al., Studies of Raman Spectra of Water Solutions of Adenosine tri-, di-, and monophosphate and some Related Compounds, Biophys. J., 9, 1969, 320-392.
(28) E.D. Eanes, D.R. Lundy and G.N. Martin, X-ray Diffraction Study of the Mineralization of Turkey Leg Tendon, Calcif. Tiss. Res., 6, 1970, 239-248.
(29) M. Fukae and M. Shimizu, Studies on the Proteins of Developing Bovine Enamel, Arch. Biol., 19, 1974, 381-386.
(30) M.J. Glimcher and P.T. Levine, Studies of

the Proteins, Peptides and Free Amino Acids of Mature Bovine Enamel, Biochem. J., 98, 1966, 742-753.

(31) S.M. Weidmann and D.R. Eyre, The Protein of Mature and Foetal Enamel, in Tooth Enamel, Its Composition, Properties and Fundamental Structure, Vol. II, R.W. Fearnhead and M.V. Stack (ed.), John Wright (Bristol), 1971, 72-78.

(32) F.M. Eggert, G.R. Allen and R.C. Burgess, Purification and Partial Characterization of Proteins From Developing Bovine Dental Enamel, Biochem. J., 131, 1973, 471-484.

(33) J.M. Seyer and M.J. Glimcher, Evidence for the Presence of Numerous Protein Components in Immature Bovine Dental Enamel, Calcif. Tiss. Res., 24, 1977, 253-257.

(34) C. Robinson, N.R. Lowe and J.A. Weatherall, Changes in Amino-acid Composition of Developing Rat Incisor Enamel, Calcif. Tiss. Res., 23, 1977, 19-31.

(35) J.M. Seyer and M.J. Glimcher, Isolation of Phosphorylated Polypeptide Components of the Organic Matrix of Embryonic Enamel, Biochem. Biophys. Acta, 236, 1971, 279-291.

(36) D.J. Carmichael, A. Chovelon and C.H. Pearson, The Composition of the Insoluble Matrix of Bovine Predentin, Calcif. Tiss. Res., 17, 1975, 263-271.

(37) W.T. Butler, J.E. Finch and C.V. DeSteno, Chemical Character of Proteins in Rat Incisors, Biochem. Biophys. Acta, 257, 1972, 167-171.

(38) A.R. Spector and M.J. Glimcher, The Extraction and Characterization of Soluble Anionic Phosphoproteins from Bone, Biochem. Biophys. Acta, 263, 1972, 593-603.

Discussion with Reviewers

F. Adar: Can the structure in any of the Raman bands of hydroxyapetite shown in Figure 4 be accounted for by either site group or factor group effects?
Authors: Yes, this is fairly well documented (text reference 22, 23) and the $\nu_4(E)$, $\nu_3(F)$ and ν_4 splittings can be accounted for by factor group effects.

R.A. Condrate: You have been able to identify the huntite carbonate phase in various mineral-containing tissues using Raman spectra. Have you also been able to identify the Raman bands associated with the normal modes of carbonate ions substituted into the lattice of hydroxyapatite, and obtain structural information from them?
Authors: The Raman band assignments for carbonate in mature dental enamel have been made by Fowler (B.O. Fowler, Polarized Raman spectra of Apatites. II. Raman bands of carbonate ions in human tooth enamel, Miner. Tiss. Res. Commun. 3, Oct., 1977, No. 68). These assignments indicate that in human dental enamel carbonate ion substitution for hydroxyl ions (equivalent to 5-10% of the total carbonate content of enamel) results in a Raman band (ν_1) at 1105 cm^{-1} and carbonate ion substitution for the phosphate ion (equivalent to approximately 90-95% of the total carbonate content of enamel) results in a Raman band at 1070 cm^{-1}, coincident with the ν_3 phosphate mode. In the regions of early mineralization and low mineral content, the intensity of the ν_3 phosphate mode is very weak in our spectra and of course superimposed on any bands associated with the protein which may occur in this spectral region. Therefore, to determine this carbonate component would require observing an increased intensity in the region of 1070 cm^{-1} over that of the ν_3. We have not attempted such a study at this stage of enamel development.

R.A. Condrate: You mentioned that high irradiance of your tissues by a laser, generated fluorescence which interfered with its Raman spectra. Does this mean that Raman microprobe techniques could not be used to investigate the effects of high intensity laser damage on related tissues?
Authors: Irradiance generated fluorescence in the Raman microprobe technique is not greater than that generated in the usual laser Raman technique. The fluorescence component which we have observed on initial exposure of the sample to the laser decreases to low background levels in 15-20 minutes. Other authors (text reference 26) have indicated similar effects with biological samples.

We have observed that samples fixed with glutaraldehyde in sodium cacodylate [$NaAsO_2(CH_3)_2$] buffer do not defluoresce. This we have attributed to the arsenic ion.

Fluorescence may also be eliminated by the use of longer wavelength exciting radiation (568.2 nm, yellow or 647.1 nm, red).

Beam focusing to any spot size with high irradiance levels will damage biological samples. The samples in this work are exposed to irradiance levels well below this threshold.

The Raman microprobe techniques could, therefore, be used to study the effect of laser damage to tissues.

J.C. Elliott: Did the authors find the same instability for the mineral huntite as was observed for the line at 1123 cm^{-1}?
Authors: No, we did not observe this instability. We are suggesting that this difference may be due to two factors 1) the difference in size of the mineral form (approximately 1 μm) and that of the biological form at this stage of development (approximately 10 x 200 A^o) and 2) the presence of water in the biological form.
J.C. Elliott: Under what conditions might huntite be expected to form?
Authors: P. Moller and F. Kubanek (Role of magnesium in nucleation processes of calcite, aragonite and dolomite, N. Jb. Miner. Abh., 126(2), 1976, 199-220) have suggested that huntite is a tertiary product of secondary nucleation (after calcite and dolomite) from solutions with an Mg/Ca molar ratio greater than 40.

R.L. Hayes: At the mineralizing front of chick tibia and rat incisor enamel, a prominent and unstable band appears at 1123 cm^{-1}. This band coincides with that of the carbonate ion in huntite. Would you comment on the significance of such a magnesium-calcium carbonate at the mineralizing front of these tissues?
Authors: As mentioned previously, the 1123 cm^{-1} band alone does not identify the mineral huntite,

however, its spectral characteristics, narrow band width and instability on exposure to the laser both suggest an association with carbonate may be correct.

Additionally, the unit cells of huntite and apatite have some interesting characteristics. The carbonate positions in the huntite unit cell (Graf and Bradley, The crystal structure of huntite, $Mg_3Ca(CO_3)_4$, Acta Cryst., 15, 1962, 238-242) nearly duplicate the phosphate and hydroxyl positions (Posner, Perloff and Diorio, Refinement of the hydroxyapatite structure, Acta Cryst., 11, 1958, 308-309) and therefore the positions which carbonate substitutes for phosphate and hydroxyl ions in the biological apatites (text reference No. 16 and 18). These similarities in unit cell structure suggest that epitaxial growth of phosphatitic apatite on a carbonate nucleus may be an important step in the mineralization process. In enamel, especially, such may explain the anomalous dissolution charateristics of enamel crystallites (Swancar, Scott, Simmelink et al., The Morphology of the enamel crystal, in Tooth Enamel II, Its Composition, Properties and Fundamental Structure, R.W. Fearnhead and M.V. Stack Eds., John Wright, Bristol, 1971, 233-234).

This evidence suggests that the huntite or carbonate structure may be the precursor of phosphate mineral in vertebrate mineralization.

R.L. Hayes: In Figure 4, mineralizing enamel gives a Raman spectrum with numerous low intensity bands. Most of these signals are absent from mature enamel. Is this difference attributable to sample quality or a function of chemical changes coincident with maturation or mineralization of the enamel?

Authors: The difference is not due to sample quality, but indicates the difference between immature or mineralizing enamel and mature enamel. The spectra reflect the loss of organic matrix which occurs as enamel mineralizes.

Additional discussion with reviewers of the paper "Coccolith Morphology and Paleoclimatology...." by P.L. Blackwelder et al, continued from page 420.

S. Margolis: Have you cultured the coccolitho- phorids at different temperatures and noticed any change(s) in size and number of coccoliths?
Authors: We are now culturing these species in the 12°C - 28°C temperature range. Our findings thus far indicate that G. oceanica and C. leptopora have quite different growth rates and isotopic ($^{18}O/^{16}O$ ratios) composition when grown under the same experimental conditions. Our experiments are not as yet complete with regard to possible effects of temperature on coccolith morphology.

R. Weiss: What is meant by "weak attachment area" in your description of the coccoliths of C. leptopora?
Authors: We have observed individual dis- articulated coccoliths (proximal and distal shields separated). The area of attachment of these plates does not appear to be fully calcified. We have described this as a "weak attachment area."

J. Matthews, W. Davis, R. Jones: How do the large intracellular coccoliths get out without extensive cell disruption? Do extracellular coccoliths have an outer membrane?
Authors: Coccolith extrusion involves fusion of the Golgi-derived vesicle (which contains the coccolith) membrane with the plasma mem- brane. This process , as viewed ultrastruc- turally, results in the extrusion of the coccolith to the extracellular surface without extensive cell disruption. Transmission electron microscopic observations of Cricosphaera carterae indicate the presence of a non-crystalline cov- ering which conforms to the shape of the cocco- lith. These coverings have been observed by several workers after decalcification of these structures, but their nature has not been described.

SCANNING ELECTRON MICROSCOPY/1979/II
SEM Inc., AMF O'Hare, IL 60666, USA

ESTIMATION OF THE SIZE OF RESORPTION LACUNAE IN MAMMALIAN CALCIFIED
TISSUES USING SEM STEREOPHOTOGRAMMETRY

A. Boyde and S. J. Jones

Department of Anatomy
University College London
Gower Street, London WC1E 6BT
U.K.

Abstract

The main aim of this paper is to report
quantitative stereophotogrammetric measurements
of the sizes of osteoclastic resorption lacunae
in enamel, bone and dentine. Surfaces for SEM
were exposed by removing surface cells mechanically with washing, or chemically with trypsin or
NaOCl. Negative replicas were made by casting
Araldite against a dried specimen, and dissolving
the specimen with 6N HCl and 5% NaOCl used alternatively. Anisotropy was more generally present
in the bone lacunae, whereas the resorbing tooth
surfaces had a higher proportion of more isotropic lacunae. This may indicate a high degree of
translocatory movements of osteoclasts resorbing
bone. Anisotropy was most easily studied in the
replicas or pseudoscopic and negative inversions
of the stereopairs.

Stereopair images of resorbed surfaces were
measured in EMPD1 or an SB 185 stereometer.
Three points, one at each rim of a resorption
lacuna and one at the deepest point on the
straight line between them, were measured so that
the widths and depths of the lacunae could be
determined and the radius of curvature of the
lacuna calculated. Minimum radii of curvature
were remarkably similar in all tissues: maximum
values reflected the anisotropy of the lacuna,
that is the lateral shift of the cell and/or the
ruffled border. By adding a measurement of the
length in the direction of the greater radius of
curvature for each resorption lacuna, it was possible to derive volumes and surface areas. The
present results indicate that bone resorption
lacunae are smaller in every respect than tooth
resorption lacunae. This method enables minimum
values of the volume of mineral that could be
removed by one osteoclast to be measured.

KEY WORDS: Osteoclast, Ruffled Border, Resorption Lacunae, Enamel, Dentine, Bone, Stereophotogrammetry

Introduction

Removal of hard tissue substance involving
the intervention of special cells (most commonly
called osteoclasts[1], whatever the substrate that is
being digested) occurs under both normal and pathological circumstances. Osteoclasis in man normally occurs during growth-remodelling of bone[2] and
calcified cartilage, and during the removal of the
roots and a small portion of the crown of deciduous teeth[3]. Portions of the roots of developing,
functional human permanent teeth are resorbed
prior to the deposition of cellular cementum[4].
Bone is also resorbed to maintain blood calcium
and phosphate levels, and this occurs excessively
in some conditions which are associated with
abnormally high bone turnover rates such as Paget's
disease[5]. In addition to the normal resorption
processes found in man, mammals such as the horse
resorb portions of the enamel of the crowns of
the teeth prior to the attachment of cementum to
the surface of the enamel[6].

Conventional light microscopic (LM) histology
using 5-15 μm thick demineralized sections gives
a poor overview of the process of hard tissue
resorption because (a) the dimensions of the
special cells involved - the osteoclasts - exceed
the thickness of the sections; (b) we have to
reckon with differential dimensional changes
during the histological processing of the osteoclasts and the demineralized matrix and (c), in
the case of enamel, the organic matrix of mature
enamel is entirely removed by demineralization.
Thus, although osteoclasts are known to be large
cells, satisfactory data concerning their overall
shape and size and details of the shape and size
of the actively resorbing territories of each cell
are lacking from the literature.

Conventional transmission electron microscopy
of ultra thin sections of resorbing bone and other
hard tissues shows that only a relatively small
area of each osteoclast is involved in the active
removal of bone substance[7,8]. This portion of the
cell may be delineated by a clear zone of cytoplasm and shows a ruffled border.   Previous estimates of the area of ruffled border zones from
TEM histomorphometric studies have been based on
single ultra-thin sections which may represent
less than 0.005% of the total cell volume, for
cells which typically cover about   0.5% of bone
surfaces[9,10,11].

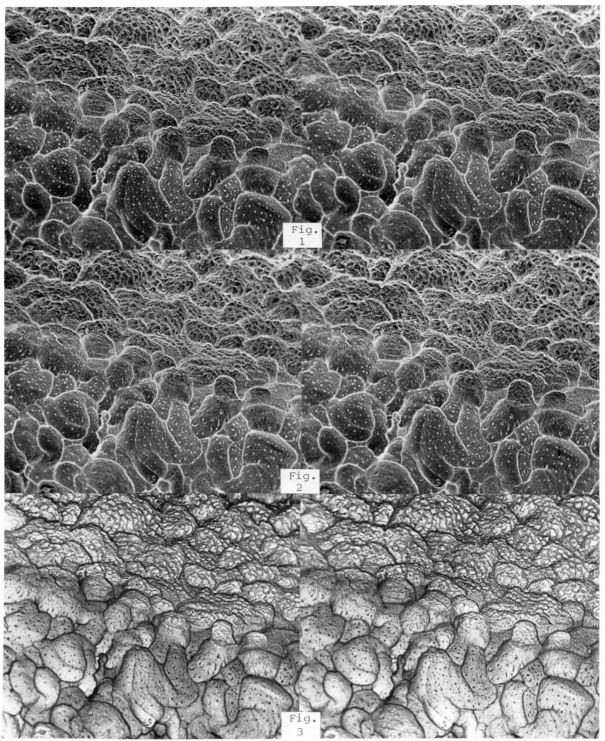

Figures 1-3: Resorbing surface of crown of human deciduous molar; enamel top, dentine below. Stereo pair, tilt angle 10°. Width of Field 270 μm.

Figure 1: "Printed and mounted the right way round".

Figure 2: Pseudoscopic inversion - viewed stereoscopically, it appears as if one is viewing the surface from within the tissue.

Figure 3: Pseudoscopic and negative inversion. The images are printed as negatives, as well as being mounted so that the wrong image is presented to each eye. It now appears as if one is looking at the surface of the osteoclasts that had been resorbing this surface. BUT it must be borne in mind that this is not so. Values for the width W; depth D; and radius of curvature R for lacunae in Figure 1 are given in Table 1.

Scanning electron microscopy (SEM) has been shown to be a satisfactory means of studying the plan extent of osteoclasts on surfaces which are good hunting grounds for these cells, such as the juxta-sutural remodelling zones of the endocranial aspect of the young rat skull[12]. SEM shows these cells to be remarkably diverse in shape, and suggests that each cell may possess more than one active resorption site, such sites being located at considerable distances apart on branched, spindly cells. However, the problems of differential shrinkage are maximal for the case of cells in contact with a mineralized tissue[13], so that measurements of CPD osteoclasts need to be interpreted with caution and their shapes will have been modified during preparation.

On the other hand, the surface of a hard tissue which was undergoing resorption is one of the most undeformable which presents in biological SEM. The hard tissues contain so little water that they do not shrink seriously on drying, and can be examined directly in the SEM. Such surfaces are also easily replicated with, for example, epoxy resin casts, which are also relatively dimensionally stable. The present investigation makes use of both procedures in the morphological and morphometric study of normal osteoclasis in bone, cementum, dentine and enamel of man, the rat and the horse.

Materials and Methods

Hard tissue surfaces were prepared by simple dissection and washing with a stream of water to remove surface cells. In some cases this was aided by ultrasonication and trypsination. Such surfaces were freeze-dried (FD), critical point dried (CPD) or air-dried from ethanol (AD). Anorganic preparations were made by treatment of the tissues with 5% NaOCl, followed by washing in water and air drying.

Negative replicas of surfaces cleaned as above were made by casting Araldite CY 212 (EM-Scope, 13 Bedford Road, London SW4 7SH, U.K.) against the dried sample. After this had set (2 days) the Araldite was trimmed and the hard tissue dissolved by alternate treatment with 6N HCl and NaOCl.

All samples, native or replica, were coated with carbon and gold by evaporation or with gold only by sputtering in argon, and examined in a Cambridge Stereoscan M I or S410 operated at 10 kV.

Photogrammetric methods

Stereo-pair images were recorded with 10° tilt angle difference[14-16]. One inch square contact prints of the photographic negatives were measured at 8-fold enlargement in EMPD1[15,16]: height differences at the plot scale are read directly from this instrument. The X,Y coordinate data was measured from marks made using the contouring pencil of this instrument.

Alternatively, prints made to the original CRT magnification were measured using an SB 180 stereometer[14]. Parallaxes for 10° tilt pairs were multiplied by 5.74X to give the height difference[17]. X,Y co-ordinate data were measured from marks made with a contouring pen attached to the plattens of this instrument[17], or measured

directly using a transparent millimeter scale over the left photograph[16]. With both measuring instruments, three points on a straight line across each resorption loculus were measured; one at each rim and one at the deepest point (or point of maximal elevation in the case of Araldite replicas) of each depression.

In obtaining the results shown in Table 1, stereopair photographs were used in which the mean tilt of the local specimen surface facet had not been controlled. Each lacuna was measured in the direction which permitted a simple 3 points-in-one-line analysis. The data recorded were the parallax screw readings at the three points (A1, B1, C1) and the linear distance measured in the left photo between A1 and B1 (A) and between A1 and C1 (B). These data were processed using a Hewlett-Packard 9830A programmable calculator. The width, W, between the two rims and the depth, D, normal to the line connecting the two rims (i.e. a perpendicular from B1 to the line joining A1 and C1 in true space) were calculated. From observations on many enamel and dentine lacunae it was concluded that it would be justifiable to assume that they had nearly spherical outlines. Using W and D, the radius of curvature, R, of the sphere was calculated from the expression

$$R = W/(2.\sin(180 - (2.\arctan(0.5.W/D))))$$

The volume, V, of each resorption loculus was calculated as the segment of a sphere from the expression

$$V = ((\pi/6) . D) . ((3.((\tfrac{1}{2}W)^2)) + (D^2)).$$

When more bone resorption loculi were measured, it became clear that the working assumption of near sphericity was not justified, since the shapes were closer to segments of cylinders or ellipsoids. It was therefore either necessary to measure a largish number of X,Y,Z co-ordinates around the rims and in the central portions of resorption bay, which would be very time-consuming with the existing system for manual transfer of the data, or to make an alternative simplifying assumption.

It was therefore decided to measure the resorption lacunae in the line of the least radius of curvature. To make this selection possible, the stereo pairs were recorded with the specimen surface facet nearly normal to the beam, i.e. as near as possible to the condition of symmetrical tilt about the beam axis. A sixth measurement of each lacuna was then recorded, namely, the working length, L, of a line along the direction of the greatest radius of curvature: for bone resorption tracks, this is often nearly flat and is usually the longest axis of the lacunae. This working length, L, was recorded with the knowledge that it was to be used primarily to multiply the cross sectional area, A, in the direction of the least radius of curvature in calculating the surface area, S, and the volume, V, of the resorption lacuna using expressions for the segment of a cylinder.

The sector angle,

$$T = \arctan (0.5W)/\sqrt{1 - ((0.5W)/R^2)}.$$

The cross sectional area,

$$A = ((\pi.R^2) . 2 . T)/360) - (((R^2) \times \sin (2.T))/2).$$

The length of the arc in the least-radius direction which we designate as the perimeter,

$$P = \pi . 2 . T/90.$$

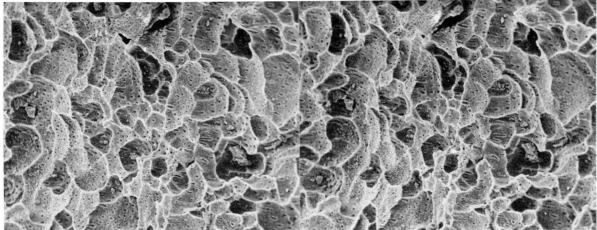

Figure 4: Human deciduous molar showing resorption of dentine in the crown region. Field 270 µm.

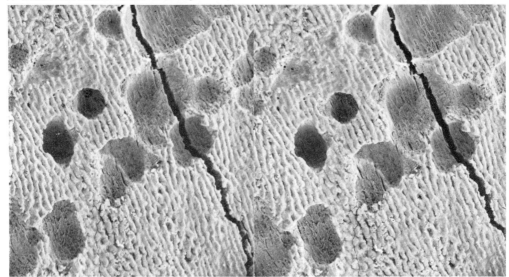

Figure 5: Horse molar enamel undergoing superficial resorption prior to deposition of coronal cementum. Stereo-pair, tilt angle 10°. Field 170 µm.

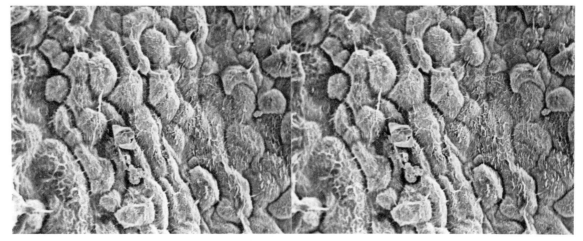

Figure 6: Araldite replica of resorbing horse molar enamel. Stereo-pair tilt angle 10°.
Field 300 µm.

Table 1.   Dimensions of resorption lacunae in figures of this paper
Standard deviations in brackets below each value.

Figure No.	N	Tissue	Width	Depth	0.5W/D	Radius	Spherical Volume
1-3 11	11	Human dec. enamel.	44.7 (10.0)	13.6 (2.9)	1.7 (0.4)	26.2 (8.0)	12944 (6643)
1-3	20	Human dec. dentine	39.0 (14.1)	14.8 (5.3)	1.5 (0.8)	23.5 (11.8)	12379 (9282)
4	28	Human dec. dentine	39.5 (14.5)	13.0 (4.2)	1.6 (0.5)	22.9 (10.5)	11601 (10244)
5 & not figd.	10 / 30	Horse enamel	27.4 (9.7)	11.2 (4.5)	1.4 (0.5)	15.1 (5.5)	5447 (5308)
6	50	" " replica	25.4 (6.9)	8.7 (2.9)	1.6 (0.5)	14.4 (4.7)	3036 (2212)
9	12	Human perm. root dentine	24.2 (9.9)	6.9 (4.3)	2.0 (0.9)	16.9 (7.8)	3058 3348
10	19	Paget's bone	23.2 (10.2)	5.0 (3.2)	2.6 (1.3)	20.3 (12.6)	2184 (3709)
11	20	Osteopetrotic bone	18.3 (6.9)	5.9 (2.4)	1.5 (0.7)	11.3 (5.7)	1389 (1108)
13 & not figd	10 / 8	Human femur endosteal	11.0 (4.2)	2.2 (1.0)	2.4 (1.0)	8.3 (3.9)	156 (139)

Widths, depths and radiuses in microns; volumes in cubic microns

The surface area, $S = L . P$.
The volume, $V = L . A$.
Table 2 shows results processed in the above fashion. It is important to note certain features of the most extrapolated values, viz: the surface area and the volumes of the lacunae computed in this way. Firstly, we measured 50 lacunae in two directions so as to compute the major and minor radii of an ellipsoid. The surface areas derived from the measurements made in the two extreme directions were in very close agreement. The volumes obviously agreed only if the ends of the resorption loculi were at the same height as the base. Generally, however, the resorption loculi are shaped like boats with a good sheer to the gunwhales. We decided that the relevant volume was, to use the boat analogy, that which lay beneath a longitudinally curved plane resting on the gunwhales (not, that lying beneath a flat plane touching only at the stem and the stern).

Processing measurements from dentine replicas using the "cylindrical-segment" model (Table 2) gave volumes which were close to those calculated from the spherical model assumption (Table 1). However, a qualitative difference was noted between resorbing bone and dentine surfaces in that L in dentine was often less than W, i.e. the larger radius of curvature was often present for a length less than the width measured in the direction of the least radius of curvature.

The elongation ratio shown in Table 2 is the ratio of L/W or W/L, whichever was the greater.

In bone, dentine and enamel, resorption loculae often occur in lines, rows or strings which strongly suggest that one cell was responsible for several loculae; the present system of measurements and calculations was designed to establish the minimum dimensions of an individual resorption loculus.

Results and Discussion

SEM examination of resorbing hard tissue surfaces (Figures 1-14) shows the classical resorption lacunae, which, it is commonly supposed, are each occupied by, or had been caused by, one osteoclast. The diameters of these bays in a resorption front in bone do not match the known dimensions of osteoclasts measured in histological sections by LM or as seen by SEM. In both cases the cells are usually seen to be several times as wide as the diameter of a resorption

Table 2.   Dimensions of resorption lacunae with volumes calculated according to cylindrical model: see Fig. 15.

N	Tissue	Width	Depth	0.5W/D	Rad.	cyl. vol.	P	surf. Area	elongn. ratio	L
135	Human decid. dentine replica	29.3 (9.1)	7.9 (3.3)	2.0 (0.8)	19.3 (9.2)	5184 (4078)	35 (11)	1010 (539)	1.5 (0.5)	28 (12)
165	Human mandible buccal bone	7.1 (2.8)	1.7 (0.8)	2.4 (1.2)	5.4 (3.9)	119 (134)	8 (3)	102 (86)	1.7 (0.7)	11 (7)

Width, depth, radius & perimeter in microns; areas in square & vols. in cubic microns

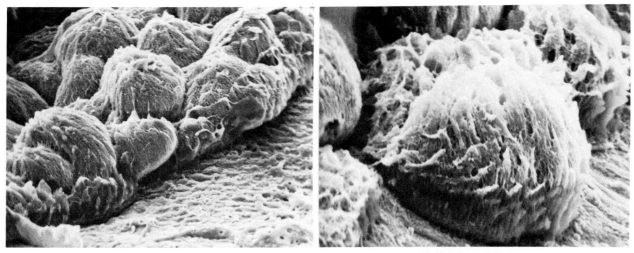

Figures 7 and 8: High tilt angle images of araldite casts of resorbing horse enamel.
Field width for Figure 7 = 105 μm and Figure 8 = 40 μm.

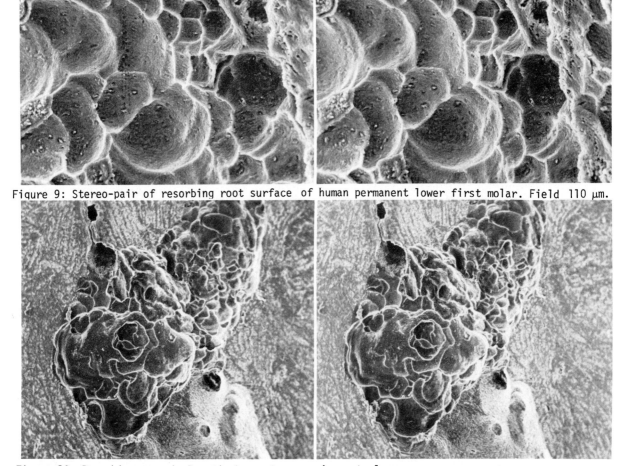

Figure 9: Stereo-pair of resorbing root surface of human permanent lower first molar. Field 110 μm.

Figure 10: Resorbing area in Paget's bone. Lacunae in central area were measured at a higher
magnification (see Table 3). Field 350 μm. Stereo-pair, tilt 10°.

loculus, and this discrepancy is greater if a correction is applied for the shrinkage of the cells during tissue processing. The discrepancy is simply explained by assuming that only small portions of osteoclasts are actively engaged in resorption at any one time[7,8,12] (see Figure 12). Dynamic data also suggest that some cells may be without active ruffled membranes for a proportion of the time that they are in the vicinity of skeletal sites known to be resorbing[18,19].

The data of this study make a contribution to knowledge of osteoclasis by indicating a range of sizes for the active areas of osteoclasts (Tables 1 and 2) in different hard tissues. There was a striking contrast between values for odontoclasis and osteoclasis. Thus resorption loculi in bone are smaller in all parameters. The sizes of resorption loculi in bone agree with our previous identification of active resorbing areas of osteoclasts[12], representing only a small fraction of the cell's territory. In enamel and dentine resorption, the actively resorbing area is much greater (Tables 1 and 2) and constitutes a major part of the territory of the cell. A greater proportion of nearly spherical lacunae were found in resorbing tooth, both root and crown surfaces (Figures 1-9), whereas anisotropy is more the rule than the exception in bone (Figures 9-11, 13 and 14). Thus in bone, many lacunae bear more resemblance to paths than pits, and we use the descriptive term snail track lacunae to signify this type (Figures 13 and 14).

Photogrammetry in the SEM will probably continue to find more applications in the measurement of indentations rather than elevations, since the latter can usually be usefully assessed by combining information available from different projections. In the present study, we have been successful in replicating resorbing areas using Araldite epoxy resin erosion casts (Figures 6-8). Measurement of resorption "bays" in low angle SEM images of casts of the lacunae agree with and provide confirmation of, the photogrammetric measurements. Low angle mono-images of replicas are easier to interpret correctly than mono-images of the original pits (Figures 7 and 8).

Deviations from roundness (sphericity) of the resorption loculi are also better seen in mono-images of the resin casts than in monoimages of the original surface. However, if the stereo-pair images of the original surface are viewed pseudoscopically (with the wrong images presented to the left and right eyes, so that the surface is seen from inside out) then such deviations are seen with equal, if not greater, ease (Figures 1-3 and 6). Whether or not such minor features are appreciated in a simple stereoscopic analysis does not affect the accuracy of 3-D coordinate reconstructions, which merely depend on precise distance measurements[15-17].

We infer that the extent of translocation of osteoclasts resorbing enamel and dentine did not approach that of osteoclasts resorbing bone. It is possible that the orientation of the collagen in the lamellae of lamellar bone may guide the direction of progress of the osteoclast. However it may also be the case that the collagen, or the mineral which it contains - since the mineral is usually dissolved first[3] - may be preferentially dissolved along the long axis, analogous to the situation with enamel crystals[20,21]. A similar guidance or preferential axial dissolution velocity mechanism could not exist in dentine, where the collagen fibril orientation is essentially random. In enamel, the prism and crystallite orientation is normal to the mean direction of progress of the osteoclasts, whether these are approaching from without, as in horse enamel (Figures 5-8), or within, as in resorbing human deciduous teeth (Figures 1-3). The fine structure of the etched surface shows perhaps a slight preference for prismatic core/head crystals rather than for prism-tail/interprismatic crystals. Indeed, in some instances, particularly in the smallest resorption loculi at the surface of the horse enamel, it would seem that the resorption progresses faster along, than across, the prism/crystallite axis (Figures 7 and 8)[6].

It would be of considerable interest to compare the shapes of resorption lacunae in primary (foetal, woven) bone with those in adult lamellar bone, having regard to the important differences which exist in the orientation of the collagen and the composition of these two tissues: foetal bone possesses a more random collagen arrangement, and a higher degree of mineralization of the matrix, both features being partly analogous with dentine. We feel that, with some small evolution, the present methodology would be adequate to study such problems in the future. However, the resorption of calcified cartilage - about which it is debated to what extent osteoclasts are involved at all - would not be sensibly studied by our approach, because this tissue contains spaces (the hypertrophic chondrocyte lacunae) which occupy a high proportion of the the volume and which bear a strong natural resemblance to osteoclastic resorption lacunae. The natural spaces in dentine and bone - the dentinal tubuli and the osteocytic lacunae and canaliculi - are easily distinguished on morphological grounds from resorption lacunae. For example, there is a high concentration of canaliculi in the bone matrix immediately adjacent to an osteocyte lacuna (Figure 11), so that, even if a resorption lacuna involves this area, one can suspect that the shape may perhaps be that of a modified osteocyte lacuna. Such useful clues are lacking in cartilage.

The present study has concentrated on methodological problems. In future work we hope to examine the details of the shape and size of the resorption lacunae in different types of bone in more detail, paying particular attention to associations with age and disease processes.

The present results indicate the very minimal volumes of mineral which can be removed by a single resorptive cell: the single-bite resorption loculi in the surface of horse enamel could only be interpreted in this way. This interpretation is supported by our own SEM studies of isolated osteoclasts among ameloblasts on completed enamel surfaces. Such resorption lacunae penetrating a surrounding intact enamel surface in horse enamel are frequently larger than 3500 μm^3 (the volume of the larger, dark pit in Figure 5). Another site at which we can be sure that each loculus is a primary "bite" into horse enamel is at the

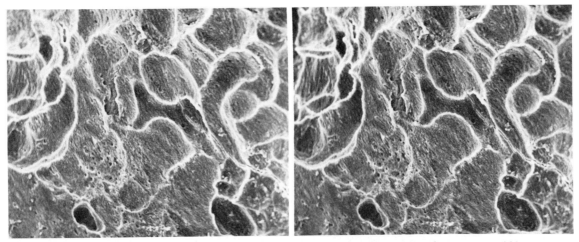

Figure 11: Resorbing area in bone from Albers-Schönberg disease (osteopetrosis). Field 380 μm.

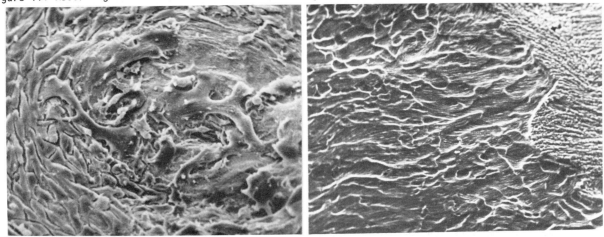

Figure 12: Osteoclast on endocranial aspect of rat skull. Field width 210 μm.

Figures 13 Human femur, endosteal resorbing surface. Field width 210 μm.

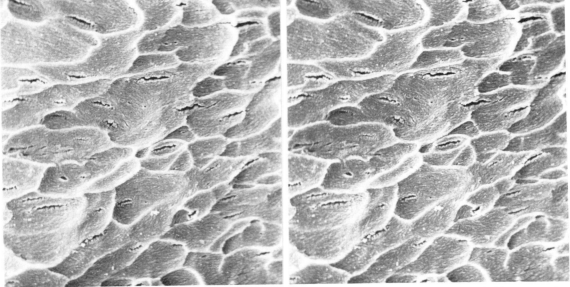

Figure 14. Buccal surface of human alveolar Sharpey fiber bone. Measurements of 28 resorption lacunae in this field are included in Table 2. Stereo-pair, tilt 10°. Fieldwidth 90 um.

periphery of the spreading resorbing patches. Here, a row of one or two pits is followed by cementum deposition which fills the pits in. Apart from these instances, the computed volumes of resorption lacunae with interrupted rims must be less than the volumes of hard tissue originally removed by the ruffled membrane resorptive organs to form the original loculi. We cannot estimate the maximum volume of calcified tissue that can be removed by one cell, both because most resorption loculi are partly removed by others and because each osteoclast may be operative at several sites and in several lacunae[12].

References

1. A.V.Koellicker, Handbuch der Gewebelehre, W.Engelmann, Leipzig, Germany, 1889, 6th edn.Vol.I.
2. Z.F.G.Jaworski, Parameters and indices of bone resorption, In: Bone Histomorphometry, P.J. Meunier, Ed., Armour Montagu, 75011 Paris, France, 1977, 193-234.
3. A.Boyde and K.S.Lester, Electron microscopy of resorbing surfaces of dental hard tissues. Z. Zellforsch. 83, 1967, 538-548.
4. S.J.Jones and A.Boyde, A study of human root cementum surfaces as prepared for and examined in the SEM. Z.Zellforsch. 130, 1972, 318-337.
5. F.R.Singer, Paget's disease of bone, Plenum Medical Book Co., New York, U.S.A., 1977.
6. S.J.Jones and A.Boyde, Coronal cementogenesis in the horse. Archs.oral Biol. 19, 1976, 605-614.
7. M.E.Holtrop, Quantification of the ultra-structure of the osteoclast for the evaluation of cell function. In: Bone Histomorphometry, P.J. Meunier, Ed., Armour Montagu, 75011 Paris, France, 1977, 133-156.
8. M.E.Holtrop, L.G.Raisz and G.J.King, The response of osteoclasts to prostaglandin and osteoclast activating factor as measured by ultra-structural morphometry. In: Mechanisms of localized bone loss, J.E.Horton, T.M.Tarpley and W.F. Davis, Eds., Information Retrieval Inc., Washington D.C., Spec.Suppl.to Calcified Tissue Abstracts 1978, 13-20.
9. R.K.Schenk, Histological estimates of bone resorption surface parameter in iliac bone. In: Bone Morphometry, Z.F.G.Jaworski, Ed., Univ.of Ottawa Press, Ottawa, Canada, 1976, 153-155.
10. P.Meunier, C.Edouard and P.Courpron, Morphometric analysis of trabecular resorption surfaces in normal iliac bone. In: Bone Morphometry, Z.F.G. Jaworski, Ed., Univ.of Ottawa Press, Ottawa, Canada, 1976, 156-160.
11. A.Schulz and G.Delling, Age-related changes in bone resorption parameters in iliac crest trabecular bone. In: Bone Morphometry, Z.F.G.Jaworski, Ed., Univ.of Ottawa Press, Ottawa, Canada, 1976, 161-162.
12. S.J.Jones and A.Boyde, Some morphological observations on osteoclasts. Cell Tiss.Res. 185, 1977, 387-397.
13. A.Boyde, E.Bailey, S.J.Jones and A.Tamarin, Dimensional changes during specimen preparation for SEM. SEM/1977/I, IIT Research Institute, Chicago, Il. 60616, 507-518.
14. A.Boyde, Height measurements from stereo-pair scanning electron micrographs, BEDO, 1, Remy Verlag, Münster, Germany, 1968, 97-105.
15. A.Boyde, A stereo-plotting device for SEM micrographs: and a real time 3-D system for the SEM. SEM/1976/I, IIT Research Institute, Chicago, Il. 60616, 93-100.
16. A.Boyde and H.F.Ross, Photogrammetry and the SEM. Photogrammetric Record 8, 1975, 408-457.
17. A.Boyde, Stereo tutorial. SEM/1979/II, SEM Inc., AMF O'Hare, Il. 60666.
18. P.Goldhaber, Behaviour of bone in tissue culture. In: Calcification in biological systems, R.F.Sognnaes, Ed., A.A.A.S., Washington D.C., U.S.A., 1960, 349-372.
19. N.M.Hancox and B.Boothroyd, Motion picture and electron microscope studies on the embryonic avian osteoclasts. J.biophys.biochem.Cytol. 11, 1961, 651-661.
20. A.N.Sharpe, Influence of crystal orientation in human enamel on its reactivity to acid as shown by high resolution microradiography. Archs oral Biol. 12, 1961, 583-591.
21. J.C.Voegel and R.M.Frank, Ultrastructural study of apatite crystal dissolution in human dentine and bone. J.Biol.Buccale 5, 1977, 181-194.

Discussion with Reviewers

J.Eurell: Do you feel a technique could be developed to measure sites on trabecular bone occupied by osteoblasts? If so, how might you propose to do this?

Authors: We have measured osteoblast territories on endocranial surfaces of rat parietal bones using methods that could be adapted for trabecular bone (S.J.Jones. The secretory territories and rate of matrix production of osteoblasts. Calcif. Tiss.Res. 14, 1974, 303-315; S.J.Jones and A.R. Ness. A study of the arrangement of osteoblasts of rat calvarium cultured in medium with, or without, added parathyroid extract. J.Cell Science 25, 1977, 247-263; S.J.Jones, A.Boyde and A.R.Ness SEM studies of osteoblasts: size, shape and anisotropy in relation to hormonal status in organ culture. Bone Histomorphometry. 2nd International Workshop, Lyon 1976, Ed. P.J.Meunier, pp.275-289, Distributor: Armour Montagu, 183 rue de Courcelles 75017 Paris, France, 1977).

J.L.Matthews, W.L.Davis and R.Jones: These SEM's show osteoclasts to have more than one site for brush border activity, each brush taking a "bite". Without a combination of cells followed by removal of cells and inspection of the calcified matrix how is it possible to determine the resorption capability per osteoclast?

Authors: We are accumulating data on the volumes of resorption lacunae below individual, well characterized osteoclasts. The number of nuclei per cell is counted by light microscopy, the cell is dissected off in the SEM and the surface recorded, and the surface than made anorganic and recorded again in the SEM. This approach confirms our previous SEM observations[12] showing that osteoclasts may have several ruffled borders in as many separate resorption loculi. However this report gives information concerning the amount of calcified matrix removed per "bite", where one "bite" corresponds to one ruffled border zone. It is possible to get the "greatest minimal" value from an inspection of the continuity of the rim of a lacuna, but not a maximum value. Beside the possibility of more than one brush border active in an osteoclast at one time[12], the site of a brush border relative to the whole cell and to the bone surface may change and several resorption loculi, often in a line, be the work of one moving resorptive organ (see Figure 15 below).

J.L.Matthews, W.L.Davis and R.Jones: Granted that shrinkage during processing may reduce cells by 15%, do SEM's of tooth resorbing cells show the numbers of cells to correspond one to one with the individual oval lacunae shown? Do osteoclasts of teeth have more than one "active site"?
Authors: Firstly, processing shrinkage of cells is greater than 15% even on a linear basis (see, for example, Figure 5 in J.L.Matthews, R.V.Talmage, J.H.Martin and W.L.Davis, Osteoblasts, bone lining cells and the fluid compartment, pp.239-247 in Bone Histomorphometry, P.J.Meunier, Ed., Armour Montagu, 183 rue de Courcelles, 75017, Paris, France; and see list of references in A.Boyde, Pros and Cons of CPD and FD for SEM. SEM/SEM Inc /1978, Vol.2, pp.303-314).
    We know from our own unpublished SEM studies that many odontoclasts only have one ruffled border. We cannot, however, exclude the possibility that some cells may have more than one. Certainly, we interpret more than one resorption loculus in a continuous short string of two or three as being the work of one cell; the loculi in this situation might be also the result of one moving ruffled border.

S.C.Miller: What is meant by "active areas of osteoclasts"? Does this refer to the resorption pits or perhaps the ruffled border areas on the osteoclast?
Authors: The active area of an osteoclast is the ruffled border.

S.C.Miller: Are you implying that small resorption pits are occupied by small cells? Please clarify.
Authors: No. Small resorption bays are occupied by small ruffled border zones. We are presently correlating cell size with the sizes and numbers of resorption lacunae per osteoclast.

S.C.Miller: I was not aware that woven bone was more mineralized than adult lamellar bone. Please provide a primary reference for this.
Authors: R.Amprino, A.Engstrom, Studies on Xray absorption and diffraction of bone tissue. Acta Anat. 15, 1952, 1-22. We have confirmed this in our microradiographic studies with M.H.Hobdell (1967 - unpublished).

S.C.Miller: I presume that the doubt you raise concerning the involvement of osteoclasts in resorbing cartilage has reference to calcified, intact hypertrophic cartilage. For example at the growth plate. Certainly the osteoclast can resorb calcified cartilage cores, for example in the metaphyseal regions of growing bones.
Authors: Exactly. But R.K.Schenk, D.Spiro and J. Wiener (Cartilage resorption in the tibial epiphyseal plate in young rats, J.Cell Biol. 34, 1967, 275-291) show that the resorption of the transverse non-mineralized bars of cartilage which

they show to be identical with the capsule of the chondrocyte are destroyed by perivascular cells and endothelial cells. Later remodelling resorbtion at the metaphysis involves osteoclasts. This work has been confirmed recently by G.Silvestrini, M.E.Ricordi and E.Bonnucci in a study of: The resorption of uncalcified cartilage in the diaphysis of the chick embryo tibia. Cell Tiss.Res. 196, 1979, 221-235.

S.C.Miller: Since only a small percentage of resorption pits have been found to have osteoclasts adjacent to them in normal human bone (see reference 9 of text), are there differences in the resorption pits that are actively being resorbed by osteoclasts from those that are not?
Authors: Yes, there are. When we dissect osteoclasts off the bone surface, and match the site of the ruffled membrane on the cell surface with the underlying bone, the bone adjacent to the ruffled membrane is seen to be 'frayed' compared with the surrounding surface.

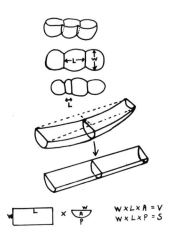

Figure 15: The three upper diagrams illustrate conjoined resorption lacunae. The width, W, is in the line of the smallest radius of curvature of each loculus, and the length, L, in the line of the largest radius. Hence L may be larger, or smaller, than W. The measured volume per loculus is smaller than the volume of tissue removed by a ruffled border of the osteoclast at each resorptive incident.
    The lower diagrams illustrate a snail track lacuna, such as may occur in bone. These lacunae were usually boat-shaped, gently curved along their length, L. Measurements of volume, V, and surface area, S, were calculated as shown, assuming the shape of the lacuna to be part of a straight cylinder. A is the area of the section, and P the perimeter of the arc. The length has been exaggerated for ease of representation.

SCANNING ELECTRON MICROSCOPY/1979/II
SEM Inc., AMF O'Hare, IL 60666, USA

ULTRASTRUCTURAL STUDIES ON CALCIFICATION IN VARIOUS ORGANISMS

N. Watabe and D. G. Dunkelberger

Electron Microscopy Center and
Department of Biology
University of South Carolina
Columbia, SC 29208

Abstract

Recent studies of ultrastructure and formation of selected types of calcareous skeletons
are summarized. The structures discussed are
echinoderm skeletons, normal and regenerated
mollusc shells, egg shells, and calcareous spicules and axial skeletons.

Morphological characteristics of $CaCO_3$
minerals in those structures and the processes of
their formation are reviewed.

KEY WORDS: Biological Calcification,
Calcareous Structures, $CaCO_3$ Crystals

Introduction

Mineral formation in living systems is a
widespread phenomenon occurring throughout the
animal and plant kingdoms. The minerals produced by the organisms include carbonates, silicates, phosphates, oxides, and sulphates. Of
these, calcium carbonate has the widest distribution and is found in almost all the phyla from
protozoa through chordata and in some groups of
algae[1] and higher plants[2]. The minerals of $CaCO_3$
are present in those organisms as one or more of
the four polymorphic modifications, i.e., calcite,
aragonite, vaterite, or amorphous state. The
size, morphology, and the state of aggregate of
these minerals show great diversity between different types of tissues or different groups of
organisms[3].

During the past decades, the ultrastructure
of calcified tissues and the processes of calcification have been studied extensively with the
transmission electron microscope (TEM) in a
variety of organisms. However, calcium carbonate
crystals are extremely difficult to section even
with the diamond knife for TEM observation, and
except for a handful of examples, most of the
TEM studies of calcified structures were carried
out by replica methods. Although this technique
gave us abundant information on the structure,
especially of molluscan shells (see for example,
Grégoire[4], Wilbur[5]), it has some disadvantages.
It is almost impossible to replicate extremely
irregular surfaces; the area of observation is
limited to very small (up to about 10 mm^2) dimensions; and the correlation of the orientation
of the observed images to actual structures is
often difficult. The introduction of scanning
electron microscopy (SEM) in the mid- to late
sixties has added a new dimension to calcified
tissue studies and has now become an indispensable research tool[6,7,8] along with the TEM and
x-ray microanalysis methods.

This paper attempts to summarize some recent knowledge in the ultrastructure and formation of several selected calcified tissues.
Since excellent reviews have been recently published for various aspects of calcification,[4,9,]
[10,11,12] the discussion will be concentrated on
those findings by SEM, but pertinent information
from TEM and/or x-ray microanalysis works will
also be included.

Due to the limitation in time and space, it

is not possible to cover all the calcified structures, and we will primarily deal with those related to our field of interest.

Echinoderm Skeleton

Echinoderm spines and plates are generally single crystals of magnesium calcite,[13,14] or as Towe[15] suggested for the plates, single crystals in the inner portion but polycrystalline aggregates with preferred orientation in the exterior. Exceptions seem to be in the cortex of primary spines of Cidaridae in which calcite is in the polycrystalline state.[16]

The spines indicate complex fenestrated structures. The central region of the solid spines such as in Strongylocentrotus or Arbacia is a spongy lattice of anastomosing crystal, and is continuous with an outer zone of radial wedges which thickens towards the margin of the spine.[17,18] In the hollow spines of Diadematidae (Diadema antillarum and Echinothrix diadema), the central region is a perforated tube with or without inner meshwork (Fig. 1).[19]

Fracturing the spines induces their regeneration. Regeneration of solid spines is initiated at the central region of the fractured surface with the appearance of microspines growing distally (Fig. 2). Crossbridges are formed to connect the microspines and continued growth completes the meshwork (see Heatfield[17] for details). In contrast, new microspines first appear on the margin of the microtube in the hollow spines of Diadema and Echinothrix (Fig. 3).[19] The bridge formation was similar to that of the solid spines but the rate of regeneration was about 4 times faster.

Electron probe analysis of the spines during regeneration of Arbacia punctulata showed that the calcite was relatively low in magnesium and high in strontium compared to that of the old non-regenerate part.[18] As the regeneration proceeded and the spine matured, the composition changed to the high-magnesium, low-strontium calcite similar to the old spine. Thus, the regenerated spines were not different from the normal in mineralogy or in morphology. The magnesium content of the regenerate was directly proportional to the temperature, but the temperature control on the calcite chemistry was considered to be indirect and it merely affected the rate of growth.

The chewing apparatus (lantern of Aristotle) including the teeth are also made of high magnesium calcite. Extensive SEM studies have been carried out by Märkel on these structures.[20,21,22] The echinoid lantern and teeth are classified into 1) the "clypeastroid type" present in Cassiduloida, Clypeastroida, and probably in Oligopygoidia and 2) the "regular type" present in all recent regular echinoids and probably in Holectypoida. The regular type teeth are highly complex in structure. Each tooth is composed of two rows of calcareous elements, each consisting of primary plates, side plates and prisms. Each of the primary plates is a single crystal of calcite, which is connected with each other by a polycrystalline calcitic disc. There is a narrow area called "stone zone" in the tooth containing short fibrous calcite. X-ray microanalysis indicates that the magnesium content in this zone was extremely high (38 Mol % MgCO$_3$ versus about 15 Mol % in the other area). In the "clypeastroid type" tooth, the primary plates are small, the prisms are flattened or reduced to a small area, and replaced by side plates.

The sites of formation of echinoderm skeletons have been controversial; however, tooth development[24] as well as test regeneration are considered to be intracellular and associated with matrix material at least during early stages.[23] Spine regeneration seems to take place extracellularly [12] within a membrane-bound area in which amorphous material is present.[25]

Molluscan Shell Structure

Mollusc shells consist of several calcified layers exhibiting a variety of configurations. Since these structures have recently been reviewed rather extensively[4], only a few aspects will be discussed here.

Fig. 1. Fracture surface of a hollow spine of the sea urchin Echinothrix diadema. (From Mischor[19]).
Fig. 2. Regenerating microspines of the sea urchin Arbacia punctulata. The width of the photograph represents 2,000 μm. (From Davies et al.[18]).
Fig. 3. Regenerating microspines at the periphery of the perforated tube of Echinothrix diadema. (From Mischor[19]).
Fig. 4. Brickwall pattern of the nacreous layer in the pelecypod (Aetheria elliptica typica). The width of the photograph represents 20 μm. (From Grégoire[29]).
Fig. 5. Step pattern of the nacreous layer of the pelecypod (Aetheria elliptica typica). The width of the photograph represents 20 μm. (From Grégoire[29]).
Fig. 6. Vertical stack pattern of the nacreous layer of the gastropod (Clanculus margaritarium). The width of the photograph represents 40 μm. (From Iwata[46]).
Fig. 7. Vertical stacks of aragonite in the nacreous layer of the gastropod (Turbo castanea). The width of the photograph represents 45 μm. (From Wise[27]).
Fig. 8. Parallel growth bandings on the surface of aragonite tablets (Tegula excavata). (From Erben[31]). The bar represents 10 μm.
Fig. 9. Adoral surface of lamella from the inner part of the last septum of Nautilus pompilus showing indented radial pattern. (20 min. treatment with sodium hypochlorite). The width of the photograph represents 25 μm. (From Crenshaw and Ristedt[41]).
Fig. 10. The spherulitic prismatic layer (Nautilus pompilus). (From Wise[7]).
Fig. 11. Stellate crystal aggregates of early stage development of sherulitic prismatic layer (Nautilus macromphalus). The width of the photograph represents 85 μm. (From Meenakshi et al.[50]).
Fig. 12. Spherulites from the stellate aggregates as shown in Fig. 11. The width of the photograph represents 110 μm. (From Meenakshi et al.[50]).

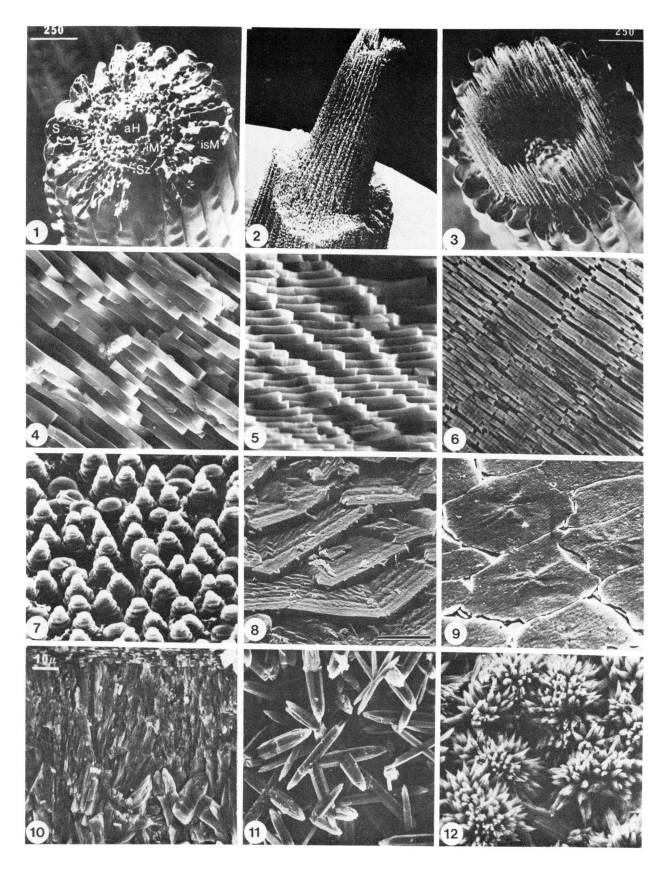

See page 404 for captions.

Aragonite crystals in the nacreous layer

The nacreous layer is the innermost portion of the shell and the constituent crystals are always aragonite. The layer is composed of numerous lamellae each of which is a horizontal aggregate of tabular crystals deposited on an organic sheet of interlamellar matrix, and intercrystalline matrix separates the individual crystals. In most pelecypods, the arrangement of crystals in the successive lamellae is either a "brickwall" type in which the intercrystalline spaces are offset in the adjacent lamellae (Fig. 4), or a "stair step" pattern (Fig. 5). In the gastropods and cephalopods, the crystals are deposited in tall vertical stacks (Fig. 6, 7). Another type of stack-up pattern called "lenticular nacre" by Taylor, Kennedy and Hall[26] is found in some pelecypods.[27,28] Usually, each group of molluscs indicates one of these charcteristic patterns of layer accumulation; however, all of those "brickwall", "stair step", and "lenticular" patterns were found in a single shell of the freshwater mussel (Aetheria and Bartlettia.[29]

Each aragonite tablet of the nacreous layer is polygonal to rounded, and is considered to be a single crystal or twins.[28,30,31] Each single crystal may contain continuous dendritic components of lamellae or blocks,[30] but it is by no means a mosaic aggregate of microcrystallites. Mutvei[32] showed by artificial etching that each polygonal crystal in Mytilus, Nucula, and Unio was composed of four crystal individuals occurring in two structurally different pairs which probably represented cyclical twins. Interestingly, nucleation of new nacreous tablets was always observed on the top surface of the pair less soluble in a glutaraldehyde-acetic acid etching solution.

Eight different forms of aragonite crystals were reported in the nacreous layer of the septal neck of Nautilus, each form being associated with different amounts of organic matrix.[33] For example, crystals which grew in a region low in conchiolin were either dentritic or tended to be porous, and those in conchiolin-rich regions often had a granular structure. Variations in

the structure of nacreous conchiolin were observed in different parts of the septal neck, which substantiated the relationship between the organic matrix and crystal morphology modifications.[34]

The outer surfaces of the aragonite crystals show many structural features. One type is the concentric or parallel lamellae representing growth lines of the crystals (Fig. 8).[7,27,28,31] Similar structures were observed after artificial etching which revealed internal crystal structure.[31,35] Although acicular components were revealed by etching,[31,35] these seem to represent small growth faces, ridges, and intercrystalline matrix rather than individual aragonite needles.[31]

Occurrence of natural decalcification of the nacreous layer (also of other shell layers) is quite common,[7,28,36,37,38] and is considered to be the results of anaerobic respiration during valve closing.[5] It is proposed that insoluble matrix left after decalcification is reincorporated within the shell and forms subdaily growth lines.[38,39] However, Koike[40] reports that the growth lines are rich in calcium and do not represent organic matrix.

The surface of crystal tablets in the nautiloid septa revealed indented radial patterns (Fig. 9) after treatment with sodium hypochlorite.[41,42] This was shown to be the replica of calcium binding sites rich in sulfated acid mucopolysaccharides of the interlamellar matrix. The penetration of portions of interlamellar membranes onto the crystal surface was also noted in gastropods.[31] The sites are proposed to have two functions: 1) initiation of mineral deposition, and 2) control of the thickness of the crystal tablet (i.e., the nacreous lamella). According to Crenshaw and Ristedt,[41,42] the calcium-binding sites (or structure) secreted by the mantle epithelium is selectively adsorbed to the active growing center of the crystal tablet such as screw-dislocation in which calcium ions are less tightly fixed in the crystal lattice. Once the binding site is adsorbed, further crystal growth is inhibited. However, the free surface of the binding site remains exposed to the extrapallial fluid and binds additional calcium, and vertical crystal growth continues until the

Fig. 13. A dumbbell-shaped aggregate of aragonite needles in a regenerating shell of Pomacea paludosa. The width of the photograph represents 20 μm. (From Blackwelder and Watabe[54]).

Fig. 14. A mixture of rhombohedral calcite crystals and spherulites of aragonite developed during shell regeneration of Cepaea nemoralis. (Watabe, unpublished). The bar represents 10 μm.

Fig. 15. Aragonite tablets developed on the organic matrix membrane from Haliotis nacreous layer during shell regeneration in Cepaea nemoralis. (Watabe, unpublished). The bar represents 1 μm.

Fig. 16. Egg shell layer of the land snail Anguispira alternata. O--outer layer; M--middle layer; I--inner layer. The crystals are all calcite. The width of the photograph represents 30 μm. (From Tompa[67]).

Fig. 17. Egg capsule of the freshwater snail Pomacea paludosa. O--outer layer of small spherulites; I--inner layer of acicular elements. The crystals are all vaterite. The width of the photograph represents 30 μm. (From Meenakshi et al.[66]).

Fig. 19. Vesicle inclusions in the hen's egg-shell. (From Erben and Kriesten[75]).

Fig. 20. Crystalline units of the turtle egg-shell Geochelone elephantopus ephippium. The width of the photograph represents 400 μm. (From Krampitz and Erben[79]).

Fig. 21. Single crystal triradiate spicules from sea urchin embryos. The width of the micrograph represents 30 μm. (From Inoue and Okazaki[86]).

Fig. 22. Rhombohedral calcite crystals epitaxially grown on a spicule which was isolated from the sea urchin embryo and incubated in a solution of calicum chloride and sodium bicarbonate. The width of the photograph represents 70 μm. (From Inoue and Okazaki[86]).

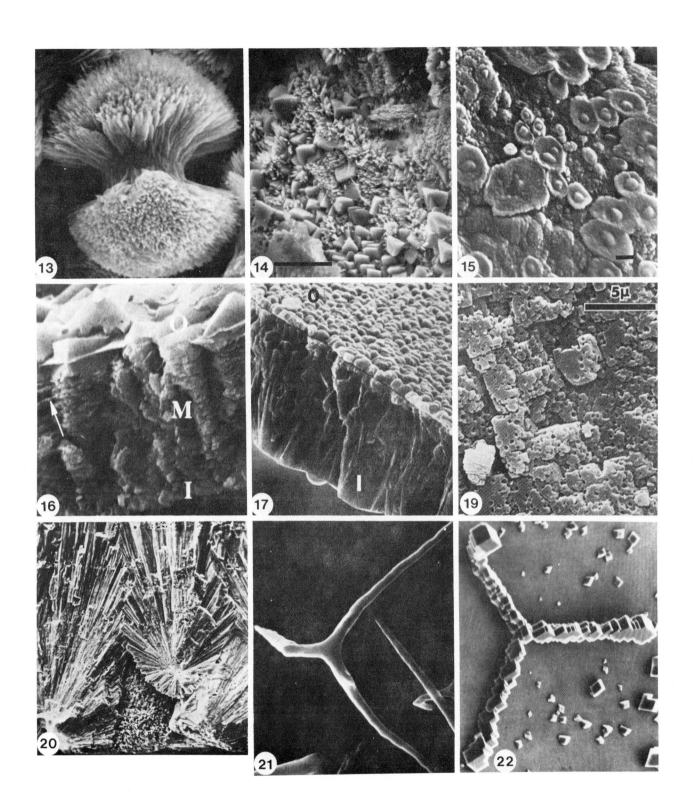

See page 406 for captions.

next episodic secretion of calcium-binding sites. In this manner, the thickness of the lamellae is controlled. In this connection, it is an interesting observation that all the lamellae at the surface of the nacreous layer are terminated by screw dislocation.[43] The orderly arrangement of the nacre is suggested to be related to this dislocation.[43,44] On the other hand, Mutvei[32] reported that the interlamellar membrane covering the top surface of the nacre tablet does not determine the nucleation sites because the membrane has a uniform structure on each side of the tablet.[45,46]

Spherulitic prismatic layer

This layer consists of aggregates of aragonite spherulites and the prismatic extension of their acicular crystallites (Fig. 10). It is present in mollusc shells such as archaeogastropods (Cittarium pica),[37,47,48] cephalopods (Nautilus),[33,49,50] tegmentum and shell eyes of polyplacophorans,[51,52] or in some bivalve ligaments.[53] The spherulites are present in those structures assuming the typical form or polyhedral or granular groups of acicular crystallites. The process of development of this layer was observed during shell regeneration.

Doubly pointed needle crystals develop first, which grow into stellate groups (Fig. 11) and finally into the spherulites (Fig. 12).[50] Alternatively, the needle crystals form spindle-shaped aggregates, at each end of which additional needles develop radially to form a dumbbell structure (Fig. 13); spherulite is completed by further development of needles in the middle region of the dumbbell.[54] Competitional growth of closely apposed spherulites leads to fan shaped prisms.

Shell regeneration

When a portion of the shell is removed, the animal secretes shell material to repair the damaged portion. The structure of regenerated shells may be similar to the normal shell such as seen in the freshwater snail Helisoma duryi duryi,[55] or the cephalopod Nautilus macromphalus.[50] In Mytilus edulis the regenerated shell was also similar to the normal, but very often and additional shell layer (complex prismatic or crossed-lamellar-type) not found in the normal animal was present.[56] In the experiment mentioned above, the regeneration was induced at the shell edge in Helisoma, but at the mid-shell region in Nautilus and Mytilus. The results imply that depending on the animal, and/or the region of shell removal, the mantle can create an environment of shell formation quite similar to the normal conditions. However, in many cases, the regenerated shells differ from the normal in: 1) the overall structure; 2) type of shell layers; 3) the type and form of the crystals; and 4) the chemical nature of the organic components (see review by Wilbur).[57]

The freshwater snails Pomacea paludosa,[54,58] Viviparus intertextus,[59] or terrestrial snails Otala lactea,[60] Helix pomatia,[61,62] or Cepaea nemoralis (Watabe, unpublished) have shells composed completely of aragonite, but their regenerated shells contain calcite, aragonite, and in some cases vaterite. Except for Pomacea in which calcite crystals were present as a discrete foliated layer, both aragonite and calcite were present in a mixed state at least in the initial stage of "normal" regeneration. (The "normal" shell regeneration implies here the regeneration induced in the normal environmental conditions with the portion of shell removal being left uncovered or covered with a plastic sheet). In these instances, the calcite crystals were rhombohedral, and aragonite crystals were the needle-stellate - dumbbell - spherulitic sequence of forms. This coexistance of polymorphic crystals apparently indicates the control of mantle was not working to provide a condition favoring the formation of each polymorph of $CaCO_3$ and specific shell layers. Therefore, it may be possible to change the crystal patterns by altering the environment, and infact, temperature changes caused a change in the aragonite - calcite ratio.[59] It was also reported that microtopography of the surface of the substrata put in contact with the mantle reflected in the morphology of the regenerate crystals.[60] The fact that this mold and cast relationship was produced is remarkable in view of the situation that usually a new organic matrix membrane is formed initially over the substrata. Thus, the influence of the substrata should be reduced. It may be that the matrix membrane was so thin that it reflected the substrata topography, which in turn, controlled the regenerate morphology. In similar experiments, the organic matrix substrata from an aragonitic shell failed to control the crystal type of the regenerates. In this case, even if the matrix substrata could have influenced the crystal type (see Wilbur),[63] they might have failed to do so because the regenerate crystals developed without direct contact with them. In our recent experiment, almost all the shell regenerates of the snail Cepaea nemoralis were a mixture of aragonite and calcite (Fig. 14), regardless of whether the substrata used was the matrix from aragonitic shell (nacreous layer of Haliotis) or from calcitic shell (prismatic layer of Pinna). (The organic matrix substrata were prepared following the methods of Crenshaw and Ristedt.)[41,42] However, in one out of 10 cases, aggregates of predominantly aragonite (Fig. 15) were found to be formed on the matrix from aragonite shell. These crystals were not spherulitic and were very similar to the tabular form normally found in the nacreous layer. Irregular and small calcite crystals were also formed on the calcitic matrix. Although the results were not conclusive, they may suggest the positive role of organic matrix in affecting crystal type in molluscs.

Small calcareous spherules similar in structure to those in many invertebrate tissues[64,65] were found in the mantle and foot of Pomacea paludosa. These spherules were formed intracellularly and were made of amorphous calcium carbonate or vaterite and organic matrix. They were shown to be utilized as calcium source for shell regeneration.[65]

Calcified Egg Shells

The egg shells or capsules of most snails are made of conchiolin or gelatinous materials, but are calcified in certain snails.[66] Tompa[67] found that the egg shell was calcified in at least 36 of the 65 known families of land snails (Stylommatophora). X-ray diffraction analyses indicated that egg shells in all of the 22 families examined were made of only calcite, or of a combination of calcite with smaller amounts of aragonite. These shells showed an enormous range of structural diversity suggesting that egg shell structure may have taxonomic significance.[67] For example, the land pulmonate _Anguispira alternata_ indicated 3 calcitic layers (Fig. 16):[67,68] the outer surface layer composed of large (rhombohedral) crystals about 10-15 μm in maximum diameter; the thick middle layer composed of alternating layers of crystals and organic matrix; and the thin inner layer of small crystals. On the other hand, the crystals of the egg shell of the freshwater snail _Pomacea paludosa_ were found to be present exclusively as vaterite.[66,69] The shell consisted of an outer proteinaceous covering underlain by an outer calcified layer formed of small and closely packed spherules, and an inner calcified layer of acicular elements elongated perpendicularly to the shell surface (Fig. 17). In regions where two or more shells were in contact, relatively flat and disk-like crystals were present.

Egg shell structure of reptiles, amphibians, and birds is more complex. It was shown that the crystals of avian egg shells of 32 species from 21 orders including the hens were calcite with traces of aragonite.[70] Basically the avian egg shell is composed of 2 layers of membrana testacea (inner and outer shell membrane), eisospherites, mammillary layer, the spongila, and the cuticle (Fig. 18).[71,72] The mammillary layer is composed of mammillary knobs of spicular to prismatic crystallites surrounding an organic matrix core, and cunei of tabular crystallites. The spongia is an aggregate of tabular crystallites in a fishbone arrangement (palisade layer), and the external zone was of crystallites oriented perpendicularly to the shell surface. The mammillary knobs are composed of aragonite crystals,[70] and calcite is localized in the other calcified portions of the shell. However, in the egg shells of the ostrich[72] and the Japanese quail,[73] the spherules around the mammillary core were formed by aggregates of calcite (rhombohedrons) in contrast to the spicular elements of aragonitic core. It is not known whether these differences in mineralogy in avian egg shells are controlled by genetic and/or environmental factors. Their taxonomic or phylogenetic significance is also unknown.[72] Although amorphous type deposition was reported to replace "definite crystalline growth" at the early stage of mammillae formation in hens' eggshell,[74] no such amorphous aggregates were seen in the Japanese quail[73] or were reported in the hen studied by Erben and Kriesten.[75] Fractured surface[70,76] or thin sections[73,77] of the spongy layer revealed many small globular inclusions or vesicles about 0.5μ in diameter within the crystals (Fig. 19). the nature of this structure is unknown; but they could be dendritic voids.[76]

Structure and composition of organic components differ according to the shell layer and the type of egg shells.[70,71,78] X-ray microanalysis of semi-thin sections of the Japanese quail egg shells indicated that the shell membrane contained Ca, P and S. The spherulitic concretions found in the cuticle were rich in Ca, P, and S, which were believed to be present in the form of amorphous calcium and magnesium phosphate.[73]

Egg shells of turtles are aragonitic; crocodiles and geckos have calcitic shells; and dinosaurs have calcitic egg shells with a trace of aragonite.[70,78] In turtles, and probably in squamates, the entire shell wall is built up by the zone of radial aggregates (Fig. 20), which seems to be a phylogenetically ancestral structure. In contrast, the crocodile egg shells are composed of tabular aggregates of calcite.[70] Traces of octacalcium phosphate were found in turtles, crocodiles, dinosaurs as well as in avian egg shells,[70] but later investigation could not confirm this.[78]

The insecticides DDT and its derivatives, Kepone, or cadmium caused extreme fragility of the egg shells in pelicans and Japanese quails by decreasing the shell thickness mainly in the spongy layer.[80,81] There were no changes in shell ultrastructure except for increase in the vesicles of the spongy layer. (Vesicles were rather reduced by cadmium.)

Pathological shell thickening was observed in some dinosaur eggs in southern France. These shells were characterized by the double mammillary layers causing the blockage of the pore canals for air passages. The mammillary tip, which is normally resorbed, was left unetched, showing evidence that the embryo perished without utilizing shell calcium.[80,81,82] Apparently these

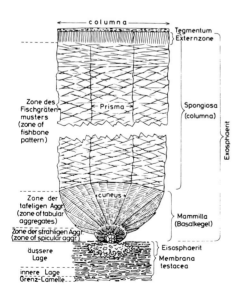

Fig. 18. Schematic drawing of avian egg shell structure. (From Erben[70]).

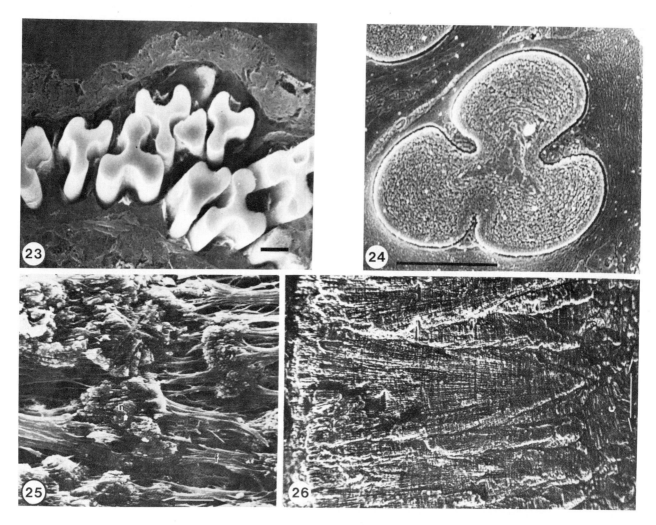

Fig. 23. Trilobed spicules of Renilla reniformis developed in the mesoglea. SEM photograph of the tissue processed by oxygen plasma etching after embedded in resin. (Watabe and Dunkelberger, unpublished).
The bar represents 10 μm.
Fig. 24. A Renilla spicule showing concentric lamellar structure. The surface was treated briefly with 1/10 N HCl prior to plasma etching. (Watabe and Dunkelberger, unpublished). The bar represents 10 μm.
Fig. 25. Partially fused nodules of calcite crystals at an early stage of calcification of the axial rod in the pennatulid, Veretillum cynomorium. f-- collagen fibers. The width of the photograph represents 180 μm. (From Ledger and Franc).[93]
Fig. 26. Calcite columns radiating from the axial core (c) of the axial rod of Veretillum cynomorium.
The width of the photograph represents 30 μm. (From Ledger and Franc).[93]

eggshells represent the "ovum in ovo" condition which is fatal to the embryo.

Calcareous Spicules and Axial Skeletons

The calcareous spicules of microscopic size are found in many groups of invertebrates and assume a variety of forms. In sponges belonging to the class Calcarea, each spicule is composed of a single crystal of magnesium calcite with $MgCO_3$ content ranging from 5.2 to 12.0 Mol % (7 species).[83] When fractured, the spicules indicate no regular cleavage patterns characteristic of calcite. However, some of their etched and unetched sections did show concentric rings, which Jones and James[84] interpreted as growth patterns.

The triradiate spicule (Fig. 21) of the sea urchin larva is also a single crystal of high magnesium calcite, the ratio of $Mg^{2+}/Ca^{2+} + Mg^{2+}$ being about 5/100.[85] Similar to the sponges, no fractured or etched surfaces of these larval spicules indicated a texture of calcite. Interestingly, Okazaki and Inoue showed that when normal living gastrulae with triradiate spicules were transferred into low Ca^{2+} and Mg^{2+} medium, the spicules thickened without elongating, and the thickened spicules exhibited calcite-like faces.[85] Concentric lamination with a central hole was found at the top of some spicules. It was also demonstrated that rhombodehral calcite crystals grew epitaxially over the isolated triradiate or pluteus spicules in a solution containing $NaHCO_3$ and $CaCl_2$ (Fig. 22).[86] Examination of the morphology and orientation of these rhombo- dehrons showed that the larval sea urchin spicule is not made up of an array of microcrystals, but the crystal lattice is uniform throughout the entire spicule.

The colonial pennatulid Renilla reniformis contains trilobed calcite spicules (Fig. 23) in the mesoglea. SEM and TEM as well as polarizing light microscope observations indicated that the spicules are not single crystals but are aggre- gates of small needles with their c-axes oriented in the direction of elongation. Etching of the transversely fractured surface (Fig. 24) as well as thin sectioning revealed many concentric growth rings composed of alternating crystals and organ- ic lamellae (intercrystalline matrix).[87]

The sponge spicule (Sycon ciliatum) is form- ed in an intercellular cavity containing organic material, whereas initial development of the spicule of sea urchin larva and Renilla takes place intracellularly within a vacuole in as- sociation with matrix material. Although organic materials are found to be associated with the spicule,[85,88] intraspicular matrix has not been confirmed except for Renilla.[87] Intracellular formation of spicules in the Didemnidae (colonial ascidians) were studied by Lafargue and Kni- prath,[89] and an extensive survey of spicule oc- currences in species of Turbellaria and meiofauna in general were carried out by Rieger and Ster- rer.[90] Detailed ultrastructural information on the morphology and the processes of formation of these spicules is needed.

Skeletal rods, or axial skeletons of el- lisellid gorgonian and a pennatulid Vertillum cynomorium are calcified. The rods of V. cyno- morium consist of 10% organic material and 90% calcite. Roughly 50% of the organic material was found to be collagen.[91] Goldberg[92] also re- ported appreciable amounts of collagen (42% of proteins in Nicella schmitti) in the axial skel- tons of gorgonians containing calcareous rods and suggested that collagen is a fundamental com- ponent of all "axiferous" octocorals. The as- sociation of collagen with calcium carbonate crystals is unique in mineralized tissues in that in all other known cases the minerals associated with collagen are calcium phosphate. (Sponge and echinoderm exoskeletons were reported to contain collagen fibers, but these are not regarded to be of skeletal origin.)[12]

The structure and processes of formation of collagenous axial rod of Veretillum cynomorium were studied by Ledger and Franc.[93] The rod is enveloped by a thin layer of endodermic epithe- lium from which fibrous proteins and calcium salts are secreted. The formation of calcite crystals takes place in the matrix region at the inferior tip of the rod, sometimes being associated with

the vesicular and membranous remains of cell fragments. Collagen fibers in the matrix are not responsible for the nucleation of mineral. The calcite crystals were initially in the form of nodules (Fig. 25) which come to fuse together to form the core of the axis. Irregular columns develop radially from the core and increase the diameter of the rod. The transverse section of the rod (Fig. 26) indicates a structure very much similar to the spherulitic sector of shells of eggs or molluscs (cf. Figs. 10,20). The c- axes of the calcite were found to be parallel to the direction of elongation of radial units. Collagen fibers were reported to be embedded in the calcite but never impregnated with it. The direction of elongation of collagen fibers is not parallel to that of the calcite but almost at right angles to it.

In Nicella schmitti, the rod consists of: 1) central chord of poorly crystalline or amor- phous minerals and organic fibers; and 2) peri- pheral axial cortex with highly calcified radial columns.[92] The crystal type of calcium carbo- nate and its structural relationships with the collagen fibers are not known. These calcified axes are distinguished from other non-calcified axes by the presence of acidic and sulfated poly- saccharides,[92] which were also present in Vere- tillum.[91]

Summary Comment

It is clear that the biogenic $CaCO_3$ crystals often indicate common morphological characteris- tics even if they are formed by different tis- sues or organisms. At the same time the charac- teristics may differ significantly within a given organism. For example, spherulites and their related structures are found in mollusc shells, egg shells, axial rods and exoskeletons of coelenterates (see also Sorauf[94]), bryozoan exoskeltons,[95]; aragonite tablets are present in the nacreous layer of mollusc shells. On the other hand, the siphon of the Nautilus shell con- tains both aragonite tablets and spherulites.[33,34]

Many factors have been suggested to control the crystal polymorphism of $CaCO_3$ in biological systems. These factors include the structure and/or composition of organic matrix, inorganic and organic components of the mother fluid, and environemntal temperature.[3,63,96] However, not much attention has been given to the problems of crystal habit modifications. As shown by Mutvei[33] and Wada,[97] the form of the aragonite tablets appears to be altered by the amount of organic matrix or secretory activity of the or- ganism. The perfection of spherulitic formation in the coral exoskeleton is thought to be re- lated to the rate of crystallization, and the nature of the organic matrix.[94]

In non-biological crystallization, the crystal habit is known to be affected by: 1) the type of solvent; 2) the pH of the solution; 3) the presence of impurities; 4) the degree of supersaturation or supercooling; 5) the rate of cooling; 6) the temperature of crystalliza- tion; 7) the degree of agitation, etc.[98] The calcite or aragonite formed by the silica gel technique exhibited each distinct morphologic type at three different pH ranges.[99] However, very little information is available at present on the conditions covering crystal morphology in living organisms. The shell regeneration may be an ideal system to investigate these conditions since the past studies have shown that crystal formation could be altered by experimental methods.

The morphology of the single-crystal spicules, spines, and tests of the echinoderms or spicules of sponges are obviously under strict cellular control,[85,88] but we have no knowledge concerning the conditions under which those crystals can grow as single crystals. A comparison with a system such as Renilla spicules which are polycrystalline aggregates may give us some clues.

The occurrence of collagen fibers in association with $CaCO_3$ crystals in the gorgonians and the pennatulid is significant in that the crystals associated with collagen in other mineralized tissues are calcium phosphate. Further detailed investigation of these organisms is needed to explore this unique process of calcification.

Acknowledgement

Most of our work discussed here was supported by PHS grant DE 03424 from the National Institute of Dental Research; NSF grant GB-39886, PCM 77-01528, PCM 77-27866; and the funds of the Electron Microscopy Center, University of South Carolina.

The shell regeneration studies on Cepaea were carried out by N. Watabe at the Institute of Paleontology, University of Bonn, Germany in 1976-1977 as a recipient of the U.S. Senior Scientist Award from Federal Republic of Germany, and he is most grateful to Prof. Dr. H.K. Erben, Institute of Paleontology, Bonn and the staff of Alexander von Humboldt Stiftung for their hospitality, encouragement and advice throughout the study.

We thank Tom Bargar, Andrea Blake, and Carm Finneran for their assistance in preparing the manuscript.

References

1. Lowenstam, H.A. Biologic problems relating to the composition and diagenesis of sediments. In: The Earth Sciences. T.W. Donnelly, Ed., Univ. Chicago Press, Chicago, 1963, pp. 137-195.
2. Arnott, H.J., and Pautard, F.G.E. Calcification in Plants. In: Biological Calcification: Cellular and Molecular Aspects. H. Schraer, Ed., Appleton-Century-Crofts, New York, 1970, pp. 375-446.
3. Watabe, N. Crystal growth of calcium carbonate in biological systems. J. Cryst. Growth, 24/25, 1974, 116-122.
4. Grégoire, C. Structure of the molluscan shell. In: Chemical Zoology, Vol. VII, Mollusca, M. Florkin and B.T. Scheer, Eds., Academic Press, New York, 1972, pp. 45-102.
5. Wilbur, K.M. Shell formation in mollusks. In: Chemical Zoology, Vol. VII, Mollusca, M. Florkin and B.T. Scheer, Eds., Academic Press, New York, 1972, pp. 103-145.
6. Sandberg, P.A., and Hay, W.W. Application of the scanning electron microscope in paleontology and geology. SEM/1968, IIT Research Institute, Chicago, Ill., 60616, 29-38.
7. Wise, S.W., Jr. Study of molluscan shell ultrastructure. SEM/1969, IIT Research Institute, Chicago, Ill., 60616, 205-216.
8. Erben, H.K. Application of the SEM in paleobiology. SEM/1971/I, IIT Research Institute, Chicago, Ill, 60616, 233-240.
9. Arnott, H.J. Calcification in higher plants. In: The Mechanisms of Mineralization in the Invertebrates and Plants, N. Watabe and K.M. Wilbur, Eds., Univ. of South Carolina Press, Columbia, S.C., 1976, pp. 55-78.
10. Pautard, F.G.E. Calcification in single cells: With an appraisal of the relationship between Spirostomum ambiguum and the osteocyte. ibid., pp. 33-53.
11. Simkiss, K. Cellular aspects of calcification. ibid., pp. 1-31.
12. Wilbur, K.M. Recent studies of invertebrate mineralization. ibid., pp. 79-108.
13. Nissen, H.-U. Crystal orientation and plate structure in echinoid skeleton units. Science, 166, 1969, 1150-1152.
14. Donnay, G., and Pawson, D.L. X-ray diffraction studies of echinoderm plates. Science, 166, 1969, 1147-1150.
15. Towe, K.M. Echinoderm calcite: single crystal or polycrystalline aggregates. Science, 157, 1967, 1048-1050.
16. Markel, K., Kubanek, F., and Willgallis, A. Polykristalliner Calcit bei Seeigeln (Echinodermata, Echinoidea). Z. Zellforsch., 119, 1971, 355-377.
17. Heatfield, B.M. Growth of the calcareous skeleton during regeneration of spines of the sea urchin Strongylocentrotus purpuratus (Stimpson): A light and scanning electron microscope study. J. Morph. 134, 1971, 57-90.
18. Davis, T.T., Crenshaw, M.A., and Heatfield, B.M. The effect of temperature on the chemistry and structure of echinoid spine regeneration. J. Paleontol., 46, 1972, 874-883.
19. Mischor, B. Zur Morphologie und Regeneration der Hohlstracheln von Diadema antillarum Philippi und Echinothrix diadema (L.) (Echinoidea, Diadematidae). Zoomorphol., 82, 1975, 243-258.
20. Märkel, K. Das Wachstum der "Laterne Aristoteles" und seine Anpassung an die Funktion der Laterne (Echinodermata, Echinoidea). Zoomorphol., 86, 1976, 25-40.
21. Märkel, K. On the teeth of the recent cassiduloid Echinolampas depressa Gray, and on some Liassic fossil teeth nearly identical in structure (Echinodermata, Echinoidea) Zoomorphol., 89, 1978, 125-144.
22. Märkel, K., Gorny, P., and Abraham, K. Microarchitecture of sea urchin teeth. Fortsch. Zool., 24, 1977, 103-144.
23. Shimizu, M., and Yamada, J. Light and Electron microscope observations of the regenerating test in the sea urchin, Strongylocentrotus intermedius. In: The Mechanisms of Mineralization in the Invertebrates and Plants, N. Watabe, and K.M. Wilbur, Eds., Univ. South Carolina Press, Columbia, S.C., 1976, pp. 261-281.

24. Kniprath, E. Ultrastructure and growth of the sea urchin teeth. Calcif. Tiss. Res., 14, 1974, 211-228.
25. Heatfield, B.M., and Travis, D.F. The fine structure of calicoblasts in the regenerating sea urchin spine. Proc. 30th Am. EMSA Mtg., 1972, 162-163.
26. Taylor, J.D., Kennedy, W.J., and Hall, A. The shell structure and mineralology of the Bivalvia. I. Introduction. Nuclacea - Trigonacea. Bull. Brit. Mus. (Natural History), Zool. Suppl 3, 1969, 1-125.
27. Wise, S.W., Jr. Microarchitecture and mode of formation of nacre (Mother-of-pearl) in pelecypods, gastropods, and cephalopods. Ecol. geol. Helv., 63, 1970, 775-797.
28. Erben, H.K. Uber die Bildung und das Wachstum von Perlmutt. Biomineralization, 4, 1972, 15-46.
29. Grégoire, C. On the organic and mineral components of the shell of Aetheriidae (Mollusca, Bivalvia, Unionacea). Rev. Zool. afr., 4, 1974, 847-869
30. Watabe, N. Studies on shell formation. XI. Crystal-matrix relationships in the inner layers of mollusk shells. J. Ultrastr. Res., 12, 1965, 351-370.
31. Erben, H.K. On the structure and growth of the nacreous tablets in gastropods. Biomineralization, 7, 1974, 14-27.
32. Mutvei, H. The nacreous layer in Mytilus, Nucula, and Unio (Bivalvia). Calcif. Tiss. Res., 24, 1977, 11-18.
33. Mutvei, H. Ultrastructural studies on cephalopod shells. Part I. The septa and siphonal tube in Nautilus. Bull. geol. Instn. Univ. Upsala. N.S. 3, 8, 1972, 237-261.
34. Grégoire, C. On the submicroscopic structure of the organic components of the siphon in the Nautilus shell. Arch. intern. physiol. bioch., 81, 1973, 299-316.
35. Mutvei, H. Ultrastructural relationships between the prismatic and nacreous layers in Nautilus (Cephalopoda). Biomineralization, 4, 1972, 81-86.
36. Watabe, N. Electron microscope observations of aragonite crystals on the surface of cultured pearls. Rep. Fac. Fish. Pref. Univ. Mie, 1, 1954, 449-454.
37. Erben, H.K. Anorganische und organische Schalenkomponenten bei Cittarium pica (L.) (Archaeogastropoda). Biomineralization, 3, 1971, 51-64.
38. Lutz, R.A., and Rhoads, D.C. Anaerobiosis and a theory of growth line formation. Science, 198, 1977, 1222-1227.
39. Gordon, J., and Carriker, M.R. Growth lines in a bivalve mollusk: subdaily patterns and the dissolution of the shell. Science, 202, 1978, 519-521.
40. Koike, H. Estimation of paleoenvironment by shell structure (In Japanese), Suri-Kagaku, 170, 1977, 1-7.
41. Crenshaw, M.A., and Ristedt, H. Histochemical and structural study of nautilid septal nacre. Biomineralization, 8, 1975, 1-8.
42. Crenshaw, M.A., and Ristedt, H. The histochemical localization of reactive groups in septal nacre from Nautilus pompilus L. In: The Mechan-

isms of Mineralization in the Invertebrates and Plants, N. Watabe, and K.M. Wilbur, Eds., Univ. South Carolina Press, Columbia, S.C., 1976, pp. 355-367.
43. Wise, S.W., Jr. and deVilliers, J. Scanning microscopy of molluscan shell ultrastructure: screw dislocations in pelecypod nacre. Trans. Amer. Micros. Soc., 90, 1971, 376-380.
44. Blackwell, J.F., Gainey, L.F., Jr. and Greenberg, M.J. Shell ultrastructure in two subspecies of the ribbed mussel, Geukensia demissa (Dillwyn, 1817). Biol. Bull., 152, 1977, 1-11.
45. Mutvei, H. On the micro- and ultrastructure of the conchiolin in the nacreous layer of some recent and fossil molluscs. Stockholm Contr. Geol. 20, 1969, 1-17.
46. Iwata, K. Ultrastructure of the conchiolin matrices in molluscan nacreous layer. J. Fac. Sci. Hokkaido Univ., 17, 1975, 173-229.
47. Wise, S.W., Jr., and Hay, W.W. Scanning electron microscopy of molluscan shell ultrastructures. I. Techniques for polished and etched sections. Trans. Amer. Microscop. Soc. 87, 1968, 411-418.
48. Ibid. II. Observations of growth surfaces. Trans. Amer. Microscop. Soc. 87, 1968, 419-430.
49. Erben, H.K., Flajs, G., and Siehl, A. Die fruhontogenetische Entwicklung der Schalenstruktur ectocochleater Cephalopoden. Palaeontographica. 132, A, 1969, 1-54.
50. Meenakshi, V.R., Martin, A.W., and Wilbur, K.M. Shell repair in Nautilus macromphalus. Mar. Bio., 27, 1974, 27-35.
51. Haas, W. Untersuchungen über die Mikro- und Ultrastrukturder Polyplacophoren Schale. Biomineralization 5, 1972, 3-52.
52. Haas, W., and Kriesten, K. Die Astheten mit intrapigmentarem Schalenauge von Chiton marmoratus L. (Mollusca, Placophora). Zoomorphol., 90, 1978. 253-268.
53. Mano, K., and Watabe, N. Scanning electron microscope observations on the calcified layers of the ligament of Mercenaria mercenaria (L.) and Brachydontes exustus (L.) Pelecypoda. Biomineralization, 10, 1979, 25-32.
54. Blackwelder, P.L., and Watabe, N. Studies on shell regeneration. II. The fine structure of normal and regenerated shell of the freshwater snail Pomacea paludosa. Biomineralization, 9, 1977, 1-10.
55. Wong, V., and Saleuddin, A.S.M. Fine structure of normal and regenerated shell of Helisoma duryi duryi. Can. J. Zool., 50, 1972, 1563-1568.
56. Meenakshi, V.R., Blackwelder, P.L., and Wilbur, K.M. An ultrastructural study of shell regeneration in Mytilus edulis (Mollusca: Bivalvia). J. Zool., Lond., 171, 1973, 475-484.
57. Wilbur, K.M. Mineral regeneration in echinoderms and molluscs. Ciba Found. Symp., 11 (New ser.), 1973, 7-33.
58. Meenakshi, V.R., Blackwelder, P.L., Hare, P. E., Wilbur, K.M., and Watabe, N. Studies on shell regeneration. I. Matrix and mineral composition of the normal and regenerated shell of Pomacea paludosa. Comp. Physiol. Biochem., 50A, 1975, 347-351.
59. Wilbur, K.M., and Watabe, N. Experimental studies on calcification in molluscs and the alga

Coccolithus huxleyi. Ann. N.Y. Acad. Sci., 109, 1963, 82-112.

60. Meenakshi, V.R., Donnay, G., Blackwelder, P.L., and Wilbur, K.M. The influence of substrata on calcification patterns in molluscan shell. Calcif. Tiss. Res., 15, 1974, 31-44.

61. Saleuddin, A.S.M., and Wilbur, K.M. Shell regeneration in Helix pomatia. Can. J. Zool., 47, 1969, 51-53.

62. Saleuddin, A.S.M., and Chan, W. Shell regeneration in Helix. Shell matrix composition and crystal formation. Can. J. Zool., 47, 1969, 1107-1111.

63. Wilbur, K.M. Shell formation and regeneration. In: Physiology of Mollusca, Vol. I, K.M. Wilbur and C.M. Yonge, Eds., Academic Press, New York, 1964, pp. 243-282.

64. Simkiss, K. Intracellular and extracellular routes in biomineralization. Symp. Soc. Exp. Biol., 30, 1976, 432-444.

65. Watabe, N., Meenakshi, V.R., Blackwelder, P.L., Kurtz, E.M., and Dunkelberger, D.G. Calcareous spherules in the gastropod, Pomacea paludosa. In: The Mechanisms of Mineralization in the Invertebrates and Plants. N. Watabe and K.M. Wilbur, Eds., Univ. South Carolina Press, Columbia, S.C., 1976, pp. 283-308.

66. Meenakshi, V.R., Blackwelder, P.L., and Watabe, N. Studies on the formation of calcified egg-capsules of Ampullarid snails. I. Vaterite crystals in the reproductive system and the egg capsules of Pomacea paludosa. Calcif. Tiss. Res., 16, 1974, 283-291.

67. Tompa, A.S. A comparative study of the ultrastructure and mineralogy of calcified land snail eggs (Pulmonata: Stylommatophora). J. Morph., 150, 1976, 861-888.

68. Tompa, A.S. Calcification of the egg of the land snail Anguispira alternata (Gastropoda: Pulmonata). In: The Mechanisms of Mineralization in the Invertebrates and Plants. N. Watabe, and K.M. Wilbur, Eds., Univ. South Carolina Press, Columbia, S.C., 1976, pp. 427-444.

69. Meenakshi, V.R., and Watabe, N. Studies on the formation of calcified egg capsules of ampullarid snails. II. Calcium in reproductive physiology with special reference to structural changes in egg capsules and embryonic shell. Biomineralization, 9, 1977, 48-58.

70. Erben, H.K. Ultrastrukturen und Mineralization rezenten und fossilen Eischalen bei Vögeln und Reptilien. Biomineralization, 1, 1970, 2-66.

71. Wilbur, K.M., and Simkiss, K. Calcified shells. In: Comprehensive Biochemistry, 20A. M. Florkin and E.H. Stotz, Eds., Elsevier, New York, 1968, pp. 229-295.

72. Sauer, E.G.F., Sauer, E.M., and Gebhardt, M. Normal and abnormal patterns of struthious eggshells from South West Africa. Biomineralization, 9, 1975, 32-54.

73. Quintana, C., and Sandoz, D. Coquille de l'oeuf de caille: étude ultrastructuralle et cristallographique. Calcif. Tiss. Res., 25, 1978, 145-159.

74. Creger, C.R., Phillips, H., and Scott, J.T. Formation of egg shell. Poultry Sci., 55, 1976, 1717-1723.

75. Erben, H.K., and Kriesten, K. Micromorphologie der Frühstadien bei der Kristalbildung in normalen und anormalen Hühner-Eischalen. Biomineralization, 7, 1974, 28-36.

76. Heyn, A.N.J. The calcification of the avian egg shell. Ann. N.Y. Acad. Sci., 102, 1963, 246-250.

77. Simons, P.C.M., and Wiertz, G. Notes on the structure of membranes and shell in the hen's egg. Z. Zellforsch., 59, 1963, 555-567.

78. Erben, H.K., and Newesely, H. Kristalline Bausteine und Mineralbestand von Kalkigen Eischalen. Biomineralization, 6, 1972, 32-48.

79. Krampitz, G., and Erben, H.K. Über Aminosaurenzusammensetzung und Struktur von Eischalen. Biomineralization, 4, 1972, 87-99.

80. Erben, H.K. Ultrastrukturen und Dicke der Wand pathologischer Eischalen. Abh. Akad. Wiss. Lit. Mainz, Math.- Naturw. Kl. 6, 1972, 195-216.

81. Erben, H.K., and Krampitz, G. Eischalen DDT-Versuchter Vögel: Ultrastruktur und organische Substanz. Abh. Akad. Wiss. Lit. Mainz, Math.- Naturw. Kl. 2, 1971, 3-24.

82. Erben, H.K. Dinosaurien: Pathologische Strukturen der Eischale als Letalfaktor. Umschau, 17, 1969, 552-553.

83. Jones, W.C., and Jenkins, D.A. Calcareous sponge spicules: A study of magnesian calcite. Calc. Tiss. Res., 4, 1970, 314-329.

84. Jones, W.C., and James, D.W.F. Examination of the large triacts of the calcareous sponge Leuconia nivea Grant by scanning electron microscopy. Micron, 3, 1972, 196-210.

85. Okazaki, K., and Inoue, S. Crystal property of the larval sea urchin spicule. Develop., Growth, Differ., 18, 1976, 413-434.

86. Inoue, S., and Okazaki, K. Biocrystals. Scient. Amer., 236, No. 4, 1977, 82-92

87. Dunkelberger, D.G., and Watabe, N. An ultrastructural study on spicule formation in the pennatulid colony Renilla reniformis. Tissue & Cell, 6, 1974, 573-586.

88. Ledger, P.W., and Jones, W.C. Spicule formation in the calcareous sponge Sycon cilatum. Cell Tiss. Res., 181, 1977, 553-567.

89. Lafargue, F., and Kniprath, E. Formation des spicules des Didemnidae (ascides composées), 1. L'apparition apres la spicules chez l'oozoide après la métamorphose. Mar. Biol., 45, 1978, 175-184.

90. Rieger, R.M., and Sterrer, M. New spicular skeletons in Turbellaria and the occurrence of spicules in marine meiofauna. I, II. Z. Zool. Syst. Evolut.-forsch., 13, 1975, 207-278.

91. Franc, S., Hue, A., and Chassagne, G. Etude ultrastructurale et physicochemique de taxe squelettigue de Veretillum cynomorium Pall. (Cnidaire, Anthozoaire): Cellules, calcite, collagene. J. microscopie, 21, 1974, 93-110.

92. Goldberg, W.M. Comparative study of the chemistry and structure of gorgonian and antipatharian coral skeletons. Mar. Biol., 35, 1976, 253-267.

93. Ledger, P.W., and Franc, S. Calcification of the collagenous axial skeleton of Veretillum cynomorium Pall (Cnidaria: Pennatulacea). Cell Tiss. Res., 192, 1978, 249-266.

94. Sorauf, J.E. Observations on microstructure and biocrystallization in Coelenterates. Biomineralization, 7, 1974, 37-55.

95. Ristedt, H. Personal communication, 1979.
96. Kitano, Y., Kanamori, N., and Yoshioka, S. Influence of chemical species on the crystal type of calcium carbonate. In: The Mechanisms of Mineralization in the Invertebrates and Plants, N. Watabe and K.M. Wilbur, Eds., Univ. South Carollina Press, Columbia, S.C., 1976, pp. 191-202.
97. Wada, K. Nucleation and growth of aragonite crystals in the nacre of some bivalve molluscs. Biomineralization, 6, 1972, 141-159.
98. Mullin, J.W. Crystallization, 1961, pp. 268 + ix, Butterworths, London.
99. McCauley, J.W., and Roy, R. Controlled nucleation and crystal growth of various $CaCO_3$ phases by the silica gel techniques. Amer. Mineral., 59, 1974, 947-963.

Discussion with Reviewers

H.J. Arnott: Can you make a generalized statement as to the importance of living cells to the various calcification systems you have discussed?
Authors: It is generally accepted that, regardless of whether the sites of calcification are extracellular or intracellular, the skeletogenic cells synthesize and secrete organic matrix and pump a fluid of calcium and other ions into these sites. Mineralogy, composition, and morphology of the calcified structures are directly or indirectly under cellular control. However, it is not known to what extent each of these characteristics is controlled by organic matrix or composition of the fluid.

H.J. Arnott: In a physical sense what does the replacement of calcium by magnesium mean? For example, is this mineral harder or softer? More or less soluble? or more or less brittle?
Authors: Generally, Ca Mg (CO_3) or dolomite is harder, more brittle, and less soluble than $CaCO_3$ crystals. The stone zones in the echinoid teeth are the hardest skeletal component of echinoderms and contain a high amount of magnesium. Märkel et al. (text reference 16) attributed this enormous hardness to the "strong interlacing of microcrystalline (high magnesium) calcite and organic matter."

H.J. Arnott: In shell regeneration what is the role of the mantle?
A.S.M. Saleuddin: Is there any evidence that potential difference in the mantle cells during shell regeneration can cause the deposition of crystals other than normal type?
Authors: In Pomacea paludosa, when the mantle is damaged by shell removal, some of the damaged epithelial cells become degenerated and detached from the epithelium. Further degeneration of these cells results in a deposition of cellular debris in the extrapallial space. Calcite crystals not found in normal shell develop in close association with those debris. At about the same time, amoebocytes migrate out into the extrapallial space and form cell aggregates, which eventually degenerate and leave fibers, vesicles, and other cell fragments. Aragonite spherulites, also an abnormal type in the shell, are formed in the fragments. However, the bulk of substances for regenerated shell may be secreted by the intact

mantle epithelium and mix with the debris and form conditions favoring the development of those abnormal type crystals. In Anodonta, Tsujii[100] observed that epithelial cells change their structure and function to form abnormal type shells. These cells return to the normal type eventually.

H.J. Arnott: Other than spherules in Pomacea, are there other cases where amorphous calcium carbonate is important in animal mineralization?
Authors: Yes. The examples include: the bivalve Anodonta;[101] gastropods Helix pomatia,[102] Helisoma duryi endiscus;[103] and arthropods Orchestia, and Niphargus.[104]

I.S. Johnston: One of the biggest frustrations in the study of biomineralization has been the failure to perceive unifying concepts which might apply to the fundamental process of mineral deposition in all calcifying tissues. In considering the extremely divergent systems alluded to in this review, do you have any hopes that such unifying concepts might ever be discovered, particularly on the basis or morphological studies?
Authors: Calcification is a complex phenomenon involving physiological, biochemical, morphological or physico-chemical process. We believe it is rather unfruitful and dangerous to seek for unifying concepts of these processes which occur in such a diversified system. Our approach is to address our questions to some specific aspects of calcification which are common to various calcifying tissues. We hope that detailed morphological observations in different organisms and experimental studies will enable us to identify the factors controlling the formation of such specific calcified structures.

I.S. Johnston: Could you please briefly describe some of the pitfalls and some of the better preparative techniques for SEM investigation of mineralizing systems, especially of newly-formed deposits at a growth surface?
Authors: One of the problems involved in TEM or SEM studies of mineralized systems, especially of those containing relatively large $CaCO_3$ crystals, is of course the difficulty in observing crystals and soft tissues simultaneously. We have applied the oxygen plasma etching techniques to embedded tissues. These techniques enabled us to see fairly large polished-etched surfaces of tissues with minerals under SEM, and seem to have promising applications. Details will be sent for publication soon.

References

100. Tsujii, T. An electron microscopic study of the mantle epithelial cells of Anodonta sp. during shell regeneration. In: The Mechanisms of Mineralization in the Invertebrates and Plants. N. Watabe and K.M. Wilbur, Eds., Univ. South Carolina Press, Columbia, S.C., 1976, pp 339-353.
101. Istin, M. Role du manteau dansle metabolism du calcium chez les lamellibranches. Bull. Infs. Scient. tech. Commit. Energy. atom., 144, 1970, 53-80.
102. Abolins-Krogis, A. The histochemistry of

the hepatopancreas of Helix pomatia (L.) in relation to the regeneration of the shell. Arkiv
Zool., 13, 1961, 159-201

103. Kapur, S.P., and Gibson, M.A. A histochemical study of calcium storage in the foot of the
freshwater gastropod, Helisoma duryi eudiscus
(Pilsbry). Can. J. Zool., 10, 1968, 987-990

104. Graf, F. De stockage de calcium avant la
mue chez les Crustaces Amphipodes Orchestia (Tali-
tride) et Niphargus (Gammaride hypoge). Arch.
orig. Centre Document C.N.R.S. No. 2690, 1969,
1-216.

SCANNING ELECTRON MICROSCOPY/1979/II
SEM Inc., AMF O'Hare, IL 60666, USA

COCCOLITH MORPHOLOGY AND PALEOCLIMATOLOGY...II.   CELL ULTRASTRUCTURE AND
FORMATION OF COCCOLITHS IN CYCLOCOCCOLITHINA LEPTOPORA (MURRAY AND BLACKMAN)
WILCOXON AND GEPHYROCAPSA OCEANICA KAMPTNER

P. L. Blackwelder, L. E. Brand* and R. L. Guillard*

Nova Univ. Ocean Sciences Center   *Woods Hole Oceanographic Inst.
8000 N. Ocean Drive   Woods Hole, MA 02543
Dania, FL 33004

Abstract

Current interest in utilization of coccoliths for paleoclimate reconstruction
necessitates background information on
environmental limits for growth and coccolith
production as well as examination of cell
ultrastructure in specimens collected in the
field and in cultured representatives.
Successful isolation of the two geologically
important species Gephyrocapsa oceanica
(strain A674) and Cyclococcolithina leptopora
(strain A650) allows investigation of ultrastructure in cultured forms. Fine structure
of cells and coccoliths was observed in the
SEM using critical point dried preparations
and ultrastructure was examined with the
transmission electron microscope.
Cell diameters in G. oceanica ranged from
2.5 - 4.0 microns and in C. leptopora from
7.5 - 9.5 as measured in thin section. In
both species, a single layer of interlocking
coccoliths surrounds each cell. Coccoliths are
formed intracellularly and appear to form within
Golgi-derived vesicles located near the nuclear
membrane. Neither coccolithosomes nor scales
were associated with the coccoliths. Arrangement and morphology of cell organelles was
typical of the coccolithophorid group.
Formation and development of coccoliths in
the two species resemble these processes in
Emiliania huxleyi but differ from those of
Cricosphaera carterae, notably in the absence
of coccolithosomes and scales and in the fact
that coccoliths are produced intracellularly
one at a time.

KEY WORDS: Coccolithophores, Coccolith Formation,
Cell Ultrastructure, Gephyrocapsa oceanica,
Cyclococcolithina leptopora

Introduction

Use of fossil coccolith assemblages for
paleoclimate study is an established technique
for reconstruction of paleoenvironments.[1]
Relatively less is known of the ultrastructure
of coccolith producing species or of changes in
the calcification process taking place under
different environmental regimes. Much of what
is now known has been obtained using either
Emiliania huxleyi[2,3] or Cricosphaera carterae.[4,5,6,7,8] E. huxleyi is certainly of
interest to geologists in that it is relatively
ubiquitous in the world's oceans, but
C. carterae, although convenient for physiological studies, is not a common open ocean
form.[5]
Successful isolation of Gephyrocapsa
oceanica and Cyclococcolithina leptopora now
permits comparative study of calcification in
two coccolithophores common in today's oceans
as well as in the geologic record. Sketches
of Cyclococcolithina leptopora were drawn from
light microscope observations made by
Shiller,[9] but no work on its ultrastructure has
been attempted as far as we know. In nature
C. leptopora is common in Pacific Central and
Transitional Zones, and is abundant in the
Equatorial Zone. It has been found in plankton
samples down to the 2°C isotherm.[10] In the
Atlantic it is common from the Equator to the
poles.[11] G. oceanica is common in Pacific
Central and Transitional areas, abundant near
the Equator, and common in the Atlantic in
Transitional and Subtropical areas.[11] It has
been found in plankton samples only down to the
14°C isotherm.[10] In the geologic record
G. oceanica and C. leptopora are common in
sediments from late Pliocene to the present.[12]
In this study we describe the ultrastructure of Gephyrocapsa oceanica and
Cyclococcolithina leptopora with emphasis on
coccolith production.

Materials and Methods

Gephyrocapsa oceanica (clone A674;
(Figure 1) was isolated from a 17.5°C water
sample collected on III-3-78 at 34°00.5N,
68°00W. Cyclococcolithina leptopora (clone
A650) was isolated from 21.0°C water collected

on III-1-78 at 27°15'N, 66°43'W. Isolation techniques will be discussed elsewhere.

Cells were prepared for electron microscopy after centrifugation to a pellet and fixation in 2% glutaraldehyde solution in 0.5 M sodium cacodylate buffered sea water for 2-4 hr. For transmission microscopy, cells were washed in buffer, post-fixed in 2% osmium solution made up with the buffered sea water dehydrated in a series of ethanols, and embedded in Spurr embedding resin.[13] Thin sections were cut on a Porter-Blum MT-2 microtome fitted with a diamond knife and examined with a Philips EM 201 transmission microscope. For scanning electron microscopy, fixed cells were dehydrated in a series of ethanols, then passed through a graded series of Freon TF and ethanol.[14] The cells were dried using a Bomer SPC-900 critical point dehydrator, placed on an aluminum stub with a piece of double-stick tape, coated with gold palladium in a vacuum evaporator, and examined with a JEOL JSM-U3 scanning electron microscope.

Results

Coccoliths of Gephyrocapsa oceanica (strain A674) cultured at 20°C range in diameter of from 2.0 - 2.5 microns (Figure 2) consist of a single layer of approximately 20 - 30 coccoliths. The plates interlock and the coccosphere often maintains its structure even after removal of the cellular material (Figure 3), while some coccospheres become disarticulated and are thus observed as individual plates in preparations (Figure 4).

Cyclococcolithina leptopora (strain A650) cultured at 20°C is covered with a single layer of as many as 20 - 40 coccoliths per cell. Coccolith diameters ranged from 7.0 - 10.0 microns (Figure 5). Less frequently than in G. oceanica, coccoliths occasionally maintained a tightly packed interlocking configuration around each cell (Figure 6). Individual coccolith shields are often observed as disarticulated plates (Figure 7) due to a weak attachment area (Figure 8).

Cell Ultrastructure

Cell diameters in G. oceanica (strain A674), as estimated from transmission electron micrographs, ranged from 2.5 to 4.0 microns in cultures maintained at 20°C. Cells of C. leptopora (strain A650) grown under similar culture conditions ranged in cell diameter from 7.5 to 9.5 microns. The organelle arrangement and ultrastructure of both species was similar to that of other coccolithophores with a nucleus containing dense chromatin and one nucleolus and a prominent Golgi apparatus with closely packed cisternae in a parallel arrangement. Cells have chloroplasts with the usual array of parallel laminations. Filamentous and granular material was observed between the plasma membrane and the cell wall. Reticular bodies were observed in close proximity to intracellular coccoliths. A large chrysolaminarin vesicle[5] was present in C. leptopora, but was not observed in G. oceanica.

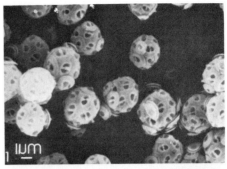

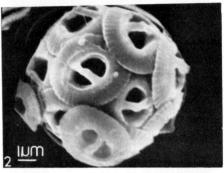

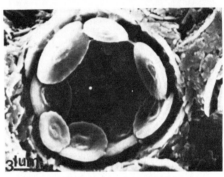

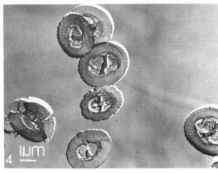

Figure 1. Coccospheres of Gephyrocapsa oceanica (strain A674) from isolate. Bar = 1 micron.

Figure 2. Geohyrocapsa oceanica cell and surrounding coccoliths. Bar = 1 micron.

Figure 3. The inner surface of a coccosphere of Gephyrocapsa oceanica coccosphere showing configuration of interlocking coccoliths. Bar = 1 micron.

Figure 4. Individual coccoliths of G. oceanica from disarticulated coccosphere. Bar = 1 micron.

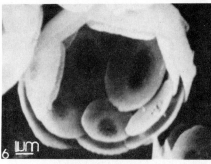

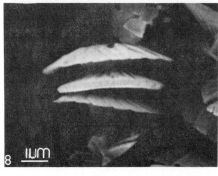

Figure 5. Coccosphere of a cell from an isolate of Cyclococcolithina leptopora. Bar = 1 micron.

Figure 6. The inner surface of a coccosphere of Cyclococcolithina leptopora. Bar = 1 micron.

Figure 7. Disarticulated coccoliths of Cyclococcolithina leptopora showing distal and proximal plates. Bar = 1 micron.

Figure 8. Coccolith Cyclococcolithina leptopora showing delicate connection between distal and proximal plates. Bar = 1 micron.

Coccolith Formation

In both G. oceanica and C. leptopora coccoliths appeared to be produced intracellularly one at a time (Figure 9). Coccoliths were invariably observed associated with Golgi-derived vesicles and in fact appear to form in a homogeneous fibrous material within the vesicles. No organic scales were observed beneath intra- or extra-cellular coccoliths, but scale-like structures were observed in C. leptopora which were not associated with coccoliths (Figure 9). A reticular body similar to that observed in E. huxleyi[2,15] was present near the calcifying vesicle in G. oceanica and C. leptopora. Intracellular mineralization in G. oceanica was observed to form the base of a central cylinder which calcified upwards and outward to form the proximal and distal shields. In C. leptopora, the pattern differed in that several centers were observed and no distinct continuity in sequence was discernable.

Insufficient data now exist to speculate on the complete sequence of coccolith formation in these species, but further study is planned.

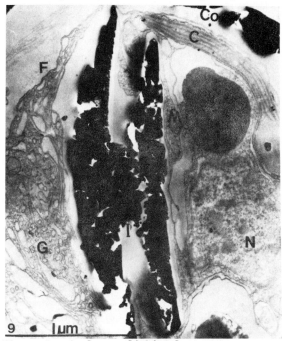

Figure 9. Cyclococcolithina leptopora (strain A650). Cell organelles include nucleus (N), chloroplasts (C), fibrous material between plasma membrane and cell wall (F), single layer of extracellular coccoliths (Co), prominent Golgi apparatus (G), intracellular coccolith (I). No stain. Bar = 1 μ.

Discussion

Isolation and culture of Gephyrocapsa oceanica and Cyclococcolithina leptopora demonstrates the feasibility of study of living coccolithophores of geologic importance. Aspects of coccolith production observed in culture have direct impact on interpretation of fossil distributions. For example, published estimates of species abundances based on coccolith numbers in sediments should take into account the fact that G. oceanica is covered by 20-30 coccoliths per cell while C. leptopora has 25-40 coccoliths per cell.

Gephyrocapsa oceanica cell size was considerably smaller than Cyclococcolithina leptopora. The ultrastructure of both species was typical of most other coccolithophores, especially Emiliania huxleyi,[2,15] but differed considerably from the neretic Cricosphaera carterae in many respects.[2,15] In both our species, as in E. huxleyi,[2,15] coccoliths are formed one at a time within each cell, while C. carterae often contains several intracellular coccoliths and under certain experimental conditions may contain as many as eleven per cell.[16] Organic scales are always associated with coccolith production in Cricosphaera carterae,[4] but were not observed beneath coccoliths in G. oceanica or C. leptopora, nor have they been reported in E. huxleyi.[2,15]

In G. oceanica and C. leptopora thin sections of cells in the process of cell division revealed that each new cell after division retained some of the coccoliths of the original cell. This has also been observed in Cricosphaera carterae[4] and in Emiliania huxleyi.[15]

Our present culture techniques will permit work with these species on effects of environmental factors on coccolith production, variation in coccolith morphology, on nutrition, life cycles, and growth physiology.

Acknowledgements

We thank Dr. N. Watabe for permitting the first author use of the SEM at the Electron Microscopy Center of the University of South Carolina. Jan Witte provided aid in preparation of the final manuscript. This study was supported by National Science Foundation (Geological Oceanography) grant OCE78-09643 (PLB), National Science Foundation (Biological Oceanography) grant OCE77-10876 (RRLG) and an NSF pre-doctoral fellowship to L.E. Brand.

This is Contribution Number 4338 from the Woods Hole Oceanographic Institution.

References

1. B.U. Haq, "Calcareous Nannoplankton," Introduction to Marine Micropaleontology, Elsevier, New York, U.S.A. (1978),79-107.
2. K.M. Wilbur and N. Watabe, "Experimental Studies on Calcification in Molluscs and the Alga Coccolithus huxleyi," N.Y. Acad. Sci. 109 (1963), 82-112.
3. P.L. Blackwelder, "Temperature Relationships in Coccolith Morphology and Dimension in Fossil and Living Emiliania huxleyi" (1976), Ph.D. Dissertation, University of South Carolina, Columbia, 106 pp.
4. S.F. Levi, "Effects of Salinity on Growth, Calcification, and Ultrastructure in a Calcareous Alga, Hymenomonas carterae" (1972), Masters Thesis, University of South Carolina, Columbia, 148 pp.

5. P.L. Blackwelder, R.E. Weiss and K.M. Wilbur, "Effects of Calcium, Strontium, and Magnesium in the Coccolithophorid Cricosphaera carterae. I. Calcification," Marine Biol.34 (1976), 11-16.
6. R.E. Weiss, P.L. Blackwelder and K.M. Wilbur, "Effects of Calcium, Strontium, and Magnesium in the Coccolithophorid Cricosphaera carterae. II. Cell Division," Marine Biol. 34 (1976), 17-22.
7. M.L. Blankenship and K.M. Wilbur, "Cobalt Effects on Cell Division and Calcium Uptake in the Coccolithophorid Cricosphaera carterae," J. Phycol. 11 (1975), 211-219.
8. J.L. Dorigan and K.M. Wilbur, "Calcification and its Inhibition in the Coccolithophorides," J. Phycol. 9 (1974), 450-456.
9. J. Schiller, "Coccolithinae. In: L. Rabenhorst, Kryptogamen-Flora 10. Akad. Verlagsgesellschaft, Leipzig, 89-263.
10. A. McIntyre and A. Be, "Modern Coccolithophoridae of the Atlantic Ocean. I. Placoliths and Crytoliths," Deep-Sea Res. 14 (1967), 561-597.
11. H. Okada and A. McIntyre, "Modern Coccolithophores of the Pacific and North Atlantic Oceans," Micropaleontology,23 (1977), 1-55.
12. K.R. Geitzenauer, "The Pleistocene Calcareous Nannoplankton of the Subantarctic Pacific Ocean," Deep-Sea Res., 19 (1972), 45-60.
13. D. Dunkelburger and N. Watabe," An Ultrastructural Study of Spicule Formation in the Pennatulid Colony Renilla reniformis," Tissue and Cell,6 (1974), 573-586.
14. A.L. Cohen, D.P. Marlow and G.E. Garner, "A Rapid Critical Point Method Using Fluorocarbons ("Freons") as Intermediate and Transitional Fluids," J. Microscopie 7-3 (1968),331-342.
15. D. Klaveness, "Coccolithus huxleyi I.- Morphological Investigations on the Vegetative Cell and the Process of Coccolith Formation," Protistologica (1972), 335-346.
16. R.E. Weiss and K.M. Wilbur, "Effects of Cytochalasin B on Division and Calcium Carbonate Extrusion in a Calcareous Alga," Exp. Cell Res. 112 (1978), 47-58.

Discussion with Reviewers

J. Matthews, W. Davis, R. Jones: What is the crystalline nature of the intracellular and extracellular material?
Authors: Most coccolithophores studied by X-ray or electron diffraction have been found to contain intracellular and extracellular coccoliths composed of the calcium carbonate polymorph calcite. In certain instances under experimental culture conditions coccoliths have been found to contain varying amounts of calcite, aragonite, and vaterite. We plan to perform electron and X-ray diffraction on the species we have described under normal culture conditions as well as those grown at different experimental temperatures.

For additional discussion see page 392.

SCANNING ELECTRON MICROSCOPY/1979/II
SEM Inc., AMF O'Hare, IL 60666, USA

THE ORGANIZATION OF A STRUCTURAL ORGANIC MATRIX WITHIN THE SKELETON
OF A REEF-BUILDING CORAL

I. S. Johnston[*]

Department of Biology
University of California
Los Angeles, CA 90024

Abstract

Although a variety of organic compounds can be isolated from the skeletons of reef-building corals, attempts to implicate them in the processes which control and direct mineralization are hampered by a lack of data concerning their structural organization. This review paper describes two different structural organic matrix components in the skeleton of *Pocillopora damicornis*. One of these components is a transient crystal sheath present only at the growth surface of the skeleton. The other is a boundary lamella separating individual bundles of aragonite crystals. The identity of these elements of a structural matrix is reviewed with respect to previous reports on matrix materials, and their potential role within the general concepts of biomineralization is also considered.

*Present address:
Dental Research Institute
Center for the Health Sciences
University of California
Los Angeles, CA 90024

KEY WORDS: Reef-Coral, Skeleton, Organic-Matrix, Mineralization, Calcium Carbonate

Introduction

Chemical analyses of biogenic mineralized structures almost invariably reveal the presence of organic compounds in addition to the inorganic mineral(s). Whereas some of these organic compounds may represent random contaminants fortuitously incorporated into these structures or secreted materials which modify the mechanical properties of these structures, other secreted compounds which are usually more uniformly distributed throughout are presumed to have some direct role in the process of biomineralization. These latter compounds constitute the organic matrix in any given mineralized structure. The most widespread and generalized statement of a presumed role for a matrix is the so-called organic matrix concept i.e. the implication that organic matter serves as a template in crystal nucleation and the oriented growth of mineralized structures.[1,2] This concept of a physical template, although historically predominant, is rather restrictive and ought to be expanded, for example, to include the possibility that dynamic metabolic processes may be involved.

The aragonitic calcium carbonate exoskeletons of reef-building scleractinian corals contain a variety of protein, carbohydrate and lipid compounds.[3-7] These findings are based on gross chemical analyses of whole coral skeletons. Attempts to implicate these organic materials in the processes which control and direct mineralization[2,8] have largely been made in ignorance of this material's spatial distribution and microarchitecture within the skeleton, particularly at sites close to the skeletal growth surface. However, if the organic matrix concept is to be invoked in any way at all, it must be demonstrated that the organic material displays precise structural organization that correlates in some way with the organization of the mineral elements of the skeleton, as for example, in the structural model proposed by Wainwright.[4] This paper will present the results of an investigation of the morphology of some components of the skeletal organic matrix of the colonial reef coral, *Pocillopora damicornis*, while a subsequent paper will deal with the functional morphology of the skeletogenic epithelial cells. An account of some preliminary aspects of these studies has already been published.[9] The problems associated with the functional interpretation of ultra-structure in this system are no different than those posed by other invertebrate mineralizing systems or even by vertebrate bone and tooth formation. If there are unifying concepts in the field of biomineralization i.e. fundamental processes common to all biogenic mineralizing systems, then they might reasonably be sought at the lower phylogenetic level of the coelenterates.

Materials and Methods

Small, rapidly growing P. *damicornis* colonies (up to 10cm diameter) were collected in Kaneohe Bay, Oahu. 3-5mm long tips from apical branches were excised with side-cutting pliers. The tips were immediately anaesthetized for 5-8 mins in a 1:1 solution of 0.36M $MgCl_2.6H_2O$ and seawater. Following anaesthetization, the tips were transferred to the primary fixative, which was a modification of that developed by Futaesaku et al.[10]: 3% glutaraldehyde, 7.5% sucrose, and 0.5% tannic acid in 0.1M sodium cacodylate at pH 7.4. The material was fixed for one hour at 24°C followed by 7-14 hours at 4°C, and then post-fixed in 1% osmium tetroxide in 0.1M cacodylate buffer for one hour at 20°C. Some of the material was immediately dehydrated in a graded series of ethanol, while the rest was partially demineralized and then dehydrated. Demineralization was carried out in a 1:1 solution of freshly-made 2% ascorbic acid and 0.3M NaCl for 48 hours.[11] This process was carried out in the dark with constant agitation of the samples and the demineralizing solution was changed every 6-8 hours.

For scanning electron microscopy (SEM), samples of both intact and partially-demineralized fixed adult material were subjected to ethanol cryofracture[12] to produce cross-sectional surfaces through tissue layers and skeleton. This technique seemed to be superior to simple mechanical fracture of critical-point-dried material.[13] These specimens were subsequently critical-point-dried from liquid carbon dioxide after the complete removal of ethanol by exchange with liquid carbon dioxide. For examination of "naked" skeletal material, tips were removed from apical branches of living colonies and immersed in fresh water for 4-5 hours. They were then sprayed with a jet of fresh water to remove most of the soft tissue and finally treated for 15 mins in warm 1N NaOH to remove all organic materials from the surface of the skeleton. The tips were then rinsed with distilled water to remove all the NaOH, then allowed to air-dry. Dried SEM specimens were glued to appropriate mounting stubs with a silver conducting paste and then were sputter-coated with an approximately 10-15nm thick layer of palladium and gold. Scanning microscopy was carried out on an ETEC "Autoscan" microscope.

For transmission electron microscopy (TEM), dehydrated specimens of both intact and partially-demineralized fixed adult material were embedded in Spurr plastic (hard formula). The mineralized specimens were infiltrated in 100% unpolymerized plastic for 72 hours under a vacuum of approximately 0.1 torr, during which time the plastic was changed every 12 hours. Polymerized tissue blocks were sectioned using glass knives or a Dupont diamond knife. Sections through undemineralized tissue were cut on knife edges wetted with a saturated solution of calcium carbonate and the sections were picked out of the knife-boats with naked 200 mesh copper grids within 10 seconds of being cut. These precautions were taken in an attempt to prevent the dissolution of calcareous structures. Prior to sectioning, adult-tissue blocks were trimmed so that sections passed through or close to costal spines situated around the rim of an apical polyp calyx. Demineralized-tissue sections were stained with aqueous 2% uranyl acetate or with saturated 50%-ethanolic uranyl acetate, while mineralized-tissue sections were examined unstained. Transmission microscopy was carried out on a Phillips "EM 200" microscope operated at 60 or 80 KV.

Results

Surface features of NaOH-treated P. *damicornis* skeleton are illustrated in Fig. 1-6. This coral has no single dominant terminal polyp, as in the family Acroporidae and some other families. Thus, the rim of the most terminal polyp calyx (Fig. 1&2) from an apical branch was chosen for the most extensive study since branch tips represent zones of maximal skeletal accretion.[14] The individual calyxes have only rudimentary septa but prominent extra-thecal vertical costal spines (Fig. 2&3). Except for the walls of the costal spines, the surface of the skeleton is nodular (Fig. 3). Each 4-7μm diameter nodule (Fig. 4&5), or fasciculus, represents the termination of a continuous bundle of aragonite crystals.[15,16] Within these bundles, the crystals are all aligned with their 'c' axes parallel to each other and perpendicular to the skeletal surface.[17,18] In the walls of the costal spines, the crystals are similarly aligned with respect to each other, but at these sites the 'c' axes terminate tangential to the growth surface, and from a consideration of skeletal surface topography, their organization

Fig. 1. NaOH-cleaned apical branch tip from a colonial skeleton of P. *damicornis*; each large depression or calyx formerly housed a single coral polyp.

Fig. 2. The rim of the terminal calyx bears prominent costal spines (cs); the rudimentary septa are represented by shallow ridges (s).

Fig. 3. The smooth surface features of the costal spines contrast with the fasciculate surface of the skeleton between the costal spines.

Fig. 4. An NaOH-cleaned fasciculate growth surface between the bases of the costal spines, with a single fasciculus outlined by arrows; the depression (des) marks the site of attachment to the skeleton of a desmoidal process which is a polypal-tissue anchoring structure.

Fig. 5. A single fasciculus with details of the termination of individual aragonite crystals.

Fig. 6. A non-fasciculate growth surface on the walls of a costal spine.

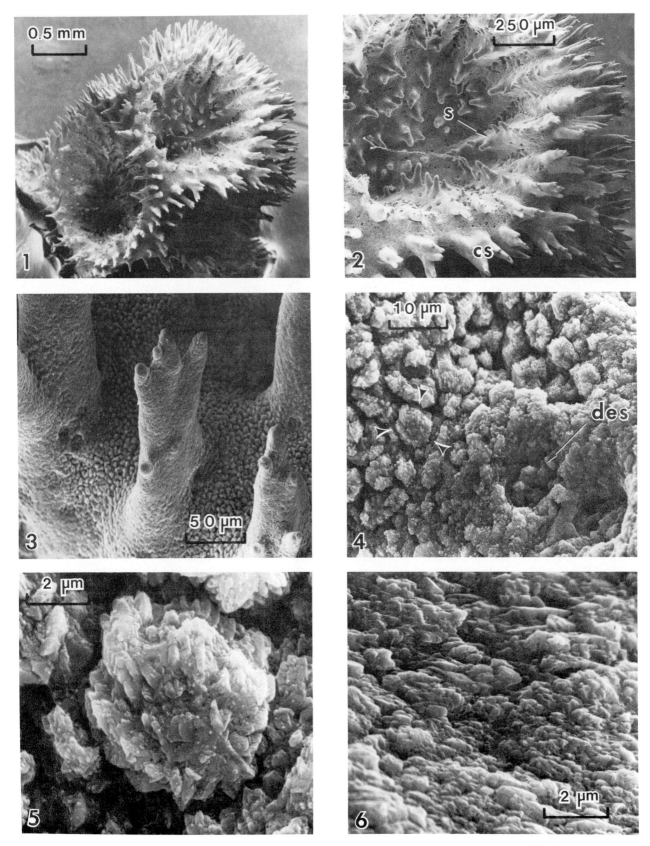

Figure 1 contains the labels: 0.5 mm, 250 µm, s, cs, 50 µm, 10 µm, des, 2 µm, 2 µm

See page 422 for captions.

into crystal bundles is not immediately apparent (Fig. 6).

Dissolution of the mineral component of the skeleton under the experimental conditions described reveals an organic remnant in the space formerly occupied by the skeleton. This remnant is a structural skeletal organic matrix (Fig. 7). Cryofractured specimens show that the matrix has two principal components (Fig. 8). At the growth surface, a meshwork of tiny compartments forms an exact replica or ghost of the intact skeletal surface (Fig. 8&9). This is most clearly seen in the outline of individual fasciculi e.g. compare specimens treated in three different ways: a) an intact skeletal surface with the epithelial cell layer removed (Fig. 10); b) NaOH-treated naked skeleton (Fig. 5); c) the surface of a demineralized skeletal specimen (Fig. 11). This surface matrix (fine mesh) component penetrates into the skeleton to a variable depth. The penetration is much deeper in the faster growing areas of the skeleton where it may be found as far as 5-8μm from the growth surface (Fig. 8), whereas, in slower-growing coenosteal areas along the sides of apical branches, the penetration may be less than 1μm (Fig. 12). The second component of the structural matrix appears as sheets of material fused together to form the walls of 3-6μm diameter chambers (Fig. 8&12). The long axes of these chambers run in a plane which is perpendicular to the general plane of the skeleton surface. The chambers correspond in dimension to that of the individual crystal bundles. The demineralization process probably accounts somewhat for the tearing and collapse of these chamber walls at depths greater than about 10-15μm into the skeleton (Fig. 7). However, this component does eventually degenerate at depth in the skeleton i.e. with age. At the cut end of a branch taken from a basal part of a colonial skeleton, the skeletal mass would have been deposited at least 1-2 years previously. At such a site, even though the outline of the chambers may still be apparent (Fig. 13), the chamber walls are very ragged and they reveal a granular and fibrous substructure (Fig. 14). Webs of 30-40nm diameter fibers are present on, and penetrating through, the chamber walls. They may

or may not represent an invading microflora, although there is ample evidence that coral skeletons are bored into by algae; see for example, the 1μm diameter algal thallus in Fig. 13.

Transmission electron microscopy of thin sections through the tissue-skeleton interface gives further information on the structural matrix. The surface component of the matrix consists of a series of partially fused, adjacent envelopes forming small compartments (Fig. 15&16). At the growth surface, these compartments are more discretely outlined than deeper in the skeleton, i.e. deeper in the skeleton the envelope wall material appears to be teased out with many breaks and discontinuities (Fig. 16). The compartments often appear irregular in shape, but there is a general trend towards them being elongated and in some planes of section, the long axes of adjacent compartments are parallel to each other and generally perpendicular to the skeletal growth surface (Fig. 16). The compartments are, approximately, 0.15-0.25μm wide with some over 1μm in length and they are generally oriented with their long axes parallel to each other. The wall that divides two adjacent compartments is made up of two partially fused 2-3nm thick sheets, indicating that each compartment has a complete and independent envelope (Fig. 17).

There is a subtle transition between the predominance of one matrix component and the other, and there is no apparent difference in ultrastructure between the surface compartment walls and the deeper chamber walls (Fig. 15). However, there is a difference in the composition or chemical state of these matrix components since, when tannic acid was omitted from the prefixative, the surface component was no longer preserved while the deeper matrix component was still visible (see ref. #19, plate 45).

TEM examination of the skeleton from undemineralized specimens demonstrates that the growth surface matrix compartments are always completely filled with aragonite mineral so that no unfilled or partially filled compartments protrude above the mineralized skeleton (Fig. 18). There is a population of 50-75nm diameter vesicles with electron lucent contents which are

Fig. 7. A cryofracture section through a demineralized specimen with coral soft tissues intact; an organic remnant (sk) marks the space formerly occupied by the base of a costal spine; the ectoderm (ec) and the endoderm (en) are the principal tissue layers of the animal; the spherical bodies within the endodermal cells are symbiotic algae.

Fig. 8 Detail of the tissue-skeleton interface in Fig. 7; the calicoblastic ectoderm (cal) is narrowly separated from the skeletal organic matrix by the subectodermal space (between opposing arrows). The matrix has a fine-mesh surface component (me) and a deeper lamellar component (la).

Fig. 9. Cryofracture section through a demineralized specimen showing the growth surface component of the matrix (me) separated from the calicoblastic ectoderm (cal) by the subectodermal space (s).

Fig. 10. The skeletal growth surface of an undemineralized specimen from which the coral soft tissues were fortuitously removed by mechanical shearing during preparation for cryofracture. The vesicles present on the surface of the fasciculi probably are vesicles normally found in subectodermal space.

Fig. 11. The skeletal growth surface of a demineralized specimen with the organic ghosts of some fasciculi intact and others cryofractured; the fractured faces reveal the meshwork characteristic (me) of the surface component of the structural matrix.

Fig. 12. Oblique cryofracture section through a tissue-skeleton interface from a coenosteal area at the base of a demineralized apical skeletal branch; the fasciculus outlined by arrows shows only a very thin layer of the growth surface matrix component; the calicoblastic ectoderm (cal) intimately envelopes the skeletal surface.

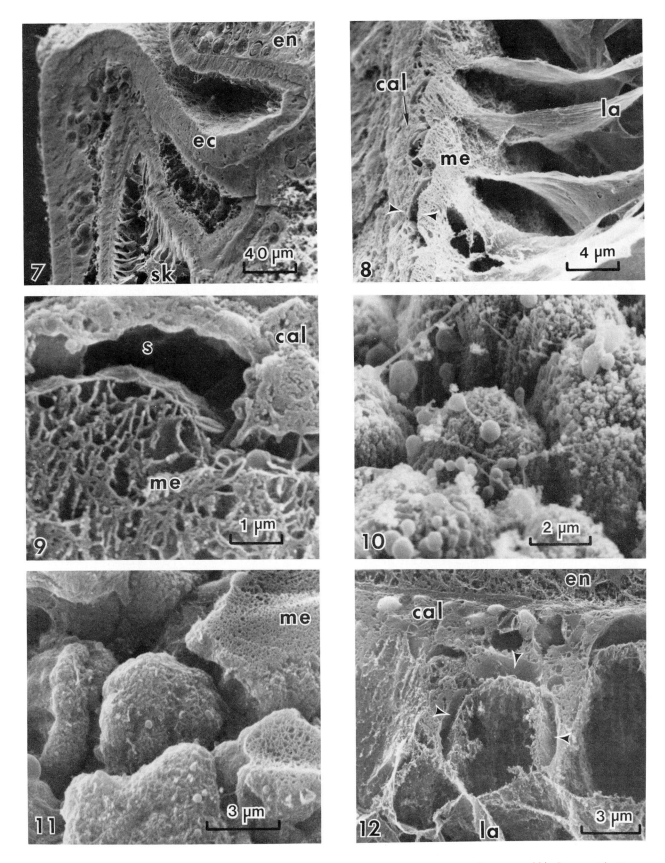

See page 424 for captions.

found in the subectodermal space i.e. in that space separating coral soft tissue and skeleton (Fig. 9,10&15). These vesicles are occasionally seen in close proximity or possibly in process of fusing to the skeletal surface (Fig. 15&18), however, they contain no preformed aragonite mineral.

Discussion

The exoskeletons of reef corals are precisely sculpted in a species-specific manner. The complex and spectacular microarchitecture of both the outer surfaces and the interiors of these skeletons is readily studied using SEM techniques. These techniques have principally been used to investigate the inorganic mineral elements, and they have provided important crystallographic and taxonomic data for paleontologists, geologists and biologists studying both Recent and fossil coral skeletons.[15,16,18,20-26] These same data beg questions about tissue-skeleton relationships, and in particular about the process of mineralization in which the coral soft-tissues exert genetically defined cellular control over the initial formation and subsequent accretionary growth of the skeleton. An organic matrix secreted by the skeletogenic epithelium might mediate this kind of cellular control over mineralization.

Whereas there has been much speculation about the function of the various organic materials found in coral skeleton, there has also been some controversy about their identity, for example, as to whether they have some structural integrity[4,27-29] or are merely present as a homogeneous or colloidal ground substance.[30,31] This present study indicates that within the mineralized adult skeleton of *P. damicornis* is an organic matrix with at least two different structural components: 1) At, or close to, the growth surface of the skeleton, matrix compartments seen in demineralized specimens correspond in size and orientation with the individual acicular aragonite crystals which make up the inorganic phase of the skeleton [17,18,32], thus it appears that the matrix provides a complete and independent envelope or sheath for each aragonite crystal. It is possible that this matrix component is identical with skeletal organic material previously described by Bourne[27] as an "external limiting membrane" and by Goreau[33] as a "mucopolysaccharide-like lamella". Both of these observations were also made on demineralized specimens but it was not clear if prior to demineralization the matrix was within the skeleton or merely enveloping its surface. However, the surface matrix component described in this present study certainly does correspond with structures previously identified by Sato[34] and characterized as an "intercrystalline organic matrix made up of reticular fibers". Since his observations were made on unfixed skeletons, Sato's reticular fibers probably represent a stage in the degeneration of whole crystal-sheaths. 2) Deeper in the skeleton the structural matrix persists as sheets of material lying between the individual crystal bundles (= Wainwright's "crystal fibers"[17]). I propose calling this matrix component the crystal fiber boundary lamellae. These lamellae are almost certainly equivalent to skeletal organic structures previously observed by light microscopy[27-29], although their relationship to the crystal bundles and to the surface fasciculi was formerly unknown. In demineralized specimens, cross sections through the chambers formed by this matrix component reveal cell-like profiles with dimensions similar to those of the overlying calicoblastic epithelial cells. This is probably what prompted Von Heider[35] and later workers[36-39] to believe, incorrectly, that the skeleton was made up of transformed calicoblast cells.

A relationship between the two matrix components is not immediately obvious. They were differentially preserved when tannic acid was omitted from the prefixative indicating that they are either composed of different chemical compounds or of the same compounds present in different chemical states.[19] However, the sheaths of the most terminal aragonite crystals are structurally indistinguishable from the crystal fiber boundary lamellae at a depth of 10μm or so below the growth surface. Both components show structural degradation with age, although these diagenetic trends occur on very different time scales. The crystal sheaths become disorganized and then disappear completely at depths greater than 2-8μm below the skeleton surface, being consistently present at greater depths beneath the most rapidly growing areas than elsewhere on the corallum. It seems unlikely that this is due solely to differential rates of fixative penetration into the skeleton since, when deeper areas of the skeleton were exposed to fresh fixative e.g. at the cut-end of a branch, no crystal sheaths were apparent. Therefore, a more likely explanation is that the crystal sheaths naturally degrade and disappear after new skeletal material has been deposited on top of them. If this process takes a finite time, then the deeper presence of the sheaths in the faster growing areas is due to the fact that more skeleton is deposited over the top of them during that finite period. The time over which the sheaths apparently degrade may be very short. Using a conservative estimate of linear skeletal extension rate, calculated from the laboratory experiments of Clausen and Roth as 6.1mm/year[14], and assuming equal daily growth rates throughout the year (which is certainly not the case), then the mean daily extension rate would be about 16.7μm/day. Therefore, it seems unlikely that any of the intact crystal sheaths persist for more than a few hours after subsequent calcium carbonate deposition.

On the other hand, the crystal-fiber boundary lamellae are still apparent after months of subsequent skeletal growth, although they do eventually become disorganized, exposing a granular and fibrous substructure that is not apparent at earlier stages. It is not known if the webs of 30-40nm fibrils at the surface of the degenerating lamellae are a diagenetic product of the lamellar material or if they originate independently of the lamellae. For example, they might be formed from degradation products of the crystal sheaths. If each fibril has a 10nm coating of gold-palladium, then the true diameter is probably about 20nm. These fibrils, therefore, correspond in size and morphology to the fibrils that Wainwright isolated from *P. damicornis* skeletons and which gave the same X-ray diffraction pattern as chitin[4], although their organization does not correspond with Wainwright's structural model.

It seems quite possible that diagenesis of the structural matrix is closely, and may be even causally, linked with incipient diagenesis of the inorganic elements of the skeleton. Diagenetic events, such as the complete calcitization of an aragonite coral skleton, may not occur for hundreds of years depending on the environmental conditions. On the other hand, syntaxial coating or cementation of the original aragonite crystals occurs very soon after initial deposition[13,40] and could well coincide with the disappearance of the crystal sheaths.

Each of the two principal matrix components described in this study shares a common structural dimension with one of the two basic organizational units of the mineral phase of the skeleton i.e. the individual aragonite crystal and the crystal fiber. This satisfies a very minimal criterion for implicating one or both of these matrix components, but especially the crystal sheaths because of their location at the growth surface, within the organic matrix concept. Therefore, it seems worthwhile to pursue a possible role for these crystal sheaths in the process of mineralization.

Within the general field of biomineralization, the extracellular organic matrix template hypothesis, which in vertebrates has centered principally on a role for collagen[41], is opposed by the hypothesis that mineralization first occurs within intracellular organellar enclosures, from which the crystals are subsequently "planted out" in an extracellular environment for further oriented crystal overgrowth.[42] The discovery, in the extracellular space of some developing hard tissues, of membrane enclosed vesicles in which mineral nucleation is first apparent has led to the matrix vesicle concept.[43] The matrix vesicle concept incorporates elements of the first two alternatives, since the vesicles each potentially represent a small enclosed intracellular environment, and they also contain specific organic compounds which might function within the organic matrix template concept.

The same kinds of alternative explanations have been proposed for coral calcification. Goreau proposed that skeletal growth was controlled by an extracellular, extraskeletal mucopolysaccharide template.[44] Sorauf indicated that he may have seen this template[22] but it is not clear if he actually saw components of the structural matrix as described in this present study, rather than a structural component from the subectodermal space. Hayes and Goreau propose an intracellular intravacuolar site for crystal nucleation, followed by secretion of the seed crystal and subsequent extracellular crystal overgrowth.[45,46] Neither alternative seems correct in light of the data derived from this present study.

Although the mode of synthesis and assembly of the crystal sheaths is not known, I tentatively suggest that the vesicles in the subectodermal space may be, or may contain, sheath precursor material.[9,19] Therefore, since the vesicles are

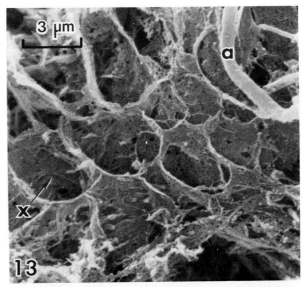

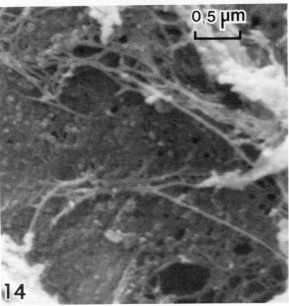

Fig. 13. Partially demineralized cut-surface of an excised basal branch; the deep lamellar organic component is still apparent although no longer completely intact; (a) the thallus of a boring alga ; (x) marks an area shown in greater detail in Fig. 14.

Fig. 14. Detail of the lamellar matrix component in an older area of a colonial skeleton; the lamella has a granular and fibrous substructure and there are webs of 30-40nm fibrils lying on and apparently penetrating through the lamella.

membrane-bound, it is possible that each terminal crystal sheath may also consist in part of a true biological membrane. As such, each sheath would be analogous to the matrix vesicles in which hydroxylapatite nucleation occurs in vertebrate calcifying cartilage.[43] Once again, the inside wall of the crystal sheath may still operate within the template hypothesis e.g. a presumed role for the calcium-binding, membrane lipid, phosphatidyl serine[47], and the sheath may also have the ability to modify the composition of the enclosed space by active transport mechanisms e.g. matrix vesicles from epiphyseal chondrocytes concentrate calcium to levels 25 to 50 times higher than adjacent cells.[48] These possibilities remain to be tested.

The presence of sheaths enveloping individual inorganic crystals in biogenic mineralized structures is not uncommon in the invertebrates, especially in molluscan hard tissues e.g. bivalve nacre[49], and even in coelenterate tissues e.g. alcyonarian calcite spicules.[50] The unusual feature of the crystal sheaths in this reef-coral skeleton, however, is the speed with which incipient diagenesis occurs. This observation once again emphasizes the importance of looking at the growth surface i.e. at the site of mineral deposition, to study matrix-calcification relationships. Concomitantly, the value of attributing a mineralizing role to skeletal organics isolated from "aged" skeletons is highly questionable. Not only are these aged compounds diagenetically altered, but they are also almost invariably contaminated by a variety of endolithic algae[51] and bacteria.[52]

Acknowledgments

The advice and encouragement of Dr. Leonard Muscatine is gratefully appreciated. Financial support came from NSF grant #PCN76-21390. Field work was supported by a grant from the Edwin Pauley Fund.

References

1. K.M. Towe. Invertebrate shell structure and the organic matrix concept. Biomineralization 4: 1-14. 1972.
2. E.T. Degens. Molecular mechanisms on carbonate, phosphate and silica deposition in the living cell. Topics Curr. Chem. 64: 1-112. 1976.
3. D. Lester and W. Bergmann. Contributions to the study of marine products. VI. The occurrence of cetyl palmitate in corals. J. Org. Chem. 6: 120-122. 1941.
4. S.A. Wainwright. Skeletal organization in the coral *Pocillopora damicornis*. Quart. J. Microsc. 104: 169-183. 1963.
5. S.D. Young. Organic material from scleractinian coral skeletons. I. Variation in composition between several species. Comp. Biochem. Physiol. 40B: 113-130. 1971.
6. S.D. Young, J.D. O'Connor and L. Muscatine. Organic material from scleractinian coral skeletons. II. Incorporation of ^{14}C into protein, chitin and lipid. Comp. Biochem. Physiol. 40B: 945-958. 1971.
7. K. Mopper and E.G. Degens. Aspects of the biogeochemistry of carbohydrates and proteins in aquatic environments. Woods Hole Oceanographic Institution Technical Report, WHOI-72-68. 1972.
8. R.M. Mitterer. Amino acid composition and metal binding capability of the skeletal protein of corals. Bull. Mar. Sci. 28: 173-180. 1978.
9. I.S. Johnston. Aspects of the structure of a skeletal organic matrix, and the process of skeletogenesis in the reef-coral *Pocillopora damicornis*. Proc. III Int. Coral Reef Symp. Miami 2: 447-453. 1977.
10. Y. Futaesaku, B. Mizuhira and H. Nakamura. A new fixation method using tannic acid for electron microscopy and some observations of biological specimens. Histochem. Cytochem. (Proc. 4th Int. Cong. Histochem. Cytochem.) 4: 155-156. 1972.
11. F. Dietrich and A.R. Fontaine. A decalcification technique for ultrastructure of echinoderm tissue. Stain Tech. 50: 351-354. 1975.
12. W.J. Humphreys, B.O. Spurlock and J.S. Johnson. Critical point drying of ethanol-infiltrated, cryofractured biological specimens for scanning electron microscopy. Scanning Electron Microscopy/1974, IIT Research Institute, Chicago, IL, 60616; pp. 275-282.
13. J.A.E.B. Hubbard. Life and afterlife of reef corals: A timed study of incipient diagenesis. IX International Sedimentological Congress pp. 75-80. 1975.
14. C.D. Clausen, and A.A. Roth. Estimation of coral growth-rates from laboratory ^{45}Ca-incorporation rates. Mar. Biol. 33: 85-91. 1975.
15. S.W. Wise. Scleractinian coral skeletons: Surface microarchitecture and attachment scar patterns. Science 169: 978-980. 1970.
16. S.W. Wise. Observations of fasciculi on developmental surfaces of scleractinian coral exoskeletons. Biomineralization 6: 160-175. 1972.
17. S.A. Wainwright. Studies of the mineral phase of coral skeleton. Exp. Cell Res. 34: 213-230. 1964.
18. J.H. Vandermeulen and N. Watabe. Studies on reef corals. I. Skeleton formation by newly-settled planula larva of *Pocillopora damicornis*. Mar. Biol. 23: 47-57. 1973.
19. I.S. Johnston. Functional ultrastructure of

Fig. 15. Thin section through a demineralized tissue-skeleton interface at a rapidly calcifying area of the animal; the growth surface (me) and deeper lamellar (la) components of the matrix are apparent; within the subectodermal space (s) is a population of 50-75nm vesicles (ve).

Fig. 16. Oblique section through the growth surface matrix component (me) showing the gradual degeneration of the individual crystal sheaths with increasing distance from the subectodermal space (s).

Fig. 17. Detail of the terminal compartments, or crystal sheaths, of the growth surface matrix component; the sheaths of adjacent crystals are normally fused together but in places they may be resolved into their individual components (between opposing arrows).

Fig. 18. Unstained thin section through the tissue-skeleton interface of an undemineralized specimen; the open spaces in the skeleton (sk) represent places from which the aragonite mineral has been chipped out during sectioning; the terminal compartments of the growth surface matrix component are all completely filled with aragonite; there are some 50-75nm vesicles (ve) with electron-lucent contents close to the skeletal surface.

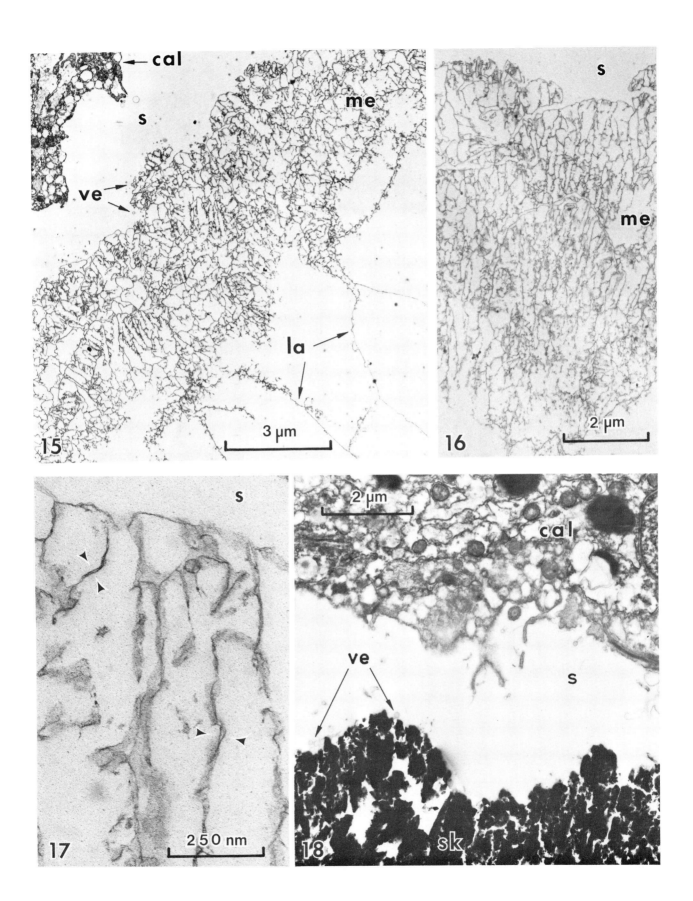

the skeleton and the skeletogenic tissues of the reef coral *Pocillopora damicornis*. PhD. dissertation, University of California at Los Angeles. 1978. University Microfilms International, order # 7811356.

20. D.J. Barnes. Coral skeletons: An explanation of their growth and structure. Science 170: 1305-1308. 1970.

21. J.E. Sorauf. Microstructure and formation of dissepiments in the skeleton of the Recent Scleractinia (hexacorals). Biomineralization 2: 1-22. 1970.

22. J.E. Sorauf. Skeletal microstructure and microarchitecture in Scleractinia (Coelenterata). Paleontol. 15: 88-107. 1972.

23. J.E. Sorauf. Observations on microstructure and biocrystallization in coelenterates. Biomineralization 7: 37-55. 1974.

24. J.S. Jell. The microstructure of some scleractinian corals. in Proc. II Int. Symp. Coral Reefs, Gt. Barrier Reef Comm. Brisbane 2:301-320. 1974.

25. K. Oekentorp. Electron-microscope studies on skeletal structures in Coelenterata and their systematic value. in Proc. II Int. Symp. Coral Reefs, Gt. Barrier Reef Comm. Brisbane 2: 321-326. 1974.

26. B.F. Spiro. Ultrastructure of *Acropora*. Bull. Geol. Soc. Den. 23: 76-78. 1974.

27. G.C. Bourne. Studies on the structure and formation of the calcareous skeleton of the Anthozoa. Quart. J. Microsc. Sci. 41: 499-547. 1899.

28. A. Krempf. Sur la formation du squelette chez les hexacorallaires a polypier. Acad. Sci. Paris, Comptes Rendus 144: 157-159. 1907.

29. E. Hayasi. On the detection of calcium in the calicoblasts of some reef corals. Palao Trop. Biol. Sta. Stud. 2: 169-176. 1937.

30. J.E. Duerden. Recent results on the morphology and development of coral polyps. Smithson. Misc. Coll. 47: 93-111. 1904.

31. W.H. Bryan and D. Hill. Spherulitic crystallization as a mechanism of skeletal growth in the hexacorals. in Proc. R. Soc. Queensland 52: 78-91. 1941.

32. J. Vahl. Sublichtmikroskopische Untersuchungen der kristallinen Grundbauelemente und der matrix Beziehung zwischen Weichkorper und Skellett an *Caryophyllia* Lamark 1801. Z. Morph. Okol. Tierre 56: 21-38. 1966.

33. T.F. Goreau. Histochemistry of mucopolysaccharide-like substances and alkaline phosphatase in Madreporaria. Nature 177: 1029. 1956.

34. T. Sato. Electron microscope observation on the minute structures of the skeletons of some scleractinians. Earth Sci. (Chikyu Kagaku) 66: 9-14. 1963.

35. A. von. Heider. Die Gattung *Cladocora*, Ehrenb. Sitzungsber. Akad. Wiss. Wien. 84: 634-667. 1881.

36. M.M. Ogilvie. Microscopic and systemmatic study of madreporarian types of corals. Phil. Trans. Roy. Soc. Lond. B 187: 83-345. 1896.

37. M.M. Ogilvie-Gordon. The lime-forming layer of the madreporarian polyp. Quart. J. Microsc. Sci. 49: 203-211. 1906.

38. M.M. Ogilvie-Gordon. Note on the formation of the skeletons in the Madreporaria. Quart. J. Microsc. Sci. 51: 473-482. 1907.

39. S. Kawaguti and K. Sato. Electron microscopy on the polyp of staghorn corals with special reference to its skeleton formation. Biol. J. Okayama Univ. 14: 87-98. 1968.

40. N.P. James. Diagenesis of scleractinian corals in the subaerial vadose environment. J. Paleont. 48: 785-799. 1974.

41. M.J. Glimcher. Composition, structure, and organization of bone and other mineralized tissues and the mechanism of calcification. Handbook Physiol. Sect. 7 Endocrinology 7: 25-116. 1976.

42. A.L. Lehninger. Mitochondria and calcium ion transport. Biochem. J. 119: 120-138. 1970.

43. H.C. Anderson. Matrix vesicle calcification. Fed. Proc. 35: 105-108. 1976.

44. T.F. Goreau. The physiology of skeleton formation in corals. I. A method for measuring the rate of calcium deposition by corals under different conditions. Biol. Bull. 116: 59-75. 1959.

45. R.L. Hayes and N.I. Goreau. Intracellular crystal-bearing vesicles in the epidermis of scleractinian corals, *Astrangia danae* (Agassiz) and *Porites porites* (Pallas). Biol. Bull. 152: 26-40. 1977.

46. N.I. Goreau and R.L. Hayes. Nucleation catalysis in coral skeletogenesis. in Proc. III. Int. Coral Reef Symp. Miami 2: 439-445. 1977.

47. R.E. Wuthier. Lipids of matrix vesicles. Fed. Proc. 35: 117-121. 1976.

48. R.E. Wuthier. Electrolytes of isolated epiphyseal chondrocytes, matrix vesicles, and extracellular fluid. Calc. Tiss. Res. 23: 125-133. 1977.

49. H.K. Erben and N. Watabe. Crystal formation and growth in bivalve nacre. Nature 248: 128-130. 1974.

50. D.G. Dunkelberger and N. Watabe. An ultrastructural study on spicule formation in the pennatulid colony *Renilla reniformis*. Tiss. Cell 6: 573-586. 1974.

51. K.J. Lukas. Two species of the chlorophyte genus *Ostreobium* from skeletons of Atlantic and Caribbean corals. J. Phycol. 10: 331-335. 1974.

52. L.H. DiSalvo. Isolation of bacteria from the corallum of *Porites lobata* (Vaughan) and its possible significance. Amer. Zool. 9: 735-740. 1969.

Discussion with Reviewers

J.E. Sorauf: You speak of subectodermal space but illustrate vesicles and other matter "floating" within something that must have filled the space (Fig. 15&18). Do you have an idea of what material might have filled this space, either as to consistency or composition?

Author: I presume that the subectodermal space is an extracellular space much like the lateral intercellular spaces within the calicoblastic epithelial cell layer.[9,19] As such it may contain a fluid of composition very distinct from that of the intracellular environment, although the cells bounding this space probably exert an influence over its composition. For example, there is evidence for the secretion of materials, other than the presumptive matrix precursor vesicles, into this space.[19]

From ultrastructural studies, little can be inferred about the consistency of the contents of this space. Further more, soft-tissue shrinkage during fixation has undoubtedly distorted the original dimensions of the subectodermal space in the living coral, so the absolute dimensions of this space are also unknown.

J.E. Sorauf: I'm sure that in Ref. #22, Plate 11 Fig. 1 does not illustrate "matrix" as you have shown it in this present study. I now think that it shows an ectodermal layer, whereas I formerly supposed that I was seeing the "amorphous layer" supposed to be matrix by Tom Goreau.

Author: It seems quite unlikely that any of the exposed calicoblastic ectodermal soft tissues were still present after an unfixed coral was exposed to 5% NaOCl for three days. On the other hand, if diffusion of the NaOCl into the mineralized skeleton was a slow process, then some of the matrix materials might have persisted, to be subsequently exposed by EDTA decalcification.

R.S. Blanquet: Is there any information regarding the role of zooxanthellae in providing materials for, or affecting the rate of formation of, organic matrices in corals?

Author: Young et al.[6] demonstrated that isotopically labelled carbon, photosynthetically fixed by the zooxanthellae, could very rapidly be detected in coral skeletons particularly in a skeletal lipid fraction. Algal photosynthetic products are, therefore, certainly available to the animal for the synthesis of skeletal matrix materials, and they may even represent a principal or limiting source for certain matrix precursor materials.[4] However, until more is known about the chemical composition of the structural matrix at the growth surface of the skeleton, it will be difficult to speculate about an algal role in controlling matrix formation. If the algae are involved in matrix formation and if the matrix mediates cellular control over the rate of mineralization, then this would provide an adequate explanation for algal enhancement of coral calcification.[44]

R.L. Hayes: The extracellular matrix vesicle described in vertebrate mineralizing systems is not only crystal-bearing, but is bounded by a "unit" membrane. If the coral vesicles in the subectodermal space are analogous to the vertebrate matrix vesicles, they would be expected to reveal the typical configuration of a cell membrane. Has unit membrane structure been resolved in the coral vesicles and if so what is the thickness of that membrane?

Author: The vesicles in the subectodermal space and in the lateral intercellular spaces of the calicoblastic ectoderm have a unit membrane which is 5-6nm thick. The crystal sheaths however lack the configuration of a typical lipid bi-layer structure.

R.L. Hayes: The use of tannic acid in your fixative solution has led to the proposal that two chemical constituents or configurations are found in the organic matrix. Do you have any other evidence to suport chemical differentiation of the matrix? As an alternative interpretation of your

results, perhaps the cross-linking influence of glutaraldehyde is not sufficient to fix both matrix components and tannic acid, as a coagulating fixative, supplements the retentive capacity of the aldehyde solution.

Author: The differential fixation of the crystal sheaths in the presence and absence of tannic acid is the only evidence that these sheaths are different from the crystal fiber boundary lamellae. Tannic acid is able to precipitate proteins at concentrations 1000X less than the minimum protein concentration needed for aldehyde fixatives to cause precipitation[10], therefore the presence of tannic acid is quite likely to supplement the retentive capacity of glutaraldehyde in the prefixative.

Additional discussion with reviewers of the paper "Statolith Synthesis and Ephyra Development..." by D.B. Spangenberg continued from page 438.

R. S. Blanquet: Do you know if the total amount of calcium incorporated into gypsum differed in your experiments? That is, though decreased levels of calcium cause increased numbers of statoliths, the total amount of gypsum in greater numbers of smaller statoliths might be the same as smaller numbers of larger ones?

R. M. Dillaman: Is the absolute amount of calcium deposited per animal as statoliths any different in 7.5 mM Ca versus 0.7 mM Ca?

Author: I do not know whether the total amount of gypsum in the greater number of smaller statoliths (0.7 mM Ca) might be the same as smaller numbers of larger ones (7.5 mM Ca). This has not yet been determined quantitatively. I use the term hyper-mineralization in regard to the low calcium condition because more mineralization sites are formed.

R. S. Blanquet: Do you know how much strontium is incorporated into statoliths compared to calcium?

Author: Quantitative studies have not been done.

R. M. Dillaman: Were you able to get a strontium peak in the analysis of the soft tissue?

Author: We did not get a strontium peak in the analysis of soft tissue adjacent to statolith material using identical analytical techniques.

Reviewer I: Are there any calcium storage sites in the ephyra? Calcium storage sites are known in many invertebrates (Molluscs, Nematodes, Platyhelminthese).

Author: There are no calcium storage sites in the ephyrae to my knowledge. At one time I wondered whether statoliths could serve as calcium storage sites, but the low calcium studies suggest that calcium is not transferred from statoliths to the ephyra tissue since the ephyrae were demonstrating calcium deficiency symptoms while the statolith material was not depleted.

SCANNING ELECTRON MICROSCOPY/1979/II
SEM Inc., AMF O'Hare, IL 60666, USA

STATOLITH SYNTHESIS AND EPHYRA DEVELOPMENT IN AURELIA METAMORPHOSING
IN STRONTIUM AND LOW CALCIUM CONTAINING SEA WATER

D. B. Spangenberg

Department of Pathology
Eastern Virginia Medical School
Norfolk, VA 23501

Abstract

Statolith synthesis and ephyra development were studied in Aurelia strobilating in artificial sea water (ASW) deficient in Ca^{2+}, ASW deficient in Ca^{2+} but supplemented with Sr^{2+} or PO_4^{3-} and in standard ASW. Stobilae developing in low Ca^{2+} ASW (0.7-1.4 mM) give rise to ephyrae with statolith numbers which are significantly higher than normal. Supplementation of the low Ca^{2+} ASW with either Sr^{2+} or PO_4^{3-} further enhances the statolith numbers in animals developing in these ASWs.

SEM investigations of the form of statoliths from organisms from the various ASWs revealed that they had different shapes. Energy dispersive X-ray analysis studies of statoliths from organisms developing in low Ca^{2+} ASW supplemented with Sr^{2+} revealed the presence of Sr in the statoliths. Incorporation of Sr^{2+} into the statoliths of Aurelia is especially of interest because it suggests the possibility that statoliths of jellyfish medusae from nature might be used to readily detect small amounts of radioactive Sr^{2+} contamination in marine environments.

Ephyrae exposed to a low Ca^{2+} environment synthesized excessive numbers of statoliths even though their morphological development is severely disturbed. These ephyrae are ruffled in appearance, cannot pulse, and often have abnormal numbers of arms and rhopalia. Further investigations into the specific mechanisms whereby calcium deficiency causes ephyra abnormalities could lead to a better understanding of the specific actions of Ca^{2+} in normal and abnormal mineralization and in organismal development.

KEY WORDS: Calcium, Strontium, Phosphate, Statoliths, Aurelia

Introduction

Regulatory mechanisms involved in the strobilation and mineralization processes of Aurelia aurita have been under investigation in this laboratory for many years. It was found that metamorphosis is induced by iodide and/or thyroxine in artificial sea water (ASW) so that routinely large numbers of mineralizing ephyrae (young medusae) can be obtained in a period of six days [1,2]. Special attention has been given to the factors which may control or directly affect statolith synthesis.

A major factor of importance in statolith synthesis is the ionic balance of the sea water in which the organisms live and metamorphose. Spangenberg and Beck [3] established that the chemical composition of jellyfish (Jf) statoliths is calcium sulfate dihydrate (gypsum). Spangenberg later found that statolith synthesis cannot proceed if the sulfate content of the ASW is reduced below 4.2 mM although strobilation proceeds at a normal rate. The pulsing rate of the ephyrae is somewhat reduced and the nematocyst numbers are lower in these organisms, but the overall external appearance of the ephyrae is normal. Phosphate treatment of the low sulfate containing ASW causes reversal of abnormalities of ephyrae which had developed in it. There is a restoration to normal of the statolith numbers and a marked improvement of nematocyst numbers [4].

Having determined that phosphate supplementation of low sulfate ASW improved uptake of both sulfate and calcium into the ephyrae and especially into the statoliths, it became of interest to determine whether low calcium ASW could also cause statolith deficiency and, if so, whether phosphate would reverse the low calcium effects.

Groups of polyps were induced to strobilate in the ASW of various calcium content and the number of statoliths synthesized per rhopalium was recorded. The general morphology of the ephyrae developing in the low calcium ASW was recorded with the SEM and pulsing activity was observed under the light microscope. Phosphate and strontium were added to the low calcium medium in an effort to counteract the negative effects of the low calcium ASW on the organisms. Energy dispersive X-ray analysis studies were made of the statoliths from organisms from the low calcium plus strontium ASW, low calcium plus phosphate ASW, as well as ASW plus strontium.

The extraordinary capacity of rhopalia to synthesize statoliths in a calcium deficient medium which causes severe morphological abnormality in the ephyrae is subsequently described in this paper as is the influence of strontium and phosphate supplementation on these effects.

Methods

Two different strains of Aurelia aurita were used for these studies. One culture was collected by the author in 1961 from Corpus Christi, Texas and are called the Texas Aurelia (TA)[5]. The other strain was collected in the summer of 1978 from Norfolk, Virginia and are termed the Norfolk, Aurelia (NA). Large numbers of polyps were maintained in the laboratory according to previously described methods[5]. Groups of 10 organisms were induced to strobilate in ASW containing 1×10^{-5} M sodium iodide at 30°C. This concentration of iodide was used because strobilation is more rapid and better synchronized at this dosage than at lower concentrations. Free ephyrae were found released from the strobilae 6 days after the administration of iodide. Ephyrae were placed in a wet film and the number of statoliths formed per rhopalium was counted. The average number of statoliths made per rhopalium per test group was calculated and an analysis of variance was done in each series of tests. Since in all test series, highly significant differences were found between the groups at the p<0.05 level, the Duncan's multiple range test[6] was applied to each test sequence with results discussed in the text. The organisms used for testing were chosen from cultures at random and each test (including controls) was done from a single culture. Ten polyps were used per test group and at least 3 tests were performed per experimental condition usually with 2 tests using the TAs and one test using the NAs. The ASW containing the monobasic sodium phosphate was adjusted to pH 7.8 with NaOH but the strontium-containing ASW did not require adjustment.

The Cambridge SEM was used for most of these studies although a few recent observations were made with the JEOL JMS 35. Technics for the preparation of the Jf tissue were done according to previously reported methods.

Results

Statolith synthesis: Three groups of ten polyps were exposed to ASW with calcium concentrations of 7.5 mM (standard ASW concentration) 5.4 mM, 3.4 mM, 2.0 mM, 1.4 mM, and 0.7 mM (added as $CaCl_2 \cdot 2H_2O$) during strobilation. The organisms were placed in the altered ASW at the time of iodide induction of strobilation. Table 1 shows the numbers of statoliths found microscopically per rhopalium per organism in the various concentrations of calcium. Organisms developing in ASW containing 0.7 mM, 1.4 mM and 2.0 mM calcium developed significantly higher numbers of statoliths than organisms developing in ASW with higher amounts of calcium. Microscopic examination of the statoliths from the low Ca^{2+} ASW (0.7 mM) reveal that they are approximately one third smaller in size than those from control organisms. (Fig. 1)

Supplementation of the low Ca^{2+} ASW (0.7 mM) with 0.75 mM of Sr^{2+} significantly increased the statolith numbers in the organisms developing in this modified ASW (Table 2A). At the light microscope level, the statoliths of the Sr^{2+} supplemented organisms were found to be more rounded in shape (Fig 1D) than control statoliths. Energy dispersive X-ray analysis of the statoliths from 1.4 mM Ca^{2+} plus 0.75 mM Sr^{2+} and of ephyrae from ASW plus 0.75 mM Sr^{2+} revealed the presence of small Sr peak in these statoliths (Fig. 1F). This indicates that Sr^{2+} is incorporated into the statoliths of these organisms.

Supplementation of the low Ca^{2+} ASW (1.4 mM) with 0.72 mM of PO_4^{3-} given as NaH_2PO_4 caused a significant increase of statolith numbers (Table 2B). Supplementation of standard ASW with PO_4^{3-} likewise

Table 1. Effects of calcium concentration range in ASW in statolith synthesis.

	Duncan's Multiple Range Test*	
ASW	5.4 mM Ca^{2+}	3.4 mM Ca^{2+}
24.3	25.6	22.4
2.0 mM Ca^{2+}	1.4 mM Ca^{2+}	0.7 mM Ca^{2+}
30.5	36.9	46.1

Values not underscored are significantly different. P = 0.05

*Mean number of statoliths per rhopalium.

Table 2. Statolith synthesis in ephyrae developing in low Ca^{2+} ASW supplemented with (A) strontium or (B) phosphate (330 Jf in 6 tests)

	Duncan's Multiple Range Test*	
A. ASW	1.4 mM Ca^{2+}	0.75 mM Sr^{2+} 1.4 mM Ca^{2+}
24.3	36.9	34.3
ASW	0.7 mM Ca^{2+}	0.75 mM Sr^{2+} 0.7 mM Ca^{2+}
24.3	46.1	62.1
B. ASW	1.4 mM Ca^{2+}	0.72 mM PO_4^{3-} 1.4 mM Ca^{2+}
20.0	40.3	51.5
ASW	0.7 mM Ca^{2+}	0.72 mM PO_4^{3-} 0.7 mM Ca^{2+}
23.3	51.9	62.2

All values are significantly different p = 0.05

*Mean number of statoliths per rhopalium

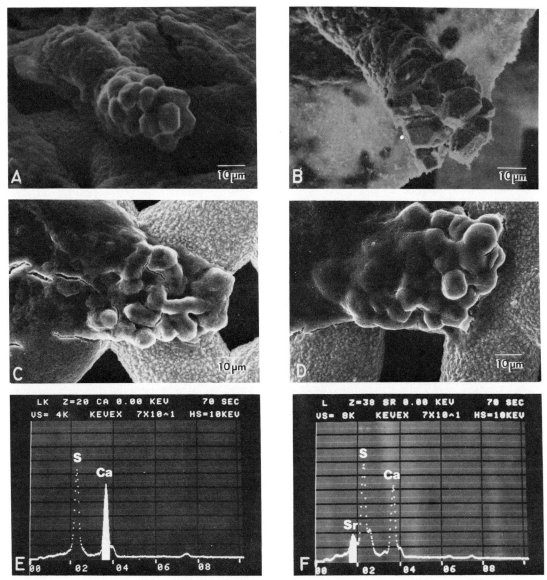

Figure 1. SEM and energy dispersive x-ray analysis of statoliths from ephyrae which has developed in: A. ASW; B. 1.4 mM Ca^{2+} + PO_4^{3-}; C. 1.4 mM Ca^{2+}; D. 1.4 mM Ca^{2+} + 0.75 mM Sr^{2+}; E. ASW; F. ASW + 0.75 mM Sr^{2+}.

causes an increase in statolith numbers in ephyrae which had developed in it[8]. Microscopic examination of statoliths of organisms from low Ca^{2+} supplemented with phosphate reveals that these statoliths are more pointed and somewhat larger than controls (Fig. 1B).

Ephyra development: General morphology of ephyrae which had developed in low Ca^{2+} ASW is normal in those organisms from the ASW containing 5.4 mM and higher concentrations of calcium. Organisms from the 3.4 mM Ca^{2+} ASW have arms which are slightly ruffled in appearance. Ephyrae from the ASW containing 2.0 mM of Ca^{2+} have a slow pulsing rate and they cannot swim. Morphologically the ruffling of the arms and lappets is more pronounced as compared with those of organisms from ASW of higher Ca^{2+} concentration. This deformity is more severe in organisms which had developed in ASW containing 1.4 mM Ca^{2+}. These organisms are extremely ruffled (Fig. 2C) and often exhibit folding in half either toward the oral region or

backwards (Fig. 2D), and they cannot pulse or swim. Large clusters of microbasic heterotrichous eurytele nematocyts are found on the umbrella side of these ephyrae (Fig. 2D). The ephyrae from ASW containing only 0.7 mM of Ca^{2+} are nearly unrecognizable as being ephyrae. These organisms look like rumpled balls with recognizable mouths and collections of statocysts containing large numbers of statoliths around the bell. Some of these statolith sites have recognizable rhopalia but many do not. Some organisms are found from the 1.4 mM and the 0.7 mM groups with fewer than normal arms (Fig. 2B) and some with as many as three rhopalia per arm.

Supplementation of the low Ca^{2+} ASW with Sr^{2+} or with PO_4^{3-} failed to improve the general morphology of the ephyrae developing in it nor did it enable the ephyrae to pulse. The general rate of development and the size of the ephyrae, however, was improved in the organisms which had developed in the phosphate supplemented ASW.

Discussion

Statolith Synthesis: Exposure of Aurelia polyps to low Ca^{2+} ASW during strobilation and ephyra development yielded some interesting and unexpected experimental results. It was not possible to prevent synthesis of gypsum statoliths by reducing the number of available calcium ions in the ASW. In fact, in a very low concentration of Ca^{2+} (0.7 mM) in the ASW, ephyrae developed with dramatically high numbers of statoliths. These statoliths were usually smaller than those of animals developing in sea water with standard amounts of calcium. Statoliths do not grow to a normal size in animals developing in a low calcium environment and therefore, perhaps, the animals synthesize more statoliths as a compensatory measure. Formation of high statolith numbers in these organisms suggest that either more calcifying vesicles are synthesized or more calcifying vesicles are mineralized than in control organisms but this aspect has not yet been investigated.

Addition of Sr^{2+} or PO_4^{3-} to the low Ca^{2+} ASW caused further enhancement of statolith synthesis in these organisms. The fact that Sr^{2+} is incorporated into the statoliths and phosphate apparently is not (based on the survey of statoliths from organisms exposed to Sr^{2+} and PO_4^{3-} using energy X-ray dispersive analysis), suggests that their mechanisms of statolith enhancement may differ. Stontium incorporation into the statoliths of ephyrae indicates that Aurelia medusae in nature may also incorporate radioactive strontium into their statoliths. These organisms, therefore, may be good indicator organisms for readily ascertaining the levels of radioactive strontium fallout in marine environments.

Low calcium diets in a variety of higher organisms have been reported to cause osteoporosis[9]. This effect is especially dramatic in lactating rats according to deWinter and Steendijk[10]. De Luca[11] found that a low calcium diet markedly stimulates intestinal calcium transport. This transport effect is reported to result from the stimulation of parathyroid hormone which in turn stimulates the production of $1-25-(OH)_2$ D_3 (vitamin D metabolite) which causes increased calcium transport. The low calicum ASW may be stimulating an increased calcium uptake in the Jf rhopalia through a similar type of hormonal stimulus.

Ephyra Development: Ephyrae which had developed in low Ca^{2+} ASW revealed a bizarre morphology (Fig. 2). Lappets and arms were often curled to give organisms a ruffled appearance. Some ephyrae developed with reduced numbers of arms and multiple rhopalia per arm (Fig. 2). Large clumps of microbasic heterotrichous euryteles were found on the umbrella surface of the organisms. In many, the bodies were bent in the oral direction or backward to give a butterfly appearance.

Morphological aberrations of ephyrae developing in low Ca^{2+} ASW could have several causes. Cell migration and adhesion are probably important factors in the reorganization of strobilae necessary for ephyra development. Calcium is necessary for normal cell motility and adhesion[12,13]. When Jf are unable to obtain sufficient calcium intracellularly during strobilation (as in a low calcium ASW) aberration of form could result. Indeed, the formation of unusually large clusters of nematocysts in the ephyrae may have resulted from an inability of these nematocysts to migrate from sites of synthesis into the body at large. Another possible cause of aberration of form in the ephyrae could be an abnormal synthesis of mesogloea due to the calcium deficiency.

Pulsing inability of the ephyrae in the low Ca^{2+} ASW could be caused by the lack of sufficient calcium for neuro-muscular functioning or for muscular contraction. Calcium is reported to be essential as a current carrier for excitation-contraction coupling in cardiac and smooth muscle and for excitation secretion coupling at presynaptic endings[14]. It has been known for some time that calcium controls muscular contraction and relaxation of the sarcoplasm[15]. Although chemical contraction mechanisms have not been studied in Aurelia striated muscle, it is likely that calcium is required for the highly active contractile muscle system. Insufficient availability of calcium, therefore could result in the inability of this organism to pulse.

The ability of rhopalia to synthesize abnormally high numbers of statoliths in organisms which have severe morphological abnormalities due to calcium deficiency emphasizes that calcium uptake and/or concentration mechanisms are different in rhopalia than in the rest of the organism and it also indicates that rhopalia do not serve as calcium storage sites from which stored calcium is distributed throughout the organism when needed.

Efforts to reverse the morphological abnormalities caused by calcium deficiency in the ephyrae with supplementation of low Ca^{2+} ASW with Sr^{2+} and PO_4^{3-} were unsuccessful. Failure of PO_4^{3-} to alleviate low Ca^{2+} morphological effects on ephyrae was particularly interesting since phosphate was found to improve morphological defects in Aurelia caused by low sulfate deficiency wherein statolith numbers and nematocyst numbers were improved[4,3] In other systems, phosphate has been reported to enhance calcium uptake. Sherman and Pappenheimer[16] found that the addition of phosphorus as K phosphate to a lower calcium diet accelerated calcium uptake in rats, and Shohl[17] found that phosphorus uptake is a limiting factor in the control of the calcium content of the body in the rat. In cardiac sarcoplasmic reticulum, Wasserman and Kallfelz[18] report that the amount of calcium accumulated is directly related to the external calcium concentration and is facilitated by oxalate or inorganic phosphate. More recently, the importance of phosphate in promoting calcium uptake into mitochondria has been emphasized by Lehninger et al [19] and LeBlanc et al [20].

In conclusion, this research demonstrates that the simple developmental model of metamorphosing Aurelia offers a unique system for the study of basic roles of calcium in mineralization and in organismal development. The fact that strobilation is able to proceed in a severely calcium deficient environment is remarkable even though the resultant ephyrae are abnormal in form and in their mineralizing ability. Further investigations into the specific mechanisms whereby calcium deficiency causes ephyra abnormalities could lead to a better understanding of the specific actions of calcium in normal and abnormal mineralization and in organismal development.

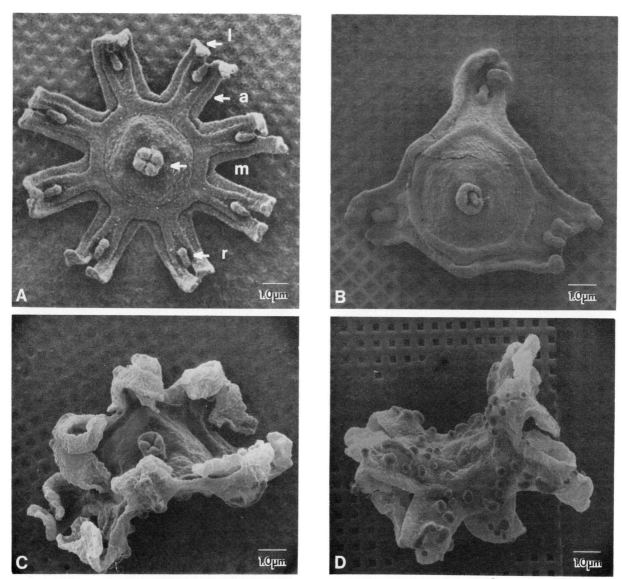

Figure 2. SEM of normal ephyra (A) and ephyrae which had developed in low CA^{2+} ASW (1.4 mM). Note lack of arms and extra rhopalium of B; ruffling of arms and lappets of C; and ruffling and large nematocyst clusters of D. Structures of A: a-arm; l-lappet; m-mouth; r-rhopalium.

ACKNOWLEDGEMENTS: The author gratefully acknowledges the financial support of the National Institute for Dental Research, Grant #DE 05048-01 and the Department of Energy (Contract #EE-77-05-5593). She is also very grateful to the following people for their technical assistance: Tim Aardrup, Kirk Ives, Bill Kuenning, Michal Patten, Lloyd Rhodes, and Vera Wang.

BIBLIOGRAPHY

1. Spangenberg, D.B. Iodine induction of metamorphosis in Aurelia. J. Exp. Zool. 165 (3), 1967, 441-450.
2. Spangenberg, D.B. Thyroxine induced metamorphosis in Aurelia. J. Exp. Zool. 178, 1971, 183-194.
3. Spangenberg, D.B., and Beck, C.W. Calcium sulfate dihydrate statoliths in Aurelia. Trans. Am. Microsc. Soc. 87, 1968, 329-335.
4. Spangenberg, D.B. Statolith differentiation in Aurelia aurita. J. Exp. Zool. 169, 1968, 487-500.
5. Spangenberg, D.B. New observations on Aurelia. Trans. Am. Microsc. Soc. 83, 1964, 448-455.
6. Duncan, D.B. Multiple range and multiple F tests. Biometrics 11, 1955, 1-42.
7. Spangenberg, D.B. Intracellular statolith synthesis in Aurelia aurita. In International symposium on the mechanisms of mineralization in invertebrates and plants. U. of South Carolina Press, Columbia, USA 1976, 231-248.
8. Spangenberg, D.B., and Kuenning, W. SEM studies of strobilating Aurelia. In Coelenterate Ecology and Behavior. Plenum Press: New York, USA, 1976, 377-386.
9. Sevastikoglou, J.A., Thomaidis, V. Th., and Lindholm, T.S. Reversibility of osteoporosis induced in adult rats by calcium deficiency. Long and short-term observations. Calcif. Tiss. Res. 22(S), 1977, 260-265.

10. deWinter, F.R., and Steendijk, R. The effect of a low-calcium diet in lactating rats; observations on the rapid development and repair of osteoporosis. Calcif. Tiss. Res. 17, 1975, 303316.

11. Deluca, H. Vitamin D and calcium transport, in Calcium Transport and Cell Function. Annals of N.Y. Acad. Sci. Scarpa, A., and Carafoli, E. (Eds) 307, 1978, 356-375.

12. Taylor, D.L. Dynamics of cytoplasmic structure and contractility. In: International Cell Biology, 1976-1977. Rockefeller U. Press: New York. USA, 1978, 367-377.

13. Loewenstein, W.R. On the genesis of cellular communication. Devel. Biol. 15, 1967, 503-520.

14. Brown, A.M., Ahaihe N., and Lee K.S. The calcium conductance of neurons, in Calcium Transport and Cell Function. Annals of N.Y. Acad. Sci. Scarpa, A., and Carafoli, E. (Eds) 307, 1978, 330-344.

15. Rodan, G.A. Cellular functions of calcium in Calcium and Phosphorous Metabolism. Academic Press, New York, USA, 1973, 187206.

16. Sherman, H.C., and Pappenheimer, A.M. Experimental rickets in rats; diet producing rickets in white rats and in prevention by addition of an inorganic salt. J. Exp. Med. 34, 1921, 189.

17. Shohl, A.T., Brown, H.B., Chapman, E.E., Rose C.S., and Saurwein, E.M. Evaluation of phosphorous deficiency of rickets-producing diet. J. Nutr. 6(27), 1933, 271-284.

18. Wasserman, R.H., and Kallfelz, F.A. Vitamin D_3 and unidirectional calcium fluxes across the rachitic chick duodenum. Am. J. Phys. 203, 1962, 221-224.

19. Lehninger, A.L., Carafoli, E., and Rossi, C.S. Active ion transport by mitochondria. Protoplasma 63, 1967, 90-94.

20. Leblanc, P., Bourdain, M. and Clauser, H. A specific ADP requirement in the course of Ca^{++} and phosphate accumulation in mitochondria. Biochem. Biophys. Res. Commun. 40, 1970, 754-762.

Discussion with Reviewers

N. Watabe: What was the salinity of each experimental ASW? Could the effects of low Ca ASW be due to the salinity and not necessarily due to low Ca?

Author: We routinely use ASW of a total salt concentration of 435.5 mM. In the past, we have lowered the total salinity (by reducing the NaCl concentration) to 271.5 mM without causing abnormality of ephyra morphology or of statolith number or form. The lowest salinity modification in our low calcium experiments using a calcium concentration of 0.7 mM concentration resulted in a total salinity concentration of 428.7 mM, not a significant reduction in total salt concentration and certainly not enough of a change to affect ephyra morphology or statoliths.

N. Watabe: You stated in the DISCUSSION that low calcium ASW is perhaps stimulating an increased calcium uptake in the jellyfish. Doesn't this statement contradict with the later discussion that morphological and pulsing abnormalities of the animals developed in the low Ca ASW are the results of the inability of the animals to obtain sufficient calcium? Do you have the data of calcium concentration in the organisms in each experiment?

Author: The statement that low calcium ASW is stimulating an increased calcium uptake in the jellyfish is in reference to the rhopalium and not the total organism. The total organisms have symptoms of calcium insufficiency while the rhopalia and prerhopalia regions show evidence of increased calcium uptake and hypermineralization because of the excessive numbers of statoliths formed. We did not quantitate calcium concentrations in these experiments.

N. Watabe: You discussed the roles of phosphate in calcium uptake. What is your speculation on why phosphate was effective in recovering morphological abnormalities in sulphate deficient media but not in calcium deficient media?

Author: My response to this question must indeed be speculative as I do not know the answer. Perhaps phosphate stimulates sulfate uptake and the calcium is taken into the organism simultaneously with the sulfate (through ion pairing or some other mechanism). In the low sulfate-phosphate effect, there is plenty of available calcium to enter the organisms with the sulfate and to effect statolith synthesis. In the low calcium-phosphate situation, the low calcium concentration is rate limiting and insufficient calcium is taken in for statolith growth. A sufficient amount is available however for increased numbers of small statoliths to form.

R. S. Blanquet: In your statocyst experiments, I noticed that the proportions of calcium to strontium or phosphate differ for each run. Could you comment on possible competitive or enhancement effects between these ions and how these might relate to your results?

Author: The proportions of calcium to strontium or to phosphate did not differ for each run. We ran three identical tests in which the proportions were: 1.4 mM Ca: 7.2 mM PO_4; 0.7 mM Ca: 7.2 mM PO_4; 1.4 mM Ca: 7.5 mM Sr; and 0.7 mM Ca: 7.5 mM Sr. I would not expect competitive effects between calcium and phosphate but it is possible that Sr competes with Ca. Phosphate has shown a statolith enhancement effect in organisms developing in ASW, indicating a calcium enhancement uptake. Strontium may compete with Ca^{2+} in being incorporated into the statolith crystal but it apparently does not cause a reduction of statoliths formed in the low calcium environment.

For additional discussion see page 432.

SCANNING ELECTRON MICROSCOPY/1979/II
SEM Inc., AMF O'Hare, IL 60666, USA

COMPARATIVE SHELL MICROSTRUCTURE OF THE MOLLUSCA, BRACHIOPODA AND BRYOZOA

J. G. Carter

Department of Geology
University of North Carolina
Chapel Hill, NC 27514

Abstract

Molluscs differ from brachiopods and bryozoans in their greater diversity of aragonitic and calcitic shell microstructures. This greater diversity may be related to their higher level of integration between shell microstructure and shell mechanical design. Microstructural evolution may be limited in brachiopods by their retention of more or less unicellular controls over the size, shape, and orientation of their major microstructural units. For this reason, brachiopods may have lacked the evolutionary potential to form extremely large diameter microstructures, such as crossed lamellae, that are uniformly oriented with respect to directional shell mechanical properties.

Because molluscs can form smaller diameter and relatively unoriented as well as larger diameter and directionally oriented microstructures, there is little basis for regarding any brachiopod shell microstructure as diagnostic of non-affinity with the molluscs. Thus, considered alone, the presence of crossed bladed structure in fossil tentaculitids cannot be regarded as conclusive evidence for closer phylogenetic affinity with brachiopods than with molluscs. The occurrence of uniformly directionally oriented crossed lamellar structure in fossil hyolithids supports evidence of larval development suggesting their close relationship with the Mollusca. No other mantled invertebrate phylum except the Mollusca shows comparable evidence for a high level of integration between microstructure orientation and shell mechanical design.

KEY WORDS: Shell Microstructure, Mollusca, Brachiopoda, Bryozoa, Tentaculitids, Hyolithids, Evolution

Introduction

Despite the considerable attention paid to shell microstructure in the Mollusca, Brachiopoda and Bryozoa in recent years, few attempts have been made to evaluate these data for application to problems of phylogenetic affinity of poorly understood fossil groups. Of the two recent articles concerning phylum-level phylogenetic significance of shell microstructure, one concluded that crossed lamellar structure is non-diagnostic of membership in the Mollusca,[1] and the other concluded that crossed bladed structure suggests close phylogenetic affinity with the Brachiopoda.[2] The present paper compares shell microstructures among the Mollusca and the two shelled lophophorate phyla (Brachiopoda and Bryozoa) and interprets some of the differences from the perspective of different modes of shell secretion.

Materials and Methods

The data for this paper are based largely on a comprehensive review of molluscan, brachiopod, and bryozoan shell microstructures recently undertaken by the writer for the purpose of defining a uniform classification and nomenclature of shell microstructures. Most of the present microstructure terms have been proposed by previous investigators, but they are restated or redefined slightly here on the basis of uniform micromorphologic criteria. Each microstructure is considered to have, at least theoretically, both aragonitic and calcitic mineralogical analogs. The present microstructure definitions differ from some previous definitions in their independence from genetic and optical crystallographic criteria. The accompanying figures are based on SEM of acetate-peeled surfaces of embedded, sectioned, polished and acid-etched specimens, or are simple fracture surfaces. Surfaces prepared by acetate peeling were initially imbedded in Epon 815 epoxy resin, sectioned, polished, etched for 3 seconds in 5% HCl, air dried, flooded with acetone, covered with an acetate slab, and the dried slab was then peeled off the surface to reveal the etched shell section. These and other surfaces were coated with gold-palladium using a Polaron E5100 sputter coater, and then examined with an ETEC Autoscan

SEM. The reader is referred to Carter[3] and Carter and Tevesz[4] for more complete definitions of the microstructure terms used here.

Explanations of Symbols in Figures

In Figs. 2-21 the thick and thin arrows indicate the direction toward the depositional surface and toward the shell margin, respectively. The letters indicate the viewing perspective, i.e., radial (R), transverse (T), horizontal (H), and oblique (O). The abbreviations USNM, UNC and YPM represent the United States National Museum, the University of North Carolina, and the Yale University Peabody Museum of Natural History, respectively.

Results

The most striking conclusion of the present survey is the microstructural conservatism of brachiopods and bryozoans. The only major difference in calcitic microstructure noted between brachiopods and bryozoans is the apparent restriction of well developed crossed bladed structure to strophomenid brachiopods[5,6] and the slightly more regular development of regularly foliated structure among certain bryozoans.[7] Bryozoans show a few aragonitic microstructures not presently known in brachiopods (e.g., spherulitic prismatic and irregularly nacreous), but data are unknown for comparison with the rare aragonitic brachiopods[8]. A highly irregular nacreous structure occurs in certain laminar inarticulate brachiopods (see below), but this is calcitic rather than aragonitic.

Whereas bryozoans and articulate brachiopods are predominantly calcitic, molluscs are predominantly aragonitic, and they have evolved largely calcitic shells generally through the suppression of ancestral aragonitic shell layers. Most molluscs retain some aragonite secretion, if only in their larval shell or at sites of shell-muscle attachment (i.e., in myostracal deposits).[9,10,11] Molluscan aragonitic microstructures include analogs of virtually every aragonitic and calcitic microstructure known in lophophorate shells, with the exception of well developed calcitic crossed bladed and calcitic regularly foliated structures. But aragonitic analogs of these two structures are approximated in the lamello-fibrillar structure of certain endocochleate cephalopods,[12] and in the "fibrous structure" described for certain bivalves by Kobayashi,[13] respectively. In addition, calcitic crossed bladed structure is not uncommonly approximated in certain bivalves as a local variation in their calcitic crossed lamellar and calcitic regularly foliated structures.

The greater diversity of molluscan microstructures cannot be attributed solely to a mineralogic effect. Relatively few molluscs secrete calcitic shells, but their calcitic microstructures are still more diverse than those in the largely calcitic Brachiopoda and Bryozoa combined (Fig. 1). Other differences between molluscan and lophophorate shell secretion become apparent when we compare their major categories of shell microstructure in detail.

Prismatic structures are similarly developed in molluscs, brachiopods and bryozoans in the form of smaller irregularly shaped simple prisms, smaller spherulitic prisms, and fibrous prisms (Figs. 2-4). Aragonitic irregular simple prisms are characteristic of molluscs at their sites of shell-muscle attachment (Fig. 2), but comparable layers in brachiopod and bryozoan shells are generally calcitic, and vary from prismatic to more or less structureless (i.e., homogeneous).[7,14-16]

Brachiopods and bryozoans typically lack larger diameter simple prisms as well as composite prisms radially aligned with respect to their shell margins, despite the widespread occurrence of these structures in calcite or aragonite (respectively) among certain bivalved molluscs (Figs. 5, 6). Brachiopod and bivalve shell calcitic fibrous prisms generally differ in their variable versus uniform orientation with respect to the shell margins. The variable orientation of brachiopod fibrous prisms has been attributed to the path of migration of their mantle secretory cells over the shell interior.[17,18]

Nacreous structures consisting of persistent laminae and regular tablets and tablet stacking modes are restricted to the Mollusca. The unusual row stack nacre of *Pinna* shows elongate tablets uniformly oriented with respect to the shell margins (Fig. 7), probably for the purpose of producing directional flexibility in the shell posterior.[3] Other varieties of molluscan nacreous structure, e.g., sheet nacre and columnar nacre, are morphologically more or less nondirectional in the plane of the depositional surface (Figs. 8, 9). The reader may note that nacreous luster occurs in many molluscs and brachiopods, and not always in association with nacreous microstructure.[19]

A form of nacreous structure occurs in certain "laminar" bryozoans and brachiopods,[20,21] but never with the regularity characteristic of most molluscan nacre. Aragonitic sheet nacreous structure in the bryozoan *Cupuladria* consists of flat-lying orthorhombic "prisms" with a poorly defined laminar arrangement, and occasionally showing spiral growth scarps (see Pl. 13, Fig. 45 in Tavener-Smith and Williams[7]). Calcitic sheet nacreous layers in certain bryozoans and inarticulate brachiopods show similar irregular horizontal laminae with relatively small, randomly stacked tablets commonly with spiral growth scarps.[7,14,22]

Crossed bladed structure is well developed as a calcitic layer in certain strophomenid brachiopods (Fig. 10).[5,6] This microstructure, like fibrous prismatic structure, has been described by previous workers as formed by unicellular secretory activity in brachiopods.[5] Like lophophorate fibrous prisms, the blades in crossed bladed structure are not uniformly oriented with respect to the shell margins, and they may change their directions from one part of the shell to the next.

Approximations of crossed bladed structure occur as local variations within other microstructures in the Mollusca. The closest molluscan analog occurs in the inner part of the outer

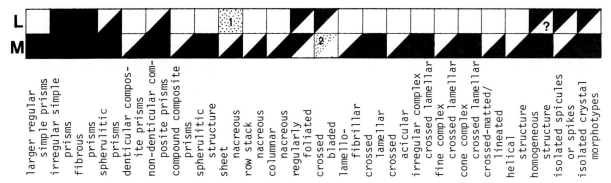

Figure 1. Distribution of shell microstructures by mineralogy in the lophophorates (L) and molluscs (M) in aragonitic (▟) and calcitic (◨) shell layers. Microstructures not defined in the text are described by Carter.[3,4]

layer of the bivalve Ctenoides, where crossing of its bladed microfabric is apparent (Fig. 11). Faint ridges and grooves occur on and between the calcitic blades in the regularly foliated structure of the bivalve Plicatula (Fig. 12). This lacks the strong alternation of blade directions typical of well developed crossed bladed structure, but it may be noted that many crossed bladed shell layers in brachiopods show a very low blade angle between successive laminae. A change of blade direction occurs in the crossed lamellar structure of Propeamussium (Fig. 13). In this case the blades lack the superficial ridges characteristic of well developed crossed bladed structure. But such ridges are not invariably well developed in brachiopod crossed bladed structure. Crossing of elongate structural units occurs in adjacent laminae in the lamello-fibrillar structure of certain cephalopods,[12] but this microstructure is aragonitic and consists of fibers rather than blades. These comparisons indicate that crossed bladed structure is approximated in modern molluscs, but that it is not well developed here because the associated calcitic crossed lamellar and calcitic regularly foliated structures are simply mechanically more adaptive.

Regularly foliated structure occurs in several calcitic bivalve and gastropod molluscs,[11,23] but only rarely occurs in calcitic bryozoans and brachiopods (Figs. 116, 117, 122 in Tavener-Smith and Williams[7]) (Fig. 12). Brachiopod regularly foliated structures are generally limited to local variations within fibrous prismatic or crossed bladed shell layers. As with the other microstructures so far considered, regularly foliated structure commonly shows a uniform dip direction with respect to the shell margins in molluscs, but not in brachiopods and bryozoans.

Crossed lamellar structure has evolved several times in the Mollusca in aragonite and calcite, but this structure is without parallel in the lophophorates. Unlike crossed bladed structure, the major structural units in crossed lamellar structure (the first order lamellae) are more or less vertically oriented, and their structural subunits are weakly to strongly dipping relative to the surface of deposition. By definition, crossed lamellar structures show only two predominant dip directions of their subordinate structural units relative to the shell margin (Figs. 13-

15). As observed on the surface of deposition, the first order lamellae may vary considerably in size and shape, and in the case of larger first order lamellae, these are clearly much longer and wider than the cells of the adjacent mantle epithelium.

Complex crossed lamellar structure is analogous to crossed lamellar structure except for the occurrence of three or more predominant dip directions of its subordinate structural units. These structural subunits may be arranged in parallel or as stacked cones. Fine complex crossed lamellar structure consists of individual crystallites or aggregations of only a few parallel structural subunits with three or more uniform dip angles. The aggregations are large and sometimes clearly visible to the naked eye in the case of certain irregular and cone varieties of complex crossed lamellar structure (Figs. 16-18). All of these varieties of complex crossed lamellar structure are presently unknown in the Brachiopoda and Bryozoa.

Homogeneous structure, or an irregular aggregation of minute elongate, rounded, or irregularly flattened crystallites, occurs only infrequently and sporadically in the Mollusca, Brachiopoda and Bryozoa. This structure is presently known to comprise a thicker shell layer only in certain molluscs (Fig. 19). Homogeneous structure occurs in a few molluscs and lophophorates as local variations in other structures.

Spherulitic structure differs from spherulitic prismatic structure in the more or less equidimensional shape of its constituent spherulites. Aragonitic spherulitic structure seldom comprises a major shell layer in molluscs. Aragonitic and calcitic spherulitic layers are presently unknown in the Brachiopoda and Bryozoa.

Helical structure consists of nested, helically coiled elongate structural units arranged with their helix axes more or less perpendicular to the surface of deposition. This structure occurs in certain aragonitic pteropod gastropods,[24] but is unknown in other aragonitic or calcitic molluscs, brachiopods or bryozoans. Calcitic helical structure is only roughly approximated in certain brachiopods where their fibrous prisms twist in a partial spiral as a result of migration of the mantle epithelium over the surface of deposition.[18]

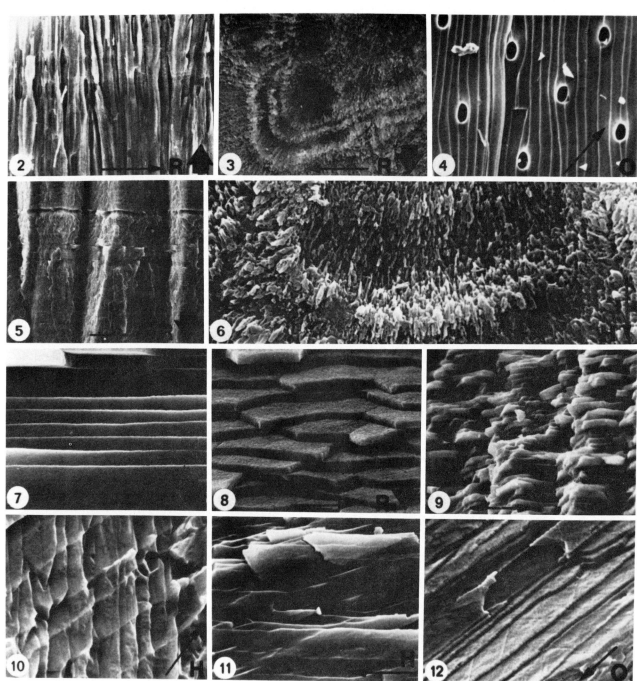

Figure explanations. All of these figures (2-21) are scanning electron micrographs of fracture surfaces except Figs. 2, 3, 6, 14-17, which are acetate peeled sections, and Fig. 20, which shows spikes freed from the periostracum by dissolution in 5% NaOCl. Fig. 2: aragonitic irregular simple prismatic structure; inner layer of Lithophaga nigra (Orbigny), Bivalvia, Florida, YPM 9989. Photograph width = 75 μm. Fig. 3: aragonitic spherulitic prismatic structure; outer layer of Spisula solidissima Dillwyn, Bivalvia, New Jersey, UNC 7034. Photograph width = 20 μm. Fig. 4: calcitic fibrous prismatic structure; inner layer of Laqueus sp., Brachiopoda, California, UNC 5134. Photograph width = 120 μm. Fig. 5: aragonitic regular simple prismatic structure; outer layer of Parapholas californica (Conrad), Bivalvia, California, YPM 10258. Photograph width = 110 μm. Fig. 6: aragonitic composite prismatic structure; outer layer of Dosinia troscheli Lischke, Bivalvia, North China coast, USNM 217994. Photograph width = 16 μm. Fig. 7: aragonitic row stack nacreous structure; inner layer of Pinna bicolor Gmelin, Bivalvia, Philippines, YPM 6964. Photograph width = 7 μm. Fig. 8: aragonitic sheet nacreous structure; inner layer of Pinctada radiata Leach, Bivalvia, Bahamas, YPM 6889. Photograph width = 7 μm. Fig. 9: aragonitic columnar nacreous structure; middle layer of Neotrigonia gemma Iredale, Bivalvia, Australia, UNC 5427. Photograph width = 40 μm. Figure explanations continued below.

442

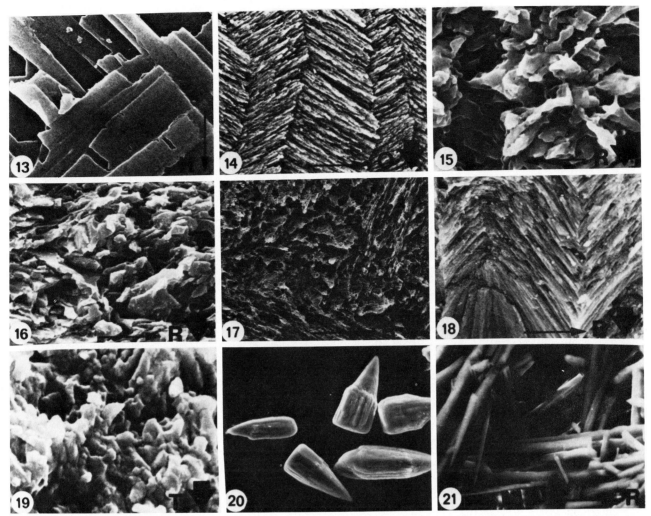

Figure explanations, continued. Fig. 10: calcitic crossed bladed structure, inner layer of <u>Strophomena</u> <u>planumbona</u> Hall, Brachiopoda, Ordovician of Ohio, UNC 7173. Photograph width = 18 μm. Fig. 11: approximation of crossed bladed structure in the calcitic largely regularly foliated portion of the outer layer of <u>Ctenoides scabra</u> (Born), Bivalvia, Florida, YPM 9702. Photograph width = 14 μm. Fig. 12: calcitic regularly foliated structure showing inter-blade grooves and crossing by faint oblique grooves; outer layer of <u>Plicatula gibbosa</u> Lamarck, Bivalvia, West Indies, YPM 9631. Photograph width = 8 μm. Fig. 13: calcitic crossed lamellar structure; left valve of <u>Propeamussium dalli</u> (Smith), YPM 8387. Photograph width = 12 μm. Fig. 14: aragonitic crossed lamellar structure; right valve of <u>Propeamussium dalli</u> (Smith), YPM 8387. Photograph width = 60 μm. Fig. 15: transitional structure between aragonitic crossed acicular and aragonitic crossed lamellar; middle layer of <u>Arctica islandica</u> Linnaeus, Bivalvia, New Jersey, UNC 7033. Photograph width = 8 μm. Fig. 16: aragonitic fine to irregular complex crossed lamellar structure; middle layer of <u>Geloina</u>? sp., Bivalvia, YPM 9701. Photograph width = 8 μm. Fig. 17: aragonitic irregular complex crossed lamellar structure; inner layer of <u>Geloina</u>? sp., Bivalvia, YPM 9701. Photograph width = 80 μm. Fig. 18: aragonitic cone complex crossed lamellar structure; inner part of the outer layer of <u>Spisula solidissima</u> Dillwyn, Bivalvia, New Jersey, UNC 7034. Photograph width = 30 μm. Fig. 19: aragonitic homogeneous structure; inner layer of <u>Cooperella subdiaphana</u> (Carpenter), Bivalvia, California, YPM 9645. Photograph width = 5 μm. Fig. 20: aragonitic periostracal spikes freed from the posterior periostracum of <u>Spengleria rostrata</u> (Spengler), Bivalvia, Florida, YPM 9473. Photograph width = 200 μm. Fig. 21: aragonitic isolated crystal morphotypes from between the periostracum and the outer layer of the shell proper in the posterior of <u>Spengleria rostrata</u> (Spengler), Bivalvia, Florida, YPM 9473. Photograph width = 70 μm.

Isolated spicules or spikes are variably shaped minute structures sparsely distributed within a cuticle, periostracum, or connective tissue layer. Aragonitic spikes or spicules are secreted in the cuticle of aplacophoran molluscs, in the girdle cuticle of chitons, and in the periostracum of certain bivalves (Fig. 20).[25,26] Similar cuticular or periostracal structures are presently unknown in the Brachiopoda and Bryozoa. Calcitic spicules occur in the connective tissues of the lophophore and mantle of certain brachiopods.[27] Calcified spicules (supporting rods) occur in the gills of certain unionacean and trigoniacean bivalves, but their mineralogy is presently unknown.[28]

Several malacologists have suggested that the spicule-like cuticular calcification of modern aplacophoran molluscs represents a primitive mode of calcification.[25,29] If this view is correct, then the earliest shelled molluscs may have resembled brachiopods in their largely unicellular controls on the size, shape and arrangement of their shell microstructural units. According to Hoffman,[30] cuticular spicule secretion in the aplacophoran Proneomenia is essentially a unicellular process. It may also be noted that shell plates in the chiton Middendorffia are apparently initiated by secretion of several rows of more or less isolated granules.[29] But post-larval shell plate formation in this and other chitons is probably a tissue-grade activity, judging from the large size and uniform orientation of their crossed lamellar structures. It is presently unknown whether the aragonitic periostracal spikes in bivalves (Fig. 20) represent unicellular or tissue-grade mantle secretory activity.

Isolated crystal morphotypes are sparsely distributed and irregularly oriented crystal forms occurring as single crystals, as twins, or as mutually isolated spherulites (Fig. 21). These structures are sporadically developed in pockets and other voids filled with pallial fluid but isolated from direct contact with the mantle epithelium.[31] Although presently known only in a few molluscs, they never comprise entire shell layers, so their occurrence is of little phylogenetic significance at lower taxonomic levels.

Discussion

The fact that all shell microstructures presently known in brachiopods are also developed in molluscs, although sometimes only poorly so, indicates that no lophophorate microstructure is diagnostic of non-affinity with the Mollusca. Only one microstructure is found in certain strophomenid brachiopods that is only poorly developed in mollusc shells, i.e., calcitic crossed bladed structure. Crossed bladed structure is presently known in the Mollusca only as a local variation within calcitic crossed lamellar and calcitic regularly foliated shell layers. However, these local variations show some of the basic features of crossed bladed structure as it appears in tentaculitids, including adjacent calcitic blades separated by grooves and crossed by ridges. The apparent absence of major crossed bladed shell layers in modern molluscs may simply reflect the evolution of more directionally oriented microstructures, such as calcitic crossed lamellae, which are mechanically superior to crossed bladed structure. However, there are insufficient data to generalize about the presence or absence of crossed bladed structure in early Paleozoic molluscs, because too little is presently known about the range of their microstructural adaptations.

It might be argued that well developed crossed bladed shell layers characterize certain brachiopods because this microstructure is relatively mechanically adaptive in the context of their relatively low level of integration of (1) cellular level microstructure controls and (2) coordination between microstructure orientations and shell mechanical design. Williams[32] demonstrated that the size, shape, and the orientation of the fibrous prisms in modern terebratulid and rhynchonellid brachiopods reflects the size, shape, and migration direction of secretory mantle epithelial cells. He also suggested that the morphology of the exposed portions of these calcitic fibers can be used to identify modifications in the mantle epithelium in fossil brachiopods. According to Armstrong[5] blade secretion in strophomenid crossed bladed structure is likewise controlled by more or less unicellular secretory activity. Evidence for this hypothesis is indirect, because no modern brachiopod is known to secrete crossed bladed structure. The indirect evidence lies in the fact that some crossed bladed shell layers grade laterally into nearly parallel structural units which resemble the fibrous prisms of other brachiopods. In addition, the paleontological record is compatible with the hypothesis that brachiopod crossed bladed structure evolved through rearrangement of ancestral fibrous prisms structurally like those in modern brachiopods. Brunton[6] indicated that some Devonian Productacea show variations in their fibrous prismatic shell layer that are transitional toward the crossed bladed structure of later Paleozoic productaceans. The fact that the blades in crossed bladed structure show superficial ridges that span several adjacent blades suggests that limited integration of cellular secretory activity may be involved in the secretion of this structure. The relatively large diameters of certain brachiopod fibrous prisms are likewise suggestive that some species show locally coordinated secretory activity of their mantle cells to form larger diameter structural units. However, it may be noted that extremely large diameter microstructural units (i.e., comparable to most simple prisms in the Bivalvia) are unknown in the Brachiopoda, despite the large size of many Brachiopod shells. The absence of larger diameter microstructures does not require that brachiopods show more or less unicellular shell secretory activity. However, considering their long and rich evolutionary history, and considering also their diverse shell morphologies and life habits, one might suspect that , had brachiopods evolved advanced levels of integration of secretory activity, they might have shown much larger microstructural units in at least a few representatives. It may be emphasized here that our knowledge of Paleozoic brachiopod microstructure is relatively extensive in comparison with Paleozoic molluscan microstructure. Even assuming that brachiopods are comparable to molluscs in their potential for the evolution of extremely large diameter microstructural units, they nevertheless show no sign of integration between the orientation of their microstructural units and adaptive shell microstructural design. Rather, their non-vertical microstructural units appear to be oriented haphazardly relative to the shell margins, even in crossed bladed structure. Uniformly oriented concentric and radial microstructures such as crossed lamellae are entirely unknown in any lophophorate, despite their well studied fossil record. It is also important to note that brachiopods and bryozoans do not even approximate the extremely large size and uniform orientation in the shell of molluscan crossed lamellar structure in any of their presently known microstructures. I regard this difference as suggestive that brachiopods do in fact lack a level of integration between shell microstructure and shell mechanical design comparable to modern molluscs. If the earliest molluscan shells evolved through direct modification of more or less unicellular mechanisms of shell secretion, such as occurs in modern aplacophorans,[26,29,30] then it is possible that the most primitive molluscan shell microstructures were likewise subject to largely unicellular controls. In this initial evolutionary context, crossed bladed structure may have been relatively more adaptive than it apparently is in modern shelled molluscs. In any event, further examination of Cambrian molluscan shell microstructure is clearly required before conclusions can be drawn about the range of their microstructural adaptations.

Based on our present knowledge of microstructural variability in the mantled phyla, the poorly developed crossed bladed structure described by Towe[2] in Tentaculites cannot be regarded as indicative of their phylogenetic affinity with the Mollusca or Brachiopoda. It may be noted here that portions of the crossed bladed structure in Tentaculites resemble certain molluscan foliated structures more than they do well developed brachiopod crossed bladed structure. For example, the microstructure illustrated by Towe[2] in his Fig. 1d is indistinguishable from portions of the regularly foliated shell layer presently illustrated for the mollusc Plicatula gibbosa (Fig. 12, this paper). Information regarding the mineralogy of the myostracal deposits in Tentaculites might be helpful for assessing their affinities, because molluscs typically show aragonitic myostracal deposits. However, Towe[2] has indicated that Tentaculites lacks undoubted muscle scars. The condition in Tentaculites may reflect evolutionary loss of the major shell musculature, and with it, either brachiopod calcitic or molluscan aragonitic myostracal microstructures. In this regard it may be noted that certain modern bivalves are known to have lost at least part of their ancestral aragonitic pallial myostracum associated with the evolution of a largely calcitic shell.[33]

Inasmuch as the directionally oriented crossed lamellar structure in hyolithids[1] reflects a high level of integration of shell microstructure with shell mechanical design, this group of aragonitic shelled invertebrates shows evidence for a close relationship with the Mollusca. However, it must be kept in mind that the crossed lamellar microstructure per se probably originated independently in the Hyolitha, as it has apparently done in the molluscan classes. The conclusion of close affinity between molluscs and hyolithids

is compatible with their likewise similar larval shell development.[34]

Acknowledgement

This study was supported by NSF grant DEB 77-00022.

References

1. B. Runnegar, J. Pojeta, Jr., N. J. Morris, J. D. Taylor, M. E. Taylor, and C. McClung, "Biology of the Hyolitha," Lethaia 8 (1975) 181-191.
2. K. M. Towe, "Tentaculites: evidence for a brachiopod affinity?," Science 201 (1978) 626-628.
3. J. G. Carter, Ph.D. Dissertation, Yale University, New Haven Conn. (1976).
4. J. G. Carter, and M. J. S. Tevesz, "Shell microstructure of a Middle Devonian (Hamilton Group) bivalve fauna from central New York," J. Paleontol. 52 (1978) 859-880.
5. J. D. Armstrong, "The crossed bladed fabrics of the shells of Terrakea solida (Etheridge and Dun) and Streptorhynchus pelicanensis Fletcher," Paleontology 12 (1969) 310-320.
6. C. H. C. Brunton, "The shell structure of chonetacean brachiopods and their ancestors," Bull. Br. Mus. (Nat. Hist.), Geol. 21 (1972) 1-26.
7. R. Tavener-Smith and A. Williams, "The secretion and structure of the skeleton of living and fossil Bryozoa," Phil. Trans. R. Soc. London B 264 (1972) 97-159.
8. V. Jaanusson, "Fossil brachiopods with possible aragonitic shell," Geol. Foren. Stockholm Foerh. 88 (1966) 279-281.
9. H. B. Stenzel, "Aragonite and calcite as constituents of adult oyster shells," Science 142 (1963) 232-233.
10. H. B. Stenzel, "Oysters: composition of larval shell," Science 145 (1964) 155-156.
11. J. D. Taylor, W. J. Kennedy, and A. Hall, "The shell structure and mineralogy of the Bivalvia. Introduction, Nuculacea-Trigonacea," Bull. Br. Mus. (Nat. Hist.) Zool., Suppl. 3 (1969) 1-125.
12. H. K. Erben, "Über die bildung und das Wachstum von Perlmutt," Biomineralisation 4 (1972) 15-40.
13. I. Kobayashi, "Internal shell microstructure of Recent bivalvian molluscs," Sci. Rept. Niigata Univ., Ser. E. Geol. and Min. 2 (1971) 27-50.
14. A. Williams and A. D. W. Wright, "Shell structure of the Craniacea and other calcareous inarticulate Brachiopoda," Palaeont. Assoc. Sp. Pap. in Palaeont. 7 (1970) 1-51.
15. D. I. Mackinnon, "The formation of muscle scars in articulate brachiopods," Phil. Trans. R. Soc. London B 280 (1977) 1-27.
16. J. D. Armstrong, "Microstructure of the shell of the brachiopod Wyndhamia clarkei," J. Geol. Soc. Aust. 17 (1970) 13-19.
17. A. Williams, "Growth and structure of the

shell of living articulate brachiopods,"
Nature (London) 211 (1966) 1146-1148.

18. M. Jope, "Constituents of brachiopod shells,"
in M. Florkin and E. H. Stotz, eds.,
Comprehensive Biochemistry, 26 C (1971)
749-782.

19. K. M. Towe and C. W. Harper, Jr., "Pholido-
strophid brachiopods: origin of the nacreous luster," Science 154 (1966) 153-
154.

20. D. Schumann, "Inaequivalver Schalenbau bei
Crania anomala, Lethaia 3 (1970) 413-421.

21. A. Williams, "Spiral growth of the laminar
shell of the brachiopod Crania," Calcif.
Tissue Res. 6 (1970) 11-19.

22. A. Williams, "Scanning electron microscopy
of the calcareous skeleton of fossil and
living Brachiopoda," in Haywood, ed.,
Scanning Electron Microscopy, Chapter 3
(1971) 37-66.

23. C. MacClintock, "Shell structure of patel-
loid and bellerophontoid gastropods
(Mollusca)," Peabody Mus. Nat. Hist. Yale
Univ. Bull. 22 (1967) 1-140.

24. A. W. H. Bé, C. MacClintock, and D. C.
Currie, "Helical shell structure and
growth of the pteropod Cuvierina colum-
nella (Rang) (Mollusca, Gastropoda),"
Biomineralisation 4 (1972) 47-79.

25. G. E. Beedham and E. R. Trueman, "The cuticle of the Aplacophora and its evolutionary significance in the Mollusca," J.
Zool., Lond. 154 (1968) 443-451.

26. J. G. Carter and R. C. Aller, "Calcification
in the bivalve periostracum," Lethaia 8
(1975) 315-320.

27. A. Williams, "A history of skeletal secretion among articulate brachiopods,"
Lethaia 1 (1968) 268-287.

28. D. Atkins, "On the ciliary mechanisms and
interrelationships of lamellibranchs.
VII. Latero-frontal cilia of the gill
filaments and their phylogenetic value,"
Q. J. Micros. Sci. 30 (1938) 345-436.

29. L. v. Salvini-Plawen, "Zur Morphologie und
Phylogenie der Mollusken: Die Beziehungen der Caudofoveata und der Solenogastres
als Aculifera, als Mollusca, und als
Spiralia," Z. Wiss. Zool., Leipzig 184
(1972) 205-394.

30. S. Hoffman, "Studien über das Integument der
Solenogastren nebst Bemerkungen über die
Verwandtschaft zwischen den Solenogastren
und Placophoren," Zool. Bidr. Uppsala 27
(1949) 293-427.

31. F. H. Wind and S. W. Wise, Jr., "Organic vs.
inorganic processes in archaeogastropod
shell mineralization," in N. Watabe and
K. M. Wilbur, The Mechanisms of Mineralization in the Invertebrates and Plants,
Univ. South Carolina Press, Columbia,
S. C. (1976) 369-387.

32. A. Williams. "Evolution of the shell structure of articulate brachiopods," Spec.
Pap. Palaeontol., Palaeontol. Soc. 2
(1968) 1-55.

33. H. B. Stenzel, "Oysters, composition of the
larval shell," Science 145 (1964) 155-156.

34. J. Dzik, "Larval development of hyolithids,"
Lethaia 11 (1978) 293-299.

Discussion with Reviewers

K. M. Towe: If, as you suggest, primitive mollusks had the ability to secrete cross-bladed
structure why have such molluscan fossils never
been found although many brachiopods with this
structure are commonly preserved?
Author: I suggested that well developed crossed
bladed structure may be mechanically superior to
modern molluscan calcitic structures only in the
evolutionary context of largely unicellular controls over the orientation of microstructural
units within the shell. Inasmuch as some molluscs had evolved uniformly oriented crossed
lamellar structure at least by Devonian time, I
would expect for well developed crossed bladed
structure to be found, if at all, only among
certain Lower Paleozoic molluscs. I am assuming
here that the earliest shelled molluscs were
evolutionarily transitional between aplacophorans
and modern monoplacophorans in their reliance
upon a largely unicellular mechanism of shell secretion. Obviously, we presently lack sufficient shell microstructure data on early Paleozoic molluscs to critically evaluate this hypothesis. I may point out, in this regard, that
crossed bladed structure is taxonomically restricted at the ordinal level among brachiopods.
Thus, had it not been for the fact that certain
Paleozoic strophomenid shells were described
microstructurally in the past 10 years, the apparent absence of crossed bladed structure in all
five orders of inarticulate brachiopods, in the
other six orders of articulate brachiopods, and
in the phyla Phoronida and Bryozoa might have
been taken to suggest that well developed crossed
bladed structure is absent from the Brachiopoda
as well as from the Mollusca.

N. Watabe: What is your definition of nacreous
structure? Why do you include calcitic layers
in this category?
Author: Nacreous structures are defined as showing laminae consisting of polygonal to rounded
tablets lying essentially parallel to the general
depositional surface. Spiral growth of the tablets may locally disrupt the laminar arrangement
of the structure. In the classification I employ
(reference #3, above) all microstructures are considered to have, at least potentially, both aragonitic and calcitic varieties. This arbitrarily
imposed independence of microstructure nomenclature from shell mineralogy facilitates comparisons
between phyla that may be strongly convergent in
some of their shell microstructures, yet generally
distinct in their shell mineralogy, such as the
Mollusca and Brachiopoda.

For additional discussion see page 456

SCANNING ELECTRON MICROSCOPY/1979/II
SEM Inc., AMF O'Hare, IL 60666, USA

THE EFFECTS OF MARINE MICROPHYTES ON CARBONATE SUBSTRATA

K. J. Lukas

Department of Geology and Geography
Vassar College
Poughkeepsie, NY 12601

Abstract

Marine microphytes are involved in constructive and destructive processes with respect to calcified tissues and to inorganic carbonate substrata. By changing the pH of seawater, and possibly by means of secreted metabolites, some pit and corrode carbonate surfaces while others bore into them. These processes increase the porosity of and weaken the surface layers of inhabited substrata. The microphytes provide food for herbivores, and in that way are involved in carbonate removal by abrasion. The resulting biokarst surfaces can be extremely rugged. Endolithic microphytes are also involved in producing "constructive" and "destructive" micrite envelopes around carbonate particles.

The rate at which endolithic algae bore into their substrata has been measured at between 0.3 and 36 μm/day. Filament densities within substrata may range from $150,000/mm^3$ to $500,000/cm^2$. Because they are adapted to life with very little light, endolithic algae may be found to surprisingly great depths within oceans. Plectonema, a cyanophyte, has been found to 370 meters on the Florida continental slope. Fungi, which are not light-limited, have been reported from 780 meters.

Most genera of endolithic microphytes have a worldwide distribution, though we are increasingly aware of regionality at the species level. They are identified, and their growth forms studied by 1) removing them from their substrata, 2) observing them in situ within transparent substrata, 3) preparing resin casts of their borings for study by means of reflected light and SEM, and 4) observing serial sections of doubly embedded material. The mechanism by which endoliths bore remains unknown, though it is thought to be by chemical rather than by mechanical means.

Introduction

Microorganisms, particularly cyanophytes, chlorophytes, rhodophytes, fungi and lichens, are involved in constructive and destructive processes with respect to calcified tissues and to carbonate rocks and sediment. At times these processes occur simultaneously. Constructive processes include the trapping and binding of sedimentary particles by mat-forming microphytes[1,2], and the precipitation of skeletal Mg-calcite by encrusting rhodophytes which cements particles together. They also include the encrustation of freshwater (and a few marine) microphytes by inorganically precipitated carbonate the crystal orientation of which is influenced by these organisms[1]. Destructive processes include the pitting of rock surfaces and crevices by metabolic products of epilithic and chasmolithic (cavity-dwelling) microphytes[3], and boring by endolithic microphytes[4,5]. Boring removes carbonate directly and leaves holes which conform to the shapes of the thalli. Boring also weakens the substrata so that they are more easily eroded by physical processes. In addition grazing animals abrade rock while feeding on all three types of microphytes[6]. Although the microorganisms involved are very small, their impact is enormous. Their activities result in the alteration and long-term evolution of coastal morphology, in the acceleration of the global carbon cycle[7], and at times, in the destruction of man's architectural and art treasures.

In addition to the summaries cited above, a brief treatment of constructive and destructive processes involving endolithic and epilithic algae is provided in Schneider[8], and extensive bibliographies are available in Pia[9], Lukas[4] and Kobluk and Kahle[10].

This paper is concerned primarily with destructive processes which involve endolithic microphytes. It builds on the earlier work of Golubic et al[5]; however some of the older literature is also considered. It deals with the following aspects: Methods for studying endoliths, the identity of endolithic microphytes, their geographic and depth distribution, the mechanism by which endoliths bore, the rate at which they inhabit substrata, and bioerosion.

Methodology

Endolithic microphytes are usually removed from their substrata for taxonomic work by dissolving the matrix with dilute acid, or, less commonly, by means of the Ca-chelating agent EDTA[5]. The organisms are then mounted in

KEY WORDS: Bioerosion, Carbonate Erosion, Endolithic Microphyta, Microboring Algae, Microboring Fungi glycerin or water for examination with a light microscope. Observation is enhanced tremendously by the use of Nomarski Interference Phase Contrast or similar optics. However study of spatial relationships among microphytes and of their growth habits requires in situ observation or casting of their boreholes.

Translucent crystals of Iceland Spar calcite are often provided as a substratum for colonization since the organisms can be seen within them[11,12]. Glover[13] presents a rather unique method for enhancing in situ visibility within other carbonate substrata. He converted the $CaCO_3$ of ooids into calcium fluoride (CaF_2) by immersing them in 10% hydrofluoric acid. When micropores within the resulting pseudomorph, 33.5% by volume[14], are filled with a fluid having the same refractive index, the grains are rendered transparent. The technique allows one to identify and to assess the abundance of live vs dead colonies in calcitic and aragonitic grains up to 0.4 mm in diameter which are relatively free of non-carbonate impurities.

Direct observation of boreholes can be made by means of SEM following treatment of specimens with clorox or peroxide to remove organic matter[1]. Quantitative measures of boring densities have been made from SEM photomicrographs of cast material[15] and by light microscope examination of resin embedded material which has been slabbed, polished and etched with acid[16,17,18].

Casts generally are made using the method pioneered by Golubic et al[19] and summarized in Figure 1. It provides three dimensional casts of borings, those containing endolith filaments as well as unfilled or partially filled boreholes. Casts of short filaments are usually sturdy enough to remain in growth position during drying and coating for SEM examination. However long, thin filaments may collapse during drying and hence may not provide accurate renditions of the organism's life form. This is a problem particularly with Ostreobium Bornet & Flahault and Conchocelis Batters whose filaments frequently are millimeters long, but average only 5 μm in diameter. Partial rather than complete removal of the substratum helps prevent this problem. The identification of an endolith from its cast is difficult, and requires examination of serial sections cut from re-embedded (doubly embedded) material (see Figure 1). In such sections the organisms can be seen within their borings. Their visibility within sections is enhanced by post-fixation in osmium tetroxide prior to metal coating. This also makes it easier to assess the relative number of occupied vs empty borings. The casting/double embedding method is the only reliable way to demonstrate the relationships between specific taxa and their borings. This approach is necessary as a basis for comparison between Recent and fossil material in which the borings are usually all that is preserved. Natural casts of fossil endoliths are known. They have been found composed of aragonite[20], limonite[21,22,23], phosphatized chalk[24] and pyrite[25]. Occasionally the organism itself is preserved, but this appears to be rare[26].

Taxonomy and Distribution

The incorrect or uncertain identification of endolithic microphytes has diminished the value of many of the papers written about them[16,17,23,27, 28,29,30,31]. For example the endolith which grew into Kobluk and Risk's Iceland Spar chips was Phaeophila dendroides (Crouan) Batters and not Ostreobium as they state[13,28,29,30]. Their Figure 3[30] clearly shows an eukaryote having cross walls and hyaline hairs. Ostreobium is a siphonaceous alga (lacks cross walls) which has no such hairs. Perkins and his early co-workers have improperly identified several endolithic taxa, mistaking Ostreobium and Conchocelis for fungi (their Figures 4 & 5)[16], an unidentified organism or its trace for a siphonaceous green alga (their Figures 10-12)[17], and Hyella for an endolithic red alga (their Figure 6)[17]. The identification in later papers appears to be largely correct[18,32]. A cast identified by Gatrall and Golubic[23] as that of an endolithic fungus has proved to be that of Gomontia, a green alga (their Figure 3b). These comments are made to caution people who are new to the field to be critical of the identity attributed to endolithic microphytes and of the ecological and paleoecological interpretations presented in early papers. These errors can largely be avoided by studying sections of doubly embedded as well as decalcified material and casts.

Although many endolithic algae are difficult to identify at the species level, most may be recognized quite readily at the generic level. Several papers contain descriptions with drawings or photographs of decalcified material. The more reliable of these include: Geitler[33], Ercegovic[34,35,36,37,38,39], Golubic[40], Nielsen[41], Golubic & LeCampion-Alsumard[42], Golubic et al[5], Schneider[15], Lukas[43,44], and in part Perkins & Tsentas[32] (except organisms identified in their Figures 9,10,11,15 and possibly also Figures 28, 29,30). Fewer people have worked on the endolithic fungi and lichens, so they present a more difficult taxonomic problem. What is known about the taxonomy and ecology of these groups is summarized in Lukas[4], Golubic et al[5] and Schneider[15].

The majority of algal borers are cyanophytes, and most of the genera, at least, are distributed worldwide. These include Hormathonema Ercegovic, Hyella Bornet & Flahault, Mastigocoleus Lagerheim, Plectonema Bornet & Flahault and Solentia Ercegovic. We have become increasingly aware of regionality at the species level, and a revision of the taxonomy of several endolithic cyanophyta, as well as the introduction of new genera, is in progress. The genera of boring chlorophytes, including Eugomontia Kornmann, Gomontia (Lagerheim) Bornet & Flahault, Ostreobium and Phaeophila and the one boring rhodophyte, Conchocelis also occur worldwide[5]. Regionality at the species level has also been reported for the chlorophytes[43]. At least one author[45] has suggested that Ostreobium quekettii, an endolith which is found worldwide, is but one stage in the life cycle of the siphonaceous chlorophyte Pseudobryopsis myura (Ag.) Berthold.

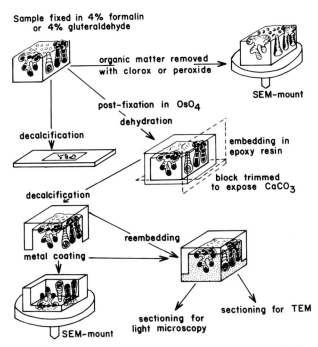

Figure 1. Summary of the casting/double embedding method used to prepare endolithic microphytes for light microscopy, SEM and TEM. Adapted from Golubic et al[19] and Schneider[15].

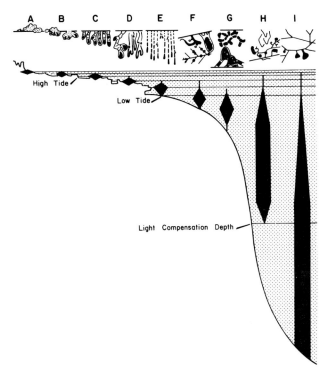

Figure 2. Relative vertical distribution of common marine microboring algae superimposed on a coastal profile. The upper limits are controlled by water supply; the lower limits by light penetration into the water column. A. coccoid epilithic cyanophytes, B. *Hormathonema luteobrunneum*, *H. violaceo-nigrum*, C. *Hormathonema paulocellulare*, D. *Solentia foveolarum*, *Kyrtuthrix dalmatica*, E. *Hyella tenuior*, F. *Mastigocoleus testarum*, *Gomontia polyrhiza*, *Phaeophila dendroides*, *Conchocelis*-stages of various rhodophytes, G. *Hyella caespitosa*, *Eugomontia sacculata*, H. *Plectonema terebrans*, *Ostreobium quekettii*, I. fungi. Revised from Golubic et al[5].

The maximum reported depths of occurrence of endolithic microphytes are 370 meters for the cyanophyte, Plectonema[44], 200 meters for the chlorophyte, Ostreobium quekettii[44,46] and 780 meters for fungi[16,17]. A summary of the depth distribution of the most common taxa is presented in Figure 2. Absolute depths below or elevations above the intertidal zone are not given since these will vary with the species involved and with water clarity and the maximum reach of wave spray.

Boring Rate, Boring Mechanism and Bioerosion

Speculation about the importance of endoliths as agents of limestone bioerosion began almost as soon as their borings and thalli were recognized within rocks and calcified tissue[15,47-59]. For example, Bornet and Flahault[60] suggested that they were responsible for the breakdown and removal of mollusk shells in calm embayments where agents of mechanical erosion are not effective. It has, however, proved difficult to devise experiments or to make observations which quantify their effectiveness.

Microborings weaken the surface layers of carbonate substrata and increase their porosity, making them susceptible to dissolution and to mechanical erosion in turbulent water[16,50,51,61, 62,63]. Their presence, and the presence of epilithic and chasmolithic algae as well, render the substrata liable to removal by grazing vertebrates such as parrotfish and surgeon fish as well as by grazing invertebrates such as gastropods, chitons and urchins. Several authors have reported that endoliths preferentially attack certain substrata, and may selectively remove them from the sediment.

This has been observed for corals[47,50,51] and for mollusk shells[17,50,51]. Similar preferences have been reported from fossil material[21,56].

The majority of endolithic algae are near-surface phenomena[5], that is they never penetrate more than a few millimeters into their substrata. They are prevented from boring deeper by a lack of light. Golubic and Schneider[6] speculate that most endoliths bore until the distal ends of their filaments reach light levels approaching the compensation light intensity for that species, then stop unless the proximal end of the filament is removed. Since that is usually accompanied by the removal of calcium carbonate as well, the organism bores deeper until it once again reaches its compensation light intensity. Bioerosion appears to occur primarily in this way[6,15]. Fungi, which are not light-limited, must have a carbon source, however. That source may be the organic lamellae within calcified tissues or the filaments of endolithic algae which their hyphae have been observed to penetrate[15,44,64].

The rate at which endolithic microphytes do bore may be rapid for such small organisms. Reported growth rates vary from 0.3 μm/day[65] to

36 μm/day[66] under field and experimental conditions. Of course this growth is not necessarily in a direction perpendicular to the substrate surface. In independent experiments carried out in Marseille by LeCampion-Alsumard[12], in St. Croix, Virgin Islands by Perkins and Tsentas[32] and in Jamaica by Kobluk and Risk[27-30], endolithic algae occupied provided substrata within 8-9 days from the start of the experiments. These substrata were Iceland Spar calcite chips and fragments of coral and conch shell. No difference in the time of initial colonization was recorded among the substrata. The results of the experiments were analyzed in different ways by the three teams. LeCampion-Alsumard found that endolithic algae (Mastigocoleus testarum, Hyella caespitosa and Plectonema terebrans) covered 90% of the surfaces of spar chips at the end of 90 days, and that within 28 days Hyella had penetrated the chips to a depth of 50 μm[12]. Kobluk and Risk obtained similar results. Endolithic algae covered 76.7% of the surfaces of their spar chips at the end of 95 days; 100% of the surfaces were covered to a depth of 30 μm in 213 days. The alga involved was Phaeophila dendroides. When examined on day 129 of the experiment, the algal filaments had penetrated to depths exceeding 300 μm. Perkins and Tsentas did not calculate the amount of surface area occupied by borers, but found that 95% of their experimental chips were inhabited by at least some filaments within 120 days. They did find that the conch shell became more heavily infested than did the spar chips in their study area (St. Croix). Perkins and Tsentas reported the greatest variety of endoliths, including both algae and fungi, but found that the fungi were much less abundant than were the algae. Their substrata were placed in the greatest range of depths (intertidal zone to 30 meters) compared to the others (60 cm for Kobluk and Risk; intertidal zone and "shallow subtidal zone" for LeCampion-Alsumard). They concluded that at the rate of boring observed during their study period, a layer of carbonate 20-30 μm thick would be removed in two years.

That endolithic algae may grow prolifically within their substrata is illustrated in Figure 3. Published estimates of filament density within substrata vary from 1500/mm^3 [67] to 500,000/cm^2 [15].

The mechanism by which microphytes bore remains uncertain. Two pieces of evidence suggest that it is by chemical rather than by mechanical means: 1) endolithic microphytes are found only within carbonate substrata or within the carbonate cements of clastic sandstones, and 2) the crystals located around the margins of the borings look as though they have been etched by acid[23]. See Pia[9] for references to suggested boring mechanisms. Alexandersson[59] observed that the etch pattern on the surfaces of borings produced by the cyanophyte Hormathonema is bilaterally symmetrical: Two zones of grooves oriented perpendicularly to the axis of the boring are separated by two zones of pyramids. He states that "the pattern of grooves shows no relationship to the internal ultrastructure or to the crystallographic parameters of the bored substrate (p. 237)[59]. His illustration (Figure 1b)[59] shows, however, that the angle of intersection at the summits of the pyramids conform to the crystallography of calcite.

Alexandersson believes that the borings are produced by some externally located boring "organelle" of subcellular size. Indeed he has observed "small threads" 0.2 μm in diameter and 2-6 μm long on the surfaces of Hormathonema filaments. His material was air dried, however, so what he observed was most likely a shrivelling of the algal sheath. The sheaths of cyanophytes are extracellular polysaccharide secretions which have not been observed to contain any cellular products. Furthermore the cyanophytes are prokaryotes which do not have organelles.

Although it has not been measured, the amount of carbonate removed directly by endolithic microphytes is probably less than the volume removed by herbivores which feed on them and on other components of the rock-dwelling flora or by the endolithic fauna, particularly the sponges, bivalves and worms. For a survey of endolithic invertebrates see Warme[68]. For example the clionid sponges bore into substrata by a combination of chemical and mechanical means[69] which result in the production of small characteristically shaped carbonate chips[70]. As a result of studies of clionids from Bermuda, Reutzler[71] has estimated that the sponges produce 250 mg of chips per square meter each year, an amount equal to about 40% of the interreef sediment. Fütterer[70] estimated that clionid chips amount to 30% of the total lagoonal sediment of Fanning Island (but only to about 2% of the total in the Persian Gulf and northern Adriatic Sea).

In 1974 Folk et al[72] described a particularly rugged type of karstic surface from Hell, Cayman Islands, which they termed "phytokarst." It is described as "a landform produced by rock solution in which boring plant filaments are the main agent of destruction, and the major morphological features are determined by the peculiar nature of their mode of attack"(p. 2351)[72]. The algae involved cover the surface of the rock giving it a black color, though some filaments were observed to penetrate as much as 1 mm below the surface. Whether the algae actually are borers is unclear from their photographs. It is possible that they produce the overlapping concavities and lacy pinnacles typical of phytokarst by etching their way down into the substratum rather than by truly boring into it. Thus they would be the "algues cariant" of Chodat[73] rather than "algues perforants." In any case the treacherously ragged surface is truly impressive with razor-sharp pinnacles having a relief of as much as 3 meters. Folk et al do not report on the presence of grazers.

Biokarst produced jointly by microphytes and grazing animals also occurs in the shallow subtidal, intertidal and supratidal zones[15]. The particular plants involved vary according to moisture, illumination and the characteristics of the substratum, producing variously colored rock zones[74]. In her study location near Marseille, which has steeply sloping cliffs, LeCampion-Alsumard found that variations in salinity were less important to algal distribution[74]. However Hyella, Solentia, Mastigocoleus and Kyrtuthrix were absent from rocks near river mouths where the salinity may be as low as 4‰[75]. It appears, therefore, that short, repetitive fluctuations in salinity have a different effect on algal borers

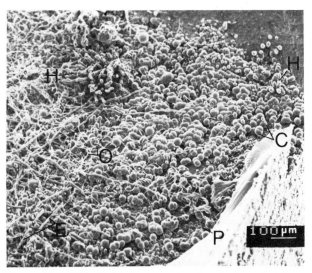

Figure 3. SEM photomicrograph of a "micro-land-scape created by resin casts of microboring algae. C-Cyanosaccus (gen. nov.), E-Eugomontia, H-Hyella, O-Ostreobium, P-Plectonema. The substratum was a Pecten shell from 11 meters from the Florida continental shelf.

than does prolonged exposure to water whose salinity is lower than that of normal seawater. Hyella, Kyrtuthrix and Mastigocoleus were also absent from tide pools where evaporation may raise the salinity to 283‰ during low tide[76]. However other factors, such as changes in oxygen content and organism interactions, must also be involved since salinities may be that high on a microscale due to the evaporation of seawater in small crevices where endoliths live in the intertidal zone.

Following a study of the water chemistry and the floras and faunas of rock pools, Schneider[15] concluded that the water was always saturated with respect to calcium carbonate in all but the highest pools, and that, therefore biological corrosion and abrasion rather than inorganic dissolution must be responsible for most of the coastal erosion in his study areas. He further concluded that the microflora, epilithic as well as endolithic, contribute to coastal erosion primarily by providing food for grazers and by preventing anaerobic conditions from occurring in the rock pools thereby making them suitable habitats for animals. Corrosion of limestone by the microphytes is a continuous, but slow, process.

Microphytes contribute to carbonate destruction in subtidal environments in similar ways. That the algal microphytes, particularly those which are endolithic, can survive with very little light has been shown experimentally[77,78,79]. These studies involve light penetration into live coral colonies which contain two genera of endolithic algae, Ostreobium and Plectonema[44]. The siphonaceous filaments of Ostreobium are found throughout the coralla of all of the corals which I have examined from the Atlantic and Pacific Oceans. They are generally most abundant within one or more concentric green or brown bands the outermost of which is located 0-15 mm beneath the coral surface[4]. Only the outermost band appears

to contain living Ostreobium filaments. Above this band lie a varying thickness of calcium carbonate containing fewer numbers of Ostreobium filaments and a layer of living animal tissue which contains millions of single-celled plant protists, zooxanthellae. The zooxanthellae make the animal tissue brown, and absorb light of the same wavelengths as are required by endoliths for photosynthesis[77,79].

Halldal[77] estimated that from 0.1-2% of the light incident upon a coral head penetrated to the top of a green band located 13 mm beneath the coral surface. This corresponds to 100-1000 lux on a sunny day in the tropics. This quantity is reduced to 0.1-1 lux, 1 mm beneath the top of the green band. Kanwisher and Wainwright[67] estimated that 1-2% of the incident light reached a band 10 mm beneath the surface of an Atlantic coral, about 25-50 lux. Franzisket[78] found even less light (15 lux) penetrating to 1 mm in Hawaiian corals.

That the plants are adapted to live at low light intensities is shown by the fact that filaments from the top of a band located 10 mm beneath the surface of a coral reached a maximum rate of photosynthesis (saturation) at a light intensity of $1000 \text{ erg/cm}^2/\text{sec}$ (200 lux) at 440 nm, while samples from the center of the same band reached saturation at $100-700 \text{ erg/cm}^2/\text{sec}$ (20-140 lux) at the same wavelength[77]. Photooxidation occurred at light intensities above $1800 \text{ erg/cm}^2/\text{sec}$ (360 lux). Samples from the bottom of the band showed photooxidation at all light intensities at 440 nm. Since Halldal did not include a baseline curve for the reduction in oxygen level, the cause of the oxygen depletion is uncertain. Franzisket[78] found that the maximum rate of photosynthesis in algae from Hawaiian corals occurred at 10 lux with photooxidation occurring at 500 lux.

Micritization and Intragranular Cementation in the Marine Environment

Since 1975 the greatest number of papers written on endoliths have been concerned with their involvement in the formation of micrite envelopes and intragranular cements. In 1966 Bathurst[80] suggested that micrite envelopes around carbonate particles are produced gradually by the filling of abandoned algal borings with 4-5 μm sized aragonite. As defined by Friedman and Sanders[81] (p. 565), a micrite envelope is: "An opaque, cryptocrystalline or microcrystalline surficial layer bordering a carbonate particle and formed by either 1) the filling of closely spaced, overlapping algal bores, which penetrate inward from the exterior of a skeletal particle; or 2) the concentric enlargement of a particle whose outer parts may or may not have been bored."

Micrite envelopes are important geologically since they are more resistant to dissolution than are the grains which they surround. Thus they preserve the original outline of the grain even after the original carbonate has been removed. Bathurst's hypothesis is widely accepted today with a few additions and qualifications as one of the more important ways in which micrite is formed in shallow tropical water (see 82,83,84 for other mechanisms). We now know that micrite envelopes may be composed of Mg-calcite or low Mg-calcite as well as of aragonite which is precipitated inorganically within vacated boreholes[23,84,85]. Thus particles from deep water as well as those from shallow water may become micritized[85].

Kobluk[27] and Kobluk and Risk[28,29,30] have designated this type of micrite "destructive micrite" to distinguish it from a type produced by the coalescence of calcified algal filaments which are epilithic with respect to the grain. This type, which forms on the grain's exterior, they call "constructive micrite." In experiments carried out in Jamaica they found both types of micrite in association with filaments they identified as those of endolithic algae. According to their results, endolithic filaments grew out of the substratum, thus assuming a partly epilithic habit. Dead filaments eventually became encrusted as well as filled by micrite-sized crystals of low Mg-calcite. This phenomenon was first noticed and described by Schroeder[85], though he found that the crystals were composed of high Mg-calcite, as was the substratum in which the filaments were found. Eventually a sufficient quantity of calcified filaments was present to form a continuous micritic envelope around the grains. At the same time vacated boreholes became filled by high Mg-calcite (destructive micrite). Both processes take place rapidly. Kobluk and Risk[28] also cited the presence of etched microporous patches on the surfaces of spar chips located under and produced by epilithic algae. Kahle[31] reported micritization associated with borings in grains and sparry cements of the Miami Limestone of Pleistocene age. However his photographs of the algae and fungi lack sufficient clarity and detail to distinguish between them, and one photograph (his Figure 2e) is most likely of a sponge spicule[31].

Harris et al[86] observed intragranular cements of acicular aragonite in borings of endoliths which had died after colonizing ooids from Joulter's Cay, Bahamas. The aragonite grew radially inward from the margins of the boreholes as well as radially outward from the centers of the boreholes, apparently from collapsed algal sheaths. Decalcification of the ooids showed that they contained an abundance of moribund cyanophyte colonies, some containing shrivelled cells, and some lacking cells.

Alexandersson[83] found that calcium carbonate did not precipitate within vacated boreholes in skeletal sands from Sweden where the seawater is undersaturated with respect to calcium carbonate. Similarly vacated borings in skeletal sands from the coast of western Ireland remained empty even though the water there is substantially supersaturated with respect to calcium carbonate, at least at a local level[87]. This led Gunatilaka to conclude that micritization is a biochemical process. This contradicts the conclusions of others who have found consistently that micritization is associated with dead, but not live, algal filaments[28,29,30,86].

Conclusions

During the last five years progress has been made in the study of the role which microphytes play in bioerosion. Their most important role appears to be an indirect one as a primary food source for herbivores. These animals remove carbonate by abrasion while freeing the enclosed microphytes, a process which takes place most actively in the supratidal, intertidal and shallow subtidal zones. Although it has proved difficult to measure the amount of carbonate which the microborers alone remove by corrosion under natural conditions, it probably is small when compared to that removed by abrasion.

Endolithic algae colonize provided substrata within 8-9 days, then grow and increase in numbers very rapidly until they cover their surfaces. It has not been possible to confirm the mechanism(s) by which they bore. Micritization of carbonate particles, with which endolithic algae are involved in shallow tropical water, has been shown to occur more rapidly than was previously thought (weeks or months rather than years). Micritization takes place by means of calcium carbonate precipitation around or within dead algal filaments or vacated borings.

There have been continued difficulties with identifying Recent endolithic taxa even though a number of papers are available which provide good illustrative and descriptive information. Experience has proved the value of examining sections made from doubly embedded material along with a study of decalcified preparations and resin casts to avoid making errors in identification and interpretation. This approach is also needed for comparison between Recent and fossil microborers. In situ observations of endoliths can be made by providing transparent substrata, such as Iceland Spar calcite, for colonization, or by converting natural carbonate substrata into calcium fluoride and mounting the pseudomorph in a medium which has the same refractive index.

The maximum depth of occurrence of algal borers varies with water clarity, and thus with depth of light penetration into ocean water. However they are found much deeper than was previously thought. Two endoliths, Ostreobium and Plectonema, have been found alive and in quantity to 200 meters depth in the Atlantic Ocean off Florida. Ostreobium has also been reported from 200 meters in the Mediterranean Sea. Plectonema occurs to at least 370 meters off Florida, though most endolithic algae are found to only 100 meters. This means that the presence of algal borers cannot be taken as an indicator of shallow water.

Acknowledgments

I acknowledge support from NSF Grant #EAR-76-84233 to S. Golubic and B. Cameron, Boston University, and thank Dr. Golubic for reading the first draft of the manuscript, and for discussing his unpublished results.

References

1. S. Golubic, "The relationship between blue-green algae and carbonate deposits," in The Biology of Blue-green Algae, N.G. Carr & B.A. Whitton (eds)., University of California Press, Los Angeles, CA, 1973, Ch. 21.

2. M. Walter (ed)., "Stromatolites," Devel-

452

opments in Sedimentology, 20, Elsevier Scientific Publishing Company, New York, NY 1976, 790p.

3. W.D. Krumbein (ed)., Environmental Biogeochemistry and Geomicrobiology, 3 vols., Ann Arbor Science Publishers, Inc., Ann Arbor, MI, 1978, 1054p.

4. K.J. Lukas, Taxonomy and Ecology of the Recent Endolithic Microflora of Reef Corals with a Review of the Literature on Endolithic Microphytes, Unpubl. Ph.D. Dissertation, University of Rhode Island, 1973, 154p.

5. S. Golubic, R.D. Perkins & K.J. Lukas, "Boring microorganisms and microborings in carbonate substrates," in The Study of Trace Fossils, R.W. Frey (ed)., Springer-Verlag, New York, NY, 1975, Ch. 12.

6. S. Golubic & J. Schneider, "Carbonate dissolution," in Biogeochemical Cycling of Mineral-forming Elements, P.A. Trudinger & D.J. Swaine (eds)., Elsevier Scientific Publishing Company, New York, NY, 1979, Ch. 2.4.

7. S. Golubic, W. Krumbein & J. Schneider, "The carbon cycle," in Biogeochemical Cycling of Mineral-forming Elements, P.A. Trudinger & D.J. Swaine (eds)., Elsevier Scientific Publishing Company, New York, NY, 1979, Ch. 2.1.

8. J. Schneider, "Carbonate construction and decomposition of epilithic and endolithic microorganisms in salt- and freshwater," in Fossil Algae, Results and Developments, E. Flugel (ed)., Springer-Verlag, Berlin, FDR, 1977, Ch. 27.

9. J. Pia, "Die Kalklösenden Thallophyten," Arch. Hydrobiol., 31, 1937, 264-328 & 341-398.

10. D.R. Kobluk & C.F. Kahle, "Bibliography of the endolithic (boring) algae and fungi and related geologic processes," Bull. Canadian Petrol. Geol., 25, 1977, 208-223.

11. S. Golubic, "Distribution, taxonomy and boring pattern of marine endolithic algae," Am. Zool., 9, 1969, 747-751.

12. T. LeCampion-Alsumard, "Étude expérimentale de la colonisation d'éclats de calcite par les Cyanophycées endolithes marines," Cah. Biol. Mar., 16, 1975, 177-185.

13. E.D. Glover, "Organic remains seen in oolites after conversion to fluorite," J. Sed. Petrol., 48, 1978, 795-798.

14. E.D. Glover & R.F. Sippel,"Experimental pseudomorphs: replacement of calcite by fluorite," Am. Mineral., 47, 1962, 1156-1165.

15. J. Schneider, "Biological and inorganic factors in the destruction of limestone coasts," Contributions to Sedimentology, 6, 1976, 1-112.

16. R.D. Perkins & S.D. Halsey, "Geologic significance of microboring fungi and algae in Carolina shelf sediments," J. Sed. Petrol., 41, 1971, 843-853.

17. W.S. Rooney & R.D. Perkins, "Distribution and geologic significance of microboring organisms within sediments of the Arlington Reef Complex, Australia," Geol. Soc. Amer. Bull., 83, 1972, 1130-1150.

18. B.D. Edwards & R.D. Perkins, "Distribution of microborings within continental margin sediments of the southeastern United States," J. Sed. Petrol., 44, 1974, 1122-1135.

19. S. Golubic, G. Brent & T. Lecampion, "Scanning electron microscopy of endolithic algae and fungi using a multipurpose casting-embedding

technique," Lethaia, 3, 1974, 203-209.

20. P.M. Harris, K.J. Lukas & R.B. Halley, "A comparison of endolith floras from Holocene-Pleistocene (Bahama-Florida) ooids," (abs), Amer. Assoc. Petrol. Geol. Bull., 61, 1977, 793-794.

21. I. Hessland, "Investigations of the Lower Ordovician of the Siljan District, Sweden, II. Lower Ordovician penetrative and enveloping algae from the Siljan District," Geol. Inst. Univ. Uppsala, Bull., 33, 1949, 409-428.

22. J. Wendt, "Stratigraphie und Paläogeographie des Roten Jurakalks im Sonnwendgebirge (Tirol, Österreich), Neues Jahrb. Geol. Paläontol., Abh., 132, 1969, 219-238.

23. M. Gatrall & S. Golubic, "Comparative study on some Jurassic and Recent endolithic fungi using a scanning electron microscope," in Trace Fossils, T.P. Crimes & J.C. Harper (eds)., Geol. J. Spec. Paper 4, 1970, 167-178.

24. R.G. Bromley, "Borings as trace fossils and Entobia cretacea Portlock, as an example," in Trace Fossils, T.P. Crimes & J.C. Harper (eds)., Geol. J. Spec. Paper 4, 1970, 49-90.

25. D.R. Kobluk & M.J. Risk, "Algal borings and framboidal pyrite in Upper Ordovician brachiopods," Lethaia, 10, 1977, 135-144.

26. J. Kazmierczak & S. Golubic, "Oldest organic remains of boring algae from Polish Upper Silurian," Nature, 261, 1976, 404-406.

27. D.R. Kobluk, "Calcification of filaments of boring and cavity-dwelling algae, and the construction of micrite envelopes," in Geobotany, R.C. Romans (ed)., Plenum Press, New York, NY, 1977, 195-207.

28. D.R. Kobluk & M.J. Risk, "Micritization and carbonate grain-binding by endolithic algae," Amer. Assoc. Petrol. Bull., 61, 1977, 1069-1082.

29. D.R. Kobluk & M.J. Risk, "Calcification of exposed filaments of endolithic algae, micrite envelope formation and sediment production," J. Sed. Petrol., 47, 1977, 517-528.

30. D.R. Kobluk & M.J. Risk, "Rate and nature of infestation of a carbonate substratum by a boring alga," J. Exp. Mar. Biol. Ecol., 27, 1977, 107-115.

31. C.F. Kahle, "Origin of subaerial Holocene calcareous crusts: role of algae, fungi and sparmicritization," Sedimentology, 24, 1977, 413-436.

32. R.D. Perkins & C.I. Tsentas, "Microbial infestation of carbonate substrates planted on the St. Croix shelf, West Indies," Geol. Soc. Amer. Bull., 87, 1976, 1615-1628.

33. L. Geitler, "Cyanophyceae," in Kryptogamen-Flora von Deutschland, Österreich, und der Schweiz, 14, L. Rabenhorst (ed)., 1932, 1196p.

34. A. Ercegović, "Tri nova roda litofitiskim cijanoficeja sa jadranske obale (Trois nouveaus genres de Cyanophycées lithophytes de la côte adriatique), Acta Bot. Inst. Bot. Univ. Zagreb, 2, 1927, 78-84.

35. A. Ercegović, "Dalmatella, nouveau genre de Cyanophycées lithophytes de la côte adriatique," Acta Bot. Inst. Bot. Univ. Zagreb, 4, 1928, 35-41.

36. A. Ercegović, "Sur la valeur systématique et la ramification des nouveau type d'algue perforante," Ann. Protistol., 2, 1929, 127-135.

37. A. Ercegović, "Sur quelques nouveau types de Cyanophycées lithophytes de la côte adriatique,"

Arch. Protistenk., 66, 1929, 164-174.

38. A. Ercegović, Sur quelque types peu connus de Cyanophycées lithophytes," Arch. Protistenk., 71, 1930, 361-376.

39. A. Ercegović,"Sur la valeur systématique de quelques algues perforantes recemment decrites" Acta Bot. Inst. Bot. Univ. Zagreb, 9, 1934, 34-40.

40. S. Golubic, "Scanning electron microscopy of recent boring Cyanophyta and its possible paleontological application," in Proceedings of the Symposium on Taxonomy and Biology of Blue-green Algae, T.V. Desikachary (ed)., Univ. Madras, Madras, India, 1972, 167-170.

41. R. Nielsen, "A study of the shell-boring marine algae around the Danish Island Laeso," Bot. Tidsskrift., 67, 1972, 245-269.

42. S. Golubic & T. LeCampion-Alsumard, "Boring behavior of marine bluegreen algae, Mastigocoleus testarum Lagerheim and Kyrtuthrix dalmatica Ercegovic, as a taxonomic character," Schweiz. Zeitschr. Hydrol., 35, 1973, 157-161.

43. K.J. Lukas, "Two species of the chlorophyte genus Ostreobium from the skeletons of Atlantic and Caribbean reef corals, J. Phycol., 10, 1974, 331-335.

44. K.J. Lukas, "Depth distribution and form among common microboring algae from the Florida continental shelf," Geol. Soc. Amer. Abstracts with Programs, 10, 1978, 448.

45. M.H. Mayhoub, "Reproduction sexuees et cycle du developpement de Pseudobryopsis myura (Ag.) Berthold (Chlorophycees, Codiale), C.R. Acad. Sci. Paris, 278, 1974, 867-870.

46. G. Fredj & C. Falconetti, "Sur la présence d'algues filamenteuses perforantes dans le test des Gryphus vitreus (Born)(Brachiopodes, Térébratulidés) de la limite inférieure du plateau continental méditerranéen," C.R. Acad. Sci. Paris, 284, 1977, 1167-1170.

47. J. Duerden, "Boring algae as agents in the destruction of corals," Carnegie Inst. Washington Pub., 20, 1902, 1-130.

48. J.S. Gardiner, The Fauna and Geography of the Maldive and Laccadive Archipelagos, 1, 1901-1903, 333-341.

49. J.S. Gardiner, "Photosynthesis and solution in formation of coral reefs," Linn. Soc. (London), Proc., 143, 1930/31, 65-71.

50. R.N. Ginsburg, "Intertidal erosion in the Florida Keys," Bull. Mar. Sci. Gulf and Caribbean, 3, 1953, 55-69.

51. R.N. Ginsburg, "Early diagenesis and lithification of shallow water carbonate sediments in South Florida," in Regional Aspects of Carbonate Deposition, R.J. LeBlanc & J.G. Breeding (eds)., Soc. Econ. Paleontol. Mineral. Spec. Pub., 5, 1957, 80-99.

52. N.D. Newell, "Reefs and sedimentary processes of Raroia," Atoll Res. Bull., 36, 1954, 1-35.

53. E.C. Purdy & L.S. Kornicker, "Algal disintegration of Bahamian limestone coasts," J. Geol., 66, 1958, 96-99.

54. G. Ranson, "Erosion biologique des calcaires côtiers et autres calcaires d'origine animale," C.R. Acad. Sci. Paris, 249, 1959, 438-440.

55. C.D. Neumann, "Observations on coastal erosion in Bermuda and measurements of the boring rate of the sponge, Cliona lampa," Limnol.

Oceanogr., 11, 1966, 92-108.

56. M. Rioult & L. Dangeard, "Importance des Cryptogames perforantes marines en geologie," Botaniste, 50, 1967, 389-413.

57. C.G. St. C. Kendall & P.A. d'E Skipwith, "Holocene shallow-water carbonate and evaporite sediments of Khor al Bajam, Abu Dhabi, Southwestern Persian Gulf," Amer. Assoc. Petrol. Geol. Bull., 53, 1969, 841-869.

58. E. Driscoll. "Selective bivalve shell destruction in marine environments," J. Sed. Petrol., 40, 1970, 898-905.

59. E.T. Alexandersson, "Marks of unknown carbonate-decomposing organelles in cyanophyte borings," Nature, 254, 1975, 212 and 237-238.

60. E. Bornet & C. Flahault, "Sur quelques plantes vivant le test calcaire des mollusques," Bull. Soc. Bot. (France), 36, 1889, CXLVII-CLXXVI.

61. J.P. Swinchatt, "Significance of constituent composition, texture and skeletal breakdown in some Recent carbonate sediments." J. Sed. Petrol., 35, 1965, 71-90.

62. K.W. Klement & D.F. Toomey, "Role of the blue-green alga Girvanella in skeletal grain destruction and lime mud production in the lower Ordovician of west Texas," J. Sed. Petrol., 37. 1967, 1045-1051.

63. W. Klepal & H. Barnes, "Further observations on the ecology of Chthalamus depressus (Poli)," J. Exp. Mar. Biol. Ecol., 17, 1975, 269-296.

64. O. Jaag, "Untersuchungen über die Vegetation und Biologie der Algen des nacktengesteins in den Alpen, im Jura und im Schweizerischen Mittelland," Beitr. Kryptogamenflora Schweiz., 9, 1945, 1-560.

65. M. Parke & H.B. Moore, "The biology of B. balanoides. II. Algal infection of the shell," J. Mar. Biol. Assoc. (UK), N.S., 20, 1935, 49-56.

66. K.M. Drew, "Studies in the Bangioides. III. The life-history of Porphyra umbilicalis (L.) Kutz. var. laciniata (Lightf.) J. Ag." Ann. Bot., N.S., 18, 1954, 183-211.

67. J.W. Kanwisher & S.A. Wainwright, "Oxygen balance in some reef corals," Biol. Bull., 133, 1967, 378-390.

68. J.E. Warme, "Borings as trace fossils, and the processes of marine bioerosion," in The Study of Trace Fossils, R.W. Frey (ed)., Springer-Verlag, New York, NY, 1975, Ch. 11.

69. W. Cobb, "Penetration of calcium carbonate substrates by the boring sponge Cliona," Am. Zool., 9, 1969, 783-790.

70. D.K. Fütterer, "Significance of the boring sponge Cliona for the origin of fine grained material of carbonate sediments," J. Sed. Petrol., 44, 1974, 79-84.

71. K. Reutzler, "The role of burrowing sponges in bioerosion," Oecologia, 19, 1975, 203-216.

72. R.L. Folk, H.H. Roberts & C.H. Moore, "Black phytokarst from Hell, Cayman Islands, British West Indies," Geol. Soc. Amer. Bull., 84, 1973, 2351-2360.

73. R. Chodat, "Sur les algues perforantes d'eau douce. Études du biologie lacustre," Bull. Herbier Boissier, 6, 1898, 431-476.

74. T. LeCampion-Alsumard, "Contribution à l'étude des Cyanophycées lithophytes des étages

supralittoral et médiolittoral (région de Marseille)," Tethys, 1, 1969, 119-172.

75. A. Ercegović, "Sur la tolérance des Cyanophycées vis-à-vis des variations brusque de la salinité de l'eau de mer," Acta Bot. Inst. Bot. Univ. Zagreb, 5, 1930, 48-55.

76. A. Ercegović, "Ekološke i sociološke studije o litofitskim cijanoficejama sa jugoslavenske obale jadrana (Études écologique et sociologique des Cyanophycées lithophytes de la côte yougoslave de l'Adriatique)", (Croatian, French summary), Jugoslav Akad. Znanosti Umjetnosti Rad., 244, 1932, 129-220.

77. P. Halldal, "Photosynthetic capacities and photosynthetic action spectra of endozoic algae of the massive coral Favia," Biol. Bull., 134, 1968, 411-424.

78. L. Franzisket, "Zur Ökologie der Fadenalgen im Skelett lebender Riffcorallen," Zool. Jahrb. Physiol., 74, 1968, 246-253.

79. K. Shibata & F.T. Haxo, "Light transmission and spectral distribution through epi- and endozoic algal layers in the brain coral, Favia," Biol. Bull., 136, 1969, 233-241.

80. R.G.C. Bathurst, "Boring algae, micrite envelopes, and lithification of molluscan biosparite," Geol. J., 5, 1966, 15-22.

81. G.M. Friedman & J.E. Sanders, "Principles of Sedimentology, John Wiley & Sons, New York, NY, 1978, 792p.

82. E.C. Purdy, "Carbonate diagenesis: an environmental survey," Geologica Romana, 8, 1968, 183-228.

83. E.T. Alexandersson, "Micritization of carbonate particles: processes of precipitation and dissolution in modern shallow-water sediments," Bull. Geol. Inst. Univ. Uppsala, N.S., 3, 1972, 201-236.

84. G.M. Friedman, C.D. Gebelein & J.E. Sanders, "Micritic envelopes of carbonate grains are not exclusively of photosynthetic algal origin," Sedimentology, 16, 1971, 89-96.

85. J.H. Schroeder, "Calcified filaments of an endolithic alga in Recent Bermuda reefs," Neues Jahrb. Geol. Palaontol. Mh., 1, 1972, 16-33.

86. P.M. Harris, R.B. Halley & K.J. Lukas, "Endolith microborings and their preservation in Holocene-Pleistocene (Bahama-Florida) ooids," Geology, in press.

87. A. Gunatilaka, "Thallophyte boring and micritization within skeletal sands from Connemara, western Ireland," J. Sed. Petrol., 46, 1978, 548-554.

Discussion with Reviewers

Reviewer I: Do these organisms produce carbonic anhydrase?
Author: Although the presence of carbonic anhydrase is suspected, it has not been demonstrated that these microphytes do produce it. Most recent work on endoliths has been done by taxonomists and geologists--the biochemists seem not to have discovered them yet.

Reviewer I: Are the organisms sufficiently closely adherent to the substrate that a microenvironment conducive to resorption could be compartmentalized preventing dilution by sea water?
Author: That is certainly true for the endoliths, and is probably true for the epiliths and chasmoliths (cavity dwellers) as well.

Reviewer I: How do the microphytes change the pH of sea water (second sentence of abstract)?
Author: They lower the pH of sea water by increasing the partial pressure of carbon dioxide during respiration. They may also produce organic acids. Several, e.g. carbonic acid, have been suggested as the means by which endoliths bore (see Pia[9]).

Reviewer I: What do SEM's of the eroded tunnels and substrates look like?
Author: Unfortunately I do not have photographs available for publication to show that. There are excellent photographs of endolith borings in Golubic et al[5] and in Schneider[15]. Biokarst is illustrated in Folk et al[72].

*K. Lukas: Is there any evidence that microboring algae are facultative heterotrophs?
Author: The possibility of heterotrophy among the microboring algae has not, to my knowledge, been investigated. My experience from dives made in a research submersible to 200 meters off Florida is that there is sufficient light for photosynthesis to at least 100 meters (the maximum depth of occurrence of most endoliths there). I think it is significant that the two endoliths having the greatest depth range, Ostreobium (200 m) and Plectonema (370 m) are also the only algae found deep within skeletons of live corals where light levels are very low (see text). I think they are autotrophs to the limits of their occurrence.

*Question rephrased by the author based on the comments of one of the reviewers of her paper.

REVIEWER I: J.L. Matthews, W.L. Davis and R. Jones

455

Additional discussion with reviewers of the paper "Comparative Shell Microstructure of the Mollusca..." by J.G. Carter et al, continued from page 446.

K. M. Towe: You have noted that crossed bladed structure is absent from modern molluscs. It is also absent from fossil molluscs. These observations are perfectly consistent with the belief that no mollusc ever developed this structure. Why then should any speculations about primitive molluscan unicellular control be considered a more sound basis for phylogenetic comparison if no evidence exists to support it?

Author: I do not maintain that crossed bladed structure is absent from modern molluscs. Both modern bivalves and tentaculitids closely approximate crossed bladed structure in their calcitic shells layers. Furthermore, there are insufficient data to warrant any generalization that crossed bladed structure is absent from fossil molluscs. We know very little about the range of microstructural adaptations in the early Paleozoic molluscs. Finally, there is no need to rely only upon the fossil record for verification of the possibility of unicellular controls on calcification in primitive molluscs, because ample evidence is provided by histological work on modern aplacophorans (text references 29, 30). If aplacophorans are considered to represent a primitive grade of molluscan calcification (text references 25, 26), it is only natural to assume that their largely unicellular controls on spicule formation might be inherited by the most primitive shelled molluscs, to be replaced eventually by more advanced levels of integration of secretory activity. This hypothesis is compatible with the apparent lack of directionally oriented microstructures (e.g., concentric crossed lamellae) in the modern Monoplacophora.

I do not rely upon speculations about primitive unicellular controls on shell secretion as a basis for phylogenetic comparison. On the contrary, I suggest that such considerations reinforce data of microstructural convergence between brachiopods and molluscs to indicate that there is no sound basis for phylogenetic comparisons when dealing with brachiopod and the more primitive molluscan microstructures, at least at the phylum level.

N. Watabe: Please explain why the lack of larger diameter microstructures indicates a low level of integration of mantle secretory activity?

Author: The lack of larger diameter microstructures does not, in itself, indicate a low level of integration of mantle secretory activity, especially when comparisons are made between closely related species showing both larger and smaller microstructural diameters. However, when an entire phylum, such as the Brachiopoda, consistently shows microstructural diameters that are generally similar to its likely mantle cell diameters, despite the fact that ecologically and morphologically similar shells in other phyla have evolved much larger diameter microstructures in response to similar requirements for shell mechanical design, then the data argue for some limitation on integration of mantle secretory activity. Some bivalve shells lack larger units microstructures, but the fact that many of their close relatives also secrete larger unit microstructures suggests that, in this case, the lack of larger diameters merely reflects natural selection for specific shell mechanical properties.

SCANNING ELECTRON MICROSCOPY/1979/II
SEM Inc., AMF O'Hare, IL 60666, USA

ON THE INTERNAL STRUCTURES OF THE NACREOUS TABLETS IN MOLLUSCAN SHELLS

H. Mutvei

Swedish Museum of Natural History
104 05 Stockholm, and
Paleontological Institute
Box 558
751 22 Uppsala

Abstract

The aragonite crystals in molluscan nacreous tablets show complicated twinning patterns when etched with a glutaraldehyde-acetic acid solution. In the bivalves, Mytilus, Nucula and Unio, each tablet is composed of two layers. The outer layer consists of four, radially-arranged, crystal individuals which are probably cyclically twinned. The surface of this layer exhibits concentric growth lamellae. The surface of the inner layer shows numerous parallel crystalline laths which indicate a lamellar, polysynthetic twinning of this layer. These laths are associated with an organic matrix which remains after demineralization.

In gastropods and in the cephalopod Nautilus, only one layer has been hitherto distinguished in the nacreous tablets. This layer is composed of a varying number of radially-arranged crystal individuals which form either cyclic or interpenetrant twins. The surface of each crystal individual exhibits parallel laths which are similar to the surface of the inner layer of the bivalve nacreous tablets.

By using other etching solutions, the tablets of the molluscan nacreous layer become subdivided into numerous vertical angular crystalline plates. The plates from consecutive tablets are fused together, probably through the inter-trabecular bridges of the inter-lamellar organic membranes. The plates are associated with an organic matrix which remains after demineralization as vertical cords.

KEY WORDS: Shell Structure, Etching, Demineralization, Molluscs, Nacre, Crystal Twinning, Organic Matrices

Introduction

The nacreous layer of the molluscan shells has been one of the most intensely studied mineralized tissues. Several of these studies have been devoted to elucidating the mechanism of biomineralisation.

The nacreous layer is composed of regularly arranged mineral and organic components. Aragonite tablets, commonly with pseudo-hexagonal outlines in early stages of growth, form thin mineral lamellae that are separated by inter-lamellar organic membranes. As recently demonstrated, the nacreous tablets in bivalves are composed of four crystal individuals, which occur in two different pairs, and which are probably cyclically twinned.[1] One pair of crystals is considerably less soluble in the glutaraldehyde-acetic acid solution than the other pair. The nucleation of new nacreous tablets invariably takes place on the top surface of the less-soluble pair of crystal individuals.

In contrast to bivalves, the nacreous tablets in gastropods and in the cephalopod Nautilus consist of a varying number (2 to 50) of crystal individuals.[2] These individuals form cyclic and interpenetrant twins. Poly-synthetic (lamellar) twins also occur in each individual. The central part of each nacreous tablet contains a significant amount of calcified organic matrix which is insoluble in either a chromium sulphate solution or a 25% glutaraldehyde solution. In most cases, the crystal individuals are separated by radial organic membranes which extend from the central portion to the periphery of the tablet. The nucleation of new tablets takes place in the central portion of the tablet. A similar type of crystal twinning to that of gastropods and Nautilus, has been described in non-biogenic aragonite crystals.

In the present paper, additional information on the crystalline components and their twinning in the nacreous tablets of bivalves is given. The structural relationship between the consecutive nacreous tablets and the inter-lamellar organic membranes is described in bivalves and gastropods.

Materials and Methods

Dry shells of the following molluscs were studied: the bivalves <u>Mytilus</u> <u>edulis</u> L. and <u>Nucula</u> <u>sulcata</u> (Bronn) from the Gullmar Fjord on the west coast of Sweden, and <u>Unio</u> <u>tumidus</u> Retzius from the Lake Rörsjön, north of Stockholm; the gastropod <u>Gibbula</u> <u>cineraria</u> (L.) from the Gullmar Fjord on the west coast of Sweden; and the cephalopod, <u>Nautilus</u> <u>pompilius</u> L. from the Solomon Islands.

The organic components within the nacreous tablets are soluble in EDTA and several other etching solutions. It is therefore essential that the solution used for etching and demineralization should also provide a good and rapid fixation of these organic components.[1,2,3,4] The following solutions were used: (1)* A 12% glutaraldehyde solution to which various amounts of 1% acetic acid, and alcian blue, dissolved in ethanol, were added. The duration of the etching varied from 5 to 30 min. (2) A saturated cetylpyridinium chloride solution at pH 8.5. (3) A 25% glutar-aldehyde solution at pH 6.2 to 7.2. (4) A 1:1 mixture of the saturated cetylpyridinium chloride solution at pH 8.5 and a 25% glutar-aldehyde solution at pH 6.2. The duration of the etching with the solutions (2), (3) and (4) varied from 2 to 30 days at +32°C.

The uncalcified parts of the inter-lamellar organic membranes were dissolved with a concentrated sodium hypochlorite solution containing 8% to 12% active chlorine. The duration of this treatment varied from 15 min to 2 hours.

Most preparations were dehydrated in ethanol, transferred into amyl acetate, and critical point dried in CO_2. The preparations were sputter coated with evaporated gold and studied with a SEM (Stereoscan, Cambridge Instrument Ltd.) at the Swedish Museum of Natural History, Stockholm.

Results

The structure of the nacreous tablets in the bivalves, <u>Mytilus</u>, <u>Nucula</u> and <u>Unio</u>, has been recently studied by using glutaraldehyde-acetic acid etching.[1] According to these studies, each tablet is composed of four crystal individuals, which are probably cyclically twinned. The etching also revealed numerous concentric lamellae on the surface of the tablet. These lamellae are parallel to the edges of the tablet, and probably mark the successive growth stages.

My further studies on the nacreous tablets in <u>Mytilus</u>, <u>Nucula</u> and <u>Unio</u>, have modified the results referred to above. When the preparations are briefly etched by the glutar-aldehyde-acetic acid solution, a subdivision of the tablet into two horizontal layers is clearly visible. The surface of the outer layer, directed towards the shell-secreting epithelium, shows the characteristics of the bivalve nacre - four crystal individuals and concentric growth lamellae (Figs. 1,2,3).

Conversely, the surface of the inner layer exhibits parallel laths which are about 0.1 to 0.2 millimicrons in diameter (Figs. 1,2,3,4). These mineral laths have been described in my previous papers, and they may indicate a lamellar, polysynthetic twinning.[2,5] With further etching, the laths become completely demineralized and elastic, without losing their original shape (Fig. 5). This indicates that the mineral laths are associated with an organic matrix that remains after the demineralization. The demineralized laths decrease only slightly in diameter.

The outer layer of the nacreous tablets in <u>Nucula</u> is considerably thicker than that in <u>Mytilus</u> (compare Figs. 1 and 2). However, further studies are required to know whether this difference is constant or influenced by environmental factors.

In gastropods and in the cephalopod <u>Nautilus</u>, only one layer has been hitherto distinguished in the nacreous tablets. This layer is composed of a varying number of radially-arranged crystal individuals which form either cyclic or interpenetrant twins.[2] The surface of each crystal individual exhibits similar parallel laths as those on the surface of the inner layer of the bivalve nacreous tablets.

The relationship between the consecutive nacreous tablets and the inter-lamellar organic membranes was studied in <u>Mytilus</u> (bivalve), <u>Gibbula</u> (gastropod) and <u>Nautilus</u> (cephalopod). Vertical fracture planes of the nacreous layer in the shells of these molluscs were etched for several days with one of the following three solutions: (1) a saturated aquaeous solution of cetylpyridinium chloride at pH 8.5, (2) a 25% glutaraldehyde solution at pH ranging from 6.2 to 7.2, and (3) a 1:1 mixture of the saturated cetylpyridinium chloride and 25% glutaraldehyde solutions.

Etching with these three solutions gives almost identical results. The nacreous tablets in <u>Mytilus</u>, <u>Gibbula</u> and <u>Nautilus</u> become subdivided into numerous crystalline plates which have an angular shape and which extend vertically through the tablet. The acute angles of these plates are projected into the adjacent inter-lamellar organic membranes (Fig. 6). In certain places of these preparations, and in the preparations in which the inter-lamellar membranes have been dissolved by a treatment with sodium hypochlorite solution, it is clearly seen that the angular plates from the consecutive nacreous tablets are fused to each other along their acute angles (Figs. 6,7,8). Thus, in contrast to what was previously thought, the consecutive nacreous tablets are partially separated by the inter-lamellar organic membranes, and partially continue through these membranes.

---

*The range of proportions was: 1 to 5 parts of 1% acetic acid, 0.1 to 0.5 parts of alcian blue, and 10 parts of 12% glutaraldehyde. The best results were obtained with 2 parts of 1% acetic acid, 0.2 part of alcian blue, and 10 parts of 12% glutaraldehyde.

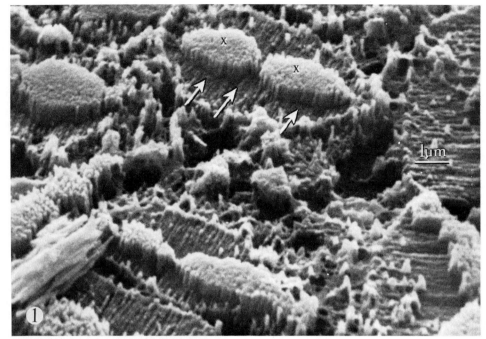

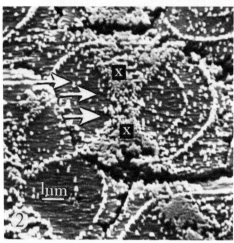

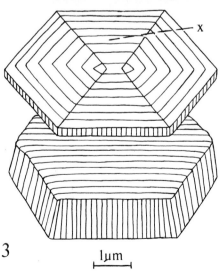

Fig. 1. Inner shell surface of <u>Nucula sulcata</u> (Bronn), showing the two-layered structure of the nacreous tablets. The less-soluble crystal individuals (x) of the outer layer are partially preserved, whereas the highly-soluble individuals are completely dissolved. The orientation of the crystalline laths on the exposed surface of the inner layer is indicated by arrows. The preparation was treated with sodium hypochlorite solution for 30 min, and etched with glutaraldehyde-acetic acid solution for 15 min. It was dehydrated with acetone and air dried.

Fig. 2. Horizontal fracture plane of the nacreous layer in <u>Mytilus edulis</u> L., to show the two-layered structure of the tablets. Two less-soluble crystal individuals (x) and two concentric growth lamellae remain of the outer layer, whereas the rest of the layer is completely dissolved. On the exposed surface of the inner layer, the orientation of the crystalline laths is indicated by arrows. The preparation was treated in the same way as that in Figure 1.

Fig. 3. Schematic diagram of a nacreous tablet in <u>Mytilus edulis</u> L., to show the two structurally different layers. The thin outer layer is composed of two pairs of crystal individuals which probably are cyclically twinned. This layer also shows concentric growth lamellae. The surface of the inner layer exhibits parallel crystalline laths which indicate a lamellar, polysynthetic twinning of this layer. These lamellae, as well as the outer layer, seem to be composed of vertical acicular crystalline elements.

459

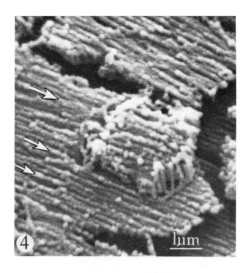

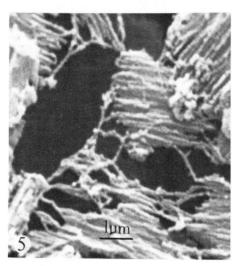

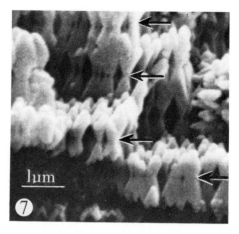

Fig. 4. Inner shell surface of <u>Mytilus edulis</u> L., to show the crystalline laths, indicated by arrows, on the surface of the inner layer of the nacreous tablets. The outer layer of the tablets is completely dissolved. The preparation was treated with sodium hypochlorite solution for 20 min, and etched with glutaraldehyde-acetic acid solution for 10 min. It was dehydrated in ethanol, transferred into amyl acetate, and dried in a critical point apparatus with CO_2.

Fig. 5. Similar preparation as that in Figure 4, but etched 10 min longer with glutaraldehyde-acetic acid solution. Note the elastic organic residue which remains after the demineralization of the crystalline laths.

Fig. 6. Etched vertical fracture plane of the nacreous layer in <u>Mytilus edulis</u> L., to show that the vertical angular plates continue through the horizontal inter-lamellar organic membranes. The preparation was etched with a 1:1 mixture of saturated cetylpyridinium chloride solution at pH 8.5 and 25% glutaraldehyde solution at pH 6.2 for three days. It was dehydrated in acetone and air dried.

Fig. 7. Etched vertical fracture plane of the nacreous layer in <u>Mytilus edulis</u> L., to show that the consecutive nacreous tablets are continuous with each other. The positions of the dissolved inter-lamellar organic membranes are indicated by arrows. The preparation was treated with sodium hypochlorite solution for 2 days, and etched with 25% glutaraldehyde solution at pH 6.2 for 2 days at a temperature of +32°C.

The inter-lamellar organic membranes are composed of trabeculae and inter-trabecular membraneous bridges.[3,6,7] The inter-trabecular bridges are nearly always soluble in EDTA (sodium ethylene diamine tetra-acetate).

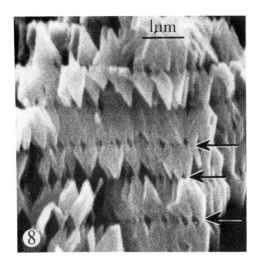

Fig. 8. Vertical fracture plane of the nacreous layer in _Gibbula cineraria_ (L.), etched in a similar way as the preparation in Figure 7. The arrows indicate the positions of the dissolved inter-lamellar organic membranes.

Conversely, the trabeculae are insoluble in EDTA. With all probability, the continuation between the consecutive nacreous tablets takes place at the inter-trabecular membraneous bridges, whereas at the trabeculae, the consecutive tablets are separated. This means that at the membraneous bridges, the inter-lamellar membrane is calcified. This conclusion is in agreement with my observations that by dissolving the inter-lamellar organic membranes with the sodium hypochlorite solution, the consecutive mineral lamellae usually do not become separated. However, the adhesion between the lamellae is weak and can be easily disrupted by gentle pressure with a needle.

Several vertical fracture planes of the nacreous layer in _Mytilus_ were demineralized with the glutaraldehyde-acetic acid solution. In these preparations, there remains a well defined organic residue. This residue forms numerous vertical cords which extend between the horizontal inter-lamellar organic membranes (Fig. 9). It is probable that these cords are remnants of organic membranes, which originally enveloped the vertical crystalline components of the tablets. Similar "intra-crystalline membranes" have been figured by Iwata in demineralized ultra-thin sections of the nacre in _Nautilus_.[8]

Discussion

As pointed out in previous papers, the structure of the nacreous tablets is complicated, mainly because of the twinning of the aragonite crystals.[1,2] In the present paper, new features of this twinning are described. The nacreous tablets in bivalves seem to consist of two layers, each of which shows a different twinning pattern. The outer layer is composed of four crystal individuals, which probably are cyclically twinned, whereas

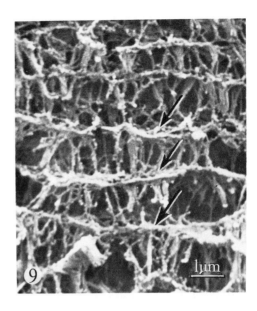

Fig. 9. Demineralized vertical fracture plane of the nacreous layer in _Mytilus edulis_ L., to show numerous vertical cords of organic matrix which extend between the inter-lamellar organic membranes (indicated by arrows). The preparation was treated with dilute sodium hypchlorite solution for 2 hours, and etched with glutaraldehyde-acetic acid solution for 3 hours at a temperature of $+32^{\circ}C$. It was dehydrated in ethanol, transferred into amyl acetate, and dried in a critical point apparatus with CO_2.

the inner layer is probably polysynthetically twinned and forms crystalline laths. The arrangement of the soluble organic matrix seems to be different between these two layers.

It should be emphasized that our knowledge of the structure of the nacreous tablets is still imperfect and that many structural details cannot be fully explained. This is true with regard to the angular crystalline plates, which appear in the etched nacreous tablets. These plates seem to correspond to the "acicular crystallites" described in my earlier paper.[9]

It was previously thought that the inter-lamellar organic membranes are uncalcified in their entire extension. However, the results forwarded in the present paper indicate that only the trabeculae are uncalcified, whereas the inter-trabecular membraneous bridges are calcified. The consecutive nacreous tablets continue through these membraneous bridges. Wada also assumed that there is a continuation between the consecutive tablets through "pores" of the inter-lamellar organic membranes, and that the new tablets grow epitaxially with the underlying tablets.[10]

Acknowledgement

This study was supported by grants 287-17, 287-18 and 287-101 from the Swedish Natural Science Research Council.

References

1. H.Mutvei, "The nacreous layer in Mytilus, Nucula and Unio (Bivalvia). Crystalline composition and nucleation of nacreous tablets", Calcif. Tiss. Res. 24 (1977) 11-18.
2. H.Mutvei, "Ultrastructural characteristics of the nacre in some gastropods", Zool. Scr. 7 (1978) 287-296.
3. G.Goffinet, Ch. Grégoire and M.-F. Foss-Foucart, "On the ultrastructure of the trabeculae in the interlamellar membranes of the nacre conchiolin of the Nautilus shell", Arch. intern. Physiol. Biochim. 85 (1977) 849-863.
4. M.A.Crenshaw and H.Ristedt, "Histochemical and structural study of nautiloid septal nacre", Biomineralisation 2 (1975) 1-8.
5. H.Mutvei, "Ultrastructure of the mineral and organic components of molluscan nacreous layers", Biomineralisation 2 (1970) 48-61.
6. H.Mutvei, "On the micro- and ultrastructures of the conchiolin in the nacreous layer of some recent and fossil molluscs", Stockholm Contr. Geol. 20 (1969) 1-17.
7. Ch. Grégoire, "Ultrastructure des composants organiques des coquilles de mollusques", Haliotis 2 (1972) 51-79.
8. K.Iwata, "Ultrastructure of the conchiolin matrices in molluscan nacreous layer", J. Fac. Sci. Hokkaido Univ. 17 (1975) 173-229.
9. H.Mutvei, "Ultrastructural studies on cephalopod shells. Part 1: The septa and siphonal tube in Nautilus", Bull. geol. Instn Univ. Uppsala (N.S.) 3 (1972) 237-261.
10. K.Wada, "Nucleation and growth of aragonite crystals in the nacre of some bivalve molluscs", Biomineralisation 6 (1972) 141-159.

Discussion with Reviewers

T. Watabe: Etch figures represent early stages of crystal dissolution, and they vary by the nature of solvent and duration of etching. The structures shown by etching do not necessarily mean that the crystals are actually composed of such structures. What evidence do you have to suggest that the lath and rhombic plates revealed by your etching method are 1) the real components of the crystal, and 2) represent polysynthetic or other twins?
Author: Both the horizontal and vertical crystalline elements of the nacreous tablets are associated with an organic matrix. They seem therefore to be real components of the tablets. However, the actual shape of these components is still poorly known. The horizontal components appear as laths or lamellae, whereas the vertical components are acicular or plate-like in shape. I still believe that the vertical components in the nacreous tablets correspond to the acicular components in the aragonite spherulites.[5,9] The interpretation that the horizontal components (laths, lamellae) correspond to the polysynthetically twinned lamellae is based on the common occurrence of such lamellae in non-biogenic aragonite.

K. M. Towe: Figures 1-3 are offered in support of the idea that each nacre tablet is "composed of two layers". Yet the two layers are not seen in the cross-sectional views of figures 5-8. Why? In your previous paper [1] you studied the same mollusks with the same etchants. But instead of two layers you claimed that the tablets were made of numerous horizontal lamellae. How do you account for this difference?
Author: One of the main differences between the two layers of the nacreous tablets is the different arrangement of the organic matrix. It is extremely difficult to preserve this matrix during the process of etching and demineralization because it is highly sensitive to slight changes in the composition (pH, aldehyde concentration, etc.) and temperature of the etchant, and the duration of the etching. Also, the methods used for dehydration and drying have a great influence on the state of preservation of this matrix. The above mentioned conditions may explain the slight discrepancy which exists between my previous and present results. Special studies are required to show the two-layered structure of the nacreous tablets in vertical sections. For example, it is possible that the two-layered structure is present in the preparation in Figure 9 but that it is obscured by the shrinkage of the organic matrix.

R. M. Dillaman: How do you relate the vertical rhombic plates (Figs. 7-8) to the horizontal lamellae seen in the less soluble crystal individuals treated only with sodium hypochlorite?
Author: The horizontal lamellae seem to be restricted to the side faces of the incipient nacreous tablets of Mytilus edulis. These lamellae have not been observed in fully grown tablets used for CPC-glutaraldehyde etching. The interspaces between these lamellae probably become filled in.

T. Watabe: How thick are the outer and inner layers of the crystal tablets?
Author: In Mytilus edulis, the inner layer is about 0.8 to 1.0 millimicrons, and the outer layer about 0.2 millimicrons in thickness. In Nucula sulcata and Unio tumidus, the inner and outer layers both are about 0.4 millimicrons in thickness.

K. M. Towe: The laths in a given tablet all appear to be parallel to one another (Figs.1,4). Can you make it consistent with twinning of the crystals?
Author: It is probable that the laths represent parallel lamellae, which are the result of polysynthetic (lamellar) twinning.

K. M. Towe: How can the rhombs so clearly seen in Figure 8 be related to the crystallography of orthorhombic aragonite?
Author: On closer examination these plates were seen not to be regularly rhombic in shape. In the text, I now call them the angular plates. The etch figures between the plates are not clear enough to permit definite conclusions about the crystallographic orientation.

SCANNING ELECTRON MICROSCOPY/1979/II
SEM Inc., AMF O'Hare, IL 60666, USA

SHELL DISSOLUTION: DESTRUCTIVE DIAGENESIS IN A METEORIC ENVIRONMENT

B. M. Walker

Department of Geography
University College London
Gower Street
London WC1 6BT
U.K.

Abstract

An in depth study of Quaternary aragonitic mollusca from South East Iran was carried out prior to Carbon-14 dating. SEM comparison with present-day examples has revealed a number of ultrastructural diagenetic features. The majority of these features are a product of carbonate dissolution in a meteoric environment, that is to say as a result of exposure to rain water. The process has affected these marine carbonates during the last 6,000-25,000 years since their emergence.

Dissolution features which have occurred to varying degrees in the shell material include: (1) etching of the outer (external and growth) surfaces accentuating the underlying ultrastructure; (2) surface etching of the individual internal ultrastructural units. Fracture sections show that these features are the result of a single inward moving gradational zone of marginal dissolution and are related to the shell's organic framework and its mineralogical structure. Features and factors involved in their production are discussed and illustrated for the ultrastructural types found in the dated Bivalvia and Gastropoda examined in this study.

SEM has proved to be a valuable tool in the detection of Carbon-14 contamination and has provided an example of submicroscopic destructive carbonate diagenesis under surface conditions.

KEY WORDS: Quaternary, Shell, Mollusca, Ultra-structure, Carbon-14, Dissolution, Diagenesis, Carbonate, Meteoric, Subaerial

Introduction

During the last decade the SEM has revealed many ultrastructural details of shell material which could not be detected using optical microscopes. The study of present-day shell structures with the SEM was begun by Wise[1], who commented on the wide variety of scientific disciplines interested in the topic. The work described in this paper stems from the use of fossil shell carbonate as a dating material.

Sedimentological and tectonic studies in the Makran of South East Iran have emphasized the importance of establishing an accurate chronology[2]. As a result, a large number of aragonitic and calcitic molluscan samples from a series of raised beaches along this coastline are being dated using the Carbon-14 method. This dating is being carried out by a number of international Carbon-14 laboratories. The results will establish rates of uplift and help in an understanding of tectonic history in an unstable region.

The use of shell carbonate for Carbon-14 dating is considered to be less reliable than other sources for a number of reasons. The main problem is that of contamination[3], particularly carbonate alteration and/or cementation. As a result, the author thought it desirable to investigate samples with the SEM as well as with other established geological and geochemical techniques. The use of the SEM to detect post-depositional alteration has apparently not previously been carried out in dating work, although the method was used recently in a study of oxygen-isotope ratios in shell material[4].

An extensive SEM study of the fossil aragonitic material in question has revealed a number of ultrastructural diagenetic features. Comparison with modern Iranian specimens and with published observations suggests that these features are largely the result of natural etching or carbonate dissolution. As a result of uplift the shells have spent most of their post-depositional history (up to 25,000 years) near the ground surface exposed to meteoric water but above the groundwater zone.

The purpose of this paper is to describe and discuss the features of dissolution observed in the dated fossil species Anadara uropigmelana,

Asaphis deflorata, Purpura rudolphi and Oliva sp. Similar destructive features have been described on shells and shell fragments from present-day marine environments[5,6], but not in detail from meteoric environments. There are also some features of constructive diagenesis in the form of neomorphic alteration and surface cements. Although the area of study may not be a typical meteoric environment, the processes of submicroscopic destruction which are descibed here are of importance both in Carbon-14 dating and carbonate diagenesis[7].

Location and Environment

Shell samples were collected from a number of fossil raised beaches along the Makran coast of South East Iran. As they were intended for dating, unabraded (little post-depositional movement) and, if possible, only weakly cemented shells were selected. Some of these source fossiliferous deposits displayed slight calcitic cementation.

This area of Iran receives very low, infrequent rainfall (mean annual total 15.0 cm or less). When rain does occur there is rapid infiltration through the recent porous deposits, and the fossil beach surfaces are unaffected by permanent groundwater conditions. Temperature on the other hand is very high (mean annual temperature 27.0°C).

Preparation and Method

Bivalve and gastropod shell samples were cleaned by dry mechanical brushing and grinding to remove superficial sand and carbonate prior to possible dating. Representative specimens were broken to provide fracture sections for SEM study. They were not artificially etched as has been recommended for present-day shells[8], to avoid the introduction of any preparational artefacts. The sections were then mounted and coated with approximately 50nm of gold in a Diode Sputtering unit. The stub-mounted specimens were viewed in a Cambridge Stereoscan Mk II. Observations were made on both the fracture surfaces and the outer surfaces of these shells. Micrographs were taken at a number of magnifications (from 10 x to 20,000 x) to facilitate reference to the original specimen.

Species and Ultrastructural Types

The following aragonitic species provided possible dating material and therefore were studied under the SEM.

Bivalvia

    (a) Anadara uropigmelana

    (b) Asaphis deflorata

Gastropoda

    (a) Purpura rudolphi

    (b) Oliva sp.

It should be noted that Purpura is thought to have a mixed mineralogy, but the observations made were in aragonitic structures.

Comparisons between samples can be made as some beaches provided two or more species and the same species was found on beaches of different ages.

The following observations were made on two ultrastructural types: crossed-lamellar and composite prismatic. They were studied purely because of their abundance in the species mentioned previously. The reader is referred to MacClintock[9], Taylor, Kennedy and Hall[10,11] and Wise[1,12] for definitions of these structures and the terminology used in their description.

Observations

The observations have been divided into groups depending where the features occur in the shells but this is not to imply any difference in the processes involved as will be shown later.

Features of the Outer Surfaces

External and growth surfaces showed varying degrees of corrosion of the shell which accentuated the underlying ultrastructure. Figure 1 shows severe corrosion of the crossed-lamellar structure on the external surface of a specimen of Anadara uropigmelana which is 7,000 years old. A considerable number of the smallest structural units, the third-order lamellae, appear to have been completely removed. Those that are present have pointed terminations presumably due to etching. This surface corrosion has produced a considerable microporosity following the loss of the original organic films between lamellae, as well as the etching and even the total dissolution of individual aragonite lamellae.

Features of the Fracture Surfaces

Shell fracture sections show that corrosion is not confined to the outer surface but also occurs internally. The third-order lamellae (Fig.2) 100 um from the exterior surface in a fossil Anadara uropigmelana show no differences to those seen in present-day examples. In comparison the same structure (Fig.3) in an Anadara uropigmelana from another location but of similar age (20,000 - 25,000 years) shows signs of etching 200 - 300 μm from the exterior surface. The lamellae, originally lath shaped units with planar surfaces and straight edges, have changed. The surfaces are pitted and the edges notched, resulting in a generally more rough appearance. It appears that the smallest shell structural units are being progressively dissolved.

The sequence of micrographs in figures 4, 5 and 6 with increasing magnification provides an interesting view of internal corrosion of an area in Purpura rudolphi 200 μm from the exterior surface. In this case the sample is about 6,000 years old. The crossed-lamellar structure in figure 4, when viewed at higher magnification (Fig.5), is seen to have two distinct zones of possible dissolution. Fissures occur along the boundary between first-order lamellae and more irregularly across these lamellae. These latter fissures are most clearly seen in first-order lamellae where the long axis of the third-order lamellae are emerging almost directly from the micrograph (i.e. central and outer two lamellae

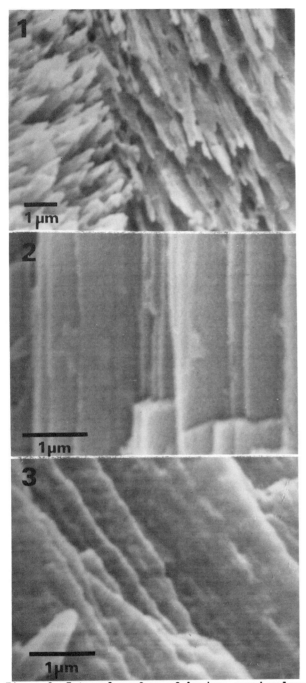

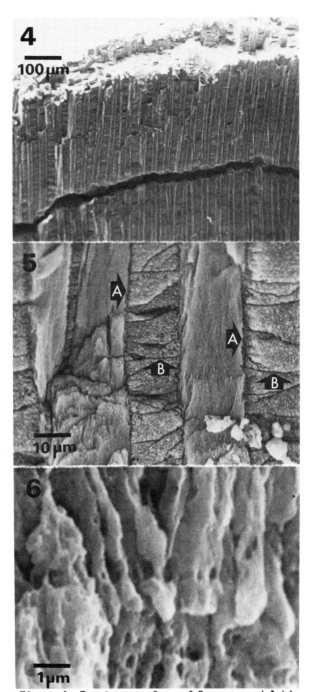

Figure 1. External surface of <u>Anadara uropigmelana</u> (7,000 years old). Deep surface etching of the underlying crossed-lamellar structure.

Figure 2. Fracture surface of <u>Anadara uropigmelana</u> (20,000-25,000 years old). Unaltered third-order lamellae in crossed-lamellar structure 100 μm from the exterior surface.

Figure 3. Fracture surface of <u>Anadara uropigmelana</u> (20,000-25,000 years old). Etched third-order lamellae in crossed-lamellar structure 200-300 μm from the exterior surface in a similar position within the shell of Fig.2 but from another site.

Figure 4. Fracture surface of <u>Purpura rudolphi</u> (6,000 years old). First-order lamellae in crossed-lamellar structure 200 μm from the exterior surface.(Crack due to preparation).

Figure 5. As Fig.4, dissolution along boundary between first-order lamellae (A) and across first-order lamellae (B).

Figure 6. As Fig.4, severe corrosion of third-order lamellae producing a large internal microporosity.

465

in Fig.5).This may suggest that the fissures have a similar plane of orientation. In figure 6 at a much higher magnification the third-order lamellae are themselves heavily corroded and a large internal surface area has been produced.

Similar evidence for internal etching is found in the composite prismatic structure seen in Asaphis deflorata (Fig.7)which is also about 6,000 years old. This exterior structural layer has been etched and is therefore porous, but both these characteristics decrease away from the outer surface (base of Fig.7). Sixty μm from the exterior, etching is still evident at higher magnification (Fig.8). In this case internal corrosion has taken place in a zone up to 100 μm thick. This marginal zone of dissolution is gradational, being strongest where it meets the surface.

Features of the Original Tubules

Tubules are cylindrical pores which pass through the whole shell surrounded by solid shell structure,and they are a primary feature of Anadara. Figure 9 shows that dissolution can take place in a narrow zone (less than 10 μm) adjacent to the original tubule wall. In this fossil Anadara uropigmelana 2 mm from the exterior surface and 1-2 mm from the interior growth surface the crossed-lamellar structure beyond this tubule dissolution zone is unaltered and so lies inside any marginal zones of corrosion or dissolution. In detail (Fig.10) it is again clear that third-order lamellae have been attacked and selectively removed.

Discussion

The results which have so far been outlined all provide evidence that shell carbonate is being dissolved as a result of destructive diagenesis in a meteoric environment. The observed features are not unlike those described from a present-day marine environment[5,6], where the etching or dissolution took place because the marine waters were assumed to be undersaturated with calcium carbonate. Alexandersson[5] noted the similarities between the two environments and showed some shell material from Barbados which had suffered diagenesis under surface conditions.

Subaerially leached shells have commonly been described as "chalky"[13]. This is clearly a description of the visual appearence after submicroscopic dissolution. The Iranian samples sometimes displayed a superficial "chalky" appearence, although totally "chalky" shells would not have been collected for dating as it was previously thought that this meant that all such shells were contaminated. This visual superficial loss of normal shell lustre corresponds to the SEM observed severe marginal zones of dissolution.

Surface etching and marginal zones of dissolution or corrosion should not be regarded in the cases reported here as separate areas of dissolution or what have been defined in the marine examples[6] as stages in order of increasing attack. Both sets of features are the result of a single inward moving gradational zone of dissolution (Fig.7). The outer surface is the most exposed. It is also where the maximum removal has taken place with time. This marginal zone in the specimens studied was commonly between 100 μm and 300 μm in thickness. The etching adjacent to the original tubules was clearly similar to the marginal dissolution in that it is a zone of attack away from a free surface. In this case the depth of attack is an order of magnitude less than that in the marginal zones.

All the dissolutional features are reflections of the underlying ordered biogenic framework in each shell and ultrastructural type. The selective nature of some of this dissolution (Fig.5) probably results from differences in the organic network. Organic films play an important role as it is generally held that their breakdown produces an agent of carbonate dissolution. Their removal then provides a suitable pathway for further corrosive agents, in this case meteoric water[14]. This organic breakdown is poorly understood but in the fresh parts of these shells organic material is still present. It was not possible to show any relationship between time and the extent or severity of dissolution in these samples. It was possible to suggest that the position and situation of individual shells within the original sediments (source deposits) may be as important as time.

There is some visual evidence of constructive diagenesis,and under SEM some examples of calcitic neomorphic replacement are found (Fig.11) with aragonite relics[15]. The prior dissolution and resultant microporosity help to explain the process by which these ultrastructural units are preserved within this calcite[16]. The weak calcitic cements which lightly bind these deposits can only have come from the shells themselves. This autochthonous source is only revealed fully by the dissolution features described, a factor which may have been underestimated in the past[7].

As this study demonstrates, the SEM provides an extra valuable tool in the detection of internal shell alteration. Carbon-14 contamination should not be a problem when shells have suffered only dissolution. As concluded previously[16], and confirmed in these examples, x-ray diffraction shows that the etching process does not alter the shell's aragonitic mineralogy. But where open system replacement has occurred dating would be pointless without further pretreatment. In the example of Oliva sp. (Fig.11) this would require the further mechanical removal of neomorphic calcite if practical. This study also provides another example of subaerial carbonate diagenesis and highlights the appreciation of possible submicroscopic destructive processes.

Acknowledgements

The author would like to thank Dr. C. Vita-Finzi who collected the samples, made available the Carbon-14 dates and read a number of drafts of this paper. Also Mr. S. Phethean and Mr. C. Cromerty for technical assistance, and Mr. N.

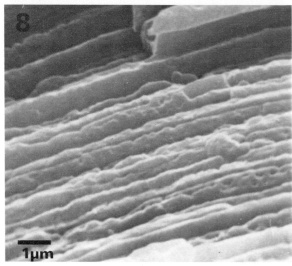

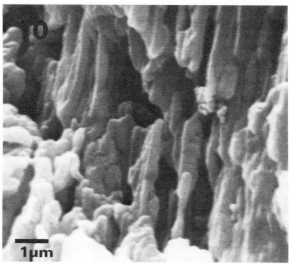

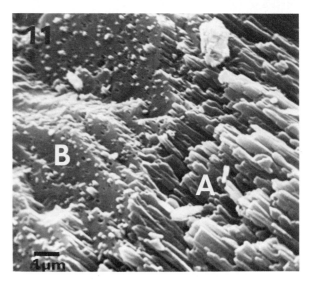

Figure 7. Fracture surface normal to external surface of <u>Asaphis deflorata</u> (6,000 years old). Marginal zone of corrosion in composite prismatic structure. Dissolution decreasing from the outer surface (just off base of figure) towards the centre of the shell (top of figure).

Figure 8. As Fig.7, detailed etching of composite prismatic structure 60 μm from the exterior surface.

Figure 9. Fracture surface of <u>Anadara uropigmelana</u>. The original tubules (A) passing through crossed-lamellar structure, 2-3 mm from the exterior surface and 1-2 mm from the growth surface.

Figure 10. As Fig.9, dissolution around original tubule producing etching and some selective removal of third-order lamellae.

Figure 11. Fracture surface of <u>Oliva</u> sp. Boundary between etched crossed-lamellar structure (A) and neomorphic calcite with relic aragonite lamellae (B).

Moore and the Department of Crystallography, Birkbeck College, London for the use of the SEM.

References

1. Wise, S.W., Study of Molluscan Shell Ultrastructures. SEM/1969, IIT Research Institute, Chicago, IL, 60616, 205-216.
2. Vita-Finzi, C., Quaternary Deposits in the Iranian Makran., Geogrl. J., 141, 1975, 415-420.
3. Olsson, I.U., Some problems in connection with the evaluation of C^{14} dates., Geol. Foren. Stockh. Forh., 96, 1974, 311-320.
4. Buchardt, B., Oxygen isotope ratios from shell material from the Danish Middle Paleocene (Selandian) deposits and their interpretation as paleotemperature indicators., Palaeogeogr., Palaeoclimatol., Palaeoecol., 22, 1976, 209-230.
5. Alexandersson, E.T., Micritization of Carbonate Particles: Processes of Precipitation and Dissolution in modern shallow-marine sediments., Bull. geol. Instn Univ. Upsala N.S., 1972, 201-236.
6. Alexandersson, E.T., Petrographic Saturometry in marine carbonate sediments., SEM/1978/I, SEM. Inc., AMF O'Hare, IL, 60666, 503-509.
7. Bathurst, R.G.C., Carbonate Sediments and their Diagenesis., Elsevier, Amsterdam, The Netherlands, 1971, Chapter 8.
8. Wise, S.W. and Hay, W.W., Scanning Electron Microscopy of Molluscan Shell Ultrastructures I. Techniques for polished and etched sections., Trans. Amer. Microsc. Soc., 87, 1968, 411-418.
9. MacClintock, C., Shell Structure of Patelloid and Bellerophontoid Gastropods (Mollusca)., Peabody Mus. Nat. Hist. Yale Univ. Bull., 22, 1967, 1-140.
10. Taylor, J.D., Kennedy, W.J. and Hall, A., The shell structure and mineralogy of the Bivalvia. Introduction, Nuculacea-Trigonacea., Bull. Br. Mus. Nat. Hist., Zool., Suppl., 3, 1969, 1-125.
11. Taylor, J.D., Kennedy, W.J. and Hall, A., The shell structure and mineralogy of the Bivalvia. II, Lucinacea-Clavagellacea, conclusions., Bull. Br. Mus. Nat. Hist., Zool. 22(9), 1973, 225-294.
12. Wise, S.W., Shell ultrastructure of the Taxodont Pelecypod Anadara notabilis. Eclogae geol. Helv., 64(1), 1971, 1-12.
13. Land, L.S., Diagenesis of skeletal carbonates., J. Sedim. Petrology, 37, 1967, 914-930.
14. James, N.P., Diagenesis of Scleractinian corals in the subaerial vadose environment., J. Paleontology, 48, 1974, 785-799.
15. Sandberg, P.A., Schneidermann, N. and Wunder, S.J., Aragonitic ultrastructural relics in calcite-replaced Pleistocene skeletons. Nat. Phys. Sci., 245, 1973, 133-134.
16. Pingitore, N.E., Vadose and phreatic diagenesis: Processes, products and their recognition in corals., J. Sedim. Petrology, 46, 1976, 985-1006.

Discussion with Reviewers

M.R. Carriker: Your case for "internal corrosion" is well made, but I wonder, if rather than internal corrosion, your fracture sections may not be demonstrating a weakening of the shell structure which on parting by fracturing resembles internal corrosion? To test this hypothesis, you might have to examine opposing matching surfaces of a fracture with SEM.
Author: A "non-fit" of these surfaces could result from slight breakage and disruption along the fracture plane. However, I think my case is justified by the gradational severity observed in these zones of internal corrosion at the margins of the Iranian shells, a feature consistently found both in many of the shells studied and from different parts of the same shell, which is independent of ultrastructural type and shows maximum loss at the exposed free surface. Secondly similar features are associated with the zone around original tubules, in particular at the back of these tubules (Fig.10) where the features could not be artefacts of fracturing. The extent to which weakening is only a result of the loss of organic material is hard to assess. As a precursor of carbonate dissolution, organic weakening or loss may precede aragonite removal and so give a zone wider than the zone of true dissolution. Therefore it may be a matter of opinion to what extent the features in Figure 8 for instance are the result of dissolution and/or the loss of intracrystalline organic material.

S.W. Wise: Could the fissures across the first-order lamels shown in Figure 5 have possibly been induced by fracturing of the shell?
Author: They may partially be the result of the fracturing but I have not observed this feature in crossed-lamellar structure from the other bivalve species studied. I feel that they may reveal the presence of some more major intercrystalline organic films which are inherent lines of weakness or have been weakened by the process of dissolution, in this gastropod species.

M.R. Carriker: Is there any information on the pH of water percolating through the soil into ground water, or of the rainfall, on the Makran coast? Such information could help in explaining dissolution of the mineral components of your shells.
Author: Unfortunatly I know of no such data for this part of Iran.

M.R. Carriker: If carbonate specimens are not dry, considerable charging can occur in the SEM; how much did you dry your specimens before coating?
Author: Shells were collected dry and no further systematic drying was carried out. Only slight charging was found to be a problem in a few of these specimens. However, in the future I will make a point of checking that moisture is not present before coating.

SCANNING ELECTRON MICROSCOPY/1979/II
SEM Inc., AMF O'Hare, IL 60666, USA

FORMATION OF A DISSOLUTION LAYER IN MOLLUSCAN SHELLS

D. A. Wilkes and M. A. Crenshaw*

Curriculum in Marine Sciences
University of North Carolina
Chapel Hill, NC 27514

*Curriculum in Marine Sciences
Dept of Pedodontics and
Dental Res. Center
University of North Carolina
Chapel Hill, NC 27514

Abstract

Dissolution layer formation in the shell of Geukensia demissa was induced by removing the mussel from water and compared to the dissolution layer of Mercenaria mercenaria. The dissolution layer developed only on the inner shell surface inside the pallial line in both bivalves. The amount of shell solution by the mussel was not related to time out of water. The pattern of the development of the dissolution layer in the mussel was more complex than in the clam. The complexities of this pattern and its repair were related to the periodic opening of its valves by the mussel when it was exposed to air. It is suggested that the dissolution layer in some intertidal molluscs may not be a good tidal cycle marker.

KEY WORDS: Shell Dissolution, Decalcification, Nacre, Prismatic Aragonite, Anaerobiosis, Intertidal Adaptations, Molluscs, Geukensia demissa, Mercenaria mercenaria

Introduction

When bivalves close their shells they become anaerobic.[1,2] During the anaerobic periods, acidic metabolic end products are formed[3] which dissolve previously deposited shell.[1,2] This shell solution produces a chalky layer on the inner shell surface that is readily distinguished from the normal crystal surface. This dissolution layer is covered by normal crystal structure during subsequent aerobic periods when the valves are opened.[4] In this manner, a record of anaerobic periods to which the bivalve has been subjected is preserved in the shell structure.

An intertidal mollusc would be expected to retain such a record of tidal cycles because a dissolution layer would be produced at each low tide when the animal was anaerobic. However, the Atlantic ribbed mussel, Geukensia demissa, obtains oxygen from air by periodically opening its valves when it is exposed during low tide.[5] This adaptation may alter the formation of the solution layer.

Here, we report the patterns of dissolution layer formation and its repair in Geukensia demissa when it was removed from water and the subsequent formation of normal nacre when the mussel was returned to water.

Materials and Methods

The mussels were collected near Morehead City, North Carolina in mid-November and divided into two groups. Each group included equal numbers of three shell-length classes: small - 1 to 3 cm, medium - 3 to 6 cm, large - 6 to 10 cm. The animals were left overnight in a running sea water table at 20°.

The next morning, the first group of animals was removed from the water, placed on a dry bench for four hours and, then, returned to the water. At this time, the second group was removed from the water and placed on the dry bench. Note was taken of valve movements throughout the duration of the experiment. Dissolution layer formation was studied in the second group. The subsequent formation of normal nacre was followed in the first group.

At intervals from 5 minutes to 20 hours,

one small, two medium and one large mussel from each group were opened by cutting the adductor muscles. The mantle and extrapallial fluids were collected for calcium determination by atomic absorption spectrophotometry.

The shells were cleaned free of tissue, rinsed in tap water and air dried. Representative samples were broken from several areas of each shell. The organic matrix was removed by soaking the shell fragment in 2.5% sodium hypochlorite for two hours. The fragments were rinsed with water, air dried and sputter coated with gold. Micrographs were taken with an ETEC Autoscan microscope at 20kV.

Shells of Mercenaria mercenaria that were out of the water for two hours were prepared for SEM observation in an identical manner. The clam shells served as reference material for dissolution layer formation. The shells from several specimens of G. demissa that had been left, undisturbed for 24 hours, in the sea water table were also examined by SEM.

Results

Formation of the Dissolution Layer

The chalky dissolution layer extended over the entire inner surface inside the pallial line in the M. mercenaria shells. Dissolution was evidenced by the irregular edges of the crystals, by their poor organization and by the large voids in the crystal layers (Fig. 1). The area outside the pallial line did not undergo dissolution. Here, the crystals had sharp edges, they were well organized and filled the available space (Fig. 2).

In Geukensia demissa, shell dissolution was also always confined to the area inside the pallial line. At five minutes, solution from the center of the nacreous tablets was evident (Fig. 3). Small crystals having sharp boundaries were scattered over the innermost layer of dissolving nacreous tablets. The small crystals did not have the normal hexagonal outline of normal incipient nacreous tablets, and they seemed to be the result of surface reprecipitation.

Dissolution of the nacreous tablets did not take place outside the pallial line. In the same shell showing evidence of dissolution inside the pallial line (Fig. 3), nacreous tablet growth outside the pallial line was evident. Here, the incipient tablets were solid and had regular hexagonal outline. Numerous small tablets were present, and growth at screw dislocations was observed (Fig. 4).

As dissolution progressed in G. demissa, a chalky dissolution layer much like that observed in M. Mercenaria (Fig. 1) was formed (Fig. 5). However, the dissolution layer in the mussel never developed over the entire surface inside the pallial line as it did in the clam. In areas not covered by the dissolution layer, normal nacreous tablets were present (Fig. 6). The transition zone from the fully developed dissolution layer (Fig. 5) to normal nacre (Fig. 6) spanned a distance of 50 to 100 μm.

The dissolution layer developed in G. demissa from two foci. The first and larger focus was located ventral to the posterior adductor muscle, and it advanced anteriorly and, then, dorsally. The second focus started at the inner border of the pallial line posterior to the anterior adductor muscle and advanced dorsally and laterally.

The area of the inner shell surface that was covered by the dissolution layer was not related to the time an animal was out of water. Small mussels, generally, had a larger portion of the inner shell surface covered by dissolution layer. A dissolution layer was also formed by undisturbed and submerged mussels. The animals would periodically close their valves and form a limited dissolution layer.

The absence of a correlation between the area of the dissolution layer and time out of water coincided with the lack of a correlation between the calcium concentration in the fluids and time out of water. The calcium concentration in G. demissa was 8.4 ± 0.7 mM (mean $\pm$ standard deviation), and there was no significant difference between any of the time periods. M. mercenaria showed evidence of shell solution in that the fluid concentration in these animals was 9.2 ± 0.2 mM. The water from the water table had a calcium concentration of 8.1 ± 0.1 mM.

Changes in the Dissolution Layer

All of the mussels that were out of water opened their valves at irregular intervals and for variable times. Generally, the fraction of time that the valves were open increased with exposure time. The longer periods of valve opening were associated with structural changes in the dissolution layer. These changes began at the margins of the dissolution layer and progressed centrally. The earliest change observed was the appearance of small, sharp-edged crystals on the surface of the dissolution layer (Fig. 7). Later, a layer of prismatic aragonite was formed (Fig. 8). In some cases, the entire dissolution layer appeared to have been reorganized into prismatic aragonite. In others, the reorganization was confined to the inner surface. Often, a second chalky dissolution layer was formed over the prismatic aragonite. Normal nacre was deposited over the margin of the dissolution layer in a few cases although the animal was out of water (Fig. 9). When the animals were returned to water, normal nacre was deposited over the inner shell surface.

One of the simpler sequences is shown in Figure 10. The dissolution layer that was formed the night before and which was used as a reference is in the lower right corner. Above this is the normal nacre that deposited overnight. The next layer is the unmodified dissolution layer that was formed while the mussel was out of water for four hours. This is covered by normal nacre that was deposited after the animal was returned to water for one hour.

Discussion

Mercenaria mercenaria has been an excellent experimental animal for demonstrating that the shell is dissolved by acid produced when the

Fig. 1. Dissolution layer in <u>Mercenaria mercenaria</u>. Bar = 1 μm.

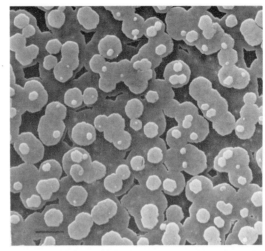

Fig. 4. Nacreous tablet growth outside the pallial line. Same valve as Fig. 3. Bar = 2 μm.

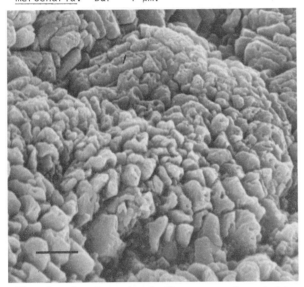

Fig. 2. Growing edge in <u>M. mercenaria</u> outside the pallial line. Bar = 0.1 μm.

Fig. 5. Advanced (chalky) dissolution layer in <u>G. demissa</u>. Bar = 1 μm.

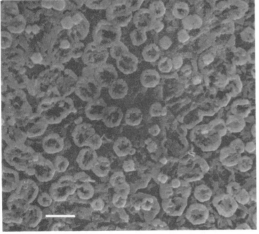

Fig. 3. Initial dissolution of nacre inside the pallial line in Geukensia demissa. Bar = 2 μm.

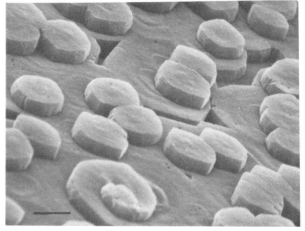

Fig. 6. Nacreous tablets inside the pallial line where dissolution layer was not developing. Same valve as Fig. 5. Bar = 2 μm.

animal metabolizes anaerobically.[1,2] When the clam is removed from water, its valves remain closed with sufficient tightness to prevent significant gas exchange with air.[1,2,6] Therefore, there is a high and positive correlation between time out of water with the amount of shell solution and with the extent to which the dissolution layer is developed.[1,2] Conclusions drawn from data obtained from this model may not be generally applicable to bivalves, however.

When Geukensia demissa is taken out of water, it periodically obtains oxygen from air by frequently opening its valves.[5] Therefore, it alternates between anaerobic and aerobic metabolism when out of water. This adaptation to the intertidal environment accounts for our observations that there was no correlation between time out of water and the amount of shell solution or the limited extent to which the dissolution layer was developed. In fact, some mussels were able to obtain sufficient oxygen from air to deposit nacre over portions of the anaerobically-formed dissolution layer.

However, shell solution is associated with anaerobic metabolism, and shell deposition occurs during aerobic metabolism. This does not imply that mineral deposition is an aerobic process. Net calcification may be prevented by the decrease in the pH that is caused by anaerobic metabolism.[1,2,3,6] After a chalky dissolution layer was formed and the animals gained limited access to oxygen, the chalky dissolution layer was usually replaced by prismatic aragonite. This formation of prismatic aragonite may not be an active biological process. It may be caused by a gradual increase in pH as the acid, produced anaerobically when the valves are closed, is metabolized aerobically when the valves later open. When animals were provided an ample supply of oxygen by returning them to water, the dissolution layer was covered by freshly deposited normal nacre.

The two foci of shell solution were located under areas of greatest tissue mass in the muscle. Therefore, one would expect the tissue in these areas to become anaerobic sooner, when the valves closed, than in other areas where there was less tissue. Conversely, the tissues overlying the solution foci would also be supplied with proportionally less oxygen when the valves opened in air.

The inner shell surface outside the pallial line does not appear to be part of the alkali reserve that buffers anaerobically produced acid. This difference in the two zones of the shell was present in the clam as well as the mussel. It may reflect a functional zonation in the mantle. Alternatively, sufficient oxygen may diffuse between the closed shell margins to allow the mantle edge to continue aerobic metabolism.

Observations reported here, show that mussels close their valves and begin dissolution layer formation even when they are submerged. Similar observation of shell solution by submerged and undisturbed bivalves has been previously reported.[3,6] These facts make any cor-

relation of dissolution layers with tidal cycle tenuous.[4]

Acknowledgements

This research was supported, in part, by PHS grants DE 02668 and RR 05333. The authors wish to thank Mr. Douglas Wilson for his aid in preparing the micrographs. We should also like to thank the reviewers for their comments.

References

1. L. P. Dugal, "The use of calcareous shell to buffer the product of anaerobic glycolysis in Venus mercenaria," J. Cell Comp. Physiol. 13, 1939, 235-251.
2. M. A. Crenshaw and J. M. Neff, "Decalcification at the mantle-shell interface in molluscs," Am. Zoologist 9, 1969, 881-885.
3. A. deZwaan and T. C. M. Wijsman, "Anaerobic metabolism in bivalvia," Comp. Biochem. Physiol. 54B, 1976, 313-342.
4. R. A. Lutz and D. C. Rhoads, "Anaerobiosis and the theory of growth line formation," Science 198, 1977, 1222-1227.
5. C. M. Lent, "Adaptations of the ribbed mussel Modiolus dimissus, to the intertidal habitat," Am. Zoologist 9, 1969, 283-292.
6. M. A. Crenshaw, "The inorganic composition of molluscan extrapallial fluid," Biol. Bull. 143, 1972, 506-512.

DISCUSSION WITH REVIEWERS

S. J. Jones: There seems to be no evidence from the SEMs to support the identification of the "dissolution layer" as material remaining after acid dissolution of a previously formed layer with "normal" crystal structure, rather than a layer formed under different conditions with a different crystal structure/size and distribution; i.e. that decalcification or, indeed, recrystallization has occurred. Surely the degree of calcification in the "dissolution layer" is so much less than in the other layer, that the alteration from 8.1 to 9.2 mM calcium could not account for the change?

Authors: That bivalves become anaerobic when they close their shells and that molluscs dissolve previously deposited shell while they are anaerobic are well documented.[1,2,3,4,6,7] Evidence is also available showing that the specific structures that we termed the "dissolution layer" are formed by the solution of previously deposited shell mineral.[8,9,10] The change in the concentration of calcium in the fluids of M. mercenaria was equivalent to 0.1% of the total shell.[2,3] This was a significant amount of dissolution, considering the limited area from which the mineral was removed.

We have used the term, dissolution layer, to describe the overall process. It seems likely that some reprecipitation of the dissolved mineral may take place especially when the valves open to air. It is our hypothesis that reprecipitation is the process by which

Fig. 7. Surface changes in the chalky dissolution (arrow) layer after the valves opened. G. demissa. Bar = 3 μm.

Fig. 8. Fully formed prismatic aragonite. G. demissa. Bar = 2 μm.

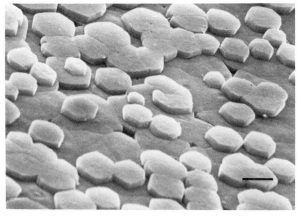

Fig. 9. Nacre has grown over the dissolution layer while the mussel was out of water. G. demissa. Bar = 2 μm.

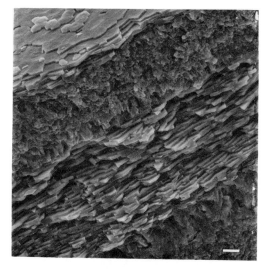

Fig. 10. Sequence of dissolution layer (lower right) formed when the animal was collected, normal nacre deposited overnight, chalky dissolution layer formed during four hours out of water and normal nacre deposited when the mussel was returned to water for 30 minutes. G. demissa. (Fracture surface.) Bar = 3 μm.

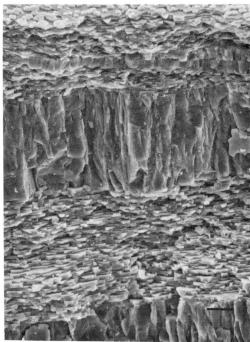

Fig. 11. More common sequence of prismatic dissolution layer (bottom), normal nacre deposited overnight, prismatic dissolution layer formed while the mussel was out of water for four hours, a layer of normal nacre deposited when the animal was returned to water, a prismatic dissolution layer formed when the submerged mussel closed its valves after 30 minutes, and normal nacre deposited during the subsequent period after the valves were reopened. G. demissa. (Fracture surface.) Bar = 10 μm.

the initial chalky layer is replaced by prismatic aragonite.

K. M. Towe: As I understand it, dissolution results from the succinic acid formed anaerobically combining with the shell calcium. If oxygen stimulates dissolution layer repair out of water then the calcium involved in this repair can only come from reversing this reaction placing succinic acid back in solution. This seems contradictory. Can you clarify?

Authors: It is our hypothesis that oxygen reaching the mantle tissue when subaerially exposed bivalves open their valves allows the tissue to metabolize the succinic acid, thereby raising the pH and freeing calcium for dissolution layer repair. We are now testing this hypothesis.

Authors Late Addition: In Figure 10, normal nacre was deposited directly over the chalky dissolution layer when the animal was returned to water. This sequence was unusual. The most common sequence is shown in Figure 11. The dissolution layer formed the night before is shown in the bottom of the micrograph, and it has the form of prismatic aragonite. The next layer is the normal nacre that was formed overnight. The prismatic layer above this is the dissolution layer that was modified while the mussel opened its valves while it was in the air. The layer of normal nacre covering the dissolution layer was formed during the first hour that the animal was returned to water. A third prismatic layer was formed when the valves were closed for about thirty minutes while the animal was submerged. The top layer of normal nacre was deposited when the valves were open again.

S. W. Wise, Jr.: How prevalent are natural dissolution layers in G. demissa shells freshly collected from intertidal environments?

Authors: Our observations indicate that they are very common. Figure 12 shows a fracture surface of a freshly collected shell in which the natural dissolution layers occupy a greater fraction of the aragonitic portion of the shell than does normal nacre. This mussel was collected from the same bed and at the same time as our experimental animals.

K. M. Towe: If as you state no dissolution takes place outside of the pallial line, would you agree that the formation of growth lines in the prismatic layer of Mercenaria cannot therefore be related to anaerobic decalcification as has been suggested by some authors?

Authors: We most definitely agree.

K. M. Towe: In one instance G. demissa formed normal nacre while out of the water (Fig. 9), but in another instance, it formed the disturbed dissolution layer shown in Figure 10. What causes this distinction between the markedly different modes of calcification out of water?

Authors: We do not know. Dissolution layer formation was the overwhelming process that occurred while the animals were out of water.

The extent to which G. demissa deposited normal nacre while out of water was very limited, occurred only at the margins of the dissolution layer and only after the valves had been open for extended periods.

Reviewer II: I would like to see the authors change the title of the paper, the text and the conclusions, so that they eliminate the use of the term "dissolution layer." As they introduce the use of this term, they imply that the shell mineral is partly dissolved during the anaerobic respiration phase. However, their own data presented in this paper clearly shows that a lot of shell is being formed during the anaerobic phase--only the shell is of different texture. The latter is the important finding of this paper which should be radically edited to take account of this criticism.

Authors: Our evidence does not indicate that net shell formation took place while the animals were out of water. The only source of calcium for these animals to deposit mineral was their shells. If the dissolution layer were new shell formation, then dissolution of shell in another region would have had to occur. The structure of the nacreous tablets in the areas when the dissolution layer was not forming did not indicate that these tablets were undergoing solution.

Additional References

7.   K. M. Wilbur, "Shell formation in mollusks" in Chemical Zoology VII Mollusca, M. Flokin and B. T. Scheer, eds., Academic Press, New York, 1972, 103-145.

8.   S. W. Wise, Jr., "Study of molluscan shell ultrastructures" in SEM, O. Johari, ed., IIT Research Institute, Chicago, Illinois, 1969,   205-216.

9.   T. T. Davies, "Effect of environmental gradients in the Rappahannock River estuary on the molluscan fauna" in Environmental Framework of Coastal Plain Estuaries, B. W. Nelson, ed., Geol. Soc. Amer., Boulder, Colorado, 1972, 263-290.

10. K. Wada, "Nucleation and growth of aragonite crystals in the nacre of some bivalve molluscs," Biomineralization 6, 1972, 141-159.

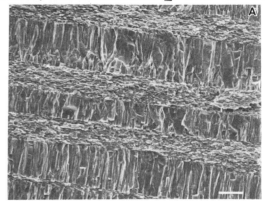

Fig. 12. Fracture surface of freshly collected G. demissa. Inner shell surface is at A. Bar = 50 μm.

SCANNING ELECTRON MICROSCOPY/1979/II
SEM Inc., AMF O'Hare, IL 60666, USA

ULTRASTRUCTURAL RELATIONSHIPS OF MINERAL AND ORGANIC MATTER
IN AVIAN EGGSHELLS

A. S. Pooley

Peabody Museum of Natural History
Yale University
New Haven, CT 06520

Abstract

Avian eggshells are composed of calcite with a
small amount of organic matrix. Questions concerning the existence of single or multiple crystals traversing the main layer or the nature of the organic-
mineral relationships have not been completely
settled. Despite evidence for large crystals from
X-ray diffraction, polarized light, and histological
studies of the organic matrix within the crystals,
reports of multiple crystals and of organic matter
surrounding but not penetrating small crystals have
persisted.

Eggshells of the American White Pelican (Pelecanus erythrorhynchos) were studied in detail, using
EDTA with glutaraldehyde and sodium hypochlorite
to etch the calcite and organic phases respectively.
Eggshells of 70 species representing other families
of birds, and representing 24 orders were studied
less extensively, usually without etching, to establish
the validity of the results for birds in general.

The eggshells of all species of birds examined
have three principal layers: 1) a spherulitic cone
layer interpenetrating the outer egg membrane but
otherwise usually having less organic matrix than the
rest of the shell. 2) A layer of radially oriented, large,
continuous crystals. Each of these crystals has a
single orientation of etch patterns and no sign of polycrystalline aggregates or twinning within it. Each
crystal contains a network of matrix, the two materials interpenetrate each other. The matrix exists as
hollow spheroidal films lining the vesicles, flat sheets
of matrix and thin fibrils and granules extending between the vesicle linings. The calcite etches preferentially along the C axis producing lozenge and pyramidal etching features. 3) The outer layers of the
eggshell, the cover and/or cuticle are usually rich in
organic matter which in many species has small
crystals of calcite embedded in it.

KEY WORDS: Avian Eggshells, Mineralized Tissue,
Calcite Etching, Organic Matrix, Avian Taxonomy

Introduction

An examination of the structure of eggshells of
species from different families of birds was originally
made to obtain information possibly useful for the
verification and improvement of the classification of
birds. Interest in the structure of the eggshell itself
developed and a more extensive study of the shells of
a few species, including the development of methods
to visualize the structure and orientations of the
mineral and organic phases, was made. Data on the
eggshell of the American White Pelican, Pelecanus
erythrorhynchos, will be presented in detail.

Avian eggshells have been studied most extensively by Tyler who has reviewed the situation up
until the last decade.[1] Of major interest has been the
nature of the inter-relationships of organic and mineral materials. Debate has centered particularly on
the size of the crystals of calcite and whether the
structure of the shell can be described as a crystal
phase embedded in organic matter or an organic
phase embedded in mineral.[1,2,3] Schmidt[2] presented evidence for large, crystallographically continuous
crystals extending nearly through the shell and having
dimensions of up to several hundred micrometers. A
number of other workers have presented evidence for
aggregates of small (under 0.1 μm) crystals having a
common orientation and embedded in a micro-net of
organic matter.[4,5,6] X-ray diffraction data for the
presence of large crystals with orientations of their
C axes approximately perpendicular with the shell
surfaces were produced by Cain and Heyn.[7] More
recently eggshells have been examined with the SEM,
confirming earlier findings concerning the external
shape of the crystals in the cone layer, and the presence of vesicles and cleavage surfaces on the broken
shell.[8-14]

Materials and Methods

Eggshells from 70 species representing as many
families and representing 24 orders of birds were
examined with the SEM. The specimens were from
the collections of the Peabody Museum of Natural
History and the Deleware Museum of Natural History.

Dry shells were broken to produce fracture surfaces roughly parallel with the radial crystals. For the American White Pelican specimens from five localities collected prior to DDT pesticide introduction were compared. Pieces from one of these shells was subjected to a variety of etching conditions. A solution of 0.01 M ethylenediaminetetraacetic acid (EDTA), 2 percent glutaraldehyde and 0.1 M sodium phosphate buffer pH 7.4 was used to etch the calcite of the freshly fractured pieces of eggshell for $\frac{1}{2}$, 2 and 4 hours and 5 days. Prior to, following, or without EDTA etching organic matter was removed from shell pieces with 2.5 percent sodium hypochlorite for 16 hours or by Soxhlet extraction with ethyelenediamine for 24 hours.[15]

The pieces were washed, dehydrated in ethanol, and dried by the critical point technique. Specimens were coated with about 0.02 μm of gold-palladium with a Polaron diode sputter coater or by vacuum evaporation. Specimens were examined with an ETEC Autoscan U-1 SEM. Shell fragments were prepared for polarized light examinations by standard mineralogical embedding, sectioning and polishing. Shell pieces demineralized in EDTA were prepared by standard histological techniques including staining with amido black and eosin for examination with the light microscope and, after critical point drying, with the SEM.

A piece of Iceland spar calcite was examined to locate the C axis and small fragments of known orientation were subjected to etching with EDTA under the conditions used for eggshells and were examined with the SEM. A mineralogical section was also subjected to EDTA-glutaraldehyde etching, coated with gold-palladium, the orientation of preferred etching of a few crystals was established with the SEM and the orientations of the C axes of the same crystals were determined by polarized light.

Results

Eggshells of birds representing 70 out of 160 families and 24 out of 28 orders show a consistent basic pattern with many variations of detail. All avian eggshells examined consisted of large, closely packed columnar shaped, radially oriented crystals originating from spherulitic centers in the inner part of the shell. These spherulites are anchored to the outer eggshell membrane by penetration of the calcite into the membrane. The major thickness and strength of the shell is in the columnar crystals. These terminate at a fairly consistent level and are sometimes covered by a layer of small, radially oriented crystals. They are next covered by a layer of organic material sometimes containing small spheroidal calcite crystals, usually surmounted by a thin, purely organic layer. Pore channels, usually straight and unbranched except in the Ostrich (Struthio camelus)[1] and the extinct Elephant Bird (Aepyornis) pass through the layer of columnar crystals and are often covered and plugged with material of the outer layers. The shells of all species have vesicles, which are air filled spaces

within the columnar calcite crystals. These are usually spheroidal, closely packed and about 1-5 μm in diameter. The Anatadae (ducks and geese), Ratites (ostrich, emu, rhea), and Galliformes (pheasants, grouse, chicken) have smaller, more irregular and less frequent vesicles. The structure of the eggshells of the members of 70 families were examined for characteristics useful for establishing taxonomic relationships between the families. Few characters or groups of characters were unequivocally associated with the accepted classification of birds and it appears that at the familial level and above, the structure of eggshells is not stable enough in evolution to provide useful taxonomic information.

The eggshell of the American White Pelican (WP) was chosen as being representative of the structure found in many eggshells.[16] The eggshell of the domestic chicken on which most of the previous work has been done, is not typical because it lacks a mineralized cover, has few, small vesicles and a fairly dense organic matrix throughout the shell. The eggshell of the White Pelican has four principal layers plus the membranes below it. At the surface is the cuticle, a layer of dried mucous (Fig. 1, 2, 3, layer A), next comes the cover (layer B), a porous layer of small spheroidal calcite crystals embedded in organic material. This material also fills the openings of pore channels (PC) in the column layer (Fig. 2). The column layer (layer C) consists of radially arranged columnar crystals which originate in spherulitic centers in the cone layer (D). The spherulites of the cones surround the upper layer of fibers of the membrane (E), anchoring the two together.

Histologically the organic matrix of the column and cone layers are a unit, alike and continuous except for a lower density of the matrix in the cone layer. The cuticle and cover are similar but show histochemical differences,[16] and both differ from the column and cone layers. The membrane stains a distinctive color. Thus the continuous matrix of the column and cone layer does not extend into the cover or into the membrane.

A mineralogical section of the eggshell viewed between crossed polarizers (Fig. 1B) transmits light only through those calcite crystals whose C axes are not aligned with one of the polarizers. The cover (B) consists of small crystals in many orientations. The cones (D) are spherulitic, the crystals radiating outward are oriented in many directions. The column layer consists of large crystals that darkens as a single unit when the sample is rotated between the polarizers. The cuticle (A) and membrane (E), having no calcite, do not transmit light.

Details of the structure of the layers described above are visible in fractured cross sections with the SEM. Untreated cross sections do not show the matrix or the boundaries and orientation of the crystals, therefore shells treated to remove some of the organic or mineral phases were also studied. The results will be presented layer by layer.

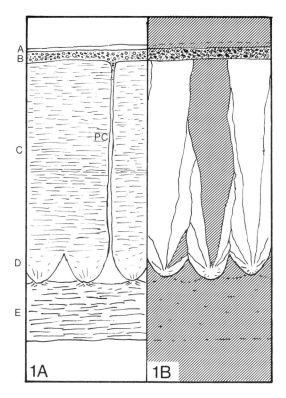

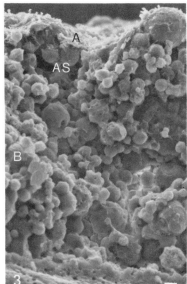

Fig. 1A. Diagram of histological section of demineralized, stained White Pelican (WP) eggshell matrix and membranes.
Fig. 1B. Diagram of mineralogical section of WP eggshell viewed with crossed polarized filters.
Fig. 2. WP eggshell. Bar = 100 um.
Fig. 3. Outer part of WP eggshell. Bar = 1 um.
Fig. 4. Outer part of WP eggshell after 1/2 hour EDTA - glutaraldehyde. Bar = 1 um.
Symbols: A = cuticle, B = cover, C = column layer, D = cones, E = membrane, AS = air spaces, PC = pore channel.

Fig. 2

The cuticle appears as a thin film conforming to the particles of the cover (Fig. 3). The cover consists of spheroidal particles of calcite surrounded by organic material which leaves air spaces (AS) between them. The cuticle and cover are both completely removed by sodium hypochlorite which dissolves the organic material and allows the calcite particles to be washed away. EDTA- glutaraldehyde removes the calcite of those particles that were exposed at the fractured surface (Fig. 4), some small particles disappear, larger broken ones are hollowed out but the purely organic cuticle and organic portion of the cover are unaffected.

The cones partially penetrate the membrane below them. A fractured cone reveals the junctions (J) of the spherulitic crystals (Fig. 5) radiating outward from the center of the cone. Etching the calcite causes the organic matrix to stand out in relief (Fig. 6); prior to treatment it could not be distinguished. The crystals radiate outward from a concentration of organic matrix called the core of the cone (CC). [17] No sign of any aragonite crystals as reported by Erben, [10] were seen in the cones of any bird studied. The organic matrix of the cones consists of fibrils, granules (FG) and sheets (S) of matrix (Fig. 7) which are completely removed by sodium hypochlorite (Fig. 8). Large organic structures such as membrane fibers leave large visible spaces (FS) after their removal, but the

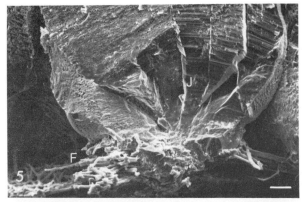

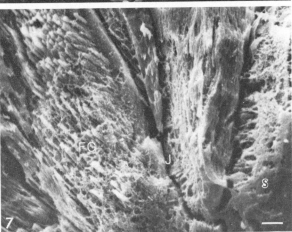

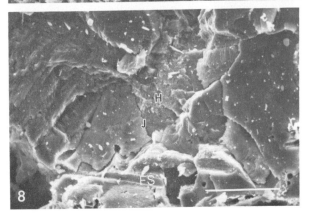

Fig. 5. WP eggshell: a cone fractured in half. Bar =
10 um.
Fig. 6. Cone after 1/2 hour EDTA - glutaraldehyde.
Bar = 10 um.
Fig. 7. Detail of cone after 1/2 hour EDTA -
glutaraldehyde. Bar = 1 um.
Fig. 8. Cone after 1/2 hour EDTA - glutaraldehyde
followed by sodium hypochlorite. Bar = 10 um.
Symbols: J = junction between crystals, F = fibers of
the membrane, CC = core of the cone, FG = fibrils
and granules of the matrix, S = sheet of matrix, FS =
space left by removal of membrane fiber, H = holes
left by removal of matrix.

---

finer elements of the matrix leave only a few visible
holes (H) in the calcite. This is not surprising since
the fibrils are only 0.01 micrometer thick. [3]

The column layer is a continuation of one or two
crystals, now columnar in shape, from each spherulite. In some areas the calcite fractures with
regular cleavage (R, Fig. 9), with the characteristic
interfacial angles of calcite. Vesicles (V), gas filled
spaces, and areas of regular cleavage and irregular
fracture occur throughout the layer. Sodium hypochlorite removal of organic matrix does not change
the appearance of the surface; the spaces left by the
matrix are seldom visible. EDTA etching in the
columnar calcite occurs preferentially along the C
axis (perpendicular to the plane of the carbonate ions
(0001) plane), etching outward from each vesicle.
This results in rounded lozenge shaped spaces (Fig. 10)
around the vesicles and pyramidal projections of
calcite (P) aligned with the C axis. Each vesicle was
lined with a film of organic material that stands in
relief after etching (VL). Shrinkage of the organic
linings during critical point drying results in distortion into flattened structures of the previously
spheroidal linings. Sheets of organic matter (S) pass
through the calcite without changing its orientation.

That the EDTA etching occurs in the direction of
the C axis was verified by etching the surface of a
mineralogical thin section and comparing the orientation of the lozenge shaped spaces in the SEM with the
C axes of the crystals using crossed polarizers. The
C axes of crystals of the column layer are inclined at
angles from about 0 to 40° from perpendicular with
the shell surface. It was possible to trace individual
crystals having a single orientation of the lozenge
shaped spaces from the cone layer to the top of the
column layer.

Sodium hypochlorite treatment of EDTA etched
column layer fracture surfaces results in loss of the
organic vesicle linings and sheets at the exposed
surface (Fig. 11), the calcite is not affected. The
junction between crystals (J) did not have as thick a
space left by a layer of organic matter as existed
within the crystal.

Etching the column layer with EDTA for a longer
time completely removes the calcite from the surface

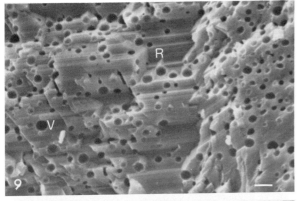

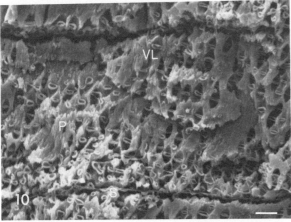

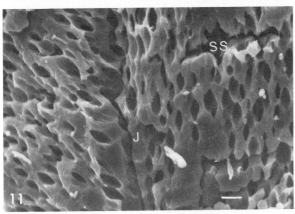

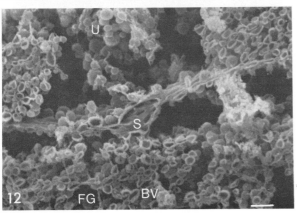

Fig. 9. Unetched column layer of WP eggshell. Bar = 1 um.

Fig. 10. Column layer after 1/2 hour EDTA - glutaraldehyde. Bar = 1 um.

Fig. 11. Column layer after 1/2 hour EDTA - glutaraldehyde followed by sodium hypochlorite. Bar = 1 um.

Fig. 12. Column layer after 5 days EDTA - glutaraldehyde. Bar = 1 um.

Symbols: V = vesicles, R = area of regular cleavage, VL = vesicle linings, P = pyramidal projections of calcite, S = sheets of matrix, SS = space left after removal of sheet of matrix, J = junction between 2 crystals with different C axis orientation, U = unbroken vesicle linings, BV = broken vesicle linings, FG = fibrils and granules of matrix.

---

and leaves the matrix standing free (Fig. 12). The linings of the vesicles that were broken open on the fracture surface before etching are open hemispheroidal structures (BV) whereas the unbroken vesicle linings (U) are closed, spheroidal and have fine fibrils and granules (FG) on their surfaces. If the broken vesicle linings are removed by sodium hypochlorite treatment before EDTA etching, only the unexposed, unbroken linings (U) are left (Fig. 13) along with fibrils and granules of matrix (FG) with a background of calcite etching features (P). The sheets of organic material are sometimes continuous with the vesicle linings (S, FIG. 12). It is not known whether the fibrils and granules are the same substance as the vesicle linings and sheets. The fibrils and granules tend to be washed off and lost from the etched surface. Etching followed by critical point drying and then refracturing of the dried specimen shows that extensive fibrils (F , Fig. 14) surround the vesicle linings. In the deeper, unetched zone (UZ) the vesicle linings and matrix are present but do not stand in relief and are not visible. In the partially etched zone (P) the vesicle linings stand in relief but the fibrils are still embedded in calcite and are invisible or have been pulled off the surfaces of the vesicle linings.

Previous studies on the etching of calcite [18] did not produce etching features similar to the lozenges and pyramidal shapes of calcite found in the eggshell. However they were done on the cleavage surface (1011) with concentrated acid or saturated EDTA producing surfaces at roughly 45° to the C axis. Here the etching proceeded along the C axis producing surfaces approximately parallel to the rhombohedral unit cell of calcite (4041).

A crystal of Iceland spar calcite with a known orientation of the C axis was etched with the EDTA solution used on the eggshells. At rare places that had fractured perpendicular to the C axis ((0001) plane), pyramidal shapes of calcite (Fig. 15) similar to those produced in the eggshell were consistently found after etching with EDTA but not before. Compression of calcite results in twinning, a change in the orientation of the C axis, [18,19] and in a change in

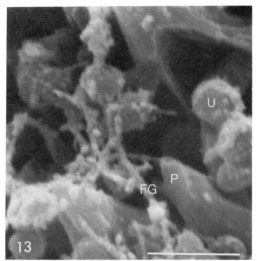 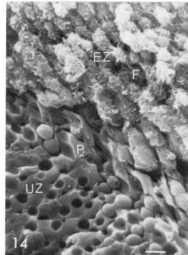 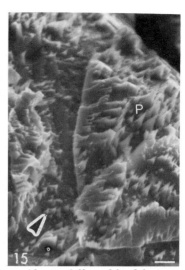

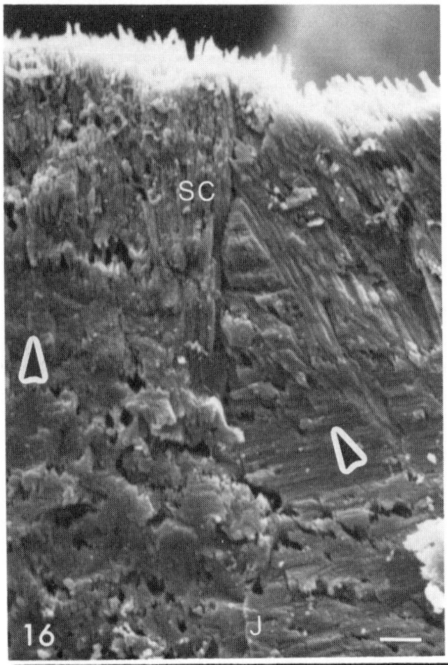

Fig. 13. Column layer after sodium hypochlorite followed by 2 hours EDTA - glutaraldehyde. Bar = 1 um.

Fig. 14. Column layer after 5 days EDTA - glutaraldehyde, critical point dried, then refractured. Bar = 1 um.

Fig. 15. Calcite (Iceland Spar) of known C axis orientation (arrow) after 1/2 hour EDTA - glutaraldehyde. Bar = 1 um.

Fig. 16. Domestic chicken eggshell, top of column layer after 1/2 hour EDTA - glutaraldehyde followed by sodium hypochlorite. Bar = 1 um.

Symbols: FG = fibrils and granules of matrix, P = pyramidal projection of etched calcite, U = unbroken vesicle lining, UZ = unetched zone of shell, SC = small crystals, J = junction between large crystals, arrow EZ = etched zone of shell

the orientation of the pyramidal shapes on naturally twinned calcite as observed in the SEM but this was not observed in the eggshell crystals.

The eggshell of the Domestic Fowl (Gallus gallus) shows features not shared with the American White Pelican, one of which is briefly described. Above the column layer a thin layer of small, radially oriented crystals is found in the Domestic Fowl and many other species [17] but not in the American White Pelican. In mineralogical sections with crossed polarizers the C axes of these small crystals are aligned with those of the larger crystals on which they rest and they appear continuous with them. EDTA etching of this layer followed by sodium hypochlorite treatment to remove the exposed matrix and cuticle (there is no cover) reveals many small crystals (SC) having the same C axis orientation as the larger crystals on which they

rest (Fig. 16). This is a true polycrystaline aggregate with preferred orientation of the C axes and it can be seen that EDTA etching reveals the multiple nature of the crystalline units. Further etching causes these to dissolve rapidly and be washed away.

Discussion

Previous studies have shown that the avian egg-shell is an unusual vertebrate mineralized tissue.[1-14] It is not a living tissue but is connected physiologically to the growing embryo which dissolves some of the mineral of the shell to form its skeleton.[20] The mineral of the shell is in the form of calcite,[1,7] unlike the mixed apatite of vertebrate bone or the aragonite of many invertebrate shells.[21] Small amounts of other minerals are also present.[1,10] The content of organic matter is much lower in the egg-shell than in other mineralized systems, often only 3 to 5 percent of the mass of the true shell.[1,16,17] Studies with the TEM and SEM have described the organization of demineralized matrix,[3,4,6,8-14] the orientation of calcite crystals in thin sections,[14] and the distribution of vesicles and cleavage features on fractured shells.[8,13]

This study concerns the size and orientation of the individual calcite crystals, the form and organization of the matrix elements and the relationship of the two components with each other. Neither the matrix nor the crystal boundaries are visible on fractured shells. Total demineralization does not permit study of the matrix in situ within the mineral.

Etching the calcite with EDTA dramatically reveals both the boundaries of the crystals and their orientation and also shows the distribution and organization of the matrix within the crystals. Etching is preferential approximately to the (4041) plane producing lozenge shaped pits around the vesicles and pyramidal projections on edges. The fact that the vesicle linings continue to cling to the sides of the lozenge shaped pits shows that little etching occurs perpendicular to the C axis. The single orientation of etching features on crystals whose junctions with other crystals are now visible allows the dimensions of the individual, single unit crystals to be determined. They are large, up to several hundred micrometers in dimension and are continuous from the cones through the column layer in agreement with other studies. [10,14] The fact that polycrystalline aggregates, such as those in the outer layer of the chicken eggshell, [22] are easily revealed by EDTA etching, is further evidence that the continuous units of the column layer are in fact single crystals. Eggshells are therefore unlike the calcitic shells of some invertebrates which are composed of polycrystalline aggregates, sometimes with shared orientation. [21] The absence of dislocations or twinning in undamaged calcite in this study is in agreement with Tyler [19] but not with Quintana and Sandoz who found possible twinning defects in Quail eggshell. They discuss the possibility that the defects may be artifacts of thin sectioning. [14] Twinning changes in C axis orientation have been observed after damage to shells. [19]

Unlike vertebrate bone, where the crystals of apatite are surrounded and enclosed in compartments of organic matter [17] the calcite crystals of the eggshell contain and surround the matrix but are not surrounded by it; in fact thicker sheets of matrix occur within the crystals than between them. The cover differs; there, small crystals are enclosed in organic matter.

The matrix of the column layer consists of fibrils, granules, sheets and vesicle linings. The fact that the fibrils are largely lost from the surface of the vesicle linings during etching (Fig. 12) and are usually preserved where they are protected below the surface being etched (Fig. 14) suggests a loose attachment. The calcite crystallizes around and between the matrix elements, not usually forcing the matrix into a pattern except sometimes in the cones.

A close association of mineral and matrix arises during growth of the shell. Erben and Kriesten have suggested that the matrix of the cores of the cones somehow initiates nucleation and determines the average distance between cones and hence the size of the columnar crystals. They suggest the crystals are initiated by epitaxy on organic templates but continue to grow by accretion. [11] The matrix appears to be passively engulfed in mineral which grows through it without deflection of its crystalline axes. The situation

is analogous to that of sand-calcite where calcite crystals form in sand, growing between the grains. Even though the sand occupies 60 percent of the volume, the calcite forms regular crystals up to several centimeters in dimension, engulfing the sand without losing its own crystalline continuity. [23] The picture of eggshell structure that emerges is one of large, radially oriented, continuous crystals filled with a nearly continuous network of matrix and many vesicles whose organic linings are loosely attached to the matrix.

Acknowledgements

I wish to thank Dr. Keith S. Thomson for use of facilities and materials, Dr. Charles G. Sibley for materials, advice and reading the manuscript, Mr. John E. duPont, Deleware Museum of Natural History, for the loan of eggshells and Dr. Horace Winchell for mineralogical advice.

References

1. C. Tyler. Avian egg shells: Their structure and characteristics. Int. Rev. of Gen. and Exp. Zool. 4, 1969, 81-130.
2. W. J. Schmidt. Polarisationoptik und Bau der Kalkschale des Hühnereies. Z. Zellforsch. 57, 1962, 715-729.
3. P. C. M. Simons and G. Wiertz. Notes on the structure of membranes and shell in the hen's egg: a TEM study. Z. Zellforsch. 59, 1963, 555-567.
4. W. Massehoff and H. J. Stolpmann. Licht-und elektronenmikroscopische Untersuchungen an der Schalenhaut und kalkschale des Huhneries. Z. Zellforsch. 55, 1961, 818-832.
5. A. R. Terepka. Structure and calcification in avian egg shell. Exp. Cell Res. 30, 1963, 171-182.
6. A. N. J. Heyn. The crystalline structure of calcium carbonate in the avian egg shell: a TEM study. J. Ultrastruct. Res. 8, 1963, 176-188.
7. C. J. Cain and A. N. J. Heyn. X-ray diffraction studies of the crystalline structure of the avian eggshell. Biophys. J. 4, 1964, 23-39.
8. R. Bellairs and A. Boyde. SEM of the shell membranes of the hen's egg. Z. Zellforsch. 96, 1969, 237-249.
9. H. K. Erben. Ultrastrukturen und Mineralisation rezenter und fossiler Eischalen bei Vögeln und Reptilien. Biomineralization 1, 1970, 1-66.
10. H. K. Erben and H. Newesely. Kristalline Bausteine und Mineralbestand von kalkigen Eischalen. Biomineralization 6, 1972, 32-48.
11. H. K. Erben and K. Kriesten. Mikromorphologie der Kristallbildung in normalen und anormalen Hühner-Eischalen. Biomineralization 7, 1973, 28-36.
12. J. H. Becking. The ultrastructure of the avian eggshell. The Ibis 117, 1973, 143-151.
13. S. Henstra, P. C. M. Simons and G. Wiertz. Untersuchungen von Hühnereierschalen in Raster-Elektronenmikroskop. Beitr. Elektronenmikroskop.

Direktabb. Oberfl. 3, 1970, 343-350.

14. C. Quintana and D. Sandoz. Coquille de l'oeuf de caille: étude ultrastructurale et cristallographique. Calcif. Tiss. Res. 25, 1978, 145-159.

15. C. W. Skinner, E. S. Kemper and C. Y. C. Pak. Preparation of the mineral phase of bone using ethylenediamine extraction. Calcif. Tiss. Res. 10, 1972, 257-268.

16. C. Tyler. A study of the egg shells of the Gaviiformes, Procellariiformes, Podicipitiformes and Pelecaniformes. J. Zool. Lond. 158, 1969, 395-412.

17. K. Simkiss and C. Tyler. A histochemical study of the organic matrix of hen egg-shells. Quart. J. Microscop. Sci. 98, 1957, 19-28.

18. R. E. Keith and J. J. Gilman. Dislocation etch pits and plastic deformation in calcite. Acta. Metallur. 8, 1960, 1-10.

19. C. Tyler and D. Moore. Types of damage associated with measuring egg shell strength. Brit. Poult. Sci. 6, 1965, 175-182.

20. C. Tyler and K. Simkiss. Studies on egg shells. XII. Some changes in the shell during incubation. J. Sci. Food Agric. 11, 1959, 611-615.

21. K. M. Towe. Invertebrate shell structure and the organic matrix concept. Biomineralization 4, 1972, 1-14.

22. C. Tyler. A study of the egg shells of the Anatidae. Proc. Zool. Soc. Lond. 142, 1964, 547-583.

23. A. F. Rogers and R. D. Reed. Sand-calcite crystals from Monterey County, California. Am. Mineralogist 11, 1926, 23-28.

DISCUSSION WITH REVIEWERS

I. S. Johnston: You claim that the column layer above each spherulitic cone is made up of one or two individual crystals. However, there are multiple horizontal organic sheets interrupting each column unit and the etching pattern of Fig. 10 would indicate that the mineral phase is not continous through these sheets. Can you suggest how crystal orientation might be conserved across such an organic barrier even though the barrier completely disrupts the continuity of the crystal lattice?
Author: The organic sheets are not completely continuous though they do appear so in a small area such as Fig. 10. Calcite as it grows has the ability to establish a continuity of crystalline organization even though a large part of the volume of the crystal is occupied by matrix, or by sand grains in the case of sand-calcite.[23]

Reviewer IV: Fig. 10 seems to indicate that etching occurs around the vesicles. Is that always the case? If so, how do you interpret the phenomenon?
Author: Etching occurs on all exposed features, whether covered by matrix or not. All vesicles etch into lozenge shapes until pits coalesce into larger shapes during further etching. The differential rate of dissolution on different planes explains the shapes produced. The pyramidal projections are produced by the intersection of two or more lozenge shaped pits.

I. S. Johnston: From Fig. 12 it is not obvious that the vesicle linings and the organic sheets are continuous. Could the circular depression in the sheet at "S" represent the site of fortuitous close juxtaposition of vesicles and sheets, each being structurally independent of each other?
Author: This may be true but the sheet and vesicle lining must have coalesced during formation of the shell and this suggests they are of the same substance and are merely laid down differently as the shell grows around them.

I. S. Johnston: The eggshells which you studied were certainly not living tissues, but isn't it quite likely that during shell biogenesis the organic inclusions may have been metabolically active? The organic inclusions which you describe are likely to have been diagenetically altered in both structure and composition following biogenesis, however based on their final structure and location do you think that they may have had some role other than merely modifying the physical properties of the deposited mineral?
Author: I have no evidence nor do I know of anything in the literature bearing on these questions.

A. Boyde: What is the gas in the "vesicles" (gas bubble spaces) in the eggshell?
Author: Tyler and Geake found that the shell became opaque as it dried out and that this was reversible with rewetting (The effect of water on egg shell strength including a study of the translucent areas of the shell. Brit. Poultry Sci. 5, 1965, 277-284.) The opacity presumably arises from gas filling the previously water filled vesicles, increasing the scattering of light. As water diffuses out either air or water vapor must replace it.

Reviewer IV: What methods were used for identification of crystals? For example, it is stated without proof that the small, spheroidal crystals of the cover are calcite; and that no sign of any aragonite was seen in the cones.
Author: Calcite, identified originally in eggshells by X-ray diffraction,[7] is homogenous in structure from the cones through the column layer. Nothing comparable to the structures reported to be aragonite in incubated eggs by Erben and Newesely[10] was seen in this study. The mineral particles of the cover each extinguish as a single unit with polarized light and are therefore crystalline, presumably calcite.

Reviewer IV: You mentioned that the cuticle and membrane do not transmit polarized light. However, many organic fibers and cuticles are birefringent. What magnification did you use to observe the extinction?
Author: Magnifications up to 1000 X were used. The organic layers transmit a little light but the intensity does not change as the specimen is rotated. Possibly this is light scattered rather than rotated by birefringence in the specimen.

SCANNING ELECTRON MICROSCOPY/1979/II
SEM Inc., AMF O'Hare, IL 60666, USA

SEM OF THE ENAMEL LAYER IN ORAL TEETH OF FOSSIL AND EXTANT
CROSSOPTERYGIAN AND DIPNOAN FISHES

M. M. Smith

Department of Oral Anatomy
Royal Dental Hospital
St. George's Hospital Medical School
Cranmer Terrace
London, SW17 ORE
U.K.

Abstract

The intention was to investigate simple
enamel structure in crossopterygian oral teeth
and establish whether it is homologous with
reptilian non-prismatic enamel, or with enameloid
as found in actinopterygian teeth. Enamel on
tooth plates of extant dipnoans was also
investigated for the same reasons, as previous
histological studies had not resolved the
question.

The oral teeth of Onychodus were examined in
the SEM after removal of the specimens from the
fossil matrix and the preservation of microstructure was found to be excellent. Additional
criteria have been obtained from the microstructure of the surface layer which fully substantiate the histological observations that it is
enamel of the type found in all tetrapods. The
surface of the enamel is sculptured by a system
of ridges, possibly a generic pattern. Each
ridge of the enamel surface is further characterized by a regular ribbing in a herring bone
arrangement. The organization of the crystal
components within the enamel ridges, and the
regular surface pattern of each ridge is such
that the enamel can be said to be of the same type
as that of porolepid and osteolepid crossopterygians.

It is concluded that examples of simple
enamel, of a type leading to pseudoprismatic
enamel in reptiles, can be found in the extinct
forms of crossopterygian fishes.

The glossy, hypermineralized layer around
the margins of the tooth plate of the extant
dipnoan, Lepidosiren paradoxa, has been compared
in microradiographs, polarized light and the SEM.
It is concluded that the layer has many of the
features of an extremely thin layer of enamel.
In some regions it forms over an extensive
resorption surface of bone and it is suggested
that although it is clearly not enameloid, its
homologies with reptilian enamel must be in
question.

KEY WORDS: Enamel, Crossopterygian Teeth,
Dipnoan Teeth, Fossil Teeth, Fish,
Onychodus, Lepidosiren

Introduction

The enamel of the oral teeth in mammalian,
reptilian and amphibian vertebrates is secreted
by specialized cells of the dental epithelium and
grows by apposition on the surface of the dentine.
The morphology of the enamel surface, the thickness of the enamel, the arrangement of the
enamel crystallites and the pattern of incremental
growth are all under the control of the epithelium. It has been proposed (1,2,3) that enamel-
covered teeth had evolved in the Devonian period
in the crossopterygians, independently of
enameloid in the teeth of actinopterygians and
elasmobranchs. This investigation forms part of
a study to test this hypothesis.

Of the crossopterygian fishes, the
rhipidistians are all fossil forms and only a few
have simple conical teeth implanted in shallow
sockets of a type that may represent an ancestral
form for the tetrapods.[4] One example belongs to
the genus Onychodus; the oral teeth have been[1,5,6]
reported from histological studies to possess a
relatively thick layer of enamel, although this
is of the type interpreted by others as enameloid.[7]
The purpose of the present study was to examine
the microstructure of the enamel layer in the
teeth of Onychodus with the SEM. It was intended
from this to comment on the competence of the
dental epithelium to influence the microstructure
of the enamel, at this level of organization in
the evolution of reptilian and mammalian enamel.

The only extant forms considered to be
either crossopterygians or closely related,
belong to two widely divergent groups, the
dipnoans, or lungfish which have an extremely
specialized form of dentition, and within the
actinistians, the coelacanth, Latimeria chalumnae,
Smith.[2] Two recent SEM investigations[2,3] of the
oral teeth in the coelacanth have confirmed
previous histological studies that a form of
simple enamel covers these teeth. The reviews
of the histological investigations of dipnoan
tooth plates have shown[8,9] that the conclusions
on the homologies of the outer layer in dipnoan
tooth plates are equivocal. Correlation of
histology with SEM of the enamel in the extant
dipnoan Lepidosiren paradoxa Natterer was undertaken to resolve this problem.

Materials and Methods

Fossil teeth of Onychodus sp. were from the Devonian Gogo formation of Western Australia,[10] and also from the Devonian Silica formation in Ohio, U.S.A. The specimens had been extracted from the calcareous rock nodules or from the silica bed by the standard acetic acid technique.[11] The protective coating, polymethyl methacrylate, was completely removed from each specimen using acetone before air drying from acetone and applying approximately 30nm of gold in a Polaron Cool Sputter Coating Unit. Specimens were mounted on $\frac{1}{2}$" aluminium stubs using either double-sided adhesive tape to facilitate removal after examination, or Epoxy resin (Devcon, Danvers, Mass. 01923, U.S.A.) for complete specimen support. Colloidal silver was used between the specimen and the stub to improve conduction. Specimens were either mounted whole to allow examination of natural surfaces or fractured longitudinally through the cone of dentine to examine fracture surfaces through the dentine and enamel layer. The teeth prepared for the SEM in this way were on average 5mm in height and typical of oral teeth from the margins of the jaws, considerably smaller than the large 15-20mm symphyseal teeth. The specimens were all uncrushed with exceptionally well preserved morphology. The interpretation of microstructure of the dental tissues is based on the assumption that the organization of the inorganic crystals is faithfully maintained as in the living state. The principal changes in calcified tissues, caused by fossilization, are the loss of organic material and the conversion of hydroxyapatite into fluorapatite.[7]

The posterior third of one side of the prearticular tooth plate of Lepidosiren paradoxa was removed from a specimen previously fixed in 10% neutral formal saline and stored in 70% ethanol. Sections were cut from the unembedded piece of tooth plate in a labio-lingual plane, vertical to the tritoral surface using a diamond impregnated copper wire saw (Lastec, Model 206A, Laser Tech. Inc.) Sections 100-200 μm thick were ground and polished down to 70 μm for microradiography and for examination by polarized light and phase contrast microscopy. Adjacent thick slices were prepared for SEM. The labial part of one tooth plate ridge was prepared as an anorganic surface by treating with 10% NaOCl, followed by dehydration in a graded series of alcohols before critical point drying using Freon 113/CO_2.[12]

Contact microradiographs were made on Kodak Spectrographic Plates 649-0 with 20-30 minutes exposure time. The X-rays were generated by a Machlet O-2 diffraction tube, with a Cu-target and beryllium filter, operating at 20KV, 16 mA, and a target-film distance of 30 cm.

SEM of all specimens was carried out in either a Cambridge Steroscan IIa at 20-28KV accelerating voltage, or a Coats and Welter Field Emission Scanning Microscope at 15KV.

Results

Onychodus sp. (Crossopterygii)

Within the sample of teeth isolated from the fossil matrix, two distinct stages were represent-ed in the formation of the dentition. Those with worn flat tooth tips, extensively resorbed bases and a thickened cone of secondary dentine inside the primary dentine were mature teeth in the process of shedding from the jaws. The second type of tooth was younger and newly formed within the dentition because none of these features of the mature tooth was found. The tooth tip was intact and pointed (Fig 1) and there was a thin tapering edge at the base of the primary dentine cone, where the pulpal dentine had the typical appearance of a forming calcospheritic surface. Comparisons between these two types of teeth have shown differences in the microstructure of the surface layer which are interpreted as develop-mental changes. The conical tip of the tooth had a relatively smooth and featureless surface (Fig 1) whereas the entire surface of the tooth shaft was sculptured by a system of ridges and furrows. These ridges were arranged in the longitudinal direction of the tooth, adjacent ridges joining and then separating in a random pattern (Fig 1). This surface relief could be identified at 50x magnification and was a feature noted in previous studies as typical of the genus Onychodus.[13] With increasing magnification add-itional microstructure was apparent as a ribbing in herring bone arrangement along each ridge (Figs 2,3,4). The ribs were regularly spaced and par-allel to each other but each set of ribs diverged from a common point along the crest of the ridge. In the mature tooth the ribs from each side of the ridge meet each other along the crest (Fig. 4), but in the young tooth they did not meet but merged with an area of unribbed surface (Fig 3). The difference in these two surfaces suggested that it was a developmental pattern not completed on the younger tooth. Also, although in the mature tooth the surface was smooth and untextured (Fig 4), in the young tooth the surface was rough and granular in texture (Fig 2,3).

In fracture surfaces through this outer layer of the teeth there was always a distinct junction between the dentine and the sculptured layer (Figs 5,6). In the fracture surface the predominant direction of the crystal groups in the outer layer was perpendicular to the surface. Comparing a fracture surface through a ridge (Fig 5) with one through the smooth conical tip of the tooth (Fig 6), the crystal groups of the ridge diverged slightly from the mid-line axis through the ridge crest, whereas in the region of the tip they were predominantly parallel to each other and to the normal to the surface. The thickness of the layer was also greater in the tooth tip (Fig 6) than the tooth shaft and tapered towards the base of the tooth shaft. Incremental lines have been observed running parallel to the surface and were approximately 2-5μm apart (Fig 7). In the strongly ridged sur-face layer these lines were not obvious but the arrangement of the crystal groups seemed to be staggered to correspond with the ribs running

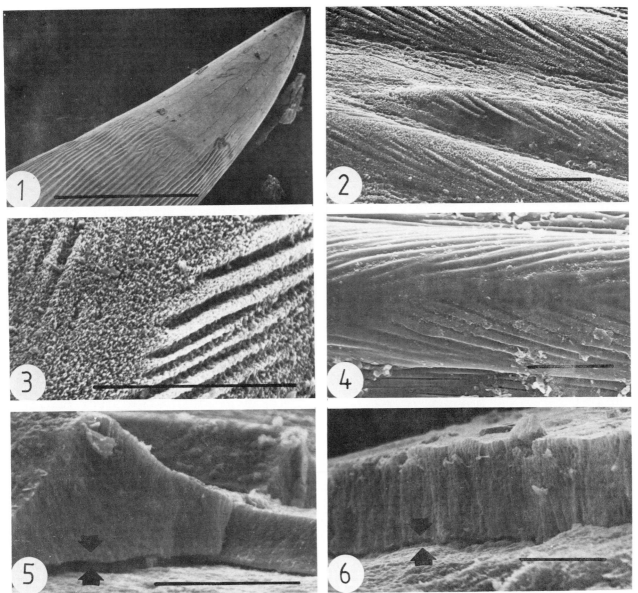

Figures 1-6 Natural surfaces and fracture surfaces of oral teeth of _Onychodus_ extracted from Devonian calcareous nodules.

1) Tip of young tooth, smooth enamel surface gives way to characteristic ridged enamel layer. Bar - 1 mm.

3) Part of ridge from same surface as in Fig. 2, it shows the unribbed part of the ridge crest. Bar - 10 μm.

5) Fracture edge of ridged enamel layer in mature tooth, the enamel dentine junction is smooth and cleaves naturally at this point (arrows). Crystallite orientation is normal to this surface but changes gradually throughout the ridge. Bar - 10 μm.

2) Enamel surface with ridges from young specimen in Fig. 1. Parallel ribs of enamel meet those on the opposite side of the ridge at an acute angle, as in an 'arrow head' with the point away from the tooth tip (left). Bar - 10 μm.

4) Part of ridge from surface of a mature tooth; it shows the ribs from opposite sides of the ridge joining on the ridge crest in contrast with Fig 3. Bar - 10 μm.

6) Fracture edge through smooth enamel of the tooth tip, crystallites are all perpendicular to the surface and radially arranged. Enamel-dentine junction (arrows). Bar - 10 μm.

along the surface of the ridges (Fig 8). Tubules were confined to the dentine and were not observed in the surface layer (Figs 5,6,7).

Lepidosiren paradoxa (Dipnoi)

The labial and lingual surfaces of the non-triturating part of the tooth plate are covered by a glossy layer of tissue, enamel or enameloid. In situ this part of the tooth plate is surrounded by a cuff of epithelial-lined tissue. This enamel surface is shown on the labial margin of one ridge of the tooth plate (Fig 9), it extends from below the tritoral surface to the basal growing margin where it overlaps a bone surface. The cut surface of this block of tissue is adjacent to the sections used for microradiographs (Fig 10) and optical microscopy (Fig 11). The central part of the ridge of the tooth plate is a column of hypermineralized petrodentine, embedded in trabecular dentine and bone. Also on the lingual surface there is an extremely thin, hypermineralized, glossy layer, opaque to X-rays (Fig 10), strongly refractile and birefringent (Fig 11). The sign of birefringence is +ve with respect to the surface tangent, although not all regions show extinction in the parallel position but require an angle of approximately 20° from the parallel position before extinction is achieved. These features indicate that not all the birefringent elements of the layer are parallel to each other, although from the sign of birefringence the crystalline elements are predominantly perpendicular to the surface. Radial striations can be seen in both phase contrast and polarized light, also incremental lines are arranged obliquely to the dentine surface and taper towards the basal margin of the tooth plate (Fig 11). There is a narrow band of +ve birefringence juxtaposed to the enamel layer, although it is not part of the hypermineralized region. The enamel layer readily fractures away from the dentine-bone surface and SEM's of the vertical fracture surface demonstrate lines parallel to the outer surface (Fig 12). Comparing the surface features of the glossy enamel layer seen in the SEM with those features described in the sections, incremental lines are found running parallel to the basal margin (Fig 9) and also, a series of lines or grooves running approximately perpendicular to these. On the lingual surface (Fig 13) the lower margin of the hypermineralized, glossy layer overlays a resorption surface of bone (Figs 13, 14). This is an extensive area of resorbing bone, identified in the microradiographs by the presence of cell lacunae, and by numerous concave pits at the surface, each with raised edges. These edges identify the margins of resorption[14] formed at the peripheries of the sites previously occupied by the bone resorbing cells (Fig 15). From the sections and microradiographs it is apparent that the enamel layer, together with an extremely thin layer of dentine or bone, has been deposited on top of the resorption surface of the bone, where a reversal line between the tissues can   frequently be observed.

Discussion

Enamel in the teeth of Onychodus sp. (Teleostomi: Crossopterygii, Sensu Miles [15]

The principal criteria for the conclusion that the surface layer to the tooth is enamel, are those which indicate that the layer is formed by appositional growth on the surface of the dentine in a regular, controlled, incremental pattern and that the crystallite orientation is not dependent upon a pre-existing collagen fibre matrix. The reduction of a collagen-fibre matrix from the developing to the mature tissue cannot be directly demonstrated in fossil forms, particularly using only the SEM. However there are certain observed features from which this information can be extrapolated. One is the natural fracture line between enamel and dentine with no continuity of structures across the junction. Another is the contrast between the regular and highly organized pattern of the crystal groups of the enamel, with the irregular arrangement in the underlying dentine. The spatial arrangement of the crystal groups in the enamel layer in the teeth of Onychodus, as described from the SEM, shows them to be of a similar orientation and all perpendicular to the surface. There is a slight but regular variation of the orientation of the crystal groups within each ridge on the sculptured part of the tooth surface. These are both features of enamel, with crystallite orientation[16] controlled by the cells at the formative surface, such as the enamel epithelium, rather than by a collagen-fibre matrix. The radial arrangement of the crystal groups is similar to that described in histological studies of enamel in porolepid and osteolepid rhipidistians. This, together with the fact that the enamel junction is not similarly shaped into ridges, implies that the entire thickness of the layer and the organization of the crystallites is under the control of the enamel epithelium; a prerequisite for deriving the pseudoprismatic enamel of the mammal-like reptiles.[17] The enamel in the teeth of rhipidistians has been described as pseudoprismatic and it is believed that this type of enamel is not very different from nonprismatic pre-mammalian enamel.

Incremental lines in which the slope of the line indicates appositional growth from the dentine have been demonstrated in the smooth, unsculptured enamel of the tooth tip.[3] This feature has been used previously,[3] as one of the criteria for enamel in the teeth of Latimeria chalumnae. This, together with an increase in thickness of the enamel layer towards the tooth tip, is indicative of enamel growing from the dentine surface. Peyer has illustrated incremental lines in the enamel of Onychodus sigmoides in one of the larger symphyseal teeth and interpreted them as similar to striae of Retzius in mammalian teeth.

A further feature used as evidence that the outer layer is enamel, is the observation that the surface microstructure changes with maturity of the tooth. This has been described as a fine herring bone pattern superimposed on the ridged surface. The regularity and change with maturity

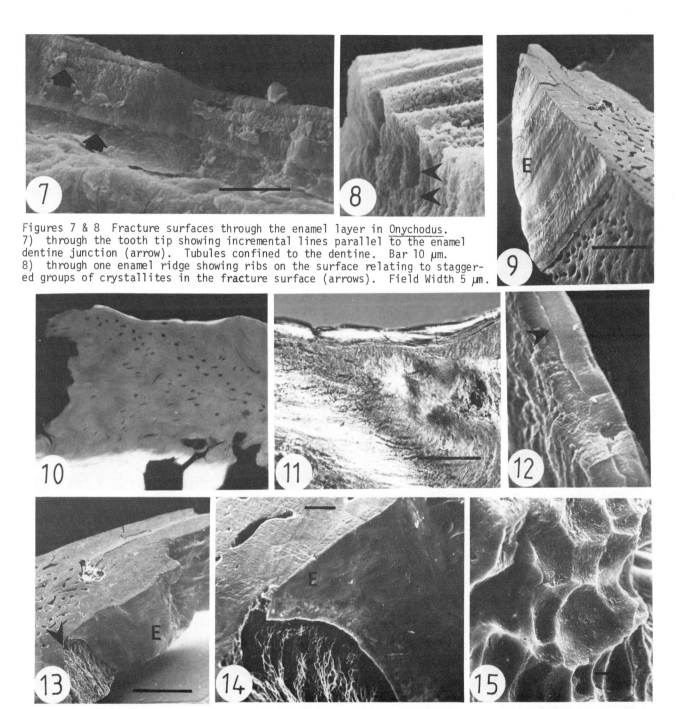

Figures 7 & 8 Fracture surfaces through the enamel layer in <u>Onychodus</u>.
7) through the tooth tip showing incremental lines parallel to the enamel
dentine junction (arrow). Tubules confined to the dentine. Bar 10 μm.
8) through one enamel ridge showing ribs on the surface relating to staggered groups of crystallites in the fracture surface (arrows). Field Width 5 μm.

Figures 9-15 <u>Lepidosiren paradoxa</u>. The SEM block (9 & 13) is part of the tooth plate showing the glossy
enamel (E) on both labial and lingual surfaces. Figs 10 & 11 are sections parallel to the cut surface
and show the enamel layer as hypermineralized in the microradiograph and strongly birefringent in
polarized light. In Fig 12 the SEM shows an incremental line in the fracture surface of the same enamel
layer (arrow). Fig 9) Enamel layer (E) covers bone at the lower margin, incremental lines are parallel
to this. Bar 1mm. Fig 10) Density of enamel layer is similar to that of the hypermineralized dentine,
in contrast to the bone between with lacunae. Height - 1mm. Fig 11) Birefringent enamel seen as overlapping increments. Bar 100 μm. Fig 12) SEM showing an incremental line in the fracture surface of the
same enamel layer (arrow). Width 10μm. Fig 13) Lingual enamel (E) covering a resorbing bone surface
(arrow). Bar 1mm. Fig 14) SEM of the same region showing enamel (E) overlapping the resorbing bone surface. Incremental lines are parallel to this lower border. Bar 100 μm. Fig 15) SEM of the same resorbing bone surface with prominent margins around each resorption pit. Bar - 10 μm.

of the surface ribbing are considered to be of significance because they imply that a great degree of control of the crystallite orientation has been achieved by the dental epithelium in this Devonian group of vertebrates. It may be possible to demonstrate generic or species differences from this pattern on the enamel ridges. It is of considerable interest that, in this group of crossopterygian fishes, there is enamel homologous with reptilian enamel as they are thought to be most closely related to the osteolepids, the group of fishes with the closest relationship to the ancestors of the tetrapods. It is concluded that the ability to secrete and organise the components of the enamel to give characteristic patterns did not suddenly evolve within the mammal-like reptiles but already existed within Devonian fishes, the crossopterygians.

Enamel in adult tooth plates of Lepidosiren paradoxa (Teleostomi: Dipnoi)[15]

In most respects the glossy, superficial layer covering the labial and medial aspects of the tooth plate ridges, satisfies criteria used previously[2,3] identify a similar, extremely thin layer of enamel in the oral teeth of the coelacanth, Latimeria chalumnae. It is hyper-mineralized, of comparable microradiographic density to the petrodentine, removed by dilute acid, fractures perpendicular to the surface, cleaves from the dentine, exhibits radial striations in phase and polarized light and is positively birefringent relative to the surface tangent. Both the histological and SEM appearances suggest that the crystalline components are arranged predominantly perpendicular to the surface. It is unlikely that this arrangement is due to hypermineralization of radially arranged collagen fibres of the dentine, as it would be in enameloid, for two reasons; one, the tissue immediately below has predominantly longitudinal collagen fibres, and two, the enamel layer has in many places formed by apposition directly onto a bone surface. The arrangement of the incremental lines indicates that the layer grows away from the underlying surface and it must be assumed from this and from the surface microstructure in the SEM that the epithelial cells of the 'gingival' cuff have retained the ability to secrete an extremely thin layer of enamel. To this extent I agree with Peyer[1] that enamel is present in the tooth plates of dipnoans. It is contrary to the views of Schmidt and Keil,[7] Lison [18] and Kemp [19] who all regard this layer as enameloid. The observation that the enamel layer forms on top of a resorption surface is interesting because Schmidt and Keil,[7] did in one part of their account describe the tooth plates of Lepidosiren paradoxa as covered by an enamel layer formed after a resorption process had occurred in this region. It may be more realistic to consider this layer of enamel as comparable with the ganoin of dermal bones and scales of actinopterygians, and the outer layer of cosmine of the dermal bones of dipnoans, rather then enamel homologous with that in the oral teeth of crossopterygians. Ørvig[20] comments in two recent papers that ganoin in the dermal skeleton of fossil actinopterygians is similar in its microstructure to the outer layer of the

cosmine described[8] in the Devonian dipnoans. The only part of the tooth in dipnoans, which could be considered strictly homologous with enamel described in crossopterygian oral teeth, is that covering the larval denticles in the first stages of tooth plate formation. This has been discussed in a previous paper.[9]

In view of these conclusions it would seem anomolous to consider that the enamel layer in adult dipnoan tooth plates has much relevance to the understanding of simple enamel structure in pre-reptilian vertebrates. This can only be achieved by investigating the microstructure of enamel in fossil forms such as Onychodus, and others among the rhipidistians. It is apparent from the observations with the SEM of resorption surfaces and forming or mature surfaces, that much relevant information can be obtained from fossil material of the type preserved in the Gogo formation in Western Australia.

Acknowledgements

Gogo Fossil specimens from Dr. S.M. Andrews (Royal Scottish Museum) and from the British Museum (Natural History). Lepidosiren paradoxa specimen from Dr. Adam Locket of the Institute of Opthalmics, University of London. Cambridge Steroscan Mk IIa in the Clincial Research Labs., Coats & Welter SEM in the Department of Structural Biology, St. George's Hospital Medical School. Microradiographic equipment in the Department of Morbid Anatomy, Royal National Orthopaedic Hospital.

References

1. B. Peyer: Comparative odontology. (Trans. and ed. by R. Zangerl) University of Chicago Press, Chicago, U.S.A., 1968, 114-118.
2. Moya M. Smith: Enamel in the oral teeth of Latimeria chalumnae: an S.E.M. study. J. Zool. Lond. 185, 1978, 355-369.
3. R.P. Shellis and D.F.G. Poole: The structure of the dental hard tissues of Latimeria chalumnae Smith. Archs oral Biol, 1979, (In Press).
4. H.P. Schultz: Folded teeth and the mono-phyletic origin of tetrapods. Am. Mus. Nov. 2408, 1970, 1-10.
5. H.P. Schultz: Die Faltenzahne der rhipidistiiden Crossopterygier, der Tetrapoden, U.S.W. Palaeontogr. ital, O.S. 65, N.S. 35, 1969, 63-136.
6. W. Gross: Über Crossopterygier und Dipnoer aus dem baltischem oberdevon im Zusammenhang einer Vergleichenden ......K. Svenska Vetensk. Akad. Handl. (4) 5, (6), 1956, 3-140.
7. W.J. Schmidt and A. Keil. Polarizing micro-scopy of dental tissues (Trans. by D.F.G. Poole and A.L. Darling), Pergamon Press, Oxford, England, 1971, 240-249.
8. Moya M. Smith: The microstructure of the dentition and dermal ornament of three dipnoans from the Devonian of Western Australia. Phil. Trans. R. Soc. Lond. B 281, 1977, 29-72
9. Moya M. Smith: Structure and histogenesis of tooth plates in Sagenodus inaequalis Owen

considered in relation to the phylogeny of post-Devonian dipnoans. Proc. R. Soc. Lond. B 199, 1979, (In Press).

10. B.G. Gardiner and R.S. Miles: Devonian fishes of the Gogo formation in Western Australia. Colloques int. Cent. natn. Rech. scient. 218, 1975, 73-79.

11. H.A. Toombs and A.E. Rixon: The use of acids in the preparation of vertebrate fossils. Curator 2, 1959, 304-312.

12. A. Boyde and Sheila J. Jones: Bone and other hard tissues. In: Principles and techniques of scanning electron microscopy, M.A. Hayat (ed.) Van Nostrand Reinhold Company, New York, U.S.A. 2, 1974, 123-149.

13. W. Gross: Onychodus jackeli Gross (Crossopterygii, Oberdevon), Bau des Symphysenknochens und seiner Zahne. Senckenberg. leth. 46a, 1965, 123-131.

14. A. Boyde and M.H. Hobdell: Scanning electron microscopy of lamellar bone. Z. Zellforsch. mikrosk. Anat. 93, 1969, 213-231.

15. J.A. Moy-Thomas and R.S. Miles: Palaeozoic Fishes, Chapman and Hall Ltd., London, England 1971.

16. A. Boyde: The structure and development of mammalian enamel. Ph.D. Thesis, University of London, 1964.

17. D.F.G. Poole: The structure of the teeth of some mammal-like reptiles. Quart. J. micr. Sci. 97, 1956, 303-312.

18. L. Lison: Rescherches sur la structure et l'histogenese des dents des Poissons Dipneustes. Archs Biol. Paris 52, 3, 1941, 279-320.

19. A. Kemp: The pattern of tooth plate formation in the Australian lungfish. Zool. J. Linn. Soc. 60, 1977, 223-258.

20. T. Ørvig: Microstructure and growth of the dermal skeleton in fossil actinoptergyian fishes. (a) Zool. Scr. 7, 1978, 33-56. (b) Zool. Scr. 7, 1978, 125-144.

Discussion with Reviewers

R.P. Shellis: The surface sculpturing of the enamel shows, as far as I can see, only that the inner dental epithelium could fold in this fish, but this is common throughout vertebrates. If as Boyde[16] concluded, enamel crystals form roughly perpendicular to the epithelium when Tomes processes are absent, could not the pseudoprismatic structure be generated simply by the altered profile of the epithelium without any change in activity of the epithelial cells? In this respect, the surface shapes of elasmobranch teeth are determined before matrix is formed, so that here also folding of the epithelium would be the morphogenetic factor.

Author: The question you raise is of course, a very important point. Clearly any discussion of cell arrangement or cell activity is based on very few facts, the cells themselves cannot be investigated only the product of their secretion. Interpretation of the role of the inner dental epithelium, in this case, is based only on the arrangement of the mineral components of the enamel layer. I do not believe that the surface details described in Figs 2,3 and 4 superimposed on both ridges and the inter-ridge area could be achieved simply by folding of the epithelial layer over a collagenous matrix. Also, the junction between the dentine and the enamel is not similarly shaped but is smooth. If you are wishing to suggest that it is enameloid formed by hypermineralization of a collagenous matrix then this would imply a modifying influence by the inner dental epithelium which was capable of penetrating to different depths of the forming outer layer to produce a thicker layer of enamel along the ridges and a thin layer between the ridges. Folding of the epithelium would not, in my opinion, produce the staggered arrangement of crystal groups related to the surface ribbing demonstrated in Fig 8. Whether or not it is justified to call this pseudoprismatic enamel can possibly only be decided by examination of ground sections of the same material and by reference to previous findings by Schultze[5] and Poole[17]. I believe that the morphogenetic factors reside in the secretory activity of the inner dental epithelial cells, rather than simply folding of the epithelium.

R.E. Weiss: The regular ribbing of the enamel surface ridge in the oral teeth of Onychodus has been clearly demonstrated in your report. Is there any structural similarity between this ribbing and the "saw-tooth" pattern of enamel crystallites seen in mammals of the late Triassic and succeeding Jurassic periods as described by M.L. Moss.

Author: Mammalian enamel is prismatic and, therefore, the "saw-tooth" pattern is of a different order of organization than the ribbing described on the enamel surface of the oral teeth of Onychodus.

D.G. Gantt: What is the sample size? Were any of the specimens etched prior to analysis? Could the rough, granular texture in the young teeth be lost or worn away, thus resulting in the smooth, untextured surface of the older specimens?

Author: Between 8-10 teeth or parts of teeth have been examined in the SEM, none of the specimens has been etched prior to analysis. No, this is not likely to be an explanation of the surface differences because the ribbing is still present and more complete in the older teeth. I believe that the two surfaces represent genuine differences due to a change with maturity of the surface layer.

Shiela J. Jones: This paper demonstrates an interesting application of SEM to unravel the evolutionary history of enamel. Enamel is by definition an epithelial secretion. Have you been able to correlate your SEM findings of the fossil material with biochemical investigations?

Author: No, I have not attempted any biochemical investigation. The amount of sample available for analysis is alarmingly small and traces of any organic material in the extremely thin enamel layer would predictably be in minute quantities.

Additional discussion with reviewers of the paper "Asbestos Fibre Counting by Automatic Image Analysis" by R.N. Dixon and C.J. Taylor continued from page 366

P.W. Hawkes: How easy would it be to implement your strategy (and hence software) on the various systems reviewed?

Authors: As indicated in the text, we have implemented analysis of detected image structure in a general-purpose minicomputer. Thus, any system which includes a minicomputer could be used to implement this part of the analysis. The critical processing phase is, however, the initial detection and subsequent skeletonisation. A minicomputer is simply too slow to execute these routines in realistic times. We have implemented the detection operator in the Nova minicomputer, for development purposes, and it runs about 80 times more slowly than in the image processor. Thus, some additional processing power is essential, though it must be programmable to allow a software solution. Most instruments lack this facility but we understand that the Bausch and Lomb systems have an image processor that operates on the post-detection binary representation of the image. High-speed skeletonisation is therefore a possibility. However, this processor appears to lack the flexible access to grey-level image data that is necessary to implement the detection operator in reasonable times.

SCANNING ELECTRON MICROSCOPY/1979/II
SEM Inc., AMF O'Hare, IL 60666, USA

A METHOD OF INTERPRETING ENAMEL PRISM PATTERNS

D. G. Gantt

Section on Palaeoanthropologie
Senckenberg Museum   and
6000 Frankfurt am Main, W. Germany

Dept. Anthropology
Florida State Univ.
Tallahassee, FL 32306

Abstract

The importance of enamel structure as a phyletic indicator is not a new concept. In 1849, Tomes conducted studies on the enamel of marsupials and suggested that taxonomic affinities could be determined. However, it was not until the advent of the scanning electron microscope that researchers could begin to understand the development and ultrastructural features of enamel, especially the enamel prism.

The problem which must be resolved is the variety of prism patterns reported for the same species. It has been suggested that a number of different patterns can be obtained from the same specimen. However, the methods of analysis are ill defined leading to the possibility of an incorrect pattern or an artifact. An attempt to solve this problem of interpretation has recently been aided by the work of Boyde on acid etchants and their effects on tooth enamel.

Human and nonhuman primate teeth were prepared by placing a 1cm.^{2}, highly polished (~6μm) facet on the mid-cervical crown. The facet was then etched, using a 0.074M solution of phosphoric acid for 60 seconds. The specimens were then coated with gold and analyzed with a Cambridge S-4 stereoscan at various magnifications. Stereo-pairs were taken to determine, and correctly describe, the enamel prism pattern for each specimen.

The results clearly document a "keyhole" pattern in the human teeth and a "hexagonal" pattern in the teeth of *Pongo*. No other patterns were obtained using this technique. Therefore, in a study of enamel prism patterns the critical factors are (1 etching, (2 polishing, and (3 use of stereo-analysis.

KEY WORDS: Enamel, Fossils, Prisms, Patterns, Etching, Artefacts, Phylogeny, Taxonomy

Introduction

The importance of enamel structures as relevant taxonomic characters has been known for a number of decades[1-7], although it is only recently that researchers have focused on this aspect of dental anatomy to assist in phylogenetic reconstruction and identification[8-16]. This disparity of interest was, perphas, due to the lack of understanding of enamel structures in man and other mammals.

It was not until the development of such instruments as the scanning electron microscope (SEM), transmission electron microscope (TEM), and X-ray diffraction that researchers could begin to understand the developmental and ultrastructural features of enamel, especially the enamel prism. Poole and Brooks have identified the arrangement of crystallites within human enamel prisms and Meckel has defined the "keyhole" human concept of enamel microstructure[8,9]. However, it has been Boyde who has done much to provide a more complete understanding of enamel structures in mammals[10,16-19].

Recent studies have documented that specific enamel prism patterns do exist and can be used as a taxonomic indicator [10-14,16]. Although, a few researchers have reported different prism patterns in particular species[14], while others have even suggested that prism morphology contains no information on phylogenetic relationships[15].

The purpose of the present study is to develop a standardized technique to study the morphology of the enamel prism, and to define and discuss those factors contributing to the variety of patterns reported in the literature. Three variables are considered to be responsible for the present confusion in the literature. First, the main problem in studying enamel is that its structure must be seen in three dimensions[16]. The importance of stereo-analysis of enamel structures can not be overemphasized. Boyde, among others, has published extensively on this subject[17].

A second problem is the differential effects of acid etchants. Etching is of potential interest in the field of mammalian taxonomy, because enamel undergoes few post mortem changes. Recently, Boyde and his colleagues have conducted an intensive study of the quantitative and qualitative aspects of enamel etching with acid and EDTA[18]. The results have brought to light certain inadequacies and the presence of artifacts as the result of specific etchants and/or lengths of etching times. They concluded that a dilute H_3PO_4 solution (0.074M = 0.5% vol./vol. of concentrated, 85%, acid) is most suitable for etching polished enamel to study crystallite orientation and prism patterns, with etch times of 30 to 60 seconds. More prolonged etching with any of the agents tested gave rise to artifacts of one sort or another[18].

A third problem arises when attempting to study fossil material. Fossil remains are often subjected to significant post mortem changes due to a vast variety of chemical and physical agents. These agents change the surface chemical and physical composition. Therefore, treatment of the tooth surface with acid etchants will produce greatly differing prism patterns due to the remineralization products; the differences in mineral content of the crystals.

Materials and Methods

Twenty molars and premolars were prepared from ten individuals of *Homo sapiens sapiens* and compared to the patterns obtained from ten molars from the pongid genus *Pongo*. Each specimen was cleaned in alcohol and acetone and air dried. A highly polished facet of approximately 1cm^2 was prepared in successive stages.

Polishing

The facet is placed on the mid-cervical crown, on either the buccal or linual surfaces. A facet is prepared first by fine grinding on 600 grit carbinet paper using distilled and deionized water as an extender on a low speed polisher (AB Low Speed Polisher/Grinder-Buehler LTD.,Illinois). The specimen is then washed and the facet is rough polished, using 6 micron Metadi on nylon cloth with Automet lapping oil (Buehler, LTD.) as an extender. Polishing continues until all the scratches from the 600 grit are removed. Again, the specimen is washed, but in soap and water to remove the oil and is then allowed to air dry. A final polish is applied using Micropolish B (Aluminum Oxide 0105 micron) on Microcloth (Buehler, LTD.) until all scratches from the 6 micron rough polishing stage are removed. Distilled or deionized water is used as an extender for the alumina polishing. After washing the specimen is ready to be etched.

This polishing procedure is designed to be used on rare, fossil specimens. The mid-cervical crown provides a large area of the tooth where generally the prisms have the same orientation with minimal prism decussation. This region is the most often preserved part of a fossil tooth. In addition, this area provides little morphological data. Casts and photos can be taken to provide a permanent record prior to polishing and etching.

Etching

Once the facet was prepared the specimen was again washed in distilled water, dehydrated in ethanol, and allowed to air dry, although it is suggested that when conducting etching studies the specimens should be critical point dried[18]. Etching was conducted in 50mm diam. dishes using 10mls of 0.074M H_3PO_4 with etching times of 30 to 60 seconds. At the end of the etching procedure, the specimens were washed in distilled water, dehydrated in ethanol, and air-dried. The specimens were then mounted on 1/8" diameter aluminum rivets and coated with approximately 10nm C and 10nm Au by vacuum evaporation. The specimens were then kept under vacuum until analysis in a Cambridge Scientific Instrument-*Stereoscan MK-S4* scanning electron microscope operated at 10kV beam voltage. Stereopair micrographs were recorded in every case and further analyzed with the aid of the Hilger and Watts Stereometer as described by Boyde[17].

Results

The differential affects of acid etching are clearly seen in Plates 1-3. Differences in prism orientation, surface topology, and specific etchants are clearly documented. These differences have led to a variety of reported prism patterns. It can be observed that in Plates 1-3, the prism patterns vary from archade shape to hexagonal. From this evidence it is impossible to determine what the prism pattern is or if there is really such a thing as an enamel prism pattern. Such studies have led researchers to disregard prism patterns as important taxonomic indicators[15]. Also, Boyde *et al* (1978) has documented the variety of artifacts which can be obtained due to acid etchants[18].

By producing a highly polished facet and subsequently etching with a dilute solution of H_3PO_4 as previously described, reproducible and consistent patterns were obtained. The "keyhole" prism pattern was observed in all human teeth (Plate 4). In *Pongo* the typical "hexagonal" pattern of the pongids was obtained (Plate 5).

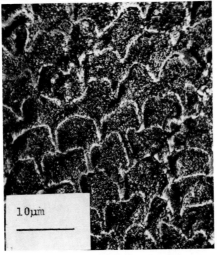
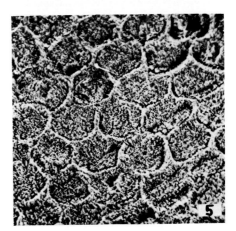
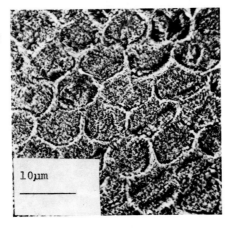

Plates 1 to 3 document the effects of simply etching the tooth; plates 1 to 3 show
   hydrochloric acid etched pongid molar, surface enamel.
Plate 1. Reveals a number of different patterns across the tooth surface.
Plate 2. A higher magnification of one area (A) showing a hexagonal to open archade
   pattern.
Plate 3. A higher magnification of one area (B) showing irregular shaped prism.

Plate 4. 0.074M H₃PO₄
   polished and etched
   human molar.

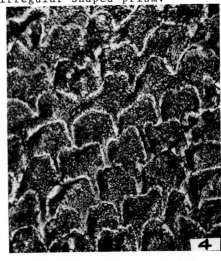

Plate 5. 0.074M H₃PO₄
   polished and etched
   molar of *Pongo*.

   (Stereo-pairs).

A fossil hominoid which may be related to the genus *Pongo* was also studied to determine if this methods of analysis could be applied to fossil material. The specimen, a lower molar of *Sivapithecus*, was prepared using the same procedures previously described. Analysis reveals a pattern which is "hexagonal", like that of *Pongo* (Plate 6). In plate 7 the hexagonal prism is clearly presented.

Discussion

Enamel prism patterns are specific and can be accurately described if the following conditions are met: (1 a highly polished facet is prepared on the mid-cervical crown; (2 the use of a dilute solution of H_3PO_4, with etching times of approximately 60 seconds; and (3 the use of stereopairs to interpret accurately the specimen's prism pattern. The polished surface must be deep enough into the surface of the tooth to penetrate the non-prismatic layer.

If fossil specimens are to be studied the surface replacement layer, which may be of considerable thickness, must be penetrated. Present studies of fossil primate teeth indicate that the problems of fossilization can be solved *if* the polishing and etching methods are continued until the crystallites are clearly visible within each prism as they are in plates 6 and 7.

The problems of interpretation of enamel prism patterns are basically due to the lack of understanding of the ultrastructure of enamel and in the correct applications of SEM techniques to the study of ultrastructural features. Numerous studies, especially those by Boyde, on (1 enamel structures, (2 the effects of acid etchants, and (3 the use of the SEM in the study of mineralized tissues, have provided the framework for this an analysis. Researchers wishing to study the enamel prism patterns of mammals must apply ridgid standards of analysis, if they are to correctly describe and discuss enamel prism patterns. At present, the methods described are currently being used in a study of fossil hominoids. I have applied the procedures described in this paper with success. The patterns are reproducible and consistent between individuals and within the genus. However, it is often necessary to repolish and etch the specimen several times before the remineralized zone is removed. This polishing and etching procedure is continued until the crystallites are clearly visible, at which time I consider the observed pattern to be representative of the species' prism pattern.

The importance of enamel histology, especially enamel prism patterns in mammalian phylogeny still awaits a comprehensive study of not one but several taxa. Those who wish to study prism patterns must be aware of the differential effects of acid etching as well as the other variables discussed before a species' prism pattern is described. Future studies will provide a more complete understanding of enamel prism patterns and their phyletic affinities.

Acknowledgements

I wish to thank Dr. Alan Boyde for his assistance and comments, also Mr. Dan Cring and Mr. Bill Miller for their technical assistance.

Financial support was received from the L.S.B.Leakey Foundation, the American Philosophical Society, faculty research grant Florida State University, and the Alexander von Humboldt-Stiftung Foundation.

References

1.   J.T. Carter: The microscopical structure of the enamel of two Sparassodonts, Cladosictis and Pharsophorus, as evidence of their marsupial character together with a note on the value of the pattern of the enamel as a test of affinity. J. Anat. (Lond.) 54, 1920, 189 - 195.

2.   J.T. Carter: On the structure of the enamel in the primates and some other mammals. Proc. Zool. Soc. Lond., 1922, 599 - 608.

3.   V.A. Korvenkontio: Mikroskopische Untersuchungen an Nagerincisiven unter Hinweis auf die Schmelzstruktur der Backenzähne. Histologisch-phyletische Studie. Annal. Zool. Soc. Zool.-Bot. Fenn. Vanamo (Helsinki), 2, 1934-5, 1 - 274.

4.   M. Shobusawa: Vergleichende Untersuchungen über die form der Schmelzprismen der Säugetiere. Okajimas Folia Anat. Japan, 24, 1952, 371 - 392.

5.   C.S. Tomes: On the minute structure of the teeth of creodonts, with special reference to their suggested resemblance to marsupials. Proc. Zool. Soc. Lond., 1906, 45 - 58.

6.   J. Tomes: On the structure of the dental tissues of marsupial animals, and more especially of the enamel. Phil. Trans. R. Soc. Lond., 139, 1849, 403 - 412.

7.   J. Tomes: On the structure of the dental tissues of the order Rodentia. Phil. Trans. R. Soc. Lond., 140, 1850, 529 - 567.

8.   D.F.G. Poole and A.W. Brooks: The arrangement of crystallites in enamel prisms. Arch. Oral Biol., 5, 1961, 14 - 26.

9.   A.H. Meckel: The keyhole concept of enamel microstructure; pp. 25 - 42, in Chemistry and Physiology of Enamel, U. of Mich. 1971.

10. A. Boyde: Development of the structure of the enamel of the incisor teeth in the three classical subordinal groups of

the Rodentia. pp. 43 - 58, in Development Function and Evolution of Teeth (P.M. Butler and K.A. Joysey eds.) Academic Press, London, 1978.

11. D.G. Gantt: Enamel of primate teeth: Its thickness and structure with reference to functional and phyletic implications: Ph.D. Thesis, 1977, Washington University.

12. D.G. Gantt, D.R. Pilbeam and G.P. Steward: Hominoid enamel prism patterns. Science, 198, 1977, 1155 - 1157.

13. W. von Koenigswald: Mimomys cf. reidi aus der villafranchischen Spaltenfullung Schambach bei Treuchtlingen. Mitt. Bayer. Staatslg. Palaont. hist. Geol. 17, 1977, 197 - 212.

14. C.L.B. Lavelle, R.P. Shellis and D. F.G. Poole: Evolutionary Changes to the Primate Skull and Dentition. (Primate Dental Tissues) pp. 157 - 219, Springfield, 1977.

15. E.S.Vrba and F.E. Grine: Australopithecine enamel prism patterns. Science, 202, 1978, 890 - 892.

16. A. Boyde: The structure and development of mammalian enamel: Ph.D. Thesis, 1964, University of London.

17. A. Boyde: Some aspects of the photogrammetry of SEM images. Photogrammetric Record, 8, 1975, 408 - 445.

18. A. Boyde, S.J. Jones, and P.S. Reyonlds: Quantitative and qualitative studies of enamel etching with acid and EDTA. SEM/1978/II, SEM Inc. IL, 991-1002.

19. A. Boyde: Electron microscopic observations relating to the nature and development of prism decussation in mammalian dental enamel. Bull. Group. Int. Rech. Sc. Stomat., 12, 1969, 151 - 207.

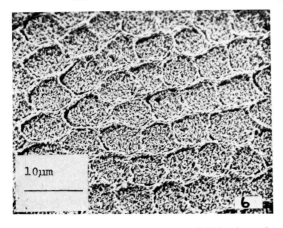

Plate 6. 0.074M H_3PO_4 polished and etched molar of a fossil hominiod *Sivapithecus*.

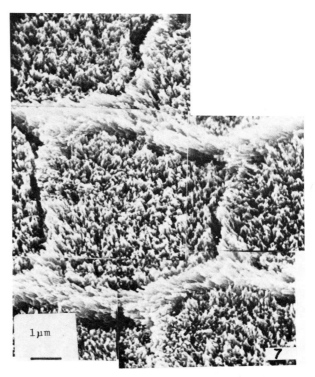

Plate 7. A composite photo revealing the crystallite arrangement of the "hexagonal" prism.

Discussion with Reviewers

K.A. Galil: *In your Materials and Methods you mentioned that "it is suggested that when conducting etching studies the specimen should be critical point dried". The question is have you done any comparisons between air dried and critical point dried specimens?*

A.J. Gwinnett: *It is not entirely clear why such a highly mineralized tissue like enamel should be critical point dried.*

Author: The suggestion that enamel etching studies using the SEM should make use of critical point drying and/or freeze drying is not mine, but Boyde et al[18]. The authors suggest that drying artifacts might be prevented due to the demineralized state[18]. I have not done any comparisons between air dried and critical point dried specimens, for I do not get drying cracks with the described techinque as can be seen in Plate 7.

J.C. Rose: *Though the micrographs indicate, in most cases, a distinct difference in prism shape between Homo and Pongo, are they consistently distinct for all individuals in the samples or is there some gradation between the two shapes?*

Author: I did not obtain any type of gradation between the two shapes. The only problem was in the quality of the final polish which can be seen between the hexagonal pattern of *Pongo* and *Sivapithecus*.

R.P. Shellis: *Could you give any indication on how deeply into the enamel the effects of fossilization are likely to reach, and on how fossilization might be expected to alter the etch pattern?*
Author: The process of fossilization subjects the enamel to a vast variety of chemical and physical agents. The agents may cause the specimen to already be highly etched, or affect its acid resistance. Acid resistance may be high, related to a high fluoride content, or lowered due to a high carbonate content, than the natural state. Concerning the depth of the effects of fossilization I can not give any specific measurements. Studies of fossil hominoids, dating from 3 to 25 million years reveal that some specimens appear to have undergone little if any post mortem changes while others have undergone extensive changes.

Analysis of the South African australopithecines reveal a variety of prism patterns. These fossils have been recovered by limestone caves and are highly etched with a high content of carbonate, contributing to the variety of patterns obtained by Vrba and Grine[15]. On the other hand, etching of extant primate teeth by acid etchants other than 0.074M H_3PO_4 may result in pattern descriptions which are the result of artifacts. Although, previous studies[11,14,16] have provided data which are comparable at present.

SCANNING ELECTRON MICROSCOPY/1979/II
SEM Inc., AMF O'Hare, IL 60666, USA

THE ARRANGEMENT OF PRISMS IN THE ENAMEL OF THE ANTERIOR
TEETH OF THE AYE-AYE

R. P. Shellis and D. F. G. Poole

MRC Dental Unit
Dental School
Lower Maudlin Street
Bristol BS1 2LY
U.K.

Abstract

The Aye-aye, an unusual primate, has large, pro-
cumbent, anterior teeth which, like rodent
incisors, have enamel on the anterior surface
only. The structure of this enamel has been
examined, in various section planes, by SEM. In
the plane transverse to the tooth axis, two types
of structure were observed: 1. In the midline
region, undulating rows of oblique prism sections
passed outwards from the enamel-dentine junction
to the enamel surface. 2. In lateral enamel,
to each side of the midline enamel, highly
regular HUNTER-SCHREGER bands were present,
making an angle with the enamel surface of about
60°. In alternate bands, groups of prisms were
cut tangentially (PARAZONES) or transversely
(DIAZONES). The SEM showed a gradual, rather
than abrupt, change in prism orientation from
parazone to diazone. Tangential sections of
lateral enamel appeared similar to transversely
sectioned midline enamel, and tangential sections
of midline enamel showed Hunter-Schreger banding
similar to transversely sectioned lateral enamel.
Therefore, the banding seen in the latter cannot
be due to longitudinally (with respect to the
tooth axis) disposed sheets of prisms with
alternating orientations. Instead, it is
proposed that each prism follows a uniform
helical course, the helix being sheared or
inclined to the transverse plane. Diazones
would be regions where the section plane cuts
the spiralling prisms transversely, and parazones
where it cuts them tangentially, in a close-
packed mass of such helices. The observed
absence of bands in the midline enamel would
occur because a change of pitch or slope of the
helices would not allow the prisms to be
sectioned tangentially. A three dimensional
model based on this concept, demonstrating some
of the features of the enamel, has been constructed, and helical regions in some prisms
have been demonstrated by SEM using deep etching
techniques.

Introduction

The aye-aye, Daubentonia, is a primate related to the lemurs but so specialised that it
is placed in a separate family[1]. The dentition
consists of a short row of small cheek teeth in
each quadrant separated by a diastema from a
large, powerful anterior tooth which is often
regarded as an incisor but may be a canine.
These anterior teeth resemble rodent incisors in
being of continuous growth and in being covered
with enamel only on the anterior surface; differential wear of the incisal surface leads to the
formation of a sharp gouge-like morphology. As
part of a survey of primate dental hard tissues
we have described the principal features of the
enamel[2].

The enamel is characterised by the presence
in the lateral regions of a strikingly regular
system of Hunter-Schreger bands. Observations
with polarising microscopy and SEM suggested a
model for the structure based on the concept
that the course followed by each prism was that
of a regular cylindrical helix, sheared so that
its long axis was oblique to the long axis of
the tooth. The enamel could be regarded as a
large number of such prisms closely packed together. On this hypothesis, Hunter-Schreger
bands would not be discrete regions in which the
prisms as a whole are inclined at an angle to
those in neighbouring bands. Instead, diazones
would occur where the spiralling prisms were cut
transversely and parazones where they were cut
tangentially. The grouping of tangential and
transverse sections to give the bands would be
generated by the mode of packing of the spirals
and by their orientation to the section plane.
This concept will be more fully explained with
the aid of a model, the construction of which is
detailed below.

In the present paper we present a description of the enamel structure more complete than
our previous, necessarily brief report, and to
report new observations with SEM on an additional
specimen. We begin by describing the polarised
light appearance, so that the light and electron
microscopic levels of structure can be correlated.

KEY WORDS: Aye-Aye, Enamel, Hunter-Schreger bands,
Polarising Microscopy, Primates, Prisms

Materials and methods

The material consisted of an anterior tooth from the left mandible of a dried specimen of _Daubentonia madagascariensis_. Three transverse slices were cut with a diamond-impregnated annular blade in a Microslice 2 cutting machine (Metals Research Ltd., Royston, Herts., U.K.) from the region of the tooth where the enamel was fully formed but not yet erupted (Fig.1).

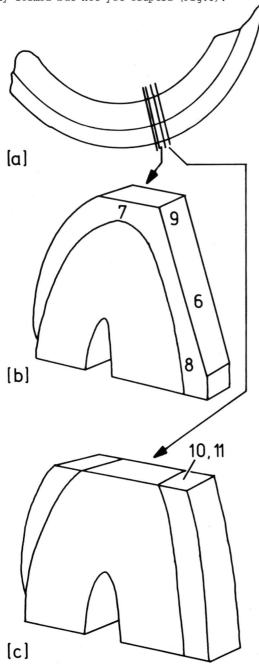

Fig. 1 - (a) Specimens in relation to tooth; (b) Specimen 1; (c) Specimen 2; with planes prepared for SEM study. Numbers refer to areas shown in later figures.

One slice was polished with alumina powder to a thickness of about 70 μm for examination by light microscopy. The others, each about 1 mm thick, were prepared for examination by SEM. Each slice was divided in two and the portion covered with enamel polished on both faces. The specimens were then embedded in cold-cure methacrylate ('Simplex' clear: Howmedica Ltd., London N16 OBP, U. K.) and facets in defined planes prepared as in Fig. 1, by grinding on wet-and-dry carborundum paper grade 320 with water as lubricant and polishing with alumina powder. After removal of the acrylic resin with chloroform, the specimens were mounted on stubs with the orientation shown in Fig. 1, and etched with molar hydrochloric acid. Specimen 1 was etched for 10 sec, Specimen 2 for 30 sec. The specimens were washed thoroughly in water, then in 50 % ethanol and allowed to dry. They were coated with gold-palladium alloy by evaporation and examined in a Cambridge Stereoscan 600 SEM using an accelerating voltage of 15 kV. Stereo-pair photographs were prepared by rotation of the specimen through 6-10° between successive exposures.

The model of the proposed structure was prepared in wax as follows. Cylindrical profile wax, 3 mm in diameter (Chaperlin & Jacobs Ltd., 591 London Road, North Cheam, Surrey, U.K.), was cut into 300 mm lengths, simulating individual prisms, which were softened in a water bath at 50°C. These were wound in groups of 20 around a cylindrical former of glass tubing 15 mm in diameter, to produce batches of close-packed spirals. Since the same number of lengths was used for each batch and care was taken to cover the same length of the glass tube (240 mm) each time, all the helices had the same pitch. After cooling at 4°C, each batch was slid from the tube and the helices carefully detached from each other. About 60 helices were then assembled together one at a time into a close-packed array, each new helix being fixed in place with touches of molten wax. Finally, the mass was sectioned with a hot blade in three planes: transverse, longitudinal and oblique to the long axis of the helix on which the model was based. The cut surfaces, analogous to sections through the enamel, could then be examined.

Results

Polarising microscopy

In transverse section, there were two patterns of banding in the enamel; one in the region of the midline of the tooth (midline enamel) and one in the regions on either side (lateral enamel). (Fig.2).

In the midline enamel, broad birefringent bands ran parallel with the enamel surface, the sign of birefringence reversing between successive bands. In the transitional zone between the midline and the lateral enamel the bands diverged from this orientation towards the enamel surface. The lateral enamel itself was traversed by straight, parallel bands, almost equal in width, which ran obliquely outwards from the enamel-dentine junction to make an angle of about 60° with the enamel surface. The bands

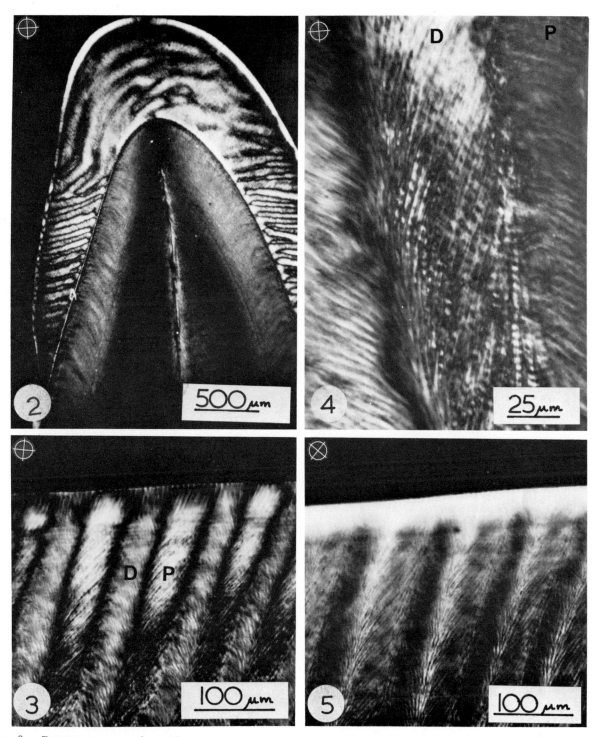

Fig. 2 - Transverse ground section in polarised light: different structure of midline enamel (top) and lateral enamel (sides). Symbol at top left shows polarisation axes.

Fig. 3 - Same section. Lateral enamel, near-parallel position: parazones, P. and diazones,D.

Fig. 4 - Part of field of Fig. 3: prisms with cross-striations in parazones, indistinct striations in diazones.

Fig. 5 - Same field as Fig. 3, but diagonal to polarisation axes: uniform negative bire-fringence of surface layer.

were most distinct when viewed with the enamel surface 10-15° from the parallel position when they became separated by isotropic lines (Figs. 3 and 4). In this position there was alternation of strongly and weakly birefringent bands, accompanied by reversal of sign. In the brighter bands (parazones) prisms with pronounced cross-striations were observed, running obliquely and with a marked curvature across the bands (Fig.4). In the less bright bands (diazones) there were striations running obliquely so as to make an angle of about 65° with the prisms in the parazones but these striations could not be identified as prisms (Fig. 4)

Over the whole outer surface of the enamel there was a layer, about 30 μm thick, with a structure different from that of the inner enamel (Figs. 2,3,5). In the diagonal position it appeared homogeneous and was positively bire-fringent with respect to the surface tangent whether mounted in balsam (RI 1.54), water (RI 1.33) or Thoulet's solution (RI 1.62) (Fig.5). At the near-parallel orientation, however, this layer was broken up by the appearance of bire-fringent 'caps' surmounting the parazones, separated by regions of opposite sign of bire-fringence (Fig. 3).

Scanning electron microscopy

The prisms were arranged in rows (Figs. 6-9). Since our previous report it has been found that the rows are separated by distinct sheets of interprismatic enamel (Fig. 6), showing that the prism arrangement is that described as pattern 2 by Boyde[3]. The crystalline texture of the enamel was very marked and it was clear that the crystal orientation in the sheets was approximately perpendicular to that of the prisms (Fig. 6).

In the transverse plane the pattern of band-ing observed by polarising microscopy was re-produced. In the midline region obliquely and transversely cut prism profiles formed rows following an undulating course between the inner and outer enamel surfaces; registration of the rows produced the effect of bands running parallel with the outer enamel surface (Fig. 7).

In the lateral enamel (Fig. 8), parazones appeared as bands of prisms cut more or less tangentially, and diazones as regions where the prisms were cut more transversely. Inter-row sheets were prominent where the constituent crystals were roughly parallel with the etched surface, i.e. in the diazones, but where the crystals were more perpendicular to the surface, i.e. in the parazones, the sheets were removed by the acid. Examination of stereo pairs (Fig.8) showed that, after acid etching, the parazones were elevated above the diazones and that in the latter regions, the etching affected the interrow sheets less than the prisms, again as a result of variations in crystal orientation with respect to the surface.

The boundaries between parazones and diazones, although appearing distinct at low magnifications, did not involve a sudden change in prism orientation. At higher magnifications it was seen that the prism profiles curved obliquely across the parazones. There were slight changes in orientation from one prism to

the next, this resulting in considerable differences between prisms on opposite sides of the parazone. For instance, in Fig. 8, prism segment A is concave while prism segment B, on the other side of the band and nearer the dentine, is convex. Moreover in the transitional region between parazone and diazone the prisms changed progressively from being obliquely sectioned to being transversely sectioned, although differ-ences in the level of etching of differently orientated prisms often made the transition appear more abrupt.

In the tangential facet through the lateral enamel, the central regions, which were both closer to the dentine and most nearly tangential with the enamel-dentine junction, were character-ised by undulating rows of prism ends forming bands like those in transverse sections of mid-line enamel. Towards the edge of this facet near the midline, there appeared diazones and parazones somewhat similar to those in transverse sections of lateral enamel (Fig. 9). Here, how-ever, the inter-row sheets lay more nearly parallel with the surface in the parazones and survived the etching. It was observed that the inter-row sheets were continuous between diazones and parazones, the continuity apparently involv-ing a bending of the sheets. Diazones and parazones, running across the long axis of the tooth, were also present in the tangential facet prepared in the midline enamel proper.

In longitudinal facets the bulk of the enamel was opened up by etching but the 30 μm layer near the natural outer surface retained most of the interprismatic material (Fig. 10). Superficially this material had a rough texture but near the deep surface, where inter-row sheets ran into it, the structure was more regular.

Banding occurred in longitudinal facets, whether the facet was prepared parallel with the bands in transverse section or oblique to them. The bands in the parallel and oblique facets sloped in different directions, however, those in the parallel facets sloping anteriorly from the dentine-enamel junction, those in the oblique facet posteriorly. The longitudinal facets were etched more deeply than facets in other planes; this removed the inter-row enamel preferentially and allowed observation of the course of the prisms to a greater depth of the tissue. The prisms were markedly sinuous and in many places crossed from diazone to parazone (Fig. 11). In Fig. 12 is shown the terminal portion of a diazone which extended only part of the way through the enamel thick-ness. In this field, one prism can be seen to descend below the surface, cross over another prism and then ascend to cross over other prisms. The stereo-pair shows the prism con-cerned to have an unmistakably spiral course.

The model

Fig. 13 shows the cut surface of a model constructed of helices of wax, sectioned obliquely. The surface shows diagonal bands consisting of profiles of the rods. In successive bands the profiles are alternately transverse/oblique and tangential sections.In

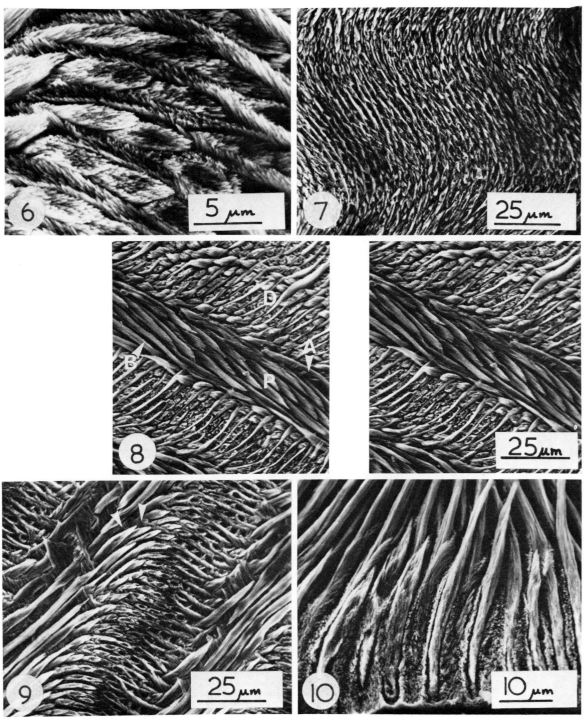

Fig. 6 – Tangential facet in lateral enamel
(6, Fig. 1): prism profiles in rows separated
by inter-row sheets with different crystal
orientation.

Fig. 7 – Transverse facet, midline enamel
(7, Fig. 1): transverse/oblique prism profiles
in undulating rows running between dentine
(below) and outer surface (above).

Fig. 8 – Transverse facet, lateral enamel
(8, Fig. 1). Stereo pair: parazones and diazones.

Fig. 9 – Tangential facet through side of tooth
but in region of midline enamel (9, Fig. 1):
parazones and diazones, with signs of bending of
inter-row sheets (arrow).

Fig. 10 – Longitudinal-parallel facet
(10. Fig. 1): surface zone of enamel.

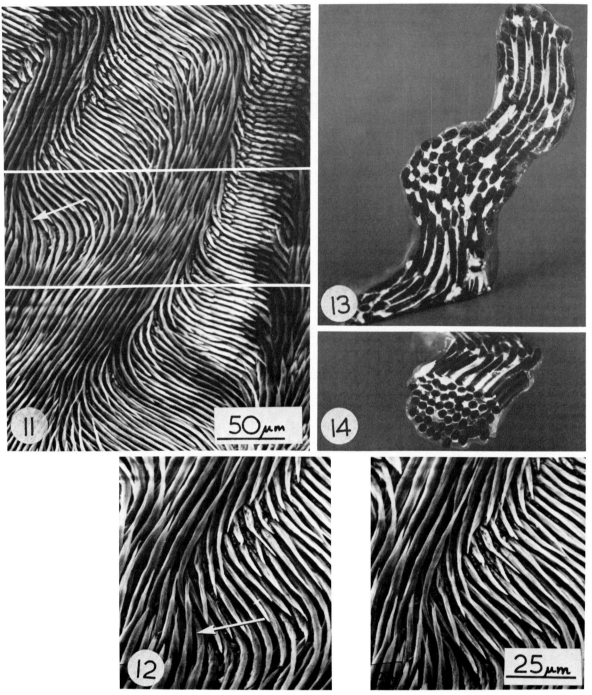

Fig. 11 - Longitudinal parallel facet (11,Fig.1).
Montage. Crossing-over of sinuous prisms between
parazones and diazones. Arrow shows prism in
Fig. 12.
Fig. 12 - Part of field of Fig. 11: arrow
indicates prism with spiral course.
    (Fig. 12- Stereo-pair).

Fig. 13 - Wax model sectioned obliquely: alternating bands of transverse and tangential
profiles.
Fig. 14 - Wax model sectioned transversely:
transverse profiles only.

surfaces exposed by cutting the model longitudinally, i.e. parallel with the helix axes, only curved, somewhat elongated, oblique profiles of the rods were exposed, these forming undulating rows following the axis of the model (not illustrated). Transverse cuts produced surfaces containing only transverse or slightly oblique profiles (Fig. 14).

DISCUSSION

The present study helps to clarify certain aspects not properly dealt with in our previous report. First, it is clear that the prisms have a genuine pattern 2 (row) arrangement[3], as found in ungulate and marsupial enamels, with distinct inter-row sheets of crystals orientated at an angle to those in the prisms. In all other primates, the prism shape is always a variant of the open, pattern 3, type[2]. It is common to find in other primates areas of enamel where the prisms form rows [2,3,4] but this does not constitute a pattern 2 arrangement; the crystals within the prisms are not parallel and the rows are not separated by sheets with crystals orientated at right angles to the prisms. In our previous report[2], we had wrongly concluded that the aye-aye enamel had such a pseudo-pattern 2 structure. In this connection, it is worth noting that the surface texture in the SEM suggests that the prisms are bundles of parallel crystals without the deviations associated with pattern 3 prisms, as are found in man and other primates[2-5].

Secondly, the polarised light and SEM appearances can be better correlated. It is quite clear that, whereas the striping of the parazones in polarised light is due to the prisms themselves, that in the diazones is due to the inter-row sheets, not to oppositely orientated prisms, as we stated previously[2]. Finally the birefringent surface zone has been found to consist of prisms embedded in highly mineralized interprismatic enamel. The uniform birefringence in the diagonal position indicates that this material consists of crystals standing largely perpendicular to the outer enamel surface but in the near-parallel position, the deviations of the prisms from this orientation cause the formation of a complex image. Further evidence of the density of the outer layer comes from the finding that it always has the same birefringence, irrespective of the refractive index of the medium; the intercrystalline spaces must thus be extremely narrow.

It has long been recognised that Hunter-Schreger banding is an effect resulting from periodic variations in prism orientation which, in turn, produce variations in light reflection, birefringence and other properties[3,4,6-8] The possible arrangements of prisms to produce such variations have been the subject of much work. A widely accepted view is that the prisms form zones in which the prisms are inclined with respect to those in neighbouring rows[6,9-11]. Appropriate section planes (usually longitudinal) would cut at right angles through these rows and reveal bands of alternating prism directions.

We believe that such a hypothesis cannot explain the banding in aye-aye enamel. It does not account for several features:

1) The oblique orientation of prism profiles across the parazones.
2) The replacement of bands by undulating rows of prism cross-sections in the transverse sections of midline enamel.
3) The banding seen in tangential sections of the midline enamel.
4) The appearance of banding in both longitudinal and transverse sections of the lateral enamel.

We suggest that, for aye-aye enamel, a better model, suggested initially by the obliquity of the prism profiles in the parazones and the gradualness of the diazone-parazone transition, would be one in which it was assumed that all prisms followed a similar path through the enamel and that this path was some form of helix. This geometrical figure is capable of generating a variety of profiles according to the section plane and the point where the section intersects the helix. Since both tangential and transverse profiles are necessary for Hunter-Schreger banding, it is clear from Fig. 15 that the helices must be orientated obliquely with respect to the section plane, as in d-h.

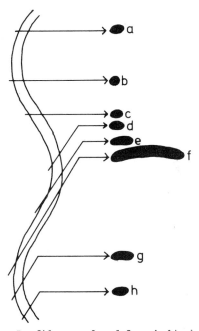

Fig. 15 - Profiles produced from helix by sections at two orientations.

When a model was constructed as outlined and sectioned obliquely, the predicted alternating bands of transverse and tangential profiles appeared (Fig.13). Comparison of three section planes through the model established that, when the cut surface appeared banded (oblique section only), each helix appeared only once in that surface and that in other, parallel sections, it would appear with a different profile, i.e. would be found in diazones or parazones according to the level of the section.

Examination of deeply etched longitudinal surfaces clearly showed prisms coursing between diazones and parazones and the impression of helically orientated prisms was reinforced (Figs. 11,12) It has only been possible to find direct evidence of a helical course in areas such as that in Fig.12, which are limited because there is no way of examining the course of a prism for more than a short distance below the surface.

The rather crude model illustrated is intended only to show that the proposed structure is possible, and represents only a very limited section area. The width and orientation of the bands would obviously depend on the plane of section and the pitch and diameter of helices. Any attempt at a more accurate representation of the enamel would necessitate systematic variation of these parameters and also other manipulations; for instance, the prism helices might be elliptical rather than cylindrical in section. Such a study could only be made with a computer. An interesting feature of the wax model (Fig. 13), however, is that it suggests that, for the bands in the lateral enamel to be oriented as they are (Fig. 2), the prism helices would have to slope in the opposite direction to the bands, i.e. towards the midline, as well as to lie oblique to the transverse plane of the tooth.

It has been suggested previously[12-15] that prisms follow a spiral course (in human, rodent and lagomorph enamels). However, these authors either still retained the concept of discrete zones of prisms or gave no clear idea of the relationship between the spirals and the Hunter-Schreger bands. On the present hypothesis, the enamel would be homogeneous and the banding an effect produced by the mode of packing of the helices in relation to the section plane. Prisms would not 'belong' to particular zones.

We would suggest that in the midline enamel the helices are directed more nearly in the long axis of the tooth and subtend a smaller angle with the dentine surface such that, in transverse sections of the tooth tangential prism profiles were no longer produced (Fig. 16b). This change of orientation would at the same time cause tangential profiles, and hence banding, to appear in tangential sections of this part of the enamel (Fig. 16b). In the lateral enamel, because of the less acute inclination of the helices, the appearances in the two section planes would be reversed (Fig. 16a). There may be a functional reason for the change of structure in the midline region. The anterior teeth of the aye-aye are more compressed laterally than those of rodents and have a pointed, rather than a rounded, gouge-shaped tip. It may be that the altered structure of the midline enamel makes this region more resistant to wear, so that the point, which is used in breaking up tree bark[1], rather than for gnawing, is maintained.

The production of helical prisms standing perpendicular to the dentine surface would necessitate each ameloblast following a circular path. However, to account for the proposed obliquity of the helices, relative to the transverse and longitudinal planes of the tooth, this would be translated into a gyratory or, more

probably, a sinusoidal path.

The suggested structure might be applicable to the enamel of other species. The enamel most similar in appearance to that of the aye-aye is the 'multi-serial' enamel of lagomorphs and many rodents[3,6,11], in which it is known that prisms pass from one band to the next[3,6]. However, given appropriate modifications of the parameters of the helices, it could, we believe, account for banding in superficially dissimilar types of enamel, such as that found in higher primates[2].

Acknowledgements

We are indebted to Dr. K.A. Joysey of the Cambridge University Museum of Zoology for aye-aye material. We thank Mr. M. S. Gillett and Miss J. Poole for assistance, and are most grateful to Mrs. D. Jelfs for typing the paper.

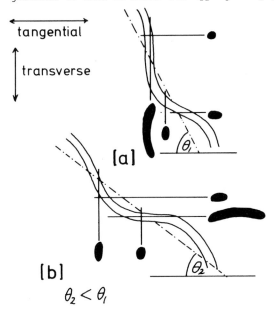

Fig. 16 - Diagram showing how appearances in transverse and tangential section planes can be reversed by change in slope of helices. (a) Lateral enamel; (b) Midline enamel.

References

1. J.R.Napier and P.H. Napier: A handbook of living primates. Academic Press - London, 1967.
2. R. P. Shellis and D. F. G. Poole: Calcified dental tissues of primates. pp. 197-279 in Evolutionary changes to the primate skull and dentition (C. L. B. Lavelle, R. P. Shellis and D. F. G. Poole) Charles C. Thomas - Springfield, 1977.
3. A. Boyde: The structure and development of mammalian enamel. PhD Thesis - University of London, 1964.
4. J. W. Osborn: Variations in structure and development of enamel. Oral Sci. Rev. 3, 1973, 3-83.
5. A. H. Meckel, W. J. Griebstein and R. J. Neal: Ultrastructure of fully calcified human dental enamel. pp. 160-162 and 190-192 in Tooth enamel (M. V. Stack and R. W. Fearnhead-eds) Wright - Bristol, 1965.

6. A. Boyde: Electron microscopic observations relating to ... prism decussation in mammalian dental enamel. Bull. Group. int. Rech. Stomat. 12, 1969, 151-207.

7. R. Hoffman and L. Gross: Microstructure of dental enamel I...J. dent. Res. 46, 1967, 1444-1455.

8. B Sundstrom: Schreger bands and their appearance in microradiographs...Acta odont. Scand. 24, 1966, 179-194.

9. J. W. Osborn: Directions and interrelationships of enamel prisms... J. dent. Res. 47, 1968, 223-232.

10. J. Erausquin: The aspect of the bands of Schreger in the horizontal sections of enamel. J. dent. Res. 28, 1949, 195-200.

11. W. J. Schmidt and A. Keil: Polarising microscopy of dental tissues. Pergamon - Oxford, 1971.

12. C.S. Tomes; A manual of dental anatomy. Churchill - London, 1904.

13. H. P. Pickerill: The structure of enamel. Dent. Cosmos 55, 1913, 969-988.

14. J. L. Williams: Disputed points ...in the normal and pathological histology of enamel. J. dent. Res. 5, 1923, 27-107.

15. J. G. Helmcke: Form and Verlauf von zehn Schmelzprismen der Hunter-Schreger-Zone ... Dtsch. Zahnartzl. Z. 21, 1966, 1065-1070.

Discussion with Reviewers

W. von Koenigswald & D. G. Gantt: Could not the model illustrated below explain the structure of aye-aye enamel? In this model each prism rises from the dentine-enamel junction at a steeper angle than the Hunter-Schreger band. Subsequently the prisms would have to move from one band to the next and this could be accomplished by bending through approximately 90° at the junctions between the bands, as shown in the figure below and in your Fig. 9. This model involves a planar zigzag instead of a helix.

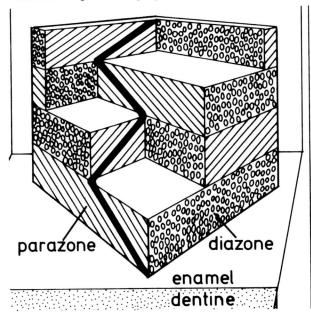

parazone   diazone

enamel
dentine

Authors: This alternative model has some features in common with ours; thus, the prisms are inclined to the bands and pass from one band to the next. However, we do not think that it explains the structure, for the following reasons: 1. We have not observed straight lengths of prisms, let alone right angle bands, which seem unlikely in terms of crystal growth; 2. Tangential sections of lateral enamel would, on this model, give bands of oblique prism sections, those in one band pointing uniformly at right angles to those in the next. In fact (text ref. 2, Fig. 73), sinuous rows of oblique prism sections are found with no abrupt changes of angle. The appearance of midline enamel in transverse section is similar (Fig. 7) so the proposed model does not account for the appearance in Fig. 7 either; 3. Transition from parazone to diazone is gradual (Figs. 8,9). You may have misinterpreted Fig. 9. Here there is a superficial appearance of right angle bending but close inspection shows that the linear structures in the parazones are prisms, those in the diazones are inter-row sheets. We included polarised light work partly to have the chance of correcting a similar misinterpretation in our previous work.

Z. Skobe: I observed similar bending of enamel rods in Hunter-Schreger bands of human molar teeth (J. dent. Res. 56, 1977, B133 Abstr.). However, I interpreted the course of enamel prisms to be sinusoidal ... through the deeper enamel because the prisms alternated bends in clockwise and counterclockwise directions. Do the authors have comments comparing sinusoidal and helical models of enamel rod pathways?
Authors: We cannot comment in detail because of the lack of illustrations but it seems to us that if the helices in human enamel were elliptical in section, they could resemble planar sinusoidal structures, especially if stereo pairs were not used.

K. A. Galil: In your last statement, you mentioned that 'given modifications of the parameters ...'. Could you elaborate on these modifications?
J. C. Rose: In your discussion you suggest a functional explanation for the complex enamel prism arrangement in the midline. Could you discuss your model of the aye-aye enamel structure in relation to the gnarled enamel frequently observed in the cuspal region of human enamel?
Authors: It is probably unnecessary to invoke a helical model for many types of enamel. For instance, most insectivores and prosimians have straight prisms roughly vertical to the surface, which is probably primitive. The 'uniserial' enamel of rat and other rodents may have developed directly from this pattern. In the anthropoid primates, the absence of banding in the outer enamel indicates that here, any helical structure would be lost. In the inner enamel of these animals, the banding is usually less distinct than in the aye-aye; this could be due to the helices having a greater pitch,

giving more gradual transitions between parazones and diazones. Specimens of some enamels, e.g. that of rabbit, suggests the presence of spirals with the diameter increasing inwards the outer enamel. The wavy appearance of 'gnarled' cuspal enamel resembles that of our model when sectioned longitudinally, suggesting that in the cuspal region the helices are not inclined to the longitudinal section plane.

D. K. Whittaker: You have commented briefly upon the ameloblast movements necessary to produce the helical prism pattern which you have described. Have you any suggestions as to the forces involved in producing this complex movement which on occasions (Fig. 12) results in cross over of individual prisms?

W. von Koenigswald & D.G. Gantt: Spirals tend to build cone-shaped units but not long bands. How could the ameloblasts change their direction by rotating 360° and not produce large areas of disturbance? We see our model (above) as a more simplistic model of cellular movement and continuity.

Authors: We were careful to use the word helix, as suggesting a cylindrical rather than a conical spiral. We must refer you to the description of our model and emphasise that inclined helices of uniform diameter are envisaged, the bands being generated by the packing of the helices and the orientation of the section plane. Since the spirals are probably of rather high pitch and are inclined, we do not imagine gyration of ameloblasts but rather a sinusoidal movement. We would suggest that such a motion is no more complex in terms of organisation of the sheet of ameloblasts than any other motion. As to mechanism, we are sorry to say that we can proffer no suggestions; presumably there is interaction between forces on the epithelium generated by the growing prisms, and intra-epithelial factors such as functional asymmetry of Tomes processes and patterns of intercellular adhesion.

J. C. Rose: Based on your observation of primate enamel structure, could you state your opinions of Daubentonia's evolutionary and taxonomic position within primates?

Authors: While enamel structure has some value as a taxonomic feature, we believe its use in this respect is limited; quantitive analysis of our material (unpubl.) indicates that prism shape and arrangement may be dependent on enamel thickness and the pattern of amelogenesis, along the lines indicated by Boyde (text ref 3). However, the aye-aye shares pattern 2 enamel with the flying lemur and elephant shrew (unpubl.); these animals are widely regarded as related to the ancestral primates (W. E. LeGros Clark: The antecedents of man. Edinburgh University Press, 1959). The aye-aye's line of descent may thus have been separate from that of other lemuroids for a long time, the others having acquired pattern 3 enamel (text ref. 2). This would support the status recognised in some classifications (LeGros Clark) by assigning the aye-aye to a separate superfamily.

D. K. Whittaker: To be a little pedantic, is it correct to use the term Hunter-Schreger bands for arrangements seen under the SEM? Should we not refer to prism orientations which result in the production of Hunter-Schreger bands under the light microscope?

Authors: This is a useful concise term to specify the banding and although we take your point we do not think it wrong to use the term thus, as Hunter-Schreger bands were described originally from the appearance of polished surfaces in reflected light, which has analogies with SEM.

K. A. Galil: Do you think that the results of this investigation - which are based on one tooth only - justify extrapolating, in view of the fact that you are dealing with the anterior teeth of a highly specialised primate?

Authors: W. von Koenigswald sees the same appearances in this enamel as we do, but interprets them differently (see above). It was precisely because of the unusually regular structure that we became interested in explaining it. Justification of the hypothesis depends on its power of explaining prism arrangement in other species and this in turn must await fresh studies of enamel structure.

K. A. Galil: In your discussion you mentioned that the birefringent zone has been found to consist of prisms embedded in highly mineralized interprismatic substance. Are you really looking at interprismatic substance or at prisms in other section planes? Should not the interprismatic substance be more organic in nature than highly mineralized?

Authors: The sign of birefringence shows the outer layer to be predominantly mineral and it is currently accepted that interprismatic regions are not particularly rich in organic material. Fig. 10 shows continuity between inter-row sheets in deep enamel and interprismatic enamel of the surface layer. We see only longitudinal prism profiles in the layer, either by SEM or polarised light.

K. A. Galil: Why did you etch with 1M HCl rather than the conventional acids used in etching currently, e.g. 37% or 50% phosphoric acid?

Authors: The etchants you mention seem more suitable for clinical application. Brief etching with M HCl seems to produce minimal reprecipitation artifact and we have found it gives good differentiation of structure on a wide range of enamels and similar tissues. It works particularly well if, as in the aye-aye, there is a marked difference of crystal orientation between prisms and interprismatic enamel.

SCANNING ELECTRON MICROSCOPY/1979/II
SEM Inc., AMF O'Hare, IL 60666, USA

COMPARATIVE INVESTIGATIONS ON FLUOROSED ENAMEL

M. Triller, A. Bouratbine, D. Guillaumin[*], and R. Weill

Laboratoire d'Histologie
Faculté de Chirurgie Dentaire
(Paris V)
1, rue Maurice Arnoux
92120 - Montrouge
France

[*]Laboratoire d'Evolution
des Etres Organisés
(Paris VI)
105, blvd. Raspail
75006 - Paris
France

Abstract

Structural observations upon fluorosed enamel were made on human teeth extracted from Nabeul (Tunisia). Longitudinal sections of enamel were made and microradiographed. Topographical observations in the S.E.M. showed alterations of the pattern of mineralization. Fissures and clefts filled with plaque material were observed to reach to the surface of the teeth. They penetrated into enamel along the prisms. Acid etching of the surface by 30% H_3PO_4 revealed enlarged acid-resistant sheaths while the core of the prisms was irregularly etched. Inner portions of the enamel were observed on fractured fragments. The structure was not significantly perturbated except for areas of porosity. After demineralization of the sections by 10% EDTA at pH 7 and critical point drying, the organic matrix was observed in the S.E.M. : It appeared as a fibrous network. Compared with sound teeth the matrix was dense and irregular. Histochemical stainings showed the presence of organic material. Alcian blue and PAS positive contents were very low except near the enamel-dentine junction. Preliminary observations were made using SMI Cameca ion microanalyser and histochemical studies on undecalcified sections were made in order to compare the observations made with light microscopy with S.E.M. findings.

KEY WORDS: Tooth, Enamel, Fluorosis, Histology, Mineralization, Enamel Structure, Enamel Matrix, Etching Pattern

Introduction

Numerous reports of the structure of fluorosed enamel agree in describing alterations as a pitted surface and areas of hypomineralization which appear principally in the subsurface layer beneath a well mineralized surface zone (1). Ingestion of high levels of fluoride during amelogenesis causes the enamel to develop a porous structure. There is a general lowering of the mineral content, up to 25% in severe cases (2), while the amount of organic component is increased compared to sound teeth (3,4). However, the mechanisms by which fluoride inhibits proper mineralization are still not clearly understood. During experimental fluorosis in rats, morphological changes were observed in the structure of ameloblasts suggesting disturbances in cell metabolism (5,6). Secretion and maturation of the matrix are altered and modifications occur in the incorporation of different amino-acids (7). In a study on amino-acids of enamel matrix protein, fluoride administration inhibited mineral deposition and crystal growth and/or the withdrawal of the matrix during maturation of enamel. Fluorosed enamel subsequently showed disturbed structure and yet, in spite of its poorly mineralized pattern, it appeared remarkably resistant to acid attack and caries.

The present study aimed to observe nondecalcified and decalcified specimens of human fluorosed enamel by scanning electron microscopy and to correlate these observations with histochemical light microscopical and microradiographical findings from adjacent sections prepared from the same tooth.

Material and Methods

The fluorosed teeth were collected from Nabeul (Tunisia) from areas with endemic fluorosis. They were free of caries and showed macroscopically characteristic alterations such as scattered opacities and chalky white or yellowish areas. Six teeth were used for this study. Their structure was compared to controls with caries-free premolars extracted in Paris (France) for orthodontic purposes. Fluorosed and control teeth were stored in 10% Formalin.

The teeth were cut in half in longitudinal direction. One half was prepared for ground sections for microradiography and light microscopy. Some sections were studied using a SMI 300 Cameca ion microanalyser.

The second half of each tooth was divided in two parts :

a) in one part fragments of the surface and fractured specimens were mounted on aluminium stubs, coated with a 300 Angstroms thick layer of gold-palladium and observed with a Cameca M.E.B./07 at 20 Kv. Some fragments, prior to the gold-palladium coating were etched with 30% H_3PO_4 for 60 sec.

b) the second part was demineralized by 10% EDTA at pH 7, the samples embedded in paraffin and cut at 7 um. After removal of the paraffin the enamel matrix could be observed 1) using light microscopy after histochemical staining for proteins by the tetrazoreaction of Danielli, for G.A.G. by Alcian blue and for glycoproteins by P.A.S. 2) after critical point drying and gold-palladium coating for S.E.M.

Results

The untreated external surfaces of fluorosed enamel, observed under S.E.M. were irregularly roughened and pitted (fig. 1) compared to sound teeth (fig. 2). At a higher magnification the fluorosed surfaces appeared porous (fig. 3). These porosities were located either on buccal or occlusal surfaces. This hypomineralized structure was confirmed by microradiographs taken from adjacent sections (fig. 4). On occlusal surfaces, irregular clefts could also be seen (fig. 5). When observed by light microscopy, these clefts appeared to be filled with plaque which penetrated into the enamel along the prism borders.

The acid-treated external surfaces revealed after H_3PO_4 treatment for 60 sec. an irregular pattern of etching in adjacent areas : Generally the border of the prisms appeared very resistant while the core was irregularly etched (fig. 6).
Sometimes there was a reverse pattern of etching with an accentuated destruction of the sheaths (fig. 7). When the acid was applied for a longer time (5 min), the core of the prisms was always deeply etched while thick acid resistant sheaths remained (fig. 8). Surface of sound enamel etched for 60 sec. by 30% H_3PO_4 revealed a regular pattern. (fig. 9).

The broken fragments which allowed observations of the inner regions of fluorosed enamel did not differ significantly from corresponding areas of sound enamel, especially when the prisms were fractured longitudinally (fig. 10). When they were transversally cut, they appeared more irregular (fig. 11). Usually this aspect is also observed on sound enamel, due to the fact that the plane of section was seldom strictly transverse. On these fragments, the pattern of etching varied similarly in the same way as that observed at the surface.

Heterogeneity of the mineralization of fluorosed enamel is known to be associated with a high content of remaining organic matrix. Histochemical observations using light microscopy after demineralization by 10% EDTA at pH 7 showed a high content of proteins by the method of tetrazo/reaction of Danielli. The amount of alcian blue and P.A.S. stainable material was very low except near the enamel-dentine junction Similarly prepared matrix was observed by SEM after critical point drying. The matrix then appeared as a reticular network in the sheath areas (fig. 12). At a higher magnification, this network generally demarcated each prism (fig.13) without any material in the core. Sometimes, however, it appeared more diffuse : the sheaths were not so well defined and the cores were almost completely filled with a fibrillar material (fig. 14), particularly in the most inner parts of enamel. The matrix of sound enamel did not present this irregular and reticular appearance (fig. 15).

Near the enamel-dentine junction areas of hypomineralization were visualized by microradiography (fig. 16), and globular disturbances were observed with the ion microanalyser (fig.17) This appearance did not occur on sound teeth (fig. 18).

Discussion

Most previous reports have described fluorosed enamel as a porous, hypomineralized structure. Our observations confirm these earlier studies (8,9). In this study, we did not attempt to correlate the observations with the severity of fluorosis, according to clinical criteria, since the histological features were very similar from tooth to tooth.

The rough, irregularly pitted surface corresponded to areas of hypomineralization as previously reported (1). The variations in the pattern of etching of the acid-treated specimens might be due partly to differential crystallite orientation (10,11). However, it seems more likely related to the heterogeneity of the structure itself.

According to our histological observations the matrix is thicker in fluorosed than in sound enamel. Histochemical studies revealed a high level of organic components. When observed in a S.E.M. the matrix appeared as a dense reticular network, mostly located at the periphery of the prisms, but extending sometimes into the core, particularly in the inner parts of enamel. When observed with the ion microanalyser, it has been previously reported (12) that fluorosed enamel demonstrated abnormally large sheaths and hypocalcified prisms. Disturbances in the structure and mineralization occurred near the enamel-dentine junction. It seems from the foregoing that fluoride acts very early during amelogenesis. In experimental animal fluorosis, biochemical data (13,14,15) suggested that fluoride might bind firmly to newly deposited protein (16) thus causing a disturbance of maturation and resulting in hypocalcification of enamel.

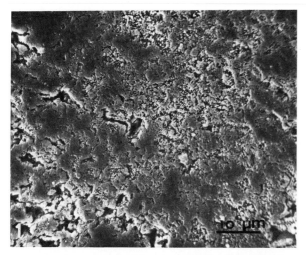

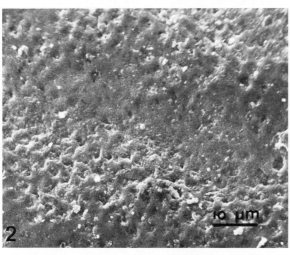

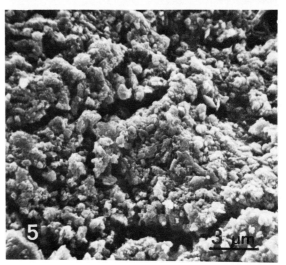

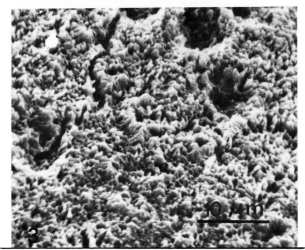

Fig. 1 : External surface of untreated fluorosed enamel showing roughened areas or irregular pits

Fig. 2 : External surface of untreated sound enamel in perikymata area.

Fig. 3 : Higher magnification of roughened area of untreated fluorosed enamel showing the porous aspect of the surface.

Fig. 4 : Microradiograph of fluorosed enamel section showing areas of hypomineralization (arrows) of the surface.

Fig. 5 : Sinuous and irregular clefts reaching the occlusal surface of fluorosed enamel.

The fact that the main disturbances in enamel fluorosis were observed at the surface and on the most inner parts of enamel, near the enamel-dentine junction, may indicate that enamel fluorosis has a dual origin (17,18).

In conclusion the data point to
1) early disturbances affecting amelogenesis and
2) post eruptive secondary damages affecting the surface zone.

We suggest this to be the case in the present study, since the teeth we observed were extracted from adult patients who were born and had spent their life in areas of endemic fluorosis.

This investigation has been supported by A.T.P. INSERM 55 77 85 - N° 13.

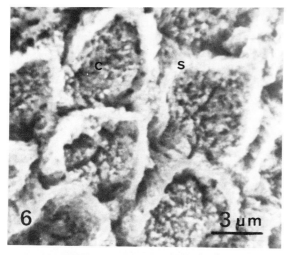

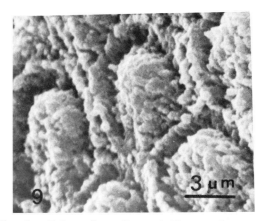

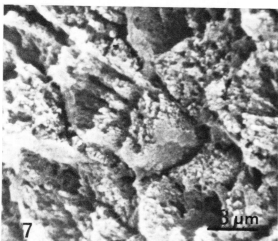

Fig. 6 : External surface of fluorosed enamel after 60 sec. etching by 30% H_3PO_4 : the sheaths (S) are acid resistant while the cores (C) of the prisms are irregularly etched and present a crystalline aspect.

Fig. 7 : Reverse pattern of etching : the periphery of the prisms is deepened. The cores are more deeply etched than in fig.6 and have a pitted aspect.

Fig. 8 : After 5 min. etching by 30% H_3PO_4 the core of the prisms is deeply etched. Thick acid-resistant sheaths remain.

Fig. 9 : Sound enamel surface after 60 sec.etching by 30% H_3PO_4 shows regular pattern.

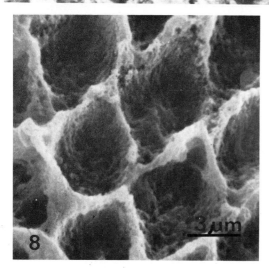

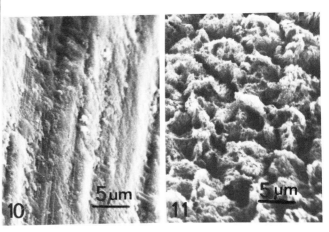

Fig. 10 : Broken fragment of fluorosed enamel : the prisms are cut longitudinally. Their course does not differ from the situation in normal enamel.

Fig. 11 : Broken fragment of fluorosed enamel : the prisms are cut transversly : their mineralization is irregular. The center of the core is often porous or pitted.

References

1. Fejerskov O., Johnson N.W. and Silverstone L.M. : The ultrastructure of fluorosed human dental enamel. Scand. J. Dent. Res. 82 : 1974, 357-372.

2. Fejerskov O., Silverstone L.M., Melsen B. and Moller I.J. : The histological features of fluorosed human dental enamel. Caries Res. 9 : 1975, 190-210.

3. Darling A.I. and Brooks A.W. : Some observations on the mottled enamel of fluorosis. J.Dent. Res. 38 : 1959, 1226-1227.

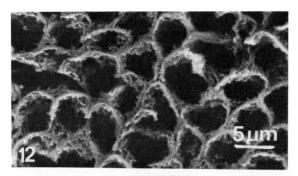

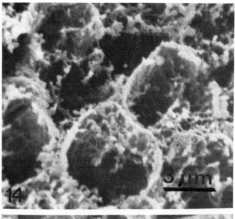

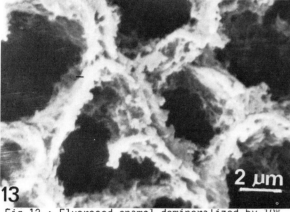

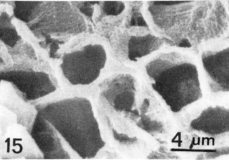

Fig.12 : Fluorosed enamel demineralized by 10% EDTA at pH 7 : The matrix appears as a thick reticular network.

Fig.13 : Higher magnification of fig. 12 : the matrix is limited to the sheath areas. It looks irregular and fibrous. The cores of the prisms are free from any organic material.

Fig.14 : In the inner parts of enamel, near the enamel-dentine junction, the sheaths are not so well defined as they were in fig.13 : the cores of the prisms are invaded by a fibrous organic material.

Fig.15 : Sound enamel matrix has not the reticular and irregular appearance as observed in fluorosed one.

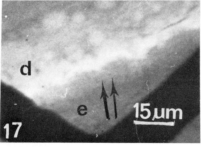

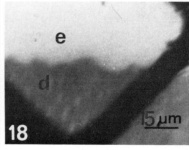

Fig. 16 : Microradiograph of the enamel-dentine junction (EDJ) shows globular areas of hypomineralization. (arrows)

Fig. 17 : Enamel-dentine junction of fluorosed enamel observed in ion microanalyser showing globular alterations in the enamel (arrows). Dentine (D) - Enamel (E).

Fig. 18 : Enamel-dentine junction of sound enamel observed in ion microanalyser. Dentine (D) Enamel (E).

4. Silness J. and Gustavsen F. : Some observations of the fine structure of fluorosed dental enamel. Acta Odont. Scan. 28 : 1970, 701-720.

5. Kruger B.J. : The effects of different levels of fluoride on the ultrastructure of ameloblasts in the rat. Archs.Oral Biol. 15 : 1970, 109-115.

6. Walton R.E. and Eisenmann D.R. : Ultrastructural examination of various stages of amelogenesis in the rat following parenteal fluoride administration. Archs. Oral Biol. 19 : 1974, 171-183.

7. Basford K.E., Patterson C.M. and Kruger B.J. Multivariate analyses of the influence of mottling doses of fluorid on the amino-acids of enamel matrix protein of rat incisors. Archs. Oral Biol. 21 : 1976, 121-129.

8. Gustafson A.G. : The histology of fluorosed teeth. Archs Oral Biol. 4 : 1961, 57-69.
9. Sundstrom B., Jongebloed W.L.and Arends J : Fluorosed human enamel. A SEM investigation of the anatomical surface, outer and inner regions of mildly fluorosed enamel. Caries Res. 12 (6) 1978, 329-338.
10. Johnson N.W., Poole D.F.G. and Tyler J.E. : Factors affecting the differential dissolution of human enamel in acid and EDTA, a SEM study. Archs. Oral Biol. 16, 1971 : 385-396.
11. Silverstone L.M., Saxton C.A., Dogon I.L. and Fejerskov O : Variation in the pattern of acid etching of human dental enamel examined by SEM. Caries Res. 9 : 1975, 373-389.
12. Lefevre R. and Frank R.M. : Etude de la fluorose dentaire au microanalyseur ionique. Jour. Biol. Buccale 4 : 1976, 29-41.
13. Deutsch D., Weatherell J. and Robinson C.: Distribution of fluoride in developing bovine enamel. J. Dent. Res. 53 : 1974, 1053.
14. Hallsworth A.S., Weatherell J. and Deutsch A.D. : An ultrasensitive method of determining fluoride ion in mineralized tissue. J. Dent.Res. 53 : 1974, 1088.
15. Patterson C.M., Basford K.E. & Kruger B.J.: The effect of fluoride on the immature enamel matrix protein of the rat. Archs. Oral Biol. 21 : 1976, 131-132.
16. Crenshaw M.A., Wennberg A. and Bowden J.W. : Fluoride binding by the organic matrix of developing bovine enamel. Archs. Oral Biol. 23, 1978 285-287.
17. Shinoda H : Effect of long-term administration of fluoride on physico-chemical properties of rat incisor enamel. Calc. Tis. Res. 18 : 1975, 91-100.
18. Sundstrom B and Myhrberg H : Light and scanning electron microscopy of fluorosed enamel from human permanent teeth. Caries Res. 12 (6) 1978, 320-328.

Discussion with reviewers

Reviewers I and II : What fixation technique was utilized ? This aspect seems critical as demineralization in 10% EDTA exposes organic enamel matrix.
Authors : Both fluorosed and sound teeth were stored in 10% formaline before histological investigations and critical point drying.

Reviewer I : Do the authors suggest that if fluorosed teeth of a patient who moves away from the endemic area at an early age were observed, only disturbances of amelogenesis could be observed ?
Authors : If it were possible to obtain recently erupted or unerupted teeth, we might be able to determine whether the second phase disturbance (which we hypothesize exists) occurred during pre-eruptive or post-eruptive maturation.

Reviewer II : How do the histochemical reactions you reported for fluorosed enamel differ from normal enamel ?

Authors : According to histochemical stainings, we found an increase of organic components in fluorosed enamel compared to sound teeth and an heterogenic distribution of alcian blue and PAS stainable material near the enamel-dentine junction (Triller et al : Preliminary histochemical observations of fluorosed matrix, Caries Res. 12 (2), 122, 1978.)

Reviewer III : From which part of the tooth were the fragments taken and in what relation to the anatomical surface; this in regard to the findings of Sundström et al. (Caries Res. 12, 329-338, 1978): a changing morphology going from the outer surface to the dentine-enamel junction.
Authors : We have observed all parts of the teeth from longitudinal sections.

Reviewer III : In the paragraph "Results" it is mentioned that the general appearance of the untreated external surface was rough and pitted. Was this observed morphology related to hypomineralization and/or were there signs of in vivo demineralization-remineralization and strongly acid-resistant areas.
Authors : Comparing the histochemical appearances of the fluorosed teeth with the normal controls, we found a greater protein concentration in the fluorosed teeth which we presume to be a primary effect, but progressive data could not be obtained from our material. However, as we said in the conclusion, we think that high levels of fluoride in the drinking water may induce secondary effects at the surface of the teeth.

Reviewer III : It has been stated that heterogeneity of the mineralization is being associated with a high content of remaining matrix. Can you give some figure about matrix content in fluorotic and sound enamel. Do you agree that the heterogeneity in mineralization in the surface and subsurface layer to a large extent is being induced by mineral concentrations in the oral cavity.
Authors : Illustration of organic matrix of control sound enamel shows a very different appearance to the one observed in fluorosed enamel. We do agree that alterations of the surface and subsurface layer are largely due to mineral concentration in the oral cavity. In our conclusion we suggest therefore a dual origin for the disturbances observed : preeruptive alterations occurring during amelogenesis and secondary damages .

Reviewer I   : C.H. PAMEIJER
Reviewer II : W.R. COTTON
Reviewer III : W.L. JONGEBLOED

512

SCANNING ELECTRON MICROSCOPY/1979/II
SEM Inc., AMF O'Hare, IL 60666, USA

SCANNING ELECTRON MICROSCOPY OF RACHITIC RAT BONE

D. W. Dempster, H. Y. Elder[*] and D. A. Smith

Department of Medicine
*Department of Physiology
University of Glasgow
Glasgow, G12 8QQ
Scotland

Abstract

The SEM appearances of lateral endosteal surfaces of anorganic tibial diaphyses from rachitic, vitamin D-treated and normal rats are described. Different types of surface could be recognised in all three groups. These included irregular, incompletely mineralized surfaces corresponding to mineralization fronts (I) and partially mineralized collagen fibers (II), smooth resting surfaces (III) and scalloped surfaces which were undergoing, or had previously undergone, osteoclastic erosion. Percentage areas occupied by each surface type was measured using a computer-linked planimeter. In normal bones 38% of the sampled surface was in a resting phase. The bulk of the remaining surface consisted of mineralization fronts (47.3%) and partially mineralized collagen fibers (10%). There was little resorbed surface (4.7%) on normal bones and all of this had the appearance of active rather than non-active resorption. Compared with normal bone, rachitic and D-treated bones had a significantly smaller (P < 0.05) proportion (<10% in each case) of surface occupied by resting bone matrix. In the rachitic bones 84.7% of the endosteal surface was incompletely mineralized matrix. Of this, 39.7% was mineralization fronts. The amount of partially mineralized collagen fibers in both rachitic (45.0%) and D-treated bones (37.0%) was greater (P <0.05) than that present on normal bones. The amount of resorbed surface on D-treated bones (35.3%) was greater (P <0.05) than on both rachitic (10.6%) and normal bones. The results are interpreted as indicating a slowing in the rate of mineralization between stages II and III in the rachitic and D-treated groups. This may be a hypophosphataemic effect. The D-treated rats may show an accelerated transition between stages I and II. The increased bone resorption is probably an indirect effect of low phosphate, an effect blocked in the rachitic animals by the deficiency of vitamin D.

KEY WORDS: Rat, Bone, Rickets, Mineralization, Phosphate Deficiency, Resorption, Vitamin D, Osteoid, Morphometry, Anorganic

Introduction

The processes of bone formation and destruction are surface phenomena. The scanning electron microscope (SEM) is therefore a very useful tool for studying these processes. In 1968, Boyde and Hobdell[1], introduced the practice of removing the organic component of a hard tissue matrix to expose the mineral or 'anorganic' component for examination in the SEM. The following year Boyde and Hobdell[2,3] made a study of free surfaces in anorganic lamellar and primary membrane bone and gave a detailed description of the mineralizing fronts exposed in the anorganic specimens. They also demonstrated that on the basis of the morphological appearance of a surface before and after it was made anorganic, it was possible to determine whether the surface was undergoing formation or resorption or if it was in a 'resting' phase. In 1970 Jones and Boyde[4] investigated the changes induced in the surface topography of anorganic rat parietal bones by either calcitonin (CT) or parathyroid hormone (PTH) administration. In this study these authors were able to further differentiate the state of activity on the bone surface. Actively resorbing zones could be distinguished from previously resorbed surfaces which were in a resting phase and surfaces which had only recently entered the latter phase could be distinguished from prolonged resting surfaces. CT was found to suppress bone resorption whilst PTH enhanced it.

To date there appears to be only one study in which rachitic bone has been examined in the SEM[5]. These authors described the appearance of anorganic bone specimens from patients with hypophosphataemic rickets. Whilst the appearance of resting free bone surfaces was indistinguishable from those in normal bone, the authors detected a textural difference in the nature of the mineralizing front (forming surface) and a difference in its shape in the pathological bones. In rachitic bone, mineralizing fronts displayed a wider range of orientation of the unjoined mineral particles than is seen in normal bone.

The object of the present study was to examine bone from rats with experimentally induced rickets and to quantify the distribution of the different types of anorganic surface.

Materials and Methods

Eighteen female Sprague - Dawley rats (~6 weeks old) of mean weight 93.6 ($\pm$ 2.4) grams were divided into 3 equal groups each of which was fed a different experimental diet for a period of 7 weeks:
(1) Rachitic group (R): Vitamin D deficient, low phosphate diet[6]. (2) Vitamin D-treated group (D): The same diet as group R with a vitamin D supplement (160 I.U. Vit. D per 100 g. of diet for the first 4 weeks of the experiment, 320 I.U. thereafter). (3) Normal group (N): Laboratory animal diet No. 41b (Oxoid Ltd., Basingstoke, England).

The rats in the vitamin D-treated group (D) were pair-fed with those in the rachitic group (R). The rats in the normal group (N) were fed ad libitum. Water was provided ad libitum. At the end of the experimental period, the rats were anaesthetised by an intra-peritoneal injection of Nembutal (0.5 mg/kg body wt.) and were exsanguinated. The right tibiae were prepared for examination of the intact bone matrix and the left tibiae were prepared for examination of anorganic matrix using the methods of Boyde and Jones[7].

Preparation for Examination of Intact Bone Matrix in the SEM

(1) The epiphyses of the tibiae were sawn off using a dental saw and the bone marrow was flushed out with a 0.5% trypsin solution containing a small amount of Lissapol detergent (Union Chemical Co.) (2) The bones were immersed in this solution, subjected to a few seconds of ultrasonication to aid mixing and were then left in the solution for 10 minutes. (3) The bones were washed with distilled water.

The above treatment is just sufficient to remove the endosteal lining cells without apparent disruption of the underlying trypsin-resistant collagenous matrix. (4) A length of the diaphysis was cut from the bones using a rotary saw (Metals Research Ltd. Macrotome II) with a very thin cutting blade (0.5 mm thick) electroplated with diamond grit. Distilled water was used as coolant and lubricant. The length of diaphysis ran from the point of attachment of the fibula to the point where the crest of tibia meets the shaft (Fig. I). This shaft segment was then cut longitudinally into medial and lateral halves. (5) The lateral half was then dehydrated in ascending concentrations of acetone and was dried from liquid CO_2 in a Polaron E3000 critical point drier.

Preparation for Examination of Anorganic Bone Matrix in the SEM

(1) The left tibiae were fixed in neutral formol saline for 48 hours. (2) A lateral shaft segment was prepared as described above and was immersed in a 5 - 7% sodium hypochlorite solution overnight. This treatment removes the organic material in the bone, leaving the mineral component of the extracellular matrix. (3) The specimens were gently washed in distilled water for 15 minutes and dehydrated in ascending concentrations of ethanol. The ethanol was substituted by ascending concentrations of ether (in ethanol) and the specimens

were finally air-dried from absolute ether.
When dry, the lateral shaft segments of both left and right tibiae were glued with the endosteal surface upwards onto aluminium stubs (Agar Aids Ltd.) using electrodag 915 high conductivity paint (Acheson Celloids Co.). The specimens were then given a 50 nm coating of gold in an Edwards S150 Sputter Coating Unit and were viewed and photographed on a Philips PSEM 500 microscope using an accelerating voltage of 12 kV. The magnification of the microscope was calibrated by photographing a 1500 mesh silver grid and a diffraction grating replica with 30,000 lines/inch, both supplied by Agar Aids Ltd.

Morphometric Analysis

The percentage area occupied by different types of surface was measured on anorganic tibiae from 3 rats in each of the experimental groups. The surface of each bone was 'sampled' as follows:

Starting at the proximal cut edge, a series of 25 micrographs was taken along the middle of each bone segment (Fig.2). The micrographs were all taken at a nominal magnification of X640. To avoid overlap, each frame was separated by a distance of half the C.R.T. width. In order to minimise errors due to foreshortening, the specimen surface was aligned normal to the electron beam before each micrograph was taken. The micrographs were analysed using a computer-linked planimeter[8].

Results

(1) Bone with Intact Organic Matrix Component

The endosteal surface of the rachitic bone (Group R) was lined by a thick layer of unmineralized or poorly mineralized organic matrix (osteoid) (Fig. 3a). This appeared lighter than the adjacent mineralized matrix and individual lamellae could readily be discerned. Up to 12 lamellae were present on the endosteal surface of rachitic bones. In bones from D-treated (Group D) and normal rats (Group N), the osteoid lining was very thin (only 1 lamella) (Fig. 3b and c). The vascular channel entrances in the rachitic bone were also lined by a thick layer of osteoid (Fig. 4).

(2) Anorganic Specimens

(a) Low-power views of the anorganic endosteal surface: The appearance of bones from the three groups is shown in Fig. 5. The vascular channel entrances in rachitic bones were wider ($\leqslant$ 150 μm across) than those in bones from D-treated and normal rats ($\leqslant$ 45 μ m across). Moreover, adjacent vascular channel entrances in the rachitic rats were often confluent. This was never seen in bones from D-treated and normal rats. Clearly, the appearance of the rachitic bones is due to the fact that they have been rendered anorganic. In life, the vascular channels were lined by a thick layer of osteoid.

The endosteal surfaces of rachitic bones were characterised by stacked and overlapping layers of mineral (Fig. 6a). This appearance was not seen in bones from D-treated or normal rats. (Fig. 6b and c).

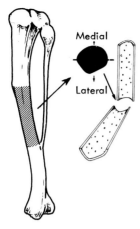

Fig. 1: The lateral segment of tibial shaft used for SEM.

Fig. 2: Endosteal surface of left tibia from a normal rat illustrating the sampling procedure for morphometric analysis. Anorganic. The rectangles representing each frame are drawn to scale. 700 μ m bar.

(b) High-power views of the anorganic endosteal surface: As has been described by Boyde and his co-workers[2,3,4,9],several different types of anorganic surface can be distinguished. In this study, surfaces with similar appearance to those already described have been found. We are in agreement with the interpretation given by Boyde and his co-workers of the functional activity represented by these surfaces. The different types of surfaces which were seen are as follows:- (I) Surfaces composed of small, fusiform mineral particles approximately 500 nm long (Fig. 7a). In some areas the particles displayed a random orientation. In others they were aligned in rows. This type of surface is taken to correspond to mineralization fronts, representing an early phase of mineralization of the matrix. (II) Surfaces consisting of rows of larger mineral nodules resembling strings of beads (Fig. 7b). The nodules were usually spindle-shaped and the range of their short axis

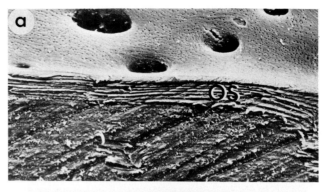

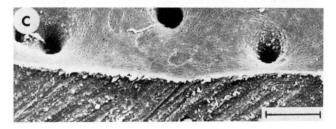

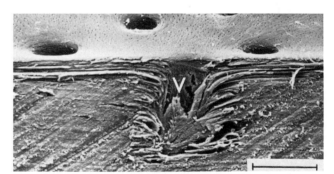

Fig. 4: Cut edge of the right tibia from a rachitic rat. Endosteal surface. A vascular channel (V) is cut in oblique section and is lined by a thick layer of osteoid. 60 μm bar.

---

diameters (∿ 0.5 to ∿ 3.0 μm) was comparable with those of intact collagen fibers in specimens which had not been rendered anorganic. These areas are taken to represent a later stage in the calcification of the collagen matrix. The mineral nodules correspond to the mineralized zones along the length of what were partially calcified collagen fibers before the specimens were made anorganic. Surfaces I and II therefore

515

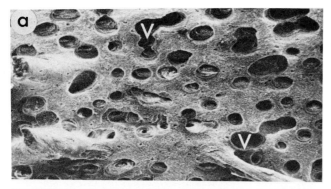

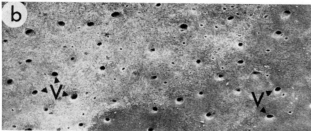

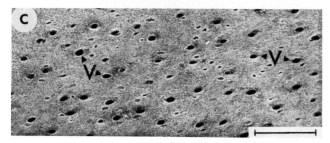

Fig. 5: Endosteal surface of left tibiae from rachitic (a), D-treated (b) and normal (c) rats. Anorganic. The vascular channel entrances (V) in the rachitic bone are wider than those in bone from D-treated and normal rats and many of them are confluent. 250 μm bar.

both represent incompletely mineralized matrix. (III) Smooth surfaces consisting of a continuum of collagen fiber skeletons (Fig. 7c). (The term 'collagen fiber skeleton' is used to describe the mineral which impregnated an individual collagen fiber). These areas are taken to represent completely mineralized, 'resting' bone matrix. (IV) Surfaces displaying numerous depressions or pits with bright, scalloped edges (Fig. 8). These surfaces were always at a lower level than the adjacent bone matrix and their borders were usually clearly demarcated by well-defined edges. These surfaces correspond to bone matrix which has undergone osteoclastic resorption. Two types of resorbed surface can be distinguished: (a) In some areas (Fig. 8a), individual collagen fiber skeletons were visible and their orientation could readily be distinguished. The orientation was seen to vary between successive, exposed lamellae. These surfaces represent areas of recent osteoclastic activity and were never overlain by mineralization fronts. (b) In other resorbed areas (Fig. 8b), the surface has a much

smoother texture and individual collagen fiber skeletons cannot be distinguished. These surfaces represent resorbed surfaces which have been partially remineralized and will be referred to as 'resting resorbed surfaces.' Occasionally, these surfaces are overlain by mineralization fronts (Fig. 8c).

(3) Distribution of Different Surface Types

The percentage area of endosteal surface occupied by each of the different surface types is given in Fig. 9. The data were subjected to analysis of variance and all significant differences are given below.

Incompletely mineralized matrix (Surfaces I and II): In the rachitic bones, 85% of the endosteal surface was occupied by incompletely mineralized matrix (Surface I and II). This compares with the 57% which was present on bones from D-treated and normal rats. The proportion of surface occupied by rows of mineral nodules (Surface II) on bones from both rachitic (45.0%) and D-treated rats (37.0%) was significantly ($P<0.02$ and $P<0.05$ respectively) greater than that present on normal rat bones. There were no significant differences between groups in the proportion of surface which was occupied by mineralization fronts (Surface I).

Fully mineralized, resting matrix (Surface III): There was a significantly ($P<0.05$) greater proportion of resting matrix present on bones from normal rats (38%) than on those from rachitic (5%) and D-treated rats (7%).

Resorbed matrix (Surface IV): There was a significantly ($P<0.05$) greater amount of resorbed matrix (Surfaces IVa + IVb) present on bones from D-treated rats (35%) than on those from rachitic (11%) and normal rats (5%). The ratio of actively resorbing surface (Surface IVa) to resting resorbed surface (Surface IVb) was 0.2 in the rachitic group and 0.6 in the D-treated group. All of the resorbed surface present on the normal bone had the appearance of active resorption (Surface IVa).

Discussion

The rat, unlike man, is very resistant to a dietary deficiency of vitamin D. Thus, in order to induce rickets, it is necessary not only to restrict the vitamin D intake but also to interfere with the mineral balance of the diet[10]. In the present experiment rickets was achieved by feeding a diet deficient in both vitamin D and phosphorus.

Although the surface types observed in all three groups conform to those previously described,[2,3,4,9] extreme caution must be exercised in the interpretation of the functional activity represented by these surfaces, particularly in the rachitic group. The thin osteoid layer present in the D-treated and normal groups would suggest that the underlying mineral surface was of recent origin. By contrast, the endosteal surface was lined by a thick layer of osteoid in the rachitic group, and therefore most of the mineral surface examined was probably formed early in the experimental period. Indeed if, as seems likely, that early period was a time

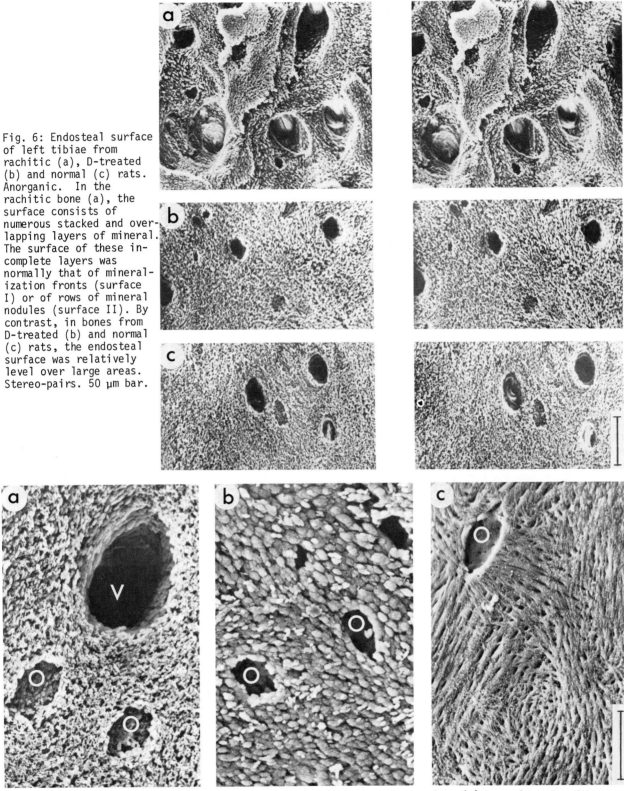

Fig. 6: Endosteal surface of left tibiae from rachitic (a), D-treated (b) and normal (c) rats. Anorganic. In the rachitic bone (a), the surface consists of numerous stacked and overlapping layers of mineral. The surface of these incomplete layers was normally that of mineralization fronts (surface I) or of rows of mineral nodules (surface II). By contrast, in bones from D-treated (b) and normal (c) rats, the endosteal surface was relatively level over large areas. Stereo-pairs. 50 μm bar.

Fig. 7: Endosteal surface of the left tibia from a normal rat. Anorganic. (a) mineralization fronts (b) rows of larger mineral nodules corresponding to partially mineralized collagen fibers (c) resting, fully mineralized bone matrix. V = vascular channel entrance. O = forming osteocyte lacuna. See text for more details. 15 μm bar.

of low phosphate but more slowly falling vitamin D levels in the R group, then the similarity in the distribution of surfaces I, II and III between the R and D groups is perhaps of no surprise.

The relative areas of the different types of anorganic surface must be a function of the rates of formation or transformation of these surfaces. We assume that surfaces I, II and III described above represent three stages in the process of mineralization of the organic matrix (Fig. 10).

The larger proportion of partially mineralized matrix in the rachitic group than in the other two groups could be due to an increased mineralization rate at the time of deposition. Alternatively, it could be due to a decreased rate of conversion to the fully mineralized matrix. These two possibilities are not mutually exclusive. When the rats were killed the serum was assayed for alkaline phosphatase levels and significantly higher levels were found in both the rachitic (13.48±2.31 K.A. Units/100 ml.) and D-treated groups (10.42±1.74 K.A.U./100 ml.) than in the normal (5.68±0.63 K.A.U./100 ml.). Serum alkaline phosphatase is considered to be roughly correlated with the rate of bone matrix formation[11]. If formation was increased in the rachitic rats at a time prior to inhibition of initial mineralization this may account for the larger proportion of partially mineralized matrix found in the rachitic bones. However, the significantly smaller amount of fully mineralized matrix in the rachitic rats in comparison to that in normal rats would indicate that there was a reduction in the rate of conversion to fully mineralized matrix in the rachitic bone. When we consider the distinction between the early (Stage I) and late (Stage II) phases in the proposed mineralization sequence the rachitic bone is seen to have a significantly greater area in stage II than the normal bone. This suggests that there is a decrease in the rate of conversion of stage II to stage III.

In the D-treated group the relative amounts of stages II and III were similar to those in the rachitic bone and were again significantly different from those in the normal group. This suggests that in the D-treated bones there is also a slowing in the rate of conversion of stage II to stage III. At the time of sacrifice, the mean serum phosphorus levels in the rachitic (0.72±0.11 mmoles/1) and the D-treated (1.71±0.30 mmoles/1) groups were both considerably lower than in the normal (2.68±0.39 mmoles/1) group. Thus, the failure in completion of mineralization of the matrix in these rats may be due to the low serum phosphorus levels. The D-treated group have a smaller proportion of bone surface in stage I than the normal group. However, the serum alkaline phosphatase was significantly higher in the D-treated than in the normal rats. Since initial mineralization in the D-treated group would not appear to be inhibited, as evidenced by the small amount of osteoid present, and the serum enzyme levels indicate an increased bone formation rate, the small proportion of bone surface in stage I of mineralization could be due

to an accelerated transition from stage I to stage II.

There was a significantly greater amount of resorbed surface in the D-treated group than in the rachitic and normal groups. This finding is in agreement with that of Baylink et al.[12] using a microradiographic technique. Recent studies on the role of phosphate upon vitamin D metabolism would indicate that part of the effect of low phosphate levels may be indirect. Tanaka and DeLuca[13] have shown that in vitamin D deficient, thyroparathyroidectomised rats, the rate of conversion of 25-(OH) D$_3$ to 1,25-(OH)$_2$D$_3$ in the kidney varies inversely with the plasma phosphate concentration. 1,25-(OH)$_2$D$_3$ has been shown to be a potent stimulator of bone resorption both in vivo[14] and in vitro[15]. Therefore the increased bone resorption in hypophosphataemia may be mediated by an increase in the circulating level of 1,25-(OH)$_2$D$_3$. The serum phosphorus levels in the rachitic rats were even lower than those in the D-treated group. However, significantly less resorbed surface was found in the rachitic group than in the D-treated group. This finding is also in accordance with those of Baylink et al.[12] and would suggest that the increased resorptive activity associated with hypophosphataemia is inhibited by vitamin D deficiency.

The above interpretations are based upon the surface morphology at one point in time. Their confirmation or refutation awaits further time dependent studies.

Acknowledgements

The authors are grateful to the National Fund for Research into Crippling Diseases (Grant No. G/A/8/915) for generous financial support and to Professors D. Newth and K. Vickerman for the use of the SEM. We also wish to thank Miss Aileen Smith, Mrs. Moira McRae and Mr. I.Ramsden for photographic printing, typing and art work.

References

1. Boyde, A., and Hobdell, M.H. A scanning electron microscope study of mammalian bone surfaces. J. Dent. Res. 47, 1968, 1006 (Abstract No. 120).
2. Boyde, A., and Hobdell, M.H. Scanning electron microscopy of lamellar bone. Z. Zellforsch. 93, 1969, 213-231.
3. Boyde, A., and Hobdell, M.H. Scanning electron microscopy of primary membrane bone. Z. Zellforsch. 99, 1969, 98-108.
4. Jones, S.J., and Boyde, A. Experimental studies on the interpretation of bone surfaces studied with the SEM. SEM/1970 IIT Research Institute, Chicago, Il. 60616, 193-200.
5. Steendijk, R., and Boyde, A. Scanning electron microscopic observations on bone from patients with hypophosphataemic (vitamin D resistant) rickets. Calc. Tiss. Res. 11, 1973, 242-250.
6. Numerof, P., Sassaman, H.L., and Rodgers, A. The use of radioactive phosphorus in the assay of vitamin D. J. Nutr. 55, 1955, 13-21.

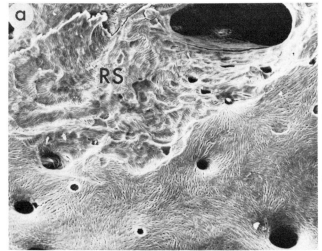

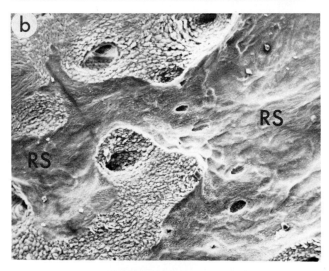

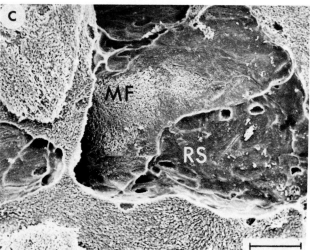

Fig. 8: Resorbed surfaces (RS) on the endosteum of anorganic tibiae from normal (a) and rachitic (b and c) rats. MF = mineralization front overlying resorbed surface. See text for further details. 50 μm bar.

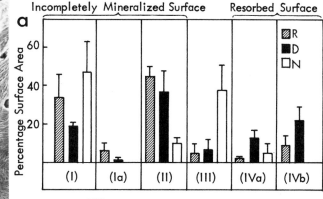

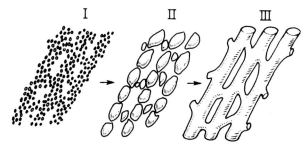

Fig. 9: Percentage area occupied by: (a) all of the different surface types and (b) total incompletely mineralized, resorbed and resting surfaces on the tibial endosteum of rachitic (R), D-treated (D) and normal (N) rats. Values are means ± S.E.M. (n = 3).

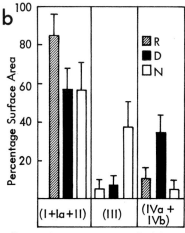

Fig. 10: Proposed stages in the mineralization sequence. See text for further details.

7. Boyde, A., and Jones, S.J. Bone and other hard tissues. In Principles and Techniques of Scanning Electron Microscopy, Vol.2. Ed. M.Hayat: Pub. Van Nostrand-Reinhold Co., N.Y., London, 1974. (pp. 123-149).
8. Biddlecombe, W.H., Dempster, D.W., Elder, H.Y., et al. Morphometric analysis by computerized planimetry. J. Physiol. 273, 1977, 22-23P.
9. Boyde, A. Scanning electron microscope studies of bone. In: The Biochemistry and

Physiology of Bone, Edn. 2, Vol. I, Ed. G.H. Bourne. Pub. Academic Press, N.Y., U.S.A., 1972. (pp. 259-310).

10. McCoy, R.H. Dietary requirements of the rat. In The Rat in Laboratory Investigation. Eds. J.Q. Griffith and E.J. Ferris. Pub. J.B. Lippincott & Co., Philadelphia, Montreal, London, 1942. (pp. 67-101).

11. Rasmussen, H., and Bordier, P. The Physiological and Cellular Basis of Metabolic Bone Disease. The Williams and Wilkins Co., Baltimore, U.S.A., 1974. p.183.

12. Baylink, D., Wergedal, J., and Stauffer, M. Formation, mineralization and resorption of bone in hypophosphatemic rats. J. clin. Invest. 50, 1971, 2519-2530.

13. Tanaka, Y., and DeLuca, H.F. The role of inorganic phosphate in the control of vitamin D metabolism. Arch. Biochem. Biophys. 154, 1973, 566-574.

14. Haussler, M.R., and Rasmussen, H. Metabolism of vitamin D3 in the chick. J. Biol. Chem. 247, 1972, 2328-2335.

15. Reynolds, J.J. Bone remodelling: in vitro studies on vitamin D metabolites. In Hard Tissue Growth, Repair and Remineralization. Eds. K. Elliott and D.W. Fitzsimons. Pub. Associated Scientific Publishers, Amsterdam, London, N.Y., 1973. (pp. 315-330.)

Discussion With Reviewers

J.L. Matthews, W.L. Davis and R. Jones: Why was a D-deficient, phosphate sufficient control group omitted from the experimental protocol?
Authors: This paper reports the results from the first study in a planned series of investigations into the effects of serum phosphorus and vitamin D levels on bone morphology as seen in the SEM. D-deficient, phosphate sufficient rats are included in an experiment currently in progress.

J.L. Matthews, W.L. Davis and R. Jones: Why did the authors use vitamin D supplementation when 1,25 dihydroxycholecalciferol would have bypassed the PO_4 effect on the kidney?
Authors: 1,25 DHCC was not available to us when this experiment was performed.

J.L. Matthews, W.L. Davis and R. Jones: What was the spatial relationship of forming, resting, and resorbing fronts? Was the distribution suitable to account for reutilization of mineral?
Authors: Resorbed surfaces were found adjacent to both resting and forming surfaces. No particular pattern emerged from our computer-linked planimetry which might suggest a spatial relationship associated with possible mineral reutilization.

R.M. Dillaman: Would you care to comment on the surface morphology one might expect to encounter on the periosteal surface of the bone?
Authors: Each of the different surface types encountered on the endosteum was also seen on the sub-periosteal surface of anorganic specimens. Additionally the surface described as a "prolonged resting surface" by Boyde and colleagues[2,3,4,9] was observed. This type of surface was not seen on the endosteum in the present study. As for the endosteal surfaces, the vascular channel entrances in the rachitic bones were of larger diameter than those in D-treated and normal bones. The rachitic bones displayed the stacked and overlapping mineral layers, as seen on the endosteum. However, these did not extend through as great a depth. We have yet to quantify the distribution of the various surface types on the anorganic periosteum.

R.M. Dillaman: What might one expect to see if examining the metaphysis or epiphysis rather than the diaphysis?
Authors: These sites were specifically avoided because of the abundance of trabecular bone. The small diameter cylindrical form of trabecular bone does not lend itself to a quantitative morphological study using the present method. We might, however, anticipate a quantitatively different distribution of surface types at these sites for the following reasons. Firstly the turnover rate in trabecular bone is generally held to be higher than in cortical bone. Secondly it is probable that the dynamic balance between formation and erosion is different in these sites from that in either the diaphyseal endosteum or periosteum.

SCANNING ELECTRON MICROSCOPY/1979/II
SEM Inc., AMF O'Hare, IL 60666, USA

THE ARCHITECTURE OF METAPHYSEAL BONE HEALING

K. Draenert and Y. Draenert

Orthopädische Klinik und Poliklinik
rechts der Isar der
Technischen Universität
Ismaningerstrasse 22
D-8000 München 80
W-Germany

Abstract

A simple technique for the processing of embedded, non-decalcified bone specimen is described. This method allows
three-dimensional visualization of the
cancellous bone trabeculae and assessment
of the surfaces of the trabeculae. A healing osteotomy of metaphyseal bone is used
as an example for demonstration of the
technique, which consists of elution of
the resin from ground sections with 100%
acetone.

In the study of fracture healing the
structure and arrangement of the bone trabeculae is of particular interest. Both
the trabecular framework and pre-injected
blood vessels are revealed by maceration
techniques and can be documented by scanning electron microscopy (SEM). However,
the samples are very fragile and are
easily destroyed.

The preparative technique described
complements the conventional methods for
the processing of non-decalcified bone
samples. Apart from stereoscopic visualization of the trabeculae it is possible
to demonstrate accretion, resorption, and
resting zones on the active surfaces of
the bone. The collagen fibers are also
well visualized.

The results of this technique are
exemplified by specimens of direct metaphyseal bone union in the proximal tibia
of the dog.

KEY WORDS: Bone, Cancellous Bone, Fracture
Healing, Biological Specimen Preparation, Metaphyseal Bone, Internal Fixation, Bone Histology

Introduction

The histological processing of non-
decalcified bone usually involves embedment in synthetic resin. Resin blocks
have the advantage of allowing relatively
large pieces of bone to be stored inde-
finitely. This makes it possible to collect
serial sections and evaluate the tissue
specimen systematically.

The study of fracture healing(1,2,3,
4) with the aid of conventional preparative techniques is hampered by the lack
of three-dimensional information. Serial
preparations for the SEM cannot be prepared in unlimited numbers, since they
have to be examined immediately. Specimens
for scanning electron microscopy which
are stored show increasing charges, reduced
resolution, and irregular contrast.

Preparations which have been embedded
for light microscopy (LM) and which require further processing for electron
microscopy frequently provide cytological
information of disappointing quality. LM
techniques may cause one or more of a variety of artefacts. On the other hand, it
is possible to achieve good visualization
of the micro-and ultrastructures of preparations which were originally processed
for transmission electron microscopy (TEM)
and embedded in epoxy resin (5).Good results have also been obtained with SEM
specimens which were re-embedded and then
examined by TE-microscopy(6,7).

Three-dimensional visualization provides valuable information about the architecture of the bone trabeculae. A preparative technique which allows systematic
three-dimensional investigation of the
structure of embedded bone specimens represents a valuable addition to the conventional techniques for the processing
of non-decalcified bone samples.

Materials and Methods

Bilateral hemiosteotomies of the proximal tibiae of 15 female Beagle dogs were performed. The osteotomy was stabilized in each case with an ASIF (Association for the Study of Internal Fixation) plate (Figs.01 & 02). Four and eight days, and

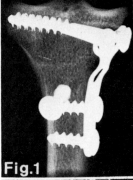

Fig.1 Roentgenogram 4 weeks following hemiosteotomy of the proximal tibia of a dog. The osteotomy was stabilized by compression fixation with an ASIF plate.

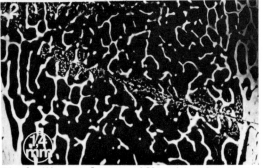

Fig.2 View at low magnification of a hemiosteotomy in the proximal tibia of a dog. 2 weeks after the operation the osteotomy gap has already been bridged by a framework of woven bone which is reinforced in a lamellar fashion. Microradiogram.

2,3,4,5,6,7,7.5,8,9, and 10 weeks after the operation general anaesthesia was induced, and the animal was sacrificed by administration of a lethal dose of Nembutal. The perfusion was carried out by injection of Ringer solution and 40% alcohol in the abdominal aorta. The non-decalcified bone samples were then prepared for histological examination. Following block staining with basic fuchsin(8) the bone specimens were embedded in methyl methacrylate (Merck), 2.5ml of Plastoid (Röhm), and 0.2g of benzoyl peroxide. Polymerization took place over a period of 4 weeks at room temperature. The hardened resin blocks were sawn in the frontal plane and the resulting slices were ground. Ground slices 200-400 µm in thickness were selected for SE-microscopy. This thickness was considered necessary to allow adequate 3-dimensional visualization of the honeycomb-cells of the cancellous bone, which are 300-500 µm broad and deep in the proximal tibia of the dog.

The resin was dissolved out with 100% acetone. The rate of elution of the resin was determined with a 400 µm ground bone slice in the following manner. The slice was divided into 8 segments of equal size which were removed from the acetone after immersion periods of 6, 12, 18,24,30,36,42, and 48 hours, dried by the critical point technique, and sputtered with a 200 Å gold layer. The acetone bath was changed 4 times in each case.

The ground bone slices which were used for the investigation of fracture union were eluted for 48 hours in 100% acetone. The cellular constituents were removed with dry nitrogen or with the water jet (WaterPik®) and the sections were dehydrated over acetone. They were then dried in a desiccator, mounted on aluminium stubs, sputtered with a gold layer of 200 Å thickness, and photographed stereoscopically in the PSEM 500 with 12 kV. The angle of tilt was 10° in each case (9,10).

Results

Within 6 hours the methacrylate had been dissolved out of the ground surfaces of the specimens. Residual methacrylate was easily recognizable, since it desintegrated in the electron beam and thus caused the gold layer to break up. After 36 hours remnants of methacrylate were still present in the slices with a thickness of 400 µm. After 48 hours the samples were free of methacrylate. Rinsing out of the cellular components of the bone marrow with the water jet fully revealed the architecture of the bone trabeculae. In many cases this latter procedure left the points of adhesion of the fat cell membranes intact. Stretches of the arterial vessels were also revealed (Fig.03). A striking feature of the cancellous bone trabeculae below the surface of the joint was the formation of vault-shaped structures with weight-bearing columns, the majority of which contained blood vessels and avascular mural segments of the cancellous "honeycomb" (Fig.03 & 04). In many of the medullary spaces the straight arteries were still intact, and the sites of adhesion of the yellow marrow cells on their adventitia were clearly recognizable. The vascular network was spread out under the shield of the stable cancellous trabecular arches. The medullary sinuses and venous system were completely removed by the preparative procedure (Fig.05).

The investigation of fracture healing showed that within 8 days the hemiosteotomy gap had been bridged by a framework of woven bone with trabeculae ranging from 10 µm to 35 µm in thickness. The

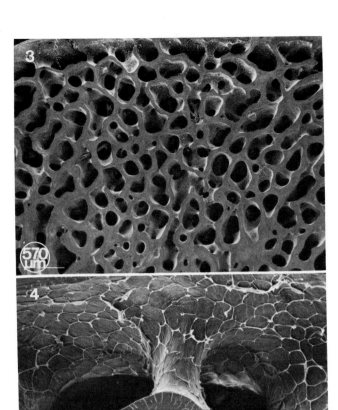

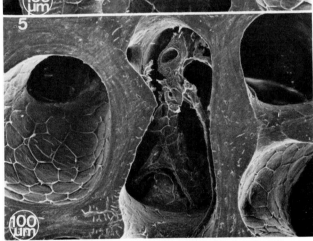

Fig.3 The proximal tibial epiphysis above the osteotomy. Ground section, 400 μm thick,MMA eluted in 100% acetone. The marrow was removed with the water jet and the ground section was dehydrated in acetone, dried in the desiccator, and sputtered with a 200 Å gold film. PSEM 500, 12 KV. Fig.4 The weight-bearing columns of the trabecular arches are often vasiferous cancellous bone trabeculae. Preparation as Fig.3. Fig.5 The arterial vessels have remained partially intact. Preparation as Fig.3.

framework of woven bone was irregularly round. No continuous layers of osteoblasts were seen (Fig.06 & 07). The gap was 80-300 μm in width. On each side of the osteotomy the interstices of the cancellous bone had also become filled with a scaffolding of woven bone. Further away from the osteotomy signs of incipient lamellar reinforcement of the cancellous bone trabeculae were visible. Here more or less intact layers of osteoblasts were found. At this stage the supporting framework was still only slightly calcified, as was shown in the microradiographs. In those samples in which broad osteoid seams were visible under the light microscope, SE-microscopy revealed irregular round trabeculae whose disordered fiber bundles were still clearly visible. The woven bone trabeculae first appeared in the form of garlands along the original cancellous bone trabeculae and then grew out across the osteotomy gap. Gaps of 50-100 μm in width were filled by one osteon, and gaps of 200-300 μm were crossed by a 2-or 3-stage framework of woven bone. The sizes of the enclosed medullary cavities within the osteotomy varied very little and ranged from 70-80 μm. The trabeculae were always aligned at right angles to the axes of the in-growing vascular buds.

The formation of the woven bone framework was followed in the second week by concentric lamellar reinforcement of the network of trabeculae. In the microradiographs a marked increase in the degree of mineralization was apparent.

Examination at low magnification (Fig.8) revealed thickening of the woven bone trabeculae and additional filling of the medullary cavities with bone. The trabeculae formed arch-like structures, the domes of which were directed proximally. This lamellar reinforcement of the bone trabeculae did not take place concentrically with respect to the courses of the vessels but took the form of concentric layers around the framework of woven bone (Fig.9). In the vicinity of the plate the concentric lamellar reinforcement of the bone trabeculae proceeded until the medullary cavities had been filled in and concentric lamellae were then laid down around the central vessels. This stage of conversion to compact bone was reached at the end of the third week (Fig.10).

At this point in time the union of the osteotomy was complete. The networks of cancellous bone on each side of the osteotomy were stably supported by stress-oriented bone trabeculae (Fig.11).

In the third week resorption of the supporting framework began at the end of

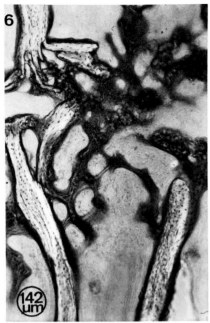

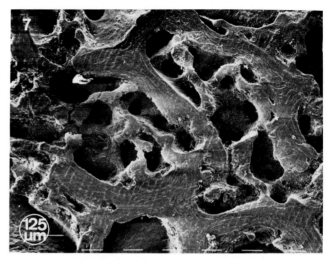

Fig.6 Osteoto-
my gap with a
distally conti-
guous cancellous
bone cavity one
week after the
operation. At
this stage the woven bone framework has already
taken on an arch-like orientation.
Leitz Orthoplan, polarized light,10/0.45 oil immers.

Fig.7 Osteotomy gap 8 days after the operation.
Irregular round trabeculae with randomly oriented
bundles of fibers which are arranged in garlands
along the original cancellous bone trabeculae. The
trabeculae are always oriented at right angles to
the axes of the ingrowing vascular buds.
Preparation as Figure 3.

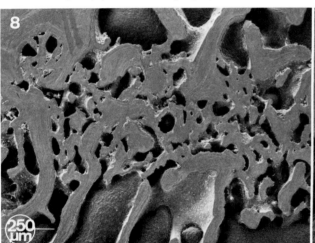

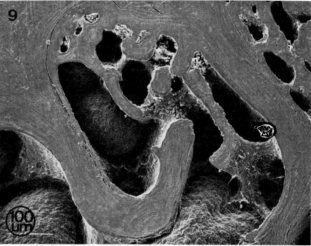

Fig.8 Hemiosteotomy gap 2 weeks after the operation. Thickening of the woven bone trabeculae is
seen, together with further filling-in of the medullary spaces with bone.
Preparation as Fig.3.

Fig.9 Supporting framework in a cancellous bone
cavity above the osteotomy. 2 weeks after the
operation. The trabeculae are more rounded-off and
form arches. The lamellar reinforcement conforms
to the supporting framework of the woven bone.
Preparation as Fig.3.

the hemiosteotomy and continued until
there was only a residual thickening of the
resynthesized cancellous bone trabeculae.
This remodeling was complete at the end
of the sixth week. The only remaining traces of the osteotomy were the thickening
of the cancellous bone trabeculae and
small residual fragments of the previous
generation of supporting trabeculae.
     After 9 weeks, all but traces of the
old supporting trabeculae had been resor-

bed. The bone marrow up to the vicinity
of the osteotomy had been transformed into yellow marrow (Fig.12). Zone of resorption were still found in the osteotomy itself.
Some of the lacunae had sharp edges,
whilst others were rounded off and covered with deposits of new bone.
The reconversion of the cellular bone
marrow into yellow marrow continued until the end of the twelfth week.

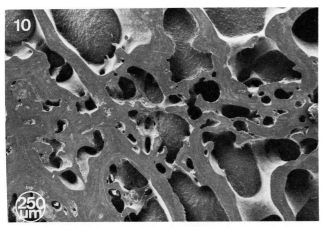

Fig.10 Hemiosteotomy gap 3 weeks after the operation. In the vicinity of the plate the concentric lamellar reinforcement is filling in the medullary spaces (conversion to compact bone). Preparation as Fig.3.

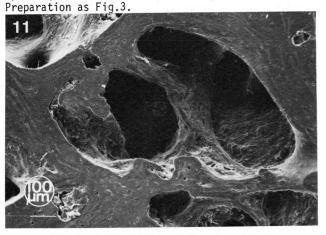

Fig.11 Supporting framework of a cancellous bone cavity above the osteotomy. SE-microscopy reveals the three-dimensional shape of the arches. Preparation as Fig.3.

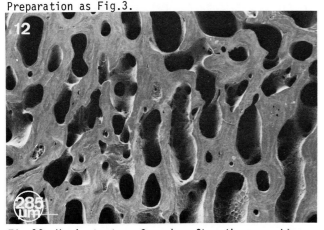

Fig.12 Hemiosteotomy 9 weeks after the operation. The only residual signs of the osteotomy are the thickened cancellous bone trabeculae and the reduction in size of the medullary cavities. Only traces of the original supporting trabeculae remain. Preparation as Fig.3.

Discussion

A range of tested preparative techniques is available for the processing of non-decalcified bone. These techniques allow precise assessment of normal and pathologically altered bone. The standard methods include the Frost method of block staining(8), embedment in methyl methacrylate or Epon synthetic resin, the preparation of sawn slices and grinding with wet carborundum paper (11), examination of the mineralized bone by microradiography, and the preparation of non-decalcified microtome sections by staining with the Goldner, Movat and von Kossa techniques (12,13,14). The staining characteristics are mainly determined by the degree of mineralization of the tissue. However, no technique is available for the specific detection of calcium salts. Staining with alizarin red and counterstaining with toluidine blue are not specific for calcium. The von Kossa reaction is unspecific and takes place in the presence of phosphate groups. The combination of microradiography with light microscopy allows differentiation of the mineralized matrix from the non-decalcified ground substance. If further differentiation of the osteoid structure is required the ultrastructure of the tissue has to be investigated by transmission electron microscopy (15).

The constant turnover of the bone can be visualized by fluorochrome staining(16,17,18,19) or by SE-microscopy of the inner and outer surfaces of the hard tissue. The stage of activity of the bone cells can only be assessed by investigation of the ultrastructure of the cut surfaces and outer surfaces of the cells. Here, the maximum amount of information can be obtained by combination of TEM and SEM. The freeze-drying methods are best for the preparation of the cells for SEM (20,21,22,23).

The study of fracture healing involves not only investigation of the ultrastructural processes of bone formation but also that of the stress-oriented architecture of the bone trabeculae. The first three-dimensional images of fracture healing were obtained with macerated preparations (24,25,26). Valuable information can also be derived from macerated specimens subjecting them to SE-microscopy (27). However, these specimens are very fragile and are easily destroyed.

Various etching and decalcification procedures have been described for the demonstration of fiber orientation, surface activity and crystal arrangement by SEM (22,28,29,30,31,32,33).

Boyde und Hobdell (28,30)

used 1,2-ethanediamine or 5-10% hypochlorite. Lindenfelser and others (34)
described a method using potassium hydroxide. Swedlow et al(32)and Lindenfelser et al
(35)used hydrazine.Anderson and Kennett (36)
prepared samples for SEM with a 5% anti-
formin solution described by Ludwig. However, none of these methods can be combined with the conventional light microscopic examination of non-decalcified
ground sections.

Boyde predicted that SE-microscopy
would replace the conventional techniques
for the assessment of the bone surfaces
(37). According to Whitehouse, preparation
for SE-microscopy produces inferior results when a bone sample is subjected
to statistical evaluation (38).

The maceration techniques which are
used by various workers for SE-microscopy clearly reveal the surface of the
mineralized bone. However, removal of the
organic components decreases the available information about the surface of the
bone. Furthermore, combination with the
proven techniques of light microscopy is
not possible.

Embedded preparations have the great
advantage that they can be stored for unlimited periods and can be systematically
examined by sectioning them serially
with the saw or microtome. An embedded
bone can be processed in any direction.
For light microscopy the bone is routinely embedded in resin and further processed
as described above. Thus by the addition
of a simple preparative procedure the information provided by light microscopy
can be complemented by that of a three-dimensional image. The latter reveals the
active surface of the bone, the trabecular
architecture, and the structure of those
vessels which remain intact.

The processing of non-decalcified
bone samples for SEM without extracting
the organic constituents requires considerable time and effort. For this reason, the number of fresh preparations
which can be processed is, irrespective
of the technical difficulties, limited.
Resin blocks, on the other hand can be
stored for unlimited periods of time and
can be systematically processed.

The preparative technique described
is simple and suitable for routine use.
It produces beautiful images of the structure of the fibers and trabeculae, and,
in addition, long sections of the arteries
and arterioles of the medullary cavities
are often displayed. Area of bone accretion and resorption and the quiescent
zones with the sites of adhesion of the
adipocytes can also be visualized. Furthermore, the osteoblast layer remains
intact if care is taken when rinsing out

the marrow cells. To be sure, the epithelial (i.e.contiguous) structure of the
groups of cells is usually lost, and
the surfaces of the cells lose their pri-
stine appearance in the majority of cases.
However, the main feature is the stereographic documentation of the trabecular
architecture.

Acknowledgements

The study was funded by the M.E.
Müller Foundation, Bern, the Ciba-Geigy
Jubily Foundation, Basel, and the Emil-
Barell Foundation of Hofmann-La Roche,
Basel. The microphotographs were taken on
a PSEM 500, donated by the Stanley Thomas
Johnson Foundation, Bern.

References

01.Schenk R.,H.Willenegger: Histologie
der primären Frakturheilung,Langenbecks
Arch.klin.Chir. 308,440-452(1964).
02.Geiser M.: Beiträge zur Biologie der
Knochenbruchheilung.Z.Orthop.Supplementum
97(Ferdinand Enke Verlag Stuttgart 1963).
03.Krompecher St.: Ueber den Spongiosakal-
lus.Z,Orthop.112,1196-1201(1974).
04.Oberdahlhoff H.: Der Einfluss mecha-
nisch-funktioneller Kräfte auf die fein-
eren Vorgänge der Knochenbildung. Dtsch.
Med.Wschr.73,291(1948).
05.Barber V.C.:Preparative techniques for
the successive examination of biological
specimens by light microscopy, SEM, and
TEM.SEM/1972   ,IIT Research Institute,
Chicago, IL 60616,321-326.
06.Erlandsen S.L., A.Thomas,G.Wendel-
schafer: A simple technique for correlating SEM with TEM on biological tissue
originally embedded in epoxy resin for
TEM. SEM/1973   ,IIT Research Institute
Chicago,IL 60616, 349-356.
07.Barber V.C.,A.Boyde:Scanning electron
microscopic studies of cilia.Z.Zellforsch.
84,269-284(1968).
08.Frost H.M.: Preparation of thin undecalcified bone sections by rapid manual
method.Stain Technol.33,273(1958).
09.Boyde A.: Photogrammetry and the
scanning electron microscope.Photo-
grammetric Rec.8,408-457(1975).
10.Boyde A.:Quantitative photogrammetric
analysis and qualitative stereoscopic
analysis of SEM images.J.Microsc.98,
452-471(1973).
11.Schenk R.: Zur histologischen Bearbei-
tung von unentkalkten Knochen.Acta anat.
(Basel)60,3(1965).
12.Goldner J.:A modification of the Masson
trichromtechnique for routine laboratory
purpose.Amer.J.Path.14,237-243(1938).
13.Movat H.Z.:Demonstration of all connective tissue elements in a single section.

Arch.Pathol.60,289-295(1955).
14.Kossa v.: Nachweis von Kalk.Ziegl.Beitr.
29,163(1901).
15.Fornasier V.L.:Osteoid:an ultrastructural study.Hum.pathol.8,243-254(1977).
16.Milch R.A.,D.P.Rall,J.E.Tobie:Fluores-
cence of tetracycline antibiotics in bone.
J Bone Jt.Surg40.A,897-910(1958).
17.Rahn B.A.,S.M.Perren:Calcein blue as a
fluorescent label in bone.Experentia 26,
519(1970).
18.Rahn B.A.,S.M.Perren: Xylenolorange, a
fluorochrome useful in polychrome sequential labeling of calcifying tissue.
Stain.Technol.46,125-129(1971).
19.Rahn B.A.,S.M.Perren:Alizarin Komplexon
Fluorochrom zur Markierung von Knochen-
und Dentinanbau.Experentia 28,180(1972).
20.Boyde A.,C.Wood:Preparation of animal
tissues for surface scanning electron
microscopy.J.Microsc.90,221-249(1969).
21.Boyde A.,P.Vesely:Comparison of fixation and drying procedures for preparation
of some cultured cell lines for examination in the SEM.SEM/1972   ,IIT Research
Institute,Chicago,IL,60616,265-272.
22.Boyde A.,V.C.Barber:Freeze-drying methods for the scanning electron microscopical study of the protozoon Spirostomum
ambiguum and the statocysts of the cephalopod mollusc Loligo vulgaris.LCell Sci.4,
223-239(1969).
23. Draenert Y.K.Draenert: Freeze-drying of
articular cartilage.SEM/1978/II,SEM Inc.
AMF O'Hare,IL 60666,759-766.
24.Helferich H.:Frakturen und Luxationen.
(Lehmann, München 1914).
25.Matti H.:Die Knochenbrüche und ihre Be-
handlung.II.Auflage(Springer Berlin 1931).
26.Block W.: Die normale und gestörte
Knochenbruchheilung.Neue Deutsche Chirurgie, Band 62(Sauerbruch Berlin 1942).
27.Whitehouse W.J.:Cancellous bone in the
anterior part of the iliac crest.Calc.
Tiss.Res.23,67-76(1977).
28.Boyde A.,M.H.Hobdell:SEM of lamellar
bone.Z.Zellforsch.93,213-231(1969).
29.Hobdell M.H.,A.Boyde: Microradiography
and SEM of bone sections.Z.Zellforsch.94,
487-494(1969).
30.Boyde A.,M.H.Hobdell:SEM of primary
membrane bone.Z.Zellforsch.99,98-108(1969).
31.Piekarski K.:Fracture of bone.J.Appl.
phys.41,215-223(1970).
32.Swedlow D.B.,R.A.Harper, J.L.Katz: Evaluation of a new preparative technique
for bone examination in the SEM. SEM/1972
IIT Research Institute, Chicago, IL,60616
335-342.
33.Jones S.J.,A.Boyde: Experimental studies
on the interpretation of bone surfaces
studied with the SEM.SEM/1970,IIT Research
Institute,Chicago,IL 60616,193-200.
34.Lindenfelser R.W.Krönert,H.Orth: Strukturen der Knochenspongiosa. Untersuchung-
en mit dem Raster-Elektronenmikroskop.
Virchows Arch.5,201-208(1970).
35.Lindenfelser R.,HP.Schmitt,P.Haupert:
Vergleichende rasterelektronenmikroskopische Knochenuntersuchungen bei prim.
und sek.Hyperparathyroidismus.Virchows
Arch.360,141-154(1973).
36.Anderson C.,D.Kenneth: Scanning electron microscopic observations on bone.
Arch.Pathol.Lab.Med.101,19-21(1977).
37.Boyde A.: SEM studies on bone in burn.
In the biochemistry and physiology of
bone II.Ed.(Academic Press New York 1972)
pp.259-301.
38.Whitehouse W.J.:Scanning electron
micrographs of cancellous bone from the
human sternum.J.Pathol.116,213-224(1975).

Discussion with Reviewers

S.Marks: Have you quantitated the contributions of bone formation and resorption
to fracture healing by measuring the relative areas of forming, resorbing and
resting surfaces during repair of the
fracture site? While this is no easy
task, your method of specimen preparation
combined with the use of an appropriate
grid should make the job much easier and
give quantitative support to your observations that formation predominates the
first three weeks and resorption for the
next three.
Authors: Yes. This work is reported in a
paper entitled "Direct Metaphyseal Bone
Healing" which has not yet been published.
For morphometry we prefer the stereological techniques applied to microtome
sections (literature reference 38).
R.W.Rice: Since many labs may be un-
equipped to grind bone, how suitable is
the specimen preparation using only a
high speed saw?
Authors: Carborundum paper costs us 0.20
DM (approximately 10 US cents) per sheet
and we use 2-3 sheets for each ground
section. However, sawn sections can also
be used.
M.Zimny: Were any other synthetic resin
experimented with in addition to methacrylate?
Authors: No
M.Zimny: Do you feel that blowing out
the marrow with the water jet may have
been detrimental to the trabeculae?
Authors: No.
M.Spector: Rather than used the proposed
procedure for preparing resin embedded
specimen for SEM, can't adjacent specimens
of trabecular bone be processed for
histology and SEM (one specimen for SEM
and the adjacent specimen for histology)?

Authors: It is also possible to examine a ground section by light microscopy and then process it for scanning electron microscopy.

M.Spector: What information can three-dimensional visualization (including stereo visualization) of trabeculae provide that histological evaluation of a few serial sections cannot? How has three-dimensional visualization helped the authors to better understand cancellous bone healing?

Authors: Complete visualization of the three-dimensional structure of bone is not possible using serial sections. Approximately 100 serial sections with thicknesses of 5 μm are needed for the three-dimensional visualization of the honeycomb-like structure of cancellous bone (assuming that none of the sections are destroyed).Three-dimensional visualization provides the only means of examining the stress-oriented trabecular structure of a fractured and stably fixed bone.

M.Spector: It is implied that when trabecular bone specimens are excised and processed for SEM there is some resulting distortion of the fragile cancellous network. Why can't trabecular bone specimens be excised using a sharp razor blade and carefully processed in such a manner as to minimize distortion? Did the authors ever directly compare the SEM appearance of bone processed using their etching procedure with trabecular bone they processed using conventional techniques?

Authors: Bone which has been macerated cannot be further processed. Fresh bone can be sawn without difficulty but the preparative possibilities are limited, since the methods are laborious and bone cannot be stored for unlimited periods unless it is embedded.

M.Spector: How were the hemiosteotomies made in precisely the same transverse plane in each tibia? Since the cancellous network varies quite a bit from point to point in the proximal tibia could differences in the position of hemiosteotomy in different dogs produce inconsistent results.

Authors: There is little variation in the structure of the cancellous bone trabeculae. In the proximal tibia of the dog the lamellar trabeculae are 90-110μm thick and those bearing vessels are 250-270μm thick. The pattern of healing was dependent on the width of the osteotomy gap and on the stability of the fixation but was independent of the cancellous bone network.

M.Zimny: Have comparable studies been done on osteotomies stabilized by immobilization? How did the results compare?

Authors: Similar experiments were performed by Krompecher (Krompecher literature reference 3). In his experiments, unlike ours, lamellar concentric reinforcement did not occur and the framework of woven bone was more like a fracture callus. We conclude from this that his immobilization was unstable.

M.Spector, M.Zimny: What basis was used to estimate or measure stress as mentioned in this study and what constitutes stress oriented trabeculae?

Authors: Stress-oriented trabeculae are cancellous bone trabeculae which have a lamellar structure. They are aligned with the lines of force which act on the physiologically loaded bone. Typical examples are the cancellous trabeculae at the upper end of the femur.

SCANNING ELECTRON MICROSCOPY/1979/II
SEM Inc., AMF O'Hare, IL 60666, USA

COLONIZATION OF VARIOUS NATURAL SUBSTRATES BY OSTEOBLASTS IN VITRO

S. J. Jones and A. Boyde

Anatomy Department
University College London
Gower Street, London WC1E 6BT
England

Abstract

The ability of osteoblasts to transfer from normal rat growing skull bone to small fragments of various biological materials was tested in an organ culture system. The substrates were tested as overlays and included intact or anorganic cut, polished or fractured bone, osteoid, cementum, dentine and enamel. Osteoblasts migrated on to and over all these materials, whether collagenous or not and whether mineralized or not. The culture medium affected the cell shape, elongation, alignment and dorsal ruffling activity of the cells similarly on the original bone surfaces and on the overlays. The position of the cells on the overlay was influenced by constraints due to space available for each cell, sharp edges and corners, and fine textural patterns, in that order. It is concluded that the ability of such substrates to support new bone formation cannot rest with the readiness of the osteoblasts to coat them. The procedures outlined constitute a valuable, simple and rapid means of testing one aspect of the bio-compatibility of potential prosthetic bone implant materials.

KEY WORDS: Bone, Osteoblasts, Cell Migration, Colonization of Substrates, Biocompatibility of Bone Implant Materials

Introduction

It is sometimes clinically desirable to provide a framework for bone formation[1]. If the framework has a high capacity for osteogenic induction, repair and the recovery of function will be speeded. Substrata have been found to differ in their capability to support or induce new bone formation[2]: allogenic bone or dentine which have been decalcified in hydrochloric acid are currently most highly rated in this respect[3]. Osteogenesis in vivo follows vascularization and the differentiation of new osteoblasts at the site of the implant. This sequence of events has emphasized the importance of the stimulatory effect that the substratum may have on vascular tissue, and hints at a blood-borne or vascular origin of the osteoblast precursor cells [4]. The differentiation to osteoblasts rather than chondroblasts appears to be dependent on vascularization[5,6], probably due to the higher oxygen tension prevailing.

By using an in vitro model it is possible to exclude the role of the vascular tissue and to examine the response of differentiated bone cells to different substrata. Previous studies have shown that osteoblasts retain the ability to repopulate denuded bone surfaces in vitro and will migrate on to new non-biological substrata[7]. This study was undertaken to investigate whether the composition or the structure and texture of the substratum influence the once functional, fully-differentiated osteoblasts. Because the geometry of the system studied was three-dimensionally complex, and the substrata for the cells were opaque, this work necessitated the use of scanning electron microscopy (SEM).

Materials and Methods

Biological substrata of various compositions and textures were prepared in the following ways: (1) Flat osteoid surfaces were obtained from rat calvaria which were divested of fibrous tissue by dissection under tissue culture medium and then swept completely free of cells on the endocranial aspect with a wedge of silicone rubber. Specimens of different shapes were cut, and these were frozen in liquid nitrogen and thawed three times to kill any cells within the tissue, and then stored frozen at -4oC. The osteoid provided a

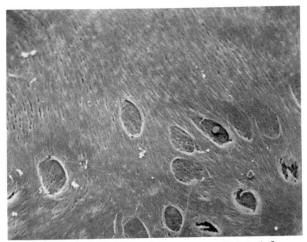

Fig. 1: Flat, osteoid surface on endocranial aspect of rat parietal bone from which all cells have been swept away. CPD. Field width 180 μm.

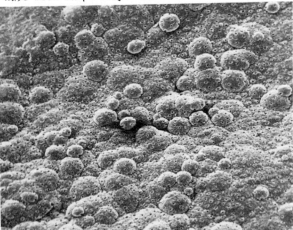

Fig.3: Anorganic mineralizing front of dentine showing calcospherites. The dentine tubules are about 2 μm across. CPD. Field width 330 μm.

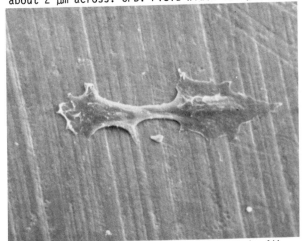

Fig.4: Smeared, grooved tooth surface cut with a diamond wheel. The elongation and the ruffles seen at the leading edges of the osteoblast show that it was moving across the grooves. CPD. Field width 80 μm.

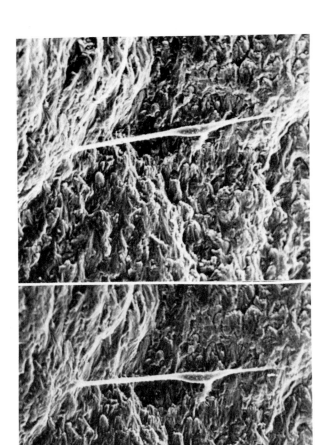

Fig.2: A cell is stretched across a crevasse in this very rough fractured enamel surface. CPD Stereopair: field 110 μm.

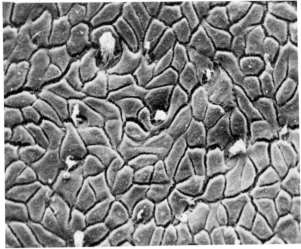

Fig.5: Rat endocranial osteoblasts lying in a monolayer over the bone surface immediately following removal of overlying tissue. The overlays were placed directly on such a surface for the culture period. Field width 160 μm.

finely grained, flat surface with occasional depressions at sites of lacunae and holes where blood vessels had passed into the bone (Fig.1). Anorganic bone was prepared using a 5% sodium hypochlorite solution.

(2) Fragments of air-dried human or bovine teeth were prepared by fracturing to provide surfaces of dentine, cementum and enamel (Fig.2). These specimens provided extremely rough, irregular surfaces. Any effect of the collagenous substrata could also be compared with that of the non-collagenous, highly mineralized enamel in the same culture. Free formative dentine surfaces were made anorganic using NaOCl to provide irregular mineral surfaces penetrated by small canals, i.e. the mineralizing front of dentine (Fig.3).

(3) Transverse slices of roots and crowns of dried human teeth were cut to a thickness of approximately 50 μm using a diamond wheel: a smeared, grooved surface resulted (Fig.4). Some of these specimens were treated with a 5% solution of sodium hypochlorite to remove most of the organic matrix and all the cellular remnants.

Calvaria were excised from anaesthetized Albino Wistar rats, aged from 2 to 8 days, and immersed in Dulbecco's modification of Eagle's minimal essential medium (DMEM:k'13 Gibco Bio-cult Ltd., Paisley, Scotland). The fibrous layers were removed from the bone, leaving a single layer of cells covering the bone matrix (Fig.5). The specimens were oriented with the endocranial aspect uppermost, and one or more overlays of the prepared substrata were placed at sites known from previous studies to be appositional and therefore covered by osteoblasts (Figs.5,6,7). The calvaria were then cultured in DMEM alone, or in DMEM with added 10% foetal calf serum (Gibco Bio-cult Ltd.), or in DMEM plus 10% serum together with 0.5U/ml of added parathyroid extract (Lilley). A few specimens were cultured with or without added parathyroid extract in Hank's MEM (Cat.157 Gibco Bio-Cult Ltd.) with 10% foetal calf serum. The organ cultures were continued for 1, 2 or 3 days, at 37OC, then fixed in 3% glutaraldehyde in 0.15 M cacodylate buffer (pH 7.2), dehydrated in graded ethanols, substituted with Freon 113, critical point dried from carbon dioxide and coated with gold in a sputter coater. They were examined in a Cambridge Stereoscan Mark IV operated at 10 kV. Stereo pairs were recorded with 10° tilt angle difference.

Results

Transference of cells to the overlay.

Osteoblasts migrated on to all the types of overlay used. Neither the constitution, nor the shape, size, nor gross roughness of the overlay deterred the cells from moving on to and over it (Figs.8,9). Cells were separated from one another in the leading ranks (Fig.9), but formed a more continuous sheet behind (Fig.10). Because the cells were able to form confluent sheets over very irregular surfaces it was not possible to guess the type of substratum underlying the osteoclasts from their habit (Fig.11). The appearance of the cells reflected the hormonal and nutritional levels in the medium rather than the nature of the substratum.

Effect of different culture media.

The shape of the cells on the overlay, particularly when confluent, was the same as that shown by the osteoblasts that remained on the surrounding bone. Thus, when cultured in the nutritionally rich, serum-supplemented DMEM, cells on the overlays exhibited dorsal ruffling and a low elongation ratio although their polarity was obvious (Fig.12). The addition of parathyroid extract (PTE) to this medium was accompanied by a reduction of dorsal ruffling and an increase in elongation and alignment of the cells on the overlay so that they appeared more fibroblastic (Fig. 13), just as did their peers on the surrounding normal substratum (Figs.13,14). The same effect on the cells as elicited by adding parathyroid extract to DMEM could be achieved just by using serum-supplemented Hank's MEM instead, in which the calcium content is lower and the amino acid levels approximately half that of DMEM (Fig.15). The addition of PTE to Hank's MEM did not further alter the overall appearance of the cells. Osteoblasts transferred from the original bone to the overlays in the absence of serum, but the rate of colonization of the overlay was markedly lower than that in serum-supplemented media.

Effect of the texture of the overlay.

The distribution or crowding of the cells on the new surface had a greater influence on the arrangement of the cells than did the grain of the substratum. The locomotory cells were more influenced by the overall direction of spread over new territory and the restriction for movement imposed by near neighbours than by the fine grain of the enamel prisms, or of the dentine tubules that were fractured open longitudinally leaving hemicylindrical grooves of about 2 μm in diameter (Fig.16). However, if the directions of spread of the cells and the grain were nearly the same, cells aligned parallel to the grain (Fig.17). The even finer grooving of the aligned collagen in the osteoid surface and the parallel grooves and ridges made in enamel by the diamond grit of the cutting wheel did not give guidance to the cells that took priority over the general pattern of their spread over the overlay, even when the cells were cultured in media that promoted an elongated form for the cells.

Abrupt changes in the orientation of the surface at the corners and sides of overlays which were shaped like flat, angular tablets did affect the order of the cells, particularly where these were elongated (Fig.18). This edge-effect was dependent on the crowding of the cells: spaced cells tended to align so that their long axes were parallel to the edge, and spread along the edges preferentially. However, where the cells reached the edge at right angles to it and in close array, the edge effect was nullified; then cells were observed in transit of the edge, bent at right angles with their long axis at right angles to the edge.

Finally, away from areas exhibiting the edge effect, where the elongated cells were in a confluent monlayer and not thought to be translocating but overlying a bone surface with fine collagen fibres in a parallel pattern, they were usually true to that pattern (Fig.19).

The sizes of the gaps that the cells bridged

Fig.6: One day culture of human enamel on rat bone. Osteoblasts coat the vertical fractured faces and were spreading over the enamel surface at left. CPD. Field width 1660 μm.

Fig.7: Five fragments of devitalized unfixed rat bone on endocranium. Three day culture in medium and serum and PTE. The overlays are entirely covered by colonizing osteoblasts. CPD. Field width 4160 μm.

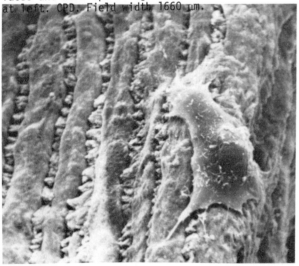

Fig.8: Osteoblast migrating over fractured enamel. One day culture in DMEM and serum. CPD. Field 42 μm.

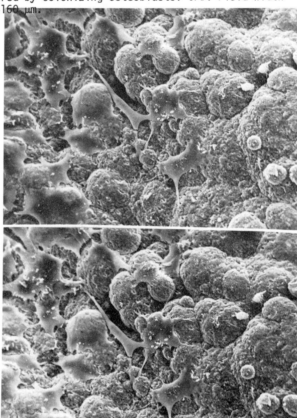

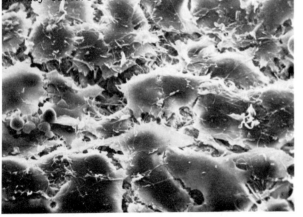

Fig.10: Confluent sheet of osteoblasts over mineralized front of dentine (same specimen as Fig.9). CPD. Field width 90 μm.

Fig.9: Osteoblasts were spreading across a rough anorganic dentine surface from lower left. Two day culture in DMEM and serum. CPD. Field of stereopair 180 μm.

varied with the substratum presented. The openings of vascular channels into the bone were up to 50 μm wide and were successfully bridged by one or more cells. Cross-cut or longitudinally cut dentine tubules (of 2 μm in diameter and at intervals of about 5 - 10 μm) were not heeded, although occasionally it appeared that a leading edge of the cell had lapped around a transversely fractured tubular opening. There was no evidence that the cells had extended processes down these holes even in cases where the locations of holes or semicylindrical grooves in the dentine could be determined below thin parts of the cells due to backscattered electrons from the dense substrate (Fig.16).

Effect of the constitution of the overlay.

The composition of the overlay made no difference to whether or not it was colonized by the osteoblasts. All overlays were populated, whether uncalcified, as was the osteoid, or almost entirely inorganic, as were the untreated enamel or anorganic bone, dentine, cement and enamel. The osteoblasts seemed not to be concerned by the substance that supported them: in one case a rat scalp hair that rested on the surface was also coated with cells.

Migration of osteoclasts on to overlays.

There were a few cells on the overlays that were similar in size and shape to osteoclasts as observed by SEM in situ on bone (Fig.20). One (Fig.21) was proved to be multinucleated by dissecting it off the enamel substrate using a micromanipulator and collecting it, in the SEM, on a coverslip in a drop of glycerol which contained some methylene blue. This was then removed from the SEM and examined by light microscopy.

Discussion

It is apparent from these experiments that migratory osteoblasts will coat any firm substrate presented in vitro, whether these substances are likely to be encountered and populated in vivo, such as enamel and calcified connective tissues, or unnatural substrates such as glass[7] or hairs. The lack of specificity for, or tolerance of, all the various substrates can explain such phenomena in vivo as the deposition of bone, as coronal cementum, on enamel when the excluding layer of reduced enamel epithelium is breached, and the reattachment of fibres of the periodontal ligament via the deposition of cementum (bone on a tooth) to scaled or planed root surfaces. It is not known, however, whether osteoblasts will coat a substrate that is not firm, nor what conditions are necessary for bone formation to follow on from their establishment as a continuous sheet. The local factors that control the migration of differentiated osteoblasts and their function may not be those that encourage the differentiation of new osteoblasts. However, it is certainly unnecessary to provide substrates of a particular texture or smoothness in an effort to establish a bone cell layer. The dimensions of the substrate particles would be critical at the lower end of the range of sizes, since cells have a stretch-dependency for cell division (fibroblasts[8,9]; osteoblasts[5]).

The present results provide additional weight to the suggestion that osteoblasts may migrate in vivo when the continuity of the cell layer is

disrupted[7]. They also show that new bone formation on implants of whatever nature in vivo is unlikely to be influenced by the ability of osteoblasts to coat the substrate. If this is the case, then the difference in the capabilities of such implants to support new bone formation must be solely related to the factors controlling the provision and survival of osteogenic precursor cells and their differentiation into osteoblasts. The survival of the osteoblast precursors will depend upon problems of toxicity and antigenicity. The provision of these cells is likely to be directly related to the local environment that stimulates angiogenesis, the vascular penetration of the implant and stem cell invasion. The observations on osteogenesis within semi-enclosed substrates such as compact cortical bone[2,4] or tooth [5], where decalcification seems to be necessary for success, suggest that the nutritional levels are affected by diffusion through the matrix. The demineralized inductive bone matrix[2] can only be effected if appropriate stem cells are present [10]. Macerated bone[11,12] may be a successful host for new bone formation for the same reason: the partly anorganic bone would be more permeable than intact bone matrix.

It was not useful to measure the rates of population of the different substrates because they were of such varied forms, and the ease and extent of access to the overlay from the cell sheet of the calvarium differed. No obvious differences were noted, however, apart from those cultures in which serum was absent. Here cells did transfer, but to a small extent: this may have been due merely to residual capacity from the in vivo condition of the cells which would be used up early in the culture period. Altering the hormonal or nutritional levels of serum-supplemented media changed the form of the cells but not their ability to populate the overlays.

The positive identification of osteoclasts that have migrated on to new substrates is not easy because some osteoclasts may not be very much greater than extended, migratory osteoblasts. The form of the cells can be very different but identification then depends on the skill and experience of the investigator. Perhaps only two morphological characteristics are fool-proof: the presence of a resorptive organ, the ruffled border, together with the possession of more than one nucleus. Resting osteoclasts do not have ruffled borders, however. The demonstration by SEM of a ruffled border in an actively resorbing cell necessitates upturning the cell from its substrate [13]. We have recently developed simple methods of demonstrating the number of nuclei in an osteoclast in situ on bone substrates. It is sometimes possible to identify the nuclei of loosely attached cells in BSE images. Caution is required in designating any multinucleate cell derived from a mixed cell population as an osteoclast[14], even when the occurrence of these cells seems to be dependent upon the presence of a bony substrate.

In conclusion, we believe that the organ culture, specimen preparation and SEM examination procedures outlined here constitute a valuable, simple and rapid means of testing the non-immunological aspects of the problems of biocompatibility of potential new bone prosthetic implant materials.

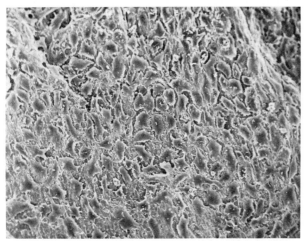

Fig.11: Confluent sheet of osteoblasts over fractured vertical enamel surface. The elongation
ratio of the cells mimics that of osteoblasts in
vivo. One day culture in DMEM and serum. CPD.
Field width 435 µm.

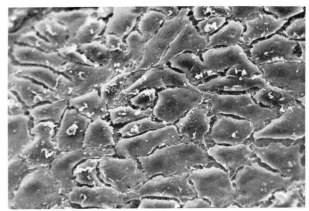

Fig.12: Same specimen, showing osteoblasts on
smooth external enamel surface. Ruffling is
present on the dorsal cell surface. CPD. Field
width 180 µm.

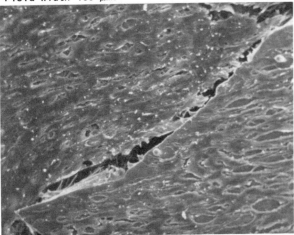

Fig.13: Three day culture in DMEM and serum and
PTE. Cells on the tooth slice overlay (top right)
and the bone surface are elongated and parallel.
CPD. Field width 165 µm.

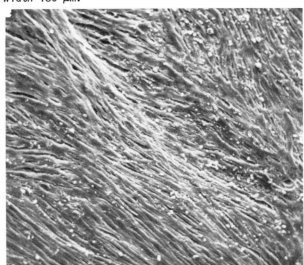

Fig.14: Three day culture of osteoblasts on bone
in DMEM and serum and PTE: the cells are elongated and parallel over the entire surface. CPD.
Field width 430 µm.

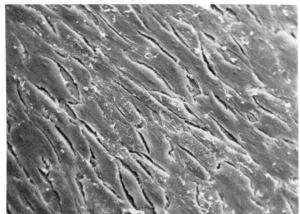

Fig.15: Three day culture in Hank's MEM and serum
of osteoblasts with dentine overlay. The cells
look similar to those of PTE-treated cultures.
CPD. Field width 190 µm.

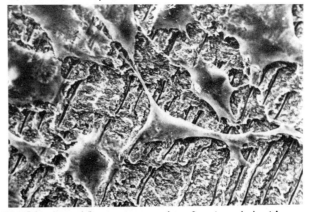

Fig.16: Osteoblasts traversing fractured dentine
surface: two day DMEM and serum. Note that the
fractured dentine canals and texture of the
matrix can be "seen" through the cells. CPD.
Field width 140 µm.

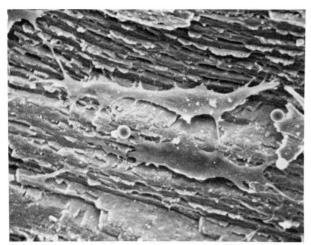

Fig.17: Three day culture in DMEM, serum and PTE. Elongated osteoblasts migrating along grain of fractured enamel. CPD. Field width 85 μm.

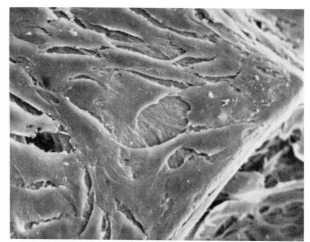

Fig.18: Cells on bone overlay are aligned parallel to the edges. Two (bottom right) span a vascular opening. Three day DMEM, serum and PTE culture. CPD. Field width 160 μm.

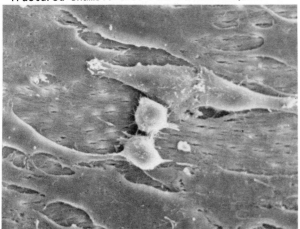

Fig.19: Same specimen as Fig.18. Elongated osteoblasts are more sparse in centre of overlay and lie parallel to the collagen fibre orientation. One cell spans an empty lacuna. Mitosing cells are also present. CPD. Field width 80 μm.

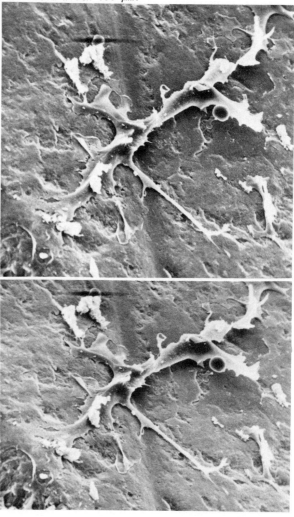

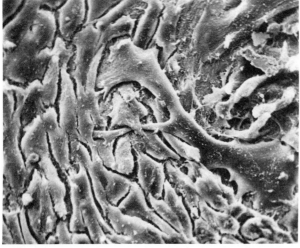

Fig.20: Osteoclast on bone surface immediately following dissection. CPD. Field width 105 μm.

Fig.21: Osteoclast that had migrated from bone on to an enamel overlay. One day culture in DMEM and serum. CPD. Stereopair: field 90 μm.

References

1. S.Nade and R.G.Burwell, Decalcified bone as a substrate for osteogenesis. J.Bone and Joint Surg. 59B, 1977, 189-196.
2. M.R.Urist, T.A.Dowell, P.H.Hay and B.S.Strates, Inductive substrates for bone formation. Clin. Orthop.Rel.Res. 59, 1968, 59-96.
3. S.M.Tuli and A.D.Singh, The osteoinductive property of decalcified bone matrix. J.Bone and Joint Surg. 60, 1978, 116-123.
4. J.A.Bombi, D.Ribas-Mujal and J.Truetta, An electron microscopic study of the origin of osteoblasts in implants of demineralized bone matrix. Clin.Orthop.Rel.Res. 130, 1978, 273-284.
5. A.H.Reddi, In: Advances in Biological and Medical Physics, Eds. L.H.Lawrence and J.W.Gofman. Academic Press, London, 1974, pp.1-18.
6. A.H.Reddi, Changes in collagen and proteoglycan types during matrix-induced endochondral bone differentiation. Abstr. 3rd Int.Workshop on Calc. Tiss., Kiriat Anavim, Israel, 1978.
7. S.J.Jones and A.Boyde, The migration of osteoblasts. Cell Tiss.Res. 184, 1977, 179-193.
8. N.G.Maroudas, Anchorage dependence: correlation between amount of growth and diameter of bead, for single cells grown on individual glass beads. Exp.Cell Res. 74, 1972, 337-342.
9. J.Folkman and H.P.Greenspan, Influence of geometry on control of cell growth. Biochimica et Biophysica Acta 417, 1975, 211-236.
10. A.J.Khan and D.J.Simmons, Chondrocyte-to-osteocyte transformation in grafts of perichondrium-free epiphyseal cartilage. Clin.Orthop.Rel. Res. 129, 1977, 229-304.
11. R.Maatz and A.Bauermeister, A method of bone maceration, J.Bone and Joint Surg. 39A, 1957, 153-166.
12. R.Salama and S.L.Weissman, The clinical use of combined zenografts of bone and autologous red marrow. J.Bone and Joint Surg. 60B, 1978, 111-115.
13. S.J.Jones and A.Boyde, Some morphological observations on osteoclasts. Cell Tiss.Res. 185, 1977, 387-397.
14. A.J.Altman, J.G.Bandelin, J.H.Domingues and G.R.Mundy, Differentiation of isolated calvarial cells into a mature heterogenous bone cell population in culture. Metab.Bone Dis.and Rel.Res. 1, 1978, 75-79.

Discussion with Reviewers

R.E.Weiss: One must be extremely careful in drawing conclusions based on the addition of PTE to culture medium and observing cell morphology. The heterogeneity of this extract along with non-specific responses (also seen with serum supplemented Hank's Medium) leave one with little to say about the specific effect of PTE on osteoblasts.
Authors: We don't know what it is in (Lilley) PTE that changes the osteoblast shape, but we do know that (Lilley) PTE reproducibly causes elongation of osteoblasts (S.J.Jones and A.Boyde, Experimental study of changes in osteoblastic shape induced by calcitonin and parathyroid extract in an organ culture system, Cell and Tissue Res. 169, 1976, 449-465; S.J.Jones and A.R.Ness, A study of the arrangement of osteoblasts of rat calvarium cultured in medium with, or without added para-thyoid extract, J.Cell Sci. 25, 1977, 247-263). The purpose of using PTE in the present context was as a tool to cause cell elongation and to determine whether this influenced the colonization of substrates - it was not to study the effect of parathyroid hormone. We concluded that there was no evident effect of elongation (due to PTE or Hank's MEM) on colonization.

R.E.Weiss: Do you have any evidence to rule out the possibility that the presumptive osteoclast you describe (Fig.21) is in fact a fusion product of two, or more, osteoblasts?
Authors: No, we do not: except that we have never seen osteoblasts looking like this.

R.E.Weiss: Have you tested osteoblasts from diabetic subjects or osteosarcoma cells to see if they exhibit the same ability to migrate and colonize regardless of the substrate?
Authors: No, we have not.

R.E.Weiss: Your report on the migration and colonization of osteoblasts on a variety of hard tissue substrates raises an important point: In order for these cells to migrate they must be able to attach to the substratum. In part, this can be due to an electrical charge attraction between the substrate and the cell surface, but do you have any evidence (either biochemical or otherwise) to implicate the role of a "fibronectin-like" component in this process? Your results in the absence of serum might reflect this possibility.
Authors: Recapitulating, migration did occur in the absence of serum, although to a markedly reduced extent. Unfortunately, we have no data from other types of study.

K.& Y. Draenert: What was the size of the rat calvarial fragments?
Authors: The cultured calvarial bones were mostly 4 to 7 mm across, square or oblong in shape. The overlaid, denuded and devitalized bone fragments were about 250 μm to 1 mm across, and of various shapes (see Fig.7).

K.& Y. Draenert: Do you have any ideas that would explain the edge effect?
Authors: The cells may possibly align themselves with the nearly right-angled edges of the substrates because this position allows the cytoskeletal elements, particularly the microfilament bundles, to be in a position of least strain. The work of G.A.Dunn and J.P.Heath (A new hypothesis of contact guidance in tissue cells, Exp.Cell Res. 401, 1976, 1-14) and Y.A.Rovensky and I.L.Slavnaya (Spreading of fibroblast-like cells on grooved surfaces, Exp.Cell Res. 84, 1974, 199-206) with fibroblasts is relevant to this point. However, whereas these authors argued that the filament-bundle-strain problem might be the explanation why fibroblasts would not cross sharply radiussed edges, we have shown here that osteoblasts do when they approach the edge in close array.

M.M.Smith: Could you explain how you establish that the cells are fully differentiated osteoblasts prior to translocation of the cells onto the overlay : is it on their elongation ratio and dorsal ruffling?

Authors: The single layer of cells in intimate contact with the bone matrix surface of the endocranial aspect of the rat calvarium are producing that matrix, and are therefore osteoblasts (S.J. Jones, Secretory territories and rateof matrix production of osteoblasts. Calcif.Tiss.Res. 14, 1974, 309-315). These are the only cells, apart from osteoclasts, that are present on a good preparation of the type which we have used. There can be no doubt that these are the cells which migrate off the bone and on to another substrate.

M.M.Smith: You suggest in a previous paper (Jones & Boyde, 1977) that reduction in synthetic activity of osteoblasts may lead to a decrease in cell-substrate adhesivity and suggest that mobility of cells may be increased. Do you think that the procedure for preparation of the osteoblasts as explants would predispose the cells to locomotion, and if so, would all cells be equally affected?

Authors: It is likely, as you imply, that the rate of synthetic activity of these osteoblasts would be reduced under our culture conditions. We do not know yet if reduced synthetic activity is linked with reduced adhesiveness of these cells - that is still speculative. But there is circumstantial evidence suggesting that less active osteoblasts are less adhesive. For example, from our own experience, we recognize that osteoblasts (surface osteocytes, if you prefer) near resorptive areas, and therefore likely to be the less active cells, are more easily washed away during specimen processing. The osteoblasts in the rat endocranial system that are known to be active are always well attached. You may, therefore, be correct in supposing that the procedure of culturing the osteoblasts would increase their migratory tendencies, and we might suppose that most of the osteoblasts in the endocranial sheet, which appear to be alike, would behave equally.

M.M.Smith: Do you consider that it is important to distinguish between the mobility of fully differentiated osteoblasts and that of osteoblast progenitor cells?

Authors: We believe that our dissection procedure is adequate to reduce the osteoprogenitor cell number to an insignificantly small proportion of the population which might colonize the substrates. Therefore, it is not that we are not interested in osteoprogenitors, but we have excluded them from the system. It should further be borne in mind that these osteoblasts can mitose, and do so markedly after release from treatment with PTE in our culture system (S.J.Jones and A.Boyde, Experimental study of changes in osteoblastic shape induced by calcitonin and parathyroid extract in an organ culture system. Cell Tiss.Res. 169, 1976, 449-465).

M.M.Smith: Have you any idea what proportion of the active osteoblasts become osteocytes and therefore what percentage become resting osteoblasts susceptible to changes inducing locomotion?

Authors: The number of cells incorporated as osteocytes obviously increases as the bone thickness

increases. Assuming a volume proportion of about 4% for the osteocytes in the neonatal/young rat endocranially formed bone, that the layer of osteoblasts make up to $\simeq$ 100 µm in thickness of bone, and that the mean surface covered by each cell is 154 µm², then 100 osteoblasts would overlay 1,540,000 µm³ of bone. A 20 x 10 x 5 µm three-axis ellipsoid (the osteocyte lacuna) has a volume of $\simeq$ 520 µm³, so that there are $\simeq$ 118 cells in this volume, i.e. we calculate that as many as 118% of the osteoblasts may eventually be incorporated as osteocytes. Therefore, 100% replacement (by mitosis or recruitment from osteoprogenitors) must have occurred by the time that 85 µm thickness of bone had been formed by a layer of osteoblasts, assuming that all parameters above are correct and remain constant. Now, $\simeq$ 85 µm of this young bone would take $\simeq$ 21 days to construct, so that each day $\simeq$ 5% of the cells would be incorporated in the bone matrix. By the time that one half of the cell was incorporated (that is, its lacuna was half-formed), it would probably be covered by neighbouring osteoblasts: that is, after 2½ µm of bone (a little more than one half-day's production). We might, therefore, predict that about 2-3% of the cells visible in the osteoblast sheet are about to be included in the matrix.

M.M.Smith: Will it be possible to predict from the procedures described here what factors might induce active bone formation after colonization of the substrata?

Authors: It would be possible to remove the cells from the substrata and examine whether they have deposited bone matrix on the colonized substrate, but estimates of the thickness and volume of this new matrix would be better derived from sectioning techniques for TEM or light microscopy. We know that they make a matrix when they colonize glass (S.J.Jones and A.Boyde, SEM of bone cells in culture, pp.97-104 In: D.H.Copp and R.V. Talmage, eds., Endocrinology of calcium metabolism. Exerpta Medica Int.Congr.Ser.No.421,Elsevier /North Holland Inc., Amsterdam, 1978), but we have not studied matrix production on enamel, dentine and bone as yet.

S.C.Miller: This study takes advantage of an interesting observation made by the author that osteoblasts will apparently leave their natural substrate and migrate onto other, even foreign, substrates. This is certainly a unique opportunity to study cell-substrate interactions and leaves one wondering why an osteoblast would want to wander away from its natural environment. In any case, there are some interesting questions raised by this observation.

One aspect of the work that I am a little uneasy about is the assumption that these cells might be properly called osteoblasts which have migrated off the bone. In the classical sense an osteoblast is a functional cell secreting collagen and regulation mineral deposition (see specific question below). It may be possible that these migratory cells are 'pre-osteoblasts' or even post-osteoblast "osteoprogenitor cells" (as proposed by Young). These apparent precursors to osteoblasts are not active in the secretion of

collagen (at this time they would be called osteoblasts) and are also migratory in vivo. It seems to
me very essential to eventually establish the functional capabilities of these cells before they are
strictly classed as a given cell type.
     Is there any evidence that the cells which
migrate onto the foreign substrates retain or regain
functional capabilities of osteoblasts, for example,
collagen synthesis or mineral deposition?
Authors: This question is similar to one raised by
M.M.Smith which we have answered above. There are
very few cells other than active osteoblasts on the
endocranial surface of our preparations. We really
show that osteoblasts are osteoprogenitor cells in
that they can beget more osteoblasts by mitosis (see
references in answer to Smith).

S.C.Miller: The developing calvaria contains some
red marrow between the diploe. Could it be possible
that some of the larger cells observed in this study
might be marrow derived macrophages and giant cells?
Certainly these types of cells develop very rapidly
in culture, especially in the presence of serum.
The authors have, however, pointed out that these
distinctions would be difficult to make using the
limited techniques employed in this study.
Authors: There is no red marrow, there are no diploe
in rat skulls of the age that we study. In another
paper in this volume (I.M.Shapiro, S.J.Jones, N.M.
Hogg, M.Slusarenko and A.Boyde) we show that macrophages on bone do not resemble osteoblasts and
neither do osteoclasts.

S.C.Miller: I also have reservations as to the
authors claim that this procedure is a way to test
biocompatibility of implant material. Certainly
the system could eventually be used for these types
of determinations, but since the cells coated all
substrates and foreign materials (even hair), the
real diagnostic value of the test eludes me as no
negative controls were demonstrated. If the cells
will always coat a foreign material, then what is
the value of the test?
Authors: We only claim that it tests one aspect of
the complex of biocompatibility, which we might
express as acceptability to osteoblasts. Although
we have not tried the experiment, we believe that
toxic substances would soon be sorted out via this
approach.

SCANNING ELECTRON MICROSCOPY/1979/II
SEM Inc., AMF O'Hare, IL 60666, USA

USE OF SEM FOR THE STUDY OF THE SURFACE RECEPTORS OF OSTEOCLASTS IN SITU

I. M. Shapiro[**], S. J. Jones, N. M. Hogg[*], M. Slusarenko[*] and A. Boyde

Department of Anatomy
[*]Department of Zoology
University College London
Gower Street
London WCIE 6BT, England

Abstract

The monocyte macrophage has been suggested as a possible precursor to the osteoclast. In an attempt to determine if these two distinct cell types are related, a study was conducted to ascertain whether Fc and C3 receptors, which are characteristically present on the macrophage cell surface, could also be demonstrated on bone cells. Using SEM as a means of detecting sensitized red blood cells rosetting to cell surface ligands, the presence of Fc and C3 receptors on macrophages grown in culture and adherent to both glass cover slips or freeze-thawed cell-free surfaces of bone was noted. The receptors on the macrophages could still be detected by the assay following freeze drying and glutaraldehyde fixation, but the results indicated that the Fc was less stable than the C3 receptor.

Similar immunological techniques were applied to endocranial bone cells in situ and SEM was used to identify osteoclasts and osteoblasts. The rosetting assay was performed immediately after dissection, or following culture for 1 day in medium in the presence or absence of serum. However, no rosettes were observed. The results of these experiments speak against a common ancestry of osteoclasts and macrophages. In addition, the absence of any rosette-forming cells on bone suggests that under normal growth conditions, macrophages are not members of the cell population of the bone surface.

[**]On study leave from the Department of Biochemistry, School of Dental Medicine, Univ. of Pennsylvania, Philadelphia, PA.

KEY WORDS: Fc and C3 Receptors, Osteoclasts, Bone, Macrophages

Introduction

Despite the characteristic morphology of the osteoclast and its reactivity to a number of physiological stimuli, ideas concerning its origin remain divided and the geneology of the cell line is disputed[1-5]. Evidence from studies of bone resorbing systems suggests that macrophage precursors may comprise an important source of osteoclasts[6-16]. If this concept is valid and osteoclasts are related to macrophages through a common ancestor, then the Fc and C3 receptors that are present with high frequency on macrophages[17] might also be present on osteoclasts[*].

From an experimental point of view, the conventional immunological methods that are used to study surface receptors on isolated cells or cell lines cannot be applied to the cells of bone. In this tissue a mixed cell surface population exists on an optically opaque substrate and the pro-osteoclast cannot be differentiated morphologically from other cells.

SEM has been used successfully to identify osteoclasts and osteoblasts of isolated calvaria in situ[18,19]. This morphological technique obviates the need for methods, such as trypsinization, harvesting and cell culture, that could damage or remove membrane proteins. Furthermore, osteoclasts cannot be characterized morphologically after the latter procedures. It should be emphasized that the great advantage of SEM of intact bone fragments is that the cells are maintained within their normal environment. Their morphology can also be matched with that of the underlying bone matrix by dissecting off the individual cells[20].

In this report, we describe the application of SEM and immunological techniques for the study of surface receptors on positively identified bone cells and macrophages. In addition to providing information on the occurrence of these receptors on bone cells, we have used the receptors to assess the relationship between macrophages and osteoclasts.

***

[*] The Fc receptor binds the Fc portion of the IgG molecule, the C3 receptor binds fragments of the third component of complement.

Methods

Isolation of Macrophages

Male Balb/C mice were injected intraperitoneally with 2% starch. After 4 days, the peritoneal exudate was removed by flushing with 5 ml cold phosphate buffered saline. The cells were plated on to 13 mm glass cover slips at a density of 200,000 cells/cover slip. The non-adherent cells (lymphocytes and granulocytes) were removed after 12 hours by washing the cover slip with saline. Non-specific esterase staining[21] showed that better than 99% of the cells on the cover slip were macrophages.

Preparation of Calvarial Cells in situ for Receptor Assay

Half calvaria were excised from 2 or 3 day old Albino Wistar rats under anaesthesia and washed free of superficial blood in buffered Eagles medium at 37°C. The overlying connective tissue, the dura mater, was dissected off the endocranial aspect of the bone so that the surface cells were exposed as a single layer. It was known from previous studies that, at this age, with the exception of the sutural areas which are rich in osteoclasts, much of the cranial vault surface is covered with osteoblasts. Some calvaria were maintained briefly at 37°C in the medium prior to the Fc and C3 receptor tests. Others were cultured for 30 hours in 5% CO_2:95% O_2 in Dulbecco's modification of Eagle's medium, with and without 10% foetal calf serum, and then assayed for receptors.

Preparation of Macrophages on Bone

Half calvaria were excised from 2 day old rats under ether anaesthesia. All the overlying fibrous connective tissues were removed from the bone with fine forceps using a dissecting microscope. Particular care was taken to ensure that only bone cells remained on the endocranial surface. These cells were then swept away using a flexible silicone rubber wedge and the bone surface rinsed with fresh medium. The specimens were frozen in liquid nitrogen and thawed four times to ensure that any residual cells died. These devitalized cell-free bone surfaces formed the substrata for macrophages, prepared as described above except that the concave endosteal surfaces of the calvaria formed 'wells' for the attachment of the macrophages. The macrophages were plated at a density of 200,000 cells/half calvarium, and were then cultured for 24 hours before testing for surface receptors.

Fc and C3 Receptor Assays

Fc(EA) rosettes. 0.5 ml of a 10% solution of twice washed sheep red blood cells (SRBC) in 0.9% saline were added to 0.5 ml of a 1:100 dilution of rabbit anti-sheep red blood cell serum (Flow Labs Inc., Rockville MD USA Cat.No.8-744R) in 0.9% saline which was heat inactivated at 56°C for 30 min. This was incubated at 37°C for 30 mins and then washed twice with 5 ml 0.9% saline. The pellet (erythrocyte-antibody (EA)) was resuspended in 2 ml serum-free Minimum Essential Medium (MEM). A control suspension of 2.5% SRBC in 0.9% saline was also prepared (erythrocyte(E)).

Control peritoneal macrophages, grown overnight on glass coverslips and rat calvaria, were incubated with 200 μl of the EA or E suspensions for 30 mins at 4°. The samples were then washed three times in cold MEM and examined microscopically.

C3 (EAC) rosettes. 1 ml of a 5% solution of twice washed SRBC in 0.9% saline was added to 1 ml of a 1:500 dilution of rat anti-SRBC serum. The rat anti-SRBC serum was made by injecting old Sprague-Dawley rats intravenously with 10^9 SRBC and collecting serum four days later in order to ensure maximum IgM antibody levels. This SRBC - anti-SRBC serum mixture was incubated for 30 mins at 37°C. After one wash, with 5 ml 0.9% saline, the pellet was resuspended in 1 ml of 1:10 dilution of normal mouse serum, as a source of complement. This was incubated for 30 mins at 37° and then washed with saline before making up to 10 ml in cold MEM. These were the EAC (erythrocyte antibody complement) cells. Control EA cells were prepared by omitting the mouse serum. The incubation of EAC and EA cells with samples proceeded as outlined for Fc rosettes except that the incubation took place for 15 mins at room temperatures.

Preparation of Specimens for SEM

After the receptor assay, the specimens were fixed in 3% glutaraldehyde in 0.15 M cacodylate buffer, pH 7.2. Specimens to be critical point dried were washed in distilled water and dehydrated with graded ethanols. Freon 113 was substituted in steps for the absolute ethanol and the specimens were critical point dried from CO_2. Other specimens were, after fixation, quenched in liquid nitrogen cooled Freon 12, transferred to liquid nitrogen and freeze dried at -70°C (Edwards Speedivac-Pearse Tissue Dryer).

All specimens were coated with gold in a sputter coater equipped with a Peltier cooled cold stage (0°C) and magnetic electron deflectors (Polaron 5000) and examined in a Stereoscan S4-10 operated at 10 kV.

Effects of Fixation and Freeze Drying on Macrophage Receptors

Macrophages, established on 13 mm glass coverslips were either fixed with cacodylate buffered 3% glutaraldehyde for four days or quenched in liquid Freon 12 at its melting temperature, transferred to liquid nitrogen and freeze dried at -70°C. These cells were then rehydrated as they were subjected to the receptor assays and checked by light microscopy for rosetting.

Results

Demonstration of Rosetting on Macrophages using SEM Techniques

Macrophages on glass. The presence of rosetting of red blood cells (RBC) on macrophages can be reliably assessed by light microscopy when the cells are on a glass coverslip. Because of the opaque nature of the bone substratum, the glass rosettes were used as controls for time and conditions for the preparations of cells on bone. While the morphological characteristics of

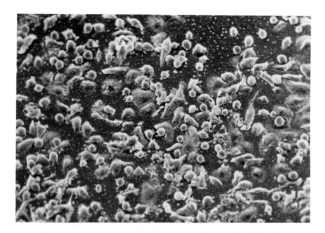

Fig. 1   Mouse macrophages (peripheral blood mono-cytes) grown on glass coverslip for 24 h.   Field width 410 μm.

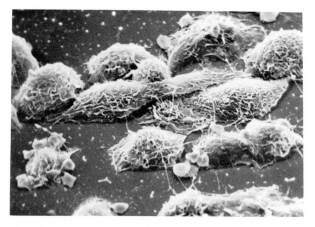

Fig. 2   Some sheep red blood cells (SRBC) lie amongst the macrophages, but no rosettes are present.   Field width 75 μm.

Fig. 3   Macrophages on glass following Fc recep-tor assay.   All cells have rosettes of SRBC attached.   Field width 400 μm.

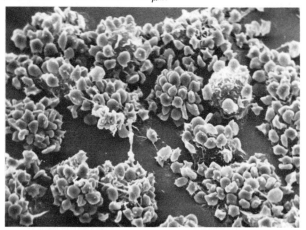

Fig. 4   Higher magnification of macrophages on glass showing rosetting.   Field width 80 μm.

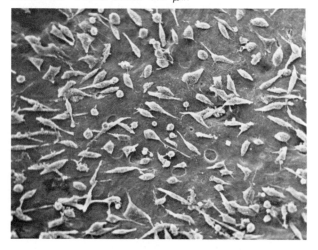

Fig. 5   Mouse macrophages grown on endocranial surface of rat parietal bone, previously swept free of cells.   24 h culture: the macrophages have attached and spread.   Field width 410 μm.

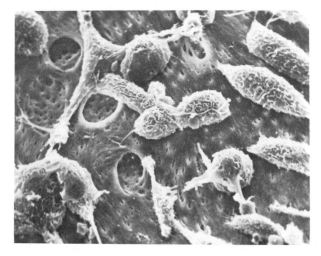

Fig. 6   Higher magnification of macrophages on bone.   The collagenous surface is penetrated by canalicular openings and contains empty, half-formed osteocyte lacunae.   Field width 80 μm.

macrophages will vary with different culture conditions, in these experiments the morphology of the cells was similar on glass and bone. The SEM morphology of the macrophages on glass is seen in Figures 1 and 2; figures 3 and 4 show the Fc rosetting of RBC on macrophages on glass. Rosetting is seen both after freeze drying and critical point drying. Hence RBC bound to macrophage cell surfaces are not displaced by either of these routine SEM preparative techniques.

Conversely, the receptor sites were still present when macrophages had been first fixed with glutaraldehyde or freeze dried and subsequently tested for Fc and C3 rosetting. Rosettes were formed on all macrophages tested for C3 receptors after fixation or freeze drying, but decreased RBC binding occurred in the Fc receptor assay. These results suggest that the C3 receptor is more stable than the Fc receptor of the cell membrane.

Macrophages on bone. Macrophages attach readily to bone matrix surfaces (Figs 5,6 & 7) within a 24 hour culture period and demonstrate the same morphological characteristics that they show when they have attached to glass (Figs 5-6). They can be easily distinguished from bone cells (Figs 8 & 9). Fc and C3 rosetting of RCB to macrophages on bone occurred to the same extent as to macrophages on glass (Figs 10 & 11). The results indicate that macrophages on bone contain receptors on their free surface. It was therefore concluded that the rosetting techniques could be applied to bone cells in situ to determine the presence or absence of similar receptor sites.

Examination of Bone Cells for Fc and C3 Receptors

Receptor assays were conducted on (1) freshly prepared calvaria, (2) calvaria cultured for 24 hours in medium, and (3) calvaria cultured for 24 hours in medium with added serum. Culturing and the addition of serum were used to stimulate the expression of cell surface receptors[22]. While the osteoclasts and osteoblasts could be reliably identified following all procedures, no rosettes were observed (Fig. 12).

Discussion

This study has shown that, during normal growth and remodelling, the surface cells of bone do not contain Fc or C3 receptors on their free surfaces. This finding is of particular relevance to the hypothesis concerning the origin of the osteoclast. Current experimental evidence supports the view that the osteoclast precursor is haematogeneous in origin and derived from the monocyte-macrophage cell line[7-10]. As osteoclasts and macrophages do not appear to express similar receptors, doubt must now be thrown on the putative common ancestry theory.

Failure to demonstrate receptors that are similar to those of the macrophage may be due to (1) conditions under which the experiment was performed and (2) incomplete knowledge of the life history of the osteoclast. Firstly, with respect to the conditions of the experiment, care was taken to ensure that, when grown on a bone surface and when subjected to SEM preparative procedures, macrophage Fc and C3 receptors would bind to antibody-coated red cells. However, it may be argued that the distribution of receptors on macrophages and osteoclasts differ . There is evidence to favour this view. Thus, bone forming and bone resorbing cells are polarized with respect to the bone surface; the possibility therefore exists that the receptors are present only on the surface facing the bone. Using the technique described in this paper, such receptors would not be available for binding such comparatively large markers as RBC . Experiments are now in progress to free individual osteoclasts from their matrix in order to perform the rosetting assay on the surface that has the ruffled membrane.

A second reason for failing to demonstrate receptors may be related to the dearth of knowledge concerning the osteoclast life history. For example, there is little information available describing membrane turnover kinetics and the possibility exists that receptor development may be transitory. In this case, receptors that may be present at an early developmental stage may be lost as the osteoclast matures. As fully differentiated and resorptive cells, they might have lost surface receptors that were present at an earlier time in their life history. If this is the case, then it might be expected that culturing the calvaria in the presence of serum would stimulate receptor formation. It should be noted that macrophages that have been cultured for extended periods exhibit an increase in the number of Fc and C3 receptors[22]. However, as this did not occur it must be presumed that the absence of receptors is not simply an age-dependent phenomenon. Of course, this observation does not exclude the possibility that the presence of osteoclast surface receptors is modulated by endocrinal factors. It is known that membrane ruffling and bone cell surface polarity are increased by parathyroid hormone[23]. Parathyroid hormone has the additional property of increasing osteoclast number, possibly by facilitating osteoclast recruitment from extra-osseous sources[4,23]. These newly recruited cells might have receptors that are lacking in the osteoclasts that are well established on the bone. Thus, before definitely concluding that macrophages and osteoclasts are not related, it will be necessary to perform receptor assays on osteoclasts newly recruited to bone following the administration of parathyroid hormone in vivo and in vitro.

Finally, a comment should be made concerning the importance of macrophages in physiological bone resorption. Mundy et al.[15] and Kahn et al.[16] have shown that monocytes will resorb bone in vitro; however, it was not established whether these cells were present on the bone surface under normal conditions in vivo. In this investigation, it was found that none of the cells on the endocranial surface had Fc or C3 receptors. It can therefore be assumed that there are no macrophages in the osteoblast plus osteoclast monolayer. This result strongly supports the view that macrophages do not play a direct role in normal bone turnover. Whether they resorb

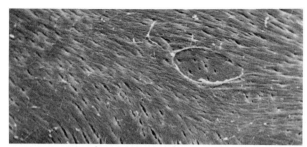

Fig. 7 Endocranial aspect of rat parietal bone from which all the surface cells have been removed mechanically. Field width 90 μm.

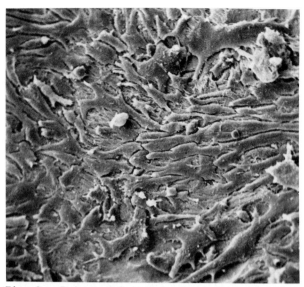

Fig. 8 Endocranial aspect of rat parietal bone, showing osteoblasts (the smaller cells) and osteoclasts (larger, irregular cells). Field 180 μm.

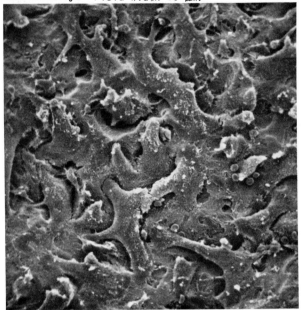

Fig. 9 Inner surface of the cranial vault of the newborn rat showing more extensive zones of osteoclasts. Field width 190 μm.

Fig. 10 SRBC rosettes on mouse macrophages cultured for 24 hours on the bone matrix surface and then assayed for Fc receptors. Field width 410 μm.

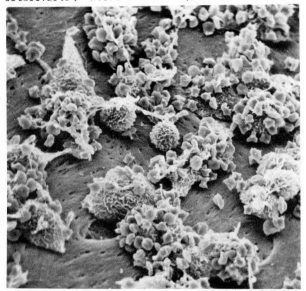

Fig. 11 Macrophages can be distinguished from bone cells by surface topography. Field width 80 μm.

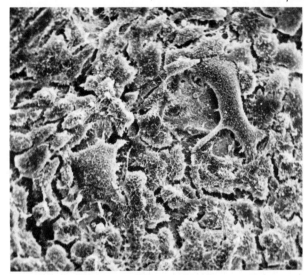

Fig. 12 Osteoclasts and osteoblasts maintained in organ culture in serum-supplemented medium for 24 hours and then tested in situ for Fc receptors. No rosettes are present. Field width 200 μm.

bone in pathological states has yet to be determined.

Acknowledgements

This study was supported by N.I.H. grant DE 02623, a Fogarty International Fellowship L F06 TWOO 274-01 (I.M.S.), the Leukemia Research Fund of Great Britain (N.M.H.), and the Science Research Council (M.S.). The authors wish to thank Miss Elaine Maconnachie for her help.

References

1. N.F.Kember, Cell division in endochondral ossification. A study of cell proliferation in rat bones by the method of tritiated thymidine autoradiography, J.Bone & Joint Surg. 42B, 1960, 824-839.
2. M.Owen, Cell population kinetics of an osteogenic tissue, 1, J.Cell Biol. 19, 1963, 19-32.
3. B.L.Scott, Thymidine-3H electron microscope autoradiography of osteogenic cells in the fetal rat, J.Cell Biol. 35, 1967, 115-126.
4. P.J.Bingham, I.A.Brazel, M.Owen, The effect of parathyroid extract on cellular activity and plasma calcium levels in vivo, J.Endocrinol. 45, 1969, 387-394.
5. E.A.Tonna, E.P.Kronkite, Use of tritated thymidine for the study of the origin of the osteoclast, Nature(Lond.) 190, 1961, 459-460.
6. W.S.S.Jee, P.D.Nolan, Origin of osteoclasts from the fusion of phagocytes, Nature(Lond.) 200, 1963, 225-226.
7. D.A.Fischman, E.D.Hay, Origin of osteoclasts from neuromuscular leucocytes in regenerating newt limbs, Anat.Rec. 143, 1962, 329-337.
8. K.Buring, On the origin of cells in heterotopic bone formation, Clin.Orthopaed.Rel.Res. 110, 1975, 293-301.
9. G.Gothlin, J.L.E.Ericksson, On the histogenesis of the cells in fracture callus, Virchows Arch.Abt.B.Zellpath. 12, 1973, 318-329.
10. S.J.Jones, A.Boyde, Experimental study of changes in osteoblastic shape induced by calcitonin and parathyroid extract in an organ cell culture system, Cell Tiss.Res. 109, 1976, 449-465.
11. D.G.Walker, Spleen cells transmit osteopetrosis in mice, Science 190, 1975, 785-787.
12. D.G.Walker, Bone resorption restored in osteopetrotic mice by transplants of normal bone marrow and spleen cells, Science 190, 1975, 784-785.
13. S.C.Luk, C.Nopajaroonsri, G.T.Simon, The ultrastructure of endosteum, a topographic study in young adult rabbits, J.Ultra.Res. 46, 1974, 165-183.
14. A.S.Kahn, D.J.Simmons, Investigation of cell lineage in bone using a chimaera of chick and quail embryonic tissue, Nature(Lond.) 258, 1975, 325-327.
15. G.R.Mundy, A.J.Altman, M.D.Gondek et al., Direct resorption of bone by human monocytes, Science 196, 1977, 1109-1111.
16. A.J.Kahn, C.C.Stewart, S.L.Teitelbaum, Contact-mediated bone resorption by human monocytes in vitro, Science 199, 1978, 988-989.
17. H.Huber, M.Weiner, Binding of immune complexes to human macrophages: the role of membrane receptor sites, In: Activation of Macrophages, W.Wagner, H.Hahn, R.Evans (eds.), American Elsevier Publishing Co.Inc., New York, USA, 1974, p.54-75.
18. S.J.Jones, A.Boyde, Some morphological observations on osteoclasts. Cell Tiss.Res. 185, 1977, 387-397.
19. S.J.Jones, A.Boyde, Morphological changes in osteoblasts in vitro, Cell Tiss.Res. 166, 1976, 101-107.
20. S.J.Jones, A.Boyde, J.B.Pawley, Osteoblasts and collagen orientation, Cell Tiss.Res. 159, 1975, 73-80.
21. I.R.Koski, D.G.Poplack, R.M.Blaese, A monospecific esterase stain for the identification of monocytes and macrophages, In: In vitro Methods in Cell Mediated and Tumor Immunity, B.R.Bloom, J.R.David (eds.), Academic Press, New York, USA, 1976, 359-362.
22. J.Rhodes, Macrophage heterogeneity in receptor activity, The activation of macrophage Fc receptor function in vivo and in vitro. J.Immunol. 114, 1975, 976-981.
23. M.E.Holtrop, L.G.Raisz, H.A.Simmons, The effects of parathyroid hormone, colchicine, and calcitonin on the ultrastructure and the activity of osteoclasts in organ culture, J.Cell Biol. 60, 1974, 346-355.

Discussion with Reviewers

S.C.Marks,Jr.: Have you been able to compare cell surface receptors of osteoclasts with a less differentiated cell of the macrophage lineage such as a circulating monocyte or resident (as opposed to elicited) peritoneal macrophage? Do your results really argue against a common ancestry for osteoclasts and macrophages? Is there evidence that cells derived from a common precursor must express similar surface receptors? The evidence from lymphocytes, for example, favours the view that surface receptors are a manifestation of function. The functional differentiation of osteoclasts and macrophages from a common progenitor, if one exists, could be sufficiently divergent to produce different surface receptors.
S.C.Miller: There is an assumption in this study that the surface receptors remain relatively constant throughout the differentiation pathway of the monocyte-macrophage line. Has in fact this been established for the known pathways of differentiation of this cell line? For example, are there differences in surface receptors in the pathway of promonocytes to monocytes to macrophages to epitheloid cells to giant cells? Since the osteoclast morphologically and functionally resembles a foreign body giant cell, at least more than a peritoneal macrophage does, wouldn't it be better to compare the surface receptors of a macrophage-derived foreign body giant cell to an osteoclast than a macrophage?
Authors: Fc and C3 receptors are present on both monocytes and on some resident peritoneal macrophages. Recent studies have indicated that receptors are also present on macrophage-derived giant cells (J.M.Papadimitriou, T.A.Robertson and M.N-I.Walters, An analysis of the phagocytic potential of multinucleate foreign body giant cells, Am.J.Path. 78, 1975, 343-358; T.J.Chambers, Studies of the phagocytic capacity of macrophage

polykaryons, J.Path., 123, 1977, 65-77), fully differentiated microglial cells (N.M.Hogg and M. Slusarenko, unpublished observations), Langerhan's cells (G.Stingl, E.C.Wolf-Schreiner, W.J.Pichler, F.Gschnait, W.Knapp and K.Wolf, Epidermal Langerhans cells bear Fc and C3 receptors, Nature(Lond.) 268, 1977, 245-246), and liver (Kupffer) cells (H.Huber, S.D.Douglas and H.H.Fudenberg, The IgG receptor: an immunological marker for the characterization of mononuclear cells, Immunol. 17, 1969, 7-21). These latter cell types are also considered to have common ancestry with the macrophage.

As far as promonocytes are concerned, it has been shown that receptors are present not only on promonocytes, but also on monoblasts, the earliest and most immature cell recognizable in mononuclear /phagocytic colonies (T.J.L.M.Goud, C.Schotte, R.van Furth, Identification and characterization of the monoblast in mononuclear phagocyte colonies grown in vitro, J.Exp.Med. 142, 1975, 1180-1199). Hence it can be concluded that the Fc and C3 receptors are present on the primitive as well as the most differentiated cells of the mononuclear series.

Of course, our results do not negate the possibility of a common ancestry for monocytes/macrophages and osteoclasts because, as we have noted in the discussion, receptors that are present at one early developmental state may be lost as the osteoclast matures. Arguably, the divergence of the monocyte/macrophage line from the osteoclast line could have occurred prior to the monoblast stage and before the Fc and C3 receptors are expressed.

S.C.Miller: Is there any logical reason to believe that a functional osteoclast may need immunoglobulin or complement receptors on its surface?

Authors: We do not know whether the presence of Fc and C3 receptors would be advantageous, but it is possible that these receptors would facilitate phagocytic activity which might be a functional asset, particularly in resorption associated with inflammatory processes.

P.A.Farber: Did the authors do blocking experiments in which macrophages were first incubated with serum prior to incubation with the treated red cells?

Authors: The macrophages had been incubated with serum prior to the Fc assay. In addition, we have cultured calvaria for 24 hours in medium supplemented with serum prior to the Fc test. In this latter case, no rosetting was seen on the calvarial bone surface cells. We must point out that the reason for culturing the bone in the presence of serum was to enhance the expression of receptors (see text, ref. 22).

P.A.Farber: Why did the authors compare macrophages from mice with osteoclasts from rats? Wouldn't a comparison of these two cell types from the same species be more appropriate?

Authors: This question addresses itself to the specificity of the rosette assay. As far as plasma membrane receptors for the Fc portion of the IgG molecule are concerned, there is little species specificity. Antibodies raised in rabbits to sheep erythrocytes can be used to demonstrate the presence of macrophage Fc receptors in hamster, guinea-pigs, mice and humans. We have also tested

rat macrophages for Fc receptors using the same system and, as might be expected, they also rosetted. In this study, we merely used the mouse macrophage to show that the rosette assay was working; we did not compare the abilities of mice macrophages and rat osteoclasts to form rosettes with antibody-covered RBC.

J.A.Eurell: Are there any other cell types besides cells of the same ancestry as macrophages that you might consider investigating as osteoclast precursors?

Authors: Osteoprogenitor cells, osteoblasts, and osteocytes have all been implicated as osteoclast precursors. Some workers have even suggested that osteoblasts and osteoclasts are modulations of the same cell type (R.W.Young, Cell proliferation and specialization during endochondral osteogenesis in young rats, J.Cell Biol. 14, 1962, 357-370; H.Rasmussen and P.J.Bordier, The physiological and cellular basis of metabolic bone disease, Williams & Wilkins, Baltimore, USA, 1974). Both TEM and SEM studies, however, indicate that neither osteoblasts nor osteocytes fuse with osteoclasts. Osteocytes rather protect their territory from osteoclastic action (text ref. 12: A.Boyde, SEM studies of bone, In: The Biochemistry and Physiology of Bone, G.H.Bourne (ed.), Academic Press Inc., New York, USA, 1972, pp.259-310). One current view is that it is most likely that the osteoclast precursor belongs to the hemopoietic stem cell series rather than the bone stromal stem cell series (W.S.S.Jee and D.B. Kimmel, Bone cell origin at the endosteal surface, In: Bone Histomorphometry, P.J.Meunier (ed.), Armour Montagu, 75017 Paris, France, 1976; M.Owen, Histogenesis of bone cells, Calc.Tiss.Res. 25, 1978, 205-207). We must also take note of work implicating cells of the lymphoid series, both thymus and spleen derived, in restoring deficient osteoclastic function (G.Milhaud, M.L.Labat, B.G. Graf and M-J.Thillard, Guerison de l'osteopétrose congénitale du rat (op) par greffe de thymus, C.R.Acad.Sci.Paris D. 283, 1976, 531-533; S.C. Marks Jr., Osteopetrosis in the ia rat cured by spleen cells from a normal littermate, Am.J.Anat. 146, 1976, 331-338; S.C.Marks and G.B.Schneider, Evidence for a relationship between lymphoid cells and osteoclasts: Bone resorption restored in ia (osteopetrotic) rats by lymphocytes, monocytes and macrophages from a normal littermate, Am.J.Anat. 152, 1978, 331-342). However, these studies do not indicate whether the lymphoid cells differentiate directly into osteoclasts themselves or activate existing non-functional osteoclasts, or cause other cells to differentiate into osteoclasts. Experiments are currently under way to investigate the role of lymphoid cells in the resorption process.

Since submitting our manuscript, T.J. Chambers (St Bartholomew's Hospital Medical College, London) has sent us a copy of his paper in press in the Journal of Pathology, entitled "Phagocytosis and trypsin resistant glass adhesion by osteoclasts in culture". He also reports no Fc or C3 receptors in osteoclasts derived from rabbit bone using collagenase.

Additional discussion with reviewers of the paper "Structural Convergences Between Enameloid of Actinopterygian Teeth and of Shark Teeth" by W.-E. Reif, continued from page 554.

W.L. Davis, R. Jones and J.L. Matthews: Are there other lines of evidence other than the enameloid to support the precept that these species had a common ancestor?
Author: All recent discussions of teleost systematics regard the families Sphyraenidae and Characidae as natural groups, i.e. ichthyologists assume that each of the two families has a common ancestor not shared with any other fish species. Compagno (Amer. Zool. 17, 1977, 303-322) bases his extensive discussion of the phyletic relationships of living sharks and rays on a large number of anatomical characters. His conclusions concur with my own, namely that the euselachians (sensu 21; which he calls "neoselachians") form a monophyletic group and share a common ancestor. The significant contribution of my "enameloid method" is, that incomplete fossil skeletons or even isolated fossil teeth can be identified as belonging to the euselachians. Problems arise with the rays, which Compagno includes in the "neoselachians". If this is correct, it would mean that the rays secondarily evolved rather simple and often woven enameloid structures, and in many genera reduced the thickness of the enameloid cap (21). With the enameloid structure alone, fossil ray teeth would be very difficult to identify.

W.L. Davis, R. Jones and J.L. Matthews: As the crystals described are extraordinarily large, is it reasonable to attribute their nucleation and formation to the matrix collagen? What type of collagen is this? Isn't some other factor a more likely candidate? What plays this role in enamel?
Author: These are very important questions, but I think the problems of crystal nucleation have not yet been solved in enamel or enameloid. The mineralization of enameloid seems to be of a different type from the mineralization of dentine and bone, because, as mineralization occurs collagen fibers revert to a labile form and become extruded as crystal growth takes place (15, 20; Poole & Gwinnet, J. Dent. Res. 48, 1969, 1119). TEM pictures, however, show that the crystals have exactly the same orientation as the collagen fibers previously had! According to Poole & Gwinnet (1969) the banding periodicity of the collagen in immature dogfish enameloid is 450-500 Å which resembles that of fetal pulp fibers.

W.L. Davis, R. Jones and J.L. Matthews: If collagen existed in early enameloid and was subsequently resorbed by cells of inner epithelium, did they have collagenase? Is this another example of cells losing the use of a specific enzyme during evolution?
Author: Shellis & Miles (5) assume that the removal of collagen from the enameloid matrix is due to a non-enzymatic mechanism. "The epithelial protein in enameloid might interact with the matrix collagen in such a way that, during mineralization, the latter becomes labile and can be removed, so that crystal growth is not impeded as it is in dentine" (5, p. 67). If this is the case, inner epithelium cells of tetrapods which produce enamel have not lost the use of a specific enzyme (collagenase). Rather, the difference between enameloid and enamel is that the enameloid matrix is a mixed matrix, where epithelial protein and collagen can interact; whereas in a developing tetrapod tooth, enamel matrix consisting of epithelial protein and the mesenchymal collagen of the dentine matrix are spatially separated and hence cannot interact at all.
M.M. Smith: Do you think that throughout the extent of the enameloid the degree of lability of the 'degraded' collagen is the same? Could some differences in this property of the matrix explain some of the structural complexities, perhaps dependent upon distance of matrix from the inner dental epithelium?
Author: This could very well be the case, but the available data to answer this question is not sufficient. According to the hypothesis of Shellis & Miles (5; see last answer) the lability of the 'degraded' collagen depends on its interaction with the epithelial protein. It is unknown whether the amount of epithelial protein is the same throughout the uncalcified matrix.

SCANNING ELECTRON MICROSCOPY/1979/II
SEM Inc., AMF O'Hare, IL 60666, USA

STRUCTURAL CONVERGENCES BETWEEN ENAMELOID OF ACTINOPTERYGIAN
TEETH AND OF SHARK TEETH

W. -E. Reiß

Department of Geology and Palaeontology
University of Tübingen
D-7400 Tübingen
W-Germany

Abstract

An enameloid coating of dermal armour and teeth is a primitive feature
in Vertebrata. Originally the microstructure of this enamel-like hard tissue had
a very low degree of order. Highly ordered
microstructures evolved secondarily in
the teeth of two teleost families, Characidae and Sphyraenidae, and in the
Euselachii (modern sharks). Certain aspects of the microstructure of this
highly ordered enameloid ("parallel-
structured enameloid") of the Euselachii
point to the fact that this type of hard
tissue evolved only once in the sharks.
This conclusion supports the assumption
that the Euselachii form a natural group.
The common ancestor of all Euselachii
must have lived in the late Triassic
(200-210 million years b.p.).

A. Introduction

The histogenesis of the enamel-like
cap (=enameloid) of fish teeth has posed
difficult problems for a long time (see
Poole, 1, for a literature review).
Transmission electron microscopy studies,
histochemical studies and experiments
with radioactive tracers (2-5) show that
the matrix of the enameloid of teleosts
is secreted by the inner dental epithelium (ectoderm) and by the odontoblasts
(mesenchyme). It is hence a "mixed" matrix with a high amount of collagen. The
inner dental epithelium degrades the collagen and resorbs the organic matrix from
the calcifying enameloid and it also provides the mineral salts (4). (This histogenesis is markedly different from that
of true enamel- of crossopterygians,
dipnoans and tetrapods -, where the matrix of the enamel is secreted solely by
the inner dental epithelium, 6, 7). The
data which are available so far, point
to the fact that the enameloid of sharks
is formed in the same way as the enameloid of teleosts (4, 8-15). This similarity of hard tissue formation is probably

Fig. 1 Astraspis sp. (agnathan), Ordovician. Dermal tubercle, vertical section, etched for 2 sec in 2N HCl, SC=
single-crystallite enameloid.

KEY WORDS: Enameloid, Teeth, Actinopterygii,
Euselachii, Elasmobranchii, Sphyraenidae,
Characidae, Apatite Crystals, Agnathi

not due to convergence between teleosts and sharks, but to common ancestry. *Astraspis desiderata,* an agnathan from the Ordovician, which belongs to the oldest vertebrates known so far, has dermal plates with tubercles, which have a cap of hypermineralized tissue. To all intents and purposes this tissue is enameloid and not enamel (see discussion in 16-18). Like enameloid types of all sharks and many actinopterygians it consists of crystals which are approximately 10 times as large as the crystals of enamel (thickness 0.1 µm; length 0.5 µm); the crystals are randomly arranged (fig. 1). According to the terminology developed in (19), this enameloid type can be called "single-crystallite enameloid". It can be concluded that a common ancestor of the agnathans and the gnathostomes evolved enameloid (which originated from a mixed matrix and which had a low degree of structural complexity). A change in timing of the activities of inner dental epithelium and odontoblasts led to a separation of the two components of the matrix and thus to an evolution of true enamel in crossopterygians, tetrapods and dipnoans (4). (For an alternative hypothesis see 20).

The purpose of this paper is to describe new microstructures in actinopterygian teeth and thus contribute to the knowledge of the evolution of the enameloid in sharks on one hand and in teleosts on the other hand. Both groups have evolved enameloid with a high structural complexity.

B. Material and Methods

a) Dermal tubercles of *Astraspis desiderata* (Agnathi, class Pteraspidomorphi, subclass Heterostraci), Ordovician, Colorado Canyon, U.S.A.

b) Teeth of *Sphyraena barracuda, Sph. novaehollandiae* and *Sph. obtusata,* Recent. (Teleostei, order Perciformes, family Sphyraenidae).

c) Teeth of *Bryconamericus* sp., *Leporinus* sp., *Tetragonopterus* sp., *Astyanax* sp., *Serrasalmus* sp., *Moenkhausia* sp., *Hoplias malabaricus, Hoplias* sp.; Recent (Teleostei, order Cypriniformes, family Characidae sensu lato).

d) Teeth of *Pyrrhulina* sp., *Metynnis* sp., *Curimatus* sp., *Hemiodus* sp., *Prochilodus* sp., *Ctenolucius* sp., *Casteropelecus* sp., *Characidium* sp., *Charax* sp., *Acestiorhynchus* sp., *Chilodus* sp., *Distichodus* sp., *Saccodon* sp., *Alestes* sp.; Recent (Teleostei, order Cypriniformes, family Characidae sensu lato).

e) Additionally 30 genera of Recent and fossil actinopterygians of the following families were studied: Saurichthyidae, Colobodontidae, Birgeriidae,

Eugnathidae, Pachycormidae, Lepisosteidae, Esocidae, Mormyridae, Anarhichadidae, Sparidae, Cichlidae, Cyprinidae, Balistidae, Tetraodontidae, Diodontidae, Molidae. A full account of these results will be given in another paper.

The microstructures of shark enameloid, summarized in the text, were fully described in (19).

The fossil teeth are either from bone beds or from articulated specimens. Wherever necessary they were freed from the matrix with diluted acetic acid. Recent teeth were either not treated at all with chemicals before etching or they were cleaned in diluted H_2O_2 or NaOCl.

Natural tooth surfaces, sectioned and polished surfaces, and fresh fracture surfaces were etched with 2N HCl for 3 to 7 seconds. (Before sectioning small teeth were imbedded in polyester resin).

After washing in distilled water, the specimens were dried, mounted with cement on stubs and coated with carbon and gold-palladium. The specimens were studied under a "Stereoscan" mark IIA of the Cambridge Instrument Co. Ltd. Acceleration voltages of 10 KV and 30 KV were used.

C. Observations and Discussion

The enameloid of all fossil sharks, except for the Euselachii (Triassic to Recent) is "single crystallite enameloid". In this tissue the crystals are not aligned to form crystal-bundles, but they remain independent. They are either randomly oriented or are oriented more or less perpendicular or more or less parallel to the tooth surface (19, 21, 22). During evolution of the Euselachii a three-layered type of enameloid evolved (figs. 2 and 3), with (a) a shiny layer at the surface; this layer is formed on a very late stage of tooth morphogeny; one has thus to assume that it is derived solely from the ectoderm; (b) a thick layer of parallel-structured enameloid below and (c) a layer of woven enameloid of varying thickness at the enameloid-dentine junction (19). Layers (b) and (c) consist of slender apatite crystals which are aligned parallel to each other, thus forming crystal-bundles. There is almost no collagen in the mature hard tissue (1,4); but there is some evidence that the direction of the crystal-bundles reflects the direction of the original collagen fibers in the matrix (5,20). In the parallel-structured enameloid the crystal-bundles have only two directions: (a) the majority of the bundles are oriented parallel to the tooth surface and have a more or less basal-apical direction, with respect to the tooth ("surface-parallel bundles") and (b) the rest of the fibers run in a centripetal direction from the tooth surface to the enameloid-dentine junction ("radial bundles"). The angle between surface-parallel bundles and radial bundles is 90°. (Some authors e.g. Mörnstad,13 , assume that the radial bundles are unmineralized collagen fibers. This, however, is not true for the mature tissue). The threelayered enameloid occurs in <u>all</u> fossil and Recent euselachians (except for the posterior, crushing, teeth of <u>Heterodontus</u>, 19,21). It can be shown that the parallel-structured enameloid povides a high bending strength to the teeth (23); this is especially important, as all euselachians have more or less well developed fangs

and cutting teeth with sharp cutting edges. Four types of bundle architecture occur in the different shark teeth (19). Experiments show that the surface parallel bundles in all four types have the direction of the main stress-lines (23). This explains the bending resistance of this type of enameloid.

Much less is known about the structure of the enameloid of actinopterygians. Generally, two types of enameloid occur on the same tooth, cap enameloid and collar enameloid (4, 24, 25); they can be distinguished with respect to their position on the tooth and, to a certain degree, with respect to their microstructure. In many teeth collar enameloid is strongly reduced in thickness, and it will not be discussed here. Instead of a shiny surface layer,

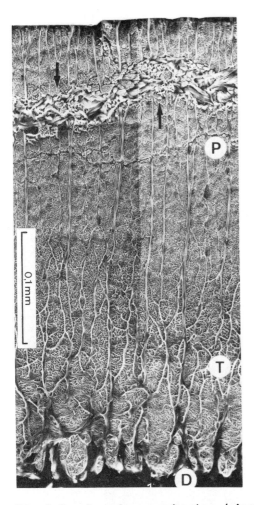

Fig.2 <u>Carcharodon carcharias</u> (shark), tooth; Recent. Horizontal section, etched for 2 sec in 2N HCl. Surface of the tooth on top, enameloid-dentine junction at the bottom. The shiny layer has been etched away; P= parallel-structured enameloid, T= woven enameloid; D= dentine. An arrow marks a disturbed part of the structure. (After 19, modified).

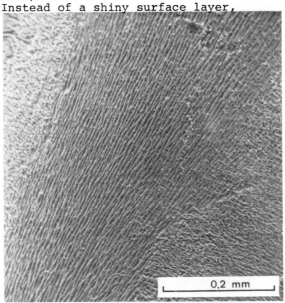

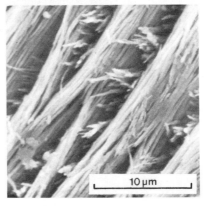

Fig. 3a,b <u>Scapanorhynchus raphiodon</u>, (shark), middle Cretaceous, England. Etched for 5 sec in 2N HCl. Surface view, the shiny layer was etched away; both components of the parallel-structured enameloid, the surface-parallel bundles and the radial bundles, are clearly visible.

actinopterygian teeth have a mineralized or unmineralized acid resistant cuticle (26). With few exceptions (see below), in all fossil and Recent actinopterygian teeth the enameloid has a woven structure throughout (fig. 4); it should be mentioned, however, that the enameloid can be differentiated in an inner layer with very low degree of structural order and an outer layer with a higher degree of order. If we assume that single crystallite enameloid (as in agnathans and in fossil sharks) is the most primitive enameloid, the actinopterygians already show a derived condition. The woven enameloid probably has a high compression strength, rather than a high bending strength (23). It forms a very thick layer in blunt crushing teeth, but sometimes a thin layer in cone-shaped fangs, which are used in prehension of prey.

Fangs and cutting teeth with <u>sharp cutting edges</u> are very rare in actinopterygians. They have been found in two Recent teleost families, Characidae sensu lato (<u>Piranha</u> and related genera) and Sphyraenidae (<u>Barracuda</u> and related species). Shellis & Berkovitz (24) found that the enameloid cap of Piranha teeth consists of a parallel-structured layer and of a woven layer, similar to the shark teeth. This finding could be confirmed for the three sphyraenids available (see Material and Methods) and for the characid genera listed under (c) in Material and Methods (figs. 5-9). It was not yet possible to find parallel-structured enameloid in the characids listed under (d). The reason for this is that the teeth are very small and the cuticle is very thick and resistant to acids; thus the cuticle could not be removed with acid without considerable damage to the enameloid layer.

Several important observations were made:
a) The bundle architecture in the teleost teeth (fig. 6,7) resembles the simplest type of the shark teeth (19,27). Serration denticles are not well developed (fig. 6) or do not occur at all. There is never more than one layer of parallel-structured enameloid (fig.8,9). In contrast to this, the sharks <u>Isurus</u>, <u>Lamna</u>, <u>Carcharodon</u> and <u>Odontaspis</u> have two to three layers of parallel-structured enameloid of different bundle directions. The functional reason for this is probably that the teleost teeth hardly ever become as big as the teeth of these shark genera.

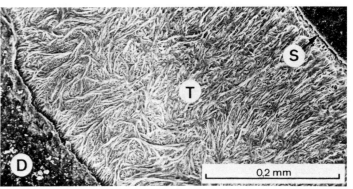

Fig. 4 <u>Pagrus</u> sp., (teleost), Recent. Vertical section, etched for 6 sec in 2N HCl. D= dentine, S= tooth surface, T= woven enameloid.

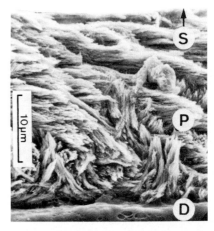

Fig. 5 <u>Sphyraena barracuda</u> (sphyraenid teleost), Recent. Horizontal section, etched for 4 sec in 2N HCl. D= dentine; P= parallel-structured enameloid, S= tooth surface; surface-parallel and radial bundles of the parallel-structured enameloid can be clearly distinguished.

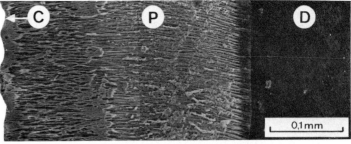

Fig. 6 <u>Sphyraena barracuda</u> (sphyraenid teleost), Recent. Surface picture of the cutting edge, etched for 6 sec in 2N HCl; the cuticle has been etched away. D= dentine; P= parallel-structured enameloid; C= cutting edge.

b) In the sphyraenids the enameloid is restricted to the cutting edge itself and its vicinity (fig. 6). This probably helps the animals save phosphate. In the characids the (cap-) enameloid covers the whole cusps or even the major part of the crown (fig. 7a, b).

c) The internal structure of the cutting edges differs significantly from that of the shark teeth (27).

d) Like in the sharks, parallel-structured enameloid in actinopterygians has surface-parallel bundles and radial bundles. Horizontal sections through the teeth, however, show that the internal structure of the parallel-structured enameloid is very irregular and varies from species to species. This also is a significant difference to the shark teeth. In shark teeth there is a clear separation of aggregates of surface-parallel bundles from aggregates of radial bundles. The boundary plane between two aggregates has always a radial-vertical orientation (fig. 2). Towards the center of the tooth the parallel-structured enameloid grades into woven enameloid, which has still a high degree of order. In the Sphyraenidae the distribution of radial and of surface-parallel bundles is very irregular (fig. 5). In Hoplias malabaricus (Characidae) the horizontal sections show that in addition to the surface-parallel (vertical) bundles there are bundles of many other directions. All bundles are closely interwoven (fig.9). In Serrasalmus sp. (Characidae) thick strands of bundles which originate in the dentine and run more or less radially towards the tooth surface dominate the picture (fig. 8).

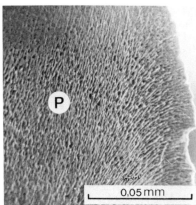

Fig. 7 Astyanax sp. (characid teleost), Recent, etched for 7 sec in 2N HCl; (a) two lateral cusps of the tooth crown; all cusps have well developed cutting edges; P= parallel-structured enameloid; D= dentine; the cuticle has been etched away; (b) central tooth cusp.

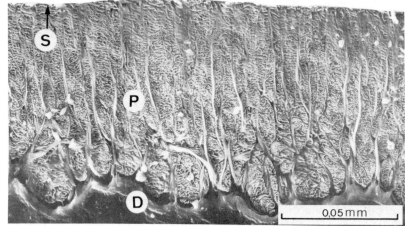

Fig. 8 Serrasalmus sp. (characid teleost), Recent, etched for 2 sec in 2N HCl; horizontal section. D= dentine; P= parallel-structured enameloid, S= tooth surface.

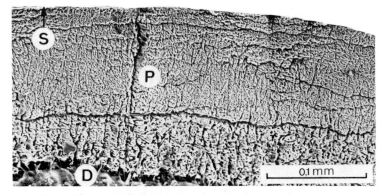

Fig. 9 Hoplias malabaricus (characid teleost), Recent, etched for 7 sec in 2N HCl; horizontal section. D= dentine; P= parallel-structured enameloid; the cracks resulted from sectioning; S= tooth surface.

It is doubtful whether this lower degree of order in the enameloid of teleost teeth has any biomechanical significance, i.e. decreases the strength of the hard tissue. The more important conclusion to be drawn is, that if only a small fragment of a tooth cap is available, it is still possible to tell whether the fragment is from a teleost or a shark.

D. Conclusions

This study shows that the bending-resistant parallel-structured enameloid evolved at least 3 times convergently, in the Characidae, the Sphyraenidae, and the Euselachii. The structural peculiarities of the parallel-structured enameloid of shark teeth however point to the fact that this particular structure evolved only once, namely in the last common ancestor of all Euselachii. This enameloid is a synapomorphic character of the Euselachii with respect to all other sharks. The study of the microstructure of the tooth enameloid under the SEM thus provides a tool to distinguish Euselachii from other elasmobranchs. This tool can be applied in Recent animals as well as in fossils, no matter whether a whole fossil skeleton is available or only a single tooth.

Acknowledgements

The teeth were provided by the late W. Gross and by C. Liem (Harvard University, Cambridge, Mass.), A. Seilacher (Department of Geology, Tübingen) and G.v. Wahlert (Naturkunde-Museum, Ludwigsburg). R. Klett and H. Hüttemann assisted at the "Stereoscan". The study was supported by the Sonderforschungsbereich 53 "Palökologie". This paper is Konstruktions-Morphologie no. 103 (No 102, see Seilacher, A.: Constructional morphology of Sand-Dollars.- Paleobiology, in press).

Literature

1. Poole, D.F.G.: Phylogeny of tooth tissues: Enameloid and enamel in recent vertebrates, with a note on the history of cementum.- In: Structural and Chemical Organization of Teeth (ed. A.E.W. Miles) 1, 111-149, Academic Press, New York and London (1967).
2. Herold, R.C.: Ultrastructure of odontogenesis in the Pike (Esox lucius). Role of dental epithelium and formation of enameloid layer.- Archs. oral Biol., 19 (1974) 633-644.
3. Shellis, R.P.: Histological and histochemical studies on the matrices of enameloid and dentine in teleost fishes.- Archs. oral Biol. 20 (1975) 183-187.
4. Shellis, R.P.: The role of the inner dental epithelium in the formation of the teeth of fish.- In P.M. Butler and K.A. Joysey (eds.): Development, Structure and Evolution of Teeth, 31-42, Academic Press, London 1978.
5. Shellis, R.P. & Miles, A.E.W.: Autoradiographic study of the formation of enameloid and dentine in teleost fishes using tritiated amino acids.- Proc. R. Soc. Lond. B. 185 (1974), 51-72.
6. Eastoe, J.E.: Recent studies on the organic matrices of bone and teeth.- In H.J.J. Blackwood (ed.): Bone and Tooth, 269-281, Pergamon Press, Oxford 1965.
7. Reith, E.J.: The stages of amelogenesis as observed in molar teeth of young rats.- J. Ultrastruct. Res. 30 (1970) 111-151.
8. Garant, P.R.: An electron microscopic study of the crystal-matrix relationship in the teeth of the dogfish Squalus acanthias L.- J. Ultrastructure Res. 30 (1970) 441-449.
9. Goto, M.: Development of shark teeth and phylogeny of teeth in vertebrates.- Earth Sciences (Chikyu no Kagaku) 30 (1976) 206-221. In Japanese, with English summary.
10. Kemp, N.E. & Park, J.H.: Ultrastructure of the enamel layer in developing teeth of the shark Carcharhinus menisorrah.- Archs. oral Biol. 19 (1974) 633-644.
11. Kérébel, B. & Daculsi, G.: Ultrastructure de l'émailloide des dents de Prionace glauca L.- Biol. Buccale 3 (1975) 3-12.
12. Kérébel, B., Daculsi, G. & Renaudin, S.: Ultrastructure des améloblastes au cours de la formation de l'émailloide des Sélaciens.- Biol. Cellulaire, 28 (1977) 125-130.
13. Mörnstad, H.: On the histogenesis of shark enamel.- Odont. Revy 25 (1974) 317-326.
14. Moss, M.L.: Skeletal tissues in sharks.- Amer. Zool., 17 (1977) 335-342.
15. Ripa, L.W., Gwinnett, A.J., Guzman, C. et al.: Microstructural and microradiographic qualities of lemon shark enameloid.- Archs. oral Biol., 17 (1972) 165-173.
16. Denison, R.H.: Ordovician vertebrates from Western United States.- Fieldiana, Geology, 16 (1967) 131-192.

17. Halstead, L.B.: Calcified tissues in the earliest vertebrates.- Calc. Tiss.Res. 3 (1969) 107-124.
18. Ørvig, T.: Pycnaspis splendens, new genus, new species, a new ostracoderm from the Upper Ordovician of North America.- Proc. U.S. Nat. Mus. 108 (1958) 1-23.
19. Reif, W.-E.: Morphologie und Ultrastruktur des Hai- "Schmelzes".- Zool. Scr., 2 (1973) 231-250.
20. Poole, D.F.G.: An introduction to the phylogeny of calcified tissues.- In A.A. Dahlberg (ed.): Dental Morphology and Evolution, 65-79, University of Chicago Press, Chicago 1971.
21. Reif, W.-E.: Tooth enameloid as a taxonomic criterion: 1. A new Euselachian shark from the Rhaetic-Liassic boundary.- N.Jb.Geol. Paläont.Mh. 1977, 565-576.
22. Reif, W.-E.: Tooth enameloid as a taxonomic criterion: 2. Is "Dalatias" barnstonensis Sykes, 1971 (Triassic, England) a squalomorphic shark? - N.Jb.Geol.Paläont. Mh 1978, 42-58.
23. Preuschoft, H., Reif, W.-E. & Müller, W.H.: Funktionsanpassungen von Haifischzähnen in Form und Struktur.- Z. Anat. Entwicklungsgesch. 143 (1974), 315-344.
24. Ørvig, T.: Fossila fisktänder i svepelektronmikroskopet: gamla frägeställningar i ny belysning.- Fauna och Flora 68 (1973) 166-173.
25. Ørvig, T. (MS): Studies by light and scanning electron microscopy of hypermineralized dental hard tissues in some fossil actinopterygian fishes, with an appendix containing remarks on the investigated material from the point of view of taxonomy and comparative biology. 58 pp., 16 pl., Museum of Natural History, Department of Paleozoology, Stockholm.
26. Shellis, R.P. & Berkowitz, B.K.B.: Observations on the dental anatomy of Piranhas (Characidae) with special reference to tooth structure.- J. Zool. Lond., 180 (1976), 69-84.
27. Reif, W.-E.: Bending-resistant enameloid in carnivorous teleosts.- N.Jb.Geol.Paläont.Abh. 157 (1978), 173-175.

Discussion with Reviewers

R.P. Shellis: Indications of a two-layered enameloid structure (parallel-structured superficial to woven) have been reported in a number of teleosts, e.g. Anquilla, Gadus, by light microscopy (Kerr, T., Proc. zool. Soc. Lond. 133, 1960, 401-422; Schmidt, W.J., Z. Zellforsch. 93, 1969, 447-450) and TEM (Shellis, R.P. & Miles, A.E.W., Proc. R. Soc. Lond. B 194, 1976, 253-269). In other species, including Polypterus, the longitudinal, but not the radial elements of the outer layer were described (Kerr, 1960).
Has the author examined the species in question?
Author: I have not studied the species in question, nor has any other author studied them in the SEM. From the published evidence I am not sure whether the outer layer in these species really is parallel-structured enameloid as defined in the text. My own studies lead to the assumption that in actinopterygians generally the outer layer of the enameloid can have some structural regularity; a typical parallel-structured enameloid, however, only occurs in teeth with well developed cutting edges (Sphyraenidae and Characidae among the Recent teleosts).

M.M. Smith: Ørvig, T. (Zool. Scripta, 7, 1978, 33-56) has published Sem's of actinopterygian jaw-teeth in which apparently parallel-structured 'acrodin' /= enameloid_7 and woven 'acrodin' are featured. Can you relate this to your own observations, as they seem to show two layers of parallel-structured enameloid similar to those in shark teeth.
Author: The species in question is the Triassic birgeriid Birgeria groenlandica, which I have not studied. According to Ørvig it has teeth with sharp cutting edges. It would hence not be surprising to find parallel-structured enameloid in these teeth. According to Ørvig's very limited data and a single figure (loc.cit., fig. 13) this tissue seems to occur only in the apex of the tooth. (Birgeria mougeoti, which I studied, has conical teeth without cutting edges).

R.P. Shellis: Could the author comment on the relationships between the SEM image of enameloid structure and that provided by other techniques?
Author: Ideally, in a study of hard-tissues light microscopy, polarization microscopy, SEM and TEM are used in combination. SEM studies could confirm the basic assumptions of earlier authors, who used polarization microscopy. The advantage of the SEM is that

many more details, including individual crystals can be made visible and that it can be applied to the study of natural or fracture surfaces or to the study of extremely thin layers, where the light microscopy fails. Many more TEM studies would be needed, however, to determine more exactly size and shape of individual crystals and the presence or absence of collagen fibers in the mature tissue.

M.M. Smith: In the text you describe figure 1 as representing typical "single crystallite enameloid", with separate crystals 10 times as large as enamel. This is not as convincing as previously published Sem's. As this is used as evidence in the introduction that a common ancestor of the agnathans and the gnathostomes possessed this type of enameloid and later that it is the most primitive type, I feel it would be helpful to give more information on the structural arrangement of this tissue.
Author: To the best of my knowledge figure 1 is the first SEM-picture of enameloid of the Ordovician Astraspis or any other Paleozoic agnathan. (In the contemporary Eryptichius I could not find enameloid at all.) I should mention that the preservation of these dermal tubercles of Astraspis is not as favourable as it could be, and preparation is very difficult. Throughout the enameloid cap, which is only up to 80 μm thick, the structure is uniform. Traces of tubules, which were found in thin-sections (18), can also be seen in the SEM. In vertical sections the line of demarcation between enameloid and dentine is smooth and fairly sharp. It seems important to mention that I found enameloid with the same structure also in the upper Silurian heterostracan Oniscolepis dentata (but not yet in the contemporary Traquairaspis and Tolypelepis). It also occurs in the scales of all fossil sharks studied so far (19, fig. 25c).

M.M. Smith: Could you state how you can demonstrate from an acid etched surface that the groups of crystal bundles are entirely mineral and do not contain in the center any collagen fibres. Is it necessary to refer to TEM of similar regions to clarify this point?
Author: Yes, it is necessary to refer to TEM of similar regions, because it is not possible to identify collagen under the SEM. It is however possible to tell from SEM pictures whether the hard tissue contains a high amount or a low amount of organic matrix.

M.M. Smith: You have mentioned collar enameloid in the discussion. Could you make it clear what your own observations are on this structure. I feel it is important to establish this information if you are going to mention collar enameloid.
Author: My own information comes mainly from the Recent sparid Pagrus and from the Triassic semionotid Sargodon. As far as I know, no SEM pictures of collar enameloid have been published so far (see however, Ørvig, 25). Ørvig (24, 25, Zool. Scripta, 7, 1978, 33-56 and 125-144) uses the term ganoin for collar enameloid. Without discussing terminology problems here, I agree with Ørvig's statement that the structure of collar enameloid is strikingly similar to genuine ganoin in scales. The apatite crystals are never aligned to form bundles. They are more or less parallel to each other and vertical to the tooth surface. They can also have a fan-shaped arrangement (see e.g. Ørvig 1978, p. 51, fig. 29-34). In Sargodon tomicus more than 100 very thin growth lines are developed in the collar enameloid. The tissue seems to grow outward (like true enamel), rather than inward. (In this context it is important to note that Shellis & Miles, Proc. R. Soc. Lond. B, 194, 1976, 253-269 found in Anguilla teeth a layer which may be homologous to enamel and which covers cap enameloid as well as collar enameloid. The terminology of different kinds of "enameloids" in fossil fish teeth may thus become a difficult problem!)

For additional discussion see page. 546.

SCANNING ELECTRON MICROSCOPY/1979/II
SEM Inc., AMF O'Hare, IL 60666, USA

APPLICATION OF ELECTRON PROBE X-RAY MICROANALYSIS TO CALCIFICATION STUDIES
OF BONE AND CARTILAGE

W. J. Landis

Department of Orthopaedic Surgery
Harvard Medical School
The Children's Hospital Medical Center
300 Longwood Avenue
Boston, MA 02115

Abstract

The use of electron probe x-ray microanalysis in previous studies of bone and cartilage has been reviewed with emphasis on the results which have contributed to some of the current concepts of the mechanism of mineralization in these tissues. A number of investigations continuing in the author's laboratory utilizing high spatial resolution x-ray microanalysis and anhydrous methods of specimen preparation are described, including aspects concerning the derivation of calibration curves from synthetic calcium phosphate solids, qualitative and quantitative analyses of calcium and phosphorus in bone from embryonic chicks and in growth plate cartilage from rats, and the role of organically-bound phosphorus in mineralizing tissues. The data obtained have helped identify brushite, $CaHPO_4 \cdot 2H_2O$, as the major crystalline solid phase of calcium phosphate in the earliest mineral deposits of bone tissue, brushite and poorly crystalline hydroxyapatite in bone mineral of increasing age, and poorly crystalline hydroxyapatite in the most mature mineral portions of the tissue. Growth plate cartilage examination has revealed calcium and phosphorus in single mitochondrial granules within chondrocytes and in certain extracellular particles distinct from matrix vesicles. These results have provided important information about the possible roles of cells, extracellular components, and the organic matrix in the regulation of mineralization and about the composition, structure, and organization of the mineral phase as a function of progressively increasing age and maturation of the tissues studied.

KEY WORDS: X-Ray Microanalysis, Mineralization, Bone, Cartilage, Calcium, Phosphorous, Brushite, Hydroxyapatite, Mitochondrial Granules, Matrix, Vesicles

Introduction

Electron probe x-ray microanalysis has been particularly applicable to certain studies of calcification in the vertebrate tissues of bone, cartilage, dentin, enamel and cementum. The calcification process in these tissues occurs by incompletely understood, complex biological and physical-chemical processes which are influenced by the regulatory activities of specialized cells, by the various properties of the organic constituents synthesized by those cells, and by specific interactions occurring between certain of the organic components and ions of calcium and inorganic orthophosphate present in the extracellular tissue spaces. The organic matrices of bone, dentin, and embryonic enamel are known to constitute approximately 30% of their respective dry weights and are largely acellular, consisting of collagen, proteoglycans, acidic proteins, and phosphoproteins. A solid inorganic mineral phase of a calcium phosphate salt, of which hydroxyapatite, $Ca_{10}(PO_4)_6(OH)_2$, is the final form, contributes the remaining 70% dry weight.

The majority of the inorganic solid phase of mineralized tissues is located in the extracellular tissue spaces and at the ultrastructural level is arranged in a highly ordered manner, in intimate association with the various components of the organic matrix which quite likely strictly specify the close relationship. In addition, solid phase calcium phosphate is also found to be deposited in mitochondria and other intracellular organelles.

Much of the present knowledge of some of the complex mechanisms in calcification is based on the well-documented results of light and electron microscopy, correlated with x-ray and electron diffraction, and more recently on the data provided by electron probe x-ray microanalysis. This laboratory has been applying high spatial resolution x-ray microanalysis to study the mineralization principally of bone and cartilage, and, for purposes of review, the remarks following will therefore focus on these two among the vertebrate calcified tissues. For the literature concerning x-ray microanalysis of dentin, enamel, cementum and other calcifying tissues, the reader is referred to the recent publications and collected papers of Boyde, Höhling, and others.[1-8]

X-Ray Microanalysis
of Thick Specimen Sections

Identification and distribution of specific elements: calcium and phosphorus

Because the mineralized tissues were known to contain elements which were fundamental to normal physiological processes and which were present in sufficiently high concentrations to be accessible to detection by early instrumentation, they were among the first biological specimens to be examined by electron probe x-ray microanalysis. Indeed, the initial biological and medical applications of x-ray microanalysis were directed toward the identification and localization of calcium, studied in specimens which had been embedded and sectioned, sometimes highly polished,[9,10] to thicknesses on the order of 0.5 µm or greater. The work in dentin and enamel by Boyde, Switsur, and Fearnhead,[11] in mature bone by Tousimis[12-14] and by Baud et al.,[15] and in epiphyseal cartilage by Brooks, Tousimis, and Birks[16] showed that calcium was disposed in the extracellular matrices of the respective tissues and that its concentration varied qualitatively in different areas of the sections. These first investigations indicated the utility of the electron probe in identifying elements and determining their distribution at the histological (micron) level in mineralized tissues, and accordingly the subsequent pattern of research followed these established directions. Phosphorus was also found to be distributed in the matrix of the calcifying regions of epiphyseal cartilage in rat tibiae,[10,17] and, like calcium, its concentration, as deduced from x-ray maps, varied according to the degree of mineralization in different areas of measurement. The first quantitative determinations of a calcium-to-phosphorus molar ratio made by Mellors[18-20] in human bone and by Tousimis[21] in mineralized epiphyseal cartilage from guinea pig tibia gave results consistent with the stoichiometry of hydroxyapatite, the accepted principal inorganic component of the tissues. Mellors[19] also showed the presence of an age-dependent minor component whose properties corresponded to calcium carbonate. In further work, Höhling et al.[22] reported that the Ca/P ratios for newly synthesized bone regions of the tibial and femoral diaphyses from chick embryos were slightly lower than the ratios for hydroxyapatite and that, prior to nucleation, calcium is associated with an excess number of weakly bound phosphate groups, some of which become more strongly bound as mineralization progresses.

In a more recent and extensive study concerned with measurement of bone mineralization rates in the tibial diaphyses from young growing rats, Wergedal and Baylink[23] determined from microanalysis-derived Ca/P molar ratios that mineral deposition in bone consisted of two phases, the first of a rapid deposition of an amorphous calcium phosphate (Ca/P = 1.35) followed by a deposition at an exponentially decreasing rate of hydroxyapatite. The maximum Ca/P ratio (1.60) measured in the second phase was found to be below the theoretical value predicted for hydroxyapatite alone (1.67), the results indicating that the conversion of amorphous calcium phosphate to hydroxyapatite was incomplete. Although this was the first quantitative description of calcium and phosphorus deposition in vivo relative to time, the identification and characterization of the exact mineral species present in bone (as well as in the other calcified tissues) as a function of the degree of mineralization and maturation of the tissue have remained controversial. Other studies by x-ray microanalysis, utilizing alternative specimen preparation methods and instrumentation, have yielded results different from those just described (see below).

Other investigations[10] of the phosphorus content in cartilage revealed that it appeared in the chondrocytes throughout the epiphyseal plate, as well as in the extracellular matrix. The cellular phosphorus, presumably representing nucleic acid, qualitatively decreased toward the calcifying cartilage zones where the chondrocytes were thought to degenerate and where alkaline phosphatase, an enzyme considered important for promoting mineral deposition,[24,25] coincidentally was found to occur in the extracellular matrix at the same time calcification was first observed. At the time these results were published, the role of alkaline phosphatase in mineralization was being elaborated by Fleisch, Neuman, and their colleagues,[26-29] who showed that certain phosphate esters which inhibit calcification in vitro are degraded by the action of alkaline phosphatase or a more specific inorganic pyrophosphatase. The concept of inhibitors of mineral deposition represents one mechanism currently thought to be involved in mineralization control.

Silicon and magnesium

In other studies of elements and ions present in mineralizing tissues and possibly involved in the regulation of calcification, silicon has been found in tibiae from young mice and rats by Carlisle.[30] It increased directly with calcium in unmineralized tissue regions where calcium concentrations were relatively low, and it decreased in calcifying regions where calcium stoichiometry approached that of hydroxyapatite. The results were interpreted to imply that silicon is associated with calcium in early stages of calcification and may be involved in the initiation of mineral deposition in the extracellular tissue spaces.

Magnesium, which is contained in large proportion in bone mineral (approximately 65% of the total body magnesium is present in the skeleton of vertebrates), was measured by Green et al.,[31] using mandibular bone from the rhesus monkey, and by Wergedal and Baylink,[23] using rat tibiae. The latter study showed that the magnesium and calcium concentrations rose in a similar manner during mineralization, to a point at which crystal perfection of hydroxyapatite became dominant. Silicon was also analyzed but was found only at very low, constant concentrations, inconsistent with the prior results of Carlisle,[30] showing variable concentrations, the highest of which occurred near the mineralizing front.

Sulfur

Sulfur has been examined by x-ray microanalysis at the histological level. Andersen[10] localized and analyzed the sulfur content in epiphyseal cartilage from rat tibia in an effort to determine whether a correlation existed between the sulfate of protein polysaccharides and the calcium and phosphorus previously found in the extracellular matrix. Glimcher[32] had already noted that the sulfate and carboxyl groups of such molecules are capable of selectively binding considerable amounts of calcium ions, thereby inhibiting mineral deposition. The results of x-ray microanalysis[10] showed some sulfur confined to chondrocytes with the principal concentration located in the extracellular matrix; the newly calcifying regions had a larger sulfur content than more mature regions of the matrix. Sulfur was also examined in tibiae from embryonic chicks by Höhling et al.,[33] who found the element in both the bone matrix and the surrounding unmineralized cells of the periosteum. In the bone a sulfur-rich macromolecule was assumed to bind calcium.

More recently, in a critical and extensive study of the rat tibial cortex utilizing in part x-ray microanalysis of sulfur, calcium, and phosphorus, Baylink et al.[23,34] significantly extended the previous results correlating protein-polysaccharide content in uncalcified and calcified bone matrix with the presence of the mineral phase. Their observations showed that the sulfur (proteinpolysaccharide) concentration in uncalcified portions of the bone matrix was relatively much higher than in the newly mineralized bone matrix and it declined abruptly as calcium concentration increased at the mineralizing front. Further, they found that the concentration of sulfur varied within the uncalcified bone matrix, itself, and was greater in peripheral regions adjacent to the cells synthesizing matrix components (osteoblasts) than in the regions nearer the mineralizing front. In addition, they found that mineralization was initiated in bone matrix only when the sulfur concentration of the matrix was lost. These results have provided evidence that proteinpolysaccharides are in some way involved in the initiation of bone mineralization, either by inhibiting the formation of a solid phase of calcium phosphate in the extracellular tissue spaces or by acting directly as nucleation catalysts, possibly serving as a template for nucleation.[34-36]

X-Ray Microanalysis of Thin Specimen Sections

Initial studies

In 1971 Sutfin et al.,[37] using a scanning electron microscope modified for high spatial resolution of thin (~2000 Å) biological sections, were the first to demonstrate that x-ray spectra could be obtained from certain ultrastructural components of cells. The tissue analyzed was the costal epiphyseal plate from mice, from which calcium and phosphorus were detected in electron-dense mitochondrial granules of hypertrophic chondrocytes. The results confirmed the evidence deduced from earlier work with

microincineration methods,[38-40] electron microscopy,[38,40-42] electron diffraction,[40] and biochemical techniques[43-46] that these elements were sequestered in the cells of various tissues and provided additional support to yet another aspect of calcification, possible direct cellular participation in mineralization control. Other results[37] also demonstrated a number of limitations to x-ray microanalysis of thin sections, related to specimen contamination and sample preparation, difficulties which were investigated further in subsequent work.[47-49]

At the same time that experimental qualitative microanalysis of ultrastructural components was shown to be successful, Hall[50-52] published a basic method for the quantitative determination of elemental composition from thin biological specimens. In the field of calcified tissue research, these reports were followed by new investigations, now conducted at the cellular and subcellular level, again intended to localize individual elements of critical interest to determine their distributions and to provide some quantitative or semi-quantitative estimation of their concentrations. Such examinations included additional studies localizing calcium and phosphorus in mitochondrial granules and other structures within cells from bones, cartilage, and teeth from mice and rats[2,53-56] and in a matrix vesicle (see below) from the calcifying epiphyseal cartilage of the rabbit.[57] The latter was also used as an example of the application of quantitative x-ray microanalysis to measure a calcium mass fraction.

Specimen preparation

It became increasingly apparent in studies with thin biological sections, and especially those obtained from the calcified tissues, that the method of specimen preparation demanded serious consideration for meaningful applications of x-ray microanalysis. As x-ray diffraction, electron microscopy, and other techniques had already demonstrated,[58-63] the mineral phase of bone and other calcified tissues could be artifactually changed by exposure to aqueous solvents during its treatment for electron optical analysis. These alterations in chemical composition, which included translocation of diffusible ions, dissolution and reprecipitation of mineral salts, and mineral phase transformations, were even more convincingly shown in direct x-ray microanalysis studies such as those reported by Morgan et al. using rat aorta,[64] by Thorogood and Craig Gray using rat bone,[65] and by Landis et al. utilizing a variety of synthetically prepared calcium phosphate solids.[66,67] Besides illustrating the application of microanalysis for evaluating specimen preparation techniques, the data emphasized the necessity for alternative approaches to the conventional use of aqueous solvents, and as a result a number of new developments in specimen treatment were applied to ultrastructural and x-ray microanalysis studies of calcified tissues.

The current methodology for thin section preparation of calcified tissues involves the use of organic solvents such as ethylene glycol[68,69] as well as ultracryomicrotomy.[66,70-77] These anhydrous techniques as applied by Landis et al.[68,71,78] have been documented by quantitative analytical measurements,[36,61] x-ray and electron diffraction,[78] electron microscopy,[68,70,71] and electron probe microanalysis[66,67,78,79] to eliminate or at least significantly reduce the artifacts otherwise apparent with aqueous solvents, although it must be mentioned that the anhydrous methods may themselves introduce some particular changes in tissue integrity related to the mineral phase. The use of these techniques, their merits and disadvantages have been described in detail elsewhere as applied to investigations of calcified tissue.[68,71,78]

Calcified cartilage

There are currently a number of studies concerned once again with cartilage structure, directed in part at examining the role of matrix vesicles in mineral deposition. These membrane-bound structures, or bodies similar to them, were first described in the extracellular matrix of guinea pig cartilage by Bonucci[80] and have since been reported in other calcifying tissues. Supported by many electron microscopic observations that particles of mineral appear within matrix vesicles before calcification occurs in the remainder of the extracellular tissue spaces, the concept has developed that these structures are causally related to the onset of mineralization and in some manner initiate the formation of the solid phase of mineral in the extracellular matrix.[73,81-83] There are, however, serious objections to this view based upon a number of significant conceptual and technical considerations.[36,68,84]

Electron probe microanalysis has been used to examine matrix vesicles found in thin sections prepared with aqueous solvents and more recently by ultracryomicrotomy. Earliest work by Hall et al.[57,85] demonstrated the presence of both calcium and phosphorus in vesicles from guinea pig cartilage but accurate elemental concentrations could not be made because specimen preparation included the use of aqueous solutions. The same tissue prepared by freeze drying without fixation or staining, but resin-embedded, briefly floated on water and sectioned, was studied much later by Barckhaus and Höhling,[86] who obtained Ca/P molar ratios by microanalysis of extracellular regions containing clusters of mineral phase particles and distinguished by differences in electron density. On the basis that the ratios were found to correspond to that of hydroxyapatite and from other measurements, it was conjectured that the particle clusters might possibly represent mineralized matrix vesicles. Ali et al.[73,83] examined the calcifying cartilage from the rabbit and found on analysis of unstained thin sections prepared conventionally that matrix vesicles accumulated calcium and phosphorus in concentrations which varied in the different zones of the growth plate, the highest concentrations corresponding to hydroxyapatite occurring in regions in which calcification of the extracellular tissue spaces was clearly observed (lower hypertrophic zone). As it was recognized that these results could be made more reliable by the use of ultracryomicrotomy to avoid changes in the mineral phase, Ali et al.[72,87] subsequently reexamined rabbit cartilage and were able to prepare unfixed, unembedded, and unstained frozen thin sections of the tissue in which chondrocytes and certain cellular organelles could be identified. Unfortunately, although the method of specimen preparation circumvented the artifacts inherent in the previous studies, it did not impart contrast sufficient to identify large numbers of extracellular matrix vesicles. For those few vesicles which were observed, they were characterized by a high Ca/P ratio, again similar to that of hydroxyapatite.[87] On the other hand, the same study[87] showed many mitochondrial granules present in the chondrocytes and on microanalysis these were found to have relatively low Ca/P ratios which were attributed to the presence of amorphous calcium phosphate.

Bone tissue

With respect to the application of electron probe microanalysis to previous ultrastructural studies of bone, Landis and Glimcher[78] reported extensive results obtained from unstained thin sections of embryonic chick tibiae, prepared by anhydrous means with either organic solvents or dry ultracryomicrotomy. Their data showed calcium, phosphorus, and magnesium present compared to background in mitochondrial granules of bone cells, in the most recently deposited clusters of a mineral phase located in the extracellular tissue spaces, and in the regions of the tissue containing progressively more mature dense mineral masses. Sulfur was detected as well in unmineralized portions of the extracellular matrix but not in densely calcified areas, a result consistent with the work of Baylink et al.[34] Molar ratios of Ca/P were determined and found to be 0.8 - 1.1 for the mitochondrial granules, 1.2 - 1.3 for the recently deposited clusters of mineral particles, and 1.4 - 1.5 for the dense mineral masses. Correlative electron microscopy and electron diffraction of the same sites of mineral deposition examined by x-ray microanalysis established that no electron diffraction patterns of a specific calcium phosphate solid phase were generated from mitochondrial granules or from the smallest clusters of the mineral phase first deposited in the extracellular matrix. The older, more mature extracellular deposits consisted progressively of a more mature crystalline phase which diffracted as hydroxyapatite. The changes in the electron diffraction characteristics accompanied by the increases in the Ca/P ratios of the solid mineral phase suggested either that the mineral phase which is deposited later in the calcification of the tissue as a whole is chemically and structurally different from that which is deposited early or that the mineral phase initially deposited undergoes a phase transformation. As also found and discussed by Wergedal and Baylink,[23] the observation that the maximum Ca/P molar ratios determined (~1.5) did not correspond to that of hydroxyapatite (1.67) was attributed to a possibility that an initially deposited solid phase having a low Ca/P ratio may remain in significant proportions in the mature mineral phase. In this context, the presence of brushite, $CaHPO_4 \cdot 2H_2O$, must be considered; it has been identified [88,89] in the youngest (most recently deposited) mineral fractions obtained from embryonic chick bone, has a Ca/P molar ratio of 1.0, and fails to generate any electron diffraction pattern when prepared by the methods of Landis and Glimcher.[78]

High Spatial Resolution X-Ray Microanalysis

The work in this laboratory has been concerned for some time with various aspects of the mechanism of calcification, including (1) the identification of the exact physical and chemical nature of the initial mineral phase deposited in calcified tissues, (2) the characterization of the changes in the chemical and structural nature of the solid phase with progressively increasing degrees of mineralization and maturation, (3) the analysis of the possible role of protein- or peptide-bound organic phosphorus in nucleation and the regulation of mineralization, and (4) the elaboration of the spatial and temporal events which possibly exist between the mineral deposits already identified and described in mitochondria, matrix vesicles, and the extracellular tissue spaces. Each of these areas of concern is uniquely suitable for investigations extending past results utilizing electron probe x-ray microanalysis in conjunction with proper methods for the preparation of thin specimen sections.

Recently a JEOL JSM 50A scanning electron microscope has been highly modified for scanning transmission capability and high spatial resolution x-ray microanalysis, in the range of 100 Å or less.[49] The instrument incorporates a finely focused electron beam currently generated from a standard hot tungsten filament; resolutions of 100 Å in SEM and 70 Å in STEM have already been achieved. Additional features include an 80 mm^2 solid state Si(Li) x-ray detector located beneath the specimen stage so as to measure x-rays on the emergent side of the sample, a mechanically stable stage to minimize specimen drift, and a pumping system maintaining column vacuum on the order of 1×10^{-8} Torr or better. Some of the advantages of these characteristics for thin section microanalysis have been discussed.[47,49] Together with several procedures[68,71,78] which permit the processing of calcified tissues without introducing the significant artifacts in the mineral phase attributable to the use of aqueous solvents, the instrumentation has made it possible to follow in much greater detail the sequential changes which occur in the mineral phase and in its relationship to other components of the tissues as a function of time and maturation. A few of these studies will be described below to illustrate the application of x-ray microanalysis to calcified tissue research in this laboratory.

Materials and methods: Tissues and synthetic mineral standards

For the variety of experiments reported here, the principal tissues examined included the periosteal cuff of bone (primary center of ossification) from the tibiae of normal 17-day old embryonic chicks and the cartilaginous portion of the epiphyseal growth plate from the proximal tibiae of normal 4-week old rats. The standards for electron probe microanalysis included a number of calcium phosphate solids prepared in vitro and characterized by powder x-ray diffraction[78] and by wet chemical analysis,[90,91] from which molar Ca/P ratios were calculated. The group of calcium phosphate solids consisted of calcium metaphosphate [$Ca(PO_3)_2$, experimentally determined Ca/P = 0.50], monetite [$CaHPO_4$, Ca/P = 1.00], brushite [$CaHPO_4 \cdot 2H_2O$, Ca/P = 1.03], octacalcium phosphate [OCP, $Ca_8H_2(PO_4)_6 \cdot 5H_2O$, Ca/P = 1.34], an amorphous calcium phosphate solid phase prepared at neutral pH [ACP, Ca/P = 1.42], a poorly crystalline calcium phosphate containing some poorly crystalline hydroxyapatite [PCHA, overall Ca/P = 1.45], tricalcium phosphate [$Ca_3(PO_4)_2$, Ca/P = 1.50], and a well-crystallized hydroxyapatite [HA, $Ca_{10}(PO_4)_6(OH)_2$, Ca/P = 1.62].

Materials and methods: X-ray microanalysis

Carbon-reinforced, Parlodion-coated copper grids were used for mounting specimens and standards, prepared as thin sections ($\sim$800 Å) or as dry particles ($\sim$200 Å to 3 μm in their largest dimension) by a variety of techniques described below. Grids were examined in the JEOL JSM 50A and x-ray data were acquired at an accelerating voltage of 25 keV and a beam current of 1×10^{-11} amp, measured with a Faraday cage located below the specimen stage. Instrumental working distance used was 6 mm. The electron beam position was adjusted with manually-controlled deflection coils at magnifications of $1-5 \times 10^4$ so that x-ray spectra obtained using a static spot mode for analysis were precisely correlated spatially with the ultrastructure from which they were generated. All x-ray spectra were recorded during 100 sec. integrated detecting time periods. Contamination of the specimen was not observed during data acquisition. Indeed, the high vacuum characteristics of the instrument are such that a focused static electron beam of 10^{-11} amp can be positioned on a collodion film for a period of 15 minutes without evidence of contamination.[49]

X-ray spectra were monitored on a Tektronix model 4010-1 computer display terminal and recorded on a Hewlett-Packard model 7202A X-Y plotter for qualitative analysis of the particular elements present in an excited volume of the specimen. For quantitative estimations of calcium-to-phosphorus intensity ratios in specimens, the characteristic K_α emissions for each element were used. The number of counts under the elemental peak of interest (E) was integrated over an appropriate counting interval, and from that figure was subtracted the value of background counts ($\bar{B}$). $\bar{B}$ was determined from the mean of the integrated counts from two small intervals, one on either side of the interval for the elemental peak E, using windows of five channels each with 20 ev/channel. The quantity ($E - \bar{B}$) was obtained for calcium and phosphorus counts generated on peak and off peak, the latter by positioning the electron beam just away from the ultrastructural site of interest. The intensity ratio of the two elements was then calculated as:

$$I_{Ca/P} = \frac{[(E - \overline{B})_{on\ peak} - (E - \overline{B})_{off\ peak}]_{Ca}}{[(E - \overline{B})_{on\ peak} - (E - \overline{B})_{off\ peak}]_P}$$

Calibration curves

From a calibration curve relating Ca/P intensity ratios and Ca/P molar ratios for the calcium phosphate solid phase standards given above, a Ca/P molar ratio can then be determined for an unknown once its Ca/P intensity ratio has been determined. The method has been used previously[67] for similar calcium phosphate standards and specimens of bone examined in other instruments and under conditions different from those described here. The preparation of the standards for x-ray microanalysis has been detailed elsewhere[67,68] but will be briefly recounted. The techniques were designed to follow the procedures used for the preparation of calcified tissue specimens and included (1) sprinkling dry powders of the individual calcium phosphate solids onto carbon-reinforced, Parlodion-coated grids; (2) processing the solids anhydrously in organic solvents as follows: the dry powders were placed in Beem capsules filled with 100% ethylene glycol at room temperature and then into a vacuum desiccator on a shaker for 24 hours at 4°C. The capsules were centrifuged to sediment the powders and the glycol was replaced with Cellosolve, the monoethyl ether of ethylene glycol. Cellosolve was changed twice at 12 hour intervals, then replaced with a mixture of propylene oxide-epon (1:1). This mixture was replaced by 100% epon after one week. The embedded calcium phosphate solids were thin-sectioned (∿800 Å) with diamond knives containing a trough fluid of 100% ethylene glycol and then collected on grids; and (3) processing the solids in a manner simulating anhydrous ultracryomicrotomy by freezing untreated powders at -196°C then transferring them to a holding temperature of -70°C for six hours. They were collected by gently touching them to precooled grids. All samples were dried at room temperature, during either storage over P_2O_5 in a desiccator or transfer to the electron probe subsequent to preparation. All the standard calcium phosphate solids, prepared by the three methods described, were examined unstained and with no further treatment; evaporation of a carbon film onto the grids to stabilize the specimens in the electron beam was found to be unnecessary. The data obtained from x-ray microanalysis at 25 keV are summarized in Table 1 and Figure 1. It is apparent that the anhydrous treatments utilizing organic solvents or ultracryomicrotomy do not significantly change the Ca/P ratios compared to dry powders. This result has also been obtained using x-ray and electron diffraction to evaluate potential alterations in these samples induced by the anhydrous preparative techniques.[78]

X-ray microanalysis of bone mineral from embryonic chicks

This laboratory has recently developed a method by which the macroscopic mineral component of embryonic chick tibiae can be separated into individual fractions using a differential density centrifugation technique.[88,89] The method is extremely powerful because it provides a series of relatively homogeneous samples representing the successive stages of mineral development and maturation. This is extremely important because many previous investigations have been severely limited by the mineral phase heterogeneity inherent in the calcified tissues selected for study. The fact has frequently been overlooked that, biologically, macrosamples of bone tissue represent a system which is continuously developing in growing animals and remodeling in growing and mature animals. Consequently, such samples contain a population of mineral phase constituents which vary widely with respect to their "age" and degree of maturation, depending, for example, on the source of the tissue and the age of the animal, the precise location from which the tissue was obtained, the region examined, and other factors. Because of this heterogeneity in age and maturation of the mineral phase, macrosamples, and even microsamples, of bone reflect only the composite or average properties of the bone mineral present at a particular stage of maturation. This problem may be circumvented by the selection of appropriate tissues at specified ages, such as the long bones (primary center of ossification) from embryonic chicks,[36,78] or by the current technique of fractionating whole tissue by differential density centrifugation. By separating the mineral in organic solvents, the additional problems of mineral phase changes are avoided, and relatively small quantities of mineral, especially those representing initial phases (lowest densities), available from single animals can be pooled and thereby increased substantially with multiple centrifugations.

Fractionation of bone powder, obtained from the central portions of the tibial diaphyses from 17-day old embryonic chicks, was accomplished[88,89] using stepwise (0.1 g/cm³ increments) centrifugation in bromoform-toluene mixtures spanning a range of densities from 1.4 to 2.3 g/cm³. The isolated fractions were then examined by x-ray microanalysis in order to determine their chemical composition. Bone fractions and synthetic calcium phosphate standards (see above) were prepared either dry on individual grids or by anhydrous techniques utilizing organic solvents, embedding, and thin-sectioning. Samples were analyzed in the JEOL JSM 50A at 15 keV with a beam current of 1×10^{-11} amp. Counting times of 100 sec were used for recording x-ray data, shown in Table 2. The number of particles analyzed on a grid is given by n. Errors represent twice the standard deviation of mean Ca/P x-ray intensity ratios, calculated as described above. Ca/P molar ratios for the various bone fractions were determined by interpolation from the calibration curve (15 keV) plotting Ca/P x-ray intensity ratio as a function of Ca/P molar ratio for the synthetic calcium phosphate standards. The relative Ca/P ratios given in Table 2 are consistent with other results[89] from x-ray and electron diffraction studies of the same bone mineral fractions which have shown that brushite is the major crystalline solid phase of calcium

Table 1

Electron Probe X-Ray Microanalysis (25 keV) of Calcium Phosphate Standards

Treatment	Ca/P Intensity Ratio							
	$Ca(PO_3)_2$	Monetite	Brushite	OCP	ACP	PC HA	$Ca_3(PO_4)_2$	HA
Dry Powder	0.569 ± 0.022 (20)	1.283 ± 0.047 (24)	1.263 ± 0.045 (22)	1.680 ± 0.034 (25)	1.743 ± 0.038 (23)	1.850 ± 0.046 (26)	1.752 ± 0.031 (25)	1.983 ± 0.038 (25)
Organic Solvents	0.576 ± 0.012 (25)	1.221 ± 0.026 (25)	1.129 ± 0.057 (25)	1.613 ± 0.050 (25)	1.662 ± 0.049 (25)	1.819 ± 0.076 (23)	1.793 ± 0.050 (25)	2.096 ± 0.073 (25)
Freeze-Thaw	0.519 ± 0.029 (25)	1.312 ± 0.036 (25)	1.181 ± 0.026 (25)	1.645 ± 0.040 (24)	1.729 ± 0.051 (24)	1.807 ± 0.040 (22)	1.801 ± 0.037 (25)	2.095 ± 0.055 (24)
Molar Ratio (Ca/P)	0.50	1.00	1.03	1.34	1.42	1.45	1.50	1.62

Legend: Ca/P intensity ratios were determined in a JEOL JSM 50A operated at 25 keV with a beam current of 1×10^{-11} amp. X-ray data were collected during 100 sec counting times. The number of particles analyzed on a grid is given in parenthesis below the intensity ratio. Errors represent twice the standard deviation of mean x-ray intensity ratios. Molar ratios were determined by wet chemical means.

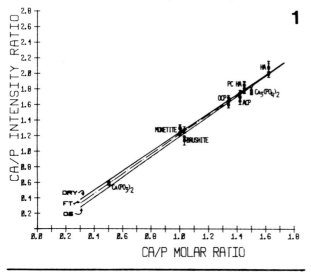

1

Fig. 1. Calibration curves [dry; organic solvents (OS); freeze-thaw (FT)] for the calcium phosphate standards in Table 1 were obtained by plotting Ca/P intensity ratios as a function of Ca/P molar ratios. Least squares regression curves with 95% confidence bars on each mean. Accelerating voltage 25 keV, beam current 1×10^{-11} amp.

phosphate present in the earliest mineral deposits of embryonic chick bone, brushite and poorly crystalline hydroxyapatite are present in bone mineral of increasing age and maturation (increasing density), and only poorly crystalline hydroxyapatite is present in the most mature mineral portions of the tissue. In this instance, the application of electron probe x-ray microanalysis has provided information concerning the nature of the solid phase deposited in calcified tissues and has helped characterize some of the progressive changes which occur in the mineral during maturation.

X-ray microanalysis of organic phosphorus

Organically-bound phosphorus has been identified in reconstituted collagens from a variety of tissues,[92] in decalcified bone and dentin,[41,93,94]

in highly purified α-chains of bone and dentinal collagen,[94,95] and in enamel proteins.[96,97] From these results, and from theoretical, structural, physicochemical, and biological considerations,[36] organic phosphorus has been implicated as serving an important role in mineralization,[36,61,92-97] and its detection by electron probe microanalysis and correlation spatially with specific components or regions within the cellular or extracellular matrix of calcified tissues would be highly significant in studies of the mechanism of calcification.

In a series of recently completed preliminary experiments, it was possible to demonstrate the presence of organic phosphorus in a variety of samples which had been extensively decalcified, dialyzed, or otherwise freed of inorganic phosphorus contaminants. The samples included the phosphoproteins casein and phosvitin (Figure 2), the allantoic and vitelline membranes from eggs, and purified phosphoproteins from dentin and enamel. Examination of calcified tissues is now being followed by high spatial resolution x-ray microanalysis in order to compare the organic

Table 2

Electron Probe X-Ray Microanalysis of Bone Powder from 17-Day
Embryonic Chick Tibiae and of Calcium Phosphate Standards

	Ca/P Intensity Ratio		Ca/P Molar Ratio
	Dry Preparation	Embedded and Thin-Sectioned	
$d < 1.4$ g/cm^3	0.863 ± 0.215, n = 8	*	0.90 ± 0.30
$d = 1.4-1.5$ g/cm^3	1.065 ± 0.092, n = 9	*	1.04 ± 0.02
$d = 1.6-1.7$ g/cm^3	0.928 ± 0.027, n = 8	0.925 ± 0.043, n = 10	0.95 ± 0.02
$d = 1.8-1.9$ g/cm^3	1.499 ± 0.137, n = 7	1.491 ± 0.059, n = 9	1.44 ± 0.11
Whole bone powder	1.517 ± 0.046, n = 9	1.539 ± 0.091, n = 10	1.46 ± 0.06
Brushite	0.912 ± 0.018, n = 20	0.953 ± 0.025, n = 25	1.03
Monetite	0.960 ± 0.021, n = 25	0.979 ± 0.016, n = 25	1.00
Poorly crystalline HA	1.497 ± 0.016, n = 21	1.598 ± 0.050, n = 25	1.45
Well crystallized HA	1.679 ± 0.033, n = 20	1.659 ± 0.062, n = 23	1.62

*Material available in these density fractions was insufficient to permit anhydrous preparations for thin-sectioning.

Legend: Specimens were examined at 15 keV with a beam current of 1×10^{-11} amp. Counting times were 100 sec. The number of particles analyzed on a grid is given by n. Errors represent twice the standard deviation of mean ratios. Statistical analyses of data show no significant differences in the Ca/P intensity ratios obtained for the three fractions of density $d < 1.4-1.7$ g/cm^3 and synthetic brushite or monetite (see text).

2

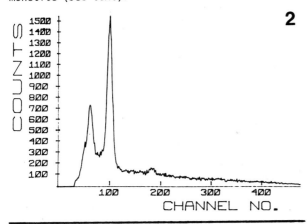

Fig. 2. X-ray spectrum of phosvitin, extensively dialyzed free of inorganic phosphorus, deposited dry on a carbon-reinforced, Parlodion coated copper grid. Analysis was performed at 10 keV using a static beam spot with on-screen magnification of 10,000x and 100 sec counting time. The multichannel analyzer in this system stores 20 eV/channel. Peaks identified are magnesium (maximum at channel 63) and phosphorus (maximum at channel 101).

phosphorus content of each and its specific location within the respective specimens.

X-ray microanalysis of rat epiphyseal cartilage

High spatial resolution x-ray microanalysis has also been used recently to identify characteristic elements of certain ultrastructural components in the unstained thin sections from the growth plate cartilage of tibia in normal 4 week old rats. The tissue sections used were prepared by ultracryomicrotomy,[71,78] examined first in a JEOL 100C transmission electron microscope and then in the JEOL JSM 50A operated in the scanning transmission mode at 25 keV with a beam current of 1×10^{-11} amp. X-ray spectra recorded during 200 sec counting times were obtained from single mitochondrial granules (500-1000Å in diameter) and from the mitochondrial matrices directly

adjacent to the granules (within 100-300Å of the granules) in proliferating and hypertrophic chondrocytes. Spectra from mitochondrial granules were typically characterized by the presence of sodium, silicon, phosphorus, sulfur, chlorine, potassium, and calcium, as shown in Figure 3. In spectra from the mitochondrial matrices, Figure 4, calcium and much of the phosphorus peak intensity was absent, the amount of the latter remaining most likely because of the presence of organically-bound phosphorus. Matrix vesicles, which in these preparations could be identified by their size (800-3500Å in diameter), shape, relative electron density, and location in the thin sections, were also examined, as well as the adjacent extracellular regions surrounding them. Matrix vesicles characteristically generated no calcium signals and phosphorus was not

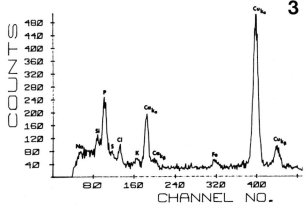

3

Fig. 3. Spectrum of a single mitochondrial granule in a proliferating chondrocyte from rat cartilage (tibia) prepared by ultracryomicrotomy. Data were obtained at 25 keV using a static beam spot, beam current of 1×10^{-11} amp., and 200 sec counting time. On-screen magnification was 30,000x. The Fe signal originates from the objective lens pole piece. The same tissue was examined for generating the spectra in Figures 4-8 and the same conditions of analysis were maintained (MCA storage 20 eV/channel).

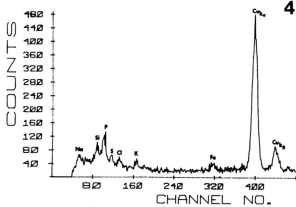

4

Fig. 4. Spectrum of the mitochondrial matrix directly adjacent to the granule analyzed in Figure 3. A phosphorus signal is still apparent most likely because the mitochondrial matrix contains some of the element in an organically-bound state rather than because of beam spreading or inaccurate beam placement. Phosphorus can be detected in regions of the matrix well away from any granules.

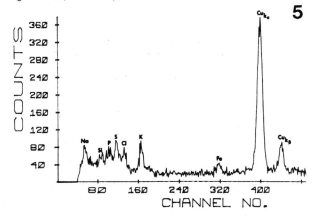

5

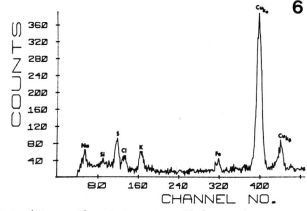

6

Figs. 5-6. Spectra obtained from a single matrix vesicle (Figure 5) and the extracellular region (Figure 6) adjacent to it (hypertrophic zone of the cartilaginous growth plate). Little or no calcium and only small phosphorus peaks could be detected in vesicles and analyses showed no significant differences in the phosphorus within or outside the vesicles in many cases.

significantly greater than that detected from the adjacent extracellular matrix. These results are shown in Figures 5-6. On the other hand, small, widely distributed extracellular structures (300-800Å in diameter), which appeared to be distinct from matrix vesicles on the basis of high resolution, high magnification electron microscopy, generated strong calcium and phosphorus intensities compared to background (Figures 7-8). Quantitative analysis of the x-ray spectra, based on the calibration curves derived for synthetic calcium phosphate solids prepared by freezing and thawing to simulate the experimental conditions for ultracryomicrotomy (Table 1 and Figure 1), gave Ca/P molar ratios of approximately 1.0 for mitochondrial granules and 1.2 for the small extracellular particles. The presence of such extracellular particles containing calcium and

phosphorus in the growth plate may be an important new consideration in the mechanism of calcification in cartilage. They may have quite possibly escaped detection in instruments less sensitive than those used in this study, but their identification and characterization, as well as that of individual mitochondrial granules and matrix vesicles, emphasize yet another application of x-ray microanalysis in studies of mineralizing tissues.

Conclusion

Electron probe x-ray microanalysis has been used extensively in a variety of previous studies concerning calcified tissues. In general, its applications have been directed toward identifying characteristic elements and ion species and

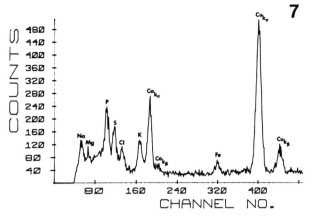
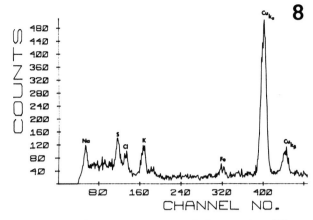

Figs. 7-8. Spectra from a small, non-vesicular structure (Figure 7) and the extracellular region (Figure 8) surrounding it (in the same field of view at 30,000x as the matrix vesicle analyzed in Figures 5-6). Calcium and phosphorus are present and above background levels. There is little difference in the spectra from the extracellular regions shown in Figures 6 and 8.

elaborating their distribution in the tissues, localizing such elements or ions within structural components and compartments of the cells and extracellular spaces, and assessing by quantitative or semi-quantitative means the chemical composition within a given region of interest. With bone and cartilage, the results already have provided important information concerning the nature of the mineral phase, its relationship to cellular and extracellular components, and the changes it undergoes both spatially and temporally in the sequence of progressively increasing degrees of mineralization and maturation of the tissue. Correlated with other electron optical methods, biochemical and physicochemical investigations, the appropriate choice of standards, specimens, and specimen preparative techniques, electron probe x-ray microanalysis should provide even more significant data in additional studies of the normally mineralizing tissues as well as of the many disease states in which abnormalities of calcification occur. In this manner x-ray microanalysis should offer one approach to obtaining a more comprehensive understanding of the structural, chemical, and biological events of mineralization.

Acknowledgments

For their invaluable technical assistance with the studies reported here, the author is especially grateful to Mr. Manuel Casiano, Mrs. Mary C. Paine, Miss Margaret Hammett, Miss Judy Blech and Miss Beatrice Lefteriou. For their helpful discussions and encouragement, appreciation is also expressed to Dr. Melvin J. Glimcher and Dr. Albert H. Roufosse. The purified phosphoproteins from dentin and enamel were a gift of Dr. Sandra L. Lee, Rush Medical College, Chicago.

This work was supported by Grant AM 15671 from the National Institutes of Health and by the New England Peabody Home for Crippled Children, Inc.

References

1. A. Boyde and E.J. Reith. Qualitative electron probe analysis of secretory ameloblasts and odontoblasts in the rat incisor. Histochemistry 50, 1977, 347-354.
2. W.A.P. Nicholson, B.A. Ashton, H.J. Höhling, P. Quint, J. Schreiber, I.K. Ashton, and A. Boyde. Electron microprobe investigations into the process of hard tissue formation. Cell Tiss. Res. 177, 1977, 331-345.
3. B. Ashton, H.J. Höhling, W.A.P. Nicholson, U. Zessack, W. Kriz, and A. Boyde. Quantitative analysis of Ca, P and S in mineralizing and non-mineralizing tissues. Naturwissenschaften 60, 1973, 392-393.
4. H.J. Höhling, R.H. Barckhaus, E.R. Krefting, and J. Schreiber. Electron microscopic microprobe analysis of mineralized collagen fibrils and extracollagenous regions in turkey leg tendon. Cell Tiss. Res. 175, 1976, 345-350.
5. J.W. Edie and P.L. Glick. Dynamic effects on quantitation in the electron probe analysis of mineralized tissues. In Proc. 11th Annual Conf. of Microbeam Analysis Soc., 1976, 65A-65E.
6. R.M. Frank, M. Capitant, and J. Goni. Electron probe studies of human enamel. J. Dent. Res. 45, 1966, 672-682.
7. A.R. Johnson. Strontium, calcium, magnesium and phosphorus content of rat incisors as determined by electron microprobe analysis. J. Dent. Res. 51, 1972, 115-121.
8. M.E. Neiders, J.D. Eick, W.A. Miller, and J.W. Leitner. Electron probe microanalysis of cementum and underlying dentin in young permanent teeth. J. Dent. Res. 51, 1972, 122-130.
9. A. Boyde and V.R. Switsur. Problems associated with the preparation of biological specimens for microanalysis. In X-Ray Optics and X-Ray Microanalysis, H.H. Pattee, V.E. Cosslett and A. Engström (Eds.), Academic Press, New York, 1963, 499-506.
10. C.A. Andersen. An introduction to the electron probe microanalyzer and its application

to biochemistry. In Methods of Biochemical Analysis, Vol. XV, D. Glick (Ed.), Interscience, New York, 1967, 147-270.

11. A. Boyde, V.R. Switsur, and R.W. Fearnhead. Application of the scanning electron-probe x-ray microanalyzer to dental tissues. J. Ultrastruct. Res. 5, 1961, 201-207.

12. A.J. Tousimis. Electron probe analysis of biological tissues. Proc. Instr. Soc. Am. 8, 1962, 53.

13. A.J. Tousimis. Elemental analysis at the subcellular level in biological tissue sections with the electron probe. Appl. Spectroscopy 16, 1962, 66.

14. A.J. Tousimis. Electron probe microanalysis of biological specimens. In X-Ray Optics and X-Ray Microanalysis, H.H. Pattee, V.E. Cosslett, and A. Engström (Eds.), Academic Press, New York, 1963, 539-557.

15. C.A. Baud, S. Kimoto, and H. Hashimoto. Etude de la distribution du calcium dans l'os haversien avec le radioanalyseur à microsonde électronique. Experientia 19, 1963, 524-525.

16. E.J. Brooks, A.J. Tousimis, and L.S. Birks. The distribution of calcium in the epiphyseal cartilage of the rat tibia measured with the electron probe x-ray microanalyzer. J. Ultrastruct. Res. 7, 1962, 56-60.

17. A.J. Tousimis. Applications of the electron probe x-ray microanalyzer in biology and medicine. Amer. J. Med. Electronics, First Quarter, 1966, 15-23.

18. R.C. Mellors. Electron probe microanalysis: calcium and phosphorus in normal human cortical bone. Lab. Invest. 13, 1964, 183-195.

19. R.C. Mellors and T.N. Solberg. Electron microprobe analysis of human trabecular bone. Clin. Orthop. 45, 1966, 157-167.

20. R.C. Mellors, K.G. Carroll, and T. Solberg. Quantitative analysis of Ca/P molar ratios in bone tissue with the electron probe. In The Electron Microprobe, T.D. McKinley, K.F.J. Heinrich, and D.B. Wittry (Eds.), John Wiley and Sons, New York, 1966, 834-840.

21. A.J. Tousimis. Scanning electron probe microanalysis of biological specimens. In Proceedings of the First National Biomedical Sciences Instrumentation Symposium, Vol. 1, F. Alt (Ed.), Plenum Press, New York, 1963, 249.

22. H.J. Höhling, H. Schöpfer, R.A. Höhling, T.A. Hall, and R. Gieseking. The organic matrix of developing tibia and femur, and macromolecular deliberations. Naturwissenschaften 57, 1970, 357.

23. J.E. Wergedal and D.J. Baylink. Electron microprobe measurements of bone mineralization rate in vivo. Amer. J. Physiol. 226, 1974, 345-352.

24. R. Robison. The possible significance of hexosephosphoric esters in ossification. Biochem. J. 17, 1923, 286-293.

25. R. Robison, M. McLeod, and A.H. Rosenheim. The possible significance of hexosephosphoric esters in ossification. IX. Calcification in vitro. Biochem. J. 24, 1930, 1927-1941.

26. C.C. Solomons and W.F. Neuman. On the mechanisms of calcification: the remineralization of dentin. J. Biol. Chem. 235, 1960, 2502-2506.

27. H. Fleisch and W.F. Neuman. Mechanisms of calcification: role of collagen, polyphosphates and phosphatase. Amer. J. Physiol. 200, 1961, 1296-1300.

28. H. Fleisch, R.G.G. Russell, and F. Straumann. Effect of pyrophosphate on hydroxyapatite and its implications in calcium homeostasis. Nature 212, 1966, 901-903.

29. H. Fleisch, R.G.G. Russell, S. Bisaz, J.D. Termine, and A.S. Posner. Influence of pyrophosphate on the transformation of amorphous to crystalline calcium phosphate. Calcif. Tiss. Res. 2, 1968, 49-59.

30. E.M. Carlisle. Silicon: a possible factor in bone calcification. Science 167, 1970, 279-280.

31. L.J. Green, J.D. Eick, W.A. Miller, and J.W. Leitner. Electron microprobe analysis of Ca, P, and Mg in mandibular bone. J. Dent. Res. 49, 1970, 608-615.

32. M.J. Glimcher. Specificity of the molecular structure of organic matrices in mineralization. In Calcification in Biological Systems, R.F. Sognnaes (Ed.), American Association for the Advancement of Science, Washington, D.C., 1960, 421-487.

33. H.J. Höhling, T.A. Hall, B. Boothroyd, C.J. Cooke, P. Duncumb, and S. Fitton-Jackson. Untersuchungen der Vorstadien der Knochenbildung mit Hilfe der normalen und elektronenmikroskopischen Electron Probe X-Ray Microanalysis. Naturwissenschaften 54, 1967, 142-143.

34. D. Baylink, J. Wergedal, and E. Thompson. Loss of proteinpolysaccharides at sites where bone mineralization is initiated. J. Histochem. Cytochem. 20, 1972, 279-292.

35. J.W. Smith. The disposition of proteinpolysaccharide in the epiphyseal plate cartilage of the young rabbit. J. Cell Sci. 6, 1970, 843-864.

36. M.J. Glimcher. Composition, structure, and organization of bone and other mineralized tissues and the mechanism of calcification. In Handbook of Physiology: Endocrinology, Vol. 7, R.O. Greep and E.B. Astwood (Eds.), American Physiological Society, Washington, D.C., 1976, 25-116.

37. L.V. Sutfin, M.E. Holtrop, and R.E. Ogilvie. Microanalysis of individual mitochondrial granules with diameters less than 1000 Ångstroms. Science 174, 1971, 947-949.

38. J.H. Martin and J.L. Matthews. Mitochondrial granules in chondrocytes, osteoblasts, and osteocytes. Clin. Orthop. 68, 1970, 273-278.

39. E.C. Weinbach and T. Von Brand. Formation, isolation, and composition of dense granules from mitochondria. Biochim. Biophys. Acta 148, 1967, 256-266.

40. R.S. Thomas and J.W. Greenawalt. Microincineration, electron microscopy, and electron diffraction of calcium phosphate loaded mitochondria. J. Cell Biol. 39, 1968, 55-76.

41. J.H. Martin and J.L. Matthews. Mitochondrial granules in chondrocytes. Calcif. Tiss. Res. 3, 1969, 184-193.

42. J.L. Matthews, J.H. Martin, H.W. Sampson, A.S. Kunin, and J.H. Roan. The mitochondrial granules in the normal and rachitic rat epiphysis. Calcif. Tiss. Res. 5, 1970, 91-99.

43. F.D. Vasington and J.V. Murphy. Ca^{++}

uptake by rat kidney mitochondria and its dependence on respiration and phosphorylation. J. Biol. Chem. 237, 1962, 2670-2677.

44. A.L. Lehninger, C.S. Rossi, and J.W. Greenawalt. Respiration dependent accumulation of inorganic phosphate and Ca^{++} by rat mitochondria. Biochem. Biophys. Res. Commun. 10, 1963, 444-448.

45. G.W. Engström and H.F. DeLuca. The nature of Ca^{++} binding by kidney mitochondria. Biochemistry 3, 1964, 379-383.

46. A.L. Lehninger. Mitochondria and calcium ion transport. Biochem. J. 119, 1970, 129-138.

47. L.V. Sutfin. High spatial resolution x-ray microanalysis of thin specimens in the scanning electron microscope. In SEM/1972 , O. Johari (Ed.), IIT Research Institute, Chicago, IL., 1972, 65-72.

48. L.V. Sutfin. Electron probe analysis of subcellular structures in biological tissues. In Proc. 8th Natl. Conf. on Electron Probe Analysis, 1973, 59A-59E.

49. L.V. Sutfin. An ultra-high vacuum SEM for microanalysis of ultrastructure of thin specimens. Proc. 11th Annual Conf. of Microbeam Analysis Soc., 1976, 61A-61E.

50. T.A. Hall and H.J. Höhling. The application of microprobe analysis to biology. In Proc. Vth International Conference on X-Ray Optics and Microanalysis, G. Möllenstedt and H. Gaukler (Eds.), Springer, Berlin, 1969, 582-591.

51. T.A. Hall and P. Werba. Quantitative microprobe analysis of thin specimens: continuum method. In Proc. 25th Anniversary Meeting of the Electron Microscopy and Analysis Group of the Institute of Physics, The Institute of Physics, London and Bristol, 1971, 146-149.

52. T.A. Hall. The microprobe assay of chemical elements. In Physical Techniques in Biological Research, Vol. IA, Second Edition, G. Oster (Ed.), Academic Press, New York, 1971, 157-275.

53. L.V. Sutfin and P.R. Garant. Analysis of intracellular structures in cells of calcifying tissues for the presence of inorganic elements. J. Bone Joint Surg. 55A, 1973, 659.

54. L. Sutfin and W.J. Landis. High spatial resolution microanalysis of ultrastructural features of calcifying tissues. J. Dental Res. 53A, 1974, 76.

55. M. Vannier and L. Sutfin. Identification and localization of minerals at the subcellular level. Clin. Res. 22, 1974, 251A.

56. H.J. Höhling and W.A.P. Nicholson. Electron microprobe analysis in hard tissue research: specimen. J. Microscopie 22, 1975, 185-192.

57. T.A. Hall, H.C. Anderson, and T. Appleton. The use of thin specimens for x-ray microanalysis in biology. J. Microscopy 99, 1973, 177-182.

58. J. Termine. Quoted in Chapter VI, Biophysical properties of connective tissues. In The Comparative Molecular Biology of Extracellular Matrices, H.C. Slavkin (Ed.), Academic Press, New York, 1972, 443-450.

59. B. Boothroyd. The problem of demineralisation in thin sections of fully calcified bone.

J. Cell Biol. 20, 1964, 165-173.

60. B. Boothroyd. Sources of artefact in preparations of bone for electron microscopy. In Electron Microscopy, Vol. 2, D. Steve-Bocciarelli (Ed.), Tipografia Poliglotta Vaticana, Rome, 1968, 429-430.

61. M.J. Glimcher and S.M. Krane. The organization and structure of bone and the mechanism of calcification. In A Treatise on Collagen, Vol. 2B, B.S. Gould and G.N. Ramachandran (Eds.), Academic Press, New York, 1968, 68-251.

62. R.C. Moretz, C.K. Akers, and D.F. Parsons. Use of small angle x-ray diffraction to investigate disordering of membranes during preparation for electron microscopy. I. Osmium tetroxide and potassium permanganate. Biochim. Biophys. Acta 193, 1969, 1-11.

63. R.C. Moretz, C.K. Akers, and D.F. Parsons. Use of small angle x-ray diffraction to investigate disordering of membranes during preparation for electron microscopy. II. Aldehydes. Biochim. Biophys. Acta 193, 1969, 12-21.

64. A.J. Morgan, T.W. Davies, and D.A. Erasmus. Changes in the concentration and distribution of elements during electron microscope preparative procedures. Micron 6, 1975, 11-23.

65. P.V. Thorogood and J. Craig Gray. Demineralization of bone matrix: observations from electron microscope and electron-probe analysis. Calcif. Tiss. Res. 19, 1975, 17-26.

66. W.J. Landis, M.C. Paine, B.T. Hauschka, and L.V. Sutfin. High resolution electron probe microanalysis of bone prepared by inert dehydration and ultracryomicrotomy. Trans. Orthop. Res. Soc. 1, 1976, 111.

67. W.J. Landis. X-ray microanalysis of calcium phosphate solids prepared anhydrously as calibration standards for mineralized tissues. Proc. 8th Internatl. Conf. on X-Ray Optics and Microanalysis, 1977, 174A-174F.

68. W.J. Landis, M.C. Paine, and M.J. Glimcher. Electron microscopic observations of bone tissue prepared anhydrously in organic solvents. J. Ultrastruct. Res. 59, 1977, 1-30.

69. A.L. Miller and H. Schraer. Ultrastructural observations of amorphous bone mineral in avian bone. Calcif. Tiss. Res. 18, 1975, 311-324.

70. W.J. Landis, B.T. Hauschka, and M.C. Paine. An examination of bone ultrastructure using inert dehydration and ultracryomicrotomy for tissue preparation. J. Bone Joint Surg. 57A, 1975, 571.

71. W.J. Landis, B.T. Hauschka, C.A. Rogerson, and M.J. Glimcher. Electron microscopic observations of bone tissue prepared by ultracryomicrotomy. J. Ultrastruct. Res. 59, 1977, 185-206.

72. S.Y. Ali and A. Wisby. Mitochondrial granules of chondrocytes in cryosections of growth cartilage. Amer. J. Anat. 144, 1975, 243-248.

73. S.Y. Ali. Analysis of matrix vesicles and their role in the calcification of epiphyseal cartilage. Fed. Proc. 35, 1976, 135-142.

74. C. Gay and H. Schraer. Frozen thin-sections of rapidly forming bone: bone cell ultrastructure. Calcif. Tiss. Res. 19, 1975, 39-49.

75. C. Gay and H. Schraer. Ultrastructure of rapidly forming bone prepared by frozen thin-sectioning. Fed. Proc. 34, 1975, 936.

76. H. Schraer and C. Gay. Matrix vesicles in newly synthesizing bone observed after ultracryotomy and ultramicroincineration. Calcif. Tiss. Res. 23, 1977, 185-188.

77. C. Gay. The ultrastructure of the extracellular phase of bone as observed in frozen thin sections. Calcif. Tiss. Res. 23, 1977, 215-223.

78. W.J. Landis and M.J. Glimcher. Electron diffraction and electron probe microanalysis of the mineral phase of bone tissue prepared by anhydrous techniques. J. Ultrastruct. Res. 63, 1978, 188-223.

79. W.J. Landis and M.J. Glimcher. X-ray microanalysis of molar Ca/P in subcellular structures of bone tissue prepared anhydrously. Trans. Orthop. Res. Soc. 3, 1978, 113.

80. E. Bonucci. Fine structure of early cartilage calcification. J. Ultrastruct. Res. 20, 1967, 33-50.

81. H.C. Anderson. Vesicles associated with calcification in the matrix of epiphyseal cartilage. J. Cell Biol. 41, 1969, 59-72.

82. H.C. Anderson. Matrix vesicles of cartilage and bone. In The Biochemistry and Physiology of Bone, Vol. IV, G.H. Bourne (Ed.), Academic Press, New York, 1976, 135-157.

83. S.Y. Ali, A. Wisby, L. Evans, and J. Craig Gray. The sequence of calcium and phosphorus accumulation by matrix vesicles. Calcif. Tiss. Res. (Suppl.), 1977, 490-493.

84. J. Thyberg and U. Friberg. The lysosomal system in endochondral growth. In Progress in Histochemistry and Cytochemistry, Vol. 10, No. 4, W. Graumann, Z. Lojda, A.G.E. Pearse, and T.H. Schiebler (Eds.), Gustav Fischer Verlag, Stuttgart and New York, 1978, 1-45.

85. T.A. Hall, H.J. Höhling, and E. Bonucci. Electron probe x-ray analysis of osmiophilic globules as possible sites of early mineralization in cartilage. Nature 231, 1971, 535-536.

86. R.H. Barckhaus and H.J. Höhling. Electron microscopical microprobe analysis of freeze dried and unstained mineralized epiphyseal cartilage. Cell Tiss. Res. 186, 1978, 541-549.

87. S.Y. Ali, J. Craig Gray, A. Wisby, and M. Phillips. Preparation of thin cryo-sections for electron probe analysis of calcifying cartilage. J. Microscopy 111, 1977, 65-76.

88. A. Roufosse, W.T. Sabine, W.J. Landis, and M.J. Glimcher. X-ray diffraction identification of brushite in newly mineralized embryonic chick bone. Trans. Orthop. Res. Soc. 2, 1977, 83.

89. A.H. Roufosse, W.J. Landis, W.K. Sabine, and M.J. Glimcher. Identification of brushite in newly deposited bone mineral from embryonic chicks. J. Ultrastruct. Res., in press.

90. E.P. Clark and J.B. Collip. A study of the Tisdall method for the determination of blood serum calcium with a suggested modification. J. Biol. Chem. 63, 1925, 461-464.

91. R.L. Dryer, A.R. Tammes, and J.I. Routh. The determination of phosphorus and phosphatase with N-phenyl-p-phenylene-diamine. J. Biol. Chem. 225, 1957, 117-183.

92. M.J. Glimcher, C.J. Francois, L. Richards, and S.M. Krane. The presence of organic phosphorus in collagen and gelatins. Biochim. Biophys. Acta 93, 1964, 585-602.

93. A. Veis and A. Perry. The phosphoprotein of the dentin matrix. Biochemistry 6, 1967, 2409-2416.

94. A. Veis and R.J. Schlueter. The macromolecular organization of dentine matrix collagen. I. Characterization of dentine collagen. Biochemistry 3, 1964, 1650-1657.

95. C.J. Francois, M.J. Glimcher, and S.M. Krane. The organic phosphorus content of the α chains of chicken bone collagen. Nature 214, 1967, 621-622.

96. M.J. Glimcher and S.M. Krane. The identification of serine phosphate in enamel proteins. Biochim. Biophys. Acta 90, 1964, 477-483.

97. P.T. Levine, M.J. Glimcher, and S.M. Krane. The identification and isolation of serine phosphate in the developing proteins of rodent enamel. Arch. Oral Biol. 12, 1967, 311-313.

Discussion with Reviewers

Reviewer I: There are certain inherent limitations to x-ray microanalysis. For example, the method cannot distinguish between any substances which, although quite different in their structure, have the same ratio of elements being analyzed, such as calcium pyrophosphate, $Ca_2P_2O_7$, and calcium hydrogen phosphate, $CaHPO_4$, which have the same Ca:P ratio. Would the author mention some of the other limitations to the use of the method for thin section examination.

Author: Theoretical and practical considerations for microanalysis have been well documented. These include (A) the nature of x-ray production in thin sections and the influence of extraneous signals (see, for example, text reference 52 and J.J. Goldstein and H. Yakowitz (Eds.), Practical Scanning Electron Microscopy, Plenum Press, New York, 1975), (B) instrumental limitations such as the effects of accelerating voltage, beam current, and brightness as well as the location and position of the x-ray detector (see J.I. Goldstein and D.B. Williams, X-ray analysis in the TEM/STEM, in SEM/1977/I, O Johari (Ed.), IIT Research Institute, Chicago, IL, 651-662), (C) specimen limitations including the important problems associated with specimen preparation, specimen contamination, radiation damage, mass loss, and spatial resolution (see P. Echlin and A.J. Saubermann, Preparation of biological specimens for x-ray microanalysis, in SEM/1977/I, 621-637, and H. Shuman, A.V. Somlyo, and A.P. Somlyo, Theoretical and practical limits of ED x-ray analysis of biological thin sections, in SEM/1977/I, 663-672), and (D) considerations for quantitation of x-ray signals, including multiple least squares fitting, spectral deconvolution of overlapping peaks, spectral filtering and smoothing, and other techniques to maximize the spectrum information (see text reference 47 and H. Shuman, A.V. Somlyo, and A.P. Somlyo, Quantitative electron probe microanalysis of biological thin sections: methods and validity, Ultramicroscopy 1, 1976, 317-339).

Reviewer II: Could the author elaborate on the tissue processing methods he uses himself.
Author: Specific details of anhydrous specimen preparative techniques are given in text references 68 and 71. These deal respectively with material prepared in organic solvents or by ultracryomicrotomy. The essential features of the former include dissection of fresh tissue in 100% ethylene glycol; replacement of glycol with anhydrous Cellosolve, the monoethyl ether of ethylene glycol; a change from Cellosolve to propylene oxide-epon (1:1), and final embedding in epon. Thin sections are cut with diamond knives onto an ethylene glycol trough and staining may be accomplished using salts dissolved in ethylene glycol.

Ultracryomicrotomy involves rapid freezing of small tissue pieces (< 1 mm^3) by immersion in isopentane (melting point -160°C) cooled by liquid nitrogen or by immersion directly into liquid nitrogen (melting temperature -196°C). Sectioning is performed at temperatures in the range of -60° to -80°C with fresh dry glass or diamond knives. Sections are collected with an eyelash and gently flattened on grids with a cold, highly polished copper rod; they are then dried in the nitrogen vapors of the ultracryomicrotome unit (an LKB Cryokit). Alternate grids may be exposed to osmium tetroxide crystal vapor to enhance ultrastructural features. Grids not examined immediately are stored over P_2O_5 in a desiccator.

Reviewer I: What are the advantages and disadvantages of x-ray microanalysis as compared to electron diffraction?
Author: While both methods provide a means for characterizing in situ certain ultrastructural features observed in tissue thin sections prepared by a variety of techniques, the essential differences between x-ray microanalysis and electron diffraction are the following: X-ray microanalysis is capable of identifying and localizing the elements present in specimen volumes on the order of 10^{-3} μm^3 or greater. Generally, all the periodic table elements above beryllium can be detected with good sensitivity. Detection limits in the range of 10^{-17} to 10^{-18} grams of an element are possible although the limit is poorer for the lower atomic number elements such as carbon, nitrogen, and oxygen. With careful consideration of background, statistical noise, spectral peak interferences, and other limitations specific to x-ray microanalysis, the elemental analysis may be made quantitative or semi-quantitative to provide elemental concentration ratios or absolute elemental mass or concentration values for the excited region under examination.

On the other hand, electron diffraction can identify elements as well as their chemical form as compounds in the specimen; and, whereas x-ray microanalysis cannot distinguish a particular structure from an admixture of several different species (phases, compounds, and the like), electron diffraction is capable of such an analysis. Further, electron diffraction can provide information on atomic structure and may in certain instances determine other general structural features of the specimens at that level of

resolution, which is superior to the level obtainable by x-ray microanalysis. It should be noted, however, that in usual practice the specimen volume examined by electron diffraction is relatively large, on the order of 10^{-1} μm^3 or more, compared to that for x-ray microanalysis, so that the ease of correlating the presence of a particular chemical species with the ultrastructure observed is not so great as with x-ray microanalysis. Moreover, and importantly, ultrastructural features amorphous to electrons, that is, those that do not generate a coherent electron diffraction pattern, may yet be characterized in terms of their element composition and distribution by x-ray microanalysis.

Regardless of their relative merits, however, it is clear that, when used to advantage, these two methods can provide valuable complementary information concerning the critical properties of unknown samples.

Reviewer IV: Using the formula for calculating Ca/P ratios, was there a significant difference if the off peak counts were not subtracted from the on peak counts?
Author: The answer depends upon the particular sample being examined as well as the number of counts generated from a given excited volume. In the case of microanalysis of the various calcium phosphate solids directly sprinkled dry onto grids, for example, the number of Ca or P counts on peak (from an individual particle) is high and the background contribution (Ca or P from the carbon-reinforced, Parlodion substrate) is very low; the off peak effect, then, is negligible in such a situation. On the other hand, the Ca/P ratios obtained from embedded (or frozen) and thin-sectioned tissue samples may change significantly if off peak counts are ignored. This is the result of the detection of organic phosphorus frequently present in the matrix (background) of regions also containing inorganic P (on peak). Since the net Ca counts generated from the specific volume are not necessarily large compared to the net P counts, even small changes in the net P counts extensively influence the Ca/P ratios calculated.

Reviewer III: Although the calibration curves (Figure 1) show a linear relationship between Ca/P intensity ratio and Ca/P molar ratio for various calcific species, the Ca/P intensity ratios obtained by x-ray microanalysis are generally higher than the chemically determined Ca/P molar ratios. What factors are considered to be responsible for this result?
Author: The relation between Ca/P intensity ratio and molar ratio is derived empirically. While the molar ratios are values fixed by the chemical nature of the compound under examination, theoretical considerations predict that characteristic x-ray production from a sample changes under different conditions. To a first approximation, x-ray intensity varies according to the number of atoms of the specific element (Ca or P in this case) encountered per cubic centimeter for the interaction between ionizing electrons and each element, the excitation energy of the ionizing electrons, and the ionization

cross section of each element. The cross section in turn exhibits a dependence on both the excitation energy of the ionizing electrons and the absorption edge energy of the characteristic x-ray of the element of interest. In a series of preliminary experiments in this laboratory, a variation in x-ray intensity of Ca, P, and Ca/P was observed for calcium phosphate compounds as a function only of excitation energy in the JEOL JSM 50A. The Ca/P intensity ratios were somewhat higher than the corresponding Ca/P molar ratios at 25 keV (Figure 1) but lower when calculated for 10 and 15 keV. A few such values may be directly compared in Tables 1 (25 keV) and 2 (15 keV) of the text.

Reviewer I: What role, if any, does pyrophosphate have in early calcification?
Author: There is considerable evidence that inorganic pyrophosphate may act in the regulation of mineral formation and dissolution through its participation in a number of physical chemical reactions principally inhibitory in their effect. These include inhibition of the formation of a calcium phosphate solid phase from a solution, inhibition of the growth of solid phase particles, inhibition of secondary nucleation, inhibition of the phase transformation of amorphous calcium phosphate to poorly crystalline hydroxyapatite, and inhibition of the dissolution rate of calcium phosphate solid phase particles. On the other hand, a stimulatory effect on the rate of mineral formation in tissue culture by low concentrations of inorganic pyrophosphate has also been reported.

Pyrophosphate in the form of hydrated calcium pyrophosphate has been identified in a number of pathological conditions such as pseudogout, chondrocalcinosis, and other joint diseases; the mechanism of calcification occurring in these cases is unclear.

A few articles pertinent to this topic include text references 26-29 and
N.W. Alcock and M.E. Shils. Association of inorganic pyrophosphatase activity with normal calcification of rat costal cartilage in vivo. Biochem. J. 112, 1969, 505-510.
H.C. Anderson and J.J. Reynolds. Pyrophosphate stimulation of calcium uptake into cultured embryonic bones. Fine structure of matrix vesicles and their role in calcification. Dev. Biol. 34, 1973, 211-227.
S.M. Krane and M.J. Glimcher. Transphosphorylation from nucleotide di- and triphosphates by apatite crystals. J. Biol. Chem. 237, 1962, 2991-2998.
D.J. McCarty. Calcium pyrophosphate dihydrate crystal deposition disease. Arthritis Rheum. 19, 1976, 275-285.
D.J. McCarty, D.C. Silcox, F. Coe, S. Jacobelli, et al. Diseases associated with calcium pyrophosphate dihydrate crystal deposition. Amer. J. Med. 56, 1974, 704-714.
K.P.H. Pritzker, P.-T. Cheng, M.E. Adams, and S.C. Nyburg. Calcium pyrophosphate dihydrate crystal formation in model hydrogels. J. Rheum. 5, 1978, 469-473.
R.G. Russell. Metabolism of inorganic pyrophosphate (PPi). Arthritis Rheum. 19, 1976, 465-478.

R.G.G. Russell, S. Bisaz, A. Donath, D.B. Morgan, and H. Fleisch. Inorganic pyrophosphate in plasma in normal persons and in patients with hypophosphatasia, osteogenesis imperfecta, and other diseases of bone. J. Clin. Invest. 50, 1971, 961-969.

Reviewers I and IV: Could you provide a photograph showing both the matrix vesicles and the smaller extracellular particles analyzed in the rat growth plate cartilage? What is the nature of these extracellular particles?
Author: The vesicles and small particles are frequently located within the same regions of the extracellular tissue spaces. While a limiting membrane distinguishes the matrix vesicles, no such structure appears to enclose the small dense particles. The particles are generally irregularly-shaped and are much more numerous and widely distributed in the extracellular spaces than are the rounded vesicles.

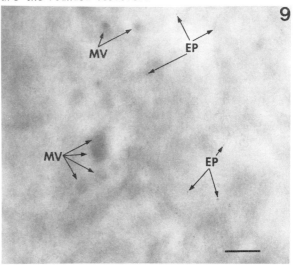

Fig. 9. Electron micrograph of an unstained thin section showing a region of the extracellular tissue space from the proliferating zone of rat growth plate cartilage, prepared by dry ultracryomicrotomy (-80°C). The matrix vesicles (MV) are membrane-bound, relatively large rounded structures, while the small extracellular particles (EP) appear as loci of intrinsic electron density with no detectable limiting membrane. (Bar = 0.5 μm).

Reviewer IV: The author has labeled a phosphorus peak in a matrix vesicle, Figure 5. Since this peak is not significantly different from no peak, Figure 6, is the labeling justified?
Author: Channel-by-channel subtraction of Figure 6 from Figure 5 clearly indicates a phosphorus peak at the position labeled for these spectra.

Reviewer I: Are the Ca/P ratios different in the collagen adjacent to the calcifying extracellular particles as compared to collagen adjacent to matrix vesicles?
Author: No significant difference is apparent (compare, for example, text Figures 6 and 8).

Reviewer III: What is the author's conclusion regarding the nature of the calcium phosphate species in the mitochondrial granules of rat chondrocytes? Does organic phosphorus of the matrix contribute to the observed Ca/P ratio of 1.0 obtained from the mitochondrial granules?

Author: Selected area electron diffraction has been used in conjunction with x-ray microanalysis to examine these particular structures. No coherent electron diffraction pattern could be generated from the mitochondrial granules, a result consistent with that reported for mitochondrial granules observed in bone tissue (text reference 78). The failure of mitochondrial granules in rat chondrocytes to generate a coherent diffraction pattern, together with the observed low Ca/P ratio determined by microanalysis, may be attributable to a number of plausible explanations, a situation preventing specific identification and characterization of the chemical composition of the granules. An elaboration of this problem may be found in text reference 78.

The presence of organic phosphorus from sources other than those associated with the granules, themselves, has been considered in the calculation of the Ca/P ratios. The contribution of organic phosphorus is sampled in spectra generated adjacent to the granules (off peak, see text Figure 4, for example) and then subtracted from the phosphorus signal obtained directly from the granules (on peak, text Figure 3, for example). It is possible, however, that ATP molecules, which may be bound to the mitochondrial granules, and phosphorus-containing components of mitochondrial membranes, which appear in close contact with granules, may provide an extraneous phosphorus source measured on peak. Their effect, then, would be to lower the Ca/P ratios obtained from the mitochondrial calcium phosphate solid phase.

REVIEWERS: I K. P. H. Pritzker
II A. Boyde
III L. M. Buja
IV W. G. Banfield

SCANNING ELECTRON MICROSCOPY/1979/II
SEM Inc., AMF O'Hare, IL 60666, USA

REPLICATION TECHNIQUES WITH NEW DENTAL IMPRESSION MATERIALS IN
COMBINATION WITH DIFFERENT NEGATIVE IMPRESSION MATERIALS

C. H. Pameijer

Boston University Goldman School of Graduate Dentistry
100 East Newton Street
Boston, MA 02118

Abstract

New materials and new techniques have enabled the fabrication of more reliable and more accurate replicas. Not only is the reproduction of detail of importance, but the expertise required from the operator and the time involved to produce a replica are considered key factors. For various reasons a reliable and reproducible replication technique for scanning electron microscopy offers many advantages. Recently a new dental precision impression material has been introduced, which in combination with low viscosity resins has produced superior results over other techniques. This combination processed by means of a centrifugal casting machine has produced replicas which could easily be compared to a standard test die at magnifications up to 3000 X. More in depth testing will have to be performed to establish whether these materials can be universally applied to a broad spectrum of replication problems. The combination Reprosil/ Spurr low viscosity imbedding medium yielded replicas of high quality which can be made with simple equipment and without possessing special skills. Centrifugation of the positive replication material into the negative impression virtually eliminated the entrapment of airbubbles.

KEY WORDS: Replica, Synthetic Elastomers, Silicone Rubber, Polyethers, Vinyl Polysiloxane, Spurr Resin, Epoxy Resin, Dentistry, Biological Specimen Preparation, Centrifugation, Archaeological Artifacts.

Introduction

The use of replica techniques for scanning electron microscopy (SEM) has opened avenues of research hereunto impossible. This method has been employed in laboratory studies and *in vivo* experiments and has found application in numerous other areas in which archeology assumes an important place. Replication of precious artifacts offers several advantages; i.e.:
a) The object is too valuable for direct examination,
b) The object is too large for SEM examination,
c) The object can not be sampled,
d) The object is in the field.
Obviously certain precautions have to be taken into account and the conditions to which the materials used for replication are subjected as reported by Pfefferkorn[1] and Pameijer[2] cannot be ignored. This presentation offers techniques which can easily be acquired utilizing materials which can readily be obtained, most frequently from a dental supplier. For a complete review of the literature, the reader is referred to two articles published in the past[1,3]

Materials and Methods

Accuracy of reproduction tests were conducted on a machine fabricated stainless steel cylinder measuring 9mm in diameter and 10mm high. On the top side (experimental side) Knoop hardness indentations were made at subsequently 25, 50, 100, 300, 500 and 1000 gr. A similar approach has been reported by Ayer[4]. These indentations were clearly visible under a stereo dissecting microscope and had the advantage that they allowed for an easy orientation in the SEM. The test cylinder was mounted in the center of an aluminum base which measured 50 x 50 x 3mm, with a threaded bolt. Thus the die could be handled easily without danger of tipping over. In addition it could be observed directly in the SEM. The test cylinder was not metallurgically polished and scratches remained clearly visible creating a surface comprising of Knoop indentations and numerous grooves, ridges and fine scratches. In this particular study, all comparisons were made on the 25 gr indentation which measured 28 μm in

Table I: This table presents the ratings on a scale from 1-5,in which the with an * indicated combinations were not tested.Ratings were based on magnifications in the order of 1000X and 3000X. ———►

Table I

Pos. Impression Materials / Neg. Impression Material	Impregum	Citricon	Acetate tape	Polysulfide	Reprosil	Xantopren
Oxy-dent	5	5	1	5	4	4
Spurr	4	5	5	5	1	5
Stycast	3	5	5	5	2	3
Citricon	*	*	5	*	*	*
Reprosil	*	*	5	*	*	*
Xantopren	*	*	3	*	*	*

length and 4.6 μm in width in the center of the indentation.

Based on past experience a selection amongst the most promising materials for negative and positive replication was made. Table I lists the materials tested and their evaluations reflecting ·the quality of reproduction by comparing the micrographs of the replicas to those of the original die and rating them as follows: excellent = 1, very good = 2, good = 3, fair = 4, and poor = 5 at standard magnifications of 560 and 1000 X. Optimum resolution of replicas was examined at 3000 X.

The following techniques were employed to obtain negative replicas. With exception of the acetate tape, which technique has been described previously[2], all materials were mixed according to the manufacturers instructions and syringed on the test cylinder. Once polymerization was completed the impression was separated. The materials used for positive replication were prepared and either inserted directly with a spatula or pipetted or centrifuged in the mold with an automatic electric centrifuge operating at 500 revolutions / minute*. After polymerization all positives were separated and coated with gold palladium at a thickness of approximately 150 Å in a.Hummer II**. In addition half the elastomeric positives were coated in a vacuum evaporator***. Analysis of the micrographs made from positive silicone rubber replicas has indicated that the method of coating is of paramount importance to the stability of the replicas. An earlier report has clearly demonstrated the surface distortion caused by metal coating in a sputtering machine[3].

Observations

Table I demonstrates the quality of replicas rated on a scale from 1 - 5. This evaluation was somewhat subjective and was therefore separately carried out by an independent observer. The similarity in judgement was remarkable and without error certain replicas were favored over others. At low magnifications (up to 200 X) most methods produced acceptable results. At magnifications of 560 X, however, differences were becoming noticeable and some replicas were already considered inadequate. At 1000 X only a few combinations in particular demonstrated excellent reproduction of detail which can be seen in Figures 1-3. Emphasis should be placed on the combina-

---

* Oxy-dental Products, Irvington, New Jersey
** Techniques, Alexandria, Virginia
*** JEOL, Medford, Massachusetts

tion Reprosil and Spurr being only slightly better than Reprosil/Stycast (Figures 2 and 3). Reprosil is an addition cured vinyl polysiloxane elastomeric impression material which is non-toxic and non-irritant. The material complies with the highest requirements of the American Dental Association Specification No. 19 and American National Standard Z 156. 19-1971 for elastomeric impression materials with regard to dimensional stability. Mixing time is 45 seconds allowing approximately 3 minutes to syringe the material on the object to be replicated. As a rule the faster this is done the least risk one takes to lose fine detail. Setting time is 3 1/2 minutes at 37°C. Usually more time is required at room temperature (5-8 minutes). The material will remain dimensionally stable for many days. Shelf life of the material is quite good for one year at room temperature. An added advantage is the fact that separation of the materials presents no difficulties at all, thus reducing the risk, as is so common with other combinations, that the positive replica is contaminated with negative impression material.

Without exception the replicas cast by means of the centrifugal casting machine demonstrated very few and usually no air bubbles. The advantage of this method cannot be overemphasized since air bubbles have been one of the most frequently reported artifacts.

Although often condemned elastomeric positive replicas yielded good results if prepared properly. The use of a vacuum evaporator rather than sputter coating bypassed all problems of distortion usually caused by the latter method. The recommendation to immediately coat and observe these replicas after preparation still has to be maintained since polymerization shrinkage over a period of time does take place. The resin

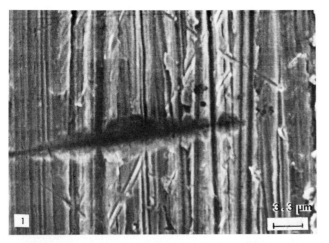

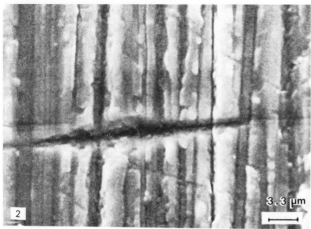

replicas offer therefore a distinct advantage over this method. It is also advised that elastomeric positives are either coated with a thicker layer of metal or observed at a kV of 10 or below. Both methods diminish resolution, however too thin a coating and a higher kV can cause damage to the specimen.

It should also be pointed out that the versatility of application of elastomeric impression materials far exceeds that of applying acetate tape. Especially heavily contoured surfaces and undercut areas present limitations to the use of acetate tape. On the other hand the reproduction of detail is unsurpassed and the dimensional stability is excellent over any period of time.

Conclusions

1. A combination of Reprosil and Spurr low viscosity embedding medium resulted in the best reproduction of detail of all combinations tested.

2. The combination Acetate tape/Oxy-dent although very accurate lacked the versatility of Reprosil/Spurr.

3. The combination Spurr/Reprosil appeared to be the most versatile method.

4. The dimensional stability of Reprosil allows several days before pouring the positive replica without causing changes.

5. Centrifugal casting greatly improved the quality of replicas by eliminating air bubbles contaminating the surface.

Appendix

Impregum, Premier, Philadelphia, PA. USA;
Citricon, Kerr, Romulus, MI. USA;
Acetate tape, Ladd, Burlington, VT. USA;
Polysulfide, Experimental rubber base, Healthco, Boston, MA. USA.;
Reprosil, De Trey AG, Zurich, Switzerland;
Xantopren Blue, Unitek, Monrovia, CA. USA;
Oxy-Dent, Oxydental Products, Irvington, NJ, USA;
Spurr low viscosity embedding medium, Polysciences, Warrington, PA. USA;
Stycast 1266, Emerson & Cuming, Canton, MA. USA.

Acknowledgement

The author wishes to acknowledge Wayne Sharaf for his assistance in the technical phase of this project.

References

1. G.E. Pfefferkorn and A. Boyde, Review of replica techniques for scanning electron microscopy, Scanning Electron Microscopy/1974, IIT Research Institute, Chicago, IL., 75-82.
2. C.H. Pameijer, Replica techniques, in Principles and Techniques of Scanning Electron Microscopy, Vol. 4, Van Nostrand, NY., 1975, Chapter 3.
3. C.H. Pameijer, Replica techniques for scanning electron microscopy - A review. Scanning Electron Microscopy/1978, Vol. II. SEM Inc., AMF O'Hare, IL 60666, USA. 831-836,810.
4. H.D. Ayer, A detail duplication test for dental materials. N.Y. State Dent. J., 1959, 25:

Fig. 1. Knoop rhomboidal pyramid indentation of 25 gr. in stainless steel test specimen.

Fig. 2. Replica specimen of Reprosil/Spurr combination depicting 25 gr. indentation. Positive replica was made by centrifugal casting.

Fig. 3. Less detail was achieved with the Reprosil/Stycast combination. Positive replica was made by centrifugal casting.

82-84.

DISCUSSION WITH REVIEWERS

N.A. Nielson: Comparison of Figure 2, the Reprosil/Spurr replica, with Figure 1, the original metal surface, shows that the replicated surface detail is unsharp. A major diagonal scratch in the stainless steel does not show in the replica. On this basis why was the replica system rated 1 or excellent, and do you believe that a superior negative/positive replication system remains to be found or needs to be found?
Author: Since at all times a two step impression method will have to be utilized to produce a positive, it is unreasonable to expect a replica to exhibit the same features at increasingly high magnifications.   The best system was rated excellent in comparison to the other replicas. In the meantime the search to find better materials should continue.

M. Arteaga: Is it possible that impression materials may be too accurate such that tearing or distortion may occur when severe undercuts are encountered?
Author: One of the advantages of elastomeric impression materials is their elastic memory, i.e. they will return to their original shape when withdrawn from undercuts. Of course when tearing occurs the impression should be discarded. If this happens when the positive is separated from the negative cast the result may still be a usable positive replica.

G. Pfefferkorn: How is the shrinkage of Reprosil with the time? There seems to be a contradiction in the text between good stability of silicone material on the one hand and the recommendation to immediately prepare silicone replica positives on the other hand.
Author: According to the manufacturer Reprosil is dimensionally stable for several days. This is an extra advantage and means that the time of pouring the positive replica is not critical. Reprosil is classified as a vinyl polysiloxane. None of the silicone impression materials (Citricon, Xantopren) or polysulfides exhibit this feature. The polyethers however, possess the same dimensional stability. Therefore when any of these elastic materials are utilized as positive replicas it is of utmost importance to observe them immediately.

M. Arteaga: Could you provide more information regarding the replicating materials you used: Identification of the type of compound would be helpful even though the complete formulation is unknown.
Author: A variety of negative impression materials were used: Polyethers (Impregum), Silicone rubbers (Citricon, Xantopren), Acetate tape, Polysulfide rubber materials, Oxy-dent, Spurr and Sty-cast are epoxy resins. It would carry too far to present a complete description of each individual material. For more information I refer to "Dental Materials, Properties and manipulation". R.G. Craig, W.J. O'Brien and J. M. Powers. The C.V. Mosby Co., St. Louis, 1975.

M. Arteaga: I am sure you will agree that the faithfulness of the replication of ultra-fine structure is inversely related to the viscosity. Have you experimented with a less viscous starting material or a slower polymerization to determine the effect on resolution of detail in the sub-micron range?
Author: I do agree with your statement relative to viscosity. It is important however to add to this the wetting capability of a material which relates to the surface tension. Reprosil has a low viscosity and sets fast. This is for the type of research on patients in which I am mainly involved very important.

C. Pearson: Is there any exothermic reaction during polymerisation of the replica materials?
Author: No. All materials tested for the first step replication polymerize without generating heat. Materials for positive replication were Oxy-dent and Stycast (both polymerize at room temperature) and generate no heat and Spurr, which in contrast requires heating to 70° C and a cure for 8 hours.

C. Pearson: These techniques have only been used on a hard solid surface. In the archaeological field this would mean restriction to bone, ceramics, glass and stone (and then only if not painted or glazed). Can the negative replica materials be used on more friable material such as wood, painted objects, organic fibres and glazed (deteriorated) objects?
Author: I don't think that the use of Reprosil is restricted to hard surfaces only. Since I have not as yet applied this technique to soft specimens and to the materials you refer to a word of caution seems appropriate. All negative impression materials contain dyes to render the material color specific. It was observed that in certain combinations the positive cast was stained. Since for the tests reported in this article, a stainless steel die was used the effect of staining could not be observed. However the possibility exists that porous surfaces may be stained by the negative impression material, thus causing irreversible damage.

SCANNING ELECTRON MICROSCOPY/1979/II
SEM Inc., AMF O'Hare, IL 60666, USA

INLAYED TEETH OF ANCIENT MAYANS: A TRIBOLOGICAL STUDY USING THE SEM

A. J. Gwinnett   and   L. Gorelick*

Dept. Oral Biology and Pathology *Department of Dentistry
Health Science Center   Long Island Jewish/Hillside
S.U.N.Y.   Medical Center
Stonybrook, NY 11794   New Hyde Park, NY 11040

Abstract

A study has begun of inlayed teeth of Meso-American Indian skulls using scanning electron microscopy and modelling techniques. Round cavity preparations 2-3 mm in diameter and 1-2 mm deep had been cut through the enamel and just into the underlying dentin of the teeth. The vertical walls of the preparation met the floor in either a square, rounded or undercut form. Towards its center, the floor was occasionally elevated, sometimes depressed and commonly rounded. Closer examination showed abrasion anomalies as concentric, shallow grooves cut into the tooth tissue. A modification of the Semenovian principle was employed to determine the tool: 1) from the marks registered in the cavity and 2) the outline form of the preparation itself. Preparing cavities experimentally in teeth using wood and stone drills and sand as an abrasive produced certain characteristics consistent with those in the Meso-American teeth in which wooden drills created a variety of cutting patterns which included flat, elevated and depressed floors in the preparations. We have tentatively concluded that suggestions for the use of a tubular drill does not adequately explain the variety of forms encountered and that the cutting patterns were more consistent with the use of a wooden drill and sand.

KEY WORDS: Inlayed Mayan Teeth, Abrasion Patterns, Cutting Anomalies, Tribology, Archaeological Applications.

Introduction

At the time of the Spanish conquest of Meso-America, chroniclers observed that the native peoples filed their front teeth. The many geometric configurations caused by filing have since been classified.[1] Unnoted at the time, but later revealed by archeological excavation, were teeth which had been inlayed with jade and other stones. Confined to the anterior maxillary and mandibular teeth,[2,3] this practice is believed to have existed several centuries before Christ.

The purpose of inlaying teeth is considered by many to have religious and esthetic connotations and clearly inferred the existence of a fascinating, sophisticated but as yet unknown technology. Fastlicht[4] in his beautifully illustrated treatise on dentistry among pre-Columbian Indians, analyzed Knoblock's classical work[5] on Indian drilling. He concluded that because there was an elevation at the base of the cavities, characteristic of tubular drilling, the cavities were made with a round, hard perforating stone tube of jade or later of copper. An abrasive slurry such as powdered quartz or obsidian was probably used. It is important to note that Knoblock used a variety of drills of which only one was tubular. He drilled holes in a stone pebble, some of which had a slightly raised central portion in the floor. This corresponded with the same characteristic seen in the inlayed teeth. However, this elevation was caused not only with a tubular drill but also with a solid hickory peg. There is no doubt that tubular drills were used by Meso-Americans since evidence exists in trephined skulls.[4]

In order to better understand the possible means of drilling human teeth for stone inlaying we chose to modify the Semenovian[6] principle of "Functional Analysis". In essence, such a principle states that the wear pattern on a tool can be used to identify its use if the wear pattern on the tool can be duplicated experimentally. Our hypothesis was that the tool marks, indelibly engraved into the walls of a cavity together with its configuration, can be used to determine the tools used in its preparation. Using scanning electron microscopy, a study was conducted to develop standards of the outline form of the cavities and the cutting patterns (science of tribology) engraved in their walls. These standards were then reproduced experimentally on extracted teeth using wood and stone drills (microliths) with fine sand as an abrasive. Wood and stone are logical choices since both would have been readily available. In addition, changes in tool shape and cavity form with time were monitored using a sequential drilling and replication method.[7]

Material for Study

A Mayan skull with an intact, inlayed dentition was placed at our disposal through the courtesy of Dr. Gordon Ekholm of The American Museum of Natural History in New York.

Nine anterior teeth were available for study in the skull (Fig. 1). Eight of the teeth were inlayed either with jade or hematite. While the inlays were retained by mechanical fit they were readily and carefully displaced with the tyne of a dental explorer. The unfilled cavity preparations were occluded with sand which was readily and carefully removed with a small, blunt plastic instrument.

After removal of the inlays, brownish, relatively smooth, chalky deposit was found lining the floor and part of the walls of the cavity. This deposit, possibly a cement, was readily detached from the tooth with light application of a plastic instrument. Some of the larger deposits were recovered for later analysis. Clean preparations were now ready for examination.

Method of Examination

Impressions were made of the teeth and the cavity preparations using a commercially available silicone material (Unitek Corp. California). Models were cast from the impressions using an epoxy resin (Epoxydent, New Jersey) in which the original shapes, forms and abrasion patterns were accurately reproduced. Individual models of teeth and cavities were mounted on aluminum stubs with conductive cement and coated with a conductive layer of gold in a water-cooled sputtering device (Technics Inc.). The coated samples were then examined in an AMR scanning electron microscope. Observations were recorded photographically.

Findings

Cavities

All the cavity preparations were circular and located on the labial face of the teeth (Fig. 2). The diameters were different for different preparations. The long axis of the cavity was almost perpendicular to the tooth surface. This produced a cavity whose incisal wall depth frequently exceeded its cervical wall depth. This, coupled with the location of the cavity in the incisal half of the tooth, produced relatively deep preparations without exposing the vital pulp of the tooth. In the skull, four preparations measured 2.5 mm in diameter, one was 3.0 mm, two were 3.5 mm and two were 4.0 mm. With one exception, all these preparations were approximately 1.5 mm deep and all into dentin. The 2.0 mm

exception had a floor perforation which appeared to expose the pulp chamber. The two widest preparations contained hematite inlays whereas the smaller ones were inlayed with jade. In three of the preparations the floors were slightly raised (Fig. 3) in three others, slightly depressed (Fig. 4 A & B), with the remaining four being slightly rounded (Fig. 5 A & B). The dentin of the floor and walls of the preparations showed occasional concentric cutting patterns (Fig. 6), while the angle between the wall and the floor was usually slightly rounded (Fig. 7). The margin of the cavity was often chamfered (Fig. 8).

Inlays

The hematite inlays were flat, almost planoparallel, circular disks measuring only slightly less in diameter than the cavity in which they were found. They were 0.5 - 1.0 mm thick and 3-4 mm across. The size and shape of the jade inlays varied with the oral surface commonly appearing convex, (Fig. 9) a shape known in lapidary as cabochon. The stones were seldom circular and varied in width from 1-3 mm and 1-2 mm in height. The oral surface was always polished while the remaining unexposed surfaces were rough. The stones lacked any cutting or abrasion marks.

Experimental Drilling

The experimental drilling was conducted on extracted human teeth using a "sequential replication" procedure. This procedure involved frequent interruptions in the cutting process to take silicone impressions of the cavity and the tool. Epoxy resin models were then cast from the impressions and when placed in order showed the sequential changes in shape of the cavity and the tool as drilling progressed.

The cavities were cut in extracted human teeth suitably clamped, using wood and flint drills. Each drill was approximately 2 mm in diameter with the wood being either birch or lignum vitae shaped into cylindrical, solid rods. The flint rods were fashioned using a rotating, heatless stone wheel.

The rods were held in the chuck of a hand powered drill and driven alternately in a clockwise and counter-clockwise manner similar to the discontinuous motion of a bow drill. Fine sand was used as an abrasive with water to bind the sand when using the flint drill. Sand only was used with the wood drill since this material softened in water to significantly reduce its cutting efficiency.

Wood rod

Birch, as well as the hard lignum vitae, wore rapidly as drilling progressed. Commencing with a flat, circular end, the rods soon began to round off (Fig. 10) with the center changing from flat, to depressed, to a slightly elevated configuration. The experimental cavities conformed, in reverse, to the shape of the drill. For example, a central depression in the drill matched a central elevation in the cavity (Fig. 11). The shape of the tool and also the cavity changed continuously as drilling progressed. The cavities were 1-2 mm deep and approximately 0.5 mm wider than the 2 mm rod. Abrasion or cutting

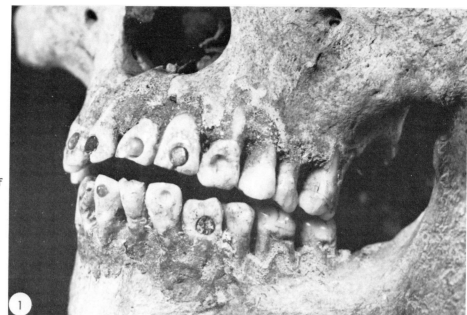

Fig. 1. Inlayed teeth of ancient Mayan skull.

Fig. 2. Composite showing configuration of the cavity preparation.

2

Fig. 3. (Below) Cavity with elevated floor. Photo width-- 4.3 mm.

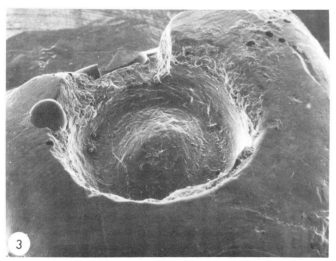

3

patterns appeared as concentricities both on the walls and floor of the cavities located predominantly in dentin. These characteristics, as well as cavity configuration, were remarkably similar to those seen in the Mayan teeth (Fig. 12 A & B).

Flint rod

The flint in comparison to wood showed remarkably little wear during cavity preparation. The cavities were basically similar in dimension to those created with the wood, except that the floor of the preparation was flat, producing a more simple topography than that seen with wood.

Discussion

The depth and diameter of the jade inlayed cavities and those inlayed with hematite were generally similar. However, in all cases the topography of the floor showed significant variation. The floor represents the face upon which the abrasive works and upon which it is maintained by pressure exerted through the drill. It was postulated that the topographical variations in the floor (tool or abrasive marks) reflect the nature of the tool and concomitant change in the

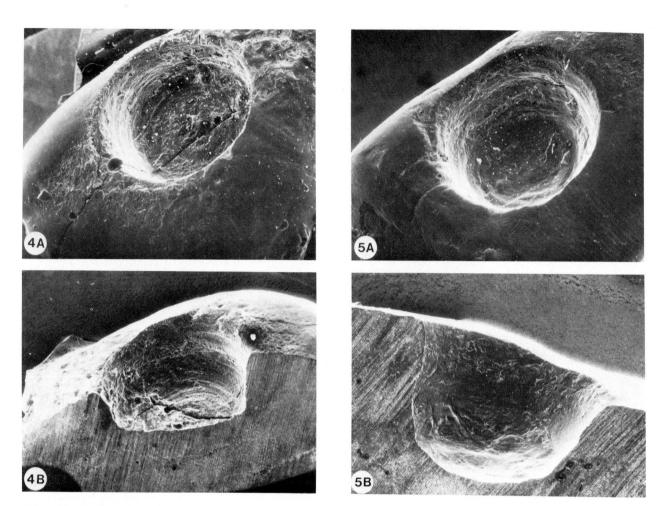

Fig. 4A Cavity with slightly depressed floor.
   4B Sagittal section through cavity
       P.W. - 5.5 mm.

Fig. 5A Cavity with rounded floor.
   5B Sagittal section through cavity.
       P.W. - 5.1 mm.

P.W.= Photo Width.

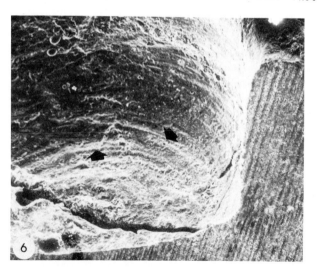

Fig. 6 Concentric abrasion lines (arrow)
P.W. - 0.8 mm.

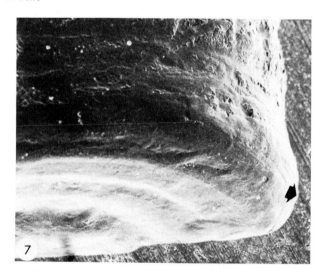

Fig. 7 Rounded inner line angle of cavity
(arrow). P.W. - 0.6 mm.

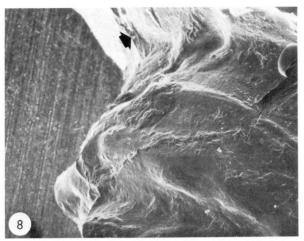

Fig. 8   Beveled cavo-surface margin (arrow) of cavity. P.W. - 0.8 mm.

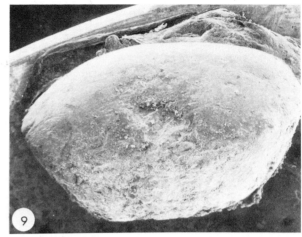

Fig. 9   Jade inlay with rounded (cabochon) oral surface. P.W. - 3.6 mm.

Fig. 10   Wear configuration on wooden drill. P.W. - 3.8 mm.

Fig. 11   Wear configuration on drill matching cavity outline. P.W. - 5.1 mm.

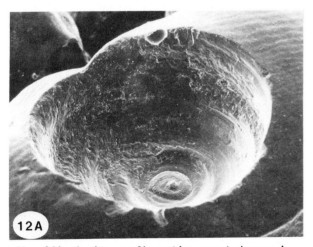

Fig. 12A   Cavity configuration created experimentally. P.W. - 3.2 mm.

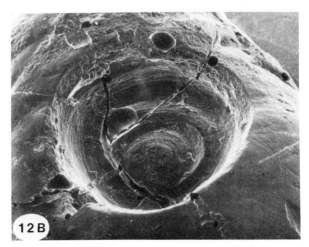

Fig. 12B   Cavity configuration in Mayan tooth. P.W. - 4.1 mm.

tool due to its wear. This was supported by observations from the experimental drilling sequence. While flint maintained its geometric integrity, the wood constantly wore. Elevations and depressions and rounded corners were found in the floor of the cavity even at different times in the preparation of the same cavity. The shape at any particular time represented a reciprocal shape in the tool at that time (see Fig. 11). It is conceivable, therefore, that different floor patterns could exist in any number of teeth in the same skull as was the case in the skull examined. In addition, all inlays were not necessarily done at the same time. At present, some of the subtle variables in floor topography can be accounted for only with the use of wooden drills. The concentric, abrasion patterns were consistent with those produced through rotation of a drill using fine sand as an abrasive.

It was presently concluded that at a time before metal tool technology, native indians drilled teeth with wooden drills rotated with a bow. There is evidence that bow drills are still used in parts of Central and South America. Dry sand was probably among the abrasives used. How the cavity was begun is not known but the angle of the bevel at the margin of the cavity suggests the use of a relatively large starter tool used to create a depression which would contain and confine the abrasive. Keeping the abrasive at the working face of the drill is a key to cutting efficiency. Using a drill of relatively smaller diameter, the bevel remained as evidence of cavity initiation. Probably supine, facing the sun and suitably euphoric from chewing on coca leaves the subject, by our estimates, must have had to endure the slow, traumatic drilling process for several hours.

References

1. D.F.R. Borbolla. Types of tooth mutilations found in Mexico. Am. J. Phys. Anthrop. 26; 1940, 30-34.
2. S. Fastlicht. Tooth mutilations in Pre-Columbian Mexico. J. Am. Dent. Assoc. 36; 1948, 197-203.
3. P.A. Sweet, M.G. Buonocore and I.F. Buck. Pre-Hispanic Indian dentistry. Dent. Radiog and Photog. 36; 1963, 3-8.
4. S. Fastlicht. Tooth mutilations and dentistry in Pre-Columbian Mexico. Quintessence Books, Los Angeles, 1976, 50-64.
5. B.W. Knoblock. Banner stones of the North American Indian. Published by author, La Grange Illinois, 1939, 29-48.
6. S.A. Semenov. Prehistoric Technology. Barnes and Noble. New York, 1964, 2-7.
7. L. Gorelick and A.J. Gwinnett. Ancient seals and modern science. Expeditions. 20; 1978, 38-47.

Discussion with Reviewers

D.K. Whittaker: In view of the valuable nature of the material were the teeth radiographed, and if so could you comment upon the post operative vitality of the teeth and secondly upon the often debated question as to whether these types of inlays were carried out pre- or post mortem.

V.C. Barber and C. Emerson: What evidence is there that holes were drilled into the teeth of living people? Is it possible that they were later decoration of the skull for some sort of burial ritual?

Authors: Radiographic examination was not part of our study design. However, studies[4] have shown pathological change associated with the roots of inlayed teeth suggesting that their preparation was done pre-mortem in some cases.

V.C. Barber and C. Emerson: Can the authors provide representative total times that were taken to produce the appropriate holes in the experimental drilling programme?

Authors: The question is a valid one but we have not yet timed the procedure except to note that preparation of a single cavity took several hours.

D.K. Whittaker: In your paper you imply that the cementing material was discarded--would it not have been of interest to compare its analysis with that carried out by Linne on similar material in 1950?

V.C. Barber and C. Emerson: Has any attempt been made to analyze the material that the authors propose might be the remains of cement? One presumes that the fit of the inlays in the teeth cavities was not sufficiently close to preclude the use of an adhesive. Is this correct?

N.A. Nielsen: Would you know if anyone has attempted to confirm that a cement was used, if so, what analytical work may have been done?

Authors: We have not had an opportunity to analyze the cement used to retain the inlays. Analyses have been conducted by Linne's (Dental Decoration in Ancient Mexico. Ethnos Stockholm (1950) pp. 166-173) and Romero, J. and Romano, A. (quoted by Fastlicht, F. Tooth Mutilations and Dentistry in Pre-Columbian Mexico. Quintessence Books. Los Angeles, 1976, p. 81). The principle elements were calcium, phosphorus, silicon and aluminum, iron and sodium.

H. Newesely: The need for use of the SEM in this project is not convincing--could you comment?

Authors: This report presents preliminary findings and is not intended to show justification for the use of scanning microscopy in such a study. Further research will no doubt answer the question more fully.

SCANNING ELECTRON MICROSCOPY/1979/II
SEM Inc., AMF O'Hare, IL 60666, USA

OPTIMIZATION OF AN ANALYTICAL ELECTRON MICROSCOPE FOR X-RAY MICROANALYSIS:
INSTRUMENTAL PROBLEMS

J. Bentley, N.J. Zaluzec, E.A. Kenik, and R.W. Carpenter

REMAG, Metals and Ceramics Division,
Oak Ridge National Laboratory,
P.O. Box X
Oak Ridge, TN 37830

Abstract

The addition of an energy dispersive x-ray spectrometer to a modern transmission or scanning transmission electron microscope can provide a powerful tool in the characterization of materials. Unfortunately this seemingly simple modification can lead to a host of instrumental problems with respect to the accuracy, validity, and quality of the recorded information. This tutorial reviews the complications which can arise in performing x-ray microanalysis in current analytical electron microscopes. The first topic treated in depth is fluorescence by uncollimated radiation. The source, distinguishing characteristics, effects on quantitative analysis and schemes for elimination or minimization as applicable to TEM/STEMs, D-STEMs and HVEMs are discussed. The local specimen environment is considered in the second major section where again detrimental effects on quantitative analysis and remedial procedures, particularly the use of "low-background" specimen holders, are highlighted. Finally, the detrimental aspects of specimen contamination, insofar as they affect x-ray microanalysis, are discussed. It is concluded that if the described preventive measures are implemented, reliable quantitative analysis is possible.

1. Introduction

Over the past several years Transmission Electron Microscopes (TEMs), Scanning Transmission Electron Microscopes (STEMs), and combination TEM/STEMs equipped with Energy Dispersive X-Ray Spectrometers (EDS) have become very popular. One can regard such an instrument as an Analytical Electron Microscope (AEM) of the simplest type. Their popularity is due to the attractive possibility of performing elemental analysis with high spatial resolution (<100 nm) in a straightforward manner [i.e., without many of the often gross corrections which have to be made in bulk analyses performed using Scanning Electron Microscopes (SEMs) or Electron Probe Micro Analyzers (EPMAs) — the ZAF (atomic number, absorption, fluorescence) correction procedures]. The ability to obtain image and diffraction analysis simultaneously with the elemental analysis is also often a very desirable feature. It was expected, several years ago, that all that was necessary to perform elemental analysis was a rather simple (in most cases) mechanical interface of a suitable EDS detector to the microscope and then easily interpreted, observed spectra originating solely from the small area illuminated by the electron probe could be recorded. However more careful investigations have shown the situation in practice is more complex. Most problems arise from evolution of the AEM from the TEM or STEM rather than being developed from its inception as a "true AEM."

This paper will deal with all the major effects reported to date and solutions to problems caused by these effects. In sections 2 and 3 the various problems associated with the so-called "systems background"[1] are discussed, and in section 4 the effects of contamination, as they enter into x-ray microanalysis, will be treated. Here we define the term "systems background" in its more general sense to be those effects resulting from interactions with the hardware of the instrument, or in other words, the "system." Section 2 discusses effects originating primarily in hardware remote from the specimen, and section 3 with effects more concerned with the immediate environment of the specimen. This division is somewhat arbitrary, and is made solely for clarity since all the effects associated with

KEY WORDS: Analytical Electron Microscopy, STEM, High Votage Electron Microscopy, X-ray Microanalysis, System Background, In-hole Signal, Specimen Holders, Contamination, Field Emission Electron Gun, Artefacts "systems background" are interrelated and in any classification there is considerable overlap.

The most serious errors in elemental analysis are usually associated with uncollimated fluorescing radiation produced in the illumination system. The presence of this radiation is revealed when the electron probe is positioned in a hole in the specimen, circumstances under which no signal should be detected. However, invariably x-rays characteristic of the specimen and/or the specimen environment are observed — the so-called "In-Hole" signal. The effect is quite general. Over the last few years there has been (sometimes heated) controversy with, for example, one faction claiming x-rays as the responsible radiation, another faction claiming the effects were all due to electron "tails," and some even protesting that there were no such problems in their particular microscope. As we shall see, this situation arose mainly because of (1) differences in the detailed design of the variable second condenser aperture and/or (2) differences in specimen type. However, these two reasons were not among the most frequently proposed reasons which included accelerating voltage, instrument manufacturer, operator incompetence et cetera.

Two aspects of optimization will be briefly mentioned for the sake of completeness: detector positioning and collimation. Although it is common practice for detectors to be mounted orthogonally to the electron beam, considerable increases in peak-to-background (P/B) ratio result from orienting the detector at an acute angle to the beam on the electron entrance side of the specimen.[2] Optimum detector collimation can certainly help to reduce some of the undesirable effects considered in this paper. However, we will assume that the collimation is already adequate, since the guidelines are well established (see, for example, ref. 3). The single most important consideration is to locate the "entrance" of the collimator as close as possible to the specimen.

Artifacts arising during the physical process of x-ray detection in the Si(Li) detector and those associated with the signal processing chain (e.g., detector efficiency, silicon escape peaks, peak broadening, peak distortion, pulse pile-up, deadtime correction, and microphonics) are not discussed in this paper. The interested reader is referred to Reed,[4] Woldseth,[5] or Fiori and Newbury.[6]

Although the emphasis in this paper is toward the materials sciences, many of the topics covered are applicable to biological research. Unless otherwise stated we will be dealing with instruments having accelerating voltages of ~100 kV, since these are most popular. X-ray analysis performed at up to 1 MeV in high voltage electron microscopes will also be discussed.

2. Systems Background (I): Remote Production

In this section, possibly the most important, we will deal with various aspects of uncollimated fluorescing radiation produced remote from the specimen in the illumination system. The effects are generally undesirable, but many of the principles discussed have been turned to advantage in at least one application.[7] A study of similar effects in the SEM was made by Bolon and McConnell[8] and many of the same processes have been found to be important on transmission instruments.[1,9—22]

2.1 Sources and Characteristics of Uncollimated Radiation

Uncollimated electrons may arise as a result of scattering around the condenser apertures, if their design allows this, or scattering from the bore of thick or contaminated apertures. Even if these scattered electrons are subject to the focusing action of a final probe-forming lens, there is the possibility of the formation of electron "tails" — in other words, appreciable current at large radial distances from the small probe used in microanalysis. The energy of these scattered electrons will be a large fraction of the incident energy, so their range in solids will be on the order of a few tens of μm. These electron tails can be very effective in fluorescing both the specimen and its environment.

The current emitted from the filament is reduced by many orders of magnitude in the process of "condensing" the illumination. This focusing and collimating action necessitates the interception of the beam with various fixed and variable condenser apertures. The electron current absorbed by each aperture depends, of course, on the precise geometric and electron-optical configuration employed in the particular instrument. In conditions appropriate for microanalysis it is doubtful that the second condenser (C_2) aperture is ever primarily responsible for reducing the probe current by >10^5 times as implied by Goldstein and Williams[18] and considerable current can be absorbed by the fixed condenser apertures.[1] The range of the electrons in the aperture material is typically a few tens of μm, so that they are absorbed within the top part of the aperture. Both bremsstrahlung and x-rays characteristic of the aperture material are produced (for a complete treatment see Dyson[23]) and although considerable absorption may occur, large fluxes of both may be transmitted through the aperture if it is thin enough. The energy spectrum of the bremsstrahlung exhibits a monotonic decrease with increasing photon energy, but the low energy photons are absorbed most strongly and so the resulting energy spectrum of x-rays incident towards the specimen tends to exhibit a maximum at intermediate energies.[18] The large flux of x-rays incident onto the specimen or its environment may cause considerable fluorescence, and because of the nature of the energy spectrum the fluorescence may be quite selective (the photoelectric cross section is greatest for incident photons of energy just greater than the absorption edge).

The nature of the x-ray energy spectrum can be judged experimentally by (Compton) scattering from a specimen in the form of a sheet of low atomic number material with the electron probe arranged so as to pass through a central small hole in the sheet.[14,22]

2.2 Effects of Uncollimated Radiation

The presence of uncollimated fluorescing radiation can have a number of effects detrimental to x-ray microanalysis. The first is the loss of spatial resolution. This is a special problem in inhomogeneous specimens where even qualitative analysis may be precluded. The fluoresced area can easily be several mm diameter at the specimen position as evidenced by the presence of characteristic peaks of the specimen holder material in the in-hole signal. The second related effect is that the specimen environment may be detected. This will be covered fully in section 3, below. Thirdly and most important, accurate quantitative analysis, the primary purpose of an AEM, is prevented.

Two types of experiment where the effects of uncollimated radiation can be most insidious for the unwary are: (1) quantitative analysis of homogeneous specimens and (2) determination of small compositional changes (either qualitative or quantitative analyses) in basically homogeneous specimens. Because of its importance the remainder of this section will deal with the effect on quantitative analysis. At the outset we wish to point out that the type of specimen will be important and that the details will be dependent upon the particular fluorescing species. We will consider a metallurgical specimen of the self-supporting thinned-disk type, with high ($\geqslant$6 keV) and low ($\leqslant$3 keV) energy characteristic lines. This may be a medium- or high-atomic number element or a suitable homogeneous alloy. The uncollimated radiation causes fluorescence, particularly in the "bulk" part of the specimen, which is detected simultaneously with that induced by the electron probe in the thin region of interest. The problem is that the "bulk" fluorescence signal may contribute so significantly that quantitative analyses based on thin film electron excitation are incorrect.

Figure 1 shows schematically the type of behavior that is observed as a function of foil thickness. In fig. 1(a) the intercepts at zero thickness represent the in-hole signal. The intensities then increase linearly with increasing foil thickness until absorption effects become significant (affecting the low energy line first). The resulting high energy/low energy intensity ratio is plotted in fig. 1(b) together with the value characteristic of thin foil electron excitation [i.e., the dotted line of fig. 1(b) corresponding to the ratio of the two lines in fig. 1(a) after subtraction of the in-hole signal]. Initially this ratio decreases with increasing sample thickness as the (collimated) electron probe induced signal begins to increase and eventually dominate that produced by the uncollimated radiation. In general, this ratio will approach but not reach the value characteristic of thin foil electron excitation. The amount of this discrepancy between the experimental and theoretical thin foil values will depend on the intensity of the in-hole signal and can be as high as several orders of magnitude. Eventually the ratio goes through a relative minimum and begins to increase as preferential absorption by the specimen of the lower energy characteristic line occurs. This type of behavior has been observed by a number of authors.[10,11,14,15,17,24] The same general behavior should be followed regardless of whether electrons or x-rays are the uncollimated fluorescing radiation. In particular, the in-hole high/low energy intensity ratio normally will be greater

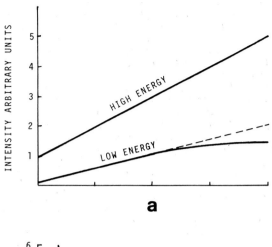

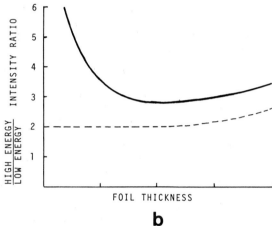

Fig. 1   Schematic behavior of (a) x-ray intensities and (b) the resulting high/low energy intensity ratio in the presence of uncollimated fluorescing radiation. See text for details.

than the ratio induced by the electron probe in a thin foil.

If x-rays are the dominant uncollimated fluorescing species then the in-hole signal exhibits very small low energy peaks, possibly hardly detectable. This is the result of two effects: (1) Low energy transitions will not be excited to the same extent as higher energy transitions because of the characteristics of the incident x-ray flux described above (i.e., an absence of low energy photons with a maximum in the energy spectrum at intermediate photon energies), and (2) greater absorption of low energy x-rays in the thick regions of the specimen where their predominant production occurs.

If electron tails are the dominant fluorescing radiation then the low energy peaks may be small but certainly detectable, since only absorption (effect 2, above) is operative.

We mentioned above that specimen type (thickness and composition) will be an important variable. For uniformly thin specimens such as evaporated films or extraction replicas, the effects will not be as pronounced. For x-rays in particular, the in-hole signal may be vanishingly small because of their weak interaction with matter but electron tails may still cause a significant, if greatly reduced, signal. Specimens of low atomic number will also produce lower in-hole signals if x-rays are the fluorescing species because of less absorption of the incident flux by the specimen and a decreased contribution of the photoelectric cross section to the total absorption cross section (the inelastic scattering cross section tending to dominate). Fluorescence by electron tails may still be fairly extensive even in low atomic number materials.

2:3 Distinguishing between Electron and X-Ray Fluorescence

Electron tails *may* be present at significant levels in some microscopes, but unless special precautions and design features have already been incorporated into the instrument it is almost certain that x-rays *will* be present at significant fluxes. As a first step in the elimination of uncollimated radiation, identification of the specific species present is necessary. Here we simply list several techniques which may help in the distinction between electrons and x-rays. However, some are rather qualitative and should be applied with care. Generally it is best to use several or all in combination.

1. Examine the design of the illumination system in order to determine if it is physically possible for scattered electrons to travel to the specimen other than through the apertures. In particular, examine whether the variable C_2 aperture holder fills the whole cross section of the bore of the microscope at that point.

2. With the variable C_2 aperture blocked off, with, for example, a suitably thick blank disk in place of the normal C_2 aperture, and with the aperture drive in position, attempt to measure any current below the C_2 aperture or at the specimen position, using a Faraday cup or special specimen holder and electrometer. The measuring device should be extremely sensitive since the currents involved may be $\leqslant 10^{-13}$ A.

3. With a perforated bulk specimen of "suitable" material perform an "in-hole" experiment. If high energy x-rays are the only fluorescing species, then the continuum background will be very low and the peak-to-background ratio should be very high compared to the P/B ratio for a spectrum measured with the probe on the specimen. By a "suitable" material, we mean an element or homogeneous alloy of medium- or high-atomic number with characteristic lines well separated from possible systems peaks arising from the specimen environment or illumination system. Considerable caution should be exercised in this test since the continuum background can arise through several processes other than generation by stopping of electrons in the specimen (e.g., Compton scattering of continuum x-rays by the specimen or specimen holder, particularly if these are of low atomic number).

In addition, the energy spectrum of the continuum resulting from electron excitation of a bulk specimen has definite characteristics which can also be used to help identify the fluorescing species. An indication of the characteristics of the energy spectrum due to excitation by electron "tails" can be obtained from an analysis made with the electron beam spread over the entire specimen.

4. Perform an "in-hole" experiment on a perforated bulk specimen of an element with high (>6 keV) and low energy (<3 keV) characteristic lines or a homogeneous alloy with similar characteristics. In this case the interpretation is based on the previous section 2.2 and is difficult to quantify. Therefore, great care in its application should be exercised. However, in general, high-energy x-rays fluorescing the specimen will result in negligibly small low-energy characteristic peaks in the in-hole signal, but the presence of electron tails will cause easily detectable, if much reduced, low-energy peaks.

Points 3 and 4 are probably the easiest to perform (they can be made at the same time on a suitable specimen) and when used together the risk of misinterpretation is minimized. They therefore comprise a logical first choice but points 1 and/or 2 should not be neglected.

2.4 Elimination or Minimization of Uncollimated Radiation

The precise methods employed to eliminate or minimize the amount of uncollimated radiation will, of course, depend upon the particular instrument. An important first step is to confer with the instrument manufacturer. Several now offer modifications or accessory devices available both for new and older microscopes as retro-fit items. For those unable to pursue this course of action and also to help in providing a basis for critical evaluation of solutions offered by manufacturers, we will now outline in a systematic way a procedure developed from various users solutions.[1,10—12,16,18,19,22,25,26] The basic scheme is first to eliminate electron tails and then reduce the x-ray flux by modifications to the condenser apertures, followed by the installation of extra non-beam-defining apertures or collimators.

2.4.1 Elimination of Electron Tails. The elimination of electron tails is a fairly straightforward task. Basically, the entire bore of the column, except for the apertures themselves, must be completely filled at or near the aperture planes. The most important aperture to be considered is the variable C_2. Simple modifications, such as the insertion where needed of spray apertures above and/or below the variable C_2, will usually intercept and virtually eliminate any electrons scattered around the fixed and variable apertures.

The only other precaution needed to prevent or minimize electron tails is to ensure that the aperture edges do not become badly contaminated. This point needs most careful attention when using thick (see below) rather than the more common thin or semi-thick self-cleaning types.

2.4.2 Fixed Condenser Apertures. As mentioned earlier, considerable x-ray fluxes may be generated at fixed apertures in the illumination system.[1] An obvious solution is to replace the standard apertures with thick ones of high atomic number. Tantalum or, more usually, platinum has been used. The purpose, of course, is to absorb within the aperture itself as much as possible of the x-ray flux created near the top (electron entrance) surface. Thicknesses of ~0.5 mm should be used, if possible, and the aperture should be uniformly thick to as close as possible to the beam-defining edge. This is because the region near the aperture hole intercepts the highest flux of electrons and is therefore the most prolific source of x-rays. However, it has been observed that electron tails may be produced by scattering from the sides of the bore of thick apertures; therefore, a compromise solution was employed.[26] A conventional molybdenum condenser aperture was aligned concentrically with a 0.5-mm-thick Pt aperture of slightly ($\sim$50 μm) larger bore, the molybdenum being on the electron entrance side. With this arrangement, as shown schematically in fig. 2, the molybdenum serves as the beam-defining edge but presents only a very small annulus as an effective x-ray source. The thin edge of the molybdenum also remains cleaner for longer times (see above), presumably due to the greater heating effects experienced.

2.4.3 Variable Second Condenser Aperture. As with fixed condenser apertures, an obvious suggestion for the reduction of the x-ray flux produced by the impingement of the beam is to employ thick, high atomic number apertures. Many of the considerations above (section 2.4.2) also apply. A thick variable C_2 aperture also provides a further reduction of the flux of high-energy x-rays produced at the fixed apertures located above the variable aperture. Thick ($\sim$0.5 mm) platinum sheet added as part of the aperture clamping device on the end of the aperture rod may also aid in this absorption action.[26]

2.4.4 Non-Beam-Defining Apertures and Collimators. The final and most innovative of the modifications to reduce the high-energy x-ray flux onto the specimen is the introduction of non-beam-defining apertures or collimators between the second condenser aperture and the specimen.[10] Although simple thick apertures have been used and there is some merit in the associated operational ease, more conventional collimators are considerably more effective in reducing the high-energy x-ray flux. The extent of both the umbra and penumbra is reduced as simple ray-tracing will confirm. Two radiation opaque non-beam-defining apertures, one just below the C_2 aperture and the other just above the specimen, act in the same way as a long collimator (but reduce scattering from the bore) and probably constitute the optimum choice. With this arrangement the source and irradiated area of the specimen are limited to areas just greater than the size of the apertures used.

The interference of the aperture or collimator with the beam in certain electron optical configurations is an important practical consideration. For example, rocking beam modes, low magnification scanning and secondary electron imaging may all be precluded or limited by incorrectly placed apertures. Removable apertures/collimators may allow positioning at optimum shielding locations without restricting special operating modes. Further, an optimum design should also incorporate a mechanism which allows fine adjustments in position and orientation, this being particularly useful for a collimator where the most precise location and orientation relative to the beam is required for correct operation.[25]

2.4.5 Illustrative Example. The spectra in fig. 3 illustrate the effect of sequentially

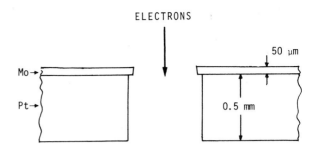

Fig. 2 Arrangement of conventional molybdenum and thick platinum apertures in combination for use as fixed condenser apertures.

performing some of the above procedures for the reduction of uncollimated radiation. The results are for a JEM 100 C AEM operated at 120 kV in the STEM mode. The specimen, mounted in a double-tilt beryllium specimen holder, was a Ni-20 at.% Mo thinned disk. A 400 μm C_2 aperture was used producing a probe current of $\sim$1 nA. In comparing the spectra of fig. 3, note the changes in vertical scale. Figure 3(a) is the signal obtained with the probe on the specimen at a foil thickness of $\sim$50 nm and fig. 3(b) is the in-hole signal before any special preventative actions were made. The aperture design in the instrument allows little possibility of electron tails and no special measures were taken. [Note that in all in-hole spectra, fig. 3(b) to (f) the high P/B ratios and the extremely small Mo L/Ni K and Mo L/Mo K and large Mo K/Ni K ratios compared to fig. 3(a)]. Figure 3(c) shows the effect of replacing the fixed Mo, C_1 and C_2 condenser apertures with ones made from 0.5-mm-thick Pt and fig. 3(d) of using the Mo and Pt apertures in combination. Figure 3(e) illustrates the further reduction obtained by also using a thick (0.5 mm) Pt variable C_2 aperture, fixed on the aperture rod using 0.5-mm-thick Pt sheet. Figure 3(f) shows the final in-hole spectrum obtained after further adding a brass non-beam-defining collimator of 1 mm diameter and 15 mm long just above the objective lens. A reduction of about an order of magnitude in the in-hole signal is seen between fig. 3(b) and fig. 3(f). Figure 3(g) is from the same area as fig. 3(a) but after the modifications and subtraction of fig. 3(f). For comparison, fig. 3(h) shows a spectrum produced by electron excitation of a "bulk" part of the self-supporting disk specimen.

Again it is necessary to realize that this example is specific for a JEM 100 C and that different effects can occur in different instruments.

2.4.6 Special Problems in Some Dedicated STEMs. The dedicated STEM, in addition to sharing many of the problems of a conventional TEM-based AEM, may have its own unique problem in that the objective or illumination aperture can be very near ($\sim$5 mm) the specimen. There is no effective way to reduce the x-ray flux produced at this aperture, but two solutions to this special problem exist:

1. Simply perform x-ray microanalysis with the aperture removed. This greatly reduces the problem of uncollimated radiation, but

unfortunately results in a severe deterioration in spatial resolution due to an increase in probe size.

2. Installation of a probe-defining aperture nearer the electron gun. This may involve major design changes to the instrument (H. L. Fraser, University of Illinois; V. G. Microscopes Ltd., private communication).

2.5 Corrections for Quantitative Analysis

Although the above procedures can greatly reduce the flux of uncollimated fluorescing radiation, complete elimination is impossible. It is still necessary to subtract the in-hole signal from the signal obtained with the probe on the specimen. An important assumption is that if the in-hole signal is collected for the same period of (machine) time as the specimen signal the correction procedure will be fairly accurate. However, an implied assumption is that the probe/emission current is the same for the two experiments. This assumption is probably good for a stable thermionic electron gun, but for an instrument using a cold FEG*it may be quite inaccurate. Commonly, the emission current from a cold FEG shows a continual decrease with time following flashing of the tip. Therefore, a method is needed for measuring the integrated electron dose to use in determining the correct time of acquisition.[27] Another implied assumption is that the dead-time correction is accurate since the signal from the specimen and hole may be different by orders of magnitude.

We emphasize that it is important to first reduce the in-hole signal before subtraction; otherwise large statistical errors may arise. A suitable in-hole signal which one could specify as "acceptable" might be one of the same order as the average specimen bremsstrahlung[1] in specimens of interest at foil thicknesses typically used or, as used by the authors recently in a performance specification for the purchase of an AEM, 1% of the signal from areas of thickness ≤100 nm for several specified materials covering a wide range of atomic numbers.

2.6 X-Ray Analysis in the HVEM

There is considerable interest in AEMs employing accelerating voltages greater than 100 kV. As far as x-ray microanalysis is concerned, the important advantage is an increase in mass sensitivity. Although the ionization cross section decreases with increasing electron energy, the bremsstrahlung cross sections decrease more quickly so that the peak/background ratio shows a monotonic increase with electron energy. For example, the calculated P/B for nickel at 1 MeV is ∿2.5 times that at 100 keV.[19]

As expected, however, the uncollimated fluorescing radiation problems (particularly high-energy x-rays) become more severe at higher accelerating voltages.[10,19,25] At an accelerating voltage of 200 kV the general behavior is rather similar to that discussed above for 100 kV instruments. An acceptably low in-hole signal was obtained after the installation of a non-beam-defining collimator.[10,25] Serious attempts at x-ray analysis on an HVEM have only been reported recently.[19,20] In the more systematic and complete investigation severe problems were encountered.[19] In addition to the installation of a non-beam-defining lead

---
*Field emission gun.

collimator above the objective lens, large amounts of lead shielding around the detector were found to be necessary. However, in this case a usable, if not optimum, system was achieved and the predicted increase in P/B ratio was confirmed experimentally. Some development is needed in order to optimize x-ray microanalysis in an HVEM, not the least of which being an electron optical system for small high-current density probes.[28] Research on this topic is being continued by the present authors.

3. Systems Background (II): Specimen Environment

It is quite possible that for the unwary investigator the extent and even existence of uncollimated fluorescing radiation, dealt with in the previous section, could be difficult to estimate. However, it is uncommon to find that any operator is unaware of some problems regarding the local specimen environment. Characteristic x-rays from the specimen support grid and/or specimen holder are almost always present. The problem may seem trivial, as often indeed it is, but significant detrimental effects, sometimes quite subtle, can result in problems for x-ray microanalysis which are just as important and limiting as those caused by uncollimated fluorescing radiation.

3.1 Sources of Near-Specimen Systems Background

Two types of fluorescing radiation contribute to the excitation and possible subsequent detection of the specimen environment: remotely produced uncollimated fluorescing radiation, and what might be called specimen generated fluorescing radiation. The former was dealt with extensively in section 2 where the effects of direct fluorescence of the specimen were emphasized. However, the specimen holder and other components of the instrument hardware located near the specimen may also be fluoresced.

Specimen generated fluorescing radiation includes backscattered electrons, diffracted high energy electrons, characteristic and bremsstrahlung x-rays produced by the electron probe, and characteristic x-rays produced in the specimen by uncollimated high energy x-rays. The characteristic x-rays are emitted isotropically but the other forms of radiation all exhibit anisotropic distributions. The distribution of scattered electrons depends upon the diffracting conditions in the specimen and on the electron energy. The continuum or bremsstrahlung x-rays from the specimen generally exhibit a forward peaked distribution which moves more toward the forward direction as the accelerating voltage increases.[2,23]

The two contributions to the fluorescence of the specimen environment can be demonstrated and evaluated from data similar to that shown in fig. 4. An aluminum specimen in a conventional copper specimen holder was used, and the probe current was varied by changing the size of the variable C_2 aperture. The copper signal was measured with the probe at the same position on the specimen and "in-hole" for each size of condenser aperture. As expected, there are two contributions to the signal — the approximately constant in-hole signal, and one due to specimen generated fluorescing radiation which is directly

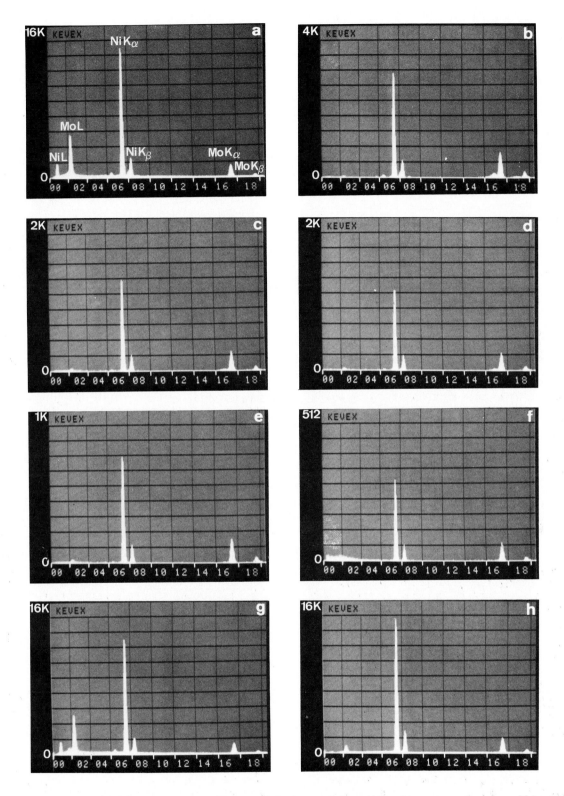

Fig. 3 X-ray spectra illustrating the effects of instrumental modifications to reduce uncollimated radiation. Acquisition time = 100 s. Note the changes in indicated vertical scales. (a) Specimen, before modifications, (b) In-hole, before modifications, (c) In-hole, thick platinum fixed condenser apertures, (d) In-hole, Mo-Pt combination fixed condenser apertures, (e) In-hole, thick platinum variable C_2 aperture, (f) In-hole, brass non-beam-defining collimator, (g) Specimen after modifications, and subtraction of in-hole spectrum, (h) "bulk" region of specimen, electron excitation. See text for details.

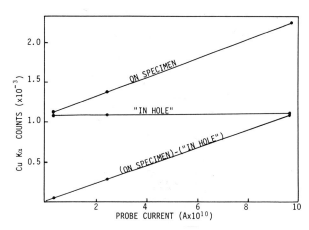

Fig. 4 CuK$_\alpha$ intensity versus probe current measured with the probe positioned on a high-purity aluminum specimen and also passing through a hole in the specimen.

proportional to the probe current.

3.2 Effects of Near-Specimen Systems Background

The most obvious effect of detecting systems background from the specimen environment is that the acquired spectrum no longer represents the area of specimen analyzed. This may be more of an aesthetically displeasing effect than a serious problem, in which case most investigators would ignore it. However, subtle effects may be present together with more obviously undesirable problems for accurate quantitative analysis.

3.2.1 Overlap of Specimen and Systems Peaks. The problem of overlap of systems peaks with characteristic lines in the desired specimen signal may take one of two forms. First, overlapping of neighboring peaks may occur. In obtaining quantitative data it may be necessary to use complex peak stripping or profile fitting procedures for which computer processing is required. The data usually can be obtained but often only after expending considerable effort. In particular the most severe problems may arise when small specimen peaks are overlapped by large systems peaks.

The second form of peak overlap arises when an analysis is required for an element also detected in the systems background signal. In this situation quantitative and possibly even qualitative analysis may be precluded. Particularly common elements are copper and iron. Magnesium and aluminum in "modified" specimen environments are other possibilities.

3.2.2 Fluorescence by Near-Specimen Systems Background. Besides fluorescence of the specimen by uncollimated radiation originating in the illumination system of the microscope, radiation originating in the specimen itself or in its local environment can cause substantial problems for quantitative analysis. In particular, systems background radiation from the local specimen environment can fluoresce the specimen. Although bremsstrahlung x-rays will surely have an effect, strong characteristic lines in the systems background have the most potential for pronounced detrimental consequences. This is because of the potential for highly selective fluorescence by

characteristic peaks. One reported example of this effect concerned analysis of a Ni—20 at. % Mo specimen in copper and beryllium specimen holders.[14] A further example is shown in fig. 5. The spectra were obtained from the same area of a specimen of type 316 stainless steel examined in beryllium [fig. 5(a)] and copper [fig. 5(b)] double-tilt specimen holders. Note in particular the magnitude of the NiK$_\alpha$ peak compared to the FeK$_\beta$ peak. In the copper holder the CuK$_\alpha$ systems background causes increased fluorescence of iron and chromium (only CuK$_\beta$ radiation strongly fluoresces the nickel). The apparent nickel content of the analyzed region is reduced from 12.8 wt % in the beryllium holder to 10.8 wt % in the copper holder.

The magnitude of the effect in fig. 5 appears at first sight to be surprisingly large, considering the rather moderate magnitude of the copper systems peak. However, it was observed from the diffraction pattern that the lower part of the copper specimen holder was intercepting a considerable fraction of the diffracted electrons. It is therefore probable that the large copper signal being generated below the specimen was responsible for the majority of the fluorescence, but that only a small part was actually being detected. This example thus also serves to illustrate some of the more subtle consequences of the generation of systems background.

As far as fluorescence of the specimen (or support grid) by the bremsstrahlung produced in the specimen itself is concerned, one point is worth emphasizing. As the specimen is tilted (to a suitable specimen-detector orientation) the lower half of the specimen will intercept a larger flux of bremsstrahlung and this may lead to increased fluorescence problems. The magnitude of this effect is proportional to the amount of bremsstrahlung radiation produced by the electron probe-specimen interaction and is therefore proportional to the specimen thickness being analyzed. The effect can be minimized by the use of high take-off angle detectors which permit the specimen to remain more nearly normal to the electron probe.

3.2.3 Continuum Generation in the Near-Specimen Environment. The above effects of systems background have come primarily from characteristic peaks. However, the bremsstrahlung component, produced by absorption of backscattered and diffracted electrons in the solid materials around the specimen, must also be considered under conditions where its presence may be important. A case in point is the quantitative analysis of biological specimens.[21] In addition, the decrease in P/B ratio will result in a decrease in mass sensitivity. Both of these points are also discussed in section 4.2.3.

3.3 Elimination or Minimization of Local Systems Background

In the elimination or minimization of local systems background, we will first assume that beryllium support grids are used, near specimen apertures are withdrawn, and that, following the procedures in section 2, electron tails have been eliminated and the flux of high energy x-rays significantly reduced. Also, as mentioned in section 1, effective collimation on the x-ray detector can also help to reduce significantly the

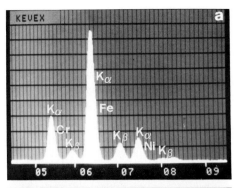

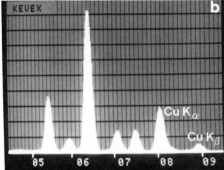

Fig. 5 X-ray spectra from the same area of a type
316 stainless steel specimen examined in (a)
beryllium and (b) copper double-tilt specimen
holders. Note the change in the relative intensities of chromium, iron, and nickel due to fluorescence of iron and chromium by CuK_α systems
background.

detected local systems background.[9] On the other
hand, the only really effective solution for many
of the effects described above is to incorporate
low-atomic-number materials in the construction of
the specimen environment and in particular the
specimen holder. We consider beryllium to be the
optimum choice for such a material, but its potentially hazardous nature, which requires special
precautions to be taken in machining processes,
reduce its availability. Although reduced bremsstrahlung production may result from the use of
materials such as aluminum or magnesium alloys,
the presence of detectable characteristic peaks
can still leave some problems unsolved (such as
peak overlap or fluorescence). Carbon coating
has been extensively used in the SEM in an attempt
to overcome systems background, but this is not
necessarily compatible with the "clean" vacuum
requirements discussed in the next section in connection with the minimization of contamination.
Some special high-strength graphites, however, may
be a reasonable second choice to beryllium.

The extent of necessary modifications may
vary from one instrument to another and depends
to some degree on individual operational procedures.
Among those reported are: (1) shortening of the
objective aperture holder,[16] (2) use of low-atomic
number anti-contamination devices,[16,29] (3) use of
low-atomic-number shields over the lower objective
pole-piece,[16] and (4) use of special specimen
holders constructed from low-atomic number
materials.[14,16,19,20,29-33]

Most manufacturers now supply "low-background"
EDS specimen holders and there have been a large
number of indepednent designs. It is not possible
to give a universal solution, but we feel that a
list of criteria or important design considerations
for low-background specimen holders is appropriate,
particularly in view of the fact that many of those
available are far from optimum. Desirable features
and their consequences include:

1. The absence of detectable characteristic
lines in the systems background. This requires the
use of materials with atomic number <11, and also
determines the amount of the holder replaced near
the specimen position.

2. Double-tilting capability. This is required for diffraction analysis prior to x-ray
microanalysis, achievement of specific diffracting
conditions during x-ray microanalysis and optimization of the specimen-detector geometry.

3. Sufficient mechanical strength and
robustness.

4. Compatibility of specimen holder material
with "clean" vacuum requirements and resistance to
damage by the electron probe. A nonporous material is required to minimize the surface area
available for contamination pick-up outside the
microscope.

5. A wide range of available take-off angles
to complement the double-tilting capability and to
eliminate shadowing effects.

6. The possibility for the incorporation of
other special features such as: a specimen current monitor, a Faraday cup, and specimen heating
or cooling capabilities.

Compton scattering of any remaining uncollimated x-rays produced in the illumination system
is much more pronounced with the use of low-atomic
number specimen holders [e.g., the Compton scattered MoK x-rays in fig. 3(b–d)]. This problem
may require special attention.

Finally, even in an extensively modified
system, it may be necessary to incorporate a final
subtraction procedure in order to more accurately
obtain the correct continuum intensity, if this is
required in the analysis.[21]

4. Specimen Contamination

Specimen contamination during electron microscope investigations is not a new phenomenon;
however, the development of STEM and its consequent high current density small electron probes
have increased contamination rates to a point at
which detrimental effects on microanalysis can
sometimes occur. This accumulation of foreign
material on the specimen will, in general, degrade
the analytical information from a specific region
of interest. It is beyond the scope of this
paper to consider the mechanisms of the contamination process; however, there is discussion of this
in the literature.[34-38] This section is concerned
with the problems related to performing microanalytical experiments in the presence of
contamination.

4.1 Effects of Contamination on Image Quality and Spatial Resolution

Figure 6 is a TEM micrograph showing rather
severe contamination formed during a series of
x-ray measurements on a thin metal foil using an

electron probe approximately 10 nm in diameter. Although the contamination may be convenient from the point of view that the region which was analyzed is now clearly indicated and an estimate of the foil thickness may be obtained following tilting of the specimen,[39] it is also immediately apparent that all further imaging experiments on those particular areas are now impossible. There is, however, a more serious aspect of the formation of these contamination zones — namely, their effect on spatial resolution. The unique feature of performing microanalysis on an AEM instrument, particularly those equipped with STEM capability, is a direct consequence of the availability of small electron probes which can be directed to specific regions of the specimen. Any buildup of a contamination layer on the electron entrance surface of the specimen will adversely affect the spatial resolution afforded by these small probes as a result of beam broadening occurring within the contaminant before the probe interacts with the specimen. The time dependence of the spatial resolution can also produce its own set of problems. Furthermore, there is some evidence that charging effects, due to the poor conductivity of the contamination, can cause the probe to wander which further decreases the effective spatial resolution.[15]

4.2 Effects of Contamination on Quantitative Analysis

4.2.1 Systems Peaks from Contaminant.
A potential problem associated with contamination is related to the generation of systems peaks in the volume of contaminant material accumulating on the specimen. If this material were purely hydrocarbon, then during x-ray analysis using standard EDS detectors, no additional characteristic emission will be observed. However, if vacuum pump oils or greases containing silicon are used in the instrument the contaminant may also contain silicon. The silicon will be detected in the measured x-ray emission spectrum. An example is shown in fig. 7 which is an expanded region of an x-ray spectrum measured from a thin, single crystal specimen of β-NiAl.[40] Three peaks are clearly resolved: Ni L ($\sim$0.85 keV), Al K ($\sim$1.49 keV), and Si K ($\sim$1.74 keV). The magnitude of the silicon line measured in this case correlated directly with the accumulation of observable contamination. The elimination of such a systems peak can in some cases be accomplished by using silicon-free mechanical and diffusion pump oils and greases, following of course, a thorough cleaning of the entire microscope column and vacuum system.

4.2.2 Absorption by Contamination.
In cases where large contamination rates occur, it is sometimes possible to detect preferential absorption effects of low-energy x-rays due to the accumulating mass of material on the specimen surface. This was demonstrated by Zaluzec and Fraser[15] who monitored the characteristic intensity ratio of a medium-energy x-ray line (Ni K_α $\sim$7.48 keV) to a relatively low-energy line (Al K_α $\sim$1.48 keV) as a function of time (and thus contamination). Their results shown in fig. 8 indicate that in some cases measurable absorption effects can be observed, and thus caution should be exercised in interpreting such data quantitatively. This preferential absorption effect is most pronounced for low-energy x-ray lines and would be negligible for higher energy x-rays or for intensity ratios of lines whose difference in mass absorption coefficients is small.

4.2.3 Contamination Generated Continuum Background.
As the incident electron probe interacts with the material deposited on the specimen surfaces, bremsstrahlung x-rays (as well as undetected characteristic lines) are produced. The bremsstrahlung x-rays are detected in addition to the specimen related x-ray spectrum. Two problems with respect to x-ray microanalysis arise as a result. The first arises in quantitation procedures which require the measurement of the continuum background intensity, in order to obtain estimates of the mass thickness of the analyzed region of the specimen.[41] These procedures are particularly useful in analyses of biological specimens. Here the contaminant will have the largest detrimental effect because of the inherent low mass thickness of the specimens.

The second problem which arises because of an increase in continuum background is a decrease in the ultimate mass sensitivity. The mass sensitivity is directly related to the characteristic P/B ratio for a particular element in the specimen. This is lower in the presence of contamination due to the anomalously high continuum levels. A simple model of a contaminating specimen would predict a monotonic decrease in the P/B ratio with analysis time and, although a decrease in the P/B ratio eventually occurs, some experimental results indicate that the process may be much more complex.[40]

4.3 Reduction of Contamination Rate

Although the elimination of systems peaks such as silicon may be overcome by using appropriate oils and greases in the vacuum system without necessarily reducing contamination buildup, the other effects described in this section do require reduction of the contamination rate. This is not a simple task. Two distinct approaches are necessary. The first is the minimization of contributions from the instrument vacuum system and the second is the reduction of the specimen-borne contribution. Manufacturers, as well as many microscopists, are constantly attempting improvements to instrument vacuum systems. The use of oil-free vacuum pumps, effective cryo-pumping around the specimen area, dry O-rings, low vapor pressure greases, column baking, leak checking and prevention of backstreaming by traps are well known, successful approaches. Once a "clean" vacuum system is realized, the contamination appears to be controlled by specimen-borne material. Several techniques to reduce specimen-borne contamination have been suggested during the last few years.[12,37,38,42-44] These include in situ specimen heating[37,44] and cooling[12,37,42,43] and immobilization of surface contaminants by electron flooding.[37] At present there is no general method applicable to all situations. It is prudent to be aware of possible adverse effects and take appropriate steps to minimize them.

5. Conclusions

Although many detrimental effects arising from various instrumental problems have been discussed, none of them appear to be insurmountable.

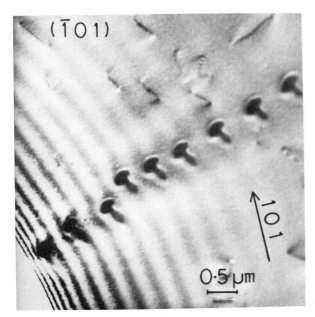

Fig. 6 An example of severe contamination formed during 100 s analyses of a thin foil of β-NiAl using a ∿10 nm electron probe at 200 keV.

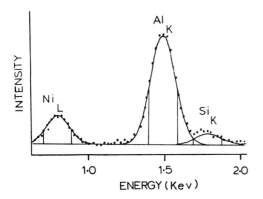

Fig. 7 Portion of the x-ray spectrum from β-NiAl illustrating the detection of a SiK peak originating from the contamination deposited during x-ray microanalysis.

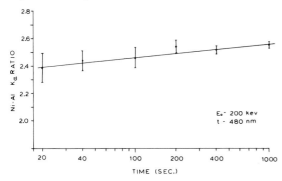

Fig. 8 Experimental variation of the Ni/Al K_α intensity ratio as a function of time (hence contamination) illustrating absorption effects of the low-energy Al K_α line.

When appropriate preventative measures are implemented, the achievement of reliable quantitative analysis is a more realistic proposition.

Acknowledgments

This research was supported by the Division of Materials Sciences, U. S. Department of Energy, under contract No. W-7405-eng-26 with the Union Carbide Corporation. One of us (NJZ) appreciates the support of the E. P. Wigner Fellowship at ORNL. The authors would like to thank Dr. P. S. Sklad for a helpful review. The continued encouragement and support of Dr. J. O. Stiegler are appreciated. Finally, we would like to thank Mrs. Frances Scarboro and Mrs. Judy Young for manuscript preparation.

References

1. D. C. Joy and D. M. Maher, *Optimization of an X-Ray Energy-Dispersive System* in Analytical Electron Microscopy (1976) 111-113.*

2. N. J. Zaluzec, *Optimizing Conditions for X-Ray Microchemical Analysis in Analytical Electron Microscopy*, Ninth Int. Cong. on Electron Microscopy (1978) Vol. 1, 548-549.†

3. E. F. Sturcken, *SEM and XES in High Beta-GAMMA Radiation Fields*, SEM/1976/I, 247-256.‡

4. S.J.B. Reed, *Electron Microprobe Analysis*, Cambridge University Press, Cambridge, England (1975).

5. R. Woldseth, *X-Ray Energy Spectrometry*, Kevex Corporation, Burlingame, California (1973).

6. C. E. Fiori and D. E. Newbury, *Artifacts Observed in EDS in the SEM*, SEM/1978/I, 401-422.‡

7. L. M. Middleman and J. D. Geller, *Trace Element Analysis Using X-ray Excitation with an EDS on a SEM*, SEM/1976/I, 171-178.‡

8. R. B. Bolon and M. D. McConnell, *Evaluation of Electron Beam Tails and X-Ray Spatial Resolution in the SEM*, SEM/1976/I, 163-170.‡

9. R. H. Geiss and T. C. Huang, "Quantitative X-Ray Energy Dispersive Analysis with the TEM," *X-Ray Spectr.* 4 (1975) 196-201.

10. N. J. Zaluzec and H. L. Fraser, *X-Ray Absorption Effects in Thin Metal Foils* in Analytical Electron Microscopy (1976) 118-120.*

11. N. J. Zaluzec and H. L. Fraser, *Microchemical Analysis of Thin Metal Foils* in 34th Ann. Proc. of EMSA (1976) 420-421.§

12. H. Shuman, A. V. Somlyo, and A. P. Somlyo, "Quantitative EPMS of Biological Thin Sections," *Ultramicroscopy* 1 (1976) 317-339.

13. J. I. Goldstein and D. B. Williams, *X-Ray Analysis in the TEM/STEM*, SEM/1977/I, 651-662.‡

14. E. A. Kenik and J. Bentley, *Influence of X-Ray Induced Fluorescence on Energy Dispersive X-Ray Analysis of Thin Foils* in 35th Ann. Proc. of EMSA (1977) 328-329.§

15. N. J. Zaluzec and H. L. Fraser, *Contamination and Absorption Effects in X-Ray Microchemical Analysis of Thin Metal Films*, 8th Int. Conf. on X-Ray Optics and Microanalysis, Boston (1977) ed. Beaman, Ogilivie, Wittry, Science Press, Princeton, New York (in press).

16. W.A.P. Nicholson, B. W. Robertson, and J. N. Chapman, *The Characteristics of X-Ray Spectra from Thin Specimens in the TEM* in Developments in

Electron Microscopy and Analysis, 1977, ed.,
D. L. Misell, Institute of Physics, Bristol,
England, 373—380.

17. P. L. Morris, N. C. Davis, and
J. S. Treverton, "Effects of a Surface Film Upon
Thin Foil Microanalysis" in *Developments in Electron Microscopy, 1977*, ed. D. L. Misell, Institute
of Physics, Bristol, England, 377—380.

18. J. I. Goldstein and D. B. Williams,
Spurious X-Rays Produced in the STEM, SEM/1978/I,
427—434.‡

19. N. J. Zaluzec, E. A. Kenik, and
J. Bentley, "X-Ray Microanalysis Using an HVEM"
in *Analytical Electron Microscopy* (1978) 179—182.‖

20. G. Cliff, M. J. Nasir, G. W. Lorimer,
and N. Ridley, "X-Ray Microanalysis of Thin Specimens in the TEM at Voltages up to 1000 kV" in *Ninth
Int. Cong. on Electron Microscopy* (1978) Vol. I
540—541.†

21. B. W. Robertson, J. N. Chapman,
W.A.P. Nicholson, and R. P. Ferrier, "A Quantitative Investigation of Thin Specimen X-Ray Spectra
in the TEM," ibid, 550—551.†

22. T. J. Headley and J. J. Hren, "Sources
of Background X-Radiation in AEM" ibid, 504—505.†

23. A. Dyson, *X-Rays in Atomic and Nuclear
Physics*, Longman, London, England (1973) 7—135.

24. H. L. Fraser, N. J. Zaluzec,
J. B. Woodhouse, and L. B. Sis, "On the Feasibility of Quantitative Microchemical Analysis of Thin
Metal Foils" in *33rd Ann. Proc. of EMSA* (1975)
106—107.§

25. N. J. Zaluzec, *An Analytical Electron
Microscopy Study of the Omega Phase Transformation
in a Zirconium-Niobium Alloy*, Ph.D. Thesis, University of Illinois, 1978; also published as Oak Ridge
National Laboratory Report ORNL/TM-6705. Copies
available from NTIS, U.S. Department of Commerce,
Springfield, VA 22161.

26. E. A. Kenik, J. Bentley, and N. J. Zaluzec,
*Modifications of Analytical Electron Microscopes for
Improved Performance*, Oak Ridge National Laboratory
Report, ORNL/TM-6857 (1979).

27. H. L. Fraser and J. B. Woodhouse, "Energy
Dispersive X-Ray Analysis Using a Field Emission
Gun: A Precaution," in *Analytical Electron Micros-
copy* (1978) 191—194.‖

28. A. F. Moodie, C. J. Humphreys, D. Imeson,
and J. R. Sellar, "Convergent-Beam Diffraction in
the Oxford HVEM," in *Electron Diffraction 1927-1977*,
ed., P. J. Dobson, J. B. Pendry, and C. J. Humphreys,
Institute of Physics, Bristol, England (1977)
129—134.

29. J. J. Hren, P. S. Ong, P. F. Johnson,
and E. J. Jenkins, "Modification of Philips EM 301
for Optimum EDX Analysis," in *34th Proc. EMSA*
(1976) 418—419.§

30. J. W. Sprys, "Specimen Holder for EDS
X-Ray Analysis in the TEM," *Rev. Sci. Instr.* 46
(1975) 773—774.

31. N. J. Zaluzec and H. L. Fraser, "Modified Specimen Stage for X-Ray Analysis in a TEM,"
J. Phys. E 9 (1976) 1051—1052.

32. B. Liljesvan and G. M. Roomans, "The
Use of Pure Carbon Specimen Holders for AEM of
Thin Sections," *Ultramicroscopy* 2 (1976) 105—107.

33. P. F. Johnson, S. R. Bates, and
J. J. Hren, "EDS Spectral Contamination in the
EM 301," in *9th Int. Cong. on Electron Microscopy*

Vol. I (1978) 502—503.†

34. A. E. Ennos, "The Origin of Specimen
Contamination in the Electron Microscope," *Brit.
J. Appl. Phys.* 4 (1953) 101—111.

35. W. A. Knox, "Contamination Formed
Around a Very Narrow Electron Beam," *Ultramicroscopy* 1 (1976) 175—180.

** 36. J. T. Fourie, *Contamination in Cryo-
pumped TEM and STEM*, SEM/1976/I, 53—60.‡

37. G. Rackham, J. A. Eades, and
H. H. Willis, "Specimen Contamination in the
Electron Microscope," *Optik* 47 (1977) 227—232.

38. L. Reimer and M. Wächter, "Contribution
to the Contamination Problem in TEM," *Ultramicroscopy* 3 (1978) 169—174.

39. G. W. Lorimer, G. Cliff, and J. N. Clark,
"Determination of the Thickness and Spatial Resolution for the Quantitative Analysis of Thin
Foils," in *Developments in Electron Microscopy
and Analysis*, Academic Press, London (1976)
153—159.

40. N. J. Zaluzec and H. L. Fraser, "On the
Effects of Contamination in X-Ray Microanalysis,"
in *Analytical Electron Microscopy* (1978) 122—129.‖

41. D. J. Marshall and T. A. Hall, "Electron
Probe X-Ray Microanalysis of Thin Films," *Brit.
J. Appl. Phys.* 1 (1968) 1651—1656.

42. J. Wall, "Contamination at Low Temperature," in *Analytical Electron Microscopy* (1978)
112—113.‖

43. R. Egerton and C. J. Rossouw, "Measurement of Contamination Rate Using ELS," *J. Phys.*
D 9 (1976) 659—663.

44. L. M. Brown, A. J. Craven, G. P. Jones,
et al., *Applications of a High Resolution STEM
to Materials Science*, SEM/1976/I, 353—360.‡

*Report of a Specialist Workshop, August 3—6, 1976,
Materials Science Center, Cornell University,
Ithaca, NY 14853.

†Ed., J. M. Sturgess, pub. Microscopical Society
of Canada, Toronto, Ontario M5S 1A1.

‡1968—77: IIT Research Institute, Chicago, IL
60616. 1978: SEM Inc., AMF O'Hare, IL 60666.

§Claitor's Publishing Division, Baton Rouge,
LA 70821.

‖Report of a Specialist Workshop, July   , 1978,
Materials Science Center, Cornell University,
Ithaca, NY 14853.

Discussion with Reviewers

R. H. Geiss: *Would you explain the concept of
fluorescence, bremsstrahlung, and continuum
radiation, especially in the sense that this is
a tutorial?*
Authors: Fluorescence is the emission of radiation by a substance during exposure to external
radiation. In the context of this paper it is
the emission of characteristic x-rays which
occurs following ionization by incident electrons
or x-rays.

Bremsstrahlung is electromagnetic radiation
produced by the interaction of incident charged
particles with target nuclei. The acceleration
imparted to the incident particle results in
emission of radiation with all possible energies

** See also, Fourie, this vol.

between zero and the energy of the incident particle.

Continuum radiation is synonymous with bremsstrahlung but may also be used to describe the "background" of energy-dispersive x-ray spectra.

See text reference 5 for a simple but more complete discussion and text reference 23 for an advanced treatment of these three concepts. Information on scattering mechanisms and some cross-section data may also be found in these references.

P. L. Ryder: *Can the authors quote experimental measurements of the magnitude (fraction of the primary beam current) and lateral dimensions of "electron tails"? In which instruments have they been observed to cause trouble?*
Authors: Goldstein and Williams (text reference 18) reported a current of $\sim 3 \times 10^{-13}$ A in electron tails under conditions which produce a probe current of $\sim 1.5 \times 10^{-10}$ A. However, it is important to realize that the effects of electron tails are disproportionate to the magnitude of the ratio of electron tails current/probe current. This is because the electron tails can interact with the "bulk" parts of the specimen and produce many more x-ray quanta per electron than the primary probe, which interacts with the very thin regions of the specimen. The lateral dimensions of "electron tails" have not been determined experimentally but typically may be several mm at the specimen position as inferred from the observation that x-rays from the specimen holder may be detected in the "in-hole" signal when the specimen is totally absent. Electron tails have been observed to be a particularly severe problem in Philips EM 300, 301, and 400 series instruments. However, the manufacturers are aware of the problem and modifications are available.

R. H. Geiss: *Have there been any experiments published which support the idea that one may distinguish between electron and x-ray bremsstrahlung excited spectra, based on a comparison of relative strengths of short and long wavelength x-ray photons?*
Authors: To our knowledge there have been no definitive experiments reported. These conclusions are based on the relative cross-section ratios for fluorescence by electrons and x-ray photons. Cross-section ratios for Mo K_α/Mo L_α and Ni K_α/Ni L_α characteristic x-rays for electron excitation are shown in table D1, while table D2 gives the values for fluorescence by x-rays. Clearly x-ray fluoresence favors high energy emission while electron

Table D1. Cross-section ratios for electron excitation

Incident electron energy keV	Cross-section ratio: $\sigma_{K_\alpha}^{electron}/\sigma_{L_\alpha}^{electron}$	
	Mo	Ni
25	0.002	0.010
50	0.011	0.012
75	0.014	0.013
100	0.016	0.015

Table D2. Cross-section ratios for x-ray excitation

X-ray photon energy keV	Cross-section ratio: $\sigma_{K_\alpha}^{x-ray}/\sigma_{L_\alpha}^{x-ray}$	
	Mo	Ni
25	205	625
50	230	610
75	235	590

fluorescence favors the low energy lines. The actual measured intensity ratios will be different because the effects of detector efficiency, x-ray fluorescence yield, etc., must be included (text reference 25).

J. I. Goldstein: *Can you suggest specific test specimen(s) which can be used to identify the presence of spurious* x-rays and their source(s)?*
Authors: There are a vast number of possibilities. We feel that the general requirements given in section 2.3 are sufficient to allow an appropriate choice to be made. To reiterate: a homogeneous alloy or element with low (<3 keV) and high (>6 keV) characteristic lines in a specimen with some "bulk" character, i.e., a self-supporting thinned disk. Characteristic lines of energy >20 keV are less desirable than those with energy <20 keV.

R. H. Geiss: *There is a small peak at about 6.4 keV in figures 3(a) to 3(e). Is this an Fe K_α peak? If not, what is it, and in either case where does it come from?*
Authors: It is the Fe K_α peak. We suppose its origin is the objective lens polepiece material.

E. L. Hall and J. B. Vander Sande: *Fraser has shown quite convincingly that the primary hole count problem in a D-STEM (VG HB-5) arises due to uncollimated electrons scattered by the probe-forming aperture, rather than x-ray flux. Our own work with the HB-5 has shown that the use of special thick apertures in this position has no effect on hole count, which is consistent with Fraser's conclusion.*
Authors: H. L. Fraser has never attributed the in-hole signal in the VG HB-5 to scattered electrons (H. L. Fraser, University of Illinois, private communication). The observed characteristics are consistent with production by high energy x-rays. Electrons scattered from the probe-forming aperture have been surmised to be the source of the in-hole signal in the Siemens ST 100F (Oppolzer and Knauer, these proceedings) but here the electron-optical configuration and aperture position are different. The fact that the use of thick apertures has no apparent effect on the in-hole signal may not be surprising considering the proximity of the aperture and specimen.

D. B. Williams: *The comments concerning spurious* x-rays in a dedicated STEM make this instrument appear to be relatively useless for high resolution x-ray microanalysis. Have the instrument* *manufacturers taken steps to correct the problem? How do these comments affect data from such instruments already published in the literature?*

Authors: Although severe problems of fluorescence by uncollimated x-rays produced at the final probe-forming aperture exist in some D-STEMs, the situation is not hopeless, as we clearly indicated in the text. It is the nearness of the aperture to the specimen which makes the conventional solutions impossible to implement. Furthermore, not all D-STEMs have an aperture near the specimen (e.g., Oppolzer and Knauer, these proceedings). The instrument manufacturers are aware of the problem and have begun to implement design changes. Problems with equally severe consequences for x-ray microanalysis have existed in TEM based AEMs. Even with the probe forming aperture removed, the analytical spatial resolution is still higher than in many TEM based AEMs. Most users are aware of the problems and generally, published data from D-STEMs is in no worse a condition than that from TEMs.

J. I. Goldstein: *Could you give the readers some criteria to define a minimum level of spurious* x-rays which can be tolerated and which will not interfere with a quantitative analysis?*

Authors: There is no simple answer. Such an amount would depend on the accuracy required in the particular analysis. The selective nature of the effects and the sensitivity to specimen type (atomic number, geometric shape, etc.) also cause problems. It is our opinion that subtraction of the in-hole signal is a better way to achieve more accurate analyses. The experiment is easy to perform and the subtraction procedure is trivial. In fact, it is part of the computer program used for quantitative analysis at ORNL. As emphasized in the text, it is also important to reduce the in-hole signal as much as possible to avoid the introduction of excessive statistical errors.

D. B. Williams: *The possible effect of carbon contamination on the absorption of low energy x-rays is an important point. However, in the conventional (at present) analytical set-up, the detector is normal to the beam and thus to the contamination peak. Therefore, detected x-rays have only passed through the specimen, and very little carbon contamination, prior to entering the collimator. Have the authors done any absorption calculations to explain the data in fig. 8 and compared the results with observed carbon thickness? For this particular problem may not a high take-off angle detector be less desirable? (See diagram below for clarification.)*

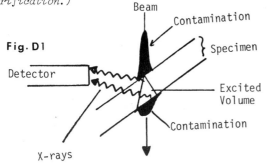

Fig. D1

Beam

Contamination

Specimen

Detector

Excited Volume

Contamination

X-rays

Authors: Although fig. D1 may illustrate a typical arrangement, the contamination that formed during the experiment for fig. 8 (text reference 15) covered considerably larger areas of the specimen as shown in the representative secondary electron micrograph of fig. D2. X-ray path lengths in the contamination of ~1 μm are required to explain the data of fig. 8. This is consistent with the size of the contaminant in fig. D2.

High take-off angle detectors would appear to be less desirable for this particular problem. However, it is our opinion that on modern instruments the contamination rate can be kept sufficiently small that the problem is not severe enough to nullify the other advantages of high take-off angle detectors.

E. L. Hall and J. B. Vander Sande: *Could the authors outline briefly specimen stage stability problems and include acceptable drift rates and possible procedures for minimizing drift?*

Authors: The main problem associated with specimen drift is the decrease in spatial resolution. Acceptable drift rates depend upon the spatial resolution required and the time for analysis. Generally, top-entry stages have greater stability than side-entry stages but the former have less desirable geometrical arrangements for x-ray microanalysis and eucentric tilting is usually not available. Even with side-entry specimen holders, however, the guaranteed drift rates rarely exceed 3 nm/min. For most analyses this is much less than the limits to spatial resolution imposed by the probe size necessary to achieve desired probe currents, beam spreading within the specimen (which may be enhanced by contamination build-up) or electrical stability of the electron-optical system. For the highest spatial resolution (<5 nm), e.g., when using FEGs and very thin specimens, then specimen drift may turn out to be the limiting factor in spatial resolution. The procedures for minimizing drift are well recognized and include: maintenance of backlash-free stage drives; sufficient lubrication of moveable "o-rings;" close temperature control of (objective) lens cooling water; and avoidance of poorly clamped specimens and drastic changes in illumination of the specimen, which result in fluctuations in beam heating effects.

*Since the term "spurious x-rays" does not help to identify the source of this undesirable signal, the present authors prefer to use <u>systems background</u>.

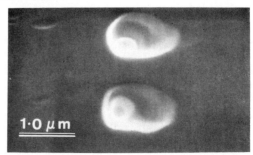

1·0 μm

Fig. D2. Secondary electron micrograph of contamination formed during the experiment[15] of fig. 8.

SCANNING ELECTRON MICROSCOPY/1979/II
SEM Inc., AMF O'Hare, IL 60666, USA

PRINCIPLES OF X-RAY MICROANALYSIS IN BIOLOGY

J. A. Chandler

Tenovus Institute for Cancer Research
Welsh National School of Medicine,
Heath Park,
Cardiff, CF4 4XX, U. K.

Abstract

Progress in biological X-ray micro-analysis has been limited in recent years by specimen preparation procedures. Instrumentation and high levels of sensitivity for low concentrations are well developed for a large number of biological applications but reliable preparation of samples in the thick or thin form has yet to be established. In addition the possible damage of specimens by the electron probe has been incompletely studied in the majority of applications. In this review the basic principles of X-ray production, design and operation of X-ray detectors, quantitative procedures for thick and thin samples, analytical operation of the microscope, and current specimen preparation techniques are briefly described with the aim of highlighting weaknesses and discussing recent developments in each field.

KEY WORDS: X-Ray Microanalysis, Biology, Biological Specimen Preparation, Analytical Electron Microscopy, Sensitivity, Quantitative Analysis

1. Introduction

Rapid progress has taken place in the last few years in the development of instrumentation and techniques for biological X-ray microanalysis. Not least of these developments has been the introduction of the solid state energy dispersive X-ray detector in 1968[1]. The comparative ease with which solid state detectors can be used to obtain multi-elemental information within seconds has, however, led the biologist into a false position of confidence. Whereas technological advances in the solid state technology and the design of electron optical instrumentation have been very great, the biologists' understanding of processes affecting the tissue during preparation and electron irradiation have been somewhat limited. New, improved analytical instruments have become available each year promising greater sensitivity, higher energy resolution, smaller probe diameters and more flexibility in stage design and specimen handling. The biologist, however, working from relatively little background information, has struggled to understand the processes occurring during freezing of tissue or immersion of samples in fixatives and solvents. In addition the effects of the electron beam on frozen or fixed and embedded tissues are poorly understood so that an element of uncertainty often clouds the analyses even when there is optimism about the integrity of the specimen during preparation.

The purpose of this paper, in addition to reviewing the basic theoretical principles of biological X-ray analysis, is briefly to highlight the areas of weakness together with recent technical developments in the fields of instrumentation, quantitative theory, specimen preparation, and analytical operation required for maximising sensitivity of trace element analysis in organic material. It is hoped that further papers in this Workshop will enlarge on specific aspects of these topics such that the present state of X-ray technology for biological application will be fully elucidated. A more thorough review of X-ray analysis of inorganic samples, particularly with regard to prevailing quantitative methods, is to be found elsewhere[2].

2.   X-ray analysis - basic theory

The basic principles of X-ray analysis have been comprehensively described in a number of papers[2,3,4,5]. Briefly, the atoms comprising the specimen may be considered as having a nucleus of protons and neutrons surrounded by electrons in various energy states. These electrons occur in orbits with well defined energies, the energy of each orbit increasing with greater distance from the nucleus. The orbits or 'shells' are designated K, L, M and N shells, the K shell being of lowest energy. Heavier atoms, with greater numbers of electrons, fill more shells and each shell may contain more than one orbit of a discrete energy level.

An incident electron may interact with an orbital electron, say from the K shell, causing it to be ejected from the atom which then becomes ionised. Since the K shell has the least energy, this vacancy is immediately filled with an electron falling from a higher energy orbit, say the L shell. This L shell electron loses energy during the transition (E_L-E_K) which is then emitted as an X-ray photon. This transition may in turn give rise to an electron jump from an M shell to the new vacancy in the L shell, causing the emission of another X-ray photon of energy E_M-E_L, and so on. Thus one ionisation may give rise to a whole spectrum of X-ray emissions. Each element in the periodic table contains electrons in orbits with particular discrete energy levels and it is on the basis of these specific X-ray energies that each element can be identified. Thus if a number of elements in a sample are simultaneously irradiated by the electron beam a spectrum of X-ray energies is emitted. X-ray analytical systems are designed to unravel this spectrum by the identification of specific X-ray energies, or wavelengths (where the energy is proportional to the reciprocal of the wavelength). The generation of X-rays from the specimen, which may be a bulk sample or a thin or ultrathin section, depends on the initial energy E_O of the electron probe. This must be sufficiently high to ionise the atom being detected, the critical ionisation potential being designated E_C. The choice of probe voltage is an important parameter in analysis of biological material for it determines X-ray yield as well as depth of electron penetration and hence spatial resolution of

analysis in the sample.

In addition to the characteristic X-ray radiation described above a continuous radiation is also emitted from the specimen. This radiation, otherwise known as 'continuum', 'bremsstrahlung', or 'white radiation', arises from the interaction of the primary electron beam with atomic nuclei in the sample, producing an X-ray spectrum with continuously varying energy from E_O down to zero. The continuum forms much of the background underlying the characteristic spectrum and is primarily detected for use in quantitative procedures.

3.   X-ray detectors

Gas flow proportional counters and wavelength dispersive crystal spectrometers were the earliest types of X-ray detectors to be employed but, in biological work at least, these have largely given way to lithium drifted silicon Si(Li) energy dispersive solid state detectors (SSD), mainly because of the ease with which these detectors can be interfaced without great geometric problems to scanning and transmission microscopes, and because of the relative simplicity of their operation. Analysis with crystal spectrometers and Si(Li) solid state detectors should, however, be considered complementary rather than competitive, especially for analysis of inorganic materials. Analysis with Si (Li) detectors is now routine in SEM's, TEM's and STEM's. When the SSD first appeared it had an energy resolution not much better than that of the gas flow proportional counter (about 600 eV), but with careful attention to the quality of lithium diffused silicon, reduction of thermal noise, and associated signal processing, this figure has been steadily reduced until detector resolutions of 150 eV or less are routinely obtained. This is still far from the typical energy resolution of 1-10 eV obtained with a crystal spectrometer but with electronic methods of enhancing resolution the advantages of the SSD predominate for the majority of biological analyses. The main advantages of the solid state detector are the high efficiency of X-ray detection that can be achieved and the large solid angle of collection that can be covered compared with the crystal spectrometer.

The combination of relatively poor energy resolution and peak to background ratios is the greatest limitation to the use of solid state detectors in biological analyses. One is frequently working with samples where a complex spectrum is produced and intensities of the spectral lines from elements of interest are in the same order of magnitude as the minimum level of detection. Hence very small peaks in the spectrum can easily be swamped by adjacent peaks or background noise. Peak to background ratios can be improved by increasing the peak or decreasing the background.[6] Most contributions to background arise before the detector, i.e. in the microscope, often due to incorrect choice of analytical conditions, spurious scatter of electrons out of the sample, and inadequate collimation of X-rays towards the detector. Enhancement of the peak signal in the SSD may also be more greatly affected by the instrumental and analytical conditions than by small improvements in detector design. There has been a trend towards bringing the SSD closer to the specimen in the microscope to increase the solid angle of collection and hence the collection efficiency. Both peak and background signals are likely to be affected in the same way by this manoeuvre, however, and the main advantage would lie in the reduced electron beam current required to produce a given collected signal. This in turn allows a smaller probe diameter to be employed, which, for thin specimens at least, would provide better spatial resolution of analysis. In addition the lower beam current reduces the risk of specimen damage.

Another disadvantage of the solid state detector (although also a problem with other detectors) lies in the very rapid fall off in detection efficiency at the lower energies through X-ray absorption in the protective beryllium window in front of the silicon crystal. This has been improved by the use of thinner windows and even windowless detectors but the latter are difficult to use and, because of the greater risk of damage, have not become widely employed.

Because of the relatively poor energy resolution in the spectra obtained from biological specimens the identification of small peaks often becomes a problem. With a wavelength dispersive crystal spectrometer the excellent peak to background ratios obtained and high energy resolution make the detection of such peaks theoretically more easy than with the SSD. Indeed with a very high precision spectrometer and completely stable analytical system this would be the case, but in practice the crystal spectrometer detection method often suffers from problems of non-reproducibility and instability which for the very low signals may make detection difficult. Here the solid state detector has the further advantage of being capable of measuring the whole spectrum simultaneously, peak and background intensities, such that variations in probe current are less important and no mechanical movement is required. Some shifts in energy calibration may occur through changes in ambient temperature or high X-ray signal intensities however, which could give rise to errors in computing background subtraction and in deconvolution of overlapping peaks[7,8,9]. It is in the field of electronic deconvolution of complex energy spectra and the reliable determination of low intensity peaks that continuous progress needs to be made if the biologist is to enjoy all the other advantages afforded by the SSD.

4. Instrumentation and analytical operation

Almost all SEMs and TEMs are now designed with the facility for adding an X-ray detection system. In the case of the SEM this is largely just a matter of providing another port in the specimen chamber so that the snout of the SSD can be moved in close to the specimen. In the TEM stricter considerations govern the design of the interface because the specimen sits inside the objective lens. The most recent advances in instrumental design have been to provide a STEM system to combine with X-ray analysis. In the SEM this is again possible without too much difficulty whereas in TEM's some modification is required to the objective lens to allow entry of the X-ray detector. The true STEM mode of operation, however, has its maximum value when small diameter probes, such as those provided by field emission guns (FEG), are employed. It is more difficult to operate this kind of electron source in a conventional microscope because of the relatively poor vacuum systems available, although some current models do provide better specification in this respect. Custom built STEM instruments, however, operate routinely at the higher vacuums and employ FEG's as a matter of course. Such guns can provide probe diameters, and hence potential spatial resolutions of analysis, of just a few nanometres or less for ultrathin specimens compared with the tens of nanometres available from conventional hairpin filaments. The increased brightness available from these small diameter probes can be of great use to the biologist determining trace elements in small areas of the specimen since, in thin specimens at least, the actual resolution of analysis is improved and, because of the reduced total mass being analysed, background readings are lower, thus enhancing sensitivity. In the analysis of bulk specimens, however, such small diameter probes are of less value because of the extensive diffusion of electrons into soft tissue.

In pursuit of greater sensitivity and the analysis of smaller and smaller masses the biologist may choose between bulk (thick) specimens where the electrons are totally stopped by the sample, and thin or ultrathin sections where diffusion problems become of less significance. A great advantage of thin samples is found in the increased spatial resolution and enhanced peak to background ratios obtainable, but this is partly offset by the relatively lower signals obtained. With such low peak intensities great attention has to be paid to the reduction of factors contributing to the relatively high background. As described below much of this background arises from scattered electrons striking parts of the microscope and specimen support and must be accounted for in quantitative correction procedures. The use of more coherent electron sources together with adequate shielding of aperture systems in the microscope[9] has helped to reduce some of this spurious background radiation. Another major source of X-ray background arises from electrons scattering from the specimen itself, and while this may be reduced by the correct choice of analytical conditions (accelerating voltage, current, specimen tilt), the use of carbon and beryllium to create a light element specimen environment, together with strict X-ray collimation and efficient electron traps before the X-ray detector have all contributed to a much enhanced sensitivity[4,6,7]. In a transmission microscope, for example, using a conventional thermionic gun, single molecules of ferritin have been analysed to demonstrate just 10^{-19}g of iron[10], while Shuman et al,[6] have calculated that, with proper control of all sources of background and maximisation of signal peak intensities it should be possible, in thin section at least, to detect quantities as low as 10^{-22}g. In such cases the limitation becomes due to statistical uncertainty in the data obtained and to specimen damage caused by the electron beam.

Another major problem affecting sensitivity and reliability of analysis is the possible damage of the specimen during electron beam irradiation. This is still poorly understood for two main reasons. Firstly, no one tissue type will satisfactorily act as a model for the behaviour of all samples under the electron probe since elements may exist in tissue with varying degrees of binding. Secondly, whatever losses or changes occur during irradiation may take place in a very short time interval at the start of the irradiation and before monitoring is possible. The long term losses of specimen components have been measured for a number of elements under various analytical conditions but the immediate changes are virtually impossible to detect during the analysis. Typical losses that may occur in the initial exposure to the electron beam have been demonstrated using model systems[11]. In a real sample this is far less predictable. Instead, the approach

of most workers has been to minimise these losses in as many ways as possible. This has been achieved by reducing the total electron dose imparted to the specimen, (hence the desirability of larger detector solid angles); by providing conducting coatings on specimens; and, probably most effectively, by analysing specimens maintained at very low temperatures [12,13,14]. Because of the obvious errors such elemental and mass losses will introduce into an analysis, this problem is an area of major weakness in the analytical operation and needs to be given much attention.

5. Quantitation

For the quantitative assessment of elemental concentrations in biological tissue it is necessary to distinguish thick, thin and ultrathin samples since different theories exist in each case. In thick samples none of the incident electrons can reach the far surface. Strictly, thick samples for a given X-ray line are those in which all of the electrons eventually fall below an energy level E_C required to produce that line. This may be calculated from the electron range equation derived by Reed[15],

$$D = 0.048 \, \frac{(E_O^{1.5} - E_C^{1.5})}{\rho} \qquad --(1)$$

where D is the electron penetration in μm, E_O and E_C are the probe energy and critical excitation energy in keV, and ρ is the specimen density in g/cm^3. Thus for Zn Kα X-rays, for example, where E_C = 9.66 keV, at an accelerating voltage of 30kV, in a tissue of density $1g/cm^3$, $D \simeq 7\mu$m. At 100kV, however, D = 49μm by this relationship, while at 5kV the Kα line would not be excited at all and for Zn Lα (1.04 keV), D = 0.5μm. Thus the definition of thickness depends on the probe energy, on the line being excited, and on the specimen density.

A more thorough way of approaching specimen thickness is to use the well known Bethe formula to calculate the gradual loss of electron energy during penetration into the sample. This has been adapted by Hall[3] to provide an estimate of the electron penetration into soft tissue for a given energy loss:

$$\rho s = \int_E^{E_O} \frac{E de}{4.\ln.16E} \qquad --(2)$$

were ρ is the specimen density (g/cm^3), s is the electron path (μm), and E is the electron energy in keV. Thus calculations for specimen thickness, s, may be made according to the tolerable energy loss, E_O-E, of the electrons. By this criterion, for a thick specimen E would have a value of E_C.

Thin samples are those in which most

of the electrons reach the far side before their energy falls below E_C. Ultra-thin samples are those in which almost no effects of inelastic electron scattering or X-ray absorption occur, although some energy must be imparted to the specimen to produce X-rays, and generally covers the range up to $0.2\mu m$. In these specimens lateral diffusion of the electron beam is negligible and the area of analysis is that defined by the probe at the surface.

In order to understand the inherent advantages and weaknesses of the current methods employed for each sample type, a brief account of the basic principles is given here. A more thorough treatise is given elsewhere[2,3].

5.1. Thick samples

During electron radiation the intensity I_{xs} of X-rays from an element x in a standard sample is proportional to the concentration C_{xs} of that element. Thus

$$I_{xs} = k_s \, C_{xs} \qquad --(3)$$

where k_s is a constant depending, among other things, on the overall composition of the sample. By simple comparison the concentration C_x of the same element in another specimen may be calculated from a ratio of the X-ray intensities under the same analytical conditions, i.e.

$$C_x = \frac{I_x}{I_{xs}} \cdot C_{xs} \cdot \frac{k_x}{k_x} \qquad --(4)$$

and if the specimen and standard have similar overall compositions, then

$$C_x = \frac{I_x}{I_{xs}} \cdot C_{xs} \qquad --(5)$$

This would be the simplest approach and for a number of biological samples and elements actually works. There are, however, certain errors involved here which, in some cases, must be corrected by the well known ZAF correction procedures. In this case the equation becomes:

$$C_x = \frac{I_x/I_{xs} \cdot C_{xs}}{C_Z \, C_A \, C_F} \qquad --(6)$$

where C_Z is a correction factor to account for the mean atomic number of the matrix in which the electrons are scattered and gradually lose their energy; C_A is a correction factor to account for absorption of X-rays as they leave the specimen from varying depths within the specimen; and C_F is a correction factor to account for the possibility that emerging X-rays will ionise other atoms with a resulting change in I_x. Each of these factors needs to be examined in turn and such a lengthy exposition would be out of

place here. It is necessary to say, however, that the theory was developed primarily for metallurgical and mineralogical samples where the specimen often presents an inhomogeneous matrix and complex mathematical procedures are required to account for variations in atomic number and absorption, while fluorescence can be a very real problem. With the majority of soft biological tissues, however, we are dealing with an organic matrix whose mean atomic number often does not deviate very greatly and in which the elements to be detected represent a small (usually < 1%) contribution to the total mass. By the choice of an appropriate standard having a similar matrix composition to the specimen itself it is possible to revert back to the original simple relationship shown in equation 5 since the ZAF effects will be comparable in both sample types under the same analytical conditions. An exception will be in harder biological tissues such as teeth or bone and in heterogeneous samples containing areas with densities different to the overall matrix. In these cases estimates of the mean atomic number and relative absorption become a real problem, especially for X-rays of lower energies. For accurate quantitation, and where non-biological standards are employed, one must resort to the more complex correction procedures. Since some of these correction parameters are functions of concentration and require a prior estimate of specimen composition, the iterative method of successive approximations must be used. This is usually far too difficult for hand calculations and complex computer correction procedures are required[16].

The ZAF correction procedures were originally developed for metallurgical and mineralogical use where mass absorption coefficients and fluorescent yields for the heavier elements were well established. For biological samples, however, consisting mainly of C,H,O, and N, such parameters are less well known and, for heterogeneous samples in particular, allow a lower accuracy ($\sim 10\%$) in quantitation compared with that of inorganic samples ($\sim 1\%$)[16].

In addition, the electron beam will penetrate a thick sample according to the electron voltage and the composition of the matrix. Thus the resolution of analysis is always worse than the initial probe diameter. This may introduce errors into the calculations in a number of ways. If the specimen is heterogeneous, i.e. the determined element is confined to a volume smaller than the total volume analysed, while the standard is homogeneous, fewer of the scattered electrons will produce X-rays from that element than in the standard and a simple comparison will not be sufficient. Alternatively, if the element is being analysed in a discrete volume near to a volume containing the same element but of a greater concentration, part of the scattered electron beam or even fluorescent X-rays may produce emission from the latter region and give falsely elevated readings. These are really problems of spatial resolution rather than of quantitation but they may seriously affect the final results. Attention must then be given to improvement of the spatial resolution in the thick sample by the correct choice of accelerating voltage and possibly by the use of smaller and more coherent electron probes. The best alternative would be to use thinner samples.

An interesting but disturbing phenomenon encountered in the analysis of frozen hydrated thick specimens has been described by Fuchs[17]. Since ice contained within the frozen specimen is non-conducting, the electron probe builds up an internal space charge field which significantly changes the shape of the X-ray source volume from that predicted for normal ZAF correction procedures. Only by obtaining information on the depth of analysis by experimentation with thin layered samples and monitoring the ice content by measuring oxygen $K\alpha$ intensities can accurate quantitative analysis be performed in this kind of situation.

Recent developments   The equations and theory presented above, although devised for non-biological samples a number of years ago, have been shown to be workable for most kinds of biological quantitation in thick samples required till now. Indeed, the relatively homogeneous nature of organic material compared with many metallurgical samples makes many calculations quite uncomplicated. Accuracy of quantitation required from most biological samples is generally much less than that needed in inorganic problems due to the innate biological variation that occurs between cells and tissues, while the magnitude of errors from other sources such as specimen preparation is likely to mask any small approximations made in quantitative theory. In certain conditions, however, such as in hard tissues where ZAF effects become more serious, or in soft tissues containing certain discrete volumes of material with high concentrations of some elements, correction procedures may become very necessary and the use of computer correction programmes is of some value[16].

Since the introduction of energy dispersive spectrometry there has been further progress in quantitative analysis because of the compatibility of these detectors to computer interfacing. Methods for determining the shape of the continuum and correcting for background have allowed the determination of small intensity peaks to be made in the range 0.1% or less. Unfortunately, this is still far higher than the concentrations of many elements in soft tissues and the uncertainties associated with statistical substraction of background limit the accuracy of such quantitative determinations. It may be that for the majority of biological analysis, in thick soft tissue samples at least, the old hand calculated comparison of intensities from specimen and standard is proving to be just as reliable as the more sophisticated methods.

5.2. Thin samples

Samples are considered thin if some incident electrons emerge from the far side before falling below the critical excitation potential E_C for the elements being analysed. The definition therefore depends on the initial probe voltage and on the elements of interest but for 30kV electrons and for Zn $K\alpha$ emission, for example, this would correspond to about 7μm or less of fully hydrated or embedded tissue. For lighter elements the electrons could travel further into the tissue before being brought below the threshold energy for ionisation, but one must also consider the possibility of absorption of the lower energy X-rays.

The principle of quantitation, first developed by Marshall and Hall[18] and further elaborated by Hall[3], makes use of the fact that while the intensity of characteristic emission from an element is proportional to the number of atoms of that element irradiated, the continuum generated in the sample is proportional to the total mass of the matrix being irradiated by the electron probe. Thus, simply, the relative weight fraction of that element is given by the ratio of these intensities.

$$C_x = k\,\frac{I_x}{W} \qquad \text{--(7)}$$

where W is a measured fraction of the continuum or white radiation and k is a constant. Thus, again by comparison with a standard of similar composition and having a known concentration of the element C_{xs}, we have

$$C_x = \frac{I_x/W}{I_{xs}/W_s}\,C_{xs} \qquad \text{--(8)}$$

The thickness of the specimen or standard need not be known provided both are technically thin, but the matrix composition of the standard should be similar to that of the specimen, and analytical conditions must be the same for both specimen analyses.

The method works quite well for most biological applications and has the advantage that for many purposes the ZAF correction procedures may be largely avoided. The rather simplistic formula above, however, is dependent on a number of assumptions which contribute to some weaknesses in the technique.

The method is only valid when the specimen is truly 'thin'. Once the sample thickness reaches a stage where a large fraction of the electrons are emerging with energies lower than E_C then the specimen is neither wholly thin nor wholly thick. There is a more complicated procedure for dealing with this 'semi-thin' situation by taking account of the X-ray emission from the substrate on which the sample is mounted[16,19].

Although elaborate ZAF correction procedures are not always required, absorption of lower energy X-rays may well occur as the sample thickness increases. This may be partly offset by appropriate monitoring of continuum in an energy band close to the chosen spectral line since both will be absorbed to a similar extent. With increasing thickness, however, fluorescence may in certain cases again become a problem.

A further problem with increasing specimen thickness is that of spatial resolution due to lateral diffusion of the electron probe in the section. Thus the possibility increases that spurious X-ray emissions may arise from heterogeneous samples as in thick samples. In addition, local changes in sample mass thickness close to the area of interest may mask the true mass thickness of the region being analysed if electrons are scattered in this way, especially if the outer regions have a very different density.

Recent developments   The quantitative theory described by Hall[3] for thin samples is widely used and has been improved very little since its development. It has been applied to embedded, frozen hydrated, and frozen dried sections by a number of investigators with some success. The most useful development has been the design of suitable standards which truly represent the specimen being analysed. In the case of heterogeneous specimens (as most biological tissues really are) this is not always possible but some standards have been developed in the form of embedding resin containing accurately known concentrations of elements[20] to compare with thin embedded sections. For unembedded tissue the methods of incorporating salts into tissue homogenates or into protein gels before freezing and sectioning have been available for some years and are still used[4]. A most useful technique for frozen hydrated specimens described by Gupta et al,[12] incorporates a peripheral standard into the specimen before freezing and sectioning. The additional organic standard, in the form of a physiological medium of known salt concentration together with 20% Dextran as a cryoprotectant, not only provides an adjacent

specimen area for comparative standard measurements, but is subjected to exactly the same environmental and analytical conditions as the specimen itself.

5.3. Ultrathin samples

The theories for thin and ultrathin samples are identical in most respects and for ultrathin specimens are even more applicable because of the virtual complete absence of ZAF correction procedures, electron diffusion in the sample, and limited electron energy loss during irradiation. Ultrathin samples present a number of difficulties for quantitative considerations, however, mainly because of the low X-ray signals obtained and the relatively high spurious signals which occur from electron scatter. Limitations in the quantitation of elements in ultrathin sections have been examined by Shuman et al,[6].

The same equation may be employed as that shown for C_x in thin samples (equation 7) but corrections to W must be made for continuous radiation arising from other parts of the microscope and specimen support[4,6]. These spurious signals often turn out to contribute a high proportion of the total continuum signal and can lead to serious errors. Contamination or erosion can also seriously affect the results.

The method has been applied to frozen dried sections as well as to embedded sections and air dried samples. In frozen dried sections local variation in matrix density due to the removal of water and subsequent collapse of tissue structure make it difficult to interpret the data quantitatively on a wet weight basis unless the section thickness is known to be constant. No satisfactory method has been described for producing such sections routinely from a range of soft tissues apart from striated muscle[13] and frog skin[21]. With embedded sections the method relies on the assumption that embedding resin has completely replaced the original water content of the tissue. The errors in this assumption are likely to be far less than the dangers of removing elements from the tissue during the preparative procedures. So far very little work has been described for the successful production and analysis of frozen hydrated ultrathin sections of soft tissue.

Recent developments   For embedded sections standards have been developed that closely represent the specimen composition and contain known concentrations of given elements. Such elements are incorporated as salts into Araldite or epoxy resins as for the thin specimens at concentrations from 0.01 to several percent and may be analysed and treated in exactly the same way as the true specimens[4,20,22]. Shuman et al,[6] have indicated that a limitation of these standards may be that elements are dissolved in resin while in real specimens the matrix is largely protein. For embedded thin sections, however, where resin is assumed to have replaced the subcellular water content, the composition will not be very dissimilar from that of the standard. Errors are more likely to occur when using these standards to compare with frozen sections but even then the errors are only in the order of 10%. Shuman et al,[6] have also proposed, as an alternative to resin standards, the use of proteins such as BSA, insulin and phosvitin which contain elements (S,P,Na,Mg) in known stoichiometric concentrations. For cryo-sections Roomans and Seveus[23] have developed ultrathin frozen dried standards incorporating salts into albumin and gelatin with good homogeneity of elemental distribution.

By the careful use of light elements (e.g. C,Be) supporting and surrounding the specimen and by suppression of electron scatter, strict X-ray collimation, and electron filtering, the errors involved in determination of specimen mass thickness have been greatly reduced and the accompanying reduction of X-ray background has provided enhanced sensitivity in a number of laboratories[24].

6. Specimen preparation

Of all the areas of development in biological X-ray analysis specimen preparation is undoubtedly the weakest. Early attempts at analysis using conventional procedures of fixation and embedding made it clear that some elements were being lost or at least displaced from their in vivo sites and that others were being introduced. However, a number of apparently successful analyses have been reported using conventional techniques[25,26]. In a general sense all elements could be considered diffusible since their rigid retention in subcellular locations only depends on their insolubility and tissue-binding in the various solutions used. In practice it is the electrolytes (K,Na,Cl and Ca) which are most difficult to retain, and various alternative methods of specimen preparation have been adopted in an attempt to retain these elements for analysis.

Any preparative procedure must satisfy at least two criteria: a) it must retain elements in situ, and b) it must provide sufficiently good image detail in the microscope to allow the microanalysis to take place. Unfortunately, the difficulty in satisfying these two conditions together in the same sample has proved to be the biggest obstacle to soft tissue analysis.

It soon became clear that freezing tissue was the primary approach to maintaining specimens near to their in vivo states. Difficulties arose, however, with ice crystal damage through insufficiently rapid freezing techniques which not only destroyed the image but caused the subcellular displacement of elements. Continuous attempts by a number of workers have so far failed to completely overcome this problem but some success has been reported in specific tissues. Unfortunately, the success of any particular cryo-technique is nowhere near good enough to be called a routine method. Thus, extensive and costly laboratory experiments, which may take long times to complete, cannot be entirely trusted to a method which will be unable to guarantee the final production of specimens for analysis.

A wide range of preparative techniques has been employed for analysis of bulk specimens and thin and ultrathin sections. A number of them have been compared together[4,26,27], the latter authors providing a list of those papers dealing with elemental losses during wet preparative techniques. The results of such wet methods are disappointing to say the least in that very high and variable losses of elements from tissues occur in almost all instances. Such losses may not be due to the displacement of elements alone but also to the extraction of macromolecules to which the elements are bound to the tissue.

Wet methods include the use of glutaraldehyde and osmium tetroxide together with precipitating agents such as potassium pyroantimonate, potassium or sodium oxalate and gold and silver salts. Of all the methods involving solvents or aqueous solutions, however, the most promising, in that it produces the least total loss of elements from the tissue, appears to be that of freeze substitution[27]. This method, first devised for location of electrolytes in plant cells[28], falls halfway in between methods involving fixation and methods employing only cryo-techniques since it involves the rapid freezing of the tissue followed by the slow removal of ice by anhydrous solvents at low temperatures. The dehydrated specimen is usually embedded with resin. Images produced in the microscope from such preparations are superior to those obtained from purely frozen material while several workers have indicated that for certain elements, including electrolytes, losses may be quite minimal. There is no guarantee by this method, however, that translocation of elements within subcellular locations is eliminated.

As mentioned previously, the nearest one can get to the in vivo situation is to quench freeze tissue for analysis. A number of workers have developed this technique for a range of tissues such that samples may be examined in the thick or thin form, and may either be freeze dried or hydrated. For the latter, samples are maintained at low temperatures on a cold stage during the analysis. Again, maintaining the water in the tissue during examination is as near as one can get to the original tissue state. Gupta et al[12] have made extensive studies of thin sections prepared in this way and have concluded that, in order to satisfy the two conditions of image resolution and elemental retention, a slight (10%) degree of dehydration of the sample must be allowed to provide image contrast. Rigorous attention to freezing and mounting techniques and the prevention of dehydration or frosting of specimens during transfer of sections to the microscope has allowed the successful analysis of Na, K and Cl in 1μm frozen hydrated sections in the SEM. Other workers[13] have successfully analysed ultrathin frozen-dried sections of striated muscle in the TEM, again after paying careful attention to freezing, mounting and transfer techniques. Such a periodic structure as presented by striated muscle is not, however, typical of most soft tissues and the good quality of the images of these sections cannot yet be repeated on other material.

For analysis of bulk material, cryotechniques are again used to avoid translocation of elements, involving the rapid freezing of tissue and the examination of freeze fractured surfaces while maintaining full hydration during the analysis[29]. Compared with the preparation of frozen sections such a method is relatively easy and reproducible.

A further advantage of thick samples over sections is the greater ease with which the frozen water may be retained during preparation and handling, this being due to the reduced surface area (one half) for evaporation and the greater thermal mass of the sample. The two main disadvantages of the method, however, are that one is limited to a surface rather than a transmitted image, and that the spatial resolution of analysis is far worse than with thin sections. In addition, the quantitation of data obtained from such samples may, as previously mentioned, be more complicated due to the need for ZAF correction procedures and the space charge effects produced by internal charging[17].

Conclusions

Instrumentation is not now a major limiting factor in the analysis of biological material. The main areas of weakness lie in the preparation of specimens that truly represent the original tissue in both elemental content and structural detail, and the analysis of these samples under conditions that avoid wasting all the hard work of preparation by irradiation damage to the sample. No one method of preparation will be suitable for all tissues, and no set of analytical conditions will satisfy the stability requirements of all elements in all samples. Many investigators, working on a wide range of tissue types will, for the technology to advance further, have to produce data that can be supported by stringent correlative analytical techniques. Whereas cryotechniques offer the most obvious method for studying physiological processes, the methods will have to become more reliable, reproducible and less cumbersome than at present. For sensitivity in analysis to improve, both instrument manufacturers and analysts will need to give particular attention to those factors inherent in their microscopes which produce unwanted background and limit peak detection and accurate quantitation.

Advances in the fields of specimen preparation have been slower than those seen in instrumental design. Perhaps with greater commitment to studying the basic events occurring during these preparative procedures a fuller understanding and more reliable interpretation of the data obtained from soft tissue analysis will be possible.

Acknowledgements

The generous financial support of the Tenovus Organisation is gratefully acknowledged.

References

1. R. Fitzgerald, K. Keil & K.F.J. Heinrich. Solid state energy dispersion spectrometer for electron microprobe X-ray analysis. Science 159, 1968, 528-530.
2. S.J.B. Reed. Electron Microprobe Analysis. Cambridge Univ. Pr. 1975.
3. T.A. Hall. The microprobe assay of chemical elements. In: Physical Techniques in Biochemical Research, 2nd ed. Ed. G. Oster, Academic Press N.Y., 1971, 1A, 158-275.
4. J.A. Chandler. X-ray Microanalysis in the Electron Microscope, Elsevier/North Holland, 1977.
5. J.C. Russ. Electron probe X-ray microanalysis - principles. In: Electron Probe Microanalysis in Biology, Ed. D.A. Erasmus, Chapman & Hall, London, 1978, 5-36.
6. H. Shuman, A.V. Somlyo. Quantitative electron probe microanalysis of biological thin sections : methods and validity. Ultramicroscopy 1, 1976, 317-339.
7. P.J. Statham. A comparative study of techniques for quantitative analysis of the X-ray spectra obtained with a Si(Li) detector. X-ray Spectrometry 5, 1976, 16-28.

8. F.H. Schamber. A new technique for deconvolution of complex X-ray energy spectra. Proc. 8th Natl. Conf. on Electron Probe Analysis, New Orleans, 1973, paper 85.

9. D.B. Williams & J.I. Goldstein. A study of spurious X-ray production in a Philips EM 300 TEM/STEM. Proc. 8th Int. Conf. on X-ray Optics and Microanalysis. Eds. B. Ogilvie & D. Wittry, Boston, USA, 1977, 113A-113E.

10. H. Shuman & A.P. Somlyo. Electron probe X-ray analysis of single ferritin molecules. Proc. Nat. Acad. Sci. USA, 73, 1976, 1193-1195.

11. T. Bistricki. Quantification of losses during X-ray microanalysis of mercury. Proc. 9th Int. Cong. on Electron Microscopy, Toronto, Canada, 2, 1978, 12-13.

12. B.L. Gupta, T.A. Hall & R.B. Moreton. Electron probe X-ray microanalysis. In: Transport of Ions and Water in Animals, Eds. B.L. Gupta, R.B. Moreton, J.L. Oschman and B.J. Wall, Academic Press, 1977, 83-143.

13. A.V. Somlyo, H. Shuman & A.P. Somlyo. Elemental distribution in striated muscle and the effects of hypertonicity : electron probe analysis of cryo sections. J. Cell Biol. 74, 1978, 828-857.

14. T.A. Hall & B.L. Gupta. Beam induced loss of organic mass under electron microprobe conditions. J. Microsc. 100, 1974, 177-188.

15. S.J.B. Reed. Spatial resolution in electron probe microanalysis. Proc. 4th Cong. on X-ray Optics and Microanalysis. Eds. R. Castaing, P. Deschamps & J. Philibert, Orsay, France, 1966, 339.

16. R.R. Warner & J.R. Coleman. Quantitative analysis of biological material using computer correction of X-ray intensities. In: Microprobe Analysis as Applied to Cells and Tissues, Eds. T. Hall, P. Echlin & R. Kaufmann. Academic Press, N.Y. & London, 1974, 249-268.

17. W. Fuchs. Electron probe microanalysis of frozen hydrated bulk specimens : basic experiments. Proc. 8th Int. Conf. on X-ray Optics and Microanalysis. Eds. B. Ogilvie & D. Wittry, Boston, USA, 1977, 163A-163C.

18. D.J. Marshall & T.A. Hall. Electron probe X-ray microanalysis of thin films. Br. J. Appl. Phys. 1, 1971, 1651-1658.

19. T.A. Hall. Methods of quantitative analysis. J. de Microscopie et de Biologie Cellulaire, 22, 1975, 271-282.

20. A.R. Spurr. Choice and preparation of standards for X-ray microanalysis of biological materials with special reference to macrocyclic polyether complexes. J. de. Microscopie et de Biologie Cellulaire, 22, 1975, 287-302.

21. A. Dorge, K. Gehring, W. Nagel & K. Thurau. Intracellular Na^+ and K^+ concentration of frog skin at different states of Na-transport. In: Microprobe Analysis as Applied to Cells and Tissues, Eds. T. Hall, P. Echlin and R. Kaufmann. Academic Press, N.Y. & London, 1974, 337-350.

22. G.M. Roomans & H.L.M. Van Gaal. Organometallic and organometalloid compounds as standards for microprobe analysis of epoxy resin embedded tissue. J. Microsc. 109, 1977, 235-242.

23. G.M. Roomans & L.A. Seveus. Preparation of thin cryosectioned standards for quantitative microprobe analysis. J. Submic. Cytol. 9, 1977, 31-35.

24. B.J. Panessa, J.B. Warren, J.J. Hren, et al. Beryllium and graphite polymer substrates for reduction of spurious X-ray signal. SEM/1978/II, SEM Inc., AMF O'Hare, Ill 60666, 1055-1062.

25. A.J. Saubermann. The application of X-ray microanalysis in physiology. J. de Microscopie et de Biologie Cellulaire, 22, 1975, 401-414.

26. J.A. Chandler. The application of X-ray microanalysis in TEM to the study of ultrathin biological specimens - a review. In: Electron Probe Microanalysis in Biology. Ed. D.A. Erasmus, Chapman & Hall, London, 1978, 37-93.

27. A.J. Morgan, T.W. Davies & D.A. Erasmus. Specimen preparation. In: Electron Probe Microanalysis in Biology. Ed. D.A. Erasmus, Chapman & Hall, London, 1978, 94-147.

28. A. Spurr. Freeze substitution additives for Na and Ca detection in cells studied by X-ray analytical electron microscopy. Bot. Gaz. 133, 1972, 263-270.

29. W. Fuchs, B. Lindemann, J.D. Brombach & W. Trosch. Instrumentation and specimen preparation for electron beam X-ray microanalysis of frozen hydrated bulk specimens. J. Microsc. 112, 1977, 75-87.

Discussion with Reviewers

Reviewer III: You have described sources and reduction of spurious signals in some detail. Can you add comments on sources of continuum radiation and its suppression as well as a brief description of methods that are employed for subtraction of this portion of an X-ray spectrum.

Author: The major source of background in the X-ray spectrum arises from bremsstrahlung due to electron interaction with both the specimen and the supporting material. Ideally the majority of this radiation should come from the specimen region selected for analysis but electrons outside of the focused beam may strike other regions of the sample as well as the relatively heavy specimen support and even pole pieces and other parts of the instrument in the specimen region[9]. In addition a high background signal may arise from the primary electron beam producing an X-ray source from the condenser aperture, this in turn irradiating the whole of the specimen region with consequent spurious characteristic and continuous radiation being produced. The latter source of background has, in current instruments, been largely eliminated by the careful screening of the condenser aperture with lead shields, while spurious signals from outside the focused probe have been greatly reduced by careful collimation of X-rays and by the extensive use of light elements such as beryllium and carbon in the specimen region. In thick specimens the very large background produced by penetration of the electrons into the sample limits the sensitivity of analysis. In order to accurately measure the intensity of small peaks on such a varying background computerised methods are employed to iteratively fit a synthetically generated bremsstrahlung distribution to that actually produced by the sample, this background being subtracted from the whole spectrum, leaving just the spectral lines for examination. Methods of computing this background have been described in detail by Statham[7].

Reviewer III: Is it not true that smaller probe provides better spatial resolution regardless of specimen thickness despite more extensive diffusion of electrons into a bulk of specimens? Are there any negative consequences associated with irradiating thick specimens with a probe of a small diameter?
Author: Spatial resolution with smaller probes in thick specimens would be improved primarily in the surface layers of the sample. The probe would then diffuse laterally with increasing depth of penetration to excite the well known pear shaped volume of the sample. The actual extent of the diffusion depends on the accelerating voltage of the electron beam and on the specimen mean atomic number. The main problem with using a very small probe (say 10nm) on a thick sample would be the error of concluding that this represented the resolution of analysis. In a soft tissue sample such a probe at an accelerating voltage of, say, 15kV would excite a volume of about $7\mu m$ in depth and $3.5\mu m$ in diameter for Ca $K\alpha$ X-rays. Other consequences, such as irradiation damage to the sample, are dependent on the total electron dose in the probe itself.

Reviewer III: When analysing thin sections, you said, the main limitation is a relatively high spurious signal caused by the electron scatter. How does the carbon coating influence this phenomenon?
Author: Since an ultrathin specimen produces such a low X-ray signal, other spurious signals not directly arising from the sample have a relatively great effect. Thus, in a sample of thickness 100nm, for example, coated with a carbon film of thickness 20nm, approximately 20% of the bremsstrahlung may arise from the coating film alone (less actually since carbon has a lower atomic number than average soft

tissue). In addition, if contamination is allowed to build up the resultant bremsstrahlung will also be elevated while the characteristic signal from the specimen may remain unchanged. An increase in overall mass thickness may also cause an increase in electron scatter from the specimen resulting in a higher spurious signal from electron interaction outside of the specimen itself.

Reviewer III: A brightness of an electron source can be used as a relative measure of a beam energy. Can you give examples of maximum current versus spot size comparing FEG and conventional sources. A SEM in my laboratory yields 50 pA current in a beam diameter of 4nm at 30kV. How does this compare with FEG under the same operating conditions?
Author: It is not easy to exactly compare conventional thermionic filaments with field emission guns unless they are operated in the same electron microscope under similar focusing conditions. However, it is generally agreed that while tungsten thermionic filaments produce, typically, current densities of $1A/cm^2$ at the gun, field emission sources can produce current densities of greater than 10^4 A/cm^2. For example Bovey et al[30] have obtained beam currents of 10^{-8} A in a 5nm diameter probe operated at 100 keV in a STEM, equivalent to 4×10^4 A/cm^2 current density, from a cold field emission source. In an early study Broers[31] obtained 10^{-10} A in a 1nm probe diameter from a tungsten field emission source compared with 8×10^{-13} A in a 5nm probe from a tungsten hairpin filament.

Reviewer III: How to determine identical analytical conditions when comparing X-ray intensities of an unknown and a standard. What is the most appropriate parameter to monitor if a device to measure a beam current is not available? What are usual deviations in beam currents found in a typical SEM?
Author: If there is no device to measure beam current in the microscope, i.e. a Faraday cage, the best advice is to make one and install it! It is very difficult to perform reproducible analyses unless one can accurately measure beam current. The current may alternatively be collected by the viewing screen if insulated from the microscope, or the exposure meter may monitor beam current intensity. In some microscopes gun bias current can be related to probe current. Fluctuations of probe current usually arise through incorrect saturation of the thermionic gun but even under saturated conditions it is possible for filament drift to take place and for changes of a few percent to occur. Under normal stable conditions, over a typical analysis time of a hundred seconds,

the beam current should vary no more than 2-5%. Since most methods of quantitation employ the simultaneous measurement of characteristic and continuous radiation, this fluctuation should not prove troublesome.

Reviewer II: The author suggests that 10% accuracy may be achieved in thin section microanalysis. This sort of number is also used by other authors, and seems to be based primarily on the quantifiable errors in intensities and the various quantitative models. Can the author give meaningful estimates of errors in preparation, mass loss, etc., even if only for specific cases?
Author: Unfortunately the errors due to specimen preparations and mass loss are less easily defined and likely to be much larger than instrumental errors. Errors due to specimen preparation can only be estimated by correlative techniques of analysis, e.g. atomic absorption of tissues during each procedure. These vary enormously from tissue to tissue and for different preparation methods[27]. In some analyses it is possible to compare frozen tissues (the most reliable) with other methods and estimate losses in subcellular regions. For example, in a study on sperm cells[32] a comparison was made of freeze dried, freeze substituted, air dried, glutaraldehyde fixed and histochemically fixed tissue. Whereas freeze drying and air drying produced identical analyses, large changes took place with other methods. Thus it is not possible to give a general figure for accuracy of analysis but rather to quote a comparative value against that of a reliable preparation. Mass and element loss depend on analytical conditions and on elemental binding in the tissue. Again, using sperm cells, it has been shown[33] that the wrong choice of analytical conditions can introduce large (~50%) errors into the analysis due to specimen damage.

Reviewer I: Do you define a thin sample as one of thickness below the range given by equation 1?
Author: A thin sample is one in which almost no energy loss of electrons takes place by the time they emerge from the sample. There is a 'no man's land' at the region where a thick specimen becomes a thin specimen and this is approximately defined by equation 1 since some of the emergent electrons will have an energy lower than E_c. Strictly speaking one should choose the thickness of a thin specimen to be well below that defined by equation 1. The thinness of a specimen is also defined by the degree of absorption (and fluorescence) that occurs, this again depending on the energy of the emitted X-ray line.

Reviewer I: Does the quality of the vacuum affect the analysis?
Author: The analysis of biological specimens may be affected in a number of ways by the quality of the vacuum. An unclean vacuum may cause the build up of contamination on the specimen giving rise to changes in specimen mass thickness and in extreme cases affecting the penetration of electrons into and the emission of X-rays from the sample. In addition certain impurities may be deposited on the specimen, possibly arising from diffusion pump oils, dirty internal surfaces, etc, which could cause unwanted peaks in the energy spectrum. Water vapour present in the vacuum of the microscope, for example from a frosted anticontaminator or cold stage, has been known to cause erosion of the specimen in the area of the electron probe.

Reviewer I: Can cations such as Na^+ and K^+ move in ice under electron bombardment?
Author: There have, to my knowledge, been no reported measurements of this phenomenon in the literature. It is known that complete loss of elements may occur through volatilisation and that this depends on element binding, conductivity, etc. The crystallisation of the ice may play an important role in lateral diffusion of cations in frozen hydrated specimens, this crystallisation depending on the method of quench freezing the specimen[12]. Diffusion or displacement of ions in the electron beam does occur in certain minerals and in glass. It would seem likely that such changes may also take place in ice but analysis of frozen hydrated specimens reported elsewhere[12] show the subcellular analyses of these cations to give meaningful results implying that no losses had occurred.

Reviewer II: J. Russ
Reviewer III: T. Bistricki

References

30. P. Bovey, I. Wardell & P.M. Williams. X-ray microanalysis in the STEM. Proc. 8th Int. Conf. on X-ray Optics and Microanalysis. Eds. B. Ogilvie & D. Wittry, Boston, USA, 1977, 117A.
31. A.N. Broers. Factors affecting resolution in the SEM. SEM/1970, IITRI, 8-13.
32. J.A. Chandler & S. Battersby. X-ray microanalysis of diffusible and non-diffusible elements in ultrathin biological tissues using human sperm cells as a model for investigating specimen preparation. In: Biological X-ray Microanalysis by Electron Beam Excitation. Academic Press, N.Y., 1979. (in press).
33. J.A. Chandler & S. Battersby. X-ray microanalysis of ultrathin frozen and freeze dried sections of human sperm cells. J. Microsc. 107, 1976, 55-65.

For additional discussion see page 618

SCANNING ELECTRON MICROSCOPY/1979/II
SEM Inc., AMF O'Hare, IL 60666, USA

GENERAL CONSIDERATIONS OF X-RAY MICROANALYSIS OF FROZEN HYDRATED
TISSUE SECTIONS

A. J. Saubermann

Department of Anaesthesia,
Harvard Medical School at Beth Israel Hospital
Boston, MA 02215

Abstract

X-ray microanalysis of frozen hydrated sections can be applied for analysis of diffusible or highly mobile elements which might be displaced or lost when conventional preparatory techniques are used. Once frozen the specimen must be handled in such a way that it becomes vacuum compatible for subsequent analysis. Analysis cannot be accomplished without compartment recognition. Initial freezing can be considered successful if ice crystal damage does not exceed the dimension of the compartment to be analyzed. As compartment size decreases ice crystal artifacts become a significant problem. The application of special freezing methods including use of non-penetrating cryoprotectants is an important advance in freezing methodology for X-ray analysis. Cryosectioning is a fracturing process at low temperatures which can affect morphology and may affect analysis. An understanding of this process can aid in optimizing cryosectioning for use in preparing frozen hydrated and frozen dried section for X-ray microanalysis. Special equipment is necessary for both freeze dried as well as frozen hydrated specimens; while special transfer devices are necessary for frozen hydrated specimen transfers. Because of the design of cold stages special consideration for reducing extraneous background radiation is an intrinsic part of their application to biological X-ray analysis.

KEY WORDS: Cryotechniques, Biological Specimen Preparation, X-ray Microanalysis, Frozen Hydrated, Cryosectioning

Introduction

Cryo-techniques applied to biological specimens for the purpose of immobilizing and retaining water soluble or diffusible elements or ions has held great promise for biological application of X-ray analysis. Yet cryo-techniques have been used successfully to obtain useful biological data in only a limited number of cases (see for example 1-5). The general preparation of biological specimens for X-ray analysis has been reviewed[6] and illustrates the plethora of techniques and their variations applied, sometimes successfully and sometimes not. This review will discuss one of these techniques, namely frozen hydrated section analysis including freezing, cryosectioning, specimen transfer and analysis. Hopefully, by understanding the limitations, as well as the advantages, of frozen hydrated tissue section analysis the interested reader may be able to adapt or use these methods to solve a physiological problem.

Freezing

Much has been written about freezing biological specimens. Still the process, except in certain special cases, is poorly understood; however this does not preclude cryofixation from use. Ice crystal formation remains the major problem since it is regarded as both a cause of morphological artifact and a potential cause of elemental displacement or distortion. Freezing techniques vary depending upon the end point one is interested in, such as survival or morphology free of ice crystal artifacts. It is abundantly clear that good morphology is not a sign of a surviving cell.[7] How successfully must freezing be to apply X-ray analysis as a tool for physiological studies? Can we, in fact, accomplish our ends with techniques available today?

To use X-ray analysis for diffusible elements (usually ions) our major concern is to retain those elements in their original position within some compartment and therefore obtain a sample representing that compartment at the moment the specimen was originally sampled. Secondly we must be able to identify, with reasonable certainty, that compartment we wish to analyze. It is probably possible to freeze tissue and cells adequately for these ends, providing the nature and extent of ice crystal damage is known; and as long as the ice crystal damage does not exceed the spatial limits of the volume which is to be analyzed.

The size of ice crystals is the main spatial limit to analysis from which useful data can be obtained from cryoprepared specimens. In many applications compartment size is relatively large, such as cytoplasm, vacuoles, spaces, lumens, gradients either side of a boundary, groups of cells, or tissue layers. Such compartments are visible with light microscopy being several microns in diameter. For physiological problems requiring analysis of such a compartment, tissue can be frozen rapidly in any number of cryogenic liquids such as Freon 22, 13, or 14, nitrogen slush, isopentane or propane. There is probably little advantage, one over another, since the major limit to freezing rate is the poor thermal conductivity of the tissue itself. The least toxic and cheapest would be liquid nitrogen slush which can be simply prepared by placing LN_2 in a styrofoam cup and exposing it to a rotary pump vacuum (40 torr) for 30-60 seconds. To optimize freezing, just as for chemical fixation, sample size should be minimized. Ice crystal artifacts formed during freezing are likely to be sufficiently small to permit use of the sample for the majority of problems requiring analysis of larger compartments (several microns). One to two mm cubes of tissue are a good size for conventional freezing. Special methods such as spring loaded or specially propelled plungers[8,9] or highly polished conductive metals may not be necessary for large compartment analysis. Detection of elemental displacement artifacts caused by ice crystal damage in large compartments can be compensated for during analysis by using a raster of sufficient size to include "matrix" and ice crystal. By thus enlarging the volume analyzed, elemental displacement within that volume is not detected providing, of course, that the average mass fraction of the element within the compartment is unchanged. For example if the specimen has larger ice crystal artifacts (>1μ) it is necessary to use a raster several times that size. Obviously the usefulness of information obtained from such a specimen would depend upon the original homogeneity of the compartment.

If the morphological boundaries of the compartment are sufficiently large to apply this technique the ice crystals are inconsequential to analysis. However if on the other hand the compartment size is similar in size to ice crystal artifacts this technique is obviously invalid. The presence of ice crystal damage, therefore, does not mean that a specimen cannot be analyzed, it does mean that caution should be exercised and that a raster be chosen which is much larger than the ice crystal damage. To always attempt to achieve "perfect" crystal free freezing for many applications, although desirable, may not, in fact be necessary.

Compartment identification (morphological recognition) can be difficult or impossible with extensively damaged tissue. This fact may be the limiting factor in judging the usefulness of specimens for analysis. Normally we identify structures by two criteria, first the position of the structure of interest in relationship to other structures and secondly the peculiar characteristics of that structure itself. Structures which are not generally visible or easily identifiable at the light microscope level demand greater care in freezing so as they can be identified. Comparison techniques have been useful as aids to structure identification. Somlyo et al[3] have used, for example, OsO_4 vapor fixation of sections cut specifically for morphological reasons and not for analysis. Another useful technique is to prepare "freeze substituted" - alcohol fixed sections in series with those used for analysis. These sections can be stained with Toluidine blue to aid in structure and cell identification in those cryosections submitted to X-ray analysis. A simple method is to place a section into a drop of cold 90% ethanol placed on a glass slide cooled to cryostat temperatures. The section and slide are then brought to room temperature, the ethanol evaporated and the section stained. This technique is obviously best suited to low resolution compartment identification.

Cryoprotectants have been used to reduce ice crystal size to below the size of the analytical compartments. Penetrating cryoprotectants have been extensively studied and used to preserve conventional morphology. However they are of little use in X-ray analysis as they can cause extensive loss of elements. Franks et al,[10] Echlin et al,[11] and Skaer et al[12] in a series of papers have reported on the use of three nonpentrating cryoprotectants: polyvinylpyrrolidone (PVP-molecular weights 44,000 and 700,000), hydroxethyl starch (HES-molecular weight 450,000), and dextran (molecular weight 68,500). These large polymers are presumed to act by suppressing external nucleation allowing subcooling of the cell to a homogeneous nucleation temperature thereby promoting formation of a microcrystalline state within the cell. The high water binding capacity of these agents prevents osmotic gradients and minimizes water loss.[10] To be effective they must be used in concentrations exceeding 25%.[10,11,12] These non-penetrating cryoprotectants offer a number of advantages over penetrating cryoprotectants for use in X-ray microanalysis. These include (1) minimal interference with cell function (2) morphological preservation (3) easier freezing (4) minimal effects on elemental loss (5) they are nontoxic and (6) easy to use. Echlin et al[11] have demonstrated that normal cellular function continues in a number of organisms when placed in solutions of PVP, HES, and dextran as judged by several criteria including movement, appearance, secretion and growth. However they point out that some systems can be affected, possibly by binding of cellular transmitters or drugs to the cryoprotectants.[11] They therefore caution against use without thoroughly testing the specimen in the cryoprotectant. The absence of diffusion across cell boundaries has been demonstrated by examining plant cells sectioned after being placed in PVP.[11]

The morphology of specimens treated with

Table 1

Work of Cryosectioning

Angle (-80°C)	Work µJ	(n)	Heat[†] (cal/gm)	Cutting Patterns[*]
2°	558.9+284.4	(8)	274+136	1
4°	225.6+256.9	(8)	107+ 68	3
6°	470.7+294.2	(7)	224+142	2
Temperature				
-30°C	127.5+ 58.8	(6)	59+ 27	1
-80°C	431.5+284.4	(7)	206+137	2
-120°C	264.8+156.9	(4)	125+ 74	3+2
-30°C to -80°C[#]	78.5+ 19.6	(6)	35+ 11	3
Thickness (-80°C)				
0.5 µ	470.7+294.2	(7)	224+142	2
1.0 µ	431.5+186.3	(7)	102+ 44	2
2.0 µ	411.9+127.5	(6)	99+ 30	2

[*]refers to pattern type in Fig. 1.

[†]Worst case calculation assumes all the work of sectioning is in the form of heat and all the heat is transferred to section. For full details see Ref. 14.

[#]Block warmed to -30°C then recooled to -80°C and sectioned.

these nonpenetrating cryoprotectants has been studied in single cells as well as tissues using high resolution techniques. The preservation of morphology is generally in excess of most X-ray analysis requirements.[10,11,12] Furthermore neither special high speed freezing nor cryogenic liquids need be used for freezing since good freezing can be obtained by direct immersion in liquid nitrogen.

Sectioning

Cryosectioning has been a major problem largely because of unpredictable consistency and is perhaps one of the most problematic steps in cryospecimen preparation. Unfortunately cryosectioning is a poorly understood process. Historically cryosectioning has been considered a variant of conventional sectioning thus instruments available for cryosectioning are basically refrigerated conventional microtomes. A number of questions have been raised regarding whether tissue which has been cryosectioned can be used for microanalysis. Does the cryosectioning process cause elemental displacement or morphological changes? Furthermore, and this is a strictly practical consideration, can consistent reproducible sections be produced? Appleton[13] has pointed out that each tissue seems to have an optimal temperature at which to section. He uses as an end-point the temperature at which his sections form ribbons.[13] He considers this presumptive evidence of melting. Saubermann et al[14] measured the work of cryosectioning by placing strain gauges on a load

cell placed in the drive arm of a Sorvall-MT2 microtome fitted with a low temperature Christensen type cryochamber. Forces occurring during cryosectioning were recorded on a chart recorder. Since the speed of sectioning was known and the forces measured, the work could be calculated. Sections were photographed through a calibrated eyepiece on the microtome microscope so that dimensions could be measured. Assuming the density of the cryosection to be unity an estimate of the work, converted to calories per gram, could be made. Thus the calculations, based upon direct measurement of work, estimated the worst possible case, namely, that all work was converted to heat and that all the heat was available for distribution to the section. Table 1 summarizes these data which indicate that there is generally sufficient heat available, were it all to be distributed to the section (taking into consideration latent heat of fusion) that the section could transiently melt, except for sections cut at -30°C. The unanswered questions are where is the heat distributed, does melting actually occur, and, if so, under what conditions? Furthermore, is cryosectioning an acceptable method for specimen preparation for X-ray microanalysis? A clue to these questions can be seen from observing the force pattern during sectioning. These patterns (Fig. 1) which confirm observations made by others suggest there are three types of "sections" which form a spectrum of patterns reflecting the processes of cryosectioning. Thus the patterns reflect fracturing at low temperatures while cutting (or shearing) at warmer temperatures. Observation of the surfaces

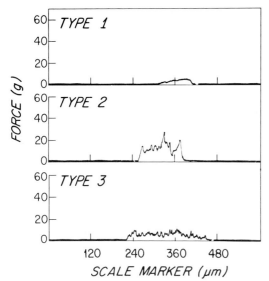

Fig. 1. Three distinct force patterns can be obtained during cryosectioning depending upon temperature, knife clearance angle, and other factors.[15] Type 2 and Type 3 patterns represent a fracturing process during sectioning, while Type 1 is seen with smooth sections and requires the least work.

of frozen hydrated sections fractured and cut support this hypothesis (Fig. 2). At very low temperatures (-100°C or lower) the size of fractures decreases making the sections more "powdery" in appearance. As the chamber temperature warms the fractures become large (-80°C) until the fractures disappear (-40°C) and the sections appear smooth and glossy.

There are several important and practical corollaries regarding the effects of controlled variables during cutting. Clearance angle is very important to cryosectioning, as very low clearance angles < 2° result in drag of the block against the knife and may affect the sectioning properties of the remaining block. The fracturing process is not affected by section thickness between 5 and .5 μm thicknesses, nor is the work per unit mass. This may indicate that if a "fixed" percent of heat available does go into the section, the absolute heat may not be sufficient to cause transient melting until very thin (<100 nm for example) sections are cut. It is these very thin sections which are generally required for analysis of the smallest compartments. For thicker sections, providing the structural deformities resulting from the fracture segment overlapping do not disrupt or prevent compartment identification nor compartment integrity, cryosectioning is a reasonably safe step. An unanswered question as yet is, of course, does any transient melting which may occur adversely affect elemental distribution and compartment? The answer from the point of view of X-ray analysis will probably rely upon the size and nature of compartment to be analyzed.

In a recent study[15] we have examined the

effect of cryosectioning 0.5 μm thick sections at warmer (-30°C) temperature to determine if smearing or displacement of diffusible elements occur. This study was prompted by the observation that at -30°C flat smooth sections are produced with the least amount of work and therefore the smallest potential for transient melting were all the work converted to heat and transferred to the section (Table 1). In an initial series of experiments a standing gradient was prepared with 15% KI and NaCl, on a 30% gelatin substrate. A block was frozen and sectioned at -80°C and the section analyzed for a gradient over 20 μm areas at distances of 60 μm between centers beginning at the leading edge of the block surface. The same block was then warmed to -30°C and another 0.5 μm thick section cut and analyzed in a similar fashion. The results showed two identical exponential curves. This gradient remained unchanged regardless of the direction of the knife. Tissue sections (0.5 μm thick) examined at -30° and -80°C showed no difference in ice crystal size after freeze drying in the microscope, suggesting that sectioning at -30°C produces no additional significant ice crystal damage when examined at lower magnification (1,000 - 5,000 x). Distinct compartments in kidney can be observed and repeatedly measured[16] strongly suggesting that these warmer temperatures may be a reasonable compromise for X-ray analysis at low magnification with raster sizes in the range of 1-2 μm in width. One of the major advantages of cryosectioning at -30°C is the relatively good morphology one can obtain which is normally obscured by the overlapping chips produced during the fracturing occurring at lower temperatures (Fig. 3).

Handling, picking up, and mounting sections on a suitable support is a weak link in the whole process of frozen hydrated section analysis. Generally we have found that sections can be manipulated onto a specimen support placed close to the knife by using an eyelash attached to an applicator stick. For our specimen support we use a specially designed Be holder which minimizes extraneous background radiation but provides good thermal and electrical conductivity.[17] The annular stub has a 3 mm hole which is covered by a 75 mesh Be grid over which is placed a nylon film[18] which is then carbon coated. Forces holding the section onto the surface are probably friction and some weak charge attraction. The Be grid provides support for the nylon film, and helps keep the specimen hydrated in the SEM. Since the carbon coating in the nylon film is too thin to act as a significant pathway for thermal conductivity, heat transfer between the specimen and the Be stub occurs principally via the supporting Be grid.

Special Equipment

Most analytical systems can be adapted for cryo-preparatory techniques, and most investigators in this field now generally agree that

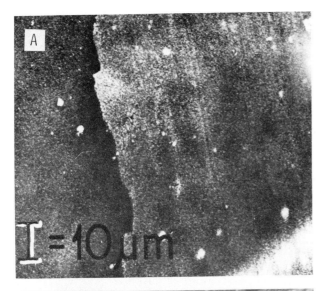

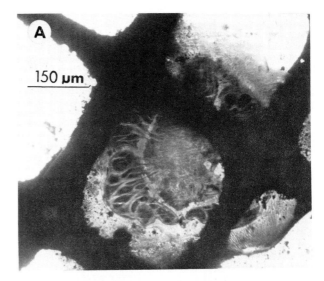

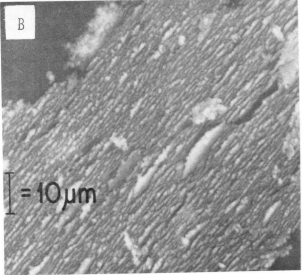

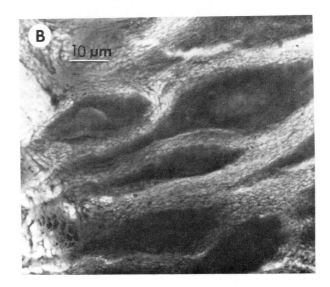

Fig. 2. Secondary images of the surfaces of frozen hydrated sections of mouse liver cut at -30°C (A) and -80°C (B). The section cut at -80°C shows the overlapping chips formed during sectioning which obscures tissue morphology when viewed in the STEM mode.

Fig. 3. Frozen dried 0.5 μm thick sections of leech ganglia cut at -30°C. STEM images demonstrate that good morphology can be obtained by preventing the overlapping of chips formed during sectioning at lower temperature (see Fig. 2).

a cold stage is essential for analysis of biological specimens whether examined frozen dried or frozen hydrated. The major reason for examining frozen dried material at low temperatures is to reduce matrix and volatile elemental losses. The extent of these losses may be different for different specimens and obviously varies with radiation dosage. Cl seems to be easily lost and this has been nicely demonstrated by Roinel[19] in the specialized technique for microdroplet analysis. Mass loss from the matrix can also be decreased significantly by analyzing at lower temperatures.[20] For frozen hydrated specimens the temperature requirements of the stage are quite stringent since the temperature must be

sufficiently low to reduce sublimation loss rate to undetectable rates. This means a stage which is able to operate in the temperature range below -160°C.

A number of specially built coldstages have been reported.[18,21-25] Recently a stage using a Joule-Thompson refrigerator was commercially built.[23] Good, high resolution secondary scanning images of frozen hydrated plants have been obtained which would suggest that one of the major theoretical limitations of the Joule-Thompson stage, namely vibration, can be overcome. Efficient stages, cooled by copper braid have been adapted to a number of SEM's. Taylor et al[21] reported a design for the JXA-50A which operates at -175°C. Their stage cannot be used for STEM imaging. Saubermann and Echlin,[18] and Saubermann et al[22] have also used a copper stage cooled by copper braid, which also operates in the -180° - 175°C range. The basic principle in designing such a stage is to minimize connections which can be disconnected. That is to braze as many connections as possible along the cold pathway. Also by keeping the stage mass at a minimum and isolating it from the substage assembly by minimizing the number and size of contacts, and by thermally insulating those connections, one can build a fairly efficient cold stage. Major deficiencies of the copper braid type stage, when compared with Joule-Thompson type stages are the difficulties in rapidly warming the cold stages and some compromise in range of stage motion. By comparison J-T type stages can be easily cooled and warmed by adjusting gas flow. Low temperature stages are also available for "conventional" TEM's which can be equipped with X-ray spectrometers and probe forming lenses. While it is possible to do frozen hydrated tissue section analysis using a TEM, transfer systems have remained a problem. Hutchinson et al[24] have been working with a TEM to obtain routine thin frozen hydrated section analysis. Their system uses sections sandwiched between Cu grids.

One word of caution which, although seemingly obvious, should be kept firmly in mind, is that the stage temperature is not a measure of the specimen temperature. So far, attempts at biological specimen temperature measurement have not been successful. However a reasonably safe estimate is to assume that specimen temperature is likely to be 10° to 15° warmer than the specimen holder which is likely to be 10° - 15° warmer than the stage. If one is attempting to work with frozen hydrated specimens this rough rule of thumb is generally sufficient to keep specimens frozen hydrated by observing stage temperature.

Anticontamination devices are cold plates placed adjacent to the cold specimen which act to prevent potentially contaminating vapors from being trapped by the specimen. These devices seem to be most useful in higher resolution problems. The kind of contamination which is of greatest significance to the biologist wishing to utilize cryotechnique is contamination which significantly alters specimen mass. Such contamination would result in a decrease in apparent mass fraction of an elemental species in the specimen, rather than the contamination seen with high resolution microscopy related to surface migration of volatile, or beam induced mobile elements. Larger compartment (several micrometer) analysis at low temperature does not seem to be terribly sensitive to the usual level of residual "contamination" in the vacuum of the microscope chamber. Very small compartments, usually analyzed in the TEM, are more sensitive. As a simple measure of contamination rate, the white radiation generation rate from a thin (with respect to the beam) film cooled to stage operating temperature (-175°C) can be monitored. Under stable beam conditions and clean vacuum conditions, increases in white radiation generation rate are not observed nor are they observed over the surface of a frozen hydrated section.[22]

Transfer systems for loading the microscope with frozen hydrated specimens have been of two kinds, the vacuum and non-vacuum type. The vacuum type is a type of system in which the microscope chamber remains at high vacuum during the transfer and the non-vacuum type is where the microscope chamber is brought to atmospheric pressure and the transfer made at atmospheric pressure. The vacuum transfer system has the advantage of reducing the likelihood of microscope and specimen contamination from atmospheric water vapor while insuring that sublimation rate is likely to be extremely small. The major advantage of the non-vacuum transfer system is its relatively low cost since additional prechambers and pumps are not necessary. It does however require special care that the transfer is done as rapidly as possible.[18,23] A vacuum transfer system has been reported[22] and a commercial system is also available.[23] The prepumped airlock of the JXA has been used by adapting the transfer rod for low temperature transfers.[5] Gupta et al have suggested that sections can be stored for several days in a dry N_2 atmosphere on a LN_2 cooled "parking" stage before transferring these sections into the STEM.[5] The "heart" of all these transfer systems is a rod which can be withdrawn into a shroud or tube.[24] The principle is to make a small portable "cold stage" which has sufficient mass to keep the specimen cold during the transfer to the microscope cold stage and to keep the specimen from being contaminated with atmospheric water vapor. Details of such transfer rods can be found in the literature.[5,18,22-25]

The high thermally conductive metals, having relatively high mass, used in cold stage construction pose a peculiar problem for low temperature biological X-ray microanalysis since these metals can add considerable extraneous background radiation. Thus special care is required to "clean" the system, using collimation and low density thermally conductive metals (such as Be and C whenever possible to optimize peak to background ratios.)[17] This is extremely important since biologically important elements are often present in quantities less than 1%, are low atomic numbers, and small differences are likely to be significant physiologically. Electron scatter within the chamber can cause spurious background which can obscure small peaks, and the Copper L line, if an EDS system is used, can obscure Na detection unless great efforts are made to reduce these extra peaks through use of colloidal carbon application and collimation. With higher accelerating energies (> 30 keV) backscattering of electrons into the Si(Li) detector can cause a spurious shape to the continuum. Hall[26] has reported that this artifact can be corrected by placing small ferro-magnets around the detector nose to deflect these backscattered electrons.

Morphology and Analysis

In order to analyze a compartment it is essential to be able to observe it. A useful technique for obtaining morphology in frozen hydrated sections is scanning transmission microscopy. There has been some controversy as to what one can or should be able to observe, using what mode, in the frozen hydrated state.[18,5] Gupta et al[5] have maintained that one cannot observe structures in frozen hydrated tissue in the STEM mode but when dried, or partially dried, a contrast enhancement occurs. The presence of an observable image, they feel, is an indication of partial or total dehydration. They show STEM images in which the dried specimen is darker than the same hydrated specimen prior to drying. This suggests that the total mass of the dried specimen is greater than the hydrated specimen which is in direct conflict with their data on white radiation generation in the hydrated and dried state. If the microscope brightness and contrast controls are kept unchanged as a section dries there is a general lightening as the beam penetrates more easily and we have not observed a general darkening but it is clear that drying a section can improve morphology through enhanced contrast. Thinner frozen hydrated sections, when viewed in STEM mode, will often have better morphology than thicker sections yet have similar elemental mass fractions. Consequently, an observable image with STEM is not an absolute criterion for dehydration. Obviously the particular instrumentation's ability to amplify and differentiate small signal losses plays an important part in image formation.

Observing the surface of a section viewed in the secondary mode is useful in determining if drying has occurred. If an etched surface, quite distinct from the surface features of the cryofractured low temperature section, is observed then drying has occurred. We have found that drying tends to occur primarily in the cryostat prior to transfer to the cold stage. The difference in secondary and STEM images between frozen hydrated sections and frozen dried sections can be seen in Fig. 4. In this example both sections were transferred at the same time, but one section had dried in the microtome prior to transfer.

We have found the Hall[27] method of quantitative analysis to be the easiest and most useful for quantitation. This method has been used successfully by others.[3,5] Briefly, this method uses the characteristic X-ray counts normalized to a portion of white radiation which is proportional to mass of the compartment irradiated. This ratio can be compared to a standard having a similar composition to the specimen. The difficulty involves extracting these data from the spectrum produced by energy dispersive detectors (EDS). Wave length dispersive detectors (WDS) make it easier to obtain the peak minus background since the P/B ratio is so good. However, a Si(Li) detector has ~ 100 times the efficiency of a WDS, consequently, the electron dose can be correspondingly reduced. Commerical

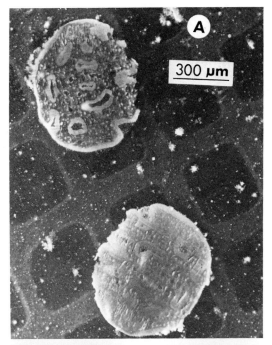

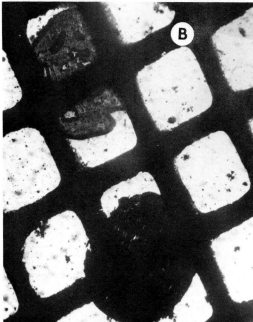

Fig. 4. Secondary (A) and STEM (B) images of two rat renal papilla, one of them frozen hydrated (lower) and the other freeze-dried (upper). Structural detail is nearly absent in the hydrated section, 0.5 μm thick, cut at -40°C.

software is available for data reduction using EDS as well as extensive literature (see for example[28,29,30]).

By using the white radiation counts as a measure of mass it is possible to obtain numbers reflecting mass/volume rather than direct mass fraction (mass/Σmass). To do this one must first analyze a compartment in the frozen hydrated state, dry the section in the microscope by allowing the stage to warm, and reanalyze the same compartment. The difference in white radiation counts is a measure of the H_2O originally present. The characteristic counts obtained in the hydrated state can then be normalized to the white radiation difference as a measure of the mass/mass of water which can be used as an approximation of mass/volume of water. A major difficulty still remains is how to determine a truly fully hydrated section from a partially hydrated section in any single particular instance. Although hydrated standards can be used it is obvious that some difference in hydration can be observed in the same sections as well as in adjacent sections.

Summary and Suggestions

Frozen hydrated sections can be used for biological specimen preparation and X-ray analysis of biological specimens. While the potential user of these techniques may find the difficulties and problems a formidable barrier upon first encounter, we would like to summarize this brief review by making several practical suggestions and by offering some hopefully useful guidelines.
1. Choose a freezing method which is the simplest for obtaining adequate morphology for compartment identification. This will probably mean that ice crystal artifacts are smaller than the compartment to be analyzed (Table 2).
2. The object of cryosectioning is to produce a sample of a specimen where compartment identification (size and morphology) is possible and compartment contents are retained. To achieve these ends section at as low a temperature possible without distorting morphology by the fracture pattern. Cut at hand speeds, using low angle knives with > 6° clearance angles.
3. Comparative microscopy of alternate serial sections may be helpful in compartment identification.

4. Carefully weigh the relative benefits of frozen hydrated vs. frozen dried specimens (Table 3).
5. Use collimation and colloidal carbon to reduce extraneous background as much as possible to optimize detection capabilities.

Table 2

Minimal Freezing Methods for
Compartment Identification

I. Light microscopy, single cells
   a) Spray freezing
   b) Direct immersion into cryogenic liquid

II. Intermediate magnification 2000x - 1500x
   a) Propelled immersion into cryogenic liquid
   b) Freezing against cooled polished metal surface
   c) 15-20 μm depth limit from freezing front
   d) Non-penetrating cryoprotectants

III. High magnification 15000x and greater
   a) Very small sample
   b) Propelled immersion
   c) Non-penetrating cryoprotectants
   d) Less than 15 μm depth limit from freezing front without cryoprotectants

---

Application of frozen hydrated section X-ray microanalysis requires considerable effort which exceeds the efforts normally needed for analyzing dried tissue sections. In Table 3 we have attempted to summarize some of the pros and cons regarding a choice between using dried or hydrated sections. In general hydrated sections can give direct measurements of elemental concentration in the wet weight, and in vacuoles or lumens, dried sections give measurements in terms of dry weight fractions. Attempts have been made to calculate wet weight fractions on the basis of dry weight fraction[31] but these calculations can at best be only approximations or estimates based upon as yet unproven assumptions[4].

Table 3
Frozen Hydrated Sections vs. Frozen Dried Sections

	Advantages	Disadvantages
Frozen hydrated section	a) Wet weight concentration b) Probability of elemental displacement minimal c) Can be used to analyze liquid filled compartments	a) Additional mass from H_2O limits minimum detection limits (MDL). b) Poor inherent contrast makes compartment identification difficult c) Difficult to handle and transfer
Frozen dried section	a) Better contrast for compartment identification b) Reduced minimum detection limits (MDL) • c) No special transfer system required	a) Dry weight concentration b) Possibility of elemental displacement during drying c) Cannot use for liquid filled compartments d) Specimen shrinkage

That is not to imply that analysis of frozen dried sections cannot give useful or valuable data, but rather to caution the reader to assess his needs carefully and then choose the method best suited for his particular problem.

Acknowledgment

The author wishes to thank Ms. Joan Krier and Ms. Linda McAulay for their help in preparation of this manuscript. I am also indebted to Mr. William Riley, Dr. Patricia Peters, Dr. Ruth Bulger, Dr. Rein Beeuwkes, and Dr. Patrick Echlin whose participation in this methodology has contributed greatly to its development.
This work was supported by NIH GM 15904.

References

1. R.G. Kirk, C. Bronner, W. Barba, et al. Electron probe microanalysis of red blood cells. Am. J. Physiol. 235, 1978, C245-C250.

2. P. Echlin, A. Saubermann, J.F. Franks, et al. The use of polymeric cryoprotectants in the preparation of biological material for X-ray microanalysis. Microscopica Acta (Supp. 2), 1978, 64-78.

3. A.V. Somlyo, H. Shuman, and A.P. Somlyo. Elemental distribution in striated muscle and the effects of hypertonicity. J. Cell Biol. 74, 1977, 828-857.

4. R. Rick, A. Dörge, E. Van Arnim. X-ray microanalysis of frog skin epithelium: evidence for a syncitial Na transport compartment. Microscopia Acta (Supp. 2), 1978, 156-165.

5. B.L. Gupta, T.A. Hall, R.B. Moreton. Electron Probe X-ray Microanalysis. In: Transport of Ions and Water in Animals. Academic Press, London, 1977. Chapt. 4.

6. P. Echlin, and A.J. Saubermann. Preparation of biological specimens for X-ray microanalysis. SEM/1977/I, IIT Research Institute, Chicago, Ill., 60616, 621-637.

7. J. Farrant, C.A. Walter, H. Lee, et al. Structural and functional aspects of biological freezing techniques. J. Microsc. 111 1977, 17-34.

8. P. Echlin, and R. Moreton. Low temperature techniques for scanning electron microscopy. SEM/1976/I, IIT Research Institute, Chicago, Ill., 60616, 753-762.

9. B.F. Trump, I.K. Brezesky, R.E. Pendergrass, et al. X-ray microanalysis diffusible elements in scanning electron microscopy of biological thin sections. SEM/1978/II, SEM. Inc., AMF O'Hare, Ill., 60666,1027-1039.

10. F. Franks, M.H. Asquith, C.C. Hammond, et al. Polymeric cryoprotectants in the preservation of biological ultrastructure. I. J. Microsc. 110, 1977, 223-238.

11. P. Echlin, H. le B. Skaer, B.O.C. Gardiner, et al. Polymeric cryoprotectants in the preservation of biological ultrastructure. II. J. Microsc. 110, 1977, 239-255.

12. H. le B. Skaer, F. Franks, M.H. Asquith, et al. Polymeric cryoprotectants in the preservation of biological ultrastructure. III. J. Microsc. 110, 1977, 257-270.

13. T.C. Appleton. A cryostat approach to ultrathin 'dry' frozen sections for electron microscopy. J. Microsc. 100, 1974, 49-74.

14. A.J. Saubermann, W.D. Riley, R. Beeuwkes, III. Cutting work in thick section cryomicrotomy. J. Microsc. 111, 1977, 39-49.

15. P.D. Peters. Frozen hydrated sections. In Proceedings of Microbeam Analysis Society 13th Annual Conference 1978, available from K.F. Heinrich, NBS, Washington, D.C., W8A-W8B.

16. A.J. Saubermann, R. Beeuwkes, P.D. Peters, et al. Definition of tissue compartments in renal papilla by direct X-ray microanalysis of frozen specimens. Kidney International 14, 1978, 779.

17. P.D. Peters, A.J. Saubermann, W.D. Riley. Improvement of peak-background ratios for analysis of frozen-hydrated tissue sections. In Proc. of Microbeam Analysis Soc. 13th Ann. Conf. 1978, available from K.F. Heinrich. NBS, Washington, D.C., 22A.

18. A.J. Saubermann, and P. Echlin. The preparation, examination and analysis of frozen hydrated tissue sections by scanning transmission electron microscopy and X-ray microanalysis. J. Microsc. 105, 1975, 115-191.

19. N. Roinel. Elementary quantitative analysis of lyopholized 10^{-10}l volume solutions. Proc. EMSA/1977. ed. G.W. Bailey. Claitor's Publish. Baton Rouge, Louisiana, 1977, 362-365.

20. T.A. Hall, and B.L. Gupta. Beam-induced loss of organic mass under electron microprobe conditions. J. Microsc. 100, 1974, 174-188.

21. P.G. Taylor, and A. Burgess. Cold stage for electron probe microanalyzer. J. Microsc. 111, 1977, 51-44.

22. A.J. Saubermann, W. Riley, and P. Echlin. Preparation of frozen hydrated tissue sections for X-ray microanalysis using a satellite vacuum coating and transfer system. SEM/1977/I, IIT Research Institute, Chicago, Ill., 60616, 347-356.

23. J.B. Pawley, and J.T. Norton. A chamber attached to the SEM for fracturing and coating biological samples. J. Microsc. 112, 1978, 169-182.

24. T.E. Hutchinson, M. Bacaner, J. Broadhurst, et al. Instrumentation for direct microscopic elemental analysis of biological tissue. Rev. Sci. Instrum. 45, 1974, 252-255.

25. W. Fuchs, and B. Lindemann. Electron beam X-ray microanalysis of frozen biological bulk specimen below 130K. I. Instrumentation and specimen preparation. J. de Microsc. et de Biol. Cell. 22, 1975, 227-232.

26. T.A. Hall. Reduction of background due to backscattered electrons in energy-dispersive X-ray microanalysis. J. Microsc. 110, 1977, 103-106.

27. T.A. Hall. The microprobe assay of chemical elements. In Physical Techniques in Biological Research, Vol 1A 2nd ed. (ed G. Oster). Academic Press, N.Y., 1971, 157-275.

28. H. Shuman, A.V. Somlyo, A.P. Somlyo. Quantitative electron probe microanalysis of biological thin sections methods and validity. Ultramicroscopy 1, 1976, 317-339.

29. P.J. Statham. Computer techniques for analysis of energy dispersive X-ray spectra. In Proc.. of MAS 13th Ann. Conf. 1978, available from K.F. Heinrich, NBS, Wash. D.C. 47A-47I.

30. F.H. Schamber. A modification of the linear least-squares fitting method which provides continuum suppression. In X-ray Fluorescence Analysis of Environmental Samples. Ed. T. Dzubay. Ann Arbor Science Publishers Inc.,Ann Arbor, Mich. 1977, 241-257.

31. Gupta, B.L., and T.A. Hall. Quantitative electron probe X-ray microanalysis of electrolyte elements within epithelial tissue compartments. Federation Proc. 38, 1979, 144-153.

Discussion with Reviewers

T.A. Marshall: Can you make any suggestion which might account for sections (100 to 200 nm thick) cut in a cryostat (see Appleton, T.C., 1978. In Electron Probe Microanalysis in Biology. Chapman and Hall, London, p. 148) appearing "smooth and glossy" at -70°to -80°C whilst your sections do not achieve this state until -40°C?

Authors: Other investigators have also confirmed Dr. Appleton's observation using systems similar to ours, and it would seem smooth 'thin' sections are easier to produce at lower temperatures than 'thick' sections.[3] One possible explanation for this observation may relate to the energy (possibly in the form of heat) actually transferred to the section (and block) during cutting. From Table 1 it can be seen that the amount of work (at -80°C for example) required for sectioning is the same for 0.5 μm, 1.0 μm, 2.0 μm (and 5.0 μm) thick sections, however the ratio of energy to mass (cal/gm) is obviously increased as a function of the smaller mass of the thinner sections. If one hypothesizes that a fixed quantity of energy is transferred to the section, it is conceivable that the brittleness of the block and section is changed to produce conditions at -70°or -80°C which are similar to those at -30° or -40°C for thicker sections.

T.A. Marshall: With respect to section collection, why not use Appleton's vacuum technique for drawing off a ribbon? I have used this technique quite successfully in a modified LKB 'cryokit'.

Authors: We have not used Appleton's vacuum technique for the simple reasons that a) we began with the eyelash pick-up method, b) it works, c) we cut only one or two sections before transferring and have not tried to produce ribbons. I can see no reason for anyone wishing to use Dr. Appleton's method not to do so. The choice between the two methods seems to me a matter of personal choice rather than particular merit other than for the fact that Appleton has had success with picking up ribbons.

W. Fuchs: Cryogenic liquid adhering to the specimen may impair cryosectioning and analysis. Could you comment on its removal?

Authors: The majority of useful cryogenic liquids have boiling points and melting points which will generally mean that these agents do not adhere during sectioning at -80°C or warmer. There is sufficient choice in cryogenic liquids so that their interference need not be a problem.

W. Fuchs: You do not mention the curling of the cryosections. How do you prevent it? Do you find any dependence on temperature and section thickness?

Authors: To prevent curling we use an antiroll plate. There are two major effects of temperature on our system. The first, and most important, effects the size of the chips and amount of chip overlap during sectioning (see Fig. 2b). The second effects section thickness through thermal changes in dimensions of microtome components at -40°C, for example, in 4-5 sections taken from the same block we may see a variation in white radiation of up to ± 50% with the same relative mass fractions. This observation, we feel, reflects a variation in thickness probably thermally induced.

W. Fuchs: Do you assume that the cutting patterns shown in Fig. 1 are independent from the microtome in use?

Authors: The patterns, until repeated on other systems, must be considered specific for our system and the conditions we used. However our observations with our system are consistent with the hypothesis that cryosectioning is a process which more closely approximates metal cutting (fracturing and shaving) than plastic cutting. Therefore we would expect that the patterns we observed would appear in general with other systems.

W. Fuchs: How did you prepare the KI and Na Cl standing gradient on the gelatin substrate?

Authors: A 30% gelatin solution was prepared and poured into a petri dish solidified, sealed, and refrigerated. A solution of 15% KI and Na Cl was mixed with a drop of green food coloring. A plug of solidified gel, 1 cm x 2 cm, was removed with a clean cork borer and a drop of the KI, NaCl solution placed on the surface for 1 minute, after which the surface was momentarily blotted and the plug frozen in Freon 12.

T.E. Hutchinson: Why was the standing gradient of KI and NaCl formed rather than a sharp discontinuity? It would seem that any smearing which occurred would be more noticeable at a sharp interface.

Authors: We felt that a standing gradient would be a sensitive indication of displacement since any deviation from an exponential curve would indicate redistribution. Although we can demonstrate distinct compartments either side of a cell boundary (for example K in erythrocytes and plasma) quantitation of the compartment composition is critical for demonstration of the absence of movement or smearings as well as a prior certainty of the original distribution. For the erythrocyte we can obtain values for plasma K which are consistent with those measured by flame photometer but the validity of our intracellular K measurement is less certain although it reflects a 20/1 ratio one might expect to see for K_I/K_0. As a generalization, we have as yet been unable to demonstrate smearing or displacement at -30°C within the resolution and compartment size we use, however we cannot exclude the possibility that nondetected displacement occurred over shorter distances in small compartments.

T.A. Marshall: You make a point about the danger of sections dehydrating in the N_2 atmosphere of the microtome chamber. This will obviously depend upon temperature but can you give any guide to permissible "storage" times?
Authors: At -30°C we tried to transfer within 5 minutes after placing the section on the film at -40°C we tried to transfer within 10 minutes after placing the section on the film. At -50°C we have not had difficulty with 20 to 30 minutes. These are of course only rough guidelines and generally include a generous margin of safety.

T.A. Marshall: Is it not possible to attribute the increase in contrast in a dehydrated frozen section to a) an increase in average atomic number due to water removal and b) an increase in density due to collapse of the specimen as it dries? Because of these two effects, electron scattering and therefore contrast should be enhanced.
Authors: The average atomic number of a biological specimen does not increase appreciably with water loss since the average atomic number of water is similar to the remaining matrix. Although it is conceivable that, under certain circumstances, one could encounter a situation where this factor may be of significance, it is generally less important than the differential loss of mass which occurs during drying. With freeze drying of sections there is relatively little shrinkage (we've measured a 6% decrease in width). Instead the water as it sublimates leaves a lace-like honeycomb pattern reflecting the ice crystal damage which occurs during freezing. Compartments which have less water have less ice crystal damage and will retain a higher total mass than compartments having greater amounts of water. It is conceivable that some of this increase in density results from differential shrinkage but I would think that this would be a relatively minor effect compared to the differential loss of mass (primarily water) which occurs during drying.

C.E. Fiori: What was the current density of your electron probe?
Authors: Our usual operating conditions are 30 KeV accelerating voltage, 0.1 nA probe current (measured at the stage in a Faraday cup). For analysis we use a small raster, whose size varies with the compartment, but is frequently in the range of 1 to 5 μm in width. Thus the current density would be in the general range of 10 μA/cm^2 to .4 μA/cm^2.

C.E. Fiori: In addition to collimation and the use of low Z coating materials (such as carbon dag) to reduce the effects of stray electrons and X-rays should you not also attempt to eliminate or reduce these before they occur. This subject is discussed by Bolan and McConnell (SEM/1976/I P. 163) and Fiori and Newbury (SEM/1978/I P. 401).
Authors: I agree completely with your point that every effort should be made to eliminate the stray electrons and X-rays before trying to eliminate their effects. The two references you suggest are recommended to the readers.

Additional discussion with reviewers of the paper "Principles of X-Ray Microanalysis in Biology" by J.A. Chandler, continued from page 606.

Reviewer III: You have been involved in applications of X-ray microanalysis since its early days. To your opinion do we biologists lack interest, knowledge or financial support to conduct studies in the basic events governing processes in X-ray generation and analysis as you recomended in the conclusion?

Author: Whereas there can be no doubting the great enthusiasm with which X-ray microanalysis has been applied to biological problems in the last decade it is also painfully true from a perusal of the literature that relatively little effort has been made to understand the factors affecting elemental concentrations in soft tissue during preparative procedures. That this is so may be concluded from the observation that 50% of all analyses so far reported in the literature, for thin specimens at least, have been performed on tissues fixed and processed by conventional electron microscope procedures. The use of cryoultramicrotomy is not yet routine and much progress is still required to develop alternative methods which will yield reliable analyses as well as providing satisfactory morphological information.

SCANNING ELECTRON MICROSCOPY/1979/II
SEM Inc., AMF O'Hare, IL 60666, USA

QUANTIFICATION OF ELECTROLYTES IN FREEZE-DRIED CRYOSECTIONS BY
ELECTRON MICROPROBE ANALYSIS

R. Rick, A. Dörge, R. Bauer, K. Gehring and K. Thurau

Department of Physiology
University of Munich
12 Pettenkoferstrasse
D-8000 Munich 2
West Germany

Abstract

The preparation of freeze-dried cryosections
and quantitative energy dispersive X-ray microanalysis of thin biological specimen is described.
The problems encountered during the tissue
preparation are discussed, and the preconditions
for quantitative analysis are outlined. It is
shown that the preparation, as well as the
quantification procedure used, are feasible for
quantitative determination of electrolyte
concentrations on a cellular level in biological
soft tissues.

The tissue preparation is characterized by
shock-freezing in liquid propane (- 188° C), dry
cryosectioning into 1 µm thick serial sections
(- 90° C), which are mounted on thin films and
freeze-dried (- 80° C). Energy dispersive
analysis is performed in a scanning electron
microscope at 17 kV acceleration voltage and
0.5 nA probe current. The energy dispersive
X-ray spectra are analyzed with the aid of a
computer programme. Cellular wet weight
concentrations are evaluated by comparing the
characteristic X-ray intensities obtained from
the cell with those from an internal albumin
standard.

KEY WORDS: Electron Probe Microanalysis, Energy
Dispersive X-ray Analysis, Biological Soft Tissue,
Electrolytes, Cryofixation, Cryosectioning,
Freeze-Drying, Irradiation Damage, Internal
Standard

Introduction

Electron microprobe analysis (EMA) provides
a method capable of detecting elements in such
small quantities that analysis of element
concentrations in individual cells or even in
subcellular structures is in principle possible.
The major difficulties for the application of
this method to biological specimen lie in the
nature of living material. Biological soft
tissues consist of various compartments which,
in the living state, have completely different
element compositions. The in vivo concentration
gradients of diffusible elements within the
tissue must be anticipated to be changed during
preparation of the specimen for analysis.
Furthermore, biological electron microprobe
analysis is complicated by the sensitivity of
biological material to electron irradiation,
resulting in loss or dislocation of elements
and specimen contamination. Therefore, the main
objectives of quantitative EMA of biological
specimen are the adequate preparation of the
tissue and the control of the factors influencing
the quantitative interpretation.

In the present paper the use of thin
freeze-dried cryosections for electron microprobe
analysis of water soluble elements in biological
soft tissues is discussed. The main problems
encountered during freezing, cryosectioning and
freeze-drying of the tissue, and the preconditions for quantitative X-ray microanalysis of
thin biological specimen are outlined. A
quantification method is described, which allows
the determination of cellular element concentration in units per wet weight. Details of the
method have already been reported in a previous
paper[1].

Cryofixation

Quantitative EMA of electrolytes requires
that the in vivo distribution of water soluble
elements is preserved during the preparation.
It is generally accepted that "physical fixation"
by rapid freezing of the tissue is the only way
to keep readily diffusible elements such as Na,
K and Ca in place. However, due to ice crystal
formation during freezing inevitably some
dislocation of electrolytes and distortions of
the tissue structure occur. Even under optimal

freezing conditions only a superficial zone of 2 - 3 μm is free of ice crystals (vitrification[2]) and only in a surface layer of 5 - 8 μm[3], or 10 - 15 μm[4,5], the cellular ultrastructure is well preserved. Cellular analysis on a slightly coarser scale can be performed in a region extending up to 50 μm from the surface.

Fig. 1 shows a scanning transmission electron micrograph of a 1 μm thick freeze-dried cryosection of frog skin, which had been snap-frozen in liquid propane (-188°C). The structural preservation of the epithelium is good enough to allow discrimination between different epithelial cell types or nucleus and cytoplasm of the same cell. In addition, the original distribution of water soluble elements seems to be well maintained[1]. In principle, the structural preservation can be improved by cryoprotectives, however, all cryoprotectives are likely to affect the intra-/extracellular electrolyte distributions. Therefore, analysis of deeper structures in biological tissues is only possible, when the specimen is cut prior to freezing[6].

Cryosectioning

Measurements of intracellular element concentrations require that the volume excited by the incident electrons does not exceed the size of an individual cell. As shown in Fig. 2 this can be achieved by using sections which are at least thinner than the cell diameter. The attainable spatial resolution power is mainly a function of the section thickness the lateral resolution was found to be about 0.5 and 0.1 μm in 1 and 0.1 μm thick freeze-dried cryosections, respectively[1,7,8].

In order to maintain the original distribution of water soluble elements the material has to be kept frozen during sectioning. However, dry cutting of thin frozen section[3,9,10,11,12] is beset with difficulties. Ultrathin sections can only be obtained at very low temperatures (less than -100°C), at which already fragmentation of the sections occurs[3]. In addition, recrystallization of ice and transient thawing during sectioning have been described[13,14]. Yet, measurements of the element distribution at a cellular border have shown that gradients of electrolytes between extracellular and intracellular spaces are well preserved[1].

In principle, cryosectioning can be avoided by freeze-drying and vacuum embedding[15,16,17] or by precipitation of labile compounds followed by conventional tissue preparation[18,19]. However, the applicability of these techniques for the analysis of diffusible elements in biological soft tissues is not yet well established.

Fig. 3 depicts schematically the sectioning of frozen tissue pieces in a modified cryo-microtome (Reichert OmU2) at -80 to -100°C. Using the edges of the steel knife, the specimen is trimmed into a cube with a sectioning area of about 0.2 x 0.3 mm. The material is then cut into serial sections approximately 1 μm thick. The curling up of the sections is prevented by a small glass plate, positioned a few μm above the edge of the knife. The sections are picked up by a formvar or collodion film suspended over a Ni grid, which is then placed on a second film, mounted over a hole in the storage table. In this way, the sections are sandwiched between two thin films, suspended over a 1 x 2 mm slot in the Ni grid.

Freeze-drying

Cryosections can be analyzed either in the frozen hydrated state[20,21,22,23,24], or after freeze-drying in the dehydrated state[12,25,26,27]. Freeze-drying offers the advantage that redistributions of water soluble substances such as electrolytes is still prevented even when the specimen is allowed to warm up. Furthermore, compared to analysis in the frozen hydrated state, freeze-dried tissue sections allow a much better electron optical visualization and, since almost 80% of the mass is removed during drying, X-ray microanalysis of freeze-dried specimen yield much higher signal to noise ratios. On the other hand, freeze-drying may lead to redistributions of electrolytes, when drying is performed at insufficiently low temperatures, and some shrinkage of the tissue may occur[17,28]. In addition, the applicability of freeze-dried sections for EMA is limited to biological soft tissues with sufficiently high matrix content. In large watery compartments containing almost no biological matrix such as the lumen of the kidney tubules, the electrolytes were found to be dislocated during freeze-drying [29].

Spatial arrangement of analysis

EMA of thin specimen entails a very low X-ray yield. Therefore, the shielding of the detector against X-rays generated in the specimen support or in other parts of the specimen chamber becomes a crucial point for quantitative analysis especially when EDS-systems are employed. Extraneous signals can be reduced by supporting the specimen only by thin films, which are suspended over large openings[1,30], or by the use of grids with low background properties[31,32] and by additional shielding of the detector. Fig. 4 shows the spatial arrangement of specimen, specimen stage and detector used in the present study. The specimen is sandwiched between two thin films, which are suspended over the 1 x 2 mm wide opening of a Ni grid. The grid is placed over a bore in a specimen stage. The stage is made of carbon, thus reducing X-ray excitation of the specimen stage by scattered electrons. Furthermore, using this spatial arrangement stray radiations originating from the final aperture as well as X-rays excited by transmitted electrons are shielded from the detector.

Fig. 5 shows a typical energy dispersive X-ray spectrum obtained from a frog skin epithelial cell. Aside from a small Ni and Fe peak originating from excitation of the grid and specimen chamber, all peaks of the spectrum (P, Cl, K) correspond to elements which are present only in the biological specimen. The fraction of the white radiation not originating

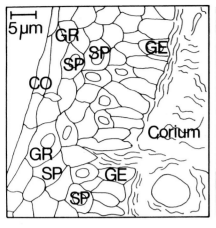

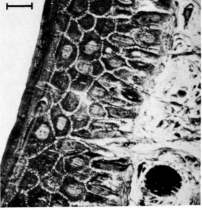

Figure 1. Freeze-dried cryosection of frog skin. CO = cornified, GR = granular, SP = spiny and GE = germinal layer.

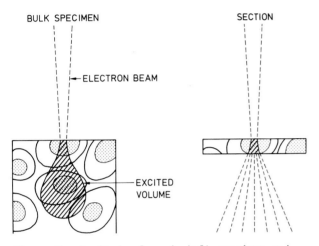

Figure 2. Excited volume in bulk specimen and section.

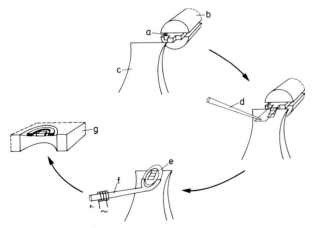

Figure 3. Cryosectioning (for details see text). a = specimen, b = specimen holder, c = steel knife, d = anti roll device, e = Ni grid, f = electromagnetic rod, g = storage table.

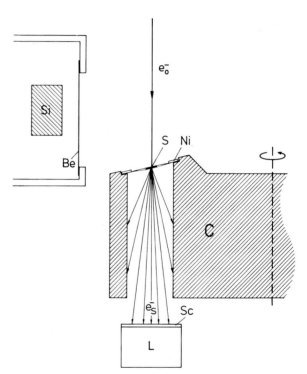

Figure 4. Spatial arrangement during analysis. e_o^- = incident electrons, e_s^- = transmitted electrons, S = specimen, Ni = Ni grid, Si = Si detector, Be = Be entrance window, C = carbon specimen stage, Sc = scintillator, L = light guide.

from the specimen itself has been shown to be less than 10%[33].

The low X-ray yield obtained from thin specimen further requires a high efficiency of X-ray detection. Using an EDS system, the detector can be positioned very close to the specimen (see Fig. 4), resulting in a large solid angle of detection. At 30 mm^2 surface area of the detector (Link/Kevex) and at 7 mm distance between detector and specimen, a solid angle of 0.6 steradian, equal to a detection efficiency of almost 5% for X-rays between 2 and 30 keV, can be obtained.

Irradiation damage

EMA of biological specimen is further complicated by the sensitivity of organic materials to electron bombardement. During irradiation, elements may be lost or dislocated. This artifact will become critical for the analysis of thin specimen at high spatial resolution, since mass loss largely depends on the applied current density[1,34]. Mass loss can be minimized by coating the section with a thin metal layer[35] or by cooling the specimen[36]. Fortunately, the mass loss appears to affect only constituents of the biological matrix, whereas electrolytes are stable, as shown in Fig. 6. During mass loss the element characteristic peaks for Na, Cl and K are unchanged, whereas the S peak as well as the white radiation is reduced. Despite a 10,000 fold increase in the applied current dose, the Na, K and Cl countrate is stable, whereas the S and white countrate shows a significant drop. A further possible irradiation artifact is the formation of a contamination layer at the site of analysis, largely depending on the cleanness of the specimen chamber vacuum[37].

Evaluation of EDS spectra

The quantification of energy dispersive X-ray spectra is made difficult by the inherently low energy resolution, resulting in peak overlapping and low peak to background values. Furthermore, EDS spectra are distorted by additional X-ray absorption in the Be entrance window and Si dead layer of the detector. However, during recent years a number of computer programmes have been developed, allowing sufficiently accurate analysis of EDS spectra from biological specimen[12,38,39].

Fig. 7 illustrates the fitting of a computer generated spectrum to a measured one. The X-ray spectrum is considered to be a linear combination of characteristic peaks and a background curve, taking into account the physical basis of X-ray excitation, X-ray absorption and X-ray detection by an EDS system[38]. Already after 4 repetitive adaptations, the agreement between artificially generated and measured spectrum is excellent ($x^2 \approx 1.0$). In an individual X-ray spectrum obtained from a thin freeze-dried section of a biological soft tissue, peaks corresponding to 0.4 mmole/kg wet weight of K, or 2 mmole/kg wet weight of Na can be detected.

Quantification

The use of thin sections greatly simplifies the interpretation of EMA data, provided that the energy of the exciting electrons is sufficiently high[40,41]. In a 1 µm freeze-dried section of biological soft tissue - virtually equivalent to a carbon layer of 0.2 µm thickness - the deceleration of 17 keV electrons is negligibly small, so that constant ionization cross sections for all elements can be expected to exist throughout the specimen thickness.

Furthermore, X-ray absorption and secondary fluorescence within the specimen are insignificantly small for X-ray energies greater than 1 keV. For Na ($E_K = 1.041$ keV) the self-absorption is only about 5% (at 1 µm section thickness and 30° takeoff angle). Thus, under constant measuring conditions the detected X-ray intensities should depend only on the respective elemental mass per area. Fig. 8 shows the relative Na intensities obtained from 1 µm thick freeze-dried section of dextran standards (25 g%) containing 100 mmole/kg wet weight K and varying concentrations of Na. Over the entire concentration range (5.3 to 186 mM) the intensity of the Na radiation is linearly dependent on the Na concentration. Similar calibration curves were obtained for all biologically relevant elements from Na through Ca.

Quantification of element concentrations in thin biological specimen can be achieved by comparing the ratio of element characteristic intensity and continuum intensity with that obtained from a thin standard[41]. Preconditions of this type of analysis are that specimen and standard are of similar gross elemental composition and that no differential mass loss occurs. In freeze-dried section his method provides dry weight concentrations; conversion into wet weight concentrations is only possible, when the local dry weight content is known.

A second quantification method for thin specimen directly compares the element characteristic intensities with that of an element present in the specimen in known concentrations, the internal standard[42]. However, since in biological soft tissues the local concentration of no element is actually known, the specimen has to be supplied with an artificial internal standard. One way to achieve this is to immerse the tissue immediately prior to freezing in a solution of known elemental composition. For the analysis in the freeze-dried state, the standard should contain a sufficiently high concentration of a macromolecular compound, e.g. albumin or dextran, to avoid dislocations of water soluble elements during freeze-drying.

Fig. 9 illustrates this method. The scanning transmission image in the lower half shows the outer region of a freeze-dried frog skin section and an adherent albumin standard layer. The spectra A and B were obtained by scanning the indicated areas in the albumin layer and in the nucleus of an epithelial cell. The cellular concentrations can be evaluated by a direct comparison between the respective characteristic peaks in the cell and in the standard, as shown for Na. For elements which are not present in the internal standard in sufficiently large concentrations, a comparison with a different standard element can be performed as formulated for K, taking into account the differences in the X-ray yields by a correction factor α. The numerical value of α for any given pair of elements can be determined experimentally, as shown in Fig. 8.

Since the thickness of standard layer and cell are the same, the computed values yield

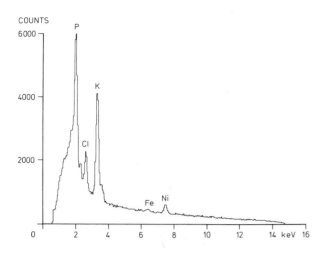

Figure 5. Energy dispersive X-ray spectrum of a frog skin epithelial cell (50 eV/channel).

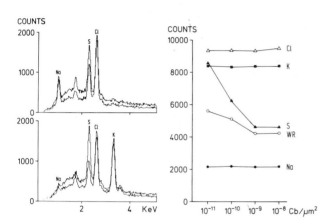

Figure 6. Mass loss of 1 μm thick freeze-dried cryosections of different albumin standard solutions (at room temperature). X-ray spectra and countrates of Cl, K, S, Na and white radiation (WR) after administration of varying current doses. The respective upper spectra were obtained after 10^{-11}, and the lower ones after 10^{-8} Cb/μm2.

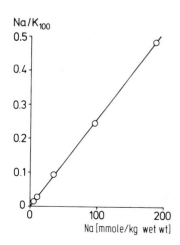

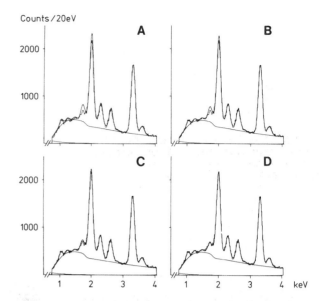

Figure 7. Comparison of measured (histogram) and computer generated spectrum (smooth line). A: after evaluation of starting values; B-D: after 1, 2 and 4 iterations.

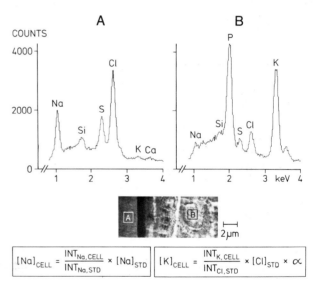

$$[Na]_{CELL} = \frac{INT_{Na,CELL}}{INT_{Na,STD}} \times [Na]_{STD}$$
$$[K]_{CELL} = \frac{INT_{K,CELL}}{INT_{Cl,STD}} \times [Cl]_{STD} \times \propto$$

Figure 9. Illustration of the quantification method (spectra with 20 eV/channel).

Figure 8. Calibration curve for Na, obtained from dextran standards containing different Na concentrations and a constant K concentration of 100 mmole/kg wet weight. The ratios of Na and K intensities are plotted against the Na concentrations.

concentrations per unit of volume, which for soft tissues with a specific weight of about 1 represent wet weight concentrations. However, possible small differences in the thickness within an individual cryosection might lead to errors. This artifact can be expected to be minimal if the analysis is performed on a cellular level and, using serial sections, it should cancel out. A further precondition of this type of quantification is that during the preparation of the specimen no differential swelling or shrinkage of the tissue and the standard has occurred. The small possible shrinkage of biological soft tissues during freeze-drying is very likely to be almost the same as in the albumin standard solution[17]. Furthermore, it can be expected that using thin sections which are supported by films, a differential shrinkage in volume will result primarily in an uneven reduction of the section thickness[28].

Table 1 compares the cellular concentration of Na and K obtained by this method with those obtained by other techniques. The cellular Na and K concentration obtained from isolated frog skin epithelial cells, which have been jet-washed in Na-free solution in order to reduce the amount of adherent extracellular Na[43], are in good agreement with the mean cellular Na and K concentrations obtained by X-ray microanalysis[25]. Moreover, X-ray microanalysis of

	Na	K
	mmole/kg wet wt	
Frog skin		
X-ray microanalysis	9.4	118.4
Chemical analysis[a]	12.1	109.1
Rat kidney		
X-ray microanalysis	20.2	139.5
K-selective electrode[b]	--	113.0

[a]data from washed isolated cells[43]
[b]minimal estimate of free K concentration[44]

Table 1. Cellular electrolyte concentrations of frog skin epithelium and rat kidney proximal tubule.

rat kidney proximal tubules (see accompanying paper by Thurau et al.) revealed a K concentration, which is only about 20% higher than the minimal estimate of the free K concentration calculated from measurement with a K-sensitive microelectrode[44].

Acknowledgement

This investigation was supported by the Deutsche Forschungsgemeinschaft.

References

1. Dörge, A., R. Rick, K. Gehring, and K.Thurau: Preparation of freeze-dried cryosections for quantitative X-ray microanalysis of electrolytes in biological soft tissues. Pflügers Arch. 373, 85-97 (1978)
2. Moor, H.: Recent progress in the freeze-etching technique. Philos. Trans. R. Soc. Lond. (Biol.) 261, 121-131 (1971)
3. Seveus, L.: Preparation of biological material for X-ray microanalysis of diffusible elements. I. Rapid freezing of biological tissue in nitrogen slush and preparation of ultrathin frozen sections in the absence of trough liquid. J. Microscopy 112, 269-280 (1978)
4. Heuser, J.E., and T.S. Reese: Freeze-substitution applied to the study of quickfrozen synapses. J. Cell Biol. 70, 357a (1976)
5. Van Harreveld, A., J. Trubatch and J. Steiner: Rapid freezing and electron microscopy for the arrest of physiological process. J. Microscopy 100, 189-198 (1974)
6. Van Harreveld, A., and E. Fifkova: Rapid freezing of deep cerebral structures for electron microscopy. Anat. Rec. 182, 377-386 (1975)
7. Appleton, T.C., and P.F. Newell: X-ray microanalysis of freeze-dried ultrathin frozen sections of a regulating epithelium from the snail Otala. Nature 266, 854-855 (1977)
8. Somlyo, A.P., A.V. Somlyo, H. Shuman, B. Sloane and A. Scarpa: Electron probe analysis of the sarcoplasmic reticulum and mitochondria in muscle. In: Microprobe Analysis in Biology and Medicine. S. Hirzel Verlag, Stuttgart, p. 79-91 (1978)
9. Appleton, T.C.: A cryostat approach to ultrathin 'dry' frozen sections for electron microscopy: A morphological and X-ray microanalytical study. J. Microscopy, 100, 49-74 (1974)
10. Bernhard, W., and A. Viron: Improved techniques for the preparation of ultrathin frozen sections. J. Cell. Biol. 49, 731-746 (1971)
11. Christensen, A.K.: Frozen thin sections of fresh tissue for electron microscopy, with a description of pancreas and liver. J. Cell Biol. 51, 772-804 (1971)
12. Shuman, H., A.V. Somlyo and A.P. Somlyo: Quantitative electron probe microanalysis of biological thin sections: Methods and validity. Ultramicroscopy 1, 317-339 (1976)
13. Hodson, S., and J. Marshall: Ultracryotomy: A technique for cutting ultrathin sections of unfixed frozen biological tissues for electron microscopy. J. Microsc. 91, 105-118 (1970)
14. Hodson, S., and J. Marshall: Evidence against through-section thawing whilst cutting on the ultacryotome. J. Microsc. 95, 459-466 (1972)

15. Edelmann, L.:
Visualization and X-ray microanalysis of
potassium tracers in freeze-dried and plastic
embedded frog muscle. In: Microprobe Analysis
in Biology and Medicine. S. Hirzel Verlag,
Stuttgart, p. 166-174 (1978)
16. Höhling, H.J., and W.A.P. Nicholson:
Electron microprobe analysis in hard tissue
research: Specimen preparation. J. Microsc.
Biol. Cell. 22, 185-204 (1975)
17. Ingram, F.D., M.J. Ingram and C.A.M.
Hogben:
An analysis of the freeze-dried, plastic embedded electron probe specimen preparation.
In: Microprobe Analysis as Applied to Cells
and Tissues. Academic Press, London and
New York, p. 119-146 (1974)
18. Läuchli, A.:
Precipitation technique for diffusible
substances. J. Microsc. Biol. Cell. 22,
239-248 (1975)
19. Yarom, R., and J.A. Chandler:
Electronprobe microanalysis of skeletal
muscle. J. Histochem. Cytochem. 22,
147-154 (1974)
20. Echlin, P.:
The preparation of frozen-hydrated biological
material for X-ray microanalysis. J. Microsc.
Biol. Cell. 22, 215-226 (1975)
21. Gupta, B.L., M.J. Berridge, T.A. Hall,
and R.B. Moreton:
Electron microprobe and ion-selective microelectrode studies of fluid secretion in the
salivary glands of Calliphora. J. Expl. Biol.
72, 261-284 (1978)
22. Hutchinson, T.T., M. Bacaner, J. Broadhurst
and J. Lilley:
Elemental microanalysis of frozen biological
thin sections by scanning electron microscopy
and energy selective X-ray analysis. In:
Microanalysis as Applied to Cells and Tissues.
Academic Press, London and New York, p. 191-
200 (1974)
23. Marshall, A.T.:
Electron probe X-ray microanalysis of frozen-hydrated biological specimens. Microscopica Acta
79, 253-266 (1977)
24. Saubermann, A.J., and P. Echlin:
The preparation, examination and analysis of
frozen hydrated tissue sections by scanning
transmission electron microscopy and X-ray
microanalysis. J. Microscopy, 105, 155-191
(1975)
25. Rick, R., A. Dörge, E. v.Arnim and K.Thurau:
Electron microprobe analysis of frog skin
epithelium: Evidence for a syncytial sodium
transport compartment. J. Membrane Biol. 39,
313-331 (1978)
26. Sjöström, M.:
X-ray microanalysis in cell biology,detection
of intracellular elements in the normal muscle
pathology. J. Microsc. Biol. Cell. 22, 415-425
(1975)
27. Somlyo, A.V., H. Shuman and A.P. Somlyo:
Elemental distribution in striated muscle and
the effects of hypertonicity. Electron probe
analysis of cryosections. J. Cell. Biol. 74,

828-857 (1977)
28. Appleton, T.C.:
Physiology and pharmacology at a sub-cellular
level. In: Microprobe Analysis in Biology and
Medicine. S. Hirzel Verlag, Stuttgart, p. 1-10
(1978)
29. Dörge, A., R. Rick, K. Gehring, J. Mason,
and K. Thurau:
Preparation and applicability of freeze-dried
sections in the microprobe analysis of biological
soft tissue. J. Microsc. Biol. Cell. 22, 205-
214 (1975)
30. Rick, R., A. Dörge, and E. v.Arnim:
X-ray microanalysis of frog skin epithelium:
Evidence for a syncytial Na transport compartment.
In: Microprobe Analysis in Biology and Medicine.
S. Hirzel Verlag, Stuttgart, p. 156-165 (1978)
31. Packwood, R.H., E.E. Laufer, and W.N.
Roberts:
Graphite support grids for X-ray analysis in the
electron microscope. 8th Int. Conf. X-ray Opt.
Microanal., Boston, Massachusetts, p. 115A-D
(1977)
32. Panessa, B.J.:
Beryllium and graphite polymer substrates for
reduction of spurious X-ray signals. SEM, (II)
SEM Inc. AMF O'Hare, Il 60666, USA, p. 1055-1061
(1978)
33. Rick, R., A. Dörge, K. Gehring, R. Bauer,
and K. Thurau:
Quantitative determination of cellular electrolyte
concentrations in thin freeze-dried cryosections
using energy dispersive X-ray microanalysis.
Workshop on biological X-ray microanalysis by
electron beam excitation, Boston, USA, 1977
(to be published in book form by Academic Press,
London and New York)
34. Bahr, G.F., F.B. Johnson, and G. Zeitler:
Elementary composition of organic objects after
electron irradiation. Lab. Invest. 14, 1115-1127
(1965)
35. Höhling, H.J., T.A. Hall, W. Kriz, A.P.
v.Rosenstiel, J. Schnermann, and U. Zessack:
Loss of mass in biological specimens during
electronprobe X-ray microanalysis. In: Modern
Techniques in Physiological Sciences. Academic
Press, London and New York, p. 335-343 (1973)
36. Hall, T.A. and B.L. Gupta:
Measurement of mass loss in biological specimen
under an electron microbeam. In: Microprobe
Analysis as Applied to Cells and Tissues.
Academic Press, London and New York, p. 147-158
(1974)
37. Sutfin, L.V.:
An ultra-high vacuum SEM for microanalysis of
ultrastructure of thin specimen. Proc. 11th Ann.
Conf. Microbeam Analysis Society, Miami Beach,
Florida, USA, p. 66A-E (1976)
38. Bauer, R. and R. Rick:
Computer analysis of X-ray spectra (EDS) from
thin biological specimens. X-ray Spectrometry 7,
63-69 (1978)
39. Russ, J.C.:
A basic language program for dedicated energy
dispersive microanalysis of thin samples in TEM
or STEM. In: Microprobe Analysis in Biology
and Medicine. S. Hirzel Verlag, Stuttgart,

p. 217-227 (1978)

40. Hall, T.A.:
The microprobe assay of chemical elements. In:
Physical Techniques in Biological Research.
Academic Press, London and New York, p. 157-275
(1971)

41. Hall, T.A.:
Methods of quantitative analysis. J. Microsc.
Biol. Cell. 22, 271-281 (1975)

42. Russ, J.C.:
Evaluation of the direct element ratio calculation method. J. Microsc. Biol. Cell. 22,
283-286 (1975)

43. Zylber, E.A., C.A. Rotunno, and M.Cereijido:
Ion and water balance in isolated epithelial
cells of the abdominal skin of the frog
Leptodactylus ocellatus. J. Membrane Biol. 13,
199-211 (1973)

44. Edelmann, A., S. Curci, I. Samarzija and
E. Frömter:
Determination of intracellular K^+ activity in
rat kidney proximal tubular cells. Pflügers
Arch. 378, 37-45 (1978)

Discussion with Reviewers

J. Chandler: How often can one obtain areas of
tissue with the good morphology shown in
figure 1. How does their morphology compare with
the preparation, by the same methods of other
soft tissues apart from muscle which is well
known to give good pictures?
Authors: Cryosections of similar quality can
be obtained in epithelial tissues, such as frog
skin, toad urinary bladder, rat kidney or liver.
It is our experience that cryosectioning of soft
tissue composed of largely different structural
elements, such as the axon and Schwann cell of
peripheral nervous tissue, is extremely difficult.
J. Chandler: How can one be sure that while
preparing internal standards diffusion of these
elements does not occur between the standard
and the subcellular regions. The distance
between the standard layer and the cell nucleus
shown in figure 9 is only a few micrometres.
Couldn't such a movement be rapid over this
distance. In addition, does the introduction
of the tissue into the standard containing
medium not change the whole electrolyte
distribution anyway since it would be different
to the in vivo situation. Doesn't this destroy
all the advantages of the freezing procedures?
Authors: The albumin standard closely mimics
the conditions found in vivo because, apart from
the albumin, the solution contains electrolytes
in concentrations, normally found in extracellular fluid. It is unlikely that the use
of the standard leads to alterations in the
intracellular electrolyte concentrations, for it
was applied to the tissue for only a few seconds
before freezing. Furthermore, tissue pieces
from a single skin prepared both with and
without the standard albumin layer revealed
identical X-ray spectra.
T.E. Hutchinson: The Bremsstrahlung spectrum
seems to reflect very little of the absorption
edges which would occur after the large peaks

shown. Does your computer program take these
into account in generating the X-ray spectrum
prior to substraction of the Bremsstrahlung
background?
Authors: The X-ray spectra do not show absorption edges, since the specimens are thin and,
therefore, self-absorption of X-rays within the
specimen is negligibly small. For thicker
specimens,the absorption can be taken into
account by multipying the primary X-ray spectrum
with a transmission function (see reference 38).
T.E. Hutchinson: In figure 5 the spectrum which
you show has an unusually large Bremsstrahlung
background beneath the P peak. It is clearly
not due to heavy elements at least in the
region shown. This might suggest that is is due
to the lower lying elements, such as C within
your system?
Authors: In our system, the Bremsstrahlung
originates almost exclusively from excitation
of the specimen itself (see reference 33).
Therefore, the light elements, which constitute
the biological matrix (C,N,O and P), account for
most of the Bremsstrahlung background.
T.E. Hutchinson: While it is reasonable to
assume that a conversion from dry weight to wet
weight concentrations is relatively accurate
on a cellular level, such conversion in subcellular regions will depend upon the amount of
local water lost. Would you agree that it is
necessary to obtain the data from frozen hydrated tissue at least in representative cases in
order that local concentrations on a subcellular
scale may be obtained?
Authors: The applied quantification method
directly provides wet weight concentrations,
presuming that the section thickness is uniform
and no differential swelling or shrinkage of
standard and tissue had occurred during the
preparation. As long as these preconditions are
fulfilled, quantitative analysis of subcellular
structures is, in principle, possible.

SCANNING ELECTRON MICROSCOPY/1979/II
SEM Inc., AMF O'Hare, IL 60666, USA

ORIGIN OF ARTIFACTUAL QUANTITATION OF ELECTROLYTES
IN MICROPROBE ANALYSIS OF FROZEN SECTIONS OF ERYTHROCYTES

J. M. Tormey and R. M. Platz

Dept. of Physiology
UCLA School of Medicine
Los Angeles, CA 90024

Abstract

Frozen sections of erythrocytes have been used to validate microprobe X-ray analysis of diffusible elements in biological samples. At this meeting last year we reported that intracellular Na concentrations measured by microprobe were much higher than those measured by bulk chemical methods. It was suggested that this might be due to the movement of extracellular material over the cells by microtomy.

We now present evidence that such results are better explained by electron scattering within the sample during analysis. This evidence includes observations that the excess measured Na varies directly with section thickness and inversely with accelerating voltage. At 80 kV accelerating voltage there is excellent agreement between microprobe and chemical analyses.

It may be concluded that specimen preparation techniques such as used in our laboratory are satisfactory for reliably localizing diffusible elements in small volumes, but that instrumental factors can seriously affect their quantitation.

KEY WORDS: X-ray Microanalysis, Quantitative Analysis, Biological Specimen Preparation, Freeze-Drying, Electrolytes, Erythrocytes, Electron Scattering, Artifact

Introduction

Microprobe X-ray analysis has the potential to become as well established in the biological as in the materials sciences. For this to happen, its reliability as a quantitative tool first needs to be demonstrated on well characterized biological specimens.

With this in mind, we have been using the microprobe to quantitate electrolyte concentrations in freeze-dried frozen sections of erythrocytes. These small cells have the advantage not only of structural homogeneity, but also of the ease and accuracy with which their electrolyte composition can be determined by chemical analysis. Validation of biological microprobe methods requires that microprobe measurements of such specimens agree with their known chemical composition.

Last year at this meeting we reported unsatisfactory agreement between microprobe and chemical estimates of erythrocyte electrolyte concentrations.[1] Even though results with K were in good agreement, microprobe estimates of Na and Cl were consistently too high. The concentration of each element was elevated by a constant fraction of that element's concentration in the extracellular matrix that surrounded each cell. This phenomenon was referred to as "smearing". For instance, an extracellular Na concentration of 150 mM ($mM/10^3cm^3$) and a "smearing factor" of 0.3 (30%) would cause intracellular Na to be too high by 0.3 X 150 = 45 mM, a serious error in a cell that actually contains only 10 mM Na; whereas the same smearing factor operating on an extracellular K of 5 mM would elevate intracellular K by only 1.5 mM, hardly significant in a cell that contains 100 mM K.

The problem was, and is, to understand the physical basis of this artifact. The possibility that ionic gradients were being partially dissipated by diffusion was definitively ruled out. The two remaining possibilities are either bulk movement of a portion of the extracellular matrix over the cells during microtomy, or excitation of the extracellular phase by stray radiation when the cells are being probed. Since a number of tests for stray radiation turned out negative, microtomy was implicated as the more likely source.

The present paper reports a series of additional experiments on the origin of the artifact. It now appears that it arises not from microtomy but from radiation scattered within the sections during analysis. Furthermore we have found that it can be prevented by proper choice of instrumental conditions.

Methods

Specimen Preparation

The method of sample preparation was virtually identical to that described here last year.[1] Suspensions of human high-potassium (HK) and sheep low-potassium (LK) erythrocytes were incubated at 37°C in a Ringer's solution containing 20% dextran. Thin layers of suspension containing a mixture of HK and LK cells were formed on Millipore filters and frozen in liquid propane. Separate suspensions of each cell type were incubated in parallel with the cells to be frozen, and, at the time of freezing, these were centrifuged for chemical analysis, with H[3]-mannitol included as a marker for trapped extracellular volume. Intracellular Na and K concentrations were determined by flame photometry, and Fe concentration by a standard method for hemoglobin Fe.

Frozen sections were cut on glass knives in a Sorvall Frozen Thin Sectioner. Except as otherwise noted, microtomy was carried out at -80°C, as measured by a thermocouple at the knife edge. The specimens were mounted in vise chucks, and the resulting sections were cross sections through the filter with adhering layers of suspended cells. Section thickness was usually a nominal 0.5 μm. In most cases, the sections were sandwiched between two 250 Å carbon films mounted on 100 mesh folding Cu grids.

The sandwiched sections were freeze-dried for two days at -70°C. Following warmup to room temperature, specimens were handled and stored in dry nitrogen atmospheres.

Analytical Conditions

Sections were analyzed at room temperature with either an ETEC Autoscan SEM or a JEOL 100C TEM operated in the STEM mode. The former was used at a beam current of 4×10^{-10} A at either 10 or 20 kV, and the latter at currents up to 4×10^{-9} A at either 20 or 80 kV. Both instruments were equipped with energy dispersive X-ray detectors in concert with either a Northern 880 or a Kevex 7000 analyzer.

Specimen tilt was varied between 0° and 45° on the SEM without significant effect on the results, while most measurements on the TEM/STEM were at 30° tilt. Cells were analyzed by small rectangular rasters for 150-400 secs. Data was usually obtained from pairs of HK and LK cells within a few μm of each other.

Except where specifically indicated otherwise, results were obtained on the SEM at 20 kV.

Quantitation

The results to be reported were obtained by a quantitation scheme introduced here last year.[1] It takes advantage of the fact that erythrocytes contain a substantial amount of hemoglobin Fe whose concentration is nearly constant from cell to cell and which can be measured by chemical analysis. This is used as an internal standard against which other cellular concentrations can be estimated. The X-ray intensity ratios of Na and K with respect to Fe were corrected to concentration ratios by means of correction factors; the concentration ratios were then multiplied by the known Fe concentration to obtain Na and K concentrations.

The correction factors were determined at each accelerating voltage by means of a standard. This consisted of a dextran solution containing salts of Na, K, and Fe that was frozen and sectioned in the same way as the suspensions of erythrocytes.

The Fe ratio quantitation scheme has the advantage of being particularly easy to implement with our samples. Its validity was established by the fact that it gave results comparable to two other methods: One is the principal method employed in last year's paper. It equates cellular-to-extracellular concentration ratios with cellular-to-extracellular X-ray intensity ratios for each element, and then uses the known extracellular electrolyte concentrations as an internal standard. "Smearing" was then estimated by multiple linear regression.[1] The other is the Hall peak to continuum method, which yields concentrations per unit mass rather than per unit volume. This method employed Spurr plastic standards and required use of a -170°C cold stage to minimize mass change. All three methods when applied to X-ray data from the same cells led to virtually identical estimates of the degree of smearing.

All concentrations in this paper are expressed in $mM/10^3 cm^3$ of cell volume.

Results

Microprobe measurements of intracellular Na and K gave the same pattern of results reported previously.[1] While K concentrations were close to those measured chemically, those of Na were consistently too high. The absolute value of the Na excess was independent of whether the cells contained large or small concentrations of Na (see for instance Tables 1 and 2); the excess was typically about 30% of the extracellular Na concentration.

Such results are an example of the phenomenon we have termed "smearing".[1] In other words, the intracellular measurements of each element are contaminated by additional X-ray signals that are proportional to the concentration of that element in the extracellular matrix. The effect is thus negligible for K, but quite large for Na. Although measurements of other elements such as Cl are also affected,[1] the rest of this paper concerns itself primarily with smearing as exemplified by Na.

The following experiments were carried out to determine whether the difference between microprobe and chemical measurements of Na (excess Na) could be due to movement of extracellular material over the cells during microtomy.

Effect of Cutting Temperature

It has been frequently questioned whether frozen sectioning in the -70 to -80°C range involves transient melting at the surface of the sections. One estimate suggests melting zones on

Table 1
Effect of Cutting Temperature on Excess Na

	$[Na]_{Probe}$	$[Na]_{Chem}$	$[Na]_{Excess}$
-80°C			
HK	78.5±5.5	13.2	65.3±5.5
LK	162.7±3.8	105.0	57.7±3.8
-110°C			
HK	74.2±3.5	13.2	61.0±3.5
LK	174.5±6.1	105.0	69.5±6.1

All concentrations are mM/10^{3}cm^3 cell volume.
n = 13 to 27.
Standard errors of mean are reported.

the order of 600 Å in thickness.[2] However, it is also known that proceeding to lower temperatures alters the nature of the cutting process from melting to cleaving.[3] Therefore, if surface melting is the explanation for the discrepancy in Na measurements, results should be improved by sectioning at lower temperatures.

Alternatively, we have wondered whether the inclusion of large concentrations of dextran in the extracellular matrix could give it a plastic quality that would allow it to be deformed during cutting and perhaps carried over the cells by the knife. If so, such a property ought to be affected by temperature.

Therefore, results from sections cut at -80°C were compared with those from sections cut at -110°. Table 1 demonstrates the results. Varying cutting temperature over this range obviously had no effect on the excess Na signal.

Effect of Section Thickness

In spite of the above result, let's suppose again that cutting causes a layer of extracellular material to be displaced over the cells. If so, each section will consist of a central region that has remained undisturbed and a pair of surfaces on which smearing has occurred. The thickness of the surface layers should be relatively independent of section thickness. On

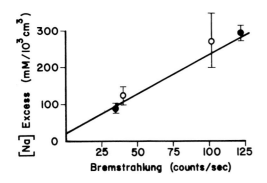

Fig. 1. Effect of section thickness on excess Na. Open circles are LK cells, closed are HK. Bremstrahlung levels are an index of section thickness which varied between approximately 0.3 and 0.9 µm. Each point represents 7 to 22 measurements. Bars indicate standard errors. Regression coefficient, 0.988.

this basis, the ratio of undisturbed to displaced material should increase as section thickness increases, and an inverse relation between section thickness and excess Na would be expected.

Figure 1 shows the result of an experiment in which section thickness was varied over a three-fold range from approximately 0.3 to 0.9 µm. Excess Na is plotted against bremstrahlung intensity, which is an index of relative section thickness. (Collodion rather than carbon support films were used in this experiment; these films made negligible contribution to bremstrahlung, since irradiation caused them to lose most of their mass in our instrument.) Instead of the anticipated inverse relationship, there was a direct relation between section thickness and excess sodium. (The correlation coefficient was 0.988.)

These two experiments make it unlikely that the excess Na comes from smearing during microtomy. In addition, the pattern of results rules out the possibility that it results from a partial melting of the sample that allows ions to diffuse down their concentration gradients. (If this were the case, the cells would lose K as well as gain Na; furthermore, since the cells with low Na have a much steeper gradient favoring Na entry, they would gain more excess Na than cells containing high Na.) Therefore, it is unlikely that we are dealing with an artifact of specimen preparation.

Microprobe as Source of Artifact

These negative results lead to a reconsideration of the possibility that excess Na results from stray microprobe radiation that excites the extracellular matrix.

We had already ruled out two sources of stray radiation, namely electron beam "tailing" and hard X-rays generated in the electron column.[1] When the electron beam was placed 0.5 µm from the edge of a 5 µm diameter Pt aperture, the intensity of the Pt M peak was only 0.5% of that when the Pt was directly irradiated. This amount of stray radiation is nearly two orders of magnitude too small to explain the excess Na signal.

We have also made calculations which indicate that X-rays generated within the cells cannot produce significant numbers of fluorescent X-rays from the extracellular matrix.

However, another possibility is electron scattering within the section. A fraction of the electrons passing through the cells will be elastically scattered over sufficiently wide angles to enter the extracellular matrix. Those which do so will have much longer trajectories through the section than those in the unscattered primary beam, and hence will have a higher probability of exciting X-rays. The question is whether such scattering is sufficiently large to explain the excess Na. On the one hand, simple application of a formula based on the Rutherford model for elastic scattering suggests that too little scattering would occur.[1,4] On the other, we have two pieces of evidence that implicate scattering:

(1) The data in Fig. 1 show a direct relationship between section thickness and excess

Table 2
Effect of Accelerating Voltage on Excess Na

	$[Na]_{Probe}$	$[Na]_{Chem}$	$[Na]_{Excess}$
20 kV			
HK	53.7±2.7	9.2	44.5±2.7
LK	136.6±7.0	102.2	34.4±7.0
10 kV			
HK	95.4±7.5	9.2	86.2±7.5
LK	200.1±20.2	102.2	97.9±20.2

All concentrations are mM/10^3cm^3 cell volume.
n = 15 to 18.
Standard errors of mean are reported.

Table 3
Effect of Type of Microprobe and Accelerating Voltage on Excess Na

	$[Na]_{Probe}$	$[Na]_{Chem}$	$[Na]_{Excess}$
TEM, 80kV			
HK	10.1±1.9	9.2	0.9±1.9
LK	91.0±8.3	102.2	-11.2±8.3
TEM, 20kV			
HK	12.2±2.5	9.2	3.0±2.5
LK	76.8±10.5	102.2	-25.4±10.5
SEM, 20kV			
HK	70.5±5.6	9.2	61.3±5.6
LK	152.5±5.9	102.2	50.3±5.9

All concentrations are mM/10^3cm^3 cell volume.
n = 11 to 28.
Standard errors of mean are reported.

Na. Though incompatible with a sectioning arti-fact, this is entirely compatible with scatter-ing, since the Rutherford model predicts that scattering though a given angle is a direct linear function of target thickness.

(2) The Rutherford model also predicts that scattering will vary inversely as the square of the accelerating voltage. Therefore, we examined the effect of accelerating voltage. Table 2 shows the result when the same sections were probed at 10 and 20 kV. Lowering the accelera-ting voltage caused the excess Na to increase by a factor of at least 2. This clearly indicates something happening within the microprobe. Since electron beam tailing is virtually the same at both accelerating voltages, the effect is almost certainly related to scattering by the section.

Actually, two forms of scattering need to be considered. One is scattering from cells direc-tly into the extracellular phase. The other is scattering into other parts of the specimen chamber followed by scattering back to the sec-tion. However, it is unlikely that enough radi-ation could be scattered back to explain the ex-cess Na signal. Assuming for sake of argument that the electrons which are scattered back have the same average trajectory through the section as those in the primary beam, 70% of the elec-trons in the primary beam would have to return to the section to explain the excess Na signal at 10 kV. This is improbable, not only because the backscattering coefficient for specimen chamber material is 0.3, but also because, on geo-metrical grounds, only a small fraction of elec-trons scattered by the specimen chamber could return to the sections. Therefore, it is more likely that the excess Na arises from scattering within the section itself.

Use of TEM/STEM vs. SEM

All the foregoing results were obtained with a conventional SEM operating in the STEM mode with a maximum accelerating voltage of 20 kV. Although 30 kV was also available, we did not use it, because backscattered electrons at this potential flooded the X-ray detector. However, a TEM/STEM has the double advantage of accelera-ting voltages up to 100 kV and the presence of a strong magnetic field around the specimen that prevents scattered electrons from entering the X-ray detector.

Therefore, we used a TEM/STEM in the STEM mode to see whether higher accelerating voltages would reduce or eliminate smearing. Table 3 com-pares results obtained at 80 and 20 kV from a TEM/STEM with those at 20 kV from the SEM. While excess Na was typically high on the SEM, it was virtually non-existant on the TEM/STEM. How-ever, results at 20 kV on the latter instrument were surprising, since they were similar to those at 80 kV rather than those at 20 kV on the SEM.

Two additional comments on Table 3 need to be made. (1) Although we have consistently found that results at 20 kV are better with the TEM/STEM than the SEM, we have seen statistically significant differences between 20 and 80 kV in some cases. To cite our worst experience with the TEM/STEM we found excess Na as high as 20 mM at 20 kV. However this is still a factor of 2 less than the lowest excess Na we have measured on the SEM. It is obviously desirable to work at higher accelerating voltages. (2) Although the table indicates some negative values of Na excess, we do not take them seriously. Not only are the differences between microprobe and chemi-cal analyses of dubious statistical significance in this instance, but also we have found nearly perfect correspondence (within 2%) between types of analysis in other experiments.

Since the sections examined in the SEM were either the same as those examined in the other instrument or were prepared at the same time, the results must be attributed instrumental effects. The improved results at 80 kV are to be expected, but there is no satisfactory explana-tion for the difference between the two instru-ments at 20 kV. Both had primary electron beams with negligible beam tailing and hard X-rays. Furthermore, since backscattering does not satis-factorily explain the excess Na of the SEM, differences in the configuration of their speci-men chambers do not provide a likely cause. How-ever, there is a very strong magnetic field around the specimen in the TEM, but virtually none at all in the SEM. This magnetic field will alter the trajectories of scattered electrons, and might thereby minimize smearing. It remains to be seen whether this fundamental difference between probe forming instruments is truely significant for minimizing the artifact we have described.

Conclusion

This paper shows that the artifact we have described arises from the interaction of the electron beam with the specimen in the microprobe. By the same token, the specimen preparation techniques we have employed are shown to be adequate for valid quantitation of electrolytes.

Erroneously high concentrations are liable to be measured whenever one is studying an element at low concentration in one phase that is surrounded by another where that element is at high concentration. Biologists who are making analyses in such a situation should be aware of this effect and make suitable alterations in their instrumental conditions as necessary.

Clearly this artifact can be controlled. It is reduced by decreasing section thickness and by increasing accelerating voltage. The ability of a TEM type of instrument to work at high voltages with an energy dispersive X-ray detector make it preferable to a conventional SEM in this regard.

Nevertheless, successful biological microprobe analysis has been carried out on SEM's at low accelerating voltages. The difference between our results and those of others might lie in differences in sample geometry. Ours consists of quite small cells surrounded by a large extracellular phase that represents 75% of the volume of the sample. Stray extracellular signals should be less of a problem when one is working with, say, a multilayered epithelium with a relatively small amount of extracellular fluid. Also, they should prove no problem when probing areas where an element is accumulated at much higher concentration than in its surroundings.

Some theoretical issues are left unresolved. Although scattering within the section appears to explain smearing, we lack a model that predicts this quantitatively. Monte Carlo calculations should be carried out to learn how theoretically reasonable this explanation is. It also remains to be seen whether immersion of the specimen in a strong magnetic field provides a satisfactory basis for the difference that we have observed between instruments. Comparisons between other instruments of both types might illuminate this question.

Acknowledgements

This project was supported in part by Research Award #510 from the American Heart Association, Greater Los Angeles Affiliate, and by Program Project Grant #HL 11351 from the National Institutes of Health. Special thanks to Mr. Roland Marti and JEOL USA, and to Dr. Thomas Hutchinson and his colleagues at the Center for Bioengineering, University of Washington, for allowing us to make extensive use of the TEM/STEM used in this project.

References

1. Tormey, J.M. Validation of methods for quantitative X-ray analysis of electrolytes using frozen sections of erythrocytes. In, Scanning Electron Microscopy/1978/II (O. Johari, ed.) SEM, Inc., AMF O'Hare, Il, 60666, pp. 259-266 (1978).
2. Thornburg, W., and Mengers, P.E. An analysis of frozen tissue sections: I. Sectioning of fresh-frozen tissue. J. Histochem. Cytochem., 5:47-52 (1957).
3. Kirk, R.G., and Dobbs, G.H. Freeze fracturing with a modified cyro-ultramicrotome to prepare large intact replicas and samples for X-ray microanalysis. Science Tools, LKB Instru. J., 23:28-31 (1976).
4. Goldstein, J.I., et al. Quantitative X-ray analysis in the electron microscope. In, Scanning Electron Microscopy/1977/I (O, Johari, ed.) IIT Research Inst., Chicago, IL, 60616, pp. 315-324 (1977).

DISCUSSION WITH REVIEWERS

Reviewer I: This paper is essentially a retraction of a similar paper at last year's meeting.
Authors: Not really. Both last year's data and our methods of analyzing it were basically sound, and we stand behind them. Furthermore, our interpretation was the most reasonable, based on the data available.

We have now presented new types of data designed to test critically the hypothesis we suggested. Since these are incompatible with that hypothesis, we have rejected it in favor of another that fits the data better.

Reviewer IV: I am not convinced by your argument about "scattering" within the samples for two major reasons. The first is that you calibrate your Na signal within the cells by comparison with a standard frozen and sectioned in the same way as the suspension of erythrocytes. If "scattering" exists, it will contribute by the same factor in both the standard and erythrocyte suspension. Thus it will not affect the final result within the erythrocytes.
Authors: Scattering will certainly occur in both the standard and the erythrocyte preparation. However, we disagree with your conclusion.

The standard is used to obtain constants that allow calculation of Na/Fe concentration ratios from Na/Fe X-ray intensity ratios. Scattering will not affect the experimental determination of this constant for two reasons: (1) Since wide angle scattering is elastic in nature, the scattered electrons will have the same energy as the primary beam, and hence will produce the same X-ray intensity ratio. (2)

Since the standard has only one phase, the scattered electrons will "see" the same relative concentrations of Na and Fe as the primary beam.

The situation with the erythrocyte suspension is fundamentally different, since this is a multiphase system. The cells are so small that wide angle scattering will allow electrons to excite the extracellular matrix. Since this has a different composition from the cells, the intensity ratio from it will differ correspondingly. The observed intensity ratio will therefore be the weighted sum of two different ratios, one from the cell, the other from the matrix.

Clearly, then, scattering out of the erythrocytes will alter the observed intensity ratios, whereas scattering within the standard will not.

Reviewer IV: I am also not convinced about scattering, because you worked with two microscopes equipped with two kinds of energy dispersive systems: Northern and Kevex. The sodium peak is very difficult to investigate with Si(Li) detectors as the characteristic peak is located over the hump of the bremsstrahlung curve. When sodium concentration is low, the way the background is calculated and subtracted may dramatically affect the result, while at higher concentrations this effect is not so great; this would explain why at 20 kV you can achieve good results with one system, and not with the other.
Authors: Your point is well taken and has been a matter of concern to us. Therefore, we have used both the Kevex and Northern systems on both types of microscope. Such differences between the two analyzers as we do observe are much too small to change any of our conclusions. For example, at 80 kV on a JEOL 100CX, one system measured a Na concentration of 9.1 ± 1.8 mM/10^3cm^3 while the other measured 16.1 ± 1.4. These values bracketed the chemical measurements.

Reviewer I: You have shown that use of the TEM/STEM gives correct results for Na concentrations and speculate that it is due to the magnetic field preventing scattered electrons entering the detector. Have you tried placing a simple magnetic shield around the detector in the SEM?
Authors: Immersing the specimen in a strong magnetic field will alter the trajectories of electrons. This has one undoubted advantage, namely, keeping scattered electrons out of the X-ray detector. However, electrons scattered into the detector do not contribute to any of the characteristic peaks and so would not affect our results. The reason why these electrons are so undesirable is that they produce a huge increase in background that overwhelms the counting circuitry, resulting in poor counting statistics for the peaks. The second advantage is very highly speculative, namely, that the field might reduce scattering within the section itself and thereby explain better results in one instrument as opposed to the other.

If we placed a magnetic shield around the detector of an SEM operating at 20 kV, it would have no effect. It would not reduce scattered electrons entering the detector, since at this voltage they do not have sufficient energy to penetrate the Be window of the detector. And it would not materially affect scattering within the section, since it would be placed too far away from it.

Reviewer I: You have specifically defined "smearing" as due to the cutting and "...smearing of extracellular material over cells..." (Text reference 1). Since you now show this not to be the case, don't you think it would be appropriate to drop the term?
Authors: The term smearing is imprecise and confusing. Therefore, we agree it would best be dropped.

Part of the confusion stems from the fact that we have used the same term to describe both a phenomenon and a mechanism. The phenomenon is the contamination of intracellular measurements by X-ray signals related to extracellular concentrations (Text reference 1, p. 262, and our present paper). One mechanism by which this could occur is bulk smearing of extracellular material over the cells during microtomy. A second is beam scattering from one part of the section into another, which is the smearing of radiation. According to my dictionary, anything can be smeared. However, the more common connotation of smearing appears to be the spreading of material, hence the imprecision.

T. Barnard: Have you considered whether "smearing" could be related to local charging? This has been shown to alter the interaction volume for hydrated samples (J.D. Brombach, J. de Microscopie, 22:233-8, 1975).
Authors: Yes, we have. Our sections are usually sandwiched between two pre-formed films, and it is possible that many parts of the section might be in poor electrical contact with the substrates. Therefore, we have also worked with sections that were situated on a single film and then directly evaporated with either aluminum or carbon. Although this should considerably reduce any local charging, it had no effect on smearing.

Reviewer IV: Why didn't you suspend your cells in a Na-free isotonic medium, which could be an efficient way to measure a "smearing" effect if it exists?
Authors: We have thought of doing this experiment in a solution in which choline is substituted for Na, but we have not gotten to it. Since there would be no Na in the extracellular matrix, there should be no smearing regardless of whether it were caused by movement of extracellular material or by stray radiation.

J. Shelburne: It would help if you would give some raw data (uncorrected) as well as corrected data to illustrate how the correction factors are used.
Authors: Let's illustrate by showing how we obtained a Na concentration of 16.5 mM/10^3cm^3 from a Na/Fe intensity ratio of 0.125 at 80 kV.

We had previously determined from a standard at 80 kV that the Na/Fe intensity ratio must be multiplied by 5.26 to equal the Na/Fe concentration ratio. We had also measured an intracellular Fe concentration of 25 mM/10³cm³ in this particular batch of cells.

Therefore, the intensity ratio of 0.125 was multiplied by 5.26 to equal 0.658, the concentration ratio of Na/Fe. This ratio times the Fe concentration equals the Na concentration, 16.5 mM/10³cm³.

J. Shelburne: Was background iron ever a problem in either microscope? If so, could its presence have affected your quantitation?
Authors: We had a small Fe background in the SEM that was negligible compared to the size of the cell Fe peak. Proof that background Fe does not affect the quantitation lies in the fact that the Fe ratio approach yielded results similar to two independent quantitation schemes.

J. Shelburne: In your studies on the JEOL 100C did you try turning the objective lens on and off to affect the magnetic field?
Authors: Yes, we did. However, when the resultant beam was focused on the section by the condenser, it was too large to visualize the cells. Nevertheless we found that the Cu signal due to scattering from the support film to the grid was significantly increased when the objective was turned off.

J. Shelburne: Can you give more details on the composition of the dextran solution?
Authors: This was a Ringer's solution containing 143 mM NaCl, 6 mM KCl, 2.125 mM Na_2HPO_4, .375 mM NaH_2PO_4, and 10.1 mM glucose per liter H_2O. 20 g of Sigma Type 60C dextran (average MW 83,000) was added to 100 ml of this solution to make "20%" dextran Ringer's

J. Shelburne: Why do you dry for two days? Have you experimented with shorter times?
Authors: Shorter times could possibly be used. All we know is that considerably longer drying times have no effect on results.

J. Shelburne: Have you experimented with handling and storing you sections in ordinary room atmospheres? In other words, is it really necessary to use dry nitrogen atmospheres?
Authors: We are not certain how careful one has to be. We do know that storage for several days in a Petri dish in room air causes gradients between cells and extracellular matrix to be partially wiped out.

J. Shelburne: Have you had time yet to probe red cell ghosts that still pump Na and K but which (of course) lack hemoglobin? Presumably less scattering would occur with sections of these cells.
Authors: No we haven't, but this would be an interesting experiment for the reason you suggest.

A. Saubermann: To the best of my knowledge, no one has ever shown that sectioning at any temperature causes melting. Therefore, it is somewhat suprising that you would expect differences when you vary cutting temperature between -80 and -110°C.

Authors: We agree that there is no unequivocal demonstration that cutting of cryosections involves melting. However, there is enough evidence for the possibility to make it an a priori concern that needs to be ruled out.

For instance, Thornburg and Mengers (Text reference 2) presented a physical model that predicts melting and showed that their experience with frozen sections was consistent with it. They also predicted that at sufficiently low temperatures the character of cutting would abruptly change to cleaving. Kirk and Dobbs (Test reference 3) made replicas of block faces from which frozen sections had been cut. Their appearance suggested that freeze fracturing occurred at -110°C, but that "cutting" took place at -65°C. The two processes appear to be different, although it remains problematic whether "cutting" entails melting.

A. Saubermann: Does the "smearing" artifact you describe occur with other low Z elements such as Mg?
Authors: Not in our preparations. Smearing appears to involve only elements which are present in the extracellular matrix at relatively high concentrations. Thus, we found (Text reference 1) both Na and Cl smeared in proportion to their extracellular concentrations, while it was not measurable for elements outside the cell in low concentrations, such as K. We made no attempt to assess smearing of Mg, since microprobe measurements indicated low intracellular concentration, and since Mg was absent from the extracellular solution.

A. Saubermann: How do you eliminate the Cu L line which can interfere with Na?
Authors: The Cu L peak does partially overlap the Na K peak. However the Cu-Na complex was satisfactorily deconvoluted using two different types of spectral analysis programs. One made use of multiple least squares fitting with digital filtering to fit Cu and Na reference peaks to the unknown spectrum. The other subtracted a visually fitted background followed by a "simplex" deconvolution routine.

We have considerable evidence that smearing is real, rather than a result of poor deconvolution. (1) Both deconvolution routines gave essentially the same results. (2) Excess Na had the same value regardless of whether sections were on Cu grids or Cu-free supports (such as the 3 mm diameter collodion films on carbon tubes that were the basis for the data we presented last year). (3) In experiments in which accelerating voltage or section thickness was varied, there was poor or non-existent correlation between the intensity of Cu L and the amount of excess Na.

A. Saubermann: Was the magnitude of the artifact you describe affected by the position of your section on the Cu grid? Was there significant additional scattering from the Cu grid bars?
Authors: No, although Cu signals increased close to the grid bars. Your question seems to suggest a concern that our results could be due to some kind of interaction involving the Cu grids. We believe this is ruled out by points (2) and (3) of our reply to your previous question.

Additional discussion with reviewers of the paper "Evaluation of Tissue-Response to Hydrogel Composite Materials" by R.T. Greer et al, continued from page 878.

B. D. Ratner: No attempt was made in the discussion section of this paper to compare the tissue responses noted for the various non-porous materials. Do you feel that significant differences were observed between these materials? Can you attribute these differences to monomer composition, crosslinking density, water content, graft level, etc.?
Authors: The non-porous materials, both the hydrogel coated and uncoated silicone rubber, showed no significant differences in the categories of tissue response which were evaluated and presented in Table 1. No definitive conclusions were stated as to particular factors which may contribute to the tissue response for these implanted materials since the evaluated response to the coated and non-coated materials were very similar.

V. F. Holland: Please indicate which SEM micrographs are of sections, thin films or bulk samples.
Authors: Specimens for the light micrographs (Figures 3 and 4) were sectioned at six micrometers. Bulk samples of uncoated silicone rubber, hydrogel prepared by the radiation method, and Dacron velour impregnated with hydrogel were used for the SEM micrographs (Figures 1, 10, and 12). Figure 8 is an SEM micrograph of a thin film coating of hydrogel on a silicone rubber sample. Figures 5, 6, 7, 9, 11, 13, and 15 are SEM micrographs of samples prepared by 10μm sectioning.

J. J. Rosen: What precautions and/or controls have been employed to account for shrinkage, cutting angle, section thickness variation?
Authors: The histological measurements were taken with a light microscope on samples prepared by standard light microscopy techniques. Minimal shrinkage artifact is believed to occur in the preparation of these glass slides. The samples observed in the SEM do undergo some shrinkage because they are not covered with a glass coverslip after staining and hence undergo air drying before they are coated with gold for SEM observation. The priorities of this work were first of all to determine the general level of biocompatibility and secondly, to assess the range of responses as regards the degree of encapsulation and the amount of ingrowth. In this context, the qualitative nature of the SEM work was sufficient to demonstrate the nature of the collagenous portion of the capsule.

Samples were visually aligned in the microtome to achieve a perpendicular cut. The section thickness depends on the cosine of the cutting angle. Therefore, an alignment error of up to 30° amounts to only about 15% error in the reported capsule thickness. Actual alignment errors are expected to be on the order of 15° or less giving a capsule thickness measurement error of only <4% or less.

Section thickness did vary from about 5 micrometers up to 8 micrometers but was not considered to be a critical effector of artifact.

Other sample preparation techniques, such as freeze-fracturing, freeze-etching, or freeze-drying could be employed to further pursue the microstructural relationship of implant to tissue.

J. J. Rosen: Based on your observations of the in vivo response to these materials, can you suggest a model that will describe the specific interactions between the inflammatory cellular response and the materials parameters varied at the implant site?
Authors: The in vivo response to these implanted materials was used to determine the extent of possible chemical interactions between host and implant as a result of possible cytotoxic moieties released to contiguous tissue. All the materials tested showed no acute cytotoxic activity. Implant stability (mechanical integrity and properties) was determined to depend on substrate and hydrogel polymerization methods. The Silastic composite materials were surrounded by a thin, dense encapsulating membrane of collagen with little suggestion of tissue ingrowth. The cellular reaction to the Dacron composite materials suggests that fibroblasts, macrophages, and small blood vessels can advance into the Dacron-hydrogel composite material due to the filamentous nature of the Dacron velour and the relative porosity of 10-50 micrometers and low mechanical strength of the hydrogel filling the Dacron interstices.

SCANNING ELECTRON MICROSCOPY/1979/II
SEM Inc., AMF O'Hare, IL 60666, USA

NON-FREEZING TECHNIQUES OF PREPARING BIOLOGICAL SPECIMENS FOR
ELECTRON MICROPROBE X-RAY MICROANALYSIS

A. J. Morgan

Zoology Department
University College, P.O. Box 78
Cardiff, CF1 1XL,
U. K.

Abstract

In this paper a review is presented of the
many ways the fluids involved in conventional wet
chemical E.M. preparative techniques (including
precipitation procedures) disturb the in vivo
chemical integrity of biological tissues. The
artifacts discussed are: (i) bulk element loss,
(ii) phase transformations in mineralized
tissues, (iii) redistribution of endogenous
tissue elements; (iv) introduction of extraneous
elements. Finally, the range of biological
applications to which wet chemical methods have
been successfully applied by various authors is
briefly discussed to provide a realistic working
perspective. It is stressed that in most of
these applications the authors restrict
themselves to a comparison of the chemistry of
cells or their constituent organelle systems and
do not attempt to define the in vivo chemical
composition of cells. The author concludes that
since all preparative procedures available to the
microprobe analyst suffer from some form of
restrictive compromise, it may not be advisable
to think of a general microprobe preparative
technique but, alternatively to apply different
preparative techniques according to the specific
biological answers sought.

KEY WORDS: Biological Specimen Preparation,
X-Ray Microanalysis, Fixation, Dehydration,
Embedding, Elemental Loss, Extraneous Elements,
Precipitation

Introduction

In general, electron microprobes are
composite instruments consisting of an X-ray
detector/analyzer system attached to the column
of a relatively unmodified electron microscope.
Early workers often considered the analytical
facility of the microscope as a rather convenient
adjunct to the basic imaging facility of their
electron microscope and, consequently, prepared
their specimens for analysis by the well
established wet chemical regimes involving
fixation, dehydration, resin embedding etc. By
and large the frailties of the biological specimen
as a complex highly ordered chemical entity were
ignored. However, it soon became clear that these
wet chemical preparative procedures, though
obviously providing specimens with excellent and
easily-interpretable fine-structural detail,
yielded specimens whose chemical integrity was
grossly disturbed.

During the last few years, the above
practical perspective has been almost reversed.
The microprobe is probably now considered to be
primarily an analytical instrument, although the
imaging capacity is still frequently of a very
high order. This change of perspective has led
to the rapid evolution of cryo-techniques, whereby
biological specimens are 'fixed' by quench
freezing. Bulk frozen specimens or frozen
sections may either be transferred to and analyzed
in the microprobe in the hydrated state, or after
prior vacuum drying. In any event, cryo-
techniques have been developed and exploited in an
attempt to preserve in vivo electrolyte
concentration gradients in intracellular and
intercellular domains.

However, all available methods of preparing
biological specimens for microprobe analysis
involve some sort of restrictive compromise.
Indeed, cryo-techniques though offering the best
(and probably the only) means of maintaining
electrolyte and ion gradients in cells, do not
yield specimens where the quality of the
ultrastructural information discernible is
adequate for all applications, and especially in
highly cellular tissues. Thankfully, the
biologist is not always interested in the
distribution of highly exchangeable cell
components. One may advance the thesis, therefore,
that no single preparative technique can be

recommended for all microprobe applications to the complete exclusion of all others; the method of choice in a given situation often depends on the specific questions under investigation.

In this paper I shall attempt to survey in some detail the reasons why wet chemical preparative techniques are generally unsuitable for microprobe applications. Secondly, I shall outline with reference to specific applications those areas where wet chemical procedures can provide viable biological information in a microprobe context.

Conventional (Wet Chemical) Preparative Procedures

First of all I would like to discuss the chemical artifacts that tend to be introduced by the exposure of biological specimens to aqueous media:
1. Element loss
2. Phase transformations in mineralized tissues
3. Redistribution of elements within the specimen
4. Introduction of extraneous elements.

1A. Element loss during fixation, dehydration and embedding.

The bulk loss of endogenous elements or of radioisotopes from various plant and animal tissues during chemical fixation has been measured by many authors (Table 1)[1,2,3,4,5,6,7,8,9]. The recorded losses during initial fixation often exceeded 50% of the original element concentration in the tissue. Predominantly structural elements, such as sulphur and phosphorus, are less readily lost,[5] although Hayat [10] reviewed several papers showing that S- and P- containing macromolecules, including proteins, lipids and polysaccharides can be lost during E.M. processing.

Several general conclusions may be drawn from the results summarized in Table 1:

(a) Some authors have described interesting differences in the kinetics of electrolyte loss during fixation in various aldehydes[2] and in osmium tetroxide[8]. However, these differences between fixatives are of but theoretical interest, since all losses are highly restrictive.

(b) Losses can occur during all stages of processing[3,9,11,12,13] i.e. during fixation, buffer washing, dehydration and resin embedding (Figs. 1,2). The rapid loss during initial fixation probably reflects the loss of the more readily exchangeable electrolyte compartments. DeFilippis and Pallaghy (1975)[13] have measured the rate of loss of bound + unbound and unbound Hg and Zn compartments from plant tissue (Fig. 2). The chemical state of binding of an element within a tissue compartment quite obviously affects its rate of loss during processing. Furthermore, just because a late stage in a wet chemical regime - alcohol dehydration (say) - does not remove significant quantities of a given element, it does not mean that potentially alcohol dehydration does not affect the element concentration; it may simply mean that the alcohol-soluble fractions have already been leached during earlier processing stages.

(c) The rate of element loss from a specimen during exposure to histological fluids can be influenced by specimen size[3,12] (Figs. 1C and 1D), the processing temperature[8], the type of fixative used[7,8], and the concentration of the fixative[2,7,8] (Fig. 3). The actual rates of element loss during processing have seldom been measured, but the data presented by Pentilla et al. (1974)[8] indicate that loss rates can be high e.g. during the first 2 minutes of fixation in 3% glutaraldehyde at $0^{\circ}C$ the rate of loss was $\simeq$ 15% per min. of the original K^+ conc. in Ehrlich ascites tumor cells; at $37^{\circ}C$ the rate of loss during the first 2 minutes was $\simeq$ 40% per min. Thus the reduction of fixation time is not a viable practical proposition for the retention of mobile electrolytes.

1B. Loss of elements during flotation and staining

Embedded specimens are often sectioned onto and retrieved from water baths. A discussion of element losses during section flotation and staining is of wide significance since many workers, who prepare their specimens by techniques which (theoretically at least) offer a better prospect of element retention e.g. precipitation procedures, freeze-substitution, freeze drying/ resin embedding/sectioning, often section onto aqueous media, and occasionally also stain their specimens.

Mineralized structures are readily leached during water flotation[14,15]. These losses are often due to an incomplete infiltration by resin and can be largely avoided by cutting onto dry knives and by avoiding staining (Fig. 4). A major problem associated with the loss of mineralized structures during flotation and staining is that the specimen is immediately surrounded by a medium containing increasingly high concentrations of the extracted elements. Non-specific precipitation[11] and secondary binding of ions to high-affinity sites with the section are likely further artifacts.

Agostini and Hasselbach (1971)[16] performed systematic studies of ^{45}Ca loss from sarcoplasmic reticulum vesicles mounted directly on E.M. grids, and exposed to on-grid fixation and staining (Table 2). Similar findings were reported by Harvey et al.[17] on ^{36}Cl and ^{22}Na losses from freeze-substituted plant tissues embedded with Spurr resin. Others[18] have shown considerable Ca losses during uranyl acetate and lead citrate staining of myocardial tissue stabilized by osmium + pyroantimonate precipitation.

2. Phase transformations in mineralized tissues

Apart from the losses induced during exposure to histological fluids[14,19,20,21,22,23,24], these preparative procedures can also induce a significant transformation of the mineral phase. Amorphous calcium phosphate is readily transformed to apatite crystalline phase during exposure to histological fluids e.g. Termine[24] showed that in a mineral sample where 25% of the amorphous phase was leached during processing, 40% of the amorphous phase remaining in the sample was transformed to apatite crystals. These changes are obviously of considerable consequence in mineralization studies, and can be partially circumvented by the adoption of freeze-substitution in organic solvents[25], or by dry cryo-microtomy and freeze-drying[26].

Table 1   Summary of element losses from plant and animal tissues during various conventional (wet chemical) procedures[*]

Element	Tissue	Preparative Technique	% Loss	Reference
K	Uterus	1% OsO_4 in water:	75	1
		3% glutaraldehyde in water:	95	
	Erythrocytes	2% formaldehyde in PO_4 buffer, pH 7.2:	70[**]	
		4% formaldehyde in PO_4 buffer, pH 7.2:	"	
		1.65% glutaraldehyde in PO_4 buffer, pH 7.2:	"	2
		3.3% glutaraldehyde in PO_4 buffer, Ph 7.2:	"	
		2% acetaldehyde in PO_4 buffer, pH 7.2:	"	
Rb	Leaf tissue	1% OsO_4 in cacodylate-acetate buffer:	90	3
Na	Smooth muscle	5% glutaraldehyde in physiological saline, pH 7.2, 36°C or 4°C, 4h:	increased by 200% over controls	4
Ca	Aorta	3% glutaraldehyde in PO_4 buffer, pH 7.4, 2h:	47	
		3% glutaraldehyde in PO_4 buffer, 1% OsO_4, PO_4 buffer, alcohols, propylene oxide:	51	5
	Tendon	3% glutaraldehyde in PO_4 buffer, pH 7.4, 2h:	64	
		3% glutaraldehyde in PO_4 buffer, 1% OsO_4, PO_4 buffer, alcohols, propylene oxide	57	
^{45}Ca	Pancreas	0.5%, 1% or 3% glutaraldehyde	42-58 (depending on conc. of fixative)	6
Ca	Smooth muscle	5% glutaraldehyde in physiological saline, pH 712, 36°C or 4°C, 4 hr:	increased by 35% over controls	4
^{86}Sr	Isolated mitochondria	6.25% glutaraldehyde in 0.1 M PO_4 buffer, pH 7.4:	25-35	
		12.5% glutaraldehyde in 0.1 M PO_4 buffer, pH 7.4:	25	
		12.5% glutaraldehyde + 1 mM succinate:	13	
		12.5% glutaraldehyde + 3 mM ATP:	26	
		12.5% glutaraldehyde + 10 mM succinate + 3 mM ATP:	28	7
		12.5% glutaraldehyde + 1% OsO_4 in 0.1 M PO_4 buffer, pH 7.4:		
		1% OsO_4 in veronal-acetate buffer, pH 7.4:	50	
		10% formaldehyde ± succinate and ATP:	40	
Mg	Ehrlich ascites tumor cells	5% glutaraldehyde + 1% OsO_4 in Krebs-Ringer PO_4 buffer, pH 7.45:	55	8
	Smooth muscle	5% glutaraldehyde in physiological saline pH 7.2, 36°C or 4°C, 4 hr:	0	4
^{31}Si	Diatoms	2% glutaraldehyde, wash, 1% OsO_4, alcohols	25	
	Spleen	"	35	
	Lung	"	55	
	Liver	"	70	9
	Kidney	"	90	
	Mitochondria (liver & diatom)	"	40-60	
^{68}Ge	Diatoms	2% glutaraldehyde, wash, alcohols:	60	
	Spleen	"	35	
	Liver mitochondria	"	60	
	Diatom nuclei	"	80	9
	Diatom vesicles	"	65	
	Diatom microsomes	"	60	

*   See also Table 5
**   The erythrocyte potassium content equilibrated at approximately similar levels during prolonged fixation in the three aldehydes, but the kinetics of potassium loss displayed considerable differences.

3.   Redistribution of elements

The translocation of elements during E.M. processing to loci they do not occupy in vivo is an especially serious artifact that cannot be overlooked even in specimens prepared by cryo-techniques. The seriousness of the artefact is amplified and not diminished by the fact that its presence is difficult to demonstrate[27]. Coleman and Terepka (1972)[11] proposed 4 criteria which may prove useful in identifying redistribution artifacts:

(a) The occurrence of unusual intracellular crystals, especially of calcium phosphate salts The shortcomings of this criterion is that it is often difficult to distinguish between pathologically-mediated intracellular calcium precipitation, and calcium precipitation in (apparently) normal cells induced by their exposure to aqueous preparative media. A good example of this difficulty, which also happens to represent an excellent example of the way wet chemical preparative techniques can, or should, be applied to yield useful bio-medical information, was reported by Maunder et al.[28] These authors recorded elevated Ca:P ratios in the myonuclei of patients suffering from Duchenne muscular dystrophy (Table 3), but proposed that the elevation was probably a secondary development i.e. by the entry of Ca^{2+} into cells during processing as a consequence of the pre-disposing influences of the disease on membrane properties. Maunder et al. were interested only in a comparison of the properties of diseased and non-diseased cells and in the use of the probe to identify non clinical states of the disease. The redistribution artifact which they have monitored, therefore provides an indirect but valid reflection of a more fundamental disease-mediated change in cellular chemistry.

(b) Distinct chemical boundaries and electrolyte gradients may reflect the absence of marked redistribution. In practice this is probably the most easily applied criterion, and has been used by many workers for the validation of a wide range of preparative procedures, including cryo-techniques.[29,30,31,32] The reasoning behind the use of this criterion is that diffusion artifacts would ways tend to diminish and not to create concentration gradients and discrete distribution boundaries.[32] However, distinct but false chemical boundaries can be produced in cells whose chemical integrity has been severely disturbed during exposure to preparative fluids. For example, we[5] have consistently recorded higher calcium concentrations in the nuclei of osmium-fixed arterial smooth muscle cells than in nuclei of unfixed cells prepared by cryo-microtomy, even though the wet-chemical regime was known to extract at least 50% of the total tissue calcium content. Yarom et al.[33] similarly detected cytoplasmic calcium in myocardial cells only After osmium tetroxide/pyroantimonate fixation, but not after glutaraldehyde or pyroantimonate fixation. These false calcium localizations after osmication may be due initially to a passive influx of calcium into cytoplasm,[4] followed by an interaction between Ca^{2+}, osmic acid, and osmic acid reactive groups within specific cellular domains.[34]

(c) The use of gelatine models loaded with elements to test whether the elements are displaced during processing. Their extreme simplicity, and especially the absence of membrane boundaries renders such models of little practical significance even for the comparison of different wet chemical regimes.

(d) Translocation obviously occurs if different preparative procedures yield different element distribution patterns. Surprisingly, few workers have systematically compared elemental distribution in cells prepared by wet chemical procedures and by cryo-techniques. Thus, the study by Erasmus and Davies[35] of calcium distribution in the vitelline cell of Schistosoma mansoni is of considerable technical interest since it shows unequivocally the loss of an

element (calcium) from one organelle and the uptake of calcium by another organelle during osmium fixation. This particular cell contains two structures (Fig. 5) with significant inherent electron density which facilitates their identification in unfixed, unstained cryo-sections. The vitelline droplets contain no calcium when prepared by cryo-microtomy, whereas the dense bodies within the dilations of the endoplasmic reticulum yield significant calcium and phosphorus signals. In contrast, the dense bodies in glutaraldehyde and osmium double fixed cells are extremely pale and possess little detectable calcium. However, the vitelline droplets in double-fixed cells contain a significant calcium content. Erasmus and Davies[35] have also shown that the vitelline droplets in glutaraldehyde single-fixed cells contained no detectable calcium.

4.   Introduction of elements
     Elements may be incorporated into cells during processing either by diffusion from extracellular fluid[4,34,36], or from truly exogenous sources by virtue of being chemical constituents or contaminants of given processing fluids (Table 4).

Wet chemical regimes introduced in conjunction with ion localization applications
     Two general categories of wet-chemical procedural variants are worthy of discussion in the present context:

(a) Simplified embedding procedures. Firstly, the direct embedding of fixed specimens in a polymerizable mixture of 50% glutaraldehyde and urea[41] removes the necessity to dehydrate. Yarom et al.[42] found that up to 10x more calcium is retained in the sarcomeres of myocardial tissue embedded in glutaraldehyde/urea than in samples dehydrates and subsequently embedded in Epon 812.

Secondly, the direct embedding of glutaraldehyde-fixed tissues in Epon 812, after sequentially increasing the glutaraldehyde concentration from 2.5% to 50%, has also been shown to reduce the loss of calcium and zinc (from rat tongue muscle) during alcohol dehydration.[48]

Both the above procedures undoubtedly reduce the loss of elements that otherwise occur in regimes that include dehydration. Nevertheless, irrevocable major losses and redistribution artifacts are already present as a result of chemical fixation (Table 1).

(b) Precipitation procedures. These procedures represent attempts to reduce diffusion artifacts that occur during wet-chemical processing by the sequestration of endogenous ions (hopefully) in situ with introduced ions, often heavy metals, to produce an insoluble precipitate e.g. Cl^- ions have been precipitated with gold[49,50] and silver[12,51]; Na^+ with pyroantimonate ions[52]; PO_4^{3-} with lead ions[53]; Ca^{2+} with oxalate ions[11].

The success of these various precipitation reactions, and especially of the pyroantimonate reaction, in reducing ion loss and drift from in vivo locations has been the subject of considerable controversy.[54] I would simply wish to make the following three observations which

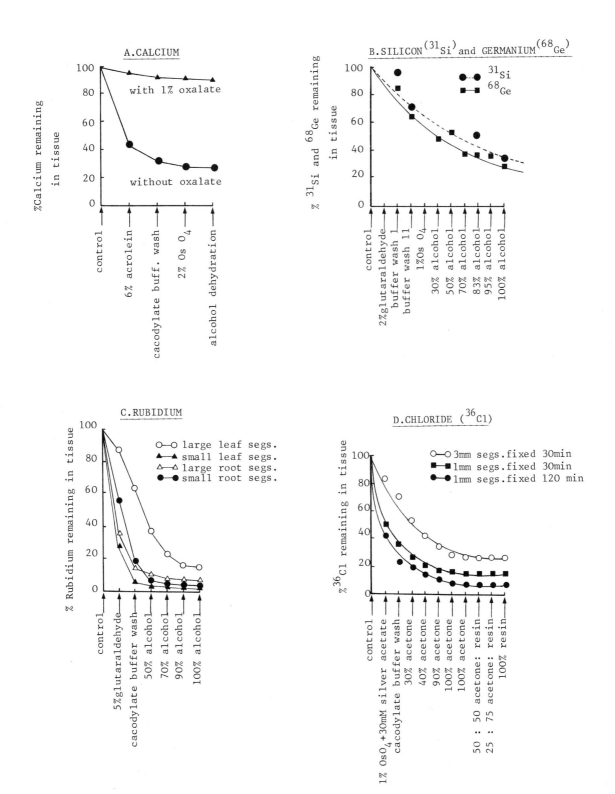

Figure 1 Loss of divalent and monovalent ions from plant and animal tissues during the various stages of preparation for electron microscopy. A, ^{45}Ca loss from labelled chloroallantoic membranes: B, ^{31}Si and ^{68}Ge loss from double-labelled rat liver: C, loss of rubidium from plant and leaf segments; D, loss of ^{36}Cl from labelled leaf segments. In 'A' and 'D' precipitation agents are included in the preparative fluids. A redrawn from Coleman & Terepka (1972)[11]; B redrawn from Mehard & Volcani (1975)[9]; C redrawn from Hall et al. (1974)[3]; D redrawn from Harvey et al. (1976)[12].

Table 2 is at bottom of
the facing page.

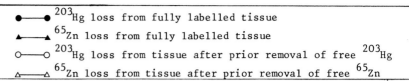

Table 3 Ca: P ratios (derived from the relative mass
        fractions of calcium and phosphorus x 10³) in
        myonuclei of structurally undamaged fibres in
        male patients suffering from Duchenne muscular
        dystrophy (DMD) and in controls.

	Case No.*	Age	Ca : P x 10
Non-diseased Controls	1	17 days	4.7
	2	6 years	3.6
	3	7 years	4.1
	4	16 years	2.5
	5	32 years	2.6
	6	34 years	4.3
			$\bar{x}$ =3.6
			s.d.=± 0.8
Patients with diagnosed DMD	1	3 months	5.0
	2	4 years	8.8
	3	5 years	11.5
	4	6 years	10.4
	5	8 years	5.6
	6	9 years	6.5
			$\bar{x}$ =8.0
			s.d.=± 2.4

* Average number of fibres analyzed per patient = 7
(From Maunder, Yarom and Dubowitz, 1977 - J. of the
Neurological Sciences 33, 323-334)[28]

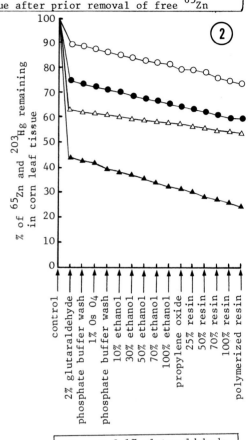

Figure 2
Percentage loss of ^{65}Zn and ^{203}Hg
(before and after exchange of labels
occupying the tissue free space) from
labelled corn leaf segments during
the course of conventional
preparative procedure for electron
microscopy.
Redrawn from De Filippis and Pallaghy
(1975) - Micron 6, 111-120[13]

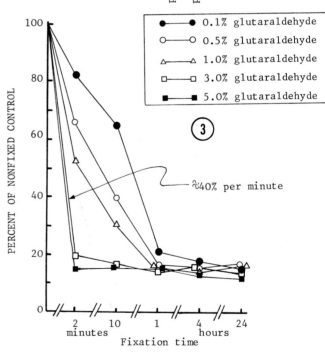

Figure 3
The effect of fixation in different
concentrations of glutaraldehyde
at 37°C on the rate of loss of
potassium from Ehrlich ascites tumor
cells.
Redrawn from Penttila et al., 1974 -
J.Cell Biol. 63, 197-214[8].

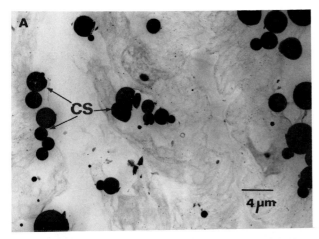

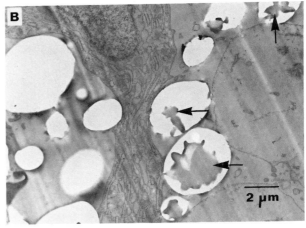

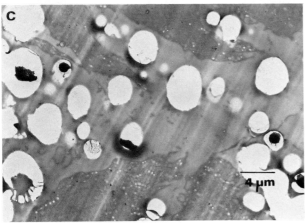

Figure 4. Electron micrographs of sections through the calciferous gland of the earthworm Lumbricus terrestris. All specimens were double fixed in 3% glutaraldehyde and 1% osmium tetroxide, and embedded in Spurr's resin. A. Section cut onto a dry knife, unstained. Note the relatively intact mineralised (calcitic) spherules, CS, in the intracellular spaces which have survived exposure to the fixatives, alcohols and embedding media. Some cellular morphological detail is discernible. B. Section cut onto a dry knife and stained in uranyl acetate and lead citrate. Mineral has been leached from the spherules, leaving characteristically shaped holes some of which have a resin 'core' (arrow) suggesting that some spherules normally possess hollow centres. C. Section cut onto water, but not stained. Mineral has again been almost completely removed from the spherules-compare with 'B'.

Table 2 ^{45}Ca loss from ^{45}Ca-loaded sarcoplasmic reticulum vesicles mounted on E.M. grids during various staining procedures

A. ^{45}Ca loss from unfixed specimens

Controls*	Water treated grids	2% Ammonium molybdate, pH 7.0	1% phospho-tungstic acid (PTA), pH 7.0	1% Ammonium molybdate + 0.5% PTA, pH 7.0	1% Uranyl acetate, pH 4.5
0% (n=32)	41% (n=32)	37% (n=32)	62% (n=32)	65% (n=32)	45% (n=32)

B. ^{45}Ca loss from sarcoplasmic specimens during fixation on grids, and from the fixed samples during various staining procedures

Controls*	2.5% glutaraldehyde + 10mM potassium oxalate + 30mM KCl + 20mM imidazole, pH 7.0	2% Ammonium molybdate, pH 7.0	1% PTA, pH 7.0	1% Uranyl acetate, pH 4.5
0% (n=15)	53% (n=15)	67% (n=15)	76% (n=15)	59% (n=15)

* Controls represent untreated, ^{45}Ca-loaded samples mounted on grids. Droplets of the reagents were placed directly onto individual grids, and removed after standardized periods (0.5-3 min for stains, 10-30 min for fixative) with blotting paper.
From Agostini and Hasselbach, 1971 - Histochemie 28, 55-67.

Table 3 is on the facing page.

Table 4 An inexhaustive list of some of the elements that may enter tissues from various sources during typical wet chemical processing procedures

Processing Stage	Introduced Element	Implications and Comment
FIXATIVES		
Glutaraldehyde	Ca (contaminant)	Some commercial grades of glutaraldehyde contain various amounts of calcium; the presence of electron-opaque deposits in a variety of tissues can be correlated with the presence of calcium in the fixative. The implications are that whilst calcium-uptake pathways may be localized, the distribution of calcium in the living state cannot. This has led some workers to use re-distilled glutaraldehyde [37,38,39].
Osmium tetroxide:	Os	The Mα emission line of Os at 1.914 keV overlaps the Kα line of P at 2.013 keV. Can be resolved with crystal spectrometer.[40]
Potassium pyroantimonate	Sb	The Sb (Lα), emission line at 3.604 keV and the Ca (Kα) line at 3.690 keV. Can be resolved with crystal spectrometer.[18]
BUFFERS		
k-phosphate) s-collidine) Na-cacodylate)	contaminants and normal constituents	All three buffers were shown to contain significant quantities of contaminants: calcium and potassium or sodium, s-collidine buffer contained the least amount of Ca.[37] Cacodylate buffer also contains As as a normal constituent - the As (Lα) emission at 1.282 keV overlaps the Mg (Kα) emission at 1.253 keV in energy-dispersive spectra and may create sufficient background signal to swamp weak Na (Kα) emissions at 1.041 keV.
RESINS		
Araldite) Epon 812) Spurr's medium)	Cl	Workers studying endogenous chlorine distribution have developed or exploited various low-chlorine[41,42,43] embedding media e.g. 50% glutaraldehyde-urea mixture does not contain Cl and has advantage that no dehydration and long infiltration periods are necessary; 50% glutaraldehyde-carbohydrazide mixture[43,44] contains no Cl; Epon 826 contains 30mM, kg^{-1} organically bound Cl compared with 1400mM.kg^{-1} in Epon 812;[45] Pallaghy's modification[46] of Spurr's resin contains no Cl and retains the low viscosity feature of Spurr's.[47]
CHLOROFORM VAPOUR	Cl	Chloroform vapour is routinely used for stretching resin-infiltrated sections on a water bath. The vapour is readily absorbed by sections, thus producing significant Cl signals with the specimen.[46] Pallaghy (1973) suggested that methyl cyclohexane can be a useful alternative; although flotation on aqueous media is generally not to be recommended for sections destined for microprobe analysis (see above)
STAINS		
Uranyl acetate	U	The major emission lines of the heavy metals commonly used for staining E.M. sections arise within the energy range of many of the biological elements e.g. Pb(Mα-2.342keV and Mβ-2.442keV) emissions are adjacent to S(Kα-2.307keV) and Cl (Kα-2.307keV) and Cl (Kα-2.621keV) lines; U(Mα-3.165keV) and Mβ-3.336keV) emissions are adjacent to the K(Kα-3.312 keV) line. In general, sections to be analyzed should not be stained with heavy metal salts.
Lead citrate	Pb	
Potassium permanganate	Mn	
Ammonium molybdate	Mo	

seem to indicate that the inclusion of a precipitation agent in the media for wet chemical processing does not fundamentally reduce the seriousness of ion diffusion artifacts, and in no way should precipitation procedures be considered alternatives to cryo-techniques:

(i) A number of authors have independently shown that gross element losses may still occur during processing even in the presence of a 'specific' precipitation agent. (Table 5).

(ii) Significant element losses during dehydration, embedding (Fig. 1[11,12]), flotation, and staining[18,33] have been recorded even after ion immobilization by precipitation.

(iii) The formation of discrete precipitates implies that elements are mobilized from their in vivo locations and sequestered by the precipitating ions at focal points.

(ii) and (iii), above, indicate that considerable ion diffusion may occur during precipitation reactions. The details of the mechanisms involved in the reaction of precipitating ions with endogenous tissue ions are essentially not known. The microprobe does not distinguish between the various bound and unbound element compartments. In theory, some

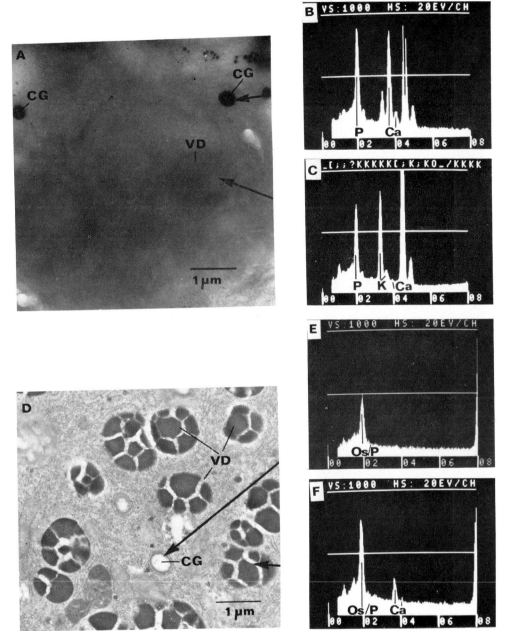

Figure 5.   Redistribution of calcium from the dense calcareous granules of the vitelline cell of Schistosoma mansoni to the vitelline droplets during osmication (either after glutaraldehyde fixation or during osmium textroxide single-fixation).   Fixed sections were mounted on copper grids, cryosections were mounted on titanium grids.   A. Electron micrograph of an unfixed, freeze-dried cryosection through the vitelline cell.   Note the dense spherular calcareous granules (CG) and the vitelline droplets (VD).   B. Spectrum obtained from the analysis of a calcareous granule prepared as in 'A'.   Note especially the approximately equal P and Ca signals.   C. spectrum obtained from the analysis of a single vitelline droplet prepared as in 'A'.   Note the absence of a significant Ca signal.   D. Electron micrograph of a section through a glutaraldehyde and osmium fixed vitelline cell.   Note the vitelline droplets (VD), and calcareous granules (CG) in the distended cisternae of the granular endoplasmic reticulum.   E. Spectrum obtained from analysis of a calcareous spherule prepared as in 'D'. Note especially the absence of detectable calcium.   F. Spectrum obtained from the analysis of a vitelline droplet prepared as in 'D'.   Note especially the presence of a very significant Ca signal (cf. with Fig. 5C).
(Reproduced by courtesy of Dr. D.A. Erasmus and Mr. T.W. Davies).

Table 5 Summary of quantitative measurements of the loss
of given elements from biological specimens during
exposure to certain precipitation reactions

Precipitation Reaction	Endogenous ion/ radioactive tracer	Tissue	Processing regime	% loss With precipitant	% loss Without precipitant
Pyroantimonate[1] for monovalent and divalent cations	Na^+	Uterus	3% glutaraldehyde + 2% pyroantimonate:	90	95
			1% OsO_4 + 2% pyroantimonate:	75	75
	Ca^{2+}	Uterus	3% glutaraldehyde + 2% pyroantimonate:	55	30
			1% OsO_4 + 2% pyroantimonate:	85	0
	Mg^{2+}	Uterus	3% glutaraldehyde + 2% pyroantimonate:	30	20
			1% OsO_4 + 2% pyroantimonate:	35	10
Oxalate for[11] Calcium	Ca^{2+}	Chorio-allantoic membranes	6% acrolein in 0.1 M cacodylate-HCl + 1% sodium oxalate, buffer + oxalate wash, 2% OsO_4 + 1% oxalate, graded alcohols:	12	75
Sulphide for[13] Heavy Metals	^{65}Zn	Corn leaf, Barley root, Chlorella	0.3% or 1% glutaraldehyde in PO_4 buffer + 30% Na_2S:	7.6 - 9.5	62 - 70
[13]	^{203}Hg	Corn leaf, Barley root, Chlorella	0.3% or 1% glutaraldehyde in PO_4 buffer + 30% Na_2S:	72 - 81	34 - 41
[55]	Cd*	Bivalve tissues	Glutaraldehyde + H_2S:	14	28
			Glutaraldehyde + H_2S, embedded in glutaraldehyde/urea:	20	(-)**
			Glutaraldehyde + H_2S, alcohols:	33	(-)
			Glutaraldehyde + H_2S, alcohols, embedded in Spurr resin:	41	(-)
Silver Ions[49] for Chloride	^{36}Cl	Root tissue	1% OsO_4 in 0.1 M cacodylate-acetate buffer, pH 6.5 + either 0.5% Ag acetate or 1% Ag lactate, buffer wash, acetone (+ brief wash in 0.05 N HNO_3), propylene oxide, Spurr resin:	4	(-)**
[12]		Leaf	1% OsO_4 in cacodylate-acetate, buffer wash, acetone, Pallaghy modification of Spurr's resin:	70 - 95 (depending on fixation time and sample size)	(-)**

* Details of processing regime not provided in this communication
** not measured

precipitation reactions may be used, possibly in conjunction with cryo-procedures, for the identification of ion compartments.

Biological applications of wet chemical procedures

The above discussion has shown how each stage of a typical wet chemical procedure, traditionally used for the preparation of biological specimens for ultrastructural study, can grossly alter the chemical integrity of the specimen. Having comprehensively illustrated this fact, I shall now briefly attempt to balance the practical perspective, for there can be no doubt that much useful bio-medical information has been derived from specimens prepared by exposure to aqueous media. The localization of readily diffusible components is obviously precluded, but there are certain types of applications that fairly readily lend themselves to wet-chemical treatment, e.g.:

1. Analysis of natural inclusion bodies[56,57]
2. Pharmacology - distribution of deliberately introduced compounds[58]
3. Pathology[28,59]
4. Identification of histochemical reaction products[60,61,62]
5. Analysis of bound elements[63,64]

Almost invariably the success of the studies quoted above are due to their limited terms of reference: they do not attempt to define the in vivo chemistry of a cell or cell part, but essentially compare the chemistry of cells or their constituents (diseased v normal, young v old, etc.).

Conclusion

The papers presented earlier in this volume have argued strongly and convincingly in favour of cryo-procedures of one type or another. My comments on wet-chemical preparative procedures serves but to amplify their comments. However, I believe that we are not yet at the stage where a single technique can be considered universal or general - each and every technique, in different measure, can contribute to our chemical knowledge of biological structures and entities. Regardless of their serious deficiencies, this I feel applies to wet-chemical techniques, and also to other non-freezing preparative techniques that have been omitted from this discussion e.g. whole mounting and air drying.[65]

Acknowledgements

I would like to sincerely thank: Mrs. N. West and Mrs. J. Rees for painstakingly typing the manuscript, and the British Science Research Council for financial support of our microprobe laboratory.

References

1. Garfield, R.E., Henderson, R.M. and Daniel, E.E. Evaluation of the pyroantimonate technique for localization of tissue sodium. Tissue and Cell 4, 1972, 575-589.
2. Vassar, P.S., Hard, J.M., Brooks, D.E., Hagenberger, B. and Seaman, G.V.F. Physicochemical effects of aldehydes on the human erythrocyte. J.Cell Biol. 53, 1972, 809-818.
3. Hall, J.L., Yeo, A.R. and Flowers, T.J. Uptake and localization of rubidium in the halophyte Suaeda maritima. Z. Pflanzenphysiologie 71, 1974, 200-206.
4. Schoenberg, C.F., Goodford, P.J., Wolowyk, M.W. and Wootton, G.S. Ionic changes during smooth muscle fixation for electron microscopy. J.Mechanochem. Cell Motility 2, 1973, 69-82.
5. Morgan, A.J., Davies, T.W. and Erasmus, D.A. Changes in the concentration and distribution of elements during electron microscope preparative procedures. Micron 6, 1975, 11-23.
6. Howell, S.L. and Tyhurst, M. 45-Calcium localization in islets of Langerhans, a study by electron-microscope auto-radiography. J.Cell Sci. 21, 1976, 415-422.
7. Greenawalt, J.W. and Carafoli, E. Electron microscope studies on the active accumulation of Sr++ by rat-liver mitochondria. J.Cell Biol. 29, 1966, 37-61.
8. Penttila, A., Kalimo, H. and Trump, B.F. Influence of glutaraldehyde and/or osmium tetroxide on cell volume, ion content, mechanical stability, and membrane permeability of Ehrlich ascites tumor cells. J.Cell Biol. 63, 1974, 197-214.
9. Mehard, C.W. and Volcani, B.E. Evaluation of silicon and germanium retention in rat tissues and diatoms during cell and organelle preparation for electron probe microanalysis. J.Histochem. Cytochem. 23, 1975, 348-359.
10. Hayat, M.A. Principles and Techniques of Electron Microscopy. Biological Applications. Van Nostrand Reinhold Company, New York, Cincinnati, Toronto, London & Melborne, 1970, Vol. 1.
11. Coleman, J.R. and Terepka, A.R. Electron probe analysis of the calcium distribution in cells of the embryonic chick chorioallantoic membrane. I. A critical evaluation of techniques. J. Histochem. Cytochem. 20, 1972, 401-413.
12. Harvey, D.M.R., Flowers, T.J. and Hall, J.L. Localization of chloride in leaf cells of the halophyte Suaeda maritima by silver precipitation. New Phytol. 77, 1976, 319-323.
13. De Filippis, L.F. and Pallaghy, C.K. Localization of zinc and mercury in plant cells. Micron. 6, 1975, 111-120.
14. Boothroyd, B. The problems of demineralization in thin sections of fully calcified bone. J.Cell Biol. 20, 1964, 165-173.
15. Morgan, A.J. Strontium metabolism in the earthworm, Lumbricus terrestris. I. Morphology of calciferous glands and chloragogenous tissues; preliminary electron microprobe analysis. Cell Tiss. Res. 1979. (In press).
16. Agostini, B. and Hasselbach, W. Electron cytochemistry of calcium uptake in the fragmented sarcoplasmic reticulum Histochemie 28, 1971, 55-67.
17. Harvey, D.M.R., Hall, J.L. and Flowers, T.J. The use of freeze-substitution in the preparation of plant tissue for ion localization studies. J.Microsc. 107, 1976, 189-198.
18. Yarom, R., Peters, P.D., Scripps, M. and Rogel, S. Effect of specimen preparation on intracellular myocardial calcium. Electron microscopic X-ray microanalysis. Histochemistry 38, 1974, 143-153.
19. Boothroyd, B. The adaptation of the technique of microincineration to electron microscopy. J.R. Microsc. Soc. 88, 1968, 529-544.
20. Renaud, S. Superiority of alcoholic over aqueous fixation in the histochemical detection of calcium. Stain Technology 34, 1959, 267-271.
21. Glimcher, M.J. and Krane. The organisation and structure of bone, and the mechanism of calcification. In: Treatise on Collagen, Vol. 2, Part B, Biology of Collagen (Gould, B.S., ed.), Academic Press, London & New York, 1968, 67.
22. Dudley, H.R. and Spiro, D. The fine structure of bone cells, J. Biophys. Biochem. Cytol. 11, 1961, 627-649.
23. Posner, A.S. In: The Comparative Molecular Biology of Extracellular Matrices (Slavkin, H.C., ed.), Academic Press, New York & London, 1972, 437.

24. Termine, J.D. In: The Comparative Molecular Biology of Extracellular Matrices (Slavkin, H.C., ed.), Academic Press, New York & London, 1972, 444.

25. Landis, W.J., Paine, M.C. and Glimcher, M.J. Electron microscopic observations of bone tissue prepared anhydrously in organic solvents. J. Ultrastruct. Res. 59, 1977, 1-30.

26. Landis, W.J., Hauschka, B.T., Rogerson, C.A. and Glimcher, M.J. Electron microscopic observations of bone tissue prepared by ultracryomicrotomy. J. Ultrastruct. Res. 59, 1977, 185-206.

27. Morgan, A.J. Preparation of biological specimens for microprobe analysis; changes in chemical integrity. In: X-Ray Microanalysis in Biology (Hayat, M.A., ed.), University Park Press, Baltimore, 1979 (In press).

28. Maunder, C.A., Yarom, R. and Dubowitz, V. Electron-microscopic X-ray microanalysis of normal and diseased human muscle. J.Neurol. Sci. 33, 1977, 323-334.

29. Spurr, A.R. Freeze substitution additives for sodium and calcium retention in cells studied by X-ray analytical electron microscopy. Botanical Gazette 133, 1972, 263-270.

30. Appleton, T.C. A cryostat approach to ultrathin 'dry' frozen sections for electron microscopy; a morphological and X-ray analytical study. J. Microsc. 100, 1974, 49-74.

31. Ali, S.Y., Craig Gray,J., Wisby, A. and Phillips, M. Preparation of thin cryosections for electron probe analysis of calcifying cartilage. J. Microsc. 111, 1977, 65-76.

32. Gupta, B.L., Naftalin, R.J. and Hall, T.A. Microprobe measurements of concentrations and gradients of Na, K and Cl in epithelial cells and intercellular spaces of rabbit ileum. Nature (Lond.) 272, 1978, 70-73.

33. Yarom, R., Hall, T.A. and Peters, P.D. Calcium in myonuclei: electron microprobe X-ray analysis. Experientia 31, 1975, 154-157.

34. Krames, B. and Page, E. Effects of electron microscopic fixatives on cell membranes of the perfused rat heart. Biochim. Biophys. Acta 150, 1968, 23-31.

35. Erasmus, D.A. and Davies, T.W. Schistosoma mansoni and S. haematobium: aspects of the calcium metabolism of the vitelline cell. Expt. Parasitol. 1979 (In press).

36. Elbers, P.F. Ion permeability of the egg of Limnaea stagnalis L. on fixation for electron microscopy. Biochim. Biophys. Acta 112, 1966, 318-329.

37. Oschman, J.L. and Wall, B.J. Calcium binding to intestinal membranes. J. Cell Biol. 55, 1972, 58-73.

38. Oschman, J.L., Hall, T.A., Peters, P.D. and Wall, B.J. Microprobe analysis of membrane-associated calcium deposits in squid giant axon. J.Cell Biol. 61, 1974, 156-165.

39. Skaer, R.J., Peters, P.D. and Emmines, J.P. The localization of calcium and phosphorus in human platelets. J. Cell Sci. 15, 1974, 679-692.

40. Chandler, J.A. and Battersby, S. X-ray microanalysis of ultrathin frozen and freeze-dried sections of human sperm cells. J. Microsc. 107, 1976, 55-65.

41. Pease, D.C. and Peterson, R.G. Polymerizable glutaraldehyde-urea mixtures as polar, water-containing embedding media. J. Ultrastruct. Res. 41, 1972, 133-159.

42. Yarom, R., Peters, P.D. and Hall, T.A. Effect of glutaraldehyde and urea embedding on intracellular ionic elements. X-ray microanalysis of skeletal muscle and myocardium. J. Ultrastruct. Res. 49, 1974, 405-418.

43. Van Stevenick, R.F.M., Van Stevenick, M.E., Hall, T.A. and Peters, P.D. A chlorine-free embedding medium for use in X-ray analytical electron microscopic localization of chlorine in biological tissue. Histochimie 38, 1974, 173-180.

44. Heckman, C.A. and Barrnett, R. GACH: A water-miscible, lipid-retaining embedding polymer for electron microscopy. J. Ultrastruct. Res. 42, 1973, 156-179.

45. Ingram, M.J. and Hogben, C.A.M. Procedures for the study of biological soft tissue with the electron microprobe. In: Developments in Applied Spectroscopy (Baer, W.K., Perkins, A.J. and Grove, E.L., eds), Plenum Press, New York, 1968, 43.

46. Pallaghy, C.K. Electron probe microanalysis of potassium and chloride in freeze-substituted leaf sections of Zea mays. Australian J. Biol. Sci. 25, 1973, 1015-1034.

47. Spurr, A.R. A low-viscosity epoxy resin embedding medium for electron microscopy. J. Ultrastruct. Res. 26, 1969, 31-43.

48. Yarom, R., Maunder, C.A., Scripps, M., Hall, T.A. and Dubowitz, V. A simplified method of specimen preparation for X-ray microanalysis of muscle and blood cells. Histochemistry 45, 1975, 49-59.

49. Lauchli, A., Stelzer, R., Guggheim, R. and Henning, L. Precipitation techniques as a means of intracellular ion localization by use of electron probe analysis. In: Microprobe Analysis as Applied to Cells and Tissues (Hall, T., Echlin, P. and Kaufmann, R., eds), Academic Press, London & New York, 1974, 107-118.

50. Lauchli, A. Precipitation technique for diffusible substances. J. Microsc. Biol. Cell. 22, 1975, 239-246.

51. Van Steveninck, R.F.M., Chenoweth, A.R.F. and Van Steveninck, M.E. Ultrastructural localization of ions. In: Ion Transport in Plants (Anderson, W.P., ed), Academic Press, London & New York, 1973, 25-37.

52. Tandler, C.J., Libanati, C.M. and Sauchis, C. A. The intracellular localization of inorganic cations with potassium pyroantimonate. J.Cell Biol. 45, 1970, 355-366.

53. Tandler, C.J. and Solari, A.J. Nucleolar orthophosphate ions. Electron microscope and diffraction studies. J.Cell Biol. 41, 1969, 91-108.

54. Bowen, I.D. and Ryder, T.A. The application of X-ray microanalysis to histochemistry. In: Electron Probe Microanalysis in Biology (Erasmus, D.A., ed), Chapman and Hall, London, 1978, Ch. 6, 183-205.

55. George, S.G., Nott, J.A., Pirie, B.J.S. and Mason, A.Z. A comparative quantitative study of cadmium retention in tissues of a marine bivalve during different fixation and embedding procedures. Proc.R.Microsc. Soc. 11, Part 5, Micro 76 Suppl, 1976, 42.

56. Sohal, R.S., Peters, P.D. and Hall, T.A. Fine structure and X-ray microanalysis of mineralized concretions in the Malpighian tubules of the housefly, Musca domestica. Tissue & Cell 8, 1976, 447-458.

57. Humbert, W. Cytochemistry and X-ray microprobe analysis of the midgut of Tomocerus minor (Insecta, Collembola) with special reference to the physiological significance of the mineral concretions. Cell Tiss. Res. 187, 1978, 397-416.

58. Erasmus, D.A. The application of X-ray microanalysis in the transmission electron microscope to a study of drug distribution in the parasite Schistosoma mansoni (Platyhelminthes). J.Microsc. 102, 1974, 59-69.

59. Chandler, J.A. Application of X-ray microanalysis to pathology. J. Microsc. Biol. Cellulaire 22, 1975, 425-432.

60. Ryder, T.A. and Bowen, I.D. The use of X-ray microanalysis to investigate problems encountered in enzyme cytochemistry. J. Microsc. 101, 1974, 143-151.

61. Bowen, I.D., Ryder, T.A. and Winters, C. The distribution of oxidizable mucosubstances and polysaccharides in the planarian Polycelis tenuis Iijima. Cell Tiss. Res. 161, 1975, 263-275.

62. Lever, J.D., Santer, R.M., Lu, K.S. and Presley, R. Electron probe X-ray micro-analysis of small granulated cells in rat sympathetic ganglia after sequential aldehyde and dichromate treatment. J. Histochem. Cytochem. 25, 1977, 275-279.

63. Jessen, H., Peters, P.D. and Hall, T.A. Sulphur in different types of Keratohyaline granules. A quantitative assay by X-ray microanalysis. J.Cell Sci. 15, 1974, 359-377.

64. Jessen, H., Peters, P.D. and Hall, T.A. Sulphur in epidermal Keratohyalin granules: a quantitative assay by X-ray microanalysis. J. Cell Sci. 22, 1976, 161-171.

65. Morgan, A.J., Davies, T.W. and Erasmus, D.A. Specimen preparation. In: Electron Probe Microanalysis in Biology (Erasmus, D.A., ed). Chapman and Hall, London, 1978, 94-147.

Discussion with Reviewers

T. Bistricki: In spite of your justified criticism of conventional preparative procedures can you perhaps recommend some that are more suitable for analytical applications than others? Cryo-techniques are still far from being a routine.

Author: My personal view is that it is probably ill-advised to unreservedly "recommend" any single preparative procedure (including the various cryotechniques). However, within the limited context of a specific biological problem, one preparative procedure may be more suitable than others. For example, the cryosectioning (freeze-drying) regime is probably the technique of choice for the microprobe localization of diffusible substances in matrix-filled intra-cellular loci; whilst ions and electrolytes can only be validly localized in frozen-hydrated specimens. In contrast, wet chemical procedures are useful only where the investigated elements reside in relatively insoluble reaction products or endogenous states, and where a high degree of ultrastructural information is crucial to the study (e.g. text reference, 28). In future the so-called conventional wet chemical procedures may be increasingly replaced by improved freeze-substitution regimes. Freeze-substitution under rigorously maintained anhydrous conditions, and with an optional simultaneous fixation in >20% acrolein (Van Zyl et al., 1976- Micron 7, 213), offers the dual advantages of good, easily-interpretable structural preservation and of >95% total retention of diffusible electrolytes.

W.C. de Bruyn: The tables provided with this paper give situations where at least appreciable amounts of elements are preserved at the end of the procedure. I prefer the view that such elements in one way or another might occur as bound to the tissue and survive. Could the author comment on the idea of element binding to tissue structures.

Author: Apart from the well known fact that preparative fluids can extract significant proportions of some of the various tissue macro-molecules (text reference, 10), the data presented by De Filippis and Pallaghy (text reference, 13) illustrate how the 'bound fraction' can be seriously depleted during the preparative sequence from fixation to resin infiltration (Fig. 2) e.g. approximately 25% of the non-exchangeable ^{203}Hg fraction is lost gradually during the fixation, dehydration and embedding. Clearly, therefore, what survives at the end of wet chemical processing is often but an unknown proportion of the bound fraction.

C.M. Fenoglio: Are there any techniques that can be applied to the specimens prior to wet preparative techniques or cryo-preparative techniques which would help bind the material in question to a given site in the cell, thus reducing its subsequent loss by either type of procedure?

Author: The precipitation reactions discussed in the text are examples of ion immobilization, though their success in reducing element losses during wet preparative procedures is variable (Table 5). Histochemical reactions in combination with quantitative microprobe analysis may, however, enable us to measure bound and unbound element compartments. Chandler and Battersby (1976- J.Histochem. Cytochem.24, 740-748) hinted that by careful scrutiny of Ca: Sb ratios obtained by EMMA-4 analysis of pyroantimonate-treated sperm that they were able to qualitatively distinguish differences in the proportion of bound to 'free' calcium in various subcellular locations. In pursuit of this important principle, it may well be advantageous in the future to combine the technique of ion precipitation (bearing) in mind that the electron microprobe does not require that the reaction product be electron opaque, thus broadening the histochemical horizon) with cryosectioning, for example, where the possibility of loss and translocation of the reaction product is minimized

P. Echlin: Why was no mention made of the anhydrous sectioning procedures where sections are cut dry or onto troughs containing non-aqueous liquids?
Author: Text Fig 4 deals exclusively with the problem of element losses during section 'flotation' on a water bath. The discussion was also extended to include a summary of Agostini and Hasselbach's quantitative data on ^{45}Ca loss from fixed and unfixed specimens during various staining procedures (Table 2).
    Similar information may be obtained from Boothroyd (J.Cell Biol. 20, 1964, 165-173), Harvey et al. (J. Microsc. 107, 1976, 189-198) and Yarom et al. (Histochemistry 38, 1974, 143-153). Interestingly, Harvey et al. found that flotation on saturated $MgSO_4$ and Ca $(NO_3)_2$ solutions did not significantly reduce the major losses of ^{22}Na and ^{31}Cl from freeze-substituted plant tissues.

W.C. de Bruyn: I am not convinced that the present cryo-fixation procedures are adequate, though the elements are preserved inside the cells, dislocation still might occur during too "slow" freezing.
C.M. Fenoglio: It is obvious from the discussion that the author favours cryoprocedures of one sort or another over wet chemical techniques. Might it not also happen that shifts of intracellular materials can occur during the freezing process? What guarantees are there that highly mobile substances such as electrolytes or sugars do not transfer from one cellular compartment to another during the freezing procedure?
Author: Cryoprocedures (i.e. those which do not include intervention by fluid phases) are the best techniques at our disposal. Thus it is virtually impossible to establish by the analysis of specimens prepared by other methods whether translocation of diffusible substances occurs across distances which the microprobe can resolve? It may also be pertinent to consider whether changes occur in the superficial layers of the tissue (i.e. the 'useful', structurally well-preserved zone) during pre-freezing delays. Rapid freezing is an obvious necessity; equally important perhaps is a simple freezing arrangement that minimizes the pre-freezing interval.

C.K. Pallaghy: I am not entirely convinced from your presentation that there is a redistribution of calcium from the dense calcareous granules of the vitelline cell of Schistosoma mansoni to the vitelline droplets during osmication. Although the Ca peak in Fig. 5F looks higher than that in 5C, there is no indication whether this is not simply due to differences in section thickness since the background in Fig. 5F is markedly higher. It is also evident that Figures 5E and 5F were obtained under different conditions as these show a strong copper peak at 8keV. Variations in spot size during analysis can be critical. Have the differences in calcium levels been verified by background stripping techniques?

Author: Apart from the difference in grid material analytical parameters were identical for both specimens. It should be noted that the vitelline droplet in the cryosection yields a major potassium signal (with a Kα emission at 3.312 keV and a Kβ emission at 3.589 keV. The intensity of the Kβ emission is approximately 1% of the Kα, and cannot be resolved by EDS from the Ca (K) emission at 3.690 keV. Background stripping by a pre-programmed EDAX 707B/ MICROEDIT system (Edax Inc.) showed that there was no significant calcium emission in the cryosectioned vitelline droplets.

T. Bistricki: Occasionally one has need to analyze stained sections that have not been prepared specifically for microanalysis. Can you comment on the value of digital stripping techniques to resolve unwanted interference?
Author: My experience of digital stripping techniques is limited, but for a discussion see Russ (X-ray Spectrometry 6, 1977, 37). My response to the question, however, falls into two parts. Firstly one must be absolutely certain that the comprehensive wet chemical regime to which the specimens have been exposed has not seriously disturbed the chemistry of the structure of interest. Secondly, the signals from the tissue elements will inevitably be considerably less than the signals emanating from the heavy-metal stains. I question whether stripping techniques are of much value under such circumstances.

SCANNING ELECTRON MICROSCOPY/1979/II
SEM Inc., AMF O'Hare, IL 60666, USA

STANDARDS FOR X-RAY MICROANALYSIS OF BIOLOGICAL SPECIMENS

G. M. Roomans

Wenner-Gren Institute
University of Stockholm
Norrtullsgatan 16,
S-11345, Stockholm,
Sweden

Abstract

Quantitative X-ray microanalysis of biological specimens present some special problems :
one of these is the development of suitable
standards. It has been attempted to make standards that are comparable to the specimen with
regard to its chemical and physical properties,
so that correction procedures, which in biological microanalysis may introduce inaccuracies,
are minimized. In addition, mass loss under the
electron beam would be comparable in specimen
and standard. Much progress has been made in
preparing suitable standards during the last few
years. The present paper will discuss the choice
of standards, give a survey of the development
of standards for analysis of biological specimens
and discuss some problems related to the manufacturing of the standards.

For quantitative analysis of specimens, embedded in epoxy resin, a standard dissolved in
resin is a logical choice. To dissolve an element homogeneously into an epoxy resin, this element has to be part of a molecule with sufficient
organic properties to make it soluble in the
resin. For quantitative analysis of frozen-dried
or frozen-hydrated sections, often standards are
used that consist of mineral salts in an organic
matrix such as gelatin or albumin. The principal
difficulty is here that ice-crystal formation
during freezing has to be avoided.

Relatively few methods have been tested for
quantitative analysis of bulk specimens. Blocks
of frozen-dried or air-dried gelatin containing
mineral salts may be used. Also, crystals of
mineral or organic salts, e.g. of potassium,
have been used; with these standards, a ZAF-
correction may be necessary.

KEY WORDS: X-Ray Microanalysis, Biological
Specimens, Quantitative Analysis, Standards,
Organometallic, Epoxy Resin, Cryotechniques,
Microdroplets, Crystals, Bulk Specimens

Introduction

Recent developments in X-ray microanalysis
of biological specimens show a clear trend to a
more quantitative approach. Although there still
are some fields, for instance pathology, where
localization and identification of a xenobiotic
or otherwise unusual element may be the primary
goal, the biologist will no longer be satisfied
with qualitative data only. In physiological
studies of ion transport, reliable quantitation
is necessary to combine results obtained by X-ray
microanalysis with results obtained by other
techniques in a meaningful way [1].

An obvious requirement, though not the only
one, for reliable quantitation is a good standard.
In X-ray microanalysis of metallurgical specimens,
where the sum of all detectable elements adds up
to 100%, standardless quantitation may be carried
out since only the ratio in which the elements
occur has to be determined [2]. In addition, for
this type of specimen, the pure element, either
as bulk specimen or in the form of thin metal
foils, may be used as a standard. Preparing
standards for analysis of biological specimens
presents more difficulties, and this is one of
the factors that has delayed quantitation in
biological microanalysis. Spurr [3] has reviewed,
some years ago, the attempts that had been made
at preparing suitable standards for biological
X-ray microanalysis, and discussed the principles
underlying the preparation of such standards. In
subsequent work, it has been attempted to overcome some of the practical difficulties encountered in the early stages of quantitative biological microanalysis, and to extend existing
methods developed for a limited group of elements,
to include more elements. Various types of standards have now become available. In the present
paper, attention will be given to the general
principles governing the choice of standard. The
preparation of various types of commonly used
standards, and practical problems that may be
encountered, will be discussed.

Choice of Standard

The ultimate aim of quantitation in X-ray
microanalysis is to obtain data on local in vivo
concentrations in a biological specimen. The
choice of standard should be seen in this perspective. The ideal standard should be chemically

well-defined and homogeneous at the level of resolution used. These requirements are self-evident and do not need any further discussion.

A more intricate point is, whether the standard should resemble the specimen in its chemical and physical properties, and if so, to what extent. Two aspects are of interest here: one associated with the 'correction factors' used in quantitative analysis, one associated with the mass loss of the specimen during irradiation by the electron beam.

In thick specimens, the relation between elemental concentration and observed X-ray intensity is basically non-linear, due, among other factors, to absorption of X-rays in the specimen. In thin sections, the peak-to-background ratio is generally used for quantitation[4] and this introduces a non-linearity since the continuum ratio is dependent on the composition of the specimen. The factors causing deviations from linearity may be negligible in some biological specimens under some conditions[5] but otherwise they have to be accounted for by the introduction of correction factors[4,6,7]. Considerable errors may be introduced by neglecting these corrections and assuming a linear relationship between peak intensity and elemental concentration for comparison of an unknown with a standard. The magnitude of the error depends on the concentration difference, and on the atomic number of the element under consideration. An example is given in fig. 1.

If the composition of the specimen differs markedly from that of the standard, the correction factors may become unproportionally important in the calculation and in view of the uncertainties in some of these correction factors[8,9] this is generally not desirable. Especially in the case of bulk specimens in which extensive absorption occurs, this problem may be serious.

During irradiation of the specimen by the electron beam, chemical bonds within the specimen are broken, which leads to the formation of small molecules like CO_2, CH_4 and C_2H_6 [10], that are subsequently lost from the specimen. Due to this selective mass loss of light elements the concentration of elements detectable by energydispersive X-ray microanalysis is increased during analysis. The extent to which mass loss occurs seems to be dependent on the chemical nature of the specimen. Shuman, Somlyo and Somlyo[11] reported a loss of about 45% with sucrose, compared with only 13% with albumin. Others have, however, reported higher values for mass loss from proteinaceous specimens [12-14]. It is not yet clear to which extent mass loss is reduced at lower specimen temperatures [11-13]. It is assumed, that if specimen and standard resemble each other closely in their chemical and physical properties, mass loss will occur to about the same extent in specimen and standard, and corrections for mass loss will cancel. Since an easy correction for mass loss seems, at least at this stage, not possible, this is an attractive assumption; although the assumption seems reasonable, it cannot yet be said that it is fully supported by experimental evidence.

It is also unknown, to what extent, or in what respect, standard and specimen may differ, without unacceptable differences in selective mass loss. Spurr [3] showed in an interesting experiment

that two types of standard which differed markedly in their chemical and physical properties gave different results on the same specimen. This observation, quite likely due to differences in selective mass loss, shows that the problem of selective mass loss is indeed serious enough for careful consideration.

As a general rule, therefore, it has been accepted that specimen and standard should resemble each other in their physical and chemical properties.

Standards for Thin Sections - Embedded Specimens

If quantitative analysis of sections of tissue embedded in epoxy resin is to be carried out, a standard dissolved in resin seems to be a logical choice. Mineral salts cannot be dissolved homogeneously into epoxy resins, only dispersed; elements to be dissolved in the resin must form part of a molecule with organic properties. In a systematic search for such a compound, a list of properties of the 'ideal compound' may be used.

This 'ideal compound' should be:
(1) an organic or organometallic compound containing the element(s) of interest
(2) easily soluble in the epoxy resin; this means that the compound should have a sufficiently large apolar part, while, on the other hand, a reasonably high concentration of the element in the resin should be attainable. The chance that a certain compound can be dissolved in epoxy resin can be judged from its solubility properties in organic solvents like chloroform and ether. The compound should be insoluble in water.
(3) stable during analysis under the electron beam; no reaction should occur that involves the formation of a volatile compound containing the element of interest, whereas loss of organic parts of the molecule containing C, H,N and O may be unavoidable.
(4) non-reactive towards the constituents of the resin mixture, since a chemical reaction may delay or inhibit polymerization.
(5) non-reactive towards other compounds with which it may come into contact, such as oxygen (air) and water
(6) it should have a sufficiently high boiling point (and a low vapour pressure) to avoid appreciable evaporation during the polymerization process.

Bearing these requirements in mind, one will find that the choice of compounds to prepare standards for alkali or earth alkali elements is rather limited. For transition metals there are much more possibilities and for the non-metallic elements in the periodic system it is usually not very difficult to find a suitable compound.

Alkaline and earth-alkaline metals may form complexes with - usually rather large - organic molecules, which contain, apart from carbon and hydrogen, also oxygen and/or nitrogen. These molecules form a 'cavity' in which the metal is 'hidden'; also the counterion forms part of the (neutral) complex. The use of macrocyclic polyethers (crown-ethers) [15] has been suggested by Spurr [3,16]. The use of Na- and K- complexes of one of these crown ethers, dicyclohexyl-18-crown-6,

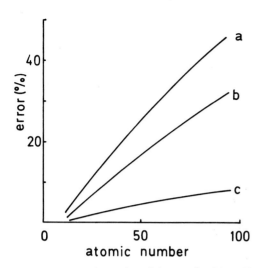

Figure 1. Errors introduced by neglecting the correction of the background intensity according to Hall [4,7] as a function of the atomic number. In the calculation an organic matrix was simulated that contained one element with atomic number higher than 11. In case (a) the standard contained 5% of this element, and the specimen 0.5%; in case (b) the standard contained 5% and the specimen 2.5% and in case (c) the standard contained 1% and the specimen 0.2%.

is now well documented [3,11,17,18]; homogeneous standards with a Na or K content of 200 meq/kg can be prepared. At higher concentrations problems are encountered [11], but although this is a limitation of these standards, especially for model studies, the concentration range that can be obtained seems to be sufficient for most biological problems. This type of standard is prepared as follows [16-18]: equimolar amounts of the crown ether and a suitable salt of the element of interest are dissolved in methanol, where complex formation takes place. The methanol is removed by heating the solution to 70°C and the complex, a honey-coloured viscous mass is dried at 70°C for some hours. The dried complex can then be dissolved in epoxy resin by stirring. Mixing the complex and the epoxy resin is made easier, if the resin mixture (without accelerator) is gently heated to make it less viscous. The accelerator is then added later. Not all salts give stable metal-crown ether complexes: cyanides and thiocyanides are recommended and also iodides and bromides can be used; cyanoferrates and permanganates give less stable complexes and cannot be recommended, chlorides and phosphates cannot be used at all [15]. Polymerization of the final mixture is reported to be slightly delayed, but otherwise without problems [16].

A possible alternative for crown ethers is the use of cryptates of alkali and earth alkali metals. Cryptates are complexes of alkali or earth alkali elements with 4,7,13,16,21,24 - hexaoxa - 1,10 - diazabicyclo - (8,8,8)-hexacosan, marketed as Kryptofix 222 (Merck, Darmstadt, German Federal Republic) or with related compunds [19,20]. Cryptates are somewhat more stable than crown ether complexes and also somewhat easier to prepare:

Kryptofix and a suitable salt of the element of interest (e.g. thiocyanide) are dissolved in chloroform/methanol by gentle heating. Hexane/benzene is added and the crystalline complex is precipitated and filtered. Preliminary experiments indicate similar solubility properties in epoxy resin as crown ether complexes; it seems, however, to be very difficult to prepare Mg-complexes.

Transition metals have a richer coordination chemistry, providing more convenient alternatives for the preparation of resin-soluble standards. Roomans and Van Gaal [21] tested a number of organometallic compounds containing transition metals for possible use as standards. Various groups of compounds were found to be potentially useful for this purpose. Among these were cyclopentadienylderivatives, acetylacetonates, and dithiocarbamates. Cyclopentadienylderivatives contain one or two cyclopentadienyl rings; as an example the structure of ferrocene (fig. 2) is given. Standards for Mn and Fe from commercially available compounds could easily be prepared (fig. 3) [21].

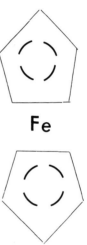

Figure 2. Structure of ferrocene ($Fe(C_5H_5)_2$).

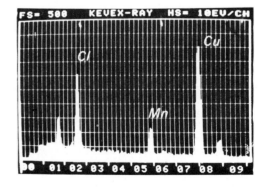

Figure 3. Spectrum of a Mn-containing standard (tricarbonylmethylcyclopentadienylmanganese, 1.25% Mn). For experimental details see the legend of table 1.

Nickelacetylacetonate was also easily soluble in epoxy resin, but other acetylacetonates (e.g. cobalt) did not always give good results. Dithiocarbamates are complexes of metals with the general formula $Me(R_2 dtc)_m$, where Me is the metal, R is an alkyl group (C_nH_{2n+1}), dtc is dithiocarbamate and m varies from 1 to 5. The structure of a dithiocarbamate complex with m = 2 is given as an example (fig. 4). Nearly all transition metals have been shown to form dithiocarbamate complexes; of some transition metals (Hf,Tc,Os) the dithiocarbamate chemistry is unexplored [22]. Dithiocarbamate complexes are not commercially available, but are not very difficult to make. The preparation is carried out in two steps: first, sodium di-N-n-butyl-dithiocarbamate is made by adding, under cooling with ice, 1 mol CS_2 to a stirred aqueous solution of NaOH and the secondary di-n-butylamine (1 mol of each). This solution can be used as it is for the preparation of the dithiocarbamate complexes: the required stoichiometric amount of metal salt (chloride) dissolved in water is added and the complex precipitates and is isolated by filtration. Recrystallisation can be carried out from various solvents (acetone,ethanol) and the complex can be dried in vacuum above P_2O_5. Generally, the longer the alkyl chain is, the better is the solubility of the complex in epoxy resin; for this reason butylanalogues are preferred to ethylanalogues[21].

Standards, using these organometallic compounds, can be easily prepared. Some compounds dissolve directly into epoxy resin, others dissolve rapidly upon gentle heating (for instance in a water bath). In this latter case, the accelerator was added after the organometallic compound had been dissolved. The compounds tested were more soluble in Spurr's resin [23] than in Epon 812, possibly due to the lower viscosity of Spurr's resin [23]. Polymerization of the blocks can be carried out overnight.

Not only transition metals, but also metals from groups IVb and Vb of the periodic system such as Pb and Bi may form dithiocarbamate complexes which can be used to prepare standards [21].

Standards for phosphorus can be prepared by dissolving triphenylphosphine in epoxy resin; the analogous As and Sb compounds can be used for the preparation of standards for these elements.

The dithiocarbamate complexes discussed here can also serve as a standard for sulfur. Another possibility is the use of the sulfur-containing flexibilizer DY 041 in Araldite as a standard [24]. However, the number of theoretically satisfactory sulfur-containing organic compounds is so large that preparation of standards for sulfur should not be a very serious problem. The same applies to the halogens: single or multiple halogen-substituted benzenes or benzene-derivatives are theoretically possible candidates. A standard for Br using 3,5-dibromo-4-hydroxybenzaldehyde was prepared (fig. 5). In a similar way, standards for Cl and I can be prepared.

From table 1 an impression of the homogeneity and reproducibility of this type of standard can be obtained. The standard deviation in the relative peak intensity is about 4-6%, which is only somewhat higher than the optimal standard devia-

tion calculated on theoretical grounds (the reasons for this discrepancy have been discussed by Shuman and coworkers [11]). The minimal detectable concentration (at 95% confidence level) can, for the metal standards be estimated at about 0.05-0.1% under the experimental conditions of table 1; the sensitivity can, however, be somewhat improved at higher count rates.

Calibration curves give a linear relationship between observed X-ray intensity and concentration [16,24]. The metal content of the standards may be independently determined by dissolving the resin in ethanol containing NaOH or KOH and analyzing the solution by atomic absorption spectrometry or flame spectrophotometry [11].

In conclusion, it can be said that in principle it is possible to prepare standards for (nearly) all elements using suitable organic or organometallic compounds and dissolving these in epoxy resin. This type of standard has the following advantages:
(1) the standards are easy to prepare; the most difficult step generally is the preparation of the organometallic complex.
(2) the standards are stable and can be kept for a very long time
(3) sections can be made in the same range of thickness as the specimen

Problems associated with this type of standards are:
(1) the maximal concentration of an element that can be dissolved into the resin seems to be about 1% (for metals) [11,21]; this may be sufficient for most biological problems but limits the use of these standards as model systems.
(2) the standard resembles only part of the specimen in its chemical and physical properties, namely the embedding medium, but not the tissue itself.

A type of standard that would theoretically overcome this latter difficulty was suggested by Ingram and coworkers [25]. Drops of a 20% albumin solution containing the elements of interest in a known concentration were shock-frozen, freeze-dried, osmium-fixed and embedded. However, practical problems are still reported with this type of specimen preparation [26].

Standards for Thin Sections - Cryosections

Since elements of eminent biological importance are lost during the conventional preparation procedure, cryotechniques are increasingly used in the preparation of biological specimens for X-ray microanalysis, especially if one is interested in diffusible elements such as Na, K, Cl, Mg and Ca. For this type of problem, dry sectioning of frozen specimens at very low temperatures has become the method of choice. The sections are either dried before analysis, or examined in the frozen-hydrated state. In accordance with the notion that the standard should resemble the specimen as much as possible, several groups have turned to standards which consist of mineral salts in an organic matrix.

In this case, the problem is not the choice of the salt to be dissolved, since most water-soluble salts can be used. Just as in the case of

Table 1.   Analysis of thin standards prepared by dissolving organic or organometallic compounds in epoxy resin.

exp	standard		R	SD	SD%
1	1.26% Mn	Mn	0.709	0.037	5.3
		Cl	2.065	0.109	5.3
		Mn/Cl	0.343	0.014	4.1
2	0.75% Fe	Fe	0.426	0.019	4.4
		Cl	2.033	0.087	4.3
		Fe/Cl	0.210	0.005	2.5
3	0.57% Ni	Ni	0.328	0.017	5.3
		Cl	2.019	0.107	5.3
		Ni/Cl	0.162	0.005	2.9
4	0.69% Cu	Cu	0.388	0.023	5.9
		Cl	2.060	0.119	5.8
		Cu/Cl	0.188	0.013	6.9
5	2.86% Br	Br	1.264	0.061	4.8
		Cl	0.897	0.049	5.4
		Br/Cl	1.410	0.059	4.2

Thin (100-150 nm) sections of the standards [21] were placed on copper grids (titanium grids in experiment 4) and coated with carbon layer. Analysis was carried out using the spot mode of the STEM at an accelerating voltage of 80 kV. Live counting time ranged from 200-400 s; the number of counts in the metal peak was about 800, and the number of background counts 1000-2000 (experiments 1-4); in experiment 5 the number of counts in the Br peak was about 1600, the number of background counts was about 1250. The relative peak intensity R was calculated according to Hall et al. [7]. Ten measurements (on five different sections) were carried out on each standard, and the standard deviation (SD) and its relative value (SD% = SD/R x 100%) were calculated. The table shows data for the metal and the Br peak, and for the chlorine peak due to the resin (Spurr's resin was used in all experiments). In addition, the ratio between the relative intensity of the 'element of interest' and the chlorine was determined. Experiment 5 is not comparable to the other experiments since a different peak width was used.

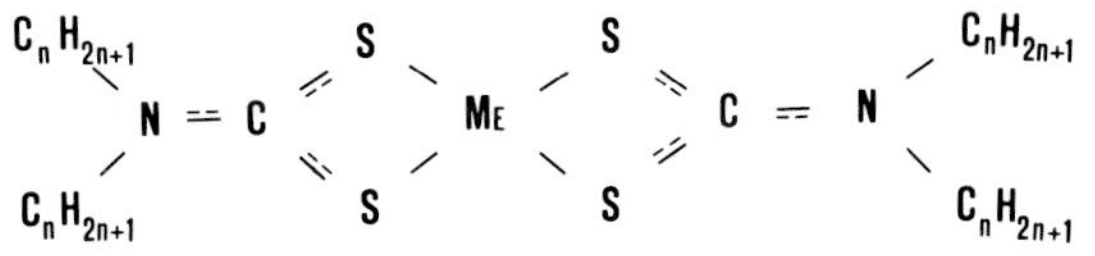

$$C_nH_{2n+1} \quad\quad S \quad\quad S \quad\quad C_nH_{2n+1}$$
$$N = C \quad\quad Me \quad\quad C = N$$
$$C_nH_{2n+1} \quad\quad S \quad\quad S \quad\quad C_nH_{2n+1}$$

Figure 4. Structure of a metal bis(dialkyldithio-carbamate) complex (Me = metal).

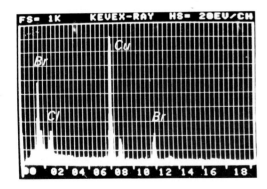

Figure 5. Spectrum of a Br-containing standard, 2.86% Br. For experimental details see the legend of table 1.

specimen preparation by cryofixation, the problem is how to avoid ice-crystal formation during freezing, recrystallisation and diffusion during sectioning and drying.

Various substances have been used as an organic matrix: gelatin [26-30] and albumin [26,31] are the most commonly used, generally in concentrations of 20-30%, resembling the conditions in tissue. One should be aware, that commercially available gelatin and albumin may contain impurities [5,26]; in that case it may be advisable to dialyze the matrix against distilled water [28,30]. Sometimes, cryoprotectants are added [30]. The standard is shock-frozen and sectioned at low temperature under the same conditions as the specimen. Roomans and Sevéus [30] reported good results with a standard consisting of 20% gelatin to which 5% glycerol was added as a cryoprotectant; in addition, mineral salts in known concentrations were added. Thin (100-150 nm) sections were cut at very low temperature (-100°C) and freeze-dried. An accuracy of better than 5% could be obtained with these standards.

Standards for various elements have been made by this method. Naturally, Na, K and Cl have been the elements most frequently introduced in this kind of standard, but the method is equally well suited for other elements like P, Ca and also heavier elements like Rb and Cs, of which high concentrations (up to 10%) can be accommodated [30]. Very homogeneous standards can be prepared mainly from the outermost part of the block, where ice-crystal formation is minimal [30]. Plots of peak intensity versus concentrations show (at least for light elements) a linear relationship [26,31]. For heavier elements corrections for background effects will have to be applied [29].

Although proteinaceous standards resemble

the specimen to a quite considerable extent, it is, of course, possible that mass loss is not exactly the same in a gelatin standard and in biological tissue. Hall and Peters [12] used kidney homogenate as an organic matrix for the standard. It remains to be seen, however, whether the advantage of the increased likeness is greater than the disadvantage of the inferior homogeneity [3].

In some cases, standard and specimen are joined. Gupta and coworkers [32] dissect the specimen in saline solution containing 10-20% dextran, and this solution, containing various elements in known concentration, is then used as an internal standard. Dörge and coworkers [31] cover the specimen before freezing with a layer of albumin standard solution of known ionic composition, which is preserved during freezing and freeze-drying and is sectioned with the specimen. Internal standards of this kind offer the advantage that specimen and standard can be analyzed under exactly the same conditions, without any need for specimen exchange. In addition, one is sure that specimen and standard have been subjected to exactly the same preparative procedure.

Freezing artifacts may be circumvented by using the protein standards suggested by Shuman and coworkers [11]: bovine serum albumin and phosvitin contain, respectively, covalently bound sulfur or phosphorus in known concentrations. Standards may be prepared by either dusting bare copper grids with the crystals or by placing a drop of 3% aqueous protein solution on a carbon foil covered grid and then drying at 60°C for 1 h.

Microdroplets

X-ray microanalysis offers a unique tool for the determination of the elemental composition of extremely small volumes of fluid. A considerable number of publications in which this technique is used is now available. Specimen and standard droplets are deposited from a specially constructed micropipette onto a support and either freeze-dried [33,34] or heated [35,36]. The need to know the droplet volume accurately may be circumvented by adding an element in a known amount to both specimen and standard, so that the analyzed quantity can be compared [36].

The Ratio Model and Quantitative Analysis of Thin

Specimens

It is possible to circumvent the need to have standards for all elements present in the specimen by using the so-called 'ratio model'. The intensity ratio of two elements (I_1/I_2) in a thin specimen in which no absorption occurs, may be converted to the concentration ratio (C_1/C_2) by:

$$C_1/C_2 = p \ (I_1/I_2)$$

where p is a proportionality constant. There is a theoretical approach to determine p [37], but often p is determined experimentally for a given accelerating voltage and for a given spectrometer configuration by measuring the relative intensities obtained from thin specimens with known com-

position. From the resulting ratios between elements or pairs of elements a complete 'efficiency curve' may be constructed [11,17,38]. It should be noted that new efficiency curves have to be made for any change in instrumental parameters. Theoretically, the use of the ratio method should not affect the accuracy of the quantitative determination. It is, however, recommended to restrict the use of the ratio model to elements which do not differ too much in atomic number. In addition it has been found that the ratio model may be less accurate in the case of low energy X-rays.

Bulk Specimens

Bulk specimens are less commonly used in quantitative microanalysis of biological specimens, mainly because of their inferior spatial resolution. Relatively few methods have been tested for quantitative analysis of bulk specimens. Three types of standards for use with biological bulk specimens will be shortly discussed: (1) frozen-dried or air-dried gelatin blocks, (2) crystals of mineral or organic salts and (3) thick sections of epoxy resin.

Frozen-dried [39] or air-dried [5] gelatin blocks containing the elements of interest (either as mineral or as organic salts) have been used in conjunction with frozen-dried specimens. Such standards resemble the specimen closely, and omission of ZAF-correction procedures is often permitted. In frozen-dried blocks, homogeneity may be impaired; even if cryoprotectants are used, ice-crystal formation will take place at some μm depth. On the other hand, since spatial resolution is less important in bulk specimens, the demands on homogeneity are not quite as high as for thin sections. Instead of gelatin, also other types of organic matrix may be used [40].

Crystals of mineral and organic salts are very easy to prepare and very homogeneous [39,41]. On the other hand, they differ considerably from the biological specimen. It was found that, especially in the case of those crystalline standards that contained, in addition to the elements C, H, N, O and K, one other element of atomic number higher than 10, a ZAF-correction was necessary (fig. 6) [39].

The use of organic and organometallic compound dissolved in epoxy resin has been extended to use with bulk specimens, although this type of standard was originally designed for use with thin sections. A ZAF-correction may sometimes be necessary, but is usually small. Wroblewski and coworkers [42] give a method to prepare a carbon specimen holder containing standards for several elements which may be used repeatedly. Several holes are drilled in blocks of pure carbon; the holes are filled with epoxy resin containing an element of interest in a known concentration (e. g. a solution of a dithiocarbamate complex in resin). After polymerization, plates are cut from the blocks and used as specimen holders, on which the specimen (in this case thick sections [42]) may be placed in between the standards.

Of these three types of standard, the gelatin standard has the advantage that it resembles

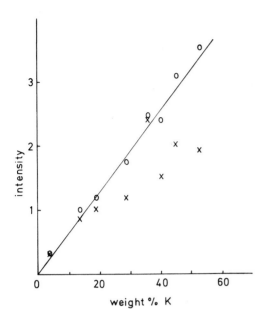

Figure 6. Plot of the measured intensity of the K K_α line against the known concentration of K in the standard. The standards are, (in order of ascending K concentration) a gelatin standard, sodium-potassium tartrate, potassium biphthalate, potassium biphosphate, potassium hexacyanoferrate, potassium thiocyanide, potassium sulfate, potassium carbonate. The spectra were determined at 25 kV [39]. (x) before ZAF-correction, (o) after ZAF-correction. Only after ZAF-correction a linear regression line through the origin can be obtained.

the specimen to a much greater extent. Corrections for mass loss and ZAF-corrections generally will be negligible. It is not clear, however, whether this advantage completely outweighs the disadvantage of inferior homogeneity, and more work remains to be done.

Discussion

For nearly all elements, standards can now be made that meet many of the requirements for a correct and reliable standard. The standards developed for biological microanalysis are not yet perfect in the sense that they do not add at all to the total accuracy of the quantitative determination. On the other hand, one can no longer say that the preparation of standards is the major bottleneck in quantitative biological microanalysis. Other problems, residing in instrumental factors, such as excessive external background, and in spectrum handling [11] are certainly of equal, if not of more, importance.

The choice between various types of standards will, as we have seen, mainly depend on the kind of specimen that is analyzed. Arguments based on convenience and ease of preparation are as yet of secondary importance.

Acknowledgements

I would like to thank Dr. A. Boekestein (Department of Submicroscopical Morphology, University of Nijmegen, The Netherlands) for permission to report unpublished observations on the use of cryptates, and Dr. H.L.M. Van Gaal (Department of Inorganic Chemistry, University of Nijmegen) for valuable discussion. Financial support was obtained from the Netherlands Organization for the Advancement of Pure Research (Z.W.O.) through the Netherlands Foundation for Biophysics, during the time I was working at the University of Nijmegen, and from the Swedish Natural Science Research Council.

References

1. B.L. Gupta, T.A. Hall and R.B. Moreton "Electron probe X-ray microanalysis" in Transport of ions and water in animals, Academic Press, London, 1977, 83-143.
2. M.J. Nasir "X-ray analysis without the need for standards", J. Microsc. 108, 1976, 79-87.
3. A.R. Spurr "Choice and preparation of standards for X-ray microanalysis of biological materials with special reference to macrocyclic polyethercomplexes", J. Microsc. Biol. Cell. 22, 1975, 287-302.
4. T.A. Hall "The microprobe assay of chemical elements" in Physical techniques in biochemical research, vol. 1A, Academic Press, New York, 157-275.
5. A.T. Sumner "Quantitation in biological X-ray microanalysis with particular reference to histochemistry", J. Microsc. 114, 1978, 19-30.
6. P.M. Martin and D.M. Poole "Electron probe microanalysis: the relation between intensity ratio and concentration" Metallurgical Reviews 150, 1971, 19-46.
7. T.A. Hall, H. Clarke Anderson and T. Appleton "The use of thin specimens for X-ray microanalysis in biology", J. Microsc. 99, 1973, 177-182.
8. S.J.B. Reed "Electron microprobe analysis", Cambridge University Press, Cambridge, 1975, 240-260.
9. A.G.S. Janossy and D. Neumann "Quantitative X-ray microanalysis: microcrystal standards and excessive background", Micron 7, 1976, 225-229.
10. G.F. Bahr, F.B. Johnson and E. Zeitler "The elementary composition of organic objects after electron irradiation", Lab. Invest. 14, 1965, 1115-1132.
11. H. Shuman, A.V. Somlyo and A.P. Somlyo "Quantitative electron probe microanalysis of biological thin sections: methods and validity", Ultramicroscopy 1, 1976, 317-339.
12. T.A. Hall and P.D. Peters "Quantitative analysis of thin sections and the choice of standard" in Microprobe analysis as applied to cells and tissues, Academic Press, London, 1974, 229-238.
13. J.M. Tormey "Validation of methods for quantitative X-ray analysis of electrolytes using frozen sections of erythrocytes" SEM/1978/II, SEM Inc., AMF O'Hare, IL, 60666, 259-266.

14. T.A. Hall and B.L. Gupta "Measurement of mass loss in biological specimens under an electron beam" in Microprobe analysis as applied to cells and tissues, Academic Press, London, 1974, 147-158.

15. C.J. Pedersen "Cyclic polyethers and their complexes with metal salts", J. Am. Chem. Soc. 89, 1967, 7017-7036.

16. A.R. Spurr "Macrocyclic polyether complexes with alkali elements in epoxy resin as standards for X-ray analysis of biological tissue", in Microprobe analysis as applied to cells and tissues, Academic Press, London, 1974, 213-228.

17. J.A. Chandler "A method for preparing absolute standards for quantitative calibration and measurement of section thickness with X-ray microanalysis of biological ultrathin specimens in EMMA", J. Microsc. 106, 1976, 291-303.

18. J.A. Chandler "X-ray microanalysis in the electron microscope" in Practical methods in electron microscopy, vol. 5, North-Holland Publishing Company, Amsterdam, 1977, 317-547.

19. B. Dietrich, J.M. Lehn and J.P. Sauvage "Diaza-polyoxa-macrocycles et macrobicycles", Tetrahedron Letters 34, 1969, 2885-2888.

20. B. Dietrich, J.M. Lehn and J.P. Sauvage "Les Cryptates", Tetrahedron Letters 34, 1969, 2889-2892.

21. G.M. Roomans and H.L.M. Van Gaal "Organometallic and organometalloid compounds as standards for microprobe analysis of epoxy resin embedded in tissue", J. Microsc. 109, 1977, 235-240.

22. J. Willemse, J.A. Cras, J.J. Steggerda and C.P. Keyzers "Dithiocarbamates of transition group elements in unusual oxidation states", Structure and Bonding 10, 1976, 83-126.

23. A.R. Spurr "A low viscosity epoxy resin embedding medium for electron microscopy", J. Ultrastruct. Res. 26, 1969, 31-43.

24. H. Jessen, P.D. Peters and T.A. Hall "Sulphur in different types of keratohyalin granules: a quantitative assay by X-ray microanalysis", J. Cell Sci. 15, 1974, 359-377.

25. F.D. Ingram, M.J. Ingram and C.A.M. Hogben "An analysis of the freeze-dried, plastic embedded electron probe specimen preparation" in Microprobe analysis as applied to cells and tissues, Academic Press, London, 1974, 119-146.

26. B.F. Trump, I.K. Berezesky, R.E. Pendergrass "X-ray microanalysis of diffusible elements in scanning electron microscopy of biological thin sections. Studies of pathologically altered cells", SEM/1978/II, SEM Inc., AMF O'Hare, IL 60666, 1027-1039.

27. G.M. Lehrer and C. Berkeley "Standards for electron probe microanalysis of biological specimens", J. Histochem. Cytochem. 20, 1972, 710-715.

28. B.L. Nichols, H.A. Soriano, D.J. Sachen "Electron probe localization of electrolytes in immature muscle", Hopkins Med. J. 35, 1974, 322-335.

29. G.M. Roomans and L.A. Sevéus "Subcellular localization of diffusible ions in the yeast Saccharomyces cerevisiae: quantitative microprobe analysis of thin freeze-dried sections", J. Cell Sci. 21, 1976, 119-127.

30. G.M. Roomans and L.A. Sevéus "Preparation of thin cryosectioned standards for quantitative microprobe analysis", J. Submicr. Cytol. 9, 1977, 31-35.

31. A. Dörge, R. Rick, K. Gehring and K. Thurau "Preparation of freeze-dried cryosections for quantitative X-ray microanalysis of electrolytes in biological soft tissues", Pflügers Arch. 373, 1978, 85-97.

32. B.L. Gupta, T.A. Hall, S.H.P. Maddrell and R.B. Moreton "Distribution of ions in a fluid transporting epithelium by electron probe X-ray microanalysis", Nature 264, 1976, 284-287.

33. C. Lechene "Electron probe microanalysis of picoliter liquid samples" in Microprobe analysis as applied to cells and tissues, Academic Press, London, 1977, 351-368.

34. K. Baumann, C. De Rouffignac, N. Roinel "Renal phosphate transport: inhomogeneity of local proximal transport rates and sodium dependence", Pflügers Arch. 356, 1975, 287-292.

35. R. Beeuwkes III, J.M. Amberg and L. Essandoh "Urea measurement by X-ray microanalysis in 50 picoliter specimens", Kidney Intern. 12, 1977, 438-442.

36. A.J. Morgan, T.W. Davies and D.A. Erasmus "Analysis of droplets from iso-atomic solutions as a means of calibrating a transmission electron analytical microscope (TEAM)", J. Microsc. 104, 1975, 271-280.

37. J.C. Russ "X-ray microanalysis in the biological sciences", J. Submicr. Cytol. 6, 1974, 55-79.

38. G.M. Roomans, B.A. Afzelius, H. Kollberg and B. Forslind "Electrolytes in nails analyzed by X-ray microanalysis in electron microscopy. Considerations on a new method for the diagnosis of cystic fibrosis", Acta Paediatr. Scand. 67, 1978, 89-94.

39. G.M. Roomans and A. Boekestein "Distribution of ions in Neurospora crassa determined by electron microprobe analysis", Protoplasma 95, 1978, 385-392.

40. V.S. Sottiurai, R.L. Malvin, L.F. Allard and W.C. Bigelow "Biological standard for electron microprobe analysis of intracellular sodium concentration", J. Histochem. Cytochem. 24, 1976, 749-751.

41. I. Zs. Nagy, C. Giuli "Energy-dispersive X-ray microanalysis of the electrolytes in biological bulk specimen", J. Ultrastruct. Res. 58, 1977, 22-33.

42. R. Wroblewski, G.M. Roomans, E. Jansson and L. Edström "Electron probe X-ray microanalysis of human muscle biopsies", Histochemistry 55, 1978, 281-292.

Discussion with Reviewers

A.R. Spurr: In addition to the selective mass loss of small molecules like CO_2, CH_4 and C_2H_6 isn't it possible that elements like Na, K and even heavier elements would also be lost due to beam damage?

Author: Also other elements than C, H, N and O can be lost from the specimen. Especially Cl and other halogens appear to be easily removed if high beam currents are used; this is the type of radiation damage most extensively described in the literature. It cannot be excluded, that under some experimental conditions, i.e. extremely bad thermal conductivity of the specimen, also other elements may be removed.

A.R. Spurr: In listing the properties of an ideal compound for addition to the resin in the preparation of a standard, wouldn't one expect that if the organic part of the molecule is lost because of the action of the beam, the element of interest would also be disrupted or volatilized and possibly redeposited elsewhere on the specimen?
Author: In many cases, loss of small organic molecules from the organic part of the organometallic complex will not necessarily lead to loss of the metal from the standard. In the case of nonmetal elements, the risk of loss is, for various reasons, greater. Since much depends on the experimental conditions, it is hardly possible to give a general rule.

T. Jalanti: Organometallic compounds are sometimes used in activators (for instance cobalt naphthenate for polyester resins). Have you noticed any modification of the hardening properties and of the sectionability of the epoxy resins after introduction of organometallic compounds?
Author: In my experience, the standards are easy to section and no obvious differences in sectionability have been noticed.

H.K. Hagler: What results has the author obtained with the preparation of an epoxy resin embedded calcium standard which could be used for the quantification of biological and pathological calcific deposits observed in epoxy embedded thin sections?
Author: Epoxy resin embedded calcium standards can be made by dissolving Ca-crown ether-complexes into resin. The maximal concentration of calcium that can be reached (if one wishes to obtain homogeneous standards) is, however, rather low - in my experience about 0.5%. This might be considerably lower than local calcium concentrations found in calcific deposits. Sectioned microcrystal standards (text references 7 and 9) might be a good alternative for this special case. Use of cryosectioned specimens or standards (instead of epoxy embedded) should, also for other reasons, be seriously considered.

H.K. Hagler: When the peak to continuum measurements of Hall are used for quantitation, how are local instrument and grid bar contributions to the continuum measured and taken into account before a final concentration is computed?
Author: Hall (text reference 7) gives detailed instructions how to proceed. The beam is placed onto an empty part of the grid, in a position relative to the grid bars corresponding to that where the measurement on the section is carried out. The thus obtained background intensity is subtracted from the total background generated by specimen plus grid and instrument. The peak intensity of the elements of interest is thus related to the background generated by the specimen only. Janossy and Neumann (text reference 9) have pointed out that there is a risk for underestimation of the contribution of the grid bars due to differences in lateral spreading of the electron beam. It can be calculated, however, that

this will often have only a very minor effect on the calculation. It is clear that the overall accuracy of the method is improved by reducing the background due to instrument and grid.

Additional discussion with reviewers of the paper "Quantitative Methods in Biological X-Ray Micro-analysis" by N.C. Barbi, continued from page 672.

J.C. Russ: The error due to ignoring absorption should be calculated before using large particle standards or thick sections. Would the author please show the expected absorption (and error if it is neglected) for the 1μm sulfate particles used as standards? Also the extension of "thin" to cover "1 or even 5μm thick" organic sections should be critically examined. Please show the expected absorption for representative elements such as Na, P, K, etc., for such sections, and again if they have been stained or fixed with heavy elements. Note that in a Hall calculation the background is usually measured at a higher energy, so any argument that the background absorption is the same and cancels cannot be invoked.

Author: I agree that the error due to ignoring absorption should be estimated before application of the thin film assumption. It was cautioned in the text that for the sulfates of Na, Mg and Al, an effort should be made to stay near the 0.1μm particle size range when using the described technique, in order to obtain proper thin section results. It was also recognized, however, that such particles are usually difficult to locate in the SEM. For the sulfates of the heavier metals, particles up to several μm's are acceptable.

Using Beer's law for x-ray transmission, one can approximate the particle size necessary to limit absorption to some reasonable value, e.g., 10% (thereby accepting a 10% error in the results obtained):

$$\frac{I}{I_o} = e^{-\mu_m \rho x} = 0.9$$

where
$\frac{I}{I_o}$ = fractional x-ray transmission through a material of density (ρ) and thickness (x)

μ_m = mass absorption coefficient

Applying the above equation, and assuming that the x-ray is generated in the center of the particle, the particle diameter which would give 10% absorption is given by:

$$(x)_{10\%} = \frac{2.1 \times 10^3}{\mu_m \rho} \ (\mu m)$$

The particle sizes that are then "acceptable" for the various sulfates are as follows:

SULFATE	$(x)_{10\%}$ (μm)
Na	0.5
Mg	0.6
Al	0.9
Ca	4
Cu	13

For 1μm particles (path length = 0.5μm), one calculates ≈10% absorption for Na and ≈1% absorption for Cu.

For the cases of 1μm and 5μm carbon films, the expected absorption for selected elements is given below:

Absorption (%)

Element	1μm Film	5μm Film
Na	7.5	32
Mg	4.5	20
P	1	6
Ca	-	1

If heavy elements are present from fixation or staining, the absorption will significantly increase, the degree of which depends on the concentration of these elements.

The usefulness of the above exercise, I think, has been to demonstrate that "thinness" should not be hastily presumed. The same sample can be thin for some elements, and not thin for others. Heavy metal additions to the tissue will degrade the thinness assumption severely.

One other point could be raised concerning the thinness of samples in the 1μm and greater range. As the reviewer noted (text reference 23), the sample may be thin from an absorption consideration, as calculated by the above method, but perhaps not from an electron stopping power consideration.

I.K. Berezesky: In your experiments to determine thin section relative pure element intensities, was any other accelerating voltage used besides 15 kV?

Author: Yes, 25 kV was also used, with comparable results. Goldstein, et al (text reference 7) show similar curves theoretically calculated for several voltages (up to 200 kV), with some measured data for comparison. However, in this reference the inverse of the intensity is plotted as a "k-value" rather than the intensity directly.

C.E. Fiori: Is there a documented computer program which uses primarily the "Hall method" with its various adaptations. If so, where can this program be obtained, which computer language is it in and how big is it?

Author: Most x-ray system manufacturers offer a Hall-type program for use on their system. Our program, for example, is written in BASIC, but several assembly language sub-routines are called by the program to perform such tasks as digital filtering and least squares fitting.

I am not aware of a generally available documented program written in a high level language.

SCANNING ELECTRON MICROSCOPY/1979/II
SEM Inc., AMF O'Hare, IL 60666, USA

QUANTITATIVE METHODS IN BIOLOGICAL X-RAY MICROANALYSIS

N. C. Barbi

Princeton Gamma Tech
P.O. Box 641
Princeton, NJ 08540

Abstract

Quantitative methods in biological x-ray microanalysis have been developed primarily for thin sections for the following reasons: 1) x-ray intensity and concentration are linearly related in thin sections, thereby simplifying the quantitative algorithms compared to those required for thick samples; and 2) spatial resolution is vastly improved in thin sections compared to that attainable in the bulk material.

The method developed by Hall and co-workers relates the mass fraction of an element in the sample to the ratio between the characteristic peak intensity and the sample-generated continuum. Local changes in mass thickness of the section are accounted for in the ratio and need not be known explicitly. Standards can be either biological "type" standards or thin stoichiometric compounds or minerals. The non-sample background must either be negligible or subtracted from the total continuum intensity for proper application of the technique.

The elemental ratio method determines the concentration ratio between elements in the sample from their intensity ratio, and is therefore also insensitive to local mass thickness changes in the section. The proportionality constant is determined by measurement, calculation, or a combination of both.

Quantitative techniques have also been developed for other specific sample configurations, such as thin sections on infinitely thick substrates and samples prepared from the liquid. However, quantitative techniques in bulk biological samples have not been thoroughly developed because of poor x-ray spatial resolution in such samples and difficulties in applying a conventional ZAF correction procedure in the light matrix.

KEY WORDS: Biological Specimens, X-Ray Microanalysis, Quantitative Analysis, Energy Dispersive X-Ray Spectroscopy, Solid State Detectors

Introduction

Electron beam x-ray microanalysis, now routinely applied by the physical scientist to the investigation of microstructure and microprocesses in materials, is beginning to offer similar promise to the biologist. Cautious acceptance of x-ray microanalysis in the life sciences has occurred because of 1) problems in sample preparation, generally associated with the transition from the living state to vacuum; 2) poor x-ray spatial resolution in thick samples compared to that achieved in materials due to the low atomic number matrix of the biological specimen; and 3) specimen damage or alteration due to electron beam effects. Over recent years, however, techniques to solve, alleviate, or circumvent these difficulties have been sufficiently developed to encourage application and stimulate an accelerated rate of development of the analytical method.

The most prominent difficulty to overcome before meaningful x-ray microanalysis can become routine, particularly for the mobile ions such as Na, K, and Ca, is that of sample preparation. Conventional preparations involving chemical fixation, dehydration, embedding and sectioning result in the loss or redistribution of these elements. Their location and concentration no longer reflect the living state. Furthermore, the presence of heavy metals introduced in fixation or staining, such as osmium, lead or uranium, interfere with the analysis of the lighter elements by virtue of the absorption of the lower energy x-rays or by a direct overlapping of their lines with the less intense lines of the lighter elements. In order to preserve the location of the diffusible elements, cryogenic preparation techniques must be employed. Coleman[1] and Echlin and Sauberman[2] have recently reviewed sample preparation problems and techniques suitable for microanalysis.

Symbol Table

d : Bulk sample quantitative spatial resolution

E_o : Electron beam accelerating voltage, keV

E_c : Critical excitation energy for x-ray, keV

ρ : Density (g/cm^3)

b : Thin section spatial resolution, cm

Z : Atomic number

A : Atomic weight

t : Specimen thickness, cm

C_i : Concentration of element i in the sample

C_i^o : Concentration of element i in the standard

I_T : Total x-ray intensity, peak + background

I_B : Background intensity

n : Number of x-ray measurements

k : k-ratio = I_i/I_i^o

W_i : Window intensity for peak of element i

W_c : Window intensity for continuum measurement

F : Overlap factor

k´ : Proportionality constant

I_i : Net peak intensity

I_c : Continuum intensity from sample

I^o : Intensity from standard

I^t : Intensity from thin sample of thickness t

I^{ot} : Intensity in thin standard of thickness t

n_i : Number of atoms of element i in chemical formula

$\overline{\dfrac{Z^2}{A}}$: Average Z^2/A for matrix

q_i : (I_i/I_c) specimen/(I_i/I_c) standard

G_i : $n_i/\Sigma(n_i/Z_i^2)$ standard

C_i' : $C_i/\sum_m C_i$

m : As subscript, refers to matrix elements

u : As subscript, refers to non-matrix elements

Q : Ionization cross section

ω : Fluorescence yield

R : Relative line intensity (e.g., $\alpha/\alpha+\beta$)

T : Detector efficiency

o : As superscript, refers to standard

The problems associated with electron beam damage are of concern in any material, but are more pronounced in biological specimens or other materials which exhibit low thermal and electrical conductivity. Further problems arise because mobile ions can migrate, and organic matter can actually be vaporized under the influence of the electron beam. Hall described observations of mass loss in biological samples due to the electron beam[3], Echlin discussed other associated problems and coating techniques employed to alleviate them[4] and Shuman, Somlyo and Somlyo discussed the implications of radiation damage on minimum detectability limits and spatial resolution[5].

This review will focus on the techniques for quantitation applicable to biological specimens. A properly prepared sample and minimal electron beam effects will be presumed, assumptions which must be critically evaluated in practice. The remaining problem noted, that of spatial resolution, will be discussed in detail, since the analyzed volume is intimately bound to the quantitative result.

Spatial Resolution

If a measure of the concentration of an element is to be given as a microanalysis result, then the volume and location over which that concentration applies must be specified. The "diameter" of the analytical volume will always exceed the beam diameter due to the scattering and diffusion of electrons within the sample, a phenomenon often called "beam spreading". For samples which are infinitely thick (with respect to electron beam penetration), the analytical volume has been estimated by Reed and others. Reed proposed the following x-ray spatial resolution, d, defined as the diameter of a sphere which would contain 99% of the x-ray production[6]:

$$d = 0.231 \, (E_o^{1.5} - E_c^{1.5})/\rho \qquad (1)$$

For typical soft tissue, $\rho = 1g/cm^3$, and the quantitative spatial resolution for Ca or K at 15 kV would exceed 10μm. At 20 kV, the diameter would be near 20μm. These analytical volumes are quite large relative to the size of most structural features of interest in the sample and may, therefore, render the analysis useless.

The only means to improve the spatial resolution achieved for a given x-ray line in a given sample of infinite thickness, according to Equation (1), is to lower the accelerating voltage. This can be done effectively only to a point, since eventually, the overvoltage (E_o/E_c) becomes too low to efficiently excite the line of interest.

In thin sections, however, Equation (1) does not apply. If one examines in more detail the shape of the analytical volume for low atomic number materials, it is shown to be "tear drop" shaped rather than spherical. If one now assumes a section thickness less than the length of the neck in the tear drop, it is easy to visualize the improvement in spatial resolution (Figure 1). The section is thinner than the depth at which the beam would spread significantly in the sample.

Equation (2) has been proposed by Goldstein, Costley, Lorimer and Reed to describe the amount of beam broadening as a function of accelerating voltage and sample density and thickness[7].

$$b = 6.25 \times 10^2 \frac{Z}{E_o} \left(\frac{\rho}{A}\right)^{1/2} (t)^{3/2} \qquad (2)$$

Equation (2) predicts that at 100 kV and a specimen thickness of 500nm, the expected broadening is only 50nm. Thinner sections or higher voltages would reduce the broadening still further.

Unfortunately, in beam sensitive materials, one must be mindful of the maximum radiation dose permitted if sample integrity is to be retained. To obtain high spatial resolution with good minimum detectability limits, a finely focused intense probe is necessary. However, this condition imparts a large radiation dose to the sample (in number of electrons per unit area). If the dose is too large, mass loss can occur and quantitation is defeated. To reduce the dose, the beam diameter must be increased, with corresponding degradation in spatial resolution. According to Shuman et al, who have evaluated this problem in detail, the radiation damage limitation on spatial resolution for many biological materials appears to be quite severe[5]. Unless methods to reduce radiation damage are successfully developed, the promised resolutions of Equation (2) may not be realized.

Data Collection and Processing

Statistics of X-ray Analysis

A familiarity with the statistics of x-ray analysis is a pre-requisite to the performance of quantitative analysis. The subject has been concisely reviewed by Goldstein[8]. Counting statistics are of particular concern to biologists who are often dealing with low total counts imposed by a beam-sensitive sample (which limits counting time and permissible probe current) and low concentrations (which yield low count rates). Ziebold has shown that the analytical sensitivity for a 95% probability level can be approximated by:[9]

$$\Delta C = C_1 - C_2 \geq \frac{2.33 \, C_1 \sqrt{I_T}}{\sqrt{n} \, (I_T - I_B)} \qquad (3)$$

To obtain a general feeling for the value of ΔC, assume that the concentration (C_1) of an element at point 1 in the sample is 10%; the integrated peak to background ratio for the pure element is 50 to 1; the intensity and concentration are linearly related; and only one x-ray measurement is made on each analysis point (n = 1).

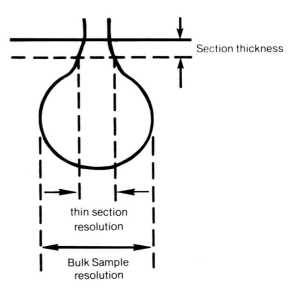

Figure 1: Schematic Comparison of Spatial Resolution Attainable in Bulk Samples and Thin Sections

For a total window intensity of 120 counts (Peak = 100, Background = 20):

$$\Delta C(\%) = \frac{23.3 \, (10.95)}{100} = 2.55$$

For a total window intensity of 1200 counts, $\Delta C = 0.74\%$; and for 12000 counts, $\Delta C = 0.26\%$.

The difference between the concentrations measured at two points in the sample is not significant unless that difference exceeds ΔC. For the case of 100 peak counts in the example, even a concentration of 7.5% cannot be considered a statistically valid difference in concentration from the 10% value measured at point 1. Since most quantitation techniques in biological samples are considered to be capable of 10-15% relative accuracy for this concentration level, the 1000 count statistics (7.4% relative) are just narrowly acceptable if 10% accuracy is to be realized. Note that the precision calculated by Equation (3) is based on counting statistics only and assumes that the background is accurately known; it therefore represents a theoretical best value. Errors in background subtraction, peak overlap correction, or instrumental or experimental contributions to the total precision of the analysis can only degrade the actual value obtained.

Intensity Determination

The following considerations will pertain to energy dispersive X-ray systems (EDS). The EDS technique is particularly suited to the low probe currents required for biological samples, since the efficiency of detection is much greater than the crystal spectrometer counterpart.

The measured intensity of an X-ray line is determined by integration of the counts in the central portion of the peak. In a modern EDS system, an X-ray line at 6 keV will be represented by a near-Gaussian peak with a FWHM of approximately 150 eV, comprised of several discrete channels of data. The characteristic X-ray line, which is emitted at a single energy, is therefore, spread over a broad range by the detecting system. In order to make optimum use of the accumulated data, the counts in the peak are integrated over a window or band of energy centered on the centroid of the peak, having a width equal to or slightly greater than the FWHM (1-1.5 FWHM). Since the width of a peak increases with energy, the number of channels comprising the window will increase accordingly in order to maintain a constant fraction of the peak width.

Background Subtraction

The integrated count in the peak, or window intensity, obtained from a collected spectrum includes background. The background in the electron-excited X-ray spectrum is primarily due to the continuum (white radiation or Bremsstrahlung) emitted as a result of the deceleration of electrons in the sample. The continuum is emitted over an energy range from 0 to the accelerating voltage and does not directly relate to the presence or concentration of a particular element. The background must be subtracted from the total window intensity in order to obtain the net intensity due to the characteristic line of the element of interest, the parameter which is related to concentration.

There are several methods for background subtraction or suppression currently in use for biological samples.

a) The Linear Method - A simple method, which can be applied without the aid of a computer, is to set windows on either side of the peak of interest and interpolate to obtain the background under the peak. If the background windows are not of the same width as that on the peak, the intensities must be normalized to the peak window (i.e., by determining the intensity per channel in the background window and multiplying by the number of channels comprising the peak window).

If a computer is available, the method is simplified by permitting the selection of background "points" (actually single channels). The background is then constructed by interpolation between these points over the whole spectrum (Figure 2) The accuracy of this method is generally good for X-ray lines greater than 3 keV, but is often poor in the 1 to 3 keV region, where the background shape is rapidly varying and absorption edges are lying underneath the peaks.

The linear technique is also subject to the short-coming that it requires a few points between 1 and 3 keV to approximate the shape of the background in this region.

If the selection of points here is prohibited by the presence of Na, Mg, Al, Si and P in the spectrum, then a much greater error in the estimation of background under these peaks will result.

b) The Background Spectrum Method - Another simple technique for background subtraction, which overcomes the point-selection problem noted above, is to collect a "background spectrum", which inherently contains the approximate background shape information. This spectrum can be collected from pure C or an organic resin containing no heavy elements which would give rise to measured X-ray peaks. The "background standard" should be of approximately the same thickness as the samples to be analyzed, and the spectrum should be collected at the same voltage and geometric conditions as the working samples. The background spectrum can then be scaled to the working spectrum and subtracted. The background spectrum method will provide reasonable results if the detectable elements in the tissue are relatively low in atomic number and of low concentration.

Figure 3 shows this result for a background spectrum obtained from bulk carbon at 15 kV and a spectrum from a Ti-bearing plastic. The background shapes match well because of the similar average atomic number of the two materials, and because the detector efficiency information is inherently contained in the background spectrum. This method can be used without a computer by setting a scaling window at a spectrum location that is usually free from peaks (a window centered at ≈3 keV is often suitable). The window intensities representing the background for the peaks of interest are then multiplied by the ratio of the scaling window intensities from the sample and background spectrum.

There are weaknesses inherent in either of these simple techniques described. However, the effect of the error can be minimized if the procedure is consistent from sample to sample. The errors will then generally be the same for all samples studied, and the noted trends in intensity will be valid.

c) Digital Filter Method - A more sophisticated method, which requires a computer and extensive software, incorporates a digital filter to effectively suppress background. It can be viewed as a mathematical band-pass filter, eliminating the slowly varying background and rapidly varying channel-to-channel statistical fluctuations, while retaining the peaks, which are of intermediate frequency. However, with this method the peaks become distorted in shape and cannot be directly integrated. The digital filter technique is generally used in conjunction with a standard spectrum, also filtered. The filtered spectrum from the standard is then

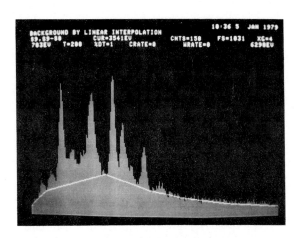

Figure 2: Linear Background Subtraction.
Spectrum from Rudimentary Kidney
in Salamander Embryo, 25 kV

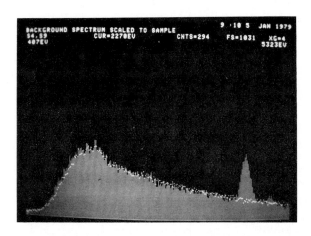

Figure 3: "Background Spectrum" Method
For Subtraction of Background

fit, usually by a least squares fitting
technique, to the filtered spectrum from
the sample, and a k-ratio is obtained.
This k-ratio is actually the fraction of the
standard spectrum that is required to match
the intensity (and shape) of the peak in
the unknown and provides an estimate of
the concentration in the sample:

$$C_i = k \, C^o_i \qquad (4)$$

Statham has described the digital filter
mathematics in detail[10] and Shuman, Somlyo
and Somlyo have described the fitting
mathematics in detail[11].

The filter and fit approach, although
it involves extensive software, has the
advantages that it requires no operator
intervention and includes an inherent peak
overlap correction, as discussed below.

Overlap Correction

Since the peaks in an X-ray energy
spectrum are quite broad, overlap of one
peak on another in some region of a complex
spectrum is likely. The intensity in a
channel shared by overlapping peaks is the
summation of the intensity in that channel
contributed by each of the individual peaks.
Therefore, a window intensity will not only
contain the counts from the desired peak,
but will contain additional counts from
the overlapping peaks as well. The task is
to separate a composite peak into its component parts, a process called spectrum
deconvolution.

a) The Window Method - An example of
peak overlap is given in Figure 4, which
shows the proximity of the K Kβ to the Ca
Kα peak. The K window intensity, determined
on the Kα peak, is unaffected by the overlap; the Ca Kα window, however, contains
the undesired contribution from the K Kβ.

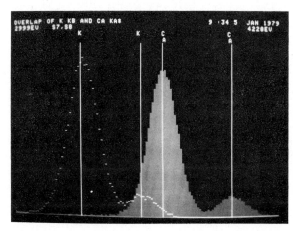

Figure 4: Display of K Spectrum Overlayed with Ca Spectrum

The Ca window intensity must be corrected
for the K Kβ overlap before that intensity
can be used in quantitation. The problem
is most pronounced when a small amount of
Ca is present with a relatively high concentration of K. A technique to determine
the degree of overlap, or overlap factor,
for K on Ca without the aid of a computer
is as follows:

1. Set the K and Ca windows as previously described. It is best to initially
set the windows on "standard" spectra which
contain no overlap (e.g., K_2SO_4 for potassium
and $CaSO_4$ for calcium).

2. On the K standard spectrum, determine the ratio of the counts in the Ca window
to the counts in the K window $(W_{Ca}/W_K)_K$.
This quantity should be determined after
background subtraction, and is the overlap
factor, F, to subsequently be used in

correcting the Ca window intensity from the sample for K Kβ overlap.

    3. Subtract background from the spectrum obtained from the sample. The intensity values to be used in quantitation are:

$$I_K = W_K$$

$$I_{Ca} = W_{Ca} - (F \cdot W_K) \qquad (5)$$

The deconvolution process described above is illustrated on a K-Ca overlap in a silicate glass. Figure 5a shows the K spectrum (of Figure 4) after background subtraction, having used the linear interpolation method of Figure 2. The K and Ca windows and their intensities are displayed. The overlap factor for K on Ca, determined from the K standard spectrum, is thus $W_{Ca}/W_K = 3828/51652 = 0.075$. Figure 5b is the glass spectrum, after background subtraction, displaying the K and Ca window intensities. The corrected Ca intensity, following Equation (5) is thus, $6052 - (0.075 \cdot 10831) = 5240$.

To further illustrate overlap and its correction, Figure 5c shows the K standard spectrum scaled by computer to the K Kα in the glass spectrum. The "excess" counts at the K Kβ location are due to Ca. Figure 5d shows the net spectrum after subtracting the scaled K spectrum from the glass. The Ca window (5248) is essentially the same result as previously calculated for the corrected Ca intensity, but utilizing the computer to perform the arithmetic.

b) Computer Least Squares Fitting - Least squares fitting of reference spectra, whether used in conjunction with the digital filter or applied directly to background subtracted spectra, performs an accurate overlap correction in most practical cases. Assume that a standard spectrum exists for each element in the sample and that all spectra are background subtracted. In the fitting technique, the appropriate fractions of standard spectra are added together to build a composite spectrum which most closely matches that from the sample. In so doing, the computer has effectively deconvoluted the spectrum into its component parts. The fraction of each standard used in constructing the match to the sample is the k-ratio, described by Equation (4) which can be used directly in quantitation.

In the illustration using the K and Ca overlap, there would be little advantage of least squares fitting over the manual window method; with more complex spectra involving several overlaps, computer fitting would have obvious advantages. Both least squares fitting and the overlap factor methods rely on the basic premise that the relative line intensities for a given element are the same in the standard and sample. In the K-Ca or similar overlap

situations, the α to β ratios must be the same in standard and sample for the correction to be accurate. Since α and β lines of a series are close in energy, their relative absorption will usually be similar in both sample and standard, and the assumption will be valid. However, if one hopes to use an L-line at high energy (e.g., Os Lα at 8910 eV) to determine an overlap factor for the M-line at much lower energy (e.g., Os M at 1914 eV on P at 2013 eV), then the relative absorption between the L and M-line in standard and sample must be critically evaluated. If they differ significantly, a more appropriate standard must be prepared, or a second-order correction for differential absorption applied.

c) Overlap Factor Method - Another method for overlap correction which appears promising is the computerized overlap factor method. These factors are conceptually similar to the measured overlap factor previously described, and are mathematically tailored to individual spectrometer characteristics. The factors have been measured for K and L-lines and stored for use in the program FRAME C, a ZAF program specifically developed for EDS by the National Bureau of Standards[12]. Even though the program is best suited to the analysis of metals and minerals, the overlap correction is generally applicable and can be viewed as a subroutine adaptable to any computer program. The advantage of this technique over least squares fitting is the elimination of actual stored library spectra as a requirement for analysis. This simplifies implementation of standard intensity calculation schemes (described in a later section), since actual spectra need not be generated in order to perform overlap corrections, and reduces significantly the amount of required computer mass storage.

Converting Intensity to Concentration

For present purposes, it is convenient to classify biological samples in the following categories:
    1. Thin sections
    2. Thick sections
    3. Sections on infinitely thick substrates
    4. Infinitely thick (or bulk) samples
    5. Liquid samples

It is assumed that the "Thin" and "Thick" sections are either unsupported or supported by thin low Z substrates which contribute little to characteristic or continuum x-ray signals. Bulk, hard samples, such as bone and teeth will not be discussed explicitly since standard microanalysis techniques developed for metals and minerals are applicable. However, the data reduction schemes to be discussed for thin sections are in fact applicable to hard thin samples as well as soft tissue.

Because the exact composition of the biological matrix is often unknown, and

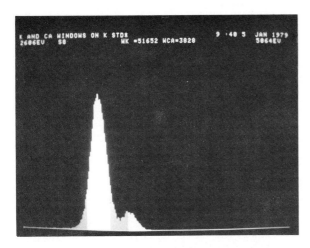

Figure 5a: K Reference Spectrum After
Linear Background Subtraction

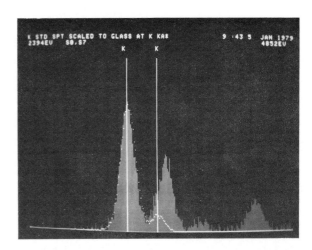

Figure 5c: K Reference Spectrum Scaled
to K Kα Intensity of Sample

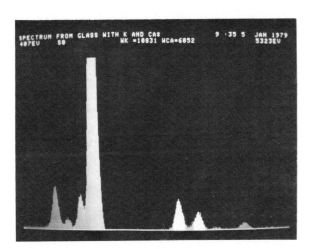

Figure 5b: Glass Spectrum After Background
Subtraction, Showing K and Ca
Window Intensities

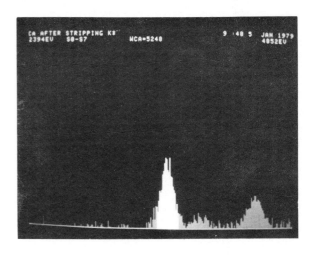

Figure 5d: Net Spectrum After Subtracting
the Scaled K Reference from the
Working Spectrum

intensities from the light elements (Z < 9)
cannot be measured with conventional EDS
spectrometers, a ZAF type analysis, routinely applied to infinitely thick metals
and minerals, is not readily applicable
to bulk biological materials. Additionally,
there has been a lack of enthusiasm in performing microanalysis in bulk organic samples due to poor spatial resolution. As a
result, there is a general lack of good
experimental data to fit the associated
constants for fundamental equations in the
ZAF program for the low Z matrix. These
difficulties have all contributed to dis-
courage development of quantitative techniques in bulk biological samples. These
comments do not exclude, however, the application of calibration curves derived from

rigorously prepared standards, nor do they
exclude quantitation in bulk samples where
linear relationships over narrow concentration ranges can be shown to apply.

The most promising sample form for
quantitative analysis is the thin section.
The term "thin" is used here in the x-ray
sense: a sample is thin if the thin film
assumption is applicable, which holds that
interelement effects of absorption and
fluorescence are minimal and can be ignored.
Intensity and concentration are then linearly related. It is generally assumed that
the thin section techniques to be described
are applicable to section thicknesses up to
1μm, and possibly up to 5μm. The applicability of the thin film assumption should
be evaluated for the particular

analytical situation.[13]

Thick sections represent a departure from the validity of the thin film assumption. Therefore, analytical accuracy will degrade as the thin film assumption becomes less valid.

There are a variety of other sample configurations encountered. For example, Warner and Coleman[14] and Coleman[1] described the development of a quantitation scheme applicable to sections on infinitely thick substrates (silicon wafers). A computer program called BASIC (not to be confused with the programming language of the same name) was written for an IBM 360 computer and can handle up to 12 elements simultaneously. This technique uses the substrate x-ray intensity as a basis for internal compensation for changes in mass thickness from point-to-point on the sample. The investigators report ± 12% relative accuracy using readily available bulk microprobe standards (pure metals or compounds).

Lechene has reviewed the development of a technique to analyze picoliter liquid samples[15]. A known volume of liquid containing known concentrations of the elements of interest is allowed to dry on a Be substrate. The salt residue is analyzed by scanning the entire droplet area and calibration curves are prepared. Unknowns are then analyzed under precisely the same experimental conditions and x-ray intensities are converted to concentrations by reference to the standard curve.

Both of the above techniques have been well developed and thoroughly described by the investigators. Readers who are interested in such sample configurations are encouraged to study the references noted.

Before beginning a detailed description of thin section quantitation techniques, some comprehensive papers on the subject of biological microanalysis will be noted. Coleman reviewed the subject in detail covering the basic physics of microanalysis in biology[1]. Shuman, Somlyo and Somlyo thoroughly described the procedures and mathematics of their approach to biological quantitation[11] and estimated minimum detectability limits for thin section analysis[5]. Trump, et al described thin section preparation techniques and presented results of microanalysis of diffusible elements in normal and pathologically altered cells[16].

Thin Section Quantitation

The Hall Method

The quantitation method proposed by Hall and colleagues is based on the following premises:[17-19]

1) the intensity of a characteristic line of an element is proportional to its mass present in the analyzed volume.

2) the intensity of the continuum radiation is proportional to the total mass of the analyzed volume.

Therefore, the ratio of the two intensities is proportional to the mass fraction or concentration of the element of interest:

$$C_i = k \frac{I_i}{I_c} \qquad (6)$$

If a biological "type" standard is available with concentration of element i in the range of the unknown, then C_i can be calculated:

$$C_i = \frac{(I_i/I_c)}{(I_i/I_c)^o} \cdot C_i^o \qquad (7)$$

It is not necessary to have an organic standard to use the Hall method. A thin sample of a stoichiometric compound or mineral can be a useful standard as well, but the equation must change to reflect the dissimilar matrices between sample and standard:

$$C_i = \frac{I_i/I_c}{(I_i/I_c)^o} \left(\frac{n_i}{\Sigma\, n_i Z_i^2}\right)^o \frac{\overline{Z^2}}{A} A_i \qquad (8)$$

where Z^2/A is taken as 3.28 for soft biological tissue. The quantity $\dfrac{n_i}{\Sigma n_i Z_i^2}$ is calculated from the stoichiometry of the mineral standard. For example, using apatite ($Ca_3P_2O_8$) as a standard for Ca,

$$\Sigma\, n_i Z_i^2 = (3*20^2)+(2*15^2)+(8*8^2) = 2162 \qquad (9)$$

and

$$n_{Ca}/\Sigma\, n_i Z_i^2 = 3/2162 \qquad (10)$$

Equation (8) is applicable to analytical situations where the concentration of the heavy element is not sufficient to significantly alter Z^2/A. If Z^2/A does not have a constant known value, then an iterative procedure could be invoked, in which a value of Z^2/A is assumed and concentrations calculated; a new value of Z^2/A is then calculated, and concentrations redetermined; the procedure continues until new values of Z^2/A no longer significantly change the calculated concentrations. Hall, however, considered the case where one or more elements are present in an otherwise simple organic matrix in sufficient concentration to affect Z^2/A, and proposed a non-iterative equation[17] as reformulated by Reed[20]:

$$C_a = \frac{q_A\, A\, G_A}{\dfrac{1+\sum\limits_u (q_i\, G_i\, Z_i^2)}{\sum\limits_m (C_i'\, Z_i^2/A_i)} + \sum\limits_u (q_i\, A_i\, G_i)} \qquad (11)$$

Selection of Region for Continuum Measurement

The region selected for the continuum measurement should be free from peaks from any source. Preferrably, it should also be a region where the sample continuum is relatively intense and that from the non-sample components is minimal. The optimum region will therefore depend on the instrument and sample type. For example, the second criterion stated above will generally be met in the 1-3 keV range. The specific choice will depend upon the elements present in the family of samples being studied. The presence of Na(1.04 keV) and K(3.31 keV) would suggest a narrowing of the range to 1.3 - 3.1 keV. If a significant Si peak (1.74 keV) were present in the particular EM-EDS system, the region might be further narrowed to 2.0 - 3.1 keV. If S (2.31 keV) and Cl(2.62 keV) were also present, the region must be chosen at higher energy (perhaps in the 4-7 keV range).

Standards

Biological type standards can be made by the addition of known quantities of salts to an organic material (albumin agar, 20% albumin, and tissue homogenates have been used as matrices). Spurr described in detail, the preparation of standards for Na and K in organic material[21]. Mineral standards can be made by grinding stoichiometric minerals into fine particles and dispersing them onto carbon coated grids. A simple modification of the latter method is to purchase high purity stoichiometric compounds and disperse them using familiar particle dispersion techniques such as described in a later section of this manuscript.

Hall and Peters conclude that both types of standards have application in biological analysis[22]. They consider that the organic standards may suffer radiation damage while the mineral standard will be more stable. If the mass loss in the standard is approximately equivalent to that in the sample, a first approximation correction for mass loss may be inherent in the method. However, unless experimental verification of the equivalence of mass loss between sample and standard is made, they caution that it would be dangerous to invoke such an assumption.

Removal of Non-Sample Background - One of the experimental difficulties in application of the Hall method is the separation of the sample background (that which is used in the quantitative equation) from the non-sample background. The latter contribution can come from "stray" or scattered electrons which strike the specimen holder, the support grid, etc. Assuming that all spectral components are due to electrons, the individual spectra that would come from the grid or holder can be scaled by the characteristic peak and subtracted from the working spectrum. This provides a reasonable

first approximation correction of the background intensity, provided that x-ray excitation has not contributed significantly to the intensities of the "non-sample" peaks. Since x-rays do not produce continuum, the ratio of background in the continuum window to characteristic peak counts coming from the construction or support material may not be the same when the beam is on the sample as when the beam is made to strike the foreign material directly. X-ray excitation is a problem in low voltage instruments where thin film appertures are used and in high voltage instruments even when conventional appertures are used.

Illustration of the Non-Sample Background Correction - Sixteen CaSO4 particles were examined in the size range of 0.3 to 8μm diameter. The particles were dispersed on a carbon film, coated with a thin evaporated layer of Cr on a Cu grid. The continuum window was set between 2700 eV and 3400 eV. The spectra, as collected, show extraneous peaks of Cu from the grid and Cr from the substrate (Figure 6). Although it is prudent to analyze regions of the sample away from grid bars, sample permitting, two particles very near grid bars were analyzed.

Figure 7 shows one of the two spectra and the particularly large Cu peak (compared to Figure 6). The region selected for the continuum measurement is shown in cross-hatch. A spectrum from pure Cu was scaled to the net Cu peak intensity exhibited in the sample spectrum and is overlayed in the Figure. It is apparent that a significant fraction of the counts contained in the continuum window will be due to Cu rather than the sample. The contribution to the continuum from the substrate was negligible. A first order correction to the continuum window due to an "overlap" from the Cu continuum, can be made by subtracting the scaled Cu spectrum from the sample spectrum. Alternatively, if a computer is not available, one may use an overlap factor, F, given by the ratio of the continuum window counts to the Cu peak integral, both taken from the pure Cu spectrum. The continuum window counts (W_c) and the net Cu window count, after background subtraction, (W_{Cu}) are determined on the sample spectrum. The corrected continuum intensity (I_c) is then given by:

$$I_c = W_c - F \, W_{Cu} \qquad (12)$$

The Ca and S ratios to the total continuum and to the corrected continuum were calculated for the particles exclusive of the two collected near grid bars, and an average value determined. Table I shows the deviation from the mean for these two spectra, with and without the continuum correction for Cu. It can be seen that the calculated peak to continuum ratio is significantly improved when the correction is made. The experimental relative standard

TABLE I
CORRECTION FOR BACKGROUND CONTRIBUTION FROM Cu GRID
(Two Particles)

	I_S/I_c	MEAN*	% DEV FROM MEAN	I_{Ca}/I_c	MEAN*	% DEV FROM MEAN
NO CORRECTION	1) 4.22	5.61	24.8	3.96	5.17	23.4
	2) 4.73	±0.52	15.7	4.45	±0.50	13.9
CORRECTION	1) 5.35	5.88	9.0	5.03	5.42	7.2
	2) 5.49	±0.49	6.6	5.12	±0.46	5.5

*MEAN calculated from 14 particles exclusive of the two tabulated.

deviation (as % of the mean) also improved when the correction was applied to the data.

Elemental Ratio Method

Because of the many sample preparation and beam effect problems associated with biological materials, particularly as they are related to mass changes, calculated concentrations often only apply to the sample as examined, and not to the material as part of the parent system from which it was taken. Furthermore, it is general in microprobe analysis that changes in concentration (that is, relative enhancement or depletion at particular structural features in the sample are more meaningful and more accurately quantitated than absolute values of concentration. Therefore, it is often preferred and more justified to report quantitative results in terms of elemental concentration ratios.

The Hall method can of course be used in a relative mode. However, the elemental ratio method is used here to indicate quantitation based on an expression derived from the thin film assumption and the well-known microprobe relationship between concentration and k-ratio:

$$C_i = k_i = I_i^t/I_i^{ot} \qquad (13)$$

In Equation (13) the thickness of the standard must be equal to that of the unknown. Equation (13) can be rewritten in ratio form:

$$\frac{C_1}{C_2} = \frac{I_1}{I_2} \cdot \frac{I_2^{ot}}{I_1^{ot}} \qquad (14)$$

Since the efficiency of the EDS spectrometer for a given photon energy and analysis geometry is constant over long periods of time, the relative pure element intensities can be determined once and retained as analytical proportionality constants for use in Equation (14).

Determining Relative Pure Element Intensities for Thin Sections

Russ proposed that the pure element intensity for a thin section can be calculated by an expression of the form:[23]

$$I^o = \frac{k' Q \cdot \omega \cdot R \cdot T}{A} \qquad (15)$$

In ratio form, the proportionality constant cancels and the quantity for application of Equation (14) can be computed.

Unfortunately, spectrometer efficiency will vary with the spectrometer and with sample-detector geometry. Even for a given system and geometry, the effective efficiency can change with time due to window contamination or changes in detector or electronic characteristics. Therefore, it has been suggested that an intensity equation of the form of Equation (15) be fit to experimental data points, essentially producing a calibration of the pure element intensity response for a given set of analytical conditions[24,25]. The intensity calibration can then be easily redone periodically or established for new conditions as they are required.

Illustration of Calculation

Shuman, Somlyo and Somlyo prepared solutions of known concentrations of salts and allowed small droplets of the solution to evaporate on a thin carbon film. The result was a thin crystal of the compound, from which relative pure element intensities for thin sections could be measured.[11] Alternatively, small particles of stoichiometric compounds can be used. The calculation of the relative pure element intensities based on the small particles, is illustrated below. The example given is for a low voltage SEM case, but the technique is equally applicable to high voltage instruments.

High purity metal sulfates were obtained: Na_2SO_4, $MgSO_4$, $Al_2(SO_4)_3$, $CaSO_4$ and $CuSO_4$. Particles of a given compound were sprinkled on a glass slide and then covered by a few drops of amyl acetate. A second glass slide was placed over the top of the suspension and the slides moved back and forth in a reciprocating motion, thereby rubbing the particles between them to

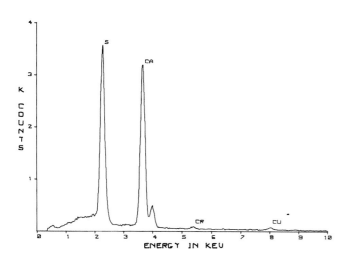

Figure 6: Typical Spectrum from Calcium
Sulfate Particle, 15 kV

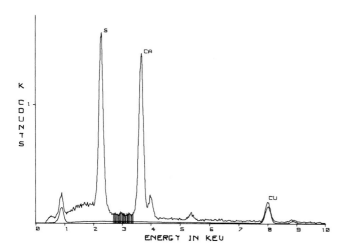

Figure 7: Spectrum showing "overlap" of
Cu Continuum with Sample
Continuum

obtain a high degree of dispersion. A drop
of the dispersion was allowed to fall on a
small area of a smooth graphite stub. (A
carbon film on a grid for TEM work is an
equivalent substrate). In this manner,
droplets of dispersed particles from each
of the compounds were placed on the stub
surface. The amyl acetate was allowed to
dry for several minutes before the stub
(or grid) was placed in the microscope.
No coating of any kind was applied, although
a thin carbon coating would be recommended.
     Particles less than 1μm in diameter
were then located and analyzed. It would
be preferable to restrict the analysis to
particles near 0.1μm in size for the Na,
Mg and Al sulfates, but this is experimen-

tally difficult on most SEMs. On TEMs,
small particles can be readily found.
Since the small particles yield relatively
low count rates, the time of analysis was
extended to give at least 1000 net counts
in the integral (window) of the peak of
interest. The actual time of analysis
(as well as probe current and exact particle size) is unimportant, since all spectra
will be normalized on the basis of expected
S intensity. A scanning beam was used at
high magnification rather than a spot probe,
at 15 kV accelerating voltage. Spectra
from at least two particles of each type
were collected. The spectra were background
subtracted using the linear interpolation
method and those from each type of particle
were averaged together to form the standard
spectrum.

Normalization Based on Metal Concentration

Each standard spectrum was then scaled
to the intensity expected from the pure element by dividing by the weight fraction of
the metal in the sulfate. For example, the
weight fraction of Ca in $CaSO_4$ is 0.294;
therefore, the intensity from the pure
element based on the thin film assumption,
is 1/0.294 times the measured Ca intensity
in the sulfate.

Normalization Based on Sulfur Concentration

Each spectrum was then normalized to
the appropriate S intensity expected from
the S concentration in the compounds, again
based on the thin film assumption. This
step compensates for various times of
analysis, particle sizes and probe currents
that may have been used. Table II below
provides the required data:

TABLE II

NORMALIZATION OF COLLECTED STANDARD SPECTRA

COMPOUNDS	$\underline{S}$	$\underline{X}$	"EXPECTED" S INTENSITY*
Na_2SO_4	.226	.324	22600
$MgSO_4$	.266	.202	26600
$Al_2(SO_4)_3$	.281	.158	28100
$CaSO_4$	.236	.294	23600
$CuSO_4$	.201	.398	20100

S=S Mass Fraction; X=Metal Mass Fraction
*Based on 100,000 Counts for Pure S

     For example, in the $CaSO_4$ spectrum,
the S window intensity was 13355 counts.
Table II shows a desired intensity of
23600. Therefore, the spectrum was multiplied by the factor 23600/13355. In this
manner, all other standard spectra were

scaled to the tabulated S integral.

The intensities of the peaks from the metals in the compounds now represent proper relative intensities from pure element thin films. They can be used directly in Equation (14), or if necessary, more compounds can be analyzed to provide the intensities for the missing elements. Alternatively, a pure element intensity equation can be fit to the Na, Mg, Al, Ca and Cu data points. Figure 8 shows the calculated relative pure element intensities after fitting. As a check, S in the compounds (which was not used in fitting the intensity equation) and $NiCl_2$ particles were analyzed. The elemental ratios of S to metal in the sulfates and Ni/Cl were calculated to within 10% of the stoichiometric values.

Summary

Quantitative methods in biology have been developed extensively for thin sections in the form of the Hall and elemental ratio methods. Attainable accuracies of 10-15% relative for concentrations in the range of 1-2% have been reported[11]. Because of counting statistics and errors in background subtraction when the background is high compared to the net peak intensity, accuracies generally degrade rapidly for lower concentrations (and would degrade still further if overlap corrections were necessary).

The development of cryogenic thin section preparation techniques has greatly enhanced the credibility of the quantitative result. Improvements in these techniques will be necessary if quantitation in the electron microscope is to be applicable to the material in the living condition.

Use of thin sections has also greatly increased the attainable x-ray spatial resolution. Resolutions on the order of 50nm for a 500nm thick section are predicted for Ca in soft tissue at 100 kV. However, when high intensity probes are focused to a fine spot on the sample, radiation damage to the sample may result. To limit the dose in radiation-sensitive samples, the probe diameter would have to be increased, with a consequent degradation of spatial resolution, perhaps significantly, over that predicted.

Quantitative methods have also been developed for other, more specialized, sample configurations such as thin sections on infinitely thick substrates (Warner and Coleman) and samples prepared from the liquid (Lechene). However, quantitation in the bulk sample has not been enthusiastically pursued because of poor x-ray spatial resolution and difficulties in applying a ZAF correction procedure in the light matrix material.

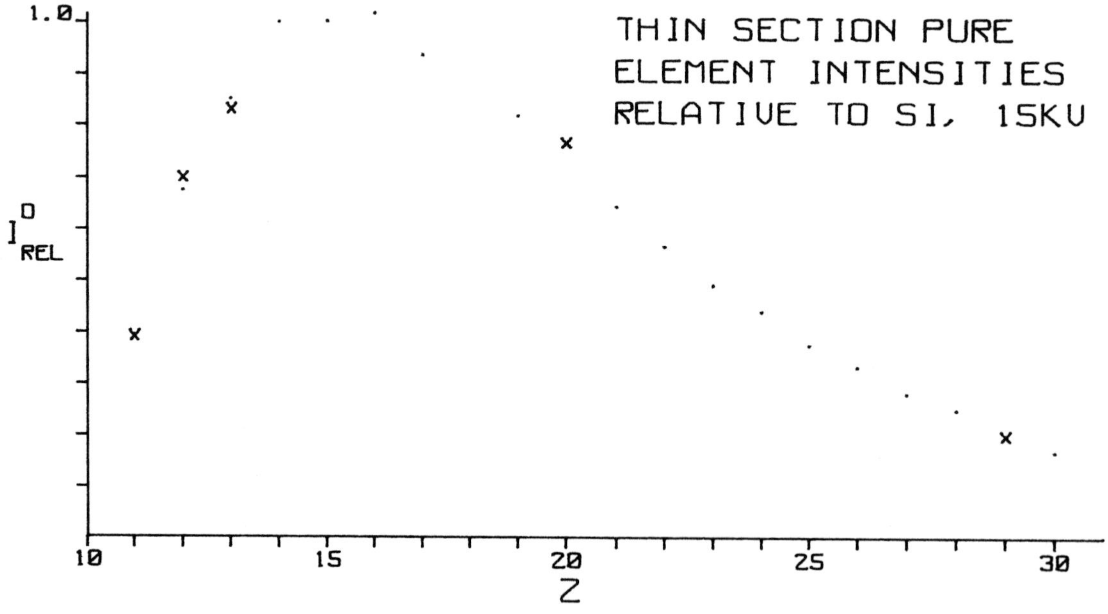

Figure 8: Relative Thin Section Pure Element Intensities, Back-Calculated from Sulfates

References

1. J.R. Coleman, "Biological Applications: Sample Preparation and Quantitation" in Practical Scanning Electron Microscopy (J.I. Goldstein and H. Yakowitz, eds.), Plenum Press, NY, 1975, 491-527.

2. P. Echlin and A.J. Sauberman, "Preparation of Biological Specimens for X-ray Microanalysis", SEM/1977/I, IIT Research Institute, Chicago, IL 60616 621-637.

3. T.A. Hall and B.L. Gupta, "Measurement of Mass Loss in Biological Specimens Under an Electron Microbeam" in Microprobe Analysis as Applied to Cells and Tissues (T. Hall, P. Echlin and R. Kaufman, eds.), Academic Press, London, 1974, 147-158.

4. P. Echlin, "Coating Techniques for Scanning Electron Microscopy and X-ray Microanalysis", SEM/1978/I, AMF O'Hare, ILL. 60666, 109-131.

5. H. Shuman, A.V. Somlyo and A.P. Somlyo, "Theoretical and Practical Limits of ED X-ray Analysis of Biological Thin Sections", SEM/1977/I, IIT Research Institute, Chicago, IL. 60616, 663-670.

6. S.J.B. Reed, Electron Microprobe Analysis, Cambridge Univ. Press, Cambridge, England, 1975, 216-218.

7. J.I. Goldstein, J.L. Costley, G.W. Lorimer, and S.J.B. Reed, "Quantitative X-ray Analysis in the Electron Microscope", SEM/1977/I, IIT Research Institute, Chicago, IL, 60616, 315-323.

8. J.I. Goldstein, "Statistics of X-ray Analysis", Proceedings, 11th Annual Conference of MAS., 1976, 1-8, available from K. Heinrich, Anal. Chem. Div., NBC, Washington, DC. 20234.

9. T.O. Ziebold, "Precision and Sensitivity in Electron Microprobe Analysis", Anal. Chem., 39, 1967, 858.

10. P. Statham, "Reliability in Data-Analysis Procedures for X-ray Spectra", Proceedings 8th International Conference on X-ray Optics and Microanalysis and 12th Annual Conference of the Microbeam Analysis Society, 1977, 95, available from K. Heinrich, Anal. Chem. Division, NBS, Washington, DC 20234.

11. H. Shuman, A.V. Somlyo and A.P. Somlyo, "Quantitative Electron Probe Microanalysis of Biological Thin Sections: Methods and Validity". Ultramicroscopy 1, 1976, 317-339.

12. R.L. Myklebust, C.E. Fiori, and K.F.J. Heinrich, "FRAME C: A Compact Procedure for Quantitative Energy Dispersive Electron Probe X-ray Analysis", Proceedings 8th International Conference on X-ray Optics and Microanalysis and 12th Annual Conference of the Microbeam Analysis Society, 1977, 96, available from K. Heinrich, Anal. Chem. Division, NBS. Washington, DC.

13. R. Tixier and J. Philibert, "Analyse Quantitative d'echantillons," Proc. 5th Int'l Cong. on X-ray Optics and Micro-analysis, eds. G. Mollenstdet and K.H. Gaukler, Springer-Verlag, Berlin, 1969, 180-186.

14. R.R. Warner and J.R. Coleman, "A Computer Program for Quantitative Micro-analysis of Thin Biological Material", Proceedings of the 7th National Conference on Electron Probe Analysis, 1972, 41., available from K. Heinrich, Anal. Chem. Div., NBS, Washington, DC 20234

15. C. Lechene, "Electron Probe Microanalysis of Picoliter Liquid Samples", in Microprobe Analysis as Applied to Cells and Tissues, (T. Hall, P. Echlin and R. Kaufman, eds.) Academic Press, London, 1974. 351-367.

16. B.F. Trump, I.K. Berezesky, R.E. Pendergrass, S.H. Chang, R.E. Bulger, and W.J. Mergner, "X-ray Microanalysis of Diffusible Elements in Scanning Electron Microscopy of Biological Thin Sections. Studies of Pathologically Altered Cells", SEM/1978/II, AMF, O'Hare, IL 60666, 1027-1039.

17. T. Hall, "Some Aspects of the Microprobe Analysis of Biological Specimens" in Quantitative Electron Probe Microanalysis (K.F.J. Heinrich, ed.) NBS Special Pub. 298, U.S. Govt. Printing Office, Washington, DC. 20402, 1968. 269-299.

18. D.J. Marshall, T.A. Hall, "A Method for the Microanalysis of Thin Films, in Optique Des Rayons X et Microanalyse (R. Castaing, J. Descamps and J. Philibert, eds.), Hermann, Paris, 1966. 374-381.

19. T.A. Hall and P.R. Werba, "Quantitative Microprobe Analysis of Thin Specimens; Continuum Method", in Proc. 25th Anniversary Meeting of EMAG, Institute of Physics, London, 1971. 146-149

20. op. cit. S.J.B. Reed, 330-332.

21. A.R. Spurr, "Macrocylic Polyether Complexes with Alkali Elements in Epoxy Resin as Standards for X-ray Analysis of Biological Tissues", in Microprobe Analysis as Applied to Cells and Tissues, (T. Hall, P. Echlin and R. Kaufman, eds.) Academic Press, London, 1974, 213-227.

22. T.A. Hall and P.D. Peters, "Quantitative Analysis of Thin Sections and the Choice of Standards", in Microprobe Analysis as Applied to Cells and Tissues (T. Hall, P. Echlin and R. Kaufman, eds.), Academic Press, London, 1974. 229-238.

23. J.C. Russ, "The Direct Element Ratio Model for Quantitative Analysis of Thin Sections", in Microprobe Analysis as Applied to Cells and Tissues (T. Hall, P. Echlin, and R. Kaufman, eds.), Academic Press, London, 1974. 269-276.

24. N.C. Barbi, D.P. Skinner and S. Blinder "The Calculation of Pure Element X-ray Intensities from Empirically Derived Expressions and its Application to Quantitative SEM/EDS Analysis", Proc. 11th Annual Conference of MAS., 1976, 8, available from

K. Heinrich, Anal. Chem. Div., NBS,
Washington, DC 20234
25. N.C. Barbi, M. Foster and L. Goldman,
"Semiquantitative Energy Dispersive
Analysis in STEM using Generated Reference
Spectra and Empirical Scale Factors"
SEM/1977/I., IIT Research Institute,
Chicago, IL 60616. 307-314.

Discussion With Reviewers

C.E. Fiori: What is "radiation damage"
when < 20 keV electrons are used in an
SEM? Are we talking more about heat or
actual molecular destruction and dislocation due to the kinetic energy of a beam
electron. The kinetic energy of a 20 keV
electron is quite low. What are the rec-
commended methods to reduce radiation
damage? Are there preferred preparation
techniques?
Author: I am not aware of references on
radiation damage with voltages less than
20 kV. Hall and Gupta (text reference 3)
observed mass loss in tissue sections at
30 kV, which they presumed to be due to the
"escape of volatile products after the beam-
induced rupture of chemical bonds". They
reference a paper by K. Stenn and G. Bahr
(J. Ultrastruct. Res., 31, 1970, 526-550)
in which it is suggested that chemical
bonds are broken and radiation products
including CO_2, H_2 and N_2 are formed and
escape. Use of a cold stage to maintain
the sample at low temperature should not
reduce radiation damage to bonds but does
seem to inhibit mass loss.

Specimen damage which is due to gradual heating of the specimen caused by the
electron beam can be reduced significantly
by use of conductive coatings and low beam
currents.

J.C. Russ: The discussion of statistically
introduced error in net intensities is fine,
but would the author please extend the discussion to the consequences for P/B ratio
errors. This is especially critical for
low count spectra where the background
intensity becomes very small and is further
affected by the subtraction of extraneous
background. Representative cases and error
magnitudes would be useful.
Author: The Hall method depends on the
measurement of the peak: continuum ratio.
The statistical uncertainty in determining
the ratio of the two intensities will
therefore be increased relative to the determination of the peak intensity alone.
For example, consider that the peak intensity = 1000 and the continuum intensity
= 400. Using 2σ statistics,

$$\frac{I_i}{I_c} = \frac{1000 \pm 63}{400 \pm 40} \Rightarrow 2.13 < \frac{I_i}{I_c} < 2.95$$

The uncertainty due to counting statistics
for the peak intensity alone is 6.3%. For
the peak: continuum ratio, the uncertainty
is increased to 15-20%. Generally, the
situation is worsened by the inaccuracy in
computing background under the peak, which
increases the uncertainty in the peak intensity. The severity of the statistical
problem in quantitating low count spectra
must be realized.

I.K. Berezesky: It is stated in the paper
that "quantitation in bulk samples has not
been enthusiastically pursued because of
poor x-ray spatial resolution and difficulties in applying the ZAF correction procedure in light matrix material". In view
of this statement, please comment on the
recent work of Zs.-Nagy et al. (J. Ultrastr.
Res. 58:22, 1977) who proposed, according
to their data, an acceptable quantitation
of electrolyte concentrations in bulk
biological specimens and the work of
Wroblewski et al. (Histochem. 55:281, 1978)
who used a ZAF correction for quantitation
of 16μm thick muscle cryosections.
Author: I did not mean to imply that bulk
sample quantitation has not been attempted
or even successfully performed. However,
there has not been strong emphasis in
performing bulk sample quantitation, compared to the effort placed in thin section
work, because of the circumstances described
in the text. Zs.-Nagy, et al in the reference cited state in their abstract: "At
present one must be satisfied with the low
spatial resolution obtainable and with the
relative exactness of quantitation, since
the validity of the known ZAF-correction
methods is doubtful for light elements in
an organic matrix". In the work reported
in that investigation, the authors
established a linear relationship between
the quantity $(I_i/I_c \cdot Z^2/A)$ and concentration of selected elements in crystals whose
average atomic number, exclusive of the
element being analyzed, was similar to the
biological matrix. The calibration curves
established from these crystal standards
then provided acceptable quantitation in
their study.

In the investigation by Wroblewski, et
al, a ZAF correction was, in fact, applied
to bulk biological samples. The authors
in this study utilized "type" standards
prepared by dissolving appropriate compounds in Spurr's resin, which provides
some degree of normalization of the ZAF
procedure: if concentrations are low, the
Z correction will be unnecessary and the
A correction minimized due to the similar
matrix of sample and standard; the
authors suggest that for these samples,
the F correction is negligible. Thus, the
degree of ZAF correction becomes minimal
and its relative inaccuracy for this
matrix becomes less important (as does its
benefit).

For additional discussion see page 658.

672

SCANNING ELECTRON MICROSCOPY/1979/II
SEM Inc., AMF O'Hare, IL 60666, USA

ANALYSIS OF ELEMENTAL RATIOS IN "THIN" SAMPLES

J. C. Russ

Edax Laboratories
P.O. Box 135
Prairie View, IL 60069

Abstract

Elemental microanalysis of thin or semi-thin sections in the Transmission Electron Microscope (TEM) and Scanning Transmission Electron Microscope (STEM) is often used to determine ratios of elements at the same analysis point rather than absolute mass fractions. The use of ratios avoids the need to consider gain or loss of water or organic components in preparation (or under the electron beam) and makes the measurement relatively insensitive to extraneous background produced by stray electrons. The latter problem is common in many instrumental configurations and introduces errors in the methods using characteristic-to-continuum ratios to measure weight fraction. With the ratio method, it is also comparatively easy to transfer standardization from instrument to instrument, to use a variety of types of standards, or to work without standards.

The principal standardization methods include measurement of intensity ratios from mineral chips or other chemical compounds of known stoichiometry, measurement of "pure" intensities from single element standards (for instance prepared with a droplet technique) and the calculation of pure intensities. The latter are in principle simply the product of the probabilities of the element being excited by the electron (cross-section), the excited atom emitting an X-ray (fluorescence yield), the X-ray leaving the sample (absorption) and its being detected (spectrometer efficiency). Several different mathematical models for each term, ranging from simple to elaborate, are shown along with the assumptions and magnitude of error for each.

Key words: X-Ray Microanalysis, Quantitative Analysis, Thin sections, Ratio Model

Introduction

Several types of quantitative data interpretation can be applied to thin sections analyzed in the Transmission Electron Microscope (TEM) or Scanning Transmission Electron Microscope (STEM). Briefly, these may be classified according to whether they are: 1) absolute or relative methods, depending on whether the actual amount of an element is determined or whether only ratios to other elements or locations are obtained; and 2.) mass or concentration methods, depending on whether the result is expressed as the amount of the element or its weight fraction in the excited volume. In addition, combinations of these methods can be used. Standards can be intentionally similar to the unknowns, as in the case of resins doped with elements for biological tissues, or quite dissimilar, including the use of particles, crystals, and even metal blocks. Standards may also be replaced by quasi-theoretical calculation in some cases.

Faced with so many choices in technique, it is necessary for researchers to select one or a few that match their needs, and then perhaps adapt the standards, calculations, and so on. It is the purpose of this paper to review the major methods used for the analysis of elemental ratios. Generally, any of the ratio methods can be made absolute if at least one of the elements in the sample is either known or can be quantified by reference to appropriate absolute standards. However, in many circumstances it is the ratios themselves which are the desired information.

"Thin" Sections

Computation of the absolute mass of an element on a truly "thin" section is straightforward, since the number of X-rays from the element counted is linearly proportional to the number of atoms of the element for a given analytical line. Since the constant of proportionality depends on the element and the voltage, it is usually measured by using tiny particles of known composition, whose mass can be estimated from observations of particle size. The ratio of mass in the unknown to that in the standard is then just the ratio of intensities. (In all of these descriptions, the intensities will be assumed to be counts per second per electron, and measurement and/or control of beam current will be assumed unless otherwise noted).

This linear relationship is in fact an operational definition of a "thin" section. As the product of density times thickness increases, two factors can cause deviation from linearity: 1.) the probability that the incident electron interacts with the atom of the element is a function of the electron energy. In the original beam, this is highly monoenergetic, and in a thin section those electrons which interact with atoms (and lose energy thereby) leave the sample without having a second interaction. If the sample mass thickness is great enough to cause that portion of the TEM image to be dark (for example a typical stained particle in a tissue section) it is likely that multiple electron - atom interations are occurring.This changes the relative yield of X-rays per electron, and makes comparison to other samples (or standards whose electron "stopping power" is different) incorrect.

2.) The absorption of generated X-rays is usually very small and may be ignored in a truly thin section. As mass thickness increases, X-ray absorption increases. Absorption is exponential with distance following the expression

$$I_{meas.}/I_{gen.} = e^{-\mu \rho x} \qquad (1$$

The x is not the section thickness, but the distance the X-rays travel (which depends on takeoff angle and will be discussed further). The mass absorption coefficient μ is itself a function of X-ray energy and specimen composition. For organic matrices, μ is generally fairly low; but it rises for lower energy X-rays, so that a section which may be considered "thin" for iron X-rays (6.4 keV) or even calcium X-rays (3.7 keV) may not be for sodium X-rays (1.0 keV).

Corrections for the absorption of generated X-rays can be made provided the mass thickness (ρt) can be estimated, although since the entire specimen composition must be used, the computation is iterative and may require a computer. Correction for slowing down of the incident electrons is generally impractical in the conventional models, and requires the use of Monte-Carlo calculations. Fortunately, the levels of accuracy thus far required and justified in biological "thin" section analysis have not necessitated making this correction.

Sources of Error

Accuracy of results in biological thin section analysis rarely approaches the level that can be obtained in conventional "microprobe" analysis of bulk samples. The sources of the errors can for most purposes be found in three places: 1.) difficulty in obtaining, making, measuring or adapting standards; 2.) uncertainties in relating the measured spectrum to the sample, because of extraneous X-rays generated by scattered electrons or X-rays (from components in the electron column) striking the sample or its supports, and because of possible sample damage or contamination under the beam; 3.) statistical or counting errors due to low count rates. The errors due to the use of approximations in the quantitative models or in the treatment of measured spectra to obtain net element intensities, separating peaks from background and from each other, are often much smaller than these three; they will be discussed, but the user is warned that using the more elaborate (and costly in terms of computer and time) calculations may not be justified if the major source of error cannot be reduced.

Elemental Ratios

From the simple linear relationship between elemental mass and intensity mentioned before, it follows that the ratio of intensities of two or more elements should be proportional to their relative weight fractions. The change from number of atoms to concentration is justified since the different elements are all included in the same excited volume, even though we may not know exactly how large that volume is nor what its shape is. Incidentally, none of the methods for correcting for non-linearities such as absorption can compensate for unequal distribution of the elements within the excited volume; we must always assume that the beam excites a small enough region in the sample for us to assume it is homogenous.

The classic method to obtain absolute concentration is to use the element characteristic intensity as the measure of the element mass and the background (bremsstrahlung) in the spectrum as a measure of the total mass of the excited volume, so that the Peak-to-Background (P/B) ratio is proportional to concentration[1]. The background has been measured either under the peak or elsewhere in the energy spectrum. Of course, if ratios of the P/B values for different elements are used to obtain elemental ratios, the background values should cancel out (perhaps leaving a constant factor if backgrounds at different energies are used) so that we again find elemental concentration ratios are proportional to the characteristic intensity ratios.

The P/B method encounters two main problems in practical application. The first is the need to carefully separate the true background due to bremsstrahlung generation in the specimen from the extraneous background already mentioned. This is usually done by measurements on blanks after careful modification of the instrument and selection of operating conditions to minimize such extraneous signals. The second is the difficulty in relating the weight fraction of the element in the section as it exists in the microscope to the original specimen. Addition or removal of mass in the form of water or organic molecules can occur during preparation; usually this can be determined at least on the average, but not necessarily at the point of analysis. Loss of mass under the beam[2] and contamination can also be reduced to acceptable levels by the use of specimen cooling and careful cold trapping around the specimen.

However, if only elemental ratios are needed, these problems can be largely ignored, so long as the extraneous background (not of course any extraneous characteristic X-rays) does not alter the spectrum so much that separation of peaks from background is frustrated, and so long as the mass loss or gain of the sample does not involve the elements being analyzed nor seriously change the specimen ρt to affect X-ray absorption. In practice these criteria are usually met. This does not mean that extraneous background should be tolerated in the microscope, or that mass loss or gain can be completely ignored; rather, the wise researcher will first attempt to control those factors and estimate their magnitude. Only then can he safely decide to ignore their remaining contribution.

The Proportionality Constant

The simple ratio model thus far proposed is simply

$$\frac{C_1}{C_2} = k_{12} \frac{I_1}{I_2} \qquad (2$$

where C's are concentrations and I's intensities, and k is a proportionality factor for the particular elements.

Since the intensities here are measured at the same time and with the same beam current, it is possible to relax the earlier condition of measuring and compensating for beam current. It is even possible to use intensities collected with a varying beam current. It may be dangerous, however, to use intensities from spectra collected while the beam scans over a large inhomogeneous area. Not only may the elements so measured not really be in the same place, so that the implication of an elemental "ratio" loses meaning, but if the flux of all X-rays (including real and extraneous background, and characteristic X-rays of other elements) varies, it will change the system dead-time of the counting electronics. This will cause the loss of X-rays from the high-count-rate regions and their replacement by extra X-rays from the low-count-rate regions, so that a biased result may be obtained.

The factor k_{12} depends on the two elements in the equation, and so rather than have a table of all possible binary pairs it is more convenient to rewrite the equation as

$$C_1 : C_2 : C_3 : ... = \frac{I_1}{P_1} : \frac{I_2}{P_2} : \frac{I_3}{P_3} : ... \qquad (3$$

where the factors P_i correspond to each of the elements in the ratio. Regardless of which way the factors are expressed, they describe the relative number of X-rays emitted from equal concentrations of each element, for excitation by electrons of a particular voltage. Before considering this process step-by-step with a view to calculating the proportionality factors, it is useful to consider the ways they have been determined by measurement. This raises the need to measure spectra and obtain net intensities for the elements, and so that will be taken up next.

Processing Spectra

The X-ray intensity versus energy spectrum measured by the analytical system consists primarily of characteristic peaks, whose energy location and height identify the element and its concentration, and continuum or background, which as mentioned before can be used to determine the total excited region of the sample. In addition to the peaks and background coming from the sample, there may also be peaks and background from extraneous sources such as specimen holders, grids, or portions of the sample far from the point of analysis. If we are using a ratio method, the only information we need from the spectrum is the size of the characteristic peaks for the elements of interest.

For the peak "size", we may use the height (above background), the area of the peak, the sum of areas of several peaks such as $K_\alpha + K_\beta$, or the ratio of the peak to the size of a "standard" peak measured on a standard. Any of these methods is suitable if consistently applied, and each has been used in some cases. The peaks themselves are nearly Gaussian in shape. The principal deviation from this simple shape is a tail on the low energy side of the peak which results from incomplete charge collection in the silicon detector. The size of the tail is energy-dependent, since as X-ray energy varies the average depth of penetration into the detector changes and the amount of charge trapping changes. One way to overcome this problem is to use measured and stored peaks for each element, another is to calculate a modified Gaussian shape in generating peaks.

A suitable model for the peak tailing uses a shape calculated as:[3]

$$y = e^{-0.5(x-E)^2/s_o} + a_o \, e^{4(x-E)}(1-e^{0.4(x-E)^2/s_o}) \qquad (4$$

where y is the channel height, x is the energy, E the centroid energy and the peak width is defined by

$$s_o = \left\{ (\text{Res'n @ Mn } K_\alpha)^2 + 2735 (E-5.894) \right\} / 2.77259 \cdot 10^6$$

the first term is the normal Gaussian, the remainder is the tail. The coefficient for the size of the tail can be calculated from peak energy as:

$$a_o = 0.01 + 0.0025E \qquad (5$$

With this modification excellent fits between measured and calculated lines have been shown[4]. In principle, the calculated peak shape should offer an advantage over the use of measured library spectra since the ratios of peaks (eg. $L_\alpha : L_{\beta_1} : L_{\beta_2}$) are not fixed and therefore may be adjusted to fit the spectrum. Since these ratios are affected by matrix absorption, they should vary from sample to standard. Also, they can vary somewhat (for L and M shells) with accelerating voltage.

In practice, the refinements of modifying peak shape or accommodating different peak ratios are rarely important in analyzing biological thin sections. This is due to the poor counting statistics in the spectrum (because of low count rates) and to the limiting effect of other errors on total results. The matrix absorption in biological thin secitons is rarely great enough to alter peak ratios significantly (in fact in many cases we ignore absorption altogether) for two different lines of an element. The accelerating voltage is usually high enough that the relative excitation of the three L edges or five M edges is nearly constant. Since the peaks are small - often only a few hundred or few thousand counts high in the highest channel - the small portion (usually less than 2%) of the total peak area in the tail may not even be statistically separable from background.

In any case, these factors can introduce in general no more than a few per cent error, and we will see that other sources of error are much greater. Hence the fitting of peaks to the spectrum to ascertain their size may be done equally well for our purposes using either generated or stored peaks. It is most common to use a linear least-squares technique to do this fitting. There are cases of very close overlaps where this may give mathematically "best" answers that are incorrect[5,6] but these are not often encountered in biological thin sections unless the M lines of heavy elements added as stains are present, and it is recommended that this be avoided in any case. Quantitatively useful separation of such overlaps as Sb(L)-Ca(K) can be carried out by such fitting[7].

Background Fitting

Separation of the characteristic peaks from the background must also be performed. Sometimes this is included in the linear least-squares fitting procedure by using library spectra for peaks which include background, but this can introduce substantial errors if the background is highly curved, as it is in the 1-3 keV range, or if the P/B ratios on the standards are dramatically greater than on the unknowns, as they may be if high concentration standards are used. It is more common therefore, to discriminate against the background in these spectra by fitting not the measured spectra, but their second derivatives (often produced by a digital filtering method)[8]. This is quite suitable for obtaining useful results, especially considering the limits on overall accuracy coming from other sources, and requires little effort on the part of the user once his standard spectra are on file.

The major potential sources of error in this approach are: 1) Shifts in the energy calibration (gain and zero) of the spectrum so that the library spectral peaks do not line up with the unknown. This kind of shift is a problem in any fitting procedure, but can be aggravated by the use of the second-derivative method. Calibration of peaks so that they always lie within a few eV, typically much less than one channel in the multichannel analyzer, is essential. 2) The presence of significant amount of extraneous background, as mentioned before, in either the standard or the unknown spectra, or both. Since this is not linearly additive and may have a quite different shape from the true background, it can distort the results. Because of these problems, we recommend that fitting of library spectra (either original or second derivative) be used only if the extraneous background is very small, or has been separately determined and fit as an additional component[9].

For routine work where these conditions cannot be assumed, we find it better to make a separate, discrete subtraction of background before fitting the peaks. This can be done using any shape that adapts itself to the general form of the background (which may be altered if extraneous background is present). Even straight line segments could be used if they were drawn between enough points in the spectrum. However, a more generally useful shape for interpolation is obtained by approximating background with the equation:

$$B \propto \frac{V - E}{E} \quad T \tag{6}$$

where V is the accelerating voltage, E is the energy, and T is the efficiency of the X-ray detector (which drops off dramatically at low energy). This general shape (Figure 1) matches that of the background rather well, and in fact differs from the theoretically predicted shape only in ignoring any matrix absorption effects. If line segments with this shape are successively fit to background points in the measured spectrum, the only assumption regarding absorption is that it is not so high as to vary significantly from the energy of one fitting point to the next. This condition is easily met even in samples with significant total absorption, provided enough points are selected.

The drawback of this method is that the user is usually required to select several background points in his spectrum, which bracket all of the characteristic peaks of interest. Note that this requires in general more points than are needed for background fitting to spectra measured on bulk specimens[10], where the complete shape of the continuum can be quite accurately calculated from only two points. In the case of a thin section (and especially if extraneous background coming from thin and bulk materials excited by electrons of different energies and by stray X-ray photons is present) the shape cannot be predicted in detail, so fitting over short energy ranges, using multiple background points, is necessary. However, this method will in most cases fit very adequately.

The need for the user to select the background points can also be relaxed by the use of a simple algorithm to search the spectrum for suitable points. One such method[11] is to move in steps of a few hundred electron volts, testing each location to see that it does not exceed (by more than statistics would allow) the intensity of either neighboring location, and that it does not lie below both neighboring locations. The first test eliminates any point that could lie on a peak, and the second assumes that the point is not the valley between two peaks. In a computerized system, such searching and fitting requires a fraction of a second and produces results such as shown in Figure 1.

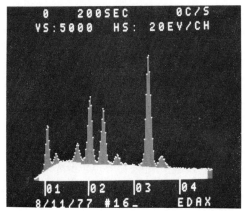

Figure 1: Background (bright area) fitted to a spectrum measured from a thin section of freeze-dried muscle tissue.

Errors in Intensity Values

If the intensity values at a given photon energy obtained are expressed as counts per second per unit of beam current it is obviously necessary to know the time and current accurately to obtain accurate results. Since the ratio method allows these terms to cancel, we can consider only the counts (or peak size) in discussing accuracy. The external sources of error - miscalibration of the energy scale, poor choice of peak shape or background shape, etc. - can all be controlled by operator care. There remains a limiting error due to counting statistics that frequently assumes major importance in the final results.

Counting of truly random events, such as X-rays, gives a number of counts N whose standard deviation is given (assuming as usual that N is the best estimate of the true number) as $\sqrt{N}$. This means that on a percentage basis, we would assume an error from N counts of $100/\sqrt{N}$, or 1% for 10^4 counts, 3% for 10^3, 10% for 10^2 and so on. However, that is not quite the situation we actually face. The counts in one channel, or a few channels covering a peak in our spectrum, are not all X-rays of one type, coming from one process. Instead, we may have N_E counts in the peak of the element of interest, N_B counts in the true background, N_X extraneous background counts (and there may be several sources of these) and N_O counts from overlaps of peaks of other elements (again there may be several). Each N has its own statistical uncertainty, and even if we do a proper job of separating the spectrum into its various parts, the errors all influence the uncertainty in our value for intensity. The total number of counts (which is all we have really measured) is the sum of all the individual ones. We obtain the net element intensity by estimating each other signal and subtracting it, to leave behind the value we want, but this adds to the error. Since there are independent sources of X-rays, the errors add in quadrature. The equation for the percentage error in our net intensity is

$$\Delta N(\%) = 100 \cdot \sqrt{N_E + 2(N_B + N_X + N_O)} / N_E \quad (7$$

For the case of a large peak with small background and no overlaps, this approaches $100/\sqrt{N}$, but for the case often encountered in analysis of relatively low concentrations of elements in biological thin sections these terms are not likely to be negligible. Obtaining 10^3 counts for a minor element, with a background under the peak at least as great (a P/B ratio of about 1:1) and perhaps a similar number of counts from an overlapping peak will about double the predicted inherent statistical error, from 3% (considering just $100/\sqrt{N}$) to more than 5%. Obviously, many worse cases can be found. Consideration of the magnitude of the statistical error will often lead to 1) a realistic error estimate for the entire analysis; 2) a realization that some of the refinements already mentioned for more accurate peak fitting, and to be discussed for more accurate calculation, are not justified; and 3) discouragement.

Standards

Standards are used chiefly in the ratio method to determine the interelement proportionality factors (k_{12} or P.), although they may also be used to record elemental peak profiles for fitting if each standard contains only a few elements with well separated peaks. Different standards for these two purposes may also be used. Our concern at this point is with standards containing elements in known concentration ratios used to determine P factors (or k factors, which are simply ratios of P factors).

"In-type" standards (standards similar in composition to the unknowns) are attractive because it is intuitively clear that they will minimize any need for correction factors or any error from ignoring them. This type of standard has proved quite satisfactory in other fields; in the analysis of metal foils, for example, if the average composition of the bulk metal is known, then spreading the TEM beam, or scanning the STEM raster, to cover a large area of the specimen gives a standard spectrum (provided that absorption is negligible and no local counting rate paralysis occurs) that can be used to obtain factors for use in analyzing smaller areas or points in the same sample[12].

Another use of in-type standards is found in the analysis of asbestos-like minerals. There samples are often not thin enough to assume that absorption is negligible[13] but provided the standards are similar in size to the unknowns, the factors will include the "not-so-thin" effects. Several authors[14,15] have measured factors for a variety of these minerals; Figure 2 shows a typical curve of relative intensity (P factor).

There have been attempts to produce standards similar to biological material but doped with a homogeneous distribution of several elements. These have included frozen albumin[16], methacrylate[17], epoxies[18], and other materials to which compounds containing elements such as Na, K and Cl have been added. The main criteria for such standards are: 1) homogeneity at a scale of at least tens of nanometers; 2) stability in vacuum and under the electron beam. The most successful materials reported so far are the macrocyclic polyether complexes reported by Spurr[19], which because of their cyclic structure will trap a variety of elements if introduced as appropriate salts, and then in turn may be introduced into an epoxy resin.

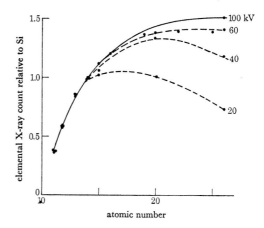

Figure 2: Sensitivity factors relative to silicon measured at various accelerating voltages using mineral standards (ref. # 15).

Hall[20] has discusssed the relative merits of organic standards and finely ground mineral chips. The latter are stable, and have homogeneous, well known composition because of their stoichiometry. To use the intensities from such chips in an absolute concentration calculation based on P/B ratios it is necessary to adjust the measured data for the different continuum production in the higher atomic number of the matrix, as compared to an organic sample. For ratio calculations this is not the case, and even absorption corrections may usually be ignored. For elements that can be obtained in suitable mineral form, such standards are quite suitable.[21,22]

Another approach to making standards for ratio factors is to spray droplets of solutions containing known (often equal) concentrations of many elements onto a thin film[23,24,25]. Measurement of intensities using one such standard containing all elements of interest can provide a master curve of P factors such as shown in Figure 3.

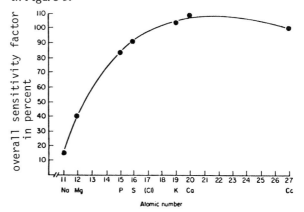

Figure 3: Sensitivity or P-factors measured from droplet standards (ref. # 23).

These three latter methods are all in fairly widespread use at this time. It should be noted that an experimentally determined set of P factors may include instrument-dependent variables as well as fundamental physical effects, and so should not be transferred blindly from one instrument or class of specimens to another. Also, they apply only to one accelerating voltage, since this affects the probability of an electron ionizing each element. Finally, there may be elements of interest that are inconvenient or difficult to prepare in these ways (for instance, loss of Cl from droplets has been reported[25], particularly where many different elements are encountered from time to time. In such cases it may be more suitable to measure a curve of P factors for a few elements and mathematically or graphically fit a curve to interpolate for other elements, or to calculate such a curve from theoretical terms.

Calculated Proportionality Factors

In principle, the P_i factor for each element (or the relative intensity measured from equal concentrations) is simply the product of several probabilities: 1) the probability Q that in passing through the section, an electron strikes and ionizes a particular shell of an atom of a particular element (the ionization cross-section); 2) the probability ω that the ionized atom emits an X-ray (the fluorescence yield) 3) the probability L that the X-ray is in the line or lines being measured (line α: β:γ ratios); 4) the probability F that the X-ray is not absorbed in leaving the sample (the absorption correction, which is often ignored); 5) the probability T that the X-ray spectrometer detects and counts the X-ray (the spectrometer efficiency). It is also convenient to divide this P factor by the atomic weight of the element A if results are desired in weight fraction ratios rather than atomic ratios. We can write this equation as

$$P \propto Q\omega \; LFT/A \qquad (8$$

where the order of the terms corresponds to the definitions given above. For each of these terms, several relatively simple models and equations are available. The entire calculation, with the possible exception of the absorption correction if this changes significantly from sample to sample, can be done once and for all for a given accelerating voltage. The calculation results can also be fitted to however many factors have been measured from standards, so that better interpolations or extrapolations are possible. Note that the P_i factors enter only as ratios into the calculation of relative concentration. This means that such terms as beam current, sample thickness (except for any absorption term), and detector solid angle, cancel and need not be considered. Also any fundamental constants can be left out, and are in the expressions shown.

Ionization Cross-Section

An expression derived from the original Bethe[26] equation can be written as

$$Q_i = N_{el} \left(\log_e (E_o/E_i) \right) / (E_o E_i) \qquad (9$$

where N_{el} is the number of electrons in the shell being excited (2 for the K shell, 8 for the L shell, and 18 for the M shell) and E_o is the accelerating voltage and E_i the absorption edge energy of the element. The ratio E_o/E_i is the "overvoltage", usually written as U_i.

Several improvements or modifications have been suggested to this basic equation. Since at the high voltages used in STEM the electrons have relativistic speeds, the energy can be adjusted to correct for the change in electron wavelength[27]. This gives

$$U_i^* = U_i (1 + 9.875 \cdot 10^{-4} \; E_o) \qquad (10$$

which is roughly a 10% change for a 100 kV accelerating voltage. When the value of U_i then is put into the equation for Q_i, the change in Q_i is somewhat smaller (about 7% for potassium at 100 kV), and since this affects all elements, the ratios of P_i factors are only slightly changed (less than 1% for K relative to Na, for example).

The poorest performance from equation # 9) is found when intensities from different shells are ratioed to each other. This can be experimentally checked using K/L or L/M ratios from one element, and it is seen that the simple use of the number of electrons is inadequate. Of the various modifications to compensate for this effect, we find most satisfactory the inclusion of two adjustment terms A_1 and A_2[28].

$$Q_i = A_1 \; N_{el} \left(\log_e (A_2 U) \right) / (E_o E_i) \qquad (11$$

where A_1 = 1.0 for K shell and 0.714 for L or M shell (this is effectively the same as using the numbers 2, 5.714, 12.857 for the number of electrons instead of 2, 8, 18) and A_2 = 4/(1.65 + 2.35 exp (1 − U)).

The effect of A_2 is to decrease the cross section at low overvoltages, by as much as 10%. This may be important for K lines of moderately high energy, if the accelerating voltage is below 50 kV, but can be practically ignored for very low energy lines at high accelerating voltages.

With or without the modifications, the calculated values of Q are very good for relative element ratio calculations where all elements are analyzed by lines of the same shell, and acceptable in most cases if mixed shells are used.

Fluorescence Yield

Three commonly used expressions for fluorescence yield are based on the atomic number Z.

$$\omega = \exp (A \log Z - B)$$

Shell	A	B	(12
K	2.373	8.902	
L	2.946	13.94	

$$\omega = Z^4/(A + Z^4)$$

Shell	A		(13
K	10^6		
L	10^8		
M	7.5 10^8		

$$\omega = Z^{*4}/(1 + Z^{*4})$$
$$Z^* = A + BZ + CZ^3$$

Shell	A	B	C	(14
K	−.03795	−.11107	−.00036	
L	+.03426	+.01368	+.00386	
M	−1.163 10^{-6}	−2.177 10^{-7}	+2.01 10^{-7}	

The first expression is not recommended by the authors[29] outside the 1-12 keV range, as it diverges to values greater than 100%. The second[30] is easy to calculate, but if adequate computational capability is at hand should be replaced by the third[31], which is superior.

Relative Line Intensity

In principle, tables of relative intensities are available[32], but in practice the values reported in those tables have been determined using other means of excitation, and show a tendency toward rounding off, that introduces potential errors. Furthermore, the ratio we require, if we use the α line, is $(\alpha/\Sigma$ all) where other minor lines may underly the α peak and alter the intensity. Finally, the ratios are not readily fit to simple analytic expressions and so large tables would be required. We have found it more satisfactory to eliminate this factor by using not just the intensity in one peak, but the combined intensity for all peaks. This is provided automatically if fitting of library spectra is used, and is also readily obtained if discrete peak generation is used since the minor peaks are normally fit to resolve any interferences with other elements. The L term is then 1.0 and can be eliminated.

Absorption Correction

As has already been mentioned, this term is most often ignored (set equal to 1, meaning that there is no absorption). Even if absorption is not insignificant, it is only in cases where the difference in absorption for the different elements being ratioed is significant that the term needs to be included.

The simple correction for absorption is[27]

$$F = e^{-m_o/2} \qquad (15$$

where $m_o = \mu \rho t \cos (tilt)/\sin(takeoff)$, and

μ is the matrix absorption coefficient which must be estimated from a presumed composition as $\Sigma C_i \mu_i$ using concentrations C_i for each major element and their individual mass absorption coefficients μ_i for the element being calculated

ρt is the density times thickness of the sample (the mass thickness)

tilt is the angle of tilt of the specimen

takeoff is the takeoff angle of the X-rays

The factor of 2 is an implicit assumption that the X-rays originate, on the average, at the center of the section thickness. If we instead assume that generation is distributed through the section thickness uniformly, the expression becomes

$$F = (1 - e^{m}o)/m_o \qquad (16$$

but the difference is less than 1% for practical cases of biological matrices and viewably thin sections. Neither this term nor the use of the more exact form may be ignorable if the values of μ, ρ, or t become large, as they may in the case of heavily stained particles or other regions in tissue sections. When the absorption correction is used, the major error may still lie in the difficulty of measuring t for the specimen[33,34]

Spectrometer Efficiency

Energy-dispersive detectors have constant efficiency approaching 100% in the middle of the normal energy range (5-15 keV), as each X-ray directed toward the detector will be detected and counted. At very high energies there is a finite probability that the X-ray may penetrate completely through the detector without being absorbed, and at low energies, it may be absorbed in the beryllium window, the surface layer of dead silicon, or the thin gold contact layer This latter term is practically always negligible for our purposes; the spectrometer efficiency T can therefore be expressed as

$$T = e^{(-\mu_{Be}\rho_{Be}t_{Be} - \mu_{Si}\rho_{Si}t_{Si})}(1 - e^{(-\mu_{Si}\rho_{Si}t_{Det})}) \quad (17$$

where μ, ρ, and t have their usual meanings, and the subscripts indicate the Be window, silicon dead layer, and total detector thickness respectively. Since μ is the energy-dependent variable which can be readily calculated, and we have available reasonable values for density, it remains only to specify the thickness of the window, dead layer and detector. Typical values are 7-10μm for the window (more of the X-rays pass through it at an angle), 0.1 - 0.3μm for the dead layer, and 2-4 mm for the detector. In many cases these values can best be determined on the individual unit, particularly the first two, by adjusting them to match measured intensities from known standards or by calculating them from measurements of continuum spectra[29,35]

If the calculation method being used ignores the absorption term, then it is likely that there is not a routine handy for computing the mass absorption coefficients. In that case, a reasonably useful fit to the more accurate spectrometer efficiency curve is provided by the simple expression[36]

$$T = e^{-C_1/E^{2.8}}(1 - e^{-C_2/E^{2.8}}) \qquad (18$$

where typical values for C_2 are about 10^4 (describing the detector thickness) and for C between 0.9 and 1.5 (describing the combined absorption of the beryllium window and dead layer). Figure 4 shows the comparison of this simple curve of T as a function of energy E with the more exact expression. The deviation does not exceed 10% for energies from Na up, and the constant C_1 (C_2 is important only for very high energy lines, not often used) can be determined from a single measurement on a standard.

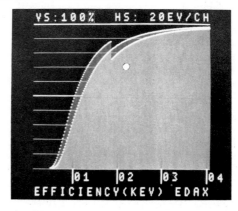

Figure 4: Plot of spectrometer efficiency versus energy. The broken line is calculated exactly from equation (16 for a 7.5μm Be window and 0.2μm Si dead layer. The smooth line is calculated from equation (18 with a constant of 1.1.

Atomic Weight

Of course, atomic weights are normally used in

table form, but for those cases in which a calculation is being performed in minimal memory, and other limitations permit some error, the atomic weight can be calculated from atomic number using

$$A = 1.428 \, Z^{1.1289} \tag{19}$$

with fair accuracy (except for a few specific elements such as Argon and nickel, the values for elements above $Z = 10$ are generally within 4%).

Conclusion

The calculation of elemental ratios in thin sections is an attractive analytical method because it makes the results largely independent of mass loss or gain in the sample, extraneous background, and beam current instability. The ratio is easy to compute from measured intensities, which can themselves be obtained from spectra in a variety of ways with accuracies limited primarily by inherent statistical considerations. Standards can be made in various ways, or calculations of relative factors can be used to supplement or eliminate the measurements.

Regardless of the details of the procedure, the general accuracy of the method is generally of the order of 10%. It may be worse in specific instances where an oversimplified term or terms are used, or absorption is ignored, and could in principle be better, approaching perhaps 5%, if all of the refinements mentioned are used. However, in the analysis of biological thin sections the greatest and therefore limiting errors arise in counting statistics, specimen preparation variables, difficulty in determining density & thickness of the section, etc. This means that accuracies of 10% may be considered as a practical limit for most cases at this time, which means in turn that the crude models and simpler equations can be used.

References

1. Hall, T.A., 1971, In Physical Techniques in Biological Research, 2nd Edition, Vol. IA, G. Oster, Editor, Academic Press, New York and London, p. 157-275.
2. Hall, T.A. and Gupta, B.L., Measurement of Mass Loss in Biological Specimens under an Electron Microbeam, Microprobe Analysis as Applied to Cells and Tissues, edited by T. Hall, P. Echlin and R. Kaufmann, Academic Press, London, 1974, p. 147.
3. W.M. Sanders & D.M. Holm, An Analytical Method for Unfolding Gamma-Ray Spectra, Los Alamos Scientific Laboratory, Los Alamos, N. Mex., Rept. LA-4030, 1969.
4. J.C. Russ, Resolving Spectrum Interferences Using Non-Gaussian Peaks, Canadian Spectroscopy, February, 1978, Vol. 23 p. 22.
5. Statham, P.J., X-Ray Spectrometry, Volume 7 Number 3 page 132-137, (1978).
6. J.C. Russ, Multiple Least Squares Fitting for Spectrum Deconvolution, EDAX EDITor, Volume 8 Number 4, 1978, p. 3.
7. J.C. Russ, Analyzing Calcium in the Presence of Antimony, EDAX EDITor, Vol. 7 No. 4, 1977, p. 18.
8. F. Schamber, X-Ray Fluorescence Analysis of Environmental Samples, T. Dzubay (ed.) Ann Arbor Press, 1977, p. 241-257.
9. Schuman, H., Somlyo, A.V. and Somlyo, A.P., Quantitative Electron Probe Microanalysis of Biological Thin Sections: Methods and Validity, Ultramicroscopy 1(1976) 317-339.
10. Fiori, C.E., Myklebust, R.L. and Heinrich, K.F.J., Prediction of Continuum Intensity in Energy-Dispersive X-ray Microanalysis, Analytical Chemistry Vol, 48, No. 1, Jan. 1976, p. 172-176.
11. Russ, J.C., Automatic Fitting of Calculated Backgrounds in Energy Dispersive X-Ray Spectra, Proc. 8th Annual MAS Meeting, 1977, p. 102.
12. Cliff, G. and Lorimer, G.W., Quantitative Analysis of Thin Metal Foils Using EMMA-4 - The Ratio Technique, Proc. Fifth European Congress on Electron Microscopy, Inst. Phys., London, 1972, p. 140.
13. Russ, J.C., Variation in Intensity Ratios Used to Identify Asbestos Fibers, Proc. Annual Denver X-Ray Conference, Plenum, 1978, p. 121.
14. Beaman, D.R., Quantitative Determination of Asbestos Fiber Concentrations, Analytical Chemistry, Vol. 48, No. 1, p. 101-110, January, 1976.
15. Pooley, F.D., The Use of an Analytical Electron Microscope in the Analysis of Mineral Dusts, Phil. Trans. R. Soc. Lond. A. 286, 625-638 (1977).
16. Ingram, M.J. and Hogben, A.M., 1968, In: Developments in Applied Spectroscopy, 6, 43-54.
17. Sawhney, B.L. and Zelitch, I., 1969, Plant Physiol., 44, 1350-1354.
18. Roomans, G.M. and vanGaal, H.L.M., Organometallic and Organometalloid Compounds as Standards for Microprobe Analysis of Epoxy Resin Embedded Tissue, Journal of Microscopy, Vol. 109, Pt 2, March 1977, p. 235-240.
19. Spurr, A.R., Macrocyclic Polyether Complexes With Alkali Elements in Epoxy Resin as Standards for X-Ray Analysis of Biological Tissues, Microprobe Analysis as Applied to Cells and Tissues, edited by T. Hall, P. Echlin and R. Kaufmann, Academic Press, London, 1974, p. 213-227.
20. Hall, T.A. and Peters, P.S., Quantitative Analysis of Thin Sections, and the Choice of Standards, Microprobe Analysis as Applied to Cells and Tissues, op.cit. p. 229-237.
21. Janossy, A.G.S, Czaran, E. and Papp, J., X-Ray Microanalysis of Zeolites, I. A natural Mineral Zeolite: Mordenite, EDAX EDITor, Vol. 6 No. 4, p. 16.
22. Rowse, J.B., Jepson, W.B., Bailey, A.T., Climpson, N.A. and Soper, P.M., Composite Elemental Standards for Quantitative Electron Microscope Microprobe Analysis, Journal of Physics E: Scientific Instruments, Vol. 7, p. 512-514, 1974.
23. Morgan, A.J., Davies, T.W. and Erasmus, D.Z., Analysis of Droplets from Iso-Atomic Solutions as a Means of Calibrating a Transmission Electron Analytical Microscope (TEAM), Journal of Microscopy, Vol. 104, 1975, p. 271-280.
24. Davies, T.W. and Morgan, A.J., The Application of X-Ray Analysis in the Transmission Electron Analytical Microscope (TEAM) to the Quantitative Bulk Analysis of Biological Microsamples, Journal of Microscopy Vol. 107, 1976, p. 47-54.
25. Marshall, A.T., Iso-Atomic Droplets as Models for the Investigation of Parameters Affecting X-Ray Microanalysis of Biological Specimens, Micron, 1977, Vol. 8: 193-200.
26. Green, M. and Cosslett, V.E., Proc. Phy. Soc., 78, 1206 (1961).
27. Goldstein, J.I., Lorimer, B.W. and Cliff, G.,

Conf. Aug. 1976, p. 25A-C.
28. Worthington, C.R. and Tomlin, S.G., Proc. Phys. Soc. (London), A69, 401 (1956).
29. Myklebust, R.L. and Heinrich, K.F.J., FRAME: An On-Line Correction Procedure for Quantitative Electron Probe Microanalysis, NBS Technical Note 796, Superintendent of Documents, Government Printing Office, Washington, D.C. 20402.
30. Wentzel, G., Z. Phys., 43, 524 (1927).
31. Burhop, E.H.S., J. Phys. Radium, 16, 625 (1955).
32. Johnson, G.G. and White, E.W., X-Ray Emission Wavelengths and keV Tables for Nondiffractive Analysis, ASTM Data Series DS46, ASTM, Phila., 1970.
33. Chandler, J.A., A Method for Preparing Absolute Standards for Quantitative Calibration and Measurement of Section Thickness with X-Ray Microanalysis of Biological Ultrathin Specimens in EMMA, Journal of Microscopy (April 1976) Vol. 106, Pt. 3, p. 291-302.
34. Misell, D.L. and Burdett, I.D.J., Determination of the Mass Thickness of Biological Sections from Electron Micrographs, Journal of Microscopy, Vol 109, Pt. 2, March 1977, p. 171-182.
35. Russ, J.C., Measuring Detector Entrance Windows, Proc. 8th Annual MAS Meeting, 1977, p. 105.
36. Russ, J.C. Quantitative Results with X-Ray Fluorescence Spectrometry Using Energy-Dispersive Analysis of X-Rays, X-Ray Spectrometry, Vol. 1, No. 3, July, 1971, 119.

Discussion with Reviewers

Barbi: In your discussion of errors, you should mention that there is only a 68% probability of obtaining results within one standard deviation. A more conservative view would be 3σ, or 30% expected error, where you suggest 10%. Do you feel that the 1σ criterion used in the paper gives an overly optimistic view of the problem, particularly with regard to counting statistics?

Newbury: In comparison to quantitative X-ray microanalysis of bulk samples, how does the distribution of errors compare for thin specimens?

Statham: Several sources of error contribute to your estimate of 10% for overall accuracy. In a typical analysis situation, what are your estimates of the relative magnitude of each of these?

Bentley: The author seems too willing to accept sloppy results and too content with low accuracy. Does he really believe that there have been no reported determinations of elemental ratios with accuracy better than about 5%?

Author: In most practical analyses of biological thin sections, the error due to counting statistics (small peaks with significant background or overlaps) will usually be the dominant one. The other two significant terms will usually be uncertain ratios between shells, if no standards are used, and the error in ignoring absorption of low energy X-rays from locally dense areas (eg. because of staining) in otherwise "thin" samples.

The estimate given in the paper of 10% is not intended as either an acceptance of poor accuracy on the one hand, nor a guarantee on the other that such results are readily or automatically obtained. There are comparatively few results available showing analyses of thin biological samples which contain accurately known local elemental ratios or concentrations to serve as a check. Instead, we must extrapolate from work with metal thin sections or mineral particles. In these cases

the better counting statistics, more complete knowledge of the matrix, and availability of either good standards or at least well-measured fundamental parameters, make it practical and worthwhile to perform the more exact corrections. Indeed, it is the existence of these other application areas which has stimulated the development of most of the more exact corrections described in the paper.

Achieving a level of accuracy of about 10% clearly requires attention to the principal sources of error already mentioned, and if the counting statistics are poor a much larger error estimate is indicated. Considering 2σ, or even 3σ, based on counting statistics in such cases is quite appropriate. Similarly, of course, any estimate of error does not preclude a particular measurement lying much closer to the "truth". Some such data almost surely must result. For the types of samples we are discussing, in which the "truth" is rarely known and confirmation of errors is difficult, I believe it unlikely that you can expect results to consistently approach 5%.

Janossy: Did you try "tailing functions" other than equation 4? Would a second Gaussian shifted to lower energy work as well?

Statham: The formula for peak tailing does not contain any adjustable parameters, implying that the degree of tailing does not depend on the particular detector being used. Would this hold even for different detectors and ones made by different manufacturers?

Author: The expression for peak tailing is one of several that have been proposed, all of them largely empirical. In applications where peaks large enough to have statistics in the tail region adequate to distinguish the effects of the parameters (a_0 and the constant 0.4) we see no significant change from one detector to another, but have not obtained comparative data on units from other manufacturers. All of the peak-tailing expressions available seem to give adequate fitting for applications with large peaks. For applications to biological thin sections, the statistics are not adequate to even tell if the tails are present, let alone to distinguish between different models.

Barbi: You describe the modification of the Gaussian shape for tailing, but not the use of the function. How does the analyst use these shapes?

Author: They are fit to the measured spectrum using a linear least-squares method. Since the shape is defined for each energy, and the centroid position is presumed to be known for each element's lines, only the height must be determined for best fit.

Fiori: What is the significance of the factor 2 in equation 7?

Author: The measured intensity value is actually the sum of the net counts N_E plus the other photons of the same energy ($N_B + N_X + N_O$). If this total is written as N_T and the total to be subtracted is N_S then the standard error in N_E is $\sqrt{N_T + N_S}$, which can be rewritten as the equation shown. The factor of 2, in other words, is due to the fact that those counts are first added in and must later be subtracted out.

Newbury: Please provide the justification for the choice of constants A_1 and A_2 in equation 11. How do calculated and experimentally measured values of the generated K/L and L/M ratios compare?

Janossy: The numerical values of constant A_1 in equation 11 are not the same for L or M shell. Instead

A_K = 0.7, A_L = 2, A_M = 4.5 should be used. We have tested and published (A.G.S. Janossy, K. Kovacs, I. Toth, Anal. Chem. 51 to be published March 79) results using weighted averages for E_i and compared intensity ratios K/L and L/M finding much better agreement between calculated and measured cross-sections by using weighted averages for the L and M-absorption edge energies.

Author: One of the greatest outstanding problems in the ratio method has been shell/shell ratios. The constants used in all models are essentially empirical adjustments to measured data, often covering only a small range of elements, conditions, etc. The new publication by Janossy is welcomed as an important contribution in the field.

Bentley: By using the combined intensity from a series of peaks in order to avoid relative line intensity problems, the overall P/B ratio and hence statistical accuracy will be decreased. Since this often is the largest source of error, please comment on the desirability of this procedure.

Author: It is true that the use of minor lines will degrade the statistical precision. Generally, the addition of those lines large enough to require stripping anyway, in the event of peak overlaps, will provide a consistent estimate of total shell intensity with a minimal effect on statistical precision (for example, at least the $L\alpha_1$, $L\beta_1$ and $L\beta_2$). If a minor peak is so small that it can be ignored from a peak overlap standpoint (and this will depend on the overall counting statistics and the elemental concentration) then it should certainly not be included in the summation. Since any fitting method using library spectra automatically fits all of the peaks, there have been methods proposed to use weighted fits so that the major peaks control the result.

Statham: If the absorption correction is ignored, at what section thickness would absorption produce errors of 5% for sodium? How can one be sure this thickness is not exceeded?

Author: This depends of course on the matrix composition. If we assume a concentration of elements representative of tissue in the hydrated state we obtain absorption coefficients (μ/ρ) ρ = 0.25±0.1μm^{-1}. These would predict a 5% absorption of Na X-rays in about 0.2μm, and considering that on the average the X-rays come from the center of a thin section and emerge at an angle of, say 30°, this corresponds to a section thickness of 0.2μm. Since the X-rays of elements to which Na is ratioed will also be absorbed, a slightly greater thickness could be tolerated; however, as higher energy X-rays are absorbed less, the effect is small and the 0.2μm value serves as a reasonable estimate. Section thickness measurement can be carried out in several ways, including measurement of electron backscattering or continuum production. The most straightforward method is a measurement of displacement of contamination spots on top and bottom of the section as the sample is tilted.

Statham: In calculating P factors, the detector efficiency varies rapidly with energy below 2 keV and becomes a critical factor in the overall accuracy. Would it not be best to use known standards in this energy range? How do you determine the constant C_1 in equation 18 from a single measurement?

Author: Standard samples are always preferred when they are available. However, at very low keV, the X-ray absorption within the sample (or the standard) can also be very significant. The calculation of efficiency using either equation 17 or 18 is an adequate substitute for standards, and sometimes preferable if the standards are not truly thin, provided that the constants (C_1 in equation 18 or the various thicknesses in equation 17) are measured on the actual system. The constant C_1 can be determined by measuring a line ratio from a known sample, computing the other terms in expression 8, and thus obtaining T. A particularly simple method is to use the copper L and K lines from a grid. Equation 8 must be modified for bulk samples and terms for effective current, stopping power, and self absorption included. The copper K line is practically speaking not absorbed in the window, so the absorption of the L line can be used to directly calculate T and thus C_1. A similar method using continuum from a bulk sample, typically a low-Z material, permits determining the thicknesses of the beryllium and dead Si separately, the latter from the height of the Si absorption edge in the spectrum. In both these cases the use of a bulk sample will yield satisfactorily high count rates and make the contribution of stray radiation small or negligible.

Janossy: Would not the use of specimen cooling and cold trapping near the sample be redundant?

Author: The purpose of specimen cooling is to reduce damage to the sample, either in the form of mass loss or ion movement. Cold trapping is designed to reduce contamination on the sample surface, and indeed the cold trap must be colder than the cooled specimen in order to protect it.

Johari: You have several references to the EDAX EDITor. Is this a publicly available journal, and if so how may it be obtained?

Author: The EDITor is published quarterly and has a total subscription of about 1600, including many libraries. The scope is generally limited to energy dispersive analysis applications, using electron or other means of excitation. Submitted papers are reviewed by an independent board and accepted without regard to what type of equipment was used. Inquiries regarding papers, submission, reprints or subscriptions may be addressed to the author.

SCANNING ELECTRON MICROSCOPY/1979/II
SEM Inc., AMF O'Hare, IL 60666, USA

PREVENTION OF LOSSES DURING X-RAY MICROANALYSIS

T. Bistricki

National Water Research Institute
Canada Center for Inland Waters
Burlington, Ontario,
Canada   L7R 4A6

Abstract

Not many years ago, chemical analysis at a cellular level became a reality by a successful merging of Energy Dispersive Spectroscopy with Electron Microscopy. The rapid development of ancillary electronics and computerized data correction techniques contributed largely to a massive utilization of Analytical Electron Microscopy. A familiar electron microscope, in a new edition, became an efficient analytical instrument at an ultrastructural level. The new technology, although readily mastered from the technical point of view, has brought with it numerous problems previously not encountered. Nowadays, after the initial excitement of finally being able to analyze at a cellular level, one of the main problems remaining is the loss or displacement of ions during preparatory processes and analyses. The preservation *in situ* of inorganic tissue components is being investigated by a number of authors. The appropriate preparative methods have been devised to minimize losses and instrumental modifications to allow microanalysis at liquid nitrogen temperature have been designed for the same reason. On the contrary, optimization of instrumental working parameters has been often neglected. Reduced beam current with correspondingly smaller beam diameter will, in many experiments, yield more information in comparison to a high dose excitation. Conditions that are optimal for high resolution imaging, such as slow scan and relatively high beam current, are not always equally suitable for analytical work. This review presents some recent results of efforts to reduce radiation and thermal damage, associated mass loss and loss of x-ray intensities.

KEY WORDS.   X-Ray Microanalysis, Biological Specimens, Specimen Damage, Mass Loss, Beam Diameter, Scan Rate, Analytical Electron Microscopy

Introduction

Theoretical and experimental studies in determining the effect of radiation are usually associated with high resolution transmission or scanning transmission electron microscopy. It is in those kinds of studies where the radiation and thermal damage is restricting the fulfillment of an atomic resolution not otherwise limited by instrumental technology[1-3]. Consequently, investigations in that field have been performed on organic macromolecules with prime consideration being the preservation and presentation of ultrastructure[4-7]. Theoretical models were designed and experimental considerations were analyzed by complicated relationships derived from physical theory[8-10]. Possible consequences of electron beam radiation have been predicted for thin or moderately thick biological specimens in the hostile environment of an electron microscope. According to one author[2], conditions of high resolution electron microscopy are so severe that they resemble the situation in close proximity to the centre of a nuclear explosion! A calculation somewhat easier to comprehend evaluates the amount of energy absorbed by an average sized bulk specimen exposed for one second to 100 kV electron beam as sufficient to bring 20g of ice to steam[11].

A number of specimen-electron beam interactions are useful in obtaining evidence about physical and chemical characteristics of a specimen under investigation; electron emission, x-ray emission or emission of light are some of these signals generated as a secondary effect of excitation and ionization of an atom. However, there are effects of the beam irradiation that are detrimental to effective data collection. Electrostatic charging, rise of temperature, contamination, chemical bond breaking and mass loss of the material are all the symptoms of a phenomenon generally known as specimen damage. Such damage was recognized in the early stages of transmission electron microscopy[12] and, for some purposes, usefulness of the technique has been, and still is, occasionally questioned.

The high resolving power of conventional transmission electron microscopes (CTEM) and transmission microscopes utilizing the scanning transmissive mode (STEM) that allow recognition of minor changes have been used to perform the majority of measurements for assessing specimen damage. For a comprehensive bibliography in that field, a reader is encouraged to consult the excellent reviews of Isaacson[13] and Cosslett[14]. In the scanning electron microscope (SEM) employing secondary electron emission for imaging of surfaces, the energy absorbed by a specimen is considerably greater in comparison to the above mentioned cases. Despite this, small changes caused by beam damage are not readily discernible. The resolution level of SEM is by orders of magnitude lower in comparison with CTEM and STEM. The damage is nonetheless induced and large changes in total mass or loss of a specific element have been theoretically assessed[15] and experimentally demonstrated[16]. The mass loss in SEM is the most damaging structural degradation caused by specimen irradiation. Such losses, when manifested in the process of x-ray microanalysis will certainly contribute to erroneous quantitative analyses or, in extreme cases, a loss of a specific element can lower its concentration beyond the detection limit of the technique.

Unlike thin transmission specimens that are absorbing only a small portion of incident energy, a poorly conductive bulk sample absorbs all of the energy impinging upon its surface. It might be considered unwise to perform x-ray microanalysis on poorly conductive samples and to do so at room temperature. However, due to reasons of economy or necessity to examine unaltered specimens (forensic, occupational health, environmental), such analyses are being conducted. Fortunately, if certain precautions are considered, the specimen damage can be considerably reduced even on bulk specimens and significant improvement in data collection can be achieved. As expressed rather poetically by Prof. Robert M. Glaeser "...this topic would be a melancholy task if it were not that several avenues still exist for attempting to overcome the limitations that are presently associated with radiation damage"[3].

Indicators of Losses and Their Quantification

In relation to numerous efforts to measure and minimize preparation losses[17-19], very little is known about the magnitude of losses that are occurring during electron excitation. The effects of specimen damage, including losses of mass have been measured on a variety of organic specimens. In the majority of cases, these measurements are derived from methods not applicable to the problems of x-ray microanalysis in SEM.

Changes of crystalline structures, for example, could only be applied to thin specimen exhibiting regular spacing of atomic layers by using TEM. Moreover, a concurrent alteration, or even a complete loss, of a diffraction pattern does not necessarily affect the intensity of x-ray emission. Measurement of changes in contrast (bright field[5] or dark field[20]) also requires thin film specimens and high resolution TEM or STEM instrumentation. The method is not applicable to x-ray microanalysis in SEM. Very interesting determinations of radiation damage can be performed by measurements of characteristic energy loss spectra[5]. Losses in very low energy regions, difficult to measure by other methods, reflect changes in masses of C, N and H, but as an unidentified participant in a discussion at the 1973 Battelle Conference described: "We could probably live with C, N and H going away if we knew that things like Na are not being lost"[21]. The measurements are performed on thin films in the microscopes (TEM, STEM) equipped with an energy loss analyzer. Results can be compared with optical absorption experiments at different wavelength bands (IR, UV)[2,4]. The energy loss method requires expensive instrumental arrangements and is not applicable to SEM microanalysis. Direct weighing of organic compounds before and after exposure to the electron beam has also been used to determine total mass loss[22]. Such a method has an advantage that initial values can be obtained without having a specimen exposed to the damaging irradiation. Physical restrictions of the method require specimens of large dimensions susceptible to heat damage. Although applicable to SEM, the method does not permit detection of losses smaller than 5 μg.

In 1974, Hall and Gupta[21,23] recognized the need for a sensitive method capable of detecting rapid changes of mass loss under conditions of analytical electron microscopy. They recorded continuum x-ray radiation which is linearly proportional to the total mass in the volume excited by a static electron beam. Sensitivity of their method could be seriously affected by the background radiation generated from sources other than the specimen itself. In analyses of small particles, for example, a large portion of the background can be generated by excitation of a specimen support. The advantage of the method is that monitoring of the mass loss can be performed simultaneously with analysis of specific elemental radiations. The method does not allow measurement of losses that are occurring in the first moments of irradiation and requires excitation of relatively large volume (beam diameter greater than 1 μm). A contamination effect, if high, can mask the actual mass loss. Similarly, it is possible to measure diminution of a specific element's x-ray intensity[24]. However, Höhling[25] reported no changes in intensity of measured Na, Cl and K intensities but observed large variations in the intensity of x-ray continuum, a measure of a total excited mass. Changes in characteristic x-ray intensity of Hg have been observed in this author's laboratory[26], although not in all samples analyzed under identical conditions. The variability has been attributed to the chemical form of Hg present in the samples. This method too, does not provide information on the initial concentration (mass) of an element before exposure.

It became obvious that some other method of determining elemental loss must be developed. A method should be sufficiently sensitive to detect minute losses from an irradiated specimen and should allow rapid and accurate determination of mass or concentration before and after specimen exposure to an electron beam. Furthermore, such a method should be independent of possible mass increase due to contamination effects.

The above mentioned requirements were partially satisfied by using an independent (Ge)Li detector and associated electronics for measurement of specific gamma radiation from $^{203}Hg^{16}$. The sensitivity of the method as originally described is in the nanogram (ng) range. The sensitivity of the method can be considerably increased by using safe radioactive compounds of higher specific activity. The total mass under examination can be contained in any suitable volume and dried on the surface of a specimen support. A disadvantage of this method is that a specimen must be removed from an SEM between consecutive measurements. This, and the fact that the beam does not follow the same path in consecutive scans, contributes to yet undetermined degree of inaccuracy. Some measurements, to determine initial specimen losses and the effects of operational conditions during x-ray microanalysis, are described in the following paragraphs.

Effect of Operational Conditions

The instruments used in x-ray spectrometry using an electron beam as a source of excitation energy can generally be separated into two distinctly different categories. The first category referred to as 'microprobes' (the expression often extended to all microanalytical systems) are designed primarily as devices for microchemical analysis of bulk specimens with the prime requirement on efficient x-ray generation[27]. They operate at relatively high beam currents of 10^{-6} to $10^{-10}A$ with a typical beam diameter in a range of 0.1 - 1.0 μm and are equipped with mechanically-operated crystal spectrometers. These instruments, designed specifically for applications in materials sciences, are not suitable for ultrastructural examinations of delicate biological material. Instruments from this category are seldom employed in biomedical applications.

The second category of instruments are those with primary functions being visual presentation of structures. They operate with beam currents from 10^{-10} to $10^{-12}A$ with a correspondingly smaller beam diameter (.002 to 0.1 μm). Relative inefficiency in x-ray production necessitates the usage of the efficient and sensitive Si(Li) detector[28].

It is well known that an electron beam entering a bulk specimen excites a volume whose size and shape is determined by the beam diameter, accelerating voltage and atomic number of the element in a specimen[29]. Since all qualitative or quantitative changes in the specimen depend on the size of that excited volume, it is extremely important to learn how to efficiently control its size and the speed over the exposed specimen area. This last parameter has been a subject of several discussions but, to the best of the author's knowledge, the importance of the scan rate has not been previously experimentally qualified.

Figure 1 shows several curves of mass losses typical for various conditions of scanning rate and beam diameter. All measurements have been effected at 30kV accelerating potential with an emission current of 150 μA. Magnification was adjusted to such a level that the entire surface of air dried, 1 μl droplet was exposed to the scanning raster. Tilt angle and working distance, 0^0 and 12 mm, respectively, were kept constant for all measurements. All samples were coated with a 20 nm layer of carbon. Some measurements, on samples coated with Al, are not included in this diagram.

In these experiments, the concentration of stock solution (aqueous mercuric chloride, 0.81 mg/ml, specific activity 1 mCi/ml) was adjusted to contain 0.81 ng of Hg in the volume of one microlitre. The microscope (AMR 1000) was not equipped with a device to monitor the beam current. The current was adjusted by focussing the beam on a small fragment of Al foil while recording the characteristic x-ray emission (EDAX x-ray spectrometer, TN-11 data acquisition and analysis system). Variations in Al x-ray intensities were found to be negligible for intervals not exceeding one minute. When irradiating for longer times, the intensities were recorded before and after the measurements and appropriate corrections were applied to normalize data. The data points on the diagram represent a mean value of three readings. Corrections for decay of radioactivity were applied as well for measurements not performed on the same day.

Operational conditions can not be separated entirely from certain preparation procedures. Three curves, at the lower left-hand side of the diagram were constructed from the measurements on non-conductive and poorly conductive supports. The loss of almost 50% was recorded within the first minute of exposure to the beam. This period actually represents a very short beam residence (dwell time) at any given point on the specimen surface. A total cumulative time does not exceed more than a score of microseconds in a typical analysis.

Under conditions of poor conductivity, temperature rise is probably the most important factor contributing to the mass loss, undoubtedly aggravated by radiation induced rupture of chemical bonds[11]. Theoretically, if a specimen of 1μm³ is irradiated by a beam of $10^{-8}A$, the temperature can rise to 2690K in 300 μs from an initial 300K[15].

The middle trace in the upper part of the figure represents a set of measurements obtained under identical irradiating conditions but the specimen was deposited on a conductive support. The reduction of loss was considerable but still intolerable in any kind of quantitative analysis. Although illuminating conditions have been rather exaggerated to illustrate the point, it is obvious that the importance of conductivity cannot be overemphasized.

Three curves on the top of Figure 1 illustrate variability in mass losses when the scanning rate or the beam diameter is altered. In the illustrated example, a beam of 0.1 μm diameter, when travelling slowly along the exposed area, resides on any given "point" of specimen surface for 0.25 μs. This time can be defined (Appendix) as an Apparent Dwell Time (ADT). If the beam diameter is kept constant but the scan rate is increased[30] the ADT decreases(.08μs) resulting in appreciable reduction of losses. An important factor to note here is that faster scanning leaves some unirradiated area between lines that are spread further apart in comparison with a slow scan. The mass loss of a total sample indicates a reduction although locally, along the scan lines that are exposed more frequently, the opposite may be true. The ADT, however, may still give a relative indication of absorbed energy if the above mentioned uncertainty, caused by inhomogeneous irradiation, is disregarded.

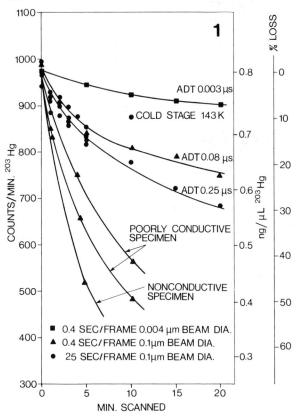

Figure 1. Diagram of mass losses at various operating conditions and specimen conductivities.

The energy deposited per discrete point on the specimen surface is proportional to ADT which can be used thus as a relative measure of absorbed energy for a given beam diameter and accelerating potential. Further improvement in reduction of mass losses can be achieved by using the electron beam of a smaller diameter. The uppermost curve on the diagram illustrates results of such action. It must be noted that reduction of beam diameter will increase the time required to obtain statistically significant x-ray counts. A compromising solution must be chosen in dependence with specimen sensitivity.

Effect of Low Temperature

The main objective of this presentation was to discuss operating conditions of SEM in relation to the possible outcome of x-ray microanalysis. That would not be possible without touching, however briefly, on the role of operating at low temperatures.

It is conceivable that very low specimen temperatures might prevent radiation damage and accompanying molecular disorder[3]. If disorder is promoted by the increased motion of ionized molecular fragments then, by lowering the temperature, a molecular motion could be restricted and disorder prevented or delayed[11,13]. Consequently, an improvement in reducing mass loss could be realized by keeping specimens at low temperature during irradiation.

Talmon and Thomas[31] have theoretically predicted the possible temperature rise and sublimation rate from frozen hydrated specimens, but there is only a limited number of experimental data available either to support or disprove their calculations. Theoretically, the temperature of the frozen specimen will rise only by a few degrees when irradiated by an electron beam. However, Fuchs and Lindemann[32], exposing a bulk specimen to a stationary beam, have observed a depression being produced on its surface. By scanning a small area instead, the effect of local heating was decreased and surface charging prevented. These authors also noted a reduction of x-ray intensities at low temperatures (-145°C). The x-ray count rate increased considerably after allowing for some ice sublimation at -90°C for a short time. Salih and Cosslett[33] described reduction of damage by a factor of 3-4, as a result of lowering the temperature of a sample from room temperature to 4K. Although a considerable improvement in radiation resistance can be achieved at the temperature of liquid nitrogen, the full effect, the authors claim, requires a stage cooled by liquid helium. From experiments of Hall and Gupta[23] it is obvious that mass loss is dependent on the specimen temperature at the time of excitation. Even the initial effect of radiation damage attributed to chemical bond rupture is greatly reduced at -180°C. Further discussion on the subject of low temperature specifically pertaining to specimen preparation, interpretation of results and morphology can be found in a recent review of Echlin[34].

In this author's experience, while exposing a specimen to adverse beam irradiation at 143K (1μA beam current in .1 μm beam at 30 kV) the mass loss as compared to room temperature excitation was reduced by a factor of 2-3. It is hoped that using a somewhat less energetic beam will result in a reduction of loss below the level obtained at room temperature. An interesting point to note is that, for example, an $HgCl_2$ specimen deposited, as described in a previous section, on an aluminum surface or coated with a layer of Al, did not show any statistically significant reduction in loss at 143K. This is regarded as occurring in a consequence of amalgamation of metallic surface resulting in formation of an extremely volatile product. In such a case, it does not seem possible to reduce the damaging effect of impinging radiation; by using an inert carbon support and carbon coating, the loss of specimen mass was reduced to a level indicated in Figure 1. Experiments at low temperature have been performed by placing a specimen on a microscope stage at room temperature and then cooling the stage after the microscope was evacuated below 10^{-3} torr. The stage, (Figure 2) was designed to replace a standard microscope stage and to retain the possibility of X, Y and Z adjustments. The specimen elevation angle was fixed at 35° but, with freedom to adjust the working distance and variable geometry of an x-ray detector, the selection of optimal take-off angles was readily facilitated. Cooling the stage to 143K from room temperature was achieved within 3 minutes by delivering liquid nitrogen through flexible steel bellows from a pressurized container. Temperature has been measured on the specimen support using an iron-constantan thermocouple. The stage was manufactured using local resources at a cost of less than $500.

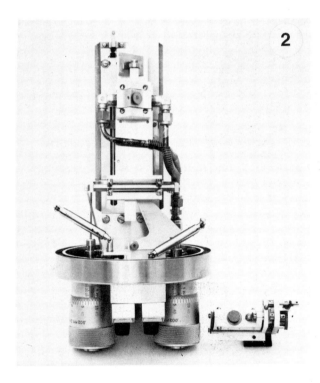

Figure 2. A simple liquid nitrogen cooled stage

Discussion and Conclusions

Specific requirements of x-ray microanalysis are frequently disregarded and counts may be collected under conditions that are generally useful for morphological examinations but inappropriate for x-ray microanalysis. It is of course possible in many experimental situations to obtain results without taking precautions regarding the amount of irradiating energy. In scanning beam instruments, the beam is usually stopped and positioned at the area of interest. If the concentration of an element in an excited volume is low and an excitation energy is high, it is certain that the loss of mass or a loss of a specific element can occur before the time required to collect a statistically significant number of counts. Situations of that kind may be rare in some scientific disciplines utilizing x-ray microanalysis with electron beam instruments (materials sciences) but there are a number of instances where a correct answer to an analytical problem must be given in a first attempt. Examples of this kind are frequent in forensic or in occupational hazard laboratories where samples can be "one of a kind", unique in their conditions, composition and heterogeneity. Erroneous analytical results may have unforeseeable consequences in some occasions, while in environmental surveillance a consequence may be of a financial nature; it may cost thousands of dollars to collect another sample from the same location, which may not be the same as the previous one. When selecting operational conditions for microanalysis, it must be remembered that conditions for high resolution imaging and conditions for error-free, quantitative analyses are not identical. This presentation, it is hoped, has given evidence that it is possible to analyze with minimal or no losses, even at room temperature, if operational conditions are carefully selected.

The method for assessing mass loss using radioactive compounds is potentially powerful but requires further refinements. A gamma spectrometer should ideally be attached to a microscope column and measurements performed without removing the specimen or turning the beam off. Rotation of a model specimen during exposure would ensure homogeneous irradiation and better accuracy of measurements. Experiments, utilizing the above are presently in progress in the author's laboratory, but results are yet inconclusive.

Acknowledgments

The author is indebted to Dr. R.W. Durham of NWRI for permission to use a gamma spectrometer, to Mr. R.J. Goble and Dr. S.R. Joshi for their assistance in operating the equipment, to Mr. H. Savile for engineering the cryo-stage, to Dr. G.G. Leppard and reviewers of the paper for helpful suggestions and discussions during preparation of the manuscript.

References

1. P. Echlin and J.A. Fendley: "The future of electron microscopy in biology"; Nature, 244 (1973), 409-414.
2. L. Reimer: "Review of the radiation damage problem of organic specimens in electron microscopy"; Chapter 13, in Physical Aspects of Electron Microscopy and Microbeam Analysis, (B.M. Siegel and D.R. Beaman, eds.), John Wiley and Sons, New York 1975.

3. R.M. Glaeser: "Radiation damage and biological electron microscopy"; Chapter 12, in Physical Aspects of Electron Microscopy and Microbeam Analysis, (B.M. Siegel and D.R. Beaman, eds.), John Wiley and Sons, New York 1975.
4. M. Isaacson, D. Johnson and A.V. Crewe: "Electron beam excitation and damage of biological molecules: Its implication for specimen damage in electron microscopy"; Rad. Res. 55, (1973), 205-244.
5. S.D. Lin: "Electron radiation damage of thin films of glycine, di-glycine and aromatic amino acids", Rad. Res. 59, (1974), 521-536.
6. M. Isaacson, J. Langmore and J. Wall: "The preparation and observation of biological specimens for the scanning transmission electron microscope"; in SEM/1974, pp. 19-26, (O. Johari and I. Corvin, eds.), IITRI, Chicago 1974.
7. R.M. Glaeser: "Limitations to significant information in biological electron microscopy as a result of radiation damage"; J. Ultrastruct. Res. 36 (1971) 466-482.
8. D. Misell: "Conventional and scanning transmission electron microscopy: image contrast and radiation damage"; J. Phys. D: Appl. Phys. 10, (1977), 1085-1107.
9. Y. Talmon and E.L. Thomas: "Beam heating of moderately thick cold stage specimen in the SEM/STEM"; J. Microsc. 111, (1977), 151-164.
10. Y. Talmon and E.L. Thomas: "Electron beam heating temperature profiles in moderately thick cold stage STEM/SEM specimens"; J. Microsc. 113, (1978), 69-75.
11. K. Stenn and G.F. Bahr: "Specimen damage caused by the beam of the transmission electron microscope, a correlative reconsideration"; J. Ultrastruct. Res. 31, (1970), 526-550.
12. J. Hillier, S. Mudd, A.G. Smith and H. Beutner: "The 'fixation' of electron microscopic specimen by the electron beam"; J. Bacteriol. 60, (1950), 641-654.
13. M.S. Isaacson: "Specimen damage in the electron microscope"; Chapter 1 in Principles and Techniques of Electron Microscopy, Biol. applications, Vol. 7, (M.A. Hayat, ed.), Van Nostrand Reinhold Co., New York 1977.
14. V.E. Cosslett: "Radiation damage in the high resolution electron microscopy of biological materials: a review"; J. Microsc. 113, (1978), 113-129.
15. R.W. Shaw and R.D. Willis: "X-ray emission analysis; Sample losses during excitation"; in Analytical Electron Microscopy and X-ray Applications to Environmental and Occupational Health, Vol. 1, 1978, (P.A. Russell and A.E. Hutchings, eds.), Ann Arbor Science Publ. Inc., Ann Arbor, Michigan, 1978, pp. 51-64.
16. T. Bistricki: "Quantification of losses during x-ray microanalysis of mercury"; in Electron Microscopy 1978, Vol. 2, pp. 12-13; Proc. 9th Int. Congress on Electron Microscopy, Toronto 1978, (J.M. Sturges, ed.), Microscopical Society of Canada, Toronto 1978.
17. P. Echlin and A.J. Sauberman: "Preparation of biological specimens for x-ray microanalysis"; in SEM/1977, Vol. I, pp. 621-637, (O. Johari, ed.), IITRI, Chicago 1977.

18. J.R. Coleman and R.A. Terepka: "Electron probe analysis of the calcium distribution in cells of the embryonic chick chorioallantoic membrane, I. A critical evaluation of techniques"; J. Histochem. Cytochem. 20, (1972), 401-413.
19. J. Gulash and R. Kaufmann: "Energy-dispersive x-ray microanalysis in soft biological tissues: relevance and reproducibility of the results as depending on specimen preparation (air drying, cryofixation, cool-stage technique)", pp. 175-190, in Microprobe Analysis as Applied to Cells and Tissues, (T. Hall, P. Echlin and R. Kaufmann, eds.), Academic Press, New York 1974.
20. J.P. Langmore: quoted in text reference 13., "Studies of unstained and selectivity stained biological molecules using scanning transmission electron microscopy"; Ph.D. dissertation, The University of Chicago, 1975.
21. T.A. Hall and B.L. Gupta: "Measurements of mass loss in biological specimens under an electron microbeam"; pp. 147-158 in Microprobe Analysis as Applied to Cells and Tissues, (T. Hall, P. Echlin and R. Kaufmann, eds.), Academic Press, New York 1974.
22. K.S. Stenn and G.F. Bahr: "A study of mass loss and product formation after irradiation of some dry amino acids, peptides, polypeptides and proteins with an electron beam of low current density"; J. Histochem. Cytochem. 18, (1970), 574-580.
23. T.A. Hall and B.L. Gupta: "Beam induced loss of organic mass under electron-microprobe conditions"; J. Microsc. 100, (1973), 177-188.
24. A.R. Spurr: "Freeze-substitution systems in the retention of elements in tissues studied by x-ray analytical electron microscopy"; pp. 49-61 in Thin Section Microanalysis, (J.C. Russ and B.J. Panessa, eds.), Edax Laboratories, 1972.
25. H.J. Höhling et al.: quoted in text reference 23.
26. S. Ramamoorthy, A. Massalski and T. Bistricki: "Energy dispersive x-ray microanalysis of mercury in environmental samples"; pp. 247-254 in Analytical Electron Microscopy and X-ray Applications to Environmental and Occupational Health, Vol. 1., 1978, (P.A. Russell and A.E. Hutchings, eds.) Ann Arbor Science Publ. Inc., Ann Arbor, Michigan, 1978.
27. R. Woldseth: X-ray Energy Spectrometry, 1st edition, Chapter 4, 1973; Kevex Corp., Burlingame, Ca., 1973.
28. R. Fitzgerald, K. Keil and K.F.J. Heinrich: "Solid-state energy-dispersion spectrometer for electron probe x-ray microanalysis"; Science, 159, (1968), 528-529.
29. J.I. Goldstein: "Electron beam-specimen interaction"; Chapter III in Practical Scanning Electron Microscopy, (J.I. Goldstein and H. Yakowitz, eds.), Plenum Press, New York 1975.
30. D.E. Newbury: "Image formation in the scanning electron microscope"; Chapter IV in Practical Scanning Electron Microscopy, (J.I. Goldstein and H. Yakowitz, eds.), Plenum Press, New York 1975.
31. Y. Talmon and E.L. Thomas: "Temperature rise and sublimation of water from thin frozen hydrated specimens in cold stage microscopy"; pp. 265-272 in SEM/1977/I, (O. Johari, ed.), IITRI, 1977.

32. W. Fuchs and B. Lindemann: "Electron beam microanalysis of frozen biological bulk specimens below 130°K: Instrumentation and specimen preparation"; J. Microsc. Biol. Cell 22, (1975), 227-232.

33. S.M. Salih and V.E. Cosslett: "Radiation damage in E.M. of organic materials: effect of low temperatures"; J. Microsc. 105, (1975), 269-276.

34. P. Echlin: "Low temperature scanning electron microscopy: a review"; J. Microsc. 112, (1978), 47-61.

35. J.C. Wiesner: "Measurements of optical parameters for the scanning electron microscope"; pp. 675-682, in SEM/1976/I, (O. Johari, ed.), IITRI, Chicago 1976.

Discussion with Reviewers

H.K. Hagler: What materials are used to fabricate the various conductive, poorly conductive and non-conductive specimen supports? What characteristics did you measure to get these three categories?

Author: Specimen supports of various conductivities were prepared by mixing graphite powder with a low viscosity epoxy resin in different proportions. These mixtures were cured on surfaces of standard Al stubs and categorized by measuring a resistance between graphite-epoxy surface and the base. Non-conductive: infinite resistance; poorly conductive: $\gtrsim$ 3kΩ resistance; conductive: $\lesssim$ 300Ω resistance.

A.R. Spurr: As oxygen and carbon usually make up a very large part of the dry weight of animals and plants, wouldn't one expect losses in mass due to beam damage to include O as well as C, N and H?

J.L. Costa: The experiments presented refer specifically to the mass loss of ^{203}Hg following electron bombardment. Does the author think it is wise to extrapolate from the data general principles about the problem of electron-associated mass loss for other elements (especially in biological specimens)?

J.S. Wall: Most biological samples are non-volatile and react quite differently from volatile mercury during electron beam excitation. Why did you choose mercury as your experimental model for comparison to biological specimens?

Author: The difference between experimental model here and other elements that may be analyzed in biological specimens is only in a degree of possible losses. It would be perhaps more realistic to incorporate a tracer into organic matrix, but it is my opinion that general principles of beam energy management would hold in any situation. Hg in oxidized mercuric chloride is not more volatile than many other compounds. HgCl$_2$ decomposes well above 600K. It was selected because of a need in the author's laboratory to quantitatively determine Hg in specific biological components of aquatic environments.

H.K. Hagler: What happens to the analysis time required as a function of beam conditions, i.e.,

how much longer do you have to analyze under beam conditions that are gentle and achieve acceptable statistics for peak identification?

Author: X-ray emission varies directly with the beam current. A 1000 count spectrum accumulated at 30 pA would require ten times longer acquisition than the same spectrum accumulated at 300 pA. However, the more energetic beam in a static mode or at slow scanning rate may (depending upon the sensitivity of a specimen) give an incorrect quantitative result as illustrated on Figures 3 and 4. Lower spectrum on Figure 5 was obtained in 100 sec. using the smallest beam diameter (.004 µm on AMR 1000) with a current of 30 pA. Although increased current (200 pA) produces a statistically "better" spectrum in the same time (upper trace, Fig. 5) the gentle excitation yields a readily identifiable spectrum.

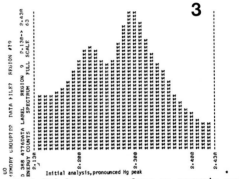

Figure 3. Initial spectrum from an organic sediment component.

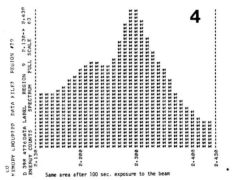

Figure 4. Spectrum from the same area after 100 sec irradiation.

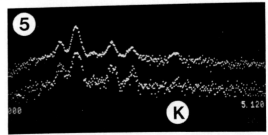

Figure 5. Comparison of spectra from the same specimen using 30 pA (lower) and 200 pA (upper) current. K(Kα) peak represent a concentration of 0.2%.

A.R. Spurr: Is there any evidence from your work with reduced beam current, that the beam traverses less redeposited material from the prior scan line and therefore fewer spurious x-ray signals are generated?
Author: During the experimental work presented here this factor was not important since the specimen was of homogeneous composition. It is conceivable, however, to believe that improvement in this respect can be realized by using low energy irradiation.

J.S. Wall: How did you ascertain that the diminished signal obtained from the mercury preparations was not due to the mercury 'flaking' away as it charges up?
Author: Each specimen was coated with a relatively thick ($\simeq$ 20nm) layer of carbon. Except on the surface of a fully non-conductive support, charging symptoms were not observed.

B.J. Panessa-Warren: Would you explain why you chose to build a cold stage rather than buy a commercially-available model. Are there advantages to your cold stage design?
Author: The objective of this design was to achieve an effective and rapid cooling of thick specimens. The major concern was to prevent a mass loss and to optimize the x-ray signal collection. It can be used in morphological studies of frozen hydrated material in a bulk form. The main advantage in comparison to commercial models is in the economy of construction.
B.J. Panessa-Warren: You stated that local heating can be decreased and surface charging prevented by scanning a small area of specimen. Would you explain how reducing the area scanned can make a specimen appear to be more thermally and electrically conductive?
Author: In the work of Fuchs and Lindemann (text ref. 32.) on frozen hydrated specimens, the scan of a small area (4 μm x 4 μm) has shown reduced heating effect in comparison to an excitation with a static beam. In this case a small area scan extensively increased exposed surface. Charging was reduced due to sublimation of ice which is a good thermal insulator at -145^0 C.

J.L. Costa: What was the efficacy of the EDAX detection as compared to that of a gamma counter?
Author: This was not compared. The main concern was to develop a method for measurement of initial loss which cannot be monitored using x-ray signal as an indicator.
J.L. Costa: It is difficult to compare the data given, since different scan rates were apparently used to produce different curves. It would be helpful, for example, to have presented the "cold stage" data in the same scan terms as were used to derive the remainder of the data.
Author: One of the objectives of this paper was to compare the effect of different beam speeds over the exposed area. Cold stage data are produced by using high excitation energy and are directly comparable to other data indicated by circles.

Appendix

The ADT can be determined by the following calculations:

First, the length of a beam path (Σd) is defined by

$$\Sigma d = t_{\delta} \cdot d / t_{\ell} \qquad (1)$$

where t_{δ} is a time to scan one full faster from the top to the bottom of an image in seconds; t_{ℓ} is a time required to scan one line on the x axis of a cathode ray tube (CRT) in seconds and d is a length of a single x sweep (t_{ℓ}) in μm.

t_{δ} is readily determined by observing and timing the Y sweep of a beam on CRT while t_{ℓ} is found from the microscope service manual. To determine d, a length of X in mm is measured on the CRT and divided by a magnification (on the screen) x 10^{-3}. Knowing the distance which the beam has to sweep in a given time, the ADT is calculated as follows:

$$ADT = t_{\delta} \cdot \phi / \Sigma d \qquad (2)$$

where ϕ is beam diameter in μm. The beam diameter must be measured[35], if the "spot size" control on the instrument is not already calibrated in such units.

SCANNING ELECTRON MICROSCOPY/1979/II
SEM Inc., AMF O'Hare, IL 60666, USA

IDENTIFICATION AND PREVENTION OF ARTIFACTS IN BIOLOGICAL X-RAY MICROANALYSIS

B. J. Panessa-Warren

Department of Orthopedic Surgery
Health Sciences Center
State University of New York,
Stony Brook, NY 11794

Abstract

Although X-ray microanalysis is considered a "non-destructive" analytical technique (when compared to analytical methods such as atomic absorption spectrophotometry, proton activation analysis, secondary ion mass spectometry etc..), all organic samples, whether biological or non-living, are affected by the vacuum and beam of the electron microscope, as well as by the physico-chemical trauma which may be experienced during specimen preparation. With the advent of sophisticated cryotechniques and the recent advances in specimen cold stage design, it is possible to examine freshly frozen hydrated cells by X-ray microanalysis. Unfortunately, many investigators do not have cryoequipment available to them, and therefore they are limited to using routine electron microscopy preparation procedures. Although these preparation methods are not ideal for X-ray microanalysis, valid elemental information can be obtained by observing certain precautionary measures and carefully scrutinizing the resultant X-ray data. This paper is designed to aid the researcher who wishes to utilize X-ray microanalysis but is limited to conventional preparation methods or modest instrumentation. Artifacts due to specimen preparation (the introduction of exogenous elements, translocation and loss of materials, apparent concentration of elements due to drying), the electron beam (radiation and heat damage, contamination, electron scatter, fluctuations in current and accelerating voltage) and artifacts resulting from detection and display of X-ray events (spurious signal, signal detected from areas other than the area directly under the beam), are discussed and illustrated in terms of their identification and prevention.

KEY WORDS: X-Ray Microanalysis, Biological Specimens, Artifacts, Artifacts Prevention, X-Ray data Comparision

Introduction

X-ray microanalysis offers the biological investigator an accurate, relatively rapid technique by which bulk specimens, thin and thick sectioned material, powdered samples and liquids may be analyzed qualitatively and quantitatively for their elemental content. This is a relatively non-destructive technique in that the specimen remains intact during the analysis. However, it must be noted that the electron beam does interact with the specimen and may cause minor to severe alterations in the sample. Using either wavelength or energy dispersive spectrometry the specimen can be imaged by scanning or transmission electron microscopy, and the sample examined at the gross or ultrastructural level. Immediately thereafter, the beam can be positioned directly over a desired sample area or organelle, and the analysis begun. This ability to visually examine a specimen at high magnifications and perform elemental analysis of areas as small as 5 - 10 nm in diameter without appreciably destroying the morphology of the sample is unique in microanalytical methods.

However, like all scientific tools this attractive technique is only as good as the investigator's ability to interpret the resultant data. This paper attempts to discuss the identification and prevention of some of the artifacts encountered when biological samples are subjected to X-ray microanalysis.

Redistribution and Loss of Elements During Specimen Preparation

During tissue processing for microscopy, various constituents of biological cells may be redistributed or lost entirely from the specimen. Therefore, the specimen preparation methods used most successfully for TEM imaging may prove totally useless for the preparation of cells for X-ray microanalysis. Low molecular weight soluble substances, ions and electrolytes[1] are easily lost from tissues during fixation and dehydration. In actuality, any material which is not securely bound within the cell is capable of transcellular movement if water is removed from the sample, whether by vacuum desiccation or by chemical substitution.[2] For this reason, fixatives have been used to try to immobilize cellular constituents, and retain at least those elements that are linked to cellular protein moieties.

Fixation and Elemental Precipitation

Ideally one would prefer not to introduce any exogenous elements into a specimen. For this reason many investigators have turned to examining unfixed, frozen tissue on a cold stage. [3,4] These exciting cryotechniques, however, are not universally available, or applicable to all tissues or research projects, and therefore, many investigators, with care, successfully continue to use fixation techniques. [5,6,7,8]

In order to retain diffusible ions, the biological sample should be examined frozen or freeze dried. [1,4,9] If these two methods are not possible, precipitation techniques have been developed which can immobilize an otherwise diffusible substance by histochemical reactions. Such techniques include precipitating Ca^{++} with oxalate[5,6] and recently with 2% pyroantimonate[8]; Cl^- with Ag^+ salts in animal and plant tissue[10]; PO_4^{3-} with lead salts[11]; and carbonic anhydrase with a cobalt reaction product. [8] These are merely a few of the myriad of histochemical reactions that may be used to immobilize and "elementally tag" diffusible cellular materials. It must be remembered, however, that using such heavy metal stains does introduce exogenous elements into the specimen, also increasing the amount of X-ray continuum generated.

Some controversy exists concerning the use of fixatives for X-ray microanalysis. Many investigators feel that the use of fixatives, such as glutaraldehyde, causes alterations in the "semipermeability" of cellular membranes, resulting in some redistribution of cellular constituents. [4,9] This phenomenon is well documented when cellular membranes are treated with osmium tetroxide alone. It should also be mentioned here that some tissues are more profoundly altered by fixation than others, and each specimen must be considered separately.

If a fixative must be used, the key to reducing fixation artifacts is to determine experimentally the type of fixative to be used (glutaraldehyde, paraformaldehyde, formaldehyde, acrolein, etc.), the optimal tonicity of the buffer (whether it is cacodylate, veronal, phosphate, Hepes, etc.) and the proper concentration of fixative-to-buffer for each tissue to be studied. The pH of the buffer solution is an equally important factor to consider. This is markedly demonstrated in the preparation of T7-coliphages which remain intact in 1.5% glutaraldehyde at pH 7.2 but extrude their DNA when the pH is raised to 7.4. [12] Boyde et al. [9] noted an increase in specimen volume when embryonic tissues were transferred from a fixative to a buffer wash (the same buffer used to dilute the fixative). If care is taken to make the buffer isotonic to the inherent tonicity of the cells comprising the specimen, this swelling should not occur. Usually Na^+, Ca^{++}, or K^+ salts are used to alter the tonicity of a buffer. The choice of salts would be made with regard to the analysis of the desired characteristic X-ray line. For example, calcium should not be added

to a fixative or buffer solution if calcium, or even potassium, X-ray signal is to be detected. All fixative and buffer washes should be monitored for the presence of elements which may complicate the experimental spectrum. Cacodylate buffer, for example, contains arsenic and samples fixed or washed in solutions containing this buffer exhibit arsenic X-ray signal, as well as a slight increase in X-ray continuum.

Dehydration. Materials may also be washed out of tissues during dehydration and drying. For this reason a dehydrating agent should be sought which is less apt to remove cellular constituents that are to be studied. For example we prefer to use acetone as a dehydrating agent because it shows negligible leaching of lipids from the articular nerves studied in our laboratory, as compared to ethanol. To prevent osmotic shock, tissues should be dehydrated gradually. We prefer to use an incremental drop-wise dehydration with constant agitation, over a period of 15 min. to 2 hours. For most tissues this allows for adequate solvent penetration without the dissolution of lipid moieties.

Shrinkage is always to be avoided in preparing samples for X-ray microanalysis because it is usually accompanied by elemental translocation. Shrinkage has been shown to occur in varying degrees in tissues dehydrated in acetone, ethanol, methanol and propyl alcohol. [9]

It is not accurate to assume that tissues subjected to solvent dehydration will always show elemental redistribution, especially if the elements in question are immobilized within the cells. However, to rule out this potential source of artifact, fresh tissues analyzed by atomic absorption spectrophotometry (or an equally sensitive form of elemental analysis) can be compared to the same type of tissue following dehydration. If the element in question shows a decrease in concentration in the dehydrated tissues, it may be assumed that elemental loss has occurred.

Drying Artifacts. Critical point drying, air drying and freeze drying have also been shown to produce some shrinkage of tissues. However, this distortion will vary according to the tissue processed, technique used and the ability of the operator. [13,14] Boyde et al[9] found that critical point drying produced far less distortion than air drying. The stage at which the most evident distortion occurred was during decompression, particularly when the pressure bomb approaches atmospheric pressure. This is thought to be due to the retention of unsubstituted intermediate fluid (or water) in the specimen, which evaporates causing surface tension associated shrinkage. Consequently, this problem can be minimized by using smaller pieces of tissue which have had adequate dehydration, and thorough CO_2 or Freon replacement of the dehydration solvent. Cohen[13] suggests that turbulence in the critical point bomb can also produce severe tissue damage, which may ultimately result in elemental loss. The two periods of greatest turbulence are when the bomb is being filled with the transition fluid, and as the critical temperature is approached. During the latter period, convection currents become extremely strong and care must be taken to avoid too rapid heating of the bomb. Some of these deleterious conditions may be ameliorated by placing the specimens in a porous container held off the floor of the bomb.[13] For the past eight years we have been using a simple stainless steel wire mesh, cylindrical basket measuring 12mm x 9mm to protect our specimens during critical point drying. These cylindrical baskets are placed in a larger wire mesh basket within the critical point bomb (Fig. 1).

If samples are dried by any method, there is always a chance that materials may be relocated within the body of the specimen. For this reason some other form of analysis should be used to determine the elemental composition of a tissue, and serve as a comparison for the X-ray data. We have used atomic absorption spectrophotometry in the past to check the electrolyte content within fresh, fixed, dehydrated and dried samples, and used this information to determine if the content or distribution of the element in question was altered at any stage of specimen preparation.[15] Radioactive labeling followed by scintillation counting also has been used as a comparative method for elemental determination.[5]

We also have found that translocation may occur due to desiccation of the specimen during transferral from the dehydrating solvent to the critical point bomb. To eliminate this problem, the tissues were placed in a wire mesh basket to retard desiccation, and the basket (filled with tissue), plunged immediately from the solvent into CO_2 "snow." By surrounding the tissue with the snow, and immediately placing it into the bomb and flushing it with liquid CO_2, desiccation was prevented, as was rehydration from atmospheric water vapor.

With improved instrumentation and expertise in freeze drying, this method has now proven to be equally good, and in many cases better, than critical point drying.[4,9] If the freezing step is not done properly, ice crystal damage may alter tissue morphology and elemental distribution. Since water is drawn from the cytoplasm of the cells to the forming ice crystals, elements may follow the water movement causing elemental redistribution. Therefore, great care must be taken during the initial freezing procedure to minimize ice crystal formation.

Concentration of elements after drying

Samples treated to remove all cellular water may show a somewhat "non realistic" concentration of elements. It must be remembered that electrolytes and elements in a living cell are suspended in an aqueous environment. After the cellular water is removed, the residual elements will be deposited onto the adjacent cellular structures. When these dried cells are subjected to X-ray microanalysis as much as a 10% increase in elemental concentration may be evident. Figure 2 shows a schematic representation of this phenomenon. This occurrence is far more frequent in plant and animal cells possessing extensively vacuolated cytoplasm, in comparison to cells having dense cytoplasm with fewer vacuoles.

Figure 1. To decrease the physical trauma experienced by tissues during critical point drying, we place specimens in a 12mm x 9mm stainless steel wire mesh basket with a mesh snap-cap.

hydrated plant cell   dried cell

Figure 2. This schematic representation of a hydrated plant cell shows intracellular substances suspended in the tonoplast vacuole (V). When the cell is dried, these materials can no longer be suspended in an aqueous vacuole, and therefore, are deposited on adjacent membranes giving the appearance that there has been an increase in the concentration of the vacuolar elemental constituents.

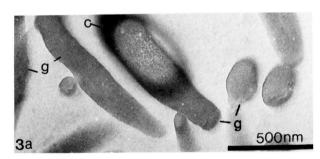

Figure 3a. This plastic embedded, unstained section (120 nm thick) of frog retinal pigment granules bombarded with an 80 keV beam (32 μAmp beam current) for 100 sec. at 32° tilt. The incident elliptical spot is approximately 470 nm in length. A pigment granule (g) in this unstained section, after analysis exhibits a burn mark (C). Although a liquid nitrogen cold finger was used, an elliptical contamination ring appears around a central lighter core showing some sample degradation.

Artifacts due to the Electron Beam

Elemental loss may also result from electron bombardment of the specimen surface. The use of high beam currents (especially when using STEM at high accelerating voltages) can cause a rapid loss of organic material from plastic embedded samples, thereby resulting in substantial elemental losses.[16] Usually this type of artifactual elemental loss can be significantly reduced by decreasing the beam current, which, unfortunately will also simultaneously reduce the count rate.

Radiation Damage

Radiation damage as a source of artifact can be classified into two categories; loss of mass, and heat damage. The radiation dose deposited in a sample subjected to high resolution electron microscopy is about 10^{10}-10^{11} rads. A dose of 10^9 rads, according to Glaeser[17] is sufficient to destroy the chemical properties of most organic materials, while 10^6 rads will inactivate most enzymes. Nucleotide bases have been found to be 100 times more resistant to electron beam exposure, as are extensively conjugated crystalline molecules. This may explain why aliquots of biological secretions and similar fluid samples which have been frozen at $-160^{\circ}C$ (the ice sublimed in vacuum), and crystallized, have been reported to withstand extensive electron beam bombardment (sequential 3 min. counting periods) without any alteration in elemental detection.[18]

Stenn and Bahr[19] demonstrated that chemical bonds which are in an excited or ionized state may often break, forming highly reactive free radicals. Intermolecular cross linking and the formation of double bonds (from single bonds) may also occur. The end result of these phenomena is the release of low molecular weight species from the sample, such as NH_3 (from amines and amides), CO_2 (from carboxylic acids), H_2, and ultimately the alteration of the specimen's chemical composition.

Figure 3 shows a micrograph of a 120 nm thick epon embedded unstained section of frog retina pigment epithelial melanin granules analysed for 100 seconds at 80 keV in a Philips 301 TEM (32 μAmp beam current). Although a liquid nitrogen cold finger was used at the time of the analysis, what appear to be extensive burn marks can be seen surrounding the damaged central region. During the analysis of this sample the count rate was initially high, but decreased by 50% within 20 seconds after the analysis was begun. Apparently the plastic and associated tissue were so damaged by the beam that material has actually been lost from the section. This could have been prevented in large part by reducing the beam current.

Heat Damage

To a certain extent, Figure 3 may be attributed to some local heat damage. Heat damage can be especially troublesome when examining biological tissues containing metallic inclusions. When we examined thin sections (100-150 nm thick) containing large electron dense aggregations of hemosiderin-like inclusions by X-ray microanalysis, a significant FeK_α peak was initially observed (greater than 2 standard deviations above background).[20] In many similar preparations this strong iron signal would dissipate within 20-40 seconds after the analysis was begun. After the analysis, when the section was subsequently viewed, the area formerly containing the electron dense inclusion was replaced by a hole in the plastic section. We assumed that the iron inclusion became heated during electron bombardment, which loosened the attachment of the iron inclusion to the surrounding plastic. When bulk samples (critical point dried) of pitcher plant leaves, fed horse spleen ferritin, were subjected to wavelength dispersive X-ray microanalysis in spot mode (0.5 μm beam diameter) the same type of loss phenomenon was observed. Initially a strong iron X-ray signal was detected, follwed by a cessation of iron X-ray signal above background. Microscopy of the leaf surface after analysis revealed that the surface layer of cells had collapsed during the analysis. Although contamination artifact may also play a role in this type of count rate diminution, it is minimal in comparison to the effects of tissue collapse.

Heat damage can be readily reduced in severity by applying a coating of conductive material to the specimen surface prior to analysis. By providing a conductive pathway, not only is the specimen rendered thermally stable (which will eliminate local heating artifacts), but electrically conductive as well. The latter is especially significant in X-ray microanalysis because an accumulation of electrons on the surface of the specimen (charging) can lead to specimen movement and instability. In extreme cases it may also lead to the destruction of the sample.[4] Heat damage may also be minimized by using a scanning mode of analysis rather than a spot mode. Since the incident probe is always moving to a different part of the specimen at any given time, the radiation dose is minimized.

Mass loss and heat damage may also be reduced by bombarding a biological specimen with an intermitent incident beam. This beam blanking "strobe" effect switches off the electron beam as soon as an X-ray pulse arrives at the detector and remains off until another pulse can be individually accepted. Using this method, ideally "no pulses are rejected, no current is wasted on rejected pulses and there is less spectrum distortion due to reduced mean input pulses."[21] In this case, the specimen is also spared excessive electron radiation, thereby reducing specimen damage.

Conductive coatings. The accumulation of charge on the specimen surface will also result in deflection of the incident beam from the desired analysis site on the specimen surface (Fig. 4), producing X-rays from unknown areas of the sample. Whether the specimen is examined as a section, single cell, or bulk tissue, it should be coated with a uniform layer of conductive, low atomic number material. Although aluminum can be used to coat specimens, it does produce a characteristic X-ray line which may obscure some softer (low energy) X-ray signal from the specimen. Aluminum will also contribute to the continuum, although its contribution is minimal. Carbon is the material of choice because of the ease of vacuum deposition, and with most X-ray

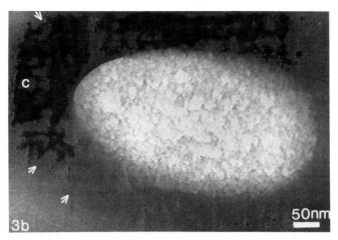

Figure 3b. Upon higher magnification of the un-stained plastic embedded section described in Figure 3a, an adjacent granule, subjected to electron beam bombardment for 100 sec., shows definite destruction.

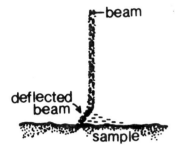

4.

Figure 4. When electrons are allowed to accumulate on the specimen surface, the incident beam may be deflected away from the area of interest. Electrons are represented as dashes.

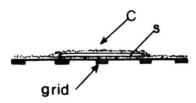

5.

Figure 5. To prevent charging, sections (s) can be placed on a formvar-carbon coated grid and given an additional surface coating of carbon (C) forming a 'carbon sandwich.'

systems carbon does not produce any significant X-ray signal. Although carbon is an excellent coating material for X-ray microanalysis, it does not produce appreciably high secondary electron emission and is, therefore, a poor coating material for specimens which need to be imaged by scanning electron microscopy as well. The samples should be coated in such a manner that the coating is uniform and continuous with the specimen holder (bulk analysis) or grid (STEM or TEM analysis). Some sectioned samples need more than a 2-3 nm surface coating of carbon to prevent charging. These samples may be placed on a grid which has been coated with formvar followed by a thin layer of carbon (2-3 nm).[2] If the tissue to be examined is embedded in plastic and is relatively thick (120-300 nm), an additional layer of carbon may be needed on the surface of the section, forming a carbon sandwich (Fig. 5). Some bulk specimens may continue to charge after coating due to their geometry. By painting a continuous line from the tissue surface onto the specimen stub with a conductive graphite paint (paste or cement), conductivity is usually improved (Fig. 6) without increasing the X-ray continuum.

Specimen contamination

By their very nature, most biological specimens subjected to electron bombardment will exhibit contamination. This contamination is not only due to hydrocarbon deposition resulting from the presence of residual gases and vapors in the electron optical column,[22] but also to the condensation of substances (on the specimen surface) which are released from the tissues themselves during electron bombardment.[17,23] During electron beam radiation C-H, -COH, -COOH, and $-NO_2$ groups have been reported to leave organic specimens.[23] When these gaseous constituents are combined with the deposition phenomenon previously mentioned, the resultant contamination layer at the specimen surface can be considerable, ultimately causing a reduction in count rate of softer X-rays emitted from the sample.[22,24] Using a liquid nitrogen cold finger around the specimen can significantly reduce specimen contamination by providing an area for vapor condensation which is out of the 'line-of-sight' of the incident beam and the X-ray detector. Heating the specimen can increase its stability in the beam,[25,26] but in biological samples it will cause some compositional alterations that may obscure the X-ray data. Liquid nitrogen traps on the diffusion pump, molecular sieves in the roughing line and the use of certain diffusion pump oils can collectively reduce contamination due to oil vapor.[22] As a general rule, when working with delicate tissues, it is often better to focus over the entire specimen surface, at a slow raster speed, rather than use a reduced raster. This decreases the specimen's contribution to contamination, and eliminates the characteristic "burn mark" seen after reduced raster focusing. Directly cooling a specimen will also decrease the specimen contamination,[17,23] however a cold finger at a lower temperature than the specimen should be used adjacent to the sample, in order to prevent the specimen surface from acting as a cold trap and becoming contaminated.

Where is the X-ray Signal Generated?

There is a tendency when looking at EDS or WDS X-ray data to interpret all of the X-ray signal which is 2 standard deviations above background as significant. Due to the fact that most biological specimens are not homogeneous (and that they are usually mounted on a support having a markedly different elemental composition), it is important to critically evaluate all X-ray data in respect to the X-ray excitation volume for that sample.

Placement of the X-ray excitation volume

Because biological samples are primarily composed of carbon, hydrogen, nitrogen, and oxygen, the electron beam tends to readily penetrate these specimens. Bulk samples subjected to EDS or WDS analysis may show signal, or increased continuum, derived in part from the specimen stub.[2] The electron scatter volume may penetrate beyond the desired depth and excite the underlying adhesive or specimen holder. If a graphite cement and/or graphite holder are used, there may be a marked increase in the continuum which may obliterate signal derived from trace amounts of elements (Fig. 7). This can be corrected by lowering the incident probe voltage, which then produces a smaller, more superficial (to the surface of the specimen) electron scatter volume. When examining bulk biological samples, it is also important to know where the majority of electron scattering events are occurring in relation to tissue morphology. Since this 'teardrop' shaped region will also produce most of the characteristic X-ray signal, it is essential to know what cells or tissues are being excited during the analysis. For this reason a knowledge of the size and depth of the X-ray excitation volume is of significant importance.

A recent paper by Reuter et al[27] offers a relatively simple formula for calculating the penetration depth in a specimen. Although this equation was developed for use with solid, non-biological materials, it is useful for estimating the penetration depth in moderately uniform, bulk biological samples.

$$\rho X = 5.2 \cdot (E_0^{1.3} - E_c^{1.3}) \frac{\text{av. atomic wt.}}{(\text{av. atomic number})^{0.9}} \quad (1)$$

In this case an estimate of the density of the tissue (ρ) must be initially made. The unknown (X) represent the deepest depth at which X-ray signal would be generated (in μgcm^{-2}). E_0 is the accelerating voltage in keV and E_c the absorption edge energy of the X-ray line of interest.

When using thin sections, the problems encountered with producing X-ray signal from the specimen holder are significantly reduced due to the decrease in the specimen electron scatter volume. However, it is very common for biologists to observe X-ray signal which is generated from the specimen grid even though the grid is not visible in the field of view. For example, a thin section placed on a copper grid will frequently show a relatively high continuum, as well as characteristic CuKα and Kβ signal even when the

beam is placed directly in the center of a grid square. This phenomenon may occur because the specimen is tilted at such an angle that there is some excitation of a grid bar, even though the grid bar is not readily visible in TEM or STEM modes (Fig.8).

This may be corrected by examining the sample at 0° tilt. With the geometry of most microscopes this orientation unfortunately also causes a drastic reduction in characteristic X-ray signal. By using a beryllium, graphite or graphite polymer grid[15] the possibility of generating spurious X-ray signal is reduced without necessitating a change in specimen orientation. It should be noted here that specimens placed on Be grids should be oriented in relation to the incident beam in such a way as to maximize the characteristic X-ray signal, rather than the overall count rate. Although Be is a low atomic number material, it does elastically scatter some of the incident electrons and, therefore, some secondary flourescence has been observed in X-ray spectra of Be mounted, plastic embedded biological thin sections.[15] For this reason, the specimen should be oriented so that grid bars are not directly bombarded by the beam. Figure 9 shows two EDS spectra taken of the same plastic embedded unstained section of amphibian eye on a formvar carbon coated beryllium grid at 30° tilt and 36° tilt using the same beam current and spot diameter. The incident probe was positioned in the center of a grid square, away from all grid bars. The arsenic X-ray signal is from the cacodylate buffer used. The sulfur is endogenous to the tissue, and the cadmium was introduced in situ. The copper signal which is generated from the parts of the microscope is observed to increase with the tilt angle, suggesting that backscattered electrons may be responsible for this artifactual peak.

Signal may also be generated from the specimen support, specimen holder and/or specimen chamber as a result of uncollimated electrons.[28] Needless to say, the microscope should be carefully and routinely aligned before each analysis.

Often Fe, Zn, Cr and Cu signals are seen in biological X-ray spectra, which cannot be attributed to the specimen holder, stub or grid. Since these metals are of significant interest to many biologists and biomedical research scientists, it is paramount to determine if the detected signal is produced by the specimen. In the TEM, X-ray spectra exhibiting these spurious elemental lines have been demonstrated to arise from electron scatter around the second condenser (C2) aperture, and from X-rays excited from the thin foil C2 apertures.[28,29] By using thicker condenser apertures (which will not permit penetration of X-rays), and by inserting a heavy metal collar (lead) in the electron optical column, spurious characteristic and continuum X-rays [28,29] coming down the column can be reduced.

Interpretation of X-ray Data

How can each X-ray line in a spectrum be analyzed to determine if it is an artifact or a characteristic X-ray from the specimen? As stated before, there should be some preliminary knowledge of the elemental content and elemental distribution in the biological specimen before X-ray microanalysis is attempted. If an X-ray peak appears at an unusual X-ray energy (for example, characteristic of vanadium Kα) when the sample should only exhibit Cl, Na, Ca, S and P, the data should be carefully examined to determine if the unexplained X-ray signal could be artifactual. The first thing that should be done, if energy dispersive spectrometry is available, is to expand the energy scale and look at all of the X-ray lines produced by the specimen. Frequently what appears to be a questionable low energy X-ray K line may in actuality be a rather ordinary L or M line from a higher atomic number element in the tissue. For example, what initially appeared to be a vanadium Kα line could in reality be a cesium Lα line. Occasionally when uranyl acetate stained thin sections are analyzed, and the counting statistics are low, the peak produced at approximately 3.3 keV is misinterpreted to be potassium (K Kα 3.312 keV) when it is probably a uranium M line at 3.239 keV.

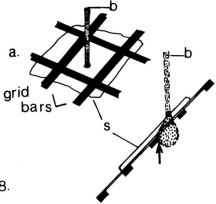

8.

Figure 8. In thin section X-ray microanalysis, it may appear (a) that the beam is being positioned away from the grid bars, but in actuality at high tilt angles incident electrons and X-rays may excite them due to the increased X-ray excitation volume in the section (s).

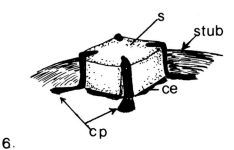

6.

Figure 6. Conductivity of bulk samples coated with carbon may be improved by painting lines of colloidal graphite paint (cp) from the stub to the specimen (s) surface. Graphite cement (ce) used to attach specimen to stub further increases conductivity.

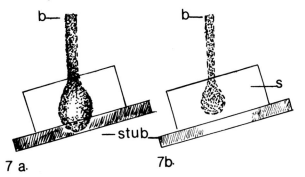

7 a.   7 b.

Figure 7. In order to prevent the X-ray excitation volume from entering the region of attachment of the specimen (s) to the stub (7a), the accelerating voltage is reduced, thereby decreasing the penetration depth of the electron scatter volume as shown in 7b. Incident electron beam (b).

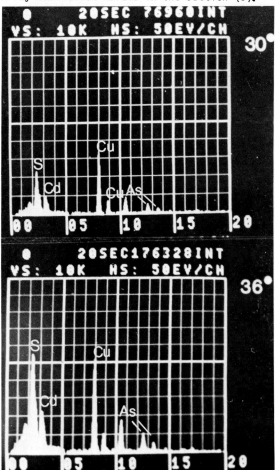

Figure 9. X-ray spectra taken of the same section (100 nm thick) on a formvar-carbon coated beryllium grid, using the same beam current, accelerating voltage (80keV), counting time (20 sec.) and spot diameter (200nm), but different tilt angles.

697

Spurious Peaks

Escape peaks (EDS detectors). The second way to check a biological spectrum is to subtract the energy of the silicon K α line (1.74 keV) from the energies of the largest X-ray peaks. For example, if large Cu k α signal is observed in a spectrum, then:

$$Cu\ K\alpha(8.04\ keV) - Si\ K\alpha(1.74) = 6.20\ keV \qquad (2)$$

The difference between these two lines produces an escape peak at 6.20 keV. If a peak is observed around this energy range, it may initially be mis-interpreted as iron signal (Fe K α 6.398 keV). If no Fe K β signal is seen in the spectrum (Fe K β 7.057 keV), and the counting statistics are low, it may be assumed that the signal at 6.2 keV is an artifact called an escape peak. In the EDS detector bombarded with incoming photoelectrons, the silicon atoms of the crystal which are in an ionized state may emit a Si X-ray. When this X-ray escapes, it robs the cascade being measured by the detector, resulting in the formation of an escape peak at 1.74 keV less than the parent line.[24]

Biological specimens usually produce low X-ray counting rates; however, it is important to note that when the count rate is high, escape peaks may be produced by any significant X-ray peak, K, L and M lines. For example, when Cu K α and Cu K β X-ray signals are statistically well defined with respect to the background, escape peaks may appear at 6.2 keV and 7.06 keV, mimicking Fe K α and Fe K β X-ray lines. To distinguish whether the recorded X-ray events are representative of iron or copper escape peaks, the counting period should be extended to permit the accumulation of more X-ray events. The data should then be checked to see if the X-ray peaks are shifted away from their designated X-ray energy. The X-ray intensity ratios should also be checked to see if the ratios of the various X-ray lines for a given element are in agreement. For example, the Fe K β X-ray peak should be approximately 12% of the FeK α signal. Using these X-ray intensity ratios will provide a way to identify discrepancies of X-ray peak height for the same element. However, it is far better to prepare another specimen, or make another analysis where the questionable element is eliminated. If copper grids may be producing spurious X-ray signal, it would be advisable to make another tissue preparation and mount it on a beryllium or graphite grid.

Pulse pile-up (EDS detector). When the count rate is high, X-ray photons may enter the EDS detector almost simultaneously. Due to the time it takes the detector to process a received pulse, additional incoming pulses may be combined with the initial one. The detector is unable to discriminate between individual X-ray energies and an integer multiple of the same energy X-ray signal. Therefore, X-ray signal may appear at summations of the most significant X-ray lines. This pulse pile-up is usually not a serious problem with biological specimens because the count rates are usually low, and the incoming signal can be easily discriminated by the detector. Occasionally, when a copper grid is excited during thin section X-ray microanalysis, high Cu K α, K β and L α X-ray events are observed. If the count rate is high, summation peaks may appear at double the spectral energy of each primary X-ray line (at 16.08 keV

and far smaller peaks at 17.6 keV and 1.86 keV).

Almost all EDS systems are equipped with pile-up rejection electronics that reduce the artifactual peaks from the X-ray spectrum. Most pulse pile-up rejection systems cannot eliminate summation peaks, especially at X-ray energies less than approximately 1.3 keV, but they do reduce the associated pile-up continuum.

Comparing X-Ray Data

The specimen preparation procedures for X-ray microanalysis, and the delicate mutable nature of the biological sample, make it necessary for certain tissue controls to be employed if X-ray data from different samples are to be compared with any validity.

Samples that will be subjected to fixation and dehydration should be prepared with the same stock solutions, or at least with solutions having the same elemental composition and tonicity. Occasionally reagents may become contaminated, producing one batch of fixative or buffer with a slightly different elemental composition from subsequent batches.

Samples that are to be dried, should preferably be dried at the same time, under the same conditions. This eliminates variations arising from inconsistencies in operator performance of the procedures, as well as mechanical fluctuations in machinery. Once a sample has been dried, great care should be taken to keep it free of moisture. The samples should not be exposed to room air, but can be safely stored in a vacuum desiccator. We do not use chemical desiccants to keep the specimens dry for fear of silicon (silica gel), calcium ($CaCl_2$) or other elemental contamination.

Current and voltage fluctuations

All X-ray data to be compared should be obtained with the same and similar specimen current for biological X-ray microanalysis, as the X-ray excitation volume is dependent on these factors. Variations in either functions will alter the number and location of cells producing the detected signal. If a significant difference in the number of X-ray events (height of the background) is observed in comparative spectra of the same specimen, or if there is a large difference in the X-ray signal detected from a calibration standard counted at the beginning and end of an examining period, fluctuation in the beam current may be suspect. This is especially true when a TEM is used for X-ray microanalysis of a thin specimen.

Validity of X-ray Maps

By setting a voltage window on an energy or wavelength dispersive detector and scanning the incident beam over the specimen surface, it is possible to accumulate signal from the generated X-ray events falling within the acceptable energy range of the window. By displaying these X-ray events cumulatively as dots, it is possible to demonstrate the localization of specific elements in a sample as a distribution map. Frequently an investigator will make several maps of a sample (each for a specific element) and compare these maps to determine the presence or absence of an element in different cellular structures. Before this can be done, it is important to determine if a map is actually representing an element in question or really is a distribution map of the continuum signal (i.e. a map of the average atomic number of the area).

Biological samples prepared with solutions containing arsenic, osmium, chromium or zinc will produce X-ray maps representative of the continuum produced by the presence of such high atomic number materials in a relatively low atomic number matrix. For this reason it is important to check all X-ray maps by setting a window no wider than the full-width-half-max (FWHM) of the characteristic X-ray line to be mapped.[22] To insure that the distribution observed is really from the element in question, another map should be made that is "off peak" (the voltage window is set so that the energy of interest is not selectively accepted by the detector).

Figure 10 shows an X-ray map of a carbon coated plastic embedded 1½ µm section of human articular fat-pad tissue treated with a histochemical procedure that deposits a silver reaction product in the associated neurological cells. This X-ray map shows the X-ray signal measured between 2.96 keV and 3.04 keV; this energy spread is FWHM of the silver Lα line. A few regions are clearly seen to show higher dot density than the surrounding fairly uniform background. The window was reset (2.64 - 2.80 keV) and another map was made of the same tissue using the same experimental conditions (20 keV, 1.7 x 10^{-9} Ampere current, 250 sec., 30° tilt). This window is just below Ag Lα and adjacent to Cl Kα. The X-ray map taken off peak bears no resemblance to the Ag map, indicating that the observed distribution of the Ag map was not merely a background phenomenon.

Once windows are set for X-ray mapping, the calibration should be checked periodically for drift to insure the validity of subsequent X-ray maps.

Due to the low concentrations of many elements in biological specimens, it is often necessary to use long counting periods when making X-ray maps. Background radiation or X-ray noise becomes prominent in X-ray images with long exposure times, and sometimes it is not possible to produce a clear map because the background is contributing too much signal, in comparison to the specimen. Therefore a good rule-of-thumb to consider before attempting an X-ray map is that at least a 5:1 (or better) signal to noise ratio is required.

Conclusions

X-ray microanalysis of biological tissues is an exciting analytical tool which can reveal compositional information with a minimum of specimen preparation and instrument time. Due to the complex chemical and morphological nature of biological tissues, and the way in which X-ray signal is produced, detected and displayed, the interpretation of biological X-ray data is not as straightforward as it seems. However, a body of knowledge does exist with which spectral artifacts can be identified and prevented. And, with the daily progress in instrument design and quantitative analytical computer programs, many of the problems faced by the biological investigator attempting to use X-ray microanalytical techniques are minimized.

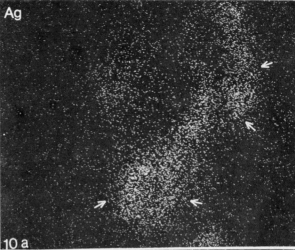

Figure 10. (a) This Ag Lα X-ray map was taken of a 1.5 µm thick plastic embedded section of human articular tissue treated with a histochemical procedure which deposits a silver reaction product in neurofilaments. The map shows an accumulation of X-ray events (white arrows), which are totally lacking when compared to an off-peak map of the same area under the same experimental conditions (b): 20 keV, 1.7x10^{-9} Amp specimen current, 250 sec. counting time, 30° tilt, same magnification, window for silver map, 2.96-3.04 keV; window for off-peak map, 2.68-2.86 keV.

Acknowledgement

The author would like to sincerely thank Dr. John B. Warren and Dr. Hobie Kraner at Brookhaven National Laboratory, Department of Instrumentation, for their advice, assistance, and instrument time. Gratitude is also expressed to Mr. Frank Cardone and Mr. John Kuptsis of IBM, T. J. Watson Research Center, for their time and willingness to discuss data and methods.

References

1. P. Echlin and R. Moreton, The Preparation of Biological Materials for X-ray Microanalysis. Microprobe Analysis as Applied to Cells and Tissues, ed. T. Hall, P. Echlin and R. Kaufmann, Academic Press, N.Y. 1974: 159-173.

2. B. J. Panessa, X-ray Microanalysis of Biological Samples: Considerations and Problems, J. of Submicrosc. Cytol. 6, 1974: 81-94.

3. A. Marshall and A. Wright, Detection of Diffusible Ions in Insect Osmoregulatory Systems by Electron Probe X-ray Microanalysis Using Scanning Electron Microscopy and a Cryoscopic Technique, Micron 4, 1973: 31-45.

4. P. Echlin, Low Temperature Biological Scanning Electron Microscopy, J. Microsc. 112, 1978: 47-61.

5. J. Coleman and A. Terepka, Electron Probe Analysis of the Calcium Distribution in Cells of the Embryonic Chick Chorioallantoic Membrane. I. A Critical Evaluation of Techniques. J. Histochem. Cytochem. 20, 1972: 401-413.

6. L. M. Popescu and I. Diculescu, Calcium in Smooth Muscle Sarcoplasmic Reticulum in Situ. J. Cell Biol. 67, 1975: 911-918.

7. S. Suzuki and H. Sugi, Ultrastructural and Physiological Studies on the Longitudinal Body Wall Muscle of Dolabella auricularia. J. Cell Biol. Vol. 79, 2, 1978: 467-478.

8. P. Gambetti, S. Erulkar, A. Somlyo and N. Gonatas, Calcium-Containing Structures in Vertebrate Glial Cells, J. Cell Biol. 64, 1975: 322-330.

9. A. Boyde, E. Bailey, S. Jones and A. Tamarin, Dimensional Changes During Specimen Preparation for Scanning Electron Microscopy; in SEM/1977/I, ITT Research Institute, Chicago: 507-518.

10. A. Läuchli, R. Stelzer, R. Guggenheim and L. Henning, Precipitation Techniques as a Means of Intracellular Ion Localization by Use of Electron Probe Analysis. Microprobe Analysis as Applied to Cells and Tissues. ed. T. Hall, P. Echlin and R. Kaufman. Academic Press, 1974: 107-118.

11. C. Tandler and A. Solari, Nucleolar Orthophosphate Ions: Electron Microscope and Diffraction Studies. J. Cell Biol. 41, 1969: 91-108.

12. P. Serwer, A Technique for Observation of Bacteriophage T7 Capsid-DNA Complexes. J. Ultrastruc. Res. 65, 1978: 112-118.

13. A. Cohen, A Critical Look at Critical Point Drying--Theory, Practice and Artifacts; in SEM/1977/I, ITT Research Institute, Chicago: 525-534. See also Cohen, in this volume.

14. H. Burstyn and A. Bartlett, Critical Point Drying: Application of the Physics of the PVT Surface to Electron Microscopy. Amer. J. Physics, 43, 5, 1975: 414-419.

15. B. J. Panessa, J. B. Warren, J. J. Hren, J. A. Zadunaisky and M. Kundrath, Beryllium and Graphite Polymer Substrates for Reduction of Spurious X-ray Signal, in SEM/1978/I, SEM, Inc.: 1054-1062.

16. P. Echlin, Coating Techniques for Scanning Electron Microscopy and X-ray Microanalysis, in SEM/1978/I, SEM Inc. O'Hare, Illinois: 109-117.

17. R. Glaeser. Radiation Damage and Biological Electron Microscopy, Physical Aspects of Electron Microscopy and Microbeam Analysis. ed. B. Siegel and D. Beaman, J. Wiley and Sons, N.Y., 1975: 205-209.

18. C. LeChene, Electron Probe Microanalysis of Picoliter Liquid Samples, Microprobe Analysis and Applied to Cells and Tissues. ed. T. Hall, P. Echlin and R. Kaufmann. Academic Press, N.Y. 1974: 351-368.

19. K. Stenn and G. Bahr, Specimen Damage Caused by the Beam of the Transmission Electron Microscope J. Ultrastruct. Res. 31, 1970: 526.

20. B. J. Panessa, J. D. Kuptsis, L. Piscopo-Rodgers, and J. Gennaro Jr., Iron and Uranium Transport Studied by SEM and X-ray Microanalysis. in SEM/1976/I, ITT Research Institute, Chicago: 461-468.

21. J. Pawley, P. Statham and T. Menzel, Use of Beam Blanking and Digital Scan-Stop to Speed the Microanalysis of Particles. in SEM/1977/I, ITT Research Institute, Chicago: 297-306.

22. D. Beaman and J. Isasi, Electron Beam Microanalysis, Amer. Soc. for Testing and Materials, 506, 1972: 2-80.

23. L. Reimer, Review of the Radiation Damage Problem of Organic Specimens in Electron Microscopy, Physical Aspects of Electron Microscopy and Microbeam Analysis, J. Wiley & Sons, New York, 1975: 231-345.

24. C. E. Fiori and D. E. Newbury, Artifacts Observed in Energy Dispersive X-ray Spectrometry in the Scanning Electron Microscopy. in SEM/1978/I, SEM Inc., O'Hare, Ill: 401-426.

25. J. Rouberol, M. Tong and C. Conty, Application of Soft X-ray Spectrometry in the 20$\overset{o}{A}$ to 80$\overset{o}{A}$ Region by the Electron Probe Microanalyzer to Elements having Low Atomic Number, in Proc. Conference for the Advancement of Spectroscopic Methods (GAMA), Paris, June 8, 1966.

26. A. N. Broers, B. J. Panessa and J. F. Gennaro Jr., Scanning Electron Microscopy of Bacteriophage in Proc. of the Microbeam Analysis Society, Las Vegas Nevada, 1975: 11A-11F. Copies obtainable from Dr. J. Goldstein, Leheigh University, Bethlehem, Penn. 18015.

27. W. Reuter, J. D. Kuptsis, A. Lurio and D. F. Kyser, X-ray Production Range in Solids by 2-15 KeV Electrons, J. of Physics D, App. Physics 11, 1978: 2633-2642.

28. J. Goldstein and D. B. Williams, Spurious X-rays Produced in the Scanning Transmission Electron Microscope, In SEM/1978/I: 427-444.

29. J. J. Hren, P. S. Ong, P. F. Johnson and E. J. Jenkins, Modification of a Philips EM301 for Optimum EDX Analysis, in Proc. 34th EMSA Meeting. ed. G. W. Bailey, Claitor's Publishing Division, 1976: 418-419.

For availability of past SEM volumes:

1968-77: IIT Research Institute, Chigago, IL 60616
1978-   SEM Inc., AMF O'Hare, IL 60666

Discussion with Reviewers

P. Echlin: Has the author any experience with dehydration using buffered alcohol or acetone?
Author: I am not sure what is meant by "buffered alcohol or acetone." Several years ago I attempted to dehydrate specimens in a mixture of acetone and either cacodylate or phosphate buffer. In both cases a precipitate resulted, and this procedure was abandoned. Although other buffers are now available, I have not tried to use them for fear of introducing exogenous elements into the specimens. I am very interested however in any techniques that you have found to be successful.

P. Echlin: Has the author any evidence that the formula derived by Reuter et al. is of use on biological material?
Author: This formula is simply used for estimating the penetration depth in a metal or organic specimen. It is based on experimental, rather than theoretical results, and has been successfully applied to thin film and bulk samples where the specimen composition and density can be assumed. Since this equation was only published a few months ago, I doubt that any biologists have had the chance to use it yet. It is proposed here because it offers a very quick way to estimate the penetration depth using only a pocket calculator, and can, therefore, be done while sitting at the microscope.

T. Bistricki: Can you evaluate a rapid dimethoxypropane (DMO) dehydration in respect to retention or leaching of elemental constituents? I find that such dehydration of small monocellular organisms yields superb results as compared to gradual alcohol or acetone dehydration.
Author: I cannot evaluate a rapid DMP dehydration without setting up an experiment in which a known specimen containing the element in question is divided into 2 or 3 equal segments. One segment would be analyzed fresh (without fixation or dehydration) by atomic absorption spectrophotometry or frozen-hydrated X-ray microanalysis for the element in question. The other 2 segments would be treated with a fixative and dehydrated rapidly in DMP or dehydrated incrementally over a short period of time (as described in this review). Following the dehydration, each piece of tissue, as well as the remaining dehydration-fluids would be analyzed in order to determine if the element in question was lost to the dehydrating solvents. The fresh tissue analysis is used as the control. Each tissue is different, and therefore, the optimal specimen preparation for elemental analysis will vary. The most important concept to keep in mind is that no preparation method should be accepted without some initial experimentation to see if elements are lost or translocated.

T. Bistricki: Is it possible that solid CO_2-solvent mixtures can cause some adverse effect to tissues exposed to this thermal shock?
Author: If a tissue is completely dehydrated, and contains no residual water, the thermal shock of liquid CO_2 cannot cause any ice crystal formation. This is the primary cause of damage in frozen specimens, and results in elemental translocation as well as morphological distortions. Again, I would like to stress that each type of sample may respond differently to a preparation procedure. I have not seen any damage from thermal shock in the cells and viruses I have studied.

T. Bistricki: The aluminum coating, you said, does not contribute significantly to an increase of continuum. Could you explain why the carbon support, which is of lower atomic number, markedly increases this signal?
Author: I think that we are thinking about two different things. An aluminum coating is usually thin (5-20 nm), and will produce characteristic X-ray signal at 1.486 keV. Because of aluminum's low atomic number and the fact that, in this case, it is a thin film, the continuum is not appreciable. Graphite planchets and stub's, however are not thin films. A graphite stub will produce a sizeable X-ray excitation volume, which in turn will result in Bremsstrahlung. Fig. A shows a spectrum and the resultant continuum of a pyrolytic graphite planchet. The second problem with graphite is that it may contain impurities. Therefore, it is important to run a control spectrum on graphite stubs or grids. The latter procedure should be routine for any specimen supports employed.

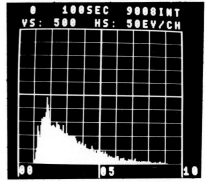

C. Fiori: Varying the current and voltage in an SEM will alter the X-ray excitation volume. How does this volume change when STEM or TEM are employed and the current and voltage are changed?
Author: In the SEM the primary way to alter the excitation volume in bulk specimens is by adjusting the accelerating voltage. When sections, especially thin sections, are examined, the X-ray excitation volume is much reduced (Figures 7 and 8). Since 2/3's of the X-ray excitation volume is lost (in thin sections), changes in the accelerating voltage do not cause significant alterations in the area producing the X-ray signal. However, the beam current plays an important role in TEM and STEM X-ray microanalysis. Since increasing the current also increases the probe size, the volume producing the X-rays also increases, as does the specimen damage.

<u>C. Fiori</u>: When one fixes or stains biological
materials the average atomic number and density
of the analytical volume is altered. How does
this alteration effect the quantitative results?
<u>Author</u>: To avoid confusion when analyzing thin
sections, I cut two sets of sections. One section
is placed on a copper TEM grid, stained and used
for orientation purposes. The second set of
sections are placed on carbon or beryllium grids
and analyzed unstained. If contrast is poor with
the unstained section, I drop the accelerating
voltage to 60 or 40 keV for viewing. The accelerating voltage is raised to 80 keV for the analysis. To insure that the beam has not shifted off
the area of interest, after the analysis is complete the cold finger is removed and a burn spot
allowed to form.

SCANNING ELECTRON MICROSCOPY/1979/II
SEM Inc., AMF O'Hare, IL 60666, USA

MICROANALYSIS IN BIOLOGY AND MEDICINE. A REVIEW OF RESULTS OBTAINED
WITH THREE MICROANALYTICAL METHODS

P. Galle, J. P. Berry and R. Lefevre[*]

Laboratoire de Biophysique
de la Faculté de Médecine
de Créteil
8, rue du Général Sarrail
94000 - CRETEIL, France

[*]Laboratoires de Physique
des Solides et de Géologie
Université Paris-Sud
91405 - ORSAY, France

ABSTRACT

Three microanalytical methods have been used : X-ray emission, secondary ion emission and electron energy loss analysis.

Although X-ray microanalysis is the most used method, many biological problems cannot be solved because of its relatively low sensitivity and inability to analyze light elements. New possibilities are offered by Secondary Ion Mass Analysis and Electron Energy Loss Analysis. Analysis by secondary ion emission permits the study of elements at low and even trace element concentration, and even the lightest elements such as hydrogen and beryllium are detected. Electron Energy Loss Analysis makes possible the study of very small volumes, less than 500 Å in diameter.

These methods have been applied to the study of a variety of biological tissues (mostly lungs, kidneys and biomineralized tissues) and typical applications to pathology and cell biology will be presented.

KEY WORDS: Secondary Ion Emission Microanalysis, Electron Energy Loss Analysis, Analytical Electron Microscope, X-ray Microanalysis, Cell Biology, Biomineralisation, Pathology, Submarine Polymetallic Nodules, Geology

INTRODUCTION

Three microanalytical methods have been used in the Biophysics Department of the Creteil Medical School : X-rays microanalysis, secondary ion emission microanalysis and electron energy loss microanalysis. A review of results obtained in different applications (human pathology, cell physiology and biomineralisation) is presented.

OBJECTIVES

The objectives sought in biology are generally quite different from those sought in studying inert matter. In most cases the relation between the structure and the function of a very small intracellular organelle (mitochondrion, lysosome) is sought in cellular biology ; the exact intracellular localization of an abnormal mineral element is sought in pathology, and the visualization system adapted to the microanalytical instrument consequently becomes extremely important. The ideal and most convenient system for this is the conventional transmission electron microscope and the ideal preparation is an ultrathin tissue section. Such visualization can be obtained in X-ray[1] and electron energy loss[2] microanalysis. When ultrathin tissue sections cannot be obtained, the best system of visualization in most cases is a simple optical transmission microscope. With some exceptions, the scanning electron microscope has appeared to be less useful than an optical microscope, despite its better resolution power, as it does not allow us to observe the interior of tissues.

On the other hand, quantitative analysis, among the most important preoccupations of physicists studying inert matter, often appears in biology as a secondary problem. Considering the very small size of observed mineral deposits and their irregular distribution within a tissue, a highly localized absolute quantitative study often loses meaning. When the distribution of an element is homogeneous in a given volume (diffusible ions), a quantitative analysis can be useful ; in such a case we shall see that the difficulties are very great and still not completely resolved.

Main problems posed in biological applications

These problems are generally associated either with the very small volumes to be analyzed (often much less than 1μm³) or with the low concentration of elements sought. Despite the method used, the analytical system must always be as sensitive as possible. At this level possible artifacts are very frequent in X-ray microanalysis[3], which is not a method adapted to the study of elements at a trace concentration, and the greatest precaution must be taken in ion mass analysis and in electron energy loss analysis.

In practice, the problems posed are different depending on the applications and, schematically, 3 cases can be differentiated: the study of non-diffusible elements in soft tissue, the study of diffusible elements in soft tissue and the study of hard tissue.

The study of non-diffusible elements in soft tissue presents no difficulty. The choice of a method to be used basically depends on 3 factors: the element sought, its local concentration and the volume to be studied. Whenever possible, X-ray microanalysis is preferable, considering its rapidity and its simple manner of interpreting results. This method, however, cannot be used or is poorly adaptable for the study of light elements ($Z < 11$), elements with low concentrations (< 100 ppm) or very small inclusions (< 0.05 μm). Secondary ion emission analysis is highly interesting in studying elements at a very low concentration or of a low atomic number but then the relation between the ultrastructure and microanalysis cannot any longer be established. Electron energy loss analysis is especially well adapted to the study of very small particles or elements with a low atomic number.

The study of diffusible elements presents many difficulties, some of which are associated with the preparation of the tissues: the problem is to avoid during such preparation the diffusion of these elements through peri or intracellular membranes, the thickness of which is less than 0.01 μm. Many authors call for this on freezing techniques - A. Tousimis[4], T. Hall[5], P. Echlin[6], A.J. Saubermann[7], A. Dorge[8], F.D. Ingram[9], W. Fuchs[10] - or on precipitation techniques - A. Lauchli[11]. Experiments show that the diffusion of elements is very difficult to avoid. Another difficulty is associated with the fact that the concentration of a large number of these elements is not very high (the intracellular concentration of an element such as sodium is approximately 10^{-15} g/μm³). Under such conditions, qualitative analysis using X-ray microanalysis is barely possible, and any attempt at quantitative analysis is often illusory due to the fact that preparations do not support high intensity electron microprobes without damage. Moreover, it is often impossible to differentiate in the signal observed between the characteristic X-rays emitted under the probe and the same characteristic rays emitted at a given distance from the probe[3]. Under such conditions, X-ray microanalysis often fails. Secondary ion emission analysis, however, will appear as the best method,

once the problems associated with the preparation of samples have been solved, considering its very high sensitivity in studying electrolytes.[12]

The study of hard tissue presents difficulties associated with the preparation of tissues, most of which have been solved in different ways- H.J. Höhling[13], R. Lefevre[14]. Considering the frequent need in this field of working on thick sections or on bulk samples, important difficulties appear in X-ray microanalysis in which diffusion of electrons in the specimen prevents the study of very small volumes, and obtaining distribution images for major elements (Ca, P) or minor elements (F, Na, K, Mg) is impossible with a sufficiently high resolution. In most cases completely uniform images are obtained or no structure appears. In this field, secondary ion emission analysis becomes extremely interesting[15], considering the fact that resolution in depth is excellent (< 0.01 μm).

Basic features of the instrumentation and analytical conditions

Four instruments have been used: 1) a CAMECA M.S.46 microprobe equipped with a transmission light microscope and 4 wave length dispersive spectrometers (Quartz, LiF, PET and KAP crystals); 2) a CAMEBAX microprobe equipped with a conventional transmission electron microscope, an energy dispersive spectrometer (Tracor) and wave length dispersive spectrometers (LiF, PET, TAP and ODPb crystals); a secondary ion emission microanalyzer SIMS 300 equipped for high resolution mass analysis with an electrostatic filter; electron energy loss analysis has been made in the "laboratoire d'Optique Electronique du CNRS de Toulouse" with a high voltage electron microscope equipped with an energy filter[2].

Specimens for X-ray or energy loss analysis are generally Epon ultrathin sections (0.08 μm) obtained with the standard method used for normal conventional electron microscopy, and deposited on aluminum or cupper grids; the deposition of a thin carbon film is sometimes necessary to avoid the destruction of the section under the beam. Specimens for Ion mass analysis are, for soft tissues, thick paraffin sections, frozen hydrated sections or Epon 1 μm thick sections ; for hard tissues, we use bulk specimen the surface of which has been carefully polished ; in this case, the evaporation of a metallic grid is necessary to avoid charging effects.

The conditions of analysis in X-ray spectrometry vary depending on the element to be detected ; generally, we use an accelerating voltage as low as possible even on ultrathin sections for a better sensitivity and to avoid artifacts ; in our application we generally use wave length dispersive spectrometers rather than energy dispersive spectrometers for a better sensitivity and to avoid peak interferences. Energy loss analysis has been made with a million volt electron beam and in secondary ion analysis we use O_2^+ in the primary ion beam.

Results

Results obtained by us in the field of human pathology, cell physiology and structure of bio-mineralized tissues will be discussed below.

Human pathology

a) Lung disease. X-ray microanalysis combined with electron microdiffraction appears as the ideal method for studying normal and pathological lung dust[16]. The chemical composition and the eventual crystalline structure of each bit of dust can be individually defined as well as its exact site within a macrophage or another cellular variety. A wide variety of intra-cellular dust has consequently been identified[16]. Figures 1 and 2 show, as examples, the identification of vermiculite and biotite, varieties of dust frequently found in the atmosphere. The most striking result observed in this study has shown that the lung macrophage can dissolve atmospheric dust in its phagolysosomes[17]. Most of the mineral deposits observed, except in pneumoconiosis, in the macrophages of human lungs, contain the same elements as those observed in the atmosphere (Si, Ti, Al, Fe, Ca) but, contrary to the latter, they are not found in a crystalline state (figures 3 and 4), which indicates a dissolution of such dust within phagolysosomes. This dissolution phenomenon has been experimentally

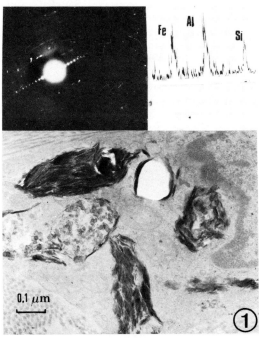

Figure 1. Crystalline atmospheric mineral dust in a human lung macrophage:ultrastructure, x-ray microanalysis and electron diffraction pattern. Identification : vermiculite

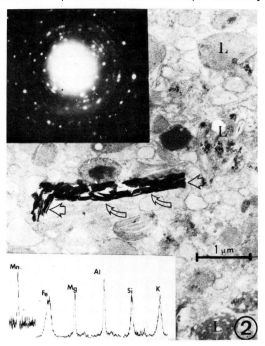

Figure 2. Crystalline atmospheric mineral dust in a human lung macrophage:ultrastructure, x-ray microanalysis and electron diffraction pattern. Identification : biotite

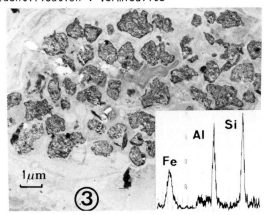

Figure 3. Amorphous mineral deposit frequently observed in human lung macrophages and containing Si, Al and Fe.

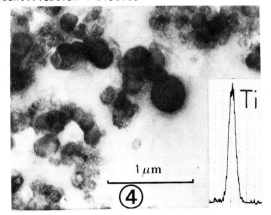

Figure 4. Another variety of amorphous mineral deposit frequently observed in human lung macrophages and containing Ti.

confirmed by J.P. Berry[17]. Figure 5 shows an illite particle within a lung macrophage of a rat 3 days after inhalation of this variety of dust. The ultrastructure, microanalysis and diffraction diagram show that non-modified illite crystals are observed at this stage. Figure 6 shows the mineral deposits as they are observed in the macrophages 3 months after inhalalation of the same illite particles. No diffraction diagram can be obtained, the structure is no longer crystalline and the respective proportion of the different illite elements (Si, Al, Fe, K) is no longer the same.

In some cases, x-ray microanalysis cannot be used -lithium, for example, has too low an atomic number to be detected by this method. Secondary ion emission analysis, however, readily studies the distribution of this element. Figure 7 shows the distribution of lithium in a human lung (pneumoconiosis of unknown origin).

When the mineral particle is too small (<0.05 µm), x-ray microanalysis cannot also be used in practice as the characteristic signal received by the detector is too weak. Under such conditions electron energy loss analysis is quite interesting. Figure 8 shows the characteristic K loss of beryllium which is obtained on a particle present in a macrophage, the size of which is approximately 0.01 µm.

b) Brain pathology. A variety of abnormal mineral deposits in the human brain has been studied in different encephalopathies: Sturge-Weber disease[18], Alzheimer disease[19], Parkinson disease and striatonigral disease[20] ; this work has been directed by S. Duckett. The most striking result has undoubtedly been the discovery of the role of aluminum in the genesis of some human encephalopathies, the cause of which is unknown[21].

These results can be compared with those later reported by Alfrey in cases of encephalopathies in dialyzed patients.

c) Renal pathology. A great variety of mineral deposits have been studied by x-ray microanalysis in the kidneys of patients with nephropathies (Ca, Mg, P, S, Ag, Au, Fe, Si, Se). This method appears to be very useful in determining the cause of a nephrocalcinosis:hyperparathyroidy, hypervitaminosis D, Burnett's disease[22], or in finding the cause of a nephrotic syndrome. Figure 10 shows, as an example, the analysis of a dense inclusion containing gold in a renal mitochondrion (experimental intoxication in rats). Identical inclusions can be observed in some extra-membranous glomerulonephritis ; the cause of this disease can thus be suspected.

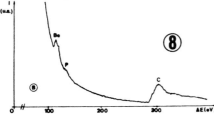

Figure 8. Electron energy loss spectrum obtained from a very small particle (about 0.015 mm in diameter) in a macrophage ; thickness of the section: 0.1 mm (with the courtesy of Mrs. Jouffrey and Sévely, Toulouse,France.)

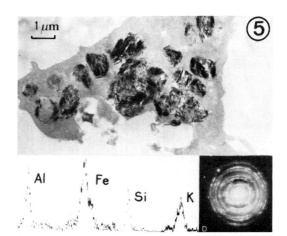

Figure 5.Illite crystal in a rat macrophage 3 days after inhalation of Illite particles;ultrastructure and diffraction pattern of the crystal is not modified

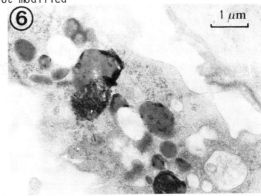

Figure 6.Mineral dust in a rat macrophage 3 months after inhalation of Illite particles;no diffraction pattern can be obtained and the relative concentrations of Si, Al, Fe and K are different from the Illite crystal.

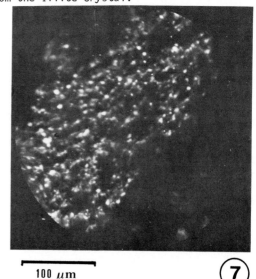

Figure 7.Lithium distribution (7Li) at the surface of a thick section of a human lung (secondary ion emission microanalysis)

Cell physiology

a) In this field, secondary ion emission microanalysis will appear in the future as an exemplary method considering its very high sensitivity. Elements present in very low concentrations and free or associated with macromolecules can be locally detected and distribution images can be readily obtained. Figure 9 shows the distribution of mass 56 (^{56}Fe) at the surface of a nucleated red blood cell. Such images are impossible to obtain by x-ray microanalysis. In some cases, unexpected elements have been detected in the structure of the retina such as barium[23].

b) X-ray microanalysis, however, is still interesting if we want to analyze elements in an average or high concentration in very small volumes. The role of mitochondria or lysosomes in the intracellular concentration phenomenon of some elements can thus be revealed[24]. Figure 10 shows the role of the mitochondrion of the proximal tubule cell of the kidney in the concentration of gold. Figure 11 shows the concentration of calcium and phosphorus in a mitochondrion of a human synovial cell. Figures 12,13,14,15, respectively, show the concentration of gold in the lysosome of a macrophage in a rat, the concentration of chromium and uranium in the lysosomes of the proximal tubule cells of rats, and the concentration of aluminium in a phagolysosome of a rat's brain neurone.

Biomineralization

In this field, R. Lefevre has shown the interest of secondary ion emission microanalysis in human dental or bone pathology[25,26,27] and in the study of marine biomineralization[28,29].

Figure 16 shows the distribution of Sr, Mg and Na in a triassic red Algae (200 millions years old) ; the two varieties of $CaCO_3$ can be distinguished : Aragonite (orthorhombic) with hyperconcentration of Sr, and Calcite (rhomboedral) with hyperconcentration of Mg.

Submarine nodules are well known examples of polymetallic accumulation in the sea floor, in relation with volcanic activity of mid-oceanic ridges. Analytical Ion Microscopy shows that biomineralizations (Radiolaria for example) are present in these nodules. The skeleton of living Radiolaria consists of amorphous silica. Inside the nodules, the chemical composition is modified and different elements such as Na, Al, Mn, Sr and Ca are observed instead of silica (figure 17).

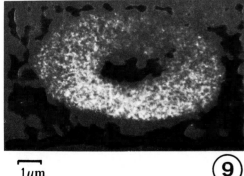

1μm

(9)

Distribution of iron (^{56}Fe) at the surface of a nucleated red blood cell(secondary ion emission microanalysis)

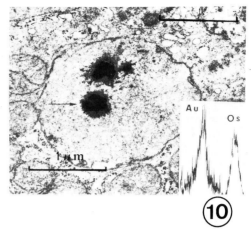

(10)

Figure 10. Concentration of gold in a mitochondrion of a proximal tubule cell of the kidney.

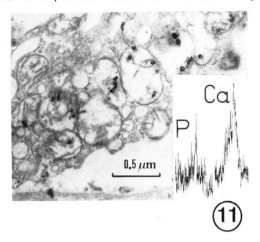

(11)

Figure 11. Concentration of Ca and P in a mitochondrion of a human synovial cell.

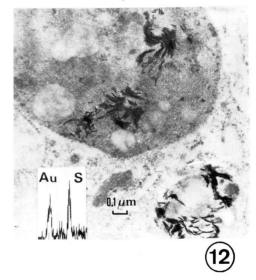

(12)

Figure 12. Concentration of gold with sulphur in a lysosome of a macrophage.

References

1. P. Galle. Cytochimie sur coupes ultrafines par spectrographie des rayons X. Sixth International Congress for Electron Microscopy. Maruzen ed., Nihonbashi, Tokyo Japan 1966, 79-80.
2. C. Zanchi, J. Sevely and B. Jouffrey. An energy filter for high voltage electron microscopy. J, Microscopy Spectr. Electr., 1, 1976, 95-104.
3. P. Galle. Les artéfacts en microanalyse par sonde électronique. J. Micr. Biol. Cell., 22, 1975, 315-334.
4. A.J. Tousimis. Electrolytes in cells of biologic tissues. Proceedings Seventh National Conference on Electron Probe Analysis (San Francisco, USA 1972) paper 47.
5. T.A. Hall. Physical Techniques in Biological Research, 2nd edition, vol. 1, A. Academic Press New-York and London, 1971, The microprobe assay of chemical elements.
6. P. Echlin. The preparation of frozen hydrated biological material for X-ray microanalysis. J. Micr. Biol. Cell., 22, 1975, 215-226.
7. A.J. Saubermann. The application of X-ray microanalysis in physiology. J. Micr. Biol. Cell., 22, 1975, 401-414.
8. A. Dörge, R. Rick, K. Gehring ... Preparation and applicability of freeze-dried sections in the microprobe analysis of sort tissue. J. Micr. Biol. Cell., 22, 1975, 205-214.
9. F.D. Ingram and M.J. Ingram. Quantitative analysis with the freeze-dried, plastic embedded tissue specimen. J. Micr. Biol. Cell., 22, 1975, 193-204.
10. W. Fuchs and B. Lindemann. Electron beam x-ray microanalysis of frozen biological bulk specimen below 130 K. J. Micr. Biol. Cell., 22, 1975, 227-232.
11. A. Lauchli. Precipitation technique for diffusible substances. J. Micr. Biol. Cell., 22, 1975 239-246.
12. P. Galle. Sur une nouvelle méthode d'analyse cellulaire utilisant le phénomène d'émission ionique secondaire. Ann. Phys. Biol.Med., 1970, 84-94.
13. H.J. Höhling and W.A. Nicholson. Electron microprobe analysis in hard tissue research: specimen preparation. J. Micr. Biol. Cell., 22, 1975, 185-192.
14. R. Lefevre. Physico Chimie et Cristallographie des Apatites d'Intérêt Biologique. CNRS Paris, France, 1975, 143-148.
15. P. Galle and J.P. Berry. Microanalyse in biology: a review of some specific problems. Twelfth Annual Conference on the Microbeam Analysis Society. Boston, U.S.A. 1977 (paper 167).
16. J.P. Berry, P. Henoc, P. Galle ... Pulmonary mineral dust. Am. J. Path., 83, 1976, 427-438.
17. J.P. Berry, P. Henoc and P. Galle. Phagocytosis by cells of the pulmonary alveoli. Am. J. Path., 93, 1978, 27-36.
18. S. Duckett, G. Lyon and P. Galle. Mise en évidence du Magnésium dans la maladie de Sturge-Weber. C.R. Acad. Sc. Paris, 282, 1976, 113-115.
19. S. Duckett and P. Galle. Mise en évidence de l'aluminium dans les plaques séniles de la maladie d'Alzheimer. C.R.Acad. Sc. Paris, 282, 1976, 393-396.
20. S. Duckett and P. Galle. X-ray microanalysis of pallidal arteries in Parkinson Disease. Sixth European Congress on Electron Microscopy, Jerusalem, Israel, 1976, 210-211.
21. J. Lapresle, S. Duckett, L. Cartier. Documents cliniques, anatomiques et biophysiques dans une encéphalopathie avec présence de dépôts d'aluminium. C.R.Soc.Biol. 169, 1976, 282-285
22. P. Galle and J.P. Berry. Ultrastructure et microanalyse des néphrocalcinoses. Physicochimie et cristallographie des Apatites d'Intérêt Biologique. CNRS Paris, 1975,51-66
23. M.B. Bellhorn and R.K. Lewis. Localization of ions in retina by secondary ion mass spectrometry. Exp. Eye Res., 1976, 22, 505-518.
24. P. Galle. Rôle des lysosomes et des mitochondries dans les phénomènes de concentration et d'élimination des éléments minéraux par le rein. Journ. Microscopie, 19, 1974, 17-24.
25. R. Lefevre, R.M. Frank and J.C. Voegel. The study of human dentin with secondary ion microscopy and electron diffraction. Calcified Tissue Res., 19, 1976, 251-261.
26. R. Lefevre. Analyse par émission ionique secondaire. J. Micr. Biol. Cell., 22, 1975, 335-347.
27. R. Lefevre and R.M. Frank. Etude de la fluorose dentaire au microanalyseur ionique. J. Biol. Buccale, 4, 1976, 29-41
28. J.P. Cuif et R. Lefevre. Etude par microanalyse ionique de l'évolution diagénétique d'algues solénopores triasiques. C.R. Acad. Sc. Paris, 278, 1974, 2263-2270.
29. R. Lefevre, J. Marcoux et J.P. Cuif. Microscopie ionique avec optique de transfert expérimentale de radiolaires fossiles: premiers résultats. J. Microsc. Spectrosc. Electron., 3, 1978, 469-476.

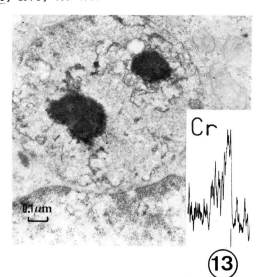

Figure 13. Concentration of chromium in a lysosome of a tubular proximal tubule cell of the kidney.

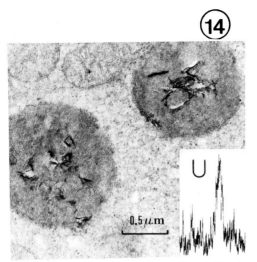

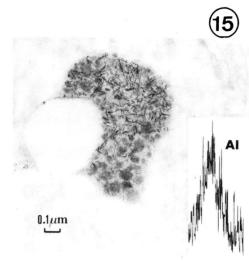

Figure 14. Concentration of uranium with phosphorus in a lysosome of a tubular proximal tubule cell of the kidney

Figure 15. Concentration of aluminium in a lysosome of a neuron of the brain cortex of the rat

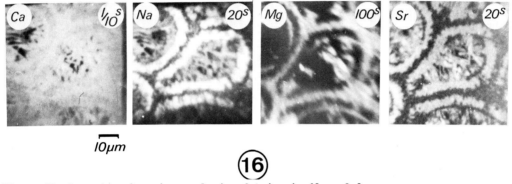

Figure 16. Secondary ion microanalysis of triassic Algae Solenopora. Distribution of four elements.

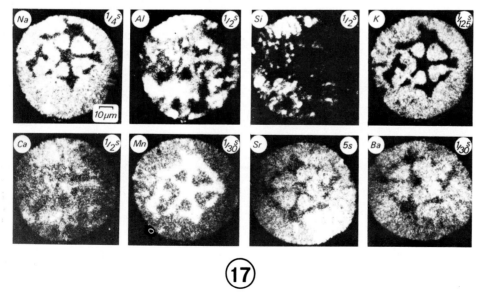

Figure 17. Secondary ion microanalysis ; Radiolaria included into triassic polymetallic concretions : substitution of original silica by Na, Al, Ca, Mn and Sr.

Discussion with reviewers

Reviewer I: What are the basic features of the instrumentation and conditions used by the authors for x-ray microanalysis, electron microdiffraction, secondary ion emission microanalysis and electron energy loss microanalysis ?

Authors: The basic features of the instrumentation and conditions have been added to the final manuscript.

Reviewer I: What are the lower limits of elemental detection for the three microanalytical methods presented ?

Authors: The lower limits of elemental detection for the three microanalytical methods vary depending on many factors such as the nature of the element, the conditions of analysis (accelerating voltage in electron probe analysis, nature of primary ions in secondary ion mass analysis) and the nature of the specimen itself ; thus, only some typical examples can be given here. In our experience, on epon sections 0.1 μm thick, the lower limit in x-ray analysis is between 10^{-15} and 10^{-17}g for many elements of Z > 11 ; the major factor of limitation is due here to the destruction of the Epon section under intense electron beams. In secondary ion mass analysis, with O_2^+ in the primary beam, less than 10^{-19}g of elements such as Na or K can be easily detected under a surface of 1 square μm. On Epon sections 0.1 μm thick, 10^{-18}g of beryllium has been easily detected by electron energy loss analysis. Better limits of detection may be obtained in x-ray or energy loss analysis in thinner specimens (molecules of ferritin deposited on an ultrathin carbon layer).

Reviewer I: Have the authors attempted semi-quantitative analysis of differences in elemental content using peak-to-continuum ratios as a more precise means of differentiating between the various types of pulmonary mineral inclusions ?

Authors: A semi-quantitative analysis using peak-to-continuum ratios has been attempted. These results are necessary for the interpretation of the electron diffraction pattern.

Reviewer I: Have the authors explored potential limitations of energy loss microanalysis of biological specimens including problems related to section thickness and large carbon peaks which overlap peaks of biological elements of interest such as calcium ?

Authors: We have a limited experience in energy loss microanalysis but it appears that the section thickness of an even "ultrathin" epon section is an important limiting factor. A standard "ultrathin" section of 0.1 μm is too thick for analysis of many inclusions of 0.01 μm in diameter ; there are many interferences with the organic matrix.

Reviewer II: Please indicate the preparation techniques of the tissues and at what temperature the preparation was done for each analysis (Fig.1-17).

Authors: The preparation techniques have been added to the final manuscript. The temperature of the specimen under the electron beam varies depending on the thickness and the nature of the section, and on the intensity and the energy of the electron beam. On a thick specimen of low thermal conductivity such as Epon the temperature can easily reach several 100°C (Ref. 3). On an ultra-thin Epon section, I cannot give the temperature of the specimen under the beam. This temperature raising makes it impossible to work with a very intense electron probe current, and is probably a major limiting factor in x-ray analysis of "relatively thick" Epon sections (0.1 μm). In secondary ion mass analysis, the temperature of the first atomic layers of the specimen is impossible to accurately predict.

Reviewer II: What kind of ion did you use for the primary beam of secondary ion emission ?

Authors: O_2^+ for the applications presented here.

Reviewer II: Did you find any tissue damage during secondary ion emission analysis ?

Authors: Yes, secondary ion emission analysis is a destructive method ; scanning electron microscopy of the specimen surface shows many irregularities after the ion bombardment. These irregularities do not affect the quality of the image but they may be at the origin of many difficulties in quantitative analysis.

Reviewer II: What was the thickness of the samples for electron energy loss analysis ?

Authors: Of the order of 0.1 μm.

Reviewer II: Which method did you use for x-ray microanalysis, wave length dispersive x-ray spectroscopy (WDS) or energy dispersive x-ray spectroscopy (EDS) ?

Authors: Generally, wave length dispersive spectrometry. Energy dispersive spectrometry has been sometimes used for a rapid screening of relatively large inclusions containing elements at a high concentration but the results obtained have always been checked by wave length dispersive spectrometry to avoid the very frequent peak interferences.

Editor: Can you put your work in perspective of other work using these techniques ?

Authors: In an overall sense, the work of our laboratory oriented towards the study of non-diffusible elements and to the detection of these elements on the ultrastructural scale. We have studied the problem of diffusible elements very little. Whereas, this problem has been extensively studied in most other laboratories using these techniques. As regards instrumentation we use principally CTEM as a method of visualization and WDS for x-ray analysis, while most other laboratories use SEM with EDS.

REVIEWERS: I   L.M. Buja
            II E. Arima

SCANNING-ELECTRON MICROSCOPY/1979/II
SEM Inc., AMF O'Hare, IL 60666, USA

ELECTRON PROBE ANALYSIS OF MUSCLE AND X-RAY MAPPING OF BIOLOGICAL SPECIMENS
WITH A FIELD EMISSION GUN

A.P. Somlyo, A.V. Somlyo, H. Shuman and M. Stewart*

Pennsylvania Muscle Institute   *C.S.I.R.O.
Presbyterial-Univ. of Pennsylvania Division of Computing Research
    Medical Center   Canberra, Australia
51 N. 39th Street
Philadelphia, PA 19104

Abstract

Recent electron probe analytic studies of
freeze-dried cryosections of vascular smooth
and vertebrate striated muscle are reviewed.
The results show that the sarcoplasmic reticulum of striated muscle is not in ionic communication with the extracellular space. Vacuolation by hypertonic solutions and fatigue involves the
T-tubule system. The high calcium content of
the terminal cisternae of the resting muscle has
been quantitated in situ.

In smooth muscle, the high Cl content is
distributed in the cytoplasm, and mitochondria
in rabbit portal vein smooth muscle cells do not
contain high concentrations of calcium. Mitochondrial calcium loading in the form of
granules is generally due to fiber damage.
Nuclear and mitochondrial composition in situ
has been quantitated and compared to the composition of the cytoplasm of the same cells.

Preliminary phosphorus x-ray maps of
smooth muscle show the feasibility of this
approach in defining the composition of organelles in thin cryosections. The use of x-ray
maps at intermediate resolution is illustrated
with tropomyosin paracrystals labelled with Hg-
containing dye at the thiol residues. Mercury
x-ray maps of such paracrystals show the 40nm
periodicity of the thiol groups and their Fourier
transforms contain information to a spatial
resolution of 10-20nm.

KEY WORDS: Muscle, Nucleus, Mitochondrion,
X-ray Microanalysis, Field Emission Source STEM,
Sarcoplasmic Reticulum, Calcium, X-ray Mapping

Introduction

This review article summarizes the
results of electron probe analysis of smooth and
striated muscle conducted in our laboratory and
our preliminary experience with x-ray mapping
of biological material with a field emission
source. Detailed accounts of our studies on
muscle have been published elsewhere.[1-5]

Electron probe analysis has directly
resolved major questions about the composition
of the sarcoplasmic reticulum (SR) in normal,
resting striated muscle, about the changes in
subcellular composition induced by hypertonic
solution or severe fatigue, and it is beginning to
reveal the changes that occur in the SR of
activated muscle. In smooth muscle, the distribution of the high cytoplasmic chloride and the
question whether mitochondria play a significant
role in accumulating calcium during relaxation
have been particularly amenable to electron
probe analysis that also yielded information of
general interest about the composition of nuclei
and mitochondria in situ.

We have used freeze-dried cryosections
for electron probe analysis of muscle, because
in this tissue there are, at least normally, no
large aequeous domains such as found in some
epithelial structures[6-7] that would require the
use of frozen hydrated specimens.[8] Furthermore, the analyses reported were performed on
unstained specimens: negatively stained preparations (or those cut onto a wet trough), while
valuable for structural studies[9-11], are not suitable for the analysis of diffusible elements.[12-14]
Freeze-substitution is another ancillary preparatory technique particularly suitable for the
structural study of muscles in which vacuolation
has been experimentally induced[3, 15, 16], under
special conditions it is also useful for electron
probe analysis of non-diffusible precipitates of
calcium[17] but not for the study of diffusible
elements.[18] In our experience, the use of
trapping agents (e.g., pyroantimonate for cal-

cium or silver for chloride) does not yield reliable localization or quantitation, and results obtained with these methods will not be dealt with in this communication.

Methods

The experimental animals (frog, toad-fish and rabbit) and the preparation of the muscles have been described in our detailed publications. We wish to reemphasize the importance of careful dissection in preparing tissues for electron probe analysis, lest the use of cut or otherwise damaged surface fibers result in erroneously high mitochondrial calcium content (granules) and abnormally low cytoplasmic K and high NaCl. Tissues were frozen in supercooled Freon 22[1], the rates of freezing obtained by this method have also been published.[19] This method avoids the explosive hazard associated with the use of liquid propane.

Analyses were done on a Philips EM400 electron microscope interfaced with a $30mm^2$ Kevex Si (Li) energy dispersive detector and Kevex 7000 multichannel analyzer, using Hall's approach[20] of determining concentrations in ultrathin sections through the relationship of the characteristic x-ray/continuum ratio. Details of the multiple least square fitting technique and the sensitivity of our instrumentation have been published.[21-23] Two modifications have recently been added to the instruments previously described: 1) the microscope is now fitted with a field emission gun normally operated at 80kV and separated from the main column by a differentially pumped aperture and 2) the multiple least squares fitting routine has been rewritten in our laboratory (by Mr. K. McGinnis) and is now executed in less than 5 seconds on a PDP11/34 computer equipped with hard wired multiplier and dual RKO5 discs, as compared to the approximately 200 second computation required by the software multiplication based Tracor Northern NS880 system.

X-ray mapping was performed with the field emission source by scanning the beam; x-ray counts from the appropriate characteristic window were entered into the computer while simultaneously accumulating the scanning transmitted electron image. For x-ray maps and STEM images of the paracrystals, three to twelve frames of 256 lines (32 or 64 msec. or to 1/16 or 1/8 sec. line time) were summed in the computer memory to minimize the effects of source noise. The majority of the maps were obtained with a total scanning time of between 200 and 300 seconds. The images and x-ray maps were stored on RKO5 discs and when appropriate, have been subjected to two dimensional Fourier analysis by conventional

methods. The thiol residues in tropomyosin paracrystals were stained with a mercuric dye containing 4Hg atoms per molecule: this staining produces a characteristic 40nm period banding pattern in the paracrystal localizing the thiol residues.[24]

Results

Vascular smooth muscle

A freeze-dried cryosection of portal anterior mesenteric vein (PAMV) vascular smooth muscle used for analysis is illustrated in Fig. 1 and typical spectra of a mitochondrion, cytoplasm and nucleus are shown in Fig. 2.

The cytoplasm of smooth muscle, as contrasted to frog striated muscle, had a relatively high Cl content of 278mmol/kg dry wt. (168 fibers, 8 animals). Sampling along a length of smooth muscle cell with $1\mu m$ diameter probes showed homogeneous distribution of Cl within the cytoplasm of a given fiber.[12]

Nuclear and cytoplasmic composition was compared by analyzing the nucleus and cytoplasm of the same fiber with identical probe parameters, and the results are shown in Table 1. Neither the Na nor the Ca content of the nucleus was significantly different from that of the adjacent cytoplasm. The continuum counts obtained in the two regions were identical, suggesting that they also have similar degrees of hydration. There was, as expected, a large excess of P in the nucleus, and also a somewhat higher K and lower S and Cl concentration than in the cytoplasm.

Mitochondria in normal PAMV smooth muscle did not contain high Ca concentrations (Fig. 2 and Table 2). The transmitochondrial distribution of K, Na and Cl did not show a large gradient even assuming that these ions were distributed in mitochondrial water and correcting for the lower degree of mitochondrial than cytoplasmic hydration. Incidentally, the mitochondrial hydration calculated on the basis of x-ray continuum counts obtained showed good agreement with the measured water content of isolated mitochondria (for discussion see reference 5).

Mitochondrial granules containing Ca and P (Figs. 3-4) were only found in fibers that were damaged as indicated by their high Na and low K content[5].

Striated Muscle

An unstained dry cryosection of frog striated muscle used for electron probe analysis is shown in Fig. 5. It is evident, that, in spite of minor ice crystal formation in this

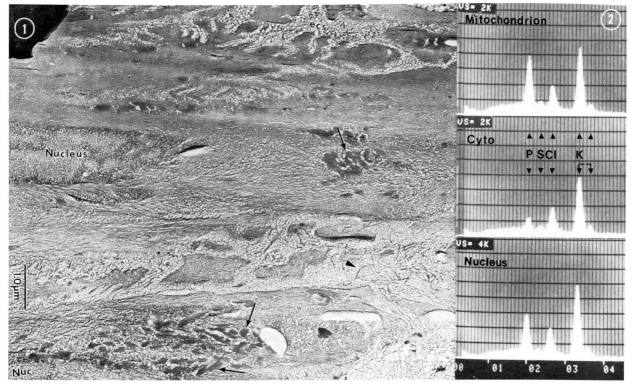

Fig. 1 Unstained dried cryosection of PAMV smooth muscle bundle showing gradation from vitreously frozen region (cell on the left) to one showing fine ice crystal formation. Muscle contracted for 30 mins. with high K solution prior to freezing. Arrowhead=collagen; arrow=mitochondria; e.c.s.=extracellular space. From reference 5.

Fig. 2 X-ray spectra of mitochondrion, cytoplasm and nucleus of a depolarized PAMV smooth muscle fiber illustrated in Fig. 1. Extraneous peaks (Cu L and Si) have been subtracted by the computer. From reference 5.

particular muscle, the sarcomeres and their components are well identifiable, as are the triadic regions and mitochondria.

During the validation of the methods used in these studies, we have attempted to determine whether the preparatory techniques used preserve a sharp demarcation of the in vivo ionic distribution across the surface membrane. X-ray spectra and the results of quantitative analysis of frog striated muscle fibers and adjacent extracellular space (e.g.: Fig. 6) show preservation of the physiological gradients. Furthermore, in vacuoles induced in the T-tubule system by hypertonic solutions[1,15] or by fatigue[3] there was a clear demarcation between the extracellular composition (high NaCl) of vacuoles, and the adjacent cytoplasm analyzed within 100nm of the vacuole. Therefore, these findings indicate that within the precision of our quantitation (estimated to be approximately 15%), the in vivo distribution of diffusible elements is maintained at least across such membranous barriers.

The terminal cisternae (TC) and cytoplasm have been analyzed in frog[1] and toadfish[2]

striated muscle, and in neither case was their content comparable to that of the extracellular space. The Ca content of the TC in resting muscles is approximately 66-100mmol/kg dry wt.,[1-3] and the TC also contain somewhat higher K than the adjacent cytoplasm.

X-ray mapping

A phosphorus x-ray map of a freeze-dried cryosection of rabbit portal anterior mesenteric vein smooth muscle is shown in Fig. 7. The nucleus and the condensation of nuclear chromatin at the nuclear membranes are well resolved and the (mostly mitochondrial) structures of the nuclear pole are also evident. A tear in the section (arrow) shows the extent of extraneous background contributed to the map in this example. However, this specimen was supported on copper grid, and the extraneous background can be further reduced by the use of beryllium grids.

Mercury x-ray maps of tropomyosin paracrystals showed a well-resolved periodicity of 40nm (Fig. 8A) corresponding to the thiol

Table 1

Paired comparison of nuclear and cytoplasmic composition in
rabbit portal-anterior mesenteric vein smooth muscles

	K	Na	Cl	Mg	Ca	P	S	Continuum
Nucleus	649+4.2	177+6	256+2.2	39+2.7	2.1+1.0	593+4.2	95+1.7	1668+11
Cytoplasm	592+3.8	168+5.9	282+2.3	36+2.6	1.6+1.0	250+2.4	139+1.8	1620+11
Nucleus - cytoplasm (paired)	70+5.8	18+8.5	-25+3.1	1.5+3.8	0.6+1.5	375+5	-47+2.5	3+15

28 paired analyses (nucleus and adjacent cytoplasm each) obtained from 6 animals, mmole/Kg dry weight $\pm$ S.E.M., except for continuum given as the number of counts. From reference 5.

Table 2

Paired comparison of mitochondrial and cytoplasmic composition in
rabbit portal-anterior mesenteric vein smooth muscles

	K	Na	Cl	Mg	Ca	P	S	Continuum
Mitochondria	464+1.7	193+3.5	220+1.1	38+1.5	0.8+0.5	571+2.1	142+1.0	1789+7
Cytoplasm	565+2.6	246+4.6	308+1.7	43+1.9	0.7+0.7	261+1.7	125+1.2	1071+6
$\dfrac{[X_i]\ \text{mito}}{[X_i]\ \text{cyto}}$	1.2	1.1	1.1					

68 pairs of mitochondria and adjacent cytoplasm (each) from 9 animals were analyzed; values are mmol/kg dry weight except continuum, given as the number of counts. From reference 5.

groups of tropomyosin, observed as regions of increased scattering in the scanning transmitted electron image (Fig. 8B) and previously described by one of us in transmission electron micrographs.[24] Two dimensional Fourier analysis of the x-ray maps showed very strong second, and occasional weak third and fourth order reflections. These findings indicate that an x-ray spatial resolution of 20nm can be readily obtained in biological material and that our current instrumentation is suitable for attaining a spatial resolution of at least 10nm. Inspection of the STEM images of the mercury labelled tropomyosin paracrystals suggests that they may contain no further high resolution information due to the globular nature of the deposit that can be approximated closely by a Fourier spectrum of a few sinusoids, without significant high frequency components (sharp edges). We have not determined whether this is due to

radiation damage or to biochemical artefact introduced during the preparation of the reagents for labelling the thiol groups in tropomyosin.

Discussion

The high Cl content of smooth muscle, previously detected by wet chemical methods,[25-28] has been definitely localized through electron probe analysis to the cytoplasm. The implications of this finding and of the apparent deficit of detectable cellular anions have been discussed elsewhere in detail.[5] Cytoplasmic analysis of frog striated muscle has also suggested that Cl is not completely Donnan distributed in this tissue.[1]

Analysis of mitochondria in both relaxed (Table 2) and in contracted vascular smooth muscles,[5] shows that these organelles do not sequester significant concentrations of calcium,

714

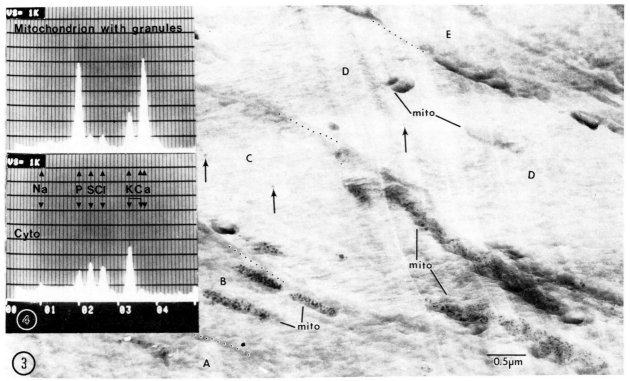

Fig. 3 Unstained dried cryosection of PAMV smooth muscle including several damaged fibers containing mitochondria with numerous Ca and P containing granules. Cells A, B, C and E contain mitochondrial granules. Mitochondria not shown in A and E. Fiber C also contains Ca deposits (arrows) in the SR. In fiber D the mitochondria do not contain granules, but one small deposit of Ca was present in (presumably) the SR. Fibers containing mitochondrial granules had high cytoplasmic Na and low K concentrations. Dots: cell borders. From reference 5.

Fig. 4 X-ray spectrum of mitochondrion containing granules and of adjacent cytoplasm. Note the high mitochondrial Ca and P and the Na peak in the cytoplasm. Extraneous peaks (Cu, L and Si) subtracted by computer. From reference 5.

Fig. 5 Unstained dried cryosection from a control bundle of frog semitendinous fibers. Brackets mark a sarcomere. A nucleus is shown in the lower portion. Pairs of terminal cisterna (arrows) occur at the Z lines.

at least in rabbit portal anterior mesenteric vein smooth muscle. This finding is in agreement with the apparent K_m ($\sim 17\mu M$) of isolated mitochondria for $Ca^{29, 30}$ being significantly higher than the free cytoplasmic Ca required for contractile activation (for review, see 5). However, only the in situ measurements afforded by electron probe analysis of cryosections enabled us to exclude the possibility that the apparent K_m in a true cytoplasmic environment may be significantly lower than that measured in vitro. The low concentration of mitochondrial calcium in rabbit smooth muscle contrasts with the concentrated deposits of Ca found in the sarcoplasmic reticulum of the same tissues.[5]

Mitochondrial calcium granules were found only in cells that were damaged as evidenced by their high cytoplasmic Na and low K content. Therefore, at least in non-calcifying tissues, mitochondrial calcium granules are generally associated with cell damage (for heart, see reference 4, for epithelium, see 31). The possibility that such granules are due to cell damage induced during tissue preparation, such as cubing of otherwise normal tissues[32, 33] must be considered and, in our experience, is very likely.[12] Moderately heavy granule formation in mitochondria represents a load of approximately $0.1-1\mu$ mol Ca/mg mitochondrial protein[34] and unpublished observations. Since even maximal rates of calcium uptake by isolated mitochondria are only about 10nmoles/sec. mg protein, the accumulation of such large quantities of calcium must require a mitochondrial energized calcium transport of at least 10 sec. duration. The presence of relatively high concentrations of calcium in isolated bovine mitochondria, increased in atherosclerotic blood vessels,[34] does raise the possibility of pathological and/or species related differences in mitochondrial calcium content.

The finding that the sarcoplasmic reticulum of striated muscle does not have a composition similar to that of the extracellular space resolves a long-standing controversy based on compartmental analysis of Na and Cl fluxes, and excludes the possibility of significant ionic communication between the intracellular compartment involved in contractile activation and the extracellular space (for review, see 1, 11, 35). The presence of high concentrations of calcium in the terminal cisternae (TC) of the SR of resting muscle is in obvious agreement with a large body of earlier evidence based on other techniques (for review, see 1, 34, 35-39); more recently, work in progress on activated muscles has shown that approximately 60% of the calcium can be released from the TC during a maintained tetanus.[40, 41] There is not an equivalent increase in the Mg and K content of the TC of

activated muscle, suggesting that charge neutrality is not maintained by the uptake of these cations, but may be due to proton movements or changes in the charges of organic constituents. Our finding of the large quantity of Ca released during tetanus is in agreement with physiological measurements suggesting that during a caffeine contracture the total calcium in the frog fiber is raised to mmolar level [42] and does not contradict by any means the general conclusion that free calcium in activated striated muscle is in the micromolar range (43, and for review, see 37-39). Since frog striated muscle containing at least 0.5mmolar parvalbumin with two high affinity calcium sites/molecule[44] in addition to troponin, a relatively large quantity of total calcium has to be released to attain the free Ca concentration required to fully activate muscle. However, we have no information on how much, if any, of the calcium released from the TC has already been taken up by the longitudinal tubules of the sarcoplasmic reticulum[45] at the time our measurements were made, or whether there is a sufficient rise in total (bound and free) cytoplasmic Ca to account for the amount released from the TC.

Electron probe analysis has also been valuable in determining the composition of abnormal compartments such as those created by incubation of muscle in hypertonic solutions[1] or by extreme fatigue.[3] The finding of considerable interest in the latter study was the observation that the Ca content of the TC is not depleted in severely fatigued muscles that no longer respond to electrical stimulation,[3] revealing that some earlier step in excitation-contraction coupling, rather than inavailability of activator calcium, is involved in the mechanism of fatigue. These findings, as well as recent electron probe studies of diseased human skeletal muscle[46] and immature muscle[47] clearly indicate the enormous potential of electron probe analysis in studying the cellular physiology and pathology of muscle.

X-ray mapping has been extensively employed in materials sciences but only rarely, and then for the analysis of thick sections and/or concentrated elemental deposits,[48] in biology. We have been interested in using the method for study of the distribution of naturally occurring elements and labels in thin cryosections and in protein paracrystals, taking advantage of the potentially greater spatial resolution afforded by a field emission source. Our preliminary results[40] suggest that with a field emission gun and a commercially available energy dispersive detector, x-ray maps can be obtained of physiologically occurring natural elements, such as phosphorus, without destruction of the specimen and at a spatial resolution of at least 10 to 20nm. We anticipate further improvement in spatial resolution with better detector efficiency and more suitable specimens. Recently, electron energy loss maps of cell membranes and organelles have also been obtained with a field emission source.[49,50] The combination of the two methods with image processing of computer stored signals should be an extremely powerful approach in "biological microchemistry" at electron microscopic resolution.

(Supported by HL15835 to the Pennsylvania Muscle Institute and GM00092.)

References

1. Somlyo, A.V., H. Shuman and A.P. Somlyo: Elemental distribution in striated muscle and effects of hypertonicity: electron probe analysis of cryosections. J. Cell Biol. 74: 828-857, 1977.
2. Somlyo, A.V., H. Shuman and A.P. Somlyo: The composition of the sarcoplasmic reticulum in situ: electron probe x-ray microanalysis of cryosections. Nature 268: 556-558, 1977.
3. Gonzalez-Serratos, H., A.V. Somlyo, G. McClellan et al.: The composition of vacuoles and sarcoplasmic reticulum in fatigued muscle: electron probe analysis. Proc. Natl. Acad. Sci. 75: 1329-1333, 1978.

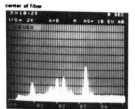

Fig. 6

ML FOR FILE 29 TIME=300
CONT AREA = 1763 ± 63

ELEMENT	CONC (mM/KG)	PEAK CNT	PEAK ERROR
Na	290.9 ± 45.7	918	± 140
Mg	-16.5 ± 14.3	-120	± 103
P	45.3 ± 11.5	561	± 142
Cl	344.8 ± 16.8	5566	± 183
K	17.4 ± 8.4	290	± 141
Ca	6.6 ± 8.1	*116	± 142

CHISQD = 2.97

ML FOR FILE 30 TIME = 300
CONT AREA = 2631 ± 76

ELEMENT	CONC (mM/KG)	PEAK CNT	PEAK ERROR
Na	33.8 ± 33.2	159	± 156
Mg	33.2 ± 11.2	360	± 121
P	350.1 ± 14.6	6475	± 193
Cl	75.6 ± 7.8	1821	± 180
K	496.6 ± 17.0	12359	± 224
Ca	2.9 ± 6.6	77	± 173

CHISQD = 3.74

ML FOR FILE 33 TIME = 300
CONT AREA = 3134 ± 79

ELEMENT	CONC (mM/KG)	PEAK CNT	PEAK ERROR
Na	27.9 ± 26.7	156	± 150
Mg	31.9 ± 9.7	411	± 125
P	308.1 ± 12.0	6785	± 201
Cl	58.3 ± 6.5	1672	± 182
K	475.9 ± 14.3	14102	± 227
Ca	-9.9 ± 5.4	-307	± 169

CHISQD = 3.78

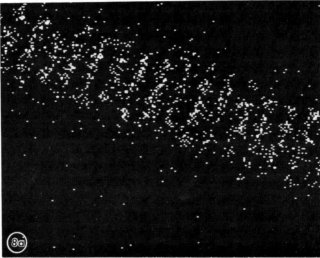

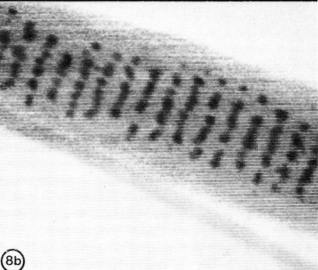

Fig.6 (at the top) X-ray spectra and elemental concentrations from two areas of the same fiber and adjacent extracellular space. From reference 1.

Fig.7 (above) Phosphorus x-ray map of freeze-dried section of rabbit portal-anterior mesenteric vein smooth muscle. Nuc: nucleus, M: mitochondria, arrow: tear in section. Scan time: 160 sec.

Fig.8 (right) Tropomyosin paracrystal with H-labelled thiol residues at 40nm repeats. A. Hg (L,M) x-ray map. B. STEM image. Each repeat=40nm.

*4. Somlyo, A.V., J. Silcox and A.P. Somlyo: Electron probe analysis and cryoultramicrotomy of cardiac muscle: mitochondrial granules. Proc. 33rd Ann. EMSA Meetings, 532, 1975.

5. Somlyo, A.P., A.V. Somlyo and H. Shuman: Electron probe analysis of vascular smooth muscle: composition of mitochondria, nuclei and cytoplasm. J. Cell Biol. 81, 1979.

6. Lechene, C.P.: Electron probe microanalysis: its present, its future. Amer. J. Physiol. 232: F391-F396, 1977.

7. Gupta, B.L., M.J. Berridge, T.A. Hall and R.B. Moreton: Electron microprobe and ion-selective microelectrode studies of fluid secretion in the salivary glands of Calliphora. J. Exp. Biol. 72: 261-284, 1978.

8. Hutchinson, T.E., D.E. Johnson and A.P. Mackenzie: Instrumentation for direct observation of frozen hydrated specimens in the electron microscope. Ultramicroscopy 3: 315-324, 1978.

9. Sjöström, M. and J.M. Squire: Fine structure of the A-band in cryosections. J. Mol. Biol. 109: 49-68, 1977.

10. Somlyo, A.P. and A.V. Somlyo: Ultrastructure of the contractile apparatus: controversies resolved and questions remaining. In: Excitation-Contraction Coupling in Smooth Muscle, eds. R. Casteels, T. Godfraind and J.C. Rüegg (Elsevier/North-Holland Biomedical Press, Amsterdam), pps. 317-322, 1977.

11. Somlyo, A.V.: Bridging structures spanning the junctional gap at the triad of skeletal muscle. J. Cell Biol. 80: 743-750, 1979.

12. Somlyo, A.V. and J. Silcox: Cryoultramicrotomy for electron probe analysis. In: Microbeam Analysis in Biology. C. Lechene and R. Warner, eds. Academic Press, New York, in press, 1977.

*13. Somlyo, A.V., H. Shuman and A.P. Somlyo: Electron probe analysis of vertebrate smooth muscle: distribution of Ca and Cl. Proc. 34th Ann. EMSA Meetings, 334, 1976.

14. Sjöström, M. and L.E. Thornell: Preparing sections of skeletal muscle for transmission electron analytical microscopy (TEAM) of diffusible elements. J. Micros. 103: 1-12, 1975.

15. Franzini-Armstrong, C., H.E. Heuser, R.S. Reese et al.: T-tubule swelling in hypertonic solutions: a freeze substitution study. J. Physiol. (Lond.) 283: 133-140, 1978.

16. Sommer, J.R., N.R. Wallace and W. Hasselbach: The collapse of the sarcoplasmic reticulum in skeletal muscle. Z. Naturforsch 33c: 561-573, 1978.

17. Ornberg, R.L. and T.S. Reese: A method for localizing divalent cations in freeze-substituted thin sectioned tissue. J. Cell Biol. 79: 257a, 1978.

*18. Bonventri, J.V. and C. Lechene: A method for electron probe microanalysis of organic components in picoliter samples. Proc. MAS 9th Ann. Conf., Ottawa, pps. 8a-8d, 1974.

19. Costello, M.J. and J.M. Corless: The direct measurement of temperature changes within freeze-fracture specimens during rapid quenching in liquid coolants. J. Micros. 112: 17-38, 1978.

20. Hall, T.A.: The microprobe assay of chemical elements. In: Physical Techniques in Biological Research. G. Oster, ed., Academic Press, Inc., N.Y., 1A, 1971.

21. Shuman, H., A.V. Somlyo and A.P. Somlyo; Quantitative electron probe microanalysis of biological thin sections: methods and validity. Ultramicroscopy 1: 317-339, 1976

22. Shuman, H., A.V. Somlyo and A.P. Somlyo: Theoretical and practical limits of Ed x-ray analysis of biological thin sections. Scan. Elect. Micros., IIT Res. Inst. 1: 663-672, 1977.

23. Shuman, H. and A.P. Somlyo: Electron probe x-ray analysis of single ferritin molecules. Proc. Natl. Acad. Sci. 73: 1193-1195, 1976.

24. Stewart M. and V. Diakiw: Electron microscope location of protein thiol residues. Nature (Lond.) 274: 184-186, 1978.

25. Jones, A.W., A.P. Somlyo and A.V. Somlyo: Potassium accumulation in smooth muscle and associated ultrastructural changes. J. Physiol. (Lond.) 232: 247-273, 1973.

26. Jones, A.W. and L.A. Miller: Ion transport in tonic and phasic vascular smooth muscle and changes during deoxycorticosterone hypertension. In: Blood Vessels, Molecular and Cellular Aspects of Vascular Smooth Muscle in Health and Disease. D.F. Bohr and F. Tanenaka, eds. S. Karger, Switzerland, pps. 83-92, 1978.

27. Casteels, R.: The distribution of chloride ions in the smooth muscle cells of the guinea pig's taenia coli. J. Physiol. (Lond.) 214: 225-243, 1971.

28. Kao, C.Y. and A. Nishiyama: Ion concentrations and membrane potentials of myometrium during recovery from cold. Amer. J. Physiol. 217: 525-531, 1969.

29. Vallieres, J., A. Scarpa and A.P. Somlyo: Subcellular fractions of smooth muscle. I. Isolation, substrate utilization and Ca^{++} transport by main pulmonary artery and mesenteric vein mitochondria. Arch. Biochem. Biophys. 170: 659-669, 1975.

30. Wikström, M., P. Ahonen and T. Tuukkainen: The role of mitochondria in uterine contractions. FEBS Lett. 56: 120-123, 1975.

31. Gupta, B.L. and T.A. Hall: Electron

microprobe x-ray analysis of calcium. In: Ann. N.Y. Acad. Sci., Calcium Transport and Cell Function. A. Scarpa and E. Carafoli, eds., N.Y. Acad. Sci., N.Y. 307: 28-51, 1978.

32. Saetersdal, T.S., R. Myklebust, N.-P. Berg Justesen et al.: Calcium containing particles in mitochondria of heart muscle cells as shown by cryoultramicrotomy and x-ray microanalysis. Cell Tiss. Res. 182: 17-31, 1977.

33. Seveus, L., D. Brdiczka and T. Barnard: On the occurrence and composition of dense particles in mitochondria in ultrathin frozen dry sections. Cell Biol. Intl. Rep. 2(#2), 1978.

34. Somlyo, A.P., A.V. Somlyo, H. Shuman et al.: Electron probe analysis of calcium compartments in cryo sections of smooth and striated muscles. In: Ann. N.Y. Acad. Sci., Calcium Transport and Cell Function. A. Scarpa and E. Carafoli, eds., N.Y. Acad. Sci., N.Y. 307: 523-544, 1978.

35. Costantin, L.L.: Contractile activation in skeletal muscle. Prog. Biophys. Biol. 29: 197-224, 1975.

36. MacLennan, D.H. and P.C. Holland: Calcium transport in sarcoplasmic reticulum. Ann. Rev. Biophys. Bioeng. 4: 377-404, 1975.

37. Ebashi, S. Excitation-contraction coupling. Ann. Rev. Physiol. 38: 293-313, 1976.

38. Endo, M.: Calcium release from the sarcoplasmic reticulum. Physiol. Rev. 57: 71-108, 1977.

39. Fabiato, A. and F. Fabiato: Calcium release from the sarcoplasmic reticulum. Circ. Res. 40: 119-129, 1977.

40. Somlyo, A.P., H. Shuman and A.V. Somlyo: Mitochondrial and sarcoplasmic reticulum contents in situ: electron probe analysis. In: Frontiers of Biological Energetics: Electrons to Tissues, A. Scarpa, P.L. Dutton and J. Leigh, eds., Academic Press, N.Y. Vol. 1: 742-751, 1978.

41. Somlyo, A.V., H. Gonzalez-Serratos, H. Shuman et al.: Changes in sarcoplasmic reticulum composition in activated frog skeletal muscle. Biophys J. 25: 25a, 1979.

42. Moisescu, D.G. and R. Thieleczek: Calcium and strontium concentration changes within skinned muscle preparations following a change in the external bathing solution. J. Physiol. 275: 241-262, 1978.

43. Blinks, J.R., R. Rüdel and S.R. Taylor: Calcium transients in isolated amphibian skeletal muscle fibres: Detection with aequorin. J. Physiol. 277: 291-323, 1978.

44. Gosselin-Rey, C. and C. Gerday:

Parvalbumins from frog skeletal muscle (Rana temporaria L.). Isolation and Characterization. Structural modifications associated with calcium binding. Biochemica et Biophysica Acta 492: 53-63, 1977.

45. Winegrad, S.: Intracellular calcium movements of frog skeletal muscle during recovery from tetanus. J. Gen. Physiol. 51: 65-83, 1968.

46. Wróblewski, R.: Healthy and diseased human skeletal muscle studied by electron microscopy and x-ray microanalysis with special reference to fibre types. University of Stockholm, Sweden. Thesis 1978.

47. Nichols, B.L., H.A. Soriano, D.J. Sachen, et al.: Electron probe localization of electrolytes in immature muscle. Hopkins. J. 135: 322-335, 1974.

48. Warner, R.R. and J.R. Coleman: Electron probe analysis of calcium transport by small intestine. J. Cell Biol. 64: 54-74, 1975.

49. Hainfeld, J. and M. Isaacson: The use of electron energy loss spectroscopy for studying membrane architecture: a preliminary report. Ultramicroscopy 3: 87-95, 1978.

*50. Costa, J.L., D.C. Joy, D.M. Maher et al.: Subcellular localization of fluorinated serotonin in human platelets by electron energy-loss spectroscopy. Proc. 35th Ann. EMSA Meetings, pps. 238-239, 1977.

Discussion with Reviewers

R. Kaufman: You mention the use of a field emission gun but not of a cooling device for your specimen. What about the mass loss of your specimen which is expected to occur, particularly, if you want to take advantage of the higher brightness of your gun? Can you exclude that some of your data are influenced by mass loss phenomena?

Authors: Quantitative analysis is done at approximately -165°C. Mass loss in fibrous proteins is about 13% (ref. 21). See also answer to Question 2 of H. Jessen.

R. Kaufman: Can you comment on the ultimate detection limits you may reach with your deconvolution techniques, particularly with respect to low Ca in the presence of high K-signals? Can you give an example of how long you have to collect counts (beam current and irradiated volume given) in order to have the contour of your continuum smooth enough for the extraction of very small Ca-peaks?

Authors: The minimal detectable concentration, (see also ref. 21) for a 100 sec. count with 1000cps originating from the specimen, is 10mmol/kg dry weight. Therefore, reduction of the minimal detectable concentration to 1mmol/kg dry weight in the same specimen

*Availability Information: EMSA Proceedings: Claitor's Publishing, Baton Rouge, Louisiana. MAS Proceedings: K.F.J. Heinrich, Analytical Chemistry, N.B.S., Washington, D.C., 20234.

would require a 100-fold increase in counting time (or increase in probe current), or the averaging of 100 comparable measurements. Errors due to faulty deconvolution arise when detector calibration is incorrect and/or the detector resolution is different from that used for the collection of primary standards. These considerations, and specifically the uncertainty associated with measuring low concentrations of calcium in the presence of physiological intracellular concentrations of potassium are discussed in reference 21. In 100nm thick dried cryosections of muscle, given the number of analyses reported in the publications cited, we consider the uncertainty of absolute cytoplasmic calcium concentrations measured to be approximately 2mmol/kg dry weight (see also ref. 5).

R. Kaufman: Can you comment on your statement that in an activated skeletal muscle fiber, myoplasmic Ca rises to mmolar levels which is conflicting with every evidence obtained by other techniques (aequorin technique, birefringence signals, pCa-tension curve, model calculations)?

Authors: Thank you for pointing out the apparent, but not real, conflict between the two sets of data. We are naturally referring to total, rather than ionized, calcium and have emphasized this point in the Discussion. As noted there, the parvalbumin content of frog muscle is at least 0.5mM, and this protein has two high affinity calcium sites/molecule.

I.K. Berezesky: In recent years, many investigators have published substantial evidence that mitochondria in various types of tissue are the principal Ca stores and that the Ca content of these organelles far exceed those levels found in the cytoplasm. Your x-ray data on normal PAMV smooth muscle show that the mitochondria do not contain high Ca concentrations and, in fact, that the mitochondrial Ca levels are about the same as those found in the cytoplasm. Do you believe that this might also be the case in other types of tissue beside muscle, a finding which is in contradiction to the currently held theory?

Authors: We cannot rule out completely the possibility of there being a "relatively" small ($<10^5$) mitochondrial/cytoplasmic free calcium gradient in muscle, in view of the approximately 2mmol/kg dry weight uncertainty of the present measurements (see answer to question 2 of R. Kaufman), and our inability to distinguish free from bound calcium. Recent studies on squid axons (for review, see Brinley, F.J., Ann. Rev. Biophys. Bioeng. 7: 363-392, 1978) also show relatively low mitochondrial calcium content and mitochondria isolated from most tissues (e.g., liver, heart) have calcium contents of less than 10mmol/kg dry weight. The most recent evidence suggests that in non-

muscle cells such as nerve (McGraw, C.F., A.V. Somlyo and N.P. Blaustein: Neuroscience. Abst. 4: 332, 1977; Henkart, M.P., T.S. Reese and F.J. Brinley Jr.: Science. 202: 1300-1303, 1978; Gambetti, P., S.E. Erulkar, A.P. Somlyo, N.K. Gonatas: J. Cell Biol. 64: 322-330, 1975), fibroblasts (Moore, L. and I. Pastan: Ann. N.Y. Acad. Sci. 307: 177-194) and endothelium (Somlyo, A.V., and A.P. Somlyo: Science. 174: 955-958, 1971), there is a smooth endoplasmic reticulum (SER) that can sequester calcium. Increasing attention is being given to the possibility that the SER, rather than mitochondria, is the site of physiological calcium sequestration in non-muscle cells, and that mitochondria take up large amounts of calcium only when cytoplasmic concentrations rise to abnormally high levels.

I.K. Berezesky: The authors state that "mitocondrial granules containing Ca and P were found only in fibers that were damaged as indicated by their high Na and low K content." Such granules have been observed ultrastructurally following routine electron microscopic preparation by several investigators in different types of normal tissue. In addition, these granules have recently been observed even more frequently in unfixed cryosections leading to the belief that osmium leaches out most of the calcium. Some investigators, using x-ray microanalysis, have found that these granules contain Ca and P. Rather than the possibility that such granules are due to cell damage induced during tissue preparation such as cubing, is it not possible that their formation in normal tissue may represent a type of reaggregation of Ca and P which possibly develops during the initial freezing of tissue and subsequent cryosectioning?

Authors: In cells in which the cytoplasmic composition has also been determined (ref. 5, 7), mitochondrial granules were found only in damaged cells, while other reports showing the presence of such granules provide no information about cytoplasmic composition (see also Discussion in ref. 5). We do not think that the major issue is whether the granules are present in mitochondria in living cells or whether they form in a eutectic phase during freezing, but that such granules represent a mitochondrial calcium content of minimally 70mmol/kg dry weight and up to 1.8mole/kg. Although some living, possibly pathological cells may contain such high concentrations of mitochondrial calcium (e.g., ref. 34), simple calculation (taking into account mitochondrial mass) suggests, for example, that a mitochondrial calcium concentration of 100mmol/kg dry weight is completely inconsistent with the known total calcium content of cardiac muscle.

I.K. Berezesky: It is stated that "the ML

routine was rewritten so that execution now takes 5 sec. as compared to 200 sec. required by the software multiplier based Tracor Northern system." Is this rewritten program the same as the original, commercially available routine or has it been changed in any other manner than execution time?

Authors: A Fortran program (based on F. Schamber, Proc. 8th Natl. Conf. on Electron Probe Analysis (1973) 85) was written for our PDP 11/34. The change in execution time is mainly due to the improved hardware (floating point processor). The only software improvement was to store the filtered reference spectra, rather than filtering both reference and unknown for each pass of the ML routine.

I.K. Berezesky: Other than the x-ray mapping, were any of the probe data obtained using the field emission gun?

Authors: Our more recent quantitative analyses of tetanized striated muscle were also obtained with the field emission gun at approximately -165°C.

I.K. Berezesky: Please expand on the two dimensionsal Fourier analysis reduction scheme used for the x-ray mapping.

Authors: A 256 x 256 array was stored on RKO-5 discs and conventional Cooley-Tukey FFT algorithm was performed on a PDP 11/34 computer. The matrix transposition necessary for a 2-dimensional FFT was optimized for the 11/34's memory size and disk.

D. Joy: Was there any change in your "x-ray hole count" when you switched from thermionic to the field emission source?

Authors: Yes, there was a decrease in the "x-ray whole count" since installation of the field emission source. However, in addition to the source, the spray aperture, condenser optics and objective are also different. Therefore, any estimate of a change in hole count due to the source itself would be meaningless. While the hole counts currently obtained are tolerable with quantitative analysis, we believe that manufacturers should be further encouraged to make improvements in collimating the beam and reducing parasitic scattering.

D. Joy: Have you tried any dynamic background stripping techniques for x-ray mapping, and do you believe that they offer any advantage?

Authors: We have no experience with dynamic background stripping techniques. For some of the tropomyosin x-ray maps, although not all of them, we have used beryllium grids to reduce the background. In general, we plan to obtain a "background x-ray map" simultaneously with the characteristic maps and subsequently subtract, if necessary. Given 512 x 512 array and 16 bits available storage, a STEM image and several x-ray maps can be obtained and stored

simultaneously.

H. Jessen: Please, elaborate on the instrumental parameters, using the field emission gun for biological x-ray mapping. What are the probe parameters (current density, spot size, scanning rate etc.)?

Authors: We tried a range of instrumental parameters to optimize the x-ray maps, and those most frequently used are now given in the text. The manufacturer's specifications on the field emission gun are: brightness 10^9 amps/cm^2 STER; 0.2×10^{-9} amps in a 0.3nm, and 0.5×10^{-8} amps in a 5nm diameter spot. Good tropomyosin x-ray maps were obtained with measured probe currents of approximately 0.3×10^{-9} amps.

H. Jessen: To what extent is radiation damage of a biological specimen limiting the usefulness of a field emission gun? Have you any comparative data illustrating the benefits of a field emission gun (vs. a conventional gun) in quantitative, biological x-ray microanalysis?

Authors: The total radiation dose delivered with a field emission source need not exceed that associated with the use of conventional guns, because the duration of the scan can be limited. Most radiation damage studies (ref. 21 and for review, see Isaacson, M. in Principles and Techniques of Electron Microscopy, Biological Application; Glaeser, R.M. in Physical Aspects of Electron Microscopy and Microbeam Analysis) suggest that, within limits, the total dose, rather than the dose rate, determines radiation damage; the maximal damage dose for most damage parameters is of the order of $10^{-2}C/cm^2$, a value generally exceeded during electron probe analysis with conventional sources. Hence, theoretically at least, the use of field emission source should not materially increase radiation damage, and our preliminary experiments merely verify the theoretically expected result.

Regarding the second part of your question, we do not believe that a field emission gun has any advantages over a LaB_6 source for quantitative, biological x-ray microanalysis with stationary probes of 5nm or greater diameter. Sufficient current for analysis of thin sections can be obtained in a 50nm probe with an LaB_6 source, and for 100 to 200nm diameter probes (and larger), the current produced by a conventional source is actually greater than that produced by a field emission gun. For our purposes, the great advantage of the field emission gun is for obtaining relatively high resolution x-ray maps and for electron energy loss analysis of EXAFS type-fine structure that requires the inherent lower energy spread of field emission sources. The conventional use of field emission sources for high resolution, dark field STEM imaging has been discussed in several publications by A. V. Crewe, who introduced this approach (e.g., Crewe, A. V.: Electron Microscopy $\underline{3}$: 197-204, 1978).

H. Jessen: Can a "spatial resolution of at least 10 to 20nm" be obtained in mapping elements, which are not present in a crystalline or paracrystalline context?

Authors: Our results show an instrumental resolution measured on a test object having a regular lattice, just as the optical resolution of an electron microscope can be tested, for example, with a gold lattice. The Fourier transform is merely a convenient way of defining the spatial frequencies that still contain information, and has been used for similar purposes in conventional transmission electron microscopy. We expect that attainment of a "point-to-point" resolution of 10-20nm will also be possible, although more difficult, in noncrystalline specimens, and we need different test objects having such intrinsic detail for an experimental demonstration.

SCANNING ELECTRON MICROSCOPY/1979/II
SEM Inc., AMF O'Hare, IL 60666, USA

ANALYTICAL ELECTRON MICROSCOPIC STUDIES OF ISCHEMIC AND HYPOXIC
MYOCARDIAL INJURY

H.K. Hagler, K.P. Burton, L. Sherwin, C. Greico, A. Siler, L. Lopez and
L.M. Buja

Department of Pathology, Southwestern Medical School
University of Texas Health Science Center
5325 Harry Hines Blvd.
Dallas, TX 75235

Abstract

Our work has been directed to the application of analytical electron microscopy to the study of hypoxic and ischemic cell injury. Abnormal mitochondrial inclusions were produced in canine myocardium after 40 minutes of temporary coronary occlusion and 20 minutes of reflow. Energy dispersive x-ray spectroscopy (EDS) showed that some mitochondrial inclusions had small Ca peaks in tissue fixed directly in phosphate-buffered osmium, but that the inclusions uniformly showed large Ca and P peaks in tissue fixed in alcohol or in unfixed, freeze-dried cryosections. These results indicate that the initial stage of calcification in ischemic mitochondria occurs in the form of readily soluble calcium phosphate. EDS of unfixed cryosections also showed changes in other electrolytes in ischemic versus control mitochondria. An ionic lanthanum (La) probe technique was developed to evaluate the role of altered plasma membrane functional integrity in the evaluation of electrolyte alterations. In an isolated cat papillary muscle preparation, La was selectively localized to extracellular sites of control oxygenated muscles in both unfixed cryosections and conventionally fixed specimens. Abnormal intracellular La deposition was demonstrated at a transitional stage in the evolution of irreversible hypoxic injury in the papillary muscle model. These results suggest that an alteration in functional membrane integrity associated with altered movement of polyvalent ions plays an important role in the pathogenesis of irreversible myocardial injury.

KEY WORDS: Cell Injury, Ischemia, Hypoxia, Myocardium, Membranes, Calcium, Calcification, Analytical Electron Microscopy, X-ray Microanalysis, Cryoultramicrotomy, Fixation, Biological Specimen Preparation

Introduction

Evidence is accumulating that analytical electron microscopy can provide important new information for studies of cell injury. Areas of interest include evaluation of cellular and subcellular electrolyte shifts, documentation of pathological calcium accumulation and calcification and localization of membrane probes and other cytochemical markers.[1-10] Recent work in our laboratory has been devoted to the evaluation of optimal methods of tissue preparation and analytical techniques for the study of cell injury, with emphasis on myocardial ischemic and hypoxic injury. Also, cryoultramicrotomy of fresh unfixed cat papillary muscles was used to further validate the use of ionic lanthanum as a probe of altered membrane integrity.[5]

Materials and Methods

Experimental Models

Mongrel dogs were anesthetized with sodium pentobarbital, ventilated and the heart exposed following a left thoracotomy. In five dogs, the proximal left circumflex coronary artery was temporarily occluded for 40 minutes followed by release and reflow for 20 minutes. The hearts were then excised and placed in ice cold saline prior to tissue sampling.

An isolated cat papillary muscle preparation was used for the studies involving ionic lanthanum as a probe of membrane integrity.[5] Cats were anesthetized with sodium pentobarbital, the heart rapidly excised and a suitable papillary muscle removed from the right ventricle. Muscles were stimulated and maintained in a physiological salt solution with various concentrations of glucose as substrate and bubbled with 100% O_2. Muscles were subjected to progressive intervals of hypoxia by replacing the 100% O_2 with 100% N_2, with and without reoxygenation. The muscles were then exposed to lanthanum for one hour by replacement of the standard medium with one containing 5mM $LaCl_3$.

Tissue Fixation

Samples of control and ischemic tissue taken from the coronary occlusion animals were immersed in fixative and diced into 1mm^3 cubes. The following fixatives were employed: 1) 1% osmium tetroxide (OsO_4) in 0.1M phosphate, 0.1M cacodylate or veronal acetate buffers with or without 5% sucrose for 2 1/2 hours; 2) 3% glutaraldehyde in 0.1M phosphate or cacodylate buffers for 1 hour with or without buffer wash for 1/2 hour and post-fixation in buffered osmium for 1 hour; or 3) absolute alcoholic (80%) aldehyde for 3 1/2 hours. All fixatives and buffers were at pH 7.2-7.4. The fixation was performed at room temperature.

Control and ischemic papillary muscles were initially fixed by immersion of the muscles into the fixative while held in the clips of the muscle bath. Initial fixation was achieved with a solution of 2.5% glutaraldehyde and 2% formaldehyde in 0.1M cacodylate buffer. More recently a cold solution of 3% purified glutaraldehyde in 0.1M phosphate buffer has been used to improve retention of lanthanum. After initial fixation the muscles were diced up and placed in fixative at room temperature. Tissue from each muscle was then rinsed in buffered sucrose and either post-fixed in buffered 1% OsO4 or left unosmicated.

Tissues fixed in aqueous solutions were taken through a graded series of alcohols to 100% alcohol. The tissues fixed in aqueous solutions or in alcohol were processed through several changes of absolute alcohol, followed by propylene oxide and embedded in epon-araldite. Ultramicrotomy was performed using an LKB Ultrotome III. Semithin (1 micron) sections were mounted on glass and stained with toluidine blue. Thin sections were mounted on copper or nylon grids and either stained with lead citrate and ethanolic uranyl acetate or left unstained for x-ray microanalysis.

Cryoultramicrotomy

Travenol biopsy needles were used to obtain ribbons of control and ischemic tissue from the coronary occlusion hearts. The ribbons were rapidly diced into small cubes of less than 1mm^3 and mounted on LKB silver pin tissue holders. Control papillary muscles were removed from the muscle bath, diced and mounted on silver pin tissue holders. The tissue holders were placed in a spring-loaded quick freezing device used to thrust the samples beneath the surface of the freezing solution. A slurry of liquid and solid nitrogen at -208°C was used to freeze the tissue. The elapsed time from excision of the heart to sample freezing was always less than 10 minutes and for the papillary muscles was less than 2 minutes. The sampling for cryoultramicrotomy was always performed prior to obtaining tissue for the other tissue preparation procedures.

It is possible to judge the adequacy of the freezing technique used on the tissue by observing the appearance of the tissue before and after freezing. If the specimen has undergone virtually no change in color then a good freeze has been obtained. Tissue poorly frozen will take on a more white or frosted appearance. After a good freeze of the specimen, cryoultramicrotomy was performed using an LKB Ultrotome and cryokit. The kit was modified in our laboratory to obtain optimal performance at the low temperature required for sectioning unfixed fresh tissue. The modifications were similar to those described by Sevéus.[11,12]

We have developed two additional modifications which make it easier to collect and retain specimens on the grids. Due to the extremely low temperatures within the chamber, a great deal of static electricity develops which makes the sections "jump" in unpredictable directions. A radioactive alpha source or "static master" has been mounted within the cryochamber in the area of section collection and where the grids are transferred to the drying container. The second device which was extremely useful is a vacuum pipette for moving sections from the knife to the grid. This consists of a glass pipette pulled by hand to a very small diameter tip. The tip just covers one grid square of a 300 mesh grid. The micropipette is connected to a small vacuum pump and the suction is controlled using a finger over one end of a Y connector in the line. The tip may be immersed in liquid nitrogen to prevent section melting.

Ultramicrotomy was performed with the specimen at -120°C and the glass knife at -100°C. Sections were cut in the 100-200nm range from the surface of the block to a depth of a few μm and dry mounted on copper or nylon grids. Formvar or carbon support films were used to help retain sections on the grids. Section quality can be evaluated at this step by observing how the sections come off of the block. If curling of the specimen was observed, the sections were invariably damaged by remelting or were not frozen well initially. Good sections come off flat down the base of the knife. The grids with sections are next pressed flat using a 5 gram weight for 5 minutes and placed in a vial filled with desiccant within the cryokit chamber. The sections are allowed to warm up within the cryochamber for about 2 hours. The grids are coated with 10-20nm of carbon in a Denton vacuum evaporator and then stored in a desiccator prior to analysis.

Analytical Electron Microscopy

A JEOL 100C transmission electron microscope with a high resolution scanning attachment and Kevex 30mm^2, 158eV energy dispersive x-ray detector was used. The x-ray spectra were collected using a Tracor Northern TN 2000 multichannel analyzer operated at 20eV/channel. The x-ray spectra were sent to and stored in a Tracor Northern NS 880 analyzer. Spectral plots were obtained using a Hewlett-Packard 2648 graphics terminal attached to the NS 880. Microscope conditions consisted of scanning transmission mode at 80KV, 50μamp emission current (above a dark current of 52-54μamp), 30° specimen tilt and a beam diameter estimated to be around 40-60nm in diameter. The detector was placed within 20mm of the specimen area of interest.

Spectra were collected using the summation of five to ten ultrastructurally similar areas using brief 10 to 30 second analysis times for each area in order to avoid deterioration in peak detection due to progressive mass loss and specimen contamination. This technique was described previously in this symposia series.[2] Stability of beam position was confirmed by the localization of contamination spots and by comparing spectra obtained on and off of

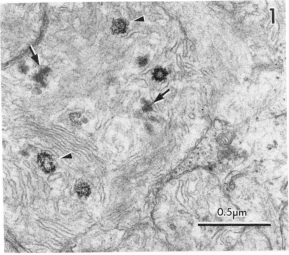

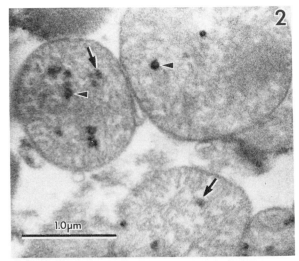

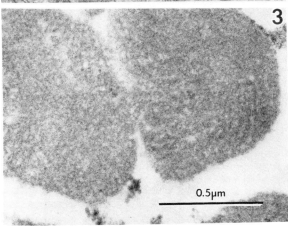

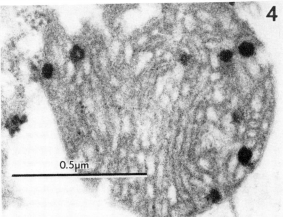

Fig. 1. There are two types of ischemic mitochondrial inclusions: moderately electron dense,
osmiophilic, amorphous matrix (flocculent) densities (arrows) and very electron dense, granular,
calcium-containing inclusions which often have
lucent centers (arrowheads).

Fig. 3. TEM of normal myocardium. Alcohol
fixation. Epoxy section stained with uranyl
acetate and lead citrate. The mitochondria are
devoid of inclusions.

Fig. 2. Scanning transmission electron micrograph (STEM) of an unstained epoxy section of
osmium-phosphate-fixed ischemic myocardium
after temporary coronary occlusion. This preparation used for analytical data collection
also shows the amorphous matrix densities
(arrows) and the very electron dense, granular,
calcium-containing inclusions (arrowheads).

Fig. 4. TEM of ischemic myocardium after temporary coronary occlusion. Alcohol fixation.
Epoxy section stained with uranyl acetate and
lead citrate. The mitochondria contain numerous
very electron dense, granular, calcium-containing
inclusions, but do not exhibit any of the
moderately electron dense, amorphous matrix
densities observed in osmicated tissue. Compare
with Figs. 1 and 2.

characteristic structures. Elemental reference
spectra were collected from thin crystals of
various salts (NaCl, CaCl$_2$, etc.) which have
elements of interest with spectral peaks well
removed from each other. The unknown spectra
were analyzed using the multiple least squares
fitting routine available on the NS 880.[13]
Analysis of thin biological specimens was based
on the thin film criteria described by Hall
et al[14] and as applied by Shuman et al.[15,16] The
continuum region of 5.5 to 6.0 KeV was chosen for
the data reduction programs. This region was
always free of interfering peaks and proved more
reliable than was 1.34 to 1.64 KeV reported by
Shuman et al.[15] Elemental peaks were considered
real for the purposes of detectability when they
were shown to be present at the two sigma confidence limit.

Absolute quantitation of elemental concentration using spectra obtained from salt crystals
was not reproducible due to the unknown degree
of hydration present in most salt crystals. The
present work only involves use of the salt
crystal spectra to model the detector response
and to identify the presence of elements for
comparison with other spectra. We have experienced considerable difficulty with the preparation of reproducible standard curves from thin
film biological materials for purposes of

absolute quantitative analysis of test spectra from tissue sections.

Results

In Vivo Myocardial Ischemia

The temporary ischemia model with reflow resulted in a variety of ultrastructural alterations compared to control myocardium, including the appearance of two types of mitochondrial inclusions when the tissue was fixed directly in osmium-phosphate and stained with uranyl acetate and lead citrate (Figs. 1 and 2). These inclusions consisted of small, moderately electron dense, homogeneous, amorphous matrix (flocculent) densities and small, very electron dense, granular inclusions which usually showed lucent centers resulting in an annular-granular appearance (Figs. 1 and 2). Both types of inclusions had a size range of approximately 50 to 100nm. Analytical electron microscopy of unstained sections collected on nylon grids resulted in the association of a small calcium peak with the granular type density but not with the amorphous (flocculent) density (Fig. 2, Table 1). Analytical studies were typically performed on unstained sections because of the known propensity of acidic uranyl acetate solution to dissolve calcium deposits.[17]

Table 1. Effect of Tissue Preparation on the Preservation of Early Mitochondrial Calcification

Tissue Preparation	Ca Peak/Continuum Ratio*
Glu-PO$_4$ and Os-PO$_4$	N.D.**
Os-PO$_4$	0.14±0.43
Alcohol	1.15±0.04
Unfixed cryosections	0.91±0.05***

  * Each value represents the peak intensity and its confidence limit (1 sigma) from a summed spectrum of ten microareas.
 ** Not detected.
*** Relatively reduced due to increased continuum from copper grid compared to nylon grids for other sections.

When tissue was initially fixed in glutaraldehyde in phosphate or cacodylate buffers or in osmium buffered with cacodylate or veronal acetate, the granular type inclusions were not preserved and only after the tissue was osmicated were the amorphous (flocculent) densities apparent. Analytical electron microscopy failed to reveal mitochondrial inclusions with detectable calcium.

The tissue fixed directly in alcoholic aldehyde or absolute alcohol showed generally poor ultrastructure, but the mitochondria were preserved (Figs. 3-5). Both unstained and stained sections of ischemic tissue exhibited numerous very electron dense granular mitochondrial inclusions which typically exhibited a homogeneous, solid appearance. The small, moderately electron dense, amorphous matrix (flocculent) densities were not observed. Analytical electron microscopy

of the unstained alcohol-fixed samples of ischemic myocardium revealed prominent Ca and P peaks in the mitochondrial inclusions (Fig. 6, Table 1). These elements were not present in adjacent sarcoplasm.

Unfixed, frozen-dried sections of normal and temporarily ischemic myocardium are shown in Figs. 7 and 8. The normal tissue contains rows of dense ovoid structures which resemble mitochondria. Myofibrils with A, I and Z bands are also visualized in the tissue. In the sections of ischemic myocardium, only ovoid mitochondrial profiles with dense inclusions are readily visualized. Variation in ultrastructural appearance may be related to the quality of freeze-drying with increased visualization of ultrastructure in areas with slight ice crystal damage.[11,12] Analytical electron microscopy of the sections provided the x-ray spectra shown in Figs. 9 and 10. The spectrum collected from mitochondria of normal tissue reveals multiple elements of biological interest including a prominent K peak and small Ca peak (Fig. 9). The spectrum obtained from ischemic mitochondrial inclusions shows that the mitochondrial calcium has increased and the potassium level has decreased with ischemia and reflow (Fig. 10). The elemental peak-to-continuum ratios derived from the x-ray spectra from the cryosections are shown in Table 2.

Tables 1 and 2 summarize the effects of tissue preparation on the preservation of calcium in the form of small electron dense inclusions in ischemic dog myocardium following 40 minutes occlusion and 20 minutes reflow. The calcium peak-to-continuum ratio for the cryosections is reduced due to an increased continuum from the copper grids in this preparation as compared to the nylon grids used for the other sections. We have generally used copper rather than nylon grids for cryoultramicrotomy since flatter sections have been obtained with the copper grids.

In Vitro Lanthanum Probe Studies

Analytical electron microscopy has provided data regarding the localization of lanthanum in isolated cat papillary muscles after control and hypoxic conditions. In control oxygenated papillary muscles, electron dense deposits consistent with lanthanum accumulation were confined to extracellular regions involving the basal lamina-plasma membrane complex of muscle cells in both fixed, epoxy embedded sections and unfixed, frozen-dried cryosections (Figs. 11 and 12). Analytical electron microscopy confirmed the selectively extracellular localization of detectable lanthanum (Figs. 13 and 14). Selective extracellular lanthanum deposition also was observed after 30 minutes of hypoxia. When the hypoxic interval was maintained for 75 minutes or longer, abnormal intracellular localization of lanthanum occurred in the form of electron dense deposits on the myofibrils and in the mitochondria (Figs. 15 and 16). The onset of abnormal intracellular lanthanum accumulation correlated with the transition from reversible to irreversible contractile depression in the isolated muscles subjected to progressive intervals of hypoxia and reoxygenation.[5]

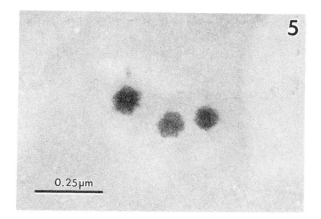

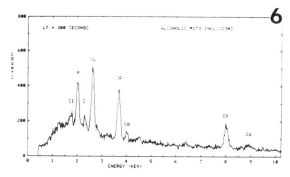

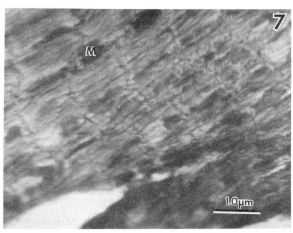

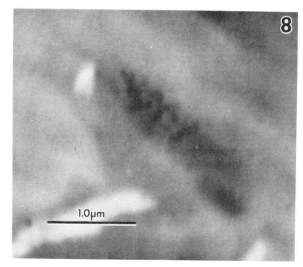

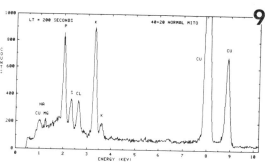

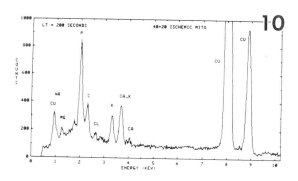

Fig. 5. STEM of unstained, alcohol-fixed epoxy section of ischemic myocardium. The inherent electron density of these mitochondrial inclusions correlates with a high calcium content on energy dispersive x-ray analysis.

Fig. 6. Summed x-ray spectrum (200 seconds) from mitochondrial inclusions in unstained epoxy section of alcohol-fixed ischemic myocardium. The mitochondrial inclusions exhibit prominent Ca and P peaks and small Si and S peaks. Cl is from the embedding medium and Cu from the microscope.

Fig. 7. STEM of unfixed, unstained, freeze-dried cryosection of normal myocardium shows myofibrils with Z, I and A bands and oval dense structures which appear to represent mitochondria (M).

Fig. 8. STEM of unfixed, unstained, freeze-dried cryosections of temporarily ischemic myocardium exhibits numerous electron dense inclusions in mitochondrial profiles.

Fig. 9. Summed x-ray spectrum (200 seconds) from mitochondria of unfixed cryosection of normal myocardium exhibits prominent K peak which obscures a small Ca peak. Other elements detected include Mg, P, S, Si and Cl. Na is obscured by the peak from the copper grid.

Fig. 10. Summed x-ray spectrum (200 seconds) from mitochondria of unfixed cryosection of temporarily ischemic myocardium exhibits an increased Ca peak and reduced K and Cl peaks compared to control mitochondria. Other peaks detected include P, Mg, Si and S. Compare with Fig. 6.

Table 2. Elemental Content (Peak-to-Continuum Ratios) of Unfixed, Frozen-Dried Cryosections of Control and Ischemic Myocardium Following Forty Minutes Temporary Coronary Occlusion and Twenty Minutes Reflow

	Mg	P	S	Cl	K	Ca
Control Mitochondria*	0.14±0.2	2.44±0.04	0.72±0.07	0.88±0.06	2.97±0.03	0.15±0.2
Ischemic Mitochondrial Inclusions*	0.17±0.16	1.77±0.04**	0.54±0.08**	0.17±0.18**	0.64±0.05**	0.86±0.05**

* Each value represents the peak intensity and its confidence limits (1 sigma) from a summed spectrum of ten microareas.
** These values are significantly different from the control mitochondrial values since the differences in the values exceed the confidence limits.

Discussion

Analytical electron microscopy has provided direct information regarding the elemental content of various mitochondrial inclusions observed in ischemic and infarcted myocardium. The data indicate that the small, very electron dense, granular inclusions represent an early stage of pathological calcification which is associated with electrolyte shifts, including potassium efflux, involving muscle cells damaged by temporary coronary occlusion and reflow. The small calcific inclusions appear to represent a readily soluble, chemically amorphous (non-crystalline) form of calcium phosphate.[18-20] Formation of these calcific mitochondrial inclusions is associated with excess calcium influx from the blood.[19] It appears likely that cell injury produced by permanent coronary occlusion is associated with electrolyte shifts between extracellular and intracellular compartments and mitochondria and cytosol, although minimal changes in total tissue electrolytes are evident.[6-8,18,19] Some supportive analytical electron microscopic data for the latter statement has been presented by others.[6-8]

Electrolyte shifts and pathological calcium accumulation appear to represent consequences of altered membrane integrity induced by ischemic damage. Our in vitro lanthanum probe studies have provided direct cytologic evidence of altered movement and intracellular accumulation of polyvalent ions at a transitional stage in the evolution of irreversible hypoxic injury. Selective extracellular localization of lanthanum in unfixed freeze-dried cryosections indicates that fixation does not produce major alterations in lanthanum distribution and provides supportive evidence for the use of the lanthanum probe technique for the study of membrane pathophysiology.

Studies in our own and other laboratories of the amorphous matrix (flocculent) densities have shown that these inclusions are osmiophilic rather than inherently electron dense, have low or non-detectable calcium content, are rich in lipid and protein, and form in mitochondria of severely damaged cells regardless of the mechanism of injury and level of perfusion.[1-4,6-10, 18,19,21] These findings suggest that the amorphous matrix densities represent aggregates of denatured organic material formed as a result of mitochondrial damage.

Our findings indicate that the method of tissue preparation significantly affects the amount of calcium preserved in early stages of pathological calcium accumulation. Optimal preservation of the calcific inclusions was obtained using alcoholic fixation or cryoultramicrotomy of fresh tissue. Others have reported similar results.[6-8,20] The annular-granular inclusions observed with osmium-phosphate fixation appear to represent sites of calcification in which much of the calcium is replaced by osmium. Other aqueous fixation methods resulted in complete dissolution of these soluble calcific deposits. Staining with heavy metals, particularly acidic uranyl acetate solutions, also has a deleterious effect on calcium preservation.[17] Later stages of biological and pathological calcification are associated with progressive enlargement of deposits and conversion from soluble to insoluble calcium phosphate in the form of hydroxyapatite-like material.[1-4,20] More advanced calcific deposits are at least partially preserved by all methods of aqueous fixation tested.[1-4]

Acknowledgements

The authors wish to thank Ms. Kathy Handrick for typing this manuscript and Mrs. Linda Bolding and Mr. Stacy Bartus for the photographic reproductions.

This work was supported in part by the Charles and Kathryn Moore Foundation, the Moss Heart Center, NIH Specialized Center of Research (SCOR) Grant HL-17669 and NIH Research Training Grant in Cardiovascular Pathophysiology HL-07360.

References

1. L.M.Buja, J.H.Dees, D.F.Harling and J.T. Willerson, "Analytical electron microscopic study of mitochondrial inclusions in canine myocardial infarcts", J. Histochem. Cytochem. 24, 1976, 508-516.
2. H.K.Hagler, K.P.Burton, R.H.Brown, R.C. Reynolds, G.H.Templeton, J.T.Willerson and L.M. Buja, "Energy dispersive x-ray spectroscopic (EDS) analysis of small particulate inclusions in

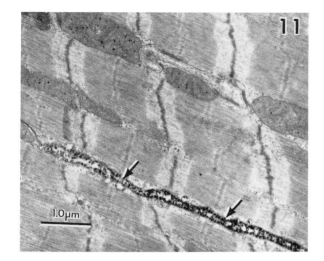

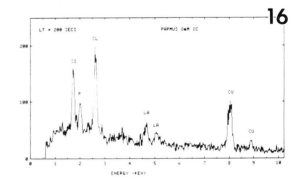

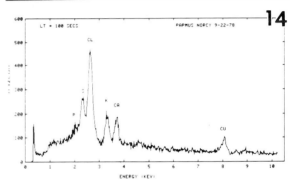

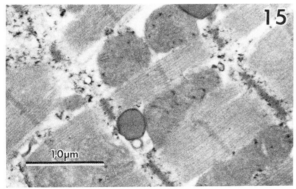

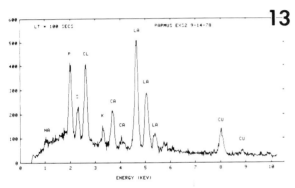

Fig. 11. TEM of control, oxygenated cat papillary muscle exposed to 5mM LaCl3 prior to aldehyde and osmium fixation. Ultrastructure is normal. La deposition is limited to the plasma membrane-basal lamina region of the cells (arrows).

Fig. 13. Summed x-ray spectrum from interstitial deposits observed in unfixed cryosections of control papillary muscle exposed to LaCl and shown in Fig. 12. Note the prominent La peaks.

Fig. 15. TEM of cat papillary muscle subjected to 75 minutes of hypoxia prior to exposure to 5mM LaCl3; aldehyde and osmium fixation. Abnormal intracellular accumulation of La on the myofibrillar I bands and in the mitochondria is consistent with hypoxia-induced alteration in functional membrane integrity.

Fig. 12. STEM of unfixed, unstained cryosection of control, oxygenated cat papillary muscle shows linear electron dense deposits consistent with La deposition in the interstitial space. Compare with Fig. 11.

Fig. 14. Summed x-ray spectrum from low contrast, intracellular regions observed in unfixed cryosections of control papillary muscles exposed to LaCl3 and shown in Fig. 12. Note the absence of detectable La and the prominent K peak consistent with an intracellular localization.

Fig. 16. Summed x-ray spectrum of abnormal mitochondrial La deposits observed in aldehyde and osmium fixed epoxy sections of hypoxic papillary muscle shown in Fig. 15. Prominent La and P peaks suggest that the deposits represent precipitates of La phosphate.

hypoxic and ischemic myocardium", SEM/1977/II, O. Johari (ed.), IIT Research Institute, Chicago, Illinois 60616, 145-152.

3. H.K.Hagler, L.Sherwin and L.M.Buja, "Analytical electron microscopy of mitochondrial inclusions in infarcted myocardium: Effect of tissue preparation", (abstract), Circulation 58 (Suppl. II), 1978, II-241.

4. H.K.Hagler, L.Sherwin and L.M.Buja, "Effect of different methods of tissue preparation on mitochondrial inclusions of ischemic and infarcted canine myocardium: Transmission and analytical electron microscopic study", Lab. Invest., May, 1979, in press.

5. K.P.Burton, H.K.Hagler, G.H.Templeton, J.T. Willerson and L.M.Buja, "Lanthanum probe studies of cellular pathophysiology induced by hypoxia in isolated cardiac muscle", J. Clin. Invest. 60, 1977, 1289-1302.

6. B.F.Trump, W.J.Mergner, M.W.Kahng and A.J. Saladino, "Studies on the subcellular pathophysiology of ischemia", Circulation 53 (Suppl. I), 1976, I-17-I-26.

7. B.F.Trump, I.K.Berezesky, R.E.Pendergrass, S.H.Chang, R.E.Bulger and W.J.Mergner, "X-ray microanalysis of diffusible elements in scanning electron microscopy of biological thin sections. Studies of pathologically altered cells", SEM/1978/II, R.P.Becker and O.Johari (eds.), SEM, Inc., AMF O'Hare, Illinois 60666, 1027-1039.

8. M.Hoffman, W.J.Mergner, J.Hickey, J.Rudert and G.W.Mergner, "Electrolyte shifts following permanent coronary artery occlusion: Microprobe analysis and bulk determination", (abstract), Fed. Proc. 37, 1978, 495.

9. M.Ashraf and C.M.Bloor, "X-ray microanalysis of mitochondrial deposits in ischemic myocardium", Virchows Arch. B Cell. Path. 22, 1976, 287-297.

10. M.Ashraf, H.D.Sybers and C.M.Bloor, "X-ray microanalysis of ischemic myocardium", Exptl. Molec. Path. 24, 1976, 435-440.

11. L.Seveus, "Preparation of biological material for x-ray microanalysis of diffusible elements. I. Rapid freezing of biological tissue in nitrogen slush and preparation of ultrathin frozen sections in the absence of trough liquid", J. Microsc. 112, 1978, 269-279.

12. T.Barnard and L.Seveus, "Preparation of biological material for x-ray microanalysis of diffusible elements. II. Comparison of different methods of drying ultrathin cryosections cut without a trough liquid", J. Microsc. 112, 1978, 281-291.

13. F.H.Schamber, "A modification of the linear least-squares fitting method which provides continuum suppression", Workshop on X-Ray Fluorescence Analysis of Environmental Samples, T. Dzubay (ed.), Ann Arbor Science Publications, Ann Arbor, U.S.A., 1976, 1-26.

14. T.A.Hall, H.Clarke-Anderson and T.Appleton, "The use of thin specimens for x-ray microanalysis in biology", J. Microsc. 99, 1973, 177-182.

15. H.Shuman, A.V.Somlyo and A.P.Somlyo, "Quantitative electron probe microanalysis of biological thin sections: Methods and validity", Ultramicrosc. 1, 1976, 317-339.

16. H.Shuman, A.V.Somlyo and A.P.Somlyo, "Theoretical and practical limits of EDS x-ray analysis of biological thin sections", SEM/1977/I, O. Johari (ed.), IIT Research Institute, Chicago, Illinois 60616, 663-672.

17. A.P.Somlyo, A.V.Somlyo, C.E.Devine, P.D. Peters and T.A.Hall, "Electron microscopy and electron probe analysis of mitochondrial cation accumulation in smooth muscle", J. Cell Biol. 61, 1974, 723-742.

18. A.C.Shen and W.B.Jennings, "Myocardial calcium and magnesium in acute ischemic injury", Amer. J. Path. 67, 1972, 417-440.

19. A.C.Shen and W.B.Jennings, "Kinetics of calcium accumulation in acute myocardial ischemic injury", Amer. J. Path. 67, 1972, 441-452.

20. W.J.Landis and M.J.Glimcher, "Electron diffraction and electron probe microanalysis of the mineral phase of bone tissue prepared by anhydrous techniques", J. Ulstr. Res. 63, 1978, 188-223.

21. R.B.Jennings, A.C.Shen, M.L.Hill, C.E. Ganote and P.B.Herdson, "Mitochondrial matrix densities in myocardial ischemia and autolysis", Exptl. Molec. Path. 29, 1978, 55-65.

Discussion with Reviewers

T. Makita: When it is almost self-evident that the precipitates in Figs. 11, 12 and 15 are lanthanum and nothing else, why should analytical electron microscopy be used except to obtain more quantitative information?
Authors: Analytical electron microscopy was used to provide more definitive information regarding the intracellular and extracellular localization of La in control and hypoxic muscle preparations.

T. Makita: As you show in Figs. 11 and 15, lanthanum on the surface of the cell membrane and also over intracellular structures could be beautifully localized by conventional methods of tissue preparation. Why then do you attempt to analyze it over vague images in cryosections such as Fig. 12?
Authors: The demonstration of selective extracellular localization of La in unfixed cryosections of control oxygenated tissue provides important validation of the La probe technique for the study of altered membrane integrity in cell injury.

T. Barnard: Please comment on the apparent decrease in phosphorus signal between control mitochondria and dense inclusions in experimental mitochondria, as shown in Table 2.
Authors: The total phosphorus signal represents the sum of all organic and inorganic phosphorus species preserved by a given combination of biological state of the tissue and the method of tissue preparation. The data from the alcohol fixed tissue indicate that the inclusions of ischemic mitochondria are rich in calcium phosphate (Fig. 6). The ischemic mitochondria, however, undoubtedly became depleted in organic phosphate species, including ATP. Therefore, the total phosphorus peak of the ischemic mitochondria is reduced compared to that of control mitochondria when unfixed cryosections are used for analysis.

T. Barnard: Why do the spectra from La-treated controls have such high calcium signals both extra- and intra- cellularly?
Authors: La treatment may alter the properties of the plasma membrane and thereby allow some intracellular leakage of Ca. This intracellular Ca is highly soluble and only detectable in unfixed frozen sections of La treated muscle.

T. Barnard: Would you comment on the relationship between the amorphous calcium phosphate granules and the intramitochondrial matrix granules seen in most mitochondria after osmium tetroxide fixation?
Authors: The amorphous calcium phosphate granules described in this report represent a pathological finding related to excess calcium accumulation induced by ischemia or other forms of cell injury.[4,6,7,18,19] These pathological inclusions were demonstrable in tissue fixed in osmium-phosphate or alcohol as well as in unfixed cryosections. The normal osmiophilic granules probably are composed of organic material rather than inorganic calcium phosphate.[22-24] After brief periods of hypoxia or ischemia, the normal osmiophilic granules are lost prior to the onset of pathological calcification.[4,6]

T. Makita: A vacuum pipette to collect cryosections was discussed by Dr. Appleton of Cambridge University some years ago.[25] Is yours different?
Authors: The vacuum pickup device described by Dr. Appleton is a macrotool for collecting whole ribbons of sections. In our experience, however, cryoultramicrotomy does not usually yield ribbons of sections. The device described in our report is a micropipette which enables the collection of individual sections and the accurate placement of the sections on grids in the cryochamber.

T. Makita: An early publication of Sutfin et al claimed that individual mitochondrial granules less than 100nm in diameter could be analyzed.[26] From my own experience, using a JEOL 100C and EDAX system, however, it is hard to believe how you could collect significant signals of La (shown in spectrum in Fig. 16) from such a fine precipitate as in Fig. 15. Is there any special mode of operation? If so, could you not detect elements over tiny densities such as shown in Figs. 1, 2, 4 and 5 instead of analyzing cryosections? Why wasn't osmium evident in spectrum of Fig. 16?
Authors: For routine microscopy, we used osmicated tissue sections on copper grids. For analytical electron microscopy, however, we frequently collected spectra, such as the one shown in Fig. 16 from unsmicated, unstained tissue mounted on nylon grids. Elemental detection from such small deposits is achieved because of the reduction in background radiation with nylon grids and unosmicated tissue as well as the use of multiple brief analyses of several microareas to produce a summed spectrum.[2] In fact, Ca was detected in the inclusions of Figs. 1, 2, 4 and 5 as shown in Fig. 6 and Table 1.

T. Barnard: What methods were used to evaluate if the beam was kept on target during analysis since this is of crucial importance for confidence in the results? This question is related to the fact that the spot size is stated to be 40-60nm whereas the resolution shown in Fig. 2 indicates a beam diameter of less than 10nm (e.g. cristae details).
Authors: Beam position was evaluated by checking the localization of the contamination spot after analysis and by comparing spectra obtained on and off of structures of interest. Instrument conditions resulting in a finer beam diameter were used for enhanced photographic resolution, whereas a larger beam diameter was used for analytical work.

G. L. Todd: Were Ca or P peaks observed over any sarcoplasmic components other than mitochondria with any of the procedures employed?
Authors: In the cryosections of canine myocardium, detectable Ca peaks were limited to mitochondria whereas P and other elements were identified in mitochondria and cytosol.

G. L. Todd: Fig. 11 shows numerous electron dense deposits in the mitochondria of control tissue. What is the nature of these deposits?
T. Barnard: Did you analyze the intramitochondrial granules in La-treated control preparations, such as those shown in Fig. 11? If so, what did they contain?
Authors: These structures are typical osmiophilic granules of normal mitochondria.[5,6] These granules are not prominent in aldehyde-fixed, unosmicated tissue and do not exhibit detectable Ca peaks. Specifically, analysis of these granules in fixed specimens of control muscles exposed to La revealed prominent Os peaks but no detectable La or Ca.[5] The normal osmiophilic granules do not appear to be related to the small Ca-containing deposits which may develop in unfixed cryosections of control tissue, probably as a result of movement and aggregation of calcium and phosphorus under certain conditions of tissue preparation.[22-24]

M. Ashraf: Are all electron dense inclusions shown in Fig. 8 granular or agranular? Was there any variation in the Ca contents of individual granules or altogether absent in a few (shown in Figs. 9 and 10)?
Authors: All electron dense mitochondrial inclusions observed in unfixed cryosections of temporarily ischemic myocardium contained calcium (Figs. 8 and 10). The substructure of these inclusions was not apparent in frozen sections examined by scanning transmission electron microscopy. Mitochondria in unfixed cryosections of normal myocardium did not exhibit prominent inclusions or show calcium peaks (Figs. 7 and 9).

M. Ashraf: Your work with La (Figs. 15 and 16) shows that it is not possible to combine elemental analysis with La probe studies. Is there any other way to confirm whether fluctuations in intracellular Ca content takes place or not and where Ca goes upon displacement?
Authors: Our experience indicates that the La

probe technique can serve as a useful marker of altered movement and intracellular accumulation of polyvalent ions in appropriate models of myocardial hypoxic and ischemic injury.[5] Analytical electron microscopy of unfixed cryosections can be used to directly evaluate local changes in Ca and other electrolytes.

T. Barnard: In Table 2, why compare control "mitochondria" and experimental "granules" when it would be more appropriate to compare granules to organic matrix within the same profile?
Authors: The data in Table 2 are presented because comparisons between control and ischemic mitochondria are important and because the ischemic mitochondria contained large numbers of closely packed inclusions (Fig. 8) making the alternative method of analysis difficult.

T. Barnard: What criteria are used to assign an intramitochondrial location to the particles in Fig. 5?
Authors: Although no structural detail is apparent in the transmission electron micrograph, scanning transmission electron microscopy provided more contrast and structural detail and confirmed an intramitochondrial location similar to that shown in Fig. 3. Fig. 5 is presented to stress the inherent electron density of the inclusions.

G. L. Todd: Was there any difference in the distribution of the two types of mitochondrial deposits in cells containing hypercontracted sarcomeres and "contraction bands" and in those cells with a normal striation pattern?
Authors: In the temporary coronary occlusion model, damaged muscle cells generally showed myofibrillar hypercontraction as well as both calcific and amorphous matrix inclusions in the mitochondria. With infarcts produced by permanent coronary occlusion, damaged muscle cells with hypercontracted myofibrils and both types of mitochondrial inclusions were localized to peripheral infarct regions with some collateral blood flow. In contrast, severely ischemic central infarct regions showed damaged muscle cells with relaxed myofibrils and amorphous matrix densities but no calcific inclusions in the mitochondria.[1-4]

M. Ashraf: Do you think an irreversible injury is associated with changes in mitochondria only? What about changes in the nucleus and contractile elements or other organelles? Is there any relevance of these to the pathogenesis of irreversible injury?
Authors: Irreversible cellular injury should be regarded as a complex phenomenon involving changes in multiple subcellular systems.[6] Nevertheless, our work has been directed toward an evaluation of the relationship between the progression of cell injury, mitochondrial calcium alterations and plasma membrane permeability changes.

Discussion References

22. T.S.Saetersdal, R.Myklebust, N.P.Berg Justesen and H.Engedal, "Calcium containing particles in mitochondria of heart muscle cells as shown by cryoultramicrotomy and x-ray microanalysis", Cell Tiss. Res. 182, 1977, 17-31.
23. T.Barnard and B.A.Afzelius, "The matrix granules of mitochondria: A review", Sub. Cell. Biochem. 1, 1972, 375-389.
24. L.Sevéus, D.Brdiczka and T.Barnard, "On the occurrence and composition of dense particles in mitochondria in ultrathin frozen dry sections", Cell Biol. Internatl. Rep. 2, 1978, 155-162.
25. T.C.Appleton, "A cryostat approach to ultrathin 'drug' frozen sections for electron microscopy: A morphological and x-ray analytical study", J. Microsc. 100, 1974, 49-74.
26. L.V.Sutfin, M.E.Holtrop and R.E.Ogilvie, "Microanalysis of individual mitochondrial granules with diameter less than 1,000 angstroms", Science 174, 1971, 947-949.

SCANNING ELECTRON MICROSCOPY/1979/II
SEM Inc., AMF O'Hare, IL 60666, USA

INTRACELLULAR ELECTROLYTE CONCENTRATIONS IN EPITHELIAL TISSUE DURING
VARIOUS FUNCTIONAL STATES

K. Thurau, A. Dörge, R. Rick, Ch. Roloff, F. Beck, J. Mason & R. Bauer

Department of Physiology,
University of Munich,
12 Pettenkoferstrasse,
D-8000, Munich 2, West Germany

Abstract

The scanning electron microscope together with an energy dispersive system was used to quantify intracellular elemental concentrations of transporting epithelial cells under a variety of experimental conditions.

In the frog skin, the Na transport pool is comprised of the intracellular compartments of all vital cells in the different cell layers, except the mitochondria rich cells. The Na content of this transport pool exchanges easily with the epithelial (outer) bathing solution. Vasopressin increases the Na permeability of the corial cell barrier.

Proximal and distal tubular cells of the rat kidney show differences in the pattern of intracellular element concentrations. The distal tubular cell is more resistant to 20 min of ischemia than the proximal cell. The changes in intracellular electrolyte concentrations following 60 min of ischemia are reversible after reperfusing the kidney with blood for 60 min.

In the cells of the frog skin and rat kidney, Na and K were equally distributed between the cytoplasm and the nucleus. Differences exist for P, Cl and dry weight.

KEY WORDS: X-Ray Microanalysis, Transepithelial Electrolyte Transport, Ouabain, Amiloride, Vasopressin, Frog Skin, Tubular Epithelium, Renal Ischaemia

Introduction

Extensive studies have been performed in the past to understand the transport of water and solutes across epithelial cell layers such as in the intestinal tract, the gall bladder, the nephron, the urinary bladder or frog skin. A variety of methodological approaches, such as the measurement of electrical potential, uni- and bidirectional fluxes and transepithelial chemical gradients have disclosed some of the characteristics of transepithelial transport phenomena. However, most of the intracellular events participating in transepithelial transport, in particular the route of transport, the location of the transport pool, the intracellular concentration and distribution of the various elements and their changes at altered transport rates, are still incompletely understood.

We have applied electron microprobe analysis to quantify intracellular elemental concentrations in a variety of transporting epithelia at different rates of transepithelial transport. This method is particularly suitable for such an analysis because (1) several elements, such as Na, K, Cl and P, can be measured simultaneously and (2) measurements can be performed in individual cells which is of utmost importance in tissues which are composed of different cell types. This paper summarizes some of the data recently obtained from various epithelial structures in our laboratory [1, 2, 3].

Methods

A detailed description of the methodology of electron microprobe analysis and of the preparation of the tissue specimen, as applied in our laboratory, is discussed by Dr. Rick in this Symposium in a separate paper. Hence, the methods employed are mentioned here only briefly.

The analysis was performed on about 1 μm thick freeze-dried cryosections with an energy dispersive X-ray detector (EDAX) in a scanning electron microscope (Cambridge S4). The acceleration voltage was 17 kV, the beam current 0.5 nA and the scanning area during analyses approximately 1 μm^2. Quantification of element concentrations and dry weight content was achieved by comparing the element characteristic radiations and the white radiation from the specimen with those of an internal standard. The separation of the element characteristic peaks from the background radiation was performed using a computer programme, especially designed for this purpose.

Frog skin

The abdominal skin of Rana temporaria and Rana esculenta, kept in running tap water, was mounted into an Ussing-type chamber. The skin was short circuited except for short (10 sec) intervals every 20 min during which the transepithelial PD was measured. Initially, both sides of the skin were bathed with Ringer's solution before the Na concentration was changed or ouabain (10^{-4}M), amiloride (10^{-4}M) or vasopressin were added. When Na-free Ringer's solution was used, Na was replaced by choline. The pH of all solutions was about 8.3.

Before the skin was shock-frozen in propane (-180 °C) the epithelial surface was blotted and covered with a thin layer of standard albumin-Ringer's solution containing 25g/100 ml albumin and of known electrolyte composition. This layer, attached to the tissue, served as an internal standard in each tissue section with which the X-ray emissions from the cells could be quantified.

The various cell types and cell layers of the frog skin were identified from the transmission image of the freeze-dried section of the skin (see Fig. 1).

Kidney

The experiments were performed on male Sprague Dawley rats anaesthetized by intraperitoneal injection of 100 mg/kg b.w. inactin. The left kidney was exposed by a flank incision, freed from adhering tissue and placed in a lucite cup. In some experiments the renal artery was occluded by a weak sprung clip for periods of up to 60 min (ischaemic model of acute renal failure).

The application of the standard albumin-Ringer's solution and the procedure used for shock-freezing was the same as for the frog skin.

Results and Discussion

Frog skin

The outer, epithelial layer of the frog skin and the cornified cells of the stratum corneum showed electrolyte concentrations almost identical with those of the bathing solution. Ouabain, amiloride, vasopressin and Na-free solution corially produced no alterations (see Fig. 2). In all the layers of the frog skin composed of living cells (i.e. stratum granulosum, spinosum and germinativum) the intracellular electrolyte concentrations under control conditions were similar, Na was 9.4 and K was 118.4 mmole/kg wet weight. For each of the elements Na, Cl and K, the concentration in cytoplasm and the nucleus were found to be practically identical. In the nucleus, P concentration was significantly higher than that in cytoplasm, whereas dry weight was significantly lower. These data are summarized in table 1.

The effects of removing Na from the epithelial (outer) bathing solution and of applying ouabain and amiloride is shown in fig 2. When the active transport step for Na, located at the corial side of the cell membranes, is inhibited by ouabain, the intracellular Na concentration of all epithelial layers increases by about 100 mmole/kg wet weight. K, not shown here, showed an almost identical drop. This effect of ouabain was abolished when the epithelial (outer) bathing solution was free of Na or contained amiloride.

These data demonstrate that the various living cell layers of the frog skin have almost identical intracellular concentrations of Na and K under a variety of experimental conditions and at various rates of transepithelial transport. Hence, the intracellular space of these living cells appears to represent a single distributional space for these ions. The bulk of the intracellular Na can exchange easily with the epithelial (outer) bathing medium, indicating a considerably higher Na permeability of the outer-facing than the inner-facing membranes.

It is, therefore, concluded that the major quantity of intracellular Na originates from the outer bathing solution. This conclusion is in contradistinction to earlier results, which indicated that the corial (inner) side is the major source of intracellular Na [4, 5]. If a transcellular pathway for Na transport is assumed, then the cells comprising the Na transport compartment should be identifiable by the typical behaviour of their intracellular Na concentrations after inhibition of transport either at the epithelial (outer) barrier by amiloride or at the corial (inner) barrier by ouabain. This assumption could be confirmed experimentally, thus supporting the view of a syncytial Na transport compartment, as suggested earlier [6, 7]. The coupling of the individual cells is thought to occur at the level of the intercellular junctions, which have a high Na permeability. Since the tight junctions between the outermost living cells abolish the extracellular shunt pathway, it has been postulated that the Na of the deeper epithelial layers originates mainly from the outer bathing solution and is distributed among these cells via the intercellular junctions.

Vasopressin is known to increase water permeability and transepithelial Na transport. The stimulatory effect upon Na transport may result from an increased Na influx across the epithelial (outer) cell barrier or from a direct stimulation of the active transport. The simultaneous measurement of intracellular Na concentration together with the short circuit current (SCC) as a measure of transepithelial net Na transport should provide information about the site of action of vasopressin. Fig. 3 depicts the data from such an experiment. During control incubation the intracellular Na concentration in the various cell layers is approx. 11 mmole/kg wet weight and the SCC is 40 µA/cm². Vasopressin increases the intracellular Na concentration to 30 mmole/kg wet weight and the SCC to 75 µA/cm². Note that the mitochondria-rich cells (MRC) do not participate in the

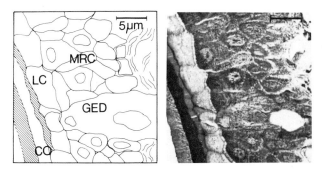

Figure 1. Scanning transmission electron micrograph of a freeze-dried cryosection of frog skin epithelium (about 1 μm thick), together with a sketch of the epithelium. MRC = mitochondria-rich cell; LC = light cell; GED = gland excretion duct; CO = stratum corneum.

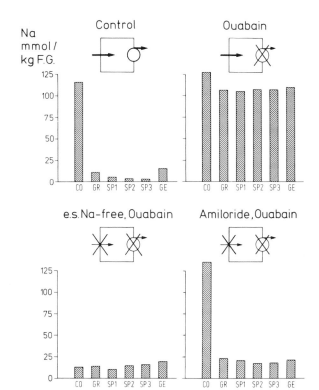

Figure 2. Cellular Na and K concentrations in the different layers of frog skin.
Upper left: Control
Upper right: After ouabain (10^{-4} M)
Lower left: After ouabain and when the outer bathing solution is replaced by Na-free Ringer's solution
Lower right: After ouabain and when amiloride (10^{-4} M) was added to the outside bathing medium
CO = stratum corneum; GR = stratum granulosum; SP1, SP2 and SP3 = stratum spinosum 1, 2 and 3; GE = stratum germinativum

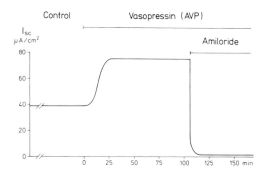

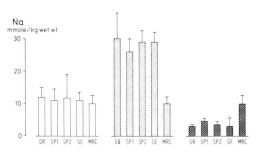

Figure 3. Short circuit current (I_{SC}) and intracellular Na concentration in the different epithelial layers of the frog skin in control, during vasopressin (AVP, 150 mU/ml, inner bathing solution) and during vasopressin and amiloride (10^{-4} M, outer bathing solution). GR = stratum granulosum; SP1 and SP2 = stratum spinosum 1 and 2; GE = stratum germinativum; MRC = mitochondria-rich cell.

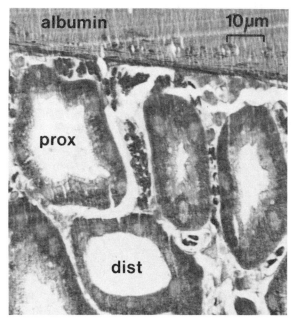

Figure 4. Scanning transmission electron micrograph of a freeze-dried cryosection of the superficial layer of the rat kidney. The homogeneous layer attached to the renal surface is the albumin standard layer.

	Na	K	P	Cl	dry wt
	mmole/kg wet wt				g/100g
FROG SKIN					
CYTOPLASM	7.4	118.8	98.8	36.5	25.4
	±4.5	±12.1	±19.8	±5.4	±2.3
NUCLEUS	5.1	115.0	144.4	32.9	21.8
	±3.9	±11.4	±19.1	±3.7	±1.9
RAT KIDNEY, PROXIMAL TUBULES					
CYTOPLASM	22.4	135.9	177.8	33.1	33.0
	±5.4	±17.4	±25.9	±8.9	±5.0
NUCLEUS	20.2	139.5	145.2	20.9	23.1
	±4.6	±17.4	±20.9	±6.6	±2.2
RAT KIDNEY, DISTAL TUBULES					
CYTOPLASM	12.5	142.3	201.0	16.9	27.8
	±2.8	±20.9	±36.2	±3.7	±3.9
NUCLEUS	10.7	138.7	183.9	11.3	22.8
	±3.1	±21.2	±23.2	±4.3	±3.2

Table 1. Cytoplasmic and nuclear element concentrations and dry weight of frog skin and rat kidney epithelia cells.

increase in intracellular Na. This finding suggests that the increase in SCC during vasopressin results, at least in part, from an increase in Na influx into the cell and, hence, from an increase in intracellular Na concentration.

When the epithelial Na influx is blocked by amiloride, intracellular Na decreases below control to approx. 4 mmole/kg wet weight, excepting the mitochondria-rich cells, and SCC falls almost to zero indicating a stop in transepithelial net Na transport.

Both findings, the increase in Na concentration in all cell layers during AVP and the decrease after amiloride, again demonstrate a syncytial connection between all cell layers in respect to the Na transport pool. Excluded from this syncytium are the few mitochondria-rich cells.

Tubular epithelium of the rat kidney

A scanning transmission electron micrograph of a freeze-dried section of the rat kidney cortex is shown in Fig. 4. The nuclei can easily be distinguished from the slightly darker-appearing cytoplasm. The characteristic feature which differentiates distal from proximal tubules is the pronounced appearance of microvilli in the latter. Analysis of intracellular elemental concentrations in tubular epithelial cells is complicated by the occurrence of the brush border structure and the basal infoldings. This cytological architecture implies that X-ray signals obtained from these regions result from both extra- and intracellular compartments. The lowest Na and Cl concentrations and the highest K and P concentrations were found in cytoplasmatic areas in juxtaposition to the nucleus. These data obtained in the proximal and in the distal tubular epithelium are included in Fig. 5 and Table 1, which also contains the values for the nuclei. It is apparent that the ratio nucleus/cytoplasm for Na and K is practically 1, whereas Cl, P and dry weight are higher in the cytoplasm. This holds both for proximal and distal tubular cells. There are systematic differences between proximal and distal cells: in the proximal tubular cells Na and Cl concentrations are higher, P is lower and K is identical compared to the distal cells. Dry weight of proximal and distal cells is similar.

During ischaemia the kidney surface was either exposed to air or to 100% nitrogen gas. For ischaemic periods up to 60 min no significant

PROXIMAL TUBULE   DISTAL TUBULE

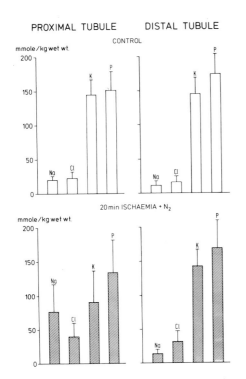

PROXIMAL TUBULE

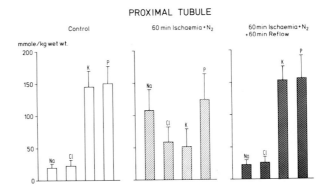

Figure 5. Intracellular concentrations for Na,
Cl, K and P of the epithelial cells in the
proximal and distal tubule of the rat kidney
during control and following 20 min of renal
artery occlusion. During the ischaemic period
the renal surface was maintained in a N_2-
atmosphere.

Figure 6. Intracellular concentrations for Na,
Cl, K and P of the proximal tubular cells of
the rat kidney during control, following 60 min
of renal artery occlusion and following reper-
fusion with blood for 60 min. During the
ischaemic period the renal surface was maintained in a N_2-atmosphere.

alteration of the cellular electrolyte concentrations in superficial tubules were detected when
the kidney was kept in air.

Fig. 5 shows the Na, Cl and K concentrations
in proximal and distal tubular cells after 20 min
of ischaemia in a nitrogen atmosphere. In the
proximal tubules the Na and Cl concentrations
increased to 76 and 35, whereas K concentration
decreased to 89 mmole/kg wet weight. In contrast,
in the distal tubule almost no effect upon the
cellular electrolyte concentrations was observed
after 20 min of ischaemia. Fig. 6 shows the
cellular electrolyte concentrations after 60 min
of ischaemia in a nitrogen atmosphere and after
60 min blood reflow. Since, under this condition,
the electrolyte concentrations of the proximal
and distal tubules showed no systematic differences, only the data obtained in the proximal
tubule are considered. After ischaemia the Na
concentration increased to 108, while the K concentration decreased by almost the same amount
to 52 mmole/kg wet weight. At the same time the
Cl concentration increased to 60 mmole/kg wet
weight. The cell swelling during ischaemia
known from morphological studies [8], [9] is reflected in the present investigation by a decrease in the P concentration of about 25%. In
addition, the observed increase in the cellular
Cl concentration agrees well with the view of
cellular uptake of extracellular fluid.

In order to determine whether the changes
in cellular electrolyte concentrations resulting
from ischaemia were reversible, reflow was permitted for 60 min after the ischaemic period.
As can be seen from Fig. 6, 60 min after opening
the arterial clamp, the Na, K and Cl values had
almost returned to the control levels.

The observed differences in the electrolyte
concentrations in the proximal and distal tubule
in the control and after a shorter period of
ischaemia may result from functional differences
between these nephron segments. On the other
hand, the impairment of renal functions following
ischaemia is apparently not reflected in a disturbance of the cellular electrolyte distribution,
since electrolyte concentrations were found to
be almost completely restored after 60 min of
blood reflow.

Acknowledgement

This investigation was supported by the
Deutsche Forschungsgemeinschaft.

References

1. Dörge, A., Rick, R., Gehring, K., Thurau, K.
Preparation of freeze-dried cryosections for
quantitative X-ray microanalysis of electrolytes
in biological soft tissues. Pflügers Arch. 373 :
85 - 97. (1978).
2. Rick, R., Dörge, A., Macknight, A.D.C.,
Leaf, A., Thurau, K. Electron microprobe
analysis of the different epithelial cells of
toad urinary bladder. J. Membrane Biol. 39:
257 - 271. (1978).
3. Rick, R., Dörge, A., v.Arnim, E., Thurau, K.
Electron microprobe analysis of frog skin
epithelium: Evidence for a syncytial sodium
transport compartment. J. Membrane Biol. 39:
313 - 331. (1978).

4. Cereijido, M., Rotunno, C.A.
Fluxes and distribution of sodium in frog skin.
A new model. J. Gen. Physiol. 51: 280 - 289
(1968).
5. Nagel, W., Dörge, A.
A study of the different sodium compartments and
the transepithelial sodium fluxes of the frog
skin with the use of ouabain. Pflügers Arch.
324: 267 - 278 (1971).
6. Ussing, H.H., Windhager, E.
Nature of shunt path and active sodium transport path through frog skin epithelium. Acta
Physiol. Scand. 61: 484 -504 (1964).
7. Farquhar, M.G., Palade, G.E.
Cell junctions in amphibian skin. J. Cell.
Biol. 26: 263 - 291 (1965).
8. Flores, J., DiBona, D.R., Beck, C.H.,
Leaf, A.
The role of cell swelling in ischaemic renal
damage and the protective effect of hypertonic
solute. J. Clin. Invest. 51: 118 - 126 (1972).
9. Frega, N.S., DiBona, D.R., Guertler, B.,
Leaf, A.
Ischaemic renal injury. Kidney Intern. 10:
suppl. 6 S17 - S25 (1976).

Discussion with Reviewers

S.L. Kimzey: You state that the cell layers have
identical intracellular concentrations of Na and
K under various rates of transport. What is the
sensitivity of your measurement technique with
respect to ions?
Authors: In an individual measurement the minimal detectable concentration is about 1 mM for
Na and 0.5 mM for K.
S.L. Kimzey: How do you separate the possibility
of overlying cytoplasm affecting your nuclear
signal when you are comparing nuclear/cytoplasm
ratio?
Authors: Usually, the measurements were performed on serial sections, so that overlying of
cytoplasm could be checked by looking at the
neighbouring sections.
S.L. Kimzey: Have you observed any difference
in intracellular Na or K distribution other than
that reported for the nucleus and cytoplasm?
Authors: Only average values of the nucleus and
cytoplasm could be obtained, as the excited
volumes during analysis were about 1 μm^3. These
conditions were chosen, since it can be expected
that concentration gradients which might exist
within small subcellular regions are disturbed
during freezing of freeze-drying.
J.A.V. Simson: In terms of interpretation of
data, do the authors believe that their data
rule out substantial paracellular contribution
to transepithelial Na flux in the frog skin?
Authors: In principle, the contribution of an
extracellular component to net Na transport in
frog skin cannot be excluded from the present
experiments. However, since all the changes in
cellular electrolyte concentrations observed
are in accordance with a cellular route of Na
transport, an extracellular route seems
unlikely.

J.A.V. Simson: What are the authors' conclusions
concerning the apparent paradox that although the
mitochondria-rich cell is the most likely candidate for a "pumping" cell, since it has ready
energy supply for driving the pump, its intracellular ion concentration seems not to be
affected by drugs such as vasopressin and amiloride, which markedly influence transepithelial
Na transport?
Authors: The finding that amiloride, which
blocks the transepithelial net Na transport
almost completely, has no effect upon the cellular electrolyte concentrations of the mitochondria-rich cells argues against a substantial
contribution of this cell type to transepithelial
Na transport. It is conceivable that the mitochondria-rich cells provide energy to those cells
involved in transepithelial Na transport.
However, since they are apparently not well
coupled to the neighbouring cells it remains
obscure how the energy transfer is accomplished.
J.A.V. Simson: In the kidney experiments, how
was intravascular coagulation in the ischaemic
kidney prevented (so that reflow could be
obtained)?
Authors: No special precautions were taken to
prevent vascular coagulation, for it has been
shown that temporary ischaemia does not lead
to fibrin deposition and that heparinization
has no influence upon the renal blood flow
after an ischaemic episode (see reference 9).

SCANNING ELECTRON MICROSCOPY/1979/II
SEM Inc., AMF O'Hare, IL 60666, USA

X-RAY ANALYSIS OF PATHOLOGICAL CALCIFICATIONS INCLUDING URINARY STONES

K. M. Kim

Department of Pathology (151)
V.A.Medical Center Research Svc.
3900 Loch Raven Blvd.
Baltimore, MD 21218

and also

Department of Pathology
University of Maryland
School of Medicine
22 South Greene Street
Baltimore, MD 21201

Abstract

Electron-probe microanalysis (EPM) is an ideal technique with which to study biological calcification. It is particularly effective in identification of crystalline or non-crystalline deposits of minute size in tissues, and in detecting artifacts which may occur during tissue processing.

Human aorta, aortic valves, tumours, joint fluid, and calculus specimens were analyzed via scanning electron microscopy (SEM), EPM, transmission electron microscopy (TEM), and selected-area electron diffraction (ED). Crystals found in the specimens were definitively identified by combined SEM-EPM. It is apparent that EPM is an invaluable tool that will potentially improve the acuity of 'in house' laboratory diagnosis of many pathological calcifications.

Introduction

Electron-probe microanalysis (EPM) is one of the most powerful methods available for analyzing of minute quantities of samples at the picoliter level.[1] It has been utilized extensively in the field of cell biology to elucidate compartmentalization and distribution of inorganic elements in tissues and cells.[2] The major limitation has been diffusibility of the ions in tissue; it appears this obstacle has not yet been fully overcome. EPM of non-diffusible deposits in tissue, however, is relatively simple, non-destructive, and accurate. The technique, therefore, has been applied in the various biological sciences including: forensic science,[3] microparticle identification in the lungs,[4] studies of calcium-turnover in bone,[5] and applied histochemistry.[6] Despite these potential clinical applications, relatively few attempts have been made to utilize EPM as a diagnostic tool.

The mechanisms of biological calcification have been studied extensively. Relatively little progress has been made, due mainly to technical difficulties in elucidating the early stages of calcification. In vitro systems have demonstrated the initial phase separation from solution occurs as amorphous calcium phosphate followed by a spontaneous transformation into hydroxyapatite crystals.[7] The amorphous calcium phosphate has been observed in human bone and aortic valve.[8,9] Morphologically similar structures have long been recognized in mitochondria, i.e. mitochondrial granules, and the amorphous calcium phosphate nature of these granules has been determined by EPM.[10] EPM is, therefore, an invaluable tool with which to study biological calcification. This paper reports findings concerning a number of pathological calcifications studied via a combination of scanning electron microscopy (SEM), EPM, transmission electron microscopy (TEM), and selected-area diffraction (ED).

Materials and Methods

Human aorta, aortic valves, and tumours were fixed in 4% glutaraldehyde, postfixed in osmium, and embedded in Epon 812. Thin sections

KEY WORDS: X-Ray Microanalysis, Correlative Microscopy, Electron Diffraction, Pathology, Calcification, Urinary Stones were examined either unstained or double stained with uranium and lead. In the study of joint fluid crystals, the fluid was diluted in tris buffer, pH 7.4, and centrifuged at 2000 r.p.m. for 15 minutes. The crystals were pipetted onto aluminum stubs for SEM and formvar-coated grids for TEM studies. Salivary and urinary stones were cracked with sharp razor blades and ground into a powder with an acid-washed mortar and pestle. The resulting gritty fluid was transferred dropwise to coated grids and received an additional carbon-coat prior to TEM and ED analyses. Portions of stones were mounted, cracked surface up, onto aluminum stubs with silver conducting paint for SEM and EPM. All samples for SEM and EPM were carbon-coated in a Hummer (Technics) prior to examination.

SEM and EPM analyses were performed with an ETEC Autoscan electron microscope, equipped with a transmission module and a computerized energy dispersive microanalytic system, KEVEX-TRACOR NS880. Standard operating voltage was 20 or 30 KeV. TEM and ED analyses of calculus samples were made with a JOEL 100B transmission electron microscope, equipped with a goniometer. Standard operating voltage was 60 KeV or 80 KeV. Diffraction patterns were generated with a camera length of 40 cm.

Results and Discussion

Calcification of human aorta and aortic valves is a common age-related phenomenon. The calcific deposits generally consist of needle-shaped hydroxyapatite crystals. There are, however, morphologically different depositions, including amorphous calcium phosphate.[9] Porous or spongy deposits are particularly prevalent in human aortic calcification; the precise nature of these porous densities has not been determined.[11] Upon EPM, these porous densities contained calcium and phosphorus, and ED yielded the pattern of carbonate apatite, which is believed to be a common component of tissue calcifications, including urinary stones.[12,13] (Fig. 1a and 1b) The porosity of the crystals apparently reflects a high carbonate content.[14]

A majority of tissue calcifications occur in the form of calcium orthophosphate. In joint cartilage and synovium of certain patients, the precipitate is calcium pyrophosphate dihydrate (pseudogout); a common clinical condition which needs to be distinguished from urate crystal deposition disease (gout). Pseudogout crystals contain calcium and phosphorus, while gout crystals do not yield these elements but do contain sodium. EPM is a rapid and highly sensitive method to obtain differential diagnosis of joint crystal deposition diseases (Fig. 2 and 3).

Perhaps the most beneficial area for application of EPM in diagnostic electron microscopy is the elemental analysis of urinary and other stones. EPM is particularly effective for this purpose because of its simplicity, accuracy, and efficiency in time and expense. Examination of a salivary stone by SEM revealed fibrillar crystals with pointed or rounded tips. The crystals were frequently curved and tended toward parallel arrangement although loose, haphazard

arrangements were also observed (Fig. 4a).

EPM yielded calcium and phosphorus peaks and produced a powder pattern characteristic of whitlockite (Fig. 4b). Other areas of the same stone demonstrated laminated plate-shaped crystals and only yielded calcium peaks upon EPM. This suggested the presence of calcium carbonate.

SEM of urinary weddelite crystals revealed tetrahedral bipyramids and spongy spherules.[15] (Fig. 5) They appeared as rectangular, triangular and spongy spherular silhouettes upon TEM. Whewelite consisted of laminated tetragonal crystals producing brick wall-like appearances by SEM (Fig. 6). TEM morphology of whewelite is similar to that of the weddelite; these two crystals were distinguished by selected-area ED. EPM of both weddelite and whewelite yielded only a tall calcium peak. The characteristic SEM morphology, however, warrants distinction between these two crystals. Weddelite and whewelite often coexisted and calcium apatite was frequently mixed with the oxalates.

Calcium (carbonate) apatite was usually found intermixed with other crystals, particularly oxalates and struvite. Pure apatite stones were not encountered in this study. Apatite formed aggregates of radially arranged rod-shaped crystals, usually visualized as irregular matted spherules. At high magnifications needle or rodlet-shaped crystals were apparent (Fig.7). The apatite crystals can be identified by TEM, which revealed fine rodlet-shaped crystals. EPM demonstrated calcium and phosphorus. It can be distinguished from other forms of calcium phosphate by its morphology or by ED.

Struvite crystals were rhombohedral or pinacoid by SEM and invariably coexisted with calcium apatite (Fig. 8a). TEM demonstrated round or oval particles approximately 30-50nm in diameter along the periphery of rectangular or spherular silhouettes. EPM revealed the presence of magnesium, phosphorus, and frequently calcium, due to coexistent calcium phosphate (Fig. 8b). ED patterns were highly characteristic.

SEM of uric acid stones demonstrated orderly aggregates of square columns, equants, and prismatic crystals (Fig. 9). The crystals appeared in clusters of granular, rod, or plate shapes via TEM.[16] These silhouettes rapidly disintegrated into homogeneous spherules upon exposure to the electron beam. ED is, therefore, difficult to perform. Trace amounts or no inorganic elements were detected by EPM.

Brushite, like whitlockite, is rare in urinary stones. The crystals formed long rectangular columns and elongated arrow-head shapes in parallel arrangements (Fig. 10). EPM yielded calcium and phosphorous peaks. Both whitlockite and brushite appeared as rod-shaped silhouettes via TEM, but ED defined the nature of the crystals.

The cases reviewed illustrate the successful utilization of x-ray microanalysis in diagnosis of pathological calcifications. Advantages include: observation of tissue morphology, simultaneous evaluation of elemental contents and structure of discrete crystal deposits at desired

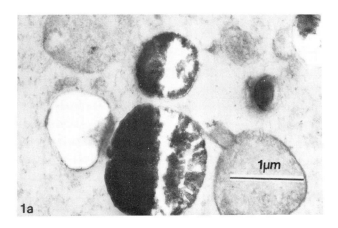

Fig. 1a. Porous crystal deposition in human aortic medial calcification.

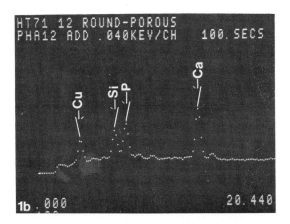

Fig. 1b. EPM of the porous crystal showing copper, silicon, calcium, and phosphorus peaks. Silicon, and copper are from the grid and its preparatory membrane.

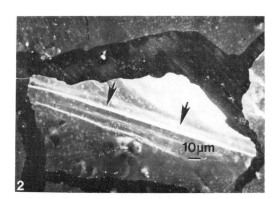

Fig. 2. SEM of a crystal from joint fluid. EPM of the crystal did not show detectable elements.

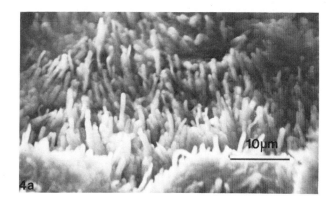

Fig. 3. TEM of crystals in synovial tissue from a patient with pseudogout. EPM showed calcium and phosphorus.

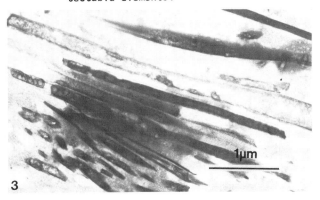

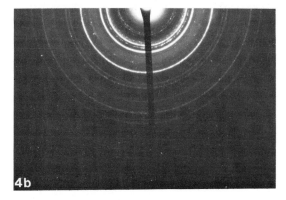

Fig. 4a. SEM of round rod-shaped crystals in a sialolith. EPM demonstrated calcium and phosphorus.

Fig. 4b. ED of the crystals showed a pattern of whitlockite.

---

magnifications, and use of small sample sizes.

Urinary calculi are especially intriguing pathological calcifications, demonstrating variation in structure and frequency of occurrence. They have been classified by x-ray diffraction as weddelite (calcium oxalate dihydrate), whewelite (calcium oxalate monohydrate), calcium carbonate apatite, struvite (ammonium magnesium phosphate), uric acid, brushite (monocalcium phosphate), whitlockite (tricalcium phosphate) and rare organic or metabolic stones. X-ray diffraction is undoubtedly the reference method of crystal identification; however, it is not without limitations since, as in chemical analysis, it deals with the totality of samples analyzed. In addition, the instrumentation is expensive and available only at certain referral centers. Crystal components of uroliths appear to possess specific morphological characteristics and elemental contents via combined SEM and EPM. It is likely definitive stone identification can be accomplished with these rapid and simple methods. If available, EPM is an invaluable tool that will potentially improve the acuity of 'in house' laboratory diagnosis of many pathological calcifications.

Fig. 5.   SEM of weddelite showing tetrahedral bipyramids and spherular particles.
Fig. 6.   SEM of whewelite with a brick wall-like appearance. Irregular particles on the surface are calcium apatite.
Fig. 7.   SEM of calcium apatite. Rodlet-shaped crystals form spherular aggregates.
Fig. 8.   A pinacoid crystal of struvite. Irregular spherules in the background are calcium apatite. EPM yielded magnesium and phosphorus peaks.
Fig. 9.   SEM of uric acid crystal showing large square-shaped or cubic blocks. Smaller crystals on the surface suggest epitactic growth.
Fig. 10. SEM of whitlockite crystals appear as rectangular columns.

Acknowledgements

This work was supported, in part, by the Medical Research Service of the Veterans Administration. The author thanks Mrs. Helen Spencer for her patient clerical assistance.

References

1. R. Beeuwkes, III, J.M. Amberg and L. Essandoh, "Urea measurement by x-ray microanalysis in 50 picoliter specimens", Kidney Int. 12, 1977, 438-442.
2. C.P. Lechene, "Electron probe microanalysis: its present, its future", Am. J. Physiol., 232, 1977, F391-F396.
3. D. Samale, "The examination of paint flakes, glass and soils for forensic purposes, with special reference to electron probe microanalysis", J. Forensic Sci. Soc., 13, 1973, 5-15.
4. J.P. Berry, P. Henoc, P. Galle, et al, "Pulmonary mineral dust. A study of ninety patients by electron microscopy, electron microanalysis, and electron microdiffraction", Am. J. Pathol., 83, 1976, 427-456.
5. J.E. Weergedal and D.J. Baylink, "Electron probe measurements of bone mineralization rates in vivo", Am. J. Physiol., 226, 1974, 345-352.
6. T. Daimon, V. Mizuhira and K. Uchida, "Ultrastructural localization of calcium around the membrane of the surface connected system in the human platelet", Histochemistry, 55, 1978, 271-279.
7. E.D. Eanes and A.S. Posner, "Kinetics and mechanism of conversion of non-crystalline calcium phosphate to crystalline hydroxyapatite", Trans. N.Y. Acad. Sci., 28, 1965, 233-241.
8. J.D. Termine and A.S. Posner, "Amorphous/crystalline interrelationships in bone mineral", Calcif. Tiss. Res., 1, 1967, 8-23.
9. K.M. Kim and B.F. Trump, "Amorphous calcium precipitations in human aortic valve", Calcif. Tiss. Res., 18, 1975, 155-160.
10. L.V. Sutfin, M.E. Holtrop and R.E. Ogilvie, "Microanalysis of individual mitochondrial granules with diameters less than 1,000 Angstroms", Science, 174, 1971, 947-949.
11. M.B. Gardner and D.H. Blankenhorn, "Aortic medial calcification. An ultrastructural study", Arch. Pathol., 85, 1968, 397-403.
12. R.M. Biltz and E.D. Pellegrino, "The nature of bone carbonate", Clin. Orthoped., 129, 1977, 279-292.
13. E.L. Prien, "Crystallographic analysis of urinary calculi: A 23 year survey study", J. Urol., 89, 1963, 917-924.
14. R.Z. LeGross, O.T. Trautz and W.P. Shirra, "Apatite crystallites: Effects of carbonate on morphology", Science, 155, 1967, 1409-1411.
15. W.H. Boyce, "Some observation on the ultrastructure of 'idiopathic' human renal calculi", in Urolithiasis, Physical Aspects, National Acad. Sci., Washington, D.C., 1972, 97-114.
16. B. Finlayson and A.S. Meyer, "Stone ultrastructure", in Urolithiasis, Physical Aspects National Acad. Sci., Washington, D.C., 1972, 115-123.

Discussion with Reviewers

F.E. Dische: The reviewer has noted discrepancies in results for calcium identification in renal biopsy material of human nephrocalcinosis by light microscopical histochemistry and EPM carried out on osmium postfixed araldite-embedded thin sections: routine EM processing appears to have removed calcium compounds from the less dense deposits. Has the author encountered this problem and how has he overcome it?

Author: The only precaution we take is to limit exposure of the tissue to aqueous solutions as much as possible. Thin sections are routinely screened by TEM at magnification of 25,000 times or higher. Any crystals detected are then studied via SEM. The ideal method of preserving crystals in tissue would be ultrathin cryostat sectioning. By use of STEM, crystal detection can be facilitated.

K.G. Gould: Would you please elaborate on the use of ED in crystal identification in pathology, and specifically provide ED patterns discriminating between whitlockite and brushite.

Author: I would like to refer the first question to: D.F. Parsons, "The examination of mineral deposits in pathological tissues by electron diffraction", Int. Rev. Exp. Pathol., 6, 1968, 1-54. For the second question, densitometric tracings of the ED patterns of whitlockite and brushite are shown below.

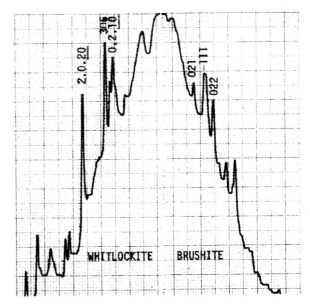

H. Harasaki: A qualitative chemical analysis for urinary stones, such as Simmons and Grentzkow's method, is a simple and well established method requiring as much as 10 to 20mg of test material. Could you outline the unique advantages of SEM, TEM, ED and EPM for daily clinical utilization in identifying the constituents of urinary stones?

Author: The advantages of EM methods in stone analysis are: minimal sample preparation without production of artifacts, use of small specimen sample to perform a complete analysis, simultaneous evaluation of elemental content, crystalline structures and morphology of the individual crystals; and the ability to observe spatial relationships and growth patterns of the crystals at the molecular level. The author believes that once the detailed criteria are compiled, crystal identification can be performed expeditiously by SEM morphology alone.

K. Pritzker: In your methods, the stones were ground. How can a nidal crystal which is different from the bulk of calculus be distinguished?

Author: This can be done by careful dissection of the stone under a dissecting microscope. ED data are then correlated with SEM morphology and EPM data of discrete crystals.

K. Pritzker: What technical problems were encountered with the electron diffraction of urinary calculi?

Author: As stated in the text, organic crystals are difficult to diffract because of 'melting' under the electron beam. Inorganic crystals also transform occasionally upon electron beam bombardment. Proper exposure to obtain analyzable patterns is not always easy. In addition, the intensity of ED patterns do not coincide with those from x-ray diffraction.

K. Pritzker: Is it possible to do electron diffraction on the same crystal as observed and analyzed on the scanning electron microscope?

Author: The specimen used for SEM can be powdered and prepared for ED. In general, crystals in urinary stones have zonal distribution. Sampling from different areas of the stones, therefore, is not difficult.

K. Pritzker: What is the significance of calcium hydrogen phosphate dihydrate (brushite) in urinary calculi?

Author: No significance is known to my knowledge.

SCANNING ELECTRON MICROSCOPY/1979/II
SEM Inc., AMF O'Hare, IL 60666, USA

APPLICATION OF SCANNING AND TRANSMISSION ELECTRON MICROSCOPY, X-RAY ENERGY SPECTROSCOPY, AND X-RAY DIFFRACTION TO CALCIUM PYROPHOSPHATE CRYSTAL FORMATION IN VITRO

S. A. Omar, P.-T. Cheng, S. C. Nyburg,* and K. P. H. Pritzker

Department of Pathology
Mount Sinai Hospital
600 University Avenue,
Toronto, Canada, M5G 1X5

*Department of Chemistry
University of Toronto
Toronto, Canada

Dr. Pritzker and Dr. Cheng also affiliated
with University of Toronto

Abstract

This paper describes the correlated application of polarized light microscopy, scanning electron microscopy using both the secondary emitted and back-scattered electron imaging modes, x-ray energy spectroscopy, transmission electron microscopy and micro x-ray diffraction analysis to the study of calcium pyrophosphate crystal formation in an in vitro model system. The model consisted of hyaline cartilage into which two parallel troughs were cut and filled with powdered calcium chloride and sodium pyrophosphate respectively. Ions diffused through the cartilage from the troughs forming a precipitin band. Subsequently, the cartilage was fixed in cacodylate buffered 10% formalin, freeze dried and analyzed by the above techniques. The precipitin band was localized by scanning electron microscopy, the localization enhanced by back-scattered electron imaging. X-ray energy spectroscopy showed peaks of Ca, Na, and P within the precipitin band. X-ray mapping revealed the distribution of Ca and P not only in the band but in the matrix between the precipitin line and the troughs. Transmission electron microscopy of the precipitin line showed crystals with bubbly morphology typical of hydrated calcium pyrophosphates. Micro x-ray diffraction analysis on plastic embedded sections identified crystals within the line as monoclinic calcium disodium pyrophosphate tetrahydrate, α form. This study demonstrates that multimodal analytic techniques are essential for characterization of crystal deposits and for localization of ions in the matrix of this model system.

KEY WORDS: Correlative Microscopy, Calcium Phosphate Crystals, X-Ray Microanalysis, X-Ray Diffraction, Pseudogout, Arthritis, Cartilage

Introduction

Calcium pyrophosphate crystal deposition disease, more commonly known as pseudogout is a form of arthritis in which crystals of the sparingly soluble substances monoclinic and triclinic calcium pyrophosphate dihydrate (CPPD) are found in synovial fluid, articular cartilage, fibrocartilage and synovium.[1] Although the morphology of CPPD crystal deposits has been described by transmission[2,3,4,5] and scanning electron microscopy[5,6,7,8,9] the mechanism of CPPD crystal formation in tissues remains unknown. Recognizing that hyaline cartilage may be considered as a gel, we have hypothesized that calcium pyrophosphate crystal formation is regulated by the physical/chemical state of the gel[10]. This paper describes the application of correlative polarized light microscopy, scanning electron microscopy, x-ray energy spectroscopy, transmission electron microscopy and micro x-ray diffraction to an in vitro model system for the study of calcium pyrophosphate crystal formation in postmortem human articular cartilage.

Materials and Methods

Tibial plateaus were obtained at autopsy from patients without calcium pyrophosphate crystal deposition disease or other forms of arthritis. Parallel troughs 2.5 cm. apart were cut in the articular cartilage to the depth of the subchondral bone on both the lateral and medial faces. Powdered calcium chloride dihydrate was placed in one trough and powdered sodium pyrophosphate dihydrate placed in the other. The system was incubated in air-tight plastic containers for 48 hours either at 10°C to retard ionic diffusion or at 37°C to simulate physiologic conditions. Within 48 hours, a sharply defined white precipitin band formed 3 mm. from the pyrophosphate trough in the 10°C experiment; the precipitin band was equidistant from the troughs in the 37°C experiment. Samples, cut coronally from the precipitin bands were fixed in sodium cacodylate-buffered 10% formalin. The samples were then thoroughly rinsed with dionized double distilled water before being freeze dried onto aluminium stubs for scanning electron microscopy. The stubs were coated with carbon and examined in a Cambridge Stereoscan 180 scanning electron microscope equipped with an Ortec 6230 x-ray energy spectrometer. The x-rays were collected at a fixed take-off angle of 45°. The x-ray spectra in each case were recorded with count time of 40 seconds at an accelerating voltage of 30 KV with a specimen magnification ranging from 33X to 460X. The data were corrected for background by subtraction of the Brehmsstrahlung contribution calculated through analysis of single channels before and after each elemental window. The x-ray dot maps were made with a specimen magnification ranging from 33X to 460X.

A sample of fixed tissue was alternatively dehydrated through graded concentrations of alcohol and embedded in low viscosity embedding medium (LVM)[11]. Sections 80 μm. thick were examined with a Leitz Orthoplan II polarizing microscope fitted with polarizer, analyzer and lambda first order retardation plate. Areas of sections showing birefringent material were further analyzed by micro x-ray powder diffraction using nickel-filtered CuKα x-ray radiation 100 μm. collimator, and recording on a flat plate 75 mm. from the specimen with an exposure time of 48 hours or more. Subsequent to micro x-ray diffraction, the same sections were partially deplasticized by immersion in propylene oxide for 3-5 hours. Sections were then mounted on aluminium stubs with double-backed adhesive tape, coated with carbon and examined in the scanning electron microscope operated at 30kV as described for the bulk specimen.

For transmission electron microscopy, the fixed dehydrated material was embedded in Epon-Araldite epoxy resin embedding medium. Thin sections were cut with a diamond knife, stained with uranyl acetate and lead citrate or alternatively left unstained and viewed with a Philips 301 transmission electron microscope.

Results

The precipitin band cut coronally and viewed with the scanning electron microscope appeared as a ridge running from the cartilage articular surface to the subchondral bone (Fig. 1A). The definition of this ridge was enhanced by imaging the specimen with the back-scattered electron mode (Fig. 1B). X-ray mapping (Fig. 1C) showed peaked concentrations of calcium and phosphorus corresponding to the precipitin band.

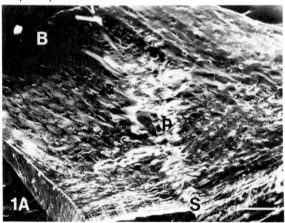

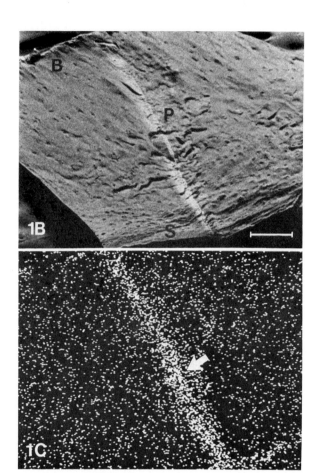

Fig. 1: Scanning electron micrographs of tibial plateau articular cartilage. The troughs of calcium chloride, and sodium pyrophosphate, are to the left and the right. Precipitan band P appears as a ridge running from the surface (S) of the cartilage to the subchondral bone (B). Bar 300 μm.

A. Specimen imaged in secondary emitted electron mode.
B. Specimen imaged in back-scattered electron mode. Note the increased definition of the precipitin band.
C. X-ray dot map for calcium and phosphorus. The elements are concentrated along the precipitin band (arrow).

However, calcium in particular was seen on the pyrophosphate side of the band indicating that the precipitin line was not a complete barrier to ionic diffusion. X-ray energy spectroscopy (XES) of the band showed a Ca:P ratio of 1:0.9 with the detection of a trace of sodium. XES analysis of cartilage adjacent to the pyrophosphate trough showed a similar Ca:P ratio while analysis of cartilage adjacent to the calcium trough and of control cartilage showed low amounts of Ca and P.

Thin sections of the epoxy resin embedded precipitin band examined with the transmission electron microscope revealed minute elongated crystals (Fig.2) 0.1 x 0.25 μm. with a bubbly ultrastructure, situated amongst the collagen fibers.

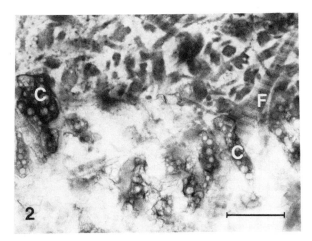

Fig. 2: Transmission electron micrograph of resin embedded precipitin band showing 0.1 x 0.25 μm. elongated crystals (C) amongst the collagen fibers (F). Bar 0.25 μm.

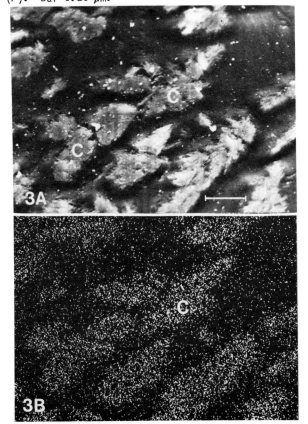

Fig. 3: Scanning electron micrographs of partially deplasticized 80 μm. LVM - embedded sections. With the precipitin band cut en face, the islands of crystalline material (C), 30 to 45 μm. in diameter, within the cartilage can be seen. Bar 30 μm.

A. Specimen imaged in conventional secondary emitted electron mode.
B. Calcium and phosphorus x-ray dot map showing elemental concentration corresponding to crystalline island areas.

Scanning electron microscopy of the 80 μm. LVM - embedded sections, partially deplasticized with propylene oxide showed islands of crystalline aggregates (Fig. 3A) within the articular cartilage. X-ray mapping (Fig. 3B) confirmed the concentration of calcium and phosphorus in these crystalline areas. XES (Fig. 4) of the crystalline aggregates showed a Ca: Na: P ratio of 1:0.1: 1.4. Micro x-ray powder diffraction analysis of these crystals in thick Spurr embedded sections in situ definitively identified the crystals in the precipitin band as monoclinic calcium disodium pyrophosphate tetrahydrate, α form (CaNa$_2$P$_2$O$_7$·4H$_2$O) (Fig. 5).

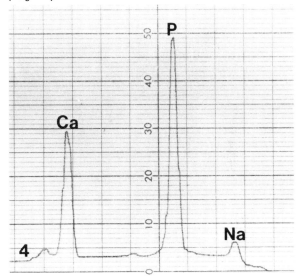

Fig. 4: X-ray energy spectroscopy chart of crystals seen in Fig. 3. Ca, Na and P are present in a 1:0.1:1.4 ratio.

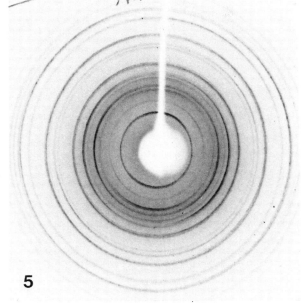

Fig. 5: Micro x-ray powder diffraction pattern of the precipitin band visualized in Fig. 3. This pattern identifies the crystals as monoclinic calcium disodium pyrophosphate tetrahydrate α form.

Discussion

Correlative microanalytic techniques centered upon the scanning electron microscope and/or x-ray energy spectroscopy have been increasingly used to investigate crystals associated with arthritis[5-9,12] as well as inorganic particles associated with other pathological processes[13-17].

In the investigation described above, each analytic mode contributed valuable information; this information in turn could be applied to the data from the other modes so as to strengthen our knowledge of the model system. Compensated polarizing light microscopy of the embedded sections showed that the precipitin band contained birefringent particles i.e. anisotrophic crystals rather than amorphous material. Low magnification scanning electron microscopy of the bulk cartilage specimen, particularly in the back-scattered electron imaging mode, defined the size and shape of the precipitin band. Mapping of this specimen confirmed the high concentration of Ca and P within the band and the distribution of calcium ions on the pyrophosphate side of the band. This latter information indicated that calcium ions can diffuse through the precipitin band. Further, as the Ca and P patterns were present in tissues previously washed with deionized double distilled water, the Ca and P present outside of the precipitin line must be bound to the tissues. In the sections, x-ray mapping showed that the elements were present in the precipitate in discrete angular particles i.e. the crystals. X-ray energy spectroscopy revealed the Ca:P ratios (1:0.9 and 1:1.4) and the presence of sodium indicating that the crystals contained these elements. Transmission electron microscopy of the crystals showed a morphology common to hydrated calcium pyrophosphates. With this background information, the lengthy procedure of micro x-ray diffraction analysis was performed until an identifiable crystalline pattern was obtained. This powder diffraction pattern was identical to a crystal species which we have previously designated in other model systems as Unknown II[10] and which we have recently characterized by single crystal x-ray diffraction analysis as monoclinic calcium disodium pyrophosphate tetrahydrate, α form ($CaNa_2P_2O_7 \cdot 4H_2O$), a substance with a Ca:Na:P ratio of 1:2:2. With this knowledge and with the ratios of the elements obtained by x-ray energy spectroscopy, we can state that some of the calcium and phosphorous both inside and outside the precipitin band exists in forms other than this particular crystal.

It is of considerable interest that this model system which forces calcium pyrophosphate crystal formation in cartilage results in a non-physiological product, α monoclinic calcium disodium pyrophosphate tetrahydrate rather than the physiological monoclinic or triclinic calcium pyrophosphate dihydrate. This observation indicates that normal cartilage strongly inhibits CPPD crystal formation. That the product of these experiments varies from the crystal product formed in similar experiments with gelatin is in accord with the hypothesis that formation of CPPD crystals in affected patients takes place in a matrix with special physical/chemical characteristics. From the analytical point of view this study illustrates how each mode of this multimodal analytical approach contributes complementary information to our understanding of the matrices and products involved in calcium pyrophosphate crystal formation.

Acknowledgements

This work is supported by the Canadian Arthritis Society, the Canadian Geriatrics Research Society, the National Research Council of Canada, the Ontario Provincial Lottery Health Awards Program and the Mount Sinai Institute. We thank Mrs. M. Khoury for technical assistance, Mr. E. Lin, Department of Zoology, University of Toronto for assistance with the scanning electron microscope studies, Ms. E. Parks, Department of Instructional Media, Mount Sinai Hospital for photography, and Ms. E. Sinclair for typing the manuscript.

References

1. McCarty, D.J. Calcium Pyrophosphate Dihydrate Crystal Deposition Disease - 1975. Arthritis and Rheumatism, 19(3), 1976, 275-285.
2. Bjelle, A.O. Morphologic Study of Articular Cartilage in Pyrophosphate Arthropathy (Chondrocalcinosis articularis or calcium pyrophosphate dihydrate crystal deposition disease). Ann. Rheum. Dis. 31, 1972, 449-456.
3. Reginato, A.J., Schumacher, H.R., Martinez, V.A. The Articular Cartilage in Familial Chondrocalcinosis, Light and Electron Microscopic Study. Arth. and Rheum., 17(6), 1974, 977-992.
4. Bjelle, A.O., Sundstrom, B.K.G. An Ultrastructure Study of the Articular Cartilage in Calcium Pyrophosphate Dihydrate (CPPD) Crystal Deposition Disease (Chondrochalcinosis Articularis). Calcif. Tiss. Res. 19, 1975, 63-71.
5. Pritzker, K.P.H., Phillips, H., Luk, S.C., Koven, I.H., Kiss, A., Houpt, J.B. Pseudotumour of Temporomandibular Joint: Destructive Calcium Pyrophosphate Dihydrate Arthropathy. J. of Rheum. 3(1), 1976, 70-81.
6. Cameron, H.U., Fornasier, V.L., MacNab, I. Pyrophosphate Arthropathy. Am. J. of Clin. Path. 63(2), 1975, 192-198.
7. Doyle, D.V., Dieppe, P.A., Crocker, P.R., Ibe, K., Willoughby, D.A. Mixed Crystal Deposition in an Osteoarthritic Joint. J. of Path. 123(1), 1977, 1-5.
8. Gaucher, A., Faure, G., Netter, P., Pourel, J., Duheille, J. Identification Des Cristaux Observes Dans Les Arthropathies Destructrices De La Chondrocalcinose. Revue du Rhum. 44(6), 1977, 407-414.
9. Gaucher, A., Faure, G., Netter, P., Malaman, B., Steinmetz, J. Identification of Microcrystals in Synovial Fluids by Combined Scanning Electron Microscopy and X-ray Diffraction: Application to Triclinic Calcium Pyrophosphate Dihydrate. Biomedicine, 27, 1977, 242-244.
10. Pritzker, K.P.H., Cheng, P-T., Adams, M.E., Nyburg, S.C. Calcium Pyrophosphate Dihydrate Crystal Formation in Model Hydrogels. J. Rheum. 5, 1978, 469-473.

11. Spurr, A.R. A Low-viscosity Epoxy Resin Embedding Medium for Electron Microscopy. J. Ultrastruct. Res. 26, 1969, 31-43.

12. Pritzker, K.P.H., S.C. Luk. Apatite Associated Arthropathies: Preliminary Ultrastructural Studies. SEM/1976/1, IIT Research Institute, Chicago, IL, 60616, 493-499.

13. Landis, W.J., Glimcher, M.J. Electron Diffraction and Electron Probe Microanalysis of the Mineral Phase of Bone Tissue prepared by Anhydrous Techniques. J. Ultrastruct. Res. 63, 1978, 188-223.

14. Brody, A.R., Dwyer, D.M., Vallyathan, N.V., Visco, G.P., Craighead, J.E. The Elemental Content of Granulomata: Preliminary Studies of Pulmonary Sarcoidosis and Hypersensitivity Pneumonitis. SEM/1977/II, IIT Research Institute, Chicago, IL, 60616, 129-136.

15. Abraham, J.L. Scanning Electron Microscopy as an Aid in Diagnosis - An Overview. SEM/1977/II, IIT Research Institute, Chicago, IL, 60616, 119-128.

16. Brody, A.R. Use of Scanning Electron Microscopy and X-ray Energy Spectrometry to Determine the Elemental Content of Inclusions in Human Tissue Lesions, SEM/1978/II, SEM Inc., AMF O'Hare, IL, 60666, 615-622.

17. Dieppe, P.A., Crocker, P.R., Willoughby, D.A. Microanalysis of Particulate Material Involved in Inflammation, in Perspectives in Inflammation; Future Trends and Developments. MTP Press Ltd., Lancaster, England, 1977, 233-235.

Discussion With Reviewers

Reviewer II: The amount of sodium detected by x-ray analysis appears much too low to explain the formula $CaNa_2P_2O_7.4H_2O$. Is this due to substitution of atoms in the crystal?
Authors: No. The low ratio of sodium is related to the insensitivity of XES for analysis of low molecular weight elements.

Reviewers II & IV: Did you measure the K x-ray peaks for Ca, Na and P on control (known) $CaNa_2P_2O_7.4H_2O$ crystals to determine the relation between the ratios of the control crystals and those of the crystals forming in the cartilage model?
Authors: Yes. XES analysis was performed on pure crystals as determined by single crystal diffraction analysis. The results are as follows:

Crystal	Ca:Na:P Ratio
α $CaNa_2P_2O_7.4H_2O$ monoclinic	1:0.1:1.2
β $CaNa_2P_2O_7.4H_2O$ monoclinic	1:0.1:1.1
Cartilage crystals in section	1:0.1:1.4

Reviewer IV: How were the dot maps of calcium and phosphorous (Fig. 1C & 3B) generated?
Authors: Rather than superimposing calcium and phosphorous dot maps or assuming that unwindowed scalar outputs were predominantly Ca and P

signals, we used double windows for K x-ray peaks of phosphorous and calcium to generate the maps.

Reviewer V: What x-ray diffraction methods were employed to obtain Fig. 5?
Authors: We used a micro x-ray powder diffraction method in a transmission mode using a 100 um. collimator. (Micro x-ray diffraction: definitive crystal identification in tissue sections, K.P.H. Pritzker et al, Lab. Invest. 40, 1979, 42-43.)

Reviewer II: Were any other diffraction patterns obtained from the precipitin band?
Authors: No.

Reviewer I: Why was this model system chosen to study calcium pyrophosphate crystal formation?
Authors: Presently, there is no known natural or experimental animal model for this disease. This model forces crystal formation in the natural matrix therefore it is appropriate to understanding the conditions under which calcium pyrophosphate crystals form in cartilage.

Reviewer III: With respect to different matrices for mineralization, what new information about the mechanism of CPPD crystal formation in affected patients is provided by this in vitro model system?
Authors: The fact that in this system, $CaNa_2P_2O_7.4H_2O$ forms rather than $Ca_2P_2O_7.2H_2O$ indicates that cartilage normally inhibits CPPD crystal formation. What has not been completely sorted out by this model is the influence of the high ambient sodium ion concentration on the crystal product. It should be noted, however, that at physiological extracellular sodium concentration, 0.14M, one would predict the formation of $CaNa_2P_2O_7.4H_2O$, a crystal product which has been found in diseased human cartilage.

Reviewer III: Do you have any information concerning the nature of calcium and phosphorus present in the cartilage outside the limits of the precipitin band?
Authors: No, not yet.

REVIEWERS: II   K.M. Kim
III J. Ruffolo
IV   T. Okagaki
V   V.K. Berry

Additional discussion with reviewers of the paper "X-ray Microanalysis of Pyroantimonate-Precipitable Cations" by J.A.V. Simson, continued from page 792.

Reviewer I: Although the water of hydration in various pyroantimonate salts is held by much weaker forces than normal chemical bonds, it is conceivable that the salts may hold water to different degrees. Some calcium salts are notably quite hygroscopic or deliquescent, and it may be that calcium pyroantimonate is more difficult to dehydrate than other antimonate salts. In the weight determination experiments and in particular in view of the high Sb (Lα): Sb (Lβ) peak height ratios achieved with the calcium precipitate, was any attempt made to take possible differences in degree of hydration into account?

Authors: No, we did not attempt oven-drying the precipitates. Water of hydration should not affect the specific emission peaks except possibly as a non-specific "quenching" agent.

Reviewer I: Unless more washing is achieved through the use of a graded series of ethyl alcohol, wouldn't it have been just as effective to utilize only absolute alcohol in the dehydration of the precipitates?

Authors: Probably, but in order to follow, as nearly as possible, the type of protocol that is used with tissue samples, we ran the precipitates through graded alcohols. We have found quite a striking concordance in the analyses of water-washed, graded alcohol dehydrated, Epon-embedded, thin-sectioned precipitates (*i.e.*, routine tissue-processing protocol) and water-washed absolute-alcohol-dehydrated, planchette-mounted crystalline precipitates.

REVIEWER I: A.R. *Spurr*

SCANNING ELECTRON MICROSCOPY/1979/II
SEM Inc., AMF O'Hare, IL 60666, USA

DOCUMENTATION OF ENVIRONMENTAL PARTICULATE EXPOSURES IN HUMANS
USING SEM AND EDXA

J. L. Abraham

Department of Pathology, M-012
University of California San Diego,
School of Medicine,
La Jolla, CA 92093

Abstract

There is increasing awareness of health hazards from environmental and occupational exposures to particulates. Scanning electron microscopy (SEM) and energy dispersive x-ray analysis (EDXA) can document these exposures by analysis of small portions of cells, tissues and environmental samples. Previous work is briefly reviewed and special attention is given to discussion, with examples, of the various types of particulates which may be found in tissues (exogenous, endogenous, inhaled, injected, ingested, inorganic, organic), the different tissues in which they may be found (lung, heart, liver, skin, brain, kidney, lymph nodes, etc.), methods of tissue sampling (e.g. pulmonary lavage, transbronchial biopsy, open biopsy, percutaneous biopsy, autopsy), specimen preparation (fixation, embedding, sectioning, choice of substrate), SEM and EDXA data collection (backscattered electron imaging, etc.) data interpretation (artefacts, limitations of SEM and EDXA) and other new techniques (ion microprobe, laser Raman microprobe).

KEY WORDS: Inorganic Particulates, Backscattered Electrons, X-ray Microanalysis, Specimen Preparation Methods, Lung, Pathology, Etiologic Agents, Environmental Exposures, Occupational Exposures, Human Diseases

Outline

Introduction

Review and Previous Studies

Limits of SEM/EDXA
   Case Selection for Microanalysis

Tissue Samples

Sample Preparation
   Sections
   Embedding Media
   Contamination
   Artefacts

One Approach to Analysis of Particulates
   Polarized Light Microscopy
   Finding Particles with the SEM

Method of Analysis

Interpretation of Findings
   Sources of Particles in Tissues

Conclusion

Introduction

In this review I will deal with an approach to the study of inorganic particulates which may be found in humans. Following a brief review, I will describe the methods I have found most useful, with examples of important points including some pitfalls, artefacts and limitations of the scanning electron microscope (SEM)/energy dispersive x-ray analysis (EDXA) approach.

Most environmental exposures to inorganic particulates of medical significance are via inhalation. Thus, the lung is the most common tissue sampled for documentation of such exposure. The size of particulates which penetrate deeply into the lung is at the limits of resolution for light microscopy. The SEM is ideally suited for studying such particulates, and coupled with an EDXA system, is able to morphologically and chemically characterize most particles to a degree sufficient for identification. When combined with appropriate sampling of the environment and history of exposure, not only the exposure, but its source and consequences (the in situ tissue reaction) may be documented.

The increasing importance of documenting environmental exposures and the need for more efficient and sensitive techniques has been referred to in a recent report on macrophage biology by a committee of the National Heart, Lung and Blood Institute.[1]

This report emphasizes the analysis of particulates in tissue. However, the preparation of environmental samples is critical to the documentation of sources of possible particulate exposure. Methods for preparing particulates themselves for SEM examination have been recently reviewed by DeNee.[2] This will not be repeated here, except to point out that one should attempt to obtain as complete a sample of the environmental exposure(s) as possible and to prepare it in such a way as to retain all the particulates in a sufficiently dispersed condition so that the analysis will be possible. In some cases one may wish to separate respirable sized particles from others to facilitate relevant analysis.[3]

The importance of a detailed lifetime environmental and occupational exposure history should be stressed. This is vital to documentation of exposures. Detail is important because significant exposures may have been brief and may have occurred many years ago, such as during World War II. This is illustrated by a recent example from a patient with end-stage interstitial lung disease.[4] Multiple dusty jobs were held by the patient during his 30 plus year work experience. The last (and earliest) entry in the occupational history (truly the "bottom line") mentioned that the patient worked in the Navy during World War II, at which time he almost certainly had significant asbestos exposure.

Review and Previous Studies

The use of SEM and EDXA in diagnostic pathology and in identification of particulates in the lung has been reviewed.[5,6] Only selected more recent reports will be reviewed here.

Most of the analyses of particulate material in tissue and in environmental samples fall into the qualitative or semi-quantitative category. It is usually not possible to perform true quantitative analysis of either the individual

particles themselves or of their components in the tissue analyzed at low magnification. Microanalytical capabilities of SEM and EDXA are best suited to individual particle analysis. The sensitivities for "bulk" analyses of tissues are inadequate for most real situations. Furthermore, the important associations of elements within individual particles provide clues as to their composition and origin. These clues are lost when numerous particles, usually of mixed types, are analyzed in the aggregate.[6-10] The theoretical techniques for quantitative analysis of particles in the size range of those likely to be found in tissues (less than $5\mu m$ in thickness in standard sections) have been dealt with by several workers and reviewed by Armstrong.[11]

The potential of microanalytical techniques for true quantitative analysis of biologically occurring elements is being developed using cryotechniques for those more plentiful ions such as sodium, potassium and chlorine. Trump, et al. review the work in microanalysis of diffusible elements in biological thin sections.[12] Tormey reported meticulous studies of quantitative analysis of frozen sections of erythrocytes as a model for x-ray microanalysis of frozen sections of tissue.[13] Quinton reviewed the results of microdroplet analysis of approximately 10^{-10} liters of fluid.[14] By mounting the droplets on a thin film the Bremsstrahlung background is reduced so that satisfactory peak-to-background ratios may be obtained using EDXA.

Quantitative analysis of particles supported on a thick substrate is difficult. Some attempts have been made to deal with the theory and practical application of quantitative particle analysis.[11,15]

The sensitivity of x-ray microanalytical techniques should be carefully defined and understood by those using it. For most materials, detection limits are on order of 0.1% by weight in the volume analyzed. While very small masses of material can be detected (approximately 10^{-16} to 10^{-18} gms), these materials are detected in local regions of high concentration. Particulates such as those discussed and illustrated in this report are examples of such high local concentrations in very small volumes. For example, a particle approximately 0.05 micrometers in diameter would have a volume of approximately $5x10^{-4}\mu m^3$, which equals $5x10^{-16}cm^3$. If one can detect an element present at the level of 1% by weight within such a particle, assuming a density of 2 gms per cm^3, one is detecting 10^{-17} gm of material. However, this mass of material would not be detectable using SEM and EDXA microanalysis if it were distributed more diffusely throughout the tissue.

There is less of a problem if one can locate the particles or regions of high concentration of materials to be analyzed. The backscattered electron BSE) image is the tool in the SEM which facilitates location of regions of differing composition for subsequent microanalysis.[5,6,16,17] The use of more sensitive techniques such as ion microprobe mass analysis (IMMA)[18] make it possible to perform more quantitative analysis of tissue sections for elements present in much lower concentration. The sensitivity of the IMMA is on the order of parts per billion for many elements. It is especially sensitive for the very light elements which cannot be seen with EDXA and only with poor sensitivity even with wavelength dispersive x-ray spectrometry. While the IMMA does allow high sensitivity and spatial resolution of approximately two micrometers, it does not provide morphological images of particulates being analyzed, but only ion distribution images. Thus, optimally it needs to be combined with study by some other technique such as SEM and/or light microscopy.[6]

Quantitative bulk analyses of lung tissue in inorganic dust induced lung disease have been relatively few. For example, in beryllium disease, the levels required for acceptance of cases into the U.S. Beryllium Case Registry[19] are greater than 0.1 micrograms Be per gram dry weight (0.1ppm) (0.00001%) (G. Boylen, personal communication). The highest Be levels recorded have been on the order of 10-15ppm (0.0015%). The sensitivity required to detect Be is within the capability of the ion microprobe (IMMA). Quartz in normal lungs exposed to industrial dust is 0.18 to 1.5% of dry lung tissue, and 0.23 to 0.6% in normal hilar lymph nodes. The total quartz content of silicotic lungs is 0.2%, commonly 2-3%. Actual free silica is usually less than total quartz.[20] For EDXA purposes, it is the silicon concentration rather than the silica (SiO_2) concentration which is important. The elemental silicon concentration in SiO_2 is 47 weight %. One can calculate that the concentration of silicon even in some cases of silicosis is near the limits of meaningful SEM-EDXA analysis when a "bulk" or low magnification approach is used. The information on silicates is even less certain, as the low magnification approach will not reveal which silicates are present and whether there is significant silica present. Also, the percent silicon by weight in most silicates is less than that in silica.

DeNee[21] and others[6,22-26] have presented and reviewed methods for studying particulates in tissues, including both in situ methods and particle isolation methods. The important correlations of particles with specific reactions in tissues are lost when particle isolation methods are used. Also, one cannot be certain that all the particles recovered from a tissue sample were in fact genuine particles in the tissue rather than contamination.

Echlin reviewed the coating techniques suitable for SEM and x-ray microanalysis.[27] He points out that while heavy metal coating is easily applied with a sputter coater, vacuum evaporators are still the best method for applying thin films of low atomic weight elements for coating samples for x-ray microanalysis. Echlin's negative findings about carbon coating have not been justified in my experience (see below). Murphy reviewed techniques other than coating to render samples conductive.[28] Most of these apply to bulk samples of tissue and involve heavy metals. These are not recommended for microanalytical studies. More importantly, microanalysis of bulk tissues is quite difficult to interpret due to complex geometry and x-ray absorption problems.

Brody et al. reported the application of SEM and EDXA to lesions in the lung and oral cavity.[29] The cases they present are useful illustrations of the application of these techniques. One controversial aspect of their analytical method in some cases, and also that used by another group of investigators[10] is the use of low magnification analysis rather than individual particle analysis. The sensitivity of x-ray microanalysis in a scanning electron microscope or electron probe has been discussed above and recently reviewed by Coleman.[15] An illustrative example of the difference in sensitivities was given in a case with granulomatous lesions in the lungs of a foundry worker.[6] In this case low magnification analysis of the granuloma revealed the presence of silicon, but analysis of individual particulates revealed a mixture of particles, some of which were silica, some silicates, and others zirconium silicates. The element zirconium, present in high concentration in the zirconium silicate particles, was not even detected with low magnification analysis, despite the presence of numerous such particles in the lesion. Also not detected by the low magnification analysis were numerous fume-sized metal particles containing various metals in combination with iron.

Kaszynski et al. showed a comparison and advantages of SEM and EDXA analysis of tissue sections versus a light optical histochemical fluorescence technique, which was less specific in identifying the elements in the animals exposed to zirconium and aluminum compounds.[30]

Stettler et al. reported a careful study involving analysis of large numbers of particles from lung tissue and air filter samples to document exposure to stainless steel welding fume particles.[31] This is a model study showing that large numbers of particles need to be individually analyzed to draw correlations between environmental samples and tissue samples.

Bonin and Abraham presented a scientific exhibit demonstrating the use of SEM and EDXA in diagnostic pathology.[32] Among other cases, they demonstrated the presence of thorium dioxide (thorotrast) in the liver of a patient with angiosarcoma. Similar reports of identification of thorium in liver[33] and brain[34] have been recently published.

Sherwin, Barman and Abraham[3] have recently studied a group of cases in which severe fatal interstitial fibrosis developed in nonsmoking residents of agricultural areas. Numerous plate-like birefringent particles were found by light microscopy associated with areas of fibrosis. Analysis of the tissue sections revealed silicates and silica, with the majority of the silicates being of the mica group containing mostly aluminum, silicon and potassium. Approximately 10% of the particles were silica. The source of this was presumed to be soil and confirmed by analysis of respirable particles isolated from the soil by sedimentation in water. The analysis of the soil particles showed similar distribution of particle types and composition.

Chatfield,[26] Mah and Boatman[35] and Beaman and File[36] have discussed some of the special problems and successes in using SEM and transmission electron microscopy (TEM) for analysis of asbestos fibers.

In the Symposium on Inhaled Particles held in 1975,[37] numerous clinical and pathological studies concerned with various types of particulates are reported.

Limits of SEM/EDXA

There are limits to the usefulness of SEM and EDXA in diagnostic situations. The sensitivity for trace elements and light elements is poor and the technique is not suitable for bulk analysis. The technique is capable of revealing associations of several elements in single particulates, which may be of significance in suggesting their possible origin. However, this technique does not provide exact chemical identification. Finally, these techniques give virtually no information regarding organic materials. Therefore, other techniques should be used when appropriate. A few merit mention here. Histochemical techniques can be quite useful for organic and some inorganic materials.[38,39] Electron diffraction provides crystallographic information, such as with asbestos studies.[25] Secondary ion mass spectrometry (SIMS) can reveal isotopic information down to the lightest elements including hydrogen, and into the parts per billion range for many of these elements.[18,40] The ion microprobe has the potential of quantitative microanalysis of tissue sections[41] and has been used to analyze a number of human lesions.[42] A very recently developed and highly promising technique for investigative, diagnostic and forensic work is the laser Raman microprobe. This yields molecular information to aid in identification of complex organic as well as inorganic materials.[43,44]

Chatfield[26] has beautifully documented the inability of SEM to detect small asbestos fibers and very thin plates, even in a best-possible situation of isolated particles on a feature-less substrate. There is very little chance of finding many individual asbestos fibers in situ in tissue sections using BSE imaging. The TEM is necessary for this important area.

Case selection for microanalysis should be made only after routine pathologic examination, on the basis of clinical history, light microscopy and appropriate microbiological studies.[6,45]

Tissue Samples

Tissue samples obtained by various methods are useful for SEM and EDXA. The increasingly popular technique of transbronchial biopsy[46] can be a good source for documentation of environmental particulate exposure. Although the tissue sample is small (usually less than 5 millimeters in diameter), it almost always contains tissue from near the airways. The peribronchiolar and peribronchial regions are the location of lymphatics through which pass macrophages and associated particulates being cleared from the lung. Thus, even a tiny transbronchial biopsy may give important clues as to environmental exposures. An example of this is finding of typical asbestos bodies in a transbronchial biopsy. (The importance of the iron stain for light microscopy cannot be over emphasized.) One often will note opaque fine particulate material with black or brownish color in the peribronchial interstitial regions. From light microscopy, all one can say is that this is opaque material. One can use polarizing light microscopy to see if there is associated birefringent crystalline material, but the identification cannot be made by light microscopy. These tiny biopsies can be serially sectioned with one section being placed on a carbon substrate. Alternately, a section may be transferred from the glass slide to the carbon disc. An instructive example was a case in which there was abundant fine black particulate material which one would normally assume to be "anthracotic" pigment. In this case, SEM and EDXA revealed that most if not all the particulates were in fact mainly composed of molybdenum. There may also have been sulfur present in these particles but the energy dispersive spectrometer was not able to resolve the sulfur K lines from the molybdenum L lines. Wavelength dispersive spectrometry would be able to resolve this. Although the transbronchial biopsy may not always obtain representative or sufficient tissue for a pathologic diagnosis of interstitial lung disease, it usually contains material reflecting the environmental exposures of the patient.

These transbronchial biopsies will not necessarily reflect recent exposure, but the distribution in an open lung biopsy may give some clue as to recent versus remote exposure. The very recent exposure may result in a certain particulate being found in the air spaces (free or in exudates or epithelium), with more remotely respired particulates being found in the dust clearance pathways, e.g. peribronchial and perivascular locations. Draining lymph nodes are also useful sites to examine for evidence of previous exposure.

Bronchial lavage and sputum may also contain evidence of environmental exposures. However, the finding of particulates in the lavage or sputum does not necessarily indicate recent exposure. For example, in the case recently examined in our laboratory,[47] a lady with occupational exposure to tungsten carbide and associated metals in a tool grinding factory showed presence of these materials in an open lung biopsy taken at the time of her work. A repeat (transbronchial) biopsy and lavage taken one year after cessation of work also showed similar particulates in the tissue and in lavaged macrophages. These similar particles were correlated with the environmental sample collected at the workplace. Identical size and composition particles were observed in the lung tissue, the bronchopulmonary lavage and the environmental sample. The only difference between the environmental sample and the patient's samples was the presence of the element cobalt in the environmental sample but not in the tissue samples. Cobalt is commonly associated with these materials in usage[20] but is quite soluble and thus not retained in the tissues at levels high enough for electron microprobe an-

alysis.

Bronchopulmonary lavage is the treatment of choice in cases of the rather uncommon pulmonary disease, pulmonary alveolar proteinosis.[48,49] Since several studies have suggested that alveolar proteinosis in most, if not all cases, may be due to environmental exposure to toxic particulates and other chemicals,[50,51] the examination of lavage fluid from these cases, as well as biopsy tissues, may be most informative.

Sample Preparation

Sample preparation can be identical with that of light microscopy with standard paraffin sections picked up on carbon substrates instead of glass slides. No staining nor special drying is necessary, and the morphology is preserved for correlation with the microanalytical findings. This important correlation is lost when destructive or bulk analytical techniques are used.[5,6,16,52]

It is not usually necessary to coat the tissue when sections are supported on conductive carbon substrates. The majority of the electron beam (at reasonable accelerating voltages, e.g. 20keV and above) goes through the thin tissue into the carbon. One should recall that approximately 90% of the tissue weight is water, which is removed during any dehydration and drying process prior to placing the specimen in the SEM vacuum chamber. This elimination of one more step in the preparation reduces the preparation time and eliminates another possible source of contamination. If one does carbon coating, it is important to use clean carbon rods which have not been left in an evaporator during its use for metal coating -- carbon rods will get coated with metal and flakes of metal coating may then be deposited on the sample which was intended for carbon coating only.

Paraffin sections are easily deparaffinized whereas the embedding material is only removed from plastic sections with difficulty. When one examines the surface of the tissue section, the cellular morphology and topography is available in the secondary electron image of deparaffinized sections, however, the smooth surface of plastic sections is all the topography available.[53] The size of sections readily cut from the paraffin is also greater than that with plastic embedding materials.

The use of serial sections, with one being used for light microscopy and the next for SEM, suffices for most cases for correlation between SEM and light microscopy. In the rare instance in which there is either a unique feature noted in one glass slide or in which there is absolutely no additional source of tissue available, the entire section or a small specific portion of it may be readily transferred from the glass slide to the carbon disc for x-ray microanalysis. Convenient plastic materials such as diatex (available from Scientific Products) can be used after removal of the cover slip from the selected tissue section. The plastic is allowed to form a thin film and to harden, after which the section is easily removed from the glass slide and transferred to the carbon disc. The plastic is then removed using xylene. Other materials could be used for the same purpose such as the water soluble polyvinyl alcohol, although I have not had experience with the latter for this purpose. There is thus a vast library of tissue sections available in most pathology departments which are suitable for SEM and x-ray microanalysis of interesting or otherwise unresolved particulate related diseases.

Perhaps the best example illustrating the durability of tissue specimens and the availability of information on particulate exposures in tissue stored for many years is the demonstration of silica and silicate exposure in 3,000 year-old tissues from Egyptian mummies.[54] Although the cellular detail was lost from the tissue section prepared 3,000 years after mummification, the connective tissue framework of the lung was well enough preserved to easily recognize alveoli, vessels and airways. In the characteristic peribronchial and perivascular locations, dust particles were found which showed a distribution of silica and silicates remarkably similar to that found in current exposures to soil dust.[3,55]

In transferring sections from glass slides to carbon discs for microanalysis there are a few minor problems, most of which can be avoided. The selection of the stained slide for transfer should be carefully made. For example, certain stains require treatment with chemicals which may alter particulates or add extraneous materials which may interfere with microanalysis. The usual Prussian blue ferric-ferrocyanide stain[38] leaves innumerable precipitates of potassium and iron-containing salts in the tissue. These appear, of course, as particulates and greatly impair any microanalysis or even searching for significant particles. Treatments with acid and bases may be necessary for certain stains; however, these treatments may alter partially soluble compounds such as some calcium-containing salts. Metal stains such as a silver stain may be quite useful for histological and histochemical visualization of specifically stained components using backscattered electron imag-

ing.[5,16] However, these do not usually help when one is searching for particulates and, in fact, the chromic acid used as part of the pre-treatment in the silver staining in some cases provides a generally increased level of backscatter yield from the tissue, reducing the contrast between foreign particulates and the tissue itself. There is generally no problem with hematoxylin and eosin stained sections, or elastic stained sections. On occasion, traces of bromine contained in the eosin may be detected. Occasionally small fragments of glass may be transferred with the section. These probably result from fragments of glass which are created during etching of glass slides with a carbide marking pen or from breakage of a cover slip during its removal. These usually are easily identified as contaminants on the surface of the section, and their analysis can be compared with samples of glass slides and cover slips from which the sections have been taken. Most glass slides show silicon as the major element on EDXA, with lesser amounts of calcium and sodium frequently present.

The choice of embedding media for tissue sections for microanalysis is an important one. Many workers used to looking at transmission electron microscopy routinely think that any kind of electron microscopy tissue sections should be embedded in plastic. This is not necessarily the case. There are many advantages of paraffin embedding for particulate analysis in tissue sections. The size of most particles in tissues will be in the range of 10 microns or less in maximum diameter. Most plastic section techniques are one micron or less in thickness. For transmission electron microscopy sections are routinely less than 100 nm. The sectioning of inorganic particulates is usually difficult and sections may only show holes where particles have been torn from the sections. Also, if the particles are successfully sectioned, only the random cross-section is available for determining particle size and morphology. In contrast, in paraffin sections, most of the particles will remain intact in a 5 micrometer section. Those which are intracellular will be easily seen as such by comparison of the secondary electron and backscattered electron images. The intact particle will be available for determination of its morphology and true size may be determined by stereological techniques. There is little point in using cryosectioning techniques for particles in tissue unless one wishes to analyze soluble materials or semi-liquid materials which are difficult to section at room temperature or are lost during conventional fixation and processing. An example of this would be the presence of organic crystals such as the amino acid cystine. In a case of cystinosis I have studied, numerous crystals of the sulfur rich amino acid were demonstrated in the frozen section but were absent after routine formalin fixation and paraffin sectioning.

To avoid problems of contamination and interference with the analysis of particles in tissue requires careful handling of all tissues and sections, use of clean, filtered solutions, avoidance of heavy metal buffers, such as cacodylate, and avoidance of heavy metal fixatives and stains, such as Zenker's, osmium, lead or uranium. Representative sample substrates chosen for support of the tissue sections should be examined before any sections are added. This should be repeated with each lot of substrates obtained for use. Several different sources of carbon planchets or stubs are available. They vary in the quality of the surface and in the contamination by small particles either inherent in graphite or used for polishing. The most common contaminants visualized with backscattered imaging and analyzed by EDXA, in my experience, have been silica or silicon monoxide, iron or alloys, presumably from cutting equipment used to prepare the discs, and other small particles such as diamond, which do not interfere with the x-ray microanalysis or backscattered imaging. Those substrates treated with high temperature to volatilize metals seem to be generally cleaner for the purposes of particle analysis in situ in tissue sections. The production of an ultra clean surface with a thin layer of pyrollized carbon may look fine with secondary electron imaging; however, the polishing particles of silica, for example, may still be evident in backscattered imaging and may interfere with in situ particle analysis. The interference is greater, of course, with analysis of environmental samples or particles isolated from tissues because one loses the criteria for identifying particles in situ when the tissue is removed. All particles on the substrate appear to be suitable candidates for analysis.

Small biopsies such as needle biopsies and transbronchial biopsies are especially likely to have contamination around their edges where they have been in contact with supporting materials such as gauze or gloves or paper during their transfer or fixation. These materials may be carried across the tissue during sectioning but will usually be much larger than the in situ particles and will lie on top of the section, extracellularly. They should be excluded from the analysis of significant in situ particles.

A few artefacts to be recognized include contaminating particles on or under the section or in the substrate. By comparison of the secondary image and the backscattered electron image, those particles which are not obviously in the tissue or in cells should be considered artefacts.

One Approach to Analysis of Particulates

The value of polarized light microscopy cannot be emphasized too strongly. This provides a most valuable and readily available screening technique for tissue sections. It enables one to survey large areas of tissue much more rapidly than SEM. However, only a fraction of the inorganic particulates will be birefringent, whereas most will be detectable by BSE imaging in the SEM. Its use is not restricted to the lung, as illustrated by some examples recently seen of unsuspected drug abuse being documented by a percutaneous liver biopsy. In one case a moderate inflammatory infiltrate in the portal areas of the liver, plus moderate elevation of serum enzymes suggested a low grade of liver injury. Examination of the biopsy with polarized light revealed numerous tiny strongly birefringent particles in the portal areas. It is important to distinguish these from contaminants, which would not likely show any selective distribution in the tissues. SEM examination revealed the plate-like morphology of these particles and their analysis showed magnesium and silicon, confirming their identity as talc. The history of drug abuse (intraveneous heroin had been used twice ten years prior to the biopsy) was obtained only after the confirmation of the situation by SEM and EDXA.

Among the artefacts encountered in polarized light microscopy, formalin pigment is the most common. This has fine needle-like organic crystals appearing brown to opaque with transmitted brightfield light and moderately birefringent in polarized light. The distribution of these crystals in association with blood in the tissue sections and their extracellular location in most cases should easily avoid confusion with inorganic crystals worthy of microanalysis. These are easily removed from sections by treatment with picric acid.[38] Other crystals with strong birefringence can be analyzed with x-ray microanalysis. Some organic crystals such as urate may give negative results, but usually these have characteristic morphology. Some forms of calcium crystals may be lost from the tissues during acid treatment or even during storage in inappropriately buffered fixative solutions.

Finding Particles with the SEM

The theory and applications of backscattered electron (BSE) imaging have been reviewed elsewhere.[16,17] A few technical points are worth emphasizing here, however. The type of detector necessary for useful backscattered electron imaging when searching for the small regions of higher atomic number in tissue or environmental samples is the solid state detector.[52] One previous limitation of earlier solid state detectors was their inability to function well at rapid scan rates. With appropriate amplifier circuitry, this problem has now been overcome and good sensitivity can be maintained even at TV rates in the SEM.[56] With the appropriate detector it should be possible to see particles down to 0.05 micrometers diameter in an organic low Z matrix such as a tissue section. Small differences of a few atomic numbers can produce sufficient contrast for separation of materials with this kind of a system. The calculation of the backscattered coefficient and of predicted contrast between two materials was discussed by Newbury[57] and examples of some pairs of very light elements and silicates are given in another paper.[58] Other methods of locating materials in the tissues are generally less efficient in both sensitivity and time than backscattered electron imaging. The sensitivity of cathodoluminescence detectors and limitation of finding materials with cathodoluminescence properties makes this less useful than backscattered electron imaging. The comparative sensitivity of solid state BSE detector with the conventional scintillator detector for secondary electrons used for BSE detection is at least an order of magnitude less with the latter.[52] A comparison of x-ray distribution mapping with BSE imaging reveals that only one or at most a few elements may be mapped at any one time and that the time required for mapping is usually several minutes per map.[59] This can be compared with BSE imaging which can operate on line at TV rates.[56]

The morphology of the particles found in examining tissue sections may also be an important clue to the source material. An example of this is illustrated in the following case in which a desquamative interstitial pneumonia (DIP) histologic pattern[60] was seen in a patient with multiple exposures. The examination of the tissue by SEM and EDXA revealed no asbestos particles as was questioned in the clinical history, but identified numerous plate-like particles composed of silica, aluminum silicates and magnesium silicate. A few of the silica particles had the appearance of fragments of diatoms. Diatomaceous earth is itself a form of silica formed by the minute skeletal remains of diatoms. This is widely used in industrial applications because of its high surface area. Figure 1 shows a fragment with the delicate radially symmetrical structure often found in diatoms. This helped document the exposure to diatomaceous earth as at least one of the materials to which the patient had been exposed. Other cases of DIP have also been shown to be related to inorganic particulate exposure including asbestos[61] and talc in another case seen by this author.

Method of Analysis

Two approaches have proven successful in my experience. The first applies when there are areas of obvious particulate concentration in tissue observed by light microscopy. These areas can be directly searched for and particulates within them analyzed in the SEM. To avoid bias it is important to analyze all the particles detected, documenting the location, size and analytical results for each particle. If only selected particles are analyzed, a possible bias may result. One is already working with a small sample in analyzing a tissue section, so that additional selection should be avoided if possible. Within the individual sections and accumulations of particulates, it seems that there is a representative concentration and admixture of particulates in most non-fibrotic accumulations of macrophages. Reactions typical of silicosis, with formation of dense concentric layers of collagen, may distort the blend of particulates and favor or reflect concentration of certain types of particles, e.g. silica, in specific locations. The association of specific particles with specific pathologic reaction is most important to preserve and to recognize.

A second type of analysis is done when one is searching for particulates in a tissue and attempting to get a semi-quantitative estimate of the number and kind of particultes. In this approach one must select an appropriate instrument operating configuration (specimen tilt, working distance, detector-to-specimen distance, accelerating voltage and beam current). To compare x-ray results from one particle to the next, or one sample to the next, it is essential to have standardization of analytical conditions. Most SEM's equipped with x-ray detectors also have means to measure specimen current or beam current. In the absence of this a standard sample (such as a blank area of the supporting carbon substrate) can be analyzed and the x-ray count rate standardized to a reasonable value for

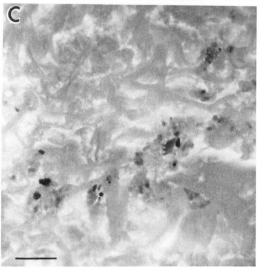

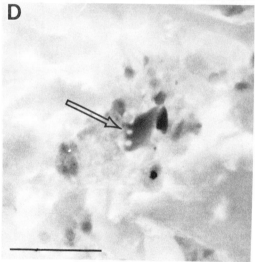

Fig. 1 Autopsy lung, formalin fixed, par-affin embedded, 5µm section on carbon disc, deparaffinized, uncoated, examined in SEM at 20kV, 14-15mm working distance, 1.0 nanoamp beam current, 45° tilt. A. SE image, low magnification showing sec-tion on disc. Marker=100µm. B. SE image of perivascular area illustrating moder-ate charging problems evident with some uncoated sections when high beam current is used (bright areas). Inorganic par-ticles in tissue are not immediately evi-dent. Marker=10µm. C. Same area as in B, BSE image (negative polarity, with darker areas representing regions of greater backscatter yield, i.e., higher atomic number). More than 50 individual particles can be recognized and subse-quently analyzed with EDXA. The loca-tion of the particles in the tissue can be confirmed by comparing the SE and BSE images. Note that charging evident in B does not interfere with BSE image. 130 seconds recording time (twice the standard) was used to improve signal-to-noise ratio. Marker=10µm. D. BSE image (negative polarity) of one macrophage near center of B and C, showing several particles with varying brightness. One (arrow) shows morphology suggestive of a fragment of a diatom. EDXA analysis of this particle revealed 756cps for Si, and no other peaks above background, con-sistent with SiO_2. Some other particles detected include: Ti; Fe; Fe,Cr,Mn; Sn; Si; Al,Si,K; Al; Al,Si; and Mg,Si. Marker=10µm.

repeating on subsequent analyses. The parameters and physics of x-ray microanalysis have been carefully reviewed by Coleman.[15] Once the instrument conditions have been standardized a standard magnification should be selected at which the smallest particles one wishes to analyze can be visualized. A final magnification on a CRT screen of 6000X has proven to be the approximate minimum magnification acceptable for such searching on line. At lower magnifications, one may overlook the tiny fume sized particles frequently found in lung tissues. In searching the sample one follows some standard scan pattern which will traverse the entire sample or randomly selected fields in the sample. Each field is of known area and thickness (correction for tilt must be considered) and the number of fields examined is counted. Within these fields the numbers of particulates detected are also counted. These are categorized by size, shape, location and composition. One end result is a semi-quantitative estimate of the number of particulates per volume of tissue. If only sections with certain criteria are counted, then the calculation of volume becomes meaningless as standard stereological sampling techniques need to be observed.[62]

In the actual analysis of particles sometimes one encounters macrophages containing numerous particles very close together. The problems of fluorescence must be considered in such cases. One tries to analyze background areas of adjacent tissue, particles on which other particles may be deposited, and various edges of particles at high magnification. By appropriate analysis of different regions of the sample one can usually determine which particles have which major components. The ideal would be to have particles separated by a few micrometers at least to avoid fluorescence effects. Coleman has considered many of these factors in his review paper.[15]

It is necessary to have reproducible conditions for observation and microanalysis, but it is often necessary to vary the operating parameters of the SEM to obtain optimal images and data. The large beam current (spot size) needed for sufficient signal for BSE imaging and x-ray microanalysis may be reduced to produce better secondary electron (SE) images. In microscopes designed to collect x-ray data at zero degrees specimen tilt, the topographic contrast in the SE image of uncoated or carbon coated sections may be low. Tilting the specimen will yield better surface morphology, but one must exercise care in keeping regions and particles of interest in perspective as the tilt changes.

In our system using a 30mm^2 detector and a 3cm specimen-to-detector distance it is possible to obtain useful spectra from most particles in 10 or 20 seconds analysis time. This means that with a reasonable sample one can analyze between 20 and 50 particles per hour, with complete documentation of each particle by appropriate micrographs and data printout. The majority of the time in most cases is spent in searching for particles and in quantitating the number of particles during searching of random or consecutive fields of view.

Coleman reviews the theory for dealing with particles or sections supported on thick and thin substrates and also methods of background subtraction.[15] The method which I find most useful is straightforward background subtraction. A blank spectrum from the carbon substrate with no tissue in the region analyzed is collected and stored. This is subtracted from the spectrum of particles subsequently analyzed. This is more rapid than complex mathematical means of generating theoretical backgrounds for given substrates. It also comes closer to the actual background in a given system than any of the calculation methods which I have seen to date. This is especially useful for small particles which contribute negligibly to the general background. Larger particles will create more of a continuum x-ray background and give a generally higher background than the carbon disc blank spectrum. This would give rise to significant problems only when quantitative analysis is being attempted.

As a result of the nonspecificity of EDXA analysis in identifying specific silicate crystal structures and compounds, a grouping of particles by general type such as that shown by Stettler et al.[31] has been found most useful by this author as well as others.[24] Particles can be separated into silica (showing only the element silicon), silicate (showing silica and one or more other elements), and other (particles showing no silicon). Within the silicate and "other" groups, particles can be classified by the type of materials found such as magnesium silicates, iron, calcium phosphate, steel (iron plus chromium, manganese or nickel), etc.

The limits of detection for light elements are commonly reported as sodium being the lightest element detectable. Recently I have examined a biopsy from a neck mass in a lady with previous injection of teflon particles to reinforce a paralyzed vocal cord. In this case the particles of teflon were easily recognized in a backscattered electron image, and the x-ray spectrum revealed a strong peak for the element, fluorine, at 0.68 keV. Approximately 85cps were detected in the fluorine peak under conditions for which a similar sized particle of silica, for example, would have given approximately 3500cps in the silicon peak. Thus, although there is great absorption of the very soft x-rays of fluorine, the yield from teflon (C_2F_4) is high enough so that it can be detected (teflon is 75% fluorine by weight). Likewise, on rare occasions a small peak for oxygen can be seen from some silicates with 50 or 60% oxygen by weight. One must be careful to exclude escape peaks when unusual peaks are found. Fiori and Newbury[63] discuss artefacts observed in x-ray microanalysis.

It is worth noting that the analysis of silicates with EDXA (or even with WDXA) can be incomplete because lithium-containing silicates are common in the soil. In the lungs of a patient with beryllium disease reported elsewhere[6] three silicate particles were shown by IMMA ion distribution mapping, and two of these were shown to be rich in lithium. This element is, of course, not detectable by any x-ray microanalytical technique at present. SEM study of the particles after ion microprobe analysis showed three respirable sized particles containing silicon and aluminum, but there was no way to distinguish between those with lithium and the one without.

Interpretation of Findings

The finding of inorganic particulates in the lung does not, by itself, prove the inhalation route of exposure. There are other mechanisms by which inorganic particulates could reach the lung. Ingested materials can be regurgitated and aspirated. For example, barium sulfate used as radiographic dye can be aspirated into the lung. It may also reach the lung directly in the case of abnormal connections between the alimentary and respiratory tract, e.g. tracheo-esophageal fistula.

Another mechanism for ingested material to reach the lung is by absorption and re-crystallization or precipitation within the lung. An example of this could be ingested metal such as lead which becomes partially soluble and may end up depositing in the cells of the lung as it does elsewhere in the body. This is probably not a common mechanism by which inorganic materials reach the lung in particulate form.

Injected materials may often be found in the lung. Intravenously injected particulate matter, most commonly associated with intraveous drug abuse,

is first filtered from the blood when it reaches the lung. Here the particles too large to traverse the capillary bed are trapped in various sized arteries and arterioles.[64]

Injected material may also reach the lung by intentional therapeutic embolization, as in a radioactive lung perfusion scan. Particulate matter with varying size particles and varying radioisotopic labels can also be used experimentally to assess blood flow in the lung.[65]

Injected soluble drugs may be also found in the lung. An example of this sort of inorganic particulate would be the lysosomal accumulation of intramuscularly administered gold salts as an anti-inflammatory agent in rheumatoid arthritis.[66,67] These gold salts may accumulate at the site of inflammatory reactions and are thought in some cases to be the cause of an injurious reaction on a hypersensitivity basis.[68]

When one is searching tissue for foreign material based on increased backscattered electron signal, a possible major source of interfering material can be endogenous inorganic or metallo-organic deposits. The most prevalent type of deposits in this category would be hemosiderin and calcification. Hemosiderin droplets in cells may appear morphologically identical with fume sized particles when small, and when larger, their round shape will usually suggest their endogenous origin, as such large heavy particles would not likely get into the distal lung parenchyma. Calcification may show quite a range of particle sizes from the submicron up to many tens of microns, well beyond the respirable particle size range. Sometimes these calcifications may be intracellular and others extracellular, especially in perivascular connective tissue.

The composition of the questioned endogenous material may be helpful in its identification. For example, hemosiderin iron is usually accompanied by a significant content of phosphorus. The form taken by calcification may be most commonly calcium phosphate with nearly equal amounts of calcium and phosphorus. Often there is associated small amounts of magnesium. Pure or seemingly pure calcium containing particulates may represent carbonate or oxalate.

Another source of particulate inorganic material in tissues is illustrated by the presence of wear particles from prosthetic materials. Wear particles from plastic and metallic joints have been studied by light microscopy and analysis of particles done with SEM and x-ray microanalysis by Semlitsch.[69] I have reviewed a case[70] in which unusual foreign material was associated with foreign body reaction in an axillary lymph node. Analysis of the material using EDXA revealed only the presence of the element silicon. Morphologically, the material did not appear to be crystalline silica and by light microscopy it did not have the appearance of silica but showed many irregular fragments. This has been subsequently shown to be silicone rubber using laser Raman microprobe analysis (Abraham and Etz, unpublished results). This example illustrates one limitation of EDXA, that is, its inability to analyze compounds. It also points out the great potential of the recently developed laser Raman microprobe in pathology and other fields.[44]

Conclusion

There is ample evidence in the literature and in the above discussion that SEM and EDXA can play an important role in documenting particulate exposure in humans. Increasing experience with methods of specimen preparation and analytical methods which overcome some of the limitations of SEM and EDXA can be expected to lead to rapid growth of this field of investigation.

Acknowledgements

I thank Mrs. C. Lloyd for painstaking manuscript preparation, Ms. N. Tyler and Mrs. J. Douglass provided excellent technical assistance. Partially supported by USPHS Grant HL 19619. Some of the cases studied are part of the Averill A. Liebow Pulmonary Pathology Collection which is supported in part by the Division of Lung Diseases, NHLBI, NIH, Bethesda, Md.

References

1. J.D. Brain, D.W. Golde, G.M. Green: Biologic potential of pulmonary macrophages. Am. Rev. Resp. Dis. 118, 1978, 435-443.
2. P.B. DeNee: Collecting, handling and mounting of particles for SEM. In Scanning Electron Microscopy/1978, Vol. 1, pp. 479-486. SEM, Inc., AMF O'hare, IL, 1978.
3. R.P. Sherwin, M.L. Barman and J.L. Abraham: Silicate pneumoconiosis of farm workers. Lab. Invest. (in press)
4. A.A. Liebow: Definition and classification of interstitial pneumonias in human pathology. Prog. Resp. Res. 8, 1975, 1-33.
5. J.L. Abraham: SEM as an aid in diagnosis - An overview. In Scanning Electron Microscopy/1977, Vol. 2, pp. 119-128. I.I.T. Research Institute, Chicago, IL, 1977.
6. J.L. Abraham: Recent advances in pneumoconiosis - The pathologist's role in etiologic diagnosis. In The Lung, Williams and Wilkins, Co., Baltimore, 1978, pp. 96-137.
7. K. Pintar, A. Funahashi and K.A. Siegesmund: A diffuse form of pulmonary silicosis in foundry workers. Arch. Pathol. Lab. Med. 100, 1976, 535-538.
8. A.R. Brody, N.V. Vallyathan and J.E. Craighead: Distribution and elemental analysis of inorganic particles in pulmonary tissue. In Scanning Electron Microscopy/1976, Vol. 1, pp. 477-484. I.I.T. Research Institute, Chicago, IL, 1976.
9. A.R. Brody, D.M. Dwyer, N.V. Vallyathan: The elemental content of granulomata: Preliminary studies of pulmonary sarcoidosis and hypersensitivity pneumonitis. In Scanning Electron Microscopy/1977, Vol. 2, pp. 129-136. I.I.T. Research Institute, Chicago, IL, 1977.
10. A. Funahashi, K.A. Siegesmund, R.F. Dragen: Energy dispersive x-ray analysis in the study of pneumoconiosis. Br. J. Ind. Med. 34, 1977, 95-101.
11. J.T. Armstrong: Methods of quantitative analysis of individual microparticles with electron beam instruments. In Scanning Electron Microscopy/1978, Vol. 1, pp. 455-468. SEM, Inc., AMF O'Hare, IL, 1978.
12. B.F. Trump, I.K. Berezeski, R.E. Pendergrass: X-ray microanalysis of diffusible elements in scanning electron microscopy of biological thin sections. - Studies of pathologically altered cells. In Scanning Electron Microscopy/1978, Vol. 2, pp. 1027-1039. SEM, Inc., AMF O'Hare, IL, 1978.
13. J.M. Tormey: Validation of methods for quantitative x-ray analysis of electrolytes using frozen sections of erythrocytes. In Scanning Electron Microscopy/1978, Vol. 2, pp. 259-266. SEM, Inc., AMF O'Hare, IL, 1978.
14. P.M. Quinton: SEM-EDS x-ray analysis of fluids. In Scanning Electron Microscopy/1978, Vol. 2, pp. 391-398. SEM, Inc, AMF O'Hare, IL, 1978.
15. J.R. Coleman: X-ray analysis of biological samples. In Scanning Electron Microscopy/1978, Vol. 2, pp. 911-926. SEM, Inc., AMF O'Hare, IL, 1978.
16. J.L. Abraham and P.B. DeNee: Biomedical applications of backscattered electron imaging -- One year's experience with SEM histochemistry. In Scanning Electron Microscopy/1974, pp. 251-258. I.I.T. Research Institute, Chicago, IL, 1974.
17. O.C. Wells: Scanning Electron Microscopy, 421 pages. McGraw-Hill, New York, 1974.
18. C.A. Andersen and J.R. Hinthorne: Ion microprobe mass analyzer. Science 175, 1972, 835-860.
19. F.M. Hasan and H. Kazemi: Progress Report-U.S. Beryllium Case Registry, 1972.

Am. Rev. Resp. Dis. 108, 1973, 1252-1253.
20. W.R. Parkes Occupa. Lung Disorders. Butterworths, London, 1974.
21. P.B. DeNee: Identification and analysis of particles in biological tissue using SEM and related techniques. In Scanning Electron Microscopy/1976, Vol. 2, pp. 461-468. I.I.T. Research Institute, Chicago, IL, 1976.
22. A.M. Langer, R. Ashley, V. Baden: Identification of asbestos in human tissues. J. Occup. Med. 15, 1973, 287.
23. T. Ehrenreich, A.D. Mackler, A.M. Langer: Identification and characterization of pulmonary dust burden in pneumoconiosis. Ann. Clin. Lab. Sci. 3, 1973, 118.
24. F.D. Pooley: The use of an analytical electron microscope in the analysis of mineral dusts. Phil. Trans. R. Soc. Lond. A.286, 1977, 625-638.
25. I.B. Rubin and C.J. Maggiore: Elemental analysis of asbestos fibers by means of electron probe techniques. Environ. Health 9, 1974, 81-94.
26. E.J. Chatfield and M.J. Dillon: Some aspects of specimen preparation and limitations of precision in particulate analysis by SEM and TEM. In Scanning Electron Microscopy/1978, Vol. 1, pp. 486-496. SEM, Inc., AMF O'Hare, IL, 1978.
27. P. Echlin: Coating techniques for scanning electron microscopy and x-ray microanalysis. In Scanning Electron Microscopy/1978, Vol. 1, pp. 109-132. SEM, Inc., AMF O'Hare, IL, 1978.
28. J.A. Murphy: Noncoating techniques to render biological specimens conductive. In Scanning Electron Microscopy/1978, Vol. 2, pp. 175-193. SEM Inc., AMF O'Hare, IL, 1978.
29. A.R. Brody, N.V. Vallyathan and J.E. Craighead: Use of scanning electron microscopy and x-ray energy spectrometry to determine the elemental content of inclusions in human tissue lesions. In Scanning Electron Microscopy/1978, Vol. 2, pp. 615-622. SEM, Inc., AMF O'Hare, IL, 1978.
30. E. Kaszynski, E.O. Bernstein and R.M. Sauer: Zirconium and aluminum location in lung by Morin fluorescence and energy dispersive x-ray analysis. Bull. Soc. Pharmacol. & Environ. Pathologists 6, 1978, 8-11.
31. L.E. Stettler, D.H. Groth and G.R. Mackay: Identification of stainless steel welding fume particulates in human lung and environmental samples using electron probe microanalysis. Am. Ind. Hyg. Assoc. J. 38, 1977, 76-86.
32. M.L. Bonin and J.L. Abraham: Combined scanning electron microscopy and x-ray microanalysis: Diagnostic and investigative uses in pathology. Lab. Invest. 38, 1978, 374 (Scientific exhibit)
33. A. Odegaard, E.M. Ophus and A.M. Larsen: Identification of thorium dioxide in human liver cells by electron microscopic x-ray microanalysis. J. Clin. Pathol. 31, 1978, 893-896.
34. M. Meyer, R.M. Wagner, G. Niyawama: Thorotrast induced adhesive arachnoiditis associated with a meningioma and a Schwannoma. Hum. Pathol. 9, 1978, 366-369.
35. M. Mah and E.S. Boatman: Scanning and transmission electron microscopy of new and used asbestos cement pipe utilized in the distribution of water supplies. In Scanning Electron Microscopy/1978, Vol. 1, pp. 85-92. SEM, Inc., AMF O'Hare, IL, 1978.
36. D.R. Beaman and D.M. File: Quantitative determination of asbestos fiber concentrations. Analytical Chem. 48, 1976, 101-110.
37. W.H. Walton (ed.): Inhaled Particles IV. Pergamon Press, Oxford, 1977.
38. L.G. Luna (ed.): Manual of histologic staining methods of the Armed Forces Institute of Pathology, 3rd edition. Blakiston, New York, 1968.
39. J.D. Bancroft: Histochemical techniques, 2nd edition. Butterworths, London, 1975.
40. K.F.J. Heinrich and D.E. Newbury: Secondary ion mass spectrometry. Proc. of Workshop on Secondary Ion Mass Spectrometry and Ion Microprobe Mass Analysis. Special publication 427. National Bureau of Standards, Washington, D.C. 1975.
41. J.L. Abraham and T.A. Whatley: Ion microprobe analysis of beryllium-containing particles in situ in human lungs. Fed. Proc. 36, 1977, 1090.
42. J.L. Abraham: Microanalysis of human granulomatous lesions. Proc. of 8th International Conference on Sarcoidosis, Cardiff, Wales (in press).*
43. G.J. Rosasco and E.S. Etz: The Raman microprobe: a new analytical tool. Res. Devel. 28, 1977, 20.
44. E. Etz: Raman microprobe analysis - Principles and applications. In Scanning Electron Microscopy/1979. Johari, O. ed., SEM, Inc., AMF O'Hare, IL (In press) (these proceedings).
45. C.B. Carrington and E.A. Gaensler: Clinical pathologic approach to diffuse infiltrative lung disease. In The Lung, Williams and Wilkins, Co., Baltimore, 1978, pp. 58-87.
46. S.K. Koerner, A.J. Sakowitz, R.I. Appelman: Transbronchial lung biopsy for the diagnosis of sarcoidosis. New Engl. J. Med. 293, 1975, 268-270.
47. J.L. Abraham and R.G. Spragg: Documentation of environmental exposure using open biopsy, transbronchial biopsy and bronchopulmonary lavage in giant cell interstitial pneumonia (GIP). Am. Rev. Resp. Dis. (in press).
48. S.H. Rosen, B. Castleman and A.A. Liebow: Pulmonary alveolar proteinosis. N. Engl. J. Med. 258, 1123-1142.

---

*Contact author for more information.

49. J. Ramirez-R: Pulmonary alveolar proteinosis: treatment by massive bronchopulmonary lavage. Arch. Intern. Med. 119, 1967, 147-156.

50. H. Spencer: Pathology of the Lung, 3rd edition. Pergamon Press, Oxford, 1977.

51. D.D. McEuen and J.L. Abraham: Particulate concentrations in pulmonary alveolar proteinosis. Environ. Res. 17, 1978, 334-339.

52. J.L. Abraham, R.M. Wagner, K. Miyai: Choice of imaging modes in SEM study of respirable radiographic contrast medium: Tantalum particles in the lung. In Scanning Electron Microscopy/1976, Vol. 2, pp. 691-698. I.I.T. Research Institute, Chicago, IL, 1976.

53. G.M. Wickham, R. Rudolph and J.L. Abraham: Silicon identification in prosthesis associated fibrous capsules. Science 199, 1978, 437-439.

54. A. Cockburn, R.A. Barraco, T.A. Reyman, and W.H. Peck: Autopsy of an Egyptian mummy. Science 187, 1975, 1155-1160.

55. J. Bar-Ziv and G.M. Goldberg: Simple siliceous pneumoconiosis in Negev Bedouins. Arch. Environ. Health 29, 1974, 121-126.

56. D.A. Gedcke, J.B. Ayers and P.B. DeNee: A solid-state backscattered electron detector capable of operating at Tv scan rates. In Scanning Electron Microscopy/1978, Vol. 1, pp. 581-594. SEM, Inc., AMF O'Hare, IL, 1978.

57. D.E. Newbury: Fundamentals of scanning electron microscopy for the physicist: contrast mechanisms. In Scanning Electron Microscopy/1977, Vol. 1, pp. 553-568. I.I.T. Research Institute, Chicago, IL, 1977.

58. J.L. Abraham: Multiple imaging and microanalytical modes for biological specimens - Electron and ion beam studies. In Transactions of the 1977 International Conference on X-ray Optics and Microanalysis, D. Beaman, R. Ogilvie and D. Wittry, eds. (In press).

59. J.B. Pawley and G.L. Fisher: Using simultaneous three color x-ray mapping and digital-scan-stop for rapid elemental characterization of coal combustion by-products. J. Microsc. 110, 1977, 87-102.

60. A.A. Liebow, A. Steer and J.G. Billingsley: Desquamative interstitial pneumonia. Am. J. Med. 39, 1965, 369-404.

61: B. Corrin and A.B. Price: Electron microscopic studies in desquamative interstitial pneumonia associated with asbestos. Thorax 27, 1972, 324.

62. E.R. Weibel: Morphometry of the Human Lung. Springer-Verlag, Berlin, 1963.

63. C.E. Fiori and D.E. Newbury: Artefacts observed in energy-dispersive x-ray spectrometry in the scanning electron microscope. In Scanning Electron Microscopy/1978, Vol. 1, pp. 401-422. SEM, Inc., AMF O'Hare, IL, 1978.

64. J.L. Abraham and C. Brambilla: Particle size for differentiation between inhalation and injection pulmonary talcosis. Am. Rev. Resp. Dis. (In press).

65. J.H. Reed and E.H. Wood: Effect of body position on vertical distribution of pulmonary blood flow. J. Appl. Physiol. 28, 1970, 303-311.

66. J.S. Lawrence: Comparative toxicity of gold preparations in treatment of rheumatoid arthritis. Ann. Rheum. Dis. 35, 1976, 171-174.

67. G.B. Bluhm: The treatment of rheumatoid arthritis with gold. Semin. Arthritis Rheum. 5, 1975, 147-166.

68. R.H. Winterbauer, K.R. Wilske and R.F. Wheelis: Diffuse pulmonary injury associated with gold treatment. New Engl. J. Med. 294, 1976, 919-921.

69. H.G. Willert and M. Semlitsch: Reaction of the articular capsule to plastic and metallic wear products from joint endoprotheses. Sulzer Technical Review 57(2), 1975, 1-15. (Available from Sulzer Brothers, Ltd, CH-8401 Winterthur, Switzerland).

70. A.J. Christie, K.A. Weinberger and M. Dietrich: Silicone lymphadenopathy and synovitis. JAMA 237, 1977, 1463.

DISCUSSION WITH REVIEWERS

Reviewer I: You have pointed out the disadvantages of epon embedding but failed to point out its advantages. To balance the review please include the following: 1) TEM or STEM can prove the cell type plus the intracellular and subcellular localization of particles. Sometimes the presence of mucous, fibrin and cell debris might make a particle appear intracellular using SEM and BSE. 2) TEM provides information on the health of the cell. 3) Since plastic sections are flat, geometric problems common with a $5\mu m$ deparaffinized section are not a problem, and quantitation is more feasible. 4) Epon thin sections provide thin regions of crystals ideal for electron diffraction and x-ray microanalysis of the same particle. With a diamond knife one can cut thin sections of particles such as fly ash and show their lysosomal location (Ingram, P. and Shelburne, J.D. X-ray ultrastructural studies in cadmium coated fly ash particles. Available from the authors) 5) Very thick (e.g. $80\mu m$) epon sections can be effectively studied with a dissecting microscope to aid selection of specific areas for further analysis. 6) Epon can be removed just as with paraffin for SEM/EDXA, although it takes longer. 7) The problem of particle overlap and fluorescence can be partially solved

with thin sections. 8) It is not necessary to fix or stain with heavy metals for TEM or STEM. The inherent electron density of particles can be evaluated in unstained sections. Serial stained sections can reveal other features. When mounted on beryllium grids the only spurious peak from the sample should be the chlorine in epon. 9) Just as there is a library of paraffin embedded tissues, so to is there a library of epon embedded tissues (Shelburne, J. O. and Ingram, P. Use of epon embedded tissues for X-ray microanalysis; Available from authors). These can be especially useful if the routine fixative for light microscopy uses a heavy metal fixative, e.g. Zenker's.
Author: These are important points and welcome additions to this review. To select a method, one must consider the preparation, alteration, volume of tissue sampled in a given sample and time, and the level of ultrastructural detail required for a particular investigation.

Reviewer I: The greater sensitivity of X-ray fluorescence (XRF) over electron induced X-ray emision should be included in your review.
Author: This is adequately discussed elsewhere (Middleman and Geller, SEM/1976 1:171-176). Since XRF is unable to analyse single particles it was not discussed in this review.

Reviewer I: Are there particles which contain only lithium, silicon and oxygen and no other elements (detectable by EDXA)? If so could these be confused with SiO_2 in an EDXA study? Or is the demonstration of Si alone by EDXA fairly good evidence of SiO_2?
Author: I have not encountered such particles, but they would not be separable from SiO_2 either with EDXA or WDXA. Only SIMS or Raman spectroscopy could prove the lithium component in a particle. The finding of Si alone usually indicates SiO_2, but Si metal, silicone particles or SiC would show only Si by EDXA (text ref. 53).

Reviewer I: Can you recommend specific suppliers of carbon stubs or planchets which are free of interfering contaminants (in your experience)?
Author: This has been a frustrating problem. Many planchets or stubs which appear suitable by secondary electron imaging reveal few or many inorganic particles when examined with sensitive backscattered electron imaging. I have found none which are free from contaminants. The most common are silicon (SiO_2 or SiO used for polishing or inherent in the graphite source material), steel (Fe,Cr,Ni

used probably in cutting the discs) and tungsten (used in polishing or cutting). These are a potentially great interference in SEM analysis of particulates, and I can only recommend that a few sample discs or stubs be examined by the potential user before purchasing many. There is also variation from lot to lot.

Reviewer I: What buffers are appropriate and what buffers inappropriate for preserving calcium crystals?
Author: Generally, acid buffers will solubilize many calcium containing compounds. If one can avoid aqueous solutions altogether, such as using alcohol or acetone fixation, the water soluble crystals will be preserved. Care should also be taken to avoid floating sections on water if one wishes to preserve soluble crystals. Other workers such as yourself have recently reviewed cryotechniques and the use of complexing salts to preserve soluble components in tissues and cells.

Reviewers II & V: Considering the range of instruments and analytical capabilities available with electron beam instruments, would you still recommend the use of a scanning microscope and energy dispersive x-ray spectometry techniques for the analysis of particles in tissues? Would you recommend existing alternatives? That is, if you had an analytical electron microscope available for use, in which scanning-transmission microscopy, selected area electron diffraction, and energy dispersive analysis was available (with appropriate programs for "spectrum stripping" etc.), would you still see advantages in your present techniques?
Author: These important questions emphasize that the SEM/EDXA approach is resolution limited, as I mentioned in the text. It is a fact that most laboratories do not have all state of the art techniques available simultaneously, consequently, each lab tends to develop the available methods to best advantage and, to a degree, to consider the disadvantages of less available techniques. This has been especially true of earlier SEM's with poorer resolution and unavailable solid state backscattered electron detectors. The optimum system and approach to these studies has yet to be established. I am aware of no study in which samples of a group of tissue or other specimens were divided and distributed to various laboratories utilizing varying techniques. Such a cooperative study would be welcome and would serve both a quality control and optimizing function.
In answer to the specifics of the question I feel SEM/EDXA offers the best first approach to non-destructive analysis of particulates in tissues because of the ease of specimen preparation, the ability

to correlate particle with pathology in situ, the relative volume of tissue sampled (5μmx1cmx1cm) compared with TEM (0.1 μmx0.5mmx0.5mm) in a section, and the chance to examine intact rather than sectioned particles. I do not think any one instrument answers all questions, and one should seek other information whenever the most available instrumentation is not the best for the particular problem. Obviously, all your own pioneering asbestos studies would not have gone far if only SEM/EDXA had been available.

Reviewer III: In what percentage of cases would you estimate that SEM and EDXA give significant etiological clues to the disease rather than just confirming associations of particles with lesions?
Author: This would depend on how cases were selected for analysis. In my experience (limited to approximately 100-200 cases) I would estimate that more than 50% of cases suggest etiologic agents. These are best confirmed by careful history taking and subsequently (if possible) by animal studies. The problem with the latter is that many of the diseases in humans take decades to develop or manifest symptoms and result from impure exposures.

Reviewer III: Do you think that wavelength spectrometry can be used to quantitate levels of certain elements in particles of rather uniform size and shape?
Author: Both EDXA and WDXA can quantitate major elements in particles of uniform size and shape (text ref. 11). However, the only such particle populations of which I am aware are those artificially produced. It is exceedingly unlikely that human exposures to such particles will be found.

Reviewer IV: Which techniques have proved to be legally acceptable in court cases? Without such indications an investigator may select quite the wrong technique and waste everyone's time. Would you list references discussing legal acceptability of tests for some important particulates?
Author: I have had limited experience with the courts but the forensic sciences workshop at this symposium should contain some information. See also text references 6, 11 and 23.

Reviewer V: Does the author believe that every university medical center should have an SEM with EDXA?
Author: The equipment should be secondary to the personnel. If the university has interested medical scientists then I believe they can make productive use of the equipment. I do not believe the reverse is necessarily true.

Reviewer V: Discuss standards for this type of work.
Author: The selection and use of standards for particulate analysis has been discussed by others (see text references 11, 15, 26, 63). Theoretically, since particles found in tissues vary in size, one would need standards of the same size for the most accurate quantitative analysis. This is circumvented by some rather elaborate computations (ref. 11). More of a problem than standards are controls for the study of particulates in human tissues. The multiple variables make exact control mearly impossible (see ref. 51). The best way at present to document a given particulate exposure is to find the same particles (composition and size) in both the tissue and environmental samples (see ref. 31).

Reviewers: I   J. Shelburne
             II   A. Langer
             III   A. Brody
             IV   D. F. Parsons
             V   B. F. Trump

SCANNING ELECTRON MICROSCOPY/1979/II
SEM Inc., AMF O'Hare, IL 60666, USA

TECHNIQUES FOR QUANTITATIVE ORGANIC ANALYSIS IN MICRODROPLETS

R. Beeuwkes III

Department of Physiology
Harvard Medical School
25 Shattuck Street
Boston, MA 02115

Abstract

Techniques are described for the chemical analysis of urea by means of X-ray microanalysis. A selective labelling reagent, thioxanthen-9-ol, is used to precipitate urea together with a sulfur label. Reagents are added to and removed from microdroplets of 50 picoliter volume by means of oil-water partition. The sensitivity of the method is less than 1 picomole. The possibility of extension of these techniques to other analyses is discussed.

KEY WORDS: X-ray Microanalysis, Urea, Microdroplets, Partition Techniques.

Introduction

Electron probe X-ray microanalysis has been routinely used for more than 10 years for elemental characterization of aqueous droplets of less than 10^{-10} liter volume.[1-3] Presently, such specimens usually consist of droplets of fluid obtained in renal micropuncture experiments. However, studies of the secretion of small glands such as sweat glands or the secretory organs of insects may provide other areas of application. In addition, the ability to analyze such small droplets may allow analysis of the fluid surrounding single cells cultured in droplets under oil. More recently, techniques have been developed for the quantitative X-ray microanalysis of an organic compound, urea, by the application of chemical techniques to microdroplets.[4-6] Urea, containing only carbon, hydrogen, oxygen and nitrogen, is not elementally distinguishable from other organic compounds. However, by means of a chemical precipitation procedure, it may be selectively and quantitatively linked with sulfur in an insoluble precipitate, and this sulfur may be readily measured. Although developed for the analysis of one particular compound, certain methodological features of this technique may prove useful for ultra-micro analytical chemistry in general, and may extend the range of applicability of X-ray microanalysis.

Assuming that an appropriate labelling reaction scheme can be devised, most of the problems arising in microdroplet chemical analysis derive from the difficulty of performing chemical manipulations in specimens of 50 picoliter volume. Although specimens and standards may be pipetted onto a supporting surface using conventional microdroplet techniques, means must be found for the quantitative addition of reagents and buffer to each of many microdroplets; for ensuring the formation of a precipitate of small crystal size; and for the removal of unreacted labelling reagent at the completion of the reaction. In the method to be described, reagent addition and removal is achieved by partition between aqueous micro-droplets and the protective oil surrounding them. By precoating silicon sample supports with a thin layer of the unlabelled form of the reaction product, multiple nucleation sites are provided for the growth of product microcrystals.

Method

During both chemical and analytical steps, specimens are supported on wafers of pure silicon (Semiconductor Processing, Inc., Hingham, MA). Such wafers are extremely pure, low in cost, non-toxic, and have appropriate wettability. To ensure uniform surface properties, silicon wafers are first placed in 70% nitric acid for 10 minutes, then thoroughly rinsed in distilled water. If this step is omitted, microdroplets may subsequently fail to adhere to the surface. After drying on lint-free cloth, wafers are marked with a diamond scribe in a grid pattern to aid specimen relocation. Silicon dust is then removed by wiping with a lint-free cloth wet with acetone, and the wafer again allowed to dry. Then the wafers are briefly heated to bright red color in a Bunsen flame to remove any organic residue. This step apparently also provides a controlled amount of oxide formation which promotes proper sample adhesion.

Wafers so prepared are coated with a thin layer of unlabelled product microcrystals (di-xanthydrol urea, Fig. 1) by means of a simple atomizer (Fig. 2A). A dilute solution of this compound in solvent is forced by means of a syringe pump through the tip of a capillary tube drawn to a coarse point. A jet of air is directed across this point by means of a Pasteur pipette. The resulting spray is made visible by means of a strong light from a fiber optic bundle. The wafer is held about 4 inches from the atomizer and moved about rapidly to form an even coating. Good microcrystals show a blue interference color.

Specimens and standards are deposited under oil onto the silicon surface by means of ultra micropipettes of classic form[7] prepared by drawing capillary tubing in an electrode puller adjusted to form a segment about 2 cm long of nearly parallel wall tubing, with an inside diameter of 20-40 microns. The tip is squared by nipping off jagged edges with extra fine forceps. Using a loop of 0.002 inch platinum wire connected to an adjustable low voltage power supply, a region behind the tip is heated to form a constriction, and the tip is partially closed. Volumes are measured between the tip and the midpoint of the constriction. Finished pipettes are flushed with distilled water, ethanol and chloroform, and silicone coated. A length of polyethylene tubing connects the pipette to a syringe filled with air, and the pipette is mounted in a holder having an O-ring compression clamp (Brinkman).

The prepared silicon wafer is placed under water saturated paraffin oil (Fisher O-119) in a shallow dish formed by a bead of silicone rubber on the surface of a microscope slide (Fig. 2B). Droplets of standards and unknowns are drawn up to the constriction of the micropipette and followed by oil spacers. Thus many specimens may be contained within the straight segment behind the tip. These are then deposited in a regular array on the silicon surface (Fig. 2C). Because the droplets have an affinity for the metallic surface, they assume roughly hemispherical form. This procedure is performed under a stereomicroscope at a magnification of 64. Because both standards and unknowns are measured with the same volumetric pipette, precise calibration of the volume is not necessary.

After pipetting is complete, the wafer carrying the specimens is placed in oil containing the specific reagent, thioxanthen-9-ol, methanol, and acetic acid for pH adjustment. When the oil containing these reagents comes in contact with an aqueous droplet, the reagents diffuse into the droplet until a stable equilibrium is reached, defined by the oil-water partition coefficient of each. Thus it is not necessary to add reagents to each drop individually, and the concentration in each droplet is identical. As the reaction proceeds, thioxanthen-9-ol is precipitated and more reagent diffuses into the drop so as to maintain the equilibrium. Because the pool of reagent in the oil is effectively infinite, it is impossible to deplete the reagent source during the reaction. The reaction is allowed to go to completion during a waiting period of 60 minutes, during which the sample support is maintained at 30°C in a closed chamber.

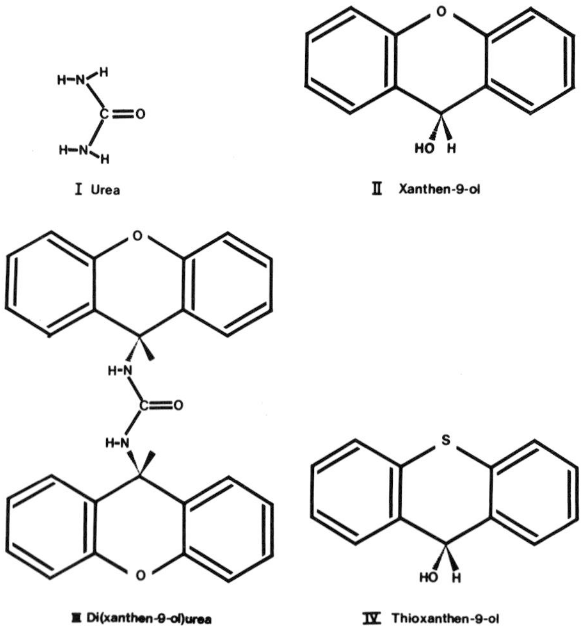

Figure 1. Structural formulae of compounds used in this analytical procedure. Urea (I) lacks a distinctive elemental label. Xanthen-9-ol (II) reacts with urea to form the insoluble complex di-xanthen-9-ol urea (III). By substitution of sulfur for oxygen, thioxanthen-9-ol (IV) is formed. This compound forms an insoluble complex with urea and labels each urea molecule with two sulfur atoms.

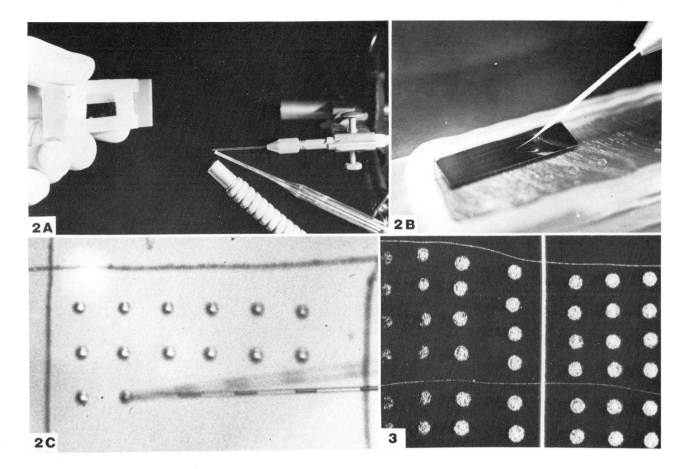

2A

2B

2C

3

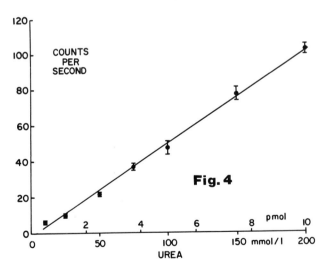

Fig. 4

COUNTS
PER
SECOND

pmol

UREA

mmol/l

Figure 2A. A silicon wafer, mounted to a microscope slide with tape, is coated with microcrystals
of unlabelled reaction product by means of a simple
atomizer to the right. A coarse micropipette fed by
a syringe pump delivers solutions at a controlled
rate into a jet of air emerging from a Pasteur
pipette. A fiber optic light source (below) makes
the atomized stream visible.

Figure 2B. The precoated silicon wafer is placed
under oil on a microscope slide. Droplets of standards and unknowns are placed on its surface by
means of an ultra micropipette.

Figure 2C. Droplets, here of about 50 picoliter
volume, are placed in a regular array on the silicon surface. Successive specimens are held within
the bore of the pipette and are separated by oil.

Figure 3. After removal of the covering oil and
unreacted reagent the reaction product appears as
patches of small crystals, each about 100 μm in
diameter (dark field illumination).[6]

Figure 4. Sulfur counting rate observed during
X-ray microanalysis of the reaction products shown
in figure 3. Each point represents the mean ± SD
of 6 replicate specimens of a single standard
solution. The line was calculated by linear regression. The picomole scale is based on a pipette
volume of 50 picoliters.[6]

During the reaction, microcrystals may be
observed to form within the droplets. The reaction conditions employed must be adjusted to
yield crystals of very small size so as to avoid
problems with beam penetration and X-ray absorption. After reaction the partition technique is
again used in order to remove excess reagent.
By rinsing the sample support in reagent-free
organic solvent, reagents partition outward to
an effectively infinite compartment and are
washed away. In order to perform this step
without displacing the droplets, the silicon

wafer is picked up by a forceps, lowered into isopentane, and gently agitated. It is then lifted into the air above the rinse container and fresh isopentane is allowed to flow over its surface from a pipette while the slide is held nearly horizontal. The wafer is then tilted, allowing the isopentane to drain towards an unused area, and dried in air. The reaction product then consists of small patches (about 100 microns diameter) of very fine crystals (about 1 to 2 microns thick) (Fig. 3). Because the reaction procedure has required no exposure to aqueous solvents, these crystals also contain the salts (Na, Cl, etc.) present in the original specimens. For this reason the product is hygroscopic and should be stored in a desiccator until analyzed.

Electron probe X-ray microanalysis has been performed using a Cameca MS 46 instrument and a 20 KeV, 200 nA beam, defocussed to cover the largest specimen spot. Crystal spectrometers were used; tuned on the reaction product itself. Because of the presence of the silicon support, high resolution is necessary. (If an energy dispersive spectrometer is used, it appears necessary to form the product on carbon supports. On such a support reasonable peaks have been obtained for S, Cl, and K.[5]) Each spot was counted for 10 seconds, although for greater accuracy longer counting time may be used. No change in counting rates for any of the elements studied was observed in repeated counts made over a period of 100 seconds. Background was defined on-peak, off-specimen. Sulfur counting rate (background subtracted) was found to be quite linear between urea concentrations of 10 to 200 millimoles per liter. Between 400 and 1000 millimoles per liter the slope of the standard curve decreases by a factor of two.[5] The signal was equal to background at 10 millimolar concentration, corresponding to 0.5×10^{-12} moles of urea in a 50 picoliter specimen (Fig. 4). Simultaneous analysis of sodium, potassium, and chlorine, together with urea has been shown to be practical.[6]

Potential sources of error in this technique include the presence of significant quantities of sulfur in the solution to be analyzed, possible loss of the unknown compound into the covering oil, and absorption of low energy X-rays (i.e. from sodium) in the precipitate during simultaneous analysis. Renal collecting duct fluid, for which this assay was developed, may contain about 20 millimolar sulfur together with 5-50 times as much urea. If error from this source is expected, duplicate analyses may be performed on reacted and unreacted specimens. Although urea is often regarded as oil soluble, experiments have shown that even after 2 hours under oil, no loss of urea can be detected.[6] Finally, in order to avoid artifacts due to X-ray absorption, standard solutions must be used whose total composition closely approximates that of the unknowns.

Some features of this analytical technique may be useful in the development of methods for other compounds. The labelling reagent employed was modified from a compound long known in the literature[8] by the substitution of sulfur for oxygen in a non-functional site. Non-functional oxygens are common in organic reagents and may often be replaced with sulfur or selenium by relatively simple synthetic procedures. The partition technique has wide potential applicability. Many organic reagents have appropriate partition coefficients or can be given them by substitutions of non-polar or polar groups. Organic acids or bases may be used for buffering. In the technique described, the partition technique allowed simultaneous addition and removal of thioxanthen-9-ol, acetic acid, and methanol to hundreds of droplets simultaneously. And, because of the equilibrium nature of the partition process, reagent concentrations remain constant within the droplets during the reaction. It is by virtue of the partition technique that this analytic procedure requires fewer manipulations than other techniques involving much larger samples.[9, 10] Because ionic constituents are neither added nor removed, simultaneous multi-element analysis is possible.

Acknowledgement

This work was supported by the National Institute of Arthritis and Metabolic Diseases (grant AM 18249), NIH Research Career Development Award AM 00224, and a gift from R. J. Reynolds Industries. The author acknowledges the important contributions of J. M. Amberg and L. Essandoh to the development of this technique, and thanks S. Gilbert and the personnel of the Biotechnology Resource in Electron Probe Microanalysis, Harvard Medical School, for their assistance.

References

1. M. J. Ingram and C. A. M. Hogben, "Electrolyte analysis of biological fluids with the electron microprobe." Anal. Biochem. 18, 1967, 54-57

2. F. Morel and N. Roinel, "Application de la microsonde electronique a l'analyse elementaire quantitative d'echantillons liquides d'un volume inferieur a 10^{-9} L." J. Chim. Phys. 66, 1969, 1084-1091.

3. C. Lechene, "Electron probe microanalysis of picoliter liquid samples." In Microprobe Analysis as Applied to Cells and Tissues. T. Hall, P. Echlin, and R. Kaufmann, editors. London, New York, Academic Press, 1974, pp. 351-368.

4. R. Beeuwkes III, J. M. Amberg, and L. Essandoh, "Urea and electrolyte measurement in picoliter specimens." Federation Proc. 36(3), 1977, 605.

5. R. Beeuwkes, "Quantitative organic analysis by electron probe: Urea." In 35th Ann. Proc. Electron Microscopy Soc. Am., Boston, 1977, G.W. Bailey, editor. pp. 358-361.
6. R. Beeuwkes III, J. M. Amberg, and L. Essandoh, "Urea measurement by X-ray micro-analysis in 50 picoliter specimens." Kidney Int. 12, 1977, 438-442.
7. P. de Fonbrune, Technique de Micromanipulation. Paris, Masson et Cie, Editeurs, 1949.
8. R. Fosse, "Origine et distribution de l'Uree dans la nature: Application de nouvelles methodes d'analyse de l'uree, basees sur l'emploi du xanthydrol." Ann. Chim. 6, 1916, 13-95.
9. F. Roch-Ramel, "An enzymatic and fluorophotometric method for estimating urea concentrations in nanoliter specimens. Anal. Biochem. 21, 196/, 373-381.
10. D. J. Marsh, C. Frasier, and J. Decter, "Measurement of urea concentrations in nanoliter specimens of renal tubular fluid and capillary blood." Anal. Biochem. 11, 1965, 73-80.

Discussion with Reviewers

J.B. Pawley: You show a calibration curve but no results on actual biological samples. Could you include such results along with analyses of the ionic compounds present and relevant bulk analyses for comparison?

Author: We have reported direct comparison of the electron probe technique and the standard photometric urea analysis in text reference 6. The data below for urea in urine are from that report.

Probe, mM	Photometer, mM	Difference %
25 + 3	25	0
36 ∓ 3	37	-3
72 ∓ 8	68	+6
117 ∓ 3	104	+13
133 ∓ 5	124	+7

J.B. Pawley: It would seem that you could increase your sensitivity still further if you were to deposit your drops on a thin substrate such as a formvar film possibly coated with SiO or SiO_2 to improve adherence (e.g.: Quinton SEM/78, II, 391 or Stratham and Pawley SEM/78, I, 469). Do you see any serious disadvantages to such a procedure, and, if not, do you plan to try it? Would this modification not make it possible to use EDS to get useful results?

R. Rick: The sensitivity of the method for detecting urea is sufficiently high to enable physiological concentrations of this compound to be determined in urine. However, for samples of other biological fluids or for other organic substances the sensitivity may be inadequate and might impose limitations on the applicability of this technique. Do you think, that thin films or Be supports could be employed for improving the sensitivity?

Author: The only serious disadvantage is that such sample supports are much more difficult to handle than the solid silicon wafer. We have tried it and found that the major difficulty results from the mechanical weakness of the thin film. Beryllium wets well, but we avoid using it due to its toxicity and chemical reactivity. Solid carbon supports have wetting problems but potentially offer lower background together with good mechanical strength. For our purposes 10^{-12} molar sensitivity was sufficient, so we have not followed this up.

J.B. Pawley: Would this technique work with even smaller droplets? What would limit it ultimately?

Author: Presumably yes. We can pipette 5 picoliter droplets and they react. The fact that water is very slightly soluble in the covering oil seems likely to define the ultimate limit.

N. Roinel: The sensitivity of the method is limited by the use of silicon holders which have a higher continuum background than beryllium holders. In the former case, did you try to improve the peak to background ratio for the sulfur signal by decreasing your beam high voltage? This would both increase the slope of the response curve and decrease the background. Or, are the crystals in your samples too thick to be fully excited by a high voltage of 15 kV? What is the average value of sample mass thickness of the more concentrated standard samples? Do you think that you could reduce the crystal size by a rapid freezing followed by lyophilization?

Author: As stated above, we felt that 10^{-12} moles sensitivity was sufficient for our purposes and have not made extensive efforts to improve sensitivity. At 21 KeV we feel secure that the samples are fully excited, given their estimated mass thickness of 1 to 4 times 10^{-4} g/cm^2. Because the crystals are formed as a precipitate during the reaction phase, their size is based upon the number of nucleation sites available and the reaction conditions. Thus, the crystal size cannot be reduced by a change in the postprecipitation handling procedures.

N. Roinel: How do you explain that the curve relating the intensity of sulfur signal to the urea concentration in figure 4 does not seem to pass through the origin? If this is so, is the reason that the background cannot be estimated accurately enough and in this case is overestimated, or that the precipitation reaction requires some minimum amount, thus concentration, of urea to exceed the solubility product?

Author: With the short counting times ordinarily employed neither the signal nor the background is very accurately defined in the lower concentration range. However, we have noted that product does not seem to be formed at concentrations lower than 5 mM urea. This obviously limits the application of the technique

to relatively concentrated solutions. It is
reasonable to suppose that alteration of the
reaction conditions might improve this lower
level of sensitivity.

J.D. Kuptsis: You stated what material the
technique was developed to analyze but did you
actually perform these analyses? If so what
were the results?
Author: This analytical technique is now being
used routinely to measure urea, sodium, potassium, and chlorine in 5×10^{-12} L specimens of
ice from frozen sections. This work is still in
progress, and only a very preliminary note has
been published (R. Beeuwkes and L. B. Kinter,
"The micropunch: A new method for study of
medullary tubular fluid composition." Abstracts
VIIth International Congress of Nephrology,
Montreal, Q4, 1978).

J.D. Kuptsis: You stated that you defocussed
the beam to cover the largest spot. Since a
crystal spectrometer was used was there any
problem with part of the area being outside X-ray
focus?
Author: The beam was defocussed to a size not
exceeding 100 microns diameter. Given the take-
off angle and Rowland circle radius, this would
result in a relative wavelength off-set of less
than 5×10^{-4} at the very edge of the spot.

J.D. Shelburne: What other organic compounds -
especially those of possible clinical relevance
- could be (or have been) quantitatively analyzed in microdroplets using similar procedures?
Please speculate.
Author: To my knowledge no other organic compound is presently being analyzed in microdroplets by chemical techniques. To speculate, one
might possibly measure reducing sugars such as
glucose by adaptation of the Fehling copper
reduction technique or measure protein or amino
acids such as tyrosine by iodination or other
labelling reactions. By adaptation of histochemical precipitation techniques, one might perform in droplets many of the reactions now
conducted in tissue sections. We have not had
time to explore these, but would be pleased to
assist others wherever we have relevant
experience.

SCANNING ELECTRON MICROSCOPY/1979/II
SEM Inc., AMF O'Hare, IL 60666, USA

ELECTRON PROBE STUDIES OF Na+-K+-ATPase

S. Rosen and R. Beeuwkes[*]

Department of Pathology
Beth Israel Hospital
330 Brookline Avenue
Boston, MA 02215

[*]Department of Physiology
Harvard Medical School
25 Shattuck Street
Boston, MA 02215

Abstract

Use of appropriate reaction conditions allows direct quantitative measurements of the activity of Na-K-ATPase to be made with the electron probe microanalyzer. In additon, the demonstration by microanalysis of the linearity of the complete cytochemical reaction sequence is the first detailed analysis of the elemental sequence in a histochemical method and provides the basis for a visual semi-quantitative analysis of enzyme activity in tissue sections. This technique indicates the potential of electron probe microanalysis as a basic tool in enzymology.

Introduction

Hale[1] was the first investigator who applied the electron probe to the identification of cytochemical reaction products, utilizing X-ray elemental maps to depict the alkaline phosphatase ion sequence. A similar type of study was done by Engel[2] who examined the elements in the intermediate step of a method used to demonstrate EDTA-activated adenosine triphosphatase. In tissues from patients with Wilson's disease[3] the presence of lysosomal reaction product, Pb, in acid phosphatase preparations was correlated with the occurrence of Cu. Rosen[4] studied the surface topography of turtle bladder mucosa in conjuction with electron probe analysis of such tissue incubated for the demonstration of carbonic anhydrase activity. None of these studies did fulfill the promise that Hale[1] foresaw for electron probe analysis in histochemistry: "It cannot be used to study the rate of deposition of reaction products directly,but may provide a means of doing so by analysis of a series of sections prepared under controlled circumstances." However, the methods for the cytochemical study of sodium-potassium activated adenosine triphosphatase (Na-K-ATPase) all contain probe detectable elements, and a recent technique [5,6] provides appropriate circumstances for <u>quantitation</u> of enzyme reaction product.

Na-K-ATPase is thought to play a fundamental role in the handling of sodium and potassium in the kidney but it has been established that its activity shows wide variations throughout the renal parenchyma.[7] Because an understanding of the tubular location of the enzyme is critical for the elucidation of its role, multiple histochemical approaches have been attempted, and such studies of the cytochemical localization of ATPase are devisable into two phases. The first, primarily dealing with Mg-activated ATPase, stems from Wachstein and Meisel's work in 1957.[8] The substrate was ATP, lead was included in the incubation medium to stabilize product phosphate, and the ensuing compound was converted to visible lead sulfide by exposure to ammonium sulfide. However, the validity of this procedure has been questioned by many investigators for reasons that include lead-induced

KEY WORDS: Enzyme, Histochemistry, ATPase, Electron Probe

non-enzymatic catalysis and direct enzyme inactivation[9, 10] and those modifications used to demonstrate Na-K-ATPase have been subjected to similar criticism.

In 1972, Ernst[10, 11] proposed a new approach, the "second" phase of ATPase histochemistry, in which phosphate was the substrate and strontium the phosphate stabilizing ion. This method was based on the understanding that the hydrolysis of ATP by Na-K-ATPase occurs in two phases: an initial Mg and Na dependent phosphorylation of the phosphorylated intermediate. P-nitrophenyl phosphate (PNPP) can serve as phosphate donor in this enzyme system. As would be expected if the PNPPase were part of the ATPase system, PNPP competitively inhibits ATP driven transport[12] and its hydrolysis is both potassium-dependent and ouabain-sensitive. Thus Ernst utilized PNPP as a substrate for the sodium-potassium activated ATPase system. Although there was less PNPPase inhibition by strontium than found with ATPase and lead, the need for a phosphate stabilizing ion in the incubation medium still resulted in partial inactivation of the system under study. Furthermore, the PNPPase activity demonstrated by Ernst[11] was substantially but not completely inhibited by ouabain addition or potassium deletion.

The basis for a major improvement in the Ernst technique was established by the biochemical studies of Albers and Koval.[13] These workers observed that the pH optimum of the K-dependent, ouabain-sensitive PNPPase was shifted from neutral to 9.0 by addition of dimethylsulfoxide (DMSO) to the incubation medium, and that in such media the PNPPase was selectively activated. When Guth and Albers[5] used these conditions for the histochemical demonstration of ATPase activity, they found the initial product phosphate remained within the tissue without the need for a foreign stabilizing ion. The activation ions of the enzyme double as primary phosphate capture ions.[6] This reaction was both potassium-dependent and ouabain-sensitive. A dark visible product could then be formed by subsequent coupling of the insoluble phosphate with cobalt, and later formation of cobalt sulfide.

Methods

Section preparation

These studies were performed in kidney tisuse from humans, rabbits, and rats.[6, 14] Blocks 3-4 mm thick, frozen by immersion in liquid nitrogen, were mounted using Tissue-Tek and sections of 10 μm thickness cut in an International cryostat at -20°C. Sections were picked up on room temperature glass slides or pure silicon wafers and incubated for periods ranging from 2 minutes to 3 hours at 37°C in medium prepared according to Guth and Albers.[5] The medium contained 30 mM KCl, 5 mm $MgCl_2$, 5 mM P-nitrophenyl phosphate (PNPP), and 25% DMSO in 70 mM 2-amino-2-methyl-1-propanol buffer (2-2-1) adjusted to pH 9 with hydrochloric acid. Following incubation, slices were either dehydrated immediately for electron probe analysis

or, for formation of visible product, placed in 2% cobalt chloride for 5 minutes, rinsed briefly in distilled water, washed in 3 changes of buffer for 30 seconds and placed in 0.6% ammonium sulfide for 3 minutes. Sections were then washed in distilled water, dehydrated and mounted.

Electron probe analysis

For electron probe analysis, sections were picked up on pure silicon substrates (7 x 24 mm) obtained from Semiconductor Processing, Inc., Hingham, Massachusetts at relatively low cost allowing them to be used on disposable basis. THe resistivity of the silicon was low (from 0.1 to 25 Ohm - cm), preventing charging effects, and the highly polished surface of the silicon supports permitted optical observation using vertical illumination. The sections mounted on silicon wafers, stored in a desiccator, may be reanalyzed at any time since the reaction product is stable.

The sections were analyzed in a Cameca type MS46 electron probe microanalyzer. This instrument employs four fully focussing spectrometers of 250 mm Rowland circle radius, all of which may be used simultaneously. An accelerating voltage of 20 KV, beam current of 50 nA, and beam diameter of 5 um were employed. Spectrometer crystals were potassium acid phthalate (KAP) for phosphorus and magnesium, pentaerythritol (PET) for sulfur and potassium and lithium fluoride (LIF) for cobalt. Probe operating conditions were standardized by obtaining a phosphorus counting rate of 4000 counts per second (cps) on a fluorapatite crystal (Tousimis Research Inc.) at the beginning of each run. The large sample area of the instrument permitted up to 16 kidney sections to be accommodated at one time together with the standard, thus all of the sections required for any experimental group were introduced into the instrument at one time. Only sections free from tearing artifacts and showing uniformly good adhesion to the sample support were analyzed. Because the electron probe includes an optical microscope whose objective is co-axial with the electron beam, tissue areas for analysis were selected by optical observation. For consistency, only thick ascending limb segments located within 100 μM of the junction between inner and outer medulla were analyzed. X-ray counts for four elements were recorded directly from the counters by an Hewlett Packard 2100 computer and sample current was continuously recorded within the computer so that corrections could be made if necessary. Data obtained from the electron probe are generally presented directly as counting rates, without background subtraction, since the peak to background ratios were very high. Peak to background ratios were determined by detuning each spectrometer above and below the line of interest and interpolating to find the background beneath each peak.

Results

Optical observations

High activity is present in the medullary

thick ascending limbs, less but still prominent activity in the distal convoluted tubule, moderate activity in cortical thick ascending limbs and very low activity in proximal tubules.[6] High magnification study in 4 micron thick sections shows a predominately basal product distribution (Fig. 1).

X-ray microanalysis

Measurements of Product phosphorus

Sections dehydrated directly after incubation showed high levels of phosphorus in cells of outer medullary thick ascending limbs (Fig. 2). Counting rates on the order of 500 to 1000 counts per seconds were recorded and these were stable during exposure to the electron beam. The peak to background ratio observed at a counting rate of 500 per second was typically more than 100 to 1, and interference from the peaks of silicon or sulfur was not detectable. The phosphorus signal was found to vary quite widely from point to point within a given tubule and from one tubule to another. Accordingly, for estimates of mean enzyme product, measurements were generally made of approximately 50 to 100 thick ascending limbs in each section. The product increased linearly with incubation time up to 10 minutes, the incubation period used for probe analysis; the phosphorus signal at two hours was about twice that observed at this 10 minute period. Detailed studies of this reaction product indicate that it is composed of a mixture of magnesium phosphate, $Mg_3(PO_4)_2$ and potassium magnesium phosphate, $KMgPO_4$, each together with water of hydration.[6] Analysis of sections incubated in medium without substrate showed that about 6% of the phosphorus could be attributed to the endogenous sources.

Reproducibility

The reproducibility of the method was studied by analyzing groups of serial sections prepared under identical conditions. Table 1

Table 1

SECTION	C.P.S. $\pm$ S.D.
1	544 $\pm$ 133
2	469 $\pm$ 112
3	526 $\pm$ 132
4	598 $\pm$ 174
5	606 $\pm$ 151
6	563 $\pm$ 168
7	573 $\pm$ 128
8	473 $\pm$ 123
9	473 $\pm$ 132
10	470 $\pm$ 106
11	560 $\pm$ 148
All	536 $\pm$ 146

Measurements of product phosphorus (mean $\pm$ standard deviation) obtained by electron probe analysis of 49 or 50 thick ascending limbs in each of 11 identically prepared sections. The mean and standard deviation of all 546 measurements treated as a single group appears as the last entry in the table. Incubation time: 5 minutes (modified from 6).

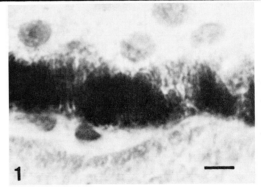

Figure 1

Micrograph showing localization of optically visible histochemical product in kidney. A high magnification photograph of segment of a rabbit distal tubule shows activity apparently associated with basal membranes (counterstain: Masson trichrome; 1 cm = 5 μm).

Figure 2

A silicon mounted section of rat kidney dehydrated immediately after incubation. Thick ascending limbs at the transition between inner and outer medulla contain phosphate product which appears dark under incident illumination. Black carbon deposits identify tubule regions subjected to electron probe microanalysis. The carbon deposits are larger than the analyzed regions (5 micron diameter) which lie at the center of each spot. Incubation time: 10 min (1 cm = 20 μm).

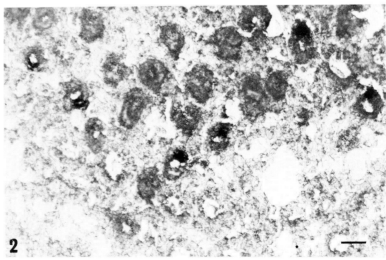

shows the result of a typical experiment including a group of 11 sections, each incubated for 5 minutes. The single section means, each based on 49 to 50 measurements, were found to lie within one half of one standard deviation of the grand mean. The entire group of measurements appeared normally distributed, with 70% of the values lying within one standard deviation of the grand mean. These observations indicate that for many experiments, measurements made on single sections will be sufficiently precise.

Ouabain sensitivity and potassium dependence

Sections incubated in the presence of 1 and 5 millimolar ouabain showed 70 and 90% inhibition of product formation respectively. When potassium or substrate was deleted from the incubation medium, the counting rate fell to 6% of control. The peak to background ratio in the phosphorus channel remained approximately 10:1 even at these low counting levels. The analysis of larger numbers of ouabain concentrations allows the estimation of K_I (50% activity) (Fig. 3). Each point is based on comparison of product accumulation in 50 tubular segments of control and treated sections. The K_I under these conditions in the rat and rabbit is 1 mM and 0.05 mM respectively. Both the range and difference in sensitivity of rat and rabbit corresponds with the known sensitivity of kidneys from these species using microsomal preparations. Studies are now underway in which various diuretics are being analyzed for their effects on transport ATPase.[15]

Visible product formation

Simultaneous electron probe measurements, made of cobalt and phosphorus at 100 points in each of 10 sections dehydrated after cobalt treatment of the phosphate product, revealed a linear relationship without loss of product phosphorus. No cobalt was lost during ammonium sulfide treatment for visible product formation. Simultaneous measurements of cobalt and sulfur in sections dehydrated at this stage also showed a linear relationship with loss of half of the original product phosphorus. The stoichiometric relationship observed between cobalt and phosphorus after incubation and cobalt treatment, the retention of cobalt during sulfide treatment, and the final relation between sulfur and cobalt, indicate that the amount of visible product formed was linearly related to the original enzymatic product. Furthermore, the continuity of spatial relationship[14] during the reaction sequence can be also shown by X-ray elemental mapping techniques (Fig. 4). Thus, we may conclude that both in the spatial distribution and quantity the visible product accurately reflects the initial enzymatic reaction.

Acknowledgements

This research was supported by NIH grants RRO5479 (S. Rosen), RRO0679, HL15552, AM18249, and HLO2493 (R. Beeuwkes).

References

1. J. A. Hale, "Identification of cytochemical reaction products by scanning X-ray emission microanalysis." J. Cell. Biol. 15, 1962, 427-435.

2. W. K. Engel, J. S. Resnick, and E. Martin, "The electron probe in enzyme histochemistry." J. Histochem. Cytochem. 16, 1968, 273-275.

3. S. Goldfischer and J. Moskal, "Electron probe analysis of liver in Wilson's disease." Am. J. Pathol. 48, 1966, 305-316.

4. S. Rosen, "Surface topography and electron probe analysis of carbonic anhydrase-containing cells in the turtle bladder mucosa." J. Histochem. Cytochem. 20, 1972, 548-551.

5. L. Guth and R. W. Albers, "Histochemical demonstration of (Na^+-K^+)-activated adenosine triphosphatase." J. Histochem. Cytochem. 22, 1974, 320-326.

6. R. Beeuwkes III and S. Rosen,, "Renal sodium-potassium adenosine triphosphatase: optical localization and X-ray microanalysis." J. Histochem. Cytochem. 23. 1975, 828-839.

7. U. Schmidt and U. C. Dubach, "Activity of (Na^+K^+)-stimulated adenosine triphosphatase in the rat nephron." Pfluegers Arch. 306, 1969, 219-226.

8. M. Wachstein and E. Meisel, "Histochemistry of hepatic phosphatases at a physiologic pH. With special reference to the demonstration of bile canaliculi." Am. J. Clin. Pathol. 27, 1957, 13-23.

9. H. L. Moses and A. S. Rosenthal, "Pitfalls in the use of lead ion for histochemical localization of nucleoside phosphatase." J. Histochem. Cytochem. 16, 1968, 530-539.

10. S. A. Ernst, "Transport adenosine triphosphatase cytochemistry. I. Biochemical characterization of cytochemical medium for the ultrastructural localization of ouabain-sensitive, potassium dependent phosphatase activity in the avian salt gland." J. Histochem. Cytochem. 20, 1972, 13-22.

11. S. A. Ernst, "Transport adenosine triphosphatase cytochemistry. II. Cytochemical localization of ouabain-sensitive, potassium dependent phosphatase activity in the secretory epithelium of the avian salt gland." J. Histochem. Cytochem. 20, 1972, 23-38.

12. P. J. Garrahan and A. F. Rega, "Membrane phosphatase and active transport of cations.: Nature (New Biol), 232, 1971, 24-25.

13. R. W. Albers and G. J. Koval, "Sodium-potassium-activated adenosine triphosphatase. VII. Concurrent inhibition of Na^+-K^+-adenosine triphosphatase and activation of K^+-nitrophenylphosphatase activities." J. Biol. Chem. 247, 1972, 3088-3092.

14. S. Rosen and R. Beeuwkes III, "Renal Na-K-ATPase: Quantitative X-ray microanalysis." In 1st International Workshop on Biological Electron Probe Microanalysis, 1977. C. Lechene, editor (in press).

15. R. Beeuwkes III, J. Shahood, M. Chirba and S. Rosen, "Renal Na-K-ATPase inhibition by ethacrynic acid and ouabain: optical and electron probe analysis. Abstracts 22nd International Congress of Physiological Sciences, Paris, 1977.

See p778 for Discussion With Reviewers.

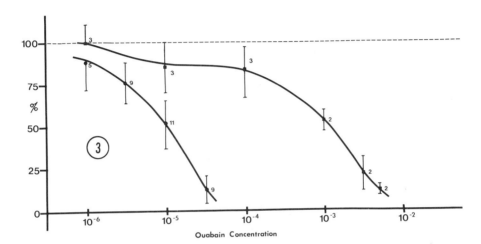

Figure 3
    Ouabain dose-response curves in rat and rabbit. Quantitative electron probe microanalysis of phosphate product in thick ascending limbs after incubation of sections in ouabain-containing media showed the rabbit enzyme (left) to be much more sensitive to ouabain than the rat (right). The K_I calculated for rabbit is 10^{-5} M and for rat about 10^{-3} M. These values are comparable with those observed for Na-K-ATPase in tissue homogenates.

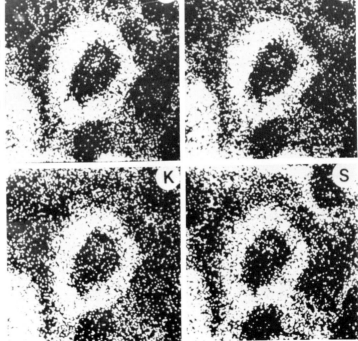

Figure 4
    X-ray element maps (15,000 counts/element). Thick ascending limb in section dehydrated immediately after incubation to demonstrate K-dependent ouabain-sensitive transport ATPase. The highly insoluable initial product is a mixture of $Mg_3(PO_4)_2$ and $KMgPO_4$, each with water of hydration. Thus the enzyme activating ions serve as capture ions.

Discussion with Reviewers

S.L. Kimzey: What governed the selection of a 5 micron probe diameter? Was this the smallest size possible under your operating conditions?
Authors: The 5 micron diameter spot was chosen to cover the full height of basal infoldings of the cells of the thick ascending limb. The spot size can be made smaller without difficulty, but there is no advantage in making it much less than the section thickness because of beam spreading.

S.L. Kimzey: You used no background correction method for contamination or the presence of nonspecific elements. Do you not consider this necessary?
Authors: With the crystal spectrometer used, overlap of elemental peaks does not occur. Experiments were performed in which background was defined by detuning the spectrometer and no contribution from contamination or non-specific elements was found. Since the background from all sources was less than 10% even at very low counting rates, we feel that such corrections are unnecessary.

R.J. Barrnett: If the analyzed regions are 5 microns in diameter (and accepting the counts recorded for P), what were the P counts in similar cells incubated without substrate? What is the rate of precipitation of $Mg_3(PO_4)_2$ in comparison to the rate of production? If $Mg_3(PO_4)_2$ precipitates at predilected sites, efforts at localization by counts are at best brave.
Authors: As stated in the text, the P count (reported in more detail in text reference 6) in which thick ascending limb cells incubated without substrate was 6% of that in cells incubated with substrate. Given that the solubility product of $Mg_3(PO_4)_2$ is 8.0×10^{-24}, it appears unlikely that PO_4 radicals would diffuse far before combining with Mg.

R.J. Barrnett: Are you trying to indicate that under the circumstances of your experiment that after incubation and washing, etc., followed by bath in 2% $CoCl_2$, that Co reacted only with Pi produced from PNPP? No loss of product phosphorus is only half the information. Where is the control that no cobalt binds to cell moieties in the absence of precipitates derived from Pi?
Authors: The quantitation by X-ray microanalysis is performed on precipitates not subjected to washing or Co treatment. As is true of all histochemical techniques involving heavy metals there is a low level of non-specific tissue binding particularly to nuclear PO_4.

J.W. Mills: Can the "quantitative" aspect of this study be refined so that differences in product formation between tubule segments can be directly related to some value of enzyme specific activity.
Authors: The method as used defines relative activity in a standardized volume defined by section thickness and beam diameter. These measurements could be considered analogous to relative activity per unit wet weight. It is conceivable that some estimate of protein content within the analyzed region could be made on the basis of electron probe analysis. Direct measurement of real specific activity, i.e., activity per mg protein, could only be made by complicated microdissection and ultramicrochemical techniques (see text reference 7).

J.W. Mills: Is there a known minimum level of enzyme activity below which this technique will not reveal the presence of Na^+-K^+-ATPase?
Authors: The minimal level of Na^+K^+ATPase activity detectable with this technique has not as of yet been defined. We have focussed our efforts on cells with high activity.

SCANNING ELECTRON MICROSCOPY/1979/II
SEM Inc., AMF O'Hare, IL 60666, USA

X-RAY MICROANALYSIS OF PYROANTIMONATE-PRECIPITABLE CATIONS

J. A. V. Simson, H. L. Bank* and S. S. Spicer*

Department of Anatomy
*Department of Pathology
Medical University of South Carolina
Charleston, SC 29403

Abstract

Modifications of the Komnick potassium (pyro)antimonate precipitation method have been widely used for the subcellular localization of a variety of cations. The identity of cations precipitated with this method has often been controversial, and it is therefore important to establish definitive criteria for identifying precipitated cations *in situ*. In the present study, we have precipitated antimonate salts *in vitro* and examined the salts both in crystalline powder form and after embedment in Epon, using energy-dispersive X-ray microanalysis, in an attempt to identify biologically important antimonate-precipitable cations. We have found that the cations sodium, magnesium, and calcium, if present in physiological concentrations, will precipitate antimonate under "standard" conditions (2.5% antimonate, pH 7.2 - 7.4, in the presence of 1% OsO_4). Characteristic X-ray emissions were observed for sodium and magnesium, as well as for lead similarly precipitated, but $L\alpha$ and $L\beta$ emission peaks from antimony interfered with calcium identification and necessitated complex computerized deconvolution or peak stripping to determine the presence of a calcium peak. Precipitates of sodium did not contain appreciable potassium, whereas variable amounts of potassium were present in precipitates of calcium and lead, depending upon the extent of washing prior to dehydration. Sizeable potassium peaks were consistently present in even well washed magnesium precipitates. X-ray spectra of standardized precipitates were found useful as an aid in interpreting the more complicated spectra obtained from tissue samples.

KEY WORDS: Pyroantimonate, X-ray Microanalysis, Nucleolus, Cations Microanalysis

Introduction

Cation cytochemistry and cell function

The critical importance of local ambient cation concentration in the functioning of cellular enzymes, as well as in the integrity and activity of cytoskeletal and cytomuscular systems is well known. Understanding the role of cations in cell function requires knowledge of cation distribution within the cell and its organelles. However, the subcellular localizations of cation pools and the redistribution patterns of cations that occur in stimulus-response coupling are poorly understood.[1] An exception is the striated muscle cell in which a sarcoplasmic reticulum calcium pool that is released upon cell membrane depolarization has been convincingly demonstrated.[2]

It is not yet clear to what extent tissue cations captured with electron microscopic cytochemical techniques reflect the distribution or the quantities found in living cells. Difficulties in determining the subcellular distribution of cations by morphological or biochemical methods include: (1) loss of many tissue cations during routine preparative procedures for electron microscopy;[3] and (2) extensive redistribution of all but the most tightly bound cations occurs during cell fractionation, making routine biochemical localization procedures impractical.[4]

Several studies have shown that the pyroantimonate method for cation precipitation *in situ* yields heavy cytochemical deposits in selected cellular sites, and have indicated that physiologically-induced cation redistribution can be detected with this method.[5-9] Identification of precipitated cations is particularly important to an understanding of the role of cation redistribution in modulating cellular physiology. To date, no systematic study has been undertaken applying microanalytical techniques to antimonate salts of known composition. The present study utilizes energy-dispersive X-ray microanalysis in the identification of cations in precipitates produced *in vitro* with pyroantimonate.

The antimonate method for subcellular cation
localization

The pyroantimonate (antimonate) anion forms
an insoluble sodium salt and has been used in
analytical chemistry for the precipitation of
sodium from solution after the removal of other
cations. Since the antimony in this anion has
an atomic weight of 122, Komnick and Komnick[10,11]
considered it a good candidate for fine
structural cytochemistry, and utilized an
antimonate-containing fixative in an attempt to
determine the subcellular distribution of sodium
in herring-gull salt gland. They suggested that
other inorganic cations such as Ca^{2+} and Mg^{2+}
would precipitate antimonate, but concluded that
sodium was probably responsible for most of the
tissue precipitates in salt-water-adapted glands
since it was present in 100-fold greater concentration than the other potentially precipitating
cations.[10,11]

The pyroantimonate technique has been
extensively utilized with the intent of localizing sodium fine-structurally.[12-29] However,
concern about the specificity of the technique
arose when it was unequivocally demonstrated that
several additional cations, including Mg^{2+}, Ca^{2+}
and H^+, readily precipitated antimonate.[30-38]
Thus, the ability of the antimonate technique to
precipitate and localize sodium selectively was
challenged. More recently, the antimonate
method has been used primarily to localize
sites of calcium accumulation, for example, in
muscle,[35,39-47] secretory cells,[4,6,48-51] and
other known sites of calcium concentration.[52,53]
It is clear, however, that several tissue cations
other than calcium are also precipitated by
antimonate in biological samples.[54-56]

The variable precipitability of different
cations with antimonate, and the effects of
buffers and pH on cation precipitability have
been elucidated in *in vitro* model systems.[31,32,
57] In a series of precipitation experiments,
we have demonstrated that the major ions forming
insoluble precipitates with either buffered or
unbuffered antimonate were Na^+, Ca^{2+}, and
Mg^{2+}.[1,57] The formation of precipitates with
potassium was highly pH-dependent, and these
precipitates, either in the test tube or in
tissues, were readily dissolved by adequate
rinsing with distilled water (pH > 7.0).
Exposure of potassium pyroantimonate solutions
to either alcohol or acidic pH results in the
formation of antimonate precipitates. On the
other hand, both 0.1 M phosphate and 0.02 M EGTA
inhibit the *in vitro* precipitation of Na^+ and
Ca^{2+} with antimonate. Inclusion of certain
buffers in the pyroantimonate-containing
fixative solution has been found to alter the
density and distribution of antimonate deposits
in tissue samples.[57] It is now clear that the
source of much of the variability in precipitate
distribution reported with the pyroantimonate
procedure has been the frequent failure of
investigators to employ a standardized method or
to recognize the effects of methodological
variants of technique on the pattern of cation
localization.

X-ray microanalysis

The pyroantimonate anion precipitates a
wide range of cations found in biological tissue
and hence has been useful for preservation of
elemental ions which are generally eluted during
routine preparative procedures of tissue samples
for electron microscopy.[58]

Wave-length dispersive techniques. Most
analyses have been performed using wave-length
dispersive techniques. The most commonly
occurring cellular cations, including Na, K, Ca,
and Mg,[33,48,59-61] were detected. Many recent
analyses have focused primarily upon calcium
localization because of its emerging importance
in the stimulus-response coupling of contractile
and secretory cells.[6,7,46,62] Other recent
studies have explored the precipitation of less
commonly occurring cations such as zinc[54,55,63]
and cadmium.[64,65]

Energy-dispersive analysis. An alternate
approach for detecting spectra of X-rays emitted
from small areas of tissue bombarded with
electrons utilizes solid-state detectors and
multichannel analyzer systems for energy-dispersive analysis. These analytical systems can
be adapted to scanning and transmission electron
microscopes widely utilized for elemental
analysis. Energy-dispersive spectroscopy
has been used in a few studies for analysis
of pyroantimonate-precipitable cations,[44,49,
51-53] and in these, calcium has been
reported as the major cation associated
with antimonate precipitates. Several difficulties exist, however, in the analysis and
interpretation of data from tissue containing
antimonate-precipitable cations using energy-
dispersive techniques.[58] Characteristic
emissions by elements such as sodium and
magnesium may be obscured by the high background
(Bremsstrahlung) in this low energy region
(< 2 KEV). Secondly, resolution available with
energy-dispersive detectors is limited ($\sim$ 160 EV
at 5.9 KEV) compared with < 5 EV for wave-length
detectors.[66] Therefore, overlapping peaks such
as those of manganese (Mn, $K\alpha,\beta$) vs. iron (Fe,
$K\alpha,\beta$) or osmium (Os, $M\alpha$) vs. phosphorus (P, $K\alpha$)
cannot be resolved without computerized
processing. Of particular concern in this
regard is the fact that calcium $K\alpha$ and $K\beta$ peaks
(3.69 and 4.02 KEV) are obscured by the major
antimony ($L\alpha$ and $L\beta$) emission peaks (3.60, 3.84
and 4.10 KEV), thus complicating identification
of this important cation.

Another limitation on the utility of
techniques such as energy-dispersive analysis,
for the determination of elemental composition
of tissue samples, is the spatial resolution
of the system. Despite the fact that modern
scanning electron microscopes, operated in the
STEM mode, are capable of focusing on an $\sim$ 5 nm
spot point on the sample, the effective spatial
resolution depends not on the area of illumination but rather the area of excitation of the
sample. In general, the area of excitation is
equal to about half the section thickness.

In order to achieve reasonable counting statistics, it is desirable to utilize relatively thick sections (150 - 200 nm); therefore, the minimum effective resolution would be between 75 and 100 nm. The ionic composition of areas adjacent to the illuminated area cannot be ignored in the analysis of chemical deposits such as antimonate precipitates.

The approach taken in the present study has been to examine antimonate precipitates by X-ray microanalytical techniques in a model system in which the precipitated cation and the buffer composition were known. These *in vitro* results were used as a reference to aid in the interpretation of the complex spectra obtained from precipitates in tissue samples fixed in the presence of pyroantimonate.

Materials and Methods

In vitro studies

In vitro precipitation of cations was performed as described previously[57] with serial dilutions of cations in order to determine whether or not the original cation concentration influenced the final composition of the precipitate. A 2.5% pyroantimonate (Fisher, lots #730550 and #730312) solution with 1% OsO_4 was used. Precipitates were formed by addition of salt to antimonate solutions with or without additives commonly used for electron microscopy (*e.g.*, 0.1 M phosphate, collidine or sucrose). All reactions were performed in acid cleaned, distilled water-soaked tubes; 0.5 ml of salt solution (NaCl; $MgCl_2$; $CaCl_2$; Pb acetate, Fisher reagent grade, in distilled water) was added to 4.5 ml of the antimonate-containing "fixative" solution. The final concentrations of the salts were: sodium, 10^{-1} M, 10^{-2} M and 10^{-3} M; magnesium, 10^{-2} M, 10^{-3} M and 10^{-4} M; calcium, 10^{-2} M, 10^{-3} M and 10^{-4} M; and lead, 10^{-2} M, 10^{-3} M and 10^{-4} M. Precipitated salts were rinsed twice with distilled water (at least 10 min per rinse), then dehydrated through the graded series of alcohols employed for dehydrating tissue specimens. Some samples were decanted carefully prior to dehydration, then dehydrated with ethyl alcohol (beginning with 50%) without prior rinsing in distilled water (unwashed samples; see Table 1).

Powder samples

After dehydration, a portion of the precipitate was either air dried or lyophilized and then attached to a spectroscopically pure carbon planchette with carbon dag for analysis of the crystalline samples. Both the carbon planchette and the carbon dag were analyzed prior to use to insure that they would make a minimal contribution to the spectrum. The samples were analyzed in a Coates and Welter Model #106 electron microscope operated at 12 KEV. The emitted X-ray photons were detected with a cryogenically cooled Nuclear Electronics Corporation, silicon-drifted lithium detector. For energy-dispersive analysis, the signals were amplified, shaped, and digitized in an analog to digital converter (Tracor

Northern Model #623); the resulting amplitude histogram was stored in a Northern Scientific NS 880 analyzer. Subsequent data manipulation, including background subtraction and quantitative analysis, utilized a PDP-11/05 computer. Background subtraction utilized a nonlinear Spline procedure to approximate the continuum (Forst, Bank and Lam, in preparation). After background subtraction, the resulting spectra over the energy range of 0-10 KEV were plotted on a Hewlett Packard XY recorder. Since the precipitated crystal surfaces were irregular, it was impossible to calculate the precise takeoff angle of the X-rays. In order to minimize the effect of tilt angle, the samples were analyzed at low magnification (250 diameters), and the tilt angle was altered from 45° to 85° approximately every 150 seconds. The samples were counted for 500 seconds with a beam current of approximately 20 μa. Under these conditions counting rates of 500 to 1500 cps were obtained.

We utilized two methods to determine if calcium was present in samples known or suspected to be calcium antimonate precipitates. In the most straightforward type of analysis, reference spectra were accumulated for sodium antimonate precipitates. The reference file was then utilized to mathematically "strip" away the pure antimonate spectrum from the spectrum of "calcium antimonate" precipitates after normalizing the reference spectrum to the observed antimony Lα peak heights. The residual spectrum constituted the calcium contribution to the spectrum. The second method of deconvolution of calcium from antimony utilized a Simplex optimization procedure to estimate the heights, means, and standard deviation of the peaks (Lam, Forst and Bank, Applied Spectroscopy, in press). From initial estimates of the peak locations, heights, and standard deviations for the overlapping Gaussian curves, the Simplex procedure iteratively adjusts those parameters to find the best fit of a series of Gaussian curves. Although this procedure is somewhat indirect, the results typically yield a residual error of < 5%.

Epon-embedded samples

Portions of the washed salt precipitates were dehydrated through ethanol and xylene, and placed in the bottom of BEEM #2 embedding capsules in a slurry of Epon 812. The salts were covered with Epon and the blocks oven-hardened, sectioned with an ultramicrotome by glass knives at a thickness of about 0.1 to 0.2 μm and recovered on plastic grids (Fullum #2550). Grids containing sections of Epon-embedded antimonate deposits were inserted in a specially designed transmission holder and observed in the Coates and Welter microscope operated in the STEM mode. The precipitates were observed at 80,000 diameters in order to minimize the contribution of the epoxy resin to the spectrum. In those samples which contained small flocculent precipitates, several separated areas were analyzed for 150 to 200 seconds each, and the results were summed to obtain a composite spectrum. To

Table 1

Weights of *In Vitro*-Precipitated Antimonate Salts:
Effects of Buffer and Washing

Cation: Concentration and Washing[2]		Calculated Weight (mg)[1]	Buffer or Additive (0.1 M)				
			None	Sucrose	Cacodylate	Collidine	Phosphate
Na$^+$ (10^{-1} M)	Washed: Exp. A	127.5	78.4	73.0	-	85.2	NP
	Unwashed: Exp. A		103.4	67.7	-	99.2	NP
	Exp. B		-	NP	118.7	120.8	45.2
Mg^{2+} (10^{-2} M)	Washed: Exp. A	33.2	22.4	17.3	-	36.3	35.9
	Unwashed: Exp. A		25.8	21.3	-	40.3	48.0
	Exp. B		-	22.5	81.7	37.9	72.6
Ca^{2+} (10^{-2} M)	Washed: Exp. A	34.0	26.2	24.3	-	27.8	21.1
	Unwashed: Exp. A		41.7	41.3	-	45.2	60.5
	Exp. B		-	58.1	91.8	52.3	103.9
Pb^{2+} (10^{-2} M)	Washed: Exp. A	42.3	26.6	23.9	-	39.9	48.4
	Unwashed: Exp. A		46.9	55.9	-	50.6	53.5
	Exp. B		-	53.1	77.8	69.0	90.0

NP = No Precipitate

[1]Based on a cation:antimonate stoichiometry of 1:1 for Na and 1:2 for the divalent cations. The final volume of solution was 5 ml; the cation concentration is indicated.
[2]Experiment A: Precipitates were allowed to settle 1½ hours, then centrifuged and the supernatants were carefully decanted prior to ethanol dehydration and lyophilization (unwashed precipitates). Dehydrated, lyophilized precipitates were resuspended in 5 ml distilled water (2 x 20 min @) prior to addition of ethanol and air-drying (washed precipitates). Experiment B: Precipitates were allowed to settle overnight in the tube and the supernatant was removed carefully with a pipet prior to ethanol dehydration and air-drying (unwashed precipitates).

facilitate analysis and increase the counting rate, the detector was inclined 7 degrees from the horizontal.

Tissue samples

Tissue samples were fixed for one hour in 2.5% pyroantimonate and 1% OsO$_4$ (unbuffered, pH 7.2 - 7.4). Tissues were dehydrated, embedded, sectioned with a diamond knife, and picked up on plastic grids. These were analyzed as described for the embedded, *in vitro*-precipitated crystals. The major precipitates analyzed were those from nucleoli and cytoplasm of rat trigeminal neurons, which have been previously demonstrated to possess abundant antimonate-precipitable cation.[67,68]

Data analysis

After correcting for background, peaks were identified according to K, L, or M lines of any of the elements known or suspected to be present in solutions used in these studies. Background-subtracted peak height counts were then recorded for all detectable peaks. The ratio of the peak height of the element of interest vs. the antimony Lα peak was used to normalize the data between spectra. These figures were expressed as specific element:Sb Lα peak height ratios (*e.g.*, Na (Kα,β):Sb (Lα) peak heights).

Weight determinations (Table 1)

Two experiments were performed to obtain approximate weights of precipitated antimonate salts in order to assess, with greater accuracy than has been previously obtained, the extent of cation precipitation in the presence of buffers and the effect of washing on the quantity of precipitate formed. In one experiment (B), precipitated antimonate salts were allowed to settle overnight, and the supernatant was carefully removed by pipet. The precipitates were then dehydrated directly through graded alcohols, beginning with 50% and ending with absolute ethanol. The contents of the tubes were air dried and weighed. In the other experiment (A), precipitates were allowed to settle for 1½ hours, then centrifuged, and the supernatant was carefully decanted. These precipitates were dehydrated as above, lyophilized and weighed. The above salts are referred to as "unwashed precipitates." In experiment (A), the precipitates were then resuspended in 5 ml distilled water for 20 minutes, centrifuged and the supernatant decanted. This step was repeated once and the remaining precipitates were air dried and weighed. These salts are referred to as "washed precipitates." The weights of precipitated salts were compared with weights calculated from the final cation concentration in 5 ml of solution, assuming precipitation of all cation present by an excess of antimonate ions.

Results

Precipitation studies

Precipitates were formed when a 2.5% antimonate and 1% OsO_4 "fixative" solution (pH 7.4 to 7.6) contained the following salts: NaCl ($\geq 10^{-2}$ M); $MgCl_2$ ($\geq 10^{-4}$ M); $CaCl_2$ ($\geq 10^{-4}$ M); and $Pb(C_2H_3O_2)_2$ ($\geq 10^{-4}$ M). Sodium formed precipitates which readily settled to the bottom of the tube, whereas the other (divalent) cations formed copious, flocculent precipitates which required centrifugation in order to decant the liquor without loss of precipitate. Certain buffers and additives, when present in the fixative solution, inhibited the precipitation of selected cations with antimonate.[57] Weights of washed precipitates were generally somewhat less than the calculated weights (Table 1). In the presence of phosphate the amounts of precipitated Na^+ and Ca^{2+} salts were substantially decreased. Sucrose sometimes also appeared to inhibit sodium precipitation with antimonate. Unwashed precipitates frequently were heavier than calculated, presumably because additional potassium pyroantimonate was coprecipitated in the presence of alcohol during the initial dehydration step, confirming observations by Torack and LaValle[32] that alcohol alone can precipitate potassium pyroantimonate.

Analysis of unembedded precipitates

Residual potassium was present in precipitated salts of divalent cations that had not been well washed, but was either not detectable or present near background levels in precipitates of sodium (Table 2, Fig. 1). This confirmed other evidence, shown in Table 1, that

little or no potassium was trapped in precipitates of sodium pyroantimonate. On the other hand, magnesium precipitates consistently exhibited a high potassium peak (Fig. 2), whether or not the precipitates were washed. Spectra from calcium precipitates revealed variable amounts of potassium, depending upon the extent of washing (Fig. 3). One feature of calcium spectra which was notable and quite consistent was the diminished trough between the antimony $L\alpha$ and $L\beta$ peaks. Lead precipitates exhibited a clearly identifiable lead $M\alpha$ peak (Fig. 4).

The relative amount of sodium in the sodium antimonate precipitates, as determined by Na ($K\alpha,\beta$):Sb ($L\alpha$) peak height ratio, was similar whether or not the precipitates were washed thoroughly (Table 3). The peak ratios of specific elements to antimony ($L\alpha$) were generally greater in washed precipitates of divalent cations than in unwashed precipitates. In our system, it was not possible to determine directly the relative amount of calcium present in the calcium pyroantimonate precipitates, since the antimony $L\alpha$, $L\beta_1$, and $L\beta_2$ emissions (3.60 - 4.10 KEV) completely overlapped the calcium emission peaks (3.69 - 4.01 KEV).

In an effort to find some reproducible criterion for determining whether or not calcium was present in antimonate-containing precipitates, we calculated ratios of antimony $L\alpha$:$L\beta$ peak heights (Table 4) on the assumption that a calcium contribution would affect this ratio, since calcium emissions should be additive to the observed antimony peaks. In general, these ratios averaged between 1.9 and 2.5 with calcium antimonate exhibiting slightly higher $L\alpha$:$L\beta$ ratios than salts of the other cations. However, Mg and Pb also frequently exhibited high Sb $L\alpha$:$L\beta$ ratios. Moreover, potassium pyroantimonate alone gave a high $L\alpha$:$L\beta$ peak height ratio (Table 5 and Fig. 5). Because the potassium $K\beta$ peak (3.59 KEV) is inseparable from the antimony $L\alpha_1$ peak (3.60 KEV) with our energy-dispersive detection system, we postulated that some of the high Sb ($L\alpha$):Sb ($L\beta$) peak height ratios were the result of a significant contribution in the K ($K\beta$) emission, particularly in the unwashed precipitates and the potassium pyroantimonate salt. In order to substantiate this supposition, a correlation curve was plotted comparing K ($K\alpha$):Sb ($L\alpha$) peak height ratios against Sb ($L\alpha$):Sb ($L\beta$) peak height ratios (Fig. 6). A high positive correlation (0.92) was found between these two parameters for the cations Na, Mg and Pb. However, two of the calcium antimonate data points deviated significantly from the curve in having high Sb ($L\alpha$):Sb ($L\beta$) peak height ratios but little or no detectable potassium. Thus, a high Sb ($L\alpha$):Sb ($L\beta$) peak height ratio may be indicative of calcium, but *only in the absence of coprecipitated potassium*.

Analysis of Epon-embedded precipitates

These analyses were performed on Epon-embedded precipitates, sectioned and mounted on plastic grids and lightly carbon-coated. In general, the analyses of embedded, thin-sectioned precipitates correlated well with data obtained from analyses of the dried, mounted powder (Tables 2 through 6; Figs. 1-4). The most notable features of the analyses were as follows. (1) Sodium was readily detectable in sodium-antimonate precipitates but not in precipitates of divalent salts. (2) Potassium was scarce or non-detectable in sodium-antimonate precipitates. (3) Calcium and lead precipitates generally exhibited a lower potassium to antimony ratio than did magnesium precipitates (Table 2). (4) Magnesium peaks were detectable in virtually all of the samples analyzed with the exception of a few sodium antimonate samples (Table 6). (5) The ratios of the Mg Kα to the Sb Lα peak heights generally ranged between 0.05 and 0.20, with magnesium antimonate salts generally having Mg/Sb peak height ratios above 0.10. However, overlap with ratios from other precipitates (especially Pb) precluded the establishment of criteria for the unequivocal presence of "endogenous" magnesium in the precipitates. This observation could, however, indicate contamination with magnesium of salts used for precipitation or of other reagents used for processing of the precipitates. An alternative explanation for the presence of the magnesium peak is that it is an artifact of the system resulting from an escape peak of the potassium (Kα) emission, since the difference between the energies of K (Kα) and Mg (Kα,β) is about 1.7 KEV, the energy differential at which an escape peak would be predicated. (6) Hydrogen

Figure 1. X-ray spectra (background subtracted) of SODIUM-antimonate precipitates. (*1a*) Powder from antimonate-osmium solution containing 0.1 M sucrose and 0.1 M NaCl. These were washed precipitates. Note sizeable sodium peak (Na) and virtual absence of potassium peak (K). A silicon peak (Si) is a contaminant of all the spectra. The two major peaks are the antimony Lα (Sbα) and antimony L β (Sbβ) peaks. (*1b*) Powder from antimonate-osmium solution containing 0.1 M collidine and 0.1 M NaCl. Despite the fact that these were unwashed salts, the features are very similar to those of the previous spectrum, that is, a sizeable sodium peak (Na) with little or no potassium (K) present. (*1c*) Epon section of sodium antimonate precipitated from an antimonate-osmium solution containing 0.1 M collidine and 0.01 M NaCl. Although there is a small peak in the region of aluminum (Al) as well as silicon, the essential features of the spectrum, *i.e.*, substantial sodium (Na) and barely detectable potassium (K), are the same as those of the powders.

Figure 2. X-ray spectra (background subtracted) of MAGNESIUM-antimonate precipitates. (*2a*) Powder from antimonate-osmium solution containing 0.1 M sucrose and 0.01 M MgCl$_2$. These were unwashed precipitates. Two features of this spectrum should be noted: (1) the high potassium peak (K) relative to antimony (Sb) Lα(α) and Lβ (β) peaks, and (2) the presence of a magnesium peak (Mg). (*2b*) Powder from antimonate-osmium solution containing 0.1 M phosphate and 0.01 M MgCl$_2$. Despite the fact that these were washed precipitates, the magnesium peak (Mg) is accompanied by a sizeable potassium peak (K). In addition, a distinct phosphorus peak (P) is present in this sample. (*2c*) Epon section of magnesium antimonate precipitated from an antimonate-osmium solution containing 0.1 M phosphate and 0.02 M MgCl$_2$. As in the other spectra from sections (*e.g.*, Fig. 1c), aluminum (Al) and chlorine (Cl) are detectable contaminants. As with the spectra from powder precipitates (Figs. 2a, b), magnesium (Mg) is also present, as is a sizeable potassium peak (K).

Figure 3. X-ray spectra from CALCIUM-antimonate precipitates. (*3a*) Powder from antimonate-osmium solution containing 0.1 M sucrose and 0.01 M CaCl$_2$. Despite the fact that these were washed precipitates, the potassium peak (K) is present, although not as high (relatively) as was seen with magnesium precipitates (Fig. 2). Note also the high antimony Lα (α) peak relative to the Lβ (β) peak, and the shallow trough separating them. This shallow trough was characteristic of the calcium spectra (*cf.* Fig. 2). (*3b*) Powder from antimonate-osmium solution containing 0.1 M collidine and 0.01 M CaCl$_2$. This is an unwashed precipitate, showing that the potassium peak (K) is relatively higher than in the previous spectrum. (*3c*) Epon section of calcium antimonate precipitates from antimonate-osmium solution containing 0.1 M sucrose and 0.01 M CaCl$_2$. As with other sections, the low energy regions (esp. Al and Si) show contaminating peaks. There is also a sizeable chlorine (Cl) peak. In contrast to the powders, this sectioned material exhibited virtually no potassium peak, but the antimony Lα peak (α) to Lβ peak (β) is higher than the 2:1 ratio normally obtained for well-washed samples, and the trough separating them is shallow. (opposite page)

Figure 4. X-ray spectrum (background subtracted) of LEAD-antimonate precipitates from an antimonate-osmium solution containing 0.1 M sucrose and 0.01 M lead acetate. This is a spectrum from an unwashed powder, showing a sizeable potassium peak (K) as well as a sizeable lead peak (Pb).

Figure 5. X-ray spectrum (background subtracted) of a section of embedded potassium pyroantimonate (lot #730312). There is very little contribution to the spectrum of elements other than potassium (K) and antimony (Sb). Note, also, the clean separation between the antimony Lα (α) and Lβ (β) peaks.

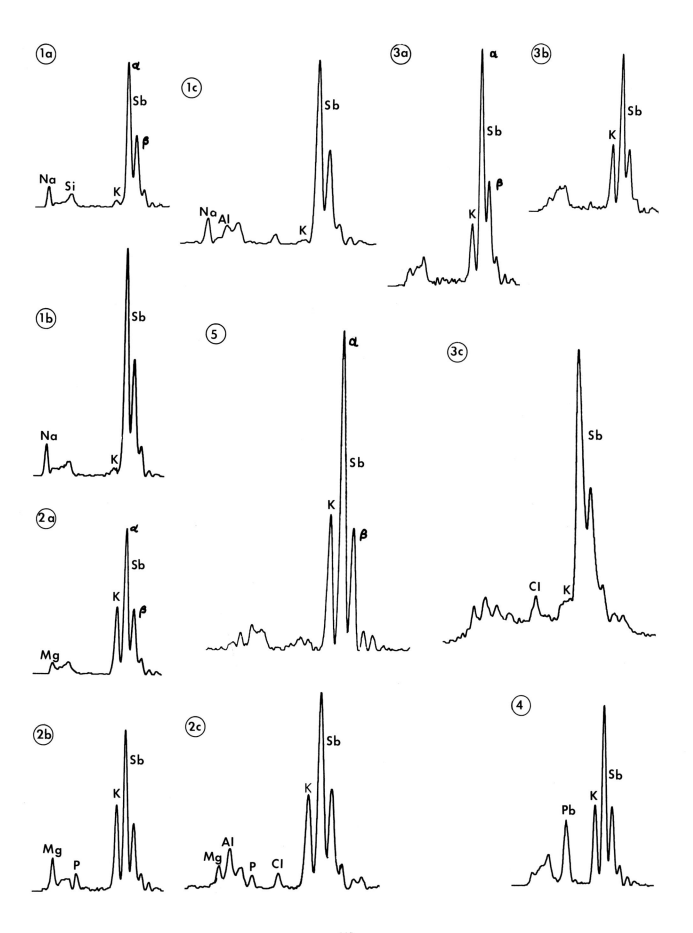

785

Table 2

Residual Potassium in Precipitates of Various Cations with Antimonate: $\dfrac{\text{Potassium }(K\alpha)}{\text{Antimony }(L\alpha)}$ Peak Height Ratios

Cation		Sucrose	Collidine	Cacodylate	Phosphate[1]
Na+	Powder: washed	-	0.05	-	-
	Powder: unwashed	0.05	0.04	-	ND
	Epon(washed)	ND	ND	ND	-
Mg2+	Powder: washed	0.11	-	-	0.53
	Powder: unwashed	0.46	0.48	-	1.27
	Epon(washed)	0.17	-	0.23	0.48
Ca2+	Powder: washed	0.26	-	-	-
	Powder: unwashed	0.55	0.43	-	1.24
	Epon(washed)	ND	0.04	0.15	-
Pb2+	Powder: washed	-	-	-	-
	Powder: unwashed	0.45	0.47	-	1.01
	Epon(washed)	-	0.07	-	-

ND = Not Detected
[1] The buffer was potassium phosphate

Table 3

$\dfrac{\text{Specific Cation}}{\text{Antimony }(L\alpha)}$ Peak Height Ratios

Cation		Sucrose	Collidine	Cacodylate	Phosphate
Na+ (Kα,β)	Powder: washed	-	0.13	-	-
	Powder: unwashed	0.15	0.11	-	0.12
	Epon(washed)	0.04	0.19 / 0.14	0.10	-
Mg2+ (Kα,β)	Powder: washed	0.15	-	-	0.20
	Powder: unwashed	0.08	0.07	-	0.08
	Epon(washed)	0.19	-	0.08	0.12
Pb2+ (Mα)	Powder: washed	-	-	-	-
	Powder: unwashed	0.36	0.23	-	0.22
	Epon(washed)	-	0.40	-	-

Table 4

$\dfrac{\text{Antimony }(L\alpha)}{\text{Antimony }(L\beta)}$ Peak Height Ratios

Cation		Sucrose	Collidine	Cacodylate	Phosphate
Na+	Powder: washed	-	1.82	-	-
	Powder: unwashed	2.0	1.94	-	1.99
	Epon (washed)	2.03	1.92 / 1.96	2.01	-
Mg2+	Powder: washed	1.96	-	-	2.40
	Powder: unwashed	2.20	2.43	-	3.01
	Epon (washed)	2.17	-	1.99	1.98
Ca2+	Powder: washed	2.26	-	-	-
	Powder: unwashed	2.49	2.49	-	3.24
	Epon (washed)	2.26	2.99	1.89	-
Pb2+	Powder: washed	-	-	-	-
	Powder: unwashed	2.28	2.16	-	2.78
	Epon (washed)	-	2.13	-	-

Table 5

Analysis of Spectra from Sections of
Epon-embedded Antimonate Salts or Precipitates

Salt	Ratios		
	$\frac{Sb\ (L\alpha)}{Sb\ (L\beta)}$	$\frac{K\ (K\alpha)}{Sb\ (L\alpha)}$	$\frac{Mg\ (K\alpha,\beta)}{Sb\ (L\alpha)}$
K-Ant.[1] (Lot 730312)	2.62	0.43	0.06
H-Ant.[2]	1.91	0.08	0.12

[1]This was embedded directly from the bottle without dehydration.
[2]Precipitated from 4.5 ml antimonate-osmium solution containing 0.1 M sucrose, by the addition of 0.5 ml glacial acetic acid.

Table 6

$\dfrac{Magnesium\ (K\alpha,\beta)}{Antimony\ (L\alpha)}$ Peak Height Ratios

		Buffer or Additive			
		Sucrose	Collidine	Cacodylate	Phosphate
Na+	Powder: washed	-	0.06	-	-
	unwashed	ND	0.04	-	ND
	Epon 10^{-1} M	-	0.09	-	-
	10^{-2} M	ND	ND	0.11	-
Mg2+	Powder: washed	0.15	-	-	0.20
	unwashed	0.08	0.07	-	0.08
	Epon	0.19	-	0.08	0.12
Ca2+	Powder: washed	0.08	-	-	-
	unwashed	0.05	0.09	-	0.08
	Epon	0.09	0.07	0.04	-
Pb2+	Powder: washed	-	-	-	-
	unwashed	0.07	0.06	-	0.10
	Epon	-	0.12	-	-

ion appears not to coprecipitate much potassium (Table 5) since the potassium to antimony peak height ratios are very low (less than 1/5 that of the pure salt). (7) Neither the buffers nor sucrose seemed to have much influence on the ratios of the peak heights of various elements to the antimony $L\alpha$ peak height. However, a phosphorus peak was always present in antimonate salts precipitated in the presence of phosphate buffer (Fig. 2b, c), suggesting some phosphate coprecipitation.

Thin-sectioned tissue samples

The tissue examined initially with X-ray microanalysis was trigeminal ganglion. The nucleoli of trigeminal neurons possessed abundant antimonate precipitates (Fig. 7) when fixed in an unbuffered osmium tetroxide fixative containing potassium pyroantimonate (pH 7.4).[67,68] The precipitates in nucleoli of trigeminal neurons exhibited high antimony counts but very low potassium (Fig. 7). A sodium peak was also present in this spectrum. Both of these observations indicate that at least part of the precipitate at this site is probably sodium antimonate. Another interesting peak observed in this spectrum was tentatively identified as iron, although manganese has not been ruled out. Iron has

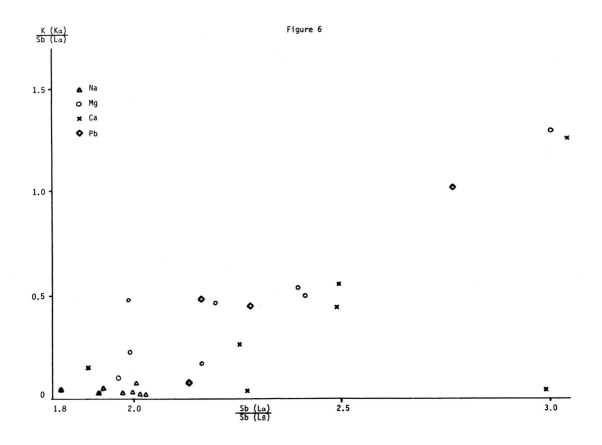

Figure 6. Comparison of relative potassium concentration, expressed as K (Kα):Sb (Lα) versus antimony peak height ratios, expressed as Sb (Lα):Sb (Lβ). The values for sodium, magnesium and lead exhibit a high coefficient of correlation (0.92) and can be fitted to a line with an X intercept (theoretically no potassium) at about 1.9 and a slope of about 1.1.

been reported to be present in nucleoli of ashed tissue sections[69] and in plant nuclei.[70]

Studies currently underway are intended to provide data on the precipitability and X-ray spectra of iron, zinc, and manganese. Such studies will provide information on the precipitability of these biologically important cations with pyroantimonate, as well as analyses of the X-ray spectra of their antimonate salts which will supplement the data obtained in the present study for sodium, magnesium and calcium.

Discussion

By utilizing data from an *in vitro* model of the pyroantimonate cation capture technique, certain conclusions can be drawn concerning specificity of the reaction and the identification of antimonate precipitates. (1) Sodium is the most readily detected and consistently identified pyroantimonate-precipitable cation of the three common antimonate-precipitable cations (Na^+, Mg^{2+} and Ca^{2+}) despite the low counting efficiency for sodium with X-ray microanalytical systems. Two major reasons for the identifiability of sodium in pyroantimonate salts are: [a] sodium was *not* detectable in salts of antimonate precipitates with other cations and [b] other cations (*e.g.*, K and Mg) were essentially absent from sodium antimonate precipitates. (2) A magnesium peak appeared to be present in precipitates of virtually all other cations except sodium. Whether this peak is an artifact caused by a potassium escape peak or whether magnesium is a contaminant of many reagent-grade salts remains to be determined. (3) Magnesium-containing precipitates also possessed substantial coprecipitated potassium,

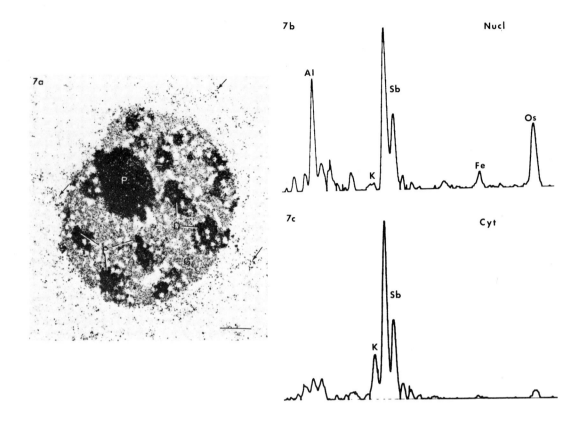

Figure 7. Rat trigeminal neuron, fixed in unbuffered antimonate osmium tetroxide solution, pH 7.4. (*7a*) A nucleolus exhibits heavy coarse antimonate deposits in the pars amorpha (P), less abundant, coarse precipitates in the dense component (D), and light fine deposits in the granular component (G). Intervening spaces or channels lack deposits. The highly dispersed heterochromatin characteristic of neuronal nuclei contain moderate antimonate deposits (arrows). Bar represents 1 µm. (*7b*) X-ray spectrum of <u>nucleolar</u> antimonate precipitates. Note the virtual absence of a potassium peak (K), and the presence of iron (Fe) and osmium in this spectrum. The reason for the high aluminum (Al) peak is not clear. (*7c*) X-ray spectrum of cytoplasmic antimonate precipitates. There is a much higher potassium (K) contribution, and the osmium, iron, and aluminum peaks of the nucleus (Fig. 6b) are markedly diminished or absent in the cytoplasmic precipitates.

whereas precipitates formed by simply lowering the pH did not (*cf*. Tables 2 and 4). These data suggest that the precipitation reaction of pyroantimonate with cations is complicated and that precipitates of antimonate salts contain varying amounts of potassium and/or water of hydration, depending upon the precipitating cation.

The present studies confirm our previous conclusions[57] that phosphate buffer should be avoided in antimonate precipitation reactions if capture of sodium or calcium ions is intended. In addition, cacodylate should probably also be avoided if Mg localization is desired, since the As Lα peak is very close to the Mg Kβ peak.

The overall conclusion is that, except in the case of sodium antimonate, the identification of any given intracellular antimonate precipitate (or precipitate aggregate) with energy-dispersive X-ray microanalysis is not unequivocal. Although it is possible to deconvolute complex spectra using reference standards or a Simplex curve fitting procedure, such procedures are indirect and time-consuming. For the direct separation of overlapping calcium and antimony peaks in thin sections of biological material, wave-length dispersive X-ray analysis, although time-consuming and cumbersome, remains the analytical tool of choice. However, the capability of the energy-dispersive system to scan a broad spectrum and pick up unexpected peaks, as the iron in the trigeminal nucleolus, makes it an important tool in the analysis of biological samples. It is of interest that iron has also been detected in certain plant cell nuclei.[70]

Efforts to determine cellular ion localization have been hindered by the diffusibility of small ions and the unavailability of appropriate technology to immobilize these ions *in situ*. Ultracryotomy and ion-capture techniques have begun to provide a handle on cellular ion localization and, in conjunction with X-ray microanalysis,[58,71] may also permit identification and quantitation of these important and ubiquitous cellular constituents. The older notion that cellular ions were fully diffusible is currently being supplanted by the realization that diffusible ions are bound by fixed tissue macromolecules to a greater or lesser extent.[1,72] Their localization by ion-capture techniques thus is feasible, since ion concentrations at given cellular sites are modulated by the relative binding affinities of the cations and cellular macromolecules. In order to interpret, realistically, the results of ion-capture cytochemistry, one must take into consideration the many factors which can rapidly alter ion distribution, such as temperature, pH, and ambient concentrations of competing ions. If proper controls are performed, however, it is possible to gain important information concerning the subcellular localization of ion binding sites, as with the Oschman and Wall technique,[73] and of cellular ions themselves, as with antimonate,[74] oxalate,[75] and silver precipitation techniques.[46]

Acknowledgments

We would like to acknowledge the technical expertise of Mr. David Emerson and the skilled editorial assistance of Mrs. Cindy Gue. We would also like to thank Dr. Hurshell Hunt of the Biometry Department for performing an analysis of the data in Figure 6. Supported by: NIH #AM 18115 (HLB), NIH #AM 11028 (SSS), and #RR-5767 to the College of Dental Medicine, Medical University of South Carolina (JVS).

References

1. J. A. V. Simson, S. S. Spicer and T. Katsuyama, "Cell Membrane Cation Localization by Pyroantimonate Methods: Correlation with Cell Function", In: Pathobiology of Cell Membranes (B. J. Trump and A. Arstila, eds.), Academic Press, New York City, New York, U.S.A., in press.
2. W. Hasselbach, "Relaxation and the Sarcotubular Calcium Pump", Federation Proc. 23, 1964, 909-912.
3. I. D. Bowen and T. A. Ryder, "The Application of X-ray Microanalysis to Histochemistry", In: Electron Probe Microanalysis in Biology (D. A. Erasmus, ed.), Chapman Hall, Ltd., London, England, 1978, 183-211.
4. F. Clemente and J. Meldolesi, "Calcium and Pancreatic Secretion. 1. Subcellular Distribution of Calcium and Magnesium in the Exocrine Pancreas of the Guinea Pig", J. Cell Biol. 65, 1975, 88-102.
5. J. A. V. Simson and S. S. Spicer, "Cytochemical Evidence for Cation Fluxes in Parotid Acinar Cells Following Stimulation by Isoproterenol", Anat. Rec. 178, 1974, 145-167.
6. T. Sato, L. Herman, J. A. Chandler, A. Stracher and T. C. Detwiler, "Localization of a Thrombin-Sensitive Calcium Pool in Platelets", J. Histochem. Cytochem. 23, 1975, 103-106.
7. R. Yarom, D. Ben-Ishay and O. Zinder, "Myocardial Cationic Shifts Induced by Isoproterenol. Electron Microscopic and Electron Probe Studies", J. Molec. Cell. Card. 4, 1972, 559-570.
8. R. Yarom, P. D. Peters, T. A. Hall, J. Kedem and S. Rogel, "Studies with EMMA-4 on Changes in the Intracellular Concentration and Distribution of Calcium in Heart Muscle of the Dog in Different Steady States", Micron 5, 1974, 11-20.
9. L. Boquist, "Ultrastructural Changes in the Parathyroids of Mongolian Gerbils Induced Experimentally *In Vitro*", Acta Pathol. Microbiol. Scand. 85-A, 1977, 203-218.
10. H. Komnick, "Elektronenmikroskopische Lokalization von Na+ and Cl− in Zellen und Geweben", Protoplasma 55, 1962, 414-418.
11. H. Komnick and U. Komnick, "Elektronenmikroskopische Untersuchungen zur Funktionellen Morphologie des Ionentransportes in der Salzdrüse von *Larus argentatus*", Z. Zellforsch. Mikrosk. Anat. 60, 1963, 163-203.
12. G. I. Kaye, J. D. Cole and A. Donn, "Electron Microscopy: Sodium Localization in Normal and Ouabain-Treated Transporting Cells", Science 150, 1965, 1167-1168.
13. G. I. Kaye, H. O. Wheeler, R. T. Whitlock and N. Lane, "Fluid Transport in the Rabbit Gallbladder. A Combined Physiological and Electron Microscopic Study", J. Cell Biol. 30, 1966, 237-268.
14. L. W. Tice and A. G. Engel, "The Localization of Sodium Pyroantimonate in Frog Muscle Fibers", J. Cell Biol. 31, 1966, 118A.
15. J. F. Hartmann, "High Sodium Content of Cortical Astrocytes", Arch. Neurol. 15, 1966, 633-642.
16. V. Mizuhira and T. Amakawa, "Detection of Electrolytes in Tissues at the Electron Microscopic Level with Special Reference to Sodium Ion Transport Mechanism in Rat Kidney", J. Histochem. Cytochem. 14, 1966, 770-771.
17. A. Nolte, "Elektronenmikroskopische Untersuchungen zum Natrium-und Chloridionentransport in der Proximalen Tubuluszelle der Rattenniere", Z. Zellforsch. Mikrosk. Anat. 72, 1966, 562-573.
18. J. A. Zadunaisky, "The Location of Sodium in the Transverse Tubules of Skeletal Muscle", J. Cell Biol. 31, 1966, C11-C16.
19. R. J. Grand and S. S. Spicer, "Preliminary Studies on the Electron Microscopic Localization of Sites of Sodium Transport in the Human Eccrine Sweat Gland", Mod. Probl. Pediat. 10, 1967, 100-106.

20. E. Yamada, "Sodium Localization in the Plasma Membrane of the Intestinal Absorptive Epithelial Cell", Arch. Histol. Jap. 28, 1967, 419-423.

21. J. Ochi, "Elektronenmikroskopischer Nachweis der Natriumionen in den Schweissdrüsen der Rattenfussohle", Histochemie 14, 1968, 300-307.

22. H. F. Edelhauser and K. A. Siegesmund, "The Localization of Sodium in the Teleost Cornea", Invest. Ophthal. 7, 1968, 147-155.

23. S. S. Spicer, J. H. Hardin and W. B. Greene, "Nuclear Precipitates in Pyroantimonate-Osmium Tetroxide-Fixed Tissues", J. Cell Biol. 39, 1968, 216-221.

24. J. Villegas, "Transport of Electrolytes in the Schwann Cell and Location of Sodium by Electron Microscopy", J. Gen. Physiol. 51, 1968, 61s-71s.

25. E. Tani, T. Ametani and H. Handa, "Sodium Localization in the Adult Brain. I. Normal Brain Tissue", Acta Neuropath. (Berlin) 14, 1969, 137-150.

26. R. M. Torack, "Sodium Demonstration in Rat Cerebrum following Perfusion with Hydroxyadip-aldehyde-Antimonate", Acta Neuropath. (Berlin) 12, 1969, 173-182.

27. Y. A. Vinnikov and K. Koichev, "Sodium Localization in the Spiral Organ during Relative Quiet and after Exposure to Sound", Nature (London) 223, 1969, 641-642.

28. R. Henrikson, "Ultrastructure of Ovine Ruminal Epithelium and Localization of Sodium in Tissue", J. Ultrastruct. Res. 30, 1970, 385-401.

29. P. Satir and N. B. Gilula, "The Cell Function in a Lamellibranch Gill Ciliated Epithelium. Localization of Pyroantimonate Precipitate", J. Cell Biol. 47, 1970, 468-487.

30. R. E. Bulger, "Use of Potassium Pyroantimonate in the Localization of Sodium Ions in Rat Kidney Tissue", J. Cell Biol. 40, 1969, 79-94.

31. S. Shiina, V. Mizuhira, T. Amakawa and Y. Futaesaku, "An Analysis of the Histochemical Procedure for Sodium Ion Detection", J. Histochem. Cytochem. 18, 1970, 644-649.

32. R. M. Torack and M. LaValle, "The Specificity of the Pyroantimonate Technique to Demonstrate Sodium", J. Histochem. Cytochem. 18, 1970, 635-643.

33. L. Herman, T. Sato and B. A. Weavers, "An Investigation of the Pyroantimonate Reaction for Sodium Localization Using the Analytical Electron Microscope, EMMA-4", In: 29th Ann. Proc. Electron Microscopy Soc. Amer. (C. J. Arceneaux, ed.), Claitor's Publishing Division, Baton Rouge, Louisiana, U.S.A., 1971.

34. S. M. Sumi and P. D. Swanson, "Limitations of the Pyroantimonate Technique for Localization of Sodium in Isolated Cerebral Tissues", J. Histochem. Cytochem. 19, 1971, 605-610.

35. R. L. Klein, S.-S. Yen and A. Thureson-Klein, "Critique on the K-pyroantimonate Method for Semiquantitative Estimation of Cations in Conjunction with Electron Microscopy", J. Histochem. Cytochem. 20, 1972, 65-78.

36. R. E. Garfield, R. M. Henderson and E. E. Daniel, "Evaluation of the Pyroantimonate Technique for Localization of Tissue Sodium", Tissue and Cell 4, 1972, 575-589.

37. S. S. Spicer and A. A. Swanson, "Elemental Analysis of Precipitates Formed in Nuclei by Antimonate-Osmium Tetroxide Fixation", J. Histochem. Cytochem. 20, 1972, 518-526.

38. C. C. Tisher, B. A. Weavers and W. J. Cirksena, "X-ray Microanalysis of Pyroantimonate Complexes in Rat Kidney", Am. J. Pathol. 69, 1972, 255-270.

39. R. Yarom and U. Meiri, "N Lines in Striated Muscle: A Site of Intracellular Ca^{2+}", Nature New Biol. 234, 1971, 254-256.

40. R. Yarom and K. Braun, "Ca^{2+} Changes in Myocardium Following Scorpion Venom Injections", J. Molec. Cell. Card. 2, 1971, 177-179.

41. R. Yarom and U. Meiri, "Ultrastructural Cation Precipitation in Frog Skeletal Muscle", J. Ultrastruct. Res. 39, 1972, 430-442.

42. B. K. Yeh, "Localization of Calcium Antimonate in the Atrial and Ventricular Muscle Fibers of the Cat Heart", J. Mol. Cell. Card. 5, 1973, 351-358.

43. R. Yarom and U. Meiri, "Pyroantimonate Precipitates in Frog Skeletal Muscle Changes Produced by Alterations in Composition of Bathing Fluid", J. Histochem. Cytochem. 21, 1973, 146-154.

44. T. S. Saetersdal, R. Myklebust, N.-P. B. Justesen and W. C. Olsen, "Ultrastructural Localization of Calcium in the Pigeon Papillary Muscle as Demonstrated by Cytochemical Studies and X-ray Microanalysis", Cell Tiss. Res. 155, 1974, 57-74.

45. W. L. Davis, J. L. Matthews and J. H. Martin, "An Electron Microscopic Study of Myofilament Calcium Binding Sites in Native EGTA-chelated and Calcium Reloaded Glycerolated Mammalian Skeletal Muscle", Calc. Tiss. Res. 14, 1974, 139-152.

46. R. Yarom, P. D. Peters and T. A. Hall, "Effect of Glutaraldehyde and Urea Embedding on Intracellular Ionic Elements. X-ray Microanalysis of Skeletal Muscle and Myocardium", J. Ultrastruct. Res. 49, 1974, 405-418.

47. G. Debbas, L. Hoffman, E. J. Landon and L. Hurwitz, "Electron Microscopic Localization of Calcium in Vascular Smooth Muscle", Anat. Rec. 182, 1975, 447-472.

48. L. Herman, T. Sato and C. N. Hales, "The Electron Microscopic Localization of Cations to Pancreatic Islets of Langerhans and Their Possible Role in Insulin Secretion", J. Ultrastruct. Res. 42, 1973, 298-311.

49. M. E. Stoeckel, C. Hindelang-Gertner, H.-D. Dellmann, A. Porte and F. Stutinsky, "Subcellular Localization of Calcium in the Mouse Hypophysis. I. Calcium Distribution in the Adeno- and Neurohypophysis under Normal Conditions", Cell Tiss. Res. 157, 1975, 307-322.

50. J. E. Schechter, "Cations in the Rat Pars Distalis Ultrastructural Localization", Am. J. Anat. 146, 1976, 189-206.

51. E. B. Cramer, C. Cardasis, G. Periera, L. Milks and D. Ford, "Ultrastructural Localization of Cations in the Rat Pars Distalis under Various Experimental Conditions", Neuroendocrinology 26, 1978, 72-84.
52. C. T. Brighton and R. M. Hunt, "Mitochondrial Calcium and Its Role in Calcification: Histochemical Localization of Calcium in Electron Micrographs of the Epiphyseal Growth Plate with K-pyroantimonate", Clin. Orthopaed. and Rel. Res. 100, 1974, 406-416.
53. L. Boquist and E. Lundgren, "Effects of Variations in Calcium Concentration on Parathyroid Morphology in vitro", Lab. Invest. 33, 1975, 638-647.
54. J. A. Chandler and S. Battersby, "X-ray Microanalysis of Zinc and Calcium in Ultrathin Sections of Human Sperm Cells Using the Pyroantimonate Technique", J. Histochem. Cytochem. 24, 1976, 740-748.
55. J. A. Chandler, F. Sinowatz, B. G. Timms and C. G. Pierrepoint, "The Subcellular Distribution of Zinc in Dog Prostate Studied by X-ray Microanalysis", Cell Tiss. Res. 185, 1977, 89-103.
56. J. A. Chandler and B. G. Timms, "The Effect of Testosterone and Cadmium on the Rat Lateral Prostate in Organ Culture", Virchows Arch. B Cell Path. 25, 1977, 17-31.
57. J. A. V. Simson and S. S. Spicer, "Selective Subcellular Localization of Cations with Variants of the Potassium (Pyro)antimonate Technique", J. Histochem. Cytochem. 23, 1975, 575-598.
58. J. A. Chandler, In: Electron Probe Microanalysis in Biology (D. A. Erasmus, ed.), Chapman Hall, Ltd., London, England, 1978, 37-93.
59. B. P. Lane and E. Martin, "Electron Probe Analysis of Cationic Species in Pyroantimonate Precipitates in Epon-embedded Tissue", J. Histochem. Cytochem. 17, 1969, 102-106.
60. C. J. Tandler, C. M. Libanati and C. A. Sanchis, "The Intracellular Localization of Inorganic Cations with Potassium Pyroantimonate. Electron Microscope and Microprobe Analysis", J. Cell Biol. 45, 1970, 355-366.
61. R. Yarom and J. A. Chandler, "Electron Probe Microanalysis of Skeletal Muscle", J. Histochem. Cytochem. 22, 1974, 147-154.
62. C. N. Hales, J. P. Luzio, J. A. Chandler and L. Herman, "Localization of Calcium in the Smooth Endoplasmic Reticulum of Rat Isolated Fat Cells", J. Cell Sci. 15, 1974, 1-15.
63. R. Yarom, E. Wisenberg, P. D. Peters and T. A. Hall, "Zinc Distribution in Injured Myocardium. EMMA-4 Examinations of Dogs' Hearts after Coronary Ligation", Virchows Arch. B Cell Path. 23, 1977, 65-77.
64. B. G. Timms, J. A. Chandler, M. S. Morton and G. V. Groom, "The Effect of Cadmium Administration in vivo on Plasma Testosterone and the Ultrastructure of Rat Lateral Prostate", Virchows Arch. B Cell Path. 25, 1977, 33-52.
65. J. A. Chandler, B. G. Timms and M. S. Morton, "Subcellular Distribution of Zinc in Rat Prostate Studied by X-ray Microanalysis. I. Normal Prostate", Histochem. J. 9, 1977, 103-120.
66. E. Lifshin, "X-ray Generation and Detection in the SEM", In: Scanning Electron Microscopy (O. Wells), McGraw-Hill Book Co., New York City, New York, U.S.A., 1974, 243-276.
67. J. H. Hardin and S. S. Spicer, "An Ultrastructural Study of Human Eosinophil Granules: Maturational Stages and Pyroantimonate Reactive Cation", Am. J. Anat. 128, 1970, 283-310.
68. J. H. Hardin, S. S. Spicer and G. E. Malanos, "Quantitation of the Ultrastructural Components of Nucleoli of Rat Trigeminal Ganglia", J. Ultrastruct. Res. 32, 1970, 274-283.
69. D. M. Steffensen, "Chromosome Structure with Special Reference to the Role of Metal Ions", Int. Rev. Cytol. 12, 1961, 163-197.
70. M. E. van Steveninck, R. F. M. van Steveninck, P. D. Peters and T. A. Hall, "X-ray Microanalysis of Antimonate Precipitates in Barley Roots", Protoplasma 90, 1976, 47-63.
71. B. F. Trump, I. K. Berezesky, S. H. Chang and R. E. Bulger, "Detection of Ion Shifts in Proximal Tubule Cells of the Rat Kidney Using X-ray Microanalysis", Virchows Arch. B Cell Path. 22, 1976, 111-120.
72. G. N. Ling, C. Miller and M. M. Ochsenfeld, "The Physical State of Solutes and Water in Living Cells According to the Association-Induction Hypothesis", In: Ann. N. Y. Acad. Sci. 204, 1973, 6-50.
73. J. L. Oschman and B. J. Wall, "Calcium Binding to Intestinal Membranes", J. Cell Biol. 55, 1972, 58-73.
74. R. Yarom, P. D. Peters, M. Scripps and S. Rogel, "Effect of Specimen Preparation on Intracellular Myocardial Calcium. Electron Microscopic X-ray Microanalysis", Histochemistry 38, 1974, 143-153.
75. J. R. Coleman and A. R. Terepka, "Electron Probe Analysis of the Calcium Distribution in Cells of the Embryonic Chick Chorioallantoic Membrane. II. Demonstration of Intracellular Location during Active Transcellular Transport", J. Histochem. Cytochem. 20, 1972, 414-424.

Discussion with Reviewers

A.R. Spurr: Rather than periodically alter the tilt angle of the samples from 45° to 85°, wouldn't it have been more advantageous to select the tilt angle that gave optimum counts for all analyses?

Authors: While empirically optimizing the tilt angle does increase the number of X-ray photons processed, it also may result in the majority of the counts originating from the edge of a crystal. We found that the variability of spectra obtained from a single sample could be substantially decreased by collecting data at low magnification, relatively low accelerating voltage (~ 12 KEV) and by varying the tilt angle periodically. Such procedures minimize the effects of micro-inhomogeneities in the sample.

For additional discussion see page 750.

SCANNING ELECTRON MICROSCOPY/1979/II
SEM Inc., AMF O'Hare, IL 60666, USA

CYTOCHEMICAL LOCALIZATION OF CATIONS IN MYELINATED NERVE USING TEM, HVEM,
SEM AND ELECTRON PROBE MICROANALYSIS

M. H. Ellisman, P. L. Friedman and W. J. Hamilton*

Department of Neurosciences
University of California San Diego
School of Medicine
La Jolla, CA 92093

*Hasler Applications Laboratory,
Applied Research Laboratory
9545 Wentworth Street,
Sunland, CA 91040

Abstract

By combining the techniques of the direct osmium-pyroantimonate fixation procedure for precipitation of cations, with high voltage electron microscopy (HVEM), and electron probe wavelength spectroscopy we have localized elevated concentrations of Na^+ to the cytoplasmic compartments of the Schwann cell paranodal loops in myelinated axons. We have also localized high concentrations of Ca^{++} to areas of compact myelin using both the osmium-pyroantimonate procedure and the Oschman and Wall technique for visualizing bound Ca^{++}. With HVEM it was possible to determine the location of precipitates within semi-thick (1/4 - 1/2 μm) unstained tissue sections.

Electron probe wavelength spectroscopy was used to detect the presence of Na^+ and Ca^{++} in the Schwann cell paranodal loops, the axoplasm at the node, the compact myelin, and the extracellular matrix.

Na^+ concentrations within cytoplasmic compartments of the Schwann cell paranodal loops in close proximity to nodes of Ranvier are as yet without a physiological explanation. A role for the paranodal organ as a source or sink for Na^+ involved in impulse conduction is suggested. The presence of high concentrations of Ca^{++} within compact myelin is consistent with the role calcium is thought to play in the mechanism of adhesion between the lipid bilayers of multilammelar liposomes.

KEY WORDS: Sodium, Calcium, Myelinated Axons,
Node of Ranvier, Paranode, Myelin, Cytochemistry,
Potassium-Pyroantimonate, X-ray Microanalysis,
High Voltage Electron Microscopy, Correlated
Microscopy

Introduction

This communication concerns the localization of ions in relation to the glial wrappings in the vicinity of nodes of Ranvier. In their study of the distribution of ions in myelinated axons, Rick and co-workers[1] showed that elevated levels of sodium (Na^+) exist at the nodes of Ranvier, particularly in the paranodal region. They studied cryosections of rapidly frozen sciatic nerves (of Rana esculenta) by electron probe microanalysis after freeze-drying. It has been shown that this procedure maintains a distribution of diffusible ions representative of the in vivo condition[2-3]. Although Rick et al.[1] were able to identify and circumscribe the location of Na^+ at the node, they were not able to specify whether Na^+ was located extracellularly or cytoplasmically in respect to the paranodal windings. The following studies were therefore aimed at determining whether Na^+ is localizable to particular compartments within the complex structure of the node of Ranvier. By correlating the approaches of the osmium-pyroantimonate cation precipitation technique[4], high voltage electron microscopy (HVEM) of thick sections, and electron probe wavelength spectroscopy, we were able to identify specific regions of elevated Na^+ and calcium (Ca^{++}) concentrations at the node of Ranvier.

Materials and Methods

A. Cytochemical Procedures
    1. Osmium-pyroantimonate fixation without pre-wash of tissue.[4] All glassware used in these pyroantimonate procedures was pre-rinsed in 2 N HCl to eliminate contaminants. The osmium-pyroantimonate solution was prepared by adjusting 200 ml of 0.01 N glacial acetic acid to pH 7.4 by addition of 0.1 N KOH, then adding 4 grams of osmium-pyroantimonate (EM Laboratories, Inc., affiliate of Merck). This was dissolved in the acetic acid solution using a small water bath on a heater/stirplate (temperature should not exceed 80°C). After 1.5 hours the Merck solution thus prepared was quite clear. One gram of Sigma grade OsO_4 was then added to 100 ml of this warm solution which we continued to stir until the osmium dissolved. The final pH of this osmium-pyroantimonate acetic acid

fixative solution was then adjusted to 7.8 with 0.05 N acetic acid. This was then allowed to stand well sealed overnight in the cold. Ten to fifteen minutes before use on the following day we spun 15 ml of this fixative at 15,000 G.

Five-week old rats were anesthetized with nembutal (50 mg/kg) after which the spinal cord and roots were carefully exposed by dorsal laminectomy after which the dura was removed. The tissue was then flooded with the cold osmium-pyroantimonate fixative. Spinal roots were dissected out and placed into the cold fresh fixative for·60 minutes. The tissue was washed with triple distilled water made slightly alkaline with the addition of 0.1 N KOH (1 ml/100 ml) and dehydrated in slightly alkaline ethanols. Tissues were embedded in Epon-Araldite.

2. EDTA/Osmium-pyroantimonate.[4] Five-week old rats were anesthetized with pentobarbital (50 mg/kg) and the spinal cord similarly exposed by dorsal laminectomy. The dissected area was then flooded with Ca^{++} and Mg^{++} free rat Ringers solution (see Ellisman, et al.[5] for composition) containing heparin (200 units/100 ml) and xylocaine (30 mg/100 ml). The Ringers solution was bubbled with 5% CO_2 - 95% air prior to use. The dissected anesthetized rats were then perfused via the left ventricle for 15 minutes passing a volume of 400 ml of Ca^{++} and Mg^{++} free Ringers containing 3 mM EDTA-dipotassium salt (Eastman) to chelate Ca^{++} and Mg^{++}. The tissues were simultaneously flooded, in situ, with the K-EDTA Ringers. The dissected area was then quickly rinsed with Ca^{++}, Mg^{++} free Ringers containing no EDTA and flooded with the osmium-pyroantimonate solution. The roots were then removed and placed into fresh, cold 2% osmium-pyroantimonate fixative for 1 hour. Subsequent treatments were identical to those of section 1 above.

3. 5 mM Ca^{++} inclusion during tissue preparation.[6] Four-week old rats were anesthetized with pentobarbital (50 mg/kg) and perfused via the left ventricle with 400 ml rat Ringers which contained 5 mM $CaCl_2$, heparin (200 units/100 ml), and xylocaine (30 mg/100 ml). After the Ringers rinse, fixative was substituted containing 2.5% glutaraldehyde in 0.08 M S-collidine at pH 7.2 with 5 mM $CaCl_2$ and 5% sucrose added. The spinal roots were then removed by dorsal laminectomy and placed in fresh cold fixative for an additional 30 minutes, then washed in 0.08 collidine, pH 7.2 also containing 5 mM $CaCl_2$, postfixed for 1 hour in 1% OsO_4 buffered with 0.08 M collidine to which 5 mM $CaCl_2$ had been added. Subsequent buffer washes and dehydrations (in ethanols) also contained 5 mM Ca^{++}. Tissues were embedded in Epon-Araldite as above.

4. Osmium tetroxide fixation only without pre-wash of tissue. Five week old rats were anesthetized with pentobarbital (50 mg/kg) after which dorsal laminectomy was performed as in procedure 1. The spinal cord and roots were then flooded with a cold 1% solution of OsO_4 prepared according to Klein[4] but without the addition of the osmium-pyroantimonate. Spinal roots were removed and processed according to procedure 1.

5. EDTA-osmium only. Five-week old rats were prepared in the same manner as described in procedure 2 with the single exception that the fixative consisted of 1% solution of OsO_4 without the addition of osmium-pyroantimonate. All other aspects of the preparation of the tissue remained the same.

B. Electron Microscopic Procedures

1. Conventional transmission and high voltage electron microscopy (EM & HVEM). Thin and semi-thick sections (1/4-1/2 μm) of representative regions from each experimental condition were prepared. These were picked up on uncoated grids and then lightly coated with carbon for stability and examined without poststaining at 1,000 KV on a JEM 1000 HVEM in Boulder, Colorado. Stereo micrographs were taken by tilting the specimen to angles of +6 and -6°.

2. Combined scanning EM and wavelength spectroscopic analysis. Additional unstained 1000 A sections were prepared from each of the cytochemical conditions and mounted on unfilmed nickel finder grids. A multi-crystal computer assisted spectrometer system, the SEMQ (provided by Applied Research Laboratories) allowed for the simultaneous detection of Na^+ and Ca^{++}. To minimize beam damage to the specimen, we used the optical imaging capabilities of the SEMQ to first locate the known areas of interest on the finder grids. The nodes of Ranvier were most effectively imaged by the Backscatter electron (BSE) mode. The microprobe was operated at an accelerating voltage of 30 KV with a beam current of 80 nanoamps which provided a probe size of approximately 0.5 μm. Individual spectroscopic analyses were done for 20 seconds. The following crystals were tuned to each element: thallium acid phthalate (TAP) for Na^+ and pentaerythritol (PET) for Ca^{++}. The following areas of interest were comparatively analyzed: Schwann cell paranodal loops, axoplasm at the node, compact myelin and extracellular matrix.

The two independent crystal spectrometers were focused or tuned to their respective element by obtaining a peak value (P) with the appropriate standard. This set position was thereafter used to monitor (in counts/sec) the characteristic x-rays from that element within the area of the specimen being analyzed. For each area analyzed a background (B) value was obtained with the same detector by shifting the position of the crystal off the peak value by a pre-determined amount for each element: B was P±300 for Ca^{++} and P±400 for Na^+. This background value or noise was then subtracted from the peak value to obtain a more accurate measure (P-B) of the number of characteristic x-rays/sec.[7]

Results

A. Cytochemistry and Wavelength Spectroscopic Analysis

1. Osmium-pyroantimonate fixations. Within the Schwann cell paranodal loops (PN) of Figure 1, the pyroantimonate precipitate is seen as an electron dense, small-grain deposit. This precipitate may be contrasted with the coarser and

larger-grain precipitate located within the axoplasm (Ax). When HVEM stereopairs were used to examine these precipitates, the grains in the paranodal region were found to be localized to the cytoplasmic compartments of the paranodal loops. It was also noted that grains of precipitate formed in areas of compact myelin.

The data obtained from x-ray microanalysis of the various ultrastructural regions of the node of Ranvier are presented in Tables 1 and 2. The first of each triad of numbers represents the Mean (P-B) in counts/sec. The second number is the Standard Deviation, followed by a third value which represents the number of different areas analyzed with the microprobe within that specific region. For each of these areas analyzed (within a given anatomical region) three 20 second analyses were done. To interpret accurately Tables 1 and 2, comparisons between values should be made along the horizontal axis in each experimental condition only. Direct numerical comparisons between different nodes is not valid because of variation in section thickness. However, trends such as the elevated values for Ca^{++} in myelin over Ca^{++} in the extracellular matrix apparent from comparisons of individual samples (Table 2) are significant as they represent aggregates of significant differences.

In the osmium-pyroantimonate fixed material, the relative concentration of Na^+ localized to the paranode (for both nodes) exceeded that found in the extracellular matrix, the nodal axoplasm and in the compact myelin (Table 1). X-ray microanalysis for Ca^{++} in this same material (Table 2) shows that the greatest relative concentrations of Ca^{++} were found in the paranodal loops and in compact myelin. Significantly smaller concentrations of Ca^{++} were localized to the extracellular matrix and to the axoplasm. The high concentration of Ca^{++} within compact myelin was significantly larger than that found in all other ultrastructural regions examined for both nodes.

2. _EDTA-osmium-pyroantimonate fixations._ Klein et al.[4] have shown that osmium-pyroantimonate precipitates the cations Ca^{++}, Mg^{++} and Na^+ when these ions exceed concentrations of 10^{-6} M, 10^{-5} M, and 10^{-2} M respectively. This cytochemical approach lacks the sensitivity necessary for reliable localization of Na^+. However, if the tissue is pre-washed with K-ethylenedinitrilo-tetraacetic acid (K-EDTA) to chelate much of the Ca^{++} and Mg^{++}, the resulting precipitate in the tissue will more largely reflect the presence of Na^+ where it is in high concentration.[4]

When spinal nerves were prepared according to the EDTA-osmium-pyroantimonate protocol,[4] a small-grained, closely-packed precipitate was formed within the paranodal loops (PN) seen in Figure 2. A similar pyroantimonate reaction was localized to various areas of compact myelin. HVEM stereo examination of semi-thick sections from this tissue revealed diminished amounts of precipitate along the axolemma and within the axoplasm (Ax) when compared to precipitates in Figure 1.

X-ray microanalysis of the EDTA-osmium-pyroantimonate preparation (Table 1) revealed a localized and comparatively concentrated amount of Na^+ within the paranodal loops. When compared to all other anatomical regions analyzed, the paranodal loops exhibited significantly higher levels of Na^+. All other areas analyzed retained only negligible amounts of Na^+.

Despite the K-EDTA pre-wash, the x-ray microanalysis for Ca^{++} (Table 2) revealed significant concentrations of this cation within the paranodal loops, the axoplasm at the node, and in the compact myelin. An analysis of the data from node #1 shows an example of substantially larger relative concentrations of Ca^{++} localized to the paranode whereas node #2 exhibits an almost equal relative concentration of Ca^{++} localized in both the paranode and to the compact myelin. These results suggest that the EDTA pre-wash may not remove the Ca^{++} from the tissue but may compete with pyroantimonate to bind it.

3. _Glutaraldehyde-osmium with 5 mM Ca^{++} inclusion fixation._ The inclusion of Ca^{++} in the fixatives results in the appearance of regions of electron opacity not seen when Ca^{++} is removed from conventional glutaraldehyde-osmium fixation protocols.[6] These sites of elevated electron density have been demonstrated by x-ray microanalysis to represent Ca^{++} deposits. In the semi-thick (1/4-1/2 μm) tissue sections of nerve prepared for HVEM examination, a presumed Ca^{++} precipitate was visualized along the inner leaflet of the axolemma in both the nodal and paranodal loops, and in regions of compact myelin. Most notable, however, was the enhanced density seen at the paranodal end feet directly adjacent to the axolemma. Unlike the axolemma at the paranodal glial axonal junction, the interparanodal axolemma exhibited no enhanced density. It was of interest to note that the Schwann cell microvilli which surround the bare nodal segment of the axon, although directly adjacent to both the precipitate positive paranode and compact myelin, displayed no enhanced contrast. We also observed that the Ca^{++} related precipitation occurred more consistently in the paranodal loops of larger myelinated axons than small ones. The compact myelin exhibited enhanced electron density regardless of the extent of myelination.

To test for a correlation between the distribution of precipitate and presence of Ca^{++}, 1000 Å thick sections of this tissue were examined with the x-ray spectrometer (Table 2). It was confirmed that a significantly large amount of Ca^{++} was present in the paranodal loops as well as in the axoplasm at the node. However, the greatest comparative concentration of Ca^{++} was found to be localized within the compact myelin, with relatively little Ca^{++} present in the extracellular matrix. The concentrations of Na^+ found in these specimens was negligible (Table 1).

4. _Osmium tetroxide fixation only._ Conventional electron microscopic (EM) examination of this tissue revealed no evidence of any electron dense precipitates. However, when tis-

TABLE   1

CRYSTAL SPECTROMETER; Sodium Kα, X-ray counts from myelinated nerves.

NERVE TREATMENT	EMISSION MEASUREMENTS	PARANODAL GLIAL LOOPS	AXOPLASM AT NODE	COMPACT MYELIN	EXTRACELLULAR MATRIX
I. OsO_4/PYRO					
Node #1	P-B	26±4 *(18)*	6±3 *(18)*	3±3 *(18)*	5±4 *(18)*
Node #2	P-B	16±6 *(18)*	11±7 *(18)*	0±6 *(18)*	5±4 *(18)*
II. EDTA/OsO_4/PYRO					
Node #1	P-B	82±9 *(18)*	16±7 *(18)*	19±16 *(18)*	15±11 *(18)*
Node #2	P-B	99±16 *(18)*	9±6 *(18)*	12±7 *(18)*	12±4 *(18)*
III. GLUT/OsO_4 w/5 mM Ca^{++}					
Node #1	P-B	15±8 *(18)*	0±2 *(18)*	0±7 *(18)*	4±4 *(18)*
Node #2	P-B	2±3 *(18)*	0±2 *(18)*	0±3 *(18)*	0±5 *(18)*

TABLE   2

CRYSTAL SPECTROMETER; Calcium Kα, X-ray counts from myelinated nerves.

NERVE TREATMENT	EMISSION MEASUREMENTS	PARANODAL GLIAL LOOPS	AXOPLASM AT NODE	COMPACT MYELIN	EXTRACELLULAR MATRIX
I. OsO_4/PYRO					
Node #1	P-B	42±8 *(36)*	40±15 *(12)*	263±32 *(36)*	4±3 *(12)*
Node #2	P-B	177±33 *(27)*	46±13 *(27)*	246±29 *(27)*	40±13 *(27)*
II. EDTA/OsO_4/PYRO					
Node #1	P-B	911±86 *(48)*	118±39 *(12)*	211±25 *(15)*	38±6 *(15)*
Node #2	P-B	185±26 *(18)*	146±82 *(18)*	187±20 *(18)*	22±4 *(18)*
III. GLUT/OsO_4 w/5 mM Ca^{++}					
Node #1	P-B	157±22 *(27)*	106±32 *(27)*	295±34 *(27)*	31±21 *(27)*
Node #2	P-B	164±20 *(27)*	62±5 *(27)*	298±32 *(27)*	0±12 *(27)*

Instrumental conditions: Sodium data, Thallium acid phthalate - diffracting crystal; Calcium data, Pentaerythritol - diffracting crystal; voltage, 30kV, at beam current of 80 nAmp; probe diameter 0.5 μm, counting time 20 seconds per point.

P, peak; B, background; P-B, characteristic emission. Number in parentheses are the number of different analyses within the specified nerve structure. Three, twenty second integrations were made per beam position, multiple beam positions within each structure were examined. Small numbers not in parentheses are mean standard deviation of P-B.

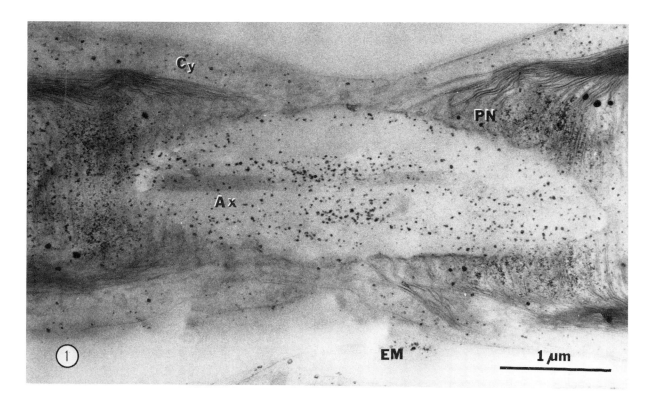

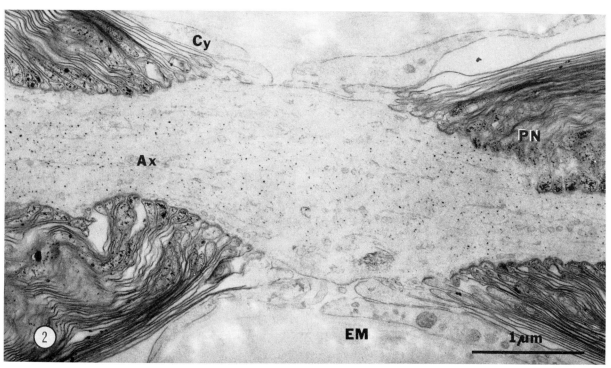

Fig. 1. In this unstained section of a node of Ranvier, prepared according to the direct osmuin-pyro-antimonate technique, numerous particles of electron dense precipitate are localized within the para-nodal glial loops (PN). The axoplasm (Ax) contains comparatively fewer grains of precipitate, with even less precipitate seen in both the Schwann cell cytoplasm (Cy) and the extracellular matrix (EM).
Fig. 2. This unstained node of Ranvier was pre-washed with K-EDTA before being fixed by the osmium-pyro-antimonate technique. Note the highly localized precipitate within the cytoplasmic compartments of the paranodal glial loops (PN), the reduced staining of the axoplasm (Ax), and the absence of precipitate in the Schwann cell cytoplasm (Cy).

sue sections were examined by wavelength spectroscopy, it was found that the Ca^{++} distribution was similar to that of the 5 mM Ca^{++} procedure and for both the osmium-pyroantimonate procedures. For example, the paranode and the compact myelin contained mean P-B counts/sec of 658 and 741 respectively (standard deviations of ±25 and ±29) for six different locations examined. These far exceeded the mean Ca^{++} x-ray counts/sec from the axoplasm and the extracellular matrix of 57 and 0 respectively (standard deviations of ±13 and ±9). The x-ray analysis of these same areas for Na^+ revealed no detectable ionic concentrations.

5. <u>K-EDTA osmium fixation</u>. The results from this experimental condition parallel those found in condition #4. No precipitate was localized in the EM, and the distribution of Ca^{++} within the Schwann cell paranodal loops and the compact myelin was elevated with mean P-B counts/ sec of 480 and 614 respectively (standard deviations of ±25 and ±24). The axoplasm yielded a relatively lower mean P-B counts/sec for Ca^{++} of 68 (standard deviation of ±13), and a mean of 6 counts/sec eminated from the extracellular matrix (standard deviation ±9). Na^+ again was not detected in any of the ultrastructural regions in these preparations.

Discussion

By combining the results of the direct osmium-pyroantimonate fixation technique for precipitating cations, with HVEM and electron probe wavelength spectroscopy, we have been able to localize elevated concentrations of Na^+ to the cytoplasmic compartments of the Schwann cell paranodal loops in myelinated axons. These data, therefore, extend and define the spacial resolution for the localization of Na^+ as presented by Rick et al.[1] using frozen sections. In addition, we have localized high concentrations of Ca^{++} to areas of compact myelin, thus providing new information about the <u>in vivo</u> composition of myelin.

Criticisms of the osmium-pyroantimonate technique for localizing Na^+ are generally of two types. The first, is that ion mobility is not arrested after primary aldehyde fixation.[8] The second is that osmium-pyroantimonate acts as a non-specific cytochemical marker, since at physiological levels, pyroantimonate precipitates with K^+, Ca^{++}, and Mg^{++} as well as with Na^+.[4,9] The techniques used in this study circumvent these problems. Klein et al.[4] and Simson and Spicer[10] have shown that if primary fixation is carried out using osmium-pyroantimonate, diffusion of small ions may be kept to a low level, approximating that of the <u>in vivo</u> condition. In addition, if the protocol as developed by Klein and co-workers[4] is used, the cation K^+ does not appear to be precipitated by the pyroantimonate. The electron dense precipitates obtained by this procedure may however, still result from Ca^{++}, Mg^{++} or Na^+. The identity of precipitant species has been accomplished with the aid of x-ray microanalysis.[7,11] Thus, it is possible to determine whether Ca^{++}, Mg^{++} or Na^+ are giving rise to precipitates in a specific region of a cell. In this regard it is noteworthy that, in experiments where the osmium-pyroantimonate fixative was not used to precipitate Na^+ (Ca^{++}-glutaraldehyde fixations or osmium fixations without pyroantimonate) Na^+ was not found with the microprobe.

The localization of Na^+ concentrations to the cytoplasmic compartment of the paranodal loops lacks a physiological explanation at this time. A role for the paranodal organ as a source of Na^+ for impulse conduction has been suggested.[12,13] The intimate apposition of the paranodal loops to the axolemma forming the glial-axonal junctions[14,15] may provide an ultrastructural pathway for the transcellular movement of these ions.

That comparatively high concentrations of Ca^{++} are localized to areas of compact myelin is not surprising in view of the model of adhesion between bilayers of multilammellar liposomes proposed by Newton et al.[16] In this simple system, Ca^{++} is presumed to act as a divalent ligand linking the polar head groups of adjacent bilayers. With a similar location in myelin a similar role for Ca^{++} is to be expected.

Acknowledgements

This work was supported by research grants to M.H. Ellisman from the N.I.N.C.D.S. #NS14718 and the Muscular Dystrophy Association of America. The HVEM work was in part supported by a grant from the N.I.H. to K.R. Porter (#RR-00592). The authors would like to express appreciation to K. Porter, G. Wray and R. Livingston for encouragement during these studies as well as to Applied Research Laboratories for allowing us to use the SEMQ system.

References

1. R. Rick, A. Dörge and A. Tippe. Elemental distribution of Na, P, Cl, and K in different structures of myelinated nerve of Rana esculenta. Experientia 32 (1976) 1018-1019.
2. A. Dörge, R. Rick, K. Gehring and K. Thurau. Preparation of freeze-dried cryosections for quantitative x-ray microanalysis of electrolytes in biological soft tissues. Pflügers Archive. 373 (1978) 85-97.
3. B.F. Trump, I.K. Berezesky, et al., X-Ray microanalysis of diffusible elements..., in SEM/ 1978/II, SEM Inc., AMF O'Hare, IL. 1027-1040.
4. R.L. Klein, S.-S. Yen and A. Thureson-Klein. Critique on the K-pyroantimonate method for semiquantitative estimation of cations in conjunction with electron microscopy. J. Histochem. Cytochem. 20 (1972) 65-78.
5. M.H. Ellisman, M.H. Brooke et. al. Appearance in slow muscle sarcolemma of specializations characteristic of fast muscle after reinnervation by a fast muscle nerve. Exper. Neurol. 58 (1978) 59-67.
6. J.L. Oschman and B.J. Wall. Calcium binding to intestinal membranes. J. Cell Biol. 55 (1972) 58-73.
7. R. Yarom and J.A. Chandler. Electron probe microanalysis of skeletal muscle. J. Histochem. Cytochem. 22 (1974) 147-154.

8. R.E. Bulger. Use of potassium pyroantimonate in the localization of sodium ions in rat kidney tissue. J. Cell Biol. 40 (1969) 79-88.
9. R.M. Torack and M. LaValle. The specificity of the pyroantimonate technique to demonstrate sodium. J. Histochem. Cytochem. 18 (1970) 635-647.
10. J.A.V. Simson and S.S. Spicer. Selective subcellular localization of cations with variants of the potassium (pyro)antimonate technique. J. Histochem. Cytochem. 23 (1975) 575-598.
11. B.P. Lane and E. Martin. Electron probe analysis of cationic species in pyroantimonate precipitates in epon-embedded tissue. J. Histochem. Cytochem. 17 (1969) 102-106.
12. H. Müller-Mohnssen, A. Tippe et. al. Is the rise of the action potential of the Ranvier node controlled by a paranodal organ? Naturwissenschaften 61 (1974) 369-370.
13. A. Tippe and H. Müller-Mohnssen. Further experimental evidence for the synapse hypothesis of Na$^+$ current activation and inactivation at the Ranvier node. Naturwissenschaften 63 (1975) 490-491.
14. R.B. Livingston, K. Pfenninger et. al. Specialized paranodal and interparanodal glial-axonal junctions in the peripheral and central nervous system: A freeze-fracture study. Brain Res. 58 (1973) 1-24.
15. K. Akert, C. Sandri et. al. Extracellular spaces and junctional complexes at the node of Ranvier. Actual. Neurophysiol. Paris, France (1974) 9-22.
16. C. Newton, W. Bangborn et. al. Specificity of Ca^{2+} and Mg^{2+} binding to phosphatidylserine vesicles and resultant phase changes of bilayer membrane structure. Biochim. Biophys. Acta 506 (1978) 281-287.

Discussion with Reviewers

J.A.V. Simson: Why was xylocain used in the perfusion fluid?
Authors: Xylocain is used in the Ringers wash to anesthetize the smooth muscles of small arteries (preventing them from constricting) providing for a more complete perfusion.

J.A.V. Simson: Does xylocain effect the calcium distribution or membrane permeability of the cells being analyzed?
Authors: Like other local anesthetics, xylocain stabilizes excitable membranes with respect to activation of the voltage dependent Na$^+$ fluxes and should thereby help to stabilize the in vivo ionic concentrations.

R.J. Barrnett: Why use the pyroantimonate method according to Klein et al. instead of a different technique?
Authors: We tried several pyroantimonate methods which involved prefixation with glutaraldehyde. These alternative techniques produced highly variable results. The method used in our study yielded very repeatable results.

R.J. Barrnett: Can the pyroantimonate method be used as a means for quantitation?
Authors: We do not feel that the pyroantimonate technique can yield any reliable quantitative information. In this study we have used pyroantimonate-cation precipitates as a means to see where Ca^{++} and Na$^+$ ions are compartmentalized at nodes of Ranvier.

R.J. Barrnett: I would have expected severe beam damage resulting from instrument conditions of 30 kV at a beam current of 80na. How did you avoid this?
Authors: As mentioned in the text the current was measured on each grid. Perhaps the fact that we used thick sections coated on both sides with carbon provided for adequate dissipation of the current. We were able to make multiple (20-50) 20 second peak measurements and an equal number of background measurements with the sample remaining intact.

R.J. Barrnett: What is the merit of your method for determining background versus keeping the beam at peak and moving the spot?
Authors: In our study, both methods of background (B) measurement were used. The "background" as produced by the continuum fluorescence at the location in the specimen being analyzed may best be estimated in a wavelength spectrometer by "detuning" the spectrometer. Background due to the non-specific staining can best be determined by measurement at the peak wavelength in the extracellular matrix.

J.R. Coleman: If the K-EDTA wash removed extracellular Ca^{++} and Mg^{++}, why is there so much Ca^{++} left?
Authors: First, we are aware of no study in which high concentrations of Ca^{++} (or Mg^{++}) have been shown to reside within the extracellular matrix of peripheral nerves. Second, we know of no evidence which clearly shows K-EDTA pre-wash removes Ca^{++} from a biological system; rather, it is well recognized that K-EDTA will chelate Ca^{++}, thus preventing it from participating in any biological function or reaction, such as its precipitation with pyroantimonate. Consequently, a marked reduction in precipitate occurred in the K-EDTA pre-washed tissue (Fig. 2) as compared to the osmium-pyroantimonate fixed tissue (Fig. 1). However, the wavelength spectrometer was able to detect the presence of this chelated Ca^{++} since detection is not dependent upon a Ca^{++}-pyroantimonate precipitate.

J.R. Coleman: Why does fixation in the presence of 5 mM Ca^{++}, which is claimed to show calcium localizations, have so little effect on the ratio of Ca^{++} in the paranodal loops compared to extracellular spaces?
Authors: We do not know the answer to this question. We do however suspect that Ca^{++} inclusion in fixatives is not necessary to retain a significant amount of Ca^{++} in these tissue sites. This is supported in our data on the Ca^{++} distribution after osmium fixation alone.

G. Pappas: What do you conclude about Ca^{++} at the node?
Authors: Since Ca^{++} appears in the region of the paranodal junction it may be involved in the adhesion of the Schwann cell loops to the axolemma and/or be involved in the active transport of Na^+ between Schwann cell and axon.

G. Pappas: What could the "capturing" of Na^+ in the paranodal region mean?
Authors: It is possible that the voltage dependant Na^+ channels are localized in the paranodal region of the axon. A regulated reservoir of Na^+ closely apposed to this region may function to stabilize or modulate the impulse conduction related process.

A.R. Lieberman: In view of the differences in nodal organization between peripheral and central myelinated fibers, it would be interesting to compare the findings presented here with findings on central nodes. Are such observations available?
Authors: Yes, we have examined central nodes of the optic nerve after pyroantimonate fixation. There are precipitates formed in the glial loops however we have not yet tested these with the spectrometer.

A.R. Lieberman: Have the authors any comment on the possible relationship between their findings and other histochemical studies of nodes of Ranvier, particularly those of Langley and Landon and Waxman?
Authors: Langley and Landon have noted the polyanionic nature of the extracellular matrix above the nodal membrane away from the region of the paranodal glial loops. The presence of this matrix may be important for the absorption of Na^+ by the Schwann cell. Waxman's findings pertain to the staining of the nodal and paranodal membrane with $KFCN_6$. The functional significance of this stainability is obscure.

A.R. Lieberman: Does the region of increased subaxolemmal electron density seen after the Oschman and Wall procedure correspond in location and extent to the subaxolemmal undercoating of conventionally prepared material?
Authors: No, the subaxolemmal (cortical) density seen in conventional material is beneath the nodal membrane only. The Ca^{++} related density is associated with the paranodal glial-axonal-junction area.

A.R. Lieberman: Have the authors any observations to report on the precise localization of K-pyroantimonate precipitates within the axoplasm (at the node or elsewhere) and in particular, do they confirm the findings of Duce and Keen with respect to the SER?
Authors: It may be noteworthy in this regard that the pyroantimonate precipitates occur periodically along neurofilaments. This distribution may relate to the cross-bridges interlinking these elements. There are also precipitates formed in SER cisternae however, since this material was first fixed with osmium, much vesicu-lation of these labile cisternae has probably occurred. We have made no attempts to determine the elemental composition of these precipitates as others have determined these to be Ca^{++} rich sites.

SCANNING ELECTRON MICROSCOPY/1979/II
SEM Inc., AMF O'Hare, IL 60666, USA

METAL BINDING BY INTESTINAL MUCUS

J. R. Coleman   and   L. B. Young

Department of Radiation Biology and Biophysics
School of Medicine and Dentistry
University of Rochester
Rochester, NY 14642.

Abstract

Electron probe microanalysis offers distinct advantages for the study of intestinal mucus. This technique permits analysis of metal binding in situ, requires only a small amount of tissue, allows several experiments to be performed with one animal, and can resolve variations in binding that may occur in different portions of the intestine.

We have used electron probe microanalysis to examine the metal binding capacity of intestinal mucus in situ. We have exposed portions of excised intestine to various concentrations of several metals, rapidly frozen the tissue and freeze dried it. After anhydrous embedding, thick sections were cut and analyzed on silicon discs or carbon coated copper grids. Qualitative analysis shows two distinctive patterns of distribution. The results of this work show clearly that at least three divalent cations are bound by mucus, that mucus exhibits different affinities for different metals, and that binding of metals is not uniform throughout mucus.

Introduction

The role of intestinal mucus has not been clearly defined. It is commonly felt that mucus serves to lubricate the intestinal surfaces and may prevent microorganisms and some food products from making direct contact with the lumenal surfaces of intestinal cells[1,2]. Previous electron probe investigations of calcium absorption by rat small intestine had shown that intestinal mucus could bind a significant amount of calcium and that calcium binding capacity was not uniform throughout mucus[3]. It was not possible to determine whether the calcium bound to mucus might reflect the presence of the vitamin D dependent calcium binding protein (CaBP) isolated and studied by Wasserman and colleagues[4], or the presence of the goblet cell mucus component that has been isolated and studied by the Forstners and their colleagues[5-7]. This latter molecule, a glycoprotein with a polydisperse hexose content, is estimated to bind as many as 280 moles of calcium per mole of glycoprotein (MW 2×10^6) while CaBP binds tightly only about 4 moles of calcium per mole of CaBP (MW 28,000) with a few more low affinity sites available[4,6].

We wished to know whether mucus could also bind other divalent ions and, if so, whether the ions were bound to the same mucus components that bound calcium. This paper reports the distribution of some divalent ions within mucus.

Materials and Methods

Duodenum from white Charles River rats which had been deprived of food overnight were removed and cut into everted gut slices, and incubated as previously described[3]. The medium for incubation contained 120 mM NaCl, 10 mM KCl, 20 mM HEPES, pH 7.2 and divalent ions as in Table 1. Incubations were usually at 22° C for 10 min after which the tissue slices were briefly rinsed in cold incubation medium containing no divalent ions. The tissue was immediately flash-frozen and freeze dried according to the procedure of Halloran[8]. When dry, the tissue was exposed to osmium vapor under anhydrous conditions and embedded in Spurr's medium[9]. After polymerization, sections (2μ-0.1μ) were cut on dry glass knives and mounted on either silicon

KEY WORDS: Intestinal Mucus, Calcium, Metal Binding, Divalent Ions, Intestinal Absorption, X-Ray Microanalysis discs or carbon coated copper grids for analysis. Some tissue was removed from the animals, rinsed briefly in incubation medium with no added divalent ions, and prepared for analysis. Analysis was carried out with an ARL EMX-SM electron probe equipped with crystal spectrometers under conditions previously described[3,10], using the BASIC quantitation technique[11,12]; a Philips EM 200 was used for transmission electron microscopy.

Results

Each of the ions tested was found localized in mucus. Two types of distribution were found. Figure 1 shows the first type of distribution in which calcium and cadmium showed essentially equivalent distribution patterns. Of course, X-ray images of this sort are not suitable to determine whether the distributions are truly equivalent; quantitative analysis of many sites within the mucus will be required, and these are now underway. Similar patterns of equivalent distribution for calcium and the other divalent ion being tested were found in each preparation.

The second pattern of distribution is seen in Figure 2. These images show that Zn is distributed throughout the mucus in the lumen while calcium is distributed unevenly throughout the same mucus. While these distributions, in which the non-calcium divalent ion was more extensively distributed than was calcium, were common, the converse situation, in which calcium was more extensively distributed than the non-calcium ion, was not seen.

Tissue prepared directly after only brief rinsing or after incubation with calcium showed mucus localizations of only calcium. Therefore, localizations of other metals, after incubation with those metals, represented binding only during the incubation and were not endogenous.

In order to test whether these metal localizations were really mucus associated or whether they might represent adherent drops of incubation medium, the ratios of X-ray intensities (P-B) of each element in the localizations were compared

to the same ratios in samples of medium freeze dried or rapidly dried onto silicon discs. In each case, the media produced X-ray intensity ratios markedly different from the mucus localizations indicating that they were not dried medium. An example is given in Table 2.

Quantitative analyses of these localizations are now in progress. Examples of quantitative analyses of two regions in the mucus of Figure 2 are presented in Tables 3 and 4. Table 3 represents the analysis of a calcium "poor" region (#2) and Table 4 that of a calcium "rich" region (#1). This shows clearly that the two regions differ quite markedly in the amounts of each element. Further quantitative analyses are now in progress.

Discussion

The glycoprotein isolated by Forstner et al.[5] is 10% weight sialic acid and contains about 300 potential binding sites per molecule. Consequently, the mucus must act as a "buffer" to influence the concentration of calcium free in the lumen, so that the concentration of free calcium will be determined by the amount of mucus and the binding constant of mucus for calcium, the association of calcium with food or other materials in the lumen, and the rate of removal of calcium from the lumen by absorption.

$$\begin{array}{c} Ca \cdot food \\ Ca \cdot mucus \end{array} \rightleftarrows Ca^{2+} \longrightarrow absorption$$

It is appropriate to consider what the calcium binding capacity of mucus might mean for the overall absorption process. First, because mucus components are only slowly degraded in the small intestine, any calcium associated with them is likely to remain with them until they reach the large intestine where calcium absorption is low or until they are excreted. If so, then mucus may play a role in calcium excretion.

Table 1

INCUBATION MEDIA ADDITIONS

1. Rinse only, no divalent cations

2. 10' incubation, no divalent cations

3. 10' incubation, 1 mM $CaCl_2$

4. 10' incubation, 1 mM Ca + 10 mM $HgCl_2$

5. 10' incubation, 1 mM Ca + 1 mM Pb acetate

6. 10' incubation, 1 mM Ca + 10 mM Pb acetate

7. 10' incubation, 1 mM Ca + 1 mM $CdCl_2$

8. 10' incubation, 1 mM Ca + 10 mM $CdCl_2$

9. 10' incubation, 1 mM Ca + 1 mM $ZnCl_2$

Table 2

X-RAY INTENSITIES (P-B) CPM

	Medium	Mucus Localization
Hg	1958	5340
Si	20318	53895
Na	83392	708
S	4389	431
Cl	84637	8975
K	12065	1142
Na/K	6.912	0.620
Hg/Na	0.023	0.763

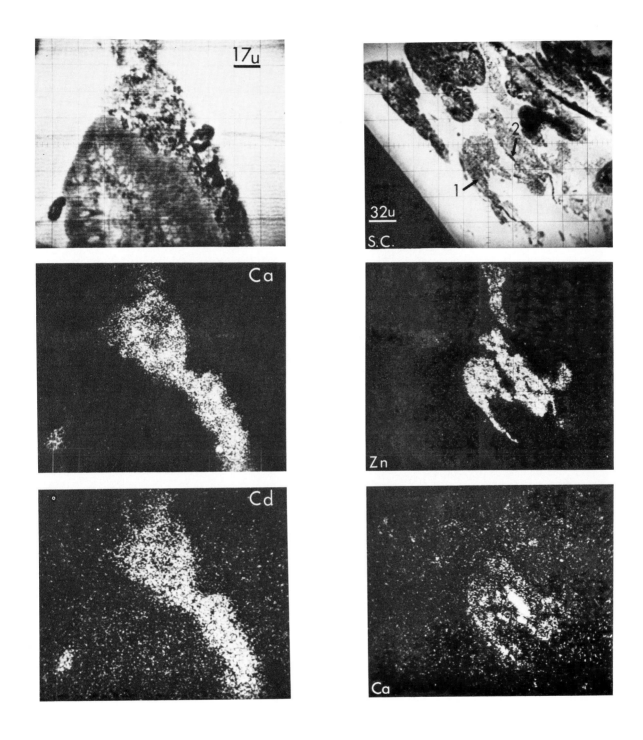

Figure 1. Sample current image (top) of section from duodenum showing the tip of a villus and associated mucus in tissue which had been exposed to 1 mM Ca and 10 mM Cd. Middle image is Ca K$_\alpha$ X-ray image. Bottom is Cd K$_\alpha$ image.

Bar=17 μm

Figure 2. Sample current image (top, S.C.) showing portion of villi in section of duodenum which had been exposed to 1 mM Ca and 10 mM Zn. Middle image is Zn K$_\alpha$ image. Bottom is Ca K$_\alpha$ image.

Bar=32 μm

Table 3

ANALYSIS OF AREA #2 IN FIGURE 2

Element	Weight Percent	Atomic Percent
Ca	2.33	13.31
P	1.46	10.78
K	1.07	6.28
S	0.21	1.48
Cl	4.57	29.49
Na	2.32	23.06
Zn	4.46	15.60

Accelerating Voltage	20.0 kev
X-Ray Emergence Angle	52.5 degrees
Density Times Thickness	2.903 mg/cm^2
E sub L	16.2 kv
Original Substrate Intensity	0.4470
Corrected Substrate Intensity	0.7565

Matrix of Elemental Ratios

	Ca	P	K	S	Cl	Na	Zn
Ca	1.000	0.810	0.472	0.111	2.217	1.733	1.172
P	1.234	1.000	0.582	0.138	2.736	2.139	1.447
K	2.120	1.717	1.000	0.236	4.699	3.673	2.485
S	8.973	7.269	4.233	1.000	19.899	15.549	10.519
Cl	0.451	0.365	0.213	0.050	1.000	0.782	0.529
Na	0.577	0.468	0.272	0.064	1.279	1.000	0.677
Zn	0.853	0.691	0.402	0.095	1.891	1.478	1.000

Column is numerator, row is denominator, ratios in atomic percent.

Table 4

ANALYSIS OF AREA #1 IN FIGURE 2

Element	Weight Percent	Atomic Percent
Ca	0.21	0.90
P	2.81	16.05
K	0.59	2.67
S	0.17	0.94
Cl	5.45	27.17
Na	1.75	13.48
Zn	14.35	38.79

Accelerating Voltage	20.0 kev
X-Ray Emergence Angle	52.5 degrees
Density Times Thickness	2.307 mg/cm^2
E sub L	17.1 kv
Original Substrate Intensity	0.4450
Corrected Substrate Intensity	0.8084

Matrix of Elemental Ratios

	Ca	P	K	S	Cl	Na	Zn
Ca	1.000	17.754	2.948	1.038	30.049	14.908	42.905
P	0.056	1.000	0.166	0.058	1.693	0.840	2.417
K	0.339	6.023	1.000	0.352	10.195	5.058	14.556
S	0.963	17.104	2.840	1.000	28.949	14.362	41.333
Cl	0.033	0.591	0.098	0.035	1.000	0.496	1.428
Na	0.067	1.191	0.198	0.070	2.016	1.000	2.878
Zn	0.023	0.414	0.069	0.024	0.700	0.347	1.000

Column is numerator, row is denominator, ratios in atomic percent.

The fact that mucus binds other divalent ions as well as calcium indicates that it may accomplish the same buffering actions with these metals as well. An additional role for mucus then might be to prevent uncontrolled or excessive absorption of toxic divalent metals. The fact that chronic ingestion of Cd produces a tropical sprue-like condition marked by goblet cell hyperplasia and an excessive amount of mucus in the intestinal lumen[13], may even be an indication that mucus is being produced in large quantities to affect the excretion of cadmium.

Our findings indicate that there are portions of mucus that have higher affinities for one divalent ion than another (Figure 2). If so, then it may mean that mucus could be synthesized to handle overloads of specific divalent ions. This will be examined in greater detail.

Finally, the observations raise the question of whether mucus in other epithelial tracts, including the respiratory tract, might function in a similar fashion to trap ions and thereby control the concentration of free divalent ions.

Acknowledgements

This paper is based on work performed partially under contract with the U. S. Department of Energy at the University of Rochester Department of Radiation Biology and Biophysics and partially under NIH grant #AM-14272 and has been assigned Report No. UR-3490-1539.

References

1. Trier, J.S. Morphology of the Epithelium of the Small Intestine, in Handbook of Physiology, Vol. 3 (C.F. Code, ed.) American Physiology Society, Washington, D.C., p. 1151, 1968.
2. Hendrix, T.R. The Secretory Functions of the Alimentary Canal, in Medical Physiology (V. Mountcastle, ed.) C.V. Mosley, p. 1178, 1974.
3. Warner, R.R. and J.R. Coleman. Electron Probe Analysis of Calcium Transport by Small Intestine, J. Cell Biol., 64: 54-68, 1975.
4. Fullmer, C.S. and R.H. Wasserman. Vitamin D-dependent Calcium Binding Proteins, in Calcium Binding Proteins and Calcium Function (R.H. Wasserman, R.A. Corradino, E. Carafoli, R.H. Kretsinger, D.H. McLennan, and F.L. Siegel, eds.) North-Holland, New York, p. 293, 1977.
5. Forstner, J.F., I. Jabbal, and G.G. Forstner. Goblet Cell Mucin of Rat Small Intestine. Chemical and Physical Characterization, Can. J. Biochem., 51: 1154-1166, 1973.
6. Forstner, J.F. and G.G. Forstner. Calcium Binding of Goblet Cell Mucin, Biochim. Biophys. Acta, 386: 283-292, 1974.
7. Forstner, J.F. and G.G. Forstner. Effects of Calcium on Intestinal Mucin: Implications for Cystic Fibrosis, Pediat. Res., 10: 609-613, 1976.
8. Halloran, B.P. Calcium Transport in the Intestine. Ph.D. Thesis, University of Rochester, 184 pp., 1976.
9. Spurr, A. A Low Viscosity Embedding Medium for Electron Microscopy, J. Ultrastruc. Res., 26: 31-43, 1969.
10. Coleman, J.R. and A.R. Terepka. Electron Probe Analysis of the Calcium Distribution in Cells of the Embryonic Chick Chorioallantoic Membrane. I. A Critical Evaluation of Techniques, J. Histochem. Cytochem., 20: 401-413, 1972.
11. Warner, R.R. and J.R. Coleman. A Biological Thin Specimen Microprobe Quantitation Procedure that Calculates Composition and ρ_X, Micron, 6: 79-84, 1975.
12. Coleman, J.R. X-Ray Analysis of Biological Samples, in SEM/1978/II, SEM Inc., AMF O'Hare, IL 60666, 911-926, 1978.
13. Richardson, M.E. and M.R. Spivey-Fox. Dietary Cadmium and Enteropathy in the Japanese Quail, Lab. Invest., 31: 722-731, 1975.

Discussion with Reviewers

Forstner: It seems possible that some of the material in the lumen to which ions are bound might not be mucus. If so, then differences in distribution of various ions may not be related to mucus heterogeneity. What evidence do you have that the major ion-binding component in your duodenal slices is in fact mucin?

Authors: Even in fasted animals, the intestinal lumen may contain non-mucus materials such as bacteria or desquamated epithelial cells, and it is quite conceivable that these could bind metals. Bacteria and senescent cells or cell fragments might not be readily identified in sample current images such as in Figures 1 and 2. In order to eliminate this possibility, we analyze thin sections using coordinated transmission electron microscopy/electron probe microanalysis (J.R. Coleman, S. Davis, B. Halloran, and P. Moran. A Simple Transmitted Electron Detector (TED) for Thin Biological Samples. Proc. X Ann. Conf. Microbeam Analysis Society [available from Dr. J.I. Goldstein, Dept. Metallurgy and Materials Science, Lehigh University, Lehigh, Penn.] 45A-45G, 1975) and this permits us to determine whether mucus or other materials occur at the metal binding sites. In every case, metal binding has been by the fibrous material identified as mucus in transmission electron micrographs. We will continue to check our samples for this possibility.

Forstner: Have you noted any differences in metal binding between mucus in crypt and villus locations; or have you investigated whether differences occur in other sections of the small intestine, or even in the large intestine?

Authors: The analyses are rather time consuming because we use wavelength dispersive analysis; consequently, our studies have been limited mainly to duodenum. The few analyses we have performed in other small intestinal regions did not reveal any striking differences. We hope to have access to an energy dispersive X-ray spectrometer which will greatly reduce the amount of time required for analysis.

Forstner: Why did you choose the incubation conditions you used? Do you know if varying these conditions affects binding? During the rinsing, is it possible to wash away surface mucus and find a parallel decrease in metal binding to intestine?

Authors: The incubation conditions are essentially those we use to study intestinal calcium absorption. We have been studying the effects of these metals on calcium absorption and vice versa so that we wish to keep conditions uniform. We have varied the buffer, the incubation times, and the temperature in a few cases, but we do not have sufficient information to draw any firm conclusions from these variations. We think that the rinsing procedure is very important. When we study calcium absorption, we incubate similar intestinal slices with ^{45}Ca and measure absorption by scintillation counting. Consequently, any ^{45}Ca bound by surface mucus attached to the slice of intestine will be measured as having been absorbed. Thus, it is necessary to have a measure of how much of the total ^{45}Ca is due to mucus and how much is really within absorptive cells. If the rinsing procedure is variable, then the amount of mucus left clinging to the tissue may also be variable, and it is likely that this will influence the proportion of total ^{45}Ca which bound to mucus and the remainder which will be within cells. For these reasons, we have tried to keep the rinsing procedure constant.

M. and S. Tannenbaum: Are there physiological or pathological variations in mucus and, if so, have you examined these?

Authors: We have not studied such variations yet, but a good deal of information is available from studies of mucoviscidosis (cystic fibrosis); for example: Mucus Secretions and Cystic Fibrosis (G.G. Forstner, ed.) Modern Problems in Pediatrics, XIX, S. Karger, Basel, 1977.

Carter: Have you compared fasted to non-fasted animals or have you examined the effects of oral administration of metals?

Authors: We have examined only a few non-fasted animals and have noted no difference. Our studies of calcium absorption employ fasted animals, thus we have chosen to study binding by mucus in these too. We plan oral administration of metals but have chosen in vitro exposures to begin this work because of the convenience of performing control experiments on the same animal and because of the reproducibility of this technique.

Reviewer III: Have you attempted to examine the interaction of iron with intestinal mucus?

Authors: The intestinal absorption of iron is an interesting and important problem. Absorption occurs mainly as the divalent ferrous form rather than the trivalent ferric form, but the nature of the absorption process is not clear. However, we have not specifically looked for iron in our studies.

Reviewer III: Have the authors given any thought to the use of this technique as a diagnostic tool for malabsorption problems?

Authors: The potential for this technique to be employed as a diagnostic tool in malabsorption problems seems good. Only a small tissue sample is required and this can be taken by biopsy. Techniques are currently available which permit biopsy of various portions of the intestine, and it seems reasonable to compare the metal binding properties of normal mucus with those from patients exhibiting malabsorption symptoms. Our immediate aim is to use the technique to determine the role of mucus in the absorption of various metals in animal models.

Forstner: How do you distinguish calcium "rich" from calcium "poor" regions as in Figure 2?

Authors: This is simply a qualitative and rather arbitrary judgment based on estimating the X-ray intensity at the sites in the X-ray image.

REVIEWERS: H.W. Carter
                    J. Forstner
          III A.L. Jones
               M. Tannenbaum
               S. Tannenbaum

SCANNING ELECTRON MICROSCOPY/1979/II
SEM Inc., AMF O'Hare, IL 60666, USA

PROGRESS IN THE DEVELOPMENT OF THE PEAK-TO-BACKGROUND METHOD FOR THE
QUANTITATIVE ANALYSIS OF SINGLE PARTICLES WITH THE ELECTRON PROBE

J.A. Small, K.F.J. Heinrich, D.E. Newbury and R.L. Myklebust

Gas and Particulate Science Division,
Building 222
National Bureau of Standards
Washington, DC 20234

Abstract

A method is described for the quantitative analysis of particles with electron excitation. For this method, the ratios of the characteristic x-ray intensities to the continuous x-ray intensities of the same energy are used to eliminate major particle effects. The procedure consists of scaling the peak intensity of the particle up to a value appropriate to a bulk material. The scaled intensity can then be compared to a standard and the concentration determined by a ZAF routine.

In order to use the analytical procedure it is necessary to calculate a value of the continuous x-ray intensity for a hypothetical bulk material with the same composition as the unknown. To date, two methods have been used to calculate this value. The first method uses Kramers' law with Green's corrections to Kramers' constant. The results from the analysis of both glass and talc particles with this version of the program are in good agreement with the actual concentrations. The second method uses an empirical equation derived at NBS to calculate the continuous x-ray intensity of the bulk. The results on FeS_2 particles with this version of the program decrease the relative error in the Fe concentration from 19 percent for Kramers' equation to six percent.

In addition to calculating the values for the hypothetical bulk material, work is also being done with a Monte Carlo program in an effort to identify those conditions under which the correspondence of the peak-to-background ratio between particles and bulk material of the same composition breaks down. Calculations show that there is a significant deviation in the peak-to-background ratios from particles compared to bulk due to two effects: 1) the anisotropy of the generation of the continuous x-radiation and 2) differences in the energy dependence of the cross-section for characteristic and continuous x-radiation coupled with differences in the energy distribution of electrons backscattered from bulk and particle targets.

KEY WORDS: Electron Probe Microanalysis, Particle, Quantitative Analysis, Peak-to-Background Ratios, Monte Carlo Methods, Glass Particles, Mineral Particles, X-Ray Microanalysis

Introduction

For the purpose of quantitative particle analysis with the electron probe, particles can be divided into three size ranges. The first size range is the large particles which have dimensions in excess of 20 μm. The electron interaction volume in these particles at a beam energy of 20 keV or less is completely contained within the particle and is virtually equivalent to the interaction volume generated within a flat, bulk specimen. The surface topography across the interaction volume of the particle may lead to slight differences in the average x-ray absorption path between particle and standard, but for an x-ray detector placed at a high take-off angle, the deviation from the flat bulk standard leads to relatively small errors. As a result, large particles can be analyzed with conventional ZAF procedures providing: 1) the electron excitation volume is contained within the particle volume. 2) the detector take-off angle is relatively high and the particle orientation is such that the difference between the average absorption pathlength in the particle and the path-length in a flat standard is minimized.

The second size range is the small particles which have dimensions less than 0.2 μm. The particles in this range are small enough that x-ray absorption and fluorescence is relatively insignificant. In addition, at high beam energies, 100 keV or more, elastic scattering effects are minimized. This effect reduces the need for an atomic number correction. The primary effect remaining is a mass effect (explained in the next section), and as a result small particles can be analyzed by the techniques of analytical electron microscopy which have been developed for thin films and which are based on relative sensitivity factors.[1]

The third size range consists of particles which have dimensions from 0.2 to 20 μm. For the particles in this size range there are two major effects which make conventional ZAF analysis of these particles incorrect and which therefore must be taken into account in developing a quantitative analysis scheme.

The first major effect arises from the finite size (mass) of a particle. This mass effect is important when the interaction volume

of the primary electron beam approaches and
becomes larger than the volume of the analyzed
particle. As the particle volume decreases, the
mass of material participating in x-ray generation is less in a particle than it is in a thick
specimen since the beam electrons escape through
the sides and bottom of the particle. This
effect results in a reduction of x-ray intensities from the particles relative to the intensities from bulk materials and the deviation
from bulk behavior increases in magnitude with
decreasing particle size. The mass effect can be
demonstrated by comparing the x-ray emissions
from a particle to the emissions from a bulk target of the same composition. The effect is more
evident when the comparison is done for a relatively hard energy x-ray line which has minimal
absorption. Such a case is shown in Figure 1 by
the decrease in the Ba intensity from the particles relative to the bulk material for particles smaller than 3 μm in diameter.

The second major particle effect is the
result of absorption and is important when the
average x-ray absorption path length for a particle is different from the average path length
for a bulk material. The greatest effect occurs
when there is high absorption as is typically
observed for low energy x-ray lines. Such a
situation is common in many samples containing
elements such as aluminum and silicon. The
difference in average path length can result in
widely different values of x-ray intensities
between particle and bulk. For the situation
shown in Figure 2 the path length A-B in the particles is less than the path length A-C in the
bulk. The reduced average path length results in
a greater x-ray emission from the particle compared to bulk, despite the opposing action of the
particle mass effect. The absorption effect is
demonstrated in Figure 3 where the silicon intensity ratio between particle and bulk is above
unity for spheres greater than 2.0 μm in diameter.

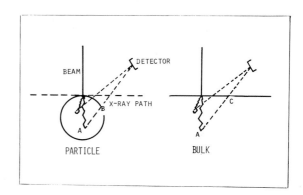

Figure 2. Diagram of particle absorption effect.

Various analytical techniques have been
developed for the analysis of the intermediate-
size particles, 0.2 μm to 20 μm in diameter.
These procedures range in complexity from the
simple scaling of the peak intensities obtained
by conventional ZAF analysis[2] to the extensive
treatment based on shape determination proposed
by Armstrong.[3] The main problem with these
existing techniques are that they require either
a large computer and extensive analyst training
or they are limited to a small number of particle
compositions for which the particle effects are
minimized.

At the 1978 Scanning Electron Microscopy
meeting, a new approach to quantitative analysis
of the intermediate-sized particles based on peak-
to-local-background measurements was proposed
independently by Small et al. for the determination of concentrations and by Statham and Pawley
for the determination of concentration ratios.[4,5]
This approach is based on the use of the ratio
between the characteristic x-ray intensities and
the continuous x-ray intensities of the same
energies (peak-to-local-background ratios) to

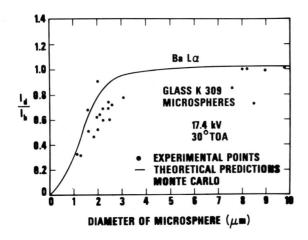

Figure 1. Ba Lα x-ray intensity for glass spheres
ratioed to bulk intensity and plotted as a function of sphere diameter. I_d=sphere intensity,
I_b=bulk intensity and TOA= take-off angle.

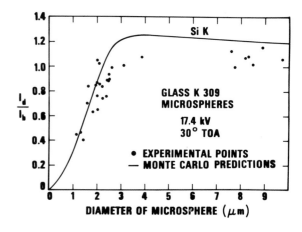

Figure 3. Si K x-ray intensity for glass spheres
ratioed to bulk intensity and plotted as a function of sphere diameter. I_d=sphere intensity,
I_b=bulk intensity and TOA= take-off angle.

eliminate the major particle effects. It is a modification of the method used by Hall to cortect for the excitation volume in thin films.[6] In this paper, the current status of the analytical procedure being developed at NBS will be discussed and the results obtained from the analysis of glass and mineral particles will be compared to the results obtained from conventional ZAF analysis.

Analytical Procedure and Results

Unlike the procedure being developed by Statham in which the results are expressed as concentration ratios, the method under development at NBS expresses results as elemental concentrations. The NBS procedure requires that the particles are mounted on a thin carbon foil. This is necessary in order to minimize the interference of x-rays produced by the mounting substrate. A discussion of sample preparation procedures is given in reference 7.

Equation 1 is the foundation for the particle procedure and resulted from the experimental observation that the peak-to-local-background ratios from particles are, to a first approximation, equivalent to the peak-to-local-background ratios from a bulk sample of the same composition.[4]

$$(P/B)_{particle} = (P/B)_{bulk} \qquad (1)$$

Rearranging equation 1, the peak intensities for the particle can be scaled up to values appropriate to a hypothetical bulk material of the same composition, equation 2.

$$P^*_{particle} = P_{bulk} = \frac{P_{particle} \cdot B_{bulk}}{B_{particle}} \qquad (2)$$

The values of the scaled intensities, P*s, can then be used to form x-ray intensity ratios which would then be used as input for a standard bulk analysis, ZAF, routine such as FRAME C as if no particle effects existed. In practice, for a typical particle of unknown composition, bulk material with the same compositions will not be available to measure the values of B(bulk). These values, however, can be determined from the first estimate of composition obtained from the hyperbolic iteration loop in the ZAF program. These values of B(bulk) can then be used in equation 2 to obtain the first estimates of P*'s. A set of intensity ratios is calculated from equation 3 and used as input for the ZAF routine. The set of concentrations from each iteration is used to calculate new values of B(bulk) and the sequence is repeated. A block diagram for the particle procedure is shown in Figure 4.

$$k = P^*_{particle}/P_{standard} \qquad (3)$$

Equations for Calculating B(bulk)

In the development of the analytical procedure, two equations have been used to calculate the values of B(bulk). Equation 4 which predicts B(bulk) from the continuum on a standard, B(std)

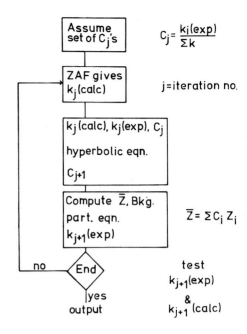

$$C_j = \frac{k_i(exp)}{\Sigma k}$$

j=iteration no.

$$\overline{Z} = \Sigma C_i Z_i$$

test $k_{j+1}(exp)$ & $k_{j+1}(calc)$

Figure 4. Flow chart showing the interface of the particle corrections with a ZAF analysis scheme.

used the atomic number dependence of Kramers' law as modified by Green.[8,9]

$$B_{bulk} = \frac{\overline{Z}_{bulk}}{\overline{Z}_{std}} \cdot B_{std} \times \frac{f_\chi(bulk)}{f_\chi(std)} \cdot \frac{F(\overline{Z}_{bulk})}{F(\overline{Z}_{std})} \qquad (4)$$

The f_χ's are the absorption terms from the ZAF routine for the current estimate of bulk composition and standard, $\overline{Z}$s refer to the average atomic numbers of the bulk and standard, and the $F(\overline{Z})$'s are the Green corrections for the Z dependence of Kramers' constant. Substituting equation 4 into equation 2 the scaled particle intensity can be expressed by equation 5.

$$P^*_{(part)} = (P/B)_{part} \times B_{std} \times \frac{F_\chi(bulk)}{F_\chi(std)} \cdot$$

$$\frac{\overline{Z}_{bulk}}{\overline{Z}_{std}} \times \frac{F(\overline{Z}\,bulk)}{F(\overline{Z}\,std)} \qquad (5)$$

In order to test the particle analysis scheme it was necessary to obtain particles of known composition. For this purpose, glass particles were made by crushing bulk glass from the set of NBS Research Material Glasses.[10] Results from the analysis with Kramers' law of NBS glass K-309 particles are shown in Table 1. Both particles were irregularly shaped and are shown in Figures 5a 5b. Glass K-411 (see Table 2) was used as the standard for Si, Ca, and Fe and bulk glass K-309 was used as the standard for Al and Ba. The results from the particle program are substantially closer to the nominal values than the

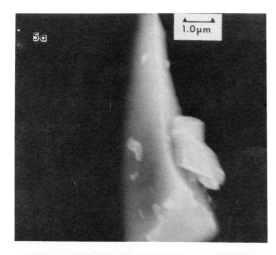

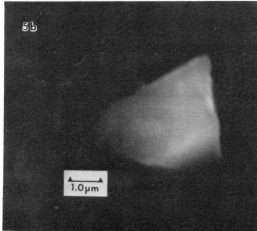

Figure 5a and 5b. Micrographs of K-309 glass particles

results from a conventional ZAF routine such as FRAME C. The relative errors for the particle results are all under 10 percent except for Si which is 11 percent. In comparison, the results from FRAME C have relative errors on the order of 65 percent.

Table 1. Analysis of K-309 Particles

Element	True Composition	Particle 5a	FRAME 5b	FRAME C 5a	5b
Al	0.079	0.070	0.084	0.014	0.034
Si	0.187	0.178	0.208	0.035	0.079
Ca	0.107	0.099	0.101	0.020	0.039
Ba	0.134	0.124	0.128	0.026	0.050
Fe	0.105	0.093	0.104	0.021	0.043

K-411 Glass, $\overline{Z}$-13 was the standard for Si, Ca, Fe. K-309 Glass, $\overline{Z}$-19 was the standard for Al, Ba. Accelerating Voltage was 20 kV.

Table 2. Composition of Glass K-411

Element	Wt. Fraction*
Si	.2539
Fe	.1120
Mg	.0884
Ca	.1106
O	.4351

*Determined by wet chemical analysis.

The second method used to calculate the values of B(bulk) from a standard is given by equation 6, which was derived empirically at NBS. This equation takes into account an experimentally observed nonlinear relationship between generated continuum intensity and atomic number. The equation was determined by fitting, at 3, 5, and 7 keV, a curve to the plot of generated continuum versus atomic number for some 40 pure elements. All elements were run with a normal beam incidence. The various points of the plot were calculated from a background fit of the experimental data.[11] The values from the fit were then converted to generated continuum by dividing by the absorption and detector efficiency terms used in FRAME C.

At the present time, equation 6 is being tested in order to evaluate its applicability to different detector and electron probe systems. In addition, the equation is being modified to improve the accuracy in predicting the continuous x-ray emission of a multielement system from a pure element standard.

$$B_{bulk} = \frac{B_{(std)} \times [(0.006E + 0.0642)\overline{Z}^n_{bulk} - 0.012E - 0.088]}{(0.006E + 0.0642)\overline{Z}^n_{(std)} - 0.12E - 0.088} \quad (6)$$

E is the energy in keV of the intensity to be calculated. n is a term which is dependent on the beam energy and ranges from about 1.0 for 10 keV to 0.44 for 25 keV.

The results of the analysis with this equation for talc particles are shown in Table 3. The calculated values for Al and Mg compare favorably between the methods based on Kramers' equation and the NBS equation. The analyses by both methods are well within a 10 percent relative error while the values from FRAME C are about 63 percent low.

A more dramatic effect can be seen in Table 4 which reports the analysis of FeS_2 particles. The results show a decrease in the relative error of Fe from 19 percent for the method based on Kramers' law to approximately five percent for the method based on the NBS equation. The relative error on the sulfur analyses is somewhat greater for particle #2 using the NBS equation but the value is still within six percent.

Table 3. Analysis of Talc Particles

	Z	Stoic.*	Frame C	Kramers (Eq. 5)	NBS (Eq. 6)	Aver. dimension
#1	12	.193	.087	.190	.191	3 μm
	13	.298	.134	.312	.301	
#2	12	.193	.056	.187	.191	2 μm
	14	.298	.085	.307	.298	
#3	12	.193	.166	.179	.175	6 μm
	14	.298	.254	.209	.272	

Accelerating voltage was 10 kV.
K-411 Glass (Table 2) was used as the standard
   for Mg and Si.

---

*Stoichiometric concentration

Table 4. Analysis of FeS$_2$ Particles

	Z	Stoic.*	Frame C	Kramers (Eq. 5)	NBS (Eq. 6)	Aver. dimension
#1	16	.534	.313	.532	.535	5 μm
	26	.466	.308	.557	.497	
#2	16	.534	.460	.577	.562	5 μm
	26	.466	.397	.546	.483	

Accelerating voltage was 15 kV.
ZnS and Fe were used as standards.

---

*Stoichiometric concentration

Monte Carlo Calculation of Peak-to-Background Ratio

In addition to the two major particle effects mentioned earlier, an additional source of error in the analysis of particles 0.2 to 2 μm in size may result from deviations in the peak-to-background ratios for particles compared to the values for bulk specimens. In an effort to predict these deviations, a Monte Carlo electron trajectory simulation technique has been employed to study the behavior of the peak-to-background ratio as a function of the beam and particle parameters.[12,13] In particular, the Monte Carlo technique has been used to investigate those conditions under which the correspondence of the peak-to-background ratio between particles and bulk material of the same composition breaks down.

The hybrid Monte Carlo simulation technique used for this work has been described in detail elsewhere.[14,15,16] For the present application, the cross-section for inner shell ionization proposed by Green and Cosslett[17] and the cross-section for continuum x-ray generation described by Kirkpatrick and Wiedmann[18] were used for the calculation of the peak-to-background ratios. All calculations were made for an x-ray emergence angle of 40 degrees (50 degrees from the beam axis) since previous calculations indicated that anomalous behavior of the x-ray intensities is observed for low x-ray emergence angles (less than 20 degrees) due to absorption effects.[12] The electron beam was assumed to be of constant current density and the beam diameter was set equal to the particle diameter (for spherical particles). These beam conditions closely approximate the conditions used in experimental practice, when raster scanning has been employed.[4]

Comparison of Peak-to-Background Ratios in Particle and Bulk Material

In order to theoretically test the experimental observation (mentioned earlier) that, for materials of identical composition, the peak-to-background ratio for a particle is the same as that for the bulk, Monte Carlo calculations have been made for the multielement targets with the compositions of NBS glasses K-309 and K-227 (Table 5). A spherical particle target with a diameter of 6 μm was considered, and the beam energy was taken as 20 keV. The calculated results, shown in Table 5, confirm that the peak-to-background ratios are indeed similar for bulk and for particle targets over a wide range of x-ray energy. For comparison purposes, the ratio of the characteristic intensities for particle

Table 5. Comparison of Calculated X-ray Emissions from Spherical Particles and Bulk Targets

Element	Concentration (wt.%)	X-ray Line Energy keV	$\frac{(P/B)particle}{(P/B)bulk}$	$\frac{P_{particle}}{P_{bulk}}$
NBS Glass K-309 (Particle diameter 6 μm)				
Al	7.9	1.49	1.03	1.12
Si	18.7	1.74	1.00	1.06
Ca	10.7	3.69	1.02	0.86
Ba	13.4	4.45	1.06	0.86
Fe	10.5	6.40	1.05	0.87
NBS Glass K-227 (Particle diameter 1.5 μm)				
Si	9.4	1.74	1.07	0.72
Pb L	74.3	10.5	1.25	0.64
Pb M	same	2.35	1.08	0.81

Accelerating voltage was 20 kV

X-ray emergence angle 40° (detector axis 50° from beam axis)

and bulk targets are also given. This ratio shows a significantly larger deviation from unity, generally in excess of 10 percent, than the ratio of peak-to-background values, which deviate from unity 0-6 percent.

Effect of Particle Size

Monte Carlo calculations for x-ray emission from small particles have been carried out for K-309 glass as a function of particle diameter. The results, shown in Table 6, reveal that the peak-to-background ratio for the particle deviates from that of the bulk, and becomes larger than the bulk value as the particle diameter decreases. For a K-309 particle with a diameter of 1 μm, the deviation from unity is in the range 30-55 percent depending on the energy of the x-ray line considered.

Effect of Particle Shape

To assess the effect of the particle shape on the peak-to-background ratio, calculations were made comparing the x-ray production of K-309 particles of spherical and cylindrical shape

Table 6. $\dfrac{(Peak/BG)_{Sphere}}{(Peak/BG)_{Bulk}}$ Experimental

Element	X-ray Line Energy keV	Diameter (μm)				
		6	3	2	1	0.5
Al	1.48	1.03	1.10	1.21	1.39	1.51
Si	1.74	1.00	1.05	1.14	1.29	1.38
Ca	3.69	1.02	1.10	1.22	1.43	1.54
Ba	4.46	1.06	1.14	1.30	1.55	1.67
Fe	6.40	1.05	1.11	1.31	1.54	1.67

For K-309 Glass at 20 kV
X-ray emergence angle = 40°

(Table 7). The results indicate that there is a shape effect, with the cylindrical particles showing less deviations from bulk behavior than the spherical particles. For example with a diameter of 1 μm, the deviation from unity of the peak-to-background ratios for several elements is 14-33 percent for a cylinder and 29-55 percent for a sphere.

Effect of Beam Energy

The effect of changing the beam energy of the peak-to-background ratio of particles compared to bulk targets has also been calculated with the Monte Carlo simulation for K-309 glass. The results for two different energies and three particle sizes, Table 8, reveal that lowering the beam energy causes a reduction in the deviation of the peak-to-background ratio from that of a bulk target.

Origin of the Deviation

Two effects have been identified as contributing to the deviation of the peak-to-background ratio of the particles from the bulk value. These effects are: (a) the anisotropy of the generation of the continuous x-radiation and (b) differences in the energy dependence of the cross sections for characteristic and continuous x-radiation coupled

Table 7. Comparison of Spheres (SPH) and Cylinders (CYL) of Glass K-309 at 20 kV (Emergence angle = 40°)

$\dfrac{(Peak/BG)_{particle}}{(Peak/BG)_{bulk}}$ Calculated

Element	X-ray Line Energy keV	Diameter (μm)			
		6		1	
		SPH	CYL	SPH	CYL
Al	1.48	1.03	1.01	1.39	1.21
Si	1.74	1.00	0.98	1.29	1.14
Ca	3.69	1.02	1.00	1.43	1.24
Ba	4.46	1.06	1.02	1.55	1.31
Fe	6.40	1.05	1.01	1.54	1.33

with differences in the energy distributions of electrons backscattered from bulk and particle targets.

Anisotropy of the Continuum. The generation of the x-ray continuum is anisotropic; its directional distribution is related to the trajectory of the beam electrons, and as a result it peaks in the forward direction.[18] The generation of the characteristic radiation, to the contrary, is isotropic, since the characteristic x-rays are produced during the de-excitation of an ionized atom, a process which is decoupled from the direction of flight of the ionizing electron. However,

Table 8. Effect of Acceleration Potential on Peak-to-Background Ratios

$\dfrac{(Peak/BG)_{sphere}}{(Peak/BG)_{bulk}}$

Element	X-ray Line Energy keV	Diameter (μm)					
		3		1		0.5	
		20kv	10kv	20kV	10kV	20kV	10kV
Al	1.48	1.10	1.01	1.39	1.06	1.51	1.21
Si	1.74	1.05	1.01	1.29	1.05	1.38	1.18
Ca	3.69	1.10	1.01	1.43	1.08	1.54	1.27
Ba	4.46	1.14	1.01	1.55	1.11	1.67	1.33
Fe	6.40	1.11	1.01	1.54	1.08	1.67	1.31

K-309 Glass, Emergence Angle = 40°.

in a bulk target, elastic scattering randomizes the trajectories of different electrons. Although the continuous radiation generated in a single electron trajectory segment is anisotropic, the integrated continuum generated from the entire interaction volume is randomized and appears to be nearly isotropic. In a thin foil or particle target, the electrons undergo fewer elastic scattering events before escaping, and hence, most of the trajectory segments are nearly aligned. The anisotropy of the generation of the continuum will thus be more apparent from a thin target than from a bulk target.

In order to assess the portion of the deviation of peak-to-background ratios of particles from bulk values which could be ascribed to anisotropic generation effects, Monte Carlo calculations were carried out with the proper anisotropic continuum cross section and with a postulated isotropic continuum cross section. The results for several x-ray energies in Figure 6 show that, at any particle diameter, about half of the deviation observed is due to the anisotropy of continuum generation. However, deviations of 10 to 30 percent from the bulk value still would be observed for small particles even if the generation of the continuum were isotropic.

Energy Dependence of the Cross Sections.
The remainder of the deviation observed in Figure 6 (curves marked "isotropic") arises from differences in the behavior of the processes of generation of characteristic and continuous x-radiation as a function of energy. The cross sections for ionization of the iron K-shell[17] and for the generation of 6.4 keV continuum[18] are plotted as a function of electron energy in Figure 7. As the electron energy decreases, the cross section for continuum generation increases continually, while the cross section for inner shell ionization goes through a maximum at an overvoltage U (U=E/Eq, where Eq is the critical ionization energy for the shell) of about two to three keV and then sharply decreases to zero. In the case of iron, electrons in the energy range of 6.4 to 12 keV make significant contributions to the 6.4 keV continuum but are relatively ineffective at generating characteristic FeK x-radiation. In

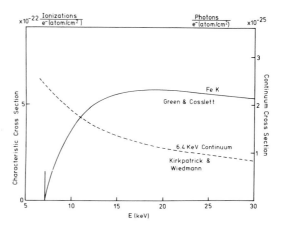

Figure 7. The cross sections for ionization of the iron K shell and generation of 6.4 keV continuum plotted as a function of electron energy.

the energy range of 15-20 keV, the situation is reversed.

Electrons scattering within a small particle are much more likely to escape ("backscatter") due to the proximity of the sides and bottom surface of the particle as compared with a bulk target. This effect leads to a greater value of the electron backscatter coefficient, η, for particles as compared to bulk targets. In addition, the energy spectrum of the backscattered electrons is also altered significantly. Figure 8 shows a

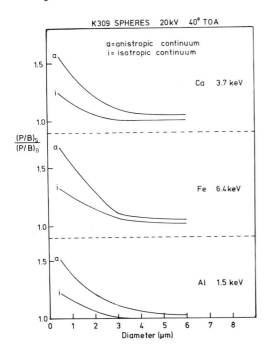

Figure 6. The peak-to-local-background ratios for glass spheres divided by the ratios for bulk glass and plotted as a function of sphere diameter.

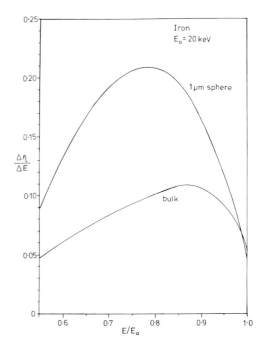

Figure 8. Monte Carlo calculations of the energy distribution of backscattered electrons from bulk iron and from a 1 μm diameter iron sphere.

Monte Carlo calculation of the energy distribution of backscattered electrons for bulk iron and for a 1 μm diameter iron sphere. The distribution has a higher and sharper peak for the sphere. Relatively few electrons reach low energies in a particle target before escaping. Thus, the electron energies at which the most efficient continuum generation is achieved are not reached, while the efficiency of characteristic generation is virtually unchanged. The peak-to-background ratio therefore tends to be higher for a particle or film than for a bulk target.

Correction for the Deviation

In order to accurately analyze small particles by the peak-to-background method, a secondary correction must be applied to compensate for the deviation of the value of (P/B) particle from (P/B) bulk. According to Figure 6, the deviation follows a monotonic function with particle diameter. A variation with x-ray energy is also observed, with the highest energy line showing the largest deviation. From Tables 5-8, it appears that the important parameter is the particle size relative to the size of the interaction volume, as measured by an electron or an x-ray range. For the anisotropic curves in Figure 6, an expression of the following form approximates the observed behavior:

$$\frac{(P/B)_{part.}}{(P/B)_{bulk}} = \exp\ (Ax) + B \qquad (7)$$

where A and B are constants and X=Fx/D, where Rs is the x-ray range. An evaluation of the constants A and B is currently being made for a wide range of target compositions and beam energies.

Additional Areas of Investigation

At the present time there are two known sources of relatively minor errors in the particle analysis scheme, which need to be studied in more detail. The first of these is a decrease in the relative intensities of x-rays due to characteristic and continuum fluorescence. Both of these fluorescence effects are long-range effects, with the fluorescence-induced characteristic x-ray production in bulk samples occurring over a range of distance greater than 10 μm from the beam-sample interaction volume.[4] As a result, for the micrometer-sized particles virtually all the fluorescence-induced x-rays will be lost. This effect is most significant for heavy elements in a light matrix, and could cause deviations on the order of 5-10 percent for particles relative to bulk targets.

The second is a breakdown in the method used to fit the background on small particles. Currently the continuum for the particles is predicted by modification of Kramers' equation which incorporated the second order term proposed by Lifshin.[11] This equation was derived for bulk specimens but is in good agreement with the continuum for particles down to 10 μm in size. For particles less than 10 μm the predicted continuum is higher than the true value by about five percent.

Work is underway to derive a more exact method for the prediction of the continuum emitted by small particles.

Conclusions

The proposed method for the quantitative analysis of particles is designed to obtain elemental concentrations rather than concentration ratios. The method involves minor alterations to a conventional ZAF routine such as FRAME C and can therefore be run on-line with a small computer. In addition, the method is to a first approximation, independent of particle shape and requires no additional input other than that required for a conventional analysis.

The results from the analyses on synthetic and natural mineral particles are in good agreement with the known compositions and are substantially better than the results from a conventional ZAF routine. In the near future additional corrections will be incorporated to account for the various second order effects.

References

1. J. I. Goldstein, G. W. Lorimer, S. J. B. Reed, "Quantitative X-ray Analysis in The Electron Microscope", SEM/ 1977/1, IIT Research Institute, Chicago, Il. 60616, pp. 315-324.

2. J. Gravrilovic, "Quantitative Analysis of Small Particles Using Wavelength and Energy Dispersive Systems in an Electron Beam Instrument", Proceedings of the 13th Annual Conference of the Microbeam Analysis Soc., Ann Arbor, Mi., June 1978, paper #60.

3. J. T. Armstrong, P. R. Buseck, "Quantitative Chemical Analysis of Individual Microparticles Using the Electron Microprobe: Theoretical", Anal. Chem., 47, 1975, pp. 2178-2192.

4. J. A. Small, K. F. J. Heinrich, C. E. Fiori, R. L. Myklebust, D. E. Newbury, and M. F. Dilmore, "The Production and Characterization of Glass Fibers and Spheres for Microanalysis", SEM/1978/1, SEM Inc., AMF O'Hare, Il. 60666, pp. 445-454.

5. P. Statham and J. Pawley, "A New Method for Particle X-ray Microanalysis Based on Peak to Background Measurements", SEM/1978/1, SEM Inc., AMF O'Hare, Il. 60666, pp. 469-478.

6. T. Hall, "Some Aspects of the Microprobe Analysis of Biological Specimens", in Quantitative Electron Probe Microanalysis, K. F. J. Heinrich ed., U. S. Government Printing Office, Washington, D.C., 1968, pp. 269-299.

7. J. A. Small, K. F. Heinrich, C. E. Fiori, D. E. Newbury, and R. L. Myklebust, "Progress in Quantitation of Single-Particle Analysis with the Electron Probe", in Proc. 13th Ann. Conf. Microbeam Analysis Soc., Ann Arbor, Mi., 1978, paper #56.

8. H. A. Kramers, "On the Theory of X-ray Absorption and of the Continuous X-ray Spectrum", Phil. Mag., 48, 1923, p. 836.

9. M. Green, Ph.D. Thesis, University of Cambridge, Chapter 4, 1962, pp. 60-75.

10. C. E. Fiori, et al., "An Overview of the Glass Standards Program for Microanalysis at the National Bureau of Standards", Proceedings of the 11th Annual Conference of the Microbeam Analysis Society, Miami Beach, Fl., 1976, paper #27.

11. E. Lifshin, The Use of Solid State X-ray Detectors for Obtaining Fundamental X-ray Data, in Proc. 9th Ann. Conf. Microbeam Analysis Soc., OTTAWA, Canada, 1974, paper #53.

12. H. Yakowitz, D. E. Newbury, and R. L. Myklebust, "Approaches to Particle Analysis in the SEM with the Aid of a Monte Carlo Program", SEM/1975, IIT Research Institute, Chicago, Il. 60616, pp. 93-102.

13. R. L. Myklebust, et al., "Monte Carlo Electron Trajectory Simulation - An Aid for Particle Analysis", in Proc. 13th Ann. Conf. Microbeam Analysis Soc., Ann Arbor, Mi., 1978, paper #61.

14. R. L. Myklebust, D. E. Newbury, H. Yakowitz, "The NBS Monte Carlo Electron Trajectory Calculation Program", Use of Monte Carlo Calculations in Electron Probe Microanalysis and Scanning Electron Microscopy, eds. K. F. J. Heinrich, D. E. Newbury, and H. Yakowitz, NBS Special Publication 460, U. S. Government Printing Office, Washington, D.C., 1976, pp. 105-125.

15. D. E. Newbury, R. L. Myklebust, and K. F. J. Heinrich, "A Monte Carlo Procedure Employing Single and Multiple Scattering", in Proc. 12th Ann. Conf. Microbeam Analysis Soc., Boston, Ma., 1977, paper #27.

16. L. Curgenven, P. Duncumb, "Simulation of Electron Trajectories in a Solid Target by a Simple Monte Carlo Technique", Tube Investments Research Laboratories, Report No. 303, Hinxton Hall, Suffron Walden, Essex, England, 1971.

17. M. Green and V. E. Cosslett, "The Efficiency of Production of Characteristic X-radiation in Thick Targets of a Pure Element", Proc. Phys. Soc. of London, 78, 1961, pp. 1206-1214.

18. P. Kirkpatrick and L. Wiedmann, "Theoretical Continuous X-ray Energy and Polarization", Phys. Rev., 67, 1945, pp. 321-339.

Discussion with Reviewers

C. Maggiore: You state that cylindrical particles show less deviation from bulk behavior than spherical particles; this appears to be based on analysis of a single particle with no indication of the orientation of the cylinder relative to the beam and detector. I agree that shape will have an effect, but it is not obvious that such a sweeping generalization is justified. Would you please comment?
Authors: The statement that cylindrical particles show less deviation from bulk behavior than spherical particles is not based on experimental measurements but on theoretical Monte Carlo calculations. For these studies, the fiber orientation was perpendicular to the detector.

C. Maggiore: Many microscopes have the detector at 90° relative to the beam, do you think appropriate modifications to equation (6) will make this technique universally applicable?
Authors: Although we have not tried the method with data collected at a 90° take-off angle, I think it would be possible to make the necessary changes so that the method will work with the standard and sample tilted.

R. B. Bolon: Were the particles in fig. 5 measured by using the beam in a raster or fixed spot mode? If in the fixed spot mode, were measurements obtained from the various faces? What was the magnitude of the associated error?
Authors: The glass particles in fig. 5 were measured with a rastered beam. Although we have not done a systematic measurement on different faces, the results obtained for particles measured with a spot have been insensitive to the position of the beam. A study of the effect of the beam position on the peak-to-background ratios was done by Statham and Pawley in reference 5. The results from this study also indicate that the analysis is insensitive to the beam entry point.

R. B. Bolon: What kind of substrate and mounting techniques did you use? How significant is the bremsstrahlung from the substrate (caused by either transmitted or incident electrons) in comparison to the component from the smaller particles? How would you correct for this problem?
Authors: The particles are mounted on a thin, approximately 15 nm, carbon film which is supported by a Be TEM grid. The grid is mounted over a hole in a graphite block. The bremsstrahlung generated from the substrate is less than three counts per second which is sufficiently low enough that to a first approximation it can be ignored when analyzing particles in the 0.2-20 μm size range.

R. B. Bolon: Did you coat the glass particles? If so, with what? Would you expect a coating to contribute significantly to the continuum, especially in the case of the smaller particles, thereby introducing more error?
Authors: The glass particles are coated with approximately 5 nm of carbon. I would not expect a 5 nm coating on a 200 nm or larger particle to contribute significantly to the continuum. This however is certainly an area which should be investigated particulary, as you suggest, for the smaller particles.

P. Statham: Both Rao-Sahib and Wittry (J. Appl. Phys. 45, 5060, 1974) and Smith et al. (X-Ray Spectrom. 4, 149, 1975) have studied the atomic number dependence of continuum intensity after correction for backscatter and absorption so how do your experimental results compare with those of these investigators? Do you have a more complete set of values for the exponent, n?
Authors: We have not as yet compared our results with the values determined with the equations of Rao-Sahib and Wittry, and Smith et al. At this time we do not have a more complete set of values for n.

P. Statham: Equation 4 contains terms for both absorption of the continuum and departure from Kramers' law yet these refinements are missing in similar formulae in references 4 and 7. Were the good results in this earlier work obtained through fortuitous cancelling of the $F(\chi)$ and atomic number terms or is there some implicit normalization which makes the result rather insensitive to formula chosen for the background intensity?

Authors: The results reported in references 4 and 7 were for glass spheres analyzed against the parent glass as the standard. Since B_{std} and B_{bulk} are equivalent, the various terms in equation 4 cancel. It should be noted that in the analysis of the glass spheres we were primarily interested in demonstrating the ability of the peak-to-background method to correct for particle effects and as a result we chose a system where the prediction of B_{bulk} was trivial. Kramers' law was used in its original form only as a first approximation.

P. Statham: There is some disagreement in the literature concerning the formulae which should be used to describe the continuum cross section. In the Monte Carlo calculations, what, if any, relativistic correction did you use, and would you expect use of different cross-section formulae to give significantly different results? Experiments with different sample/detector geometries would certainly help to verify the existence of a large anisotropy effect, as shown in figure 6, but does your present choice of theory predict a geometry where the anisotropy effect will be minimal?

Authors: We used the cross-section proposed by Kirkpatrick and Wiedmann. Since a ratio of backgrounds is used, the actual value of the cross-section is not that important. As yet, we have not studied the effect of different geometries on the anisotropy of the continuum.

D. R. Beaman: Referring to Table 4. Why do the concentrations calculated for two 5 μm FeS_2 particles differ so markedly when using the FRAME program? The concentrations from FRAME on particle #1 are surprisingly low.

Authors: In this instance, the variance in the concentrations determined by FRAME is probably due to the position of the beam during analysis.

D. R. Beaman: Is the NBS particle program available on request?

Authors: The NBS particle program is still being developed and is not yet available.

SCANNING ELECTRON MICROSCOPY/1979/II
SEM Inc., AMF O'Hare, IL 60666, USA

PROGRESS IN THE QUANTITATION OF ELECTRON ENERGY-LOSS SPECTRA

D. C. Joy, R. F. Egerton,* and D. M. Maher

Bell Laboratories,
Murray Hill, NJ 07974

*University of Alberta
Edmonton, Canada, T6G 2J1

ABSTRACT

This paper describes recent progress that has been made in the quantitation of electron energy-loss spectra using characteristic inner-shell excitations for elemental analysis. Three quantitation approaches are outlined, namely Efficiency Factor, Calculated Cross-Section and Standards methods. The results obtained from these three methods have been examined in terms of their stability, relative accuracy and absolute accuracy. The variables considered in this examination are the energy window over which integrations are carried out, the scattering angle accepted by the spectrometer and instrumental factors. Possible errors are discussed and it has been concluded that under suitable conditions: i) a stability of $\pm$ 5% is attainable; ii) a relative accuracy of 10% or better can be expected when analyzing compounds (i.e. MgO, BN etc.); and iii) an absolute accuracy of $\pm$ 20% or better should be possible. The latter two figures are presently limited by partial cross-section determinations and/or specimen related variables.

KEY WORDS: Electron Energy-Loss Analysis, STEM, Quantitative Analysis

1. INTRODUCTION

Over the past few years it has been recognized that electron-energy losses caused by inner-shell excitations and detected in the transmission mode provide a sensitive method of microanalysis for elements in the first two rows of the periodic table.[1-4] Unlike Auger microanalysis, the energy-loss method can detect elements distributed within the interior of a specimen, provided the latter is sufficiently thin. Both X-ray and Auger techniques rely on secondary processes which involve higher level shells and the inner-shell vacancies created by incident electrons, whereas electron energy-loss spectrometry measures the inner-shell excitation directly through its effect on the transmitted electrons. Therefore, relatively simple equations can be used for quantitative analysis and corrections for absorption or yield of the secondary process are not required. Unlike Auger and X-ray microanalysis, electron energy-loss spectroscopy can be employed as a *standardless* technique - that is, the absolute amount of a particular light element within a given region of a specimen (defined, for example, by the incident beam) is measured. This measurement requires that one record the low-loss region of the spectrum, in addition to the regions containing characteristic K-, L- or M-edges and that an appropriate partial cross section is known[5] for the inner-shell excitation. However, before placing undue reliance on the answers obtained, we believe it is desirable to test both the experimental procedures and the assumptions concerning cross sections by means of "standard" specimens which can be exchanged between different laboratories (to check for differences in operating procedure or possible instrumental effects) and whose thickness and/or composition can be measured by independent means.

2. ANALYSIS METHODS

The characteristic losses due to excitations of inner-shell electrons from a particular element occur in the energy spectrum of transmitted electrons as a rise in electron intensity at some threshold energy loss E_k (k representing the type of shell: K, L, etc.), followed by a more gradual decay (see Fig. 1). The characteristic signal is superimposed on a background which arises from the excitation of outer shell (e.g. valence) electrons and can generally be fitted over a moderate energy range by the function: $A \cdot E^{-r}$, where E represents the energy loss of a transmitted electron, A and r being constants. If the spectrum is stored electronically in a micro- or minicomputer[6,7], the values of A and r can be found from a digital fit to the background over an energy range of typically 50-100 eV preceding the threshold energy E_k. The modeled background can then be extrapolated under the edge for a specified energy window Δ and the inner-shell intensity obtained by stripping this

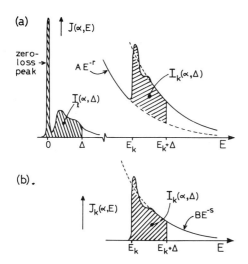

Figure 1   Schematic electron energy-loss spectra of the signal intensity $J(\alpha,E)$ versus energy loss E: (a) the zero-loss peak, low-loss intensity $I_l(\alpha,\Delta)$ and characteristic-loss intensity $I_k(\alpha,\Delta)$ superimposed on a background of the form $A \cdot E^{-r}$; and (b) the characteristic-loss intensity after background subtraction and of the limiting form $B \cdot E^{-s}$.

fitted background from the edge. Although the procedure has been found to be very accurate in practice, deviations between the modeled background and experimental data ahead of the edge usually being less than the channel to channel noise fluctuations, it is clearly sensible to use the smallest energy window Δ so as to avoid the necessity of having to extrapolate the fitted background over too wide a range. By measuring the intensity $I_k(\alpha,\Delta)$ above this extrapolated curve (over an energy range Δ above E_k) and also the intensity $I_l(\alpha,\Delta)$ under the low-loss region of the spectrum (representing $O \leqslant E \leqslant \Delta$), the quantity N of a given element within the area of the beam can be obtained to a good approximation[5] from the equation

$$N \doteq \frac{1}{\sigma_k(\alpha,\Delta)} \frac{I_k(\alpha,\Delta)}{I_l(\alpha,\Delta)} \qquad (1)$$

Where $\sigma_k(\alpha,\Delta)$ is a "partial" cross section for transitions from shell k which cover a range Δ of energy loss and a range of scattering angles between zero and α. α is normally defined by means of an aperture (e.g. the objective aperture in CTEM mode) placed between the specimen and the electron spectrometer. If $\sigma_k(\alpha,\Delta)$ is expressed in cm^2 per atom, then equation (1) gives N in atoms per cm^2 of surface within the incident beam.

In the data systems that we have used, the values of I_l and I_k are found through interactive computer programs which read the spectrum from the memory of a multi-channel analyzer and allow the regions of interest (such as the low-loss, background and edge regions) to be selected, for example, by "painting" them with a cursor controlled by the operator. The time required to model and extrapolate the background under an edge and to derive the values of I_k and I_l is about 15 sec, the limiting factor being the time it takes the user to mark the regions of interest. The other experimental parameters required, i.e. the scattering angle α accepted by the spectrometer and the accelerating voltage E_o, are also entered into the computer at this time. The last step is to calculate $\sigma(\alpha,\Delta)$ and three methods are described below.

Method 1: Efficiency Factors

The partial cross-section $\sigma(\alpha,\Delta)$ in equation (1) is used

because only electrons scattered through angles equal to, or less than, α and suffering energy losses between the threshold energy E_k and $E_k+\Delta$ are collected by the spectrometer and made use of in the analysis. However electrons are scattered through all angles up to π, and they can lose any amount of energy between E_k and E_o. The cross-section $\sigma(\pi,\Delta \to \infty)$ which takes account of all of these events is the total cross-section σ_T and is the same quantity as the ionization cross-section used in quantitative X-ray analysis. If we restrict our discussion to K-edges then it can be shown[3] that the partial and total ionization cross-section are related, to a good approximation, by

$$\sigma(\alpha,\Delta) \doteq \sigma_T \cdot \eta_\alpha \cdot \eta_\Delta \quad . \qquad (2)$$

That is the variables α and Δ can be treated separately in their action on $\sigma(\alpha,\Delta)$. If appropriate forms for the functions η_α and η_Δ can be found then $\sigma(\alpha,\Delta)$ can be calculated because σ_T is available from standard X-ray references (for a comprehensive review see Powell[8]). The quantities η_α and η_Δ are usually referred to as "efficiency factors" since η_α is the fraction of electrons collected by the spectrometer as a result of its finite acceptance angle α compared to the total scattered flux, and η_Δ is the fraction of the signal lying in the energy window Δ compared to the true integral which stretches to E_o. Analytical forms can be found for both efficiency factors if K-edges are used and some reasonable approximations are made.[3] η_α is obtained by noting that the inelastically scattered electrons have an angular distribution $I(\theta)$ about the incident direction $(\theta = O)$ of the form

$$I(\theta)/I(O) = (\theta^2+\theta_E^2)^{-1} \qquad (3)$$

where θ_E is the characteristic scattering angle for an energy loss E, and is equal to $E/2E_o$.[9] Using this expression one readily finds that the fraction η_α of the total signal lying within the angle α is

$$\eta_\alpha = \frac{ln\,[1+(\alpha^2/\theta_E^2)]}{\ln\,(2/\theta_E)} \qquad (4)$$

where now θ_E will be set equal to $(E_K+\Delta/2)/2E_o$, that is the average scattering angle over the energy window from E_K to $E_K+\Delta$. Measurements show that this expression does represent the variation of the collected signal with α quite well.[10] Since the quantities α,E_o,E_K and Δ are all fed into a computer program, the value of η_α can be calculated directly when required.

The value of η_Δ is found by noting that, after stripping away the background, the K-edge intensity I(E) has the form

$$I(E) = B \cdot E^{-s} \quad (E > E_K) \qquad (5)$$

where B and s are constants for any set of experimental conditions.[3,9,10] The ratio η_Δ of the signal integrated over the window Δ to that extending over the energy-loss range from E_K to E_o, then is found simply by integrating equation (5) and the result is

$$\eta_\Delta = \left[1 - \left(\frac{E_K}{E_K + \Delta}\right)^{s-1}\right] \quad . \qquad (6)$$

For given values of E_K, Δ and s, therefore η_Δ can be calculated. The value of s is obtained by fitting the expression $B \cdot E^{-s}$ to the edge profile after the background has been stripped and using a least-squares criterion to find the best value of s. As in the case of the background fit, the portion of the edge to be modeled can be "painted-in" using the cursor. The computer then reads all the intensity values and calculates the best fit value of s and then η_Δ. Figure (2) shows the various steps in the application of these procedures to a carbon K-edge and demonstrates in turn: the original form of the edge; the background fit and its extrapolation under the edge; the stripped edge; and finally the edge and fitted form. For that part of the

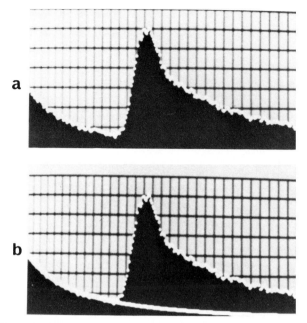

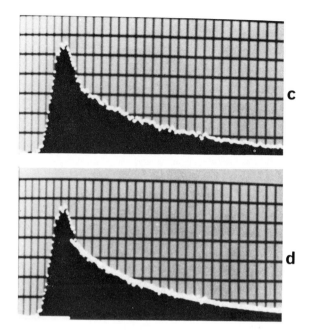

Figure 2   Illustration of the basic steps to process a characteristic energy-loss peak:

    a)   spectral intensity for a carbon K-edge ($E_K = 284\ eV$);

    b)   background fit and extrapolation - solid line;

    c)   stripped edge; and

    d)   limiting edge fit - solid line.

edge immediately following its onset at E_K, the shape is very dependent on such variables as the specimen thickness and even the chemical bonding of the atoms concerned. As a result the accuracy of the fit in this region is not very good. However from 20 eV or more away from the edge, the stripped edge is accurately described by $B \cdot E^{-s}$, as can be seen from Fig. 2d. Consistent values of the parameter s are obtained for windows of 80 to 100 eV or wider. The value of $\sigma(\alpha,\Delta)$ now can be found using the computed values of η_α and η_Δ and the appropriate total cross-section σ_T. In the program described here the value of σ_T is computed using the Bethe[11] cross-section but with the constants selected in accord with the recommendations of Powell.[8] The accuracy of σ_T for light elements at 100 kV is not well established and the literature contains other values for σ_T which can differ by up to a factor of 2 in either direction.[10] Until better experimental and theoretical data is available, it is therefore necessary to treat these computed σ_T values with some caution as far as their absolute value is concerned, although the ratio of the σ_T values for different elements is probably fairly accurate.

Method 2: Calculated Cross-Sections

A second way to find the required partial cross-section $\sigma(\alpha,\Delta)$ is to calculate it directly from first principles. The ionization cross-section can be computed using quantum mechanical methods[11,12] and this has been shown to give results which are in detailed agreement with the best experimental data, although considerable computing time is required. However, the simplest procedure is to approximate the initial and final-state wave functions within the inner-shell region by Coulombic wave functions, so that the theoretical situation becomes similar to that of the hydrogen atom, for which the generalized oscillator strength (for a given energy and momentum transfer) is known analytically.[13] For the conditions normally used in transmission electron spectroscopy, this hydrogenic approximation gives results which are in good agreement with experimen-

tal data for K-shell losses.[14] With some correction in the region of the excitation threshold, a similar approximation can be used for L-shells. Therefore values of $\sigma_k(\alpha,\Delta)$ can be calculated for K- and L-edges in less than 1 sec by means of a short Fortran program,[14] using the experimental values of α and Δ.

A complete quantitation program incorporating all the required integration, background stripping and computational steps can be put into about 13 kilo-byte of store on a conventional 16 bit word mini-computer. This is about the same storage as is required for the program described in method 1.

The only inputs required for the hydrogenic cross-section programs (SIGMAK and SIGMAL) are the threshold energy of the edge E_k, the energy window Δ, the accelerating voltage E_o, the scattering angle α and the atomic number Z. The required absolute partial cross-section is then produced directly and so it can be inserted straight into equation (1). Because this method does not require a second curve fitting operation, it is faster in use than method 1 and it has the further advantage that, since

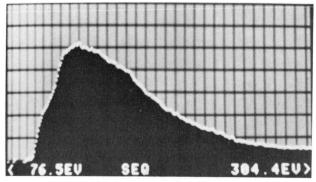

Figure 3   Experimental silicon L-edges after background stripping: $E_o = 100$ keV, $\alpha = 3$ mrad, spectrometer resolution $= 5$ eV and a dwell time of 200 msec/channel at 0.9 eV/channel.

it does not rely on an edge fitting procedure, smaller energy windows can be used without loss of accuracy. This is of special benefit when consecutive elements are being analyzed (e.g. boron nitride on a carbon substrate).

Method 3: Standards

A third method of performing a practical quantitation based on equation (1) is to use a separate experiment to find $\sigma(\alpha,\Delta)$. This implies the use of a standard, a procedure which clearly introduces a number of additional problems into the quantitation since effects like the specimen thickness are liable to have a pronounced influence on the observed form of the edge. However this approach would have the compensating advantage that it could be applied to any edge for which a suitable standard could be found. We have applied this technique to the quantitation of L-edges, in particular the silicon L_{23} at 99 eV.

Figure 3 shows this edge after the background has been stripped using the procedures described previously. The variation of the integral under the edge now can be found as a function of the energy window Δ. If this quantity is divided by the corresponding integral from the zero-loss region of the spectrum over the same energy window, then the result is directly proportional to the partial cross section $\sigma(\alpha,\Delta)$, as can be seen from equation (1). Figure (4) shows the experimentally determined variation for the silicon L-edges with $\alpha = 3$ mrad at 100 kV. For convenience the curve has been normalized to the value obtained at an energy window of 100 eV, and it thus represents the function $f(\Delta)$ where

$$\sigma(\Delta) = \sigma(100) \cdot f(\Delta) \quad . \qquad (7)$$

By finding an absolute value for $\sigma(100)$ at 3 mrad, the partial cross-section for any other energy window can thus be determined. The value of $f(\Delta)$ can either be read from the graph or calculated from a polynomial fit to this curve. Experimentally the form of $f(\Delta)$ has been found to be substantially independent of the value of α. Therefore the cross-section can be scaled, in the same way as was done in method 1 using equation (4), as α is varied from 3 mrad.

The absolute value for the partial cross section at $\alpha = 3$ mrad and $\Delta = 100$ eV was obtained by simultaneously taking an energy-loss spectrum and a convergent-beam

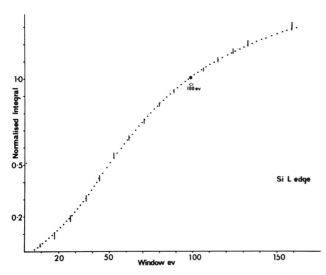

Figure 4   Normalized silicon L-edge intensity, i.e. $f(\Delta) = \sigma_L$ (3 mrad,Δ)/σ_L (3 mrad, 100 eV), versus energy window.

diffraction pattern from the same area of a silicon crystal. The thickness t of the crystal can be derived directly from the convergent-beam diffraction pattern to an accuracy of about 3%[15] and then the number N of atoms-cm^{-2} calculated from t and the bulk density of silicon. Since N is now known, the partial cross-section can be found using equation (1) and the intensity data from the energy-loss spectrum. The result is that σ_L (3 mrad, 100 eV) = 5.23×10^{-20} cm^2 per atom at 100 kV.

The spirit of this method is in essence the same as that described in method 1, except that the correction (or efficiency) factors are determined experimentally. It has general applicability, and is very easy to carry out in practice since the necessary correction with Δ can be found graphically. Although the experimental form of the result is slightly dependent on the thickness of the crystal, the net effect is small for energy windows greater than $\sim$ 20 eV. Apparently this is because the major consequence of thickness is to redistribute the intensity immediately after the edge.

In the absence of an on-line computer capable of performing cross-section calculations of the type described in methods 1 and 2, this approach would appear to be the only way to carry out L- and M-edge analyses. Furthermore cross-sections obtained in this way can be compared directly to those derived from model calculations.

3. TESTS OF THE QUANTITATION METHODS

Because the techniques of both data collection and analysis in electron energy-loss spectroscopy are relatively new, it is important to test the results as fully as possible. This implies not only comparing results obtained from energy-loss measurements with those obtained from other techniques, but also checking the results from energy-loss measurements for self-consistency and even performing "round-robin" type of experiments in which the same sample is examined on different microscopes by different workers in an attempt to look for instrumental effects.

Three test criteria, in ascending order of severity, therefore have been used to examine the quality of the quantitation routines described in the preceding sections:

1) The measurement of N for an element of interest must remain constant when the experimental parameters of the system (i.e. accelerating voltage, acceptance angle or energy window) are varied. This is a test of self consistency for the data reduction scheme.

2) Measurements of the composition of compounds of known stoichiometry (such as BN, MgO etc.) should agree with the expected results and likewise should be independent of the operating conditions. This is a test of the relative accuracy of the quantitation schemes when applied to two or more elements. And

3) The value of N should agree with measurements made on the same material by independent quantitation schemes and with similar measurements performed on different instruments. This is a test of the absolute accuracy of the procedures employed and of the effects of instrumental artefacts on the data.

Any technique satisfying all three of these conditions will qualify as an absolute, quantitative micro-analytical method. If only the first two conditions are met then although the method will be useful for compositional studies of compounds, standards will be necessary to produce absolute answers.

3.1 The Stability of Quantitation Result

The most basic requirement of any quantitation is that the answer produced be independent of the choice of experimental variables. In the case of electron energy-loss spectroscopy,

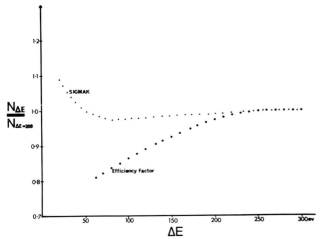

Figure 5   Typical results of an N determination as a function of Δ for an amorphous-carbon film $\sim$ 40 nm thick using the calculated cross-section method (Δ) and efficiency factor method (o): E_o = 100 keV and α = 3 mrad.

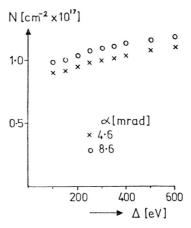

Figure 6   Variation of N for copper (evaporated onto carbon) with the energy window Δ and α from an L-loss analysis using hydrogenic cross-sections with a modified correction for outer screening: E_o = 80 keV.

these are the accelerating voltage, the spectrometer acceptance angle and the energy window. In general the choice of the voltage and the angle is governed by factors such as the mode of microscope operation and as a consequence these two parameters are not varied frequently. For that reason their influence on the results will be considered when other instrumental variables are discussed in a later section. Therefore the remaining parameter to consider here is the energy window Δ, as it can be varied over a wide range when the data is being analyzed. Unless the variation of N with Δ is small there is the possibility that two separate analyses of the same spectrum will yield different results. It is also desirable that the quantitation method used is reasonably tolerant of changes in the edge shape due to multiple scattering, as in many practical cases it is not possible to produce a specimen which is thin enough to ensure an ideal single-scattering profile. These aspects of quantitation have been extensively investigated for both the efficiency factor and the calculated cross-section methods by studying the variation in N as a function of the energy window for a suitable edge, such as the carbon K-edge. Figure 5 shows typical data using the two methods to quantitate a spectrum obtained from a carbon film about 40 nm in thickness and at E_o = 100 kV. For clarity in presentation the answers have been normalized to the value each method gives when Δ = 300 eV.

The efficiency factor method which relies on the curve fitting approach described in Section 2 gives a value of N which varies by about 25% over the energy window range 50 to 300 eV. No results are given for windows of less than 50 eV because the value of the exponent "s", derived from a fit to the stripped edge profile, varies rapidly in this region and the resultant changes in the calculated energy efficiency factor produce large swings in N. Beyond this cut off, however, the results are smoothly varying out to the largest window used. The observed variation is mainly due to the fact that the value of "s" changes slowly as the window is widened. Above Δ = 175 eV, stability of better than 5% is achieved. Clearly the larger the energy window the less sensitive the result is to noise on the spectrum. In the case where substantial multiple scattering is present on the edge this procedure is rather less satisfactory, as the shape of the edge is distorted. This leads to significant changes in "s" as the window is varied and poor stability of the result. In summary, this approach satisfies the stability condition adequately provided that a window of sufficient width is chosen and multiple scattering is minimal.

The stability of N values derived using the calculated (SIGMAK) cross-section method is seen to be excellent even for relatively narrow windows. From about 30 to 300 eV the variation is within 5%, and some of the variation that is observed for large Δ is probably due to cumulative errors in the integral as a result of inaccuracies in the background extrapolation. Even on profiles showing some multiple scattering good stability is still achieved since, as discussed in Section 2, the form of the integrated edge is relatively insensitive to the exact shape of the edge profile.

Figure 6 shows the corresponding result when using the hydrogenic model (with an edge correction) to quantitate the L-shell of copper (in this example evaporated on to a carbon substrate). Over the Δ range 50 to 500 eV, the variation in N, for either of two acceptance angles, is less than 15%. Some of this variation is attributed to cumulative errors from the background subtraction. This is a very satisfactory result, and the conclusion is therefore that the hydrogenic (SIGMAK) technique meets the first test criterion for both K- and L-edges.

The standards method has not been applied to K-edges but when used on L edges (for which a correction curve of the type shown in Fig. 4 is available) the stability of the N value is good, 10% or better being obtained. Since the experimental correction curve is derived from an edge integral, the result again is tolerant of limited amounts of multiple scattering.

In summary, therefore, all three methods adequately meet the first test criterion provided that they are applied under properly chosen conditions.

3.2 The Accuracy of Elemental Ratios

When a quantitation routine is employed to measure the composition of a known compound we are, in effect, testing the relative accuracy of the partial cross-section calculations for each of the elements concerned. Several possible results can be obtained from a test of this kind: the elemental ratios could vary with the choice of Δ, or the ratios could be reasonably constant but not in agreement with the expected value, or they could be both constant and correct within reasonable experimental error.

One of the problems in this type of experiment is to find test samples, as many compounds containing light elements with suitable K-edges tend to suffer differential mass-loss under the electron beam. However, two satisfactory specimens have

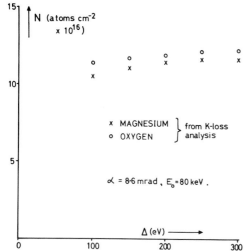

Figure 7   Compositional analysis of MgO showing the variation of N with energy window Δ using the calculated cross-section method: Mg:O = 0.93 ± 0.03.

been found, namely BN and MgO, both of which are readily available in a pure form suitable for micro-analysis. Alternatively standard specimens can be obtained by an ion-implantation technique.[16] An additional difficulty is that two (or more) edges in the same spectrum will be affected differently by multiple scattering, with the lowest energy-loss edge being more severely affected than a higher energy-loss edge. This means that the samples should be as thin as possible and that the program routines used should be able to deal effectively with limited multiple scattering artefacts.

Both methods 1 and 2 have been found to give good results when analyzing compounds. Figure 7 shows a typical result when using the SIGMAK routine to analyze MgO. Both the N values for Mg and O remain stable with Δ, and the resultant atomic ratio Mg:O = 0.93 ± 0.03 is in good agreement with the expected value. This indicates that the SIGMAK calculation of the cross-sections correctly accounts for the variation with edge energy. It does not, of course, prove that the absolute value is correct since in any ratio the multiplying constants will cancel. Equally good results have been obtained on the BN system for which a ratio of B:N = 0.90 ± 0.03 has been found for energy windows in the range 40 to 100 eV. Although the absolute value of the ratio is in 10% error with the expected value, the agreement is to be considered satisfactory at this stage of development, particularly as no standards were employed to obtain these numbers. It is also worth noting that in the case of BN a carbon K-edge (from the support film) is interposed between the two edges used for the quantitation. The success of the programs in dealing with this situation indicates that the background stripping routines are sufficiently accurate at present.

The efficiency factor method has also been found to give good results for BN. Although the concentration of either element varies with the choice of Δ in the same way as shown in Fig. 5, their ratio is sensibly constant for windows of 100 eV or greater. In this method the value of the ratio obtained is determined by the constants inserted into the Bethe formula[13] used to calculate σ_T for each element. Although the absolute value of σ_T for any element can be made to vary over a range in excess of 2:1 by an appropriate choice of the constants published in the literature,[8,10] the ratio of the values obtained is surprisingly constant indicating that the functional variation of any of the "standard" X-ray cross-section formulae is correct.

For both methods the ratios obtained are essentially independent of the choice of the spectrometer acceptance angle. Thus the angular efficiency factor, or the angular dependence of the SIGMAK cross-section, correctly account for this variation. The only cases where this might not be so are those in which strong diffraction occur, or where the spectrometer acceptance angle was less than the convergence angle of the incident beam, resulting in a systematic error.[9] The third quantitation method is referenced to a standard specimen of known N. The accuracy of elemental ratios derived from such a technique will therefore be governed by the accuracy of the original standard determination. The probable accuracy of such an approach is discussed in the next section.

In summary, with the currently available methods, elemental ratios using the K-losses can be obtained with an accuracy that is probably better than 10% with suitable choice of the energy window and spectrometer acceptance angle. Either method can therefore be considered suitable for routine micro-analysis.

3.3 The Absolute Accuracy of Quantitation

A final requirement for any program is that the N value agrees, in absolute terms, with some independently determined value. This is not a pre-requisite for a micro-analytical technique since, in general, it is compositional ratios that are most often required. Thus energy dispersive X-ray spectroscopy is widely used even though the result is in no way an "absolute" measurement. However, because the quantitation relation is so simple in electron energy-loss spectroscopy, there is a strong incentive to achieve data reduction programs which do provide correct, absolute values of N.

We have attempted to examine this aspect of the methods discussed previously by comparing the answers to those obtained, from the same specimen area, by independent methods of measurement. Further we have followed up these experiments with others in which the results were taken from the same specimens but on different microscopes and under varied operating conditions in an attempt to look for instrumental contributions.

The specimens chosen for this purpose were a sequence of amorphous-carbon films (prepared by E. J. Fulham Inc.) with nominal values of thickness ranging from 20 nm to 160 nm. These films were examined both as received and after an anneal at 300 °C. Although no elements other than carbon were detected in these films considerable changes in the characteristics of the films were observed after annealing, indicating the possible presence of hydro-carbons mixed with the amorphous carbon. The experiments were conducted on two different microscopes; the first was a JEOL 100B, operated at 80 kV in the conventional TEM mode and equipped with a magnetic prism spectrometer;[17] the second was a JEOL 100B, operated at 100 kV in the STEM mode and also using a magnetic prism spectrometer.[18] For both instruments the resolution used was between 5 and 10 eV, as measured from the zero-loss peak. The spectra were computer processed using the programs described in Section 2.

Independent microscopic determinations of N are possible using several techniques. The best known of these is that based on the *Plasmon Peaks*. Provided that the mean-free path for a plasmon excitation λ_p is known then the thickness of the sample can be determined from measurements of the intensity of the plasmon peaks in the low-loss region of the spectrum. Strictly speaking in the case of amorphous carbon the prominent peak observed at 24 eV loss is not definitely classed as a plasmon but may contain some contribution from single-electron excitations. However for this purpose it can be treated in an identical way to a normal plasmon. The thickness can be derived from the relation

$$t/\lambda_p = \log_e(I_T/I_o) \qquad (8)$$

where I_o is the integrated intensity under the zero-loss peak, I_T is the total spectrum intensity (measured by integrating the spectrum over an energy window of 200 eV or more) and t is the specimen thickness. This expression has been found to be more reliable than measurements based on the ratio of the heights or areas of individual plasmon peaks. Using equation (8) a simple determination of two intensities thus gives t. Two warnings about the accuracy of this procedure, particularly in the case of carbon, need to be made. Firstly, the λ_p values tabulated in the literature[19,20] are experimental values calculated from films whose thickness was assumed to be known and secondly, the thickness uniformity of the films is usually not given. Therefore the literature values must be treated with caution. An estimate of 10% probable error would seem likely. Once the film thickness is estimated, its density is required in order to obtain N. Densities quoted for carbon films vary between 1.8 and 2.2 gm/cm^3, the exact value certainly being sensitive to the prior history of the film. Thus a further uncertainty of at least 10% in the determination of N is introduced.

An alternative technique would involve the use of the *Annular Dark-Field Signal* collected on a suitably large detector placed beneath the specimen. This signal is linearly proportional to the mass-thickness of the sample[21] and thus, in principle, N can be estimated if a suitable calibration is devised. Attempts to obtain a calibration from measurements on latex spheres were unsuccessful because of the high rate of mass loss from these objects. Another possible way of obtaining a calibration using the dark-field signal would be by comparison with the plasmon losses. Since this would not yield an independent estimate of N, the use of the dark-field signal for the purposes of assessing the accuracy of quantitation was not pursued at this time.

In the case of a crystalline specimen an absolute thickness determination can be made from the *Convergent-Beam Diffraction Pattern Technique,*[15] as outlined in Section 2. Because no data other than the density of the specimen is required in this approach, it should be capable of high accuracy. However it is limited to crystals and thus cannot be used in the case of carbon. Samples containing a known concentration of a given elemental species can be produced by *Ion Implantation*[16] but insufficient data have been obtained at present on these specimens to make use of them as an absolute standard.

Thus the quantitation data obtained from the carbon K-loss edges has been compared with values of N derived from plasmon intensities and the bulk density of carbon. Figure 8 summarizes the results of one set of experiments, the specimen being a film of nominal thickness 20 nm. The 80 kV data were obtained in the CTEM mode on a 100B microscope at The University of Alberta and the 100 kV data were obtained from a 100B instrument at Bell Laboratories with the same specimen being examined in both cases. The data processing routines were run, as described in Section 2, on spectra without any collaboration between the two sites until the results were finalized. From equation 8, N values of 1.27 and 1.38×10^{17} $atoms/cm^2$ at 100 and 80 kV, respectively, were obtained assuming a density of 2 gm/cm^3. This agreement is judged excellent and justifies the choice of this technique as a suitable standard of comparison.

The value of N obtained from the K-edge at 80 kV (with $\alpha = 8.6$ mrad) is $1.22 \pm 0.02 \times 10^{17}$ $atoms/cm^2$ over the range 30 $eV < \Delta < 250$ eV using the SIGMAK routine. This compares with a K-edge value at 100 kV (for $\alpha = 3$ mrad) of $1.74 \pm 0.03 \times 10^{17}$ $atoms/cm^2$ for 30 $eV < \Delta < 250$ eV using the SIGMAK routine and analyzing the same data using the

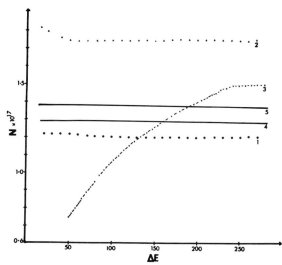

Figure 8   Quantitative analyses of an amorphous carbon film (~ 20 nm thick):

1   CTEM, $E_o = 80$ keV, K-loss analysis using SIGMAK, $\alpha = 8.6$ mrad;

2   STEM, $E_o = 100$ keV, K-loss analysis using SIGMAK, $\alpha = 3$ mrad;

3   STEM, $E_o = 100$ keV, K-loss analysis using efficiency factor method, $\alpha = 3$ mrad;

4   STEM, $E_o = 100$ keV, plasmon analysis; and

5   CTEM, $E_o = 80$ keV, plasmon analysis.

efficiency factor method, resulted in a value of $1.48 \pm 0.05 \times 10^{17}$ $atoms/cm^2$ (200 $eV < \Delta < 250$ eV). Although not shown on this figure, N values were obtained at both kV's for a range of α about the particular values given. The measured spread in N was of the order $\pm$ 5%. Similar experiments were done on other film thicknesses and in every case the comparative values of N exhibited the same relative order as those given in Fig. 8. After annealing the carbon films, the values of N determined under all conditions decreased by about 40%, with a small reduction in the spread between the various techniques used to evaluate N.

If an error of $\pm$ 10% in the plasmon determination of N is assumed (see previous discussion), then the 80 kV SIGMAK value and the limiting value from the 100 kV efficiency factor method, are in good agreement with the "absolute calibration". However the 100 kV SIGMAK value is about 25% higher than this. Since this ordering is maintained over the entire range of film thicknesses, the 100 kV SIGMAK value is clearly showing some systematic discrepancy compared to the other data. This general level of agreement is very encouraging, particularly bearing in mind the uncertainties in obtaining a reliable calibration technique. Moreover if the density of the carbon film had been taken as 2.2 gm/cm^3 rather than 2 gm/cm^3, the errors would have been reversed, with the 100 kV value being in better agreement than the 80 kV ones. Thus for any operating condition, on either microscope and using either data reduction technique, the value of N obtained is probably within $\pm$ 15% of the "true" value. Since in the STEM case only a few tens of thousands of carbon atoms were being analyzed, these results are encouraging.

There are several possible explanations for the systematic differences between the results obtained at 100 kV (using the two data reduction methods) and at 80 kV (using only SIGMAK).

1) The difference between the 100 kV values obtained when using SIGMAK and the efficiency factor method are due to our limited knowledge of cross-sections at high voltages. If the values in the Bethe expression, as chosen by Powell,[8] are selected to give agreement with cross-sections obtained from either SIGMAK[14] or a full quantum-mechanical approach,[12] then better agreement should be possible. The spread in N as a function of Δ when using the efficiency factor method, is an inherent drawback of the approach since small amounts of multiple scattering, or even noise, can affect the value of the "s" exponent obtained by curve fitting and hence the energy efficiency factor.

2) The discrepancy between the 80 kV and 100 kV SIGMAK values could be due to several factors:

a) There is a difference in operating mode since the 80 kV data was taken in CTEM while the 100 kV data was obtained in STEM. The convergence of the incident illumination is convoluted with the scattered distribution and this has the result of leading to a smaller effective value of α.[9] In the case where $\alpha_o \sim \alpha$ this effect can give an error of up to 30% unless an appropriate correction is made. However in both microscopes, the value of α_o was of the same order of magnitude ($\sim 1\times10^{-3}$ rads) because in CTEM focussed illumination was used in order that a small area could be selected for analysis. It is therefore unlikely that this effect explains the discrepancy observed here. However, it is worth noting that a quantitation performed on a dedicated STEM will usually be in the regime for which $\alpha < \alpha_o$ (since α_o will be of the order 10^{-2} rads) and in this case proper account must be taken of the convolution between the incident beam and the scattering profile.

b) The values of α used in either case could be in error. The magnitude of α needs to be known to about 5% if significant errors are to be avoided in the calculation of the partial cross-section using SIGMAK. However α is measured directly from a diffraction pattern[19] and the values used here were rechecked to verify their correctness. It is important to keep in mind that 5% accuracy can only be achieved if the variation of α with objective lens current or specimen height is known.

c) An instrumental artefact, such as stray scattered background in the spectrometer, could produce a systematic error in the value of I_K and hence in N. This possibility is a likely one if precautions are not taken in the spectrometer design in order to minimize stray scattering. However in both designs used here, the stray background around the carbon edge is very small ($<< 1\%$) and any substantial contribution of this type would lead to variations in the value of N with Δ, and these were not observed. And

d) The actual N values for the two experiments may have been different, either because different regions of the specimen were sampled in the two laboratories or possibly because of changes which occurred during transit (e.g. absorption of O_2 and H_2O). There is also a possibility that microscope induced changes, due to hydrocarbon contamination or mass loss, could have contributed to differences in N. In the absence of any other major source of error, such sample variations seem to be the most probable explanation.

4. CONCLUSION

The three methods of quantitation which have been described in this paper all show considerable promise. In particular the two computer-based schemes for K- and L-loss quantitation perform well under a wide variety of conditions, accounting correctly for changes in both Δ and α. When used for compositional analysis an accuracy of 10% or better without recourse to standards or arbitrary corrections is obtainable. Because of the problem in finding suitable independent standard methods, the absolute accuracy is less easy to determine. However results obtained under different conditions (kV, α etc.) and in different laboratories still show agreement to within $\pm$ 20%. Until thin-film standards become available, it is difficult to see how this figure can be improved upon. Nevertheless this degree of agreement for these methods at this early stage of development is encouraging.

REFERENCES

1. D. B. Wittry, R. P. Ferrier and V. E. Cosslett, *Selected-Area Electron Spectrometry in the Electron Microscope,* Brit. J. Appl. Phys. **2** (1969) 1767-73.

2. R. F. Egerton and M. J. Whelan, *High-Resolution Microanalysis of Light Elements by Electron Energy Loss Spectrometry,* Proc. 8th Int. Cong. on Electron Microscopy (ed. J. V. Sanders and D. J. Goodchild, Australian Academy of Science, Canberra, 1974), Vol. 1, 384-5.

3. M. Isaacson and D. Johnson, *The Microanalysis of Light Elements Using Transmitted Energy-Loss Electrons,* Ultramicroscopy **1** (1975) 33-52.

4. C. Colliex, V. E. Cosslett, R. D. Leapman and P. Trebbia, *Contribution of Electron Energy Loss Spectroscopy to the Development of Analytical Electron Microscopy,* Ultramicroscopy **1** (1976) 301-315.

5. R. F. Egerton, *Formulae for Light-Element Microanalysis by Electron Energy-Loss Spectrometry,* Ultramicroscopy **3** (1978) 243-251.

6. D. M. Maher, P. Mochel and D. C. Joy, *A Data Collection and Reduction System for Electron Energy-Loss Spectroscopy,* Proc. 13th Ann. Conf. Microbeam Analysis Society (available from K. F. J. Heinrich, NBS, Washington, 1978) 53A-G.

7. R. F. Egerton, *Minicomputer Software for Energy Loss Analysis,* Proc. of a Specialist Workshop on Analytical Electron Microscopy, Cornell University Materials Science Center (available from Prof. J. Silcox) Report 3082 (1978) 232-5.

8. C. J. Powell, *Cross Sections for Ionization of Inner Shell Electrons by Electrons,* Rev. Mod. Phys. **48** (1976) 33-47.

9. M. Isaacson, *All You Wanted to Know About ELS and Were Afraid to Ask,* Proc. 11th Ann. SEM Symposium (Chicago: SEM Inc.) **1** (1978) 763-76.

10. R. F. Egerton and D. C. Joy, *Cross Sections for K-Shell Excitations by Fast Electrons,* Proc. 35th Ann. Meeting EMSA (Baton Rouge: Claitors Press, 1977) 252-3.

11. S. T. Manson, *Inelastic Collision of Fast Charged Particles With Atoms,* Phys. Rev. **A6** (1972) 1013-24.

12. R. D. Leapman, P. Rez and D. Mayers, *A Quantitative Approach to Inner Shell Losses,* Proc. 9th Int. Cong. on EM (Toronto: Imperial Press and available from the Canadian Microscopical Society) **1** (1978) 526-7.

13. H. Bethe, *Zur Theorie des Durchgangs schneller Korpuskularstrahlen durch Materie*, Ann. Physik **5** (5) (1930) 325-400.

14. R. F. Egerton, *K-Shell Ionization Cross Sections for Use in Microanalysis*, Ultramicroscopy **4 (1979) No. 2 (in press)**.

15. S. Amelinckx, *Direct Observation of Dislocations*, (Academic Press, London 1964) 193.

16. D. M. Maher, D. C. Joy and P. Mochel, *A Standard For Transmission Electron Spectroscopy*, Proc. 9th Int. Congress on Electron Microscopy (Toronto: Imperial Press and available from the Canadian Microscopical Society) **1** (1978) 528-9.

17. R. F. Egerton, *A Simple Electron Spectrometer for Energy Analysis in the Transmission Microscope*, Ultramicroscopy **3** (1978) 93-67.

18. D. C. Joy and D. M. Maher, *A Practical Electron Spectrometer for Chemical Analysis*, J. of Microscopy **114** (1978) 117-129.

19. R. F. Egerton, *Inelastic Scattering of 80 keV Electrons in Amorphous Carbon*, Phil. Mag. **31** (1975) 199-215.

20. Y. Kihn, J. Sevely and B. Jouffrey, *Excitation des Niveaux Atomiques K du Carbone...*, Phil. Mag. **33** (1976) 733-41.

21. J. P. Langmore, J. Wall and M. Isaacson, *The Collection of Scattered Electrons in Dark-Field Electron Microscopy*, Optik **38** (1973) 335-50.

DISCUSSION WITH REVIEWERS

Reviewer I: Since equation 2 is derived assuming $\Delta \ll E_k$, how do you explain the fact that it seems to be a good approximation even for $\Delta \approx E_k$?

Authors: It is true that in the original, theoretical, derivation of equation (2) (Isaacson and Johnson (1975)) a necessary assumption is that $\Delta \ll E_k$. However many experimental measurements have confirmed that, over a wide range of values of α, the shape of the edge is independent of α and hence equation (2) can be treated as an established experimental result without regard to the approximations required to derive it theoretically.

Reviewer I: You mention that, unlike X-ray spectroscopy, electron energy loss spectroscopy is a standardless method. Isn't it true that, for example, with measured beam current, and measured or calculated cross-sections, fluorescence yields and geometrical collection and detection efficiencies, X-ray spectroscopy can be just as standardless?

Authors: No. For one reason there is no simple relationship between the volume of the specimen sampled by the beam and the actual volume over which any given X-ray line can be generated. Thus an absolute standardless X-ray measurement is impossible. Secondly, although in principle all the necessary terms can be calculated the accuracy with which some key parameters can be evaluated under the relevant conditions is questionable. Thus even so basic a quantity as the ionization cross-section at 100 kV is not yet known for most lines to better than probably a factor of 2 (see Powell 1976). While this uncertainty is reduced in a ratio measurement it is not eliminated. Thirdly, while the use of data "measured" from the microscope can be used to cover gaps in our ability to calculate the equivalent parameter effectively, the method cannot then qualify as *standardless* since some

assumptions have been made about the sample used to obtain the data, and even in the most recent, careful experiments (Goldstein - private communication) it has proved difficult to get reliable data for the elements Na and Mg because of irreproducible factors such as X-ray absorption from contamination build-up on the sample or detector window.

Reviewer II: Would you please justify expression (6) for η_Δ?

Authors: The intensity in the ionization edge extends from $E = E_k$ to $E = E_o$, and η_Δ represents the fraction of this that falls in the window $E = E_k$ to $E = E_k + \Delta$. For a K-edge we find experimentally $I(E) = B \cdot E^{-s}$ ($E \geq E_k$). By direct integration, and assuming s to be constant over the range Δ.

$$\int_{E=E_k}^{E=E_k+\Delta} I(E)\,dE = \left[\frac{B(s-1)}{E^{s-1}} \right]_{E=E_k}^{E=E_k+\Delta}$$

Hence

$$\eta_\Delta = \int_{E_k}^{E_k+\Delta} I(E)\,dE \,/\, \int_{E_k}^{E_o} I(E)\,dE = 1 - \left\{ \frac{E_k}{E_k+\Delta} \right\}^{s-1}$$

assuming that $E_o \gg E_k$.

Reviewer II: How was the influence of thickness considered in your experiments?

Authors: There is no need to consider thickness directly. Our requirement was that the quantitation programs should give stable results on any sample likely to be used in the microscope. As we show, the three approaches differ in the success with which they met this condition, but under suitably chosen conditions any of the methods give acceptable stability ($\sim$ 5%) and accuracy ($\sim \pm 10\%$) over the range of mass-thicknesses considered. There will, of course, be a thickness beyond which the multiple scattering becomes so severe as to cause a substantial error. We estimate this to start occurring at about one mean free path for inelastic scattering (typically 800-1500 Å at 100 keV) and we obtain this condition from calculations based on the low-loss (plasmon) data. In cases where the plasmon ratio is $\geqslant$ 1 we would treat any quantitation result with caution. This conclusion is the same for any atomic number Z.

Reviewer II: Can you give some examples of r and s values measured on your experimental spectra?

Authors: For a carbon film at 100 keV, r varies from 4.0 at $\alpha = 0.75$ mrad to 3.5 at $\alpha = 25$ mrad, whereas s goes from 5.2 at $\alpha = 0.75$ mrad to 3.75 at $\alpha = 25$ mrad. The s values are very reproducible as would be expected since they come from the fundamental physics of the ionization process. The r values are more variable since they represent the sum of a number of processes (plasmon tails, noncharacteristic single-electron excitations etc) and they are influenced by the prior history of the film. The values of s measured for carbon, and other elements, agree with theoretical values - see for example Leapman et al. 1978, Egerton 1979. A more detailed discussion of r and s values can be found in Joy, Maher and Mochel (Proc. Specialist Workshop on Analytical Electron Microscopy, Cornell University, Materials Science Center Report 3082 (1978) pp. 236-41 - available from Prof. J. Silcox, Cornell Univ.) and Maher, Joy, Egerton and Mochel (J. of Applied Physics - in press).

Reviewer II: What kind of precautions are taken in order to determine, and to prevent, stray scattering in the spectrometer?

Authors: The precautions taken include the use of large, graphite coated, drift tubes, the removal of all scattering surfaces in the flight path, and the use of collimators in the slit region. Under all operating conditions the stray scattering background, on both spectrometers, is between 2 and 3 orders of magnitude less than the typical signal of interest at any energy loss up to ~ 2 keV.

Reviewer II: Under what conditions do you estimate equation (1) isvalid and, for instance, sufficiently accurate as compared to the N value obtained from the deconvoluted spectra?

Authors: Equation (1) has been shown to be valid for most conditions likely to be encountered in the microscope (see Egerton, Proc. 11th Ann. SEM Symp. (ed. O. Johari) **1**, 13 (1978)) with the exception of the case where strong diffracted beams are present. While the expression is not exact, it is a sufficiently close approximation for the error to be insignificant unless *substantial* multiple scattering is occurring. In that case there will be other problems with the interpretation of the data as well. While it would be desirable to always quantitate from a deconvoluted spectrum to yield a single-scattering profile, this is not feasible at present, for routine studies, because of the time required for the deconvolution, and experimentally (see Ref. 5) no significant improvement in accuracy has been found.

Reviewer III: In what language are the computer programs written? Are they available to other workers in the field, and how should they go about getting copies of the programs?

Authors: The programs are written in Fortran as provided in the usual RT-11 compiler of the DEC 11/03. The implementation of the programs, in the interactive mode that we use, requires also the use of machine-code subroutines derived from the operating software of the multichannel analyzer (here a KEVEX-7000). Versions of these programs have been supplied to several laboratories, and more information can be obtained from the authors at the addresses given on the title page.

Reviewer III: In the examples of analyses given, ~ 1×10^{17} atoms/cm^2 were determined. What was the beam diameter, beam current, and sample thickness? How many atoms need to be in the beam to obtain quantitation with the accuracies and precisions described?

Authors: The STEM data was typically obtained with a 200 Å diameter probe, containing 10^{-10} amp incident on a 500-1000 Å thick sample. The analyzed volume therefore contains about 10^6 atoms. At this level the figures we quote are valid. The precision can be evaluated from the fact that the typical $P/\sqrt{P+B}$ figure in this condition is ~ 500 (assuming Δ = 100 eV, and dwell time = 200 ms/channel).

Reviewer III: While quantitation of BN, MgO and C films are useful for this paper, how practical is it to expect quantitation of a microprecipitate in a silicon matrix, or in an iron matrix?

Authors: We have experience in the quantitation of elements in a silicon matrix, and provided that the specimen is properly prepared no problems are encountered. Several examples of "practical" quantitations have already been published by other groups of workers as well (e.g. Leapman, pp. 203-219, and Kokubo et al., pp. 242-3, both in "Proceedings of a Specialist Workshop on Analytical Electron Microscopy", Cornell University Materials Research Laboratory Report 3082 (1978) - available from Prof. J. Silcox).

Reviewer III: Don't the experimental problems associated with sample preparation make quantitation of microprecipitates in any matrix of questionable value?

Authors: No. Provided the mass-thickness of the sampled volume falls below the value we suggest, there are no special problems. There are no specimens that cannot be prepared properly if adequate care is taken.

REVIEWERS: I   D.E. Johnson
            II   J. Sevely
            III   G.B. Larrabee

SCANNING ELECTRON MICROSCOPY/1979/II
SEM Inc., AMF O'Hare, IL 60666, USA

ELECTRON ENERGY ANALYSIS IN A VACUUM GENERATORS HB5 STEM

M. T. Browne

Physics Department
University of London,
Queen Elizabeth College,
Campden Hill Road
London, W8 7AH, U.K.

Abstract

The design and construction of a deceleration-dispersion - acceleration energy analyser is described. In a theoretical treatment the dispersion and focusing properties are assessed. In addition the stability against HT fluctuations is considered and is found to be three orders of magnitude better than a sector magnet analyser. The analyser has been fitted to a Vacuum Generators Ltd. HB5 STEM. The measured dispersion curve corresponds closely to the theoretically predicted one. The energy resolution using small apertures has been measured as 0.15 eV. Using larger apertures which correspond to a collecting semi-angle (α_A) of 0.5 mrad from a source of diameter (S_A) of 200 μm a resolution $\Delta E < 0.5$ eV at an electron energy (E) of 60 KeV has been obtained. In this mode very low dose rate (10^{-3} electrons $\text{Å}^{-2}\text{s}^{-1}$) energy loss measurement of beam sensitive material can be made. A pre-analyser lens system can be used to present to the analyser a source of apparent diameter S_A and angle α_A from a specimen of diameter S and angle α where $\alpha S = \alpha_A S_A$. The collecting angle from the specimen can thus be modified without affecting the resolution if S is correspondingly adjusted. For K shell energy loss measurements where large collecting angles are required, the specimen can be placed before the objective lens which can be used as a pre-analyser lens to collect up to 30 mrad semi angle from a source of 3 μm diameter while maintaining a resolution of 0.5 eV. Examples of various modes of operation are given. The working properties of the present analyser (using a 'figure of merit' based on $\alpha_A, S_A, \Delta E$ and E) are formally compared with those of the sector magnet analyser used on other STEM instruments.

KEY WORDS: Electron Optics, STEM, Image Analysis, Electron Energy Loss Analysis

Introduction

Electron energy analysers are used on scanning transmission electron microscopes to produce energy gated images or to obtain selected area energy loss spectra. The important analyser properties are the energy resolution (ΔE) and the electron collecting semi-angle (α). The resolution requirements vary with the application. A high resolution is required for fine structure spectral observations but for energy gated images the resolution requirements are generally less stringent. The ultimate resolution of the analyser system is limited by the energy half width of the electron source which for a field emission tip is of the order of 0.2 eV. The collecting angle requirements also vary with the application. The angular dependence of inelastic scattering depends on the energy loss. For example for carbon at 80 KeV, 70% of the K shell loss is scattered >10 mrad, while for the low loss range 10 - 50 eV < 2% is scattered > 10 mrad.[1] Thus for K shell loss measurements, where the spectral intensity is low, when high collection efficiency is required collecting semi angles >10 mrad are necessary. In the low loss range collecting semi-angles ~1 mrad are used.[2] An additional parameter of importance is the area of specimen (diameter S) from which the spectrum is obtained as this should be large if low beam damage measurements are required.

Analysers usually have some focusing properties so that a monoenergetic electron beam of energy E incident on the entrance aperture with a collecting semi angle α is brought to a focus, at least in the direction of dispersion. The resolution of the analyser with focusing properties will depend on the dispersion ($\delta\mu$m/eV) and the size of the image of the source. For a point source the size of the image will depend on the aberrations which are α dependent. For a given analyser a particular resolution ΔE_A can be obtained provided α is limited to say α_A. If the source is finite this will increase the size of the aberrated image and thus affect the resolution. Provided S is limited to say S_A the resolution is little affected. (For most analysers the image is produced at near unity magnification and thus to a reasonable approximation $S_A \approx \delta \Delta E_A$.) Thus both α_A and S_A are fixed for a given resolution ΔE_A. These values may not be suitable for the experiment at hand. A pre-analyser lens system can be used to present to the analyser a source of apparent diameter S_A and angle α_A from a specimen of diameter S and angle α where $\alpha S = \alpha_A S_A$. The collecting angle from the specimen can thus be modified without affecting the resolution if S is correspondingly adjusted.

A variety of energy analysers has been designed[3] which would be suitable to couple to a STEM or CTEM. They can broadly be classified into those which first decelerate the electrons before dispersion and those which do not. Of the latter type the sector magnet analyser is an example which has been used extensively with STEMs but this has the disadvantages of a very small dispersion ($\sim 1\mu m/ev$). Examples of the former type are the Wien filter[4] and the Möllenstedt analyser[5] These have large dispersions but accept only a small value of α. They focus only in one place unless additional astigmatic elements are incorporated into the system. In considering alternative designs the two directional focusing and the dispersive properties of off-axis electrons in an axial magnetic field have not been exploited as far as the author is aware. The incorporation of this as the dispersive element in a deceleration dispersion-acceleration system was therefore considered and developed. As the system showed promise the choice of an analyser suitable for attachment to an HB5 STEM was reduced to that between the sector magnet and the new design. The main advantage of the sector magnet was the simplicity of construction and the relatively large collecting angle, while the disadvantages included the small dispersion and the need for high stability of the magnetic field and HT supply. The new design would give a large dispersion, high stability against HT variations and, because the energy loss scan could be obtained by varying a bias voltage on the central electrode, energy losses would be directly read without the need for calibration. An additional advantage over the sector magnet, which deflects the beam through about $90°$, is that it was a 'straight through' design so that if the analyser was switched off the detector would be in the normal bright field position. For these and other reasons the new design was selected.

General description and theoretical treatment

A schematic diagram of the analyser is shown in Fig.(1). It consists of a symmetrical three electrode electrostatic lens. The central electrode has an internal cylindrical bore with rounded ends. A coil wound around the internal bore provides an axial magnetic field. The central electrode is held at the field emission tip supply voltage (HT) plus a variable bias voltage. The electrons are decelerated as they travel towards the central electrode where they encounter the magnetic field. Off axis electrons spiral around in the magnetic field. When the electrons emerge from the region of the central electrode they are reaccelerated and finally form an image in the detector plane. The combined effect of the electrostatic and magnetic fields produce both focusing and dispersive properties. If the field distributions are known, solutions of the paraxial ray equations are possible and the focusing and dispersive properties may be obtained. In the consideration of the type of decelerating field to use it was realised that for a potential of the form $\phi(z)=A\exp(-kz)$, $\phi'(z)/\phi(z)$ and $\phi''(z)/\phi(z)$ are constants and this property will give a smooth deceleration. The electrodes were therefore designed to approximate to this field. Thus for the electrode configuration used it was assumed that the electric potential could be expressed in the form

$$\phi(z)=A_1\exp(-kz)+A_2\exp(kz) \qquad (1)$$

where z is the axial distance from the analyser entrance electrode and A_1, A_2 and K are constants which give $\phi(o)$ and $\phi(Zmax)$ equal to (HT) and $\phi(Zmax/2)$ equal to the saddle potential (P). A value for P is obtained from measurements of the minimum energy of electrons which are transmitted by the analyser.

Electron trajectories of electrons entering the analyser with initial energy (HT + L) electron volts have been computed from the paraxial ray equations[6]

$$\qquad \qquad \qquad \qquad \qquad \qquad ...(2)$$
$$r''(z)=-\left[\frac{r'(z)}{2}\phi'(z)+r(z)\left\{\frac{\phi''(z)}{4}+\frac{10^{-9}}{8}\frac{e}{m}(B(z))^2\right\}\right]/(\phi(z)+L)$$
$$\Theta(Z)=\Theta(o)+\int\left[\frac{10^{-9}}{8}\frac{e}{m}(B(z))^2/(\phi(z)+L)\right]^{\frac{1}{2}}dz. \qquad (3)$$

where $B(z)$ is the axial magnetic field, and r is the perpendicular distance from the z axis and Θ is the angle of the 'guiding plane' in which r is measured. The method of computation involved setting initial values of $r(o)$, $r'(o)$ and dividing the analyser path length (Zmax) into 100 equal elements. Assuming a constant value of $r(z)$ and $r'(z)$ over each element the value of $r''(z)$ was calculated for that element. This allowed estimates of the values of $r(z)$ and $r'(z)$ for the next element to be obtained. This process was repeated to give a value of $r(z\ max)$ and $r'(z\ max)$. The number of elements was then doubled and the process repeated. The new values of $r(z_{max})$ and $r'(z_{max})$ were compared with the previous values. If there was a significant difference the number of elements was again doubled to obtain a more accurate value. Up to eight iterations were allowed in the programme but in general four or five were sufficient. It should be noted that the electrons are rapidly decelerated on entering the analyser and thus a relativistic correction can be ignored as it is only applicable to a very small region where the dispersion is very small anyway.

From the values of $r(o)$ $r'(o)$ $r(z_{max})$ and $r'(z_{max})$ at different values of L the focusing and the dispersion properties of the analyser were obtained, allowance having been made for

the aperture lens effect at Z=o and Z max.
Typical electron trajectories for various
values of L are plotted in Fig. (2). The
reciprocal of the focal length (1/F_2) and the
position of the principal plane (Z_{H_2}) together
with the angle of the 'guiding plane' Θ are
plotted in Fig. (3). The dispersion curve is
in the form of a spiral and is shown in Fig.
(4). The source is focused once on each loop
of the spiral with approximately unity magnification. The entrance aperture is also
focused once on each loop at a magnification
greater than unity.

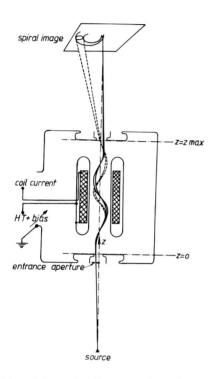

Fig. (1). Schematic diagram of Analyser

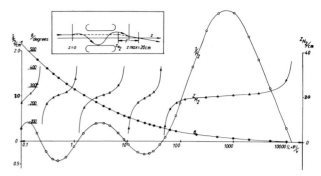

Fig.(3). 1/F_2, Z_{H_2} and Θ v (L+P) for HT = 60kV
and B (10cm)= 7×10^{-4}T.
Inset. Schematic diagram indicating
F_2 and Z_{H_2}

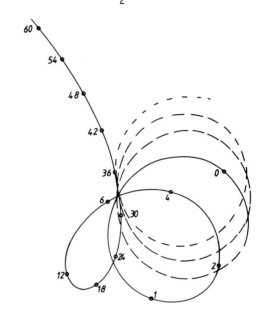

Fig. (4). Dispersion curve for HT = 60KV and
B($Z_{max}/2$)= 7×10^{-4}T

Values of L are marked along curve.

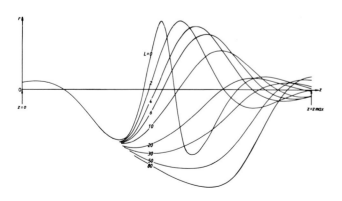

Fig. (2). Electron trajectories for HT = 60KV
and B($Z_{max}/2$) = 7×10^{-4}T

If a bias voltage v is applied to the
central electrode the effect on the electron
trajectories is similar to that resulting from
an energy loss of the same magnitude. This is
confirmed by comparing the arrival point on the
spiral of electrons with L=50 V , V=0 and HT=60 kV
with L=50 V , V=0 and HT=60.3KV, which is
exactly equivalent to L=350V, V=300V and HT=60KV.
The error was found to be equivalent to a ΔV of
0.15V. For smaller values of V the error was
found to be proportionately less. The percentage error is 0.05%. Thus by varying the bias
voltage V energy loss electrons can be made to
scan across an aperture placed on the spiral.
If the bias is adjusted so that the zero is
aligned with the aperture a change of bias of X
volts will bring the energy loss (X$\pm$0.05%) eV to
the aperture. Thus the value of the bias can be

used to directly read the energy loss.

The above analysis also indicates that the analyser would be very stable against H.T. fluctuations. Thus at 60KV a 1 V change during a scan would result in only a 0.5 meV error in energy loss measurements. For the sector magnet the error would be 1eV.

Analyser Design

The analyser was designed to fit onto an HB5 STEM. The high vacuum requirements of the STEM made it necessary to couple the analyser via a large pumping port. To allow mechanical alignment a 15cm diameter bellows was used to couple the analyser to the top of the STEM as shown in Fig. (5). The tilt mechanism is of a kinematic design which allows the analyser to be tilted about a point in the specimen plane. To minimise HT instability which would affect the STEM performance an HT insulator similar to that used in the STEM was selected and the radii of curvature of surfaces and electrode separations were made as large as possible. The shape of the central electrode of length 15cm and internal diamter 2.5cm was designed to give an axial electric field which approximates to the simple exponential form given in equation (1).

The detector system mounted above the analyser consisted of a plastic scintillator (NE 160) and variable detector apertures inside the vacuum column with the photomultiplier placed outside a vacuum window. To facilitate the alignment of the analyser and the selection of the best position for the detector aperture a set of x-y scan coils and a set of alignment coils were placed between the analyser and the detector. In a later modification the vacuum window was replaced by a fibre optic vacuum window and the sheet of plastic scintillator was pressed against the window. With this arrangement a detector aperture inside the column was unnecessary because an area of the scintillator could be selected from outside the fibre optic window. In fact more than one area could be selected at a time by using a configured light pipe arrangement to couple to more than one photomultiplier.

The analyser bias voltage has to be super-imposed on the HT voltage. The coil is also held at HT potential. The bias voltage and coil current are supplied from rechargable batteries, which, together with the associated electronic circuits are housed in a cylindrical container mounted at the centre of a supplement-ary HT tank. Perspex and oil are used as mixed dielectric insulators in the HT tank and the gap between the container and the tank wall made large enough so that the stability of the STEM HT supply is not affected. The bias voltage can be linearly scanned in a 40V range, shifted by a variable 0-60v or stepped by twelve 30v steps and one 240v step. The scan generator of approximately one second period produces synchronising pulses which are optically coupled to the outside of the HT tank.

Analyser Performance

A photograph of the analyser dispersion curve is shown in Fig. (6). This was obtained by applying the STEM X and Y scan voltages to the analyser X and Y scan coils thus causing the dispersion curve to be scanned across a small detector aperture. The resultant image could be viewed on the STEM video monitor. The photograph is a composite of several carbon energy loss spectra shifted by 6v steps. The main features are the zero loss peaks (shifted by 6v steps) which are the bright regions of the image. The dispersion curve can be seen to closely correspond to Fig. (4) which is the theoretically predicted curve.

The spectrum source for this photograph was a few μm in diameter. The spectral image of this source can be seen to be elongated due to astigmatism in the analyser. The elongation is perpendicular to the direction of dispersion in one region of the curve. This is the region where the best energy resolution is obtained and is thus the position chosen for the detector aperture. For high collection efficiencies at the detector, slit apertures of dimensions 1500μm x 200μm and 1500μm x 500μm have been used.

To test the resolution the zero loss peak can be scanned across the detector aperture by scanning the bias voltage and observing the photomultiplier signal. Using a 50μm detector aperture the f.w.h.m. was measured as 0.26eV as shown in Fig. (7). As the f.w.h.m. of the field emission tip was calculated to be 0.22 eV using the Fowler-Nordheim plot the analyser resolution was approximately 0.15eV. By using the larger slit detector aperture (1500μm x 500μm), a source size of up to 200μm diameter and a 200μm diameter analyser entrance aperture (corresponding to α =0.5mrad) the resolution was found to be $<$0.5eV. These measurements were made at E=60keV.

Data Collection

The photomultiplier signal obtained when the energy loss spectrum is scanned across the detector aperture is sampled at 256 points per scan and digitised with 8 bit resolution. The digital information is stored in a PDP8E minicomputer which is interfaced to the STEM. The average of the last eight or sixty four scans can be displayed on a video monitor screen. To calibrate the display a 30v bias step is applied and the resultant zero loss shift can be measured in terms of the 256 display points. Every eighth point can be brightened to aid this process.

Modes of operation

A STEM fitted with an energy analyser can be used to obtain electron energy loss spectra from very small areas of specimen as limited by the probe size. However some specimens are beam sensitive and in order to minimise beam damage larger areas must be used. The STEM is very adaptable and can be used for such measurements. Isaacson[2] in measurements on nucleic acid bases defocused the electron probe to 2-5μm diameter (5μm would limit the resolution to ≈ 1eV) and each spectrum consisted of the sum of spectra from different areas of the specimen. The present analyser has been used for such low-dose measurements. Three modes of operation have been used which are illustrated in Fig. (8).

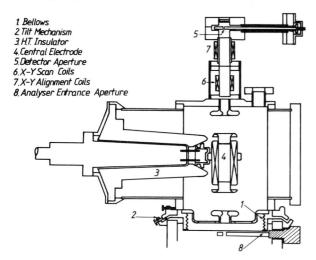

1. Bellows
2. Tilt Mechanism
3. H.T. Insulator
4. Central Electrode
5. Detector Aperture
6. X-Y Scan Coils
7. X-Y Alignment Coils
8. Analyser Entrance Aperture

Fig. (5). Cross section of analyser (1/4 scale)

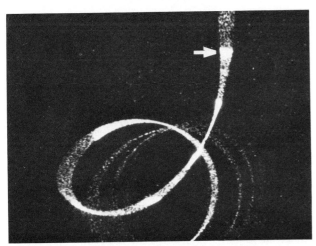

Fig. (6).   Analyser dispersion curve showing 6v steps of the zero loss (A suitable position for the aperture is indicated ⇨)

In mode A with the STEM set in 'diffraction' mode the condenser lens is focused into the back focal plane of the objective lens which in turn is focused into the analyser entrance aperture. The diffraction aperture is conjugate with the specimen plane and thus defines the area of specimen illuminated. The largest diffraction aperture which could be used was limited to approximately 450μm diameter due to astigmatism

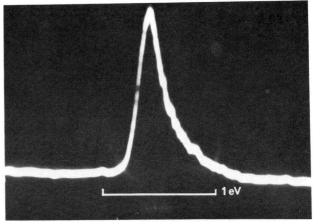

Fig. (7.) Energy distribution from the field emission source with measured fwhm. of 0.26eV. (Calculated source f.whm. = 0.22eV indicating an analyser resolution of 0.15eV)

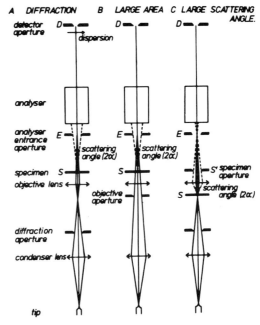

Fig. (8). STEM energy analyser modes of operation

in the condenser and gun. As there is a magnification of approximately x 30 between the diffraction aperture and the specimen plane the largest area of specimen which can be used is 15μm in diameter. Doses of ≈ 1 electron $Å^{-2}$ per spectrum can be obtained.

In order to reduce the dose rate it is necessary to illuminate a larger area of the specimen. This is possible if the objective lens is turned off and the large area mode B is used. In this mode the objective aperture is used to define the specimen area. Using a 200μm diameter objective aperture dose rates of 10^{-3} electrons $Å^{-2}s^{-1}$ can be obtained.[7]

For K shell losses it is necessary to collect a large scattering angle. In mode A and B the collecting angle is limited to ~0.5mrad for <0.5eV resolution. In mode C the specimen is lowered to the objective aperture position. The objective lens can be used as a pre-analyser lens to modify the collecting angle which is determined by the aperture S placed in the normal specimen position. This can have a value of up to 200μm diameter without reducing the resolution giving a collecting semi angle up to 30mrad. The area of the specimen from which the spectrum is obtained is now determined by the analyser entrance aperture which is conjugate with the specimen and is demagnified x 60 into the specimen plane. As the entrance aperture is limited to 200μm for <0.5eV resolution this corresponds to 3μm diameter. An example of this mode of operation is shown in Fig. (9) which is the K shell carbon loss for Thymine. The spectra are the means of 64 scans each one obtained from a fresh area of specimen. A damaged energy loss spectrum is also shown.

If the STEM is used in the imaging mode an energy gated image can be obtained. Using a 200μm diameter analyser entrance aperture to give a beam semi angle of 0.5mrad a resolution of <0.5eV is obtained for a field of view at the specimen of 200μm diameter. As the electron probe incident on the specimen subtends a beam semi-angle of ~10mrad the collection efficiency would be poor. If the resolution requirements are not so stringent a larger analyser entrance aperture can be used. Fig. (10a) shows a dark field image of thorium atoms on a carbon substrate while (b) is an energy gated image with an energy window of approximately 10 to 35eV and (c) shows the subtraction (b) from (a). In this example an analyser entrance aperture of 1000μm diameter was used corresponding to a collecting semi angle of 2.5mrad.

The present STEM performance can be compared to that of the Castaing and Henry[8] arrangement where for a resolution of 1.5eV a 1 mrad beam semi angle was used. The field of view was limited to 1.4μm. However if the present analyser was properly matched to the STEM a resolution of <0.5eV could be obtained for a beam semi angle of 30 mrad. The field of view would then be limited to 3μm diameter.

Discussion

Many factors must be considered in the choice of an energy analyser for attachment to a STEM. The performance parameters are α_A, S_A, and $\Delta E_A/E$. However α_A and S_A can be modified to α and S at the specimen (where $\alpha S = \alpha_A S_A$) by the use of a suitable lens system to better match the application in hand. Thus, providing a suitable lens system is available the analyser

with the largest $\alpha_A S_A$ value for a given $\Delta E_A/E$ has the greatest potential. As both α_A and S_A increase with ΔE_A it is found that $(\alpha_A S_A)^{\frac{1}{2}}/\frac{\Delta E_A}{E}$ is reasonably constant for a given type of analyser and can thus be used as a 'figure of merit' for the comparison of one analyser performance with another. In table (1) values are given for a typical sector magnet analyser[9], a median plane aberration corrected analyser[10] and the present design.

Table 1

	$\frac{E}{KeV}$	$\frac{\Delta E_A}{eV}$	$\frac{\alpha_A}{mrad}$	$\frac{S_A}{\mu m}$	$(\alpha_A S_A)^{\frac{1}{2}}/\frac{\Delta E_A}{E}$ $m^{\frac{1}{2}}rad^{\frac{1}{2}}$
Sector magnet	100	3.5	1	5	2.0
Median plane aberration corrected sector magnet	20	0.5	$\sqrt{25 \times 2.25}$ = 7.5	2.5	5.5
Browne	60	0.5	0.5	200	38

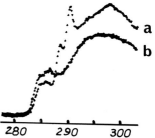

Fig. (9). Characteristic energy loss spectra of Thymine showing fine structure in the region of the carbon K-shell excitation edge. a) undamaged b) damaged.

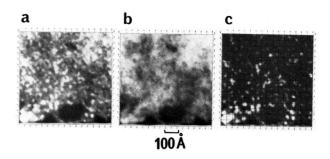

100Å

Fig. (10). Small clumps of Thorium atoms on a thin carbon film.
(a) dark field image.
(b) energy gated (10-35eV) image
(c),(b) subtracted from (a).

There is a strong case for using a lens system to match a given analyser to the STEM to allow large collecting angles[11]. The present system has been used in mode C where the objective lens is used as a pre-analyser lens to collect up to 30 mrad at 0.5eV resolution. The use of a high excitation objective lens in an 'objective-condenser' arrangement would allow collection angles of 30mrad in the imaging mode.

The ideal detecting system for a STEM would collect all the transmitted electrons. The transmitted electrons should be separated into the inelastically scattered electrons which can be divided up into various energy gates and the non-loss electrons which should be divided up spatially[12] This may be very difficult to realise. The present analyser has a certain capability which might be exploited. The analyser entrance aperture is imaged at one region of the dispersion spiral in the detection plane. If the zero loss electrons are brought to this region a zero loss image which retains the spatial integrity of the electrons in the entrance aperture could be obtained while the energy loss electrons would be dispersed around the spiral.

Acknowledgement

I am grateful for financial support to the Science Research Council. Dr. D. L. Misell kindly provided the computer program. I also thank Dr. Lackovic and Professor Burge for discussion. The analyser was made in the Department Workshop.

References

1. R. F. Egerton. Inelastic scattering of 80KeV electrons in amorphous carbon. Phil. Mag. 31 (1) 1975 199-215.
2. M. Isaacson. Interaction of 25KeV electrons with the nucleic acid bases, Adenine, Thymine and Uracil. J. Chem. Phys. 56, 1972 1803-1818.
3. O. Klemperer. Electron beam spectroscopy. Rep. Prog. Phys. 28,(1965) 77-110.
4. H. Boersch, J. Geiger and W. Stickel. The resolving power of an electrostatic-magnetic energy analyser for fast electrons. Z.Phys. 180, (1964) 415-424.
5. A.J.F. Metherell. Energy analysing and energy selecting electron microscopes. Adv. Optical Electron Microscopy. 4, 1971. 263-360.
6. B. Paszkowski. Electron optics. ILIFFE LONDON 1968 Ch. 5.
7. R. E. Burge et al. Image analysis and beam damage. Inst. Phys. Conf. Ser. No. 36. 1977, p.103. Inst. Physics, Bristol, U.K.
8. R. Castaing and L. Henry. Filtrage Magnétique des vitesses en microscopie électronique. J. Microscopie 3,1964, 133-152.
9. R. F. Egerton and C. E. Lyman. An electron spectrometer system for use with a STEM. Proc. EMAG 75, J. A. Venables Academic Press London, 1975, 35-38.
10. A. V. Crewe, M. Isaacson and D. Johnson. A high resolution electron spectrometer for use in STEM. Rev. Sci. Inst. 42,1971, 411-420.
11. A. V. Crewe. Post-specimen optics in the STEM. Optik, 47, 1977, 299-311.
12. R. E. Burge et al. STEM imaging at high resolution: the influence of detector geometry. SEM/]979/I, this volume.

Discussion with Reviewers

D. Johnson:
Since you incorporate both deceleration and dispersive elements in your analyzer, would a "figure of merit" comparison with a deceleration magnetic sector system (e.g. Y. Kokubo and M. Iwatsuki, J. Electron Micros. 25 123 (1976)) be more meaningful, and have you done this?
Author: The 'figure of merit' table shows values for analysers which have been fitted to STEMs but the 'figure of merit' benefit is only obtained if the analyser is properly matched. Of course other designs could be considered. Insufficient information is available in the paper referred to to derive a 'figure of merit.' The effect of the deceleration will increase dispersion and a 'figure of merit' comparable with the present design might result. A deceleration system is probably necessary to obtain a high'figure of merit.'

D. Johnson:
Ultimately, a parallel detection system for electron energy loss spectrometry is desirable. Do you have any thoughts as to how such a system might be implemented with the spiraling, non-linear dispersion figure shown?
Author: Ultimately parallel detection is desirable both for electron energy loss spectrometry and for 'total electron' collection in imaging where parallel detection of the zero loss spatial information is also required. The ultimate aim would be to perform both tasks.

In order to implement parallel detection with the present analyser it would be necessary to use a pre-analyser lens system to match a large (~30mrad) semi angle scattering cone from the specimen into the analyser entrance aperture. A sketch of the dispersion curve at the detector plane which would result is shown in Fig. (11). Note that the analyser entrance aperture is focused at one point on the dispersion curve and the bias is adjusted to bring the zero loss to this point. K-shell losses would not be transmitted by the analyser but would be reflected and imaged on the input side of the analyser. The reason for using ~30mrad entrance cone would be to make sure the K shell losses were collected with reasonable efficiency.

Parallel detectors could be arranged to
collect the zero loss spatial information and
the low (5-45 eV) energy loss on the output side
of the analyser and the K shell losses on the
input side. This arrangement would be best
suited to imaging.

For parallel energy loss measurements high
resolution (<0.5eV) would only be obtained in
a limited range (~15eV) around the 30eV point
in Fig. (11). To examine fine structure in say
the plasmon (p) or a K shell (K) region that
part of the spectrum could be shifted to the
high resolution region by applying a bias step.
The time required to collect K spectra is
considerably longer than for p spectra because
of the low K signal. If both p and K spectra
are to be obtained very little extra time is
required to take p first and then K rather than
both together.

D. Johnson: All the data shown are for 60kV.
Are there reasons for not running at 100kV?
Author: We have the HB5 STEM prototype instrument which was built to operate at 20, 40, 60
and 80kV only. We found certain instabilities
in the STEM itself at 80kV and generally work at
60kV. The analyser and the associated HT bias
tank have been taken up to 100kV on a separate
supply as part of a conditioning process for
80kV working and should therefore run at 100kV.

D. Johnson: Could you estimate a time and
material cost for this type of analyser?
Author: The materials would cost approximately
$6,000 including made up cables and insulators
and labour would be one man year.

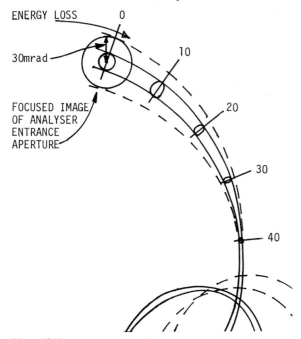

Fig. (11). Sketch of dispersion curve showing
position of focused image of the analyser
entrance aperture as might be used for parallel
detection.

SCANNING ELECTRON MICROSCOPY/1979/II
SEM Inc., AMF O'Hare, IL 60666, USA

VISUALIZATION OF SUBSURFACE STRUCTURES IN CELLS AND TISSUES BY
BACKSCATTERED ELECTRON IMAGING

R. P. Becker[+]   and   M. Sogard[*]

Department of Anatomy, Coll. Medicine   *Physics Dept. and Enrico Fermi Inst.
University of Illinois at the Med. Ctr.   University of Chicago
808 South Wood Street   5640 S. Ellis Avenue
Chicago, IL 60612   Chicago, IL 60637

Abstract

A fraction of the beam electrons which interact with a specimen scatter back. The number of
backscattered electrons (BE's) increases with
the atomic number of the elements encountered.
Cell and tissue structures lacking a heavy metal
content yield few BE's compared to structures
affixed with heavy metals, either vitally or by
means of staining methods applied after fixation.
The BE imaging mode of a scanning electron
microscope (SEM) provides an intensity map of
the BE yield from the specimen. BE imaging of
selectively stained structures in cells and
tissues renders these structures visible in
contrast to the unstained surround. Since BE's
can emerge from a significant depth within the
material, BE imaging can be used to view such
heavy metal stained structures beneath intact
cell surfaces. The microcontours of the overlying surface can be viewed concurrently by using
the surface scanning (i.e., the secondary
electron imaging; SEI) mode of the microscope.
Methods for selectively contrasting subsurface
structures can be adapted from existing light
microscope (LM) and transmission electron microscope (TEM) methods. Staining methods have been
devised for subsurface viewing of cell organelles, including nuclei, mitochondria, peroxisomes, lysosomes, and phagosomes.
A physical model is presented which describes
these observations and suggests future possible
trends in this subject. Specifically the image
contrast and resolution are described in terms of
the physical properties of the stain and specimen
and of the SEM operating conditions of energy and
current. Finally a summary of instrumentation considerations describes present and potential BE detectors, their ancillary electronics, and image
processing.

[+]R.P. Becker also affiliated with:
Department of Anatomy, University of Chicago
Chicago, IL 60637

KEY WORDS: Cell Nuclei, Mitochondria, Lysosomes,
Peroxisomes, Enzyme Cytochemistry, Heavy Metal
Staining, Biological Specimen Preparation, Contrast,
Resolution, SEM Accessories, Backscattered Electron
Imaging, Electron-Specimen Interactions

I. Introduction

The scanning electron microscope (SEM), in
its secondary electron (SE) imaging mode, is
unparalleled in its ability to inform about the
microcontour of natural surfaces. As such, it
has become a "must" tool in many biological
investigations. Unfortunately, natural surfaces
have, with some notable exceptions, proven to be
remarkably uninformative and, indeed, unreliable
as to the identity of cells, their specific
functions, their stages in the cell cycle, and
their possible neoplastic state.[1] This circumstance has led investigators to devise alternative means to extract such information from
their SEM preparations through light microscopic
(LM) correlation[2-4] or transmission electron
microscopy (TEM) of areas previously scanned[5-7]
(for a review of methods ancillary to SEM,
consult Albrecht and Wetzel[8]). Additionally,
however, investigators have recognized and begun
to exploit alternative sources of information
available from within the SEM itself, i.e., by
using SEM imaging modes which map the
distribution over the specimen of characteristic
X-ray emission,[9-14] cathodoluminescence
(CL)[15-23] (for a general bibliography on
cathodoluminescence, consult Brocker and
Pfefferkorn[24]) and backscattered electrons
(BE)[9-11,13,14,25-34] (for a general
bibliography on the backscattered electron
image, consult Wells[35]). Of these, backscattered electron imaging (BEI) is particularly
attractive because it offers the visualization
of cellular organelles beneath intact cell
surfaces concurrent with the imaging of the
surface, efficiently providing direct correlation of natural surfaces, as seen by
collecting secondary electrons and sub-surface
components, as revealed by BEI. Our purpose
here is to put forth a practical understanding
of BEI theory, instrumentation and methodology
for the purpose of visualizing subsurface
structures in cells and tissues.

II. Physical Considerations

In this section we will briefly review the
properties of secondary electron and backscattered electron imaging as it relates to our
topic.

Secondary Electron Imaging

When a high-energy electron beam passes through a specimen it produces many low-energy electrons within the material through electron - electron scattering. The range of these low-energy electrons is quite short. However, if they are produced within a few nm of the surface of the specimen, they have a chance of escaping. Such "secondary" electrons (SE's) have energies typically less than 20 ev. They are easily collectable by a detector maintained at a positive potential in order to attract them (usually a scintillation counter). The number of secondaries collected depends on the local surface properties of the material, mainly its topography. This number varies as the beam is scanned over the specimen and leads to the contrast necessary for producing an image of the surface. The secondary electron coefficient, the number of secondary electrons produced per incident electron, has little dependence on Z, the atomic number of the material forming the surface, increasing by about a factor 2 from carbon to gold. It decreases slowly as the incident electron energy increases, changing by a factor of between 3 and 4 as the incident energy goes from 5 to 30kV[36]. This is why the contrast in a secondary electron image decreases as the beam energy is raised. Because the secondary electrons have a finite range within the specimen, their intensity really represents an average of the surface properties over a small region surrounding the beam spot. The size of this region is of the order 4-5 nm and represents the ultimate resolution attainable with this imaging technique.

Backscattered Electron Imaging

In addition to electron - electron interactions, which act mainly to simply reduce the incident electron's energy, the electron may also suffer collisions with the atomic nuclei. Little energy loss is suffered in these collisions, but the direction of the electron can be appreciably altered. Indeed a single collision can occasionally reverse the electron's direction. More frequently, the same effect is achieved by a succession of smaller angle collisions. If the direction-reversed electron regains the surface of the specimen it is referred to as a backscattered electron (BE). Figure 1 describes this process graphically. It is obtained from a computer Monte Carlo simulation of electron interactions in a slab of carbon. A narrow beam of electrons enters the carbon from above. The higher the energy the more deeply the beam penetrates the material. In so doing it spreads out and some fraction of the electrons eventually regain the surface. The energy of these backscattered electrons can be considerable, approaching the incident beam energy. Figure 2 shows the energy spectrum of BE's from various elements. For high atomic number (Z) materials a considerable fraction of highly energetic BE's are produced. This qualitative change from low Z to high Z materials arises from the growing importance of elastic, relative to inelastic, electron collisions as Z increases.

The production of BE's is strongly Z dependent. The BE coefficient η, defined as the number of BE's produced per incident electron increases with Z, ranging from about 0.05 for carbon to 0.50 for gold. It is this Z dependence which makes possible the technique described in this paper. If a biological structure is selectively stained with a heavy metal, internal details of the specimen can be imaged by means of the increased backscattering from the high Z regions. Figure 1d shows this effect graphically. Below the upper (we shall call it the surface) layer of carbon is a region which has been stained with silver. The enhanced backscattering from the lower silver-stained region is evident. Such an image is referred to as a Z contrast (or atomic number contrast) image. Normally, when working with Z contrast, we would like to minimize surface topographic detail which tends to obscure the internal structure of the specimen. Since exiting BE's travel in a straight line, topographic detail tends to be enhanced when the BE detector is located to one side of the beam and near the specimen's "horizon". We should, therefore, try to collect electrons with a detector concentric and near to the beam. In addition, for maximum penetration into the specimen, the specimen should be oriented approximately perpendicular to the beam. Figure 1e illustrates the case of an obliquely incident beam. The effective depth of the beam within the specimen is reduced. Figure 3 shows the BE angular distribution for the above desired conditions. Because of their higher energy, BE's, unlike SE's, cannot be attracted to a detector biased at a moderate positive voltage, so the detector must intercept a significant solid angle in order to achieve reasonable collection efficiency.

In order to understand the potential and limitations of this technique we must answer a number of basic questions. We must know how far into the specimen we can look with this technique. We must know what kind of resolution we can achieve, and we must know the amount of contrast as a function of beam energy. We must also know the beam currents which are required and whether they will put constraints on the attainable resolution. To answer these questions requires information about the specimen: its density and the heavy metal stain concentrations.

These questions will be discussed further and answered below.

III. Biomedical Applications

Early Studies

Priority for recognizing the potential of BE imaging in biomedical studies and for the development of early methodology is shared by Watanabe,[9] Watanabe and Ohishi,[10] Abraham and DeNee,[25,37] and DeNee et al.[26] In pioneering studies, these investigators established that heavy metal staining methods, adapted from classical light microscopic (LM) and TEM methodology taken together with BE imaging, could be used to visualize structures at or below the surface of SEM preparations, including cell nuclei, collagen and reticular fibers, basement membranes, sites of phagocytic

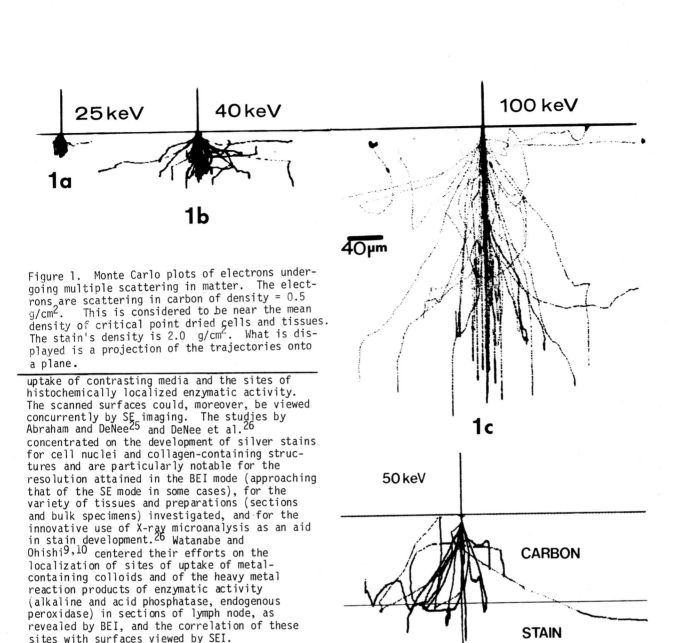

25 keV   40 keV   100 keV

1a

1b

40 μm

1c

50 keV

CARBON

STAIN

1d

10 μm

45°   50 keV

1e

20 μm

Figure 1. Monte Carlo plots of electrons under-going multiple scattering in matter. The electrons are scattering in carbon of density = 0.5 g/cm^2. This is considered to be near the mean density of critical point dried cells and tissues. The stain's density is 2.0 g/cm^2. What is displayed is a projection of the trajectories onto a plane.

uptake of contrasting media and the sites of histochemically localized enzymatic activity. The scanned surfaces could, moreover, be viewed concurrently by SE imaging. The studies by Abraham and DeNee[25] and DeNee et al.[26] concentrated on the development of silver stains for cell nuclei and collagen-containing structures and are particularly notable for the resolution attained in the BEI mode (approaching that of the SE mode in some cases), for the variety of tissues and preparations (sections and bulk specimens) investigated, and for the innovative use of X-ray microanalysis as an aid in stain development.[26] Watanabe and Ohishi[9,10] centered their efforts on the localization of sites of uptake of metal-containing colloids and of the heavy metal reaction products of enzymatic activity (alkaline and acid phosphatase, endogenous peroxidase) in sections of lymph node, as revealed by BEI, and the correlation of these sites with surfaces viewed by SEI.

Subsequent biomedical applications of Z contrast in BEI have for the most part followed the initiatives of these early studies (a notable exception was the early use of BEI in SEM autoradiography[31,38]), and can be grouped conveniently into: (1) studies dealing with the viewing of structures at or contiguous with cut or fractured surfaces;[27,37,39-41] and (2) studies, which are of primary interest here, dealing with the viewing of structures wholly beneath intact natural surfaces, including cell nuclei, peroxidase containing leukocyte granules and phogosomes[13,14,28-30,32,33,34].

Below we give methods of staining these (and other) structures for subsurface viewing in a variety of SEM preparations. It will become readily apparent that these methods for BEI represent simple modifications of established LM and TEM staining methods adapted to standard SEM biological sample preparation.

Methodology

Our approach will be to present a sampling, from our own experience, of reliable protocols

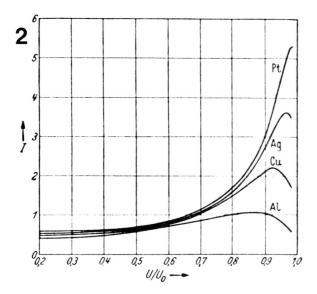

Figure 2. BE intensity is plotted against BE energy U normalized to the primary energy Uo. Figure from H. Kulenkampff and W. Spyra, Z. Phys. 137, (1954), 416-425.

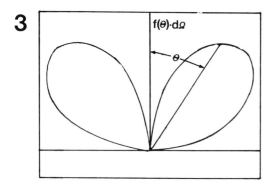

Figure 3. The angular distribution $F(\theta) = \cos \theta$ multiplied by the solid angle $d\Omega = 2\pi \sin \theta d\theta$ is plotted. This is proportional to the angular intensity which would be measured at various angles θ.

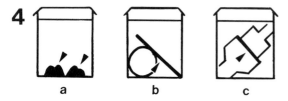

Figure 4. Staining with GMS is carried out in snap cap vials. (a) Bulk tissue rests on vial bottom (arrowheads); (b) a cover slip or other substrate is positioned against a glass bead such that the attached cells (arrowhead) face but do not contact the bottom of the vial; (c) a Millepore filter (arrowhead) with adhering cells remains in a disposable Swinnex holder, the whole unit being immersed in the stain.

for contrasting specific structures in a particular cell or tissue preparation intended for BEI. The examples presented have been selected to demonstrate the breadth of possible applications, whether the preparation be bulk tissue, tissue culture, or cell suspension or whether the contrasting method be direct staining with heavy metals, localization of heavy metal products of enzymatic reactions, or marking with ingested heavy metal particulates.

Cell nuclei - bone marrow, lymph node - silver reduction staining (Figs. 4-10). The femoral bone marrow and popliteal lymph nodes of black-hooded Long-Evans rats were fixed by vascular perfusion with 1.5% formaldehyde (prepared from paraformaldehyde) and 2.5% glutaraldehyde together in 0.08 M sodium cacodylate buffer, pH 7.2, with 0.15% $CaCl_2$ added. The osmolality of this fixative solution was 1285 $\pm$ 10 mOsmol/kg. The marrows and lymph nodes were excised, divided into small (about mm^3) blocks, fixed further by immersion for 2h at 4 C, and washed thoroughly with several changes of the same buffer through 24h.

Just prior to staining, the tissue blocks were rinsed free of buffer with several changes (5 min each) of distilled water. The tissues were then stained by exposure to a warm methenamine silver solution[42] as follows: tissue blocks were placed in distilled water contained in clean (need not be acid clean) 16 ml, Wheaton, snap cap, "800" glass, specimen bottles (No. 225536) (Fig. 4). The water was then decanted and replaced with about 10 ml of a methenamine silver solution (GMS).[42] This solution was prepared fresh just before use by mixing 20 ml of GMS stock (5 ml of 5% silver nitrate added dropwise with stirring to 100 ml of 3% hexamethylene tetramine) with 20 ml of distilled water, and adding 1.6 ml of 5% sodium borate. The pH of the resulting soluton is 8.8 to 9.0. The GMS stock solution can be stored, if kept in the dark with refrigeration, for at least one week without obvious loss of staining potency or reproducibility. After allowing the GMS solution to infiltrate the tissues for a few minutes, the solution was decanted and the remaining 10 ml of the GMS solution was added to the vial. The vial was then placed in a preheated 50 C water bath, covered, and warmed with gentle shaking until the solution became just perceptibly gray (usually 30-45 min). The tissue itself should by this time be brown. The vial was then removed, the tissue blocks removed by drawing up into a medicine dropper, and reimmersed in a large excess (100 ml) of distilled water. The blocks were thoroughly rinsed by reimmersion in several changes of distilled water, again placed in a clean 16 ml vial, and restained as above.

Following the second staining, the tissues were briefly rinsed free of GMS solution with distilled water and placed in a 5% sodium thiosulfate solution for 10 min. The hypo was rinsed free with distilled water, and the tissues were dehydrated by means of a graded ethanol series and critical point dried from CO_2, with Freon TF as an intermediate fluid. The dried preparations were affixed to aluminum stubs or carbon planchettes with Coates and Welter conductive specimen cement (also available as "Televison Tube Koat", GC Electronics, Rockford, IL, 61101, Cat. No. 49-2) and coated with 20 to 40 nm of carbon by evaporation.

These preparations (and others illustrated in the presentation) were viewed in Hitachi HFS-2 or JEOL 35 C scanning microscopes. The Hitachi HFS-2 was equipped with a field emission source and either an overhead (with respect to the specimen) annular silicon diode surface barrier detector[43] (ASSD) or an overhead scintillator light guide - photomultiplier detector (OSD) of original design. This latter detector is described in detail below.

The JEOL 35 C was equipped with a LaB_6 source and overhead annular silicon diode detector (Ortec Inc., Oak Ridge, TN).

The microscope, detector, specimen current, and accelerating voltage employed to obtain each image are specified in the appropriate figure caption below. The specimen stage tilt was 0° unless otherwise stated.

Figures 5-8 are illustrative of images which may be obtained by SEI and BEI of cells in tissue blocks stained as above. Figure 5a and b are, respectively, SE and BE images of the same area of the luminal surface of a high endothelial venule (HEV) in rat lymph node. Subsurface cell nuclei are visualized as highlights (conventional signal polarity: full white areas correspond to areas from which the maximum BE signal is collected; full black areas correspond to areas from which the minimum BE signal is collected). The highlights displayed represent incident beam electrons, which having entered the cells, encounter the heavy metal (silver; Z = 47) selectively deposited on chromatin by the staining procedure, and then scatter backwards through the cell surface, to strike the BE detector overhead and contribute to the image highlight at that point. Such highlights stand out in contrast to areas containing little or no silver deposits, which do not cause many beam electrons to scatter back on to the BE detector and which, therefore, appear as degrees of blackness.

Figures 6-8 show comparably stained areas of rat bone marrow. The nuclei of sinusoidal endothelial cells in bone marrow can be quite close to the cell surface, and hence also to the vessel surface, due to the extreme thinness of these cells. The resolution attainable in the BE image in this circumstance can be sufficient to view aspects of nuclear morphology such as folds and nuclear pores to good advantage (Fig. 6,7). A similar situation prevails for viewing the nuclei of small lymphocytes, which have a large nuclear to cytoplasmic ratio (Fig. 8). Figure 8 further demonstrates the utility of BE stereo imaging in the demonstration of nuclear morphology.

The selectivity of the GMS staining can, and should, be checked by TEM of the SEM preparations. This can be accomplished simply by delicately scraping the scanned sample (or a comparably prepared sample) off of the substrate into propylene oxide, and thereafter processing the tissue through plastic embedding, sectioning

and viewing, as one would for routine TEM.

TEM images of the preparation seen in Fig. 8 verify the impression obtained from Figs. 5-8 that the GMS procedure detailed above is an effective stain for chromatin (Figs. 9,10). The GMS is not, however, a specific stain for chromatin, as evidenced by the presence of some cytoplasmic staining, especially in cells of the erythrocyte series.

Cell nuclei - rabbit cornea fibroblasts - silver reduction staining (Fig. 11). Cell nuclei in cultured cells can be stained with GMS and otherwise prepared for SEI and BEI scanning by employing methods analogous to those given above for bulk tissue.

Serum Institute rat cornea fibroblasts infected with a virus isolated from human embryo lung (SIRC-HEL 1)[44] and growing in Falcon plastic dishes were washed twice with phosphate-buffered saline, pH 7.3 and fixed by flooding the dishes with 2% glutaraldehyde contained in sodium cacodylate buffer, pH 7.3 (300 mOsmol/kg) for 1 h at 4 C. Discs were then punched out of the bottom of the plastic dishes, immersed in 0.1 M sodium cacodylate buffer, pH 7.2, and, following several changes of buffer, stored in this vehicle until staining. Prior to staining, discs were rinsed free of buffer by immersion in several changes of distilled water, and placed together with a clean glass bead in a 16 ml Wheaton snap cap vial (as above) containing a freshly prepared GMS solution. The disc was positioned in the vial such that the cell-covered surface faced the bottom of the vial, while yet supported by the glass bead (Fig. 4). This simple expedient insured that particles of reduced silver did not "rain" on the cells during the staining process. The vial was then warmed in a 50 C bath until the GMS solution became just perceptibly gray. Further processing (including a restaining cycle) and viewing was performed as for bulk tissue above.

Figure 11 shows correlating SE (a) and BE (b,c) images of the same site in a SIRC-HEL 1 culture stained for BEI of cell nuclei. Several features of these images are worthy of note. The BE images (11 b,c) have been obtained in both conventional signal polarity (11b), as in previous figures (Figs. 5-8) and in a reversed polarity signal, which we shall designate BE (-). The BE (-) image is perhaps more readily interpreted by biologists, because it resembles familiar LM and TEM bright field images.[25]

The surface of the expansive cell which occupies most of the field of view is very smooth; there is no surface topography in the SE image which locates the nuclei or which reveals that the cell is multinucleated. Yet the location of these nuclei can be seen in the SE image. The viewing of these nuclei by SEI is attributable to the collection of SE's produced by emerging BE's. The outward-going BE's are, in a fashion, projecting images of the nuclei on the surface of the cell. The overall effect is that of seeing a structure dimly, as through a veil. This viewing is nevertheless useful. It allows one, up to a point, to realize some of the advantages of Z contrast BEI, without the

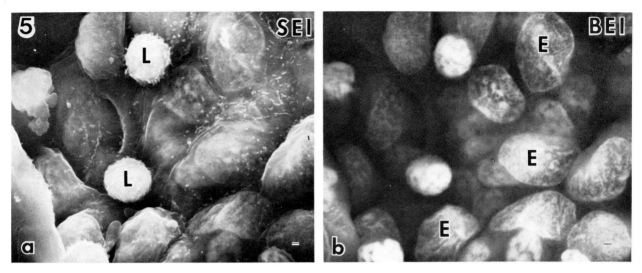

Figure 5. Correlating SE (a) and BE (b) images of the same surface area of a high endothelial venule in rat lymph node. Subsurface cell nuclei of HEV endothelial cells (E) and lymphocytes (L) adhering to the luminal surface are evident. GMS stain. Hitachi (HFS-2), Overhead Scintillator Detector (OSD), 2×10^{-11}A, 25 kV. Bar = 1µm.

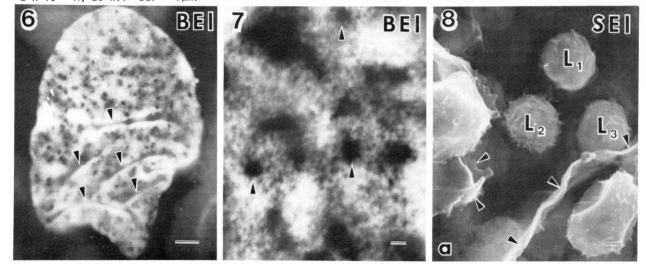

Figure 6. BEI of the luminal aspect of the flattened nucleus of a sinusoidal endothelial cell in rat bone marrow. Certain aspects of nuclear morphology are evident, including folds (arrowheads) and chromatin-free areas opposite nuclear pores, which give the profile a "measled" appearance. A portion of this image is seen at higher magnification in Fig. 7. GMS stain. HFS-2, Annular Solid State Detector (ASSD), 2×10^{-11}A, 25 kV. Bar = 1µm.
Figure 7. Higher magnification BEI of a portion of Fig. 6. The granular nature of the stain deposits is clearly seen. Several areas which presumably represent nuclear pore areas are seen to good advantage (arrowheads). GMS stain. HFS-2, ASSD, 2×10^{-11}A, 25 kV. Bar = 100 nm.
Figure 8. Correlating SE (a) and BE (b) images of the same site in rat bone marrow. Visualization of cell nuclei within the cells present in the field, as aided by the BE stereo image pair, allows the observer to identify three small lymphocytes (L_{1-3}) adhering to the luminal surface of the sinusoidal endothelium (vessel margin indicated by arrowheads). The lobed configuration of the nuclei of two of the lymphocytes (L_{1-3}) is clearly revealed in the BE stereo images. N, nucleus of sinusoidal endothelial cell. GMS stain. HFS-2, OSD, 2×10^{-11}A, 25 kV. Stereo image tilt: 0°/4°. Bar = 1µm.

need of a BE detector.

Cell nuclei - dissociated tumor cells - silver reduction staining. Black-hooded rats of Long-Evans descent, bearing subcutaneous tumors of the Shay transplantable leukemia, were anesthetized with ether and the tumor exposed. One to four grams of tumor tissue were excised from non-necrotic areas of the tumor and placed, together with 15 ml of 0.9% sterile, pyrogen-free NaCl in a glass tissue homogenizer. Tumor cells were dissociated with three plunges of a loosely fitting teflon pestle. The resulting suspension was filtered twice through a funnel, containing glass wool, for removal of vascular

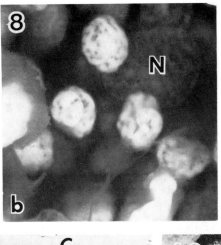

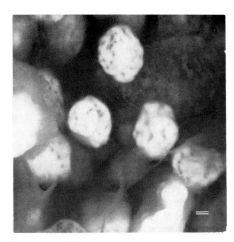

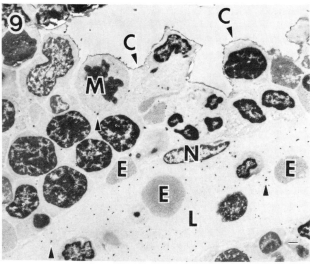

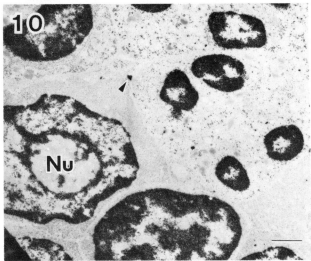

Figures 9 and 10. TEM images of the same SEM preparation stained with GMS for the BEI visualization of nuclei, as seen in Fig. 8. The chromatin of these bone marrow cell nuclei is heavily stained with silver. Chromatin-free portions of the nucleolus (Nu) are unstained, while light deposition is seen in the cytoplasm. Some silver precipitate is also present (e.g., arrow-heads). C, carbon coat; L, lumen of sinusoidal vessel. N, nucleus of sinusoidal endothelium; M, mitotic figure; E, erythrocyte. GMS stain. Bar = 1μm.

and fibrous tissue components, washed (two sequences of centrifugation at 750 RPM for 5 min and resuspension in fresh sterile, pyrogen-free 0.9% NaCl) and fixed, following a final centrifugation for 5 min at 750 RPM, by resuspension in 2.5% glutaraldehyde in 0.08 M HEPES buffer, pH 7.2.

Fixed cell suspensions were washed by twice decanting the fixative from over the sedimented cells and resuspending in Hanks' balanced salt solution (HBSS), pH 7.2. Just prior to staining the resedimented cells were resuspended by gentle agitation. A volume of the suspension containing 0.5 to 1.0 X 10^6 cells was withdrawn, and the cells collected by expressing the volume through a 0.22 μm pore size Millipore filter contained in a disposable plastic Swinnex holder (Millipore Corp.). Care was taken not to void the fluid over the cells. Just prior to staining, the HBSS was replaced by gently flushing the holder with 6 ml of distilled water. The distilled water in the Swinnex holders was then replaced

with a freshly prepared GMS solution, and the holder, now containing leukemic cells on the filter, was placed in a 16 ml snap cap Wheaton vial such that the GMS solution in the holder was confluent with the GMS solution in the vial. The vial was then warmed to 50 C in a water bath until a slight gray turbidity was attained (as above). The holders were then removed, the filters taken out and rinsed by immersion in distilled water, followed by immersion for 10 min in 5% sodium thiosulfate. The filters were then reimmersed in distilled water, dehydrated and processed as for bulk tissue and cells in culture (above).

Correlating SE and BE images of this preparation are shown in Figures 12-14. It should be noted at the outset that the BE images (Figs. 12b, 13b, 14b) have been obtained in reversed contrast with respect to conventional imaging. It can be seen that, with this method, cells of a heterogenous population, as viewed by SEI, can be sorted on the basis of nuclear

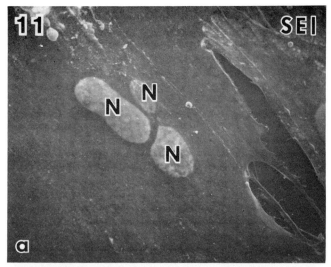

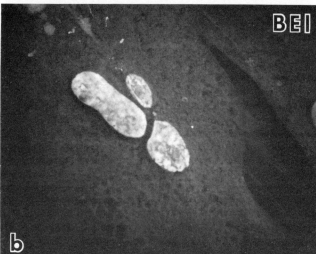

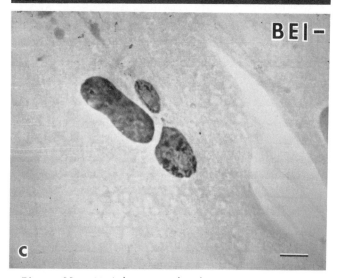

morphology, as seen in the BE images. Interphase cells can be separated from cells in mitosis; both of these types can be distinguished from pycnotic cells. Furthermore, if one knows the TEM nuclear morphology of a particular cell expected to be in the population, as was the case here for the myeloblast-like leukemic cells,[45] one can then identify these cells in the SEI on the basis of BEI nuclear morphology.

It should also be noted that in this application the SE images, except for Fig. 14a, were obtained at 15 kV in order to optimize the topographic imaging, while the BEI were obtained at 39 kV, in order to increase the number of beam electrons which reach the deeper nuclei in these cells; in many of these cells, especially the tumor cells, the nucleus is considerably further below the surface than for the squamous endothelial cells and cultured cells viewed above. Figure 14a shows a SEI obtained at 39kV. The image is composed of SE's produced by the beam electrons striking the cell surface, and also by beam electrons which have penetrated to strike the stained nuclei, and then have produced SE's upon being scattered back through the surface. The physical interactions here yield a faint image of the nuclei in the SEI in much the same manner as seen previously, (Fig. 11a). The companion stero BEI (Fig. 14b.) is again useful in demonstrating nuclear lobulation or other three-dimensional aspects of nuclei within such cells (see also Fig. 8.).

Sperm accessory fibers - silver reduction staining.[41,46] The testes and epididymis of adult Wistar rats were excised and dissociated in a pool of fixative consisting of 1.5% depolymerized paraformaldehyde together with 2.5% glutaraldehyde in 0.08M sodium cacodylate buffer, pH 7.2. After two hours testis tissue was washed free of fixative by immersion in the same buffer (several changes in 24 hours), rinsed (3 changes in 15 minutes) with distilled water to remove buffer salts, and then stained and further processed for BEI as previously described for bulk tissue. Some testis blocks were cryofractured in ethanol[47] to expose the lumens of seminiferous tubules.

Figures 15-18 show that the GMS reagent, as used here, can also be used to contrast the accessory or outer dense fibers of mammalian sperm. These nine fibers surround the axoneme of the sperm tail for the major part of its length. They are presumed to serve a function which is accessory to the doublet microtubules of the axoneme.[48]

In cryofractured preparations, BEI of cross-fractured tails yields profiles of the nine fibers which correlate well with the familiar TEM cross-sectional appearance.

While the basis for the staining affinity of GMS for chromatin is not yet known, it seems clear that the staining of the fibers is due to the reducing power of sulfur-containing proteins in the fibers.[49,50] Swift[51-54] has amassed

Figure 11. SE (a) and BE (b,c) images of the same area of a tissue culture dish overgrown with SIRC-HEL 1 cells. The BE image contrast in 11c was obtained in reversed polarity, BEI (-), yielding an image which resembles the LM and TEM bright field images. Visualization of the cell nuclei by GMS staining and BE imaging allows one to determine that the large squamous cell in the center is multinucleated. The nuclei are also weakly imaged in the SEI due to the enhanced production of SE's at surface areas through which BE's emerge. N, nucleus. GMS stain. HFS-2, ASSD, 2×10^{-11}A, 25 kV. Bar = 10µm.

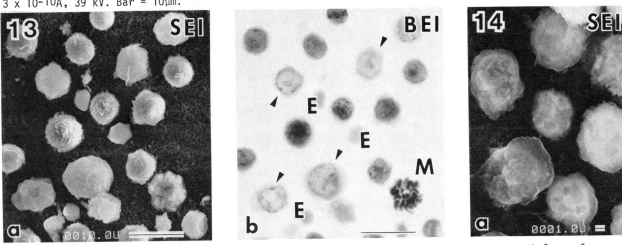

Figure 12.   SE and BE (-) images of the same area of a Millipore filter, upon which rest cells obtained from a dissociated tumor (subcutaneous rat myelogenous leukemia).   Nuclear contrast is sufficient in the BEI (b) to survey the variety of profiles at low magnification, prior to viewing at greater detail as in Fig. 13 a, b.   M, mitotic figure; P, pycnotic nuclei.   GMS stain.   JEOL 35C (35C), ASSD.   a: 3 x 10-10A, 39 kV. Bar = 10μm.

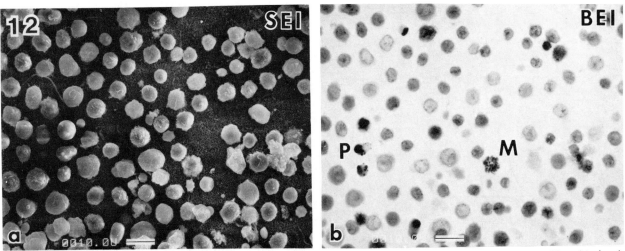

Figure 13.   SEI and BEI (-) of a portion of the field seen in Fig. 12.   The nuclear morphology of certain cells (arrowheads) in the BEI (b) correlates well with the known TEM and LM morphology of the leukemic myeloblast-like cells in the tumor.[45] M, mitotic figure; E, erythrocyte.   GMS stain.   35C, ASSD.   a: 3 x 10-11A, 25 kV.   b: 3 x 10-10A, 39 kV.   Bar - 10μm.

Figure 14.   SEI and BEI (-) of the same preparation as seen in Figs. 12 and 13. At 39 keV the topographic signal in the SEI (a) is somewhat "washed out" by the SE signal due to emerging BE's.   The three-dimensional nuclear morphology of isolated cells can be appreciated by obtaining a BEI stereo pair (b).   GMS stain.   35C, ASSD, 3×10-10A, 39 keV. Stereo image tilt: 3°/10°.   Bar = 1μm.

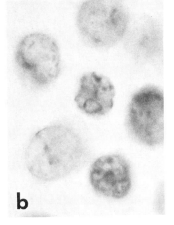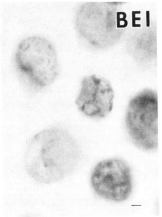

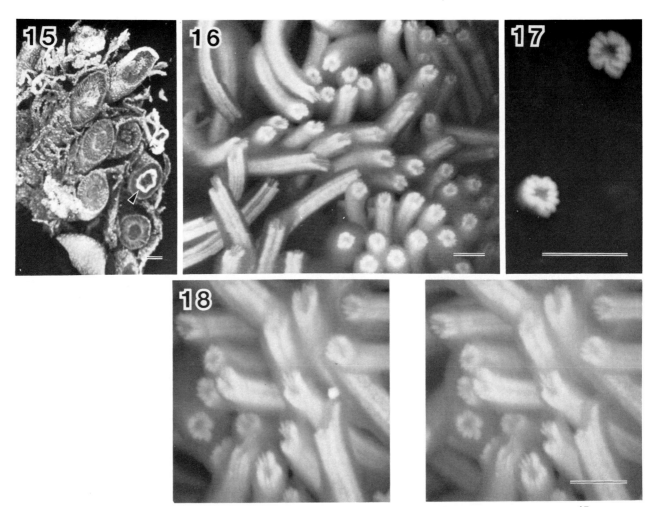

Figure 15. SEI of a block of rat testis seminiferous tubules and interstitium, cryofractured[47] and stained with GMS as described. A portion of the lumen of one tubule (arrowhead) is seen at higher magnification in Fig. 16. GMS stain. HFS-2, 2×10^{-11}A, 25 kV. Bar = 100µm.
Figure 16 and 17. BEI of sperm tails, cryofractured[47] and stained with GMS. The outer dense fibers of the tails are viewed within the tails and at the fracture surface, due to the increased collection of BE's in the silver stained fibers. The profiles of the nine fibers, as seen in cross-fractured tails by BEI, is readily correlated with the familiar TEM cross sectional appearance. GMS stain. HFS-2, ASSD, 2×10^{-11}A, 25 kV. Bars = 1µm.
Figure 18. BEI stereo pair of cryofractured, GMS stained sperm tails. The nine accessory fibers of the tails are seen to good advantage. GMS stain. HFS-2, ASSD, 2×10^{-11}A, 25 kV. Stereo image tilt: 5°/10°. Bar = 1µm.

evidence which suggests that sulfhydryl groups and disulfide bonds, particularly the latter, are demonstrated by GMS. Reduction with benzyl-thiol followed by alkylation with iodoacetate, moreover, blocks the GMS staining of these structures.[41] Burr[55], on the other hand, has reported that GMS reacts with sulfhydryl and 1,2 glycol groups.

Based on the probable demonstration of disulfide bonds by GMS, as used here, it is to be expected that other structures rich in S-S bonds might also be readily contrasted by GMS for BEI. Such structures would include elastic fibers[56] and the insulin producing β cells of the endocrine pancreas.

Peroxidase-containing leukocyte granules - cytochemical localization with DAB-osmium.[28]
Among the most widely used of LM and TEM cytochemical methods are methods demonstrating oxidases and peroxidases. For TEM localization of these enzymes, the Graham-Karnovsky[57] method employing diaminobenzidine (DAB) and osmium (Z = 76) as the heavy metal is, perhaps, the most widely used. The two applications which follow demonstrate the adaptation of this method for the purpose of supplying contrast for BEI.

Rat femoral bone marrow was fixed by perfusion. The fixative consisted of 1.5% depolymerized paraformaldehyde together with 2.5% glutaraldehyde in 0.08 M HEPES (N-2-Hydroxy-

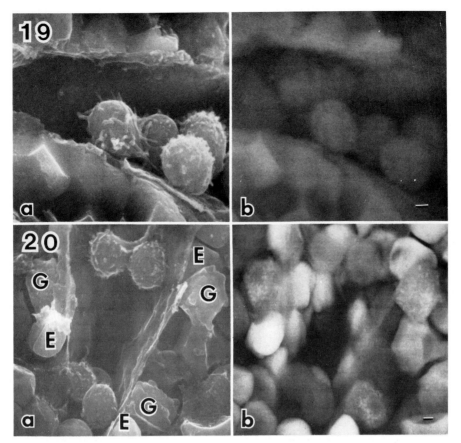

Figure 19. SEI (a) and BEI (b) images of the same site in rat bone marrow. The control was incubated without substrate for the demonstration of endogenous peroxidase (End PO). The BE image contrast is due to both Z contrast, imparted by a uniform nonspecific staining with osmium (Z=76), and specimen topography. No selective staining for End PO is seen (compare with Fig. 20b). HFS-2, ASSD, 2×10^{-11}A, 25 kV. Bar = 1μm.

Figure 20. SEI (a) and BEI (b) of the same site in rat bone marrow stained to localize End PO activity as described in the text. The highlights in the BEI are due to the increased collection of BE from structures with an enhanced osmium content due to the cytochemical staining procedure for endogenous peroxidase. Even at this low magnification, certain cells (G) can be identified as granulated myelocytes due to End PO (+) granules, while others (E) may be identified as belonging to the erythrocyte series, owing to the homogenous staining resulting from the peroxidase activity of hemoglobin. DAB-osmium cytochemical stain. HFS-2, ASSD, 2×10^{-11}A, 25 kV. Bar = 1μm.

ethylpiperazine-N'-2-ethanesulfonic Acid) buffer, pH 7.4, or in 0.08 M sodium cacodylate buffer, pH 7.4, plus 0.15% $CaCl_2$. The femurs were excised and the marrows removed. The marrows were then fixed further by immersion in the same fixative (2 hours at 4C) and washed (12-24 hours at 4C) in the same 0.08 M buffer, pH 7.4, as was used for fixation, plus 7% sucrose. The marrow tissue was divided into blocks with razor blades or by fracturing after freezing in isopentane cooled with liquid nitrogen. Cryofractured tissue was thawed in the same sucrose buffer. Marrow tissue blocks were incubated for 30-60 minutes at 37C in a modified Graham-Karnovsky medium[57] for the demonstration of peroxidase activity. The medium was freshly prepared by addition (with stirring) of 10 mg of 3,3'diaminobenzidine and 10 drops of 0.3% H_2O_2 to 10 ml of 0.08 M HEPES buffer, pH 7.4. Control tissue preparations were incubated in media lacking DAB or H_2O_2. Following

incubation, marrow tissue was washed thoroughly with distilled water and immersed in 0.5% aqueous osmium tetroxide for two minutes. The osmicated tissue was rapidly rinsed free of excess osmium tetroxide with several brief changes of distilled water, dehydrated to absolute ethanol, transferred through a graded ethanol/Freon PCA series to Freon PCA and critical point dried from CO_2. Dry marrow blocks were mounted on aluminum discs with conductive specimen cement (as above) and coated with carbon by evaporation to a thickness of 20-40 nm.

Figures 19-23 demonstrate the BEI visualization of endogenous peroxidase in cells of bone marrow preparations together with images of control preparations. In LM or TEM histochemical localizations it is necessary to perform the proper controls. It should be no less requisite for SEM histochemistry. Figures 19a and b are SE and BE images of the same site in a control

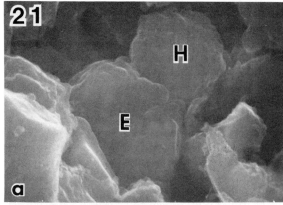

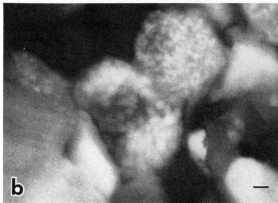

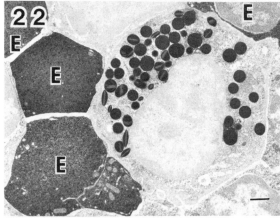

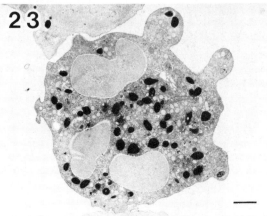

preparation of marrow. The BEI image contrast
is due to the nonspecific binding of the heavy
metal osmium (Z = 76) to cell and tissue
components; no selective cytochemical localization is present. This appearance is to be
compared with Figures 20a and b, which show, in
the BEI of the SEI/BEI pair, the selective
enhancement in contrast of certain structures
due to the selective concentration of osmium at
sites of endogenous peroxidase activity. As is
readily apparent, these structures include most
prominently leukocyte granules of various sizes
(Figs. 20b, 21b). Additionally, however, cells
of the erythrocyte series can be distinguished
and identified in the SEI, owing to the
peroxidatic activity of hemoglobin.[58] That
this is so can be verified by TEM of the preparations scanned, i.e., by embedding in plastic
and further processing in the conventional
manner. Such control TEM images are seen in
Figures 22 and 23.

The above procedure offers a direct means
through BEI of identifying certain leukocytes in
SEM preparations where identification on the
basis of surface features may be ambiguous.

Mitochondria - sperm - cytochemical
localization with DAB - osmium.[46] The
following procedure for BEI of mitochondria
depends on the localization of a specific
membrane-bound mitochondrial enzyme, cytochrome
oxidase. The enzyme is localized by a
DAB-osmium incorporating method,[59-61] not
unlike that used for the BEI visualization of
sites of endogenous peroxidase activity (above).

The testes and epididymis of adult Wistar
rats were exised, dissociated and immersed for
20 minutes in a 4 C pool of 4% formaldehyde
(prepared from paraformaldehyde) in 0.05 M
sodium phosphate buffer, pH 7.4, to which 5%
sucrose had been added. Following fixation, a
crude suspension of sperm was prepared by gently
pelleting the larger tissue and cell aggregates.
Serial dilutions of the supernatant were then
prepared by addition of 0.05 M sodium phosphate
buffer, pH 7.4, to which 5% sucrose had been
added. A 5 ml volume of each dilution was then
passed through a 0.22 m pore size Millipore
filter contained in a 13 mm Swinnex holder, as
for the collection of tumor cells (above). The
filters, plus collected sperm, were removed from

Figure 21. SEI (a) and BEI (b) of the same site
in bone marrow stained with osmium for End PO.
Two cells, which could not be identified or distinguished from one another on the basis of their
surfaces in the SEI (Fig. 21a) can be tentatively
identified as eosinophilic (E) and heterophilic
(H) myelocytes on the basis of the characteristics
of their peroxidase (+) granules (compare with
Figs. 22, 23). DAB-osmium cytochemical stain.
HFS-2, ASSD, 2 x 10^{-11}A, 25 kV. Bar = 1μm.
Figures 22 and 23. TEM images obtained from rat
bone marrow which was prepared to demonstrate End
PO and viewed by SEI/BEI. End PO activity is
present in granules of an eosinophilic myelocyte
(Fig. 22) and a heterophil (Fig. 23). The cytoplasm of four cells of the erythrocyte series (E)
also shows End PO (+) staining. DAB-osmium cytochemical stain. Bar = 1μm.

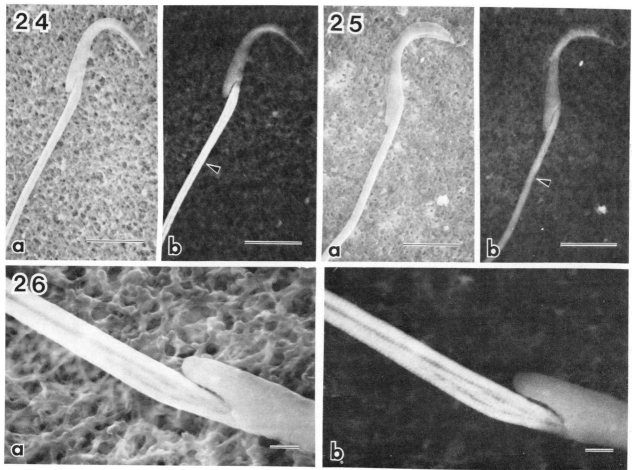

Figure 24. SEI (a) and BEI (b) of the same sperm on a Millipore filter. The sperm has been stained to demonstrate mitochondria through the cytochemical localization of cytochrome oxidase.[59-61] The middle piece (arrowhead) of the sperm is bright in the BE image, as compared to the head, due to the collection of more BE from the osmium concentrated by the staining method in this structure, than in the head. Compare with the control image incubated with the addition of the cytochrome oxidase inhibitor, cyanide (Fig. 25). DAB-osmium cytochemical stain. HFS-2, OSD, $2 \times 10^{-11}A$, 25 kV. Bar = 10µm.
Figure 25. SEI (a) and BEI (b) of the same sperm in a control preparation incubated for the cytochemical localization of cytochrome oxidase activity as in Fig. 24, except for the addition of the specific inhibitor, cyanide. The level of BE yield from the middle piece (arrowhead) is not greater than that from the head, both of which show only a low background level of general osmium staining. DAB-osmium cytochemical localization. HFS-2, OSD, $2 \times 10^{-11}A$, 25 kV. Bar = 10µm.
Figure 26. SEI (a) and BEI (b) images of the same sperm prepared for BEI visualization of mitochondria/cytochrome oxidase. The helical coil of elongated mitochondria in the middle piece is seen in the BE image (26b). DAB-osmium cytochemical stain. HFS-2, OSD, $2 \times 10^{-11}A$, 25 kV. Bar = 1µm.

the holder, washed overnight in the same buffer plus sucrose and incubated for one hour at 37 C in a medium modified from Seligman et al.[59] for the demonstration of cytochrome oxidase. The medium consisted of 5 mg of 3,3´ - diaminobenzidine (Sigma Chemical Co., St. Louis) and 750 mg of sucrose in 9 ml of 0.05 M sodium phosphate buffer, pH 7.4. Some control incubations were performed with the cytochrome oxidase inhibitor, potassium cyanide (6.5 mg, 0.01 M), added to the incubation medium. Following incubation, sperm were washed with several changes of the same sucrose buffer, followed by a rinse with distilled water, and then immersed in 0.5% OsO_4 for 10 minutes at 20 C. Excess OsO_4 was quickly washed from the

filters with distilled water. The filters were dehydrated in graded ethanols and critical point dried from CO_2 with Freon TF as an intermediate fluid. Dry filters were mounted and coated with carbon for viewing as detailed for the GMS stained tissue above.

Some pelleted testis tissue was also incubated for the demonstration of cytochrome oxidase and subsequently osmicated and prepared for TEM.

BE images of mitochondria in the middle piece sheath of sperm, plus correlating SE and TEM images and SEI/BEI views of the cyanide control incubation are shown in Figures 24-29. The selective concentration of osmium in the middle piece of sperm incubated for cytochrome

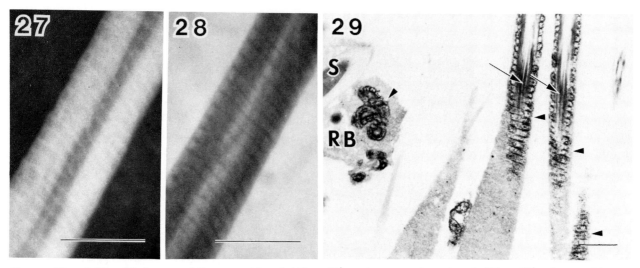

Figure 27 and 28. BE images with conventional (Fig. 27) and reversed contrast (Fig. 28) of a portion of the middle piece sheath of sperm incubated to localize cytochrome oxidase. The staining localized between the inner and outer mitochondrial membranes (compare with the TEM image of comparably stained sheath mitochondria in Fig. 29) yields a good visualization of the lateral margins of the elongated mitochondria and the pitch of the mitochondrial gyres. DAB-osmium cytochemical stain. HFS-2, OSD, 2×10^{-11}A, 25 kV. Bar = 1µm.

Figure 29. TEM image of sperm prepared to demonstrate cytochrome oxidase in mitochondria, as in Figs. 24, 26, 27 and 28. The staining procedure has concentrated osmium selectively in the mitochondrial intracristal space and between the inner and outer mitochondrial membranes, effectively outlining the mitochondria of sperm tails (arrowheads) and of a residual body (RB). S, sperm head; arrow, outer dense fibers. DAB-osmium cytochemical stain. Bar = 1.5µm.

oxidase causes the sheath to "light up" in the BE image (Fig. 25b), as compared to the lesser nonspecific level of osmium density seen in the head of the same sperm. In control preparations incubated in the presence of the specific inhibitor of the cytochemical reaction, BEI shows the middle piece to lack selective osmium staining (Fig. 26b). Only the BE yield due to nonspecific cytoplasmic staining, at a level equivalent in intensity to that of the head portion, is viewed. At an increased magnification (Figs. 27-28), one is able to view the helical coil of mitochondria to good advantage.

In order to compare the localization observed by BEI with that found by TEM of scanned, embedded and sectioned preparations (Fig. 29), it is useful to obtain a BE image with reversed contrast (Fig. 28b). Both BE and TEM images show the proper distribution of reaction product in the mitochrondrial intracristal space and between the mitochondrial inner and outer membranes.56-61

It is expected that this methodology for the viewing of mitochondria in SEM preparations of sperm can also be applied to other cell types, where it may be of use, for example, in studies quantifying the mitochondrial population.

Phagosomes - bone marrow sinusoidal endothelial cells - Thorotrast uptake. Means for contrasting phagosomes for BEI can be divided into two categories: 1) supplying heavy metal particulates to cells in vivo for ingestion, and 2) supplying a tracer substance, e.g., horseradish peroxidase, the location of which can be marked by heavy metal staining subsequent to fixation. The following method is illustrative of the former category.

One ml of Thorotrast (Testagar and Co., Detroit, MI) containing 250 mg of stabilized colloidal thorium or stet dioxide was injected into the saphenous vein of anesthetized adult Wistar rats. The incision prepared for the injection was closed and the rat was returned to the animal care facility where it was housed for a specified period (in the case of images shown here, 3 mo).

The femoral bone marrow of the injected rat was fixed by perfusion with 1.5% depolymerized paraformaldehyde plus 2.5% glutaraldehyde together in 0.08 M sodium cacodylate buffer, pH 7.4, with 0.15% $CaCl_2$ added. The marrows were removed, washed free of fixative with the same buffer, dehydrated to absolute ethanol and cryofractured according to the method of Humphreys.47 Thereafter, fractured marrow blocks were critical point dried from CO_2, with Freon TF as an intermediate fluid, mounted on specimen supports and viewed as described above.

Figures 30-32 show correlating SEI, BEI and TEM images of bone marrow prepared as above. In the BE image (Fig. 30b), phagosomes related to the sinusoidal endothelial cells are viewed in good contrast due to the high yield of BE from concentrated thorium dioxide particles. The intracellular location of the tracer in sinusoidal endothelial and adventitial cells can be demonstrated by TEM of scanned preparations (Figs. 31,32). One can in this manner identify phagocytic cells (here a cell of the reticulo-endothelial system) and, perhaps, in appropriate protocols, accurately quantify the phagocytic capability of such cells. This BEI approach to investigating macrophage function is, perhaps,

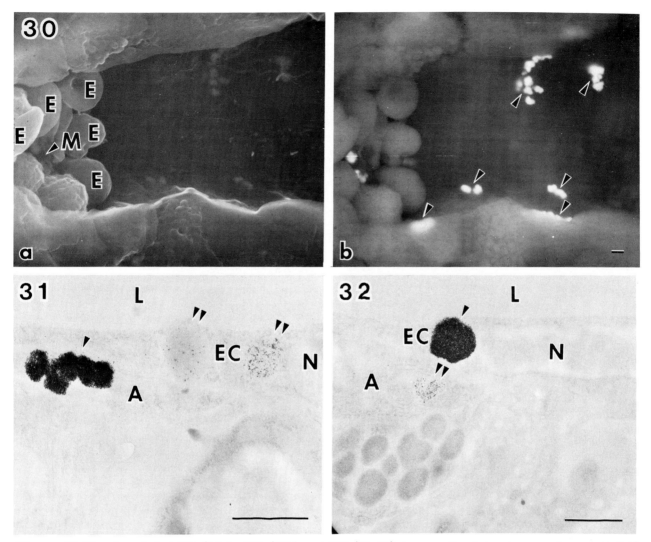

Figures 30-32. Correlated SEI (30a), BEI (30b) and TEM (31,32) images of rat bone marrow from an animal injected intravenously with colloidal thorium dioxide (Thorotrast) 3 months prior to sacrifice. The heavy metal particulate (Z=90) has been sequestered in phagosomes (arrowheads) of endothelial cells (EC) and adventitial cells (A) of the sinusoidal blood vessels. In some phagosomes (double arrowheads) the concentration of thorium dioxide is too low to render them visible by BEI. E, erythrocyte; L, lumen of sinusoidal vessel; M, circulating phagocytic cell; N, nucleus. Thorotrast injection. Unstained. Fig. 30: HFS-2, OSD, 2×10^{-11}A, 25 kV. Bars = 1µm.

the most widely used of BEI methods to date (see, for example,[13,14,30,33,62,63]).

Phagosomes - bone marrow sinusoidal endothelial cells - DAB-osmium localization of ingested horseradish peroxidase (HRP).[28] Adult Wistar rats received 1 ml of Hanks balanced salt solution containing 10 mg of HRP (Sigma VI) by injection into the saphenous vein. The enzyme was allowed to circulate three minutes prior to initiation of bone marrow fixation by perfusion. Perfusion fixation and subsequent tissue preparation for the cytochemical localization of exogenous peroxidase activity were performed as for localization of endogenous peroxidase in bone marrow (above). Control tissue preparations were incubated in media lacking DAB or H_2O_2. Following incubation for the BEI demonstration of HRP, the material

was prepared for SEM and TEM as described above.

Correlating SEI and BEI of the same luminal surface area of the sinusoidal endothelium in bone marrow are shown in Figures. 33a and b. Subsurface phagosomes containing osmium, deposited at sites of sequestered HRP, are evident as luminescent spheres and shells in the BEI (Fig. 33b). A consideration of the TEM image (Fig. 33c) yields a verification of this localization in components of the endocytic vacuolar system of these cells.[64]

Lysosomes/cell nuclei -tissue culture macrophages - cytochemical lead stain for acid phosphatase/silver reduction stain.[65] Cultured macrophages were fixed and incubated for the localization of acid phosphatase by deposition of lead (Z = 82) at active sites.[66] Following incubation, the medium was washed free with

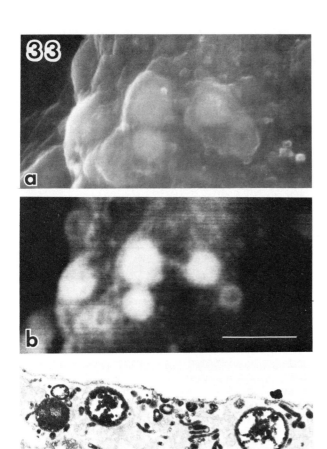

Figure 33. Correlating SE (a), BE (b) and TEM (c) images of the sinusoidal endothelium in bone marrow of a rat which had been injected intravenously with horseradish peroxidase (HRP). Osmium (Z=76) has been deposited cytochemically at sites of uptake of HRP into phagosomes and other components of the vacuolar apparatus of these cells.[64] DAB-osmium cytochemical stain. Fig. 33 a, b: HFS-2, ASSD, 2×10^{-11}A, 25 kV. Bars = 1µm.

distilled water and stained with GMS, as described above. The doubly stained cells were then prepared for SEI/BEI SEM as above, except for coating of the cells with about 5 nm of gold instead of carbon. This staining innovation, which was supplied by H. Carter and M. Tannenbaum,[65] is notable in that it demonstrates the feasibility of combining cytochemical staining methods, where Pb is the heavy metal, with GMS reduction. The result in this case is BEI of the lysosomal distribution in cells within which nuclear morphology can also be appreciated (Figs. 34,35). This application is also interesting in that it demonstrates that a thin layer of heavy metal (gold; Z = 79) does not preclude the BEI visualization of the structures stained. It can be expected, however, that the gold coat does impair BEI to some degree.[13]

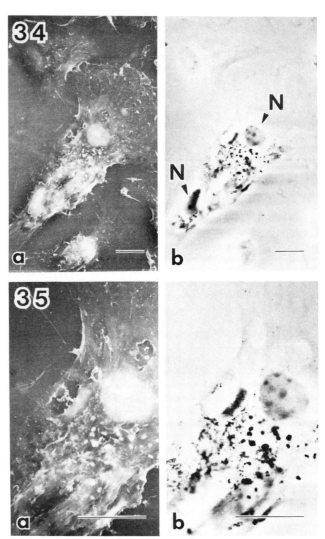

Figure 34. SE (a) and BE (b) images (contrast reversal in BE image) of the same area of a macrophage culture stained cytochemically with lead (Z=82) for sites of acid phosphatase activity (e.g., arrowheads), followed by GMS staining with silver (Z=47) to reveal nuclear (N) morphology.[65] Lead cytochemical stain - silver reduction stain. 35C, ASSD, 3×10^{-10}A, 25 kV. Bar = 10µm

Figure 35. Correlating SE (a) and BE (b) images of a portion of the macrophage culture in Fig. 34. Sites of acid phosphatase activity, as contrasted by the BE yield from lead (Z=82) can be related to surface micro-contour in the SEI, and to nuclear position and morphology.[65] Lead cytochemical stain - silver reduction stain. 35C, ASSD, 3×10^{-10}A, 25 kV. Bar = 10µm

IV. Predicting the Feasibility of BEI in Biological Investigations

In this section we will describe the physical model of backscattering with which we will address the questions posed in Section II. We will then summarize the results in a form which can be used by biologists to determine the feasibility of this technique for a specific application.

A physical model for electron backscattering.

Our model is shown in Figure 36. We describe a stained specimen as consisting of a layer of stained material of thickness x_2 sandwiched between an upper layer of carbon of thickness x_1 and a lower layer of carbon thicker than the range of the electrons. As shown there are three basic types of BE events. Events of type 1 and 3 arise from backscattering in the two carbon layers. Only event type 2 arising from BE's produced in the stained layer will provide contrast. In moving through the specimen both before and after scattering, the electrons lose energy. In addition, some of them are absorbed. The transmission factors describing this absorption are strong functions of the electron's energy. Since BE's possess a continuous energy spectrum (see, e.g., Figure 2) the absorption taking place after backscattering must be integrated over the BE energy spectrum.

Finally as we will see in section V some BE detectors do not respond efficiently, or at all, to BE's below a certain threshold energy. Also there may be circumstances where it is desirable to reject low energy BE's. Therefore, in calculating the contrast, BE's with energy E_3 or E_3' below some value may not be included.

Putting all of this together leads to the following expression for the total BE coefficient $\eta \equiv \eta(x_1, x_2, E_0)$:

$$\eta = \eta_C(x_1, E_0) + T_C(x_1, E_0) \int_0^{E_1} \frac{d\eta_S}{dE_1'}(x_2, E_1, E_1') \cdot$$

$$T_C(x_1, E_1') dE_1'$$

$$(1)$$

$$+ T_C(x_1, E_0) T_S(x_2, E_1) \cdot \int_0^{E_2} \frac{d\eta_{oC}}{dE_2'}(E_2, E_2') \cdot$$

$$T_S(x_2, E_2') T_C(x_1, E_1') dE_2'$$

where $\eta_i(x, E)$ is the BE coefficient for material i of thickness x for incident electrons of energy E; $T_i(x, E)$ is the transmission coefficient for a film i of thickness x and electron energy E ($T = 1.0$ if no electrons are lost in traversing the film); $d\eta_S(x_2, E_1, E_1')/dE_1'$ is the BE energy spectrum from the stained layer (it is normalized such that $\int_0^{E_1} \frac{d\eta_S}{dE_1'} dE_1' = \eta_S(x_2, E_1)$); and $d\eta_{oC}(E_2, E_2')/dE_2'$ is the BE energy spectrum from the bulk carbon layer ($\int_0^{E_2} \frac{d\eta_{oC}}{dE_2'} dE_2' = \eta_{oC}(E_2)$). Because the latter is thicker than the BE range, η_{oC} is independent of thickness.

To proceed further we need to know more about the specimen's properties. As indicated we will regard dehydrated biological material as being composed solely of carbon. This is a reasonable simplification based on available chemical analysis of tissue. Marshall[67] has determined the density of air dried tissue to be

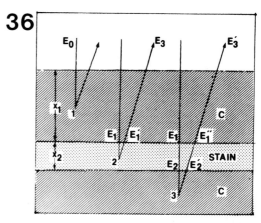

Figure 36. Schematic illustrating the physical model of contrast formation. Electrons with incident energy E_0 enter the material. The electrons lose energy as they penetrate the material: at depth X_1 of the stain they are reduced to energy E_1. If they penetrate through the stain, they are reduced to energy E_2. If they backscatter within a layer they emerge from that layer with a variable energy E_1 or E_2. They continue to lose energy as they pass through layers of "overburden", being reduced to E_1'' in passing through the stain and E_3 or E_3' in emerging from the surface.

about 0.5 g/cm^3. We have obtained similar results in our laboratory with critical point dried material. The density is a ratio of mass to volume, both of which shrink considerably during the dehydration process. As reported by Boyde and others[68] shrinkage may be considerably reduced if freeze drying is used instead of critical point drying. The dry density in this case might be significantly lower. However, we will assume a density $\rho = 0.5$ g/cm^3 in this work. The physical properties we will be discussing - energy loss, transmission, backscattering and beam broadening due to multiple scattering - all depend on the mass thickness of the material, i.e., how thick it is in terms of number of atoms. Two films of the same material, one thin with a high density, the other thick with a lower density will yield identical results for the above properties so long as the product of their density and thickness is the same. By specifying a density we can describe our system unambiguously in terms of lengths (μm) instead of mass thicknesses (μg/cm^2), which are more commonly used by physicists.

We now need to know the physical properties of the stained region. We will work with the silver stain which has been used for much of the work described in section III. This stain is "fixed" by a reduction process somewhat similar to that in photographic development which leads to the accretion of silver into many microcrystallites. Fig. 37a shows a TEM micrograph of a thin ($\sim$120 nm) section taken from a silver stained specimen originally prepared for SEM. Three regions can be distinguished: a stain excluding region (mainly intercellular boundaries and parts of the nucleus), lightly stained areas

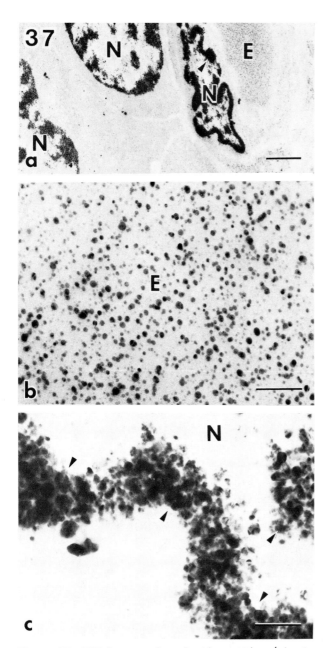

Figure 37. TEM images of a plastic section (about 120 nm in thickness) obtained from the SEM preparation stained with silver as described above (Section III) and illustrated in Fig. 8. Chromatin (arrowheads) in cell nuclei (N) is heavily stained (a,c), while an erythrocyte (E) is lightly stained (a,b). Scale bar in (a,b) = 1μm. Scale bar in (c) = 100nm.

(the cytoplasm) and heavily stained areas in the nucleus (the chromatin). Fig. 37b shows a lightly stained region at higher magnification. The crystallites are apparent. X-ray microprobe analysis and selected area diffraction confirm that they are composed of pure silver. By treating them as spherical and determining their size distribution and concentration, the amount of silver per unit volume can be obtained. This

was determined to be about 4.6×10^{20} Ag atoms/cm^3. Fig. 37c shows a chromatin stained region at high magnification. The concentration of Ag particles is now so high that counting and sizing become difficult. We estimate the Ag concentration in this region to be at least 8.4×10^{21} Ag atoms/cm^3. It could be 2-3 times larger. Using this value for the Ag concentration we calculate the density of the stained layer in our model at about $\rho_s = 2.0$ g/cm^3.

The energy lost in traversing a layer of the specimen is calculated from the Thomson-Whiddington law[69]. Straggling effects are ignored. This leads to the following expressions for the electron energies defined in Figure 36:

$$E_1 = (E_0^2 - 7.6\sqrt{E_0}\ x_1)^{\frac{1}{2}}$$

$$E_2 = (E_1^2 - 22.5\sqrt{E_1}\ x_2)^{\frac{1}{2}}$$

$$E_1'' = (E_2^2 - 22.5\sqrt{E_2}\ x_2)^{\frac{1}{2}} \qquad (2)$$

$$E_3 = (E_1''^2 - 7.6\sqrt{E_1''}\ x_1)^{\frac{1}{2}}$$

$$E_3' = (E_1''^2 - 7.6\sqrt{E_1''}\ x_1)^{\frac{1}{2}}$$

where the units are keV and μm.

Expressions for the backscattering coefficient from a thin layer of C or a mixture of C - Ag are needed as a function of electron energy and film thickness. These were obtained by fitting functions of the form

$$n/n_0 = \tanh(a(E)x + b(E)x^2) \qquad (3)$$

to the data of Niedrig and Sieber[70] who have measured η as a function of energy E and thickness x for Al, Cu, Ag, Au and Bi in the energy range 20 - 53 keV. n_0 is the BE coefficient for bulk specimens. It was observed during the fitting procedure that the relation

$$\frac{\rho Z^2}{n_0 A}\ x_{0.5} = constant \qquad (4)$$

was approximately true independent of atomic number Z. The quantities ρ and A are the mass density and atomic weight for the material of atomic number Z. $x_{0.5}$ is that thickness for which n/n_0 achieves half of its full bulk value. This relationship was used to obtain an expression for C (which was not measured) from that for Al.

The expressions for the BE coefficients are as follows:

$$\eta_C(x,E) = n_{0C}(E)\tanh(\frac{5.55x}{E^{1.36}} + \frac{66000x^2}{E^{4.86}})$$

$$(5)$$

$$\eta_{Ag}(x,E) = n_{0Ag}(E)\tanh(\frac{1812x}{E^{2.11}} + \frac{2230x^2}{E^{2.12}})$$

where the units are μm and keV.

Energy dependences for the bulk BE coefficients for C and Ag have been included and were obtained from fits to data[71,72,73]:

$$\eta_{0C}(E) = \begin{cases} 0.070(1-0.0059E) & E \gtrsim 50 \text{ keV} \\ \dfrac{0.049}{1+0.928(E/511)^{0.823}} & 50 \text{ keV} < E \leq 1 \text{ MeV} \end{cases} \tag{6}$$

$$\eta_{0Ag}(E) = \begin{cases} 0.402(1+0.00013E) & E \gtrsim 50 \text{ keV} \\ \dfrac{0.404}{1+0.053(E/511)^{1.47}} & 50 \text{ keV} \leq E \leq 1 \text{MeV} \end{cases}$$

The following expression was used to represent the BE coefficient for the C - Ag stain mixture:

$$\eta_S(x,E) = f_C \eta_C(x/f_C,E) + f_{Ag}\eta_{Ag}(x/f_{Ag},E) \tag{7}$$

where f_i is the mass fraction of element i in the mixture:

$$f_C = \rho_C/\rho_S \qquad f_{Ag} = \rho_{Ag}/\rho_S \tag{8}$$

This expression can be shown to reduce to the known forms of the BE coefficient for a mixture in the two limits of very thin films[74], where single scattering dominates, and bulk specimens.[75]

The transmission coefficients were obtained from the same Monte Carlo program which generated the plots in Fig. 1. Fits to the program's output led to analytic expressions for the coefficients:

$$T_C(x,E) = e^{-[(x/\lambda_C)^2 + 1.29(x/\lambda_C)^4]/3.3}$$

$$\lambda_C = 0.0222 \, E^{1.67} \tag{9}$$

$$T_S(x,E) = e^{-(x/\lambda_S)^2}$$

$$\lambda_S = 0.0776 \, E^{1.85}$$

where again the units are μm and keV.

Few measurements of BE energy spectra from thin films exist, so a theoretical expression derived from the Everhart theory was used.[76] Define the dimensionless quantities

$$y = x/R \quad , \quad \varepsilon = E/E_B \tag{10}$$

where E_B is the energy before backscattering and E after reemerging from the layer within which the backscattering took place. R is the electron range which we will take here as[77]

$$R = 0.025 \, E^2/\rho \quad (\mu m) \tag{11}$$

where E is in keV. For thin films y < 0.5; for a bulk specimen y = 0.5. We also define

$$y_0 = (1-\varepsilon^2)/2 \tag{12}$$

We then write the energy spectrum as

$$\frac{d\eta}{dE}(x,E_B,E) = \begin{cases} \varepsilon/y_0 \, [\eta(\varepsilon,y)-1+(1-y_0)^a]; y_0 \leq y \\ \varepsilon/y_0 \, [\eta(\varepsilon,y)-1+(1-y)^a]; y_0 > y \end{cases} \tag{13}$$

where a = 0.045Z

and $\eta(\varepsilon,y)$ is the spectrum integrated between $E = \varepsilon E_B$ and E_B:

$$\eta(\varepsilon,y) = \begin{cases} \dfrac{(a+1)y_0 - 1 + (1-y_0)^{a+1}}{(a+1)y_0} & ; y_0 \leq y \\[2mm] \dfrac{1}{(a+1)y_0}\Big\{(a+1)y_0 - 1 + (1-y)^a[(a+1)(1-y_0) \\ \qquad -a(1-y)]\Big\} & ; y_0 > y \end{cases} \tag{14}$$

The expression for dη/dE is normalized to the Everhart value for the BE coefficient for a thin film:

$$\eta(0,y) = \frac{(a-1)+(1-a+2ay)(1-y)^a}{a+1} \tag{15}$$

We really want to include the presumably more accurate expressions, eqs. 5-7, for η from the fits, so we multiply by $\eta_{fit}/\eta(o,y)$. This gives us the fitted BE coefficient times an energy spectrum weighting factor normalized to 1.0.

From the form of Everhart's theory it can be shown that corresponding expressions for the C - Ag stain are obtained simply by replacing a by $f_C a_C + f_{Ag} a_{Ag}$ where again the f's are mass concentrations.

We are now in a position to calculate the contrast for a given thickness and depth of a stained region in a specimen. We define the contrast as the increase in signal from a stained region over that from an adjacent unstained region, divided by the latter signal. Let the beam current be I_0. Then the signals from the two areas of the specimen are given by

$$I_S = \eta(x_1,x_2,E_0)kI_0$$
$$I_C = \eta_{0C}(E_0)kI_0 \tag{16}$$

where k is an efficiency factor associated with the BE detector. The contrast C is therefore

$$C = \frac{I_S - I_C}{I_C} = \frac{\eta(x_1,x_2,E_0) - \eta_{0C}(E_0)}{\eta_{0C}(E_0)} \tag{17}$$

To get some idea of the range of practical values of contrast, values as low as about 1% have been used for magnetic contrast.[78]

In general, contrast will depend on the thickness of the stained layer x_2, its depth x_1 below the surface of the specimen and the electron energy E_0. In addition, there may be an energy threshold for the BE detector below which it does not respond efficiently or at all

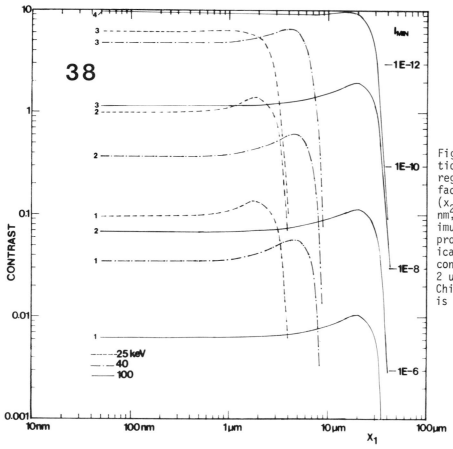

Figure 38. Contrast as a function of depth (x_1) of stained region below the specimen surface is shown for 4 thicknesses (x_2) of stain: 1) 10 nm, 2) 100 nm, 3) 1 μm, 4) 10 μm. The minimum beam current necessary to produce a statistically significant contrast, I_{min}, (for the conditions of the Hitachi HFS-2 used in the University of Chicago SEM User's Laboratory) is also shown.

to BE's. Fig. 38 shows a plot of contrast as a function of x_1, x_2 and Eo. A detector threshold of 10 keV has been used and we have assumed k=0.25. The contrast is typically flat over a considerable range of depth for a given thickness of stain, finally dropping off as the depth increases still further. The contrast, not too surprisingly, tends to increase with stain thickness and decreasing energy. Fig. 39 shows this clearly for an intermediate depth. The detailed shape of the curve arises from the different behavior of the transmission functions and BE energy spectra from C and Ag.

The basic shape of these curves comes from the transmission functions. The BE energy spectra shapes only modify the steeply falling region.

Fig. 39 also shows the effect of removing the 10 keV detector threshold. The contrast decreases, more for the thinner layers of stain. This occurs because the Ag stain BE energy spectrum exceeds that from C more at higher energies, as is shown in Fig. 40. Contributions from C are much more important relative to the stain at lower BE energies, especially for thinner layers of stain.

For given values of x_1 and x_2 Fig. 38 implies that there is an optimum beam energy which provides maximum contrast. For too low an energy, the electrons cannot reach the stained layer and still backscatter with enough energy

to trigger the detector. Once an adequate energy is reached, the contrast drops monotonically as the energy is increased further. Fig. 41 illustrates this behavior for particular values of x_1 and x_2. The shape of this curve is not very sensitive to their values, however. Thus, for given values of x_1 and x_2, the range of energies within which the best contrast is achieved, is only about 10 keV.

The maximum depth at which the stain can be seen at a given energy occurs at the sharp knee where the plateau in contrast ends as shown in Fig. 38. Fig. 42 shows the depth at which the contrast has decreased to 50% of its plateau value, as a function of energy. This curve does not vary much as x_2 is changed.

We can draw a number of important conclusions from these figures. First, it is clear that there is more than adequate contrast available for many situations of biological interest. This is already clear from the experimental results of section III. Secondly this technique should enable us to look into the interior of biological specimens to some considerable depth, depending on the energy. Again, the results of section III verify the general concept; however, quantitative evidence is hard to come by. Abraham et al.[13] used stereometry to determine the depth at which they could see Ta particles within a block of lung tissue. They found the particles to be visible to a depth of 1.6 μm at

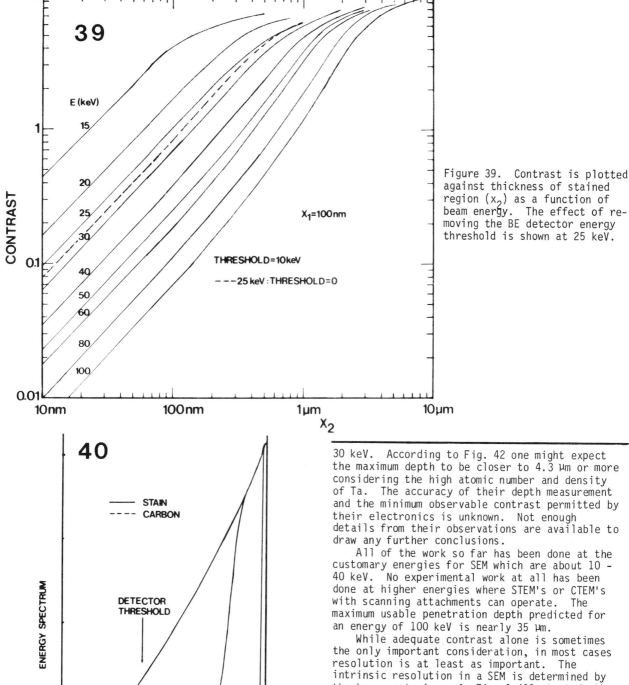

Figure 39. Contrast is plotted against thickness of stained region (x_2) as a function of beam energy. The effect of removing the BE detector energy threshold is shown at 25 keV.

Figure 40. BE energy spectra predicted by the Everhart theory.

30 keV. According to Fig. 42 one might expect the maximum depth to be closer to 4.3 μm or more considering the high atomic number and density of Ta. The accuracy of their depth measurement and the minimum observable contrast permitted by their electronics is unknown. Not enough details from their observations are available to draw any further conclusions.

All of the work so far has been done at the customary energies for SEM which are about 10 - 40 keV. No experimental work at all has been done at higher energies where STEM's or CTEM's with scanning attachments can operate. The maximum usable penetration depth predicted for an energy of 100 keV is nearly 35 μm.

While adequate contrast alone is sometimes the only important consideration, in most cases resolution is at least as important. The intrinsic resolution in a SEM is determined by the beam spot size. As Fig. 1 illustrated, the beam spreads out as it penetrates the specimen, so the resolution will deteriorate with increasing depth. At a given depth the resolution will improve with increasing beam energy. To estimate the resolution attainable in this technique the Monte Carlo program used to produce Fig. 1 was employed. A beam of zero width was assumed to impinge on the specimen and the median radius of the beam, within which half of it was contained, was determined as a function of depth for several energies. The

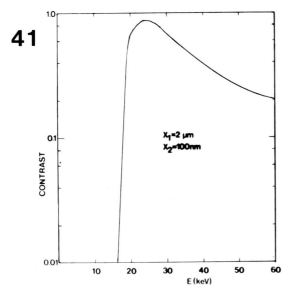

Figure 41. The contrast from a given stained area is shown as a function of beam energy.

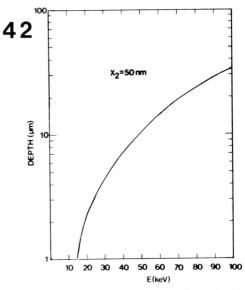

Figure 42. The approximate maximum depth at which stained regions can be seen at a given energy (thickness of stain = 50nm).

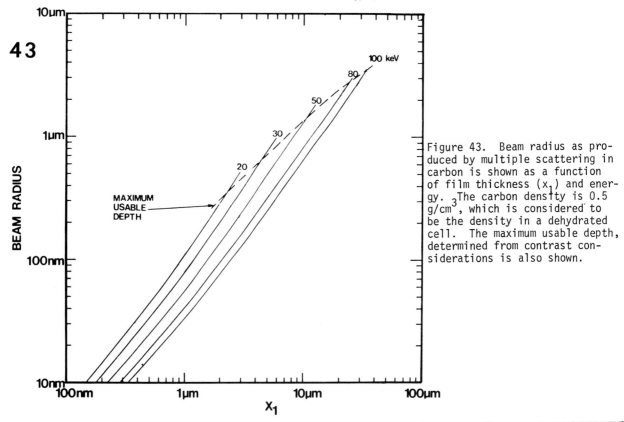

Figure 43. Beam radius as produced by multiple scattering in carbon is shown as a function of film thickness (x_1) and energy. The carbon density is 0.5 g/cm³, which is considered to be the density in a dehydrated cell. The maximum usable depth, determined from contrast considerations is also shown.

results are shown in Fig. 43. The curve from Fig. 42 representing maximum usable depth is also drawn. At a depth of about 3.5 μm at 30 keV the spot size (median diameter) is about 0.9 μm. At a depth of 1.6 μm, where as noted above, Abraham et al[13] were able to see Ta particles, the resolution should be about 0.3 μm. The smallest Ta particles visible in their

Fig. 5 are of the order 0.4-0.5 μm. These particles may have an appreciable intrinsic size. At 100 keV the resolution at a depth of 5μm should be about 0.5μm. At the maximum usable depth at 100 keV of about 35 μm the resolution is about 5μm. At a depth of 1μm the resolution is about 66 nm at 100 keV.

At great depths the resolution is inferior to that available in a light microscope. However the depth of focus is enormously greater, and the possibility of using stereometry to aid in image interpretation exists. Moreover the capability always exists of correlating these deep structures with surface features at higher resolutions than is possible with a light microscope. One could also correlate these images with X-ray microprobe studies (The large BE coefficient makes BE imaging many times more efficient than X-ray imaging.) This could be of great utility for example in the clinical study of heavy metal diseases.

These numbers apply assuming the beam is properly focused at the depth of interest. The beam in the SEM is convergent, and optimal resolution is obtainable only at the crossover height. This is illustrated in Fig. 44. While particle 1 is optimally focused, particle 2 is seen (if at all) with much poorer resolution despite its being closer to the surface. Clearly if one is viewing structure at some depth a decision must be made on whether to maximize resolution of the deep structure or to maintain a sharp focus of the SE produced surface image.

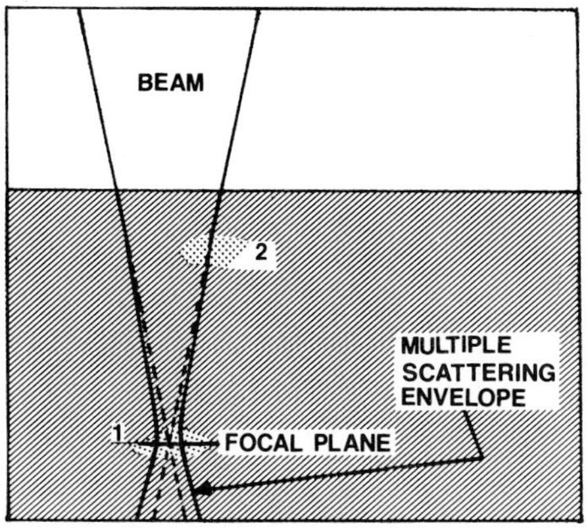

Figure 44. The effect of the finite depth of focus is shown.

Returning to the subject of contrast for a moment, one other consideration is the beam current required to produce a statistically significant contrast. A certain minimum number of BE's are required per picture element in order to reduce the statistical noise to a level where specimen detail can be discerned. For a micrograph with a given number of picture elements taken in a given amount of time this puts a lower limit on the required beam current. Using Rose's criterion[79] we derive in the Appendix an expression for the minimum beam current required to make a given level of contrast statistically significant. The minimum current is plotted in Fig. 38 for picture conditions appropriate for the Hitachi HFS-2 at the Universiy of Chicago SEM User's Laboratory. This calculation ignores other sources of noise, such as detector amplifier noise, beam current fluctuations etc., so the real minimum current is certainly higher. It should probably be good to better than an order of magnitude, however.

As is well known, the beam spot size increases with increased current[80]. Table I shows some spot sizes arising from spherical aberration and diffraction as a function of beam

Table 1

a) Tungsten Hairpin

V_o	$I = 10^{-7}$	10^{-8}	10^{-9}	10^{-10}	10^{-11} Amp
10 kV	0.5 μm	0.21 μm	87.0 nm	36.8 nm	15.4 nm (19.1)
20	0.36	0.16	64.0	27.0	11.4
30	0.32	0.13	56.0	23.8	10.0
40	0.28	0.12	50.0	21.2	8.9
100	0.20	83.9 nm	35.3	14.9	6.3

b) LaB_6

V_o	$I = 10^{-7}$	10^{-8}	10^{-9}	10^{-10}	10^{-11} Amp
10 kV	81.5 nm	34.4 nm (36.4)	14.5 nm (20.5)	6.1 nm (11.6)	2.6 nm (6.9)
20	60.0	25.3	10.7 (12.2)	4.5 (6.9)	1.9 (4.1)
30	52.7	22.2	9.4	4.0 (5.0)	1.7 (3.0)
40	47.0	19.8	8.4	3.5 (4.1)	1.5 (2.5)
100	33.1	13.9	5.9	2.5	1.0 (1.2)

Beam diameters are plotted as a function of beam current and voltage for two types of emission systems. Objective lens parameters are assumed to be C_S = 2 cm, C_C = 0.8 cm. The minimum beam size from spherical aberration is tabulated. When less than that from chromatic aberration, the latter is included in parentheses.
  a) Brightness ratio: β/Vo=2.0, energy spread = 2 ev.
  b) Brightness ratio: β/Vo=200, energy spread = 2.5 eV, tip radius = 1 μm.

current and energy for a W hairpin filament and a LaB_6 tip. Other parameters are adopted from Wells[81]. In most cases, the limiting resolution is likely to arise from multiple scattering rather than the effect just described. However, when examining relatively thin, shallow stained areas at high energy the W tip may become the limiting factor. The specimen displayed in Fig. 7 might be such an example. In that case a LaB_6 tip, which, because of its greater brightness β, produces a smaller spot size for a given beam current, would be required for optimum resolution. For the same reason even higher resolution could be obtained with a field emission tip.

Feasibility flow chart

We can now write down a prescription for putting this information to use. A flow chart describing this procedure is also shown in Fig. 45. We assume a Ag stain of approximately the concentration employed here. Some initial information about the biological system is needed: the approximate size of the selectively stained regions (x_2) and their location relative to the surface of the specimen (x_1). These dimensions must be for the specimen in a dehydrated state, since considerable shrinkage may have occurred during preparation. Also, the density should be known approximately. If the density of the material ρ_m differs significantly from the 0.5 g/cm^3 we have assumed here, all lengths quoted in the graphs and tables must be scaled by the factor (0.5/ρ_m.

Using the value of x_1 determined from a priori knowledge of the system, and Fig. 42, the minimum value of E_0 can be determined. As we have seen in Fig. 41 the maximum contrast occurs close to this energy.

From x_1, x_2 and E_0 the contrast can now be determined from Figs. 38 and 39. If the contrast is inadequate, stop here; BE imaging will not solve your problem.

If the contrast is adequate use x_1 and E_0 to estimate the resolution permitted by multiple scattering from Fig. 43. If contrast is high but resolution marginal, Fig. 43 shows that resolution can be improved by increasing E_0. However, contrast falls at the same time. Somewhere is a "breakeven" point whose location will depend on the details of your problem.

Finally using, the expressions in the appendix, the minimum current I_{min} corresponding to your contrast can be determined for the conditions of your SEM. The intrinsic beam size corresponding to I_{min} for your SEM can then be determined and compared with the beam size determined above from multiple scattering.

While this technique is fairly general, it is presently restricted to systems stained only with a particular concentration of Ag. We hope in the future to measure concentrations of other heavy metal stains of interest to biologists. Adopting these results to other concentrations of Ag stain is not straightforward. Doubling the stain concentration would not double the contrast, in general. First, the BE coefficient for the stain is not, in general, a linear function of the concentration. Secondly, and more fundamentally, the contrast depends strongly on absorption, and the transmission functions which determine absorption are not simple functions of the Ag concentration. Further work is needed here.

The model presented here has been limited in energy to about 100 keV, because it is presently non-relativistic. Also, some of the physical information used as input (e.g. the BE coefficients for thin films) has only been measured in this range. There is no reason to suppose that this model could not be modified and extended to higher energies. In general, at higher energy

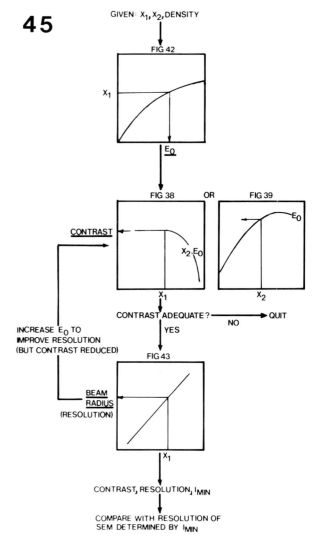

Figure 45. Flow chart for determining the feasibility of using Z contrast BEI in biological systems.

we would expect the penetration to increase, and the resolution at a given depth to improve, both at the expense of reduced contrast. The BE coefficients will also slowly decrease with energy up to about 1 MeV or more, above which they drop off precipitously.[73]

V. Instrumentation Considerations

We will attempt to summarize the present state of instrumentation relating to this application of backscattering. We will deal primarily with detectors. We will also attempt to suggest future instrumentation trends implied by our conclusions.

Detectors

The most common BE detectors currently employed are scintillator and solid state detectors. Scintillator detectors have been used for many years to detect secondary electrons[82]. Characteristically, they are mounted about 45° above the "horizon" of the specimen (or the specimen is tilted to that orientation). SE's are attracted to a grid maintained at several hundred volts positive potential. Beyond the grid they are accelerated to a scintillator by a potential of 10-12 kV. Striking the scintillator they produce light which is detected and amplified by a photomultiplier tube.

Occasionally, this detector is used to detect BE's by putting a negative bias on the grid to repel SE's. The resulting signal now arises mostly from BE's which emerge from the specimen in the direction of the scintillator. The collection efficiency is poor because, as mentioned in section II, the BE angular distribution is broad and the solid angle at the specimen intercepted by the detector is usually quite small (if the collection efficiency were not small the BE's would appreciably affect the SE signal when the detector is operated normally). The resulting BE signal, therefore, tends to be noisy and of marginal quality. This has sometimes led to the erroneous conclusion that the intrinsic BE signal is too weak to provide adequate contrast.

Because of the BE angular distribution (Fig. 3) the best detector geometry requires the detector to intercept a significant fraction of the back hemisphere above the specimen and be roughly axially symmetric. The latter requirement reduces topographic contrast, which arises mainly from surface detail, to a minimum relative to Z contrast, which is generally a bulk property.

Most BE scintillator detectors so far have been homemade. Fig. 46 is a sketch of the detector currently in use at the SEM User's Laboratory at the University of Chicago. The beam passes through a hole in the light pipe and scintillator. The scintillator is a CaF (Eu) crystal[83] which is highly radiation resistant and about twice as efficient in terms of light output as plastic scintillators which have been commonly used in SE detectors in the past. Its response time, however, is of the order of 1μsec which makes it marginal for video rate scanning.

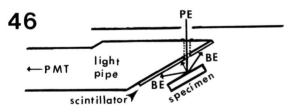

Figure 46. Schematic drawing of the scintillator-light guide-photomultiplier detector currently installed in the Hitachi HFS-2 SEM employed in this presentation. The electron beam (PE) passes through the hole in the detector's light pipe and scintillator. BE's from the specimen strike the scintillator and produce light, which is transmitted to the photomultiplier tube (PMT) by the light pipe. Drawing taken from Becker and Vogel.[46]

Other radiation resistant scintillators with faster response times are available but typically at the cost of reduced light output[83]. The scintillator is epoxied to a lightpipe of optical grade quartz and the whole assembly aluminized to prevent charging and to prevent the photomultiplier from responding to any ambient sources of light. Enough space exists between the BE detector and the specimen that a SE signal can be obtained simultaneously with the BE signal. The BE detector can also be removed from the beam area by means of a high vacuum bellows.

A commercial scintillator detector designed by Robinson exists.[84,85] The scintillator-lightpipe is a single piece of plastic scintillator which covers a large fraction of the back hemisphere above the specimen. Plastic scintillator is not very radiation resistant[86], so the lifetime of the detector is questionable.

As mentioned earlier, scintillator detectors do not respond to electrons of arbitrarily small energy. The electron must possess enough energy to penetrate the aluminized coating on the scintillator and produce enough light for the photomultiplier to respond efficiently. Empirically the threshold is typically around 5-10 keV (above this energy the signal/noise ratio of the detector signal remains constant). Above threshold, the response from the detector increases linearly with electron energy. This may not be a desirable effect for BEI of subsurface structures, since higher energy BE's tend to come from near the specimen surface and the response of the detector will tend to be intrinsically weaker for BE's coming from deeper within the specimen.

The photomultiplier tube is a very efficient, relatively noise free amplification device and the resulting signal is of high quality.

Solid state detectors are also used in detecting BE's. These are usually silicon surface barrier diodes. They have the advantage of being quite compact and requiring nothing but electrical feedthroughs to the vacuum. Until recently, these detectors consisted of discrete chips mounted on opposite sides of the beam[87]. Adding the signals together gave a reasonable amount of Z contrast. Taking their difference produced considerable topographic contrast. Presently, annular detectors, similar to those developed by Wolf and Everhart[88] which are mounted concentric with the beam, are available and can operate at video rates.[89]

Their energy response is similar to that for scintillator detectors. The threshold depends mainly on the thickness of the conduction coating applied to the face of the detector and can be 5 keV or less.[43,89] As with scintillator detectors, this coating serves a dual purpose because these detectors are also somewhat photosensitive.

These detectors are not too resistant to radiation damage. In the Crewe laboratory significant deterioration in the detector's properties has been observed in less than a year's use (when used as dark field transmission detectors in STEM's). This is manifested by increased noise and slow drifts in the DC signal level. Some of this damage may arise from local heating effects caused by inadvertently placing the very intense direct beam on the detector. This is not possible in the case of the BE detector, so the situation may be more hopeful there.

Specimen current (absorbed electron current) is another means of obtaining a BE signal. The beam current I_0 must be equal to the other currents entering and emerging from the specimen in the absence of charging. These are the SE current I_{SE}, the BE current I_{BE}, and the specimen current I_{SC}:

$$I_0 = I_{SE} + I_{BE} + I_{SC} \qquad (18)$$

If we treat the beam current as constant in time the change in the other currents must be related as follows:

$$\Delta I_{SC} = -(\Delta I_{SE} + \Delta I_{BE}) \qquad (19)$$

Thus the specimen current should provide the same magnitude, but opposite polarity and contrast as the sum of the SE and BE signals.[90] It will also provide the same resolution. The energy response in this case is just the energy dependence of the SE and BE signals.

A pure BE signal can be obtained by covering the specimen with a grid biased about 50V negative with respect to specimen ground. This will prevent the SE's from leaving the specimen's surface. Because all BE's contribute to the SC signal topographic contrast is minimized.

The SC signal is very small, so very high gain amplification is required. The specimen must also be well insulated from ground except through the SC amplifier. In such a situation considerable care is required to avoid pickup from ambient electrical noise sources. Furthermore, there are limits to the frequency response range available when the amplifier gain becomes very high. A video rate image is possible only if the specimen current exceeds approximately 10nA.[91]

A fairly simple means of detecting BE's has been proposed recently by Moll et al.[92,93] It has been observed that SE's produced by BE's can contribute significantly to the SE signal.[94] This is evident in many of the micrographs of section III, e.g., Figs. 11, 14, 33-35. There are two sources for these. First, when a BE emerges from the specimen surface it can produce SE's, typically with greater efficiency than the primary electron. The BE can again produce SE's when it strikes a surface within the microscope, usually the objective lens pole piece. These SE's are then collected along with the rest by the SE detector. If a negatively biased grid is placed over the specimen to suppress SE's produced at the specimen surface, as described earlier in connection with the specimen current, only SE's produced by BE's at surfaces other than the specimen's will be collected. Thus a signal proportional to BE production is obtained.

Since this is a multiple step process the efficiency might be expected to be low. Moll et al. measured this signal to be about 20% of the standard SE signal. However, the latter signal contains sizable contributions from the two BE processes described above, so in fact this signal is probably roughly comparable to the "true" SE signal. The basic reason for this lies in the energy dependence of the SE coefficient described in section II. It increases as energy decreases, reaching a maximum typically at several keV. Thus the lower energy BE's are more efficient at producing SE's than the primary electrons. The energy response of this system is, therefore, to some extent complementary to that of the scintillator detector and solid state detector, with it responding most efficiently to BE's of several keV and declining steadily above that.

The frequency response of this technique is determined by the SE detector and associated electronics. One cannot simultaneously observe the standard SE image, but the latter can be produced simply by turning off the bias on the grid.

A possible future candidate for a BE detector is the microchannel plate.[95] It can detect UV photons, X-rays, electrons and ions. In shape and size it resembles the annular solid state detector. It consists of a very fine honeycomb (hole diameters can be less than 15 μm) of glass with a resistive secondary electron emission coating in the channels. A voltage of approximately 1kV is maintained across the plate and electrons entering the channels are multiplied by a factor of 10^3-10^4 and collected at an anode on the other side. The detector is sensitive to single electrons, has low intrinsic noise and excellent frequency response. It can detect secondary electrons (and so must be negatively biased to repel them) and its energy response is fairly uniform, varying by less than 60% from below 500eV to 50keV.[96] Venables et al.[97] report employing one to measure electron channeling patterns. Above a beam current of about 1nA the plate saturated, reducing the contrast. Reducing the high voltage across the plates corrected the problem. Clean vacuums of 10^{-6} Torr or better are required for stable operation. The gain of the microchannel plate slowly decreases over time.[98] Over a limited range, increasing the high voltage could presumably compensate for this decrease. The noise level appears to be unaffected by this aging process.

As we have seen, there is some relation between the energy of BE's and the depth from which they emerge. This relation is particularly true for high Z thin films, as Figs. 2 and 40 show. To the extent that this relationship holds, the variation of the BE detector's response with respect to BE energy may be either a help or a hindrance. For example, if one is trying to look deep within a specimen, the lower energy BE's from that depth will produce less of a response from the scintillator or solid state detector than a comparable number of higher energy BE's from shallower depths. Better contrast might be obtained using one of the other detectors whose response does not increase with BE energy. The relative energy response of all the detectors mentioned here is shown in Fig. 47.

Their absolute collection efficiency must also be considered, of course. Only the specimen current detector's response does not vary appreciably with energy. The importance of these considerations is difficult to assess at present because of the very limited experience with these detectors in the application considered here. As long as we are working in a relatively narrow energy range of say 10-30keV

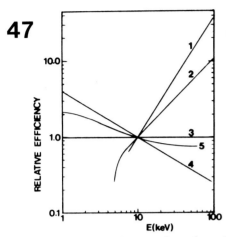

47

Figure 47. The efficiency of various BE detectors, normalized so that they are all equal at E=10 keV; 1) scintillator detector, 2) solid state detector, 3) specimen current detector, 4) Moll et al[92,93] detector, 5) channelplate detector.

the detector's energy response may not be too important. If it proves useful to work at higher voltages, however, problems might arise.

The existence of an energy dependent response in the BE detector will, of course, also change the measured contrast from that expected from Figs. 38 and 39, because, unlike most signals used in electron microscopy, the BE signal involves a wide range of electron energies. For example, consider a solid state detector with a response which increases linearly with increasing electron energy. As we consider deeper stained regions, the mean BE energy will decrease, and consequently, instead of a broad plateau in contrast, as shown in Fig. 38, there will be a steady decrease as the depth x_1 increases, for fixed stain thickness x_2.

Electronics

We will make a few comments concerning the electronics ancillary to the BE detectors. This mainly reflects experience gained with the BE detector at the University of Chicago SEM User's Laboratory.

It is very convenient to have a simultaneous display of both SE and BE images, so that identifying surface and subsurface features can be carried on continuously. Being able to switch electronically from one image to the other quickly is tolerable, but not quite as convenient. Having to physically remove a detector to change images, however, is quite inconvenient.

A further advantage of several simultaneous displays is that one can then consider electron-

ically mixing signals from two detectors to enhance various features.[13] The subsurface structure visible in some SE images, as mentioned in Section III, is sometimes strong enough to obscure important surface detail. Subtracting the BE signal, appropriately scaled, from the SE signal can eliminate this interference. Crewe and Lin[29] have further suggested that removing BE produced SE's from the SE image can enhance the intrinsic resolution of that image, since this species of SE's is less locally produced than the "true" SE's produced by the incident beam. The most sensitive way to make this subtraction is to examine line scans from the "mixed" and BE images simultaneously and adjust the gain of the BE signal so that prominent features in that signal are just removed from the "mixed" signal.

Other combinations of signals may also prove useful. For example, a BE signal restricted to lower energy BE's might be obtained by subtracting the signal from a scintillator or solid state BE detector (which does not respond to electrons below an energy threshold, which could be deliberately increased, if desired) from the signal from a specimen current detector, biased to suppress SE's (which detects all BE's).

Finally, it is probably worth repeating that it is frequently convenient to reverse the contrast of the BE image, so that regions of strong back-scattering are dark instead of bright. This sometimes aids in interpreting the image, as mentioned in Section III.

Appendix

Contrast and minimum beam current

From eq. 17 in section IV we know that contrast is proportional to the difference in collected BE currents from stained and unstained regions of the specimen. The number of BE's, n, contribution to the signal in a picture element ("pixel") is just proportional to this current times the time the beam is scanned on this pixel. Thus the contrast could also be written as

$$C = \Delta n / n \qquad (A1)$$

where Δn is the difference in number of collected BE's from two pixels, one stained, the other unstained. Rose[79] found that in order for contrast in a noisy picture to be statistically reliable, the difference in number of electrons had to exceed the statistical uncertainty in number of electrons by about a factor of 5. For small contrast, where these consideration are important the numbers of electrons contributing to both stained and unstained pixels are approximately equal. We can then state Roses' criterion as

$$\Delta n / \sqrt{2n} \geq 5 \qquad (A2)$$

Combining eqs. A1 and A2 we get that in order for the contrast to be statistically significant we must have enough electrons n contribution to each pixel so that

$$n \geq 50/C^2$$

(So far a contrast of 1% to be meaningful we need about 5×10^5 electrons/pixel). If the average BE coefficient for the specimen is n_{ave} and the overall BE collection efficiency is k, then the required number of electrons in the beam, n, is

$$n_0 = n/kn_{ave} > 50/C^2 kn_{ave} \tag{A3}$$

For a picture of N^2 elements in time t this leads to a condition for the minimum beam current I_{min}:

$$I_{min} = 50N^2 e/C^2 kn_{ave} t \tag{A4}$$

where e is the charge of the electron.
For the User's Laboratory Hitachi HFS-2 we have taken N = 1000, t = 60 sec, k = 0.25, and $n_{ave.} = 0.06$. This leads to

$$I_{min} = 9 \times 10^{-12}/C^2 \quad \text{(Amp)}. \tag{A5}$$

Acknowledgements

This investigation was supported in part by NCI grant CA 05493, NSF grant BNS77-28493, NIH grant 5-P41-RR-00984, and the Louis Block Fund of the University of Chicago.

The Hitachi HFS-2 scanning electron microscope, located in the University of Chicago Scanning Electron Microscope User's Laboratory in the Enrico Fermi Institute, was established by a grant from the Alfred P. Sloan Foundation and is currently supported by a grant from the Biotechnology Resources Branch of the National Institutes of Health.

We wish to thank T. Huber, R. Steiner and A. Kabaya for their assistance and for the use of the facilities at the Application Laboratory of JEOL, U.S.A., Inc. in Medford, Mass.

We wish also to thank Professor A.V. Crewe for support and encouragement.

We wish further to thank A. Cibavicius and V. Kriho for technical contributions and proof reading, and the personnel, particularly Bridget Archer and Gail Whitaker of the Word Processing Center of the College of Medicine, University of Illinois at the Medical Center for the typing and editing of this manuscript.

References

1. Wetzel, B. Cell kinesics: an interpretive review of the significance of cell surface form. In: Scanning Electron Microscopy/1976/II, Johari, O., and Becker, R.P. (Eds.), IIT Res. Inst., Chicago, 135-144.

2. Wetzel, B., Erickson, B., Jr., and Levis, W. The need for positive identification of leukocytes examined by SEM. In: Scanning Electron Microscopy/1973, Johari, O., and Corvin, I (Eds.), IIT Res. Inst., Chicago, 535-542.

3. Wetzel, B., Jones, G.M., and Sanford, K.K. Cell cycle and topography in non-synchronized monolayers: the use of autoradiography and time-lapse for more rigorous SEM studies. In: Scanning Electron Microscopy/1977/I, Johari, O. (Ed.), IIT Res. Inst., Chicago, 545-552.

4. Noonan, J.M., and Riddle, J.M. Dynamic surface activities of exudative leukocytes. In: Scanning Electron Microscopy/1977/II, Johari, O. and Becker, R.P. (Eds.), IIT Res. Inst., Chicago, 53-58.

5. Geissinger, H.D., Yamashiro, S., and Ackerley, C.A. Preparation of skeletal muscle for intermicroscopic (LM, SEM, TEM) correlation. In: Scanning Electron Microscopy/1978/II, Becker, R.P., and Johari, O. (Eds.), SEM Inc., AMF O'Hare, IL, 60666, U.S.A., 267-274.

6. Wickham, M.G., and Worthen, D.M. Scanning and transmission electron microscopy of single tissue specimens. In: Principles and Techniques of Scanning Electron Microscopy, Vol. 2, Hayat, M.A. (Ed.), Van Nostrand Reinhold, New York, 1974, 60-71.

7. Becker, R.P., and DeBruyn, P.P.H. The transmural passage of blood cells into myeloid sinusoids and the entry of platelets into the sinusoidal circulation; a scanning electron microscope investigation. Am. J. Anat. 145, 1976, 183-206.

8. Albrecht, R.M., and Wetzel, B. You can usually tell a cell beneath its cover: ancillary methods for biological SEM. In: Scanning Electron Microscopy/1979/III, Johari, O. (Ed.), SEM Inc., AMF O'Hare, IL, 60666, U.S.A., 1979, 203-222.

9. Watanabe, S. The scanning electron-microscope study of the lymph node. Acta Haematol. Jap. 35, 1972, 483-505.

10. Watanabe, S., and Ohishi, T. Labelling methods of lymph node cells for a scanning electron microscopy. In: Histochemistry and Cytochemistry, 1972, Takeuchi, T., Ogawa, K., and Fujita, S. (Eds.), Japan Society of Histochemistry and Cytochemistry, Kyoto, Japan, 1972, 167-168.

11. DeBruyn, W.C., VanMourik, W., and Bosveld, I.J. Cell border demarcation in the scanning electron microscope by a silver stain. J. Cell Sci. 16, 1974, 221-239.

12. DeNee, P.B., Abraham, J.L., and Gelderman, A.H. Methods for a SEM study of coal workers' pneumoconiosis. In: Scanning Electron Microscopy/1973, Johari, O., and Corvin, I. (Eds.), IIT Res. Inst., Chicago, 411-418.

13. Abraham, J.L., Wagner, R.M., Miyai, K., Bullock, R., and Friedman, P.J. Choice of imaging modes in SEM study of respirable radiographic contrast medium: tantalum particles in lung. In: Scanning Electron

Microscopy/1976/I, Johari, O., and Becker, R.P. (Eds.), IIT Res. Inst., Chicago, 691-698.

14. Abraham, J.L. Scanning electron microscopy as an aid in diagnosis: an overview. In: Scanning Electron Microscopy/1977/II, Johari, O., and Becker, R.P. (Eds.), IIT Res. Inst., Chicago, 119-128.

15. Pease, R.F.W., and Hayes, T.L. Scanning electron microscopy of biological material. Nature 210, 1966, 1049.

16. Manger, W.M., and Bessis, M. Cathodo-luminescence produced in cells and proteins by paraformaldehyde as seen with the scanning electron microscope. In: Microscopie Electronique 1970, Favard, P. (Ed.), Societe Francaise de Microscopie Electronique, Paris, 483-484.

17. Falk, R.H. Cathodoluminescence - its potential for biology. In: Proceedings of the 5th Annual Stereoscan SEM Colloquium, Kent Cambridge Scientific Company, Morton Grove, U.S.A., 1972, 35-41.

18. Padawer, J. Scanning electron microscopy: identification of mast cells by indirect cathodoluminescence. Histochem. 38, 1974, 351-360.

19. Soni, S.L., Kalnins, V.I., and Haggis, G.H. Localization of caps on mouse B lymphocytes by scanning electron microscopy. Nature 255, 1975, 717-719.

20. Bröcker, W., Schmidt, E.H., Pfefferkorn, G., and Beller, F.W. Demonstration of cathodoluminescence in fluorescein marked biological tissues. In: Scanning Electron Microscopy/1975, Johari, O., and Corvin, I. (Eds.), IIT Res. Inst., Chicago, 243-250.

21. Schmidt, E.H., Bröcker, W., and Pfautsch, M. Cathodoluminescence properties of frequently used fluorescent stains in vaginal smears. In: Scanning Electron Microscopy/1976/II, Johari, O., and Becker, R.P. (Eds.), IIT Res. Inst., Chicago, 327-334.

22. Haggis, O., Bond, E.F., and Fulcher, R.G. Improved resolution in cathodoluminescent microscopy of biological material. J. Micros. 108, 1976, 177-184.

23. Bröcker, W., and Pfefferkorn, G. Applications of the cathodoluminescence method in biology and medicine. In: Scanning Electron Microscopy/1979/II, Johari, O. (Ed.), SEM Inc., AMF O'Hare, IL, U.S.A., 125-132.

24. Bröcker, W., and Pfefferkorn, G. Bibliography on cathodoluminescence. In: Scanning Electron Microscopy/1976/I, Johari, O. (Ed.), IIT Res. Inst., Chicago, 725-737.

25. Abraham, J.L., and DeNee, P.B. Biomedical applications of backscattered electron imaging. One years experience with SEM histochemistry. In: Scanning Electron Microscopy/1974, Johari, O., and Corvin, I. (Eds.), IIT Res. Inst., Chicago, 252-258.

26. DeNee, P.B., Abraham, J.L., and Willard, P.A. Histochemical stains for the scanning electron microscope. Qualitative and semiquantitative aspects of specific silver stains. In: Scanning Electron Microscopy/1974, Johari, O., and Corvin, I. (Eds.), IIT Res. Inst., Chicago, 259-266.

27. Ashraf, M., Livingston, L.H., and Bloor, C.M. SEM of T tubules in myocardial cells. In: Scanning Electron Microscopy/1976/II, Johari, O., and Becker, R.P. (Eds.), IIT Res. Inst., Chicago, 179-186.

28. Becker, R.P., and DeBruyn, P.P.H. Backscattered electron imaging of endogenous and exogenous peroxidase activity in rat bone marrow. In: Scanning Electron Microscopy/1976/II, Johari, O., and Becker, R.P. (Eds.), IIT Res. Inst., Chicago, 171-178.

29. Crewe, A.V., and Lin, P.S.D. The use of backscattered electrons for imaging purposes in a scanning electron microscope. Ultramicroscopy 1, 1976, 231-238.

30. DeNee, P.B., and Abraham, J.L. Backscattered electron imaging (applications of atomic number contrast). In: Principles and Techniques of Scanning Electron Microscopy, Vol. 5, Hayat, M.A., (Ed.), Van Nostrand Reinhold, New York, U.S.A., 1976, 144-180.

31. Hodges, G.M., and Muir, M.D. Scanning electron microscope autoradiography. In: Principles and Techniques of Scanning Electron Microscopy, Vol. 5, Hayat, M.A. (Ed.), Van Nostrand Reinhold, New York, U.S.A., 1976, 78-93.

32. Nopanitaya, W., and Grisham, J.W. Comparative observations on fatty liver using secondary and backscattered electrons. In: Proceedings, 35th Annual Meeting, Electron Microscopy Society of America. Bailey, G.W. (Ed.), Claitor's Publ. Division, Baton Rouge, LA, U.S.A., 1977, 490-491.

33. Albrecht, R., Jordan C., and Hong, R. Indentification of monocytes, granulocytes and lymphocytes: correlation of histo-logical, histochemical and functional properties with surface structure as viewed by scanning electron microscopy. In: Scanning Electron Microscopy/1978/II, Becker, R.P., and Johari, O. (Eds.), SEM Inc., AMF O'Hare, IL, U.S.A. 60666, 511-523.

34. Tannenbaum, M., Tannenbaum, S. and Carter, H.W. SEM, BEI and TEM ultrastructural characteristics of normal preneoplastic and neoplastic human transitional epithelia. In: Scanning Electron Microscopy/1978/II, Becker, R.P., and Johari, O. (Eds.), SEM Inc., AMF O'Hare, IL, U.S.A., 949-958.

35. Wells, O.C. Backscattered electron image (BSI) in the scanning electron microscope (SEM). In: Scanning Electron Microscopy/1977/I, Johari, O. (Ed.), IIT Res. Inst., Chicago, 747-771.

36. Wittry, D.B. Secondary electron emission in the electron probe. In: X-ray Optics and Microanalysis, Castaing, R., Deschamps, P., and Philibert, J. (Eds.), Hermann, Paris, 1966, 168-180.

37. Abraham, J.L., and DeNee, P.B. Scanning-electron-microscope histochemistry using backscatter electrons and metal stains. Lancet 1, 1973, 1125.

38. Hodges, G.M., and Muir, M.D. Autoradiography of biological tissues in the scanning electron microscope. Nature 247, 1974, 383-385.

39. Carr, K.E., and McGadey, J. Staining of biological material for the scanning electron microscope. J. Microscopy 100, 1974, 323-330.

40. Miyai, K., Abraham, J.L., Linthicum, D.S., and Wagner, R.M. Scanning electron microscopy of hepatic ultrastructure. Secondary, backscattered and transmitted electron imaging. Lab. Inves. 35, 1976, 369-376.

41. Vogel, G., Becker, R.P., and Swift, H. Backscattered electron imaging of nuclei in rat testis. In: Scanning Electron Microscopy/1976/II, Johari, O., and Becker, R.P. (Eds.), IIT Res. Inst., Chicago, 1976, 409-415.

42. Gomori, G. A new histochemical test for glycogen and mucin. Am. J. Clin. Pathol. 16, 1946, 177-179.

43. Lin, P.S.D., and Becker, R.P. Detection of backscattered electrons with high resolution. In: Scanning Electron Microscopy/1975, Johari, O., and Corvin, I. (Eds.), IIT Res. Inst., Chicago, 61-67.

44. Kindly provided by Dr. Sandra Panem, Department of Pathology, University of Chicago, Chicago, IL.

45. DeBruyn, P.P.H., Becker, R.P., and Michelson, S. The transmural migration and release of blood cells in acute myelogenous leukemia. Am. J. Anat. 149, 1977, 247-268.

46. Becker, R.P., and Vogel. G. Visualization by backscattered electron imaging of the accessory (outer dense) fibers and mitochondria of sperm. In: Proceedings of the Fourteenth Annual Conference of the Microbeam Analysis Society. San Francisco Press, Inc. In press.

47. Humphreys, W.J., Spurlock, B.O., and Johnson, J.S. Critical point drying of ethanol infiltrated cryofractured biological specimens for SEM. In: Scanning Electron Microscopy/1974, Johari, O., and Corvin, I. (Eds.), IIT Res. Inst., Chicago, 275-282.

48. Fawcett, D.W. The mammalian spermatozoon. Devel. Biol. 44, 1975, 394-436.

49. Baccetti, B., Pallini, V., Burrinil, A.G. Accessory fibers of the sperm tail. I. Structure and chemical compositon of the full "coarse fibers." J. Submicr. Cytol. 5, 1973, 237-256.

50. Baccetti, B., Pallini, V., and Burrini, A.G. The accessary fibers of the sperm tail. III. High sulfur and low sulfur components in mammals and cephalopods. J. Ultrastr. Res. 57, 1976, 289-308.

51. Swift, J.A. The electron histochemical demonstration of sulphydryl and disulphide in electron microscope sections, with particular reference to the presence of these chemical groups in the cell wall of the yeast Pityrosporum ovale. In: Electron Microscopy/1966, Vol. II, Uyeda, R. (Ed.), Maruzen Co. Ltd., Tokyo, 63-64.

52. Swift, J.A. The electron histochemistry of cystine - containing proteins in thin transverse sections of human hair. J. Roy. Micr. Soc. 88, 1968, 449-460.

53. Swift, J.A. The electron histochemical demonstration of cystine - containing proteins in the guinea pig hair follicle. Histochemie. 19, 1969, 88-98.

54. Swift, J.A. The electron cytochemical demonstration of cystine disulphide bonds using silver - methenamine reagent. Histochemie. 35, 1973, 307-310.

55. Burr, F.A. Staining of a protein polysaccharide model with silver methenamine. J. Histochem. Cytochem. 21, 1973, 386-387.

56. Ross, R., and Bornstein, P. The elastic fiber. I. Separation and partial characterization of its macromolecular components. J. Cell Biol. 40, 1969, 366-381.

57. Graham, R.C., and Karnovsky, M.J. The early stages of absorption in injected horseradish peroxidase in the proximal tubules of the mouse kidney: ultrastructural cytochemistry by a new technique. J. Histochem. Cytochem. 14, 1966, 291-302.

58. Pietra, G.G., Szidon, J.P., Leventhal, M.M., and Fishman, A.R. Hemoglobin as a tracer in hemodynamic pulmonary edema. Science 166, 1969, 1643-1646.

59. Seligman, A.M., Karnovsky, M.J., Wasserkrug, H.L., and Hanker, J.S. Nondroplet ultrastructural demonstration of cytochrome oxidase activity with a polymerizing osmiophilic reagent, diaminobenzidine (DAB). J. Cell Biol. 38, 1968, 1-14.

60. Cammer, W., and Moore, C.L. Oxidation of 3,3'-diaminobenzidine by rat liver mitochondria. Biochemistry 12, 1973, 2502.

61. Roels, F. Cytochrome c and cytochrome oxidase in diaminobenzidine staining of mitochondria. J. Histochem. Cytochem. 22, 1974, 442-446.

62. Albrecht, R., and Bleier, R. Histochemical, functional, and structural features of isolated adherent supraependymal cells: characterization as resident mononuclear phagocytes. In: Scanning Electron Microscopy/1979/III, Becker, R.P., and Johari, O. (Eds.), SEM Inc., AMF O'Hare, IL, U.S.A., 60666, 55-61.

63. Adler, R.B. The human pulmonary alveolar macrophage: two distinct morphological populations. In: Scanning Electron Microscopy/1979/III, Becker, R.P., and Johari, O. (Eds.), SEM Inc., AMF O'Hare, IL, U.S.A., 60666, 921-927.

64. DeBruyn, P.P.H., Michelson, S., and Becker, R.P. Endocytosis, transfer tubules, and lysosomal activity in myeloid sinusoidal endothelium. J. Ultrastr. Res. 53, 1975, 133-151.

65. This application, specimen preparation, and Figures courtesy of Dr. Harry Carter, Department of Pathology, Saint Barnabas Medical Center, Livingston, New Jersey and Dr. Myron Tannenbaum, Department of Pathology, Columbia University, New York, New York.

66. Brunk, V.T., and Ericsson, J.L.E. The demonstration of acid phosphatase in in vitro cultured tissue cells. Studies on the significance of fixation toxicity and permeability. Histochem. J. 4, 1972, 349-363.

67. Marshall, A.T. Electron probe X-ray microanalysis. In: Principles and Techniques of Scanning Electron Microscopy, Vol. 4, Hayat, M.A. (Ed.). Van Nostrand Reinhold, New York, 1975, 103-173.

68. Boyde, A., Bailey, E., Jones, S.J., and Tamarin, A. Dimensional changes during specimen preparation for scanning electron microscopy. In: Scanning Electron Microscopy/1977/I, Johari O. (Ed.), IIT Res. Inst., Chicago, 507-518.

69. Wells, O.C., Boyde, A., Lifshin, E., and Rezanovich, A. Scanning Electron Microscopy, McGraw-Hill, New York, 1974, Chapter 3, 37-68.

70. Niedrig, H., and Sieber, P. Rückstreuung mittelschneller Elektronen an dünnen Schichten. Z. Angew. Phys. 31, 1971, 27-31.

71. Bishop, H.E. Some electron backscattering measurements for solid targets. In: X-ray Optics and Microanalysis, Castaing, R., Deschamps, P., and Philibert, J. (Eds.), Hermann, Paris, 1966, 153-158.

72. Heinrich, K.F.J. Electron probe microanalysis by specimen current measurement. In: X-ray Optics and Microanalysis, Castaing, R., Deschamps, P., and Philibert, J. (Eds.), Hermann, Paris, 1966, 159-167.

73. Tabata, T., Ito, R., and S. Okabe. An empirical equation for the backscattering coefficient of electrons. Nuel. Instrum. and Meth. 94, 1971, 509-513.

74. Niedrig, H. Film-thickness determination in electron microscopy: the electron backscattering method. Optica Acta 24, 1977, 679-691.

75. Castaing, R. Electron probe microanalyses. In: Advances in Electronics and Electron Physics, Vol. 13, Marton, L.L., and Marton, C. (Eds.), Academic Press, New York, 1960, 317-386.

76. Everhart, T.E. Simple theory concerning the reflection of electrons from solids. J. Appl. Phys. 31, 1960, 1483-1490.

77. Cosslett, V.E., and Thomas, R.N. Multiple scattering of 5-30 keV electrons in evaporated metal films II: Range-energy relations. Brit. J. Appl. Phys. 15, 1964, 1283-1300.

78. Newbury, D.E. Fundamentals of scanning electron microscopy for physicists: contrast mechanisms. In: Scanning Electron Microscopy/1977/I, Johari, O. (Ed.), IIT Res. Inst., Chicago, 553-568.

79. Rose, A. Television pickup tubes and the problem of vision. Adv. Electron. 1, 1948, 131-166.

80. Smith, K.C.A. Scanning electron microscopy: the next ten years. In: Scanning Electron Microscopy/1972, Johari, O., and Corvin, I. (Eds.), IIT Res. Inst. Chicago, 1-8.

81. Wells, O.C., Boyde, A., Lifshin, E. and Rezanovich, A. Scanning Electron Microscopy, McGraw-Hill, New York, 1974, Chapter 4, 69-88.

82. Wells, O.C., Boyde, A., Lifshin, E., and Rezanovich, A. Scanning Electron Microscopy, McGraw-Hill, New York, 1974, Chapter 5, 89-107.

83. Nuclear Enterprises, Inc., 931 Terminal Way, San Carlos, CA, 94070.

84. ETP Semra, 47 Alexander St., Crows Nest, Sydney, Australia, 2065.

85. Robinson, V.N.E. Backscattered electron imaging. In: Scanning Electron Microscopy/1975, Johari, O., and Corvin, I. (Eds.), IIT Res. Inst., Chicago, 51-60.

86. Pawley, J. Performance of SEM scintillation materials. In: Scanning Electron Microscopy/1974, Johari, O., and Corvin, I. (Eds.), IIT Res. Inst., Chicago, 27-34.

87. Kimoto, S., and Hashimoto, H. Stereoscopic observations in scanning microscopy using multiple detectors. In: The Electron Microprobe, McKinley, T.D., et al. (Eds.), Wiley and Sons, New York, 1966, 480-489.

88. Wolf, E.D., and Everhart, T.E. Annular diode detector for high angular resolution pseudo-Kikuchi patterns. In: Scanning Electron Microscopy/1968, Johari, O., and Corvin, I. (Eds.), IIT Res. Inst., Chicago, 41-44.

89. Gedcke, D.A., Ayres, J.B., and DeNee, P.B. A solid state backscattered electron detector capable of operating at T.V. scan rates. Scanning Electron Microscopy/1978/I, Johari, O. (Ed.), SEM INC., AMF O'Hare, IL, U.S.A., 60666, 581-592.

90. Newbury, D.E. The utility of specimen current imaging in the SEM. In: Scanning Electron Microscopy/1976/I, Johari O. (Ed.), IIT Res. Inst., Chicago, 111-118.

91. Quoted by G.W. Electronics Inc., 22 Perimeter Park, Suite 101, Atlanta, Ga., 30341, for their model 103 specimen current amplifier.

92. Moll, S.H., Healey, F., Sullivan, B., and Johnson, W. A high efficiency, nondirectional backscattered electron detection modes for SEM. In: Scanning Electron Microscopy/1978/I, Johari, O. (Ed.), SEM Inc., AMF O'Hare, IL, U.S.A., 60666, 303-308.

93. Moll, S.H., and Healey, F. Further development of the converted backscattered electron detector. In: Scanning Electron Microscopy/1979/II, Johari, O. (Ed.), SEM Inc., AMF O'Hare, IL, U.S.A., 60666, 149-154.

94. Seiler, H. Einige aktuelle Probleme der sekundärelektronen emission. Z. Angew. Phys. 22, 1967, 249-263.

95. Galileo Electro-Optics Corp., Galileo Park, Sturbridge, MA, 01518.
96. Archuleta, R.J., and DeForest, S.E. Efficiency of channel electron multipliers for electrons of 1-50 keV. Rev. Sci. Instrum. 42, 1971, 89-91.
97. Venables, J.A., Harland, C.J., and Bin Jaya, R. Crystallographic orientation determination in the SEM using electron backscattering patterns and channel plates. In: Developments in Electron Microscopy and Analysis, Venables, J.A. (Ed.), Academic Press, London, 1976, 101-104.
98. Sandel, B.R., Broadfoot, A.L., and Shemansky, D.E. Microchannel plate life tests. Appl. Optics 16, 1977, 1435-1437.

Symbol Table

BE   Backscattered electron.
SE   Secondary electron.
η BE coefficient for a thin film.
η_0 BE coefficient for a bulk film.
T_i Transmission coefficient for material i.
ρ mass density (g/cm^3).
E_0 Primary electron energy (keV).
Z   Atomic number.
f_i Mass fraction of element i in a binary compound.
λ_i "Attenuation length" in transmission coefficient for material i (μm).
x   depth of electron penetration (μm).
R   Range of electrons (μm).
y   Electron penetration normalized by range (=x/R).
ε BE energy normalized to E_0.
a   Constant in Everhart theory(=.045Z).
I_0 Beam current on the specimen (A).
I_S BE current from stained area in specimen (A).
I_C BE current from unstained area in specimen (A).
k   BE collection efficiency.
I_{min} Minimum beam current required to provide statistically significant contrast.

Discussion with Reviewers

Editors: Please provide a plot of the BE coefficient for different elements.

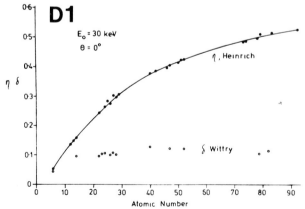

Figure D1. The BE and SE coefficients, η and δ respectively, are shown as a function of atomic number Z.

Authors: Figure D1 shows both the bulk BE and SE coefficients as a function of atomic number Z. The figure is taken from a review paper by D. Newbury (Fundamentals of scanning electron microscopy for physicists:contrast mechanisms. Scanning Electron Microscopy/1977/I,IIT Research Institute, Chicago,553-568). An empirical fit to this BE data is quoted by Newbury:
$$\eta = -0.0254 + 0.016Z - 0.000186Z^2 + 8.31 \times 10^{-7}Z^3.$$

It is perhaps worth commenting on the relative performance of heavy metal stains as a function of atomic number in BE imaging and in conventional TEM (bright field) work. In the latter the brightness of a stained area is related (among other things) to the total elastic cross section of the heavy element. This increases approximately like $Z^{1.5}$, so a stain like Au or U should be much brighter than Ag, for the same concentration. However the BE coefficient increases rather slowly with Z especially for the high atomic numbers. As a result Ag is nearly as bright as the heaviest metals, its BE coefficient being only about 23% less than that of U.

J.L. Abraham: Figures 5 and 8 show considerable blurring of the backscattered image from nuclei less than 1μm beneath the surface. This does not seem consistent with your theoretical explanation. Please comment.
Authors: A resolution from multiple scattering of $\sim$180 nm would be expected from the model at a depth of 1μm. Some details this small are resolvable in the pictures. Another problem is that the range of contrast from the nuclei is relatively small. The nuclear structures are frequently of the order 1μm or more in size (see Figs. 9 and 10) and, as Fig. 39 shows, the contrast does not change rapidly for stain thicknesses in this range at 20-25 keV. Thus it is not easy to see detail on the nuclear surface(which may be close to the specimen surface) when it sits above the main (stained) body of the nucleus.

J.L. Abraham: Is it fair to say that the density of tissue is the same as carbon? What about void volumes? The density could be much less. Clearly the assumption of dehydrated tissue as carbon is a gross overestimate, since the volume from which water has been removed will be an empty void. Please comment.
Authors: In creating a model for backscattering from biological material there are two independent considerations: 1) assuming for simplicity we want to represent the material as a single element, what atomic number do we choose to give approximately the same backscattering probability per atom as the material; 2) what density do we choose to give the same number of atoms per unit volume as the material. In answer to 1) we assumed dry tissue to be approximately represented by an average amino acid of composition $C_{5.35}H_{9.85}O_{2.45}N_{1.45}S_{0.10}$. The bulk BE coefficient of this compound differs from that of carbon by only 12%. What counts is the difference in BE coefficient between the tissue and the stain. The BE coefficient for Ag is about 0.40 while that for C is about 0.06. Clearly the BE coefficient for the tissue could have a sizable error without significantly affecting the results as far as contrast is concerned. In an-

swer to 2) we used a density of 0.5 g/cm^3 for C, because that is the observed density of critical point dried tissue (the normal density of C is about 2.0 g/cm^3). This is a bulk, average property of the tissue. Clearly it will be in error within a void volume, where the local density will be zero. It will also be in error within a region of compact protein which typically has a density of about 1 g/cm^3. But since we are concerned here with viewing a heavy metal stained region against a background of unstained tissue, it is sufficient, given the level of sophistication of the model, to deal with the average properties of the latter.

P.B. DeNee: How do your theoretical results compare with the work of DeNee (Measurement of mass and thickness of respirable size dust particles by SEM backscattered electron imaging. Scanning Electron Microscopy/1978/I, SEM Inc., AMF O'Hare, I1,60666,741-746) and Niedrig[74]? Please comment with respect to the cases treated in the appendix of the DeNee paper, e.g. for the case of a thin, low Z material on a bulk,high Z material and for a thin, high Z material on a bulk, low Z material.
Authors: We are somewhat limited in what predictions we can make because of our need of physical input data, specifically the BE coefficient for a thin film and the transmission function for a material. Because we presently lack this information for iron we can make no comparison with DeNee's results for that element. We can, however, make predictions for the BE coefficient for a metallic bilayer consisting of a thin film of Au on top of a thick film of Al. For the transmission functions of these elements we used fits similar to eqs. 9, which were obtained from Monte Carlo results of Reimer and Krefting (L. Reimer and E.R. Krefting, The effect of scattering models on the results of Monte Carlo calculations, in "Electron Probe Microanalysis and Scanning Electron Microscopy", K.F.J.Heinrich, D.E.Newbury and H.Yakowitz (eds), NBS Special Pub. 460,(1976),45-60). This prediction was compared with data from Hohn et al (F.J. Hohn, M.Kindt, H.Niedrig, B.Stuth, Elektronen-rückstreumessungen an dünnen Schichten auf massiven Trägersubstanzen, Optik 46 (1976), 491-500). We also compared an empirical expression of DeNee from his above referenced paper with the data:

$$\eta_{bilayer} = \eta_{oAl} + \eta_{oAu}(x)(1-\eta_{oAl}/\eta_{oAu}),$$

where x is the thickness of the Au film. The results are shown in Fig. D2. Both models seem to fit the data satisfactorily.

P.B. DeNee: Can the atomic number of ThO$_2$ be considered to be 90, which is the Z of Th? Or is the Z of ThO$_2$ equal to the mean of the three atoms involved? I have observed,for example,that the BE yield from Si is higher than that from SiO$_2$.
Authors: It is not a good idea to think in terms of an effective atomic number for compounds, because the BE coefficient is not a linear function of Z. One must instead determine an expression for the BE coefficient for a compound.

For a thick film of a binary compound AB, the BE coefficient is just

$$\eta_{AB} = f_A \eta_A + f_B \eta_B$$

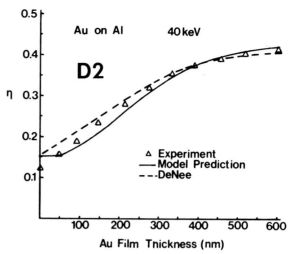

Figure D2. BE coefficient of a thin layer of Au on a thick Al substrate as a function of Au film thickness: experiment and theory.

where f_i is the mass fraction of element i present and η_i the corresponding BE coefficient. This was first suggested by Castaing[75] and has been thoroughly verified experimentally (e.g., see reference 72). For ThO$_2$ this works out to

$$\eta_{ThO_2} = 0.88\eta_{Th} + 0.12\eta_O$$

The BE coefficients of Th and O are about 0.51 and 0.09 respectively, which leads to η_{ThO_2}=0.46, about 90% of that of pure Th.

This analysis is consistent with your observation for SiO$_2$.

For thin films we suggest using Eq. 7.

O.C. Wells: Where does Eq. 3 come from?
Authors: The basic characteristics of the BE coefficient are as follows. For very thin films η increases linearly with the film thickness. For thicker films η increases more slowly and finally "saturates" at a constant value. This behaviour arises from the fact that,for thin films,BE's arise predominantly from single, large angle scatterings, and the probability of these occurrences increases linearly with the film thickness (i.e. with the number of atoms available for scattering). η saturates for thick films because of the finite range of the electrons. They reach a depth where they no longer have enough energy to backscatter and regain the surface. Increasing the film thickness beyond this point clearly can have no further effect on η.

The hyperbolic tangent tanh possesses these properties and is thus a convenient candidate as a fitting function. Adding a quadratic term in film thickness to the argument of tanh gives a needed extra degree of freedom to the fit. Fits were originally made individually to data at different energies. The energy dependence shown in Eqs. 5 was then determined. The success of this procedure is illustrated in Fig. D3 for Ag. All of the data (symbols) for that element are satisfactorily fitted by Eq. 5.

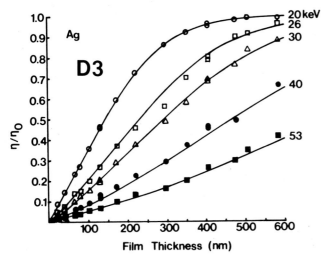

Figure D3. BE coefficients for a thin film of Ag as a function of film thickness and electron energy. The lines are the fit represented by Eq. 5.

G. Pfefferkorn: Why do you define an efficiency factor k in Eqs. 16 independent of the energy spectrum of the BE's emerging from the stained or unstained region? From Figs. 2 and 40 one might expect different efficiency factors, k_C and k_S, for instance.
Authors: The efficiency factor k refers to energy independent factors such as detector solid angle. Energy dependent efficiencies are implicitly included by the use of the 10 keV threshold, which neglects electrons with less than this energy. The threshold was included in a crude attempt to simulate the energy behaviour of scintillator or solid state detectors. As Fig. 40 illustrates, the BE energy spectra from C and the stain are different, so the 10 keV threshold will affect them differently.

G. Pfefferkorn: Comparing your calculations with the experimental results of Abraham et al[13] you doubt the accuracy of their measurements. Do you have any new experiments to verify your calculations on the maximum usable depth at which stained regions can be seen at a given energy? Do you have any experiments to show the resolution as a function of depth and beam energy? Are there any experiments of imaging subsurface detail with energies > 40 keV to test your theory?
Authors: We cannot "doubt" the accuracy of the measurements of Abraham et al, because no errors were quoted. We feel there is not enough information available to decide whether their observations represent a significant test of the model. Our theoretical work was completed quite recently, and consequently we have no experimental results to report yet.
    We are unaware of any BE imaging performed at energies exceeding 40 keV.

J.L. Abraham: In Fig. 4 you refer to SE topographic detail as "washed out" by the emerging BE's. This is more likely a function of increased penetration of the very thin tissue at 39 keV. Would not the same loss of surface detail occur with unstained, carbon coated cells?
Authors: There are two reasons why the SE contrast decreases with increasing energy. First, the SE coefficient decreases with increasing electron energy above a few keV typically, so there is less SE signal at higher energies. Second, there is a background of SE's produced by BE's emerging from the surface. For a thick homogeneous specimen this background should be roughly a constant fraction of the "real" SE signal, since the bulk BE coefficient is roughly energy independent, and the energy dependence associated with BE's producing SE's is presumably similar to that associated with producing SE's by primary electrons. However, if there is a heavy metal concentration some distance below the surface, it may be that electrons at 15 keV have insufficient energy to reach this layer and backscatter with enough energy to regain the surface (and produce SE's). At 39 keV, however, they may have enough energy, and then the higher BE coefficient of the stain leads to a fractional increase in the background SE's.

O.C. Wells: How much do BE's contribute to the SE signal?
Authors: BE's can produce SE's both when they emerge from the specimen surface and when they subsequently strike some part of the specimen chamber. The contribution from the former effect can be represented by a change in the SE coefficient δ. The effective SE coefficient δ' is given by

$$\delta' = \delta(1 + \eta B),$$

where η is the mean BE coefficient of the specimen, and B is an enhancement factor which reflects the greater efficiency with which BE's, rather than primary electrons, produce SE's. Reported values of B range from about 2 to 4 depending on the experimental situation[95], and as we have seen η can range from about 0.05 to 0.5. Thus this BE contribution to the SE's can be considerable. George and Robinson have demonstrated this by means of a Monte Carlo study (George, E.P. and Robinson, V.N.E., The dependence of SEM contrast upon electron penetration. Scanning Electron Microscopy/1976/I, IIT Research Institute, Chicago, 17-23).
    The second BE contribution has been measured by Moll et al[33] to be about 23% of the total SE signal.

G. Pfefferkorn: Under what conditions is the resolution of the BE image comparable to that of the SE image?
Authors: The resolution of the BEI can be close to that of the SEI only when viewing structures at the specimen surface. When viewing deeper structures the resolution deteriorates sharply, as shown in Figure 43.

O.C. Wells: Does Fig.3 represent the intensity I would measure, if I moved a small detector about? I have performed such measurements and have not seen a decrease in signal when I moved near the beam (i. e., for small θ).
Authors: You are correct. Fig. 3 is a bit mis-

leading. It represents the intensity as a function of θ for an annular detector and is a useful plot in deciding on the angular acceptance of such a detector (or calculating its solid angle). The intensity which would be measured by a small (ideally "point") detector would be proportional to cosθ. Thus it would increase as the detector approached the beam (θ decreased) in agreement with your observation.

J.L. Abraham: In your consideration of the density of critical point dried material you state that both the volume and mass of the tissue shrink during the dehydration process. Does mass "shrink"?
Authors: A considerable fraction of the mass of wet tissue consists of water. When the tissue is dried this mass is removed, and the mass of the dry tissue is thus significantly reduced from its original value.

J.L. Abraham: I disagree with your interpretation of Fig. 11. Could not the large multinucleated cell be seen as such in the SEI more clearly than illustrated by increasing the amplifier gain?
Authors: The contrast could be increased at the expense of the signal-to-noise ratio. The resolution of the features common to the two images would be the same.

J.L. Abraham: You indicate that the GMS stain should include elastic fibers as positive- which is not my experience for LM and SEM. Can you provide some evidence on this?
Authors: After we prepared this paper we learned that Gnepp and Green have recently demonstrated GMS staining and BE imaging of elastic fibers (Gnepp, D.R., and Green, F.H.Y., Scanning electron microscopy of collecting lymphatic vessels and their comparison to arteries and veins. in SEM/1979/III, Sem Inc., AMF O'Hare, Il 60666, 757-762.).

G. Pfefferkorn: Normally fixative solutions show an osmolality between 200 and 600 Osmol/kg. Is the amount of 1285 mentioned in Section III correct?
Authors: The osmolality of the fixative consisting of 1.5% formaldehyde, 2.5% glutaraldehyde, and 0.15% $CaCl_2$ in 0.08 M sodium cacodylate, pH 7.2 is indeed 1285 mOsmol/kg. This fixative is a so-called "Karnovsky" fixative of lesser concentration and osmolality than the even more hypertonic (2010 mOsmol/kg) original formulation (Karnovsky, M.J., A formaldehyde-glutaraldehyde fixative of high osmolality for use in electron microscopy. J. Cell Biol. 27, 1965, 137A). "Karnovsky" fixatives, albeit hypertonic, are widely used and considered to yield reliable preservation of structure and enzymatic activity.

J.L. Abraham: You have used several different buffers in your specimen preparation. Is there one you recommend over the others? Specifically, does not the arsenic in the cacodylate buffer reduce the contrast from specific stains in the BE image?
Authors: We recommend that until contraindicated one should use the buffer that one ordinarily would use for TEM or LM, as the case may be. We have no evidence which suggests that cacodylate buffer reduces the intensity of specific stains sufficiently to significantly alter the yield of BE's from stained regions.

G. Pfefferkorn: Why do you use a pH 8.8-9.0 GMS solution? This can lead to artefacts in cell ultrastructure.
Authors: The use of the GMS solution at pH 8.8 to 9.0 is recommended on the basis of our experience with GMS staining at lesser alkalinities and the studies of Swift.[52-54] Lesser alkalinities require a corresponding greater staining time and staining temperature in order to achieve the intensity suitable for BEI. Our exposure conditions (pH 9.0, 30 to 40 min, at 50 C) do not appear to result in an unacceptable level of artefact (see, for example, Figures 9 and 10).

J.L. Abraham: You indicated that the glass vials need not be acid cleaned. What % of stained samples are rendered useless by uncontrolled precipitation? How reproducible is the method in the text?
Authors: If one abides by the protocol, removing the samples from the stain when the GMS solution becomes just perceptibly gray, one can expect that no sample will be rendered useless by our method. We have encountered no problem with the reproducibility of the staining.

J.L. Abraham: What is the usual depth of penetration of your tissue block with useful silver staining?
Authors: In those cases where we have embedded stained samples for TEM, we have found nuclei in excess of 50 μm below the surface with no apparent diminution of staining intensity. This depth exceeds the useful depths as predicted by our own model (Figure 43)- at least up to an energy of 100 keV.

J.L. Abraham: You say that your histochemical staining offers a means of identifying certain leukocytes. Which ones, and what % of cells can be identified by this technique?
Authors: We suggest that our SEM method can be used to distinguish leukocytes which contain peroxidase (+) granules from those which do not, to identify the former on the basis of the number, size and shape of the peroxidase containing granules, and to distinguish among stages of immature forms of these leukocytes. To our knowledge, no one has investigated the utility of this method of identification in a comprehensive fashion.

J.L. Abraham: In quantifying the mitochondrial population of a cell by SEM, how can this be done practically when overlapping mitochondria will certainly be observed in the BE image?
Authors: Overlapping will indeed be a problem. It is to be expected that mitochondria will be more quantifiable in cells which are squamous (e.g. most endothelial cells), and within which little overlapping of mitochondria occurs, than within thick, polygonal cells, such as hepatocytes. However, in the latter case, stereology can be expec-

ted to aid in sorting out overlapping areas.

P.B. DeNee: Do you have any experience with staining of embedded material, such as that reported by DeNee et al (DeNee,P.B.,Frederickson, R.G., and Pope, R.S., Heavy metal staining of paraffin, epoxy and glycol methacrylate embedded biological tissue for scanning electron microscopy histology. Scanning Electron Microscopy/ 1977/II,IIT Research Institute, Chicago,83-92)?
Authors: We have very limited experience with this BEI application. Our own experience does indicate, however, that BEI can be usefully applied to classically prepared, sectioned material, as demonstrated in the reference which you cite.

G. Pfefferkorn: Do you have any experience with the possibilities of signal mixing with respect to BE and SE images?
Authors: Following the suggestion of Crewe and Lin[20] we have electronically subtracted a fraction of the BE signal from the SE signal to remove contributions to the latter from BE-produced SE's, arising from high Z, subsurface regions. The contribution from such inclusions can obscure surface detail. An example of this procedure is illustrated in Fig. D4. Fig D4a is the original SE image and Fig. D4b is the corresponding BE image. The "subtraction" image in Fig D4c illustrates the successful elimination of these BE effects: a presumably more faithful surface image results.

One must use this technique with care however. A constant fraction of the BE signal is subtracted over the entire image. There is no reason to assume that this will represent the correct amount of subtraction everywhere[29]. In general there may be regions of over- or under-subtraction in the image which lead to a subtraction image little better than the original SE image. In our experience however there seem to be enough practical applications of this technique which are free of this problem to make it worthwhile. One must always bear in mind these potential dangers however in applying it.

G. Pfefferkorn: Is it possible to say which type of detector will be most suitable to get maximum information of subsurface structure in cells and tissues?
Authors: So little experimental work has been done so far that it does not appear to be possible to rank the detectors.

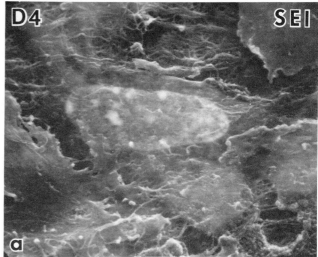

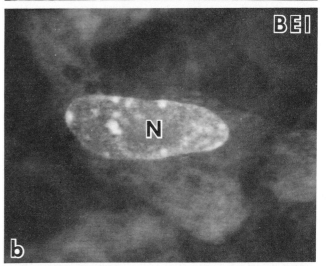

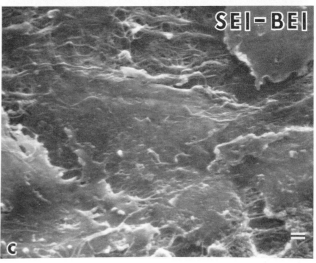

Figure D4. SE (a) and BE (b) and "subtraction" images (c) of the same area on the exposed undersurface (the surface adhering to the culture dish) of a chick limb bud in organ culture. The tissue was stained en bloc with the GMS method described in the text. The silver-stained nucleus (N) of one cell is evident in the BEI (b) and also in the SEI, due to the BE-produced SE's, but not in the subtraction image. HFS-2, OSD, 2×10^{-11} Amp, 25kV. Scale bar = 1μm.

SCANNING ELECTRON MICROSCOPY/1979/II
SEM Inc., AMF O'Hare, IL 60666, USA

EVALUATION OF TISSUE-RESPONSE TO HYDROGEL COMPOSITE MATERIALS

R. T. Greer, R. L. Knoll and B. H. Vale

Biomedical Engineering Program,
Engineering Research Institute
Department of Engineering Science and Mechanics
Iowa State University
Ames, IA 50011

Abstract

Hydrogel formulations of 10-20% HEMA (hydrox-yethyl methacrylate), 0-3% EGDM (ethylene glycol dimethacrylate) and 0-15% NVP (n-vinyl pyrroli-done) were prepared to coat silicone rubber sheets and polyethylene terephthalate (Dacron) velour sub-strates by chemical and radiation polymerization methods. The rabbit paravertebral muscle implan-tation test was used to evaluate the tissue response to the hydrogel composite materials. Each fabricated sample was implanted in ten separate rabbits. The samples and surrounding muscle tissue were examined histopathologically at ten days, one month, and three months. The samples were sectioned and stained with hematox-ylin and eosin, and trichrome stains. Samples were observed by LM (light microscope) and SEM (scanning electron microscope) techniques both prior to implantation and after removal from the muscle site.

All the materials tested showed no cyto-toxic activity. The silicone rubber composite materials were all surrounded by a thin, encap-sulating, collagenous layer in a range of 10 to 30 μm. The polyethylene terephthalate composite materials showed a considerable amount of tissue ingrowth up to 700 μm. This ingrowth was prob-ably due to the filamentous nature of the poly-ethylene terephthalate velour and the hydrogel's relative porosity of 20-50 micrometers filling the interstices. The P-HEMA bulk polymer exhibited a small amount of tissue ingrowth.

SEM microstructural details (e.g., the rela-tionship of tissue penetration to the intercon-necting pores of the hydrogel) indicate that a variety of microstructural features can be cor-related to the fabrication variables (radiation initiated polymerization compared with chemical grafting; and the composition of the monomer for various crosslink formulations).

KEY WORDS: Hydrogels, Prosthesis, Prosthetic Coatings, Tissue-Response, Silastic, Dacron, Biocompatibility, Encapsulation, Collagen

Introduction

The development of hydrogels, especially polyhydroxyethyl methacrylate (P-HEMA), as materials for medical applications has been reviewed by Ratner and Hoffman (1) and Hoffman et al (2). The P-HEMA monomer can be cross-linked with ethylene glycol dimethacrylate to form stable three-dimensional, water-imbibing gel networks. It is believed that the presence of imbibed fluid within the polymeric system is intrinsically related to its high biocompati-bility (1).

With the increasing interest in biomedical applications of hydrogels, there have been impor-tant discussions raised in the literature as to the surface and bulk properties in relation to the areas of blood compatibility, tissue compat-ibility, and cell adhesion. A better under-standing of basic interactions between material parameters and biological response is needed so that a higher level of control over the amount of biological reaction can be attained compared with current technology.

The amount of graft, water imbibed and degree of biocompatibility of the hydrogel sys-tems can be easily varied by changing substrate material, monomer concentrations, percent of crosslinking agent, and grafting methods. The degree of surface roughness of pre-implant materials, the amount of biological tissue pene-tration into the pores of the hydrogel, and organization of the fibrous capsule can be inves-tigated using light microscopy and scanning electron microscopy techniques.

Materials and Methods

Materials Fabrication

Radiation initiated polymerization was used to coat Dacron (U.S. Catheters and Instruments, DeBakey Double Velour, lot no. 015984) and sili-cone rubber (Dow Corning, Silastic, lot no. HH1404). 1 x 4 cm strips (Dacron and Silastic) were cleaned ultrasonically (5 minute wash, 3 rinses) with a nonoily soap (Ivory flakes), dried and weighed. Monomer solutions of 10-20% 2-hydroxyethyl methacrylate (HEMA, Haven Chemical Co., Philadelphia, lot no. 700-238-22) and 0-15% n-vinyl pyrrolidone (NVP, Haven Chemical Co., lot no. 700-205-2) crosslinking agent with 0-2% ethylene glycol dimethacrylate, were mixed in a solvent consisting of 25% methanol and 75% distilled water. The substrate materials were suspended in Pyrex test tubes to which the monomer solution of interest was added. The tubes were then saturated with nitrogen to minimize oxidative reactions and irradiated in a ^{60}Co radiation source to receive 0.25 Mrad.

Impregnation was achieved using an Interpenetrating Network (IPN) fabrication method (3). Silastic sheets cut into strips (1 cm x 4 cm) were washed and weighed as described above. The materials were preswelled by boiling in xylene for ten minutes and then were placed in the monomer solution of interest for two hours. Monomer solutions of 10-25% 2-hydroxyethyl methacrylate (HEMA, Haven Chemical Co., Philadelphia, lot no. 700-238-22) with 0-2% crosslinking agent added (ethylene glycol dimethacrylate, EGDM, Haven Chemical Co., lot no. 53-95) were prepared in xylene with 5% ethanol. The reaction was completed just below the boiling point of the resulting monomer-crosslinker solution which is in the range of 118-135°C, depending on the concentrations of monomer and crosslinker in the particular solution. This technique has also been used to coat Dacron with hydrogel although the resulting network is not truly an IPN.

In both fabrication methods, a bulk product was formed in the reaction vessel. The samples were stripped free mechanically from the bulk polymer. The freed substrate with residual hydrogel coating was rinsed in an ethanol/water (50/50 v.%) solution (3 rinses) and stored in distilled water.

Graft and H_2O Percent Determination. Samples to be used for graft and water imbibition measurements were dried for four hours in an oven at 100°C and then stored in a desiccator. The dry samples were weighed to determine the percent hydrogel graft acquired. Water imbibition was determined by blotting wet samples (stored in distilled water at least 24 hours) between sheets of Whatman #1 filter paper with a 100 gram weight for five seconds and weighing sample. Percent water imbibition is reported as:

$$\frac{\text{weight of water imbibed}}{\text{weight of dry hydrogel graft}} \times 100\%$$

SEM Techniques. Critical point drying of non-implanted samples involved taking the samples through a series of acetone rinses (30, 60, 90, 100, 100%), from the distilled water storage medium. Samples were critical-point dried in a Polaron Equipment drier using CO_2, mounted on aluminum stubs with colloidal silver, sputter coated with 300 Å gold by a Polaron Instruments SEM coating Unit E5100, and observed at 5-20 KeV in a JEOL-U3 SEM.

Surgical Implantation

Rabbit paravertebral muscles were used as the site of implantation for the test discs according to the method suggested by Coleman, King, and Andrade (4). Briefly summarized, the method was a sterile technique performed under general anesthesia: a one-centimeter incision was made approximately two centimeters from the vertebral spines in the midlumbar region over the right and left paravertebral muscles (iliocostalis lumborum and longissimus lumborum); each disc was implanted into a pocket formed by blunt dissection (separating the muscle fibers longitudinally) with an iris scissors; the muscle sheath, fascia, and skin were closed with 000-prolene[1] sutures. Sham surgery sites did not receive an implant. The rabbits were given a single postoperative intramuscular injection of 200,000 units of combiotic[2] (a penicillin and streptomycin combination).

Ten animals per material were used. Six sites per animal were employed, one being a control silastic, one being a sham surgical site, and four being material implantation sites.

Specimen Preparation. Tissue and material samples were harvested at 10 days, 30 days, and 90 days. 1 cm^2 blocks of tissue containing the samples were excised from freshly killed rabbits, fixed in buffered 10% formalin, and embedded in paraffin. The paraffin blocks were chilled and sectioned at six micrometers through a representative interface of tissue and implant. Sections were prepared with the Gomori one-step trichrome stain (5) to demonstrate collagen. Aniline blue was substituted for light green stain resulting in the blue staining of collagen. Other sections, intended for both light microscopy and ultimately scanning electron microscopy, were similarly prepared with the exception that no coverslip was applied.

These samples were adhered to aluminum stubs with colloidal silver, sputter coated with 300 Å of gold by a Polaron Instruments SEM coating instrument (Unit E51000), and observed at 5-20 KeV in a JEOL-U3 SEM.

Histological Evaluation

Several histological features were chosen to be evaluated for each implant specimen. These included: the number of inflammatory cells per field around the implant surface, the total depth of the encapsulation, the amount of ingrowth into the polymeric specimen, the degree of muscle cell deterioration around the implant, and the amount of fat encircling the implant. Samples stained with the Gomori trichrome stain were used for the analysis although parallel preparations of hematoxylin and eosin-stained samples were used as a basis for the identification of cell types. Then, the parameters were quantified as follows.

Inflammatory Cell Count. Monocyte derived cells were counted in a circular area of sample equal to 20,000 square micrometers, which was bisected by the interface of tissue and implant. Measurements were taken at four sites, two from both of the long sides of the implant. The disc shaped implants appear rectangular in cross section and the long sides of the rectangle are used as the characteristic interface. Both ends of the rectangle often demonstrate a higher degree of reactivity due to the stress concentration there.

---

[1]Ethicon Inc., Somerville, NJ

[2]Pfizer, Agricultural Division, New York, NY

Total Encapsulation Response. The "healing in" process of a foreign body reaction includes both the initial inflammation and the subsequent collagen deposition. In the case of the paravertebral muscle implantations done here, a course from healthy muscle to the center of the polymeric implant would pass through a layer of degenerating muscle, a collagenous layer whose cell population is primarily fibroblasts, an inflammatory cell layer and in some cases a final monolayer of cells which are of an endothelial nature. The total encapsulation response is the measured distance from the outermost edge of the collagenous layer to the innermost cell layer, which may be at the surface of the implant or may have considerably penetrated the material. Measurements were taken at four sites, two from both of the long sides of the implant.

Collagenous Portion of Capsule. As described above, the collagenous layer has primarily fibroblasts as its cellular elements and represents a connective tissue layer. This layer stains light blue in the Gomori preparation and is usually distinguishable from the inner inflammatory layer in that the latter stains dark blue. The collagen is often more diffuse than in normal granulomatous reactions (6). When viewed with a polarizing microscope, the collagen is only slightly refractile, suggesting immaturity or incomplete organization. The layer is measured at the same four sites as the total encapsulation response.

Amount of Ingrowth. The penetration of tissue, measured from the outer edge of the polymer to the innermost layer of cellular activity, was recorded for four sites, two from both of the long sides of the implant.

Muscle Cell Deterioration. A zone of muscle surrounding the implant often demonstrated an increase in the number of centralized nuclei, a proliferation of interfascicular connective tissue, and swelling or waviness of muscle fascicles. These properties are often indicative of localized trauma. The condition of surrounding muscle was graded according to the following scale:

1. Deterioration extends completely around the implant to a distance greater than 750 micrometers.

2. Deterioration distance is less than 750 micrometers, but greater than 250 micrometers for a representative amount of sample.

3. Deterioration does not exceed 250 micrometers, but more than half the sample is surrounded.

4. Less than half the sample is surrounded by less than 250 micrometers of deterioration.

Amount of Fat. A layer of fat cells filled with lipid material was often observed peripheral to the collagenous portion of the capsule. This was graded and recorded in the following manner:

1. No fat cell layers.
2. Fat cells on one side or both ends.
3. Fat cells on any three sides.
4. Fat cells completely surrounding the capsule.

Results

Sample fabrication and histological evaluation data are presented in Table 1. The information presented in Table 1 includes fabrication and tissue response information. The following emphasizes observations first for fabrication considerations (controls, chemical grafted and radiation grafted) and then for tissue response characteristics. The discussion follows the sample listing in Table 1.

Uncoated silicone rubber (shown in Figure 1) was used as the control. Like all of the samples with silicone rubber substrates, it had the smallest amount of collagen in its capsule at 90 days (the thickness of the collagenous portion of the capsule is plotted against days of implantation in Figure 2). This pattern in collagen thickness at 90 days was seen for all materials listed in Table 1. Several samples had a 30-day collagen value that was higher than either the 10-day or 90-day value.

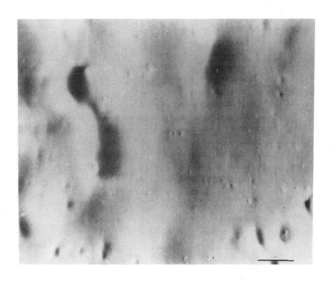

Figure 1. SEM of uncoated silicone rubber (scale bar = 10μm). 25 KeV. 30° tilt.

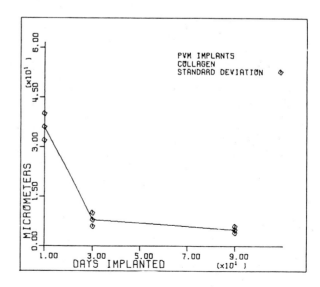

Figure 2. Collagen in capsule of control.

Table 1. Sample Fabrication and Histological Evaluation Data

Sample Fabrication	Graft	Water	Days Implanted	Collagen Thickness μm	Capsule Thickness μm	Ingrowth Thickness μm	Inflammatory Cell Count	Fat Score	Muscle Score
Silastic, Control	--	--	10	36 ± 4	59 ± 12	0 ± 0	28 ± 5	1	2
			30	8 ± 2	14 ± 3	0 ± 0	8 ± 5	2	2
			90	5 ± 1	8 ± 2	0 ± 0	7 ± 2	3	4
Silastic, 20%HEMA, 1.5%EGDM, IPN	8%	40%	10	14 ± 3	19 ± 6	0 ± 0	14 ± 7	1	3
			30	38 ± 13	44 ± 13	0 ± 0	8 ± 2	1	1
			90	5 ± 1	7 ± 2	0 ± 0	3 ± 2	3	4
Silastic, 10%HEMA, 0%EGDM, IPN	11%	38%	10	28 ± 15	39 ± 15	0 ± 0	24 ± 10	1	3
			30	17 ± 15	24 ± 15	0 ± 0	15 ± 3	2	3
			90	7 ± 4	17 ± 3	0 ± 0	17 ± 9	3	3
Silastic, 10%HEMA, 10%NVP Radiation	2%	44%	10	17 ± 3	25 ± 5	0 ± 0	20 ± 10	1	2
			30	13 ± 0	17 ± 3	0 ± 0	11 ± 4	2	2
			90	8 ± 3	17 ± 4	0 ± 0	8 ± 2	-	-
Silastic, 20%HEMA, 1.5%EGDM Radiation	3%	30%	10	14 ± 8	22 ± 8	0 ± 0	10 ± 6	1	3
			30	30 ± 17	61 ± 22	0 ± 0	2 ± 1	2	3
			90	9 ± 4	13 ± 0	0 ± 0	3 ± 2	2	4
Silastic, 20%HEMA, 2.0%EGDM Radiation	2%	35%	10	63 ± 33	78 ± 45	0 ± 0	34 ± 14	1	2
			30	27 ± 28	53 ± 21	0 ± 0	31 ± 13	2	2
			90	11 ± 6	22 ± 6	0 ± 0	17 ± 10	4	3
Silastic, 15%HEMA, 5.0%NVP Radiation	4%	45%	10	48 ± 26	66 ± 18	0 ± 0	27 ± 16	1	3
			30	86 ± 52	103 ± 39	0 ± 0	26 ± 18	2	2
			90	20 ± 13	23 ± 16	0 ± 0	13 ± 9	2	3
Bulk, 20%HEMA, 2.0%EGDM Radiation	--	--	10	23 ± 6	234 ± 94	47 ± 21	30 ± 9	1	1
			30	13 ± 7	44 ± 22	20 ± 9	19 ± 9	1	2
			90	8 ± 3	66 ± 25	38 ± 9	71 ± 17	2	2
Dacron, 20%HEMA, 1.5%EGDM Radiation	163%	201%	10	32 ± 9	220 ± 94	163 ± 83	46 ± 13	1	2
			30	13 ± 5	119 ± 47	91 ± 45	37 ± 10	2	3
			90	13 ± 7	175 ± 79	125 ± 61	71 ± 12	3	2
Bulk, 20%HEMA, 1.5%EGDM Radiation	--	--	10	55 ± 29	106 ± 65	23 ± 26	43 ± 15	1	1
			30	16 ± 12	42 ± 13	35 ± 16	15 ± 6	2	4
			90	34 ± 12	48 ± 13	42 ± 34	24 ± 12	3	2
Dacron 20%HEMA 1.5%EGDM, IPN	128%	152%	10	75 ± 76	750 ± 228	700 ± 191	54 ± 23	1	2
			30	25 ± 9	688 ± 125	606 ± 108	67 ± 7	1	2
			90	18 ± 6	644 ± 105	631 ± 83	56 ± 7	2	3

The basic character of the silicone rubber substrate as observed in the SEM was not found to change with IPN fabrication procedure (7). Similarly, the inclusion of NVP with the HEMA in the radiation technique produces a surface penetration effect and the basic microstructure was left intact (8). An example of these material types where fabrication did not alter the surface texture is the silicone rubber substrate impregnated with 20% HEMA and 1.5% EGDM by the IPN technique. Figure 3 is a low power light micrograph of the implant at 30 days showing the increased collagen deposition and tissue reactivity at the ends of the implant compared to that at the sides due to mechanical stress variations. The layers of collagen and inflammatory cells forming the capsule can be seen in Figure 4. A densification of the capsule, which may account for the decreased thickness commonly measured at 90 days, progressed toward 90 days. Figure 5 shows the capsule around a 20% HEMA, 1.5% EGDM, IPN silicone rubber sample at 30 days. At 90 days, shown in Figure 6, the capsule is of a denser arrangement. The organization of a high density capsule for the 10% HEMA, 10% NVP, radiation-grafted silicone rubber sample at 90 days is shown in Figure 7 (tissue face which was in apposition with the polymer).

The radiation treatment without NVP results in a coating of the substrate as shown in Figure 8 for 20% HEMA, 1.5% EGDM radiation grafted onto silicone rubber which has a different general surface character than in Figure 1. The resultant capsule in Figure 9 is typical of low density organization features seen for 10-day specimens.

Figure 10 shows the condition of the bulk porous hydrogel material prior to implantation. An open three dimensional network (interstices of

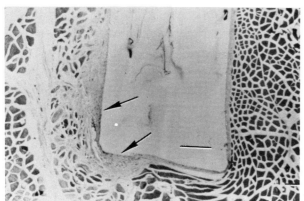

Figure 3. LM of 20% HEMA, 1.5% EGDM IPN in silicone rubber sample at 30 days (scale bar = 1mm). Lower left represents increased encapsulation due to mechanical stress (arrows).

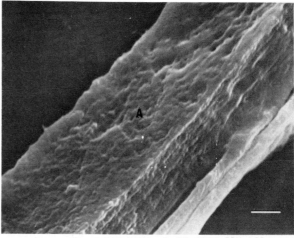

Figure 6. SEM of 20% HEMA, 1.5% EGDM IPN silicone rubber sample at 90 days (scale bar = 2μm). Face A was in contact with implant surface. 15 KeV. 20° tilt.

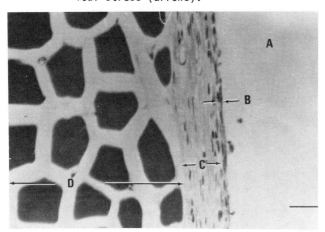

Figure 4. LM of 20% HEMA, 1.5% EGDM IPN in silicone rubber sample at 30 days (scale bar = 100μm). A is implant, B is inflammatory cell layer, C is collagen layer, and tissue is D.

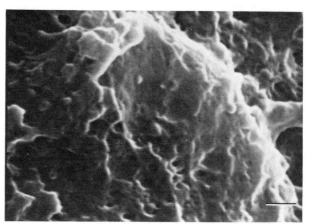

Figure 7. SEM of 10% HEMA, 10% NVP, silicone rubber radiation grafted sample at 90 days (scale bar = 1μm). 15 KeV. 20° tilt.

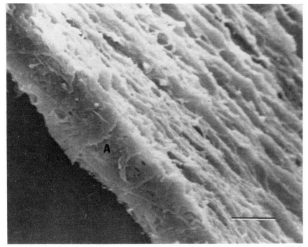

Figure 5. SEM of 20% HEMA, 1.5% EGDM IPN in silicone rubber sample at 30 days (scale bar = 5μm). Face A was in contact with implant surface. 10 KeV. 40° tilt.

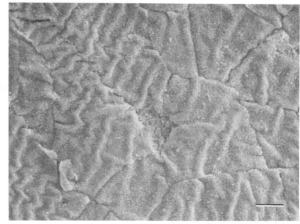

Figure 8. SEM of 20% HEMA, 1.5% EGDM, radiation grafted on silicone rubber sample before implantation (scale bar = 10μm). 10.KeV. 0° tilt.

the order of 50 micrometers and less in diameter) is present within the polymer. This sample was fabricated with 20% HEMA and 1.5% EGDM using the radiation polymerization technique. The nature of tissue ingrowth and encapsulation is presented in the SEM micrograph, Figure 11. This view shows the bulk hydrogel-collagen interface in cross section. The bulk hydrogel comprises the left half of the image, and the dense, organized collagen capsule is to the right. At the interface, there is attachment of the collagen by some ingrowth into the bulk polymer ($\sim$ 30 μm penetration).

The hydrogel resulting from the radiation polymerization of 20% HEMA, 1.5% EGDM within the Dacron velour is shown in Figure 12. The openings among Dacron fibers have been packed by hydrogel deposited during the polymerization. The nature of the encapsulating tissue at 30 days implantation period is shown in Figure 13. The field of view shows the hydrogel packed within the Dacron fiber network (top two thirds of photograph). In this cross section, the Dacron fibers ($\sim$ 15 μm in diameter) are seen as short cylinders. The collagen capsule ($\sim$ 30 μm thick) and tissue are located in the lower third of the photograph. The inflammatory cell count at the three harvesting times is graphed in Figure 14 (data in Table 1). The cell count increase at 90 days is probably due to a fragmentation of the hydrogel that fills the Dacron network.

The IPN hydrogel-Dacron composite has considerable ingrowth at 10 days ($\sim$ 700 μm) that was maintained in thickness through the 90-day implantation period. Figure 15 shows the nature of the infiltrating tissue around Dacron fibers. Tissue penetrates into the hydrogel. Small areas of hydrogel are encapsulated. The inflammatory cell count is higher for these samples compared with the radiation initiated polymerization within the Dacron fabric. The IPN system is more open than the radiation initiated polymerization system and therefore ingrowth occurs more rapidly. Since the hydrogel is more susceptible to fragmentation in the IPN case, there is an early encapsulation of hydrogel, and thus a high inflammatory cell count is observed continuously through the 90-day implant period.

Conclusions

While many approaches have been used in the attempt to define the soft tissue response to hydrogel implants (9,10,11,12), the resultant data have not conclusively demonstrated the roles of certain parameters such as water content and porosity. Barvic, et al. (12) defined three categories of response dependent on porosity (which was a function of water content of starting material). Sprincl, et al. (9) found that higher porosity produces narrower capsules. Gilding, et al. (11) report no ingrowth for samples with <90% water in the final gel.

In an attempt to investigate the range of responses of soft tissue to hydrogel implants, we have used a protocol (4) which satisfactorily demonstrated a range of degrees of encapsulation and ingrowth as shown in Table 1. The study confirmed that no chronic inflammation or other path-

ological condition that would exclude the materials from biomedical applications resulted.

The time and nature of the healing process that occurs after implantation of specific polymer materials can be affected by the choice of material substrate and hydrogel formulation. Implant studies over a 90-day period for Dacron and for silicone rubber substrates coated or impregnated with hydrogel suggest that several major parameters can be varied to meet specific site and design requirements: degree of ingrowth, thickness of capsule, and inflammation. These parameters may be controlled according to fabrication conditions.

Acknowledgments

This work was supported by a Grant-in-Aid from the American Heart Association and with funds contributed in part by the American Heart Association, Iowa Affiliate. The authors thank T. J. Cipriano, G. M. Riley, and J. L. Amenson for valuable discussions and technical assistance.

References

1. Ratner, D. B., and A. S. Hoffman, "Synthetic Hydrogels for Biomedical Applications," in Hydrogels for Medical and Related Applications, American Chemical Society Series No. 31, ed. by J. D. Andrade, A.C.S., Washington, D.C., pp. 1-36, 1976.
2. Hoffman, A. S., T. A. Horbett, and B. D. Ratner, "Interactions of Blood Components at Hydrogel Interfaces," Annals of the New York Academy of Sciences, Vol. 283, pp. 372-382, 1977.
3. Predecki, P., "A Method for Hydron Impregnation of Silicone Rubber," J. Biomedical Materials Research, 8, pp. 487-489, 1974.
4. Coleman, D. L., R. N. King, and J. D. Andrade, "The Foreign Body Reaction: An Experimental Protocol," J. Biomedical Materials Research Symposium, 5, pp. 65-76, 1974.
5. Gomori, G., "A Rapid One-Step Trichrome Stain," American Journal of Clinical Pathology, 20, pp. 661-664, 1950.
6. Rosen, J. J., D. F. Gibbons, and L. A. Culp, "Fibrous Capsule Formation and Fibroblasts Interactions at Charged Hydrogel Interfaces," in Hydrogels for Medical and Related Applications, American Chemical Society Series No. 31, ed. by J. D. Andrade, A.C.S., Washington, D.C., pp. 329-343, 1976.
7. Greer, R. T., B. H. Vale, and R. L. Knoll, "Hydrogel Coatings and Impregnations in Silastic, Dacron, and Polyethylene," in Scanning Electron Microscopy/1978/I, O. Johari (ed.), pub. by SEM Inc., AMF O'Hare, IL, 60666, pp. 633-642.
8. Ratner, B. D., A. S. Hoffman, "Radiation grafted hydrogels on silicone rubber as new biomaterials," in Biomedical Applications of Polymers, ed. by H. P. Gregor, Plenum Publishing Corporation, New York, N.Y., pp. 159-171, 1975.
9. Sprincl, L., J. Kopecek, D. Lim, "Effect of Porosity of Heterogeneous Poly (Glycol Monomethacrylate) Gels on the Healing-In of Test Implants," J. Biomedical Materials Research, 4, pp. 447-458, 1971.

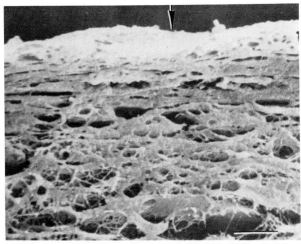

Figure 9. SEM of capsule around 20% HEMA, 1.5% EGDM, radiation grafted onto silicone rubber at 10 days implantation (scale bar = 5μm). The implant was in contact with upper surface (arrow); tissue side view. 10 KeV. 30° tilt.

Figure 10. SEM of bulk hydrogel polymer composed of 20% HEMA, 1.5% EGDM--radiation initiated polymerization (scale bar = 10μm). 10 KeV. 10° tilt.

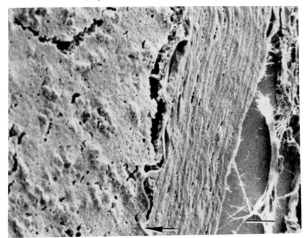

Figure 11. SEM of capsule around bulk polymer composed of 20% HEMA, 1.5% EGDM--radiation initiated polymerization (scale bar = 10μm). Arrow indicates the interface between bulk hydrogel and collagen. 10 KeV. 10° tilt.

Figure 12. SEM of 20% HEMA, 1.5% EGDM--radiation initiated polymerization into the Dacron velour network before implantation (scale bar = 50μm). 10 KeV. 30° tilt.

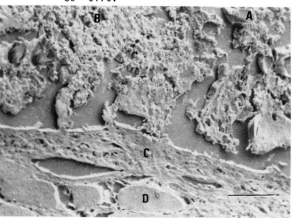

Figure 13. SEM of 20% HEMA, 1.5% EGDM--radiation initiated polymerization into the Dacron velour network at 30 days implantation (scale bar = 50μm). Example features shown as: A = Dacron, B = hydrogel, C = collagen, and D = tissue. 5 KeV. 40° tilt.

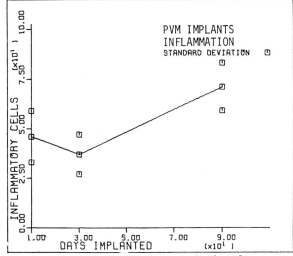

Figure 14. Inflammatory cell population for implantation of 20% HEMA, 1.5% EGDM--radiation initiated polymerization of hydrogel into Dacron velour.

10. Sprincl, L., J. Vacik, J. Kopecek, "Biological Tolerance of Ionogenic Hydrophilic Gels," J. Biomedical Materials Research, 7, pp. 123-136, 1973.
11. Gilding, D. K., G. F. Green, D. Annis, J. G. Wilson, "Soft Tissue Ingrowth into Hydrogels," Trans. American Society for Artificial Internal Organs, 24, pp. 411-414, 1978.
12. Barvic, M., Kliment, M. Zavadil, "Biologic Properties and Possible Uses of Polymer-like Sponges," J. Biomedical Materials Research, 1, pp. 313-323, 1967.

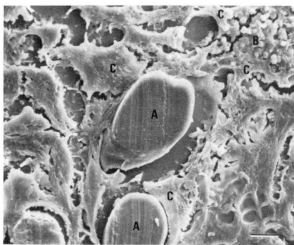

Figure 15. SEM of 20% HEMA, 1.5% EGDM, IPN in Dacron velour sample at 30 days (scale bar = 10μm). Features are designated as: A = Dacron, B = hydrogel fragment, and C = collagen. 10 KeV. 30° tilt.

Discussion with Reviewers

C. C. Haudenschild: Is there a possible release of toxic material (e.g. methanol) incorporated into the hydrogels during fabrication, which could be responsible for some of the tissue response?
Authors: With both the radiation and chemical grafting methods, the bulk hydrogel polymer and the substrates with residual hydrogel coatings were rinsed in an ethanol/water (50/50 V.%) solution (3 rinses over 24 hours) to remove any unreacted monomer, crosslinker and solvent. Samples for graft and H_2O percent determinations were stored in distilled water. Samples for implantation were stored at least 24 hours in sterile physiological saline.

C. C. Haudenschild: How are the hydrogel-impregnated implants sterilized?
Authors: Prior to implantation, 5mm disks of the test materials in physiological saline were steam autoclaved at 121°C, 15psi for 20 minutes.

C. C. Haudenschild: For practical purposes, what technique would you choose if an implant with maximum anchorage, i.e. best collagen ingrowth, is desired?
Authors: The Dacron fabric prepared by the chemical or the radiation grafting methods allowed for the maximum amount of collagen ingrowth (and thus anchorage) compared with the other materials tested in this study.

C. C. Haudenschild: How strong is the adhesion between the hydrogel and the Silastic or Dacron respectively, as compared with the adhesion between uncoated Dacron fibers and collagen?
Authors: We did not measure the adhesion properties.

B. D. Ratner: How does the healing response to the hydrogel impregnated Dacron velour compare to the healing response for non-impregnated Dacron velour materials as studied either by yourselves or by others?
Authors: In studies of both subcutaneous and intramuscular implants in rabbits, non-impregnated Dacron materials have been observed to illicit a dense fibrous capsule and some chronic inflammation (Lee, H. and K. Neville, Handbook of Biomedical Plastics, Pasadena Technology Press, Pasadena, CA, p. 14-9, 1971). Davila and coworkers (Davila, J. C., E. V. Lautsch, and T. E. Palmer, "Some Physical Factors Affecting the Acceptance of Synthetic Materials as Tissue Implants," Ann. N.Y. Acad. Sci., 146, pp. 138-147, 1968) have found that Dacron felt implants which are 13-1/2 months old show a loosely fibrous (collagen) organization of a capsule around each Dacron fiber. A few nuclei of fibroblasts are seen in close proximity to the Dacron fibers. Arterioles, venules and capillaries invaded the interstices of the felt. Arterioles no larger than the plastic fibers are abundant and in close proximity to clumps of fibroblasts which form and maintain the collagenous framework in which the Dacron is enmeshed.

In a similar fashion, the encapsulating tissue of the hydrogel-Dacron composite materials was found to be a loosely fibrous organization with fibroblasts surrounding both Dacron fibers and small areas of hydrogel. The depth of fibroblast and collagen penetration was found to depend on grafting technique. The IPN chemical grafted hydrogel-Dacron composite had considerable ingrowth at 10 days (∿ 700 μm) compared with the radiation grafted material which had ingrowth of approximately 120μm. The IPN system was more macroporous and more susceptible to fragmentation of the hydrogel than the radiation grafted system. This permits significant ingrowth and may account for the increase in inflammatory cell count. Some vascularization of 90-day implants was noted especially in the hydrogel-Dacron composite materials prepared by the IPN method which had evidence of small vessels at 10 days within the fibrous capsule layer. No serous exudate nor calcification was evident in any of the implant materials of this study.

For additional discussion see page 634.

SCANNING ELECTRON MICROSCOPY/1979/II
SEM Inc., AMF O'Hare, IL 60666, USA

PRACTICAL EXPERIENCES WITH SCANNING ELECTRON MICROSCOPY IN A
FORENSIC SCIENCE LABORATORY

J. Andrasko, S. Bendtz and A. C. Maehly

The National Laboratory of Forensic Science
Fack,
S-581 01 Linköping, SWEDEN

Abstract

The practical experiences of four years'
casework with scanning electron microscopy (SEM)
are reviewed. This review is based on the analysis of close to 2200 items of evidence. The types
of crime, the kind of material and the techniques
used are discussed. There is a trend of expanding
SEM techniques to investigations traditionally
carried out by other methods. The importance of
correct evidence collection is stressed.

Introduction

In 1974, the Swedish National Laboratory
of Forensic Science (SNLFS) acquired a scanning
electron microscope (SEM). At that time several
forensic science laboratories had access to the
SEM, however, only a few laboratories had an
instrument of their own[x].

Several articles on the application of SEM
methods to forensic science have appeared since
1970, some of them having the character of reviews. The first publications dealt exclusively
with topological applications[1-6], but later the
combination of microphotography and chemical
analysis by induced X-ray excitation was also
evaluated. A rather extensive bibliography on
SEM applications to forensic science has been
compiled by Stewart[7], it contains 63 references[xx].

The majority of publications deal with interesting single items or with groups of material investigated (paint flakes, tool marks,
glass, hair etc.). Only a few authors discuss
the routine use of the SEM in forensic science[8-13]. It was felt therefore that the experiences at our laboratory could be useful in assessing the role of the SEM in casework.

The first months after receiving the SEM
were used for familiarization and especially
for testing energy dispersive X-ray equipment
from several manufacturers. Also, an attachment
for measuring cathodoluminescence was constructed along the lines developed by Carlsson and
van Essen[14]. In 1975, the laboratory was moved
to the university town of Linköping. From August
of that year, both research work and casework
were carried out by the SEM.

---

x)
To our knowledge the Metropolitan Police
Forensic Science Laboratory (London), the
West Midlands FSL (Birmingham), the National
Research Institute of Police Science (Tokyo),
the Department of the Chief Medical Examiner
(Los Angeles), and Bundeskriminalamt (Wiesba-
den).

xx)Unfortunately the 1975 bibliography contained
a number of errors and an updated bibliography
with 84 references appeared in SEM/1976/I, p. 739-
743. See also, R.L.Taylor, SEM/1979/II, p.209-216.
                                        -- Editor.

KEY WORDS: Forensic Applications, X-ray Microanalysis, Forensic Science Laboratory

It is not the purpose of this paper to report the results of our research work which have been published elsewhere[15-22]. Rather, we should like to comment on the role played by the SEM in routine casework.

The two main functions of the SEM are the production of microphotographs with a great depth of focus (topography), and the elemental analysis by electron excited X-ray fluorescence, either by wavelength dispersive or energy dispersive techniques. In forensic science, the analytical mode is predominant and has a wide range of applications.

In 1976, there was a change of policy regarding the use of the SEM at the SNLFS for casework. Prior to this year, the instrument was used only when and if other techniques failed to give adequate results. In particular, chemical analyses were performed by the SEM when atomic absorption, flame photometry, wet chemistry or - particularly - emission spectroscopy failed. It was felt that the SEM was something of a luxury instrument, and also that research and development of SEM applications would be hampered by too heavy case loads. Reevaluation showed that SEM analysis was usually faster and less expensive than many other techniques. As a result, the number of cases handled by the SEM has increased substantially (cf Table 1).

Table 1

The number of routine cases handled by SEM from 1st July 1975 to the 31st of December 1978.

Sixth-month period	Number of cases
1975, 2nd half	40
1976, 1st "	66
1976, 2nd "	87
1977, 1st "	132
1977, 2nd "	125
1978, 1st "	109
1978, 2nd "	113
Total	672

With this case load it was found that two operators are needed for an efficient use of the instrument. One of these is mainly concerned with method development, the other with casework. It will probably be necessary to acquire a second instrument in the early 1980's.

Methods

The instrument used is a Jeol JSM-35, equipped with a Princeton Gamma Tech energy dispersive X-ray analyzer (PGT-1000) connected to a "Servogor" x/y recorder. In the first year of the SEM's operation, an energy dispersive X-ray

analyzer (Kevex-Ray System 5000A) was used. Coatings were done in a Jeol vacuum evaporator JEE 4B.

Results

The results of four years' investigations of routine cases are shown in Tables 2 and 3. A few of these cases have been described in some detail[15-22]. In Table 2 the use of SEM methods in the investigation of different types of crimes is illustrated.

In connection with burglaries, paint chips, smears of paint on tools, and glass fragments are the most common materials analyzed. The examination of individual layers of paint chips by emission spectroscopy is only feasible in the rare cases where a physical separation prior to analysis is possible. As is well known, the SEM can easily perform layer by layer analyses. Its limitations are the sensitivity: elements at concentrations below 0.1-1% usually cannot be detected. In such cases, emission spectroscopy can be used for a whole flake, unless the laboratory has access to an electron probe (which we have not). It should be mentioned that the topographic (surface) structure of paint or plastic particles can be of evidential value; the surface can have characteristic weathering features[23], or the innermost paint layer may be a replica of the surface underneath.

In traffic accidents, including so-called hit and run cases, particles of paint are often exchanged between colliding vehicles. The same consideration as above applies to the investigation of paint flakes. The evidential value of such analyses is greatly enhanced whenever repainting and/or repairing (e.g. by "plastic padding") had been carried out.

In addition, light bulbs are frequently sent in for deciding if the filaments were on (hot) or off (cold) when the glass bulb broke. It is well documented that the large depth of focus of the SEM often gives much clearer pictures of such filaments than conventional macrophotography.

It may be less well known that SEM techniques can successfully be used in the elucidation of the causes for fires. Topographically, the appearance of defects in and damage to electrical components (wiring, contacts) can be investigated. We use electron-excited X-ray analysis of soot and distilled inflammable liquids for the detection of lead and bromine. Burned matches can be analyzed topographically and chemically[20].

In the investigation of fraud by alterations in documents, counterfeit bills, coins etc. the SEM can often play a decisive role. The non-destructive mode of operation of the X-ray analysis can here be used to its full advantage[21]. Printing inks, paper, and coin surfaces can be analyzed repeatedly. In particular, the examination of very valuable objects (rare jewellery[13], antique objects, stamps) can be performed. A great number of coins, with no connection to criminal cases, were analyzed at our laboratory in parallel with their genuine counterparts.

Previously it was costumary in Sweden to drill microcores from such coins for emission spectroscopy (the holes were filled and painted over!).

In cases of murder, manslaughter, suicide and attempts of these a great variety of materials is sent to a forensic science laboratory. Some of these materials can be successfully investigated in the SEM. We analyzed paintflakes, gunshot residues[15], mineral particles[17], metal particles, electric contacts[18], bullets, and fibres[22] in connection with crimes.

Table 2 lists also a scattering of other crimes where SEM investigations were of evidential value. In one case, a car-owner, suspected of insurance fraud, could be exonerated only due to the SEM[16]. This case along warranted the cost of acquiring and operating the scanning electron microscope.

Table 2

The number of cases investigated by SEM from 1975 to 1978 grouped according to the type of crime suspected

Type of crime	1975	1976	1977	1978	total
Burglary	11	39	45	36	131
Fires (incl arson)	5	17	41	30	93
Traffic accidents (incl hit & run)	4	21	20	26	71
Murder & manslaughter (incl attempts)	4	13	21	21	59
Bombings[x]	2	4	30	21	57
Fraud & falsifications	4	14	15	9	42
Narcotics offences	0	7	19	10	36
Property damage	3	9	9	9	30
Theft	2	7	9	12	30
Armed robbery	2	7	3	7	19
Suicide	1	3	4	3	11
Other crimes	2	7	35	30	74
Crime unknown	0	5	6	8	19
Total	40	153	257	222	672

x)Includes blasting of bank service boxes

Table 3 lists the kind of objects investigated. In common with the experience of other forensic laboratories, paint chips are the most usual items. It is important to identify the individual paint layers correctly by taking micro photographs, or - more simply - by making a sketch of the chip under a lower-power microscope. If this is done with circumspection, the two chips to be compared can be mounted side by side for analysis. As a rule no coating is needed. If a semiquantitative analysis is needed one has to analyze each layer at several locations; however, modern automobile paints have very constant compositions within one and the same coat.

Table 3

Types of material investigated by SEM from 1975 to 1978

Items	1975	1976	1977	1978	total
Chips and smears of paint	93	252	291	323	959
Glass	0	13	76	65	154
Dye-stuff	0	9	73	45	127
Metal fragments	1	16	53	28	98
Explosives	0	9	59	19	87
Coins and medals	3	9	61	13	86
Gunshot residues	0	14	40	31	85
Fibres and cloth	1	13	19	39	72
Wires and contacts	8	11	15	16	50
Soot (from fires)	0	0	29	17	46
Burnt materials	0	0	23	9	32
Toolmarks	2	2	10	9	23
Mineral particles	0	8	3	11	22
Plastics	0	6	5	9	20
Hair	0	7	5	7	19
Rubber	0	10	6	0	16
Insulating material	1	4	7	3	15
Modelling clay	0	6	0	0	6
Paper	0	0	6	0	6
Jewelry	3	2	0	0	5
Miscellaneous	15	21	50	24	110
Items of unknown nature	0	19	66	40	125
Total	127	431	897	708	2163

Glass fragments are analyzed in the SEM when the samples are found indistinguishable with respect to their physical properties (refractive index, density). As a rule a semiquantitative analysis is carried out[19]. Care must be taken in mounting and analyzing the samples to insure that sample geometry and all instrument settings are kept constant. Polishing and coating of glass fragments is generally not needed. In several cases, the analysis in the SEM could distinguish between glass samples with the same density and refractive index.

Metal fragments have the advantage of being good conductors and need no coating prior to examination. For lamp filaments, a topographic investigation is usually sufficient. Coins must be photographed for subtle differences in details due to the manufacturing process (embossing or casting). In some cases, relatively common coins are "turned into" rare coins by mechanical means. Rodgers et al.[4] have described such a case for Canadian coins. In this country, the 5 öre coins of 1910 are rare collectors items. Often coins from 1916 or 1919 are filed to make a 1910 number. Also, a variety of gold coins has been investigated such as $ 20 coins. Fig. 1 gives details of a SEM analysis of such coins.

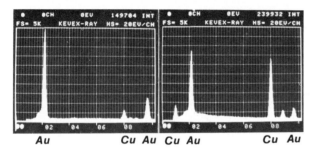

Au   Cu Au Cu Au   Cu Au

Fig. 1. Topographical details and elementary analysis of $ 20 gold coins.
    Upper part: The "0" of the text "IN GOD WE TRUST", and a part of the US coat of arms; to the left: genuine coin (made by casting). Note the fineness of detail in the genuine coin and the irregularities in the cast coins.
    Lower part: Energy dispersive X-ray analysis of the genuine coin (to the left), showing gold and a trace of copper, and of the counterfeit coin (to the right), showing approximately equal parts of gold and copper respectively.

An increasing number of cases concerning bombings and blasting of bank service boxes has occurred in Sweden during the past years. The explosives are analysed by high pressure liquid chromatography. The SEM is used to identify the inorganic constituents of the samples. As a rule, an analysis of individual compounds (i.e. crystals or other particles in the explosive mixture) is carried out.
    The detection of gunshot residues (GSR) and the analysis of glass are new methods introduced to casework in our laboratory during 1976. GSR are usually collected from the shooting hand,

cloth etc, with the help of adhesive tape[x]. The tape is coated with carbon and particles embedded in the adhesive material examined in the SEM. The detection is based on the characteristic appearance and elemental composition of GSR. So far, 85 samples of GSR have been handled by the SEM. Questionable suicides, murder/manslaughter by shooting have been the cases were the analysis for GSR is of importance.

    The enthusiasm of the SEM operator and his superiors must be curtailed by a critical attitude. It is not prudent to use the instrument in actual casework without careful consideration of the pitfalls involved. Just as with other methods introduced into forensic science, a thorough testing of new SEM-techniques is an important prerequisite. The SEM has opened up a new dimension for us which in many respects is still unfamiliar and has inherent properties of uncertainty. Many modern methods used in forensic science tend to be micro methods. But in general, the smaller the object the larger the risk of contamination or other untoward effects. The signal to noise ratio tends to decrease with decreasing size of objects and increasing sensitivity of the method. Definite difficulties enter already at the stage of evidence collection and handling. It is therefore important to inform and train the personnel in charge of scene of crime investigation in the correct assessment of the SEM technique. They have to know what new possibilities for analyses arise – and how to collect and preserve evidence of the size and nature required for a correct investigation in the SEM.

---

[x] Similar methods for the analysis of GSR were developed simultaneously, and independently by several other laboratories.

References

1. L.W. Bradford and J. Devaney, Scanning electron Microscopy Applications in Criminalistics, J. Forens. Sci., 15, 1970, 110-119.
2. H. Hantsche and W. Schwartz, Das Raster-Elektronenmikroskop als Hilfsmittel zur Identifizierung von Werkzeugspuren, Arch. Kriminol., 148, 1971, 24-32.
3. H.L. Mac Donell and L.H. Pruden, Application of the Scanning Electron Microscope to the Examination of Firearm Markings, Scanning Electron Microscopy Proc., IIT Research Inst., 1971, pp. 569-576.
4. P.G. Rodgers, J.R.G. Jakob, P. Blais and D.O. Harris, Scanning Electron Microscopy of Forged Dots on 1936 Canadian Coins, J. Forens. Sci., 16, 1971, 92-102.
5. K.A. Siegesmund and G.M. Hunter, Scanning Electron Microscopy of Selected Crime Laboratory Specimens, Scanning Electron Microscopy Proc., IIT Research Inst., 1971, pp. 577-584.
6. M.E. Taylor, Forensic Applications of Scanning Electron Microscopy, Scanning Electron Microscopy Proc., IIT Research Inst., 1971, pp. 545-552.

7. W.D. Stewart Jr., Bibliography on Forensic Applications of the SEM, Scanning Electron Microscopy Proc., IIT Research Inst., 1975 pp. 589-592.

8. J.L. Brown and J.W. Johanson, Electron Microscopy and X-ray Microanalysis in Forensic Science, J.A.O.A.C., 56, 1973, 930-943.

9. C. van Essen, The Scanning Electron Microscope in Forensic Science, Physics in Technology, 5, 1975, 234-243.

10. R. Keeley, The Economics of SEM and Electron Probe Microanalysis in Forensic Science, Int. Micro. J. Leg. Med., No. 131, 1976.

11. R.H. Keeley and M.C. Robeson, The Routine Use of SEM and Electron Probe Microanalysis in Forensic Science, Scanning Electron Microscopy Proc., IIT Research Inst., 1975, pp. 479-486.

12. V. Schneider, Die Röntgenmikroanalyse am Rasterelektronenmikroskop - ein Routineverfahren im Rahmen kriminaltechnischer Untersuchungen, Forens. Sci., 7, 1976, 223-224.

13. R.L. Williams, An Evaluation of the SEM with X-ray Microanalyzer Accessory for Forensic Work, Scanning Electron Microscopy Proc., IIT Research Inst., 1971, pp. 537-544.

14. L. Carlsson and C.G. van Essen, An Efficient Apparatus for Studying Cathodoluminescence in the Scanning Electron Microscope, J. Phys. E: Sci. Instrum., 7, 1974, 98-100.

15. L. Carlsson, A.C. Maehly and E. Rudh, Tötungsversuch mit elektrischem Strom. Eine kriminaltechnische Untersuchung, Arch. Kriminol., 157, 1976, 30-36.

16. L. Carlsson and A.C. Maehly, Autoeinbruch oder Versicherungsbetrug? Arch. Kriminol., 157, 1976, 107-110.

17. A.C. Maehly and L. Rammer, Zinkblende unter der Vorhaut. Tatortspur bei einem Sexualmord, Arch. Kriminol., 159, 1977, 139-143.

18. J. Andrasko and A.C. Maehly, Detection of Gunshot Residues on Hands by Scanning Electron Microscopy, J. Forens. Sci., 22, 1977, 279-287.

19. J. Andrasko and A.C. Maehly, The Discrimination Between Samples of Window Glass By Combining Physical and Chemical Techniques, J. Forens. Sci., 23, 1978, 250-262.

20. J. Andrasko, The Identification of Burnt Matches, J. Forens. Sci., 23, 1978, 637-642.

21. J. Andrasko, P. Kylsäter, A.C. Maehly and M. Knuuttila, Analysis of Counterfeit Gold Coins by Scanning Electron Microscopy and X-ray Diffraction, in Scanning Electron Microscopy/1979/I, SEM Inc., AMF O'Hare, IL; this publication.

22. A.C. Maehly and M.-A. von Wachenfelt, Textilfasern unter der Vorhaut - kriminaltechnischer Beweis bei einem Sexualverbrechen, Z. Rechtsmed. 82, 1979, 237-242.

23. R. Wilson, G. Judd and S. Ferriss, Characterization of Paint Fragments by Combined Topographical and Chemical Electron Optics Techniques, Twenty-fifth Annual Meeting, Academy of Forensic Sciences, Las Vegas, Nevada, USA, 1973, 30. See also G. Judd, J. Sabo and S. Ferris, SEM Applications under the FORSEM Project, in Scanning Electron Microscopy/1975, p. 487-492.

DISCUSSION WITH REVIEWERS

J.L. Abraham: In your table you only list the number of cases of different types analysed through 1977. Would you please update this through 1978?

Authors: In the final version of this paper, the results have been updated. As a consequence of this the numbers in the reviewers' questions have been adjusted.

R.L. Taylor: Of the 2163 examinations reported in Table 3, how many gave inconclusive results?

Authors: In about 9-10 % of the examinations reported in Table 3 the results obtained were inconclusive.

V. R. Matricardi Of the examinations run with the SEM/EDX would you know what percentage used imaging only, analysis only and both?

Authors: Totally, in 70 % of the examinations X-ray analysis only, in 8 % imaging only, and in 22 % both analysis and imaging respectively were used. However, in most of the examinations involving X-ray analysis only, imaging without photography is used for determining which part of the material is to be analysed.

R.L. Taylor: Of the 672 cases handled by SEM from July 1975 through December 1978, how many resulted in courtroom testimony?

Authors: Unfortunately we get rarely any information on the handling of the cases in the courts.

V.R. Matricardi: What criteria are used in determining if a piece of physical evidence is to be examined with the SEM/EDX? Administratively, how is this done in your laboratory?

Authors: Our SEM/EDX equipment is situated in the same building as all other laboratory divisions (chemistry, biology, document examination, firearms, arson and general physical sections). The instrument serves all of these divisions. The forensic scientist in charge of a given case contacts the SEM operator directly for a discussion if a given material should be examined by SEM.

M.A. Clark: Do the Swedish courts accept in testimony evidence obtained through SEM materials analysis?

Authors: We have no information to the contrary.

R.L. Taylor: Of the different types of examinations listed in Table 3, which are the most cost-effective and which the least?

If your work load was such that you had to reject half of the requests you receive, which examinations would you accept and which would you reject?

Authors: If the equipment available at our laboratory is considered, the most cost-effective types of SEM investigation were:

chips and smears of paint,
coins and medals,
gunshot residues,
toolmarks,

items of unknown nature.

The first four of the examples mentioned above would be given priority if the work load should increase.

V.R. Matricardi: Of the examinations requiring analysis by SEM/EDX, would you care to comment on what percentage relied on other analytical techniques to provide added information (Quantitative results, better minimum detection limit, etc.)?
Authors: In more than 80 % of the examinations requiring analysis by SEM/EDX other analytical techniques were used as well to provide additional information. The policy of our laboratory is to use several independent methods of examination whenever this is possible.

J.L. Abraham: Would you please give some data on the evaluation you mentioned showing that SEM was usually faster and less expensive than many other techniques?
M.A. Clark: Have you found the use of the SEM to be less costly than the conventional techniques of evidence analysis?
Authors: A typical analysis of an unknown sample by X-ray analysis in the SEM requires only few minutes including sample handling.

Similar analyses may be carried out using other techniques. We shall comment on some of them:

emission spectrography – time consuming
atomic absorption spectrophotometry – analyzes usually only one element at a time
spark source mass spectrometry and
neutron activation analysis – expensive equipment; the analysis is time consuming
polarography – not so suitable for complex mixtures; analyzes only solutions
X-ray fluorescence spectroscopy – a rapid technique.
Of the instruments mentioned above, our laboratory possesses only an emission spectrograph and has access to an atomic absorption spectrophotometer.

R.L. Taylor: With regard to the examination illustrated in Figure 1, isn't it true that the X-ray analysis provided all the evidence needed to prove the forgery and that the topographic examination and resulting photomicrographs are superfluous?
Authors: It is always valuable in forensic science to have several independent methods. Also, the loss of time was negligible.

K.F.J. Heinrich: Which characteristic features are observable in gunshot residues?
Authors: GSR particles appear to have a characteristic morphology (suggestive of melted and rapidly cooled-down material). These particles, embedded in the adhesive tape, show very high brightness when observed by SEM, particularly with backscattered electron imaging. The presence of lead, barium and antimony in one and the same particle confirms the identification of GSR.

J.L. Abraham: In examining paint chips, do you analyse individual particles visualized within the paint, such as easily discerned with backscattered electron imaging, or do you only analyse the paint at low magnification?
Authors: In examining paint chips, we analyse several regions of each layer of paint at low magnification. Individual particles are not analysed.

J.L. Abraham: You mentioned the use of a cathodoluminescence detector. Would you please give some examples of the usefulness of this detector and perhaps compare it to your experiences with the use of backscattered electron imaging as well?
Authors: A cathodoluminescence detector was constructed in our laboratory. So far, we have not used this equipment, but we hope that it will be useful in some investigations - i.e. comparison of glass samples, fibres etc.

J.L. Abraham: In Table 3 the list of material investigated does not include tissue samples. Do you not find the SEM approach useful in this important part of forensic sciences as well? If so please give a few examples of your work in this area.
M.A. Clark: Have you analysed any biologic specimens, e.g. wounds, and used the evidence in court?
Authors: Wounds and tissues are not examined in our laboratory. The examination of other biological material (hair, fibres, wood) has occasionally been performed by SEM.

D.M. Lucas: What useful information is gained by examining dye-stuff and hair in the SEM?
Authors: In 1977 numerous bank service boxes were robbed after blasting them. As a consequence, the banks added dye-stuff containers to the customers individual bags. Our laboratory had to identify spots from these dyes on bank bills (in about 200 cases). Some of these were examined in the SEM.

Investigation of hair can give information in exceptional cases, such as physical damage or surface treatments.

J.L. Abraham: Would you please illustrate the appearance of the filaments of light bulbs broken when they were on or off? If the bulb is broken shortly after being turned off, how could you differentiate this from one broken while it was on? How much time is the usual period of uncertainty of this determination?
Authors: At present we have no data on this subject.

SCANNING ELECTRON MICROSCOPY/1979/II
SEM Inc., AMF O'Hare, IL 60666, USA

SEM AND X-RAY MICROANALYSIS OF ARTIFACTS RETRIEVED FROM MARINE
ARCHAEOLOGICAL EXCAVATIONS IN NEWFOUNDLAND

V. C. Barber

Department of Biology
Memorial Univ. of Newfoundland
St. John's, Newfoundland
Canada, A1B 3X9

and

Newfoundland Marine Archaeology Society
P.O. Box 181, Station "C"
St. John's, Newfoundland
Canada A1C 5J2

Abstract

A wide selection of artifacts have been recovered in the course of the excavations of three archaeologically important shipwreck sites in Newfoundland. Prior to the conservation of these artifacts, and in attempts to assist in the identification of one of the shipwrecks, SEM and X-ray microanalytical studies were undertaken. Both a non-dispersive X-ray analyser attached to a routine SEM, and a microprobe analyser, were used to obtain analytical data on the elemental composition of certain metal artifacts. Such data were used to ensure that the proper conservation processes were undertaken on these artifacts. In addition, the SEM was used in the identification of the type of wood that formed the remains of the hull of one of the sunken vessels. The SEM and X-ray equipment was therefore used to provide a service facility for conservators, and also certain other items of useful information to assist an ongoing marine archaeological program.

KEY WORDS: Marine Archaeology, X-ray Microanalysis, Wood Structure, Metal Identification

Introduction

Since the formation of the Newfoundland Marine Archaeology Society (NMAS) in 1972, three excavations of marine archaeological sites have been undertaken by the group (Fig. 1). The latest excavation undertaken by the NMAS was in 1978 at Conche, on the site of the Marguerite, a 22 gun St Malo vessel sunk in action with the British in 1707[1]. The area dug in this trial excavation was $4m^2$, and the artifacts recovered were of the appropriate dating period. The second site worked was that of a so far unidentified shipwreck at Trinity. The artifacts retrieved in a surface collection made in 1977 suggested that the vessel dated from the mid-1700's. An excavation in 1978 of $19m^2$ of the site allowed the retrieval of over 400 artifacts. These confirmed the dating of the shipwreck. The initial site worked on was that of HMS Sapphire, a fifth rate, 32 gun, Royal Navy frigate, sunk by the French in Bay Bulls in 1696. An initial survey was followed by a trial excavation that was undertaken in 1974[2,3].

Various projects that could be assisted in their solution by certain SEM or X-ray microanalytical studies arose as a result of these operations. The present publication outlines the studies that were done by the present author and his collaborators in these projects.

Energy-dispersive Analysis of Metal Artifacts Retrieved from the Wreck of HMS Sapphire

Introduction

As a result of the trial excavation of 1974 over 300 artifacts were retrieved from the wreck of the Sapphire. At the time of writing this article only one other similar vessel has been extensively investigated, that of HMS Dartmouth, wrecked off Mull, Scotland, in 1690 [4-6]. So it can be realized that our knowledge of the implements carried on board such vessels, and also our knowledge of the construction of these ships, are still limited.

Methods used

Many metal artifacts were retrieved from the Sapphire, and it was essential in some cases that these were rapidly identified so that the appropriate conservation methods could be used. To enable this identification to be made, small samples of certain of the metal artifacts were taken prior to conservation. The surface corrosion of a small area of each artifact was removed from the object. Several very small samples of each item were then taken. These were stuck to aluminum viewing stubs with epoxy cement. To reduce "charging" the specimens were then coated with 99.9% pure gold in an Edwards Vacuum Coating Unit (Model E12E), equipped with a planetary rotation accessory. The coated specimens were then viewed in a Cambridge Stereoscan Mk. 2A scanning electron microscope operating at an accelerating voltage of 20kV or 28kV. The emitted X-rays were detected with a Kevex Corporation Model 3201 Mark AA Non Dispersive X-ray spectroscopy system with a detection area of 80mm^2 and a guaranteed resolution of 178 ev. The results of the analyses were printed graphically by a Tektronic 4015 plotter.

Results

So that the results of the analyses of the artifacts could be properly evaluated, several analyses were made of the composition of the epoxy cement in which they had been mounted (Fig. 2). The main elements detected in these analyses were silicon and phosphorous from the cement, gold from the coating process, and iron (presumably from extraneous X-rays emitted from the SEM column).

The three artifacts chosen to be presented in this article will be examined separately.

Figure 3 shows an unidentified object (part of a candlestick?), and Figure 4 presents the X-ray analysis of it. The major additional elements detected in the analysis were calcium and chlorine, from the sea water in which the object had been stored while awaiting conservation (and also possibly from remnants of encrusting calcareous seaweed), aluminum (presumably from the viewing stub), tin, lead, a small amount of copper, and iron.

A small bell (dinner bell?) is shown in Figure 5, and the results of the X-ray analysis of the object are presented in Figure 6. The major elements found in addition to those present in the epoxy cement, the gold coating, and the initial sea water, were tin, copper, and iron. Note that only the first half of the analysis was made as

opposed to the previously described artifact. The reasons for this will be commented upon in the discussion.

The last object to be discussed is a medicinal syringe (Fig. 7). The use of such items for the treatment of venereal disease on board ships of the period is well documented[7], and a similar but better preserved syringe, was found on the wreck of the Dartmouth (R. Holman, personal communication). The X-ray analysis of the syringe is shown in Figure 8. The major additional element found was tin. Only the first half of the analysis was made.

Discussion

When presented with an artifact it is generally possible with experience to decide of what metal it is made. However, appearances alone can sometimes be deceptive, particularly with artifacts that have iron compounds encrusted on them. So a method of analysis that requires only a very small sample, and hence does not damage the artifact, and is rapid, can be of use. The major metals that are likely to be encountered in addition to iron, lead, copper, silver, and gold, which can generally be visually identified, are bronzes (copper and tin), brass (copper and zinc), pewter (tin and Lead), and Brittania metal (tin, plus antimony, with some copper)[8-10]. The proportions of these metals in the various alloys are variable, and in addition to the metals indicated, several other elements are also generally present in small amounts.

It can be seen from this information that the composition of the artifacts presented in this publication can easily be determined. The unidentified artifact is made of a tin/lead alloy, although some copper is present. The artifact is therefore made of pewter. The dinner bell contains a copper/tin alloy, and hence is bronze, while the syringe contains tin, and no copper. In this latter case the presence of tin, and the absence of copper suggested that the metal was pewter. However, as the second half of the energy analysis was not run, the presence of lead was not determined. Clearly, no detailed information was obtained on the quantitative composition of the artifacts. However, this was relatively unimportant and the information that was acquired was sufficient to enable the appropriate conservation measures to be undertaken.

Microprobe Analysis of Artifacts

Introduction

Since 1976 a Microprobe Microanalyser has become freely available in the author's university, and the energy dispersive unit is not presently operational. With the appropriate control

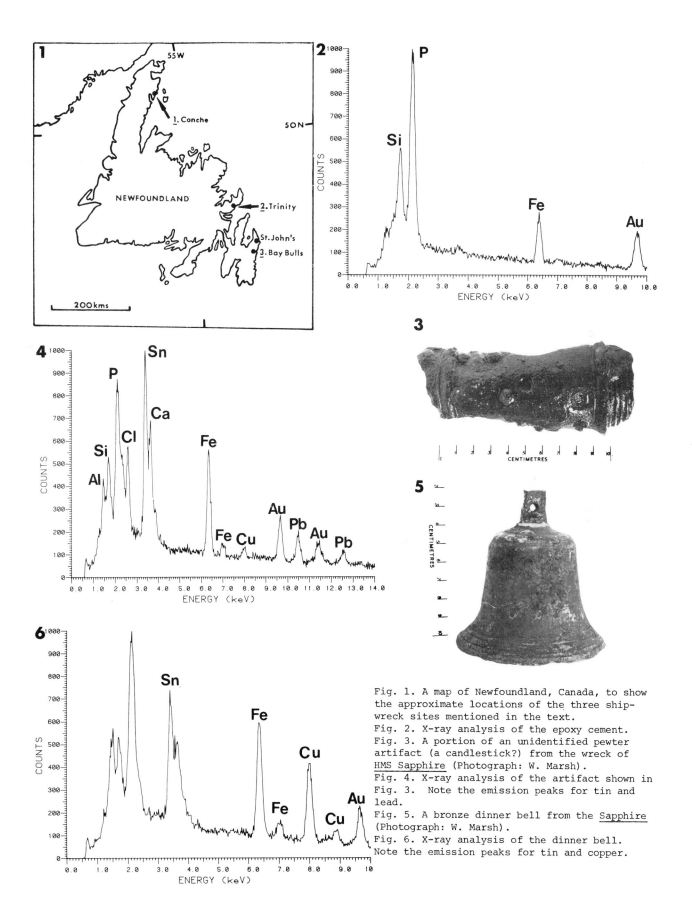

Fig. 1. A map of Newfoundland, Canada, to show the approximate locations of the three shipwreck sites mentioned in the text.

Fig. 2. X-ray analysis of the epoxy cement.

Fig. 3. A portion of an unidentified pewter artifact (a candlestick?) from the wreck of HMS Sapphire (Photograph: W. Marsh).

Fig. 4. X-ray analysis of the artifact shown in Fig. 3. Note the emission peaks for tin and lead.

Fig. 5. A bronze dinner bell from the Sapphire (Photograph: W. Marsh).

Fig. 6. X-ray analysis of the dinner bell. Note the emission peaks for tin and copper.

programs, analysis of the elemental composition of artifacts by the use of this microprobe equipment can rapidly be made, and semi-quantitative results can if required be provided by comparison with standard data. Such studies have been made on several artifacts from the Trinity wreck site.

Methods

Small samples of the artifacts were mounted on glass cover slips with Lepages epoxy cement, were carbon coated in a vacuum coating unit, and were examined in a JEOL JXA 50A Microprobe Analyzer while operating with an accelerating voltage of 15kV.

Results and Discussion

The results from one analysis will be briefly presented. The item is a dated seal (Fig. 9). This important artifact has a date on it of 1738, which gives the earliest year when the shipwreck at Trinity could have taken place. The blackened corroded exterior made it difficult to decide whether the artifact was composed mainly of copper (likely) or silver. Analysis of the item showed the presence of copper (average of two K_α readings of 344 counts/second, compared with 1,344 for a pure copper standard), and of sulphur (average of two K_α readings of 452 counts/second). Negligible amounts of silver, tin, zinc, and lead, were detected. Some iron was present, which is presumed to be a contaminant. The recorded material shows that the sample taken was composed mainly of copper sulphide, a normal corrosion product of copper in an anaerobic silt in a wreck environment.

SEM Examination of Wood from the Trinity Wreck Site

Introduction

It is important to determine the identity of the Trinity shipwreck so that a dating of the loss of the vessel, and hence the date of the "time capsule" collection of artifacts can be found. An extensive examination of contemporary records presented two vessels that could be the wreck. These records showed that, "both foundered by Means of the Ice, in Sight of Trinity, Newfoundland, where they were bound", in 1781[11]. This date is at the upper end of the dating of the site but would still be acceptable. The vessels were the Betsey, of 160 tons, known to have been built in Newfoundland[12], and the Speedwell. The fact that the Betsey was built in Newfoundland would have meant that the vessel was probably constructed from softwoods, as this was generally the case with "Canadian" built vessels[13]. If built in England the Speedwell would most likely have been built of oak[13]. An examination of

samples of wood taken from the hull of the vessel by light microscopy and SEM would allow the identification of the wood, and hence a decision as to which vessel could be the one under investigation. In addition, observations of wood that had been submerged for at least 200 years under water would be of interest.

Methods

Samples were taken by divers at several locations on the hull of the vessel, namely from the keelson, the frames (ribs), the ceiling, unidentified planking, and the outer planking. Some of these locations had been exposed by excavation. The samples were stored in sea water until they could be transported to the conservation laboratory. Comparison samples were also available from the wrecks of the Sapphire and the Marguerite but these will not be considered further. Samples were sawn off from wood taken from the various locations indicated, were washed extensively in distilled water, and were trimmed to size by the use of sharp razor blades. It might be mentioned that it is usually obvious in such specimens as to the main axes of the wood, and so appropriate orientation is relatively easy. The specimens were air dried from water, were mounted on viewing stubs with silver conducting glue, and were then coated with gold in a vacuum coating unit (see earlier). The specimens were viewed in a Cambridge Stereoscan Mk. 2A operating at accelerating voltages of from 3 - 10kV.

Results and Discussion

Various planes of a sample of wood from the Trinity wreck can be seen in Figure 10. The wood can be seen to be from a tree that was ring porous (vessels of the early wood being considerably larger than those found later in the season)(Fig. 11). The wood can also be seen to have large rays (Figs. 11 & 12). These features show the wood to be an oak. The vessels of the specimen are relatively circular in transverse section and contain tyloses (Figs. 11 & 12), showing it to be a white oak[14-16].

There are three ways that wood is likely to be degraded; by physical changes in the environment (e.g. temperature changes), by biological action of bacteria, fungi, and larger wood borers, or by chemical action[17-19]. All samples from the wreck are of white oak, are well preserved, and showed no obvious degradative changes. This has been the general experience with wood retrieved from Newfoundland waters, and it is presumed that the cold sea water temperatures that prevail for much of the year, and the fact that at many wreck sites the remaining wood is covered by the substratum, helps to slow degradation.

Authorities say that it is not

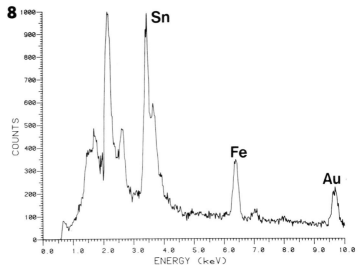

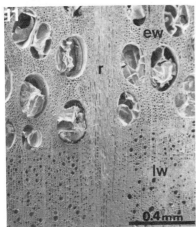

Fig. 7. A pewter medicinal syringe from the Sapphire. The bumps on the artifact are the sites of tin oxide corrosion (Photograph: W. Marsh).

Fig. 8. X-ray analysis of the syringe. Note the emission peak for tin.

Fig. 9. A dated copper seal from the Trinity shipwreck site. Note the date of 1738 (arrow). This is one half (the indented half) of a two part seal (Photograph: R. Ficken).

Fig. 10. A stereoscopic SEM view of a wood sample from the Trinity wreck. The three planes of the wood are indicated (R, radial; T, tangential; X, transverse or cross section).

Fig. 11. A cross-sectional view of a wood sample from the Trinity shipwreck. A ray (r) is clearly visible and passes through the early wood (ew) to the late wood (lw). Note the disparate sizes of the vessels in the early and late wood.

Fig. 12. A higher magnification stereoscopic view of Fig. 10 to show a ray (r), and tyloses (t) that are present in a large vessel of the early wood.

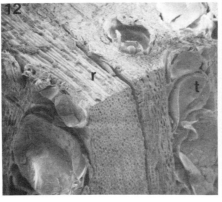

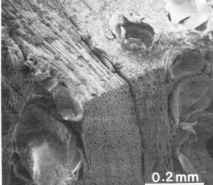

possible to distinguish the two species of English, or European oaks (Quercus robur L., and Q. petraea Liebl) from the so-called true white oak (Q. alba L) of North America[14]. However, the fact that the vessel was built of oak suggests that it was not the Betsey, which is known to have been built in Newfoundland. Therefore it is thought that the vessel was probably built of English oak (Q. robur), and that the vessel might be the Speedwell. Further literature searching is underway to exclude the possibility that so far undiscovered losses occurred at Trinity.

Conclusions

It is obvious from the studies reported here that access to SEM and X-ray analytical equipment can rapidly provide useful additional information when investigating shipwrecks of archaeological interest. Such equipment is not available in most conservation laboratories, nor do many marine archaeologists have easy access to these items, and so this emphasizes the importance of collaboration between "on site" investigators, the conservator, the historical researcher, and scientists in other disciplines. The present author is fortunate in having access to such facilities, and colleagues who have been willing to assist in such studies.

There have been several extensive studies of the composition of artifacts [8-10]. However, there has been little work done on the corrosion of such artifacts after long periods of immersion in sea water. There is clearly an area of investigation here for collaborative research. Another project that would seem worth doing is to use the dated collections of wood from shipwreck sites to extend our investigations of wood degradation. Obviously there are many worthwhile projects that could be undertaken on the retrieved artifacts or on the original shipwreck sites.

Acknowledgements

I would like to thank the following colleagues from Memorial University or the Newfoundland Marine Archaeology Society for their assistance: Ms. J.M. Barber, for advice on the artifacts and on the historical documentation; Ms. C.J. Emerson, for assisting with the SEM and with the preparation of this paper; Dr. J. Molgaard and Mr. R. Smith, for assistance with the non-dispersive X-ray analyses; Mr. M. Rayment, for developing a computer program for the printing of the non-dispersive X-ray analyses; Dr. H. Longerich for assistance with the microprobe analyses; Dr. P. Dabinett, for collecting the wood samples, and for advice on the identification of the wood; and Mr. R. Ficken, and Mr. W. Marsh, for photographic assistance.

References

1. Barber, V.C., Barber, J.M., Allston, M., and Stone B. "An initial survey of a wreck, thought to be a St Malo vessel, the Marguerite, sunk in 1707 at Conche, Newfoundland". Int. J. Naut. Archaeol., 8 (1), Published, 1979.
2. Barber, V.C. "The Sapphire, a British frigate, sunk in action in Bay Bulls, Newfoundland, in 1696". Int. J. Naut. Archaeol., 6 (4), 1977, 305-313.
3. Farmer, G., and Carter, J. "The marine dig for the Sapphire at Bay Bulls". Canadian Geographic J., Feb./March, 1979, 98 (1), 62-67.
4. Adnams, J.R. "The Dartmouth, a British frigate wrecked off Mull, 1690". Int. J. Naut. Archaeol., 3, 1974, 269-274.
5. Holman, R.G. "The Dartmouth, a British frigate wrecked off Mull, 1690. 2. Culinary and related items". Int. J. Naut. Archaeol., 4, 1975, 253-265.
6. Martin, C.J.M. "The Dartmouth, a British frigate wrecked off Mull, 1690. 5. The ship". Int. J. Naut. Archaeol. 7 (1), 1978, 29-58.
7. Trotter, T. Medicina Nautica: An essay on the diseases of seamen: comprehending the history of health in His Majesty's Fleet, under the command of Richard Earle Howe, Admiral. London, 1797, 467-469.
8. Hanson, V.F., Carlson, J.H., Papouchado, K.M., and Nielson, N.A. "The Liberty bell: composition of the famous failure". Amer. Scientist, 64, 1976, 614-619.
9. Hanson, V.F. "Quantitative elemental analysis of art objects by energy-dispersive X-ray fluorescence spectroscopy". Appl. Spectrosc., 27, 1973, 309-333.
10. Carlson, J.H. "Analysis of British and American pewter by X-ray fluorescence spectroscopy", in Winterthur Portfolio, XII, Quimby, I.M.G. (ed.), University Press of Virginia, Charlottesville, 1977, 65-85.
11. Lloyds List. No. 1299, 7th September, 1781, London.
12. Lloyds Register. 1781, London.
13. Wallace, F.W. Wooden ships and iron men. The story of the square-rigged merchant marine of British North America, the ships, their builders and owners, and the men who sailed in them. Charles E. Lauriat Co., Boston, 1937.
14. Panshin, A.J., de Zeeuw, C., and Brown, H.P. Textbook of wood technology. Vol. 1. Structure, identification, uses, and properties of the commercial woods of the United States. McGraw-Hill Book

Company, New York, 1964.

15. Core, H.A., Côté, W.A. and Day, A.C. Wood structure and identification. Syracuse University Press.
16. Jane, F.W., revised by Wilson, K., and White, D.J.B. The structure of wood. Adam & Charles Black, London, 1970.
17. Borgin, K., Faix, O., and Schweers, W. "The effect of aging in lignins of wood". Wood. Sci. Technol., 9, 1975, 207-211.
18. Borgin, K., Parameswaran, N., and Liese, W. "The effect of aging on the ultrastructure of wood". Wood. Sci. Technol., 9, 1975, 87-98.
19. Jutte, S.M., and Sachs, I.B. "SEM observations of Teredo attack of tropical hardwoods in brackish water". In: SEM/1976/II. (ed. by O. Johari & R.P. Becker) 555-562, IITRI, Chicago, IL.

Discussion with Reviewers

I.B. Sachs: Many cements contain Zn, Ca, Ti, Na, Si, and Cl, please comment on the possible elemental contaminants of the cement used to mount your specimens which might interfere with the elemental information gained from your samples.
Author: Figure 2 shows the elemental analysis of the epoxy cement used to attach the specimens for energy dispersive X-ray analysis. A comparison of such results with the analytical ones obtained from the artifact analyses, enabled one to qualitatively estimate which elements can be apportioned to the cement.

C. Pearson: How has the bell been identified as a "dinner bell"?
Author: While the bell cannot be definitely identified as a dinner bell, manifests of vessels of this type show that such a bell would have been carried by the Sapphire.

I.B. Sachs: How general are your results? How many specimens of each artifact reported were examined?
Author: It was only thought necessary to analyse artifacts that were of particular interest, and this number was a small fraction of the over 300 items recovered in the trial excavation of the Sapphire. In each case several very small samples of each were examined.

C. Pearson & Reviewer No. 4.: What did the SEM show which was not visible by light microscopy concerning the identification of the wood?
Author: It is generally possible to identify wood species solely by the use of the light microscope. However, additional structural wood features can only be observed using the increased resolution of the SEM. In addition,

minute degradative changes, and the microbiological causative agents that could be present, can only be satisfactorily studied by electron microscopy because of their small size (see text reference 19). It should also be mentioned that while the identification of the wood was one aim of the study, the wood was also collected as part of an on-going biological sampling program of wreck sites. This is being undertaken with a view to identifying microbiological wood degrading organisms that might be present in Newfoundland waters. In particular, the effect of such agents on the biodegradation of wooden wreck sites is being investigated.

C. Pearson,& Reviewer No. 4.: From extensive experience with the conservation of maritime archaeological artifacts it is rare that a metal cannot be identified by inspection. In any case, chemical spot analysis could be carried out to determine the major elements present. Also, there are routine conservation procedures that can be applied to the major metals, and these different methods also apply to alloys of these metals. Would the author like to justify his use of various X-ray analytical methods in the light of these comments?
Author: It is of course correct that an experienced conservator can usually identify metal objects solely by inspection. However, the work described on the Sapphire, was the first experience of artifact identification and conservation by those involved in the project, and additional external advice was not readily available in Newfoundland. The energy dispersive X-ray analytical apparatus was freely available, and it was found to be a rapid, non-destructive way for information to be obtained on metal composition so that appropriate conservation treatments could be undertaken. It should be noted that it is sometimes difficult, even for an experienced conservator, to visually identify the composition of certain metal objects. For fragile, and important objects, such as the dated seal referred to in the text, X-ray analysis has proved its usefulness for such compositional determinations.

SCANNING ELECTRON MICROSCOPY/1979/II
SEM Inc., AMF O'Hare, IL 60666, USA

DEHYDRATION OF SCANNING ELECTRON MICROSCOPY SPECIMENS
- A BIBLIOGRAPHY (1974-1978)

*K. S. Howard and M. T. Postek**

Department of Microbiology
*Department of Botany

Louisiana State University
Baton Rouge, LA 70803

Abstract

Research in recent years into dehydration of wet specimens for scanning electron microscopy has raised several areas of concern, including the production of shrinkage artifacts, structural collapse, elemental leaching, and the operator safety of instrumentation. As a result, questions exist as to the most appropriate approach to dehydrating samples.

The following bibliography was compiled in an attempt to assemble a collection of recent, pertinent literature on the variety of dehydration techniques used in preparing wet specimens for scanning electron microscopy. The origins of these methods date well into the light microscopic and early transmission electron microscopic literature. The year 1974 was chosen arbitrarily as a cut-off point for the present bibliography. The references included are listed alphabetically by senior author's last name and numbered sequentially. A cross-reference key is provided at the end of the list organizing the references by dehydration method.

The references found here undoubtedly represent only a portion of those appropriate for inclusion. Omissions or errors are unintentional on the part of the authors. Further, we would appreciate any reference information or reprints of articles which should have been included for future up-dates of this bibliography.

Introduction

A specimen containing or contained within a substance that is volatile in the vacuum of a scanning electron microscope (SEM) is not readily observable. Sublimation or evaporation of volatile materials may result in damage to the specimen surface and contamination of the instrument. Due to its ubiquitous nature, water is a common volatile liquid which has been a problem for scanning electron microscopists. The preparation of a specimen associated with water for SEM is often approached by one of two routes.

1. The water may be stabilized by cooling it to a solid and maintaining it in association with the specimen during observation.
2. The water may be removed in such a way as to minimize specimen damage while yielding a nearly totally dry sample.

The bibliography compiled herein will deal with the latter approach to preparing specimens for SEM observation.

Problems produced by the various dehydration methods themselves are serious and well-documented. They are known to include shrinkage, structural collapse, elemental leaching, and in certain instances, the operator safety of high pressure equipment being used. The current bibliography was established not only to serve as a reference source, but also to aid in determining the extent of use of various techniques on different specimen types and to assist in comparatively analyzing the factors affected directly by these techniques.

The list of references is arranged alphabetically by the senior author's last name and numbered sequentially. A cross-reference key is provided at the end of the listing organizing the references by number into the following five categories:

1. Chemical substitution of water
2. Chemical dehydration
3. Air drying
4. Critical point drying
5. Freeze drying

In appropriate instances, the articles dealing with each category are further subdivided into

KEY WORDS: Dehydration, Air Drying, Critical Point Drying, Freeze Drying, Chemical Substitution technical modifications versus applications of dehydration methods. Scientific articles and abstracts are included. In addition to the personal collection of the authors, computer sources were utilized to compile this bibliography, including BIOSIS, ENGINEERING COMPENDIX, and MEDLARS II. Due to the lengthiness of the topic and the authors' interest in current applications of these methods, 1974 was chosen as a cut-off point. Users should bear in mind that the development and use of these techniques date from well before 1974. The authors would certainly appreciate further information on references which should be included or corrections to those presented for possible future up-dates of the bibliography.

In view of the continued scrutiny of methods and instrumentation for dehydrating specimens for SEM observation, efforts to improve the existing techniques for complete, nondestructive dehydration of aqueous materials are underway. The following bibliography will hopefully aid in pin-pointing problem areas and serve as a reference source to elucidate the best available techniques for particular applications by individual investigators.

Acknowledgements

The authors would like to thank Bonnie Jackson, Assistant Librarian, and Sue Loubiere, Veterinary Medicine Librarian, of the Louisiana State University Library and LSU School of Veterinary Medicine Library respectively for assistance with the computer resources used in assimilating this bibliography. We would also like to thank Henry LaFleur for the accurate typing of the manuscript.

Author Index

1. Agren, L. 1975. Comparison between air drying and critical point drying for scanning electron microscopy studies of the antennae of Apis mellifera Hymenoptera Apidae. Zoon. 3: 155-158.

2. Akagawa, T. 1975. Scanning electron microscope study on facial development. Jpn. J. Plast. Reconst. Surg. 18:367-379.

3. Albrecht, R. M. and A. P. MacKenzie. 1975. Cultured and free-living cells. in Principles and Techniques of Electron Microscopy, Biological Applications, Vol. 3., M. A. Hayat (ed.), Van Nostrand Reinhold Co., New York, N.Y., USA, 109-153.

4. Albrecht, R. M., D. H. Rasmussen, C. S. Keller and R. D. Hinsdill. 1976. Preparation of cultured cells for scanning electron microscopy. Air drying from organic solvents. J. Microsc. 108:21-29.

5. Allen, D. J. and L. J. A. Didio. 1976. Hartmannella culbertsoni as revealed in scanning electron microscopy. Ohio J. Sci. 76:167-171.

6. Al-Samarrai, S. F. 1975. Studies on preparation technique of human erythrocytes for clinical scanning electron microscopy. Arch. Histol. Jpn. 38:321-334.

7. Andrews, P. M. and C. R. Hackenbrock. 1974. A scanning and stereographic ultrastructural analysis of the isolated inner mitochondrial membrane during change in metabolic activity. Exp. Cell. Res. 90:127-136.

8. Aoba, T. and Y. Moriwaki. 1977. The effect of fluoride ion on apatite formation from amorphous precursor. J. Dent. Res. 56:698.

9. Appleton, T. C. and P. F. Newell. 1977. X-ray microanalysis of freeze dried, ultrathin, frozen sections of a regulating epithelium from the snail Otala lactea. Nature 266:854-855.

10. Arakawa, M. and J. Tokunaga. 1974. Further scanning electron microscope studies of the human glomerulus. Lab. Invest. 31:436-440.

11. Arvidson, K. 1976. Scanning electron microscopy of fungiform papillae on the tongue of man and monkey. Acta Oto-Laryngol. 81:496-502.

12. Ashraf, M. and H. D. Sybers. 1974. Scanning electron microscopy of ischemic heart. SEM /1974, IIT Research Institute, Chicago, IL, 60616, 721-726.

13. Baccetti, B. 1975. Critical point drying for preservation of larvae and adult insects with soft tegument. Bull. Soc. Entomol. Ital. 107:42-47.

14. Baker, F. L. and L. H. Princen. 1975. Positive displacement holder for critical point drying of small particle materials. J. Microsc. 103:393-402.

15. Baker, J. R. J. and T. C. Appleton. 1976. A technique for electron microscope autoradiography and x-ray microanalysis of diffusible substances using freeze dried, fresh frozen sections. J. Microsc. 108:307-316.

16. Barber, T. A. and P. M. Burkholder. 1975. Relation of surface and internal ultrastructure of thymus and bone marrow derived lymphocytes to specimen preparatory technique. SEM/1975, IIT Research Institute, Chicago, IL, 60616, 369-378.

17. Barrett, L. A. and R. E. Pendergrass. 1977. A method for handling free cells through critical point drying. J. Microsc. 109:311-314.

18. Bartlett, A. A. and H. P. Burstyn. 1975. A review of the physics of critical point drying. SEM/1975, IIT Research Institute, Chicago, IL, 60616, 305-316.

19. Basu, S., N. C. Lyon and D. F. Parsons. 1975. Cell surfaces by wet replication, critical point drying and high voltage electron microscopy. J. Cell. Biol. 67:22A.

20. Bautz, A. 1977. Fine structure of epidermis in terrestial and paludal triclad planarians. Arch. Zool. Exp. Gen. 118:155-172.

21. Beju, D., T. A. Calvelli and E. de Harven. 1974. Ultrastructural comparison between critical point dried type B and type C on-cornavirus particles. Fed. Proc. 33:754.

22. Bendet, I. J. and N. I. Rizk. 1974. Thermoelectric freeze drying. J. Microsc. 101:311-316.

23. Bergstrom, S. and C. Lutwak-Mann. 1974. Surface ultrastructure of the rabbit blastocyst. J. Reprod. Fertil. 36:421-422.

24. Billings-Gagliardi, S., S. M. Pickwinse and G. B. Schneider. 1978. Morphological changes in isolated lymphocytes during preparation for scanning electron microscopy. Freeze drying vs. critical point drying. Am. J. Anat. 152: 383-390.

25. Bistricky, T. and B. A. Silverberg. 1976. Chemical dehydration with 2, 2-dimethoxypropane (DMP) for the study of red blood cells by scanning electron microscopy. in Proc. Third Ann. Mtg. Microsc. Soc. Can. III, 182-183.

26. Boublik, M., F. Jenkins and H. R. Kaback. 1975. Use of critical point drying in the preparation of Escherichia coli membrane vesicles for electron microscopy. Cytobiologie. 11:304-308.

27. Boyde, A. 1974. Freezing, freeze fracturing and freeze drying in biological specimen preparation for the scanning electron microscope. SEM/1974, IIT Research Institute, Chicago, IL, 60616, 1043-1046.

28. Boyde, A. 1975. A method for the preparation of cell surfaces hidden within bulk tissue for examination in the scanning electron microscope. SEM/1975, IIT Research Institute, Chicago, IL, 60616, 295-304.

29. Boyde, A. 1978. Pros and cons of critical point drying and freeze drying for scanning electron microscopy. SEM/1978/II, SEM. Inc., AMF O'Hare, IL, 60666, 303-314.

30. Boyde, A., E. Bailey, S. J. Jones, and A. Tamarin. 1977. Dimensional changes during specimen preparation for scanning electron microscopy. in SEM/1977/I, IIT Research Institute, Chicago, IL, 60616, 507-518.

31. Boyde, A. and E. J. Reith. 1976. Scanning electron microscopy of the lateral cell surfaces of rat incisor ameloblasts. J. Anat. 122:603-610.

32. Brown, J. A. and A. Teetsov. 1976. Some techniques for handling particles in scanning electron microscopic studies. SEM/1976/I, IIT Research Institute, Chicago, IL, 60616, 385-392.

33. Brown, J. N. 1977. A simple low cost critical point dryer with continuous flow dehydration attachment. J. Microsc. 111:351-358.

34. Brown, M. F., H. G. Brotzman and D. A. Kinden. 1976. Use of a tissue sectioner to expose internal structures of biological samples for scanning electron microscopy. Stain Technol. 51:267-270.

35. Brummer, M. E. G. and L. W. Schwartz. 1974. Scanning electron microscopy of pulmonary acariasis (Pneumonyssus sp.) in the rhesus monkey (Macaca mulatta). SEM/1974, IIT Research Institute, Chicago, IL, 60616, 753-759.

36. Bryant, A. H. 1974. Transmission electron microscopy of clay structure. Geotechnique 24:39-43.

37. Buckley, I. K. and K. R. Porter. 1975. Electron microscopy of critical point dried whole cultured cells. J. Microsc. 104:107-120.

38. Burstyn, H. P. and A. A. Bartlett. 1975. Critical point drying: Application of the physics of the PVT surface to electron microscopy. Amer. J. Physics. 43:414-419.

39. Bystricky, V., G. Stotzky and M. Schiffenbauer. 1975. Electron microscopy of T bacteriophage adsorbed to clay minerals. Application of the critical point drying method. Can. J. Microbiol. 21:1278-1282.

40. Cagle, G. D. 1974. Critical point drying rapid method for the determination of bacterial extracellular polymer and surface structures. Appl. Microbiol. 28:312-316.

41. Cagle, G. D. 1974. Improved staining of extracellular polymer surrounding Eucapsis sp. and Anabaena cylindrica. A comparative study. Can. J. Microbiol. 20:735-738.

42. Cameron, C. H. S., D. L. Gardener and R. B. Longmore. 1976. The preparation of human articular cartilage for scanning electron microscopy. J. Microsc. 108:1-12.

43. Chasey, D. 1974. The 3-dimensional arrangement of radial spokes in the flagella of Chlamydomonas reinhardtii. Exp. Cell. Res. 84:374-380.

44. Cohen, A. L. 1975. I. Definitions and abbreviations used in the critical point procedure. The Bomar Co., Tacoma, WN.

45. Cohen, A. L. 1975. II. Annotated bibliography of critical point drying applications and techniques for scanning and transmission electron microscopy. The Bomar Co., Tacoma, WN.

46. Cohen, A. L. 1977. A critical look at critical point drying--theory, practice and artifacts. in SEM/1977/I, IIT Research Institute, Chicago, IL, 60616, 525-536.

47. Cohen, W. D. 1976. Simple magnetic holders for critical point drying of micro-specimen suspensions. J. Microsc. 108:221-226.

48. Cole, G. T. and R. Ramirez-Mitchell. 1974. Comparative scanning electron microscopy of Penicillium sp. conidia subjected to critical point drying, freeze drying and freeze etching. SEM/1974, IIT Research Institute, Chicago, IL, 60616, 367-374.

49. Colvin, J. R. and G. G. Leppard. 1977. The biosynthesis of cellulose by Acetobacter xvlinum and Acetobacter acetigenus. Can. J. Microbiol. 23:701-709.

50. Costa, J. L., Y. Tanaka, K. Pettigrew and R. J. Cushing. 1977. Evaluation of the utility of air dried whole mounts for quantitative electron microprobe studies of platelet dense bodies. J. Histochem. Cytochem. 25:1079-1086.

51. Coulter, D. 1975. Current refinements of freeze drying of biological tissues for electron microscopy. Anat. Rec. 181:338-339.

52. Coulter, H. D. and L. Terracio. 1978. An all glass freeze-drier for TEM specimens with an improved design for temperature regulation, fixation and infiltration. in Proc. IX Int. Cong. Electron Microsc., Vol. II, J. M. Sturgess (ed.), Microsc. Soc. Can., Toronto, Ontario, Can., 60-61.

53. Cox, C. S., W. J. Harris and J. Lee. 1974. Viability and electron microscope studies of phage T-3 and phage T-7 subjected to freeze drying, freeze thawing and aerosolization. J. Gen. Microbiol. 81:207-215.

54. Cross, R. H. M., B. R. Allanson, B. R. Davies and D. Howard-Williams. 1977. Critical point drying as a preparative technique for scanning electron microscopy and its application in limnology. J. Limnol. Soc. South Afr. 3:59-62.

55. Crossley, A. 1976. A versatile multi-specimen holder for processing and critical point drying of materials for examination in the scanning electron microscope. J. Microsc. 108: 349-352.

56. Curry, J. C., E. E. Burns and N. D. Heidelbaugh. 1976. Effect of sodium chloride on rehydration of freeze dried carrots. J. Food Sci. 41:176-179.

57. Cutz, E., W. Chan and P. E. Conen. 1975. Surface morphology of airway epithelial lining. Lab. Invest. 32:444-445.

58. Davies, P. F., M. A. Reidy, T. B. Goode and D. E. Bowyer. 1976. Scanning electron microscopy in the evaluation of endothelial integrity of the fatty lesion in atherosclerosis. Atherosclerosis. 25:125-130.

59. Day, J. W. 1974. A BEEM capsule chamber-pipette for handling small specimens for electron microscopy. Stain Tech. 49:408-410.

60. de Estable-Puig, R. F. and J. F. Estable-Puig. 1975. Brain cyst formation. A technique for scanning electron microscopy study of the central nervous system. SEM/1975, IIT Research Institute, Chicago, IL, 60616, 281-286.

61. de Harven, E.,N. Lampen and D. Pla. 1977. Alternatives to critical point drying. in SEM/1977/I, IIT Research Institute, Chicago, IL, 60616, 519-524.

62. DeHertogh, A. A., H. P. Rasmussen and N. Blakely. 1976. Morphological changes and factors influencing shoot apex development of Lilium longiflorum during forcing. J. Am. Soc. Hortic. Sci. 101:463-471.

63. Delahunt, B., W. D. Trotter and D. D. Samarasinghe. 1975. The morphology of the pineal recess of the brush tailed possum Trichosurus vulpecula. Proc. Univ. Otago. Med. Sch. 53:63-64.

64. Demsey, A., T. A. Calvelli, D. Kawka, C. W. Stackpole and N. H. Sarkar. 1976. Surface structure of virions budding from L1210 (V) gln-mouse leukemia cells. Virology. 75:484-487.

65. Demsey, A., D. Kawka and C. W. Stackpole. 1977. Application of freeze drying intact cells to studies of murine oncornavirus morphogenesis. J. Virol. 21:358-365.

66. Demsey, A., D. Kawka and C. W. Stackpole. 1978. Cell surface membrane organization revealed by freeze drying. J. Ultrastruct. Res. 62:13-25.

67. De Vos, L. 1977. Scanning electron microscopy of the cells of the sponge Ephydatia fluviatilis. Arch. Biol. 88:1-14.

68. Dorge, A., R. Rick, K. Gehring, J. Mason and K. Thurau. 1974. Preparation of biological soft tissue for electron microprobe analysis of electrolytes. Cryobiology 11:563.

69. Dorge, A., R. Rick, K. Gehring, J. Mason and K. Thurau. 1975. Preparation and applicability of freeze dried sections in the microprobe analysis of biological soft tissue. J. Microsc. Biol. Cell 22:205-214.

70. Dorge, A., R. Rick, K. Gehring and K. Thurau. 1978. Preparation of freeze dried cryosections for quantitative x-ray micro-analysis of electrolytes in biological soft tissues. Pfleugers Arch. Eur. J. Physiol. 373:85-97.

71. Draenert, Y. and K. Draenert. 1978. Freeze-drying of articular cartilage. in SEM/1978/II, SEM. Inc., AMF O'Hare, IL, 60666, 759-766.

72. Dumas, C. and M. Lecoco. 1975. Scanning electron microscopy and stigmatic secretion observations with freeze drying-like technique. C. R. Hebd. Seances. Acad. Sci. Ser. D Sci. Nat. 280:837-840.

73. Echlin, P. and A. Burgess. 1977. Cryofracturing and low temperature scanning electron microscopy of plant material. SEM/1977/I, IIT Research Institute, Chicago, IL, 60616, 491-500.

74. Edanaga, M., Y. Masu and J. Tokunaga. 1974. Studies on procedure of critical point drying method in free cells. J. Electron Microsc. 23:73.

75. Edelmann, L. 1978. A simple freeze-drying technique for preparing biological tissue without chemical fixation for electron microscopy. J. Microsc. 112:243-248.

76. Eichler, W. and W. Sixl. 1974. Scanning electron microscopic pictures of Eomencanthus stramineus. Angew. Parasitol. 15:151-156.

77. Eranko, O. 1976. Histochemical demonstration of catecholamines in sympathetic ganglia. Ann. Histochim. 21:83-100.

78. Erlick, B. J., A. A. Fuscaldo, I. Brodsky and K. E. Fuscaldo. 1977. Critical point drying and biochemical scans from density gradients for rapid detection of oncornaviruses. Abstr. Annu. Meet. Am. Soc. Microbiol. 77:331.

79. Erlick, B. J., W. W. West and K. E. Fuscaldo. 1977. Rapid oncornavirus detection using critical point drying,biochemical and density scans. Proc. Am. Assoc. Cancer Res. 18:96.

80. Evan, A., W. G. Dail, D. Dammrose and C. Palmer. 1976. Scanning electron microscopy of tissues following removal of basement membrane and collagen. SEM/1976/II, IIT Research Institute, Chicago, IL, 60616, 203-208.

81. Fiala-Medioni, A. 1978. A scanning electron microscope study of the branchial sac of benthic filter feeding invertebrates Ascidians. Acta. Zool. 59:1-10.

82. Flechon, J. E. S. Bergstrom, S. Jaszczak and E. S. C. Hafez. 1975. Techniques for critical point drying of gametes and embryos. SEM/1975, IIT Research Institute, Chicago, IL, 60616, 325-332.

83. Flink, J. 1975. The retension of volatile components during freeze drying, a structurally based mechanism. in Food Science and Technology: Freeze Drying and Advanced Food Technology, S. A. Goldblith, L. Rey and W. W. Rothmayr (ed.'s), Academic Press. New York, N. Y. USA. 351-372.

84. Flood, P. R. 1975. Dry fracturing techniques for the study of soft internal biological tissues in the scanning electron microscope. SEM/1975, IIT Research Institute, Chicago, IL, 60616, 287-294.

85. Frederik, P. M. and D. Klepper. 1974. Freeze drying of unfixed tissue samples for the autoradiographic localization of steroid hormones in testes. J. Microsc. (Paris) 19: 11A-12A.

86. Frederik, P. M. and D. Klepper. 1976. The possibility of electron microscopic autoradiography of steroids after freeze drying of unfixed testes. J. Microsc. 106:209-219.

87. Frisch, B., S. M. Lewis, P. R. Stuart and J. S. Osborn. 1975. Further observations of the effects of ion etching on blood cells. SEM/1975, IIT Research Institute, Chicago, IL, 60616, 165-172.

88. Fukuzumi,F.M. Osumi, H. Hidaka and T. Nagatani. 1975. Examination of the critical point drying method. J. Electron Microsc. 24:201.

89. Gangemi, J. D. and F. E. Cole, Jr. 1978. Venezuelan equine encephalomyelitis virus aggregation and immunogenicity following freeze drying. J. Biol. Stand.   6:117-120.

90. Gershenbaum, M. R., J. W. Shay and K. R. Porter. 1974. The effects of cytochalasin B on BALB/3T3 mammalian cells cultured in vitro as observed by scanning and high voltage electron microscopy. SEM/1974, IIT Research Institute, Chicago, IL, 60616, 589-594.

91. Geymayer, W., F. Grasenick and Y. Hodl. 1978. Stablizing ultra-thin cryo sections by freeze drying. J. Microsc.   112:39-46.

92. Glauert, A. M. 1974. Fixation, dehydration and embedding of biological specimens. in Practical Methods in Electron Microscopy, A. M. Glauert (ed.), North-Holland Pub. Co., New York, N. Y., USA, 207 p.

93. Goldfarb, D., K. Miyai and J. Hegenauer. 1977. A hydrostatic device for tissue dehydration. Stain Technol. 52:171-176.

94. Gotjamanos, T. and D. Swedlow. 1976. Critical point drying of odontoblasts and dentin and examination by scanning electron microscopy. J. Dent. Res. 55:517.

95. Greenwood, M. F. and P. Holland. 1974. Plasma membrane ultrastructure of human leukocytes. Pediatr. Res. 8:401.

96. Grut, W., J. Edwards and E. J. Evans. 1977. Scanning electron microscopy of freeze dried aortic elastin. J. Microsc. 110:271-276.

97. Gusnard, D. and R. H. Kirschner. 1977. Cell and organelle shrinkage during preparation for scanning electron microscopy. Effects of fixation, dehydration and critical point drying. J. Microsc. 110:51-58.

98. Hafez, E. S. E. and H. Kanagawa. 1974. The endometrium and blastocyst at implantation in the rabbit as observed by scanning electron microscopy. Fertil. Steril. 25:302.

99. Haggis, G. H., E. F. Bond and B. Phipps. 1976. Visualization of mitochondrial cristae and nuclear chromatin by scanning electron microscopy.

SEM/1976/I. IIT Research Institute, Chicago, IL, 60616, 281-286.

100. Hall, J. D., E. J. Skerrett and W. D. E. Thomas. 1978. Critical point drying for scanning electron microscopy: a semi-automatic method of preparing biological specimens. J. Microsc.   113:277-290.

101. Hallman, M., K. Miyai and R. M. Wagner. 1976. Isolated lamellar bodies from rat lung. Correlated ultrastructural and biochemical studies. Lab. Invest. 35:79-86.

102. Halton, D. W. 1978. Transtegumental absorption of L-alanine and L-leucine by a Monogenean Diclidophora merlangi. Parasitology 76:29-38.

103. Hansson, H.-A., A. Linde and H. Nygren. 1975. Scanning electron microscopic studies of the rat incisor odontoblastema. Cell Tissue Res. 159:233-243.

104. Harris, J. L. 1975. Some 3-dimensional aspects of Ceratocystis ulmi as observed by high voltage electron microscopy. Mycologia. 67:332-341.

105. Hashimoto, K. and T. Kanzaki. 1975. Surface ultrastructure of human skin. Acta. Derm-Venereol.   55:413-430.

106. Hattori, A., S. Ito, A. Sugawara and M. Matsuoka. 1975. Studies on fixation and drying of blood cells for scanning electron microscopic observation. Acta Haematol. Jpn. 38:86-95.

107. Hayat, M. A. 1978. Introduction to biological scanning electron microscopy. Univ. Park Press, Baltimore, MD, U S A, 323 p.

108. Hayes, T. L. 1974. Stereo pair projection as an aid in the evaluation of critical point and freeze drying techniques. J. Ultrastruct. Res. 48:175.

109. Hayunga, E. G. 1977. A specimen holder for dehydrating and processing very small tissue samples. Trans. Amer. Micros. Soc. 96:156-158.

110. Hesse, I. and W. Hesse. 1978. Artifacts on the surface of articular cartilage due to drying methods. in Proc. IX Int. Cong. Electron Microsc., Vol. III, J. M. Sturgess (ed.), Microsc. Soc. Can., Toronto, Ontario, Can., 680-681.

111. Hicks, M. L., J. D. Brilliant and D. W. Foreman. 1976. Electron microscope comparison of freeze substitution and conventional chemical fixation of undecalcified human dentin. J. Dent. Res. 55:400-410.

112. Hiratsuka, Y. and J. Maruyama. 1973. A modified critical point drying to study germ tubes of rust fungi under scanning electron microscope. Can. For. Serv. Bi. Mon. Res. Notes. 30:5-6.

113. Hoehling, H. J., H. Steffens and G. Stamm. 1976. Transmission microscopy of freeze dried, unstained epiphyseal cartilage of the guinea-pig. Cell Tissue Res. 167:243-263.

114. Hogger, C. H. and R. H. Esty. 1976. Chamber for critical point drying of nematodes and other biological specimens. J. Nematol. 8:357-358.

115. Hogger, C. H. and R. H. Estey. 1976. Scanning electron microscopy of a plant parasitic nematode Xiphinema americanum. Phytoprotection 57:150-154.

116. Hogger, C. H. and R. H. Estey. 1977. Cryofracturing for scanning electron microscope observations of internal structures of nematodes. J. Nematol. 9:334-337.

117. Homes, J. 1974. Scanning electron microscopy of carrot cells and embryoids growing in vitro under various culture conditions. SEM/1974, IIT Research Institute, Chicago, IL, 60616, 335-341.

118. Huijing, P. A., M. Bacaner, T. E. Hutchinson, J. Lilley and J. Broadhurst. 1975. Alterations in ultrastructure of deep frozen thin sections of rabbit psoas muscle by fixation and dehydration. Fed. Proc. 34:474.

119. Humphreys, W. J. 1975. Drying soft biological tissue for scanning electron microscopy. SEM/1975, IIT Research Institute, Chicago, IL, 60616, 707-714.

120. Humphreys, W. J., B. O. Spurlock and J. S. Johnson. 1974. Critical point drying of ethanol-infiltrated, cryofractured biological specimens for scanning electron microscopy. SEM/1974, IIT Research Institute, Chicago, IL, 60616, 275-282.

121. Ingram, F. D. and M. J. Ingram. 1975. Quantitative analysis with the freeze dried, plastic embedded tissue specimen. J. Microsc. Biol. Cell. 22:193-204.

122. Iwasaki, Y. 1978. Application of the critical point dried whole cell technique to the study of animal rhabdoviruses. Intervirology 9:214-225.

123. Iwata, H. and S. Aita. 1976. Freeze drying technique for small biological materials using a modified freeze etching device. J. Electron Microsc. 25:305-306.

124. Iwata, H. and S. Aita. 1976. Freeze drying technique for small biological objects. J. Electron Microsc. 25:205.

125. Jacks, T. J. and L. L. Muller. 1975. Instant chemical dehydration for electron microscopy. Tex. Rep. Biol. Med. 33:352.

126. Jalanti, T. and G. Demierre. 1976. Chemical dehydration of microorganisms for scanning electron microscope study. Experientia 32:798.

127. Joshi, K. R., J. B. Gavin and E. E. Wheeler. 1975. The ultrastructure of Prototheca wickerhamii. Mycopathologia. 56:9-14.

128. Joshi, K. R., E. E. Wheeler and J. B. Gavin. 1975. The ultrastructure of Candida krusei. Mycopathologia. 56:5-8.

129. Kahn, L. E., S. P. Frommes and P. A. Cancilla. 1977. Comparison of ethanol and chemical dehydration methods for the study of cells in culture by scanning electron microscopy and transmission electron microscopy. SEM/1977/I, IIT Research Institute, Chicato, IL, 60616. 501-506.

130. Kawakami, S. 1975. Scanning electron microscopic observations on human placental villi. Acta Obstet. Gynaecol. Jpn. (Engl. ed.) 22:132-137.

131. Kistler, J., U. Aebi, and E. Kellenberger. 1977. Freeze drying and shadowing a 2-dimensional, periodic specimen. J. Ultrastruct. Res. 59:76-86.

132. Kistler, J. and E. Kellenberger. 1977. Collapse phenomena in freeze drying. J. Ultrastruct. Res. 59:70-75.

133. Kistler, J., B. Ten Heggeler and E. Kellenberger. 1976. Freeze drying shadowing of supramolecular crystalline structures. Experientia 32:799-800.

134. Klainer, A. S., S. Jernigan and P. Allender. 1974. Evaluation and comparison of techniques for examination of bacteria by scanning electron microscopy. SEM/1974, IIT Research Institute, Chicago, IL, 60616, 313-317.

135. Klainer, A. S. and M. Rectenwald. 1974. Surface morphology of phagocytosis of bacteria by peritoneal macrophages. SEM/1974. IIT Research Institute, Chicago, IL, 60616, 821-825.

136. Klein, H.-P. and W. Stockem. 1976. Preparation of biological specimens for the scanning electron microscope. Microsc. Acta 78:388-406.

137. Klein, M. and S. W. Applebaum. 1975. The surface morphology of locust hind gut cuticle. J. Entomol. Ser. A Physiol. Behav. 50:31-36.

138. Knutton, S. 1976. Transmission electron microscopy and scanning electron microscopy of critical point dried whole cultured cell preparations. J. Cell. Biol. 70:202A.

139. Koenhen, D. M., M. A. deJongh, C. A. Smolders and N. Yucesoy. 1975. Preparation technique for examination of wet-spun polymer fibers in a scanning electron microscope. Colloid. Polym. Sci. 253:521-526.

140. Krey, H. 1974. Scanning electron microscopy of the pars plicata of the human ciliary body. Comparative investigations with air drying and critical point drying. Albrecht von Graefes Arch. Klin. Exp. Ophthalmol. 191:127-137.

141. Kuntz, R. E., G. S. Tulloch, D. L. Davidson and T.-C. Huang. 1976. Scanning electron microscopy of the integumental surfaces of Schistosoma haematobium. J. Parasitol. 62:63-69.

142. Kuntz, R. E., G. S. Tulloch, T.-C. Huang. and D. L. Davidson. 1977. Scanning electron microscopy of integumental surfaces of Schistosoma intercalatum. J. Parasitol. 63:401-406.

143. Labuschagne, M. C. and G. C. Loots. 1975. Techniques for the preparation of ticks for scanning electron microscopy. Reeks B Natuurwet. #67:1-9.

144. Lampky, J. R. 1976. Ultrastructure of Polyangium cellulosum. J. Bacteriol. 126:1278-1284.

145. Larsson, F. and H. E. Myhrberg. 1976. Six years of experience with scanning electron microscopy in the electron microscopy laboratory in Lund. J. Ultrastruct. Res. 57:212-213.

146. Leene, W., C. Van Steeg and H. R. Hendriks. 1974. Scanning electron microscopy on critical point dried lymphocytes. An attempt to classify lymphocyte sub-populations. J. Microsc. (Paris) 19:13A-14A.

147. Lengsfeld, H. M. and W. Hasselbach. 1974. The structure of membrane preparations of the fragmented sarcoplasmic reticulum after freeze drying. Histochemistry. 40:113-127.

148. Lewis, E. R., L. Jackson, and T. Scott. 1975. Comparison of miscibilities and critical point drying properties of various intermediate and transitional fluids. SEM/1975, IIT Research Institute, Chicago, IL, 60616, 317-324.

149. Liepins, A. and E. DeHarven. 1978. A rapid method for cell drying for scanning electron microscopy. SEM/1978/II, SEM. Inc., AMF O'Hare, IL, 60666, 37-43.

150. Lin, C. H., R. H. Falk and C. R. Stocking. 1977. Rapid chemical dehydration of plant material for light microscopy and electron microscopy with 2, 2-dimethoxypropane and 2, 2-diethoxypropane. Am. J. Bot. 64:602-605.

151. Liu, K. C. and J. K. Sherman. 1975. Comparison of freeze drying and freeze substitutions in characterizing ultrastructure of frozen cells. Tex. Rep. Biol. Med. 33:337-338.

152. Liu, K. C. and J. K. Sherman. 1977. Ultrastructural comparison of freeze drying and freeze substitution in preservation of the frozen state. Cryobiology. 14:382-386.

153. Locci, R., B. Petrolini Baldan, S. Quaroni and P. Sardi. 1977. Scanning electron microscopy techniques for the mycological study of the rhizosphere. Riv. Patol. Veg. 13:49-59.

154. Locker, R. H. and G. J. Daines. 1974. Cooking loss in beef. The effect of cold shortening searing and rate of heating time course and histology of changes during cooking. J. Sci. Food. Agric. 25:1411-1418.

155. Lofberg, J. 1974. Preparation of amphibian embryos for scanning electron microscopy of the functional pattern of epidermal cilia. Zoon 2:3-11.

156. Lohnes, R. A. and T. Demirel. 1978. Scanning electron microscopy in soil mechanics. in SEM/1978/I, SEM. Inc., AMF O'Hare, IL, 60666, 643-654.

157. Ludwig, H., H. Metzger and E. S. E. Hafez. 1976. Critical point drying and gold sputtering as applied to scanning electron microscopy of human reproductive tissues. Acta Anat. 96:469-477.

158. Lung, B. 1974. The preparation of small particulate specimens by critical point drying: Application for scanning electron microscopy. J. Microsc. 101:77-80.

159. Lynch, S. P. and G. L. Webster. 1975. A new technique of preparing pollen for scanning electron microscopy. Grana. 15:127-136.

160. Mackenzie, A. P. 1975. Collapse during freeze drying. Qualitative and quantitative aspects. in Food Science and Technology. S. A. Goldblith, L. Rey and W. W. Rothmayr (ed's.).

Academic Press. New York, N.Y., U S A , 277-307.

161. Madden, P. A. and F. G. Tromba. 1976. Scanning electron microscopy of the lip dentricles of Ascaris suum adults of known ages. J. Parasitol. 62:265-271.

162. Makita, T. 1975. Histochemistry, scanning electron microscopy and x-ray micro-analysis of freeze dried whole body sections on the adhesive tape. Anat. Rec. 181:537.

163. Malech, H. L. and N. A. Wivel. 1976. Properties of murine intracisternal A particles. Electron microscopic appearance after critical point drying and platinum shadowing. Virology. 69:802-809.

164. Marchant, H. J. 1974. Scanning electron microscopy of algal cells. SEM/1974, IIT Research Institute, Chicago, IL, 60616, 351-357.

165. Marinaro, R. E., M. R. Gershenbaum, F. J. Roisen and C. M. Papa. 1978. Tinea versicolor a scanning electron microscopic view. J. Cutaneous Pathol. 5:15-22.

166. Maser, M. D. and J. J. Trimble, III. 1977. Rapid chemical dehydration of biological samples for scanning electron microscopy using 2, 2-dimethoxypropane. J. Histochem. Cytochem. 25:247-251.

167. Matsuguchi, M., K. Takeya, A. Umeda, K. Amako, T. Watabe, and S. Aita. 1977. Optimal condition for observing bacterial flagella by the scanning electron microscope. J. Electron Microsc. 26:343-344.

168. Matsumura, H., K. Komatsu and K. Ogura. 1974. Direct observation of silkworm spinning and spun cocoon filament by cryo-scanning electron microscopy. J. Electron Microsc. 23:229.

169. Maueel, J. and T. Jalanti. 1976. Phagocytosis of Leishmania enrietti by activated and normal macrophages. A scanning electron microscopy. Experientia 32:804.

170. Miller, A. L. and H. Schraer. 1975. Ultrastructural observations of amorphous bone mineral in avian bone. Calcif. Tissue Res. 18:311-324.

171. Miller, M. M. and J.-P. Revel. 1974. Scanning electron microscopy of the apical lateral and basal surfaces of transporting epithelia in mature and embryonic tissue. SEM/1974, IIT Research Institute, Chicago, IL, 60616, 549-556.

172. Miyai, K. J. L. Abraham, D. S. Linthicum and R. M. Wagner. 1976. Scanning electron microscopy of hepatic ultrastructure. Secondary, backscattered and transmitted electron imaging. Lab. Invest. 35:369-376.

173. Mizuhira, V., Y. Futaesaku, M. Shiihashi, K. Fukuda and T. Sakai. 1976. Elemental distribution in the fresh air dried or fresh freeze dried sections. J. Electron Microsc. 25:218.

174. Mizuhira, V., M. Shiihashi and H. Makabe. 1977. Quantitative elemental analysis of the fresh parietal cell and the proximal tubular cell of a mouse. J. Electron Microsc. 26:254-255.

175. Mizuno, K. and Y. Takei. 1975. Retinal glycogen by freeze drying method. Acta Histochem. Cytochem. 8:63.

176. Muller, L. L. and T. J. Jacks. 1975. Rapid chemical dehydration of samples for electron microscopic examinations. J. Histochem. Cytochem. 23:107-110.

177. Myklebust, R., H. Dalen and T. S. Saetersdal. 1975. A comparative study in the transmission electron microscope and scanning electron microscope of intracellular structures in sheep heart muscle cells. J. Microsc. 105:57-65.

178. Nadakavukaren, M. J., D. A. McCraken and B. L. Bertagnolli. 1977. Scanning electron microscopy of isolated chloroplasts. J. Submicrosc. Cytol. 9:247-250.

179. Nagata, T. and F. Murata. 1977. Electron microscopic dry mounting radioautography for diffusible compounds by means of ultracryotomy. Histochem. 54:75-82.

180. Naguro, T., K. Tanaka and A. Iino. 1975. Critical point drying method using crushed dry ice. J. Electron Microsc. 24:201.

181. Nakanishi, N., M. Takano and H. Kumito. 1977. Preparation of fine powders of MgO by means of freeze-drying processing. J. Jpn. Soc. Powder Metall. 24:1-5.

182. Nayak, R. K. 1977. Scanning electron microscopy of the camel uterine tube oviduct. Am. J. Vet. Res. 38:1049-1054.

183. Naymik, T. G. 1974. The effects of drying techniques on clay-rich soil texture. in Procedings Thirty-Second Annual Meeting Electron Microscopy Society of America, C. J. Arceneaux (ed.), Claitor's Publishing Division, Baton Rouge, U S A, 466-467.

184. Nei, T. and S. Fujikawa. 1977. Freeze drying of biological specimens observed with a scanning electron microscope. J. Electron Microsc. 26:251.

185. Nei, T. and S. Fujikawa. 1977. Freeze drying process of biological specimens observed with a scanning electron microscope. J. Microsc. 111:137-142.

186. Nermut, M. V. 1977. Freeze drying for electron microscopy. in Principles and Techniques of Electron Microscopy, Vol. 7. Biological Applications. M. A. Hayat (ed.) Van Nostrand Reinhold Co., New York, N.Y., U S A, London, England. 79-117.

187. Newell, D. G. and S. Roath. 1975. A container for processing small volumes of cell suspensions for critical point drying. J. Microsc. 104:321-323.

188. Niimi, M., M. Edanaga, H. Murakami and M. Tokunaga. 1975. Several problems in critical point drying method for scanning electron microscopy, especially shrinkage and distortion of specimens. J. Electron Microsc. 24:201.

189. Oh, S. Y. and R. T. Holzbach. 1976. Transmission electron microscopy of biliary mixed lipid micelles. Biochim. Biophys. Acta. 441:498-505.

190. Osumi, M., F. Fukuzumi, F. Saito, T. Nagatani, M. Hozumi and T. Sugimura. 1974. Ultrastructure of some biological specimens after the critical point drying. J. Electron Microsc. 23:228.

191. Pameijer, C. H. 1975. Replica techniques. in Principles and Techniques of Scanning Electron Microscopy, Vol. 4. Biological Applications. M. A. Hayat (ed.), U S A, London, England. 45-93.

192. Pankhurst, C. E. and J. I. Sprent. 1975. Surface features of soybean root nodules. Protoplasma 85:85-98.

193. Pappalardo, G. 1974. Scanning electron microscopy for the identification of spermatozoa on difficult substrates in forensic science. J. Submicrosc. Cytol. 6:433-434.

194. Parsons, E., B. Bole, D. J. Hall and W. D. E. Thomas. 1974. A comparative survey of techniques for preparing plant surfaces for the scanning electron microscope. J. Microsc. 101:59-75.

195. Petzold, H. and M. Ozel. 1976. A simple addition to a critical point apparatus for drying of serial sections. Microsc. Acta. 78:292-294.

196. Pfaller, W., E. Rovan and H. Mairbaeurl. 1976. A comparison of the ultrastructure of spray frozen and freeze etched or freeze dried bull and boar spermatozoa with that after chemical fixation. J. Reprod. Fertil. 48:285-290.

197. Polliack, A., V. Hammerling, N. Lampen and E. deHarven. 1975. Surface morphology of murine bone marrow derived lymphocytes and thymus derived: A comparative study by scanning electron microscopy. Eur. J. Immunol. 5:32-39.

198. Polliack, A., N. Lampen, B. Clarkson and E. deHarven. 1974. Scanning electron microscopy of human leukocytes. A comparison of air dried and critical point dried cells. Isr. J. Med. Sci. 10:1075-1085.

199. Polliack, A., N. Lampen and E. deHarven. 1974. Scanning electron microscopy of lymphocytes of known B and T derivation. SEM/1974, IIT Research Institute, Chicago, IL, 60616, 673-679.

200. Pomeranz, Y. 1976. Scanning electron microscopy in food science and technology. in Advances in Food Research, Vol. 22. C. O. Chichester (ed.), Academic Press, New York, N.Y., U S A 205-307.

201. Pool, R. R. 1975. Scanning electron microscopy of primary bone tumors. Lab. Invest. 32:455.

202. Pool, R. R. 1975. Scanning electron microscopy of the remodeling unit of endosteum. Låb Invest. 32:455.

203. Portch, P. A. and A. J. Barson. 1974. Scanning electron microscopy of neurulation in the chick. J. Anat. 117:341-350.

204. Postek, M. T. 1975. Techniques for the preparation of phytoplaktonic organisms for scanning electron microscopy. Tex. Rep. Biol. Med. 33:361.

205. Postek, M. T., W. L. Kirk, and E. R. Cox. 1974. A container for the processing of delicate organisms for scanning or transmission electron microscopy. Trans. Amer. Micros. Soc. 93:265-268.

206. Postek, M. T. and S. C. Tucker. 1977. Thiocarbohydrazide binding for botanical

specimens for scanning electron microscopy. A modification. J. Microsc. 110:71-74.

207. Prentø,P., 1978. Rapid dehydration-clearing with 2, 2-dimethoxypropane for paraffin embedding. J. Histochem. Cytochem. 26:865-867.

208. Revel, J.-P. 1975. Elements of scanning electron microscopy for biologists. SEM/1975, IIT Research Institute, Chicago, IL, 60616, 687-696.

209. Reymond, O. and T. Jalanti. 1976. Scanning electron microscopy of unicellular alga Treubaria. Simplified critical point drying apparatus using solid carbon dioxide. Experientia 32:808.

210. Robards, A. W. 1974. Low temperatures in ultrastructural research. Cryobiology 11:561.

211. Rosowski, J. R. and R. L. Willey. 1977. Development of mucilaginous surfaces in Euglenoids. Part 1. Stalk morphology of Colacium mucronatum. J. Phycol. 13:16-21.

212. Rostgaard, J. and P. Christensen. 1975. A multipurpose specimen-carrier for handling small biological objects through critical point drying. J. Microsc. 105:107-113.

213. Rovan, E. and P. Simonsberger. 1974. The agar tube method for electron microscopical preparation of cell suspensions and small tissue fragments. Mikroskopie 30:129-134.

214. Rovenskii Yu, A., F. S. Sokolovskii and S. V. Chuiko. 1975. A simple apparatus for drying cytological objects by the critical point technique. Tsitologiya 17:864-866.

215. Ruffolo, J. J., Jr. 1974. Critical point drying of protozoan cells and other biological specimens for scanning electron microscopy: Apparatus and methods of specimen preparation. Trans. Am. Micros. Soc. 93:124-131.

216. Ryder, T. A. and I. D. Bowen. 1977. The use of x-ray microanalysis to demonstrate the uptake of the molluscicide copper sulfate by slug eggs. Histochemistry. 52:55-60.

217. Sakamoto, K. and Y. Ishii. 1976. Fine structure of schistosome eggs as seen through the scanning electron microscope. Am. J. Trop. Med. Hyg. 25:841-844.

218. Sakamoto, K. and Y. Ishii. 1977. Scanning electron microscope observations on adult Schistosoma japonicum. J. Parasitol. 63:407-412.

219. Sale, F. R. 1974. Fine tungsten and nickel powders by the freeze-drying technique. in Fine Part, 2nd Int. Conf., Electrochem Soc., Inc., Princeton, N. J., U S A , 283-295.

220. Sale, W. S. and P. Satir. 1977. Direction of active sliding of microtubules in Tetrahymena cilia. Proc. Natl. Acad. Sci. U S A, 74:2045-2049.

221. Sanders, S. K., E. L. Alexander and R. C. Braylan. 1975. A high yield technique for preparing cells fixed in suspension for scanning electron microscopy. J. Cell Biol. 67:476-480.

222. Sarkar, N. H., W. J. Manthey and J. B. Sheffield. 1975. The morphology of murine on-cornaviruses following different methods of preparation for electron microscopy. Cancer. Res. 35:740-749.

223. Schneider, G. B. 1976. The effects of preparative procedures for scanning electron microscopy on the size of isolated lymphocytes. Part 1. Am. J. Anat. 146:93-100.

224. Schneider, G. B., S. M. Pockwinse, and S. Billings-Gagliardi. 1978. Morphological changes in isolated lymphocytes during preparation for SEM: A comparative TEM/SEM study of freeze drying and critical-point drying. SEM/1978/II, SEM. Inc., AMF O'Hare, IL, 60666, 77-84.

225. Sitte, H., H. Fell, W. Hölbl, H. Kleber, and K. Neumann. 1977. Fast freezing device. J. Microsc. 111:35-38.

226. Sjoestroem, M., R. Johansson and L. E. Thornell. 1974. Cryo-ultra microtomy of muscles in defined functional states. Histochem. J. 6:100.

227. Sjostrand, F. S. and F. Kretzer. 1975. A new freeze drying technique applied to the analysis of the molecular structure of mitochondrial and chloroplast membranes. J. Ultrastruct. Res. 53:1-28.

228. Sjostrom, M. and L-E. Thornell. 1975. Preparing sections of skeletal muscle for transmission electron analytical microscopy of diffusible elements. J. Microsc. 103:101-112.

229. Small, E. B. and T. K. Maugel. 1978. Observations on the permanence of protozoan preparations for scanning electron microscopy. SEM/1978/II, SEM. Inc., AMF O'Hare, IL, 60666, 123-127.

230. Sommerlad, B. C. and J. M. Creasey. 1975. A technique for preparing human dermal and scar specimens for scanning electron microscopy. J. Microsc. 103:369-376.

231. Somogyi, E. and P. Sotonyi. 1977. On the possibilities of application of scanning electron microscopy in forensic medicine. Z. Rechtsmed. 80:205-220.

232. Spicer, R. A., P. R. Grant and M. D. Muir. 1974. An inexpensive, portable freeze drying unit for scanning electron microscope specimen preparation. SEM/1974. IIT Research Institute, Chicago, IL, 60616. 299-304.

233. Straeuli, P. and C. Lunscken. 1974. Tumer invasion into the diaphragm. A scanning electron microscopic analysis. Experientia. 30:710.

234. Swanson, J. and B. Zeligs. 1974. Studies on gonococcus infection. Part 6. Electron microscopic study on in-vitro phagocytosis of gonococci by human leukocytes. Infect. Immun. 10:645-656.

235. Sweney, L. R., N. J. Laible and B. L. Shapiro. 1976. Some methods for more efficient processing of scanning electron microscopy specimens. J. Microsc. 108:335-338.

236. Takagi, A. and T. Katsumoto. 1976. Studies on cell arrangement in bacterial colonies by scanning electron microscopy. Application of carbon dioxide critical point drying technique. Jpn. J. Bacteriol. 31:637-648.

237. Takawa, K. 1977. Electron microscopy of human melanosomes in unstained fresh air dried hair bulbs and their examinations by electron probe microanalysis. Cell Tissue Res. 178:169-174.

238. Takaya, K. 1974. Preparation of the fresh frozen dried ultra thin sections and the interpretation of their figures using the unstained Epon sections and freshly air dried tissue spread. J. Electron Microsc. 23:219.

239. Takaya, K. 1975. Energy dispersive x-ray microanalysis of neurosecretory granules of mouse pituitary on fresh, air dried tissue spreads. Cell Tissue Res. 159:227-231.

240. Takaya, K. 1975. Energy dispersive x-ray microanalysis of zymogen granules of mouse pancreas using fresh, air dried tissue spread. Arch. Histol. Jpn. 37:387-394.

241. Takaya, K. 1975. Intranuclear silicon detection in a subcutaneous connective tissue cell by energy dispersive x-ray microanalysis using fresh, air dried spread. J. Histochem. Cytochem. 23:681-685.

242. Takaya, K. 1975. Mitochondrial granules in fresh, air dried tissue spreads. Electron microscopy and x-ray microanalysis. J. Electron Microsc. 24:203-204.

243. Takaya, K. 1976. Electron microscopy of unstained, fresh, air dried spreads of mouse pancreas acinar cells and energy dispersive x-ray microanalysis of zymogen granules. Cell Tissue Res. 166:117-123.

244. Takaya, K. 1976. Melanosomes of human plucked hair follicle. Electron microscopy and energy dispersive x-ray microanalysis using fresh air-dried spreads. J. Electron Microsc. 25:218-219.

245. Takaya, K. 1977. Electron microscopy of unstained, fresh, air dried spreads of brain and application to electron probe microanalysis. J. Electron Microsc. 26:265.

246. Takaya, K. 1977. Electron probe x-ray microanalysis of animal tissues using unstained, fresh, air dried tissue spreads. J. Electron Microsc. 26:65.

247. Tanaka, K. and A. Iino. 1974. Critical point drying method using dry ice. Stain Technol. 49:203-206.

248. Thijssen, H. A. C. 1975. Effect of process conditions in freeze drying on retension of volatile components. in Food Science and Technology. Freeze Drying and Advanced Food Technology. S. A. Goldblith, L. Rey and W. W. Rothmayr (ed.'s), Academic Press. New York, N.Y., U S A, 373-400.

249. Thomas, R. J., M. G. Wolery and J. Taylor. 1974. Critical point drying of liverwort spores for scanning electron microscopy. Stain Tech. 49:261-264.

250. Thornwaite, J. T., M. L. Cayer, B. F. Cameron, S. B. Leif, and R. C. Leif. 1976. A technique for combined light microscopy and scanning electron microscopy of cells. SEM/1976/II, IIT Research Institute, Chicago, IL, 60616, 127-130.

251. Thornwaite, J. T., B. N. Thornwaite, M. L. Cayer, M. A. Hart and R. C. Leif. 1975. A new method for preparing cells for critical point drying. SEM/1975, IIT Research Institute, Chicago, IL, 60616, 387-392.

252. Tokunaga, J., M. Masu and M. Edanaga. 1974. Cryo-fractured specimen procedures for scanning electron microscopy. J. Electron Microsc. 23:239-240.

253. Tombes, A. S. 1974. Invertebrate endocrine structures as viewed with the scanning electron microscope. Assoc. Southeast Biol. Bull. 21:88.

254. Treiblmayr, K. and K. Pohlhammer. 1974. The use of a micro-filter apparatus for fixation and dehydration of small biological specimens in electron microscopy. Mikroskopie. 30:229-233.

255. Tsien, H. C., G. D. Shockman and M. L. Higgins. 1975. Structure and molecular organization of the cell wall of Streptococcus faecalis. Abstr. Annu. Meet. Am. Soc. Microbiol. 75:165.

256. Turner, R. D. and P. J. Boyle. 1974. Studies of bivalve larvae using the scanning electron microscope and critical point drying. Am. Malacol. Union Inc. Bull. 40:59-65.

257. Van Ewijk, W. and M. P. Mulder. 1976. A new preparation method for scanning electron microscopic studies of single selected cells cultured on a plastic film. SEM/1976/II, IIT Research Institute, Chicago, IL, 60616, 131-134.

258. van Steveninck, M. E. and R. F. M. van Steveninck. 1975. Evidence for structural units in chloroplast thylakoids. Protoplasma. 86:381-389.

259. Walker, G. K. 1975. The tubular cristae of protozoan mitochondria preservation by critical point drying. Trans. Am. Microsc. Soc. 94:275-279.

260. Walker, G. K., T. K. Maugel and D. Goode. 1976. Preservation of sub-cellular structures via the combined Langmuir trough critical point drying technique. Trans. Am. Microsc. Soc. 95:702-707.

261. Waterman, R. E. 1974. Scanning electron microscopic studies of early facial development in rodents and man. SEM/1974, IIT Research Institute, Chicago, IL, 60616, 533-540.

262. Weber, W. J., Jr., M. Pirbazari, and G. L. Melson. 1978. Biological growth on activated carbon: An investigation by scanning electron microscopy. Environ. Sci. Technol. 12:817-819.

263. Wendelschafer-Crabb, G., S. L. Erlandsen and D. H. Walker, Jr. 1975. Conditions critical for optimal visualization of bacteriophage adsorbed to bacterial surfaces by scanning electron microscopy. J. Virol. 15:1498-1503.

264. Westfall, J. A. and J. W. Townsend. 1977. Scanning electron stereomicroscopy of the gastrodermis of Hydra. SEM/1977/II, IIT Research Institute, Chicago, IL, 60616. 623-629.

265. Wheeler, E. E., J. B. Gavin and R. N. Seelye. 1975. Freeze drying from tertiary butanol in the preparation of endocardium for scanning electron microscopy. Stain Technol. 50:331-337.

266. Wiggins, G. B., E. Y. Lin and K. E. Chua. 1976. Preliminary scanning electron microscopy investigation of an aqueous carbohydrate material, the gelatinous matrix of caddis-fly eggs. Insecta Trichoptera. SEM/1976/II, IIT Research Institute, Chicago, IL, 60616, 605-610.

267. Williams, S. T. and C. J. Veldkamp. 1974. Preparation of fungi for scanning electron microscopy. Trans. Br. Mycol. Soc. 63:408-412.

268. Winborn, W. B. and D. L. Guerrero. 1974. The use of a single tissue specimen for both transmission electron microscopy and scanning electron microscopy. Cytobios. 10:83-91.

269. Wroblewski, R. 1976. Analytical scanning microscopy correlated with light microscopy as a tool in experimental physiology and pathology. J. Ultrastruct. Res. 57:213.

270. Wroblewski, R. and E. Jansson. 1976. X-ray microanalysis on isolated human muscle fibers. J. Ultrastruct. Res. 54:481-482.

271. Yamada, E. and H. Watanabe. 1977. High voltage electron microscopy of the critical point dried cryo-section. J. Electron Microsc. 26:253.

272. Yonehara, K., H. Ishikawa, E. Yamada and T. Asai. 1976. Some improvements for cryoultramicrotomy. J. Electron Microsc. 25:179.

273. Yoshii, Z. and S. Tanaka. 1975. A new attempt in the critical point drying of free cell materials for scanning electron microscopy. J. Electron Microsc. 24:200-201.

274. Zimny, M. L. and I. Redler. 1974. Chondrocytes in health and disease. A scanning electron microscope study. SEM/1974, IIT Research Institute, Chicago, IL, 60616, 805-811.

275. Zs-Nagy, I., C. Pieri, G. Giuli, C. Bertoni-Freddari and V. Zs-Nagy. 1976. A new method of biological specimen preparation for x-ray microanalysis by means of freeze fracturing and subsequent freeze drying. J. Submicrosc. Cytol. 8:256-257.

276. Zs-Nagy, I., C. Pieri, C. Guili, C. Bertoni-Freddari and V. Zs-Nagy. 1977. Energy dispersive x-ray microanalysis of the electrolytes in biological bulk specimen. Part 1. Specimen preparation, beam penetration and quantitative analysis. J. Ultrastruct. Res. 58:22-33.

Subject Index

I. Chemical substitution of water
4,30,61,92,93,97,101,106,109,111,118,120,129, 134,137,155,159,164,167,172,177,194,205,213, 254,255,258.

II. Chemical dehydration
25,125,126,129,150,166,176,206,207.

III. Air drying
A. Technical information
4,61,107,110,134,136,145,149,183,194,198, 204,238.

B. Specific applications
1,49,50,58,110,140,164,168,173,179,183, 193,194,198,203,204,229,237,238,239,240, 241,242,243,244,245,246.

IV. Critical point drying
A. Technical information
14,17,18,29,30,32,33,34,38,44,45,46,47, 54,55,59,61,74,82,88,97,100,107,108,109, 110,112,114,120,130,134,136,145,148,158, 180,183,187,188,190,194,195,198,204,205, 208,209,212,214,215,221,223,224,235,247, 250,251,257,260,273.

B. Specific applications
1,2,3,5,6,7,10,11,12,13,16,19,20,21,24, 26,28,31,34,35,37,39,40,43,48,57,58,60, 62,63,67,73,74,78,79,80,81,82,84,87,90, 94,95,97,98,99,105,106,107,110,112,115, 116,117,119,120,122,126,132,135,138,140, 141,142,143,144,145,146,150,155,157,159, 161,163,164,165,168,169,171,172,177,182, 183,190,192,194,197,198,199,201,202,204, 209,211,215,217,218,220,222,223,224,231, 233,234,236,249,250,252,253,256,257,259, 261,262,263,264,268,271,273,274.

V. Freeze drying
A. Technical information
15,22,29,30,32,51,52,72,75,107,108,110, 123,124,132,133,136,139,145,152,153,160, 178,183,184,185,186,191,194,196,208,210, 224,225,227,232,238,248,265,272.

B. Specific applications
3,8,9,23,24,27,28,36,41,42,48,51,53,56, 62,64,65,66,68,69,70,71,76,77,83,84,85, 86,89,91,96,102,103,104,107,110,113,119, 121,127,128,131,139,143,145,147,151,154, 155,156,160,162,164,170,173,174,175,178, 179,181,183,184,189,194,196,200,211,216, 219,224,226,227,228,230,238,248,265,266, 267,269,270,275,276.

CLASSIFICATION OF PAPERS BY MAJOR SUBJECTS:

This index includes an abridged title, name of the first author and the starting page number for each paper. It should be used in conjunction with the Table of Contents and the Author Index in SEM/1978/I.

GENERAL INTEREST PAPERS ON INSTRUMENTATION AND TECHNIQUES

KEYNOTE PAPER: HOW TO GET THE BEST FROM YOUR SEM; G.E. Pfefferkorn (1

GRID APERTURE CONTAMINATION IN ELECTRON GUNS USING DIRECTLY HEATED LaB_6 SOURCES; P.B. Sewell (221

REVIEW: IMAGE PROCESSING FOR SCANNING MICROSCOPISTS: A.V. Jones (13

A MICROPROCESSOR-BASED DIGITAL SCAN GENERATOR; B.M. Unitt (27

REVIEW; ...PERFORMANCE OF PRESENTLY AVAILABLE TV-RATE STEREO SEM SYSTEMS; J.B. Pawley (157

A METHOD FOR TAKING STEREO SEM PICTURES BY EUCENTRIC ROTATION..; J. Blödorn (283

REVIEW: NEW APPLICATIONS OF SUBMICROMETER STRUCTURES..; H.I. Smith (33

TUTORIAL: COATING TECHNIQUES FOR SEM AND X-RAY MICROANALYSIS; P. Echlin (109

SPUTTER COATING IN OIL-CONTAMINATION FREE VACUUM; E. de Harven (167

TUTORIAL: SEM VACUUM TECHNIQUES AND CONTAMINATION MANAGEMENT; D.E. Miller (513

See also: Male (643/II), Murphy (175/II), Boyde (303/II), Pameijer (831/II), Thurston (849/II), Pawley (683/II), Robards (927/II), Strojnik (319), Ward (783), Crabtree (543), Ura(747), papers under Image Interpretation and Contrast, Analytical Electron Microscopy (for papers on x-ray microanalysis etc.), Lee (677), Lyman (529)

REVIEW: BIOLOGICAL SEM FOR PHYSICISTS AND ENGINEERS; J.-P. Revel (829

REVIEW: OPTICAL TECHNIQUES FOR ANALYSING SE MICROGRAPHS; N.K. Tovey (381

SCANNING X-RAY PROJECTION MICROSCOPY USING AN ENERGY DISPERSIVE SPECTROMETER; W. Brünger (423

HEAVY ION MICROSCOPY OF HYDRATED UNFIXED CELLS..; T.L. Hayes (233

IMAGE INTERPRETATION AND CONTRAST (BSE electrons):

REVIEW: BACKSCATTERED ELECTRONS AS A TOOL FOR FILM THICKNESS DETERMINATION; H. Niedrig (841

REVIEW: ELECTRON SCATTERING IN THE SEM; V.N.E. Robinson (859

EXPERIMENTS WITH A SMALL SOLID ANGLE DETECTOR FOR BSE; L. Reimer (705

A SOLID STATE BSE DETECTOR CAPABLE OF OPERATING AT T.V. SCAN RATES; D.A. Gedcke (581

A HIGH EFFICIENCY, NONDIRECTIONAL BSE DETECTION MODE FOR SEM; S.H. Moll (303

EFFECT OF COLLECTOR POSITION ON TYPE-2 MAGNETIC CONTRAST IN THE SEM; O.C. Wells (293

NOTE ON SIGNAL-TO-NOISE RATIO IN SEM; O.C. Wells (299

See also: DeNee (741), Robinson (595)

ANALYTICAL ELECTRON MICROSCOPY, STEM, X-RAY MICROANALYSIS:

TUTORIAL: ARTEFACTS OBSERVED IN ENERGY-DISPERSIVE X-RAY SPECTROMETRY..; C.E. Fiori (401

TUTORIAL: HOW TO USE QUANTITATIVE X-RAY ANALYSIS PROGRAMS; R.B. Bolon (813

ENERGY DISPERSIVE FILTRATION FOR SIMULTANEOUS ENERGY AND WAVELENGTH DISPERSIVE ANALYSIS; J.D. Geller (201

MEASUREMENTS NEAR THE DETECTABILITY LIMITS WITH A HORIZONTAL CRYSTAL SPECTROMETER FOR THE SEM; I. Dienwiebel (807

REVIEW: METHODS OF QUANTITATIVE ANALYSIS OF INDIVIDUAL MICROPARTICLES..; J.T. Armstrong(455

A NEW METHOD FOR PARTICLE X-RAY MICROANALYSIS..; P.J. Statham (469

ESTIMATING ELEMENTAL CONCENTRATIONS IN SMALL PARTICLES..; N.C. Barbi (193

THE PRODUCTION AND CHARACTERIZATION OF GLASS FIBERS AND SPHERES FOR MICROANALYSIS; J.A. Small (445

THE EFFECT OF CHEMICAL VARIABILITY OF INDIVIDUAL FLY ASH PARTICLES..; T.L. Hayes (239

ELECTRONIC ADDING-UP AND STORING OF SEM COLOR IMAGES; F. Buschbeck (835

SPURIOUS X-RAYS PRODUCED IN THE STEM; J.I. Goldstein (427

QUANTITATIVE STEM CHEMICAL ANALYSIS; B. Bengtsson (655

STEM MICROANALYSIS OF DUPLEX STAINLESS STEEL WELD METAL; C.E. Lyman (213

DETERMINATION OF GRAIN BOUNDARY SEGREGATION..; D.R. Clarke (77

See also: Coleman (911/II), Panessa (1055/II), Quinton (391/II), Trump (1027/II), Tormey (259/II), Raymond (93), Papers under surface analysis in Semiconductor Applications.

TUTORIAL: COMPARISION OF SEM, TEM, STEM..; C.E. Lyman (529

REVIEW: STRUCTURAL ANALYSIS OF MACROMOLECULAR ASSEMBLIES WITH STEM; L.T. Germinario (69

REVIEW: FACTORS IN HIGH RESOLUTION BIOLOGICAL STRUCTURE ANALYSIS BY CONVENTIONAL TEM; W. Chiu (569

TUTORIAL: ALL YOU MIGHT WANT TO KNOW ABOUT ELS..; M.S. Isaacson (763

REVIEW: THE DESIGN OF SPECTROMETERS FOR ELS; H.T. Pearce-Percy (41

REVIEW: QUANTITATIVE ELS; R.F. Egerton (133

ELEMENTAL ANALYSIS OF SECOND PHASE CARBIDES USING ELS;   H.L. Fraser   (627

See also: Lee (677), Miller (513),   Papers under Particulates, Petroff (325)

BIBLIOGRAPHY ON CATHODOLUMINESCENCE;
        W. Bröcker,   (333

LAMMA: A NEW LASER MICROPROBE MASS ANALYSER;
        R. Wechsung   (611

PRINCIPLES AND ANALYTICAL CAPABILITIES OF SCANNING LASER ACOUSTIC MICROSCOPE;   L.W. Kessler (555

SCANNING SYSTEM FOR A 5MeV ELECTRON MICROSCOPE;
        A. Strojnik   (319

IMAGE SIGNALS AND DETECTOR CONFIGURATIONS FOR STEM;
        J.M. Cowley   (53

PRACTICAL PHASE DETERMINATION OF INNER DYNAMICAL REFLECTIONS IN STEM;   J.C.H. Spence   (61

PARTICULATE MATTER IN THE ENVIRONMENT

See also: Papers under analytical electron microscopy, particularly, Armstrong (455), Statham (469), Barbi (193), Small (445), Hayes (239)

TUTORIAL: BASIC CONCEPTS OF ELECTRON DIFFRACTION AND ASBESTOS IDENTIFICATION..; R.J. Lee (677

TUTORIAL: COLLECTING, HANDLING AND MOUNTING OF PARTICLES FOR SEM;   P.B. DeNee   (479

See also: Papers under surface analysis under Semiconductor applications

MEASUREMENT OF MASS AND THICKNESS OF RESPIRABLE SIZE DUST PARTICLES BY SEM BSE IMAGING;
        P.B. DeNee   (741

SOME ASPECTS OF SPECIMEN PREPARATION AND LIMITA-TIONS OF PRECISION IN PARTICULATE ANALYSIS..;
        E.J. Chatfield   (487

SIZING OF PARTICULATES FOR ENVIRONMENTAL HEALTH STUDIES; J.R. Millette   (253

PREPARATION LOSSES AND SIZE ALTERATIONS FOR FIBROUS MINERAL SAMPLES;   K. Bishop   (207

SEM/TEM OF NEW AND USED ASBESTOS-CEMENT PIPE..;
        M. Mah   (85

SEDIMENTS INCLUDING COAL

· PREPARATION OF GEOLOGICAL SAMPLES FOR SEM;
        D.A. Walker   (185

SAND GRAIN SELECTION PROCEDURES FOR OBSERVATION IN THE SEM; N.K. Tovey   (393

OBSERVATIONS ON SMALL SEDIMENTARY QUARTZ PARTICLES ..;   P.A. Bull   (821

AN SEM EXAMINATION OF QUARTZ GRAINS FROM SUB-GLACIAL AND ASSOCIATED ENVIRONMENTS..;
        W.B. Whalley   (353

CATHODOLUMINESCENCE IN QUARTZ SAND GRAINS;
        D.H. Krinsley   (887

CATHODOLUMINESCENCE AND MICROSTRUCTURE OF QUARTZ OVERGROWTHS ON QUARTZ;   P.R. Grant   (789

MICROCRACKS AND MATRIX DEFORMATION IN STRESSED GAS-BEARING SANDSTONES;   L.A. Dengler   (603

REVIEW: PETROGRAPHIC SATUROMETRY IN MARINE CAR-BONATE SEDIMENTS;   E.T. Alexandersson (503

DIAGENETIC FEATURES OBSERVED INSIDE DEEP-SEA Mn NODULES..;   V.M. Burns   (245

A MOUNTING MEDIUM FOR COAL PARTICLES;
        A. Moza   (289

REVIEW: ... ORGANIC SULFUR ANALYSIS IN COAL..;
        R. Raymond, Jr.   (93

PYRITE DISTRIBUTION IN COAL; R.T. Greer   (621

DETERMINATION OF TRACE ELEMENT SITES..IN COAL..;
        R.B. Finkelman   (143

CORRELATED MOSSBAUER-SEM STUDIES OF COAL..;
        R.J. Lee   (561

...STUDIES OF A SOLVENT REFINED COAL PILOT PLANT CARBONACEOUS PLUG;   L.A. Harris   (537

MICROSTRUCTURE STUDIES OF CHEMICALLY DESULFURIZED COALS;   T.V. Rebagay   (669

SEM APPLIED TO THE STUDY OF ASPHALTENES..;
        T.V. Rebagay   (663

See also: Tovey(381), Lohnes(643), Lewin (695), Ross (53/II), LeFurgey (579/II), Papers under General Interest, Analytical Electron Microscopy, Particulates, and Materials Characterization. Also Surface Analysis under Semiconductor Applica-tions, Petroff (325).

MATERIALS CHARACTERIZATION

A TECHNIQUE FOR FULL SURFACE EXAMINATION OF SMALL SPHERES..;   C.M. Ward   (783

DEVELOPMENT OF HIGH TEMPERATURE SEM..;
        D.N.K. Wang   (777

MATERIALS CHARACTERIZATION IN A SEM ENVIRONMENTAL CELL; V.N.E. Robinson   (595

QUANTITATIVE IMAGE ANALYSIS OF BRITTLE AND TOXIC CERAMICS; H.W. Arrowsmith   (311

REVIEW: METALLURGICAL APPLICATIONS OF KOSSEL DIFFRACTION IN THE SEM;   D.J. Dingley   (869

HYDROGEL COATINGS AND IMPREGNATIONS IN SILASTIC..;
        R.T. Greer   (633

REVIEW: SEM APPLICATIONS IN SOIL MECHANICS;
        R.A. Lohnes   (643

SEM IN THE DIAGNOSIS OF "DISEASED" STONES;
        S.Z. Lewin   (695

...CHARACTERIZATION OF BIODEGRADATION OF AN ALUMINUM BEARING ROCK BY FUNGI;   A. Mehta   (171

...MORPHOLOGICAL AND LEACHING CHARACTERISTICS OF THERMOPHILIC MICROORGANISMS..; V.K. Berry   (177

See also: Papers under general interest, image formation and contrast, analytical electron microscopy, particulates, sediments, surface analysis (under semiconductor applications) and Petroff (325)

SEMICONDUCTOR APPLICATIONS:
SURFACE ANALYSIS

See papers under analytical electron microscopy.

TUTORIAL: QUANTITATIVE AUGER ELECTRON SPECTROSCOPY - PROBLEMS AND PROSPECTS;   P.H. Holloway (361

BEAM BRIGHTNESS MODULATION IN AUGER ELECTRON MICROSCOPY;   T.J. Shaffner   (149

CHARACTERIZATION OF SEMICONDUCTOR WAFER DEFECTS BY
ASLEEP, SEM AND SCANNING AES;
    A. Christou   (273

TUTORIAL: SURFACE ANALYSIS FROM SCATTERED AND
SPUTTERED IONS; A.W. Czanderna   (259

EXTENDING SEM CAPABILITIES USING A SIMS ATTACHMENT
..;   G.R. Sparrow   (711

SEMICONDUCTOR CHARACTERIZATION AND QUALITY CONTROL

REVIEW: PLASMA ETCHING TECHNIQUES IN SEMICONDUCTOR
MANUFACTURE..;   P.N. Crabtree   (543

APPLICATION OF ELECTRON BEAM MEASURING TECHNIQUES
FOR VERIFICATION OF COMPUTER SIMULATIONS FOR LSI
CIRCUITS;   H.-P. Feuerbaum   (795

AUTOMATIC MEASUREMENTS BY MEANS OF A PROBE FORMING
ELECTRON-BEAM SYSTEM; D.E. Davis   (497

ELECTRON OPTICAL DESIGN OF PICOSECOND PULSE
STROBOSCOPIC SEM;   K. Ura   (747

SUBMICRON ELECTRON BEAM PROBE TO MEASURE SIGNAL
WAVEFORM..;   H. Fujioka   (755

THE SEM AS A DIAGNOSTIC TOOL FOR RADIATION HARDNESS
DESIGN OF MICROCIRCUITS; C.G. Thomas   (735

TUTORIAL: ELECTRON BEAM INDUCED CURRENT
      H.J. Leamy   (717

RESOLUTION LIMITS OF THE EBIC TECHNIQUE IN THE
DETERMINATION OF DIFFUSION LENGTHS IN SEMICONDUCTORS;
      S.P. Shea   (435

DETERMINATION OF DIFFUSION LENGTH AND DRIFT
MOBILITY IN SILICON BY USE OF A MODULATED SEM BEAM;
    S. Othmer   (727

REVIEW: VOLTAGE CONTRAST..;
      A. Gopinath   (375

ON-WAFER DEFECT CLASSIFICATION OF LSI-CIRCUITS USING
A MODIFIED SEM;   P. Fazekas   (801

REVIEW: STEM TECHNIQUES FOR SIMULTANEOUS
ELECTRONIC ANALYSIS AND OBSERVATION OF DEFECTS IN
SEMICONDUCTORS;   P.M. Petroff (325

*See also papers under General Interest, Image
Formation and Contrast, Analytical Electron
Microscopy, Particulates and Materials
Characterization.*

SCANNING ELECTRON MICROSCOPY/1978/II

A selected list of papers of interst to
readers of SEM/1978/I:

SEM AND SURFACE ANALYTIC STUDY OF AN ISOTROPIC
VAPOR DEPOSITED CARBON FILM ON MICROPOROUS MEMBRANE;
    H.S. Borovetz   (85/II

TUTORIAL: NON-COATING TECHNIQUES TO RENDER
BIOLOGICAL SPECIMENS CONDUCTIVE;
    Judith A. Murphy   (175/II

VALIDATION OF METHODS FOR QUANTITATIVE X-RAY ANALYSIS
OF ELECTROLYTES USING FROZEN SECTIONS OF ERYTHROCYTES;
    J.M. Tormey   (259/II

TUTORIAL: PROS AND CONS OF CRITICAL POINT DRYING
AND FREEZE DRYING FOR SEM; A. Boyde (303/II

REVIEW: PHYSICS OF SEM FOR BIOLOGISTS;
    D.C. Joy   (379/II

SEM-EDS X-RAY ANALYSIS OF FLUIDS;
    P.M. Quinton   (391/II

USE OF SEM AND X-RAY ENERGY SPECTROMETRY TO DETERMINE
THE ELEMENTAL CONTENT OF INCLUSIONS IN HUMAN TISSUE
LESIONS; A.R. Brody   (615/II

PRELIMINARY STUDIES OF COATED COMPLEMENTARY FREEZE-
FRACTURED YEAST MEMBRANES VIEWED DIRECTLY IN THE
SEM;   J.B. Pawley   (683/II

FREEZE-DRYING OF ARTICULAR CARTILAGE;
    Y. Draenert   (759/II

PREPARATION OF AQUATIC BACTERIA FOR ENUMERATION BY
SEM;   T.M. Dreier   (843/II

HEALTH AND SAFETY HAZARDS IN THE SEM LABORATORY;
    E.L. Thurston   (849/II

TUTORIAL: X-RAY ANALYSIS OF BIOLOGICAL SPECIMENS;
    J.R. Coleman   (911/II

A TRANSFER SYSTEM FOR LOW TEMPERATURE SEM;
    A.W. Robards   (927/II

BERYLLIUM AND GRAPHITE POLYMER SUBSTRATES FOR
REDUCTION OF SPURIOUS X-RAY SIGNAL;
    B.J. Panessa   (1055/II

TUTORIAL: REPLICA TECHNIQUES FOR SEM;
    C.H. Pameijer   (831/II

ERRATA/SUGGESTIONS

ERRATA:

Despite the best efforts of authors, reviewers and editors, errors may remain. Please help by pointing out errors that you notice. Please provide enough information to locate each error (volume, part, page, column, line etc.) and indicate suitable correction.

If you disagree with the results, conclusions or approaches in a paper, please send your comments, typed in a column format (each column is 4-1/8 inches wide and 11½ inches long; i.e., 10.5 by 29.2 cm.). Your comments along with author's response will be published in the following issue.

SUGGESTIONS:

Based on the suggestions we receive, constant improvements in our approach, publications and programs are being made. The volumes are no longer published as proceedings but as a regular journal (at present annual, soon to be bi-annual). Even though most papers are at present based on the presentations at the Annual SEM meetings, we do accept papers for publication only. Considering the number of initial abstracts and final manuscripts we receive, almost 20% of the contributions are not finally published.

We invite your suggestions to improve our publications and meetings, consistent with the aims of our organization (see p.xxiv). You can specifically help in the areas listed below. Since the program planning for each meeting is finalized in late summer/ early fall, please send your suggestions related to the meetings early.

TUTORIAL & REVIEW PAPERS TOPICS: For a definition of these presentations in the context of our publication/meetings, see page iv. All papers are subjected to same intensive reviewing, which is a unique feature of these meetings/publications. One may suggest oneself for these contributions.

WORKSHOP THEMES: Workshops are organized for concentrated discussion and development of SEM applications and techniques in the selected theme. Workshops center around a number of contributed/review/tutorial papers, including papers for publication only.

REVIEWERS: The contribution of reviewers to the quality of this publication and our meetings is tremendous. We find suitable reviewers from the suggestions we receive from the authors, our advisors, and from our past contacts. We would welcome your suggesting your own name or other's names(along with full mailing address) as reviewers. The reviewers are requested to select papers they feel qualified to review through an invitation sent along with the list of papers.

> Important Note: The time restrictions imposed by the printing schedule requires that each reviewer returns his review along with the manuscript sent within five days from its receipt. *Please donot commit yourself if you feel that you cannot respond within this time frame; while we are grateful for your desire and efforts to help us, the reviewers who do not respond in time, in fact, seriously hamper our efforts.*

PROMOTION OF OUR PUBLICATIONS/MEETINGS: Your help is requested in bringing our publications to the attention of your library; many libraries now have standing orders with us either directly or through subscription agencies. Also please bring our meetings/ publications to the attention of your colleagues and contacts, specially editors of suitable journals (who may be interested in reviewing/abstracting our publications) or officers of specific societies and publications (who announce meetings etc. to their members/readers).

BIBLIOGRAPHIES: This publication contains two bibliographies (one on Macrophages,637/ III, and the other on drying of specimens, 892/II). A bibliography on Forensic Applications is included in part II/1979, and a Cathodoluminescence bibliography was published in SEM/1978/I. Many review papers contain extensive bibliographies also. Volunteers to prepare specific bibliographies should contact Om Johari. Also, please send us two copies of each of your publications so that they can be submitted for inclusion in appropriate bibliography.

CONTACT: OM JOHARI, SEM INC., P.O.BOX 66507, AMF O'HARE, IL 60666, USA.

Actinopterygian teeth 547
Analytical electron microscopy   1, 103, 581, 807,
    see also biolgical microanalysis   817, 827
Arthritis 745
Asbestos 361
ATPase 773
Auger electron spectroscopy (microscopy)   1, 259
Automatic focusing 47
Automatic image processing   47, 53, 61, 67, 361
Axon 793
Backscattered Electron Imaging   53, 61, 111,
                149, 155, 159, 751, 835

Bibliography on
    Dehydration 892
    Forensic applications 209
Biological microanalysis   595, 607, 619, 627, 635,
        649, 659, 673, 683, 691, 703, 711, 723,
    733, 739, 745, 751, 767, 773, 779, 793, 801
    other techniques   259, 279, 291, 383, 817, 827
Biological mineralization 383, 393, 403, 417, 421,
    433, 439, 447, 457, 463, 469, 475, 483, 491,
    497, 507, 513, 521, 529, 539, 547, 555, 575,
                                     739, 745

Biological specimen preparation:
    Air drying 892
    Ashing 345
    Charge neutralization 31
    Coatings   21, 133, 325, 355
    Critical point drying   235, 303, 892
    Cryosectioning   607, 619, 711, 723
    Cryotechnique   291, 325, 595, 607, 619
                  627, 649, 683, 711, 723
    Dehydration 892
    Etching 345
    Fixation 635
    Freeze drying   291, 619, 711, 723, 892
    Freeze fracture 325
    Microdroplets 767
    Pyroantimonate precipitation   779, 793
    Staining   125, 835
Biomaterials   529, 871
Bone   383, 393, 513, 521, 529, 539, 555
Brachiopoda shell 439
Broyozoa shell 439
Bullet fragments 167
Carbon film substrate 375
Carbonate substrata 447
Cartilage   555, 745
Cathodoluminescence   1, 125
Cell injury 723
Cell lysosome 835
Cell mitochondria   711, 835
Cell nuclei   711, 779, 835
Cell peroxisome 835
Cell process 793
Charging   31, 133
Coating   21, 133, 325, 355
Coccolith morphology 417
Computer image analysis   53, 373
Computer topography analysis   47, 61
Contamination   87, 103, 581
Contrast, atomic number   111, 149, 155, 159, 835
Contrast, discharge 133
Contrast, topographic   61, 111, 133, 149
Converted backscattered electron detector   149
Coral 417
Correlative microscopy   739, 745, 793, 835

Crossopterygian fish teeth 483
Cryotechnique see under biological specimen prep.
Cytochemistry   779, 793, 835
Dehydration, bibliography 892
Dentine   383, 393
Dental impression material 571
Dipnoan fish teeth 483
Discharge contrast 133
Dynamite 203
Eels   817, 827
Egg shells   403, 475
Electrolyte transport 733
Electron detector   149, 155, 835
Electron energy loss spectroscopy   703, 817, 827
Electron-specimen interactions   111, 133, 835
Enamel   383, 393, 483, 491, 497, 507, 547
Environmental particulate exposure 751
Ephyra development 433
Erythrocyte   133, 627
Etching 345
Evaporative coating   21, 355
Ferritin 133
Firearms identification 167
Fish teeth 483
Forensic science   155, 159, 167, 175, 179,
                185, 193, 203, 209, 879
Fossil shell 463
Fossil teeth   483, 491, 547, 575
Graphite 61
Gunshot residue   155, 159, 167, 175, 179
Hair 193
Height measurement 47
Histochemistry   773, 779, 835
Hydrogel 871
Hypoxic myocardial injury 723
Image analysis   47, 53, 361, 367, 827
Image processing   53, 61
IMMA 751
Intestinal mucosa   133, 801
Intestinal mucus 801
Ion microprobe mass analysis 751
Ion scattering spectroscopy 1
Ion sputtering deposition 355
Ischemic mycardial injury 723
Laser microprobe mass analysis   1, 279
Laser, Raman microprobe analysis   1, 383, 751
Lysosome 835
Marine archaeology 885
Marine microphyte 447
Mass Measurement 291
Mayan teeth 575
Microanalysis 1, 53, 103, 111, 167, 175, 179, 185,
        193, 203, 209, 259, 433, 555, 595, 607, 619,
        627, 635, 649, 659, 673, 683, 691, 703, 711,
        723, 733, 739, 745, 751, 767, 773, 779, 793,
                801, 807, 817, 827, 879, 885
Microdiffraction 103
Microphyte 447
Mineralized tissue 383
Mitochondrion   711, 835
Mollusc shell   403, 439, 457, 463, 469
Muscle 711
Myelinated axon 793
Nacreous tablets 457
Na$^+$-K$^+$- ATPase 773
Nervous system   373, 793
Node of Rabvier 793
Nucleus   711, 779, 835

On-line topographic analysis   47
Osteoblast   529
Osteoclast   393, 539
Peroxisome   835
Plasma ashing   185, 345
Preparation methods   21, 31, 103, 125, 133, 179
                             235, 291, 303, 325, 345, 595
Prosthesis   871
Pseudogout   745
Quarternary mollusc shell   463
Rachitic bone   513
Red blood cell   133, 627
Reef coral   421
Resorption locunae   393
Replica   133, 571
Retinal rod   133
Rickets   513
Scanning transmission electron microscopy   87,
                   291, 367, 375, 581, 711, 817, 827
Secondary ion mass spectroscopy, SIMS   1, 703
Shark teeth   547
Specimen
     Ashing   185, 345
     Charging   31, 155
     Coating   21, 133, 325, 355
     Damage   683
     Preparation, see preparation methods,   607, 835
     Staining   135, 835
     Substrate   375
Sputter coating   21, 133, 325, 355
Standards for microanalysis   649
Statolith synthesis   417
STEM see scanning transmission electron microscopy
Stereo photogrammetry   67, 79, 393
Surface contamination   87
Teaching SEM   217, 221, 225, 231, 235,
                             241, 243, 247, 253
Teeth   383, 393, 483, 491, 497, 507, 547, 575
Thin films   21, 133, 355, 375
Tissue compatibility   871
Topographic analysis   47
Transepithelial electrolyte transport   733
Urea   767
Urinary stones   739
X-Ray Microanalysis   see microanalysis
ZAF method   1, 595

Abraham, J.L.	751
Andrasko, J.	879
Baker, K.K.	217, 243
Bank, H.L.	779
Barber, V.C.	885
Barbi, N.C.	659
Bauer, R.	619, 733
Beck, F.	733
Becker, R.P.	835
Beeuwkes, R.,III	767, 773
Bendtz, S.	879
Bentley, J.	581
Berry, J.P.	703
Bistricki, T.	683
Blackwelder, P.L.	417
Bouratbine, A.	507
Boyde, A.	67, 393, 529, 539
Brand, L.E.	417
Bröcker, W.	125
Browne, M.T.	827
Buja, L.M.	723
Burton, K.P.	723
Carpenter, R.W.	581
Carter, J.G.	439
Casciani, F.S.	383
Chandler, D.B.	345
Chandler, J.A.	595
Chapman, R.L.	225
Chatfield, E.J.	53
Cheng, P.-T.	745
Chujo, M.	367
Cohen, A.L.	303
Coleman, J.R.	801
Crawford, C.K.	31
Crenshaw, M.A.	469
Crewe, A.V.	375
Crosby, P.	325
Davidson, D.L.	79
Dempster, D.W.	513
Dixon, R.N.	361
Dörge, A.	619, 733
Doty, S.B.	383
Draenert, K.	521
Draenert, Y.	521
Dunkelberger, D.G.	403
Echlin, P.	21
Edwards, R.	61
Egerton, R.F.	817
Elder, H.Y.	513
Ellisman, M.H.	793
Etoh, T.	103
Etz, E.S.	383
Flegler, S.L.	217, 243
Fourie, J.T.	87
Friedman, P.L.	793
Galle, P.	703
Gantt, D.G.	491
Gehring, K.	619
Geller, J.D.	355
Goldstein, J.I.	221
Gorelick, L.	575
Greer, R.T.	871
Greico, C.	723

Guillard, R.L.	417
Guillaumin, D.	507
Gwinnett, A.J.	575
Hagler, H.K.	723
Hamilton, W.J.	793
Harada, Y.	103
Healey, F.	149
Heinen, H.J.	279
Henk, W.G.	235, 345
Heinrich, K.F.J.	807
Hillenkamp, F.	279
Hillman, D.	367
Hogg, N.M.	539
Holburn, D.M.	47
Hooper, G.R.	217, 243
Howard, K.S.	225, 892
Humphreys, W.J.	235, 345
Hurd, D.A.	355
Isaacson, M.S.	375
Janssen, A.P.	259
Johnston, I.S.	421
Johnson, W.	149
Jones, S.J.	393, 529, 539
Joy, D.C.	817
Kaufmann, R.	279
Kaye, G.	21
Keeley, R.H.	179
Kenik, E.A.	581
Kim, K.M.	739
Knoll, R.L.	871
Landis, W.J.	555
Lebiedzik, Jana	61
Lebiedzik, Jozef	61
Lefevre, R.	703
Llinás, R.	367
Lopez, L.	723
Lukas, K.J.	447
Maehly, A.C.	879
Maher, D.M.	817
Mason, J.	733
Matricardi, V.R.	159
McConville, R.L.	231
Moll, S.H.	149
Moore, C.H., Jr.	225
Morgan, A.J.	635
Murphy, Judith A.	253
Mutvei, H.	457
Mykelbust, R.L.	807
Newbury, D.E.	1, 383, 807
Noguchi, T.T.	167, 203, 209
Nyburg, S.C.	745
Ohtsuki, M.	375
Omar, S.A.	745
Pameijer, C.H.	571
Panessa-Warren, B.J.	691
Pausak, S.	185
Peters, K.-R.	133
Pfefferkorn, G.E.	125
Phillips, B.	61
Platz, R.M.	627
Poole, D.F.G.	497
Pooley, A.S.	475

Postek, M.T.	892
Pritzker, K.P.H.	745
Reif, W.-E.	547
Reimer, L.	111
Rick, R.	619, 733
Robards, A.W.	325
Roloff, Ch.	733
Roomans, G.M.	649
Rosen, S.	773
Russ, J.C.	673
Sato, S.	193
Saubermann, A.J.	607
Scheltens, F.	231
Seta, S.	193
Shapiro, I.M.	539
Shellis, R.P.	497
Sherwin, L.	723
Shuman, H.	711
Shürmann, M.	279
Sild, E.H.	185
Siler, A.	723
Simson, J.A.V.	779
Slusarenko, M.	539
Small, J.A.	807
Smith, D.A.	513
Smith, K.C.A.	47
Smith, M.M.	483
Socolofsky, M.D.	225
Sogard, M.	835
Somlyo, A.P.	711
Somlyo, A.V.	711
Spangenberg, D.B.	433
Spicer, S.S.	779
Spitzer, R.	367
Stewart, M.	711
Stott, W.R.	53
Sullivan, B.	149
Suzuki, I.	203
Tassa, M.	155, 175
Taylor, C.J.	361
Taylor, M.S.	167
Taylor, R.L.	167, 203, 209
Thurau, K.	619, 733
Thurston, E.L.	241
Tomita, T.	103
Tormey, J.M.	627
Triller, M.	507
Ueyama, M.	203
Vale, B.H.	871
Venables, J.A.	259
Walker, B.M.	463
Wall, J.S.	291
Wallace, J.S.	179
Watabe, N.	403
Watabe, T.	103
Watanabe, H.	103
Wechsung, R.	279
Weill, R.	507
Wells, O.C.	247
Wilkes, D.A.	469
Williams, D.B.	221
Yoshino, M.	193
Yoshioka, T.	355
Young, L.B.	801
Zaluzec, N.J.	581
Zeldes, N.	155, 175